# THE OXFORD
# DICTIONARY OF
# OPERA

# THE OXFORD
# DICTIONARY OF
# OPERA

John Warrack and Ewan West

Oxford   New York
OXFORD UNIVERSITY PRESS

Oxford University Press, Walton Street, Oxford OX2 6DP

Oxford  New York
Athens  Auckland  Bangkok  Bombay
Calcutta  Cape Town  Dar es Salaam  Delhi
Florence  Hong Kong  Istanbul  Karachi
Kuala Lumpur  Madras  Madrid  Melbourne
Mexico City  Nairobi  Paris  Singapore
Taipei  Tokyo  Toronto
and associated companies in
Berlin  Ibadan

Oxford is a trade mark of Oxford University Press

First published 1992
Sixth impression 1995

British Library Cataloguing in Publication Data
Data available

Library of Congress Cataloging in Publication Data
Warrack, John Hamilton, 1928–
The Oxford dictionary of opera/John Warrack and Ewan West.
p.  cm.
Includes bibliographical references.
1. Opera—Dictionaries.  I. West, Ewan,  II. Title.
782.1'03—dc20  ML102.06W37  1992  92-6730
ISBN 0-19-869164-5

Printed in Great Britain
on acid-free paper by
Bookcraft (Bath) Ltd
Midsomer Norton, Avon

# PREFACE

OPERA—frequently pronounced ailing, senile, brain-damaged, in need of major surgery or transfusions of new ideas, even clinically dead—has never been so healthy or so active. It has achieved a breadth of appeal unprecedented in its history, commanding the patronage of the wealthy but also a popular following, generating well-filled opera-houses but also spectacular mass events, public television relays of famous occasions, and huge record sales. It has increasingly drawn the interest of designers and producers from the theatre, treating the art of the composer as no less intellectually or dramatically exacting than that of the playwright. It is now also generating, alongside a mass of musicological work, a serious body of critical thought, with some of the most valuable contributions coming from philosophers, aestheticians, literary critics, and historians of ideas. Most crucially, composers who in the past neglected it, either for economic reasons or through lact of faith, now regard it as the most exciting and important musical genre—the word 'tradition', so sustaining but also so cramping in the past, is perhaps no longer helpful when such a variety of dramatic music now claims a place as operatic.

To respond to this creative energy is a stimulating task for a reference book; it is also one demanding more than the scholarly skills which remain central. We have taken a wider view of the definition of opera than might have been thought reasonable even ten years ago, both in paying some attention to different ancient and modern dramatic forms and in taking account of some of the dramatic music that preceded the traditional starting-point of opera, the deliberations and experiments of Count Bardi's Camerata in Florence in the late 16th century. We have, however, remained chiefly attentive to the European tradition that continues to provide the material for repertories throughout the world, and so to the needs and interests of opera-goers who wish to inform themselves about what they might see, hear, or read about.

Though the book is, we hope, self-explanatory, there are certain points to which we should like to draw attention. Cross-references within the text are given by asterisks. An asterisk before a word indicates an entry that is relevant to the one being consulted, containing significant further information or comment; it does not automatically occur every time there is mention of someone or something with a separate entry in the book. Place-names present problems, as ever. We have normally given the name in most general use at the time in the country concerned, sometimes with alternatives but without the explanation that would involve a short historical essay: thus, we refer to Ljubljana, but give the previously used German name Laibach in brackets under the main entry. There are two natural exceptions. We follow general English usage in writing Munich, Warsaw, Florence, etc., not München, Warszawa, Firenze; and when a town or city has changed its name, we try to give this according to how it was widely known at the time. Thus it becomes possible to be born in St Petersburg, study in Petrograd, work in Leningrad, and now retire in St Petersburg, all without moving. It is not, however, possible to be born in Pressburg, study in Pozsony, and retire in Bratislava, since these are the German, Hungarian, and Slovak names of what is now a Slovak city. These matters are never easy to resolve, still less so in times of renewed national upheaval. We have tried to identify artists by what we believe to be the nationality they claim, but it is often difficult to know how local this should be. If we have offended, it is not through any wish to make political points.

As to dates: we retain the Old Style English dates before the adoption of the Gregorian calendar in 1752, as far as possible, but calendar changes are complex and potentially misleading (for instance, as to which year saw a work's première, when the New Year is not on 1 January). Russian dates, and those of other countries that maintained the Julian calendar into modern times, are given in New Style. Dates of operas, unless otherwise explained, are of first production; we have normally added the composition date only when it antedates production by a significant number of years.

Opera titles are usually given in the original language, and in the form reflecting the widest usage at the time of production. We have not, however, used Cyrillic, Greek, or other non-Roman Alphabets, though we do give foreign accents; and in general reference we follow conventional English usage with titles such as *The Bartered Bride*, and sometimes use familiar abbreviations, e.g. *Die Meistersinger*, where it is natural and above all space-saving. Works are, with rare exceptions, described simply as 'opera', rather than by a more specific designation such as 'opera buffa' or 'italienisches Singspiel', since it is often impossible to establish what the original should be: even so immediately popular an opera as *Figaro* was described by these and several other titles in the original librettos, score, playbill, and Mozart's catalogue of his works. We also give, where appropriate, the most important sources of librettos. Voice categories are given (sop, ten, etc.) after roles, though here, too, there is sometimes inconsistency of description; we do not include them for early operas where no definition was given by the composer and where often the range does not correspond closely with any modern category. Revivals are only listed when they are of special significance.

Under composers, we give surname and then the first name or names in conventional use; we do not list alternative names or spellings, maiden names, or Russian patronymics. In the performer entries, the seasons and appearances are only some of the most notable. We use the conventional term 'created' to mean 'first singer of'.

Under Bibliography will be found a list of some of the works to which the interested user of this book may wish to turn for further or fuller information. Among them is *The Concise Oxford Dictionary of Opera*, from which we have drawn a small amount of material. We owe an incalculable debt to an army of friendly scholars and scholarly friends, all of whom responded to our requests with unstinted generosity by reading material for us, chasing facts, answering queries, giving critical opinion, making suggestions, and otherwise checking, correcting, and advising. Among these, we should like particularly to thank Karina Asenova, William Ashbrook, Corinne Auger, Rosamond Bartlett, Lucy Beckett, Jack Belsom, Victor Borovsky, Peter Branscombe, Jonas Bruveris, Julian Budden, Bojan Bujić, Janet Caldwell, Tim Carter, David Cummings, Martin Dreyer, June Emerson, Rómulo Ramírez Esteva, Pablo Fernandez, Per Fosser, Angel Fumagalli, Elena Komarova, S. Kortes, George Leotsakos, Antonia Malloy, Arda Mandikian, the Earl of Mexborough, Rodney Milnes, Maia Museridze, Malgorzata Nawrocka, Ágnes Nyilassy, Matilda Onofrei-Voiculet, Richard Osborne, Mare Põldmäe, Caroline Rae, Patrick J. Smith, David Trendell, John Tyrrell, Richard Vendome, Jaak Viller, and Chris Walton. In a category all of his own is John M. Gehl: we owe him an exceptional debt for his enthusiastic and knowledgeable assistance over many matters particularly to do with opera in the USA, and for his tireless correspondence and his generosity with checking proofs. We also gladly acknowledge the help given by the representatives of opera-houses in many different countries.

We are grateful to a number of libraries, including the Glinka and Shalyapin Museums, Moscow, the Österreichische Nationalbibliothek, the British Library, and in Oxford the Bodleian Library (including the Music Room) and the Library of the Taylorian Institute (including its Slavonic department). Particular gratitude is due to the Library of the Oxford Faculty of Music and its knowledgeable and (fortunately) tolerant Librarian, John Wagstaff. For help over some of the day-to-day organization, we are grateful to the Faculty Administrator, Anne Roberts, and to two Secretaries, Sophie Browne and Anna Mochan. At the Oxford University Press, Melissa Spielman gave us much technical guidance in the early stages; later, Pam Coote showed us many kindnesses and much support. Rowena Anketell was a keen-eyed copy-editor who made helpful contributions of her own. As proof-readers, Bonnie Blackburn and Leofranc Holford-Strevens went far beyond the normal bounds of such a task with their scholarly correction, advice, and augmentation (including over obscure linguistic matters): we acknowledge their contribution with awe as well as gratitude. We should also like to thank our respective Colleges in Oxford, St Hugh's and Mansfield, for practical support and encouragement, and for allowing leave of absence to enable the book to be completed.

To our two research assistants we are also deeply in debt. In the early stages, Gwen Hughes helped to devise working practices, draw up lists, check much material, and write some entries. Subsequently, Theresa Lister worked particularly on the performer entries, also checking much and contributing ideas and material. Without them, the work would have been slower and less complete, not to mention less enjoyable.

JOHN WARRACK
EWAN WEST

*Rievaulx and Oxford, 1992*

# BIBLIOGRAPHY

IN a book of this size, it is clearly impossible to provide a comprehensive bibliography for every entry. We have included only a few references to further literature, and these are intended to do no more than indicate the most important work on the subject: normally this is chosen because it also includes a good, up-to-date bibliography, though some items (e.g. E. J. Dent's pioneering study of Mozart) are cited chiefly for their classic status or historical interest. With a wide English-speaking readership in mind—enthusiasts, students, scholars, and all three in one—we have shown a general preference for English-language works, though in the case of country, city, and opera-house entries we have generally found it necessary to cite foreign publications. Periodical articles have been included in exceptional cases only. For many composers and artists of lesser rank—and indeed for some significant composers of the 17th, 18th, and 19th centuries—there exists no individual study sufficiently reliable for its use to be recommended without extreme caution. The reader is, in all cases, referred to the relevant entries in such standard works as *The New Grove Dictionary of Music and Musicians* or *Die Musik in Geschichte und Gegenwart* and their specialist bibliographies. Furthermore, most countries with a claim to a musical tradition have published a dictionary or encyclopedia, sometimes general, sometimes national; we have not listed these, though of course they include coverage of opera. Autobiographies have not been included as a matter of course (those by singers, in particular, are often entertaining but highly individual), and we have restricted our choice to items of some historical importance.

While it would be equally difficult to draw up even a summary list of writings on the history, theory, and practice of opera in so brief a space, we think it may be useful to indicate some of the most important large-scale studies of the genre. There are listed below, chronologically by date of publication within each category. Caution is advised in consulting some of the catalogues, particularly the earlier ones: a common fault is the inclusion of a number of titles that prove on investigation to be ballets, masques, incidental music, or other works that cannot by the most generous interpretation be described as opera.

### CATALOGUES

Some of the works given here are lists, with or without attendant statistics; some add comment or even brief synopses.

**Allacci, L.**, *Drammaturgia* (Rome, 1666); continuation by G. Pasquali (Venice, 1755).

**Ivanovich, C.**, *Le memorie teatrali di Venezia*, in *Minerva al tavolino* (Venice, 1681, 2/ 1688).

**de Lavallière, L.-C.**, *Ballets, opéras, et autres ouvrages lyriques* (Paris, 1760, R/1967).

**Clément, F.**, and **Larousse, P.**, *Dictionnaire lyrique, ou Histoire des opéras* (Paris, 1867–9; 4 suppls. to 1881; further suppls. by A. Pougin, 1899–1904, R/1969).

**Lajarte, T.**, *Bibliographie musicale du Théâtre de l'Opéra* (Paris, 1877–9).

**Riemann, H.**, *Opern-Handbuch* (Leipzig, 1887).

**Dassori, C.**, *Opere e operisti* (Genoa, 1903, 2/1906).

**Towers, J.**, *Dictionary-Catalogue of Operas and Operettas* . . . (Morgantown, 1910, R/1967).

**Albinati, G.**, *Piccolo dizionario di opere teatrali, oratori, cantate ecc.* (Milan, 1913).

**Prod'homme, J.-G.**, *L'Opéra (1669–1925)* (Paris, 1925).

**Loewenberg, A.**, *Annals of Opera, 1597–1940* (Cambridge, 1943, 2/1955, 3/1978).

**Mooser, R.-A.**, *Opéras, intermezzos, ballets, cantates, oratorios joués en Russie durant le dix-huitième siècle* (Geneva, 1945, 2/1955).

**Michalowski, K.**, *Opery polskie* (Cracow, 1954).

**Manferrari, U.**, *Dizionario universale delle opere melodrammatiche* (Florence, 1954–5).

**Bauer, A.**, *Oper und Operetten in Wien* (Graz, 1955).

**Smith, W.**, *The Italian Opera and Contemporary Ballet in London, 1789–1820* (London, 1955).

**Mattfeld, J.**, *Handbook of American Operatic Premières* (Detroit, 1963).

**Caselli, A.**, *Catalogo delle opere liriche pubblicate in Italia* (Florence, 1969).

**Stieger, F.**, *Opernlexikon* (Tutzing, 1975–83).

**Northouse, C.**, *Twentieth Century Opera in England and the United States* (Boston, 1976).

**Gruber, C.**,ed., *Opern-Uraufführungen: Ein internationales Verzeichnis von der Renaissance bis zur Gegenwart* (Vienna, 1978– ).

**B. Shteynpress**, *Opernye premyery xx veka*, vol. 1, 1901–1940 (Moscow, 1983), vol. 2, 1941–60 (Moscow, 1986).

**Pitou, S.**, *The Paris Opéra: An Encyclopedia of Operas, Ballets, Composers and Performers* (Westport, CT, 1983–90).

**Marco, G.**, *Opera: A Research and Information Guide* (New York, 1984).

**Never, A.**, *Polish Opera and Ballet of the Twentieth Century* (Cracow, 1986).

## DICTIONARIES AND ENCYCLOPEDIAS

### General Theatrical and Other Reference Works

**Eisenberg, L.**, ed., *Grosses biographisches Lexicon der deutschen Bühne im XIX. Jahrhundert* (Leipzig, 1903).

**d'Amico, S.**, ed., *Enciclopedia dello spettacolo* (Rome, 1954–62, suppl. 1966).

**Mokulsky, S.**, ed., *Teatralnaya entsiklopedia* (Moscow, 1961–7).

**Dabrowski, S.**, ed., *Słownik biograficzny teatru polskiego* (Warsaw, 1973).

**Gammond, P.**, *The Oxford Companion to Popular Music* (Oxford, 1991).

### Operatic Reference Works

**Ehrlich, A.**, *Berühmte Sängerinnen* (Leipzig, ?1896).

**Bernandt, G.**, *Slovar oper, vperviye postavlennykh ili izdannych v dorevolyutsionnoy Rossii i v SSSR (1736–1962)* (Moscow, 1962).

**Rosenthal, H.**, and **Warrack, J.**, *The Concise Oxford Dictionary of Opera* (Oxford, 1964, 2/1979; trans. and ed.: *Friedrichs Opernlexicon*, Hanover, 1969; *Guide de* *l'opéra*, Paris, 1974, 2/*Dictionnaire de l'opéra*, Paris, 1986; *Dizionario dell'opera lirica*, Florence, 1974, 2/*Dizionario enciclopedico dell'opera lirica*, Florence, 1991).

Gozenpud, A., *Operny slovar* (Moscow, 1965).

Seeger, H., *Opernlexikon* (Berlin, 1978).

Dahlhaus, C., ed., *Pipers Enzyklopädie des Musiktheaters: Oper, Operette, Musical, Ballet* (Munich, 1986– ).

Hamilton, D., ed., *The Metropolitan Opera Encyclopedia* (London, 1987).

Kutsch, K., and Riemens, L., *Grosses Sängerlexikon* (Berne, 1987).

Anderson, J., *Bloomsbury Dictionary of Opera and Operetta* (London, 1989).

Honegger, M., and Prévost, P., *Dictionnaire des œuvres de l'art vocale* (Paris, 1991–2).

## SYNOPSES AND ANALYSES

Annesley, C. [C. and A. Tittman], *The Standard Operaglass* (London, 1888, 51/1937).

Melitz, L., *Führer durch die Oper* (Berlin, 1910).

Harewood, ed., *Kobbé's Complete Opera Book* (London, 1922, 10/1987).

Newman, E., *Opera Nights* (London, 1943).

—— *Wagner Nights* (London, 1949).

—— *More Opera Nights* (London, 1954).

Fellner, R., *Opera Themes and Plots* (London, 1958).

Martin, G., *The Opera Companion* (New York, 1961, 3/1982).

Lubbock, M., *The Complete Book of Light Opera* (London, 1962).

Zentner, W., and Würz, A., *Reclams Opern-und Operettenführer* (Stuttgart, 25/1969).

Pankratova, V., and Polyakova, L., *Opernye Libretto* (Moscow, 1970).

Martin, G., *Companion to Twentieth Century Opera* (New York, 1979, R/1984).

Reger-Bellinger, B., *et al.*, *Knaurs grosser Opernführer* (Munich, 1983).

Neef, S., *Handbuch der russischen und sowjetischen Oper* (Berlin, 1985).

Wagner, H., *Das grosse Handbuch der Oper* (Wilhelmshaven, 1987, 2/1991).

Gänzl, K., and Lamb, A., *Gänzl's Book of the Musical Theatre* (London, 1988).

Kornick, R., *Recent American Opera* (New York, 1991).

## HISTORICAL STUDIES

### General

Hanslick, E., *Die moderne Oper* (Berlin, 1875).

Dent, E., *Opera* (Harmondsworth, 1940, 5/1949).

Grout, D., *A Short History of Opera* (New York, 1947, 3/1988).

Orrey, L., *Opera: A Concise History* (London, 1972, rev. R. Milnes, 3/1987).

Donington, R., *The Rise of Opera* (London, 1981).

Sadie, S., ed., *History of Opera* ('New Grove Handbooks in Music') (London, 1989).

### Bulgaria

Bix, R., *Bălgarski operen teatr* (Sofia, 1976).

### Czechoslovakia

Eckstein, P., *Die tschechoslowakische zeitgenössische Oper* (Prague, 1967).

Tyrrell, J., *Czech Opera* (Cambridge, 1988).

## Great Britain

**Dent, E.**, *Foundations of English Opera* (London, 1928, R/1965).
**Fiske, R.**, *English Theatre Music in the Eighteenth Century* (Oxford, 1973, 2/1986).
**Hume, E.**, ed., *The London Theatre World 1660–1800* (Carbondale, IL, 1980).
**White, E.**, *A History of English Opera* (London, 1983).

## France

**Prunières, H.**, *L'opéra en France avant Lulli* (Paris, 1913).
**Crosten, W.**, *French Grand Opera: An Art and a Business* (New York, 1948, R/1972).
**Cooper, M.**, *Opéra-Comique* (London, 1949).
**Anthony, J.**, *French Baroque Music from Beaujoyeulx to Rameau* (London, 1973, 2/1978; rev. and trans. Fr., Paris, 1981).
**Pistone, D.**, ed., *Le théâtre lyrique français 1945–1985* (Paris, 1987).

## Germany and Austria

**Brockpähler, R.**, *Handbuch zur Geschichte der Barockoper in Deutschland* (Emsdetten, 1964).
**Goslich, S.**, *Die deutsche romantische Oper* (Tutzing, 1975).
**Wenzel, J.**, *Geschichte der Hamburger Oper 1678–1978* (Hamburg, 1978).
**Schusky, R.**, ed., *Das deutsche Singspiel im 18. Jahrhundert: Quellen und Zeugnisse zur Ästhetik und Rezeption* (Bonn, 1980).
**Bauman, T.**, *North German Opera in the Age of Goethe* (Cambridge, 1985).

## Italy

**Solerti, A.**, ed., *Gli albori del melodramma* (Milan, 1904–5, R/1969).
**Worsthorne, S.**, *Venetian Opera in the Seventeenth Century* (Oxford, 1954, R/1968).
**Della Corte, A.**, *Drammi per musica dal Rinuccini allo Zeno* (Turin, 1958).
**Pirrotta, N.**, *Li due Orfei* (Turin, 1969, 2/1978; trans. (Cambridge, 1982) as *Music and Theatre from Poliziano to Monteverdi*).
**Robinson, M.**, *Naples and Neapolitan Opera* (Oxford, 1972).
**Bianconi, L.**, and **Pestelli, G.**, *Storia dell'opera italiana* (Turin, 1987– ).
**Kimbell, D.**, *Italian Opera* (Cambridge, 1991).
**Rosand, E.**, *Opera in Seventeenth-Century Venice: The Creation of a Genre* (Berkeley, CA, 1991).

## Portugal

**Rebello, L.**, *História do teatro de revista em Portugal* (Lisbon, 1984).
**de Brito, M.**, *Opera in Portugal in the Eighteenth Century* (Cambridge, 1989).

## Russia

**Ginzburg, S.**, *Russkiy muzykalnyy teatr 1700–1833* (Leningrad, 1941).
**Cooper, M.**, *Russian Opera* (London, 1951).
**Gozenpud, A.**, *Muzykalnyy teatr v Rossii ot istokov do Glinki: ocherki* (Leningrad, 1959).
—— *Russkiy sovetskiy opernyy teatr (1917–1941)* (Leningrad, 1963).
—— *Russkiy opernyy teatr mezhdu dvukh revoliutsiy (1905–1917)* (Leningrad, 1975).
**Taruskin, R.**, *Opera and Drama in Russia as Preached and Practiced in the 1860s* (Ann Arbor, MI, 1981).

**Maximowitsch, M.**, *L'opéra russe: 1731–1935* (Lausanne, 1987).

*Spain*

**Cotarelo y Mori, E.**, *Historia de la zarzuela, ó sea el drama lírico* (Madrid, 1934).
**Chase, G.**, *The Music of Spain* (New York, 1941, 2/1959).
**Subirá, J.**, *Historia de la música teatral en España* (Barcelona, 1945).

*United States of America*

**Sonneck, O.**, *Early Opera in America* (New York, 1915, R/1963).
**Hipsher, E.**, *American Opera and its Composers* (Philadelphia, 1927, R/1978).
**Johnson, H.**, *Operas on American Subjects* (New York, 1964).
**Davis, R.**, *A History of Opera in the American West* (Englewood Cliffs, NJ, 1965).
**Virga, P.**, *The American Opera to 1790* (Ann Arbor, MI, 1982).

## GUIDEBOOKS

**Turnbull, R.**, *The Opera Gazetteer* (London, 1988).
**Stockdale, F.**, and **Dreyer, M.**, *The Opera Guide* (London, 1990).
**Couch, J.**, *The Opera Lover's Guide to Europe* (New York, 1991).
**Zietz, K.**, *Opera: The Guide to Western Europe's Greatest Houses* (Santa Fe, NM, 1991).
**Cowley, M.**, ed., *The Mentor Opera Handbook* (London, 1991).

# KEY TO VOCAL COMPASSES

c''' to c''''

c'' to b''

c' to b'

c to b

C to B

C, to B,

# ABBREVIATIONS

*Voice ranges*

| | |
|---|---|
| alt | alto |
| con | contralto |
| con cas | contralto castrato |
| counter-ten | counter-tenor |
| bar | baritone |
| bs | bass |
| bs-bar | bass-baritone |
| mez | mezzo-soprano |
| mezzo-sop cas | mezzo-soprano castrato |
| sop | soprano |
| sop cas | soprano castrato |
| ten | tenor |
| treb | treble |
| nar | narrator |

*Opera-Houses, Conservatories, Companies, and Venues*

| | | |
|---|---|---|
| Barcelona | L | Teatro Liceo |
| Berlin | D | Deutsche Oper |
| | H | Hofoper (Court Opera, i.e. Königliche Oper) |
| | K | Komische Oper |
| | S | Staatsoper |
| | Sch. | Schauspielhaus |
| | SO | Städtische Oper |
| | V | Volksoper |
| Bologna | C | Teatro Comunale |
| Brussels | M | Théâtre Royale de la Monnaie |
| Buenos Aires | C | Teatro Colón |
| Cardiff | WNO | Welsh National Opera |
| Dresden | H | Hofoper (Court Opera) |
| | S | Staatsoper (Semperoper) |
| Florence | P | Teatro alla Pergola |
| | C | Teatro Comunale |
| Genoa | CF | Teatro Carlo Fenice |
| Glyndebourne | Gly. | Glyndebourne |
| Hamburg | S | Staatsoper (Stadttheater) |
| Hanover | H | Hofoper (Court Opera) |
| London | BNOC | British National Opera Company |
| | C | Coliseum |
| | CG | Covent Garden |
| | COG | Chelsea Opera Group |
| | CR | Carl Rosa Opera Company |
| | DL | Drury Lane |
| | ENO | English National Opera |
| | EOG | English Opera Group |
| | GSM | Guildhall School of Music |
| | H | Haymarket 1705–14   : Queen's Theatre |
| | | Haymarket 1714–1837: King's Theatre |
| | | Haymarket 1837–      : see Her Majesty's |

| | | |
|---|---|---|
| London (*cont.*) | HM | Her (His) Majesty's |
| | J | St James's Theatre |
| | L | Lyceum (English Opera House) |
| | QEH | Queen Elizabeth Hall |
| | QH | Queen's Hall |
| | RAH | Royal Albert Hall |
| | RAM | Royal Academy of Music |
| | RCM | Royal College of Music |
| | RFH | Royal Festival Hall |
| | SW | Salder's Wells Theatre |
| | TCM | Trinity College of Music |
| Manchester | RMCM | Royal Manchester College of Music |
| | RNCM | Royal Northern College of Music |
| Milan | S | Teatro alla Scala |
| | L | Teatro Lirico |
| | N | Teatro Nuovo |
| | PS | Piccola Scala |
| | RD | Teatro Regio Ducal |
| | V | Teatro dal Verme |
| Moscow | B | Bolshoy Teatr |
| | Z | Zimin Teatr |
| Munich | N | Hof- und Nationaltheater |
| | P | Prinzregententheater |
| | S | Staatsoper |
| Naples | B | Teatro San Bartolomeo |
| | C | Teatro San Carlo |
| | N | Teatro Nuovo |
| New York | AM | Academy of Music |
| | CC | City Center |
| | CO | City Opera |
| | M or Met. | Metropolitan Opera House |
| | P | Palmo's Opera House |
| Paris | B | Opéra de la Bastille |
| | BP | Bouffes-Parisiens |
| | C | Théâtre du Châtelet |
| | CE | Théâtre des Champs-Elysées |
| | CI | Comédie-Italienne |
| | FP | Fantaisies-Parisiennes |
| | I | Théâtre Italienne |
| | L | Théâtre Lyrique |
| | O | Opéra (Académie Royale de Musique) |
| | OC | Opéra Comique |
| | SB | Théâtre Sarah Bernhardt |
| Philadelphia | AM | Academy of Music |
| Prague | C | Czech Theatre |
| | N | Národní Divadlo (National Theatre) |
| | P | Prozatímní Divadlo (Provisional Theatre) |
| Rome | Ad. | Teatro Adriano |
| | Ap. | Teatro Apollo |
| | Arg. | Teatro Argentina |
| | C | Teatro Costanzi |
| | R | Teatro Reale dell'Opera |
| | | (after 1946 Teatro dell'Opera) |

| St Petersburg | B | Bolshoy Teatr |
| | K | Kirov Teatr (when Leningrad) |
| | M | Maryinsky Teatr |
| Turin | R | Teatro Regio |
| Venice | B | Teatro San Benedetto |
| | C | Teatro San Cassiano |
| | F | Teatro La Fenice |
| | GG | Teatro San Giovanni Grisostomo |
| | GP | Teatro Santi Giovanni e Paolo |
| | L | Teatro Santa Lucia |
| | S | Teatro San Samuele |
| Vienna | B | Burgtheater |
| | H | Hofoper |
| | J | Theater in der Josephstadt |
| | K | Kärntnerthortheater (Theater nächst dem Kärntnerthor) |
| | L | Theater in der Leopoldstadt |
| | S | Staatsoper |
| | V | Volksoper |
| | W | Theater auf der Wieden (an der Wien) |

## Nationalities and Languages

| | | | |
|---|---|---|---|
| Belg. | Belgian | Hung. | Hungarian |
| Brit. | British | It. | Italian |
| Bulg. | Bulgarian | Lat. | Latin |
| Cz. | Czech | Nor. | Norwegian |
| Dan. | Danish | Pol. | Polish |
| Dut. | Dutch | Port. | Portuguese |
| Eng. | English | Rom. | Romanian |
| Finn. | Finnish | Russ. | Russian |
| Flem. | Flemish | Span. | Spanish |
| Fr. | French | Srb-Cr. | Serbo-Croatian |
| Ger. | German | Swed. | Swedish |
| Gr. | Greek | | |

## Miscellaneous

| | | | |
|---|---|---|---|
| abr. | abridged | Co. | Company |
| ad. | adapted | Coll. | College |
| AMZ | *Allgemeine musikalische Zeitung* | collab. | collaborated |
| | | comp. | composed |
| anon. | anonymous | compl. | compiled |
| arr. | arranged | cond. | conductor/conducted |
| Assoc. | Association | Cons. | Conservatory/Conservatoire |
| BBC | British Broadcasting Corporation | dim. | diminutive |
| | | dir. | director/directed |
| cap. | capacity | E | Eastern/East |
| Carn. | Carnival | ed(s). | edited |
| CBS | Columbia Broadcasting System | edn. | edition |
| | | F | French franc |
| cent(s). | centuries | facs. | facsimile |
| Fest. | Festival | pseud. | pseudonym |

| | | | |
|---|---|---|---|
| GDR | German Democratic Republic | post. | posthumous(ly) |
| gns. | guineas | pt | part (e.g. pt2) |
| H | Hall | pub. | published |
| incl. | including/inclusive | q.v. | *quod vide* (which see) |
| introd. | introduction/introduced | R | (in bibliographies) |
| *JAMS* | *Journal of the American Musicological Society* | | reprinted |
| | | (R) | has made operatic gramophone record |
| jun. | junior | recit. | recitative |
| L | Italian lire | rev. | revised/revived |
| lit. | literally | RAI | Radio Italiana |
| LPO | London Philharmonic Orchestra | RPO | Royal Philharmonic Orchestra |
| LSO | London Symphony Orchestra | S | San, Santa, Santo, or Santi (e.g. S Carlos T) |
| m | metres | S | Southern/South |
| MS | manuscript | sen. | senior |
| *MT* | *Musical Times* | Sig. | Signor |
| N | Northern | SO | State Opera (also Staatsoper) |
| Nat. | National | | |
| NBC | National Broadcasting Corporation | Soc. | Society |
| | | St | Saint |
| n.d. | no date | St. | Street |
| net. | network (e.g. net. TV) | T | Theatre (also for Teatro, Teatr, Theater, etc.) |
| NSW | New South Wales | | |
| NYPO | New York Philharmonic Orchestra | | |
| | | TH | Town Hall |
| O | Opera | trans. | translated |
| OC | Opera Company | TV | television |
| *OED* | *Oxford English Dictionary* | U | University |
| OH | Opera-House | UK | United Kingdom |
| Orch. | Orchestra | US | United States of America |
| orig. | originally | USAF | United States Air Force |
| OT | Opera Theatre | unfin. | unfinished |
| OUOC | Oxford University Opera Club | unprod. | unproduced |
| | | vol. | volume |
| perf. | performance | W | Western/West |
| pf. | pianoforte | * | see marked entry for further information |
| prod. | producer/produced | | |

# A

**Aachen** (Fr., Aix-la-Chapelle). City in North-Rhine Westphalia, Germany. Companies from Holland, later France and Italy, visited in the 18th cent., performing in the Komödienhaus auf dem Katschhof (1752, one of the first German civic theatres). A new theatre was built by Schinkel in 1825. The town has a reputation as a springboard for young talent; artists who worked there early in their careers include Leo Blech, Karajan, and Sawallisch. The present theatre (cap. 944) opened in 1951.

**Abbado, Claudio** (*b* Milan, 26 June 1933). Italian conductor. Studied Milan, Vienna with Swarowsky. Won Koussevitsky and Mitropoulos competitions (1958 and 1963). Milan, S, 1965, music director 1968–86; London, CG, 1968; New York, M, 1968. Salzburg since 1965, including first visit by a British orchestra (LSO) in 1973. Vienna, S, music director 1986–91. Although recognized as an especially fine conductor of Verdi, and Italian opera generally, Abbado's intelligent musicianship encompasses a wide stylistic range. He has conducted much contemporary opera, including works by Nono and Manzoni, with whose *Atomtod* he made his Salzburg début, and has conducted distinguished revivals in Vienna, including *Khovanshchina* (with Stravinsky's ending) and Schubert's *Fierrabras*. (R)

**Abbatini, Antonio Maria** (*b* Città di Castello, 1609 or 1610; *d* Città di Castello, 1677 or 1679). Italian composer and scholar. Held positions in Rome and Orvieto as maestro di cappella and composed a large quantity of sacred music. His first stage work, *Dal male il bene*, to a libretto by Rospigliosi based on Calderón, is a seminal work in the development of opera buffa. Composed for a wedding of the *Barberini family, it includes both comic arias and primitive versions of the *ensemble finale, while the role of the servant girl Marina is one of the earliest operatic instances of the *soubrette. Act II was provided by Marazzoli.

Abbatini's two later operas, *Ione* and *La comica del cielo*, were less influential.

WORKLIST: *Dal male il bene* (Rospigliosi, after Calderón; Rome 1653); *Ione* (Draghi; Vienna 1664); *La comica del cielo* (Rospigliosi; Rome 1668).

**Abbott, Emma** (*b* Chicago, 9 Dec. 1850; *d* Salt Lake City, 5 Jan. 1891). US soprano. Studied New York with Achille Errani and in Europe with Wartel, Sangiovanni, and Delle Sedie. Début London, CG, 1876 (Marie, *La fille du régiment*) and New York 1877 (same role). Married Eugene Wetherell 1875; together promoted Emma Abbott English Grand OC. Introduced 'specialities', such as popular ballads, into operas in which she appeared.

**Abencérages, Les, ou L'étendard de Grenade.** Opera in 3 acts by Cherubini; text by V. J. Étienne de Jouy, after J. P. Florian's novel *Gonzalve de Cordove* (1791). Prod. Paris, O, 6 Apr. 1813, in the presence of Napoleon, with Branchu, J. Armand, Nourrit, Dérivis, Lavigne, Alexandre, Bertin. Rev. Florence 1957.

The historical Abencerages were a family holding a prominent position in the Moorish kingdom of Granada in the 10th cent., their name deriving from Yusuf ben-Serragh, head of the tribe. The plot tells of the disputed triumphs of Almansor, the last of the Abencerage warriors, and his final overthrow at Granada in 1492. A Grand Opera in Spontini vein, it had little success, though Mendelssohn was to call it free, bold, and spirited.

**Abendspielleiter** (Ger.). Stage manager.

**Abgesang.** See BAR.

**Abigaille.** Nabucco's supposed daughter (sop) in Verdi's *Nabucco*.

**abonnement** (Fr.: 'subscription'). The term used in German and French opera-houses for the various subscription series, the financial mainstay of the season. The subscribers are known in Germany as *Abonnenten*, in France as *abonnées*, in Italy as *abbonati*.

**Abos, Girolamo** (*b* La Valetta, 16 Nov. 1715; *d* Naples, Oct. 1760). Maltese composer. Of Spanish descent, Abos studied in Naples, where he composed his first stage work, the opera buffa *Le due zingare simili*, in 1742. Composed 14 operas, including *La serva padrona*, *Artaserse*, *Alessandro nell'Indie*, and *Adriano in Siria*. *Tito Manlio* was performed in London in 1756, and thereafter Abos's music appeared sporadically in English pasticcios such as *The Maid of the Mill* (1765).

SELECT WORKLIST: *Le due zingare simili* (Palomba; Naples 1742); *La serva padrona* (Federico; Naples 1744); *Artaserse* (Metastasio; Venice 1746); *Alessandro nell'Indie* (Metastasio; Ancona 1747); *Adriano in Siria* (Metastasio; Rome 1750); *Tito Manlio* (?Roccaforte; Naples 1751).

**Abraham, Paul** (*b* Apatin, 2 Nov. 1892; *d* Hamburg, 9 May 1960). Hungarian composer. After study in Budapest became conductor of the Budapest Operetta T in 1927. First success with *Viktoria und ihr Husar*, which skilfully combines traditional Viennese operetta with elements of jazz. Further successes with *Die Blume von Hawai* and *Ball im Savoy*, which were soon performed throughout Germany. Forced to move initially to Vienna, and thence to Cuba, with the rise to power of Hitler, Abraham had little success in later life and died in an asylum. His works, which were once highly regarded as the successors to Lehár and Kálmán, have been neglected in recent years.

SELECT WORKLIST (all librettos by Grünwald & Löhner-Beda): *Viktoria und ihr Husar* (Leipzig 1930); *Die Blume von Hawai* (Leipzig 1931); *Ball im Savoy* (Berlin 1932).

**Ábrányi, Emil** (*b* Budapest, 22 Sep. 1882; *d* Budapest, 11 Feb. 1970). Hungarian composer, son of Emil (1851–1920), a poet, librettist, and translator of many librettos into Hungarian, including *Tristan* and *Carmen*. Emil jun. composed a number of operas, some to his father's texts. His works include *Monna Vanna* (1907), *Paolo és Francesca* (1907), *Don Quijote* (1917), and an opera on Bach, *A Tamás templom karnagya* (1947). Music director Cologne 1904–6, Hanover 1907–11, Budapest from 1911 in various theatres.

**Abreise, Die** (The Departure). Opera in 1 act by D'Albert; text by Ferdinand von Sporck, after August von Steigentesch's drama. Prod. Frankfurt 20 Oct. 1898; London, King's (Hammersmith), 3 Sep. 1925; Provo, UT, 30 Oct. 1973.

A simple anecdote about the efforts of Trott (ten) to encourage Gilfen (bar) to depart on a journey, thus leaving Luise (sop), Gilfen's neglected wife, free to receive his advances. Eventually it is Trott who is sent off, the jealousy aroused in Gilfen serving to renew his devotion to his wife.

D'Albert's most successful comic opera.

**'Abscheulicher!'** Leonore's (sop) aria in Act I of Beethoven's *Fidelio*, in which she first rages against her husband Florestan's imprisoner Don Pizarro, and then prays for a rescue.

**Abschrift** (Ger.). The MS copy of a musical score.

**Abu Hassan.** Opera in 1 act by Weber; text by Franz Carl Hiemer, after a tale in the *1,001 Nights* (added by Antoine Galland to the original collection; pub. 1712). Prod. Munich 4 June 1811; London, DL, trans. W. Dimond with music ad. T. S. Cooke, 4 Apr. 1825; Philadelphia, Chestnut St. T, 21 Nov. 1827.

Abu Hassan (ten) and his wife Fatime (sop) are being pursued by their creditors. Abu Hassan plans for them each to fake death so that the other can claim funeral money and a shroud from the Caliph (spoken). Omar (bs) agrees to settle their debts so as to further his pursuit of Fatime. She

returns with the money and Abu Hassan leaves to try his luck. During his absence, Omar woos Fatime. Seeing her husband return, Fatime hides Omar in a cupboard. She and Abu Hassan fake a jealous quarrel. One of the Caliph's men arrives to see if Abu Hassan was telling the truth about Fatime's death. She promptly pretends to be dead, and he departs, happy to have won a bet. The deception is repeated with Abu Hassan when another of the Caliph's men appears. On hearing the Caliph himself approach, both now feign death, but Abu Hassan miraculously comes to life when the Caliph offers a reward to whoever can explain the situation. The amused Caliph forgives them.

Weber's third completed opera, and his first real success.

Also opera by Hodemann (1790s).

**Abul Hassan.** In full, Abul Hassan Ali Ebn Bekar (bar), the eponymous Barber of Bagdad in Cornelius's opera.

**Académie Royale de Musique.** The name given in France, especially in the 17th cent., to the company with royal privilege to perform opera. First used 1672. See also OPÉRA.

**academy** (It., *accademia*). A type of institution popular in Italy from the Renaissance onwards devoting itself to promoting literature, science, and the arts. In essence gatherings of intellectuals and creative artists, most grew up around an individual, as well as providing a forum for the exchange of ideas, participating in publishing and disseminating the fruits of their discourse. A smaller academy was sometimes known as a *camerata: one such, based in Florence, was of crucial importance for the evolution of opera.

Many academies were involved with discussions of poetry and drama and hence, by extension, opera: almost all Italian 17th- and 18th-cent. librettists belonged to at least one academy. It was for a meeting of the **Accademia degli Invaghiti** in 1607 that Monteverdi's *Orfeo* was first performed. Most important of all was the **Accademia dell'Arcadi**: founded Rome 1692 and including Stampiglia and Metastasio, it had an important influence on the development of opera seria. Like many academies, the Accademia dell'Arcadi required its members to take names, 'Arcadian' in spirit: it espoused those virtues of pastoral simplicity and innocence which they believed to have existed in Arcadia. Of the many other academies which had some influence on the shaping of Italian opera, most important were: **Accademia della Crusca** (Rome), **Accademia degli Elevati** (founded Florence, 1607, by Gagliagno), **Accademia degli Filarmonici** (Verona), **Accademia degli Intonati** (Siena), **Accademia degli Intrepidi** (Ferrara), **Accademia**

**degli Olimpici** (Vicenza), **Accademia dei Pellegrini** (Florence).

**Academy of Music.** New York opera-house (cap. 4,600) on Irving Place and 14th St., opened 2 Oct. 1854 with Grisi and Mario in *Norma*. Succeeded the Astor Palace OH. Home of the Mapleson seasons and of the US débuts of Patti and Tietjens and US premières of many Verdi operas and other important works. Succeeded by the *Metropolitan 1883, but still used for opera until around 1900.

**accademia.** See ACADEMY.

**accento** (It.: 'accent'). The final accent in a line of the *verso. If it falls on the last syllable, it is called *tronco*, if on the penultimate syllable *piano*, if on the antepenultimate *sdrucciolo*.

**accompanied recitative.** See RECITATIVE.

**'Ach, ich fühl's'.** Pamina's (sop) aria in Act II of Mozart's *Die Zauberflöte*, lamenting Tamino's apparent indifference.

**Achilles.** In classical mythology, the greatest of the Greek warriors who laid siege to Troy. Metastasio's text *Achille in Sciro* was set by many composers, including Caldara (1736), Leo (1739), Jommelli (1749 and 1771), Sarti (1759 and 1779), Hasse (1759), Paisiello (1772), and Gazzaniga (1781). Other Achilles operas are by Lully (Act I of *Achille et Polyxène*, 1687), A. Scarlatti (*Achille e Deidamia*, 1698), Campra (*Achille et Déidamie*, 1735), Cimarosa (*Achille nell'assedio di Troia*, 1797), Paer (*Achille*, 1801), and F. E. Barbier (*Achille chez Chiron*, 1864). Also ballad opera by Gay (*Achilles in Petticoats*, 1733), and opera after Gay by Arne (1773). Tippett's *King Priam* (1962) includes a central role for Achilles (ten).

**Acis and Galatea.** Masque in 3 acts by Handel; text by John Gay, including adaptations of, or original work by, others including Pope and Dryden, after Ovid's *Metamorphoses*, xiii. 750–897. Probably written and performed at Cannons between 1718 and 1720; first complete public performance, London, Lincoln's Inn T, 26 Mar. 1731, with Rochetti, Wright, Leveridge; New York, Park T, 21 Nov. 1842.

Galatea (sop), alone among the nymphs and shepherds, is sad because Acis (ten) is absent. He returns, followed by the giant Polyphemus (bs), who also loves Galatea. Polyphemus crushes Acis under a rock, and Acis is transformed into a spring.

Other operas on the subject are by Lully (1686), Stölzel (1715), Blinzig (1716), Haydn (1762), Sépine (1786), Bianchi (1792), Naumann (1801), Cooke (1840), Hatton (1844), and Zarbo (1892).

**Ackermann, Otto** (*b* Bucharest, 18 Oct. 1909; *d* Wabern, 9 Mar. 1960). Swiss conductor. Studied Bucharest and Berlin. When 15 conducted Royal Romanian O on tour. Düsseldorf 1928–32; Brno 1932–5; Berne 1935–47. Guest at leading Italian opera-houses, Vienna, Paris, Brussels, Barcelona, 1946–53. Music director Cologne 1953–8; Zurich 1958–60. A fine Mozart conductor, also good in operetta. (R)

**Ackte** (orig. Achté), **Aïno** (*b* Helsinki, 23 Apr. 1876; *d* Nummela, 8 Aug. 1944). Finnish soprano. Daughter of Lorenz Nikolai Achté, baritone and conductor, and Emmy Achté (Strömer), soprano. Studied with her mother and then Duvernoy in Paris. Début Paris, O, 1897 (Marguerite); Paris until 1904; New York, M, 1904–5; London, CG, 1907. First British Salome, 1910. Had a voice of purity and power, and an excellent dramatic instinct and stage presence. Director of Finnish Nat. O 1938–9. Published two vols. of autobiography in 1925 and 1935. (R) Her sister **Irma Tervani** (1887–1936) was for many years leading mezzo-soprano, Dresden O. (R)

**act** (Fr., *acte*; Ger., *Aufzug*; It., *atto*). A formal subdivision of an opera, indicated as such by the composer, often further subdivided into scenes, or tableaux. Each act usually constitutes an integral dramatic whole, and often has a climax of its own. The origin of the act lay in the interventions of the chorus in Greek tragedy, which served to delineate portions of the drama. Following the plan of classical tragedy, the earliest operas (e.g. Peri, Monteverdi, Landi) were divided into five acts; this model also served for the first French operas (Lully, Rameau), which were written under the influence of Corneille and Racine. With the rise of Venetian opera in the mid-17th cent. (Cavalli) the number of acts in Italian opera contracted to three, and the progression of the drama was standardized: in the first act, the main lines of the plot were drawn; moving through the second, relationships between the characters were developed to the point of maximum confusion; in the third, the strands were untangled for a happy denouement.

The 3-act plan survived the reforms of Zeno and Metastasio and was usually employed for 18th-cent. opera seria, while that for opera buffa was more flexible, with 2-act librettos sometimes appearing. In France, the strong classical heritage meant almost unswerving loyalty to the 5-act plan, even when it had long been discarded in Italy: hence much *Grand Opera (Berlioz, Meyerbeer), even that provided by foreign composers such as Verdi (e.g. *Don Carlos*), was framed around an essentially old-fashioned structure. Opéra comique adopted the more modern

3-act plan, but in Germany much greater latitude was permitted, the rise of Singspiel in the 18th cent. paving the way for the use of anything between one and four acts. Wagner later established the 3-act scheme as a fundamental principle of music drama, but by this time a more flexible approach was generally being adopted. While many composers of the 20th cent. have continued to indicate formal act breaks, some have chosen to ignore such formal divisions, or to conceive quite different structures (e.g. Brecht in *Die *Dreigroschenoper*).

**action musicale.** See HANDLUNG FÜR MUSIK.

**act tune** (Fr., *entr'acte*, *divertissement*; Ger., *Zwischenspiel*). A term common in England in the late 17th and early 18th cents. to describe the interlude performed between the acts of a semi-opera, or spoken play containing incidental music. It was also used to refer to the music played before the curtain was raised (sometimes called the 'First Music' and 'Second Music'), which adumbrated the later operatic *overture. Sometimes also called the 'curtain tune' because it was performed while the curtain was down. Examples may be found in Purcell's *Dioclesian* and *The Indian Queen*.

**Adalgisa.** A Druid priestess, Norma's rival (sop), in Bellini's *Norma*.

**Adam, Adolphe** (*b* Paris, 24 July 1803; *d* Paris, 3 May 1856). French composer. Studied with Reicha and Boieldieu, who encouraged him in the direction of opera composition. His fluency as a composer of opéra-comique hindered his success with larger forms; but as he said himself, 'my only aim is to write music which is transparent, easy to understand, and amusing to the public'. He had some early successes, despite political difficulties that drove him briefly to London. On his return to Paris in 1832, he triumphed with *Le chalet* (which had achieved 1,000 performances at the Opéra-Comique by 1873). There followed a long list of works that won easy acceptance for the qualities Adam had said he sought. His most important opera is *Le postillon de Lonjumeau*, a lively, entertaining piece that offers a brilliant part to the central character but is also written with a vivid sense of character and situation. He also achieved a wide success with *Si j'étais roi*. In 1847, having been banned from the Opéra-Comique by the new director, Basset, he opened what was then Paris's third opera-house, the Opéra-National, but was bankrupted when it had to close in 1848. He was able to return to the Opéra-Comique in 1850.

SELECT WORKLIST (all first prod. Paris): *Le chalet* (Scribe & Mélesville; 1834); *Le postillon de Lonjumeau* (De Leuven & Brunswick; 1836); *Le roi d'Yve-*

*tot* (De Leuven & Brunswick; 1842); *Giralda* (Scribe; 1850); *Si j'étais roi* (Ennery & Brésil; 1852).

WRITINGS: *Souvenirs d'un musicien* (Paris, 1857).

BIBL: A. Pougin, *Adolphe Adam* (Paris, 1876).

**Adam, Theo** (*b* Dresden, 1 Aug. 1926). German bass-baritone. Sang in Dresden Kreuzchor, then studied with Dietrich. Début Dresden 1949 (Hermit, *Freischütz*). Bayreuth since 1952; Berlin, S, since 1953. New York, M, 1969–72. Sang his first Wotan in 1963, and has performed the role in many leading European opera-houses (incl. Bayreuth) and at the New York, M. Has also sung with success as Don Giovanni, Pizarro, Boris, and Wozzeck, and has produced opera in Berlin. A fine, powerful artist, with a well-placed voice and a warm, striking stage presence. (R)

**Adamberger, Valentin** (*b* Munich, 6 July 1743; *d* Vienna, 24 Aug. 1804). German tenor. Studied Munich with Valesi. Singer in the Bavarian Hofkapelle. Success in Italy 1776 (under name of Adamonti). An unofficial visit to London (1777–9) was also a success, but led to his dismissal from Munich. Vienna, German company 1779–89, Italian company 1789–93. Gluck rewrote Orestes for him in the 1781 revival of *Iphigénie en Tauride*. Mozart, a close friend, called him 'a singer of whom Germany can well be proud', and wrote for him Belmonte (he declared that 'O wie ängstlich' was 'expressly to suit Adamberger's voice'), Vogelsang (*Der Schauspieldirektor*), and several concert arias.

**Adami, Giuseppe** (*b* Verona, 4 Feb. 1878; *d* Milan, 12 Oct 1946). Italian playwright, librettist, and film writer. A successful light playwright, he met Puccini through Ricordi; and, having failed to interest him in the Quintero brothers' *Anima allegra* (set by Vittadini, Rome 1921), wrote for him first *La rondine* (1917), then *Il tabarro* (1918) and (with Simoni) *Turandot* (1926). Also wrote more librettos for Vittadini, and for Zandonai, among others. Published his correspondence with Puccini (Milan, 1928), and a sympathetic biography, *Il romanzo della vita di Giacomo Puccini* (Milan, 1935).

**Adams, John** (*b* Worcester, MA., 15 Feb. 1947). US composer and conductor. Studied Harvard with Del Tredici and Sessions. After the structuralism of his early works, he developed a minimalist manner, reliant on strict systems of rhythmic repetition but introducing continuous modulation, and created an easily accessible style.

*Nixon in China* (1987) breaks new ground in Western opera by its treatment of a contemporary episode. The work covers three days of President Nixon's visit to China in 1972, the first by a Western leader, each act dealing with one day.

Adams's choice of subject reveals his belief in opera as an art form of immediate relevance, seeing as 'the myths of our time . . . not Cupid and Psyche, but characters like Mao and Nixon'. Despite the techniques of musical repetition, he infuses his music with dramatic elements, writing lyrical vocal lines, scoring resourcefully, and varying the rate of harmonic change.

Another 'myth' to capture Adams's imagination was the murder of the elderly Klinghoffer during the terrorist hijacking of the cruise liner *Achille Lauro* in 1985. *The Death of Klinghoffer* is virtually staged oratorio, with dancer doubles for the main characters, but little action, and with large choruses punctuating the meditations and narrations of the leading players. The score, which includes electronics, moves away from narrow minimalism, but remains non-developmental and barely thematic. As with *Nixon*, Adams's collaboration with Peter Sellars is fundamental: both work towards opera as a new dramatic view of the present world rather than as an extension of traditional means.

WORKLIST: *Nixon in China* (Goodman; Houston 1987); *The Death of Klinghoffer* (Goodman; Brussels 1991).

**added number.** An aria or other number inserted in the text of an opera, usually for a special occasion with a distinguished guest singer. Mozart wrote several arias for insertion into other composers' works, such as the quartet 'Dite almeno' (K479) and the trio 'Mandina amabile' (K480) written in Nov. 1785 for insertion in Francesco Bianchi's *La villanella rapita* (1783). The practice was widespread in the 18th cent., when the *doctrine of affections meant that many arias expressing a certain emotion were interchangeable, and survived well into the 19th cent. Some singers would arrive with a favourite number in their luggage, hence the term aria di baule (suitcase aria). In the Lesson Scene in Act II of Rossini's *Barbiere*, the original 'Contro un cor' was often dislodged by the visiting prima donna in favour of a piece the better to flatter her talents. Many different arias, including some prophetically produced from the future, have been here laid under contribution. Melba would have a piano wheeled on to the stage and turn the scene into a ballad concert, ending with 'Home, Sweet Home'. The same song was used by Joan Sutherland in the Ball Scene of *Die Fledermaus*, an occasion appropriate, if not originally intended, for all manner of added numbers.

**'Addio del passato'.** Violetta's (sop) aria in Act III of Verdi's *La traviata*. After reading Germont's letter, she bids farewell to her happy past with Alfredo.

**'Addio fiorito asil'.** Pinkerton's (ten) aria in Act III of Puccini's *Madama Butterfly*, bidding farewell to the home where he has lived with Butterfly.

**'Addio senza rancor'.** Mimì's (sop) aria bidding farewell to Rodolfo in Act III of Puccini's *La bohème*.

**Addison, Joseph** (*b* Milston, 1 May 1672; *d* London, 17 June 1719). English essayist, poet, and man of letters. Some of the earliest musical journalism consisted of the reflections on the Italian opera he published in *The Spectator* (from 1711), and these influenced Mattheson in the publication of the first musical periodical, the *Critica Musica* (1722–3 and 1725). Addison's libretto *Rosamond* was set by Clayton (1707), Arne (1733—his first opera), and Arnold (1767).

**Adelaide.** City in South Australia. The first opera staged was *La muette de Portici* in 1840, followed by an Italian season in 1865 by the Bianchi Co., one of the touring companies upon which Adelaide was long dependent. There was also a local Intimate Opera Group. Performances were normally given in the Tivoli T, opened in 1913, reopened as Her Majesty's in 1962, and renovated in 1979: now known as the Opera T (cap. 1,009). In 1970 foundations were laid for a new Festival T on the south bank of Torrens Lake. The project was oversubscribed within a week, and the theatre (cap. 1,978) opened in 1973. The Festival Centre includes a smaller theatre (cap. 635), opened 1974, together with an experimental open-air amphitheatre, The Space (cap. 380). The New Opera South Australia, formed in 1974, pursued an adventurous policy under the musical direction of Myer Fredman (resigned 1979). The company is now the State Opera of South Australia, and has contributed some notable productions to the Adelaide Festival; its Youth Co. has premièred Graham Dudley's *The Snow Queen* (1985) and Alan John's *Frankie* (1987).

**Adele.** Eisenstein's maid (sop) in Johann Strauss's *Die Fledermaus*.

**Adelson e Salvini.** Opera in 3 acts by Bellini; text by Andreas Leone Tottola. Prod. Naples, T del Conservatorio di S Sebastiano, 12 Dec. 1825.

Set in Naples, the opera concerns Lord Adelson and the artist Salvini in their independent courtships of the beautiful Countess Fanny.

Bellini described this opera, his first, as 'a play alias a muddle'.

Also operas by Fioravanti (1816), Savi (1839).

**'Adieu, notre petite table'.** Manon's (sop) aria in Act II of Massenet's *Manon*, bidding farewell to the little table in the room she has shared with Des Grieux.

**Adina.** A wealthy landowner (sop) in Donizetti's *L'elisir d'amore*.

**Adler, Kurt Herbert** (*b* Vienna, 2 Apr. 1905; *d* San Francisco, 9 Feb. 1988). Austrian, later US, conductor and manager. Studied Vienna. Worked with Reinhardt; one of Toscanini's assistants, Salzburg 1936. Emigrated to USA 1938. Chorus-master and assistant conductor, Chicago 1938–42 and San Francisco 1943. Later also assistant to director, San Francisco O Association, Gaetano Merola; succeeded him 1953 as general director. Lengthened season from five weeks in 1953 to ten in 1972, and greatly enlarged the repertory and enterprisingly developed the range and importance of the company. (R)

**Adler, Peter Herman** (*b* Jablonec, 2 Dec. 1899; *d* Ridgefield, CT, 2 Oct. 1990). Czech, later US, conductor. Studied Prague with Zemlinsky. After appointments at Brno, Bremen, Kiev, and Prague, emigrated to USA 1939. Helped Fritz Busch to launch New OC, New York 1941. Director NBC TV O 1949, and did much to establish TV opera in the USA, commissioning Martinů's *The Marriage* and Menotti's *Amahl and the Night Visitors* and *Maria Golovin*; cond. latter Brussels World Fair 1958. Founded National Educational TV O (NET), giving first US *House of the Dead*, 1969, and commissioning Henze's *La cubana*. New York, M, 1972–3.

**Admeto, re di Tessaglia.** Opera in 3 acts by Handel; text an altered version by Haym or Rolli of an Italian libretto by Aurelio Aureli, *L'Antigone delusa da Alceste*. Prod. London, H, 31 Jan. 1727, with Bordoni and Cuzzoni. Rev. Abingdon, 1964.

**Admetus.** The King of Pherae, husband of Alcestis. Operas on Admetus are by Magni (1702), Handel (1727), and Guglielmi (1794). Appears in Gluck's *Alceste* as Admète. See also ALCESTIS.

**Adolar.** The Count of Nevers (ten), lover of Euryanthe, in Weber's *Euryanthe*.

**Adolfati, Andrea** (*b* Venice, 1721 or 1722; *d* Padua, 1760). Italian composer. A pupil of Galuppi, Adolfati held positions as maestro di cappella in Venice, Genoa (from 1748), and Padua (from 1760), and was briefly in the service of the court at Modena. He composed ten operas, of which seven are lost. A skilful if unimaginative exponent of the conventional mid-18th-cent. style of opera seria, Adolfati made six settings of Metastasian librettos, including *Artaserse*, *Didone abbandonata*, and *La clemenza di Tito*: Metastasio heard the latter and commented unfavourably.

SELECT WORKLIST: *Artaserse* (Metastasio; Verona 1741); *Didone abbandonata* (Metastasio; Venice 1747); *La clemenza di Tito* (Metastasio; Vienna 1753).

**Adorno, Theodor Wiesengrund** (*b* Frankfurt, 11 Sep. 1903; *d* Visp, 6 Aug. 1969). German critic and aesthetician. Studied Frankfurt with Sekles, Vienna with Berg. Edited the influential journal *Anbruch* 1928–31. After emigrating first to England, then to the USA, returned to Germany 1949 and became Director of the Institut für Sozialforschung in Frankfurt. Influenced by Hegel and Marx, but also by Freud, he in turn exercised a powerful influence on musical aesthetics, writing with a perceptive view on the role of the artistic individuality of, in particular, Schoenberg and Berg. Wagner, especially his relationship with his world and his social ideology, were discussed in a valuable study, *Versuch über Wagner*, in which the *Versuch* is not only 'essay' but 'quest' and even, more equivocally, 'attempt' in the sense of an attempt on his reputation, which ultimately survives.

WRITINGS: *Versuch über Wagner* (Berlin, 1952, trans. as *In Search of Wagner*, London, 1981); *Arnold Schoenberg* (Berlin, 1957); *Alban Berg* (Berlin, 1968).

**Adriana Lecouvreur.** Opera in 4 acts by Cilea; text by Arturo Colautti, from the drama *Adrienne Lecouvreur* (1849) by Scribe and Legouvé. Prod. Milan, L, 6 Nov. 1902, with A. Pandolfini, Caruso, De Luca, cond. Campanini; London, CG, 8 Nov. 1904, with Giachetti, Anselmi, Sammarco, cond. Campanini; New Orleans, French OH, 5 Jan. 1907, with Tarquini, Constantino, Fornari, cond. Conti.

The opera tells of the famous actress Adriana (Adrienne) Lecouvreur (1692–1730) (sop), of the Comédie-Française, rival of Princess Bouillon (mez) for the love of Maurice de Saxe (ten). She dies from inhaling the scent of a bunch of poisoned violets sent by the Princess. Michonnet (bar), the stage-manager of the theatre, is also in love with Adriana.

Other operas on the subject are by Vera (1856), Benvenuto (1857), and Perosio (1889).

**Adriano.** Steffano Colonna's son (mez) in Wagner's *Rienzi*.

**Aegisth.** Aegistheus (ten), lover of Klytemnestra, in Strauss's *Elektra* (1909).

**Aeneas.** In classical mythology, the Trojan hero who, escaping after the sack of Troy, led his men to Carthage. Here he became the lover of Queen Dido, but abandoned her to take his followers on to found Rome. The hero (ten) of Purcell's *Dido and Aeneas* (1689 or 1690) and Berlioz's *Les Troyens* (comp. 1856–8). Other operas on Aeneas are by Pallavicini (1675), Franck (1680), Fux (1731), Porpora (1734), Jommelli (1755), Traetta

(1760), Piccinni (1775), Righini (1793), Krauss (1799), and Topatyński (1912).

See also DIDO AND AENEAS.

**Aennchen.** Agathe's cousin (sop) in Weber's *Der Freischütz*.

**Affektenlehre.** See DOCTRINE OF THE AFFECTIONS.

**Africaine, L'.** Opera in 5 acts by Meyerbeer; text by Scribe. Prod. Paris, O, 28 Apr. 1865, with Marie Sass, Marie Battu, Naudin, Faure, cond. Haine; London, CG, 22 July 1865, with Lucca, Fioretti, Wachtel, Graziani, cond. Costa; New York, AM, 1 Dec. 1865, with Carozzi-Zucchi, Ortolani, Mazzoleni, Bellini, Antonucci, cond. Bergmann.

The opera tells how Vasco da Gama (ten) sails to find a new land beyond Africa, and is wrecked on the African coast. He returns to Portugal with two captives, Nelusko (bar) and Selika (sop) ('l'Africaine' of the title), with whom he has fallen in love. She finally sacrifices her life so that Vasco can marry his former love, Inez (sop).

Enormously popular in the 19th cent., the opera had nearly 60 performances during its first four seasons. A zarzuela by Manuel Fernandez Caballero, *El duo de la Africana* (1893), derives from the opera.

**afterpiece.** A theatrical genre popular in England in the 18th and early 19th cents.: in essence a short opera or pantomime which was given after the performance of a play or other major dramatic work.

**Agathe.** The daughter (sop) of Kuno, the Prince's head forester, and lover of Max in Weber's *Der Freischütz*.

**Agazzari, Agostino** (*b* Siena, 2 Dec. 1578; *d* Siena, ?10 Apr. 1640). Italian composer, organist, and writer on music. Little is known of Agazzari's early life, although he was possibly a pupil of Viadana. In 1602 he became maestro di cappella at the Collegio Germanico in Rome, moving to the Seminario Romano 1606. Returned to Siena 1607, and was briefly organist at the cathedral there. A prolific composer of church music and madrigals, he also made an important contribution to the early history of opera and the development of *stile rappresentativo, though his achievements have been overshadowed by those of Peri, Caccini, Cavalieri, and Monteverdi.

In 1606 Agazzari was asked to produce, at one month's notice, a theatrical entertainment for the Seminario Romano during Carnival. The resultant work, *Eumelio*, was written in only 15 days, using a libretto by Torquato de Cupis and Tirletti, two priests at the Seminario. Although called a dramma pastorale, it is effectively a sacred opera, cast in a prologue and three acts. As befits a work for a Jesuit seminary, it presents an instructive tale, in which Christian and pagan elements combine. The virtuous Eumelio is lured away from his idyllic pastoral pleasures by demons disguised as vices. Although taken to the Underworld, he is rescued by Apollo and Mercury, who return him to his former state of Arcadian bliss.

The similarity of this story to the Orpheus myth, popular in so many early operas, is obvious. For most of the work Agazzari relies upon stile rappresentativo as the main musical vehicle, but this is interspersed with choruses for the demons who transport Eumelio to the Underworld, and for the shepherds among whom he frolics. This contrast of solo and choral writing presages Monteverdi's *Orfeo* (1607), where it is used to even greater effect, as is the technique of setting slightly varied vocal strophes over a repeated bass pattern, which appears here for the first time also.

*Eumelio* is an important milestone in the early history of opera. Although (in conveying its moralizing message) it lacks some of the dramatic impetus of Monteverdi, or even Peri, it is a worthy successor to the model of Cavalieri's *Rappresentazione di anima e di corpo*, and with it represents the beginnings of Roman opera. Shortly after its composition Agazzari's treatise *Del sonare sopra 'l basso con tutti li stromenti e dell'uso loro nel conserto* (Siena, 1607; facs. edn. Milan, 1933) appeared, in which the ideals of the earliest composers of monody are cogently summarized and discussed.

WORKLIST: *Eumelio* (De Cupis & Tirletti; Rome 1606).

**Agnesi, Maria Theresa** (*b* Milan, 17 Oct. 1720; *d* Milan, 19 Jan. 1795). Italian composer, harpsichordist, singer, and librettist. One of the first female opera composers; three of her works were performed in Milan in the mid-18th cent., including *Ciro in Armenia*, for which she provided her own libretto.

SELECT WORKLIST: *Ciro in Armenia* (Agnesi; Milan 1753).

BIBL: L. Anzoletti, *Maria Gaetana Agnesi* (Milan, 1900).

**Agnes von Hohenstaufen.** Opera in 3 acts by Spontini; text by Ernst Raupach. Prod. Berlin 12 June 1829 (Act I had already been given on 28 May 1827). First 20th-cent. revival Florence 14 May 1954, with Tebaldi, cond. Serafin.

Set in Mainz, the plot tells of the love of Agnes (sop), daughter of the Countess Ermengard, for Henry of Brunswick (ten), son of the rebel Duke of Saxony, and describes the political intrigues by the Emperor Henry VI of Hohenstaufen (bs) and the French King (disguised as the Duke of Burgundy) (bar), to prevent Agnes and Henry of Brunswick marrying.

Spontini's last opera.

**Agostini, Pietro Simone** (*b* Forlì, *c*.1635; *d* Parma, 1 Oct. 1680). Italian composer. After studying in Ferrara, Agostini embarked first on a military career, but later attempted to gain a church appointment. A notoriously rakish character, his immoral conduct precluded him from serious consideration by the ecclesiastical authorities. He was forced to flee from Genoa after seducing a nun and only received his first appointment as maestro di cappella one year before his death, in Parma.

He composed seven operas, of which three survive: *Ippolita, reina delle Amazoni*, *Gl' inganni innocenti*, and *Il ratto delle Sabine*. All are representative of the Venetian style of Cavalli, which at this time was at the height of its popularity.

SELECT WORKLIST: *Ippolita, reina delle Amazoni* (Maggi; Milan 1670); *Gl' inganni innocenti* (Apolloni; Rome 1673, rev. Bologna 1675, rev. Milan 1679); *Il ratto delle Sabine* (Bussani; Venice 1680).

**Agricola, Johann Friedrich** (*b* Dobitschen, 4 Jan. 1720; *d* Berlin, 2 Dec. 1774). German composer, conductor, organist, and musicographer. Studied Leipzig with Bach and Berlin with Quantz. Appointed to the Berlin court by Frederick the Great in 1751. Here he wrote *c*.11 operas: those which survive, including *Cleofide* and *Achille in Sciro*, show him to be a generally uninspired exponent of mid-18th-cent. opera seria who rarely rises above the conventional. Much of his operatic career was hampered by Frederick the Great, who took against him after he married one of the court singers, Emilia Molteni: according to court protocol, singers were expected to remain single and both Molteni and Agricola had their salaries reduced as punishment. In 1759, following Graun's death, he was placed in overall charge of operatic productions.

Agricola made a more lasting contribution to operatic history through his writings. These included pamphlets on musical taste, contributions to Nicolai's *Allgemeine deutsche Bibliothek*, and, most importantly, a glossed translation of ★Tosi's *Opinioni*, recognized today as one of the most important documents for a study of contemporary vocal performing practice.

SELECT WORKLIST: *Cleofide* (Metastasio; Berlin 1754); *Achille in Sciro* (Metastasio; Berlin 1765).

WRITINGS: *Anleitung zur Singekunst* (Berlin, 1757: an annotated trans. of Tosi's *Opinioni*, 1723; facs. edn. Celle, 1966).

**Agrippina.** Opera in three acts by Handel; text by Vincenzo Grimani. Prod. Venice, GG, 26 Dec. 1709, with Carli, Durastanti, Valeriano, Scarabelli, Vanini-Boschi, Pasini. One of Handel's few operas produced in Italy during his three-year stay. One of the original librettos which Handel set. Also given in Naples (1713), Hamburg (1718–22), Vienna (1719); the aria 'Ho un non so che nel cor', sung by Vanini-Boschi in Scarlatti's *Pirro e Demetrio* was the first music by Handel heard in London (6 Dec. 1710). Rev. Halle 1943 with Dertil, cond. R. Kraus.

**Aguiari, Lucrezia** (*b* Ferrara, 1743; *d* Parma, 18 May 1783). Italian soprano. Also known, from her illegitimacy, as La Bastardella or La Bastardina. Studied Ferrara with Petrucci and Lambertini. Début Florence 1764. Court singer Parma 1768, when also created Teti in Paisiello's *Peleo e Teti* in Naples. She was engaged for the Pantheon, London, in 1775 and 1776 to sing two songs at £100 a night. In 1780 married Giuseppe Colla, having created leading roles in several of his operas, and retired. Her voice, which had a range of three and a half octaves, was much praised for its quality; she was also an excellent actress. Mozart heard her sing in Parma in 1770 and said that she had 'a lovely voice, a flexible throat and an incredibly high range', quoting a passage he heard her sing that went up to c''''.

**Ägyptische Helena, Die** (The Egyptian Helen). Opera in 2 acts by Richard Strauss; text by Hugo von Hofmannsthal, after various classical legends. Prod. Dresden 6 June 1928, with Rethberg, Rajdl, Taucher, cond. Busch; New York, M, 6 Nov. 1928, with Jeritza, Laubenthal, cond. Bodanzky.

The sorceress Aithra (sop), daughter of the Egyptian King, learns that Menelas (ten) plans to murder his wife Helena (sop), who was the cause of the Trojan war. Aithra wrecks the ship in which they are sailing, gives Helena a potion which erases her memories of past evils, and tells Menelas that Helena has been asleep in Egypt while he and the Greeks were at the Trojan war. Under Aithra's charms, Menelas and Helena journey to a place where the war is unknown. Menelas is invited to hunt with two Arab chiefs, Altair (bar) and his son Da-Ud (ten), who both admire Helena. Provoked by jealousy, Menelas kills Da-Ud. Helena discovers an antidote to Aithra's potion. She and Menelas drink it and, in full awareness of themselves, the couple are happily reunited.

Originally conceived as an operetta. Rev. (abr.) for the 1933 Salzburg Festival.

**'Ah! Belinda'.** Dido's (sop) aria in Act I of Purcell's *Dido and Aeneas*, in which she confesses her anguish over her love for Aeneas.

**'Ah! fors' è lui'.** Violetta's (sop) aria in Act I of Verdi's *La traviata*, in which she asks herself if she is really falling in love (then rejecting the idea in 'Sempre libera').

**'Ah, fuggi il traditor'.** Donna Elvira's (sop) aria in Act I of Mozart's *Don Giovanni*, in which she urges Zerlina to flee from Giovanni's advances.

**'Ah, fuyez, douce image'.** Des Grieux's (ten) outburst in Saint-Sulpice in Act III of Massenet's *Manon*, when he tries in vain to drive the image of Manon from his mind.

**'Ah! non credea mirarti'.** Amina's (sop) sleep-walking aria in Act III of Bellini's *La sonnambula*, which leads to the final cabaletta, 'Ah! non giunge!' as she awakens and sees her beloved Elvino beside her.

**'Ah sì, ben mio'.** Manrico's (ten) aria to his beloved Leonora, consoling her as they shelter in a fortress in Act III of Verdi's *Il trovatore*.

**'Ah taci, ingiusto core'.** The trio at the beginning of Act II of Mozart's *Don Giovanni*, in which Don Giovanni (bar) pretends to woo Donna Elvira (sop), while Leporello (bar) mocks the situation.

**Aida.** Opera in 4 acts by Verdi; text by Ghislanzoni from the French prose of Camille du Locle (1868), plot by August Mariette Bey. Prod. Cairo, OH, 24 Dec. 1871, with Pozzoni, Grossi, Mongini, Medini, Costa, Steller, cond. Bottesini; Milan, S, 8 Feb. 1872, with Stolz, Waldmann, Fancelli, Pandolfini, Maini, cond. Faccio; New York, AM, 26 Nov. 1873, with Torriani, Cary, I. Campanini, Maurel; London, CG, 22 June 1876, with Patti, S. Scalchi, Nicolini, Cotogni, Bagaggiolo, cond. Bevignani.

*Aida* was not, as is often supposed, written for the opening of the Suez Canal (1869), nor commissioned by the Khedive of Egypt to open the new *Cairo OH the same year; Mariette did suggest his story to the Khedive to celebrate the opening of the Canal, but the synopsis did not reach Verdi until 1870.

Egypt, in the age of the Pharaohs. Radamès (ten), commander of the Egyptian army, is in love with Aida, the Ethiopian slave girl of Princess Amneris (mez), daughter of the King of Egypt (bs). In a ceremony in the temple, Ramfis the High Priest (bs) consecrates Radamès's armour.

Amneris provokes Aida into admitting her love for Radamès. Knowing that Amneris also loves him, Aida offers to relinquish him in order to placate her. A great triumphal scene celebrates the return of the victorious Egyptian army. As his reward, Radamès asks that the Ethiopian prisoners be released. All are freed except Amonasro (bar), who has divulged that he is Aida's father, but not that he is King of Ethiopia. Radamès is offered Amneris's hand in marriage.

On the banks of the Nile, Aida is persuaded by her father to obtain from Radamès details of the army's proposed route on its next attack. The three are about to escape when Amneris and Ramfis appear crying 'traitor'. In order to allow Aida and her father to escape, Radamès gives himself up.

Amneris tells Radamès she will see that his life is spared if he will swear to forget Aida. He refuses and is sentenced to be entombed alive. As the stones of the tomb close above him, he laments that he will never see Aida again. But she appears, having hidden herself in the tomb so as to die with him. In the temple above, Amneris prays for her own peace of mind.

Subject of a parody by Sudessi & Zorzi (1881).

BIBL: H. Busch, ed., *Verdi's Aida: The History of an Opera in Letters and Documents* (Minnesota, 1978).

**Aiglon, L'** (The Eaglet). Opera in 5 acts by Honegger (Acts II–IV) and Ibert (Acts I & V); text by Henri Cain after Edmond Rostand's drama (1900). Prod. Monte Carlo 11 Mar. 1937, with Fanny Heldy and Vanni-Marcoux. Announced for production in Naples in Feb. 1939 but cancelled on Mussolini's orders after the dress rehearsal. Rev. Paris, O, 1952.

The Duke of Reichstadt (sop), son of Napoleon, remains loyal to the French cause, although held captive by the Austrian Prince Metternich (bar). On the day of a grand ball, the Duke escapes with his loyal footman Flambeau (bs). Arriving at the site of the battle of Wagram, the Duke becomes so deeply lost in thought that he is easily recaptured. Having taken farewell of his mother, he dies of consumption, hearing as his final sounds French songs.

**'Ai nostri monti'.** The duet for Manrico (ten) and Azucena (mez) in Act IV of Verdi's *Il trovatore*, in which they console one another with memories of their homeland.

**air.** See ARIA.

**'Air des bijoux'.** See 'JEWEL SONG'.

**Aix-en-Provence.** City in Bouches-du-Rhône, France. At the annual summer festival founded 1948, now the most important in France, opera is mostly staged in the open-air theatre of the Archbishop's palace (cap. 1,639, rebuilt 1975 and 1985), occasionally at Les Baux (*Mireille*, 1954), the Parc du Tholonet (*Carmen*, 1957), and Arles arena (*Médée*, 1976). The festival's repertory was at first largely classical, with emphasis on Mozart, and has introduced to a wider public Sciutti, Berganza, Berbié, Stich-Randall, Éda-Pierre, Troyanos, and Simoneau early in their careers. After a decline in standards in the 1960s, the repertory was extended to include more Romantic (especially Rossini, with Caballé) and modern works under Bernard Lefort, 1973–80, and standards recovered; his interest in English singers led to early appearances by Masterson, Langridge, and Robert Lloyd. Under Louis Erlo (from 1982), tradition has been preserved but extended with an interest in period instruments, in the lesser-known Mozart, and the exploration of Campra (born in Aix) and Rameau. English conductors have included Mackerras and Gardiner.

**Ajo nell'imbarazzo, L'** (The Embarrassed Tutor). Opera in 2 acts by Donizetti; text by Jacopo Ferretti, based on Giovanni Giraud's comedy. Prod. Rome, T Valle, 4 Feb. 1824; London, HM, 28 July 1846.

Don Giulio (bar) wants his sons Enrico (ten) and Pipetto (bs) to be brought up in total ignorance of women until the age of 25. However, Enrico has already married Gilda (sop), and they have a son. When the old housekeeper Leonarda (mez) discovers Gilda in the room of Don Gregorio (bs), the boys' tutor, she assumes her to be his mistress. The confusion is resolved and Pipetto's education is entrusted to Enrico.

Other operas on the subject are by Guarnacci (1810), Pilotti (1811 or 1818), Celli (1813), G. Mosca (1815), and Nicelli (1825).

**Aladdin.** (1) Fairy opera in 3 acts by Bishop; text by G. Soane, based on the oriental tale *Aladdin and his Wonderful Lamp*. Prod. London, DL, 29 Apr. 1826 as a counter-attraction to Weber's *Oberon* at Covent Garden. A failure, it had to be withdrawn. Weber was present at the first performance and noted a charming hunting chorus; but the audience greeted it by whistling the *Freischütz* huntsmen's chorus. Bishop's major opera, though containing spoken dialogue.

(2) Opera by Atterberg. Prod. Stockholm 18 Mar. 1941.

Other operas on the subject are by Gyrowetz (1819), Isouard (1822), Guhr (1830), Claudius (c.1840), Wichtl (c.1840), Hornemann (1888), Gibbons (1892), and Rota (1968).

**'A la faveur de cette nuit obscure'.** The trio in Act II of Rossini's *Le comte Ory*, in which the Count (ten) believes he may successfully woo the reluctant Countess Adèle (sop), cynically observed by the page Isolier (mez).

**Albanese, Licia** (*b* Bari, 22 July 1913). Italian, now US, soprano. Studied with Baldassare-Tedeschi. Début Milan, L, 1934 (Butterfly). London, CG, 1937 (Liù and Nannetta); New York, M, 1940–66. Broadcast and recorded Mimì and Violetta with Toscanini. A singer of great feeling. (R)

**Albani,** (Dame) **Emma** (*b* Chambly, Montreal, 1 Nov. 1847; *d* London, 3 Apr. 1930). Canadian soprano. Studied Paris with Duprez, Milan with Lamperti. Début Messina 1870 (Amina). London, CG, 1872–96; New York, M, 1890; also Paris, Germany, Australia, etc. She was a famous Elsa and Elisabeth (one of the few of her time to sing these roles in German), and her repertory encompassed Gilda, Lucia, Violetta, Valentine, and Isolde. A beautiful woman, she possessed a rich voice that lasted unimpaired until her retirement in 1911. (R)

BIBL: C. Macdonald, *Emma Albani* (Toronto, 1985).

**Albania.** During the Turkish occupation (1506–1912) Albania was unable to develop an operatic life, and it was not until the Communist Party took power in 1944 that native arts were encouraged. In 1949 all performers were grouped into the Albanian Philharmonia, and from this developed the Theatre of Opera and Ballet in 1953. The first full-length opera performed was Dargomyzhsky's *Rusalka*, and this was followed by popular works from the international repertory, among them *La traviata* and *The Bartered Bride*. At first the company was housed in a building which is now the Higher Institute of Arts (Conservatoire), but in 1964 it moved to the newly constructed Palace of Culture in Tirana (cap. 1,000).

Although Martin Gjoka (1890–1940) began work on a melodrama *Juda Makabej* in 1919, it remained unfinished, and the first true Albanian opera is *Mrika* (1958) by Prenkë Jakova (1917–69), on the building of a power-station. He followed this socialist-realist work with *Gjergj Kastrioti-Skënderbeu* (1968), an epic on the 15th-cent. national hero who in a series of battles held off the Turks. The first Albanian operetta was *Agimi* (Dawn, 1954) by Kristo Kono (1907–91), who also wrote *Lulja e Kujtimit* (The Flower of Remembrance, 1961). The 1960s saw the growth of opera, with a number of composers writing works that were of necessity usually based on patriotic and propagandist themes, drawing extensively on folk music. Tish Daija (*b* 1926) has written *Pranvera* (Spring, 1960), and *Vjosa* (1980) and Vangjo Nova (*b* 1927) *Heroina* (1967). *Zgjimi* (Awakening, 1976) by Tonin Harapi (*b* 1928) gives an unusual twist to an old operatic device by including a letter song as good news and political tracts arrive from Party headquarters. Avni Mula (*b* 1928) has written *Borana* (1984) and the patriotic *Nëna e trimave* (Mother of Courageous Sons, 1987), and Nikolla Zoraqi (*b* 1929) the propagandist *Komisari* (The Commissar, 1975) and *Paja* (1989). Also of significance is Pjeter Gaci (*b* 1939), with *Përtej mjëgulles* (Beyond the Vine, 1971) and *Toka jonë* (Our Land, 1981).

**Albéniz, Isaac** (*b* Camprodón, 29 May 1860; *d* Cambô-les-Bains, 18 May 1909). Spanish composer. After an erratic early life as a child-prodigy pianist, studied Leipzig with Jadassohn and Reinecke. Though known principally as a pianist-composer, his pioneering work for Spanish music included the composition of operas and zarzuelas. In 1891 he signed a contract with the banker Francis Burdett Money-Coutts to set the latter's librettos, but the only successful work to result was *Pepita Jiménez* (1896). He had meanwhile produced *The Magic Opal* (London 1893). The ambitious trilogy *King Arthur* (Money-Coutts, after Malory: *Merlin*, *Lancelot*, and *Guinevere*) was never performed. Of his zarzuelas, *San Antonio de la Flórida* (1894) had some success.

BIBL: G. Laplane, *Albéniz: sa vie, son œuvre* (Paris, 1956).

**Alberich.** The Nibelung (bs-bar) of the title of Wagner's *Der Ring des Nibelungen*, whose renunciation of love and theft of the Rhinegold helps to precipitate the events of the whole cycle. His name derives from the Old High German and Middle High German *alp* or *alb* = elf, and Gothic *reiks* = ruler, also the source of the name *Oberon. In *The Ring* he is the dark embodiment of this magic authority: in the answers to Mime's riddles in *Siegfried*, Wotan describes the Nibelungs as Schwarzalben and their ruler as Schwarz-Alberich, the gods as Lichtalben and their ruler, himself, as Licht-Alberich.

**Albert, Eugen d'** (*b* Glasgow, 10 Apr. 1864; *d* Riga, 3 Mar. 1932). German pianist and composer. Largely self-taught. He had early successes as a pianist, and was much encouraged and influenced by Liszt. He was also active as an opera composer in Germany from 1893, though his first real success came with his fourth opera, *Die Abreise*. This light, cheerful romantic comedy won popularity for its direct, tuneful idiom barely touched by the musical tensions of the day. He capped this with his best-known opera, *Tiefland*, a work of distinct power which owes something to verismo but also appeals to a nostalgia for the 'clean' world of the mountains in contrast with the corrupt life of the 'lowlands' of the title. *Flauto solo*, a light opera about Frederick the Great, was briefly popular; and he had a certain success with another work in highly charged verismo manner, *Die toten Augen*.

SELECT WORKLIST: *Die Abreise* (The Departure) (Sporck, after Steigentesch; Frankfurt 1898); *Tiefland* (The Lowlands) (Lothar, after Guimeras; Prague 1903); *Flauto solo* (Wolzogen; Prague 1905); *Die toten Augen* (The Blind Eyes) (Ewers; Dresden 1916).

BIBL: W. Raupp, *Eugen d'Albert* (Leipzig, 1930).

**Albert Herring.** Opera in 3 acts by Britten; text by Eric Crozier, after Maupassant's story *Le rosier de Madame Husson* (1888). Prod. Gly. 20 June 1947, with Pears, Cross, cond. Britten; Tanglewood, Lenox, MA, 8 Aug. 1949 with Lloyd, Pease, Faull, cond. Goldovsky.

The plot turns on the attempt of a small Suffolk town to elect a May Queen, and, in default of one of sufficient virtue being found, the choice of Albert Herring (ten) as May King: he dismays the community by spending his prize money on a debauch, which nevertheless serves to liberate him from his domineering mother. The Suffolk types include the formidable Lady Billows (sop).

**Albinoni, Tomaso** (*b* Venice, 8 June 1671; *d* Venice, 17 Jan. 1750). Italian composer. The son of a wealthy paper merchant, Albinoni remained a dilettante all his life: his musical activities were not dictated by any patron or employer. Though he is chiefly remembered today for his instrumental music, there is evidence that he wrote about 50 theatrical works, many to librettos by Silvani, Pariati, and Zeno.

Albinoni's earliest operas were all written for Venice, beginning with *Zenobia, regina de' Palmireni*. His *Didone abbandonata* was probably the first setting of a Metastasian libretto by a Venetian composer. The revival of *L'inganno innocente* in Naples in 1702 as *Rodrigo in Algeri* launched him on successful career outside his native city. A setting of Zeno's *Griselda* for Florence in 1703 was his first truly non-Venetian work; other cities where his operas were staged included Genoa, Bologna, and Ferrara. His fame also spread outside Italy and his works were performed, usually in adaptations or pasticcios, in several European cities, including Hamburg and London. Probably the most significant appearance of his work outside Italy came in 1722, when he was commissioned to write *I veri amici* for the Munich court, to celebrate the wedding of the Elector of Bavaria.

Albinoni himself claimed to have composed over 80 operas, but only two survive complete in score: *Il tiranno eroe* and *La statira*. It is thus impossible to make any general comments about his style and approach to operatic composition, although his skill as an instrumental composer was clearly employed to beneficial effect.

SELECT WORKLIST: *Zenobia, regina de' Palmireni* (Marchi; Venice 1694); *Didone abbandonata* (Metastasio; Venice 1725); *L'inganno innocente* (Sitani; Venice 1701, rev. Naples 1702 as *Rodrigo in Algeri*); *Griselda* (Zeno; Florence 1703); *Il tiranno eroe* (Cassani; Venice 1711); *I veri amici* (Silvani & Lalli; Munich 1722); *La statira* (Zeno & Pariati; Rome 1726).

BIBL: Michael Talbot, *Tomaso Albinoni* (Oxford, 1990).

**Alboni, Marietta** (orig. Maria Anna Marzia) (*b* Città di Castello, 6 Mar. 1823; *d* Ville d'Avray, 23 June 1894). Italian contralto. Studied Bologna with Mombelli. Taught (uniquely) by Rossini, studying his contralto roles. Début Bologna 1842 (Climene, Pacini's *Saffo*). Successes throughout Italy, then in N Europe; London, CG, 1847, in its first season as Royal It. O, making début on opening night as Arsace. With her rich voice and virtuosity, she became at CG a rival attraction to Jenny Lind at HM, and her salary was quadrupled to £2,000. Meyerbeer composed Urbain's aria in *Les Huguenots* at CG for her in 1848, when she also sang the baritone role of Carlos, rejected by Tamburini and Ronconi, in *Ernani*. Continued to appear in London until 1858. Toured France 1850; USA 1853; Paris 1852 and 1857. Officially retired 1863, after she married Count Pepoli, though made further appearances until 1872. One of operatic history's great contraltos, famous for her impeccable singing. When she became obese, Rossini called her 'the elephant that swallowed a nightingale'. She sang at his funeral.

BIBL: A. Pougin, *Marietta Alboni* (Paris, 1912).

**Alceste.** Opera in 3 acts by Gluck; text by Calzabigi, after the tragedy by Euripides (438 BC). Prod. Vienna, B, 26 Dec. 1767, with Antonia Bernasconi, Tibaldi, Poggi; London 30 Apr. 1795, with Giorgi-Banti, Killy, Baghetti; Wellesley Coll., MA, 11 Mar. 1938. French version revised by Gluck, with text by L. du Roullet, prod. Paris, O, 23 Apr. 1776, with Levasseur, Legros, Gelin.

A few years after the Trojan war. The people of Thessaly call on the gods to cure their king, Admète (ten), who is dangerously ill. His wife Alceste (sop) offers sacrifices to Apollon. As the High Priest invokes the god in his temple, the earth shakes and an oracle is heard declaring that Admète will die unless someone is sacrificed in his place. Alceste is torn between love for her children and her husband but finally offers herself.

Admète, now recovered, much to the joy of the people, is told that a stranger has promised to sacrifice her life for his. He discovers from his wife that it is she, and vows to die with her. The people implore her to change her mind, but she resolves to keep her promise to the gods.

When Hercule (bar) learns Alceste's fate, he determines to save her. Alceste reaches the entrance to Hades but, since she must wait till night before she can enter, is driven back. She is joined by Admète, who refuses to be separated from her in death. However, only one of the couple may enter Hades; Alceste and Admète vow to die for each other. Hercule then appears

and, pointing to the couple as a model of conjugal love, declares that they shall both live.

The preface to the score, one of the key documents in the history of opera, embodies Gluck's view that opera should be not merely an elegant concert in costume, but a form of music drama (text often reprinted, e.g. Eng. trans. in A. Einstein, *Gluck* (1936) and O. Strunk, *Source Readings in Music History* (1950) ). For other settings of the story see below.

**Alcestis.** In Greek mythology, the wife of *Admetus, King of Pherae. Admetus was permitted by Apollo not to die if someone would agree to take his place, and this Alcestis did. She was rescued from the grasp of Death by Hercules. Operas on the subject, mostly deriving from Euripides' drama (438 BC), are by Lully (1674), Trenta (1679), Franck (1680), Strungk (1693), Draghi (1699), Schürmann (1719), Lampugnani (1744), Gluck (1767), Guglielmi (1768), Schweitzer (1773), D'Ordonez (1778), Floquet (comp. 1784), F. Benda (1786), Wolf (1786), Gresnick (1786), Portugal (1798), W. Müller (parody, *Die neue Alceste*, 1806), Appiani (1811), Staffa (1852), Gambaro (1882), Boughton (1922), and Wellesz (1924).

**Alcina.** Opera in 3 acts by Handel; text by A. Fanzaglia, after Ariosto's *Orlando furioso* (1516). Prod. London, CG, 16 Apr. 1735, with Carestini; Dallas 16 Nov. 1960, with Sutherland. Rev. frequently in modern times.

Also operas by Martinelli (1649), Campra (1705), Albinoni (1725), Alessandri (1775), Lacepède (comp. 1786), and Bruni (1795).

**Alda,** (orig. Davies) **Frances** (*b* Christchurch, 31 May 1879; *d* Venice, 18 Sep. 1952). New Zealand soprano. Studied Paris with Marchesi. Début Paris, OC, 1904 (Manon). Appearances at Brussels, M; London, CG, and Milan, S. Here she met Toscanini, who became her lifelong friend, and Gatti-Casazza, whom she later married. New York, M, 1908–29. Created female leads in operas by Damrosch, Victor Herbert, and Henry Hadley. Famed for her volatile temperament, professional quarrels, and lawsuits. Autobiography, *Men, Women and Tenors* (Boston, 1937). (R)

**Aldeburgh.** Town in Suffolk, England. In 1948 Benjamin Britten, Eric Crozier, and Peter Pears founded a festival of music and the other arts. Britten's own music and taste were at the centre of the programmes, which rapidly won a reputation for high standards and enterprise (and the capacity to draw the greatest artists), together with a holiday atmosphere. Performances were originally given in the Jubilee Hall (cap. 300), with Britten's church parables in Orford Church (cap. 475). In 1967 the disused Maltings building at Snape was adapted as an auditorium, but burnt down on the first night of the 1969 festival. It was rebuilt for use the following summer (cap. 840). Premières given by the festival include Britten's *Let's Make an Opera!* (1949), *Noyes Fludde* (1958), *A Midsummer Night's Dream* (1960), three *church parables (1964, 1966, 1968), and *Death in Venice* (1973), Berkeley's *A Dinner Engagement* (1954), Williamson's *English Eccentrics* (1964), Walton's *The Bear* (1967), Birtwistle's *Punch and Judy* (1968), Crosse's *The Grace of Todd* (1969), Gardner's *The Visitors* (1973), and Musgrave's *The Voice of Ariadne* (1974). With Britten's death in 1976 the festival lost its importance as an operatic centre.

BIBL: R. Strode, ed., *Music of Forty Festivals* (Aldeburgh, 1987).

**Alden, David** (*b* New York). US producer. Scottish O 1979; O North; London, ENO, from 1984; New York, M; New Israeli O; Netherlands O. Productions include *Calisto*, *Fidelio*, *Rigoletto*, *Masked Ball*, *Boccanegra*, *Don Carlos*, *Werther*, *Mazeppa*, *Rake's Progress*, *Mahagonny*. Works regularly with the designer David Fielding: their unconventional productions are typically provocative and intense.

**Aleko.** Opera in 1 act by Rakhmaninov; text by Vladimir Nemirovich-Danchenko, after Pushkin's poem *The Gypsies* (1824). Prod. Moscow, B, 27 Apr. 1893, with Korsov, Klementyev, Deysha-Sionitskaya, Vlasov, Shubina, cond. Altani; New York, Jolson's T, 11 Jan. 1926, with La Touche, Saratovsky, Ignatiev; London, Collegiate T, 2 May 1972, with M. King, Robiczek, cond. Badacsonyi.

Aleko (bs) has abandoned his settled life to wander with a group of gypsies. When Zemfira (sop) tires of him and plans to run off with a young gypsy (ten), he kills her; the gypsies abandon him to his fate.

The work was composed as a graduation exercise for three students of Arensky's composition class at the Moscow Conservatory (the others being Nikita Morozov and Lev Konius). All set the same libretto; with his version Rakhmaninov graduated with the highest honours.

**Aler, John** (*b* Baltimore, 4 Oct. 1949). US tenor. Studied New York with O. Brown, Tanglewood with Malas. Début Juilliard American O Center, 1977 (Ernesto). Brussels, M, 1979; Gly. 1979; New York, City O, 1981; London, CG, 1988; also Milan, S; Lyons; etc. Roles incl. Ferrando, Don Ottavio, Tamino, Ory, Nemorino, Percy (*Anna Bolena*), Iopas (*Troyens*). A musicianly singer of poise and elegance. (R)

**Alessandri, Felice** (*b* Rome, 24 Nov. 1747; *d* Casinalbo, 15 Aug. 1798). Italian composer. Had

success in Italy with *Ezio* and *Il matrimonio per concorso* before moving to London *c*.1768, where he was employed at the King's T with his wife, the singer Maria Lavinia Guadagni (*b* Lodi, 21 Nov. 1735; *d* Padua, *c*.1790). During the 1770s and early 1780s he worked in various Italian cities; 1777–8 in Paris, where he shared the direction of the Concert Spirituel with Legros. In 1786 he tried unsuccessfully to obtain a post at the St Petersburg court, so settled in Berlin until dismissed 1792. Works written here were unsuccessful, but his last pieces, *Zemira* and *Armida*, were well received in Italy.

Alessandri wrote *c*.35 operas: though his opere serie rarely rise above the conventional, his opere buffe, which were among the most popular works of their kind, show him to be a composer of some talent. Especially successful were *La villanella rapita* and *I sposi burlati*.

SELECT WORKLIST: *Ezio* (Metastasio; Verona 1767); *Il matrimonio per concorso* (Martinelli, after Goldoni; Venice 1767); *La villanella rapita* (Bertati; Bologna 1784); *Zemira* (Sertor; Padua 1794); *Armida* (Foppa; Padua 1794); *I sposi burlati* (?; Mantua 1798).

**Alessandro.** Opera in 3 acts by Handel; text by P. A. Rolli. Prod. London, H, 5 May 1726, with Faustina Bordoni (making her début, as Rossane), Cuzzoni, Boschi, Senesino. Rev. as *Roxana, or Alexander in India*, probably with additions by Lampugnani, London, H, 15 Nov. 1743. Rev. in German edn., Stuttgart 1959. See also ALEXANDER THE GREAT.

**Alessandro Stradella.** Opera in 3 acts by Flotow; text by 'W. Friedrich' (Friedrich Wilhelm Riese) after a French *comédie mêlée de chants* by P. A. A. Pittaud de Forges and P. Duport, prod. Hamburg, 30 Dec. 1844; London, DL, 6 June 1846; New York, Niblo's Garden, 29 Nov. 1855.

Set in Venice and Rome in about 1670, the opera tells of episodes in the life of the composer Stradella. At a carnival, Stradella (ten) abducts Leonore (sop), ward of Bassi (bs), and carries her off to Rome. The lovers are pursued by Bassi's hired assassins, but first they and later Bassi himself are won over by Stradella and his art.

Other operas on Stradella's adventurous life are by César Franck (1844, piano score only), Schimon (1846), Boccaccio (1852), and Sirico (1863).

**Alexander, John** (*b* Meridian, MS, 21 Oct. 1923; *d* Meridian, MS, 8 Dec. 1990). US tenor. Studied Cincinnati with Weede. Début Cincinnati 1952 (Gounod's Faust). New York, City O, 1957; New York, M, from 1961; San Francisco from 1967; Vienna, S, 1968; London, CG, 1970. A highly versatile singer, whose repertory included Idomeneo, Ferrando, Elvino, Edgardo, Pollione,

Alfredo, Rodolfo, Pinkerton, Hoffmann, Julien (*Louise*), and Bacchus. (R)

**Alexander the Great** (356–323 BC). King of Macedon. He features in well over 100 operas, including some under the name of his second wife Roxana. The first setting is by Cesti (1651). Over 70 are to Metastasio's text *Alessandro nell'Indie*, and include works by Leonardo da Vinci (1730, the first), Galuppi (1738), Duni (1743), Araia (1743), Graun (1744), Wagenseil (1748), Piccinni (1758), Manfredini (1758), Holzbauer (1759), Jommelli (1760), J. C. Bach (1762), Sacchini (1763), Kozeluch (1769), Anfossi (1772), Paisiello (1773), Piccinni (1774), Cimarosa (1781), Cherubini (1784), and Pacini (1824).

**Alfano, Franco** (*b* Posillipo, 8 Mar. 1875; *d* San Remo, 27 Oct. 1954). Italian composer. Studied Naples with De Nardis and Serrao, then Leipzig with Jadassohn. His first operatic success came with *Risurrezione*: rapidly taken up, and a favourite work of Mary Garden's, this shows him moving away from the verismo tradition. He revealed a weightier and more personal vein with *L'ombra di Don Giovanni*, which he developed, especially under the influence of Debussy, in his most important stage work, *La leggenda di Sakuntala*. Accused over *Madonna Imperia* of having Frenchified Italian opera, he attempted a reconciliation with his origins in subsequent works, though these mark a continued decline from *Sakuntala*.

Though some of these operas have retained a hold on the Italian repertory, it is for his completion of Puccini's *\*Turandot* that he is best remembered. He had once planned an opera on the subject himself, and was chosen for the task by Toscanini, who conducted the opera's première without Alfano's ending, adding it at the next performance. Enough was left in the sketches for Alfano to work them up into an acceptable shape; and though a break in style has been observed, no major attempt has been made, despite some revision, to improve upon Alfano's work.

SELECT WORKLIST: *Risurrezione* (Hanau, after Tolstoy; Turin 1904); *L'ombra di Don Giovanni* (Moschino; Milan 1914); *La leggenda di Sakuntala* (Alfano, after Kalidasa; Bologna 1921); *Madonna Imperia* (Rossato, after Balzac; Turin 1927).

BIBL: A. Della Corte, *Ritratto di Franco Alfano* (Turin, 1935).

**Alfio.** A teamster (bar), husband of Lola and rival of Turiddu, in Mascagni's *Cavalleria rusticana*.

**Alfonso.** 1. Don Alfonso (bs), the cynical philosopher who proposes the wager that produces the plot of Mozart's *Così fan tutte*.

2. Alfonso d'Este (bar), Duke of Ferrara, in Donizetti's *Lucrezia Borgia*.

3. Alfonso XI (bar), King of Castile, in Donizetti's *La favorite*.

**Alfonso und Estrella.** Opera in 3 acts by Schubert; text by Franz von Schober. Comp. 1821–2; prod. Weimar 24 June 1854; Reading Univ., 22 Feb. 1977.

Estrella (sop) meets Alfonso (ten), son of the usurped King Troila (bar). Estrella's father, Mauregato (bar), is the usurper. When Adolfo (bs), Mauregato's general, is refused Estrella's hand in marriage, he plots against the throne. The situation is saved by Alfonso, who is welcomed as the rightful heir. Adolfo is put in prison but King Troila forgives everyone else's misdemeanours, and Alfonso marries Estrella.

**Alfred.** Rosalinde's lover (ten) in Johann Strauss's *Die Fledermaus*.

**Alfredo.** The younger Germont (ten), Violetta's lover, in Verdi's *La traviata*.

**Alfred the Great.** King of England, AD 871–99. Operas on him are by Marinelli (1815), Weinlig (*c.*1815), Mayr (1818), Reinecke (1818), Lee (1820s), Ritter (1820), Donizetti (1823: the semifictional plot includes a Danish general named Atkins), Wolfram (1826), Böhmer (1827), J. P. Schmidt (1830), Geissler (1831), Neumann (1837), Reuling (1840), Raff (1851), Chemin-Petit (1858), Duggan (comp. 1860), Stainford (*Alfred, King of Wessex*, 1864), Kistler (comp. 1876), and Gatty (*King Alfred and the Cakes*, 1930).

**Algarotti, Francesco** (*b* Venice, 11 Dec. 1712; *d* Pisa, 3 May 1764). Italian poet and writer on music. Studied Rome and Bologna. He was well acquainted with Voltaire, who greatly admired his work, and was also in the service of Frederick the Great and Augustus III, Elector of Saxony and King of Poland. At both Berlin and Dresden he took part in opera productions as a poet. His importance to opera rests chiefly on his *Saggio sopra l'opera in musica* (Essay on Opera) (1755). One of the classic documents of opera, this argues for a unified art form, with all the elements at the service of a poetic idea. The example of Quinault and Lully is upheld in contrast to Metastasio and the artificiality his texts promoted, though Algarotti accepts the stylization necessary to the genre and thus argues for exotic, fantastic, or otherwise distanced subjects. He offers his own example with a text for *Iphigénie en Aulide*; and this was to exert a strong influence on subsequent Iphigenia operas, especially on the reforms embodied in that by Gluck. His recommendations extended to such practical matters as theatre construction and the proper demeanour of audiences.

**Algeria.** The T de l'Impératrice (later Grand-T Municipal) opened in 1853 with an occasional piece, *Alger*, music by Baron Bron, a government official. Enlarged and restored 1856, 1859, 1873. Burnt down 1883, restored on model of Paris, C; a mixed repertory included opera and operetta. Restored 1935, reopened 1939.

**Alice.** 1. Ford's wife (sop) in Verdi's *Falstaff*.
2. Robert's foster-sister (sop) in Meyerbeer's *Robert le diable*.

**Allegranti, Maddalena** (*b* Venice, 1754; *d* ?Ireland, *c*.1802). Italian soprano. Début Venice 1770; then studied at Mannheim. Great success in Germany until 1778; London 1781–3. For some years then prima donna buffa at Dresden. Earlier in her career Burney had noted her 'pretty, unaffected manner', while Mozart thought her far better than Ferrarese, adding that this was not saying much. Casanova, who saw her in Bologna in 1771, found her both 'adorabile' and 'pericolosa'.

**Allen, Thomas** (*b* Seaham, 10 Sep. 1944). English baritone. Studied London with Hervey Alan. Début WNO 1969 (Rossini's Figaro), then regularly WNO. London, CG, from 1972; Gly. from 1973. Milan, S, 1976. London, ENO. Also New York, M 1981; Vienna; Munich; etc. Repertory includes Don Giovanni, Count Almaviva (Mozart), Valentin, Posa, Pelléas, Billy Budd, Doktor Faust (Busoni). Created Valerio in Musgrave's *Voice of Ariadne*. Musically and dramatically intelligent, with a virile, mellifluous voice, and assured stage presence. (R)

**allestimento** (It.: 'preparation'). The term used in Italy for the production of an opera, especially as it is announced in the *cartellone outlining the season's plans. A nuovo allestimento is a new production of a repertory work.

**Almaviva.** The Count who, as a tenor, woos and wins Rosina in Rossini's *Il barbiere di Siviglia*, and, as a baritone, has to be taught a lesson for his infidelity to her as his Countess in Mozart's *Le nozze di Figaro*.

**Almaviva, ossia l'inutile precauzione.** See BARBIERE DI SIVIGLIA, IL.

**Almeida Theatre.** Theatre in London where the enterprising choice and staging of opera grew into a festival by 1982 with the première of Jean-Jacques Dunki's *Prokrustes*. This was followed by Thomson's *Four Saints in Three Acts* (1983), Robert Ashley's *Perfect Lives (Private Parts)* (1984), Claude Vivier's *Kopernikus* (1985), Wolfgang Rihm's *Jakob Lenz* (1987), Michael Finnissy's *The Undivine Comedy* (1988), John Casken's *Golem* (1989), Smirnov's *The Lamentations of Thel* (1989), Cage's *Europeras 3 and 4* (1990), and Gerald Barry's *The Intelligence Park* (1990).

**Almira.** Opera in 3 acts by Handel; text by Friedrich Feustking, after Pancieri's Italian libretto for Boniventi. Prod. Hamburg 8 Jan. 1705. Handel's first opera, containing 41 German and 15 Italian airs. Other operas on the text are by Boniventi (1691), Fedeli (1703), Keiser (1706), and Erler (1893).

**Altani, Ippolit** (*b* S Ukraine, 27 May 1846; *d* Moscow, 17 Feb. 1919). Russian conductor. Studied St Petersburg with Zaremba and A. Rubinstein. Chorus master Kiev 1867–9, then conductor to 1882; conducted there first provincial performance of a Tchaikovsky opera (*Oprichnik*, 1874). Chief conductor Moscow, B, 1882–1906, including premières of Tchaikovsky's *Mazeppa* (1884) and Rakhmaninov's *Aleko* (1893), and first Moscow performances of works including *Boris* (1888), *Queen of Spades* (1891), *Yolanta* (1893), and *Snow Maiden* (1893). Tchaikovsky had some lessons from him before conducting the première of *Cherevichki*.

**alto** (It.: 'high'). A term for certain voices. The female alto is usually known as a *contralto. The male alto is usually a bass voice singing *falsetto. It is also one of the categories of the *castrato voice. The range is about from f to a''. See COUNTER-TENOR.

**Alva, Luigi** (*b* Lima, 10 Apr. 1927). Peruvian tenor. Studied Lima with Morales, and Milan with Ghirardini. Début Lima 1949 (in the zarzuela *Luisa Fernanda*). Sang Paolino (*Il matrimonio segreto*) at the opening night of Milan, PS, in Dec. 1955; Edinburgh 1957; Début London, CG, 1960 (Rossini's Count Almaviva), subsequently Fenton and Ferrando; New York, M, 1964–75. Has sung extensively in Europe; especially admired in Mozart and Rossini. (R)

**Alvaro.** Don Alvaro (ten), a Peruvian of Inca blood, in Verdi's *La forza del destino*.

**Alvary, Max** (orig. Maximilian Achenbach) (*b* Düsseldorf, 3 May 1856; *d* Gross-Tabarz, 7 Nov. 1898). German tenor. Studied Frankfurt with Stockhausen, Milan with Lamperti. Début Weimar 1879 (Alessandro Stradella). Weimar till 1885; New York, M, 1885–9 (début as Don José opposite Lilli Lehmann, also making her US début). First US Loge, Siegfried, and Adolar. Bayreuth 1891 (Tristan and Tannhäuser); sang in the first CG *Ring* under Mahler, 1892, which began with *Siegfried* so that Alvary could make his début in his favourite role. Retired 1897.

**Alyabyev, Alexander** (*b* Tobolsk, 15 Aug. 1787; *d* Moscow, 6 Mar. 1851). Russian composer. Fought as a hussar against Napoleon in 1812; then studied music in St Petersburg, leaving the Army in 1823. His comic opera *Lunnaya noch* appeared in this year; and he then also wrote the song 'Solovey' ('The Nightingale'), which became very popular and was often used by Patti, Viardot, and Sembrich for the Lesson Scene in Act II of Rossini's *Il barbiere di Siviglia*. He was banished from Moscow in 1828 partly for alleged murder, partly perhaps for his Decembrist associations; he returned at the end of the 1830s, after travels in the East and South of Russia. His knowledge of folk music, systematically acquired in exile, and his understanding of oriental and Caucasian subjects were significant in the development of Russian opera in the years immediately preceding Glinka.

WORKLIST: *Lunnaya noch* (The Moonlight Night) (Mukhanov & Arapov; 1822); *Burya* (The Tempest) (after Shakespeare; c.1835); *Volshebnaya noch* (The Enchanted Night) (Weltman, after Shakespeare; comp. 1838–9); *Rybak i rusalka* (The Fisherman and the Water Nymph) (comp. 1841–3); *Ammalet-Bek* (Weltman, after Bestuzhev-Marlinsky; comp. 1842–7).

**Alzira.** Opera in a prologue and 2 acts by Verdi; text by Cammarano, after Voltaire's tragedy *Alzire, ou Les Américains* (1730). Prod. Naples, C, 12 Aug. 1845, with Tadolini, Fraschini, Coletti; New York, Carnegie H, 17 Jan. 1968, with Ross, Cecchele, Quilico, cond. Perlea; London, Collegiate T, 10 Feb. 1970, with Duval, Wilcock, cond. Badacsonyi. Rev. Rome 1967, with Zeani, Cecchele, MacNeil, cond. Capuana.

Peru, c.1550. Zamoro (ten), chief of the Incas, betrothed to Alzira (sop), is presumed dead after being taken prisoner by Gusmano (bar), the son of the Governor of Peru. Gusmano takes over as Governor and Alzira is forced to marry him. Zamoro has been captured in battle but is released when Alzira's wedding is announced. On the day of the ceremony, he returns and stabs Gusmano, who pardons Zamoro and returns Alzira to him as he dies.

One of Verdi's rare failures: he later called it 'brutta'.

Other operas on the subject are by Krauss (1777), Zingarelli (1794), Roszisky (1794), Nicolini (1796), Horzizky (comp. 1800), Bianchi (1801), Manfroce (1810), and Rossi (comp. 1880); also pasticcio by Nasolini and others (1796).

**Amadei, Filippo** (*fl.* 1690–1730). Italian composer. Spent his early years in Rome, where he earned a reputation as a cellist and composer of oratorios. Arrived in London c.1719; 1720 joined orchestra of the *Royal Academy of Music, also acting as composer. Remembered for his contribution (Act I) to *Muzio Scevola*, written jointly with Handel and Bononcini.

**Amadis.** Opera in a prologue and 5 acts by Lully; text by Quinault, based on a Spanish chivalric romance by García Ordonez de Montalvo. Prod. Paris, O, 18 Jan. 1684; Cambridge, MA, 15 Mar. 1951.

Amadis (ten) falls in love with the Princess Oriane (sop), and embarks upon chivalric adventures to win her hand. During his journeys he encounters many difficulties but always succeeds in accomplishing heroic deeds, rescuing not only Oriane and her father but also his own brother and parents, whom he does not recognize since they had cast him to sea in infancy as he was illegitimate. He gains victories in many countries and is eventually united with Oriane.

A parody on the opera by Romagnesi and Riccobini was prod. Paris, Nouveau T Italien, 1740. A sequel, *Amadis de Grèce*, was set by Destouches (1699).

Other operas on the subject are by Handel (London, H, 25 May 1715), Torri (1724), Blaise (parody, 1740), Berton (1771), J. C. Bach (1779), Stengel (1798), and Massenet (comp. 1895, prod. Monte Carlo 1922).

**Amahl and the Night Visitors.** Opera in 1 act by Menotti; text by composer, after Bosch's painting *The Adoration of the Magi*. Prod. NBC TV, 24 Dec. 1951 (the first opera written for TV); stage prod., Bloomington, Indiana U, 21 Feb. 1952.

The opera describes the reception of the Magi, on their way to Bethlehem, by a little crippled boy.

**Amalia.** An orphan (sop), niece of Count Moor, in Verdi's *I masnadieri*.

**Amato, Pasquale** (*b* Naples, 21 Mar. 1878; *d* Jackson Heights, NY, 12 Aug. 1942). Italian baritone. Début Naples 1900 (Lescaut). London, CG, 1904, with Naples, C, but despite success never returned. New York, M, 1908–21, where he created Jack Rance, Cyrano de Bergerac (Damrosch), and Napoléon (*Madame Sans-Gêne*). Successes in Italy, Germany, and Odessa, 1902–13. An excellent singer, with a resonant voice and a wide range of dramatic expression. (R)

**Amelia.** 1. Boccanegra's daughter Maria, brought up as Amelia Grimaldi (sop), in Verdi's *Simon Boccanegra*.

2. The wife (sop) of Anckarstroem (Renato) and lover of Gustavus III (Riccardo) in Verdi's *Un ballo in maschera*.

3. The heroine of Menotti's *Amelia Goes to the Ball*.

**Amfortas.** The Keeper of the Grail (bar) in Wagner's *Parsifal*, where he is the son of Titurel. His name may originally derive from his battle cry in other legends, 'Amor fortis' (Love is strong).

**Amico Fritz, L'** (Friend Fritz). Opera in 3 acts by Mascagni; text by P. Suardon (N. Daspuro), after Erckmann-Chatrian's novel (1864). Prod. Rome, C, 31 Oct. 1891, with E. Calvé, Synnemberg, De Lucia, Lhérie, cond. Ferrari; London, CG, 23 May 1892, with E. Calvé, G. Ravogli, De Lucia, Dufriche, cond. Bevignani; Philadelphia, Grand OH, 8 June 1892.

A light pastoral comedy about the confirmed bachelor Fritz (ten), a rich landowner; Rabbi David (bar), the consummate match-maker; and Suzel (sop), the charming daughter of one of Fritz's tenants, with whom Fritz eventually falls in love. Also opera by Edwards (1893).

**Amina.** An orphan (sop), adopted daughter of the mill-owner Teresa, the sleepwalker in Bellini's *La sonnambula*.

**Aminta.** The wife (sop) of Henry Morosus in R. Strauss's *Die schweigsame Frau*.

**Amirov, Fikret** (*b* Ganja (Kirovabad), 22 Nov. 1922; *d* Baku, 20 Feb. 1984). Azerbaijani composer. Studied Baku with Zeydman and Hadjibekov. After war service, resumed studies and graduated with his first opera *Ulduz*. His most important work, a pioneering one for Azerbaijani opera, is *Sevil*, which incorporates local folk music into a number opera.

WORKLIST: *Ulduz* (Idayat-zade; comp. 1948); *Sevil* (Eyubov, after Jabarly; Baku 1953).

**Amneris.** The King of Egypt's daughter (mez) and Aida's rival for Radamès in Verdi's *Aida*.

**Amonasro.** The King of Ethiopia (bar), Aida's father, in Verdi's *Aida*.

**Amore dei tre re, L'** (The Love of the Three Kings). Opera in 3 acts by Montemezzi; text by Benelli, after his verse tragedy (1910). Prod. Milan, S, 10 Apr. 1913, with Villani, Ferrari-Fontana, Galeffi, De Angelis, cond. Serafin; New York, M, 2 Jan. 1914, with Bori, Ferrari-Fontana, Amato, Didur, cond. Toscanini; London, CG, 27 May 1914, with Edvina, Crimi, Cigada, Didur, cond. Moranzoni.

Fiora (sop) has been forced for political reasons to marry Manfredo (bar), son of King Archibaldo (bs). While Manfredo is absent on a campaign, the old king murders his daughter-in-law for refusing to reveal the name of her lover Avito (ten). He poisons her lips, hoping Avito will come to see her body. Avito dies from kissing the poison, as does Manfredo, who cannot live without his wife.

**Amores de la Inés, Los** (The Loves of Inés). Zarzuela by Falla, in collaboration with Amadeo Vivès; text by Emilio Dugi. Prod. Madrid, T Comico, 1902. It was the only one of Falla's five zarzuelas to reach the stage.

**'Amour, viens aider ma faiblesse'.** Dalila's (mez) aria in Act II of Saint-Saëns's *Samson et Dalila*, in which she invokes the power of love to fetter Samson.

**Amphion.** Opera in 1 act by Honegger; text by Paul Valéry. Prod. Paris, O, 23 June 1931.

Apollon (bar) gives his lyre to Amphion (ten), a favourite of the gods. The music Amphion plays is so powerful that stones form themselves into shapes of their own accord; the event is cited as the birth of architecture. Amphion builds a temple to Apollon and the city of Thebes, but is prevented from appreciating his achievements: he is led away by a veiled woman, who also casts away his lyre.

**Amsterdam.** Capital of The Netherlands. The first opera performed was *Le fatiche d'Ercole per Deianira* by P. A. Ziani on 31 Dec. 1680. Together with the pastoral play with music *De triomfeerende Min* by Hacquart (1678), it was given in a specially built theatre on the Leidse Gracht. When this burnt down in 1770, performances were given in the Stadsschouwburg (burnt down 1772), in the Operahuis (where a German-Jewish company played in 1784), and in the T Français (where a French company, having previously performed in an inn, was established 1788–1855). Some of these performances were in Dutch. In 1882 Angelo Neumann's Wagner company gave *The Ring* in the Paleis voor Volksvlijt; and in 1883 the successful Wagner-Vereniging was founded: it gave many performances of operas by composers other than Wagner. There was also a resident Italian company, 1897–1942.

During the German occupation the Municipal Opera Enterprise was formed (1941), and from this developed in 1946 De Nederlandse Opera, which gave performances in Dutch in the Stadsschouwburg. The first company with that name had been founded in 1886, and eventually gave 200 performances a year not only in Amsterdam but in N Holland, Rotterdam, Utrecht, and Hilversum. In 1956 it was reorganized as the Nieuwe Nederlandse Opera under Maurice Huismann (also administrator of the Brussels, M), performing in other Dutch cities regularly. After many delays and difficulties, the Muziektheater (cap. 1,594) opened in 1986 with a programme including Otto Ketting's *Ithaka*, and rapidly developed an enterprising policy. An opera studio was formed in the 1974–5 season, to encourage young audiences and to give chamber opera in schools. The VARA broadcasting network has run an adventurous opera-in-concert series.

**'Am stillen Herd'.** Walther's (ten) song in Act I of Wagner's *Die Meistersinger von Nürnberg*, telling the Mastersingers where and from whom he learnt to sing.

**Anckarstroem.** A secretary (bar) in the service of King Gustavus III in the 'Swedish' version of Verdi's *Un \*ballo in maschera*.

**Anders, Peter** (*b* Essen, 1 July 1908; *d* Hamburg, 10 Sep. 1954). German tenor. Studied Berlin with Grenzebach, later with Mysz-Gmeiner. Début Berlin 1931, in Max Reinhardt's production of *La belle Hélène*. Berlin, SO, 1936–49 in the lighter lyric roles; especially successful as a Mozart singer. Hamburg 1950 (Walther and Verdi's Otello). British début 1950, Edinburgh Festival, Bacchus under Beecham, with whom he also sang Walther at London, CG. An intelligent and musical singer. (R)

**Andersen, Hans** (*b* Odense, 2 Apr. 1805; *d* Copenhagen, 4 Aug. 1875). Danish writer. He is best known for his fairy-tales (168 published 1835–72), but originally hoped to become an actor or singer and worked in the Royal T, Copenhagen. He also worked as librettist, adapting Scott for Bredal's *Bride of Lammermoor* (1832) and for his friend Weyse's *Festen på Kenilworth* (1836), and Manzoni's *I promessi sposi* for Gläser's *Bryllupet ved Como-søen* (The Wedding on Lake Como, 1849), also writing the text of *Nøkken* (The Water-Sprites, 1853) for Gläser. Operas based on his fairy-tales are as follows (alphabetical order):

*The Cat and Mouse in Partnership*: T. Chanler (*The Pot of Fat*, 1955).
*The Cobbler of Delft*: Bersa (1914).
*The Emperor's New Clothes*: Kjerulf (1888); J. Clokey (1924); Kósa (*A király palástja*, 1926); Høffding (1928); Wagner-Régeny (1928); R. Kubelik (1946); E. F. Burian (1947); D. S. Moore (1948); G. Ránki (*Pomádé király új ruhája*, radio 1950, prod. 1953); Simeone (1956); Jenni (1965); Novák (1969); Wood (1972); Glaser (1973); Gingold (20th cent.).
*The Garden of Paradise*: Bruneau (1923); Němeček (*Rajská zahrada*, 1933).
*Little Christina*: Hartmann (1846); Leoni (1901).
*The Little Mermaid*: Maliszewski (*Syrena*, 1928); M. More (1951); Boyd (1985).
*The Magic Mirror*: Beckett (1955).
*The Match Girl*: Enna (1897); Veretti (1934); Nordon (20th cent.).
*The Mother*: L. Moreau (*Myrialde*, 1912); J. Wood (1942); Vlad (1951); S. Hollingsworth (1954); Koppel (1965).
*The Nightingale*: Enna (1912); Stravinsky (1914); J. Clokey (1925); Irmler (1939); Schanazara (1947); D. Lamb (1954); B. Rogers (1955); Gallois-Montbrun (1959); Windsor (1966); C. Strouse (1982).
*The Princess and the Swineherd*: Poldini (*Csavargó és királyleány*, 1903); Reutter (1928).
*The Princess and the Pea*: Enna (1910); Toch (1927).
*The Raven*: Hartmann (1832).

*The Shepherdess and the Sweep*: J. Smith (1966).
*The Snow Queen*: Asafyev (1907); Guzewski (1907); Gerrish-Jones (1917); Gaboro (1952); Zanaboni (1955); K. Thies (1957); Vander (20th cent.).
*The Ten Kisses*: Sekles (1926).
*The Tinderbox*: Fougstedt (*Tulukset*, 1950).
*The Travelling Companion*: Stanford (1926); Hamerik (1946).
*What the Old Man Does is Always Right*: Hoddinott (1977).
Also musical on his life by Loesser (1974: includes a part for Jenny Lind).

**Anderson, June** (*b* Boston, 30 Dec. 1952). US soprano. Studied New York with Leonard. Début New York, CC, 1978 (Queen of the Night). Successes in USA; Rome 1982; Munich 1983; Milan, S, and Paris, O, 1985; London, CG, 1986; New York, M, 1989. Roles incl. Donna Elvira, Desdemona (Rossini), Rosina, Semiramide, Amina, Lucia, Gilda, Isabella (*Robert le diable*). (R)

**Anderson, Marian** (*b* Philadelphia, 27 Feb. 1899; *d* Portland, OR, 8 Apr. 1993). US contralto. Studied New York with Boghetti. Most of career a concert artist; operatic début as first Black singer at New York, M, 1955 (Ulrica). (R)

**André, Johann** (*b* Offenbach, 28 Mar. 1741; *d* Offenbach, 18 June 1799). German composer and publisher. Translated opéras-comiques by Philidor and others into German for Marchand's theatrical troupe, who performed his first Singspiel, *Der Töpfer*, 1773. Its success inspired him to compose further Singspiels, and in 1775 he was commissioned by Goethe, who had praised *Der Töpfer*, to set his text *Erwin und Elmire*; he later set another of Goethe's Singspiels, *Claudine von Villa Bella*.

André's Singspiels were all composed 1773–83, mostly when he was with Doebbelin's troupe in Berlin, 1776–84. Here he produced 17 Singspiels in addition to other stage works (e.g. incidental music for *King Lear* and *Macbeth*, both 1778). His success encouraged other north German composers to write Singspiels. The Berliners' preference for music from Vienna from 1783 led to a decline in his popularity; he returned to Offenbach and devoted more of his energy to his music publishing firm.

Described by Goethe (*Dichtung und Wahrheit*) as a man of 'natural, lively talent', André made a significant contribution to the development of comic Singspiel at a time when other composers were trying to infuse a serious element into the genre. Though he kept to the pattern evolved by Hiller, André's style was more varied, blending folk-style music with elements of opéra-comique and opera buffa. Largely unadventurous, his works contain some passages of note. *Laura Rosetti* and *Erwin und Elmire* have \*melodrama scenes; the chorus is employed in an unusually dramatic manner in *Belmonte und Constanze* (a setting of the Bretzner libretto used by Mozart in *Die Entführung aus dem Serail*) and *Der Barbier von Bagdad*. There are also several examples of accompanied recitative, and of continuous ensemble movements, e.g. the abduction scene in *Belmonte und Constanze*, modelled after the early buffo ensembles of Galuppi and Piccinni. Among André's most successful incidental music was that for Beaumarchais's *Le barbier de Séville* (1776).

His son **Johann Anton** (1775–1842) composed two Italian operas, but was chiefly occupied with editing and printing Mozart's manuscripts, having purchased the Mozart *Nachlass* from Constanze Mozart in 1799.

SELECT WORKLIST: *Der Töpfer* (André; Hanau 1773); *Erwin und Elmire* (Goethe; Frankfurt 1775); *Laura Rosetti* (D'Arien; Berlin 1778); *Claudine von Villa Bella* (Goethe; Berlin 1778); *Belmonte und Constanze* (Bretzner; Berlin 1781); *Der Barbier von Bagdad* (André; Berlin 1783).

**Andrea Chénier.** Opera in 4 acts by Giordano; text by Illica. Prod. Milan, S, 28 Mar. 1896, with Carrera, Borgatti, Sammarco; New York, AM, 13 Nov. 1896; London, Camden T, by CR (in English), 16 Apr. 1903.

France, the time of the Revolution. Carlo Gérard (bar), a servant, is in love with the aristocratic Madeleine de Coigny (sop). Guests arrive for a ball at her father's house, among them the poet Andrea Chénier (ten), whom Madeleine loves. The party is interrupted by Gérard and a band of peasants, and the servant is dismissed by the Comtesse de Coigny, Madeleine's mother (mez).

Following the Revolution, Chénier is being watched by a spy of the new government. His friend Roucher (bs) advises him to flee Paris but Chénier will not leave until he has discovered the identity of the woman who has, anonymously, asked for his protection. It is Madeleine. The spy sees her talking to Chénier and reports this to Gérard, who is a member of the government. Gérard challenges Chénier, but is seriously wounded by him. When the police arrive, Gérard says that he does not know his attacker.

A year later Gérard is informed that Chénier has been arrested. Madeleine arrives and pleads for Gérard to save her beloved. He agrees when, finally, she offers herself to him. At the tribunal, however, Chénier is condemned to death despite Gérard's good word.

Awaiting his execution, Chénier reads his last poem to Roucher. Gérard makes a final plea, and Madeleine bribes the gaoler to let her take the place of a condemned female prisoner. As the lovers die together, strengthened by their unity, Gérard weeps.

**Andreini** (orig. Ramponi), **Virginia** (*b* Milan, 1 Jan. 1583; *d* Bologna, 1630). Italian singer. Originally an actress in her husband's theatrical company. Sang in the *Ballo delle ingrate* at the Gonzaga wedding in 1608, when she also created Monteverdi's Arianna. Toured Europe with the Comici Fedeli, 1613–28. Known as La Florinda from the title role of her husband's opera. **Giovanni Battista Andreini** (*b* Florence, 9 Feb. 1579; *d* Reggio Emilia, 8 June 1654), was also the author of the sacra rappresentazione, *La Maddalena*, for which Monteverdi wrote the prologue. He published the librettos of several works during the early years of the 17th cent. His second wife was the actress Virginia Rotari.

**Andreozzi, Gaetano** (*b* Avessa, 22 May 1755; *d* Paris, 21 or 24 Dec. 1826). Italian composer. Studied at the Cons. di S Maria di Loreto in Naples, though also received instruction from his uncle Nicolò Jommelli (hence his nickname 'Jommellino'). Andreozzi's operas quickly found fame throughout Europe: in 1784 he undertook an important journey to Russia, composing *Didone abbandonata* and *Giasone e Medea*, though these were not produced there. He wrote *c*.40 operatic works, mainly opere serie in the conventional Metastasian mould: most interesting, because of their subject-matter, are *Giovanna d'Arco* and *Amleto*. In 1786 Andreozzi married the famous soprano **Anna dei Santi**. He settled in Naples in 1801. After 1803 he composed little music, preferring to concentrate on his second career as a singing-teacher. Though highly acclaimed, especially at the Neapolitan court, Andreozzi was eventually forced to flee to Paris in severe financial difficulties.

SELECT WORKLIST: *Didone abbandonata* (Metastasio; Pisa 1785); *Giovanna d'Arco* (Sografi; Vicenza 1789); *Amleto* (Foppa; Padua 1792); *Giasone e Medea* (Metastasio; Naples 1793).

**Andromache.** In classical mythology, the wife of the Trojan hero Hector. Operas on Salvi's text *Andromacca* are by Torri (1716), Bioni (1729), Feo (1730), Leo (1742), Lampugnani (1748), Perez (1750), Aurizicchio and Pampani (1753), Valentini (1754), Sarti (1760), Sacchini (1761), Tozzi (1765), Bertoni (1775), Grétry (1780), Martin (1780), and Nasolini (1790). Other operas are by Caldara (1724), Scolari (1757), Pavesi (1804), Tritto (1807), Raimondi (?1815), Ellerton (1830s), and Windt (1932). She also appears in Tippett's *King Priam* (sop), and in a silent role in Berlioz's *Les Troyens*. See also RACINE.

**Andromeda.** In Greek legend, the daughter of Cepheus and Cassiopeia. To pacify the offended

Poseidon, she was chained to a rock at the mercy of a sea monster, but was rescued by Perseus, who married her.

Operas on the legend, sometimes with Perseus included in the title, are as follows: Monteverdi (1618–20; lost); Draghi (*Perseo*, 1669); Franck (1679); Lully (*Persée*, 1682); Bononcini (1702); Cocchi (1755); Paisiello (1774); Sacchini (*Perseo*, 1774); Philidor (*Persée*, 1780); M. Haydn (1787); Naumann (1792); Zingarelli (1796); Elsner (1807); Schneider (1807).

**'Andrò ramingo e solo'.** The quartet in Act III of Mozart's *Idomeneo*, in which Idamante (sop) announces his wish to leave and seek death; Ilia (sop) declares she will follow him, while Idomeneo (ten) curses Neptune and Elettra (sop) asks herself when she will be avenged.

**Anelli, Angelo** (pseud. Nicolò Liprandi, Marco Landi) (*b* Desenzano, 1 Nov. 1761; *d* Pavia, 9 Apr. 1820). Italian librettist. Attached to Milan, S, 1793–1817. A highly practical man of the theatre, with a particular gift for comedy, he wrote the texts for *L'italiana in Algeri* (Mosca, 1808; Rossini, 1813) and *Ser Marcantonio* (Pavesi, 1810: the basis of the text for Donizetti's *Don Pasquale*). The many other composers for whom he wrote include Cimarosa, Coccia, Dussek, Gazzaniga, Guglielmi, Martín y Soler, Mayr, Paer, Pacini, Pavesi, and Zingarelli.

**Anfossi, Pasquale** (*b* Taggia, 5 Apr. 1727; *d* Rome, ?Feb. 1797). Italian composer. Studied with Piccinni and Sacchini, and composed his first opera, *La serva spiritosa*, in 1763. Apart from 1782–6, when he was music director of London, H, Anfossi worked mainly in Venice and Rome. He became maestro di cappella at St John Lateran (Rome) in 1791 and thenceforth wrote only sacred music, including 20 oratorios.

Anfossi composed at least 60 operas, of which five, including *Il trionfo della costanza*, were written for London. In his opere serie he demonstrated a fondness for heroic themes, often setting Metastasian librettos, including *La clemenza di Tito*, *Alessandro nell'Indie*, *Demofoonte*, and *Lucio Silla*. But his comic operas—beginning with his first major success, *L'incognita perseguitata* in 1773—won him more lasting fame. *La finta giardiniera* is overshadowed by Mozart's more famous setting, but *La vera costanza* is a worthy companion to Haydn, while *L'avaro* and *Il curioso indiscreto* were particularly popular in their day. Gifted with the ability to write instantly memorable melodies, Anfossi also experimented in his later works with formal designs less rigid than the ubiquitous da capo aria.

SELECT WORKLIST: *La serva spiritosa* (?; Rome 1763); *La clemenza di Tito* (Metastasio; Rome 1769 or Naples 1772); *Alessandro nell'Indie* (Metastasio; Rome 1772); *L'incognita perseguitata* (?Petrosellini, after Goldoni; Rome 1773); *Demofoonte* (Metastasio; Rome 1773); *La finta giardiniera* (?Calzabigi; Rome 1774); *Lucio Silla* (De Gamerra; Venice 1774); *L'avaro* (Bertati; Venice 1775); *La vera costanza* (Puttini; Rome 1776); *Il curioso indiscreto* (?Bertati; Rome 1777); *Il trionfo della costanza* (Badini; London 1792).

**Angélique.** Opera in 1 act by Ibert; text by Nino. Prod. Paris, T Fémina, 28 Jan. 1927, with Bériza, Warnery, Ducros, Marvini, cond. Golschmann; New York, 44th St. T, 8 Nov. 1937, with Morel, Dolci, Iposvay, Perulli, Chabay; London, Fortune T, 20 Feb. 1950, with Castle, D. Craig, T. Jones, Sale, Lawrence, cond. Renton.

Boniface (bs-bar), the owner of a Paris china shop, is persuaded by his friend Charlot (ten) that the only way to rid himself of his shrewish wife Angélique is to put her up for sale. She is eventually bought by the Devil (ten), who soon returns her.

**Angelo.** Opera in 4 acts by Cui; text by Victor Burenin, based on Victor Hugo's *Angelo, tyran de Padoue* (1835). Prod. St Petersburg, M, 13 Feb. 1876, cond. Nápravník.

Padua, 1549. Caterina and Thisbe, wife and mistress of Angelo, both really love Rodolfo. Homodei, an Inquisition spy, also loves Caterina. He jealously arranges a meeting between her and Rodolfo, meaning to betray her to Angelo. But Thisbe, discovering that Caterina had once saved her mother's life, sacrifices herself for her former rival, and is killed. Rodolfo then discovers that he has misinterpreted a selfless act.

**Angiolini, Gaspero** (*b* Florence, 9 Feb. 1731; *d* Milan, 6 Feb. 1803). Italian dancer, choreographer, and composer. Studied principally Vienna with Hilverding. He collaborated with Gluck on *Don Juan* (1761), also producing a dance version of the opera *La Cythère assiégée* (1762) and choreographing *Orfeo* (1762). He worked further with Gluck, arguing a case and demonstrating with his work a belief in dance as a functional rather than merely decorative element in opera. He also worked in St Petersburg, Venice, and Milan.

**Aniara.** Opera in 2 acts by Blomdahl; text by Erik Lindegren, after Harry Martinson's epic poem (1956). Prod. Stockholm 31 May 1959, with Hallin, Djellert, Vikström, Ulfung, Sædén, Tyrén, cond. Ehrling; Edinburgh Festival, same cast, 3 Sep. 1959; Montreal, Place des Arts, by Royal Swedish O, 31 May 1967.

A spacecraft carrying refugees from earth, devastated in an atomic war, is on its way to Mars. It is knocked off course, and must continue to fly in space for ever.

A very successful work, produced in a number of countries after its première.

**Ankara.** Capital of Turkey. The first permanent theatre, the Halk Evi (cap. 638), opened in 1930. In 1934 Carl Ebert was invited to form a drama school, opened in 1939 with musical help from Hindemith. Opera was organized in 1947, when the Devlet Tiyatrosu (State T) opened; this had an opera section which began with the French and Italian repertory and in 1952 premièred the first Turkish opera, Adnan Saygun's *Kerem*. It also premièred Ferit Tüzün's *Midas'in kulaklari* (Midas's Ears) in 1978, despite a difficult decade in the company's history; the 1979–80 season saw Çengiz Tanç's *Deli Dumrul* and Cemal Resit Rey's *Çelebi*.

**Anna Bolena.** Opera in 2 acts by Donizetti; text by Romani. Prod. Milan, T Carcano, 26 Dec. 1830, with Pasta, Orlandi, Rubini, F. Galli; London, H, 8 July 1831, with Pasta, Rubini, Lablache; New Orleans, T d'Orléans, 12 Dec. 1839, with J. Calvé, Bamberger. Frequently performed, with Grisi and Tietjens, but then not revived until 1948, Barcelona, L, with Scuderi, and 1957, Milan, S, with Callas.

Loosely based on the well-known historical events, the plot tells how King Henry VIII (bs), in love with Jane Seymour (mez), wants to dispose of his wife Anne Boleyn (sop). He tries to accuse her of treason and recalls Percy, Earl of Northumberland (ten), to give evidence against her. Percy, however, professes his continuing love for Anne, as does her page Smeaton (con). Having come upon Anne alone with Percy, the King believes her guilty of adultery although Smeaton protests Anne's innocence; Henry has all three arrested.

Jane visits Anne in prison to persuade her to confess and so save her life. Although Anne refuses, she forgives Jane for supplanting her in the King's favour. When it emerges that Anne and Percy do in fact love each other, they are both condemned to death, as is Rochefort (bs), Anne's brother. The two men are offered pardons, which they refuse since Anne must still die. As she awaits her execution in the Tower of London, Anne loses her senses and imagines it is the day of her wedding to the King. As Anne and her three accomplices are led to the scaffold, a cannon and bells are heard acclaiming the new Queen.

BIBL: P. Gossett, *Anna Bolena and the Artistic Maturity of Gaetano Donizetti* (Oxford, 1985).

**Anne Trulove.** The daughter (sop) of Trulove, lover of Tom Rakewell, in Stravinsky's *The Rake's Progress*.

**Annibali, Domenico** (b Macerata, c.1705; d ?Rome, 1779). Italian soprano castrato. Earliest evidence of him in Porpora's *Germanico*, Rome 1725. After successes there, in Venice, and in Vienna, engaged 1731 at Dresden, where he sang until 1764, notably in Hasse's operas. In 1736 brought to London by Handel. Début in *Poro* (when Handel, exceptionally, let him interpolate other composers' arias). Created roles written for him in *Arminio*, *Giustino*, and *Berenice* (all 1737). He resumed his Dresden career, also singing in Italy, and retired with a comfortable pension and the title of Kammermusikus.

**Ansani, Giovanni** (b Rome, 11 Feb. 1744; d Florence, 15 July 1826). Italian tenor. With Giacomo David, helped to establish the supremacy of the tenor in Italian opera after the age of the castrato. During the 1780s and 1790s sang in the leading Italian theatres and also in London. Much admired by Burney. Created roles in operas by Paisiello, Cimarosa, Anfossi, etc. Retired from stage in 1795 and became a teacher; his pupils included Manuel García jun. and Lablache.

**Anseaume, Louis** (b ?; d Paris, Aug. 1784). French librettist. He wrote many texts for Duni and Philidor, to a lesser extent for Grétry and Monsigny, in which his light, fluent touch does not exclude some originalities which helped his composers to make formal innovations, especially in the genre of *comédie mêlée d'ariettes.

**Anselmi, Giuseppe** (b Catania, 16 Nov. 1876; d Zoagli, 27 May 1929). Italian tenor. Studied violin and composition, then joined an operetta company touring Italy and the Near East. Heard by Giulio Ricordi, who suggested that he study with Mancinelli. Début Athens 1896 (Turiddu). Italian début 1901. London, CG, 1901, and occasionally until 1909. Much admired in Buenos Aires; also appeared regularly in Warsaw, St Petersburg, and Madrid. Left his heart to Madrid's theatrical museum, where it is preserved. One of the foremost exponents of bel canto, with a beautiful voice and finished style. (R)

**Ansermet, Ernest** (b Vevey, 11 Nov. 1883; d Geneva, 20 Feb. 1969). Swiss conductor. Studied Lausanne, becoming professor of mathematics. Then studied composition with Bloch, and had advice from Nikisch. Début 1911. Conducted premières of *The Soldier's Tale* (1918), *Rape of Lucretia* (1946), and Martin's *Der Sturm* (1956). His opera performances, especially of *Pelléas*, *Boris*, *Magic Flute* in Hamburg, Geneva, and elsewhere in his latter years, were widely admired for the clarity and judgement that marked all his music-making. Wrote several books and essays on music and philosophy. (R)

BIBL: B. Gavoty, *Ernest Ansermet* (Geneva, 1961).

**antefatto** (It.: 'antecedent fact'). The device for explaining to an audience background facts or previous events which they must know in order to understand the plot from the start; it may be contained in a narration, as with *Il trovatore*, or printed in a prefatory paragraph to a libretto, as with some by Piave and Romani. See also ARGOMENTO.

**Antheil, George** (*b* Trenton, NJ, 8 July 1900; *d* New York, 12 Feb. 1959). US composer and pianist. Studied with Sternberg and Bloch. Assistant music director, Berlin Stadttheater, 1928–9. His first opera, *Transatlantic*, caused a stir on its first production: it concerns a corrupt presidential campaign, and includes prominent jazz elements. Autobiography, *Bad Boy of Music* (New York, 1945).

SELECT WORKLIST: *Transatlantic* (Antheil; Frankfurt 1930); *Helen Retires* (Erskine; New York 1934); *Volpone* (Perry, after Jonson; Los Angeles 1953).

**Antigone.** In Greek mythology, the daughter of Oedipus by his mother Jocasta. After Oedipus had put out his eyes, she accompanied him to Colonus, returning to Thebes in order to defy the tyrant Creon's ban on the burial of her brother Polyneices. Sentenced to be buried alive, she killed herself with her lover Haemon, Creon's son. Operas on the subject, mostly derived from Sophocles' drama (441 BC), are by Orlandini (1718), Galuppi (1751), Fini (1751), Casali (1752), Latilla (1753), Bertoni (1755), G. Scarlatti (1755), Ciampi (1762), De Majo (1768), Bianchini (1769), Traetta (1772), Mysliveček (1773), Mortellari (1776), Campobasso (1788), Winter (1791), Bianchi (1796), Pedrotti (1839), Honegger (1927), Ghislanzoni (1929), Liviabella (1942), Orff (1949) and Joubert (1954). Metastasio's *Antigono*, the text of many settings, takes as its subject Alexander the Great's general Antigonus.

**antimasque** (sometimes also **antemasque**). A section of the *masque featuring grotesque, comic, or vulgar characters, included as a contrast to the more elevated and refined spirit of the masque proper. First seen in *The Hue and Cry after Cupid* (Jonson, 1608) and *The Masque of Queens* (Jonson, 1609), the antimasque was often the most popular part of the masque and was usually performed by professional actors. Also sometimes used later in a wider sense to describe any masque whose characters were not the grand ones of the more elaborate entertainments.

**Antonia.** The consumptive singer (sop), one of Hoffmann's lovers, in Offenbach's *Les contes d'Hoffmann*.

**Antonio.** The gardener (bs), Barbarina's father and Susanna's uncle, in Mozart's *Le nozze di Figaro*.

**Antony and Cleopatra.** Opera in 3 acts by Samuel Barber; text by Franco Zeffirelli, based on Shakespeare's drama (1606–7). Prod. New York, M, 16 Sep. 1966, with Leontyne Price, Diaz, Jess Thomas, Flagello, cond. Schippers. Commissioned for opening of new theatre.

The opera tells of the love between Antony (bar), ruler of the eastern portion of the Roman Empire, and Cleopatra (sop), Queen of Egypt. Despite conflict between their two nations, and Antony's marriage to Octavia (sop), daughter of another member of the ruling triumvirate, the couple are united in death. Antony kills himself when he believes Cleopatra to be dead and, poisoned by the bite of an asp smuggled to her past guards, she also dies.

**Antwerp** (Fr., Anvers; Dut., Antwerpen). City in Antwerpen, Belgium. Home of the Royal Flemish O, founded 1893 by the bass Hendrik Fontaine. Performances were originally given in the Royal Dutch T, but since 1907 the Opera has had its own theatre (cap. 1,050), and the title Royal was granted to it in 1920. The company is subsidized by state, city, and province. All operas are sung in Flemish, and over 300 artists are employed each season. During the 1960s several important foreign works had their Belgian premières there, e.g. *A Midsummer Night's Dream* (1962), *Capriccio* (1964), and *The Nose* (1965). Performances of *The Ring* are a regular feature. The large repertory also includes works by Flemish composers, e.g. Blockx and August de Boeck.

The Antwerp Chamber O, founded 1960, amalgamated with the Dutch Chamber O in 1971 to form the Flemish Chamber O, with its headquarters in the Ring T.

**anvil** (Fr., *enclume*; Ger., *Amboss*; It., *incudine*). A percussion instrument, imitating a real anvil, first used in opera in Le Sueur's *La mort d'Adam* (1809). Anvils appear in both Wagner's *Rheingold* and Verdi's *Il trovatore* (1853); they normally consist of metal bars mounted on a resonator. Verdi, in the chorus 'Chi del gitano', notated his two on a higher and lower C, to fit the harmony, though the pitch is actually indeterminate. Wagner, to indicate the labouring Nibelungs, wrote for 18 anvils in three groups of sizes (3.3.3: 2.2.2: 1.1.1.), notating them on three octave Fs. A real anvil appears on the stage in *Siegfried*, to be split by the hero with the newly forged sword Nothung. Real anvils also appear on the stage in Auber's *Le maçon*, are scored for as *petites enclumes* in Berlioz's *Benvenuto Cellini*, and occur in Gounod's *Philémon et Baucis*.

**Anvil Chorus.** The chorus of gypsies 'Chi del gitano' in Act II of Verdi's *Il trovatore*, swinging their hammers in time to the music as they work at their forging.

**Aperghis, Georges** (*b* Athens, 23 Dec. 1945). Greek composer. Studied Athens with Papaioannou, moving to Paris, where he was influenced by Xenakis and Kagel. Has taken a great interest in all forms of musical theatre, including the traditional, and written works that have won him a respectful following, especially in France. *Jacques le fataliste* (after Diderot, 1974) is characteristic in its mix of genres (including here use of cinema) and an idiomatic range that includes chant and atonal music.

**Apollo.** In Greek mythology, the son of Zeus and Leto. He was the god of the sun, of agriculture, also of prophecy and song: the latter two attributes were held by the Greeks to be connected. He is as frequently represented holding a lyre as, in his manifestation as a warrior god, armed with a bow.

Operas on him, or aspects of his legend, are by Kaiser Leopold I (1669), Buini (1720), Stölzel (1729), Aliprandi (1737), Mozart (1767), Franceschini (1769), and Parma (1909). See also DAPHNE.

In late-19th-cent. German aesthetics, under the influence of Nietzsche's *The Birth of Tragedy* (1872), the Apollonian principle in art stands for order, reason, and discipline, in opposition to the Dionysian principle of the instinctive, physical, and elemental. It is this antithesis which is dramatized in Szymanowski's *King Roger* and is at the centre of Britten's *Death in Venice*.

**Apollo et Hyacinthus, seu Hyacinthi Metamorphosis.** Opera in 1 act by Mozart; text by Rufinus Widl. Prod. Salzburg 13 May 1767; London, Fortune T, 24 Jan. 1955.

The plot, a revised version of the story in Ovid's *Metamorphoses*, concerns the love of Zephyrus (sop cas) for Melia (sop) and his jealousy of Apollo (con cas). Apollo slays Hyacinthus (sop cas), turning him into a flower, has Zephyrus borne away by the winds, and himself weds Melia.

The work, comprising a prologue and nine numbers, is not strictly an opera, but an intermezzo to Widl's drama *Clementia Croesi*.

**Apolloni, Giovanni Filippo** (*b* Arezzo, *c*.1635; *d* Arezzo, ?15 May 1688). Italian librettist. From 1653–9 at the court of Archduke Ferdinand Karl of Austria in Innsbruck, collaborating with his friend Cesti on *Argia* (1655) and *Dori* (1657). The latter proved especially popular; its style was strongly influenced by \*Cicognini. Returning to Italy in 1659, Apolloni was employed by Cardinal Flavio Chigi in Rome: his later librettos include

*Circe* (Stradella, 1668) and *Amor per vendetta* (Pasquini, 1673).

**Apolloni, Salvatore** (*b* Venice, *c*.1704; *d*?). Italian composer. A barber by trade, he played the violin in the orchestra of the T di S Samuele, Venice, for which he composed three works. Written for their troupe of comedians, these were mostly witty parodies of opere serie and proved highly popular: *La fama dell'onore* was later revived in Vienna and *Le metamorfosi odiamorose* in Dresden.

SELECT WORKLIST: *La fama dell'onore* (Miani; Venice 1727); *Le metamorfosi odiamorose* (Gori; Venice 1732).

**Apostolescu, Nicu** (*b* Brăila, 31 July 1896). Romanian tenor, formerly baritone. Studied Bucharest with Popovici. Début Cluj, 1921 (Amonasro); début as tenor, 1923 (Radamès), while continuing to sing Scarpia, Rigoletto, etc. Bucharest 1926–44, where he was the first Romanian Parsifal, Tristan, Tannhäuser, Walther, and Otello (Verdi). Italy 1928–30. A fine singing actor. (R)

**appaltatore** (It.: 'contractor'). A theatrical agent, sometimes also impresario.

**Appia, Adolphe** (*b* Geneva, 1 Sep. 1862; *d* Nyon, 29 Feb. 1928). Swiss designer. Studied Geneva, Leipzig, and Dresden; also spent much time examining contemporary theatrical practice, especially at Bayreuth. His pioneering designs for Wagner are based on three-dimensional forms, sometimes stylized or abstract, intended as an extension of the music so as to express it visually, and with the singing actor giving them life. He also interested himself in lighting as part of the scenery in its own right. His intention was to realize the works' inner action, 'to enable the audience to see things the way the heroes of the drama typically see them': thus, Isolde should be surrounded not with 'the mild summer's night' but 'the terrible space that separates her from Tristan'. He designed *Tristan* for Milan, S (1923), *Rheingold* and *Walküre* for Basle (1924–5). Cosima Wagner, though interested, rejected his designs on the grounds that Wagner himself had already said all there was to be said on the subject; but Appia's influence on the ideas of Wieland and Wolfgang Wagner was crucial.

**Aprile, Giuseppe** (known as Sciroletto or Scirolino) (*b* Martina Franca, 29 Oct. 1732; *d* Martina Franca, 11 Jan. 1813). Italian contralto castrato. Studied Naples with Sciroli, whence his nicknames. Début Naples, C, 1753 (small roles); began his international career Parma 1755. In 1756 was engaged on Jommelli's suggestion for Stuttgart, where he remained for ten years, and where Schubart heard him and wrote, 'In him art and nature are marvellously combined; he sings with the purity of a bell up to E above the treble stave.' Taught from 1780; one of his pupils was Lady Hamilton.

**Arabella.** Opera in 3 acts by Richard Strauss; text by Hofmannsthal. Prod. Dresden 1 July 1933, with Ursuleac, Bokor, Jerger, cond. Krauss; London, CG, 17 May 1934, same artists; New York, M, 10 Feb. 1955, with Steber, Gueden, London, cond. Kempe.

Vienna, Carnival 1860. Graf Walder (bs) is in debt, but hopes to find a rich husband for his daughter Arabella (sop). He writes to his friends and one, Mandryka (bar), offers help if he may marry Arabella. A young officer, Matteo (ten), is in love with Arabella, though she rejects him together with her other suitors. Her sister Zdenka (sop), however, who has been brought up as a boy because Walder cannot afford to have two débutante daughters, loves Matteo and tells Arabella that it was she who wrote to him, pretending to be her sister.

Arabella realizes that Mandryka is the interesting stranger she has seen outside their hotel; she abandons her other suitors, to Matteo's despair. However, Zdenka gives Matteo a note, purportedly from Arabella, containing the key to her room. Mandryka overhears, and suspects Arabella's behaviour.

After his assignment, Matteo descends to the hall of the hotel and is astonished to see Arabella entering. In the middle of their misunderstanding, Arabella's parents and Mandryka arrive; the situation seems to confirm their worst suspicions. Zdenka then appears, declaring her love for Matteo and her intention of throwing herself into the Danube; it emerges that it was she who made, and kept, the assignment. Matteo seems happy to transfer his attentions to the younger sister. Arabella offers Mandryka a glass of water, following an old Croatian custom symbolizing a bride's purity, and the couple embrace.

Hofmannsthal's last Strauss libretto, left partly unrevised at his death.

BIBL: K. Birkin, *Richard Strauss: 'Arabella'* (Cambridge, 1989).

**Aragall, Giacomo** (*b* Barcelona, 6 June 1939). Spanish tenor. Studied Barcelona with Puig, Milan with Badiali. Début Venice, F, 1963 (Gastone, *Gerusalemme*). Milan, S, from 1963; Vienna, S, and London, CG, 1966; New York, M, 1968; Naples, C, 1972; also Chicago, Edinburgh, San Francisco, Berlin, etc. Roles incl. Don Carlos, Duke of Mantua, Rodolfo, Pinkerton, Werther, and Faust and Roméo (Gounod). (R)

**Araia, Francesco** (*b* Naples, 25 June 1709; *d* ?Bologna, before 1770). Italian composer. After writing several operas for Italian cities, he was appointed to the St Petersburg court by the Empress Anne. He began with a new version of his *La forza dell'amore e dell'odio* in 1736. After a return to Italy to form a new troupe, he staged for the Empress Elizabeth operas including *Cephalus and Procris* (1755), the first opera sung in Russian. His work thereafter became less popular, and he withdrew to Italy.

**Araiza, Francisco** (*b* Mexico City, 4 Oct. 1950). Mexican tenor. Studied Mexico City with I. Gonzalez. Début Mexico City 1973 (Rodolfo). Munich, Bayreuth 1978; Paris, O, 1981; Salzburg 1983; New York, M, 1984; also Brussels, Aix, Vienna, etc. Roles incl. Belmonte, Ferrando, Tamino, Idreno (*Semiramide*), Steersman, Des Grieux (Massenet). (R)

**Archers, The.** Opera in 3 acts by Benjamin Carr; text by W. Dunlap, based on Schiller's version of the traditional William Tell story. Prod. New York, John St. T, by Old American OC, 18 Apr. 1796. The first US opera of which parts of the music survive: there are two extant numbers, a rondo from the overture, and the song 'Why, huntress, why?'.

**Arditi, Luigi** (*b* Crescentino, 16 July 1822; *d* Hove, 1 May 1903). Italian composer and conductor. Studied Milan, incl. with Vaccai. Conducted in Milan, then in Havana, USA, and Russia, before settling in London. Conducted regularly at London, HM and CG, 1858–92 and also many touring companies, including those organized by Mapleson with Patti, as well as continuing to make many appearances on the Continent. Conducted first London performances of many operas. Composed several operas, though he is best remembered for the song 'Il bacio'. Autobiography, *My Reminiscences* (London, 1896).

**'Ardon gl'incensi'.** The beginning of the Larghetto section of Lucia's (sop) Mad Scene in Act III of Donizetti's *Lucia di Lammermoor*, in which she imagines her wedding to Edgardo.

**Arensky, Anton** (*b* Novgorod, 12 July 1861; *d* Terioki, 25 Feb. 1906). Russian composer and conductor. Studied St Petersburg, including with Rimsky-Korsakov. Moving to Moscow, he taught Rakhmaninov, Skryabin, and Glière, and made friends with Tchaikovsky and Taneyev. His greatest operatic success was with *Son na Volge*, on a text passed on to him by Tchaikovsky on abandoning *The Voyevoda*. His later operas were less successful. His style, elegant and lyrical if somewhat lacking in personality, owes much to Tchaikovsky's manner.

WORKLIST (all first prod. Moscow): *Son na Volge* (A Dream on the Volga) (Ostrovsky & Tchaikovsky; 1891); *Raphael* (Kryukov; 1894); *Nal i Damayanti* (M. Tchaikovsky, after Zhukovsky; 1904).

**Argentina.** For all its present richness, the operatic life of Argentina did not develop until the last third of the 19th cent. The country was then dependent on Italian singers visiting Buenos Aires. As in most countries, the first national operas treated local subjects (Inca stories and, to a lesser degree, popular tales of pampas life) in an Italianate style; later works drew upon folk melodies and national musical characteristics. The first important Argentinian opera was *La gata blanca* (1877) by Francisco Hargreaves (1849–1900), a work in buffo vein that proved very influential. Arturo Berutti (1862–1938) won a popular success with his *Pampa* (1897, after the gaucho drama *Juan Moreira* by Juan José Podestá); his other works include *Yupanky* (1899), the very successful *Khryse* (1902), *Horrida nox* (1908), and *Los héroes* (1909), each in some way making use of indigenous Argentinian subjects and colouring. Ettore Panizza's *Aurora* (1908; libretto by Illica) sought to reconcile folklore with a verismo manner. Other operas showing a strong European influence were by Constantino Gaito (1879–1945) (11 operas, including *Shaffras* (1910), *Caio Petronio* (1919), *Flor de Nieve* (1922), and *Ollantay* (1926)), and E. García Mansilla (a Rimsky-Korsakov pupil whose *Ivan* was produced in St Petersburg).

The first work which really sought to make fully operatic use of Argentinian musical characteristics was *Huemac* (1916), by Pascual de Rogatis (1880–1980), though it retains a European tinge. The first local opera in Spanish was by Felipe Boero (1884–1958), his *Tucumán*, which won the first local opera competition in 1918. Many Argentinian operas now began to appear, including Felipe Boero's *Raquela* (1923), set on a pampas ranch, and his *El matrero* (1929), also set in the countryside and, for its effective use of local tunes and dances, regarded as one of the most important Argentinian operas. Other significant works were written by Alfredo Schiuma (1885–1963) (*Amy Robsart* (1920), *Tabaré* (1925)), Gilardo Gilardi (1889–1963) (*Ilse* (1923), *La leyenda del Urutaú* (1934)), Raúl Espoile (1888–1958) (*La ciudad roja* (1936)), and Arnaldo D'Esposito (1907–45) (*Lin-Candel* (1941)). These works use local subjects and music drawn either from Inca scales and rhythms or from folksong and dance. Parallel with this development went a movement that kept strong links with European opera: important works were written by Carlos López Buchardo (1881–1948) (*El sueno del Alma* (1914)), Athos Palma (1891–1951) (*Nazdah* (1924)), Rafael Peacan del Sar (1884–?)

(*Chrysanthème* (1927): a Puccinian work based on Pierre Loti and set in Japan), J. B. Massa (1885–1938) (*La Magdalena* (1929)), Panizza (*Bizancio* (1939)), and Juan José Castro (1895–1968) (*Proserpina y el extranjero* (1951), a Verdi prize opera transferring classical legend to South America). The most important post-war opera composer has been Alberto *Ginastera. His pupil Antonio Tauriello (*b* 1931) has conducted at the Colón and in the USA; his operas include *Les guerres picrocholines* (after Rabelais, comp. 1971).

Outside Buenos Aires, the country has been largely dependent on touring opera, with few premières. Exceptions are Massa's *L'evaso* (1922) at Rosario and Enrique Casella's *Corimayo* (1926) and *Chasca* (1939) at Tucumán. Occasional seasons have been given at Rosario, Tucumán, Mendoza, Santa Fe, Córdoba, and Bahía Blanca, by companies based on Buenos Aires. At the T Argentino of La Plata, the capital of the province of Buenos Aires, there have been regular seasons.

See also BUENOS AIRES, COLÓN.

**Argento, Dominick** (*b* York, PA, 27 Oct. 1927). US composer. Studied Baltimore with Nabokov and Cowell; Rochester with Hovhaness and Hanson, also acting as opera coach; Italy with Dallapiccola. Music director, Hilltop O, Baltimore, also forming Center OC, Minnesota, for which he wrote the opening work, *The Masque of Angels* (1964), and *Postcard from Morocco* (1971). Opera has been at the centre of his interests, and his works show a wide range of taste and techniques, as well as a sympathetic understanding of the voice.

SELECT WORKLIST: *The Boor* (Olon-Scrymgeour, after Chekhov; Rochester 1957); *Christopher Sly* (Manlove, after Shakespeare; Minneapolis 1963); *The Masque of Angels* (Olon-Scrymgeour; Minneapolis 1964); *Postcard from Morocco* (Donahue; Minneapolis 1971); *The Voyage of Edgar Allan Poe* (Nolte; Minneapolis 1976); *The Aspern Papers* (after James; Dallas 1988).

**argomento** (It.: 'argument'). In 17th-cent. and early 18th-cent. librettos, especially Venetian, the text was usually prefaced by an argument, or summary of events that had taken place before the opera story; sometimes this included a summary of the plot itself. Although the argomento was often complex and involved, the audience was usually well acquainted with it, since copies of the libretto were widely sold or circulated before the performance. See also CERENI, ANTEFATTO.

**Århus.** City in Jutland, Denmark. Founded in 1947 at the Musikhuset (cap. 1,477), Den Jyske Opera (Jutland Opera) is Denmark's second largest company, and tours widely. A new concert-house, in which opera is given, opened in

1982 with Nielsen's *Maskarade*. A complete *Ring* was given there in 1987. Århus Sommeropera opened in the Helsingore T (cap. 300) in 1989.

**aria** (It.: 'air', from Gr. and Lat., *aer*, a term which came to mean a scheme or model, i.e. a melody which could be repeated with different words). The elaborated song-form of opera (and oratorio) for solo voice (occasionally two voices) with instrumental accompaniment. Its nature has changed substantially, with its high point of formality *c*.1680–1780 and then growing looser.

Even in the earliest operas, moments of heightened expression or deepened contemplation were emphasized, with the *stile rappresentativo becoming noticeably more melismatic: the imposition of a regular formal structure upon this *arioso style resulted in the aria, the first great example of which is 'Possente spirto' in Monteverdi's *Orfeo*. During the first half of the 17th cent. the stylistic division between the music used for narrative and reflective text widened further in Roman and Venetian opera, gradually settling into the categories of *recitative and aria, although such rigid definitions were not widely accepted until much later. The first use of the word aria, in the score of Mazzochi's *La catena d'Adone* (1626), embraces both styles of solo writing, as well as duets and other ensembles. At this point, the music of the embryonic aria was organized according to a variety of formal patterns: most popular were the *strophic aria, the aria over a ground bass (the so-called *ostinato aria), and the *da capo aria. Sometimes the forms were used in combination, as in the final duet of Monteverdi's *Poppea*, which employs both ostinato and da capo principles.

An important distinction between recitative and aria, and one of great use when deciding how best to describe passages from operas composed up to *c*.1680, when the two styles were still blurred, concerns the presence or absence of word repetition. The aria, as a point of repose, usually took a small amount of text and repeated it, whereas the recitative, not repetitive, was set more in speech rhythms, and was more narrative. This practice was adumbrated in the works of A. Scarlatti and his contemporaries and became routine in the *opera seria of Handel and Hasse. Following *Zeno and his contemporaries, and even more so *Metastasio, composers concentrated on the aria, maintaining the recitative as a narrative link between highly stylized outpourings of emotion. Most of the arias were of the **da capo** variety, cast in three parts, the second of which presented a complementary poetic image to the first, usually in a contrasting, but related key, such as relative major or minor, while the third repeated the opening section with florid ornamentation added by the singer. During the mid-18th cent. it became customary for the return of the first section to be abridged, by starting at a later point than its beginning: this variant was known as the **dal segno aria**. In accordance with the *doctrine of affections, each aria was framed around only one main emotion, although the middle section provided an element of variety, often by introducing a simile drawn from nature to which a moral would be applied. This structure involved such verbal repetition that the text was sometimes overwhelmed by the music, but by now the aria had become so formalized that the musical style alone proclaimed its prevailing character. So uniform was the treatment of emotions that arias could be (and frequently were) transposed from one opera to another, even from one composer's work to that of another.

In part, the supremacy of the da capo aria was also a response to the pre-eminence of the solo singer whose virtuosity, beginning with the works of Cavalli, had increasingly come to dominate opera. Rigid rules applied to the distribution of the arias throughout the opera and this, combined with the overriding prominence of the **aria di sortita** (entrance aria), whose conventions demanded that the singer leave the stage after its performance, served to inhibit any real sense of dramatic progression. Each singer had to have at least one aria in each act, but was not allowed to have two arias consecutively; each principal was required to have five arias, all different in character. The secondary characters had fewer, so that their subordinate status was maintained, and no two arias of a similar kind could be placed together, even if sung by different characters. To handle all these rules, an elaborate categorization of aria-types grew up, widely observed, but with local variations. Arias chiefly concerned with expressive content included the **aria di carattere**, **aria di mezzo carattere** (usually more restrained in feeling), **aria di sentimento**, **aria di strepito**, **aria agitata**, **aria infuriata**, and **aria di imitazione** (in which voice and instruments vied with each other in imitating the sounds of nature). Arias chiefly for technical display included the **aria di bravura**, **aria di agilità**, **aria di portamento** (usually dignified with a wide dynamic range, unornamented) and **aria cantabile** (gentle and sad, delicately ornamented). Arias more directly connected to the dramatic action included the **aria di lamento** (or simply lamento) and **aria del sonno** (or sleep aria, sung to the sleeping beloved). Arias of a more declamatory nature included the **aria parlante** (sometimes also called the **aria di nota e parola**) and **aria declamata** (usually for expressing powerful or passionate sentiments). Arias concentrating on formal niceties included the **aria in rondò** (a common 5-part form was a-b-a-c-a) and **aria variata** (in the form of theme and variations). There was also the elaborately accompanied **aria concertata** and the very rare unaccompanied **aria senza accompagnamento**. Miscellaneous categories included the **aria di baule** (or suitcase aria, the favourite number which a singer carried in his luggage for introduction into whichever opera he sang) and the **aria del sorbetto** (or sorbet aria, the number for a minor character, which gave the spectators the chance to retreat to enjoy an ice). Occasionally, the aria was sung by more than one performer: Scarlatti has examples of an **aria a quattro** for four people. Naturally, these categories were usually assigned to principals and to noble characters: those of humbler birth were normally confined to the **arietta** or **canzonetta**. These were much simpler in style, rather like song, neither lengthy nor difficult, and often cast in a strophic form recalling the earlier treatment of such characters in Venetian opera.

The influence of the 18th-cent. aria was considerable and extended beyond the immediate confines of the opera-house: the slow movements of many instrumental works, for example, are often described as being in 'aria style', on account of their da capo structure and expressive nature. However, the monotonous alternation of aria and recitative rapidly became a suffocating convention, and from the 1750s onwards a gradual (if often explicitly unacknowledged) reform of opera seria took place. Beginning with Jommelli and Traetta, these changes were an attempt to rekindle the spirit of genuine drama within opera. This was achieved partly by allowing the action to advance within the aria, and partly by paying less heed to the doctrine of affections, presenting a wider emotional range within individual arias. Instead of recitative-aria chains, such composers often preferred to construct longer sequences in which aria was replaced by arioso, and freely mixed with recitative, ensemble, and choral writing.

The most important reform of the mid-18th cent. came with Gluck, who found his inspiration in French opera. From its earliest beginnings under Lully, this tradition had been characterized by a more refined style of melodic writing, in which florid virtuosity was of less importance. Indeed, such displays were largely relegated to the **ariette**, a coloratura piece with Italian words, not to be confused with a later simple style of French song which formed part of the *comédie mêlée d'ariettes*. Under Rameau the contrast with opera seria became more marked, and the **air** a noble and dignified expression of emotion. In Gluck's reform operas (notably *Orfeo ed Euridice* and *Alceste*) this French style was carefully assimilated, leading to a noticeably more balanced and restrained vocal line. *Opera buffa also provided inspiration to some composers as, in German countries, did *Singspiel, for both permitted greater flexibility in the crucial respect that there was no prohibition on action taking place within the aria. In other respects, opera buffa had certain similarities with opera seria, not least in having a similarly accepted categorization of aria types, including that of the *catalogue aria. All of these strands come together in Mozart, where there is a remarkable degree of cross-fertilization between the different genres. *Die Zauberflöte*, for example, although usually considered a Singspiel, draws on all three traditions and even its serious numbers go beyond the model of da capo aria to embrace other forms as well.

After 1800 the growth of strong national traditions lent the solo vocal passages in opera a richness and diversity that prevents easy description. However, two main concerns of such writing may be identified, concerns shared with the practices of opera seria which they replaced: the expression of emotion, and the display of vocal technique. In 19th-cent. Italian opera these functions were admirably served by the *cantabile-*cabaletta pairing, a structure already adumbrated by Mozart (e.g. *Idomeneo*, *Die Zauberflöte*) and found also in many German works (e.g. Beethoven's *Fidelio*). The important differences in 19th-cent. opera were that dramatic action during the aria was now entirely accepted, while an increased use of ensemble and choral writing (which, with few exceptions, was totally absent in 18th-cent. opera seria) had the effect of throwing the appearances of the solo singers into greater prominence. This was particularly true of French *Grand Opera, where the protagonists were presented against a spectacle of epic proportions, hence rendering simple solo vocal numbers effective (e.g. 'Par une telle nuit' in Berlioz's *Les Troyens*).

The *convenienze of the Italian tradition notwithstanding, the main movement throughout the 19th cent. was not towards a development of the aria as an individual form, but in favour of its subsumption into a more genuinely dramatic framework, where solo vocal music of all kinds was telescoped together with ensemble and choral writing. The precedent for this was provided in the 18th-cent. *ensemble finale, and another important early step was taken with the construction of elaborate *scenas in Italian opera. This breakdown of formal divisions reached its apotheosis, and most cogent expression as a principle of operatic writing, in Wagner. However, even here the ancient dramatic instinct to balance narration with reflection is shown in moments whose melodic distinctiveness and contemplative intensity suggest the equivalent of an aria; e.g. Siegmund's 'Winterstürme' or Walther's 'Am stillen Herd'. In Italian opera the large number of individual passages performed from, for example, Puccini's works (where to all superficial appearances the old 'number' structure has been totally supplanted by continuous music) also suggest equal caution in proclaiming the death of the aria in the 19th cent. Subsequently, the rise of the neo-classical movement in the 20th cent. has re-kindled an interest in the more or less formal aria as capable of playing a functional role in modern opera.

**Ariadne.** In Greek mythology, the daughter of King Minos of Crete and Pasiphae. Having helped Theseus escape from the labyrinth concealing the Minotaur, she fled with him to Naxos, where he abandoned her. In some versions of the legend she killed herself, in others she became the bride of Dionysus (Bacchus). Among some 50 operas on the subject are those by Monteverdi (1608: the first setting, lost apart from the Lament), Cambert (1674), Conradi (1691), Kusser (1692), Marais (1696), Porpora (1714), Leo (1721), Keiser (1722), Feo (1728), Handel (1734), Ristori (1736), Orlandini (1739), Sarti (1756), Galuppi (1763), Benda (duodrama, 1775), Anfossi (1781), Klein (1825), Massenet (1906), Richard Strauss (1912, rev. 1916), and Tovey (1929).

**Ariadne auf Naxos** (Ariadne on Naxos). Opera in 1 act, later with a scenic prelude, by Richard Strauss; text by Hofmannsthal. Originally intended to follow Molière's comédie-ballet, *Le bourgeois gentilhomme* (1670), for which Strauss had written incidental music: in this form, prod. Stuttgart 25 Oct. 1912, with Jeritza, Siems, Jadlowker, cond. Strauss; London, HM, 27 May 1913, wih Von der Osten, Bosetti, Mařák, cond. Beecham. Second version prod. Vienna, H, 4 Oct. 1916, with Jeritza, Kurz, Lotte Lehmann, Körney, cond. Schalk; London, CG, 27 May 1924, with Lotte Lehmann, Ivogün, Schumann, F. Niemann, cond. Alwin; Philadelphia, AM, 1 Nov. 1928, with Peterson, Boykin, Williams, House, cond. Smallens.

The work's first form was impracticable, requiring both a theatrical and an operatic company for the same evening. The second, operatic, version substitutes for Molière's play a prologue in which word is sent from the 'bourgeois gentilhomme', who has engaged an opera company and a commedia dell'arte troupe, that their entertainments must take place simultaneously so as to be over in time for a fireworks display.

Vienna, a rich house. Prologue: the Composer (sop), troubled about his art, is consoled first by the old Music Master (bs), then by Zerbinetta

(sop), leader of a group of players, and lastly by his faith in music. Opera: on her island, Ariadne (sop) is the subject of much comment from the comedians. She sings of her longing for the kingdom of death, at which Zerbinetta tries to convert her to a more flirtatious philosophy. When Bacchus (ten) arrives, Ariadne greets him as Death. They ascend into the sky together, which Zerbinetta considers acceptance of her advice.

BIBL: Karen Forsyth, *Ariadne auf Naxos by Hugo von Hofmannsthal and Richard Strauss: its Genesis and Meaning* (Oxford, 1982).

**Ariane et Barbe-bleue** (Ariadne and Bluebeard). Opera in 3 acts by Dukas; text adapted from Maeterlinck's drama (1901), written with Dukas in mind. Prod. Paris, OC, 10 May 1907, with Georgette Leblanc, Vieuille, cond. Ruhlmann; New York, M, 29 Mar. 1911, with Farrar, Rothier, cond. Toscanini; London, CG, 20 Apr. 1937, with Lubin, Etcheverry, cond. Gaubert.

Barbe-Bleue (bs) has just married his sixth wife Ariane (mez). The villagers fear she will disappear mysteriously, as her predecessors have done. Ariane resolves to discover their fate: she finds them hidden in a vault in the castle and releases them. The villagers attack Barbe-bleue and force him to confront his wives. They seem to bear no grudges against him, and bathe his wounds.

Dukas's only opera, initially successful.

See also BLUEBEARD.

**Arianna in Creta.** (Ariadne in Crete). Opera in 3 acts by Handel; text ?F. Colman, after Pariati's text *Arianna e Teseo* set by Porpora (1727) and Leo (1729). Prod. London, H, 26 Jan. 1734 with Strada del Pò, Carestini, Negri, Scalzi, Waltz, Durastanti.

**arietta.** See ARIA.

**ariette** (Fr.). In 18th-cent. French opera a short vocal number, lively in character, with orchestral accompaniment. Later took on the more general meaning of 'song': used in this sense in the generic term *comédie mêlée d'ariettes.

**Ariodant.** Opera in 3 acts by Méhul; text by F. B. Hoffman, after Ariosto's poem *Orlando furioso* (1516). Prod. Paris, OC, 11 Oct. 1799.

The plot (which has some anticipations of Weber's *Euryanthe*) turns on the attempts of Othon (ten) to compromise Ina (sop) so as to force her to abandon her lover Ariodant (ten) and marry him; these are eventually frustrated. Méhul's favourite among his operas, it suffered from the production six months previously of Berton's *Montano et Stéphanie* on the same subject. Notable for its functionally dark orchestration.

See also ORLANDO.

**Ariodante.** Opera in 3 acts by Handel; text ad. from A. Salvi, after Ariosto's poem *Orlando furioso* (1516). Prod. London, CG, 8 Jan. 1735, with Strada, Negri, Young, Carestini, Beard, Waltz, Stoppelaer; New York, Carnegie H, 9 Mar. 1971, with Steffan, Wise, Caplan, Raskin, Stewart, Meredith, Solem, cond. Simon.
See also ORLANDO.

**arioso** (sometimes also **recitativo arioso**; It.: 'like an aria'). A musical style and manner of setting text falling between that of *recitative and *aria, which, lacking the essentially declamatory nature of the former and the regular phrase structure and organization of the latter, is characterized by an expressive melodic line and frequent melismatic word-setting. The practice of interrupting recitative with arioso passages for dramatic contrast was first used to prominent effect in mid-17th-cent. Venetian opera. In later opera seria, arioso appeared mostly in accompanied recitatives; its increased use (e.g. in Gluck) served gradually to weaken the rigid distinction between aria and recitative. In its widest sense, arioso style can be used to describe any music of a predominantly lyrical nature which falls outside the remit of aria proper; it is sometimes also applied (e.g. by Handel in *Orlando*) to refer to a short aria.

**Ariosti, Attilio** (*b* Bologna, 5 Nov. 1666; *d* ?England, ?1729). Italian composer. In the course of a colourful life, which is only partly documented, Ariosti composed over 20 works for the stage, and played a leading part in Handel's operatic activities. He took holy orders in his native city, but joined the service of the Duke of Mantua in 1696. The following year several operatic works, including the dramma pastorale *Tirsi*, were performed in Venice with some success. Shortly afterwards Ariosti went to the Berlin court where, though he held no formal appointment, he was effectively Kapellmeister. Several stage works were written there, including the highly successful *La fede ne' tradimenti*. But the presence of a Catholic monk at a Protestant court sparked off a diplomatic crisis, necessitating Ariosti's return to Italy in 1703. Passing through Vienna, Ariosti obtained a position at the Austrian court, for which he wrote several operas including *Il Danubio consolato*. Under Emperor Joseph I Ariosti undertook a return visit to his native country as Austrian representative to all the courts of Italy, but on Joseph's death in 1711 was banished from Vienna because of his immoderate behaviour.

Little is then known of Ariosti until 1716, when he appeared in London. Involved at an early stage in the establishment of the *Royal Academy of Music, he was formally engaged as

composer from 1722, joining the company of Handel and Bononcini in the operatic war currently raging in London. For a while he enjoyed some success: his first two operas, *Coriolano* and *Vespasiano*, were well received, but over the next few years his popularity gradually waned and his last opera *Teuzzone* was a total failure. Because none of the London operas survives complete, it is hard to assess Ariosti's mature style. But he was well known as an instrumental composer—his first recorded appearance in London was as viola d'amore player in Handel's *Amadigi* in 1716—and his skill in this field undoubtedly inspired his operatic music. Though hardly comparable to Handel's, his works have moments of great beauty, with particularly effective descriptive writing.

SELECT WORKLIST: *Tirsi* (Zeno; Venice 1696); *La fede ne' tradimenti* (Gigli; Berlin 1701); *Il Danubio consolato* (Bernadoni; Vienna 1707); *Coriolano* (Haym; London 1723); *Vespasiano* (?Haym, after Morselli; London 1724); *Teuzzone* (Zeno; London 1727).

**Ariosto, Lodovico** (*b* Reggio Emilia, 8 Sept. 1474; *d* Ferrara, 6 July 1535). Italian poet. His most important work is the poem *Orlando furioso* (first edn. 1516, third edn. 1532), written to exalt the ducal house of Este and its legendary ancestor Rogero (Ruggiero), and continuing the story of the love of Orlando (Roland) for Angelica begun in Matteomaria Boiardo's *Orlando innamorato* (1487). Operas based on the poem or its episodes, including those involving Charlemagne, Ginevra, Alcina, Angelica, Medoro, Rinaldo, and Bradamante, are as follows:

F. Caccini (*La liberazione di Ruggiero dall'isola d'Alcina*, 1625), Manelli (*La maga fulminata*, 1638), Rossi (*Il palazzo incantato*, 1642), Sacrati (*L'isola d'Alcina*, 1648), Martinelli (*Alcina*, 1649), Costa (*Ariodante*, 1655), Lucio (*Medoro*, 1658), Freschi (*Olimpia vendicata*, 1685), Lully (*Roland*, 1685), Scarlatti (*Olimpia vendicata*, 1685), Gabrielli (*Carlo il grande*, 1688), Bassani (*Ginevra*, 1690), Steffani (*Orlando generoso*, 1691), Griffini (1697), Sabadini (*Ruggiero*, 1699), Campra (*Alcine*, 1705), Lacoste (*Bradamante*, 1701), Perti (*Ginevra*, 1708), Heinichen (*Olimpia vendicata*, 1709), Scarlatti (*Orlando*, 1711), Ristori (*Orlando furioso*, 1713), Gasparini (*Rodomonte sdegnato*, 1714), Vivaldi (*Orlando finto pazzo*, 1714), Fux (*Angelica*, 1716), Pollarolo (*Ariodante*, 1716), Porpora (*Angelica*, 1720), Sarro (*Ginevra*, 1720), Fiorè (*Ariodante*, 1722), Schürmann (*Orlando furioso*, 1722), anon. (*Angelica y Medoro*, 1722), Bioni (*Orlando furioso*, 1724), Faleoni (*Ginevra*, 1724), Guerra (*Orlando furioso*, 1724), Albinoni (*Alcina delusa da Ruggero*, 1725), Polaroli (*Orlando furioso*, 1725), Vivaldi (*Orlando*, 1727), Broschi (*L'isola di*

*Alcina*, 1728), Handel (*Orlando*, 1733), Sellitto (*Ginevra*, 1733), Handel (*Ariodante*, 1735), Handel (*Alcina*, 1735), Latilla (*Angelica ed Orlando*, 1735), Vivaldi (*Ginevra*, 1736), Lampugnani (*Angelica e Medoro*, 1738), Pescetti (*Angelica e Medoro*, 1739), Latilla (*Olimpia*, 1741), Graun (*Angelica e Medoro*, 1742), Milano (*Angelica e Medoro*, 1745), Wagenseil (*Ariodante*, 1745), Scalabrini (*Angelica e Medoro*, 1746), anon. (*Orlando furioso*, 1746), Majo (*Il sogno d'Olimpia*, 1747), Mele (*Angelica e Medoro*, 1747), anon. (*Angelica e Medoro*, 1747), Bertoni (*Ginevra*, 1753), Brusa (*Angelica*, 1756), Sacchini (*Olimpia tradita*, 1758), Piccinni (*Il nuovo Orlando*, 1764), Paisiello (*Alceste in Ebuda, ovvero Olimpia*, 1768), Guglielmi (*Ruggiero*, 1769), Beecke (*Roland*, after 1770), Guglielmi (*Le pazzie di Orlando*, 1771), Hasse (*Ruggiero*, 1771), Gazzaniga (*L'isola d'Alcina*, 1772), Rust (*L'isola di Alcina*, 1772), Alessandri (*Alcina e Ruggiero*, 1775), Touchemoulin (*I furori di Orlando*, 1777), Carvalho (*Angelica*, 1778), Piccinni (*Roland*, 1778), Schuster (*Ruggiero e Bradamante*, 1779), Haydn (*Orlando paladino*, 1782), Prati (*Olimpia*, 1786), Andreozzi (*Angelica e Medoro*, 1791), Bertoni (*Angelica e Medoro*, 1791), Mortellari (*Angelica*, 1796), Paradis (*Rinaldo e Alcina*, 1797), Isouard (*Ginevra di Scozia*, c.1798), Berton (*Montano et Stéphanie*, 1799), Méhul (*Ariodant*, 1799), Guglielmi (*Alcina*, 1800), Mayr (*Ginevra di Scozia*, 1801), Tritto (*Ginevro ed Ariodante*, 1801), Mosca (*Ginevra di Scozia*, 1802), Vannacci (*Angelica e Medoro*, 1802), Portugal (*Ginevra di Scozia*, 1805), Tozzi (*Angelica e Medoro*, 1805), Reichardt (*Bradamante*, 1809), anon. (*Il ritorno del Ruggiero*, 1810), Nicolini (*Orlando*, 1811), Gandini (*Ruggiero*, 1822), Giraud (*Ruggiero e Bradamante*, 1830s), Loffredo (*Orlando furioso*, 1831), Holmes (*Ruggiero*, 1838), Thomas (*Angélique et Médore*, 1843), Schneider (*Orlando*, 1848), Noberasco (*Ginevra di Scozia*, 1852), Petrali (*Ginevra di Scozia*, 1854), Rota (*Ginevra di Scozia*, 1862), Desormes (*Roland furieux*, 1870s), Lombardi (*Ginevra di Scozia*, 1877), Jachino (*Giocondo e il suo re*, 1924).

**Arkel.** The old blind King of Allemonde (bs) in Debussy's *Pelléas et Mélisande*.

**Arkhipova, Irina** (*b* Moscow, 2 Dec. 1925). Russian mezzo-soprano. Studied Moscow with Savransky, then joining the Sverdlovsk O as soloist (1954–6). Moscow, B, 1956 (Carmen); then Eboli, Amneris, Azucena, and the classic Russian mezzo roles, as well as the first performances of many new works there. In the 1960s began to sing widely in Europe. Montreal 1967, with Moscow B; San Francisco 1972; London, CG, 1975. Possesses a rich, dramatic voice, and an imposing stage presence. (R)

**Arlecchino** (Harlequin). Opera in 1 act by Busoni; text by composer. Prod. in the same bill as his *Turandot*, Zurich 11 May 1917, with Wenck, Grunert, Moissi, cond. Busoni; London, BBC broadcast 27 Jan. 1939, TV 12 Feb. 1939; New York, Carnegie H (semi-staged), 11 Oct. 1951; Gly. 25 June 1954, with Malbin, M. Dickie, Gerster, cond. Pritchard.

Set in Bergamo in the 18th cent., it introduces such familiar commedia dell'arte figures as Arlecchino (spoken), Columbine (sop), Doctor Bombasto (bs), and Leandro (ten). The four parts of the opera are called 'Harlequin as Rogue', 'Harlequin as Soldier', 'Harlequin as Husband', and 'Harlequin as Conqueror'.

**Arlesiana, L'** (The Girl from Arles). Opera in 3 acts (originally 4) by Cilea; text by Leopoldo Marenco, after Daudet's drama (1872). Prod. Milan, L, 27 Nov. 1897, with Ricci, De Paz, Caruso, cond. Zuccari. 3-act version, same theatre, 22 Oct. 1898; Philadelphia, Society Hill Playhouse, 11 Jan. 1962; London, Camden TH, 17 May 1968, with Keetch, R. Remedios, cond. Manton.

Federico (ten), son of Rosa Mamai (sop), is in love with a girl from Arles, but discovers she is the mistress of Metifio (bar), a mule-driver; his family arrange that he marry his foster-sister, Vivetta (sop). He kills himself on the eve of the wedding.

**Arline.** Count Arnheim's daughter (sop) in Balfe's *The Bohemian Girl*.

**Armed Men.** In Mozart's *Magic Flute*, two armed men (*Zwei Gerharnischter*) (ten and bs) lead Tamino and Pamina to the ordeals they must undergo.

**Arme Heinrich, Der** (Poor Heinrich). Opera in 3 acts by Pfitzner; text by James Grun, after a medieval legend. Prod. Mainz 2 Apr. 1895, with Cruvelli, Neumann, Heydrich, cond. Pfitzner.

Heinrich (ten), a singer and a knight, is struck ill in punishment for his arrogance towards God; he will recover only if a virgin sacrifices herself for him. As Dietrich (bar), a faithful servant, and his wife Hilde (sop) fear, their daughter Agnes (sop) offers herself. At the moment of her sacrifice, Heinrich suddenly finds strength and is in time to snatch the sacrificial knife from the doctor's hand.

**Armenia.** The first Armenian opera, by Tigran *Chukhadjyan (1837–98), was *Arshak II* (1868, prod. 1945); this was a story of the liberation of Armenia in the 4th cent., composed on Italian models. Chukhadjyan followed it with a series of lighter works, some using Armenian folk music. He also directed an operetta company, 1872–8. His example encouraged other Armenian composers, including Armen Tigranyan

(1879–1950), whose Romantic opera *Anush* (Alexandropolis, later Leninakan, 1912), set in pre-revolutionary Caucasus, made fuller use of local colour and costumes, and folk instrumental techniques. It has been successfully performed in Europe and America (1989).

The first Armenian ensembles were formed in the 1860s, giving performances of the Russian and international repertory as well as of the pioneer Armenian works; but only in the 1920s did a company begin giving regular performances touring Armenian and Russian towns. It was with this group, under S. Talyan, that many Armenian artists, including Alexander Melik-Pashayev, began their careers. In 1927 the Erevan Conservatory founded an opera class. New Armenian operas included *Seda* (1922) by Anushavan Ter-Gevondyan (1887–1961) and *Almast* (1928, Moscow 1930) by Alexander Spendiaryan (Spendiarov) (1871–1928), both on historical subjects but Russian in musical tradition. The first permanent Armenian company was formed in Erevan in 1932, from the Conservatory class, and began operations on the following 20 Jan. with *Almast*; it was named after Spendiarov. The theatre (cap. 1,200) was built in 1933. Among the Armenian composers encouraged to write for the new ensemble were Aro Stepanyan (1897–1966) (the comedy *Kach Nazar* (Brave Nazar, 1935), and *David Sasunsky* (1936)) and Artemy Aivazyan (1902–75) (another comedy, *Taparnikos* (1938)), all on Armenian subjects and using folk music. A season of Armenian opera was given in Moscow in 1939.

The Spendiaryan Opera and Ballet T (cap. 1,130) opened in Erevan in 1940. Works written for it include Tigranyan's *David-bek* (on the legendary hero, 1950) and Vaagi Araratyan's (*b* 1902) *Tsar Hadjik* (comp. 1945), on patriotic themes. Many works came after the war, including Ter-Gevondyan's children's opera *V luchakh solntsa* (In the Sun's Rays, 1949) and Stepanyan's *Nune* (1947) and *Geroina* (The Heroine, on peasant collectivization, 1950). A second Moscow season followed in 1956. Later Armenian operas include *Artsvaberd* (The Eagle's Lair, 1957) by Andrey Babayev (1923–64), on the post-revolutionary struggles in Armenia, *Sos i Barditer* (1964) by Vartan Tigranyan (1906–74), *Krusheniye* (The Ruin, 1968) by Gevork Armenyan (*b* 1920), and *Skazka dlya Vzroslykh* (A Story for Grown-Ups, 1970) by Erik Arutunyan (*b* 1933). Alexander Arutunyan (*b* 1920), more strongly nationalist in his music, wrote *Sayat Nova* (on the subject of the national bard, 1969). The repertory includes Russian and Armenian works, as does that of the Musical Comedy T founded in 1942, and as well as Western classics the theatre staged the first Soviet productions of *Oedipus rex*

and *West Side Story* (both in 1963). The company made its US début in Los Angeles in 1989.

BIBL: G. Tigranov, *Armyansky muzykalny teatr* (Erevan, 1956 and 1960).

**Armide.** Opera in 5 acts by Gluck; text by Quinault, after Tasso's poem *Gerusalemme liberata* (1575). Prod. Paris, O, 23 Sep. 1777, with Levasseur, Legros, Gelin, Larrivée; London, CG, 6 July 1906, with Bréval, Laffitte, Seveilhac, Crabbé, cond. Messager; New York, M, 14 Nov. 1910, with Fremstad, Caruso, Amato, Gilly, cond. Toscanini.

Armide (sop), who has managed to seduce most of the crusader knights, carries off Renaud (ten) to a magic island, although he has not succumbed. By using supernatural powers, Hydraotes (bar) makes him fall in love with Armide and he forgets his obligations as a crusader. Two of his friends find him and make him realize that he has been under a spell. When Renaud tells Armide of his plan to resume his duties, she is distraught and calls on the gods of the Underworld to make the island sink.

Other operas are by Lully (1686), Bioni (1725), Bertoni (1746), Graun (1751), Sarti (1759), Traetta (1761), Jommelli (1770), Manfredini (1770), Anfossi (1770), Salieri (1771), Sacchini (1772), Gazzaniga (1773), Naumann (1773), Cimarosa (1777), Gluck (1777), Astarita (1777), Mysliveček (1778), Cherubini (1782), Righini (1782), Haydn (1784), Sarti (1786), Zingarelli (1786), Rossini (1817), Zajc (1896), and Dvořák (1904).

**Arminio.** Opera in 3 acts by Handel; text by A. Salvi. Prod. London, CG, 12 Jan. 1737. Rev. Oldenburg (in German) 1963.

**Armstrong, Richard** (*b* Leicester, 7 Jan. 1943). Studied Cambridge. Répétiteur, London, CG, 1966–8. Début WNO as assistant; music director WNO 1973–86, where successes included a Janáček cycle, *Midsummer Marriage*, *The Ring*, also Verdi's *Otello* with Peter Stein. WNO subsequently, incl. tours to New York, Milan, and Japan. CG from 1982; ENO and Geneva 1990; also Netherlands O; Berlin, K; Monte Carlo; Frankfurt O (principal guest conductor). Scottish O from 1993. Repertory incl. *Fidelio*, *Dutchman*, *Boccanegra*, *Salome*. (R)

**Arne, Thomas Augustine** (*b* London, ?12 Mar., *bapt.* 28 May 1710; *d* London, 5 Mar. 1778). English composer. The success of the masque *Dido and Aeneas* (1734), led to his employment at London, DL. Among a vast output there, the only work of note was the 3-act masque *Comus* (1738), his most successful stage work. The masque *Alfred* (1740), revised on several occasions, is remembered only as the original home of 'Rule

Britannia'. The mediocrity of much of Arne's music of this period led Burney to comment that 'the number of his unfortunate pieces for the stage was prodigious'. His premature experiment with opera buffa into English in *Don Saverio* (1750) proved unpopular. At the end of the season he moved to Covent Garden.

The success of *Artaxerxes* (1762) produced a favourable response even from Burney, who wrote that 'Arne had the merit of first adapting many of the best passages of Italy . . . to our own language'. He wrote recitative, and bravura arias for Tenducci, Peretti, and his pupil Charlotte Brent, but simpler arias for the other performers. It was the only English opera 'after the Italian manner' to enjoy continuing success into the 19th cent. Its popularity encouraged Arne to attempt an all-Italian opera, *L'Olimpiade* (1765), which failed. As popular as *Artaxerxes* was his pasticcio *Love in a Village* (1762), which was extensively imitated. A transition between ballad opera and a more fully composed genre, only five of its 42 numbers were specially written for the work, though most are substantial settings with orchestral preludes and postludes.

SELECT WORKLIST (all first prod. London): *Don Saverio* (Arne; 1750); *Artaxerxes* (Arne, after Metastasio; 1762); *L'Olimpiade* (Bottarelli, after Metastasio; 1765).

**Arnold.** Arnold Melcthal (ten), lover of Mathilde in Rossini's *Guillaume Tell*.

**Arnold, Samuel** (*b* London, 10 Aug. 1740; *d* London, 22 Oct. 1802). English composer, organist, conductor, and editor. Educated at the Chapel Royal. His pasticcio *The Maid of the Mill* (based on Richardson's *Pamela*) was performed to enormous acclaim at Covent Garden in 1765, launching him on a spectacular career as a theatrical composer. He produced over 100 stage works, about half of which were pasticcios, burlettas, and ballets: most were for the Marylebone Gardens, which he leased 1769–76, or for the Little T Royal in the Haymarket, where he was composer from 1777. In 1783 he became organist and composer of the Chapel Royal; in 1789 director of the Academy of Ancient Music.

Among Arnold's theatrical works are *Tom Jones*, *The Spanish Barber*, and *Macbeth*. The *Servant Mistress*, produced at Covent Garden in 1770, was an interesting adaptation, with additional music, of Pergolesi's *La serva padrona*, and was probably his most successful work. As he wrote to popular demand, and frequently to a tight deadline, the quality of his work is very variable: capable of composing music of considerable originality and flair, more often he produced uninspired banalities. Arnold was also responsible for the first attempt at a complete edition of any composer's music: his unfinished *Complete Edition of Handel's Works* represents a landmark in the history of English music. His son **Samuel James Arnold** (1774–1852) was a librettist and theatrical impresario.

SELECT WORKLIST (all first prod. London): *The Maid of the Mill* (Bickerstaffe, after Richardson; 1765); *Tom Jones* (Read, after Fielding; 1769); *The Spanish Barber* (Colman, after Beaumarchais; 1777); *Macbeth* (Shakespeare; 1778).

**Arnoldson, Sigrid** (*b* Stockholm, 20 Mar. 1861; *d* Stockholm, 7 Feb. 1943). Swedish soprano. Studied with her father, the tenor Oscar Arnoldson (1830–81), Paris with Strakosch and Marchesi, and Berlin with Artôt. Début Prague 1885 (Rosina). Moscow 1886; London, DL, 1887, and CG from 1888; New York, M, 1893. Charlotte in first US and British *Werther*, 1894. Her voice covered three octaves. Retired 1916. (R)

**Arnould, Sophie** (*b* Paris, 13 Feb. 1740; *d* Paris, 18 Oct. 1802). French soprano. Studied Paris with Marie Fel. Début Paris, O, 1757, and sang there for 20 years. Created title-role of Gluck's *Iphigénie en Aulide* and was the first Eurydice in the French version of his *Orfeo*. Also highly successful in the works of Rameau (notably in *Castor et Pollux*), Francœur (creating the title-role in *Aline*), and Monsigny. She was an impassioned actress (the only one in France of whom Garrick spoke with enthusiasm) and possessed a good voice, but her laxness led to a vocal decline before she was 40. She retired in 1788 with a state pension of 2,000 livres. Lively and intelligent, she numbered Voltaire, Rousseau, and Beaumarchais among her intellectual circle in Paris. Pierné's *Sophie Arnould* (1927) is based on events in her life. Her portrait by Greuze is in the Wallace Collection, London.

**Aroldo.** See STIFFELIO.

**Aron.** Aaron (ten), Moses' brother in Schoenberg's *Moses und Aron*.

**Arrieu, Claude** (*b* Paris, 30 Nov. 1903; *d* Paris, 7 Mar. 1990). French composer. Studied Paris with Dukas. Her stage works include the early experimental *Noé* (1934), a number of works in more traditional language such as *Cadet Roussel* (1939) and *Cymbeline* (1963), some lighter works and works for radio and for children.

**Arroyo, Martina** (*b* New York, 2 Feb. 1935). US soprano. Studied New York with Gurewich. Début New York, M, 1959 (Voice from Heaven, *Don Carlos*). Zurich 1963–8; then regular appearances all over Europe. She returned to New York to sing principal roles in 1965. London, CG, from 1968. Her repertory includes Reiza, Norma, and the Verdi dramatic soprano roles. (R)

**Arsace.** Semiramide's son (mez) in Rossini's *Semiramide*.

**Artaserse.** Metastasio's libretto has had over 100 settings including those by Vinci (1730), Hasse (1730), Gluck (1741, his first opera), Graun (1743), Terradellas (1744), Duni (1744), Jommelli (1749), Galuppi (1749), J. C. Bach (1760, his first opera), Arne (1762), Piccinni (1762), Paisiello (1771), Sacchini (1768), Mysliveček (1774), Cimarosa (1781), Anfossi (1788), Isouard (1794), Le Sueur (comp. 1811), and Nicolai (c.1830). Jommelli's setting anticipates Gluck's reforms in showing avoidance of the traditional da capo aria form.

The plot concerns repeated attempts by Artabano to gain the throne. He first of all kills King Serse and then attempts to poison Artaserse, the king's son. Arbace, Artabano's son, thwarts these efforts, since he is true to the crown. Loyalties are further confused by Arbace's love for Mandane, King Serse's daughter, and Artaserse's love for Semira, Artabano's daughter. The couples are eventually united and Artabano exiled.

**Artaxerxes.** Opera in 3 acts by Arne; text by Metastasio, trans. by the composer. Prod. London, CG, 2 Feb. 1762, with Brent, Tenducci, Peretti; Philadelphia, Chestnut St. T, 28 Dec. 1827.

See also ARTASERSE.

**artists in opera.** Painters and other visual artists who appear in opera or who have operas based on their works include the following:

Fra Angelico: Hillemacher (1924).

Hieronymus Bosch: Menotti (*Amahl and the Night Visitors*, 1951).

Benvenuto Cellini: Berlioz (1838); Lauro Rossi (*Cellini a Parigi*, 1845); Schlösser (1847); Lachner (1849); Kern (1854); Orsini (1875); Bozzano (1877); Diaz de la Peña (1890); Tubi (1906); Saint-Saëns (*Ascanio*, 1890).

Dürer: Sonnenfeld (*Albrecht Dürer w Wenecji*, c.1892); Baussnern (*Dürer in Venedig*, 1901); Baselt (c.1900); Mraczek (*Herr Dürers Bild*, 1927).

Fragonard: Pierné (1934).

Gauguin: Elizalde (1948); Gardner (*The Moon and Sixpence*, 1957).

Goya: Barbieri (appears in *Pan y toros*, 1864); Granados (*Goyescas*, 1916); Menotti (1986).

Matthias Grünewald: Hindemith (*Mathis der Maler*, 1938).

Hogarth: Stravinsky (*The Rake's Progress*, 1951).

Michelangelo: Isouard (1802); F. Ricci (*Michelangelo e Rolla*, 1841); Solera (*Genio e sventura*, 1843); Buongiorno (1903); Alfred Mendelssohn (1964).

Raphael: Raimondi (1838); C. Gurlitt (mid-19th cent.).

Rembrandt: Klenau (1937); Badings (*De Nachtwacht*, comp. 1943, prod. 1950).

Tilman Riemenschneider: Pászthory (1937).

Rubens: see Van Dyck.

Salvator Rosa: Rastrelli (1832); Bassi (1837); Bergson (c.1850); Bianchi (1855); Sobolewski (1859); Duprato (1861); Bazzoni (19th cent.); Gomes (1874); Zoboli (19th cent.).

Andrea del Sarto: Baravalle (1890); Weingartner (*Meister Andrea*, 1920); M. Rosselli-Nissem (1931); Lesur (1974).

Karel Škreta: Bendl (1883).

Wit Stwosź: Swider (1974).

Titian: Kubelík (*Cornelia Faroli*, 1972).

Van Dyck: Villent-Bordogni (1845; includes a duel between Van Dyck and Rubens); Michel (*La meunière de Savantham*, 1872); A. Müller (between 1875 and 1882).

Van Gogh: Nevit Kodalli (1957); Wilson (*Letters to Theo*, 1985); Rautawaara (1990).

Velasquez: Strauss (*Friedenstag*, 1938).

Leonardo da Vinci: Wade (*The Pupil of Da Vinci*, 1839); Rózycki (*Meduza*, 1912); Schillings (*Mona Lisa*, 1915); De Ninno (*Monna Lisa*, ?); Werle (1988).

**Artôt, Désirée** (*b* Paris, 21 July 1835; *d* Berlin, 3 Apr. 1907). Belgian mezzo-soprano, later soprano. Studied Paris with Viardot. Engaged for Paris, O, on Meyerbeer's suggestion; début 1858 (Fidès). Italy and Germany as mezzo, then sang as soprano, including in London concerts, 1859–60: HM 1863 as Marie (*La fille du régiment*), Violetta, and Adalgisa; CG 1864–6. Russia 1868, where she was briefly engaged to Tchaikovsky; in 1869 she married the Spanish baritone **Mariano Padilla y Ramos** (1842–1906). They toured together in Europe and Russia. Their daughter, the soprano Lola Artôt de Padilla (1876–1933) created Vreli (*A Village Romeo and Juliet*) Berlin, K, 1907 and was a member of the Berlin, H, 1909–27.

**Arts Florissants, Les.** See CHRISTIE, WILLIAM.

**Arturo.** Lord Arthur Talbot (ten), a Cavalier betrothed to Elvira, in Bellini's *I Puritani*.

**Arundell, Dennis** (*b* London, 22 July 1898; *d* London, 10 Dec. 1988). English actor, producer, author, and composer. Studied Cambridge with Rootham and Stanford. Prod. Rootham's *The Two Sisters* 1922, and other Cambridge performances incl. the first staged *Semele* (1925), and *The Fairy Queen* (1931). Prods. at London, SW, incl. *Shvanda* (own trans., 1948), first English *Káťa Kabanová* (1951), and *Flying Dutchman* (1958). Collaborated with Beecham on *Bohemian Girl* at London, CG (1951) and première of *Irmelin*, in Oxford (1953). Also worked in Helsinki and in Australia. His two operas are *Ghost of Abel* and *A Midsummer Night's Dream*.

WRITINGS: *Henry Purcell* (London, 1927); *The Critic at the Opera* (London, 1957); *The Story of Sadler's Wells* (London, 1965, 2/1977).

**Arvidson.** Mlle Arvidson (mez), the fortune-teller in the 'Swedish' version of Verdi's *Un *ballo in maschera.

**Ascanio in Alba.** Opera in 2 acts by Mozart; text by G. Parini, perhaps after Count Claudio Stampa. Prod. Milan, RD, 17 Oct. 1771, with Falchini, Manzuoli, Girelli, Solzi, Tibaldi.

In an Arcadian landscape near Rome, Venus (sop) tells her grandson Ascanio (sop cas) that he is to rule part of her nation in her place and is to marry Silvia (sop), a wise nymph. He is to go in search of Silvia, but in disguise so as to test her true feelings for him.

Ascanio and Silvia meet and fall in love. However, she puts herself under the protection of Aceste (ten), a priest of Venus, to await the arrival of her destined groom. Venus, convinced of the nymph's virtue, reveals Ascanio's real identity to Silvia. The couple are delighted and everyone sings Venus's praises.

Commissioned for the wedding of the Archduke Ferdinand to the Princess d'Este of Modena to accompany Hasse's opera seria *Ruggiero* (16 Oct.), it succeeded where *Ruggiero* failed, leading Hasse allegedly to comment, 'This boy will make us all be forgotten'.

Other operas are by: Bernabei (1686), Moreira (1785).

**Aschenbach.** Gustav Aschenbach (ten), the writer in Britten's *Death in Venice*.

**Assassinio nella cattedrale** (Murder in the Cathedral). Opera in 2 parts, with an intermezzo, by Pizzetti; text by the composer, after A. Castelli's translation of T. S. Eliot's dramatic poem (1935). Prod. Milan, S, 1 Mar. 1958, with Rossi-Lemeni, cond. Gavazzeni; New York, Carnegie H, 17 Sep. 1958, concert perf., with Rossi-Lemeni, cond. Halasz; Coventry, by London SW Co., 12 June 1962, with Garrard, cond. Davis.

The plot is based on the historical events leading to the assassination in Canterbury Cathedral of Thomas à Becket (bs), Archbishop of Canterbury, by four knights in the service of Henry II.

**'Assisa al piè d'un salice'.** See WILLOW SONG.

**Astarita, Gennaro** (*b* ?Naples, *c.*1745–9; *d* ?, after 1803). Italian composer. His earliest work, the opera buffa *L'orfana insidiata*, was written in collaboration with Piccinni, and followed by further successes, leading to his appointment as maestro di cappella in his native city in 1770. Little is known of his life, but his operatic activity was clearly not restricted to Naples: several operas were written for other cities, including *Il marito indolente*, *Didone abbando-*nata, and *La molinarella*. Though noted in his own country as a composer of mainly opere buffe, Astarita played an important role in the development of opera in Russia, which he first visited in 1781. In 1784 he became director of the orchestra of the Petrovsky T in Moscow, moving to St Petersburg in 1786. In 1794, at the request of the director of the Imperial Theatres in St Petersburg, Prince Nicolay Yusupov, Astarita recruited an Italian troupe, which he directed until 1799.

SELECT WORKLIST: *L'orfana insidiata* (Piccinni; Naples 1765); *Il marito indolente* (Mazzolà; Bologna 1778); *Didone abbandonata* (Metastasio; Pressburg 1780).

**Astuzie femminili, Le** (Feminine Wiles). Opera in 2 acts by Cimarosa; text by Giovanni Palomba. Prod. Naples, Fondo, 16 Aug. 1794; London 21 Feb. 1804; New York TH 31 Oct. 1966.

According to the terms of her father's will, Bellina (sop), the richest heiress in Rome, is to marry Giampolo (bs), an elderly merchant from Bergamo. Her tutor Dr Romualdo (bs), although betrothed to Leonora (mez), Bellina's governess, also wants to marry the young girl. Bellina herself is in love with the young Filandro (ten); they run off together and, returning disguised as Cossack officers, are successful in their ruse to get married.

**Atalanta.** Opera in 3 acts by Handel; text anon., after B. Valeriani's *La caccia in Etolia*. Prod. London, CG, 12 May 1736 on the occasion of the marriage of Frederick, Prince of Wales (son of George II), to Princess Augusta of Saxe-Gotha, with Conti (Gizziello), Negri, Strada, Beard. Rev. Hintlesham Festival, 1970, by Kent Opera.

This *pièce d'occasion* tells of the love of Atalanta (sop), the princess huntress-nymph, for the king shepherd, Meleager (sop cas). The opera ends with Mercury (bs) bringing Jupiter's greeting to the newly-weds, i.e. Prince Frederick and Princess Augusta.

Other operas are by Kerl (1667), Draghi (1669), Steffani (1692), Strungk (1695), Fedeli (1702), and Hasse (1737).

**Atanasov, Georgy** (*b* Plovdiv, 18 May 1882; *d* Fasano, 17 Nov. 1931). Bulgarian composer and conductor. Studied privately, in Bucharest, and (1901–3) in Pesaro with Mascagni. He worked as a bandmaster, and conducted the first opera performances in Bulgaria. Known as The Maestro, he is regarded as the founder of Bulgarian opera. In his operas, still highly regarded in Bulgaria, he sought to confer technical maturity on the emergent local art by absorbing Italian influences and later, less successfully, those of Wagner and Strauss.

SELECT WORKLIST (all first prod. Sofia): *Borislav*

(Popov, after Vasov; 1911); *Gergana* (Bobevsky, after Slaveikov; 1917); *Zapustyalata vodenitsa* (The Abandoned Mill) (Morfov; 1923); *Tsveta* (The Flower) (Chernodrinsky; 1925); *Kosara* (Danovsky; 1927); *Altsek* (Karapetrov; 1930). Also first Bulgarian operetta, *Moralisti* (1916), and five children's operettas.

BIBL: L. Sagayev, *Maestro Georgy Atanasov* (Sofia, 1961).

**Athanaël.** The coenobite monk (bar) who tries to save and then falls in love with Thaïs in Massenet's *Thaïs*.

**Athens** (Gr., Athinai). Capital of Greece. Opera was first given by a visiting Italian company in 1837 (*Barbiere*); this was immensely popular, especially with the male population. But though French operettas were performed in the summer theatre at Neon Phaleron from 1871, and there were student performances in 1876, the first professional Greek company was not formed until the first Elliniko Melodrama in 1888 (opening with Xyndas's *O ypopsifios vouleftis* on 14 Mar.), followed by the second Elliniko Melodrama 1888–90 (opening with *Betly* on 19 Dec.); the latter toured Egypt, the Near East, France, and Romania with a repertory of Italian works and some by Karrer and Xyndas. The third Elliniko Melodrama was founded by Dionysios *Lavrangas (opening with *Bohème*; last performance 6 June 1943). This company did much to establish opera in Greece, also touring Greek communities abroad; its repertory was largely Italian, and it included the baritone Ioannis Anghelopoulos (1881–1943). In 1933 *Kalomiris founded the Ethnikos Melodramatikos Omilos (Nat. OC); this closed in 1935. A national company under the auspices of the Ethnikon T was formed in 1939 (opening with *Fledermaus*), later moving to the Olympia T. In 1944 it became the Ethniki Lyriki Skini (Nat. Lyric T), under Kalomiris, opening with Samaras's *Rhea*.

After the war the theatre was renovated (cap. 952), and has a roster of 17 soloists and a chorus of 65. The season usually runs from Nov. to May, and the company normally opens the Athens Festival (first in 1955), playing in the open-air T of Herodes Atticus (cap. 5,000). It occasionally visits Salonika and Patras. A good repertory and tradition have developed, and among many distinguished singers to have begun their careers there is Maria Callas, for whom performances of *Medea* and *Norma* have been staged in the ancient theatre at Epidauros.

**Atlántida, L'.** Scenic 'oratorio' by Falla; text by the composer, after Jacinto Verdaguer's poem (1877); score completed by E. Halffter. Prod. Milan, S, 18 June 1962, with Stratas, Simionato, Puglisi, Halley, cond. Schippers; New York,

M, at Philharmonic H (concert), 29 Sep. 1962, with Farrell, Madeira, London, cond. Ansermet.

An epic Spanish saga, telling how Atlantis is submerged by flood; how Spain is saved from Hercules; how Barcelona came into being; and how Columbus discovered the New World.

Also *Atlantyda*, opera by Guzewski (comp. 1913–30, unprod.).

**Atlantide, L'.** Opera in 4 acts by Tomasi; text by F. Didelot, after Pierre Benoit's novel (1920). Prod. Mulhouse 26 Feb. 1954. Antinea (mimed role), a maenad, rules the land of Atlantis. She lures two officers of the French Foreign Legion, a Lieutenant (ten) and Captain (bs), to her domain, and they fall in love with her. One she kills with a silver hammer, the other dies in the desert pursuing a phantom that is her double.

**Atlantov, Vladimir** (*b* Leningrad, 19 Feb. 1939). Russian tenor. Studied Leningrad with Bolotin. Début as soloist Leningrad, K, 1963. Student artist, Milan, S, 1963–5. Moscow, B, since 1967. Guest appearances Milan, Vienna, Wiesbaden, New York, etc., as Hermann, Lensky, Don José, and Italian roles. Considered the outstanding tenore robusto of today by the Italians. (R)

**Attaque du moulin, L'** (The Attack on the Mill). Opera in 4 acts by Bruneau; text by Louis Gallet, after the story in Zola's *Soirées de Médan* (1880). Prod. Paris, OC, 23 Nov. 1893, with Leblanc, Delna, Vergnet, Bouret, cond. Danbé; London, CG, 4 July 1894, with Nuovina, Delna, Cossira, cond. Flon; New York, New T, 8 Feb. 1910, with Noria, Delna, Clément, cond. Tango.

The miller Morlier's (bar) daughter, Françoise (sop), is betrothed to a young Flemish peasant, Dominique (ten); but their celebrations are interrupted by a recruiting party, causing Françoise's foster-mother, Marcelline (mez), to denounce all war. Though Dominique is not in fact called up, when the mill is attacked he is seized by the Germans and condemned to be shot. When he escapes, the German captain demands Morlier's life in his stead; in order to save Dominique for his daughter, Morlier sacrifices himself deliberately.

**Atterberg, Kurt** (*b* Göteborg, 12 Dec. 1887; *d* Stockholm, 15 Feb. 1974). Swedish composer, conductor, and critic. Studied Stockholm with Hallén, privately, and in Germany. Though most admired in his own country in the earlier part of his career for his symphonic music, he had some success at home and in Germany with his operas.

WORKLIST (all first prod. Stockholm): *Härvard harpolekare* (Härvard the Harper) (1919, rev. 1954); *Bäckhasten* (The White Horse) (1925); *Fanal* (1934); *Aladdin* (1941); *Stormen* (1948).

**At the Boar's Head.** Opera in 1 act by Holst; text by the composer, after Shakespeare's dramas *Henry IV*, pt 1 and pt 2 (1597–8), with two Shakespeare sonnets. Prod. Manchester, by BNOC, 3 Apr. 1925, with Constance Willis, T. Davies, Allin, cond. Sargent; New York, Mac-Dowell Club, 16 Feb. 1935.

The plot is closely derived from the Falstaff episodes in Shakespeare's Henry plays. Falstaff (bs) arrives at The Boar's Head, where Bardolph (bar) and Pistol (bar) are drinking, after his bungled highway robbery: Poins (bs) and Prince Hal (ten) reveal that they have tricked him. News comes of civil war. Hal and Poins disguise themselves so as to observe Falstaff's pursuit of Doll Tearsheet (sop). A distant march is heard, and Bardolph brings Hal a summons to Westminster. Hal and Poins set off; Pistol quarrels with Doll, and is thrown out. Bardolph announces that a dozen captains are searching for Falstaff; they set off for the wars together, leaving the women in tears until Bardolph slyly returns to say Falstaff wants to see Doll once more.

**Attila.** Opera in prologue and 3 acts by Verdi; text by Solera, after Zacharias Werner's drama *Attila, König der Hunnen* (1808). Prod. Venice, F, 17 Mar. 1846, with Sofia Loewe, C. Guasco, Marini, N. Constantini; London, HM, 14 Mar. 1848, with Cruvelli, Gardoni, Belletti, Cuzzoni; New York, Niblo's Garden, 15 Apr. 1850, with Tedesco, Lorini, Corradi-Setti, L. Martinelli, cond. Bottesini.

Italy, AD 454. Attila's army has ransacked the town of Aquileia. Odabella (sop) is among the prisoners; her father was killed by Attila (bs) and she swears to avenge his death. Ezio (bar), a Roman general, arrives and suggests that they divide Italy between them. Attila rejects the offer, calling Ezio a traitor.

Odabella is delighted by the arrival of her lover Foresto (ten), whom she believed dead, but has to convince him that she is loyal to her country. She discloses a plan to murder Attila. He is troubled by dreams and resolves to attack Rome. Foresto and Ezio plan to overthrow Attila, who is given poison; but Odabella (determined that he shall die by her hand alone) prevents him from drinking it. The grateful Attila promises to marry Odabella and prepares to resume the attack on Rome.

As the wedding celebrations begin, Ezio and Foresto plot to ambush the Huns. Foresto once again accuses Odabella of treason: Attila arrives to find her with his enemies. He realizes he has been betrayed and Odabella stabs him.

Other operas on Attila are by Ziani (1672), Porsile (1675), Franck (1682), Farinelli (1808), Seyfried (1809), Generali (1812), Mosca (1824), Persiani (1827), Malipiero (1845), Gunkel (1895).

**Attwood, Thomas** (*b* London, *bapt.* 23 Nov. 1765; *d* London, 24 Mar. 1838). English composer and organist. Sent by the Prince of Wales to study music first in Naples, 1783–5, and then in Vienna, 1785–7. Here he took lessons with Mozart who, according to Kelly's *Reminiscences*, predicted that he would 'prove a sound musician'. Returning to England with Nancy and Stephen Storace in 1787, he took a series of court appointments, including one as composer to the Chapel Royal. His theatrical works, composed mainly 1796–1825, are chiefly afterpieces for plays and generally undistinguished. Even his more major essays, the comic operas *The Old Clothesman* (1799) and *The Castle of Sorrento* (1799), were appended to other entertainments. But the lyricism of his music shows him to be not without talent and he scored a major success with *The Prisoner* (1792), his first work, performed as an afterpiece to *The Beaux' Strategem*. Many of Attwood's stage works were pasticcios: his *The Escapes, or The Water Carrier* (1801) includes some music from Cherubini's setting and Mozart's melodies feature in almost every one of his works: *The Prisoner* includes 'Non più andrai', *The Mariners* 'In diesen heil'gen Hallen', *Caernarvon Castle* 'Sull' aria', *The Red Cross Knights* 'Bei Männern', etc.

SELECT WORKLIST: *The Old Clothesman* (Holcroft; London 1799); *The Castle of Sorrento* (Heartwell, after Colman, after Duval; London 1799).

**Auber, Daniel-François-Esprit** (*b* Caen, 29 Jan. 1782; *d* Paris, 12–13 May 1871). French composer. Studied with Ladurner, then with Cherubini, who had been impressed by his *L'erreur d'un moment*. Nevertheless, further success in opera eluded him until through Cherubini he re-established himself, at first with *La bergère châtelaine* and *Emma*. In 1823 he began his long and fruitful collaboration with *Scribe, and after a brief flirtation with an Italianate style (as in *Leicester* and *La neige*), found himself with *Léocadie*. He then gave his career as a master of opéra-comique, in succession to Boieldieu and Isouard, a sure foundation with *Le maçon*. There followed a long series of highly successful works in which to his appreciation of the wit of Rossini was added his own Gallic lightness of touch. Of the works of this period and in this vein, the most enduring has been *Fra Diavolo*: it epitomizes his lighter style with its mixture of sentimentality and wit, expressed in graceful melodies deftly accompanied. Among his most popular and enduring successes in opéra-comique were also *Le philtre*, *La marquise de Brinvilliers*, *Le cheval de bronze*, *Le domino noir*, and *Les diamants de la couronne*.

Auber's career took a new turn with *La muette de Portici*. Commissioned for the Opéra, it was the most significant of the works which inaugurated the era of French *Grand Opera. Its essential ingredients included a substantial group of important leading roles, powerful and functional choruses, increased reliance on the orchestra, the inclusion of ballet, and not least music written with an awareness of the vivid and imposing scenic effects provided by *Cicéri and *Daguerre. The work was influential on Meyerbeer and Halévy, but also (especially in *Lohengrin*) on Wagner, who wrote an enthusiastic article praising the work's swiftness and urgency of action. Later, in 1830, a performance in Brussels was to spark off the revolt of the Belgians against the Dutch. Other successful works in this vein include *Gustave III* (an exceptionally vivid and original work) and *Lestocq* (an opera also remarkable for some powerful choral writing).

After about 1840 Auber's lighter style (Wagner now compared him to a barber who lathers but does not shave) developed into a more lyrical manner, marked with a limited increase in harmonic range, as in *La part du diable*, *Haydée*, and *Le premier jour de bonheur*. He was also a distinguished director of the Paris Conservatoire, 1842–70. Shy and diminutive of stature, he was characterized by Rossini as 'piccolo musico, ma grande musicista' ('a small musician, but a great maker of music').

SELECT WORKLIST (all first prod. Paris): *L'erreur d'un moment* (Boutet de Monvel; 1805, rev. 1811); *La bergère châtelaine* (Planard; 1820); *Emma* (Planard; 1821); *Leicester* (Scribe, after Scott; 1823); *La neige* (Scribe; 1823); *Léocadie* (Scribe; 1824); *Le maçon* (Scribe; 1825); *La muette de Portici* (Scribe & Delavigne; 1828); *Fra Diavolo* (Scribe; 1830); *Le philtre* (Scribe; 1831); *La marquise de Brinvilliers* (Scribe & Castil-Blaze; 1831); *Gustave III* (Scribe; 1833); *Lestocq* (Scribe; 1834); *Le cheval de bronze* (Scribe; 1835); *Le domino noir* (Scribe; 1837); *Le lac des fées* (Scribe & Mélesville; 1839); *Les diamants de la couronne* (Scribe & Vernoy de Saint-Georges; 1841); *Haydée* (Scribe, after Mérimée; 1847); *Manon Lescaut* (Scribe, after Prévost; 1856); *Le premier jour de bonheur* (D'Ennery & Cormon; 1868).

BIBL: C. Malherbe, *Auber* (Paris, 1911).

**Auden, W(ystan) H(ugh)** (*b* York, 21 Feb. 1907; *d* Vienna, 29 Sep. 1973). English, later US, poet. He wrote librettos for Britten's *Paul Bunyan* (New York, Columbia U, 1941) and, with Chester Kallman, Stravinsky's *The Rake's Progress* (1951), Henze's *Elegy for Young Lovers* (1961) and *The Bassarids* (1966), and Nabokov's *Love's Labour's Lost* (1973). Also with Kallman translated *Don Giovanni* and *Die Zauberflöte*.

BIBL: W. H. Auden, Essays in *The Dyer's Hand* (London, 1963).

**Audran, Edmond** (*b* Lyons, 12 Apr. 1840; *d* Tierceville, 17 Aug. 1901). French composer. Studied Paris with Duprato. Success with an early operetta, *Le grand mogol* (Marseilles 1877) brought an invitation from the Paris, BP, that led to a string of light, tuneful works of which the most popular was *La mascotte* (1880). He never equalled this success despite a considerable development and expansion of his style, especially in *La poupée* (1896).

**'Au fond du temple saint'.** The duet between Nadir (ten) and Zurga (bar) in Act I of Bizet's *Les pêcheurs de perles*, in which they recall the beautiful Leïla they both once loved.

**Aufstieg und Fall der Stadt Mahagonny** (Rise and Fall of the City of Mahagonny). Opera in 3 acts by Weill; text by Brecht. Prod. Leipzig 9 Mar. 1930, with Trummer, Fleischer, Zimmer, cond. Brecher (prod. as a *Songspiel, Baden-Baden, 1927); New York, TH, 23 Feb. 1952; London, SW, 16 Jan. 1963, with Cantelo, Dowd, Te Wiata, cond. Davis; first full US perf., New York, Anderson T, 28 Apr. 1970, cond. Matlovsky.

Mahagonny is the city of material pleasure. Jenny (sop) and her colleagues move in, followed by Jim Mahoney (ten) and his lumberjack friends on the spend; there develops a love affair between the two, based upon hard cash as well as affection. A hurricane threatens the city, but is miraculously deflected. Act II illustrates Gluttony, Love, Fighting, and Drinking, all carried on to excess and without restraint. Jim cannot pay for his drinks; tried according to Mahagonny's scale of values, he is sentenced to two days for indirect murder, four years for 'education, by means of money' and, for not paying his whisky bill, death.

**Auftrittslied** (Ger.: 'entry song'). The aria in a Singspiel in which a character introduces himself to the audience, either directly or by addressing another character, e.g. Papageno's 'Der Vogelfänger bin ich ja' in Mozart's *Die Zauberflöte*.

**Augér, Arleen** (*b* South Gate, CA, 13 Sep. 1939; *d* Leusden, 10 June 1993). US soprano. Studied Los Angeles with Frey, Chicago with Errolle. Début Vienna, S, 1967 (Queen of the Night). Vienna until 1974; New York City O 1968–9; Milan, S, from 1975; New York, M, from 1978; also Moscow, B; Bayreuth; Salzburg; Munich; Frankfurt; Cologne; Aix; etc. A musical singer with a sweet-toned, flexible voice. Roles incl. Ilia, Servilia, Marzelline, Countess (Mozart), Poppea. (R)

**Augsburg.** City in Bavaria, Germany. Opera was given from 1637 in the Fechthaus (Fencing School), later Grosses Opernhaus. Johann *Kusser visited with his company to stage some of his German operas there 1697–8, though the city was largely dependent on visiting Italian troupes (notably Peruzzi's). Weber's *Peter Schmoll* was premièred there 1803; his half-brother Edmund was then conductor. The Stadt-theater (cap. 1,300) opened 1877 and was enlarged 1938–9; bombed 1944. The present theatre (cap. 1,010) opened 1956 with *Figaro*. István Kertész was music director 1958–63, Peter Ebert director 1968–73. Open-air performances are given at the Rote Tor, opened 1930 with *Fidelio*.

**Auletta, Pietro** (*b* S Angelo, Avellino, *c*.1698; *d* Naples, Sep. 1771). Italian composer. A pupil of Porpora, Auletta became maestro di cappella at S Maria la Nova in Naples *c*.1724, later serving in the same capacity to Prince Belvedere. His stage works, which include both operas and pasticcios, were performed to great effect throughout Italy: the most popular was the opera buffa *Orazio*, though his mastery of opera seria is demonstrated by *Ezio* and *Didone abbandonata*.

SELECT WORKLIST: *Ezio* (Metastasio; Rome 1728); *Orazio* (Palomba; Naples 1737); *Didone abbandonata* (Metastasio; Florence 1759).

**Aureli, Aurelio** (*fl.* 1652–1708). Venetian librettist, one of the most prolific of the second half of the 17th cent. He was established by the 1660s, and his texts were set by virtually all the important Venetian composers of the day, including Cavalli (*Erismena*), Sartorio, and Ziani. He preferred to invent his own stories, sometimes basing them on history. His contributions to the libretto include the use of fewer characters for greater clarity, a mixing of heroic figures with comic Venetian servants, and a skilful dramatic use of points of repose. Latterly he became disillusioned with the demands of singers and public for display set arias.

**Aussig.** See ÚSTÍ NAD LABEM.

**Austin, Frederic** (*b* London, 30 Mar. 1872; *d* London, 10 Apr. 1952). English baritone. Studied with Charles Lunn. Début London, CG, 1908 (Gunther, in the English *Ring* under Richter). Principal baritone, Beecham Co. Peachum in his own new version of *The Beggar's Opera*, London, Lyric Hammersmith, 1920. Artistic director, BNOC, 1924. Pupils include Constance Shacklock. (R). His son **Richard** (1903–88) conducted with the CR and at London, SW, and was from 1953 head of the RCM opera dept. (R)

**Austin, Sumner** (*b* London, 24 Sep. 1888; *d* Oxford, 9 July 1981). English baritone. CR 1919, then sang with O'Mara OC. Old Vic OC in 1920s. London, SW, until 1940. Especially distinguished in Mozart. Produced several operas at SW, including *Falstaff*, *Walküre*, *Don Carlos*,

and *The Wreckers*; also *Tannhäuser* and *Wozzeck* at London, CG, in the 1950s.

**Austral, Florence** (*b* Melbourne, 26 Apr. 1894; *d* Newcastle, NSW, 15 May 1968). Australian soprano. Studied Melbourne Cons. and New York with Sibella. Début BNOC 1922 (Brünnhilde); also Aida and Isolde, and Brünnhilde during international seasons 1924, 1929, and 1933. Guest artist London, SW, 1937–9. A generous and musical singer, who, like Leider, sang Wagner using the principles of bel canto. (R)

**Australia.** The first opera staged was Bishop's *Clari* in 1823 in Sydney. The next was *Cenerentola* by the Dramatic Company (in which the only trained singer was Vincent Wallace's sister Wallace Bushelle). In 1845 Count Carandini, a political refugee from Italy, founded a company with his wife as prima donna and gave successful performances of *Der Freischütz*, *Fra Diavolo*, and some Italian operas. The first operas written in Australia were *Merry Freaks in Troublous Times* (comp. 1843) and *Don John of Austria* (Sydney 1847) by Isaac *Nathan (1790–1864). The Anna Bishop OC gave seasons in New South Wales 1850–60. Among companies active during these years were the Lyster Grand OC., an American group that gave English-language performances until 1880; local singers were increasingly employed in their seasons. Other companies included the Montague Turner (1881–4) and Martin Simonson: the latter toured 1866 and 1886–7 with works including *Belisario* and *Roberto Devereux*, and with some Italian singers. Other groups followed in the 1890s. The first Wagner performances were under the auspices of George Musgrove. J. C. Williamson staged opera in 1893, and a season of Italian opera by the J. C. Williamson Co. followed in 1901, when Musgrove also toured with more German opera (including parts of *The Ring*); each singer used his own language. Other touring companies included the Quinlan and Gonzales. An early nationalist composer was Alfred *Hill (1870–1960).

In 1911 Nellie Melba, who had long wanted to take a company to her native land, joined Williamson in a tour with singers including John McCormack, Eleanora de Cisneros, and Rosina Buckman in a catholic repertory. The Quinlan OC toured in 1912 with Beecham among its conductors; their success led to a return in 1913 with a repertory including the first Australian *Meistersinger*. An Italian company toured in 1916; in 1921 Williamson formed a company of local artists. In 1924 the Melba-Williamson Co. was back with Toti dal Monte and Dino Borgioli; and their last season was in 1928. Melba helped to organize a further season in 1928 with a similar company strengthened by Arangi-Lombardi, Merli, and John Brownlee. New standards of presentation were set in 1932 by the Imperial Grand OC, including Lina Pagliughi and Primo Montanari.

After the Second World War, further seasons were given by the Nevin Trust. Then in 1952 the two leading groups, the National Theatre OC of Melbourne and the New South Wales National O of Sydney, pooled their resources as a first step towards a national opera; each received a grant of £A5,000. Singers included Elizabeth Fretwell, Marie Collier, and Ronald Dowd. When rivalries developed, each company continued separately until the formation of the Elizabethan Trust in 1954. In 1955 Warwick Braithwaite became director of the Sydney company, resigning in mid-season because of managerial interference. With the arrival of Hugh Hunt, director of the Elizabethan Trust, a new company was formed, the Australian O. Also in 1955, the NSW State Government launched a design competition for a new Sydney opera-house, won by Jørn Utzon. In July 1956 the Trust launched its first season at Adelaide with a Mozart season, followed by a tour until February 1957. In December 1957 Karl Rankl became music director, launching his first season in July 1958, but resigned in 1961. In 1965 Joan Sutherland toured with her own Sutherland-Williamson company. The Trust was renamed the Australian OC in 1969, with Carlo Felice Cillario as music director. He was succeeded by Edward Downes 1972–6, then Richard Bonynge. The early 1970s saw a rapid expansion in operatic activity all over the country. The new opera-house in Sydney, the first in the country specially built for opera, opened in 1973 with Prokofiev's *War and Peace*.

Composers who have worked to develop an Australian style in opera include Richard Meale (*b* 1932) with *Voss* (1985) and *Mer de Glace* (1991). Among many of foreign origin active in the country have been the Germans Felix Werder (*b* 1922) with *The Affair* (1974) and George Dreyfus (*b* 1928), and Larry Sitsky (*b* 1934 of Russian parents) with *Lenz* (1974) and *Fiery Tales* (1976). A younger generation includes Anne Boyd (*b* 1946) with *The Little Mermaid* (1985).

See also ADELAIDE, BRISBANE, MELBOURNE, PERTH, SYDNEY.

BIBL: H. Love, *The Golden Age of Australian Opera*; A. Gyger, *Opera for the Antipodes* (Paddington, 1990).

**Austria.** The history of Austrian opera is not readily separable from that of Germany; for though Austria, or Austria-Hungary, was a political entity long before Germany, the two traditions merge, overlap, and frequently coincide.

Thus Mozart is treated in this book as working in a tradition belonging to the German-speaking lands of Central Europe, while Bohemian, Moravian, Croatian, Hungarian, and other composers, especially in the 18th cent., contribute to it without losing their more local identity. Nevertheless, certain traditions and certain composers claim a more specifically Austrian identity. Viennese Singspiel is discussed under *Singspiel and *Vienna. Viennese operetta as a distinctive genre finds expression in the work of *Suppé, *Millöcker, *Heuberger, but especially the Johann *Strauss family, later Oscar *Straus, *Lehár, and *Fall. *Schoenberg and *Berg (with Anton Webern) constitute the so-called Second Viennese School, more palpably Viennese than their predecessors Mozart, Haydn, and Beethoven.

See also BREGENZ, GRAZ, INNSBRUCK, KLAGENFURT, SALZBURG, VIENNA. See further GERMANY.

**'Avant de quitter ces lieux'.** Valentin's (bar) aria in Act II of Gounod's *Faust*, calling upon God to protect his sister Marguerite as he prepares to leave for the wars. Composed by Gounod for Santley in the first English production.

**Avignon.** City in Vaucluse, France. The first staged music drama was in 1622, the *Duel de la juste rigueur et de la clémence*. A theatre was opened in the 18th cent., but it was not until 1825 that there was opened a Grand Théâtre; burnt down 1846, reopened (cap. 1,000) 1847. In recent years it has developed a reputation for enterprise, and has staged a number of important premières of French works. Now the Théâtre d'Avignon et des Pays de Vaucluse. The festival was founded by Jean Vilar in 1969 with the intention of exploring new genres of music theatre, and has won a considerable reputation in this field.

**Ayrton, William** (*b* London, 24 Feb. 1777; *d* London, 8 Mar. 1858). English writer, composer, and impresario. As director of London, H, 1817, gave English premières of works incl. *Don Giovanni*, engaging singers including Pasta, Camporese, and Crevelli. His critical work included editing and writing for *The Harmonicon* (1823–33), the most important English musical journal of the day.

**Azerbaijan.** A primitive form of Azerbaijani opera known as *mugam*, a kind of stage song-cycle, has survived since the 4th or 5th cent. Baku's first theatre was opened in 1873, and visiting opera companies performed there.

The first steps towards a formal musical theatre were taken at the turn of the century in Shusha and Baku with the staging of traditional dramas with music, including the popular *Leili we Mejnun*. This was set as the first true Azerbaijani opera (Baku 1908) by Uzeir *Hadjibekov (1885–1948), and the work remains in the repertory. Other of his operas include what is regarded as the masterpiece of Azerbaijani opera, *Kyorogly* (The Blind Man's Son, 1937). The première was an important occasion, including for the taking of the title-role by the great tenor, folklorist, and teacher known as Bül-Bül (The Nightingale: really Murtasa Meshadi Rsa-ogly Mamedov, 1897–1961: his wife Shevket was a soprano and teacher). The work makes substantial use of folk music and instruments. In May 1920 a state company was formed, the Akhundov Opera and Ballet T; the opera company became independent in 1924. For them, *Glière wrote the first large-scale national opera, *Shekh-senem* (1927), which makes use of Azerbaijani melodies in a Western harmonic context. Other operas given were by Muslim Mahomayev (1885–1937), whose *Shah Ismail* (1919) and especially *Nerghiz* (1935) have been very popular. The theatre built in 1910–11 was reconstructed in 1938 (cap. 1,281). In April 1938 a season of Azerbaijani opera was given in Moscow, including most of the above works. Further operas followed in the war, including *Khosrov i Shirin* (1942) by Niyazi (Hadjibekov) (1912–84) and *Maskarad* (1945) by Boris Zeidman (1908–81), with particular success attending *Vatan* (Fatherland, 1945) by Kara Karayev (1918–82).

Post-war operas included *Nizami* (1948) by Afrasiyab Badalbeyli (1907–76) and the very successful *Sevil* (1953) by Fikret *Amirov (1922–84), a treatment of the theme (as popular in Azerbaijani as in Russian opera) of the lovelorn, suffering heroine. In those years the repertory widened to include French and Italian operas as well as the Russian works which began being performed in the 1930s, and it has since widened further still. A new tendency has been shown by Shafiga Akhundova (*b* 1924) with *The Bride's Rock* (1974), a more sophisticated reconciliation of *mugam* with modern opera requiring improvisation and non-academic vocal techniques from the singers. The Baku Opera and Ballet T has separate Azerbaijani and Russian sections, as does the Musical Comedy T, founded in 1938. In Baku, mixed-genre works such as the children's opera-ballet *Fox and Wolf* (1963) by Ibrahim Mamedov (*b* 1928) and the experimental, parodistic *Aldanmish kevakib* (The Betrayed Stars, 1977) by Mamed Kuliyev (*b* 1936) have been given.

BIBL: N. Kerimova, *Teatr i muzyka: ocherky po istoriy teatrolnoy muzyky Azerbaidzhana, 1920–1945* (Baku, 1982).

**azione sacra** (It.: 'sacred action'). 17th- and early-18th-cent. Italian term used occasionally for an opera on a religious subject, e.g. Draghi's *L'humanità redenta*; associated particularly with the Habsburg Court in Vienna. Essentially synonymous with the terms **azione sepolcrale** and **rappresentazione sacra**, it later came to be applied almost exclusively to the *oratorio.

**azione sepolcrale.** See AZIONE SACRA.

**azione teatrale** (It.: 'theatrical action'). Term used during the late 17th and early 18th cents. for an opera or musical festival play in one act, usually for performance by a restricted body of singers and players in a private, invariably aristocratic theatre. Its reduced scale apart, such a work, which was sometimes also known as **azione scenica, componimento drammatico**, or **componimento da camera**, had many stylistic similarities with contemporary opera: indeed, the azione teatrale may be seen as the forerunner of the *chamber opera. Its subject-matter was usually mythological or historical, with a story often carefully chosen to reflect the supposed virtues of the patron before whom it was performed. Classic examples are Gluck's *Il Parnasso confuso* and Mozart's *Il sogno di Scipione*. When cast on a grander scale the azione teatrale was usually called a **festa teatrale**.

**Azucena.** The gypsy (mez) in Verdi's *Il trovatore*.

# B

**Baba the Turk.** A bearded lady from a circus (mez) in Stravinsky's *The Rake's Progress*.

**Baccaloni, Salvatore** (*b* Rome, 14 Apr. 1900; *d* New York, 31 Dec. 1969). Italian bass. Treble, Sistine Chapel. Studied Rome with Kaschmann. Début Rome, Ad., 1922 (Dr Bartolo, *Barbiere*). Milan, S, 1926, where after three seasons in normal bass repertory, he specialized in buffo roles. London, CG, 1928–9. Gly. 1936–9, Leporello, Dr Bartolo, Osmin, Alfonso, and Pasquale (his greatest role). US début, Chicago 1930; New York, M, 1940–62. Appeared in every major opera-house, and was considered by many the greatest buffo since Lablache. After his retirement from opera, appeared in comedy films. (R)

**Bacchus.** See DIONYSUS and ARIADNE.

**Bach, Johann Christian** (*b* Leipzig, 5 Sep. 1735; *d* London, 1 Jan. 1782). German composer. Youngest son of Johann Sebastian and Anna Magdalena Bach. Studied with his brother Carl Philipp Emmanuel in Berlin, where he encountered the operas of Graun, Hasse, and Agricola. In 1756 went to Italy, became acquainted with the operas of Jommelli and Traetta, and probably studied with Martini. Initially, he composed sacred music, but in 1760 wrote his first opera, *Artaserse*, for Turin, followed by *Catone in Utica* for Naples in 1761. *Alessandro nell'Indie*, performed there in 1762, confirmed his operatic abilities and led to an invitation from the impresario Colomba Mattei to come to London.

Bach arrived in England in summer 1762, replacing Gioacchino Cocchi as composer at the H. His first London opera, *Orione*, was performed in Feb. 1763 before an enthusiastic audience including George III and Queen Charlotte. Burney, who was also present, commented upon Bach's abandonment of the da capo aria, as well as the delicacy of the orchestral texture: he wrongly thought that *Orione* included the first use of clarinets in an English opera. *Zanaida*, which appeared three months later, was nearly as successful, and prompted Bach's appointment as music master to the Queen.

Now at the forefront of London musical life, Bach did not confine his activities to opera. A virtuoso keyboard performer, he was also involved with Carl Friedrich Abel in a concert series which ran for 17 years from 1764. *Adriano in Siria* was poorly received, but his popularity was partly restored with *Carattaco* in 1767. He also contributed to several pasticcios, including one based on Piccinni's *L' Olimpiade* (1769), and with Guglielmi organized the London première of Gluck's *Orfeo ed Euridice* (1770). He also wrote for Vauxhall and Ranelagh and contributed to *The *Maid of the Mill* (1765).

On arrival in London, Bach had been disappointed in the singers, except for Anna Lucia de Amicis. Although better artists were gradually attracted, including from Milan the soprano Cecilia Grassi (later his wife), the invitation to compose for Mannheim was enticing. He made two trips, in 1772 and 1774, for which he composed *Temistocle* and *Lucio Silla*. The former was especially successful, as was his last London opera, *La clemenza di Scipione*, but his final operatic work, a setting for the Académie Royale de Musique in Paris in 1779 of an old tragédie-lyrique libretto, *Amadis de Gaule*, was poorly received. By now the prevailing taste in London was for opera buffa and, despite the success of *Scipione*, he was gradually eclipsed by the popularity of Sacchini and other opera buffa composers and died relatively forgotten.

His cosmopolitan education, with its combination of German and Italian influences, perhaps explains why Bach developed such a distinctive style in an era of great uniformity. Although he followed the path of Jommelli, Traetta, and Gluck in concentrating on opera seria, he showed little interest in the dramatic or tragic aspects of the genre. But his melodic inventiveness, skilful handling of the vocal line, and colourful instrumentation lent him a distinctive voice. One of his most important influences was on the young Mozart, whom he met in London in 1764, and later in Paris. There are many points of similarity between the two composers: in particular, the Mannheim operas, which represent the summit of Bach's operatic achievements, paved the way for the imaginative approach to Metastasian opera seria manifested later in Mozart's *Idomeneo* and *La clemenza di Tito*.

His brother **Johann Christoph** (1732–95) also wrote some successful dramatic works to texts by Herder and others. Their brother **Wilhelm Friedemann** (1710–84) is the subject of an opera by Graener (1931). Their father **Johann Sebastian** (1685–1750) is the subject of an opera by Emil Ábrányi (1947).

WORKLIST: *Artaserse* (after Metastasio; Turin 1760); *Catone in Utica* (after Metastasio; Naples 1761); *Alessandro nell'Indie* (after Metastasio; Naples 1762); *Orione ossia Diana vendicata* (Bottarelli; London 1763); *Zanaida* (Bottarelli; London 1763); *Adriano in Siria* (after Metastasio; London 1765); *Carattaco* (?Bottarelli; London 1767); *Temistocle* (Verazi, after

41

Metastasio; Mannheim 1772); *Lucio Silla* (Verazi, after De Gamerra; Mannheim 1774); *La clemenza di Scipione* (?; London 1778); *Amadis de Gaule* (De Vismes, after Quinault; Paris 1779).

BIBL: C. Terry, *Johann Christian Bach* (2/London, 1967).

**Bacquier, Gabriel** (*b* Béziers, 17 May 1924). French baritone. Studied Paris. Début Nice 1950 (Landowski's *Le fou*, with José Beckmans's Compagnie Lyrique Française). Brussels, M, 1953–6; Paris, OC, 1956, O since 1958. Gly. 1962 as Count Almaviva (Mozart); London, CG, 1964 (Ford, Malatesta, Scarpia, etc.). New York, M, 1964. Aix Festivals as Don Giovanni, Falstaff (Verdi). An elegant and stylish artist. (R)

**Baden-Baden.** Town in Baden-Württemberg, Germany. Early premières at the T de Bade included Gounod's *Colombe* (1860) and Reyer's *Erostrate* (1862). The T der Stadt (cap. 512) opened 1862 with Kreutzer's *Das Nachtlager von Granada*, and Berlioz's *Béatrice et Bénédict* was premièred here in the same year. Pauline Viardot sang here 1864–6. The 1927 Chamber Music Festival included premières of Hindemith's *Hin und Zurück* and Milhaud's *L'enlèvement d'Europe*.

**Badini, Carlo Francesco** (*fl.* London, late 18th cent.). English librettist of Italian descent. Poet of the Italian opera in London, he provided both original librettos and adaptations for Bertoni (*Il duca di Atene*, 1780), Guglielmi (*Le pazzie d'Orlando*, 1771), Tarchi (*La generosità di Alessandro*, 1789), and others. Librettist of Haydn's unstaged London opera *L'anima del filosofo* (1791): if the surviving four acts constitute the complete work, a fact disputed by some scholars, Badini's libretto is virtually the only treatment after Monteverdi's *Orfeo* to retain the original tragic ending of the myth.

He was also an effective if malicious critic: according to *Da Ponte, it was Badini's 'satire and destructive gossip' which manipulated the fortunes of opera in London for many years. An implacable enemy of Da Ponte, who eventually took over his position at the Italian opera, Badini wrote an infamous and obscene attack on him after his first London opera, *The School for Husbands* (1795).

**Badoaro, Giacomo** (*b* Venice, 1602; *d* Venice, 1654). Italian librettist. A Venetian nobleman, he was a member of the Accademia degli Incogniti and together with his friends Busenello and Strozzi played an important role in the establishment of commercial opera in Venice. Although he did not (as is usually claimed) write the libretto of Monteverdi's *Le nozze di Enea con Lavinia* (1641; lost), he did provide that for *Il ritorno d'Ulisse in patria* (1640): his other extant works are *Ulisse errante*, set by Sacrati in 1644,

and *Elena rapita da Teseo,*? set by Cavalli in 1653. His skilful adaptation of classical mythology and Renaissance pastoralism to the demands of the public stage set a high standard for later Venetian librettists: of particular note was his creation of sympathetic, realistic characters displaying genuinely human emotions.

**Bagnara, Francesco** (*b* Vicenza, 1784; *d* Venice, 21 Oct. 1866). Italian scene painter. Studied Venice with Giuseppe Borsato, then worked at the Fenice 1810–23, and elsewhere in Venice and other cities, especially Vicenza. Painted over 1,100 sets, including for the premières of works by Rossini, Bellini, and Donizetti. His lively, expressive sketches (many preserved in the Museo Correr in Venice) suggest an expert and colourful style, classical in training but responding with quick instinct to the emergent Romanticism of some of his composers.

**Bahr-Mildenburg, Anna** (*b* Vienna, 29 Nov. 1872; *d* Vienna, 27 Jan. 1947). Austrian soprano. Studied Vienna with Papier, Hamburg with Pollini. Début Hamburg 1895 (Brünnhilde). Here she became the protégée and mistress of Mahler. Endowed with a magnificent voice and physique, she was musical, impassioned, and a superb actress. Soon became leading Wagner soprano. Bayreuth 1897 (Kundry); also 1909–14. London, CG, 1906 (Isolde, Elisabeth), and first London Klytemnestra 1910. Vienna 1898–1916. Darmstadt 1915–16 as actress. Vienna 1919–20. After retirement taught, but sang final Klytemnestra, Augsburg 1930. Married writer Hermann Bahr 1909, with whom she wrote *Bayreuth und das Wagnertheater* (Leipzig, 1910). (R)

**baile** (Sp.). A kind of *entremés performed after the second act of a serious play, usually featuring dancing and poetry as well as music.

**Bailey, Norman** (*b* Birmingham, 23 Mar. 1923). English baritone. Studied Vienna with Vogel, Patzak, and Witt. Début Vienna 1959 (Tobias Mill, *Cambiale di matrimonio*). Linz 1960–3; Wuppertal 1963–4; Düsseldorf 1964–7. London, SW, from 1967 (début as Mozart's Figaro), then Sachs in the memorable production of *Die Meistersinger* under Goodall 1968. London, C (Wotan, Gunther, Pizarro, Kutuzov, Gremin). Has appeared in the Wagner repertory at London, CG, and Bayreuth. Milan, S, 1967; New York, M, 1976. His Sachs, wise, human and benevolent, and his Wotan, both magisterial and moving, display an inherent musicality and feeling for the text. (R)

**Baker,** (Dame) **Janet** (*b* York, 21 Aug. 1933). English mezzo-soprano. Studied London with Helene Isepp. Début Oxford 1956 OUOC (Rosa, *The Secret*). 1959 Eduige in *Rodelinda* with Handel OC, first of a number of Handel roles. Sang regularly with EOG 1961–76, Dido, Lucretia, Nancy, Kate. Dorabella, Octavian, the Composer, and Dido (Berlioz) for Scottish O. From 1966 at London, CG, incl. Dido, Cressida, Idamante; Gly. 1965–82, incl. Penelope (*Ulisse*), Diana (*Calisto*), Orfeo (Gluck). Her repertory included Vitellia, Giulio Cesare, Maria Stuarda, and Charlotte. Her highly individual and beautiful voice, warmth, commitment, and wide dramatic range made her one of the outstanding artists of the day. Retired from opera 1982 and from all public performance 1989. Autobiography, *Full Circle* (London, 1982). (R)

**Balanchivadze, Meliton** (*b* Banodzha, 24 Dec. 1862; *d* Kutaisi, 21 Nov. 1937). Georgian composer and singer. Studied privately, and from 1880 sang in the chorus at the Tbilisi Opera, then studying St Petersburg with Rimsky-Korsakov. Part of his opera *Tamar tsbiery* (The False Tamar, 1897) was given in concert there; though not produced till Tbilisi, 1926, this is regarded as a founding work of Georgian opera. His elder son Georg became a distinguished choreographer as George Balanchine; his younger son **Andrey Balanchivadze** (*b* St Petersburg, 1 June 1906) is a composer and conductor who studied in Kutaisi with Ippolitov-Ivanov and has been very active in his country's musical life. His operas include *Mziya* (Tbilisi 1950).

**Balducci, Giuseppe** (*b* Jesi, 5 May 1796; *d* Malaga, 1845). Italian composer. Studied Naples with Tritto and Zingarelli. Probably uniquely among 19th-cent. Italian composers, he wrote a number of drawing-room operas, e.g. *Il conte di Marsico* (1839). Of his public works, the most important was his last, *Bianca Turenga* (1838).

**Balfe, Michael** (*b* Dublin, 15 May 1808; *d* Rowney Abbey, 20 Oct. 1870). Irish composer, singer, and violinist. Studied Wexford, and London with C. E. Horn. Début as violinist in London, 1823, as baritone in Norwich in an adaptation of *Der Freischütz* (Caspar). After further study in Italy with Galli, went to Paris and with Rossini's encouragement sang at the I (Figaro in *Barbiere*). He was principal baritone in Palermo, 1829–30, producing his first opera there. Back in London in 1835, he began his career as a composer of English operas with the very successful *The Siege of Rochelle* at DL, following it with the almost equally successful *The Maid of Artois* for Malibran. He also sang Papageno in the first performance in English of *The Magic Flute*. Back in Paris by 1841, having failed in an attempt to establish English opera at the *Lyceum (*Keolanthe*)*, he continued his career there, returning to England for the greatest success of his 29 operas, *The Bohemian Girl*. From

1857 to 1863 he was closely associated with the Pyne-Harrison project for an English national opera: his own chief success was *The Rose of Castille*. He continued to travel, making two successful visits to Russia, before turning gentleman farmer and retiring to an estate in Hertfordshire. His melodies are often graceful and charming, and his gift for a memorable tune led him to place much importance on ballads in his operas, but he was little interested in characterization and had less flair for dramatic development. However, he did write, in *Catherine Grey*, one of the first English Romantic operas without spoken dialogue.

SELECT WORKLIST (all first prod. London unless otherwise stated): *I rivali di se stessi* (Their Own Rivals) (?; Palermo 1829); *The Siege of Rochelle* (Fitzball, after Mme de Genlis; 1835); *The Maid of Artois* (Bunn; 1836); *Catherine Grey* (Linley; 1837); *Falstaff* (Maggione; 1838); *Keolanthe* (Fitzball; 1841); *The Bohemian Girl* (Bunn, after Saint-Georges; 1843); *The Rose of Castille* (Harris & Falconer; 1857).

BIBL: C. Kenney, *Memoir of Michael William Balfe* (London, 1875, R/1978).

**ballabile** (It.: 'suitable for dancing'). A term used to describe sung dances in 19th-cent. Italian opera. Examples include the song and dance of the witches in Act III of Verdi's *Macbeth* and the opening chorus to Act II of *Ernani*.

**ballad opera**. A popular form of English entertainment flourishing in the 18th cent., in which spoken dialogue alternated with musical numbers, usually fitting new words to old ballads, folk tunes, or (particularly in its later stages) melodies by different composers. Though developed from earlier theatrical practices, the genre may be dated from its most famous example, the Gay/Pepusch The *Beggar's Opera* (1728). This was characteristic in its satire of contemporary politicians and of the excesses of Italian opera and its prima donnas. Ballad opera resembled the contemporary French opéra-comique en vaudevilles (including its occasional parody of serious operas); the translation into German of Charles Coffey's *The Devil to Pay* (1731) and its sequel *The Merry Cobbler* (1735) influenced the development of *Singspiel. After the first flood of ballad operas in 1728–35, which rapidly became so popular as to threaten the Italian opera (even of Handel) which they lampooned, the genre waned, to be revived in a somewhat modified manner with (generally) original tunes: a famous example is Bickerstaffe's *Love in a Village, with music by Arne. Most of the early ballad operas were produced at the Little Haymarket T (where Gay had seen some opéras-comiques en vaudevilles given by the *Théâtres de la Foire between 1720 and 1728), or at Lincoln's Inn Fields T. Ballad opera was also particularly popular in Dublin, and enjoyed a certain following in America during the 1730s.

Modern operas making use of the title include Vaughan Williams's *Hugh the Drover*, in which ballad-type songs are incorporated into a fully composed opera. In Weill's *Die Dreigroschenoper* the numbers are introduced in conscious imitation of the old form in general and *The Beggar's Opera* in particular.

**ballata** (It.: 'ballad'). A term applied to numbers in 19th-cent. Italian opera which were couched in dance rhythm. Examples include 'Son Pereda, son ricco d'onore' in Verdi's *Don Carlos* and 'Questa o quella' in *Rigoletto*.

**ballet de cour** (Fr.: 'court ballet'). A popular French court entertainment of the 16th and 17th cent., whose development paralleled that of the English *masque. Its origins lay in Italy, in the traditions of the spectacular *intermedi and *balletti a cavallo* (equestrian ballets), which combined with ballroom dancing to produce a sumptuous mélange of ballet, music, and drama first seen in France in 1533 at the festivities for the wedding of the future King Henri II with Catherine de' Medici. Given the natural French love of ballet, the genre found immediate popularity, not least because of the opportunity it provided for the spectactors themselves to participate in the dancing. The earliest work with surviving music is *Circé, ou le balet comique de la royne* (1581), the first which may be considered a genuinely unified drama. The creation of the royal ballet-master Beaujoyeulx, it included two songs in monodic style; the music of most later ballets de cour, whose composers included notably Pierre Guédron (1565–1621) and Antoine Boësset (c.1585–1643), is lost.

During the early part of Louis XIV's reign, the genre underwent a significant revival at the hands of Isaac de Benserade, 'ordonnateur des divertissements' at the French court, beginning with the *Ballet de Casandre* (1651). Under his direction the ballet de cour was no longer a collective work, involving several composers and dramatists, but a genuine collaboration between just one composer and librettist. He found an ideal partner in Lully, with whom he first worked on the *Ballet de la nuit* (1653): together they fashioned the ballet de cour into the supreme expression of royal *gloire*, producing over 20 works. Louis XIV was an enthusiastic participant in ballets de cour, but from 1670, following his infirmity, the popularity of the genre waned. Already its combination of music and dancing had provided inspiration for the development of the *comédie-ballet; now it had an equally important influence on the creation of French opera. Of the many strains which came together in Lully's *Cadmus* (1673), none

was more important than ballet, which lent to the new tradition a grandiose and spectacular aura lacking in most contemporary Italian operas. Later, under Campra and Rameau, a hybrid operatic genre, in which the dramatic aspect was reduced in favour of dancing, was developed, the so-called *opéra-ballet. So crucial was the influence of the ballet de cour on the development of French opera that dance remained a staple feature of the Parisian tradition until well into the 19th cent.

**ballet-héroïque** (Fr.: 'heroic ballet'). A stage entertainment popular in France in the early and mid-18th cent., a subcategory of *opéra-ballet. Characterized by its emphasis upon the heroic and exotic, contrasting with the more worldly tone of opéra-ballet proper, the ballet-héroïque consisted of a prologue and usually three or four acts. Each of these dealt with a discrete theme, as in one of the earliest examples, François Collin de Blamont's *Les festes grecques et romaines* (1723), which originally had three acts entitled 'Les jeux olimpiques', 'Les baccanales', and 'Les saturnales'. Rameau composed five such works, including *Les Indes galantes* (1735) and *Anacréon* (1754).

**Balling, Michael** (*b* Heidingsfeld, 27 Aug. 1866; *d* Darmstadt, 1 Sep. 1925). German conductor. Studied Würzburg. Violist in Mainz, Schwerin, then Bayreuth, where he was assistant conductor 1896; cond. *Parsifal*, *Tristan*, and *The Ring*, 1904–25, and was one of Siegfried Wagner's right-hand men. Cond. *The Ring* in English for the Denhof OC, Edinburgh 1910, and *Orfeo* with Brema in London. Founded a school of music in Nelson, New Zealand, 1892. Succeeded Richter as conductor of Hallé, 1912. From 1919 music director in Darmstadt. From 1912 editor of Breitkopf und Härtel Wagner Edition.

**Ballo in maschera, Un** (A Masked Ball). Opera in 3 acts by Verdi; text by Somma, based on Scribe's text for Auber's *Gustave III, ou Le bal masqué*, in turn based on an historical event. Prod. Rome, Ap., 17 Feb. 1859, with Dejean, Scotti, Sbriscia, Fraschini, L. Giraldoni, cond. Angelini; New York, AM, 11 Feb. 1861; London, L, 15 June 1861, with Tietjens, Lemaire, Gassier, Giuglini, Delle Sedie, cond. Arditi.

Stockholm, 1789. King Gustavus III (ten) is warned of a plot against his life. The Chief Justice wants him to sign a paper banishing the fortune-teller Mlle Arvidson (con); the people consider her to be a great oracle. The King decides to go and see her in disguise.

Amelia (sop), wife of Anckarstroem (bar), the King's friend and secretary, comes to ask Mlle Arvidson's advice on how to overcome her love for the King. Officers from the court consult

Mlle Arvidson; she warns the King that he is shortly to die at the hand of a friend, the first one to shake his hand. Since this turns out to be Anckarstroem, he refuses to believe it.

The King surprises Amelia when she is collecting the necessary magic herbs at night and they confess their love. Anckarstroem comes to warn him of an imminent ambush. The King escapes and the conspirators are taken aback to find Anckarstroem. In the commotion it emerges that his veiled companion is Amelia.

Anckarstroem is convinced of Amelia's adultery and intends to kill her. When the conspirators arrive, they cast lots for who is to kill the King: it is to be Anckarstroem. He and Amelia are invited to a masked ball by the King. There, the King tells her that he has decided she and Anckarstroem must leave the court. As he says this, Anckarstroem steps between them and stabs him. The King declares Amelia's innocence and pardons Anckarstroem as he dies.

When Verdi first submitted his libretto to the San Carlo Theatre, Naples, the censor protested that the assassination of a king could not be shown on the stage, and demanded that Verdi adapt his music to a new libretto. Verdi refused, and left Naples. A Rome impresario offered to produce an altered version of the work; and so the locale was changed from 18th-cent. Stockholm to 17th-cent. Boston. Another version used at the Paris première (1861) set the story in Naples. The names of the leading characters are thus (in the order of setting, Stockholm, Boston, Naples): Gustavus III, Riccardo (Count of Warwick), Riccardo (Duke of Olivares); Anckarstroem, Renato, Renato; Count Ribbing, Samuele, Armando; Count Horn, Tomaso, Angri; Mlle Arvidson, Ulrica, Ulrica. The 'Stockholm' version has now established itself over the formerly popular 'Boston' version.

Other operas on the subject are by Auber (*Gustave III, ou Le bal masqué*, 1833) and Mercadante (*Il reggente*, 1843).

**Baltimore.** City in Maryland, USA. The Baltimore Civic O was founded in 1932 as an opera workshop and became fully professional in 1950 under the direction of Rosa Ponselle. Opera is given at the Lyric OH (cap. 2,600).

**Baltsa, Agnes** (*b* Levkas, 19 Nov. 1944). Greek mezzo-soprano. Studied Athens, Munich, and Frankfurt. Début Frankfurt 1968 (Cherubino). Frankfurt 1968–72; Berlin, D, from 1972. London, CG, from 1976 (Carmen, Octavian, etc.). Houston 1971 (Carmen). Vienna since 1976, Salzburg since 1977, New York, M, from 1979. Roles include Orfeo (Gluck), Sesto, Dorabella, Dido (Berlioz), Herodias, the Composer, Rosina, Eboli. (R)

**Balzac, Honoré** (self-styled 'de') (*b* Tours, 20 May 1799; *d* Paris, 18 Aug. 1850). French novelist. Author of over 90 novels, which he eventually had published, divided into three groups, in a collected edition entitled *La comédie humaine* (pub. 1842–8). Operas on his works are as follows:

*L'arbre rouge* (1831): Nouguès (1910)
*La peau de chagrin* (1831): Levadé (1910)
*Le colonel Chabert* (1832): Waltershausen (*Oberst Chabert*, 1912)
*Contes drôlatiques* (1832–7): Alfano (*Madonna Imperia*, 1927)
*La duchesse de Langeais* (1834): Bozza (1967)
*Peines de cœur d'une chatte anglaise* (1841): Henze (*Die englische Katze*, 1983)
*Gambara* (1837): Duhamel (1978)
*Massimilla Doni* (1839): Schoeck (1937)
*La grande Bretèche*: Tansman (*Le serment*, 1955)

**Bamberg.** Town in Bavaria, Germany. Jesuit school music dramas were given and opera was staged at the Seehof in the 18th cent. under L. Fracassini, including Grétry and Mozart as well as Singspiel. Count Soden opened a theatre 1802, and E. T. A. *Hoffmann worked here in various capacities 1808–13.

**Bampton, Rose** (*b* Cleveland, 28 Nov. 1908). US soprano, orig. contralto. Studied Philadelphia with Connell and Queena Mario. Début Chautauqua 1928 (Siebel). New York, M, 1932–50, début as Laura (*Gioconda*). Soprano début there 1937 (Leonora, *Il trovatore*). Acclaimed as Donna Anna and Alceste, and in the 1940s in several Wagner roles, including Kundry and Sieglinde. London, CG, 1937 (Amneris). A fine musician, with a rounded, velvety voice. (R)

**banda** (It., Span.: 'band'). The on-stage band was occasionally used in 18th-cent. opera, e.g. Paisiello's *Pirro* (1787), but following its use by Rossini in *Ricciardo e Zoraide* (1818), it became a staple feature of Italian opera. The 19th-cent. banda was in essence a military band, comprising about 20 woodwind and brass players, which appeared on stage for ballroom scenes, processions, etc. This ensemble was usually recruited locally by the impresario, rather than being on the strength of the theatre: e.g. in Palermo, it was provided by the garrison of Bourbon troops. Its music was written out by the composer in piano score only, with the conductor of the banda arranging it in the appropriate fashion. It makes its most marked appearance in Verdi's operas, where its stirring, frankly vulgar sound accorded well with the general Risorgimento feeling of such early works as *Nabucco* and *I Lombardi*. In his later operas he used the banda convention in a more flexible fashion, ranging from the subtlety of the ballroom scenes of *Rigoletto* and *Un ballo in* *maschera* to the stridency of the auto-da-fé scene of *Don Carlos*. While the playing standards of the banda were not generally high, some ensembles were highly regarded, and none more so than that of Kinsky's regiment of the Austrian army which performed at Venice, F.

**Bánk bán.** Opera in 3 acts by Erkel; text by Béni Egressy, after József Katona's tragedy (1814). Prod. Budapest 9 Mar. 1861, with Hollósy, Ellinger, Telek, Koszegi, cond. Erkel; London, Collegiate T, 20 Feb. 1968, by University College Opera Society, with Louis, Bateman, Jenkins, Sadler, cond. Badacsonyi.

The fortress of Visegrád in Hungary, 1213. Otto (ten), brother of Queen Gertrud (mez), pays court to Melinda (sop), the wife of Bánk bán, Count of Hungary. When he learns of this, Bánk bán (ten) joins the cause of the rebelling natives. Having gone to seek an explanation from the Queen, Bánk bán kills her in self-defence after she threatens him with a dagger. On the King's return from the Crusades, the rebellion is quashed; Bánk bán confesses his crime, and the King challenges him to a duel. However, at that moment Tiborc (bar), Bánk bán's friend, enters bearing the body of Melinda, who drowned during a rescue attempt. Distraught, Bánk bán stabs himself.

**Banská Bystrica.** Town in Slovakia, Czechoslovakia. Opera is given in the J. G. Tajovský T, normally 3–4 times a week in the season, by a company of about 20 soloists. The company, which has a reputation for energy and liveliness, revived the early Slovak composer Viliam Figuš-Bistrý's *Detvan* (comp. 1924) in 1976. It also tours the region.

**Banti, Brigida Giorgi.** See GIORGI-BANTI, BRIGIDA.

**Bantock,** (Sir) **Granville** (*b* London, 7 Aug. 1868; *d* London, 16 Oct. 1946). English composer. Apart from some early works, his only opera is *The Seal-Woman*. This makes use of some genuine Hebridean tunes collected by Marjorie Kennedy-Fraser, who took part in the first performance as an old crone spinning in the corner.

SELECT WORKLIST: *The Seal-Woman* (Kennedy-Fraser; Birmingham 1924).

**Bar.** A German verse form consisting of two *Stollen* (strophes) and an *Abgesang* (after-song). Both *Stollen* would normally be sung to essentially the same melody, with the *Abgesang* having a different but related melody. An example, suggesting the form's troubadour origins, is Adolar's 'Unter blüh'nden Mandelbäumen' in *Euryanthe*. It was taken up by Wagner, with Lohengrin's warning to Elsa not to ask his name, and occurs in various

forms in *The Ring*. His most important use of it is in *Die Meistersinger*, when Walther is lectured on the *Meistergesanges Bar* by Kothner and Sachs, eventually producing his Prize Song in the form. Suggestions that Wagner used the Bar as the basis of his form in all his mature works were put forward by Alfred Lorenz in *Das Geheimnis der Form bei Richard Wagner* (1924–33); this theory is now discredited.

**Baranović, Krešimir** (*b* Šibenik 25 July 1894; *d* Belgrade, 17 Sep. 1975). Croatian conductor and composer. Studied Zagreb with Kaiser, Vienna 1912–14. Cond. at Zagreb Opera 1915, music director at the Operas of Zagreb 1915–25, Belgrade 1927–9, Bratislava 1945–6. At Zagreb introduced wide new repertory, especially of Russian and Yugoslav opera, and gave first local *Ring*. Professor Belgrade Cons. 1946. His operas are *Striženo-košeno* (Shaven and Shorn, 1932), praised for its witty situations and skilful use of folk music, and another comedy, *Nevjesta od Cetingrada* (The Bride of Cetingrad, comp. 1942, prod. 1951). (R)

**Barbaia, Domenico** (*b* Milan, 1778; *d* Posillipo, 19 Oct. 1841). Italian impresario. Began his career as a waiter, inventing *barbaiata*, coffee or chocolate with a head of whipped cream; then ran the gambling tables at Milan, S, 1808. Director Naples, C and N, and other royal opera-houses, (1809–40, with interruptions). His productions included Spontini's *La vestale*, with his mistress Isabella *Colbran. Brought Rossini to Naples 1815 on a six-year contract to compose two operas a year. Though Colbran soon left him for Rossini, relations remained friendly. When the C burnt down in Feb. 1816, he promptly rebuilt it and reopened Jan. 1817. Also obtained the concession of the Vienna, K and W (1821–8). Here he organized brilliant seasons, introducing to Vienna Rossini and other Italian composers including Bellini and Mercadante, and commissioning *Euryanthe* from Weber. Also controlled the Scala and Cannobbiana in Milan (1826–32). A man of sharp instincts and artistic flair, he was quick to spot the talents of Rossini, Bellini, Donizetti, Mercadante, Pacini, and others, and to find effective outlets for their talents; and his practical encouragement served to set them and other composers of the day on their careers. By introducing *La vestale* and Gluck's *Iphigénie en Aulide* to Naples, he began the Italian tradition of opera seria with orchestral recitative, as in Mayr's *Medea in Corinto* (1813) and Rossini's *Elisabetta, regina d'Inghilterra* (1815). He appears as the impresario Bolbaya in Auber's *La sirène* (1844).

**Barbarina.** The gardener Antonio's daughter (sop) in Mozart's *Le nozze di Figaro*.

**Barber, Samuel** (*b* West Chester, PA, 9 Mar. 1910; *d* New York, 23 Jan. 1981). US composer. The success of his first mature opera, *Vanessa*, couched in a skilful late-Romantic style, was a prime factor in him being chosen to write the new opera by a US composer that was felt essential for the opening of the new New York, M, in 1966. However, the similar but less effectively handled idiom of *Antony and Cleopatra* proved unsuitable for the inauguration of a major new house, and even in its own terms was a failure. (R)

WORKLIST: *The Rose Tree* (Brosius; West Chester 1920); *Vanessa* (Menotti; New York 1958); *A Hand of Bridge* (Menotti; Spoleto 1959); *Antony and Cleopatra* (Zeffirelli, after Shakespeare; New York 1966).

**Barberini family**. Italian family of musical patrons whose support proved crucial for the early development of opera in Rome. Their wealth came initially from the cloth business established in Florence by Francesco Barberini (1454–1530): later the family moved to Rome, becoming one of the most powerful dynasties there. For opera their most important member was Maffeo Barberini (1568–1644), elected Pope Urban VIII 1623. For the next 21 years he placed Rome at the forefront of Italian operatic activites: composers supported by the Barberinis included Abbatini, Landi, Mazzocchi, Marazzoli, and Rossi. In 1632 an opera-house (cap. 3,000) was opened in the Barberini palace: here some of the most important Roman operas were given, including Landi's *Il Sant'Alessio* and Mazzocchi's and Marazzoli's *Chi soffre speri*.

Following the election of Pope Innocent X (1644) the Barberinis were banished from Rome. Led by Maffeo's nephew, Cardinal Antonio Barberini (1607–71), they fled to Paris, accompanied by their entourage of musicians. With Mazarin's support they were responsible for staging some of the first Italian operas to be seen in France. In 1653 a reconciliation was effected with Pope Innocent X through the marriage of Olimpia Giustiniani to Antonio Barberini's nephew: to celebrate the occasion the Barberini opera-house was reopened with Abbatini's and Marazzoli's *Dal male il bene*. It continued to be used as an operatic venue until the late 17th cent.

**Barbier, Jules** (*b* Paris, 8 Mar. 1822; *d* Paris, 16 Jan. 1901). French librettist. Collaborated with Michel *Carré on numerous librettos, providing composers with texts which derived from many authors, including Goethe, Dante, and Shakespeare, but which formed the subjects into a conventionalized, often sentimentalized, version of the original that was highly effective in the terms of most contemporary French opera. Their librettos include those for Gounod's *Faust*, *Polyeucte*, *Philémon et Baucis*, and *Roméo et Juliette*, Meyerbeer's *Dinorah*, Thomas's *Mignon*, *Francesca da Rimini*, and *Hamlet*, and Offenbach's *Les contes d'Hoffmann*. Their lighter side is shown in their text for Massé's *Les noces de Jeannette*.

**Barbiere di Siviglia, Il** (orig. *Almaviva, ossia L'inutile precauzione*) (The Barber of Seville; orig. Almaviva, or The Useless Precaution). Opera in 2 acts by Rossini; text by Sterbini, after *Beaumarchais's comedy (1775). Prod. Rome, Arg., 20 Feb. 1816, with Giorgi-Righetti (coloratura mez, though the role has often been taken by a sop), M. García, Zamboni, cond. Rossini; London, H, 10 Mar. 1818, with Fodor, García, Naldi; New York, Park T, 3 May 1819, in Bishop's version with Leesugg, Phillips, Spiller; New York, Park T, 29 Nov. 1825 (first Italian language perf. in USA), with M. García (later Malibran) and her mother, father, and brother as Bertha, Almaviva, and Figaro.

In a square in Seville, Count Almaviva (ten), masquerading as Lindoro, a poor student, woos Rosina (mez), the ward of old Doctor Bartolo (bs). The barber and general factotum, Figaro (bar), promises to help Almaviva in his courtship. By pretending to be a drunken soldier (with a forged billeting order), Almaviva gains entry to Bartolo's house; he whispers to Rosina that he is Lindoro. Don Basilio (bs), Rosina's music-master, and Bartolo are rather suspicious of the 'soldier' and try, unsuccessfully, to have him removed. They are also disturbed to hear that Count Almaviva has arrived in the town.

Almaviva visits the house again, this time disguised as a music teacher, Don Alonso, standing in for Basilio, who he says is ill. Having gained Bartolo's trust, 'Don Alonso' gives Rosina her music lesson. While shaving Bartolo, Figaro manages to steal a key which will enable the lovers to escape from the house at night. When Basilio arrives, Almaviva bribes him to support his story, but Bartolo realizes some scheme is afoot. Anxious to hasten his marriage to Rosina, Bartolo despatches Basilio to fetch a notary. Bartolo convinces Rosina that Lindoro does not really love her and is merely acting on behalf of Count Almaviva; she is so upset that she confesses she had planned to run away with the student. When Lindoro returns, he reveals his true identity to Rosina and the couple reaffirm their love. Their escape, however, is thwarted, since the ladder on to the balcony has been removed. Basilio arrives with the notary and, further bribed by Figaro, witnesses the marriage of Rosina and Almaviva. By the time Bartolo returns, the ceremony is over; after initial anger, he accepts the marriage, especially since he is not obliged to give a dowry.

The first performance in Rome was one of the greatest fiascos in the history of opera. As Paisiello's opera (see below) enjoyed great popularity, the opera was originally entitled *Almaviva, ossia L'inutile precauzione*; the title *Il barbiere di Siviglia* was first used at Bologna in 1816. The overture had already been used by Rossini for *Aurelio in Palmira* (1813) and *Elisabetta, regina d'Inghilterra* (1815).

**Barbiere di Siviglia, ossia L'inutile precauzione, Il.** Opera in 4 acts by Paisiello; text by G. Petrosellini, after Beaumarchais's comedy (1775). Prod. St Petersburg 26 Sep. 1782 in Italian: in Russian (trans. I. Vien), T Dereviany, 27 Aug. 1790; London, H, 11 June 1789, with Storace, Kelly, Benucci; New Orleans, T St Pierre, 10 Dec. 1805.

For other operas on the subject see BEAUMARCHAIS.

**Barbieri, Fedora** (*b* Trieste, 4 June 1920). Italian mezzo-soprano. Studied Trieste with Toffolo and Giulia Tess. Début Florence 1940 (Fidalma). After appearances at all leading Italian operahouses, sang at Buenos Aires, C, 1947. London, CG, 1950, with Milan, S, Co., as Quickly; 1957–8 Azucena, Amneris, Eboli. New York, M, 1950–4, 1956–7, 1967–8, making début on opening night of Bing's régime as Eboli. Among over 100 roles, much admired as Orfeo and Carmen. (R)

**Barbieri-Nini, Marianna** (*b* Florence, 18 Feb. 1818; *d* Florence, 27 Nov. 1887). Italian soprano. Teachers included L. Barbieri, Pasta, and Vaccai. After a disastrous Milan, S, début in *Belisario* (1840), and several other unfortunate appearances in which her ugliness distracted from her vocal gifts, she finally made a successful appearance at Florence, 1840–1, as Lucrezia Borgia—wearing a mask. One of the finest dramatic sopranos of her day. Created the leading roles in Verdi's *I due Foscari*, *Il corsaro*, and *Macbeth*. Retired 1856. Married the pianist L. Hackensöllner, who published her memoirs, *Le memorie di una cantatrice*.

**Barbier von Bagdad, Der** (The Barber of Baghdad). Opera in 2 acts by Cornelius; text by the composer, after *The Tale of the Tailor*, from the *1,001 Nights*. Prod. Weimar 15 Dec. 1858, with Milde, Wolf, Gaspari, Roth, cond. Liszt; New York, M, 3 Jan. 1890, with Traubmann, Kalisch, Fischer, cond. Damrosch; London, Savoy, 9 Dec. 1891, by RCM students.

The plot turns on the efforts of Nureddin (ten), abetted by the barber Abul Hassan Ali Ebn Bekar (bar), to contrive a meeting with Margiana (sop), the Caliph's daughter. Despite Nureddin's imprisonment and near-suffocation in a chest, all

is eventually resolved and the blessing of the Caliph (bar) is obtained.

The work was initially a failure owing to local intrigues against Liszt, whose departure from Weimar this occasioned; after Mottl's revision and revival (Karlsruhe 1884) it became more popular.

Other operas on the subject are by H. C. Hataš (after 1780), André (1783), and Champein (early 19th cent.).

**Barbirolli,** (Sir) **John** (*b* London, 2 Dec. 1899; *d* London, 29 July 1970). English conductor of Italian parentage. Studied London, TCM and RAM. Began career as cellist. Formed string orchestra 1924; cond. for BNOC 1926–8. London, CG, 1928–33, 1951–4; London, SW, 1934; CG, 1937 (*Tosca* and *Turandot*). Then appointed conductor of New York Philharmonic-Symphony Orch. Vienna, S, 1946–7. Cond. new prod. of *Orfeo* with Ferrier 1953, *Tristan* with Sylvia Fisher's first Isolde. Rome 1969, *Aida*. Though the centre of his career was as a concert conductor, notably with the Hallé Orchestra from 1943, he had a deep understanding of the Italian operatic tradition and a sensitive feeling for the voice. (R)

BIBL: M. Kennedy, *Barbirolli* (London, 1971).

**barcarolle**. The French, in international use, for the Italian *barcaruola* or *barcarola*, a boat-song, especially of the Venetian gondoliers. The rhythm, ostensibly imitating the motion of a gondola or other small boat, is a gentle 6/8 with alternating strong and weak beats. It has been used in opera by Hérold in *Zampa*, Auber in *Fra Diavolo* and *La muette de Portici*, Donizetti in *Marino Faliero*, Verdi in *Un ballo in maschera*, and, most famously, Offenbach in *Les contes d'Hoffmann*.

**Barcelona**. City in Cataluña (Catalunya), Spain. (Names in Catalan are given in brackets.) Opera was first given at the T de la Santa Cruz (Creu), opened in 1708 with the première of Caldara's *Il più bel nome*. The first opera by a Catalan composer was *Antigona* (1760) by José Durán (?–?after 1791). In 1838 the rival T Montesión (T Mont-Sion) opened, and the Santa Cruz changed its name to the T Principal. In 1834 opera was given at the T Nuevo (T Nou). The Gran Teatro Liceo (Gran Teatre del Liceu) (cap. 3,500) opened in 1847. Rivalry between the Principal and the Liceu tended to be between conservative and progressive. The Liceu burnt down in 1861: 1,000 shares were issued to raise funds to rebuild, and the shareholders' descendants still own the theatre with the right to individual seat occupation. The present house (cap. 2,700) opened on 20 Apr. 1862; by its charter it must give at least one Spanish opera a year. The first opera in Catalan, *A la voreta del mar* (On the Seashore, 1881) by Juan Goula (1843–1917), was premièred there. The theatre closed with the Civil War in 1939 and did not reopen until 1942. Financial and other problems in the late 1970s led to reorganization and improvement. Seasons are given from Nov. until Mar., and include visits from French, German, Italian, and even Russian companies. Singers who were born and began their careers here include Victoria de los Angeles, Montserrat Caballé, José Carreras, and Giacomo Aragall. Summer performances in the T Grec have included a Catalan *El mikado* (Sullivan) in 1986. Opera is also given in the nearby town of Sabadell at the T de la Faràndula.

**Bardi,** (Count) **Giovanni de'** (*b* Florence, 5 Feb. 1534; *d* Sep. 1612). Italian critic, playwright, poet, and composer. His importance for opera derives from the activities of the *camerata he formed *c*.1573. This attracted leading Florentine musicians and other artists: out of its discussions was born the new *stile rappresentativo, which in turn stimulated composition of the first operas. One of the most significant documents produced by the group was Bardi's *Discorso mandato a Caccini sopra la musica antica e 'l cantar bene* (*c*.1578), which paved the way for *Galilei's *Dialogo*.

**Bardi Camerata.** See CAMERATA.

**Barenboim, Daniel** (*b* Buenos Aires, 15 Nov. 1942). Argentinian, later Israeli conductor and pianist. Studied conducting Salzburg and Rome. Initially a successful pianist (début Buenos Aires, aged 7); conducting début Israel 1962. Edinburgh 1973 (*Don Giovanni*); Bayreuth 1982, 1988; Paris 1986. Appointed music director new Opéra Bastille 1988; dismissed amid much publicity over the size of his salary 1989, before the opera-house opened. Repertory incl. *Figaro*, *Così*, *The Ring*, *Tristan*. (R)

**Bari.** Town in Puglia, Italy. The T Piccinni was opened 1854 with Donizetti's *Poliuto*. The T Sediele, opened in the early 19th cent., closed when it partly collapsed in 1835, and performances were given in the grotesque T La Zuppiera ('soup-tureen'). Opera first found a permanent and worthy home in the T Petruzzelli (cap. 2,000), opened 14 Feb. 1903 with *Les Huguenots*.

BIBL: A. Giovine, *Il teatro Piccinni di Bari (1854–1964)* (Bari, 1970).

**baritone** (Gr., *barytonos*: 'heavy-tone'). The middle category of the natural male voice. Several subdivisions exist within opera-houses: the commonest in general use (though seldom by composers in scores) are given below, with examples of roles and their approximate tessitura. These divisions often overlap, and do not correspond from country to country. In general, distinction is more by character than by tessitura, especially in France: thus the examples of the roles give a more useful indication of the different voices' quality than any attempted technical definition.

French: bariton (Escamillo: c–a♭'); bariton-Martin (Pelléas: c–a♭'). In early music (Lully to Rameau), basse-taille (G–f').

German: Spielbariton (Don Giovanni: A♭–g'); Heldenbariton (Orestes in *Iphigénie en Tauride*: c–a♭'; hoher Bariton (Hans Heiling: c–a♭'); Kavalierbariton (the Count in *Capriccio*: c–a♭'). There is also the Bass-Bariton (Wotan, Sachs: A♭–f).

The Italian baritono (Dandini: c–a♭') is not generally subdivided, though professionals may speak of baritono brillante and baritono cantante.

**Barnaba.** The Inquisition spy (bar) in Ponchielli's *La Gioconda*.

**Barnett, John** (*b* Bedford, 15 July 1802; *d* Leckhampton, 16 Apr. 1890). English composer and singer of Prussian and Hungarian extraction (his father, Joseph Beer, was possibly a cousin of Meyerbeer). He appeared as a treble at the London Lyceum, 1813–18. He then studied with William Horsley and Charles Horn. From 1826 he wrote incidental music for farces and other pieces. He became music director of the Olympic T in 1832, but when Samuel Arnold reopened the Lyceum as the English Opera House in 1834, he began to write operas for it. His 'Romantic Grand Opera' *The Mountain Sylph* replaced dialogue with recitative, and is continuously composed in freer forms than were known in England at the time. It ran for 100 nights, and long remained popular. Influenced by Weber, especially by *Der Freischütz* and *Oberon*, it is in turn the immediate object of the satire of Sullivan's *Iolanthe*. Having quarrelled with Arnold, Barnett transferred to DL for *Fair Rosamond* and *Farinelli*, which returned to ballad opera in style. After abortive attempts to form a permanent English Opera House at St James's T (1838) and the Prince's T (1840), he moved to Cheltenham as a singing-teacher, and wrote no more operas. His daughters **Rosamund** and **Clara** became singers; of his brothers, **Joseph** was a singing-teacher and father of J. F. Barnett and **Zarah** wrote texts for some of his operas. Autobiography, *Musical Reminiscences and Impressions* (London, 1906).

SELECT WORKLIST (all first prod. London): *The Mountain Sylph* (Thackeray; 1834); *Fair Rosamond* (Z. Barnett & Shannon; 1837); *Farinelli* (Z. Barnett; 1839).

**Barraud, Henry** (*b* Bordeaux, 23 Apr. 1900). French composer. Studied Paris, including with Dukas, but was expelled from the Conservatoire for avant-garde work. A scrupulous attention to novel ideas, coupled with a religious attitude, shows in his operas, which have won admiration in France. He was from 1948 to 1965 an innovatory head of music in Radiodiffusion Française.

SELECT WORKLIST: *La farce de maître Pathelin* (Cohen, after medieval legend; comp. 1937, prod. Paris 1948); *Numance* (De Madariaga, after Cervantes; comp. 1950, prod. Paris 1955); *Lavinia* (Marceau; Aix 1961); *Tête d'or* (after Claudel; comp. 1980, Paris radio 1985).

**Barrault, Jean-Louis** (*b* Le Vésinet, 8 Sep. 1910). French actor and producer. Apart from his distinguished career in the theatre, he has produced opera and operetta, including *La vie parisienne* (Paris and Cologne 1958, also playing the Brazilian), Gounod's *Faust* (New York, M, 1965), and *Wozzeck* (Paris, O, 1963).

**Barsova, Valeriya** (*b* Astrakhan, 13 June 1892; *d* Sochi, 13 Dec. 1967). Russian soprano. Studied Moscow with Mazetti. Début Moscow, Z, 1917. Sang in Petrograd, then at Moscow, B, 1920–48, also at Stanislavsky Opera Studio and Nemirovich-Danchenko Moscow Arts T Studio, 1920–4. Her most famous roles included Lyudmila, Queen of Shemakha, Snow Maiden, Gilda, Violetta, Butterfly, Lakmé, Manon. After 1929 toured widely in the West. Had a light, expressive voice and a brilliant coloratura technique. In 1939 initiated operatic performances under professional guidance for workers in collectives and on the Volga-Don Canal. (R)

**Barstow, Josephine** (*b* Sheffield, 27 Sep. 1940). English soprano. Studied London with Eva Turner and Andrew Field. Joined Opera for All 1964: début as Mimì. London, SW, 1967–8 (Cherubino, Gluck's Euridice); WNO from 1968. London, CG, from 1969, creating Denise in *The Knot Garden* and Gayle in *The Ice Break*. SW, later ENO, from 1972: Natasha in first English *War and Peace*, Jeanne in *The Devils of Loudun*, and Autonoe in *The Bassarids*; also acclaimed as Emilia Marty, Octavian, Salome, Violetta, and Lady Macbeth of Mtsensk. Aix-en-Provence, Geneva, Berlin, Salzburg, and Moscow. An outstanding singing-actress with a highly individual voice. (R)

**Bartered Bride, The** (Cz.: *Prodaná nevěsta*; lit. The Sold Bride). Opera in 3 acts by Smetana; text by Sabina. Prod. Prague, P, 30 May 1866, with Gayer von Ehrenburg, Polák, Hynek, Paleček, cond. Smetana; London, DL, 26 June 1895, in German, by Ducal Court Opera of Saxe-Coburg and Gotha. New York, M, 19 Feb. 1909, with Destinn, Jörn, Didur, Reiss, cond. Mahler.

A Bohemian village, 19th cent. Mařenka (sop) loves Jeník (ten), but her parents have decided she must marry a man she has never seen, the son of Tobias Micha, as arranged by the marriage broker Kecal (bs). This unknown proves to be Vašek (ten), a simpleton whom Mařenka persuades to have nothing to do with her. Meanwhile Kecal gives Jeník 300 crowns to renounce Mařenka if she will indeed marry Micha's son. Vašek falls for the circus dancer Esmeralda, but when Jeník arrives, the bitter Mařenka decides she will after all marry Vašek—until it is revealed that Jeník has outwitted Kecal and the others, since he is himself Micha's son by a previous marriage and so can both marry Mařenka and keep the money.

Smetana made five versions of the opera: No. 1 (virtually operetta): 2 acts, overture and 20 numbers, 2 perfs.; No. 2: 2 acts, omitting a duet and adding a ballet, 27 Oct. 1866, 13 perfs.; No. 3: 2 acts, alterations incl. new Beer Chorus and Polka, 29 Jan. 1869, 4 perfs.; No. 4: 3 acts, alterations incl. new Furiant and Skočna, 1 Jan. 1869, 9 perfs.; No. 5: 3 acts, with recitatives and minor alterations, 25 Sep. 1870. After initial failure, this became Smetana's most popular opera.

**Bartók, Béla** (*b* Nagyszentmiklós, 25 Mar. 1881; *d* New York, 26 Sep. 1945). Hungarian composer. His large output includes only one opera, *Duke Bluebeard's Castle*. Influenced on the one hand by Debussy (*Pelléas*) together with Poe and the French Romantics, on the other by Schoenberg and Freud, the opera is an original and powerful piece of music drama, adapting the old Bluebeard legend as a Symbolist handling of the theme of man's essential loneliness and woman's fatal inability to penetrate it.

WORKLIST: *A Kékszakállú herceg vára* (Duke Bluebeard's Castle) (Balázs; Budapest 1918).

BIBL: T. Tallián, *Béla Bartók: the Man and his Work*, trans. G. Gulyás (Budapest, 1981).

**Bartoli, Cecilia** (*b* Rome, 4 June 1966). Italian mezzo-soprano. Sang the Shepherd in *Tosca*, Rome, aged 9, and sang in memorial concert for Callas, Paris O, aged 19. Her first recording was as Rosina, and it is with Rossini's heroines that she has been particularly associated: Rome; Milan, S; Lyons; Zurich; Cologne; Hamburg; etc. A singer with a robust chest voice, a strong, ringing high soprano range, and a brilliant technique, together with an attractive stage presence and a lively sense of characterization. (R)

**Bartolo.** See DOCTOR BARTOLO.

**Bashkiria.** Bashkir national opera began with the country's absorption into the USSR in 1919, and the foundation of a Bashkir section at the Moscow Cons. in 1932. Bashkir composers trained here included Rauf Murtazin (*b* 1910), Halik Zaymov (*b* 1914), and Zagir Ismailov (*b* 1916). A Bashkir State Opera and Ballet T was opened in

Ufa in 1938, with nine young Moscow-trained Bashkir soloists: the first opera it gave was Paisiello's *La molinara* (in Bashkiri). The ensemble was enlarged in 1941.

The first primitive attempts at Bashkir operas date from the 1930s. In 1940 *Khakmar* by Muslim Valeyev (1888–1956) was staged: regarded as the first true Bashkir opera, this is based on a contemporary folk subject and uses folk music. There followed historical operas by Anton Eichenwald (1875–1952) (*Mergen*, 1940), by Nikolay Chemberdji (1903–48) (*Karlugas* (The Swallow), 1941), and a collaboration between Zaymov and Antonio Spadavecchia (1907–87) on a folktale, *Ak buzat* (The White Stallion, 1942). After the war came Murtazin's *Azat* (1949), a story of the Civil War which won much respect as evidence of a new maturity in Bashkir opera. Further works by Ismailov include the historical opera *Salavat Yulayev* (1955), the musical comedy *Kodasa* (The Sister-in-Law, 1959), *Shaura* (1963), and *Volny Agidely* (1972).

**Basile Baroni, Adriana** (*b* Posillipo, *c*.1580; *d* Rome, *c*.1640). Italian singer. First successes in Naples, as contralto and instrumentalist. Brought to Mantua by Vincenzo Gonzaga, 1610, perhaps on the recommendation of Monteverdi, who declared her to be the finest singer of the day. Remained there, singing with great success, until 1624, when she left for Naples; settled in Rome 1633. Her beauty led to her being known as 'La bella Adriana', and her artistry was instrumental in establishing the monodic style as the new operatic manner. Her daughters **Leonora** (1611–70) and **Caterina** (1620–?) were both singers; they and their mother often performed together at her salon in Rome.

**Basilio.** See DON BASILIO.

**Basle** (Fr., Bâle; Ger., Basel). City in Basle Canton, Switzerland. Opera was first given in the Ballenhaus in 1734. The T auf dem Blömlein opened in 1834; this was rebuilt as the present Stadttheater, which opened on 20 Sep. 1909 and was completely renovated in 1975 (cap. 1,000). Basle was the scene of the abortive 1924–5 *Ring* with pioneering designs by Adolphe *Appia; other artists to be associated with the city were Walter *Felsenstein and Felix *Weingartner. The theatre performs a mixed repertory.

**Basque opera.** An early treatment of a Basque subject was *Los esclavos felices* (The Happy Slaves, 1820) by Juan Arriaga (1806–26). A few zarzuelas and operas have been written in Basque, notably by the scholar and ethnomusicologist Resurrección Maria de Azkue (1864–1951). He wrote, for the theatre of the school of Basque studies which he founded,

*Eguzkia nora?* (Where are you going, Sun?, 1896), *Viscaytik Bizkaira* (From Viscay to Biscay), and other zarzuelas. After study in Paris, Brussels, and Cologne, he wrote the Basque operas *Ortzurri* (1911) and *Urlo* (1913). Another ethnomusicologist, J. A. Santesteban, wrote *El pudor* (n.d.), consisting of a series of popular songs joined by recitative and culminating in a finale using the dance *aurresku*. Other Basque operas are *Chantón Piperri* (1899) and *La dama de Amboto* by Zapirain; *Artzai Mutilla* by F. Ortiz y San Pelayo; *Maitena* (1909) by Colin; *Mendi-Mendiyan* (High in the Mountains, 1911) by José Maria Usandizaga (1887–1915), which uses motive as well as Basque folk instruments; and *Mirentxu* (1915) (on Basque customs), *Amaya* (1920) (concerning the struggle between Christianity and Islam, using Basque instruments and dances) and the zarzuelas *El caserio* (The Village, 1926) and *Mari-Eli* (1936) by Jesús Guridi (1886–1961).

**bass** (It., *basso*: 'low'). The lowest category of male voice. Many subdivisons exist within opera-houses: the commonest in general use (though seldom by composers in scores) are given below, with examples of roles and their approximate tessitura. These divisions often overlap, and do not correspond exactly from country to country. In general, distinction is more by character than by tessitura, especially in France: thus, the examples of the roles give a more useful indication of the different voices' quality than any attempted technical definition.

French: basse-bouffe (Jupiter: F–f); basse de caractère (Gounod's Méphistophélès: G–e); basse chantante or basse noble (Brogni: F–f). In early music (Lully to Rameau), basse-contre (Eb–d').

German: tiefer Bass (Sarastro: E–e); Bass-buffo or komischer Bass (Osmin: F–f); hoher Bass (Kaspar: G–f).

Italian: basso profondo (Ramfis: D–e); basso comico or basso buffo (Don Bartolo, Don Basilio: F–f); basso cantante (Padre Guardiano: F–f).

Slavonic basses are able to achieve great depth, sometimes reaching G'.

**Bassani, Giovanni Battista** (*b* Padua, *c*.1657; *d* Bergamo, 1 Oct. 1716). Italian composer, violinist, and organist. Possibly studied with Legrenzi and Vitali. Active first in Bologna and Modena, he finally settled in Ferrara, becoming maestro di cappella at the cathedral 1686; moved to Bologna 1712. Though best known as a violinist, he wrote nine operas which were well regarded in their day.

SELECT WORKLIST: *Gli amori all moda* (?; Ferrara 1688).

**Bassarids, The.** Opera, with intermezzo, in 1 act by Henze; text by W. H. Auden and Chester

Kallman, after Euripides' *The Bacchae*. Prod. Salzburg 6 Aug. 1966, in a German trans. by Maria Basse-Sporleder, with Hallstein, Meyer, Little, Driscoll, Melchert, Paskalis, Dooley, Lagger, cond. Dohnányi; Santa Fe 7 Aug. 1968, in orig. English, with Caplan, Sarfaty, Mandac, Driscoll, Bressler, Reardon, Jamerson, Harrower, cond. Henze; London, BBC, 22 Sep. 1968, with Carlyle, Sarfaty, Watts, Young, Egerton, Bryn-Jones, Griffiths, cond. Downes; London, C, 10 Oct. 1974, with Barstow, Pring, Brecknock, Herincx, prod. and cond. Henze.

When the god Dionysus (ten) comes to Thebes, his followers, the Bassarids, go to Mount Citheron to celebrate. The new king, Pentheus (bar), forbids the cult of Semele, mother of Dionysus, and orders the arrest of all those on Mount Citheron; however, they escape back to the mountain during an earthquake. The intermezzo, a parody of 18th-cent. French courtly entertainments, reveals Pentheus' secret thoughts. Pentheus is persuaded to dress as a woman in order to observe the Dionysiac revels. He is found, and ripped to pieces by the Bassarids, who include his mother Agave (mez). On enquiring after her son, it gradually dawns on her what she has done and she bitterly rebukes Dionysus. The god commands that Thebes be burnt, and he and Semele become the object of the people's worship.

Other operas on the *Bacchae* are by Ghedini (1948), Börtz (1991), and Butler (1992).

**bass-baritone**. See BARITONE.

**bass chantante**. See BASS.

**basse taille**. See BARITONE.

**Bassi, Luigi** (*b* Pesaro, 5 Sep. 1766; *d* Dresden, 13 Sep. 1825). Italian baritone. Studied Senigallia with Morandi. Début Pesaro 1799 (in Anfossi's *Il curioso indiscreto*). After further study with Laschi, engaged for Prague 1784–1806. Mozart, impressed by his Count in the first Prague *Figaro* (1786), wrote Don Giovanni for him (1787). He is said to have persuaded Mozart to change 'Là ci darem' five times, but to have been overruled about replacing 'Fin ch'han dal vino' when he made such a success of it. Various accounts praise the flexible, somewhat tenor-like quality of his voice and the intelligence of his acting. After a period in Vienna, he returned briefly to Prague before moving to Dresden in 1815. Here, his voice now feeble, he became producer at the Italian O, where his stagings included *Don Giovanni*. He also gave a friendly welcome to Weber. Beethoven, who met him in Vienna in 1824 and 1825, called him 'the fiery Italian'.

**Bassi Manna, Carolina** (*b* Naples, 10 Jan. 1781; *d* Cremona, 12 Dec. 1862). Italian contralto. Daughter of Giovanni Bassi, a buffo bass. With her brother Nicola, she appeared in a company of *Ragazzi napoletani* founded by her father at the Naples, C, 1789. Married a noble, Pietro Manna, 1797. One of the outstanding singers of her day (Stendhal considered her 'the only singer whose genius approaches that of Mme Pasta'), she used her powerful vocal and dramatic resources to particular effect in the Rossini contralto repertory. She created leading roles in Meyerbeer's *Semiramide riconosciuta*, *Margherita d'Anjou*, and *L'esule di Granata*, Rossini's *Bianca e Faliero*, and works by Pacini and Mercadante. Retired 1828, but continued to sing in concerts. Her brother **Nicola Bassi** (1767–1825), considered by Stendhal to be the best buffo bass of the day, sang in Paris and Milan. Another brother, **Adolfo**, was a composer and for some years director of the T Nuovo, Trieste, where his five operas were produced.

**basso cantante**. See CONVENIENZE.

**Bastianini, Ettore** (*b* Siena, 24 Sep. 1922; *d* Sirmione, 25 Jan. 1967). Italian baritone. Début (as bass) Ravenna 1945 (Colline). Second début Bologna 1951 (Germont). New York, M, 1953–60, 1964–6. London, CG, 1962. Milan, Salzburg, etc. Sang Prince Andrey in first stage performance outside Russia of *War and Peace*, Florence 1953. Greatly admired as a Verdi baritone, he possessed a voice dark enough in timbre to enable him to sing Tchaikovsky roles. (R)

**Bastien und Bastienne**. Opera in 1 act by Mozart; text by F. W. Weiskern, J. Müller, and J. A. Schachtner, after M. J. B. and C. S. Favart and H. de Guerville's comedy *Les amours de Bastien et Bastienne* (1753), a parody of Rousseau's *Le devin du village* (1752). Probably prod. Vienna, in Anton Mesmer's garden theatre, Sep. 1768; Berlin, Architektenhaus, 2 Oct. 1890; London, Daly's T, 26 Dec. 1894; New York 26 Oct. 1916.

France, 18th cent. Bastienne (sop), a shepherdess, complains to Colas (bs), the village magician, that Bastien (ten) no longer loves her. He answers that Bastien has been distracted by the attentions of a grand lady. Colas tells Bastienne to hide and then informs Bastien that Bastienne has left him. At Bastien's request, Colas reads a spell and Bastienne appears. The couple are reconciled and thank Colas.

Also operas by Gaetano (1788), Kauer (1790).

**Battaglia di Legnano, La**. Opera in 3 acts by Verdi; text by S. Cammarano, after Joseph Méry's drama *La bataille de Toulouse* (1828). Prod. Rome, Arg., 27 Jan. 1849, with Giuli-Borsi, Marchesi, Fraschini, Colini; Cardiff 13 Oct. 1960 (as *The Battle*) with Harper, Ferendinos, Ronald Lewis, Alan, cond. Groves; New York, Amato T, 28 Feb. 1976.

Milan and Como, 1176. Lombard soldiers gather to form an army against the German invaders. Rolando (bar) is overjoyed to find his old friend Arrigo (ten), whom he thought dead. Lida (sop), Rolando's wife (who was once betrothed to Arrigo) spurns the attention of Marcovaldo (bar), a German prisoner released by Rolando: he declares his love for her. Rolando and Arrigo arrive. When they are alone together, Arrigo reproaches Lida for her faithlessness.

Rolando and Arrigo go to Como to ask the people to join their army. The city has already signed a pact with Barbarossa (bs), the German leader, and Arrigo accuses them of treachery. Barbarossa appears and a furious dispute ensues.

Members of the Lombard League swear to drive out the enemy. Marcovaldo shows Rolando a note Lida sent to his friend in which she declares her love for Arrigo. Lida visits Arrigo to say they must part but is surprised there by Rolando. He disowns her and denies Arrigo the honour of fighting for his country. Arrigo has to jump from a window to join the army.

The Lombards win their battle, but Arrigo is mortally wounded. When Arrigo swears to Rolando that he and Lida have not betrayed him, the two friends are reconciled and Arrigo dies a hero.

'Batti, batti, O bel Masetto'. Zerlina's (sop) aria in Act I of Mozart's *Don Giovanni*, in which she invites her lover's wrath and regains his heart.

Battistini, Mattia (*b* Contigliano, 27 Feb. 1856; *d* Contigliano, 7 Nov. 1928). Italian baritone. Studied Rome with Terziani and Persichini. Début Rome, Arg., 1878 (Alfonso, *Favorita*). Visited S America 1881, but thereafter would never cross the Atlantic. London, CG, 1883 (Riccardo, *Puritani*), with little success; DL 1887. Returned to London 1905, by when he had established himself as 'il rè dei baritoni'. Possessed a voice of exceptional range (Massenet rewrote Werther for him), reaching a', and was also masterly in florid music. In his repertory of 82 roles, he preferred noble or regal characters: his friendship with Verdi was affected by his refusal to sing Falstaff (humour was in any case not his strong point), but he was outstanding as Renato, Posa, Amonasro, and Iago. However, he was at his most characteristic in the earlier Verdi, and in Bellini and Donizetti. Other famous roles included Don Giovanni, Valentin, Wolfram, and Onegin. Sang all over Europe, including Russia and Poland. The master of a technique that included a superb *messa di voce, he continued to sing into his 70s. (R)

BIBL: F. Palmegiani, *Mattia Battistini* (Milan, 1948).

Battle, Kathleen (*b* Portsmouth, OH, 13 Aug. 1948). US soprano. Studied Cincinnati. Début New York, City O, 1976 (Susanna). New York, M, from 1977; Gly. 1979; Paris, O, 1984; London, CG, 1985; also Salzburg, Chicago, and major European opera-houses. Roles incl. Blonde, Susanna, Zerlina, Despina, Pamina, Norina, Adina, Sophie (Massenet and Strauss), Zdenka, Zerbinetta. Possesses a radiant, burnished tone and a lively temperament. (R)

Baturin, Alexander (*b* Oshmyany, 1904). Belorussian baritone. Studied Leningrad with Isachenko, Rome with Battistini. Stage début Moscow, B (Miller, Dargomyzhsky's *Rusalka*). Moscow, B, until 1958. An acclaimed William Tell, Prince Igor, Ruslan, Pimen, Dosifey, Tomsky, Escamillo. At Moscow Cons. from 1948, his students incl. Vedernikov, Ghiaurov. (R)

Baylis, Lilian (*b* London, 9 May 1874; *d* London, 25 Nov. 1937). English musician and manager. Daughter of Newton Baylis, a singer, and Liebe Cons, singer and pianist. Her aunt, Emma Cons, became lessee of the Royal Victorian Coffee Music-Hall, known as the *Old Vic, and invited her to manage it, which she did from 1898 to her death. She laid the foundations of a national English opera company, and, with Ninette de Valois, of a ballet company. Reopened London, *Sadler's Wells, 1931 as the north London equivalent of the Old Vic.

BIBL: H. Williams, *The Vic-Wells: the Work of Lilian Baylis* (London, 1935).

Bayreuth. Town in Franconia, Germany. The first opera was an anon. Singspiel *Sophia* 1661, and many other works were given in a theatre built under the Margrave Georg Wilhelm (1712–26): at one stage, 4–6 new operas were given annually. The Markgräfliches Opernhaus, designed by Giuseppe and Carlo Bibiena, opened with Hasse's *Ezio* 1748, and opera was encouraged by the Margrave Friedrich and his wife, Frederick the Great's sister Wilhelmine (composer of *Argenore*, 1740—lost). Performances were also given in the open air at the Hermitage. Under Johann Pfeiffer 1734–6 many Italians visited. Activity declined after Wilhelmine's death in 1758, and the court moved to Ansbach 1769. Opera was then given by visiting companies, especially from Bamberg, until Wagner's time.

Wagner had intended to build the festival theatre to stage his works in Munich under the patronage of Ludwig II, and plans were drawn up by Gottfried Semper. When intrigues forced the abandonment of this plan, the Bayreuth authorities in 1871 provided land for the theatre and a home ('Wahnfried'). The theatre foundation stone was laid 22 May 1872, when Wagner conducted Beethoven's Ninth Symphony in the local theatre with many of Germany's leading musicians (among them Richter as timpanist).

The Festspielhaus (cap. 1,925), originally planned as temporary, is mostly of wood and brick, with an auditorium resembling a classical amphitheatre. It has superb acoustics, largely due to the innovation of a hooded orchestra pit, and no aisles or pillars. It opened 13 Aug. 1876 with *Das Rheingold*, launching the first complete *Ring* cycle. The event, the most important in musical Europe, drew to the audience Liszt, Grieg, Bruckner, Mahler, Tchaikovsky, Saint-Saëns, Nietzsche, Kaiser Wilhelm I, and Ludwig II. The first season lost over £12,000 and there were no more until 1882, when *Parsifal* was premièred. Directors have all been members of Wagner's family: his wife Cosima 1883–1908, son Siegfried 1908–1930, daughter-in-law Winifred 1931–44, grandsons Wieland and Wolfgang 1951–66, Wolfgang from 1966. Under Cosima, the production style religiously perpetuated Wagner's own, though there was some cautious evolution under Siegfried. Winifred appointed Heinz Tietjen as artistic director. Bayreuth suffered from its associations with Hitler's patronage, but reopened after war closure in 1951. The greatest revolution in staging then came with Wieland and Wolfgang, who abandoned the strongly traditional productions in favour of simplified sets and costumes, with particular emphasis on lighting, so as to detach the works from particular associations and stress their universality. The festival has since encouraged experiment and reinterpretation (as with Patrice Chéreau's *Ring* under Pierre Boulez for the 1976 centenary), which with all its risks is seen as an essential Wagnerian tradition.

BIBL: G. Skelton, *Wagner at Bayreuth* (London, 1965, 2/1976).

**Bazin, François** (*b* Marseilles, 4 Sep. 1816; *d* Paris, 2 July 1878). French composer. Studied Paris with Berton and Halévy. In 1846 he began a long line of stage works in popular style with *Le trompette de Monsieur le Prince*, having his greatest success with *Maître Pathelin* in 1856; his most ambitious large-scale work was *Le voyage en Chine* (1865). An entrenched conservative, he refused to accept Massenet into his class in the Conservatoire, only to find his career being eclipsed by the younger man.

**Bear, The.** Opera in 1 act by Walton; text by Paul Dehn, after Chekhov's play (1888). Prod. Aldeburgh, 3 June 1967; Aspen, 15 Aug. 1968.

The widow Popova (mez) is confronted by Smirnov (bar), a creditor of her late husband. Despite quarrelling violently, they cannot shoot each other since they have, unexpectedly, fallen in love. Luka (bs), the servant, also plays a part in the confusions.

**Beard, John** (*b* ?London, *c*.1717; *d* Hampton, 5 Feb. 1791). English tenor. Studied at Chapel Royal with Gates. Sang Israelite Priest in *Esther*, 1732, while still a boy. Joined Handel's company 1734; début London, CG (Silvio, *Il pastor fido*). He won immediate acclaim and thereafter sang with Handel regularly until 1759, creating roles in *Ariodante, Alcina, Atalanta, Berenice, Samson, Judas Maccabaeus*, and *Jephtha*, among other works. Member of the London DL Co. 1737–43, and 1748–59 with Garrick; CG, 1743–8 and 1759–67. Sang in works by Boyce, Lampe, and Arne, and was a famous Macheath in *The Beggar's Opera*. Succeeded Rich as manager of CG, in 1761, resigning in 1767 chiefly on account of deafness. Considered by many the finest English singer of his day, he did much to set the tenor voice on a par with the castrato. His personal qualities and artistry won him the respect of composers and public alike.

**Beatrice di Tenda.** Opera in 2 acts by Bellini; text by Felice Romani. Prod. Venice, F, 16 Mar. 1833, with Pasta, Del Serre, Curioni, Cartagenova; London, H, 22 Mar. 1836, with Colleoni-Corti, Seguin, Winter, Cartagenova; New Orleans, St Charles T, 5 Mar. 1842, with Ober-Rossi, Marozzi, Salvatori, Perozzi. First 20th-cent. revival Catania 1935, with Arangi-Lombardi.

The castle of Binasco, near Milan, 1418. Beatrice (sop), married to Filippo Visconti, Duke of Milan (bar), is in love with Orombello (ten). Orombello is loved in vain by Agnese (mez), with whom the Duke, having tired of Beatrice, is himself in love. Agnese betrays the lovers to the Duke, and although she later pleads with him to spare Beatrice's life, the Duke refuses, as Beatrice's followers have risen in revolt against him.

Also opera by Guimaraes (to Romani's text, 1882).

**Béatrice et Bénédict.** Opera in 2 acts by Berlioz; text by the composer, after Shakespeare's comedy *Much Ado About Nothing* (1598–9). Prod. Baden-Baden 9 Aug. 1862, with Charton-Demeur, Monrose; Glasgow, 24 Mar. 1936; Washington 3 June 1964, with Dussault, Porretta, Readon, cond. Callaway.

Messina, 1700. Hero (sop) looks forward to seeing Claudio (bar) on his return from the wars, but the return of Bénédict (ten) is sardonically greeted by Béatrice (sop). Gradually the love beneath their banter is disclosed, and they agree to marry. Berlioz introduces the character of Somarone (bs), a Kapellmeister whose dull fugue is a satire on pedantry in general and probably Cherubini in particular.

**Beaumarchais, Pierre Augustin Caron de** (*b* Paris, 24 Jan. 1732; *d* Paris, 19 May 1799). French playwright and amateur musician. His fame rests upon the first two plays of the *Figaro* trilogy, *Le barbier de Séville* (Comédie-Française 1775) and *La folle journée, ou Le mariage de Figaro* (1778; Comédie-Française 1784). For the former he wrote some songs, possibly adapted from Spanish airs. The third play of the *Figaro* trilogy, *La mère coupable* (T du Marais 1792), represents a return to the *larmoyant* style of his early plays. It tells how the Countess has a child by Cherubino and how Figaro saves his master from a swindling Irish infantry officer; it was seriously considered for an opera by Grétry.

As a librettist Beaumarchais is known chiefly for *Tarare*, a 5-act opera with music by Salieri (it had been declined by Gluck) (Paris, O, 1787). The second edition of the libretto contains an interesting preface addressed 'aux abonnés de l'Opéra, qui voudraient aimer l'opéra': it claims a greater importance for words (coupled with the comment that operatic music is apt to be too dense for them), discusses the legitimate subjects for librettos, and concludes by mentioning the collaboration with Salieri. Beaumarchais's colourful and versatile personality is sketched in his own Figaro.

Operas on his works:

*Le barbier de Seville* (1775): André (1776); F. Benda (1776); Arnold (1777); Paisiello (1782); Elsperger (1783); Weigl (*Die betrogene Arglist*, 1783); Schulz (1786); Reinagle (1794); Isouard (1796); Morlacchi (1816); Rossini (1816); Dall'Argine (1868); Graffigna (1879); Giménez & Niteo (zarzuela, 1901); Cassone (1922); Torazza (1924)

*La folle journée, ou Le mariage de Figaro* (1778, prod. 1784): Shield (1784); Persicchini (1791); Mozart (1786); version of Mozart by Tarchi (1789); Dittersdorf (1789); Portugal (1799); L. Ricci (1838)

*La mère coupable* (1792): Milhaud (1966); Corigliano (1991).

Other operas making use of the character of Figaro are by Paer (*Il nuovo Figaro*, 1794); Tost (*Figaro*, 1795); Morlacchi (*Il nuovo barbiere di Siviglia*, 1816); Carafa (*Les deux Figaro*, 1827); L. Ricci (*Il nuovo Figaro*, 1832); Mercadante (*I due Figaro*, 1835); Speranza (*Les deux Figaros*, 1838); C. Kreutzer (*Die beiden Figaro*, 1840); L. Rossi (*La figlia di Figaro*, 1846); Cagnoni (*Il testamento di Figaro*, 1848); Aimon (*Les deux Figaros*, c.1850); G. Panizza (*I due Figaro*, 1926); P. H. Allen (*La piccola Figaro*, 1931); Delmas (*La conversion de Figaro*, 1931); Klebe (*Figaro lässt sich scheiden*, 1964). Also Massenet (*Chérubin*, 1905).

Opera based on *Tarare* by Mayr (*Atyr*, 1814). Also operetta *Beaumarchais* by Cools (1931). He and Marie Antoinette fall in love as ghosts in Corigliano's *The Ghosts of Versailles* (1991).

**Bechi, Gino** (*b* Florence, 16 Oct. 1913; *d* Florence, 2 Feb. 1993). Italian baritone. Studied Florence with Frazzi and Di Giorgi. Début Empoli 1936 (Germont). Rome 1937–52; Milan, S, 1939–44, 1946–53; London, CG, 1950, DL, 1958. Leading baritone of his day in Italy; at his peak celebrated as Nabucco, Amonasro, Iago (Verdi), Gérard (*Chenier*), Hamlet. Retired 1965. (R)

**Beckmesser.** Sixtus Beckmesser (bs), the town clerk in Wagner's *Die Meistersinger von Nürnberg*.

**Beecham,** (Sir) **Thomas** (*b* St Helens, 29 Apr. 1879; *d* London, 8 Mar. 1961). English conductor. Son of Sir Joseph, an industrialist and musical amateur whose wealth greatly helped his early career. Studied composition with Charles Wood and Moszkowski; self-taught as conductor. Toured with small opera company 1902–4. In 1910 launched his first season at London, CG, opening with the first English *Elektra* and operas by two other English composers he championed, Delius (*A Village Romeo and Juliet*) and Smyth (*The Wreckers*). Also introduced, surviving difficulties with the censor, *Salome*, later *Feuersnot*, *Ariadne*, and *Rosenkavalier*. Responsible for the Russian opera seasons at London, DL, 1913–14. Formed the *Beecham OC 1915. Became lessee of CG 1919, forced into liquidation 1920. Returned to CG 1932, remaining as music and artistic director, and financier, until 1939; his own performances including several of *The Ring*. Also Cologne, Munich, Hamburg, and Berlin, then at New York, M, 1941–4. Conducted little opera after the war, notable exceptions being *Ariadne* (Edinburgh 1950), *Meistersinger* and *The Bohemian Girl* (CG 1951), the première of Delius's *Irmelin* (Oxford 1953), Grétry's *Zémire et Azor* (Bath 1955), and some broadcasts. His last opera conducting was at Buenos Aires, C, 1958.

Beecham's contribution to English opera was inestimable, and in a sense tragic. He spent his fortune, as well as his talent, on heroic enterprises in days when England was unreceptive to his ideals; and at CG he was unable to achieve all of which he was capable. In part this was due to a seigneurial disregard for practicalities; but his genius burned no less brightly in the opera-house than on the concert platform, and his achievements contributed more than some superlative performances to the development of opera in England. Autobiography, *A Mingled Chime* (London, 1944). (R)

BIBL: B. Geissmar, *The Baton and the Jackboot* (London, 1944); C. Reid, *Beecham* (London, 1961).

**Beecham Opera Company.** Although Sir Thomas Beecham organized an English Opéra-Comique Company at HM in 1910, and took over the Denhof Co. in 1913, it was not until Oct. 1915 that the Beecham OC was formed for a short season at the Shaftesbury T that included the premières of Stanford's *The Critic* and Smyth's *The Boatswain's Mate*. In 1916 there followed seasons at the Aldwych T, then the Queen's T in Manchester, when Beecham declared the company to have evolved from 'a smallish troupe of Opéra-Comique dimensions into the full growth of a Grand Opera organization'. He had inherited the scenery used by Dyagilev at London, DL, in 1914, and so could add performances of various Russian operas to those of Wagner, Playfair's classic productions of Mozart, and Verdi's *Otello* and *Falstaff*. He was able to command the services of the best British singers of the day, and of the conductors Albert Coates and Eugene Goossens II and III (beside himself). The company went bankrupt in 1920, but was reorganized as the *British National OC.

**Beeson, Jack** (*b* Muncie, IN, 15 July 1921). US composer. Studied Eastman School, and with Bartók. His operas, such as *Lizzie Borden* (1965), make use of a tuneful folk-influenced idiom in setting US folk or legendary subjects.

**Beethoven, Ludwig van** (*b* Bonn, 15 or 16 Dec. 1770; *d* Vienna, 26 Mar. 1827). German composer. Beethoven's reputation as an operatic composer rests upon his only completed work *Fidelio*, but his interest in dramatic music was much wider. He composed incidental music to several stage plays, including Goethe's *Egmont* (1810) and August von Kotzebue's *Die Ruinen von Athen* (1812), and made many other attempts at opera, some of which have survived as isolated numbers, fragments, and sketches. The many revisions of *Fidelio* are proof of Beethoven's fundamental difficulty with the genre. He possessed firm ideas about the ideal libretto, and had exacting demands concerning its dramatic potential, moral tone, and linguistic content, yet none of the distinguished literary figures in his circle fulfilled his requirements.

In Bonn, Beethoven played the viola in the local opera orchestra and encountered Mozart's operas for the first time, but did not attempt to write an opera himself, nor did he apparently do so during his first ten years in Vienna—he moved there in 1792—even though his songs, notably 'Adelaide', show a new theatrical influence. It has been suggested that he was approached in 1801 by Schikaneder to set his libretto *Alexander*, as the opening work of the T an der Wien, but the earliest surviving operatic attempt is some

numbers from an abandoned setting of Schikaneder's *Vestas Feuer* (1804).

Around this time he began work on an 'old French libretto', a translation and adaptation by Sonnleithner of Bouilly's *Léonore, ou L'amour conjugal* (first set by Gaveaux, Paris 1798), and later used by Paer (Dresden 1804) and Mayr (Padua 1805). Cast in three acts, and with the overture Leonore No. 2, the opera was first performed in 1805. It was so poorly received that it was immediately revised and given in slightly altered 2-act version, with the overture Leonore No. 3, the following year. But this attempt was also unsuccessful, and only after a third, more extensive revision in 1814, made with the assistance of Treitschke, was Beethoven even moderately satisfied with his opera, by now known as *Fidelio*.

French *rescue opera had enjoyed immediate popularity with composers and audiences alike in Vienna following the performance in 1802 of Cherubini's *Lodoïska*: his *Les deux journées*, *Médée*, *L'hôtellerie portugaise*, and *Faniska*, Le Sueur's *La caverne* and Méhul's *Ariodant* were all performed there during the next four years. These models provided a style with which Beethoven could identify, for they had an emotional integrity lacking in contemporary French and Italian opera. *Leonore* was clearly fashioned after the model of Cherubini, with its wide range of aria forms, mixture of recitative with melodrama and spoken dialogue, extensive use of ensembles, and expansive finales. But, like Mozart, Beethoven also brought to opera his skills as a symphonist: the canonic quartet, Leonore's aria 'Abscheulicher!', and the finales display an unmatched complexity of musical organization and depth of expression. His supreme achievement in *Fidelio* lies in his skill in imbuing characters of a popular, almost stereotyped plot with humane feelings and passions, and an appropriate musical language: the opera's expression of faith in liberty and loathing of tyranny make it a monument both to Beethoven's artistic vision, as well as more generally to the triumph of good over evil. In this context it is not surprising that *Fidelio* was chosen to open many opera-houses after the Second World War, including the rebuilt Vienna State Opera (5 Nov. 1955).

Given his difficulties with *Fidelio*, it is strange that in 1807 Beethoven should have petitioned for a permanent post at the Viennese court theatres, undertaking to deliver at least one Grand Opera and several Singspiels annually. He was unsuccessful, but his interest in opera continued unabated. A version of *Macbeth* was abandoned when its librettist, Heinrich Joseph von Collin, gave up the work, fearing that it was becoming 'too gloomy'; musical sketches for the opening scene survive. Probably as early as 1809 Beethoven first planned a version of Goethe's *Faust*, described in 1823 as 'a work that would be the greatest achievement for me and my art'. Only a few sketches were produced, although the project occupied Beethoven intermittently up to his death. By 1815 he was making tentative sketches for *Bacchus*, a 'grand lyrical opera', to a libretto by Rudolph von Berge, as well as discussing several ideas with Treitschke, including *Les ruines de Babylon* and *Romulus*. None of these came to fruition, neither did *Die schöne Melusine*, to a libretto specially commissioned from Grillparzer. This venture followed a revival of *Fidelio* in 1822, and was an attempt to bring Vienna's most celebrated composer together with her most famous dramatist. Grillparzer finished his libretto, but Beethoven was dogged by ill-health and unable to attend to the music; the project was abandoned in 1826 and with it his operatic ambitions.

WORKLIST: *Fidelio oder die eheliche Liebe* (orig. *Leonore*) (Sonnleithner, after Bouilly; Vienna 1805; rev. Vienna 1806 with Breuning; further rev. Vienna 1814 with Treitschke).

BIBL: Max Unger, *Ein Faustopernplan Beethovens und Goethes* (Regensburg, 1952); Willy Hess, *Beethovens Oper Fidelio* (Zurich, 1953); Winton Dean, 'Beethoven and Opera', in Denis Arnold and Nigel Fortune, eds., *The Beethoven Companion* (rev. edn., London, 1973).

**Beggar's Opera, The.** Ballad opera in 3 acts arranged, adapted, and partly composed by Pepusch; text by John Gay. Prod. London, Lincoln's Inn Fields T, 29 Jan. 1728, with Fenton, Walker—run of 62 nights; New York, Nassau St. T, 3 Dec. 1750.

The Beggar (spoken), in the prologue, presents the work as his own creation. The highwayman Macheath (ten) has married Polly Peachum (sop). Her father (bs) is so irate at this that he informs against Macheath, who has been receiving stolen goods.

Macheath is arrested and sent to Newgate Prison, but the gaoler's daughter, Lucy (sop), who is in love with him, helps him escape. Macheath has promised to marry Lucy, and when she discovers he is already married to Polly, great rivalry ensues.

After further accusations by more women, Macheath finds himself back in Newgate under Lockit's (bar) care. However, the Beggar intervenes to secure release for Macheath, because 'an opera must end happily'.

This was the first, and the most famous, *ballad opera. It probably arose from a comment by Swift that a Newgate pastoral 'might make an odd pretty sort of thing'. The play has a doubly satirical purpose, poking fun not only at the contemporary conventions of Italian opera and its singers (the battle between Polly and Lucy mocks the rivalry between the sopranos Faustina Bordoni and Francesca Cuzzoni), but also lampooning politicians (Walpole as Macheath, Lord Townsend as Lockit, the gaoler); the overture, by Pepusch, contains a tune popularly known as *Walpole, or The Happy Clown*. Gay included 69 tunes in the original version of *The Beggar's Opera*; these were contemporary popular songs—including English, Irish, Scottish, and French folk-tunes—for which Pepusch provided basses. The work was a huge success; it was said to have made Gay rich, and Rich—the producer—gay. Gay himself wrote a sequel, *Polly* (pub. 1729, not prod. until 1777 for political reasons), and innumerable imitations were written. The work was revised almost every year from 1728 to 1886, when Sims Reeves sang Macheath. A version by Frederic Austin, which opened at the Lyric T, Hammersmith on 5 June 1920, ran for 1,463 consecutive performances (the longest run of any opera), and began a modern vogue for the work. Other modern adaptations have included those by E. J. Dent (Birmingham 1944), Sir Arthur Bliss (a film, with Laurence Olivier in the title role and Stanley Holloway as Lockit, 1953), and Benjamin Britten (Cambridge 1948). Brecht and Weill's *Die Dreigroschenoper* updates the story to the turn of the 20th cent.

**Begnis, Giuseppe de** and **Giuseppe Ronzi de**. See DE BEGNIS, GIUSEPPE.

**Behrens, Hildegard** (*b* Varel, 9 Feb. 1937). German soprano. Studied Freiburg with Leuwen. Début Freiburg 1971 (Countess, *Figaro*). Frankfurt 1974; Salzburg 1977; New York, M, from 1975; Bayreuth from 1983. Début London, CG, 1976 (Leonore, *Fidelio*). An intelligent, committed artist who combines lyrical singing with striking dramatic intensity. Her wide repertory includes Fiordiligi, Agathe, Rusalka, Isolde, Brünnhilde, Marie (*Wozzeck*), Salome. (R)

**'Bei Männern'.** The duet in Act I of Mozart's *Die Zauberflöte* in which Pamina (sop) and Papageno (bar) sing of the joys and power of love.

**Belarus.** See BELORUSSIA.

**bel canto** (It.: 'beautiful song', 'beautiful singing'). The traditional Italian art of singing in which beautiful tone, fine legato phrasing, and impeccable technique are emphasized, though not at the total expense of dramatic expression, as some of its greatest exponents, above all Callas, have demonstrated. The term, an imprecise one, was probably first used in a volume of *ariette da camera* by Vaccai (before 1840). However, some embryonic elements of the style can be detected in Venetian operas of the mid-17th cent., while the role which bel canto played in the stylistic formulation of opera seria has been obscured by later criticism, which has focused on its dramatic weaknesses and over-use of coloratura. Bel canto reached its apotheosis in the 19th cent.: its influence is seen most notably in *Bellini's work, though it exerted a direct influence on virtually all Italian opera composers, including Verdi and Puccini. Its influence on Wagner is often overlooked: he always acknowledged the importance of Bellini to him, and encouraged its use in his own music. Even though the importance of bel canto has diminished in the 20th cent., it is still widely taught, especially in Italy, where it is held up as a goal to which all singers should ultimately aspire.

BIBL: R. Celletti, *A History of Bel Canto* (trans. Oxford, 1991).

**Belcore.** The sergeant (bar) in Donizetti's *L'elisir d'amore*.

**Belfagor.** Opera in 2 acts with prologue and epilogue by Respighi; text by Claudio Guastalla, after Marselli's comedy (1920). Prod. Milan, S, 26 Apr. 1923, with Stabile, Sheridan, Merli, Azzimonti, Azzolini, Farrari, Gramega, cond. Guarnieri.

A Tuscan village. Candida (sop), daughter of the apothecary Mirocleto (bs), is coerced into marrying Ipsilonne (bar). Ipsilonne is, in fact, a devil, Belfagor, who has come to earth to see if it is true that marriage leads to man's downfall. When Candida's beloved Baldo (ten) returns, he refuses to believe the wedding was not of her own will. In answer to Candida's prayers, the church bells ring of their own accord to indicate her innocence.

**Belfast.** Capital of Northern Ireland. Opera has been given professionally since the late 1960s, but it was the merger of the Northern Ireland Opera Trust with the largely amateur Studio Opera Group in 1984 that formed Opera Northern Ireland. Seasons, which have included a Mozart cycle completed in 1990, are given twice a year in the Grand OH.

**Belfiore.** The Cavalier Belfiore (bar), posing as Stanislao, King of Poland, in Verdi's *Un giorno di regno*.

**Belgium.** Under the Spanish domination of the 17th cent., French taste, especially for ballet, ruled the theatre. In Brussels in 1681 the Académie de Musique was opened, and the first opera theatre, the T de la Monnaie, followed there in 1700. Few Belgian composers emerged at first in a tradition dominated by the French; opéra-comique was established with the occupation of Maréchal Saxe in 1746. The first opera by a Belgian was Pierre van *Maldere's *Le déguisement*

*pastoral* (1759). Other Belgian operas were written by the Austrian director of the Monnaie, Ignaz Vitzthumb. In Liège, Jean-Noël Hamal (1709–78) wrote some 'opéras burless' in Walloon, firstly *Li voedge di Chaudfontaine* (1757). Also from Liège came the greatest Belgian opera composer, Grétry. In the 19th cent. an international repertory began to be heard in the principal Belgian cities, as well as works by Belgian composers (e.g. Jean-Englebert Pauwels, 1795–1839, and Charles Borremans, 1769–1827). One of the first Belgian composers of opéra-comique was Albert Grisar, though the scholar-composers François-Joseph Fétis (1784–1871) and François-Auguste Gevaert (1828–1908) had opéras-comiques produced in Paris. In 1830 a famous performance of Auber's *La muette de Portici* set in motion the revolution that led to Belgian independence. A taste for Grand Opera developed, and respected works were written by Franck and Blockx. The influential Paul Gilson (1865–1942) wrote *Prinses Sonenschijn* (1903) in Flemish, but drew on an international musical language that included use of Leitmotiv. Jan Van der Eeden (1842–1913) was influenced by verismo. Eugène Samuel-Holeman (1863–1944) was an Impressionist.

See also ANTWERP, BRUSSELS, GHENT, LIÈGE.

**Belgrade** (Serb., Beograd). Capital of Serbia. The theatre was founded in 1868, though opera had been given earlier: the first was Blodek's *V studni* in 1894; the first Serbian opera was Stanislav Binički's *Na uranku* (Dawn) in 1903. A visit from Zagreb O in 1911 encouraged the foundation of a company, but not until 1919–20 was one engaged and Stanislav *Binički appointed director and conductor. He built up a principally French and Italian repertory. The company developed especially 1924–34 under Stevan Hristić (1885–1948), composer of the Expressionist *Suton* (Twilight, 1925). He engaged many foreign singers, especially Russian, and trained the first Serbian generation. The repertory was extended to many Slavonic classics and some Serbian operas. Guest singers included Shalyapin and Destinn; visits from the Milan, S, and Paris, O, encouraged standards. The National T (cap. 993) was bombed in 1941, but some operatic life continued in the war. In 1945 the first opera in freed Belgrade was *Eugene Onegin*; the enlightened director and conductor 1944–59 was Oskar Danon. The repertory was widened and tours undertaken, also recordings. The season usually lasts from Sep. to June.

**Belinda.** Dido's lady-in-waiting (sop) in Purcell's *Dido and Aeneas*.

**Belisario.** Opera seria in 3 acts by Donizetti; text by Cammarano after Jean-François Marmontel's drama *Bélisaire* (1776). Prod. Venice, F, 4 Feb. 1836 with Ungher, Vial, Pasini, Salvatori; London 1 Apr. 1837, with Giannoni, De Angioli, Sig. De Angioli, Inchiade; New Orleans, T d'Orléans, 15 Apr. 1842, with Ober-Rossi, Bamberger, Perozzi, Statutti; London, 1 Apr. 1837.

Byzantium, 6th cent. Belisario (bar), general in the Emperor Justinian's (bs) army, returns in triumph to Constantinople after defeating the Bulgarians. His wife Antonina (sop) believes him guilty of their son's murder, and denounces him on forged evidence. He is blinded and led into exile by his daughter Irene (mez). They overhear a freed captive Alamiro (ten) and Ottarino (bs) planning revenge for this on Byzantium; but Alamiro recognizes Belisario, and Irene Alamiro as her long-lost brother. Belisario is fatally wounded in the defence of Byzantium.

Initially a great success and quickly taken up by theatres all over Italy.

Other operas on the subject are by Philidor (1796) and Perretti (1799).

**'Bella figlia dell'amore'.** The quartet in Act III of Verdi's *Rigoletto*, in which the Duke (ten) woos Maddalena (mez) while Rigoletto (bar) comforts the anguished Gilda (sop) with promises of retribution.

**Belle Hélène, La** (Fair Helen). Operetta in 3 acts by Offenbach; text by Meilhac and Halévy. Prod. Paris, Variétés, 17 Dec. 1864, with Schneider and Dupuis; London, Adelphi T, 30 June 1866; Chicago 14 Sep. 1867.

Sparta and Nauplia, before the Trojan war. Vénus has decreed that Paris (ten), in the guise of a shepherd, shall win the most beautiful woman in the world in a competition: this is undoubtedly Ménélas's wife Hélène (sop). When Paris is victorious, the High Priest Calchas (bs), at Hélène's request, arranges for Ménélas (ten) to be dispatched in order that she and Paris may dine alone.

Hélène resists Paris's love but he returns later, in what she imagines to be a dream, and Ménélas arrives unexpectedly to find them kissing. Ménélas packs Paris off.

Vénus, upset at Ménélas's behaviour, causes a spate of adultery in Sparta. In an effort to restore peace, Ménélas summons the High Priest of Venus from Cythera (Paris in disguise), who promises that Vénus will forgive Ménélas if he permits Hélène to return with him to Cythera to make a sacrifice. The couple embark on their voyage.

This satirical version of the classical legend mocks the Second Empire by highlighting its preoccupation with sex and its generally lax morals. It also parodies many of the conventions of Meyerbeer's works.

**'Belle nuit, ô nuit d'amour'.** The *barcarolle in Act II of Offenbach's *Les contes d'Hoffmann*, sung by Giulietta (sop) and Nicklausse (mez) in a palace overlooking the Grand Canal in Venice, asking the night to favour their love.

**Bellezza, Vincenzo** (*b* Bitonto, 17 Feb. 1888; *d* Rome, 8 Feb. 1964). Italian conductor. Studied Naples with Martucci. Début Naples, C, 1908 (*Aida*). Toured Italy; Buenos Aires, C; New York, M, 1926–35; London, CG, 1926–30, conducting Shalyapin's London début (*Mefistofele*), Melba's Farewell, the first English *Turandot*, and the London débuts of Ponselle and Gigli; CG again 1935–6. Rome and other Italian theatres 1935–64. London, Stoll T, 1957, DL, 1958. A highly professional, effective conductor, with a reliable ability to control and deliver a successful performance. (R)

**Belli, Domenico** (*b* ?; *d* Florence, buried 5 May 1627). Italian composer. Possibly worked first at Parma court; 1610–13 tutor at S Lorenzo, Florence; 1619 joined the Medici court. He is remembered especially for *Orfeo dolente*, the intermedio he produced for a performance of Tasso's *Aminta* in Florence, 1616, and for a much-admired (lost) treatment of *Andromeda*. His work is an important link between the earliest Florentine operas and those of the Roman tradition.

WORKLIST: *Orfeo dolente* (Chiabrera; Florence 1616); *Andromeda* (Cicognini; Florence 1618: lost).

**Bellincioni, Gemma** (*b* Como, 18 Aug. 1864; *d* Naples, 23–4 Apr. 1950). Italian soprano. Daughter of the buffo bass Cesare Bellincioni, with whom she studied. Début Naples 1879 (Dell'Orefice's *Il segreto della duchessa*). Considered an outstanding Violetta by Verdi, who said 'she gave new life to the old sinner' but nevertheless rejected her for Desdemona and Alice ('intelligent but too sentimental'). The verismo repertory provided her greatest successes, of which the most momentous was the creation of Santuzza (to the Turiddu of her husband Roberto *Stagno). She was the first Italian Salome, much admired by Strauss. Her gifts lay in strength of personality and characterization rather than in the quality of her voice. Autobiography, *Io e il palcoscenico* (Milan, 1920). (R)

**Bellini, Vincenzo** (*b* Catania, 3 Nov. 1801; *d* Puteaux, 23 Sep. 1835). Italian composer. Studied with his father, Naples with Zingarelli. The success of his student opera *Adelson e Salvini* led to a commission for the Naples, C, *Bianca e Fernando* (temporarily renamed *Bianca e Gernando*, as no form of the royal name Ferdinando was permitted, and later rev. under its original title). This in turn brought a commission from Barbaia for Milan, where he at once made his name with

*Il pirata*, finding also in Felice *Romani a loyal and sympathetic librettist.

Already the essential elements of his style were formed. From Zingarelli he had learnt to give primacy to the elegant melody which came naturally to him; from his admiration for Rossini came his graceful coloratura and a feeling for orchestral drama; the grave, tender sentiment in much of the music was wholly individual. All these qualities appealed to audiences, responding to a new element of Romantic feeling in the music, no less than to singers, from whom Bellini demanded not merely beautiful singing but expressive vocal artistry. The art of *bel canto, with which he was above all composers to be associated, meant for him the expression of character and feeling through the singer's intelligent and sensitive phrasing, shading, and inflection of the melody.

His next two works, *La straniera* and *Zaira*, met with rather less success than *Il pirata*, but *I Capuleti e i Montecchi* triumphed; and with *La sonnambula* and *Norma* his gifts reached full maturity. The former is a tenderly elegiac rustic idyll, with some of his most graceful and touching melodies and at the same time a revelling in the kind of coloratura he had briefly suppressed in his effort to break free of Rossini's influence. *Norma* is a lyric drama reaching in its last act tragic grandeur. It shows his melodic mastery at its greatest, as in the famous 'Casta diva': not only does the melody develop from small phrases into a long, expressive span, but the device of introducing it at first simply on flute puts on the singer the full responsibility of giving the melody human life and expressiveness. It is not surprising that these qualities should have been admired by Verdi, especially for the ease and length of his melodies; a less likely admirer was Wagner, who praised Romani's *Norma* libretto and added, 'Bellini is one of my predilections because his music is strongly felt and intimately bound up with the words.' Nor was his influence confined to opera: Chopin's long melodies owe much to admiration for Bellini.

*Norma* was followed by the less successful *Beatrice di Tenda*, over which Bellini quarrelled with Romani. Carlo Pepoli was the less expert librettist for his last work, *I Puritani*, in which the melodic elegance is unaffected but cannot really engage fully with the drama. It was nevertheless a popular success. But Bellini was now drained in health and exhausted; and while in Paris, hoping to conclude various operatic plans, he fell ill and died, alone, in a house in the suburbs. At his requiem mass in the Invalides his shroud was held at each corner by Cherubini, Carafa, Paer, and Rossini. Reinterred in Catania, 1876.

WORKLIST: *Adelson e Salvini* (Tottola; Naples 1825); *Bianca e Gernando* (Gilardoni; Naples 1826; rev. as *Bianca e Fernando*, Romani, after Gilardoni, Genoa 1828); *Il pirata* (Romani; Milan 1827); *La straniera* (Romani; Milan 1829); *Zaira* (Romani; Parma 1829); *I Capuleti e i Montecchi* (Romani; Venice 1830); *La sonnambula* (Romani; Milan 1831); *Norma* (Romani; Milan 1831); *Beatrice di Tenda* (Romani; Venice 1833); *I Puritani* (Pepoli; Paris 1835).

BIBL: H. Weinstock, *Vincenzo Bellini* (New York, 1971).

**Belloc-Giorgi, Teresa** (*b* S Benigno Canavese, 2 July 1784; *d* S Benigno Canavese, 13 May 1855). Italian mezzo-soprano. Début Turin 1801. Italy and Paris 1802–4. Milan, S, 1804–24. London 1817–19. While successful in the works of Paer, Paisiello, and Cimarosa, she later devoted herself to Rossini's, being especially famous for her Tancredi and Cenerentola and creating Isabella (*L'inganno felice*) and Ninetta (*La gazza ladra*). One of the most prominent Italian singers of her time; Stendhal described her voice as 'pure and wonderful'. Retired 1828.

BIBL: V. Della Croce, *Una Giacobina piemontese alla Scala* (1978).

**Bell Song.** The name generally given to the soprano aria 'Où va la jeune Hindoue?' sung by Lakmé (sop) in Act II of Delibes's *Lakmé*.

**Belmonte.** A Spanish nobleman (ten), lover of Constanze, in Mozart's *Die Entführung aus dem Serail*.

**Belorussia.** During the 18th cent., when Belorussian publications were prohibited, informal theatre and music enjoyed a vigorous life, with fair theatres and school drama prominent. The history of Belorussian opera really begins, though, with *Pobór rekrutów* (Conscription, 1841: lost), to a text by Wincenty Marcinkiewicz set to music by the Minsk-born founder of Polish national opera, Stanisław Moniuszko. Opera was given in Minsk during the 1850s, but musical life was not organized until the 1920s. In 1923 Nikolay Churkin (1869–1964) wrote *Osvobozhdeniye truda* (Labour Liberated) (1924: lost); and the styles of popular music are contrasted with a more classical style in *Taras na Parnase* (Taras on Parnassus, 1927) by Nikolay Aladov (1890–1972). Attempts at staging opera in the 1920s led to the formation of opera classes in Minsk in 1927 and a state studio in 1930, thence to the foundation of the Belorussian National Opera and Ballet T (Bolshoy T) in 1933. Starting with *Carmen* (25 May 1933), this developed a national repertory. The first new Belorussian opera performed was *Tsvetok schastya* (The Flower of Happiness, 1937) by Alexey Turenkov (1886–1958). This was followed in 1939 by *Mikhas Podgorny* by Evgeny Tikotsky (1893–1970) and *V pushchakh Polesye* (In the Forests of Polesye) by Anatoly Bogatyrev (Bogatyrau) (*b* 1913), both on subjects concerning struggles for independence.

The German invasion halted progress, and the company was evacuated to Gorky and Kovrov, though Tikotsky's *Alesya* was produced in 1944. After the war there followed Aladov's *Andrey Kostenya* (1947), then *Kastus Kalinovsky* (1947) and *Pesnya o schastye* (Song of Happiness, 1951) by Dmitry Lukas (*b* 1911), and Bogatyrev's *Nadezhda Durova* (1956). A new generation of composers to emerge included Sergey Kortes (*b* 1935), with *Giordano Bruno* (1977) and Dmitry Smolsky (*b* 1937) with *Sedaya legenda* (The Grey Legend, 1978). Seasons continue to be regularly given in the Bolshoy T.

BIBL: G. Glushchenko, ed., *Istoriya belorusskoy muzyki* (Moscow, 1976).

**'Bel raggio lusinghier'.** Semiramide's (sop) aria in Act I of Rossini's *Semiramide*, in which she rejoices that a ray of love will shine into her heart now that Arsace has returned.

**Beňačková, Gabriela** (*b* Bratislava, 25 Mar. 1944). Czech soprano. Studied Zilina with Kresáková, Bratislava with Blaho. Prague 1970 (Natasha). Prague until 1980; London, CG, 1979; Munich 1981; Cologne 1983; Vienna, S, 1985; New York, M, 1991; also San Francisco, Salzburg, etc. An expressive performer with a rich, sweet tone. Roles incl. Leonore, Verdi's Desdemona, Jenůfa, Vixen, Natasha. (R)

**Benda, Jiří** (*b* Staré Benátky, *bapt.* 30 June 1722; *d* Köstritz, 6 Nov. 1795). Czech composer. He became Kapellmeister at Gotha, where he produced a number of stage works. His first opera was *Xindo riconnosciuto* (1765); despite a study trip to Italy in 1765–6, when he heard operas by Galuppi, Traetta, Piccinni, and others, it remained his only Italian work. The arrival of Seyler's theatrical troup in 1772 led to him writing German stage works, the first of which were spoken 'duodramas', *Ariadne auf Naxos* (1775) and *Medea* (1775). These works inspired many composers to attempt similar *melodramas. Mozart declared 'I like those two works so much that I carry them about with me', and said that most operatic recitatives should be treated in such a fashion 'and sung only occasionally'. In spite of his enthusiasm, Mozart himself wrote only two melodrama passages, in his unfinished Singspiel *Zaide*. Benda composed a third duodrama, *Pygmalion* (1779), modelled on Rousseau's original *scène* (1770).

Benda's most successful Singspiels include *Der Dorfjahrmarkt* (1775), *Romeo und Julie* (1776), and *Walder* (1776). The latter two were among the first Singspiels to treat serious plots. His Singspiels represent a great dramatic advancement on those of Hiller and Weisse; the tech-

niques acquired from his writing of duodramas led to a freer handling of musical forms and the creation of a musical language which was more dramatic. He achieves this by writing stronger dynamic contrasts, inserting more frequent changes of metre and tempo, and by colourful instrumentation, all of which aid deeper characterization.

His nephew **Friedrich** (1754–1814) also wrote some operas and Singspiels, including *Das Blumenmädchen* (1806); his son **Friedrich Ludwig** (1752–92) wrote songs, duets, a chorus for a translation of *Le barbier de Séville* (Dresden 1776), and Singspiels.

SELECT WORKLIST: *Xindo riconnosciuto* (Galletti; Gotha 1765); *Ariadne auf Naxos* (Brandes; Gotha 1775); *Medea* (Gotter; Leipzig 1775); *Der Dorfjahrmarkt* (Gotter; Gotha 1775); *Romeo und Julie* (Gotter; Gotha 1776); *Walder* (Gotter, after Marmontel; Gotha 1776); *Pygmalion* (Gotter; Gotha 1779).

**Bendl, Karel** (*b* Prague, 16 Apr. 1839; *d* Prague, 20 Sep. 1897). Czech composer. Studied Prague Organ School, and gained conducting experience abroad, especially in Paris. Succeeding Smetana as conductor of the Hlahol Choral Society, he did much to enlarge the repertory (e.g. with works of the young Dvořák); and his understanding of voices shows in the operas he soon began writing. His first opera, *Lejla*, is in the manner of French Grand Opera, and made a considerable impression. In 1874 he joined the Prague, P, as second conductor; but his *The Old Bridegroom* proved too similar in manner to *The Bartered Bride*, and was withheld by the composer until after Smetana's death. After a period abroad, he settled again in Prague and wrote another opera, *The Montenegrins*, on Grand Opera scale; a comedy in the Smetana manner, *Karel Šréta*, that in its treatment of the life of the Czech painter also owes something to Berlioz's *Benvenuto Cellini*; and an attempt at verismo, *Gina*. His works reveal great theatrical skill and versatility, if not a strong central personality; he was, nevertheless, one of the most important composers to emerge in Smetana's immediate wake.

SELECT WORKLIST (all first prod. Prague): *Lejla* (Krásnohorská, after Bulwer Lutton; 1868); *Starý ženich* (The Old Bridegroom) (Sabina; comp. 1871–4; prod. 1882); *Černohorci* (The Montenegrins) (Veselý; 1881); *Karel Škréta* (Krásnohorská; 1883); *Gina* (Cimino; 1884, unprod.); *Děti Tábora* (The Children of Tabor) (Krásnohorská; 1892).

BIBL: J. Polák, *Karel Bendl* (Prague, 1938).

**Benedetti, Michele** (*b* Loreto, 17 Oct. 1778; *d* ?). Italian bass. Italy and Amsterdam 1805–9; 1811 sang in first Italian *La vestale* in Naples, where he continued to work for 25 years. Created numerous roles in Rossini operas, including *Otello, Armida, Ermione, La donna del lago*, and

*Mosè in Egitto* (to which the composer added Moses' Prayer in the last act for him). Also created roles in works by Pacini, Mercadante, and Bellini. London 1822. He was at his best as strong, imposing characters and was much admired as Mosè by Stendhal.

**Benedict, (Sir) Julius** (*b* Stuttgart, 27 Nov. 1804; *d* London, 5 June 1885). German, later English, composer, conductor, teacher, pianist, and writer. Studied Weimar with Hummel, then Dresden as Weber's first pupil. In 1823 Weber took him to Vienna, where he became conductor at the K. Two years later he went to Naples as conductor of the San Carlo and Fondo theatres. Here his *Giacinta ed Ernesto* was found too German for Italian taste. In 1834 he moved to Paris, where he met Malibran: at her suggestion he visited London, where he was to remain until his death. He was at the L 1836–7, DL 1838–48, HM from 1852. At DL, as well as producing English Romantic operas including *The Bohemian Girl* and *Maritana*, he introduced operas of his own. His greatest success, however, was at CG with *The Lily of Killarney*. In the concert-hall, he was closely associated with Jenny Lind both as pianist and conductor. His music reflects the influence of Weber, and is skilfully written with a good appreciation of the voice; but in unpropitious conditions his operas were no more successful than those of his contemporaries.

SELECT WORKLIST: *Giacinta ed Ernesto* (Riciutti; Naples *c.*1827); *The Gipsy's Warning* (Linley & Peake; London 1838); *The Lily of Killarney* (Oxenford, after Boucicault; London 1862).

WRITINGS: *Carl Maria von Weber* (London, 1881).

**benefit.** In the 18th and 19th cent. these were special performances, generally at the end of the season, the proceeds of which went to a certain composer, singer, or impresario. In Italian, a distinction is made between a *beneficenza*, in support of a person or organization, and a *beneficiata*, when an artist is given an evening in which to display his special talents, usually in a mixed programme. In our day, benefit or charity performances of opera are now given in aid of an organization or institution. These are common in New York, where the Metropolitan gives several benefits, usually at increased prices.

**Bennett, Richard Rodney** (*b* Broadstairs, 29 Mar. 1936). English composer. Studied London, RAM, with Lennox Berkeley and Howard Ferguson. His excellent craftsmanship and fluent capacity for providing music for different situations are shown in his operas, where these qualities are also at the service of a quick sense of immediate dramatic effect. There is, however, a comparatively narrow range of characterization, and the more serious works, especially, tend not

to engage dramatically with the issues the texts propose. Bennett has also interested himself in music theatre and in concert works in quasi-dramatic forms, as in his series of pieces entitled *Commedia* and *Scena*.

SELECT WORKLIST: *The Ledge* (Mitchell; London 1961); *The Mines of Sulphur* (Cross; London 1965); *A Penny for a Song* (Graham, after Whiting; London 1967); *Victory* (Cross, after Conrad; London 1970).

**Benois, Alexandre** (*b* St Petersburg, 15 May 1870; *d* Paris, 9 Feb. 1960). Russian designer of French and Italian origin. Son of the architect Nikolay Benois and Camilla Cavos, daughter of the architect Alberto Cavos (1801–63), who was in turn the son of Caterino *Cavos and helped in the rebuilding of Moscow, B, 1853. After working with Bakst and other artists in Dyagilev's *Mir Iskusstva group, Benois began to design for St Petersburg, M (e.g. *Götterdämmerung*, 1902). In 1908 Dyagilev engaged him to design *Boris Godunov* for his first Paris season. Also designed the sets for the première of Stravinsky's *Nightingale* (1914). He occasionally designed for opera in the 1920s in Russia (*Queen of Spades*, 1921) and in Paris (*Golden Cockerel*, 1927), but was mostly associated with theatre and ballet until 1938. He then joined his son Nicola (see below) at Milan, S, where until 1957 his designs included *Faust, La traviata, Eugene Onegin,* and *Tosca*. An artist with a particular flair for the fantastic and spectacular, with a vivid sense of colour and form: the 1927 Paris *Golden Cockerel* was a famous example of his brilliant, contemporary, yet traditionally Russian talent. His niece **Nadia** (1894–1975) was a ballet designer, and the mother of Peter Ustinov.

WRITINGS: A. Benois, *Zhizn khudozhnika* (New York, 1955, trans. London, 1960).

**Benois, Nicola** (*b* St Petersburg, 2 May 1901; *d* Codroipo, 30 Mar. 1988). Russian designer, son of above. Worked with his father in Russia, France, and England, mostly on ballet; Milan, S, 1925 (*Boris* and *Khovanshchina* for Toscanini), chief designer 1936–70, staging Russian operas and also *The Ring* as well as other German (and Italian) works. Rome, R; Buenos Aires, C. Married to the soprano Disma de Cecco.

**Benserade, Isaac de** (*b* Paris, *bapt.* 5 Nov. 1613; *d* Paris, 19 or 20 Oct. 1691). French dramatist, poet, and librettist. Joined the court of Louis XIV in 1651 as 'ordonnateur des divertissements'; from 1674 a member of the Académie Française. Responsible for the production of about 23 *ballets de cour during the years 1651–69, providing text for both the dramatic and musical sections, as well as directing the overall production. Before Benserade the ballet de cour had been a collective effort involving several composers and dramatists: by turning it, after *Alcidiane* (1658), into a collaboration between just one librettist and one composer he made an important contribution to the eventual formulation of French opera. Particularly successful were the works he wrote with Lully, some of which approach genuine opera in the quantity and quality of their continuous music: Benserade also assisted him with the ballet interludes for the production of Cavalli's *Ercole amante* in Paris in 1662. Although he made many enemies, notably Molière and Racine, he was regarded by many as the natural heir to Corneille, and the style of his librettos had a crucial influence on Quinault.

BIBL: C. I. Silin, *Benserade and his Ballets de Cour* (Baltimore, 1940, R/1970).

**Benucci, Francesco** (also known as Pietro) (*b* ?, *c*.1745; *d* Florence, 5 Apr. 1824). Italian bass. After singing in Italy 1769–82, became member of the famous Italian company in Vienna, where he sang until 1795. Created Figaro, Leporello in the Vienna version of *Don Giovanni*, and Guglielmo. London 1788–9, when he and Nancy Storace (reputedly his mistress at one time) introduced the first piece from a Mozart opera ever heard there, 'Crudel, perchè' (*Figaro*). His voice was full, round, and deep, yet with a tenor-like quality (according to Michael Kelly), and capable of great delicacy. Described by Mozart as 'especially good', he was the most outstanding actor and singer of his generation. His last great success was as the first Count Robinson in *Il matrimonio segreto*.

**Benvenuto Cellini.** Opera in 2 acts by Berlioz; text by Léon de Wailly and August Barbier, after the autobiography of Benvenuto Cellini (1558–66) (pub. 1728). Prod. Paris, O, 10 Sep. 1838, with Dorus-Gras, Dérivis, Duprez, Stoltz, cond. Habeneck; London, CG, 25 June 1853, with Julienne, Didiée, Tamberlik, Tagliafico, cond. Berlioz; Boston, Orpheum T, 3 May 1975, with Wells, Vickers, Reardon, Beni, cond. Caldwell.

Set in Rome in 1532, the opera is based on events in the life of Benvenuto Cellini. Teresa (sop), daughter of the Papal Treasurer Balducci (bs), plans to elope with Cellini (ten), disguised as a friar, during the Roman Carnival, but their plan is overheard by Fieramosca (bar), a sculptor also in love with Teresa. In the confusion of the carnival, with Fieramosca also disguised as a friar, a fight develops in which Cellini kills Pompeo (bar). Ascanio (mez), Cellini's apprentice, brings Teresa to Cellini's studio and Cardinal Salviati (bs) comes to demand by midnight a statue he has commissioned if Cellini is not to be handed over to the law for abduction and murder. Despite a strike organized by Fieramosca, Cellini, having thrown into the crucible every piece of precious metal he possesses, succeeds in casting his Perseus and thus earning a pardon.

Berlioz's first opera, initially a failure, though later championed by Liszt in performances at Weimar (from 20 Mar. 1852 in a revised 3-act version); the London première, in Queen Victoria's presence, also failed through engineered opposition, and Berlioz immediately withdrew the work.

Other operas on Cellini are by Rossi (*Cellini a Parigi*, 1845), Schlösser (1847, unprod.), Lachner (1849), Kern (1854), Orsini (1875), Bozzano (1877), Díaz de la Peña (1890), Saint-Saëns (*Ascanio*, 1890), Tubi (1906), Courvoisier (1926).

**Bérain, Jean** (*b* Saint-Mihiel, 28 Oct. 1637; *d* Paris, 25 Jan. 1711). French designer. Assistant court painter 1671. Designer to the King 1674, and designed the costumes and sets for many court spectacles. Scenic director 1680, providing designs for the premières of all operas, notably Lully's, Paris 1680–7. Later worked with Desmarets, Campra, and Destouches. André Tessier called him 'le grand créateur du pays d'opéra'. His designs for both opera and ballet, a happy blend of formality and fluency of detail, show more concern for elegance and proportion than for spectacle, and always use central perspective. They are of the essence of the style of Louis XIV's court entertainments.

**Berbié, Jane** (*b* Villefranche-de-Lauragais, 6 May 1931). French mezzo-soprano. Studied Toulouse. Début Toulouse 1954 (Nicklausse). Milan, S, 1958, 1971; Paris, O, from 1959, B, 1990–1; Gly. 1967, 1969; London, CG, 1971–2; New York, M, 1976. A singer of great character, whose repertory includes Rosina, Dorabella, Zerlina, Marcellina, Orsini (*Lucrezia Borgia*), Ascanio (*Cellini*), Concepción. (R)

**Beregani, Nicolò** (*b* Vicenza, 21 Feb. 1627; *d* Venice, 17 Dec. 1713). Italian librettist. A famous lawyer and distinguished literary figure who spent several years exiled in Germany. He wrote librettos for Cesti (*Genserico*, 1666), Legrenzi (*Ottaviano Cesare Augusto*, 1682, and *Giustino*, 1683), and Ziani (*Heraclio*, 1671) which take after the Venetian model of Faustini and Minato. These were set by several later composers: Handel's *Giustino* (1737) is based on an adaptation of Beregani's libretto by Pariati: he also inspired Postel's *Gensericus* (1693).

**Berendey.** The Tsar (ten) in Rimsky-Korsakov's *The Snow Maiden*.

**Berenice.** Opera in 3 acts by Handel; text by Antonio Salvi (first composed by Petri in 1709). Prod. London, CG, 18 May 1737, with Strada, Gizziello, Annibali, Bertoli. Unsuccessful. There are some 20 other operas on the subject.

**Berg, Alban** (*b* Vienna, 9 Feb. 1885; *d* Vienna, 24 Dec. 1935). Austrian composer. Studied with Schoenberg, 1904–10, who deeply influenced him and to whom *Lulu* is dedicated. The 1914 Vienna première of Büchner's *Woyzeck* deeply affected Berg, and his army experiences confirmed his sense of identification with the humble, down-trodden soldier. He was also encouraged to find succinct musical forms to express the play's sequence of short, pregnant scenes, so that each act, and within that each scene, has its own self-contained form. Thus, Act I consists of five character-pieces, Act II is a five-movement symphony, and Act III a group of five inventions on ostinati. Berg also extended control over aspects of the work's production, relating lighting and other features to the motivic design. However, he was always insistent that his fugues, passacaglias, sonata movements, and so on should go unnoticed in appreciation of the compassionate musical drama which they strengthen and serve. The musical language is based on an extended diatonic idiom, and is at no time dodecaphonic. Nevertheless, the opera's première was made the occasion for veiled political attacks on Schoenbergian aesthetics and in particular on the conductor Erich Kleiber for his championship of Berg. The work in fact triumphed, and has retained a firm hold on the international repertory.

By contrast, *Lulu*, based on Wedekind's *Erdgeist* and *Die Büchse der Pandora*, draws on serial techniques and is organized in a dense network of formal relationships expressing the differing contacts between the characters; it also makes use of shorter, self-contained traditional forms such as arietta, canzonetta, etc. There is further a significant use of doubling of roles, with Lulu's victims at the beginning being identified with her clients as she sinks to becoming a prostitute and is herself eventually murdered. The rise to power of the Nazis prevented the hoped-for première, and Berg died without quite completing the full score. Musically, *Lulu* rises to greater heights than *Wozzeck*, though the latter's more easily shared compassion for an underdog, rather than for a more grotesque group of characters destroyed by raw feminine sexuality, has given it much wider popular currency.

WORKLIST: *Wozzeck* (Berg, after Büchner; comp. 1917–22, prod. Berlin 1925); *Lulu* (Berg, after Wedekind; incomplete prod. Zurich 1937, complete prod. (ed. Cerha) Paris 1979).

BIBL: H. Redlich, *Alban Berg* (incl. Berg's lecture on *Wozzeck*: Vienna, 1957, trans. 1957); D. Jarman, *The Music of Alban Berg* (London, 1979).

**Berg, Josef** (*b* Brno, 8 Mar. 1927; *d* Brno, 26 Feb. 1971). Czech composer. Studied composition and musicology in Brno. His increasing interest in dramatic music took the form, from 1962, of music theatre that owed much to Brecht's theories and, from 1968, of 'happenings' such as his *Hudebni happening* (Musical Happening) of 1970.

**Berg, Natanael** (*b* Stockholm, 9 Feb. 1879; *d* Stockholm, 14 Oct. 1957). Swedish composer. Largely self-taught. His operas, all to his own texts, are in a Romantic post-Straussian style.

WORKLIST (all prod. Stockholm): *Leila* (after Byron; 1912); *Engelbrekt* (1929); *Judith* (after Hebbel; 1936); *Birgitta* (1942); *Genoveva* (after Hebbel; 1947).

**Bergamo.** Town in Lombardy, Italy. Opera was first given when the T Riccardi opened 24 Aug. 1791 with Piccinni's *Didone*; burnt down 1797; reopened 1799. In the early 19th cent. many famous singers appeared, including the locally born Rubini and Donzelli, and great attention was given to the works of Mayr; later Rossini was more popular. Rebuilt 1897, and renamed, after Bergamo's most famous composer, the T Donizetti (cap. 2,000). Toscanini conducted the opening *Lucia di Lammermoor*, but withdrew in dissatisfaction with standards and public. In 1937 it was given a subsidiary title, T della Novità, producing each season new works by Ghedini, Barilli, Napoli, Tosatti, Malipiero, Viozzi, Ferrari, Zanon, Sanzogno, etc., and reviving many Donizetti works. The audience is considered the most difficult to please in Italy, after that of Parma.

BIBL: G. Donati-Pettini, *Teatro Donizetti* (n.d.).

**Berganza, Teresa** (*b* Madrid, 16 Mar. 1934 or 1935). Spanish mezzo-soprano. Studied Madrid with Lola Rodriguez Aragon. Début Aix-en-Provence 1957 (Dorabella). Milan, S, from 1957; Gly. 1958–9 (Cherubino and Cenerentola); London, CG, 1960 (Rosina) and 1963; New York, M, from 1967; Paris, O, and Edinburgh 1977. Repertory also includes Dido (Purcell), Octavian, Zerlina, and Carmen. A rich-voiced and sensitive singer, excellent in florid music and with a fresh stage presence. (R)

**Berger, Erna** (*b* Cossebaude, 19 Oct. 1900; *d* Essen, 14 June 1990). German soprano. Studied Dresden with Hirzel. Début Dresden 1925 (First Boy, *Zauberflöte*). Dresden until 1928. Bayreuth 1929–33; Berlin, SO, 1929 and 1932; Berlin, S, from 1934; Salzburg 1932 (Blonde) and 1953–4; New York, M, 1949–53; London, CG, 1934–47. Retired 1955. Famous for her Constanze, Queen of the Night, Gilda, and Sophie. Her voice retained its youthfulness throughout a long career. (R)

**Berghaus, Ruth** (*b* Dresden, 2 July 1927). German producer. Berliner Ensemble from 1964, director 1971–7. Director Berlin, S, 1977–9. Frankfurt from 1980; also Munich, Mannheim, Vienna, Brussels, WNO, etc. Her style, essentially alienating, symbolic, and satirical, has been admired for its stimulating qualities, and criticized for sterility, forced effects, and trivialization of great works. She has had much influence on the subsequent generation of producers. Repertory incl. *Idomeneo*, *Don Giovanni*, *Zauberflöte*, *Fierrabras*, *Tristan*, *The Ring*, *Lulu*. Produced her husband Paul *Dessau's *Lukullus*, and the posthumous première of his *Leonce und Lena*, 1979.

**Bergknappen, Die** (The Miners). Opera in 1 act by Ignaz Umlauff; text by P. Weidmann. Prod. Vienna, B, 17 Feb. 1778.

Fritz (ten) and Sophie (sop) are prevented from marrying by Sophie's guardian, Walcher (bs), who wants to marry her himself. With the help of Delda (mez), a sorceress, the couple trick Walcher, and he consents to their union. However, when he learns of the deception he withdraws his promise. The situation is saved when Fritz rescues Walcher from a landslide in the mine and in gratitude Walcher again permits the marriage.

The opera, written in a simple, popular style, was the first production of the Viennese National-Singspiel instituted by the Emperor Joseph II at the Burgtheater. It set a style followed in many works. Among some 17 other operas on the subject is one by Flotow (1835).

**Bergman, Ingmar** (*b* Stockholm, 14 July 1918). Swedish producer and film director. Early experience in opera included work at Stockholm, Royal O, 1940–2. Most of his career has been spent in film and drama, but his production for television in Swedish of *The Magic Flute* (1975) captivated a large audience with its mixture of fantasy, realism, and warm treatment of the characters. Wrote text for, and prod., Daniel Börtz's *Backanterna* (1991).

**Bergonzi, Carlo** (*b* Polisene, 13 July 1924). Italian tenor. Studied Parma with Grandini. Début Lecce 1948 (Rossini's Figaro). Sang as baritone for two years, then after further study made second début Bari 1951 (Chénier). Milan, S, from 1953; London, CG, from 1962 (Manrico, Radamès, Cavaradossi); Chicago 1955; New York, M, 1956–83. A repertory of more than 40 roles included Canio, Don José, Edgardo, and Rodolfo. His beautiful voice and musical refinement were particularly memorable in Verdi. (R)

**Berio, Luciano** (*b* Oneglia, 24 Oct. 1925). Italian composer. Berio's theatre music began with mime pieces, including *Mimusique No. 2* (1955) and then his first collaboration with Italo Calvino, *Allez Hop* (set in a flea circus). His interest in the sounds of languages led to *Passaggio*, for Milan, PS, which includes multi-lingual speaking groups in the auditorium: Berio ingeniously made the predicted audience protests material for their improvisation. *Laborintus II* (1965) was an open-ended drama, performable either on stage or in concert, using texts by Dante and others as phonetic material to create a labyrinth of meanings for the listener to explore. It draws on jazz, madrigal style, taped sounds, and traditional instruments in an intricate amalgam of styles. Berio's first full-length theatre work was *Opera* (here meaning 'Works'). A more traditional music drama, this interweaves three commentaries on human attitudes towards death (the Orpheus legend, a terminal hospital ward, and the *Titanic* disaster), and makes use of imaginative techniques that call into question the relationship between the theatre and music-making, between the music-makers and the audience. Resuming his collaboration with Calvino, Berio wrote *La vera storia*, partly influenced by Verdi's *Il trovatore* and denouncing political oppression, and his major work for the theatre, *Un re in ascolto*. Here, the different levels of reality revolve around the dying Prospero rehearsing his new musical version of *The Tempest* in an extravagant production. This includes the preparation of a musical play drawn from Gotter's Singspiel *Die Geisterinsel* (1791), and a score of great virtuosity includes some vocal writing that suggests Berio's acknowledgement of enduring Italian skills. His long preoccupation both with death and with the nature of theatrical reality, which to some extent places him in the Italian tradition of Pirandello, finds expression here in the final abandonment of Prospero to the extinction of both human and theatrical illusion.

SELECT WORKLIST: *Allez-Hop* (Calvino; Venice 1959); *Passaggio* (Sanguineti; Milan 1963); *Opera* (Berio, Eco, & Colombo; Santa Fe 1970); *La vera storia* (Calvino; Milan 1982); *Un re in ascolto* (A King Listening) (Calvino; Salzburg 1984).

BIBL: L. Berio, *Two Interviews* (trans. London, 1985).

**Berkeley, (Sir) Lennox** (*b* Oxford, 12 May 1903; *d* London, 26 Dec. 1989). English composer. He took music up seriously after leaving Oxford in 1926, and studied with Nadia Boulanger for six years. He did not write an opera until 1953, when he produced the full-scale 3-act *Nelson* for London, SW. This was not a success, in part due to a lengthy plot with some contrived situations. Berkeley found a happier vein of invention with a 1-act comedy, *A Dinner Engagement*, in which his graceful invention and wry humour are well displayed. His interest in opera led him to return twice more to the medium of 1-act opera, with the mild but well-crafted *Ruth* and a more sharply written piece, *Castaway*: another opera, *Faldon Park*, was unfinished at his death.

WORKLIST: *Nelson* (Pryce-Jones; London 1954); *A Dinner Engagement* (Dehn; Aldeburgh 1954); *Ruth* (Crozier; London 1956); *Castaway* (Dehn; Aldeburgh 1967).

BIBL: P. Dickinson, *Lennox Berkeley* (London, 1989).

**Berkshire Festival.** Annual summer music festival instituted by Serge Koussevitsky at Stockbridge, MA, USA in 1934, moving to Tanglewood in 1937. Opera provided by the opera department of the Berkshire Music Center has included the US premières of *Peter Grimes* (1946) and *Idomeneo* (1947), and premières of Lukas Foss's *Griffelkin* (1955), Louis Mennini's *The Rope* (1955), and Mark Bucci's *Tale for a Deaf Ear* (1957). The Lenox Arts Center, founded in 1971, has staged works in many different veins.

**Berlin.** City in Prussia, Germany; capital of united Germany 1871–1945; capital of German Democratic Republic 1945–90; capital of united Germany again 3 Oct. 1990. The first operas were Italian, given in 1688 during the reign of Frederick I. The first opera-house, the Hofoper (or Lindenoper), opened 1742 under Frederick the Great with *Cleopatra e Cesare* by Carl *Graun, director until his death in 1759. Opera developed chiefly at the Nationaltheater under B. A. Weber from 1795. The first German opera was Johann *Reichardt's *Brennus* (1798). The Hofoper and the Nationaltheater merged in 1814 as the Königliches Schauspiel, with August Iffland as director. The influential Count Brühl was director from 1815; he wanted C. M. von *Weber as music director, but was overruled and *Spontini held the position 1820–42. However, the first opera at the Neues Schauspielhaus, Weber's *Der Freischütz* (1821), proved a national triumph, unifying national sentiment in favour of German opera and, together with the fame of Spontini's work, giving Berlin a dominating role in a still divided Germany. The Königstädter-T opened 1824, becoming the Wallnertheater 1851; here Italian opera was given, including with Sontag and Pasta. Meyerbeer succeeded Spontini 1842, with Nicolai as a conductor. The Königliches Schauspiel burnt down 1843, reopening 1844. Opera was also given at Kroll's 'Etablissement' (1843), especially Lortzing, and at the Friedrich-Wilhelm-Städtisches-T (from 1883 Deutsches T), directed by Lortzing 1850–1. The Victoria-T gave Italian opera with singers including Artôt

and Patti. Berlin continued to increase its standing as a major centre of opera, especially German: high standards were achieved under Weingartner (1891–8) and Strauss (1898–1918).

In 1919 Max von Schillings took over the Königliche O, renamed the Staatsoper Unter den Linden. He was succeeded by Heinz Tietjen (1925–43), with as music directors Blech (1918–23, 1926–37) sharing with Kleiber (1923–34), Furtwängler (1933–4), Krauss (1935–6), and from 1936 to 1943 Heger, Schüler, Egk, Karajan, and Elmendorff. Under Kleiber premières were given of *Wozzeck* (1925) and works by Busoni, Janáček, Richard Strauss, Stravinsky, and Krenek. This adventurous policy was restrained by the Nazis. The theatre was bombed in the war, and activities were resumed under Ernst Legal at the Admiralspalast on 8 Sep. 1946 with Gluck's *Orfeo*. On 4 Sep. 1955 the Oper Unter den Linden (cap. 1,354) reopened with *Meistersinger* under Konwitschny.

In 1896 Kaiser Wilhelm II planned a new opera-house, and Kroll's T in the Königsplatz was bought. The war prevented its opening, which took place in 1924. Until May 1926, while the Oper unter den Linden was being rebuilt, the company played at the Kroll Oper (officially renamed Oper am Platz der Republik), alternating from summer 1927 with the Kroll's independent company under Klemperer with Legal as director. Under this joint direction, and when the Linden company returned home, the Kroll became the leading experimental opera theatre in Europe, with a repertory concentrating on enterprising productions and novel works, including by Hindemith, Krenek, Milhaud, Schoenberg, Weill, Stravinsky, and Janáček; conductors included Zemlinsky, Fritz Zweig, and Karl Rankl. The theatre closed in 1931, and became the home of the Reichstag in 1933.

The Städtische Oper (originally the Deutsches Opernhaus on the Bismarckstrasse, sometimes known as the Charlottenburg Opera) opened Nov. 1912 (cap. 2,100). It reached a high standard 1925–34 with Walter (1925–9) and Stiedry (1929–33) as conductors, Tietjen, Singer, and Ebert as directors. Singers included Ivogün, Lehmann, Reinhardt, Nemeth, Onegin, and Kipnis. Bombed 1943; company moved to Admiralspalast; reopened 1945 in the T des Westens (cap. 1,529), former home of the Volksoper, with Bohnen as director. He was succeeded by Tietjen 1948, and Carl Ebert 1955–61. The rebuilt house on the Bismarckstrasse opened 1961 with *Don Giovanni*, reverting to the name Deutsche Oper (cap. 1,885). Maazel was music director 1965–71. Götz Friedrich has been director since 1981.

The Komische Oper, formerly the Metropol

T, opened 1947 (cap. 1,208) and under Walter *Felsenstein became one of the most discussed opera-houses in Europe, with some famous productions, emphasizing an integrated ensemble and dramatic realism, that made brilliant use of the excellent rehearsal conditions. Felsenstein was responsible for 29 productions; he was succeeded by his disciple Joachim Herz, in turn by Harry Kupfer in 1981.

**Berlin, Irving** (*b* Temun, 11 May 1888; *d* New York, 22 Sep. 1989). Russian, later US, composer and writer. After early success as a song writer (e.g. with 'Sadie Salome, Go Home') he formed a partnership with Ted Snyder that led to stage appearances and successes. His early ragtime compositions included 'Alexander's Ragtime Band' (1911). His first Broadway show was *Watch Your Step* (1914), and this was followed by many others during the 1920s, fewer during the 1930s. His most striking musical was *Annie Get Your Gun* (1946), followed by *Call Me Madam* (1950), in which his fresh, appealing lyrics and memorable tunes easily transcend their Tin Pan Alley origins.

**Berlioz, Hector** (*b* La Côte-Saint-André, 11 Dec. 1803; *d* Paris, 8 Mar. 1869). French composer. Almost all Berlioz's music was dramatic in conception, whether ostensibly a symphony (*Symphonie fantastique*, *Harold en Italie*) or one of his hybrids—the dramatic symphony *Roméo et Juliette*, the monodrama *Lélio*, and the *légende dramatique*, *La damnation de Faust*. Though the last has been staged, it was originally entitled by Berlioz *opéra de concert* and is essentially not theatrical but a work for the listener's imagination in the concert-hall. He completed five operas, and considered other projects throughout his life: they include the offer of a libretto on *Othello* by Rouget de Lisle, who admired his setting of 'La Marseillaise'.

The early *Estelle et Némorin* is lost; and what survives of *Les Francs-juges* (included in other works) indicates his well-known admiration for Weber, and especially the influences of Méhul on Weber. He was drawn to his first complete surviving opera by the idea of setting episodes from the life of Benvenuto Cellini, whose autobiography he probably first read in 1831. Greatly attracted to a fellow-artist of heroic and Romantic individuality, he also sensed a kinship with Cellini's struggles to assert himself in the face of academicism and intrigue and to produce against all the odds a masterpiece. The project of a 2-act opéra-comique was turned down by the Opéra-Comique, and Berlioz cast the work instead for the Opéra. *Benvenuto Cellini* is an uneven opera, showing signs in its weaker moments of his attempts to match the conventions of singers and

of both the Opéra-Comique and the Opéra; but the brilliance and warmth of the music in general far outweighs these patches, and it was ironic that Berlioz's transcending of the ballet conventions of the Opéra, with the Roman Carnival scene, should have disturbed the traditionalists yet led to one of the most thrilling musical spectacle scenes in all opera.

Faced with public neglect in the field of opera after the failure of *Cellini*, Berlioz found himself cut off from the medium into which his talents should naturally have flowed; thus, when he made an opera out of Virgil, he took less account of practical considerations of performance in France. *Les Troyens* is an epic on the largest scale, and triumphs over normal operatic considerations. It embraces the spectacular and the lyrical, and combines startling originalities with a classical grandeur that looks back past Spontini to Gluck. Berlioz's love of Virgil was a typical Romantic paradox: there is the ache for a lost age of purity, balance, and order, yet equally a sense of identification with the heroes and heroines who act with Romantic urgency and passion. Traces of the conventions of French Grand Opera may readily be found, in such as the use of a large number of principals, ballet sequences, and broad panoramic effects; but these are absorbed into a grandeur of vision that dwarfs the genre. Though the work is constructed in a sequence of episodes, making no use of *Leitmotiv, there is the connecting theme of the divine command that the Trojans (with their *reminiscence motive of the Trojan March) shall found their new city of Rome; and this lends dramatic impetus and cohesion to all the skilfully contrasted scenes. With the understanding of 'open form' he had acquired from Shakespeare, Berlioz could include in his pattern familiar operatic devices together with dramatic sung and danced orchestral tableaux (the 'Royal Hunt and Storm'), a sequence of incidental entertainments (for Queen Dido at Carthage), arias for minor characters that confer new emotional dimensions on the main theme (Hylas, the sentries), dream and ghost sequences, even a scene in which the central character is silent (Andromache, a grave clarinet 'speaking' for her).

It was as a rest after the exertions of *Les Troyens*, so Berlioz said, that the comedy *Béatrice et Bénédict* was written, as well as in tribute to Shakespeare. It includes some infelicities (the leaden parody of academicism in Somarone's repeated fugue), but there is in the musical handling of the separate numbers a tenderness and an ironic wit worthy of Shakespeare's own, justifying Berlioz's description of the opera as a 'caprice written with the point of a needle'. Its German successes, before the production of the second part of *Les Troyens* (the Carthage acts), gave Berlioz false hopes of a good reception for the greater work.

In 1847 Berlioz was invited by Jullien to become music director at London, DL, as part of an ill-fated attempt to found an English national opera.

WORKLIST: *Estelle et Némorin* (comp. 1823, lost); *Les Franc-juges* (Ferrand; comp. 1826, rev. 1833 as *Le cri de guerre de Brisgaw*); *Benvenuto Cellini* (De Wailly & Barbier; Paris 1838); *La nonne sanglante* (Scribe; comp. 1841–7, unfin., unprod.); *Les Troyens* (Berlioz, after Virgil; comp. 1856–8: Pt1 (*La prise de Troie*) Karlsruhe 1890, pt2 (*Les Troyens à Carthage*) Paris 1863); *Béatrice et Bénédict* (Berlioz, after Shakespeare; Baden-Baden 1862).

CATALOGUE: D. K. Holoman, *Catalogue of the Works of Hector Berlioz* (Kassel, 1987).

SELECT WRITINGS: *Grand traité d'instrumentation* (Paris, 1843, trans. London, 1855); *Les soirées de l'orchestre* (Paris, 1852, trans. 1956); *Les grotesques de la musique* (Paris, 1859); *A travers chants* (Paris, 1862, trans. London, 1913–18); *Mémoires de Hector Berlioz* (Paris, 1870, trans. 1969); P. Citron, ed., *Correspondance générale* (Paris, 1972– ).

BIBL: J. Barzun, *Berlioz and the Romantic Century* (Boston, 1950); H. Macdonald, *Berlioz* (London, 1982); D. Cairns, *Berlioz: The Making of an Artist* (London, 1989); D. K. Holoman, *Berlioz* (Cambridge, MA, 1989).

**Bern** (Fr., Berne). Capital of Switzerland. The Stadttheater (cap. 770) opened 25 Sep. 1903, and was renovated in 1983.

**Bernacchi, Antonio Maria** (*b* Bologna, 23 June 1685; *d* Bologna, 13 Mar. 1756). Italian mezzo-soprano castrato. Studied Bologna with Pistocchi. Début Genoa 1703. Successful career throughout Italy (particularly in Venice), then in Europe generally, in works by all the major composers of the period. Sang for the Elector of Bavaria in Munich 1720–35. London, H, 1716; then 1717 as Goffredo (*Rinaldo*) and Dardano (*Amadigi*), parts previously sung by women. Also engaged by Handel 1729 to replace *Senesino; he created roles in *Lotario* and *Partenope*, but was not popular with the English (Burney found his singing artificial). After his return to Italy, continued to sing until 1736, despite a vocal decline, then founded a celebrated singing school in Bologna. A singer of great technical virtuosity, he beat *Farinelli in a famous vocal contest in 1727. His many distinguished pupils included *Guarducci, *Carestini, and *Raaff.

**Bernardini, Marcello** (*b* ?Capua, ?*c*.1740; *d* after 1799). Italian composer and librettist. Details of his life are unclear, though most of his *c*.40 operas were performed in Rome. Some of these comic works, of which *Li tre Orfei* and *La donna di spirito* are characteristic, survived into the 19th cent.

Bernardini wrote librettos both for himself and for others, notably Martín y Soler.

SELECT WORKLIST: *Li tre Orfei* (?; Rome 1784); *La donna di spirito* (?; Rome 1787).

**Bernstein, Leonard** (*b* Lawrence, MA, 25 Aug. 1918; *d* New York, 14 Oct. 1990). US conductor, composer, and pianist. Studied Harvard, including with Piston, and Curtis Institute with Reiner. Also studied with Koussevitsky, whose assistant he became; at Tanglewood, he conducted the US première of *Peter Grimes*. He also conducted the première of his own *Trouble in Tahiti* at Brandeis U in 1952; other important operatic conducting appearances included Visconti's productions of *Medea* and *La sonnambula* at Milan, S, in 1954 and 1955, and *Falstaff* at New York, M, 1964, Vienna 1966.

Bernstein's music for the theatre has ranged wide, including ballet, musical, and opera. His 1-act *Trouble in Tahiti* won respect for its serious theme, and later was given a sequel *A Quiet Place*. But his lively, irreverent score for *Candide* indicated that his real talent lay in the musical; and this was conclusively demonstrated with *West Side Story*. Its fundamentally serious theme—a modern version of the *Romeo and Juliet* story—was expressed in music that drew on jazz, Latin American music, and Broadway idioms, skilfully and wittily handled, mixing them with a more sentimental popular manner. Its enormous popularity did much to loosen distinctions between opera and Broadway musical, not least since dance scenes were an essential ingredient of the work. (R)

SELECT WORKLIST: *Trouble in Tahiti* (Bernstein; Waltham 1952); *Wonderful Town* (Green & Comden, after Fields & Chodorov; New York 1953); *Candide* (Wilbur, Latouches, Parker, & Bernstein, after Hellmann, after Voltaire; New York 1956); *West Side Story* (Sondheim, after Laurents; New York 1957); *A Quiet Place* (Wadworth; Houston 1983; rev. 1984 to include *Trouble in Tahiti* as flashback).

SELECT WRITINGS: *The Joy of Music* (New York, 1954); *The Infinite Variety of Music* (New York, 1959); *The Unanswered Question* (New York, 1976).

BIBL: P. Robinson, *Leonard Bernstein* (New York, 1982).

**Berry, Walter** (*b* Vienna, 8 Apr. 1929). Austrian bass-baritone. Studied Vienna with Gallos. Début Vienna, S, 1950. Sang small roles, then 1953 Masetto, and later Figaro (Mozart), Leporello, Guglielmo, Wozzeck, Barak, Ochs, and the Italian repertory. Salzburg since 1952; Berlin, SO, from 1961; New York, M, from 1966 (including Wotan); London, CG, from 1976; San Francisco 1985 (Alberich). Roles also include Pizarro, Escamillo, and Telramund. (R)

**Bersa, Blagoje** (*b* Dubrovnik, 21 Dec. 1873; *d* Zagreb, 1 Jan. 1934). Croatian composer. Studied Zagreb with Zajc, Vienna with Fuchs and Epstein. A leading figure in Croatian music 1900–20, he gave a new lead to opera especially with *Oganj*. Written in post-Wagnerian style, this is set in a factory and makes advanced use of music to represent machinery and its role in workers' lives. He also wrote *Jelka* on Italian models, and a comedy, *Postolar od Delfta*. His many pupils benefited from his new ideas, especially the enlarged awareness of German and Russian music. His brother **Vladimir** (1864–1927) wrote *Cvijeta* (1898) and *Andrija Čubranović* (1900). A third brother, **Josip**, sometimes acted as their librettist.

SELECT WORKLIST: *Jelka* (J. Bersa; comp. 1901, unprod.); *Oganj* (The Forge: set to German text *Der Eisenhammer*) (Willner; Zagreb 1911); *Postolar od Delfta* (The Cobbler of Delft) (Willner; Zagreb 1914).

**Bertali, Antonio** (*b* Verona, Mar. 1605; *d* Vienna, 17 Apr. 1669). Austrian composer of Italian birth. Active as an instrumentalist at the Viennese court from *c.*1623; Kapellmeister 1649. Composed *c.*600 works, including 11 operas. Important figure in the establishment of Italian opera in Vienna: most works were written for court occasions, such as *Cibele ed Atti*, staged for the engagement of Leopold I. Bertali's four extant operas, which include *La magia delusa* and *Il ciro crescente*, show that his style was greatly influenced by that of Cavalli and Cesti, to whom he is sometimes compared by his contemporaries.

SELECT WORKLIST (all prod. Vienna): *Cibele ed Atti* (?; 1666; lost); *La magia delusa* (Amalteo; 1660); *Il ciro crescente* (Almateo; 1661).

**Bertati, Giovanni** (*b* Martellago, 10 July 1735; *d* Venice, 1815). Italian librettist. His first text, for Tozzi's *La morte di Dimone* (1763), inaugurated a list of some 70, chiefly for Venice, T San Moisè, of which he was principal librettist. Most of his librettos are comic, and show the influence of Goldoni in some imitative texts and details and in their neat, rapid sense of theatre, though they are by no means void of independent and quite sharp social comment. His greatest success was his text for Cimarosa's *Il matrimonio segreto* (1792). His version of *Don Giovanni* for Gazzaniga (1787) had a direct and specific influence on Da Ponte's for Mozart (i.e. in his characteristic love of a *catalogue aria). He also wrote librettos for Galuppi, Guglielmi, Paisiello, Anfossi, Bianchi, Paer, Zingarelli, and others. He succeeded Da Ponte in Vienna, 1791.

**Berton, Henri-Montan** (*b* Paris, 17 Sep. 1767; *d* Paris, 22 Apr. 1844). French composer. Largely self-taught. He joined the Opéra-Comique at an early age as a violinist, also receiving some guid-ance from Sacchini. With the Revolution, he turned decisively to dramatic music and produced his most famous work, *Les rigueurs du cloître*, a pioneering example of *rescue opera. His most striking operas, however, were *Montano et Stéphanie* and *Le délire*, in which he shows a feeling for chromatic harmony and makes enterprising use of motive; the former, in particular, remained popular during the early 19th cent. Its success was only overtaken by *Aline, reine de Golconde*, which even reached America; it is one of his many operas to give marked extra emphasis to the orchestra, and to allow the chorus a more functional part in the drama. He was music director of Paris, I, 1807–9, then chorus-master at the Opéra. The failure of the new operas in more grandiose vein which he wrote after *Françoise de Foix* increasingly embittered him.

SELECT WORKLIST (all first prod. Paris): *Les rigueurs du cloître* (Fiévée; 1790); *Ponce de Léon* (Berton; 1797); *Montano et Stéphanie* (Dejaure; 1800); *Le délire* (Saint-Cyr; 1799); *Aline, reine de Golconde* (Vial & Favières; 1803); *Françoise de Foix* (Bouilly & Dupaty; 1809).

His son **Henri**(-François) (*b* Paris, 3 May 1784; *d* Paris, 19 July 1832) was also a composer who wrote a number of light operas of small originality, most successfully *Ninette à la cour* (1811) to a text by Favart. Taught singing at the conservatoire, 1821–7. His son **Adolphe** (*b* Paris, 1817; *d* Algiers, 28 Feb. 1857) was a tenor who had a modest career in France and Algiers.

**Berton, Pierre-Montan** (*b* Maubert-Fontaines, 7 Jan. 1727; *d* Paris, 14 May 1780). French tenor, conductor, and composer. Father of Henri-Montan Berton. Début Paris, O, 1744, immediately withdrawing as a singer and joining the cellos. Sang at Marseilles *c.*1746–8, then conducted at Bordeaux, returning to Paris as conductor at the Opéra, 1755. He helped to introduce both Gluck and Piccinni to Paris, but is best known for his arrangements of other works, e.g. by Rameau and Lully, to suit contemporary taste. Manager, Paris, O, with Trial, from 1767; director from 1770; in sole charge as director-general, 1775–8.

**Bertoni, Ferdinand** (*b* Salò, 15 Aug. 1752; *d* Desenzano, 1 Dec. 1813). Italian composer. Studied with Martini and became organist of St Mark's, Venice and maestro di cappella at the Ospedale dei Mendicanti. In 1745 had his first success in Venice with the opera buffa *La vedova accorta*, followed by a number of opere serie including *Orazio e Curiazio*. His setting of Goldoni's *Le pescatrici* led to commissions from theatres throughout Italy. After the triumph of *Quinto Fabio* in Milan, Bertoni was engaged for London, H, for the following two seasons. Here he composed three operas, *Demofoonte*, *La gover-*

*nante*, and *Il duca d'Atene*, and revived earlier works, such as *Artaserse* and *Quinto Fabio*. Returning to Venice in 1780, he produced two new operas, *Armida abbandonata* and *Cajo Mario*, though he was soon lured back to London, where he spent the 1781–2 and 1782–3 seasons; new works included *Il convito* and *Cimene*. On his return to Venice he continued to compose both opera and sacred music: his last full-scale stage work was *Nitteti*. Following Galuppi's death in 1785 Bertoni became maestro di cappella at St Mark's.

Although there is little remarkable about either their style or form, Bertoni's operas—49 in total—enjoyed considerable popularity in their day. Burney best summed up their appeal by saying that they 'would please and soothe by their grace and facility, but not disturb an audience by enthusiastic turbulence', noting that Bertoni 'never had perhaps sufficient genius and fire to attain the sublime'. His most popular opera was *Orfeo ed Euridice*, revived in London 'in the Manner of an Oratorio' (1780), and printed twice (Venice, 1776 and 1783).

SELECT WORKLIST: *La vedova accorta* (Borghese; Venice 1745); *Orazio e Curiazio* (Soggrafi; Venice 1746); *Le pescatrici* (Goldoni; Venice 1752); *Artaserse* (Metastasio; Forlì 1776); *Orfeo ed Euridice* (Calzabigi; Venice 1776); *Quinto Fabio* (Zeno; Milan 1778); *Demofoonte* (Metastasio; London 1778); *La governante* (Badini; London 1779); *Armida abbandonata* (Vitturi; Venice 1780); *Cajo Mario* (Roccaforte; Venice 1780); *Il duca d'Atene* (Badini; London 1780); *Il convito* (Andrei, after Livigni; London 1782); *Cimene* (Cassani; London 1783); *Nitteti* (Metastasio; Venice 1789).

**Bertram.** The father of Robert (bs), in reality the Devil, in Meyerbeer's *Robert le Diable*.

**Berwald, Franz** (*b* Stockholm, 23 July 1796; *d* Stockholm, 3 Apr. 1868). Swedish composer. Studied with his father Christian Berwald (1740–1825) and Dupuy, violinist of the Hovkapellet, which he then joined. He began composing seriously in 1817, beginning with an opera *Gustav Wasa* in 1827. In Berlin in 1829 he considered other opera subjects, but his first completed stage works were two operettas. Three operas followed, including *A Swedish Country Betrothal* (dedicated to Jenny Lind, who sang the first performance of excerpts), and his best-known opera, *The Queen of Golconda*.

SELECT WORKLIST: *Gustav Wasa* (after Kellgren; Stockholm 1827–8); *Eine ländliche Verlobung in Schweden* (A Swedish Country Betrothal) (Prechtler; Vienna 1847); *Drottningen av Golconda* (The Queen of Golconda) (after Vial & De Favières; Stockholm 1864).

BIBL: R. Layton, *Franz Berwald* (Stockholm, 1956; Eng. orig. 1959).

**Bess.** Crown's girl (sop), but lover of Porgy, in Gershwin's *Porgy and Bess*.

**'Bess, you is my woman now'.** The love duet between Porgy (bar) and Bess (sop) in Act II of Gershwin's *Porgy and Bess*.

**Besuch der alten Dame, Der** (The Visit of the Old Lady). Opera in 3 acts by Gottfried von Einem; text by Friedrich Dürrenmatt, after his drama of the same name. Prod. Vienna, S, 23 May 1971, with Christa Ludwig, Beirer, Waechter, Hotter, cond. Stein; San Francisco 25 Oct. 1972, with Resnik, Cassilly, Wolansky, Yarnell, cond. Peress; Gly. 31 May 1973, with Meyer, Crofoot, Bell, Garrard, cond. Pritchard.

The once impoverished Claire Zachanassian (mez), originally known as Klari Wäscher, now one of the richest women in the world, returns to her home town of Güllen. She meets Alfred Ill (bar), her former lover and father of her illegitimate child, and offers the town a huge sum if he is killed. This eventually happens; the cause of his death is declared to be heart failure. Claire claims the body and gives the mayor the cheque she promised. The people of Güllen express their happiness and satisfaction that justice has been done.

**Bettelstudent, Der** (The Beggar Student). Operetta in 3 acts by Millöcker; text by Zell and Genée, after Sardou's *Fernande* and Bulwer Lytton's *Lady of Lyons*. Prod. Vienna, W, 6 Dec. 1882, with Finaly, Schweighofer, Girardi; New York, Thalia T, 19 Oct. 1883, with Geistinger, Stebold, Schmitz, Friest, Schultze; London, Alhambra T, 12 Apr. 1884, with Fanny Leslie, Mervin, Hood. Also Singspiels by Winter (1785, after Cervantes's *La cueva de Salamanca*), and Schenk (1796).

**Betz, Franz** (*b* Mainz, 19 Mar. 1835; *d* Berlin, 11 Aug. 1900). German baritone. Studied Karlsruhe. Début Hanover 1856 (Heinrich, *Lohengrin*). Berlin, H, 1859 (Don Carlos, *Ernani*), then until retirement in 1897 (Don Giovanni, Lysiart, Hans Heiling, Amonasro, Falstaff, etc.). One of Wagner's most trusted singers, he created Hans Sachs (1868), and the Wanderer (also singing Wotan) in the first Bayreuth *Ring*, 1876. Returned to Bayreuth 1889 as Sachs, and alternated as Kurwenal and Mark. Possessed a warm, powerful, and flexible voice, excellent diction, and a sympathetic dramatic presence.

**Bevignani, Enrico** (*b* Naples, 29 Sep. 1841; *d* Naples, 29 Aug. 1903). Italian conductor and composer. Studied Naples with Albanese and Lillo. Had an early success with *Caterina Blum* (Naples 1863). Engaged by Mapleson 1864 as répétiteur for London, HM, then cond. London, CG, 1869–87 and 1890–6: cond. first London

*Aida* (1876), *Gioconda* (1883), *Amico Fritz* (1892), and *Pagliacci* (1893), and Patti's CG Farewell (1895). Much admired for his skill in accompanying singers, though Shaw had hard words for him. Moscow and St Petersburg 1871–83, receiving the Order of St Stanislas. Cond. first Bolshoy *Onegin*, 1881; Tchaikovsky declared that he owed its success mostly to Bevignani. Berlin and Vienna, 1898; New York, M, 1894–1900.

**Bianca e Fernando.** Opera in 2 acts by Bellini; text by Gilardoni after Carlo Roti's drama *Bianca e Fernando alla tomba di Carlo IV, Duca di Agrigento* (1820). Prod. Naples, C, 30 May 1826, with Méric-Lalande, Rubini. Rev. version, *Bianca e Fernando, melodramma serio* in 2 acts, altered by F. Romani. Prod. Genoa, for inauguration of CF, 7 Apr. 1828, with Tosi, David, Tamburini.

The opera tells how Fernando (ten), the son of the usurped Duke of Agrigento, overcomes Filippo (bs), the usurper, first by pretending to support him and then by rescuing his father and preventing the marriage of his sister Bianca (sop) to Filippo.

Bellini completely revised the music for the second version, retaining only two original pieces. Orig. version prod. as *Bianca e Gernando* as the Naples censorship forbade references to the royal name Ferdinando.

**Bianca und Giuseppe.** Opera in 4 acts by Kittl; text by Richard Wagner, after Heinrich König's novel. Prod. Prague 19 Feb. 1848.

Wagner wrote the scenario in 1836 for himself, but did not use it; he revised it in 1842 for Reissiger, who did not use it either.

**Bianchi, Francesco** (*b* Cremona, *c*.1752; *d* London, 27 Nov. 1810). Italian composer. A pupil of Jommelli and Cafaro, his first opera *Giulio Sabino* was followed by a succession of opere serie including *Il gran Cidde* and *Demetrio*. In 1775 he became harpsichordist at the Paris, I, staying until 1778, during which time he composed *La réduction de Paris* and *Le mort marié*. Returning to Italy, he entered on a period of enormous creative enegy, writing about 50 operas in 15 years for many of the major houses. Although his serious works included settings of heroic Metastasian librettos, such as *L'Olimpiade*, *Alessandro nell'Indie*, and *Nitteti*, he showed a preference for a newer style of text which mixed comic and serious elements, and in which the characters were drawn from the middle class or from rustic life. The most famous was *Il disertore francese*, whose thinly veiled social critique and bourgeois costumes caused an outcry at its first performance. His opere buffe, such as *La fedeltà tra le selve*, matched the light-hearted plots with music

of considerable charm and appeal, if of little profundity or memorability.

In 1782 he became vicemaestro at Milan Cathedral and in 1785 second organist at St Mark's, Venice: here his most popular opera had appeared, the dramma giocoso *La villanella rapita*. Soon taken up in many European cities, including Dresden, Paris, and Lisbon; for its performance in Vienna in 1785 Mozart wrote an additional quartet and a trio. Following the success in London of a pasticcio of *La villanella rapita*, Bianchi was invited to the H in 1794. His first work was a setting of a Da Ponte libretto, *Antigona*; in addition to revivals of earlier operas, he also composed *Il consiglio imprudente*, *Merope*, and *Cinna*. In 1798–1800 he was director of opera at Crow Street T, Dublin; returning to London, he produced *Alzira*, *La morte di Cleopatra*, and *Armida*. Although he conducted occasional revivals in London and Paris, he concentrated in later life on teaching singing, numbering Sir Henry Bishop among his pupils.

Bianchi's music has found little following in recent times, but he was one of the most widely appreciated composers of his day. Even Haydn, after hearing a revival of *Aci e Galatea* in London in 1795, commented favourably on his abilities and kept Bianchi's score close to him, apparently for solace in moments of anger. Equally at home with serious or comic opera, Bianchi had a sure understanding of the audience's expectations, which he fulfilled with suitably engaging music.

SELECT WORKLIST: *Giulio Sabino* (?; Cremona 1772); *Il gran Cidde* (Pizzi; Florence 1773); *Demetrio* (Metastasio; Cremona 1774); *La réduction de Paris* (De Rozio; Paris 1775); *Le mort marié* (Sédaine; Paris 1776); *L'Olimpiade* (Metastasio; Milan 1781); *Alessandro nell'Indie* (Metastasio; Venice 1785); *Nitteti* (Metastasio; Milan 1789); *Il disertore francese* (Benincasa, after Mercier; Venice 1785); *La fedeltà tra le selve* (?; Rome 1789); *La villanella rapita* (Bertati; Venice 1782); *Antigona* (Da Ponte; London 1796); *Il consiglio imprudente* (Da Ponte; London 1796); *Merope* (Da Ponte; London 1797); *Cinna* (Da Ponte; London 1798); *Alzira* (Rossi; London 1801); *La morte di Cleopatra* (Bonaiuti; London 1801); *Armida* (Da Ponte; London 1802).

**Biber, Heinrich** (*bapt.* Wartenburg, 12 Aug. 1644; *d* Salzburg, 3 May 1704). Bohemian violinist and composer. High steward and conductor at the Salzburg court. Remembered chiefly for his violin compositions, Biber wrote two operas: *Chi la dura, la vince* (1687), which survives intact, and *Alessandro in pietra* (1689), of which only the libretto is extant. He also composed the music for a number of the 1-act entertainments which took place at Salzburg U at the end of the academic year. (In 1767 Mozart fulfilled the same function with his *Apollo et Hyacinthus*.)

SELECT WORKLIST: *Chi la dura, la vince* (Raffaelini; Salzburg 1687); *Alessandro in pietra* (Raffaelini; Salzburg 1689; music lost).

**Bibiena.** See GALLI-BIBIENA.

**Bickerstaffe, Isaac** (*b* Ireland, 1735; *d* ?St Malo, *c.*1812). English librettist. Page to the Lord-Lieutenant of Ireland, he served in the marines before settling in London. Of his *c.*16 librettos the most famous were those for Arne's *Thomas and Sally* (1760) and Arnold's *The Maid of the Mill* (1765), a loose adaptation of Richardson's *Pamela*. He also had a notable collaboration with Dibdin; and his *Love in a Village* (1767), an adaptation of Charles Johnson's *The Village Opera* (1729), played an important role in establishing the London vogue for pasticcio. Bickerstaffe had an instinctive ability to grasp the attention of his audience, creating lively, often complex characters, whose actions and sentiments reflect the prevailing attitudes of the day. The prefaces to his librettos demonstrate a rare concern for the integration of drama and music: of the original texts written for Dibdin, the most important were *Love in the City* (1767) and *Lionel and Clarissa* (1768). In 1772 Bickerstaffe was forced to flee London to avoid arrest on a charge that also led to him being instantly cut off by all his acquaintances: there are thus few references to him in contemporary accounts, although he is generally acknowledged as the most successful English librettist of the 18th cent.

**Bielefeld.** Town in Westphalia, Germany. The Stadttheater (cap. 890) opened 1904. Among enterprising revivals was one of Weber's *Peter Schmoll* (1955).

**Bignens, Max** (*b* Zurich, 9 June 1912). Swiss designer. Studied Zurich, Munich, Florence, and Paris. Resident designer Berne 1939–45, Basle 1945–51 and 1956–61, Munich 1951–6 and 1963–5, Darmstadt 1961–3, Cologne 1965. Has also designed in many other European opera houses.

**Bilbao.** Town in the Basque province of Spain. Since 1953 the Asociación Bilbaina de Amigos de la Opera (ABAO) has mounted an opera festival initially concentrating on star casts in the French and Italian repertory and somewhat impulsive productions. In 1987 the festival moved from the Coliseo Albia to the Teatro Arriaga (cap. 1,700).

**Billington, Elizabeth** (*b* London, *c.*1765; *d* Venice, 25 Aug. 1818). English soprano. The first truly international British prima donna. Studied with J. C. Bach and James Billington (whom she married, 1783), Paris with Sacchini. Début Dublin 1783 (Polly, *Beggar's Opera*); continued to appear there for 14 years. London, CG, 1786–93. Retired and travelled in Europe 1794, but then sang with great success for two years at Naples, C; here Bianchi wrote for her *Inés de Castro*, in which she triumphed despite local fears that an eruption of Vesuvius expressed divine displeasure at a Protestant appearing at the C. Further triumphs throughout Italy. Returned to England 1801, alternating between CG and H, where she was the first English Vitellia, 1806; then sang in concerts till her final retirement. In 1817 she went to live in Italy with her fortune and her second husband. Beautiful when young (Reynolds painted her as St Cecilia), she later became obese. She was never a good actress, but had an 'exquisitely sweet' voice (Burney), whose range was a–a'''. Haydn described her as a 'great genius' and wrote 'Arianna abbandonata' for her. She was a friend of Lady Hamilton, and much admired by the Prince of Wales, whose mistress she may have been.

BIBL: Earl of Mount-Edgcumbe, *Musical Reminiscences* (London, 1825).

**Billy Budd.** Opera in 4 acts by Britten; text by E. M. Forster and Eric Crozier, after Melville's unfinished story (1891). Prod. London, CG, 1 Dec. 1951, with Pears, Uppman, Dalberg, cond. Britten; New York, NBC TV, 19 Oct. 1952. New 2-act version broadcast 13 Nov. 1960, with Pears, J. Ward, Langdon, cond. Britten; prod. London, CG, 9 Jan. 1964, with Lewis, Kerns, Robinson, cond. Solti; repeated New York, Carnegie H, 1966, with same cast and conductor. First US stage performance of rev. version Chicago 6 Nov. 1970, with Lewis, Uppman, Evans, cond. Bartoletti.

The opera, which has no female roles, is set on a man-o'-war, HMS *Indomitable*, in Napoleonic times. The ship is short of crew and Billy Budd (bar) is press-ganged into service from a merchant ship. His natural goodness and affability arouse the resentment of the master-at-arms, John Claggart (bs), who persecutes Billy and tries to bribe him into leading a mutiny.

Claggart accuses Billy of treachery to Captain Vere (ten), for whom Billy has great respect. When summoned to Vere's cabin, Billy, instead of getting the promotion he expects, is accused of mutiny; unable to defend himself because of his stammer, he becomes so frustrated that he lashes out, killing Claggart. The court of officers decides Billy must hang; Vere knows he could have altered the verdict. As he is about to be hanged, Billy shouts 'Starry Vere, God bless you!', and the cry is echoed by the crew. The opera is told as a memory of distant events by the aged Vere, haunted by his behaviour.

Also an opera by Ghedini (1949).

**Bing,** (Sir) **Rudolf** (*b* Vienna, 9 Jan. 1902). Austrian, later British, manager. Studied Vienna.

Manager of Viennese concert agency 1923. Worked Darmstadt, and Berlin under Carl Ebert, who brought him to Gly., 1934. Manager, Gly., 1936–46. Helped found Edinburgh Festival; director 1947–9. Manager New York, M, 1950–72. Played silent role of Sir Edgar in New York, C, première of Henze's *The Young Lord*, 1973. Autobiography, *5,000 nights at the Opera* (London, 1972).

**Binički, Stanislav** (*b* Jasika, 27 July 1872; *d* Belgrade, 15 Feb. 1942). Serbian composer. Studied Belgrade, Munich with Rheinberger. First music director, Belgrade National T, 1920–4. His music includes an opera, *Na uranku* (Dawn, 1903), a pioneering Serbian national opera in the manner of verismo making ingenious use of contrasting musical styles to depict the struggle between Turks and Serbs.

**Bioni, Antonio** (*b* Venice, 1698; *d* Vienna or Italy, after 1738). Italian composer. Studied with Porta. Had his first success with *Climene*. In 1723 he left Italy for Bohemia and entered the service of Count Sporck, composing several operas for his private theatre at Guckuksbade. By 1726 he had moved to Breslau, where he joined the Italian opera company as harpsichordist, becoming impresario in 1730. After the dissolution of the company in 1734 he probably returned to Italy; nothing is known of his later life. Bioni composed *c.*30 operas, but few librettos and only one score, *Issipile*, have survived. Though accounts of his music by contemporaries (e.g. Mattheson) vary, he clearly enjoyed considerable popularity, and was important for the spread of Italian opera through German-speaking territory at the beginning of the 18th cent. He occupies an especially significant position in the history of opera in Breslau (Wrocław). After the revival of *Orlando furioso*, which opened the Ballhaus T in 1725, he dominated the musical life of the city until his departure, with such works as *Lucio Vero* and *Engelberta*. He was also responsible for introducing the Breslau public to the music of many contemporary Italian composers, including Porpora and Vivaldi.

SELECT WORKLIST: *Climene* (Cassani; Chioggia 1721, lost); *Orlando furioso* (Braccioli; Guckuksbade 1724, lost); *Lucio Vero* (Burigotti; Breslau 1727, lost); *Engelberta* (Zeno; Breslau 1729, lost); *Issipile* (Metastasio; Breslau 1732).

**Birmingham.** City in Warwickshire, England. The city was long dependent on occasional and touring performances. Raybould's *The Sumida River* (1916) was followed by the première of Messager's *Monsieur Beaucaire* (1919) and then a series of productions at the Repertory T from 1920 under Sir Barry Jackson that included the première of Smyth's *Fête galante* (1923). Local

amateur companies have contributed, e.g. the Barfield OC with some lesser-known 19th-cent. works and a novelty, Margaret More's *The Mermaid* (1951), and the Midland Music Makers with some enterprising productions including *William Tell*, *Prince Igor*, *Les Troyens*, *Ivan the Terrible*, *The Jacobin*, and the British première of *Die Feen*.

The Barber Institute, part of Birmingham U, has also staged some important productions, including the British premières of Rameau's *Hippolyte et Aricie* (1965) and Haydn's *Orfeo* (1967). English Touring O and Birmingham Music T have combined forces at the Canon Hill Arts Centre. The City of Birmingham Touring O commissioned *Ghanashyam* (A Broken Branch, 1989) from Ravi Shankar.

**Birtwistle, (Sir) Harrison** (*b* Accrington, 15 July 1934). English composer. Studied Manchester, with Richard Hall; then at London RAM. Together with Peter Maxwell Davies, he founded the Pierrot Players so as to give performances of works with some element of *music theatre. The group was reorganized as the Fires of London in 1970.

Though almost all Birtwistle's music is theatrical, he distinguishes between traditional opera and his own stage works, which owe at least as much to Greek theatre. *Punch and Judy*, 'a tragical comedy or a comical tragedy', tells the old puppet-show story making use of costumed instrumentalists and a 'choregos' to introduce the plot and initiate some of the action. This emphasis on stylization, supported by the use of Baroque techniques, helps to ritualize the action and through the puppets to distance the emotions, often centred on birth, marriage, and death in terms of extreme violence. In *Down by the Greenwood Side*, 'a dramatic pastoral', the choregos has become the traditional figure of Father Christmas, and the work combines a mummers' play with a ballad, dividing singers and a speaker into two groups; again, the action is ritualized and includes some horrifying child murders in its representation of seasonal death and rebirth. *Bow Down*, described simply as 'music theatre', once more uses ballads to tell of the murder of a fair sister by her dark sister; and it repeats from earlier works the use of several versions of the same ballad not in linear narrative but so as to reflect several aspects of the story.

Much in these pieces points towards *The Mask of Orpheus*. This 'lyrical tragedy' uses a large orchestra, which carries the drama forward as much as the singing does. There is again no linear narrative, rather the juxtaposition of contrasting accounts of the same dramatic moment so as to further the exploration of the Orpheus myth.

Each of the major parts has three performers, as man (or woman) (the past), hero (the possible), and myth (the eternal). The intention is thus to create a three-layered expression of the action and its meaning. Use is made of electronic music (in six interludes), and of music constructed in Birtwistle's favourite blocks of sounds. Use of masks and of puppets again serves to ritualize immediate emotion on to the eternal.

Birtwistle followed this large-scale work with *Yan Tan Tethera*, adapted from a TV commission. The favourite ritual framework now concerns a folk-story, and use is made of the incantatory effect of the ancient sheep-counting method of the title. This is used to outline the conflict of good and evil, with the shepherd representatives of these absolutes never confronting one another but staring out at the audience: the evil Caleb has florid lines, the good Alan short, syllabic ones, with the sheep choruses using varying vocal techniques.

*Gawain* marks a departure from Birtwistle's preference for material taken directly from folklore. His treatment of the 14th-cent. romance *Sir Gawain and the Green Knight* provides recognizably human characters, and the moral problem posed by the hero's compromise of chivalric honour, accepting magic protection in order to outwit his challenger, shows a more realistic approach to the problems of human nature dealt with in earlier works. But there is still a large element of ritual, seen both in Gawain's ceremonial preparation for his quest and in the masquelike representation of the 'Turning of the Seasons'. There are also reminiscences of folk fertility figures (such as the Green Man) in the Green Knight, who miraculously survives beheading and returns at the end of the opera to claim his due and judge Gawain.

Birtwistle has also written dramatic works for performance in schools, *Music for Sheep*, *The Mark of the Goat*, and *Visions of Francesco Petrarca*, which also make use of the dramatic method of presenting different aspects of identical action. Even his purely instrumental works may incorporate a dramatic element involving players moving or perhaps opposing one another in conflict.

WORKLIST: *Punch and Judy* (Pruslin; Aldeburgh 1968); *Down by the Greenwood Side* (Nyman; Brighton 1969); *Bow Down* (Harrison; London 1977); *The Mask of Orpheus* (Zinovieff; London 1986); *Yan Tan Tethera* (Harrison; London 1986); *Gawain* (Harsent, after 14th-cent. romance; London 1991).

BIBL: M. Hall, *Harrison Birtwistle* (London, 1984).

**Bishop, Anna** (*b* London, 9 Jan. 1810; *d* New York, 18 or 19 Mar. 1884). English soprano. Studied London with Henry *Bishop, whom she married 1831. Début in concert Apr. 1831. In 1839 toured with the harpist Bochsa, with whom she then eloped to the Continent (avoiding France, where he was a wanted man). Successful concert appearances all over Europe, including St Petersburg; 1843 engaged at Naples, C, where she sang for two years, appearing 327 times in 20 operas. Though she was popular with most Italians, Donizetti said of her singing in Vienna, 'She made the very stones laugh with her tremolo [like a] tamburello', and refused to have her for Caterina Cornaro, as Verdi did for Alzira. The Pope, however, was so charmed with her singing that he conferred on her the Order of St Cecilia. She returned to England 1846, singing at DL in Balfe's *Maid of Artois*. New York début 1847 (Linda); 1852 sang first US Martha. After Bochsa's death 1856 in Sydney, she married again and continued to tour the world, surviving a shipwreck in 1866. She made her last appearance in New York, 1883. Though she possessed a brilliant technique, and eyes 'large, lustrous and full of fire' (*Morning Post*), she was said to be wanting in expression.

**Bishop, (Sir) Henry** (*b* London, 18 Nov. 1786; *d* London, 30 Apr. 1855). English composer and conductor. Started work as a music-seller when only 13, later publishing some songs and studying with Bianchi. His first opera, *The Circassian Bride*, was lost when the score was burnt in the fire that destroyed *Drury Lane the day after the first performance; but its success led to other works. Music director London, CG, 1810–24. Here he supervised and adapted a number of foreign operas, mercilessly rewriting portions of them and introducing some of his own music, as was then fashionable. Moving to DL, he produced among many other works *Aladdin*, in an attempt to rival Weber's *Oberon* at CG. He later became music director at Vauxhall. Most of his works consist of numbers interpolated into a spoken play; in many cases 'operas' credited to him include only a few numbers, or even a single number. Even *Aladdin*, his most substantial work, includes spoken dialogue. He used his song 'Home, Sweet Home' in many forms as a theme song in *Clari*. He married **Ann Riviere** (see BISHOP, ANNA) in 1831.

SELECT WORKLIST: *The Circassian Bride* (Ward; London 1809); *Clari, or The Maid of Milan* (Payne; London 1823); *Aladdin* (Soane; London 1826).

BIBL: R. Northcott, *The Life of Sir Henry Bishop* (London, 1920).

**Bizet, Georges** (*b* Paris, 25 Oct. 1838; *d* Bougival, 3 June 1875). French composer. Studied Paris with Gounod. His earliest operas are in the first period of composition which he later disavowed, too harshly in the cases of *Le docteur Miracle* and *Don Procopio*; these take Italian example (especially Rossini and Donizetti) and contain some charming music already showing a witty sense of the stage. With *Ivan IV* the model is French Grand Opera, the result awkward and most successful in the less grandiose moments.

Bizet's first important opera was *The Pearl Fishers*, where the influences are still more diverse, but are more expertly absorbed; there is a lyrical impulse that is recognizably Bizet's own, and some ambitious (occasionally over-ambitious) harmonic adventures, but the characters are alive, and the drama compelling. These gains are consolidated in *The Fair Maid of Perth*, which triumphs over a lame libretto and makes of the situations a genuine drama, with some excellent individual arias and ensembles. The surviving fragments of *La coupe du roi de Thulé* suggest that ambition was growing, especially in the handling of motive.

Bizet completed his father-in-law F. Halévy's *Noé*, unpromisingly; and then a brief but intense concentration on opéra-comique, with *Calendal*, *Clarissa Harlowe*, and *Grisélidis* (none of them completed), led to *Djamileh*. The characterization is stronger than anything in the earlier works, and the sense of the exotic stimulates some of his most original harmony and exquisite orchestral passages. With it he attains full maturity; though even this could not have led listeners to expect *Carmen*. Here his gifts are most completely realized. It was not achieved easily, and the score went through many revisions before reaching its present state. The subject suited him ideally, with its exotic setting and violent passions, its opportunity for striking, tuneful arias that fit easily into the drama (and advance it) and for brilliant orchestral colour, and its brilliant use of motive, of which he was now a master. It was notoriously a failure at first, and Bizet did not live to see it triumph. After his death, Ernest Guiraud composed recitatives so as to turn it into a Grand Opera; though they remain in use in some opera-houses, they are regrettable for their betrayal of Bizet's original opéra-comique, not least since historically the work stands at the head of the whole genre. *Carmen* was admired by a wide variety of musicians, influencing Tchaikovsky (whose favourite opera it was) in *The Queen of Spades*, and giving Nietzsche a stick with which to belabour Wagner with a demand for 'Mediterranean' values; and it has with good reason never lost its hold on audiences.

SELECT WORKLIST (all first prod. Paris unless otherwise stated): *Le docteur Miracle* (Battu & Halévy; 1857); *Don Procopio* (Cambiaggio; comp. 1859, prod. Monte Carlo 1906); *Ivan IV* (Leroy and Trianon; unfin., comp. c.1862–5, prod. Mühringen 1946); *Les pêcheurs de perles* (The Pearl Fishers) (Carré & Cormon; 1863); *La jolie fille de Perth* (The Fair Maid of Perth) (Saint-Georges & Adenis, after Scott; 1867); *Djamileh* (Gallet; 1872); *Carmen* (Meilhac & Halévy, after Mérimée; 1875).

BIBL: W. Dean, *Georges Bizet* (London, 1948, rev. 3/ 1975).

**Bjoner, Ingrid** (*b* Kråkstad, 8 Nov. 1927). Norwegian soprano. Studied Oslo with Boellemose, Frankfurt with Lohmann. Début Oslo 1957 (Donna Anna). Bayreuth 1960; Munich from 1961; New York, M, 1961–7; London, CG, 1967; also Paris, O; Milan, S; Salzburg; Berlin, SO. Roles incl. Leonore, Senta, Isolde, Brünnhilde, Tosca, Elektra, Ariadne. (R)

**Björling, Jussi** (*b* Stora Tuna, 5 Feb. 1911; *d* Stockholm, 9 Sep. 1960). Swedish tenor. Studied with his father, the tenor David Björling, and Stockholm with Forsell and Hislop. First appearance Royal Swedish O, July 1930 (Lamplighter, *Manon Lescaut*); official début there August 1930 (Don Ottavio). Sang at Stockholm O regularly until 1938, then intermittently until his death. Vienna 1936; Chicago 1937; London, CG, 1939, 1960; New York, M, 1938–60 (except during the war). A great lyric tenor of his generation, memorable in the Verdi and Puccini repertory, but also as Faust (Gounod); his voice was radiant, even, and resonant, and was used with refinement and musicianship. (R)

BIBL: B. Hagman, ed., *Jussi Björling* (Stockholm, 1960).

**Bjornson, Maria** (*b* Paris, 1949). British designer of Norwegian-Romanian origin. Studied London with Koltai. Glasgow Citizen's T 1970–2, working on 13 productions in 18 months. London from 1972: CG, ENO regularly; also WNO, Scottish O, Sydney O, Japan, etc. A dedicated and highly sought-after artist, whose large repertory contains *The Ring*, *Queen of Spades*, *Carmen*, a Janáček cycle for WNO/ Scottish O, *Toussaint* (première), and *The Phantom of the Opera*. Her very individual style, allied to a sharp eye for detail, incorporates elements of mythology, fairy-tale, and symbolic distortion; its tone encompasses the romantic, the spectacular, the harsh, and the oblique; her costumes similarly range from elaborate fullness to deceptive simplicity.

**Blacher, Boris** (*b* Newchang, China, 19 Jan. 1903; *d* Berlin, 30 Jan. 1975). German composer. Studied Berlin with Koch. His inventiveness and intelligence were applied particularly to the problems of theatre music. He wrote ballets and dance dramas, and works defined as dramatic notturno (*Die Nachtschwalbe*, 1947) and ballet-opera (*Preussisches Märchen*, 1952), as well as operas. Even in the latter, there is often a reconsideration of the conventions of opera to create, and explore sensitively, novel dramatic conditions, which in some cases anticipate the increased interest in music theatre. His *Abstrakte Oper No. 1* substitutes for narrative some considerations of fundamental human emotions and their consequences. Librettist, with *Einem, of the latter's *Dantons Tod* and *Der Prozess*.

SELECT WORKLIST: *Fürstin Tarakanowa* (Koch; Wuppertal 1941); *Romeo und Julia* (Blacher, after Shakespeare; Berlin 1947); *Die Flut* (Cramer; Dresden 1947); *Abstrakte Oper No. 1* (Egk; Mannheim 1953); *Rosamunde Floris* (Blacher, after Kaiser; Berlin 1960).

BIBL: H. Stuckenschmidt, *Boris Blacher* (Berlin, 1963).

**Blachut, Beno** (*b* Ostrava-Vitkovice, 14 June 1913; *d* Prague, 10 Jan. 1985). Czech tenor. Studied Prague with Kaděřábek. Début Olomouc 1939 (Jeník). Prague, N, from 1941, his Dalibor in 1945 establishing him as the leading Czech tenor of his day. Distinguished in the Czech repertory, he was also successful as Florestan, Don José, Otello, Hermann, and Walther. Holland 1959; Edinburgh 1964; also Venice, F; Moscow, B; Berlin, K; etc. He made skilful use of his expressive voice, and was a fine actor. (R)

**Blaise, Adolfe Benoît** (*d* ?Paris, 1772). French composer. From 1737 bassoonist at Paris, I; 1753–60 conductor; 1762–6 composer. In 1743 directed the orchestra of the T Foire Saint-Laurent, in 1744 that of the Foire Saint-Germain. A prolific composer for the I, his most important works were the collaborations with *Favart, which included *Annette et Lubin* and *Isabelle et Gertrude*. One of the most characteristic composers of the I during its heyday.

SELECT WORKLIST: *Annette et Lubin* (Favart; Paris 1762); *Isabelle et Gertrude* (Favart; Paris 1765).

**Blake, Rockwell** (*b* Plattsburgh, NY, 10 Jan. 1951). US tenor. Studied Plattsburgh. Début New York, City O, 1979 (Ory). New York, M, from 1981; Boston 1982; Aix from 1983; also Chicago, Lisbon, and Italy, where he is highly regarded. A fine bel canto singer of great virtuosity; roles incl. Don Ottavio, Almaviva, Lindoro, Rinaldo (*Armida*), Osiride (*Mosè*), Rodrigo (*Donna del Lago*), Arturo, Tonio (*Fille du régiment*), Rossini's Otello. (R)

**Blamont, François Collin de** (*b* Versailles, 22 Nov. 1690; *d* Versailles, 14 Feb. 1760). French composer. Studied with his father; after aristocratic service was appointed Surintendant of royal music in 1719. His first stage work, *Les festes grecques et romaines*, was one of the earliest *ballets-heroïques, and was followed by seven works of a similar style. During the *Guerre des Bouffons Collin de Blamont was an ardent supporter of the French faction: his views gained wide publicity following the appearance of his

*Essai sur les goûts anciens et modernes de la musique française*. In 1726 he became maître de la musique de la chambre.

SELECT WORKLIST: *Les festes grecques et romaines* (?; Paris 1723); *Les caractères de l'amour* (?; Paris 1738).

WRITINGS: *Essai sur les goûts anciens et modernes de la musique française* (Paris, 1754).

**Bland, Maria Theresa** (*b* London, 1769; *d* London, 15 Jan. 1838). English soprano. Born of Italian-Jewish parents named Romanzi; sang as Romanzini until marriage to George Bland in 1790. Appeared London, Hughe's Riding School 1773, Royal Circus 1782. Studied with Dibdin. Début, London, DL, 1786. Member of DL Co. till 1824, excelling in operas of Storace, Arnold, etc. London, HM, 1791. She was at her best in simple, naïve roles, and possessed, Leigh Hunt wrote, a voice as bland as her name. Her elder son, the tenor **Charles**, was the first Oberon in Weber's opera at London, CG, 1826, when he was found by the librettist Planché at least respectable but by Benedict 'a bad actor with an offensive voice'. Another son, **James** (1798–1861), was a buffo bass who sang at DL, latterly playing kings in Planché's pantomimes.

**Blangini, Felice** (*b* Turin, 18 Nov. 1781; *d* Paris, 18 Dec. 1841). Italian tenor and composer. A chorister at Turin Cathedral, he went to Paris in 1799 and quickly established a reputation as a singing-teacher and song composer. After completing Della Maria's unfinished opera *La fausse duègne* he took to writing stage works. Many of his operas were composed for Paris, although *Encore un tour de Caliphe* was first heard in Munich, where he was briefly Kapellmeister to the Duke of Saxe-Coburg; several works were produced for Kassel, where he was General-Musikdirektor from 1809 to 1813.

Blangini enjoyed the support of many influential patrons in Paris, including Napoleon's sister Princess Borghese, who had made him her maître de chapelle in 1806. In 1814 he became a professor at the Conservatoire, the following year Surintendant de la Chapelle du Roi. By now his smaller vocal pieces had been widely published and he was without doubt one of the most sought-after teachers in Europe. His 30 stage works were almost all opéras-comiques, although he also produced a pasticcio *Figaro ou La jour des noces* based on Mozart and Rossini. Typical of their light, attractive music and subject-matter are *Le jeune oncle* and *Le jeu de cache-cache*.

SELECT WORKLIST: *Encore un tour de Caliphe* (?; Munich 1805); *Le jeune oncle* (Advenier-Fontenille; Paris 1821); *Le jeu de cache-cache* (Artois de Bournonville; Paris 1827).

WRITINGS: *Souvenirs de Blangini*, ed. Maxime de Villamorest (Paris, 1834).

**bleat.** Vocal device whereby a single note is quickly reiterated with varied pressure of the breath. Also known as Goat's Trill (Fr., *chévrotement*; Ger., *Bockstriller*; It., *trillo caprino*; Span., *trino de cabra*). First appears in the work of the early monodists: used to good effect by Monteverdi, often as a cadential ornament, and introduced to England through his pupil Walter Porter (*c.*1595–1659). P. F. Tosi, in *Opinioni de' cantori antichi e moderni* (1723), observes that a trill is *caprino* if its two notes are less than a semitone apart, or if the trill is sung with unequal speed or force; Spohr (*Violinschule*, 1832) describes the *Bockstriller* as a trill at the unison. Not widely used in the 19th cent., though Wagner demands it of the tailors (because of the legend about the goatskin which they narrate) in *Meistersinger*, Act III.

**Blech, Leo** (*b* Aachen, 21 Apr. 1871; *d* Berlin, 24 Aug. 1958). German conductor and composer. Studied Berlin with Bargiel and Rudorff, then Humperdinck. Aachen 1893–9; Prague, Deutsches T, 1899–1906; Berlin, H, 1906–23 (music director from 1913). Toured USA 1923. Berlin, Deutsches Opernhaus (later SO), 1923 and 1926–37. As a Jew, he was prevented from returning to Berlin after conducting in Riga in 1937. Riga 1937–41, then Stockholm 1941–7. Returned to Berlin, SO, 1949. A fine Wagner and Verdi conductor, with a wide knowledge and understanding of opera. His own operas include *Das war ich* (Dresden 1902), *Alpenkönig und Menschenfeind* (Dresden 1903), and the most successful, *Versiegelt* (Hamburg 1908). (R)
BIBL: W. Jacob, *Leo Blech* (Hamburg, 1931).

**Blewitt, Jonathan** (*b* London, 19 July 1782; *d* London, 4 Sep. 1853). English composer. Studied London with Battishill and Haydn. Wrote a large amount of music especially for London, DL and SW.

**Bliss, (Sir) Arthur** (*b* London, 2 Aug. 1891; *d* London, 27 Mar. 1975). English composer and conductor. The dramatic nature of Bliss's music was first revealed in his ballet and film scores, one of which was a version of *The Beggar's Opera*: he did not turn to opera until *The Olympians* in 1949. Though he had an imaginative subject for a Romantic opera, and provided some fine music especially in Act II, the work did not make its expected impact. Autobiography, *As I Remember* (London, 1970).
WORKLIST: *The Olympians* (Priestley; London 1949); version of *The Beggar's Opera* (film, 1952–3); *Tobias and the Angel* (Hassall; BBC TV 1960, stage prod. London 1961).
CATALOGUE: L. Foreman, ed., *Arthur Bliss: A Catalogue and Critical Survey* (London, 1979).

**Blitzstein, Marc** (*b* Philadelphia, 2 Mar. 1905; *d* Fort-de-France, Martinique, 22 Jan. 1964). US composer. Studied piano with Ziloti and composition principally with Boulanger and Schoenberg. He abandoned his early experimental style under the influence of Hanns Eisler, and in the late 1930s began developing a more populist operatic style as reflection of his theories of the artist as social activist. His first significant work in this new manner was *The Cradle Will Rock*, produced by Orson Welles and John Houseman, when it was found too controversial. Here he brought into his style a feeling for US vernacular music, including blues, jazz, Broadway, and other popular musical manners, in his wish to engage a broad audience in his social message. Other operas of the period include *No for an Answer* (on the plight of immigrants). *Regina*, a version of Lillian Hellman's *The Little Foxes*, marked a return from the musical towards more traditional opera (or perhaps a modern Singspiel), though *Reuben, Reuben* resumed his Broadway manner. He had his greatest success with a version of Brecht's *Die Dreigroschenoper*. His versatility in moving between different genres, without losing his distinctive musical personality, in some ways anticipates more recent developments in music theatre.
SELECT WORKLIST: *Triple-Sec* (Jeans; Philadelphia 1928); *The Condemned* (Blitzstein, based on the Sacco & Vanzetti trial; unprod. 1932); *The Cradle Will Rock* (Blitzstein; New York 1937); *No for an Answer* (Blitzstein; New York 1941); *Regina* (Blitzstein, after Hellman; New Haven 1949); *The Threepenny Opera* (Blitzstein, after Brecht & Weill; New York 1954); *Reuben, Reuben* (Blitzstein; New York 1955); *Juno* (Stein & Blitzstein, after O'Casey; New York 1959); *Sacco and Vanzetti* (Blitzstein; unfin. 1964).

**Bloch, Ernest** (*b* Geneva, 24 July 1880; *d* Portland, OR, 15 July 1959). US composer of Swiss birth. His only opera is *Macbeth*. Written in a style that owes something to *Pelléas et Mélisande* and to Dukas, it has been highly praised for its intelligent and dramatic handling of the subject.
WORKLIST: *Macbeth* (Fleg, after Shakespeare; Paris 1910).

**Blockx, Jan** (*b* Antwerp, 25 Jan. 1851; *d* Kapellenbos, 26 May 1912). Belgian composer. Studied chiefly with Benoit and in Leipzig with Reinecke. The most gifted Belgian composer of his day, he wrote seven operas which did much to establish a native operatic tradition. Of them, *Herbergprinses* was the most significant and successful. This work, *De Bruid der Zee*, and *Baldie* constitute a trilogy of works on Flemish national life. Blockx's operas reveal his Romantic tendencies and his feeling for native folklore.
WORKLIST (all prod. Antwerp unless otherwise stated): *Iets vergeten* (To Forget Something) (Mon-

tagne; 1877); *Maître Martin* (Landoy; Brussels 1892); *Herbergprinses* (The Princess of the Inn) (Tière; 1896); *Thyl Uilenspiegel* (Cain & Solvay; Brussels 1900); *De Bruid der Zee* (The Bride of the Sea) (Tière; 1901); *De Kapel* (The Chapel) (Tière; 1903); *Baldie* (Tière; 1908; rev. as *Liefdelied*, 1912).

**Blodek, Vilém** (*b* Prague, 3 Oct. 1834; *d* Prague, 1 May 1874). Czech composer. Studied Prague, including with Dreyschock and Kittl; he also studied the flute with Eiser, becoming an influential flautist and teacher. Of his operas, the most important is *V studni*. Blodek's knowledge of *The Bartered Bride*, and his use of that work's librettist, show in this amiable 1-act village comedy, which in the 19th cent. came close to rivalling its popularity.

WORKLIST: *Clarissa* (1861; unfin.); *V studni* (In the Well) (Sabina; Prague 1867); *Zitek* (Sabina; unfin., comp. 1869, prod. Prague 1934).

BIBL: R. Budiš, *Vilém Blodek* (Prague, 1964).

**Blomdahl, Karl-Birger** (*b* Växjö, 19 Oct. 1916; *d* Kungsängen, 14 June 1968). Swedish composer. Studied Stockholm with Rosenberg. His most important opera, and his best-known work, is *Aniara*. Its novel techniques, which include a pioneering use of electronic music, aroused much interest when it first appeared, as did its wideranging eclecticism of style. The work also impressed international audiences for its exploration of serious issues, using a group of people in a doomed spaceship both as setting and metaphor.

SELECT WORKLIST: *Aniara* (Lindegren, after Martinson; Stockholm 1959).

**Blonde.** Constanze's maid (sop), sometimes known as Blondchen, in Mozart's *Die Entführung aus dem Serail*.

**Blow, John** (*b* Newark, *bapt.* 23 Feb. 1649; *d* London, 1 Oct. 1708). English composer. *Venus and Adonis* (*c.*1682), his only work for the stage, was composed as part of his duties as court composer to Charles II. Described by him as a 'masque for the entertainment of the King', *Venus and Adonis* is a miniature opera in a prologue and three acts. The King's mistress and her daughter took part in the first performance. Displaying some French influence in the forms of the dances and a few Italianate elements, it stands as an important predecessor to Purcell's *Dido and Aeneas*.

**Bluebeard.** A character in one of Perrault's *Contes de Ma Mère L'Oye* (Mother Goose's Tales, 1697) who attempted to kill his latest wife Fatima for unlocking the forbidden door behind which lay the bodies of his murdered former wives. Grétry's *Raoul Barbe-bleue* (1789) was the first of a number of Bluebeard operas, some of which

give the story new interpretations. They include Gläser (*Blaubart in Wien*, 1829), Rietz (1837), Nesvadba (1844), Offenbach's *Barbe-bleue* (1866), Thomé (*Barbe-Bleuette*, 1889), Dukas (*Ariane et Barbe-bleue*, 1907), Satso (1908), Bartók (*Duke Bluebeard's Castle*, 1918), Rezniček (*Ritter Blaubart*, 1920), and Frazzi (*L'ottava moglie di Barbablù*, 1940). According to some versions of the legend, Mélisande was one of Bluebeard's escaped wives, hence her traumatic state at the beginning of Debussy's opera. Limnander's *Château de la Barbe-bleue* (1851) is not connected with the legend.

**Bluebeard's Castle.** See DUKE BLUEBEARD'S CASTLE.

**Boatswain's Mate, The.** Opera in 1 act by Ethel Smyth; text by the composer, after W. W. Jacobs's story. Prod. London, Shaftesbury T, 28 Jan. 1916, with Buckman, Roy, Pounds, Wynn, Ranalow, cond. Smyth.

Harry Benn (ten), a former boatswain, hopes to persuade the landlady of The Beehive, Mrs Waters (sop), to marry him. He arranges that his friend Travers (bs) will pretend to attack Mrs Waters in order that Harry can rescue her. When Travers enters the house, the landlady is not frightened in the least; she takes up a gun, and locks him in a cupboard. A series of incidents then leads her to abandon Harry and marry Travers instead.

**bocca chiusa** (It.: 'closed mouth'; Fr., *bouche fermée*; Ger., *Brummstimmen*). Singing without words and with closed mouth, i.e. humming. Employed for teaching purposes, to encourage the pupil to produce tone while conserving the breath; Caruso used to study his roles by humming them before using the voice proper. A variation of the technique, in which the teeth are closed while the mouth is left open, is sometimes employed to project the sound more forcibly. The most famous uses of bocca chiusa in opera occur with the Humming Chorus in Act II of *Madama Butterfly* and in the last act of *Rigoletto*.

**Bockelmann, Rudolf** (*b* Bodenteich, 2 Apr. 1892; *d* Dresden, 9 Oct. 1958). German bassbaritone. Studied Celle, and Leipzig with Lassner and Scheidemantel. Début Leipzig 1921 (Herald, *Lohengrin*). Hamburg 1926–32; Berlin, S, 1932–45. Bayreuth 1928–42. London, CG, 1929–30, 1934–8. Chicago 1930–2. One of the finest singers of Sachs and Wotan of his generation, with a warm voice and an elegant sense of phrasing. His Nazi sympathies prevented him from resuming a full career after 1945. (R)

**Bockstriller.** See BLEAT.

**Bodanzky, Artur** (*b* Vienna, 16 Dec. 1877; *d* New York, 23 Nov. 1939). Austrian conductor.

Studied Vienna, later with Zemlinsky. Début České Budějovice 1900. Vienna 1903–4 as Mahler's assistant; Berlin, Prague, Mannheim. Cond. first stage *Parsifal* in England, London, CG, 1914. New York, M, 1915–39 (except 1928–9). A conductor of wide range and professionalism, he was known above all for his Wagner; however, he was as notorious for the cuts he made in Wagner as he was for the versions he prepared, with unsuitable additions, of operas including *Oberon*, *Freischütz*, *The Magic Flute*, and *Fidelio*. (R)

**Bogenform** (Ger.: 'bow form'). A musical form derived from a popular verse pattern of medieval German poetry, usually designated A-B-A. One of the most common and well-tried forms: in opera it appeared early on as the *da capo aria. It was proposed by Lorenz, in *Das Geheimnis der Form bei Richard Wagner* (1924–33), as one of the important structural principles underlying Wagner's music dramas. See also BAR.

**Bogianckino, Massimo** (*b* Rome, 10 Nov. 1922). Italian pianist and administrator. Studied Rome with Rossi, Casella, and Mortari. After a short career as a pianist, taught in Pittsburgh and Pesaro. Artistic director Accademia Filarmonica, Rome, 1960–3; Paris, O, 1963–8; Spoleto 1969–72; Milan, S, 1972–5; Florence, C, 1975–81; Paris, O, 1982–5.

WRITINGS: *Aspetti del teatro musicale in Italia e Francia nell'età barocca* (Rome, 1968).

**Bohème, La** (Bohemian Life). Opera in 4 acts by Puccini; text by Giacosa and Illica, after Henri Murger's novel *Scènes de la vie de Bohème* (serially, 1847–9). Prod. Turin, R, 1 Feb. 1896, with Cesira Ferrani, Gorga, cond. Toscanini; Manchester 22 Apr. 1897, with Esty, Cunningham; Los Angeles 14 Oct. 1897, with Montanari, Agostini.

The Latin Quarter of Paris, c.1830. On Christmas Eve, Marcello (bar), a painter, and Rodolfo (ten), a poet, are joined in their garret by the philosopher Colline (bs) and the musician Schaunard (bar), who brings food. When the landlord Benoit (bs) arrives, his tenants ply him with drink and he leaves without his rent. Rodolfo tells his friends he will join them in the café when he has finished the article he is writing. He hears a knock at the door and Mimì (sop) asks for a light for her candle. She and Rodolfo fall in love.

The crowds gather at the Café Momus. Mimì, wearing a new bonnet Rodolfo has just bought her, is introduced to his friends. Musetta (sop) appears with her elderly admirer Alcindoro (bs). Seeing her ex-lover Marcello, she attracts his attention and leaves with him.

A few weeks later, Mimì is seeking Marcello; she and Rodolfo have quarrelled and she asks Marcello to advise them. She overhears Rodolfo saying that he can no longer live with Mimì; he also knows she is dying from consumption. The couple agree to part when spring comes.

Back in the garret, Rodolfo and Marcello contemplate the women they have left. Spirits rise when Colline and Schaunard enter, but shortly Musetta arrives to say that Mimì has left her rich protector and is dangerously ill. She wants to spend her last hours with Rodolfo. She is brought in and Marcello is sent to pawn Musetta's earrings for medicine. While Rodolfo is covering the window, his friends notice that Mimì has died. He thinks she is asleep but realizing the truth, throws himself sobbing over her body.

The story is also the subject of an opera in 4 acts by Leoncavallo; text by composer, after Murger. Prod. Venice, F, 6 May 1897, with Frandin, Storchio, cond. Pomè; NY, Columbia U, 31 Jan. 1960, with Lo Monaco, Simeone, Polakoff, Ottaviano, cond. Rubino; London, Camden TH, 12 May 1970, with Cryer, Morgan, Collins, Baldwin, Lyons, Truleaux, cond. Gover.

Leoncavallo's opera keeps more strictly to the events and in most ways to the characters and spirit of the novel.

Also opera by Hirschmann (*La petite Bohème*, 1905).

BIBL: A. Groos & R. Parker, *Giacomo Puccini: 'La Bohème'* (Cambridge, 1986).

**Bohemian Girl, The.** Opera in 3 acts by Balfe; text by Alfred Bunn, after the ballet-pantomime *The Gypsy* by Saint-Georges (1839), based in turn on Cervantes's novel *La gitanilla* (1614). Prod. London, DL, 27 Nov. 1843; New York, Park T., 25 Nov. 1844.

The opera tells of a Polish aristocrat, Thaddeus (ten), who is in political exile, and Arline (sop), daughter of Count Arnheim (bs); she was kidnapped and brought up by gypsies. Her father recognizes her and is so overjoyed at his discovery that he permits her marriage to Thaddeus.

The opera was extremely popular in England until about 1930, and the libretto translated into several languages. Revived by Beecham, London, CG, 1951.

**Böhm, Karl** (*b* Graz, 28 Aug. 1894; *d* Salzburg, 14 Aug. 1981). Austrian conductor. Studied first law, then music in Graz, and Vienna with Mandyczewski and Adler. Répétiteur Graz 1917, second conductor 1918, first conductor 1919. Début 1917 (*Trompeter von Säckingen*). Recommended by Karl Muck to Walter, who engaged him Munich 1921. Music director Darmstadt, 1927–31; Hamburg 1931–3; Dresden 1934–42, conducting premières of *Schweigsame Frau* (1935) and *Daphne* (1938). Music director Vienna, S, 1943–5, and of rebuilt house in 1954–6; resigned after criticism of his long absences abroad. Regularly Salzburg 1938–80; Bayreuth 1962–71. London, CG, 1936, 1954, 1977, 1979; Buenos Aires, C, 1950–3; New York, M, 1957–79. With Krauss, he was the leading Strauss conductor of his day, and also outstanding in Mozart and Berg: his clarity and sense of proportion served both composers well, but were also admirably effective in the whole range of the Austro-German tradition, which lay closest to his heart. That heart was only worn on his sleeve near the end of his career; earlier, his greatest performances seemed to represent the mellowing of vast experience and ruthless professionalism into a wise maturity. (R)

**Boieldieu, (François-)Adrien** (*b* Rouen, 16 Dec. 1775; *d* Jarcy, 8 Oct. 1834). French composer. Studied Rouen with Broche. He came early to know the operas of Grétry, Méhul, Le Sueur, and especially Dalayrac, and wrote his own first opera in 1793. Moving to Paris, he quickly established himself as a successful composer of opéras-comiques, in which melodic grace and fluency, together with simple but elegant harmony, won him a wide following. His subjects ranged wider than his essential style, though he was capable of responding to the stimulus both of extremes of emotional tension, as in *Béniowski*, and of exotic settings, as in *Le calife de Bagdad*. The popularity of the latter is said to have irritated Cherubini, who then gave the modest Boieldieu some further technical instruction.

Other successes followed, especially with *Ma tante Aurore* and *Aline, reine de Golconde*, before Boieldieu left to become court composer in St Petersburg. Here he wrote about an opera a year, also greatly improving operatic standards, before returning to Paris and new triumphs with *Jean de Paris* in 1812. The assurance, vigour, and charm of this work confirmed Boieldieu's wider reputation, with several German versions being produced in the same year: Weber claimed for Boieldieu's music universal significance, praising 'the freedom and elegance of his vocal line, the skilful construction of both individual numbers and the work as a whole, his careful and excellent use of the orchestra'. Boieldieu was appointed court composer in 1815, and in 1817 succeeded Méhul as professor of composition at the Académie des Beaux-Arts.

A deepening of style, and developing awareness of Romantic influences, show in his next work, *Le petit chaperon rouge*; but Boieldieu now retreated to his country house and for seven years wrote little. It was in part the stimulus of Rossini's influence on the French musical scene that led him to return to it in 1825 with his masterpiece, *La dame blanche*. This is a richer and subtler work than anything he had composed,

one of the most important French Romantic operas of the decade, skilfully exploiting the atmosphere of mystery and suspense and showing a new freedom of construction. As Wagner found, these virtues are sustained in his next work, *Les deux nuits*, whose chorus 'La belle nuit' was (Berlioz pointed out) a source for the *Lohengrin* bridal march. In both works, the traditions of earlier Romantic opera (e.g. Gaveston as a raging villain in the line of Dourlinski, Pizarro, and Caspar) are turned to new expressive effect, especially by means of an individual vein of chromatic harmony and skilful orchestration. With them, Boieldieu reaches the peak of his achievement. His health, affected by his Russian years, then deteriorated into tuberculosis, and the 1830 Revolution removed him from his official positions. An illegitimate son, **Adrien-Louis-Victor Boieldieu** (1816–83), studied with his father and had, with his completion of *Marguerite*, a success that was not maintained.

SELECT WORKLIST: *La fille coupable* (The Guilty Girl) (J.-F.-A. Boieldieu (father); Rouen 1793); *Le calife de Bagdad* (Saint-Just, after an oriental tale; Paris 1800); *Béniowski* (Duval, after Kotzebue; Paris 1800); *Ma tante Aurore* (Longchamps; Paris 1803); *Aline, reine de Golconde* (Vial & Favières; Paris 1804); *Télémaque* (Dercy; St Petersburg 1807); *Jean de Paris* (Saint-Just; Paris 1812); *Le nouveau seigneur de village* (Creuzé de Lesser & Favières; Paris 1813); *Le petit chaperon rouge* (Little Red Riding Hood) (Théaulon de Lambert; Paris 1818); *La dame blanche* (The White Lady) (Scribe; Paris 1825); *Les deux nuits* (The Two Nights) (Scribe, after Bouilly; Paris 1829); *Marguerite* (Scribe; sketch 1830).

BIBL: G. Favre, *Boieldieu, sa vie, son œuvre* (Paris, 1944–5).

**'Bois épais'.** Amadis's (ten) aria in Act II of Lully's *Amadis*, in which he declares that the sombre woods can never be as dark as his despair, since he will never again see his lover Oriane.

**Boito, Arrigo** (orig. Enrico) (*b* Padua, 24 Feb. 1842; *d* Milan, 10 June 1918). Italian composer, librettist (sometimes as 'Tobia Gorrio'), and critic. Son of an Italian painter and a Polish countess; studied Milan with Mazzucato, and there formed a lifelong friendship with Faccio. Together they went to Paris, where Boito first met Verdi, began to consider Faust and Nero as operatic subjects, and wrote the libretto for Faccio's *Amleto* (1862). Back in Milan, he associated himself with the artistic reform movement known as the *Scapigliatura, and served under Garibaldi. In 1868 *Mefistofele* was performed at Milan, S; it failed, partly through its length and the inadequacy of the performance (to which Boito's own conducting contributed). Depressed, Boito turned to writing librettos for other composers and to translating operas (incl. *Der Frei-*

*schütz, Rienzi*, and *Ruslan and Lyudmila*). A revised version of *Mefistofele* was successful in 1875; further revisions were made and performed in 1876 and 1881.

The revision of *Simon Boccanegra* for revival in 1881 pleased the hitherto hostile Verdi, and led to work on *Otello* and hence to the friendship that Boito regarded as the most important event of his life. After *Falstaff* had followed, he began work on *King Lear* and even wrote part of the opening scene. Verdi had meanwhile urged Boito to continue his own musical career with *Nerone*; but he never completed it, and it was eventually posthumously performed in a version edited by Toscanini and Vincenzo Tommasini.

Boito's intelligence and skill as a librettist have been widely praised. He served many composers excellently, and furnished Verdi with the brilliant (though by no means flawless) text for *Otello* and the incomparable one for *Falstaff*: the occasional misjudgements in the tragedy, such as the distortion of the character of Iago, are absent from the comedy, in which Boito's verbal brilliance (he took delight in complex rhyme and arcane vocabulary) are supreme. His texts for his own two operas are ambitious, too much so for his own musical powers and his creative resolve, and the many revisions and procrastinations reveal his difficulty in realizing his lifelong vision of the conflicting attractions of virtue and evil: his greatest fascination was always for the figure of magnificent evil. His own music is capable of grandeur, but it is a quality he has difficulty in sustaining, and the idiom frequently degenerates into mechanical gestures. As a critic, he was a keen but discriminating Wagnerian, latterly unenthusiastic, and a champion of Meyerbeer and Verdi. His most celebrated non-Verdian libretto was for Ponchielli's *La Gioconda*.

WORKLIST: *Mefistofele* (Boito; Milan 1868); *Nerone* (Boito; comp. 1877–1915; ed. Toscanini & Tommasini, prod. Milan 1924).

OTHER LIBRETTOS: *Amleto* (Faccio, 1865); *La falce* (Catalani, 1875); *La Gioconda* (Ponchielli, 1876); *Ero e Leandro* (Bottesini, 1879; Mancinelli, 1897); *Simon Boccanegra* (Piave rev. for Verdi, 1881); *Otello* (Verdi, 1887); *Falstaff* (Verdi, 1893); *Basi e bote* (Pick-Mangiagalli, comp. 1919–20, prod. 1927); also others not composed. Translations include *Der Freischütz* for the 1872 Italian première.

WRITINGS: P. Nardi, ed., *Tutti gli scritti di Arrigo Boito* (Milan, 1942) (incl. all librettos).

BIBL: G. Mariani, *Arrigo Boito* (Parma, 1973).

**Bologna.** City in Emilia-Romagna, Italy. The first opera given was Girolamo Giacobbi's *Andromeda* at the T Pubblico in 1610; burnt down 1623. Opera was then given at the T Formigliari (1636–1802) and the T Malvezzi (1653–1745), and opera buffa at the T Marsigli (1710–1825).

When the T Malvezzi burnt down in 1745, the T Comunale was designed by Antonio Galli-Bibiena and opened in 1763 with Gluck directing his *Trionfo di Clelia*. Under Angelo Mariani, 1859–72, Martucci, and Toscanini, it championed Wagner, giving the Italian premières of *Lohengrin* (1871), *Tannhäuser* (1872), *Fliegender Holländer* (1877), and *Tristan* (1888), and also of Verdi's *Don Carlos* (1867). It has continued to sustain an adventurous policy and maintain high standards.

**Bolshoy Theatre.** See MOSCOW.

**Bonci, Alessandro** (*b* Cesena, 10 Feb. 1870; *d* Viserba, 9 Aug. 1940). Italian tenor. Studied Pesaro with Pedrotti, Paris with Delle Sedie. Début Parma 1896 (Fenton). Milan, S, 1897; London, CG, 1900; New York, M, 1907–10; Chicago 1919–21; Rome; St Petersburg; Berlin; Madrid. Retired 1925. With his agile, exceptionally beautiful voice and fine phrasing, a great rival of Caruso; repertory incl. Paolino (*Matrimonio segreto*), Arturo, Alfredo, Faust (Gounod), Rodolfo. (R)

**Bondeville, Emmanuel** (*b* Rouen, 29 Oct. 1898; *d* Paris, 26 Nov. 1987). French composer. Studied Rouen with Haelling and later Paris with Déré, also being advised by Dupré. His official posts included those of director of the Opéra-Comique (1948–51) and of the Opéra (1951–9). His shrewd sense of the theatre, his elegant vocal writing, and his intelligent appreciation of the operatic possibilities in works by authors as distinguished and contrasted as Molière, Flaubert, and Shakespeare, have won his operas much appreciation in France.

SELECT WORKLIST: *L'école des maris* (School for Husbands) (Laurent, after Molière; Paris 1935); *Madame Bovary* (Fauchois, after Flaubert; Paris 1951); *Antoine et Cléopâtre* (after Hugo's trans. of Shakespeare; Rouen 1974).

BIBL: A. Machabey, *Portraits de trente musiciens français* (Paris, 1949).

**Bondini, Pasquale** (*b* ?Bonn, ?1737; *d* Bruneck, 30 or 31 Oct. 1789). Italian bass and impresario. First mentioned Prague 1762–3 as buffo bass; later joined Bustelli company. Director Dresden 1777, also Leipzig 1778, presenting operas by most major Italian composers of the time. Director Prague, Italian O, 1781–8. *Figaro* was first performed there with tremendous success, 1786; as a result, Bondini requested Mozart's next opera, *Don Giovanni*, and produced it 1787. His sister-in-law Teresa Saporiti sang Donna Anna and his wife Caterina Bondini-Saporiti, Zerlina. Their daughter, the soprano **Marianna Bondini** (1780–1813), studied Bologna with Sartorini. Début Paris 1807. Sang Susanna there in the first performance in Italian of *Figaro* the same year. Married the bass **Luigi Barilli** (1767–1824), who sang with her in Paris and was manager of the Italian Co. at the Odéon.

**Boniventi, Giuseppe** (*b* Venice, ?*c*.1670; *d* ?Venice, after 1627). Italian composer. Pupil of Legrenzi; served as maestro di cappella to the Duke of Mantua from *c*.1702 and as Kapellmeister to the Margrave of Baden-Durlach from 1712. Composed 13 operas, mostly for Venice, where the first, *Il gran Macedone*, was performed in 1690. At Durlach directed operas by many composers; also revived his own *Armida al campo* in 1711. Returned to Italy *c*.1717. Almost nothing known of later life except that he wrote three operas: *Il Venceslao, Filippo re di Macedonia*, and *L'inganno fortunato. Armida* was also revived in Venice in 1742.

SELECT WORKLIST: *Il gran Macedone* (Pancieri; Venice 1690); *Armida al campo* (Silvani; Venice 1707); *Il Venceslao* (Zeno; Turin 1720); *Filippo re di Macedonia* (Lalli; Venice 1721; Act III by Vivaldi); *L'inganno fortunato* (Pavieri; after Sara; Venice 1721).

**Bonn.** City in Westphalia; capital of West Germany, 1945–90. Travelling companies performed from the late 17th cent. In 1745 Pietro Mingotti's troupe came under Locatelli, then Angelo Mingotti's 1757 and 1764. Beethoven's father organized Italian performances 1767–71, including *La serva padrona*, he and his own father taking part. The theatre opened 1778, with Neefe working there from 1779, and the repertory included opera buffa and opéra-comique as well as Singspiel. The Oper der Stadt Bonn (cap. Grosses Haus 896, Werkstättbühne 180) opened 1965 and the opera company has had sole occupation since 1986.

**Bonno, Giuseppe** (*b* Vienna, 29 Jan. 1711; *d* Vienna, 15 Apr. 1788). Austrian composer of Italian origin. Studied in Naples with Leo and Durante; returned to Vienna in 1736, where he became involved in court musical life, writing several dramatic works for imperial occasions, including *Trajano* for the birthday of Charles VI. Appointed Hofkomponist 1739; in 1741 became Kapellmeister to the Prince of Sachsen-Hildburghausen. On the death of Gassmann in 1774 he succeeded as Kapellmeister at the imperial court, where he was later assisted by Salieri. He took an active part in Viennese musical life, conducting performances by the Tonkünstlersocietät; he was a noted teacher of singing and composition, his pupils including Karl Friberth and Dittersdorf.

Bonno wrote about 30 operas, whose style was much influenced by his study in Italy and by the models of Fux and Caldara, who had dominated opera in Vienna in the previous generation. Given Metastasio's presence as *poeta cesareo it is not surprising that he, together with Pasquini, provided the librettos for most of Bonno's operas, including *Il re pastore, Didone abbandonata*, and *Il sogno di Scipione*. Though his music was soon eclipsed by a younger generation, Bonno occupied an important place in Viennese operatic history, and his occasional rudimentary use of opera buffa techniques in the framework of opera seria is noteworthy.

SELECT WORKLIST (all first prod. Vienna): *Trajano* (Pasquini; 1736); *Il re pastore* (Metastasio; 1751); *Didone abbandonata* (Metastasio; 1752); *Il sogno di Scipione* (Metastasio; 1763).

**Bononcini, Giovanni** (*b* Modena, 18 July 1670; *d* Vienna, 9 July 1747). Italian composer. Son of the church musician and theorist **Giovanni Maria Bononcini** (1642–78); studied and worked in Bologna before joining service of Filippo Colonna in Rome in 1692. His first important stage work was the pasticcio *Eraclea*, which contained about 20 of his arias, and for which he collaborated with the librettist Silvio Stampiglia. During his years in Rome he wrote five operas with Stampiglia, including *Serse*. In 1697 he joined the Viennese court, serving both Leopold I and Joseph I. In addition to operas, he composed many serenatas and pasticcios. With such works as *Euleo, Etearco*, and *Abdolomino* Bononcini became famous not only in Vienna, but throughout Europe. In 1711 he joined the entourage of the Ambassador to Rome, for whom he wrote *Astarto*, but in 1719 was engaged for the newly founded Royal Academy of Music in London.

Bononcini arrived at the peak of his fame, to join Handel and Amadei in a valiant attempt to establish Italian opera in London. From the beginning there was intense rivalry between the composers, fuelled rather than dampened by their collaboration on *Muzio Scevola*, for which each contributed one act. Although Handel already had a strong following, Bononcini initially eclipsed him in popularity: in the 1720–1 and 1721–2 seasons Bononcini had 63 performances against Handel's 28. When Ariosti joined the Academy in 1723 the Bononcini/Handel rivalry was at its height, but though the simpler style of such works as *L'odio e l'amore, Griselda*, and *Farnace* had initially made a greater impact on the audiences, Handel was already beginning to gain the upper hand. After 1724 Bononcini concentrated on a new post, organizing private concerts for the Duchess of Marlborough, although he composed a final opera for London, *Astianatte*. At the first performance the two prima donnas, Cuzzoni and Faustina, engaged in a brawl on stage, which fired their supporters in the audience to 'Catcalls, and other great Indecencies' (*British Journal*). Bononcini finally left

London in disgrace in 1731, having laid false claim to a composition of Lotti's, although the success of ballad opera had already destroyed the aspirations of the Academy. After travelling to Paris and Lisbon, he returned to Vienna, where his last operas were produced, including *Alessandro in Sidone* and *Zenobia*.

Bononcini's achievements have been eclipsed by those of his rival Handel. Though his operas appear to lack dramatic involvement and depth, this is balanced by a gentler, pastoral mood and more tuneful melodies, which perhaps reflect more accurately the general taste in opera in the early 18th cent.

SELECT WORKLIST: *Serse* (Stampiglia; Rome 1694); *Euleo* (?; Vienna 1699); *Etearco* (Stampiglia; Vienna 1707); *Abdolomino* (Stampiglia; Vienna 1709); *Astarto* (Rolli, after Zeno & Pariati; Rome 1715); *Muzio Scevola* (Rolli; 1721; comp. with Amadei & Handel); *L'odio e l'amore* (Rolli; London 1721); *Griselda* (Rolli; London 1722); *Farnace* (?; London 1723); *Astianatte* (?Haym, after ?Salvi; London 1727); *Alessandro in Sidone* (?after Zeno & Pariati; Vienna 1737); *Zenobia* (Metastasio; Vienna 1737).

His brother **Antonio Maria Bononcini** (1677–1726) was also at the Austrian court from 1704–11, having already gained a considerable reputation with his opera *Il trionfo di Camilla*, which was revived throughout Europe several times in the decades following. Returning to Italy he was impresario in Modena and from 1721 maestro di cappella at the court: among later operas are *Astianatte*, *Griselda*, and *Rosiclea in Dania*.

SELECT WORKLIST: *Il trionfo di Camilla* (?; Naples 1696); *Astianatte* (Salvi; Venice 1718); *Griselda* (Zeno; Milan 1718); *Rosiclea in Dania* (?; ? 1721).

**Bontempi, Giovanni** (*b* Perugia, *c*.1624; *d* Brufa, 1 July 1705). Italian composer, singer, and writer on music. Studied Rome with Mazzocchi; sang at St Mark's, Venice 1643–50. In 1650 joined the Dresden court, becoming joint Kapellmeister with Schütz 1656. In later life rejected music in favour of other activities, notably architecture; stage designer for the Dresden opera 1664. His *Historia musica*, the first musical history in Italian, appeared in 1695.

Remembered for *Il Paride*, probably the first Italian opera staged in N Germany, and for *Dafne*, the first German opera to survive in full score. These follow the general model of Mazzocchi; *Paride*, described by Bontempi as an 'erotopegno musicale' (pledge of love), was given to celebrate the marriage of the Princess of Saxony and is one of the most lavish of all Baroque operas.

SELECT WORKLIST: *Il Paride* (Bontempi; Dresden 1662); *Dafne* (Bontempi, after Opitz, after Rinuccini; comp. with Peranda; Dresden 1671).

BIBL: F. Briganti, *Giovanni Andrea Angelini-Bontempi (1624–1705), musicista, letterato, architetto: Perugia-Dresden* (Florence, 1956).

**Bonynge, Richard** (*b* Epping, NSW, 29 Sep. 1930). Australian conductor. Début Vancouver 1963 (Gounod's *Faust*). With his wife Joan *Sutherland as prima donna, has conducted in most leading opera-houses. Music director Vancouver O 1974–8, Australian O 1975–86. A specialist in the bel canto repertory, particularly Bellini, Donizetti; also conducts Handel, Verdi, Massenet, Gounod. (R)

**Bordeaux.** City in Gironde, France. Opera was first given in 1688, and French singers began making regular appearances soon after. Francœur's *Pirame et Tisbé* was staged 1729. Mme Dujardin built an opera-house for her company in the Jardin Municipal; burnt down 1756. In 1752 there was founded an Académie Royale de Musique, with a company including ten soloists. The Grand T Municipal, one of the finest opera-houses in France, opened 1780 (cap. 1,158); damaged in the Revolution, restored 1799. The theatre increased its reputation in the 19th cent., with a number of premières and visits from Parisian companies. It was rebuilt 1919, 1938, and 1977 (cap. 1,205), and in the 1930s strengthened its international reputation with a repertory of new works and with distinguished guest artists. Vanni Marcoux was director 1948–51, Roger Lalande 1954–70, developing an attention to enterprising productions and giving many premières and French premières of foreign works. He was succeeded by Gérard Boireau, who concentrated more on a 19th-cent. repertory, while continuing to attract international artists. The Mai Musical de Bordeaux, founded 1950, has taken more interest in novelties.

BIBL: J. Lutreyte, *Le grand théâtre de Bordeaux* (Bordeaux, 1977).

**Bordogni, Giulio Marco** (*b* Gazzaniga, 23 Jan. 1789; *d* Paris, 31 July 1856). Italian tenor. Studied with Mayr. Début *c*.1808. Milan, T Re, 1813 (Tancredi); other major Italian houses including Milan, S, until 1819, when he was engaged by Paris, I, remaining 14 years. Here he sang in the French premières of Paer's *Agnese*, Mercadante's *Elisa e Claudio*, and Rossini's *L'inganno felice*, *Otello*, *La gazza ladra*, *Cenerentola*, *Mosè*, *La donna del lago*, and *Semiramide*, among others. Though his voice was not remarkable, he became famous particularly for his refined style and technical control. Appointed professor of singing at the Paris Cons. in 1820 by Cherubini. His many pupils included Cinti-Damoreau, Sontag, and Mario.

**Bordoni, Faustina** (*b* Venice, 1700; *d* Venice, 4 Nov. 1781). Italian soprano (though Quantz describes her as a mezzo-soprano). Studied with M. Gasparini. Début Venice 1716 (Pollarolo's *Ariodante*). Known as 'the New Siren', she soon became famous throughout Italy, and also in Munich and Vienna (1725–6), where she was much admired by Fux and the Empress. In 1726 Handel brought her to London, where she scored a huge success, not only for her exquisite singing and acting, but also for her beauty, intelligence, and courtesy. The latter quality did not extend to her great rival *Cuzzoni, with whom she came to blows during a performance of *Astianatte* (*Bononcini) in 1727. (See BEGGAR'S OPERA.) Faustina created several Handel roles: Rossane (*Alessandro*), Alcestis (*Admeto*), Pulcheria (*Riccardo Primo*), Emira (*Siroe*), and Elisa (*Tolomeo*). The Royal Academy's final season ended prematurely with her illness in 1728, when she returned to Italy. In 1730 she married *Hasse; they were described by Metastasio as 'a truly exquisite couple'. In 1731 he began to work at the Dresden court with her as prima donna: she sang in many of his operas and until her retirement from the stage in 1751 remained unchallenged except by Mingotti. From Dresden she also undertook several tours to Italy, and visited Paris in 1750, with great success. On leaving in 1763, she and Hasse lived first at Vienna, then retired to Venice. One of the greatest singers of her time, she possessed a strong, agile voice, superb breath control, excellent diction, and an expressive dramatic talent well suited to contrasting roles. A fine portrait of her by Rosalba Carriera hangs in the Ca' Rezzonico, Venice. Subject of an opera, *Faustina Hasse*, by Louis Schubert (1879).

**Borgatti, Giuseppe** (*b* Cento, 17 Mar. 1871; *d* Reno, Lago Maggiore, 18 Oct. 1950). Italian tenor. Studied Bologna with Busi. Début Castelfranco Veneto 1892 (Faust). Created Andrea Chénier at short notice, Milan, S, 1896; thereafter sang with success in Italy, Spain, and Buenos Aires as Cavaradossi, Don José, Canio, Puccini's Des Grieux, etc. In 1899 sang Siegfried at Milan, S, under Toscanini, deeply impressing Richter. Italy's finest Heldentenor, he became particularly identified with the Wagner roles to which his virile, resonant voice, fine physique, and dramatic insight brought a truly heroic quality. Forced by blindness to retire in 1914.

**Borgioli, Dino** (*b* Florence, 15 Feb. 1891; *d* Florence, 12 Sep. 1960). Italian tenor. Studied Florence with Giacchetti. Début Milan, V, 1917 (*La favorita*). Milan, S, from 1918, and subsequently most of the major opera-houses, as Rodolfo, Des Grieux (Puccini), Arturo. London, CG, 1925–39 (Edgardo, Rossini's Count Almaviva, Duke of Mantua, Don Ramiro), and Gly. 1937–9 (Don Ottavio, Ernesto), at both of which he was very popular. New York, M, 1934–5. Artistic director of New London OC at Cambridge T 1949. One of the best light lyric tenors of his day, with a clear, free voice and elegant phrasing. (R)

**Bori, Lucrezia** (*b* Valencia, 24 Dec. 1887; *d* New York, 14 May 1960). Spanish soprano. Studied Milan with Vidal. Début Rome, Ad., 1908 (Micaëla). Paris, C, 1910 (Manon); first Italian Octavian, Milan 1911. New York, M, 1912–36, except 1915–20 when she withdrew with throat trouble. Also sang in Buenos Aires, C, and Monte Carlo. Her delicate, transparent tone and natural expressiveness particularly suited vulnerable heroines such as Mélisande, Mimì, Butterfly, and Violetta, though she was also a successful Norina. Created The Duchess of Towers in Deems Taylor's *Peter Ibbetson*. (R)

**Boris.** Boris Grigorievich (ten), nephew of Dikoj and lover of Káťa in Janáček's *Káťa Kabanová*. See also below.

**Boris Godunov.** Opera in 4 acts, with a prologue, by Musorgsky; text by the composer, after Pushkin's drama *The Comedy of the Distress of the Muscovite State, of Tsar Boris, and of Grishka Otrepiev* (1825) and Karamzin's *History of the Russian Empire* (1829). Orig. version comp. 1868–9, rejected by St Petersburg O 1870. Three scenes of the composer's revision (1871) prod. St Petersburg, charity perf., 17 Feb. 1873, with Petrov (as Varlaam), Komisarzhevsky, Platonova; whole opera (with further minor revisions) 8 Feb. 1874, with Melnikov as Boris and same cast, cond. Nápravník. Withdrawn after 25 performances, revived after composer's death in Rimsky-Korsakov's first rev. edn., St Petersburg 10 Dec. 1896. Rimsky-Korsakov's second version (1906–8), prod. Paris, O, 19 May 1908, with Shalyapin; Moscow, B, 17 Oct. 1908; New York, M, 19 Mar. 1913, with Didur, cond. Toscanini; London, DL, 24 June 1913, with Shalyapin. Orig. version first performed abroad London, DL, 30 Sep. 1935, with Stear. This has seven scenes—four from Pushkin, two devised by Musorgsky from indications in the play, one developed from two separate scenes in it. Other versions by Meligailis (Riga 1924) and Shostakovich (Leningrad 4 Nov. 1959).

Russia and Poland, 1598–1605 Scene 1: Boris (bs) has murdered the young Dmitry, heir to the throne, and is pretending to decline the crown himself. His agents incite the crowd to persuade him to 'relent'. Scene 2: Though plagued by guilt, Boris goes to be crowned. Scene 3: In his cell the old monk Pimen (bs) is concluding a history of Russia; with him is the novice Grigory (ten), who resolves to avenge Dmitry. Scene 4:

Grigory, who claims to be Dmitry, with two other friars Varlaam (bs) and Missail (ten), reaches an inn on the Lithuanian border. Grigory is identified but escapes the pursuing soldiers. Scene 5: To Boris in his rooms, word is brought by his councillor Shuisky (ten) of the pretender in Lithuania. To reassure him of the death of Dmitry, Shuisky recounts the murder, which throws the guilt-racked Tsar into a state of hallucination. Scene 6: Outside St Basil's Cathedral in Moscow the people begin to believe in the right of the false Dmitry. A simpleton (ten), robbed of a kopek, asks Boris to repeat his murder on the robbers; Boris prevents his arrest, and asks for prayer, but the simpleton refuses and falls to mending his shoes with a song for poor Russia. Scene 7: In the Council Hall an edict against the pretender is being read when Boris breaks in, distraught, and collapses dead.

In his second version Musorgsky made considerable additions and alterations, and arranged the scenes as follows:

Prologue: Scenes 1 and 2. Act I: Scenes 3 and 4. Act II: Scene 5. Act III (recast): Scene 1—Dmitry's lover, Marina (mez), sits in her father's Polish castle dreaming of when she will rule Russia, and Rangoni (bs) pleads the Catholic cause; Scene 2—Marina joins Dmitry in the gardens, and she persuades him not to give up his ambitions. Act IV: Scene 7, and an extra scene in the Kromy Forest, in which the people are in revolt against Boris; the pretender passes *en route* for Moscow, leaving the Simpleton singing sadly for Russia as in the now discarded Scene 6. The order of these last two scenes is frequently reversed in order to allow the protagonist the final curtain; the counter-argument holds the protagonist to be, in fact, the Russian people. A revival at London, CG, 1958, with Christoff, used all these scenes, as did that at New York, M, in 1957. Both reverted to the original orchestration.

Also opera by Mattheson (1710).

BIBL: D. Lloyd-Jones, *Boris Godunov* (performing edn., trans., & introd., Oxford, 1975).

**Borkh, Inge** (*b* Mannheim, 26 May 1921). Swiss soprano. After beginning her career as an actress, studied Milan with Muratti and at Mozarteum, Salzburg. Début Lucerne 1941. Sang in Switzerland until 1951, when her sensational Magda Sorel in Basle led to engagements in Vienna, Berlin, and London. San Francisco 1953 (Elektra), and 1955 (Lady Macbeth); New York, M, 1958; London, CG, 1959 and 1967 (Salome and Dyer's Wife); also Salzburg; Milan, S; and Bayreuth. An intense and intelligent dramatic singer. (R)

**Borodin, Alexander** (*b* St Petersburg, 12 Nov. 1833; *d* St Petersburg, 27 Feb. 1887). Russian composer. Studied music privately while a child and at the Medico-Surgical Academy, continuing to benefit from advice and guidance, including from Musorgsky, while pursuing his career as a chemist. Only one of his operatic projects was completed, a pastiche *The Bogatyrs*. Material from sketches for *The Tsar's Bride* was used in other works; his Act IV of the collective project *Mlada* includes material from his unfinished masterpiece, *Prince Igor*. This was, after 18 years of work, left unfinished and mostly unscored, in part through Borodin's self-imposed difficulties in trying to prepare his own libretto. The opera was eventually prepared for publication by Rimsky-Korsakov and Glazunov. Despite their labours, its magnificent music remains expressed in a series of tableaux rather than in an evolving drama.

WORKLIST: *Bogatyry* (The Bogatyrs) (Krylov; Moscow 1867); *Mlada* (Krylov, after Gedeonov; Act IV of collective opera-ballet, St Petersburg 1892); *Knyaz Igor* (Prince Igor) (Borodin, after scenario by Stasov; comp. 1869–70, 1874–87, unfin., prod. St Petersburg 1890).

BIBL: S. Dianin, *Borodin* (Moscow, 1955, 2/1960; rev. trans. London, 1963).

**Boronat, Olimpia** (*b* Genoa, 1867; *d* Warsaw, 1934). Italian soprano, of Spanish birth. Studied Milan with Leoni. Début Genoa or Naples, *c.*1885. Appearances in Spain, Portugal, and S America. St Petersburg, Imperial O, 1890–93, leaving to marry a Polish aristocrat. Resumed her career in Russia and Poland 1901–14, returning to Italy 1909. Opened a singing school in Warsaw, where she gave her last concert in 1922. Her voice was of an exceptional clarity and beauty, and her coloratura singing celebrated in the roles of Rosina, Elvira, and Violetta. (R)

**Borosini, Francesco** (*b* Modena, *c.*1690; *d* ?). Italian tenor. Son of the tenor Antonio Borosini, with whom he studied. Début prob. Venice 1708 (Lotti's *Il vincitor generoso*). Sang at Vienna imperial court 1712–31, in oratorio, and operas by Fux. Prague 1723 in Fux's *Costanza e Fortezza*. In 1719 he sang the title-role in F. Gasparini's *Bajazet* at Reggio, having with the poet Zanella rewritten the final scene; in London, 1724, he was the first Bajazet in Handel's *Tamerlano*, for which the composer used this revised libretto in his own revision before the first performance. Borosini also sang Sextus (*Giulio Cesare*), rewritten and enlarged for him by Handel; created Grimoaldo (*Rodelinda*), and sang in operas by Ariosti. Returned to London 1747 in works by Paradies and Terradellas. The agility and vivacity of his singing, remarked on by Quantz, were matched by his dramatic talent. He was the first important Italian tenor to sing in England.

**Bortnyansky, Dmitry** (*b* Glukhov, 1751; *d* St Petersburg, 10 Oct. 1825). Russian composer. Studied with Galuppi, whom he followed to Italy. Here he wrote his first operas, then returning with some singers to St Petersburg, where he became director of music to Paul I and turned to French opera. Though elegantly written, his operas lack any national colouring; he was concerned to learn from the best models rather than found more than the beginnings of an original Russian school. Better known for his sacred music.

> WORKLIST: *Creonte* (Coltellini; Venice 1776); *Quinto Fabio* (Zeno; Modena 1778); *Alcide* (Metastasio; Venice 1778); *Le faucon* (Lafermière; Gatchina Palace 1786); *La fête du seigneur* (Lafermière; Pavlovsk 1786); *Don Carlos* (Lafermière; St Petersburg 1786); *Le fils rival* (Lafermière; Pavlovsk 1787).

> BIBL: B. Dobrokhotov, *D. S. Bortyansky* (Moscow, 1950).

**Boschi, Giuseppe** (*fl.* 1698–1744). Italian bass. Venice 1707–9 and 1713–14 in operas by F. Gasparini, Lotti, Caldara, and Handel (*Agrippina*). London, H, 1710–11; début in *Idaspe fedele* (Mancini), and was the first Argante (*Rinaldo*). Dresden 1717–20. London, H, 1720–8, where he created roles in many Handel operas including *Floridante*, *Ottone*, *Tamerlano*, *Rodelinda*, *Scipione*, *Alessandro*, and *Siroë*. Venice 1728–9, in works by Porpora; then a member of the choir at St Mark's certainly until 1744. The power, range (G–g'), and unusual agility of his voice were inspirational: 'Handel's genius and fire never shine brighter than in the bass songs he composed for Boschi and Montagnana', wrote Burney. Bononcini also composed florid music for him in *Astarto* and *Etearco*, and Lotti in *Teofane*. His wife **Francesca Vanini** (*d* 1744) was an excellent contralto, for whom Handel wrote Otho (*Agrippina*) and Goffredo (*Rinaldo*). She was the first in London to sing an aria by Handel (interpolated into A. Scarlatti's *Pirro e Demetrio*) when she visited England 1710–11 with Boschi. Tosi, impressed by her exemplary singing, said that 'Women, who study, may instruct even Men of some Note.'

**Bostel, Luca von** (*b* Hamburg, 11 Oct. 1649; *d* Hamburg, 15 July 1716). German librettist. Trained as a lawyer, he was associated with the T am Gänsemarkt from 1679 and was its most important librettist in the early years. His works played a crucial role in the establishment of opera in Germany: they included *Crösus* (1684) and *Der unmöglichste Ding* (1684) for Förtsch, *Vespasianus* (1681) and *Der glückliche Gross-Vezier Cara Mustapha* (1686) for Franck, and *Theseus* (1683) for Strungk. Most of Bostel's work involved translating and adapting Italian, or French librettos, and

in style they were close to the prevailing Venetian model of Minato. Several were later set by Keiser and Schürmann.

**Boston.** City in Massachusetts, USA. The first opera given was Arne's *Love in a Village* in July 1769 (concert, owing to a 1750 ban on stage performances). The first staged work was a ballad opera, *The Farmer*, in July 1794. Opera in English was the rule until the 1820s, when touring companies began visiting. The Boston T opened 1854 and remained the city's opera-house for half a century. In 1855 Mario and Grisi appeared, and in 1860 Patti sang her first Rosina. The Mapleson Co. began visiting in 1878, and Grau's French Co. in 1879. During the 1880s and 1890s the Boston Ideal OC, the Theodore Thomas American Co., and the Castle Square Co. all gave seasons.

In Nov. 1909 the Boston OC, for which the Boston OH was built, began its short but spectacular career (516 perfs. of 51 operas) with *La Gioconda* (with Nordica). Under Henry Russell the company gave five seasons with Bori, Garden, Destinn, Leblanc, Dalmorès, Muratore, and Marcoux; Weingartner made his US début in 1912 (*Tristan*). After a disastrous visit to Paris in 1914, it went bankrupt. The Boston Grand Opera Co. gave two seasons 1915–16 with many of Russell's singers. Nine seasons were organized by the Boston O Association, which brought the Chicago O each spring; and the New York, M, gave annual seasons 1934–86. Local companies, performing at Boston OH (demolished 1958), included the Fleck Grand O, the Boston Civic O, and the Boston Grand O.

In 1946 Boris Goldovsky founded the New England OT (which gave the US première of a shortened *Les Troyens*, 1955). In 1958 the Boston O Group (OC of Boston, 1965) was established under Sarah Caldwell: the company has given the US premières of *Moses und Aron* (1966), *Benvenuto Cellini* (1975), and *The Ice Break* (1979). It has been based at the old Savoy T, now OH (cap. 2,605), since 1978.

**Bottarelli, Giovanni** (*fl.* 1740–80). Italian librettist. In Berlin he wrote librettos for two of Graun's operas, *Rodelinda* (1741) and *Cesare e Cleopatra* (1742): both based on Corneille, these reflect the prevailing French taste of Frederick the Great's court, while couched in a style appropriate to opera seria. Moving to England, he played an important part in the activities of the London, H, writing librettos for Sacchini (e.g. *Il Cid*, 1773 and *Montezuma*, 1775, the latter probably inspired by the Graun/Frederick the Great/Tagliazucchi collaboration of 1755), Johann Christian Bach (*La clemenza di Tito*, 1778), and others.

**Bottom.** The weaver (bs-bar), one of the 'rude mechanicals' and Tytania's illusory love, in Britten's *A Midsummer Night's Dream*.

**bouche fermée.** See BOCCA CHIUSA.

**Bouffons, Guerre des.** See GUERRE DES BOUFFONS.

**Bouffes-Parisiens.** Theatrical enterprise in Paris, opened 5 May 1855 by Offenbach at the Salle Marigny, when refused by other managements, with an entertainment including *Les deux aveugles*. He succeeded partly through the proximity of the Exposition Universelle; when this closed, he moved to a larger theatre (cap. 820), which he named Bouffes-Parisiens d'Hiver (later Bouffes-Parisiens), opening 29 Dec. 1855 with *Ba-ta-clan*. The home of light opera in general including works by Bizet, Lecocq, Chabrier, Audran, Messager, and Serpette.

**Boughton, Rutland** (*b* Aylesbury, 23 Jan. 1878; *d* London, 24 Jan. 1960). English composer. Self-taught. His attempt to establish an English school of Wagnerian music drama, drawing on Celtic and especially Arthurian myth, was pursued with remarkable single-mindedness; and although his small-scale (and partly amateur) 'Bayreuth' based on a commune at Glastonbury did not, after many efforts, establish itself permanently, he won wide popularity with *The Immortal Hour*. This had 216 consecutive performances at one revival. His music is lightly tuneful and fluently written.

SELECT WORKLIST: *The Birth of Arthur* (Buckley & Boughton; comp. 1908–9, Glastonbury 1920); *The Immortal Hour* (Macleod; Glastonbury 1914); *Alkestis* (Murray, after Euripides; Glastonbury 1922); *The Queen of Cornwall* (Hardy; Glastonbury 1926); *The Lily Maid* (Boughton; Stroud 1934).

BIBL: M. Hurd, *Immortal Hour* (London, 1962).

**Bouhy, Jacques** (*b* Pepinster, 18 June 1848; *d* Paris, 29 Jan. 1929). Belgian baritone. Studied Liège and Paris. Début Paris, O, 1871 (Gounod's Méphistophélès). Paris, OC, 1872–5, creating Escamillo in the first, disastrous *Carmen*; Paris, L, 1877; subsequently Paris, O, in *Hamlet*, *Don Giovanni*, and *La favorite*. St Petersburg 1880; London, CG, 1882. Director of New York Cons. 1885–9, and taught there 1904–7, then in Paris. Though not an outstanding singer, he brought a realistic style of characterization to his roles.

**Bouilly, Jean-Nicolas** (*b* La Coudraye, 23 Jan. 1763; *d* Paris, 14 Apr. 1842). French author and librettist. Member of the Committee for Public Instruction, friend of Mirabeau, and dedicated Jacobin. First emerged as a librettist with *Pierre le Grand* (1790) for Grétry, but his name is for ever associated with the text for Gaveaux's *Léonore, ou L'amour conjugal* (1798), which became after various vicissitudes the source of Beetho-

ven's *\*Fidelio*. He claimed that it was based on fact; his autobiography mentions some events and characters that may have provided stimulus. However, the great success in his own time was the text for Cherubini's *Les \*deux journées* (1800), which both Beethoven and Goethe regarded as the ideal opera libretto. It is a vivid and well-constructed work, fundamentally expressing Bouilly's political beliefs in natural virtue and the perfectibility of man. Other composers who made use of his dramas or librettos include Aimon, Auber, Berton, Boieldieu, Dalayrac, Gyrowetz, Isouard, Méhul, and Weigl. Autobiography, *Mes récapitulations* (3 vols., Paris, 1836–7).

**Boulevard Solitude.** Opera in 7 scenes by Henze; text by the composer and Grete Weil, after Walter Jockisch's play based on Prévost's novel *L'histoire du Chevalier des Grieux et de Manon Lescaut* (1731). Prod. Hanover, Landestheater, 17 Feb. 1952, with Clause, Zilliken, Buckow, cond. Schüler; London, SW, 25 June 1962, with Cantelo, Carolan, Glossop, cond. Lovett; Santa Fe 2 Aug. 1967, with Brooks, Driscoll Fortune, cond. Baustian. In this modern setting of the Manon Lescaut story, Des Grieux (ten) becomes a drug addict and Manon (sop) ends up wrongly accused of murder.

**Boulez, Pierre** (*b* Montbrison, 26 Mar. 1925). French composer and conductor. Studied Paris with Messiaen and Leibowitz. In 1948 acted as conductor and composer for Jean-Louis Barrault's company. Has composed some incidental music, including *Orestie* (Aeschylus, trans. Claudel; 1955), *Mon Faust* (Valéry; 1962) and *Ainsi parla Zarathoustra* (Barrault, after Nietzsche; 1974), but has never been attracted to opera: indeed his controversial utterances on the subject include a proposal that all opera-houses should be blown up. This has not prevented him from giving distinguished performances in them. Conducted first *Wozzeck* (Paris, O, 1963), and first complete *Lulu* (Paris, O, 1979). Engaged by Wieland Wagner in 1966 to conduct *Parsifal* at Bayreuth and *Wozzeck* in Frankfurt. In 1976 conducted Chéreau's centenary production of *The Ring* at Bayreuth. WNO 1992 (*Pelléas*). (R)

**Bowman, James** (*b* Oxford, 6 Nov. 1941). English counter-tenor. Studied London with De Rentz and Manén. Début Aldeburgh, EOG, 1967, Oberon (Britten). London, EOG from 1967; London, SW, and Gly. 1970; London, CG, 1972; Sydney 1978; Paris, OC, 1979, C, 1982; also Aix, Santa Fe, Dallas, Verona, Buxton, etc. An attractive, lively, and musical performer, distinguished in the second generation of counter-tenors after *\*Deller. Roles incl. Endymion, Tamerlano, Xerxes, Polinesso, Giulio Cesare,

Astron (*The Ice Break*); created Priest (Davies's *Taverner*), Apollo (*Death in Venice*). (R)

**box.** See PALCO.

**Braham, John** (*b* London, 20 Mar. 1774; *d* London, 17 Feb. 1856). English tenor. Studied London with Leoni. Concert début as treble, London, CG, 1787. Further study Bath with Rauzzini. London, DL, 1796 under Storace; then engaged by Italian O for *Zémire et Azor*. Successfully toured Continent 1798–1800 with Nancy Storace (by whom he had an illegitimate son in 1802). London, CG, from 1801, and then at the London, L and K, in contemporary English works (for which he often composed his own music) and in Italian opera, e.g. as Sesto in *La clemenza di Tito*. In 1824 he was the first English Max, and in 1826 created Huon in *Oberon*. By then he was losing the top of his voice, and caused Weber much trouble, rejecting 'From boyhood trained' and demanding the replacement 'Ah, 'tis a glorious sight'; he also made the composer write a preghiera, 'Ruler of this awful hour'. However, Weber, among others, was an admirer of the singer, who was the first great English tenor and considered without rival on the London Italian stage. His voice was strong and brilliant, ranging from A to e'', with skilful use of falsetto. Excellent in Handel and the Italian repertory, he was nevertheless capable of appalling taste and vulgar showmanship. The large fortune he amassed was lost in various unsuccessful ventures, forcing him to prolong his career; his voice having become lower, he sang Don Giovanni and William Tell at London, DL, 1838–9. A US tour, 1840–2, was unsuccessful. He sang in concerts in England until 1852. Also composed many popular songs, of which the most famous was *The Death of Nelson*.

**Braithwaite, Nicholas** (*b* London, 26 Aug. 1939). English conductor. Son of below. Studied London with Miles, Vienna with Swarowsky, and Bayreuth. Début WNO 1966 (*Don Pasquale*). London, SW, 1970, resident cond. 1971–2, associate principal cond. 1972–4. associate principal cond. ENO 1970–4; CG 1973; music director Gly. Touring O 1977–80. Repertory incl. Wagner, Verdi, Puccini, Bizet; also first British *Devils of Loudun*. (R)

**Braithwaite, Warwick** (*b* Dunedin, 9 Jan. 1898; *d* London, 18 Jan. 1971). New Zealand conductor. Father of above. Studied London, RAM. O'Mara OC 1919; Vic-Wells (later SW) Co. 1932–40, when he was responsible, with Lawrance Collingwood, for many important productions. London, CG, 1950–3; Artistic director Australian Nat. O 1954–5; music director WNO 1956–60; London, SW, 1960–8. Repertory incl.

*Fidelio, Don Carlos, Meistersinger, May Night, Mefistofele, Fledermaus.* (R)

**Brambilla.** Italian family of musicians.

1. **Paolo** (*b* Milan, 9 July 1787; *d* Milan, 1838) composed a number of operas, including *Il carnevale di Venezia* (Turin 1819). He had five children, all singers: Annibale, Ulisse, Emilia, Erminia, and Amalia, the best known:

2. **Amalia** (*b* Milan, 1811; *d* Castellamare di Stabia, Aug. 1880), soprano, whose repertory included Norma, Lucia, and Maria Padilla. She married the tenor Giambattista Verger; their son was the baritone Napoleone Verger. There were five cousins, all sisters and all singers:

3. **Marietta** (*b* Casano d'Adda, 7 June 1807; *d* Milan, 6 Nov. 1875), contralto. Studied Milan with Secchi; début London 1827 (Arsace), she singing other travesti roles. Milan, S, 1835–45, where she created Orsini (*Lucrezia Borgia*) and Vienna 1837–42, where she created Pierotto (*Linda di Chamounix*). Her range was g to g''.

4. **Teresa** (*b* Cassano d'Adda, 23 Oct. 1818; *d* Milan, 15 July 1895), soprano. Studied Milan, début there 1831. Created Gilda, and sang with success all over Italy and in Paris and St Petersburg. The three other sisters were Giuseppina, Annetta, and Laura. Their niece

5. **Teresina** (*b* Cassano d'Adda, 15 Apr. 1845; *d* Vercelli, 1 July 1921) was a soprano. Studied with her aunts Marietta and Teresa. Début Odessa 1863. In 1872 sang in opening season of Milan, V, in revised version of *I promessi sposi* by Ponchielli, whom she later married. Among her most famous roles were La Gioconda, Aida, and Elsa. Descendants of the family include Arturo, head of costumes at Milan, S, in the 1950s, and the film actor Tullio Carminato.

**Branchu, Alexandrine** (*b* San Domingo, 2 Nov. 1780; *d* Passy, 14 Oct. 1850). French soprano. Studied Paris with Garat. Début 1799, Paris, T Feydeau. Paris, O, 1799–1826, where she created leading roles in Cherubini's *Anacréon* and *Les Abencérages*, and Spontini's *La vestale, Fernand Cortez*, and *Olympie*. The greatest French dramatic soprano of her day, she profoundly affected Berlioz, for whom she was 'tragédie lyrique incarnate'. She had excellent declamation, a striking stage presence, and a rich voice, capable both of great power and of subtle pianissimo singing. Famous for her Alceste, Armide, and Iphigénie, and Piccinni's Dido. Retired 1826; after her final performance, to immense public acclaim, the chorus presented her with a diamond crown in her dressing-room.

**Brandenburgers in Bohemia, The** (Cz.: *Braniboři v Čechách*). Opera in 3 acts by Smetana; text by Karel Sabina. Prod. Prague 5 Jan. 1866.

The opera tells of the fate of the three

daughters of Volfram Obramovic (bs), Mayor of Prague, during an attack on their native Bohemia by the Brandenburgers. Tausendmark (bar), a Brandenburg spy, attempts to keep them under his control but they are eventually released thanks to the effort of Junoš (ten), a burgher of Prague, and Jíra (ten), the Bohemians' 'beggar king'.

**Brandt, Marianne** (*b* Vienna, 12 Sep. 1842; *d* Vienna, 9 July 1921). Austrian mezzo-soprano. Studied Vienna with Janda and Zeller, Baden-Baden with Viardot. Début Olomouc 1867 (Rachel, *La juive*). Berlin 1868–82; London, CG, 1872, DL, 1882; Vienna 1873–83; New York, M, 1884–8. Her powerful voice ranged from g to d' ' ', and her wide repertory included Leonore, Eglantine, Fidès, Ortrud, Brangäne, Fricka, Erda. Though she originally declined Waltraute in the first *Ring*, finding the part 'too narrative', she replaced Luise Jaide at short notice in the second cycle *Götterdämmerung* at Bayreuth, 1876, when Stanford described her as 'a genius of the first order'. Created Kundry with Materna and Malten, singing in the second *Parsifal*. Wagner thought her 'the only woman I know with the stuff in her for Kundry', possessing 'le diable au corps' (though plain, she was slim) and the 'indispensable deep notes'. She disliked Wagner, but was a friend of Liszt. Retired 1890.

**Brangäne.** Isolde's attendant (mez) in Wagner's *Tristan und Isolde*.

**Brannigan, Owen** (*b* Annitsford, 10 Mar. 1908; *d* Newcastle upon Tyne, 9 May 1973). English bass. Studied London, GSM. Début Newcastle with SW Co. 1943 (Sarastro). London, SW, 1944–9 and 1952–8. Created Swallow (*Peter Grimes*), Collatinus, Noye, and Bottom for Britten; also roles by Malcolm Williamson and John Gardner. Gly. from 1947; London, CG, from 1948. Repertory included Leporello, Don Alfonso, Don Pasquale, and Dikoj. One of the best English buffo singers, with a large, ripe voice and a warm stage presence. (R)

**Branzell, Karin** (*b* Stockholm, 24 Sep. 1891; *d* Altadena, 15 Dec. 1974). Swedish mezzo-soprano. Studied Stockholm with Hofer, New York with Rosati. Début Stockholm 1912 (Prince Sarvilaka, D'Albert's *Izeyl*). Stockholm until 1918; Berlin, S, 1918–33; New York, M, 1924–44, 1951; Bayreuth 1930–1; London, CG, 1935–8. An admired Orfeo (Gluck), Fricka, Erda, Azucena, Amneris, Klytemnestra, Kostelnička. (R)

**Bratislava** (Hung., Pozsony; Ger., Pressburg). Capital of Slovakia, Czechoslovakia. The company was founded in 1920 as a Czech theatre, and only became more distinctively Slovak under Oskar Nedbal in 1923–31; in 1945, with the region's autonomy, it became wholly Slovak. It is, with Brno, second only to Prague in size and importance. Opera is given at the Slovak Nat. T by a company of about 45 soloists. The theatre (1886) was rebuilt and enlarged in 1975 (cap. 611). It gives regular annual festivals, in which Jan Cikker's works have played a prominent part.

**Braunfels, Walter** (*b* Frankfurt, 19 Dec. 1882; *d* Cologne, 19 Mar. 1954). German composer. Studied piano with Kwast and Leschetizky, composition in Munich with Thuille and Mottl. First director, Cologne Hochschule für Musik, 1925–33, 1945–50. His operas show a strong theatrical instinct, and are cast in a late-Romantic, post-Wagnerian style that does not exclude a lightness of touch (he acquired an admiration for Berlioz from Mottl) and a feeling for the grotesque. His greatest popular success was with *Die Vögel*.

SELECT WORKLIST: *Prinzessin Brambilla* (Braunfels, after Hoffmann; Stuttgart 1909); *Ulenspiegel* (Braunfels, after De Coster; Stuttgart 1913); *Die Vögel* (Braunfels, after Aristophanes; Munich 1920); *Verkündigung* (Annunciation) (Braunfels, after Claudel; comp. 1933–5, prod. Cologne 1948).

**Braunschweig.** See BRUNSWICK.

**Brautwahl, Die** (The Marital Lottery). Opera in 3 acts by Busoni; text by the composer, after E. T. A. Hoffmann's story. Prod. Hamburg, S, 13 Apr. 1912.

Fantasy world in Berlin, *c*.1920. A young artist, Edmund Lehsen (ten), falls in love with Albertine (sop), daughter of the 'Kommissionsrat' Voswinkel (bs). She is already promised to Thusmann, but when he reports having seen her dancing at an inn, Voswinkel assumes his interest has waned. However, when an old Jew, Manasse (ten), asks for Albertine's hand for his nephew, Thusmann (bs), renews his promise of marriage. Voswinkel summons all the suitors; the one who chooses the correct one of three caskets shall win Albertine. Lehsen selects a casket containing a letter promising bliss and, eventually united with Albertine, he heads for Rome, where artistic fame awaits him.

**Bravo, Il.** Opera in 3 acts by Mercadante; text by Rossi and M. M. Marcello, based on Fenimore Cooper's novel, and on A. Bourgeois's play *La vénitienne*. Prod. Milan, S, 9 Mar. 1839, with Balzar, Benciolini, Castellan, Donzelli, Polonini, Quattrini, Marconi, Schoberlechner, Tadolini; Philadelphia, Walnut St. T, 2 Oct. 1848.

Venice, 18th cent. Foscari (bs), a Venetian nobleman, loves Violetta (sop), and has her guardian murdered. Her mother, Teodora (sop), regrets having abandoned her daughter and pays the government assassin, the Bravo (ten), to abduct Violetta. After the reunion of mother and daughter the Bravo discloses that he is Violetta's father. The government has ordered him to kill Teodora since she has offended them; Teodora seizes the dagger from the Bravo and stabs herself. A messenger then arrives with news which releases the Bravo from his dreadful occupation. Violetta is able to depart with Pisani (ten), who has returned to Venice from exile in order to arrange their elopement.

**Brazil.** The first theatres were built in the 18th cent., and the first operas were seen at the Opera Velha in the late 1770s. However, operatic life did not develop very fast until the transfer to Brazil of the Portuguese court in the Peninsular War. The Opera Nacional was founded in 1857; its repertory included zarzuela and Brazilian opera as well as Italian opera. The first important Brazilian opera composer was Carlos *Gomes (1836–96). Other significant opera composers included Joaquim de Macedo (*b* ?; *d* 1925) with *Tiradentes*, Leopoldo Migues (1850–1902) with *Saldunes*, Henrique Oswald (1854–1931), of Swiss origin, with *Croce d'oro*, and Itibere da Cunha (1848–1913). Alberto Nepumuceno (1864–1920), with *Artemis* (1898), *Abul* (1913), and *O garatuja* (1904), took a stand for Portuguese as a singing language. Also significant have been Francisco Braga (1868–1945) with *Jupira* (1900), *Anita Garibaldi* (1901), and *Contractador de diamantes* (1901), O. Lourenço Fernandes (1897–1948) with *Malazarte* (1931) and the dominating and prolific Heitor *Villa-Lobos (1887–1959). Opera is given chiefly at the T Municipal, Rio de Janeiro, and T Municipal, São Paulo, mainly by visiting companies. It has also been given at an opera-house (cap. 800) opened in 1896 at Manaus, in the Amazon jungle, to entertain wealthy rubber barons.
See RIO DE JANEIRO, SÃO PAULO.

**break.** See REGISTER.

**Brecht, Bertolt** (*b* Augsburg, 10 Feb. 1898; *d* Berlin, 15 Aug. 1956). German dramatist. He wrote the version of *The Beggar's Opera* known as *Die *Dreigroschenoper* (1928) for Kurt Weill, also the texts for *Happy End* (1929), *Mahagonny* (1930), *Der Jasager* (1930), and some of Wagner-Régeny's *Persische Episode* (*Der Darmwäscher*) (1940–50), with Caspar *Neher. *The Trial of Lucullus* by Sessions (1947) and Dessau (1951) are based on his works, as is Cezar's *Galileo Galilei* (1964).

**breeches part.** See TRAVESTI.

**Bregenz.** City in Vorarlberg, Austria. A summer open-air festival opened 1946, with opera added 1956 in the specially built lakeside theatre (cap.

4,400) with floating stage. International artists take part in large-scale productions. 18th- and 19th-cent. works are staged in the Festspielhaus (cap. 1,800) and T am Kornmarkt (cap. 633). Opera has also been given at nearby Hohenems Castle.

**Brema, Marie** (*b* Liverpool, 28 Feb. 1856; *d* Manchester, 22 Mar. 1925). English mezzo-soprano. Studied London with Henschel. Stage début London, Shaftsbury T, 1891, as first English Lola. Soon established herself as a Wagnerian, singing Ortrud, Kundry, Brangäne, and Brünnhilde: Bayreuth 1894–7; New York, M, 1895–1900; Paris 1899 and 1902. Her versatility and intelligence also made her successful as Orfeo, Amneris, and Dalila, and she created Beatrice in Stanford's *Much Ado about Nothing* (London, CG, 1901). In 1910 she organized an opera season at the Savoy T, singing Orfeo in her own production. An expressive singer and actress, memorable for her intensity and dramatic conviction.

**Bremen.** City in Lower Saxony, Germany. An early Faust opera, Ignaz Walter's *Doktor Faust* (1797), was premièred there. The Staatstheater was bombed in the Second World War; the Bremer T am Goetheplatz (cap. 900) opened 27 Aug. 1950, and maintains an enterprising repertory.

**Brent, Charlotte** (*b* c.1735; *d* London, 10 Apr. 1802). English soprano. Studied with Arne. Début Dublin 1755 (Arne's *Eliza*). London, DL, 1757, in the same. When Garrick would not engage her later, Arne arranged for her to appear at London, CG, where she sang from 1759 in a series of triumphs, notably in *The Beggar's Opera* (with John Beard). Other works included Arne's *Thomas and Sally* and *Artaxerxes*, and operas by Lampe and Linley. After appearances at the Three Choirs Festival 1765–7, she and her husband Thomas Pinto (whose name she then used) went to Edinburgh and Dublin, where she sang in M. Arne's *Cymon* (1773). Retired after a performance of *Comus* (Arne) at London, CG, 1785. A singer of great imagination, she had an appealing voice and a natural simplicity as well as a fine technique.

**Brescia.** City in Lombardy, Italy. The first opera-house was the T degli Erranti, opened 1688 with Ziani's *Tullo Ostilio*. Rebuilt, reopened as T Grande 27 Dec. 1810 with Mayr's *Ifigenia in Aulide*. The present theatre opened Aug. 1863 with *Il trovatore*. On 28 May 1904 the revised *Butterfly* scored its first success after the La Scala fiasco. Eva Turner sang her first Turandot here, 1926.

**Bresgen, Cesar** (*b* Florence, 16 Oct. 1913; *d* Salzburg, 7 Apr. 1988). Austrian composer of German origin. Studied with Hindemith, Jelinek, and Krenek. From 1939 taught at Mozarteum in Salzburg; founded young persons' conservatory there in same year. Wrote many works for youth groups showing influence of German folk-song. Among his stage works, which function as a kind of *Gebrauchsmusik*, are the children's operas *Der Igel als Bräutigam* and *Brüderlein Hund*.

SELECT WORKLIST: *Der Igel als Bräutigam* (Bresgen & L. Andersen; Esslingen, 1949); *Brüderlein Hund* (L. Andersen; 1953).

BIBL: D. Larese, *Cesar Bresgen: eine Lebenskizze* (Amriswil, 1968).

**Breslau.** See WROCŁAW.

**Bressand, Friedrich Christian** (*b* Durlach, *c.*1670; *d* Wolfenbüttel, 4 Apr. 1699). German librettist and poet. Worked in Brunswick, where he was director of court ceremonies after 1690. He was one of the most popular early German librettists and, like his contemporaries Bostel and Postel, drew much inspiration from the plots of Italian and French opera, as well as from French classical tragedy. His texts were set by Keiser (*Der königliche Schäfer*, 1694), Kusser (*Ariadne*, 1692), Mattheson (*Der edelmüthige Porsenna*, 1702), and Schürmann (*Die Plejades*, 1716), among others; though conceived in the context of an aristocratic entertainment, many were later repeated on the public stage in Hamburg.

**Bretan, Nicolae** (*b* Năsăud, 25 Mar. 1887; *d* Cluj, 1 Dec. 1968). Romanian composer, baritone, conductor, and critic. Studied Cluj with Farkas, Vienna with Geiringer, Budapest with Siklós and Sik. Principal baritone Bratislava 1913–14, 1916–17; baritone and producer Oradea 1914–16; baritone, producer, conductor, and director variously Cluj 1917–48 at both Romanian and Hungarian operas. One of the pioneers of Romanian opera, he reflects his cosmopolitan training in his musical style, and his gifts as a singer in his excellent vocal writing. His most substantial opera, *Horia*, shows some influence of Wagner, but he is better known in his native country (despite a ban on his music when he refused to join the Communist Party) for his other, 1-act operas.

WORKLIST (all prod. Cluj): *Luceafărul* (Bretan, after Eminescu; 1921); *Golem* (Bretan, after Kaczér; 1924); *Eroii de la Rovine* (Bretan, after Eminescu; 1935); *Horia* (Pop; 1937); *Arald* (Bretan, after Eminescu).

**Bretzner, Christoph Friedrich** (*b* Leipzig, 10 Sep. 1748; *d* Leipzig, 31 Aug. 1807). German librettist and dramatist. The most successful and frequently set German author of the late 18th cent. His first four librettos appeared as a set in 1779. While *Adrast und Isidore* and *Der Äpfeldieb* reflect the old tradition of Singspiel libretto in their simplicity and playful spirit, which derived partly from Molière, *Der Irrwisch* and *Das wütende Heer* were inspired by folklore and fairytale and look forward to early Romantic opera in their handling of supernatural, Gothic themes. These colourful librettos proved immensely popular and Bretzner rapidly came to dominate the Berlin stage, his collaborations with André and Kospoth enjoying great popularity.

A particular triumph was André's setting of *Belmont und Constanze* (1781); this increased Bretzner's new-found fame and, almost uniquely for a North German librettist, led to his texts being set in Vienna. When Mozart and Gottlieb Stephanie decided to collaborate on a German opera they were drawn to this latest exotic, Turkish libretto: although several important changes and additions were made for *Die Entführung aus dem Serail*, the spirit of Bretzner's original survived. Particularly important was the influence from Italian opera buffa, which was enhanced by Mozart's musical treatment, and taken further in a later set of Bretzner's librettos, comprising *Der Schlaftrank*, *Schattenspiel an der Wand*, and *Opera buffa* (pub. 1796). His other important works included *Der Talisman* and *List gegen List*.

**Bréval, Lucienne** (*b* Männedorf, 4 Nov. 1869; *d* Neuilly-sur-Seine, 15 Aug. 1935). Swiss, then French, soprano. Studied Paris with Wartot. Début Paris, O, 1892 (Selika). New York, M, 1901–2; London, CG, 1899, 1901, 1906. Created Grisélidis and Ariane (Massenet), Lady Macbeth (Bloch), Pénélope (Fauré), and Monna Vanna (Février). One of the great French singers of her time, celebrated as Brünnhilde and in the grand classical roles of Rameau and Gluck. (R)

**Brian, Havergal** (*b* Dresden, Staffs., 29 Jan. 1876; *d* Shoreham, 28 Nov. 1972). British composer. Mainly self-taught, Brian supported himself for most of his life by modest free-lance writing. Although his works were occasionally conducted by Wood, Beecham, and others, Brian never enjoyed any popular success and never held any important musical position. Apart from songs and partsongs, most of his other music, which includes several symphonies, concertos, and extended orchestral pieces, remains unpublished, and much is unperformed.

Brian composed seven operas, one of which is lost. Their subject-matter reveals the composer's catholic tastes: most important are *The Tigers*, described as a 'burlesque opera', *Turandot: Prinzessin von China*, and *Faust*. Only *Agamemnon* has ever been performed, and here the influence of Strauss is strikingly apparent.

SELECT WORKLIST: *The Tigers* (Brian; comp. 1916–36,

91

unprod.); *Turandot: Prinzessin von China* (Gozzi, trans. Schiller; comp. ?1949–51, unprod.); *Faust* (Goethe; comp. ?1954–6, unprod.); *Agamemnon* (Aeschylus, trans. J. S. Blackie; comp. 1957, prod. London 1971).

BIBL: R. Nettel, *Havergal Brian and his Music* (London, 1976).

**Bridesmaids' Chorus.** The chorus in Act II of Weber's *Der Freischütz*, in which the village girls bring a wreath to Agathe and prepare her for her wedding.

**brindisi** (It., *far brindisi*: 'to give a toast', from Span. *brindis*: 'toast', from German *Ich bring dir's*: 'I bring it to you'; cf. Juan Corominas, *Breve diccionario etimológico de la langua castellana*). A drinking or toasting song, with choral refrain. Famous examples include Alfredo and Violetta's 'Libiamo' in Verdi's *Traviata*, Lady Macbeth's 'Si colmi il calice' in *Macbeth*, 'Il segreto' in Donizetti's *Lucrezia Borgia*, and 'Viva il vino' in Mascagni's *Cavalleria rusticana*.

**Brisbane.** City in Queensland, Australia. The Lyric O of Queensland was formed in 1982 and performs in Her Majesty's T and the Lyric T (cap. 2,100). The Queensland Performing Arts Complex opened in 1985 includes the Lyric OH (cap. 2,100) and a Studio T (cap. 264).

**British Broadcasting Corporation.** Opera entered British broadcasting on 6 Jan. 1923, when the London station 2LO relayed a BNOC performance of *Hänsel und Gretel* from CG. Further BNOC broadcasts included *Figaro*, *Walküre*, and *Siegfried*. In Oct. 1923 the first studio opera was given, *Roméo et Juliette*. As well as broadcasts from CG and other theatres in the 1920s, the BBC gave an ambitious series of studio performances under Percy Pitt 1926–30, including *Roi d'Ys*, *Pelléas*, and *Louise*. The Labour government subsidized CG through the BBC 1929–31, and many broadcasts were given. Studio opera ceased until 1937, when a Music Productions Unit was formed with Stanford Robinson as music director. Concert performances at QH were given of operas including *Wozzeck*, *The Lady Macbeth of Mtsensk*, and *Mathis der Maler*, 1934–9. The war interrupted further progress, which was resumed in 1945 especially with the impetus given to music broadcasting first by the Third Programme, then the Music Programme. By now opera was established as an essential part of national and international broadcasting. The Promenade Concerts, held under BBC auspices every summer in the RAH, include visiting semi-staged performances from Gly. and CG.

**British National Opera Company.** Formed in 1922 by leading artists of the *Beecham OC forced into liquidation in Dec. 1920. Percy Pitt was the first artistic director, succeeded by Frederic Austin. The company was launched in Bradford in Feb. 1922, and in May gave *The Ring*, *Parsifal*, *Tristan*, and other works with leading artists of the day. In spring 1924 it was kept out of London, CG, by directors wanting international opera, and played at London, HM, where the première of *Hugh the Drover* was given. Other new works included *The Perfect Fool* (1923) and *At the Boar's Head* (1925). Guest artists included Melba and Dinh Gilly. By 1928 the company owed some £5,000 but in its last year had also paid £17,000 in entertainment tax. It was taken over by London, CG, in 1929, but survived only three seasons.

**Britten, Benjamin** (Lord Britten of Aldeburgh) (*b* Lowestoft, 22 Nov. 1913; *d* Aldeburgh, 4 Dec. 1976). English composer, conductor, and pianist. Studied with Frank Bridge, and at the RCM with John Ireland. His early opera *Paul Bunyan* was originally withdrawn; on its revival in 1976, the work proved (to a general surprise shared by the composer) to have a deftness of touch and flair for the stage that gave some precedent to the achievement of *Peter Grimes*. Nevertheless, the appearance of *Grimes* in 1945 was a sensation at the time and has long been seen as a historical turning-point. With it English opera seemed to come of age, and Britten was established as an opera composer of the first rank. Rooted in tradition, and owing much to the example of late Verdi, the work is individual not only in its virtuosity and dramatic instinct, but in such characteristics as a crucial use of bitonality to indicate the divided nature of the hero and his separation from the Borough. There is little doubt that this, and other of Britten's operatic themes, related to problems of acceptance by society of homosexuality: he was to show in his art an acute sympathy for the victims of intolerance and a painful sensibility towards the abuse of innocence. The opera also consolidated the position of the composer's companion Peter *Pears as an operatic artist of the first importance: Pears was to have roles in every one of Britten's operas, and his artistry lent a particular hue to Britten's writing for the tenor voice. In many other cases, though never so completely, Britten was to conceive operatic roles with particular artists in mind, and sometimes to resist alternative interpretations.

Largely owing to the uneconomic conditions of large-scale opera, Britten now turned to chamber opera, forming the *English Opera Group for the presentation of such works. *The Rape of Lucretia* shows his mastery of limited means to expressive ends, especially in his handling of the 12-piece orchestra. From this tragedy, he turned to a version of a comic Maupassant story, resituated in Suffolk, for *Albert Herring*. There is again a brilliant use of the small orchestra to point up and lend some depth to the work, though the satire of what are village types rather than characters prevents it from taking its place among the great operatic comedies. For his next two works, Britten turned to unusual forms: Gay's *Beggar's Opera* was cleverly recomposed; and *Let's Make an Opera* is in the first part a play rehearsing a cast of children and grown-ups for participation in the second, the opera *The Little Sweep*. Two years later he gave further proof of his versatility with a free realization of *Dido and Aeneas*: he was a profound admirer of Purcell, to whom he consciously owed much in his setting of English.

1951 saw the appearance of *Billy Budd*, Britten's first full-scale opera since *Peter Grimes*. As in the earlier work, there is a radical use of bitonality to embody the subject, the separation of the handsome and innocent sailor from his fellows and his destruction at the hands of evil and inaction. Set on a man-o'-war, with an all-male cast, the consequent problems were turned to creative advantage by Britten in the skilful disposition of textures both vocally and orchestrally; while the work was also formally and motivically the subtlest opera he had composed. Its successor *Gloriana* was also large in scale. Written for the Coronation of Queen Elizabeth II, it excited hostility chiefly for the libretto: this naturally drew a non-musical public's attention more than the music, whose quality has been impressively revealed in revivals.

Returning to chamber opera, Britten produced in *The Turn of the Screw* a work whose taut thematic integration is a reflection of the intense, painful story of two children possessed by ghosts: the technique includes the use of one principal theme which 'turns' in orchestral interludes as a metaphor of the tightening cruelty of the story. By contrast, *Noyes Fludde* represents Britten's freshest and most delightful vein, setting a miracle play in church and drawing on amateurs and children as well as professionals with enchanting skill and understanding. The presence of the dark and the light in his experience found new balance in one of the few Shakespeare operas worthy to stand near Verdi, *A Midsummer Night's Dream*. Aided by a skilfully arranged libretto, Britten penetrated into the heart of the play and into the mysterious world of sleep, central to the drama but given musical articulation, which had become another of his fascinations. Fairies, rustics, and nobles are expertly characterized, and the music is made a strong symbol of how, through the action of the dark dream in the wood, the lovers shed their confusion and find each other as they wake to new happiness.

Britten now followed up some of the techniques in *Noyes Fludde*, as well as his experience of Japanese \*Noh plays, with three church parables: *Curlew River*, *The Burning Fiery Furnace*, and *The Prodigal Son*. Designed to be performed in church, and taking some of their ritual nature and their dramatic and instrumental techniques from Noh, they are nevertheless individual works that absorb the Japanese influence into a form that is at least as strongly rooted in early English religious music drama. *Curlew River*, which uses a Japanese story, is not only the most oriental but also technically the strictest and most successful of the three; having mastered the disciplines he set himself, Britten felt able to use them more freely. The common framework is a dramatic fiction of performance by a group of monks who arrive to plainchant (providing musical material for the work) and withdraw to it after their performance.

A new set of problems was presented when Britten was commissioned by the BBC to write a TV opera. The theme of *Owen Wingrave* reflected Britten's own pacifism, and though dramatically it is assertion rather than argument, the work is expertly written and turns the limitations of TV to advantage in its use of special techniques such as cross-fading between characters in different places for a duet. Britten's last opera, *Death in Venice*, is in the sparer style of his final phase. It draws on ballet sequences (especially for a boy dancer) and concentrates on major roles for the central figure of Aschenbach and the figure who, in various guises, draws him away from his Apollonian stance towards a Dionysian homosexual temptation and destruction.

Britten also made an edition (1951) of Holst's *The Wandering Scholar* with Imogen Holst, his long-time assistant at Aldeburgh. As a conductor, he largely confined himself to his own works in the opera-house; but especially at the Aldeburgh Festival and on tour, he directed brilliant performances of works in the English Opera Group's repertory. (R)

WORKLIST: *Paul Bunyan* (Auden; New York 1941); *Peter Grimes* (Slater, after Crabbe; London 1945); *The Rape of Lucretia* (Duncan, after Obey; Gly. 1946); *Albert Herring* (Crozier, after Maupassant; Gly. 1947); *The Little Sweep* (second part of *Let's Make an Opera*) (Crozier; Aldeburgh 1949); *Billy Budd* (Forster & Crozier, after Melville; London 1951); *Gloriana* (Plomer, after Strachey; London 1953); *The Turn of the Screw* (Piper, after James; Venice 1954); *Noyes Fludde* (Chester miracle play; Orford 1958); *A Midsummer Night's Dream* (Britten & Pears, after Shakespeare; Aldeburgh 1960); *Curlew River* (Plomer, after Motomasa; Orford 1964); *The Burning Fiery Furnace* (Plomer; Orford 1966);

*The Prodigal Son* (Plomer; Orford 1968); *Owen Wingrave* (Piper, after James; BBC/TV 1971); *Death in Venice* (Piper, after Mann; Snape 1973).

CATALOGUE: anon., *Benjamin Britten: Complete Catalogue of Published Works* (London, 1973).

BIBL: E. White, *Benjamin Britten: His Life and Operas* (London, 1948, 3/1970); P. Howard, *The Operas of Benjamin Britten* (London, 1969); P. Evans, *The Music of Benjamin Britten* (London, 1979); M. Kennedy, *Britten* (London, 1981).

**Brno** (Ger., Brünn). City in Moravia, Czechoslovakia. Opera was given from the 1730s by Italian companies, including those of A. and P. *Mingotti, in the T in der Taffern (opened 1732 on the site of the present Reduta T). Operas included Luchini's *Vincislao* (1739) on the subject of King Václav II. In 1767 Jan Tuček's Singspiel *Zamilovaný ponocný* (The Lovelorn Nightwatchman) was given in Czech; the first opera sung in Czech was Méhul's *Joseph* (in 1839), followed by Škroup's *Dráteník* (in 1840) and the Brno composer František Kott's *Žižkův dub* (1841). The theatre burnt down in 1785. In 1786 the Nationaltheater (Redoutensaale) opened (cap. *c.*1,200); it burnt down in 1870 and was later restored as the Reduta (cap. 414) (now used for operetta). Traditions for long reflected the proximity of Vienna. However, the Czech Provisional T opened in 1884 (cap. 771), and was enlarged as the Národní Divadlo (National T) in 1894 (cap. 1,000); but standards were erratic. The German Stadttheater opened in 1882; this was the first theatre with electric light in Central Europe (cap. 1,185, later 621) (now the Mahen T). Standards were particularly high between the wars under František Neumann and after the war under František Jílek. Opera is now given in the Janáček OH, opened in 1965 (cap. 1,317), the largest in Czechoslovakia. The repertory has been notable for its enterprise, especially with contemporary works; it gave the premières of all Janáček's operas except *Mr Brouček*. There is an opera studio, the Miloš Wasserbauer Chamber O, built in 1957 by the Janáček Academy.

**Brook, Peter** (*b* London, 21 Mar. 1925). English producer of Russian-Jewish origin. Studied Oxford. Director of productions, London, CG, 1947–50. Though many of his productions were admired, notably *Figaro* and *The Olympians* (1949), he had to contend with many problems, including differences with *Rankl, and his intention to give the producer a more inventive role in opera was frustrated. He resigned after opposition to his methods came to a head with his production of *Salome* with décor by Dali (1949).

**Broschi, Riccardo** (*b* Naples, *c.*1698; *d* Madrid, 1756). Italian composer. Elder brother of castrato Carlo Broschi (*Farinelli). Little known of early life. Six operas performed in Italy 1725–35, including *La vecchia sorda*, *Idaspe*, *Merope*, and *Adriano in Siria*. 1736–7 in service of Duke of Württemberg; returned to Naples on his death and during 1740s joined brother in Spain. Chiefly remembered for his aria 'Son qual nave' written for the pasticcio on Hasse's *Artaserse* performed at London, H, in 1734, which became Farinelli's most celebrated show-piece.

SELECT WORKLIST: *La vecchia sorda* (Saddumene; Naples 1725, lost); *Idaspe* (Cande; Venice 1730); *Merope* (?; Turin 1732); *Adriano in Siria* (Metastasio; Milan 1735).

**Brouček, The Excursions of Mr.** See EXCURSIONS OF MR BROUČEK, THE.

**Brouwenstijn, Gré** (*b* Den Helder, 26 Aug. 1915). Dutch soprano. Studied Amsterdam with Stroomenbergh. Début Amsterdam 1940 (one of the Three Ladies, *Die Zauberflöte*). Netherlands O, from 1946, début as Tosca. London, CG, 1951–64 in Verdi; Bayreuth 1954–6; Vienna from 1956; Gly. from 1959 (Leonore); Chicago 1959 (Jenůfa); also at Buenos Aires, Stuttgart, and Paris. Retired 1971. A musical and intelligent singer with a firm, warm tone. (R)

**Browning, Robert** (*b* Camberwell, 7 May 1812; *d* Venice, 12 Dec. 1889). English poet. Operas on his works are as follows:

*The Pied Piper of Hamelin* (1842): Farmer (1896); Clokey (1921); Brumleau (1955); Dubbiosi (1965); Mennin (1969); Flagello (1970)

Also opera by Freer, *The Brownings Go to Italy* (1938).

**Brownlee, John** (*b* Geelong, 7 Jan. 1901; *d* New York, 10 Jan. 1969). Australian baritone. Studied Melbourne, then Paris with Dinh Gilly. Début Paris 1926 (in *Lakmé*). Brought by Melba to London, CG, where he sang Marcello at her final appearance there in 1926. Paris, O, 1927–36 as principal baritone. CG, 1930 (Golaud). Gly. from 1935 (one of its first members) as Don Alfonso, Don Giovanni, Count Almaviva (Mozart). New York, M, 1937–58, début Rigoletto. Distinguished in the French repertory and Mozart, with a well-schooled rather than outstanding voice. (R)

**Bruch, Max** (*b* Cologne, 6 Jan. 1838; *d* Friedenau, 2 Oct. 1920). German composer. Studied Bonn, later Cologne with Hiller and Reinecke.

His first opera was a setting of Goethe's *Scherz, List und Rache*, in three scenes with 16 musical numbers. He followed this with *Loreley*, his most important opera, which contains some fine and characteristic music, especially in the choral writing; this also marks his third opera, *Hermione*.

WORKLIST: *Scherz, List und Rache* (Bischoff & Brownlee, after Goethe; Cologne 1858); *Loreley*

(Brownlee, after Geibel; Mannheim 1863); *Hermione* (Hopffer, after Shakespeare; Berlin 1872).

**Brummstimmen.** See BOCCA CHIUSA.

**Bruneau, Alfred** (*b* Paris, 3 Mar. 1857; *d* Paris, 15 June 1934). French composer and critic. Studied Paris with Massenet. His early, and to conservative listeners sensational, *Le rêve* was based on Zola, whose influence on Italian literary verismo Bruneau to some extent reflects in his own veristic approach. He was to follow Zola, who wrote many of his librettos, in a number of ideas, and his *Messidor* suffered from the public opposition to Zola during the Dreyfus case. However, *L'ouragan* triumphed, as did *L'enfant roi*, a lighter work despite its serious theme on the importance of children to marriage. Zola died in 1902, mourned by Bruneau in *Lazare*. Though not fundamentally influenced by Wagner, he benefited French opera by his appreciation of Wagner's strength, without relinquishing the individually French manner he had in part acquired from his teacher Massenet. His music serves its texts faithfully and seriously, despite a certain intrusive crudity in its realism.

SELECT WORKLIST (all prod. Paris): *Le rêve* (The Dream) (Gallet, after Zola; 1890); *L'attaque du moulin* (The Attack on the Mill) (Gallet, after Zola; 1893); *Messidor* (Zola; 1897); *L'ouragan* (The Hurricane) (Zola; 1901); *L'enfant roi* (The Child King) (Zola; 1902); *Angelo* (Méré, after Hugo; 1928).

BIBL: A. Boschot, *La vie et les œuvres de Alfred Bruneau* (Paris, 1937).

**brunette** (Fr.: 'little brown one'). A kind of song popular in the late 17th and early 18th cents. in which, from the Renaissance ideal of the 'petite brune' as simple and gentle, love is affirmed in tender rather than passionate terms. It was therefore usually simple melodically, perhaps a rondo in form and cast in a few verses; found in French opera from Lully to Rameau.

**Bruni, Antonio** (*b* Cuneo, 28 Jan. 1757; *d* Cuneo, 6 Aug. 1821). Italian composer and violinist. Travelled to Paris in 1780; joined orchestra of Comédie-Italienne in 1781. His first opera, written for court at Fontainebleau, was *Coradin*. A renowned virtuoso performer who, because of Revolutionary sympathies, rose to prominence in Parisian musical life. Orchestral director of the Opéra-Comique 1799; director of the Opéra Italienne 1801. Composed *c*.20 opéras-comiques for Paris, such as *Galatée* and *L'officier de fortune, ou Les deux militaires*. These, though possessing superficial charm and very popular, were of little lasting significance. After retiring in 1802, Bruni returned to Paris in 1814 and 1815 with *La règne de douze heures* and *Le mariage par commission* in an unsuccessful attempt to recapture his previous popularity.

SELECT WORKLIST: *Coradin* (Magnitot; Fontainebleau 1785); *Galatée* (Poultier, after Rousseau; Paris 1795); *L'officier de fortune ou les deux militaires* (Patrat; Paris 1792); *La règne de douze heures* (Planard: Paris 1814); *Le mariage par commission* (Simonnin; Paris 1815).

**Brünn.** See BRNO.

**Brünnhilde.** A Valkyrie (sop), Wotan's favourite daughter, in Wagner's *Die Walküre*, *Siegfried*, and *Götterdämmerung*. Her name derives from the Middle High German *brünne* (coat of mail) and *hilt* (battle).

**Brünnhilde's Immolation.** The name generally given to the final scene of Wagner's *Götterdämmerung*, in which Brünnhilde (sop) bids farewell to life and rides on her horse *Grane into the flames.

**Brunswick** (Ger., Braunschweig). City in Lower Saxony, Germany. Performances were first given mid-17th cent. in the Rathaus; this was converted and opened 1690 with Kusser's *Cleopatra*, and saw the premières of many German operas by Kusser, Keiser, Schürmann, Hasse, and others during the late 17th and early 18th cents. The Nationaltheater opened 1818. The Landestheater opened 1861 and was bombed in the Second World War; the Staatstheater (cap. 1,370) opened 1948 with *Don Giovanni*; there is a smaller theatre (cap. 210). The company has a good history of enterprise.

**Bruscantini, Sesto** (*b* Porto Civitanova, 10 Dec. 1919). Italian baritone. Studied Rome with Ricci. Début Civitanova 1946 (Colline). Milan, S, from 1949. Gly. from 1951 in Mozart, Rossini, and Verdi. Chicago from 1961, and other major opera-houses, including London, CG; Paris; and Sydney. A thoroughly accomplished singer, whose enormous repertory encompasses lyric, dramatic, and buffo roles. (R)

**Bruson, Renato** (*b* Este, 13 Jan. 1936). Italian baritone. Studied Padua with Ceriati. Début Spoleto 1961 (Di Luna). Milan, S, 1972 in *Linda di Chamounix*. New York, M, from 1968; Edinburgh, 1972; London, CG, from 1976; Florence 1980; Orange 1984, etc. A distinguished Verdi and Donizetti singer, especially in dramatic and noble parts. (R)

**Brussels** (Fr., Bruxelles; Dut., Brussel). Capital of Belgium. The O du Quai du Foin opened in 1682, with a repertory that centred on works by Lully; this reopened in 1700 (cap. 1,200) on the site of the old Mint as the T sur la Monnaie (1763 Grand T et O de la Monnaie), with P. A. Fiocco (*c*.1650–1714) as director. French opera dominated the repertory until 1746. With the occupation by the French after the Battle of Fontenoy,

Maréchal Saxe imported Favart and the Paris, OC, who remained until 1749; the theatre was then run by a succession of directors, with French taste still prevailing. The French occupation of 1794–1814 freed the theatres and saw Belgian as well as French works. In 1815, under the Anglo-Dutch occupation, the Monnaie's title became Flemish (Schouwburg op de Munte). A new, enlarged theatre was opened in 1819, and the repertory and reputation increased. At a performance of *La muette de Portici* on 24 Aug. 1830, the call to arms in the duet 'Amour sacré de la patrie' inspired the audience to initiate forthwith the revolt that overthrew the Dutch. In 1832 gaslight was introduced; in 1855 the theatre burnt down. It was reopened in 1856 (cap. 1,170), introducing a repertory that closely mirrored Parisian taste. The fame of the theatre became international, and attracted important premières (Massenet's *Hérodiade* (1881), Chabrier's *Gwendoline* (1886)), new and great singers (Patti, Alboni, Albani, Faure, Calvé; Melba's 1887 début as Gilda), and the transfer of the original Bayreuth *Ring* (1883). The arrival in 1900 of Maurice Kufferath, an enthusiastic Wagnerian, at once improved dramatic standards and brought in a stronger German influence. After the war, under Corneil de Thoran, the theatre further widened its repertory to include a number of premières (Prokofiev's *The Gambler*, 1929) and welcomed great singers of the day (Shalyapin, Caruso, Destinn). After 1945 the Wagner influence returned; and the régime of Maurice Huisman, director from 1959, turned the emphasis on visiting foreign companies. In 1981 Gerard Mortier thoroughly, and controversially, reorganized the theatre, emphasizing music drama rather than stars and concentrating on intensive rehearsal; he left in 1991. There has been a famous Mozart series and Janáček series, and an emphasis on new works (Hans Zender's *Stephen Climax* (1990), John Adams's *The Death of Klinghoffer* (1991)). There have also been premières of works by Belgian composers (Philippe Boesmans's *La passion de Gilles* (1983), André Laporte's *Das Schloss* (1986)).

Operetta has been given in many theatres, including the T Royal du Parc (opened 1782), the Château des Fleurs (opened 1847), the Bouffes-Belges (opened 1858), and the Alcazar (opened 1867), especially the Alhambra (opened 1846 as T du Cirque); but the principal home of opéra-comique has been the T Royal des Galeries (later Fantaisies-Parisiennes) (opened 1847), and enterprising productions have been given at the Palais des Beaux-Arts from 1936.

**Bucharest** (Rom., Bucureşti). Capital of Romania. Opera was first given by Francesco's Italian company in 1787. The Theatrum Vlahicum Bucharestini opened in 1814, used by visiting companies. Johann Gerger's Viennese company came from Transylvania, a German-dominated area with a Mozart tradition, to play in Princess Ralu's theatre 'At the Red Fountain' in 1818, with a repertory of Rossini and Mozart (*Idomeneo*). Eduard Kriebig's company followed in 1823, the Fourreaux Co. in 1831–2, Theodor Müller's in 1833 on its way home from Russia; artists of the latter remained to work in a theatre built by Momolo, including Ion Wachmann (1807–63), who had directed the company in Timişoara, 1831–3, and continued in the capital until 1835. The repertory included French and Italian works, Mozart, and Wachmann's own *Braconierul* (The Poacher, 1833); 60 performances were given in 1833–4. Application for an Italian theatre in 1833 led eventually to the opening of the Teatrul Naţional in 1852 (dir. Wachmann 1852–8; the company alternated with an Italian one under Basilio Sansoni that had had a state subsidy since 1843.

A new departure came with the appointment to the T Naţional in 1877 of George Stephănescu (1843–1925), who in the face of many difficulties kept opera active: he founded Opera Română (1885–6), the Compania Lirică Română (1892–3), and the Societatea Lirică Română (1902–4). Romanian artists were still lower paid than Italians, leading to the closure of the Romanian company and the emigration of its best artists. Stephănescu gave 31 performances between 1889 and 1891; from 1897 to 1901 another company occupied the theatre. At the T Naţional between 1901 and 1910 Italian seasons were given which, though uneven, included Tetrazzini, Litvinne, Battistini, and Ruffo. French and German seasons were also given. During wartime German occupation, the Dessau O visited, and local activity included a Student O that gave 63 performances in 1914–15, a company under Stefănescu-Cerna at the Blanduzia Garden (1915), the formation of the Gabrilescu Co. and another at the Leon Popescu T (1915) and of another under Jean Atanasia (1916). In 1919 the Asociaţia Lirică Română Opera was formed under the composer Ion Nonna Otescu (1888–1940); it opened with *Aida* at the T Naţional, and as well as encouraging local artists drew famous singers (Slezak) and conductors (Weingartner, Nedbal). In 1921 this success led to the opera being made a state institution, Opera Română, and housed at the Lyric T, where the season opened with *Lohengrin* under Enescu; his successors included George Georgescu (1922–7) and Ionel Perlea (1929–30). In 1924 11 singers resigned in protest against the government's restrictive attitude, and in 1930 the company had to move to an unsuitable building and halve salaries. Nevertheless, new Romanian operas were staged. Earthquake in 1940 and bombing in 1944 wrecked the old theatre. The new Opera Română theatre (cap. 610) opened in 1953. In 1989 work began on huge arts complex to include an operahouse (cap. over 3,000).

Operetta, first given at the T Momolo in 1850 and very popular, was neglected between the wars when two revue companies dominated the stage, but is again well established in its own T de Stat de Operetă.

BIBL: I. Masoff, *Istoria Teatrului Naţional din Bucureşti* (Bucharest, 1961).

**Büchner, Georg** (*b* Godelau, 17 Oct. 1813; *d* Zurich, 19 Feb. 1837). German dramatist. Operas based on his works are as follows:

*Dantons Tod* (1835): Einem (1947)
*Woyzeck* (1836, unfin.): Berg (1925); Gurlitt (1926); Pfister (1950)
*Leonce und Lena* (1836): Merz (1926); Weismann (1925); Schwaen (1961)
*Marie Tudor* (after Hugo, 1836): Wagner-Régeny (1935).

**Budapest.** Capital of Hungary (uniting in 1872 the cities of Buda and Pest). Regular opera performances were given in German from 1787, when the Castle T opened, some also in Hungarian from 1793, the year of the first national opera, Chudy's *Pikkó Herceg* (Prince Pikkó). Most important foreign operas were performed soon after their local premières. A large German City T opened in 1812; this burnt down in 1847. The Pestí Magyar Színház (Pest Hungarian T) opened on 22 Aug. 1837; the first opera to be staged was *Il barbiere di Siviglia* on 29 Aug. This remained the country's most important musical theatre until 1884: the directors were Ferenc *Erkel (1838–74), Hans *Richter (1871–5), and Sándor Erkel (1875–84). The repertory was at first largely Italian, with Hungarian translations gradually supervening, and this helped to encourage the development of national opera. From the 1840s Erkel's operas joined the repertory, and the theatre was quick to stage important foreign works as they appeared (with the major exception of Wagner's, resisted by Erkel but encouraged by Richter). It was a mature company of international standard that moved into the Operaház (cap. 1,310), opened on 27 Sep. 1884 with acts of *Lohengrin* and *Bánk bán*. All performances (even by guest artists) were in Hungarian until the turn of the century. Performances were also given at Buda Castle.

Mahler introduced much German opera (1888–91); Nikisch (1893–5) added more Italian, French, and Hungarian works. After a slack period (1900–12), the company was revivified under Miklós Bánffy: despite the war and its aftermath, operatic life was re-established on a healthy basis, and, despite opposition, the works of Bartók and Kodály entered the repertory. With Miklós Radnai as intendant (until 1935) and Sergio Failoni conducting (1928–48), many other contemporary works followed. Damaged in the war, the theatre reopened as the Magyar Allami Operaház (Hungarian State O) (cap. 1,343) on 15 Mar. 1945, and the repertory began to include more Russian works; the company was also much enlarged. Important conductors have included Ferenc Fricsay (1945–9), Otto *Klemperer (1947–50), and János Ferencsik (1931–84). The theatre closed for renovation in 1980–4, when the company moved to the Erkel T. Before the Second World War performances were also given in the open air on Margaret Island; today they are sometimes given in the courtyard of the Budapest Hilton. Chamber opera is also given on the stage of the Odry Theatre School. The company also toured (1948–53) as Gördülő Opera (Rolling Opera), visiting towns and industrial centres in a special train.

The city's second theatre opened in 1911 as the Népopera (People's O) (cap. 2,450). In 1917 it became the Városi Színház (Civic T), and from 1925 was privately owned, giving both opera operetta. It reopened as a filial of the OH in 1951, as the Erkel T. Operetta has also been given at the Pesti Népszínház (opened 1875), the Budai Népszínház (1861–4 and 1867–70), the Kisfaludy Színház (opened 1897), and especially the Király Színház (T Royal, opened 1903).

**Buenos Aires.** Capital of Argentina. At the beginning of the 19th cent. Buenos Aires had one small and simply constructed theatre, the Coliseo (1804, renamed the T Argentino in 1838). The first operatic performances here were selections from Italian opera organized by Juan Picazzarri in 1813. From various visiting singers a regular company was formed, and gave the first complete opera, *Il barbiere di Siviglia*, in 1825. Visiting French companies came from 1831. The T de la Victoria opened in 1838. From 1848 there were many performances of Bellini, Donizetti, Mercadante, and Verdi, by leading Italian singers. Two French companies were also formed in 1852 and 1854, both of which gave an almost exclusively French repertory; these two groups, though very different in quality, managed to win an increasingly loyal public. By 1854 30 new productions were given in the city; and two years later zarzuelas were added to the repertory.

These successes led to the demand for a new and improved theatre; and the first T *Colón (cap. 2,500, with a special women's gallery) was duly opened on 25 Apr. 1857 with *La traviata*. This became the centre of the country's operatic life. The smaller T de la Opera was opened on 25 May 1872 with *Il trovatore*; it was demolished in 1935. From 1868 the Colón was directed by the powerful and influential Angelo Ferrari; though he ran a basically Italian repertory that included early performances of works by Verdi, Ferrari also saw to it that the Colón reflected the general European operatic scene of the day, and even gave a number of premières of work by Marotta, Agostini, Cavalieri, and Torrens Boqué. In 1881 a visiting French company under Maurice Grau gave a season that included *Carmen*. At the same time the T de la Opera was widening its repertory to include more Italian works; both houses were by now of sufficient repute to attract the finest Italian singers of the day. The T de la Victoria had meanwhile not ceased activity, and was responsible for the première of the first opera by an Argentinian composer, Francisco Hargreaves's *La gata blanca* (1877). At the T Politeama, local premières were given, including *Le Villi* (1886), *Der fliegende Holländer*, and *Roméo et Juliette* (1887), *Lakmé* (1888), and *Otello* (1888, a few days ahead of the Colón); one of the stars of the season was Patti.

After the Colón was closed in 1889, the rebuilt T de la Opera became the city's operatic centre in a busy period. There were regular visits from major Italian singers and conductors (Campanini, 1889 and from 1893, Mascheroni from 1894, Mugnone from 1898, Toscanini in 1901, 1903–4, and 1906), and the repertory was broadened to include more German and French and even Argentinian works, though little pre-19th-cent. opera was given.

On 25 May 1908 a performance of *Aida* opened the new T Colón (cap. 2,487, with standing room for 1,000). In 1935 the administration passed into municipal hands. The long domination of Italian opera finally began to give way to a truly international repertory that included French, German, Russian, English, Czech, and Polish, as well as Argentinian and Spanish works, and later pre-Classical opera and Mozart. The first complete *Ring* was given in 1922 under Weingartner (with Lotte Lehmann). Moreover, the rapid expansion of the Colón's activities and its emergence as one of the world's great opera-houses were reflected in the busy life of the capital: at the T Coliseo, Mascagni conducted the première of his *Isabeau* (1911), *Parsifal* was given as early as 1913, and new Argentinian works were staged; new operas were also given at the T Politeama and a more limited and sporadic repertory at the T Avenida, T Odéon, and other smaller theatres. Buenos Aires remains the richest operatic city of the continent, and one capable of challenging the greatest European and North American cities in the excellence and diversity of its operatic life.

See also ARGENTINA, COLÓN.

**buffo** (It.: 'buffoon, comedian': possibly derived from the Lat. *bufo* = toad, from the inflated gloves used for exchanging comic blows on the stage). In the theatre a *buffone* is a comic actor; in opera, a *buffo* is a singer of comic roles, hence *basso buffo*. *Opera buffa is a term for comic opera.

**Buffonistenstreit.** See GUERRE DES BOUFFONS.

**Bühne** (Ger.: 'stage or theatre'). The term is also used in the plural for a theatre, e.g. Städtische Bühnen = Civic T.

**Bühnenfestspiel** (Ger.: 'stage festival play'). Wagner's term for *Der Ring der Nibelungen*, deriving from his wish to make the work's staging a special occasion.

**Bühnenmusik** (Ger.: 'stage music'). A term used to refer both to incidental music for plays and to music played on the stage during the performance of an opera.

**Bühnenweihfestspiel** (Ger.: 'stage-consecrating festival play'). Wagner's term for *Parsifal*, deriving from his wish to restore the sacred to drama (as in classical Greece and medieval Christendom).

**Bulgaria.** Western art forms were first introduced mainly by foreign bandmasters, after Bulgaria was liberated from Turkish rule (1396–1878). The first opera was heard in 1850. The first Bulgarian orchestra was founded in Shumen in 1851 by a Hungarian, Mihály Shafran (1824–1905). His successor Dobry Voynikov (1833–78) was forced to emigrate to Romania, where he ran a small musical theatrical company. In 1890 the first Bulgarian dramatic and operatic company, the Dramatichesko-Operna Trupa, was founded in Sofia. Though two series of performances were given in 1891 and 1908, when plans were put forward for a national opera, there were no regular seasons until 1910. The lack of practical assistance held up the development of opera; many Bulgarian singers went abroad to study, to Italy, Austria, or Russia, and it was a group of these who in 1907 founded the Bulgarskata Operna Druzhba (The Bulgarian Opera Society). Performances were given in the Naroden T (National T). The first opera sung in Bulgarian was *Pagliacci* (1909).

In 1878 Emanuil Manolov (1860–1902) left to study in Moscow. Back by 1886, he set himself to found a modern Bulgarian art music on the basis of folk-song, and composed the first Bulgarian opera, *Siromakhkinya* (The Poor Woman, 1900); though it is unfinished, two acts were given in Kazanluk in Dec. 1900. It was completed through the efforts of the Operna Druzhba (prod. Sofia 1910). School performances of opera had also begun in Shumen in 1909; but it was during the next decade that opera began to spread to other towns, notably Varna, Plovdiv, and Ruse.

The next Bulgarian opera was *Kamen i tsena* (The Stone Pyramid, 1911), with melodies by Ivan Ivanov (1862–1917), harmonized and scored by the Czech conductor Václav Kaucký (1857–1917) and prod. Sofia. Next came Georgy *Atanasov's historical opera *Borislav* (1910) and Dimiter Hadjigeorgyev's (1873–1922) *Takhir-begovitsa* (Takhir Beg's Wife, 1911). Atanasov is the most important creative figure of early Bulgarian opera. Also important was Todor Hadjyev (1881–1956), who conducted most of the new operas. In 1921 opera was established more professionally under Moysey Zlatin (1882–1953), Atanasov, and Hadjyev. Prominent opera and operetta composers included Pancho Vladigerov (1899–1978), with *Tsar Kaloyan* (1936); Veselin Stoyanov (1902–69), who studied in Vienna with Franz Schmidt and came under Schoenberg's influence but reverted to folk inspiration and wrote the comedy *Zhensko tsarstvo* (The Kingdom of Women, 1935), then *Salammbô* (1940); and Lyubomir Pipkov (1904–74), with *Yaninite devet bratya* (Jana's Nine Brothers, 1937), *Momchil* (1948), and *Antigona 43* (1963). More companies were formed in the 1930s as opera gained popularity, but there was no permanent opera in Sofia until after the war. A new impetus was given to operatic life, and there are now five National Operas, companies at Vratsa (opened 1954) and Burgas (opened 1955), subsidized amateur companies based on other towns (Shumen and Sliven), as well as about 30 other amateur groups working with professional assistance. There is a festival of opera and ballet every five years to which every opera-house in the country contributes. Composers writing for these companies included Marin Goleminov (*b* 1908), with *Ivaylo* (1958); Konstantin Ilyev (1924–88), with *Boyanskiyat maystor* (The Boyana Master); and Parashkev Hadjiyev (*b* 1912, son of Todor), with *Albena* (1962) and *Maystori* (The Masters, 1966).

In 1912 a professional operetta group was formed; this encouraged native operetta, which has become extremely popular and attracts a large following. Operetta composers include (as well as Atanasov) Asen Karastoyanov (1893–1976), with *Mikhail Strogov* (1940); Yosif Stankov (*b* 1911), with *Juana* (1939); and Boris Levyev (*b* 1902).

See also PLOVDIV, RUSE, SOFIA, STARA ZAGORA, VARNA.

BIBL: R. Bix, *Bulgarsky operen teatr* (Sofia, 1976).

**Bullant, Antoine** (*b* ?nr. Amiens, ?*c*.1750; *d* St Petersburg, ?June 1821). French composer and bassoonist. Active in Paris from *c*.1771, publishing a quantity of instrumental music. Travelled to St Petersburg in 1780, where he established a reputation as a virtuoso bassoonist and joined the court orchestra of Empress Catherine II. In 1783–99 he composed five comic operas for St Petersburg and Moscow, which show influence of French opéra-comique. *Sbitenshchik* was one of the most popular Russian operas of the 18th cent.

SELECT WORKLIST: *Sbitenshchik* (The Sbiten Seller) (Knyazhnin; St Petersburg 1784).

**Bülow, Hans von** (*b* Dresden, 8 Jan. 1830; *d* Cairo, 12 Feb. 1894). German conductor and pianist. Studied Dresden with Wieck and Eberwein; then studied law in Leipzig, also music with Hauptmann. In Berlin, wrote for *Die Abendpost*, defending Liszt and Wagner. The first *Lohengrin* in Weimar (1850) so overwhelmed him that he went to Zurich to seek out Wagner, who helped him to find some conducting experience. Studied the piano with Liszt, who also admired his talent, and whose daughter Cosima he married in 1857. Conductor Munich, N, from 1865, conducting premières of *Tristan* (1865) and *Meistersinger* (1868), of which he also made vocal scores. When Cosima left him for Wagner in 1869, he abandoned Munich and toured Europe. Hanover, H, 1878; music director Meiningen 1880. A brilliant pianist and one of the great virtuoso conductors of the day, he was also a man of difficult temperament. But he behaved nobly over Cosima's desertion, and sent her a consoling message on Wagner's death; and his forceful personality exercised a strong influence on musicians of his time, including the young Richard Strauss. His compositions do not include an opera, though he had considered one on the Tristan legend.

**Bumbry, Grace** (*b* St Louis, 4 Jan. 1937). US mezzo-soprano. Studied Santa Barbara with Lotte Lehmann. Début Paris, O, 1960 (Amneris). Bayreuth 1961 (Venus; the first Black artist to appear there); London, CG, from 1963; Salzburg from 1964; New York, M, from 1965, Paris, B, 1990 (at its inauguration). Roles include Eboli, Azucena, Fricka; from 1970 has also sung soprano parts, e.g. Tosca, Salome, Lady Macbeth. Has a warm, vibrant voice and a powerful personality. (R)

**Bungert, August** (*b* Mülheim, 14 Mar. 1845; *d* Leutesdorf, 26 Oct. 1915). German composer.

Studied with Kufferath and in Cologne and Paris, later also in Berlin with Kiel. In 1889 he formed a friendship with the Queen of Romania, who as 'Carmen Sylva' wrote poems for him to set and also used her influence to found in 1911 the *Bungert-Bund*, with a magazine *Der Bund*, for the advancement of his music. His first opera was a comedy, *Die Studenten von Salamanca*. However, he is best known for his Wagnerian tetralogies (one incomplete) on Homeric subjects, for the performance of which he hoped to build a Western 'Bayreuth' at Bad Godesberg on the Rhine.

SELECT WORKLIST (texts by composer): *Die Studenten von Salamanca* (Leipzig 1884); *Homerische Welt*. I. *Die Ilias*: 1. *Achilleus* (sketch); 2. *Klytämnestra* (sketch); Pts3–5 unwritten. II. *Die Odyssee*: 1. *Kirke* (Dresden 1898); 2. *Nausikaa* (Dresden 1901); 3. *Odysseus' Heimkehr* (Dresden 1896); 4. *Odysseus' Tod* (Dresden 1903).

BIBL: M. Chop, *August Bungert* (Berlin, 1915).

**Bunn, Alfred** (*b* London, ?1798; *d* Boulogne, 20 Dec. 1860). English theatre manager and librettist. Stage manager London, DL, 1823; manager Birmingham, T Royal; joint manager London, DL and CG, 1833. He was responsible for the appearances of Malibran and Schröder-Devrient 1833–5, paying Malibran £3,375 for 19 CG appearances. He also provided poor librettos and translations for a number of operas, including some by Balfe, Benedict, and Wallace. His cheese-paring methods were legendary, and artists would even find themselves engaged for both theatres on the same evening. During the 1834–5 season, 'female dancers pattered from one house to the other six times during the evening, and underwent the operation of dressing and undressing no less than eight'. Autobiography, *The Stage* (London, 1840).

**Buona figliuola, La** (The Good Girl). Properly *La Cecchina, ossia La buona figliuola*. Opera in 3 acts by Piccinni; text by Goldoni, after Samuel Richardson's novel *Pamela, or Virtue Rewarded* (1740). Prod. Rome, T delle Dame, 6 Feb. 1760, with Borghesi, Savi, De Cristofori, Lovattini, Casaccia; London, H, 25 Nov. 1766, with Guadagni, Savi, Lovattini, Morigi, Quercioli, Piatti, Michele. It has occasionally been revived in Italy since 1945.

The opera tells of the love and eventual marriage of Cecchina (sop) and the Marquis Conchiglia (ten). The Marquis's sister opposes the match, since Cecchina is apparently of low birth, and tries to spread malicious rumours about the girl. However, when it emerges that Cecchina is the long-lost daughter of a German baron, she is quite happy for the marriage to proceed.

Piccinni's eighteenth opera and greatest success. The text had already been set by Duni (1756) and Perillo (1759); later set by Graffigna (1886). Piccinni's less successful sequel was *La buona figliuola maritata* (text by Goldoni), prod. Bologna 10 June 1761 (also set by Scolari, 1762); a further sequel, by Latilla, was *La buona supposta vedova* (text by Bianchi), prod. Venice Carn. 1766.

**Buontalenti, Bernardo** (*b* Florence, 15 Dec. 1531; *d* Florence, 6 June 1608). Italian architect designer. He was in the service of the Medici, designing for them buildings and also stage spectacles that owed much of their effect to his mechanical ingenuity. He designed some of the earliest opera productions, confirming the standards of visual splendour, in celebration of aristocratic order, which he had already set in his stagings of intermedi. These were to be influential in the early years of the opera. His last production was for Caccini's *Il rapimento di Cefalo* (1600), a highly elaborate and ingenious set with many transformations.

**Burgess, Sally** (*b* Durban, 9 Oct. 1953). English mezzo-soprano. Studied London with Alan and Studholm. Début as soprano London, ENO, 1977 (Bertha, *Euryanthe*). Sang soprano roles until 1980; further study with Salaman and Veasey. London, CG, and Gly., 1983; O North, from 1986. Roles include Orfeo, Amneris, Dido (Berlioz), Charlotte, Carmen, Octavian, Nefertiti (*Akhnaten*), Julie La Verne (*Showboat*). A highly attractive performer with a fresh voice and a strong stage presence. (R)

**Burghersh, Lord (John Fane)** (*b* London, 3 Feb. 1784; *d* Wansford, 16 Oct. 1859). English composer and musical organizer, son of the 10th Earl of Westmorland (succeeded to the title 1841). He devoted such time as he could spare from his busy military, political, and diplomatic career to music, founding and for many years acting as president of the RAM. His compositions, which he promoted during his autocratic reign at the RAM, included seven operas in the Italian manner he regarded as the only authentic musical style.

BIBL: W. Cazalet, *The History of the RAM* (London, 1854).

**Burgstaller, Alois** (*b* Holzkirchen, 21 Sep. 1871; *d* Gmund, 19 Apr. 1945). German tenor. Studied Frankfurt with Bellwidt, and Bayreuth with Kniese. Début Bayreuth 1894. Bayreuth 1896–1902 (Siegfried (his greatest role), Siegmund, Parsifal, Erik). London, CG, 1906. New York, M, 1903–9. First American Parsifal; subsequently banned from Bayreuth for infringing copyright. (R)

**Burgtheater.** Viennese theatre (cap. 920); its full title was 'National Theater nächst der kaiserlich-königlichen Burg', a reference to its location next to the *Hofburg* (imperial palace). Its repertoire included drama, opera, and ballet; following its proclamation as an 'imperial theatre' in 1776 it became the main venue for operatic productions in Vienna until the end of the century. In 1778 Emperor Joseph II placed the house at the disposal of the German company (National-Singspiel), but following the waning of its popularity, an Italian troupe returned in 1783. Opera buffa began to dominate the repertory of the theatre, but after 1788 the imperial subsidy for Italian opera was withdrawn. Gradually spoken drama began to play a more important role and in 1810 it was decided that all operatic performances should be given at the *Kärntnertortheater. The appointment of Joseph Schreyvogel in 1814 ushered in a glittering phase for the theatre, during which it gained a reputation as the finest theatre for spoken drama in any German-speaking country. Premières given at the Burgtheater included Gluck's *Orfeo* (1762), *Alceste* (1767), and *Paride ed Elena* (1770), Mozart's *Entführung* (1782), *Figaro* (1786), and *Così fan tutte* (1791), Cimarosa's *Il matrimonio segreto* (1792), and Martín y Soler's *Una cosa rara* (1786).

**Burian, Karel** (also known as Carl Burrian) (*b* Rousinov, 12 Jan. 1870; *d* Senomaty, 25 Sep. 1924). Czech tenor. Studied with Wallerstein. Début Brno 1891 (Jeník). Hamburg 1898–1902; Dresden from 1902 (creating Herod, 1905); London, CG, 1904–14; New York, M, 1906–13; Bayreuth 1908. Sang mainly Wagner roles (Tristan, Parsifal, Tannhäuser, etc.). Linguistically gifted, he possessed a powerful voice, and an impressive stage presence. Arnold Bax wrote of his 'horrifying Herod, slobbering with lust and apparently almost decomposing before our eyes'. (R) His brother **Emil** (1876–1926) was a leading baritone at Prague, N, 1906–26, and sang regularly in Germany.

His nephew was **Emil František Burian**, (*b* Plzeň, 11 Apr. 1904; *d* Prague, 9 Aug. 1959). Czech composer, stage director, and writer. Studied Prague with Foerster. He was then already working in Prague avant-garde theatre and music, later also in Brno and Olomouc and elsewhere in various roles (including as jazz band leader). In 1933 he founded the experimental theatre D34 (for *divadlo*, or theatre, of 1934), mounting many avant-garde productions that anticipate so-called 'director's theatre' (e.g. a Brecht–Weill *Dreigroschenoper*, D35) and place him as one of the most important directors of the century. He returned after the war (which he spent in a concentration camp) in 1945 to reopen it (as D46) and to work as a director in Prague

and Brno. His busy theatrical and public life included the composition of a number of operas. His earlier works are cast in a post-Romantic style, which he discarded in favour of an attachment to folk and cabaret music, jazz, and speech rhythms in a manner strongly influenced by Stravinsky and especially Janáček. He experimented with transferring essentially musical rhythms into spoken dialogue; and his work in the theatre, including the musical theatre, involved functional use of lighting and film. Janáček's manner shows particularly in his most important opera, *Maryša*.

SELECT WORKLIST: *Alladine a Palomid* (Burian, after Maeterlinck; comp. 1923); *Mastičkař* (The Quack) (Burian, after medieval text; Prague 1926); *Maryša* (Burian, after A. & V. Mrštíkové; Brno 1940). Also operettas and other stage works and films.

**Burkhard, Paul** (*b* Zurich, 21 Dec. 1911; *d* Zell, 6 Sep. 1977). Swiss composer and conductor. Studied Zurich. Cond. Berne Stadttheater (1932–4) and Zurich Schauspielhaus (1939–45). Popular for his operettas, including *Casanova in der Schweiz* (1942) and *Das schwarze Recht* (1939, rev. as *Feuerwerk* 1950).

**Burkhard, Willy** (*b* Évilard-sur-Bienne, 17 Apr. 1900; *d* Zurich, 18 June 1955). Swiss composer. Studied Berne, Leipzig, Munich, and Paris. Best known for his sacred choral music, he achieved considerable success with his only opera, *Die schwarze Spinne* (Faesi and Bonner, after Gotthelf; Zurich 1948, rev. Basle 1954).

**burla.** See BURLETTA.

**burlesca.** See BURLESQUE.

**burlesque** (Fr.) (It., *burlesca*; Ger., *Burleske*). The French term, also used in English, for a musical entertaiment parodying serious opera which developed in England during the early 18th cent. after the model of plays satirizing the content and conventions of serious theatre. Frequently coarse and vulgar in tone, it was established with Estcourt's *Prunella* (1708) and reached an early peak with Carey's *The Dragon of Wantley* (1737), with music by Lampe. Its initial development was bound up with that of *ballad opera, whose model it sometimes followed in the adaptation of folk tunes. It survived into the 19th cent. and always made musical parody its aim: satire of other matters was less important. Introduced into the USA in the 1860s, it gradually lost its connection with musical satire and came instead to be a kind of extravagant stage spectacle whose main attraction was the risqué attire of the female cast.

**burletta** (It.: 'little joke' or 'jest', sometimes also burla or burlettina). Though not used by composers, these terms were widely used in Italy in the 18th cent. for comic operas, sometimes to distinguish between such works and comic *intermezzos. Also used widely in England, where they came to refer to works which satirized serious opera; unlike the *burlesque, they did not include musical parody.

**burlettina**. See BURLETTA.

**Burnacini, Giovanni** (*b* Cesena, *c*.1605; *d* Vienna, 21 July 1655). Italian designer and architect, father of below. Worked at Venice, GP, *c*.1640, and T Santi Apostoli, justifiably claiming to have pioneered the skilful stage machinery that became a mark of Venetian opera. He may have staged some of Monteverdi's operas. His designs maintained the central perspective, with well-judged proportions that transcended the limitations of the small stage in creating an illusion of space and movement. In Vienna, with his brother Marc'Antonio from 1651, he was stage designer for the court, founding traditions that were accepted and developed by the *Bibiena family.

**Burnacini, Ludovico** (*b* ?Mantua, 1636; *d* Vienna, 12 Dec. 1707). Italian designer and architect, son of above. Went to Vienna 1651 with his father, succeeding him 1655 as architect of the court theatres and designer of all the court entertainments. Building upon his father's achievements, he developed a highly colourful, elaborate, and sensational style, often charged with symbolism, in his sets for formal court opera: the grandest of these was Cesti's *Il pomo d'oro*, but designs for other works testify to his mastery of a range of theatrical devices that included fireworks. His comic gift shows especially in his costume designs, which reflect the grotesque masks and distortions of the Venice Carnival subjected to a rougher Teutonic humour.

**Burney, Charles** (*b* Shrewsbury, 7 Apr. 1726; *d* London, 12 Apr. 1814). English writer, composer, and organist. Author of a 4-vol. *History of Music* (1776–89), and of a biography of Metastasio. Musical impressions gleaned during his travels through Europe are recorded in his two books, *The Present State of Music in France and Italy* (1771), and *The Present State of Music in Germany, the Netherlands, and the United Provinces* (1773), which contain interesting contemporary opinions of opera.

**Burning Fiery Furnace, The**. Church parable by Britten; text by William Plomer. Prod. Orford Church, Suffolk, 9 June 1966, with Pears, Tear, Drake, Shirley-Quirk, Godfrey; Caramoor Festival, New York, 25 June 1967, with Velis, Metcalf, Lankston, Berberian, Pierson.

Based closely on the biblical narrative in Daniel 3, the work is set in Babylon in the 6th cent. BC. Three young Israelite captives are appointed by Nebuchadnezzar (ten) to rule over three provinces. At a feast, they arouse the wrath of the Astrologer (bar) by refusing to eat with the courtiers, since their faith will not permit this. They also refuse to worship the Babylonian god Merodak. Nebuchadnezzar commands that a furnace be lit; the three are thrown into it but are saved by an angel. When Nebuchadnezzar sees the result of their faith he himself becomes a convert.

**Burrowes, Norma** (*b* Bangor, NI, 24 Apr. 1944). Irish soprano. Studied London with Flora Nielsen, and later with Bruce-Lockhart. Début Gly. Touring O 1970 (Zerlina). London, CG, from 1970 (Fiakermilli, Woodbird, Despina). ENO from 1971; Salzburg 1971–3 (Blonde); Gly. 1975 (Vixen); Aix 1977; Paris 1976–80. An attractive singer with a clear, appealing voice, whose early retirement was much regretted. (R)

**Burrows, Stuart** (*b* Pontypridd, 7 Feb. 1933). Welsh tenor. Studied Trinity Coll., Carmarthen. Début WNO 1963 (Ismaele, *Nabucco*). London, CG, from 1967; San Francisco, 1967; New York, M, from 1972. Guest appearances Vienna, Paris, etc. Repertory includes Elvino, Ernesto, Lensky, Alfredo; also the Mozart roles, of which, with his fine coloratura and beauty of tone, he is considered an excellent exponent. (R)

**Bury, John** (*b* Aberystwyth, 27 Jan. 1925). English designer. Studied London. After working with Joan Littlewood and Peter Hall, contributed a number of highly effective sets for opera. With Hall as producer, he designed *Moses and Aaron*, *The Magic Flute*, and *Tristan* for London CG, *Calisto*, *Ulisse*, *Don Giovanni*, and *Figaro* for Gly. His designs for Monteverdi and Cavalli in particular succeeded in catching a Baroque spirit, including in the use of modern stage apparatus to suggest the Baroque ingenuity.

**Buryat Mongolia**. Formed as an autonomous republic of the USSR in 1923, the region previously had little organized cultural life, though there flourished a primitive kind of sung and danced stage epic known as *ekhor*. In 1926 there were organized the first music courses, whose students travelled the region giving small quasi-operatic entertainments on a special railway carriage. The State T was opened in the capital, Ulan-Ude, on 7 July 1932; the director was Pavel Berlinsky (*b* 1900), who also composed the first Buryat opera, *Bair* (1938). This was followed by *Enkhe Bulat-bator* (Enkhe, the Hero of Steel, 1940), by Markian Frolov (1892–1944); both works were included in a Moscow season of Buryat works that year. In 1941 Berlinsky wrote a musical comedy, *Schaste* (Happiness), but the work's success was inhibited by difficulties with local singers used to a native musical tradition of unison singing and a pentatonic scale. Nevertheless, Berlinsky managed to stage productions of *Eugene Onegin* and *Faust* (1943).

After the war, many singers went to study in Moscow and Leningrad, and on their return they began to build up a repertory of Russian and Western classics. Lev Knipper and Sergey Ryauzov came to study the folklore and to compose, respectively, *Na Baykale* (On Lake Baikal) and *Medegmasha* (both 1948). In 1948 the opera group separated from the drama section, and in 1952 the theatre was reorganized in a new building as the Order of Lenin Opera and Ballet T (cap. 718). In 1960 the company visited Moscow with a repertory that included *Prince Igor* (in Mongolian). Other Buryat operas include the comedy *Prodelki Dyadi Morgon* (Uncle Morgon's Tricks, 1957), the influential *Prozrenie* (Insight), and *Chudesny klad* (The Miraculous Treasure, 1970) by Baudorzha Yampilov (*b* 1916), and *Pobratimy* (Blood Brothers, 1958), on a 17th-cent. subject, and *Sayan* (1967) by Dandar Ayusheyev (1910–71).

BIBL: P. Gubevich, *Operny teatr Buryati* (Ulan-Ude, 1964).

**Busch, Fritz** (*b* Siegen, 13 Mar. 1890; *d* London, 15 Sep. 1951). German conductor. Studied locally, then Cologne with Steinbach. Cond. Deutsches T, Riga, 1909. Music director Aachen 1909 and, after war service, 1918; Stuttgart 1918, giving Hindemith premières and broadening the repertory. Music director Dresden 1922–33, building a fine ensemble and giving premières of Strauss's *Intermezzo* (1924) and *Ägyptische Helena* (1928), Busoni's *Doktor Faust* (1925), Weill's *Protagonist* (1926), and Hindemith's *Cardillac* (1926). Always worked closely with designers including Appia and Kokoschka; from 1932 also collaborated with Carl *Ebert. Though not Jewish, resisted Nazism and left Germany 1933. Buenos Aires 1933–6, 1941–5. With Ebert, established Gly. 1934–9, 1950–1. New York, M, 1945–9; many European appearances. A fine representative of the best German opera-house traditions, he brought to Mozart, in particular, a care and sensitivity (extending to his work with his singers and players) that gave Gly. some of its classic years and bequeathed a legacy of treasured recordings. Autobiography, *Aus dem Leben eines Musikers* (Zurich, 1949, trans. 1953) (R)

**Busenello, Gian Francesco** (*b* Venice, 24 Feb. 1598; *d* Legnaro, 27 Oct. 1659). Italian librettist. The son of a lawyer, he spent most of his life in his native city following the same career; he was also a member of the Accademia degli Incogniti. His sharp, somewhat cynical realism, combined with a vivid awareness of sex, made him an ideal librettist for the new style of commercial opera. His greatest work is *L'incoronazione di Poppea* (1643). In this, there is not only a sense of the stage, of the nature and powers of a composer, of verbal and syllabic niceties, but above all insight into human behaviour. He possessed the ability to view even the most culpable of characters with an understanding backed by a wry tolerance and a sardonic humour, qualities which did much to enlarge the scope of the young art of opera. His other librettos are for Cavalli's *Gli amori di Apollo e di Dafne* (1640), *Didone* (1641), *La prosperità infelice di Giulio Cesare dittatore* (?1646), and *Statira* (1656). According to Livingston, he also wrote a sixth libretto, *La discesa d'Enea all'inferno* (1640).

BIBL: A. Livingston, *La vita veneziana nelle opere di G. F. Busenello* (Venice, 1913).

**Bush, Alan** (*b* London, 22 Dec. 1900). English composer. Studied RAM with Corder and Matthay, later also with Ireland, Moiseiwitsch, and Schnabel. The intellectual rigour of his idiom, which is fundamental tonal, was increasingly used to simplify his music, not least in the interests of the broadly popular approach associated with his belief in Communism. This has also dictated the choice of subjects for his operas, which had more of a following in Communist countries, especially the GDR, than in England. Nevertheless, the quality of mind shown in his music has won him many admirers among those who cannot share his sometimes simplistic operatic portrayals.

SELECT WORKLIST: *The Press Gang* (Nancy Bush; Letchworth 1947); *Wat Tyler* (N. Bush; Leipzig 1953); *Men of Blackmoor* (N. Bush; Weimar 1956).

**Busoni, Ferruccio** (*b* Empoli, 1 Apr. 1866; *d* Berlin, 27 July 1924). German-Italian composer and pianist. Studied with his parents. In his operas, he consciously sought to portray inner psychological states and scenic atmosphere and meaning, rather than the outward and visible action. He argued his position carefully in various writings, influential for their shrewd and pioneering intellectual reasoning. *Die Brautwahl* reveals his ability to create atmosphere, in this case the fantastic and sinister. Nevertheless, he also sought a light touch (as in *Arlecchino*) and a sharp use of irony (as in *Turandot*): his Italian heritage gave him an instinct for the stage that also shows in his excellent vocal writing. His masterpiece *Doktor Faust* is a work of great power that confronts, among much else, the nature of the creative quest. It is further characteristic of Busoni in its highly unusual construction, and in its capacity to draw on the long range of European music from Renaissance polyphony to extreme chromaticism. He wrote the libretto for Schoeck's *Das Wandbild*.

WORKLIST: *Sigune, oder Das versunkene Dorf* (Schanz, after Baumbach; comp. 1889, unprod.); *Die Brautwahl* (Busoni, after Hoffmann; Hamburg 1912); *Arlecchino* (Busoni; Zurich 1917); *Turandot* (Busoni, after Gozzi; Zurich 1917); *Doktor Faust* (Busoni, after Marlowe; unfin., prod. Dresden 1925).

SELECT WRITINGS: *Entwurf einer neuen Ästhetik der Tonkunst* (Trieste, 1907, trans. 1911).

BIBL: E. J. Dent, *Ferruccio Busoni* (London, 1933, 2/ 1974); A. Beaumont, *Busoni* (London, 1985); A. Beaumont, ed., *Letters of Ferruccio Busoni* (London, 1987).

**Bussani, Francesco** (*b* Rome, 1743; *d* after 1807). Italian bass. Appeared Rome 1763 as tenor in Guglielmi's *Le contadine bizzarre*. Sang throughout Italy, appearing as a buffo in works by Galuppi, Salieri, and Paisiello. By 1777 his voice was a bass-baritone, and in 1783 he went to Vienna (where he had first sung in 1772), remaining until 1794. Here he was active as stage-manager as well as singing in the Italian repertory. He was Bartolo and Antonio in the first *Figaro*, sang both Masetto and the Commendatore in the first Vienna *Don Giovanni*, and created Don Alfonso. Also created Geronimo in *Il matrimonio segreto*. Da Ponte disliked him, finding him not to be a 'gentleman', and also wrote slightingly of his wife **Dorotea Sardi** (*b* Vienna, 1763; *d* after 1810), the first Cherubino and Despina (and Fidalma in *Il matrimonio segreto*); she was popular with the public for her pretty figure, engaging voice, and vivacious acting, but not with Da Ponte's mistress Ferrarese, the first Fiordiligi. Bussani and his wife (whom he had married in 1786) left for Italy in 1795 and worked there for the next ten years. They both went to Lisbon in 1807; she appeared in London in 1809, without great success.

**Bussani, Giacomo** (*b* Venice, *c.*1640; *d* Venice, after 1680). Italian librettist. Canon regular at the Carità, he provided librettos on heroic themes for Antonio Sartorio, Pallavicino, and Pietro Agostini. His libretto for Sartorio's *Giulio Cesare in Egitto* (1676) was set by many later composers, including Jommelli, Sarti, and Piccinni.

**Büsser, Henri** (*b* Toulouse, 16 Jan. 1872; *d* Paris, 30 Dec. 1973). French composer and conductor. Studied Paris with Franck, Widor, and Guiraud, also benefiting from Gounod's advice. Prix de Rome 1893. Cond. T du Château d'Eau 1900; Paris, OC, 1902 (incl. early perfs. of *Pelléas*); Paris, O, 1905–38. Made versions of *Les Indes galantes* (1952) and *Oberon* (1954), the latter including orchestrations of Weber's piano works. Also revived Bizet's *Ivan IV* (Bordeaux 1951). Continued to conduct at Paris, O and OC, in the

1950s. His most successful operas, traditional in language, are *Colomba* (1921) and *Le carrosse du Saint-Sacrement* (1948). (R)

**Bussotti, Sylvano** (*b* Florence, 1 Oct. 1931). Italian composer. Studied Florence Cons. with Lupi and Dallapiccola and privately 1949–56. His distinctive avant-garde approach was influenced by membership of Gruppo 70 and contact with Boulez and Cage. Artistic director of Venice, F, 1975. His stage works range from music theatre to genuine opera and are characterized by a general fascination with themes of sexual decadence and political protest: typical are the 'operina monodanza' *Syron-Sadun-settimino*, the 'dramma lirico' *Nottetempo*, and the opera *Lorenzaccio*. His most notorious work is *La passion selon Sade*, a macabre psychological study in which the audience is eventually invited to participate.

SELECT WORKLIST: *La passion selon Sade* (Loulié; Stockholm 1969); *Lorenzaccio* (Bussotti, after Musset; Florence 1972); *Syron-Sadun-settimino* (Maraini & Bussotti; Royan 1974); *Nottetempo* (Amidei & Bussotti; Milan 1976) *Fedra* (Bussotti, after Racine; Rome, O, 19 Apr. 1988); *L'ispirazione* (Bussotti, after Bloch; Florence, C, 26 May 1988).

**Butt,** (Dame) **Clara** (*b* Steyning, 1 Feb. 1872; *d* North Stoke, 23 Jan. 1936). English contralto. Studied Bristol with Rootham. Début London, RAH, 1892 (Ursula, Sullivan's *Golden Legend*). Spent most of her career as concert artist, but appeared in opera as student and in 1920 at London, CG, under Beecham as Orpheus. A striking, majestic woman, with a voice to match; Elgar wrote *Sea Pictures* for her. (R)

**Buxton.** Town in Derbyshire, England. The Opera House (built 1903) was restored in 1979 (cap. 980) for the opening of the festival, which has concentrated on reviving neglected works, generally in connection with a specific theme. Outstanding productions have included the British premières of Kodály's *Spinning Room* (1982), Vivaldi's *Griselda* (1983), and Donizetti's *Pigmalione* (1987), also *Béatrice et Bénédict* (1980), and *Medée* (with Rosalind Plowright, 1984).

**Byelorussia.** See BELORUSSIA.

**Byron, George Gordon,** (Lord) (*b* London, 22 Jan. 1788; *d* Missolonghi, 19 Apr. 1824). English poet. Operas on his works are as follows:

*The Bride of Abydos* (1813): Poniatowski (1845); A. Fell (1853); F. Sand (1858); T. Dubois (1864); Barthe (1865); Lebrun (1897)
*The Giaour* (1813): Bovery (1840); Hermann (1866); N. Berg (*Leila*, 1912); Delmas (1928)
*The Corsair* (1814): Pacini (1831); Schumann

(1844, unfin.); Arditi (1847); Verdi (1848); Deffel (1871); Marracino (1900)

*Lara* (1814): Ruolz (1835); Salvi (1843); Maillart (1864); Marsick (1929)

*Parisina* (1816): Donizetti (1833); Giribaldi (1878); Keurvels (1890); Veneziani (1901); Mascagni (1913)

*The Siege of Corinth* (1816): A. Cahen (1890)

*Manfred* (1817): Petrella (1872); Bogatyryov (1926)

*Marino Faliero* (1820): Donizetti (1835); Holstein (1881, unfin.); Freudenberg (1889)

*Cain* (1821): Delvincourt (*Lucifer*, 1948); Schmodtmann (1952); Lattuada (1957)

*The Two Foscari* (1821): Verdi (1844); Bogatyryov (1940)

*Sardanapalus* (1821): Litta (1844); Alary (1852); Joncières (1867); Maître (1870); Famintsyn (1875); Libani (1880); Duvernoy (1882, prod. 1892); Grunenwald (1961)

*Don Juan* (1824, unfin.): Blaramberg (comp. 1902). Canto 2 (the Haidee episode): Polignac (1877); Fibich (*Hedy*, 1896)

Also operas *Lord Byron* by Giarda (1910) and Thomson (1972); *Lord Byron's Love Letter* by R. de Banfield (with text by Tennessee Williams, 1955).

# C

cabaletta (It., also *cabbaletta*, *cavaletta*). A term of uncertain etymology, usually applied to a short aria of simple, reiterated rhythm, with repeats, often cast in a popular style. An early example is 'Le belle imagini' in Gluck's *Paride ed Elena*. The word has also been used to describe the first section of an aria; this would on its reappearance be varied, often with triplets in the accompaniment (this has suggested a derivation from *cavallo*, a horse, from the galloping movement).

In the 19th cent. the term came to mean the final section only of an aria or duet in two or sometimes more parts, usually quick and brilliant, and preceded by a contrasting *cantabile. In its classic form the cabaletta has two quatrains which are set as a 16- or 32-bar musical unit, with an initial regularity which is gradually loosened to provide opportunity for bravura display. One of the most distinctive features of the *Code Rossini, the cabaletta was usually repeated, with the same words, after a brief orchestral ritornello, as Rossini himself described to Clara Novello. This reprise, which would include improvised embellishments, inevitably weakened the dramatic progress of the opera, and was later viewed as one of the most stultifying conventions of 19th-cent. Italian opera. Although both Bellini and Donizetti wrote many cabalettas, their most celebrated exponent was Pacini, who was known in his day as the 'master of the cabaletta'. Gradually, and largely under Verdi's influence, it was to disappear although vestiges of its influence may be traced in the final section of the mid- and late-19th-cent. aria. Famous examples of the genre include 'Ah! non giunge' in *La sonnambula*, 'Sempre libera' in *La traviata*, and 'Di quella pira' in *Il trovatore*. In *The Rake's Progress* Stravinsky included a cabaletta for Anne which is a skilful musical parody of the Rossinian form.

Caballé, Montserrat (*b* Barcelona, 12 Apr. 1933). Spanish soprano. Studied Barcelona with Eugenia Kemmeny and Annovazzi. Début Basle 1956 (First Lady). Basle 1956–9; Milan, S, 1960 (Flowermaiden) and from 1969 in major roles; New York, Carnegie H, 1965 (Lucrezia Borgia (from which dates her international fame)); New York, M, 1965; Gly. 1965 (Mozart's Countess and R. Strauss's Marschallin); London, CG, from 1972. Her wide repertory includes much

Bellini and Donizetti, Verdi, and Rossini; also the roles of Manon Lescaut, Mimì, Tosca, Tatyana, Adriana Lecouvreur, Salome. Physically strong and musically sensitive, she is, with her limpid tone, impeccable *messa di voce*, and vocal versatility, a consummate singer. (R)

Caballero, Manuel Fernández (*b* Murcia, 14 Mar. 1835; *d* Madrid, 26 Feb. 1906). Spanish composer. Studied Madrid with Eslava. One of the most successful of all zarzuela composers. His most popular works, of some 220, were *Los sobrinos del capitán Grant* (1877), *El salto del pasiego* (1878), *La viejecita* (1897), and *El señor Joaquín* (1898). His *El duo de la Africana* (1893) has as its location an Italian opera company playing Meyerbeer's *L'Africaine*.

Caccini, Francesca (*b* Florence, 18 Sep. 1587; *d* ?Florence, *c*.1626). Italian singer, composer, and teacher, daughter of Giulio *Caccini. Studied with her father and, together with her mother and sister Settimia (1591–? *c*.1638), and participated in the lavish musical events of the Florentine court, also appearing with them in Paris in 1604–5. Known as 'La Cecchina', she was, according to Monteverdi, a skilled performer on the lute, guitar, and harpsichord.

It was apparently the poet Buonarroti who encouraged her to compose. At first she concentrated on festive ballets, including *Il ballo delle zigane* (The Ballet of the Gypsies) (1615; lost) in which she herself took part. The collection entitled *Primo libro delle musiche*, published in 1618, contains skilled monodic writing after her father's style and suggests a composer of some dramatic potential. This was fulfilled in 1622, when Francesca collaborated with Gagliano on the azione sacra *Il martirio di Sant'Agata*, and in 1625 by her opera *La liberazione di Ruggiero dall' isola d'Alcina*. Written for the visit of Prince Władisław Zygmunt of Poland to the Tuscan court, its prominent use of dance suggests the influence of the intermedio, although the writing for solo voices demonstrates a sympathetic understanding of the new genre. Based on an episode from Ariosto's epic *Orlando furioso*, it is one of the earliest operas whose subject is not taken from classical mythology: it is also the first known opera by a female composer. *La liberazione di Ruggiero* clearly impressed the visiting prince: in 1628 it was performed in Cracow and published there in a Polish translation the same year. Probably the first Italian opera to be performed abroad. Its production in Cracow was a seminal influence on the development of opera in Poland; on his accession to the throne, Prince Władisław immediately established an Italian opera troupe.

Francesca Caccini married the composer **Giovanni Battista Signorini Malaspina**, by whom she had two children who themselves performed at the Medici court.

WORKLIST: *Il martirio di Sant'Agata* (Cicognini; Florence 1622, comp. with Gagliano); *La liberazione di Ruggiero dall'isola d'Alcina* (Saracinelli, after Ariosto; Florence 1625).

**Caccini, Giulio** (*b* Rome or ?nr. Tivoli, 8 October 1551; *d* Florence, 10 Dec. 1610). Italian singer, lutenist, harpist, composer, and teacher; sometimes known as Giulio Romano after his native city. Active at the Medici court from around 1564, Caccini quickly established a reputation as one of the most gifted and versatile singers of his day and in 1604–5 spent several months in Paris. He contributed towards the music for the renowned intermedi to Bargagli's *La pellegrina* in 1589 and in 1600 replaced Cavalieri as superintendant of actors and musicians at the Tuscan court.

Caccini was one of the first and most influential exponents of *stile rappresentativo. A member of the Bardi *Camerata, he participated in most of the early Florentine experiments, both as performer and composer. It is unlikely that, as sometimes claimed, he was involved with the composition of Peri's *Dafne* in 1598: his first operatic work came in 1600 with *Il rapimento di Cefalo*, the same year that he set Rinuccini's *Euridice* libretto. Peri, who used the same text, acknowledged in the preface to his score that some of the music had been provided by Caccini; although the score of Caccini's own *Euridice* appeared in 1600, and was the first opera to be published, it was not performed until 1602. Later, Caccini was to claim priority over Peri as the 'inventor' of the operatic style.

Caccini's greatest monument is the collection *Le nuove musiche*, which came to be regarded as one of the most seminal works in the new monodic style. Its very title expressed the mood of inventiveness and radicalism characteristic of Florentine musicians at the time, while its preface provided a cogent summary of their objectives. Caccini intended to provide 'a kind of music by which men might, as it were, talk in harmony', hence he 'endeavoured the imitation of the conceit of the words'. Though still partly dependent on the old techniques of madrigalian writing, some examples of which are included, Caccini followed the lead of Vincenzo Galilei in songs where the shape and pace of the music is clearly determined by the inflexion of the individual words. Unlike many of his contemporaries, he was aware that the re-creation of Greek tragedy in music—the avowed aim of the Camerata—could result in an over-rigid adherence to a dry, academic style. But by drawing on his comprehensive understanding of vocal technique and ornamentation, he produced music of considerable flexibility and eloquence, which may indeed be seen as the very beginning of *bel canto. Such was the popularity and importance of *Le nuove musiche*, which includes the only extant part of his opera *Il rapimento di Cefalo*, the final chorus, that a second collection was published in 1614.

Caccini's family was closely involved with his musical activities. Both of his wives, Lucia and Margherita, were singers, as were his two daughters, **Settimia** (1591–1638), who married the Mantuan composer Alessandro Ghivizzani (*c*.1572–*c*.1632) and created the role of Venus in Monteverdi's *Arianna* in Mantua (1608), and Francesca *Caccini.

Among his many pupils were the singers Giovanni Magli, Francesco Rasi, who created the title-role in Monteverdi's *Orfeo*, and the composer Severo Bonini.

WORKLIST: 'Io che dal ciel cader farei la luna' (from intermedio for Bargagli's *La pellegrina*) (G. B. Strozzi; Florence 1589); *Euridice* (Rinuccini; Florence 1602, printed 1600, contributed some arias and choruses for Peri's setting); *Il rapimento de Cefalo* (Chiabrera; Florence 1600, comp. with Venturi del Nibbio, Bati, & P. Strozzi); *Euridice* (Rinuccini; Florence 1600; own setting, printed 1602); *Le nuove musiche* (Florence, 1602); *Nuove musiche e nuova maniera di scriverle* (Florence, 1614).

BIBL: R. Giazotto, *Le due patrie di Giulio Caccini, musico mediceo (1551–1618)* (Florence, 1984).

**cadenza** (It.: 'cadence'). An elaborate flourish ornamenting a perfect cadence at any point in an aria. In its classic form the singer extemporized a display of vocal virtuosity over a tonic chord in its second inversion, eventually coming to rest on the dominant, which was played by the accompaniment with its succeeding tonic. The style of ornamentation had its origins in the melismas of early monody, but often resorted to conventional formulas and patterns which had little connection with the musical substance of the aria proper.

The rise in importance of the cadenza mirrored the general increase in the use of extemporized ornamentation in the 17th and 18th cents., and by the time of A. Scarlatti and Hasse the inclusion of cadenzas was standard. The ternary form of the da capo aria lent itself to three cadenzas, the last of which was always an opportunity for the singer's most brilliant acrobatics: in arias with obbligato, there was sometimes an opportunity for the instrumentalist to perform a cadenza as well. The excesses of some singers were reproached by *Tosi in 1723, but attempts to curb excessively flamboyant cadenzas were generally met with little success. By the 19th cent. many composers had taken to writing out the cadenzas, often shaping them to the characteristics of particular singers (e.g. Rossini's cadenzas for Patti in *Semiramide*) and providing a greater degree of musical integration with the aria. Now, however, singers also studied harmony, in order to prepare their own cadenzas for insertion into operas: although the results often enraged the composer, they usually pleased the audience, which revelled in the display of vocal skill. Some of these interpolations found their way into the many volumes of cadenzas which were edited by singers and voice teachers in the 19th cent. One of the most famous of such collections was Marchesi's *Variantes et points d'orgue*, which gave nine different cadenzas for 'Ah, fors'è lui' in *La traviata*.

**Cadman, Charles Wakefield** (*b* Johnstown, PA, 24 Dec. 1881; *d* Los Angeles, 30 Dec. 1946). US composer and pianist. Studied with Oehmler and Kunitz. His lifelong interest in American Indian music was reflected in a number of operas, of which by far the most successful was *Shanewis*. In them, he draws upon Indian folk legend and incorporates Indian music into his own late-Romantic idiom.

SELECT WORKLIST: *Shanewis, or The Robin Woman* (Eberhart; New York 1918); *The Garden of Mystery* (Eberhart; New York 1925); *A Witch of Salem* (Eberhart; Chicago 1926).

**Caesar.** See GIULIO CESARE.

**Cafaro, Pasquale** (*b* S Pietro in Galatina, nr. Lecce, 8 Feb. ?1716; *d* Naples 25 Oct. 1787). Italian composer. Studied Naples with Leo and Fago. In 1759 succeeded Abos at the Cons. di S Maria: from 1768 held appointments at the royal chapel including that of singing and harpsichord master for Queen Maria Caroline from 1771. Conducted performances at Naples, C, 1763–6.

As well as many oratorios and cantatas, Cafaro wrote seven operas, including *Ipermestra*, *La disfatta di Dario*, and *Creso*, which form a significant bridge in the development of Neapolitan opera between Leo and such later figures as Paisiello, Piccinni, and Cimarosa. Not an especially talented composer: perhaps the most charitable view of Cafaro and his work is that provided by the contemporary who likened him to a majestic and placid river.

SELECT WORKLIST: *Ipermestra* (Metastasio; Naples 1756); *La disfatta di Dario* (The Defeat of Darius) (Morbilli; Naples 1756); *Creso* (Pizzi; Turin 1768).

**Caffarelli** (orig. Gaetano Maiorano) (*b* Bitonto, 12 Apr. 1710; *d* Naples, 31 Jan. 1783). Italian mezzo-soprano castrato. Studied Naples with Porpora (who allegedly described him as 'the greatest singer in Europe'). Début Rome 1726 (in Sarro's *Valdemaro*). After triumphs in Venice, Milan, Rome, Genoa, and Bologna, he settled in Naples, where he sang regularly 1734–54 at court and in the T San Carlo and T San Bartolomeo. He visited London 1738–9, creating title-roles in Handel's *Faramondo* and *Serse* (whose 'Ombra mai fù' was written for him), but was not a great success, being very much in Farinelli's shadow. Sang further throughout Italy, and went to France in 1753, giving private recitals to cheer the Dauphine in her pregnancy as well as singing in Paris. Left in 1754 hurriedly, after insulting Louis XV. He was in general a highly arrogant and quarrelsome man, incurring house arrest and imprisonment for his disgraceful behaviour to other singers, and wounding the poet Ballot de Sauvot in a duel over the merits of French and Italian music. Sang in Lisbon 1755, surviving the famous earthquake of 1 Nov., and in Madrid 1756, with Farinelli. Retiring to Naples, he bought a dukedom and two palaces with the huge fortune he had amassed. One of the most famous of all the great castratos (Bartolo refers to him in the Lesson Scene of *Barbiere*), whose voice and singing at his peak were found irresistibly moving.

**Cagliari.** Capital of Sardinia. The first theatre, built partly of wood, was Las Plasas (T Regio), opened 1770; rebuilt as T Civico 1836, reopened with *Belisario*. Premières were given of various Sardinian operas, including Gonella's *Ricciarda* and Dessy's *Don Martino d'Aragona*. Bombed 1943. Opera has also been given in the T Diurno, renamed T Cerruti 1859, demolished 1895; and in the Politeama Regina Margherita, bombed in the Second World War. Opera is now given in the T Massimo, in the summer in the open air.

**Cagnoni, Antonio** (*b* Godiasco, 8 Feb. 1828; *d* Bergamo, 30 Apr. 1896). Italian composer. Studied Voghera with Moretti, then Milan. While still a student, he composed the highly successful, somewhat Donizettian *Don Bucefalo*. In 1856 he became maestro di cappella at Vigevano, then in 1863 director of the Istituto Musicale at Novara, and in 1886 maestro di cappella at Bergamo. *Michele Perrin* shows an awareness of Wagner, but the less successful *Claudia* reverts to a simpler manner. He had greater success with *Papà Martin* and especially with his last performed work, *Francesca da Rimini*, which shows a quite advanced use of *Leitmotiv.

SELECT WORKLIST: *Don Bucefalo* (Bassi; Milan 1847); *Michele Perrin* (Marcello; Milan 1864); *Claudia* (Marcello; Milan 1866); *Papà Martin* (Ghislanzoni; Genoa 1871); *Francesca da Rimini* (Ghislanzoni; Turin 1878); *Il Re Lear* (Ghislanzoni; comp. 1893; unprod.).

**Cairo.** See EGYPT.

**Calaf.** A Tartar prince (ten) in Puccini's *Turandot*.

**Caldara, Antonio** (*b* Venice, *c*.1670; *d* Vienna, 28 Dec. 1736). Italian composer. Probably a pupil of Legrenzi, he sang in the choir of St Mark's, Venice and travelled extensively throughout Europe, composing operas and oratorios for many different courts. In 1699 he joined the Mantuan court and in 1709 succeeded Handel as maestro di cappella to Prince Francesco Maria Ruspoli in Rome, where he stayed until 1716. He was then appointed Vizekapellmeister to Charles VI in Vienna, as deputy to Fux. This heralded the most fruitful and prosperous years of his career: his talents were admirably suited to the requirements of a large court with a rich tradition of grandiose musical entertainment, although pressure of time was a severe hindrance to his creative powers and many of his works fall back on well-tried formulas.

Caldara's surviving operas number about 90, and share some points of similarity with those of his near-contemporaries Vivaldi and Alessandro Scarlatti, particularly in their occasional use of the *galant* style. Most are to librettos by Zeno, Metastasio, or Pasquini, and follow the mainstream tradition of opera seria. Because of his court position, Caldara had little choice in the kind of works he wrote: usually they concern heroic figures, as in *Mitridate*, *Demetrio*, and *La clemenza di Tito*, although he was occasionally given the opportunity to handle lighter, pastoral themes, such as *Nigella e Tirsi* and *Il natale di Minerva*. During his years in Vienna Caldara also composed both sacred and secular music for the court of the Archbishop of Salzburg, Franz Anton von Harrach, including *Camaide, imperatore della China* and *Il finto Policare*.

SELECT WORKLIST (all first prod. Vienna unless otherwise stated): *Camaide, imperatore della China* (Lalli; Salzburg 1722); *Il finto Policare* (Pariati; 1724); *Mitridate* (Zeno; 1728); *Demetrio* (Metastasio; 1731); *La clemenza di Tito* (Metastasio; 1734); *Nigella e Tirsi* (Pasquini; 1726); *Il natale di Minerva* (Minerva's Birthday) (?; 1729).

BIBL: B. Pritchard, ed., *Caldara Studies* (Aldershot, 1987).

**Caldwell, Sarah** (*b* Maryville, TN, 6 Mar. 1924). US producer and conductor. Studied Boston, while still a student producing Vaughan Williams's *Riders to the Sea*. Founded Boston OC 1957, director since. Produced first US original version *Boris Godunov*, and first full US *Trojans*, *Intolleranza*, *Moses and Aaron*, and *War and Peace*. New York, City O, 1973. Has conducted most of her Boston productions, which have shown an inventive and intelligent view of opera as music theatre. First woman conductor at New York, M, 1976 (*Traviata*).

**Calegari.** Three members of this Italian family were active as opera composers.

1. **Giuseppe Calegari** (*b* Padua, *c*.1750; *d* Padua 1812). Worked as a cellist before becoming impresario of Padua, T Nuovo, 1787–1801. Wrote four operas, including a Don Juan treatment, *Il convitato di pietra*, and *Zenobia in Palmira*.

SELECT WORKLIST: *Il convitato di pietra* (Pariati; Venice 1777); *Zenobia in Palmira* (Metastasio; Modena 1779).

2. **Antonio Calegari** (*b* Padua, 17 Feb. 1757; *d* Padua, 22 or 28 July 1828). Brother of Giuseppe. Studied Venice with Bertoni. 1790–6, Music director at Padua, T Nuovo; maestro di cappella at S Antonio, Padua 1814–28. His operas, which included *Deucalione e Pirra* and *Il fanatico per gli antichi romani*, enjoyed considerable popularity in his native city.

SELECT WORKLIST: *Deucalione e Pirra* (Sertor; Padua 1781); *Il fanatico per gli antichi romani* (Palomba; Padua 1792).

3. **Luigi Antonio Calegari** (*b* Padua, *c*.1780; *d* Venice, 1849). Italian composer. Nephew of Antonio, with whom he studied. His first opera, *Il matrimonio scoperto*, was extremely well received and was followed by seven more works, of which *Amor soldato* was most highly regarded.

SELECT WORKLIST: *Il matrimonio scoperto* (The Open Marriage) (Artusi; Padua 1804); *Amor soldato* (Soldier Love) (Tassi & Rossi; Padua 1807).

**Calife de Bagdad, Le.** Opera in 1 act by Boieldieu; text by Saint-Just, after an oriental tale. Prod. Paris, OC, 16 Sep. 1800, with Gavaudan, Dugazon, Elleviou, Bertin, Paulin; New Orleans, T St. Pierre, 25 Dec. 1805; London, H, 11 May 1809.

Bagdad. The Calif Harun al Rashid (ten), wandering Bagdad disguised, rescues Zobeide (sop) from robbers. In the house of her mother Lemaide (sop) he is in turn suspected of being a robber, but privately identifies himself to the Cadi (bs) and police. Lemaide's nephew Messun (bar) appears to announce the approach of a bridal procession. All is cleared up, and the Calif claims the delighted Zobeide as the bride.

Also opera by M. García (1812).

**Calisto, La.** Opera in 3 acts by Cavalli; text by Giovanni Faustini, after Ovid's *Metamorphoses*, ii. 401–507. Prod. Venice, T San Apollinare, autumn 1651; prob. then unprod. until Leppard version Gly. 26 May 1970, with Cotrubas, Stadler, Baker, Bowman, Gottlieb, Trama, Davià, cond. Leppard; Cincinnati U, 12 Apr. 1972.

Destiny informs Nature and Eternity that Jove

has decreed that Calisto, daughter of Lycaon, King of Pelasgia, should be added to the stars as one of the immortals. Jove descends to an earth laid waste, accompanied by Mercury, to help restore nature, only to discover that Calisto has become a nymph of Diana. Calisto rejects Jove's advances, so he disguises himself as Diana and tries to seduce her. Endymion, in love with the real Diana, is taken captive by Pan and his shepherds, but eventually united with his beloved. Meanwhile Calisto, who has been changed into a little bear by Juno, is taken to Mount Olympus by Jove, now restored to his normal godhood; there she will shine eternally as Ursa Minor.

**Callas, Maria** (orig. Mary Kalogeropoulou) (*b* New York, 2 Dec. 1923; *d* Paris, 16 Sep. 1977). US, later Greek soprano. Studied Athens with Hidalgo. Début Athens, O, 1941 (Tosca). Verona Arena 1947 (La Gioconda); then throughout Italy as Turandot, Isolde, Brünnhilde, etc., but also in the coloratura repertory, e.g. Elvira (*Puritani*), evincing a Lilli-Lehmann-like versatility. Eventually deciding to concentrate on bel canto roles, she sang in (and was responsible for) revivals of many works by Bellini, Donizetti, Rossini, and Cherubini. Milan, S, 1950–8 and 1960–2; London, CG, 1952–3, 1957–9, 1964–5; Chicago 1954–6; New York, M, 1956–8, 1965; also Dallas, Vienna, Berlin, Edinburgh. A striking woman, she possessed both personal magnetism and true artistry, and was compelling in her dramatic intensity and emotional commitment. Especially memorable as Norma, Medea, and Tosca, she also brought a touching pathos and vulnerability to Violetta and Lucia. Her voice was of unique timbre, combining power with brilliance, flexibility, and a rich variety of colours. However, like Malibran (an artist with whom she bears comparison), she had a constant struggle with its imperfections, and in 1965 she withdrew from public appearances. She made a last concert tour 1973–4 with her former colleague Di Stefano. She was one of the century's greatest singing actresses, and an artist who touched every role she sang with her genius. (R)

BIBL: H. Wisneski, *Maria Callas: The Art behind the Music* (New York, 1975).

**Calvé, Emma** (*b* Decazeville, 15 Aug. 1858; *d* Millau, 6 Jan. 1942). French soprano. Studied Paris with Puget, Marchesi, and Laborde. Also influenced by Victor Maurel and the actress Eleanora Duse. Début Brussels, M, 1881 (Marguerite). Paris from 1884; Milan, S, from 1887; Rome 1891; London, CG, 1894–1904; New York, M, 1893–1904; Manhattan OC 1907–8. Then retired, but gave concerts until after 1918. One of the great actress-singers who came to the fore with verismo opera, her technique and

range enabled her to sing soprano and mezzo roles. She also possessed a high, 'disembodied' voice which Massenet employed in *Sapho*, which she created, together with *La navarraise*. She sang Pamina, Amina, and Lucia, but became especially associated with Santuzza, Ophelia, and Carmen (to which she brought a disturbing brand of realism; Shaw, though a keen admirer, was shocked by her portrayal). Her voice was rich and alluring, while her singing, according to Hahn, was 'always of the most perfect beauty'. (R)

**Calzabigi, Ranieri de'** (*b* Livorno, 23 Dec. 1714; *d* Naples, July 1795). Italian librettist. His librettos brought him in touch with Metastasio, an edition of whose works he published during a stay in Paris. Here he and his brother ran a lottery, with Casanova as partner and under the protection of Mme de Pompadour, before being expelled. He is famous as the author of three librettos for Gluck, *Orfeo ed Euridice* (1762), *Alceste* (1767), and *Paride ed Elena* (1770). These constitute a reform, away from the narrow conventions and artificiality of opera seria, in favour of 'a beautiful simplicity' and greater dramatic truth. As witness of a return both to the earliest opera and to the Greek ideal from which that had derived, he turned to the Orpheus story. There was also, in particular reaction to opera seria, an emphasis on the characters as human beings rather than stereotypes, and a renewed care for the natural declamation of poetry; music would thus play a role connected more closely to the drama than to the demands of singers. Calzabigi imposed these reforms upon himself as well as upon Gluck, for his verses are designed not as the vehicle of elaborate metaphors but as the direct, lyrical articulation of dramatic situations. This is seen at its purest in *Orfeo*, and he was the motivator, perhaps the actual author, of the famous Preface to *★Alceste*; in the other operas, especially *Paride*, more concessions are made. Calzabigi also wrote librettos for Gassmann (the comic *La critica teatrale*), Salieri, and Paisiello.

BIBL: G. Lazzeri, *La vita e l'opera litteraria di Ranieri Calzabigi* (Città di Castello, 1907).

**Cambert, Robert** (*b* Paris, *c*.1627; *d* London, Feb. or Mar. 1677). French composer. Studied with Chambonnières and for a while was active as a church musician. In 1659 he collaborated with Perrin on a dramatic musical work which became known as the *Pastoral d'Issy*, after the house of the king's goldsmith near Paris where it was first performed. This was not 'la première comédie française en musique', as its title-page claimed, though after Michel de la Guerre's *Le triomphe de l'amour et de Bacchus* it was only the second setting of a French text with genuine operatic pretensions.

Such was its success that, at Cardinal Mazarin's instigation, it was repeated before the royal court. Perrin and Cambert began work immediately on a 5-act opera, *Ariane, ou Le mariage de Bacchus*, which was not performed at this time, probably because of Mazarin's untimely death. Cambert's appointment as music-master to Queen Anne of Austria, the widow of Louis XIII, in 1666 brought him into closer contact with court life; as a result, on 28 June 1669 he and Perrin were granted a royal privilege for 'L'établissement des Académies d'opéra ou représentations en musique et en langue françoise, sur le pied de celles d'Italie'. Joining together with the Marquis de Sourdéac, a well-known stage-machinist, and the ballet-master Beauchamps they recruited a company of singers and on 3 Mar. 1671 staged their new opera *Pomone* in the Salle du Jeu de Paume de la Bouteille, which became in effect the first home of the Paris Opéra.

Though the venture was highly successful, the backers of *Pomone* retained all the profits of the performance and Perrin was imprisoned for debt. Cambert went on to write another opera, *Les peines et les plaisirs de l'amour*, in the following year, but soon afterwards Lully successfully conspired to obtain the imperial privilege. Forced to leave Paris, Cambert came to London, where he briefly ran a musical academy organized on the lines of the French Académie. *Ariane, ou Le mariage de Bacchus*, *Pomone*, and various pieces of ballet music were performed here, but otherwise little is known of Cambert's life after he left Paris. He is reputed to have been poisoned by one of his servants.

Nothing remains of the *Pastorale d'Issy* or *Ariane, ou Le mariage de Bacchus* except the librettos, while both *Pomone* and *Les peines et plaisirs de l'amour* have only come down in fragmentary form. The influence of classical French tragedy and ballet de cour are very apparent, while Cambert also made an important contribution to the development of a style of recitative suitable for the French language. Though his achievements were soon overshadowed by those of Lully, it is certain that without him and Perrin the foundation of a French operatic tradition could have not been established until much later.

WORKLIST: *La pastorale* (Perrin; Issy 1659, lost); *Ariane ou Le mariage de Bacchus* (Perrin; London 1674, lost); *Pomone* (Perrin; Paris 1671, rev. Windsor 1674); *Les peines et les plaisirs de l'amour* (Gilbert; Paris 1671).

**Cambiale di matrimonio, La** (The Bill of Marriage). Opera in 1 act by Rossini; text by Rossi, after Camillo Federici's comedy (1790). Prod. Venice, T San Moisè, 3 Nov. 1810, with Mo-randi, Ricci, Raffanelli, De Grecis; New York, 44th St. T, 8 Nov. 1937, with Morel, Ilosvay, Hollander, Zangheri, Perulli; London, SW, 23 Apr. 1954 by T dell'Opera Comica, Rome.

An English town, 18th cent. Sir Tobias Mill (bs), an English merchant, has promised the hand of his daughter Fanny (sop) to Slook (bar), a Canadian merchant. Fanny, however, is in love with Edoardo (ten), and Slook helps them outwit Sir Tobias.

Rossini's first opera to be performed. The duet 'Dunque io son' was later used in Act I of *Il barbiere di Siviglia*.

**Cambridge.** City in England. The Cambridge University Musical Society has done much for English opera (including for Handel), above all in the period 1902–41 under E. J. *Dent, lecturer at King's Coll. from 1902 and Professor of Music from 1926. A production with far-reaching consequences was the 1911 *Magic Flute* in Dent's translation, with a cast including Steuart Wilson, which paved the way for the acceptance of Mozart into the English repertory. Other works produced by Dent included Purcell's *King Arthur* (1928), *The Fairy Queen* (1931), and *The Tempest* (1938), and Handel's *Samson* (1932), *Susanna* (1935 and 1938), and *Saul* (1937). Post-war productions included Handel's *Solomon* (1948) and *Athalia* (1954), and Vaughan Williams's *The Pilgrim's Progress* (1954). In 1954 the Cambridge University Opera Group was formed, largely to stage works less 'academic' than the official ones. Works produced included *Il matrimonio segreto* (1955), *The Rake's Progress* and Vaughan Williams's *Sir John in Love* (1956), Liebermann's *The School for Wives* (1958), and Bizet's *Don Procopio* (1959). It was from the group that the *New Opera Company was formed in 1956.

**Cambridge Theatre.** Theatre in Seven Dials, London (cap. 1,285). Opened in 1930, it has staged opera including a run of *Hänsel und Gretel* (1934) and a 2-year unbroken series (1946–8) by Jay Pomeroy's New London OC under Dino Borgioli with Alberto Erede conducting an Italian repertory. The British première of *The Consul* with the original New York cast was given in 1951.

**Camden Festival.** Annual arts festival in the London Borough of Camden, founded in 1954 and devoted especially to operatic revivals and first performances, given first in St Pancras TH and after 1970 in the Collegiate T of London University. Productions (many of them British premières) have included Haydn's *Il mondo della luna* (1961), *L'infedeltà delusa* and *L'incontro improvviso* (1966), and *La fedeltà premiata* (1971); Rossini's *La pietra del paragone* (1963), *Il turco in Italia* (1965), *Elisabetta, regina d'Inghilterra*

(1968), and *Tancredi* (1971); Donizetti's *Maria Stuarda* (1966), *Marino Faliero* (1967), *Torquato Tasso* (1974), and *Francesca di Foix* and *La romanziera* (1982); Verdi's *Un giorno di regno* (1961), *I masnadieri* (1962), *Aroldo* (1964), and *Il corsaro* (1966); also Leoncavallo's *La bohème* (1970), Cilea's *Adriana Lecouvreur* (1971, 1984), Smetana's *The Secret* (1972), Rakhmaninov's *Aleko* (1972), Massenet's *La navarraise* (1972), Meyerbeer's *L'étoile du nord* (1977), Delius's *Village Romeo and Juliet* (1979) and *Margot la rouge* (1984), Grétry's *Zémire et Azor* (1980), the Riccis' *Crispino e la comare* (1981), Sacchini's *Renaud* (1981), Cavalli's *Eritrea* (1982), Vivaldi's *Juditha triumphans* (1984), Caccini's *Euridice* (1985), Boito's *Nerone* (1985), Strauss's *Friedenstag* (1985), Mozart's *La finta semplice* (1985) and *La finta giardiniera* (1986), Weill's *Protagonist* and *Tsar* (1986) and *Silbersee* and *Happy End* (1987).

**camerata** (It.: 'society'). A term in vogue in the 16th cent., often used as an alternative for *academy, but usually denoting a somewhat smaller body. Although only one of several such groups in Florence, the camerata which met during the 1580s at the house of Count Giovanni de' Bardi in Florence was of greatest importance for the development of opera, and is often known simply as the 'Florentine Camerata': its members included Jacopo Corsi, Vincenzo *Galilei (father of the astronomer), Piero Strozzi, *Peri, *Caccini, Emilio de' *Cavalieri, and *Rinuccini.

Though the Camerata was not exclusively a musical gathering, it was here that the discussions and experiments took place which led to the emergence of the new form of musical drama soon to be recognized as opera. The art of solo singing had long been popular in Italy, with polyphonic madrigals sometimes being performed by just one vocalist while the other parts were given to instruments, but the Camerata now used it as the springboard for a new practice, which simultaneously drew inspiration from theoretical discussions concerning the nature of dramatic music. Their point of departure was the ancient Greeks, who, they believed, had performed their tragedies to the accompaniment of a music in which there was a perfect union of words and melody. Under Galilei's guidance a complex new theory was evolved which owed much to his earlier correspondence with the Roman scholar Girolamo Mei, and appeared in its most cogent form in his *Dialogo della musica antica et della moderna* (Florence, 1581). Polyphonic writing was to be eschewed and, since the words must be heard clearly, performance would be by a solo singer to the simplest possible accompaniment. The text itself was to be sung with correct and natural declamation of the words, and there were

to be no picturesque madrigalian flourishes: rather, the music had to interpret the feeling of the whole passage (see STILE RAPPRESENTATIVO).

This new style made its first appearance around 1582 in Galilei's cantata *Il conte Ugolino*, a setting of lines from Dante's *Inferno*, and in his lost treatment of part of the Lamentations of Jeremiah. As it was a fundamental belief of the Camerata that the Greeks had performed their tragedies with music throughout, it was inevitable that sooner or later it would be used for a whole staged dramatic work. But though the members of the Camerata were well versed in dramatic writing—many of them, for example, participated in the lavish stage entertainments of the Medici court, such as the famous 1589 intermedi for Bargagli's *La pellegrina*—it was not until 1598 that the first opera appeared, Peri's setting of Rinuccini's *Dafne* libretto (lost). By this time Bardi had long since moved to Rome, Galilei had been dead for six years, and the Camerata was directed by Jacopo Corsi. Peri's example was soon followed by other operas: his own and Caccini's *Euridice* (both 1600) and Cavalieri's *Rappresentazione di anima e di corpo*. However, the earliest surviving examples of writing in stile rappresentativo are to be found in a non-dramatic work, Caccini's collection *Le nuove musiche*, which though published in 1602 was written during the course of the preceding decade.

Although many ideas came together for the birth of opera around 1600, it was the work of the Camerata which provided the essential catalyst for the development of the new genre. Later there were to be strong claims and counter-claims by members of the Camerata to establish primacy for the composition of the first true opera, largely in response to *Doni's researches for his *Trattato della musica scenica* (c.1635), the merits of which are now hard to assess. But it is clear that whoever may actually have written the first opera, it would never have come about without the heady intellectual debates of the Camerata and its collective originality and experimentation: ironically, it was a group fundamentally Renaissance in spirit which gave rise to the most characteristic art form of the Baroque. Although the development of opera soon passed out of the hands of the Camerata, its principles continued to exert a strong influence for some time to come. The solo passages of Monteverdi's *Orfeo*, for instance, owe much in their careful treatment of the words and discreet handling of the accompaniment, and it was not until the Roman operas of the 1620s that the principles of the Camerata were vindicated.

BIBL: C. Palisca, *The Florentine Camerata: Documentary Studies and Translations* (New Haven, CT, 1989).

**Cammarano, Salvadore** (*b* Naples, 19 Mar. 1801; *d* Naples, 17 July 1852). Italian librettist. After producing various dramas, he made his début as a librettist with *La sposa* (1834) for Vignozzi. From 1835 he was house poet of the Naples, C. He befriended Donizetti and wrote the texts for *Lucia di Lammermoor* (1835), *Belisario* (1836), *Pia de' Tolomei* (1837), *Roberto Devereux* (1837), and *Maria di Rudenz* (1838). For Verdi he wrote *Alzira* (1845), *La battaglia di Legnano* (1849), *Luisa Miller* (1849), and *Il trovatore* (1853). A careful and skilled craftsman, he was adept at fashioning an effective libretto in a robust traditional vein that would gratify composer, audience, and even censor, though often a good deal of the distinctive quality of his original source was thereby lost. He also wrote texts for Pacini and Mercadante. Two of his few comic librettos were for his composer brother **Luigi Cammarano** (*c*.1800–54).

BIBL: J. Black, *The Italian Romantic Libretto: A Study of Salvadore Cammarano* (Edinburgh, 1984).

**Campana sommersa, La** (The Sunken Bell). Opera in 4 acts by Respighi; text by Guastalla, after Gerhard Hauptmann's 'fairy-drama' *Die versunkene Glocke* (1896). Prod. Hamburg 18 Nov. 1927, with Callam, Graarud, Guttmann, cond. Wolff; New York, M, 24 Nov. 1928, with Rethberg, Martinelli, De Luca, cond. Serafin.

Heinrich (ten), a bell-founder, falls under the spell of the fairy Rautendelein (sop), and follows her to the mountains. He returns home on the death of his wife Magda (sop), but cannot forget Rautendelein, and on his own deathbed he calls for her and she returns to him.

**Campanello di notte, Il** (The Night Bell). Also known as *Il campanello dello speziale* (The Chemist's Bell). Opera in 1 act by Donizetti; text by composer, after a vaudeville by Brunswick, Troin, and Lhérie, *La sonnette de nuit*. Prod. Naples, N, 1 June 1836, with Schultz-Oldosi, G. Ronconi, Casaccia; London, L, 30 Nov. 1837; New York, Lyceum T, 7 May 1917.

Foria, nr. Naples, 19th cent. Don Annibale Pistacchio (bs), an elderly apothecary, has married young Serafina (sop). Her former lover Enrico (ten) disrupts their wedding night by presenting himself at the apothecary's door, ringing the night bell and demanding that various prescriptions are dispensed by the frustrated bridegroom.

**Campanini, Cleofonte** (*b* Parma, 1 Sep. 1860; *d* Chicago, 19 Dec. 1919). Italian conductor and violinist. Studied Parma, Scuola di Musica, but expelled; then studied with Rossi. Début as conductor, Parma 1882 (*Carmen*). Music staff New York, M, 1883–4; New York, AM, 1888 (first US *Otello* (Verdi)). In Milan from 1902: cond. pre-

mières of *Adriana Lecouvreur* (1902) at L, Giordano's *Siberia* (1903) and *Madama Butterfly* (1904) at S (1903–6). London, CG, 1904–12 (including *Manon Lescaut* with Caruso, 1904, and first European *Fanciulla del West* with Destinn, 1911). New York, Manhattan OC, 1906–9 (artistic director). Mounted centenary cycle of Verdi at his own expense, Parma 1913. Chicago 1910–19 (manager 1918–19). Married soprano Eva *Tetrazzini. His brother **Italo** (*b* Parma, 30 June 1845; *d* Corcagno, 22 Nov. 1896) was a tenor. Studied Parma with Griffini, later Milan with Lamperti. Début Parma 1863 (Vitellozzo in *Lucrezia Borgia*). Sang in Russia for three years; first Italian Lohengrin, Bologna 1871. London, DL, 1872 (Gennaro); London, HM, 1878 (Don José). New York, M, 1883 (Gounod's Faust at its opening). Idolized in his prime, he used his limited but sweet-toned voice with much feeling.

**Campiello, Il** (The Square). Opera in 3 acts by Wolf-Ferrari; text by Ghisalberti, after Goldoni's comedy (1756). Prod. Milan, S, 12 Feb. 1936, with Favero, Carosio, Tess, Baccaloni, cond. Marinuzzi.

Venice, mid-18th cent. The Neapolitan Astolfi (bar) is paying court to three girls, Gasparina (sop), Lucretia (sop), and Gnese (sop). The latter two already have lovers. Gasparina is unwilling to marry Astolfi, but her uncle Fabrizio (bs) takes a liking to him. Quarrels and confusions arise, involving three old women who live in the square, but all is eventually resolved and Gasparina consents to marry Astolfi and move to Naples.

**Camporese, Violante** (*b* Rome, 1785; *d* Rome, 1839). Italian soprano. Studied Paris with Crescentini. Sang for Napoleon; returned to Italy after his fall, and made a brilliant début at Milan, S. London, H, 1817–23 (Susanna, Donna Anna, Dorabella, Sesto, and Rossini's Desdemona and Ninetta). Milan, S, from 1818, creating Bianca in Rossini's *Bianca e Faliero*. Stendhal, while admiring her excellent singing, refers to her 'frigid virtuosity'.

**Campra, André** (*b* Aix-en-Provence, *bapt.* 4 Dec. 1660; *d* Versailles, 29 June 1744). French composer of Italian descent. Educated as a church musician, he held posts in Arles and Toulouse before moving to Paris in 1694 as maître de musique at Notre Dame. His sacred music was already highly regarded when in 1697 he composed the opéra-ballet *L'Europe galante*. Fearing that his ecclesiastical position would be compromised by an association with the theatre, he published it anonymously; other early works used his brother's name. In 1700 he was released from his duties at Notre Dame and devoted himself to opera, producing *c*.40 dramatic works, ranging from full-scale operas to shorter divertissements. He enjoyed enormous popular acclaim, receiving numerous honours and appointments, including that of Maître de la Chapelle Royale in 1722.

The majority of Campra's operas, which generally follow Lully's model, were written for the Opéra and are based on familiar stories, such as *Tancrède* and *Achille et Déidamie*. Ballet continues to constitute an important element, and in such works as *Le triomphe de l'amour* and *Les amours de Mars et Vénus* the allegorical tradition of both ballet de cour and early French opera is clearly maintained. In particular, Campra was responsible for the creation of the *opéra-ballet, a combination of music and dance in five acts, each of which presented a different comic plot. The success of *L'Europe galante* paved the way for such works as *Les fêtes vénitiennes* and *Les âges*, whose freshness and vitality proved an important ingredient for the later development of *opéra-comique.

Campra's work falls midway between Lully and Rameau, lacking the grandeur of the former and the eloquence of the latter, though its vitality, ingenuity, and pastoralism appealed to a wide audience. His background enabled him to sympathize with the prevailing styles of both French and Italian music: indeed his declared intention, expressed in the preface to his *Cantates françoises*, was to combine the best of both traditions in his vocal music. With his melodic inventiveness and skilful handling of the orchestra, it is not surprising that he should have been so highly regarded. Only the greater achievements of Lully and Rameau have caused him to be relegated to the sidelines of operatic history.

The libretto which Campra used for his highly popular *Idoménée* was modified for Mozart's opera on the same subject in 1780.

SELECT WORKLIST (all first prod. Paris): *L'Europe galante* (La Motte; 1697); *Tancrède* (Danchet, after Tasso; 1702); *Le triomphe de l'amour* (Danchet, after Quinault; 1705); *Les fêtes vénitiennes* (Danchet; 1710); *Les amours de Mars et Vénus* (Danchet; 1712); *Les âges* (Fuzelier; 1718); *Achille et Déidamie* (Danchet; 1735).

**Canada.** On 14 Nov. 1606 *Le théâtre de Neptune*, a masque by Marc Lescarbot, was played from small boats in the harbour of Port Royal on the Bay of Fundy. But this enterprise had no immediate successors. Towards the end of the 18th cent. opera performances began in Halifax, Quebec, and Montreal, and early in the 19th cent. in Toronto. By the mid-18th cent. a regular circuit of the main towns of the provinces of Upper and Lower Canada (now Ontario and Quebec) was travelled by small touring companies, and by many famous soloists from Eur-

ope and the USA, giving opera excerpts and concerts. From the 1850s full companies made regular visits to Montreal and Toronto and often to many smaller centres; by 1900 almost every town had some acquaintance with opera and the larger cities frequently saw a standard repertory, including Weber and Wagner.

Local productions were sporadic, and there were no lasting established companies until the mid-20th cent. The most successful has been the Canadian Opera Company (1950), which in addition to seasons in Toronto and visits to Ottawa has, since 1958, sent out a touring company which visits every inhabited area of Canada and many towns in the USA. Premières have included Harry *Somers's *Louis Riel* (1967) and *The Shivaree* (1982) by John Beckwith (*b* 1927). In 1961 productions by the T-Lyrique de Nouvelle-France began in Quebec. L'Opéra du Quebec was set up in 1971 to consolidate operatic production in Quebec, Montreal, and elsewhere, but closed in May 1975. A similar co-operative venture was established in 1972 in Western Canada by the Opera Associations of Vancouver, Edmonton, Calgary, and Winnipeg, all of which began regular local productions during the 1960s. Notable summer productions have been by the Stratford Festival, at the National Arts Centre in Ottawa, and at the Guelph May Festival.

The Canadian Broadcasting Corporation formed its opera company in 1948 to present Canadian performers in live studio productions of a wide range of standard, unfamiliar, and contemporary works. It has also commissioned operas including Healey *Willan's *Transit through Fire* (1942) and *Deirdre* (1946), John Beckwith's *Night Blooming Cereus* (1959), Kelsey Jones's *Sam Slick* (1967), and Murray Adaskin's *Grant, Warden of the Plains* (1967).

Despite the absence of permanent companies, numerous operas and operettas have been written by composers in Canada from the early 19th cent.

See also MONTREAL, TORONTO, VANCOUVER.

**Canadian Opera Company.** See TORONTO.

**Candeille, Pierre Joseph** (*b* Estaires, 8 Dec. 1744; *d* Chantilly, 24 Apr. 1827). French composer and singer. Came to Paris 1767, where he stayed until 1771; appeared as basse-taille in the choruses of the Académie Royale de Musique (Opéra) and the Concert Spirituels. After a period of travel returned to Paris in 1773, took up his former positions, and became involved in the production of ballets and divertissements in various theatres.

Candeille produced his first large-scale work, the heroic pastoral *Laure et Pétrarque*, in 1780. This work, with its tawdry libretto and pedestrian music, set the tone for the rest of his dramatic works, many of which were never performed. His only real success was *Castor et Pollux*, which received 130 performances 1791–1800, and after the failure of Peter von Winter's *Castor et Pollux* (1814) was revived and kept in the repertoire until 1817. As such it was one of the most enduring works of the Paris stage.

Many of Candeille's failures have been attributed to his disorganized and even unscrupulous attitude to his profession. He attempted to put his daughter (Amélie) **Julie Candeille** (1767–1834) on the stage when she was barely 15 years old, even writing an opera, *Thémire*, specially for her. Fortunately, it was not performed, and she eventually made her début in Gluck's *Iphigénie en Aulide*. In addition to her highly regarded appearances as a singer and actress, she herself later enjoyed a more modest career as a composer of opera.

SELECT WORKLIST: *Castor et Pollux* (Bertrand; Paris 1791).

**Caniglia, Maria** (*b* Naples, 5 May, 1905; *d* Rome, 16 Apr. 1979). Italian soprano. Studied Naples with Roche. Début Turin 1930 (Chysothemis). The leading Italian lyric-dramatic soprano of the 1930s. Milan, S, 1931–51; London, CG, 1937, 1939, 1950; New York, M, 1938–9. Admired as Tosca and Adriana Lecouvreur; created Roxane (*Cyrano de Bergerac*) and Respighi's Lucrezia. A warm and vivid performer with a rich, vibrant voice. (R)

**Canio.** The leader (ten) of the strolling players, husband of Nedda, in Leoncavallo's *Pagliacci*.

**cantabile** (It.: 'in a singing manner'). A term used to refer to a smooth and even style of singing or playing, but in 19th-cent. Italian opera referring more specifically to the opening section of an aria, or to the second section of a three-movement duet, where it is preceded by a \*tempo d'attacco. Invariably cast in a slow, expressive style, and often featuring a cadenza, the cantabile was usually followed by a \*cabaletta.

**cantata.** From the It. *cantare*, 'to sing', i.e. a vocal piece, as opposed to a sonata, which is performed by instruments. Usually taken to refer to a vocal work consisting of several sections or movements, based on a continuous narrative text. It first appeared as an identifiable genre early in the 17th cent. as a direct result of experiments in \*stile rappresentativo: some items in Caccini's *Le nuove musiche* may be considered embryonic cantatas, in that more than one vocal melody is used. After further development by Carissimi, the secular Baroque cantata reached a peak under A. Scarlatti, with a standardized form

comprising two or three da capo arias, connected by recitative. Though not intended for stage performance, the cantata had several points of connection with operatic music, notably through its virtuosic vocal display, use of many of the same formal structures, and generally dramatic approach to the text. The sacred cantatas of Bach, which have often been described as 'operatic', frequently approach the spirit of the genre in their colourful recitatives and vivid arias. In the late 18th cent. the cantata merged with the oratorio, but the term has continued in general usage up to the present day to refer to any non-dramatic vocal piece.

**Cantelli, Guido** (*b* Novara, 27 Apr. 1920; *d* Paris, 24 Nov. 1956). Italian conductor. Studied Milan with Pedrollo and Ghedini. Début Novara 1943. Prod. and cond. *Così fan tutte* Milan, PS, 1956. His appointment as chief conductor at Milan, S, was announced a few days before his death in an air crash. (R)

**Canterbury Pilgrims, The.** 1. Opera in 3 acts by Stanford; text by G. A. A'Beckett, after Chaucer's *The Canterbury Tales* (unfin., 1400). Prod. London, DL, 28 Apr. 1884.

2. Opera in 4 acts by De Koven; text by Percy Mackaye, after Chaucer. Prod. New York, M, 8 Mar. 1917, with Mason, Sembach, cond. Bodanzky.

**cantica** (Lat.: 'songs'). In the Roman comedies of Plautus and Terence it was customary for some sections to be sung; these were the cantica, which were interpolated within the *diverbium*, or spoken dialogue. Later the term came to mean specifically those sections sung by a soloist, as opposed to the chorus.

**canti carnascialeschi** (It.: 'carnival songs'). These part-songs, whose performance probably involved pantomimic display, were written for the spectacular carnival festivities which took place at the Medici court in Florence during the late 15th and early 16th cents.: some have texts written by Lorenzo de' Medici himself. Often containg obscene or satirical *double entendre*, they had a similar musical style and form to the frottola.

**cantilena** (It.: 'lullaby or sing-song'). Originally, the part of a choral composition with the main tune, or a small piece for one voice. Now used to describe a smoothly flowing, melodious part, or to indicate that the passage should thus be performed. Also used to describe a singer's ability to sing smoothly, flowingly, and melodiously.

**canto fiorito** (It.: 'florid song'). Melodic embellishment; used more specifically for the highly decorated style of vocal music in common usage during Rossini's era, which contrasted with the smoother lines of canto spianato.

**canto spianato** (It.: 'smooth song'). The smooth, legato style of singing often held up in contrast to Rossini's **canto fiorito**.

**canzone.** A word of Provençal origin (*canzo*) describing a certain style of song. In operatic usage it came by the 18th cent. to mean an actual song outside the dramatic situation, e.g. 'Voi che sapete' in *Le nozze di Figaro*. The diminutive canzonetta usually describes a short, simple song.

**canzonetta.** See CANZONE.

**Capecchi, Renato** (*b* Cairo, 6 Nov. 1923). Italian baritone. Studied Milan with Carrozzi. Début Reggio Emilia 1949 (Amonasro). Milan, S, from 1950; New York, M, from 1952; London, CG, from 1962; also Aix, Edinburgh, Vienna, Chicago, etc. Has a repertory of almost 300 roles, including many contemporary works. A versatile, intelligent singer, and a thoughtful artist. (R)

**Čapek, Karel** (*b* Malé Suatoňovice, 9 Jan. 1890; *d* Prague, 25 Dec. 1938). Czech novelist and dramatist. Much of his writing makes conscious use of musical techniques, including those of Leitmotiv (e.g in *Krakatit*). Operas on his works are as follows:

*Věc Makropulos* (1922): Janáček (1926)
*Krakatit* (1924): Berkovec (1961); Kašlík (1961)
*Bila Nemoc* (1937): Andrašan (1968)
*Juraj Cup* (?): Ceremuga (?)
*The Postman* (?): Feld (?).
*The Insect Play* (?): Cikker (1987)

Also the trilogy *Three Apocryphal Stories* by Podešva (??), after Čapek short stories.

**Capobianco, Tito** (*b* La Plata, 28 Aug. 1931). Argentinian producer. After working in the Argentine and Mexico, producer and technical director Buenos Aires, C, 1959–62. Director Cincinnati Summer O, 1962; New York, City O, 1965; Hartford 1972; San Diego 1976; Pittsburgh 1983. Berlin 1972; Paris and Holland, 1973; Sydney 1974 (*Hoffmann* for Sutherland).

**Cappuccilli, Piero** (*b* Trieste, 9 Nov. 1929). Italian baritone. Studied Trieste with Donaggio. Début Milan, N, 1957 (Tonio). New York, M, 1959–60; Milan, S, from 1964; London, CG, from 1967; Salzburg from 1975. His large repertory includes many Verdi roles (Boccanegra, Germont, Renato, Macbeth, etc.), to which his strong, expressive voice is well suited. (R)

**Capriccio.** Opera in 1 act by Richard Strauss; text by Clemens Krauss and the composer. Prod. Munich 28 Oct. 1942, with Ursuleac, Ranczak, Hann, Hotter, cond. Krauss; London, CG, by

Bavarian State O, 22 Sep. 1953, with Cunitz, Töpper, Kusche, Schmitt-Walter, cond. Heger; New York, Juilliard School, 2 Apr. 1954, with Davy, Mackenzie, Stewart, Rue, cond. Waldman.

A castle nr. Paris, 1775. The opera is a dramatized conversation piece, set in Paris at the time of Gluck's operatic reforms. The countess Madeleine (sop) listens to a string sextet (which also serves as the overture to the opera) written for her by one of her admirers, Flamand (ten). Her other admirer, Olivier (bar), is a tragic poet, whose play is to be staged for the Countess's birthday. A debate arises as to whether music or words are the more important in opera. When they each have the opportunity, both suitors declare their love to the Countess. In a second discussion, a theatre impresario, La Roche (bs), declares that both Olivier's characters and Flamand's music lack sufficient drama. He sets them a challenge of writing an opera, taking as a plot the foregoing discussions and using the people present as characters. The guests all depart. The Countess has made assignments with both her suitors for the same time the next morning, but still cannot decide between them; as an opera character, she needs both words and music.

The old Italian tag from which the work takes its text, *Prima la musica e poi le parole* (First the music and then the words), is also the title of an opera by Salieri (1786).

**Captain's Daughter, The.** Opera in 4 acts by Cui; text by the composer, based on Pushkin's story (1836). Prod. St Petersburg, M, 27 Feb. 1911. The opera, which takes place during the reign of Catherine the Great (mez), concerns the love of Andrey Grinyov (bs) for Maria Ivanovna (sop). He is accused of treachery but Maria secures his pardon.

**Captain Vere.** The commander (ten) of *HMS Indomitable* in Britten's *Billy Budd*.

**Capuana, Franco** (*b* Fano, 29 Sep. 1894; *d* Naples, 10 Dec. 1969). Italian conductor. Studied Naples with De Nardis. Début Brescia 1919, then conducted in many Italian towns. Turin 1929–30; Naples, C, 1930–7; Milan, S, 1937–40, 1946–52 (from 1949 music director), conducting many premières, including works by Alfano, Refice, and Ghedini, Italian premières, and revivals of neglected works. London, CG (first postwar opera), with Naples, C, company 1946, 1951–2. Outside the Italian repertory, in which he was vivid and forceful, a fine conductor of Wagner and Richard Strauss. Died while conducting *Mosè* on the opening night of the Naples, C, season. (R) His sister **Maria** (1891–1955) was a mezzosoprano. Studied Naples. Début Naples 1918 (Urbain). Possessed a firm, dark voice, and specialized in Wagner, which she sang all over Italy, in Buenos Aires, Spain, and Africa. (R)

BIBL: B. Cagnoli, *L'arte musicale di Franco Capuana* (1983).

**Capuleti e i Montecchi, I** (The Capulets and the Montagues). Opera in 4 parts by Bellini; text by Romani, modified for Bellini from his libretto for Vaccai, after Matteo Bandello's ninth *novella*, in turn drawing on an earlier libretto by Foppa set by Zingarelli (1796). Foppa's sources drew not on Shakespeare but on Gerolamo della Corte's *Storie di Verona*. Prod. Venice, F, 11 Mar. 1830, with Giuditta Grisi, Caradori-Allan, Bonfigli, Antoldi; London 20 July 1833, with Pasta, Méric, De Landey, Donzelli, Galli; New Orleans, St. Charles T, 4 Apr. 1837, with Rossi, Pantanelli, Ceresini, Candi, cond. Gabici.

Verona, 13th cent. The opera tells of the love of Romeo (mez), head of the Montagues, and Giulietta (sop), daughter of Capellio (bs), head of the Capulets, in the face of a feud between the two families. Romeo has killed Capellio's son, and Giulietta's hand in marriage is promised to Tebaldo (ten) if he can avenge this. Romeo proposes a reconciliation between the families by his marriage to Giulietta but Capellio rejects the idea. Unable to perusade Giulietta to elope with him, Romeo intends to abduct her by force, but is meanwhile arrested. The sleeping potion which causes the appearance of death is given to Giulietta by Lorenzo (bs), the Capulet family's doctor. He is prevented from telling Romeo of the plan by his own capture by Capellio. News of Giulietta's death arrives as Romeo and Tebaldo are about to fight a duel. Romeo takes poison and, waking to find her lover dying, Giulietta herself dies of grief.

Berlioz's vehement dislike of it (except for the unison passage at the end of Act II, which in his *Mémoires* he described as 'memorable') encouraged him to set Shakespeare's play as a symphony.

**Caradori-Allan, Maria** (*b* Casa Palatina, 1800; *d* Surbiton, 15 Oct. 1865). Alsatian soprano. Studied with her mother. Début London, H, 1822 (Cherubino). Sang regularly in London until 1827 and throughout England from 1834, mostly in concerts (including the first *Elijah*). Venice 1830; Paris, I; and Milan, S. Created Giulietta in *I Capuleti e i Montecchi*; also sang Amina, Zerlina, and Rosina. Mendelssohn found her singing 'pretty' but 'flat' and 'soulless'.

**Carafa, Michele** (*b* Naples, 17 Nov. 1787; *d* Paris, 26 July 1872). Italian composer. The son of the Prince of Colobrano, he was intended for the army but first studied in Mantua, in Paris with Cherubini and Kalkbrenner, and in Naples. He served in Italy and with the French on the Russian expedition of 1812, being decorated for gallantry. After Waterloo he resigned his commission to devote himself to music. He had his first major success with *Gabriella di Vergy*. In Naples, he formed a lifelong friendship with Rossini, to two of whose operas he contributed, *Adelaide di Borgogna* and *Mosè in Egitto*. From 1821 he had much success in Paris, especially with *Jeanne d'Arc à Orléans*, *Le solitaire*, *Le valet de chambre*, and particularly *Masaniello*, the last despite Auber's work on the same subject; he also had some success with *Le nozze di Lammermoor*, *La prison d'Édimbourg*, and finally *Thérèse*. He became a French citizen in 1834, from 1837 was a member of the Académie des Beaux-Arts, and from 1840 professor of counterpoint at the Conservatoire. His music was much influenced by Rossini, latterly also by Cherubini. Though excellently written, his operas lack melodic individuality. In 1860 Rossini asked him to reshape the French text of *Semiramide* from two into four acts and to compose the ballet music compulsory at the Opéra, and, knowing Carafa to be impoverished, assigned to him all the rights.

SELECT WORKLIST (all first prod. Paris unless otherwise stated): *Gabriella di Vergy* (Tottola; Naples 1816); *Elisabetta in Derbyshire* (Peracchi; Venice 1818); *I due Figaro* (Romani; Milan 1820); *Jeanne d'Arc à Orléans* (Lambert & Bournonville; 1821); *Le Solitaire* (Planard; 1822); *Le valet de chambre* (Scribe & Mélesville; 1823); *Les deux Figaros* (Turpenne; 1827); *Masaniello* (Commagny & Lafortelle; 1827); *Le nozze di Lammermoor* (Balocchi, after Scott; 1829); *La prison d'Édimbourg* (Scribe & Planard; 1833); *Thérèse* (Planard & De Leuven; 1838).

BIBL: F. Bazin, *Notice sur la vie de Michele Carafa* (Paris, 1873).

**Cardillac.** Opera in 3 acts by Hindemith; text by Ferdinand Lion after E. T. A. Hoffmann's story *Das Fräulein von Scuderi* (1818). Prod. Dresden 9 Nov. 1926, with Claire Born, Grete Merrem-Nikisch, Max Hirzel, Robert Burg, cond. Busch. Concert perf. in English, London, BBC, 18 Dec. 1936, with Eadie, Licette, McKenna, Fear, cond. Raybould. First stage version in England, New Opera Company, SW, 11 Mar. 1970, with Pashley, Robson, Wakefield, Cameron, cond. Lovett; Santa Fe 26 July 1967, with Yarick, Endich, Stewart, Reardon, cond. Craft. Rev. version (new text by composer, score changed), Zurich 20 June 1952, with Hillebrecht, Müller-Bütow, Lichtegg, Brauer, cond. Reinshagen.

Paris, 17th cent. The plot concerns a master jeweller, Cardillac (bar), who murders his customers rather than part with his creations. An officer (bs) witnesses one of the attacks but blames an innocent party, since he hopes to win Cardillac's favour and marry his daughter (sop). Finally, threatened with arson to his workshop,

Cardillac confesses and is torn to pieces by an irate crowd.

Also opera by Dautresme (1867).

BIBL: H. Schilling, *Paul Hindemith's 'Cardillac'* (Würzburg, 1962).

**'Care selve'.** Meleagro's (sop) opening arioso in Handel's *Atalanta*, when (as 'Tirsi'), he is seeking his love in the forests.

**Carestini, Giovanni** (*b* Filottrano, *c*.1705; *d* ?Filottrano, *c*.1760). Italian contralto castrato. Studied Bologna with Pistocchi and Bernacchi. Début Rome 1721 (in A. Scarlatti's *Griselda*) as soprano. His voice later developed into 'the finest, fullest and deepest counter-tenor [sic] that has, perhaps, ever been heard' (Burney); also an 'animated and intelligent actor', he was considered by many to be the best singer of his time. Vienna 1723–5; Venice between 1724 and 1731; Rome 1727–30; sang for the Elector of Bavaria 1731–41. He was engaged by Handel in London 1733–5 as the answer to Senesino and later Farinelli, stars of Porpora's rival group, and created roles in *Ariodante*, *Alcina*, *Terpsicore*, and *Arianna in Creta*. (He sang in London again, 1739–40, in operas by Hasse and Pescetti.) In 1735 he returned to the Continent, appearing throughout Italy, and in 1747–8 at Dresden. He sang for Frederick the Great 1750–4, then at Petersburg 1754–6, after which he retired. One of the foremost castratos of any age, he was admired universally for his virtuoso technique and expressive powers.

**Carl Rosa Opera Company.** English opera company founded in 1866 by Karl August Nicolaus Rose, a Hamburg violinist who came to England in 1866. After the death of his wife, the soprano Euphrosyne Parepa, in 1874, Rosa (as he became) founded the Carl Rosa Opera Company to present opera in English, opening with *Figaro* in Sep. 1875 at the Princess's T, London. Rosa was associated with Augustus Harris of Drury Lane, and a prosperous five years followed. After Harris took over Covent Garden and Rosa died (1889), the company only toured. After a Balmoral performance in 1893, Queen Victoria conferred on the company the title 'Royal'. Alfred van Noorden was director 1900–16, Arthur Winckworth and Rosa's second wife co-directors 1916–23. After the First World War the company took over the Harrison-Frewin and Alan Turner companies; the former was managed by H. B. Phillips, who became director of the Carl Rosa OC in 1923. After Phillips's death in 1950 his widow directed the company, with Arthur Hammond as music director, until 1957, when she was succeeded by H. Procter-Gregg. Financial and other difficulties ensued, leading to Procter-

Gregg's resignation and the withdrawal of Arts Council subsidy in 1958. Most of its personnel went to Sadler's Wells T. There was an unsuccessful attempt at a come-back in 1960.

The company gave the English premières of *Manon* (1885), *La bohème* (1897), and *Andrea Chenier* (1903), and the premières of British works including Goring Thomas's *Esmeralda* (1883) and *Nadeshda* (1885), Stanford's *The Canterbury Pilgrims* (1884), and Lloyd's *John Socman* (1951). Though the company's standards were sometimes rough and ready, it introduced opera to many parts of Britain and gave many artists valuable early experience.

**Carmen.** Opera in 4 acts by Bizet; text by Meilhac and Halévy, after Mérimée's novel (1845). Prod. Paris, OC, 3 Mar. 1875, with Galli-Marié, Chapuy, Lhérie, Bouhy, cond. Deloffre; Vienna 23 Oct. 1875 (in German, with recitatives by Guiraud replacing spoken dialogue); London, HM, 22 June 1878, with Hauk, Valleria, I. Campanini, Del Fuente, cond. Arditi; New York, AM, 23 Oct. 1878, with Hauk, Campanini.

Spain, *c*.1820. In a square in Seville, Micaëla (sop) is looking for Don José (ten), a corporal. He arrives and, as the girls return to work in the cigarette factory, one of them, Carmen (mez), throws him a flower. Micaëla gives him news of his mother, and he puts Carmen out of his mind. After a quarrel in the factory in which Carmen wounds another girl, she is captured and Don José is set to guard her; quickly fascinated by her, he unties her hands so that she can escape.

Carmen is dancing in Lillas Pastia's inn with Frasquita (sop) and Mercédès (sop). When the famous bull-fighter Escamillo (bar) enters, he is immediately attracted to her. Carmen, however, waits for Don José, refusing to join the others on a smuggling expedition. Soon after he arrives, he is summoned back to barracks, but Carmen tries to persuade him to desert. Zuniga (bs), Don José's captain, bursts in, hoping to find Carmen alone. He and José are about to fight when Dancairo (bar) and Remendado (ten) enter and lead Zuniga away. Guilty of having threatened a senior officer, José now has to join the smugglers.

Carmen has tired of José and he thinks, with remorse, of how he has broken the promise to his mother to marry Micaëla. Reading her fortune in the cards, Carmen foresees death. When Escamillo arrives, he and José fight over Carmen, but Escamillo is saved by the smugglers. Micaëla, who has come to find José, is discovered. He leaves with her when she tells him that his mother is dying.

On the day of Escamillo's bullfight in Seville, Carmen is warned that José is in the crowd seeking revenge. Alone, the two encounter each other; Carmen refuses to return to José and, to the sound of Escamillo's triumph in the main ring, José kills her and gives himself up.

Parodies on the opera are by Meyer-Lutz (*Carmen up to Date*, 1890). Modernized by Oscar Hammerstein II as *Carmen Jones* (New York, Broadway T, 2 Dec. 1943; Sheffield 14 Mar. 1986).

**Carnicer, Ramón** (*b* Tárrega, 24 Oct. 1789; *d* Madrid, 17 Mar. 1855). Spanish composer. Studied Barcelona with Queralt and Baguer. In 1816 he recruited an Italian company for the T de la Cruz, and in 1818 became director of the Coliseo T orchestra. He wrote three operas for Barcelona, moving to Madrid and then via Italy and France to London. Back in Madrid in 1827, he wrote four more operas. He was one of the founders of the Spanish national conservatoire, remaining a professor until his death. Much influenced by Rossini (for some of whose operas he wrote extra arias), he was nevertheless the most important Spanish opera composer of the early 19th cent.

SELECT WORKLIST: *Adele di Lusignano* (Romani; Barcelona 1819); *Elena e Costantino* (Tottola; Barcelona 1821); *Don Giovanni Tenorio* (Da Ponte, after Bertati; Barcelona 1822); *Elena e Malvina* (Romani; Madrid 1829); *Cristoforo Colombo* (Romani; Madrid 1831); *Eufemio di Messina* (Romani; Madrid 1832); *Ismailia* (Romani; Madrid 1838).

**carnival.** From It. *carnevale*, originating in the Lat. *carnem levare*, 'the putting away of flesh' as food: *OED*, which further defines it as 'the week (orig. the day) before Lent, devoted in Italy and other Roman Catholic countries to revelry and riotous amusement'. The celebrations differed widely in nature from country to country and from town to town. In Italy, the most sumptuous carnival was in Venice (and remains chiefly as a tourist attraction). The most licentious (deriving from the ancient Saturnalia) was in Rome, as Goethe distastefully described in his *Italienische Reise*; Berlioz agreed in his *Mémoires*, though for artistic purposes the Roman Carnival is represented as enjoyably riotous in Act II of his *Benvenuto Cellini*.

For operatic purposes, the season is usually dated from 26 December until the beginning of Lent (Ash Wednesday), when, especially in Italy in the 17th cent., new operas were produced. *Carnevale-quaresima* ('Carnival-Lent') was the second and most important of the three seasons into which the Italian operatic year was divided. See STAGIONE.

**Caron, Rose** (*b* Monnerville, 17 Nov. 1857; *d* Paris, 9 Apr. 1930). French soprano. Studied Paris with Masset and Obin; later with Sass. Début Brussels, M, 1883 (Alice in *Robert le Diable*). Here she created Brunehilde in Reyer's *Sigurd*, and title-roles in his *Salammbô* and Godard's *Jocelyn*. Paris, O, 1885–7, 1890–5. First French Sieglinde, Elsa, and Desdemona. Boschot portrays her 'indescribable poetic charm' as a performer. (R)

**'Caro nome'.** Gilda's (sop) aria in Act I of Verdi's *Rigoletto*, expressing her love for the 'student' Gualtier Maldé, who is in fact the Duke of Mantua.

**Carré, Marguerite** (*b* Cabourg, 16 Aug. 1880; *d* Paris, 26 Dec. 1947). French soprano. Studied Bordeaux and Paris Conservatoires. Début (semi-prof.) Nantes 1899 (Mimì); real début Paris, OC, 1901/2 (Mimì) whose director, **Albert Carré** (1852–1938) she married in 1901, divorced in 1924, and remarried in 1929. Here she created 15 roles, and was the first French Butterfly and Snegurochka. A popular singer, outstanding as Mélisande and Louise. Retired 1923. (R)

**Carré, Michel** (*b* Paris, 1819; *d* Argenteuil, 27 June 1872). French librettist. Furnished most of the leading French composers of the second half of the 19th cent. with librettos, including Meyerbeer (*Dinorah*), Gounod (*Mireille, Faust, Roméo et Juliette*, and five more), Bizet (*Pêcheurs de perles*), Thomas (*Hamlet, Mignon*), Offenbach (*Contes d'Hoffmann*), often in collaboration with Jules *Barbier. Uncle of **Albert Carré**. Sometimes wrote under the name of Jules Dubois.

**Carreras, José** (*b* Barcelona, 5 Dec. 1946). Spanish tenor. Studied Barcelona with Puig. Début Barcelona 1970 (Flavio in *Norma*). A protégé of Caballé, with whom he sang *Lucrezia Borgia* and at his London début, in concert, 1971. London, CG, from 1974 (Alfredo, Rodolfo, etc.); New York, C and M, from 1974; at San Francisco, Buenos Aires, Tokyo, etc. A gifted and popular lyric tenor with a highly appealing voice whose courageous struggle against leukaemia and comeback won him wide admiration. Autobiography, *Singing for the Soul* (Seattle, 1991). (R).

**Carte, Richard D'Oyly** (*b* London, 3 May 1844; *d* London, 3 Apr. 1901). English impresario. Abandoned ambitions as a composer of light opera for management, first of Patti, Mario, Gounod, and others, then as theatre manager, introducing to England works by Lecocq, Offenbach, Serpette and others at the Opera Comique T. Commissioned *Trial by Jury* from Gilbert and Sullivan, and its success led to his establishing the Comedy Opera C. Produced *The Sorcerer* and *H.M.S. Pinafore*, then formed his own company. Built and moved to the Savoy, opening with *Patience* 1881. Here the Gilbert and Sullivan partnership of 'Savoy Operas' flourished for ten years, and resumed with *Utopia Ltd* (1893) and *The Grand Duke* (1896). In 1891 he attempted to establish an English Grand Opera, and built the English OH (now Palace T), opening with *Ivanhoe*. The venture failed. At the Savoy he introduced electric lighting and abolished charges for programmes and cloakrooms. The D'Oyly Carte OC was revived after his death by his son Rupert then by this granddaughter Bridget. The company closed in 1982; a new company was formed in 1988 with the support of a bequest from Dame Bridget's estate and other sources, and benefited from a new artistic impetus.

**cartellone** (sometimes also **cartello**) (It.: 'a large placard'; hence a playbill). The list of operas to be performed during the season. The cartellone does not include the names of the artists appearing; they are listed under a separate heading, *elenco artistico. A leading operahouse is sometimes referred to as a **teatro di cartello**.

**Caruso, Enrico** (*b* Naples, 27 Feb. 1873; *d* Naples, 2 Aug. 1921). Italian tenor. Studied Naples with Vergine. Début Naples, N, 1894 (in Morelli's *L'amico Francesco*). Pursued an unremarkable career in Italy until 1897, when he sang with success at Palermo in *La Gioconda*. This was consolidated in the following years and in 1901 a triumph at Milan, S, in *L'elisir d'amore*, assured his future. Created Federico (Cilea, *L'arlesiana*, 1897), Loris (Giordano, *Fedora*, 1898), Florindo (Mascagni, *Le maschere*, 1901), Federico (Franchetti, *Germania*, 1902). London, CG, 1902 (Duke of Mantua), 1904–7, 1913–14; New York, M, 1903–20, where he sang in over 600 performances, among them creating Dick Johnson in *La fanciulla del West*. He possessed an astonishingly beautiful voice, rich and mellow in his youth and of an almost baritonal quality, then increasing in brilliance and smoothness in his middle years. His artistry, too, gained subtlety as he grew older, the abandoned gaiety of his earlier singing giving way to masterly phrasing and legato. His breath control was extraordinary. In 1908–9 he suffered a temporary vocal setback, and subsequently his voice darkened, but without losing its expressive qualities. His range of roles was enormous, encompassing both the lyric and dramatic, and his acting was matched by his skill in make-up, especially in the roles of Radamès, Canio, and Eléazar. The first true 'gramophone' tenor, he began recording in 1902 and made his last record in 1920. His royalties apparently reached the region of £500,000. One of his most treasured recordings is 'Una furtiva lagrima' (1904). He died of pleurisy, and was mourned as a man of immense warmth and generosity as well as one of the greatest singers of the century. (R)

BIBL: M. Scott, *The Great Caruso* (London, 1988); E. Caruso, jun., and A. Farkas, *Enrico Caruso, my Father and my Family* (Portland, 1990).

**Carvalho, Léon** (*b* Port-Louis, 18 Jan. 1825; *d* Paris, 29 Dec. 1897). French baritone and impresario. Studied Paris. Sang at the Paris, OC, 1851–5, where he met Marie *Miolan, marrying her 1853. Director Paris, L, 1856–68; producer, Paris, O, 1869–75; director, Paris, OC, 1876–87. When Paris, OC, burnt down 1887 and 131 people died, he was fined and imprisoned for negligence; reinstated on appeal 1891. Director Cairo OH 1868–72. Berlioz complained bitterly about his interference in the 1863 production of *Les Troyens*.

**Casanova de Seingalt, Giovanni Jacopo** (*b* Venice, 2 Apr. 1725; *d* Dux (Duchcov), 4 June 1798). Italian writer and adventurer. He was notorious for his amorous exploits, and these, his daring escape from imprisonment in Venice, and many other incidents are colourfully related in his 12 vols. of *Mémoires* (1791–8). He translated Rameau's *Zoroastre* for Dresden in 1752 and wrote the texts for many other works, some for the stage, with music; he also left a Utopian romance entitled *Icosaméron* in which there are thoughtful observations on the relationship of words and music in opera. He was a friend of Da Ponte, and may have been present at the first performance of *Don Giovanni*. Operas on his adventures are by Lortzing (1841), J. Wagner (*Casanova in Paris*, pub. 1889), Pulvermacher (1890), Barna (1902), Lincke (1913), Eysler (early 20th cent.), Kusterer (1921), Rózycki (1923), Benatzky (1928), Andreae (four 1-act operas, 1924), and Pick-Mangiagalli (1929).

**Casavola, Franco** (*b* Modugno, 13 July 1891; *d* Bari, 7 July 1955). Italian composer and critic. Studied in Bari, Milan, and with Respighi in Rome. An early adherent of Futurism, he wrote a number of works for the movement, turning later to traditional opera. His greatest success was with the 1-act *Il gobbo del Califfo*.

SELECT WORKLIST: *Il gobbo del Califfo* (Rossato, after the *1,001 Nights*; Rome 1929).

**Casella, Alfredo** (*b* Turin, 25 July 1883; *d* Rome, 5 Mar. 1947). Italian composer. Studied with his mother and in Paris piano with Diémer and composition with Leroux and Fauré. Returning to Italy in 1915, he associated himself with avant-garde ideas and was responsible for introducing much contemporary music to Italy. The greatest success of his three operas was *La donna serpente*. This was based on the same fable used by Wagner for *Die Feen*; however, in reaction both to Wagner and to verismo, it is a fantastic opera built out of swiftly alternating dramatic and comic scenes.

WORKLIST: *La donna serpente* (Lodovici, after Gozzi; Rome 1932); *La favola d'Orfeo* (Pavolini, after Poliziano; Venice 1932); *Il deserto tentato* (Pavolini; Florence 1937).

BIBL: F. D'Amico & G. M. Gatti, eds., *Alfredo Casella* (Milan, 1958).

**Casken, John** (*b* Barnsley, 15 July 1949). English composer. Studied Birmingham with Dickinson and Joubert, Warsaw with Dobrowolski. His first opera, the 1-act *Golem* aroused much interest at its 1989 Almeida Festival première, in particular for its concise and original dramatic sense and its rich orchestral colouring.

**Caspar.** A huntsman (bs) in Weber's *Der Freischütz*.

**Cassandra.** In classical mythology, the daughter of King Priam of Troy who received the gift of prophecy from Apollo in return for the promise of her love; when she then refused him, she was condemned to be believed by no one. Operas on her legend are by Copillo (1691), Bertin and Bouvard (1706), Faga (1713), and Gnecchi (1905). In Berlioz's *La prise de Troie*, the first part of *Les Troyens*, Cassandre (mez) leads her women to suicide rather than fall into the hands of the Greeks.

**Cassandre.** The Trojan seer (mez), daughter of Priam and Hecuba, in Berlioz's *Les Troyens*.

**Cassio.** A captain (ten), whom Iago persuades Otello is Desdemona's lover, in Verdi's *Otello*.

**'Casta diva'.** Norma's (sop) aria with chorus in Act I of Bellini's *Norma*, praying to the moon for peace between Gaul and Rome.

Also operetta in 3 acts by E. Bellini (1923).

**Castellan, Jeanne** (*b* Beaujeu, 26 Oct. 1819; *d* after 1858). French soprano. Studied Paris with Bordogni and Nourrit. Début Varese 1837 (Amina). Sang in Italy, Mexico, and USA until 1844; St Petersburg 1844–6; London, HM, 1845–7 (début Lucia); Paris, I, 1847; London, CG, 1848–53. Created Berthe (*Le prophète*), when Berlioz found her singing 'exemplary'. Though her voice was fine, her intonation and execution were not always reliable. She was evidently very beautiful.

**Castelmary, Armand** (*b* Toulouse, 16 Aug. 1834; *d* New York, 10 Feb. 1897). French bass. Paris, O, 1863–70, creating Don Diego (*L'Africaine*). Italy, 1873–9; S America, 1881; London, CG, 1889–96; New York, M, 1893–7. Was a refined and elegant Mephistopheles (Boito and Gounod); first Horatio in *Hamlet* (Thomas). Though often credited with creating the Grand

Inquisitor, he in fact sang the Monk (Verdi, *Don Carlos*).

**Casti, Giovanni Battista** (*b* Acquapendente, 29 Aug. 1724; *d* Paris, 5 Feb. 1803). Italian dramatist and librettist. Travelled widely, and visited Russia, where he wrote *Lo sposo burlato* (1779), and Vienna, where he wrote *Re Teodoro in Venezia* (1784), both for Paisiello. There followed works for Salieri, including *Prima la musica e poi le parole* (1786), a sharp satire on the low status of composers; it includes some glancing blows at Da Ponte. For Salieri he also wrote *Cublai, Gran Can de' Tartari* (1788, unprod.), and other works. He developed the dramma eroicomico, in which a famous historical character is presented in a laughable light, thus foreshadowing Offenbach. A skilled craftsman, with a cutting wit, and a vein of melancholy, he deserved, in spite of his views, greater composers.

**Castil-Blaze, François** (*b* Cavaillon, 1 Dec. 1784; *d* Paris, 11 Dec. 1857). French writer on music and composer. Studied Paris. His chief work is *De l'opéra en France* (Paris, 1820), which discusses the suitability of words for music and the components of opera, also attacking theatre managers, critics, and translators. He was himself a critic (of the *Journal des débats*, 1822–32) and a distinguished translator; but he was guilty of needlessly adapting foreign works and upsetting their proportions, not least by inserting numbers of his own. His version of Weber's *Der Freischütz* as *Robin des bois* caused the distress of the composer and the rage of his successor on the *Débats*, Berlioz. He also included parts of *Euryanthe* (which he translated, 1831) in a pasticcio, *La forêt de Senart* (1826). Father of the critic **Henri Blaze de Bury** (1813–86).

**Castor et Pollux.** Opera in a prologue and 5 acts by Rameau; text by Pierre Joseph Bernard. Prod. Paris, O, 24 Oct. 1737, with Pellisier, Tribou, Chassé, and rev. without prologue and new Act I, 1754; Glasgow (amateur), 27 Apr. 1927; New York, Vassar College, 6 Mar. 1937. Rev. Florence Festival, 1935, with Lubin, Villabella, Rouard, cond. Gaubert.

Ancient Greece. Pollux is the son of Jupiter and is immortal, Castor the son of Tyndareus and mortal. Castor has been killed, but when Pollux offers himself in Castor's place to Télaïre, she is confused and insists instead that Jupiter must be persuaded to restore Castor to life. Jupiter agrees on condition that Pollux take his place in Hades, while also revealing to him the delights of the Heavenly Pleasures on Olympus. Pollux remains firm, but at the gates of Hades he is met by Phoebe, who loves him and tries to turn him back, while Télaïre spurs him on. With Mercury's help, he enters Hades; Castor agrees to return to earth and to tell Télaïre he cannot accept Pollux's sacrifice, But Jupiter intervenes to declare that both Castor and Pollux shall be taken up to Olympus; Télaïre will follow; Phoebe is left, and kills herself.

Other operas on the legend (some with the title *I Tindaridi*) are by Traetta (1760), Bianchi (1779), Sarti (1786), Vogler (1787), Federici (1803), and Winter (1806).

**castrato** (It.: 'castrated') or evirato (It.: 'unmanned') or musico (It.: 'musician', in the 18th cent. a virtual synonym for castrato). A male soprano or contralto whose unbroken voice has been artificially preserved by means of a surgical operation on the testicles before puberty.

The practice of castrating boys so as to provide adult sopranos and contraltos was justified by the Roman Catholic Church interpreting an injunction of St Paul to the effect that women should keep silent in church (1 Cor. 14: 34). The vocal chords were prevented from undergoing the normal thickening which renders them less agile as well as causing them to produce lower notes. The voice thus created was, compared with the female equivalent, stronger and more flexible, often voluptuous in tone, and capable of the utmost delicacy and brilliance. Although one of the earliest mentions of castratos is in 1562, a time when they were employed exclusively for sacred music, the rise of opera in the 17th cent. provided them with another outlet for their talent. By the time of Metastasian opera seria in the mid-18th cent., castratos were a cardinal feature of operatic life: among the most famous were *Senesino, *Caffarelli, and *Farinelli. Admired as artists, mocked and resented for their vanity and touchiness, they formed an operatic subculture whose true nature and merit it is now difficult to appreciate. Charles D'Ancillon describes Pauluccio's tone as finer than a nightingale's and scarcely human, Jeronimo's as like the flute stop of an organ or gently falling water; Crescentini's voice was described by Vigny as seraphic, by Schopenhauer as 'supernaturally beautiful': such accounts suggest an almost inhuman quality. The castratos' abilities encouraged composers to write (both for individuals and for the castrato voice in general) music of unparalleled brilliance, through which we can obtain some notion of what their powers must have been. However, these parts must in modern revivals be sung by a counter-tenor or a woman, with their crucially different timbres, or, with transposition down the octave, by a tenor or baritone, thus distorting the balance of the music.

In Italy, though the practice was officially punishable by death, the *fait accompli* was tolerated. Parents were persuaded by the prospect of financial gain to mutilate their small sons (the operation even being performed on occasions by the mother), often long before there was any evidence of vocal or musical talent. Later it was considered more humane to wait a little longer; some castratos were even alleged to have chosen their profession. Joseph Haydn was one boy who just managed to retain his masculinity. However, the evirati were not debarred from a love life: often outstandingly handsome (on stage, their beauty in female roles could be powerfully seductive, as Casanova acknowledged), they were popular with women not only for this and their exquisite singing but also because there was no risk of pregnancy. A few even attempted marriage (e.g. *Tenducci), though this was not sanctioned by the Pope.

Following the reforms of Gluck, the art of the castrato went into a decline that continued until the demands of dramatic truth in opera, coupled with those of humanity, exterminated it: the castrato had already been banished from opera buffa. One of the greatest 18th-cent. castrato roles, Idamante in *Idomeneo*, was rewritten for a tenor by Mozart for the Vienna revival, and although castratos were still common in the early 19th cent., their activities were largely confined to Italy. Rossini composed *Aureliano in Palmira* for the last great castrato, *Velluti, but did not repeat the experiment. When the species eventually disappeared from Italian opera, it was replaced not by the tenor, but by the contralto, which also took over the title musico, and the function of the travesti, or breeches role.

The last important composer to write for the castrato was Meyerbeer, with *Il crociato in Egitto* (1824), again for Velluti. Though Wagner was interested in obtaining Domenico Mustafà (1829–1912) for the part of the self-emasculated Klingsor, he abandoned the idea. The last professional castrato (though he did not appear in opera) was Alessandro Moreschi (1858–1922), who made ten records in 1902–3. Gerald Barry's opera set in 18th-cent. Dublin, *The Intelligence Park* (1990), contains a castrato character, Serafino (counter-tenor), based on Tenducci.

BIBL: A. Heriot, *The Castrati in Opera* (London, 1960).

**Castro, Juan José** (*b* Avellaneda, 7 Mar. 1895; *d* Buenos Aires, 5 Sep. 1968). Argentinian composer and conductor, brother of musicians José Maria, Luis Arnoldo, and Washington. Studied Buenos Aires with Gaito and Fornarini, and Paris with D'Indy. Director Buenos Aires, C, 1933, giving Argentinian première of *L'heure espagnole* and other important performances. His three operas include *Proserpina y el extranjero*, a modernization of the Persephone myth set in Buenos Aires which won the International Verdi Prize;

the others reflect his admiration for his friend Lorca.

WORKLIST: *La zapatera prodigiosa* (The Marvellous Cobbler's Wife) (Lorca original; Montevideo 1949); *Proserpina y el extranjero* (Proserpina and the Stranger) (Del Carlo; Milan 1952); *Bodas de Sangre* (Blood Wedding) (Castro, after Lorca; Buenos Aires 1956).

BIBL: R. Arizaga, *Juan José Castro* (Buenos Aires, 1963).

**Catalani, Alfredo** (*b* Lucca, 19 June 1854; *d* Milan, 7 Aug. 1893). Italian composer. Studied Lucca with Magi, Paris with Bazin. In Milan he also studied with Bazzini, graduating with the promising *La falce*. Bazzini introduced him to the avant-garde *Scapigliatura group, which included Boito and Faccio, and to the music publisher G. Lucca, who commissioned *Elda*. When this was not immediately performed and *Dejanice* failed (uncharacteristically, one admirer was Puccini), he turned to Romantic subjects; but the absorption of Lucca into the firm of Ricordi placed Catalani in an inferior position to both Verdi and Puccini. His revision of *Elda* as *Loreley* was ignored by Ricordi, and when Catalani himself arranged a performance he found himself accused of Wagnerism (the work's orchestral skill in fact owes more to Weber). His fourth opera, *Edmea*, was more popular; its première was the occasion of Toscanini's Italian début. Catalani next succeeded in obtaining a performance at Milan, S, of his masterpiece *La Wally*, a work championed by Toscanini (who named his daughter after the heroine). The work has impressive and touching moments, and is carefully constructed, but its effective manner is insufficiently supported by enough strong inventive matter to give it more than occasional revivals in Italy. Nevertheless, Catalani was unlucky to make his career when Verdi was at the height of his fame, Puccini's star was rising, and the verismo movement was growing: Catalani's modest but distinct talent was of another order.

WORKLIST: *La falce* (The Sickle) (Boito; Milan 1875); *Elda* (D'Ormeville; comp. 1876, prod. Turin 1880; rev. as *Loreley*, Zanardini after D'Ormeville; Turin 1890); *Dejanice* (Zanardini; Milan 1883); *Edmea* (Ghislanzoni; Milan 1886); *La Wally* (Illica, after Hillern; Milan 1892).

BIBL: S. Pagani, *Alfredo Catalani* (Milan, 1957); M. Zurletti, *Catalani* (Turin, 1982).

**Catalani, Angelica** (*b* Senigallia, 10 May 1780; *d* Paris, 12 June 1849). Italian soprano. Studied Senigallia with her father, and Morandi. Début Venice 1797 (Mayr's *Lodoïska*). Milan, S, 1801; London 1806. The first London Susanna (taking over 'Voi che sapete') in 1812. Appeared in Riga, Lemberg, Vilna, and Brünn in 1820, and in Romania as Rosina, Angelina, etc. The last survivor of the declining years of bel canto, she possessed a 'prodigiously beautiful' voice (Stendhal) and a style lavish to the point of absurdity, both of which earned her huge fees, e.g. up to 200 guineas for singing 'Rule Britannia'. Married French diplomat Paul Valabrègue, who said that for a successful opera season it only needed 'ma femme et quatre ou cinq poupées'. This recipe, despite subsidy, failed to work at the Paris, I, where she was manager, 1814–18. She retired from opera in 1821, though continuing concert tours, and founded a school near Florence.

BIBL: M. and L. Escudier, *Vie et aventures des cantatrices célèbres* (II) (Paris, 1856).

**catalogue aria** (It., *aria di catalogo*). One of the most characteristic and popular aria forms in 18th-cent. opera buffa, consisting of a comically rapid enunciation of a list of items. Invariably featuring nimble declamation, the catalogue aria was the precursor of the *patter song: the items enumerated ranged from exotic places, or assorted grandees (e.g. Gazzaniga, *L'isola di Alcina*), to the goods in a shop (Goldoni's text for *Lo speziale*, set by Haydn among others). One of the librettists who used it most effectively was *Bertati, whose *Don Giovanni Tenorio* for Gazzaniga includes one listing the different classes of girls seduced by the hero. It is possible that Bertati and others knew contemporary and earlier collections of Anacreontic odes from the 6th cent. BC, one of which consists of a list of the poet's loves in many different lands, including 2,000 in Rhodes. Bertati's catalogue aria provided the model for Da Ponte, listing the girls by country, in his libretto for *Don Giovanni*; it is the aria sung by Leporello (bar) in Act I of Mozart's opera that is usually intended by the description 'catalogue aria'.

**Catania.** Town in Sicily. Plans for a theatre in 1812 were frustrated by Algerian pirates. The T Comunale Provisorio opened 1821 with Rossini's *Aureliano in Palmira*. The T Massimo Bellini (cap. 1,300), named after the town's most famous composer, opened 31 May 1890 with *Norma*. Closed at outbreak of Second World War, but reopened 1943 on orders of British with scenery flown from Palermo by the USAF. Rebuilt 1952.

**Catel, Charles-Simon** (*b* Laigle, 10 Jan. 1773; *d* Paris, 29 Nov. 1830). French composer and theorist. Studied Paris with Gossec. He joined the band of the Garde Nationale, also acting as répétiteur at the Opéra and as professor at the newly founded Conservatoire from 1795. His works for the Opéra, beginning with *Sémiramis*, reflect the atmosphere of the Empire and of the emergent Romanticism, and helped to confirm the success of some genres and to explore novel techniques. His most spectacular work was *Les* *bayadères*, which, though somewhat over-shadowed by Spontini, was as effective in demonstrating the power of the evolving Grand Opera. With *Zirphile et Fleur de Myrte* he explored the genre of magic opera, familiar from Grétry but handled with new effectiveness.

He was also successful as a composer of opéra-comique. *Wallace* draws on the Romantic fascination with Scotland that had absorbed opéra-comique composers including Méhul, but the others are light pieces that owe more to opera buffa. One of Napoleon's opera prize competitions in 1809 placed his *L'auberge de Bagnères* in joint second place with Cherubini's *Les deux journées*, behind Méhul's *Joseph*. He had only moderate success with *Les aubergistes de qualité*. Though his music is often somewhat slender, with enterprising ideas that are not really sustained creatively, Catel is interesting as something of a pioneer and as one of the most intelligent and perceptive composers of French opéra-comique.

SELECT WORKLIST (all first prod. Paris): *Sémiramis* (Desriaux, after Voltaire; 1802); *L'auberge de Bagnères* (Jalabert; 1807); *Les bayadères* (Jouy; 1810); *Les aubergistes de qualité* (Jouy; 1812); *Wallace* (Saint-Marcellin; 1817); *Zirphile et Fleur de Myrte* (Jouy & Lefebvre; 1818).

BIBL: F. Hellouin and J. Picard, *Catel* (Paris, 1911).

**Caterina Cornaro.** Opera in a prologue and 2 acts by Donizetti; text by Giacomo Sacchero, after Vernoy de St Georges's libretto *La reine de Chypre* for Halévy. Prod. Naples, C, 12 Jan. 1844, with Goldberg, Salvetti, Fraschini, Coletti, Arati. Concert perfs. London 1972, New York 1973. Revived Naples, C, 1972, with Gencer, Aragall, Bruson, Clabassi, cond. Cillario.

Venice and Cyprus, 1472. The wedding of Caterina (sop), daughter of Andreas (bs), to a young Frenchman, Gerardo (ten), is postponed when Mocenigo (bs) brings word that Lusignano, King of Cyprus (bs), wishes to marry her. After much intrigue, involving Lusignano being slowly poisoned by Mocenigo, Gerardo joins the Knights of the Cross to help Lusignano defend Cyprus against the Venetians. Lusignano is mortally wounded; as he dies he entrusts his people to Caterina's care. Gerardo then returns to Rhodes. (In a revised finale for the Parma production, Donizetti makes Lusignano inform Caterina that Gerardo has been killed in battle.)

Other operas on the subject are by Halévy (1841), Lachner (1841), Balfe (*The Daughter of St Mark*, 1844), and Pacini (1846).

**Catherine the Great** (Catherine II of Russia) (*b* Stettin, 2 May 1729; *d* St Petersburg, 10 Nov. 1796). Empress of Russia 1762–96. Her energies embraced the encouragement of opera, in part as court entertainment but also as an instrument of policy. She invited to Russia both Tartini and Galuppi, the first of a succession of Italians who were to stimulate and organize the Russian theatres and compose for them; she also encouraged French opera. As well as her native German, she spoke (faultily but expressively) both Russian and French, and wrote or collaborated in some librettos herself. As an admirer of the *Encyclopédistes*, she favoured opéra-comique, with its social content and comment, though of a reactionary kind; and in this vein, though with little ear for the music then provided, she produced her librettos. These consisted of a historical propaganda piece, *The Early Reign of Oleg* (music Pashkevich, Canobbio, Sarti, 1790), *Fevey* (Pashkevich, 1786), *The Novogorod Bogatyr Boyeslayevich* (Fomin, 1786), *The Brave and Bold Knight Akhrideyich* (Vančura, 1787), a piece satirizing Gustav III of Sweden, *The Woeful Bogatyr Kosometovich* (Martín y Soler, 1789), and *Fedul with his Children* (Soler and Pashkevich, 1791). She appears (mez) in Cui's *The Captain's Daughter*.

**Cavalieri di Ekebù, I** (The Knights of Ekebù). Opera in 4 acts by Zandonai; text by A. Rossato, after Selma Lagerlöf's novel *Gösta Berlings saga* (1891). Prod. Milan, S, 7 Mar. 1925, with Fanelli, Casazza, Lo Giudice, Franci, Autori, cond. Toscanini.

Ekebù, Sweden, 19th cent. The unfrocked priest Gösta Berling (ten) is given work with the 'Knights' of Ekebù by the Commander (mez), owner of the ironworks and castle of Ekebù. She is suspected of selling the knights' souls to the devil but is cleared by them. She pardons her accusers and dies, bequeathing her possessions to Gösta and his beloved, Anna (sop).

**Cavalieri, Emilio de'** (*b* Rome, *c*.1550; *d* Rome 11 Mar. 1602). Italian composer, organist, teacher, dancer, administrator, and diplomat. Organist for Lenten worship at the Oratorio del Santissimo Crocifisso in Rome *c*.1572; superintendent of musicians and actors at the Tuscan court 1588. Responsible for organizing many of the great celebrations of the Medici family, he also functioned as director of all the dramatic presentations of the Bardi *Camerata and with Peri, Caccini, and Gagliano was one of the first to experiment with the new *stile rappresentativo.

For the famous 1589 intermedi, whose co-ordination was his responsibility, Cavalieri contributed the first madrigal for the opening intermedio and a madrigal and the concluding music for the sixth intermedio, writing in the festive style of the late Renaissance appropriate to a lavish court celebration. The opening chorus of this finale, 'O che nuovo miracolo', often called the 'Ballo del Gran Duca' or 'Aria di Fiorenza', was imitated by many later composers and became one of the best-known pieces of the period. Other dramatic works composed around this time include settings of verse by Laura Guidiccioni, *Il satiro* and *La disperazione di Fileno*, and *Il giuoco della cieca*, her adaptation of Guarini's *Pastor fido*. No music survives, but the demands for stage machinery and elaborate costumes, and Cavalieri's apparent use of stile rappresentativo, suggest a link between intermedi and opera.

In 1600 Cavalieri produced another dramatic work, *La contesa fra Giunone e Minerva*, for the wedding of Henry VI of Spain. However, his plans to take control of the composition and production of *Il rapimento di Cefalo*, to be performed for the same celebration, were foiled by Caccini, and he resigned his post and returned to Rome. The same year his best-known work, *La rappresentazione di anima e di corpo*, was performed at the Oratorio della Vallicella, the church of St Philip Neri. By this means Cavalieri presented the new stile rappresentativo to a Roman audience for the first time and initiated the operatic tradition which was to flourish there in the 1620s and 1630s. Compared to the early efforts of Peri and Caccini there are some noticeable differences: the characters are not taken, like Orpheus, Eurydice, and Daphne, from classical mythology, but are allegorical figures, such as Time, Life, Intellect, and Soul. Rather than presenting a pastoral tale, *La rappresentazione di anima e di corpo* is more akin to a morality play, for there is a clear didactic purpose in the drama. For this reason, and because of its venue, it is often described as the first oratorio, but it is better called a sacred opera. Like Cavalieri's earlier dramatic works, it includes many features which belong to the older tradition of intermedio. The requirement that the actors and singers should be beautifully costumed, the need for a lavish set, the large choruses (of Blessed and Damned Spirits), and his suggestion that four intermedi be placed between the three acts all appear somewhat anachronistic, although in its skilful use of monody the work is clearly part of the new tradition as well. The preface to the score, with its detailed instructions for the playing of the figured bass and explanation of vocal ornaments, is one of the most important pieces of documentary evidence for an understanding of stile rappresentativo and confirms Cavalieri's position as a crucial figure in the early history of opera.

WORKLIST: 'Dalle più alte sfere' (Bardi), 'Godi turba mortal' (unknown), and 'O che nuovo miracolo' (Guidiccioni) (from intermedi for Bargagli's *La pellegrina*) (Florence 1589); *La disperazione di Fileno* (Fileno's Despair) (pastoral) (Guidiccioni; Florence 1590; lost); *Il satiro* (The Satyr) (pastoral) (Guidiccioni; Florence 1590, lost); *Il giuoco della cieca* (The Blind Woman's Game) (Guidiccioni, after Guarini; Florence 1595, lost); *La contesa fra Giunone e Minerva* (The Quarrel Between Juno and Minerva) (Guarini; Florence 1600, lost); *La rappresentazione di anima e di corpo* (The Representation of the Body and the Soul) (Manni; Rome 1600).

**Cavalieri, Katharina** (*b* Währing, 19 Feb. 1760; *d* Vienna, 30 June 1801). Austrian soprano. Studied with Salieri (becoming his mistress). Début Vienna, Italian O, 1775 (Sandrina in Anfossi's *La finta giardiniera*). Joined the National-Singspiel founded by Joseph II in 1778. In 1781 and 1783 created roles in two of Salieri's operas. Her voice was full and even throughout, and her technique excellent. Though not beautiful, she carried herself well and possessed 'considerable expression'. Mozart wrote Constanze (making concessions to her 'flexible throat'), and Mme Silberklang for her, as well as Elvira's 'Mi tradì' in the Vienna première of *Don Giovanni*. She also sang the Countess in 1789. One of the best in an array of very gifted singers in contemporary Vienna. She retired in 1793.

**Cavalieri, Lina** (*b* Viterbo, 25 Dec. 1874; *d* Florence, 7 Feb. 1944). Italian soprano. Of humble origins, she made a name for herself as a café singer throughout Europe. Studied Milan with Mariani-Masi; début Naples, C, 1900 (Mimì). After appearances throughout Italy, in Lisbon and at St Petersburg, she created L'Ensoleillad in Massenet's *Chérubin*, Monte Carlo 1905. New York, M, 1906–10; Paris, O, 1907 and 1909; London, CG, from 1908. Particularly successful in seductive roles (Manon Lescaut, Massenet's Salome, Giulietta, Violetta, etc.) by reason of her exceptional beauty and grace; in the USA she was known as 'the kissing prima-donna'. Much admired by the aristocracy and royalty, she boasted over £500,000 of jewels, including gifts from the Tsar and the Prince of Monaco. She had four husbands, including the French tenor Lucien *Muratore; while married to him she made several films. She died in a air-raid on Florence: on her way to a shelter, she decided to return to her villa for her jewels. A film of her life, *La donna più bella del mondo*, was made in 1957 with Gina Lollobrigida. Autobiography, *Le mie verità* (Rome, 1936). (R)

**Cavalleria rusticana** (Rustic Chivalry). Opera in 1 act by Mascagni; text by G. Menasci and G. Targioni-Tozzetti, after G. Verga's drama (1884) based on his story (1880). Prod. Rome, C, 17 May 1890, with Bellincioni, Stagno, Salassa, cond. Mugnone; Philadelphia, Grand OH, 9 Sep. 1891, with Kört-Kronold, Guille, Del Puente, cond. Hinrichs; London, Shaftesbury T, 19 Oct. 1891, with Mariani, Vignas, Brambara, cond. Arditi.

A Sicilian village, 1880. Santuzza (sop), seduced by Turiddu (ten), is distraught when she finds he has deserted her in favour of his former sweetheart Lola (mez), wife of the drover, Alfio (bar). A final appeal to Turiddu by Santuzza to return to her is of no avail, and so she tells Alfio all that has happened. Alfio swears vengeance, challenges Turiddu to a duel, and kills him. The action takes place in a small Sicilian village on Easter morning; the famous Intermezzo is played between the two scenes of the opera, while the stage remains empty.

Winner of the contest for a 1-act opera organized by Sonzogno in 1889, and a founding work of *verismo. A sequel, *Silvio*, was composed by Borch (c.1898).

Also operas by Gastaldon (*Mala Pasqua*, 1890) and Monleone (1907), and many parodies.

**Cavalli, Francesco** (orig. Pier Francesco Caletti) (*b* Crema, 14 Feb. 1602; *d* Venice, 14 Jan. 1676). Italian composer. A choirboy at St Mark's, Venice, where he probably studied with Monteverdi. In 1638, a year after the opening of the first public opera-house in Venice, Cavalli and Persiani joined forces to produce *Le nozze di Teti e di Peleo*. Over the next ten years he wrote nine operas for the T San Cassiano: his most important librettist was *Faustini, with whom he shared the management of the theatre. In *Egisto*, *Ormindo*, and *Doriclea* they consolidated the new style of Venetian opera, though a setting of a libretto by Cicognini, *Giasone*, became Cavalli's most popular work. When in 1650 financial difficulties closed the T San Cassiano, Faustini transferred to the smaller T San Apollinare, intending to produce two of his operas there annually. Four operas were written, including *Rosinda* and *Calisto*, but after Faustini's death in 1651 Cavalli was unwilling to continue his association with the theatre.

By now his fame had spread outside Venice, leading to *Orione* and *Veremonda*, while in 1653 he became associated with the T Santi Giovanni e Paolo. Here *Serse*, *Statira*, and *Artemisia* were produced, together with revivals, but in 1658 he returned to the T San Cassiano as musical director. Contracted to produce one opera each year, he was not allowed to compose for any other Venetian theatre, though was free to accept invitations from elsewhere. The commission in 1660 of *Ercole amante* for the wedding in Paris of King Louis XIV demonstrated Cavalli's status as a composer of international repute, though its unfavourable reception dealt a severe blow to his morale.

For a while he refused to compose any more operas, and though eventually persuaded to collaborate with Minato on three new works, his settings of Aureli's *Eliogabalo* in 1668 and Bussani's *Massenzio* in 1673 were never performed. The probable reason was that his style was already becoming outdated. While earlier works were revived both in Venice and elsewhere in Italy, a younger generation of composers, which oriented itself more to the demands of the singer than to the dictates of the drama was now popular with the Venetian public. By now extremely wealthy, he concentrated on his ecclesiatical duties, becoming maestro di cappella at St Mark's in 1668.

Cavalli occupies a central position in the development of opera. Together with Monteverdi he was the first to respond to the demands of the new public opera-houses, and was likewise a gifted and ingenious melodist. Rather than appealing only to aristocrats or intellectuals, opera had now to attract the bourgeoisie and even the lower classes, while commercial pressure meant that artistic considerations had frequently to yield to financial ones. The role of the chorus and size of the orchestra were reduced, while individual singers were given greater prominence. To appeal to a wider audience comic characters, such as nurses, pages, and servants, were featured, and given an appropriately simple and humorous musical style. For the protagonists, a developed *stile rappresentativo was employed, in which the later division into recitative and aria was already becoming apparent. The lament over a chromatic ground bass now became an essential feature of opera, as did the final love duet at the end of each act. The action of the opera, invariably concerned with the entanglement of two or more pairs of lovers, was complicated by elaborate misunderstandings, mistaken identity, and disguise, although a *lieto fine was essential.

Though Cavalli collaborated with a number of librettists, his partnership with Faustini proved crucial. Despite individual felicities, it is possible to see a unified style throughout these works, which is continued in the operas which he wrote with other librettists, such as *Serse* and *Erismena*. Cavalli's particular gift lay in the matching of musical progression to dramatic pace: recitative still retained some of the dramatic intensity of monody, while the aria was not yet the emotionally static display piece of the next century. The juxtaposition of serious scenes with comic interludes, in which the tension generated by the interaction of the main characters was released, engaged the attention of the audience and provided the variety of mood that had been lacking in earlier operas.

*Ercole amante* falls outside the main body of Cavalli's operas. Based on a quasi-classical story, and cast in five rather than the by now customary three acts, it featured massive choral scenes and was designed on an opulent scale which, although appropriate to a royal wedding, was far removed from that of Venetian opera. Only Lully's ballet music received public approbation, and Cavalli never fully recovered from the lukewarm reception of his work.

Although they gradually passed out of fashion, Cavalli's works exerted an important influence not only on the next generation in Italy, but also on the development of opera in France and England. The lament 'When I am laid in earth' in Purcell's *Dido* has many precursors in Cavalli's œuvre, while the recitative of Lully's operas owes much to his model. The very free realizations and recompositions of *Ormindo*, *Calisto*, and *Egisto* by Raymond Leppard for Gly. in the late 1960s began a revival of interest in Cavalli's music, and his place as a seminal figure in the development of early opera is now beyond question.

WORKLIST (all first prod. Venice unless otherwise stated): *Le nozze di Teti e di Peleo* (Persiani; 1639); *Gli amori di Apollo e di Dafne* (Busenello; 1640); *Didone* (Busenello; 1641); *L'amore innamorato* (Fusconi; 1642, lost); *La virtù de' strali d'Amore* (Faustini; 1642); *Egisto* (Faustini; 1643); *Ormindo* (Faustini; 1644); *Doriclea* (Faustini; 1645); *Titone* (Faustini; 1645, lost); *Giasone* (Cicognini; 1649); *Euripo* (Faustini; 1649, lost); *Orimonte* (Minato; 1650); *Oristeo* (Faustini; 1651); *Rosinda* (Faustini; 1651); *Calisto* (Faustini; 1651); *Eritrea* (Faustini; 1652); *Veremonda* (Strozzi; 1653); *Orione* (Melosio; Milan 1653); *Ciro* (Sorrentino; 1654); *Ipermestra* (Moniglia; Florence 1654); *Serse* (Minato; 1655); *Erismena* (Aureli; 1655); *Statira* (Busenello; 1655); *Artemisia* (Minato; 1657); *Antioco* (Minato; 1656); *Ercole amante* (Buti; Paris 1662); *Scipione affricano* (Minato; 1664); *Muzio Scevola* (Minato; 1665); *Pompeo magna* (Minato; 1666); *Eliogabalo* (Aureli; 1668); *Coriolano* (Ivanovich; Piacenza 1669, lost); *Massenzio* (Bussani; 1673, lost).

BIBL: J. Glover, *Cavalli* (London, 1978).

**Cavaradossi.** A painter (ten), lover of Tosca, in Puccini's *Tosca*.

**cavata** (It.: lit. 'something extracted, carved, or dug', hence an epigraph or inscription). In 18th-cent. opera the term was applied to the short ariosos found at the end of a long recitative, a form known as the *recitativo con cavata*. These cavate are short and epigrammatic in character, while their melody is often derived from ('dug out' of) the music of the preceding recitative.

**cavatina** (It., dim. of *cavata). In 18th- and 19th-cent. opera, a short solo song, simple in style and usually lacking the *da capo, often consisting of a short instrumental introduction to a single sentence or statement set to music. An early example is 'Godi l'amabile' in Graun's *Montezuma* (1755); more famous are the Countess's 'Porgi amor' and, in more elaborately developed form, Figa-

ro's 'Se vuol ballare' in *Le nozze di Figaro*, which is cast in a mocking minuet style. Later cavatinas include Agathe's 'Und ob die Wolke' (*Freischütz*), Euryanthe's 'Glöcklein im Thale' (*Euryanthe*) and 'Ah! che scordar non so' (*Tancredi*). In Italy, after *c.* 1830, the term is used to denote anybody's opening aria, usually in two movements. Examples include Norma's 'Casta diva' and Leonora's 'Tacea la notte placida' (*Trovatore*).

**Cavos, Caterino** (*b* Venice, 30 Oct. 1775; *d* St Petersburg, 10 May 1840). Italian composer. His father Alberto, an important figure in Venetian operatic life, was in charge of La Fenice and arranged the début there of Angelica Catalani. Caterino studied in Venice with Bianchi, and made his mark early as a composer of dramatic music. He travelled to Russia in the winter of 1798–9, and in April was made director of the Imperial Theatres. Here he flung himself energetically into musical life as a conductor, composer, arranger, organizer, and singing-teacher, playing the dominant part in the development of Russian operatic life in the first decades of the 19th cent. In 1804 he was responsible for bringing to Russia Kauer's popular *Das Donauweibchen* in an adaptation by Stepan Davydov as *Lesta, dneprovskaya rusalka*, with additional music of his own; he followed this with a second *Lesta* opera adding to Kauer's some of his own music; two more parts were composed by Davydov. These successes encouraged Cavos to abandon trivial French works in favour of a more indigenous and ambitious opera. He was responsible for the Russian premières of *Der Freischütz* (1824), *Fra Diavolo* (1831), *Robert le diable* (1834), and other works. His many pupils, some of whom formed the first generation of properly trained Russian singers, included Osip Petrov, Josephine Fodor-Mainvielle, and Anna Vorobyova. He was Kapellmeister of the Italian company (1828–31) and of the German and Russian companies (1832–40).

However, his real influence on Russian music was as a composer. From the first he took an interest in Russian subjects (having done so even while still in Italy), and beginning with a fairy opera, *Knyaz nevidimka*, he wrote a number of works which set a fashion for later Russian opera in their use of folk and patriotic, and especially legendary and magical, subjects, and also in their reliance on the fantastic and spectacular in staging. These included the successful *Ilya bogatyr* and *Kazak stichotvorets*. But his major work is *Ivan Susanin*, a *rescue opera with dialogue on the same subject as Glinka's *A Life for the Tsar* (though ending happily). This is typical of his skill in grafting Russian folk music and folk subjects onto the Italian stock as one of the first moves towards the growth of a national Russian opera, though he generously welcomed Glinka's opera as superior to his own. Cui praised his operas for their concern to assimilate Russian elements, even if they remained fundamentally Italian. His wife **Camilla Baglioni** was a singer of comic parts. Of their sons, **Giovanni** was a respected conductor in St Petersburg; and **Alberto** (1801–63) was an architect who worked on theatre design, being responsible for rebuilding the Moscow, B, after the fire of 1853 and also rebuilding the St Petersburg, M, and Mikhaylovsky; he wrote a *Traité de la construction des théâtres*. Alberto's daughter Camilla married Nikolay Benois; their son was Alexandre *Benois.

> SELECT WORKLIST (all first prod. St Petersburg): *Knyaz nevidimka* (The Invisible Prince) (Lifanov; 1805); *Ilya bogatyr* (Krylov; 1807); *Kazak stichotvorets* (The Cossack Poet) (Shakhovskoy; 1812); *Ivan Susanin* (Shakhovskoy; 1815); *Dobrynya Nikitich* (comp. with Antonolini; 1818); *Zhar-ptitsa* (The Firebird) (Antonolini; 1823).

**Cebotari, Maria** (*b* Kishinyov, 10 Feb. 1910; *d* Vienna, 9 June 1949). Bessarabian soprano. Studied Berlin with Oskar Daniel. Début Dresden 1931 (Mimì). Berlin, SO, 1936–44; London, CG, 1936, 1947; also appeared at Vienna and Salzburg. Roles ranged from Zerlina and Sophie to Turandot and Tatyana. Made several films, including *Maria Malibran*. Created Aminta (*Die schweigsame Frau*), Julia (Sutermeister's *Romeo und Julia*), and Iseut (Martin's *Le vin herbé*). An attractive performer with a clear, fresh voice. (R)

> BIBL: A. Mingotti, *Maria Cebotari* (Salzburg, 1950).

**Cecchina, La.** See BUONA FIGLIUOLA, LA.

**Cecilia.** Azione sacra in 3 acts by Refice; text by E. Mucci. Prod. Rome, R, 15 Feb. 1934, with Muzio; Buenos Aires, C, 4 Oct. 1934, with Muzio.

Based on the legend of Cecilia and Valerian, and of Cecilia's martyrdom.

Also the subject of operas by Orefice (1902), Cesi (1904), Montefiore (1905), Ryelandt (1907), and Mercuri (1910).

**'Celeste Aida'.** Radamès's (ten) aria praising Aida's beauty in Act I of Verdi's *Aida*.

**Cellini, Benvenuto.** See BENVENUTO CELLINI.

**Cenerentola, La** (Cinderella). Opera in 2 acts by Rossini; text by Jacopo Ferretti, after Étienne's text for Steibelt's opera. Prod. Rome, Valle, 25 Jan. 1817, with Giorgi-Righetti, Guglielmi, De Begnis, Verni; London, H, 8 Jan. 1820, with

Teresa Belocchi; New York, Park T, 27 June 1826, by García's company, with Malibran, Barbieri, Mme García, Milon, García jun., Rosich, Angrisani.

The plot follows the familiar fairy story by Perrault. Angelina (mez), known as Cinderella, is ill-treated by her father Don Magnifico (bs) and her two step-sisters Clorinda (mez) and Tisbe (sop). Prince Ramiro (ten), in search of a wife, changes places with his valet Dandini (bar), and falls in love with Angelina. Alidoro (bs), the Prince's tutor (the equivalent of the fairy godmother in the children's versions), helps Angelina to attend the ball at the palace. In place of the glass slipper there is a silver bracelet, by means of which the prince discovers the identity of Angelina and makes her his bride. See also CINDERELLA.

**censorship.** Opera has come into conflict with the authorities, and been restrained by censorship, in various countries, at various times, and for various reasons. It has even been banned virtually, e.g. under Paul I in Russia (1796–1801), or completely, e.g. under the Protectorate in England (1653–60).

In Britain, where censorship was formerly in the hands of the Lord Chamberlain, religious subjects were long forbidden the stage. Rossini's *Mosè* and Verdi's *Nabucco* appeared under different guises in Victorian times; *Samson et Dalila* was kept off the stage until 1909, and Massenet's *Hérodiade* appeared as *Salome* with the names of the characters changed, while Strauss's own *Salome* had to be altered for its Covent Garden première in 1910 to satisfy the Lord Chamberlain's office.

In Italy, censorship concentrated on the religious, the political, and the moral. It required composers to make a *protesta declaring their sound Roman Catholicism despite the presence in their operas of Greek or Roman deities; and Verdi encountered difficulties in 1843 over the Archbishop of Milan's refusal to admit a baptism scene in *I Lombardi*, whose incendiary patriotic choruses also alarmed the political authorities. The political censorship was most forceful during the Austrian occupation in the 19th cent., and varied from state to state. It was predominantly political in Austrian-occupied Lombardy and Venetia; in Modena and especially the Papal States it was more religious, with the Bourbon Kingdom of Naples banning mention of Luther and Voltaire and resisting anything remotely anti-monarchist. Verdi's operas were often required to appear with modified librettos and under different titles in different parts of the peninsula. Piave's libretto for *Rigoletto* was initially banned by the military governor of Venice for its inclusion of a licentious monarch and the action of a curse without the agency of the Church; and the opera played in various cities as *Viscardello*, *Lionello*, and *Clara di Perth*. This led Verdi to suggest that it be announced as by 'Don —', with the local censor's name filled in. Censorship problems also led to the relocation of *Un ballo in maschera* from Sweden to Boston.

In Vienna, censorship was systematized in 1770 by a decree of Joseph II as part of his determination to educate and enlighten his people. This included the popular extempore theatre and visits by foreign troupes, and the banning of anything thought offensive to Church and State. The wish to exclude morbid material led to the banning of *Romeo and Juliet* and the removal of the graveyard scene from *Hamlet*. A new decree under Franz II in 1795 reflected political nervousness in the wake of the French Revolution, and successive ordinances led to the strict controls confirmed under Metternich in 1810. These encouraged lightweight plays and a relaxing or uplifting musical theatre. No criticism of a monarch nor of lawful government was allowed, nor of the Roman Catholic Church, though Méhul's *Joseph* was permitted as being a serious biblical drama. All librettos were scrutinized for subversive material, with good characters not allowed too many weaknesses and the sins of bad ones equally subject to government control. The consequence for opera was to emphasize the somewhat sentimental presentation of character that typifies the Biedermeier period in Austrian life. Popular successes were not exempt: in the 1821 *Der Freischütz*, Samiel and the Hermit were both excluded, the *Wolf's Glen Scene had to be conducted within a hollow tree, and the potential subversiveness of bullets, even magic ones, caused Max and Caspar to be rearmed with crossbows.

In France, censorship, though consisting largely of the threat imposed by changing political circumstances, affected opera most strongly during Revolutionary and post-Revolutionary times. Thus, Beaumarchais's text for Salieri's *Tarare* had to change its hero from king to republican ruler to constitutional monarch as régimes changed.

Nazi Germany banned a number of operas for political or racial reasons, not only those by Jews; and the post-war German Democratic Republic exercised the Communist censorship that prevailed all over the Soviet bloc, dictating through political pressure subjects that were acceptable and finding no room for the dissident or politically contentious, even for works that did not assert 'positive' Communist values.

Russian opera was prone to the close censorship that afflicted the Russian theatre from the time of *Catherine the Great. It was firmly in place during the 19th cent., and highly experienced by the time of the 1917 Communist Revolution. Some operas were fitted with new texts (e.g. *Tosca* became *The Battle for the Commune*), and, especially once the Association of Proletarian Musicians gained control under Stalin, restraints included denial of any positive role by the Tsars (except Stalin's favourite, Ivan the Terrible). Glinka's *A Life for the Tsar*, under its original title *Ivan Susanin*, was published and performed in the USSR with a censored ending. Shostakovich was, for moral rather than political reasons, obliged to rewrite *Lady Macbeth of Mtsensk* as *Katerina Ismaylova*. Plots were approved for political content, and the music for accessibility, the situation reaching its nadir at the notorious meeting held under Stalin's associate Zhdanov in 1948.

**Central City.** City in Colorado, USA. The San Francisco OC paid visits in the 1860s, with *La Grande Duchesse de Gérolstein*. The original Montana T burnt down 1874, and the Belvedere T opened 1875 on the second floor of a brick building. The Richings-Bernard Co. gave *Trovatore*, *Martha*, and *Maritana* successfully, and in 1877 a local group gave *The Bohemian Girl*. This proved so popular that subscriptions were raised for the Teller OH (cap. 500). However, interest declined with the town's depopulation as mining interests withdrew. In 1932 a series of summer festivals began, concentrating on operetta and drama; the first opera was *The Bartered Bride*, 1940. The festival was suspended 1942–5 and resumed 1946. Works staged have included the première of Moore's *The Ballad of Baby Doe* (1956), set in the neighbouring mining town of Leadville; there is here a small opera-house, in which Oscar Wilde lectured to the miners after opening a new Oscar Shaft in the Matchless Mine in which the opera is set.

**central finale.** See 'CODE' ROSSINI.

**Cephalus and Procris.** In classical mythology, Cephalus was persuaded to test his wife Procris's fidelity by wooing her in disguise; when to his dismay he succeeded in seducing her, he drove her away, but she effected a reconciliation by the same trick. But as she was spying on his supposed meeting with a lover when he was out hunting, he hurled his spear into the bush where she was hiding and killed her. Most versions of the opera derive from the story as told by Ovid (*Metamorphoses*, vii). Operas on the legend are by Caccini (with Venturi del Nibbio, Strozzi and Bati, 1600), Hidalgo (*Celos aun del aire matan*, the first Spanish opera, 1660), Krieger (1689), Elisabeth Laguerre (the first opera by a woman at Paris, O, 1694), Bronner (1701), Bononcini (1704), Gillers (1711), Araia (the first opera in Russian, 1755),

Kerpen (1783), Reichardt (duodrama, 1771), and Benda (duodrama, 1805). See also COSÌ FAN TUTTE.

**cercar la nota** (It.: 'to seek the note'). The vocal habit of moving from a note by slightly anticipating the next one, either by linking the two with a passing note or, more commonly, by touching on the second of two notes as a kind of upbeat before it is properly sounded.

**cereni** (It.: from *cero*, 'candle'). From the 17th cent. to the early 19th cent. cereni librettos, hastily and inaccurately printed on poor paper, were often on sale outside the theatre for the use of patrons who wished to consult them by candlelight during the performance. Surviving examples are often stained with the wax drippings that gave them their name.

**Cerha, Friedrich** (*b* Vienna, 17 Feb. 1926). Austrian composer and conductor. Studied Vienna with Uhl. His association with the Second Viennese School through his work with the Viennese chamber ensemble Die Reihe led to him being invited by Berg's publishers to make a completion of *Lulu* (Paris, O, 24 Feb. 1979): this further confirmed the work's stature. He has also written *Baal* (Cerha, after Brecht; Salzburg 1981) and *Der Rattenfänger* (Graz, 1987).

**Cervantes Saavedra, Miguel de** (*b* Alcalá de Henares, *bapt.* 9 Oct. 1547; *d* Madrid, 23 Apr. 1616). Spanish writer. Operas on his works are as follows:

*La Numancia* (1582–7): J. H. van Eeden (1898); Barraud (1955)

*Don Quixote* (pt1, 1605; pt2, 1615): Sajon (1680); Förtsch (1690); Purcell, Eccles, and others (1684); Eve (1700); Gillier (*La bagatelle*, 1712); F. B. Conti (1719); Feo (1726); Treu (1727); Caldara (1727); Ristori (1727); Silva (1733); Caldara (*Sancio Panza*, 1733), Boismortier (1743); Martini (1746); Holzbauer (1755); Franchini (1757); Giorgi (*c.*1760); Telemann (1761); Philidor (*Sancho Pança*, 1762); Paisiello (1769); Bernardini (1769); Salieri (1770); Piccinni (1773); Beecke (1788); Gerl (1790); Hubaček (1791); Tarchi (1791); Chack (1792); Spindler (1797); W. Müller (1802); Generali (1805); Miari (1810); Seidel (1811); Bochsa (*Les noces de Gamache*, 1815); Mercadante (*Les noces de Gamache*, 1825); García (1827); Mendelssohn (*Die Hochzeit des Camacho*, 1827); Mercadante (1829–30); Donizetti (*Il furioso all'isola di S Domingo*, 1833); Rodwell (1833); Mazzucato (1836); Macfarren (1846); Clapisson (1847); Hervé (1848); Rispo (1859); Reparaz (*La venta encantada*, 1871); Caballero (*El loco de la Guardilla*, 1861); Arrieta (*La insula barataria*, 1864); Reparaz (*Las bodas de Camacho*, 1866); E.-H.-A. Boulanger (1869); Planas (1871); Pessard

(1874); Clay (1876); Weinzierl (1879); L. Ricci, jun. (1881); Neuendorf (1882); L. de Larra (*En un lugar de La Mancha*, 1887); Roth (1888); De Koven (1889); Santonja (1896); Jacques-Dalcroze (1897); Mohr (*Die Eifersüchtigen*, 1897); Rauchenecker (1897); Kienzl (1898); Chapí (1902); Ferrán (*Las bodas de Camacho*, 1903); Kaufmann (1903); Legouix (*Le gouvernement de Sancho Pança*, 1903); Barrera (*El carro de las cortes de Muerte*, 1907); Beer-Waldbrunn (1908); Besi (1908); Hewitt (1909); Heuberger (1910); Pasini (1910); Massenet (1910); Dall'Orso (1916); E. Ábrányí (1917); Falla (*El retablo de Maese Pedro*, 1923); Lévy (1930); Halffter (*Clavileño*, comp. 1936); Rodriguez Albert (1948); Frazzi (1952); Rivière (1960); Petrassi (1967); Halffter (1970); J. Bach (*The Student from Salamanca*, 1980); Henze (recomp. of Paisiello, 1976).

Operas using the character of Don Quixote in a new context include:

Leo (*Il Fantastico*, 1743); Gomes (*Il nuovo Don Chischiotte*, 1748); F. Bianchi (*Il nuovo Don Chischiotte*, 1788); Champein (*Le nouveau Don Quichotte*, 1789); Dittersdorf (*Don Quixote der Zweite*, 1795); Moniuszko (*Nowy Don Kiskot*, 1842); Hochberg (*Der neue Don Quixote*, 1861); Wick (*Zoribal*, 1882); Noskowski (*Nowy Don Kiszot*, 1890); Jarecki (*Nowy Don Kiszot*, comp. 1901)

*Novelas ejemplares* (1613): *El casamiento engañoso*: Anfossi (*Il matrimonio per inganno*, 1779)

*La fuerza de la sangre*: Auber (*Léocadie*, 1824)

*La gitanilla*: Kaffka (*Die Zigeuner*, 1778); Balfe (*The Bohemian Girl*, 1843); Reparaz (1861); García (1890); Gabrielson (*Gipsy Blonde*, 1934)

*El celoso extremeño*: Barrera (1908); see also *El viejo celoso* (below)

*Rinconete y Cortadillo*: Doncel (*La picaresa*, 1850)

*La ilustre fregona*: Laparra (1931)

*Ocho comedias y ocho entremeses nuevos* (1615): *La cueva de Salamanca*: Winter (*Der Bettlerstudent*, 1785); Gay (1908); Paumgartner (1923); Mojsisovics (*Der Zauberer*, 1920); Lattuada (*La caverna di Salamanca*, 1938)

*El viejo celoso* (1615, after *El celoso extremeño*): Petrassi (*Il Cordovano*, 1949).

Also operas by Grétry (*À trompeur, trompeur et demi*, 1792); Henze (*Das Wundertheater*, 1949); Orff (*Astutuli*, 1953).

Operas on Cervantes himself are by Foignet (*Michel Cervantes*, 1793), Lassen (*Le captif*, 1865), Aceves (*El manco de Lepanto*, 1867), J. Strauss II (*Das Spitzenbuch der Königin*, 1880), and Byström (1881).

**České Budějovice.** Town in Bohemia, Czechoslovakia. Opera was formerly given in the S Bohemian T about 3–4 times a week in the season by a company of about 20 soloists. The company moved to a new theatre in 1972. It tours in the region.

**Cesti, Pietro** (*b* Arezzo, *bapt.* 5 Aug. 1623; *d* Florence, 14 Oct. 1669). Italian composer: sometimes wrongly referred to as Marc' Antonio from a contraction of his later title of Marchese. A choirboy in his native city, he joined the Minorite friars in 1637 and received the monastic name Antonio. Studied Rome with Abbatini and Carissimi; by 1645 he was maestro di cappella at the seminary in Volterra and in charge of music at the cathedral. Acquainted with the Medici court at Florence, Cesti was already thinking of a career outside the Church and the production in Venice of *Orontea* in 1649 and *Cesare amante* in 1651 rapidly placed him at the forefront of operatic composers.

In 1652 he became Kapellmeister to Archduke Ferdinand Karl of Austria and moved to Innsbruck. He was manifestly unsuited to holy orders and, after a long succession of indiscreet affairs, finally resigned his vows in 1659, the same year he joined the papal choir in Rome. During his stays in Innsbruck he produced a number of operas, and *Dori*, one of his greatest successes, was given during the Florentine Carnival of 1661. Between 1666 and 1669 he was 'capelan d'honore und intendenta delle musiche teatrali' at the imperial court in Vienna, where he produced five operas, including his most famous work, *Il pomo d'oro*. He intended to settle in Venice on his retirement from the Austrian court, but was prevented from doing so by financial scandal: although his last opera, *Genserico*, was produced there in 1669, he spent his last months in Florence.

Cesti's works occupy a unique position in the early history of opera. Some were written for the Venetian public opera-houses, and aimed at a mixed audience, but his positions at Innsbruck and at Vienna obliged him to compose operas in the tradition of court entertainment established by the intermedi. Though he may be regarded as part of the general Venetian school of Manelli, Monteverdi, Ziani, and Cavalli, it is important that due recognition be given to his work outside the public opera houses. Cesti was the best-known Venetian composer after Monteverdi: *Orontea*, one of the most widely performed operas of the century, was given throughout Italy, while *Cesare amante* provoked the comment from Salvator Rosa in 1652 that Cesti was the greatest living composer active in Venice. The success of these works provided the first serious challenge to Cavalli's supremacy, and Cesti's later ability to respond more flexibly to the demands of the audience, as in *Tito*, eventually hastened the older composer's retirement. In fact, Cesti had

learnt much from Cavalli, especially concerning the choice of libretto and dramatic recitative style, as can be seen especially in *Dori*. His use of dissonance for expressive purposes is especially inventive, although the overriding concern is for lyrical vocal lines. It was this instinctive feel for aria-type writing, and his perception that the tastes of the audience were better accommodated by a concentration on genuine aria at the expense of arioso recitative, that assured his works' popularity. Such was his standing in Venice that on his death two as yet unwritten works, *Ermengarda, regina de'Longobardi* and *Giocasta*, had already been accepted for performance.

The Innsbruck court operas, such as *Argia* and *La magnanimità d'Alessandro* written for the visit of Queen Christine of Sweden, share many points of similarity with the Venetian operas, but are characterized by a concentration on visual and aural spectacle. Greater choral and orchestral resources are demanded, and there is less emphasis on the role of the solo singer. *Il pomo d'oro*, written to celebrate the wedding of Emperor Leopold I with the Infanta Margherita of Spain, is the climax of Cesti's work in this field, but is exceptional even for a court celebration. The surviving reports and drawings of the performance confirm the impression of grandeur on an almost vulgar scale suggested by the lavish musical demands and expansiveness of the score. The opening introductory sinfonia, which includes music from later scenes, was probably the first programme overture.

WORKLIST: *Orontea* (Cicognini; Venice 1649, rev. Innsbruck 1656); *Alessandro vincitor di se stesso* (Alexander his own conqueror) (Sbarra; Venice 1654); *Cesare amante* (Caesar in Love) (Varotari; Venice 1651); *Cleopatra* (Varotari; Innsbruck 1654); *Argia* (Apolloni; Innsbruck 1655); *Venere cacciatrice* (Venus the huntress) (Sbarra; Innsbruck 1659, lost); *Dori, ovvero La schiava fedele* (Dori, or The Faithful Slave) (Apolloni; Innsbruck 1657); *La magnanimità d'Alessandro* (Sbarra; Innsbruck 1662); *Tito* (Beregani; Venice 1666); *Nettuno e Flora festeggianti* (Neptune and Flora Celebrate) (Sbarra; Vienna 1666); *Le disgrazie d'amore* (The Misfortunes of Love) (Sbarra; Vienna 1667); *Semirami* (Moniglia; Vienna 1667); *Il pomo d'oro* (The Golden Apple) (Sbarra; Vienna 1668); *Genserico* (Beregani; Venice 1669, possibly finished by Domenico Partenio).

**Chabrier,** (Alexis-) **Emmanuel** (*b* Ambert, 18 Jan. 1841; *d* Paris, 13 Sep. 1894). French composer. Studied Ambert with Zaporta, later Paris with Semet and Hignard. His parents insisted that he follow a legal career, but he continued composing and his interest in poetry led him to form friendships with, among others, Catulle Mendès and Villiers de l'Isle Adam, to whom he taught music, and Verlaine, who provided him with the librettos of two operettas. He was also a close friend of Manet, and his musical gifts as pianist and improviser won him the entry to Paris salons and friendships with Fauré, Chausson, Duparc, and D'Indy. Parts of a projected opera, *Jean Hunyade*, were later used in *Gwendoline* and *Briséïs*. His vivid comic flair found expression in *L'étoile* and *Une éducation manquée*, though their liveliness does not exclude tenderness and some harmonic adventurousness. Through Duparc, he came to admire Wagner, and the experience of hearing *Tristan und Isolde* in Munich led him to resign the government post he had occupied and dedicate himself to music in 1880. The first opera to follow was *Gwendoline*, set in medieval England and filled with vigorous battles and some attractive love music; it is Wagnerian in its use of Leitmotiv and in its harmonic emancipation. He also composed quadrilles on themes from *Tristan*, in witty but appreciative vein; he was at this time a leading member of the group of French Wagnerians known as 'Le Petit Bayreuth'. He then produced an opéra-comique, *Le roi malgré lui*, the work in which his gifts for elegant melody and lively characterization are most fully displayed.

WORKLIST: *Fisch-Ton-Kan* (Verlaine; unfin., comp. 1863–4, prod. Paris 1941); *Vaucochard et fils 1er* (Verlaine; unfin., comp. 1864, prod. Paris 1941); *Jean Hunyade* (Fouquier; unfin.); *L'étoile* (The star) (Leterrier & Vanloo; Paris 1877); *Le sabbat* (Silvestre; unfin.); *Une éducation manquée* (A defective education) (Leterrier & Vanloo; Paris 1879); *Les Muscadins* (Clarétie & Silvestre; unfin.); *Gwendoline* (Mendès; Brussels 1886); *Le roi malgré lui* (King despite himself) (Najac & Burani, rev. Richepin, after A. & M. Ancelot; Paris 1887); *Briséïs* (Mendès & Mikhaël, after Goethe, unfin.; Berlin 1899).

BIBL: R. Myers, *Emmanuel Chabrier and his Circle* (London, 1969); F. Robert, *Emmanuel Chabrier: l'homme et son œuvre* (Paris, 1969).

**Chailly, Luciano** (*b* Ferrara, 19 Jan. 1920). Italian composer, critic, and administrator. Studied Bologna, Milan with Righini, and Salzburg with Hindemith. Artistic director Milan, S, 1968–71, and Turin, R, from 1972. Father of below.

SELECT WORKLIST: *Una domanda di matrimonio* (A proposal) (Fino & Vertone, after Chekhov; Milan 1957); *La riva delle Sirti* (Prinzhafer, after Gracq; Monaco 1959); *L'idiota* (Loverso, after Dostoyevsky; Rome 1970).

**Chailly, Riccardo** (*b* Milan, 20 Feb. 1953). Italian conductor, son of above. Studied Italy with Guarino, Caracciolo, and Ferrara. Assistant cond. Milan, S, 1972–4. Chicago 1974 (*Butterfly*). Milan, S, from 1978; London, CG, 1979; New York, M, 1982; Vienna, S, 1983; Salzburg from 1984; music director Bologna, C, 1986–9. Repertory incl. *Cenerentola*, *Ricciardo e Zoroaide*, *Trovatore*, *Walküre*, *Fiery Angel*. (R)

**Chalabala, Zdeněk** (*b* Uherské Hradiště, 18 Apr. 1899; *d* Prague, 4 Mar. 1962). Czech conductor. Studied Brno with Neumann, also attending Janáček's master-classes. Brno O 1925–36, 1949–52; Prague 1936–45, 1953–62; Ostrava 1945–7. With his strong leadership and expressive artistry, he did much establish a tradition of Czech and Russian opera in Brno and Prague. (R) His wife **Běla Rozumová** (*b* Příbram, 8 Feb. 1903) was a coloratura soprano; début Ljubljana 1923 (Mařenka), Brno 1925–36, Prague 1936–42.

**Chaliapin, Fyodor.** See SHALYAPIN, FYODOR.

**chamade.** See CHIAMATA.

**chamber opera.** A small-scale opera, or one of predominantly intimate character, calling for fewer vocal and instrumental resources than normally required, and often telling a simpler story than is customary. While many works of the 17th and 18th cents.—including notably Purcell's *Dido and Aeneas*, Pergolesi's *La serva padrona*, and Mozart's *Der Schauspieldirektor*—are often described as chamber operas, strictly speaking the term has no validity before the late 19th cent., for the concept of chamber music itself was not recognized until then. However, when applied to many early works, its use provides a vivid, if questionable, means of drawing comparison between large-scale court or commercial opera and the smaller genres of *azione teatrale, *serenata, and the like.

Following Wagner's expansion of both the temporal and physical dimensions, and the generally grandiose conception of *fin de siècle* opera, the genre enjoyed a certain popularity among composers who wished to explore more intimate modes of expression. Among important chamber operas of the 20th cent. are the second version of Strauss's *Ariadne auf Naxos*, Hindemith's *Cardillac*, Virgil Thompson's *Four Saints in Three Acts*, Britten's *The Turn of the Screw*, and Walton's *The Bear*. An increasing number of modern composers have shown an interest in using smaller forces, partly in reaction to the full apparatus of traditional opera, partly for the creative possibilities, and partly for purely financial reasons: at the extreme, chamber opera today merges with the growing genre of *music theatre. Among important groups which have contributed to the performance history of chamber opera have been the Intimate Opera Company and the *English Opera Group, founded by Britten and others. In recent years a number of groups have found in the smaller scale of chamber opera not only economy and convenience for touring, but a form of opera yielding rich creative results in its own right.

**Champagne Aria.** See 'FINCH'HAN DAL VINO'.

**Champs-Élysées, Théâtre des**. Paris theatre planned by Gabriel Astruc to open in the Champs-Élysées but transferred to the Avenue Montaigne while retaining the name; opened 2 Apr. 1913 with a concert, then Berlioz's *Benvenuto Cellini* on 3 Apr. Comprises three auditoriums, the Grand Théâtre (cap. 2,100) for opera, Comédie (cap. 750) for drama, Studio (added 1923, cap. 250) for experimental work. The scene of Dyagilev's 1913 season and other important guest seasons, especially Russian, in the 1920s and 1930s. After the war the theatre's tradition of sometimes belated French premières of important works continued, including visits from distinguished foreign companies. The theatre remains privately run.

**'Chanson de la puce, La'** (Song of the Flea). Méphistophélès's (bs or bar) aria in the second part of Berlioz's *La damnation de Faust*.

**Chapí (y Lorente), Ruperto** (*b* Villena, 27 Mar. 1851; *d* Madrid, 25 Mar. 1909). Spanish composer. Studied Madrid with Arrieta. His first zarzuela, *Abel y Caín* (1873), was followed by a period as a bandmaster. He then wrote a 1-act opera, *Las naves de Cortés* (1874), whose success, with Tamberlik singing Cortés, won him a study grant. He wrote more operas, but is famous as the author of over 100 popular zarzuelas.

BIBL: A. Salcedo: *Ruperto Chapí: su vida y sus obras* (Cordoba, 1929).

**Chappelou.** The postilion (ten) in Adam's *Le postillon de Lonjumeau*.

**Charakterposse.** See POSSE.

**Charlotte.** Albert's wife (mez), loved by Werther, in Massenet's *Werther*.

**Charlottenburg.** See BERLIN.

**Charpentier, Gustave** (*b* Dieuze, 25 June 1860; *d* Paris, 18 Feb. 1956). French composer. Studied Paris with Massenet. His most famous work is the opera *Louise*. Its success, partly for its startling realism and liberal social views, partly for its lyrical fluency, was as great as the rapid failure of its sequel *Julien*. Charpentier also founded, in 1902, a Conservatoire Populaire Mimi Pinson, where working girls such as his Louise could study music and dance.

SELECT WORKLIST: *Louise* (Charpentier; Paris 1900); *Julien* (Charpentier; Paris 1913).

**Charpentier, Marc-Antoine** (*b* Paris, ?1645–50; *d* Paris, 24 Feb. 1704). French composer. Studied in Rome with Carissimi. Following Lully's

rift with Molière he collaborated with the playwright on two comédie-ballets, *Le mariage forcé* (1672) and *Le malade imaginaire* (1673). This initiated a period with the Comédie-Française which continued for the next 12 years, resulting in about 26 dramatic works of various kinds.

As a dramatic composer Charpentier was inevitably overshadowed by Lully, though this was in part a result of Lully's political manœuvring. Molière clearly had a high regard for his abilities, and only his death prevented further collaborations. Many of the small-scale divertissements which Charpentier wrote for the Comédie-Française, such as *Les amours de Vénus et d'Adonis* (1678), remained in the repertoire until the middle of the 18th cent. His pastorales, such as *La noce du village* (1692) and *La couronne des fleurs* (*c*.1695), were equally popular, while the larger-scale works such as *Les plaisirs de Versailles* (early 1680s) and *La fête de Rueil* (1685) followed in the tradition of the aristocratic ballet de cour and proved equally popular.

Charpentier's training under Carissimi inspired a sympathy with the Italianate style which, like Lully, he incorporated into his dramatic works. The most important of these was the tragédie lyrique *Médée*. It received only one performance, but was highly regarded, despite a text lacking the cohesion and impact of Quinault's for Lully. Containing much original and appealing music, *Médée* follows the model of Lully's tragédies lyriques: it is cast in five acts, preceded by a prologue praising the achievements of Louis XIV, and creates similar visual splendour through the inclusion of elaborate ballet sequences.

SELECT WORKLIST: *Médée* (Corneille; Paris 1693).

BIBL: H. Wiley Hitchcock, *Marc-Antoine Charpentier* (Oxford, 1990).

**Charton-Demeur, Anne** (*b* Saujon, 5 Mar. 1824; *d* Paris, 30 Nov. 1892). French soprano. Studied Bordeaux with Bizot. Début Bordeaux 1842 (Lucia). London, DL, 1846 and HM 1852 (Amina). Paris, OC, 1849, 1853. Created Béatrice (*Béatrice et Bénédict*), Berlioz being so impressed with her 'rare beauty of style' that he asked her to be the first Dido in *Les Troyens à Carthage*. This she performed with 'grandeur and dramatic force', and he declared himself 'overwhelmed' by Dido's monologue. She became his close friend, and was with him when he died. In 1870 she sang Cassandre in the first (concert) performance of *La prise de Troie*. Married the flautist Demeur.

**Châtelet, Théâtre du**. Paris theatre (cap. 2,500) on the site of the old T du Cirque Impérial, opened 1862 for drama and concerts but also housing opera and ballet. In the 1930s an important centre for operetta. Modernized and renamed Théâtre Musical de Paris 1980, it gives a mixed repertory from visiting companies.

**Chaucer, Geoffrey** (*b* London, *c*.1340; *d* London, 25 Oct. 1400). English poet. Operas on his works are as follows:

*Troilus and Cressida* (*c*.1380–2): Walton (1954)
*The Canterbury Tales* (unfin., 1400): Stanford (1884); De Koven (1917); Hill and Hawkins (1968); Sitsky (1976)
*The Pardoner's Tale*: Sokolov (1961); E. Lubin (1966); J. Davis (1967); A. Ridout (1971)
*The Nun's Priest's Tale*: J. Barthelson (1967)
*The Prioress's Tale*: Drakeford (*The Sely Child*, 1982)
*The Knight's Tale, The Nun's Priest's Tale*, and *The Pardoner's Tale*: Crosse (*The Wheel of the World*, 1972)

**Chausson, Ernest** (*b* Paris, 20 Jan. 1855; *d* Limay, 10 June 1899). French composer. First studied law, later music with Franck and Massenet. Not needing to earn his living, he worked slowly, carefully, and diffidently at composition. He gained from his experience of Wagner a greater confidence of idiom which shows in his operas, though like many Frenchmen he was influenced by only a small part of Wagner's ideas. His early dramatic essays are slight works, though contemporaries found evidence of style and taste among the technical uncertainties. He contemplated scenarios for operas on Pushkin, Schiller, Shakespeare (*Macbeth*), Cervantes, and others, but his major work, on which he expended much time and care, is *Le Roi Arthus*. An admirer of Celtic literature as well as of Wagner, he wrote his own skilful text in prose and verse on King Arthur's renunciation of life in the wake of Guinevere's betrayal of his love with Lancelot. The work makes a discreet and individual use of motive, and is eloquently scored; but the influence of Wagner is intelligently and discreetly applied, and comparisons with *Tristan* are not really appropriate.

WORKLIST: *Les caprices de Marianne* (Chausson, after De Musset; comp. 1880–2); *Hélène* (Chausson, after Leconte de Lisle; comp. 1883–6); *Le Roi Arthus* (Chausson; comp. 1886–95, prod. Brussels 1903).

BIBL: E. Grover, *Ernest Chausson* (London, 1990).

**Chautauqua.** Town in New York State, USA. An annual summer opera festival was inaugurated in July 1929 with *Martha*. Operas are sung in English, and many young American artists have made early appearances here in a repertory that strikes a balance between the popular and the more enterprising (e.g. works by Floyd and Moore). Opera is given in the Norton Hall.

**'Che farò senza Euridice'.** Orfeo's (mez) lament at the loss of his wife Euridice in Act III (Act IV in some versions) of Gluck's *Orfeo ed Euridice*.

**'Che gelida manina'.** Rodolfo's (ten) aria to Mimì in Act I of Puccini's *La bohème*, making his first approach to her.

**Chekhov, Anton** (*b* Taganrog, 29 Jan. 1860; *d* Badenweiler, 15 July 1904). Russian writer and dramatist. A lover of music, he knew Tchaikovsky and dedicated a volume of stories to him; a plan for their collaboration on an opera on Lermontov's *Bela* came to nothing. Operas on his works are as follows:

*He Forgot!* (1882): Pribik (1921)
*Surgery* (1884): Ostroglazov(?); Ferroud (1928)
*On the Highway* (1885): Nottara (*La drumul mare*, 1934)
*Romance with a Double-bass* (1886): Dubensky (1916); Sauguet (1930); Bucchi (1954)
*The Witch* (1886): Yanovsky (1916); Hoiby (1958); Vlasov & Feré (radio 1961, prod. 1965)
*Swan Song* (1888): Chailly (1957)
*The Bear* (1888): Semenoff (1958); Jirásek (1965); Walton (1967)
*The Proposal* (1889): Chailly (1957); Röttger (1960); J. Wagner (1965); Bohác (1971)
*The Marriage* (1889): Ehrenberg (1916)
*The Seagull* (1896): Vlad (1968); Pastieri (1974)
*The Three Sisters* (1901): Pasatieri (1986)
*The Cherry Orchard* (1904): Kelterborn (1984)
*The Admiral*: Andreoli (1960)
*The Boor*: Bucci (1949); Fink (1955); Argento (1957); Kay (1960); Moss (1961)
*The Box*: C. Eakin (1961)
*The Crisis*: Chailly (1966)
*The Sneeze*: A. Mascagni (1956)
*The Joy*: Pribík (1922)

**Chelard, Hippolyte** (*b* Paris, 1 Feb. 1789; *d* Weimar, 12 Feb. 1861). French composer and conductor. Studied Paris with Fétis and Gossec, perhaps Méhul and Cherubini, later in Italy with Baini, Zingarelli, and Paisiello. He produced his first opera, *La casa da vendere* in Naples; but his first important opera, *Macbeth*, failed in Paris, despite its skilful use of Grand Opera conventions, before succeeding in Munich from 1828; and after a second failure in Paris with *La table et le logement* he settled in Germany. In Munich he again had a success with a rewriting of *La table* as *Der Student*. *Macbeth* and *Der Student* were both given during his 1832–3 season directing the German Opera (with Schröder-Devrient) at London, DL, with little success despite the contribution of Malibran. His most important work is *Die Hermannsschlacht*, in which full account is taken of the conventions of German Romantic opera. He was Liszt's predecessor in Weimar, was a loyal friend and supporter in Germany of Berlioz, to whose music his own has some resemblances, and had some influence on the young Wagner.

SELECT WORKLIST: *Macbeth* (Rouget de Lisle & Hix, after Shakespeare; Paris 1827); *La table et le logement* (Gabriel & Dumersan; Paris 1829, rev. as *Der Student*, Munich 1832); *Die Hermannsschlacht* (Weichselbaumer; Munich 1835).

**'Ch'ella mi creda libero'.** Dick Johnson's (ten) aria in Act III of Puccini's *La fanciulla del West*, in which he asks that Minnie should be told not that he has been lynched but that he has been freed.

**Chelleri, Fortunato** (*b* Parma, June 1686/90; *d* Kassel, Dec. 1757). Italian composer. The son of a German musician, Keller—a variant by which he is also known—and an Italian mother, he studied at Piacenza with his uncle Bazzani, maestro di cappella at the cathedral. In 1707 *Griselda* was performed there to great acclaim, the first of 19 operas he was to write for the Italian stage. He had an extremely itinerant life, holding court appointments in Florence, Würzburg, Kassel, and Stockholm. He was in London 1726–7 and as a close friend of Bordoni and Cuzzoni participated fully in the musical life of the city and in the activities of the Royal Academy of Music.

Chelleri occupies an interesting position between the late Venetian school, in which he was reared, and the early Neapolitan style of Leo and Porpora. That of his music which survives, such as the highly popular *L'innocenza giustificata*, suggests that he had a flair for writing appealing melodies, as well as a proper sense of dramatic pace and proportion.

SELECT WORKLIST: *L'innocenza giustificata* (Innocence Justified) (Silvani; Milan 1711, rev. Venice 1722 as *L'innocenza difesa* and as a pasticcio, with numbers by Handel and Telemann, as *Judith Gemahlin Kayser Ludewige des Frommen*, Hamburg 1732).

**Chelsea Opera Group.** A group formed in 1950 by Colin Davis, David Cairns, and Stephen Gray to give concert performances of Mozart's operas. Their successful *Don Giovanni* in Oxford was followed by other Mozart performances, and by their first London appearance with *Fidelio* at St Pancras (1953). Other performances in the 1950s and 1960s helped to arouse interest in then rarely heard works, including *Les Troyens, Benvenuto Cellini, Guillaume Tell, Euryanthe*, and *Khovanshchina*. Conductors who made appearances early in their careers include Nicholas Braithwaite, Roger Norrington, and John Eliot Gardiner, and singers who first sang favourite roles include Peter Glossop (Don Giovanni), Derek Hammond-Stroud (Beckmesser), and Elizabeth Robson (Nannetta).

**Chemnitz.** City in Saxony, Germany; from 1953–90, officially Karl-Marx-Stadt; Chemnitz

again from 1990. The Opernhaus opened 1909. Bombed in Second World War; new theatre (cap. 1,073) opened 1950.

**'Che puro ciel'.** Orfeo's (mez) arioso as he gazes on the beauties of the Elysian Fields in Act II of Gluck's *Orfeo ed Euridice*.

**Chéreau, Patrice** (*b* Lezigne, 26 Mar. 1944). French producer. Bayreuth 1976 (*Ring*); Paris, O, 1979 (*Hoffman, Lulu*). His production of *The Ring* (cond. Boulez) caused a scandal, its setting in 19th-cent. capitalist society, its emphasis on the brutal, and its reductiveness creating shock and sharp controversy at the time; nevertheless, critical opinion found much to respect in it, and its subsequent television broadcast reached a huge and appreciative audience.

**Cherevichki.** See VAKULA THE SMITH.

**Cherubini, Luigi** (*b* Florence, 8 or 14 Sep. 1760: *d* Paris, 15 Mar. 1842). Italian composer. Studied in Florence, at first with his father, later in Bologna and Milan with Sarti. He produced some church music and then his first operas, including *Quinto Fabio*. He visited London in 1784, then Paris. He had written 13 Italian operas by the time he established himself permanently in Paris in 1788; but with *Démophon* he broke away from this earlier vein and began developing the mature style that makes him a pioneer of French opéra-comique and Grand Opera, and a father figure of Romantic opera. This manner was first fully revealed in *Lodoïska*, in which there is a new, richer handling of the orchestra, a more powerful and functional use of ensemble, a stronger sense of dramatic (and moral) motivation, and the emergence of character types that were to impress subsequent composers: the villain Dourlinski, for instance, is an operatic ancestor of both Beethoven's Pizarro and Weber's Caspar.

Meanwhile, Cherubini had directed an Italian opera in Paris; and in 1795 he joined the newly founded Conservatoire as an inspector of studies. Other early traits of Romantic opera begin to emerge in the works that followed in these years, e.g. *Eliza*, in which the Alpine setting plays an important part in the action. A particular contribution was the use of natural forces to play a crucial role in the drama (a fire in *Lodoïska*, an avalanche in *Éliza*), as well as his more fully developed use of arias for secondary characters. An equally important ingredient for Romantic opera was the influence of Rousseau, not only in the handling of peasants or other simple people as embodying qualities of the highest virtue. *Éliza* also introduced into operatic currency the so-called *musique d'effet*; and the enhanced importance given to the orchestra, including the use of obbligatos especially for horn and clarinet, was

# Cherubino

another characteristic that made its mark upon the Romantics.

*Médée* is an elevated tragedy, noble and somewhat cold; it is, none the less, an opera charged with forceful statements of passion, remote from human reality though the characters, even Médée herself, may seem to be. The work is in strong contrast to *Les deux journées*, an opéra-comique making use of simple characters and consequently smaller forms such as *arietta and *couplet. Bouilly's text was regarded as exemplary by Beethoven, and as a *rescue opera the work influenced him (including its use of *melodrama): like the later *Faniska*, another rescue opera, the work includes simple, ballad-like interpolations in a dark, dramatic plot, and makes rich use of the orchestra to carry the story and the emotions of the characters. Cherubini's methods here include a looser handling of set numbers, and even a suggestion of *Leitmotiv. Above all, it was his success in embodying in a novel operatic manner the idealization of simple life, in which servants or peasants are no longer comic figures or ingénus but human beings proposing righteous moral sentiments, that won him the admiration of both Beethoven and Goethe.

Though he wrote a number of further operas (including the interesting and underrated *Pimmalione*, which uses Rousseau's text in an Italian version), and the better known *Les Abencérages*, Cherubini turned away from opera in his later years in favour of sacred music. He was greatly admired in Paris, but his career there was not smooth. After a quarrel with Napoleon, he went to Vienna with *Faniska*, returning when the French occupied the city. He was director of the Conservatoire from 1821 to 1841. Though often derided — including, unkindly, by Berlioz—for his pedantry and for a distinct and self-confessed severity in his music, he was a master of technique, and at its finest his music possesses a brilliance, purity, and originality that make him one of the most important founders of Romantic opera and Grand Opera.

SELECT WORKLIST (all first prod. Paris unless otherwise stated): *Il Quinto Fabio* (Zeno; Alessandria 1779, rev. Rome 1783); *Démophon* (Marmontel; 1788); *Lodoïska* (Fillette-Loraux; Paris 1791); *Éliza* (Saint-Cyr; 1794); *Médée* (Hoffman; 1797); *L'hôtellerie portugaise* (Aignan; 1798); *Les deux journées* (The Two Days; formerly also known as The Water Carrier (Ger., *Der Wasserträger*)) (Bouilly; 1800); *Faniska* (Sonnleithner; Vienna 1806); *Pimmalione* (Pygmalion) (Vestris, after Rousseau; 1809); *Les Abencérages* (Jouy, after Florian; 1813); *Ali-Baba* (Scribe; 1833).

BIBL: B. Deane, *Cherubini* (London, 1965).

**Cherubino.** Countess Almaviva's page (sop, sometimes mez) in Mozart's *Le nozze di Figaro*.

**chest voice.** The lowest of the three main registers of the voice, the others being the 'middle' and the 'head'. So called because the tone of the lower notes in the singer's range, when using this voice, almost seems to go from the larynx into the chest. This method of production gives the richest notes, and is essential for strength and carrying power in the lower register. García considered that the chest voice predominated in men, and it is favoured especially by the Italians. See also REGISTER.

**Cheval de bronze, Le** (The Bronze Horse). Opera in 3 acts by Auber; text by Scribe. Prod. Paris, OC, 23 Mar. 1835; London, CG, 14 Dec. 1835; New Orleans, T d'Orléans, 15 Apr. 1836, with St Clair, Heyman, Bailly.

A Chinese village and the planet Venus. Yanko (ten), Tao-chin (con), and Prince Yang-yang (bar) are all turned into wooden pagodas after disclosing the secret of the planet Venus, where they have ridden on a magical bronze horse. Tao-chin's daughter, Peki (sop), breaks the spell, and is permitted to marry her beloved Yanko, while Yang-yang marries the beautiful Stella (sop), whom he met on Venus.

**Chezy, Helmina von** (*b* Berlin, 26 Jan. 1783; *d* Geneva, 28 Jan. 1856). German poet and librettist. She read medieval romances under the influence of the Schlegels in Paris, and this led Weber, after she had settled in Dresden, to ask her to write the libretto of *Euryanthe*. Though her verses are quite well turned, she lacked (by her own admission) the dramatic experience necessary especially for such a pioneering enterprise, and the weaknesses of her text proved a damaging flaw in the opera. Autobiography, ed. B. Borngräber, *Unvergessenes* (Leipzig, 1858).

**Chiabrera, Gabriello** (b Savona, 8 June 1552; *d* Savona, 11 Oct. 1638). Italian librettist and poet. A member of the Accademia degli Alterati, he spent much time in Florence, and was one of the earliest to be involved in operatic experiments there. In addition to texts for individual monodies and court entertainments, he wrote one of the first opera librettos, *Il rapimento di Cefalo* (Caccini, 1600). This had an important influence on his contemporaries, including Rinuccini and Striggio, providing an admirable vehicle for the new musical style which did not compromise the high literary standards and classical predilections of his fellow academicians.

**chiamata** (Fr., *chamade*; 'call'). A term used to describe pieces in Venetian opera of the 17th cent. which imitate the call after the end of a hunt.

**Chiara, Maria** (*b* Oderza, 24 Dec. 1939). Italian soprano. Studied Venice with Cassinelli, Turin

136

with Carbone. Début Venice 1965 (Verdi's Desdemona). Verona Arena from 1969; Munich and Vienna from 1970; Berlin, SO, 1971; Milan, S, 1972; Naples, C, 1974; New York, M, 1978–9; also London, CG; Paris; Rome; etc. Roles incl. Violetta, Aida, Odabella, Micaëla, Tosca, Butterfly. (R)

**Chicago.** City in Illinois, USA. The first opera given was a single performance of *Sonnambula* on 29 July 1850 in J. B. Rice's Chicago T. The theatre burnt down that year but was rebuilt (cap. 1,400). Arditi's company performed, with Rosa de Vries, 1852. In 1857 James McVicker built a new theatre (cap. 2,000), where Durant's English O Troupe gave *Trovatore* and *Daughter of the Regiment*. Uranus Crosby's OH was opened 1865 with *Trovatore* (to an audience including Generals Sherman and Grant), but burnt down 1871. Opera had no permanent home until 1889, when the Chicago Auditorium (cap. 4,250) opened with *Roméo et Juliette* (with Patti) and short seasons were given by companies including the Abbey, Grau, Damrosch, Savage, Mascagni, and Conried organizations. In 1910 the Chicago Grand Opera Co. was formed from elements of the Hammerstein Co., with *Campanini as music director. Mary Garden, who sang Mélisande on the second evening (4 Nov. 1910), became the dominating figure of opera in the city, singing here 1910–31 and acting as director 1921–2; her sole season led to the première of *The Love of the Three Oranges* and a deficit of over $1m. The Chicago Civic OC's finances were guaranteed by Samuel Insull 1922–30, who before succumbing to the Depression also created the Civic OH (cap. 3,563) on the ground floor of a skyscraper. This opened 4 Nov. 1929 with *Aida*. Important Wagner seasons were given, and artists including Lotte Lehmann, Leider, Olczewska, Eva Turner, and Kipnis made their US débuts here, but the company was forced to close 1932. Other attempts to establish opera were by Paul Longone (1933–9), Fortune Gallo (1941–2), and Fausto Cleva (1944–6). Martinelli (who for a time assisted Gallo) sang his first Otello (1936) and only Tristan (1939) here.

The Lyric Theatre of Chicago was formed 1952 by Carol Fox, Nicola Rescigno, and Lawrence Kelly, and a season included Callas's US début (*Norma*); this re-formed under Fox as the Lyric O of Chicago 1956, and guests have included Tebaldi, Nilsson, Di Stefano Simionato, Del Monaco, Domingo, Pavarotti, Christoff, and especially Gobbi (début as producer, *Boccanegra*, 1956). The emphasis is now towards Italian opera. The Center for American Artists was founded 1973. Penderecki's *Paradise Lost* (1978) was commissioned for the 1976 US bicentennial. Fox was succeeded by Ardis Krainik 1981. The artistic director is Bruno Bartoletti (début here 1956).

BIBL: R. Davis, *Opera in Chicago* (New York, 1966).

**Children's opera.** The sound of children's voices and their enthusiastic and spontaneous stage presence has attracted several composers to write opera exclusively for them. One of the earliest was Cui with *The Snow Prince* (1905), *Puss in Boots*, and *Little Red Riding Hood*. Among later successful works are Respighi's *La bella dormente nel bosco* (which was first cast as a puppet opera), Milhaud's *Un petit peu de musique*, Britten's *Let's Make an Opera*, Richard Rodney Bennett's *All the King's Men*, and John Joubert's *The Prisoner*. In the former USSR children's opera proved particularly popular, where it often had a didactic purpose. Important works included Eugen Kapp's *Winter Story*, Anushavan Ter-Gevondyan's *In the Sun's Rays* and Zlata Tkach's *The Nannygoat with the Three Kids*. Though their use demands a relatively simple vocal line and a generally uncomplicated musical style, the children's vitality has usually been adjudged a sufficient recompense.

Children appear in both a singing and non-singing role in many operas of the 20th cent., among them Puccini's *Madama Butterfly* and *La bohème*, Menotti's *Amahl and the Night Visitors*, and Britten's *Death in Venice*, where the juvenile character Tadzio is portrayed by a dancer. In Britten's *The Turn of the Screw*, the part of the boy Miles is written for a boy treble, that of the girl Flora for a soprano.

**Chile.** Opera was not given until 1830, with a series of Rossini performances in Santiago by Italian singers and a Chilean orchestra. Other Italian groups began following suit. The first opera written in Chile was *La telesofra* (1846), by the German composer Aquinas Ried (1810–69). The T de la Victoria opened in Valparaiso in 1844, the T Municipal in Santiago in 1857. To a repertory of Italian opera and zarzuela were added first French and then German operas. The first opera by a Chilean was *La florista de Lugano* (1895) by Eleodoro Ortiz de Zárate (1865–1953). Other early examples included *Caupolican* (1902, on an Araucan folk subject) by Remigio Acevedo Gajardo (1868–1911).

See also SANTIAGO.

BIBL: M. Cánepa Guzmánn, *La opera en Chile (1839–1930)* (Santiago, 1976)

**'Chi mi frena'.** The sextet in Act II of Donizetti's *Lucia di Lammermoor*, in which Edgardo (ten) finds his anger checked by the sight of Lucia's distressed condition, Enrico (bar) is torn between hatred for Edgardo and pity for Lucia (sop), who deplores her misery, while Raimondo (bs),

Arturo (ten), and Alisa (mez) express their horror and pity.

**China.** Chinese opera is a mixed genre that does not match the normal European notion of opera. It is made up of four dramatic elements: singing, recitation, acrobatics, and acting or 'expression', which may include mime. The form makes minimal use of scenery and props. Chinese opera probably originated in the shaman dancing rituals of the Zhou Dynasty (1051–771 BC); in these, witch-doctors in costume and face-paint worked themselves into a trance-like state, singing and dancing to chase away evil spirits. Comic entertainments at court were also popular at this time, as were fighting demonstrations, known as Horn Butting Games which, by the time of the Han Dynasty (AD 206–20), included a plot in their action. The Emperor Ming Huan, in the Tang Dynasty (AD 618–907), founded an academy of music and dance at his court. At this time, puppet shows and short plays, based on Buddhist teachings or on myths or legends, were popular. The 'variety' drama of the Song Dynasty (960–1127) incorporated acrobatic feats, singing, and dancing. This form reached its peak in the Yuan Dynasty (1279–1368), when rigid rules developed about how a play was to be performed, particularly regarding movement on stage. By the 17th cent. increased trade links between various regions had led to a fusion of regional styles; *Peking opera evolved in the early 19th cent. by taking elements from various regional styles. There were over 300 regional styles of opera, each distinguished by local music, language, and performance techniques.

Regional opera was performed by strolling players, generally all-male troupes, whose social standing was extremely low. Most operas were aimed at the common people and made extensive use of folk-song. The category of opera known as *k'un-ch'ü*, which evolved in the mid-16th cent., however, was more sophisticated and intended for the educated classes. The subject-matter of the regional operas usually fell into one of two categories: civilian (*wen*), which recounted famous love stories, or military (*wu*), which related heroic tales and often included much in the way of acrobatics. During the Cultural Revolution (1966–76), regional opera was suppressed and much of the tradition was lost. It is now positively encouraged by the government and is performed by amateurs as well as professional companies, which are generally made up of both male and female actors.

**Chorley, Henry Fothergill** (*b* Blackley Hurst, 15 Dec. 1808; *d* London, 16 Feb. 1872). English critic and translator. Studied Liverpool. Critic of *The Athenaeum* 1833–68, in whose pages he violently opposed Wagner and, in his early days, Verdi. Translated many operas into English and wrote librettos for composers including Wallace (*The Amber Witch*, 1861) and Sullivan (*The Sapphire Necklace*, comp. 1863–4). He was a sharp, sensitive, and honest reporter of musical events rather than a major critic of the art, but his writings have a vivid flavour of London musical and especially operatic life.

WRITINGS: *Music and Manners in France and Germany* (London, 1841); *Modern German Music* (London, 1854); *Handel Studies* (London, 1859); *Thirty Years' Musical Recollections* (London, 1862, rev. E. Newman 2/1926); H. Hewlett, ed., *Autobiography, Memoirs, and Letters* (London, 1873).

**chorus** (Fr., *chœur*; Ger., *Chor*; It., *coro*). From the Greek *khoros*, a festive dance or those who performed it. The Attic drama first consisted of tales told by a single actor in the intervals of the dance; later the function of the chorus was to join in and comment upon the drama from the place before and below the stage known as the *orkhēstra* (orchestra). This function passed into opera.

During the 16th cent. the entertainments of the Medici court, such as the *intermedi for Bargagli's *La pellegrina* (1589), used large choral forces, providing an impressive, full sonority to contrast with that of the solo voices. Though early opera composers emphasized the declamatory solo voice, Peri's and Caccini's *Euridice* (both 1600) included a chorus. A body of about ten singers was on stage for most of the action, providing short ritornello-type refrains as a means of unifying the sections of monody and, as in the intermedi, offering a contrast to the solo voice. At climactic moments the chorus played a fuller role, singing longer, more lively passages, accompanied by dancing. Monteverdi's *Orfeo* (1607) is remarkable for the number and variety of its choruses, ranging from simple homophonic declamations to vigorous polyphonic numbers after the model of the madrigal. The preface to Gagliano's *Dafne* (1608) indicates a chorus of 16–18 singers, though such specifications are rarely given.

In the Roman operas of the 1620s and 1630s, produced largely under private patronage, opera grew to resemble the older spectacular tradition of the intermedi, and in the expansion of its physical dimensions the role of the chorus was very important. Although it is not known how many singers participated in such works as Landi's *La morte d'Orfeo* (1619) and Mazzochi's *La catena d'Adone* (1626), their choral sections are numerous and have a vital structural function.

Around 1640, with the rise of the public opera-houses in Venice, the importance of the chorus began to decline. Opera impresarios could not afford the lavish expenditure of earlier aristocratic patrons: rather than a chorus, it made better financial sense to engage well-known principals, thereby ensuring a good audience. For the ageing Monteverdi and for Cavalli the chorus was of little significance, as it was wherever Venetian public opera was transplanted, most notably to Hamburg. Well into the 18th cent., e.g. in the works of Handel and Scarlatti, the chorus, if present at all, had only a minor role in operas whose performances were aimed at a fee-paying public: usually the word 'chorus' refers simply to the ensemble of all the principals. In court opera the chorus retained its function as a display vehicle for the patron's wealth and power: hence, Cesti's *Il pomo d'oro* (1667), while Venetian in style, calls for lavish choral resources, since it was performed for an imperial marriage. Likewise, the Viennese court operas of Draghi and Fux often show a greater dependence on the chorus than was customary.

In France, where the birth of opera was strongly influenced by the model of Greek tragedy as reflected through the native classical tradition and by the need to reflect royal *gloire*, the chorus played an important part. It had already been a crucial element in earlier French musico-dramatic entertainments, and it is known that in 1754 the Académie Royale de Musique maintained a chorus of 38 singers, 17 of them women. For Lully and Rameau, and their many contemporaries, the chorus provided an opportunity for sumptuous display, often accompanied by dancing, and it was under the French influence that Purcell wrote a prominent part for chorus in *Dido and Aeneas* (1689). When Gluck came to reform the conventions of opera seria in the mid-18th cent., the inclusion of the chorus was central to his new dramatic conception, as in the opening scene of *Orfeo ed Euridice* (1762). Here it conforms to the practice of early opera, in which it had taken a supporting role, restricting it largely to commenting upon the events on stage. This it did not as a direct participant, but rather as a detached and neutral observer. However, the last years of the 18th cent. saw the chorus begin to participate ever more fully in the action, even if it still had little function in opera buffa. Crowds of courtiers, nobles, guests, priests, citizens, boyars, slaves, and the like made regular appearances: one of the earliest examples of their genuinely dramatic involvement occurs in the Mozart's *Idomeneo* (1781), and this dramatic use of the chorus is continued in Haydn's *L'anima del filosofo* (1791) and the Prisoners' Chorus of Beethoven's *Fidelio*. The chorus has an atmospheric, 'ghostly' role in Weber's *Der Freischütz*, which also includes a famous Huntsmen's Chorus; the Huntsmen's Chorus in his *Euryanthe* is functional to the plot. A staple feature of *rescue

opera generally, nowhere did the chorus have a more important role than in its successor, the French *Grand Opera of the 19th cent., where it made a distinctive contribution to its visual and aural splendour. Particularly fine examples of massed choral handling can be found in Berlioz's *Les Troyens* (1858) and Meyerbeer's *Les Huguenots* (1836).

Although the chorus continued to have the important function of commenting on the entanglements of the principals, as the artificial separation of musical numbers began to be supplanted by the writing of longer dramatic sequences, so the chorus gradually took part in the arias themselves: a famous instance occurs in Bellini's *Norma* (1831), during the heroine's 'Casta diva'. However, the use of the chorus to further the plot is relatively uncommon: more frequently it acts as a static point of repose, commenting on the action, summing it up, joining in to provide a bustle of activity parallel to it, and generally abetting the principals. A milestone in the history of the operatic chorus was provided by Verdi's *Nabucco* (1842), where the opposing factions of Babylon and Israel are represented by two choruses which participate so fully in the action that they rival the soloists in importance. Perhaps for the first time in opera, the most popular number became a chorus, that of the Hebrew slaves, 'Va pensiero'. Many of Verdi's early works use the chorus symbolically, to represent the oppressed Italian peoples: later, influenced by *Grand Opera, he incorporated expansive choral scenes, notably in *Don Carlos* (1867) and *Aida* (1871). The same influence can be detected in the operas of Spohr, Marschner, and others, but particularly in Wagner. *Rienzi* (1842), *Tannhäuser* (1845), and *Lohengrin* (1850) all feature grand choral scenes, although in his slightly later theoretical writings, out of which grew the concept of *music drama, Wagner argued against the inclusion of the chorus. Hence it does not appear in *The Ring* until the Gibichungs' entry in *Götterdämmerung*: by this time Wagner had already revised his early views, and made the citizens of Nuremberg intrinsic to the plot of *Die Meistersinger*.

In the 20th cent. the handling of the chorus has been as diverse as the treatment of most other aspects of opera. Britten, for example, while using the chorus as the main dramatic force pitted against the hero in *Peter Grimes*, draws on Greek example to place his Male and Female Chorus in *The Rape of Lucretia* outside the action. A still more definite reversion to Greek method occurs in Stravinsky's *Oedipus Rex*.

An opera chorus is normally divided into the conventional soprano, contralto, tenor, and bass (although 19th-cent. French choruses were

# Chorus of Hebrew Slaves

usually S.T.B.), with the numbers varying according to requirements. It is maintained on the permanent staff of an opera-house, under the direction of a chorus-master. Additional choruses of children are sometimes needed (e.g. for the Urchins' Chorus in Act I of *Carmen* and in *Porgy and Bess*), and certain schools specialize in their provision.

**Chorus of Hebrew Slaves.** See 'VA PENSIERO'.

**Christiania.** See OSLO.

**Christie, John** (b Eggesford, 14 Dec. 1882; d Glyndebourne, 4 July 1962). English operatic patron. Educated Eton, where he also taught, and Cambridge. Married the soprano Audrey *Mildmay in 1931, and in 1934 founded the *Glyndebourne Opera at his house in Sussex. A man of strong enthusiasms (originally for German music), his gift for allowing these to overrule common sense, and his innocent determination to secure the best, led to the realization of an idea that with lesser men would have remained merely an idea. His son **George** (b Glyndebourne, 31 Dec. 1934) has acted as producer's assistant 1953, chairman from 1958.

BIBL: W. Blunt, *John Christie of Glyndebourne* (London, 1968).

**Christie, William** (b Buffalo, NY, 19 Dec. 1944). US harpsichordist and conductor. Studied Tanglewood, Harvard, and Yale. First US professor at Paris Cons. Founder and director of Les Arts Florissants (Paris 1978), which devotes its activities to the revival of Baroque music (especially French) in an authentic idiom. Edinburgh 1986, Paris, OC, 1987; also Florence, Montpellier, Aix. Staged repertory incl. Purcell's *Dido and Aeneas*, Lully's *Atys*, Charpentier's *Actéon* and *Médée*, Rameau's *Les Indes galantes*. The considerable accomplishments of the group, several of whose members are well known in their own right, derive from Christie's impeccable style and lively, clear direction. (R)

**Christoff, Boris** (b Plovdiv, 18 May 1914; d Rome, 28 June 1993). Bulgarian bass. Studied Rome with Stracciari, funded by King Boris, and Salzburg with Muratti. Début Reggio Calabria 1946 (Colline). Milan, S, from 1947; London, CG, from 1949; San Francisco from 1956; Chicago 1957–63; also Paris, Salzburg, Edinburgh. Repertoire includes Rocco, King Mark, Gurnemanz, Hagen, Ivan Susanin, Dosifey, and the Verdi bass roles. A warm-voiced basso cantante with a rich variety of tonal colours, and masterly control and phrasing. Both dramatically incisive and sensitive, he is one of the great interpreters of Boris and King Philip. (R)

BIBL: A. Boshkoff, *Boris Christoff* (trans. London 1991).

**Christophe Colomb.** Opera in 2 parts (27 scenes) by Milhaud; text by Claudel. Prod. Berlin, S, 5 May 1930, in German text by R. S. Hoffmann, with Reinhardt, Scheidl, cond. Kleiber. Concert perfs. London, Queen's H, 16 Jan. 1937; New York, Winter Garden T, 30 Jan. 1957.

Spain, late 15th cent. The opera recalls the events of Columbus's (nar) life in a series of scenes unrelated by time and space. Part 1 deals with the discovery of America; part 2 is concerned with Columbus's return to Spain, where he has a hostile reception. Finally, the old man is summoned by the dying Queen Isabella (sop). He sends her his mule, and she rides to Heaven on it, traversing a carpet which depicts America.

Milhaud links the scenes by use of a narrator, a chorus, and cinematographic projection.

There are some 35 operas on Columbus, including those by Ottoboni (*Il Colombo*, 1690, lost), Morlacchi (1828), Fioravanti (1829), Ricci (1829), Carnicer (1831), Bottesini (1847), Barbieri (1848), Franchetti (1892, celebrating the 400th anniversary of the discovery of America), Egk (1933, for radio, rev. 1942 for stage), Vasilenko (1933), Zádor (1939), and Schuller (1968).

**Chrysothemis.** The daughter of Klytemnestra (sop), sister of Elektra, in Strauss's *Elektra*.

**Chukhadjyan, Tigran** (b Constantinople, 1837; d Smyrna, 23 Feb. 1898). Armenian composer. Studied Constantinople with Mangioni, and in Milan (where he was much impressed by Verdi). On returning home he worked in various theatres, then wrote the first Armenian national opera, *Arshak II*. Only excerpts were performed in his lifetime; the complete première (after rediscovery and revision of the score) took place in 1945. In his operas, Chukhadjyan was the first Armenian composer to reconcile Western operatic techniques with the characteristics of his native folk melody; the ardent nationalism of his work played a part in encouraging Armenians in their liberation movement against Turkey. His comedies (such as *Arif*, *Kyose kyokhva*, and *Leblebidji*) show a feeling for Armenian life and manners.

SELECT WORKLIST: *Arshak Erkrord* (Arshak II) (Terzyan; comp. 1868, prod. Erevan 1945); *Arif* (Atjemyan, after Gogol; Constantinople 1872); *Kyose kyokhva* (The Balding Elder) (Rshtuni; comp. 1873, prod. Istanbul 1974); *Leblebidji* (The Chick-Pea Seller) (Nalyan; Constantinople 1876, rev. 1943 as *Karine*); *Zemire* (after Arabian folk-tales; Constantinople 1891).

**Chung, Myung-Whun** (b Seoul, 22 Jan. 1953). Korean conductor and pianist. Studied conducting New York with Bamburger, and at Juilliard School. Operatic début New York, M, 1986

140

(*Boccanegra*). New York, M, 1988, 1990–1; Florence, C, 1987–9; Florence, Maggio Musicale, 1989. Also Monte Carlo, Geneva; Bologna; San Francisco; Paris, O. First music director, Paris, B, from 1989. Repertory incl. *Idomeneo, Les Troyens, Boris Godunov, Queen of Spades, Fiery Angel*, Verdi's *Otello, Samson et Dalila*. A highly accomplished and refined musician, with an intensely lyrical style. (R)

**church parable**. A termed coined by Britten for his *\*Curlew River* (1964), The *\*Burning Fiery Furnace* (1966), and The *\*Prodigal Son* (1968), works which derive from the example of the Noh Plays of *\*Japan and in which a small group of singers and players present the dramatic fiction of monks staging an opera on a moral subject to a church congregation.

**Ciampi, Vincenzo** (*b* Piacenza, 1719; *d* Venice 30 Mar. 1762). Italian composer. A pupil of Leo and Durante, he established a reputation as a composer of opera buffa. After working as a harpsichordist in Palermo and Paris, he came to London in 1748, where he remained until 1760: his pasticcio *I tre cicisbei ridicoli* was given at the Haymarket in 1749. Returning to Italy, his became Kapellmeister at the Ospedale degli Incurabili in Venice.

Ciampi's best-known works were his opere buffe, such as *Da un disordine nasce un ordine* and *Amore in caricatura*, although he also composed a number of opere serie, including *Artaserse* and *Adriano in Siria*. His most important opera was *Bertoldo, Bertoldino e Cacaseno*; it was so popular when heard in Paris in 1753 that a French version, with a text by Favart, appeared the following year under the title *Ninette à la cour*. This in turn provided the inspiration, and part of the libretto, for Hiller's Singspiel *Lotte am Hofe* (Leipzig 1767).

SELECT WORKLIST: *Da un disordine nasce un ordine* (Order born from Disorder) (Federico; Naples 1737); *Artaserse* (Metastasio; Palermo 1747); *Adriano in Siria* (Metastasio; Venice 1748); *Bertoldo, Bertoldino e Cacaseno* (Goldoni; Venice 1747, rev. London 1755); *Amore in caricatura* (Love Caricatured) (Goldoni; Venice 1761).

BIBL: C. Anguissola, *Vincenzo Legrenzio Ciampi musicista piacentino del Settecento* (2/Piacenza, 1936).

**Cibber, Susanna** (*b* London, Feb. 1714; *d* London, 30 Jan. 1766). English singer and actress. Studied with her brother Thomas Arne. Début London, Little T, Haymarket, 1732, in Lampe's *Amelia*. Appeared in Arne's *Rosamund* and *Dido and Aeneas*. Engaged at London, DL, 1734. Her marriage to Theophilus Cibber (during which she performed as an actress) ended with a sordid court case over her affair with John Sloper, at which Cibber had connived. Dublin, 1741; heard

and greatly admired by Handel, who engaged her for several oratorios, including the first *Messiah*. Her singing of 'He was despised' moved Dr Delaney to exclaim, 'Woman, for this thy sins be forgiven thee!' On her return to London, 1742, she created other Handel roles: Micah (*Samson*), and Lichas (*Hercules*). From 1744 until her death she was a great tragic actress in partnership with Garrick. Her voice, though not outstanding in an age of great singers, had 'native sweetness' (Burney), and she evidently moved her audiences deeply.

**Ciccimarra, Giuseppe** (*b* Altamura, 22 May 1790; *d* Venice, 5 Dec. 1836). Italian tenor. Naples, C, 1816–26, creating many roles in Rossini operas, including Iago, Goffredo (*Armida*), Aronne (*Mosè*), Ernesto (*Ricciardo e Zoraide*), Pilade (*Ermione*). Taught after retirement; his pupils included Joseph Tichatschek, the first Tannhäuser and Rienzi.

**Cicéri, Pierre-Luc-Charles** (*b* Saint-Cloud, 17 Aug. 1782; *d* Saint-Chéron, 22 Aug. 1868). French painter and designer. Studied singing, later turning to painting and becoming an assistant at the Paris, O, 1810; *Peintre du roi* 1825. His designs include those for *Olympie* (1819), *La muette de Portici* (1828), *Guillaume Tell* (1829), and *Robert le diable* (1833). The greatest French designer of his day, with a strong historical and architectural sense, he opened up a new expressive range by his use of three-dimensional scenery and panoramic effects. His association with Louis *\*Daguerre was epoch-making, and set new standards of accuracy in historical detail and in grandeur of pictorial effect, matching and encouraging the grandiose ambitions of French *\*Grand Opera. For *La muette*, in a typical piece of work, he designed a view of Vesuvius seen through a splendid palace, with woods and houses in the middle distance; and he ended with an eruption that included fireworks. For *Robert* he imitated the cloister of Montfort l'Amaury, building three sides as his setting for the nocturnal scene of the ghostly nuns. In this, though his style had a Romantic fluency, he showed the influence of his Baroque forebears, and he also made skilful artistic use of the latest stage machinery and equipment (e.g. gaslight in Isouard's *Aladin, ou La lampe merveilleuse*, 1822). His designs extended the expressive possibilities of the panorama (blending into the scenery), as in Isouard's *Alfred le grand*, 1822. He also pioneered the mobile panorama, revolving drums turning a backcloth so as to produce an illusion of the characters on a journey along a river in the ballet *La belle au bois dormant* (1825), and from sky to earth in Auber's *Le lac des fées* (1839) (Wagner was to make use of both ideas). Cicéri's four pupils Despléchins, Diéterle, Séchan, and

Feuchères further developed his work in the years 1833–48, satisfying the French appetite for outdoor scenes, transformations, and thrilling special effects.

**Cicognini, Giacinto** (b Florence, 1606; d Venice, 1651). Italian librettist. His father **Jacopo Cicognini** (1577–1633) was a noted dramatist and one of the first to introduce Spanish theatrical works to Italy: his one opera libretto, *Andromeda*, was set by Domenico Belli in 1618. Giacinto followed his father's example, renouncing law for the theatre, and many of his works show Spanish influences, especially in the elegance of their verse. His most famous libretto was that for Cavalli's *Giasone* (1649); his other works included *Orontea* (Cesti, 1649) and *Gli amore di Alessandro Magno e di Rossane* (Luccio, 1651, and many later settings).

**Cid, Le.** Opera in 4 acts by Massenet; text by D'Ennery, Gallet, and Blau, after Corneille's drama (1637). Prod. Paris, O, 30 Nov. 1885, with F. Devriès, J. and E. De Reszke, Plançon; New Orleans, French OH, 17 Feb. 1890, with Dauriac, Beretta, Mary, Saint-Jean, Balleroy, Geoffray, Rossi.

Spain, 11th cent. Don Rodrigue (ten) kills Don Gormas (bs), father of his betrothed Chimène (sop), to avenge a wrong done to his own father. Before he can be punished, Rodrigue leaves to fight the Moors. On his return, he is acclaimed by all as the 'Cid', the master warrior. Chimène has the right to determine his sentence, but chooses to forgive him. Racked with guilt, Rodrigue attempts suicide but Chimène prevents him, and the couple are then united.

Other operas on the character are by Batistin-Stuck (1715), Leo (1727), Sacchini (1764), Piccinni (1766), F. Bianchi (1773), Paisiello (1775), Roesler (1780), Lacchini (1783), Farinelli (1802), Aiblinger (*Rodrigo et Zimène*, 1821), Savi (1834), Pacini (1853), Neeb (1857), Cornelius (1865), Coppola (1884), Bizet (unfin.), Boehme (1887), and Wagenaar 1916.

**'Cielo e mar!'** Enzo's (ten) aria praising the sky and sea in Act II of Ponchielli's *La Gioconda*.

**Cigna, Gina** (b Paris, 6 Mar. 1900). French-Italian soprano. Studied Paris with Calvé, Darclée, and Storchio. Début (as Genoveffa Sens) Milan, S, 1927 (Freia). Reverted to her own name, and sang there 1929–43. London, CG, 1933, 1936, 1937, 1939; New York, M, 1936–8; San Francisco and Chicago. A dramatic performer with a large repertory, particularly famous as Gioconda and Turandot. (R)

**Cikker, Ján** (b Banská Bystrica, 29 July 1911; d Bratislava, 21 Dec. 1989). Slovak composer. Studied Prague with Křička and Novák, conducting

with Dědeček and in Vienna with Weingartner. Taught and worked in the opera in Bratislava from 1939. His vivid post-Romantic idiom, which owes something to Bartók and Janáček, has found its fullest expression in his operas, which have won success in his own country and in Germany.

WORKLIST: *Juro Jánošík* (Hoza; Bratislava 1954); *Beg Bajazid* (Prince Bajazid) (Smrek; Bratislava 1957); *Mr Scrooge* (Smrek, after Dickens; comp. 1958–9, prod. Kassel 1963); *Vzkriesenie* (Resurrection) (Cikker, after Tolstoy; comp. 1959–61, prod. Prague 1962); *Hra o láske a smrti* (The Play of Love and Death) (Cikker, after Rolland; Munich 1969); *Coriolanus* (Cikker, after Shakespeare; Prague 1974); *Ze Života hmyzu* (The Insect Play) (Cikker, after Čapek; Bratislava 1987).

**Cilea, Francesco** (b Palmi, 26 July 1866; d Varazze, 20 Nov. 1950). Italian composer. Studied Naples with Serrao, producing his first opera, *Gina*, with success soon after leaving the conservatory. A commission from Sonzogno led to a verismo work, *La Tilda*, but its lesser success turned Cilea towards teaching, first in Naples and then in Florence. *L'arlesiana* fared a little better, in part due to the appearance of the young Caruso (whose career it also helped: the 'Lamento' was once a favourite tenor war-horse). His most enduring success was *Adriana Lecouvreur*, more because of the work's skilful appeal to singers than for its melodic or dramatic strengths, though at its best his music has charm and style.

WORKLIST: *Gina* (Golsciani; Naples 1889); *La Tilda* (Graziani; Florence 1892); *L'arlesiana* (Marenco, after Daudet; Milan 1897); *Adriana Lecouvreur* (Colautti, after Scribe & Legouvé; Milan 1902).

BIBL: T. d'Amico, *Francesco Cilea* (Milan, 1960).

**Cillario, Carlo Felice** (b S Rafael, 7 Feb. 1917). Italian conductor. Studied Bologna and Odessa. Gly. and Chicago 1961; New York, M, 1972, 1985; Drottningholm 1982–4; San Francisco 1986; Sydney 1988 (principal guest cond.); also London, CG; Vienna; Milan, S; Buenos Aires; Paris; Berlin; Hamburg; Florence; etc. Repertory incl. *Magic Flute*, *Barber*, *Lucia*, *Aida*, *Otello*, *Falstaff*, *Werther*, *Walküre*, *Rheingold*. (R)

**Cimarosa, Domenico** (b Aversa, Naples, 17 Dec. 1749; d Venice, 11 Jan. 1801). Italian composer. A pupil of Gallo, Aprile, and possibly also Piccinni. Came to attention with *Le stravaganze del conte* in 1772 and for the next 15 years spent his time between Naples and Rome, producing a string of popular successes, including *L'italiana in Londra*, *Il convito*, and *L'impresario in angustie*, mounting a worthy challenge to the supremacy of Paisiello and Piccinni. Among other important works of this period are *Il pittor parigino*, *La villana riconosciuta*, *Il marito disperato*, and *La donna sempre al suo peggior s'appiglia*. Cimarosa moved to St Petersburg in 1787, replacing Sarti as maestro di cappella to Catherine the Great: by now his operas were known throughout Europe, including London, Paris, and Dresden. After success in Russia, where he wrote *La vergine del sole* and *Cleopatra*, he was called to Vienna in 1791 to succeed Salieri as Kapellmeister. Here in 1792 he produced *Il matrimonio segreto* which, according to legend, earned the cast a free Imperial supper before an immediate repeat performance. On the death of Leopold II, Salieri was reinstated as Kapellmeister and Cimarosa returned to Naples as maestro di cappella to the King. A happy sojourn here was curtailed by his public support for the Napoleonic cause: imprisoned in 1799, he was released, but died in Venice, *en route* for St Petersburg.

Especially during his later years, Cimarosa was attracted to opera seria as well as opera buffa. Though *L'Olimpiade*, *Penelope*, and *Gli Orazi ed i Curiazi* are carefully constructed, they show little originality and his fame rests on his opere buffe. The classic example is *Il matrimonio segreto*, where his engaging melodies, his understanding of the conventions of the genre, and his sympathetic manipulation of the characters mark him as a composer of rare dramatic sensitivity. His skilfully constructed ensembles gave his work wide popular appeal—*Il matrimonio segreto* was performed 67 times in Naples alone following his return in 1793—and a status as a composer which at times exceeded even that of Mozart, who provided an additional aria for the Viennese revival of *I due baroni di Rocca Azzurra* in 1789. *Le astuzie femminili*, one of his last works, has recently become popular and revivals both of it and of *Il matrimonio segreto* have aided our appreciation of his abilities.

There is an opera on an incident in his life by Isouard, *Cimarosa* (1808).

SELECT WORKLIST: *Le stravaganze del conte* (The Count's Eccentricities) (Mililotti; Naples 1772); *L'italiana in Londra* (Petrosellini; Rome 1779); *Il pittor parigino* (The Parisian Painter) (Petrosellini; Rome 1781); *Il convito* (The Banquet) (Livigni; Venice 1782); *La villana riconosciuta* (The Peasant Girl Unmasked) (Palomba; Naples 1783); *L'Olimpiade* (Metastasio; Vicenza 1784); *Il marito disperato* (The Desperate Husband) (Lorenzi; Naples 1785); *La donna sempre al suo peggior s'appiglia* (Women Always Go to the Bad) (Palomba; Naples 1785); *L'impresario in angustie* (The Distressed Impresario) (Diodati; Naples 1786); *La vergine del sole* (The Virgin of the Sun) (Moretti; St Petersburg 1786); *Cleopatra* (Moretti; St Petersburg 1789); *Il matrimonio segreto* (The Secret Marriage) (Bertati, after Colman & Garrick; Vienna 1792); *Penelope* (Diodati; Naples 1794); *Le astuzie femminili* (Feminine Wiles) (Palomba, after Bertati; Naples 1794); *Gli Orazi ed i Curiazi* (Sografi; Venice 1796).

**Cincinnati.** City in Ohio, USA. Visiting companies came in the 19th cent., with singers including Patti. The Cincinnati Summer Opera Association began seasons in 1920 in a bandstand next to the Zoo. Opera transferred to the Music Hall (cap. 3,630) in 1972, and became Cincinnati Opera 1975. The second oldest US company (after the Metropolitan), and known as the cradle of American opera singers from its early encouragement of artists including Peerce and Kirsten.

**Cinderella.** The English name for the heroine of Charles Perrault's story *Cendrillon* in his *Contes de Ma Mère L'Oye* (Mother Goose's Tales, 1697), known in Italian as *Cenerentola, in German as Aschenbrödel, in Russian as Zolushka, in Czech as Popelka. Operas on the story are by Laruette (1759), Isouard (1810), Steibelt (1810), Rossini (1817), García (1826), C. T. Wagner (*Popelka*, 1861), Chéri (1866), Conradi (1868), Jonas (1871), Langer (1878), Farmer (1882), Rozkošny (1885), Pitt (1899), Massenet (1899), Wolf-Ferrari (1900), Albini (*c.*1900), Forsyth (early 20th cent.), Blech (1905), Glebov (1906), Asafyev (1906), Buttykay (1912), O. Piccinni (1922), and Jarrett (1956).

**Cinti-Damoreau, Laure** (*b* Paris, 6 Feb. 1801; *d* Chantilly, 25 Feb. 1863). French soprano. Studied Paris with Plantade and Catalani. Début Paris, I, 1816 (Lilla, *Una cosa rara*). Paris, O, 1825–35; here she created leading roles in Rossini's *Le siège de Corinth*, *Moïse*, *Le Comte Ory*, and *Guillaume Tell*; also Isabelle (*Robert le diable*), and Elvira (*La muette de Portici*). At the Paris, OC, 1837–41, created roles in more of Auber's works, e.g. *Le domino noir*. London, H, 1822 and 1832. Toured USA 1844. Taught singing at Paris Cons. 1833–56, and wrote a *Méthode de chant* (Paris, 1849). She had an exceptionally pure, clear, and accurate voice, and excelled in neat ornamentation. Stendhal called her 'that charming little singer'; Fétis acclaimed her 'beau talent'.

**Cio-Cio-San.** The Japanese name of Butterfly (sop) in Puccini's *Madama Butterfly*.

**Cioni, Renato** (*b* Portoferraio, 15 Apr. 1929). Italian tenor. Studied Florence. Début Spoleto 1956 (Edgardo). Milan, S, from 1957; London, CG, from 1964; San Francisco and Chicago from 1961: Berlin, D, 1965; New York, M, 1969–70; also Verona, Vienna, Munich, Buenos Aires, etc. Repertory incl. Bellini, Donizetti, Verdi (Duke of Mantua, Gabriele Adorno), Puccini (Pinkerton, Cavaradossi). Partnered Callas and Sutherland. (R)

**Circe.** In Greek mythology, especially as related in Homer's *Odyssey*, a sorceress living on the island of Aeaea. When Odysseus visited her island, she turned his companions into swine but

# Ciro in Babilonia

was forced by Odysseus, who had evaded the spell, to release them. Odysseus stayed with her for a year. Most operas on her have as their subject the visits of \*Odysseus or of \*Telemachus. Operas on the subject are by Ziani (1665), Freschi (1679), Pollarolo (1692, lost), Sabadini (1692), Desmarets (1694), Strungk (1695), Boniventi (1711), Orefice (1713), Keiser (his last opera, 1734), Astaritta (1777), Mysliveček (1779), Cimarosa (1782), Albertini (1786), Gazzaniga (1786), Winter (1788, unprod.), Praupner (1789), Paer (1792), Romberg (1807), Perrino (1819), Chapi (1902), Hillemacher (1907), and Egk (1948).

**Ciro in Babilonia** (Cyrus in Babylon). Opera in 2 acts by Rossini; text by Francesco Aventi. Prod. Ferrara 14 Mar. 1812, with Marietta, Marcolini, Manfredini, Savinelli, Fraschi.

Amaria (sop), wife of Ciro, King of Persia (sop), is the object of the attentions of Baldassare, King of Babylon (ten). Ciro attempts to rescue Amaria and their son, who are held by Baldassare, but is captured himself. A storm seems to be a divine signal that Baldassare should sacrifice his three Persian captives. On the verge of their execution, the Persian army defeats Babylon, and Ciro is declared the new king.

Among some 20 other operas on Cyrus are those by Cavalli (1654), Caldara (1736), Hasse (1751), and Agnesi (*Ciro in Armenia*, 1753; the most important work of one of the first woman opera composers).

**City Center of Music and Drama.** Home of the \*New York City Opera 1944–65, when it moved with the New York City Ballet to the State T, Lincoln Center.

**Claggart.** John Claggart (bs), the master-at-arms in Britten's *Billy Budd*.

**Clapisson, Louis** (*b* Naples, 15 Sep. 1808; *d* Paris, 19 Mar. 1866). French composer. Son of a horn player at the Naples, C; studied Bordeaux and in Paris with Rejcha, also playing the violin in opera orchestras. He had an early success with a comic opera, *La figurante* (1838), which he followed up with a long succession of works that caught the fancy of the Paris public. He defeated Berlioz in the election to a chair at the Institut by a large majority (to the indignation of Offenbach and to Berlioz's enduring hostility); and the appeal of his *Gibby-la-Cornemuse* (1846) was partly responsible for drawing audiences away from the première of *La damnation de Faust*. His large collection of instruments was the foundation of the Conservatoire museum.

**claque** (Fr.: 'smack, clap'). A group engaged by a composer, performer, or impresario to applaud, show appreciation, demand encores, and do whatever it can as apparently ordinary members of the audience to influence a performance's reception.

Hired support of a performer (or attack on a rival) is almost as old as human vanity and deceit, and is recorded in classical antiquity. Organized groups of applauders are known to have attempted to influence the judges in the Athenian dramatic festivals. However, the father of the claque appears to be the Emperor Nero. Suetonius records that he hired young men to applaud his own performances, acting as *bombi* ('bees', who hummed loudly), *imbrices* ('roof-tiles', who clapped with hollowed hands), and *testae* ('bricks', who clapped flat-handed). There are records of organized applauders for early Venetian opera, of hired opposition bringing about the failure of an opera in Bologna in 1761, and of many similar occasions. Prudhomme in the 1790s refers to a 'M. Claque', with hands the size of a washerwoman's, who was paid 36 livres for a success and 12 livres for a failure.

The use of the term *claquer*, to applaud for money, seems to date from early 19th-cent. France. In 1820 a claque agency was opened by a M. Sauton under the title *L'assurance des succès dramatiques*. Operating under a *chef de claque* was a team of claqueurs known, from their traditional position under the chandelier in the Opéra, as the *chevaliers du lustre*. These included *tapageurs* (who clapped vigorously), *connaisseurs* (who added knowing exclamations of approval), *pleureurs* (or more often *pleureuses*) (who used concealed bottles of smelling salts to induce tears of emotion), *bisseurs* (who called for encores), *chatouilleurs* (who sustained their neighbours with witty sallies or offers of sweets), *commissaires* (who held forth on the performance's merits in the intervals), and *chauffeurs* (who 'warmed up' the house and later spread tales of fantastic triumphs). By the 1830s and 1840s the claque was playing a powerful part in French opera, and was made the subject of some vivid comments by Berlioz (in *A travers chants* and especially in the 7th Evening of *Les soirées de l'orchestre*). According to him, the claque was hired by the management, who received payment in return for the assurance of a monopoly. He reports that the *claqueurs* were also known as Romans, from the example of Nero, and in giving a lively, sardonic account of their activities he makes much play of their leader's 'Roman' name, Auguste. A late 19th-cent. tariff included plain applause, 5F; renewed applause, 15F; expressions of horror, 5F; murmurs of alarm, 15F; groans, 12.50F; guffaws, 5F; cries of 'Oh, how funny', etc., 15F. To ensure the accurate introduction of these effects, the *chef de claque* was invited to dress-rehearsals and given the score to study.

The claque has long endured in Italy, where the drama of a performance extends beyond the footlights. A tariff quoted in 1919 included applause for a gentleman's entry, 25L; for a lady, 15L; ordinary applause, 10L; insistent applause, 15–17L, pro rata; interruptions with 'Bene!' or 'Bravo!', 5L; wild enthusiasm, a special sum to be arranged. Nowadays the usual practice is for the *capo di claque* to be paid a flat sum. The most famous claque in modern Italy was that of La Scala, organized by Carmelo Alabiso (a former Toscanini tenor) and Antonio Carrara, 30–40 strong and including students, teachers, and even two barbers.

In England, the claque has not achieved the same status, though at Covent Garden in the 1890s Jean De Reszke was allowed *faveurs de claque*, free tickets for his followers. It was probably introduced into America in 1910 when Gatti-Casazza allowed the tenor Bonci's valet, so it is said, to form a claque so as to save the failing production of *Armide*. A claque flourished at the Metropolitan in the 1920s (led by Schol, an umbrella-maker). In 1954 Rudolf Bing cut the number of standing places, the traditional post of the claque, so as to restrain it; the *chef de claque* was John Bennett, who allowed his men only their tickets. Ezio Pinza used to pay them to leave him alone, a form of protection racket. Some singers have gone further and resorted to an 'anti-claque', hired to bring down a rival. They have even clashed: when a rival singer's fans tried to boo Maria Callas at La Scala, the claque responded so briskly that two people were sent to hospital as a result of the ensuing brawl.

In Germany the claque has not flourished, but one of the most famous of all time was that commanded by Schostal in Vienna in the 1920s, and vividly described by a former member, Joseph Wechsberg, in *Looking for a Bluebird*. Like many *claqueurs*, Schostal prided himself on his artistic integrity in only applauding works he admired. The Vienna claque is not alone in applauding, free of charge, for his artistry a singer who had declined to employ it. Conversely, the Parma claque once refunded a tenor his money and then booed him. It has often been the claim of *claqueurs* that they are encouragers of excellence, defenders of talent, and that they earn their money as honestly as any publicity man and with more discrimination.

**Clark, Graham** (*b* Littleborough, 10 Nov. 1941). English tenor. Studied London with Boyce. Début Scottish O 1975 (Verdi's Roderigo). London, ENO from 1978; Bayreuth 1981. Vienna, S; Munich, S; New York, M, 1985; Paris, B, 1991; also Barcelona, Zurich. Roles incl. David, Loge, Mime, Don Juan (*Stone Guest*), Gregor (*Makro-*

*poulos Case*), Captain Vere. A performer of intelligence and vitality, with a bright tone. (R)

**Clayton, Thomas** (*b* ? *c*.1665; *d* ? *c*.1725). English composer and violinist. His *Arsinoe, Queen of Cyprus*, performed at Drury Lane in 1705, was described by Clayton as 'an Opera after the Italian Manner: All Sung'; the libretto was a translation of an Italian text by Stanzani. It is the earliest full-length all-sung English opera extant, and the first wholly in the Italian style. In his preface to the work Clayton states that 'the Design of this Entertainment [is] to introduce the Italian Manner of Musick on the English Stage, which has not before been attempted'. Hawkins, in his *A General History of the Science and Practice of Music* (1776), maintained that most of the music derived from Italian sources; Clayton had studied in Italy. The anonymous author of *A Critical Discourse upon Operas in England, and a Means proposed for their Improvement* (1708) thought little of the work, saying there was nothing in it 'but a few sketches of antiquated Italian airs, so mangled and sophisticated that instead of *Arsinoe* it ought to be called the Hospital of the old, decrepit Italian operas'.

The novelty of the work, lacking the speech, masques, and stage effects usual in English entertainments, seemed to ensure it success. However, Clayton's second opera, *Rosamond*, staged at Drury Lane in 1707, enjoyed no such popularity; Hawkins described the music as 'mere noise'.

SELECT WORKLIST: *Arsinoe, Queen of Cyprus* (?Motteaux, after Stanzani; London 1705); *Rosamond* (Addison; London 1707).

**Clément, Edmond** (*b* Paris, 28 Mar. 1867; *d* Nice, 24 Feb. 1928). French tenor. Studied Paris with Warot. Début Paris, OC, 1889 (*Mireille*). Remained there until 1910, creating many roles in contemporary works, including Bruneau's *L'attaque du moulin*. New York, M, 1909–11. A stylish and elegant singer. (R)

**Clemenza di Tito, La** (The Clemency of Titus). Opera in 2 acts by Mozart; text by Mazzolà, after Metastasio's text for Caldara (1734). Prod. Prague 6 Sep. 1791, with Baglioni, Fantozzi, Antonini, Perini, Bendini, Campi; London, H, 27 Mar. 1806, with Billington, Braham; Tanglewood, Lenox, MA, 4 Aug. 1952, with Jordan, Flemming, Winston, Matheson, Lansee, Sharretts.

Rome, AD 80. Vitellia (sop) persuades Sesto (sop cas) to kill the Emperor Tito (ten) because he did not choose her as his empress. When the Emperor's wedding is postponed, Vitellia orders the assassination to be delayed. Annio (sop cas), a friend of Sesto's, wishes to marry Servilia (sop), Sesto's sister, but Tito announces that he plans to marry Servilia himself immediately. Servilia confesses to Tito that she loves Annio and he generously agrees not to come between them. Vitellia, believing she has been passed over again, has by now renewed her assassination order. When she learns of the change of Tito's plans she tries in vain to recall Sesto.

Sesto learns that Tito has not died and confesses his crime to Annio. But it emerges that Lentulus, whom Sesto stabbed in mistake, recognized his attacker and Sesto is taken before the Senate, where he is found guilty. Tito wants to save his friend but Sesto can not explain his action without incriminating Vitellia. Filled with guilt, she throws herself at Tito's feet and confesses her part in the crime. Tito pardons everyone and the people rejoice in the clemency of their emperor.

Mozart's last opera, and the first to be performed in its entirety in London.

Among nearly 50 settings of Metastasio's text are those by Caldara (1734), Leo (1735), Hasse (1735), Wagenseil (1746), Araia (1751), Gluck (1752), Jommelli (1753), Holzbauer (1757), Galuppi (1760), Jommelli (1765), Anfossi (1769), Naumann (1769), Traetta (1769), Mysliveček (1774).

BIBL: J. Rice, *Wolfgang Amadeus Mozart: 'La clemenza di Tito'* (Cambridge, 1991).

**Cleonice.** See DEMETRIO.

**Cleopatra** (69/68–30 BC). Queen of Egypt, *c.*52–30 BC. Operas on her are mostly about her love affairs with Julius Caesar or Mark Antony, and include those by Canazzi (1653), Castrovillani (1662), Anschütz (1686), Graun (1742), Monza (1775), Anfossi (1779), Cimarosa (1789), Weigl (1807), Paer (1809), Combi (1847), L. Rossi (1876), Sacchi (1877), Pedrell (1885), Morales (1891), Massenet (1914), Hazlehurst (1918), Hadley (1920), and Barber (1966).

**Cloches de Corneville, Les** (The Bells of Corneville). Opera in 3 acts by Planquette; text by Clariville and Gabet. Prod. Paris, Folies-Dramatiques, 19 Apr. 1877; New York, 5th Av. T, 22 Oct. 1877; London, Folly T, 23 Feb. 1878.

Normandy, in the time of Louis XIV. The castle of Corneville stands empty following the banishment of the Duke. His descendant Henri (bar) wishes to prove his claim by fulfilling a local legend that the bells of the castle will ring spontaneously when the rightful heir approaches. However, a letter found in the castle, and wrongly interpreted, results in the house falling into the hands of a servant girl, Serpolette (sop). It emerges that the intended recipient is Germaine (sop), niece of a wealthy tenant Gaspard (bs). She loves Henri, who once saved her life, and the couple are happily united.

Planquette's most popular opera; at the end of a run of 461 performances the manager served the last audience with 2,000 rolls and free beer.

**Cluj** (Hung., Kolozsvár; Ger., Klausenburg). City in Transylvania, Romania. Opera in Hungarian was given in the early 19th cent. The first season of Opera Romană was given in 1920 through efforts by the composer Tiberiu Brediceanu (1877–1968) and Constantin Pavel. The company prospered under the baritone Popovici, and has included in its repertory a number of new Romanian operas. The opera-house was at the centre of the fighting at the start of the Romanian revolution in 1989. There is also an opera-house giving performances in Hungarian.

**Cluytens, André** (*b* Antwerp, 26 Mar. 1905; *d* Paris, 3 June 1967). French conductor of Belgian birth. Studied Antwerp. Antwerp O 1927–32; Toulouse 1932, also conducting much elsewhere in France. Paris, O and OC, 1947–66 (music director Paris, OC, 1947). First French conductor at Bayreuth, 1955–8, 1965. (R)

**Clytemnestra.** In classical mythology, the wife and murderess of Agamemnon and lover of Aegisthus. She was in turn murdered by her son Orestes at the instigation of her daughter Electra. She appears as Klytemnestra (mez) in Strauss's *Elektra*.

**Coates, Albert** (*b* St Petersburg, 23 Apr. 1882; *d* Cape Town, 11 Dec. 1953). English conductor and composer. Studied science in Liverpool, then in commerce in St Petersburg. Studied Leipzig with Nikisch. Début Leipzig (*Hoffmann*). Assistant to Nikisch 1904; cond. Elberfeld 1906–8; Dresden 1908–10; Mannheim 1910. St Petersburg, M, 1910, then chief cond. 1911–19, incl. *Boris* (prod. Meyerhold, 1911) and Taneyev *Oresteia* (1915), returning intermittently until 1934. London, CG, 1914 (*Tristan and Ring*), 1919–24; BNOC (*Ring*, with Shalyapin) 1929, 1935–8. His operas are *Samuel Pepys* (Munich 1929); *Pickwick* (CG 1936, in season by British Music Drama OC, organized by Coates with Vladimir Rosing): scenes from the latter were the first televised opera, Nov. 1936; and *Van Hunks and the Devil* (Cape Town 1952). (R)

**Coates, Edith** (*b* Lincoln, 31 May 1908; *d* Worthing, 7 Jan. 1983). English mezzo-soprano. Studied with Clive Carey and Dino Borgioli. Début London, Old Vic, 1924 (Giovanna, *Rigoletto*). London: SW 1931–46; CG 1937–9, 1947–63. Created Auntie (*Peter Grimes*), Mme Bardeau (*The Olympians*), Grandma (Grace Williams's *The Parlour*). Particularly striking as the Countess (*Queen of Spades*). (R)

**Coates, John** (*b* Girlington, 29 June 1865; *d* Northwood, 16 Aug. 1941). English tenor. Originally a baritone; sang with D'Oyly Carte Co. 1894. After further study emerged as tenor, and sang in light opera 1899–1900. London, CG, 1901 (Gounod's Faust). First Claudio in Stanford's *Much Ado about Nothing*. Toured Germany 1902–7, and USA 1926–7. Sang with Moody-Manners and Beecham companies 1907–11. An intense, intelligent singer; one of the best British Tristans and Siegfrieds. (R)

**Coburg.** City in Bavaria, Germany. The first musical drama was the pastoral school play *Von der Zerstörung Jerusalems* (1631) by Melchior Franck (c.1579–1639). A theatre (cap. c.100) opened 1684, with elaborate stage machinery for opera-ballets. Performances were given under C. C. Schweitzelsberger 1717. The Bühnenlokal opened 1764. Opera developed in the early 19th cent. under L. Schneider. The Herzogliches Coburg-Gothaisches Hoftheater opened under A. Lübcke 1827, the Landestheater (cap. 1,000) 1840 with Auber's *Le lac des fées*. Ernst II, Duke of Saxe-Coburg-Gotha's *Diana von Solange* was premièred 1858. Malipiero's *Mysterium Venedigs* was premièred 1932.

**Cocchi, Gioacchino** (b ?Naples, c.1720; d Venice, after 1788). Italian composer, known by his contemporaries as 'Il Napoletano'. His first opera was the highly successful *Adelaide*. During the next 15 years he produced over 35 operas and other dramatic works for houses throughout Italy, embracing both opera buffa, such as *L'ipocondriaco risanato* and *Li matti per amore*, as well as opere serie on popular themes, including *Siface* and *Tamerlano*.

In 1757 he was recruited for the London, H, for whom he wrote a number of operas and pasticcios, of which the most successful were *Zenobia* and *Issipile*. He was succeeded by J. C. Bach in 1761; in 1773 he returned to Italy and to his post at the Ospedale degli Incurabili in Venice, but does not appear to have written any further operas. His greatest success was the opera buffa *La maestra*, given in Paris in 1752 as *La scaltra governatrice*.

SELECT WORKLIST: *Adelaide* (Salvi; Rome 1743); *L'ipocondriaco risanato* (The Hypocondriac Restored) (Goldoni; Naples 1746); *La maestra* (The Mistress) (Palomba; Naples 1747); *Siface* (Metastasio; Naples 1749); *Li matti per amore* (Lunatics for Love) (Goldoni; Venice 1754); *Tamerlano* (Piovene; Venice 1755); *Zenobia* (Metastasio; London 1758); *Issipile* (Metastasio; London 1758).

**Coccia, Carlo** (b Naples, 14 Apr. 1782; d Novara, 13 Apr. 1873). Italian composer. Studied Naples with Casella and Fenaroli, later Paisiello, who got him his first appointment (with the King of Naples, 1806–8). His first opera, *Il matrimonio per lettera di cambio*, was a failure, but with Paisiello's continued support he soon wrote the successful *Il poeta fortunato*. In Venice, 1809–17, he developed his characteristic genre *opera semiseria, of which the most successful example was *Clotilde*. Here he followed Mayr's lead in giving the chorus a more functional role, but he was also criticized for copying other men's music, and for the hastiness in his work that led to great unevenness. The emergence of Rossini (whom he greatly admired) forced him to seek a new arena abroad, in Lisbon (1820–3) and London (1824–7). He conducted at the London, H, and in 1827 broke a six-year silence with *Maria Stuart* (with Pasta in the title-role). In the same year he returned to Italy, where he had new success especially with *Caterina di Guisa*. A period of further study, especially of German music, during his silence led to a change in style, but the new seriousness in his work did not help him to retain popularity as fashion turned in new directions. He was director of the Turin Conservatory, 1836–40. He gave up opera in 1841. Verdi invited him to write the Lacrymosa and an Amen in the abortive Requiem for Rossini.

SELECT WORKLIST: *Il matrimonio per lettera di cambio* (Marriage by Bill of Exchange) (Checcherini; Rome 1807); *Il poeta fortunato* (Gasbarri; Florence 1808); *La donna selvaggia* (The Wild Woman) (Foppa; Venice 1813); *Carlotta e Verter* (Gasbarri; Florence 1814); *Clotilde* (Rossi; Venice 1815); *Maria Stuart* (Giannone; London 1827); *Caterina di Guisa* (Romani; Rome 1833).

BIBL: G. Carotti, *Biografia di Carlo Coccia* (Turin, 1873).

**Cocteau, Jean** (b Maisons-Laffitte, 5 July 1889; d Paris, 12 Oct. 1963). French writer and librettist. Collaborated with Honegger in *Antigone* and Stravinsky in *Oedipus Rex*, for which he also sometimes took the part of the narrator. Also provided librettos for Milhaud's *Le pauvre matelot* and Poulenc's *La voix humaine*.

**Code Rossini.** The term sometimes used in modern criticism to describe the dramatic content and formal structure of much Italian opera between c.1815 and c.1860, so called because it was largely developed by Rossini and first exploited in his operas, particularly those of his Neapolitan period (1815–22).

The opera would open either with a formal overture, which might have little dramatic relevance to the subsequent plot (Rossini's were brilliant but often interchangeable), or a short prelude setting the mood and perhaps making use of material from the opera (preferred by Verdi). The atmosphere is initially created in the introduzione e cavatina, a choral movement at the start of the opera, setting the scene and characterizing the participants, with the melodic interest largely in the orchestra. A favourite device would be then to bring on the male principal (perhaps with his companion), who would have a slow aria leading to dialogue with the chorus (initiating the plot) and then a cabaletta with choral accompaniment. The main action is then divided into a number of substantial scenes, each of which is framed around a single dramatic event involving a small group of characters. After the main issue has been stated, or suggested by the music, it is meditated upon by one of the characters; this leads to some new force being brought upon the issue, either through stage action, or through a change of mind on the part of the character. In the climax which follows the issue is resolved, usually by some demonstrative action, and the mood is thereby changed. The main drama is articulated through the musical design of the *scena, which in its musical structure mirrors the progression from meditation to action. Rossini set the seal on the convention inherited from Mozart, of ensuring that each principal singer had at least one such scena: though it originated in the bipartite (slow-fast) aria form popular in the late 18th cent., its structure was expanded by Rossini by the addition of a long introductory recitative and the interpolation of an extra section (the *tempo di mezzo) between the lyrical reflection of the *cantabile and the brilliant action of the *cabaletta. The andante tended to have two quatrains, tonally static; the cabaletta, also in simple binary form, could be in an unrelated key, tending towards the end to display. The *romanza was normally in a single movement.

The *duet was normally in three-part form. In a fast opening section, the first singer might be given a line decorating a regular melody, answered by the second singer with a similar (or the same) melody. Modulation to a new key, by way of dialogue between the singers, could then introduce a middle section in contrasting tempo (if andante, with the singers in lyrical 3rds or 6ths); and this would in turn give way (perhaps after some dramatic interruption) to a double cabaletta, with both voices singing the same music successively. There were naturally many variations on the forms of the aria and duet.

The central finale evolved out of the ensemble of perplexity in opera buffa, when the comic confusions of the plot would be at their height. Early 19th-cent. Italian opera tended to project the interest towards the end of the act, creating a multi-movement scene in which an orchestral melody might be taken through a sequence of keys and there be a Largo Concertato with complex part-writing before a final brilliant stretto.

So widely adopted was the Code Rossini in Italy that it effectively replaced the ubiquitous form of Metastasian *opera seria current in the 18th cent., back to whose exit aria convention its main features may ultimately be traced. Many attempts were made to define it, the most exhaustive and cogent of them being Carlo Ritorni's *Ammaestramenti alla composizione di ogni poema e d'ogni opera appartenente alla musica* (Milan, 1841). See also CONVENIENZE.

**Coffey, Charles** (*b* Ireland, late 17th cent.; *d* London, 13 May 1745). Irish playwright. Of his ballad operas, the most important was *The Devil to Pay* (music by Seedo, 1731). Initially a failure, it became when shortened from 42 to 16 airs a triumph, the most popular ballad opera after *The Beggar's Opera*. It was taken to Germany by the so-called English Players, or Englische Komödianten, and in a translation by Borcke, *Der Teufel ist los*, exerted a major influence on the development of *Singspiel. It was also the basis of a ballet by Adam, *Le diable à quatre* (1845).

**Colasse, Pascal** (*b* Reims, *bapt.* 22 Jan. 1649; *d* Versailles, 17 July 1709). French composer. 1677 at Paris, O, as assistant to Lully. Joined royal chapel 1683, rising to 'maître de la musique de la chambre' 1696. Wrote several works for the Paris stage: *Thétis et Pélée*, the most successful, is notable for its advanced use of the orchestra.

SELECT WORKLIST: *Thétis et Pélée* (Fontenelle; Paris 1689).

**Colbran, Isabella** (*b* Madrid, 2 Feb. 1785; *d* Bologna, 7 Oct. 1845). Spanish soprano. Studied Madrid with Pareja, Naples with Marinelli and Crescentini. Début Spain 1806. Milan, S, 1808–9. Naples 1811–22, engaged by Domenico Barbaia, who became her lover. She was also admired by the King of Naples. In 1815 she began a relationship with Rossini that led to their marriage in 1822. He wrote several parts for her: Elisabetta, Desdemona, Armida, Semiramide, and the leading soprano roles in *La donna del lago*, *Mosè*, and *Maometto II*. She possessed a fine dramatic coloratura soprano with a range of nearly three octaves; her physique was imposing and her dramatic powers considerable, though Stendhal missed the element of pathos. One of the most famous singers in Europe; however, her reputation outlasted her voice, which began to weaken in 1815. She retired in 1824 after an unsuccessful *Zelmira* in London.

BIBL: Stendhal, *La vie de Rossini* (Paris, 1824).

**Coletti, Filippo** (*b* Anagni, 11 May 1811; *d* Anagni, 13 June 1894). Italian baritone. Studied Naples with Busti. Début Naples, T del Fondo, 1834 (*Il turco in Italia*). During the next decade established himself as one of the leading exponents of Bellini and Donizetti. Lisbon 1836–9; London, HM, 1840. From 1844, when he sang Carlo in *Ernani* in Venice, F, became a prominent Verdi baritone, creating Gusmano (*Alzira*), and Francesco (*I masnadieri*) opposite Jenny Lind at London, HM, 1847. Rome 1851, the first Rigoletto there, when the opera was given as *Viscardello*. Also a distinguished Luna, Germont, and Boccanegra. Much admired by Verdi (who would have given him the title-role in *Re Lear*, had he written it). Also created roles in *Caterina Cornaro*, and Pacini's *Fidanzata corsa*. Possessed a sonorous, mellow voice capable of ringing high notes as well as subtle shading. Considered 'almost a genius' by Carlyle.

**Colin.** A village boy (ten), lover of Colette, in Rousseau's *Le devin du village*.

**Colini, Filippo** (*b* Rome, 21 Oct. 1811; *d* ? May 1863). Italian baritone. Studied Rome with Angiolini. Début Fabriano 1835, in *Aureliano in Palmira*. After appearances in Italy and Vienna, sang with great success at Parma 1843, in *Nabucco*. Verdi wrote the leading baritone roles in *Giovanna d'Arco*, *La battaglia di Legnano*, and *Stiffelio* for him. A light rather than a true Verdi baritone, he was considered rather old-fashioned in his singing but had a fine legato and a beautiful tone.

**Coliseum Theatre.** Built by Sir Oswald Stoll in 1904; London's largest theatre (cap. 2,354) with the largest stage. Once a music-hall, with appearances by opera singers and ballet dancers an important feature. In June 1913 a series of living tableaux of *Parsifal* was presented with Wagner's music arranged and conducted by Henry Wood. SW Opera first appeared there in April 1959 with *Die Fledermaus*.

In Aug. 1968 the *Sadler's Wells company moved to the theatre as its permanent home. Charles Mackerras became music director 1970. Lord Harewood was director 1972–84. In 1974 the company was renamed English National Opera. It has continued the enterprising policies of its predecessor on the larger scale which the size of the theatre and its stage invites. Among its early achievements were *The Ring* under Goodall, in a new translation by Andrew Porter, and other important productions included the British premières of Prokofiev's *War and Peace* (1972), Penderecki's *The Devils of Loudun* (1973), and Henze's *The Bassarids* (1974). Among the singers who helped to give the company its distinctive character were Josephine Barstow, Rita Hunter, Valerie Masterson, Norman Bailey, Derek Hammond-Stroud, Eric Shilling, and Alberto Remedios. Charles Groves was music director 1977–9, succeeded by Mark Elder (until 1993). Notable events during these years were the premières of

Hamilton's *Royal Hunt of the Sun* (1977) and *Anna Karenina* (1981), and Blake's *Toussaint* (1977), also the British premières of Martinů's *Julietta* (1978), and Jonathan Miller's production of *Figaro* (1978) and Goodall's *Tristan* (1981).

With the appointment of David Pountney to succeed Colin Graham as director of productions in 1982 (until 1993), and Peter Jonas to succeed Lord Harewood in 1985 (until 1993), emphasis changed towards the producer's dominant role, with results varying from the illuminating, even dazzling, to the self-indulgent and ludicrous, but ones greatly to the interest of an ever-widening operatic audience. Pountney himself produced *Rusalka* and *The Gambler* (1983), *The Midsummer Marriage* (1985), *Dr Faust* and *Carmen* (1986), *Lady Macbeth of Mtsensk* and *Hansel and Gretel* (1987), *Christmas Eve* (1988), and *Falstaff* (1989). Other producers included Jonathan Miller with *Rigoletto* (1982), *Don Giovanni* (1985), and *Tosca* and *Barber* (1987); and Nicholas Hytner with *Rienzi* (1983), *Xerxes* (1985), and *The Magic Flute* (1988). Other notable events of these years have included Goodall's performances of *Parsifal* (1986), the premières of Birtwistle's *The Mask of Orpheus* (1986) and Holloway's *Clarissa* (1990), and the British première of Reimann's *Lear* (1989). The theatre was bought by the company in 1992.

**Collatinus.** A Roman (bs), husband of Lucretia, in Britten's *The Rape of Lucretia*.

**colla voce** (It.: 'with the voice'). An instruction to keep the accompaniment synchronized with the vocal part, usually given at passages where the singer would be expected to use rubato.

**Collier, Marie** (*b* Ballarat, 16 Apr. 1927; *d* London, 8 Dec. 1971). Australian soprano. Studied Melbourne with Mme Wielaert. Début Melbourne 1954 (Santuzza). London, CG, 1956–71 (Marie (*Wozzeck*), Lisa, Chrysothemis, and created Hecuba in *King Priam*). London, SW (Emilia Marty, Minnie, etc.), San Francisco 1965; New York, M, 1967; created Christine in *Mourning becomes Electra*. A moving actress, with a highly charged voice and personality. (R)

**Colline.** A philosopher (bs), one of the four Bohemians in Puccini's *La bohème*.

**Collingwood, Lawrance** (*b* London, 14 Mar. 1887; *d* Killin, 19 Dec. 1982). English conductor and composer. Studied Oxford, then St Petersburg with Tcherepnin and Steinberg. Assistant to Coates, St Petersburg, M. London, Old Vic, 1920; chief cond. London, SW, 1931 (incl. first original *Boris* outside Russia, 1935), music director 1941–7. His operas are *Macbeth* (1937) and *The Death of Tintagiles* (concert, 1950).

**Colmar.** City in Haut-Rhin, France. The T Municipal (cap. 700) opened 1849. Opera was mostly given 1870–1972 by visiting companies from Strasburg and Mulhouse. Since 1972 it has formed part of the *Opéra du Rhin.

**Cologne** (Ger., Köln). City in North-Rhine Westphalia, Germany. The cathedral organist Caspar Grieffgens wrote a stage musical drama *Vanitas vanitatum* (1661, lost). English and French companies visited in the 18th cent. The first theatre opened 1822, with Sobald Ringelhardt as director until 1832. The company included Lortzing as singer and actor. Under Ferdinand Hiller the first local Wagner performances were given in 1853 (*Tannhäuser*) and 1855 (*Lohengrin*); Wagner himself conducted a subscription concert for Bayreuth in the Gürzenich hall 1873. The Theater in der Glockengasse opened 1872; to this was added the Theater am Habsburger Ring specially for opera in 1902. It had its greatest years under Klemperer 1917–24; these included shared première of Korngold's *Die tote Stadt* (1920) and the first German *Káťa Kabanová* (1922). The theatre was bombed in 1943; the Grosses Haus (cap. 1,346) opened 1957 with *Oberon* (there is a smaller theatre, cap. 920). Wolfgang Sawallisch was music director 1960–3, István Kertész 1964–73, John Pritchard 1978–89, James Conlon from 1991. Premières have included Fortner's *Bluthochzeit* (1957) and Zimmermann's *Die Soldaten* (1965).

**Colón, Teatro.** The leading opera-house of Argentina's capital, Buenos Aires, and of South America. For many years it gave the old-fashioned type of international opera season (May to Sep.) with French, German, and Italian works sung in their original languages by leading foreign artists. The first Colón was opened in Apr. 1857 (cap. 2,500) with *La traviata* (with Lorini and Tamberlik). Opera was performed there regularly until 1888 by Italian and French ensembles which included Bellincioni, Elena Boronat, Fricci, Borghi-Mamo, Battistini, De Lucia, Gayarre, Kaschmann, Stagno, and Tamagno. In 1887 the old Colón was sold to the National Bank for 950,000 pesos to be used for the building of a new opera-house. During the 20 difficult and controversial years before the present theatre was built, opera was given at the T de la Opera. The present building (cap. 2,478, with 1,000 standing) opened on 25 May 1908 with *Aida*. The first season, under Luigi Mancinelli, included Héctor Panizza's *Aurora*, commissioned by the government, and Shalyapin in *Mefistofele*. Cesar Ciacchi was the manager until 1914, and under his direction Serafin came, and Toscanini returned to Buenos Aires, where he had previously conducted some seasons at the T de la Opera, 1901–6. Toscanini's 1912 season sharply raised standards, and included the first performances in S America of *Ariane et Barbe-bleue*, *Germania*, and *Königskinder* by a company that included Bori, Matzenauer, Amato, De Luca, and De Angelis. Saint-Saëns conducted his *Samson et Dalila* in 1916; other composers to conduct included Strauss, Respighi, Stravinsky, and Falla.

From 1915 to 1925 the seasons were directed by Walter Mocchi; as well as bringing Italian and French artists to the Colón, he engaged Weingartner and a German ensemble including Lotte Lehmann, Wildbrunn, and Schipper for *The Ring* and *Parsifal* in 1922. From 1926 to 1930 the Colón became a municipal theatre with a general manager and artistic director. The 1931 season included several operas conducted by Klemperer (his last complete *Ring* anywhere) and Ansermet. Busch conducted regularly 1933–6 and 1941–5; Erich Kleiber, 1937–41 and 1946–9; and Albert Wolff, 1938–46.

After 1949 financial difficulties during the Perón régime resulted in a lowering of artistic standards; but after the 1955 revolution the theatre regained its former traditions. In 1958, the theatre's 50th anniversary, Beecham conducted a season including *Fidelio*, *Otello*, and *Carmen*. In 1967 six complete *Ring* cycles were given, conducted by Leitner, who had been a regular visitor since 1960, with Nilsson, Windgassen, and David Ward. The change of government in 1973 resulted in a complete upheaval of the Colón's administration and artistic policy, and standards began improving again towards the end of the decade. *Peter Grimes* was given in 1979. A good general repertory is given under the present artistic director, Carlos Montero. The theatre closed for modernization in 1987, reopening with *Aida* in 1989.

See also BUENOS AIRES.

BIBL: R. Casamano, *La historia del Teatro Colón* (Buenos Aires, 1969).

**Colonna.** Stefano Colonna (bs), head of the Colonna family, in Wagner's *Rienzi*.

**coloratura** (derived from Ger., *Koloratur*; only by adoption an Italian musical term. Coloratura of itself means elaborate ornamentation of melody; thus a coloratura soprano is one who specializes in this type of music, formerly known as *canto figurato*, distinguished by light, quick, agile runs and leaps and sparkling *fioriture* (which is the correct Italian word for flourishes or ornaments). An aria which featured extensive use of such writing was known as an **aria di coloratura** (Ger., *Koloraturarie*), or an **aria di bravura**: classic examples are the Queen of the Night's two numbers in *Zauberflöte* and Constanze's 'Martern aller Arten' in *Seraglio*.

**Coltellini, Marco** (*b* Livorno, 13 Oct. 1719; *d* St Petersburg, Nov. 1777). Italian poet and librettist. He was invited to Vienna, probably by Calzabigi, where he eventually succeeded Metastasio as court poet; but an indiscreet satire directed against Maria Theresa in 1772 led to his leaving Vienna for St Petersburg, where he became official librettist at the Imperial Theatres. He wrote the librettos for Mozart's *La finta semplice*, Haydn's *L'infedeltà delusa*, Gluck's *Telemaco*, and for operas by Traetta, Galuppi, Salieri, Sarti, and Paisiello. His daughter **Celeste** (*b* Livorno, 26 Nov. 1760; *d* Capodimonte, 24 July 1829) was a mezzo-soprano. Studied Florence with Manzuoli, Vienna with Mancini. Début Milan, S, 1780. Naples; engaged by Joseph II for Vienna in 1785. Highly regarded as an interpreter of Cimarosa, Paisiello (who wrote *Nina* for her), and Mozart, who wrote parts in his inserted arias for Bianchi's *La villanella rapita* for her. A natural actress with a pure, even voice, she excelled in the expression of sentiment. Lablache was impressed, even when she was old, by 'the perfection of her style'. Her sister **Anna** sang with success in Naples, 1782–94.

**Combattimento di Tancredi e Clorinda, Il.** Dramatic cantata by Monteverdi; text from Tasso's *Gerusalemme liberata*, xii. 52–68 (1575). Perf. Venice, Palazzo Mocenigo, Carnival 1624.

Tancredi, a Christian knight, has fallen in love with Clorinda, a Saracen maiden. He pursues a Saracen in armour, who has burnt a Christian castle, and in single combat defeats his opponent. But when he agrees to give his fallen adversary Christian baptism, 'he' reveals 'his' identity as Clorinda.

Though not strictly an opera, the work is written *in genere rappresentativo* and was intended for performance by costumed actors and a singing narrator (Il Testo) during a musical evening after madrigals had been sung 'senza gesto' (without action); a dramatic convention is thus implied. Though this did not prove fruitful for opera, in modern times there has been new interest in the musical and dramatic ideas which it embodies, and composers with a particular interest in *music theatre have turned to it for an example.

**comédie-ballet** (Fr.: 'comedy ballet'). An entertainment popular in France in the late 17th cent., in which ballet interludes were inserted into a spoken comedy. The architect of the genre was Molière, whose first such work was *Les fâcheux* (1661). With *Le mariage forcé* (1664) he began a distinguished collaboration with Lully, which lasted until *Le bourgeois gentilhomme* (1670) when, following their separation, Molière joined

with Charpentier for *Le malade imaginaire* (1673). After his death in that year, and the rise of the *tragédie en musique, the comédie-ballet lost its hold on the French public.

Molière's comedies represented the most advanced integration of music and drama on the French stage before the tragédie en musique. Their musical *entrées contained the first examples of many devices and styles, notably recitative, to be found in later French opera, while their elaborate staging and costumes looked back to the tradition of the *ballet de cour. Although the entrées were related to the main subject of the comedy, they provided an important means of introducing secondary plots, or stylized exoticism. At their most advanced, the musical elements of the comédie-ballet overwhelmed the dramatic: indeed, a famous contemporary description of *Le bourgeois gentilhomme* characterized it as a 'ballet composed of six entrées accompanied by a comedy'. The *tragédie-ballet was a related genre.

**comédie mêlée d'ariettes** (Fr.: 'comedy mixed with ariettes'). A type of French *opéra-comique, which grew out of the earlier *drame forain. It rose to popularity after the *Guerre des Bouffons and was in essence a light romantic comedy into which songs were inserted. While the text sometimes included political or social comment, the music was invariably direct, appealing, and not without a certain Italian influence. Examples include Gluck's *Le rencontre imprévue*, Philidor's *Tom Jones*, Monsigny's *Le déserteur*, and Grétry's *Zémire et Azor*.

**Comedy on a Bridge** (Cz.: *Veselohra na mostě*). Radio opera in 1 act by Martinů, text by the composer, after Kličpera. First given on Radio Prague, 18 Mar. 1937; prod. New York 28 May 1951; London, Morley College, 13 July 1977.

The scene of the opera is a bridge between two feuding villages, on which the characters become trapped. The plot recounts the many interactions which take place ranging from a schoolmaster attempting to solve a riddle to intimations of infidelity.

**'Com'è gentil'.** Ernesto's (ten) serenade in Act III of Donizetti's *Don Pasquale*. Said to have been provided by Donizetti at a late stage in the preparation of the opera for Mario to sing; he did so, with Lablache playing the lute in the wings.

**Comelli, Adelaide.** See RUBINI, GIOVANNI BATTISTA.

**'Come scoglio'.** Fiordiligi's (sop) aria in Act I of Mozart's *Così fan tutte* declaring that she will remain as firm as a rock against all temptations by her would-be seducer. In the manner of, perhaps even a parody of, an opera seria aria.

**'Come un bel dì di maggio'**. Chénier's (ten) aria in Act IV of Giordano's *Andrea Chenier*, reflecting as he awaits execution.

**comic opera**. A general name for an operatic work in which the prevailing mood is one of comedy. Under this broad definition come works from the traditions of *opera buffa, *opéra-comique, *operetta, *musical comedy, *ballad opera, and *Singspiel.

**'Comme autrefois'**. Leïla's (sop) cavatina in Act II of Bizet's *Les pêcheurs de perles*, in which she sings of the love that fills her heart.

**commedia dell'arte**. A dramatic genre of unknown origin that flourished in Italy from the 16th cent. and had a strong influence on drama and hence on opera. It consisted of improvised comedy, following a scenario rather than written dialogue, and made use of stock characters.

The most important were the *zanni*, whose *lazzi* (slapstick routines) included acrobatics and gestures as well as dialogue: often comic servants, either dull-witted or shrewd, but always ready to introduce foolery into any situation, the *zanni* were the essence of the commedia. They included the foolish, agile Arlecchino (Harlequin), the cunning, agile Brighella, the cunning, cowardly Scapino, and among others Pedrolino, Scaramuccia, Pulcinella, Mezzetino, Coviello, and Burattino. There might also be an elderly parent or guardian, Pantalone, the *magnifico*, with his pedantic, gullible crony Gratiano, the *dottore*. The swaggering Spanish captain was one of many deriving from the comedies of Plautus, in this case the *miles gloriosus*. The love-lorn *innamorata* was often named Isabella, Flaminia, Silvia, Valeria, or Olivetta; her servant or confidante Colombina, Fioretta, Violetta, or Smeraldina. The hero might be Fortunio or Fulvio, among other names. Certain of these characters were traditionally associated with various Italian cities: Pantalone was a Venetian merchant; the doctor's accent showed that he came from the great medical school at Bologna; Arlecchino and Brighella were from Bergamo, Scaramuccia and Pulcinella from Naples. Most of the characters would be masked, except for the lovers, and all would be identifiable by dress, e.g. Arlecchino's diamond-patterned motley, the doctor's black cape and hat, Pulcinella's slack white costume with pom-poms down the front.

The plots of the commedia were many and various—the confusions of disguise, mistaken identities, muddles between twins, a lover pretending to be a servant so as to win his girl from the clutches of a rich old man, a couple of unknown origin who turn out to be brother and sister, and many more. The virtuosity of the Italian troupes won the genre immense popularity, and spread its fame across Europe. It influenced many playwrights, including Shakespeare, and in France particularly Molière and Beaumarchais. Hence the influence passed into the plots of many operas, especially opera buffa, whether directly or indirectly. The characters appear in Busoni's *Arlecchino*, Ethel Smyth's *Fantasio*, Strauss's *Ariadne auf Naxos*, Mascagni's *Le maschere*, and Cowie's *Commedia* (1982); and there is a celebrated clash of reality with illusion by means of the commedia in Leoncavallo's *Pagliacci*.

More crucially, the schematism of comic plots into types provided opera librettists and composers with a comic tradition that could be varied and refreshed. The plot of Rossini's *Barbiere di Siviglia* is an instance of a commedia plot given new vividness and wit; while more than traces survive into Mozart's *Figaro* without any diminution of the characters' humanity.

**commedia in musica** (sometimes also **commedia per musica**). Strictly, a comic opera with spoken dialogue, often in Neapolitan dialect, but sometimes used more loosely as a synonym for *opera buffa.

**community opera**. A form intended to involve amateurs in the preparation and performance of an opera or musical drama. Though the idea has roots in various historical, social, and educational movements in different countries, it emerged more distinctively as a medium in Great Britain in the 1980s, when a number of opera companies began to take workshops and other projects to schools and other communities. Some of these ventures were participatory, with professional guidance, such as Opera North's Community Projects Department. This takes opera to schools but also sponsors community opera projects. Other groups with comparable aims developed, in general finding means of involving children in the preparation of an already existing opera and introducing them to the workings of an opera company, e.g. Opera North's *Girl of the Golden West*. Opera North has also sponsored opera projects involving those with no operatic experience, e.g. *Harry's Comet* (Barnsley 1986) and *Blues for Chrissie* (Huddersfield 1987).

Scottish Opera established an Education Unit in 1971, becoming full-time in 1984, and also works with children participating under minimal professional guidance in the 'Opera for Youth' group. A new work is produced each year, beginning in 1985 with *Rhyme and Reason* (Karen McIver). In some cases workpacks are sent to schools in advance and there is full participation by the children, normally during one working day and ending with a performance. In the Rock Opera Performance Project (1985) the work is developed over a longer period by means of discussion, improvisation, and rehearsal, involving all with any talent as contributors. The latter's works include *I.N.S.A.N.E.* (1986), *Cry Wolf* (1986), and *Dead of Night*, taken to Wexford in 1987.

The Welsh National Opera Company's Schools Projects have also mounted community opera in Ely (1984) and Splott (1986), and their Community and Education Department includes a 'Gentle Giant' project for Special Educational Needs Schools, involving several visits before a performance is achieved. *Poor Twm Jones* (Ely 1984) included over 400 people aged 7–75, working over ten weeks in schools, youth clubs, social clubs, and church groups; this led to the formation of a longer-lasting orchestra and choir. It led in turn to an even larger project in Splott, Tremorfa, Adamsdown, and Roath (STAR), in which about 1,000 people took part in *Song of the Streets*, working over a period of 13 months during 1985–6.

*Salomons' Dream*, by Stephen McNeff, opened David Salomons's house, near Tunbridge Wells, to opera in the summer of 1991.

**Como.** Town in Lombardy, Italy. The T Sociale opened Aug. 1813 with Portogallo's *Adriano in Siria*. Stendhal, who attended, wrote that one must go to Como to enjoy music. *Garibaldi a Como*, by the tenor Filiberti who sang the title-role, was prod. 1860. After La Scala was destroyed in 1943, the company played here.

**comparsa** (It.: 'extra actor'). Extra, walker-on: in operatic usage, it also signifies that the actor does not sing.

**componimento da camera.** See AZIONE TEATRALE.

**componimento drammatico.** See AZIONE TEATRALE.

**comprimario** (It.: 'with the principal'). Subprincipal singer. Used mostly in relation to 19th-cent. Italian opera: although he could sing a romance, or participate in an ensemble, he was not allowed an individual aria. See CONVENIENZE.

**Comte Ory, Le** (Count Ory). Opera buffa in 2 acts by Rossini; text by Scribe and Delestre-Poirson, an expansion of a 1-act comedy (1817), after an alleged old Picard legend set down by Pierre-Antoine de la Place (1785). Prod. Paris, O, 20 Aug. 1828, with Cinti-Damoreau, Nourrit, Dabadie, N. Levasseur, cond. Habeneck; in Italian, London, H, 28 Feb. 1829, prob. for Rossini's 37th birthday, with Montecillé, Castelli, Currioni, De Angelis, Galli, cond. Bochsa; New Orleans, T d'Orléans, 16 Dec. 1830, with St Clair, Paradol, Deschamps, Privat.

Fourmoutiers Castle, Touraine, 1250. The Countess Adèle (sop) is melancholy, her brother

and the menfolk of Fourmoutiers having left for the Crusades. Ragonde (con), the Countess's companion, decides to consult the hermit: he is in reality the licentious Comte Ory (ten), who has encamped at the castle gates and offers interested advice on amorous matters. He is aided in his exploits by his friend Raimbaud (bar). Ory's tutor (bs) and his page Isolier (mez), who is in love with the Countess, arrive in search of Ory. Isolier's plan to enter the castle, disguised as a nun, so appeals to the Count that he and his followers disguise themselves in nuns' habits and lay siege both to the ladies in the castle and to the wine-cellars. The return of the crusaders, just as matters are getting out of hand, results in Ory's and Isolier's defeat in their pursuit of the Countess.

The first of Rossini's two French operas and his penultimate stage work; it contains much of the music of *Il viaggio a Reims* (including the overture) with 12 additional numbers.

Also opera by Miltitz (*c.*1830).

**Concepción.** The wife (sop) of the clockmaker Torquemada in Ravel's *L'heure espagnole*.

**concertato** (It.: 'concerted', esp. in the sense of providing contrast). In the 17th cent. the *coro concertato* was a small body of singers, in contrast to the full chorus, or *coro ripieno*: from such juxtapositions of sonority the concerto eventually developed at the end of the century. In 18th-cent. opera a concertato group (i.e. a small group of solo instrumentalists) occasionally takes a prominent part in the orchestral accompaniment. Two particularly fine examples in Mozart are 'Se il padre perdei' from *Idomeneo* and 'Martern aller Arten' from *Seraglio*.

**Conchita.** Opera in 4 acts (6 scenes) by Zandonai; text by Maurizio Vaucaire and Carlo Zangarini, based on Pierre Louÿs's *La femme et le pantin*. Prod. Milan, V, 14 Oct. 1911, with Tarquini, Schiarazzi, Zinolfi, Lucca, cond. Panizza; London, CG, 3 July 1912; Chicago 2 Nov. 1912.

Seville, early 20th cent. The opera tells of the attempts of Mateo (ten), an elderly rake, to convince Conchita (sop), a poor cigar-maker and later flamenco-dancer, of the sincerity of his love for her. Eventually he succeeds, and she returns his affection.

**conductor.** In the 17th and 18th cents. opera performances were usually co-ordinated from the keyboard, although as the orchestra grew in size, it received separate direction from the principal violinist (leader). This practice was common throughout Italy and Germany, but in France, where the presence of the chorus created additional problems, time was beaten audibly, at first on stage, using a stick which was hit against a table: later the director stood near the orchestra, beating time on the floor with a large staff. According to Le Cerf, it was while performing this function that Lully hit his foot, causing it to develop the gangrene which led to his death. As the role of the continuo declined, the principal violinist took increasing responsibility. This too could involve using audible methods of keeping the ensemble together: Mendelssohn reported from Naples in 1831 that the violinist beat time throughout on a tin lampstand, producing what he called, 'something like obbligato castanets'. Gradually, as less and less of the **primo violino direttore**'s time was spent playing, and more directing the ensemble, the role of the modern conductor emerged, a musical director whose sole function was to co-ordinate the singers and chorus on stage with the orchestra. The first generation of conductors included Spohr, Mendelssohn, and above all Weber; these established the guiding principles of the new craft, especially the use of the baton (held centrally; a roll of music was also used). The conductor came to be regarded as the linchpin of the musical performance, although frequently he would be involved with the dramatic production as well. Following Weber's example, the conductor would often direct the singers from close to the footlights, with the orchestra beside and behind him; this practice at Dresden was reluctantly abandoned by *Schuch who moved the rostrum back to the familiar modern position.

**Conforto, Nicola** (*b* Naples, 25 Sep. 1718; *d* ?Madrid, after 1788). Italian composer. He came to notice as an operatic composer in Naples with *La finta vedova* and with *Antigono*. In 1755 he settled permanently in Spain and was instrumental in establishing the taste there for Italian opera. Most of his operas employ librettos by Metastasio: they include the first attempt at *Nitteti*, as well as settings of *L'eroe cinese* and *Alcide al bivio*.

SELECT WORKLIST: *La finta vedova* (The Simple Widow) (Trinchera; Naples 1746); *Antigono* (Metastasio; Naples 1750); *Nitteti* (Metastasio; Madrid 1756); *L'eroe cinese* (Metastasio; Madrid 1754); *Alcide al bivio* (Metastasio; Madrid 1765).

**congiura** (It.: 'plot, conspiracy'). The name sometimes given to scenes of conspiracy in Italian opera. Examples are 'Ad augusta!' in pt3 of Verdi's *Ernani* and 'Dunque l'onta' in Act III of his *Un ballo in maschera*.

**'Connais-tu le pays?'** Mignon's (mez) aria in Act I of Thomas's *Mignon*, the famous 'Kennst du das Land' of Goethe's *Wilhelm Meisters Lehrjahre*, trying to answer Wilhelm's questions about her origins.

**Connell, Elizabeth** (*b* Port Elizabeth, 22 Oct. 1946). Irish mezzo-soprano, then soprano. Studied London with O. Kraus. Début Wexford 1972 (Varvara). Sydney 1972–4; London, ENO, from 1975; London, CG, from 1976; Bayreuth 1980; Milan, S, 1982; New York, M, 1983; Munich, S, 1985. Roles incl. Eglantine, Ortrud, Kundry, Elisabeth, Leonore, Lady Macbeth, Vitellia, Norma. (R)

**Conradi, Johann Georg** (*b* ?; *d* Oettingen, 22 May 1699). German composer. Held court appointments in Ansbach and Römhild; went to Hamburg *c*.1690. Here he was effectively resident composer at the opera-house and dominated its repertoire during the years 1690–3. Of the nine operas he wrote there, a few numbers from *Die Verstörung Jerusalems* and *Der wunderbar vergnügte Pygmalion* survive, as well as the complete score to *Die schöne und getreue Ariadne*. This is the earliest surviving material from Hamburg's repertory and shows Conradi to be a composer of some considerable dramatic and musical abilities, much influenced by the prevailing styles of both Italian and French opera. He was highly regarded by his contemporaries and successors. *Ariadne* was revived at Hamburg in 1722 in a new version by Keiser; a similar adaptation of *Gensericus* by Telemann was given there in the same year.

SELECT WORKLIST (all texts by Postel and first prod. Hamburg): *Die schöne und getreue Ariadne* (The Beautiful and Faithful Ariadne) (1691); *Die Verstörung Jerusalems* (The Destruction of Jerusalem) (1692); *Gensericus* (1693); *Die wunderbare vergnügte Pygmalion* (The Wonderfully Happy Pygmalion) (1694).

**Conried, Heinrich** (*b* Bielitz, 13 Sep. 1848; *d* Meran, 27 Apr. 1909). German operatic impresario and manager. Began career as an actor in Vienna; after holding managerial positions in a number of German cities went to USA 1878. Managed various theatrical enterprises, then succeeded Maurice *Grau as director of New York, M, 1903–8. He engaged numerous celebrated artists, including Caruso; however, his controversial acts included the staging of the first US *Parsifal* 1903, before the expiry of the Bayreuth ban, and the first US *Salome* 1907. He resigned after dissensions with the Met. management.

BIBL: M. Moses, *Heinrich Conried* (New York, 1916).

**Constantinople.** See ISTANBUL.

**Consul, The.** Opera in 3 acts by Menotti; text by the composer. Prod. Philadelphia 1 Mar. 1950, with Neway, Powers, Lane, Marlo, C. MacNeil, Lishner, McKinley, cond. Engel; London, Cambridge T, 7 Feb. 1951, with similar cast, cond. Schippers.

The action takes place in a nameless police

state in Europe and depicts the efforts of Magda Sorel (sop) to obtain a visa for herself and her husband John (bar), a revolutionary who is being pursued by the secret police. The Consul's Secretary (mez) frustrates all attempts to see the Consul by giving out innumerable forms. John is eventually arrested and Magda commits suicide by gassing herself; as she dies, she sees in a dream her husband and all the people she has met at the Consulate who urge her to join them on the journey to Death, whose frontiers are never barred.

**Contes d'Hoffmann, Les** (The Tales of Hoffmann). Opera in 3 acts with prologue and epilogue by Offenbach; text by Barbier and Carré, founded on the stories *Der Sandmann, Geschichte vom verlorenen Spiegelbilde*, and *Rat Krespel* by E. T. A. Hoffmann. Revision, recitatives, and part of orchestration by Guiraud. Prod. Paris, OC, 10 Feb. 1881, with Isaac, Ugalde, Talazac, Taskin; New York, 5th Avenue T, 16 Oct. 1882, with Dérivis, Betty, Maire, Mangé, Ducos; London, Adelphi T, 17 Apr. 1907, with Franzillo-Kauffmann, Nadolovitch, Hofbauer, cond. Cassirer.

Prologue. In Luther's wine-cellar in Nuremberg, a letter from the prima donna Stella to Hoffmann (ten), arranging a rendezvous, is intercepted by Councillor Lindorf (bar). Hoffmann arrives and is persuaded to tell the story of his three loves; each act of the opera is one of his tales.

His first love is Olympia (sop), a doll who seems human to Hoffmann since he views her through spectacles sold to him by one of her makers, Coppelius (bs-bar). Her other maker, Spalanzani (bar), has swindled Coppelius, who then smashes Olympia.

The next is Antonia (sop), a consumptive young singer. Dr Miracle (bs-bar) forces her to sing, thus causing her death.

Hoffmann's third love is the Venetian courtesan Giulietta (mez). She is in the power of the magician Dapertutto (bs-bar), who wishes to procure Hoffmann's shadow (i.e. his soul). Giulietta attracts Hoffmann, but abandons him as soon as Dapertutto's requirements have been fulfilled.

Epilogue. Back in Luther's tavern, the 'tales' are over. Hoffmann is drunk, and when Stella arrives it is Lindorf (who has also assumed the roles of Coppelius, Dr Miracle, and Dapertutto) who leads her away.

Offenbach died during rehearsals of *Hoffmann*, and it was re-arranged (cutting the Giulietta act) by Guirand for the première. Twelve years later the missing act was restored, but before the Antonia act. Correctly, the opera is largely with spoken dialogue, and in the order given above.

**Conti, Francesco** (*b* Florence, 20 Jan. 1681; *d* Vienna, 20 July 1732). Italian composer. Active at Vienna from 1701 as a theorbo player, where had some success with *Clotilda*; appointed court composer 1713. He produced about 30 dramatic works of various kinds, of which the tragicomic opera *Don Chisciotte in Sierra Morena* was the most popular: most of his operas set librettos by Zeno or Pariati and are typical of early 18th-cent. opera seria. He showed a preference for heroic subjects, as in *Teseo in Creta* and *Issipile*, though occasionally he turned to a more pastoral libretto, as in *I sattiri in Arcadia*, a set of intermezzi and a *licenza for *Atenaidi* by Ziani, Caldara, and others. Some of his works appeared in German translation at the Hamburg Opera, including *Don Chisciotte* and *Issipile*, given under the title *Der Sieg der kindlichen Liebe*.

SELECT WORKLIST (all first prod. Vienna): *Clotilda* (Neri; 1706, lost); *I sattiri in Arcadia* (Pariati; 1714); *Don Chisciotte in Sierra Morena* (Zeno, after Pariati; 1719); *Teseo in Creta* (Pariati; 1715); *Issipile* (Metastasio; 1732).

**continuo.** Abbr. of *basso continuo* (It.: 'continuous bass'). An essential feature of early *stile rappresentativo was its orientation around a slow-moving bass line, which provided a firm foundation for the florid declamations of the solo voice. Eschewing the older polyphonic style, the early monodists stressed the need for a simple chordal accompaniment, and encouraged its performance on the lute. With the gradual separation into *aria and *recitative, the bass line developed also, adopting faster, often stereotyped, patterns for the aria, while the recitative sections retained their former generally slow motion. It became customary for these bass lines to be performed by the harpsichord, with the player (who usually directed the performance as well) realizing the harmonies, usually guided by figuration written down by the composer (**figured bass**). The harpsichord (later, sometimes a fortepiano) was normally accompanied by a bass string instrument, such as a viola da gamba or cello, or less often a bassoon: together these instruments, and their music, were known as the continuo. Heard alone in recitative, during other parts of the opera the continuo group was part of the orchestral texture. An invariable feature of opera in the 17th and 18th cents., the continuo group survived into the early 19th cent.: its presence is assumed in some of Rossini's first works.

**contralto** (It.: 'against the high voice', i.e. contrasting with the high voice). In Ger., *Alt* (the voice), *Altistin* (the singer). The lowest category of female (or artificial male) voice. Several subdivisions exist within opera-houses: the two commonest in general use (though seldom by composers in scores) are given below, with examples of roles and their approximate tessitura. In general, distinction is more by character than by tessitura: thus the examples of the roles give a more useful indication of the different voices' quality than any attempted technical definition.

German: dramatischer Alt (Erda: f–f''); komischer Alt (Frau Reich in *Die lustigen Weiber von Windsor*: f–f'').

Italian and French contralto roles are not generally further subdivided into categories, and have a similar tessitura to the German.

See also CASTRATO, MEZZO-SOPRANO.

**contratenor.** See COUNTER-TENOR.

**convenienze** (It.: 'conveniences'). The rules of etiquette governing the rank of singers in 19th-cent. Italian opera. Usually there were three principals, soprano, tenor, and baritone, known as prima donna, primo uomo, and basso cantante respectively. In earlier days, when the main male character was sung by a *castrato, the primo uomo was known as the primo musico; if there was a fourth principal, it was either a bass or a second soprano. According to the Code Rossini these singers introduced themselves in a *cavatina, and each had at least one *scena, concluding with the ubiquitous *cabaletta. The secondary characters, who acted as confidante to the heroine and hero, were known as the seconda donna and secondo uomo; they had no individual arias of their own, but were sometimes called upon to participate separately in an ensemble. Between the ranks of the primo and secondo parts came the comprimario; this intermediate character was allowed not only to sing in an ensemble, but could also perform a duet with one of the principals and even have a simple aria in one movement. With Verdi, the disposition of roles altered so that the characteristic high baritone he favoured might represent force and manly vigour, while the tenor tended to be more lyrical and ardent, with the bass dark and strong in purpose for good or ill. The soprano tended to be emotional, volatile, and strongly characterized, while the mezzo-soprano was increasingly a powerful figure, perhaps sinister, perhaps maternal (or both). See also CODE ROSSINI.

**Converse, Frederick** (*b* Newton, MA, 5 Jan. 1871; *d* Westwood, MA, 8 June 1940). US composer and teacher. Studied Harvard with Knowles Paine. After abandoning a career in commerce studied privately with Chadwick and in Munich with Rheinberger. Returning to the USA he taught at the New England Cons., Boston, and at Harvard. In 1906 *The Pipe of Desire* was given in Boston; in 1910 it became the first opera by an American composer to be staged at the New York Met. 1908–14 Converse was vice-president of the Boston OC: he wrote three further operas, two of which (*Sinbad the Sailor* (1913), *The Immigrants* (1914)) were unperformed.

SELECT WORKLIST: *The Pipe of Desire* (Barton; Boston 1906, rev. New York 1910); *The Sacrifice* (Converse and Macy; Boston 1911).

**Cooke, Tom** (*b* Dublin, 1782; *d* London, 26 Feb. 1848). Irish singer, composer, and instrumentalist. Sang in various stage works in Dublin and London; principal tenor, London, DL, 1815–35, director from *c*.1821. Produced his own *Oberon* in the same year as Weber's, in part rivalry and parody. Adapted many foreign operas, with substitutions of his own, including *Abu Hassan*, *La dame blanche*, *Le siège de Corinth*, and *La Juive*. A great wit, and a celebrated teacher.

**Cooper, Emil** (*b* Kherson, 13 Dec. 1877; *d* New York, 16 Nov. 1960). Russian violinist and conductor of English descent. Studied Odessa, Vienna with Hellmesberger and Fuchs, Moscow with Taneyev. Toured as violinist 1891–6; leader Kiev O 1896, début replacing regular conductor 1896 (*Fra Diavolo*). Kiev 1900–4; Moscow, Z (first *Golden Cockerel* and first Russian *Meistersinger*) and B. Russian seasons, Paris 1909–14. Petrograd 1919–24. Music director Riga 1925–8. Chief cond. Chicago 1929–32, 1939–44. Paris, O, and Milan, S, 1932–6. New York, M, 1944–50. His brother **Max Cooper** (1884–1959) conducted opera in Leningrad and Moscow.

**Copenhagen** (Dan., København). Capital of Denmark. Opera was given in the mid-17th cent., including Caspar Förster's *Cadmo* (1663). Poul Schindler's (1648–1740) *Der vereinigte Götterstreit* (1689, lost), the first opera by a Dane, was given at the Amalienborg T, which burnt down during the performance, killing 180 people including Schindler's wife and daughter. Foreign companies began visiting the city from the mid-18th cent.: Italian and French opera was given, and a German company under Keiser was at the Court T 1721–3, giving the première of his *Ulysses* (1722) for Frederick IV's birthday. In 1749 Prince Christian's birth was celebrated with the première of Gluck's *La contesa dei numi*. After a period of decline owing to opposition from the King on religious grounds, the opera was revived and Sarti made director of the theatre.

One of the first operas produced there was Scalabrini's *Den belønnede kjaerlighed* (Love Rewarded, translated from *L'amore premiato*, 1758). Many Italian operas were given, and a singing school founded. J. A. P. *Schulz was director 1787–95, Friedrich *Kunzen 1795–1817 (raising standards and introducing new works, especially foreign). The Kongelige T (Royal T) (cap. 1,300) opened in 1874. Johan Svendsen was director 1883–1908, introducing many new ideas. Carl *Nielsen, director 1908–14, had begun his career as second violin in the orchestra. He was succeeded by Georg Høeberg, who conducted the only complete *Ring* given in the theatre. He resigned in 1931 and was followed by Johan Hye-Knudsen, who shared the conducting with Egisto Tango (restoring the Swedish setting of *Un ballo in maschera*, introducing more modern works). During the German occupation, *Porgy and Bess* had its first production outside America (1943), but was then suppressed; there were also performances of other works banned in Nazi Germany. The Kongelige T reopened after restoration in 1986.

A second auditorium, the New Stage (cap. 1,091), opened in 1931. There is also an operetta, Nörrelzo (cap. 1,141), renamed the Scala in 1954. Opera is also given by a small but adventurous company at the Riddersalen.

BIBL: G. Leicht and M. Hallar, *Det kongelige Teaters repertoire 1889–1975* (Copenhagen, 1977).

**Copland, Aaron** (*b* New York, 14 Nov. 1900; *d* New York, 2 Dec. 1990). US composer. Studied with Wittgenstein and Rubin Goldmark, and with Nadia Boulanger in Paris (1921–4). In an attempt to encourage contemporary American music he promoted, with Roger Sessions, series of concerts in New York 1928–31. His first opera was *The Second Hurricane* (1937), a 'play opera' for high schools, which shows his early jazz-influenced style. His only other other, *The Tender Land* (1954), illustrates his efforts to create an overtly American tradition of opera; he chose a subject concerning ordinary American people (a lower-middle-class farming community in the Midwest during the Depression years), and incorporated folk melodies, as he did in his ballet *Appalachian Spring* (1944).

WORKLIST: *The Second Hurricane* (Denby; New York 1937); *The Tender Land* (Everett, after Johns; New York 1954).

BIBL: A. Copland and V. Perlis, *Copland: 1900 through 1942* (London, 1984); V. Perlis, *Copland from 1943* (London, ?).

**Copley, John** (*b* Birmingham, 12 June 1933). English producer. Studied London with Joan Cross. Appeared as the Apprentice, *Peter Grimes*, London, CG, 1950. Stage-manager, London, SW, producing *Tabarro* 1957. Resident producer CG, where he has been responsible for many revivals as well as original productions especially of Mozart; also London, C; Australian O; New York, M; Dallas; Santa Fe; and elsewhere. His productions are often highly inventive, but

always undogmatic and responsive to the music and to singers' capabilities.

**Coq d'Or, Le.** See GOLDEN COCKEREL, THE.

**Corder, Frederick** (*b* London, 26 Jan. 1852; *d* London, 21 Aug. 1932). English conductor, composer, translator, and teacher. Studied London, Cologne with Hiller, Milan. His most successful opera was *Nordisa* (Liverpool 1887, by CR); but with the death of Carl Rosa he was forced to abandon his hopes of taking part in a revival of English opera, and turned to teaching and translation. His pupils at the RAM included Bantock, Bax, and Holbrooke. Best remembered for the translations he made, with his wife Henrietta, of Wagner's *Parsifal* (1879), *Meistersinger* (1882), *Ring* (1882), *Tristan* (1882), and *Lohengrin* (1894). They were concerned to match Wagner's *Stabreim, which they did ingeniously but too often quaintly. These versions were performed at London, CG, and elsewhere.

**Corelli, Franco** (*b* Ancona, 8 Apr. 1921). Italian tenor. Studied Pesaro and Spoleto. Début Spoleto 1951 (José). London, CG, from 1957; Vienna, S, 1963; New York, M, from 1961. Milan, S, 1954–65. A personable, large-voiced heroic tenor, especially successful as Calaf, Manrico, and Dick Johnson. (R)

**Corena, Fernando** (*b* Geneva, 22 Dec. 1916; *d* Lugano, Nov. 1984). Swiss bass. Studied Milan with Romani. Zurich 1940–7; stage début Trieste 1947 (Varlaam). New York, M, from 1953, succeeding Baccaloni as leading buffo bass. London, CG, 1960, 1969; also Vienna, Berlin, Buenos Aires. An accomplished and witty buffo, memorable as Don Pasquale, Gianni Schicchi, and Falstaff. (R)

**Corfu** (Gr., Kerkyra). Capital of the Ionian island of that name, which was subject to Venice 1386–1797 and to Britain 1815–64. The first theatre, the Loggia (later T San Giacomo), opened 1691. Financed from Venice, this was the first permanent theatre in the Levant, and from 1733 until 1923 seasons were given by visiting Italian companies. Works by Spiridon Xyndas, Pavlos Karrer, and Spiridon Samaras were staged. The larger Demoticon T opened 1902, and here Lavrangas's *Dido* (1909) and other works were given; it was destroyed by shells in 1943. A festival begun in 1981 has staged opera in an open-air auditorium at the back of the Phoenix, with works including *La sonnambula* and *The Rape of Lucretia*.

**Corneille, Pierre** (*b* Rouen, 6 June 1606; *d* Paris, 1 Oct. 1684). French playwright. Operas on his works are as follows:

*Le Cid* (1637): Pollarolo (*Flavio Bertarido*, 1706);

Stuck (1715); Handel (*Flavio*, 1723); Leo (1727); Sacchini (1764); Piccinni (1766); Bianchi (1773); Paisiello (1775); Roesler (1780); Farinelli (1802); Aiblinger (*Rodrigo und Zimene*, 1821); Savi (1834); Schindelmeisser (*Der Rächer*, 1846); Pacini (1853); Neeb (1857); Gouvy (comp. 1863); Cornelius (1865); Massenet (1885); Boehme (1887); Wagenaar (1916). Debussy made sketches for *Rodrigo et Chimène*, later used in *Pelléas et Mélisande*

*Cinna* (1640): Portugal (1793); Asioli (1793); Paer (1795)

*Horace* (1640): Salieri (1786); Cimarosa (*Gli Orazi e i Curiazi*, 1796); Portugal (1798); Mercadante (*Orazi e Curiazi*, 1846)

*Polyeucte* (1642): Donizetti (1838); Gounod (1878)

*La mort de Pompé* (1643): Graun (*Cesare e Cleopatra*, 1742)

*Héraclius* (1647): Ziani (1671)

*Théodore* (1645): Papebrochio Fungoni (1738)

*Andromède* (1647): Mattiali (*Perseo*, 1664); Lully (*Persée*, 1682)

*Nicomède* (1651): Grossi (1677)

*La toison d'or* (1660): Draghi (1678)

*Sophonisbe* (1663): Caldara (1708); Traetta (1761)

*Attila* (1667): Ziani (1672)

*Psyché* (1671): Locke (1675); Champein (early 19th cent.)

**Cornelius, Peter** (*b* Mainz, 24 Dec. 1824; *d* Mainz, 26 Oct. 1874). German composer. After an attempt at an acting career, he studied with Esser and in Berlin with Dehn. In Weimar he became a close friend and supporter of Liszt and Berlioz. But the production of *Der Barbier von Bagdad* met the organized opposition to the New German School from local musicians that was the final cause of Liszt's resignation. In Vienna he entered Wagner's circle of close friends, remaining loyal but never overwhelmed. *Der Cid* shows comparable signs of respect for Wagner but independence in its attempt to write an epic on a grand scale. The unfinished *Gunlöd* draws rather closer to Wagner. However, it is for his concern, almost alone of the New German School, to give German comic opera greater maturity than he found in the works of, for instance, Lortzing, that Cornelius is important. In the entertaining *Der Barbier von Bagdad* he shows an intelligent appreciation of how the example of Berlioz might be used to this end, and there are some influences of *Benvenuto Cellini* (which he had translated for Berlioz); but it is a work of no little charm in its own right.

WORKLIST (all texts by Cornelius): *Der Barbier von Bagdad* (Weimar 1858); *Der Cid* (Weimar 1865); *Gunlöd* (comp. 1866–74, unfin.).

BIBL: C. Cornelius, *Peter Cornelius* (Regensburg, 1925).

**'Corno di Bassetto'.** See SHAW, BERNARD.

**corona.** See FERMATA.

**Corregidor, Der** (The Mayor). Opera in 4 acts by Wolf; text by Rosa Mayreder, after Alarcón's story *El sombrero de tres picos* (1874). Prod. Mannheim 7 June 1896, with Hohenleitner, Rüdiger, Krömer; London, RAM (in English), 13 July 1934, with Dawes, Rodd, Davies, Deri, Prangnell, cond. Barbirolli; New York, Carnegie H, 5 Jan. 1959 (concert), with Conner, Lipton, Kelley, Thompson, Gramm, cond. Scherman.

Andalusia, 1804. Frasquita, the miller's wife (sop), dismissing her husband Tio Lucas's (bar) jealousy as groundless, uses the advances of the elderly Corregidor (ten) to obtain a post for her nephew. When this is achieved the Corregidor arrives, soaked from having fallen into the mill stream; Frasquita defends herself with a musket. Tio Lucas has meanwhile been summoned to town by the Corregidor; he realizes that this is a false errand, but passes Frasquita in the dark on his way home as she comes to find him. When Lucas finds the Corregidor asleep in bed, he dresses in the old man's clothes; on waking, the Corregidor is beaten by the officers who have come from the town. When he reaches his own house, his wife Mercedes (sop) refuses him admission, pretending that she has mistaken Lucas for him. Lucas, taken for the murderer of the Corregidor, is also beaten; and the opera ends with everyone even.

Wolf's only opera. Also opera on the same subject by Zandonai (*La farsa amorosa*, 1933).

BIBL: P. Cook, *Hugo Wolf's 'Corregidor': A Study of the Opera and its Origins* (London, 1976).

**Corri.** Italian, later English, family of musicians.

1. **Domenico** (*b* Rome, 4 Oct. 1746; *d* London, 22 May 1825). Italian singing-teacher, composer, and publisher. Studied Naples with Porpora. His operas include *La raminga fedele* (1770), and *The Travellers, or Music's Fascination* (1806). He promoted performances in Edinburgh and London, where he established himself as a publisher in Soho in 1790.

2. **Sophia Giustina** (*b* Edinburgh, 1 May 1775; *d* after 1828). Anglo-Italian soprano. Daughter of above, with whom she studied. Married Jan Ladislav Dussek, and as Sophia Dussek appeared at London, H, 1808, in operas by Paisiello, Sarti, etc. Not to be confused with Josepha Dussek.

3. **Fanny Corri-Paltoni** (*b* Edinburgh, 1795 or 1801; *d* after 1833). Niece of (1). Anglo-Italian mezzo-soprano. Studied London with Braham and Catalani. Début London, H, 1818.

Remained until 1820. Pursued successful career in Germany, Italy, and Spain; roles included Donna Anna, Dorabella, Cenerentola. Had a fine voice, with a brilliant shake. Married the singer Paltoni.

Other members of the family include **Patrick Anthony** (1820–76), singer and conductor; **Henry** (1822–88), bass and conductor; **Haydn Jun.** (1845–76), baritone; **Ghita** (1869 or 1870–1937), soprano with London, CR; and **Charles** (1861–1941), conductor, from 1895 to 1935 chief conductor of the Old Vic and later Vic-Wells Opera.

**Corsaro, Il.** Opera in 3 acts by Verdi; text by Piave, after Byron's poem *The Corsair* (1813). Prod. Trieste, 25 Oct. 1848, with Barbieri-Nini, Rapazzini, Fraschini, De Bassini, Volpini; London, Camden T, 15 Mar. 1966, with Tinsley, Sinclair, Smith, Drake, cond. Head. New York, TH, 12 Dec. 1981, with Bergonzi, Reese, Val-Schmidt, Dietsch, cond. Lawton.

An Aegean island, early 19th cent. Corrado, the pirate chieftain (ten), plans an attack against the Pasha Seid (bar). He announces his departure to his beloved Medora (sop).

Gulnara (sop), the Pasha's favourite concubine, longs to be free from him. He summons her to a banquet to celebrate certain victory over the pirates; Corrado and the pirates launch an attack. This misfires and Corrado is captured.

Gulnara has fallen in love with Corrado, which enrages the Pasha. She bribes a prison guard to allow her access to Corrado and offers to help him escape if he will kill the Pasha. As his conscience will not permit this, she murders Seid herself and the couple flee. Back on the island, Medora is dying after taking poison and longs to see Corrado. He appears with Gulnara, who confesses that she too loves Corrado. Medora dies in Corrado's arms and he flings himself into the sea.

Also operas by Pacini (1831), Schumann (1844, unfin.), Deffei (1871), Bronsart (1875), Marracino (1900).

**Corselli, Francesco** (*b* Piacenza, c.1702; *d* Madrid, 3 Apr. 1778). Italian composer. Maestro di cappella to the Duke of Parma 1727–33. Joined Madrid court 1734; maestro di cappella 1738. Wrote two early operas for Venice and six works for Madrid. The latter, including *Alessandro nell'Indie* and *Achille in Sciro*, were among the most spectacular productions at the Spanish court during the 18th cent. and played an important part in the establishment of Italian opera there. Corselli was the most admired composer of Philip V, whose favourite castrato Farinelli performed in many of the premières.

SELECT WORKLIST: *Alessandro nell'Indie* (Val after Metastasio; Madrid 1738); *Achille in Sciro* (Metastasio; Madrid 1744).

**Corsi.** Italian family of singers.
1. **Giovanni** (*b* Verona, 1822; *d* Monza, 4 Apr. 1890). Italian baritone. Début Milan 1844 (Dandini, *Cenerentola*). Milan, S, 1845–70; Paris, I. A notable Rigoletto and exponent of Donizetti, and scrupulously attentive to characterization. Retired 1870; taught in St Petersburg.
2. **Achille** (*b* Legnano, 1840; *d* Bologna, 15 Apr. 1906). Italian tenor, brother of above. Début Milan 1859 (*Lombardi*). Milan, S, from 1860. Gifted with his pure, effortless voice; a successful Rossini singer. Retired 1882 and taught; pupils included his nephew Antonio *Pini-Corsi.
3. **Emilia** (*b* Lisbon, 21 Jan. 1870; *d* Bologna, 17 Sep. 1927). Italian soprano. Daughter of above, with whom she studied. Début 1886, Bologna (Micaëla). Her clear, supple voice brought her success as Lucia, Gilda, etc.; then she turned to the dramatic repertory and sang Manon Lescaut, Gioconda, and Sieglinde. Retired 1910. (R)

**Corsi, Jacopo** (*b* Florence, 17 July 1561; *d* Florence, 29 Dec. 1601). Italian operatic amateur. It was in his house that the group later known as the Florentine *Camerata met after Giovanni de' Bardi's departure in 1592. The discussions became more practical, and led to Peri's work completing Corsi's attempt at setting Rinuccini's *Dafne* (1598, lost, though surviving fragments include two Corsi songs). Peri's *Euridice* was also produced at his house, 1600, with him at the harpsichord.

**'Cortigiani, vil razza'.** Rigoletto's (bar) aria in Act II of Verdi's *Rigoletto*, denouncing the courtiers who have abducted his daughter.

**Cosa rara, Una.** Opera in 2 acts by Martín y Soler; text by Da Ponte, after Luis Vélez de Guevara's story *La luna della sierra*. Prod. Vienna, B, 17 Nov. 1786; London, 10 Jan. 1789, with Graziani, Elisabetta Borselli, Fausto Borselli, Forlivesi, Delicati, Fineschi, Torregiani.

A Spanish village, 18th cent. Lilla (sop) has been promised by her brother Tita (bs) as bride to Don Lisargo (bs), but she is in love with Lubino (bar), to whom she remains faithful though pursued by Prince Giovanni (ten) and his chamberlain Corrado (ten).

One of the great successes of its day, the opera drew the Viennese public away from *Figaro*, causing Mozart to quote from the Act I finale in the Supper Scene of *Don Giovanni* with Leporello's comment, 'Bravi! *Cosa rara*!'. Martín's use in the opera of a mandolin was copied by Mozart in Don Giovanni's serenade. Adapted by Stephen Storace as *The Siege of Belgrade* (1791). Sequel, music by Schack, text by Schikaneder, *Der Fall ist noch weit seltener!*, prod. Vienna, W, 10 May 1790.

**Così fan tutte** (Women are Like That). Opera in 2 acts by Mozart; text by Da Ponte, probably after Ariosto's *Orlando furioso* (1516), perhaps in turn deriving from the story of *Cephalus and Procris in Ovid's *Metamorphoses*, vii. Prod. Vienna, B, 26 Jan. 1790, with Del Bene, Villeneuve, D. Bussani, Calvesi, Benucci, F. Bussani; London, H, 9 May 1811, with Bertinotti-Radicati, Collini, Cauvini, Tramezzani, Naldi; New York, M, 24 Mar. 1922, with Easton, Peralta, Bori, Meader, De Luca, Didur, cond. Bodanzky.

Naples, late 18th cent. Ferrando (ten) and Guglielmo (bar) boast of the fidelity of the two sisters to whom they are betrothed. Don Alfonso (bs) maintains that all women are fickle and if the two men will act on his instructions for a day, he will prove it. He tells the sisters Fiordiligi (sop) and Dorabella (sop) that the two men have joined the army. There is a tearful farewell and the girls vow fidelity. Their maid Despina (sop) is persuaded to introduce the young men disguised as Albanian friends of Don Alfonso. Ferrando now courts Fiordiligi, and Guglielmo Dorabella. Next, the rejected 'Albanians' pretend to take poison. The girls weaken a little, but maintain their loyalty. Despina arrives disguised as a doctor and cures the men using magnets (a reference to Mozart's friend Dr Mesmer and his 'animal magnetism').

The sisters contemplate a little harmless flirtation. Dorabella and Guglielmo exchange love tokens but Fiordiligi still does not yield. When Ferrando learns that his Dorabella has given way, he is determined to conquer Fiordiligi, and she soon succumbs. A double wedding is arranged, with the disguised Despina as the notary. A military march signals the return of the army. The 'Albanians' flee and Ferrando and Guglielmo soon arrive. When they find the marriage contract Don Alfonso has dropped, they question the sisters angrily. They then reveal their trick. The supposed immorality of the plot led to many adaptations, often under different titles, in the 19th cent.

BIBL: N. John, ed., *Così fan tutte* (London, 1983).

**Cossotto, Fiorenza** (*b* Crescentino, 22 Apr. 1935). Italian mezzo-soprano. Studied Turin Cons., and with Campogalliani. Début Milan, S, 1957, creating Sr. Mathilde (*Carmélites*). Milan, S, until 1973. London, CG, 1959; Chicago from 1964; New York, M, from 1968; Paris; Vienna; etc. Generally considered Simionato's successor, with a warm, rich, and flexible voice. Repertory

includes Rossini, Bellini, Donizetti; also successful in dramatic Verdi roles, e.g. Eboli, Amneris. (R)

**Cossutta, Carlo** (*b* Trieste, 8 May 1932). Italian tenor. Studied Buenos Aires with Melani and Wolken. Début Buenos Aires, C, 1958 (Cassio). London, CG, from 1964; Paris, O, 1975, 1979, and B, 1991; New York, M, from 1973; also Milan, S; Chicago; Moscow, B; Vienna; Verona; etc. A singer with a heroic style, successful as Pollione, Manrico, Radamès, Otello (Verdi), Samson (Saint-Saëns), Cavaradossi, Turiddu. (R)

**Costa**, (Sir) **Michael** (orig. Michele) (*b* Naples, 4 Feb. 1808; *d* Hove, 29 Apr. 1884). Italian, later British, composer and conductor of Spanish descent. Studied Naples with Zingarelli, singing with Crescentini. Settled in London, where he became a dominant figure in English music. Engaged as maestro al piano, London, H, 1830; director and conductor 1833–46; bringing orchestra and ensemble to an unprecedented state of efficiency. Conductor Philharmonic Society from 1846; music director new Royal Italian O, London, CG, 1847–69; London, HM, 1871–9. His discipline brought high standards into the institutions he directed with despotic control; he was, however, less strict about fidelity to scores, which he unhesitatingly adapted and augmented with extraneous music. His own work was generally coolly received.

SELECT WORKLIST: *Malek Adel* (Pepoli; Paris 1838); *Don Carlos* (Tarantini; London 1844).

**Costanzi, Giovanni Battista** (*b* Rome, 3 Sep. 1704; *d* Rome, 5 Mar. 1778). Italian composer. A cellist by training, he joined the service of Cardinal Pietro Ottoboni in Rome 1721. *Carlo Magno* launched his career as an opera composer and led to important Roman musical positions, culminating in appointment as maestro di cappella at the Cappella Giulia, St Peter's. A highly regarded composer in his day; most of his works are lost.

SELECT WORKLIST: *Carlo Magno* (Ottoboni; Rome 1727).

**Cotogni, Antonio** (*b* Rome, 1 Aug. 1831; *d* Rome, 15 Oct. 1918.) Italian baritone. Studied Rome with Fontemaggi and Faldi. Début Rome, T Metastasio, 1852, Belcore (*L'elisir d'amore*). Milan, S, from 1860; London, CG, 1867–89; St Petersburg 1872–94. His vast repertory encompassed Mozart, Bellini, Donizetti, Verdi, Wagner, Puccini, and Bizet. A virtuoso singer, whose voice was brilliant, richly coloured, and extended to a'. Particularly outstanding as Posa, and said to have made Verdi weep. After his retirement (1898) he taught in Rome. His pupils included Battistini, J. De Reszke, Lauri-Volpi, Galeffi, Viglione-Borghese, Stabile, and Gigli. (R)

**Cotrubas, Ileana** (*b* Galaţi, 9 June 1939). Romanian soprano. Studied Bucharest with Elenescu and Stroescu. Début Bucharest 1964 (Yniold). Frankfurt 1968; Gly. from 1969; London, CG, from 1971; Chicago 1973; Paris, O, 1974; Milan, S, from 1975. New York, M, 1977–84. Repertory included Susanna, Gilda, Violetta, Mélisande, Mimì, Tatyana. An intense performer with a strong, lyrical voice particularly well suited to conveying pathos. Retired 1989. (R)

**Count.** Among many operatic counts, the best known is Count *Almaviva (bar) in Mozart's *Le nozze di Figaro*.

**counter-tenor.** The term originally denoted the 'contratenor' line in Renaissance polyphony lying directly above the tenor. Later it came to mean a high male falsetto voice, also known as 'male alto', with a contralto or mezzo-soprano range, occasionally extending to soprano. (Some very high counter-tenors today have described themselves as 'male soprano'.) The term has sometimes been used to mean *haute-contre, which is more generally taken, however, to be a very high, light tenor. The counter-tenor declined during the 18th and 19th cents., apart from in church choirs, to be revived in modern times by Alfred Deller, initially for singing English 16th- and 17th-cent. music, e.g. Dowland, Purcell. Others following him, including James Bowman and Jochen Kowalski, have sung in operas by Handel and Gluck, in roles conceived for castratos. Nowadays composers are increasingly interested in writing for the counter-tenor voice, often for other-worldly or ambiguous characters, e.g. Britten's Oberon, his Apollo (*Death in Venice*), and Glass's Akhnaten.

**Countess.** Among many operatic countesses, three are especially well known.

1. Countess Almaviva (sop) in Mozart's *Le nozze di Figaro*.
2. The Countess (mez) in Tchaikovsky's *The Queen of Spades*.
3. The Countess (sop), loved by Flamand and Olivier, in Strauss's *Capriccio*.

**coup de glotte** (Fr.: 'glottal stop'). A method of attacking a note whereby the false vocal chords (two membranes above the true vocal chords) are closed and then quickly opened to release the tone. It should not be an abrupt release from pressure, when a sharp cough or click results, but a start of pressure from stopped breath.

**couplets.** The French term, originally meaning 'stanzas', in general use in 18th- and 19th-cent.

opera, operetta, and Singspiel for a strophic song, generally witty in character. Examples are found in Grétry, Auber (*Fra Diavolo*), and especially in operettas by Hervé, Offenbach, and Lecocq. The last line is often echoed by the chorus, e.g. 'Di tu se fedele' in Verdi's *Un ballo in maschera*. In Germany the word is generally used in the singular, e.g. by Johann Strauss II for the couplet in *Fledermaus*, 'Mein Herr, was dächten Sie von mir' and elsewhere, also by Millöcker, Lehár, Heuberger, Fall, and others.

**Covent Garden.** The name of three theatres that have occupied roughly the same site in Bow Street, London, since 1732 on what was originally a convent garden.

The first was opened on 7 Dec. by John Rich with Congreve's *The Way of the World*; the first musical work was *The Beggar's Opera*. Most events in the first 100 years were dramatic, though some *Handel operas and Arne's *Artaxerxes* were premièred there. Burnt down 19 Sep. 1808. The second theatre opened 1809. The Old Price riots followed as a result of increased ticket prices. In 1826 the manager, Charles Kemble, invited Weber to compose *Oberon*. Important events 1833–47 included Alfred Bunn's 1833 season with Malibran and Schröder-Devrient; 1835 with Malibran in *Sonnambula* and *Fidelio* (same night); the artistically successful but financially disastrous 1839–42 seasons with Adelaide Kemble in *Norma* and other works; the Brussels company in 1845 with 'unmutilated and meritorious performances' (Chorley) of *William Tell*, *Les Huguenots*, and *La muette de Portici*. In 1847 the theatre became the Royal Italian O, with a company including artists from Her Majesty's including *Costa and opening with *Semiramide*. Frederick Gye was the influential manager 1851–79, introducing Verdi, including *Aida* with Patti, and *Tannhäuser* with Albani (1876). The theatre burnt down in 1856, and the present building opened 15 May 1858 with *Les Huguenots*.

From 1858 to 1939 (except during the First World War) opera was given annually and virtually every artist of international repute appeared. Opera was first sung in the original language (as opposed to Italian) under the management of Augustus Harris (1888–96). In 1892 Covent Garden became the Royal Opera, and Mahler conducted the Hamburg company in the first CG *Ring*. Until 1924 the theatre was controlled by the Grand Opera Syndicate: Grau managed both the New York, M, and London, CG, 1897–1900, Messager the Paris, OC, and London, CG, 1901–4; and Percy Pitt, Richter, and Neil Forsythe were in partial or full control until 1914. Beecham first conducted there 1910, introducing the major Strauss operas to London. Closed 1914–18. The BNOC occupied Covent Garden for seasons 1922–4. German opera prospered 1924–31 under Walter with Wagner and Strauss performances including Leider, Lehmann, Schumann, Olczewska, Melchior, Schorr, Jansen, and Kipnis. Beecham was artistic adviser then director 1932–9, staging Rossini with Supervia, Mozart with Lemnitz, Berger, and Tauber, and Wagner under Beecham, Furtwängler, Reiner, and Weingartner, with Flagstad, Thorborg, Bockelmann, and Weber. Strauss conducted *Ariadne* with the Dresden O, 1936.

The theatre was a dance hall 1939–45, reopening 1946 (cap. 2,250) as the home of a permanent company with a government subsidy. Karl Rankl was music director 1946–51, Rafael Kubelik 1955–8, Georg Solti 1961–71, Colin Davis 1971–86, then Bernard Haitink. General directors have included David Webster 1946–70, John Tooley 1970–88, Jeremy Isaacs from 1988. English operas receiving premières have included *Billy Budd* (1951), *Gloriana* (1953), *Troilus and Cressida* (1954), *The Midsummer Marriage* (1955), *The Knot Garden* (1970), *The Ice Break* (1977), *Taverner* (1972), and *Gawain* (1991). British premières have included *Wozzeck* under Erich Kleiber (1952), *Jenůfa* under Kubelik (1956), a complete *The Trojans* under Kubelik (1957); other important productions have included Visconti's prod. of *Don Carlos* under Giulini (1958), *Fidelio* under Klemperer (1961), Zeffirelli's prod. of *Tosca* with Callas and Gobbi (1964), *Moses und Aron* under Solti (1965), *Pelléas et Mélisande* under Boulez (1969), Friedrich's prod. of the centenary *Ring* under Davis (1976). Singers whose careers grew with the company include Joan Sutherland, Geraint Evans, Michael Langdon, Jon Vickers, and Donald McIntyre.

BIBL: H. Rosenthal, *Two Centuries of Opera at Covent Garden* (London, 1958); F. Donaldson, *The Royal Opera House in the 20th Century* (London, 1988).

**covered tone.** The tone quality produced when the singer's voice is pitched in the soft palate. It is gentler, more veiled in timbre, than open tone, and gives greater intensity and beauty to the transitional notes, i.e. c to e in baritone and bass register, the f' to a' in the tenor and soprano range. It is also a help in placing high notes, especially on awkward syllables.

**Coward,** (Sir) **Noël** (*b* Teddington, 16 Dec. 1899; *d* Blue Harbour, 26 Mar. 1973). English playwright, actor, and composer. Though self-taught as a composer, and requiring musical assistants for the notation and orchestration of his music, Coward had an individual vein of invention which shows not only in his brilliant cabaret songs but in his operetta *Bitter-Sweet* (1931).

**Cowen,** (Sir) **Frederic** (*b* Kingston, Jamaica, 29 Jan. 1852; *d* London, 6 Oct. 1935). English composer, conductor, and pianist. When only 8 he composed an operetta, *Garibaldi*, to a text by his 17-year-old sister. Studied with Goss and Benedict, later Leipzig with Moscheles and Reinecke, Berlin with Tausig. He joined Mapleson's Italian OC in 1871, also working at London, HM, under Costa. Among his operas, *Signa* was composed as a potential successor to Sullivan's *Ivanhoe* in D'Oyly Carte's project for an English National Opera; when this failed, Cowen secured a production in Milan, but there were only two performances owing to sharp dissensions (which nearly led to a duel between Boito and Sonzogno), and the unfortunate opera was finally produced in a third version by Augustus Harris (London, CG, 1894).

SELECT WORKLIST: *Garibaldi, or The Rival Patriots* (R. Cowen; comp. 1860, unprod.); *Pauline* (Hersee, after Bulwer Lytton; London 1876); *Signa* (A'Beckett, Rudall, & Weatherly, after Ouida; Milan 1893).

**Cox, Jean** (*b* Gadsden, AL, 16 Jan. 1922). US tenor. Studied Alabama, Boston, and Rome with Luigi Ricci. Début Boston 1951 (Lensky). Spoleto 1954 (Rodolfo). Kiel 1954–5, Brunswick 1955–9, Mannheim 1959. Bayreuth 1956, Steersman, later Lohengrin, Parsifal, Walther, Siegfried. Chicago 1964, 1970, 1973 as Bacchus, Erik, Siegfried. Repertory includes Števa, Apollo (*Daphne*), Sergey (*Lady Macbeth of Mtsensk*), Cardinal (*Mathis der Maler*). (R)

**Cox, John** (*b* Bristol, 12 Mar. 1935). English producer. Studied Oxford, where he produced the British stage première of *L'enfant et les sortilèges*, 1958; professional début, same work, London, SW, 1965. Assistant to Rennert Gly. 1959, returning 1970; director of productions 1972. London, SW, from 1969, and ENO; Frankfurt 1971; Australian O 1974; Munich; San Francisco; Scottish O 1981–6; London, CG, from 1988; Wexford; New York, M; Milan, S; Vienna, V; and many other cities. With Alexander Goehr, Cox was director of the Music T Ensemble. A highly gifted producer, with a quick eye for detail and a lively sense of drama in a wide repertory of composers incl. Strauss, Mozart, Massenet, Rossini, Wagner, Cavalli, and Puccini, etc.

**Cracow** (Pol., Kraków). City in Poland. The first opera libretto in Polish, a translation of Francesca Caccini's *La liberazione di Ruggiero dall'isola d'Alcina* (1625), was published here in 1628. The first private opera performances were given in 1628, the first public ones by visiting companies, mostly Italian, from about 1780. The first local staging was *Zémire et Azor* in 1782, and the first Polish opera was *Rywale, czyli Capstryk w Krakowie* (The Rivals, or A Tattoo in Cracow, 1786), an opera of local colour by Wawrzyniec Zagorski (?–?). In 1799 a Polish opera company was formed by Jacek Kluszewski, and survived until the early 19th cent. Occasional efforts to organize opera regularly were made 1844–50 and 1865–6, but from then until 1914 there were only guest performances by companies from Lvov and elsewhere. In 1915 the Opera was founded by Bolesław Wallek-Walewski; this existed until 1923, and was revived 1931–9. After the Second World War, the Society of Friends of Opera formed a company 1946–8, under Walerian Bierdiajew. In 1954 the repertory opera was organized by the Operatic Society, and a Municipal Music T of Opera and Operetta was formed in 1958.

**Craig, Edward Gordon** (*b* Stevenage, 16 Jan. 1872; *d* Vence, 29 July 1966). English designer, producer, and actor. Son of designer E. W. Godwin and actress Ellen Terry. He was, with Appia, among the first to create designs intended not as conventional settings but to give the performer a planned role in the total theatrical experience. For the Purcell Operatic Society, 1900–3, he produced *Dido and Aeneas* using only a huge skycloth against which the singers moved, and Handel's *Acis and Galatea*, in which lighting effects and projections removed the action from the particular to the universal. He was an isolated figure, but a profound influence on the succeeding generation of producers.

**'Credo in un dio crudel'.** Iago's (bar) aria in Act II of Verdi's *Otello*, in which he declares his belief in a cruel god who has fashioned him in his own likeness. There is no Shakespeare original for the words. Boito's most substantial and personal addition to the opera.

**Creon.** In classical mythology, the King of Corinth, whose daughter married Jason: he appears (bs) in M. A. Charpentier's and Cherubini's *Médée*, and (bar) in Mayr's *Medea in Corinto*. Also the King of Thebes, brother of Jocasta: he appears (bs-bar) in Stravinsky's *Oedipus rex*.

**crescendo** (It.: 'growing'). Increasing in loudness. Now a basic part of music, the crescendo was once a controversial novelty. Its use in opera largely mirrors its gradual acceptance in instrumental music, where new methods of construction permitted greater dynamic flexibility. Baroque practice was to use 'terraced' dynamics, i.e. marked shifts from one dynamic level into another contrasting one. It was the famous Mannheim orchestra which made the crescendo

popular and, significantly, it is around this time that its appearance in vocal music is first noted. According to Burney, the first opera using it was Terradellas's *Bellerofonte* (1747), which refutes Mosca's claim for his *I pretendi delusi* (1811). The innovation was often credited to Rossini, with whom it bordered on a mannerism. His finest use of it is in *Barbiere* to suggest the rising gale of slander in Don Basilio's 'La calunnia'. Generally popular in 19th-cent. Italian opera as an expressive device, the crescendo also fulfilled an important psychological function by generating excitement in the audience, leading finally to applause at the end of the aria.

**Crescentini, Girolamo** (*b* Urbania, 2 Feb. 1762; *d* Naples, 24 Apr. 1846). Italian mezzo-soprano castrato. Studied Bologna with Gibelli. Début Padua 1782 in Sarti's *Didone abbandonata*. Created Romeo in Zingarelli's *Giulietta e Romeo*, Milan, S, 1796; Cimarosa also wrote an opera for him, *Gli Orazi ed i Curiazi*. Though not a success in London (1785), he was fêted throughout Europe, and sang in all the major opera-houses. The many admirers of his 'heavenly voice' and 'noble art' (*AMZ*, 1804) included Schopenhauer, E. T. A. Hoffmann, and Napoleon, who, having heard him sing in Vienna, 1805, engaged him as singing-teacher to his family in Paris (1806–12). Here he also sang at court, and was one of the few castrati ever to be well received by the French. Returning to Italy, he then settled in Naples, where he taught. Colbran was one of his pupils.

**Crespin, Régine** (*b* Marseilles, 23 Mar. 1927). French soprano. Studied Paris with Jouatte and Cabanel. Début Mulhouse 1950 (Elsa). Paris, O, from 1950; Bayreuth from 1958; London, CG, from 1960; New York, M, from 1962; Chicago 1964. Equally at ease in French and German repertory, with particular success as Sieglinde and Marschallin. Created New Prioress in *Dialogues des Carmélites*. From 1975 began to sing mezzo-soprano roles, e.g. Carmen. A sensitive and eloquent singer, with a voice both voluptuous and dramatic. (R)

**Cressida.** The Trojan High Priest Calkas's daughter (sop, in later version mez), lover of Troilus, in Walton's *Troilus and Cressida*.

**Crimi, Giulio** (*b* Paternò, 10 May 1885; *d* Rome, 29 Oct. 1939). Italian tenor. Studied Catania with Adernò. Début Palermo 1910 (Manrico). London, CG, 1914; Milan, S, 1916; Chicago and New York, M, 1916–24. Created Luigi (*Il tabarro*), Rinuccio (*Gianni Schicchi*), and Paolo (Zandonai's *Francesca da Rimini*). Retired 1928. His good looks and warm, generous voice were ideal for romantic, and particularly verismo, roles. His pupils included Gobbi.

**Crispino e la comare** (Crispino and the Fairy). Opera in 3 acts by Federico and Luigi Ricci; text by Piave. Prod. Venice, B, 28 Feb. 1850, with Cambaggio, Pecorini; London, J, 17 Nov. 1857; New York, AM, 24 Oct. 1865.

Venice, 17th cent. The poor cobbler Crispino (bs) acquires fame and wealth thanks to the amazing healing powers given him by a fairy. He becomes so arrogant that she reveals to him that she is Death and has now come to claim him. However, Crispino repents and she allows him to return to his family.

**Critic, The, or An Opera Rehearsed.** Opera in 2 acts by Stanford; text by Lewis Cairns James, after Sheridan's comedy (1779). Prod. London, Shaftesbury T, 14 Jan. 1916, with Hatchard, Maitland, Barrigar, Mullings, Heming, Langley, Ranalow, cond. Goossens.

**Croatia.** Opera played an important part in expressing a national identity during the 19th cent. and the years of Austrian and Hungarian domination, with the Italian influence proving fruitful along the Adriatic coast. Though a translation of Peri's *Euridice* was published as early as 1617, there was little organized opera during the 17th and 18th centuries through lack of patronage. The first Croatian opera was *Ljubav i zloba* by Vatroslav *Lisinski (1819–54); his early death and the unfavourable political climate halted further development. In 1870 Ivan *Zajc left Vienna to become director of the Zagreb Opera, where his repertory was based on the standard Italian and German works; his own operas, including *Nikola Šubić Zrinski* (1876, the first Croatian Romantic opera), matched the effect of Verdi's Risorgimento operas in arousing national fervour. His principal successors were Blagoje *Bersa (1873–1934), with *Oganj* (1911), and Josip Hatze (1879–1959) with the 1-act verismo *Povratak* (1911) and with *Adel i Mara* (1932), who both showed greater willingness to accept newer, cosmopolitan influences.

In reaction, attempts were made after the First World War to create a style deriving from folk music. Antun Dobronić (1878–1955) avoided the term 'opera' in favour of, for instance, a 'musical stage mystery' *Mara* (1928) and a 'stage symphony' *Ekvinocij* (1928). Božidar Širola (1889–1956) and Jakov *Gotovac (1895–1982) favoured comedy; the latter's *Ero the Joker* (1935) achieved international success. Other significant composers were Krešimir *Baranovič (1894–1975) and Fran Lhotka (1883–1962).

'Socialist Realism', imposed after 1945, was abandoned in the 1950s, and individual styles began to emerge. Ivo Tijardović (1895–1976) and

Ivo Brkanović (1906–87) continued in pre-war styles, but Boris Papandopulo (*b* 1906), Ivo Lhotka-Kalinski (1913–87), and Natko Devčić (*b* 1914) moved towards more modern idioms. A number of Shakespeare operas included *Romeo and Juliet* (1954) by Krešimir Fribec (*b* 1908), in which only the two lovers act and the chorus provides a commentary; *Coriolanus* (1958) and *Oluja* (The Tempest, 1969) by Stjepan Šulek (1914–86); and *Richard III* (1987) by Igor Kuljerić (*b* 1938). Milko Kelemen's (*b* 1924) *Der neue Mieter* (1963) and *Der Belagerungszustand* (1970) were commissioned for Germany. Operetta is also popular.

See also ZAGREB.

BIBL: D. Cvetko, *Musikgeschichte der Südslawen* (Kassel, 1975).

**Crociato in Egitto, Il** (The Crusader in Egypt). Opera in 2 acts by Meyerbeer; text by Rossi. Prod. Venice, F, 7 Mar. 1824, with Velluti; London, H, 3 June 1825, with Velluti and Malibran.

Damietta, c.1250. Armando d'Orville (orig. cas, later mez), a knight of Rhodes, is assumed dead in Egypt during the Sixth Crusade; but under the name of Elmireno, he has become the confidant of the Sultan Aladino (bs). He falls in love with Palmide (sop), Aladino's daughter, and converts her to Christianity. Adriano (ten), Armando's uncle, arrives to sue for peace, but Emireno's true identity is discovered and he and the other captured Christians are sentenced to death. Armando, however, saved the Sultan's life after a plot to overthrow him; Armando and Palido are reunited, and a peace treaty is signed.

The first Meyerbeer opera in London, and the last and most successful of his Italian works.

**Cross, Joan** (*b* 7 Sep. 1900). English soprano. Studied London with Freer. Joined chorus of Old Vic 1924, soon singing Cherubino and First Lady. London: SW, 1931–46; CG, 1931, 1934–5, 1947–54. Founder member of the EOG 1945. Created Britten's Ellen Orford, Female Chorus, Lady Billows, and Queen Elizabeth (whose birthday she shares) in *Gloriana*. Other roles included Mimì, Snegurochka, Micaëla. After 1955 devoted herself to teaching, especially at the National School of Opera, which she founded with Anne Wood in 1948. Has also produced opera for CG and SW. One of Britain's great operatic artists, an exceptionally gifted musician, with a pure, true voice that she used with great sensitivity and feeling for line; also a fine, sympathetic actress. As director of SW during the Second World War, she did much to sustain the company and was instrumental in securing the première of *Peter Grimes*. (R)

**Crosse, Gordon** (*b* Bury, 1 Dec. 1937). English composer. Studied Oxford with Wellesz, Rome with Petrassi. His first opera was a powerful, strongly atmospheric setting of Yeats's *Purgatory*, using a freely declamatory style closely connected to the orchestral expression. *The Grace of Todd* confirmed this dramatic instinct, and an exceptional musical intelligence, though the work was hampered by a somewhat laboured comic text; his qualities were more richly revealed in *The Story of Vasco*. Crosse has also interested himself in music theatre, as with *Potter Thompson*; this was written to involve children, for whom he has also composed the Nativity opera *Holly from the Bongs*.

WORKLIST: *Purgatory* (Yeats text; Cheltenham 1966); *The Grace of Todd* (Rudkin; Aldeburgh 1969); *The Wheel of the World* (Cowan, after Chaucer; Aldeburgh 1972); *The Story of Vasco* (Hughes, after Schehadé; London 1974); *Potter Thompson* (Garner; London 1975); *Holly from the Bongs* (Garner; Manchester 1974).

**Crown.** A tough stevedore (bar) in Gershwin's *Porgy and Bess*.

**Crown Diamonds.** See DIAMANTS DE LA COURONNE, LES.

**Crozier, Eric** (*b* London, 14 Nov. 1914). English writer and producer. With Benjamin Britten and John Piper, co-founder of the EOG (1947). Librettist of Britten's *Albert Herring* and *Let's Make an Opera*, and (with E. M. Forster) *Billy Budd*; also of Berkeley's *Ruth*. Produced new version of *The Bartered Bride* (made by him with Joan Cross), London, SW, 1943, and premières of *Peter Grimes*, SW 1945, and *Rape of Lucretia*, Gly. 1946; first US *Grimes*, Tanglewood 1946.

**'Cruda sorte'.** Isabella's (con) cavatina in Act I of Rossini's *L'italiana in Algeri*, lamenting her separation from her lover Lindoro when she is captured by pirates, but then realizing she can use her charm as a woman to deal with her captors and master the situation. The original version, with its amorous *double entendres*, was censored after the 1813 Venice première for Milan.

**Cruvelli, Sofia** (*b* Bielefeld, 12 Mar. 1826; *d* Monaco, 6 Nov. 1907). German soprano. Studied Paris with Bordogni, Milan with Lamperti. Début Venice 1847 (Odabella, *Attila*). London, HM, 1848, Elvira (*Ernani*), Abigaille, and Mozart's Countess, which brought her little success; Milan 1850 (Luisa Miller); Paris, I, 1851–3 (Norma, Semiramide); London, CG, 1854, this time evoking enthusiasm as Donna Anna, Leonore, Desdemona (Rossini). Paris, O, 1854–6, Valentine, Rachel, and Hélène, which she created in *Les vêpres siciliennes*. (Her disappearance during rehearsals for this caused a sensation; it transpired that she had taken a 'honeymoon' with Baron Vigier, whom she later married.) One

of the finest dramatic sopranos of her time, especially in the operas of Verdi, whose music Fétis considered 'faite pour la cantatrice'. Her voice was powerful and rich, though rather uneven, and she possessed great dramatic flair. Her sister Friederika Marie Crüwell (1824–68), mezzo-soprano, studied with Roger and sang in France and Italy.

**Cuba.** The first opera heard in Cuba was based on Metastasio's *Didone abbandonata* (1776; composer unknown). The T Coliseo, later renamed T Principal, opened in Havana 1803. The T Tacon opened in 1838, and was for many years the home of visiting opera companies from Europe (especially France and Italy). It became the T Nacional in 1915, and was inaugurated under Serafin. In 1920 Caruso sang there, receiving his highest fee ever, $90,000 for ten performances.

Despite plans for a national opera *Antonelli*, set in Havana, by Manuel Samuell Robredo in 1838, the first Cuban opera was the 1-act *La hija de Jefté* (Jephtha's Daughter, 1875) by Laureano Fuentes Matous; this was recomposed as the 3-act *Seila* (1917). Other Cuban composers have included Guillermo Tomás (1868–1933), with *Sakuntala* (n.d.); Eduardo Sánchez de Fuentes (?–?), with *Yumuri* (1898), *El náufrago* (1901), *El caminante* (1921, with Schipa), and *Kabelia* (1942); and Gaspar Villate (1851–91), whose operas are Europeanized: they include *Angelo, tirano de Padua* (1867), *Las primeras armas de Richelieu* (1871), and *Zelia* (1877), all prod. Europe before Havana. An opera drawing on Cuban sources was *La esclava* (1921) by J. M. Estévez (?–?). Operetta and zarzuela (notably Gonzalo Ruig's *Cecilia Valdés*) are also popular. An international opera festival was held in 1987.

See also HAVANA.

**Cuberli, Lella** (*b* Austin, TX, 29 Sep. 1945). US soprano of Italian origin. Studied Dallas and Italy. Début Budapest 1975 (Violetta). Milan, S, from 1977; Pesaro 1984; Aix 1985; Paris, O, 1985–6; Salzburg 1986; New York, M, 1990. Repertory incl. Mozart's Countess, and roles in many revivals of Rossini, Bellini, and Donizetti. (R)

**Cuénod, Hugues** (*b* Corseaux-sur-Vevey, 26 June 1902). Swiss tenor. Studied Basle, and Vienna with Fr. Singar-Burian. Originally a baritone, trained as tenor, as which made début Paris 1928 in *Jonny spielt auf*. Milan, S, 1951; Gly. from 1954; London, CG, 1954, 1956, 1958. Tall, humorous, and energetic, a character tenor of great individuality and intelligence. His career spanned over 60 years and the friendship of many composers, particularly Poulenc and Stravinsky, for whom he created Sellem (*The Rake's Progress*) and subsequently many major concert works. Memorable for his Erice (*Ormindo*), Linfea

(*Calisto*), Triquet, Don Basilio (Mozart) and Vašek; at an indefatigable 85, made his début at New York, M, as the Emperor in *Turandot*. (R)

**Cui, César** (*b* Vilnius, 18 Jan. 1835; *d* Petrograd, 26 Mar. 1918). Russian composer and critic of French descent. Studied with Moniuszko, later moving to St Petersburg and studying engineering. In 1856 he met Balakirev, and became one of the five composers of the 'moguchaya kuchka' (mighty handful). Encouraged (and helped with his orchestration) by Balakirev, he wrote two operas before embarking upon what was to be his greatest success, *William Ratcliff*. Its lengthy gestation is in part responsible for its great stylistic unevenness, which embraces some fairly conventional adaptation of the text for operatic purposes and, in the second half, musical diction close to that of Dargomyzhsky. It also makes some enterprising use of reminiscence motive. Not surprisingly, the work divided critics: among those who greatly admired it were Tchaikovsky and Rimsky-Korsakov, in sharp contrast to Serov. *Angelo*, Cui's next opera, was an adaptation of Hugo's drama to match the contemporary interest in setting personal drama against a vividly depicted social background, though the intended influence of Musorgsky is less evident than that of Parisian grand opera. His next opera was actually French, *Le flibustier*, and its successor, *The Saracen*, was to a Dumas subject. *A Feast in Time of Plague* is another setting of one of Pushkin's 'little tragedies', one of which had inspired Dargomyzhsky's *The Stone Guest*, and attempts to follow that work in its manner of melodic declamation, while also imposing standard musical forms upon the poetry. His one full-length opera on a Russian subject was a setting of Pushkin's *The Captain's Daughter*. Thereafter, disillusioned, he turned to children's operas. As a critic, he was a fervent upholder of nationalist ideals, using his notoriously sharp tongue and pen in defence of his friends and to attack those unsympathetic to his aims. His fiercest attacks were intended to kill Wagnerism in Russia, and thereby wound Serov.

SELECT WORKLIST (all first prod. St Petersburg unless otherwise stated): *Kavkazky Plennik* (The Captive of the Caucasus) (Krylov; comp. 1857–8, prod. 1883); *William Ratcliff* (Pleshcheyev; 1869); *Angelo* (Burenin; 1876); *Le flibustier* (Richepin; Paris 1893); *Saratsen* (The Saracen) (after Dumas; 1899); *Pir vo Vremya Chumy* (A Feast in Time of Plague) (after Pushkin; Moscow 1900); *Kapitanskaya Dochka* (The Captain's Daughter) (after Pushkin; 1911).

WRITINGS: *La musique en Russie* (Paris, 1880, R/ 1974); Y. Kremylov, ed., *Ts A Kyui: Izbranniye Staty* (Leningrad, 1952).

BIBL: M. D. Calvocoressi, 'César Cui', in M. Calvo-coressi & G. Abraham, *Masters of Russian Music* (London, 1936).

**Cunning Little Vixen, The** (Cz.: *Příhody Lišky Bystroušky*: lit. The Tales of Vixen Sharpears). Opera in 3 acts by Janáček; text by the composer, after Rudolf Těsnohlídek's verses for drawings by Stanislav Lolek, published serially in the Brno newspaper *Lidové Noviny* during 1920. Prod. Brno 6 Nov. 1924, with Hrdličková, Snopková, Flögl, Pour, Pelc, cond. Neumann; London, SW, 22 Mar. 1961, with Bronhill, Easton, cond. Davis; New York, Hunter College, 7 May 1964, with Killmer, Fiortito, cond. Bamberger.

A Moravian village and surrounding woods. Sharpears the vixen (sop) is caught by the Forester (bar) and kept as a pet. She laments her lot to the dog Lapák (mez or ten) and, having incited rebellion among the hens against their cock (sop), she manages to escape.

Having dispossessed the badger (bs) by fouling his lair, she settles in. In the world of men, the Priest (bs) is gloomy because he is to be transferred, while the Forester sighs for the mysterious girl Terynka, who has also fascinated the Schoolmaster (ten); and though they play cards together in the inn, disorder threatens. They leave, rather drunk, and confused about the allure of Terynka; Sharpears watches them. She takes a fox, Goldenmane (sop or ten) as mate; they are married by the woodpecker.

Harašta the poacher (bs) comes upon the fox family, and shoots Sharpears. Back in the inn, the Forester is sad that he cannot now find the vixen whose image haunts him; and the marriage of Terynka to Harašta does nothing to cheer the men up. They feel their ageing emphasized by the coming of spring. Alone in the woods, the Forester falls into a doze; waking, he observes the animals, among them a foxcub exactly like her mother, and taking new comfort from this vision of nature always renewing itself, he lies back at peace.

The German version by Max Brod is at pains to emphasize an identity between the men and animals (e.g. Priest and Badger), and between the Vixen and Terynka, which Janáček deliberately left oblique.

**Curlew River.** Church parable in 1 act by Britten; text by William Plomer, after Jūro Motomasa's Noh play *Sumida-gawa* (early 15th cent.). Prod. Orford Parish Church 13 June 1964, with Pears, Shirley-Quirk, Drake, Garrard; Katonah, New York, Caramoor Festival, 26 June 1966, with Velis, Clatworthy, Berberian.

The first of Britten's *church parables, it is cast in the dramatic metaphor of monks processing into church to enact a mystery play. The Ferryman (bar) is waiting to convey passengers across the river. A Madwoman (ten) arrives, on a search for her lost son; unwillingly, the Ferryman agrees to take her across. During the crossing, he relates a story of a young boy who, fleeing from robbers, crossed the river a year previously but was so exhausted that he died on reaching the far bank. The Madwoman realizes that this was her son and, taking pity on her, the Ferryman leads her to the boy's grave.

Also opera by Raybould (1916).

**curtain tune.** A term used to refer to the music played during 17th-cent. English theatrical entertainments while the curtain was being raised: it usually followed the prologue.

**Curtin, Phyllis** (*b* Clarksburg, WV, 3 Dec. 1922). US soprano. Studied Wellesley Coll. with Regneas. New England OT 1946 (Lisa, Lady Billows); New York, City O, from 1953; New York, M, 1961; also Milan, S; Vienna; Buenos Aires; Scottish O. Created Cathy in Floyd's *Wuthering Heights*, title-role in his *Susannah*, and the three female roles in Einem's *Der Prozess*. Successful in contemporary works, but also as Salome, Mistress Ford, and in Mozart. (R)

**Cuzzoni, Francesca** (*b* Parma, *c*.1698; *d* Bologna, 1770). Italian soprano. Studied with Lanzi. First known appearance Parma, 1716. After successes in Bologna and Venice, engaged London 1722 at H, at £2,000 p.a. When she failed to arrive, Heidegger sent his harpsichordist Sandoni to bring her to England, and on the way she married him. She made a sensational début 1723, creating Teofane in Handel's *Ottone*; also created several other Handel roles, including Cleopatra (*Giulio Cesare*), Asteria (*Tamerlano*), Rodelinda, Lisaura (*Alessandro*). The engagement of Faustina *Bordoni in 1726 led to the notorious public rivalry between the two, culminating in a brawl on stage during Bononcini's *Astianatte* in 1727. Left London for Italy, but returned in 1734–6, engaged by Porpora for Lincoln's Inn Fields. Made a final visit in 1750, but was by now nearly voiceless, and heavily in debt; Walpole says she was bailed out of prison by the Prince of Wales. After further imprisonment in Holland, she returned to Bologna, where she became a button-maker and died in extreme poverty. Though physically unprepossessing ('short and squat, with a doughy cross face'— Walpole) and a bad actress, she was by all accounts a singer of the highest quality. Quantz describes her 'clear and agreeable soprano, pure intonation, and a fine shake' and declares that her singing 'took possession of the soul of every auditor'. Her range was c'–c'''.

**cycle.** A group of works telling a more or less consecutive story, but partly for reasons of length, partly for dramatic effect, separable and performable on different nights. The earliest cycle is Sartorio's *La prosperità di Elio Seiano* and *La caduta di Elio Seiano*, both in 3 acts with text by Nicolò Minato (Venice 1667). The best-known, and greatest, cycle remains Wagner's *Der *Ring des Nibelungen*. Upon this others were modelled, e.g. Bungert's *Homerische Welt* (based on the Odyssey) and Holbrooke's *The Cauldron of Annwyn*.

**Cyrano de Bergerac.** 1. Opera in 4 acts by Alfano; text by Henry Cain, after Rostand's drama (1897). Prod. Rome, R, 22 Jan. 1936, with Caniglia, Luccioni, cond. Serafin.

Paris and Arras, 1640. Cyrano (ten), the poet-soldier, is disfigured by a huge nose that makes him repulsive to women. He falls in love with the beautiful Roxane (sop) and pays court to her through the mouth of Christian (ten), who also loves her. Only when Cyrano is on his deathbed does Roxane learn the truth and realize that she loves him.

2. Opera in 4 acts by Damrosch; text by W. J. Henderson, after Rostand. Prod. New York, M, 27 Feb. 1913, with Alda, Amato, cond. Hertz.

Also operas by Tamberg (1974) and Turandowski (1962), and operetta by Herbert (1899).

**Czechoslovakia.** The name under which the countries of Bohemia, Moravia, and Slovakia were united in 1918. Opera was first given in 1627, when Italian singers came from Mantua to perform a *commedia pastorale cantata* by G. B. Buonamente and Cesare Gonzaga for the coronation of Ferdinand III. Italian opera was thereafter given by visitors, including in 1702 Bartolomeo Bernardi's *La Libussa* (the first work on a Czech subject in Czech lands). In 1723 Fux's *Costanza e Fortezza* was produced for the 4,000 visitors to the coronation of Charles VI. Later, opera (including premières of Vivaldi's works) was given in a theatre built by Count František Špork in his Prague residence in Na Poříčí in 1701, and open to the public. From 1739 Italian companies played in the new Kotzen T, which was adapted from the old town's medieval cloth market and remained the chief public theatre. The Gräflich-Nostitzsches Nationaltheater (1783), then the Königliches Ständestheater in the Old Town (Staré Město), and Count Thun's theatre in the Malá Strana (1781) replaced it.

The expansion of Italian opera in the 18th cent. led to the building of theatres in nobles' castles, where opera was given by Italian companies augmented by domestic musicians. Important castle centres included Špork's Kuks, Český Krumlov, and in Moravia Rottal's Holešov, Schrattenbach's Vyškov and Kroměříž, and especially Questenberg's Jaroměřice, where in

# Czechoslovakia

1730 *L'origine di Jaromeriz in Moravia*, by František Míča, was staged (revived there 1980). This, whether or not sung in Czech, was the first opera to be given a printed Czech translation. The Italian opera buffa and baroque drama influenced the simple Hanatic operas (from the district of Haná), composed in dialect, and *Pargamotéka* (1747) by Alanus Plumlovský (1703–59), performed in the Premonstratensian monastery at Hradiště from 1747; they also influenced the *Opera de rebellione Boëmica rusticorum* (Opera on the Bohemian Peasants' Revolt, 1775–7) by the East Bohemian teacher Jan Antoš (*fl.* 2nd half 18th cent.) and *Opera Bohemica de Camino* (The Bohemian Chimney Opera) by Karel Loos (1724–72), one of a number of 'craft' operas written in Czech. The Italian travelling companies visited Prague and elsewhere: Denzio settled in Prague and presented opera, the Mingottis brought opera buffa, and G. B. Locatelli introduced Gluck conducting his own works. In the second half of the 18th cent. these were gradually replaced by German companies, producing Singspiels, vaudevilles, and plays using songs and dances in Czech and German. As elsewhere, Italian opera appealed mainly to connoisseurs, Singspiel (often very simple) to a wider public. The castle productions died out during the first half of the 19th cent.; some were given in Náměšť, where works by Naumann, Gluck, and Handel were staged, mostly in German translations by Count Haugwitz and performed by his domestic musicians.

During these years, most Czech musicians lived abroad and worked in the main operatic languages of the day. Those working chiefly in Italian included Florian Gassmann (1729–74), Josef *Mysliveček (1737–81), and J. A. Koželuh (Kozeluch) (1738–1814); in German, Jiří Benda (1722–95), Pavel Vranický (Paul *Wranitzky) (1756–1808), and Vojtěch Jírovec (Adalbert *Gyrowetz) (1763–1850); in French, Josef Kohaut (1738–?1793) and Anton Rejcha (1770–1836). Ferdinand *Kauer (1751–1831) and Wenzel *Müller (1767–1835) worked mainly in Viennese Singspiel. Jan Rösler (1771–1813) and Václav *Tomášek (1774–1850) remained in Prague but wrote to German or Italian texts. Despite the earlier amateur operas in Czech, the first real Czech opera composer was František *Škroup (1801–62), with *Dráteník* (The Tinker, 1826). There is little influence of Weber (who had worked in Prague, 1813–16, introducing a wider repertory, and whose *Der Freischütz* was given there in 1824), and the work is closer to pre-Romantic Singspiel; whereas *Žižkův dub* (Žižka's Oak, 1841), by František Kott (1808–84), despite Italian influence, is in various ways Weberian. Škroup's lead was not decisively

followed, and his own later operas were mixed in language and genre. The most successful German-language Czech opera was *Bianca und Giuseppe* (1848; text by Richard Wagner) by Jan Kittl (1806–68).

It took Bedřich *Smetana (1824–84) to recognize that a more mature method than a simple musical play using folk-songs was demanded as a basis for the development of a true national opera. In 1868–74 he was conductor of the Provisional T (opened 1862). He impressed his audiences with *Braniboři v Čechách* (The Brandenburgers in Bohemia, 1866), the prize-winner in Count Harrach's 1861 competition for the best Czech historical and comic opera; but he won their hearts with *Prodaná nevěsta* (The Bartered Bride, 1866). He further set an example of how to treat heroic as well as folk themes in a Czech idiom with *Dalibor* (1868). In the same year building began of the National T, intended as a symbol of the nation's independent spirit. Smetana's *Libuše* was produced there in June 1881, before the completion of the building, which shortly burnt down. Of *Dvořák's operas (by which he set considerable store), only *Rusalka* has entered the international repertory; other of his works, especially *Jakobín* (The Jacobin, 1889) and *Čert a Káča* (The Devil and Kate, 1899), have remained in Czech repertories, but though wholly Czech in spirit and of much charm in their handling of history, character, and legend, they lack Smetana's essentially dramatic qualities.

Smetana's immediate successor was Zdeněk *Fibich, who absorbed some Romantic, including Wagnerian, influences into a Czech national idiom and drew upon European literature for his subjects. Others who successfully built upon Smetana's example included Richard Rozkošný (1833–1913), Vilém *Blodek (1834–74), Karel *Bendl (1838–97), Josef Nešvera (1842–1914), Karel *Šebor (1843–1903), and Karel Weiss (1862–1944). The first outstanding Czech conductor was Karel *Kovařovic (1862–1920), who introduced *Rusalka* and ran the National T 1900–20. Here he introduced much French opera, Strauss, Musorgsky, and more Wagner. Of his own operas, the most successful was *Psohlavci* (The Dogheads, 1898).

Outstanding among composers of the period before the emergence of Janáček, however, was J. B. *Foerster (1859–1951), whose *Eva* made a great impression in 1899. By now opera was securely established in Czechoslovakia, and a flourishing, independent genre: 102 new operas were given between 1900 and 1925. It still held no interest for some leading composers: Suk wrote no operas, Vítězslav *Novák (1870–1949) few, though his *Lucerna* (The Lantern, 1923) is impressive. Work of greater significance came from Otakar *Ostrčil (1879–1935), who was influenced by Fibich, with *Honzovo království* (Johnny's Kingdom, 1934); by Otakar Zich (1879–1934), with *Malířský nápad* (The Painter's Whim, 1910); and later by Rudolf Karel (1880–1945), with *Smrt kmotřička* (Godmother Death, 1933). But the period is dominated by Leoš *Janáček (1854–1928), although *Jenůfa*, performed in Brno in 1904, did not gain proper recognition until the Prague production of 1916. He is one of the greatest figures in the history of opera, and the first Czech composer to absorb completely what his country had to offer and to make of it a mature, modern, truthful, and flexible idiom at the service of profound human understanding.

He has somewhat overshadowed his successors. Alois *Hába (1893–1973) experimented with microtonal music, notably in *Matka* (The Mother, 1931). Otakar *Jeremiáš (1892–1962) wrote *Bratři Karamazovi* (1928). Jaromir *Weinberger (1896–1967) had an international success with *Švanda dudák* (Shvanda the Bagpiper, 1927). Emil František *Burian (1904–59) had a more local success with *Maryša* (1940). But the most important composer to win international fame was Bohuslav *Martinů (1890–1959), a substantial part of whose work was done outside Czechoslovakia. A later generation of important composers included Iša Krejčí (1904–68),

especially successful in his *Comedy of Errors* setting, *Pozdvižení v Efesu* (1946), Jan Hanuš (*b* 1915), Zbyněk Vostřák (1920–85), Jiří Pauer (*b* 1919), Jan F. Fischer (*b* 1921), Ilja Hurník (*b* 1922), and Ivo Jirásek (*b* 1920). The Brno composer Josef Berg (1927–71) made some interesting experiments in the 1960s with his minioperas.

Semi-independently, Slovak opera had been developing since the formation of the Czechoslovak state in 1918. The first Slovak opera composer is Ján Bella (1843–1936) with *Kováč Wieland* (Wieland the Smith, 1890, prod. 1920; libretto after Wagner), followed by Viliam Figuš-Bistrý (1875–1937) with *Detvan* (1928). The most significant has been Eugen *Suchoň (*b* 1908), whose *Krútňava* (The Whirlpool, 1949) has been widely staged, together with Ján *Cikker (1911–89), whose most important work has been *Vzkriesenie* (Resurrection, 1962). Among a younger generation, Juraj Beneš (*b* 1940) has attracted attention with more advanced techniques.

After the war the number of permanent opera companies increased to 13, some of them replacing touring companies.

See also BANSKÁ BYSTRICA, BRATISLAVA, BRNO, ČESKÉ BUDĚJOVICE, KOŠICE, LIBEREC, OLOMOUC, OPAVA, OSTRAVA, PLZEŇ, PRAGUE, ÚSTÍ NAD LABEM.

BIBL: J. Tyrrell, *Czech Opera* (Cambridge, 1988).

# D

**Dabadie, Henri-Bernard** (*b* Pau, 19 Jan. 1797; *d* Paris, May 1853). French baritone. Studied Paris Cons. Début Paris, O, 1819 (Cinna, *La vestale*). Remained at Paris, O, until 1835; here he created roles in numerous operas, including Auber's *Le philtre*, *La muette de Portici*, and *Gustave III*, and Rossini's *Le comte Ory*, *Moïse*, and *Guillaume Tell* (title-role). The first Belcore in *L'elisir d'amore*, Milan, though not to Donizetti's satisfaction. Considered rather a constrained performer; however, Rossini's influence improved his singing. He married the soprano **Louise Zulmé Leroux** (*b* Boulogne, 20 Mar. 1804; *d* Paris, Nov. 1877), who sang at the Paris, O, 1825–35. Created Sinaide (*Moïse*) and Jemmy (*Guillaume Tell*).

**da capo aria.** Although 18th-cent. arias could embrace a wide variety of moods and emotions, they were usually constructed according to a uniform tripartite scheme, in which the third section was a repetition of the first. This form is thus generally referred to as the da capo aria, a term deriving from the Italian for 'from the head' (i.e. beginning).

The da capo form appeared in the earliest operas, e.g. in the opening recitative for the shepherd in Monteverdi's *Orfeo* (1607), but its first important use was in mid-17th-cent. Venetian opera, e.g. the final duet from Monteverdi's *Poppea* (1643). Later composers, including Cavalli, Cesti, and Steffani, further developed and exploited the form, and with the rise of Neapolitan opera in the early 18th cent. established its prominent role. With A. Scarlatti, the da capo aria became the main vehicle for the expression of the characters' emotions; these moments of repose were linked by *recitative, at first secco, but later also accompanied, which advanced the dramatic action.

The da capo aria's primacy was assured by the libretto reforms of *Zeno and *Metastasio, for the structure of their verses, influenced by the *doctrine of the affections, lent itself admirably to the musical form. The text of each aria divided into two four-line stanzas. In the grand 18th-cent. da capo aria, at its most developed in Hasse's operas, the first quatrain would be set in two sections, introduced, divided, and concluded by a ritornello; before its last appearance

there would be opportunity for a cadenza. The middle part, setting the second quatrain, was in a new key with different, though usually not sharply contrasting, material; this corresponded to a new poetic idea, sometimes a simile or metaphor. It was usually concluded by another cadenza and followed by a repeat of the ritornello, leading into the reprise of the opening part. Here the singer would embellish the vocal line with ornamentation in a display of bravura technique culminating in the final, most expansive cadenza. Hence the da capo form particularly suited an era in which the solo singer's contribution to opera was highly regarded.

The da capo form was universal in the 18th cent., though the prohibition on action within the aria made it dramatically restrictive; but especially in the hands of Handel and Hasse, it became a vehicle of considerable flexibility. Partly under the influence of opera buffa, it waned in importance after 1750, as composers began to use more varied formal procedures, especially that of the *dal segno aria, though it appeared in opera seria up to the end of the century and, as in Mozart, was sometimes combined with the principles of sonata form. See also ARIA.

**Dafne.** Opera in Prologue and 6 scenes by Peri; text by Ottavio Rinuccini, after Ovid, *Metamorphoses*, i. 453–567. Prod. in Jacopo Corsi's Florence house, prob. Carnival 1598.

In the prologue, Ovid addresses the audience. Apollo has just killed the monster Pitone. He declares that he is insusceptible to love, so Amore leads him to Dafne, a beautiful nymph, and he falls in love with her and pursues her. In answer to Dafne's plea for help, Amore changes her into a laurel tree, which is ever after sacred to Apollo. The opera ends with the chorus extolling love, and expressing the hope that the feeling between men and women will always be mutual.

Generally regarded as the first opera; described as *favola pastorale*. Only fragments of the music survive, though the text (the first printed opera libretto) is complete. For other operas on the subject see DAPHNE.

**Dagestan.** Over thirty nationalities inhabit the Republic of Dagestan, founded as part of the USSR in 1921 and taking its name from the mountains of the region. For many years the music of these peoples was orally transmitted, but a concert tradition began to develop after the Second World War. In 1952 Nabi Dagirov (*b* 1921) wrote the opera *Aygazi*. In 1970 the first Dagestani opera, *Gortsy* (The Mountain Dwellers) by Shirvani Chalayev (*b* 1936), was given in Leningrad.

**Daguerre, Louis** (*b* Corneilles, 18 Nov. 1787; *d* Bry-sur-Marne, 10 July 1851). French physicist, painter, and designer. Though widely known as a pioneer of photography and inventor of the Daguerrotype (1839), he was also an influential scenic designer. After working in the theatre, he moved to the Paris Opéra in 1822, where in his first designs (with *Cicéri), for Isouard's *Aladin, ou La lampe merveilleuse*, he exploited the theatre's installation of gaslight with his own gift for lighting effects. At the Diorama, the theatre he also opened 1822, he developed a number of new pictorial and mechanical effects, and he made use of these at the Opéra to support Cicéri in giving French *Grand Opera a new scenic realism and splendour. He was largely responsible for banishing the old Italian system of wings, with their formation of a series of corridors and denial of perspective; and his contribution to the heightened realism of Cicéri's sets included the creation of the illusion of moving clouds, perhaps covering and then revealing the moon (which might move slowly during a scene), trees casting shadows when the sun came out, and many similar devices. He was responsible for the introduction of the panorama, especially exploited by Cicéri.

**Dahlhaus, Carl** (*b* Hanover, 10 June 1928; *d* Berlin, 13 Mar. 1989). German musicologist. Studied Göttingen with Gerber, Freiburg with Gurlitt. A critic of exceptional range, insight, intelligence, and energy. He was chief editor of the Wagner Collected Edition, and some of his most valuable work was on Wagner and 19th-cent. issues. Some of these (e.g. *Richard Wagners Musikdramen*, Veber 1971, trans. Cambridge, 1979) are valuable as perceptive and original introductions to their subject; some explore particular aesthetic issues in depth; some address themselves to more technical matters. His *Die Musik des 19. Jahrhunderts* (Wiesbaden, 1980, trans. 1989) provides a synoptic and sometimes idiosyncratic view, relating the music of the period to its intellectual and aesthetic world.

**Daland.** A Norwegian sea captain (bs), father of Senta, in Wagner's *Der fliegende Holländer*.

**Dalayrac, Nicolas-Marie** (*b* Muret, 8 June 1753; *d* Paris, 26 Nov. 1809). French composer. Studied Toulouse and in Paris with Langlé. His first stage works date from 1781, but his real success came with the production of *Nina* in 1786; this set a fashion for sentimental comedy, and already contains the seeds of the greater human and musical subtlety he was to bring into opéra-comique. It influenced Paisiello in his setting of the same story (1789), which in turn influenced the development of *opera semiseria. During the 1790s his works were among the most significant to reflect the perilous and alarming times in their plots, their handling of character, settings frequently gloomy or sinister, and in the new attention to moral issues. *Camille* concerns a wife's unjust imprisonment. *Léon* and especially *Léhéman* similarly treat imprisonment and freedom in the *rescue opera genre, and are also significant for their pioneering use of *reminiscence motive to draw music and plot into a closer dramatic relationship. This new development, with its consequent greater attention to the role of the orchestra in opera, helped to win Dalayrac a significant following in Germany, where his works figured in many repertories and helped to provide example and stimulus to the development of German Romantic opera. As well as writing memorable short romances, he was able to use ensemble and chorus with a new dramatic fluency. In these matters, especially in his intelligent and inventive use of motive, he influenced Méhul and hence the next generation of French and German opera composers, Berlioz and Weber among them.

SELECT WORKLIST (all first prod. Paris): *Nina, ou La folle par amour* (Nina, or The Girl made Mad through Love) (Marsollier; 1786); *Azémie, ou Le nouveau Robinson* (Azémia, or The New Robinson Crusoe) (Lachabeaussière; 1786); *Sargines, ou L'élève d'amour* (Sargines, or Love's Pupil) (De Monvel; 1788); *Les deux petits Savoyards* (The Two Little Savoyards) (Marsollier; 1789); *Raoul, sire de Créqui* (Monvel; 1789); *Camille, ou Le souterrain* (Camille, or The Cavern) (Marsollier; 1791); *Gulnare, ou L'esclave persanne* (Gulnare, or The Persian Slave) (Marsollier; 1797); *Léon, ou Le château de Monténéro* (Léon, or The Castle of Monténéro) (Hoffman; 1798); *Adolphe et Clara, ou Les deux prisonniers* (Adolphe and Clara, or The Two Prisoners) (Marsollier; 1799); *Léhéman, ou La tour de Neustadt* (Léhéman, or the Tower of Neustadt) (Marsollier; 1801); *Gulistan* (Étienne & Lachabeaussière; 1805).

BIBL: R. Guilbert de Pixérécourt, *Vie de Dalayrac* (Paris, 1810).

**Dalibor.** Opera in 3 acts by Smetana; text by Josef Wenzig, after the legend, translated from German into Czech by Ervín Špindler. Prod. Prague, Novoměstské T, on the occasion of the laying of the foundation stone of the Prague, N, 16 May 1868, with Benevicová-Milcová, Ehrenberg, Lukas, Barcal, Lev, Šebesta, Paleček, cond. Smetana; Chicago, Sokol H, 13 Apr. 1924; Edinburgh, King's T, by Prague, N, 17 Aug. 1964, with Domanínská, Přibyl, cond. Krombholc.

Prague, late 15th cent. Dalibor (ten) has been captured by his enemies and imprisoned for his assassination of the Burgrave in revenge for the killing of his friend Zdeněk. The Burgrave's daughter Milada (sop) begins to feel pity for him. Disguising herself as a boy, she apprentices herself to the gaoler Beneš (bs), and takes Dalibor's beloved violin to his dungeon. She is wounded in the attempt to rescue Dalibor and dies in his arms; he stabs himself.

In Czechoslovakia the work has been regarded as a *rescue opera with a significance for national liberty comparable to that of *Fidelio*. Also opera by Kott (1846).

**Dalila.** The Philistine temptress (mez) in Saint-Saëns's *Samson et Dalila*.

**Dallapiccola, Luigi** (*b* Pisino, 3 Feb. 1904; *d* Florence, 19 Feb. 1975). Italian composer. One of the leading dodecaphonist composers of his country, he successfully used the formal demands of the technique to give expression to an innate lyricism and love of the voice. *Volo di notte* is a skilful attempt to put contemporary ideas and incidents on the operatic stage without severing connections with tradition. *Il prigioniero* is a more powerful work in which Dallapiccola's intense sympathy with prisoners finds strong expression in the theme of the Inquisition captive almost allowed to escape. Dallapiccola declared that his operas concern 'the struggle of man against some force much stronger than he', and this finds final expression in his philosophical quest opera *Ulisse*.

WORKLIST (all texts by Dallapiccola): *Volo di notte* (Night Flight) (after Saint-Exupéry; Florence 1940); *Il prigioniero* (The Prisoner) (after Villiers de L'Isle Adam; Florence 1950); *Ulisse* (after Homer; Berlin 1968).

BIBL: R. Schackelford, *Dallapiccola on Opera* (London, 1987).

**Dalla Rizza, Gilda** (*b* Verona, 12 Oct. 1892; *d* Milan, 5 July 1975). Italian soprano. Studied Bologna with Ricci and Orefice. Début Bologna 1912 (Charlotte). Created Magda (written for her by Puccini) in *La rondine*, and leading roles in many other operas, including *Il piccolo Marat* (Mascagni), *Giulietta e Romeo* (Zandonai). Rome 1919 (Suor Angelica and Lauretta); London, CG, 1920; Milan, S, from 1923. With her clear, agile voice and versatile dramatic skills, she was notable in Verdi (especially as Violetta) and the verismo repertory. Retired 1939; taught at Venice Cons. 1939–55. (R)

**Dallas.** City in Texas, USA. The first Opera House opened 1883 with *Iolanthe*, and was for long host to touring companies (especially from Chicago and the New York, M). Dallas Civic O was formed in 1957 as an extension of the Symphony Orchestra's activities, when Lawrence Kelly came from the Chicago O. Nicola Rescigno was music director 1957–89. He opened with Zeffirelli producing Simionato in *L'italiana in Algeri*, following this in 1958 with Callas as Violetta. Subsequent seasons included important US premières and débuts, e.g. Sutherland (*Alcina*), Caballé, Domingo. The company became The

Dallas Opera in 1981, and performs in the State Fair Park Music School (cap. 3,420). Rescigno's vigorous, generally Italianate, policies included the première of Argento's *The Aspern Papers* (1988) as well as the engagement of top singers made possible by local wealth. The orchestra separated from the opera in 1989.

**'Dalla sua pace'.** Ottavio's (ten) aria in Act I of Mozart's *Don Giovanni*, declaring that his own joy or sorrow depends on that of Donna Anna.

**Dal Monte, Toti** (*b* Mogliano Veneto, 27 June 1893; *d* Treviso, 26 Jan. 1975). Italian soprano. Studied Venice with Marchisio, later with Pini-Corsi. Début Milan, S, 1916 (Biancofiore, *Francesca da Rimini*). Appearances in the provinces, then regularly at Milan, S, from 1922; Chicago 1924–8; New York, M, 1924–5; London, CG, 1925; Melbourne, Tokyo, Berlin. After retiring in 1943, she appeared in several films. Her voice was pure and remarkably agile, and she possessed a touching quality in both her singing and characterization. Celebrated as Gilda, Lucia, and Amina, she later sang Butterfly and Mimì. Autobiography, *Una voce nel mondo* (Milan, 1962). (R)

**Dalmorès, Charles** (*b* Nancy, 1 Jan. 1871; *d* Hollywood, 6 Dec. 1939). French tenor. Studied Rouen with Vergnet. Stage début Rouen 1899 (Siegfried). Brussels, M, 1900–6; Paris, T du Château d'Eau, 1902 (Tristan); London, CG, 1904–11; New York, Manhattan OC, 1906–10; Bayreuth 1908 (Lohengrin); Chicago 1910–18. Famous for his Faust (Gounod), Pelléas, Samson, and Herod. A good musician and actor, he possessed a large, resonant and sometimes plangent voice. (R)

**dal segno aria** (It.: lit. 'from the sign' aria). A modified version of the *da capo aria, increasingly popular in the mid-18th cent., when such arias began to increase in length. After the middle section, the singer would not return to the very beginning of the aria for the (ornamented) repeat, but only sing from a sign placed in the score part-way through the first section, thus reducing the overall length.

**Damase, Jean-Michel** (*b* Bordeaux, 27 Jan. 1928). French composer. Studied Paris with Busser. His skilfully written, eclectic, but also effective works include a number of operas, several to texts by Jean Anouilh, including *Eurydice* (1972).

**Dame blanche, La** (The White Lady). Opera in 3 acts by Boieldieu; text by Eugène Scribe, after Scott's novels *The Monastery* (1820) and *Guy Mannering* (1815). Prod. Paris, OC, 10 Dec. 1825, with Rigaut, Boulanger, Desbrosses, Henry, Ponchard, Féréol, Frimin, Belnie; Lon-

don, DL, 9 Oct. 1826; New Orleans, T d'Orléans, 6 Feb. 1827, with Moutonnier, Alexandre.

Count Avonel's castle, Scotland, 1759.
George Brown (ten), an English officer, asks
the Scottish farmer Dickson (ten) for lodging,
and relates his half-forgotten origins and his life
as a soldier. Jenny (sop), Dickson's wife, tells
him that the neighbouring castle is to be sold
since the owners are Jacobites who must flee to
France: the farmers hope to combine to buy
the castle, and Dickson has sworn to serve a
mysterious white lady said to be in the castle.
She has summoned Dickson, but George offers
to go instead. In the castle Gaveston (bs), the
former steward, enters with his ward Anna
(sop), who has promised to reveal the family
secret. George is admitted, and eventually is
left alone. He falls into a doze, and sees the
white lady: it is in fact Anna, who has recognized him as a young officer she once nursed,
but he believes her to be a ghost. Next morning, at the sale of the castle, Anna approaches
George, and on her instructions from the 'white
lady' he outbids his rivals. Later, in the castle,
he is led to the treasure hidden in the pedestal
of the statue of the white lady by Anna, who
has taken the statue's place: he turns out to be
the rightful heir, and when Anna is unveiled
they fall into one another's arms.

Also parodies by A. Müller, *Die schwarze Frau*
(The Black Lady, 1826) and Lally-Tollendal (*La
dame blanche de Blacknels*, 1827).

**damigella** (It.: 'damsel, maid'). Developing
from the young servant or confidante of the
*commedia dell'arte, the damigella was in turn
the ancestress of the *soubrette. Usually pretty,
impertinent, flirtatious, and considerably
shrewder than many of the upper-class characters, she often took a crucial hand in the manipulation of the plot. A classic instance of the
damigella is in Monteverdi's *Poppea*, where
Ottavia's (unnamed) maidservant explains
something of the nature of love to the agitated
valetto (Ottavia's page), providing a light interlude before the drama of Seneca's death. See
also INGÉNU.

**Damnation de Faust, La.** *Légende dramatique* in
4 parts by Berlioz; text by the composer, after
Goethe's *Faust*, ptI (1773–1808), in Gérard de
Nerval's translation (1828 and 1840), including
contributions from Almire Gandonnière. Completed 1846, incorporating *Huit scènes de Faust*
(1828) in altered form. Concert perf. Paris, OC, 6
Dec. 1846, with Duflot-Maillard, Roger, Her-
mann-Léon, cond. Berlioz. Prod. Monte Carlo
18 Feb. 1893, with D'Alba, J. De Reszke, Mel-
chissédec, cond. Jehin; Liverpool 3 Feb. 1894,
by Carl Rosa OC; New York, M, 7 Dec. 1906.

The plot makes use of episodes from Goethe,
with Méphistophélès (bs) tempting Faust (ten) to
bargain his soul in exchange for renewed youth.
Faust grows sated with the pleasures provided by
Méphistophélès, which include drunken com-
panionship in Auerbach's Cellar, military glory
(Berlioz introduces his version of the Rákóczy
March), nature, and love for Marguerite (sop),
whom he betrays: finally he is borne off to Hell in
a Ride to the Abyss, while Marguerite is saved.

There have been many subsequent attempts at
staging a work that was originally entitled 'opéra
de concert' and not conceived for the theatre
(though Berlioz had the intention of rewriting it
for staging), e.g. by Béjart at the Paris, O, in
1964, and by Geliot at the London, C, 1969.

**'D'amor sull'ali rosee'.** Leonora's (sop) aria in
Act IV of Verdi's *Il trovatore*, in which she sings
of her love for Manrico as she stands beneath the
tower in which he is imprisoned.

**Damrosch, Leopold** (*b* Posen, 22 Oct. 1832; *d*
New York, 15 Feb 1885). German violinist and
conductor. Studied Berlin with Ries, Dehn, and
Böhmer, and gained experience as concert conductor. In 1871 moved to New York, where his
many activities included, after the disastrous
1883–4 Italian season at the Metropolitan, organizing a German season 1884–5. He conducted all
the performances, which included the first M
*Walküre*. His elder son **Frank** (*b* Breslau, 22 June
1859; *d* New York, 22 Oct. 1937) studied with his
father and Moszkowski; chorus-master, New
York, M, 1885–91. His younger son **Walter** (*b*
Breslau, 30 Jan. 1862; *d* New York, 22 Dec.
1950) studied with Bülow. Assistant cond. to his
father 1884–5, continuing at the Metropolitan
under Seidl. Cond. many US premières,
especially of Wagner, 1885–91. Formed his own
opera company 1894 (see below), returning to
Metropolitan, 1900–3. Autobiography, *My Musical Life* (New York, 1923). (R)

SELECT WORKLIST (Walter): *The Scarlet Letter*
(Latrop, after Hawthorne; Boston 1896); *The Dove
of Peace* (Irwin; Philadelphia 1912); *Cyrano de Ber-
gerac* (Henderson, after Rostand; New York 1913);
*The Man without a Country* (Guiterman, after Hale;
New York 1937).

**Damrosch Opera Company.** New York company formed and financed by Walter Damrosch
in 1894. Singers in first three seasons included
Gadski, Sucher, Klafsky, Ternina, Brema, Alvary, Bispham, and Fischer. The first season's
profits exceeded $50,000 despite coast-to-coast
Wagner touring. Melba offered to join 1898, and
her manager, C. A. Ellis, also joined what
became the Damrosch-Ellis OC; other singers
now included Calvé and Lilli Lehmann. Disbanded through debt 1899.

**Damse, Józef** (*b* Sokołów, 26 Jan. 1789; *d*
Rudno, 15 Dec. 1852). Polish composer. During
a long career working in various capacities in the
Warsaw theatre, he composed, arranged, or compiled a very large number of operas, operettas,
and other musical pieces, all of them serving a
general public and making few artistic pretensions.

**Danaïdes, Les.** Opera in 5 acts by Salieri; text by
F. L. G. Lebland du Roullet and L. T. Tschudi,
partly trans. from a libretto by Calzabigi. Prod.
Paris, O, 26 Apr. 1784.

Greece, mythological times. All except one of
the 50 Danaïdes, the daughters of Danaus (bs),
obey their father and kill their husbands, the sons
of Aegyptus, Danaus' brother, on whom he seeks
vengeance. The exception, Hypermnestra (sop),
flees to Memphis with her husband Lynceus
(ten). A flash of lightning destroys Danaus'
palace; he himself is chained to a rock in Tarta-
rus, with a vulture gnawing at his entrails. The
Danaïdes are also tormented, by evil spirits and
fire.

**Dance of the Apprentices.** Dance by the Nuremberg apprentices and the girls from Fürth in
Act III of Wagner's *Die Meistersinger von Nürnberg.*

**Dance of the Blessed Spirits.** Dance in Act II of
Gluck's *Orfeo ed Euridice.*

**Dance of the Comedians.** Dance in Act III of
Smetana's *The Bartered Bride* in which the clowns
and dancers of the travelling circus are put
through their paces.

**Dance of the Hours.** The entertainment put on
by Alvise Badoero for his guests in Act III of Ponchielli's *La Gioconda*, in which the eternal
struggle between dark and light is symbolized.

**Dance of the Seven Veils.** Salome's dance
before Herod in Strauss's *Salome*, after he has
promised her anything: she chooses the head of
John the Baptist.

**Dance of the Tumblers.** Dance of the *skomorokhi* (itinerant dancers, tumblers, and singers)
before Tsar Berendey in Act III of Rimsky-Korsakov's *The Snow Maiden.*

**Dance round the Golden Calf.** Act II of Schoenberg's *Moses und Aron*, in which the people
indulge in an orgy at Aaron's instigation.

**Danchet, Antoine** (*b* Riom, 7 Sep. 1671; *d* Paris,
20 Feb. 1740). French librettist. Professor of rhe-
toric at the Jesuit College in Chartres, after 1696
he lived in Paris, where for a while he acted as
censor. In 1698 he began his famous collaboration with Campra on many opéras-ballets and
tragédies lyriques, the most famous of which

were *Tancrède* (1702), *Les amours de Mars et
Vénus* (1712), *Les fêtes vénitiennes* (1710), and
*Achille et Déidamie* (1732). In a partnership comparable to that of Lully and Quinault, Danchet
and Campra conquered the French stage, creating spectacular and forceful works which
vividly reflect the spirit of French classical tragedy. Danchet's libretto for *Idomenée* (1712) was
adapted by Varesco for Mozart's *Idomeneo*
(1781).

**Danco, Suzanne** (*b* Brussels, 22 Jan. 1911). Belgian soprano. Studied Brussels Cons. and Prague
with Carpi. Début Genoa 1941 (Fiordiligi). Success throughout Italy, incl. at Milan, S, 1947–8;
Naples, C, 1949; Gly. 1948–9; London, CG,
1951; also Vienna, Aix, Edinburgh, etc. A versatile singer of great musicality, successful as
Donna Elvira, Mimì, Mélisande, Marie (*Wozzeck*), Jocasta, Ellen Orford. (R)

**Dandini.** Prince Ramiro's valet (bar) in Rossini's
*La Cenerentola.*

**D'Angeri, Anna** (*b* Vienna, 14 Nov. 1853; *d*
Trieste, 14 Dec. 1907). Austrian soprano. Studied Vienna with Marchesi. Début Mantua 1872
(Selika). Vienna, H, 1873–80; Milan, S,
1878–81. Verdi's first Amelia in the revised *Boccanegra* (1881). London, CG, 1874–7 (Ortrud,
Venus). A handsome woman with a beautiful,
even voice, she was at her best in Verdi roles
(Aida, Elisabeth, Hélène), but refused Verdi's
offer of the first Desdemona, having retired in
1881.

**Daniel, Paul** (*b* Handsworth Wood, 5 July 1959).
English conductor. Studied London GSM, Italy
with Ferrara, London with Boult and Downes.
Staff conductor London, C, 1981–5; London,
Opera Factory, founder conductor 1981, and
music director 1987–90; music director O North
from 1990; also Nancy (Fr. première *King
Priam*). Repertory incl. *Calisto*, *Return of
Ulysses*, *Così*, *Don Giovanni*, first British *Jerusalem*, *Attila*, *Tosca*, *Mahagonny*, *Stone Guest*, *Akhnaten*, *Punch and Judy*, *Mask of Orpheus.*

**D'Annunzio, Gabriele** (*b* Pescara, 12 Mar. 1863;
*d* Gardone, 1 Mar. 1938). Italian writer, His
multifarious activities included work in the musical world. He was editor (more in name than in
fact) of the *Raccolta nazionale delle musiche ita-
liane*, together with Malipiero, Casella, Pizzetti,
and others; and the series of 36 vols. *I classici
della musica italiana* (from 1917) included several
operas. A qualified admirer of Wagner, he was
opposed to verismo and sought to encourage classical and pre-classical Italian opera. During his
military occupation of Fiume he drew up a con-
stitution that included two sections giving a central importance to music in the life of the state,

and he planned to build an arena for 10,000 spectators who would be admitted free of charge to musical events. Music, especially the operas of Wagner, Verdi, Monteverdi, and Marcello, plays an important part in his creative writing. He wrote the libretto of *Parisina* for Mascagni and of *Fedra* for Pizzetti, with whom he formed a close artistic association. Operas on his works are as follows:

*Francesca da Rimini* (1901): Zandonai (1914)
*La figlia di Jorio* (1904): Franchetti (1906); Pizzetti (1954)
*La fiaccola sotto il moggio* (1905): Pizzetti (*Gigliola*, unfin.)
*Il sogno d'un tramonto* (1905): Malipiero (unprod., 1913)
*La nave* (1908): Montemezzi (1918)
*Fedra* (1909): Pizzetti (1915)
*Parisina* (1913): Mascagni (1913)

BIBL: A. Casella and others, *Gabriele d'Annunzio e la musica* (Milan, 1939).

**Danon, Oskar** (*b* Sarajevo, 7 Feb. 1913). Serbian conductor. Studied Prague with Křička and Dĕdeček. Conductor Sarajevo 1938–41. Director Belgrade O 1945–59, doing much to establish the company's national and then international reputation with performances of, in particular, the Russian repertory. Though himself a composer, he was less committed to building up a modern repertory. From 1960 he concentrated on conducting, and appeared in most European countries, the USA, and Japan. (R)

**Dante Alighieri** (*b* Florence, May 1265; *d* Ravenna, 14 Sep. 1321). Italian poet. Operas based on *L'inferno* from *La divina commedia* (*c*.1307–21?) are as follows:

The Francesca da Rimini episode, v. 116–42: Carlini (1825); Mercadante (1828); Generali (1829); Staffa (1831); Fournier-Gorre (1832); Aspri (1835); Borgatta (1837); Morlacchi (unfin., 1839); Devasini (1841); Canetti (1843); Brancaccio (1844); Marcarini (1871); Goetz (1877); Moscuzza (1877); Cagnoni (1878); Thomas (1882); Nápravník (1902); Rakhmaninov (1906); Mancinelli (1907); E. Ábrányi (1912); Zandonai (1914)
The Gianni Schicchi reference, xxx. 32–3: Puccini (1918)
The Ugolino episode, xxxiii: Dittersdorf (1796)

Operas on Dante himself are by Carrer (*Dante e Beatrice*, 1852); Philpot (*Dante and Beatrice*, 1889); Godard (*Dante*, 1890); Foulds ('concert opera' *The Vision of Dante*, 1904); Gastaldon (*Il sonetto di Dante*, 1909); Nouguès (*Dante*, 1931).

**Dantons Tod** (Danton's Death). Opera in 2 parts by Gottfried Von Einem; text by the composer and Blacher, after Büchner's drama (1835). Prod. Salzburg 6 Aug. 1947, with Cebotari, Pat-

zak, Schoeffler, cond. Fricsay; New York, City O, 9 Mar. 1966, with Grant, Lampi, Dupree, Reardon, Beattie, cond. Märzendorfer.

Paris, 1794. Danton (bar) is disillusioned by the French Revolution but sees little hope of changing the present situation. He and Desmoulins (ten), whose views accord with those of Danton, are arrested by Robespierre (bs) after disagreeing with him. Before the revolutionary tribunal, Danton makes a speech censuring the imminent dictatorship; he is sentenced to death. The mob greets his execution with singing and dancing.

**Danzig.** See GDAŃSK.

**Dapertutto.** The evil sorcerer (bar) in the Giulietta episode of Offenbach's *Les contes d'Hoffmann*.

**Daphne.** In Greek mythology, the daughter of the River Peneios. She fled from Apollo's advances, and when he caught her she was saved by being changed into a laurel tree. Apollo made the plant sacred to him, its leaves a reward for success in the arts. The most familiar version of the story is that told by Ovid in his *Metamorphoses*, i. 453–567. Operas on the legend are by Peri (1598), Caccini (lost, perhaps prod. 1600), Gagliano (1608), Schütz (1627, music lost), Barbazza (1634), Pascatti (1635), Ariosti (1696), Aldrovandini (1696), A. Scarlatti (1700), Pollarolo (1705), Astorga (1709), Fux (1714), Caldara (1719), Reutter (1734), Porta (1738), Hensel (1798), Ferron (1892), Bird (1897), R. Strauss (1938).

**Daphne.** Opera in 1 act by Richard Strauss; text by Josef Gregor, after the classical legend. Prod. Dresden 15 Oct. 1938, with Teschemacher, T. Ralf, Kremer, cond. Böhm (in a double bill with *Friedenstag*); Santa Fe 29 July 1964, with Stahlman, Shirley, Petersen, cond. Crosby; Leeds, Grand T, 2 May 1987, with Field, cond. Lloyd-Jones.

By the hut of Peneios, mythological times. Daphne (sop), daughter of Peneios (bs) and Gaea (mez), is reprimanded by her mother for rejecting the shepherd Leukippos (ten), who loves her. The god Apollo (ten) comes down to earth disguised as a shepherd; he falls in love with Daphne, but she refuses to yield to him. Having revealed himself as the sun, Apollo kills Leukippos. Daphne is distraught, and Apollo begs Zeus to transform her into one of her laurel trees.

**Da Ponte, Lorenzo** (orig. Emanuele Conegliano) (*b* Ceneda, 10 Mar. 1749; *d* New York, 17 Aug. 1838). Italian librettist and poet. He studied for the priesthood, but was expelled from his seminary for adultery. In 1783 he settled in Vienna and, despite having written no librettos, was appointed poet to the Imperial Theatres.

After making a successful Italian adaptation of Guillard's *Iphigénie en Tauride*, his first original work came with *Il rico d'un giorno*, written for Salieri in 1784. For several years he played a major role in Viennese operatic life: in addition to his librettos for Mozart, he wrote for all the leading Viennese composers, including Gazzaniga (*Il finto cieco*, 1786), Martín y Soler (*L'arbore di Diana*, 1787), and Salieri (*Il pastor fido* and *La cifra*, both 1789). In 1791 he was ordered to leave Vienna, having fallen foul of Leopold II. He became poet at London, H, and had a notorious feud with *Badini. He wrote *Zaira* and *Il ratto di Proserpina* (both for Winter, 1804) and established himself as a printer, but was forced to flee secretly to America in 1805 because of debt. From 1826 to 1837 he held a chair in Italian at Columbia U and also wrote his entertaining, Casanovesque autobiography. In 1825 he and Manuel García were among the first to give opera in the USA, and in 1833 he was responsible for the establishment of the Italian OH in New York.

Although he wrote a total of 36 librettos, Da Ponte is rightly immortalized by the three masterpieces he produced for Mozart, *Le nozze di Figaro*, *Don Giovanni*, and *Così fan tutte*, which represent a partnership unrivalled in operatic history. In each case, pre-existing sources were carefully sifted, assimilated, and reshaped into librettos in which the focus falls firmly on the interrelationship of the characters. These are distinguished by their expression of genuine emotion, rather than the stereotyped sentiments typical of contemporary opera buffa. Perhaps most importantly, although Da Ponte's librettos are firmly rooted in the conventions, social distinctions, and attitudes of the day—many of which he satirized—the characters invariably convey a timeless message as relevant to a modern audience as to that of Enlightenment Vienna. Recently, Da Ponte has also been suggested, on stylistic grounds, as the librettist of Mozart's uncompleted opera buffa *Lo sposo deluso* (1783).

BIBL: L. Da Ponte, *Memorie* (4 vols., 1823–7, 2/ 1829–30, trans. 1929, R/1967); S. Hodges, *Lorenzo da Ponte* (London, 1985).

**Darclée, Hariclea** (*b* Brăila, 10 June 1860; *d* Bucharest, 10 Jan. 1939). Romanian soprano. Studied Bucharest, and Paris with Faure. Début Paris, O, 1888 (Marguerite). Milan, S, from 1890; Moscow; St Petersburg, 1893–6; Buenos Aires, and New York, Academy T, 1896. Possessing physical and vocal beauty, she had success in a repertory ranging from Bellini and Donizetti heroines, Gilda, and Violetta, to Aida, Desdemona, and Valentine; she also sang many verismo roles, though her rather cold temperament was not ideally suited to the dramatic demands of the genre. She created La Wally, Tosca, and Iris. Though rich when she retired in 1918, she was forced to spend her declining years in the Verdi Home, Milan, and died penniless. (R) Her son **Ion Hartulary-Darclée** (*b* Paris, 7 July 1886; *d* Bucharest, 2 Apr. 1969), composer and conductor, studied Paris with Leroux and Widor, Milan with Amedeo and Victor de Sabata. Conducted in Italy, France, Switzerland, and Spain 1915–30. His operas are *Jaretiera* (1909), *Capriciu antic* (1911), *Amorul mascat* (1913), *Anonima potin* (1916), *La signorina sans-façon* (1920), *Marjery* (1927), *Zig-zag* (1928), *Operetta* (1929), *Atlantic-City* (1930).

**Dardanus.** Opera in 5 acts by Rameau; text by Le Clerc de la Bruyère. Prod. Paris, O, 19 Nov. 1739.

Asia Minor, Antiquity. Teucer (bs) is at war with Dardanus (ten), son of Jupiter, and has promised to marry his daughter Iphise (sop) to Antenor (bs-bar), a neighbouring king. With the help of the magician Ismenor, Dardanus meets Iphise; they reveal their love for each other. He is saved from being killed through the intervention of Venus (sop); and other gods join her. A monster is sent to ravage Teucer's coasts, but Dardanus now comes to his enemy's aid and destroys the monster just as it is attacking Teucer. Peace is made between them, and Dardanus and Iphise are united.

Other operas on the subject are by Stamitz (1770) and Sacchini (1784).

**Dargomyzhsky, Alexander** (*b* Troitskoye, 14 Feb. 1813; *d* St Petersburg, 17 Jan. 1869). Russian composer. Originally destined for the Civil Service, he won himself a reputation in St Petersburg as a pianist and attempted his first opera, a setting of Hugo's *Lucrèce Borgia*. Abandoning this, he turned to another Hugo novel and completed *Esmeralda* in 1840. Though he later dismissed it as weak and trivial, over-influenced by Halévy and Meyerbeer (he might have added Auber), he was depressed by its failure to find a production and by the growing reputation of Glinka. In the 1850s he turned to the study of Russian folk-song, interesting himself especially in its dramatic and comic aspects, where Glinka had been drawn by its lyrical qualities. This led to a striking vividness in the recitative passages of *Rusalka* alongside the more immediately captivating songs: the comic role of the Miller was to become a favourite of singers including Shalyapin, and helped to win the opera wide popularity. However, dissatisfied with his artistic aims, he found himself drawn to the artists and writers associated with the satirical journal *Iskra* who were influenced by the Realist critics Belinsky and Chernyshevsky. Attempting to follow their precepts of an art that directly reflected truth in human behaviour, with as little intervention by artistic forms and formulas as possible, he began work on *Rogdana*, only to abandon it in dissatisfaction. For a time he travelled, and concentrated on orchestral music; but returning to Russia he found that some of his songs, each of which offers a vivid and immediately recognizable representation of a particular human character, had attracted the interest of Balakirev and his circle. These helped to give him a technical basis when he turned for his last opera to one of Pushkin's 'little tragedies', *The Stone Guest*. Though already fatally ill, he worked at the opera with much support from his friends (at regular meetings to try out what he had written, Musorgsky would sing Leporello and the composer Don Juan, with Rimsky-Korsakov's future wife Nadezhda accompanying). He all but finished the work on his deathbed, leaving Cui to complete the last 60 bars of vocal score and Rimsky-Korsakov to orchestrate it.

Apart from two interpolated songs, *The Stone Guest* is a virtually complete setting of Pushkin's poem, composed with the aim of enhancing the verbal music and expressing the poetry musically, rather than subjecting it to compositional forms of any kind. In practice, some kind of compromise inevitably results; but Dargomyzhsky is ingenious in avoiding the risk of monotony in his unending recitative, and can move from simple parlando to a more lyrical manner, choosing his verbal emphases subtly and providing discreet but well-judged orchestral support. This, together with the use of a whole-tone scale motive for the Stone Guest, led to misjudged accusations of Wagnerism; but Dargomyzhsky's manner is radically different. *The Stone Guest* directly colours Rimsky-Korsakov's *Mozart and Salieri*, as well as the declamatory manner of Cui's *Feast in Time of Plague* and Rakhmaninov's *The Miserly Knight*. It has won a lasting place in Russian music as the classic of the Realist genre.

SELECT WORKLIST: *Esmeralda* (Hugo, trans. Dargomyzhsky; comp. 1838–40, prod. Moscow 1847); *Rusalka* (Dargomyzhsky, after Pushkin; St Petersburg 1856); *Kammeny Gost* (The Stone Guest) (Pushkin; St Petersburg 1872).

BIBL: M. Pekelis, *A. S. Dargomyzhsky i evo okruzheniye* (A. S. Dargomyzhsky and his Circle) (Moscow, 1966–71).

**Darmstadt.** City in Hesse-Darmstadt, Germany. A 'Singballet' *Die Befreiung des Friedens* was given for a royal christening 1600. A theatre opened 1670, giving hybrid works, later many French ones. The Ducal theatre opened 1810, and the Grosses Haus (cap. 1,370) opened 1819.

In 1919 the theatre came under the control of Hesse. Karl Böhm was music director 1927–31. The theatre was bombed 1944, and performances were given in the Orangerie, then the Stadthalle, 1945–72. The Grosses Haus (cap. 956) opened 1972 with *Fidelio*. The company has a striking record of enterprise and innovation.

**Daudet, Alphonse** (*b* Nîmes, 13 May 1840; *d* Paris, 16 Dec. 1897). French writer. Operas on his works are as follows:
*L'arlésienne* (1872): Cilea (1897)
*Sapho* (1884): Massenet (1897)
*Tartarin sur les Alpes* (1885): Pessard (1888)

**Dauvergne, Antoine** (*b* Moulins, 3 Oct. 1713; *d* Lyons, 11 Feb. 1797). French violinist, composer, and theatrical director. Joined the orchestra of the Paris, O, 1744; from 1755 composer and master of the *chambre du roi*. A staunch supporter of Gluck, and an admirer of Italian music, he was director of the Paris, O, on four occasions between 1766 and 1790, for a total of 14 years.

His earliest dramatic work was the ballet *Les amours de Tempé* (1752); the following year he composed his most famous opera, the 1-act *Les troqueurs*. Performed at the Foire Saint-Laurent in the aftermath of the \*Guerre des Bouffons, it was presented as a translation of an Italian opera buffa: such was its success that it effectively ended the operatic contretemps by exposing its underlying ridicule. Historically, *Les troqueurs* is important as the first French work on the model of the Italian intermezzo, using recitative instead of spoken dialogue, and it enjoyed popularity in many European countries.

Dauvergne's other important works include *La coquette trompée*, *Enée et Lavinie*, *Canente*, *Hercule mourant*, *Polyxène*, and *La vénitienne*. He also arranged many operas by other French composers, including Campra.

SELECT WORKLIST (all first prod. Paris): *Les troqueurs* (Vade, after La Fontaine; 1753); *La coquette trompée* (Favart; 1753); *Enée et Lavinie* (Fontenelle; 1758); *Canente* (La Motte; 1760); *Hercule mourant* (Marmontel; 1761); *Polyxène* (Joliveau; 1763); *La vénitienne* (La Motte; 1768).

**Davenant,** (Sir) **William** (*b* Oxford, *bapt.* 3 Mar. 1606; *d* London, 7 Apr. 1668). English dramatist and theatre manager. His *The Siege of Rhodes* (1656), given at Rutland House with music (now lost) by various composers including Locke, is often cited as the first English opera, despite being preceded by Flecknoe's *Ariadne* (1654). By announcing *The Siege of Rhodes* as 'music and instruction' Davenant managed to evade the contemporary ban on stage plays. The work was notable for including the first appearance of a woman on stage in England. With Dryden, Davenant adapted many of Shakespeare's plays to suit contemporary taste; some of these adaptations were performed with a substantial musical component as was the fashion of the day (e.g the semi-opera *The Tempest*, ?1695, attrib. Purcell, but possibly by Weldon).

**David.** Hans Sachs's apprentice (ten) in Wagner's *Die Meistersinger von Nürnberg*.

**David.** Opera in 5 acts by Milhaud; text by Lunel. Prod. (concert version in Hebrew) Jerusalem 1 June 1954 for 3,000th anniversary of foundation of Jerusalem. First stage perf. Milan, S, 2 Jan. 1955, with Pobbe, Ratti, Gardino, Colzani, Rossi-Lemeni, Tajo, cond. Sanzogno; Hollywood Bowl, 22 Sep. 1956, with Nelli, Dixon, Presnell, Harrell, Tozzi, cond. Solomon.

The opera is based on the events chronicled in Sam. 1 and 2. David (bar), a warrior and poet, is chosen to succeed Saul (bar) as King of Israel. The giant Goliath (bs) challenges the Israelites to send a champion to fight him. David slays Goliath, and Saul's jealousy is such that he attempts to murder David. Saul and his son Jonathan are killed in battle shortly afterwards. As the new King of Israel, David chooses the site for the new city of Jerusalem. He takes another wife, Bathsheba, and their second son Solomon eventually succeeds David as King, thus fulfilling the prophecy.

Also operas by Sutor (1812) and Galli (1904).

**David, Félicien** (*b* Cadenet, 13 Apr. 1810; *d* Saint-Germain-en-Laye, 29 Aug. 1876). French composer. Studied Aix-en-Provence with Michel and Roux, Paris with Millault and Fétis. He espoused Saint-Simonism, and when the sect was dispersed in 1832 he left for Egypt. Here he found inspiration for his 'oriental' works, which included operas. *Herculanum* is in Grand Opera vein, and naturally ends with the eruption of Vesuvius; but he won greater popularity with the more modestly atmospheric *Lalla-Roukh*. He also stimulated more distinguished contemporaries (among them Gounod, Bizet, and especially Delibes) in their oriental operas. His delicate, tuneful music was much admired by Bizet, whose most immediate predecessor he was.

SELECT WORKLIST: *La perle du Brésil* (Gabriel & Saint-Étienne; Paris 1851); *Herculanum* (Méry & Hadot; Paris 1859); *Lalla-Roukh* (Lucas & Carré, after Moore; Paris 1862); *La captive* (Carré; comp. 1860–4, unprod.); *Le saphir* (Carré, Hadot, & De Leuven, after Shakespeare; Paris 1865).

BIBL: R. Brancour, *Félicien David* (Paris, 1909); D. Hagan, *Félicien David 1810–1876: a Composer and a Cause* (Syracuse, NY, 1985).

**David, Giacomo** (*b* Presezzo, 1750; *d* Bergamo, 31 Dec. 1830). Italian tenor. Studied Naples with Sala. Début Milan 1773. His long career was spent mostly in Italy, but he also sang in St Petersburg (1783) and London (1791). He sang at the openings of several major theatres: Venice, F, 1792; Trieste, T Nuovo, 1801; Milan, T Carcano, 1803. Created roles in numerous operas by Sarti, Cimarosa, Paisiello, Mayr, etc., and sang with Farinelli, Pachierotti, Catalani, and Banti. He was one of the first to challenge the supremacy of the castrati, and was, like them, a virtuoso singer. His voice was really a baritone with a falsetto extension, powerful yet capable of sensitive expression; his diction was clear and his style graceful. Mount-Edgcumbe considered him the best tenor of the day. His pupils included his son Giovanni and Nozzari.

**David, Giovanni** (*b* Naples, 15 Sep. 1790, *d* St Petersburg 1864). Italian tenor. Son of above, with whom he studied. Début Siena 1808 (Mayr's *Adelaide di Guesclino*). Milan, S, 1814, created Narciso in *Il turco in Italia*; Rossini subsequently wrote other parts for him, e.g. in *Otello, Ermione, La donna del lago, Zelmira*. Also created roles in operas by Pacini, Mayr, Donizetti, etc. London 1829. Sang until 1839, then manager of the St Petersburg O. Stendhal refers to his having 'something like true genius', and thought him the greatest tenor of his generation. His voice was full-toned, extremely flexible, and by virtue of a skilful falsetto covered three octaves. Though his singing was later sometimes in bad taste, at his peak it aroused universal admiration. His daughter **Giuseppina** (1821–1907) sang in Rome and St Petersburg.

**Davies, Cecilia** (*b* London, ?1756; *d* London, 3 July 1836). English soprano. Appeared Dublin 1763–4; London 1767. Studied Vienna 1768–9 with Hasse; here she also sang with her sister Marianne (a virtuoso glass-harmonica player), and taught Maria Theresa's daughters singing and acting. Milan 1771, then Naples and Florence, where she created Sacchini's Armida. Much admired by Hasse, Metastasio, and the Italians, who called her 'L'Inglesina', she was one of the first English singers to compete successfully with Italian prima donnas, and was considered second only to Adriana Gabrieli. Returned to London 1774–7; Florence 1784–5. Retired 1791; died in obscurity.

**Davies,** (Sir) **Peter Maxwell** (*b* Manchester, 8 Sep. 1934). English composer. Studied Manchester, Rome with Petrassi. Early in his career he became interested in the dramatic presentation of music, sometimes using a single costumed performer (as with the startlingly Expressionist *Revelation and Fall* and *Eight Songs for a Mad King*) or a dancer (*Vesalii Icones*). These brilliantly effective pieces draw on a wide range of sources, medieval music and Lehár waltzes in the first, free and controlled improvisation in the second (whose published score reflects the cage which on the stage partly represents the King's madness), Victorian hymns in the last. These diverse kinds of music are brought together sometimes for dramatic contrast or shock, sometimes for the shedding of mutual light, sometimes curdled by parody. There is throughout a powerful feeling for dramatic effect that by no means excludes doing violence to the music parodied as well as to the audience's sensibilities. There is even an underlying note of blasphemy, disclosed also in the opera *Taverner* (Christ set against Antichrist), where despite some excellent and characteristic music, drawing on Taverner's own, the stage convention did not seem ideal for Davies's imagination.

After Davies moved to Orkney in the early 1970s, the religious element that he had been contesting came to the fore, and the challenge of working with amateurs and children (which he had brilliantly met early in his career at Cirencester Grammar School) led to the highly effective chamber opera *The Martyrdom of St Magnus*, which contains elements of mystery play and also dislocation of music as well as of time and place; also to a charming piece for children, *The Two Fiddlers*, requiring the young soloists both to play and to sing. The experience of living in a remote, bare landscape stimulated much in his art (even though his compelling personality drew audiences from far and wide to take part in his musical events), and one of his strongest stage works resulted in the taut, compelling opera *The Lighthouse*. As before, diverse musics, including banjo songs and Victorian ballads, are used in a score of frightening claustrophobic power as the drama of vanished lighthousemen becomes a parable of extremes of metaphysical experience.

SELECT WORKLIST: *Revelation and Fall* (Davies, after Trakl; London 1968); *Taverner* (Davies, after 16th-cent. documents; comp. 1962–70, prod. London 1972); *Eight Songs for a Mad King* (Stow; London 1969); *Vesalii Icones* (Davies, after Vesalius; London 1969); *The Martyrdom of St Magnus* (Davies, after Mackay Brown; Kirkwall 1977); *The Two Fiddlers* (Davies, after Mackay Brown; Kirkwall 1978); *The Lighthouse* (Davies, after fact; Edinburgh 1980); *The No. 11 Bus* (Davies; London 1984).

BIBL: P. Griffiths, *Peter Maxwell Davies* (London, 1982).

**Davies, Ryland** (*b* Cwm, Ebbw Vale, 9 Feb. 1943). Welsh tenor. Studied Manchester with Freddie Cox. Début WNO, 1964 (Rossini's Count Almaviva). Gly. 1968; London, CG, from 1969; Salzburg 1970; Paris 1974; New York, M, 1975. Repertory includes Belmonte, Don Ottavio, Ernesto, Hylas, Lensky. Possesses a pleasing light tenor, and a sympathetic stage presence. (R)

**Davies, Tudor** (*b* Cymmer, 12 Nov. 1892; *d* Penault, 2 Apr. 1958). Welsh tenor. Studied London with Gustave García. Joined BNOC at its inception; London, CG, début 1921 (Rodolfo). Created Hugh the Drover, and leading roles in Holst's *At the Boar's Head* and Smyth's *Fête galante*. San Francisco 1928 (Lohengrin). Old Vic and SW 1931–41; London, CR, 1941–6; then retired. Possessed a fresh, warm voice that later deteriorated through singing the wrong roles. (R)

**Davis, Andrew** (*b* Ashridge, 2 Feb. 1944). English conductor. Studied Rome with Ferrara. Début Gly. 1973 (*Capriccio*). Music director, Gly., 1988. New York, M, from 1981; also Milan, S; London, CG. Repertory incl. *Figaro*, *Barber*, *Salome*, *Ariadne*, *Káťa Kabanová*, *Jenůfa*. (R)

**Davis, (Sir) Colin** (*b* Weybridge, 25 Sep. 1927). English conductor. Studied clarinet at London, RCM. Début as conductor with Chelsea Opera Group, London 1949 (*Impresario*, concert). After appointment with BBC Scottish Orch., London, SW, 1958, music director 1961–4. Gly. 1960. London, CG, 1965 (*Figaro*), music director 1971–86. New York, M, 1967. Bayreuth 1977 (first British conductor). He first made his name in Mozart, and his clarity of line but passionate feeling have been no less manifest in the composer with whom he became even more closely associated, Berlioz. His performances of Tippett won the trust of the composer, as well as the admiration of the audiences, for their lucidity in articulating textures and meanings that had often proved confusing. (R)

BIBL: A. Blyth, *Colin Davis* (London, 1972).

**Davydov, Stepan** (*b* ?, 1777; *d* St Petersburg, 22 May 1825). Russian composer. Studied St Petersburg with Sarti. He worked in the St Petersburg Drama School 1800–4 and 1806–10. He was music director of Count Sheremetyev's private theatre from 1815, and also taught singing at the Moscow Drama School. He is best known for his work on *Lesta*. Basically a German Singspiel, which had been brought to Russia by *Cavos, the piece was reworked and augmented by Davydov. Making use of both town and country songs, he gave it a distinctive Russian flavour and at the same time a distinctive Romantic colouring. He is one of the most important Russian opera composers before Glinka, who knew *Lesta* and may have been influenced by it.

SELECT WORKLIST (all first prod. St Petersburg): *Rusalka* (additional numbers for Kauer's *Das Donauweibchen*: Krasnopolsky, after Hensler; 1803); *Lesta, dneprovskaya rusalka* (Lesta, The Dnyepr Water Nymph) (Krasnopolsky; 1803); *Rusalka* (Shakhovskoy; 1807).

# De Amicis

**De Amicis.** Italian family of singers.
  1. **Antonio Domenico** (*b* Fermo, *c*.1716; *d* ?). Italian tenor. Sang at Naples 1744–51. Appeared in Dublin 1762, with company which he had taken over from Antonio Minelli; then at London, H. He married the soprano Rosalba Baldacci (*b* Atri, *c*.1716; *d* ?).
  2. **De Amicis Buonsollazzi, Anna Lucia** (*b* Naples, *c*.1733; *d* Naples, 1816). Italian soprano. Daughter of above, with whom she studied. Début Bologna 1755 (Galuppi's *La calamità dei cuori*). Pursued highly successful career in London, where she was J. C. Bach's favourite singer, and in Italy, retiring in 1779. Much admired by Mozart, who wrote Giunia in *Lucio Silla* for her, and his father; also by Metastasio, who thought her an excellent actress, and Burney, who noted that she was the first to sing staccato divisions. Her voice went up to e♭''', and was clear and powerful.

**De Angelis, Nazzareno** (*b* Rome, 17 Nov. 1881; *d* Rome, 14 Dec. 1962). Italian bass. Studied with Ricci and Prati. Début Aquila, 1903 (Podestà, *Linda di Chamounix*). S America 1909–25; Chicago 1911–12. Repertory included *La vestale*, *Medea*, *Barbiere*, *Don Carlos*, *Tristan*, and *Walküre*. Created Archibaldo in Montemezzi's *L'amore dei tre re*. A moving and noble actor, he was the best Italian bass of his generation. His most famous role was Mefistofele (Boito), which he sang 987 times. (R)

**Death in Venice.** Opera in 2 acts by Britten; text by Myfanwy Piper, after Thomas Mann's novella *Der Tod in Venedig* (1912). Prod. Snape 16 June 1973, with Pears, Shirley-Quirk, Bowman, Bergsma, Huguenin, choreography Ashton, prod. Graham, cond. Bedford; New York, M, 18 Oct. 1974, with Pears, Shirley-Quirk, cond. Bedford.

Munich and Vienna. Aschenbach (ten), a famous writer, leaves Germany for Venice. Here he succumbs to the allure of a boy, Tadzio (dancer), who represents the Dionysiac side of his character which he has hitherto suppressed beneath a cool Apollonian view of life. He tries to leave the cholera-stricken city, but is frustrated by an accident to his luggage, and returns to indulge his solitary and self-devouring obsession. Watching Tadzio playing on the beach with other boys, he collapses and dies.

The opera is cast in scenes connected by interludes, and makes prominent use of dance; the roles of the figures (such as an Elderly Fop) who contribute to Aschenbach's downfall are all taken by the same baritone.

BIBL: D. Mitchell, *Benjamin Britten: 'Death in Venice'* (Cambridge, 1987).

**De Bassini, Achille** (*b* Milan, 5 May 1819; *d* Cava dei Tirreni, 3 July 1881). Italian baritone. Studied Milan with Perelli. Début possibly Voghera 1837 in *Belisario* (Donizetti) and then *Norma*. Padua 1838 and 1842; Milan, S, from 1842. Rome 1844, creating Francesco in *I due Foscari*. Repeated this success at Milan, S, where he also sang in *Ernani*, *Attila*, and *Alzira*, as well as many new operas by Rossi, Ricci, etc. Created Seid in *Il corsaro* (Trieste 1848); Miller in *Luisa Miller* (Naples 1849). Verdi, seeing in him 'a certain humorous vein', despite his talent for noble and suffering characters, wrote Fra Melitone in *La forza del destino* for him; he had also thought first of him when conceiving *Rigoletto*. St Petersburg 1852–63. London, CG, 1859 (Germont, Luna, etc.). His wife, the soprano (really mezzo-soprano) **Rita Gabussi** (*b* Bologna, *c*.1815; *d* Naples, 26 Jan. 1891), made her début Milan 1830 (Rosina). Successful throughout Italy, particularly as Giovanna in Ricci's *La prigione di Edimburgo*, and Mercadante's Medea (roles she created). Her gift for portraying violent, dramatic women caused Verdi to request her, unsuccessfully, as the first Azucena. Her voice declined prematurely through singing roles too high for her. Their son **Alberto** (*b* Florence, 14 July 1847; *d* ?) was a tenor, later baritone. Début Venice 1869 (*Belisario*). After success in Rossini and Donizetti, specialized in French repertory, 1880–90. Emigrated to USA 1898. (R)

**De Begnis, Giuseppe** (*b* Lugo, 1793; *d* New York, Aug. 1849). Italian bass. Début Modena 1812 (Pavesi's *Ser Marcantonio*). Became leading buffo; engaged by Rossini 1817 as first Dandini. Paris, I, 1819–22. London, H, 1821–3, with his wife, in *Il turco in Italia*, and *Mosè*. Directed opera at Bath 1823–4, and Dublin 1834–7. A distinguished Dr Bartolo (*Barbiere*). His wife was the soprano **Giuseppina Ronzi de Begnis** (*b* Milan, 11 Jan. 1800; *d* Florence, 7 June 1853). Début Bologna 1816. After successes in Italy (including Nina in *La gazza ladra* under Rossini), Paris, I, 1819–22, where she was a notable Donna Anna. London 1821–3, in Rossini. Separating from De Begnis in 1825, she returned to Italy and sang at Naples, and Milan, S. Created Donizetti's Fausta, Sancia di Castiglia, Gemma di Vergy, Elisabeth in *Roberto Devereux*, and Maria Stuarda (during the rehearsals for this she attacked her rival Anna del Serre, who had to retire to bed for a fortnight). London, CG, 1843 (Norma). An admirable singer, she was also 'a superb figure of a woman' (Cottrau), and was reputedly the mistress of several younger men, including Donizetti (who compared her, as an artist, favourably with Malibran), and Ferdinand II of Sicily.

**Debora e Jaële.** Opera in 3 acts by Pizzetti; text by the composer, after Judges 16. Prod. Milan, S, 16 Dec. 1922, with Tess, Casazza, Pinza, cond. Toscanini.

Israel, *c.*1100 BC. Jaële (sop) is accused by the Hebrews of friendship with their enemy, Sisera (ten). Debora (mez) promises them victory and persuades Jaële to go to the enemy's camp and kill Sisera; but when the moment comes, Jaële cannot bring herself to do this. The Hebrews launch a successful battle and Sisera seeks refuge with Debora, who kills him to save him from more horrible tortures at the hands of his captors.

**Debrecen.** City in Hajdú-Bihar, Hungary. Touring companies first performed in 1799, and the first permanent opera company was formed in 1860. The Csokonai T (cap. 698) opened in 1865. The present company was formed in 1952, and tours the region.

**Debussy,** (Achille-) **Claude** (*b* Saint-Germain-en-Laye, 22 Aug. 1862; *d* Paris, 25 Mar. 1918). French composer. Studied Paris with Lavignac, Guiraud, and others including, briefly, Franck. Despite plans for operas on the subject of Tristram and Yseult and on *As You Like It*, among other projects, Debussy's only completed opera is *Pelléas et Mélisande*. His attraction to Maeterlinck (he had previously unsuccessfully sought permission to set *La Princesse Maleine*) was in part for the opportunity its shadowy, Symbolist world offered as a counter to Wagner and verismo. The score is certainly non-Wagnerian in its discretion, its understatement, and its very French attention to niceties of language-setting; but the shadow of *Parsifal* falls (constructively) across the work, and Debussy's very personal use of motive (generally evoking atmosphere or states of mind) would not have evolved as it did without the example of Wagner to resist. No less strong an influence was *Boris Godunov*, whose declamatory style and episodic structure also impressed Debussy. His command of atmosphere lends strength to the rather slender Symbolist structure of Maeterlinck, who was unwise enough to oppose the opera and even threaten Debussy with assault when the role of Mélisande was removed from his wife Georgette Leblanc, and given to Mary Garden by Carré, director of the Opéra-Comique. Further problems during the work's preparation for production arose when it was discovered that more time was needed for the scene changes: from this arose the interludes now generally regarded as a vital part of the musical fabric.

Of Debussy's other operatic projects, the two that most occupied him were both attempts at setting Poe, who greatly influenced his ideas. *Le diable dans le beffroi* suggests a novel approach to operatic choral writing (and has a whistling part for the Devil); *La chute de la maison Usher* was a further exploration of a world of morbid melancholia implicit in *Pelléas* and here made the emotional centre of what survives of the work.

WORKLIST: *Axël* (Villiers de l'Isle Adam; 1 scene comp. *c.*1888); *Rodrigue et Chimène* (Mendès, after Castro & Corneille; vocal score of parts, comp. 1890–2; completed Denisov, Lyons 1993); *Pelléas et Mélisande* (Debussy, after Maeterlinck; Paris 1902); *Le diable dans le beffroi* (Debussy, after Poe; part composed 1902–11); *La chute de la maison Usher* (Debussy, after Poe; part composed 1908–17, reconstruction by W. Harwood prod. New Haven 1977).

CATALOGUE: F. Lesure, *Catalogue de l'œuvre de Claude Debussy* (Geneva, 1977).

WRITINGS: *Monsieur Croche antidilettante* (Paris, 1921, trans. 2/1962).

BIBL: R. Orledge, *Debussy and the Theatre* (Cambridge, 1982).

**Decembrists, The.** Opera in 4 acts by Shaporin; text by Vsevolod Rozhdestvensky, using verses by A. N. Tolstoy and based on events of the Dec. 1825 rising. Prod. Moscow, B, 23 June 1953, with Ivanov, Pirogov, Petrov, Selivanov, Verbitskaya, Pokrovskaya, Kositsina, cond. Melik-Pashayev.

Russia, 1825. Dmitri Shchepin-Rostovsky (ten) is discontented with the system of ownership and serfdom, and joins forces with young aristocrats who also want to see an end to the system. The Decembrists plan an attack to capture the new Tsar, but the leaders, including Dmitri, are arrested, and sentenced to exile in Siberia. He is followed there by Yelena (sop), who loves him.

Also opera by Vasily Zolotaryov (1925, rev. as *Kondraty Ryleyev*, 1957).

**declamation.** The relationship, in the setting and/or delivery of a text, between verbal stress and musical accent. Used in a musical context to imply that the delivery of the words is nearer to spoken drama than singing, and that the text, rather than the vocal melody, is of primary importance. Much early music in *stile rappresentativo is declamatory, as was *recitativo secco. Reserved for moments of high emotion, declamation is an effective weapon in the composer's dramatic armoury.

**'Deh, vieni alla finestra'.** Don Giovanni's (bar) serenade to Elvira's maid in Act II of Mozart's *Don Giovanni*, in which he sings to his own mandolin accompaniment (often played pizzicato on an orchestral violin).

**'Deh, vieni non tardar'.** Susanna's (sop) aria in Act IV of Mozart's *Le nozze di Figaro*, in which she tantalizes Figaro.

**Deidamia.** Opera in 3 acts by Handel; text by P. Rolli. Prod. London, Lincoln's Inn Fields,

10 Jan. 1741, with Andreoni, Francescina, Edwards; Hartt Coll., Elmwood, CT, 25 Feb. 1959.

Handel's last opera; originally unsuccessful, being given for three nights only.

Other operas on the subject are by Campra (1735), Maréchal (1893), and Rasse (1906).

**De Lara** (orig. Cohen), **Isidore** (*b* London, 9 Aug. 1858; *d* Paris, 2 Sep. 1935). English composer. Studied Milan with Mazzucato, Paris with Lalo. Maurel suggested that he should rewrite his cantata *The Light of Asia* as an opera, which was performed as *La luce dell'Asia* at London, CG, 1892. His *Amy Robsart* was given a single performance there the following year. His compositions owe much to Massenet and Saint-Saëns. He constantly campaigned for the establishment of a permanent British opera company.

**De L'Épine, Margherita.** See ÉPINE, MARGHERITA DE L'.

**Delibes, Léo** (*b* Saint-Germain-du-Val, 21 Feb. 1836; *d* Paris, 16 Jan. 1891). French composer. Studied Paris, with Tariot and Adam, whose opéras-comiques much influenced him. His first stage work, described as an 'asphyxie lyrique', was *Deux sous de charbon*, staged at the Folies-Nouvelles. This initiated a long series of operettas that continued with the hugely successful *Deux vieilles gardes*. He first attempted opéra-comique with *Le jardinier et son seigneur*; this was given in 1863 at the Lyrique, where he worked as chorus-master before moving on to the Opéra, first as accompanist and then as chorus-master under Massé. This turned his interests towards ballet, and he won fame with *Sylvia* and *Coppélia*. But opera drew him back, and he wrote three works for the Opéra-Comique, *Le roi l'a dit*, *Jean de Nivelle* (a work close to Grand Opera style), and his masterpiece *Lakmé*. In keeping with the current vogue for the East, an opéra-comique tradition refreshed by the Romantic poets' fascination, and stimulated further by the example of Félicien ★David, Delibes set his tale of the English lieutenant and the Indian girl with a wealth of delightful mock-oriental melody. It is chiefly for the charming tunefulness and the decorative scoring that *Lakmé*, which owes much to Bizet, and the ballets retain an appeal which led Tchaikovsky to prefer Delibes to Brahms and Wagner.

SELECT WORKLIST (all first prod. Paris): *Deux sous de charbon* (Moineaux; 1856); *Deux vieilles gardes* (De Villeneuve & Lemonnier; 1856); *Le roi l'a dit* (Gondinet; 1873); *Jean de Nivelle* (Gondinet & Gille; 1880); *Lakmé* (Gondinet & Gille; 1883); *Kassya* (Meilhac & Gille; 1891, orch. Massenet, prod. 1893).

BIBL: A. Coquis, *Léo Delibes: sa vie et son œuvre (1836–1891)* (Paris, 1957).

**Delius, Frederick** (*b* Bradford, 29 Jan. 1862; *d* Grez-sur-Loing, 10 June 1934). English com-

poser. Studied privately during his time working on an orange grove in Florida, later in Leipzig (less beneficially) with Sitt, Reinecke, and Jadassohn. Delius's operas belong to the earlier part of his career. *Irmelin* has few of his mature characteristics apart from harmonic felicity and a feeling for orchestral colour; but more personal experiences, especially those deriving from his Florida period, show in the more vivid score of *Koanga*. His most striking work, however, is *A Village Romeo and Juliet*. Here his acute sense of chromatic harmony (deriving in part from Grieg) and his expressive orchestration are at the service of a pessimistic tale of doomed young love that owes something to aspects of Wagner. *Margot-la-rouge* was a verismo attempt to win an international Sanzogno competition. His last opera, *Fennimore and Gerda*, reflects his love of Scandinavia and is a lighter-textured work than *A Village Romeo and Juliet*, whose techniques it in other ways reflects. In part through Beecham's advocacy, the characteristic short orchestral pieces that are a feature of Delius's operas have entered the concert repertory.

WORKLIST: *Irmelin* (Delius; comp. 1890–2, prod. Oxford 1953); *The Magic Fountain* (Delius; comp. 1893–5, unprod.); *Koanga* (Keary, after Cable; comp. 1895–7, prod. Elberfeld 1904); *Romeo und Juliet auf dem Dorfe* (A Village Romeo and Juliet) (Keary, after Keller; comp. 1900–1, prod. Berlin 1907); *Margot-la-rouge* (Rosenval; comp. 1902, prod. St. Louis 1984); *Fennimore and Gerda* (Delius, after Jacobsen; comp. 1909–10, prod. Frankfurt 1919).

CATALOGUE: R. Threlfall, *A Catalogue of the Compositions of Frederick Delius* (London, 1977).

BIBL: A. Jefferson, *Delius* (London, 1972); C. Redwood, 'Delius as a Composer of Opera', in id., ed., *A Delius Companion* (London, 1976); I. Carley, ed., *Delius: A Life in Letters* (London, 1983–8).

**Della Casa, Lisa** (*b* Burgdorf, 2 Feb. 1919). Swiss soprano. Studied Zurich with Haeser. Début Solothurn-Biel 1941 (Butterfly). Zurich 1943–50; Gly. 1951 (Mozart's Countess); Bayreuth, 1952 (Eva); Vienna, S, 1952–74; New York, M, 1953–68; London, CG, 1953, and 1965 in her most famous role, Arabella. Created roles in Burkhard's *Die schwarze Spinne*, and Von Einem's *Der Prozess*. Retired 1974. An outstanding Richard Strauss and Mozart singer with a controlled legato and finely spun vocal line. (R)

BIBL: D. Debeljević: *In dem Schatten ihre Locken* (Zurich, 1975).

**Deller, Alfred** (*b* Margate, 31 May 1912; *d* Bologna, 16 July 1979). English counter-tenor. Self-taught. While a member of Canterbury Cathedral choir, was heard by Tippett, who arranged his London début, Morley College 1943. Founded Deller Consort, 1950. Influential in the revival of the counter-tenor voice, he performed neglected music of the 16th and 17th cents., but also contemporary works. Inspired many composers, including Rubbra, and Britten, who was drawn by the other-worldly quality of his voice to write Oberon in *A Midsummer Night's Dream* for him. Highly musical and intelligent, he used his pure tone to lyrical and expressive effect. (R)

BIBL: M. and M. Hardwick, *Alfred Deller: A Singularity of Voice* (London, 1968).

**Delle Sedie, Enrico** (*b* Livorno, 17 June 1822; *d* La Garenne-Colombes, 28 Nov. 1907). Italian baritone. Studied with Galeffi and Persanola. Début San Casciano 1851 (Nabucco). Milan, S, 1859; Vienna 1859; London, L, 1861, first Renato. Paris, I, 1861–74, in Donizetti, Rossini, and Verdi. London, CG, 1862 (Don Giovanni, Germont, Malatesta). Though possessing a limited voice, he was a sensitive artist, exceptional in style and taste. After his retirement, 1874, taught at Paris Conservatoire, and wrote a significant and comprehensive singing method, *L'estetica del canto e dell'arte melodrammatica* (4 vols., Paris, 1886).

**Dello Joio, Norman** (*b* New York, 24 Jan. 1913). US composer. Studied New York with Wagenaar, also with Hindemith. His most important dramatic music has been concerned with Joan of Arc. His opera *The Triumph of St Joan* was withdrawn after its première, and a second opera written for TV as *The Trial at Rouen* (1955); this was revised as *The Triumph of St Joan* (1959).

**Delmas, Francisque** (Jean-François) (*b* Lyons, 14 Apr. 1861; *d* Saint-Alban de Montbel, 29 Sep. 1933). French bass-baritone. Studied Paris with Bussine and Obin. Début Paris, O, 1886 (Saint-Bris, *Les Huguenots*). Sang there until 1927, creating more than 50 roles, including parts in *Thaïs*, *Monna Vanna*, and D'Indy's *L'étranger*. The first French Wotan (his most famous role), Hagen, and Gurnemanz, and first Paris Sachs. A noble actor with a large, smooth, and even voice, his versatility encompassed Gluck, Mozart, and Weber. (R)

**Del Monaco, Mario** (*b* Florence, 27 July 1915; *d* Mestre, 16 Oct. 1982). Italian tenor. Studied Pesaro with Melocchi; later taught himself, studying records of the great singers of the past. Official début Pesaro 1941 (Pinkerton). London, CG, 1946, 1961. San Francisco 1950; New York, M, 1950–9; also Milan, S; Buenos Aires, C; Verona. Gifted with a powerful and vigorous voice, he was popular in the verismo and Verdi repertory, particularly as Canio, Chénier, and Otello. (R)

**De Los Angeles, Victoria.** See LOS ANGELES, VICTORIA DE.

**Del Prato, Vicenzo** (*b* Imola, 5 May 1756; *d* Munich, *c.*1828). Italian soprano castrato. Studied with Gibelli. Début Fano 1772. After singing in Italy, was engaged for Stuttgart 1779. Munich 1780–1805, where he created Idamante. Mozart refers ironically to him as 'mio molto amato castrato'; he found the singer monotonous, short-breathed, uneven-voiced, and 'wretched' on stage.

**De Luca, Giuseppe** (*b* Rome, 25 Dec. 1876; *d* New York, 25 Aug. 1950). Italian baritone. Studied Rome with Persichini and Cotogni. Début Piacenza, 1897 (Valentin). Milan, L, 1902, creating Michonnet (*Adriana Lecouvreur*); Milan, S, 1903–4, creating Sharpless, and Gleby in Giordano's *Siberia*. London, CG, 1907, 1910, 1935; New York, M, 1915–35, 1939–40, 1945–6. Here he appeared over 700 times in 80 operas, creating Paquiro in *Goyescas* and Gianni Schicchi. A classic exponent of the bel canto style, his repertory covered Malatesta, Rossini's Figaro, and Alberich; he was also an admired Rigoletto. Retired 1947. (R)

**De Lucia, Fernando** (*b* Naples, 11 Oct. 1860; *d* Naples, 21 Feb. 1925). Italian tenor. Studied Naples. Début Naples, C, 1885 (Gounod's Faust). London, DL, 1887. Rome 1891. Created Amico Fritz 1891, and Osaka (*Iris*) 1898. Florence 1892, and Milan 1895, in first *I Rantzau* and *Silvano* (Mascagni). London, CG, 1892–6, 1900 (Fritz, Canio, Cavaradossi, Turiddu); New York, M, 1893–4. Retired Naples 1917, and taught; pupils included Nemeth, Pederzini, and Thill. An excellent actor, and a master of bel canto, he was a distinctive and memorable singer, renowned for his Rossini, Bellini, and Verdi, but also enormously successful in the verismo roles. (R)

BIBL: M. Henstock, *Fernando de Lucia* (London, 1990).

**De Lussan, Zélie** (*b* New York, 21 Dec. 1861; *d* London, 18 Dec. 1949). US soprano. Studied with her mother. Début Boston 1885. London, CR, 1889; London, CG, 1895–1902; Beecham OC, 1910. New York, M, 1894–5 (first US Nannetta), and 1900–1. Highly acclaimed as Carmen (which she sang over 1,000 times), Mignon, Cherubino, and Nedda. (R)

**Demetrio.** Opera in 3 acts by Hasse; text by Metastasio. Prod. Venice, GG, Jan. 1732. Rev. Vienna 1734 as *Cleonice*; also given as *Cleonice* Dresden 8 Feb. 1740.

Demetrio (sop cas) is brought up not knowing that his father was the exiled King of Syria; Cleonice (sop), daughter of the usurper Alessandro, falls in love with him. Alessandro is killed in an uprising and Cleonice, now heir to the throne, must find a husband. Eventually, Demetrio re-emerges, having been fighting on Alessandro's side; his identity is revealed and he marries Cleonice, regaining possession of his father's throne.

Among over 50 other settings of Metastasio's text are those by Gluck (1742), Wagenseil (1746), Galuppi (1748), Jommelli (1749), Paisiello (1765), Piccinni (1769), Mysliveček (1773), Mayr (1824), and Coppola (1877).

**'De' miei bollenti spiriti'.** Alfredo's (ten) aria which opens Act II of Verdi's *La traviata*, in which he sings of his happiness with Violetta.

**Demofoonte.** Opera in 3 acts by Jommelli; text by Metastasio. Prod. Stuttgart 1764.

Demophoön (ten), King of Chersonnesus, seeks to put an end to the annual sacrifice of a virgin demanded of him by the gods. The oracle of Apollo (bs) replies that the practice may cease 'when the innocent usurper of a kingdom is recognized by himself'. Demophoön selects Dircea (sop) for sacrifice, not realizing that she is secretly married to his own son Timante (sop). When Demophoön discovers this he at first condemns them both to death, but then pardons them.

Among over 70 other settings of Metastasio's text are those by Sarro, Mancini, and Leo (1 act each, 1735), Duni (1737), Lampugnani (1738), Jommelli (1743, 1753, 1764, 1770), Hasse (1748), Galuppi (1749), Traetta (1758), Piccinni (1761), Mysliveček (1769), and Paisiello (1775).

**Demon, The.** Opera in 3 acts by Anton Rubinstein; text by Pavel Viskovatov, after Lermontov's poem (1841). Prod. St Petersburg, M, 25 Jan. 1875, with Krutikova, Petrov, Raab, Komissarzhevsky, Melnikov, cond. Nápravník; London, CG, 21 June 1881, with Albani, Lassalle, E. De Reszke, the first Russian opera in London, and when given in 1888 the first opera to be sung in Russian in London; prob. Boston, 1891, by Hebrew OC, also New York, 1903.

Caucasus, legendary times. The Demon (bar), a human with satanic traits, falls in love with Tamara (sop), daughter of Prince Gudal (bs) who is to be married the following day. In the hope of winning her, he has her lover Synodal (ten) killed. Tamara hides herself away in a convent but is pursued there by the Demon. As she begins to yield, the Angel (sop) intervenes, and she is borne off to Heaven.

**Denhof Opera Company.** English opera company formed in 1910 by Ernst Denhof, a German living in Edinburgh, to give performances of The

*Ring* in English in the provinces. The first series, under Balling, was followed in 1911–12 by appearances in other cities with a repertory enlarged to include *Elektra* (first in English), *Orfeo, Fliegender Holländer, Tristan*, and *Meistersinger*. In 1913 *Pelléas, Rosenkavalier* (both first in English), and *The Magic Flute* were added; but the company was now making a serious loss. Beecham, one of the conductors, took it over with many of its artists, including Marie Brema, Agnes Nicholls, Walter Hyde, Frederic Ranalow, Frederic Austin, and Robert Radford, who soon became part of the *Beecham Opera Company.

**Denmark.** Opera was first given in the mid-17th cent., and attempts were made to initiate Danish opera with librettos by N. H. Bredal set by Sarti. These and other works were Italianate. The first opera by a Dane was *Der vereinigte Götterstreit* (1698; lost) by Poul Schindler (1648–1740). The most important composers of the 18th cent. who tried to establish Danish opera were J. E. Hartmann (1726–93) and Friedrich *Kunzen (1761–1817); the latter's *Holger Danske* (1789, after Wieland's *Oberon*) won respect. The German J. A. P. *Schulz (1747–1800) attempted a form of Danish Singspiel based on the popular French opéra-comique; he had some success with *Host-geldet* (Harvest Festival, 1790) and *Peters Bryllup* (Peter's Wedding, 1793). Romanticism was slow to influence the young tradition, but is apparent in some of the six operas by Christoph *Weyse (1774–1842); these include some to texts by Oehlenschläger and Hans *Andersen. The influence of Cherubini, and through him Weber, shows in the operas of Friedrich *Kuhlau (1786–1832).

The real influence of Romanticism is revealed in the work of J. P. E. Hartmann (1808–1900; grandson of J. E. Hartmann). He collaborated with Oehlenschläger on *Korsarerne* (The Corsairs, 1835) and with Hans Andersen on *Liden Kirsten* (Little Kristina, 1846). Niels Gade (1817–90) wrote the somewhat Mendelssohnian *Mariotta* (1850) (and left a fragmentary *Siegfried og Brunhilde*), and Peter *Heise (1830–79) *Drot og Marsk* (King and Marshal, 1878), in which there is an influence of Wagner; Berlioz is suggested in *La vendetta* (1870) by Asger Hamerik (1843–1923). Folklore and folk music are drawn upon by Peter Lange-Müller (1850–1926) in *Vikingeblod* (The Blood of the Vikings, 1900). A more eclectic composer was August *Enna (1859–1939). The dominating talent of the later period was Carl *Nielsen (1865–1931), with *Saul og David* (1902) and *Maskarade* (1906). Other successful works have included *Den kongelige gæst* (The Royal Guest, 1919) by Hakon Børresen

(1876–1954); *Fête galante* (1931) by Paul Schierbeck (1888–1949), also composer of *Tiggerens opera* (1936, a version of *The Beggar's Opera*); and *Marie Grubbe* by Poul Olsen (1922–82); *Labyrinten* (1963) and *Gilgamesh* (1973) by Per Nørgård (b 1932); and *Sandheddns Hævn* (The Revenge of Truth, 1985) by Ib Nørholm (b 1931).

See also ÅRHUS, COPENHAGEN.

**Dent, E**(dward) **J**(oseph) (*b* Ribston, 16 July 1876; *d* London, 22 Aug. 1957). English musicologist, critic, and teacher. Studied Cambridge with Wood and Stanford. His influence on English operatic life, and its relationship to the European operatic scene, was profound, in particular by reason of his practical scholarship. His work on Baroque opera led to a reappraisal of, in particular, Alessandro Scarlatti; and at a time when Mozart's operas were still in need of advocacy, he provided that with scholarship in his book on them, translations that set new and still enduring standards of dramatic liveliness, and a pioneering production of *The Magic Flute* in Cambridge in 1911 that set in motion the English Mozart revival. He also arranged and produced several early English operas, supporting his work in this area with another scholarly study. He made a neat version of *The Beggar's Opera*, and did some original composition. A brilliant, inspirational teacher and lecturer, he influenced several generations of scholars; some of his lectures were reprinted as *The Rise of Romantic Opera*.

WRITINGS: *Alessandro Scarlatti* (London, 1905, rev. 2/1960); *Mozart's Operas* (London, 1913, rev. 2/1947, rev. 3/1955); *Foundations of English Opera* (Cambridge, 1928, R/1965); *Feruccio Busoni* (London, 1933, 2/1966); *Handel* (London, 1934); *Opera* (1940, rev. 5/1949); *A Theatre for Everybody: The Story of the Old Vic and Sadler's Wells* (London, 1945, rev. 2/1946); W. Dean, ed., *The Rise of Romantic Opera* (Cambridge, 1976).

BIBL: P. Radcliffe, *E. J. Dent: A Centenary Memoir* (Rickmansworth, 1976).

**Denzler, Robert** (*b* Zurich, 19 Mar. 1892; *d* Zurich, 25 Aug. 1972). Swiss conductor. Studied Zurich with Andreae. Répétiteur Zurich and Bayreuth. Music director, Zurich O, 1915–27, 1934–47. Conducted first *Lulu* (1937) and *Mathis der Maler* (1938). (R)

**'Depuis le jour'.** Louise's (sop) aria in Act III of Charpentier's *Louise*, recalling the day when she first yielded to Julien.

**De Reszke, Édouard** (*b* Warsaw, 22 Dec. 1853; *d* Garnek, 25 May 1917). Polish bass, brother of Jean. Studied Warsaw with Ciaffei, Italy with Steller and Coletti. Début Paris, O, 1876 as the King (*Aida*), under Verdi. Milan, S, 1879–81, Fiesco in revised *Boccanegra*, and created roles in works by Ponchielli and Gomez. London, CG,

1890–4; Chicago 1891. With his brother Jean was a key member of the Paris, O, London, CG, and New York, M, companies during the last decade of the 19th cent.; they often appeared together, e.g. in *Don Giovanni*, *Faust* (Gounod), *Hérodiade*, *Roméo et Juliette*, and first *Le Cid*. A distinguished Méphistophélès (Gounod), Frère Laurent, and Leporello, he later added Wagner to his vast repertory, including Sachs (one of his greatest interpretations), Gurnemanz, Mark, and Hagen. Retired 1903, and after unsuccessful attempts to teach in London and Warsaw, withdrew to his Polish estate. Here, after the outbreak of war in 1914, he lived in extreme poverty and seclusion, first in his cellar and then in a cave. One of operatic history's great basses: a huge man with an imposing dramatic presence, possessing a voluminous, rich, and flexible voice. (R)

**De Reszke, Jean** (*b* Warsaw, 14 Jan. 1850; *d* Nice, 3 Apr. 1925). Polish tenor. Brother of Édouard. Studied Warsaw with Ciaffei, Rome with Cotogni. Début as baritone, Venice, F, 1874 (Alfonso, *La favorite*). London, DL, 1874 (Alfonso, Don Giovanni, Valentin). Paris 1876 (Fra Melitone and Rossini's Figaro). After further study with Sbriglia, made début as tenor, Madrid 1879 (Robert le diable), with little success. Paris, I, 1884 re-emerged triumphantly as John the Baptist (*Hérodiade*). Paris, O, 1885–1902, Radamès, Vasco da Gama, etc., and Roméo (under Gounod) and Rodrigue (*Le Cid*). London, DL, 1887 (Radamès, Lohengrin). London, CG, 1888–1900 (Raoul, Gounod's Faust, Walther, Siegfried, Werther). Chicago from 1891; New York, M, 1891–1901. His last appearance was his only one as Canio, Paris 1902. After retiring, he taught in Paris and Nice, his pupils including Edvina, Sayão, Slezak, and Teyte. Vocally in direct descent from Mario, and with a handsome physique, he was ideal for French and Italian romantic roles, but later also achieved huge success in the heavy Wagner repertory (acclaimed for his impeccable declamation of the German text as much as for the beauty of his singing). His astonishing versatility enabled him to sing such disparate roles as Lohengrin, Roméo (Gounod), Don José, Faust (Gounod), Radamès, and Tristan, all within days of each other. A unique figure in history, he stands at the pinnacle of operatic achievement. A record label exists, suggesting that he may have recorded. Sister to the De Reszke brothers was **Joséphine** (*b* Warsaw, 4 June 1855; *d* Warsaw, 22 Feb. 1891), soprano. Studied with Nissen-Salomon. Début Venice 1874 (*Il Guarany*). Madrid 1879 (Alice (with Jean as Robert)). Paris, O, 1874–84, very successfully as Marguerite, Rachel, Valentine, etc., and created Sita in *Le roi de Lahore*. Sang with Édouard and Jean in *Hérodiade*, Paris 1884; then retired, having married Baron Leopold de Kronenburg, and sang for charities in Warsaw.

BIBL: C. Leiser, *Jean De Reszke and the Great Days of Opera* (London, 1934).

**'Der Hölle Rache'.** The Queen of the Night's (sop) aria in Act II of Mozart's *Die Zauberflöte*, swearing revenge upon Sarastro. Famous for the top Fs it requires of the singer.

**Dérivis, Henri Etienne** (*b* Alby, 2 Aug. 1780; *d* Livry, 1 Feb. 1856). French bass. Studied Paris with Richer. Début Paris, O, 1803 (Zoroastre, *Les mystères d'Isis*, a version (or perversion) by Lachnith of *Die Zauberflöte*). Principal bass there 1808–28, creating roles in Spontini's *La vestale*, *Fernand Cortez*, and *Olympie*, Cherubini's *Les Abencérages*, and Rossini's *Le siège de Corinth*. A stentorian singer with a grand manner (his admirer, the young Berlioz, refers to his 'imposing ruggedness'), his style was more suited to Gluck than to Rossini.

**Dérivis, Prosper** (*b* Paris, 28 Oct. 1808; *d* Paris, 11 Feb. 1880). French bass, son of above. Studied Paris with Pellegrini and Nourrit. Début Paris, O, 1831 (*Moïse*). Remained until 1841, creating De Nevers (*Les Huguenots*), Balducci (*Benvenuto Cellini*), Félix (*Les martyrs*), and other roles. Milan, S, 1842–3 (first Zaccaria, *Nabucco*; Pagano, *I Lombardi*). Vienna, K, 1842 (first Prefect, *Linda di Chamounix*). Retired 1857. With *Levasseur, the best French bass of his time. His voice, a true basso profundo, was clear, vigorous, and effective in both coloratura and lyrical passages.

**D'Erlanger, Frédéric.** See ERLANGER, FRÉDÉRIC D'.

**Dermota, Anton** (*b* Kropa, 4 June 1910; *d* Vienna, 22 June 1989). Slovene, later Austrian, tenor. Studied Vienna with Rado. Début Cluj 1934. Vienna 1936–80 (First Armed Man, then Alfredo, Lensky, Rodolfo, Jeník, Des Grieux (Puccini), David, Flamand, Florestan); later sang character parts. Salzburg from 1935 (Ferrando, Belmonte, Don Ottavio). London, CG, 1947 (Ferrando, Narraboth). Also Milan, Paris, etc. A fine artist, with a sweet-timbred, distinctive voice; memorable as a Mozart singer. (R)

**Dernesch, Helga** (*b* Vienna, 3 Feb. 1939). Austrian soprano. Studied Vienna Cons. Début Berne 1961 (Marina, *Boris*). Wiesbaden 1963–5; Bayreuth 1965–8; Scottish O from 1969; Salzburg 1969–74; London, CG, from 1970. Roles include Leonore, Marschallin, Cassandre, Isolde, Brünnhilde, Sieglinde. Created Goneril (Reimann's *Lear*), title-role in Fortner's *Elisabeth Tudor*. From 1979 sang mezzo-soprano parts,

e.g. Fricka, Herodias, Dyer's Wife. An intense performer with a warm, full tone. (R)

**'Der Vogelfänger bin ich, ja'.** Papageno's (bar) aria introducing himself as the bird-catcher in Act I of Mozart's *Die Zauberflöte*.

**De Sabata, Victor** (*b* Trieste, 11 Apr. 1892; *d* S Margherita Ligure, 11 Dec. 1967). Italian conductor and composer. Studied Milan with Saladino and Orefice. His *Il macigno* (The Boulder) was prod. Milan, S, 1917. Cond. Monte Carlo 1918 (*Traviata*); conductor 1919–29, incl. first *Enfant et les sortilèges* (1925). Milan, S, 1930–9, 1941–3, 1947–53, artistic director 1953–7. Bayreuth 1939 (*Tristan*, very widely praised). London, CG, 1950 with Milan, S. Was working on a second opera at the time of his death. A brilliant, masterful conductor, whose incandescent and exciting readings suggested comparisons with Toscanini's; they were perhaps less instinctive, but more intellectually penetrating. (R)

BIBL: G. M. Gatti, *Victor de Sabata* (Milan, 1958).

**Désaugiers, Marc-Antoine** (*b* Fréjus, 1742; *d* Paris, 10 Sep. 1793). French composer. A friend of Gluck and Sacchini, he first came to attention in 1776 with a translation of a famous treatise on vocal technique by Giovanni Battista Mancini, and three years later had his first stage success with *Le petit Œdipe*, a verse comedy with ariettes and vaudevilles. This paved the way for the production of about ten successful operatic works 1780–92. Most popular were *Erixène, ou L'amour enfant* and the 1-act *Les deux jumeaux de Bergame*.

SELECT WORKLIST: *Erixène, ou L'amour enfant* (Gaillard, after Voisenon; Paris 1780); *Les deux jumeaux de Bergame* (The Twins of Bergamo) (Florian; Paris 1782).

His son **Marc Antoine Madeleine Desaugiers** (1772–1827) was a writer of popular vaudevilles, who provided librettos for Gaveaux and Piccinni among others. He also adapted Molière's *Le médecin malgré lui* as an opéra-comique libretto for his father in 1792.

**Deschamps-Jehin, Blanche** (*b* Lyons, 18 Sep. 1859; *d* Paris, June 1923). French mezzo-soprano. Studied Lyons and Paris. Début Brussels, M, 1879 (Mignon). Sang there until 1885, creating Massenet's Hérodiade, and Uta in Reyer's *Sigurd*. Paris: OC, 1885–91, O, 1891–7, then OC until 1902. London, CG, 1891–2 (Carmen and Fidès). Created Margared (*Le roi d'Ys*), and the Mother (*Louise*). A particularly successful Amneris and Dalila; possessed a rich and unusually homogeneous voice. (R)

**Desdemona.** Otello's wife (mez) in Rossini's and (sop) in Verdi's *Otello*.

**Des Grieux.** The Chevalier Des Grieux (ten), lover of Manon in Massenet's *Manon* and Puccini's *Manon Lescaut*.

**Deshayes, Prosper-Didier** (*b* ?; *d* Paris, 1815). French composer and dancer. He worked as dancer and ballet-master at the Comédie-Française, later at the Opéra. His greatest success among his opéras-comiques was *Zélia* (1791, after Goethe's *Stella*). He also contributed to the multi-authored *Le congrès des rois* in 1794.

**Desmarets, Henry** (*b* Paris, Feb. 1661; *d* Lunéville, 7 Sep. 1741). French composer. His first opera, *Endymion*, is now lost: of other early works the most successful was *Vénus et Adonis* (still in repertoire in Paris 1717 and given Hamburg 1725). He began a successful career at the Paris, O, in 1693 with *Didon*, but in 1699 was forced to leave France because of a secret marriage. Fleeing to Spain, he joined the court of Philip V, and in 1707 that of the Duke of Lorraine. Not until 1722 was he permitted to return to France.

Desmarets was one of the most important figures of the generation which preceded Rameau, although he has divided opinion: Rousseau sharply criticized his music, while Voltaire placed it on a level with Lully's. His operas, which show a generally good understanding of the tragédie lyrique, at times rank with those of Destouches and Campra in their dramatic and expressive qualities, though they are usually marred by weak librettos. The most successful was *Iphigénie en Tauride* (Paris 1704), completed by Campra, which was given 197 times up to 1763.

SELECT WORKLIST (all first prod. Paris): *Didon* (De Saintonge; 1693); *Vénus et Adonis* (Rousseau; 1697); *Iphigénie en Tauride* (Duché & Danchet; 1704).

**Désormière, Roger** (*b* Vichy, 13 Sep. 1898; *d* Paris, 25 Oct. 1963). French conductor. Studied Paris with Koechlin. Début Paris 1921. Paris: T Pigalle, 1931; OC, 1937, director 1944–6; O, assistant cond. 1945–6. London, CG, 1949 (*Pelléas*). An intelligent, versatile artist much of whose work lay outside opera, where his clear mind and enterprising taste were also manifest. Retired through ill health 1952. (R)

**Despina.** Fiordiligi's and Dorabella's maid (sop) in Mozart's *Così fan tutte*.

**Desprez, Louis-Jean** (*b* Auxerre, 9 Jan. 1742; *d* Stockholm, 17 Mar. 1804). French architect and designer. Studied Paris. Worked in Rome, then Stockholm for Gustav III at Royal OH. In his vivid, strongly charged, architecturally detailed designs, e.g. for Naumann's national opera *Gustaf Wasa* (1786), he anticipates some of the manner developed by *Cicéri and his followers in Paris in the 1820s and 1830s.

**Dessau.** City in Anhalt, Germany. The first opera given was Dittersdorf's *Das rote Käppchen* in the royal stables. The Hoftheater (cap. 1,000) opened 26 Dec. 1798 with *Bathmendi* by the director, Karl von Lichtenstein (1767–1845). This was the largest theatre after Berlin and Bayreuth, and quickly established a good reputation with Singspiel and opéra-comique, though standards had slipped by 1808; burnt down 1855 and 1922. Standards improved again under Friedrich Schneider from 1821, and especially under Eduard Thiele from 1856, building a strong Wagner tradition. The Opernhaus (cap. 1,245) opened 1885; rebuilt 1938, 1949. Hans Knappertsbusch was music director 1920–2.

**Dessau, Paul** (*b* Hamburg, 19 Dec. 1894; *d* Berlin, 28 June 1979). German composer and conductor. Studied violin in Berlin, conducting and composition (with Loewengard) in Hamburg. Music director Hamburg Kammerspiele T 1912; from 1919 under Klemperer in Cologne; Mainz 1923; Berlin, SO, from 1925. Left Germany 1933 for USA, where he began working with Brecht, returning to (East) Berlin 1948. The collaboration led to his most important stage works, in which he sought to support and interpret Brecht's ideas in dramatic music. Though highly eclectic, his style has a forcefulness and a sense of purpose that won him a keen following especially in his own country and society.

SELECT WORKLIST (all first prod. Berlin): *Die Verurteilung des Lukullus* (The Condemnation of Lucullus) (Brecht; 1951, several revisions); *Puntila* (Palitzsch, Wekwerth, after Brecht; 1966); *Lanzelot* (Müller, after Schwarz; 1969); *Einstein* (Mickel; 1973).

**Dessauer, Josef** (*b* Prague, 28 May 1798; *d* Mödling, 8 July 1876). Bohemian composer. Studied Prague with Tomašek and D. Weber. Admired as a song composer by Liszt and Berlioz, he was also a friend of Chopin. His operas, which make use of various Romantic conventions, include *Lidwina* (1836), *Ein Besuch in St Cyr* (1838), *Paquita* (1851), *Dominga* (1860), and an unprod. *Oberon*. In Paris in 1842 he persuaded Wagner to draft him a libretto on Hoffmann's *Die Bergwerke zu Falun*, which was turned down by the Opéra.

**Destinn** (orig. Kittlová), **Emmy** (*b* Prague, 26 Feb. 1878; *d* České Budějovice, 28 Jan. 1930). Czech soprano. Studied Prague with Marie Loewe-Destinn, adopting her name for professional purposes. Début Dresden 1897 (Santuzza). London, CG, 1904–14, 1919; New York, M, 1908–16, 1919–21. Bayreuth, 1901 (Senta). Berlin, 1898–1908. Prague, 1914 (Libuše). After internment during the war, her powers declined, and she retired after 1921. Created Minnie. Her wide repertory included Donna Anna, Aida,

# Destouches

Salome, Lisa, Mařenka, Butterfly. A fine actress, especially in tragic roles, while her very individual voice was well produced, even, and wide-ranging in colour. (R)

**Destouches, André Cardinal** (*bapt.* Paris, 6 Apr. 1672; *d* Paris, 7 Feb. 1749). French composer. Studied Paris, then served as a soldier in Siam, returning for further study with Campra. The pastoral *Issé* (1697) was well received and launched him on a successful theatrical career. Much admired by Louis XIV, who saw him as a worthy successor to Lully, Destouches succeeded Francine as director of the Académie Royale de Musique in 1728, where he had been inspector-general since 1713, but only held his new post for two years.

He composed *c.*12 large-scale dramatic works, of which the most highly regarded was the opera *Callirhoé*. An important figure in the post-Lully generation, Destouches is sometimes compared favourably with Rameau on account of his carefully crafted melodic lines and richly expressive harmonic vocabulary. Though Destouches concentrated on tragédie lyrique, early works such as *Amadis de Grèce* and *Omphale* contain a certain element of pastoral relief which suggests a close acquaintance with Campra's lighter *opéra-ballet style: Le carnaval et la folie* shows the influence of the older composer even more clearly, for it was based on his *Carnaval de Venise* (1699) and in turn inspired Campra's *Les fêtes vénitiennes* (1710). Later ballets, such as *Les éléments* (1721, rev. 1725), a collaboration with Lalande, and *Les stratagèmes de l'amour* (1726), also show traces of Campra's style.

Destouches's reputation rests on his trio of grand tragic works: *Callirhoé, Télémaque et Calypso,* and *Sémiramis.* It is here that he shows himself to be no mere imitator of Lully's tragédies lyriques, but an intelligent composer with a unique and distinctive voice: even to describe him as the most important forerunner of Rameau does not do justice to the originality of his style and approach. He was perhaps the first French composer to appreciate the breadth of operatic structure and its potential, supplanting the independence of the individual acts with a bond of common dramatic purpose. In his portrayal of the actions and emotions of genuinely tragic heroes he writes in a manner worthy of comparison with Gluck. Some measure of his popularity may be gleaned from the considerable number of revivals of his work, both in Paris and elsewhere. After the cool reception of *Les stratagèmes de l'amour* he wrote no new stage works, but concentrated on reviving earlier successes and on his court duties.

SELECT WORKLIST (all first prod. Paris unless stated):

*Amadis de Grèce* (La Motte; Fontainebleau 1699); *Omphale* (La Motte; 1701); *Le carnaval et la folie* (La Motte; 1703); *Callirhoé* (Roy; 1712); *Télémaque et Calypso* (Pellegrin; 1714); *Sémiramis* (Roy; 1718).

**Destouches, Franz Seraph von** (*b* Munich, 21 Jan. 1772; *d* Munich, 9 Dec. 1844). German composer. Studied Vienna with Haydn. From 1799 he worked as a theatre musician in Weimar under Goethe's direction, writing incidental music for Schiller's plays as well as several operas, including *Die Thomasnacht* (1792), *Das Missverständnis* (1805), *Die blühende Aloë* (*c.*1805), and *Der Teufel und der Schneider* (1843, prod. 1851).

**Detmold.** City in North-Rhine Westphalia, Germany. Singspiels were given in the 1770s, and the Komödienhaus opened 1778. The Hoftheater opened 1825 under A. Pichler, and Lortzing and his wife were in the company 1826–33. It closed 1848, reopened 1852, burnt down 1912, reopened 1919 as the Landestheater (cap. 730) with Lortzing's *Undine.*

**Detroit.** City in Michigan, USA. Seasons were given in the 19th cent. by companies including the Pyne-Harrison, De Vries, and Arditi, later from the New York, M, and elsewhere. The Detroit Civic OC was founded 1928, linking with the Symphony Orchestra to give seasons. The Overture to Opera Co. was founded 1920, renamed Michigan O 1973. It performs in the Fisher T (cap. 2,100) and at the Masonic Temple T (cap. 4,300), and has developed high standards and an enterprising artistic policy.

**deus ex machina** (Lat.: 'the god from the machine'). The theatrical device, originating in Greek drama, whereby a god is lowered on a cloud or triumphal car in order to resolve the complexities of the plot. The deus ex machina was a stock feature of many intermedi in the 16th cent. and was consequently adopted as a constituent part of much early opera. A classic instance occurs in the second finale of Monteverdi's *Orfeo* (1609), when Apollo descends singing on a cloud machine to remove his son Orpheus to Heaven.

As the use of elaborate stage machinery fell into disuse, the deus ex machina became an operatic rarity. By extension the term is used of any arbitrary last-minute solution to cut the tangled threads of a plot and bring about a happy ending, or *lieto fine.

**Deutsche Oper am Rhein.** See DÜSSELDORF-DUISBURG.

**Deux aveugles de Tolède, Les** (The Two Blind Men of Toledo). Opera in 1 act by Méhul; text by Benoît Marsollier. Prod. Paris, OC, 28 Jan. 1806; New Orleans, T St. Philippe, 8 May 1811. Also operas *Les deux aveugles de Franconville* (Ligon, c.1780), *Les deux aveugles de Baghdad* (Fournier, 1782), *Les deux petits aveugles* (A.-E. Trial, 1792), *Les deux aveugles* (Offenbach, 1855).

**Deux journées, Les** (The Two Days; formerly known in England as *The Water Carrier*). Opera in 3 acts by Cherubini; text by Jean Nicolas Bouilly. Prod. Paris, T Feydeau, 16 Jan. 1800; London, CG, 14 Oct. 1801; New Orleans, T St Phillippe, 12 Mar. 1811.

Paris, c.1640. Count Armand (ten) and his wife Constance (mez) are hidden in the house of an old water-seller, Daniel Mikéli (bar); Mikéli recognizes Armand as the man who once rescued him from starvation. When a policeman arrives, Mikéli pretends Armand is his sick father. Constance succeeds in crossing the barricade to the next village by using a pass belonging to Mikéli's daughter; Armand is smuggled through in a water-barrel. Once safely across, the fugitive Armand is helped to hide by Antoine (ten) (Mikéli's son) and his betrothed; unfortunately, Constance reveals her husband's whereabouts and the couple are arrested. However, they are freed when Mikéli arrives with a royal pardon.

One of Cherubini's most important works. Beethoven regarded the libretto as the best of any contemporary opera, and kept a copy on his desk. An outstanding example of *rescue opera.

Another opera on the subject is by Mayr (*Le due giornate*, 1801). A sequel, *Michelli und sein Sohn*, is by Clasing (1806).

**Devil and Daniel Webster, The.** Opera in 1 act by Douglas Moore; text by composer, after Stephen Vincent Benét's story (1937). Prod. New York, Martin Beck T, 18 May 1939.

New Hampshire, 1840. At his wedding to Mary (sop), Jabez Stone (bs) is rescued from a pact selling his soul to the devil by Daniel Webster's (bar) brilliant defence before a jury of famous villains of history.

**Devil and Kate, The** (Cz.: *Čert a Káča*). Opera in 3 acts by Dvořák; text by Adolf Wenig, after the folk-tale included in Božena Němcová's *Fairy Tales* (1845). Prod. Prague, N, 23 Nov. 1899; Oxford 22 Nov. 1932, with Turner, Ward, Manning, Wild, Beard, cond. Jacques; St. Louis, Loreto Hilton T, 14 June 1990, with Pancella, cond. Buckley.

Bohemia and Hell, legendary times. Kate (con), fails to secure a dancing partner at a country fair, so says she would dance with the Devil himself, Marbuel (bar), who now carries her off. Her garrulity oppresses even hell, which gladly relinquishes her to a rescuer. Marbuel returns to earth to capture the domineering lady of the manor, but vanishes at the sight of Kate.

**Devils of Loudun, The** (Pol.: *Diabły z Loudun*). Opera in 3 acts by Krzysztof Penderecki; text by

the composer, after John Whiting's drama *The Devils* (1961) based on Aldous Huxley's historical study *The Devils of Loudun* (1952), examining a celebrated 17th-cent. case. Prod. Hamburg, S, 20 June 1969, with Troyanos, Hiölski, cond. Czyz; Santa Fe 14 Aug. 1969, with Davidson, Reardon, cond. Skrowaczewski; London, C, 1 Nov. 1973, with Barstow, Chard, cond. N. Braithwaite.

Loudun, 1734–5. The plot describes how the priest Urbain Grandier (bar) engenders a diabolic obsession in a convent of Ursuline nuns, above all in the prioress Jeanne (sop). He is eventually tortured and burnt at the stake.

**Devil's Wall, The** (Cz.: *Čertova stěna*). Opera in 3 acts by Smetana; text by E. Krásnohorská. Prod. Prague, C, 29 Oct. 1882, with Fibichová, Reichová, Sittová, Vávra, Lev, Krössing, Hynek, Chlumský, cond. A. Čech; London, University College O, 12 Feb. 1987, with Smith, Cowboy, Hall, Hargreaves, Hamilton.

Bohemia, 12th cent. The Devil, Rarach (bs), persuades the Bohemian ruler Vok (bar) to enter a monastery, in order that a vow made by Vok's friend Jarek (ten) will be broken. He has sworn not to marry before Vok. Despite the arrival of Hedvika (sop), with whom Vok has fallen in love, he departs for his monastic life, since he believes she cannot return his love. However, Hedvika follows him, much to the annoyance of the Devil, who summons his spirits to build a dam to drown the monastery. The Abbot, Beneš (bs), drives the Devil away and destroys the wall by making the sign of the cross, and Vok and Hedvika are free to marry.

**Devin du village, Le** (The Village Soothsayer). Opera in 1 act by Jean-Jacques Rousseau; text by the composer. Prod. Fontainebleau 18 Oct. 1752; public prod. Paris, O, 1 Mar. 1753, with Laporte, Leroy, Troy; London, DL, 21 Nov. 1766; New York, City Tavern, 21 Oct. 1790.

A French village, 18th cent. Colette (sop) learns that her beloved Colin (ten) has been unfaithful to her and seeks the aid of a Soothsayer (bar) who advises her to display utter indifference to Colin's advances. At the same time, the Soothsayer tells Colin that Colette has fallen in love with someone else. The trick works and the lovers are reunited.

One of the most successful and influential works of its age, and Rousseau's most famous composition: see GUERRE DES BOUFFONS. Of its innumerable parodies, imitations, and adaptations, the most famous is Mozart's *Bastien und Bastienne* (1768).

**Devisenarie.** See MOTTO ARIA.

**Devrient, Eduard** (*b* Berlin, 11 Aug. 1801; *d* Karlsruhe, 4 Oct. 1877). German baritone, librettist, and theatre historian. Nephew of the actor Ludwig Devrient; brother of Karl Devrient, who married Wilhelmine *Schröder. Studied Berlin with Zelter. Début Berlin Royal O 1819 (Thanatos, *Alceste*). Sang there until 1822, then Dresden, Leipzig, Frankfurt 1822; later Prague and Vienna. A refined performer with excellent declamation, he was successful as Papageno, Figaro (Mozart and Rossini), Orestes, etc. Created Bois-Guilbert in Marschner's *Der Templer und die Jüdin*, and title-role in *Hans Heiling*, for which he wrote the libretto. Lost his voice 1834. Actor and producer Dresden Court T until 1846; director Karlsruhe Court T 1856–70, initiating many reforms. His writings include two other librettos (set by Taubert), plays, and several works on his ideal of a national theatre, e.g. *Geschichte der deutschen Schauspielkunst*. Responsible for (and sang in) the performance of St Matthew Passion, Berlin 1829, that began the Bach revival under his close friend Mendelssohn.

**Devriès**. Dutch family of singers.

1. **Rosa Van Os de Vries** (*b* Deventer, 1828; *d* Rome, 1889). Soprano. Début The Hague 1846 (Rachel). New Orleans 1849–51; London, L, 1856; Milan, S, and Naples, C, 1855–65. Her four children Gallicized their name to Devriès:

2. **Jeanne** (*b* ?, *c*.1849; *d* 1924). Soprano. Studied Paris with Duprez. Début Paris, L, 1867 (Amina), also creating Catherine in *La jolie fille de Perth* (Bizet described her as 'altogether splendid'). Paris, O, 1879–85; London and Lisbon 1880. Possessed a pleasing and flexible voice. In 1875 married the tenor Étienne Dereims (1845–1904).

3. **Fidès** (*b* New Orleans, 22 Apr. 1850; *d* ? 1941). Soprano. Also studied Paris with Duprez. Paris: L, 1869–71; O, 1871–4 (Marguerite, Ophelia, Elvira). Created Chimène in *Le Cid*; first Paris Elsa. Madrid and Lisbon 1884. Retired 1889. Charmingly modest, an affecting actress, and considered comparable to Calvé as a florid singer.

4. **Maurice** (*b* New York, 1854; *d* Chicago, 1919). Baritone. Début ?Liège 1874. Brussels, M, 1884, creating Gunther (Reyer's *Sigurd*). London, CG, 1884; Monte Carlo 1886; New York, M, 1885–7. Roles included Mercutio, Telramund, Nevers, William Tell. An intelligent actor, with a fine, resonant voice. Later taught in Chicago.

5. **Hermann** (*b* New York, 25 Dec. 1858; *d* Chicago, 24 Aug. 1949). Bass. Studied Paris with Faure. Début Paris, O, in *L'Africaine*. Brussels, M, 1885–6; New York, M, 1898–9. Taught in Chicago from 1910, and became a music critic.

Roles included Don Pasquale, Saint-Bris, Don Basilio (Mozart), Capulet.

6. **David** (*b* Bagnères-de-Luchon, 1881; *d* Neuilly, 1934). Tenor, son of (4). Studied Paris with Lhérie and Duvernoy. Début Paris, OC, 1903 (Gérald, *Lakmé*). Sang regularly there for 20 years. London, CG, 1910 (Pelléas); New York, Manhattan OC, 1910–11; Brussels, M, 1920–1. A notable lyric tenor in the French repertory. (R)

**Dexter, John** (*b* Derby, 2 Aug. 1925; *d* London, 23 Mar. 1990). English producer. After directing films and theatre, turned to opera with *Benvenuto Cellini*, London, CG, 1966. Hamburg from 1969. Prod. British première of Penderecki's *The Devils of Loudun*, London, C, 1973. Director of productions, New York, M, 1974–85. A gifted and stimulating producer, with original responses to many different works.

**Dezède, Nicolas** (*b* ?, *c*.1740–5; *d* Paris, 11 Sep. 1792). French composer. Came to attention with his opéra-comique *Julie* at the Paris, CI, 1772; over the next 20 years composed six operas and 15 opéras-comiques for the Parisian stage. Given his popularity, it is strange that he is rarely mentioned in contemporary writings and that little is known of his life. The suggestion that he was the bastard son of Frederick II may probably be discounted, though it is likely that he was of either German or Slavonic parentage.

The success of *Julie* was repeated with its two sequels, *L'erreur d'un moment* and *Le stratagème découvert*. In his opéras-comiques, including *Les trois fermiers*, *Blaise et Babet*, and *Les trois noces*, he proved himself the natural successor to Monsigny and Philidor, writing in an appealing and undemanding style, though moments of tenderness and expressiveness did much to enhance the quality of the genre during this period. Dezède's last opéra-comique, *Le véritable Figaro*, was banned by the censor. His serious operas, including *Péronne sauvée* and *La fête de la cinquantaine* appear anaemic when viewed against the rich tradition of Lully, Campra, Destouches, and Rameau, and have been overshadowed by his opéras-comiques.

SELECT WORKLIST (all first prod. Paris unless otherwise stated): *Julie* (Monvel; 1772); *L'erreur d'un moment* (The Mistake of a Moment) (Monvel; 1773); *Le stratagème découvert* (The Stratagem Discovered) (Monvel; 1773); *Les trois fermiers* (The Three Farmers) (Monvel; 1777); *Péronne sauvée* (Billardon de Sauvigny; 1783); *Blaise et Babet* (Monvel; Versailles 1783); *Les trois noces* (The Three Weddings) (Dezède; 1790); *Le véritable Figaro* (Billardon de Sauvigny, after Beaumarchais; comp. 1784, unprod.); *La fête de la cinquantaine* (Faur; 1796).

**Diaghilev, Serge.** See DYAGILEV, SERGEY.

**dialogue** (It., *dialogo*). 1. The spoken portions of a Singspiel, opéra-comique, or ballad opera, linked together by arias, songs, etc.

2. A 17th-cent. vocal composition with text as question and answer, invariably written for two singers with alternating parts, and frequently using echo and similar musical devices. Dialogue technique soon found its way into opera, and was exploited by many early composers.

**Dialogues des Carmélites** (Dialogues of the Carmelites). Opera in 3 acts by Poulenc; text by Ernest Lavery, after the drama by Georges Bernanos, adapted from Gertrude von Le Fort's novel *Die Letzte am Schafott* (1931) and a film scenario by Fr. Brückberger and Philippe Agostini, in turn on the recollections of one of the original nuns of Compiègne, Sister Marie of the Incarnation. Prod. Milan, S, 26 Jan. 1957, with Zeani, Gencer, cond. Sanzogno (trans. Testi); first perf. original French version Paris, O, 21 June 1957, with Duval, Crespin, Gorr, Scharley, cond. Dervaux; San Francisco 20 Sept. 1957, with L. Price, Turner, cond. Leinsdorf; London, CG, 16 Jan. 1958, with Morison, Sutherland, Watson, cond. Kubelik.

Paris, 1789. As a consequence of the Revolution, Blanche (sop), the daughter of the Marquis de la Force (bar) decides to enter a Carmelite convent. There she meets Constance (sop), who foresees that the two will die together. The convent is attacked by a rabble and all are captured and sentenced to death except Blanche, who has already fled to her home. However, when she sees Constance approaching the guillotine, she is inspired with divine strength and goes to join her in martyrdom.

**Diamand, Peter** (*b* Berlin, 8 June 1913). Austrian, later Dutch, administrator. Studied Berlin. Assistant director Netherlands O 1934–8. Secretary and artistic director Holland Festival 1947–65, also Artistic Adviser Netherlands O. Director Edinburgh Festival 1965–78.

**Diamants de la couronne, Les** (The Crown Diamonds). Opera in 3 acts by Auber; text by Scribe and Vernoy de Saint-Georges. Prod. Paris, OC, 6 Mar. 1841; New York, Niblo's Garden, 12 July 1843 with J. Calvé, Cossas, Bailly; London, Princess's T, 2 May 1844.

Portugal, 1777. Don Enriquez (ten) is captured by a gang of forgers, but is saved by Theophila (sop), heiress to the crown of Portugal; she has set one of the forgers, Rebolledo (bs), the task of copying the crown jewels, in order that she may sell the original gems to pay off her debts. News breaks out that the jewels have been stolen, but Enriquez, who has fallen in love with Theophila, protects her, aided by his cousin Diana (sop). When Theophila's identity is disclosed,

she chooses to marry Enriquez; Rebolledo is appointed head of the secret police.

Also subject of opera *Jadwiga* by Dellinger (1901).

**Dibdin, Charles** (*b* Southampton, *bapt.* 15 Mar. 1745; *d* London, 25 July 1814). English composer, singer, and musical factotum. His prodigious output of afterpieces and operas is representative of the type of operatic entertainment popular in England during the last quarter of the 18th cent. Most had spoken dialogue interspersed with songs which were at first Italianate in style but from *The Waterman* onwards aimed more towards a simple English ballad style. His significant deviation from the standard pattern was in the introduction of the *ensemble finale: he was one of the first English composers to attempt this.

Dibdin was himself a character actor, and many of his works contain tailor-made comic parts. His best works are his collaborations with Bickerstaffe, notably *Love in the City* (an unsuccessful parody of Arne's *Love in a Village*), *Lionel and Clarissa*, which originally included a substantial proportion of music by Galuppi and other Italians, and *The Padlock*. After amassing huge debts, he had to flee to France in 1776: in his autobiographical *Professional Life* he pretends that the visit was 'in order to expand my ideas, and store myself with theatrical materials'. He did actually gather librettos by Sedaine, Favart, and Marmontel which he adapted to his own use. Back in England in 1778, he quarrelled with the manager of Covent Garden, leaving him without a theatre in London which would employ him. Having managed to raise sufficient capital to construct his own theatre, he mismanaged this, too, and found himself in a debtors' prison. He decided to emigrate to India and made a one-man tour of England, recorded in *The Musical Tour of Mr Dibdin*, in order to raise money for his passage. However, having sailed as far as Torbay, he decided the sea was not to his liking (despite his fame as a composer of sea songs) and disembarked. He continued with his shows—which he called Table Entertainments—in London, and eventually made enough to build his own small theatre, Sans Souci.

SELECT WORKLIST (all first prod. London): *Love in the City* (Bickerstaffe; 1767); *Lionel and Clarissa* (Bickerstaffe; 1768); *The Padlock* (Bickerstaffe; 1768); *The Waterman* ( Dibdin; 1774).

**'Dich, teure Halle'.** Elisabeth's (sop) aria opening Act II of Wagner's *Tannhäuser*, in which she greets the Hall of Song in the Wartburg where the song contest is to take place.

**Dickens, Charles** (*b* Landport, 7 Feb. 1812; *d* Gadshill, 9 Jan. 1870). English writer. His solitary excursion as a librettist, made just before the first instalment of *The Pickwick Papers*, was for Hullah's *The Village Coquettes* (1836). Operas on his works are as follows:

*The Pickwick Papers* (monthly, 1836–7): E. Solomon (1889); C. Wood (1922); Coates (1936); N. Burnand (1956), Shartyr (1975)

*Barnaby Rudge* (1841): J. Edwards (*Dolly Varden*, 1902)

*A Christmas Carol* (1843): Herrmann (1954); Cikker (*Mr Scrooge*, comp. 1958, prod. 1963); D. Gray (1960); Kalmanov (*Mister Scrooge*, 1966); J. Cohen (1970), Musgrave (1981)

*The Cricket on the Hearth* (1846): Gallignani (1873); Goldmark (1896); Zandonai (1908); Mackenzie (1914)

*Great Expectations* (monthly, 1860–1): Argento (*Miss Havisham's Wedding Night*, 1981)

*A Tale of Two Cities* (1889): Benjamin (1949–50, prod. 1957)

**Dick Johnson.** The alias of the bandit Ramerrez (ten), lover of Minnie, in Puccini's *La fanciulla del West*.

**Diderot, Denis** (*b* Langres, 5 Oct. 1713; *d* Paris, 31 July 1784). French writer, philosopher, and critic; an important figure of the Enlightenment. Together with D'Alembert, Condillac, and Rousseau, he compiled and edited the *Encyclopédie* (pub. 1751–80). *Le neveu de Rameau* (?1761–74, never published by Diderot, later trans. Goethe) is an imagined conversation between the author and the nephew of Jean-Philippe Rameau, covering the topics of politics, philosophy, literature, and opera. It has been interpreted as a musical tract, referring to the contemporary *Guerre des Bouffons. As in a chapter of his *Les bijoux indiscrets* (1748)—a satire on the French court—Diderot sides with the 'Italian' faction, since their music seems to him to create a greater degree of verisimilitude, and to portray real emotions more successfully. He also admits that the French language distorts the human singing voice, making it unsuitable for expressive, lyrical music; he uses this as one line of his attack on librettos written by French poets. In the essays written to accompany his plays, Diderot expresses his views for a need for reform in every aspect of the theatre; these ideas were later put into practice in the reforms of Noverre, Jommelli, Traetta, and Gluck.

BIBL: B. Durand-Sendrail, ed., *Denis Diderot:écrits sur la musique* (Paris, 1987).

**Dido and Aeneas.** Opera in a prologue and 3 acts by Purcell; text by Nahum Tate, after Book 4 of Virgil's *Aeneid* (unfin. 19 BC). Prod. Chelsea, Josias Priest's School for Young Gentlewomen, spring 1689; public prod. *c*. Feb. 1700 and 9 Feb. 1704, then not revived on stage until 20 Nov.

1895, by RCM at London, L, for bicentenary of Purcell's death; New York, Plaza Hotel, 10 Feb. 1923.

Carthage, after the Trojan war. Dido (sop) cannot bring herself to declare her love for Aeneas (ten). Belinda (sop) and the court succeed in making her yield, and a chorus celebrates the triumph of love and beauty.

The Witches assemble in their cave to plot the downfall of Dido and of Carthage. They agree to conjure up a storm so as to force the royal lovers, out hunting, to take shelter in the cave; here one of the Witches (mez), disguised as Mercury, will remind Aeneas of his duty to go on to Italy. Dido and Aeneas are entertained with a masque of Diana and Actaeon: the storm drives Aeneas alone to the cave, where he is deceived by the false Mercury.

At the harbour, Aeneas plans to leave, while the Witches exult in their success: they plan to sink his fleet and fire Carthage. When Dido appears she silences Aeneas's explanations, and in the famous lament 'When I am laid in earth' she takes her leave of life; Cupids mourn above her tomb.

Various version of the opera survive, the one most generally accepted by scholars being the MS now in the Bodleian Library, Oxford (previously at St Michael's College, Tenbury). The libretto includes the text of an elaborate mythological Prologue, not in the Tenbury score: this was included in the original production, when an Epilogue by D'Urfey was spoken. It is not known whether Purcell set the Prologue: we know the opera only as an adaptation, 'A Mask in Four Musical Entertainments', made at the beginning of the 18th cent. The libretto has a different arrangement of the acts.

Dido's betrayal has been treated operatically some 90 times. Among almost 40 settings of Metastasio's libretto *Didone abbandonata* are those by Albinoni (1725), Porpora (1725), Galuppi (1741), Hasse (1742), Jommelli (1747, 1749), Galuppi (1752), Traetta (1757), Araia (1758), Jommelli (1763), Galuppi (1764), Piccinni (1770), Paisiello (1794), Paer (1810), Mercadante (1823). Operas on other texts include those by Haydn (1778), Piccinni (1783), Storace (1792), Berlioz (comp. 1856–8), and Lavrangas (1909).

BIBL: E. Harris, *Henry Purcell's 'Dido and Aeneas'* (Oxford, 1987).

**Didur, Adam** (*b* Wola Sękowa, 24 Dec. 1874; *d* Katowice, 7 Jan. 1946). Polish bass. Studied Lwów with Wysocki, Milan with Emmerich. Début Rio 1894 (Gounod's Méphistophélès). Warsaw O 1899–1903; Milan, S, 1903–6; London, CG, 1905 (Colline, Leporello); Buenos Aires 1905–8; New York, M, 1908–32, succeeding Édouard De Reszke. Created Ashby (*La fanciulla del West*) and the Woodcutter (*Königskinder*); also first US Boris. His sonorous, dark-timbred voice was suited to both Rossini and Wagner, to whose music he brought great energy and subtlety, vocally as well as dramatically. He was much admired by Mahler. After retiring, he returned to Poland and formed an opera company in Katowice, where he also taught. (R)

**'Die Frist ist um'.** The Flying Dutchman's (bar) soliloquy in Act I of Wagner's *Der fliegende Holländer*, in which he reflects that once again the term of seven years' wandering, to which he is sentenced by a curse, is at an end and he may land to seek the redeeming love of a woman who will sacrifice herself for him.

**'Dies Bildnis ist bezaubernd schön'.** Tamino's (ten) aria in Act I of Mozart's *Die Zauberflöte*, declaring his love for Pamina at first sight of her portrait.

**Dieter, Christian Ludwig** (*b* Ludwigsburg, 13 June 1757; *d* Stuttgart, 15 May 1822). German composer. Studied Stuttgart with Boroni and Poli. Though he had a success with his first Singspiel, *Der Schulz im Dorf*, his attempt to escape from the confines of the Ducal court led to imprisonment. Released, he wrote many works for the court in which he moved away from the example of his powerful predecessor Jommelli towards a lighter, more typically Germanic Singspiel. These include *Belmonte und Constanze* (1784) and a number of other light pieces popular in their own day and place.

**Dietsch, Louis** (*b* Dijon, 17 Mar. 1808; *d* Paris, 20 Feb. 1865). French conductor and composer. Studied Paris with Rejcha. Played double bass at T-Italien and Opéra. Chorus master at Opéra on Rossini's recommendation, 1840; conductor, 1860–3; left after an argument with Verdi. Also quarrelled with Wagner, who wrongly accused him of plagiarism for the text of *Le vaisseau fantôme* (Foucher and Révoil, after Marryat; Paris 1842). Conducted the notorious 1861 Paris première of *Tannhäuser*, against Wagner's wishes and incompetently though not, as Wagner asserted, maliciously so as to encompass the disaster that occurred.

**Dikoj.** Savel Prokofjevič Dikoj (bs), uncle of Káťa's lover Boris, in Janáček's *Káťa Kabanová*.

**Di Luna.** The Conte di Luna (bar), a nobleman of Aragon, in Verdi's *Il trovatore*.

**Dimitrova, Ghena** (*b* Pernik, 16 Nov. 1940). Bulgarian soprano. Studied Sofia with Dyakovich, Zagreb with Lunzer. Début Sofia 1966 (Abigaille). Belgrade, Zagreb from 1969; suc-

cesses in Italy from 1970, incl. Milan, S; Dallas 1981; London, CG, 1983; New York, M, 1988; also Buenos Aires, C; Salzburg; Moscow; Verona; Berlin, D; etc. With her full, ample tone, an acclaimed Lady Macbeth, Odabella, Leonora (*Trovatore*), Aida, Tosca, Turandot, Santuzza, Leonora (*Trovatore* and *Forza*). (R)

**Di Murska, Ilma** (*b* Zagreb, 4 Jan. 1836; *d* Munich, 14 Jan. 1889). Croatian soprano. Studied Vienna and Paris with Marchesi. Début Florence 1862 (Martha). Budapest, Berlin, Vienna, then London, HM, 1865 (Lucia). Regularly in London till 1873, including Senta (DL, 1870) in the first Wagner opera in England. New York 1873 (Amina). Possessing a brilliant technique and a three-octave range, she was a celebrated Queen of the Night.

**D'Indy, Vincent.** See INDY, VINCENT D'.

**Dinner Engagement, A.** Comic opera in 1 act by Lennox Berkeley; text by Paul Dehn. Prod. Aldeburgh 17 June 1954, with Sharp, Hooke, Cantelo, Nielsen, Young, cond. Tausky; Washington, U of Washington, 1958.

The modern buffo plot concerns the attempts of a *nouveau-pauvre* couple to marry their daughter to a Prince—with ultimate success, though not through their schemes.

**Dinorah, ou Le pardon de Ploërmel.** Opera in 3 acts by Meyerbeer; text by Barbier and Carré. Prod. Paris, OC, 4 Apr. 1859, with Cabel and Faure; London, CG, 26 July 1859, with Miolan-Carvalho, and Graziani; New Orleans, French OH, 4 Mar. 1861, with Patti, Melchisedech, Carrier, cond. E. Prévost.

Brittany, 19th cent. The goatherd Hoël (bar) puts off his wedding to Dinorah (sop) and goes in search of treasure, which will enable them to live more comfortably. At the specified site, Dinorah appears and, after a mad fit, is revived by Hoël. Her mind is cleared of all memories of his desertion and the couple are reunited.

**'Dio, mi potevi scagliar'.** Otello's (ten) monologue in Act III of Verdi's *Otello*, the equivalent of 'Had it pleased heaven to try me with affliction' in Act IV of Shakespeare's *Othello*.

**Dionysus.** In Greek mythology, the god of wine; known to the Romans as Bacchus. The son of Zeus and Semele, he was taken to Naxos, and later led expeditions at the head of a group of Bacchae (Bacchantes, Bassarids, or Maenads), who were inspired by his divine frenzy and held orgiastic feasts (Bacchanalia or Dionysia). Those who submitted to his conquests were introduced to the use of the vine and other benefits, but those who resisted him, like King Pentheus of Thebes, were savagely destroyed. Operas on the subject of

Bacchus and the Bacchae, often based on Euripides' drama (*c*.406 BC), are by Krieger (1700), Paer (1813), Dargomyzhsky (unprod., 1845), Fontana (1847), Bandini (1889), Massenet (1909), Wellesz (1931), Henze (1966), and Börtz (1991). Dionysus later married *Ariadne after she had been abandoned by Theseus on Naxos. In this role he appears as the god Bacchus (ten) in Strauss's *Ariadne auf Naxos*. He also appears in Bliss's *The Olympians*. In late 19th-cent. German aesthetics, under the influence of Nietzsche's *The Birth of Tragedy* (1872), the Dionysian principle in art stands for the instinctive, physical, and elemental, however destructive, in opposition to the Apollonian principle of order, reason, and discipline. It is this antithesis which is dramatized in Szymanowski's *King Roger*, and is at the centre of Britten's *Death in Venice*.

**'Di Provenza il mar'.** The elder Germont's (bar) aria in Act II of Verdi's *La traviata*, in which he tries to persuade his son Alfredo to return to the parental home and give up his life with Violetta.

**'Di quella pira'.** Manrico's (ten) aria in Act III of Verdi's *Il trovatore*, a rousing call to his followers to join him in saving his mother Azucena from being burnt alive. The high C which many tenors interpolate in the line 'O teco almeno, corrir a morir' is not in the score, but was introduced by Enrico Tamberlik, with Verdi's permission.

**'Di scrivermi ogni giorno'.** The quintet in Act I of Mozart's *Così fan tutte*, in which the sisters Dorabella (sop) and Fiordiligi (sop) exchange promises with their lovers Ferrando (ten) and Guglielmo (bar) that they will write to one another every day during their absence, while Don Alfonso (bar) fears he will die of laughter.

**Di Stefano, Giuseppe** (*b* Motta Santa Anastasia, 24 July 1921). Italian tenor. Studied Milan with Montesanto. Début Reggio Emilia 1946 (Massenet's Des Grieux). Milan, S, from 1947. New York, M, 1948–52, 1955–6, 1964–5. London, CG, 1961 (Cavaradossi). In his early days, as Nadir, Elvino, Fritz, etc, his voice had a pure yet velvety tone, and a ravishing pianissimo. It deteriorated when he began to sing heavier roles (Turiddu, Don José, Alvaro), and he appeared less frequently, although he partnered Callas on her final tour, 1973–4. (R)

**'Di tanti palpiti'.** Tancredi's (con) love song in Act I of Rossini's *Tancredi*. It was perhaps the most popular aria of its day, sung and whistled in the streets all over Europe. In Venice it was known as the *aria dei risi* (Rice Aria) as Rossini was alleged to have composed it in four minutes while waiting for the rice to cook—an impossible

time for either event, as the gastronome Rossini would have been the first to appreciate.

**'Dite alla giovine'.** Part of the duet between Violetta (sop) and the elder Germont (bar) in Act II of Verdi's *La traviata*, in which she begs him to make her sacrifice known to his daughter.

**Dittersdorf, Karl Ditters von** (*b* Vienna, 2 Nov. 1739; *d* Neuhof, 24 Oct. 1799). Austrian composer. His Singspiels, of which he wrote about 40 (mostly for Vienna), form the most significant part of his wide-ranging output. They enjoyed great success in their day and were popular for their lively, tuneful music and largely comic plots.

Dittersdorf's first Singspiel, *Doktor und Apotheker*, led to the commission of two further Singspiels, *Betrug durch Aberglauben* and *Die Liebe im Narrenhause*, and of an Italian opera, *Democrito corretto*. *Doktor und Apotheker* proved especially popular and was successful throughout Europe: Stephen Storace made an English adaptation of it.

Dittersdorf contributed to the development of Singspiel by establishing the practice of combining elements from both North German Singspiel and opera buffa. He had himself written at least 11 Italian operas before embarking on his first Singspiel, and among the techniques he incorporated are those of ensemble and finale writing. Particularly significant is the Act I finale of *Doktor und Apotheker*: this was the first time such a long multi-sectional finale movement had been introduced into Singspiel, replacing the more conventional *vaudeville ending. Mozart later adopted the practice, and used finale movements in both acts of *Die Zauberflöte*. However, Dittersdorf continued to use the vaudeville in *Hieronymus Knicker* and *Das Gespenst mit der Trommel*.

SELECT WORKLIST (all first prod. Vienna unless otherwise stated): *Doktor und Apotheker* (Stephanie; 1786); *Betrug durch Aberglauben* (Deceit through Superstition) (Eberl; 1786); *Die Liebe im Narrenhause* (Love in the Madhouse) (Stephanie; 1787); *Hieronymus Knicker* (Dittersdorf; 1789); *Das Gespenst mit der Trommel* (The Ghost with the Drum) (Dittersdorf, after Goldoni; Oels 1794); *Die lustigen Weiber von Windsor* (Dittersdorf, after Shakespeare; Oels 1796); *Die Opera buffa* (Bretzner; 1798).

BIBL: K. von Dittersdorf: *Lebensbeschreibung* (Leipzig, 1801; ed. N. Miller, Munich, 1967; trans. 1896, R/1970).

**diva** (It., from Lat. *divus*: 'divine'). A highly celebrated female singer, or *prima donna.

**divertissement** (Fr.: 'diversion; entertainment'). The term's use in 17th- and 18th-cent. France is so varied as to make a single definition impossible. Strictly, it refers to those parts of French opera which are inessential to the plot; their function was to entertain, rather than to advance the drama. Vocal solos, ensembles, and above all ballet played a major role in the divertissement and lent to early French opera some of the spectacle of the ballet de cour and comédie-ballet. In Lully the divertissement is sometimes woven into the plot, as in the wedding scene of *Roland*, whose placing is crucial to the overall structure of the opera. Frequently the divertissement was used to conclude the opera, as in *Armide* and *Amadis*. In later French opera, e.g. Campra and Destouches, the divertissement assumed even greater importance, to the extent that its spectacle sometimes overwhelmed the opera itself.

Divertissements were sometimes found between the acts of pastorales and opéras-comiques, acting as diverting interludes; such pieces are properly called *intermèdes. Wider applications of the word link it to genres as different in size and nature as the small-scale pastorales themselves and the elaborate court entertainments sponsored by Louis XIV in honour of royal occasions, the so-called grands divertissements, of which *Les plaisirs de l'île enchantée* (1666) was a famous example.

**'Divinités du Styx'.** Alceste's (sop) aria in Act I of Gluck's *Alceste*, invoking the gods of the underworld and defying them to do their worst.

**Djamileh.** Opera in I act by Bizet; text by Louis Gallet, after Alfred de Musset's poem *Namouna* (1832). Prod. Paris, OC, 22 May 1872, with Prelly (described by Gauthier-Villars as 'the voiceless Venus'), Duchesne, Pontel, cond. Deloffre; Manchester 22 Sep. 1892; Boston, OH, 24 Feb. 1913, with Marcel, Laffitte, Giacone, cond. Weingartner.

A palace in Cairo, 19th cent. Haroun (ten) pensions off his old mistress once a month and is bought a new one in the market by his servant Splendiano (bar). Djamileh (mez), the current mistress, falls in love with Haroun, and bargains with Splendiano to be readmitted disguised: this ruse wins Haroun's heart.

**Dmitry.** The Pretender to the throne of Russia (ten) in Musorgsky's *Boris Godunov* and Dvořák's *Dmitrij*.

**Dobroven, Issay** (*b* Nizhny-Novgorod (Gorky), 27 Feb. 1891; *d* Oslo, 9 Dec. 1953). Russian, later Norwegian, conductor, pianist, and producer. Appeared as pianist aged 5. Studied Moscow with Yaroshevsky and Taneyev, Vienna with Godowsky. Début Moscow, Kommisarzhevsky T, 1919 (*Hoffmann*); Moscow, B, 1921–2. Left Russia 1923. Dresden 1923; Berlin, V, 1924–5, and SO as producer, 1928–9; Frankfurt 1930. Sofia 1927–8, where he did much for the opera. Frequent visitor to Budapest 1936–9. Settled in

Norway 1932. Stockholm 1940–9 as cond. and prod. (incl. première of Sutermeister's *Raskolnikoff*, 1948). From 1949 cond. and prod. Russian works at Milan, S. London, CG, 1952–3 (*Boris*). (R)

**Docteur Miracle.** The evil doctor (bs) in the Antonia episode of Offenbach's *Les contes d'Hoffmann*.

**Docteur Miracle, Le.** 1. Operetta in 1 act by Bizet; text by Battu and Halévy. Prod. Paris, BP, 9 Apr. 1857; London, Park Lane Group, 8 Dec. 1957.

2. Operetta in 1 act by Lecocq; text by Battu and Halévy. Prod. Paris, BP, 8 Apr. 1857.

Padua, 19th cent. In order to win the hand of Laurette (sop), daughter of an anti-military official (bs), a young officer, Silvio (ten), disguises himself as a chef and cooks a wonderful omelette. It tastes so amazing that the official believes himself to be poisoned and summons a doctor (again Silvio in disguise) who proclaims that the only cure is to let him marry the official's daughter.

Both works were entries in a competition arranged by Offenbach (who offered 1,200 F. and a gold medal), being awarded the joint first prize among 78 entrants by a jury headed by Auber, Halévy, Scribe, and Gounod.

**Doctor Bartolo.** The old doctor of Seville (bs) in Paisiello's and Rossini's *Il barbiere di Siviglia* and Mozart's *Le nozze di Figaro*.

**doctrine of the affections** (Ger., *Affektenlehre*). The prevailing aesthetic theory of the 18th cent., expounded most cogently in the writings of Werckmeister, Marpurg, and Mattheson, although the term itself was only coined by German musicologists in the early 20th cent. Emotions were defined in terms of specific 'affections' (sorrow, hate, hope, etc.), and within each self-contained musical entity—be it a movement, aria, or other individual number—only one of these affections could be encompassed. The procedures of the da capo aria, in which, the complementary middle section notwithstanding, only one main emotional state was explored, accorded well with this theory; because each affection was linked to a particular, readily recognizable, musical style or device, it also became easy to transplant arias from one opera to another.

**Dodon.** King Dodon (bs) in Rimsky-Korsakov's *The Golden Cockerel*.

**Dohnányi, Christoph von** (*b* Berlin, 8 Sep. 1929). German conductor. Studied Munich, later US with his grandfather, Ernő Dohnányi, and Bernstein. Frankfurt 1953, chorus-master and conductor. Music director of Lübeck 1957–63, of

Kassel 1963–6, of Frankfurt 1968–75, and of Hamburg 1975–84. Chicago 1969; London, CG, 1974; New York, M, 1972; also Munich, Vienna, San Francisco. Repertory incl. *Fidelio, Fliegender Holländer, Meistersinger, Falstaff, Rosenkavalier, Salome, Wozzeck, Moses und Aron*, and premières of Henze's *Bassarids* and *Der junge Lord*. Married soprano Anja *Silja. (R)

**Doktor Faust.** Opera in 2 prologues, an interlude, and 3 scenes by Busoni (completed by Jarnach); text by the composer after the Faust legend and Marlowe's treatment of it in *Dr Faustus* (1589), also the 16th-cent. German puppet play. Prod. Dresden 21 May 1925 with Seinemeyer, Strack, Burg, Bader, Correck, cond. Busch; broadcast concert perf. London, QH, 17 Mar. 1937, with Blyth, Jones, Noble, cond. Boult; New York, Carnegie H, 1 Dec. 1964, with Bjoner, Shirley, Fischer-Dieskau, cond. Horenstein (concert); prod. Reno, Nevada, 25 Jan. 1974; London, C, in new version by Antony Beaumont, 25 Apr. 1986, with Hannan, Davies, Allen, Connell, Rivers, Newman, cond. Elder.

Wittenberg and Parma, 16th cent. Faust (bar) invokes the help of Mephistopheles (ten) to help him regain his lost youth. Faust elopes with the Duchess of Parma (sop) shortly after her marriage to the Duke (ten). At an inn in Wittenberg, Faust hears of the Duchess's death, and he too longs to die. He meets a beggar woman, who turns out to be the Duchess, clasping the corpse of a child. Faust offers his own life if the child can be revived. As Faust dies, a young man rises from the body of the dead child.

Also opera by H. Reutter (*Dr Johannes Faust*, 1936).

See also FAUST.

**Doktor und Apotheker.** Opera in 2 acts by Dittersdorf; text by G. Stephanie, after a French drama *L'apothicaire de Murcie* by 'le Comte N . . .'. Prod. Vienna, B, 11 July 1786. Adaptation by S. Storace, London, L, 25 Oct. 1788; this version also Charleston, City T, 26 Apr. 1796. Orig. version, New York, Terrace Garden, 30 June 1875.

A small German town, 18th cent. Gotthold (ten) and Leonore (sop) are forbidden to marry because of the hatred between their fathers, the doctor Krautmann (bs), and the apothecary Stössel (bs) who wants his daughter to marry an invalid soldier, Sturmwald (ten). Leonore's cousin Rosalie (sop) and her lover Sichel (ten) help the couple in a series of intrigues which involve disguises as Sturmwald and as a lawyer, and result in the drawing up of a marriage contract. The fathers are at first furious, but all is finally resolved by a miraculous change of heart on the part of Stössel's wife, Claudia (mez), who ensures that the lovers are united.

Dittersdorf's most successful Singspiel.

**Dolukhanova, Zara** (*b* Moscow, 5 Mar. 1918). Armenian mezzo-soprano. Studied Moscow with Belyayeva-Tarasevich. Début Erevan 1939. Engagements throughout USSR. Though she then spent most of her career as a concert singer, her championing of Rossini did much to establish his mezzo-soprano roles in the repertory, with broadcasts as Rosina, Cenerentola, Arsace, etc. A highly accomplished singer with a wide stylistic range. (R)

**Domgraf-Fassbaender, Willi** (*b* Aachen, 19 Feb. 1897; *d* Nuremberg, 13 Feb. 1978). German baritone. Studied Berlin with Stückgold, Milan with Borgatti. Début Aachen 1922 (Mozart's Count Almaviva). Berlin, Düsseldorf, and Stuttgart, then Berlin, SO, 1928–46. Gly. 1934 (opening night) (Mozart's Figaro), 1935, 1937 (Guglielmo, Papageno). Salzburg 1937. After the war, Vienna, Munich, and Nuremberg, where he was Oberspielleiter 1953–62, and also taught. Pupils included his daughter Brigitte *Fassbaender. A fine Mozart singer, with a warm, attractive voice; also an impressive Wozzeck. (R)

**Domingo, Plácido** (*b* Madrid, 21 Jan. 1941). Spanish tenor. Studied piano, conducting, and singing, Mexico City. Début as baritone Mexico City 1957 in Caballero's zarzuela *Gigantes y cabezudos*. Monterrey, O, 1960 (first major tenor role, Alfredo); Dallas, 1961. Israeli Nat. O, 1962–5. New York, M, from 1966; Milan, S, from 1969; London, CG, from 1971. Also San Francisco, Paris, Vienna, Berlin, etc. A leading lyric-dramatic tenor of great gifts, with a strong dramatic personality. His large repertory includes Cavaradossi, Don José, Turiddu, Samson, Aeneas (Troyens), Radamès, and Otello (Verdi), and, more recently, Heldentenor roles: Lohengrin, Parsifal, Siegmund, and his first Russian role, Hermann. Conducted *Attila*, Barcelona 1973, and *Die Fledermaus*, London, CG, 1983. Autobiography, *My First Forty Years* (London, 1984). (R)

BIBL: D. Snowman, *The World of Placido Domingo* (London, 1985).

**Domino noir, Le** (The Black Domino). Opera in 3 acts by Auber; text by Scribe. Prod. Paris, OC, 2 Dec. 1837, with Cinti-Damoreau; London, CG, 16 Feb. 1838; New Orleans, T d'Orléans, 11 Dec. 1838, with J. Calvé, Couériot, Bailly, cond. E. Prévost.

Two friends, Horatio (ten) and Juliano (ten),

attend the Christmas ball given annually by the Queen in the hope of seeing once again a beautiful woman who comes dressed in a black domino. She is Angela (mez). After many deceptions and confusions, Angela chooses Horatio as her husband, and regains a large inheritance.

Also opera by L. Rossi (1849).

**Donalda, Pauline** (*b* Montreal, 5 Mar. 1882; *d* Montreal, 22 Oct. 1970). Canadian soprano. Studied Montreal, and Paris with Duvernoy. Born Lightfoot, she took her stage name from the 'Donaldas', the first women allowed on the McGill Campus, named after Sir Donald Smith. Début Nice 1904 (Manon). London, CG, 1905, sponsored by Melba, singing Micaëla, Zerlina, Gilda, Violetta, etc. First London Concepción (*Heure espagnole*), 1919, delaying her retirement at Ravel's request for the purpose. New York, Manhattan O, 1906, Paris. Taught Paris 1922–37, then Montreal, founding the Opera Guild of Montreal 1942. (R)

**Don Alfonso.** The philosopher (bar) who engineers the plot of Mozart's *Così fan tutte*.

**Donath, Helen** (*b* Corpus Christi, TX, 10 July 1940). US soprano. Studied Corpus Christi with Dapholl, New York with Novikova. Début Cologne 1960 (Inez). Hanover 1963; Frankfurt 1965–7; Munich from 1967; Vienna from 1970; Salzburg from 1971; Milan, S, 1972. New York, M, 1990. Roles incl. Pamina, Marzelline, Micaëla, Sophie, Eva. (R)

**Donauweibchen, Das** (The Danube Sprite). Opera in 3 acts by Ferdinand Kauer; text (*Ein romantisches komisches Volksmärchen mit Gesang nach einer Sage der Vorzeit*) by K. F. Hensler. Prod. Vienna, L, 11 Jan. 1798. Kauer's most popular work, and one of the most popular of all operas in the first half of the 19th cent. all over Central and Eastern Europe. A second part followed, Vienna, L, 13 Feb. 1798, and a third, *Die Nymphe der Donau*, text by Berling, music by Bierey, Altona, 25 July 1801. Also a *Seitenstück*, *Das Waldweibchen*, by Kauer and Hensler, Vienna, L, 1 Apr. 1800. *Das Donauweibchen* was translated into Russian by N. S. Krasnopolsky and given in St Petersburg, 1803, set as *Lesta, dneprovskaya rusalka* (Lesta, the Dnyepr Rusalka), after Kauer, by Davydov (1803), Cavos (1804), Davydov (1805), and (to a new text by Shakhovskoy) Davydov (1807). These works influenced Pushkin's *Rusalka*, originally intended as a Singspiel, and hence Dargomyzhsky's *\*Rusalka*. Davydov's settings also influenced Glinka, especially in their orchestration and their use of folk music.

**Don Basilio.** The priest and music-master (bs) in Paisiello's and Rossini's *Il barbiere di Siviglia* and (ten) in Mozart's *Le nozze di Figaro*.

**Don Carlos.** Opera in 5 acts by Verdi; text by Méry and Du Locle, after Schiller's drama (1787), Cormon's drama (1849), and Prescott's history (1856). Prod. Paris, O, 11 Mar. 1867, with Sass, Guéymard, Morère, Faure, Obin, David, cond. Hainl; London, CG, (Italian version) 4 June 1867, with Lucca, Fricci, Naudin, Graziani, Petit, cond. Costa; New York, AM, 12 Apr. 1877, with Palmieri, Rastelli, Celada, Bertolasi, Dal Negro, cond. Maretzek. Rev. in 4 acts (omitting Fontainebleau act) 1884 by Du Locle, trans. Zanardini, based on orig. version by De Lauzières: prod. Milan, S, 10 Jan. 1884, with Silvestri, Tamagno, Lhérie, Navarini, Bruschi-Chiatti, Pasqua.

France and Spain, 1568. In the Forest of Fontainebleau, Don Carlos (ten), the son of Philip II of Spain, meets Elisabeth de Valois (sop), to whom he is betrothed by arrangement. The couple declare their instant love for each other; Elisabeth's father then decrees that she is to marry King Philip (bs) instead.

In the monastery of San Yuste in Spain, Carlos confesses his hopeless love for Elisabeth to his friend Rodrigo (bar). Rodrigo advises him to try to forget his passion by taking up the cause of Flanders. Rodrigo bids farewell to Elisabeth, declaring his undying love for her. King Philip returns to find the Queen alone; he is suspicious about the relationship of Elisabeth and Carlos and asks Rodrigo to watch them.

Having received an anonymous note, Carlos goes to meet a veiled woman. He assumes it will be Elisabeth but it is in fact the Princess Eboli (mez), who is in love with him. Realizing the object of Carlos's passion, she decides on revenge. Rodrigo arrives and persuades Carlos to hand over all documents incriminating him as a supporter of Flanders. At the *auto-da-fé* (a burning of heretics), Flemish officials, led by Carlos, arrive to ask for an end to the persecutions in Flanders. When Philip orders their arrest, Carlos threatens him with a sword but Rodrigo intervenes and saves the King.

Philip, compelled by the Grand Inquisitor, resolves to put his son to death and agrees to hand over Rodrigo as a heretic. Elisabeth interrupts, agitated because her jewel casket has been stolen. Philip produces it and makes her open it in his presence; it contains a portrait of Carlos. Eboli confesses to Elizabeth that it was she who stole the casket and also that she has been the King's mistress. Repentant, she vows to enter a convent. Rodrigo visits Carlos in his cell. He has implicated himself as the leader of the Flemish rebels and hopes that Carlos will be set free. Two of the King's men enter and Rodrigo is shot. Before he dies, he manages to tell Carlos that the Queen will wait for him at the monastery of San Yuste the next day.

At San Yuste, Carlos and Elisabeth bid each other a final farewell. Philip and the Grand Inquisitor have entered silently; they now seize Elisabeth and Carlos and it seems that all is lost. However, a figure who seems to be Charles V appears at the gates of his tomb and takes Carlos with him to safety.

**Don Carlos.** The lover of Laura (bar), and rival of Don Juan, in Dargomyzhsky's *The Stone Guest*. See also above.

**Don Carlos di Vargas.** Leonora's brother (bar) in Verdi's *La forza del destino*.

**Don Giovanni** (orig. *Il dissoluto punito, ossia Il Don Giovanni*). Opera in 2 acts by Mozart; text by Da Ponte, after the Don Juan legend and particularly after Bertati's libretto for Gazzaniga's *Don Giovanni Tenorio* (1787). Prod. Prague, Nostic T, 29 Oct. 1787, with Saporiti, Micelli, Bondini, Baglioni, Bassi, Ponziani, cond. Mozart; London, H, 12 Apr. 1817, with Camporese, Hughes, Fodor, Naldi, Ambrogetti, Crivelli, cond. Weichsel; New York, Park T, 23 May 1826, in Da Ponte's presence, with four members of the García family, Manuel sen. (Giovanni) and jun. (Leporello), Joaquina (Elvira), and Maria (Zerlina), cond. Étienne.

Seville, c.1600. Leporello (bs) is waiting outside a house where his master Don Giovanni (bar) is trying to seduce Donna Anna. She calls for help and her father, the Commendatore (bs), appears; Don Giovanni kills him. Donna Anna fetches her betrothed, Don Ottavio (ten), and they find the Commendatore dead; Don Ottavio vows to avenge the murder.

Donna Elvira (sop) arrives in search of Don Giovanni. He recognizes her as a former lover, and escapes, leaving Leporello to show her the catalogue of his master's conquests. She swears vengeance. Don Giovanni happens upon a peasant wedding and begins to flirt with the bride Zerlina (sop). She is about to succumb when Elvira intervenes. Don Ottavio and Donna Anna arrive and, not recognizing Don Giovanni, ask for his assistance in finding the Commendatore's murderer. As Don Giovanni departs, Anna recognizes him as the culprit. Three masked figures arrive and are invited to join the party with which Giovanni is celebrating the wedding of Zerlina and Masetto (bs); they are Elvira, Anna, and Ottavio. They discover him attempting further seduction and reveal their identities, vowing to disclose his behaviour to everyone.

Don Giovanni has exchanged costumes with Leporello and serenades Elvira's maid outside an inn. Elvira herself comes down to the street and departs with 'Don Giovanni' (actually Leporello). Masetto comes in search of Don Giovanni, who pretends to be Leporello but then beats Masetto. In a dark courtyard, Zerlina and Masetto encounter Leporello, whom they take for Don Giovanni. So as to escape, he sheds his disguise. Don Giovanni and Leporello meet in a cemetery, where the statue of the Commendatore seems to speak. Don Giovanni invites him to supper. The statue arrives and orders Don Giovanni to repent of his crimes. When he refuses, the flames of hell engulf him. Elvira, Anna, and Don Ottavio enter to confront Don Giovanni and are told of the events by Leporello. The opera ends with the cast addressing the audience with the moral of the story.

BIBL: J. Rushton, *W. A. Mozart: 'Don Giovanni'* (Cambridge, 1981).

**Doni, Giovanni Battista** (*b* Florence, *bapt.* 13 Mar. 1595; *d* Florence, 1 Dec. 1647). Italian theorist. His work on ancient Greek music includes investigations into modes and instruments, and the important *Trattato della musica scenica* (1633–5): one of the most important chronicles of early operatic activity, and fullest discussion of *stile rappresentativo.

**Donington, Robert** (*b* Leeds, 4 May 1907; *d* Firle, 20 Jan. 1990). English musicologist. Studied Oxford. His writings include a study *Wagner's 'Ring' and its Symbols* (London, 1963, 3/ 1974) that states a case for the symbolism in the work being best understood in terms of Jungian analysis. He also published *The Opera* (London, 1978) and returned to his earlier interest in *Opera and its Symbols* (Yale, 1990).

**Donizetti, Gaetano** (*b* Bergamo, 29 Nov. 1797; *d* Bergamo, 8 Apr. 1848). Italian composer. Studied Bergamo with Mayr and others, Bologna with Mattei. His first student operas were unproduced, but at Mayr's instigation he wrote some for Venice. The success of *Zoraide di Granata* led to a contract from Barbaia for Naples. His early successful comedies, which include *L'ajo nell'imbarazzo*, *Olivo e Pasquale*, and *Il giovedì grasso*, belong, despite glimpses of individuality, to the world of Italian opera dominated by Rossini. With the next works, among them *L'esule di Roma*, *Alina*, *Il paria*, and *Elisabetta*, the influence of Rossini's French style is felt, as well as that of Bellini, though Donizetti's own mature style is also beginning to emerge. However, this is fully revealed in 1830 with *Anna Bolena*, his first great tragedy and the work with which he won international recognition.

In 1832 he broke his contract with Naples so as to broaden his field of activity, signing a new and more favourable contract there in 1834. Of works written in the 1830s, *L'elisir d'amore* remains a classic of sentimental comedy, while *Torquato Tasso* and *Lucrezia Borgia* broke new ground as Romantic melodramas. A visit to Paris and the discovery of Grand Opera in 1835 influenced *Lucia di Lammermoor*, which, for all its familiarity as an example of Donizetti's style, is in fact uneven, including both original Romantic elements and pages (some of them the most famous) of routine post-Rossini vocalization. Donizetti was always happy to write so as to give the best expression to the leading vocal talents of the day; hence the virtuosity of Fanny Tacchinardi-Persiani is reflected in parts of *Lucia*, while the dramatic gift of Pasta marks *Anna Bolena* and that of Ronzi di Begnis *Roberto Devereux*, the directness of Rosine Stolz *La favorite*, and the noble example of Ronconi the writing for baritone in *Maria di Rohan* and *Maria Padilla*.

By the mid-1830s Donizetti was in full command of his dramatic gift, and though always willing to amend the content of his works to suit conditions, or to write to accommodate singers, he had begun to subordinate mere display to the needs of the drama. A striking example from this period is *Roberto Devereux*, a powerful work to contrast with the flimsiness of *Pia de' Tolomei* or the awkward dramatic effect of *Maria di Rudenz*. Commissions from Vienna and Paris encouraged him to develop his style to the needs of these new audiences. When in 1838 he was denied the directorship of the Naples Conservatory in favour of Mercadante, and *Poliuto* was banned by the Neapolitan censor, he left for Paris to produce the opera as *Les martyrs*.

In Paris he met with mixed success. He reworked some of his Italian operas and wrote *La fille du régiment*, a highly successful work, Italian in essence and only lightly coloured with Gallicisms; however, also in 1840 he was obliged to withdraw *Lucrezia Borgia* when Hugo objected to Romani's adaptation of his tragedy. The initially unsuccessful *La favorite* of the same year is a French Grand Opera. *Linda di Chamounix* (though written for Vienna) manages by contrast to reconcile the Italian *opera semiseria genre with that of French opéra-comique. Declining health progressively limited his ability to travel, though he kept up a frantic round of activity and travel. In Milan in 1841 he was depressed by censorship troubles with *Maria Padilla*. In Vienna in 1842 he was appointed Kapellmeister to the Austrian court. Other works of his last years suggest further developments: *Maria Padilla* is much more freely composed, like the later stricter and more compact *Maria di Rohan* varying formal aria and recitative in a manner that virtually achieves continuous composition. *Caterina Cor-*

*naro* is concentrated and intense in manner, whereas *Don Pasquale*, his comic masterpiece, is a relaxed and witty piece that almost wholly conceals his failing health. *Maria di Rohan* proved on its Vienna production in 1844 to be one of his most powerful Romantic dramas. He was disappointed by the failure of his last work, *Dom Sébastien*, a French Grand Opera of sombre grandeur. His illness, syphilitic in origin, caused eventual paralysis, and he was moved to Bergamo, where he was looked after by friends until his death.

A versatile and practical musician, Donizetti ranged widely in his subjects. The early works are typically comedies, his most characteristic later ones Romantic tragedies; and it was the latter genre that he regarded as his most important. Here, though always prone to relapse into triviality, he developed a gift for characterization and for dramatic momentum that can still be strongly effective in the theatre. This was partly achieved by loosening the old firm divisions of recitative and aria, and interspersing arioso sections, perhaps varied with choral comments, that allow a much freer expressive range; yet even his most original moments could be followed by a scintillating but cold display of soprano fireworks. Aware of the problem, he evolved the slow cabaletta, which allowed the singer more effective dramatic expression, with coloratura that remained satisfying to the singer's vanity but was more respectful of the audience's intelligence. Towards the latter part of his career, he also found ways of enriching his accompaniments (which can be feeble) and of chromatically inflecting his vocal lines subtly for expressive purposes without loss of tunefulness or singability. At the same time, he explored a richer vein of orchestration, and experimented with novel combinations of voices to vary the conventional ensemble patterns. In these matters, as in a good deal else, Donizetti is the most important direct forerunner of Verdi. The tunefulness of his 70 stage works, not to mention their superiority, even at their most erratic, over most contemporary operas, long ensured their place in the repertory; and in modern times they have won new audiences for their dramatic power as well as for their melodic charm and skilful stagecraft.

WORKLIST: *Il Pigmalione* (?, after Sografi, after Rousseau, after Ovid; comp. 1816, prod. Bergamo 1960); *L'ira d'Achille* (The Wrath of Achilles) (?; comp. 1817, unprod.); *Enrico di Borgogna* (Merelli, after Kotzebue; Venice 1818); *Una follia* (A Folly) (Merelli; Venice 1818); *Le nozze in villa* (Marriage in a Villa) (Merelli; Mantua 1820–1); *Il falegname di Livonia* (The Carpenter of Livonia) (Bevilacqua-Aldovandrini, after Romani, after Duval; Venice 1819); *Zoraide di Granata* (Merelli, rev. Ferretti, after Romanelli & Florian; Rome 1822); *La zingara* (The Gypsy) (Tottola, after Caigniez, after Kotzebue; Naples 1822); *La lettera anonima* (Genoino, after Corneille; Naples 1822); *Chiara e Serafina* (Romani, after Pixérécourt; Milan 1822); *Alfredo il grande* (Tottola, ?after Merelli; Naples 1823); *Il fortunato inganno* (The Fortunate Deceit) (Tottola; Naples 1823); *L'ajo nell'imbarazzo* (The Tutor in a Quandary) (Ferretti, after Giraud; Rome 1824); *Emilia di Liverpool* (Checcherini, rev. anon., after anon., after Scatizzi, after Kotzebue; Naples 1824); *Alahor in Granata* (M.A., after Florian, after Jouy, & Romani; Palermo 1826); *Elvida* (Schmidt; Naples 1826); *Gabriella di Vergy* (Tottola, ?Donizetti, after Tottola, after 14th-cent. romance; comp. 1826, prod. Naples 1869; 2nd version, comp. *c.*1838, prod. Belfast 1978); *Olivo e Pasquale* (Ferretti, after Sografi; Rome 1827); *Otto mesi in due ore* (Eight Months in Two Hours) (Alcozer, rev. Gilardoni, after Marchioni, after Pixérécourt, after Cottin; Naples 1827); *Il borgomastro di Saardam* (The Burgomaster of Saardam) (Gilardoni, after Mélesville, Merle, & Boirie; Naples 1827); *Le convenienze ed inconvenienze teatrali* (Theatrical Ups and Downs) (Donizetti, after Sografi; Naples 1827); *L'esule di Roma* (The Exile of Rome) (Gilardoni, after Marchionni, after Caigniez, Debotière; Naples 1828); *Alina, regina di Golconda* (Romani, after Sedaine, after De Boufflers; Genoa 1828); *Gianni di Calais* (Gilardoni, after Caigniez; Naples 1828); *Il paria* (Gilardoni, after Delavigne, Rossi, & Taglioni; Naples 1829); *Il giovedì grasso* (Shrove Tuesday) (Gilardoni, after Scribe & Delestre-Poirson; Naples 1828); *Elisabetta al Castello di Kenilworth* (Tottola, after Barbieri, after Scribe, after Scott; Naples 1829); *I pazzi per progetto* (The Deliberate Madmen) (Gilardoni, after Cosenza, after Scribe, Poirson, & Bertini; Naples 1830); *Il diluvio universale* (The Flood) (Gilardoni, after Ringhieri, after Byron & Moore; Naples 1830); *Imelda de' Lambertazzi* (Tottola, after Sgricci & Sacchi, both after Bombaci; Naples 1830); *Anna Bolena* (Romani, after Pindemonte, after Chénier, after Pepoli; Milan 1830); *Gianni di Parigi* (Romani, after Saint-Just; Milan 1839); *Francesca di Foix* (Gilardoni, after Bouilly & Mercier-Dupaty; Naples 1831); *La romanziera e l'uomo nero* (The Novelist and the Negro) (Gilardoni, ?after Scribe & Dupin; Naples 1831); *Fausta* (Gilardoni, Donizetti; Naples 1832); *Ugo, conte di Parigi* (Romani; Milan 1832); *L'elisir d'amore* (The Love Potion) (Romani, after Scribe, after Malaperta; Milan 1832); *Sancia di Castiglia* (Salatino; Naples 1832); *Il furioso all'isola di San Domingo* (The Madman on the Island of San Domingo) (Ferretti, after anon., after Cervantes; Rome 1833); *Parisina* (Romani, after Byron; Florence 1833); *Torquato Tasso* (Ferretti, after Rosini, after Goldoni, Goethe, & Byron; Rome 1833); *Lucrezia Borgia* (Romani, after Hugo; Milan 1833); *Rosmonda d'Inghilterra* (Romani; Florence 1834); *Maria Stuarda* (Bardari, after Schiller; Milan 1833; 2nd version as *Buondelmonte*, Salatino; Naples 1834); *Gemma di Vergy* (Bidèra, after Dumas; Milan 1834); *Marino Faliero* (Bidèra, after Delavigne & Byron; Paris 1835); *Lucia*

*di Lammermoor* (Cammarano, after Scott; Naples 1835); *Belisario* (Cammarano, after Marmontel, after Schenk; Venice 1836); *Il campanello di notte* (The Night Bell) (Donizetti, after Brunswick, Troin, & Lhérie; Naples 1836); *Betly* (Donizetti, after Scribe & Mélesville, after Goethe; Naples 1836); *L'assedio di Calais* (The Siege of Calais) (Cammarano, after Marchionni, after Belloy; Naples 1836); *Pia de' Tolomei* (Cammarano, after Sestini, Bianco & Marenco, after Dante; Venice 1837); *Roberto Devereux* (Cammarano, after Ancelot & Romani; Naples 1837); *Maria di Rudenz* (Cammarano, after Bourgeois, Cuvelier & Mallian, after Lewis; Venice 1838); *Poliuto* (Cammarano, after Corneille; Naples 1838; 2nd version *Les martyrs*, Scribe; Paris 1840); *La fille du régiment* (The Daughter of the Regiment) (Saint-Georges & Bayard; Paris 1840); *L'ange de Nisida* (Royer & Vaëz: unprod., material rev. in *La favorite*); *La favorite* (Royer & Vaëz, with Scribe; Paris 1840); *Adelia* (Romani & Marini, after anon.; Rome 1841); *Rita* (Vaëz; comp. 1841, prod. Paris 1860); *Maria Padilla* (Rossi & Donizetti, after Ancelot; Milan 1841); *Linda di Chamounix* (Rossi, after D'Ennery & Lemoine; Vienna 1842); *Caterina Cornaro* (Sacchero, after Saint-Georges; Naples 1844); *Don Pasquale* (Ruffini & Donizetti, after Anelli; Paris 1843); *Maria di Rohan* (Cammarano & Lillo, after Lockroy & Badon; Vienna 1843); *Dom Sébastien* (Scribe, after Foucher; Paris 1843).

BIBL: G. Zavadini, *Donizetti: vita, musiche, epistolario* (Bergamo, 1948); W. Ashbrook, *Donizetti* (London, 1965); W. Ashbrook, *Donizetti and his Operas* (Cambridge, 1982).

**Don José.** The corporal of dragoons (ten), lover of Micaëla and of Carmen, in Bizet's *Carmen*.

**Don Juan.** The legendary seducer and blasphemer, the nature of whose crime lies in his flouting of the conventions of man, by taking any woman for his pleasure, and of God, by mocking a dead man with the invitation of his statue to dinner. The tale was first set down in full dramatic form by Tirso de Molina (*El burlador de Sevilla y el convidado de piedra*, 1630). Later developments of the legend assert not so much the wickedness and punishment of a villain as the Romantic grandeur of a hero seeking an ideal and disdaining the pettiness of society. Operas on the legend are by Melani (*L'empio punito*, 1669), Righini (1777), Calegari (1777), Tritto (1783), Albertini (1784), Gazzaniga (1787), Gardi (*Il nuovo convitato di pietra*, in hurried imitation of Gazzaniga, 1787), Fabrizi (1787), Reeve (1787), Mozart (1787), Paisiello (1790), Federici (concocted by Da Ponte, music also by Gazzaniga, Sarti, & Guglielmi, 1794), Gardi (1802), Dibdin (music by many other hands also, 1817), Raimondi (1818), Carnicer (1822), Sanderson (1820s), Pacini (1832), D'Orgeval (*Don Juan de village*, 1863), Barbier (*Don Juan de fantaisie*, 1866), Dargomyzhsky (1872), Manent (1875),

Polignac (*Don Juan et Haidée*, 1877), Anthiome (*Don Juan marié*, 1878), Delibes (*Le Don Juan suisse*, 1880), Palmieri (1881), Vietinghoff-Scheel (1888), L. André (*Don Juan auf Reisen*, c.1890), Linan and Videgain (1900), Fayos (zarzuela, *El Don Juan de Mozart*, 1899), Helm (1911), Graener (*Don Juans letzte Abenteuer*, 1914), Benatzky (*Fräulein Don Juan*, 1915), Simon (1922), Lattuada (1929), Haug (1930), Casavola (1932), Collet (*Le fils de Don Juan*, 20th cent.), Malipiero (1963), Palester (1965), Taranu (1970), and Slater (1972). A variant of the legend, showing ultimate repentance, is derived from Prosper Mérimée's story *Les âmes du purgatoire* (1834), based on the notorious Miguel de Mañara (*d* 1697): operas by Alfano (1914), Enna (1925), Goossens (1937), and Tomasi (1956).

**Don Magnifico.** Cenerentola's father (bs) in Rossini's *La Cenerentola*.

**Donna Anna.** The Commendatore's daughter (sop), lover of Don Ottavio, in Mozart's *Don Giovanni*.

**Donna del lago, La** (The Lady of the Lake). Opera in 2 acts by Rossini; text by Tottola, after Walter Scott's poem (1810). Prod. Naples, C, 24 Sep. 1819, with Pisaroni, Colbran, David; London, H, 18 Feb. 1823, with De Begnis, De Vestris, Curioni; New Orleans, T d'Orléans, 25 June 1829, with C. and R. Fanti, Ravaglia, Bordogni, Fabj, De Rosa. Rev. Florence 1958.

Scotland, first half of 16th cent. James V of Scotland—Giacomo in the opera (ten)—loses his way and seeks shelter in the house of his enemy Douglas (bs), whose daughter Ellen (Elena, sop) has been forced to marry Roderick Dhu (Rodrigo, ten). In an uprising against the King, Roderick is killed, and Elena pleads with the King for her father's life and obtains his permission to marry her lover Malcolm (mez).

**Donna Elvira.** A lady of Burgos (sop), one of Don Giovanni's deserted lovers, in Mozart's *Don Giovanni*.

**'Donna non vidi mai'.** Des Grieux's (ten) aria in Act I of Puccini's *Manon Lescaut*, in which he voices his feelings of rapture as he sets eyes on Manon for the first time.

**Donne curiose, Le** (The Inquisitive Women). Opera in 3 acts by Wolf-Ferrari; text by Luigi Sugana after Goldoni's play of the same name. Prod. Munich, N, 27 Nov. 1903, in a German translation by H. Teibler as *Die neugierigen Frauen*, with Bosetti, Tordek, Breuer, Huhn, Koppe, Brodersen, Sieglitz, Bender, cond. Reichenberger; New York, M (first perf. in Italian), 3 Jan. 1912, with Alten, Farrar, Fornia, Mauborg, Jadlowker, De Segurola, Pini-Corsi, Didur, cond. Toscanini; Milan, S, 16 Jan. 1913, with Ferraris, Villani, Lollini, Solari, Navia, Govoni, Vannuccini, cond. Serafin; London, Peter Jones T, 28 Apr. 1952.

Set in 18th-cent. Venice, the opera tells of the suspicions of two married women and their friend, who believe that their husbands and lover are taking part in regular orgies at their men's club; they gain admission to the club and discover that all their menfolk are doing is indulging in gourmet meals.

**Donner.** The God of Thunder (bs-bar) in Wagner's *Das Rheingold*.

**Don Ottavio.** Donna Anna's lover (ten) in Mozart's *Don Giovanni*.

**Don Pasquale.** Opera in 3 acts by Donizetti; text by Ruffini and the composer after Anelli's libretto for Pavesi's *Ser Marc'Antonio* (1810). Prod. Paris, I, 3 Jan. 1843 with Grisi, Mario, Tamburini, Lablache; London, HM, 29 June 1843 with same cast but Fornasari replacing Tamburini; New Orleans, T d'Orléans, 7 Jan. 1845, with J. Calvé, Douvry, Garry, Couériot, cond. E. Prévost.

Rome, 19th cent. Don Pasquale (bs), a rich and crusty old bachelor, has decided to marry so that his nephew Ernesto (ten) will not inherit his fortune as he disapproves of his choice of a bride, the attractive young widow Norina (sop). The young couple, helped by Doctor Malatesta (bar), a friend of them both as well as of Pasquale, arrange for the old man to go through a mock marriage with 'Sofronia', Norina in disguise, who is presented as Malatesta's sister, fresh from a convent. No sooner is the wedding ceremony over, than Sofronia reveals herself in her true colours and makes the old man's life a misery. Malatesta suggests that the marriage be annulled; Pasquale agrees and, admitting he has been fooled, forgives them all.

**Don Procopio.** Opera in 2 acts by Bizet; text by Paul Collin and Paul de Choudens, translated from a libretto by Carlo Cambiaggio. Written 1859. Prod. (posth.) Monte Carlo, T du Casino, 10 Mar. 1906.

A country house in Spain, c.1600. To general disapproval, Don Andronico (bs) has agreed to marry his niece Bettina (sop) to Don Procopio (bar), a rich old miser. Her brother and aunt convince Don Procopio that Bettina would spend all his money and so he cancels the wedding, enabling Bettina to marry her beloved Eduardo (ten).

**Don Quichotte.** Opera in 5 acts by Massenet; text by Henri Cain, after Jacques Le Lorrain's play *Le chevalier de la longue figure*, based in turn on Cervantes's novel (1605, 1615). Prod. Monte

# Don Quixote

Carlo, T du Casino, 19 Feb. 1910, with Arbell, Shalyapin, Gresse, cond. Jéhin; New Orleans, French OH, 27 Jan. 1912; London, CG, 18 May, 1912.

Don Quixote (bs) is a naive champion of goodness, charged by the sophisticated Dulcinea (con) to recover her necklace from brigands. He sets off with Sancho Panza (bar) and, after attacking the windmills, is seized by the bandits. His innocence wins his release with the necklace, but, rejected by Dulcinea, he dies broken-hearted.

**Don Quixote.** See CERVANTES.

**Don Rodrigo.** Opera in 3 acts by Ginastera; text by Alejandro Casona. Commissioned and prod. Buenos Aires, C, 24 July 1964, with Bandin, Cossutta, Mattiello, De Narké, cond. Bartoletti; New York, State T, 22 Feb. 1966, to mark the opening of City Opera's new home there, with Crader, Domingo, Clatworthy, Malas, cond. Rudel.

The opera tells of the brief reign of Don Rodrigo (ten), the last Visigoth King of Spain, whose defeat at Guadalete led to the fall of Spain to the Saracens.

**Don Sanche.** Opera in 1 act by Liszt; text by Emmanuel Guillaume Théaulon de Lambert and De Rancé after a tale by Florian. Prod. Paris, O, 17 Oct. 1825 with Nourrit, cond. R. Kreutzer; London 1977.

Don Sanche (ten) wishes to enter a moated castle, owned by the magician Alidor (bar), but is told that only those who love and are loved may be admitted. He loves Princess Elzire (sop), but she does not love him. Alidor promises to help him, and raises a storm, enabling him to come to Elzire's aid. Don Sanche prevents her abduction by the knight Romualde, but is wounded: Elzire's indifference towards him is transformed into love. It emerges that 'Romualde' was Alidor in disguise; he cures Don Sanche's injuries and the happy couple may now enter the castle of love.

Liszt's only completed opera, written when he was 12. The score was lost until 1903.

**Donzelli, Domenico** (*b* Bergamo, 2 Feb. 1790; *d* Bologna, 31 Mar. 1873). Italian tenor. Studied Bergamo with Bianchi. Début Bergamo 1808 (comprimario in Mayr's *Elisa*). Further study Naples with Viganoni and Crivelli. After increasing success throughout Italy, sang in Rossini's *Tancredi*, Venice 1815, the first *Torvaldo e Dorliska*, and *L'inganno felice*, becoming the composer's friend. After *La Cenerentola*, Milan, S, 1821, his large voice, always baritonal in quality, grew heavier; he sang his famous Otello (Rossini) using full-voiced high notes (up to a), one of the first dramatic tenors to do so. (Americo Sbigoli,

trying to copy him, burst a blood-vessel and died.) London, H, 1829, and in 1830 in the first Bellini opera performed in London, *Il pirata*. Created roles in operas by Pacini, Halévy, and Bellini (Pollione in *Norma*, very successfully); also Mercadante (*Il bravo*, *Elisa e Claudio*), and Donizetti (*Maria Padilla*). An expressive and dramatic singer, he was admired by Chorley, who refers to his 'mellifluous, robust' voice, Stendhal, who pronounced him 'very fine', though he disliked his 'yell', and Rossini, who declared simply, 'That's a singer!'

**Dorabella.** Fiordiligi's sister (sop) in Mozart's *Così fan tutte*.

**Dorati, Antal** (*b* Budapest, 9 Apr. 1906; *d* Gerzensee, 13 Nov. 1988). Hungarian, later US, conductor. Studied Budapest with Bartók, Kodaly, and Weiner. Début Budapest O 1924. Dresden, assistant cond. to Busch, 1928–9; Münster 1929–33. London, CG, 1962 (*Golden Cockerel*). Autobiography, *Notes of Seven Decades* (London, 1979) (R).

**Dorfbarbier, Der** (The Village Barber). Opera in 1 act by Schenk; text by J. and P. Weidmann, originally prod. as comedy 1785. Prod. Vienna, B, 30 Oct. 1796; New York 15 Dec. 1847.

A small village, late 18th cent. The village barber and local quack, Lux (bs), wishes to marry his ward Suschen (sop). However, she loves Josef (ten), and the couple plan a trick on Lux, aided by the schoolmaster, Rund (bs). Suschen pretends to spurn Josef, who consequently takes poison. Since she has been named as Josef's beneficiary, Lux agrees to a deathbed marriage. The instant he has given his consent, Josef leaps up 'cured'.

Schenk's most popular work, it was seen on virtually every German stage and in 1816 alone received no fewer than 203 performances. Seidel composed a sequel, *Die Schmiedswitwe* (1807). Also operas by Hiller (1771) and Wernhammer (1790s).

**Dorliak, Xenia** (*b* St Petersburg, 11 Jan. 1882; *d* Moscow, 8 Mar. 1945). Russian soprano. Studied St Petersburg with Gladkaya, later Iretskaya. Sang Paris, Prague, Berlin. Paris, O, 1911 (Elisabeth, Elsa, Brünnhilde). St Petersburg and other major Russian houses. A prominent Wagner singer, she was also admired as Aida, Amelia, Tosca, Tatyana. Retired from stage 1914. Her pupils incl. her daughter **Nina Dorliak** (*b* 1908), later wife of Sviatoslav Richter.

**Dorn, Heinrich** (*b* Königsberg, 14 Nov. 1804; *d* Berlin, 10 Jan. 1892). German conductor and composer. Studied Berlin with Berger, Klein, and Zelter. Königsberg 1828; Leipzig 1829–32; Riga 1832–43; Cologne 1844–7; Berlin 1849–69.

His Riga friendship with Wagner turned to hostility; and he anticipated Wagner by composing an opera *Die Nibelungen* (Weimar 1854) to a text rejected by Mendelssohn. Much admired as conductor and teacher.

**Dorset Garden Theatre.** One of London's earliest homes for music and drama. Built by Sir William Davenant's widow on plans by Wren; opened 1671. Also known as the Duke's, later the Queen's (1698). Several of Purcell's stage works were premièred there, including *King Arthur* (1691), *The Fairy Queen* (1692), and *The Indian Queen* (1695). Closed 1706.

**Dortmund.** City in North-Rhine Westphalia, Germany. The first theatre opened 1837. The Stadttheater (cap. 1,200) opened 1904 with *Tannhäuser*; bombed 1943; the T Dortmund-Opernhaus (cap. 1,160) opened 1966 with *Der Rosenkavalier*. Premières have included Steffen's *Eli* (1957) and Gurlitt's *Nana* (1958).

**Dorus-Gras, Julie** (*b* Valenciennes, 9 Sep. 1805; *d* Paris, 9 Feb. 1896). Belgian soprano. Studied Paris with Henri and Blangini, then Paer and Bordogni. Début Brussels, M, 1826. Début Paris, O, 1830 (Angèle, *Le Comte Ory*), where she succeeded Cinti-Damoreau as prima donna 1835–45. (Her departure from the Opéra was precipitated by the victimization of Rosine Stoltz.) London, DL, 1847 (Lucia); London, CG, 1849, Alice (*Robert le diable*), and Marguerite de Valois (*Les Huguenots*), two roles which she had created. Others included Eudoxie (*La Juive*), Pauline (Donizetti's *Les martyrs*), and Teresa (*Benvenuto Cellini*). Also a famous Elvira (*Muette de Portici*), which she sang in the performance in Brussels, Sep. 1830, that sparked off the revolt of the Low Countries. Like Cinti-Damoreau, she was a brilliant and accurate singer, with a pure voice. She had to endure lampooning for the sobriety of her private life.

**Dosifey.** Dositheus (bs), leader of the Old Believers in Musorgsky's *Khovanshchina*.

**Dostoyevsky, Fyodor** (*b* Moscow, 11 Nov. 1821; *d* St Petersburg, 9 Feb. 1881). Russian writer. Operas on his works are as follows:
*Christ, the Boy, and the Christmas Tree* (1848): Rebikov (*Yolka*, 1903)
*White Nights* (1848): Tsvetayev (1933); Cortese (1970); J. Buzko (1973)
*From the House of the Dead* (1862): Janáček (1930)
*The Gambler* (1866): Prokofiev (1929)
*Crime and Punishment* (1866): Pedrollo (1926); Sutermeister (*Raskolnikoff*, 1948); Mortimer (1967); Petrovics (1969)
*The Idiot* (1869): Bogdanov-Berezovsky (*Nastasya Filipovna*, 1968); L. Chailly (1970); J. Eaton (*Myshkin*, 1973)

*The Brothers Karamazov* (1880): Jeremiáš (1928); Ruyneman (1928)
*Uncle's Dream* (1859): Krasa (*Verlobung im Traum*, 1933)
*The Crocodile* (n.d.): T. Wagner (1963)
*Poor Folk* (n.d.): Sedelnikov (1974)

**dotazione** (It.: 'endowment'). The term formerly used for the repertory of sets maintained by small theatres unable to afford new, specially designed sets for each production. Economic necessity has latterly brought about the practice of sharing sets even between major opera companies.

**Dourlinski.** Baron Dourlinski (bs), owner of the castle in which Lodoïska is imprisoned, in Cherubini's *Lodoïska*. He was a model for Beethoven's *Pizarro.

**'Dove sono'.** The Countess's (sop) aria in Act III of Mozart's *Le nozze di Figaro*, mourning the loss of her husband's love.

**Downes, (Sir) Edward** (*b* Birmingham, 17 June 1924). English conductor. Studied Birmingham, and with Scherchen. Début with CR 1950. CR until 1952; London, CG, 1952–69 (Solti's assistant from 1966), and subsequently; Australian O 1970–6; WNO 1975. Principal cond. CG 1992. Repertory incl. *Freischütz*, *Attila*, *Flying Dutchman*, *The Ring*, *Katerina Ismaylova*, *Turandot*, *Salome*, *War and Peace*, *Fiery Angel*. Completed and conducted first performance of Prokofiev's *Maddalena* (BBC 1979). Translations incl. *Khovanshchina*, *Katerina Ismaylova*, *Jenůfa*. (R)

**Down in the Valley.** Opera by Kurt Weill; text by Arnold Sundgaard. Prod. Bloomington, Indiana Univ., 14 July 1948 with Bell, Aiken, Welsh, Campbell, Carpenter, Jones; Bristol 23 Oct. 1957.

In a series of flashbacks, the opera tells the story of a young couple, Brack Weaver (ten) and Jennie Parsons (sop). Brack is awaiting execution for the killing, in self-defence, of Thomas Bouché (bs), an unpleasant rival for Jennie's affections.

**D'Oyly Carte Opera Company.** See CARTE, RICHARD D'OYLY.

**Draghi, Antonio** (*b* Rimini, 1634/5; *d* Vienna, 16 Jan. 1700). Italian composer. Began his career as opera singer in Venice; went to Vienna 1658, joining the chapel of the Dowager Empress Eleonore. Vizekapellmeister (in succession to Ziani) 1668; Kapellmeister 1669. That year he began to compose operas for the imperial court: in 1673 he became intendant of theatrical music and in 1682 Kapellmeister to the imperial chapel.

Draghi was one of the most prolific of 17th-cent. composers, producing *c*.175 dramatic works for Vienna. His first opera, *La mascherata*, used his own libretto, but most of his later works set Minato. Like Cesti and Ziani before him, he proved adept at transforming the typical musical and dramatic devices of Venetian public opera into a style appropriate to a courtly entertainment and his works represent the culmination of the first period of Italian opera. He was naturally most attracted to heroic subjects, as in *Turia Lucrezia*, *Rodogone*, *La magnanimità di Marco Fabrizio*, and *Gundeberga*, one of the first operas to be based on a German historical subject. In the smaller serenatas, trattenimenti, and commediette, such as *Fileno e Clori* (1674) and *Lo studio d'amore* (1686), he usually turned to pastoral themes. Draghi's patron, Emperor Leopold I, himself a composer of dramatic music, provided additional numbers for some of his works.

During his early years in Vienna he was also active as a librettist and wrote texts for Bertali, Sances, Tricarico, and Ziani. His son **Carlo Draghi** (1669-1711) was court organist in Vienna from 1698 and composed additional music for *Sulpizia* (Minato; 1672) and *La forza dell'amor filiale* (The Force of Filial Love) (Cupeda; 1698). It is possible that Giovanni Battista *Draghi an important figure in early English opera, was his brother.

SELECT WORKLIST (all first prod. Vienna): *La mascherata* (The Masquerade) (Draghi; 1666); *Gundeberga* (Minato; 1672); *Turia Lucrezia* (Minato; 1675); *Rodogone* (Minato; 1677); *La magnanimità di Marco Fabrizio* (Cupeda; 1695).

**Draghi, Giovanni Battista** (*b* ?, *c*.1640; *d*. London, 1708). Italian harpsichordist. Possibly the brother of Antonio *Draghi, he was active in London immediately after the Restoration, appears as the composer of many songs and instrumental pieces, and was apparently a sought-after harpsichordist. In 1667 Pepys records having heard Draghi play through an Italian opera which he had written; appointed organist to Queen Catherine of Braganza, 1677. The dances which Draghi provided for Shadwell's adaptation of *The Tempest* in 1674 have been lost, but his music for Shadwell's *Psyché* (1675), for which he collaborated with Matthew Locke, represents some of the earliest operatic writing in England. He later contributed the music for a number of stage plays, including Tate's *A Duke and No Duke* (1684) and Mountford's *The Injured Lovers* (1688).

Though there is little that is distinctive about Draghi's music, he was one of the earliest composers in England involved with the new operatic genre. He is sometimes mistakenly credited with having written the music for the comic opera *Wonder in the Sun* (1706).

**Dramaturg** (Ger.). The member of staff of a German opera-house who combines the duties of adapter of librettos, editor of programmes, and press officer, sometimes also producer and even actor. The term (usually spelt dramaturge) is now also in wide use in the UK.

**drame forain** (Fr.: 'fair-drama'). The comedies staged at the Paris fairs in the 18th cent., known as drames forains, are an ancestor of *opéra-comique and came into being following the expulsion of the Comédie-Italienne in 1696. Various companies played at the Foire Saint-Germain (Feb.–Apr.) and the Foire Saint-Laurent (Aug.–Sept.), the so-called **Théâtres de la Foire**, giving light, irreverent, satirical comedies, often mocking the Opéra and Opéra-Comique. Nimbly evading all attempts at banning them, they used music and songs, though only incidentally, and the singing was by actors rather than by professional singers; indeed, at first the drames forains were performed between spectacular acrobatic feats. The success of these highly popular pieces attracted some very skilled writers; such was René Le Sage (1668–1747), author among much else of the novel *Gil Blas*, who gave the form of these musical comedies a new control. Efforts by the Comédie-Française to restrict the Théâtres de la Foire included banning any use of spoken dialogue, a prohibition which was circumvented by the actors writing their lines on placards which were held up to the audience. This prose text was soon set to *vaudevilles which the audience sang, and soon the actors were themselves performing the music, having obtained permission to do so from the Opéra. With the composition of entire pieces around vaudevilles, an early version of opéra-comique was born. Such was its success that in 1719 the Comédie-Française banned all presentations at the Théâtres de la Foire except for acrobatics, although in 1721–3 a new Italian company performed at the Foire Saint-Laurent. It was by writing for the fairs that *Favart began his career: starting from the drame forain he assisted in the development of the *comédie mêlée d'ariettes, and so advanced the direction of *opéra-comique.

**drame lyrique** (Fr.: 'lyric drama'). A standard term for opera in the late 19th- and early 20th-cent. French opera which, growing out of opéra-comique, combined an intimacy of plot with an opulence of musical style. The first appearance of the term is probably in connection with Massenet's *Werther* (1892): although he did not use it again himself, many of his later operas present similar treatments, providing a model for the generation which followed.

**dramma comico** (It.: 'comic drama'). An 18th-cent. synonym for *opera buffa.

**dramma di sentimento.** See OPERA SEMISERIA.

**dramma eroicomico** (It.: 'heroic-comic drama'). A term used in the late 18th and early 19th cent. for an opera buffa whose plot included some heroic elements: e.g. Haydn's *Orlando paladino*. Its invention is often credited to *Casti.

**dramma giocoso** (It.: 'jocular drama'). A development of *opera buffa, the dramma giocoso was created by *Goldoni around 1750 through the addition of serious roles (*parti serie*) to the stock figures of the genre. Classic examples are provided by the librettos set in Galuppi's *Il filosofo di campagna* and Piccinni's *La buona figliuola*, where the addition of serious characters and episodes lends a more poignant aspect to the drama. Most of Haydn's comic operas, including *Le pescatrici*, *Il mondo della luna*, and *La vera costanza*, were also of this variety. The most famous dramma giocoso, and the only such work regularly performed today, is Mozart's *Don Giovanni*.

**dramma lirico** (It.: 'lyric drama'). A modern term sometimes used for opera, though not with any specific stylistic connotation.

**dramma pastorale** (It.: 'pastoral drama'). A term sometimes used to describe the earliest operas (e.g. Agazzari's *Eumelio*), reflecting the strong influence of the pastoral tradition on the shaping of their plot and characterization.

**dramma per musica** (It.: 'drama for music'). The term used by many 17th-cent. Italian librettists to describe their text, reflecting the fact that it had been written with the express purpose of being set to music. By extension the resultant opera, invariably serious, could also be known as a dramma per musica, or sometimes a dramma in musica, a term preferred by composers, since it stressed the primacy of the music. These terms continued into the 18th cent., when they effectively became interchangeable with that of *opera seria.

**dramma tragicomico.** See OPERA SEMISERIA.

**Dreigroschenoper, Die** (The Threepenny Opera). Opera in a prologue and 3 acts by Weill; text a modern interpretation of The *Beggar's Opera*, based on a translation by Hauptmann, with lyrics, drawn also from Kipling and Villon, by Brecht, with Elisabeth Hauptmann. Prod. Berlin, T am Schiffbauerdamm, 31 Aug. 1928, with Bahn, Lenja, Paulsen, Gerron, Valetti, Ponto, Busch; New York, Empire T, 13 Apr. 1933; London, Royal Court T, 9 Feb. 1956, with D. Anderson, cond. Goldschmidt.

London c.1900. The prologue presents the scene of thieves, beggars, and prostitutes at a Soho fair.

J. J. Peachum (bs) hires costumes out to beggars to arouse the public's pity. When he discovers that his daughter Polly (sop) has secretly married the robber Mack the Knife (Macheath in *The Beggar's Opera*), Peachum threatens to hand him over to the police.

Mack flees but, through bribery from Mrs Peachum (sop), is captured and taken to prison. Mack has pretended to marry Lucy (sop), the daughter of Tiger Brown, the commissioner of the police; an angry scene ensues when both girls visit him. He denies having married Polly and, with Lucy's help, escapes.

The Peachums are again informed of Mack's whereabouts. He is arrested and sentenced to hanging; he tries, unsuccessfully, to escape by bribing a policeman. As Mack goes to the gallows, a royal pardon arrives for him, and the Queen makes him a peer.

BIBL: S. Hinton, *Kurt Weill: 'The Threepenny Opera'* (Cambridge, 1990).

**Drei Pintos, Die** (The Three Pintos). Unfinished opera begun by Weber; text by Theodor Hell, after Seidel's story *Der Brautkampf* (1819). Completed and scored by Mahler, with additions, using other music by Weber and with an interlude composed on themes in the opera by Mahler. Prod. Leipzig 20 Jan. 1888 with Baumann, Artner, Rothhauser, Hedmont, Hübner, Grengg, Schelper, Köhler, Proft, cond. Mahler; London, John Lewis T, 10 Apr. 1962, with Tinsley, Brandt, Maurel, cond. Lloyd-Jones.

Madrid, 19th cent. Don Pinto (bs) is intercepted by Don Gaston (ten) on his way to Seville to marry Clarissa (sop). Gaston, with the help of his servant Ambrosio (bar), first tries to teach the inane Pinto the art of wooing and then makes him drunk. Taking his papers off him, thus becoming Pinto no. 2, Gaston sets off for Seville. When he finds that Clarissa is secretly in love with Gomez (ten) but cannot marry him, he passes on the Pinto papers: Gomez thus becomes the third Pinto. Eventually the real Don Pinto turns up and discloses himself to Clarissa's furious father Don Pantaleone (bar); but all ends happily.

Abandoned by Weber in favour of *Euryanthe*. The original scenario ordered events rather differently, and was arranged in its present form by Mahler and Carl von Weber, the composer's grandson.

**Dresden.** City in Saxony, Germany. In 1617 Heinrich *Schütz reorganized the Kapelle on Italian models, and his *Dafne*, the earliest German opera (music lost), was produced for a court wedding nearby at Torgau in 1627. After the Thirty Years War he was succeeded by Giovanni

Bontempi and Vincenzo Albrici, and the line of opera that began with the opening of the Kurfürstliches Opernhaus in 1667 with *Teseo* (Moniglia) was Italian. Under the directorship of C. *Pallavicino, Dresden court opera developed in the late 17th cent. into a generally lavish and opulent manner, following the main musical precepts of the Venetian style. In 1707 this theatre was converted by Friedrich August into the Hofkirche, and performances were given in the Riesensaal of the Castle; the enlarged Grosses Opernhaus (cap. *c*.1,500–2,000: largest in Europe) opened 1719 with *Giove in Argo* by Antonio *Lotti. Admission was to the public, and free. Under Friedrich August II, Dresden became one of the most important operatic centres in Germany: characteristic at this time were lavish productions calling for the very best musical resources. Particularly significant was Johann *Hasse, and his wife Faustina *Bordoni worked here 1731 and 1734–56. Pietro *Mingotti's company performed 1747 (with Gluck as conductor). The opera-house was altered by *Galli-Bibiena 1750. The Komödienhaus (Kleines Hoftheater) was completed 1755 and lasted until 1841; a small wooden theatre opened in the court of the Zwinger for more popular opera (burnt down 1748): operatic performances were also staged at the summer palace Schloss Moritzburg. Three Dresdeners, Johann *Naumann, Joseph Schuster, and Franz Seydelmann, visited Italy 1765–8 to study. The most popular and established form with court and nobility was Italian, and companies under Locatelli, Moretti, and Bustelli made successful visits. Singspiel was also popular; from 1770 works by Hiller, Benda, Schweitzer, and others were given by J. C. Wäser. On his appointment in 1786 Naumann did much to influence taste in a German and French direction, but unsuccessfully proposed a German opera. Opera semiseria began to appear. Joseph Seconda opened his theatre on the Linckesches Bad 1790, and E. T. A. *Hoffmann was director of the company run by the Seconda brothers in Leipzig and Dresden 1813–14.

In 1817 Carl Maria von *Weber became royal Kapellmeister; Francesco *Morlacchi was in charge of Italian opera 1810–41. Weber instituted searching reforms in the style of opera presentation, making a crucial shift towards the concept of music drama. He was succeeded by Karl *Reissiger in 1827. The Semper Opernhaus opened 1841. After the success of *Rienzi* here in 1842, Wagner succeeded Morlacchi as Hofkapellmeister, and produced his *Fliegender Holländer* and *Tannhäuser*. He had to flee after his part in the 1848 Revolution. In 1868 the opera-house burnt down; rebuilt on Semper's plans, it reopened 1878. Ernst von *Schuch was conduc-

tor from 1882, artistic director 1889–1914. A distinguished ensemble included Malten, Schuch-Proska, Sembrich, Schumann-Heink, Burian, Scheidemantel, later Siems, Von der Osten, Perron, and Plaschke. Fifty-one premières included Strauss's *Feuersnot* (1901), *Salome* (1905), *Elektra* (1909), and *Rosenkavalier* (1911). Schuch was succeeded by Fritz *Reiner. In 1918 the theatre became the Staatsoper. Fritz *Busch was appointed 1922: his régime saw the premières of more Strauss and Hindemith's *Cardillac* (1926), and the beginning of the German Verdi revival. Karl *Böhm succeeded 1934–42, Karl Elmendorff 1943–5; the company had included Seinemeyer, Cebotari, Rethberg, Höngen, Ralf, Tauber, Schoeffler, Boehme, and Frick.

On 14 Feb. 1945 the theatre was gutted during the bombing of Dresden. The company was housed in the Town Hall 1945–8, under the direction of Joseph Keilberth, then moving to the rebuilt Schauspielhaus (cap. 1,131). Rudolf Kempe was music director 1950–2, Franz Konwitschny 1953–5, Lovro von Matačić 1956–8, Otmar Suitner 1960–4. The rebuilt Semper opera-house reopened 13 Feb. 1985 with *Freischütz*. The ensemble included Schreier, Adam, and Bär.

**Drottningholm.** Palace near Stockholm, Sweden. The theatre was built by Queen Lovisa Ulrika in 1754; burnt down 1762; reopened 1766 (cap. 400). It still preserves stage machinery, a curtain, and some 30 18th-cent. stage sets used in the present annual summer seasons. Period instruments are regularly used in the performance.

**Drum-Major.** Marie's seducer (ten) in Berg's *Wozzeck*.

**Drury Lane, Theatre Royal.** There have been four theatres on the site in London. The first opened in 1663, the second, designed by Wren, in 1674; this saw the premières of several of Dryden's plays and masques with Purcell's music, and of many ballad operas. Arne, Dibdin, Linley, and Storace were all connected with the theatre. The third theatre (1794–1809) had Bishop as music director 1806–9. The fourth opened Oct. 1812. Tom Cooke was principal tenor 1813–35 and music director from 1821. Alfred Bunn took over in 1831, and presented Malibran singing *Fidelio* and *Sonnambula* in English. Bunn tried to establish English opera on a permanent basis 1835–47, producing works by Balfe, Benedict, and Wallace. In the mid-1850s seasons of English and Italian opera were given by E. T. Smith, including the English première of *Les vêpres siciliennes* (in Italian). The theatre was the home of Mapleson's Her Majesty's Opera 1867–77, and in 1870 saw the first English première of a Wagner opera, *Der fliegender Holländer* (as *L'olandese dannato*).

In 1882 Richter and a German company gave performances including the first London *Tristan* and *Meistersinger*, and from 1883 Drury Lane saw London seasons of the Carl Rosa OC. Augustus Harris began his Italian opera revival 1887, transferring to Covent Garden 1888 but using Drury Lane for some German opera 1892–3. There were English seasons 1894–1913, and Sir Joseph Beecham's Russian seasons 1913–14; these included the English premières of important Russian works including *Boris Godunov*, *Khovanshchina*, *Prince Igor*, *Ivan the Terrible*, *May Night*, *The Golden Cockerel*, and *The Nightingale*. Thomas Beecham's English OC gave some London seasons there during 1914–18. Opera was not heard again until Gorlinsky's Italian season in 1958, including *William Tell* (English première of an arrangement there 1830).

**Dryden, John** (*b* Aldwinkle, 9 Aug. 1631; *d* London, 1 May 1700). English poet, playwright, and critic. Most noted for his heroic drama (e.g. *Almanzor and Almahide*, 1670–1), Dryden also wrote tragedy and comedy; many of his plays are prefaced by critical essays on the theatre. His later dramas include two operas, *Albion and Albanius* (1685) and *King Arthur* (1691): Purcell (semi-opera, 1691). *The Indian Queen* (1664), which Dryden wrote in collaboration with Sir Robert Howard, was also set as a semi-opera by Purcell (1695).

**Dublin.** Capital of Republic of Ireland. Opera was first given in 1705–6 at the Smock Alley T (D. Purcell's *The Island Princess*). Italian opera was first performed in 1761 by a company under Antonio Minelli that included the *De Amicis family, beginning with Scolari's *La cascina*. After seven years of English opera (1770–7), Italian opera returned, in the Fishamble Street T and again Smock Alley T, where works by Paisiello, Anfossi, etc. were given. Gluck's *Orfeo* (in English) had its first Dublin performance in 1784 with Mrs Billington as Euridice: she appeared regularly for the next 13 years. Many performances were described by Michael *Kelly, who appeared throughout the 1780s. In the 19th cent. the T Royal was the scene of outstanding seasons with artists from the Italian opera-houses in London, including Persiani, Grisi, Rubini, Mario, and Pauline Viardot, who sang Lady Macbeth in the only production of Verdi's opera in the British Isles before 1938. In 1871 the Gaiety T opened, and was the home of seasons given by touring companies including the Carl Rosa, Mapleson, Harris, and O'Mara. From 1928 to 1938 the Irish Opera Society performed at the Gaiety T. In 1941 the Dublin Grand Opera Society was

formed, giving two seasons each year; normally chorus and orchestra are local, with soloists from England and the Continent, or sometimes whole visiting ensembles.

BIBL: T. Walsh, *Opera in Old Dublin, 1819–1838* (Dublin, 1952); T. Walsh, *Opera in Dublin, 1705–1797* (Dublin, 1973).

**Dubrovnik** (It., Ragusa). City in Dalmatia. Spectacular open-air performances are given as part of the annual summer festival.

**Duca d'Alba, Il** (The Duke of Alba). Opera in 4 acts by Donizetti; text by Scribe. Originally written for Paris, O, in 1840 but not produced; the libretto was later altered by Scribe and became the text of *Les vêpres siciliennes*. Donizetti's score was recovered at Bergamo in 1875 and completed by Matteo Salvi, prod. Rome, Ap., 22 Mar. 1882, with Bruschi-Chiatti, Gayarre, Paroli, Giraldoni, Silvestri, cond. M. Mancinelli; rev. Spoleto 1959, prod. Visconti, when the original scenery was reproduced, with Tosini, Cioni, Quilico, Ganzarolli, cond. Schippers; New York, TH (concert, Spoleto cast), 20 Oct. 1959.

Brussels, 1573. In this complicated tale about the oppressed Flemings suffering under Spanish tyranny, Amelia (sop), daughter of Egmont, is in love with Marcello (ten), who is revealed to be the missing son of the Duke of Alba (bar). Amelia, in an attempt to murder the Duke, instead kills Marcello, who has intervened to save his father's life. As he dies he begs the Duke to forgive Amelia.

Also opera by Pacini (1842).

**Due Foscari, I.** Opera in 3 acts by Verdi; text by Piave, after Byron's drama *The Two Foscari* (1821). Prod Rome, Arg., 3 Nov. 1844, with Barbieri-Nini, Ricci, Roppa, Bassini, Pozzolini; London, HM, 10 Apr. 1847, with Montenegro, Fraschini, Coletti; Boston, Howard Athenaeum, 10 May 1847, with Rainieri, Perelli, Vila, Battaglini.

The Doge's Palace, Venice, 1457. The Council of Ten assemble for the trial of Jacopo Foscari (ten) for murder. Jacopo's wife Lucrezia (sop) is dissuaded from going to his father Francesco (bar), the Doge, to beg for his life. The Council condemns Jacopo to exile.

Lucrezia visits Jacopo in his cell to tell him his sentence. The Doge feels unable to help his son because of the laws of Venice and tries to explain this to Lucrezia. Loredano (bs), an enemy of the Foscaris, insists that Jacopo should leave at once.

News emerges of a confession which clears Jacopo's name. The Doge is delighted, but Lucrezia rushes to say that Jacopo has collapsed and died as he was boarding the prison galley. At the instigation of Loredano, the Council of Ten 'invite' the Doge to abdicate. He agrees reluctantly. With Loredano exulting, the Doge dies as the bells announce his successor.

**Due Litiganti, I** (The Two Litigants). Correctly, *Fra due litiganti il terzo gode*. Opera in 3 acts by Sarti; text an altered version of Goldoni's *Le nozze*. Prod. Milan, S, 14 Sep. 1782; London, H, 6 Jan. 1784 (as *I rivali delusi*).

The Count and Countess each have in mind a different husband for their maid Dorina. After many confusions and surprises, she marries a third suitor, but all is resolved satisfactorily.

In its day immensely popular; Mozart not only used the aria 'Come un agnello' for a set of piano variations (K460), but quoted it, with certainty of recognition, in Don Giovanni's supper scene as the wind band's second piece—whereupon Leporello exclaims, 'Evvivano *I Litiganti!*' Also operas by Pescetti (1749), Astaritta (1766).

**Duenna, The, or The Double Elopement.** 1. Opera in 3 acts composed and compiled by Thomas Linley, sen. and jun.; text by Sheridan. Prod. London, CG, 21 Nov. 1775; Savannah, GA, 21 Feb. 1786. One of the most successful English comic operas of the 18th cent., surpassing even *The Beggar's Opera* on its first run (75 performances as against 63). In 1915 staged in Calcutta (trans. into Bengali), and in 1925 in Bombay (trans. into Marathi).

2. Opera in 4 acts by Prokofiev (sometimes known, from the translation of Prokofiev's Russian title, as *Betrothal in a Monastery*); text by Prokofiev with verses by Mira Mendelson, after Sheridan. Prod. Leningrad, K, 3 Nov. 1946 with Ulyagov, Solomyak, Khalileyeva, Velter, Bugayev, Grudina, Freidkof, Orlov, cond. Khaykin; New York, Greenwich Mews Playhouse, 1 June 1948 (pf. duet acc. only); Edinburgh Playhouse, 14 Aug. 1990, by Bolshoy OC, cond. Lazarev.

Seville, 18th cent. Louisa (sop), daughter of Don Jerome (ten), has been promised to old Mendoza (bs), but she wants to marry Antonio (ten). Disguised as her duenna Margaret (con), Louisa runs away; she encounters her brother Ferdinando (bar) eloping with Clara d'Almanzo (mez). Clara agrees to hide in a convent so that Lousia may disguise herself and take Clara's place. Meanwhile, Margaret, disguised as Louisa, has managed to persuade Mendoza to elope with her. Ferdinando and Antonio have also been taken in by the deceptions and when all parties meet at the convent, the two propose to fight a duel. Eventually all is resolved and Don Jerome allows his children to make the marriages of their choice.

3. Opera in 3 acts by Roberto Gerhard; text by the composer, after Sheridan. Prod. Madrid 21 Jan. 1992, with Cooper, Palmer, Moore, Rendall, Leggate, Van Allan, cond. Marbà;

Leeds, Grand T, 17 Sep. 1992, with Knight, Roberts, Shore, Chilcott, cond. Ros-Marbà.

**duet.** A composition for two performers, with or without accompaniment, in which the interest is equally shared. The *duetto da camera* (chamber duet) for two voices was a popular form in the 17th cent.: it entered opera in the Venetian works of Monteverdi and, above all, Cavalli, who usually ends each of the three acts of his operas with a duet. Its role was somewhat reduced following the libretto reforms of *Metastasio, but it has always since been a familiar ingredient of opera, not least because of its appropriateness to amorous situations.

**Dufranne, Hector** (*b* Mons, 25 Oct. 1870; *d* Paris, 4 May 1951). Belgian bass-baritone. Studied Brussels. Début Brussels, M, 1896 (Valentin). Paris, OC, 1899–1909, creating Marquis de Saluces (Massenet's *Grisélidis*), and Golaud, among other roles. Paris, O, from 1909, where he was 'un Jokanaan superbe' (*L'éclair*). Chicago 1913–22, where he created the Magician in *Love of Three Oranges*. London, CG, 1914. His sonorous, wide-ranged voice enabled him to sing a variety of parts. Retired 1939.

**Dugazon, Louise** (*b* Berlin, 18 June 1755; *d* Paris, 22 Sep. 1821). French mezzo-soprano. Began life as a dancer, then studied with Favart. Grétry composed an air for her to sing in his *Lucille* (1769), but her mature début was as Pauline in his *Sylvain*, 1774. Sang at the Comédie-Italienne, and later the Opéra-Comique, until 1804, creating more than 60 roles, especially in operas by Boieldieu, Grétry, and Isouard; her most famous was Dalayrac's Nina. Though her voice was unexceptional, and she was untrained musically, she was astonishingly versatile, possessing an instinctive feel for drama, and a moving intensity on stage that inspired odes, paintings, and general adoration. The genres in which she excelled came to be known as 'jeunes Dugazons', and 'mères Dugazons'.

Her son **Gustave** (1782–1826) composed 5 opéras-comiques, 3 ballets, and an anthology of arias by Rossini.

BIBL: H. and A. Leroux, *La Dugazon* (Paris, 1926); J. J. Olivier, *Madame Dugazon de la Comédie-Italienne, 1755–1821* (Paris, 1917).

**Dukas, Paul** (*b* Paris, 1 Oct. 1865; *d* Paris, 17 May 1935). French composer and critic. Studied Paris with Dubois and Guiraud. Both his first attempts at composition and his early critical writings were much influenced by Wagner. He wrote himself a libretto for *Horn et Rimenhild*, but composed no more than one act (1892). He abandoned another operatic project, *L'arbre de science* (1899), when Maeterlinck gave permission for use of *Ariane et Barbe-bleue*. The first major opera to be influenced by *Pelléas et Mélisande*, it differs in its broader structural scheme, its scope for more elaborate musical development, its more extrovert scoring, and its simpler emotional appeal: some qualities of *Pelléas* are present but dispersed into a more conventional kind of opera. Other unrealized operatic projects included a setting of *The Tempest*. Dukas also helped Saint-Saëns to complete Guiraud's *Frédégonde*, and edited several of Rameau's operas.

WORKLIST: *Ariane et Barbe-bleue* (Maeterlinck; comp. 1899–1906, prod. Paris 1907).

BIBL: G. Favre, *L'œuvre de Paul Dukas* (Paris, 1969).

**Duke Bluebeard's Castle** (Hung.: *A Kékszakállú herceg vára*). Opera in 1 act by Bartók; text by Béla Balázs. Prod. Budapest 24 May 1918 with Haselbeck, Kálmán, cond. Tango; New York, CC, 2 Oct. 1952, with Ayars, Pease, cond. Rosenstock; London, Rudolf Steiner T, 16 Jan. 1957, by Cape Town University Opera Club with Talbot, Fiasconaro, cond. Chisholm.

Bluebeard (bs) is not the Gilles de Rais monster of the *Ma Mère L'Oye* fairy-tale and Dukas's *Ariane et Barbe-bleue*, but a sorrowing, idealistic man. When his newest bride, Judith (sop), discovers his innermost secret and joins his other hidden wives, he is left with his loneliness once again.

**Duke of Mantua.** The licentious duke (ten) in Verdi's *Rigoletto*.

**Dulcamara.** The quack (bs) in Donizetti's *L'elisir d'amore*.

**Du Locle, Camille** (*b* Orange, 16 July 1832; *d* Capri, 9 Oct. 1903). French librettist and theatre director. His first libretto was for Duprato's *M'sieu Landry* (1856). In 1869 he was joint director of the Paris, O, 1870–4 of the Paris, OC, with Leuven, 1874–6 sole director. He continued the text of Verdi's *Don Carlos* after Méry's death, and made a preliminary draft of *Aida* with Verdi before it was passed to Ghislanzoni; he also translated *Simon Boccanegra*, *La forza del destino* (with Nuittier), and the first two acts of *Otello*. His other librettos include those for Reyer's *Sigurd* and *Salammbô*. He gave encouragement to Bizet at the time of *Carmen*, though he disliked the work.

**Dumas, Alexandre** (1) (père) (*b* Villers-Cotterets, 24 July 1802; *d* Dieppe, 5 Dec. 1870). French writer. Operas on his works are as follows:

*Henri III* (1829): Flotow (*Le comte de St Mégrin*, 1838, rev. as *La duchesse de Guise*, 1840); Josse (*La lega*, 1876); Hillemacher (1886)
*Charles VII chez ses grands vassaux* (1831): Donizetti (*Gemma di Vergy*, 1834): Cui (*The Saracen*, 1899)

*Don Juan de Marana* (1836): Enna (1925)
*Kean* (1836): Sangiorgi (1855)
*Mademoiselle de Belle-Isle* (1839): Samara (1905)
*Othon l'archer* (1840): Reiss (1856); Minchejmer (*Othon łucznik*, 1864)
*Pasqual Bruno* (1840): Hatton (1844)
*Ascanio* (1843): Saint-Saëns (1890)
*Les demoiselles de Saint-Cyr* (1843): Dellinger (1891); Humperdinck (*Heirat wider Willen*, 1905)
*Les trois mousquetaires* (1844): Visetti (1871); Xyndas (1885); Dionesi (1888); Somerville (1899); De Lara (1921); Benatzky (1931)
*Le comte de Monte Cristo* (1844): Dell'Aquila (1876); Caryll and others (1886)
*Les frères corses* (1845): Fox (1888)
*La dame de Monsoreau* (1846): Salvayre (1888)
*Joseph Balsamo* (1848) Litolff (*La mandragore*, 1876)
*Le tulipe noir* (1850): Flotow (*Il fiore di Harlem*, 1876)
*Le chevalier d'Harmental* (1853): Messager (1896)

Also *Gloria Arsena*: Enna (1917).

Also wrote *Piquillo*, opéra-comique, with Gérard de Nerval, music by Monpou, prod. Paris, OC, 1837.

**Dumas, Alexandre** (2) (fils) (*b* Paris, 28 July 1824; *d* Paris, 27 Nov. 1895). French writer, son of the above. Operas on his works are as follows:

*La dame aux camélias* (1848): Verdi (*La traviata*, 1853); H. Forrest (*Camille*, 1930)
*Le femme de Claude* (1873): Cahen (1896)

**Duni, Egidio Romualdo** (*b* Matera, *bapt.* 11 Feb. 1708; *d* Paris, 11 June 1775). French composer. Studied Naples with Durante. In the 1735 Rome Carnival his first opera, *Nerone*, proved a greater success than the established Pergolesi's *L'Olimpiade*. After some travels, including to London for *Demofoonte*, he had various successes in Italy, concluding his career as an opera seria composer with his own *L'Olimpiade* in 1755. At Parma, where he was maestro di cappella from 1749, he came to appreciate French music, and may then have set or contributed music to two Favart librettos (including *Ninette à la cour*). His first opera for Paris was *Le peintre amoureux de son modèle*, popular partly for its refutation of Rousseau's denial of the French language's suitability for music. With the series of works that followed, Duni established himself as one of the most important founding composers of opéra-comique. The introduction of Italian arias and recitatives into French ensembles, dances, and divertissements proved a brilliantly successful mix, his Italian liveliness and tunefulness helping to give the French *comédie mêlée d'ariettes* a substance it hardly possessed in the days of the *drame forain*. In the best of his work, the characters have a pathos and eloquence that greatly contributed to elevating the genre above the merely frivolous. Diderot praises him effusively in *Le neveu de Rameau*. He lived to see the genre he had done so much to establish growing beyond his taste or abilities.

SELECT WORKLIST: *Nerone* (after Silvoni; Rome 1735); *Demofoonte* (after Metastasio; London 1737); *L'Olimpiade* (after Metastasio; Parma 1755); *La buona figliuola* (Goldoni; Parma 1756); *Le peintre amoureux de son modèle* (The Painter in Love with his Model) (Anseaume; Paris 1757); *La fille mal gardée* (The Ill-protected Girl) (Favart & Santerre; Paris 1758); *Nina et Lindor* (Richelet; Paris 1758); *L'isle des foux* (The Island of Madmen) (Anseaume & Marcouville, after Goldoni; Paris 1760); *La fée Urgèle* (Favart, after Voltaire & Chaucer; Fontainebleau 1765); *La clochette* (The Little Bell) (Anseaume; Paris 1766); *Les moissonneurs* (The Reapers) (Favart, after the Book of Ruth; Paris 1768).

**Dunn, Geoffrey** (*b* London, 13 Dec. 1903). English tenor, producer, and librettist. Studied London, RAM. Sang with Nigel Playfair's Co., Lyric T Hammersmith, in hitherto little-known works, e.g. *Idomeneo*. Formed Intimate OC 1930 (with Margaret Ritchie and Frederick Woodhouse), for which he revived several 18th-cent. English operas by Arne, Dibdin, and Carey. Translated many foreign librettos, including *Poppea*, *Serse*, *Der Corregidor*, *La vie parisienne*, and Rimsky-Korsakov's *Mozart and Salieri*. Also wrote librettos for Malcolm Williamson, e.g. *Dunstan and the Devil*. Producer for London, SW, *Don Giovanni* (1949) and *Dido and Aeneas* (1951). After giving up singing, turned to acting.

**'Dunque io son'.** The duet between Rosina (mez) and Figaro (bar) in Act I of Rossini's *Il barbiere di Siviglia*, in which he spins a story about his cousin, a poor student, who is in love with a certain girl, and eventually is given a letter already written for the disguised Almaviva.

**duodrama.** The name given, by Mozart among others, to *melodramas for two actors. He referred with enthusiasm to the best-known works of the genre, by Jiří Benda, e.g. *Ariadne auf Naxos*, and himself contemplated writing one, *Semiramis*.

**Duparc, Elisabeth** (*b* ?; *d* ?1778). French soprano. Known as 'La Francesina'. Studied in Italy. Engaged for Opera of the Nobility, London 1736, making her début at London, H, in Hasse's *Siroe*. Sang for Handel 1738–46, creating Clotilda (*Faramondo*), Rosilda (*Serse*), the title-roles of *Semele* and *Deidamia*, and parts in most of the oratorios he composed at this time, e.g. *Belshazzar* (Nitocris) and *Hercules* (Iole). Handel admired her singing, which according to Burney was 'lark-like'. Mrs Delany commented, 'There

is something in her running divisions that is quite surprizing.'

**Duprez, Gilbert**(-Louis) (*b* Paris, 6 Dec. 1806; *d* Paris, 23 Sep. 1896). French tenor. Studied Paris with Choron. Début Odéon 1825 (Rossini's Count Almaviva). Further study in Italy, where he sang 1829–35, creating Edgardo (*Lucia*) at Naples. During this time his voice developed from a small, light tenor to one of considerable power, though with the sacrifice of agility and range of colour. Returned to Paris, replacing Nourrit as leading tenor at the Opéra, 1837–49. Here he created Benvenuto Cellini, Fernando (*La favorite*), and Polyeucte (*Les martyrs*), and took over much of Nourrit's repertory. Credited with being the first tenor to produce the *ut de poitrine*, or chest voice top C (to the delight of Paris audiences, but not of Rossini), he strongly influenced succeeding generations of tenors in this respect. Caricatured for his small size and gaping mouth, he performed with unrelenting commitment. Chorley praised his 'dramatic truth', and Roger described his singing, even at the end, as 'molten ore'. His voice declined early; Berlioz also regretted the coarsening of his style. After retiring, he was a notable teacher. Among his pupils were Miolan-Carvalho, Nantier-Didier, and Albani. He also wrote several operas, and two books on singing. His wife **Alexandrine Duperron** (1808–72) sang in Paris 1827–37, often with Duprez; their daughter **Caroline Vandenheuval-Duprez** (1832–75) was a light soprano, who sang 'exquisitely' (Berlioz) with Viardot, and also appeared with her father.

BIBL: G. Duprez, *Souvenirs d'un chanteur* (Paris, 1880); *Récréations de mon grand âge* (Paris, 1888).

**Durante, Francesco** (*b* Frattamaggiore, Aversa, 31 Mar. 1684; *d* Naples, 30 Sep. 1755). Italian composer and teacher. Studied in Naples and in Rome with Pasquini and Pitoni; in 1710 took the first of several posts in one or other of the Naples conservatories. All Durante's dramatic works were sacred, but although he wrote no opera, he had an inestimable impact on the development of the genre in 18th-cent. Italy through his activities as a teacher. Among his many pupils were Duni, Jommelli, Logroscino, Paisiello, Pergolesi, Piccini, Sacchini, Traetta, and Vinci.

A slovenly individual, according to Paisiello he died of a severe attack of diarrhoea caused by eating a surfeit of melons.

**Durastanti, Margherita** (*b* ?Venice, *fl.* 1700–34). Italian soprano, later mezzo-soprano. First known appearance Venice, 1700, in a pasticcio. From 1707, engaged by Prince Ruspoli at Rome, where she worked with Caldara, and Handel, who wrote several cantatas for her. Venice 1709–12, in operas by Lotti, etc.; created Handel's Agrippina. Dresden 1719. London, Royal Academy of Music, 1720–1, creating Handel's Radamisto, and Clelia in *Muzio Scevola*. Returned there 1722–4 (creating Sextus in *Giulio Cesare*, Gismonda in *Ottone*, and Vitige in *Flavio*), and again 1733–4, to help Handel in his struggle against Porpora (creating Tauride in *Arianna*). Though overshadowed by Senesino, Faustina, and Cuzzoni, she was a fine singer, and dramatically effective despite her size (Rolli called her 'an elephant'). She earned extremely high fees.

**Durazzo,** (Conte) **Giacomo** (*b* Genoa, 27 Apr. 1717; *d* Vienna, 15 Oct. 1794). Italian diplomat, theatre manager, and librettist. Went to Vienna in 1749 as extraordinary ambassador from the Genoese Republic; subsequently married into the Esterházy family, which led to his appointment as assistant director of Viennese theatrical affairs in 1752. With the further support of Kaunitz he progressed rapidly through the theatrical hierarchy, assuming overall control in 1754, as 'directeur des spectacles'. His reign had far-reaching consequences: the same year he engaged Gluck as Hofkapellmeister, and with him set about the reform of opera seria, in the face of stiff opposition from the *poeta cesareo, Metastasio, and from Hasse, who visited Vienna in 1760–2. In 1764 he was forced to resign from his post, but was given a diplomatic appointment in Venice, which he made the base for numerous travels during his later years.

Although Durazzo's only libretto for Gluck, *L'innocenza giustificata* (1755, partly based on Metastasio), displays some traits of reform, his influence on opera seria came more as animateur than as participant. His most important contribution was the support of the Gluck-Calzabigi collaboration *Orfeo ed Euridice* (1762): also influential was his recruitment of outstanding artists, including the singer Catterina Gabrielli and the ballet-master Gasparo Angiolini. He also encouraged the performance of French opéra-comique in Vienna, importing several vaudevilles from Paris: from 1759 he corresponded regularly with Favart. Under his regime opera buffa was also given, notably Paisiello's *Il finto pazzo* (1758).

**'Durch die Wälder'.** Max's (ten) aria in Act I of Weber's *Der Freischütz*, in which he reflects how once he wandered happily through the countryside before misfortune befell him.

**durchkomponiert** (Ger.: 'through', i.e. continuously, composed'). The term used for an opera in which there is no spoken dialogue or recitative, but continuous music into which separate numbers merge to a greater or lesser degree. See also NUMBER OPERA.

**'Durch Zärtlichkeit und Schmeicheln'.** Blonde's (sop) aria in Act II of Mozart's *Die Entführung aus dem Serail*, in which she declares that a girl's heart may be won by tenderness, never by force.

**Dussek, Sophia.** See CORRI.

**Düsseldorf-Duisburg.** Two cities in North-Rhine Westphalia, Germany, which since 1956 have together housed the Deutsche Oper am Rhein, one of the strongest German operatic ensembles. Alberto *Erede was music director 1958–62 (first Italian in such a post this century); he was succeeded by Günter Wich. It has given premières of Klebe's *Die Räuber* (1957), Zimmermann's *Die Soldaten* (1972), and Goehr's *Behold the Sun* (1985), and specialized in 20th-cent. opera, Monteverdi revivals, and cycles of Mozart, Janáček, and Rossini. Düsseldorf first saw opera in the 1680s, including works by Sebastiano Moratelli; then the première of Steffani's last opera, *Tassilone*, 1709. The Opernhaus opened 1875, rebuilt 1956 (cap. 1,342). The Duisburg Stadttheater was built 1912, rebuilt 1950 (cap. 1,118).

**Duval, Denise** (*b* Paris, 23 Oct. 1921). French soprano. Début Bordeaux 1941 (Lola). Paris, O and OC, 1947–65, when she retired. The first Thérèse (*Les mamelles*), and Blanche in first production of the original *Dialogues des Carmélites*, of which she is also the dedicatee. Created Elle (*La voix humaine*), written specially for her, with great success. Edinburgh 1960; Gly. 1962–3. A beautiful and impressive singing actress, also notable as Thaïs and Mélisande. (R)

**Dvořák, Antonín** (*b* Nelahozeves, 8 Sep. 1841; *d* Prague, 1 May 1904). Czech composer. Studied Zlonice with Liehmann, Prague with Pitsch and Krejčí. His interest in opera was first aroused when a violist in the Prague Provisional Theatre Orchestra, where he had been impressed with Wagner. Though some of his operas are popular in Czechoslovakia, they have not won acceptance in foreign repertories. Even in their own country, the early works have languished, though the sentimental humour of *Jakobín* is appreciated, as are the folk charms of *Tvrdé palice* and *Šelma sedlák*. *Čert a Káča* has had some success abroad, as to a lesser extent has *Dmitrij*, whose plot begins where *Boris Godunov* ends. Dvořák's greatest success has always been *Rusalka*, a work of much charm which is as popular with Czech children as *Hänsel und Gretel* is with Germans. In general, Dvořák lacked a strong sense of drama and characterization, so that it is for incidental numbers or general expressions of atmosphere that his operas can be appreciated. His earliest works show the influence of Wagner, to which he returned in his later works after allowing the nationalist spirit of Smetana to be felt.

WORKLIST (all first prod. Prague unless otherwise stated): *Alfred* (Körner; comp. 1870, prod. Olomouc 1938); *Král a uhlíř* (King and Charcoal-Burner) (Lobeský; 1st version comp. 1871, prod. 1929; 2nd version, 1874; 3rd version, 1887); *Tvrdé palice* (The Stubborn Lovers) (Štolba; comp. 1874, prod. 1881); *Vanda* (Šumavský, after Šurzycki; 1876); *Šelma sedlák* (The Cunning Peasant) (Veselý; 1878); *Dimitrij* (Červinková-Riegrová; 1st version, 1882; 2nd version, 1894); *Jakobín* (The Jacobin) (Červinková-Riegrová; 1889); *Čert a Káča* (The Devil and Kate) (Wenig; 1899); *Rusalka* (Kvapil; 1901); *Armida* (Vrchlický, after Tasso; 1904).

CATALOGUE: J. Burghauser, *Antonín Dvořák: thematický katalog, bibliografie, přehled života a díla* (Prague 1960, R/1990).

BIBL: J. Clapham, *Antonín Dvořák* (London, 1966); J. Clapham, *Dvořák* (Newton Abbot, 1979).

**Dvorsky, Peter** (*b* Partizánske, 25 Sep. 1951). Czech tenor. Studied with Černecká. Début Bratislava 1972 (Lensky). Vienna from 1976; New York, M, and Munich 1977; Milan, S, 1978; London, CG, 1979; New York, M, 1977, also Moscow, B; Verona; Prague; etc. Roles incl. Edgardo, Alfredo, Duke of Mantua, Rodolfo, Cavaradossi, Jeník. (R)

**Dyagilev, Sergey** (*b* Selichev, 31 Mar. 1872; *d* Venice, 19 Aug. 1929). Russian impresario. Though renowned as founder of the Russian Ballet Company that bore his name, he was also an important figure in opera, especially in Paris 1907–9, Paris and London 1913–14, and Paris and Monte Carlo 1922–4. In 1907 he presented scenes from *Sadko*, *Boris*, and *Ruslan* in concert at the Paris, O, and in 1908 the first fully staged *Boris* outside Russia, with Shalyapin. He gave *Ivan the Terrible* in Paris 1909; the first *Khovanshchina* outside Russia, Paris, CE, 1913; the first Russian season at London, DL, 1913. With Beecham, he gave seven Russian operas (*Boris*, *Ivan*, *Igor*, *Golden Cockerel*, *Nightingale*, *May Night*, and *Khovanshchina*). In Paris gave the première of *Mavra* (1922) and *Oedipus Rex* (1927); in Monte Carlo gave Gounod's *Médecin malgré lui* with Satie recitatives (1924), *La colombe* with

Poulenc recitatives, and *Philémon et Baucis* and Chabrier's *Une éducation manquée* with Milhaud recitatives.

BIBL: R. Buckle, *Diaghilev* (London, 1979).

**Dybbuk, The.** 1. Opera in a prologue and 3 acts by Lodovico Rocca (*Il Dibuk*); text by R. Simoni, after Shelomoh An-Ski's drama (1916). Prod. Milan, S, 24 Mar. 1934, with Oltrabella, Palombini, Lo Giudice, Paci, Bettoni, cond. Ghione; Detroit, Masonic Temple Auditorium, 6 May 1936.

Chanon (ten) is fascinated by the mysteries of the Kabala and tries to discover the means of winning riches so that he can marry Leah (sop). He dies, but his spirit enters Leah's body.

2. Opera in 3 acts by David Tamkin; text by Alex Tamkin. Comp. *c*.1931; prod. New York, CC, 4 Oct. 1951, with Neway, Russell, Bible, Rounseville, Raisa, cond. Rosenstock. Also rejected, partly sketched, opera by Gershwin, abandoned when he learnt that the rights belonged to Rocca. Other operas on the subject are by Füssel (1970) and Di Giacomo (1978).

**Dyck, Ernest van.** See VAN DYCK, ERNEST.

**Dyer's Wife, The.** The (anonymous) wife (sop) of Barak the Dyer in Strauss's *Die Frau ohne Schatten*.

**Dzerzhinsky, Ivan** (*b* Tambov, 9 Apr. 1909; *d* Leningrad, 18 Jan. 1978). Russian composer. Studied Moscow with Gnesin, Leningrad with Ryazanov and Asafiev. When still a student he entered his opera *The Quiet Don* in a competition organized by the Bolshoy T and the journal *Komsomolskaya Pravda*, unsuccessfully. Revised, with help from Shostakovich and others, it was produced in Leningrad under Samosud, when it was praised by Stalin as an example of an immediately comprehensible patriotic opera and hence became an instant paragon of Socialist Realism. The music's simple-mindedness ensured that this success was particular and temporary.

SELECT WORKLIST: *Tikhy Don* (The Quiet Don) (L. Dzerzhinsky, after Sholokhov; Leningrad 1935).

# E

**Eames, Emma** (b Shanghai, 13 Aug. 1865; d New York, 13 June 1952). US soprano. Studied Paris with Marchesi. Début Paris, O, 1889 (Juliette). London, CG, 1891–1901, arousing Melba's jealousy with her successes as the Countess (Mozart), Elsa, Sieglinde, Eva, Aida, etc. New York, M, 1891–1909; Boston 1911 (Tosca, Verdi's Desdemona), after which she appeared only in concert. Her voice, thought by some more beautiful than Melba's, was full, even, flexible, and accurate. She was handsome, but somewhat cold: Shaw refers to her 'casting her propriety like a Sunday frock over the whole stage'. (R)

**'È amore un ladroncello'.** Dorabella's (sop) aria in Act II of Mozart's *Così fan tutte*, in which she reflects on how love gives both imprisonment and release.

**Easter Hymn.** The Resurrection Hymn for the chorus led by Santuzza (sop), 'Inneggiamo, il Signor non è morto', in Mascagni's *Cavalleria rusticana*.

**Easton, Florence** (b Middlesborough, 25 Oct. 1882; d New York, 13 Aug. 1955). English soprano. Studied London, RCM, and Paris with Haslam. Début Newcastle 1903 (Shepherd, *Tannhäuser*). London, CG, 1903, 1909, 1927 (Turandot), 1932 (Brünnhilde). Henry Savage Co., 1905. Berlin, H, 1907–13; Hamburg 1913–15. New York, M, 1917–29, 1935–6: created Lauretta (*Schicchi*) and Aelfrida in (Deems Taylor's *The Henchman*). Had a repertory of over 100 roles in four languages, and was able to take over a part, new or old, at only hours' notice. In one week in 1922 she sang Isolde, Kundry, Fiordiligi, and Sieglinde. Her voice, pure-toned and powerful, was still magnificent at her farewell in 1936. (R)

**Ebers, John** (b London, c.1785; d London, c.1830). English impresario. Originally a bookseller; Manager London, H, 1820–7. His first season, with William Ayrton as music director, included the London premières of Rossini's *Gazza ladra* and *Turco in Italia*, but ended in ruin through financial demands by the singers. Took four-year lease 1822, but was again ruined through an absconding stage-manager to whom he had sublet. Lost a lawsuit over chorus pay-

ment, 1826. Rental and other problems forced him into bankruptcy, and he returned to bookselling. His company included De Begnis, Colbran, Camporese, Pasta, Vellutti, and Vestris. He gave the first London *Donna del lago*, *Crociato in Egitto*, and *Vestale*. A spirited account of his ventures, valuable as an account of contemporary theatrical life, is provided in his memoirs, *Seven Years of the King's Theatre* (London, 1828, R/ 1970).

**Ebert, Carl** (b Berlin, 20 Feb. 1887; d Santa Monica, 14 May 1980). German producer and manager. Studied acting with Reinhardt, and after engagements in Berlin and Frankfurt became general director of the Darmstadt State T 1927. General director and producer, Berlin, SO, 1931, where his productions came as a revelation to critics and public. Left Germany 1933. Produced opera Florence, Buenos Aires. With Fritz Busch founded *Glyndebourne Opera, where together they established its reputation for the highest international standards. Organized Turkish National T and O, 1936–47; director opera dept. U of S California 1948–56. Gly. 1947–59. Returned to former Berlin post 1956–61. New York, M, 1959–62. His productions of Mozart and Verdi set particularly high standards of intelligent dramatization and faithfulness to the composers' intentions. Staged première of *The Rake's Progress*, Venice 1951. His son **Peter** (b Frankfurt, 6 Apr. 1918) has also worked at Glyndebourne. Scottish O: director of productions 1963–75, general administrator 1977–80; Augsburg 1968–73; Wiesbaden 1975–7.

**Eberwein, Traugott** (b Weimar, 27 Oct. 1775; d Rudolstadt, 2 Dec. 1831). German composer. After achieving early fame as a violinist he joined the Rudolstadt court 1810, where he did much to raise musical standards. Kapellmeister 1817. His dramatic works include two settings of Singspiel texts by Goethe, *Claudine von Villa Bella* and *Der Jahrmarkt zu Plundersweilen*.

His younger brother **Carl Eberwein** (b Weimar, 10 Nov. 1786; d Weimar, 2 Mar. 1868) was a flautist (later violinist) in the Weimar court orchestra. A protégé of Goethe, who sent him to study with Zelter in Berlin, he later held various positions in Weimar, including that of music director for the poet's household; became director of court opera 1826–49. Made settings of most of Goethe's Singspiel librettos and produced copious amounts of incidental music for his plays.

**Eboli.** Princess Eboli (conceived as low mez, written as high mez for Lauters-Gueymard), former mistress of King Philip and lady-in-waiting to Elisabeth de Valois in Verdi's *Don Carlos*.

**Eccles, John** (*b* ?London, *c.*1668; *d* Hampton Wick, 12 Jan. 1735). English composer. He rapidly established himself as one of the leading theatre composers of the day. After a quarrel with Rich, Eccles, and other members of the United Companies, left Drury Lane and followed Betterton to the Lincoln's Inn Fields T, where he composed two 'all-sung' masques (*The Loves of Mars and Venus*, 1696, and *Acis and Galatea*, 1700) and a full length 'dramatic opera', *Rinaldo and Armida* (1698, lost). Betterton developed a policy of interpolating masques on mythological subjects into dramas; as well as the two listed above, Eccles composed a masque on Hercules. A setting of Congreve's masque *The Judgement of Paris* (1701) won him second prize in a famous competition.

Eccles's significance as an opera composer rests, ironically, on a work which was not performed. *Semele*, a setting of the Congreve text later used by Handel, appears as a sole example of an attempt to fuse English and Italian styles of opera. The text is divided into sections of aria, arietta, arioso, and recitative—specified by Congreve—and while there are three arias in traditional da capo form, they are juxtaposed with simple airs, and the style of the recitative is decidedly un-Italian, closely following natural English declamation. The work also contains the conventional masque elements of dances and incidental instrumental music. It may be supposed that Congreve wished the work to open the new theatre he and Vanbrugh were building in the Haymarket; but Eccles did not complete *Semele* until 1707, two years later, by when Congreve had left the theatre, and the best singers had moved to Drury Lane. *Semele* was given a belated première in London in 1972. Perhaps disillusioned by the apparent monopoly of Italian opera in London, Eccles left the city in 1710 and spent the rest of his life at Kingston-upon-Thames writing court odes and fishing.

**'Ecco l'orrido campo'.** Amelia's (sop) aria opening Act II of Verdi's *Un ballo in maschera*, expressing her fear and horror as she comes to the deserted spot where she is to pick the herb that will subdue her love for the King.

**'Ecco ridente in cielo'.** Count Almaviva's (ten) aria in Act I of Rossini's *Il barbiere di Siviglia*, in which, disguised as Lindoro, he serenades Rosina (or, strictly, sings her a morning song).

**Ecuador.** Opera is of comparatively recent development. It began with the efforts of a musician of Italian origin, Pedro Pablo Traversari (1874–1956). Luis Salgado (*b* 1903) wrote two operas, *Cumandá* (1954) and *Eunice* (1957) (both unprod.). Opera is given by visiting European and S American companies at the T Sucre in Quito.

**Eda-Pierre, Christiane** (*b* Fort-de-France, Martinique, 24 Mar. 1932). Martinique soprano. Studied Paris Cons. Début Nice, 1958 (Leïla). Paris, O, from 1962; London, CG, 1966; Chicago, Berlin, Moscow, etc. Repertory includes Elettra, Teresa (*Cellini*), Zerbinetta; also Rameau, Wagner, and contemporary works (creating the Angel in Messiaen's *François d'Assise*). A musical, often vocally dramatic, singer with a limpid tone. (R)

**Edelmann, Otto** (*b* Brunn am Gebirge, 5 Feb. 1917). Austrian bass-baritone. Studied Vienna with Lierhammer and Graarud. Début Gera 1937 (Mozart's Figaro). Vienna, S, from 1947; Salzburg 1948; Bayreuth 1951–2; New York, M, 1954–70; Gly. 1965. Also Milan, S; Berlin; Munich; San Francisco. Roles incl. Leporello, Rocco, Sachs, Amfortas, Ochs, Kecal. (R)

**Edgar.** Opera in 3 acts by Puccini; text by Ferdinando Fontana, after Alfred de Musset's verse drama *La coupe et les lèvres* (1832). Prod. Milan, S, 21 Apr. 1889, with R. Pantaleoni, Cattaneo, Gabrielesco, Magini-Coletti, cond. Faccio; London, Hammersmith TH, 6 Apr. 1967, with Rubini, Doyle, Byles, Rippon, cond. Vandernoot; New York, Carnegie H (concert), 13 Apr. 1977, with Scotto, Bergonzi, cond. Queler.

A village in Flanders, 1302. Fidelia (sop) and Tigrana (mez) both love Edgar (ten). Fidelia's brother Frank (bar) loves Tigrana but she spurns him. While the peasants gather in worship, Tigrana sings a profane song and they condemn her behaviour. Edgar defends her and flees with Tigrana, having set fire to his house. Frank tries to prevent them, but is injured in the fight.

Edgar sings of his disillusionment with his new life of pleasure and longs to be back in his village with Fidelia. A platoon of soldiers approaches; their captain turns out to be Frank, and Edgar soon joins the ranks. He and Frank turn on Tigrana.

Edgar is believed killed in battle and is mourned by Fidelia. As Frank delivers the eulogy, a monk reminds the people of how Edgar set his home on fire and chose a life of debauchery. Fidelia defends his memory. Tigrana enters and, having been bribed with jewels, proclaims in front of the people that Edgar planned to betray his country. The crowd storms the coffin, but they find only an empty suit of armour. The monk reveals himself to be Edgar; he denounces Tigrana, who then stabs Fidelia. As Edgar throws himself on Fidelia's dead body, Tigrana is led away.

**Edgardo.** Edgar of Ravenswood (ten), lover of Lucia, in Donizetti's *Lucia di Lammermoor*.

**Edinburgh.** Capital of Scotland. Apart from touring companies, Edinburgh had little opera before the First World War, though it was the birthplace of the *Denhof OC. In 1947 the Gly. Society Ltd. founded the annual summer festival at the instigation of Audrey *Christie and Rudolf *Bing (who was artistic director 1947–50). Despite the city's enduring resistance to providing an opera-house, opera has played a central part in the festival; it has had to be staged in the King's T and various even less suitable venues. Visiting companies first were Gly. (1947–55), Hamburg (1952, 1956), Milan, PS, (1957), Stuttgart (1958), and Stockholm (1959). Lord Harewood's régime (1961–5) saw visits by London, CG, Belgrade (1962), Naples, C (1963), Prague (1964), Munich (1965), and Holland Festival O (1965). Companies to have visited since include Stuttgart (1966), Scottish O regularly since 1967, Hamburg (1968), Florence (1969), Frankfurt and Prague (1970), Berlin, D (1971), Palermo and Deutsche O am Rhein (1972), Budapest (1973), Stockholm (1974), Berlin, D (1975), Deutsche O am Rhein (1976), Zurich (1978), Kent (1979), Cologne (1980–1), Milan, PS, WNO, and Dresden (1982), Hamburg and St Louis (1983), Washington (1984), Lyons, Connecticut, and Les Arts Florissants (1985), Leningrad Maly (1986), Stockholm Folk O (1986–9), Helsinki (1987), Houston (1988), Jutland (1989), Moscow, B, and Bratislava (1990), Moscow, and Leningrad, M (1991). The festival has itself staged Rossini and Mozart, and a successful *Carmen* (1977). The many British premières have included *Mathis der Maler* (1952), *The Rake's Progress* (1953), *Oedipus Rex* and *Mavra* (1956), *La vida breve* (1958), *Aniara* (1959), *La voix humaine* (1960), *Love of Three Oranges* and *The Gambler* (1962), *Dalibor*, *From the House of the Dead*, and Cikker's *Resurrection* (1964), *Intermezzo* (1965), Malipiero's *Sette canzoni* (1969), Reimann's *Melusine* (1971), Zimmermann's *Die Soldaten* (1972), Szokolay's *Blood Wedding* (1973), Werle's *Vision of Thérèse* (1974), Orr's *Hermiston* (1975), Musgrave's *Mary, Queen of Scots* (1977), Maxwell Davies's *The Lighthouse* (1980), Merikanto's *Juha* (1987), Adams's *Nixon in China* and Turnage's *Greek* (1988). Other outstanding events have included the first version of *Ariadne auf Naxos* under Beecham (1950), Callas in *La sonnambula*, replaced when she left by Scotto (1957), Janet Baker in *The Trojans* (1973), *Moses und Aron* (1976), and the Zurich Monteverdi cycle (1978). The Fringe has also seen a variety of events of many shapes and sizes.

From 1963 Scottish O have given annual spring seasons at the King's T and 1969–74 short winter seasons. ENO also pay regular visits. The University O Club has staged enterprising productions in various locations.

**Éducation manquée, Une** (A Defective Education). Operetta in 1 act by Chabrier; text by Leterrier and Vanloo. Prod. Paris, Cercle de la Presse, 1 May 1879, in private performance, with Hading, Réval, Morlet; first public perf. Paris, T des Arts, 9 Jan. 1913, cond. Grovlez; Tanglewood, Lenox, MA 3 Aug. 1953; London, Fortune T, 22 May 1955, with Zeri, Dickerson, Turgeon, cond. Jacob.

France. The reign of Louis XVI. Count Gontran de Boismassif (sop) arrives home with his new bride Hélène (sop), totally uninstructed by his tutor Pausanias (bs) in the facts of life. Aged relatives are reluctant to enlighten them but the problem is resolved when a violent storm drives the two into each other's arms.

**Edvina, Louise** (*b* Montreal 28 May 1878; *d* London, 13 Nov. 1948). Canadian soprano. Studied Paris with J. De Reszke. Début London, CG, 1908 (Marguerite). Sang there until 1914; also 1919–20, 1924. First London Louise, Thaïs, Francesca da Rimini. Boston 1911–13; Chicago 1915–17; New York, M, 1915 (Tosca). A 'born actress' (*The Globe*), and an affecting singer, she was particularly successful as Louise and Mélisande. (R)

**Edwards, Sian** (*b* West Chiltington, 27 Aug. 1959). English conductor. Studied London with Groves and Del Mar, Holland with Järvi, Leningrad with Musin. Début Scottish O 1986 (*Mahagonny*). Gly. from 1987 (*Traviata*); London, CG, from 1988 (*Knot Garden*); Munich and Edinburgh 1988; London, ENO, 1990 (*The Gambler*). Repertory incl. *Rigoletto*, *Trovatore*, *Carmen*, *Káťa Kabanová*, *New Year*, and first *Greek* (Turnage). Music Director, ENO, from 1993.

**Edwin and Angelina, or The Banditti.** Opera in 3 acts by Victor Pelissier; text by E. H. Smith, after Goldsmith's ballad (1764). Prod. New York, John St. T, 19 Dec. 1796.

One of the earliest US operas.

**Egisto.** Opera in 3 acts by Cavalli; text by Giovanni Faustini. Prod. Venice, C, 1643; Glasgow, Theatre Royal, 12 Jan. 1982, with D. Jones, Wallis, Rosenshein, Dalton, Maxwell, Egerton, cond. Brydon; Santa Fe, 1 Aug. 1974.

The island of Zakynthos, mythological times. Egisto relates to Climene how he and Clori were captured by Corsican bandits; Climene says she and Lidio endured a similar experience. Both are upset at the unfaithfulness of their respective lovers and go to Amor to seek revenge. When

Venus sees that Egisto now has the means of release, she begs Amor to drive him mad.

Egisto meets Clori, who wishes to have nothing more to do with him. He advises Lidio to break with Clori since she brings nothing but sorrow. Climene's brother Hipparco also loves Clori. He encourages his sister to stab Lidio, but she hesitates, asking the gods to forgive him. Lidio then appears, declaring his love for Climene is unaltered; the couples are reunited by Amor.

**Egk, Werner** (*b* Auchsesheim, 17 May 1901; *d* Inning, 10 Jul. 1983). German composer. Studied Frankfurt, and Munich with Orff. His first stage opera was *Die Zaubergeige*, using numbers based on popular tunes. *Peer Gynt* became successful after receiving Hitler's approval. His operas have been widely staged in Germany, where his dramatic flair and his capacity for drawing on the most readily acceptable aspects of his predecessors, notably Stravinsky, have made him popular.

WORKLIST: *Columbus* (Egk; radio 1933, Frankfurt 1942); *Die Zaubergeige* (The Magic Fiddle) (Egk, after Pocci; Frankfurt 1935); *Peer Gynt* (Egk, after Ibsen; Berlin 1938); *Circe* (Egk, after Calderón; Berlin 1948; rev. as *17 Tage und 4 Minuten*, Stuttgart 1966); *Irische Legende* (Egk, after Yeats; Salzburg 1955); *Der Revisor* (The Government Inspector) (Egk, after Gogol; Schwetzingen 1957); *Die Verlobung in San Domingo* (The Betrothal in San Domingo) (Egk, after Kleist; Munich 1963).

BIBL: E. Krause, *Werner Egk: Oper und Ballet* (Wilhelmshaven, 1971).

**Eglantine.** Eglantine von Puiset (mez), a rebel's daughter and Lysiart's lover, in Weber's *Euryanthe*.

**Egypt.** The first Cairo OH was commissioned by the Khedive Ismail Pasha in 1869 and opened that year, to celebrate the opening of the Suez Canal, with *Rigoletto*. He then commissioned *Aida*, premiered there in 1871. The first opera composed by an Egyptian was *Antony's Death* by Hassan Rasheed (1896–1969) (part prod. 1942 by the Egyptian O Troupe). Opera was performed by visiting companies until 1961, when Arabic performances took place (beginning with *The Merry Widow*). In 1971 the theatre burnt down; President Hosni Mubarak secured the gift of a new opera-house from Japan at the cost of £17.5m, and this opened in 1988 (cap. 1,200).

**'E il sol dell'anima'.** The Duke's (ten) aria declaring his love to Gilda in Act I of Verdi's *Rigoletto*, leading to the love duet.

**Einem, Gottfried von** (*b* Berne, 24 Jan. 1918). Swiss, later Austrian composer. Studied Plön. Répétiteur Berlin, S, 1938. Further study with Blacher during the war, when he also suffered arrest and persecution for aiding escapes from the Nazis. Composer in residence Dresden O, 1944. He won fame with his first opera, *Dantons Tod*, consolidating this with *Der Prozess*. In these he made use of separate numbers and a light, allusive style that includes the influence of jazz by way of Stravinsky and Blacher. The sentimental comedy *Der Zerrissene* is an emptier work, and Einem's eclecticism proved inadequate to the content of *Kabale und Liebe*. He has also achieved popularity with *Der Besuch der alten Dame*, in which he confirms his ability to confront sharp moral problems in music whose popular appeal need not weaken the issues. *Tulifant*, a parable about the enslavement and destruction of the world, makes use of a small orchestra to create lucid textures and atmospherics as well as to provide rhythmic dance sections.

WORKLIST: *Dantons Tod* (Danton's Death) (Blacher & Einem, after Büchner; Salzburg 1947); *Der Prozess* (The Trial) (Blacher & Cramer, after Kafka; Salzburg 1953); *Der Zerrissene* (The Man Torn in Two) (Blacher, after Nestroy, after Duvert & Lausanne; Hamburg 1964); *Der Besuch der alten Dame* (The Old Lady's Visit) (Dürrenmatt; Vienna 1971); *Kabale und Liebe* (Intrigue and Love) (Blacher & Ingrisch, after Schiller; Vienna 1976); *Jesu Hochzeit* (Ingrisch; Vienna, 18 May 1980); *Tulifant* (Ingrish; Vienna, 1990).

**'Ein Mädchen oder Weibchen'.** Papageno's (bar) aria in Act II of Mozart's *Die Zauberflöte*, in which he sings of his longing for a wife.

**'Einsam in trüben Tagen'.** Elsa's (sop) aria, sometimes referred to as 'Elsa's Dream', in Act I of Wagner's *Lohengrin*, in which she recounts her dream of a rescuing hero.

**'Ein Schwert verhiess mir der Vater'.** Siegmund's (ten) narration in Act I of Wagner's *Die Walküre*, in which he relates how he has been promised a sword in his hour of need by his father (Wotan).

**Einstudierung** (Ger.). Production.

**Eire.** See IRELAND.

**Eisenstein.** Gabriel von Eisenstein (ten), husband of Rosalinde, in Johann Strauss's *Die Fledermaus*.

**Elder, Mark** (*b* Hexham, 2 June 1947). English conductor. Début Melbourne 1972 (*Rigoletto*). Sydney O 1973; London, ENO, 1974, music director 1979–92; London, CG, 1976 (*Rigoletto*); Bayreuth 1981 (*Meistersinger*); New York, M, 1988 (*Figaro*). His period at the Coliseum was characterized by the exploration of opera as contemporary entertainment, in conjunction with David Pountney, and the revival of neglected works (*Rusalka*, *Lady Macbeth of Mtsensk*, *The Stone Guest*), as well as the introduction of new ones (British premières incl. *Mazeppa, Osud, Doktor Faust, Le Grand Macabre, Lear, Pacific Overtures, Akhnaten*; world premières incl. *The Mask of Orpheus*). A brilliant and versatile conductor, who provides sympathetic leadership to both singers and orchestra. (R)

**Eléazar.** The Jewish goldsmith (ten), supposed father of Rachel, in Halévy's *La Juive*.

**Electra.** In classical mythology, the daughter of Agamemnon and Clytemnestra. She incited her brother Orestes to avenge their father's death by killing his murderess, their mother, who had become the lover of Aegisthus. Operas on the subject are by Lemoyne (1782: a work that includes an early use of *reminiscence motive), Grétry (1780 or 1781; passed over in favour of Lemoyne, now lost), Häffner (1787), and Strauss (1909). She also appears in Mozart's *Idomeneo* as *Elettra.

**Elegy for Young Lovers.** Opera in 3 acts by Henze; text by W. H. Auden and Chester Kalman. Prod. Schwetzingen 20 May 1961, by Bavarian State Opera, with Bremert, Benningsen, Rogner, Lear, Fischer-Dieskau, Kohn, cond. Bender; Gly., 19 July 1961, with Söderström, Dorow, Meyer, Turp, Alexander, Hemsley, cond. Pritchard; New York, Juilliard School, 29 Apr. 1965, with Haywood, Shane, Wagner, Jones, Davison, Evans, cond. Henze.

An alpine hotel, 1910. Gregor Mittenhofer (bar), the self-centred and egoistic poet, goes every year to the 'Schwarzer Adler' in the Austrian Alps, to seek inspiration for his work from the hallucinations of Hilda Mack (sop). She is forever awaiting the return of her husband, who disappeared 40 years earlier on their honeymoon. When his body is recovered from the glacier, the poet has to look elsewhere for inspiration and turns to the young lovers, Toni, his stepson (ten), and Elisabeth (sop). He sends them to the Hammerhorn to gather edelweiss, and there lets them die. Their death inspires his greatest poem 'Elegy for Young Lovers', which he reads silently before the audience in the closing scene of the opera.

**Elektra.** Opera in 1 act by Richard Strauss; text by Hugo von Hofmannsthal, after his drama (1903), in turn after Sophocles' tragedy (411 or 410 BC). Prod. Dresden, H, 25 Jan. 1909, with Krull, Siems, Schumann-Heink, Perron, cond. Schuch; New York, Manhattan OC, 1 Feb. 1910 (in French), with Mazarin, Baron, Gerville-Réache, Huberdeau, cond. De la Fuente; London, CG, 19 Feb. 1910, with Walker, Rose, Bahr-Mildenburg, Weidemann, D'Oisly, cond. Beecham.

Elektra (sop) mourns the death of her father Agamemnon; she tries to persuade her sister

Chrysothemis (sop) to help her avenge it. Her mother Klytemnestra (mez), Agamemnon's murderer, is plagued by nightmares and fears, but finds some comfort in news of the death of Elektra's absent brother Orest. Alone, Elektra begins to dig up the axe that had slain Agamemnon, but is interrupted by a stranger in the courtyard: it is Orest (bs-bar), not dead but secretly returned. He enters the palace and kills Klytemnestra; when Klytemnestra's lover Aegisth (ten) appears, Elektra lights his way into the palace and to his own murder. She dances in triumph, and collapses dead.

Other operas on the subject are by Lemoyne (1782), Häffner (1787), and Gnecchi (*Cassandra*, 1905). The similarity between certain motives in the latter work and in Strauss's caused some stir at the time, and there were suggestions of both telepathy and plagiarism.

See also IDOMENEO.

BIBL: D. Puffett, *Richard Strauss: 'Elektra'* (Cambridge, 1989); B. Gilliam, *Richard Strauss's 'Elektra'* (Oxford, 1991).

**elenco artistico** (It.: 'catalogue of artists'). Properly, the names of the performers in a company or a special production or season, published on a poster. More recently, the complete list of artists, including producers, designers, conductors, etc., engaged by an Italian opera-house. In Germany the information is contained in the season's *Spielplan.

**Eletsky.** A prince (bar), in love with Lisa and Hermann's rival, in Tchaikovsky's *The Queen of Spades*.

**Elettra.** The Greek princess Electra (sop), in love with Idamante, in Mozart's *Idomeneo*.

**Elgar,** (Sir) **Edward** (*b* Broadheath, 2 June 1857; *d* Worcester, 23 Feb. 1934). English composer. Elgar's only engagement with opera came late in his career, with a projected setting of a libretto by Barry Jackson, *The Spanish Lady*, on Jonson's *The Devil is an Ass*. Copious fragments survive, but do not indicate how the work would have been developed.

BIBL: P. Young, *Elgar O.M.* (London, 1955, 2/1973).

**Elisabeth.** The Landgrave's niece (sop), Tannhäuser's 'sacred' love in Wagner's *Tannhäuser*.

**Elisabeth de Valois.** King Philip II of Spain's Queen (sop), loved by his son Carlos, in Verdi's *Don Carlos*.

**Elisabetta, regina d'Inghilterra** (Elizabeth, Queen of England). Opera in 2 acts by Rossini; text by Giovanni Schmidt, after Carlo Federici's drama (1814). Prod. Naples, C, 4 Oct. 1815, with Colbran, Dardanelli, Manzi, Nozzari, Manuel

García, Chizzola; London, H, 30 Apr. 1818, with Fodor, Corri, Crivelli, García.

London, late 16th cent. Queen Elizabeth I (sop) learns from the Duke of Norfolk (ten) that her favourite, Leicester (ten), has secretly married Mathilde (sop). Unable to persuade Leicester to give up Mathilde, she has him imprisoned; when Norfolk's intrigues eventually lead to his arrest, the Queen shows her magnanimity by pardoning Leicester and Mathilde.

The overture and finale were taken over from *Aureliano in Palmira*, and the former now replaces the lost overture to *Il barbiere di Siviglia*. A special performance was recorded by Italian Radio for the coronation of Queen Elizabeth II in 1953.

Other operas on Elizabeth I are by Pavesi (1810), Carafa (1818), Donizetti (*Roberto Devereux*, 1837), Giacometti (1853), Klenau (1939), Walter (1939), Britten (*Gloriana*, 1953), and Fortner (1972).

**Elisir d'amore, L'** (The Love Potion). Opera in 2 acts by Donizetti; text by Romani, after Scribe's libretto for Auber's *Le philtre* (1831). Prod. Milan, T della Canobbiana, 12 May 1832, with Sabine Heinefetter, Dabadie, Genero, Frezzolini; London, L, 10 Dec. 1836; New York, Park T, 18 June 1838, with Caradori-Allan, Jones, Morley, Placide.

A Tuscan village, early 19th cent. Inspired by the story of Tristan and Isolde, the shy young Nemorino (ten) buys potion from a quack, Dulcamara (bs), in the hope of winning Adina (sop). She seems to prefer Belcore (bar), a sergeant of the garrison, and, irritated by Nemorino's now apparent indifference towards her, she promises to marry Belcore immediately. Convinced that the elixir will take effect within 24 hours, Nemorino begs her to wait, but she refuses.

Nemorino joins the army, using his enlisting bonus to buy another bottle of elixir. Finding himself suddenly surrounded by village girls, he presumes the elixir is working, unaware of the rumour that his uncle has died leaving him a large fortune. Adina feels intense envy and realizes that it is Nemorino she really loves. Learning of his sacrifices, she redeems him from the army and the couple are married. Dulcamara capitalizes on the apparent success of his elixir and does a roaring trade.

Also opera by Kate Loder (*c*. 1850).

Subject of a burlesque by W. S. *Gilbert, *Dulcamara, or The Little Duck and the Great Quack* (1866).

**'Ella giammai m'amo'.** Philip II's (bs) aria in Act IV of the Italian version of Verdi's *Don Carlos*, grieving over his loneliness as a man, having never been loved by his wife, and as a king.

**'Elle a fui, la tourterelle'.** Antonia's (sop) aria at the piano, in the 'Antonia' act of Offenbach's *Les contes d'Hoffmann*.

**Ellen Orford.** The schoolmistress (sop) who befriends Peter in Britten's *Peter Grimes*.

**Elleviou, Jean** (*b* Rennes, 14 June 1769; *d* Paris, 5 May 1842). French tenor. Début Paris, CI, 1790, in Monsigny's *Le déserteur*, as bass. Second début, as tenor, in première of Dalayrac's *Philippe et Georgette*, with great success. After difficulties with the Revolutionary authorities, he rejoined the CI 1797–1801. Paris, OC, 1801–13, retiring when Napoleon refused to increase his already considerable salary. A versatile and refined actor, he made intelligent use of a pleasing voice, and was enormously popular, especially with women. Created 40 leading roles, e.g. in works by Grétry (including *Richard Cœur de Lion*), Dalayrac, Boieldieu (including *Le Calife de Bagdad*, *Jean de Paris*), Monsigny, and Méhul (with particular success in *Joseph*). Also the librettist of Berton's *Délia et Verdikan*.
BIBL: H. de Curzon, *Elleviou* (Paris, 1930).

**Elmendorff, Karl** (*b* Düsseldorf, 25 Oct. 1891; *d* Hofheim, 21 Oct. 1962). German conductor. Studied Cologne with Steinbach and Abendroth. Début Düsseldorf 1916. Berlin and Munich 1925–32. Bayreuth 1927–42. Wiesbaden 1932–5; 1952–6; Mannheim 1935–42; Dresden 1942–5; Kassel 1948–56. A distinguished Wagnerian; cond. Wagner widely in Italy and also introduced Italian operas to Germany. (R)

**Elsa.** Elsa von Brabant (sop), bride of Lohengrin in Wagner's *Lohengrin*.

**Elsner, Józef** (*b* Grodków, 1 June 1769; *d* Warsaw, 18 Apr. 1854). Polish composer and teacher of German origin. After schooling in Breslau, where he sang in the opera chorus, studied Vienna. Conductor Brno 1791–2; Lwów 1792–9. From 1799 he was director of the Warsaw O, where he developed the repertory and improved standards. His teaching activities included a period as Rector of the Conservatory from 1821; his pupils included Chopin. Among the most important of his many operatic pieces are *Andromeda* (written in honour of Napoleon, who attended the second performance) and *Leszek Biały*. The latter was encouragingly reviewed by his friend and colleague E. T. A. Hoffmann in the *AMZ*, to which Elsner also contributed. His operas are characteristic of an early phase of nationalism in that they graft local plots, characters, and folk-songs on to an established tradition, in his case Viennese Singspiel. His activities helped to provide a creative base for the succeeding generation of Polish composers.
SELECT WORKLIST: *Andromeda* (Osiński; Warsaw

1807); *Leszek Biały* (Dmuszewski; Warsaw 1809); *Król Łokietek* (King Łokietek) (Dmuszewski; Warsaw 1818).

BIBL: A. Nowak-Romanowicz, *Józef Elsner* (Cracow, 1957).

**'E lucevan le stelle'.** Cavaradossi's (ten) aria in Act III of Puccini's *Tosca*, in which he reflects on his love for Tosca as he writes his last letter before his execution.

**Elvino.** A young farmer (ten), bridegroom of Amina, in Bellini's *La sonnambula*.

**Elvira.** 1. A lady of Burgos (sop), deserted by Giovanni in Mozart's *Don Giovanni*.
2. The Bey of Algiers's about-to-be-discarded wife (sop) in Rossini's *L'italiana in Algeri*.
3. The daughter of Gualtiero Valton (Lord Walton) (sop) in Bellini's *I Puritani*.
4. Silva's kinswoman (sop), lover of Ernani in Verdi's *Ernani*.

**Emilia.** Iago's wife and Desdemona's companion (mez), in Rossini's and in Verdi's *Otello*. Also Emilia Marty, the singer (sop) in Janáček's *The Makropoulos Case*.

**Emmanuel, Maurice** (*b* Bar-sur-Aube, 2 May 1862; *d* Paris, 14 Dec. 1938). French composer and musicologist. Studied Paris with Delibes, who, outraged by his experiments, forbade him to enter for the Prix de Rome; privately with Guiraud. His scholarly interest in Greek and medieval modes is reflected in his writings and in his music, including in his two operas.

WORKLIST: *Prométhée enchaîné* (Prometheus Bound) (Emmanuel, after Aeschylus; comp. 1916–18, concert Paris 1919, prod. Liège 1923); *Salamine* (The Persae) (Reinach, after Aeschylus; Paris 1929).

**Emperor Jones, The.** Opera in 2 acts by Louis Gruenberg; text by Kathleen de Jaffa, after Eugene O'Neill's drama (1921). Prod. New York, M, 7 Jan. 1933, with Tibbett, cond. Serafin; also Amsterdam (1934), and Rome (1952) with Rossi-Lemeni, cond. Gavazzeni.

Brutus Jones (bar), an ex-Pullman porter and escaped convict, rules his island in the Caribbean, with all the outward trappings of royalty. He has systematically fleeced his people and is making preparations to flee, warned by Henry Smithers (ten) that the tribesman are about to revolt. Jones flees into the jungle and is confronted (in hallucinations) by his past victims. He begs God for forgiveness; but, discovered by the witch-doctor who has summoned the tribesmen, he kills himself with his last bullet, a silver one he has saved for this purpose.

**encore** (Fr.: 'again'). The word called by members of a British audience to demand the repeat of a number, hence the word applied to the actual repeat. On the Continent, the word *bis* (Fr. and It., from Lat.: 'twice, again') is used, and can similarly be formed into a noun (It., *chieder un bis*: 'to call for an encore') or a verb (Fr., *il fut bissé*: 'he was encored'). Though known in the ancient theatre, the practice originally began in opera with the rise of the virtuoso singer in the 17th-cent. Venetian tradition, and has been very controversial; audiences have come into conflict with artists (almost invariably conductors) who insist on artistic continuity. Toscanini was a leading example of a conductor who quarrelled with audiences over his refusal to allow encores. The practice became at one time virtually automatic in the D'Oyly Carte Opera Company, with the encore point printed into the orchestral parts and singers ready with new 'business' for each comic repeat. Caruso was obliged to repeat 'La donna è mobile' five times in Rio de Janiero in 1903; a whole scene of Verdi's *Otello* was encored at Florence in 1951; however, the longest encore on record was that of an entire opera, *Il *matrimonio segreto*, at its première in Vienna in 1792. It is said, on the other hand, that a tenor surprised to find his aria repeatedly encored by the notorious Parma audience was eventually told that he was being kept there until he sang it correctly.

**Enescu, George** (*b* Liveni Vîrnav (now George Enescu), 19 Aug. 1881; *d* Paris, 3 or 4 May 1955). Romanian composer, conductor, violinist, and teacher. Studied Vienna with Fuchs. His only opera was *Œdipe*, a work neglected outside Romania but one which includes some of his most powerful, somewhat Expressionist music in an ambitious treatment of the whole of Oedipus' life.

WORKLIST: *Œdipe* (Fleg; Paris 1936).

BIBL: M. Voicana, ed., *George Enescu: monografie* (Bucharest, 1971).

**Enfant et les sortilèges, L'** (The Child and the Enchantments). Opera in 2 parts by Ravel; text by Colette. Prod. Monte Carlo 21 Mar. 1925, with Gauley, Orsone, Lafont, Warnery, cond. De Sabata; San Francisco, Civic Auditorium, 19 Sep. 1930, with Q. Mario, Gruninger, Farncroft, Picco, cond. Merola; Oxford TH, by OUOC, 3 Dec. 1958, with C. Hunter, cond. Westrup.

A country house in Normandy. A child (mez) is punished by his mother (con) for not doing his homework. In a fury, he tears the wallpaper and pulls the cat's tail. Armchair, clock, toys, and other objects come to life and express their scorn at the child's behaviour; even Arithmetic (ten) taunts him.

At night in the garden, the child finds the animals and trees equally hostile. A squirrel is accidentally wounded in the rumpus and, when the animals see how gently the child cares for its wounds, they relent. The child is returned home to his mother and the disturbing enchantments end.

**'En fermant les yeux'.** Des Grieux's (ten) aria in Act II of Massenet's *Manon*, in which he describes how he saw Manon in a dream.

**English Music Theatre Company.** Formed in 1947 as the English Opera Group by Benjamin Britten, John Piper, and Eric Crozier to encourage new operas, especially on a chamber scale, bringing poets and playwrights together with composers. Its enterprise also lay behind the foundation of the Aldeburgh Festival, and the Opera Studio in London, which in turn became the Opera School, National School of Opera, and London Opera Centre. The group commissioned and performed works including several by Britten, also Berkeley, Williamson, Birtwistle, Gardner, Crosse, and Musgrave. It also revived works by Purcell, Mozart, Handel, Tchaikovsky, and Holst. Steuart Bedford and Colin Graham became music director and director of productions in 1971. In 1975 it was renamed to reflect its wider range of activities. It toured widely, and was able to draw upon the services of the best English singers of the day, including Baker, Ferrier, Harper, Vyvyan, Pears, Tear, and Brannigan. Disbanded 1980.

**English National Opera.** See COLISEUM, SADLER'S WELLS.

**English Opera Group.** See ENGLISH MUSIC THEATRE COMPANY.

**Enna, August** (*b* Nakskov, 13 May 1859; *d* Copenhagen, 3 Aug. 1939). Danish composer of Italian descent. Studied Copenhagen with Rasmussen. His first success as an opera composer came with *Heksen*, which won a certain international reputation, and this was confirmed with *Kleopatra*. His natural skill and graceful, late-Romantic melodic manner ensured him popularity in his own country, though his works have not generally exported well.

SELECT WORKLIST: *Heksen* (The Witch) (Ipsen, after Fitger; Copenhagen 1892); *Kleopatra* (Christiansen, after Rider Haggard; Copenhagen 1894); *Princessen på ærten* (The Princess on the Pea) (Rosenberg, after Andersen; Århus 1910); *Komedianter* (The Actors) (E. & O. Hansen, after Hugo; Copenhagen 1920).

**Enrico.** Henry Ashton (bar), Lucia's brother in Donizetti's *Lucia di Lammermoor*.

**ensemble** (Fr.: 'together'). As a noun, commonly used in English as well as French and German for a group of musical performers, although it properly refers to their musical unanimity. The earliest genuine ensembles in opera, as opposed to choral scenes, were the comic duets included in works of the 17th-cent. Venetian tradition. With the rise of opera seria in the 18th cent., the ensemble was usually relegated to an appearance at the end of the act, when it was often misleadingly referred to in the score as a chorus. The new genre of *opera buffa made much use of the ensemble, particularly in the finale (see ENSEMBLE FINALE), as did opéra-comique and the later style of Singspiel. In the 19th cent. the most popular kind of ensemble in Italian opera was the duet, but larger ensembles were found in abundance in French rescue opera and Grand Opera, and in those German operas which mirrored their style. In one such work, *Fidelio*, Beethoven writes a masterly quartet around a canon. Wagner made much use of the ensemble in his earliest operas, but it formed no part of his original conception of *music drama. Thus ensembles do not appear in the first parts of *The Ring*—such sections as the 'Ride of the Valkyries' do not qualify for the description, since the individual vocal lines are rarely sung together—although after the composition of *Die Meistersinger*, with its dramatically vital quintet, the principles of such writing found greater consideration. Many ensembles are included in the works of Puccini (e.g. the Act I love duet in *Madama Butterfly*) and Strauss (e.g. the concluding trio of *Der Rosenkavalier*), while in the 20th cent. generally, with its emphasis upon a more realistic presentation of the drama, the ensemble has been greatly favoured, not least because of the final breakdown of the *number opera.

**ensemble finale.** The closing scene of an act, usually of an opera buffa, during which the characters assemble gradually on stage, until the full or near-full cast is present.

The origins of the ensemble finale may be traced back to Venetian opera of the mid-17th cent., in which each act usually concluded with a duet. This provided a model for later composers such as A. Scarlatti, in whose only comic opera, *Il trionfo dell'onore* (1718), the first act ends with a humorous duet between two servants, while second and third acts finish with quartets. Typically, although there is an internal contrast of thematic material, none of the singers is characterized musically.

At this time the ensemble functioned as vehicle to convey the feelings and thoughts of the characters, but rarely served to advance the plot. But with the consolidation of *opera buffa as a separate genre in the mid-18th cent., the ensemble began to adopt a more dramatic role. The obligatory comic duets at the end of the first two acts were gradually extended backwards; rather than being a mere commentary on past events, they now encompassed dramatic action, and featured serious and comic characters together. In the ensemble finale the story thus progressed with the entry of each new figure, while concurrently the music changed mood or style in an attempt at characterization. At first this new approach was hampered by an over-strict adherence to closed musical forms, but the gradual use of rondo and less rigid chain-like structures eventually permitted greater dramatic flexiblity.

Leo's *Semmegliana* (1726), *Lo matrimonio annascuso* (1727), and *L'ambizione delusa* (1742) represent important stages in the expansion of the ensemble finale, as do Vinci's *Le zite* (1722), Logroscino's *Il governatore* (1747) and Galuppi's *Il mondo della luna* (1750) and *Il filosofo di campagna* (1754). But all these concluding ensembles lack significant internal dramatic development. More advanced are the ensemble finales of Pergolesi: *Lo frate 'nnamorato* (1732) has an especially fine quintet, in which formal considerations are relaxed for the sake of dramatic momentum. The most significant composers of ensemble finales before Mozart were Piccinni, Paisiello, and, to a lesser extent, Sarti. The finale to the first act of Piccinni's *La buona figliuola* (1760) runs to over 300 bars, incorporating several changes of metre and style, while similarly advanced inventions are to be found in Sarti's *Gelosie villane* (1776) and Paisiello's *Credulo deluso* (1774) and *Il barbiere di Siviglia* (1782).

No other composer approached Mozart in length, complexity, or dramatic handling. In the formal design of his ensemble finales he turned to the models of symphonic music, adapting features of sonata form in elaborate structures whose dramatic pace is superbly matched and heightened by the onward progress of the music. Of all his ensemble finales—which also feature in *Don Giovanni*, *Così fan tutte*, and *The Magic Flute*—that to the second act of *Figaro* (1786) is the most complex; during more than 600 bars of music (lasting well over 20 minutes) the plot advances to a state of bewildering complexity. At each new happening on stage there is a corresponding change in the style of the music, yet such is Mozart's craftsmanship that the various musical and dramatic elements are welded together into a perfectly balanced, homogeneous whole. Significantly, Haydn made rather little use of the ensemble finale, preferring instead to create larger dramatic blocks in the middle of acts.

Although the ensemble finale *per se* remained a feature of 18th-cent. opera buffa, it exerted an important influence on the subsequent development of opera. By freely juxtaposing distinct musical sections, and mixing both reflective and narrative styles, it had a crucial, liberating effect on current conventions, leading eventually to the continuous acts of 19th-cent. opera.

**ente autonomo** (It.: 'autonomous being, independent organization'). The name given to the independent, self-governing corporations which control leading Italian opera-houses (e.g. Milan, S, and Naples, C), as opposed to the commercial impresario who is given a concession to manage short seasons in smaller provincial theatres. The ente autonomo idea was formulated during Toscanini's 1922–9 reign at Milan, S, and was soon adopted by the major Italian theatres.

**Entführung aus dem Serail, Die** (The Abduction from the Seraglio). Opera in 3 acts by Mozart; text by Gottlieb Stephanie the Younger, altered from Christoph Friedrich Bretzner's libretto for André's *Belmont und Constanze* (1781), in turn adapted from one of various English and Italian plays and comic operas on the subject. Prod. Vienna, B, 16 July 1782, with Cavalieri, Teyber, Adamberger, Dauer, Fischer, Jautz, cond. Mozart; London, CG, 24 Nov. 1827, in a version with extra characters by Kramer, with Hughes, Vestris, Sapio, Wrench, Benson, cond. Smart; New York, Brooklyn Athenaeum, 16 Feb. 1860, with Johanssen, Rotter, Lotti, Quint, Weinlich.

The palace of the Pasha Selim, Turkey, mid-16th cent. Constanze (sop) is a captive of the Pasha (spoken), together with her servant Blonde (sop) and Pedrillo (ten). Belmonte (ten), a young Spanish nobleman and Pedrillo's original master, comes in search of his beloved Constanze. In the garden of the palace, he encounters Osmin (bs), the Pasha's steward, who flies into a rage at the mention of Pedrillo. Osmin leaves, and Pedrillo forms a plan to introduce Belmonte to the Pasha as an architect. A chorus of Janissaries signals the arrival of the Pasha with Constanze. Once alone, he pleads with her to consider giving him her love, but says will not yet force her. She reminds him of her pledge to Belmonte and departs sadly. Pedrillo introduces Belmonte.

Osmin is trying, without success, to force Blonde to love him. She stubbornly stands up for the rights of an Englishwoman to choose. Constanze resolves to refuse the Pasha and to die rather than be unfaithful to Belmonte. Pedrillo manages to inform Blonde of the escape plans for that night. He gets Osmin out of the way by making him drunk. The four lovers are reunited.

It is midnight and, as a signal to the women to start the escape, Pedrillo sings a serenade. Osmin catches them and the captives are brought before the Pasha. Belmonte turns out to be the son of the man who was responsible for sending the Pasha into exile. Instead of exacting vengeance on the son of his enemy, the Pasha decides to be magnanimous and to set all four captives free.

BIBL: T. Bauman, *W. A. Mozart: 'Die Entführung aus dem Serail'* (Cambridge, 1987).

**entr'acte** (Fr.: 'between acts'). In opera, a piece of orchestral music played between the acts or scenes; also called an *intermezzo or interlude. It sometimes erroneously appears, in writings about music, as a synonym for *intermède.

**entrée** (Fr.: 'entry'). In the *ballet de cour the entrée divided the individual acts into scenes, and featured elaborately costumed, usually masked figures who performed characteristic dances, as well as songs. Its plot and character were carefully matched to the occasion and the most lavish ballets de court could include as many as 30 entrées. The *comédie-ballet was likewise divided into entrées: a famous contemporary comment on Lully and Molière's *Le bourgeois gentilhomme* described it as a 'ballet accompanied by six entrées and a comedy'. In this sense it thus approaches the meaning of *intermède. Within Lully's operas and those of his contemporaries the term was usually applied to the music which marked the beginning of the *divertissement of songs and dances. For the *opéra-ballet and *ballet-héroïque, the term entrée was used in the sense of 'act', but may be distinguished from it by virtue of each such subdivision having a separate plot (as in Rameau's *Les Indes galantes*) rather than developing a continuous dramatic narrative. In this respect they reflect their origin in the **ballet à entrées**, an offshoot of the *ballet de cour in which each section had its own discrete plot.

**entremés** (Sp.: 'interlude'). A short stage entertainment popular in Spain in the 17th cent.; its function, character, and plot were similar to that of the Italian *intermezzo. Usually performed after the first act of a serious play: an entremés performed after the second act was known as **baile** or **jácara**, that after the third as *mojiganga. Popular in style, even distinguished playwrights were attracted to writing them, including Tirso de Molina.

**Enzo.** A Genoese noble (ten), Laura's lover, in Ponchielli's *La Gioconda*.

**Épine, Margherita de l'** (*b c*.1683; *d* London, 8 Aug. 1746). Italian soprano, known as 'La Margherita'. Went to London 1702 with the composer Greber, and appeared at Lincoln's Inn Fields in his works in 1703. London, DL, 1704–18, when, a great rival to Tofts, she sang in nearly every opera put on in London, including Handel's *Rinaldo* (Goffredo), and created Agilea (*Teseo*) and Eurilla (*Il pastor fido*). The first Italian to make a successful career in England, she

retired in 1719 with a large fortune. In *c*.1718 she married Pepusch, with whom she had worked since 1707, singing many of his odes, masques (e.g. *The Death of Dido*, *Venus and Adonis*), and cantatas. He called her 'Hecate' because she was 'so destitute of personal charms' (Hawkins); nevertheless, she attracted several peers, and lived with the Earl of Nottingham, as well as with Greber. She appears in the caricature 'Rehearsal of an Opera' (attrib. Marco Ricci).

**'Era la notte'.** Iago's (bar) narration in Act II of Verdi's *Otello*, in which he pretends to substantiate his accusations against Desdemona by alleging that he once heard Cassio talking in his sleep about her and their love.

**Erb, Karl** (*b* Ravensburg, 13 July 1877; *d* Ravensburg, 13 July 1958). German tenor. Selftaught. Début Stuttgart 1907 (Kienzl's *Der Evangelimann*). Stuttgart 1910–12. Munich, N, 1913 (Lohengrin), and subsequently member until 1925, created *Palestrina* (title-role), 1917. London, CG, 1927 (Belmonte, with his wife Maria Ivogün as Constanze). Retired from opera 1930, but continued career as recitalist. The model for Erbe in Mann's *Dr Faustus*, the intelligent 'tenor with the voice of a eunuch'. (R)

**Ercole amante** (Hercules in Love). Opera in a prologue and 5 acts by Cavalli; text by Francesco Buti. Prod. Paris, T des Tuileries, 7 Feb. 1662.

In the allegorical prelude, important royal families, including the French, are presented by the goddess Diana.

Diana promises Ercole Beauty as his bride if he will continue his tasks. The reminder of the opera is concerned with the love of both Ercole and his son Hyllo for Iole. The action involves extensive use of machines in 'heaven' and at sea (Act IV). At the conclusion of the opera, Ercole is awarded Beauty, and Hyllo marries Iole.

Louis XIV himself participated in the original production, taking the part of the Sun King. The performance lasted over six hours. Written to celebrate Louis XIV's marriage (1660), its première had to be postponed and Cavalli's *Serse* was performed in its place. The obligatory ballet music for *Ercole amante* was provided by Lully.

**Erda.** The Earth Goddess (con) and mother, by Wotan, of the Valkyries in *Rheingold* and *Siegfried* in Wagner's *Der Ring des Nibelungen*.

**Erede, Alberto** (*b* Genoa, 8 Nov. 1909). Italian conductor. Studied Genoa, Milan, Basel with Weingartner, Dresden with Busch. Début (concert) Rome 1930. Gly. music staff 1934; cond. 1938–9. Director Salzburg O Guild 1935–8. Music director New London OC 1946–8. New York, M, 1950–5. Deutsche O am Rhein 1956, music director 1958–61. A fine trainer of young singers, and a strong ensemble leader. (R)

**Erhardt, Otto** (*b* Breslau (Wrocław), 18 Nov. 1888; *d* San Carlos de Bariloche, 18 Jan. 1971). German producer. Studied music and history of art; began career as violinist. Staged German première of Monteverdi's *Orfeo* (Breslau 1913). Düsseldorf 1918–20; Stuttgart 1920–7, where his *Boris* was an early Expressionist production and he also prod. *Doktor Faust* (1927) using film sequences; Dresden 1927–31, incl. première of Strauss's *Ägyptische Helena* (1928). Chicago 1930–2. London, CG, 1933-5, incl. British premières of *Arabella* and *Shvanda* (1934). Buenos Aires, C, 1939–56. New York, CC, 1954–6, 1967; also Vienna, Milan, Munich.

**Erik.** A huntsman (ten), rejected lover of Senta, in Wagner's *Der fliegende Holländer*.

**Erinnerungsmotiv.** See REMINISCENCE MOTIVE.

**Erismena.** Opera in a prologue and three acts by Cavalli; text by Aurelio Aureli. Prod. Venice, S Apollinare, 26 Dec. 1655; rev. 1670. Rev. in version made by Lionel Salter, London, BBC, 13 May 1967, with Robinson, Neville, Sinclair, Esswood, Herincx, cond. Salter; first English stage perf. London, King's Coll., 1 Apr. 1974, cond. Bedford. Version by Alan Curtis prod. The Hague, Kon. Schouwburg, 2 July 1974, cond. Curtis.

**'Eri tu'.** Anckarstroem's (Renato's) (bar) aria in Act III of Verdi's *Un ballo in maschera*, resolving to punish not his supposedly unfaithful wife but his friend and king, Gustavus (Riccardo).

**Erkel, Ferenc** (*b* Gyula, 7 Nov. 1810; *d* Budapest, 15 June 1893). Hungarian composer, conductor, and pianist. Studied Pozsony with Klein. Moved to Kolozsvár, where he first heard some early attempts at Hungarian opera; later settled in Buda as opera conductor of the Hungarian T Co. (*c.*1835). He gave up a developing career as a pianist on hearing Liszt, and turned increasingly to composition, achieving early success with his first opera, *Bátori Mária*. He followed this up with *Hunyadi László*, the most successful of his operas in Hungary. Though drawing on Viennese and Italian elements, it achieves a genuinely Hungarian character by its use of Hungarian-inflected recitative and by the incorporation of national dances, especially the *verbunkos. Erkel also tried to develop a Hungarian equivalent of the English ballad operas known as *népszínmű*, in which were incorporated both ballads and original songs: these included *Két pisztoly*, and became very popular. Other commitments and the need to earn a living prevented him from returning to opera for some years; but in 1861 he

produced *Bánk bán*, based on a previously censored play and written in collaboration with his two sons Gyula and Sándor. Here the experiments of *Hunyadi László* are taken a stage further, with Hungarian elements more fully incorporated into fluent and original structures, and with more vivid and penetrating characterization. This was also the first composed work to make use of the cimbalom.

None of Erkel's subsequent operas achieved comparable success. He attempted comic opera on the the one hand, with *Sarolta* and *Névtelen hősök*, in which there is copious use of popular tunes and rustic scenes; and national music drama on the other, with *Dózsa György* and *Brankovics György*, in which Hungarian musical characteristics are absorbed into a continuous musical flow, with the orchestra taking a prominent role. Though there are in them anticipations of Liszt and even of Bartók, the shadow of Wagner falls more heavily across them, and deepens with Erkel's last opera, *István király* (which is mostly by Gyula Erkel). He left the National T in 1874, being succeeded by Richter, though he continued to conduct his own works and to play in public. *Hunyadi László* and *Bánk bán* have remained in Hungarian repertoris, and have occasionally been performed abroad. With them Hungarian national opera was effectively founded.

WORKLIST (all first prod. Pest): *Bátori Mária* (Egressy, after Dugonics; 1840); *Hunyadi László* (Egressy, after Tóth; 1844); *Erzsébet* (Czanyuga; 1857); *Bánk bán* (Egressy, after Katona; 1861); *Sarolta* (Czanyuga; 1862); *Dózsa György* (Szigligeti, after Jókai; 1867); *Brankovics György* (Odry & Ormai, after Obernyik; 1874); *Névtelen hősök* (Unknown Heroes) (Tóth; 1880); *István király* (King Stephen) (Váradi, after Dobsa; Budapest, 1885).

BIBL: D. Legany, *Erkel Ference művei* (Budapest, 1974).

**Erlanger,** (Baron) **Frédéric d'** (*b* Paris, 29 May 1868; *d* London, 23 Apr. 1943). English composer of French and German descent. Studied Paris with Ehmant. His *Inès Mendo* was given at London, CG, 1897 (under the anagram Ferd. Regnal). His greatest operatic success was in 1906 with *Tess*, after Hardy, with Destinn, Zenatello, and Sammarco. A banker by profession, he gave much financial support to CG and for many years was a director.

**Ernani.** Opera in 4 acts by Verdi; text by Piave, after Victor Hugo's tragedy *Hernani* (1830). Prod. Venice, F, 9 Mar. 1844, with Loewe, Guasco, Superchi, Selva; London, HM, 8 Mar. 1845, with Rita Borio, Moriani, Botelli, Fornisari, cond. Costa; New York, Park T, 15 Apr.

1847, with Tedesco, Perelli, Vita, Novelli, cond. Arditi.

Spain, 1590. Ernani (ten) (the outlawed Don Juan de Aragon, now a bandit) and Don Carlo, King of Spain (bar), both love Elvira (sop) and want to prevent her marriage to her elderly kinsman Don Ruy Gomez de Silva (bs). Ernani has vowed to avenge the death of his father at the hands of the king.

On her wedding-day, Don Carlo takes Elvira as a hostage since the laws of hospitality will not allow Silva to yield Ernani to him. Ernani tells Silva that he is willing to die at his hand, but only after he has rescued Elvira. He gives Silva a horn, promising to kill himself when it is sounded.

Conspirators meet to plot the death of Don Carlo. Ernani will not surrender his right to strike the fatal blow even in return for release from his promise to Silva. A cannon sounds, indicating the election of Don Carlo as Holy Roman Emperor, and he emerges from a hiding place having witnessed the conspiracy. The traitors are captured, but Don Carlo frees them at Elvira's intervention and offers her to Ernani in marriage.

In the middle of the wedding festivities, the horn sounds. Ernani is obliged to fulfil his promise; he stabs himself and dies.

Other operas on the subject are by Gabussi (1834), Mazzucato (1844), Laudamo (1849), and Hirschmann (1909).

**'Ernani, involami'.** Elvira's (sop) aria in Act I of Verdi's *Ernani*, in which she hopes that Ernani will flee with her.

**Ernesto.** Don Pasquale's nephew (ten) in Donizetti's *Don Pasquale*.

**Ero the Joker** (Srb-Cr. *Ero s onoga svijeta*: 'Ero from the Other World'). Opera in 3 acts by Jakov Gotovac; text by Milan Begović, after a Dalmatian folk-tale. Prod. Zagreb 2 Nov. 1935; London, Stoll T, by Zagreb Co., 28 Jan. 1955, with Tončić, Radev, Gostič, Krišžaj, Kučić, cond. Gotovac.

A village in the Dinara mountains. Mischa (ten) appears from a hayloft to land among a group of girls, who include Djula (sop). He claims to be Ero, and to have fallen from heaven because it was boring there. Djula's stepmother Doma (mez) tries to chase him away, but he convinces her that her mother is doing badly in heaven and needs financial help. Doma fetches money and gives it to Ero. With the help of the miller Sima (bar), Ero and Djula escape. They return at the time of the village fair, and Ero discloses that he is not a poor good-for-nothing but the son of a wealthy farmer, and wanted to test Djula's love. Her father gives the couple his blessing.

One of the most popular Croatian operas.

**Ershov, Ivan** (*b* Novocherkassk, 20 Nov. 1867; *d* Leningrad, 21 Nov. 1943). Russian tenor. Studied Moscow with Alexandrova-Kochetova; St Petersburg with Gabel and Paleček. Début St Petersburg, M, 1893 (Gounod's Faust). Sang there 1895–1929, in a wide repertory including Mozart, Wagner, Verdi, and many Russian roles, among them Sobinin (*A Life for the Tsar*), Golitsyn (*Khovanshchina*); also Sadko, and the first Grishka Kuterma (*Kitezh*). Told Rimsky-Korsakov he wrote awkwardly for the voice, and would do better to consult the singers. An excellent declaimer, and an actor preferred by many to Shalyapin, with a ringing tone. Taught Petrograd (Leningrad) Cons. 1916–41. His dramatic conviction was especially significant for its contribution to Wagner's popularity in Russia: 'he could claim immortality just for his interpretations of Tannhäuser, Tristan, Siegmund, and Siegfried' (Eduard Stark). (R)

BIBL: V. Bogdanov-Berezovsky: *Ivan Ershov* (Moscow, 1951).

**Erwartung** (Expectation). Monodrama in 1 act by Schoenberg; text by Marie Pappenheim. Prod. Prague, Neues Deutsches T, 6 June 1924, with Gutheil-Schoder, cond. Zemlinsky; London, SW, 4 Apr. 1960, by New OC with Harper, cond. Lovett; Washington 28 Dec. 1960, with Pilarczyk, cond. Craft.

Written for a single voice and full orchestra, the opera concerns the semi-deranged ravings of a woman (sop) searching for her lover in a forest at night; she hears weeping, mistakes a log for a corpse, imagines she is being watched, and thinks she is being pursued by the moon. In the fourth scene, the longest in the opera, she arrives at the house of her rival, who she finds has murdered her lover. Hysterically, she falls upon his body, smothering it with kisses. At length she departs.

**Escamillo.** The toreador (bar) in Bizet's *Carmen*.

**'È scherzo od è follia'.** Quintet in Act I of Verdi's *Un ballo in maschera*, in which Oscar (sop), Mam'zelle Arvidson (Ulrica) (mez), Gustavus (Riccardo) (ten), Ribbing (Sam) (bs), and Horn (Tom) (bs) react variously to the prophecy that Gustavus will die at the hand of a friend.

**Esclarmonde.** Opera in 4 acts (8 tableaux) by Massenet; text by Alfred Blau and Louis de Gramont, after a medieval romance, *Partenopeus de Blois*. Prod. Paris, OC, 14 May 1889, with Sanderson, Gilbert, Taskin; New Orleans, French OH, 10 Feb. 1893, with Bondues, Gluck, cond. Lagye; London, CG, 28 Nov. 1983, with Sutherland, Montague, Davies, Howell, cond. Bonynge.

Byzantium and France, early Middle Ages. Esclarmonde (sop), daughter of Phorcas, King of

Byzantium (bs), falls in love with a French knight, Roland (ten), despite her father's statement that her inheritance of his throne and magical powers depends on her keeping away from men. She uses her own magical powers to lure Roland who, with her assistance, accomplishes heroic deeds. When her father states that she is endangering Roland's life and her magical powers by her love, she relinquishes him. Her husband is to be the champion of a tournament organized by her father; this turns out to be Roland, and Phorcas gives his approval for their marriage.

**'Es gibt ein Reich'.** Ariadne's (sop) aria in Richard Strauss's *Ariadne auf Naxos*, in which she dreams of a land where everything is pure—the land of death.

**'E sogno'.** Ford's (bar) monologue in Act II of Verdi's *Falstaff*, reflecting upon his jealousy at the prospect of Falstaff seducing his wife.

**Essen.** City in North-Rhine Westphalia, Germany. Opera is given in the Städtische Bühnen, built 1892, rebuilt 1950 (cap. 800). Premières of works by Fortner, Klebe, and Reutter have been given.

**Estonia.** During the period of German hegemony, at the end of the 18th cent., a few operas were given by the German theatre in Tallinn, among them *Don Giovanni*. Opera performances were given by the Wanemuine music society in Tartar in 1906, and by the Estonia music society in Tallinn in 1906. With independence in 1918, new impetus was given to theatrical life, and an opera company was formed that year. Though earlier operas had been written by Estonian composers, notably *Samina* (1906), showing the influence of Russian traditions, by Artur Lemba (1885–1963), the first true Estonian opera was *Vikerlased* (The Vikings, 1928), by Evald Aav (1900–39). This was much influenced by *A Life for the Tsar*, whereas Adolf Vedro (1890–1944) modelled his *Kaupos* (1932) more closely on Wagner.

With Soviet rule, from 1944, some independent operatic life developed. New Estonian operas were written by Eugen Kapp (b 1908), notably *Tasuleegid* (The Fires of Vengeance, 1945), about the struggle against the Teutonic Knights, and a children's opera, *Talvemuinasjutt* (Winter Story, 1959). His cousin Villem Kapp (1913–64) wrote the successful *Lembitu* (1961). An important figure for Estonian musical life and opera has been Gustav Ernesaks (b 1908), with *Püha järv* (The Holy Lake, 1947), *Iormide rand* (The Edge of the Storm, 1949), *Käsi käes* (Hand in Hand, 1955), and *Vigased pruudid* (Fallible Wives, 1959). These are in a Romantic nationalistic

style, with extensive choruses. Other important opera composers are Leo Normet (*b* 1922), with *Valgus Koordis* (Light Over Koordi, 1955); Eino Tamberg (*b* 1930), with *Randne kodu* (The Iron House, 1965), *Cyrano de Bergerac* (1974), and *Lend* (Flight, 1982) about the artist's ethical dilemma; Veljo Tormis (*b* 1930), one of the few composers to use Estonian folk subjects, with *Luigelend* (The Swan's Flight, 1966) and *Eesti balaadid* (Estonian ballads, 1982) using folksongs and novel operatic forms; and Raimo Kangro (*b* 1949), with a rock opera, *Pŏhjuneitsi* (The Nordic Maid, 1980) and *Ohver* (The Victim, 1980). There is also a chamber ensemble, and a flourishing operetta tradition: a popular success has been *Kunigal on Külm* (The King is Cold, 1967) by the jazz and pop composer Valter Ojakäär (*b* 1923).

BIBL: M. Topmann, *Musik in Estland gestern und heute* (Tallinn, 1978).

**'Esultate!'** Otello's (ten) announcement of victory over the Turks in Act I of Verdi's *Otello*.

**Eszterháza.** Palace near Fertöd, Hungary. The summer residence of the Princes Esterházy, built 1762–6, it included a theatre (cap. 400) opened 1768 with *Haydn's *Lo speziale*. Opera was given twice a week during the latter part of His reign as Kapellmeister (*c.*1776–90), with a repertory that included operas by Piccinni, Sacchini, Salieri, Anfossi, Cimarosa, Dittersdorf, and others. There was also a marionette theatre on the other side of the park, at which Haydn's *Philemon and Baucis* was prod. 1773. The castle burnt down in 1779, and reopened with an enlarged theatre in 1781 with Haydn's *La fedeltà premiata*.

BIBL: M. Horányi, *Eszterházi vigasságok* (Budapest, 1959; trans. as *The Magnificence of Eszterháza* (London, 1962)).

**Etcheverry, Bertrand** (*b* Bordeaux, 29 Mar. 1900; *d* Paris, 14 Nov. 1960). French baritone. Studied Paris. Début Bordeaux 1925 (Le Bailli, *Werther*). Paris, O, 1932 (Ceprano, *Rigoletto*). Paris, OC, from 1937. London, CG, 1937, 1949. An outstanding Golaud; also popular as Boris. Created several roles, e.g. in Hahn's *Le marchand de Venise*. (R)

**Étoile, L'** (The Star). Opera in 3 acts by Chabrier; text by E. Leterrier and A. Vanloo. Prod. Paris, BP, 28 Nov. 1877 with Paola-Marié, Stuart, Daubray, Joly, Scipion, cond. Roques; New York, 18 Aug. 1890 (adapted as *The Merry Monarch*); London, Savoy T, 7 Jan. 1899 (also the adapted version, now called *The Lucky Star*).

King Ouf (ten) is seeking a victim for execution and so wanders, disguised, among his people, attempting to provoke a rebellion. He finds a victim in Lazuli (ten), a pedlar who insults the King after his suit for Princess Laoula is rejected. However, when the court astrologer, Siroco (bs), reveals that King Ouf will die 24 hours after Lazuli, the execution is cancelled, and Lazuli taken into luxurious confinement.

Lazuli and Laoula escape and news soon arrives of a shipwreck in which Lazuli has perished. King Ouf awaits his own death. The appointed hour passes, with the King still alive. Such is his rejoicing that he forgives Siroco's incompetence and permits the union of Laoula and Lazuli, whose reappearance is joyfully celebrated.

Chabrier's first stage work.

**Étoile du nord, L'** (The Northern Star). Opera in 3 acts by Meyerbeer; text by Scribe, after Rellstab's text for *Ein Feldlager in Schlesien*, based in turn on an episode in the life of Peter the Great. Prod. Paris, OC, 16 Feb. 1854 as *L'étoile du nord* with Duprez, Battaille, Lefèbvre, Herman-Léon—100th perf. on 1st anniversary; New Orleans, T d'Orléans, 5 Mar. 1855, with Pretti, Martial, Holtzem, Génibrel, Laget, Beckers; London, CG, 19 July 1855, with Bosio, Marai, Gardoni, Formes, Lablache, Zelger, cond. Costa.

Tsar Peter (bar) loves the village girl Katherine (sop). She substitutes herself for her brother George (ten) in the Russian Army, and informs the Tsar of a conspiracy. Disguised as a carpenter, the Tsar woos Katherine and makes her his Tsarina.

Originally composed as *Ein Feldlager in Schlesien* to open the new Berlin Opera, 7 Dec. 1844, with Tuczek: eight days after the première, she was replaced by Jenny Lind, who sang the work widely, in the rev. version for Vienna (1847) under the new title *Vielka*.

**Eugene Onegin.** Opera in 3 acts by Tchaikovsky; text by the composer and Konstantin Shilovsky, after Pushkin's poem (1823–31). Prod. Moscow, Maly T, by Conservatory, 29 Mar. 1879, with Klimentova, Levitskaya, Reiner, Konshina, Gilev, Medvedev, Makhalov, Tarkhov, cond. N. Rubinstein; professional première, Moscow, B, 23 Jan. 1881, with Verny, Krutikova, Yunevich, Kholkhov, Usatov, Bartsal, Abramov, cond. Bevignani; London, Olympic T, 17 Oct. 1892, with Fanny Moody, Oudin, McKay, Manners, cond. Wood; New York, M, 24 Mar. 1920, with Muzio, Perini, Martinelli, De Luca, Didur, cond. Bodanzky.

The Larins' country estate, early 19th cent. Tatyana (sop), the young and impressionable daughter of Mme Larina (mez), falls in love with Onegin (bar), a friend of her sister Olga's betrothed, Lensky (ten). Tatyana stays up all night writing an impassioned letter to Onegin. Next morning he reproves her for her forwardness and urges her to forget him.

At Tatyana's birthday party, some of the elderly women gossip about her and Onegin as they dance together. Annoyed, Onegin dances with Olga (mez), who had promised the dance to Lensky (ten). When Lensky reproaches her, she petulantly gives another dance to Onegin. Monsieur Triquet (ten), the French tutor, sings a song in praise of Tatyana. When the dancing is resumed, Lensky quarrels with Onegin and challenges him to a duel. Early next morning the two men meet beside an old mill. Lensky is killed.

Six years have passed and Onegin, who has been abroad, returns to St Petersburg. A ball is in progress at the palace of Prince Gremin (bs), now married to Tatyana. Onegin sees Tatyana again and realizes that he loves her. She agrees to meet him. He urges her to flee with him: at first she responds ardently; but then she reminds Onegin of her duty, and sends him away for ever.

**Euphrosine.** Opera in 5 (rev. to 3, finally 4) acts by Méhul; text by F. B. Hoffman. Prod. Paris, CI, 4 Sep. 1790; New Orleans, T St. Pierre, 15 Feb. 1806. The plot concerns the eventually successful efforts of Euphrosine (sop) to win the misogynist Coradin (ten) in spite of the efforts of the bitter, rejected Countess (sop). The duet 'Gardez-vous de la jalousie' became famous for its pioneering use of motive (a simple rocking figure of thirds) to express jealousy and its workings, and was much admired by Berlioz. Méhul's second opera, the first to be performed.

**Euridice.** 1. Opera in a prologue and 6 scenes by Peri; text by Ottavio Rinuccini. Prod. Florence, Pitti Palace, 6 Oct. 1600, as part of the wedding celebrations for Henri IV of France and Maria de' Medici; Saratoga Springs 9 Apr. 1941.

The plot follows the pattern of Poliziano's *Orfeo* (1480), differing only in that the opera has a happy ending. In the prologue, the spirit of Tragedy presents the subject of the opera and greets the important members of the audience.

Euridice is dancing with her companions. While Orfeo is singing of the beauties of nature, Dafne arrives to announce Euridice's death from a snake bite. In despair, Orfeo wishes to kill himself. However, Venus appears and allows him to descend to Hades to beg Pluto for his bride. He persuades Pluto and Proserpine to release Euridice, and a chorus of Shades accompanies the couple on their return to earth. The plot omits the famous condition whereby Orfeo must not look behind him as he leads Euridice back to the world.

2. Opera in prologue and 6 scenes by Caccini;

text by Ottavio Rinuccini. Prod. Florence, Pitti Palace, 5 Dec. 1602.

**Europe galante, L'** (Gallantries of Europe). Opéra-ballet by Campra; text by Antoine Houdar de la Motte. Prod. Paris, O, 24 Oct. 1697.

Each of the four entrées depicts the different romantic attitudes of a nation, France, Spain, Italy, and Turkey, allowing for much exotic colouring and a variety of dances.

**Euryanthe.** Opera in 3 acts by Weber; text by Helmina von Chezy, after a medieval French romance. Prod. Vienna, K, 25 Oct. 1823, with Sontag, Haizinger, Grünbaum, Forti, Seipelt, Teimer, Rauscher, cond. Weber; London, CG, 29 June 1833, with Schröder-Devrient; New York, M, 23 Dec. 1887, with Lilli Lehmann, Brandt, Alvary, Fischer, Elmblad, cond. Seidl.

France, 1100. Lysiart (bar), angered by Adolar's (ten) protestations of love for Euryanthe (sop), wagers that he can seduce her. She has meanwhile given away a family secret about a suicide to the evil Eglantine (mez), who has won her confidence. When Lysiart accuses her of infidelity before the court, she fails to deny it, and the miserable Adolar leads her into the desert to kill her. But when she saves his life, he abandons her instead, to be found by the King (bs) and a hunting-party. Lysiart, who has won Adolar's possessions and estate in the wager, is about to marry Eglantine when Adolar arrives; the plot is disclosed, and he is reunited with Euryanthe.

Also opera on the subject by Carafa (1828).

BIBL: M. Tusa, 'Euryanthe' and Carl Maria von Weber's Dramaturgy of German Opera (Oxford, 1991).

**Eurydice.** See ORPHEUS.

**Eva.** Opera in 3 acts by Foerster; text by the composer, after Gabriela Preissová's drama *Gazdina roba* (The Innkeeper's Daughter). Prod. Prague, N, 1 Jan. 1899.

Slovakia, late 19th cent. Mánek (ten), the son of a wealthy farmer, loves Eva (sop), a poor seamstress. Despite her love for Mánek, Eva agrees to marry Samko (bar), a tailor.

Eva grieves over the loss of her baby. Mánek is by now married, but still thinks only of Eva. They decide to abandon their families and flee.

At the harvest festival, Eva insults Zuzka (mez), and is inconsolable. Mánek's mother arrives, saying that the authorities disapprove of their relationship. Mánek is unconcerned by this, but Eva is greatly troubled and, after seeing a vision of her dead parents and child, hurls herself into the Danube.

**Eva.** Pogner's daughter (sop), lover of Walther, in Wagner's *Die Meistersinger von Nürnberg*.

**Evangelimann, Der** (The Evangelical). Opera in 2 acts by Kienzl; text by the composer, after L. F.

Meissner's story (1894). Prod. Berlin, H, 4 May 1895, with Pierson, Goetz, Sylva, Bulss, Mödlinger, cond. Muck; London, CG, 2 July 1897, with Engle, Schumann-Heink, Van Dyck, Bispham, Pringle, cond. Flon; Chicago, Great Northern T, 3 Nov. 1923, with Mörike, Metzger, Ritter, Zador.

St Othman and Vienna, 1820–50. The opera relates how jealousy leads Johannes (bar) to commit acts which ruin the love of Mathias (ten), his brother, and Martha (sop). The innocent Mathias is sent to prison; during his sentence Martha commits suicide. On his release, Mathias becomes an evangelical preacher; Johannes comes to him and confesses, not recognizing his brother, who absolves him.

**Evans, Anne** (*b* London, 20 Aug. 1941). English soprano. Studied London with Packer and Harford, Geneva and Rome with Ricci. Début Geneva 1967 (Countess Ceprano). London SW/ ENO 1968–77, and from 1986; WNO from 1978; London, CG, from 1986; Bayreuth 1989; New York, M, 1992; also San Francisco, Berlin, Paris, Turin, Zurich, etc. Repertory incl. Donna Anna, Leonore, Violetta, Elisabeth de Valois, Tosca, Elsa, Senta, Brünnhilde, Kundry, Marschallin, Chrysothemis, Milada (*Dalibor*). A musicianly singer whose attractive voice encompasses the lyrical as well as the dramatic. (R)

**Evans, (Sir) Geraint** (*b* Pontypridd, 16 Feb. 1922; *d* 19 Sep. 1992). Welsh baritone. Studied Hamburg with Hermann, Geneva with Carpi. Début London, CG, 1948 (Nightwatchman). Sang there 1949–84. Gly. 1950–61; Milan, S, 1960; Vienna 1961; New York, M, 1964. Created the roles of Flint (*Billy Budd*), Mountjoy (*Gloriana*), Antenor (*Troilus and Cressida*). Other roles include Figaro, Leporello, Beckmesser, Don Pasquale, Falstaff. Possesses a warm voice with a wide range, admirable diction, and lively dramatic skills. Has produced for WNO, and in Chicago and San Francisco. Autobiography, *A Knight at the Opera* (London, 1984). (R)

**Evans, Nancy** (*b* Liverpool, 19 Mar. 1915). English mezzo-soprano. Studied with Teyte. Début London 1938 (Sullivan's *The Rose of Persia*). London, CG, 1939. Joined EOG 1946. Gly. 1946 (sharing title-role in first *Rape of Lucretia* with Ferrier), 1957, 1959–60. Also created Nancy (*Albert Herring*), written specially for her warm, silvery voice. (R)

**Everding, August** (*b* Bottrop, 31 Oct. 1928). German producer and administrator. Studied Bonn and Munich. Director Hamburg 1973–7; Munich 1976. Vienna, San Francisco, Bayreuth (*Flying Dutchman* 1969, *Tristan* 1974). Savonlinna, *Magic Flute*, 1973–89, regularly. First

Warsaw *Ring* 1988–9. A skilful and highly practical producer, with a gift for using many different theatrical conditions to successful effect.

**evirato.** See CASTRATO.

**Ewing, Maria** (*b* Detroit, 27 Mar. 1950). US soprano, later mezzo-soprano. Studied Cleveland with Tourel and Steber. Début Ravinia Festival 1973. New York, M, 1976 (Cherubino); Milan, S, 1978 (Mélisande); Brussels, M, 1980; Salzburg 1980; Paris, O, 1981; Geneva 1982; London, CG, 1988. Gly. regularly from 1978, where she scored a memorable success as Carmen (1985). One of the most attractive and dramatic stage personalities of her generation; her roles include Susanna, the Composer, Zerlina, and Salome. (R).

**Excursions of Mr Brouček, The** (Cz.: *Výlety pana Broučka*). Opera in 2 acts by Janáček; text by the composer, with suggestions, contributions, and amendments by F. Gellner, V. Dyk, and F. S. Procházka, after Svatopluk Čech's novels (1888, 1889). Pt1, *Výlet pana Broučka do měsíce* (Mr Brouček's Excursion to the Moon); Pt2, *Výlet pana Broučka do XV. století* (Mr Brouček's Excursion to the 15th cent.). Prod. Prague, C, 23 Apr. 1920, with Štork, Miřiovská, Jeník, Zítek, Novák, Crhová, Pivoňková, Soběský, Hruška, Lebeda, Novotný, cond. Ostrčil; Edinburgh, King's T, 5 Sep. 1970, by Prague N, with Blachut, Tattermuschová, Žídek, D. Jedlička, Berman, Lermariová, Prochazková, R. Jedlička, Karpíšek, Vonásek, cond. Krombholc.

Prague and the moon. In the first of Mr Brouček's magic excursions, to the moon, he encounters a fantastic aesthetic world: though the work's ostensible purpose is to lampoon his bourgeois complacency and philistinism, he actually emerges in a more sympathetic light than the posturing moon-dwellers. His second excursion is to the 15th cent., where he behaves in a cowardly fashion before being restored to the present.

**exit aria.** See ARIA.

**extravaganza.** An entertainment combining music and drama popular in the late 18th cent., characterized by exotic or fanciful plots, and often aimed at an audience of children. The term has been applied more loosely to works with a generally larger-than-life plot, e.g. Gilbert and Sullivan's *Trial by Jury*.

**Ezio.** Opera in 3 acts by Handel; adapted from Metastasio. Prod. London, H, 15 Jan. 1732, with Strada, Senesino, Bertolli, Bagnolesi; New York, Gate T, 11 May 1959, with Caplan, Cornell, Edgar, Smith, Warwick, cond. Saffir.

Among some two dozen other settings are those by Auletta (1728), Porpora (1728), Hasse (1730), Lampugnani (1736), Jommelli (1741), G. Scarlatti (1744), Gluck (1750), Traetta (1757), J. C. Bach (pasticcio for which he provided only 2 arias (1764)), Gazzaniga (1772), and Mercadante (1827).

**Ezio.** The Roman general (bar) in Verdi's *Attila*.

# F

**Fabri, Annibale Pio** (*b* Bologna, 1697; *d* Lisbon, 12 Aug. 1760). Italian tenor. Studied Bologna with Pistocchi. Début ?1716 Bologna (Bassani's *Alarico, re dei Goti*). Sang all over Italy, 1719–29, in operas by Vivaldi, Vinci, etc. London, H, 1729–31, engaged by Handel; début with great success as Berengario (*Lotario*). Created Emilio (*Partenope*) and Alexander (*Poro*), and sang in *Giulio Cesare*, *Tolomeo*, *Scipione*, and *Rinaldo* (Goffredo, previously sung by women). Returned to Italy, where he sang successfully until 1750. Retired from stage, and was appointed cantor at the royal chapel, Lisbon. One of the first great tenors, much admired by Burney, Rolli, and Handel, who wrote demanding music for him. Mrs Pendarves declared him, with his 'sweet, clear' voice, 'the greatest master of music that ever sang upon the stage'. Composed two oratorios and some vocal exercises.

**Fabrizi, Vincenzo** (*b* Naples, 1764; *d* ?, after 1812). Italian composer. Maestro di cappella at Rome U from 1786. He wrote *c*.15 opere buffe, of which *I due castellani burlati* was most successful: these were performed throughout Europe, including Dresden, Lisbon, London, and Madrid. Other important works include the dramma giocoso *Il viaggiatore sfortunato in amore* and *Il convitato di pietra, ossia Il Don Giovanni*, which capitalized upon a contemporary vogue for operatic representations of the Don Juan legend. Little of Fabrizi's music survives: that which does suggests that he was a competent practictioner of 18th-cent. opera buffa style, though possessing few distinctive characteristics.

SELECT WORKLIST: *I due castellani burlati* (The Two Mocking Castellans) (Livigni; Bologna 1785); *Il viaggiatore sfortunato in amore* (The Traveller Unlucky in Love) (Ballani; Rome 1787); *Il convitato di pietra, ossia Il Don Giovanni* (Lorenzi; Rome 1787).

**Faccio, Franco** (*b* Verona, 8 Mar. 1840; *d* Monza, 21 July 1891). Italian conductor and composer. Studied Milan with Ronchetti-Montevito. His first opera, *I profughi fiamminghi*, led both to accusations of Wagnerism and to his friend Boito's claim in a celebratory ode that he would 'cleanse the altar of Italian opera, now stained like the walls of a brothel'—a remark taken personally by Verdi to Boito's disadvan-

tage. His greatest success was with *Amleto*, which benefits greatly from Boito's libretto and responds with some vivid scenes, but is scarcely able to rise to the demands of the subject. Taking up a career as a conductor, he toured Germany and Scandinavia with Lorini's OC 1867; Milan, T Carcano, 1868; Milan, S, 1869, assistant cond. when he made warm friends with Verdi, chief cond. 1871–90, conducting premières of *Gioconda* (1876), *Otello* (1887), and *Edgar* (1889), as well as many important Italian premières, especially of Wagner. Cond. first English *Otello* (1889) and Italian *Meistersinger* (1889). He would have been chosen by Verdi for the first *Falstaff*, but a syphilitic condition led to his mental decline and death. A pioneering figure as a conductor, he brought new quality to the Scala orchestra and thereby set new standards to which composers could respond. His sister **Chiarina** (1846–1923) was a soprano who sang in Trieste.

WORKLIST: *I profughi fiamminghi* (The Flemish Refugees) (Prague; Milan 1863); *Amleto* (Boito, after Shakespeare; Genoa 1865).

BIBL: R. de Renzis, *Franco Faccio e Verdi* (Milan, 1934).

**Fach** (Ger.: 'division', hence 'subject' or 'speciality'). The term used, strictly in Germany and more loosely internationally, to describe the range of roles that a singer may be expected to perform. Thus Mozart's Susanna, Zerlina, and Despina might be said to fall within the soubrette Fach, the Countess and Pamina within the lyric Fach. The high coloratura Fach includes the Queen of the Night and Zerbinetta. Some singers work closely within a Fach, though it may change as their voice alters; others may have the range and technique to encompass several Fachs. Examples of the latter were Lilli *Lehmann and Lillian *Nordica. See also under voice categories, SOPRANO, MEZZO-SOPRANO, CONTRALTO, TENOR, BARITONE, BASS.

**Fafner.** With Fasolt, one of the two giants (bs) in Wagner's *Das Rheingold*. He has turned into a dragon by *Siegfried*.

**Faggioli, Michelangelo** (*b* Naples, 1666; *d* Naples, 23 Nov. 1733). Italian composer. A lawyer by profession, Faggioli's importance for the history of opera lies in his one opera buffa *La Cilla* (Tullio; Naples 1706, lost), the first musical comedy to a libretto written in Neapolitan dialect. Such was its success that Faggioli's opera initiated a tradition of such works culminating in the music of Logroscino and Paisiello.

**Fair Maid of Perth.** See JOLIE FILLE DE PERTH, LA.

**Fairy Queen, The.** Semi-opera in a prologue and 5 acts by Purcell; text an anon. adaptation (?by

Elkanah Settle) of Shakespeare's comedy *A Midsummer Night's Dream* (1595–6). Prod. London, Dorset Garden T, Apr. 1692; next stage prod. Cambridge 10 Feb. 1920; San Francisco, Palace of Legion of Honor, 30 Apr. 1932. The score was lost by Oct. 1701, when an advertisement offered 20 gns. for its recovery; it was found in 1901 by J. S. Shedlock (not quite complete) in the library of the London RAM. The first work staged by London CG Opera on its formation in 1946.

**fait historique** (Fr.: 'historical act'). The name sometimes given to ephemeral opéras-comiques written at the time of the French Revolution and based on historical incidents, e.g. several on the recapture of Toulon in 1794 and the death of the young revolutionary hero Joseph Barra. The term was also used for more substantial works which took their subject matter from contemporary life, e.g. Gaveaux's *Léonore* (1798) and Cherubini's *Les deux journées* (1800).

**Falco, Michele** (*b* Naples, *c.*1688; *d* after 1732). Italian composer. Studied Naples, probably with Fago, then held various positions as maestro di cappella there. Although most of his music is lost, he was one of the earliest composers of opera buffa and played a significant role in its establishment in Naples. His most important works include *Lo Lollo pisciaportelle* and *Lo mbruglio d'ammore*.

SELECT WORKLIST: *Lo Lollo pisciaportelle* (Orilia; Naples 1709); *Lo mbruglio d'ammore* (The Love Tangle) (Piscopo; Naples 1717).

**Falcon, Cornélie** (*b* Paris, 28 Jan. 1814; *d* Paris, 25 Feb 1897). French soprano. Studied Paris with Henri and Pellegrini, later with Bordogni and Nourrit. Début Paris, O, 1832 (Alice, *Robert le diable*). Her career lasted only until 1838, when she lost her voice through overwork, and had to retire; an attempted come-back in 1840 ended badly. Created Mme Anckarstroem (*Gustave III*), Rachel (*La Juive*), Valentine (*Les Huguenots*), and many other roles; also a famous Donna Anna and Julie (*La vestale*). She possessed a full, resonant voice, and, like Nourrit, exceptional dramatic talent (they were together responsible for improving the artistic standing of the Opéra). Her name, then synonymous with the dramatic soprano parts in which she was unapproachable, still survives as a description of her voice type.

BIBL: C. Bouvet, *Cornélie Falcon* (Paris, 1927).

**Fall, Leo** (*b* Olomouc, 2 Feb. 1873; *d* Vienna, 16 Sep. 1925). Austrian composer. Studied with his father, and in Vienna with J. N. and Robert Fuchs. Conducted operetta in Hamburg 1895, Cologne, and Berlin; settled in Vienna 1906. With Lehár and Kálmán, he was one of the most popular Viennese operetta composers of his time.

His early operas were unsuccessful, but he found his métier in operetta with his light melodic grace and his neatness of invention. He visited London 1911–12.

SELECT WORKLIST: *Der Rebell* (Bernauer & Welisch); Vienna 1905; rev. as *Der liebe Augustin*, Berlin 1912); *Der fidele Bauer* (The Faithful Peasant) (Leon; Mannheim 1907); *Die Dollarprinzessin* (Willner & Grünbaum; Vienna 1907); *The Eternal Waltz* (Hurgon; London 1911); *Die Rose von Stambul* (Brammer & Grünwald; Vienna 1916); *Madame Pompadour* (Schanzer & Welsch; Berlin 1922).

BIBL: W. Zimmerli, *Leo Fall* (Zurich, 1957).

**Falla, Manuel de** (*b* Cádiz, 23 Nov. 1876; *d* Alta Gracia, 14 Nov. 1946). Spanish composer. Studied Cádiz with Odero and Broca. As a young man, he wrote a number of zarzuelas. *La vida breve* is an early work, with a weak plot; but already Falla's authentic voice is to be heard behind the different manners he adopts. Despite unevenness, it is a work of genuine dramatic power. *El retablo de Maese Pedro* handles an incident in *Don Quixote*, turning Falla's understanding of contrasting styles to skilful use in a piece of puppet music theatre treating different levels of reality and illusion. The forces are miniature, but there is a disciplined refinement and intensity now present in Falla's mature style. *L'Atlántida* is closer to dramatic oratorio than opera, and Falla did not wish it to be staged; but it is his most ambitious dramatic work, and stage performances have proved effective.

SELECT WORKLIST: *La vida breve* (The Short Life) (Shaw; Nice 1913); *El retablo de Maese Pedro* (Master Peter's Puppet Show) (Falla, after Cervantes; Paris 1923); *L'Atlántida* (Falla, after Verdaguer; unfin., completed Halffter; Milan 1962).

CATALOGUE: R. Crichton, *Manuel de Falla: Descriptive Catalogue of his Works* (London, 1976).

BIBL: E. Franco, *Manuel de Falla* (Madrid, 1977).

**falsetto** (It., lit. dim. of *falso*: 'false, altered'; or poss. from Lat. *fauces*: 'throat'). An artificial method of voice production employed by male singers, using only a partial vibration of the vocal chords. Normally used in opera only for special effects, whether for extreme refinement and softness of tone, as perhaps at the end of the romance in *Les pêcheurs de perles*, or satirically, as usually by Falstaff imitating Mistress Ford allegedly sighing, 'Io son di Sir John Falstaff'. There is a long passage in Weber's *Die drei Pintos*. In the 19th cent. a singer who specialized in a version of falsetto singing was sometimes known in Italy as *falsettone*, or *falsetto rinforzato*. The technique, when highly developed, now normally refers to *counter-tenor.

**falso canone** (It.: 'false canon'). A kind of ensemble, invariably slow, popular in 19th-cent. Italian opera. So called because it took the form of a canon or round, but abandoned the pattern after the entry of the last voice.

**Falstaff.** Opera in 3 acts by Verdi; text by Boito, after Shakespeare's comedy *The Merry Wives of Windsor* (1600–1) and *Henry IV* (Pt1, 1597; Pt2, 1598). Prod. Milan, S, 9 Feb. 1893, with Zilli, A. Stehle, Guerrini, Pasqua, Garbin, Maurel, A. Pini-Corsi, Paroli, Pelagalli-Rossetti, Arimondi, cond. Mascheroni; London, CG, 19 May 1894, with Zilli, Olghina, Ravogli, Kitzu, Beduschi, Pessina, Pini-Corsi, cond. Mancinelli; New York, M, 4 Feb. 1895, with Eames, De Lussan, Scalchi, De Vigne, Russitano, Maurel, Campanari, cond. Mancinelli.

Windsor, in the reign of Henry IV. In the Garter Inn, Sir John Falstaff (bar) defends himself and his retainers Bardolfo (ten) and Pistola (bar) against the accusations of Dr Cajus (ten), and tells his men he intends to seduce Ford's and Page's wives. When they refuse to bear letters to the women, on grounds of honour, Falstaff lectures them on the subject. In Ford's garden, Alice Ford (sop) and Meg Page (mez) read identical letters from Falstaff, with Nannetta Ford (sop) and Mistress Quickly (mez) listening. While Nannetta exchanges a quick kiss with Fenton (ten), the women resolve to teach Falstaff a lesson, and Ford (bar) another one for his jealousy.

In the Garter, Mistress Quickly encourages Falstaff to pursue his wooing of Alice. Ford, disguised as 'Fontana', pretends to be asking Falstaff to seduce Alice so as to smooth his own path, and, alone, gives vent to his jealousy. In Ford's house, Alice receives Falstaff. When Meg arrives, Falstaff hides in a laundry basket. The others arrive and begin searching the house for Falstaff. When he is tipped out of the window into the Thames, Ford realizes the unworthiness of his suspicions.

Outside the Garter, Falstaff is gloomily recovering from his ordeal when Mistress Quickly comes with the promise of a new assignation with Alice, this time in Windsor Forest at midnight. By Herne's Oak in the Forest, Nannetta invokes 'spirits', in fact the people and children of Windsor, who pinch and prick Falstaff as they enact the masque of the Fairy Queen with various marriages. When the plot is revealed, it emerges that in the confusion Fenton has actually married Nannetta. All is forgiven, and in a final fugue Falstaff suggests that all have been duped and that all the world's a jest.

BIBL: J. Hepokoski, *Giuseppe Verdi: 'Falstaff'* (Cambridge, 1983).

**Fancelli, Giuseppe** (*b* Florence, ?24 Nov. 1833; *d* Florence, Dec. 1887). Italian tenor. Début Milan, S (Fisherman, *Guglielmo Tell*). London, CG, 1866–8, 1870–2. Milan, S, 1866, and in 1872 first Italian Radamès (Verdi commented, 'a good voice but nothing else', and went so far as to assault him in rehearsal). Extraordinarily gifted vocally (with a top C unsurpassed even by Tamberlik or Caruso, according to Klein), he was musically illiterate, and declared he could not sing and act at the same time.

**Fanciulla del West, La** (The Girl of the [Golden] West). Opera in 3 acts by Puccini; text by Civinini and Zangarini, after Belasco's drama *The Girl of the Golden West* (1905). Prod. New York, M, 10 Dec. 1910, with Destinn, Caruso, Amato, cond. Toscanini; London, CG, 29 May 1911, with Destinn, Bassi, Gilly, cond. Campanini; Rome, C, 12 June 1911, with Burzio, Bassi, Amato, cond. Toscanini.

The Californian Gold Rush of 1850. Miners are playing cards at the Polka Saloon and are about to lynch one of their number for cheating when the Sheriff, Jack Rance (bar), intervenes. Rance is in love with Minnie (sop) and proposes to her. She refuses, reminding him that he already has a wife. A stranger, Dick Johnson (ten), arrives and Minnie recognizes him as the man she once knew and hoped to meet again. He is in fact the bandit leader Ramerrez, and he and his band have a plan to steal the miners' gold which Minnie guards. She shows him where it is hidden, but Johnson is so attracted to her that he ignores a signal from his band to attack.

In her cabin, Minnie receives Johnson and the couple confess their love. Outside, a blizzard rages and Johnson agrees to stay. Suddenly there is a knock at the door. Minnie hides Johnson and lets Rance and his group in. Rance shows her a photograph and Minnie realizes that Johnson is Ramerrez. When they are alone Johnson admits it and Minnie orders him to leave. A shot is heard; Johnson is injured and Minnie drags him back in, hiding him in the loft. Rance enters and finds the bandit when drops of blood fall from the loft; but Minnie persuades him to play poker, with Johnson as the prize. She wins by cheating and Rance has to abandon his pursuit.

A few days later, the miners capture Ramerrez and propose to hang him. As the noose is slipped over his head, Minnie arrives and shields him, holding the miners at bay with a gun. She persuades them to release Johnson and the couple leave together.

**'Fanget an!'** Walther's (ten) trial song in Act I of Wagner's *Die Meistersinger*, taking the marker Beckmesser's instruction to begin as words describing how spring begins everything anew.

**Faninal.** A wealthy, newly ennobled merchant (bar), Sophie's father, in R. Strauss's *Der Rosenkavalier*.

**Faramondo.** Opera in 3 acts by Handel; text adapted from Zeno. Prod. London, H, 3 Jan. 1738, with Francescina, Lucchesina, Caffarelli, Montagna, Lottini, Merighi. Unsuccessful; ran 8 nights only. Zeno's text was first set by Pollarolo (1699), later by Porpora (1719); Handel set the libretto as radically revised for Gasparini (1720).

**farewell.** The name commonly given to the last performance in public of a favourite artist. Some singers, especially sopranos, have announced their farewell seasons years before taking their actual leave of the stage—Grisi is a notable example. Melba's farewell, including her speech, at London, CG, in 1926, was recorded by HMV. German provincial houses give 'Abschied' performances for a favourite singer when he or she leaves to take up an engagement in another theatre.

**Farinelli** (orig. Carlo Broschi) (*b* Andria, 24 Jan. 1705; *d* Bologna, 15 July 1782). Italian castrato soprano. Studied with Porpora. Début Naples 1720, in his teacher's serenata *Angelica e Medoro*. Two years later he won a famous competition of skill with a trumpeter. After triumphs in Italy and Vienna, he had another brilliant contest in 1727 with Bernacchi; he lost, but learned much from the other singer. By now he was famous all over Europe for his astonishing vocal feats; however, accused in 1731 by Charles VI of seeking more to 'surprise' than 'please', he modified his style. Engaged London 1734–7 as the star attraction in Porpora's rival company to Handel's, to ecstatic acclaim. Senesino, moved by his beautiful singing, once embraced him during an opera; women frequently fainted at his performances. In Madrid, 1737, he sang to the melancholic Philip V, with such success that he was asked to stay at the court for 50,000F a year. This led to his retirement from the stage. For 25 years, he sang the same four songs every night to Philip (who rarely washed or changed his linen). Under his successor Ferdinand VI, Farinelli established an Italian Opera in Madrid, importing the best singers of the day, and his friend the librettist Metastasio. Also active in public works and as a diplomat. In 1750 he received the Order of Calatrava. On Charles III's accession, 1759, he left Spain for political reasons, with a large pension, and retired to Bologna, where he was visited by Burney in 1770. A highly cultured and agreeable man, and one of the greatest singers of all time.

Operas on him are by Barnett (1839), Espin y Guillén (1854), Zumpe (1886), and Bretón (1901).

BIBL: J. Desastre, *Carlo Broschi* (Zurich, 1903).

**Farinelli, Giuseppe** (orig. Giuseppe Francesco Finco) (*b* Este, 7 May 1769; *d* Trieste, 12 Dec. 1836). Italian composer. A pupil of Martinelli and Fago, he lived first in Naples and Turin as an independent opera composer before moving to Trieste in 1817 to become conductor at the T Grande and, from 1819, maestro di cappella of the cathedral. One of the most popular theatrical composers around the turn of the nineteenth century; his music owes much to the model of Cimarosa, and is firmly rooted in the late 18th-cent. tradition of opera buffa. Farinelli's decline came only with the rise to prominence of Rossini after 1812. At first the two composers were performed side by side—in 1813 Farinelli's *Il matrimonio per concorso* shared the season with Rossini's *L'italiana in Algeri* at the T San Moisè in Venice—but the greater talent of Rossini was soon perceived by audiences and, in an action reminiscent of that of the younger composer, Farinelli retired from opera composition in 1817.

His first opera, *Il dottorato di Pulcinella*, established the formula by which he was to achieve his success: a sparkling comedy making much use of Neapolitan dialect, its wistful humour and playful instruction is matched by an engaging and attractive musical style. It was followed three years later by *L'uomo indolente* and a string of popular triumphs which included *Teresa e Claudio, Chi la dura la vince, I riti d'Efeso, L'amico dell'uomo*, and *La contadina bizzarra*. Among Farinelli's less popular serious operas are *Pamela* and *Idomeneo*.

Although Farinelli enjoyed a tremendous following within Italy, and wrote for many of the major houses, his music was never particularly popular abroad, except in London, where both *Teresa e Claudio* and *I riti d'Efeso* were given.

SELECT WORKLIST: *Il dottorato di Pulcinella* (Pulcinella's Doctorate) (?; Naples 1792); *L'uomo indolente* (The Sluggard) (Palomba; Naples 1795); *Teresa e Claudio* (Foppa; Venice 1801); *Chi la dura la vince* (Who Endures, Wins) (Rossi, after Goldoni; Rome ?1803); *Pamela* (Rossi, after Richardson; Venice 1803); *I riti d'Efeso* (The Rites of Ephesus) (Rossi; Venice 1804); *L'amico dell'uomo* (The Man's Friend) (Foppa; Florence 1807); *La contadina bizzarra* (The Weird Peasant Girl) (Romanelli, after Livigni; Milan 1810); *Idomeneo* (Rossi; Venice 1812); *Il matrimonio per concorso* (The Marriage Contest) (Foppa; Venice 1813).

**Farlaf.** A Varangian warrior (bs) in Glinka's *Ruslan and Lyudmila*.

**Farrar, Geraldine** (*b* Melrose, MA, 28 Feb. 1882; *d* Ridgefield, CT, 11 Mar. 1967). US soprano.

Studied Boston with Mrs Long, New York with Thursby; later with Lilli Lehmann. Début Berlin, H, 1901 (Marguerite). New York, M, 1906–22; then retired. Created Goosegirl (*Königskinder*), Madame Sans-Gêne, and Suor Angelica; an outstanding Butterfly and Manon Lescaut (Puccini) and Manon (Massenet). Beautiful and a committed singer and actress, she was idolized by the public. Autobiography, *Such Sweet Compulsion* (New York, 1938). (R)

**farsa** (sometimes also **farsetta**) (It.: 'farce'). In the 18th cent. the term was usually applied to an *intermezzo, or sometimes afterpiece, in which farcical, sometimes coarse elements were particularly prevalent. Up to *c*.1780 the two terms were used interchangeably; following the establishment of opera buffa as a separate genre, the farsa came to be recognized as a 1- or 2-act comic work, but with spoken dialogue, effectively an Italian *opéra-comique. Around the turn of the 19th cent. the term was more broadly used for 1-act comic operas, of which the classic example is Rossini's *La cambiale di matrimonio*.

**Fasano, Renato** (*b* Naples, 21 Aug. 1902; *d* Rome, 3 Aug. 1979). Italian composer and conductor. Studied Naples. Founded the Collegium Musicum Italicum in Rome (1941), later reformed as the Virtuosi di Roma; and in 1957 the Piccolo T Musicale Italiano, which has appeared throughout Europe and the USA, giving performances of operas by Galuppi, Cimarosa, Paisiello, Pergolesi, Rossini, etc. (R)

**Fasolt.** With Fafner, one of the two giants (bs) in Wagner's *Das Rheingold*.

**Fassbaender, Brigitte** (*b* Berlin, 3 July 1939). German mezzo-soprano. Studied Nuremberg with her father, Willi *Domgraf-Fassbaender. Début Munich, Bavarian SO, 1961 (Nicklausse). London, CG, 1971; New York, M, 1974; also San Francisco, Salzburg, Paris, etc. Roles include Cherubino, Octavian, Amneris, Brangäne, Charlotte, Carmen. Vivid on stage, with a warm, burnished tone; also a fine Lieder singer. (R)

**Fatima.** Reiza's attendant (mez) in Weber's *Oberon*.

**Fauré, Gabriel** (*b* Pamiers, 12 May 1845; *d* Paris, 4 Nov. 1924). French composer, teacher, pianist, and organist. Studied Paris with Saint-Saëns; held various positions as an organist and teacher, before becoming director of the Paris Cons. in 1905. Though he had a consummate understanding of the potential of the human voice, Fauré composed only one true opera, *Pénélope*. This was preceded by incidental music

for several plays, including Dumas's *Caligula* (1888), Haraucourt's *Shylock* (1889), and Maeterlinck's *Pelléas et Mélisande* (1898). His most extensive dramatic work was the grand cantata *Prométhée*: as only some of the characters participate in the stage action it is scarcely an opera, though Fauré's conception of the work is at times more operatic than merely choral. In particular, it was both the first sustained adaptation of Fauré's highly individual style to the needs of dramatic action, as well as the clearest example to date of Wagner's influence on his music. *Pénélope* shows similar traits, but also has an affinity with Saint-Saëns's *Samson et Dalila* (1877) and Debussy's *Pelléas et Mélisande* (1892). Written at the request of the dramatic soprano Lucienne *Bréval, who also suggested the subject of the faithful Penelope's vigil for her husband Odysseus. Fauré's treatment of her anguish and his insight into her isolation and despair result in the creation of an operatic character worthy to stand beside the greatest of Rameau's and Gluck's tragic heroines. Though couched in Fauré's inimitably dream-like style, it also owes much to the conventions of Romantic opera, with its opening spinning chorus and final 'Gloire à Zeus'. In its seamless dramatic progression and masterful construction it is the ultimate proof that Fauré was not, as so many claim, merely a musical miniaturist: perhaps more importantly, it is in this work that he comes closest to exemplifying the true connection between the French and Greek genius—a spare, brilliant quality of thought rather than the Hellenistic languor on which the comparison generally rests.

WORKLIST: *Prométhée* (Lorrain & Hérold; Béziers 1900); *Pénélope* (Fauchois; Monte Carlo 1913).

BIBL: R. Orledge, *Gabriel Fauré* (London, 1979).

**Faure, Jean-Baptiste** (*b* Moulins, 15 Jan. 1830; *d* Paris, 9 Nov. 1914). French baritone. Studied Paris with L. Ponchard and Moreau-Sainti. Début Paris, OC, 1852 (Pygmalion, Massé's *Galathée*). Sang there until 1859. London, CG, between 1860 and 1876; also sang at London, DL and HM. Paris, O, 1861–9, 1872–6, creating Nelusko (*L'Africaine*), Posa, Hamlet, among other roles. Vienna 1878; Marseilles, Vichy, 1886, when he published his treatise *La voix et le chant*. Famous as Alphonse (*Favorite*), Guillaume Tell, Don Giovanni, Gounod's Méphistophélès. Taught at Paris Cons. 1857–60. An exceptional performer, with superb vocal control and great dramatic insight. Painted twice by Manet. (R) His wife, the soprano **Caroline Lefèbvre** (1828–1905), sang in Paris at the OC 1849–59, then at L.

**Faust.** A wandering conjuror who lived in Germany *c*.1488–1541. The legend that he had sold his soul to the Devil, in exchange for a fixed period of renewed youth and other favours, was first set down and published in the *Historia von D. Johann Fausten* (compiled Johann Spies, 1587). This inspired various writers, especially Marlowe in his *Tragicall History of Dr Faustus* (1588–93) and Goethe in his *Faust* (pt 1, 1808; pt 2, 1832).

Operas deriving from Goethe's *Faust* are as follows: Josef Strauss (*c*.1814); Lickl (1815); Seyfried (1816); I. Walter (1819); Horn (1825); Béancourt (1827); Berlioz (non-operatic, but sometimes staged: *La damnation de Faust*, 1846, incorporating and revising *Huit scènes de Faust*, 1829); Lindpaintner (1831); Bertin (1831); Pellaert (1834); Rietz (1836); Gregoir (1847); Hennebert (1853); Lutz (1855); Gounod (1859); Boito (*Mefistofele*, 1868); Hervé (1869); Lassen (1876); Zöllner (1887); Souchay (1940); Wachmann (n.d.).

Operas on Faust deriving from sources other than Goethe are by Hanke (*La ceinture du Docteur Faust*, *c*.1796), I. Walter (*Dr Faust*, 1797), Spohr (1816), W. Müller (*Dr Fausts Mantel*, 1817), Saint-Lubin (*Le cousin du Docteur Faust*, 1829), Zöllner (1887), Busoni (*Doktor Faust*, 1925), Reutter (*Dr Johannes Faust*, 1936, rev. 1955), Engelmann (*Dr Fausts Höllenfahrt*, 1951), J. Berg (*Dr Johannes Faust*, 1967), Boehmer (*Docteur Faustus*, 1985), Gifford (*Regarding Faustus*, 1988), Manzoni (*Doktor Faustus*, 1988). Also operetta by F. E. Barbier (*Le faux Faust*, 1858).

BIBL: W. Grim, *The Faust Legend in Music and Literature* (Lewiston, NY, 1989).

**Faust.** Opera in 5 acts by Gounod; text by Barbier and Carré, after Goethe's poem (pt 1, 1808; pt 2, 1832). Prod. Paris, L, 19 Mar. 1859, with M. Carvalho, Faivre, Barbot, Reynald, Balanque, cond. Deloffre: 500 perfs. by 1887, 1,000 by 1894, 2,000 by 1934; London, HM, 11 June 1863, with Tietjens, Trebelli, Giuglini, Santley, Gassier, and in every CG season 1863–1911; Philadelphia, AM, 18 Nov. 1863, with Frederici, Gross, Himmer, Steinecke, Graffi. Inaugural opera at New York, M, 22 Oct. 1883, with C. Nilsson, Scalchi, I. Campanini, Del Puente, Novara, cond. Vianesi.

Germany, 16th cent. Faust (ten) makes a bargain with Méphistophélès (bs): in return for eternal youth and the beautiful Marguerite (sop), he promises Méphistophélès his soul. Valentin (bar), Marguerite's brother, entrusts the care of his sister to the faithful Siebel (mez) while he goes off to the wars. On his return he finds Marguerite has been betrayed by Faust, and challenging him to a duel, is killed. Marguerite, in prison for killing her baby, refuses to go with Faust and Méphistophélès; as she ascends to Heaven, Faust is dragged down to Hell by Méphistophélès.

One of the most successful operas ever written, with translations into at least 25 languages. The opera inspired a poem *Fausto* by Estanislao del Campo.

**Faustina.** See BORDONI, FAUSTINA.

**Faustini, Giovanni** (*b* ?Venice, *c*.1619; *d* Venice, 19 Dec. 1651). Italian librettist and theatre manager. Wrote 11 librettos for the Venetian public opera houses 1642–51, ten set by Cavalli. As a theatre manager, Faustini understood the importance of a plot which would appeal to the widest possible audience and ensure financial success. His librettos, of which the most effective were *Egisto* (1643), *Ormindo* (1644) and *Calisto* (1652), provided considerable scope for musical characterization and are masterpieces of dramatic design. Much of Cavalli's popular success derived from the skilful construction of Faustini's librettos: their artistic collaboration, still generally underrated, ranks with that of Monteverdi-Busenello and Lully-Quinault.

**Favart, Charles Simon** (*b* Paris, 13 Nov. 1710; *d* Belleville, 12 Mar. 1792). French librettist and impresario. Stage manager *Opéra-Comique, then summoned to Brussels by the Maréchal de Saxe to organize a theatre for the troops in Flanders. Succeeded Monet as director of the Opéra-Comique, 1758, holding the position until his retirement in 1769. In 1871 the Opéra-Comique assumed the name Salle Favart. He wrote more than 150 librettos for different composers, among them Grétry, Philidor, and Gluck.

The first important French comic librettist, Favart began his career writing *vaudevilles, *drames forains, and other pieces, developing in the 1750s librettos with a continuous story, invented or at least developed from an original by himself, into which songs were fitted: this was the *comédie mêlée d'ariettes. His parodies were witty and on the whole good-natured, even constructive: his *Arlequin-Dardanus* (1739), parodying Rameau's *Dardanus* (1739), so impressed the librettist, La Bruère, that the text was revised for the next production. His shafts pierced not only fellow librettists but composers, singers, and the stage management of the Opéra-Comique. In the 1760s he turned to adaptations, in an attempt to fashion an indigenous French comic opera in the wake of the *Guerre des Bouffons. He gave the lighthearted plots a tinge of sentimentality, especially in the female lead (the *ingénue). These librettos were characterized by their nimble pace and their ingenious tangle of intrigue to which *couplets add a point of brief repose.

Favart's pastoral comedies include *Le caprice amoureux, ou Ninette à la cour*, possibly set by Duni (1755), a piece which typifies his ability to turn a parody (of Goldoni) into a work that develops the comic genre: there are in it anticipations of many subsequent ingénues in Ninette (e.g. Zerlina) and even of whole situations, and other of Favart's librettos contain suggestions of characters that were to become famous through later comic operas. *Soliman II, ou Les trois sultanes* (set by Gilbert, 1761) is an early example of the 'Turkish' plots of Mozart's *Entführung* and other works. Favart was also largely responsible for a new naturalistic treatment of peasants and rustic life, often sentimental but more accurately observed and more respectful of simple people. He helped substantially to make possible the realism of later opéra-comique.

BIBL: A. Favart and H. Dumolard, eds., *C. S. Favart: mémoires et correspondance littéraire* (Paris, 1808).

**Favart, Marie** (*b* Avignon, 15 June 1727; *d* Paris, 21 Apr. 1772). French soprano, actress, and dramatist. Wife of above. Début as 'Mme Chantilly' Paris, OC, 1745, in *Les fêtes publiques* (for the marriage of the Dauphin) by Favart, whom she married that year. Together they performed in Brussels until 1747, when they were forced to flee by their patron, the Maréchal de Saxe, who was angered by Marie's refusal to be his mistress. She sang at Paris, CI, 1749, then 1751 until her final illness in 1771. Celebrated in title-role of *La serva padrona*, and in many works with librettos by her husband, including *Les amours de Bastien et Bastienne*. Though not an outstanding singer, she was a gifted and versatile actress, possessing 'une gaieté franche et naturelle' (Favart); her truthful impersonations included a revolutionary realism in costume. She also collaborated with her husband in several works. Operetta on her life, *Mme Favart*, by Offenbach (1878).

Their son **Charles Nicolas Justin Favart** (1749–1806) was an actor and dramatist, who performed at Paris, CI, 1779–96, and wrote several vaudevilles. His son, **Antoine Pierre Charles Favart** (1780–1867) also wrote vaudevilles, including *La jeunesse de Favart*.

BIBL: A. Pougin, *Madame Favart* (Paris, 1912).

**Favero, Mafalda** (*b* Ferrara, 6 Jan. 1903; *d* Milan, 3 Sep. 1981). Italian soprano. Studied Bologna with Vezzani. Début (under name of Maria Bianchi) Cremona 1926 (Lola). Official début Parma 1927 (Liù). Milan, S, 1929–42, 1945–50; London, CG, 1937, 1939; New York, M, 1938. Created roles in Mascagni's *Pinotta*, Wolf-Ferrari's *Il campanello* and *La dama boba*. An expressive singer and actress, famous as Manon, Mimì, and Adriana Lecouvreur. (R)

**favola per musica** (It.: 'tale (lit., fable) for music'); sometimes **favola in musica**. One of the earliest designations used for the 17th-cent. Italian opera libretto, such works usually being of mythological or legendary character. The term **favola in musica** was usually preferred by the composer, as in Monteverdi's *Orfeo*, which is thus described.

**Favorite, La** (The Favourite). Opera in 4 acts by Donizetti; text by Royer and Vaëz, after Baculard d'Arnaud's drama *Le comte de Comminges* (1764) and other material. Prod. Paris, O, 2 Dec. 1840, with R. Stoltz, Elian, Duprez, Barroilhet, N. Levasseur, Wartel, cond. Habeneck; New Orleans, T d'Orléans, 9 Feb. 1843, with Lagier, Allard, Victor, Blès, cond. E. Prévost; London, DL, 18 Oct. 1843. Originally to be entitled *L'ange de Nisida*, it has been given also as *Dalila*, *Leonora di Guzman*, and *Riccardo e Matilda*.

Spain; San Giacomo and Seville, 1340. The novice Fernand (ten), though warned by the Prior Balthazar (b), abandons the monastic life because of his love for the mysterious Léonor (mez), who is actually the mistress of King Alphonse of Castille (bar). He resolves to win her through deeds of chivalry. Léonor begs for release from the King; Alphonse himself is excommunicated for abandoning his wife. Fernand, who has gained the King's favour, asks for Léonor's hand as a reward and the King agrees. The ceremony proceeds, but when Fernand learns of Léonor's relationship with the King, he is furious and storms out. Léonor seeks out Fernand, who has once again assumed his monk's habit, to beg his forgiveness. He pardons her but, before they can leave together, she dies in his arms.

**Federici, Vincenzo** (*b* Pesaro, 1764; *d* Milan, 26 Sep. 1826). Italian composer. Active in London from *c*.1780 where he wrote *L'usurpator innocente*; *c*.1790–1800 maestro al cembalo and composer at the H. Returned to Italy *c*.1802, writing operas for Milan and Turin, including *Castore e Polluce*. Joined Milan Cons. 1808; acting director 1825. An important figure in late 18th-cent. English musical life who is mentioned by many contemporary writers, usually unflatteringly, including Da Ponte, with whom he collaborated.

SELECT WORKLIST: *L'usurpator innocente* (?, after Metastasio; London 1790); *Castore e Polluce* (Romanelli; Milan 1803).

**Fedora.** Opera in 3 acts by Giordano; text by Colautti, after Sardou's drama (1882). Prod. Milan, L, 17 Nov. 1898, with Bellincioni, Caruso, cond. Giordano; London, CG, 5 Nov. 1906, with Giachetti, Zenatello, cond. Mugnone; New York, M, 5 Dec. 1906, with Cavalieri, Caruso, cond. Vigna.

St Petersburg, Paris, and Switzerland, late 19th cent. The story of the tragic love of Count

Loris Ipanov (ten), a Russian nihilist, for the Princess Fedora Romanov (sop).

**Fedra.** 1. Opera in 3 acts by Pizzetti; text by D'Annunzio. Prod. Milan, S, 20 Mar. 1915, with Krusceniski, Anitua, Di Giovanni, Grandini, cond. Marinuzzi.

The opera, based on Euripides' *Hippolytus*, tells how Phaedra (sop), the wife of Theseus (bar), falls in love with her stepson Hippolytus (ten) and when rejected by him hangs herself, leaving behind a letter falsely accusing Hippolytus of dishonouring her.

2. Opera in 1 act by Romano Romani; text by Alfredo Lenozoni. Prod. Rome, TC, 3 Apr. 1915, with Raisa; London, CG, 18 June 1931, with R. Ponselle.

Other operas on the subject by Lemoyne (1786), Paisiello (1788), Niccolini (1803), Orlandi (1820), Mayr (1821), Lord Burghersh (1824). The Phaedra legend forms the basis of the plot of Lully's *Thésée* (1675) and Rameau's *Hippolyte et Aricie* (1733).

**Feen, Die** (The Fairies). Opera in 3 acts by Wagner; text by the composer after Gozzi's comedy *La donna serpente* (1762), first used in Himmel's *Die Sylphen* (1806; the first German Gozzi setting). Prod. Munich 29 June 1888, cond. Fischer; Birmingham 17 May 1969, cond. Lee.

Pursuing a doe, Prince Arindal (ten) plunges into a river and awakes in the castle of the fairy Ada (sop). They fall in love, and marry on condition that for eight years he shall not ask her origin. At the end of this time he does ask the question, and she and her castle and her children vanish; Arindal is cast back into the world of humans. Meanwhile his father has died and the city is threatened with invasion. Arindal does not know that Ada is subject to the fairies' insistence that she must remain one of them, and may only shed her immortality for Arindal by surviving a set of ordeals; she must lay dreadful tests upon him, and he must endure them without cursing her. Arindal is driven beyond his endurance and does curse her; and when the truth is revealed he goes mad when she turns to stone. But with the help of the magician Groma, Arindal pursues her into the underworld and overcomes the spirits who defend her; with his lyre, he melts the cold stone and wins back his bride. She must remain a fairy, but he can now join her in her realm.

Wagner's first complete opera, comp. 1833–4. The first production was rehearsed by Richard Strauss, who was to have conducted it; but Perfall, the Munich Intendant, decided that 'so important a novelty cannot be left to the third conductor'. The incident precipitated Strauss's departure from Munich in the following year.

Casella used the same source for his *La donna serpente* (1932).

**Feind, Barthold** (pseud. Aristobulos Eutropius or Wahrmund) (*b* Hamburg, 1678; *d* Hamburg, 15 Oct. 1721). German librettist. Practised as a lawyer in Hamburg, where his first libretto, *Die römische Unruhe oder Die edelmüthige Octavia*, appeared in Keiser's setting in 1705. He led a tempestuous life: for a while he was banned from Hamburg for seditious libel and while visiting Denmark in 1717 was imprisoned for his pro-Swedish sentiments. His librettos, which were set principally by Keiser (*Masagniello furioso oder Die neapolitanische Fischerempörung*, 1706 and *La costanza sforzata, die gezwungene Beständigkeit*, 1706) and Graupner (*L'amore ammalato, die kranckende Liebe*, 1708), follow the Hamburg model of Bostel and Postel in mixing German and Italian text. However, Feind is noted for his greater development of comic elements and characters, particularly in the direction of satire and parody: some of his style was assimilated into the earliest *Singspiel librettos.

**Fel, Marie** (*b* Bordeaux, 24 Oct. 1713; *d* Chaillot, 9 Feb. 1794). French soprano. Studied Paris with Van Loo. Début Paris, O, 1734 (Vénus, La Coste's *Philomèle*). Sang there until 1758, creating numerous roles in operas by Rameau (e.g. *Castor et Pollux*, *Dardanus*, *Les Indes galantes*), Lully, Campra, Boismortier, Mondonville; also, particularly successfully, Colette in Rousseau's *Le devin du village*. An intelligent and graceful singer; her voice was clear, even, and supple. Voltaire found her artistry 'séduisante', the critic Friedrich Melchior Grimm fell in love with her, and she was the mistress of the librettist Cahusac, then the painter La Tour. On her retirement she was replaced at the Opéra by her pupil, Sophie Arnould. Her brother **Antoine** (1694–1771) sang at the Opéra, and was also a composer.

**Feldlager in Schlesien.** See ÉTOILE DU NORD, L'.

**Felsenstein, Walter** (*b* Vienna, 30 May 1901; *d* Berlin, 8 Oct. 1975). Austrian producer. Studied Graz, and Vienna as actor. Début Lübeck 1923. Chief producer Basle 1927; producer Freiburg 1929–32; chief producer Cologne 1932–4; Frankfurt 1934, then banned by Nazis and worked largely freelance and in drama. Director Berlin, K, 1947–75. Here he found the conditions to develop his view of opera as music expressing a complete dramatic idea, with singing actors trained individually and as a closely knit ensemble. His 'realistic music theatre' was welcomed as according with the theories of 'Socialist Realism' demanded by the Communist régime; and, freed from commercial pressures, he was able to work with very long rehearsal periods. His productions, which were also taken abroad, included *Fledermaus* 1947, *Carmen* 1949, *Freischütz* 1951, and, one of the most famous, *The Cunning Little Vixen* 1956. His pupils and assistants included Götz Friedrich and Joachim Herz.

BIBL: G. Friedrich, *Walter Felsenstein: Weg und Werk* (Berlin, 1967); P. Fuchs, *The Music Theatre of Walter Felsenstein* (London, 1991).

**Female Chorus.** The commentator (sop) who, with the *Male Chorus, introduces, comments upon, and sums up the action in Britten's *The Rape of Lucretia*.

**Fenella.** A poor Neapolitan girl, the dumb heroine of Auber's *La muette de Portici*.

**Fenena.** Nabucco's daughter (sop) in Verdi's *Nabucco*.

**Fenice, Teatro La.** See VENICE.

**Fenton.** A young gentleman of Windsor (ten), lover of Anne (Nannetta), in Nicolai's *Die lustigen Weiber von Windsor*, Verdi's *Falstaff*, and Vaughan Williams's *Sir John in Love*.

**Fenton, Lavinia** (*b* London, 1708; *d* Greenwich, 24 Jan. 1760). English singer and actress. Début London, H, 1726 (Monimia, Otway's *The Orphan*). Engaged by Rich for Lincoln's Inn Fields, she became a celebrity with her creation of Polly Peachum (1728). After her 62nd performance of the role, she left the stage for the Duke of Bolton, whom she married in 1751, having borne him three children. Clever and attractive, she was a winning actress.

**Feo, Francesco** (*b* Naples, 1691; *d* Naples, 28 Jan. 1761). Italian composer. Studied Naples with Basso and Fago; held various teaching positions there while writing operas for other Italian cities. His first success came with *L'amor tirannico*; in 1723 he set Metastasio's adaptation of *Siface*. One of the most skilled Neapolitan composers, he wrote *c*.16 operas, as well as many individual arias and scenes for other composers' works. He was one of the most popular composers of his generation and did much to maintain the position of Naples as an important operatic centre in the mid-18th cent. Among his pupils were Abos and Jommelli.

SELECT WORKLIST: *L'amor tirannico* (Lalli; Naples 1713); *Siface* (Metastasio, after David; Naples 1723).

**Ferencsik, János** (*b* Budapest, 18 Jan. 1907; *d* Budapest, 12 June 1984). Hungarian conductor. Studied Budapest with Fleischer and Lajtha. Hungarian State O; répétiteur 1927–30, cond. from 1930, music director 1953. Many guest appearances include Vienna 1948–50 and subsequently; San Francisco 1962–3; Edinburgh Festival with Budapest Co. 1963, 1973. (R)

**fermata** (It.: 'stop, pause'). The term used for a pause on a held note or chord; the Italians prefer the word **corona**.

**Fernand Cortez.** Opera in 3 acts by Spontini; text by D. Esménard and Jouy, after Alexis Piron's tragedy (1744). Prod. Paris, O, 28 Nov. 1809, with Branchu, Lainez, Laïs, Laforet, Dérivis, Bertin—*c*.250 perfs. by 1840; New Orleans, T d'Orléans, 11 Apr. 1830; never in London, initially because of Bishop's (unsuccessful) opera on same subject, text by Planché, CG, 1823.

Mexico, 1519. Alvaro (ten), brother of Cortez (ten) conqueror of Mexico, is to be sacrificed to the Aztec gods. Cortez has fallen in love with Amazily (sop), daughter of the Mexican king Montezuma (bs) and she, converted to Christianity, begs for forgiveness for the Spaniards.

Cortez puts down a mutiny, and when Telasco (bar), Amazily's brother, arrives at the camp to offer Alvaro's life in return for the Spaniard's departure, he is taken prisoner. Later he is released but refuses to free Alvaro unless Amazily is returned to the Mexicans. Cortez releases her, and Montezuma, persuaded of his sincerity, permits the marriage of the couple, thus ensuring peace.

Also opera by Ricci (1830).

**Ferne Klang, Der** (The Distant Sound). Opera in 3 acts by Schreker; text by the composer. Prod. Frankfurt 18 Aug. 1912; Leeds 14 Jan, 1992, cond. Daniel.

Germany and Venice, present-day. The musician Fritz (ten) refuses to marry his beloved Grete (sop) until he has discovered the secret 'Distant Sound'.

Ten years later Grete reigns as Queen of the demi-monde, but still thinks longingly of Fritz. She has promised to marry whichever of her suitors can best stir her heart with his song. The winner is a stranger, whom Grete identifies as Fritz. On learning that she has become notorious, he once again abandons her.

Fritz finishes his opera *The Harp*, but the public is outraged by its finale. Grete, who has sunk to prostitution, is touched by the work. Fritz meditates on how he has destroyed not only his own but also Grete's life; he wishes desperately to see her. Grete arrives and, when the couple fall into each other's arms, Fritz at last hears the Distant Sound. He rewrites the end of his opera, but then dies in his beloved's arms.

Schreker's best known opera and one of the earliest musical manifestations of Expressionism.

**Ferni.** Italian musical family.

1. **Carolina** (*b* Como, 20 Aug. 1839; *d* Milan, 4 June 1926). Italian soprano and violinist. Studied violin Paris and Brussels, and voice with Pasta. Début Turin 1862 (Léonor, *Favorite*). Milan, S,

1866–8. Repertory incl. Norma, Selika, Saffo, and Mercuri's *Il violino del Diavolo*, in which she both sang and played the violin. Retired from stage 1883 and opened a singing school in Milan, later St Petersburg. Her pupils included Caruso. She married the baritone Leone *Giraldoni; their son Eugenio *Giraldoni, who created Scarpia, was also her pupil.

2. **Vincenzina** (*b* Turin, 1853; *d* Turin, June 1926). Italian soprano and violinist. Sister of (1). After a short career, married the Spanish baritone **Manuel Carbonell Villar** (1856–1928) and retired from the stage.

3. **Virginia** (*b* Turin, 17 Dec. 1849; *d* Turin, 4 Feb. 1934). Italian soprano. Cousin of (1) and (2). Appeared aged 7 singing Spanish songs and accompanying herself on the violin. Stage début shortly after as Siebel. Successful career in Europe and USA; retired 1896. Repertory included Carmen, Mignon, Loreley, and Margherita and Elena (*Mefistofele*). Married the violinist Germano.

**Ferrandini, Giovanni Battista** (*b* Venice, *c*.1710; *d* Munich, 25 Sep. 1791). Italian composer. After study in Venice went to Munich court in 1722 as oboist; Kammermusikdirektor to the Elector 1732; Hofkapellmeister 1737. Forced to return to Italy on health grounds 1760: his house in Padua was the scene of many notable concerts, including one in 1771 at which the young Mozart performed. He composed about ten operas for Munich, mainly opere serie to Metastasian librettos, including *Demofoonte* and *Adriano in Siria*. These reflect the style of the Venetian tradition in which he was reared; he also produced a number of smaller dramatic pieces, such as the serenata *Il sacrifizio invalido* (1729). His most important works were *Catone in Utica*, written for the opening of the Residenztheater in 1753, and *Talestri, Regina delle Amazoni* (1760), whose libretto was provided by Princess Maria Antonia Walpurgis, later Princess of Saxony, one of his pupils.

Ferrandini was also a sought-after teacher: his pupils included the tenor Anton Raaff, who later created the title-role of Mozart's *Idomeneo*.

SELECT WORKLIST (all first prod. Munich): *Demofoonte* (Metastasio; 1737); *Adriano in Siria* (Metastasio; 1737); *Catone in Utica* (Metastasio; 1753); *Talestri, Regina delle Amazoni* (Walpurgis; comp. 1760).

**Ferrando.** An officer (ten), lover of Dorabella, in Mozart's *Così fan tutte*.

**Ferrani, Cesira** (*b* Turin, 8 May 1863; *d* Pollone, 4 May 1943). Italian soprano. Studied Turin with Fricci. Début Turin, R, 1887 (Micaëla). Created Manon Lescaut and Mimì. A notable Elsa and Eva, and the first Mélisande in Italy. A handsome

and sensitive singer, she was also highly cultured; after she retired in 1909, her salon in Turin became a focus of intellectual life.

**Ferrara.** City in Emilia-Romagna, Italy. The first theatre opened 1610; burnt down 1679. The T Comunale opened 1789, and gave the première of Rossini's *Ciro in Babilonia*, 1812. The Arena Tosi-Borghi opened 1856; renamed T Verdi 1913. Gatti-Casazza was director 1894–7.

**Ferrarese, La** (Adriana Gabrieli) (*b* Ferrara, *c*.1755; *d* ?Venice, after 1799). Italian soprano. Studied Venice with Sacchini. Known as La Ferrarese, and as Ferraresi del Bene after her elopement in 1783 with Luigi del Bene, whom she later married. London, H, 1785, sang in Luigi Cherubini's pasticcio *Demetrio*. Milan, S, 1787. Vienna 1788–91, where at her début she was praised for her voice (which Burney describes as 'fair, natural, and of extraordinary compass'), but not for her acting. Taken under the protection of Da Ponte, she sang in several operas with his librettos by Martín y Soler and Salieri. In 1789 she sang Susanna in the Vienna revival of *Figaro*, Mozart doubting her ability to sing the interpolated arias K577 and K579 (written specially for her) 'in an artless manner'. Though dismissive of her talent (describing Allegranti as 'far better than La Ferrarese, though I admit that is not saying much'), he wrote the role of Fiordiligi for her. Described even by Da Ponte as a trouble-maker, though 'unfailingly of the greatest use to the theatre', she was dismissed from court with him in 1791 after various scandals, and they went to Trieste (parting soon afterwards). Here in 1797 she sang in *Così*, and in 1799 she appeared at Bologna.

**Ferrari, Benedetto** (*b* Reggio Emilia, 1603 or 1604; *d* Modena, 22 Oct. 1681). Italian composer, librettist, impresario, and poet. First recognized as a theorbist—he was known as 'Benedetto della Tiorba'—Ferrari made a notable contribution to the development of the solo cantata. At the Farnese court in Modena, where he was later maestro di cappella 1619–23; served as theorbist at the Viennese court 1651–3.

He worked in Venice 1637–44 and made a crucial contribution to the development of opera there. In 1637 his *Andromeda* libretto was set by Manelli: the première of this work (at the T San Cassiano) was the first public performance of opera. He followed the model of the Roman libretto, though in his next collaboration with Manelli, *La maga fulminata* (1638) he included a comic governess: this influenced later Venetian composers, who invariably featured a humorous role for a servant. In 1653 Ferrari provided the libretto for Bertali's *L'inganno d'Amore* and also wrote the librettos for five of his own operas, being one of the earliest composers to do so. Though the texts of these works (including *Armida* and *Il pastor regio*) have survived, most of their music is lost. Ferrari's other operas include *Egisto*, a setting of a Faustini libretto first used by Cavalli, and *Gli amori di Alessandro Magna e di Rossane*. Following Alan Curtis's edition (1989), claims concerning Ferrari's involvement with the composition of Monteverdi's *Poppea* have been strengthened. (See MONTEVERDI.)

SELECT WORKLIST: *Armida* (Ferrari; Venice 1639); *Il pastor regio* (Ferrari; Venice 1640); *Egisto* (Faustini; Piacenza 1651); *Gli amori di Alessandro Magna e di Rossane* (Cicognini; Bologna 1656).

**Ferrari, Giacomo** (*b* Rovereto, *bapt.* 2 Apr. 1763; *d* London, Dec. 1842). Italian composer and theorist. A skilled performer on several instruments, he studied in Verona, Rovereto, Switzerland, and Naples, where he became friendly with Paisiello. In 1787 went to Paris, where he was accompanist at the T de Monsieur, also writing additional music for Bianchi's *La villanella rapita* and Sarti's *Due litiganti*. For the T Montansier he composed *Les évenéments imprévus* (Grétry) and *Isabelle de Salisburi* (Fabre d'Églantine). He was in London from Apr. 1792, supporting himself by teaching and writing popular vocal pieces. His first opera, *I due Svizzeri*, appeared in 1799, and was followed by a string of successes, including *Rinaldo d'Asti* and *L'eroina di Raab*, written for Catalani. He was well acquainted with a wide range of composers, including Haydn, Cherubini, and Clementi, as his operatic style suggests; he also showed an unusual concern for structural unity. His *Concise Treatise on Italian Singing* is an important source for information about the vocal technique of the period; his amusing autobiography was dedicated to King George IV.

His son **Adolfo Ferrari** (1807–70) was a singer, as were Adolfo's wife Johanna Thompson, and their daughter Sophia Ferrari.

SELECT WORKLIST (all first prod. London): *I due Svizzeri* (Buonaiuti; 1799); *Rinaldo d'Asti* (?; 1802); *L'eroina di Raab* (?; 1814).

WRITINGS: *A Concise Treatise on Italian Singing* (= *Breve trattato di canto italiano*, 2 vols., trans. Shield, London, 1818); *Studio di musica teorica pratica* (London, 1830); *Anedotti piacevoli e interessanti* (2 vols., London, 1830), ed. S. di Giacomo (Palermo, 1920).

**Ferrario, Carlo** (*b* Milan, 8 Sep. 1833; *d* Milan, 12 May 1907). Italian designer. Milan, S 1853, and T Carcano 1881; also Rome and Naples. Art director, Milan, S, 1889, incl. for Verdi on the first *Otello* and *Falstaff*. His designs for these and other Verdi productions (notably *Trovatore* 1865, *Forza del destino* 1876, *Aida* 1877, *Rigoletto* 1893)

were welcomed by the composer, and by the public, for their new realism and fidelity to the works' settings, in reaction to more Romantic interpretations. Their colourful, detailed style, with its strong sense of time and place, was widely imitated. By their practicality and effectiveness they set an influential tradition. The many premières for which he provided designs include *Mefistofele* (1868), *La Gioconda* (1876); he also designed a celebrated production of *Meistersinger* (1899).

**Ferretti, Jacopo** (*b* Rome, 16 July 1784; *d* Rome, 7 Mar. 1852). Italian librettist. After an early failure with his libretto for Fioravanti's *Didone abbandonata* (1810), a reworking of Metastasio, he made the brilliant version of the Cinderella story for Rossini (*Cenerentola*, 1817). He also wrote *Matilde di Shabran* for Rossini, and *Torquato Tasso* for Donizetti, as well as librettos for Zingarelli, Mayr, Guglielmi, Coccia, Mercadante, Pacini, L. Ricci, Coppola, and many others. His verses, especially for opera buffa, are distinguished by their verbal elegance and sharp social comment, and he was always willing to modify his own or other men's work to ensure a greater acceptability for the opera.

**Ferri, Baldassare** (*b* Perugia, 9 Dec. 1610; *d* Perugia, 18 Nov. 1680). Italian castrato. Studied Rome with Ugolini. Heard by Władysław IV (as Prince) in 1625, and went with him to Poland, remaining until 1655. Although Poland and Sweden were at war, Queen Christina begged Sigismund III to let Ferri sing in Stockholm. This permitted, his success there was commemorated with a medal. After Sweden invaded Poland, he went to Vienna, where he sang for Ferdinand III and Leopold I until 1675, becoming immensely rich. (In his will he left 600,000 scudi to build a charitable institution in Perugia.) His portrait, with the inscription 'Rè dei musici', hung in Leopold's bedroom. One of the earliest star castrati, whose art, according to Esteban Arteaga, was influential in establishing the da capo aria in opera.

**Ferrier, Kathleen** (*b* High Walton, 22 Apr. 1912; *d* London, 8 Oct. 1953). English contralto. Studied with J. E. Hutchinson, then Roy Henderson. Sang Carmen in concert, Stourbridge 1944, but declined to repeat it for Gly. Stage début Gly. 1946, creating Britten's Lucretia. Gly. 1947, Orfeo (Gluck), which she also sang in Holland. London, CG, 1953 (Orfeo), though only able to perform twice before succumbing to illness. Memorable for her generous, deep voice, noble singing, and warmth of personality. (R)

BIBL: M. Leonard, *Kathleen* (London, 1988).

**Fervaal.** Opera in a prologue and 3 acts by D'Indy; text by the composer. Prod. Brussels, M, 12 Mar. 1897, with Raunay, De la Tour, Seguin, cond. Flon.

The action, which takes place in the Midi of France at the time of the Saracen invasions, tells how Fervaal (ten), a Celtic Chief, wounded in battle, is nursed back to health by Guilhen (mez), who, thinking herself betrayed, lures the Celts to their deaths.

**Festa** (Maffei), **Francesca** (*b* Naples, 1778; *d* St Petersburg, 21 Nov. 1835). Italian soprano. Studied Naples with Aprile, Rome with Pacchierotti. Début Naples, N, 1799. Milan, S, between 1805 and 1824. Paris, Odéon, 1809–11. Munich 1821, St Petersburg 1829. Created Fiorilla (*Il turco in Italia*), and roles in operas by Paer, Mosca, etc. Though she had a voice of sensuous appeal, Stendhal classed her among 'the least impassioned of *prime donne*'.

**festa teatrale** (It.: 'theatrical festival'). A more grandiose version of the *azione teatrale, cultivated especially in Vienna and other Habsburg courts during the 17th and 18th cents.: the first important example was Cesti's *Il pomo d'oro* (1668). The festa teatrale, as the name implies, was an operatic work usually given in the context of a court celebration (marriage, name-day, etc.), leading to the use of a plot which permitted allegorical representation of royal virtues, and frequently called for lavish musical resources. The most important contributors to the genre were, inevitably, the Viennese court composers, who included Draghi, Fux, and Caldara.

**Fétis, François-Joseph** (*b* Mons, 25 Mar. 1784; *d* Brussels, 26 Mar. 1871). Belgian scholar, critic, and composer. Studied Mons, Paris with Boieldieu and Rey. Active in Paris from 1818 as composer, teacher, and critic. His compositions include a number of opéras-comiques, which had little success. He was influential as the first director of the Brussels Conservatoire from 1833, and as founder and contributor to the Paris *Revue musicale* (1827–35) and its successor (on merger with the *Gazette musicale*, founded 1834), the *Revue et gazette musicale de Paris* (1835–80). He was also the author of the *Biographie universelle des musiciens*, a dictionary which remains invaluable despite its many errors of fact. His other works include a singing manual that anthologizes other such works for use in his teaching of singers. A man of great energy and acumen, he was largely responsible for awakening an awareness in France to the importance of the music of earlier times as part of a long historical tradition. This he saw as under threat from much modern music, especially Wagner and even Verdi (whom he was still attacking in an essay in the year of his death).

WRITINGS: *Biographie universelle des musiciens et bibliographie générale de la musique* (Brussels, 1835–44, 2/1860–5, R/1963; suppl. edn. A. Pougin, 1878–80, R/1963); *Méthode des méthodes de chant* (Paris, 1869); *Histoire générale de la musique* (Paris, 1869–76).

BIBL: R. Wangermée, *François-Joseph Fétis, musicologue et compositeur* (Brussels, 1951).

**Fetonte** (Phaethon). Opera in 3 acts by Jommelli; text by Mattia Verazi, based on Ovid's *Metamorphoses*, ii. Prod. Ludwigsburg, Hoftheater, 11 Feb. 1768.

In order to prove his divine parentage to Epafo (bs), Fetonte (sop), the son of Il Sole (The Sun) (ten) and Climene (sop), asks if he may drive the chariot of the sun for a day. When he fails to keep control of the horses and sets heaven and earth alight, his father punishes him by destroying him with a flash of lightning. His mother hurls herself into the sea in order to share his final fate and his beloved, Livia (con), dies of grief.

Jommelli had written an earlier version of this opera, based on a libretto by Villati (prod. Stuttgart 1753); this had a sinfonia closely related to the opening scene, thus anticipating one of Gluck's reforms.

**Feuersnot** (Fire-Famine). Opera in 1 act by Richard Strauss; text by Ernst von Wolzogen, after a Flemish legend, *The Quenched Fires of Oudenaarde*, in J. W. Wolf's *Sagas of the Netherlands* (1843). Prod. Dresden, H, 21 Nov. 1901, with Krull, Petter Scheidemantel, cond. Schuch; London, HM, 9 July 1910, with Fay, Oster, Radford, cond. Beecham; Philadelphia, Metropolitan OH, 1 Dec. 1927, with Stanley, Salzinger, Rasely, Albert Mahler, Nelson Eddy, cond. Smallens.

Munich, 12th cent. Midsummer. Kunrad (bar), a magician, falls in love with Diemut (sop), the burgomaster's daughter. She humiliates him, so to punish the public, which laughs at him, he casts a spell which extinguishes all fires. He chides the people for their inability to comprehend the essential force of love. When Diemut accepts his love, he restores fire.

The opera pokes fun at the Munichers who rejected Richard I (Wagner), and then his disciple, Richard Strauss.

**Février, Henry** (*b* Paris, 2 Oct. 1875; *d* Paris, 8 July 1957). French composer. Studied Paris with Fauré, Leroux, Massenet, and Messager. His first opera, *Le roi aveugle*, achieved moderate success; this was followed by *Monna Vanna*, his most important work, which was produced in several countries. He wrote several more operas which were in the tradition of Massenet while showing the influence of Debussy and the Italian verismo school.

WRITINGS: *André Messager: mon maître, mon ami* (Paris, 1948).

SELECT WORKLIST (all first prod. Paris): *Le roi aveugle* (The Blind King) (Le Roux; 1906); *Monna Vanna* (Maeterlinck; 1909); *Gismonda* (after Sardou; 1918).

**Fiakermilli.** The pretty young girl who traditionally attends the annual ball of the Viennese cab drivers (*Fiaker*). She appears in Act II of R. Strauss's *Arabella*, where she sings one of the most brilliant of all coloratura arias.

**Fiamma, La** (The Flame). Opera in 3 acts by Respighi; text by Guastalla, after Hans Wiessner Jenssen. Prod. Rome, R, 23 Jan. 1934, with Cobelli, Minghetti, Tagliabue, cond. Respighi; Chicago, Civic OH, 2 Dec. 1935, with Raisa, Barova, Bentonelli, Morelli, cond. Hageman.

Ravenna, late 7th cent. Silvana (sop), second wife of Basilio (bar), is persecuted by his mother Eudossia (mez). She hides Agnes de Carvia (mez), accused of sorcery, but is unable to prevent her from being hanged; Agnes prophesies on the scaffold that Silvana will meet the same fate. She falls in love with Basilio's son Donatello (ten), and is denounced to Basilio, who dies of shock. She is accused of sorcery, and also hanged.

**Fibich, Zdeněk** (*b* Všebořice, 21 Dec. 1850; *d* Prague, 15 Oct. 1900). Czech composer. Studied privately, Leipzig with Moscheles, Richter, and Jadassohn, and Mannheim with Lachner. Apart from a period as deputy conductor and choirmaster at the Provisional Theatre in Prague (1875–8), Fibich's life was mostly devoted to teaching and composing. His interest in opera began early, and as with most of his music was strongly coloured by his German training and Romantic interests. His first surviving operas are historical works that also, in part through his choice of two of Smetana's librettists, reflect nationalist ideas.

However, a more individual voice shows in *Nevěsta mesinská*, a powerful work that achieves its grim, tragic effect in part by a dense network of motivic references, but also by its declamatory melodic style and even by such strokes as eliminating female voices from the choruses so as to darken the texture. He took his interest in declamation further with melodrama, whose effect he had admired in *Der Freischütz* and especially in the works of Benda, in his 'stage melodrama trilogy' *Hippodamia*. His later operas were written to libretto by his pupil and mistress Anežka Schulzová, and are all to varying extents studies of women, and make central use of set numbers. His most successful and enduring work is *Šárka*, a powerful and musically gripping treatment of the popular Czech legend. His second wife, **Betty Fibichová** (1846–1901) (whom he left for

Anežka), was a contralto who created several roles in his and Smetana's operas.

WORKLIST (all first prod. Prague): *Bukovín* (Sabina; 1874); *Blaník* (Krásnohorská; 1881); *Nevěsta mesinská* (The Bride of Messina) (Hostinský, after Schiller; 1884); *Hippodamie* (Vrchlický, after Sophocles & Euripides) (1. *Námluvy Pelopovy* (The Courtship of Pelops), 1890; 2. *Smír Tantalův* (The Atonement of Tantalus), 1891; 3. *Smrt Hippodamie* (The Death of Hippodamia), 1891); *Bouře* (The Tempest) (Vrchlický, after Shakespeare; 1895); *Hedy* (Schulzová, after Byron; 1896); *Šárka* (Schulzová; 1897); *Pád Arkuna* (The Fall of Arkun) (Schulzová; 1900).

BIBL: V. Hudec, *Zdeněk Fibich* (Prague, 1971).

**Fidelio oder Die eheliche Liebe** (Fidelio, or Married Love). Opera orig. in 3 acts by Beethoven; text by Josef Sonnleithner, a German version of Bouilly's *Léonore, ou L'amour conjugal*, music by Gaveaux (1798) and then set, in an Italian version, by Paer (1804) and Mayr (1805). Altered and reduced to 2 acts by Stefan von Breuning in 1806; given final form by Georg Friedrich Treitschke in 1814. First version prod. Vienna, W, 20 Nov. 1805, with Milder, Demmer, Meier, Rothe, Weinkopf, Caché, cond. Seyfried. 2nd version, Vienna, W, 29 Mar. 1806. 3rd version, Vienna, K, 23 May 1814; London, H, 18 May 1832, with Schröder-Devrient, Haitzinger, cond. Chelard; New York, Park T, 9 Sep. 1839, with Inverarity, Manvers, Giubilei, Martyn.

A prison near Seville. Florestan (ten), a Spanish nobleman, has been thrown into prison. His wife Leonore (sop) has followed him disguised as a boy, Fidelio, in the hope of rescue. The kindly jailer Rocco (bs) employs 'Fidelio', with whom his daughter Marzelline (sop) falls in love, to the annoyance of her lover Jacquino (ten). The famous quartet expresses their reactions. The tyrannical governor Pizarro (bs-bar) decides to kill Florestan to prevent his discovery at an impending inspection. Leonore persuades Rocco to allow the prisoners out for a while; they emerge groping towards the sunlight. But Florestan is not among them.

Florestan is chained in his dungeon, whither Rocco comes with Leonore to dig the prisoner's grave. Pizarro tries to kill Florestan but is prevented by Leonore with a pistol. Far-off trumpets announce the arrival of the inspecting minister. The prisoners are all released; Pizarro is arrested, and Leonore herself unshackles Florestan.

The overture now usually played is entitled *Fidelio*. Three other overtures exist, *Leonore No. 1* (composed for a projected performance in Prague), *Leonore No. 2* (actually the first, played at the première), and *Leonore No. 3* (played for the

1806 revival). There is a long-standing tradition of using the last as an interlude between the final scenes; but it fits neither harmonically, instrumentally, nor dramatically, intruding an orchestral summary of parts of the drama. At the Vienna, K, on 28 Sep. 1841 Nicolai placed it between the two acts. Carl Anschütz set it before the final scene in Amsterdam, late April 1849, and at London, DL, 19 May 1849. Balfe followed suit in London, HM, in 1851, as did Augusto Vianesi at Paris, I, in Jan. 1852 (reported by Berlioz). Levi played it after the opera in Rotterdam in 1863. Anschütz played three overtures (not known which) in Philadephia, 9 Nov. 1863, one before each act; others in USA gave *Leonore No. 3* before the final scene, incl. Seidl (New York, M, 1890–1). Mahler did so, Vienna 1904 (and has often been credited with introducing the idea). Other conductors who have done so include Bursch, Reiner, Szell, Beecham, Walter, E. Kleiber, Furtwängler, Toscanini, R. Strauss (after initial doubts), and Klemperer (later giving it up, for health rather than artistic reasons).

**Fidès.** John of Leyden's mother (mez) in Meyerbeer's *Le prophète*.

**Fieramosca.** The Pope's sculptor (bar), Cellini's rival for Teresa's love, in Berlioz's *Benvenuto Cellini*.

**Fiery Angel, The** (Russ.: *Ognenniy angel*). Opera in 5 acts by Prokofiev; text by the composer, after Valeriy Bryusov's historical novel, first published in the magazine *The Scales*, 1907–8. Concert perf. Paris 25 Nov. 1954, with Marée, Depraz, cond. Bruck. Prod. Venice, F, 29 Sep. 1955, with Dow, Panerai, cond. Sanzogno; London, SW, 27 July 1965, by New Opera Co., with Collier, Shaw, cond. Lovett; New York, CC, 22 Sep. 1965, with Schauler, Milnes, cond. Rudel.

Cologne, 16th cent. At an inn, Rupprecht (bar or bs) meets Renata (sop) in a state of possession in which she mistakes him for Heinrich, a former lover whom she associated with her guardian angel. He agrees to take her to Cologne to look for Heinrich (mute), having fallen in love with her himself. The couple try by magic to find Heinrich; Rupprecht is equally disappointed by a visit to the magus Agrippa of Nettesheim. Renata has met and been repulsed by Heinrich; she urges Rupprecht to fight him. Rupprecht loses, but wins Renata. Renata threatens to leave Rupprecht, whose obsessive physical passion for her contrasts with memories of her 'guardian angel'. In a garden by the Rhine, Renata hurls a knife at Rupprecht, accusing him of being possessed by the Devil, as Faust (bs) and Mephistopheles enter. Rupprecht watches Mephistopheles (ten) angrily devour a slow-moving serving-boy and then resurrect him from a nearby rubbish dump.

He joins the party. Rupprecht is in the suite of the Inquisitor (bs), who is investigating a story of diabolical possession in a convent. The source of the trouble is a new nun—Renata. Signs of possession appear; as the exorcism rite begins, hysteria seizes the community and the opera ends with the Inquisitor sentencing Renata to be burnt alive for having dealings with evil spirits.

**Fiesco.** A Genoese nobleman (bs) in Verdi's *Simon Boccanegra*.

**Figaro.** The barber of Seville (bar) in Paisiello's and Rossini's *Il barbiere di Siviglia*, subsequently Count Almaviva's manservant (bar) in Mozart's *Le nozze di Figaro*. See also BEAUMARCHAIS.

**Figlia del reggimento, La.** See FILLE DU RÉGIMENT, LA.

**Figlia di Jorio, La** (The Daughter of Jorio). Opera in 3 acts by Pizzetti; text, word-for-word setting of D'Annunzio's tragedy (1903). Prod. Naples, C, 4 Dec. 1954, with Petrella, Nicolai, Picchi, Guelfi, cond. Gavazzeni. One of Pizzetti's most successful operas.

The Abruzzi, long ago. The sorcerer Jorio's daughter Mila (sop), fleeing from the peasants who think she is a witch, is saved by Aligi (ten), who falls in love with her. His father, Lazaro (bar), tries to prevent their liaison and has Aligi beaten and imprisoned. He escapes, and kills his father whom he discovers about to rape Mila; but she says she is the guilty one and is burned as a witch.

Also opera by Franchetti, prod. Milan, S, 1906, with Pandolfini, De Cisneros, Zenatello, Giraldoni, Didur, cond. Mugnone.

**Figner** (Mei-Figner), **Medea** (*b* Florence, 4 Apr. 1859; *d* Paris, 8 July 1952). Italian, later Russian, mezzo-soprano, later soprano. Studied Florence with Bianchi, Carozzi-Zucchi, and Panofka. Début Sinalunga 1874 (Azucena). After successes in Italy, Barcelona, and Madrid as a mezzo-soprano, she met Nikolay *Figner and sang with him in S America 1884–6 (by now a soprano). They married in 1889. St Petersburg 1887–1912, singing with her husband until their divorce in 1904. Here she created Lisa, Yolanta, and roles in Nápravník's *Dubrovský* and *Francesca da Rimini*. Officially retired 1912, but continued to appear as Carmen and Valentine until 1916. Her roles included Tatyana, Desdemona, Mimì, and Brünnhilde. She left Russia for Paris in 1919, though making return visits and recordings there until 1930. In 1949 (released 1975) she recorded (and even sang in) an interview in which she recalled working with Tchaikovsky. A handsome woman, she was much admired for her rich voice, excellent technique, and dramatic presence. Autobiography, *Moy vospominaniya* (My Memoirs) (St Petersburg, 1912). (R)

**Figner, Nikolay** (*b* Nikiforovka, 21 Feb. 1857; *d* Kiev, 13 Dec. 1918). Russian tenor. Studied St Petersburg with Pryanishnikov and Everardi; Naples with De Roxas. Début Naples, T Sannazaro, 1882 (in Gounod's *Philémon et Baucis*). After engagements in Italy, he toured in S America 1884–6, singing in the performances which launched the young Toscanini. He later helped to secure the first Turin performance of Catalani's *Edmea* for the conductor. London, CG, 1887. St Petersburg 1887 (Raoul, with Medea Mei, his wife-to-be, as Valentine); he and Medea then became members of the Imperial O, he remaining until 1906. After singing in private companies (including his own), he became Narodny Dom O's soloist and director 1910–15. Repertory included Don José, Gounod's Faust, Turiddu, Radamès, Lohengrin, Lensky, Dmitry. He created roles in Nápravník's *Francesca da Rimini* and *Dubrovsky*; also Vaudémont (*Yolanta*) and Hermann (*Queen of Spades*). Tchaikovsky wrote, 'I liked Figner very much, and imagined Hermann to be Figner'. Though not especially gifted vocally, his eloquent singing, versatility, and dramatic commitment made him Russia's outstanding tenor for many years. (R)

BIBL: E. Stark, *Petersburgskaya opera i eyo mastera* (The St Petersburg Opera and its Stars) (Leningrad, 1940).

**figured bass.** See CONTINUO.

**filar il tuono** (or **filar la voce**, or **un filo di voce** and **filer la voix** or **filer le son**) (It. and Fr.: 'to spin the voice, or tone'). A term mentioned by G. B. Mancini in his *Pensieri e reflessioni pratiche sopra il canto figurato* (1774), and in common operatic usage with various interpretations. It is usually taken to mean an instruction to hold a long, soft note without crescendo or diminuendo. Verdi marks *un filo di voce* for the final A of Violetta's 'Addio del passato' in *La traviata*. According to other authorities, it may also require a diminuendo from piano to pianissimo, or even a swelling of tone, synonymously with *messa di voce. To be skilled in the technique, however interpreted, is in Italian *avere dei bei filati*.

**Filistri, Antonio de' Caramondani** (*fl.* 1788–1808). Italian librettist. Active in Berlin, where he wrote *c.*15 librettos, set by Alessandri, Naumann (*Medea in Colchide*, 1788) Righini, and others. His *Vasco di Gama* (Alessandri, 1792) is a rare operatic treatment of a comparatively recent historical subject.

**Fille de Madame Angot, La** (Madame Angot's Daughter). Operetta in 3 acts by Lecocq; text by Clairville, Giraudin, and Koning, after A. F. Eve Maillot's vaudeville, *Madame Angot, ou La poissarde parvenue* (1796). Prod. Brussels, Alcazar, 4 Dec. 1872, with Desclauzas, Luigini, Widmer, Joly; London, J, 17 May 1873, with Aimée; New York, Broadway T, 25 Aug. 1873.

Lecocq's most popular work, and one of the most successful post-Offenbach operettas.

**Fille du régiment, La** (The Daughter of the Regiment). Opera in 2 acts by Donizetti; text by Vernoy de Saint-Georges and Bayard. Prod. Paris, OC, 11 Feb. 1840, with Bourgeois, Boulanger, Blanchard, Marié de l'Isle, Henry, cond. Donizetti; Italian version, trans. and ad. by C. Bassi with recit. by Donizetti replacing dialogue, Milan, S, 3 Oct. 1840, with Abbadia, Salvi, Scalese, cond. Donizetti; New Orleans, T d'Orléans, 7 Mar. 1843, with Place, Blès; London, HM, 27 May 1847, with Lind, Lablache.

Tirol, 1805. Marie (sop) has been brought up in a regiment by the kindly Sulpice (bs). She loves Tonio (ten), who has been following the regiment out of love for her. When Marie is suggested to the Marquise de Birkenfeld (sop or mez) as a travelling companion, it emerges that the girl is in fact the Countess's niece. This necessitates her leaving Tonio.

Marie, somewhat reluctantly, is taught correct social etiquette. French soldiers, among them Tonio, storm the castle where she lives and the couple reaffirm their love. The Marquise discloses that she is not Marie's aunt, but her mother; despite this, Marie cannot go along with the arranged marriage and her mother permits her marriage to Tonio.

Donizetti's first French opera, and one of his most successful works: 44 perfs. in 1840. Marie Julie Boulanger (who sang the Marquise) was the grandmother of Nadia Boulanger.

**film opera.** Even in the days of the silent film, several operatic films were made. In 1903 Pathé produced a version of *Faust*, and in 1909 there were three different versions of *Rigoletto*, one made in France using Hugo's original story and Verdi's music, one in the USA entitled *The Fool's Revenge*. In 1910 a record of a bass of the Paris Opéra singing 'Vous qui faites' (*Faust*) was played behind a screen showing the singer in action. In the same year there were two *Carmen* films, one called *The Cigarette Maker of Seville*, also *Manon* and *Trovatore* in France. In 1911 there was an *Aida* and a Russian *Life for the Tsar*. Betwen 1912 and 1915 Italy produced films of *Parsifal*, *Figaro*, and *Manon*, with Lina Cavalieri and Lucien Muratore. In 1915 Cecil B. De Mille produced *Carmen* with Geraldine Farrar: subsequent *Carmen* films included *Gypsy Blood*, with Pola Negri (director Ernst Lubitsch, Germany 1919), and *The Loves of Carmen* with Dolores Del

Rio (Hollywood 1927). In 1917 Samuel Goldwyn persuaded Mary Garden to make a film of *Thaïs*. Other opera films of the time included *Madama Butterfly* with Mary Pickford, *Tosca* with Farrar, *Bohème* with Lillian Gish and John Gilbert (prod. Irving Thalberg), and Gounod's *Faust* with Emil Jannings and Yvette Guilbert. In 1918 Caruso starred in a silent film called *My Cousin*. In 1926 Strauss made a special version of his *Rosenkavalier* score for a film first shown in London with Strauss himself conducting an enlarged orchestra in the pit of the Tivoli cinema.

The first Vitaphone sound films of opera arias were made by Martinelli, Marion Talley, and Anna Case in 1926, and later Fortune Gallo produced *Pagliacci* in sound with his San Carlo Touring O. In 1932 Educational Films released a series of 18-minute 'operalogues' of *Martha*, *Carmen*, *Cavalleria rusticana*, and other works, and starred Grace Moore, Lawrence Tibbett, Lily Pons, and Jan Kiepura in feature films including operatic sequences. During the 1930s opera films made in Europe included a *Bartered Bride* with Novotná, *Louise* with Grace Moore (under Charpentier's supervision), *Butterfly* with Cebotari, *Pagliacci* with Tauber, and *The Last Rose* (based on *Martha*) with Helge Roswaenge. Pabst produced a film of *Don Quixote* with Shalyapin, music by Ibert. Other singers who appeared in films included Flagstad (who sang Brünnhilde's battle cry in *The Big Broadcast of 1938*), Lucrezia Bori, Gladys Swarthout, Risë Stevens, Gigli, and Charles Kullman.

Nevertheless, it was many years before opera received serious consideration as film entertainment, and difficulty has been experienced in overcoming the essential contradiction of the medium, namely the remoteness and formalization of the stage, with voice and personality projected to meet the audience. In addition, the cinema needs to cross the barrier of the footlights, and to vary its distances, from vast panorama to middle distance (perhaps with the camera moving among the cast) to intimate close-up. Some operas have been shot as if on stage; this merely artificializes what is in the theatre the convention. Some have attempted to compromise by using the camera's mobility of eye without taking the action beyond the confines of the stage (e.g. Paul Czinner's Salzburg *Don Giovanni* and *Rosenkavalier*, and the Zeffirelli-Karajan *Bohème*). Some have removed opera from the theatre, and made free use of distant landscape shots, panning shots, and close-ups (e.g. Stroyeva's Bolshoy *Boris*, Losey's Don Giovanni, and Zeffirelli's 1986 *Otello*). Voices are usually added after the action has been photographed: the effort in singing well usually needs the forgiving distance of the theatre.

The advent of video has extended to opera productions, and brought opera into the home in a new way. Some classic productions have been recorded, along with specially staged recordings, with the advantage that the serious opera enthusiast can now connect the visual image to his hi-fi equipment, thus overcoming one of the crucial drawbacks of ordinary *television opera.

**Filosofo di campagna, Il** (The Country Philosopher). Opera in 3 acts by Galuppi; text by Goldoni. Prod. Venice, S, 26 Oct. 1754 with three of the Baglioni family in principal roles; London, H, 6 Jan. 1761, similar cast; Boston, Cons., 26 Feb. 1960.

Campagna, 18th cent. Don Tritemio (bs) lives in the country with his daughter Eugenia (sop) and her companion Lesbina (sop), to whose charms he has succumbed. Nardo (bar), a rich farmer, has been chosen by Tritemio as the husband of Eugenia, but she is in love with Rinaldo (ten). With the help of Lesbina, who disguises herself as Eugenia, she successfully arouses Rinaldo's jealousy. He then beats Nardo, but falls in love with Lesbina. The two couples marry.

**finale** (It.: 'end'). The last movement of a piece of music, in opera of an act. From the beginning it tended to involve ensemble singing, which could be a simple chorus, as in the earliest operas, a more elaborate tableau, as in Roman opera or the works of the 17th- and 18th-cent. French tradition, or a duet, often comic, as in 17th-cent. Venetian opera. In 18th-cent. opera seria the finale was usually a simple ensemble sung by all the principals, but in *opera buffa it developed into the most complex and characteristic number, the so-called *ensemble finale. This came to influence other genres as well, and by the 19th cent. the finale was usually one of the most complex sections of the opera, involving any or all of solo aria and recitative, ensemble writing, the use of the chorus, and orchestral interludes, which were combined together in a continuous sequence.

**'Finch'han dal vino'.** Don Giovanni's aria in Act I of Mozart's *Don Giovanni*, in which he bids Leporello prepare for the party. Sometimes known as the Champagne Aria, from a tradition of the performer singing it with a glass of champagne in his hand.

**Finland** (Fin., Suomi). The first operas were given in Helsinki by amateurs, in Swedish, though *Il barbiere di Siviglia* was given by Finnish performers in 1849. Swedish long remained the official language; *Kung Carls jakt* (King Charles's Hunt, 1852) by Fredrik Pacius (1809–91) is often regarded as the first Finnish opera, but the composer was of German origin and the text was Swedish. His *Prinsessan av Cypern* (The Princess of Cyprus, 1860) inaugurated the first permanent theatre in Helsinki, and led to attempts at Finnish-language opera. But Pacius's last opera, *Lorelei* (1887), was given in German; even Sibelius's only opera, *Jungfrun i tornet* (The Maiden in the Tower, 1896), was given in Swedish. The first opera to a Finnish text was *Pohjan Neiti* (The Bothnian Maiden, comp. 1899, prod. 1908) by Oskar Merikanto (1868–1924). The first really successful Finnish opera was *Pohjalaiset* (The Ostrobothnians, 1924) by Leevi Madetoja (1887–1947); this draws on folk melodies. Madetoja also had a success with *Juha* (1935), after a setting by Aarre Merikanto had been rejected in 1922 (prod. 1963). The late 1960s saw the beginning of a remarkable new growth of opera in Finland. Among a number of striking opera composers to emerge, the most important is Aulis *Sallinen. Others include Tauno Martinen (*b* 1912) with *Poltettu oranssi* (Burnt Orange, 1976); Tauno Pulkkänen (1918–80) with *Varjo* (The Shadow, 1952), *Tuntematon sotilas* (The Unknown Soldier, 1967), and *Opri ja Oleksi* (1958); Joonas Kokkonen (*b* 1912) with *Viimeiset kiusaukset* (The Last Temptations, 1976); Einojuhani *Rautavaara; Ilkka Kuusisto (*b* 1933) with the comic *Miehen kylkiluu* (Man's Rib, 1978) and *Sota valosta* (War for Light, 1981); Paavo Heininen (*b* 1938) with *Veitsi* (The Knife, 1989); and Kalevi Aho (*b* 1949) with *Avain* (The Key, 1979). Outside Helsinki and the Savonlinna Festival, opera is given in Tampere (from 1946, and from 1991 in a conference hall, cap. 2,000), Vaasa (from 1956), and Lahti (from 1962), as well as in half a dozen other towns beginning in the 1960s and 1970s. A summer festival at Ilmajoki, from 1975, stages opera spectacularly in the open air, including in 1980 a commission, *Jaakko Ilkka*, from Jorma Panula.

See also HELSINKI, SAVONLINNA.

**Finnie, Linda** (*b* Scotland, 9 May 1952). Scottish mezzo-soprano. Studied Glasgow with Busfield. Début Scottish O, 1976. WNO 1979; Bayreuth from 1988; also London, ENO and CG; Paris, O and C; Brussels, M; Lyons; Bayreuth 1988. With her strong characterizations and firm, warm tone, successful as Ulrica, Eboli, Amneris, Brangäne, Fricka, Waltraute. (R)

**Finta giardiniera, La** (The Feigned Gardener). Opera in 3 acts by Mozart; text possibly by Calzabigi for Anfossi's opera (1774), rev. Coltellini. Prod. Munich 13 Jan. 1775, with Rosa and Teresina Manservisi, Rossi, Consoli; New York, Mayfair T, 18 Jan. 1927, with Chamberlin, Millay, Sheridan, Echolls, Rogers, Hale, Campbell, cond. Marrow; London, Scala T, 7 Jan. 1930,

with Eadie, Parry, Lemon, Leer, Wendon, Michael, Comstock, cond. Heward.

The house and garden of the Podestà of Lagonero, mid-18th cent. The Countess Violante Onesti (sop) is believed to have died at the hands of her lover Count Belfiore (ten) during a quarrel. He has fled, but she is seeking him forgivingly, and takes a post as a gardener under the name of Sandrina. Her employer, the Podestà Don Anchise (ten), has fallen in love with her, to the annoyance of his maid Serpetta (sop), who has designs on him herself. When Sandrina finds Belfiore paying court to the Podestà's niece Arminda (sop), she is determined to punish him. But when his arrest for the alleged murder of Countess Onesti is announced, she reveals her disguise; she then pretends to Belfiore that she is not Violante Onesti after all. After many fantastic confusions, all is resolved.

Other operas by Piccinni (1770) and Anfossi (1774).

**Finta semplice, La** (The Feigned Simpleton). Opera in 3 acts by Mozart; text by Coltellini, after a libretto by Goldoni, first set by Perillo (1764). Prod. Salzburg, Archbishop's Palace, 1 May 1769, with Magdalena Haydn, Fesemayer, Meissner, Spitzeder, Braunhofer, Hornung, Winter; London, Palace T, 12 Mar. 1956, with Siebert, Oravez, Küster, Maran, Jaresch, Raninger, Pernerstorfer, cond. Maedel; Boston, Jordan Hall, 27 Jan. 1961, with Dalapas.

Near Cremona, late 18th cent. Fracasso (ten), a Hungarian captain, is in love with Donna Giacinta (con); his lieutenant, Simone (bs), is in love with Giacinta's maid Ninetta (sop). It is arranged that Fracasso's sister, the Baroness Rosina (sop), who is coming to visit, will pretend to be a simpleton in order to divert the attention of Giacinta's two grumpy elder brothers, Don Cassandro (bs) and Don Polidoro (ten).

Both the brothers fall in love with Rosina. Giacinta and Ninetta disappear and the brothers, having been told that Giacinta has taken the family jewellery with her, promise Fracasso and Simone that they may marry the girls if they can bring them back, which they do.

Rosina marries Cassandro and everyone, except Polidoro, is happy.

**Finto Stanislao, Il** (The False Stanislas). Opera in 2 acts by Gyrowetz (Jírovec); text by Romani. Prod. Milan 5 July 1818. Text set by Verdi in 1840 for *Un *giorno di regno*. Another opera on the subject is by Mosca (1812).

**Fioravanti, Valentino** (*b* Rome, 11 Sep. 1764; *d* Capua, 16 June 1837). Italian composer. Studied Rome with Toscanelli and Jannacconi, Naples with Sala. He wrote some 77 operas; nevertheless, the last of them was entitled *Ogni eccesso e* *vizioso* (All Excess is Vicious). His first great success came with *Le cantatrici villane*, frequently revived including in Germany as *Die Dorfsängerinnen*. The success of *Camilla* in Lisbon led to the directorship of the S Carlos T (1801–6), whence he returned via a successful visit to Paris. His serious operas were eclipsed by his opere buffe, in which he shows a lively feeling for the manner of which Cimarosa (who admired him) was a master. His son **Vincenzo** (*b* Rome, 5 Apr. 1799; *d* Naples, 28 Mar. 1877) was obliged through parental opposition to study music in secret (with his father's teacher Jannacconi); later, after he was discovered not to have been studying medicine, his reluctant father allowed him to take up music. He experienced many difficulties in his career as a composer; but his operas, like those of his father most successful when in buffo vein, were popular in Naples. *Il ritorno di Pulcinella dagli studi di Padova* (Pulcinella's Return from Study in Padua) (1837), which includes a scene in an asylum for mad musicians, had a wider success, and was even performed in London and America, due to adaptation by the singer Carlo Cambiaggio. A second son, **Giuseppe** (*b* c.1795; *d* ?) was one of the best buffo basses of his day: he took part in a number of premières, including those of Rossini's *Bianca e Faliero* and *Matilde di Shabran* and of some of Donizetti's works. Giuseppe's two sons **Valentino** (1827–79) and **Luigi** (1829–87) were also buffo basses, specializing in the same roles as their father.

**Fiordiligi.** Dorabella's sister (sop) in Mozart's *Così fan tutte*.

**fioritura** (It.: 'flourish'). An ornamental figure, written or improvised, decorating the main line of the melody. Also known, less correctly, as *coloratura.

**First Lady.** The leader (sop) of the Three Ladies, attendants of the Queen of the Night, who save Tamino from the serpent in Mozart's *The Magic Flute*.

**Fischer, Anton** (*b* Ried, *bapt*. 13 Jan. 1778; *d* Vienna, 1 Dec. 1808). Austrian composer and tenor. Studied with his brother, the composer Matthäus Fischer, with whom he is sometimes confused; then joined chorus of Vienna, J, and moved to Vienna, W, under Schikaneder in 1800. Here he sang small roles, became assistant Kapellmeister to Seyfried in 1806 (when the theatre had moved to the Wien), and wrote many popular stage works in light and easily enjoyable vein. These include *Die Verwandlungen oder Der travestirte Aeneas*, for which Weber provided additional music in Prague in 1814, *Swetards Zauberthal*, following the vogue for magic plots,

and *Das Milchmädchen von Bercy*. Among his 11 Singspiels his great triumph was *Das Hausgesinde*, with 115 performances, sequels in 1813 and 1818: Weber wrote an admiring essay on it. He also arranged opéras-comiques by Grétry, including *Raoul Barbe-bleue* (1804; as *Raoul Blaubart*) and *Les deux avares* (1805; as *Die zwei Geizigen*). Many works are wrongly attributed to him in secondary literature.

SELECT WORKLIST (all first prod. Vienna): *Die Verwandlungen oder Der travestirte Aeneas* (The Transformations) (Baber, after Sewrin, after Blumauer; Vienna 1805); *Swetards Zauberthal* (Swetard's Magic Valley) (Schikaneder; 1805); *Das Milchmädchen von Bercy* (The Milkmaid of Bercy) (Treitschke; 1808); *Das Hausgesinde* (The Domestic Staff) (Koller; 1808).

**Fischer, Emil** (*b* Brunswick, 13 June 1838; *d* Hamburg, 11 Aug 1914). German bass-baritone. Studied with his parents, both singers. Début Graz 1857 (the Seneschal in Boieldieu's *Jean de Paris*). Danzig 1863–70; Dresden 1880–5. London, CG, 1884 (Sachs, a role in which he excelled). New York, M, 1885–91. The first American Lysiart, Sachs, King Mark, Wotan (*Rheingold*), Wanderer, Hagen.

**Fischer, Ludwig** (*b* Mainz, 18 Aug. 1745; *d* Berlin, 10 July 1825). German bass. Studied with Raaff. Début Mannheim 1772 (Salieri's *Fiera di Venezia*). Vienna 1780–3, creating Osmin, written especially for him. Mozart admired his 'excellent bass voice' and 'beautiful deep notes', and arranged the aria K512 for him. He performed with success in Paris, Italy, Prague, and Dresden, and in 1789 took up a lifelong appointment at Berlin, Königliche Oper. London 1794, 1798, and 1812. The greatest German bass of his time, he possessed a range of two and a half octaves (D–a'), a round, full tone, and unusual vocal agility. However, his Italianate training made him prone to outmoded ornamentation, and he was later criticized for continuing to embellish Sarastro's aria 'In diesen heil'gen Hallen'. He retired in 1815.

**Fischer-Dieskau, Dietrich** (*b* Berlin, 28 May 1925). German baritone. Studied Berlin with Georg Walter and Weissenborn. Début Berlin, SO, 1948 (Posa). Remained there as leading baritone; Vienna, S, and Bavarian Staatsoper, Munich from 1949; Salzburg from 1952. Bayreuth 1954–6; London, CG, from 1965; Berlin, D, 1975–6. Repertory includes Wagner (Wolfram, Amfortas, Kurwenal); Verdi (Falstaff, Rigoletto); Strauss (Mandryka, Olivier, Barak, Jochanaan); Mozart (Count Almaviva, Don Giovanni, Papageno, Don Alfonso); Wozzeck, Busoni's Doktor Faust, etc. Created Reimann's Lear, and Mittenhofer (*Elegy for Young Lovers*). The vocal mastery and sensitivity to words apparent in his Lieder singing are also impressive in his stage roles (particularly in the German repertory), to which is added an assured and commanding presence. He is married to the soprano Julia *Varady. Autobiography, *Echoes of a Lifetime* (London, 1989). (R)

BIBL: Kenneth Whitton, *Dietrich Fischer-Dieskau, Mastersinger* (London, 1981).

**Fischietti, Domenico** (*b* Naples, ?1725; *d* ?Salzburg, ?after 1810). Italian composer. Studied Naples with Durante and Leo, where his first opera, *Armindo*, was produced 1742. Having enjoyed success with *La finta sposa* in Palermo 1753, he moved to Venice, where he collaborated with Goldoni on four operas: *Lo speziale, La ritornata di Londra, Il mercato di Malmantile*, and *Il signor dottore*. These were highly acclaimed, and Fischietti's fame spread quickly beyond Italy: in 1762 he travelled to Prague, joining Bustelli's company in 1764; in 1765 he succeeded Hasse in Dresden. During this time he composed a number of opere serie, including *Vologeso* and *Nitteti*, another opera buffa to a libretto by Goldoni, *La donna di governo*, an opéra-comique, *Les métamorphoses de l'amour*, and arranged successful revivals of some earlier works. Appointed Kapellmeister to the Archbishop of Salzburg, in preference to Leopold Mozart, 1772: he held the position until 1783, visiting Italy on several occasions to produce new operas, including *La molinara*.

Fischietti's operas belong to the mainstream tradition of the mid-18th cent. and were performed in many European cities, including London. Together with Galuppi, Logroscino, and Piccinni, he is one of the most important composers of opere buffe before Mozart, who regarded him highly. The rondo finale to Act II of *Il mercato di Malmantile* constitutes an especially noteworthy contribution to the development of the *ensemble finale.

SELECT WORKLIST: *Armindo* (?; Naples 1742); *La finta sposa* (The Feigned Wife) (Fabbozzi; Palermo 1753); *Lo speziale* (The Apothecary) (Goldoni; Venice 1754; Act I by Vincenzo Pallavicini); *La ritornata di Londra* (Goldoni; Venice 1756); *Il mercato di Malmantile* (Goldoni; Venice 1757); *Il signor dottore* (Goldoni; Venice 1758); *La donna di governo* (Goldoni; Prague 1763); *Vologeso* (Zeno; Prague 1764); *Nitteti* (Metastasio; Prague 1765); *Les métamorphoses de l'amour* (Pfördten; Prague 1769); *La molinara* (The Maid of the Mill) (Livigni; Venice 1778).

**Fisher, John Abraham** (*b* Dunstable or London, 1744; *d* ?London, May 1806). English composer and violinist. First came to prominence as a violinist at the King's T, London; became member of the Royal Society of Musicians 1764. Through marriage in 1770 to the widow of one of the theatre managers, he acquired a sixteenth share in Covent Garden. Also composed the incidental music for a number of dramatic works staged in London, including the masque *The Syrens* (1776) and the pantomime *The Norwood Gypsies* (1777).

**Fisher, Sylvia** (*b* Melbourne, 18 Apr. 1910). Australian soprano. Studied Melbourne with Spivakovsky. Début Melbourne 1932 (Hermione in Lully's *Cadmus et Hermione*). London, CG, 1948–58, début Leonore. Also Sieglinde, and first British Kostelnička. Chicago 1959. London, EOG 1963–71, as Britten's Lady Billows, Female Chorus, Mrs Grose; created Mrs Wingrave (*Owen Wingrave*), BBC TV 1971, London, CG, 1973. Commanding yet warm on stage, with an intelligent approach to the study of the roles, she was particularly admired as the Marschallin. (R)

**Flagstad, Kirsten** (*b* Hamar, 12 July 1895; *d* Oslo, 7 Dec. 1962). Norwegian soprano. Studied Oslo with Jacobsen, and Stockholm with Dr Bratt. Début Oslo, 1913 (Nuri, *Tiefland*). Until 1933 she appeared only in Scandinavia, in a large repertory of opera and operetta, and was about to retire when she was engaged for small roles at Bayreuth. Here in 1934 she sang Gutrune and Sieglinde. New York, M, 1935 (Sieglinde, Isolde, Brünnhilde, Kundry), immediately establishing herself as a major artist. Remained at New York, M, until 1941, and returned there 1950–2. London, CG, 1936–7, and 1948–51 (Senta, Isolde, etc.), and Mermaid T, 1951, 1952 (Purcell's Dido). Retired in 1953, but continued to record. With her musical intuition, poise, and powerful, radiant voice, she was one of the greatest Wagner singers of the century. (R)

BIBL: E. McArthur, *Flagstad: A Personal Memoir* (New York, 1965); H. Vogt, *Flagstad* (London, 1987).

**Flamand.** A musician (ten), rival of the poet Olivier for the Countess's hand, in Strauss's *Capriccio*.

**Flavio.** Opera in 3 acts by Handel; text by Haym, partly based on Corneille's tragedy *Le Cid* (1637) and on Stampiglia's libretto *Il Flavio Cuniberti* (1696), adapted in turn from a libretto *Flavio Bertarido* by Noris (1682) set by Pollarolo (1706). Prod. London, H, 14 May 1723, with Cuzzoni, Durastini, Robinson, Senesino, Boschi, Berenstadt. Rev. Abingdon 1969.

**Fledermaus, Die** (The Bat). Operetta in 3 acts by Johann Strauss II; text by Haffner and Genée, after a French *spirituel vaudeville*, *Le réveillon* (1872), by Meilhac and Halévy, based in turn on Roderich Bendix's comedy *Das Gefängnis* (1851). Prod. Vienna, W, 5 Apr. 1874, with Charles-Hirsch, Geistinger, Szika, and Lebrecht, cond.

J. Strauss; New York, Stadt T, 21 Nov. 1874, with Cotrelly, Fiebach, Schnelle, Fritze, cond. Bial; London, Alhambra T, 18 Dec. 1876, with Cabella, Chambers, Loredan, Rosenthal, cond. Jacobi.

Vienna, 19th cent. Rosalinde (sop) is being serenaded by her former admirer Alfred (ten), an opera singer; her husband, Eisenstein (ten), is about to go to prison for having insulted the tax-collector. His friend Dr Falke (bar) comes supposedly to conduct him to prison, but suggests that they both should go to the party being given by Prince Orlofsky (mez, now often ten), and that Eisenstein should present himself at the prison the following morning. When the two men have gone, Alfred returns; his tête-à-tête with Rosalinde is disturbed by the arrival of Frank (bar), the prison governor who has come to conduct Eisenstein to jail. He mistakes Alfred for Eisenstein and takes him off to prison. Rosalinde, also invited to Orlofsky's party, arrives there masked, pretending to be a Hungarian countess; Eisenstein flirts outrageously with her. The plot is further complicated by the fact that Eisenstein's maid, Adele (sop), wearing one of her mistress's dresses, is also at the party. All misunderstandings are cleared up the following morning in the prison.

**Flensburg.** City in Schleswig-Holstein, Germany. The Städtisches Theater (cap. 610) opened 1894.

**Fleta, Miguel** (*b* Albalate, 1 Dec. 1893; *d* La Coruña, 29 May 1938). Spanish tenor. Studied Barcelona Cons., and Italy with Pierrich. Début Trieste 1919 (Paolo, *Francesca da Rimini*). New York, M, 1923–5; Milan, S, 1924 and 1926, creating Calaf. Also first Romeo in Zandonai's *Giulietta e Romeo*. Possessed a velvety, expressive voice and sang with passion, but declined early. (R)

**Fliedermonolog.** The monologue in Act II of Wagner's *Die Meistersinger von Nürnberg*, in which Sachs (bs-bar) sits under the elder tree outside his workshop reflecting on the haunting novelty of the song Walther has sung at his trial before the Masters.

**Fliegende Holländer, Der** (The Flying Dutchman). Opera in 1 act, later 3 acts, by Wagner; text by the composer, after the legend as told in chapter 7 of Heine's *Aus den Memoiren des Herren von Schnabelewopski* (1831). First used as scenario for Dietsch's *Le vaisseau fantôme* (1842). Prod. Dresden, H, 2 Jan. 1843, with Schröder-Devrient, Wächter, Reinhold, Risse, Bielezizky, cond. Wagner; London, DL, 23 July 1870, with Di Murska, Corsi, Perotti, Rinaldini, Foli, Santley, cond. Arditi (in Italian; first Wagner opera in London); Philadelphia, AM, 8 Nov. 1876, with Pappenheim, Baccei, Preusser, Sullivan, cond. Carlberg (in Italian).

A Norwegian fishing village, 1650. The Dutchman (bar) has been condemned, for his blasphemy, to sail his ship until redeemed by a faithful woman. This salvation he is allowed to attempt once every seven years. In one of these periods he is driven by storms to a Norwegian harbour. He moors beside the ship of Daland (bs), who, impressed by the Dutchman's wealth, offers him shelter. Daland's daughter Senta (sop) sings to her friends, as they spin, the Ballad of the Flying Dutchman. Her lover Erik (ten) pleads his own cause, unsuccessfully. But when the Dutchman enters, she is confused by his declaration of love. The Dutchman overhears Erik being rejected again, but concludes that he too is to be deserted by Senta. He sets sail; but Senta leaps to him from a cliff-top. Her faith redeems him, and together they are seen rising heavenwards.

Another opera on the subject is by Rodwell (1827).

**Floquet, Etienne** (*b* Aix-en-Provence, 23 Nov. 1748; *d* Paris, 10 May 1785). French composer. Moving to Paris in 1769, he first came to public attention in 1773 with his opéra-ballet *L'union de l'Amour et des arts*, which received 60 performances. The ballet which followed, *Azolan* (1774), was less successful and he left for Italy, studying with Martini. On his return he composed several stage works, including *Le seigneur bienfaisant*. Often described as the first French lyrical comedy, for its rejection of a heroic or mythological plot, it stands, like *La nouvelle Omphale*, in direct opposition to the reforms of Gluck. Floquet apparently tried to highlight this conflict by setting an adaptation by Razins de Saint-Marc of Quinault's *Alceste* libretto (1783, unprod.). Though he was regarded as one of the main supporters of the Italianate faction in the struggle between the supporters of Gluck and Piccinni, his musical style was influenced as much by his compatriots Grétry and Philidor as by Piccinni.

SELECT WORKLIST (all first prod. Paris): *Le seigneur bienfaisant* (The Benevolent Lord) (De Chabannes; 1780); *La nouvelle Omphale* (Beaunoir; 1782).

BIBL: A. Pougin, *Floquet* (Paris, 1863).

**Florence** (It., Firenze). City in Tuscany, Italy. Home of the Florentine *camerata. Peri's *Dafne* was given 1598, his *Euridice*, 1600, in the Pitti Palace. Opera was also given (and still is) in the Boboli Gardens and in the Uffizi. The Accademia degli Immobili built the wooden T della Pergola, opened 1656 with Jacopo Melani's *La Tancia*, opened to the public 22 Jun. 1718 with Vivaldi's

*Scanderbeg*; altered by Antonio Galli-Bibiena 1738; rebuilt 1857 (cap. 1,500). Scene of Italian premières of Mozart's *Figaro* (1788), probably *Don Giovanni* (1792), and *Entführung* (1935), also premières of Donizetti's *Parisina* (1833) and Verdi's original *Macbeth* (1847). The T Politeama Fiorentino Vittorio Emmanuele opened 1864 (open-air; covered 1883); renamed T Comunale 1932; rebuilt 1960; reopened (cap. 2,500) 1961.

The annual May Festival (Maggio Musicale) was established in 1933, largely due to Vittorio *Gui, who had in 1928 founded the Florence Orchestra around which the festival was built. Originally planned as biennial, it became annual in 1938 and has continued as such apart from 1943–7. Opera performed in the T Comunale, T Pergola, and the Boboli Gardens is the chief attraction. Verdi and Rossini are the composers who have contributed most to the festival's success, though the organizers have also used its renown to give a platform to lesser-known works by Bellini, Cherubini, Donizetti, and Spontini, as well as by contemporary composers, including Pizzetti, Dallapiccola, and Frazzi. There have also been many first Italian performances, and revivals of works by Cavalli, Peri, Gagliano, also Meyerbeer.

BIBL: R. and W. Wright, *A Chronology of Music in the Florentine Theatre, 1590–1750* (Detroit, 1975).

**Florentine Camerata.** See CAMERATA.

**Florestan.** A Spanish nobleman (ten), the unjustly imprisoned husband of Leonore in Beethoven's *Fidelio*. Also the Governor of Linz Castle (bs) in Grétry's *Richard Cœur de Lion*.

**Floridante.** Opera in 3 acts by Handel; text by Rolli, based on Silvani's libretto *La costanza in trionfo* set by Ziani (1696). Prod. London, H, 9 Dec. 1721, with Robinson, Salvai, Senesino, Boschi.

**Flotow, Friedrich von** (*b* Teutendorf, 27 Apr. 1812; *d* Darmstadt, 24 Jan. 1883). German composer. Studied Paris with Rejcha. He wrote part of Grisar's *Lady Melvil* in 1838 and *L'eau merveilleuse* in 1839, and his first public success was with a collaborative work, *Le naufrage de la Méduse* (Act I by Pilati): this was rewritten by Flotow as *Die Matrosen*. He wrote several other operas for Paris, notably *Alessandro Stradella*. This was followed by his most enduring success, *Martha*, which triumphed at its Vienna première and was promptly taken up by Liszt at Weimar. Few of his later operas achieved this popularity, though he had some success with *La veuve Grapin*, *Zilda*, and *L'ombre*. He was Intendant at Schwerin, 1855–63.

Flotow's operas are cosmopolitan in style, in that the early influence of French opéra-comique was to be joined by an Italian lyricism in his vocal writing and an affection for the orchestra inherited from German Romantic opera. His music is light, damagingly so when he tackles large subjects or serious issues (even in *Alessandro Stradella* it is the comic numbers which are the most memorable); but at its best his work has a grace, fluency, and charm that have continued to beguile audiences. He made a very successful Paris career by ingeniously modifying some of the more sentimental features of German Romantic opera to appeal to French taste. *Martha* has continued to survive the criticisms easily levelled at it, and the sweet Irish melody which he uses as a reminiscence motive, 'The Last Rose of Summer', is barely distinguishable from the manner of the rest of the score.

SELECT WORKLIST: *Pierre et Cathérine* (Saint-Georges, trans. Gabillon; Ludwigslust 1835); *Stradella* (Duport & Forges; Paris 1837); *Le naufrage de la Méduse* (The Wreck of the Medusa) (H. & T. Coignard; Paris 1839; rev. as *Die Matrosen*, Hamburg 1845); *L'esclave de Camoens* (Camoëns's Slave) (Saint-Georges; Paris 1843; rev. text by Putlitz as *Indra*, Vienna 1852); *Alessandro Stradella* (Riese, after Stradella; Hamburg 1844); *Martha* (Riese, after Saint-Georges, after a vaudeville; Vienna 1847); *La veuve Grapin* (The Widow Grapin) (Forges; Paris 1859); *Zilda* (Saint-Georges, Chivot, & Duru; Paris 1866); *L'ombre* (The Phantom) (Saint-Georges & Leuven; Paris 1870).

**Flower Duet.** The name sometimes given to the duet 'Scuoti quella fronda di ciliego', sung by Butterfly (sop) and Suzuki (mez) in Act II of Puccini's *Madama Butterfly* as they strew the house with flowers against Pinkerton's return.

**Flowermaidens.** The group of maidens (sop and alt) whom Klingsor sets to seduce Parsifal in Wagner's *Parsifal*.

**Flower Song.** The name usually given to Don José's (ten) aria 'La fleur que tu m'avais jetée' in Act II of Bizet's *Carmen*, telling Carmen how he has treasured the flower she threw him, and with it the hope of her love.

**Floyd, Carlisle** (*b* Latta, SC, 11 June 1926). US composer. Studied Syracuse U with Bacon. He first came to the fore with *Susannah*, which transferred the biblical story of Susannah and the Elders to Tennessee, making effective use of folk idioms as well as of more advanced musical resources. His later operas have not met with the same success, though *Of Mice and Men* was praised for its ability to use simple musical means to good theatrical effect.

SELECT WORKLIST: *Susannah* (Floyd, after the Apocrypha; Tallahassee 1955); *Wuthering Heights* (Floyd, after E. Brontë; Santa Fe 1958); *The Passion of Jonathan Wade* (Floyd; New York 1962); *Of Mice and Men* (Floyd, after Steinbeck; Seattle 1970); *Bilby's Doll* (Floyd, after Forbes; Houston, 29 Feb. 1976); *Willie Stark* (Floyd, after Warren; Houston, 24 Apr. 1981).

**Flying Dutchman, The.** See FLIEGENDE HOLLÄNDER, DER.

**Fodor-Mainvielle, Joséphine** (*b* Paris, 13 Oct. 1789; *d* Saint-Genis, 14 Aug. 1870). French soprano. Studied St Petersburg with Bianchi and Cavos. Début St. Petersburg 1808 or 1810 (Fioravanti's *Le cantatrici villane*). Paris, OC and I, 1814; London, H, 1816–18 (Fiordiligi, Countess (Mozart), Zerlina, and first London Rosina and Elisabetta (Rossini)). Successes in Paris, Naples, and Vienna, where she sang 60 performances of Semiramide in the 1824–5 season. After losing her voice (perhaps through over-use of her resonant lower register) at the end of 1825, she retired to Naples, where she attempted two come-backs in 1828 and 1831, without success. At her peak she was a brilliant and charming singer, dazzling in florid roles. Stendhal thought that her singing of Rosina approached perfection, while Weber wrote after hearing her, 'If only she would sing Euryanthe, one would go mad.'

BIBL: J. Fodor-Mainvielle, *Conseils et réflexions sur l'art du chant* (Paris, 1857).

**Foerster, Josef Bohuslav** (*b* Prague, 30 Dec. 1859; *d* Nový Vestec, 29 May 1951). Czech composer. Studied Prague Organ School. He first taught in Prague, moving with his wife Bertha Foerstrová-Lautererová to Hamburg in 1893 and Vienna in 1903, and working in both cities as a critic. In 1918 they returned to Prague, where Foerster taught at the Conservatory and elsewhere. As a critic, he wrote civilized and perceptive essays in defence of Romantic and Nationalist music. His six operas show an increasing metaphysical interest. The first two reflect real life: *Eva*, his greatest success, is a Moravian lyric tragedy that owes something to Smetana. With *Srdce* and *Bloud* the characters are increasingly identified with moral states: the expression is inward, personal, and symbolic, the drama spiritual and embodied in music that is at its finest exalted and impressive.

WORKLIST (all first prod. Prague): *Debora* (Mosenthal & Kvapil; 1893); *Eva* (Foerster, after Preissová; 1899); *Jessika* (Vrchlický, after Shakespeare; 1905); *Nepřemožený* (The Invincible) (Foerster; 1918); *Srdce* (The Heart) (Foerster; 1923); *Bloud* (The Simpleton) (Foerster, after Tolstoy; 1936).

WRITINGS: Autobiographies *Paměti poutník* (Memoirs of a Pilgrim) (Prague, 1931–2), *Poutník v Hamburku* (A Pilgrim in Hamburg) (Prague, 1938), *Poutník v cizině* (A Pilgrim Abroad) (Prague, 1947); abridged trans. as *Der Pilger* (The Pilgrim) (Prague, 1955).

BIBL: J. Bartoš, P. Pražák, and J. Plavec, eds., *J. B. Foerster* (Prague, 1949).

**Foerstrová-Lautererová, Bertha** (*b* Prague, 11 Jan. 1869; *d* Prague, 9 Apr. 1936). Czech soprano. Studied Prague with Poldková-Panachová and Tauwitz. Début Prague, N, 1887 (Agathe). Sang many important Czech roles; also first Prague Desdemona (Verdi) and Tatyana (Tchaikovsky found her excellent). Hamburg, 1893–1901, becoming one of Mahler's favourite singers. He summoned her to Vienna, where, as Foerster-Lauterer, she sang 1903–13. Described in the *Neue Freie Presse* as 'a first-rate artist' and 'a pillar of the ensemble', she was impressive for her full, ripe voice and mature dramatic sense. Her repertory embraced Mozart, Bellini, Weber, Verdi, Richard Strauss, and Wagner. Married Joseph Bohuslav Foerster, 1888.

**Foignet, Charles Gabriel** (or Jacques) (*b* Lyons, 1750; *d* Paris, 1823). French singer and composer. From 1791 he wrote opéras-comiques and vaudevilles, initially with Louis Simon, for the newly opened Paris public theatres; of one of these, the T Montansier, he and Simon became two of the joint administrators. He also founded the T des Victoires-Nationales, and directed the T des Jeunes Artistes. His son **François Foignet** (*b* Paris, 17 Feb. 1782; *d* Strasbourg, 22 July 1845) was a singer and composer who made his name at the Jeunes Artistes, especially in his own opéra-comique, *La naissance d'Harlequin* (1803). On Napoleon's closure of the small Paris theatres in 1807, he worked in Liège, Bruges, Marseilles, Nantes, Angoulême, and elsewhere.

**Foli** (Foley), **A**(llan) **J**(ames) (*b* Cahir, 7 Aug. 1835; *d* Southport, 20 Oct. 1899). Irish bass. Studied Naples. Début Catania 1862 (Elmiro, Rossini's *Otello*). After appearances in other Italian cities, London début, HM, 1865 (Saint-Bris). Sang in over 60 operas at HM, DL, and CG; Daland in the first Wagner performance in England, 1870. USA with Mapleson's Co. 1878–9; also sang in Russia and Austria. Possessed a powerful voice ranging from E to f '.

**Fomin, Evstigney** (*b* St Petersburg, 16 Aug. 1761; *d* St Petersburg, end Apr./early May 1800). Russian composer. Studied St Petersburg with Buini, Raupach, and Sartori, Bologna with Martini and Mattei. His first stage work was an operaballet, *Novgorodsky Bogatyr Boyeslavich*, to a text by Catherine the Great, in which he makes some use of folk tunes. From 1797 he was a répétiteur at the Imperial Theatres. *Yamshchiky na podstave* is remarkable for its use of Russian folk polyphony as well as its vivid depiction of the life of the post-coach drivers, and is thus an important example for the development of Russian opera.

However, his most important work is the melodrama *Orfey i Evridika*, an enterprising work that includes a number of technical and instrumental experiments (including a powerful Chorus of Shades). The most talented Russian opera composer of the end of the 18th cent., even if he did not live to develop his gifts to the full, he brought genuinely Russian elements into opera of an essentially Italian vernacular.

WORKLIST (all first prod. St Petersburg): *Novgorodsky Bogatyr Boyeslavich* (The Novgorod Bogatyr Boyeslavich) (Catherine II; 1786); *Yamshchiky na podstave* (The Post-Coach Drivers) (Lvov; 1787); *Vecherinky* (Soirées) (?Fomin; 1788); *Koldun, vorozheya i svakha* (Magician, Fortune-Teller, and Matchmaker) (Yukin; 1789); *Orfey i Evridika* (Orpheus and Eurydice) (Knyazhnin; 1792); *Amerikantsy* (The Americans) (Krylov & Klushin; 1800); *Khlorida i Milon* (Kapnist; 1800); *Zolotoye yabloko* (The Golden Apple) (Ivanov; 1803).

**Fontana, Ferdinando** (*b* Milan, 30 Jan. 1850; *d* Lugano, 12 May 1919). Italian playwright and librettist. Wrote *Aldo e Clarenza* for Massa (1878), but best known as the author of two of Puccini's earliest opera texts, *Le villi* (1884) and *Edgar* (1889), a collaboration arranged by Ponchielli. Also librettist of Franchetti's *Asrael* (1888), and the Italian translator of *Tiefland* and *Die lustige Witwe*. A member of the later *Scapigliatura.

**Foppa, Giuseppe** (*b* Venice, 12 July 1760; *d* Venice, 1845). Italian librettist. Apart from a brief period in Lisbon (1797–8), he lived largely in Venice, where he wrote over 80 works for composers including Andreozzi, Bertoni, Bianchi, Coccia, Farinelli, Fioravanti, Generali, Mayr, Nasolini, Paer, Pavesi, Rossini (*L'inganno felice*, *Il signor Bruschino*, and *Sigismondo*), Spontini (*Le metamorfosi di Pasquale*), and Zingarelli. His talent was for comedy, and his texts draw on the stock characters and situations of the 18th cent. and especially of the commedia dell'arte, somewhat under the influence of Goldoni. From the commedia dell'arte he selected and gave prominence to a number of scenes, including the *mad scene with flute obbligato. In his serious operas he showed more Romantic inclinations: he was one of the first librettists to introduce Shakespearian subjects into opera (though in the Ducis versions), e.g. *Giulietta e Romeo* for Zingarelli (1796).

**Ford.** A wealthy citizen of Windsor (bar), Alice's husband, in Verdi's *Falstaff* and Vaughan Williams's *Sir John in Love*. In Nicolai's *Die lustigen Weiber von Windsor* he is known as Herr Fluth (bar).

**Forest Murmurs** (Ger., *Waldweben*). The name given to the scene in Act II of Wagner's *Siegfried*, in which Siegfried lies basking in the beauty of the forest just before his encounter with Fafner.

**Foresto.** An Aquileian knight (ten) in Verdi's *Attila*.

**Forsell, John** (*b* Stockholm, 6 Nov. 1868; *d* Stockholm, 30 May 1941). Swedish baritone. Début Stockholm 1896 (Rossini's Figaro). Sang there regularly until 1909. London, CG, 1909 (Don Giovanni); New York, M, 1909–10. Also sang in Berlin, Vienna. Dir. Stockholm O 1924–39. His enormous repertory included Germont, Amfortas, Sachs, Onegin, and Valentin, but he was most famous as Don Giovanni, which he was still singing in 1930 (Salzburg). Possessed a beautiful voice, excellent technique, and a refreshing vigour on stage. As Prof. of Singing at Stockholm Cons., he taught J. Björling and Svanholm. (R)

**Forti, Anton** (*b* Vienna, 8 June 1790; *d* Vienna, 16 July 1859). Austrian baritone. Eszterháza 1807–11; Vienna, W 1811–13 and K 1813–35. Also sang at Prague, Hamburg, and Berlin. In 1814 he took over Pizarro from J. M. Vogl (to Beethoven's delight—he thought Forti's voice 'better adapted' to the role) for the composer's benefit on 18 July, being described as 'entirely satisfactory' in the press. Created Lysiart 1823. A handsome, intelligent man with an exceptional voice, he was influential as both actor and singer. Celebrated for his elegant Don Giovanni; also acclaimed as Sarastro, Count Almaviva (Mozart), and Figaro (Rossini). Retired after winning the State lottery. His wife **Henriette Forti** (*b* Vienna 1796; *d* Vienna 1818) was a successful light soprano until her early death; she sang at Prague, Frankfurt, and Berlin, as Zerlina, Cherubino, and the Page in Boieldieu's *Jean de Paris*.

**Fortner, Wolfgang** (*b* Leipzig, 12 Oct 1907; *d* Heidelberg, 5 Sep. 1987). German composer. Studied Leipzig with Grabner. A distinguished teacher, his many pupils have included Henze. He first came to prominence as an opera composer with his Lorca setting *Die Bluthochzeit*, following this with the more lyrical *Don Perlimplin*; both works have won considerable admiration in Germany. He has made use of a modern and personal version of Singspiel, with actors but also singers having speaking parts; and he has a particularly sensitive feeling for the orchestra.

SELECT WORKLIST: *Die Bluthochzeit* (Blood Wedding) (Fortner, after Lorca; Cologne 1957); *Corinna* (Fortner, after Nerval; Berlin 1958); *In seinem Garten liebt Don Perlimplin Belisa* (Don Perlimplin loves Belisa in his Garden) (Fortner, after Lorca; Schwetzingen 1962); *Elisabeth Tudor* (Braun; Berlin 1972).

BIBL: H. Lindlar, ed., *Wolfgang Fortner* (Rodenkirchen, 1960).

**Förtsch, Johann Philipp** (*b* Wertheim am Main, *bapt.* 14 May 1652; *d* Eutin, 14 Dec. 1732). German composer. After studying medicine in Jena, joined the Hamburg Ratschor and Oper as a singer. Hofkapellmeister to the Duke of Schleswig-Holstein 1680: court physician in 1690. Doctor and privy councillor to the bishop of Lübeck 1692.

Förtsch was one of the most influential figures in the early history of opera in Hamburg. While still a singer at the Theater am Gänsemarkt he also wrote several librettos and in 1684–90 was effectively house composer, when his works dominated the repertoire. None has survived complete, and his abilities must be assessed from excerpts from a few surviving arias. The Venetian style which he had encountered on his visits to Italy inspired many of his works, including *Croesus*, *Alexander in Sidon*, and *Xerxes in Abydus*. To celebrate the coronation of Emperor Joseph I he composed *Ancile Romanum*, while *Der glückliche Grossvesier Cara Mustapha* is one of the earliest German operas on an historical topic. A fondness for Spanish themes is seen in *Das unmöglichste Ding* and in *Der irrende Ritter D. Quixotte de la Mancha*, the first German opera based on Cervantes.

SELECT WORKLIST (all first prod. Hamburg): *Das unmöglichste Ding* (Bostel, after Vegas; 1684); *Croesus* (Bostel, after Minato; 1684); *Alexander in Sidon* (Förtsch, after Ziani; 1688); *Xerxes in Abydus* (?, after Minato; 1689); *Ancile Romanum* (Postel; 1690); *Der irrende Ritter D. Quixotte de la Mancha* (Hinsch; 1690); *Der glückliche Grossvesier Cara Mustapha* (Bostel; 1696).

BIBL: C. Weidmann, *Leben und Wirken des Johann Philipp Förtsch* (Kassel, 1955).

**Forza del destino, La** (The Force of Destiny). Opera in 4 acts by Verdi; text by Piave, after the drama *Don Álvaro, o la fuerza del sino* (1835), by Angel de Saavedra Ramírez de Banquedano, Duke of Rivas, and on a scene from Schiller's drama *Wallensteins Lager* (1799). Prod. St Petersburg 29 Oct. (new style: 10 Nov.) 1862, with Barbot, Nantier-Didiée, Tamberlik, Graziani, Bassini, Angelini rev. Milan, S, 27 Feb. 1869, with Stolz, Benzi, Tiberini, Colonnese, Rota, Junca; New York, AM, 24 Feb. 1865, with Carozzi-Zucchi, Morensi, Massimiliani, Bellini, cond. Bergmann; London, HM, 22 June 1867, with Tietjens, Trebelli, Mongini, Santley, Gassier, cond. Arditi.

Spain and Italy, mid-18th cent. The Marquis of Calatrava (bs) surprises the elopement of his daughter Leonora (sop) and Don Alvaro (bar). Professing Leonora's innocence, Alvaro throws his pistol to the ground. It goes off, killing the Marquis.

In their escape, Leonora and Alvaro are separated. Disguised as a man, Leonora encounters her brother Don Carlo (bar), posing as a student, in a village inn. Carlo relates a story showing he is determined to kill his sister's seducer. Leonora believes that Alvaro has deserted her and asks the Father Superior (Padre Guardiano, bs) of a monastery if she may become a hermit nearby.

Alvaro, now a captain in the Spanish army, believes Leonora to be dead. He rescues a man who turns out to be Carlo, though neither reveals his real identity. At a call to arms, both men rush to join the fighting. Alvaro is wounded and it emerges that he is Leonora's seducer. He recovers, and Carlo reveals his identity and challenges him to a duel. A patrol discovers them fighting; Carlo is dragged away and Alvaro decides to enter a monastery.

Alvaro is living as Father Raphael. Carlo arrives, having been seeking him for five years, and provokes him to fight. Alvaro wounds Carlo mortally and seeks absolution from the hermit, who at first refuses. Leonora and Alvaro recognize each other. Alvaro confesses that he has just injured Carlo and Leonora rushes off to her brother. The dying Carlo stabs Leonora and she dies, believing she and Alvaro will meet again in heaven.

**Forzano, Giovacchino** (*b* Borgo S Lorenzo, 19 Nov. 1883; *d* Rome, 28 Oct. 1970). Italian librettist. Studied singing and appeared in baritone roles in the Italian provinces. Provided librettos for Mascagni (*Lodoletta*, *Il piccolo Marat*), Puccini (*Gianni Schicchi*, *Suor Angelica*), Wolf-Ferrari (*Sly*), Leoncavallo (*Edipo re*), Giordano (*Il re*), and Marinuzzi (*Palla de' Mozzi*), among others. Began producing operas in 1904, and worked at the Milan, S (1922–3), and other Italian cities, as well as in London (CG), Vienna, Berlin, Budapest, and Brussels. He was responsible for staging the world premières of *Turandot* and Boito's *Nerone*, among other works. A fluent, skilful, and witty craftsman.

**Foss** (orig. Fuchs), **Lukas** (*b* Berlin, 15 Aug. 1922). US composer of German origin. Studied Berlin with Goldstein, Paris with Gallon, Philadelphia with Scalero and Thompson, later also with Hindemith. A gifted and versatile musician, he has composed music in many different manners, winning early success as an opera composer with his deft, appealingly written comedy *The Jumping Frog of Calaveras County*. His comic gifts serve more serious points in the 'miracle play' *Griffelkin*.

WORKLIST: *The Jumping Frog of Calaveras County* (Karsavina, after Twain; Bloomington 1950); *Griffelkin* (Reid; TV 1955, Tanglewood 1956); *Introductions and Goodbyes* (Menotti; New York 1960).

**Four Saints in Three Acts.** 'An opera to be sung' in 4 (*sic*) acts by Virgil Thomson; text by Gertrude Stein. Concert perf. Ann Arbor 20 May 1933; prod. Hartford, by Society of Friends and Enemies of Modern Music, Avery Memorial T, 8 Feb. 1934, cond. Smallens; Paris 1952.

The action takes place in Spain and takes the form of an allegory in which Saint Theresa (2 roles: St. Theresa 1 (sop) and St. Theresa 2 (mez)), surrounded by men, and Saint Ignatius (bar), surrounded by women, work together and help one another to become saints.

Originally performed by an all-Black cast.

**Fox, Carol** (*b* Chicago, 15 June 1926; *d* Chicago, 21 July 1981). US manager. Studied singing Chicago, New York, and in Italy. With Nicola Rescigno and Lawrence Kelly founded Lyric O of Chicago 1952; general manager 1956–80. She dominated opera in Chicago by the force of her personality, raising standards to an international level, engaging outstanding singers, conductors, producers, and designers, and enlarging the repertory. Ill health, and unwillingness to bow to budgetary pressures, led to her resignation in 1980.

**Fra Diavolo.** Opera in 3 acts by Auber; text by Scribe. Prod. Paris, OC, 28 Jan. 1830 with Boulanger, Prévost, Chollet, Féréol, Moreau-Sainti; London, DL, 1 Feb. 1831; Philadelphia, Chestnut St T (by New Orleans Co.), 16 Sep. 1831, with Saint-Clair, Milon, Saint-Aubin, Privat.

Fra Diavolo (ten), a notorious bandit, passing himself off as the Marquis of San Marco, compromises Zerlina (sop), the daughter of an innkeeper, while engaged in an elaborate plot to steal the jewels of Lady Pamela (mez), wife of Lord Cockburn (bs). Betrayed by his associates, Fra Diavolo is shot, having cleared Zerlina of any involvement and reunited her with her lover, Lorenzo (ten). An alternative ending has Fra Diavolo taken prisoner.

Also parody by Drechsler (1830).

**Fra due litiganti il terzo gode.** See DUE LITIGANTI, I.

**Fra Gherardo.** Opera in 3 acts by Pizzetti; text by the composer after the Chronicles of Salimbene da Parma (13th cent.). Prod. Milan, S, 16 May 1928, with Cristoforeanu, Stignani, Trantoul, Baccaloni, cond. Toscanini; New York, M, 21 Mar. 1929, with Müller, Claussen, Johnson, Pinza, cond. Serafin.

Parma, 1260. Gherardo (ten), a weaver, joins the order of the White Friars following a love affair with Mariola (sop). He leads a revolt of the oppressed people of Parma against the nobles and dies at the stake; Mariola is killed by a madwoman.

**'Fra gli amplessi'.** Duet in Act II of Mozart's *Così fan tutte*, in which Fiordiligi (sop) begins by trying to leave the disguised Ferrando (ten) to return to her lover and instead ends up in his arms.

**Fra Melitone.** A Franciscan friar (bar) in Verdi's *La forza del destino*.

**Françaix, Jean** (*b* Le Mans, 23 May 1912). French composer. Studied Paris with Boulanger. His lively, skilful invention has found expression principally in concert works, but he has also had success on the stage with his ballets and with an early chamber opera, *Le diable boiteux*. His most ambitious stage work is *La princesse de Clèves*.

WORKLIST: *Le diable boiteux* (The Lame Devil) (Françaix, after Le Sage; Paris 1938); *L'apostrophe* (Françaix, after Balzac; comp. 1940, radio 1947, Amsterdam 1951); *La main de gloire* (The Hand of Glory) (Françaix, after Nerval; Bordeaux 1950); *Paris à nous deux* (Paris for Us Two) (Roche & Françaix; Fontainebleau 1954); *La princesse de Clèves* (Françaix & Lanjean, after Lafayette; comp. 1953).

**France.** By the late 16th cent. various dramatic musical entertainments were fashionable at court, principally the *ballet de cour and its derivatives. These were distinguished by an emphasis on their dance content which has to varying degrees always marked the French approach to opera. During Louis XIV's minority, an attempt was made by Mazarin to establish Italian opera in Paris, probably beginning with Marazzoli's *Il giudizio* (perf. privately 1645); this was given impetus by the arrival of the exiled *Barberini family from Rome in 1644. The performances of Italian opera which were given, including Cavalli's *Egisto* (1646) and Rossi's *Orfeo* (1647) were not well received: even the work which Cavalli wrote specially for Paris for Louis XIV's wedding, *Ercole amante* (1662), failed to please, despite his attempts to accommodate French taste. Under *Benserade, ballet remained in the ascendancy; particularly significant was the series of *comédie-ballets for which *Lully collaborated with Molière. In these an embryonic form of French recitative was developed and the musical element gradually came to predominate.

The first attempts at a national genre based on the Italian example came with the foundation of the *Académie Royale de Musique (as Académie Royale d'*Opéra) by *Perrin and *Cambert in 1669. They had already collaborated on a number of small-scale stage works; their *Pomone* (1671) has the best claim to being the first French opera. Following their next work, *Les peines et les plaisirs de l'amour* (1671), the couple were declared bankrupt and their monopoly on operatic performances was seized by Lully. Beginning with *Cadmus et Hermione* in 1673, he produced a series of *tragédies en musique which established a style of French opera that survived, modified, for almost a century. Lully's basic concept owed much to classical models: a prologue (in which the monarch was usually praised) and five acts; dance functional to the entertainment, drawing in part on the tradition of the ballet de cour; an individual and expressive vein of French recitative, and a skilful use of grandiose choral forces. It was difficult for a less imposing musician than Lully to develop French opera significantly, despite the work of *Charpentier, *Campra, and *Destouches, until *Rameau, who in his *tragédies lyriques continued and developed his basic model, though placing less weight on the reflection of royal *gloire*. By now the grand formality and pomposity of the style was mocked in the *drame forain, reflecting the no less French characteristic of wit and irreverence; this eventually developed into the *opéra-comique.

The essentially French nature of this contrast, and of the love of forming factions, was shown when a troupe of Neapolitan opera buffa singers visited Paris in 1746 and 1752. There ensued the *Guerre des Bouffons. On the Italian side, *Rousseau wrote *Le devin du village* (1752). Though the liveliness of the Italians carried most Parisians with them, their style was soon absorbed into the French tradition. Having written *Orfeo* (1762) with Calzabigi under some influence of Rameau, *Gluck came to Paris in 1774 for *Iphigénie en Aulide* and French versions of *Orfeo* and *Alceste*. Gluck's success rekindled controversy, and *Piccinni was, against his will and despite his admiration for Gluck, made the subject of a rival faction. French light opera continued as plays in which music had a subsidiary if essential role, though *Philidor (e.g. *Tom Jones*, 1765) and *Duni (e.g. *Les moissonneurs*, 1768) gave naturalism and freshness greater expressive weight in a pre-Romantic manner. A third important figure was *Monsigny (e.g. *Le déserteur*, 1769), who without loss of immediacy in treatment gave opéra-comique still more substance.

With the Revolution in 1789, opéra-comique replaced the high tragédie lyrique as the most important and popular form (and was to be the ancestor of Romantic opera). *Grétry, *Dalayrac, *Catel, and others brought into opera many new ideas, from different ages and exotic countries, from different social milieus, and from fairytale and the supernatural as well as Nature and a Rousseau-inspired respect for Natural Man. The most gifted and original of this group was *Méhul, who wrote some light works but whose real gift was for the passionate and tragic (e.g. *Euphrosine*, 1790). An important consequence, especially given the French attention to

clarity of verbal expression, was the growth of interest in carrying on the action during a musical number. The orchestra took on a more functional character, and the opportunity for symphonic development led to a growth of interest in motive. Again it was an Italian, *Cherubini, who, together with Méhul, gave an important lead to opéra-comique, e.g. with *Lodoïska* (1791). Another important composer of dawning Romanticism was Le Sueur (e.g. *La caverne*, 1793). A significant genre to emerge as expressing the political and personal tensions of the period was *rescue opera.

The age of Napoleon gave France a renewed taste for grandeur and for demonstrations of *la gloire*, expressed now not to reflect aristocratic taste but, among other features, to celebrate the Hero, conquest, and the sweep of history; the audience, especially after the Restoration in 1815, was increasingly a prosperous bourgeoisie, uneducated in taste but impressed by pomp and spectacle. The style was, once again, set by an Italian, Napoleon's preferred composer *Spontini, with *La vestale* (1807); this initiated a genre of large, loud, long operas, usually on historical subjects, demanding elaborate spectacle and grandiose décor. Characteristic works which, in Spontini's wake, formed the new style of *Grand Opera in the 1820s included *Auber's *La muette de Portici* (1828) and Rossini's *Guillaume Tell* (1829). This manner was brought to its highest point by *Meyerbeer, initally with *Robert le diable* (1831), and matched by *Halévy with *La Juive* (1835).

For the first half of the 19th cent. the supremacy of Paris as an operatic centre was unquestioned, and the style of the Opéra influenced composers from Verdi and Wagner to Tchaikovsky. Not all French composers were able to match its demands: Berlioz failed to obtain a footing at the Opéra with *Benvenuto Cellini* in 1838, and wrote *Les Troyens* without hope of French production. Once again in French operatic history, the grandeur and strength that impressed so many musicians brought with it an inflexibility that excluded too much. Not surprisingly, French vivacity found expression in comic opera. Auber had (to Wagner's dismay) moved on to a series of light, rather thin works, most successfully with *Fra Diavolo* (1830). *Adam won a great success with *Le postillon de Lonjumeau* (1836). A reconciliation between the two strains of opera was one of the reasons for the outstanding popularity of *Gounod's *Faust* (1859). A more significant step came with *Bizet's *Carmen* (1875), whose passionate naturalism captivated musicians all over Europe, and was an important example for *verismo. The sentimental strain in French opera exploited by Gounod was given a distinctive new colouring by *Massenet, especially in the combination of sex and religion in *Hérodiade* (1881) and *Thaïs* (1894), and in the romantic tragedies *Manon* (1884) and *Werther* (1892). *Saint-Saëns's operas, especially *Samson et Dalila* (1877), also won wide popularity. The realistic strain found new life in the work of *Bruneau (e.g. *L'attaque du moulin*, 1893), who had Zola as his librettist, and of Gustave *Charpentier in *Louise* (1900). Meanwhile, the love of farce had been indulged with a brilliant tradition of operetta that reached its most characteristic form in the irreverent, irresistible works of *Offenbach; also popular were *Audran, *Planquette, *Lecocq, and others.

The effect of Wagner went deeper in France that in any other country outside Germany, and a number of works were written either accepting or constructively rejecting the influence. *Debussy's *Pelléas et Mélisande* (1902) both rejects Wagner and is coloured by him; it is deeply French in its delicacy, its literary awareness, its care for good declamation, its subtle use of the orchestra. Wagnerisms had crept into the work of D'*Indy, *Chabrier, *Lalo, *Dukas, and *Franck, too much so for their acceptance into the mainstream of French opera, despite many fine inventions. However, Chabrier also showed his light touch, as in his turn did *Ravel with two 1-act comedies. *Roussel, in *Padmâvatî* (1923), attempted a recreation of *opéra-ballet. In modern times, the Grand Opera tradition has proved less adaptable to new ideas, despite the example of a few works, such as *Milhaud's *Christophe Colomb* (1930), than that of light opera. *Poulenc's talent was to be characteristically employed in *Les mamelles de Tirésias* (1947) but also embraced his deeply felt *Dialogues des carmélites* (1957); and Messiaen's powerful Christian faith has taken expression in *St François d'Assise* (1983). Important contributions to the French tradition have also been made by *Sauguet and *Tomasi. The youngest generation has shown an interest in experimenting with new dramatic forms in almost all kinds and directions.

See also AIX-EN-PROVENCE, AVIGNON, BORDEAUX, COLMAR, LILLE, LYONS, MARSEILLES, METZ, MONACO, MULHOUSE, NANCY, NANTES, NICE, OPÉRA DU RHIN, ORANGE, PARIS, ROUEN, STRASBURG, TOULOUSE, TOURS, VERSAILLES.

**Francesca da Rimini.** Opera in 4 acts by Zandonai; text by Tito Ricordi, after D'Annunzio's tragedy (1902), after Dante. Prod. Turin, R, 19 Feb. 1914, with Cannetti, Crimi, Cigada, Nessi, cond. Panizza; London, CG, 16 July 1914, with Edvina, Martinelli, Cigada, Paltrinieri, cond. Panizza; New York, M, 22 Dec. 1916, with Alda, Martinelli, Amato, Bada, cond. Polacco.

Ravenna and Rimini, late 13th cent. The story, based on the fifth canto of Dante's *Inferno*, tells how Francesca (sop) falls in love with Paolo (ten), thinking him to be his brother Gianciotto (bar) whom she is to marry; he has in fact sent Paolo to bring her to him. Malatestino (ten), another brother, who has made unsuccessful advances to Francesca, betrays the lovers. They are discovered in one another's arms by Gianciotto, who kills Paolo.

There are some 30 operas on the subject. See also DANTE.

**Franchetti, Alberto** (Baron) (*b* Turin, 18 Sep. 1860; *d* Viareggio, 4 Aug. 1942). Italian composer. Studied Venice with Coccon and Maggi, Munich with Rheinberger, Dresden with Draeseke and Kretschmer. His private means enabled him to secure this thorough training and later to stage his operas under favourable conditions. His first opera, *Asrael*, proved to be a work whose apocalyptic scale drew effectively upon French Grand Opera and upon Wagner. Verdi was sufficiently impressed to suggest Franchetti to the city of Genoa for a work to celebrate the fourth centenary of the Genoese Columbus's discovery of America. In *Cristoforo Colombo* Franchetti was again able to indulge his penchant for the grandiose to good effect. *Germania* (though also widely performed, initially under the direction of his defender Toscanini) is more manufactured, and less effectively so than the popular *La figlia di Jorio*. His ambitious attempts to marry a native Italian idiom to a post-Wagnerian German manner he had learned from Draeseke foundered on a lack of real individuality in his often expert music.

SELECT WORKLIST: *Asrael* (Fontana; Reggio Emilia 1888); *Cristoforo Colombo* (Illica; Genoa 1892); *Il signore di Pourceaugnac* (Fontana, after Molière; Milan 1897); *Germania* (Illica; Milan 1902); *La figlia di Jorio* (D'Annunzio; Milan 1906); *Glauco* (Forzano, after Morselli; Naples 1922).

BIBL: B. Capobianco and others, *Ricordo di Alberto Franchetti* (Turin, 1963).

**Franck, César** (*b* Liège, 10 Dec. 1822; *d* Paris, 8 Nov. 1890). French composer of Belgian origin. Studied Liège with Dassoigne, Paris with Rejcha. Even the faithful D'Indy allowed that Franck's operas were insignificant, less operatic than some of the sacred music: blaming the librettos, he found Franck's musical style in these works less advanced than in the contemporary concert music.

WORKLIST: *Stradella* (Deschamps; comp. *c*.1844, unprod.); *Le valet de ferme* (Royer & Vaëz; comp. 1851–3, unprod.); *Hulda* (Grandmougin, after Bjørnson; comp. 1882–5, prod. Monaco 1894); *Ghisèle* (Thierry; comp. 1888–90, Monaco 1896).

BIBL: L. Davies, *César Franck and his Circle* (London, 1970).

**Franck, Johann Wolfgang** (*b* Unterschwaningen, *bapt.* 17 June 1644; *d* ?*c*.1710). German composer. At the court of the Margrave of Ansbach 1665–79, rising to be Hofkapellmeister and 'Director der Comoedie', but forced to flee after murdering one of his musicians and wounding his wife. Already known as an opera composer through e.g. *Die drey Töchter des Cecrops* (the first extant German opera in full score), he settled in Hamburg, becoming musical director of the T am Gänsemarkt. Kapellmeister at Hamburg Cathedral 1682–6, then moving to London, in whose concert life he was an active participant.

Franck's 17 operas are an important contribution to the history of opera in Hamburg, and in the development of German opera. At Ansbach he had directed many Italian operas, particularly those of Ziani. His own works, including *Semele* and *Diocletianus*, successfully reconcile the Venetian tradition of Cavalli and Cesti with that of Germany. At a time when most of his compatriots looked to Lully and to France for inspiration, Franck's preference for Italian music was noteworthy and played a crucial role in the development of a viable form of German recitative.

SELECT WORKLIST (all first prod. Hamburg): *Die drey Töchter des Cecrops* (Königsmark; 1679); *Semele* (Förtsch; 1681); *Diocletianus* (Bostel, after Noris; 1682).

**Franckenstein, Clemens von und zu** (*b* Wiesentheid, 14 July 1875; *d* Hechendorf, 22 Aug. 1942). German composer and conductor. Studied Munich with Thuille, Frankfurt with Knorr. Conducted USA 1901–2, England (Moody-Manners Co.) 1902–7, Wiesbaden 1907–8, Berlin 1908–12. Intendant Munich 1914–18, 1924–34, doing much to uphold and develop the city's operatic reputation. His brother was for many years Austrian ambassador to Britain. His wife **Maria Nežádal** (1897–?) was a soprano who sang in Munich, Vienna, and London.

WORKLIST: *Griseldis* (Mayer; Opava 1898); *Fortunatus* (Wassermann; Budapest 1909); *Rahab* (Mayer; Hamburg 1911); *Li-Tai-Pe* (Lothar; Hamburg 1920).

**Frank, Ernst** (*b* Munich, 7 Feb. 1847; *d* Oberdöbling, 17 Aug. 1889). German conductor and composer. Studied Munich with Lachner. He held conducting posts in Würzburg (1868), Mannheim (1872–8), Frankfurt (1878–9), and Hanover (1879–87). An energetic and intelligent musician, he did much to promote the works of his friend Goetz and of younger composers including Stanford.

WORKLIST: *Adam de la Halle* (Mosenthal, after Heyse; Karlsruhe 1880); *Hero* (Vetter, after Grillparzer; Berlin 1884); *Der Sturm* (The Tempest) (Widmann, after Shakespeare; Hanover 1887).

**Frankfurt** (Ger., Frankfurt am Main). City in Hesse, Germany. The first opera given was Johann Theile's *Adam und Eva* by Johann Velten's company in the Haus zum Krachbein on 4 June 1698. In the 18th cent. the city was dependent on touring companies. Lacking a court, it had no tradition of opera seria and was more drawn to opera buffa, Singspiel, and opéra-comique. Theobald Marchand was the first to give opéra-comique in German 1770–7. The popularity of Singspiel extended to the young Goethe, whose first Singspiel text, *Erwin und Elmire* was set by Johann André 1777. Other visiting companies included those of Abel Seyler (with Neefe), Böhm, and Grossmann.

In 1782 the Städtisches Comödienhaus opened in the Paradeplatz (later Theaterplatz) with a repertory centring on Mozart; also popular were Winter, Weigl, Salieri, and Paer. Friedrich Kunzen was music director 1792–4, followed by Carl Cannabich 1796. Though there were chorus difficulties, Cannabich was able to form an orchestra from the dissolved orchestras of various Rhineland cities. The repertory continued to be based on Mozart and Singspiel, with an increasing amount of opéra-comique. There was also an emphasis on décor, with the use of perspective in splendid sets by Giorgio Fuentes (1796–1805), e.g. for *Clemenza di Tito*, Weigl's *Corsar*, and Salieri's *Palmira*: Goethe tried to win him for Weimar. Weber's *Silvana* was premièred 1810; and Spohr was music director 1818–19, premièring his *Zemire und Azor* 1819. The repertory was extended to include Auber, Halévy, and Meyerbeer, also Wagner and Verdi; singers included Sontag, Schröder-Devrient, Lind, and Tichatschek. In 1842 the theatre became private, and was enlarged 1855.

In 1880 an opera-house (the present Alte Oper) was opened by Kaiser Wilhelm I, with Otto Dessof as music director, then Ludwig Rottenberg (1889–1923). Under Rottenberg there were many premières, including that of Schreker's *Der ferne Klang*. Clemens Krauss was music director 1924–9, working with singers including his wife Viorica Ursuleac; high standards were maintained, but there was less emphasis on contemporary music. William Steinberg was music director 1929, conducting the première of Schoenberg's *Von Heute auf Morgen* (1930); he was restricted by the Nazis, and was succeeded by Franz Konwitschny 1938–49; premières were given of works by Egk, Orff, and Reutter. The opera-house was bombed 1943. Opera was given in the Stock Exchange 1945–51, when the company moved to the rebuilt Schauspielhaus, with

Georg Solti as music director 1952–61. He was succeeded by Lovro von Matačić (1961–6), Christoph von Dohnányi (1968–77), Michael Gielen (1977–87), Gary Bertini (1987–90), and Sylvain Cambreling (from 1993). An arsonist destroyed the opera-house in November 1987, and performances continued in the Schauspielhaus. The new Oper Frankfurt opened 7 Apr. 1991.

**Franklin, David** (*b* London, 17 May 1908; *d* Worcester, 22 Oct. 1973). English bass. Studied Vienna with Strasser. Début Gly. 1936 (Commendatore). Sang there until 1939; roles included Sarastro, Banquo. London, CG, 1947–50 (Rocco, Pimen, Ochs (one of his finest roles)). Retired from singing 1951. Librettist of Phyllis Tate's *The Lodger*. A warm, intelligent artist with a generous voice. (R)

**Frantz, Ferdinand** (*b* Kassel, 8 Feb. 1906; *d* Munich, 26 May 1959). German bass-baritone. Studied privately. Début Kassel 1927 (Ortel, *Meistersinger*). Hamburg 1937–43; Munich 1943–59; New York, M, 1949–51, 1953–4; London, CG, 1953–4. A distinguished Sachs, Wotan, and Kurwenal; also sang Daland, King Mark, Galitsky. (R)

**Fraschini, Gaetano** (*b* Pavia, 16 Feb. 1816; *d* Naples, 23 May 1887). Italian tenor. Studied Pavia with Moretti. Début Pavia (Tamas in Donizetti's *Gemma di Vergy*). Naples, C, 1840–8, creating Gerardo in *Caterina Cornaro*, and roles in numerous operas by Mercadante, Pacini, Battista, etc. Sang in many other Donizetti works; known as the 'tenore della maledizione' from his powerful rendering of the curse in *Lucia*. London, HM, 1847; Vienna, K, 1851; Paris, I, 1864. Greatly admired by Verdi, he created the leading tenor roles in *Alzira*, *Il corsaro*, *La battaglia di Legnano*, *Stiffelio*, and *Un ballo in maschera*. He was also successful in *Simon Boccanegra*, *La forza del destino*, *La traviata*, *Rigoletto*, and *I vespri siciliani*. His voice, heroic and with a baritonal quality, sounded like 'a silver gong struck with a silver hammer' (Monaldi); yet Verdi and Donizetti appreciated his ability to sing softly and with sublety. His voice and technique were still intact at his retirement in 1873. The opera-house in Pavia is named after him.

**Frasi, Giulia** (*fl.* 1742–72). Italian soprano. Studied Milan with Brivio, and London with Burney. London, H, 1742–*c*.1761, in operas by Porpora, Galuppi, Hasse, Pergolesi, Handel, etc. Also sang at London, CG and DL. Handel himself trained 'this pleasing singer' and she sang in many of his oratorios, creating roles in *Solomon*, *Theodora*, *Jephtha*, and *The Choice of Hercules*. Her voice was 'sweet and clear', her singing 'smooth and chaste' (Burney), and in oratorio

she was particularly successful in the English provinces.

**Frate 'nnamorato, Lo** (The Brother in Love). Opera by Pergolesi; text by Gennaro Antonio Federico. Prod. Naples, T dei Fiorentini, 23 Sep. 1732.

The complicated plot tells of the sisters Nena (sop) and Nina (mez), who love Ascanio (ten) rather than their intended husbands; he is loved by Lucrezia (con). Eventually he is discovered to be the twins' brother, to Lucrezia's delight.

**Frau ohne Schatten, Die** (The Woman without a Shadow). Opera in 3 acts by Richard Strauss; text by Hugo von Hofmannsthal, after his own story (1919). Prod. Vienna, S, 10 Oct. 1919, with Lotte Lehmann, Jeritza, Weidt, Oestvig, Mayr, cond. Schalk; San Francisco, War Memorial OH, 18 Sep. 1959, with Schech, Lang, Dalis, Feiersinger, Yahia, cond. Ludwig; London, SW, by Hamburg Opera, 2 May 1966, with Tarrés, Kuchta, Dalis, Kozub, Crass, cond. Ludwig.

The Empress (sop) is a supernatural being who lacks a shadow—a symbol of her inability to bear human children. A messenger (bar) announces that the Empress must return to her own world and the Emperor (ten) will be turned into stone unless she can acquire a shadow within three days. The Nurse (mez) takes her to the house of the dyer, Barak (bar), whose wife (sop) is willing to sell her own shadow and renounce motherhood in exchange for treasure. The Nurse and Empress promise to return the following day.

The Nurse tempts the dyer's wife by offering her a handsome lover. The Empress is distressed by the anguish she is causing the dyer, although he is not aware of what is happening. Barak's wife tells him she has been unfaithful, but when he makes as if to kill her she admits that she did not actually commit the act, though she had intended to.

Barak and his wife, in an underground cave but separated by a thick wall, each try to find a solution in their own mind to the problem. The Empress enters the cave but refuses to drink the Water of Life, which would give her the woman's shadow. The spirit world rewards her for this unselfish act by giving her the long-desired shadow.

**Frazzi, Vito** (*b* S Secondo Parmense, 1 Aug. 1888; *d* Florence, 7 July 1975). Italian composer and musicologist. Studied Parma with Fano and Azzoni. A perceptive critic, he responded intelligently to the problems of setting *King Lear*, though there is a more personal voice to be heard in his response to another masterpiece, *Don Quixote*.

SELECT WORKLIST: *Re Lear* (Papini, after Shakespeare; Florence 1939); *Don Chischiotte* (Frazzi, after Cervantes & Unamuno; Florence 1952).

**Frederick II** (Frederick the Great, King of Prussia) (*b* Berlin, 24 Jan. 1712; *d* Potsdam, 17 Aug. 1786). Frederick's enthusiasm for music was aroused by a youthful hearing of Hasse's *Cleofide*, and on his accession in 1740 he was able to encourage the foundation of the *Berlin opera (opened in 1742, with Graun as Kapellmeister, though Hasse and Agricola were also performed). His tastes were for Italian opera, and not until 1771 did the artistry of Elisabeth Mara overcome a prejudice that had led him to say that he would prefer to hear his horse sing an aria rather than a German prima donna. He wrote or outlined various librettos for Graun, including *Montezuma* (1755), in which the da capo aria was displaced in primacy by the cavatina; these were versified by the court poets Villati and Tagliazucchi.

BIBL: E. Helm, *Music at the Court of Frederick the Great* (Norman, 1960).

**Freeman, David**. See OPERA FACTORY.

**Freia.** The goddess of youth (sop) in Wagner's *Das Rheingold*.

**Freiburg im Breisgau.** Town in Baden-Württemberg, Germany. A theatre opened in 1785, and opera was given regularly from 1790 with a repertory based on Mozart, Singspiel, and opera buffa. The Grosses Haus opened 1910; bombed 1944; reopened 1949 (cap. 1,133).

**Freischütz, Der** (The Freeshooter, i.e. the marksman with magic bullets). Opera in 3 acts by Weber; text by Friedrich Kind, after a tale in the *Gespensterbuch* (1811) of Johann Apel and Friedrich Laun. Prod. Berlin, Sch., 18 June 1821, with Seidler, Eunicke, Stümer, Blume, Wauer, Rebenstein, Wiedemann, Hillebrand, Gern, Reinwald, cond. Weber (500 perfs. there by 1884); London, L, 22 July 1824, with Paton, Stephens, Braham; New York 2 Mar. 1825, with Kelly, De Luce, Keene, Clarke, Richings.

Bohemia, after the Thirty Years War. The huntsman Max (ten) is in love with Agathe (sop), daughter of the head ranger Cuno (bs). He is eager to win a shooting competition to decide the next head ranger and thereby become eligible to marry Agathe, but loses in a trial. Another huntsman, Caspar (bs), has sold his soul to the evil spirit Samiel (speaker), and must bring another victim; Max agrees to get magic bullets from Samiel.

Agathe (sop), apprehensive though soothed by Ännchen (sop), is interrupted by Max, who pretends he must fetch a stag he has shot in the haunted Wolf's Glen. Amid weird sights and sounds in the Glen, Caspar forges seven bullets for Max—the last to go where Samiel wills, though Max does not know this.

Preparing to marry Max, Agathe prays for protection; she has had bad dreams, but Ännchen again comforts her. Max amazes everyone at the shooting contest with his marksmanship. The Prince orders Max to shoot a passing dove with the seventh; Agathe's voice is heard begging him not to. Max fires. Agathe falls, but it is Caspar who is fatally wounded through Samiel's treachery. Max confesses his pact with Samiel, but eventually on the intercession of the Hermit (bs) is promised forgiveness.

Other operas on the subject are by Neuner (1812), Rosenau (1816), and Roser (1816).

BIBL: A. Csampai and D. Holland, eds., *Carl Maria von Weber: Der Freischütz* (Reinbek bei Hamburg, 1981); W. Mika, ed., *Der Freischütz* (Krefeld, 1985).

**Fremstad, Olive** (*b* Stockholm, 14 Mar. 1871; *d* Irvington, NY, 21 Apr. 1951). Swedish, later US, soprano. Studied Berlin with Lilli Lehmann. Début as mezzo-soprano, Cologne 1895 (Azucena). Sang there until 1898. Vienna, Bayreuth, and Munich (1900–3); repertory at this time included Brangäne, Carmen. London, CG, 1902-3 (Ortrud, Venus). New York, M, 1903–14, by now a dramatic soprano (Sieglinde, Kundry, Brünnhilde, Isolde; also Selika, Tosca, and first American Armide (Gluck), and Salome). Chicago 1915–18, after which retired from stage but sang in concert until 1920. Intelligent and temperamental, versatile and cultivated, she was outstanding in her Wagner interpretations, to which she brought great vocal and dramatic gifts. Her popularity brought her 19 curtain calls when she left the Metropolitan in 1914. (R)

**French overture.** Beginning with the ballets *L'amore ammalato* (1657) and *Alcidiane* (1658), Lully developed a standardized overture pattern for his dramatic works which was soon recognized throughout Europe as one of the hallmarks of the French style. The Venetian opera overture, with which Lully was well acquainted, comprised a slow introduction, followed by a faster section in triple time; this provided a model for the French overture, which was characterized by its stately slow beginning, featuring dotted rhythms (*rythmes saccadés*), and by the imitative nature of the following fast section. Occasionally the slow introduction made a brief reappearance at the end of the fast section. When, beginning with Rameau's *Zoroastre* (1749), the *prologue was omitted, it became possible to make a stronger link between the overture and the drama which followed.

The influence of the French overture spread far and wide. Steffani, who probably saw the first performance of Lully's *Bellérophon* (1679), introduced it into the operas he composed for Munich and Hanover; it was, however, already known to German opera in Kusser's works. In England it was taken up in Blow's *Venus and Adonis* and Purcell's *Dido and Aeneas*; later it was also extensively used by Handel, who sometimes added a further dance movement (often a gigue) after the fast section. The French overture was also adapted to instrumental music, notably in the keyboard suites of Bach.

**Freni, Mirella** (*b* Modena, 27 Feb. 1935). Italian soprano. Studied Bologna with Campogalliani. Début Modena 1955 (Micaëla). Gly. 1960–2 (Zerlina, Susanna); London, CG, from 1961 (Nannetta, Violetta, Marguerite, Mimì). Milan, S, from 1962; Chicago, 1963; New York, M, from 1965. With the ripening of her full, limpid tone, her repertory expanded with Butterfly, Desdemona (Verdi), Elisabeth de Valois, Tatyana. A highly musical and attractive artist; married to bass Nicolai *Ghiaurov. (R)

**Frère Laurent.** The holy man (bs) (Shakespeare's Friar Lawrence) who befriends Romeo and Juliet in the operas by Gounod and Sutermeister.

**Fretwell, Elizabeth** (*b* Melbourne, 13 Aug. 1920). Australian soprano. Studied London with Hislop. Début Australian Nat. O 1947 (Senta). Dublin O 1954; London, SW and CG, and the major British companies from 1955. A warm and forthright singer, acclaimed as Violetta and Leonora. (R)

**Frezzolini, Erminia** (*b* Orvieto, 27 Mar. 1818; *d* Paris, 5 Nov. 1884). Italian soprano. Studied with her father Giuseppe Frezzolini (1789–1861), a well-known buffo, and the first Dulcamara; later with Ronconi, Tacchinardi, and M. García jun. Début Florence 1837 (Beatrice di Tenda). Milan, S, from 1841; London 1842 and 1850. Paris, I, 1853–7. Toured USA 1858–61. Her voice declined, and she retired in 1871. One of the most popular and fêted sopranos of her day. Admired by Verdi for her pure, sweet tone, legato singing, and simplicity of expression: she created Giselda (*Lombardi*, 1843), and Giovanna d'Arco, written with her in mind. Beautiful and dramatically affecting, she was also successful in *I Puritani*, *Lucrezia Borgia*, *Rigoletto*, and *Il trovatore*. She was engaged to the composer Nicolai, and married twice. Her first husband, **Antonio Poggi** (1806–75), was a tenor. He studied with Nozzari, and made his début, Paris 1827, as Giacomo in *La donna del lago*. Milan, S, 1834–6, 1845; created Carlo VII in *Giovanna d'Arco*. He and Frezzolini were married 1841–6.

**Frick, Gottlob** (*b* Ölbronn, 28 July 1906; *d* 1994). German bass. Studied Stuttgart with Neudörfer-Opitz. Début Coburg 1934 (Daland). Dresden 1940–50. Munich and Vienna from 1953. London, CG, 1951, 1957–67, 1971; Bayreuth 1957–64; New York, M, 1961. Officially retired 1970, though appeared occasionally thereafter. An expressive and powerful singer, with a large, black voice, strong and even throughout its considerable range. Repertory included Sarastro, Osmin, Caspar, Philip, and Falstaff, but especially famous for his Wagner roles: Hunding, Hagen, Daland, Gurnemanz, etc. (R)

**Fricka.** The Goddess of Wedlock (sop), Wotan's wife, in Wagner's *Der Ring des Nibelungen*.

**Fricsay, Ferenc** (*b* Budapest, 9 Aug. 1914; *d* Basle, 20 Feb. 1963). Hungarian conductor. Studied Budapest with Kodály and Bartók. Cond. Szeged 1936–44; Budapest from 1939, music director 1945. International career began 1947 after cond. première of Von Einem's *Dantons Tod* in place of the indisposed Klemperer. Music director Berlin, SO, 1949–52; Munich, Bavarian SO, 1956–8; returned to Berlin to open D, 1961 (*Don Giovanni*). Edinburgh 1950 with Gly. (*Figaro*). His clear, vigorous interpretations are preserved in a number of recordings, in the techniques of which he took special interest. (R)

BIBL: F. Herzfeld, *Ferenc Fricsay* (Berlin, 1964).

**Friedenstag** (Day of Peace). Opera in 1 act by Richard Strauss; text by Gregor, after Calderón's drama *La redención de Breda* (1625). Prod. Munich 24 July 1938, with Ursuleac, Patzak, Hotter, cond. Krauss; Los Angeles, U of S California, 2 Apr. 1957, cond. Ducloux; London, BBC, 29 May 1971.

On the day that the Peace of Westphalia is signed in October 1648, the Commandant (bar) of Breda refuses to surrender to the Commander of the besieging army (bs), despite the latter's kind words and preferred hand. Only when the Commandant's wife Maria (sop) begs her husband to acknowledge that peace really has come and that brotherhood has replaced hatred, does he throw away his sword and embrace his former enemy. The opera ends with a paean to peace.

**Friedrich, Götz** (*b* Naumburg-Saale, 4 Aug. 1930). German producer. Studied Weimar. Berlin, K, first as assistant to Felsenstein, then chief producer. Début Weimar 1958 (*Così fan tutte*). Début at Berlin, K, 1959 (*Bohème*). Guest appearances Bremen, Kassel. Holland Festival, *Falstaff* 1972, *Aida* 1973. Left East Germany 1972. Bayreuth: *Tannhäuser* 1972, *Lohengrin* 1979, *Parsifal* 1982. Chief producer Hamburg from 1973. London, CG: *Ring* 1974–6, *Idomeneo* 1978, *Elektra* 1990; chief producer from 1977. Director

259

Berlin, D, 1981, where he strengthened the ensemble, included operetta, and introduced special chamber opera performances: also *Ring* 1984–5, *Turandot* 1986, première of Henze *Das verratene Meer* 1990. Houston, *Wozzeck* 1982. Salzburg, première of Berio *Un re in ascolto* 1984. London, C, *Tristan* 1985. Munich, *Forza* 1986. Los Angeles, *Otello* (Verdi) 1986, *Fidelio* 1990. Paris, O, *Káťa Kabanová* 1988. Stockholm, *Lohengrin* 1989. One of the most talented producers of the day, bringing a powerful theatrical sense and strongly argued ideas to productions that can seem didactic (and often argue a politically left-wing case) but that are rarely less than stimulating.

**Friml, Rudolf** (*b* Prague, 7 Dec. 1879; *d* Los Angeles, 12 Nov. 1972). US composer of Czech birth. Studied Prague with Dvořák. Settled in USA 1906, making a reputation as a pianist and a teacher. He is best known for his many operettas. Among the most popular were his first, *The Firefly*, which included the popular 'Donkey Serenade', *Rose Marie* (including the 'Indian Love Call'), and *The Vagabond King*. Like Romberg, Friml at first successfully transported Central European operetta in the manner of Lehár to America. When his style was overtaken by the Broadway musical, he made a successful Hollywood career as a film composer and arranger.

SELECT WORKLIST: *The Firefly* (Harbach; Syracuse 1912); *Rose Marie* (comp. with H. Stothart, Harbach & Hammerstein; New York 1924); *The Vagabond King* (Hooker & Janney, after McCarthy; New York 1925); *The Three Musketeers* (McGuire; New York 1928).

**Fritz.** See AMICO FRITZ.

**Froh.** The God of Spring (ten) in Wagner's *Das Rheingold*.

**From the House of the Dead.** See HOUSE OF THE DEAD.

**Fuchs, Marta** (*b* Stuttgart, 1 Jan. 1898; *d* Stuttgart, 22 Sep. 1974). German mezzo-soprano, later soprano. Studied Stuttgart, Munich, and Milan. Début as mezzo-soprano, Aachen 1928. Dresden from 1931 (Octavian, Amneris, then Kundry). Bayreuth 1933–7 (Brünnhilde, Isolde). London, CG, 1936 (Marschallin, Ariadne). Berlin, SO, 1936–42; Stuttgart 1949–51. A distinguished Wagner singer, possessing a big, dark-toned voice and dramatic intensity. (R)

**Fugère, Lucien** (*b* Paris, 22 July 1848; *d* Paris, 15 Jan. 1935). French baritone. Début 1870 at the Café-Concert Ba-ta-clan. Engaged by Paris, BP, in 1873, then Paris, OC, 1877-1910, where he sang over 100 roles, creating more than 30, including Le Père (*Louise*). A famous Figaro (Mozart), Papageno, and Leporello, which he sang at London, CG, 1897. Paris, Gaîté-Lyrique 1910–19, and OC 1919. Continued to sing till he was 85. An artist of exceptional qualities, possessing a large, rich, and resonant basso cantante, excellent technique, brilliant dramatic talent, and intuitive understanding of composers' intentions. (R)

BIBL: R. Duhamel, *Lucien Fugère* (Paris, 1929).

**'Fuor del mar'.** Idomeneo's (ten) aria in Act II of Mozart's *Idomeneo*, in which he voices his despair that though now free from the dangers of the sea, he is menaced with a graver threat from Neptune as a consequence of his rash vow.

**Furtwängler, Wilhelm** (*b* 25 Jan. 1886; *d* Baden-Baden, 30 Nov. 1954). German conductor and composer. Studied Munich with Beer-Walbrunn, Rheinberger, and Schillings. Répétiteur Breslau 1905–6; cond. Zurich 1906–7; Munich, N, 1907–9; Strasburg 1910–11. Music director Lübeck 1911–15; Mannheim 1915–20. Further study with Schenker, with whom he remained in close contact: to Schenker's analysis of structural fundamentals in a work may be attributed some of Furtwängler's security of structural grasp and resulting confident freedom of phrase and tempo. His questing mind and constantly renewed experience of a work set him to many as an antithesis to the objectivity of Toscanini (who greatly respected him). After a period spent largely in the concert-hall, he began winning admiration for his Wagner in Berlin and Paris from 1924. Bayreuth 1931, 1936, 1937, 1943, 1944; reopened Bayreuth (Beethoven's Ninth Symphony) 1951, 1954. Regular guest Salzburg and Vienna, also Milan. London, CG, 1935 (*Tristan*), 1937, and 1938 (*Ring*). Though he remained politically independent, resigning all his posts over the Nazi banning of Hindemith's *Mathis der Maler* in 1934 and refusing to conduct in wartime occupied countries, he had to contend with post-war reproaches over his stance (his Jewish defenders included Schoenberg). He attached himself to what he saw as the highest German artistic and intellectual values, and these guided his music-making and gave it, at its finest, revelatory qualities. His notoriously imprecise beat was at least in part a demand for collaboration from his players and singers, deeply involving his listeners, in a creative search. (R)

WRITINGS: *Gespräche über Musik* (Zurich, 1948; trans. as *Concerning Music* (London, 1953)).

BIBL: B. Geissmar, *The Baton and the Jackboot* (London, 1944); C. Riess, *Furtwängler, Musik und Politik* (Berne, 1953; trans. as *Wilhelm Furtwängler* (London, 1955)).

**Fux, Johann Joseph** (*b* Hirtenfeld, 1660; *d* Vienna, 13 Feb. 1741). Austrian composer. Remembered for his treatise *Gradus ad Parnas-*

*sum* (1725), one of the most influential of all counterpoint tutors, Fux embraced both secular and sacred music with equal ease and was recognized as one of the most skilled composers of his generation. Having risen to be principal Kapellmeister at St Stephen's Cathedral in Vienna, he reached the height of his career in 1715, when he succeeded Ziani as court Kapellmeister.

During his long career Fux composed 19 operas for the Austrian court, including *Julo Ascanio, Re d'Alba* and *Orfeo ed Euridice*. Many of his works dealt, of necessity, with heroic themes, and by his skilful incorporation of some of the spirit of Neapolitan opera, he managed to achieve a greater depth of feeling than was customary. However, much of his works suffers from a dependence upon the heavy contrapuntal technique which was a mainstay of his religious music, thereby limiting dramatic possibilities. Such was his standing that when gout prevented him from attending the performance of his *Costanza e Fortezza*, for the coronation of Charles VI as King of Bohemia in 1723, the Emperor arranged for him to be carried all the way to Prague in a litter.

Among his many pupils were Muffat, Wagenseil, and Zelenka.

SELECT WORKLIST: *Julo Ascanio, Re d'Alba* (Bernadoni; Vienna 1708); *Orfeo ed Euridice* (Pariati; Vienna 1715); *Costanza e Fortezza* (Pariati; Prague 1723).

BIBL: J. H. van der Meer, *J. J. Fux als Opernkomponist* (3 vols., Bilthoven, 1961); R. Flotzinger, *Fux-Studien* (Graz, 1985).

# G

**Gabriele Adorno.** A Genoese nobleman (ten), lover of Amelia Grimaldi, in Verdi's *Simon Boccanegra*.

**Gabrielli, Adriana.** See FERRARESE, LA.

**Gabrielli, Caterina** (*b* Rome, 12 Nov. 1730; *d* Rome 16 Feb. or 16 Apr. 1796). Italian soprano. Daughter of Prince Gabrielli's cook; known as 'La Coghetta'. Studied Venice, probably with Porpora; later with Guadagni. Exact début unknown; sang Ermione in Galuppi's *Antigone* Venice 1754. Vienna 1755–9, 1760–61; appearances in Parma, Naples, Milan, Palermo; St Petersburg 1772–5; London 1775–6; further engagements in Italy, retiring 1782 with a large fortune. Her repertory included Piccinni, Anfossi, and Mysliveček, and she created roles in several operas by Traetta and Gluck. Metastasio, who taught her declamation, considered her voice 'unequalled in quality and quantity'; Burney found her exceptionally intelligent and well-bred, while Brydon praised her emotional delivery. Small, seductive, and daring, she was famous not only as an outstanding singer, but also for her innumerable affairs and frequent involvement in scandals (she was once imprisoned for insolence by the enraged Viceroy of Palermo). Her sister **Francesca** or **Checca** (*b* c.1735), a less gifted singer, accompanied Caterina as a *seconda donna* in many operas.

**Gadski, Johanna** (*b* Anklam, 15 June 1872; *d* Berlin, 22 Feb. 1932). German soprano. Studied Stettin with Schröder-Chaloupka. Début Berlin, Kroll, 1889 (in Lortzing's *Undine*). Berlin, K, until 1894; New York, M, with Damrosch's Co., 1895, as Co. member 1900–17 (except 1904–6); London, CG, 1898–1901; Bayreuth 1899; Munich Fest. 1905, 1906; Salzburg 1906, 1910; Chicago 1910–11. Toured USA 1929–31 in Wagner. Repertory incl. Pamina, Amelia, Aida, Valentine, Eva, Isolde, Brünnhilde, Santuzza. A fine singer and actress, with a beautiful voice and a lyrical, bel canto style; highly popular in USA until the First World War, when there was anti-German hostility to her and her husband. (R)

**Gaetano** (Kajetan) (*b* Warsaw, 1st half 18th cent.; *d* Warsaw, c.1793). Polish composer and violinist. From 1764 played in the royal orchestra in Warsaw, which he conducted 1779–93. Composed several operas, including *Żółta szlafmyca*, one of the first operas to incorporate elements from Polish folk music. *Les amours de Bastien und Bastienne* uses the libretto which Mozart had set in an adapted form 20 years earlier.

SELECT WORKLIST: *Żółta szlafmyca*, (The Yellow Nightcap) (Zabłocki, after Barré & Piis; Warsaw 1783); *Les amours de Bastien und Bastienne* (Favart; Warsaw 1788).

**Gagliano, Marco da** (*b* Florence 1 May 1582; *d* Florence, 25 Feb. 1643). Italian composer, often known as 'L'Affanato' (The Distressed One). Trained for the priesthood, but also held musical positions in Florence, including that of maestro di cappella at San Lorenzo from 1608. By 1609 was also maestro di cappella to the Grand Duke of Tuscany and soon became recognized as the leading figure of Florentine musical life. He was one of the founding fathers of opera: in 1607 he formed the Accademia degli Elevati, whose activities centred around discussion and performance of the new monodic style.

In 1607 Gagliano visited the court of Prince Francesco Gonzaga at Mantua; the following year his *Dafne* was presented there. It was an immediate success; even Peri judged it superior to his earlier setting of the same libretto. Gagliano's preface to the score (pub. 1608) is one of the most important documents in the early history of opera. Including a brief review of the development of the *stile rappresentativo, it makes telling criticism of the growing practice of melodic ornamentation. Performance instructions reveal his concern for dramatic presentation, while his command that an instrumental sinfonia should be played before the stage action is the earliest mention of the operatic overture. Although he adhered to the main precepts of the Florentine monodists, a distinction between recitative and arioso style is already noticeable.

Before leaving the Mantuan court in 1608, Gagliano contributed the third intermedio to an important performance of Guarini's comedy *La idropica*: his fellow composers included Monteverdi, who set the Prologue. In addition to a revival of *Dafne*, probably during Carnival 1610, Gagliano also composed much dramatic music, including operas, for Florence. Neither the music of *Lo sposalizio di Medoro e Angelica*, performed to celebrate the election of a Roman emperor, nor that of his two sacred operas *La regina Sant'Isola* and *La istoria di Judit*, is extant, though their librettos are. Like *Medoro*, Gagliano's last opera, *La Flora, o vero Il natal de' fiori*, was also written with Peri's help and was performed for a Medici wedding in 1628.

It is on his operatic work that Gagliano's reputation rests. One of the first to inject variety into the otherwise unvarying style of early opera, he showed much sensitivity in handling the problems of uniting drama and music.

SELECT WORKLIST: *Dafne* (Rinuccini; Mantua 1608); *Lo sposalizio di Medoro et Angelica* (Salvadori; Florence 1619, lost); *La regina Sant'Isola* (Salvadori; Florence 1625, lost) and *La istoria di Judit* (Salvadori; Florence 1626, lost); *La Flora, o vero Il natal de' fiori* (comp. with Peri: Salvadori; Florence 1628).

**Gailhard, Pierre** (*b* Toulouse, 1 Aug. 1848; *d* Paris, 12 Oct. 1918). French bass and manager. Studied Toulouse, and Paris Cons. Début Paris, OC, 1867 (Falstaff in Thomas's *Le songe d'une nuit d'été*). Paris, OC until 1870 and O 1871–84; London, CG, 1879–83. A cultured man with a warm, sonorous voice. Created roles in works by Mermet, Joncières, and Thomas; also sang Leporello, Osmin, Saint-Bris, and Gounod's and Boito's Mephistopheles, giving sharp, impressive characterizations in both comic and serious parts. In 1884 became joint manager of the Opéra, with first Ritt, then Bertrand; sole manager 1899–1906. He was responsible for many French premières (including *Otello, Lohengrin, Die Walküre, Siegfried, Tristan, Die Meistersinger*), and for the high quality of their productions. (R)

**Gál, Hans** (*b* Brünn (Brno), 5 Aug. 1890; *d* Edinburgh, 3 Oct. 1987). Austrian composer and musicologist. Studied Vienna with Mandyczewski. One of his first successes came with his opera *Die heilige Ente*, widely performed in the 1920s. His idiom owes much to the Brahms tradition he absorbed from his teacher, something to neo-classicism, nothing to the Second Viennese School. *Die heilige Ente* is a Singspiel, eclectic, folk-like, and influenced by both Johann and Richard Strauss. A stimulating writer and teacher, especially in Edinburgh's musical life, he wrote books that include a sympathetic study, *Richard Wagner* (Frankfurt, 1963, trans. London, 1975).

SELECT WORKLIST: *Die heilige Ente* (The Holy Duckling) (Levetzow; Düsseldorf 1923); *Die beiden Klaas* (Levetzow, after Andersen; comp. 1933, prod. York 1990 as *Rich Claus, Poor Claus*).

**Galeffi, Carlo** (*b* Malamocco, 4 June 1882; *d* Rome, 22 Sep. 1961). Italian baritone. Studied Paris with Sbriglia, and Rome with Cotogni. Début Fermo 1907 (Don Alfonso, *La favorita*). New York, M, 1910; Milan, S, regularly between 1913 and 1940; Chicago 1914, 1919–21. Created leading roles in Boito's *Nerone*, Montemezzi's *L'amore dei tre re*, and Mascagni's *Isabeau* and *Parisina*. First Italian Amfortas, and also famous as Rigoletto, Nabucco, and Tell. Retired 1955. Possessed an unusually round and even tone, and an expressive legato line. (R)

**Galilei, Vincenzo** (*b* S Maria a Monte, prob. late 1520s; *d* Florence, buried 2 July 1591). Italian lutenist, composer, theorist, singer, and teacher. Father of the astronomer Galileo Galilei and a prominent member of the *Camerata which met at the house of Count Bardi in Florence in the late 16th cent., out of whose discussions opera evolved. It was Galilei who initiated study of the dramatic practices of the ancient Greeks: his correspondence with Girolamo Mei formed the basis for the *Dialogo della musica antica et della moderna* (Florence, 1581), the seminal theoretical work of the Camerata. According to Doni, Galilei's lost cantata *Il conte Ugolino*, setting lines from Dante's *Inferno*, was the first work using the new *stile rappresentativo; it was followed by a similar treatment of part of the Lamentations of Jeremiah.

BIBL: V. Galilei, *Dialogo della musica antica et della moderna* (facs. R/Rome, 1934); excerpt in O. Strunk, *Source Readings in Music History* (New York, 1950).

**Galitsky.** Prince Vladimir Yaroslavich Galitsky (bs), husband of Igor's sister Yaroslavna, in Borodin's *Prince Igor*.

**Gallet, Louis** (*b* Valence, 14 Feb. 1835; *d* Paris, 18 Oct. 1898). French librettist. Of the librettos he wrote for Massenet, the most important is *Thaïs*. In the preface he sets out, with a good deal of special pleading, his ideas for so-called *poésie mélique* that would remove the French opera libretto from dependence on rhyme (at any rate, end-rhyme) in the direction of prose.

**Galli, Caterina** (*b* ?, *c*.1723; *d* London, 1804). Italian mezzo-soprano. She came to London, engaged by Italian O, appearing at H, 1742, in Brivio's *Mandane*, in the first of many travesti roles; subsequently in operas by Galuppi and Porpora. Sang for Handel 1747–54, creating roles (mostly male) in *Joshua, Solomon*, and *Jephtha*, among others. Sang again at London, H, 1747–8; in Handel; Genoa, Naples, and Venice 1754–63; London, H, 1773. Retired 1777, but forced through poverty to reappear until 1797 at London, CG. A popular singer in her time; Burney found her 'spirited and interesting'.

**Galli, Filippo** (*b* Rome, 1783; *d* Paris, 3 June 1853). Italian tenor, later bass. Début Naples 1801. Sang as tenor until 1811, when after a serious illness he re-emerged as a bass. Success in *La cambiale di matrimonio* (Padua 1811) led to an enduring association with Rossini, for whom he created the parts of Taraboto (*L'inganno felice*), Fernando (*La gazza ladra*), title-role of *Maometto II*, Mustafà (*L'italiana in Algeri*), Selim (*Il turco in Italia*), Assur (*Semiramide*), and roles in *La pietra del paragone* and *Torvaldo e Dorliska*, as well as singing many others. Paris, I, 1823; London, H, 1827–33; Milan, Carcano, 1830, creating Henry

VIII in Donizetti's *Anna Bolena*. After a vocal
decline, he taught at Paris Cons. 1842–8, and
died in poverty. Gifted with a magnificent, flexible, and expressive voice, he was also a superb
actor, exuberant and witty, yet capable of 'the
finest heights of tragic acting . . . in a style
worthy of Kean' (Stendhal, is also the source of
a story, probably apocryphal, that Rossini, furious at having lost a mistress to Galli, deliberately wrote a passage in *La gazza ladra* using
his three worst notes). His brother **Vicenzo**
(1798–1858) was a buffo bass. Début Milan, S,
1824. Sang Bellini and Donizetti, and in creations of many new works by Ricci, Rossi,
Mercadante, etc.

**Galliard, John Ernest** (Johann Ernst) (*b* ?Celle,
*c.*1687; *d* London, 1749). German oboist and
composer. Studied under Steffani; came to England 1706. Active for many years in London as
composer and performer. His *Calypso and Telemachus* was performed at the Haymarket 1712:
this was one of the most important attempts to
establish opera in English, but was unsuccessful
and only three performances were given. He
composed music to eight pantomimes by Lewis
Theobald 1723–30, including *The Necromancer*
(1723) and *Apollo and Daphne* (1726), which
proved more popular than his other operatic
attempts, *Circe* and *The Happy Captive*.

> SELECT WORKLIST (all first prod. London): *Calypso
> and Telemachus* (Hughes; 1712); *Circe* (Davenant;
> 1719); *The Happy Captive* (Theobald, after Cer-
> vantes; 1741, lost).

**Galliari.** Italian family of designers. The important members were three brothers, **Bernardino** (*b*
Andorno, 3 Nov. 1707; *d* Andorno, 31 Mar.
1794), **Fabrizio** (*b* Andorno, 28 Sep. 1709; *d* Tre-
viglio, June 1790), and **Giovanni Antonio** (*b*
Andorno, 26 Mar. 1714; *d* Milan, 1783). From
1742 they worked in Milan at the T Regio Ducale
1742–6, Itinerale 1776–8, La Scala from 1778.
The two elder brothers also worked in Turin and
abroad. Bernardino was the principal painter,
often working on designs produced by Fabrizio,
who was also a set designer; Giovanni Antonio
was a painter. Though based on the formalities of
opera seria, their designs responded to the reform
movement leading up to Gluck (*Alceste*, Vienna
1767), in particular with an increased emphasis
on naturalism. They also designed much opera
buffa. Fabrizio's sons **Giuseppino** (1742–1817)
and **Giovannino** (1746–1818) were also
designers, working in Turin, and other members
of the family worked in Italy and Germany.

**Galli-Bibiena.** Italian family of scene designers
and architects of the Baroque era, whose work
was seen all over Europe. The family name was
Galli; Bibiena was added from the birthplace of

their father, **Giovanni Galli** (*b* Bibbiena, 1625; *d*
Bologna, 1665).

The brothers (1) **Ferdinando** (*b* Bologna, 18
Aug. 1656; *d* Bologna, 3 Jan. 1743) and (2) **Francesco** (*b* Bologna, 12 Dec. 1659; *d* Bologna, 20
Jan. 1737) studied at Bologna, Ferdinando also
with Rivani, one of Louis XIV's stage machinists
at Versailles. Ferdinando was engaged by the
Duke of Parma and worked in the T Farnese and
in Bologna; then went with his brother to
Vienna, where they were responsible for the
décor for the court fêtes and theatrical performances. They were the first to exploit the *scena
per angolo*, or diagonal perspective, which
replaced the traditional central perspective of
Baroque theatre. One of the most important
designers of Baroque opera, Ferdinando interested himself particularly in architectural stage
designs, making skilful use of space and perspective, and much influencing later generations.
Francesco also worked in Bologna and other Italian cities, travelling widely and designing for
many different composers. His work has greater
imaginative range than his brother's. He
designed theatres in Bologna, also Vienna, Verona, and Nancy.

3. **Alessandro** (*b* ?Parma, 1686; *d* Mannheim,
5 Aug. 1748), son of (1), worked with his father
in Spain and Vienna, then settling in Mannheim
1720 and designing theatres there. His sets are in
the family tradition of architectural elegance and
the exploitation of asymmetry and false perspective.

4. **Giovanni Maria** (*b* Piacenza, 19 Jan. 1694;
*d* Naples, 1777), son of (1), worked first in Mannheim, then chiefly in Prague.

5. **Giuseppe** (*b* Parma, 5 Jan. 1695; *d* Berlin,
1757), son of (1), worked with his father in
Vienna, succeeding him and also working in
Prague. Designed the Bayreuth OH 1746, then
the Dresden Zwinger OH 1747; later Berlin,
under Frederick the Great from 1753. He was
associated with many important Baroque composers, including Jommelli, Hasse, and Graun,
and the librettists Zeno and Metastasio, and
expressed their opera seria ideals closely in his
work. He built upon the example of his father in
developing greater subtleties of architectural
design and perspective without losing hold of the
essential formality of the genre. He was the first
to make use of transparent scenery lit from
behind.

6. **Antonio** (*b* Parma, *bapt.* 1 Jan. 1697; *d*
Milan, 28 Jan. 1774), son of (1), worked with him
in Bologna, with (5) in Vienna, and with (2) in
Verona and Rome. He designed the Bologna
Comunale 1756–63 and many sets for it.

7. **Giovanni Carlo** (*b* ?Bologna, *c.*1720; *d* Lisbon, 20 Nov. 1760), son of (2), worked in

Bologna, then designing theatres in Portugal (none extant).

8. **Carlo** (*b* Vienna, *bapt.* 8 Feb. 1721; *d* Florence, 1787), son of (5), completed his father's Bayreuth theatre, also working in various European cities including Berlin for Frederick the Great 1765, St Petersburg for Catherine the Great 1774, and Stockholm and Drottningholm (where some of his designs are extant) for Gustavus III.

**Galli-Curci, Amelita** (*b* Milan, 18 Nov. 1882; *d* La Jolla, CA, 26 Nov. 1963). Italian soprano. Studied Milan with Carignani and Dufes, but mainly self-taught. Début Trani 1906 (Gilda). Successes in Italy, Spain, Russia, and S America; Chicago 1916–24. New York, M, 1921–30. A throat tumour (removed in 1935) eventually led to her retirement in 1937. Famous for her Rosina, Elvira (*Puritani*), Lucia, Violetta, and Lakmé, to which her pure and astonishingly agile voice was ideally suited. (R)

**Galli-Marié, Célestine** (*b* Paris, Nov. 1840; *d* Vence, 22 Sep. 1905). French mezzo-soprano. Studied with her father, Félix Mécène Marié de L'Isle. Début Strasburg 1859. After engagements in Lisbon and Rouen (in the first French *Bohemian Girl*), appeared at Paris, OC, 1862–85, where she created Carmen (giving staunch support to Bizet during the difficult rehearsal period), Mignon, and Friday in Offenbach's *Robinson Crusoé*, among other roles. Repertory also included Gounod, Massenet, Massé, etc. London, HM, 1886. 'She is small and graceful and moves like a cat', wrote a critic; she was also a penetrating actress and an intelligent singer with a voice of unusual timbre. Her Carmen was much admired by Tchaikovsky, and his brother Modest appreciated her 'unbridled passion' and 'mystical fatalism'.

**Galuppi, Baldassare** (Burano, 18 Oct. 1706; *d* Venice, 3 Jan. 1785). Italian composer. Commonly known as 'Il Buranello' after his place of birth. His first opera, *La fede nell'incostanza*, was a failure: this led him to study with Lotti. *Gl'odj delusi del sangue*, written jointly with G. B. Peschetti, was the first fruit of his studies; this, and his next collaboration, *Dorinda*, launched his successful career on the Venetian stage. Over the following years he produced many operas, sometimes five in one year. From 1738 he composed for other cities, including London, beginning with the pasticcio *Alessandro in Persia*. *Scipione in Cartagine*, *Enrico*, and *Sirbace* were widely admired and performed for many years: their champions included Burney, who avowed that Galuppi had more influence on English music than any other Italian.

Returning to Venice in 1743, he followed many of his Venetian predecessors by dividing his time

between opera and sacred music: in 1748 he became vice maestro di cappella at St Mark's, rising to the principal position in 1762. During this time he was also developing his operatic skills: crucially, in 1749 he made his first setting of a Goldoni libretto, *L'Arcadia in Brenta*, effectively the first full-length comic opera. This collaboration ushered in a string of popular successes, including *Il mondo della luna*, *Il mondo alla roversa*, *Le virtuose ridicole*, and *La calamità de' cuori*. In 1765 Galuppi went to St Petersburg at the invitation of Catherine II, where *Ifigenia in Tauride* was performed in 1768: two years later he returned to Venice, now also becoming maestro di coro at the Conservatorio degli Incurabili. After *La serva per amore* he wrote no more opera, but devoted himself to oratorio and to his duties at St Mark's.

Although Galuppi's music was gradually supplanted by that of the younger generation, at the height of his powers he was one of the leading composers of the day, with nearly 100 works to his credit. While he is remembered mainly for his comic operas, his composition of opera seria—he set many of Metastasio's librettos, including *Issipile*, *Alessandro nell'Indie*, and *Didone abbandonata*—should not be overlooked, for it highlights his position at a crossroads of operatic history. Born into an era when comic opera was in its infancy, Galuppi was of that generation which moulded it into one of the most potent vehicles of dramatic expression. To this end he found an ideal partner in Goldoni, who in his comic librettos rejected the traditional *mélange* of contrived situations and stock characters in favour of a more realistic approach, in which crude farce is supplanted by a genuinely dramatic plot where humour does not exclude expressions of pathos, sentimentality, or human emotion. This is seen at best in the most successful of the Galuppi–Goldoni collaborations, *Il filosofo di campagna*, where the charm and wit of the libretto is matched by music of a similarly appealing and carefully crafted nature. However, Galuppi continued to compose opere serie until late in his career, including *Idomeneo* and *La clemenza di Tito*.

Galuppi's most lasting contribution to the development of opera buffa lies in his expansion of the *ensemble finale. Departing from the earlier practice of Leo and Logroscino, he fashioned finales from several clearly identifiable musical sections, during which the dramatic action progresses. In this respect he foreshadows the later developments of Da Ponte and Mozart, although his chain-like structures lack their more advanced symphonic construction.

SELECT WORKLIST (all first prod. Venice unless otherwise stated): *La fede nell'incostanza* (Faith in Inconstancy) (Neri; Chioggia & Vicenza 1722); *Gl'odj delusi del sangue* (Lucchini; 1728; comp. with Peschetti); *Dorinda* (?; 1729); *Issipile* (Metastasio; Turin 1737); *Alessandro nell'Indie* (Metastasio; Mantua 1738); *Didone abbandonata* (Metastasio; Modena 1741); *Scipione in Cartagine* (Vanneschi; 1742); *Enrico* (Vanneschi; London 1742); *Sirbace* (Stampa; London 1743); *L'Arcadia in Brenta* (Goldoni; 1749); *Il mondo della luna* (The World on the Moon) (Goldoni; 1750); *Il mondo alla roversa* (The Topsy-Turvy World) (Goldoni; 1750), *Le virtuose ridicole* (The Ridiculous Virtuosos) (Goldoni; 1752); *La calamità de' cuori* (The Hearts' Disaster) (Goldoni; 1752); *Il filosofo di campagna* (The Country Philosopher) (Goldoni; 1754); *Idomeneo* (?; Rome 1756); *La clemenza di Tito* (Metastasio; Venice 1760); *Ifigenia in Tauride* (Coltellini; St Petersburg 1768); *La serva per amore* (Servant for Love) (Livigni; 1773).

BIBL: A. della Corte, *Baldassare Galuppi* (Siena, 1948); M. Muraro & F. Rossi, *Galuppiana 1985* (Florence, 1986).

**Gambler, The** (Russ.: *Igrok*). Opera in 4 acts by Prokofiev; text by the composer, after Dostoyevsky's story (1866). Orig. projected prod., Petrograd 1917, cancelled due to Revolution; second projected Leningrad prod., postponed. Prod. Brussels, M, 29 Apr. 1929, with Leblanc, Andry, Ballard, Lense, Rambaud, Yovanovitch, cond. De Thoran, in French (trans. by Paul Spaak; New York, 85th St. Playhouse, 4 Apr. 1957, with Beauvais, Bering, Lane, Falk, cond. (2 pianos) Georgette Palmer; Edinburgh, by Belgrade OC, 28 Aug. 1962, with Heybalova, Miladinović, Cakarević, Starc, Andrašević, Cvejić, cond. Danon.

A European gaming resort, *c.*1865. Alexey (ten) is in the service of a General (bs) in the German spa of Roulettenburg. In love with the coquettish Mlle Blanche (sop) and, having gambled away most of his money, in debt to a French Marquis, the General anxiously awaits the death of his rich grandmother. But she turns up, and proceeds to lose her fortune at the tables. The General's daughter Pauline (mez), whom Alexey loves, is now faced with marrying the Marquis; to prevent this, Alexey tries to win some money and succeeds in breaking the bank. But when he presents her with the money, she hysterically flings it in his face.

**Gamblers, The** (Russ.: *Igroki*). Opera by Shostakovich; text after Gogol, unfin. Leningrad, 1978; completed Meyer, prod. Wuppertal 1 June 1983, with S. Schmidt, cond. Schick.

**García**. Family of singers and teachers of singing, of Spanish origin.

1. **Manuel** (del Popolo Vicente Rodríguez) (*b* Seville, 21 Jan. 1775; *d* Paris, 10 June 1832). Spanish tenor, composer, and teacher. Father of (3), (4), (5); husband of (2). Studied Seville with Ripa. Début Cádiz 1798, in a tonadilla. After

successes in Spain (where he scandalously appeared on stage with both wife and mistress), he went to Paris in 1808, and soon became popular. Italy 1811–16, creating Norfolk (*Elisabetta, regina d'Inghilterra*), and Count Almaviva (Rossini), written for him. Paris, I, 1819–24; London, H, 1818, 1819, 1823–25. New York 1825, introduced Italian opera to USA, including *Barbiere*, *Cenerentola*, *Otello* (Rossini), and *Don Giovanni* (with himself, wife, son, and daughter in the cast, and in the presence of Da Ponte). Mexico 1827–8; Paris 1829; retired 1830. A handsome man of astonishing energy and versatility, he played the guitar, wrote a large number of popular operas and songs (major contributions to Spain's influence throughout the world), and was a gifted teacher, his pupils including Nourrit, Méric-Lalande, and his daughter Maria (Malibran). In Mexico, when much of his company's music was lost, he wrote out the entire scores of *Don Giovanni*, *Otello*, and *Barbiere* from memory. As a tenor he was best known for his Rossini, though he also sang Mozart (including Guglielmo and Don Giovanni), and his most famous role was Otello, in which 'his acting was full of fire and fury' (Stendhal), he was 'most truly the Moor', and was even praised by Edmund Kean. Some, however, found his style lacking in taste.

2. **Joaquina** (*b* Madrid, 28 July 1780; *d* Brussels, 10 May 1854). Spanish singer. A talented and vivacious actress. Second wife of (1); mother of (3), (4), and (5).

3. **Manuel** (*b* Madrid, 17 Mar. 1805; *d* London, 1 July 1906). Spanish baritone and teacher. Son of (1) and (2); brother of (4) and (5); husband of (6); father of (7). Studied with his father. Début New York 1825 (Rossini's Figaro). Retired from stage 1829. Teacher at Paris Cons. 1847–50; London, RAM, 1848–95. First to undertake study of voice production; invented the laryngoscope. With his perfection of his father's methods, he was the most influential teacher of his time; among his pupils were Lind, Frezzolini, M. Marchesi, and Santley. His writings include *Mémoire sur la voix humaine* (Paris, 1840); *Traité complet de l'art du chant* (Paris, 1840); *Observations on the Human Voice* (London, 1855).

4. **Maria Felicia** (*b* Paris, 24 Mar. 1808; *d* Manchester, 23 Sep. 1836). Spanish mezzo-soprano. Daughter of (1) and (2); sister of (3) and (5). Married E. Malibran and then C. de Bériot. See MALIBRAN.

5. **Pauline** (*b* Paris, 18 July 1821; *d* Paris, 18 May 1910). Spanish mezzo-soprano. Daughter of (1) and (2); sister of (3) and (4). Married Louis Viardot. See VIARDOT.

6. **Eugénie** (orig. Mayer) (*b* Paris, 1815; *d*

Paris, 12 Aug. 1880). French soprano and teacher. First wife of (3); mother of (7). Studied with her husband. Sang successfully 1836–58 in Italy, Paris, and London.

7. **Gustave** (*b* Milan, 1 Feb. 1837; *d* Paris, 15 June 1925). Italian baritone and teacher. Son of (3) and (6); father of (8). Studied with his father. Pursued career in London and Italy 1862–80. Taught in London from 1883.

8. **Alberto** (*b* London, 5 Jan. 1875; *d* London, 10 Aug. 1946). English baritone and teacher. Son of (7) by his second wife. Studied with his great-aunt Pauline Viardot. Sang in France and Germany, and in England, where he taught at RCM and GSM.

BIBL: J. Levien, *The Garcia Family* (London, 1932, rev. 1948 as *Six Sovereigns of Song*).

**Gardelli, Lamberto** (*b* Venice, 18 Nov. 1915). Italian conductor and composer. Studied Pesaro and Rome. Répétiteur Rome, R, under Serafin; début there 1944 (*Traviata*). Cond. Stockholm 1946–55. Berlin, Helsinki, Budapest 1961–5; Gly. 1964–5, 1968; New York, M, 1966-8; London, CG, 1968–9, 1970–1, 1975–6, 1979–80. A fine conductor of the Italian repertory, especially in Verdi. His operas include *Alba novella* (1937) and *Il sogno* (1942).

**Garden, Mary** (*b* Aberdeen, 20 Feb. 1874; *d* Inverurie, 3 Jan 1967). Scottish soprano. Studied Chicago, then Paris with Trabadelo and Fugère. Début Paris, OC, 1900 (*Louise* (taking over halfway through a performance from Rioton, its creator)). New York, Manhattan OC, 1907; Chicago O 1910–31, and its director 1919–20. Created the roles of Mélisande, Massenet's Sapho and Chérubin, and D'Erlanger's Aphrodite, among others. Repertory included Manon, Thaïs, Salome, Carmen, Ophélie, and Jean (*Le jongleur de Notre-Dame*). An intelligent, original singer-actress, remarkable for a complete identification with her roles and skilful use of vocal colour. As Mélisande, she amazed Debussy: 'That was indeed the gentle voice I had heard in my innermost soul.' (R)

BIBL: M. Garden, *Mary Garden's Story* (New York, 1951).

**Gardiner, John Eliot** (*b* Fontmell Magna, 20 Apr. 1943). English conductor. Studied Cambridge with Dart, Paris with Boulanger. After early experience as choral conductor, début London, C, 1969 (*Magic Flute*). London, CG, 1973 (*Iphigénie en Tauride*) and C, 1979 (Monteverdi's *Orfeo*). His initial experience with early music, especially with his Monteverdi Choir and Orch., led to his revival of a number of neglected Baroque works, including a concert series of Rameau operas in his own editions, 1973–5. Orange 1981; Aix-en-Provence 1982 (première Rameau *Les*

*Boréades*). Music director Lyons 1983–8, where he strengthened and widened the repertory. Has given many outstanding performances of Baroque and Classical operas, also of later French repertory incl. Chabrier's *L'étoile*, and Debussy's *Pelléas*. A demanding, highly intelligent conductor, with a gift for reanimating Baroque music through careful study of the scores and contemporary performance practice, also for sharply observed performances of Gluck, Mozart, Rossini, and Berlioz. (R)

**Gasparini, Francesco** (*b* Camaiore, 5 Mar. 1668; *d* Rome, 22 Mar. 1727). Italian composer. Studied with Corelli and Pasquini in Rome, where his first opera *Roderico* was performed in 1694. In 1702 he moved to Venice and for the next 20 years was one of the leading opera composers in the city. Among his works were *Tiberio*, *Tamerlano*, and *Gli equivoci d'amore*; of his 61 operas, many were to librettos by Zeno and Pariati. In 1720 he moved to Rome, where he produced his last dramatic work, *Tigrane*, in 1724.

Gasparini was one of the most prolific and respected composers of his day; his operas were performed throughout Italy, and were also popular in Germany and England. As products of the period immediately preceding the reforms of Metastasio, they embrace a wide variety of moods and situations, including the tragic, heroic, and tragicomic, but progress via an unvarying and static scheme of recitative/da capo aria. One of Gasparini's most successful works was *Ambleto*: this was the first Hamlet opera, though it was not based on Shakespeare and uses a different plot and secondary characters. When performed in London in 1712 it failed to repeat its Italian success and was withdrawn after six performances.

Gasparini was also important as a teacher: his pupils included Benedetto Marcello, Quantz, and Domenico Scarlatti. For a while he was engaged to Metastasio's daughter, but did not marry her.

SELECT WORKLIST: *Roderico* (?; Rome 1694); *Tiberio* (Pallavicino; Venice 1702); *Ambleto* (Zeno & Pariati; 1705); *Tamerlano* (Piovene; Venice 1710); *Gli equivoci d'amore* (Salvi; Venice 1723); *Tigrane* (?; Rome 1724).

**Gassmann, Florian Leopold** (*b* Most, 3 May 1729; *d* Vienna, 20 Jan. 1774). Bohemian composer. Studied Bologna with Martini, but had his first operatic success in Venice in 1757 with *Merope*. In 1763 succeeded Gluck as ballet composer in Vienna and also directed the opera buffa troupe at the Burgtheater; court Kapellmeister in Vienna 1772, shortly after founding the Tonkünstler-Societät. A leading figure in Viennese musical life, he also composed much sacred music and was active as a conductor and teacher:

among his pupils was Salieri, whom he brought to Vienna in 1766. His early death was caused by a fall from his carriage: he was succeeded at court by *Bonno.

Gassmann composed more than 20 operas, including both serious and comic works. Many of the former are to librettos by Metastasio, including *Issipile* and *Achille in Sciro*; while they often show little more than the ability to imitate well-tried musical and dramatic formulas, Gassmann's best works, notably *Ezio*, reflect a sympathy for Gluck's reforms and make more effective use of the orchestra than was customary. His comic works are more distinctive and include settings of Goldoni (*Filosofia ed amore* and *La notte critica*) and Boccherini (*I rovinati*). In Vienna, he also set texts by Coltellini, including *La contessina* (1770), first performed to celebrate a meeting between Joseph II and Frederick II. *L'opera seria*, a satirical commedia per musica about the conventions of serious opera, was very popular. Throughout his comic œuvre he demonstrates a talent for musical characterization: his sharply observed contrast of noble and peasant in *La contessina* has been seen as an anticipation of *Figaro*. Here, as in his other great success *L'amore artigiano*, he constructs a particularly effective *ensemble finale.

His daughters **Maria Anna Fux** (*b* Vienna, 1771; *d* Vienna, 27 Aug. 1858) and **Therese Maria Rosenbaum** (*b* Vienna, 1 Apr. 1774; *d* Vienna, 8 Sep. 1837) were singers, the latter having the more successful career in Vienna.

SELECT WORKLIST: *Merope* (Zeno; Venice 1757); *Issipile* (Metastasio; Venice 1758); *Filosofia ed amore* (Goldoni; Venice 1760); *Achille in Sciro* (Metastasio; Venice 1766); *L'amore artigiano* (Goldoni; Vienna 1767); *La notte critica* (Goldoni; Vienna 1768); *L'opera seria* (Calzigi; Vienna 1769); *Ezio* (Metastasio; Rome 1770); *La contessina* (Coltellini; Uničov 1770); *I rovinati* (The Ruined) (Boccherini; Vienna 1772).

**Gatti-Casazza, Giulio** (*b* Udine, 3 Feb. 1869; *d* Ferrara, 2 Sep. 1940). Italian impresario. Studied as an engineer, then succeeded his father as director of Ferrara, T Municipale, 1893. His success led to his appointment as director of Milan, S, 1898–1908, where with Toscanini he did much for the theatre's prestige; he gave the Italian premières of *Pelléas* (1908) and *Boris* (1909), and helped to popularize Wagner. Assistant director of New York, M, 1908, director 1910–35, when he gave more than 5,000 performances of 177 works, fostering original-language productions. He increased the theatre's international fame by the standard of his productions and the galaxy of singers he assembled. Married first Frances *Alda, then the dancer Rosina Galli.

**Gavazzeni, Gianandrea** (*b* Bergamo, 27 July 1909). Italian conductor, critic, and composer. Studied Rome with Cristiani, Milan with Pizzetti and Pedrollo. Bergamo 1930, assistant to Antonicelli. Until 1940 mostly engaged in composition, also conducting; his opera *Paolo e Francesca* (Bergamo 1935) had a success. Parma 1940; Bergamo 1941; Rome, T dell'Arte, 1942; Genoa, CF, 1942; Milan: L, with S Co., 1945, and since at S; artistic director, S, 1965–8, directing many important premières. Edinburgh Festival, with Milan, PS, 1957; Gly. 1965 (*Anna Bolena*). Chicago and New York, M, 1976. A thoughtful and acute critic.

WRITINGS: *Gaetano Donizetti* (Milan, 1937); *Musorgskij* (Florence, 1943); *La musica e il teatro* (Pisa, 1954); *I nemici della musica* (Milan, 1965); *Non eseguire Beethoven* (Milan, 1974).

**Gaveaux, Pierre** (*b* Béziers, 9 Oct. 1760; *d* Charenton, 5 Feb. 1825). French tenor and composer. Studied Béziers with Combès and Beck. He sang at the T de Monsieur in Paris, including such major roles as Floreski (*Lodoïska*), continuing at the Feydeau. Here he also began composing opéras-comiques, having immediate success with *Les deux suisses*. A genuine liberal, he found himself in trouble with 'Le réveil du peuple', the most inflammatory song before 'La Marseillaise'. His most important works include *Sophie et Moncars* and especially *Léonore*, the latter to the libretto that also inspired Beethoven. Though now completely overshadowed, the work has some vivid and affecting strokes of imagination. Gaveaux himself created his own Florestan, and received an ovation. His works are in a manner that reflected current interest in such genres as the *fait historique, and in a style that admitted ideas of some substance into an essentially light genre. They had considerable appeal for their use of local colour and their lively dramatic effects (anticipating the popular stage devices of the 1820s). Gaveaux was outstanding among opéra-comique composers of his generation for the wit and skill with which he used the conventions (admitting a certain Italian influence into his vocal writing), and he did something to extend and give substance to the finale. Latterly his voice, described as pleasant, light, and fluent, lost its timbre. His wife **Emilie Gavaudan** (1775–1837, sister of the tenor **Pierre Gavaudan** (1772–1840)) was a successful soprano at the T Feydeau, where his brother **Simon** (1759–?) was a répétiteur and souffleur; the two men also collaborated in a publishing business.

SELECT WORKLIST (all first prod. Paris): *Les deux suisses* (Demoustier; 1792; rev. as *L'amour filial*); *Sophie et Moncars* (Guy; 1797); *Léonore, ou L'amour conjugal* (Bouilly; 1798); *Un quart d'heure de silence* (Guillet; 1804); *Le bouffe et le tailleur* (Villiers &

Gouffé; 1804); *Monsieur Deschalumeaux* (Auguste; 1806).

**Gawain.** Opera in 2 Acts by Harrison Birtwistle; text by David Harsent, after the anon 14th-cent. English poem *Sir Gawain and the Green Knight*. Prod. London, CG, 30 May 1991, with Angel, Walmsley-Clark, Laurence, Greager, Le Roux, Tomlinson, cond. Howarth.

The Green Knight (bs) comes to the court of King Arthur (ten); his challenge to anyone to behead him is met by Gawain (bs), and he rides away bearing his own severed head. There follows a masque, 'The Turning of the Seasons'.

Gawain journeys to the Castle of Bertilak de Hautdesert to find the Green Knight, observed and manipulated by Morgan le Fay (sop). Bertilak's hunting expeditions and the Lady de Hautdesert's attempts to seduce Gawain are woven into another series of ritual repetitions as Gawain breaks faith with his host. Bertilak is now revealed as the Green Knight, and Gawain returns to King Arthur; but though welcomed as a hero, he is himself aware of having survived by deceit, and Morgan le Fay has succeeded in undermining the solidarity of Arthur's court.

**Gay, John** (*b* Barnstaple, *bapt.* 16 Sep. 1685; *d* London, 4 Dec. 1732). English poet, playwright, and theatre manager. He is best known as the founder of *ballad opera, of which his *The Beggar's Opera* (1728) was the first example. Its immense success encouraged Gay to write a sequel, *Polly* (comp. 1729, but not prod. until 1779, owing to a ban by the Lord Chamberlain) in which all the characters are transported to the West Indies, where Gay had never been. It was Gay himself who selected the airs for *The Beggar's Opera*, sometimes choosing tunes whose original texts bore some relation to his purpose. D'Urfey had previously included some popular melodies in his *Wonders in the Sun* (1706), but not nearly to the same extent as Gay. His third and last ballad opera was a dull, poorly constructed work, *Achilles* (1733), which enjoyed little popularity. Gay is also remembered as the librettist of Handel's *Acis and Galatea* (1718). As a theatre manager, he obtained Davenant's letters patent, which enabled him to build the first *Covent Garden Theatre in 1732.

**Gay, Maria** (*b* Barcelona, 13 June 1879; *d* New York, 19 July 1943). Spanish mezzo-soprano. Originally self-taught, later studied Paris with Adiny. Voice discovered in prison, following her arrest for singing a revolutionary song. Début Brussels, M, 1902 (Carmen). London, CG, 1906; Milan, S, 1906–7; New York, M, 1908–9; Chicago 1910–27. Roles included Dalila and Amneris, but most famous for her forceful, earthy Carmen. Married tenor *Zenatello 1913; they

founded a singing school, New York 1927, and effected the engagement of Lily Pons at the Metropolitan. (R)

**Gayarre, Julián** (*b* Valle de Roncal, 9 Jan. 1844; *d* Madrid, 2 Jan. 1890). Spanish tenor. Studied Milan with Lamperti. Début Varese 1869 (*I Lombardi*). Successes in Italy (including as first Italian Tannhäuser); St Petersburg 1873–5; Milan, S, 1876. London, CG, 1877–81, 1886–7. Paris: I, 1884; O, 1886. Created Enzo in *La Gioconda*; repertory also included Meyerbeer, Donizetti, and Verdi. Regarded by many as the supreme tenor of his time, with outstanding powers of expression; much admired by Duprez and by Gounod, who, on hearing him perform in *Faust*, said, 'I have never heard my music sung like that'. A premature vocal decline was shortly followed by his death, after withdrawing from a performance in Madrid during which his voice failed.

BIBL: F. Hernandel Girbal, *Julián Gayarre* (Madrid, 1955).

**Gayer, Catherine** (*b* Los Angeles, 11 Feb. 1937). US soprano. Studied Los Angeles and Berlin. Début Venice, F, 1961 (the Companion, Nono's *Intolleranza*). Berlin, D, from 1961. Appearances at Milan, S; Vienna; Salzburg; Edinburgh. Created Nausikaa in Dallapiccola's *Ulisse*, Christina in Orr's *Hermiston*, title-role in Reimann's *Melusine*. Other roles include Constanze, Queen of Shemakha, Zerbinetta, Mélisande, Lulu. An outstanding performer of 20th-cent. music. (R)

**Gaztambide, Joaquín** (*b* Tudela, 7 Feb. 1822; *d* Madrid, 18 Mar. 1870). Spanish composer and conductor. Studied Pamplona with Guelbenzu, Madrid with Carnicer. Chorus master Madrid, T de Santa Cruz, 1845, director T Español 1848; cond. T Real and elsewhere, introducing to Spain works by Meyerbeer and Wagner. Toured Cuba and Mexico with his own zarzuela company, but fell ill and died on his return home. The most successful of his 44 zarzuelas was *La Catalina*. One of the most important zarzuela composers of the 19th cent.

SELECT WORKLIST (all first prod. Madrid): *La Catalina* (Olona, after Scribe; 1854); *Los Magyares* (Olona; 1857); *El juramento* (Olona; 1858); *Las hijas de Eva* (Larra; 1862); *La conquista de Madrid* (Olona; 1863).

**Gazza ladra, La** (The Thieving Magpie). Opera in 2 acts by Rossini; text by Gherardini, after the comedy *La pie voleuse* (1815) by Baudouin d'Aubigny and Caigniez. Prod. Milan, S, 31 May 1817, with Giorgi-Belloc, Gallianis, Castiglioni, Monelli, Botticelli, Galli, Ambrosi, Biscottini, Rossignoli, De Angeli; London, H, 10 Mar. 1821, with Camporese; New Orleans, T d'Orléans, 30 Dec. 1828 with Mariage, Alexandre, Privat, Le Blanc.

An Italian village, early 19th cent. Ninetta (sop), a servant girl, is engaged to Giannetto (ten), the son of the farmer in whose house she works. She is accused of stealing a silver spoon, and the Podestà (bs), whose advances Ninetta has repulsed, brings her to trial and she is condemned to death. Only when she is on the way to execution is it discovered that the real thief was a magpie who flew into the farmhouse window.

**Gazzaniga, Giuseppe** (*b* Verona, 5 Oct. 1743; *d* Crema, 1 Feb. 1818). Italian composer. Studied with Porpora, and with Piccinni in Naples, where his first opera, *Il barone di Trocchia*, was performed 1768. Then moved to Venice; maestro di cappella at Crema Cathedral 1791. He wrote *c*.50 operas, mostly for Naples, Rome, and Venice, though through Sacchini he forged an important relationship with Vienna. The first of his works to be performed there, *Il finto cieco*, was a setting of a Da Ponte libretto: though the poet did not regard it highly, it proved very successful and brought Gazzaniga to public attention.

Apart from a handful of serious works, including *Ezio* and *Antigono*, Gazzaniga devoted himself to comic opera, mostly *dramma giocoso. Though conventional in approach, the lighthearted manner of *La locanda*, *La vendemmia*, *Le donne fanatiche*, and *La donna astuta* were highly popular and earned him a place in history as one of the minor forerunners of Rossini. His most influential work was *Don Giovanni Tenorio, o sia Il convitato di pietra*. Bertati's libretto was known to Da Ponte when he and Mozart began their work later that year; it is unclear whether Mozart knew Gazzaniga's music. Both versions owe much to the dual traditions of morality play and commedia dell'arte entertainment in their use of such features as the *catalogue aria. But similarities suggest a closer link than sharing a common heritage, such as the opening scene 'Notte e giorno faticar'. The clearest precedent for Mozart's alternation of serious and farcical elements, as well as his concluding buffo finale, is to be found in Gazzaniga.

SELECT WORKLIST: *Il barone di Trocchia* (Cerlone; Naples 1768); *La locanda* (The Inn) (Bertati; Venice 1771); *Ezio* (Metastasio; Venice 1772); *La vendemmia* (The Vintage) (Bertati; Florence 1778); *Antigono* (Metastasio; Rome 1779); *Il finto cieco* (The Feigned Blind Man) (Da Ponte; Vienna 1786); *Le donne fanatiche* (Bertati; Venice 1786); *Don Giovanni Tenorio, o sia Il convitato di pietra* (Bertati, Venice 1787); *La donna astuta* (Bertati; Venice 1793).

**Gazzaniga, Marietta** (*b* Voghera, 1824; *d* Milan, 2 Jan. 1884). Italian soprano. Studied Milan with Cetta and Mazzuccato. Début Voghera 1840

(Jane Seymour, *Anna Bolena*). Created Luisa Miller, and Lina (*Stiffelio*). Milan, S, 1851–3, 1862; USA 1857–69, 1871 as mezzo-soprano. Taught New York until 1876; Genoa 1881–2.

**Gdańsk** (Ger., Danzig). City in Poland, from 1918 to 1939 a Free City. Some performances were given in the early 17th cent. in the Budynek do Szermiecki (Fencing School), including a play (with music by Marcin Greboszewski), *Tragedia o Bogaczu i Lazarzu* (The Tragedy of Dives and Lazarus, 1643). Johann Meder (1649–1719) tried to introduce opera with his *Nero* (1695), but *Die wiederverehligte Coelia* (Coelia Remarried, 1698) had, owing to council opposition, to be given at Schottland. A new Schauspielhaus opened in 1801; here a German company played in about 1810. Open-air opera was given at nearby *Sopot 1929–39, including works by Wagner in some famous festivals 1924–42. The theatres were destroyed in the war, but in 1949 the T Wybrzeze began staging opera, performed by the Studio Operowe. In 1953 this joined with the symphony orchestra to form the Baltic State Opera and Philharmonic Co.

**Gebet.** See PREGHIERA.

**Gedda, Nicolai** (*b* Stockholm, 11 July 1925). Swedish tenor. Studied Stockholm with Oehmann. Début Stockholm O, 1952 (Chapelou, *Le postillon de Lonjumeau*). Milan, S, from 1953; Paris, O, 1954; London, CG, from 1955; New York, M, from 1957; also Vienna, Moscow, Chicago, etc. His wide repertory includes Don Ottavio, Huon, Cellini, Gounod's Faust, Don José, Duke of Mantua, Lohengrin, Lensky, Hermann; created the roles of the Husband (Orff's *Trionfo d'Afrodite*) and Anatol (Barber's *Vanessa*). Outstanding for his beauty of tone expert vocal control, and musical perception. (R

**Geliot, Michael** (*b* London, 27 Sep. 1933). English producer. Studied Cambridge; prod. British première of Liebermann's *School for Wives* 1958. Stage manager London, SW, 1960–1: prod. Burt's *Volpone* for New OC 1961, *Mahagonny* 1963, *Cardillac* 1970. Scottish O, Wexford; WNO resident producer 1965–8, director of productions 1969–78: *Lulu* 1971, *Greek Passion* 1981. London, CG, première of Maxwell Davies's *Taverner* 1972, *Carmen* 1973. Netherlands O, *Wozzeck* 1973. Munich, *Fidelio* 1974. Cassel, première of Cowie's *Commedia* 1979, *Figaro* 1980. Although he has produced Mozart, Verdi, and Berlioz (*Faust*, London, C, 1969), his greatest successes have been in contemporary opera, to which he brings an original intelligence. He has worked closely with Ralph *Koltai, with whom he shares an approach to opera by way of modern methods of music theatre.

**Gelsenkirchen.** Town in North-Rhine Westphalia, Germany. Until the mid-1930s, opera was generally given by companies from Düsseldorf and Duisburg. The Stadttheater opened 1935; bombed 1944. The new Stadttheater opened 1959 (cap. 1,050), and with neighbouring Bochum formed the Musiktheater im Revier, which maintains an enterprising policy.

**Gencer, Leyla** (*b* Istanbul, 10 Oct. 1924). Turkish soprano. Studied Istanbul with Arangi-Lombardi. Début Ankara 1950 (Santuzza). Naples, C, 1953; Milan, S, from 1956; London, CG, 1962; Gly. 1962–3, 1965; also Moscow, Buenos Aires, Vienna, etc. Repertory includes Donna Anna, title-roles of *Alceste*, *Medée*, *Aida*, *Gioconda*, *Maria Stuarda*, *Anna Bolena*; also early Verdi operas. Created Mme Lidoine in *Dialogues des Carmélites*. An intelligent, dramatic performer, remarkable for her pianissimo singing. (R)

**Genée, Richard** (*b* Danzig, 7 Feb. 1823; *d* Baden, 15 June 1895). German conductor, librettist, and composer. Studied Berlin with Stahlknecht. Cond. Reval, Riga, Cologne, Düsseldorf, Danzig, Mainz, Schwerin, Prague. Cond. Vienna, W, 1868–78. Of his operettas, only *Der Seekadett* (Vienna 1876) and *Nanon* (Vienna 1877) had more than local success; but he is famous for his brilliant and witty librettos, especially for Johann Strauss II (*Fledermaus*, 1874, with Haffner), Suppé, and Millöcker, often in collaboration with F. Zell (Camillo Walzel) and deriving from French farces. He also translated works by Lecocq, Offenbach, and Sullivan.

**Generali, Pietro** (*b* Masserano, 23 Oct. 1773; *d* Novara, 3 Nov. 1832). Italian composer. Studied Rome with Masi. He had a success with his first of some 50 operas, *Pamela nubile*, and other comedies, turning with less success to opera seria in 1812 with *Attila*. Overshadowed by Rossini (there is evidence of a mutual influence between the two composers), he moved to Barcelona as director of the T Santa Cruz in 1817. Even here he had some difficulty in including his own works in a Rossini-dominated repertory. He moved to Paris in 1819, to Naples in 1821; he was director of the Palermo T Carolino from 1823 until replaced by Donizetti in 1825. He became maestro di cappella at Novara in 1827 and devoted himself to sacred music. His operas have an easy brilliance which at first won him popularity, but his music lacks the individuality of Rossini's; and his work, which met with a good deal of critical reproof in his own lifetime, suffered from an indolence which often led to him finishing music when it was already in rehearsal.

SELECT WORKLIST: *Pamela nubile* (Rossi, after Goldoni; Venice 1804); *Le lagrime d'una vedova* (A Widow's Tears) (Foppa; Venice 1808); *Adelina* (Rossi; Venice 1810); *Attila* (Rossi; Bologna 1812); *I Baccanali di Roma* (Rossi; Venice 1816).

BIBL: L. Schiedemair, 'Eine Autobiographie Pietro Generalis', in *Festschrift zum 90. Geburtstage Rochus Frhrn von Liliencron* (Leipzig, 1910).

**Generalintendant** (or **Intendant**) (Ger.: 'superintendent'). The name given to the administrator of a German opera-house—often but not necessarily the artistic or musical director.

**Generalmusikdirektor** (Ger.: 'chief music director'). The title given to the senior musician in a German opera-house; his responsibilities include all decisions of musical policy, staffing, repertory, etc. and he also acts as chief conductor.

**Generalpause** (Ger.). The main interval in an operatic performance.

**Generalprobe** (Ger.: 'main rehearsal'). The final, dress rehearsal in a German opera-house. Usually opened to critics, professional colleagues, possibly opera clubs or school parties and other invited individuals or groups, it virtually constitutes the first performance of a new opera or production, and is rarely interrupted for further work. The final rehearsal at which matters may still be adjusted is the preceding *Hauptprobe. See also PROBE.

**Geneva** (Fr., Génève; Ger., Genf). City in Geneva Canton, Switzerland. The first opera given was Grétry's *Isabelle et Gertrude* in 1766. A new theatre opened in 1783; this was the home of opera until Oct. 1879, when the Grand Théâtre (cap. 1,300) opened with *Guillaume Tell*. This burnt down in 1951, and reopened (cap. 1,488) in 1952 with the French *Don Carlos* produced by Marcel Lamy, director until succeeded by Herbert Graf in 1965. Graf's régime opened with *The Magic Flute* in Kokoschka's settings under Ansermet (with the Suisse Romande Orchestra), and important revivals have included Bloch's *Macbeth* and Honegger's *Antigone*. The city's wealth has enabled it to engage international star singers and to experiment with theatre and film producers for opera, including Andrei Serban and Ken Russell.

**Gennaro.** 1. A young nobleman (ten), son of Lucrezia, in Donizetti's *Lucrezia Borgia*.
2. The blacksmith (ten) in Wolf-Ferrari's *I gioielli della Madonna*.

**Genoa** (It., Genova). City in Liguria, Italy. The first opera was Ferrari's *Il pastor regio* in the 1640s. The T Falcone opened 1650. Opera was also given at the T Sant'Agostino in the 18th cent. The T Carlo Felice (cap. 2,500) opened 7

Apr. 1828 with the première of Bellini's *Bianca e Fernando*. Angelo *Mariani was music director 1852–73. Toscanini had early successes 1891–4. In 1892, celebrating the 400th anniversary of the Genoan Christopher Columbus's discovery of America, the theatre was redecorated and Franchetti wrote his *Cristoforo Colombo* (third performance conducted by Toscanini). Wagner and Richard Strauss have been popular. Shelled 1941, bombed 1943. Opera has also been given at the T Margherita (cap. 1,800). The T Carlo Felice reopened 1991.

**Genoveva.** Opera in 4 acts by Schumann; text by Robert Reinick, altered by the composer, after Tieck's tragedy *Das Leben und Tod der heiligen Genoveva* (1799) and Hebbel's tragedy *Genoveva* (1843). Prod. Leipzig 25 June 1850, with Mayer, Günther-Baumann, Wiedemann, Brussi, cond. Reitz; London, DL, by RCM, 6 Dec. 1893.

Siegfried's castle in Strasburg, AD 730. Margareta (con) offers to help Golo (ten) woo Genoveva (sop), the wife of Count Siegfried (bar), during the Count's absence. Rejected by Genoveva, Golo plants his friend Drago (bar) in her bedroom so that she can be accused of adultery. Golo then goes to tell Siegfried, who has been injured in battle, of his wife's infidelity. Siegfried orders his men to take her to a forest, where she is to be executed. Golo makes a final attempt to persuade Genoveva to run away with him. She refuses, but her life is saved by the arrival of Siegfried, who has now learnt the truth.

Schumann's only opera. Hebbel's tragedy was also set by Natanael Berg (1947) and Eidens (20th cent.). Other operas are by Huth (*Golo und Genoveva*, 1838), Pedrotti (1854), Rogel (zarzuela, after Offenbach, 1868), Mögele (parody, 1902).

**Gentele, Göran** (*b* Stockholm, 20 Sep. 1917; *d* Sardinia, 18 July 1972). Swedish producer and administrator. Stockholm: Nat. T, 1941–52; Royal O, 1952–63 as producer, 1963–71 as director. His many productions included a famous *Ballo in maschera* in historical context (1960) and the première of Blomdahl's *Aniara* (1959). London, CG, 1961, *Iphigénie en Tauride*. Died in car crash before taking up new appointment as director New York, M.

**Gentili, Serafino** (*b* Venice, *c*.1775; *d* Milan, 13 May 1835). Italian tenor. Début Ascoli Piceno 1796. Naples 1800–3; Paris 1809; Milan, S, 1812–28; Dresden 1821–4. Created Lindoro. A much-esteemed singer in his time. His daughter **Giustina** sang with him 1820–8; Dresden début 1824 (Cenerentola).

**Gentle Shepherd, The.** Scots ballad opera in 5 acts; text by Allan Ramsay. Prod. Edinburgh,

Taylor's H, 29 Jan. 1729; London, DL, anglicized as *Patie and Peggie* by Theophilus Cibber, 20 Apr. 1729. Many other versions: Richard Tickell's prod. London, DL, 29 Oct. 1781; New York 7 Jan. 1786.

Originally a comedy without songs (1725), it was changed into a ballad opera after the success of *The Beggar's Opera* at Haddington in 1738, and quickly reached a similar standing in Scotland.

**George Brown.** A young English officer (ten), lover of Anna, in Boieldieu's *La dame blanche*.

**Georgia.** On the initiative of Anton Rubinstein, a musical society was founded in the second half of the 19th cent., and the singer Kharlampa Svaneli established some popular choir classes 1873. In charge of the society from this time was the influential Mikhail *Ippolitov-Ivanov, who maintained Georgian connections all his life. Russian opera companies visited the country from the 1880s, staging first Western operas in Russian, then Russian operas, e.g. Glinka's *Life for the Tsar* and *Ruslan* 1880–1, Rubinstein's *Demon* 1882, Tchaikovsky's *Eugene Onegin* and Rimsky-Korsakov's *May Night* 1883, Tchaikovsky's *Mazeppa* and Serov's *Rogneda* 1885.

The first Georgian opera theatre (cap. 800) (now the Paliashvili Opera and Ballet National Theatre) opened in Tiflis (Tbilisi) on 12 Apr. 1851. The first season was given by a visiting Italian company under Barbier, beginning with *Lucia* on 9 Nov. Italian opera remained popular, but French Grand Opera was also staged, including works by Meyerbeer, Halévy, and Auber. Later there was much enthusiasm for Wagner. Visiting Russian singers included Shalyapin (Brogni in *La Juive* in Kutaisi, Valentin in Tbilisi), as well as Western artists. Among the first major Georgian singers were P. Koridze and Meliton *Balanchivadze. The latter was also a composer: part of his *Tamar tsbiery* (The False Tamar) was staged in St Petersburg 20 Dec. 1897, and wholly staged Tbilisi 1925–6 (then renamed *Darejan Tsbiery*). National consciousness found an early artistic expression in opera, and pioneering works staged in 1919 were *Abessalom da Eteri* (Abessalom and Eteri) by Zachary Paliashvili (1871–1933), *The Legend of Shota Rustaveli* by Dmitry Arakishvili (1873–1953), and *Keto da Kote* by Victor Dolidze (1890–1933). Also in 1919 Arakishvili and Samuel Stolemin founded the Georgian Opera Studio (which staged *Shota Rustaveli*). Paliashvili especially was strongly influential on the course of Georgian opera with the first of these works and with *Daisi* (Sunset, 1923). These all represent the first important steps towards a Georgian opera independent of Russian example, to some extent marked by Italian vocal techniques but drawing on folk melismata, national instrumental figuration, and dance rhythms. Later important Georgian operas have included *The Legend of Tariel* (after Shota Rustaveli) (1946) by Shalva Mshevelidze (1904–84), *The Bride of the North* (1958) (after Griboyedov) by D. Toradze (1922–83), and *Mindia* (1961) by Otar *Taktakishvili (1924–89).

The main operatic centre continues to be the Tbilisi opera-house (cap. 1,061). A new opera theatre was opened in Kutaisi in 1969 as a subsidiary of the Tbilisi theatre, and since 1989 operates independently as Georgia's second opera-house.

**Gérald.** An English officer (ten), lover of Lakmé, in Delibes's *Lakmé*.

**Gérard.** A Revolutionary leader (bar) in Giordano's *Andrea Chenier*.

**Gerhard, Roberto** (*b* Valls, 25 Sep. 1896; *d* Cambridge, 5 Jan. 1970). Spanish composer. Studied Barcelona with Granados and Pedrell, Vienna with Schoenberg. His only opera, *The Duenna*, is a brilliant and intelligent setting which suggests that in more favourable circumstances he might have developed this side of his talent more fully.

WORKLIST: *The Duenna* (Gerhard & Hassall, after Sheridan; comp. 1947, prod. Madrid 1992).

**Gerl, Franz Xaver** (*b* Andorf, 30 Nov. 1764; *d* Mannheim, 9 Mar. 1827). Austrian bass and composer. Member of Salzburg choir under Leopold Mozart. After singing with the troupes of L. Schmidt and G. F. W. Grossman, he joined Schikaneder's company at Regensburg in 1787. First bass and actor with same at Vienna, T auf der Wieden, 1789–93, singing Osmin, Figaro, and Don Giovanni. Also composed (with *Schack) *Der dumme Anton*, and other successful Singspiels given by the company. Admired by Mozart, who wrote the aria K612 for him, he also created Sarastro. Brno 1794–1801; Mannheim Court T 1802–26, appearing in plays by Schiller as well as in opera. His first wife **Barbara Reisinger** (1770–1806) joined Schikaneder's company 1789 as a singer and actress, and created Papagena in 1791.

**German,** (Sir) **Edward** (orig. Jones, German Edward) (*b* Whitchurch, 17 Feb. 1862; *d* London, 11 Nov. 1936). English composer. Studied with Hay and at London, RAM. Hopes that he might prove Sullivan's successor in the field of light music were aroused by his completion of *The Emerald Isle*, and encouraged by the success of *Merrie England*; but his gift was much slenderer, though *Tom Jones* and *Fallen Fairies* were popular in their day. His music has a simple melodiousness, reflecting simple dramatic situations; it can be pretty, even dainty, or robust

and jingoistic, and is admirably written for the voice. Most of the numbers in his operettas are lyrical and decorative, and have no rhythmic or harmonic complication, though the Act I finale of *Merrie England* suggests that he had in him a vein of dramatic music that never found a true outlet.

WORKLIST: *The Two Poets* (Scott; London 1886); *The Emerald Isle* (completion of Sullivan: Hood; London 1901); *Merrie England* (Hood; London 1902); *A Princess of Kensington* (Hood; London 1903); *Tom Jones* (Courtneidge, Thompson, & Taylor, after Fielding; Manchester 1907); *Fallen Fairies* (Gilbert; London 1909).

BIBL: W. Scott, *Edward German* (London, 1932).

**Germania.** Opera in a prologue, 2 acts, and epilogue by Franchetti; text by Illica. Prod. Milan, S, 11 Mar. 1902, with Caruso, Sammarco, Pinto, Bathory, cond. Toscanini; London, CG, 13 Nov. 1907; New York, M, 22 Jan. 1910.

Set during the Napoleonic wars, the opera tells of the activities of a group of patriotic German students. Further conflict is created by Carlo's (bar) abduction of Ricke (sop) during the absence of her betrothed, Federico (ten). Eventually all are reunited in their patriotic cause but both Carlo and Federico are killed in action.

**Germany.** As a single political entity, Germany did not exist until 1871. The path of unification was tortuous, and of considerable influence for the development of German culture generally, and opera specifically. In the 17th cent. the country comprised a loose federation of about 1,700 independent states, ranging from free imperial cities (e.g. Frankfurt) and principalities and electorates (e.g. Mannheim) to minor aristocratic holdings. The larger of these courts maintained a musical establishment which was frequently involved in dramatic productions, while commercial opera was established in a number of cities. Germany's highly fragmented political structure makes the drawing of large-scale cultural lines of development very difficult. Between the 17th and 19th cents. the fortunes of opera ebbed and flowed in these cities and courts, reflecting changes in their political and economic circumstances, but not necessarily according to any general plan. A further complication is that for much of this period a common operatic heritage was shared with other German-speaking lands, principally those which combined to form the Austro-Hungarian Empire. After the Second World War, during which many opera-houses were destroyed, Germany suffered a unique blow through its partition. While operatic life in the Federal Republic (West Germany) was rebuilt and developed along Western European lines, those regions which formed the German Democratic Republic (East Germany) were forced to embrace Marxism-Leninism in their cultural life, placing political constraints on composers and producers alike. Following the reunification of Germany in 1990 it becomes possible once more to consider the country as a single whole.

The early history of German opera is, in effect, the history of Italian opera in Germany. Following the first experiments in Italy, opera was soon introduced into the South German courts: Salzburg in 1614, Vienna in 1626, and Prague the following year. The first opera by a German composer was probably *Schütz's *Dafne* (music lost), which set a translation of Rinuccini's libretto for Peri and was staged at Torgau near Dresden in 1627; the earliest German opera extant in full score is *Bontempi's and Peranda's *Dafne* (1671). Under Bontempi and C. *Pallavicino opera was firmly established at Dresden by the late 17th cent., and at Vienna, under *Cesti, *Draghi, and others. Other courts, such as those at Munich and Hanover, at both of which *Steffani worked, followed suit, and a lavish style of opera developed, taking the Venetian model as a point of departure, but with the addition of a sumptuousness and splendour appropriate to the setting.

Despite this domination, many attempts were made to foster German-language opera. The earliest work for which music survives is *Staden's *Seelewig*, a setting of a libretto by *Harsdörffer which was privately performed Nuremberg 1644: but the balance of music and text in this staged religious work is such that it barely merits description as an opera. The native German Singspiel developed slowly, initially on the basis of songs inserted into a play performed often by travelling companies who might later settle in a single place, or of music for the Reformation, and then the Counter-Reformation, school dramas. Progress towards an indigenous tradition was severely hampered by the Thirty Years War (1618–48), which spelt economic ruin for many courts, while its destruction of the vague semblance of German unity which had obtained hitherto served to encourage the influx of even stronger foreign influence. Within the realm of the court, there were few notable centres of German opera by the beginning of the 18th cent.: important exceptions were Wolfenbüttel (under G. C. *Schürmann) and Weissenfels (under Johann Philipp *Krieger).

Crucial for the development of German opera was the establishment of the first public opera-house outside Italy in 1678 in the free Hanseatic city of Hamburg. Away from courtly pressure, a repertoire developed in which elements of the German school drama were used in conjunction with features of French and Italian opera: many of the earliest Hamburg operas mixed German and Italian texts together, the title often being given in both languages. The works of the first generation, represented by *Franck, *Förtsch, and *Strungk, thus demonstrate a certain cosmopolitanism, which was carried further by *Conradi and *Kusser, and above all by *Keiser, under whose guidance German opera was for a while elevated to the standard of its rivals. Of later composers who spent time at Hamburg most important were *Handel, *Telemann, and *Mattheson: in 1738 the theatre closed and with it all hope of the establishment of serious German opera for several decades.

By the middle of the 18th cent. the supremacy of Italian opera seria in Germany seemed assured. *Agricola and *Graun in Berlin, *Caldara, and *Predieri in Vienna, and especially *Hasse in Dresden, placed their talents at the service of the *Metastasian libretto, a vehicle uniquely appropriate for courtly opera. Attempts were made, notably by *Gluck, to reform some of the more stultifying conventions of the genre, but a more serious challenge was presented by opera buffa, which proved increasingly popular. Yet more significant was the rise of a new Singspiel tradition, which drew on English *ballad opera and *opéra-comique. It began with *Der Teufel ist los* (1752), a setting of C. F. Weisse's translation of *The Devil to Pay* by J. C. Standfuss, and was followed by many works, most importantly the Weisse settings by J. A. *Hiller. Though the action is still in prose, with interpolated songs and other music, there are sometimes ensembles and finales; the subjects are generally rustic and comic, with a love interest. Other important composers were Jiří *Benda, Johann *André, J. F. *Reichardt, and Johann *Zumsteeg. Music played a significant part in early 18th-cent. Viennese Singspiel, but a crucial event was Joseph II's 1778 establishment of a German National-Singspiel designed to encourage vernacular works. This opened with Ignaz *Umlauff's *Die Bergknappen*, but did not prosper and closed 1783. Composers working in Vienna included Wenzel *Müller and especially Karl von *Dittersdorf, later Johann *Schenk and Josef *Weigl, in all of whose works there is more substantial and vital musical content. However, the condition of German opera in the late 18th cent. is seen in the attention *Mozart gave to Italian opera. His *Entführung aus dem Serail* (1782) is a Singspiel; the Viennese Singspiel provides the basis for the greatest work in the genre, *Die Zauberflöte* (1791). North German Singspiel was the basis, with the example of French *rescue opera, for *Beethoven's *Fidelio* (orig. 1805).

Early 19th-cent. German composers were concerned above all to develop a native tradition free from Italian influence. They included Peter von *Winter, Louis *Spohr, Franz *Schubert,

Johann von *Poissl, and E. T. A. *Hoffmann. However, it was Carl Maria von *Weber who grasped most fully that the real example for German Romantic opera lay not with Singspiel but with French Revolutionary opéra-comique, since it was here that a more advanced technique had developed for the handling of popular subjects, with increased use of the orchestra and hence the greater dramatic possibilities for *reminiscence motive, free from the formalities of opera seria. The performance of his *Der Freischütz* (1821) was a turning-point in the history of German opera, and Weber took his reforms further in *Euryanthe* (1823). His death in 1826 left the field principally to Heinrich *Marschner, Albert *Lortzing, Otto von *Nicolai, and Peter *Lindpaintner, who took up a number of the themes Weber had identified as distinctively German. These included the supernatural, Nature, medievalism, popular settings, melodies deriving from folk-song, and stories based on folk legend. Musically there was an increased tendency towards continuity, and to structures based on motive and whole scenes, and with this came increased depth in the portrayal of character and psychology.

These elements and much else were the basis of the growth of the art and reforms of Richard *Wagner, whose life and work culminated in the first Bayreuth Festival (1876). By now a German operatic tradition was thoroughly established, with theatres in all important towns and composers working in them or independently. Wagner's example divided musical Germany, with some consciously following it and others reacting against it, but none ignoring it. The most succesful of his followers were minor figures such as Peter *Cornelius. His inspiration fired Richard *Strauss, and continued after the First World War in the work of Franz *Schreker and Hans *Pfitzner; more characteristic was the reaction manifest in Paul *Hindemith and Kurt *Weill. Among the most important composers to emerge after the Second World War were Wolfgang *Fortner, Gottfried von *Einem, and especially Hans Werner *Henze.

Modern German operatic life continues to be as rich and solidly grounded as that of any country in the world, welcoming new ideas while acknowledging the strength of tradition: out of the contest between the two much stimulus continues to come.

See also AACHEN, AUGSBURG, BADEN-BADEN, BAMBERG, BAYREUTH, BERLIN, BIELEFELD, BONN, BREMEN, BRESLAU (under WROCŁAW), BRUNSWICK, CHEMNITZ, COBURG, COLOGNE, DARMSTADT, DESSAU, DETMOLD, DORTMUND, DRESDEN, DÜSSELDORF-DUISBURG, ESSEN, FLENSBURG, FRANKFURT, FREIBURG IM BREISGAU, GELSENKIRCHEN, GOTHA, GÖTTINGEN, HAGEN, KAISERSLAUTERN, HALLE, HAMBURG, HANOVER, KARLSRUHE, KASSEL, KÖNIGSBERG (under KALININGRAD), KREFELD, LEIPZIG, MAGDEBURG, MAINZ, MANNHEIM, MUNICH, MÜNSTER, NUREMBERG, OLDENBURG, OSNABRÜCK, PFORZHEIM, RIGA (under LATVIA), ROSTOCK, SAARBRÜCKEN, STUTTGART, TRIER, ULM, WEIMAR, WEISSENFELS, WIESBADEN, WOLFENBÜTTEL, WUPPERTAL, ZOPPOT (under SOPOT).

**Germont.** Giorgio Germont (bar), Alfredo's father, in Verdi's *La traviata*.

**Gershwin, George** (*b* Brooklyn, 26 Sep. 1898; *d* Hollywood, 11 July 1937). US composer. After an early career as pianist and song-writer for a Tin Pan Alley publisher, he wrote his first musical, *La La Lucille* (1919) and had his first success with the song 'Swanee'. During the 1920s and 1930s he wrote over 30 musicals, among them *Lady, Be Good!* (1924), *Strike Up the Band* (1927), and *Girl Crazy* (1930), many to lyrics by his brother Ira.

Gershwin's first opera, in which popular songs are connected by enterprising jazz-style recitative, was written as an interlude in George White's revue *Scandals of 1922*. Subtitled 'Opera Ala Afro-American', the Harlem setting of the 1-act *Blue Monday Blues* capitalized upon a vogue for African art and literature, but was quickly dropped from the musical comedy because of its tragic ending. In 1925 it was revived by Paul Whiteman under the title *135th Street*, with new orchestration by Ferde Grofé. Despite claims that *Blue Monday Blues* was 'the first real American opera', it was successful in neither version.

In *Porgy and Bess* (1935) Gershwin attempted to mix the idioms and techniques of jazz and classical music. At first he was attacked from both sides and forced to make extensive revisions and cuts, which were later reinstated. However, the popularity of such numbers as 'Summertime' and 'Bess, You is My Woman Now' has made *Porgy and Bess* one of the most popular operas of the 20th cent., while the skilful, almost symphonic manipulation of musical motives provides a degree of unity which refutes the charge that the opera is merely a succession of popular songs. Though Gershwin's Tin Pan Alley background is still evident, the score's musical characterization and dramatic vitality make it a worthy climax to his career. As with many of the musicals, Gershwin probably had help with the scoring of *Porgy and Bess*.

A performance of the opera in Leningrad on 26 Dec. 1955 by a travelling company was the first appearance by a US theatrical group in the USSR.

SELECT WORKLIST: *Blue Monday Blues* (De Sylva; New York 1922, rev. 1925 as *135th Street*); *Porgy and*

*Bess* (Du Bose Heyward & I. Gershwin; Boston 1935).

BIBL: E. Jablonski, *Gershwin: A Biography* (London, 1987).

**Gerster, Etelka** (*b* Kasse (Košice), 15 June 1855; *d* Pontecchio, 20 Aug. 1920). Hungarian soprano. Studied Vienna with Marchesi. Début Venice, F, 1876 (Gilda, recommended by Verdi). Berlin, Budapest 1877; London, HM, 1877, and CG, 1890. New York, Mapleson Co., 1878; tour with same 1883-4, in bitter rivalry with Patti. Berlin, Kroll O, 1889. Here set up singing school 1896–1917; taught Lotte Lehmann. A highly strung performer with a superb technique; roles included Amina, Gilda, Lucia, Queen of the Night.

**Gerster, Ottmar** (*b* Braunfels, 29 June 1897; *d* Leipzig, 31 Aug. 1969). German composer. Studied Frankfurt with Sekles. His *Enoch Arden* (after Tennyson; Düsseldorf 1936) uses popular tunes in separate numbers as well as Leitmotiv. Together with *Die Hexe von Passau* (Billinger; Düsseldorf 1941), later banned by the Nazis, this was his most successful opera.

**Gerusalemme.** See LOMBARDI, I.

**Gervais, Charles-Hubert** (*b* Paris, 19 Feb. 1671; *d* Paris, 15 Jan. 1744). French composer. Spent most of his life in Paris in the service of Philip, Duke of Orleans, himself a gifted musician, rising to be *intendant de la musique* in 1712. Of the generation between Lully and Rameau, Gervais produced three operas: *Méduse* made little impact, but its successors *Penthée*, possibly composed in collaboration with the Duke, and *Hypermnestre* were more successful, as was the ballet *Les amours de Prothée* (1720). Like his contemporaries Destouches and Campra, Gervais was concerned with developing a more flexible and lively approach to opera than that of Lully. In this he was much influenced by the Italianate tastes of his patron, as seen in his melodic writing and handling of dissonance.

SELECT WORKLIST (all prod. Paris): *Méduse* (Boyer; 1697); *Penthée* (De la Fare; 1705); *Hypermnestre* (Lafont; 1716, rev. 1728, 1746, 1765).

**Gerville-Réache, Jeanne** (*b* Orthez, 26 Mar. 1882; *d* New York, 5 Jan. 1915). French contralto. Studied Paris with Laborde and Viardot. Début Paris, OC, 1899 (Gluck's Orfeo); OC till 1903. London, CG, 1905. New York, Manhattan OC, 1907–10. Chicago, Boston, Canada 1910–15. Created Geneviève (*Pelléas*); also sang Dalila, Hérodiade, first US Klytemnestra. A true contralto with impressive dramatic qualities. (R)

**Gesamtkunstwerk** (Ger.: 'unified work of art'). A term, associated with Wagner and discussed by him in *Das Kunstwerk der Zukunft* (The Artwork of the Future, 1849) and in other places, to describe a dramatic form in which all the arts—poetry, drama, the visual arts, music, song—should be united so as to form a new and complete work of art.

**Ghedini, Giorgio** (*b* Cuneo, 11 July 1892; *d* Nervi, 25 Mar. 1965). Italian composer. Studied Turin with Cravero, Bologna with Bossi. An intelligent composer with an understanding of the Baroque, his early operas were somewhat static in nature, a condition which marks even the intense and violent *Le Baccanti*. He also had some success with *Billy Budd*.

WORKLIST: *Maria d'Alessandria* (Meano; Bergamo 1937); *Re Hassan* (Pinelli; Venice 1939); *La pulce d'oro* (The Golden Flea) (Pinelli; Genoa 1940); *Le Baccanti* (Pinelli, after Euripides; Milan 1948); *Billy Budd* (Quasimodo, after Melville; Venice 1949); *Lord Inferno* (Antonicelli, after Beerbohm; RAI 1952, rev. as *L'ipocrita felice*, Milan 1956).

**Ghent** (Fr., Gand; Dut., Gent). City in E Flanders, Belgium. The first theatre opened in 1698, when Lully's *Thésée* was given. Burnt down in 1715, this reopened in 1737 as the T Saint-Sébastien; it became the Grand T in 1803. The T-Lyrique opened in 1847. The Grand T became the T Royal in 1921. From 1925 René Coens tried to broaden the language policy from exclusively Flemish performances. A large repertory and distinguished guest appearances were features developed particularly under Vina Bovy in 1947–55.

BIBL: G. Verriest, *Het lyrisch toneel Te Gent* (Ghent, 1964).

**Ghiaurov, Nicolai** (*b* Lidjene, 13 Sep. 1929). Bulgarian bass. Studied Sofia with Brambarov; then Leningrad and Moscow. Début Moscow opera studio 1955 (Don Basilio, *Barbiere*), then Sofia 1956 (same role). Vienna, S, and Moscow, B, from 1957; Milan, S, from 1959; London, CG, from 1962; Chicago from 1964; New York, M, and Salzburg from 1965; also Paris; etc. Repertory includes Ramfis, Créon, Gremin, Pimen, and Massenet's Don Quichotte. A notable basso cantante with a rich, expressive voice, and an imposing dramatic presence, outstanding as Boris, Philip, Gounod's and Boito's Mephistopheles; married to soprano Mirella *Freni. (R)

**Ghiringhelli, Antonio** (*b* Brunello, 5 Mar. 1903; *d* Aosta, 11 July 1979). Italian administrator. Studied law in Genoa. Supervised rebuilding of Milan, S, 1945, then becoming its director till 1972. Built the Piccola Scala, and organized successful tours notably to London, Vienna, and Moscow. Established the Scala schools for young singers and stage designers, and developed the ballet school.

**Ghislanzoni, Antonio** (*b* Lecco, 25 Nov. 1824; *d* Caprino Bergamasco, 16 July 1893). Italian librettist and baritone. As a singer, he made his début in Sanelli's *Luisa Strozzi* in 1846, and sang in various Italian towns before being arrested for revolutionary activities in 1849 and deported to Corsica; resuming his career on his release, he sang Don Carlo in *Ernani* (Paris 1851). He also formed a company and began writing librettos. The loss of his singing voice compelled him to turn to writing novels and criticism as well as further librettos, and to editing periodicals. His 85-odd librettos, which earned him much respect, include works for Petrella, Ponchielli, Gomes, Cagnoni, Braga, and Catalani. He is most famous for his work on *Aida*, with which Verdi was very pleased. In it, he makes some attempt to loosen rhyme and metre.

**Giacobbi, Girolamo** (*b* Bologna, *bapt.* 10 Aug. 1567; *d* Bologna, Feb. 1629). Italian composer. Joined S Petronio, Bologna as choirboy; 1595 assistant maestro di cappella; 1604–28 maestro di cappella. He was one of the earliest non-Florentine composers to use the new *stile rappresentativo*, seen in the intermedi (pub. Venice, 1610 as *Dramatodia*) provided for Campeggi's play *Filarmindo* (1608). He also made a (lost) setting of Campeggi's *Andromeda* libretto (1610).

**Giacosa, Giuseppe** (*b* Colleretto Parella, 21 Oct. 1847; *d* Colleretto Parella, 2 Sep. 1906). Italian playwright and librettist. After practising law, he turned to writing, also holding various teaching posts including that of professor of drama at the Verdi Conservatory. As a dramatist, he is one of the most important representatives of verismo; he is also remembered as collaborator with Luigi *Illica on three operas by Puccini, for which he provided the versification, *La bohème*, *Tosca*, and *Madama Butterfly*.

**Giannini.** Italian family of singers and composers.

1. **Ferruccio** (*b* Ponte all'Arnia, 15 Nov. 1868; *d* Philadelphia, 17 Sep. 1948). Tenor. Emigrated to USA 1885. Studied Detroit with De Campi, a Lamperti pupil. Début Boston 1891. Member of Mapleson's Co. for several years. Made the first operatic records, 1896, for Berliner. (R) Married the violinist Antonietta Briglia, with whom he had six children, including:

2. **Euphemia** (*b* Philadelphia, 8 Nov. 1895; *d* Philadelphia, 15 Jan. 1979). Soprano. Studied Milan with Gigola. Début Turin, Mimì. Taught Curtis Inst., pupils incl. Frank Guarrera, Anna Moffo.

3. **Dusolina** (*b* Philadelphia, 19 Dec. 1902; *d* Zurich, 26 June 1986). Soprano. Studied with her father, then New York with Sembrich. Stage

début Hamburg 1925 (Aida). Berlin, S, 1925, 1933–4, 1937, 1949; London, CG, 1928–9, 1931; Vienna, S, 1934, 1950; New York, M, 1936–41. Repertory included Donna Anna, Aida, Santuzza, Alice, Butterfly. Created Hester in her brother's opera, *The Scarlet Letter*. Dramatically and musically a considerable interpreter, with a full, clear tone. (R)

4. **Vittorio** (*b* Philadelphia, 19 Oct. 1903; *d* New York, 28 Nov. 1966). Composer. Studied Milan and Juilliard School. Operas incl. *Lucidia* (Munich 1934), *The Scarlet Letter* (Hamburg 1938), *The Taming of the Shrew* (Cincinnati 1953), *The Harvest* (Chicago 1961).

**Gianni Schicchi.** Opera in 1 act by Puccini; text by Forzano, after an episode (perhaps based on fact) in Dante's *Inferno*, xxx. 32 (*c.*1307–21). The third part of Puccini's *Trittico*. Prod. New York, M, 14 Dec. 1918, with Easton, Crimi, De Luca, cond. Moranzoni; Rome, C, 11 Jan. 1919, with Dalla Rizza, E. Johnson, Galeffi, cond. Marinuzzi; London, CG, 18 June 1920, with Dalla Rizza, Burke, Badini, cond. Bavagnoli.

The bedchamber of Buoso Donati, Florence, 1299. Buoso has recently died. Rumour leaks out that he has left most of his money to the monks and so his relatives start a frantic search for his will. Rinuccio (ten) finds it, but will not hand it over until his Aunt Zita (con) agrees that he may marry Lauretta (sop), daughter of Gianni Schicchi (bar), if the contents are favourable. The relatives' worst fears are confirmed, and as they are deciding what can be done, Gianni Schicchi arrives. He devises a plan to help the family: he pretends to be the dying Buoso Donati and summons a lawyer to make another will. Each member of the family offers Gianni large bribes to allocate them a substantial portion of the inheritance. When Gianni dictates the will, he leaves something to each of the relatives but keeps the best things for himself. The relatives are furious but helpless. After the lawyer has left, they turn on Schicchi and carry away from the house all they can. Rinuccio and Lauretta pledge their love, and Gianni begs the audience not to think too badly of him.

**Giasone.** Opera in a prologue and 3 acts by Cavalli; text by Giacinto Andrea Cicognini. Prod. Venice, C, 5 Jan. 1649; Buxton 3 Aug. 1984, cond. Hose.

The opera, apart from the finale, follows the classical legend of Jason and the Argonauts, paying particular attention to the love of Jason and Medea. At the end, however, Jason marries Hypsipyle, Queen of Lemnos.

Cavalli's most frequently performed opera. Revised by Stradella, 1671, when a prologue and three arias were added.

**Gibert, Paul-César** (*b* Versailles, 1717; *d* Paris, 1787). French composer. Studied in Naples; made an important journey around Italy recruiting singers for the French royal chapel, including the castrato Albanese, then settled in Paris *c*.1750. Earned a reputation as a singing-teacher and composer of opéras-comiques, including *La Sybille*—a parody of Dauvergne's *Les fêtes d'Euterpe*—and *La fortune au village*. Remembered chiefly as the composer of music for Favart's *Soliman II, ou Les trois sultanes*, which was popular throughout Europe. A proponent of the Italian style, his only tragédie lyrique, *Deucalion et Pyrrha*, was so well received that he was awarded a valuable gold medal.

SELECT WORKLIST (all first prod. Paris): *La Sybille* (De Guerville; 1758); *La fortune au village* (Favart & Bertrand; 1760); *Soliman II, ou Les trois sultanes* (Favart, after Marmontel; 1761); *Deucalion et Pyrrha* (Watelet; 1772, lost).

**Gibichungs.** The tribe to whose stronghold on the Rhine Siegfried comes in Act I of Wagner's *Götterdämmerung*. Their lord is Gunther; his sister is Gutrune. Their half-brother is Hagen, son of Alberich.

**Gibson, (Sir) Alexander** (*b* Motherwell, 11 Feb. 1926; *d* Jan. 1995). Scottish conductor. Studied Glasgow, London with Austin, Salzburg with Markevitch, Siena with Van Kempen. London, SW: répétiteur, 1951–2; conductor 1954–7; music director 1957–9, conducting a wide repertory incl. première of Gardner's *Moon and Sixpence*. London, CG, 1957. Founded *Scottish Opera 1962, conducting most of its productions, including *The Trojans* and *The Ring*, and acting as an invigorating force in giving the company energy and quality. Retired from artistic directorship 1980. A versatile and sensitive musician, with a particular care in accompanying singers. (R)

**Gielen, Michael** (*b* Dresden, 20 July 1927). Austrian, later Argentinian, conductor and composer, son of producer Josef Gielen. Studied Buenos Aires with Leuchter, Vienna with Polnauer. Répétiteur Buenos Aires, C, 1947–50; Vienna, S, 1950–1. Début Vienna 1954. Chief cond. Stockholm 1960–5. Cologne 1965 (incl. première of Zimmermann's *Die Soldaten*). Chief cond. Netherlands O 1972–7. Music director Frankfurt from 1977. A musician of sharp and clear intellect, he has made a particular reputation with contemporary music. (R)

**Giesecke, Karl Ludwig** (orig. Johann Georg Metzler) (*b* Augsburg, 6 Apr. 1761; *d* Dublin, 5 Mar. 1833). German dramatist. He worked in Vienna for Schikaneder, writing opera texts among other pieces and translating librettos including Mozart's *Figaro* and *Così*. His claim to the part-authorship of Mozart's *Die Zauberflöte* is doubtful.

**Gigli, Beniamino** (*b* Recanati, 20 Mar. 1890; *d* Rome, 30 Nov. 1957). Italian tenor. Studied Rome with *Cotogni and Rosati. Début Rovigo 1914 (Enzo, *Gioconda*). Milan, S, 1918 (Faust, *Mefistofele*) in Boito celebrations under Toscanini. New York, M, 1920–32, 1938–9, being regarded as Caruso's successor. London, CG, 1930–1, 1938–9, 1946; in 1946 sang with his daughter Rina in *Bohème* and *Pagliacci* and also sang Turiddu and Canio on the same evening. Left USA for Italy in 1932 in protest against salary cuts at the New York, M.

His voice was one of the most beautiful of the century, with a technique secure well into his sixties and an exquisite cantilena. His taste was sometimes questionable, his acting often rudimentary, but his presence on the stage always assured the occasion a popular following. His best roles included the Duke of Mantua, Canio, Cavaradossi, Lionel, and Chénier. The concert tours he undertook in his last years crowned his immense popularity. His memoirs, *Le confidenze di Beniamino Gigli* (Rome, 1942), were translated as *The Memories of Beniamino Gigli* (London, 1957). (R)

BIBL: G. Pugliese, ed., *Gigli* (Treviso, 1990).

**Gilbert, W. S.** (Sir William Schwenk Gilbert) (*b* London, 18 Nov. 1836; *d* Harrow Weald, 29 May 1911). English poet, playwright, and librettist. His earliest dramatic work was a burlesque of Donizetti's *L'elisir d'amore* entitled *Dulcamara, or The Little Duck and the Great Quack* (1866). His long collaboration with *Sullivan on the *Savoy operas provided English opera with a vein of comedy which has never been consistently excelled. Idiosyncratic and ingenious, he was a versifier of matchless brilliance, with a mastery of the patter song that is the equal of anything in the whole of opera buffa. His satire of various British institutions—the Navy, the House of Lords, the Police—is as apt as that of more exotic manifestations such as the Japanese Exhibition (in *The Mikado*) and the aesthetic movement (in *Patience*), and ultimately as inoffensive. Of a number of personal quirks, the one that has caused most difficulty is his obsession with the frustrated, plain, middle-aged spinster, e.g. Katisha in *The Mikado*. His extraordinary technical proficiency was always musically stimulating to Sullivan.

BIBL: S. Dark and R. Grey, *W. S. Gilbert: His Life and Letters* (London, 1923).

**Gilda.** Rigoletto's daughter (sop) in Verdi's *Rigoletto*.

**Gillier, Jean-Claude** (*b* Paris, 1667; *d* Paris, 31 May 1737). French composer. Spent 30 years as a double-bass player in the orchestra of the Comédie-Française, for whom he also wrote incidental music for over 40 works. This mostly comprised occasional airs or a final vaudeville: among his more important contributions are those to Dancourt's *La foire des Bezons*, *La fête de village*, and *Céphale et Procris*. Also wrote divertissements and occasional works, such as *L'hyménée royale*, and some of his stage music was published as *Recueil des airs de la Comédie-Française*. From 1713 worked with the Théâtres de la Foire, for whom he wrote many works; he also appears to have made several visits to England. Like many French composers of his time, he was much influenced by the Italian style and his music is distinguished by its light, attractive, and immediately appealing temperament.

SELECT WORKLIST (all first prod. Paris): *La foire des Bezons* (Dancourt; 1695); *L'hyménée royale* (Pellegrin; 1699); *La fête de village* (Dancourt; 1700); *Céphale et Procris* (Dancourt; 1711).

**Gilly, Dinh** (*b* Algiers, 9 July 1877; *d* London, 19 May 1940). French baritone. Studied Rome with Cotogni, and Paris Cons. Début Paris, O, 1899 (Priest; *Sigurd*). Paris, O, 1902–8; New York, M, 1909–14. London, CG, 1911–14, 1919–24. Created Sonora in *La fanciulla del West*; first London Jack Rance. Also sang Athanaël (*Thaïs*), Amonasro, Germont, Rigoletto, Silvio etc. A communicative, musical singer and actor. Married the contralto Edith Furmedge, and opened a singing school in London. (R)

**Giménez, Jerónimo** (*b* Seville, 10 Oct. 1854; *d* Madrid, 19 Feb. 1923). Spanish composer and conductor. Studied Seville with his father, Paris with Alard, Savart, and Thomas. Director Madrid: T Apolo, 1885, then T de la Zarzuela (conducting first Spanish *Carmen*). One of the most important zarzuela composers of his day, he developed especially the 1-act *género chico*, making much use of Spanish song and dance. He wrote over 100 zarzuelas, including *Tannhauser el estanquero* (Tannhauser the Shopkeeper, 1890), *El barbero de Sevilla* (1901), *Los viajes de Gulliver* (Gulliver's Travels) (1911), and *Tras Tristán* (After Tristan) (1918).

**Ginastera, Alberto** (*b* Buenos Aires, 11 Apr. 1916; *d* Geneva, 25 June 1983). Argentinian composer. Studied at the Williams and Buenos Aires Cons. Held various academic posts in Argentina and USA; on several occasions was forced to leave his native country because of disagreements with the Perón régime. Came to notice with his ballet *Estancia* (1943); was subsequently regarded as the father of an Argentinian national style of composition.

He has written three operas, all of them during a period when he was experimenting with a kind of neo-Expressionism. The most important was *Bomarzo*: at first banned in Buenos Aires because of its sex-ridden plot, it was taken up elsewhere and enjoyed brief fame; musically it combines the techniques of Leitmotiv and serialism in an original fashion. *Beatrix Cenci* develops Ginastera's fascination with the macabre, drawing on serial and aleatory techniques and producing some powerful dramatic climaxes.

WORKLIST: *Don Rodrigo* (Casona; Buenos Aires 1964); *Bomarzo* (Láinez; Washington, 1967); *Beatrix Cenci* (Shand & Girri; Washington, 1971).

**Gioconda, La** (The Joyful Girl). Opera in 4 acts by Ponchielli; text by 'Tobia Gorrio' (Arrigo Boito), after Hugo's drama *Angelo, tyran de Padoue* (1835). Prod. Milan, S, 8 Apr. 1876 (two months after Cui's setting of the subject was prod. in St Petersburg), with Mariani-Masi, Biancolini, Gayarre, Aldighieri, cond. Faccio; London, CG, 31 May 1883, with Marie Durand, Stahl, Marconi, Cotogni, cond. Bevignani; New York, M, 20 Dec. 1883, with Nilsson, Fursch-Madi, Stagno, Del Puente, cond. Vianesi.

Venice, 17th cent. La Cieca (mez), the mother of the street-singer La Gioconda (sop), is accused of witchcraft, but is saved by the interception of Enzo Grimaldi (ten), a Genoese prince, whom La Gioconda loves. Barnaba (bar) is in love with La Gioconda and he offers to help Enzo elope with Laura Adorno (mez), wife of Alvise Badoero (bs). La Gioconda resolves to kill her rival, but when she realizes Laura was instrumental in obtaining the release of La Cieca, she alters her plan and helps the lovers; however, their escape attempt is foiled. When Alvise orders his wife to drink poison, La Gioconda substitutes a sleeping potion. Enzo, however, believes her to be dead and is arrested after an angry outburst. La Gioconda promises to yield to Barnaba in order to save Enzo. When he is released, Enzo disregards La Gioconda's sacrifice. As she pretends to give herself to Barnaba, she stabs herself, too late to hear Barnaba say that he has just killed La Cieca.

**Gioielli della Madonna, I** (The Madonna's Jewels). Opera in 3 acts by Wolf-Ferrari; text by Golisciani and Zangarini, German version *Der Schmuck der Madonna* by Hans Liebstöckl. Prod. Berlin, Kurfürstenoper, 23 Dec. 1911, with Ida Salden, Marák, Wiedemann; Chicago, Auditorium, 16 Jan. 1912, with Carolina White, Bassi, Sammarco; London, CG, 30 May 1912, with Edvina, Martinelli, Sammarco. Prod. Genoa 6 Feb. 1913, with Boccolini, Calleja, V. Borghese; then not in Italy again, because of the allegedly profane subject matter, until Rome, 26 Dec. 1953.

Raffaele (bar) is willing even to steal the

Madonna's jewels to prove his love for Mariella (sop); but Gennaro (ten) wins her by this very deed. Remorsefully she confesses to the enraged Raffaele and rushes away to drown herself, while Gennaro returns the jewels to an image of the Madonna, and then stabs himself.

**Giordano, Tommaso** (*b* Naples, *c*.1733; *d* Dublin, 23 or 24 Feb. 1806). Italian composer. Spent several nomadic years in his father's troupe before arriving in London in 1753; harpsichordist in its opening production, Cocchi's *Gli amanti gelosi*. Début as composer with the opera buffa *La comediante fatta cantatrice*; in 1766 his *L'eroe cinese* began his career in Dublin, where the troupe appeared regularly. Composed over 30 dramatic works, many of them collaborations and most for London or Dublin, and played an important part in the flourishing musical life of both cities. Perhaps inevitably, his works had an Italian flavour, suggesting something of the character, if not the talent, of J. C. Bach's works. Giordano also adapted the works of others, notably Gay's *The Beggar's Opera* (1765) and Hasse's *Artaserse* (1772), and composed incidental music, including for the first production of Sheridan's *The Critic* (1779). Also active as a teacher.

> SELECT WORKLIST: *La comediante fatta cantatrice* (?; London 1756); *L'eroe cinese* (Metastasio; Dublin 1766).

**Giordano, Umberto** (*b* Foggia, 28 Aug. 1867; *d* Milan, 12 Nov. 1948). Italian composer. Studied Naples with Serrao. While still a student, won sixth place in the 1889 Sonzogno competition with *Marina*, which led to the commission for *Mala vita*. The scandal caused by this outspoken piece of verismo led to it being later watered down (though it also won admiration, including from Hanslick), and Giordano turned to what he hoped would be the safer paths of Romantic melodrama with *Regina Diaz*. On its failure, he moved to Milan and achieved his most enduring success with *Andrea Chenier*. Though the post-verismo, Puccinian effects are carefully calculated, the best pages of the score have genuine strength and passion. *Fedora* was initially almost as successful; and *Siberia* had a certain initial following. Giordano continued in verismo vein with *Madame Sans-Gêne*, then tried once again to change course with *La cena delle beffe* and with a lighter work, *Il re*, which had some success due to its appeal to coloratura sopranos.

> WORKLIST: *Marina* (Golisciani; 1889, unprod.); *Mala vita* (Wicked Life) (Daspuro, after Giacomo & Cognetti; Rome 1892; rev. as *Il voto* (The Vow), Milan 1897); *Regina Diaz* (Targioni-Tozzetti & Menasci, after Lockroy; Naples 1894); *Andrea Chenier* (Illica; Milan 1896); *Fedora* (Colautti, after Sardou; Milan

1898); *Siberia* (Illica; Milan 1903); *Marcella* (Stecchetti, after Cain & Adenis; Milan 1907); *Mese Mariano* (Giacomo; Palermo 1910); *Madame Sans-Gêne* (Simoni, after Sardou & Moreau; New York 1915); *Giove a Pompei* (collab. with Franchetti) (Illica & Romagnoli; Rome 1921); *La cena delle beffe* (The Supper of Jests) (Benelli; Milan 1924); *Il re* (The King) (Forzano; Milan 1929).

> BIBL: G. Confalonieri, *Umberto Giordano* (Milan, 1958); M. Morini, ed., *Umberto Giordano* (Milan, 1968).

**Giorgetta.** Michele's wife (sop) in Puccini's *Il tabarro*.

**Giorgi, Teresa.** See BELLOC-GIORGI, TERESA.

**Giorgi-Banti, Brigida** (*b* Crema, *c*.1756; *d* Bologna, 18 Feb. 1806). Italian soprano. Began life as a street-singer. Made a spectacular début at the Paris, O, 1776, having made her way to the city by singing in cafés. London 1779–80 and 1794–1802; toured Europe 1780–94, appearing in Vienna, Italy, Warsaw, and Madrid. She was stupid, lazy, and never learned to read music, but enjoyed enormous popularity because of her glorious voice and spirited acting. Though Da Ponte found her coarse and ignorant, composers inspired to write for her included Haydn (his *Scena di Berenice*), Paisiello, Zingarelli, and Anfossi. Critics described her singing as 'charming', 'perfect', and 'impassioned'. In 1779 she married the dancer Zaccaria Banti. She left her immensely large larynx to the city of Bologna, though it has never been traced.

**Giorgi-Righetti, Geltrude** (*b* Bologna, 1793; *d* Bologna, 1862). Italian contralto. Studied Bologna. Début Bologna 1814. Married the lawyer Giorgi 1815. Created Rosina and Cenerentola, 1817, both of which reflect her gift for coloratura singing. Also highly successful as Isabella (*L'italiana*). Her voice was 'full, powerful, and of rare extension' (Spohr), and she was a good actress. Retired early in 1822; in 1823 published her well-known reply to an article by Stendhal on Rossini: *Cenni d'una donna già cantante sopra il Maestro Rossini*.

**Giorno di regno, Un** (King for a Day) (later known as *Il finto Stanislao*). Opera in 2 acts by Verdi; text by Romani, after A. V. Pineu-Duval's comedy *Le faux Stanislas* (1808). Prod. Milan, S, 5 Sep. 1840, with Rainieri-Marini, Abbadia, Salvi, Ferlotti, Scalese, Rovere, Vaschetti, Marconi; New York TH, 18 June 1960; London, St Pancras TH, 21 Mar. 1961, with Jolly, Jonic, Hallett, Hauxvell, Garrett, cond. Ucko.

Baron Kelbar's Castle, near Brest, 1733. The King of Poland has had to depart on a secret mission, leaving the Cavalier Belfiore (bar) to impersonate him for a day. At court, there is to be a double marriage; Kelbar's daughter Giulietta (mez) is to marry La Rocca (bs), though she is in love with his nephew Edoardo (ten); the Marchesa del Poggio (sop) is to marry the Count Ivrea (ten). She plans to do this because she believes she was deserted by her lover, Belfiore.

Belfiore is dismayed to learn of the Marchesa's marriage but cannot reveal his identity. After many delaying tactics and amorous complications, the situation is finally resolved: Kelbar (bs) gives his consent for Giulietta to marry Edoardo. When the King succeeds in his mission, Belfiore reveals his identity and professes his love for the Marchesa. She forgives him and they are married.

Verdi's second opera, and his only comedy apart from *Falstaff*. Also opera by Gyrowetz (1818).

**Giovanna d'Arco** (Joan of Arc). Opera in a prologue and 3 acts by Verdi; text by Solera, after Schiller's tragedy *Die Jungfrau von Orleans* (1801). Prod. Milan, S, 15 Feb. 1845, with Frezzolini-Poggi, Poggi, Colini, Marconi, Lodetti; New York, Carnegie H, 1 Mar. 1966, with Stratas, Mori, Milnes, cond. Cillario; London, RAM, 23 May 1966, with Carr, West, Charles, cond. Treacher.

France, 1429. Carlo VII (ten), King of France, is intending to abdicate. He visits a statue to the Virgin and lays his weapons beneath it. Giovanna (sop), who was asleep nearby, wakes to find them, and persuades Carlo to accompany her into battle.

The French army is victorious and Giovanna wants to return to her simple life. Carlo declares his love for her, and she in turn admits that she loves him. Angelic visions warn her against worldly love.

At Carlo's coronation, Giacomo (bar), Giovanna's father, who has been watching them, accuses Carlo of blasphemy and Giovanna of witchcraft. A thunderclap seems to confirm his statement. She is handed over to the English to be burned at the stake.

Giacomo, overhearing Giovanna's prayers, realizes that he was mistaken and frees her from her chains. She returns to battle and leads France to victory, but is mortally wounded in the fight.

See also JOAN OF ARC.

**Giraldoni, Eugenio** (*b* Marseilles, 20 May 1871; *d* Helsinki, 23 June 1924). Italian baritone. Son of below. Studied with his mother Carolina *Ferni, a Pasta pupil. Début Barcelona 1891 (Escamillo). Moscow, Odessa, Kiev, 1901–7; New York, M, 1904–5; Paris, CE, 1913. A famous Boris, and Gérard (*Chenier*), and the first Scarpia. Not vocally outstanding, but musical and impressive on stage. (R)

**Giraldoni, Leone** (*b* Paris 1824; *d* Moscow, 1 Oct. 1897). Italian baritone. Father of above. Studied Florence with Ronzi. Début Lodi 1847 (High Priest, Pacini's *Saffo*). Milan, S, 1850–85. Created Donizetti's Duca d'Alba, and Verdi's Boccanegra and Renato. Retired 1885, and taught in Italy and Moscow. Possessed a warm, even voice, and was an effective actor, especially in noble and dignified roles. Married the dramatic soprano *Carolina Ferni.

**Giraud, Fiorello** (*b* Parma, 22 Oct. 1870; *d* Parma, 28 Mar. 1928). Italian tenor. Son of the tenor **Ludovico Giraud** (1846–82). Studied Parma with Barbacini. Début Vercelli 1891 (Lohengrin). First Canio and Italian Pelléas; highly successful career in Wagner (above all as Siegfried) and in verismo opera. Possessed a robust voice and excellent declamation. (R)

**Girl of the Golden West.** See FANCIULLA DEL WEST, LA.

**Giuditta.** Operetta in 5 acts by Lehár; text by Paul Knepler and Fritz Löhner. Prod. Vienna, S, 20 Jan. 1934, with Novotná and Tauber.

Sicily and Libya, early 20th cent. Giuditta (sop) leaves her husband to accompany Octavio (ten), an army officer who has seduced her, to Africa. When he departs with his regiment, Giuditta becomes a dancer. Octavio deserts from the army and becomes a pianist in a restaurant. The two meet again in the restaurant, where Giuditta is supping with her new admirer; as they leave, Octavio remains playing the piano as a waiter turns out the lights.

Lehár's most substantial operetta. Mussolini rejected the proposed dedication on the grounds that the work's depiction of an Italian officer leaving the colours for a woman was unthinkable in Fascist Italy.

**Giuditta.** See JUDITH.

**Giulietta.** 1. The courtesan (sop or mez), one of Hoffmann's lovers, in Offenbach's *Les contes d'Hoffmann*.

2. Juliet (sop) in Bellini's *I Capuleti e i Montecchi* and other Italian *Romeo and Juliet* operas.

**Giulini, Carlo Maria** (*b* Barletta, 9 May 1914). Italian conductor. Studied Rome with Molinari. Début Rome, 1944 (concert); became music director of Italian Radio 1946–51, broadcasting many little-known operas; operatic début Bergamo 1950 (*Traviata*). Milan, S, 1951; chief cond. 1953–6, working closely with Visconti and Zeffirelli, and with Callas. Edinburgh Festival, with Gly., 1956. London, CG, 1958 (*Don Carlos*), establishing his fame in England. From 1967 turned his attention to concert work, returning to opera 1981 with *Falstaff*, Los Angeles and London, CG. One of the most thoughtful and sensitive of contemporary Italian conductors. (R)

**Giulio Cesare** (Julius Caesar). Opera in 3 acts by Handel; text by Nicola Francesco Haym, after libretto by Bussani, set by Sartorio (1677) and Kusser (1691). Prod. London, H, 20 Feb. 1724, with Cuzzoni, Durastanti, Robinson, Senesino; Northampton, MA, Smith Coll., 14 May 1927.

The Roman general Giulio Cesare (con cas) has conquered Pompey in battle; Sesto (sop cas), Pompey's son, vows to avenge his father's assassination. Achilla (bs), a captain in the Egyptian army, falls in love with Pompey's widow, Cornelia (con cas), but she is taken into the harem of Tolomeo, King of Egypt (con cas). Having reported (falsely) the death of Caesar at his hand, Achilla hopes he may marry Cornelia, but Tolomeo now plans to marry her himself. In a second battle, Achilla is fatally wounded and confesses to Sesto that it was he who killed Pompey. Caesar leads a revolt against Tolomeo during which Sesto kills the Egyptian king; Tolomeo's sister Cleopatra (sop), who is Caesar's lover, is crowned queen.

Other operas on Caesar are by Sartorio (1677), Bernabei (1680), Strungk (1694), Keiser (1710), Jommelli (1751), Perez (1762), Salieri (1800), Tritto (1805), E. Paganini (1814), Pacini (1821), Campana (1841), and Malipiero (1936). See also CLEOPATRA.

**Giuramento, Il** (The Oath). Opera in 3 acts by Mercadante; text by Gaetano Rossi, based on Hugo's drama *Angelo* (1835). Prod. Milan, S, 10 Mar. 1837, with Schoberlechner, Tadolini, Donzelli, Castellan, Balzar; London 27 June 1840; New York, Astor Place OH, 14 Feb. 1848.

Syracuse, 14th cent. Manfredo (bar) commands his wife Bianca (mez) to take poison because she has committed adultery with Viscardo (ten). However, Elaisa (sop), who also loves Viscardo, substitutes a sleeping potion. Viscardo believes Elaisa to have murdered Bianca and kills her.

**Giusti, Girolamo Aloise** (*b* ?; *d* ?, 1766). Italian librettist. Active in Venice, he produced five librettos, two of which achieved considerable attention: those for Vivaldi's *Montezuma* (1733) and for Galuppi's *L'inganno scoperto* (1735).

**Giustino.** Opera in 3 acts by Handel; text an anon. adaptation of Pariati's libretto (1711), itself based on Beregani's libretto for Legrenzi (1683). Prod. London, CG, 16 Feb. 1737, with Strada, Bertolli, Annibali, Gizziello, Beard, Negri. Rev. Abingdon 1967.

**Gizziello** (orig. Gioacchino Conti) (*b* Arpino, 28 Feb. 1714; *d* Rome, 25 Oct. 1761). Italian soprano castrato. Studied Naples with Gizzi, from whom he took his stage name. Début Rome

1730 (in Vinci's *Artaserse*). London, CG, 1736, Ariodante. Handel wrote the following parts for him: Meleager (*Atalanta*), Sigismondo (*Arminio*), Anastasio (*Giustino*), and Alessandro (*Berenice*), all of which demand exceptional virtuosity and expressiveness. His range extended to c''': the only castrato for whom Handel wrote so high. From 1738–55 he sang throughout Italy, in Lisbon, and in Madrid for Farinelli, his repertory including works by Jommelli, Hasse, and Vinci. A great rival of Caffarelli, he had the more appealing character, being modest and sensitive. Handel declared him a 'rising genius'.

**Gläser, Franz** (*b* Obergeorgenthal (Horní Jiřetín), 19 Apr. 1798; *d* Copenhagen, 29 Aug. 1861). Bohemian composer and conductor. Studied Prague with D. Weber. Worked in Vienna 1817–30 (L, 1817–18; J, 1819–27: W, 1827–30), then in Berlin. Of his many works (over 50 for the stage), most are occasional pieces of transient interest. The most important was *Des Adlers Horst*, once in most German repertories and known to Wagner; it includes 'pre-Wagnerian' features such as an apostrophe to the sun and the use of insistently repeated short figures for creating dramatic tension. In 1842 he moved to Copenhagen, being appointed court conductor in 1845. Here he wrote three more operas, two to texts by Hans Andersen.

SELECT WORKLIST: *Des Adlers Horst* (The Eyrie) (Holtei; Berlin 1832); *Brylluppet ved Como-søen* (The Wedding on Lake Como) (Andersen; Copenhagen 1849); *Nökken* (The Water-Sprites) (Andersen; Copenhagen 1853).

BIBL: N. Pfeil, *Franz Gläser* (Leipzig, 1870).

**Glasgow.** City in Lanarkshire, Scotland. Glasgow was for long dependent on visits from touring companies, such as the CR from 1877. The amateur Glasgow Grand Opera Society, formed in 1905, staged some enterprising productions, including the first British *Idomeneo* (1934), *Les Troyens* (1935), and *Béatrice et Bénédict* (1936) under Erik Chisholm. It also revived MacCunn's *Jeanie Deans* (1951). Opera was finally put on a permanent professional footing with the formation of *Scottish O in 1962.

**Glass, Philip** (*b* Baltimore, 31 Jan. 1937). US composer. Studied at the Juilliard School in New York and with Nadia Boulanger in Paris. Early conservative works were later withdrawn, as Glass adopted a more radical style of music, now termed Minimalist. The static simplicity of his writing may be seen as a reaction to the complex ideologies behind serialism, which Glass considers 'ugly and didactic'; but a primary root of its evolution was Glass's work with the Indian sitar player Ravi Shankar and tabla player Alla

Rakha. From them he learnt ideas of which the most influential was a reassessment of rhythm.

The reductive, repetitive style of his music with its rippling arpeggiations and pattering percussion accompaniment first appeared in his incidental music for Beckett's *Play*. It is from this ethos of equal collaborators in the creation of a multi-media art-form, and not from the traditional concept of opera, that Glass's works derive. *Einstein on the Beach* was created with Robert Wilson. It has four acts, interspersed with five 'knee plays' (so called because of the anatomical joining function of the knee). During scenes which represent one of three visual images, unnamed characters sing or speak words which present no narrative. Verbal communication is unimportant, and much of the text is taken up with random repetition of numbers, *solfège* names, or meaningless prose. The fundamental of the music is complex repetition; two different rhythmic patterns are superimposed and repeated in a cyclic and additive process. A three-note theme, which opens and closes the opera, later appears at strategic moments in both *Satyagraha* and *Akhnaten*.

By *Satyagraha*, Glass's approach was somewhat less radical. Some passages are drawn from the *Bhagavad-Gita*, and the opera presents, in Sanskrit, in random order, six events during Gandhi's 21 years in South Africa. 'Satyagraha' (Sanskrit: truth force) was the name given by Gandhi to his non-violent civil disobedience movement. Much of the music of *Satyagraha* is based on the chaconne structure; a four-chord harmonic pattern is repeated frequently throughout the work. *Satyagraha* makes substantial use of chorus, and here Glass adds strings to the orchestra of woodwind and synthesizer employed in *Einstein*.

The third work in his 'Portrait Trilogy' is *Akhnaten*. Its three acts concern episodes from the rise, reign, and fall of the Egyptian King Akhnaten. The work still does not present a continuous narrative. The languages of the text—ancient Egyptian, Akkadian, and Hebrew—are again alienating, though Glass now includes a scribe/narrator figure, who recites translations. *Akhnaten* does, however, contain passages in English; one is Akhnaten's 'Hymn to the Sun' (in which he declaims his ideas for religious and social reforms), the other is a passage from *Fodor's Guide to Egypt*, which forms the epilogue to the work. Glass displays an elementary concern for cohesion throughout the work; this is illustrated by key relationships and by his use of thematic devices, which are identified with particular characters or actions. He adopts this device increasingly in later works, particularly in *The Making of the Representative for Planet 8*.

The chamber opera *The Juniper Tree* (based on a story by the Grimm brothers), composed with Bob Moran, was Glass's first work written entirely in English. He aimed to write vocal lines as close as possible to the natural speaking range so that vowel sounds should not be distorted. This technique is also employed in *The Making of the Representative for Planet 8*. The opera has a definite narrative thread (though it still lacks character development). In this, together with his adoption of a conventional opera orchestra without synthesizer, and a more lyrical harmony-orientated style of music, Glass seems to have reached a point closer to traditional opera. His increasing conventionalism is indicated by his quest to infuse his works with allegorical 'meaning', where once he maintained, with *Einstein*, that his work required no objective meaning, only the subjective response of each individual in the audience. In so far as they are large scale with continuous music and, especially latterly, a narrative thread, they conform to many of opera's expectations: the composer himself continues to refer to his stage output as 'music theatre' but looks forward to a time when the opera-house will be revitalized by a breaking down of rigid barriers between traditional and contemporary forms. He is among the most prominent of a group of US composers whose work in the last decades of the 20th cent. is necessitating a reinterpretation of the term 'opera'.

WORKLIST: *Einstein on the Beach* (Wilson; Avignon 1976); *Satyagraha* (Amsterdam 1980); *A Madrigal Opera* (Amsterdam 1980); *Akhnaten* (Stuttgart 1983); *The Juniper Tree* (after Grimm; Cambridge, MA, 1985); *The Civil Wars* (Wilson; Rome 1984); *The Making of the Representative for Planet 8* (Lessing; Houston, TX, 1988); *The Fall of the House of Usher* (Yorinks, after Poe; Cambridge, MA, 1988); *1,000 Airplanes on the Roof* (Hwang; Vienna 1988); *Hydrogen Jukebox* (Ginsberg; Philadelphia (1990); *The Voyage* (Hwang; New York 1992).

WRITINGS: P. Glass, *Opera on the Beach* (London, 1988).

**Glastonbury.** See BOUGHTON, RUTLAND.

**Glière, Reinhold** (*b* Kiev, 11 Jan. 1875; *d* Moscow, 23 June 1956). Ukrainian composer. Studied Moscow with Taneyev, Arensky, and Ippolitov-Ivanov. His music, skilfully crafted and cast in a post-Romantic vein, includes several operas that make use of Asian national subjects and folk music. The first, *Shekh-senem*, was commissioned in 1923 by the Azerbaijani Composers' Union. *Leyli i Medjnun* recasts the popular folktale with a Communist slant, and uses Uzbek folk scales and rhythms with Westernized harmony. In his work Glière gave an example and much encouragement to musicians working in their own traditions.

SELECT WORKLIST: *Shekh-senem* (Jabarla & Halperin; in Russ., Baku, 1927; rev., in Azerbaijani, Baku 1934); *Leyli i Medjnun* (comp. with Sadykov, text after Navoy; Tashkent 1940); *Gulsara* (comp. with Sadykov, text Yashen & Mukhamedov; musical drama, Tashkent 1936; rev. as opera, Tashkent 1949).

BIBL: N. Petrova, *R. M. Glière* (Leningrad, 1962).

**Glinka, Mikhail** (*b* Novospasskoye (now Glinka), 1 June 1804; *d* Berlin, 15 Feb. 1857). Russian composer. Studied casually, gaining most in St Petersburg from Mayer, especially later in Berlin from Dehn. He briefly contemplated two operatic projects, the second to a text by Zhukovsky, who suggested to him instead the story of Ivan Susanin. *A Life for the Tsar* was accepted by the Imperial Theatres partly through Cavos, himself the composer of an opera on the subject, and the interest aroused in rehearsals led to the Tsar suggesting the new title to replace *Ivan Susanin*. The work has long been regarded as the foundation stone of Russian historical opera, and as well as being on a heroic national subject includes much essentially Russian music recomposed: examples are Glinka's use of *podgolosok* (folk polyphony) and of balalaika sounds woven into the musical fabric, together with his use of Russian melodic contours and a recitative based on the inflections of the Russian language. But it also depends much on examples Glinka had heard in his childhood and on his travels, especially French Grand Opera and Italian vocal writing (Bellini in particular), and the reconciliation is incomplete. It also set an example in its clear, individual orchestration.

The success of his first opera led to Glinka being appointed imperial Kapellmeister and to contemplating *Ruslan and Lyudmila*. Pushkin's death prevented their collaboration, and the confused and inadequate libretto was prepared by several hands. Despite this, the strength and individuality of the music shows in the more consistent melodic quality, the skill in developing tunes according to his famous 'varying background' technique (which influenced numerous later Russian composers), the original rhythmic and harmonic pungency (especially in the orientalisms), and the dazzling colour of the orchestration. By contrast with its predecessor, *Ruslan* is the founding work of Russian magic opera. Later 19th-cent. Russian opera composers who acknowledged Glinka's influence include Tchaikovsky, Musorgsky, Dargomyzhsky, Borodin, and Rimsky-Korsakov; and this influence was not extinct by the time Stravinsky began composing.

WORKLIST: *Rokeby* (?Bakhturin, after Scott;

sketches); *Marina Roshcha* (Zhukovsky; sketches); *Zhizn za tsarya* (A Life for the Tsar) (*Ivan Susanin*) (Rosen; St Petersburg 1836); *Ruslan i Lyudmila* (Kukolnik, Shirkov, Markevich, & Gedeonov, after Pushkin; St Petersburg 1842).

BIBL: D. Brown, *Mikhail Glinka* (London, 1974); A. Orlova, *Glinka's Life in Music* (Ann Arbor, MI, 1988).

**'Glitter and be gay'.** Cunegonde's (sop) aria in Act I of Bernstein's *Candide*, reconciling herself without much difficulty to the material consolations of her abandoned life.

**Gloriana.** Opera in 3 acts by Britten; text by William Plomer, after Lytton Strachey's study *Elizabeth and Essex* (1928). Commissioned by the Arts Council for the coronation of Elizabeth II, and prod. London, CG, at a Gala Performance in her presence, 8 June 1953, with Cross, Vyvyan, Sinclair, Pears, G. Evans, Matters, Dalberg, cond. Pritchard; Cincinnati 8 May 1956, with Borkh, Danco, Rankin, Alexander, Uppman, cond. Krips.

The plot describes the events of the later years of Queen Elizabeth I's reign. Elizabeth (sop) reassures Sir Robert Cecil (bar) of her total allegiance to the state, but in the next scene displays her affection for the Earl of Essex (ten); he pleads to be allowed to fight the rebel Tyrone in Ireland but the Queen is reluctant to let him go. However, he does depart on the mission with her blessing. When he returns defeated, she accuses him of betrayal and he falls from favour. Despite the pleas of his wife Frances (mez), he is condemned to death. The opera ends with episodes showing scenes leading up to Essex's beheading and Elizabeth's own death.

**Glossop, Peter** (*b* Sheffield, 6 July 1928). English baritone. Studied Sheffield with Mosley and Rich, later with Hislop. Début Sheffield O Soc. 1947 (Coppelius and Dr Miracle). Principal with London, SW, from 1953; London, CG, from 1961; Milan, S, 1965; New York, M, 1968; also Vienna, Buenos Aires, etc. Roles include Rigoletto, Iago, Billy Budd, Jochanaan. A vigorous performer, possessing a true Verdi baritone. (R)

**Glover, Jane** (*b* Helmsley, 13 May 1949). English conductor. Studied Oxford, where she gained experience conducting for the OUOC. Début Wexford 1975 (Cavalli's *Eritrea*). Gly. since 1982; London, CG, 1988 and C from 1989; Canadian OC 1991. Though her initial interests were in Baroque opera, she has given some outstanding performances of Mozart and Britten, in particular. Her scholarly work has centred on Baroque opera, especially with her *Cavalli* (London, 1978).

**Gluck, Christoph Willibald**, Ritter von (*b* Erasbach, 2 July 1714; *d* Vienna, 15 Nov. 1787). German composer. Having escaped from parental opposition to a musical career by running away from home, he worked and studied independently in Prague, where he also saw much Italian opera. After a brief period in Vienna, he was engaged for a nobleman's orchestra in Milan; here he probably studied with Sammartini and in 1741 made his operatic début with *Artaserse*. Its success led to seven more operas in Italy before he left for London in 1745. Both political and musical circumstances militated against a comparable success, though he formed a friendship with Handel and an increased taste for simplicity in music (it was on this visit that Handel made his famous remark that Gluck 'knows no more of contrapunto as mein cock [cook]'). Busy years of travel followed from 1746, taking him, perhaps with the Mingotti company, through Germany first to Vienna. Here *Semiramide riconosciuta* seems to have pleased everyone except the librettist, Metastasio. Resuming his travels in 1748, he toured Germany, Denmark, and Bohemia, returning to Vienna in 1750. Marriage to Maria Anna Bergin brought him not only personal happiness but good contacts with the Viennese court. He visited Prague and Naples, where he triumphed with *La clemenza di Tito*, returning to Vienna at the end of 1752.

Taking up a post as Kapellmeister to the Prince of Saxe-Hildburghausen, he consolidated the position of respect in which he was held in Vienna. When Durazzo, manager of the Vienna Imperial Theatres, sought to capitalize on the success of a visiting French company, he engaged Gluck to make suitable adaptations for the Burgtheater; but soon Gluck was responding to his own interest in French opera by composing original material. His first complete opéra-comique was the successful *La fausse esclave*. The no less successful *Le cinesi* attracted the admiration of the Emperor, and in 1755 he was commissioned for Archduke Leopold's birthday to write *La danza*, which further stimulated his natural interest in dance. Further commissions followed, including *L'innocenza giustificata* for the Emperor's birthday: collaboration with Durazzo here not only led to much shorter recitatives and consequent greater dramatic emphasis, but revealed to Gluck the importance of working directly with a librettist rather than accepting existing librettos. He also wrote *Antigono* for Rome, where he was awarded the papal title of Cavaliere. He composed operas, serenatas, and ballet music (as well as chamber music), much of it at the encouragement of Durazzo, who now also introduced him to *Calzabigi. Their first collaboration, which included the dancer Angiolini, was with the

ballet-pantomime *Don Juan* in 1761. This unified music, dance, and mime, and though a success in its own right, more significantly led to the partners collaborating on *Orfeo ed Euridice*.

This, the first of Gluck's so-called 'reform operas', elevates the fusion of different dramatic ingredients on to a new expressive plane and draws from the composer the fullest expression of his gifts so far. Though it retains conventional aspects, such as the irrelevant overture and the final *deus ex machina, and though Gluck's lack of technical sophistication continues to show, it has in ample measure both the energy and the calm sublimity that are among his greatest characteristics. It also shows the greatly increased reliance on orchestration that was to mark his mature style. The mixture of dance, song, chorus, accompanied recitative, and orchestral illustration owes much in style to contemporaries including especially *Jommelli and *Traetta, but is a natural evolution from his own previous works. Gluck was also able to work directly with the singer interpreting the role of Orpheus, Gaetano Guadagni, who had been trained as an actor in London by Garrick.

At the beginning of 1763 Gluck went to Italy with Dittersdorf, where he had a limited success with a work of mixed achievement, the Metastasian *Il trionfo di Clelia* in Bologna. Cancelling an intended visit to Paris, he then returned to Vienna for a revival of *Orfeo*, to revise *Ezio*, and to write *La rencontre imprévue*. This was to be the most successful work of his own day, and to influence even Mozart's *Entführung*. He then, early in 1764, paid his postponed visit to Paris with Durazzo, who had resigned from the Burgtheater. Among other works, there followed for Vienna two dance-dramas with Angiolini, *Semiramis* and *Iphigénie* (the latter lost). Works that followed included *Telemaco* for Joseph II's remarriage, and *La corona* for the Emperor's name-day. At some point, he also began work on *Alceste*.

The preface to *Alceste* is one of the most famous documents of operatic reform. In it, Gluck (with perhaps Calzabigi's voice also sounding) attacks the domination of the singer, and the limitations imposed by the formalities of opera seria and by its constrictive plots. He intends to 'restrict music to its true office of serving poetry by means of expression and by following the situations of the story', and to strive for 'a beautiful simplicity' and to appeal to 'good sense and reason'. Gluck was by no means the first in this field, and the terms he used well matched the ideals of the Enlightenment; and for all its quality, *Alceste* is not so consistent in its application of reform principles as its predecessor *Orfeo* or especially *Iphigénie en Tauride*. It was not immediately the success it later became. His next reform opera, *Paride ed Elena*, had a similar reception, with praise again given to Noverre's dance contribution.

Various other activities continued to keep Gluck busy. His next major work was *Iphigénie en Aulide*. With plans to set Du Roullet's text, he arrived in Paris late in 1773. Its triumphant first run was interrupted by the death of Louis XV, and during the period of mourning Gluck revised *Orfeo*, to some extent loosening the structure in the interests of adding expressive music to suit French taste. On a second visit he made a French version of *Alceste*. A third visit brought him into conflict, unsought by either composer, with the defenders of Piccinni over the latter's *Roland* and his own *Armide* (in which powerful characterization and some beautiful lyrical music overcome the drawbacks of a dated libretto). The intrigues were intended to lead to a confrontation between the settings of *Iphigénie en Tauride* by Gluck and Piccinni; but Gluck had left Paris by the time Piccinni's not unsuccessful work appeared. His own work is his masterpiece. Not only does he now completely fuse dance, action, chorus, and solo arias into a single imaginative whole, but the fluency of movement between them is guided entirely by the drama (as with the opening scene in which an actual and a psychological storm replace any formal overture). Further, the characterization is stronger and more intense than anything in his previous work.

Gluck left Paris for Vienna in 1779. He had had a stroke during rehearsals for *Echo et Narcisse*, and suffered two more while continuing work, including on a German revision of *Iphigénie en Tauride*. He was forced to abandon further plans for travel, and died after a fourth stroke.

Gluck's influence in revitalizing the traditions of tragédie lyrique shows immediately in the work of Cherubini, also in that of Méhul and Spontini. Further, if it was Gluck's grace, simplicity, and poise, and his heroic passion in treating classical legend, that deeply impressed Berlioz, Wagner (who made a not very successful version of *Iphigenia en Aulide*) admired above all the dramatic fluency and urgency of his music.

SELECT WORKLIST: *Artaserse* (Metastasio; Milan 1741); *Tigrane* (Silvani, after Goldoni; Crema 1743); *La caduta de' giganti* (The Fall of the Giants) (Vanneschi; London 1746); *Semiramide riconosciuta* (Metastasio; Vienna 1748); *La clemenza di Tito* (Metastasio; Naples 1752); *Le cinesi* (Metastasio; Vienna 1754); *L'innocenza giustificata* (Durazzo & Metastasio; Vienna 1755); *Antigono* (Metastasio; Rome 1756); *La fausse esclave* (The False Slave) (Anseaume & Marcouville; Vienna 1758); *L'ivrogne corrigé* (The Drunkard Reformed) (Anseaume & De Sarterre; Vienna 1760); *Le cadi dupé* (The Cadi

Deceived) (Le Monnier; Vienna 1761); *Orfeo ed Euridice* (Calzabigi; Vienna 1762, rev. Paris 1774); *Il trionfo di Clelia* (Metastasio; Bologna 1763); *La rencontre imprévue* (The Unforeseen Encounter) (Dancourt, after Le Sage & D'Orneval; Vienna 1764); *Telemaco* (Coltellini, after Capece; Vienna 1765); *Alceste* (Calzabigi; Vienna 1767, rev. Paris 1776); *Paride ed Elena* (Calzabigi; Vienna 1770); *Iphigénie en Aulide* (Du Roullet, after Racine; Paris 1774); *Armide* (Quinault; Paris 1777); *Iphigénie en Tauride* (Guillard & Du Roullet; Paris 1779, rev. Vienna 1781); *Echo et Narcisse* (Tschudi; Paris 1779).

CATALOGUE: A. Wotquenne, *Catalogue thématique des œuvres de Chr. W. von Gluck* (Leipzig, 1904, R/1967).

BIBL: P. Howard, *Gluck and the Birth of Modern Opera* (London, 1963); F. Lesure, ed., *Querelle des Gluckistes et des Piccinnistes* [orig. documents] (Geneva, 1984); B. Brown, *Gluck and the French Theatre in Vienna* (Oxford, 1991).

**Glückliche Hand, Die** (The Favoured Hand). Opera in 1 act by Schoenberg; text by the composer. Prod. Vienna, V, 14 Oct. 1924, with Pfundmayr, Hustiger, Jerger, cond. Stiedry; Philadelphia, AM, 11 Apr. 1930; London, RFH, 17 Oct. 1962.

The subject is the artist's isolation in the contemporary world and his quest for happiness, represented by various scenes in which a man is plagued by a monster and exhorted by a green-faced chorus to abandon his reliance on his senses. He is also frustrated in his quest for a woman.

The work is scored for a solo baritone, mixed chorus, and orchestra, and alternates use of singing and *Sprechgesang. Following Kandinsky's theories, in this work Schoenberg placed particular stress on the symbolism of colours.

**Glyndebourne.** The house and estate of the Christie family near Lewes, Sussex, where the Glyndebourne Festival Opera was launched in 1934. John Christie, encouraged by his wife, the soprano Audrey Mildmay, built an opera-house in the grounds with the intention of giving performances of opera in ideal surroundings and conditions.

The first Glyndebourne Festival opened on 28 May 1934 with *Le nozze di Figaro*, followed the next evening with *Così fan tutte*. The conductor was Fritz Busch, the producer Carl Ebert, who between them set new standards in English opera. The pre-war repertory consisted chiefly of Mozart, though *Macbeth* (the British première) and *Don Pasquale* were also staged. Although the repertory is now much wider, Mozart remains at its centre, the modest scale of the house being particularly apt for his operas. By 1939 the theatre had nevertheless been enlarged (from a capacity of 300 to 600); it was enlarged again in 1952 to its present 800. The back-stage facilities were also then extended. From the outbreak of war, when the house served as a school for evacuees, no performances were staged; however, the company toured with *The Beggar's Opera*. In 1946 the *English Opera Group gave the première of *The Rape of Lucretia* under Ansermet, and in 1947 *Albert Herring* under Britten. 1947 also saw Ebert's production of *Orfeo ed Euridice* with Ferrier, which helped to establish Gluck's place in the repertory.

Glyndebourne resumed its own annual summer festival, under the management of Rudolf Bing, in 1950, when Busch returned to work with Ebert. On Busch's death in 1951, his place was taken by Vittorio Gui, who brought several Rossini works into the repertory. In 1959 Ebert was succeeded by Rennert, who with Gui became artistic counsellor. Gui resigned this position to John Pritchard in 1963, though he continued to conduct. Rennert resigned in 1968, and Moran Caplat (assistant manager since 1945, then general manager) became general administrator, with Pritchard as music director. Pritchard was succeeded by Bernard Haitink in 1977, and Sir Peter Hall director of productions 1984–90. Brian Dickie, previously administrator of Glyndebourne Touring Opera, succeeded as general administrator in 1982, leaving in 1988.

The expansion of Glyndebourne's repertory during these years took an important turn with the performances of Venetian opera in editions by, and conducted by, Raymond Leppard. The first was *L'incoronazione di Poppea* (1962), which proved such a revelation to a new public that it was followed by Cavalli's *Ormindo* (1967) and *Calisto* (1970), and Monteverdi's *Il ritorno d'Ulisse in patria* (1972). Though Leppard's methods came in for scholarly rebuke, he and Glyndebourne were more than anything responsible for awakening a wider appreciation of Venetian opera. Another broadening of the repertory into 20th-cent. music led to Janáček's *The Cunning Little Vixen* (1975) and *Káťa Kabanová* (1988), *The Love of Three Oranges* (1982), a return of *The Rake's Progress* (orig. 1953; 1975 with designs by Hockney), and *Porgy and Bess* (1986). World premières have included Maw's *The Rising of the Moon* (1970). Since 1984 a New Opera Policy has resulted in a première every two years: this has led to Knussen's *Where the Wild Things Are* (1984) and *Higglety, Pigglety, Pop!* (1985), Nigel Osborne's *The Electrification of the Soviet Union* (1987), and the English première of Tippett's *New Year* (1990). Glyndebourne's work has become more widely known through recordings, radio and TV broadcasts, videos, Prom concerts, and foreign tours.

The number of performances each season has gradually been extended. The season now lasts from mid-May to mid-August, and since 1987 has comprised six productions (two new and two revivals). Works are generally performed in their original language, exceptions including *Intermezzo* (1974) and *The Cunning Little Vixen* (1975); *Surtitles were introduced for *Káťa Kabanová* (1988). The RPO was resident 1948–63, being replaced by the LPO in 1964. The chorus traditionally recruits from young singers likely to progress to solo careers. In 1976 a chorus scheme was launched to sponsor its financing; there are also awards for singers for study periods. Glyndebourne continues to draw all its subsidy from private and business sources. In 1953 the Glyndebourne Arts Trust was formed to relieve the Christie family of the financial burden of the festival, and now about 65% of the income derives from ticket sales, 35% from business, private, and other sources. In 1990 Sir George Christie announced plans to enlarge the opera-house to cap. 1,150 by 1994. Glyndebourne Touring Company was founded in 1967 with the intention of taking productions to the regions making use especially of younger artists. It opened in Newcastle on 25 March 1968, and continues to tour England with established singers and conductors as well as emerging talent.

Glyndebourne has been successful in preserving its unique atmosphere of opera-going in a large country house, with a famous feature being picnics in the beautiful gardens during the 'Long Interval'. Though criticized for catering too exclusively to the rich and to prosperous firms with greater social or commercial than artistic concerns, an effective reply is the maintenance of distinctive and distinguished artistic standards without which its reputation would not continue to stand so high.

BIBL: S. Hughes, *Glyndebourne* (London, 1965).

**goat's trill.** See BLEAT.

**Gobatti, Stefano** (*b* Bergantino, 5 July 1852; *d* Bologna, 17 Dec. 1913). Italian composer. Studied Bologna with Busi, Parma and Naples with Rossi. The astounding success of his first opera, *I goti*, which had 52 curtain calls, won him the honorary citizenship of Bologna and the claim by anti-Verdi circles that he was Italy's new musical laureate. The enthusiasm quickly faded, and Verdi himself called the work 'the most monstrous musical abortion ever composed'. The failure of his other operas, coupled with misfortune, debt, and slander (he was accused of having the evil eye), led to persecution mania, and after retreating to a monastery he died in an asylum.

WORKLIST: *I goti* (Interdonato; Bologna 1873); *Luce* (Interdonato; Bologna 1875); *Cordelia* (D'Ormeville;

Bologna 1881); *Masias* (Sanfelice; comp. 1900, unprod.)

**Gobbi, Tito** (*b* Bassano del Grappa, 24 Oct. 1913; *d* Rome, 5 Mar. 1984). Italian baritone. Studied Rome with Crimi. Début Gubbio 1935 (Rodolfo, *Sonnambula*). Milan, S, from 1942; London, CG, 1951–74; Chicago from 1954; New York, M, from 1956; also Vienna, San Francisco, etc. His repertory of *c*.100 roles included Rossini's Figaro, Rigoletto, Posa, Falstaff, Scarpia, Wozzeck, Jack Rance, Schicchi. Also made many films, and produced at Chicago and London, CG, 1965. Using his voice with great intelligence, he was an outstanding and unusually versatile singing actor whose characterizations were direct and humane. Autobiography, *My Life* (London, 1979) (R).

**Godard, Benjamin** (*b* Paris, 18 Aug. 1849; *d* Cannes, 10 Jan. 1895). French composer. Studied Paris with Reber. After winning an early reputation with concert music, he turned to opera in 1880, with little success until achieving it posthumously with *La vivandière*. His operatic music now survives only in the tenor Berceuse from *Jocelyn*.

WORKLIST: *Les Guelfes* (Gallet; comp. 1880–2, prod. Rouen 1902); *Pedro de Zalamea* (Détroyat & Silvestre, after Calderón; Antwerp 1884); *Jocelyn* (Capoul & Silvestre, after Lamartine; Brussels 1888); *Dante et Béatrice* (Blau; Paris 1890); *Jeanne d'Arc* (Fabre; Paris 1891); *Ruy Blas* (after Hugo; comp. 1891, unprod.); *La vivandière* (Cain; Paris 1895).

BIBL: M. Clavié, *Benjamin Godard* (Paris, 1906).

**Goehr, Alexander** (*b* Berlin, 10 Aug. 1932). German, later English, composer, son of below. Studied Manchester with Richard Hall, founding the New Music Manchester Group with *Birtwistle, *Davies, Elgar *Howarth, and the pianist John Ogdon; later studied Paris with Loriod. Professor of Music Leeds 1971, Cambridge 1976. His first opera, *Arden Must Die*, was commissioned by the Hamburg, S, and stormily received for political reasons; it is a remarkable achievement, making intelligent use of both traditional and more Brechtian means and drawing upon different musical manners without loss of artistic control or integrity. He was director of the Music Theatre Ensemble from 1967, contributing to their repertory his own sharply observed *Naboth's Vineyard*, in which the stylistic resources were from the earliest days of opera; this was joined in a triptych by *Shadowplay-2* and *Sonata about Jerusalem*. *Behold the Sun*, on the subject of the Anabaptists, continues the move away from Goehr's early Expressionism and draws upon a number of musical sources, including the Baroque oratorio but also an appreciation

of Stravinsky that had marked his music theatre pieces. A première that partly dismembered the work prevented what may well be a stronger dramatic cohesion, binding some fine and visionary music, than was permitted.

WORKLIST: *Arden muss sterben* (Arden Must Die) (Fried, after anon. 16th-cent. drama; Hamburg 1967); *Naboth's Vineyard* (Goehr; London 1968); *Shadowplay-2* (Cavander, after Plato; London 1970); *Sonata about Jerusalem* (Goehr & Freir, after Obadiah & Samuel ben Yahya; London 1970); *Behold the Sun* (Goehr & McGrath; Duisburg 1985).

**Goehr, Walter** (*b* Berlin, 28 May 1903; *d* Sheffield, 4 Dec. 1960). German, later English, conductor and composer; father of above. Studied Berlin with Schoenberg. Wrote the first radio opera, *Malpopita* (1930). Moved to England, becoming an influential conductor, teacher, and editor, including of Monteverdi's *Poppea* (1954). One of the first conductors in England to arouse wider interest in Baroque music, especially Purcell and Monteverdi, through the sympathetic, stylish quality of his performances.

**Goethe, Johann Wolfgang von** (*b* Frankfurt, 28 Aug. 1749; *d* Weimar, 22 Mar. 1832). German poet and dramatist. Goethe's deep interest and involvement in music included early experience singing bass in a choir, learning some instruments, and attending with admiration operas and Singspiels, especially those by Hiller. He was also much impressed by the visit of Marchand's company to Frankfurt, 1771–7, and came to know the composer Johann André. His own Singspiel texts were *Erwin und Elmire*, *Claudine von Villa Bella*, *Lila*, *Die Fischerin*, *Jery und Bätely*, and *Scherz, List und Rache*; he also wrote a second part to *Die Zauberflöte*. When he was director of the Weimar theatre, 1791–1817, his repertory included operas: his preferred composer was Mozart (280 perfs.) followed by Dittersdorf (139 perfs.), then French composers. Among the adaptations he made was one of Mozart's *Der Schauspieldirektor*. Beethoven began setting *Faust* several times. Operas on his works are as follows:

*Die Laune des Verliebten* (1765): Gürrlich (1812)
*Die Mitschuldigen* (1769): Riethmüller (1957)
*Götz von Berlichingen* (1773): Goldmark (1902)
*Satyros* (1773): Baussnern (1922); Bořkovec (1942)
*Der Jahrmarktfest zu Plundersweilern* (1773): Duchess Anna Amalia of Sachsen-Weimar (1775); T. M. Eberwein (1818)
*Erwin und Elmire* (1774): André (1775); Anna Amalia (1776); H. Wolf (1780); Schweitzer (1780); E. W. Wolf (1780); Vogler (1781); Agthe (1785); Stegmann (1786); Rupprecht (1790); Reichardt (1791); Bergt (*c*.1805); Schoeck (1916)

*Clavigo* (1774): Ettinger (1926)
*Die Leiden des jungen Werthers* (1774): R. Kreutzer (1792); Pucitta (1802); Benvenuti (1811); Coccia (1814); Aspa (1849); Gentili (1862); Massenet (1892); W. Müller (parody, 1830); Bose (1986)
*Claudine von Villa Bella* (1774; rev. 1779 as Singspiel): André (1779); Beecke (1780); G. Weber (1783); Reichardt (1789); Kerpen (late 18th cent.); Schneider (1807); Blum (1810); Kienlen (1810); Eisrich (1813); Schubert (part destroyed, 1815); Eberwein (1815); Kienlen (1817); Coccia (1817); Gläser (1826); Stolze (1831); Drechsler (after 1836); Müller (c.1850); J. H. Franz (Hochberg) (1864); Seidel (1871); Knappe (1882); Irmler (1932)
*Stella* (1776): Deshayes (*Zélia*, 1791; *La suite de Zélia*, 1792)
*Lila* (1776): Seckendorf (1776); Reichardt (1790); Seidel (1818)
*Die Geschwister* (1776): Rottenberg (1916); Veidl (1916); Meyerolbersleben (1932)
*Proserpina* (1777): Seckendorf (1777)
*Jery und Bätely* (1779): Seckendorf (1779); Reichardt (1789); Winter (1790); Schaum (1795); Bierey (1803); K. Kreutzer (1810); Frey (1815); Seidel (comp. 1815); Marx (1825); Bergt (early 19th cent.); Hartmann (1833); Adam (1834); Rietz (1840); H. Schmidt (before 1846); Stiehl (1868); Bronsart (1873); Bolck (1875); Pfaff (1919); Dressel (1932); Kniese (1937). Scribe & Mélesville's text for Adam's *Le chalet* (1834), reworked for Donizetti's *Betly* (1836), is derived from it
*Die Fischerin* (1782): Corona Schröter (1782: she also sang Dortchen in it, thus becoming the first composer and first singer of 'Erlkönig'); Eberwein (1818)
*Scherz, List und Rache* (1784): P. C. Kayser (1786); Winter (1790); E. T. A. Hoffmann (1801); Kienlen (1805); Bruch (1858); Wellesz (1928); Irmler (c.1930)
*Egmont* (1787): Dell'Orefice (1878); Salvayre (1886); Meulemans (1960)
*Das Märchen von der schönen Lilie* (1794): Klebe (1969)
*Wilhelm Meisters Lehrjahre* (1796): Thomas (*Mignon*, 1866)
*Der Zauberlehrling* (1797): Döbber (1907); Braunfels (1954, TV)
*Die Braut von Korinth* (1797): Devasini (1846); Duprato (1867)
*Der Gott und die Bajadere* (1797): Auber (1830)
*Hermann und Dorothea* (1798): Missa (1911)
*Faust* (ptI, 1808; pt2, 1832): Josef Strauss (c.1814); Lickl (1815); Seyfried (1816); I. Walter (1819); Horn (1825); Béancourt (1827); Berlioz (non-operatic, but sometimes staged: *La damnation de Faust*, 1846, incorporating and revising *Huit scènes de Faust*, 1829); Lindpainter (1831); Bertin (1831); Pellaert (1834); Rietz (1836); Gregoir (1847); Hennebert (1853); Lutz (1855); Gounod (1859); Boito (*Mefistofele*, 1868); Lassen (1876);

Zöllner (1887); Souchay (1940). For operas on Faust deriving from sources other than Goethe see FAUST.
*Pandora* (1808): Lassen (1886); Gerster (1949).

R. Wagner-Régeny's *Prometheus* (1959) includes a setting of Goethe's poem. Lehár's *Friederike* (1928) is an operetta on the Friederike Brion episode in Goethe's life.

**Goetz, Hermann** (*b* Königsberg, 7 Dec. 1840; *d* Hottingen, 3 Dec. 1876). German composer. Studied Königsberg with Köhler; Berlin with Stern, Bülow, and Ulrich. Tubercular from youth, he had a severe struggle in maintaining a musical career, and his most important work, *Der Widerspenstigen Zähmung*, was written when he was already seriously ill. It is one of the few significant operas of its time to resist Wagner in positive terms. The separate numbers are elegant and well constructed, especially in the finales, with which Goetz hoped to create a modern equivalent of his beloved Mozart. The work, much admired by Shaw, is one of the most successful German comedies of the mid-19th cent. He did not live to complete *Francesca da Rimini*.

WORKLIST: *Der Widerspenstigen Zähmung* (The Taming of the Shrew) (Widmann & Goetz, after Shakespeare; Mannheim 1874); *Francesca da Rimini* (Goetz, after Widmann, after Dante; unfin., completed Frank, prod. Mannheim 1877).

BIBL: E. Kreuzhage, *Hermann Goetz* (Leipzig, 1916).

**Gogol, Nikolay** (*b* Sorochintsy, 31 Mar. 1809; *d* Moscow, 4 Mar. 1852). Russian writer. Most Gogol operas are on the stories in *Evenings on a Farm near Dikanka* (1831–2):
*Christmas Eve*: Tchaikovsky (*Vakula the Smith*, 1876; rev. as *Cherevichki* (The Slippers, 1887); N. Afanasev (*Vakula the Smith*, 1875); Solovyov (comp. 1875, prod. 1886); Lysenko (operetta, 1874; opera, 1883); Rimsky-Korsakov (1895); Bogoslovsky (1929); Peysin (1929); Hartman (?); Shtzurovsky (?)
*May Night*: Sokalsky (1876); Rimsky-Korsakov (1880); Lysenko (1885); Ryabov (1937); Baumilas (c.1956)
*The Lost Charter*: Yaroslavenko (1922)
*St John's Eve*: Tikotsky (?1912)
*The Terrible Revenge*: N. Kochetov (1903)
*Sorochintsy Fair*: Musorgsky (unfin. 1881); Yanovsky (1899); Ryabov (1936); Knorr (1904); V. Alexandrov (?)

Other settings are as follows:
*The Portrait* (1835, rev. 1842): Rosenberg (1956); Weinberg (1983) Pashchenko (1968)
*The Marriage* (1834, 1 act only): Musorgsky (1868, unfin.; prod., piano acc. 1909); A. Jiránek (1945); Grechaninov (1950) Martinů (1953)
*The Coach*: Kholminov (1975)
*The Nose* (1835): Shostakovich (1930); W. Kaufmann (1953)

*The Diary of a Madman* (1835): Searle (1958); Butsko (1971); Ancelin (1975) Rajičić (1981)
*Viy* (1835): Kropivnitsky (late 19th cent.); Goryelov (1897); K. Moor (1903); Dobronić (1 act of *Vječnaja Pamjat*, 1947); Verikovsky (1946); Hubarenko (1980)
*Taras Bulba* (1835): Vilboa (19th cent.; unprod.); Afanasyev (19th cent.; unprod.); Sokalsky (1878, rev. 1905 as *The Siege of Dubno*); V. Kühner (1880); Elling (*Kosakkerne*, 1896); Lysenko (comp. 1890, prod. 1924); Kashperov (1887); Berutti (1895); Trailin (comp. *c.*1885, prod. 1914); M. S. Rousseau (1919); Richter (1935)
*The Inspector General* (1836): K. Weis (1907); Shvedov (comp. 1934); Zádor (1935); Zanella (1940); Egk (1957). Also, based on Gogol's idea, Chukhadjyan (*Arifin Khilesi*, 1872)
*The Greatcoat* (1842): W. Kaufmann (*Bashmashkin*, 1952); Marttinen (1965); Kholminov (1975) Rosenfeld (1978)
*Dead Souls* (1842): Shchedrin (1977)

**Golaud.** The grandson (bar) of King Arkel, half-brother of Pelléas and husband of Mélisande, in Debussy's *Pelléas et Mélisande*.

**Goldberg, Reiner** (*b* Crostau, 17 Oct. 1939). German tenor. Studied Dresden with Schellenberg. Début Dresden 1966 (Luigi, *Tabarro*). Dresden, S, 1969, and from 1972; London, CG, 1982; Milan, 1984; also Bayreuth 1988, New York, M, 1992, Vienna, Prague, Salzburg, Barcelona, etc. His roles include Erik, Tannhäuser, Walther, Siegfried; also as Florestan, Huon, Duke of Mantua, Hermann, Bacchus. (R)

**Golden Cockerel, The** (Russ., *Zolotoy petushok*). Opera in 3 acts by Rimsky-Korsakov; text by Belsky, after Pushkin's poem (1834), after a folk-tale and Washington Irving (1832). Prod. Moscow, Z, 7 Oct. 1909, with Dobrovolskaya, Speransky, Pikok, Rostovtseva, Zaporozhets, Ernts, Dikov, Klopotovskaya, cond. Cooper; London, DL, 15 June 1914, with Dobrovolskaya, Petrenko, Petrov, Altechevsky, cond. Cooper; New York, M, 6 Mar. 1918, with Barrientos, Didur, Diaz, cond. Monteux.

The opera tells of the miraculous golden cockerel, given to old King Dodon (bs) by the Astrologer (ten), which crows at the sign of imminent danger. Dodon brings back the beautiful Queen of Shemakha (sop) to his capital; when the Astrologer demands payments for the cockerel he is killed by Dodon, and the cockerel kills the King.

The fourteenth and last of Rimsky-Korsakov's operas. The censor refused to sanction performance during the composer's lifetime owing to the alleged reference in King Dodon's court to that of Tsar Nicholas II, and the implied criticism of the inefficient conduct of the Russo-Japanese War. The composer wanted his singers also to dance, but this was found too exhausting, and so for the Petrograd production Fokine devised the idea of having the singers seated in the theatre boxes, with the action mimed by dancers. Despite protests from the composer's family, this version introduced the work to Western Europe; but later productions in New York and London reverted to the original.

BIBL: P. Cook, *The Golden Cockerel* (London, 1985).

**Goldmark, Karl** (*b* Keszthely, 18 May 1830; *d* Vienna, 2 Jan. 1915). Hungarian composer. Largely self-taught. A staunch supporter of Wagner, he was chiefly responsible for the founding of the Vienna Wagner Verein (1872). Hanslick's enmity did not prevent the success of his first (and most popular) opera, *Die Königin von Saba*. Colourful and tuneful, it was taken up all over Germany and also in England and America. None of his later operas made a comparable effect. His eclectic style, drawing on influences as disparate as Mendelssohn and Hungarian folk music, was also influenced by Wagner. He was less successful when attempting monumental opera than when indulging his agreeable melodic vein in lighter pieces. Autobiography, *Erinnerungen aus meinem Leben* (Vienna, 1922, R/1929).

WORKLIST: *Die Königin von Saba* (The Queen of Sheba) (Mosenthal; Vienna 1875); *Merlin* (Lipiner; Vienna 1886); *Das Heimchen am Herd* (The Cricket on the Hearth) (Willner, after Dickens; Vienna 1896); *Die Kriegsgefangene* (The Prisoners of War) (Formey; Vienna 1889); *Götz von Berlichingen* (Willner, after Goethe; Budapest 1902); *Ein Wintermärchen* (The Winter's Tale) (Willner, after Shakespeare; Vienna 1908).

BIBL: M. Káldor and P. Várnai, *Goldmark Károly* (Budapest, 1956).

**Goldoni, Carlo** (*b* Venice, 25 Feb. 1707; *d* Paris, 6 Feb. 1793). Italian librettist and dramatist. Trained as a lawyer and practised in Pisa before moving to Venice, where he worked as poet at several of the opera-houses. Goldoni is credited with the invention of the opera buffa libretto, writing more than 100 texts, many under pseudonyms. In 1762 he left Venice for Paris, where he enjoyed considerable fame through his works for the Comédie-Italienne. A victim of the French Revolution, he died in poverty.

The effects of Goldoni's work were felt long after his death and profoundly influenced composers as diverse as Mozart, Rossini, Donizetti, and Wolf-Ferrari. To the traditions of the commedia dell'arte he added characters and situations drawn from everyday life, injecting a reality and earthiness into his librettos which made them popular throughout Italy. With their stock situations enlivened by rapid dialogue, often in dialect and invariably cutting and witty, and tempered by a touching sentimentality, they represented an ideal foil to the artificiality of Metastasian opera seria and proved increasingly attractive to composers. The height of Goldoni's craft is represented in the many collaborations with Galuppi, beginning with *L'Arcadia in Brenta* (1749) and embracing *Il mondo della luna* (1750) and *Il filosofo di campagna* (1754): other important librettos included *La buona figliuola* (Piccinni, 1760, based on Richardson's *Pamela*) usually regarded as the first *opera semiseria, and *Lo speziale* (Pallavicini and Fischietti, 1754). After 1748 Goldoni designated many of his librettos as *dramma giocoso, indicating the presence of serious parts alongside the stock characters of opera buffa. A particular feature of his work was the opportunity he provided for ensemble writing, especially in the finale.

Almost no composer of the late 18th cent. remained untouched by his work: particularly noteworthy were Haydn's settings, but other composers attracted to his librettos included Gassmann, Alessandro Scarlatti, and Vivaldi.

**Goldovsky, Boris** (*b* Moscow, 7 June 1908). Russian, later US, producer and conductor. Studied Moscow, Berlin with Schnabel, Budapest with Dohnányi, New York with Reiner. Director opera department Cleveland Institute, Berkshire Music Center, New England Conservatory 1942–6. Founded Goldovsky O Institute 1946 and the touring Goldovsky Grand Opera T. A versatile musician with practical and administrative skills, also a lively commentator on opera. His books include *Accents on Opera* (New York, 1953).

**Goldschmidt, Berthold** (*b* Hamburg, 18 Jan. 1903). British composer and conductor of German birth. Studied Hamburg, and Berlin with Schreker. Assistant cond. Berlin, S, 1926–7; cond. Darmstadt 1927–9; artistic adviser Berlin, SO, 1931–3. Gly. at Edinburgh 1947. His *Beatrice Cenci* (Esslin, after Shelley) was a prizewinner in the 1951 Arts Council competition for the Festival of Britain, but was not produced.

**Goldschmied von Toledo** (The Goldsmith of Toledo). Pasticcio, compiled from music by Offenbach, by Stern and Zamara; text by Zwerenz, after Hoffmann's story *Das Fräulein von Scuderi* (1818). Prod. Mannheim 7 Feb. 1919; Edinburgh 16 Mar. 1922, by BNOC, subsequently at London, CG.

The goldsmith Malaveda (bar) is driven by a desire to repossess all the jewels he has made. To this end, he commits four murders. His prize piece, a pearl necklace, is owned by the Marquesa Almedina (sop); unable to resist the demonic force driving him, Malaveda also murders her, only to discover too late that his own daughter Madalena (sop) had taken the Marquesa's place.

See also CARDILLAC.

**Goltz, Christel** (*b* Dortmund, 8 July 1912). German soprano. Studied Munich with Ornelli-Leeb. Début Fürth 1935 (Agathe). Dresden, S, 1936–50; London, CG, 1951; New York, M, 1954; also Munich, Vienna, Berlin until 1970. Created title-role of Liebermann's *Penelope*; large repertory included Leonore, Isolde, Elektra, Dyer's Wife. Impressive for her clear tone, 3-octave range, and powerful acting ability. (R)

**Gomatz**. A young Spaniard (ten), prisoner of Soliman and lover of Zaide, in Mozart's *Zaide*.

**Gomes, Carlos** (*b* Campinas, 11 July 1836; *d* Belém, 16 Sep. 1896). Brazilian composer. Studied with his father, Rio de Janeiro with Giannini, then, after the success of his first two operas, Milan with Rossi. He then developed an Italianate manner which had its most successful expression in *Il Guarany*, a work whose powerful, quasi-exotic appeal won it an international reputation. Despite Verdi's admiration, he was soon identified with foreign influences on bel canto opera, a reproach he refuted with some success with the second version of *Fosca*. Though he made use of Brazilian subjects and melodies, his music remains essentially in the Italian Romantic and veristic traditions.

SELECT WORKLIST: *A noite de castelo* (Night in the Castle) (Dos Reis; Rio 1861); *Joana de Flandres* (Mendonça; Rio 1863); *Il Guarany* (Scalvini & D'Ormeville, after Alencar; Milan 1870); *Fosca* (Ghislanzoni; Milan 1873, rev. Milan 1878); *Salvator Rosa* (Ghislanzoni; Genoa 1874); *Maria Tudor* (Praga, after Hugo; Milan 1879); *Lo schiavo* (The Slave) (Paravicini, after Taunay; Rio 1889); *Condor* (Canti; Milan 1891).

BIBL: J. Brito, *Carlos Gomes* (Rio de Janeiro, 1956).

**Gomez, Jill** (*b* New Amsterdam, 21 Sep. 1942). British soprano. Studied London, RAM and GSM. Début Cambridge University O 1967 (Mermaid, *Oberon*). Gly. 1969; London, CG, 1970 (first Flora, *The Knot Garden*). Created Countess in Musgrave's *The Voice of Ariadne*. Also Scottish O, Frankfurt O, London, ENO. Roles include Fiordiligi, Mélisande, Governess. An accomplished actress with a pure, silvery tone. (R)

**Gomis y Colomber, José** (*b* Onteniente, 6 June 1741; *d* Paris, 26 July 1836). Spanish composer. After producing *La Aldeana* in Madrid, he became music director of the royal guard but was forced for political reasons to flee to Paris. Here he was befriended by a number of important musicians, among them Rossini. He was briefly in London in 1826. Berlioz, who was to write a cordial obituary, praised, perhaps over-praised, his opéras-comiques (especially *Le revenant*) for their attempts to raise standards, in particular for their individual rhythmic and melodic nature.

SELECT WORKLIST: *La Aldeana* (?; Madrid, 1822); *Le diable à Séville* (Cavé & Hurtado; Paris 1831); *Le revenant* (The Ghost) (Calviment; Paris 1833); *Le portefaix* (The Porter) (Scribe; Paris 1835); *Rock le barbu* (Desforges & Duport; Paris 1836).

**Gonzaga, Pietro** (*b* Longarone, 25 Mar. 1751; *d* St Petersburg, 6 Aug. 1831). Italian designer. Scene painter Milan, S, 1779, then in other Italian cities. Visited Russia to build a private theatre for Prince Yusupov 1789, returning to help the inauguration of the Venice, F. Back in Russia, he worked in various theatres, where he developed his interest in architectural scenery and perspective in sets that foreshadow much in Romantic pratice.

**Gonzaga family**. Italian family of musical patrons. Rulers of Mantua 1328–1708, they supported some of the earliest operas to be staged. Their most illustrious artistic phase was under Guglielmo Gonzaga (1538–87) and Vincenzo Gonzaga I (1562–1612), when composers called to the court included Striggio, Pallavicino, and Wert. Monteverdi was engaged *c*.1589 and contributed music for several court entertainments, including *Il ballo delle ingrate* (1608) and intermedi for Guarini's *L'idropica* (1608). For a meeting of the Accademia degli Elevati in the Mantuan palace in 1607 he wrote his *Orfeo*; *Arianna* was given the next year as part of a court celebration; Gagliano's *Dafne* was also seen at this time. The accession of Ferdinando Gonzaga (1587–1626) in 1612 led to Monteverdi's departure to Venice, but he continued to write (lost) operas to commission for Mantua, including *La finta pazza Licori* (1627). Caldara was maestro di cappella to the Gonzaga court 1701–7.

**Goodall, Reginald** (*b* Lincoln, 13 July 1905; *d* Bridge, 5 May 1990). English conductor. Studied London. Début London 1936 (*Carmen*). Répétiteur with Coates in British Music Drama OC, at London, CG, 1936. London, SW, 1944–6, conducting première of *Peter Grimes* (1945; superior to Britten's own later performances, according to Pears). EOG at Gly. 1946. CG 1946, conducting *Manon*, *Meistersinger*, *Fidelio*, and much Italian repertory during Rankl's directorship, then only rarely works incl. *Walküre*, *Tristan*, *Parsifal*, *Salome*. This neglect became the more culpable when his conducting of *Meistersinger*, SW 1968, revealed to a wide public the existence of one of the great Wagner conductors of the day. It led to his no less successful *Ring* (London, C), and to further superlative Wagner performances with

the WNO and at CG (*Parsifal*, 1971). His early years of studying Wagner in Germany bore fruit in interpretations whose care of preparation—he was a long, slow, and painstaking coach and rehearser—laid the foundations of performances which at their greatest set him in the tradition of Furtwängler and Knappertsbusch, the two conductors he most admired. Very influential as a vocal coach. (R)

**Good Friday Music**. The scene (Ger., *Karfreitagszauber*) in Act III of Wagner's *Parsifal* as Parsifal, returning from his wanderings, is anointed in preparation for his entry into the Castle of the Grail.

**Goossens**. English musical family, of Belgian origin, three generations of whom named Eugene were conductors.

1. **Eugene** (*b* Bruges, 25 Feb. 1845; *d* Liverpool, 30 Dec. 1906). Studied Bruges. Came to England 1873. CR 1882, cond. first English *Tannhäuser*, 1882. Command perf. of *Fille du régiment* before Queen Victoria, 1892.

2. **Eugene** (*b* Bordeaux, 28 Jan. 1867; *d* London, 31 July 1958). Studied Bruges and Brussels. Worked under his father in CR; then chief cond. of Burns-Crotty, Arthur Rouseby, and Moody-Manners companies. Chief cond. CR 1899–1915, maintaining high standards set by his father, and making many additions to the repertory. Joined Beecham OC, Birmingham 1917, and cond. in Beecham season at London, HM. Opened Beecham season, London, DL (*Ivan the Terrible*). Married **Annie Cook**, former contralto of CR and daughter of bass T. Aynsley Cook.

3. **Eugene** (*b* London, 26 May 1893; *d* London, 13 June 1962). Studied Bruges, Liverpool, and London with Wood and Stanford. Began career as violinist. Début as conductor London 1916 (Stanford's *The Critic*). Cond. BNOC, CR. Latterly cond. little opera, apart from his own *Judith* (London, CG, 1929) and *Don Juan de Mañara* (London, CG, 1937), also performances in Philadelphia and Cincinnati. Both operas are to texts by Arnold Bennett; Goossens's operatic expertise is evident in the dramatic presentation of librettos whose merits are primarily literary, though his own highly chromatic yet chaste style is itself not essentially operatic. Director NSW Conservatory of Music 1947–56, helping many Australian singers at the start of their careers. Autobiography, *Overture and Beginners* (London, 1951). (R)

**gorgheggio** (It.: 'warbling'). A rapid decorative vocalization consisting of numerous rising and falling notes. It was already popular with composers in the 17th cent., and became in the 18th and 19th cents. one of many devices whereby singers would introduce evidence of their own virtuosity, especially in the *aria di bravura, without regard to musical or dramatic sense.

**Gorr, Rita** (*b* Ghent, 18 Feb. 1926). Belgian mezzo-soprano. Studied Ghent with Poelfiet, Brussels with Pacquot-d'Assy. Début Antwerp 1949 (Fricka). Paris, OC and O, 1952. Bayreuth from 1958; Milan, S, from 1960; New York, M, 1962–7; London, CG, 1959–71. Roles include Iphigénie, Medée, Eboli, Dalila, Amneris, Kundry, Charlotte, Mother Marie in first original *Carmélites*. A dramatic performer with a rich, ample voice. (R)

**Gorrio, Tobia.** See BOITO, ARRIGO.

**Goryanchikov.** Alexander Petrovich Goryanchikov (bar), a political prisoner, in Janáček's *From the House of the Dead*.

**Gossec, François-Joseph** (*b* Vergnies, 17 Jan. 1734; *d* Passy, 16 Feb. 1829). French composer of Belgian descent. Studied Maubeuge with Vanderbelen, Antwerp with Blavier. In Paris, he was helped by Rameau and Stamitz, and in 1762 became director of the Prince of Condé's theatre at Chantilly. His own first operas were unsuccessful, apart from an opéra-comique, *Les pêcheurs*, and rivalry with the more successful Grétry was partly responsible for him turning to instrumental music. When he resumed opera composition, having continued to work at the Opéra (and direct some of Grétry's works), he faced the still more formidable rivalry of Gluck: in 1774 his tragédie lyrique *Sabinus* was quickly forgotten in the success of *Iphigénie en Aulide*. Nevertheless, he supported Gluck in the controversy with Piccinni. In 1780 he became assistant director of the Opéra. He resigned on the outbreak of the Revolution. The works which he composed for large-scale, often outdoor Republican ceremonies gave his orchestration, always enterprising, a particular range of sonorities which was to influence French operatic composers, especially Méhul and Spontini, who were seeking to give the orchestra a more significant place in opera.

SELECT WORKLIST (all first prod. Paris): *Les pêcheurs* (The Fishermen) (D'Offémont; 1766); *Toinon et Toinette* (Desboulmiers; 1767); *Sabinus* (Maugris; 1773); *Thésée* (Chéfdeville, after Quinault; 1782); *Rosine* (Gersin; 1786).

BIBL: L. Dufrane, *Gossec* (Paris, 1927).

**Göteborg** (Ger., Gothenburg). City in Göteborgs och Bohus, Sweden. Visits from German companies in the 19th cent. encouraged the opening in 1859 of the Nya Teatern, renamed the Stora Teatern (Grand T) in 1880 (cap. 605). The city has always had international traditions, and the repertory is wide-ranging in style and genre. Opera has also been given in the Kronhuset and at the Scandinavium sports hall (cap. 14,000).

**Gotha.** Town in Thuringia, Germany. A Komödienhaus opened in the west tower of the castle 1683, and many Singspiels were given. Gottfried Stölzel (1690–1749) was music director 1719 till his death, composing about nine operas (lost), and was succeeded by Jiří *Benda, in 1778 by Anton Schweitzer (to 1787): these composers encouraged a strong Singspiel tradition. The Hoftheater (jointly with Coburg) opened 1827; a new Gotha theatre opened 1840 with *Robert le diable*. Duke Ernst II was himself a composer (*Santa Chiara*, 1854, cond. Liszt), and an enthusiastic Wagnerian. The Hoftheater became the Landestheater 1919.

**Gotovac, Jakov** (*b* Split, 11 Nov. 1895; *d* Zagreb, 16 Oct. 1982). Croatian conductor and composer. Studied Split with Dobronić, Hrazdira, and Hatze, Vienna with Marx. Cond. Zagreb O 1923–57. His most successful opera is *Ero s onoga svijeta* (known in English as *Ero the Joker*). This is a consciously folk-influenced, eclectic work transferring tried Italian methods to a local setting.

SELECT WORKLIST: *Morana* (Muratbegović; Brno 1930); *Ero s onoga svijeta* (Ero the Joker) (Begović; Zagreb 1935).

**Gotter, Friedrich Wilhelm** (*b* Gotha, 3 Sep. 1746; *d* Gotha, 18 Mar. 1797). German librettist, dramatist, and poet. Spent most of his life in Gotha as private secretary to the court, with whose Kapellmeister, Georg Benda, he collaborated on a series of works for the Seyler company. In the comic *Der Jahrmarkt* (1775) and *Der Holzhauer* (1778) he demonstrated a flair for colourful dialogue and lively farce, while *Walder* (1776) and *Romeo und Julie* (1776; the earliest operatic treatment of the Shakespeare play) show an equally assured ability to handle more serious subject-matter. Gotter also provided the text for Benda's *melodrama *Medea* (1775). His final libretto, *Die Geisterinsel* (an adaptation of Shakespeare's *The Tempest*, made with Friedrich Hildebrand von Einsiedel), was to be offered to Mozart, but was set by Reichardt (1798). Highly regarded in its day, it represents the apotheosis of the 18th-cent. German libretto and was set by several other composers, including Zumsteeg (1798).

**Götterdämmerung.** See RING DER NIBELUNGEN, DER.

**Göttingen.** City in Lower Saxony, Germany. The Stadttheater (cap. 740) opened 1890. In 1920 the revival of interest in Handel began here with *Rodelinda* under Oskar Hagen, the moving figure in establishing the Handel Festival, and editor of many of the versions staged.

**Gottlieb, Anna** (*b* Vienna, 29 Apr. 1774; *d* Vienna, 4 Feb. 1856). Austrian soprano and actress. Daughter of an actor and a singer. Studied Vienna with Mozart. Début Vienna, B, as child actress, 1779. Created Barbarina (*Figaro*), 1786. Joined Schikaneder's Co. at Freihaus T as leading soprano. Mozart admired her professionally and personally; he wrote Pamina for her, which she created and sang often, greatly contributing to the success of *Die Zauberflöte*. From 1792 a successful singer-actress at T in der Leopoldstadt. Her vocal decline began in 1808; she was finally dismissed in 1828, and passed her remaining years in poverty and neglect. At her death she was found holding the fan Mozart gave her in 1790.

**'Gott! welch' Dunkel hier'.** Florestan's (ten) aria opening Act II of Beethoven's *Fidelio*, in which he laments the silence and darkness of his imprisonment and consoles himself with the knowledge of duty done and then with a vision of Leonore.

**Gounod, Charles** (*b* Paris, 18 June 1818; *d* Saint-Cloud, 18 Oct. 1893). French composer. Studied Paris with F. Halévy and Le Sueur. He considered taking holy orders, but after early successes with church music he was encouraged by Pauline Viardot to turn to opera, in which field he acknowledged that success for a French composer had to be sought. For Viardot he wrote *Sapho*, a promisingly shapely piece which had only modest success despite Berlioz's praise (it is influenced by Gluck); and he fared no better with a less impressive Grand Opera for her, *La nonne sanglante*. He found his own voice with a witty, alert opéra-comique, *Le médecin malgré lui*, and then produced his masterpiece in *Faust*. Aware in himself of the pull between sacred and profane love, Gounod concentrates on the love story of Faust and Gretchen, and her eventual redemption. This excludes much of Goethe (and the Germans refer to the work as *Margarethe*), but Gounod's delicate touch with character, the elegant lyricism of the vocal writing, not least the melodic verve and charm gave French opera a new sense of identity in the wake of Meyerbeer. The work's success was overwhelming, such that the entrée it gave him to the Opéra obliged him to rewrite it with recitatives and a ballet. *La reine de Saba* did not clinch this success at the Opéra, and he rediscovered his real touch with the colourful, invigorating *Mireille* and especially with *Roméo et Juliette*, in which the copious love music is in his most sensuous and captivating vein. A visit to England rekindled his interest in sacred (and sentimental) choral music and extended his popularity (*Faust* was Queen Victoria's favourite opera), though this was tested by his notorious associ-ation with Georgina Weldon. Back in France, he returned to writing opera, but had little success with *Cinq-Mars*, or even with *Polyeucte*. However, the example he set was admired and imitated by an important succeeding generation of French composers, chief among them Massenet and Bizet, perhaps also Saint-Saëns and at a remove Fauré, though they did not always escape the characteristic sentimentality into which he easily fell, and by which he is too often remembered above his very real craftsmanship and sense of theatre.

WORKLIST (all first prod. Paris unless otherwise stated): *Sapho* (Augier; 1851); *La nonne sanglante* (The Bleeding Nun) (Scribe & Delavigne, after Lewis; 1854); *Le médecin malgré lui* (Doctor despite himself) (Barbier & Carré, after Molière; 1858); *Faust* (Barbier & Carré, after Goethe; 1859, rev. 1860, 1869); *Philémon et Baucis* (Barbier & Carré; 1860); *La colombe* (The Dove) (Barbier & Carré, after La Fontaine; Baden-Baden, 1860); *La reine de Saba* (The Queen of Sheba) (Barbier & Carré, after Nerval; 1862); *Mireille* (Carré, after Mistral; 1864); *Roméo et Juliette* (Barbier & Carré, after Shakespeare; 1867); *Cinq-Mars* (Poirson & Gallet, after Vigny; 1877); *Polyeucte* (Barbier & Carré, after Corneille; 1878); *Le tribut de Zamora* (D'Ennery & Brésil; 1881).

BIBL: J.-G. Prod'homme and A. Dandelot, *Gounod* (Paris, 1911, R/1973); J. Harding, *Gounod* (London, 1911); S. Huebner, *The Operas of Charles Gounod* (Oxford, 1990).

**Goyescas.** Opera in 3 scenes by Granados, amplified and scored from piano pieces after Goya's paintings; text added by Fernando Periquet. Prod. New York, M, 28 Jan. 1916—first Spanish opera at New York, M, with Fitziu, Perini, Martinelli, De Luca, cond. Bavagnoli; London, RCM, 11 July 1951. The song 'La Maja y el ruiseñor' (The Lover and the Nightingale) occurs in scene 2.

When he hears that Rosario (sop) has been invited to a ball by his rival, the toreador Paquiro (bs), Fernando (ten) decides to go too. He is fatally wounded by Paquiro in a duel, and dies in Rosario's arms.

**Gozzi, Carlo** (*b* Venice, 13 Dec. 1720; *d* Venice, 14 Apr. 1806). Italian playwright, whose comic *fiabe drammatiche*, many of them written in opposition to Goldoni, were a popular source for operas, especially in Germany since the beginning of the Romantic movement; some were translated by Schiller. Operas on his works include the following:

*L'amore delle tre melarance* (1761): Prokofiev (comp. 1919, prod. 1921)

*Il corvo* (1761): A. J. Romberg (1794); J. P. E. Hartmann (1832)

*Il re cervo* (1762): Henze (1956)

*La donna serpente* (1762): Himmel (*Die Sylphen*,

1806); Wagner (*Die Feen*, comp. 1834); Casella (1932)

*Turandot* (1762): Blumenroeder (1810); Reissiger (1835); Hoven (1839); Lövenskiold (1854); Bazzini (1867); Åkerberg (1906, unprod.); Busoni (1917); Puccini (unfin., prod. 1926); Zabel (1928)

*I pitocchi fortunati* (1764): Benda (*Das tartarische Gesetz*, 1780); Zumsteeg (*Das tartarische Gesetz*, 1780); D'Antoine (*Das tartarische Gesetz*, 1782)

Also Calandro (*I tre matrimoni*, 1756, text attrib. Gozzi)

**Grabu Louis** (*fl.* 1665–94). French composer of largely English domicile. Arrived in England 1665; appointed Master of the King's Musick under Charles II. Other court appointments included Master of the Select Band of Violins from 1667. In 1679 he retired to France, but was lured back to England in 1683. Grabu was not highly regarded, either as performer or composer, and his speedy preferment at court perhaps owes more to his French connections than to his innate talent. None of his works for the London stage found favour: his *Ariadne*, produced for the marriage of the Duke of York in 1674, was probably a mere adaptation of Cambert's work; his music for Shadwell's adaptation of *Timon of Athens* in 1678 was pedestrian, and his setting of Dryden's *Albion and Albanius* in 1685 was an almost immediate failure. However, he earns a place in music history as composer of probably the only English opera to have been given a public performance in the 17th cent.

SELECT WORKLIST: *Albion and Albanius* (Dryden; London 1685).

**Graener, Paul** (*b* Berlin, 11 Jan. 1872; *d* Salzburg, 13 Nov. 1944). German composer and conductor. Largely self-taught. London, HM, 1896, then returning to teach in Austria and Germany before concentrating on composition in the 1920s. During this decade and the next he had a wide success in Germany with his operas, which had an appeal for their direct and craftsmanlike qualities.

SELECT WORKLIST: *Don Juans letzte Abenteuer* (Don Juan's Last Adventure) (Anthes; Leipzig 1914); *Hannelies Himmelfahrt* (Graener, after Hauptmann; Dresden 1927); *Friedemann Bach* (Lothar; Schwerin 1931); *Der Prinz von Homburg* (Graener, after Kleist; Berlin 1935).

**Graf, Herbert** (*b* Vienna, 10 Apr. 1904; *d* Geneva, 5 Apr. 1973). Austrian, later US, producer. Son of critic Max Graf. Studied Vienna. Held posts in Münster, Breslau, and Frankfurt. Salzburg 1936 (*Meistersinger* and *Magic Flute* for Toscanini). Philadelphia 1936. New York, M, 1936–60. London, CG, 1958–9. Director Zurich O 1960–3; Geneva, Grand T, 1965–73. A producer with a strong sense of tradition, remembered for his encouragement of young talent.

WRITINGS: *Opera for the People* (Minneapolis, 1951); *Producing Opera for America* (Zurich, 1961).

**Graham, Colin** (*b* London, 22 Sep. 1931). English producer and librettist. Studied London. Began producing opera 1954. EOG 1958, director of productions 1963–75: staged premières of Britten's *Noyes Fludde*, church parables, *Owen Wingrave*, *Death in Venice*. London: New OC and Handel OC from 1958, SW and ENO 1961–84 (director of productions 1977–82). Staged British premières of *Cunning Little Vixen* 1961, *War and Peace* 1972, and premières of Bennett's *Mines of Sulphur* (1965) and *A Penny for a Song* (1967; also librettist) and Josephs's *Rebecca* (Opera North, 1983). London, CG, 1961–73. English Music Theatre Co. 1975–8. New York, M, 1982 (*Così*). St Louis 1982, première of Paulus, *The Postman always Rings Twice* (also librettist). Withdrew from theatre 1984 to take holy orders. Artistic director St Louis 1985: Britten cycle, *Hoffmann* 1986, first US Oliver, *Beauty and the Beast* (1987). Has staged over 200 operas, 25 for SW/ENO. One of the most gifted and musical of English producers, especially successful in his handling of the chorus and in his perceptive character realization. His close collaboration with Britten on the three *church parables was crucial in formulating their dramatic convention.

**Gramophone.** See RECORDING.

**Granados, Enrique** (*b* Lérida, 27 July 1867; *d* English Channel, 24 Mar. 1916). Spanish composer. Studied Barcelona with Pedrell, also Paris. His zarzuela *Maria del Carmen* was an immediate success on its Madrid production, and was followed by five more before he revised some of his piano pieces based on Goya to make his masterpiece, *Goyescas*. Though influenced by late 19th-cent. Central European Romantic traditions, the work is essentially Spanish both in atmosphere and in melodic techniques. His son **Edoardo** (*b* Barcelona, 28 July 1894; *d* Madrid, 2 Oct. 1928) composed 23 zarzuelas: the best known were *Bufón y hastelero* and *La Ciudad eterna*.

SELECT WORKLIST: *Maria del Carmen* (Feliu y Codina; Madrid 1898); *Goyescas* (Periquet y Zuaznabar; New York 1916).

BIBL: J. Subirá, *Enrique Granados* (Madrid, 1926).

**Grandi, Margherita** (*b* Hobart, 4 Oct. 1894; *d* ?). Tasmanian soprano. Studied Paris with E. Calvé, Milan with Russ. Début as 'Djemma Vécla' (anagram of Calvé), Paris, OC; (*Werther*). Second début Milan 1932 (Aida), under her married name of Grandi. Milan, S, 1934; Gly. 1939;

Verona Arena, 1946; London, CG, 1949; also Egypt, S America, etc. Roles included Donna Anna, Lady Macbeth, Tosca; created title-role of Massenet's *Amadis*, and Diana (Bliss's *Olympians*). A performer in the 'grand' style, with a rich vocal palette. (R)

**Grand Inquisitor, The.** The Head of the Inquisition (bs) in Verdi's *Don Carlos* and (ten) in Dallapiccola's *Il prigioniero*.

**Grand macabre, Le.** Opera in 2 acts by Ligeti; text by Michael Meschke and the composer, after M. de Ghelderode. Prod. Stockholm, Royal Opera, 12 Apr. 1978, with Slättegard, Aruhn, Vikström, Ericson, Tyrén, Söderström, Meyer, cond. Howarth; London, C, 2 Dec. 1982, with Mackay, Smith, Rigby, Howard, Hill Smith, Chard, cond. Howarth.

The opera tells of the attempts of Nekrotzar (bar) to destroy the world, having returned from the grave. The court astrologer Astradamor (bs) predicts that a comet arriving at midnight will devastate the earth. However, its arrival passes safely and Nekrotzar allows Prince Go-Go (sop) to lead him back to his grave.

**Grand Opera.** The term normally given to the genre that flourished in Paris from the early 1820s and remained influential beyond the frontiers of France throughout the 19th cent. and even into the 20th cent. Though it had its roots in the enduring aspect of French taste that responds to grandeur, which can be found in the works of Lully, and though its immediate ancestors were some of the works of Le Sueur and Spontini as expressions of Napoleonic pride, Grand Opera found its distinctive form in the 1820s. During this decade the painter and set designer Pierre *Cicéri and the lighting expert (and pioneer of photography) Louis *Daguerre contributed their joint influence to a style that became firmly established in the 1830s with the work of the librettist Eugène *Scribe. The effective organizer of these talents and of the contributing composers was the manager of the Opéra, Louis Véron.

Grand Opera typically consisted of a work in four or five acts that included ballet sequences, made functional use of sets (usually historically and geographically accurate) which would be altered for each act, and culminated in a spectacular final scene perhaps involving stage machinery to depict a natural disaster such as fire, earthquake, avalanche, shipwreck, or volcanic eruption. The scenery would involve three-dimensional sets, subtle and ambitious lighting effects, and elaborate stage machinery that could include moving backcloths to give an illusion of movement. A large number of soloists would be employed, closely linked to a chorus that was dramatically and musically essential both as a

characteristic entity and for processions, riots, battles, festivals, and other spectacular moments. Duets often replaced solos as the characteristic numbers, generally in standard A-B-A or A-B (slow-fast) form; but a feature of Grand Opera was a flexibility that largely disguised these divisions. Spoken dialogue was banished (works taken over by the Opéra had to be furnished with recitatives, as with Berlioz's for Weber's *Der Freischütz*). As well as an enlarged and augmented orchestra, stage bands became common, sometimes using unfamiliar instruments such as saxhorns, tuned bells, organ, anvils, musettes, etc.

In terms of content, the characteristic elements were drawn together and given their strongest expression by Scribe. Historical settings were valued not only for their opportunities for display, but for the chance of depicting the individual in an historical context, with the expression of character taking second pace to dramatic situations brought about by religious conflicts, dynastic clashes, warring peoples, or sudden revolution in a vividly accurate setting. The plots might involve misunderstandings, secrets belatedly revealed, affections betrayed, pulls between love and duty, between personal inclination and political necessity. Sometimes the public worlds themselves might be in a state of conflict, perhaps concerning an oppressed minority: such are Halévy's *La Juive* (1835) and Meyerbeer's *Les Huguenots* (1836). Essentially, the private drama served the function of illuminating the historical episode, with a characteristic climax involving a critical collision between private and public: this moment might concern revolution (Auber's *La muette de Portici*, 1828; Rossini's *Guillaume Tell*, 1829), religious massacre (*Les Huguenots*) or repression (Meyerbeer's *Le prophète*, 1849), political conspiracy (Halévy's *La reine de Chypre*, 1841), or military defeat (Rossini's *Le siège de Corinthe*, 1826). A large part of Scribe's success lay in his masterly varying of these ideas, and his dramatic skill in bringing about sensational confrontations of private and public, such as (a favourite device) a betrayed girl interrupting a politically expedient wedding (*La muette de Portici*) or a state ceremony (*La Juive*).

Whatever the limitations of the genre, whose success rested to a large part on a new bourgeois audience with a taste for the grandiose, its ingredients as well as this very success proved irresistible to composers from several countries. Many 19th-cent. French composers attempted Grand Opera, the greatest of them all, Berlioz, finding acceptance neither with *Benvenuto Cellini* (which failed in 1838) nor with the rejected *Les Troyens* (1856–8), whose majestic stature overwhelmed the conventions (and hence the entrenched expectations of management and audience). Gounod (*Sapho*, 1851), Massenet (*Hérodiade*, 1881), and Saint-Saëns (*Henri VIII*, 1883) were among those influenced. Composers of other nations who also wrote Grand Operas destined for Paris included Wagner (*Rienzi*, 1842), Verdi (*Don Carlos*, 1867), and Tchaikovsky (*The Maid of Orleans*, 1881). The influence endures in Verdi's *Aida* (1871); and though *Rienzi* was rejected by Paris, the influence of Grand Opera also survives in Wagner's work with such features as the interrupted wedding in *Lohengrin*, the processions in the *Festwiese* scene in *Die Meistersinger von Nürnberg*, the inundation of the Gibichungs' Hall at the end of *Götterdämmerung*, and the collapse of Klingsor's magic castle in *Parsifal*. That some of the most characteristic devices could still be effective in modern times is shown in some historical operas, especially Russian (e.g. Prokofiev's *Semyon Kotko*, 1940, and Shaporin's *The Decembrists*, 1953).

BIBL: D. Charlton, 'On the Nature of "Grand Opera" ', in I. Kemp, ed., *Hector Berlioz: Les Troyens* (Cambridge, 1989).

**Grane.** Brünnhilde's horse in *The Ring*. Seldom seen on stage since the war, but formerly ridden with effect into the flames by singers including Leider and Lawrence. The creator of the part was a gentle and nimble-footed 9-year-old, who, unlike his human colleagues, 'never whined, never stormed, never sulked' (Newman), and could be persuaded with sugar to do anything. He was so magnetic a character that Wagner cut him from the *Todesverkündigung* in *Die Walküre*, fearing that he would steal the show from Brünnhilde and Siegmund. During a production of the *Ring* in Munich, 1881, Therese Vogl rode one of the Emperor Maximilian's favourite chargers, then in retirement. His intelligence and enthusiasm were super-equine: he knew all his cues without prompting, and in the last scene of *Götterdämmerung* would wheel with alacrity and gallop into the flames, with Vogl, a fine horsewoman, vaulting on to his back. The effect was sensational. His death just before going to Berlin for Neumann's production nearly caused the performances to be cancelled; and though a last-minute replacement was found, Vogl, unable to recapture that perfect artistic rapport, could not perform her daring leap. Nicholas Nabokov records a performance during the German Depression in the 1920s in which the emaciated and starving interpreter of the role began by eating Hagen's hay-coloured beard.

**Grassi, Cecila.** See BACH, JOHANN CHRISTIAN.

**Grassini, Giuseppina** (*b* Varese 8 Apr. 1773; *d* Milan, 3 Jan. 1850). Italian contralto. Studied

Milan with Secchi and Crescentini. Début Parma 1789 (in Guglielmi's *Pastorella nobile*). Sang throughout Italy in works by Mayr, Nasolini, and Portogallo; created roles in Zingarelli's *Artaserse*, *Giulietta e Romeo*, and Cimarosa's *Gli Orazi ed i Curiazi*. Her career was enhanced by the protection of several Italian and English aristocrats; in 1800 she performed for Napoleon at Milan, S, in honour of his victory at Marengo, and went as his mistress to Paris, where she sang to great acclaim. London, H, 1804–6, appearing in Winter's *Ratto di Proserpina* with her rival Mrs Billington. Returned to Paris 1806 as a highly paid court singer to the Emperor, achieving great success in Paer's *Diddone abbandonata* (written for her), and Cherubini's *Pigmalione*. After Napoleon's fall, she sang in Italy 1817–23, then taught in Milan; her pupils included Pasta, and her nieces Giuditta and Giulia Grisi. A leading singer of her time, she was celebrated for her beauty, her 'sublimely pathetic' tones (Michael Kelly), and her superb acting, which was much admired by Mrs Siddons and Fétis.

**Grau, Maurice** (*b* Brno, 1849; *d* Paris, 14 Mar. 1907). Czech impresario. In USA organized tours of Bernhardt, Offenbach, and Kellogg OC. Joined Abbey for season at New York, M, 1890; director with Abbey and Scheffel 1891–7; Director 1897–1903. Director London, CG, 1897–1900.

**Graun, Carl Heinrich** (*b* Wahrenbrück, 7 May 1703/4; *d* Berlin, 8 Aug. 1759). German tenor and composer, brother of **August Friedrich Graun** (*b* Wahrenbrück, 1698; *d* Merseburg, 5 May 1765), a church musician, and **Johann Gottlob Graun** (*b* Wahrenbrück, 1702/3; *d* Berlin, 27 Oct. 1771), a noted instrumental composer, one of the most important figures in the formation of the pre-Classical style, and the most important exponent of Italian opera in Germany in the 18th cent. after Hasse.

Studied Dresden with Grundig, Petzold, and Johann Christoph Schmidt. His early development was much influenced by the works of Lotti and Keiser which he encountered at Dresden and by the model of Fux, in the première of whose *Costanza e Fortezza* (Prague 1723) he played as cellist. In 1725 he joined the court at Brunswick as tenor, but his skill as a composer was soon recognized and in 1731 he was appointed Vizekapellmeister under Georg Caspar Schürmann. His first opera, *Sancho und Sinilde*, was produced in 1727 and was followed by four more German operas which in style broadly followed the model of the Hamburg repertory, as well as an Italian opera, *Lo specchio della fedeltà*. This was performed for the wedding of Elisabeth Christiana, Princess of Brunswick, to the Crown Prince of

Prussia, the future Frederick the Great, who took Graun into his service at Rheinsberg in 1735. When the Crown Prince succeeded his father in 1740 he immediately made Graun Kapellmeister to the Prussian court and charged him with establishing a top-class Italian opera company.

As the musical head of one of Europe's richest and culturally most ambitious courts, Graun found ample scope to develop his talents. Having trawled Italy for singers, he launched his Berlin career with *Rodelinda* in 1741; the next year his *Cesare e Cleopatra* opened the court opera-house. Over the next 14 years Graun composed nearly 30 dramatic works for Frederick the Great, including pastorales, intermezzos, and opere serie. Almost all of the latter were to librettos by Metastasio, Zeno, and Villati, who adapted several earlier French texts, including some of Quinault's librettos for Lully, to the Italian model. Together they represent a fundamental shift of taste away from the older-style Hamburg librettos set during his Brunswick years and emphasize the primacy of Italian opera in Germany at this time, despite a certain French influence, in keeping with the taste of Frederick the Great's court.

The occasional opera by Hasse or Agricola apart, Graun's work dominated the Berlin repertory and its success amply fulfilled his patron's ambitions. It represents the pinnacle of Baroque operatic achievement, with a firm emphasis upon the importance of the solo singer, who was given considerable opportunity to display his technical command in expansive coloratura. As one of the last great composers whose reputation was secured mainly through involvement with court opera, it would be easy to characterize Graun as an anachronistic figure. Yet his later works, notably *Silla* and *Semiramide*, show some evidence of reform and most strikingly innovative of all is *Montezuma*. Unusually for this period, it is based on a modern subject; more importantly, the traditional da capo aria, on which Hasse placed great reliance, is largely avoided in favour of the emotionally concentrated and less stylized *cavatina. A similar concern for expressive writing is seen in the *accompanied recitatives and in the handling of the orchestra.

Despite their historical importance, Graun's operas have yet to receive any significant revival. Today he is best remembered for his sacred music, especially the passion cantata *Der Tod Jesu* (1755), which mirrors his operatic style in its dramatic intensity and sensitive treatment of the voices.

WORKLIST: (all first prod. Berlin unless otherwise stated): *Sancio und Sinilde* (König, after Silvani; Brunswick 1727); *Polydorus* (Müller; Brunswick 1726 or 1728); *Iphigenia in Aulis* (Postel; Brunswick 1731); *Scipio Africanus* (Fideler or Postel; Wolfen-büttel 1732); *Lo specchio della fedeltà* (The Mirror of Constancy) (Zeno; Brunswick 1733, music lost); *Pharao Tubaetes* (Müller, after Zeno; Brunswick 1735); *Rodelinda, regina de'Langobardi* (Botarelli, after Salvi; 1741); *Cesare e Cleopatra* (Botarelli, after Corneille; 1742); *Artaserse* (Metastasio; 1743); *Catone in Utica* (Metastasio; 1744); *Lucio Papirio* (Zeno; 1745); *Adriano in Siria* (Metastasio; 1746); *Demofoonte, re di Tracia* (Metastasio; 1746); *Cajo Fabricio* (Zeno; 1746); *Le feste galanti* (Villati, after De Vancy; 1747); *L'Europa galante* (Villati, after De la Motte; 1748); *Iphigenia in Aulide* (Villati, after Racine; 1748); *Angelica e Medoro* (Villati, after Quinault; 1749); *Coriolano* (Villati, after Frederick the Great; 1749); *Fetonte* (Villati, after Quinault; 1750); *Mitridate* (Villati, after ?Racine; 1750); *Armida* (Villati, after Quinault; 1751); *Britannico* (Villati, after Racine; 1751); *Orfeo* (Villati, after Du Boulair; 1752); *Il giudizio di Paride* (The Judgement of Paris) (Villati, 1752); *Silla* (Tagliazucchi, after Frederick the Great; 1753); *Semiramide* (Tagliazucchi, after Voltaire; 1754); *Montezuma* (Tagliazucchi, after Frederick the Great; 1755); *Ezio* (Tagliazucchi, after Metastasio; 1755); *I fratelli nemici* (The Hostile Brothers) (Tagliazucchi, after Frederick the Great, after Racine; 1756); *Merope* (Tagliazucchi, after Frederick the Great, after Voltaire; 1756).

**Graupner, Christoph** (*b* Kirchberg, 13 Jan. 1683; *d* Darmstadt, 10 May 1760). German composer. Studied Leipzig under Schelle and Kuhnau; began to study law in Leipzig, but was forced by the Swedish invasion to flee to Hamburg. Harpsichordist at the Hamburg Opera under Keiser 1707–9; became Vizekapellmeister 1709 and from 1712 Kapellmeister to the Landgrave of Darmstadt-Hessen. Elected Kantor of the Thomasschule in Leipzig 1722, but was persuaded not to leave Darmstadt by an enhancement in his salary, so the position was taken by Bach instead.

The majority of Graupner's ten operas date from his Hamburg years and have something in common with Handel's contemporary early efforts. They include *Dido, Königin von Carthago* and *Bellerophon*, composed for the wedding of the King of Prussia; in several of these works he was aided by Keiser and he in turn possibly contributed to some of Keiser's works, including *Der Carnaval von Venedig* (1707) and *Die lustige Hochzeit* (1708). At Darmstadt Graupner wrote three operas, including *La costanza vince l'inganno*, but these still follow the model of the Hamburg repertory, particularly Keiser, in all essential matters of style.

SELECT WORKLIST: *Dido, Königin von Carthago* (Hinsch; Hamburg 1707); *Bellerophon* (Feind; after Corneille, Fontenelle & Boileau; Hamburg 1708, lost); *La costanza vince l'inganno* (Constancy Conquers Deceit) (?; Darmstadt 1715).

**Gravenhage, 's** (Eng., The Hague). See HOL-LAND.

**Graz.** Town in Styria, Austria. The first theatre was built during seasons by the Mingotti company 1736–46. A taste for Italian opera grew, and the Nationaltheater opened 1776 with Sacchini's *La contadina in corte*. More German opera was given from 1791, later also opéra-comique and much Mozart. The theatre burnt down 1823, and the Landestheater opened 1825, with a company including Johann *Nestroy. When the censorship was abolished 1848, many new plays were introduced, also operas by Rossini, Bellini, Donizetti, and increasingly Verdi. *Tannhäuser* was staged 1854 (three years before Vienna), and the first complete local *Meistersinger* was staged 1885 under Karl Muck, who also opened the Opernhaus (cap. 1,400) in 1899 with *Lohengrin*. Conductors who worked in Graz early in their careers include Franz Schalk 1889–95, Karl Böhm 1917–21, and Karl Rankl 1932–7; and the singers Welitsch, Manowarda, Schorr, and Töpper.

**Graziani, Francesco** (*b* Fermo, 26 Apr. 1828; *d* Fermo, 30 June 1901). Italian baritone. Brother of below. Studied with Cellini. Début Ascoli Piceno 1851 (in *Gemma di Vergy*). Paris, I, 1853–61; London, CG, 1855–80; St Petersburg 1861–71. One of the best Verdi singers of the time, he successfully created Don Carlo (*Forza*), and was a notable Luna, Posa, Amonasro, Germont, and Rigoletto. Possessed an exceptionally full, mellifluous voice, and a fine technique, but was a limited actor.

**Graziani, Lodovico** (*b* Fermo, 14 Nov. 1820; *d* Fermo, 15 May 1885). Italian tenor. Brother of above. Studied with Cellini. Début Bologna 1845 (in Cambiaggio's *Don Procopio*). Paris, I, 1851; Venice, F, 1851–3; Milan, S, 1855; Vienna 1860. Created Alfredo (though Verdi, depressed by *Traviata*'s failure, referred to his singing as 'marmoreal' and 'monotonous'), and was a successful Duke of Mantua, Manrico, Riccardo (*Ballo*). His voice was clear and vibrant, but he lacked dramatic gifts. A third brother, **Vincenzo** (1836–1906), was a baritone. Début 1862 (Belcore), but despite an excellent voice had to abandon career through deafness.

**Great Britain.** Most development in opera before the 20th cent. was in England, rather than Wales, Scotland, or Ireland. For much of the 17th cent. the popular form of dramatic musical entertainment was the *masque, much influenced by the French ballet de cour. After *stile rappresentativo had been introduced into England in Dowland's lute songs, it was exploited in a number of masques, notably *Lovers made Men* (1617, music lost) by Ben Jonson and Nicholas Lanier (1588–1666). According to the libretto, it was sung throughout in *stile recitativo: if so, this must be seen as the first English opera, though it is doubtful if the music was continuous. Up to 1642 English composers, of whom the most important were Henry Lawes (1596–1662) and his brother William (1602–45), continued to develop a declamatory style, although always in the context of hybrid entertainments, such as masques. In 1639 William *Davenant (1606–68) obtained a royal patent for theatrical performances, but his plans for dramatic music were severely curtailed following the closure of the theatres during the Protectorate (1653–60). Masques continued to be produced in private: the most important was Shirley's *Cupid and Death* (1653), with music by Matthew *Locke (1621/2–77) and Christopher Gibbons (1615–76).

In 1656 Davenant staged *The Siege of Rhodes*, a 'Representation by the Art of Perspective in Scenes, and the Story Sung in Recitative Music', given at Rutland House with music (lost) by composers including Lawes and Locke. Davenant's elaborate description was an obvious attempt to circumvent the Puritan ban on theatrical performance: from the libretto it is clear that *The Siege of Rhodes* has the best claim to be the first English opera. Ten performances were given: these are notable for including the first appearance of a woman on the stage in England. At the Restoration in 1660 interest in spoken drama was so strongly renewed that it was impossible for opera to make any impact. However, several important dramatic works featuring extensive use of music were seen during this period: in 1674 *Ariadne* by Louis *Grabu (*fl.* 1665–94), an adaptation of the Cambert-Perrin collaboration, was performed for the marriage of the Duke of York, the year after Locke's music to Shadwell's *Psyche* had been given. Both of these works are eclipsed by *Venus and Adonis* (*c*.1682) which, although described by its composer John *Blow (1649–1708) as a masque, contains sufficiently continuous music to qualify as a true opera.

This served as the model for English opera's first masterpiece, *Dido and Aeneas* (1689) by Henry *Purcell (1659–95). Written for a girl's school, it is of miniature proportions and demonstrates the debt to both the French and Italian styles which was apparent in all English music of this period. Its modern fame has given it exaggerated historical importance: better known at the time was Purcell's incidental stage music, some of which, notably that for Dryden's *King Arthur* (1691), closely approaches authentic opera and is indeed known as *semi-opera. After Purcell's untimely death, Italian opera seria overtook London. Some early works set English translations of Italian texts, such as *Arsinoe, Queen of Cyprus* by Thomas Clayton (*c.*1665–*c.*1725), who later collaborated with Joseph Addison on *Rosamond* (1707). Productions of the pastoral *Gli amori d'Ergasto* (1705) (in Italian) by Jakob Greber (?–1731), *Il trionfo di Camilla* (1706) by Antonio Maria Bononcini (1677–1726) (in English, but the first Italian work to be performed in its entirety) and of *Pirro e Demetrio* (1708) by Alessandro Scarlatti (1660–1725) (in a mixture of English and Italian) prepared the way for the first performances of full-scale operas sung entirely in Italian, *Idaspe fedele* (1710) by Francesco Mancini and *Almahide* (1710) by Giovanni Bononcini (1670–1747). The first castrato appeared in London in 1706 in *Thomyris* (music adapted from Bononcini and Scarlatti), establishing a vogue which was fully exploited after the arrival of George Frideric *Handel (1685–1759) in 1710, who, beginning with *Rinaldo* (1711), produced almost 40 operas (including adaptations) for London over 30 years.

The first effective counter to the Italian domination came with *The *Beggar's Opera* (1728); its popularity triggered off a series of *ballad operas, including *The Devil to Pay* (1731) (which in turn played an important part in the growth of *Singspiel) and led to the eventual downfall of opera seria. In 1760 *Thomas and Sally* by Thomas Arne (1710–78) made use of modern characters in an all-sung opera, but his popular *Artaxerxes* (1762) did not find successors. For many years English operas, even strong examples such as *Lionel and Clarissa* (1768) by Charles *Dibdin (1745–1814) and *The Duenna* (1775) by Thomas *Linley sen. (1733–95) and jun. (1756–78), were with spoken dialogue and close to *pasticcio. But during the 18th cent. the London theatres continued to attract such Italian musicians as Gioacchino Cocchi (*c.*1715–1804), Tommaso Giordano (*c.*1733–1806), Antonio Sacchini (1730–86), Ferdinand Bertoni (1752–1813), and Francesco Bianchi (*c.*1720–*c.* 1788). The most successful of all foreign composers was probably Johann Christian *Bach (1735–82), who from his arrival in London in 1762 dominated the stage with his elegant opere serie in the galant style until the new vogue for opera buffa made popular the works of Domenico Cimarosa (1749–1801), Baldassare Galuppi (1706–85), and Niccolò Piccinni (1728–1800). Even the first complete performance of a French opera in London, Grétry's *Zémire et Azor* in 1779, was given in Italian.

By the early 19th cent. the Italian domination (especially that of Rossini) was complete, while English opera was unable to escape from pasticcio, with plot, dialogue, scenery, and stage machinery taking precedence over music. One of the most successful composers of the day was

Henry *Bishop (1786–1855), who wrote dramatic works in every vein. These conditions dismayed Carl Maria von Weber (1786–1826) when he accepted the commission for *Oberon* (1826), which is scarred by them but whose quality helped to pave the way for English Romantic opera. The first significant work in the genre was *The Mountain Sylph* (1834) by John *Barnett (1802–90). This was followed by *The Siege of Rochelle* (1835) by Michael *Balfe (1808–70): of his 30 operas, the most successful was *The Bohemian Girl* (1843). Another classic of the period was *Maritana* (1845) by Vincent *Wallace (1812–65). Also significant was Edward *Loder (1813–65), whose *Raymond and Agnes* (1855) shows an ability to build up continuous forms; and Julius *Benedict (1804–85) won a success with *The Lily of Killarney* (1862). Nevertheless, the most successful aspects of these works tended to be German- or French-influenced.

A more independent form of nationalist opera came with *Shamus O'Brien* (1896) by *Charles Stanford (1852–1924). Other works by Frederic Cowen (1852–1935) and Alexander Mackenzie (1847–1935) were staged; but the most popular figure of the latter years of the century, and enduringly in the 20th cent., was Arthur *Sullivan (1842–1900), who drew upon much in recent operatic history to fertilize his clever, Romantic idiom in the famous series of *Savoy Operas (really operettas). A serious attempt to transfer the tried methods of German opera to the stage was made by Ethel *Smyth (1858–1944); and these also affect the operas of Frederick *Delius (1862–1934). Belated attempts at English Wagnerian opera were made by Rutland *Boughton (1878–1960) and Joseph *Holbrooke (1878–1958).

However, a new attempt to discover a more authentic English idiom was made, by way of folk-song and Tudor and Elizabethan music, by Gustav *Holst (1874–1934) and Ralph *Vaughan Williams (1872–1958). Some of Holst's early operas were heavily influenced by Wagner, but with *Savitri* (1908, prod. 1916) he discovered an original vein and in particular an English musical declamation. Vaughan Williams wrote an English national folk opera in *Hugh the Drover* (1911–14, prod. 1924); and both composers also wrote comedies. If these works reveal the belated state of English opera's growth in European terms, they served to open up new vistas. Without them, the achievement of Benjamin *Britten (1913–76) in *Peter Grimes* (1945) might have been harder. With this, English opera found new maturity; and its success encouraged Britten himself, his seniors including William *Walton (1902–83), Lennox *Berkeley (1903–89), and especially the five operas of Michael *Tippett (*b

1905). The following generation included Iain *Hamilton (*b* 1922), Thea *Musgrave (*b* 1928), Alexander *Goehr (*b* 1932), Peter Maxwell *Davies (*b* 1934), and Harrison *Birtwistle (*b* 1934). Other composers who have written successful operas include Nicholas *Maw (*b* 1935), Robin *Holloway (*b* 1934), Richard Rodney *Bennett (*b* 1936), Gordon *Crosse (*b* 1937), Judith *Weir (*b* 1954), Robert Saxton (*b* 1953), Nigel *Osborne (*b* 1948), and John *Casken (*b* 1949). Though continuing economic problems have led composers to explore chamber opera or music theatre, the last half-century has seen Britain develop an operatic life creatively as rich as any in the world; and it is now the rule rather than the exception for a composer to write opera. See also ALDEBURGH, ALMEIDA FESTIVAL, BEECHAM OPERA COMPANY, BIRMINGHAM, BRITISH BROADCASTING CORPORATION, BRITISH NATIONAL OPERA COMPANY, BUXTON, CAMBRIDGE, CARL ROSA OPERA COMPANY, DENHOF OPERA COMPANY, EDINBURGH, ENGLISH MUSIC THEATRE COMPANY, GLASGOW, GLYNDEBOURNE, HANDEL OPERA SOCIETY, KENT OPERA, LEEDS, LIVERPOOL, LONDON, MANCHESTER, MOODY, OPERA NORTH, OXFORD, SCOTTISH OPERA, WELSH NATIONAL OPERA.

**Greece** (Gr., Ellada). The Ionian Islands, subject to Venice 1386–1797 and to Britain 1815–64, acquired their first knowledge of opera when visiting Italian companies gave seasons at the T San Giacomo in Corfu from 1733; seasons were also given in other islands throughout the 18th and 19th cents., mainly Cephallonia and Zakynthos, and local composers went to study in Italy. The earliest known extant Greek opera is Nikolaos Halikiopoulos-Mantzaros's *Don Crepuscolo* (Corfu 1815). Spiridon *Xyndas (1814–96) wrote his first opera, *Anna Winter* (1855), for Corfu, and also composed the first opera in Greek, *O ypopsiphios vouleftis* (The Parliamentary Candidate, comp. 1857, prod. 1867). His successors included Edouardos Lambelet (1820–1903) (three operas, lost); the Zakynthian Pavlos Karrer (1829–96), with several operas on Greek plots, including *Markos Botsaris* (Patras 1861), *Kyra Phrosyni* (Zakynthos 1868, lost), and *Marathon-Salamis* (unprod.); and Spiridon Samaras (1861–1917), whose most remarkable operas were *La martire* (1894) and *Rhea* (1908).

The mainland, occupied by Turkey from the 14th cent. until 1821, had no experience of opera until the visits of Italian companies; the first opera in Athens was *Barbiere* in 1837, and other performances followed quite frequently from the 1840s. These proved very popular, and seasons were given in Athens, Piraeus, and Patras. Operetta by French companies became popular from

1871. Local companies were also formed in Athens by Napoleon Lambelet and Dionysus Lavrangas 1888–90, the first, second, and third Elliniko Melodrama. Greek national opera began to evolve through the introduction of Greek folksongs and folk subjects and Byzantine chant into Italianate operatic forms (e.g. in *Rhea*). A successful pioneering work was *O protomastoras* (The Master Builder, 1916) by Manolis *Kalomiris (1883–1962), then *To dakhtylidhi tis manas* (The Mother's Ring, 1917). Other successful composers of the period include Theophrastos Sakellaridis (1883–1950) with *Perouzé* (1911) and at least 63 operettas; Marios Varvoglis (1885–1967), with the fluent *To apoghevma tis agapis* (The Afternoon of Love, 1944); Georgios Sklavos (1888–1976) with *Lestenitsa* (comp. 1923, prod. 1947); Petros Petridis (1892–1977) with *Zemphira* (1957, unprod.); Andreas Nezeritis (1897–1980) with *O vassilias Aniliagos* (King Aniliagos, comp. 1933, prod. 1948) and *Iro kai Leandros* (comp. 1964, prod. 1970); and Solon Mikhailidis (1905–79) with *Odysseus* (1955, unprod.). Important in a younger generation are Arghyris Kounadis (*b* 1924), who settled in Germany and whose first opera in Greece was *O yirismos* (The Return, 1991); and Perikles Koukos (*b* 1960) with the very successful *O Conroy kai i kopies tou* (Conroy's Other Selves, 1990). Opera has also been given in Crete, including the première of Butler's *The Sirens' Song* at Heraklion, 1968.

See also ATHENS, CORFU.

**Gregor, Bohumil** (*b* Prague, 14 July 1926). Czech conductor. Studied Prague with Klima. Prague, 5th May O, 1947. Brno O 1949–51; Music director Ostrava O 1958–62; Prague, N, 1962, touring with Co. to Edinburgh in first British *From the House of the Dead* and *Mr Brouček*; Royal Swedish O 1966–9; Hamburg 1969–72; San Francisco from 1969. Specialist in Janáček; repertory also incl. Verdi's *Otello*, *Bartered Bride*, *Salome*, and premières of works by Trojan, Kašlik, Pauer, and Kelemen. (R)

**Gregor, Joseph** (*b* Czernowitz (Černovice), 26 Oct. 1888; *d* Vienna, 12 Oct. 1960). Austrian librettist and writer. His extensive literary activity included the writing of librettos for Strauss's *Friedenstag*, *Daphne*, and *Die Liebe der Danae*. From 1918 he worked in the Austrian National Library, especially as archivist. His writings include *Richard Strauss* (Vienna, 1939, 3/1952) and *Kulturgeschichte der Oper* (Vienna, 1941, Zurich 3/1950).

BIBL: R. Tenschert, ed., *Richard Strauss und Josef Gregor, Briefwechsel* (Salzburg, 1955).

**Greindl, Josef** (*b* Munich, 23 Dec. 1912; *d* Vienna, 16 Apr. 1993). German bass. Studied Munich with Bender and Bahr-Mildenburg.

Début Krefeld 1936 (Hunding). Berlin since 1942, first S, then SO. Bayreuth 1943, 1951–70. New York, M, 1952–3. London, CG, 1963. Also Italy, USA, S America. Repertory included Don Alfonso, Sarastro, Hagen, Wanderer, Sachs, Boris, Ochs. A versatile and accomplished artist with a strong, warm tone. (R)

**Gremin.** Prince Gremin (bs), Tatyana's husband, in Tchaikovsky's *Eugene Onegin*.

**Gretel.** Hänsel's sister (sop) in Humperdinck's *Hänsel und Gretel*.

**Grétry, André-Ernest-Modeste** (*b* Liège, ?8 (*bapt.* 11) Feb. 1741; *d* Montmorency, 24 Sep. 1813). French composer of Walloon origin. Studied Liège with Moreau. While further studying in Rome, he had a success with a pair of intermezzos, *La vendemmiatrice* (1765, lost), but his aim was Paris and opéra-comique, which already he conceived of as serious in content. He arrived there in 1767, and soon had a success with *Le Huron*, then the very popular *Lucile* and *Le tableau parlant*, in which his Italian-trained melodic charm was placed at the service of good declamation (which he had discussed with Voltaire) and elegant taste.

He consolidated his public following (which he carefully cultivated) in a group of works in which, if there was greater variety than range, the music expressed different settings and emotional situations with unfailing grace. This period is concluded with an outstanding example, *Zémire et Azor*, a 'Beauty and the Beast' opera of pre-Romantic sensibility in which the touching quality of the story is never allowed to deteriorate into sentimentality. The librettist, as in earlier works, was Marmontel, with whom relations gradually declined despite his contribution to a group of four striking and diverse works, *Le magnifique*, *La rosière de Salency*, *La fausse magie*, and *Les mariages Samnites*. A new collaborator was Thomas d'Hèle (Hales), who provided him with texts for the witty *Le jugement de Midas* and the Italianate *Les événements imprévues*; on D'Hèle's death he turned to Sedaine for a more ambitious, Romantic work, *Aucassin et Nicolette*. *Andromaque* found him over-stretched at the Opéra with a great tragic subject; he returned to his own métier with *Colinette à la cour*, and conquered the Opéra on his own terms. He later confirmed this with *La caravane du Caire*, one of his greatest triumphs, and *Panurge dans l'île des lanternes*.

The opera which followed, *Richard Cœur-de-Lion*, was perhaps Grétry's masterpiece. Though there are suggestions of functional repetition of music in his earlier operas, it has entered history as a work which set an example in its use of a recurring melody to unify both story and score. This use of the air 'Une fièvre brulante' has been cited as an early instance of \*reminiscence motive, for which it may have set an example but from which, for all its dramatic skill, it differs in that is not worked in any symphonic manner. There followed a series of failures, from which the excellent *Raoul Barbe-bleue* stands out. Family difficulties may have been partly responsible; the Revolution further unsettled, without destroying, his position. He had some success with *Pierre le Grand*, also with *Guillaume Tell* (which includes some attractive 'Swiss' music). But he did not effectively return to the stage, though he continued to write occasional operas and to devote himself to essays and his very interesting memoirs.

Grétry's quality rests chiefly on his skill and taste in using the Italian models he had learnt in youth to express French qualities of taste, sensibility, and poetic eloquence. His clarity of thought accorded well with French expectations, as did his ability to discover the latent emotion in an apparently conventional situation. If his orchestration was generally slim, it served to present the characters sympathetically and to their best advantage; and he sought to relate the overture to the succeeding opera (strikingly in *Le magnifique*) and to write descriptive dramatic entr'actes. Moreover, without laying claim to reform, he was an ambitious innovator in his theoretical writings, looking forward to an ideal theatre with no boxes and a covered orchestra. He narrowed the old division between tragedy and comedy in opera by his perceptive and truthful treatment of human situations. Each of his best works has its distinct individuality; seen as a whole, his œuvre is an essential contribution to the maturity of opéra-comique.

SELECT WORKLIST (all first prod. Paris): *Le Huron* (Marmontel, after Voltaire; 1768); *Lucile* (Marmontel; 1769); *Le tableau parlant* (The Talking Picture) (Anseaume; 1769); *Zémire et Azor* (Marmontel, poss. after Favart, after Beaumont, Chaussée, & Villeneuve; 1771); *Le magnifique* (Sedaine, after La Fontaine, after Boccaccio; 1773); *La rosière de Salency* (The Rose-Maiden of Salency) (Pézay, after Favart, after Sauvigny; 1773); *La fausse magie* (False Magic) (Marmontel; 1775); *Les mariages samnites* (Rosoi, after Marmontel, after Montesquieu; 1776); *Le jugement de Midas* (D'Hèle; 1778); *Les événements imprévues* (Unforeseen Events) (D'Hèle; 1779); *Aucassin et Nicolette* (Sedaine, after the 13th-cent. fable; 1779); *Andromaque* (Pitra, after Racine; 1780); *Richard Cœur-de-lion* (Sedaine, after ?Paulmy, after Lhéritier; 1784); *Raoul Barbe-bleue* (Sedaine, after Perrault; 1789); *Pierre le Grand* (Bouilly, after Voltaire; 1790); *Guillaume Tell* (Sedaine, after Lemierre; 1791).

WRITINGS: *Mémoires* (Paris, 1789, rev. 2/1797, R/ 1973); *De la vérité* (Paris, 1801); ed. L. Solvay & E. Closson, *Réflexions d'un solitaire* (Paris, 1919–22).

BIBL: D. Charlton, *Grétry and the Growth of Opéra-Comique* (Cambridge, 1986).

**Grieg, Edvard** (*b* Bergen, 15 June 1843; *d* Bergen, 4 Sep. 1907). Norwegian composer. Seven numbers survive of his sole attempt at opera, *Olav Trygvason* (1873).

**Grillparzer, Franz** (*b* Vienna, 15 Jan. 1791; *d* Vienna, 21 Jan. 1872). German poet and playwright. He first contemplated writing for opera while working on his play *Sappho* (1819), and in 1823 wrote *Die schöne Melusine* as a libretto for Beethoven, although it was never set. He showed great interest in Singspiel.

**Grimaldi, Nicolò.** See NICOLINI.

**Grisar, Albert** (*b* Antwerp, 26 Dec. 1808; *d* Asnières, 15 June 1869). Belgian composer. Studied Paris with Rejcha; then, after success with his first opera, *Le mariage impossible*, production of *Sarah* in Paris and six more operas, further study with Mercadante in Naples. Returning to Paris, he produced a number of successful light works. These are melodically lively and possess a quick ear for comedy, being influenced by Grétry and Boieldieu and anticipating some of the features of Hervé, Lecocq, and occasionally even Offenbach.

SELECT WORKLIST (all first prod. Paris unless otherwise stated): *Le mariage impossible* (Mélesville & Carmouche; Brussels 1833); *Sarah* (Mélesville; 1836); *Lady Melvil* (with Flotow: Leuven; 1838); *L'opéra à la cour* (Opera at Court) (with A. Boieldieu jun.: Scribe & Saint-Georges; 1840); *Gille ravisseur* (Gille the Ravisher) (Sauvage; 1848); *Les Porcherons* (Sauvage & Lurieu; 1850); *Bonsoir, M. Pantalon* (Morven & Lockroy; 1851); *Les amours du diable* (The Devil's Loves) (Saint-Georges; 1853); *Le chien du jardinier* (The Gardener's Dog) (Lockroy & Cormon; 1855).

BIBL: A. Pougin, *Albert Grisar* (Paris, 1870).

**Griselda, La.** Opera in 3 acts by Alessandro Scarlatti; text by Zeno. Prod. Rome, T Capranica, Jan. 1721.

**Griséldis.** Opera in a prologue and 2 acts by Massenet; text by Silvestre and Morand. Prod. Paris, OC, 20 Nov. 1901, with Bréval, Tiphaine, Maréchal, Fugère, Dufranne, cond. Messager; New York, Manhattan OC, 19 Jan. 1910, with Garden, Dalmorès, Dufranne, Huberdeau, cond. De la Fuente; Wexford, 31 Oct. 1983, with Landry, Leiferkus, cond. Stapleton.

Alain (ten) loves Griséldis (sop), wife of the Marquis (bar). The Devil (bs) tries to persuade her of her husband's infidelity and to accept Alain. She resists, and he takes her child, which is eventually retrieved by the Marquis.

Also opera by Franckenstein (1898).

**Grishka Kuterma.** A drunkard (ten) in Rimsky-Korsakov's *The Invisible City of Kitezh*.

**Grisi, Giuditta** (*b* Milan, 28 July 1805; *d* Robecco, 1 May 1840). Italian mezzo-soprano, sister of below, and cousin of the famous dancer **Carlotta Grisi**. Studied Milan with her aunt, Grassini, and at Milan Cons. Début Vienna 1826 (Faliero, *Bianca e Faliero*). Appeared throughout Italy, particularly in operas by Bellini, who wrote the part of Romeo for her in *I Capuleti e i Montecchi*. Also sang in London, Paris, Madrid. Her elegant stage presence and pure, agile voice brought her success, but singing soprano roles (e.g. Norma) led to difficulties. She retired in 1838.

**Grisi, Giulia** (*b* Milan, 22 May 1811; *d* Berlin, 29 Nov. 1869). Italian soprano, sister of above. Studied Bologna with Giacomelli, Milan with Marliani, and with her sister, and aunt (Grassini). Début Bologna 1828 (in Rossini's *Zelmira*). Milan, S, 1831–2, creating Adalgisa (written for her) in *Norma*. Unhappy with conditions there, she fled to Paris, making her début with Rossini's help at Paris, I, 1832, as Semiramide. Sang there until 1846, and in 1857. London, H (later HM) 1834–46 and CG 1847–61. St Petersburg, 1849; US tour, 1854–5; Madrid 1859. Made several farewells from 1854–66. Created Elvira (*Puritani*), when according to Bellini 'she sang and acted like an angel'; Elena (*Marin Faliero*), Norina (*Don Pasquale*). Her famous tenor partners were Rubini, and Mario, with whom she had a lasting relationship, but was unable to marry. Beautiful and gifted, she was idolized by the public, including the young Queen Victoria; her well-controlled voice was remarkably even, rich, and flexible, and her dramatic ability convincing (though Chorley comments on 'a fierceness . . . which impairs her efforts'). Famous as Anna Bolena, Lucrezia Borgia, Norma, Valentine, Alice (*Robert le diable*), Leonora (*Trovatore*); also an unsuccessful Fidès, trying to eclipse its creator, Viardot.

**Grist, Reri** (*b* New York, 29 Feb. 1932). US soprano. Studied New York with Gelda. Début Santa Fe 1959 (Adèle and Blonde). Zurich 1961–4; London, CG, from 1962; New York, M, 1966; also Vienna, Salzburg, Gly., Milan, S. Roles include Susanna, Blonde, Zerbinetta, Queen of Shemakha. A lively artist with a light, sweet tone. (R)

**Grob-Prandl, Gertrud** (*b* Vienna, 11 Nov. 1917). Austrian soprano. Studied Vienna with Burian. Début Vienna, V, 1938 (Santuzza). Zurich 1946–7; Vienna, S, 1948–64; London, CG, from 1951; also Milan, S; Berlin; S America; etc. A notable Brünnhilde, Isolde, and Turandot, with a full, robust voice, and great authority on stage. (R)

**Grossi, Giovanni.** See SIFACE.

**'Grossmächtige Prinzessin'.** Zerbinetta's (sop) recitative and aria in Strauss's *Ariadne auf Naxos*, in which she tries to convert Ariadne to her fickle view of love. One of the most difficult coloratura arias in opera.

**Groves,** (Sir) **Charles** (*b* London, 10 Mar. 1915; *d* London, 20 June 1992). English conductor. Studied London. Chorus-master BBC Opera Unit 1937–9; cond. BBC Northern Orch. 1945–51, giving studio performances of operas incl. Holst's *The Perfect Fool* and Dohnányi's *The Tenor*. Music director WNO 1961–3, conducting first British *Battaglia di Legnano*, among some other distinguished Verdi performances. SW/ENO as guest from 1971, music director 1977–9; cond. première of Crosse's *Story of Vasco* (1974), and among other works *Euryanthe* (1977) and *Due Foscari* (1978). (R)

**Gruberová, Edita** (*b* Bratislava, 23 Dec. 1946). Czech soprano. Studied Prague with Medvecká, Vienna with Boesch. Début Bratislava 1968 (Rosina). Vienna, S, from 1970; Salzburg 1974–81; Gly. 1973; New York, M, from 1977; also London, CG; Milan, S; Munich; etc. An excellent coloratura singer, admired as Constanze, Queen of the Night, Lucia, Amina, Gilda, Violetta, Zerbinetta. (R)

**Gruhn, Nora.** See GRÜNEBAUM, HERMANN.

**Grümmer, Elisabeth** (*b* Niederjeutz, 31 Mar. 1911; *d* Berlin, 6 Nov. 1986). German soprano. Originally an actress. Operatic début Aachen 1940 (First Flowermaiden). Berlin, SO, from 1946; London, CG, from 1950; Gly. 1954; Bayreuth 1957–61; New York, M, 1966; also Paris, O; Vienna; Buenos Aires; etc. Retired 1972. An outstanding Countess (*Figaro*), Pamina, Eva, Marschallin. A musical and aristocratic artist with a fresh, true voice, and an unaffected style. (R)

**Grünbaum, Therese** (*b* Vienna, 24 Aug. 1791; *d* Berlin, 31 Jan. 1876). Austrian soprano. Daughter of Wenzel *Müller. Studied with her father and appeared as a child in various works. Moved to Prague in 1807, when she sang Zerlina in the first German version of *Don Giovanni*. Married the tenor J. C. Grünbaum in 1813, when she was also engaged by Weber for his theatre in Prague. She and her husband sang in his first production there, *Fernand Cortez*. Weber admired her keenly, praising her 'marvellous talent' and artistry, and describing her as 'the complete mistress of her voice'. She left his company in 1816 to be the leading soprano at Vienna, K, 1816–26;

her roles here included Rossini's Desdemona, and the first Eglantine (*Euryanthe*), written for her. Munich, N, 1827; Berlin 1828–30; here she opened a singing school. Her second husband **Johann Christoff** (1785–1870) made his début in Regensburg 1804. Prague 1808–18, then Vienna and Berlin, with his wife. Translated into German the librettos of many operas by Halévy, Rossini, Meyerbeer, and Verdi. Their daughter **Caroline** (*b* Prague, 18 Mar. 1814; *d* Brunswick, 26 May 1868) was a soprano. Studied with her mother. Début Vienna, K, 1828 (Emmeline in Weigl's *Schweizerfamilie*). Created Anna (*Hans Heiling*), Berlin 1832. Retired 1844.

**Grunebaum, Hermann** (*b* Giessen, 8 Jan. 1872; *d* Chipstead, 5 Apr. 1954). German, later English, conductor. Studied Frankfurt with Humperdinck, and Berlin. Début Koblenz 1893. Chorus-master London, CG, 1907–33: assisted Richter with English *Ring*, 1908–9. With T. C. Fairbairn, founded London School of O, conducting première of Holst's *Savitri*, 1916. Director of opera class London, RCM, 1924–46, incl. complete student *Parsifal* (1926, with Boult). His daughter **Nora Gruhn** (*b* 6 Mar. 1905) studied Munich and sang at London, CG, 1929–32 and SW 1932–6, 1945–8.

**gruppetto** (It.: 'little group'). Its most common form is the alternation of a main note with two subsidiaries, immediately above and below. Beginning as one of the myriad ornaments of Baroque composition, it survived with particularly beautiful effect in Wagner.

**Guadagni, Gaetano** (*b* Lodi, *c.*1725; *d* Padua, Nov. 1792). Italian contralto castrato, later soprano. Originally untrained. Venice, T San Moisè, 1746. London, H, 1748–9, member of Italian burletta company. Handel engaged him to sing in *Samson* and *Messiah*, and composed Didimus (*Theodora*) for him. Lisbon 1754, when he studied with Gizziello; then Paris, and London again 1755, when he performed in *The Fairies* and was trained by Garrick. Venice 1757–8; Vienna 1762–4, singing in works by Hasse and Traetta, and creating the title-roles in Gluck's *Orfeo* and *Telemaco*. London, H, 1770 and 1771 (Orfeo). Italy, Munich, Potsdam, until 1776 when he retired to Padua, singing for several years at the Basilica di Sant'Antonio. Also produced operas on a puppet stage at home, himself (now a soprano) singing the part of Orfeo. Unusual in his refusal to ornament gratuitously, he was an intelligent singer entirely to Gluck's taste. Burney admired him as an actor 'who had no equal on any stage in Europe', and for his 'impassioned and exquisite' singing. However, his artistic pride (he demanded respect for the music as well as for himself) was not appreciated in his own time. His sister **Lavinia-Maria** (*b* Lodi, 21 Dec. 1735; *d* Padua, *c*.1790) was a soprano who sang in London 1756–74 with her husband Felice *Alessandri.

**Guarducci, Tommaso** (*b* Montefiascone, *c*.1720; *d* after 1770). Italian soprano castrato. Studied Bologna with Bernacchi. Sang in Italy from 1745. Engaged by Farinelli for Spanish court 1750–70 (when he retired); Viennese court from 1752; London, H, 1766–8. Created leading roles in Gluck's *L'innocenza giustificata*, and J. C. Bach's *Alessandro nell'Indie*, among others. Burney found him, though 'inanimate as an actor', the 'plainest and most simple singer, of the first class, I ever heard'. Like Guadagni, he prized expression above virtuosity.

**Guatemala.** The first opera given in Guatemala City was Dalayrac's *Adolphe et Clara* at the T Fedriani (1830). Some Italian works were staged at the T de Oriente in 1845, and a few seasons followed, some of them by amateurs. In 1850 another season was presented including, as well as Italian works, *La mora generosa* by the Guatemalan José Escolástico Andrino. The success of Rossini's *Barbiere* in various seasons encouraged the government to arrange for regular visits of Italian and French companies. The first experience of zarzuela in turn led to regular visits by companies from other countries. Guatemalan opera has been slow to develop, but a few composers have written works based on folk or Maya music, including Jesús Castillo with *Quiché-Vinak* (1924).

**Gudehus, Heinrich** (*b* Altenhagen, 30 Mar. 1845; *d* Dresden, 9 Oct. 1909). German tenor. Studied Brunswick with M. Schnorr von Carolsfeld, Berlin with Engel. Début Berlin, 1871 (Nadori, *Jessonda*). Dresden 1880–90; Bayreuth 1882, 1886, 1888 (one of the first Parsifals, and considered a potentially ideal Tristan by the composer); London, CG, 1884; New York, M, 1890; Berlin until retirement in 1896. Contributed to the newly emerging Wagner tradition, as Tannhäuser, Walther, Siegfried, Tristan, etc. Essentially a lyric tenor, temperamentally at his best in serious roles.

**Gueden, Hilde** (*b* Vienna, 15 Sep. 1915: *d* Klosterneuburg, 17 Sep. 1988). Austrian soprano. Studied Vienna with Wetzelsberger. Début Vienna 1939 (in Stolz's operetta *Servus servus*). Munich 1941. Vienna, S, from 1947; London, CG, 1947; New York, M, 1950–60. Repertory included Zerlina, Gilda, Zerbinetta, Sophie (at composer's suggestion), Anne Trulove. A sensitive singer and actress, with a light, clear voice. (R)

**Guerre des Bouffons** (Fr: 'War of the Bouffons'; Ger: *Buffonistenstreit*). The controversy that from 1752 to 1754 divided musical Paris into two parties, those upholding French serious opera (Lully, Rameau, Destouches, etc.) and those upholding Italian opera buffa (Pergolesi, etc.). Though Pergolesi's *La serva padrona* had been given in 1746 without arousing any attention, its performance in 1752 by Eustache Bambini's troupe of Italian comic actors (*bouffons*) brought to public attention a quarrel which had long been brewing among Parisian literati. Those supporting the cause of French opera included the King, Mme de Pompadour, the court and the aristocracy; those who championed Italian opera comprised mainly intellectuals and connoisseurs. Among them was Rousseau, whose *Le devin du village* (1752) sought to utilize the best features of Italian opera buffa for the setting of a French libretto: it was one of the most important forerunners of *opera comique. Other supporters of Italian opera included Diderot, D'Alembert, and the Queen. Their argument was that Italian opera had refreshed a moribund tradition with greater simplicity and naturalness, putting less reliance on elaborate techniques. This contention was challenged by performances of works chosen to represent the best of the French tradition: particularly important was the staging of Mondonville's *Titon et l'Aurore* in 1753 and Rameau's *Castor et Pollux* and *Platée* in 1754.

The Guerre des Bouffons is often portrayed as a conflict between two national traditions, but it was in reality a rivalry between the contrasting claims of serious and comic opera. The debate which it sparked off was conducted through the publication of numerous pamphlets and articles: most famous of these was Rousseau's *Lettre sur la musique française* (1753) and François Collin de Blamont's *Essai sur les goûts anciens et modernes de la musique française* (1754). Though serious opera finally emerged triumphant, the imitation by French musicians of the Italian buffo style aided the evolution of the *comédie mélée d'ariettes. The episode is sometimes also referred to as the **Querelle des Bouffons** or the **Guerre des Coins** ('War of the Corners'), a reference to the habit of the supporters of French opera gathering by the King's box, while their opponents gathered by the Queen's.

BIBL: D. Launay, ed., *La querelle des bouffons* (Geneva, 1973).

**Gueymard, Louis** (*b* Chapponay, 17 Aug. 1822; *d* Paris, July 1880). French tenor. Début Lyons 1845. Paris, O, 1848–68. Sang the repertory of Nourrit and Duprez, and created several roles, including Henri (*Vêpres siciliennes*), Jonas (*Prophète*), and Adoniram (Gounod's *La reine de*

*Saba*). London, CG, 1852. Verdi comments on his 'fine voice'. His wife **Pauline Lauters-Gueymard** (*b* Brussels, 1 Dec. 1834; *d* ?) was a soprano. Début Paris, L, 1854; O, 1857–76. Her wide range enabled her to sing Donna Elvira, Fidès, Leonora (French *Trovatore*); she created Eboli (necessitating many changes from Verdi's original part for a low mez), the Queen (*Hamlet*), and Balkis (*Reine de Saba*). An eminent singer of her time.

**Guglielmi, Pietro Alessandro** (*b* Massa, 9 Dec. 1728; *d* Rome, 18 Nov. 1804). Italian composer. Studied with his father **Jacopo Guglielmi** (*b* ?; *d* Massa, ?1731) and with Durante in Naples. His first opera was *Lo solachianello 'mbroglione*: this was followed by a string of popular successes, including *Il ratto della sposa* and *La sposa fedele*. Engaged at London, H, 1767, where he spent five years as conductor and composer. Many of the works he directed here were pasticcios: in collaboration with J. C. Bach he produced the first performance in London of Gluck's *Orfeo*, albeit in a bowdlerized version. In 1772 he returned to Italy, where he enjoyed renewed favour with *La virtuosa in Mergellina*, *La pastorella nobile*, and *La bella pescatrice*. Maestro di cappella at San Pietro in the Vatican 1793.

Guglielmi composed about 100 operas, approximately two-thirds of them comic. Although less distinctive or influential than Paisiello's, they earned him fame throughout Europe, probably on account of what Burney called his 'Neapolitan fire'. His pedestrian opere serie are of little significance.

SELECT WORKLIST: *Lo solachianello 'mbroglione* (Pignataro; Naples 1757); *Il ratto della sposa* (Martinelli; Venice 1765); *La sposa fedele* (Chiari; Venice 1767); *La virtuosa in Mergellina* (Zini; Naples 1785); *La pastorella nobile* (Zini; Naples 1788); *La bella pescatrice* (Zini; Naples 1789).

BIBL: G. Bustico, *Un musicista massese: Pier Alessandro Guglielmi* (Barga, 1926).

**Guglielmo.** A young officer (bar), lover of Fiordiligi, in Mozart's *Così fan tutte*.

**Guglielmo Ratcliff.** Opera in 4 acts by Mascagni; text by Andrea Maffei, a translation of Heine's tragedy (1822). Prod. Milan 16 Feb. 1895, with A. Stehle, De Negri, Pacini, cond. Mascagni.

William Ratcliff (ten) is haunted by the visions of a man and a woman; in Maria (sop), daughter of MacGregor (bs), he finds the realization of the woman, and swears to kill anyone who tries to marry her; Count Douglas (bar) so intends, and William challenges him to a duel, but Douglas spares his life. Maria prepares for her marriage to Douglas and learns from her nurse Margherita (mez) that her mother and William's father had once been in love but were prevented from mar-

rying; she had, instead, married MacGregor, who had then killed William's father. William, himself wounded, bursts into Maria's room, kills her and then dies. When Douglas sees the two bodies he kills himself.

See also WILLIAM RATCLIFF.

**Gui, Vittorio** (*b* Rome, 14 Sep. 1885; *d* Florence, 17 Oct. 1975). Italian conductor and composer. Studied Rome with Settacioli and Falchi. Début Rome, Ad., 1907 (*Gioconda*). Milan, S, 1923–5 (at Toscanini's invitation), 1932–4. Founded the T di Torino 1925. Established Florence Orchestra Stabile 1928, from which developed the Florence Maggio Musicale, 1933. Here he cond. a wide and enterprising repertory, and gave the festival a strong and attractive character. Salzburg 1933. London, CG, 1938–9, 1952; Gly. 1952–65, where he conducted many distinguished performances. One of the finest Italian conductors of Gluck, Mozart, and Rossini, and responsible for revivals of a number of neglected operas in Italy. His own works include an opera, *Fata malerba* (Turin 1927). (R)

**Guillard, Nicolas Français** (*b* Chartres, 16 Jan. 1752; *d* Paris, 26 Dec. 1814). French librettist. A popular writer for the French stage; his texts were set by Désaugiers, Grétry, Lemoyne, and Paisiello among others. The quality of his best work is demonstrated in his first libretto, that for Gluck's *Iphigénie en Tauride* (1779), which treated the myth in a noble, dignified, yet human manner. The most accomplished French librettist of his day, he had an important collaboration with Sacchini: their most successful work was *Œdipe à Colone* (1786).

**Guillaume Tell** (William Tell). Opera in 4 acts by Rossini; text by Étienne de Jouy, Hippolyte Bis, and Armand Marrast, after Schiller's dramatic version (1804) of the legend first set down in a ballad (before 1474). Prod. Paris, O, 3 Aug. 1829, with Cinti-Damoreau, L. Dabadie, Mori, H. Dabadie, Nourrit, N. Levasseur, Bonel, Prévost, dancers incl. Taglioni cond. Habeneck; New York 19 Sep. 1831, with Saint-Clair, Privat, Saint-Aubin; London, HM, 11 July 1839 (orig. version), with Persiani, Rubini, Tamburini, Lablache, cond. Costa.

Switzerland, 1307. Resistance to the tyranny of the Austrian Gessler (bar) is led by the patriot Guillaume Tell (bar); he has the support of the elderly Melcthal (bs), whose son Arnold (ten) loves the Austrian Princess Mathilde (sop). Tell saves Leuthold (bar), pursued by the Austrians, and escapes with him in a boat without being recognized.

Mathilde and Arnold declare their love, interrupted by Tell with news of Melcthal's death;

Arnold agrees to join the conspirators, who now gather from different cantons.

Arnold and Mathilde realize that they must part. In the main square of Altdorf, Gessler demands obeisance to his hat. Tell alone refuses, and is condemned to shoot an apple from the head of his son Jemmy (sop). He succeeds but, when he reveals that a second bolt was reserved for Gessler in case of failure, he is arrested. Mathilde takes Jemmy into her care as Tell is removed to the Castle of Küssnacht in the middle of the Lake of Lucerne.

At the edge of the lake, Arnold rouses his companions to action and leads them into the fight. Jemmy is reunited by Mathilde to his mother Hedwig (mez), then lights a beacon to signal the start of the uprising. Tell's boat is driven ashore, and he shoots Gessler. Victory is assured, and the people hymn their new-won liberty

Other operas on the subject are by Grétry (1791), B. A. Weber (1795), Carr (1796), and Baillou (1797).

**Guiraud, Ernest** (*b* New Orleans, 23 June 1837; *d* Paris, 6 May 1892). French composer. Son of **Jean-Baptiste Guiraud** (1803–*c*.1864), who had left France when he failed to win a hearing for his operas. Ernest studied Paris with Marmontel and Halévy, where he was a classmate of his lifelong friend Bizet. He is best known (and often abused) for the recitatives he added to *Carmen*, and for his completion of the orchestration of Offenbach's *Les contes d'Hoffmann*. Most of his operas, which include *Le Kobold* (Nuittier & Gallet; Paris 1870) and *Madame Turlupin* (Cormon & Grandvallet; Paris 1872), are in light vein; the exception is *Frédégonde* (Gallet; Paris 1895), a more grandiose work which was completed after his death by Saint-Saëns and orchestrated by his pupil Dukas. His *Traité pratique de l'instrumentation* (Paris, 1892) was the first French work of its kind to evaluate Wagner's scores.

**Gulbranson, Ellen** (*b* Stockholm, 4 Mar. 1863; *d* Oslo, 2 Jan. 1947). Swedish soprano. Studied Paris with M. and B. Marchesi. Début Stockholm 1889 (Amneris). Bayreuth 1896, and from 1897 to 1914 its only Brünnhilde; also sang Kundry. London, CG, 1900, 1907; also Berlin, Vienna, St Petersburg. Retired 1915. A powerful, solid singer with fine declamation, much admired by Melba. (R)

**Gunsbourg, Raoul** (*b* Bucharest, 25 Dec. 1859; *d* Monte Carlo, 31 May 1955). French composer and impresario of Romanian birth. After managing opera in Russia (from 1881) and in Lille (1888–9) and Nice (1889–91), became director of the Monte Carlo O in 1890, a position he held until 1950. Under his long and brilliant régime Monte Carlo saw the premières of many import-

ant works, especially by Massenet, and the French premières of many more. An inventive and forceful producer, he was also a cavalier adapter, initiating his régime with his stage version of Berlioz's *La damnation de Faust*, also transposing operatic roles to different voices (e.g. the Count in *Figaro* for tenor) and in many other ways altering original texts. A gifted discoverer of talent, he was among the first to promote Caruso and Shalyapin. Autobiography, R. Gunsbourg, *Cent ans de souvenirs . . . ou presque* (n.d.).

**Gunther.** The Lord of the Gibichungs (bar), Gutrune's brother and Hagen's half-brother, in Wagner's *Götterdämmerung*.

**Günther von Schwarzburg.** Opera in 3 acts by Holzbauer; text by Anton Klein. Prod. Mannheim 5 Jan. 1777.

Though described as a Singspiel, the work (Holzbauer's only German opera) was an attempt to create a German national opera on a substantial scale, using recitatives to replace spoken dialogue. It was one of a number of overtly patriotic works, encouraging nationalistic feelings. One of the first German operas published in full score.

**Guntram.** Opera in 3 acts by Richard Strauss; text by the composer. Prod. Weimar 10 May 1894, with Pauline de Ahna, Zeller, cond. Strauss.

Germany, 13th cent. Guntram (ten) is sent by the Holy Society of Peace, a secret society of Minnesingers, to free the people under the oppressive rule of Duke Robert (bs). He saves the Duke's wife, Freihild (sop), from suicide and falls in love with her. Guntram takes part in a singing competition and, when his song is seen as threatening by the Duke, is challenged to a duel. Guntram torments himself with the thought that he wished the death of Freihild's husband and refuses to seek the judgement of the guild, punishing himself with a life of solitude instead.

**Gura, Eugen** (*b* nr. Žatec, 8 Nov. 1842; *d* Aufkirchen, 26 Aug. 1906). German baritone. Studied Munich with Herger. Début Munich 1865 (in Lortzing's *Der Waffenschmied*). Leipzig from 1870; here his Tristan d'Acunha (*Jessonda*) in 1875 impressed Wagner, who found his portrayal 'noble', and 'of a moving simplicity'. Bayreuth 1876 (Donner, and created Gunther, in first complete *Ring*) until 1892. London, DL, 1882; Munich 1882–96, when he retired. Possessing a warm tone and dramatic power, he was one of the best of the first generation of Wagner singers, especially moving as Sachs and Wotan. Also sang Lysiart, Falstaff, Iago (Verdi), Leporello. (R) His son **Hermann Gura** (1870–1944) was a baritone and producer. Début Weimar 1890 (Dutchman). Director Berlin, K, 1911. Produced (and sang Faninal in) first London *Rosenkavalier*, 1913.

**Guridi, Jesús** (*b* Vitoria, 25 Sep. 1886; *d* Madrid, 7 Apr. 1961). Spanish composer. Studied Paris with Sérieyx and D'Indy. His early success with his Basque opera *Mirentxu* (Echave; Madrid 1915) was followed by the popular *Amaya* (Jáuregui; Bilbao 1920) and, his greatest triumph, *El Caserio* (Romero & Shaw; Madrid 1928). Other zarzuelas and sainetes confirmed his position as one of the most successful composers of Basque opera.

BIBL: J. de Arozamena, *Jesús Guridi* (Madrid, 1967).

**Gurlitt, Manfred** (*b* Berlin, 6 Sep. 1890; *d* Tokyo, 29 Apr. 1973). German composer and conductor. Studied Berlin with Humperdinck and Kaun. After various appointments, music director Bremen 1914; Berlin 1924. Moved in 1939 to Japan, where he did much to make German opera known. His operas include a setting of *Wozzeck* (1926), made at about the same time as Berg's and independently successful in its day.

**Gurnemanz.** The veteran knight of the Grail (bs) in Wagner's *Parsifal*.

**Gustafson, Nancy** (*b* USA, 27 June 1956). US soprano. Studied San Francisco. San Francisco (Freia); Colorado, Minnesota, Santa Fe, Edmonton; Paris, 1984; London CG, Gly., and Chicago, 1988; Scottish O 1989. Possesses a radiant tone and a highly appealing stage personality. Roles incl. Elettra, Donna Elvira, Violetta, Marguerite, Leïla, Antonia, Amelia, Rosalinde, Rusalka, Káťa Kabanová. (R)

**Gustave III, ou Le bal masqué.** Opera in 5 acts by Auber; text by Scribe, based on historical events. Prod. Paris, O, 27 Feb. 1833; London, CG, 13 Nov. 1833, trans. Planché; New York, Park T, 21 July 1834.

See also BALLO IN MASCHERA, UN.

**Gustavus III.** The King of Sweden (ten) in the 'Swedish' version of Verdi's *Un ballo in maschera*.

**Gutheil-Schoder, Marie** (*b* Weimar, 16 Feb. 1874; *d* Weimar, 4 Oct. 1935). German soprano. Studied Weimar with Naumann-Gungl. Début Weimar 1891 (First Lady). Engaged by Mahler (who called her 'a musical genius') for Vienna, 1900; remained until 1926. London, CG, 1913; also Berlin, Munich, Dresden, etc. Though attacked by Viennese critics as 'the singer without a voice', her extraordinary dramatic gifts and artistry brought her great success. Roles included Cherubino, Donna Elvira, Carmen, Isolde, Nedda, Elektra, Octavian, and the Woman in *Erwartung*, which she created in 1924. (R)

**Guthrie, (Sir) Tyrone** (*b* Tunbridge Wells, 2 July 1900; *d* Newbliss, 15 May 1971). English producer. Connected with Old Vic from 1933, director London, SW, 1941, prod. *Figaro*. When Joan Cross became director of opera, he remained director of the Vic-Wells, also prod. *Bohème, Traviata, Carmen*. London, CG: *Peter Grimes* 1946, *Traviata* 1948. New York, M: *Carmen* 1952, *Traviata* 1957. A gifted man of the theatre, expert at creating natural stage movement though sometimes at the expense of the music.

**Gutrune.** Gunther's sister (sop), and Hagen's half-sister, in Wagner's *Götterdämmerung*.

**Gwendoline.** Opera in 2 acts by Chabrier; text by Catulle Mendès. Prod. Brussels, M, 10 Apr. 1886, with Thuringer, Engel, cond. Dupont; San Diego, 2 Oct. 1982, with Plowright, Norman, Raftery, cond. Tauriello; London, Bloomsbury T by University College, 23 Feb. 1983, with Sullivan, Hargreaves, Hetherington, cond. Fifield.

The English coast, 8th cent. The Viking King Harald (ten) loves Gwendoline (sop), daughter of his prisoner, the Saxon Armel (ten), who pretends to consent to their marriage but arranges for her to kill Harald. She refuses, and commits suicide when Harald dies.

**Gye, Frederick** (*b* London, 1809; *d* Dytchley Park, 4 Dec. 1878). English impresario. Manager London, CG, 1849–77. Introduced many operas to London, especially Verdi and Wagner, as well as some of the great singers of the day, including Patti, Lucca, Tamberlik, Faure, Maurel, and Albani, who married his son **Ernest** (manager CG 1879–85).

**Györ.** Town in Györ-Sopron, Hungary. A new theatre opened in 1978, and includes opera in its repertory.

**Gyrowetz, Adalbert** (orig. Vojtěch Matyáš Jirovec) (*b* České Budějovice, 20 Feb. 1763; *d* Vienna, 19 Mar. 1850). Austrian composer of Bohemian origin. After early successes in Vienna, Naples, and Paris, Gyrowetz arrived in London in 1789, where he was given a welcome scarcely less rapturous than that accorded Haydn. His first opera, *Semiramide*, was commissioned for the Pantheon, but the score was destroyed when the theatre burnt down in 1792, shortly after rehearsals had begun. Returning to Vienna, he spent ten years as a diplomat before becoming conductor and composer at the Hoftheater in 1804, a post he held until 1831.

He composed *c*.30 stage works, of which the most popular were *Agnes Sorel, Ida die Büssende*, and *Der Augenarzt*: his *Robert oder die Prüfung* was reputedly much admired by Beethoven. Most of them were to German librettos, although he did set some Italian texts, including Romani's *Il finto Stanislao*, later used by Verdi. *Hans Sachs im vorgerückten Alter* is of interest because of the obvious connection with Wagner's *Meistersinger*. Gyrowetz's musical diversity—he composed operas, Singspiels, and melodramas varying in length from one to five acts in a variety of styles—was matched by his eclectic taste in librettos. In addition to the common Viennese themes of magic, heroism, and exoticism, he was also drawn to American subjects, e.g. *Das Winterquartier in Amerika*. He also composed ballet music and a quantity of incidental music. He was much admired in his day and benefited from the support of many notable composers, including Mozart and Haydn, by whom he was especially influenced.

SELECT WORKLIST: *Agnes Sorel* (Sonnleithner; Vienna 1806); *Ida die Büssende* (Holbein; Vienna 1807); *Der Augenarzt* (The Eye-Doctor) (Veith; Vienna 1811); *Das Winterquartier in Amerika* (?; Vienna 1812); *Robert oder die Prüfung* (Hubet; Vienna 1813); *Il finto Stanislao* (Romani; Milan 1818); *Hans Sachs im vorgerückten Alter* (Hans Sachs in Old Age) (?; Dresden 1834).

# H

**Hába, Alois** (*b* Vizovice, 21 June 1893; *d* Prague, 18 Nov. 1973). Czech composer. Studied Prague with Novák, Vienna and Berlin with Schreker. The principal advocate of microtonal music, he wrote works, including operas, for specially constructed instruments. *Matka*, in the quarter-tone scale, was a sensation on its first production, and was given in Germany and Italy as well as at home. *Nová země* was in the normal semitonal system; but *Přijď království Tvé* requires the singers to master sixth-tones. Though much admired in his own country, Hába's works have not found wider favour, partly because of the obvious difficulty for most singers of pitching microtones accurately. Hába's brother **Karel** (*b* Vizovice, 21 May 1898; *d* Prague, 21 Nov. 1972) studied Prague with Novák, then with Alois Hába. His operas are *Jánošík* (1934), *Stará historie* (Ancient History) (1940), and *Smoliček* (1950).

WORKLIST: *Matka* (The Mother) (Hába; Munich 1931); *Nová země* (New Land) (Pûjman, after Gladkov; comp. 1935, unprod.); *Přijď království Tvé* (Hába, after Pûjman; comp. 1942, unprod.).

BIBL: J. Vysloužil, *Alois Hába* (Brno, 1970).

**habanera.** A song and dance of primitive origin, deriving from the dancing of the *ñañigos*, a people of a district of Havana (whence its name); it spread through S America, where it became the ancestor of the dance-hall tango, and reached Spain. The most familiar operatic example is Carmen's 'L'amour est un oiseau rebelle', in Act I of Bizet's opera, which reflects its characteristic rhythm and its erotic nature. Bizet first wrote another melody (in 6/8), replacing it with the familiar version based on Sebastián Yradier's song 'El Arregilito'. Also the title of an opera by Laparra (1908).

**Habsburg family.** This long-lasting dynasty, which from 1273 to 1918 ruled the greater part of Europe as the emperors of Austria-Hungary, had an influence, both direct and indirect, on the development of opera over several decades. This was strongest and most characteristic in Vienna, from where one branch of the family ruled much of present-day E Europe, and at Innsbruck, where the Habsburg archdukes maintained an impressive musical entourage in the mid-17th cent. For opera, the first important member of the family was Emperor Leopold I (1658–1705) who brought many Italian musicians to Vienna, notably Draghi, and established court opera on a firm footing. Himself a noted composer, with nine dramatic works and *c*.150 arias to his credit, Leopold I supported the production of *c*.400 new operas during his reign and increased the size of the musical establishment to over 100. His work was strengthened by his heirs Joseph I (1705–11), who appointed Ziani, Fux, and Caldara as Kapellmeister and built a new opera-house, and Charles VI (1711–40), who attracted Metastasio as *poeta cesareo and increased the number of musicians to *c*.140. He was succeeded by Empress Maria Theresa (1740–80), herself a singer, under whom the theatres were reorganized (1752) and Gluck appointed (1754). Her reign was distinguished by a gradual introduction of French opéra-comique and attempts at the reform of Italian opera seria. But it was the benevolent reign of her son Joseph II (1770–80 as Co-Regent; 1780–90 as Emperor) that coincided with the greatest era of Vienna's musical history. German opera was encouraged by the founding of the National-Singspiel (1778) and, even though Joseph II's favourite composer was Salieri, Italian influence at court was mortally challenged. Following the accession of the culturally less sophisticated Leopold II (1790–2), for whose coronation as King of Bohemia Mozart wrote *La clemenza di Tito* (1791), the Habsburg court gradually lost its importance as the main focus of Viennese musical life, with many composers preferring to work for the independent theatres or, in the early 19th cent., to participate in the flourishing public concert life. Beginning with the reigns of Franz II (1792–1835) and Ferdinand I (1835–48) the status of the Habsburg court composer began to decline, with the most highly regarded musicians choosing to work largely outside imperial patronage: by the time that Franz Joseph I (1848–1916) ascended the throne, the position of court composer had become a backwater for mediocre talent.

Of the Innsbruck archdukes, most important were Ferdinand Karl (1649–62) and Siegmund Franz (1662–5), under whose reign Cesti worked at the court.

**Hadjibekov.** Azerbaijani family of musicians.

1. **Zulfugar Abdul Hussein-ogly** (*b* Shusha, 17 Apr. 1884; *d* Baku, 30 Sep. 1950). Composer. His works include the opera *Ashug-Garib* (1916) and musical comedies.

2. **Uzeir Abdul Hussein-ogly** (*b* Agjabedi, 17 Sep. 1885; *d* Baku, 23 Nov. 1948). Composer and conductor. Studied Gori, Moscow with Ladukhin, St Petersburg with Kalafaty. Founded first Azerbaijani music school 1922. Produced his

opera *Leila and Mejnun* in Baku, 1908; others are *Sheikh Senan* (1909), *Rustum and Sohrab* (1910), *Asli and Kerem* (1912), *Shah Abbas and Hurshid-banu* (1912), *Harun and Leila* (1915), and—his most famous work—*Kyor-ogly* (1937). Also wrote many successful musical comedies.

3. **Hussein Aga Sultan-ogly** (*b* Shemakha, 19 May 1898; *d* Baku, 10 Nov. 1972). Tenor. Studied Baku with Speransky. First appeared in Azerbaijani Singspiels, then in female roles, 1916. Azerbaijan O, 1920, in operas by the two above composers. Moscow 1938, with Baku O. Roles also included Count Almaviva (Rossini), Alfredo, Lensky.

4. **Sultan Ismail-ogly** (*b* Shusha, 8 May 1919). Composer and conductor. Studied with Zeidman. Conducted Baku musical comedy company 1938–40. Wrote many musical comedies.

**Hadley, Henry** (*b* Somerville, MA, 20 Dec. 1871; *d* New York, 6 Sep. 1937). US conductor and composer. Studied Boston with Emery and Chadwick, Vienna with Mandyczewski. Held various symphonic conducting posts, including associate conductorship of NYPO (1920–7), and founded Manhattan Symphony. The easy, late Romantic appeal of his operas has not had lasting significance.

SELECT WORKLIST: *Nancy Brown* (Ranken; New York 1903); *Safie* (Oxenford; Mainz 1909); *Azora* (Stevens; Chicago 1917); *Bianca* (Stewart, after Goldoni; New York 1918); *Cleopatra's Night* (Pollock, after Gautier; New York 1920).

**Hadley, Jerry** (*b* Princeton, IL, 16 June 1952). US tenor. Studied Illinois, and with LoMonaco. Début Sarasota 1978 (Lionel). New York: CO 1979, M 1987; Vienna, S, 1982; Gly. and London, CG, 1984; also Chicago, Berlin, Munich, etc. Roles incl. Ferrando, Tamino, Duke of Mantua, Alfredo, Gounod's Faust, Werther, Nadir, Pinkerton, Tom Rakewell. (R)

**Haefliger, Ernst** (*b* Davos, 6 July 1919). Swiss tenor. Studied Vienna with Patzak, Prague with Carpi. Zurich O, 1943–52. Berlin, SO, 1952–74. Gly. 1956. Created Tiresias (Orff's *Antigonae*). Distinguished in Mozart and modern music, and also for his care with words. (R)

**Hagen.** A Gibichung (bs), half-brother of Gunther and Gutrune, in Wagner's *Götterdämmerung*.

**Hagen.** City in North-Rhine Westphalia, Germany. The theatre opened 1911 (cap. 830); bombed 1944; rebuilt 1949 (cap. 940).

**Hahn, Reynaldo** (*b* Caracas, 9 Aug. 1875; *d* Paris, 28 Jan. 1947). French composer, conductor, singer, and critic of Venezuelan origin. Studied Paris with Dubois and Massenet. Music critic of *Le Figaro* 1934–47. Cond. Salzburg

1910, Paris from 1913, incl. at O and OC. Though well known for his elegant drawing-room songs, he was also a successful opera composer. His lightness of touch and expert craftsmanship gave his operas, in particular *Ciboulette* and *Mozart*, considerable popular following.

SELECT WORKLIST (all first prod. Paris): *L'île du rêve* (The Island of Dreams) (Hahn, after Loti; 1898); *La Carmélite* (Mendès; 1902); *Ciboulette* (Flers & De Croisset; 1923); *Mozart* (Guitry; 1925); *Le marchand de Venise* (Zamacoïs after Shakespeare; 1935).

WRITINGS: *Of Singers and Singing* (trans. London, 1989).

**Haibel, (Johann Petrus) Jakob** (*b* Graz, 20 July 1762; *d* Djakovar, ?27 Mar. 1826). Austrian tenor and composer. Moved to Vienna 1789 and joined Schikaneder's company, with whom he sang Monostatos in *Die Zauberflöte*. Among his many works for the Viennese stage, the most popular was the Singspiel *Der Tyroler Wastl*, given 118 times by 1801. Most of Haibel's dramatic music is lost, but such numbers as 'Tiroller seyn aft'n so lustig und froh', the hit-song from *Der Tyroler Wastl*, suggest that he wrote in an appealing melodic style. In addition to Singspiels he also composed Possen and quodlibets of which the most successful was *Rochus Pumpernickel*; *Der Tyroler Wastl* was given Weimar 1808, under Goethe's direction. In 1806 Haibel left Vienna and became Kapellmeister at Djakovar Cathedral, where he composed much sacred music; the following year he married Sophie Weber, the youngest sister of Mozart's widow Constanze.

SELECT WORKLIST: *Der Tyroler Wastl* (Schikaneder; Vienna 1796).

**Haitink, Bernard** (*b* Amsterdam, 4 Mar. 1929). Dutch conductor. Studied Amsterdam with Hupka, later Leitner. Opera début Holland Fest. 1963 (*Flying Dutchman*). Gly. 1972–3, 1975, music director 1977–8; London, CG, 1977, music director from 1987. Repertory incl. *Entführung*, *Zauberflöte*, *Don Giovanni*, *Don Carlos*, *Lohengrin*, *Ring*, *Rake's Progress*. His sound musicianship and command of structure have, together with his sensitivity towards his singers and players, given him access to a wide range of music. His years at Covent Garden and Glyndebourne have done much to sustain and develop the companies' reputations. (R)

BIBL: S. Mundy, *Bernard Haitink* (London, 1987).

**Haizinger, Anton** (*b* Wilfersdorf, 14 Mar. 1796; *d* Karlsruhe, 31 Dec. 1869). Austrian tenor. Studied Vienna with Mozzati; later with Salieri. Début Vienna, W, 1821 (Gianetto, *La gazza ladra*). Paris, I, 1829; London, CG, 1833; St Petersburg 1835. Repertory included first Adolar (written for him); also Max, Huon, Florestan, Tamino, all opposite Schröder-Devrient. Contributed greatly to the success of Weber's operas. Later established a singing school at Karlsruhe, and published a handbook. He was much praised for his intelligence and musicianship; P. A. Wolff admired 'his sympathetic voice, his moving delivery, his excellent technique'.

**Halévy** (orig. Levy), **Fromental** (*b* Paris, 27 May 1799; *d* Nice, 17 Mar. 1862). French composer. Studied Paris with Berton, Méhul, and (most influentially) Cherubini. After some initial rejections and failures, he had his first operatic success in 1828 with *Clari* (written for Malibran), crowning this with *Le dilettante d'Avignon*. However, his real fame dates from 1835, when his masterpiece *La Juive* (his first operatic collaboration with Scribe) was produced at the Opéra. One of the key works of French *Grand Opera, it anticipates Meyerbeer's *Les Huguenots* but has a wholly individual breadth and grandeur. Berlioz praised the orchestration; Wagner admired in it 'the pathos of high lyric tragedy' and Halévy's capacity for reconstructing the atmosphere of antiquity without recourse to many circumstantial details. Halévy went on to confirm his position in Paris musical life with his opéra-comique *L'éclair*. He followed these with the powerful *Guido et Ginevra*, then *La reine de Chypre*, a work to which Wagner devoted laudatory articles: it contains suggestions of his own Tarnhelm and of the *Tristan* love potion, and is also pre-Wagnerian in its chromatic harmony and in its freedom of movement between numbers. Halévy continued with another Grand Opera, *Charles VI*, before returning to opéra-comique with *Le lazzarone*; with a sequence of works mostly in this genre, none of them matching his previous successes, he concluded his career.

Despite Berlioz's view that he was better suited to light opera, it is as one of the most important composers of Grand Opera that Halévy is remembered: especially with *La Juive*, he confirmed the pioneering work of Rossini and in particular Auber in giving the genre a European importance. He was able to work inventively within its limitations—which included a reliance on large, somewhat static choruses and grandiose effects at the expense of characterization—not least because these matched his own; but the best of his scores include originalities of harmony and orchestration that held much for his successors. His daughter Geneviève married Bizet, who completed his late opera *Noé*. He taught at the Conservatoire from 1827, his pupils including Gounod, Massé, and Bizet. His brother **Léon** (*b* Paris, 12 Feb. 1802; *d* Saint-Germain-en-Laye, 2 Sep. 1883) was an author and dramatist who helped him with revisions of librettos, and whose son **Ludovic** (*b* Paris, 31 Dec. 1833; *d* Paris, 7 May 1908) collaborated with Meilhac in writing librettos for Offenbach, Bizet, and Delibes, sometimes as 'Jules Servières' or 'Paul D'Arcy'.

SELECT WORKLIST (all first prod. Paris): *Clari* (Giannone; 1828); *La Juive* (The Jewess) (Scribe; 1835); *L'éclair* (Saint-Georges and Planard; 1835); *Guido et Ginevra* (Scribe; 1838); *Le guitarrero* (The Guitar Player) (Scribe; 1841); *La reine de Chypre* (The Queen of Cyprus) (Saint-Georges; 1841); *Charles VI* (C. & G. Delavigne; 1843); *Le lazzarone* (The Vagabond) (Saint-Georges; 1844); *La dame de pique* (The Queen of Spades) (Scribe; 1850).

WRITINGS: *Souvenirs et portraits* (Paris, 1861); *Derniers souvenirs et portraits* (Paris, 1863).

BIBL: L. Halévy, *F. Halévy: sa vie et ses œuvres* (Paris, 1862, 2/1863).

**Halka** (Helen). Opera in 2 (later 4) acts by Moniuszko; text by Włodzimierz Wolski, after K. W. Wójcicki's story *Góralka*. 2-act version concert perf. Vilnius 1 Jan. 1848, prod. Vilnius 28 Feb. 1854, with Rivoli, Dobrski, Quattrini, Ziołkowski, cond. Quattrini; 4-act version prod. Warsaw 1 Jan. 1858, with Rostowska, Nowakowski, Zamecka, Kleczyński, cond. Moniuszko; New York, People's T, June 1903; London, University Coll., 8 Feb. 1961, with Muir, Kehoe, Leftwich, cond. Addison.

Cracow and Tatras, late 18th cent. As the wealthy nobleman Janusz (bar) is celebrating his betrothal to Zofia (mez), the peasant girl Halka (sop) arrives, determined to see Janusz, who swore her undying fidelity, and whose child she is carrying. Janusz sends her back to her village, where the peasants are up in arms when they hear how Janusz has treated her. Unable to bear her sorrow, Halka throws herself into the river. News of her death puts an end to the wedding celebrations and the couple are driven from the village by the peasants.

**Hall,** (Sir) **Peter** (*b* Bury St Edmunds, 22 Nov. 1930). English producer. Studied Cambridge, where he was director Arts T 1955. Managing director Royal Shakespeare T 1960–9. London, CG: *Moses and Aaron* 1965, *Magic Flute* 1966, *Knot Garden* 1970, *Eugene Onegin* 1971, *Tristan* 1971. Appointed co-director London, CG, with Colin Davis 1969, but withdrew before taking up post. Gly. 1970–90, artistic director 1984: 15 prods. incl. *Calisto* 1970, *Ritorno d'Ulisse* 1972, *Fidelio* 1979, *Orfeo* 1982, *Poppea* 1984, *Carmen* 1985, Mozart and Verdi series; resigned over lack of consultation. New York, M: *Macbeth* 1982, *Carmen* 1986. Bayreuth 1983, *Ring*: his return to sensual and naturalistic staging, leaving interpretative options open, was not generally liked. Los Angeles 1986, *Salome* (with his then wife Maria *Ewing; also CG 1988). Houston 1989, *New Year*. His productions are notable for their keen sense of theatre and for a freshness of approach that is always musical, never wilful or doctrinaire; at his finest when working with major dramatic singers whom he admires, including Janet Baker and Hildegard Behrens.

**Halle.** City in Saxony, Germany. The first stage performances date from 1616. Duke August, who was an enthusiast for German opera, built the Comödienhaus 1654, opening with *Die Hochzeit des Thetis* by the music director, Philipp Stolle (1614–75), who remained until 1660. The city became one of the earliest operatic centres in Germany, and composers including David Pohle (1624–95), Johann Philipp Krieger (1649–1725), and Johann Beer (1655–1700) composed Singspiels there, of which only Stolle's *Charimunda* (1658) survives. When Duke August died in 1680, the court moved to Weissenfels. Handel was born Halle 1685, and his oratorios were often performed from the early 19th cent. Otherwise, Singspiel was staple fare, and Halle remained dependent on touring companies for its introduction to Romantic opera. The T des Friedens opened 1886; bombed 1945; reopened 1951 (cap. 1,035). Since 1952 there has been an important annual Handel Festival.

**Hallé, (Sir) Charles** (*b* Hagen, 11 Apr. 1819; *d* Manchester, 25 Oct. 1895). German, later English, pianist and conductor. Though most famous for the foundation of the orchestra which bears his name, he also conducted opera seasons in Manchester 1854–5, and at London, HM, 1860–1.

**Hallström, Ivar** (*b* Stockholm, 5 June 1826; *d* Stockholm, 11 Apr. 1901). Swedish composer. While studying law at Uppsala, he wrote *Hvita frun på Drottningholm* with Prince Gustav of Sweden. He is best known for his operas, among which the most popular was *Den bergtagna*: this was widely performed, and its light, easy use of folk melodies won the composer a following abroad. Of his other works, one of the most successful was the operetta *Den förtrollade katten*. Répétiteur Stockholm Royal O, 1881–5.

SELECT WORKLIST (all first prod. Stockholm): *Hvita frun på Drottningholm* (The White Lady of Drottningholm) (Prince Gustav; 1847); *Den förtrollade katten* (The Enchanted Cat) (Hedberg; 1869); *Den bergtagna* (The Girl bewitched by a Mountain Spirit) (Hedberg; 1874).

**Hamburg.** City-state in North Germany. The first independent opera-house in Germany was the T auf dem Gänsemarkt, opened against church opposition in 1678 with *Adam und Eva* by Johann Theile (1646–1724). J. W. *Franck was Kapellmeister 1682–6, and the opera gained in artistic standing under Johann *Kusser from 1693. Under Reinhard *Keiser, who was in

Hamburg from 1695, director 1703–6, and wrote over 100 stage pieces, the city developed a vigorous operatic life. Johann *Mattheson was active in various roles from the early 1690s to 1705, drawing G. F. *Handel to the city (and fighting a famous duel with him). Handel's first operas, *Almira* and *Nero*, were premièred 1705. G. P. *Telemann was in the city from 1721, and a number of his early operas were premièred here. From 1738 the city was, like others in Germany, dependent on visiting companies with an Italian repertory. The opera-house closed 1765. Under the drama director Friedrich Schröder, Mozart and Gluck were heard from 1771.

The T am Dammtor opened 1827 with Spohr's *Jessonda*. In 1851 the opera became the Grosse Oper. Bernhard Pohl (Pollini) was director 1874–97, raising standards and making the city an important Wagner centre: also given were the German premières of *Otello* (1888) and *Eugene Onegin* (1892, under Mahler, in Tchaikovsky's approving presence). The Stadttheater opened 1874, and conductors have included Bülow (1888–91), Mahler (1891–7), Klemperer (1910–12), Weingartner (1912–14), Pollak (1917–31), Böhm (1931–4), Jochum (1934-44), Ludwig (1951–71), and Dohnányi (1975–84). The theatre became the Staatsoper 1934, was bombed 1943, rebuilt (cap. 1,674) and reopened 1955. Günther Rennert was director 1946–56, Heinz Tietjen 1956–9, Rolf Liebermann 1959–72 (the latter developing a particularly adventurous policy towards modern opera), August Everding (1973–7). Premières have included Henze's *Prinz von Homburg* (1960), Goehr's *Arden muss sterben* (1967), Searle's *Hamlet* (1968), and Penderecki's *Devils of Loudun* (1969). A fire in 1975 destroyed much of the theatre's scenery and costumes. Liebermann returned 1985–8 when Dohnányi left. Gerd Albrecht became music director 1989.

**Hamilton, Iain** (*b* Glasgow, 6 June 1922). Scottish composer. Studied London with Alwyn. An early interest in opera has led to a number of large-scale works in which he is not afraid to tackle major dramatic subjects and present them with music that scrupulously serves the issues, if not always also musically enhancing them. He relinquished his earlier serial techniques, and has latterly worked in a tonal idiom; his operatic music also shows great care in writing for voices as part of his attention to the enunciation of the drama.

SELECT WORKLIST (all texts by Hamilton): *The Royal Hunt of the Sun* (after Schaffer; comp. 1967–9, prod. London 1977); *Pharsalia* (after Lucan; Edinburgh 1969); *The Catiline Conspiracy* (after Jonson; Stirling 1974); *Tamburlaine* (BBC 1977); *Anna Karenina* (after Tolstoy; London 1981).

**Hamlet.** Opera in 5 acts by Thomas; text by Barbier and Carré, after Shakespeare's tragedy (1600–1). Prod. Paris, O, 9 Mar. 1868, with C. Nilsson, Gueymard, Collin, Faure, Belval, Castelmary, cond. Hainl; London, CG, 19 June 1869, with C. Nilsson, Sinico, Carsi, Santley, Bagagiolo, cond. Arditi; New York, AM, 22 Mar. 1872, with C. Nilsson, Cary, Brignoli, Barre, James, Coletti.

The opera keeps closely to Shakespeare's tragedy. When Hamlet, Prince of Denmark (bar), learns from the ghost of his father (bs) that he was murdered by his own brother in order to usurp the throne, Hamlet vows revenge. He pretends to be mad, thus driving his beloved Ophélie (sop) to commit suicide. The usurping King poisons Hamlet's drink, but the Queen (sop) drinks it instead and dies. He then arranges a duel between Hamlet and Ophélie's brother, Laerte (ten); both are mortally wounded, but Hamlet manages to stab the King as he dies.

For other operas on the subject see SHAKE-SPEARE. Scarlatti's *Amleto* (1715) is probably derived from Shakespeare's original source-material; only one aria survives.

**Hammerstein, Oscar** (*b* Stettin, 8 May 1846; *d* New York, 1 Aug. 1919). German, later US, impresario. Emigrated to USA and made a fortune from inventing a cigar-making machine. Wrote plays and built a number of theatres. Turned to opera 1906 when he built the *Manhattan OH, establishing a brilliant company and an interesting repertory which threatened to rival the New York, M. Built Philadelphia OH 1908. In 1910 the Metropolitan bought his interests, stipulating that he should not produce opera in the USA for ten years. Built London OH (Stoll T) 1911, which failed to rival Covent Garden and collapsed after two seasons. Built Lexington OH in New York 1913, but was legally restrained from staging opera. His grandson **Oscar Hammerstein II** (*b* 12 July 1895; *d* 23 Aug. 1960) was a librettist and songwriter who collaborated with Friml (*Rose Marie*, 1924), Romberg (*The Desert Song*, 1926), Kern (*Show Boat*, 1927), and especially with Rodgers (*Oklahoma!*, 1943; *Carousel*, 1945; *South Pacific*, 1949; *The King and I*, 1951).

BIBL: V. Sheean, *Oscar Hammerstein I* (New York, 1956); D. Taylor, *Some Enchanted Evenings: the Story of Rodgers and Hammerstein* (New York, 1953).

**Hammond, (Dame) Joan** (*b* Christchurch, 24 May 1912). New Zealand soprano. Studied Sydney Cons., Vienna, London with Borgioli. Début Sydney, J.C. Williamson Grand Opera Co., 1929 (Siebel). Vienna, S, 1938; London, CR, 1942–5; CG from 1948. Also New York, Russia, Lisbon. Retired 1965. Roles included Verdi's Desde-mona, Butterfly, Tatyana, Mlada, Salome, Rusalka. Possessed a strong, clear voice and direct style.

**Hammond-Stroud, Derek** (*b* London, 10 Jan. 1929). English baritone. Studied London at TCM and with Gerhardt, Munich with Hüsch. Début London, Impresario Soc., 1957 (Publius, *Clemenza di Tito*). London: SW/ENO, from 1962; CG 1971. Gly. 1973; New York, M, 1977. Roles include Doctor Bartolo (Mozart), Albe-rich, Faninal, Beckmesser. An outstanding character actor with excellent enunciation. (R)

**Handel, George Frideric** (orig. Georg Friedrich Händel) (*b* Halle, 23 Feb. 1685; *d* London, 14 Apr. 1759). Naturalized English composer of German birth. Handel showed musical talent from an early age, but his father felt that he should pursue a more profitable career in the law. Following the intervention of the Duke of Saxe-Weissenfels, he took lessons with the organist of the Liebfrauenkirche in Halle, Friedrich Zachau. In addition to learning harpsichord, organ, oboe, and violin, he studied composition, and was also given access to Zachau's large collection of scores. In 1697 he became assistant organist of the cathedral, and first organist in 1702, the same year that he entered Halle University to read law. But with his father now dead he was able after only one year to end his law studies, resign his organist's position, and find employment at the Hamburg Opera.

At the time the Opera was directed by Keiser, who had a seminal influence on Handel's early development. As a violinist in Keiser's orchestra, he gained a useful insight into the operatic world, which was further enhanced when he was promoted to harpsichordist. In 1705 his own first operatic attempt, *Almira*, was performed at Hamburg, closely followed by *Nero*. The next year two new operas were produced, *Florindo* and *Doro*: these were originally two halves of the same large work, which was split to make it more manageable for performance. Of these, only *Almira* has survived complete: its mixture of a German libretto with occasional Italian arias is typical of the contemporary Hamburg repertory and Keiser's influence is readily discernible. More important, however, is Handel's obvious indebtedness to the Italian style, which he had first encountered while studying Zachau's scores. His sympathetic response had been fuelled by a trip to the Berlin court in 1698, when he came into contact with Ariosti and Bononcini. The visit of Prince Ferdinando de' Medici, son of the Grand Duke of Tuscany, to Hamburg in 1705 provided further encouragement for Handel to increase his knowledge of Italian music and the following year he left Hamburg and travelled southwards.

Having stopped briefly in Florence, Handel arrived in Rome towards the end of 1706, where he soon came to public attention by giving a concert on the organ at St John Lateran. Although opera had long been banned in Rome by the Church, Handel found an outlet for his creative talents in religious music, in which he was supported by the beneficence of the city's most influential cardinals. His main patron, however, was the Marchese Francesco Maria Ruspoli, for whom he wrote a number of cantatas during the period 1707–9, as well as the oratorio *La resurrezione*. Among the composers he met at this time were Corelli, who led the orchestra in *La resurrezione*, and Domenico and Alessandro Scarlatti, while the musicians of the Ruspoli retinue made such an impression that Handel later called some of them to London, including the famous soprano Margarita Durastanti.

Handel's activities were not restricted to Rome. In the autumn of 1707 his opera *Vincer se stesso è la maggior vittoria*—usually, but quite erroneously, referred to as *Rodrigo*, after the name of the principal character—was given in Florence: little of the score survives, but the work clearly drew on *Almira* and seems to have made little impact. A visit to Naples in 1708 saw the production of the serenata *Aci, Galatea e Polifemo*, while during a stay in Venice in the winter of 1709–10 Handel produced the crowning success of his Italian years, the opera *Agrippina*. Though put together hurriedly and drawing on much earlier music, it included some numbers from *Rodrigo* and is the earliest real evidence of Handel's operatic talent. According to Handel's first biographer, John Mainwaring, the audience received *Agrippina* with cries of 'Viva il caro Sassone', while the composer himself enjoyed the enthusiastic attentions of one of the leading sopranos of the day, Vittoria Tarquini ('La Bombace').

His apprenticeship over, Handel left Italy in Feb. 1710 to find a permanent appointment. His recent success and the sponsorship of the Crown Prince of Tuscany led to his being employed by the Elector of Hanover, who readily granted him an immediate year's leave of absence to visit England. By late 1710 he had arrived in London, where he already had influential friends and where the coincidental appearance of the some of his music in a popular pasticcio at the Haymarket assured him of a warm welcome. At this time the vogue was for opera seria and Handel, anxious to capitalize upon his reputation, set to work immediately. *Rinaldo*, apparently composed in just 14 days, adheres to all the main conventions, yet its invigorating music manages to inject new life into its stereotyped gestures. With its enormous orchestra and elaborate staging it was a gift to the satirists, none more so than Joseph Addison in the *Spectator*, but it served to bring Handel to the forefront of public attention. No matter that much of its score had been shamelessly lifted from earlier works: *Rinaldo* gave new impetus to the cultivation of Italian opera in England and such was its success that it was soon taken up in Ireland as well.

Handel resented having to return to Hanover in 1711, but in 1712 managed to obtain another leave of absence, on condition that he return within a reasonable time. Shortly after reaching London, his new opera *Il pastor fido* was performed in the Haymarket, with Elisabeth Pilotti in the leading female role. Pilotti also performed in Handel's third London opera, *Teseo*: like *Il pastor fido*, neither this work, nor its successor *Silla* (performed privately) achieved *Rinaldo*'s success. However, Handel's contacts amongst the nobility and his reputation ensured that he did not suffer personally: indeed, in 1713 Queen Anne took the unprecedented step of granting him a royal pension of £200 for life. Her death the following year placed Handel in an embarrassing position, for she was succeeded by his patron, the Elector of Hanover, who became King George I. Handel, having ignored the Elector's request that he should stay for only a reasonable while, was in disgrace. Whether reconciliation was effected through the composition of the *Water Music*, as some contemporary accounts claim, is uncertain: what is beyond doubt is that once Handel's patron moved to England, there was no longer any reason for him to leave the country which had so demonstrably taken him to its heart. For the next 20 years his life centred around the London stage and, whatever the fate of individual works, his position at the forefront of operatic activities was always assured. Unlike many composers, his approach to opera did not develop in an evolutionary way: indeed, his ability to revert to the practices of earlier works, often coupled with the literal reuse of existing musical material, as in *Agrippina* and *Rinaldo*, can confuse even the most perceptive and diligent observer. Handel's approach to opera seria was shaped most of all by the need to retain the interest of a fickle audience in a harsh commercial environment and to this end he demonstrated unsurpassed imagination and ingenuity.

*Amadigi di Gaula*, while unsuccessful on first performance in 1715, was well received at its revival in the following year and is particularly notable for its greater use of orchestral resources. But it was some while before Handel built on this work: in 1716 he accompanied King George I to

Hanover and returning succeeded Pepusch as chapel-master to the Earl of Carnarvon (later to become the Duke of Chandos), a post he held until 1720. As well as the famous *Chandos Anthems*, Handel reworked the Italian serenata *Aci, Galatea e Polifemo* into the English masque *Acis and Galatea* around this time, and also composed his first oratorio, *Esther*.

The new decade saw his involvement in a venture which provided some of the greatest successes, as well as some of the most spectacular failures, of his operatic career: the Royal Academy of Music. A group of noblemen, realizing the substantial financial backing necessary for regular operatic performances, as well as the potentially enormous profits which might be gained, formed themselves into a subscriptive body to raise capital. Handel was appointed musical director, Paolo Rolli librettist and secretary, and Johann Jakob Heidegger, whom Handel had already encountered at the Haymarket, manager of the new organization. Their first task was to find good singers, so Handel was sent off to Europe in May 1719. At Dresden, where Hasse was Kapellmeister, he recruited Durastanti, as well as the castrato Francesco Bernardi, known as 'Senesino'. Durastanti appeared to good effect in the Royal Academy's second production, Handel's *Radamisto*: its reception rivalled that of *Agrippina* in Italy, for the first time in several years Handel had scored a truly popular success, and the future of the Royal Academy seemed assured.

Later in 1720 its ranks were swelled by the arrival of the famous castrato Matteo Berselli and the composers Amadei and Bononcini. Almost immediately dissent broke out concerning the relative merits of these composers; the squabble was ingeniously solved by asking them each to set one act of the opera *Muzio Scevola*. Though Handel's was generally held to be the best contribution, the seeds of future disharmony had been sown. While trying to act as music director of the Royal Academy, Handel was forced to maintain a running rivalry with Bononcini who, for a while at least, gained the upper hand. Though he did not have equal the number of performances of Bononcini's works, Handel none the less produced a string of notable successes. *Floridante* was received as enthusiastically as *Radamisto*, while *Ottone* saw the first appearance with the Royal Academy of the famous soprano Francesco Cuzzoni. Such was the fever generated by this production that tickets changed hands at heavily inflated prices, causing contemporary observers to draw comparison with the South Sea Bubble. Only *Flavio* failed to excite the public, although by this time Bononcini's supremacy had finally been toppled: even the arrival of a new composer,

Ariosti, did not prove a serious challenge to Handel.

The next few seasons mark the zenith of his operatic achievements. *Giulio Cesare*, whose cast included Durastanti, Cuzzoni, and Senesino, was a resounding success, as were *Tamerlano* and *Rodelinda*. By now more singers had been attracted to the Royal Academy, notably the castrato Andrea Pacini, the tenor Francesco Borosini, and the soprano Faustina Bordoni. Rivalry between Bordoni and Cuzzoni was inevitable and was fuelled by a press eager to take any opportunity to enliven the Royal Academy's productions. Partisanship led to rowdy scenes at the opera-house, lampoons were published, and a performance of Bononcini's *Astianette* was interrupted by a fight on stage between the two prime donne. Such diversions, although serving to uphold the public interest in opera, camouflaged the true appeal of a string of masterpieces from Handel's pen, including *Scipione*, *Alessandro*, *Admeto*, and *Riccardo Primo*. While the plots of these works inevitably tend towards the conventional and contrived, the scope of Handel's originality and inventiveness, particularly in the development of dramatically engaging aria form, was without parallel at the time.

The fall of the Royal Academy was as quick as its rise. Despite the intense public interest and box office receipts they had generated, singers such as Bordoni and Cuzzoni were a substantial drain on the resources of the organization. These pecuniary problems exacerbated personal tensions within the company: Senesino proved increasingly unreceptive to Handel's management, while the Bordoni–Cuzzoni rivalry had all but destroyed the possibility of artistic co-operation. Most significant, however, was the performance of Gay's *The Beggar's Opera* in January 1728. Almost overnight this first example of ballad opera eclipsed the Italian opera of the Royal Academy, with its contrived plots and complicated arias. *Siroe* and *Tolomeo* were the last two operas which Handel wrote for the Royal Academy and after a revival of *Admeto*, the company closed down in June 1728.

But all was not lost. In 1729 Handel and Heidegger persuaded the members of the Royal Academy to fund a new venture at the King's T. Once again, the first task had been to recruit singers, but a trip by Heidegger to Italy at the end of 1728 had proved fruitless. Handel himself now travelled to Venice and Rome and engaged a number of new performers, including the castrato Antonio Bernacchi and the soprano Anna Strada. The first production of the new venture, *Lotario*, was not successful, so for his next work Handel turned to a comic, anti-heroic libretto. *Partenope* fared no better, and for a while Handel tried to recapture his former public by mounting a series of revivals, including *Tolomeo* and *Scipione*, the latter with Senesino. The limited success of these productions encouraged him to place new works before his public, but once again the venture was ill-judged. *Poro* was given a lukewarm reception, as was *Sosarme*; *Ezio* was a complete disaster and even *Orlando*, considered by many his finest opera, lasted for only a few performances.

By now another threat had arisen in the shape of a rival company, the Opera of the Nobility. With Porpora as its main composer, it engaged some of Handel's former star singers, including Senesino and Cuzzoni, and attracted a large attendance to its performances at Lincoln's Inn Fields. Despite the success of Handel's new *Arianna in Creta* and a well-received revival of *Il pastor fido*, his position had been almost fatally challenged. When his contract with Heidegger and the Haymarket expired in 1734 it was not renewed: Heidegger instead leased the theatre to the rival company, who sought to capitalize upon their success by attracting Hasse to London. He refused to come, not wishing to be drawn into direct conflict with Handel, but instead sent the score of *Artaserse*. This was the vehicle chosen by the Opera of the Nobility for its most spectacular coup to date, the introduction to London of the renowned castrato Farinelli. Handel, who had shifted his activities to Covent Garden, mounted a counter-challenge with the famous French dancer Maria Sallé and a spectacular opera, *Ariodante*, in which she performed to great acclaim. This was followed by *Alcina*, which also featured Sallé's talents.

The operatic battle proved costly to both sides and Handel, whose works had often been performed to almost empty houses, sustained enormous losses. However, a revival of *Esther* in 1732 had already suggested to him that oratorio could provide a means of recapturing his lost opera audiences, and the success of *Deborah* in the following year confirmed this. Although it was to be several years before his complete retirement from the operatic stage, the 1730s saw Handel turn increasingly to the composition of oratorio. Injecting it with the same kind of variety and melodic inventiveness as he had once opera, oratorio, especially in its dramatic choruses, proved a means for Handel to retain his position at the forefront of English musical life.

Although he attempted to keep his opera company afloat with *Atalanta* in 1736, both it and the Opera of the Nobility had entered into a terminal decline. In 1737 two new operas were produced, *Arminio* and *Giustino*, but neither captured the imagination of the public. Despite the growing success of his oratorios, Handel became increasingly depressed and ill. A visit to Aix-La-Chapelle to take the cure rejuvenated him, and he returned to England in late 1737 determined to set his operatic public alight once more. But *Faramondo*, in which the castrato Caffarelli made his London début, was not the success he had expected, while *Serse* was an unmitigated failure. In its light-hearted and entertaining plot it taps a genuinely comic vein which Handel had scarcely exploited before, although this proved no more appealing than the heroic exploits of mythological figures or the machinations of apparitions from the world of the supernatural: *Serse* did, however, provide the aria 'Ombra mai fù', which was to become the composer's single most popular piece, known inescapably today as 'Handel's Largo' (it is in fact based on Bononcini's setting of 1694). His last two operas, *Imeneo* and *Deidamia*, were both spectacular failures and many observers considered that his career was finished. As far as his operatic career was concerned this was, in fact, correct: his works quickly left the repertoire and lay unperformed until the 20th cent. However, the enormous acclaim which greeted the oratorios *Messiah* and *Samson*, produced hard on the heels of this final operatic failure, set Handel on a new path. For nearly 20 years London audiences were enthralled by the power and beauty of his oratorio writing and by the time of his death he was considered a national hero.

Handel occupies a curious position in operatic history. Ostensibly, he belongs to that large body of early and mid-18th-cent. composers whose activities centred around the production of opera seria of an increasingly stereotyped mould. However, his German background, Italian training, and English place of work and domicile impart to him a degree of cosmopolitanism lacking in most of his contemporaries, while his innately greater gifts enabled him to approach his task in a more inventive and creative way. Crucially, he turned the *da capo aria, with its rigid entrance and exit conventions, into a vehicle of great flexibility and expression for his singers, giving them every opportunity to display their vocal talents, while at the same time allowing the opera itself to unfold in a manner which was dramatically satisfactory according to the precepts of the day. At a time when the tendency was for a uniformity of style between operas, so that individual arias could be transposed from one work to another, Handel's ability, and apparent desire, to experiment with new modes of expression and declamation is remarkable. His handling of recitative drew extensively on the rich heritage of Italian opera, reaching an unsurpassed level of expressiveness, while his occasional linking of aria, arioso, and recitative—seen most notably in the

# Handel

Mad Scene from *Orlando*—enabled him to achieve a closer union between the drama and the music. His handling of orchestral resources demonstrated a similar ability to exploit to the full every facet at his disposal. His aim was, of course, to captivate the imagination of the audience and retain their interest for future productions. The choice of librettos, which were usually adapted by either Rolli or Nicola Haym, was determined on the same basis and the oscillation between the heroism of *Alessandro* and *Scipione*, the rustic beauty of *Il pastor fido*, and the magic of *Orlando* was a deliberate ploy to maintain a constant element of variety.

The failure of Handel's operatic career is perhaps as much bound up with politics as with music. Certainly, the rivalry between the Royal Academy of Music which, like Handel personally, enjoyed the support of the King, and the Opera of the Nobility, to whose colours the Prince of Wales was called, involved political partisanship on a extensive scale. In reality, the market for Italian opera had always been small and was merely increased artificially for a short while by the antics of a number of extrovert singers. With the rise of English ballad opera there came a fundamental change of taste, which even Handel was unable to arrest. He alone of those involved in the vigorous operatic activities of the 1720s and 1730s managed to retain a place in the affection of London audiences and that not without difficulty. Seeing himself as essentially a dramatic composer, he at first tried to retrieve his position through the production of pasticcios. While working on his last operas he composed a number of such works, of which *Il Parnasso in festa* was most popular (like most pasticcios, it drew extensively on pre-existing music, in this case *Athalia*, which had not yet been performed as an opera). The eventual means by which he retrieved his former standing, namely oratorio, likewise involved the manipulation of dramatic techniques. Whether Handel ever intended the works to be staged—some of the scores include such instructions—is immaterial in this respect, for it is by the nature of their music that the oratorios derive their real dramatic power.

Between 1754 and 1920 none of Handel's operas was staged, although most of the scores had been made available in the 19th cent. through Chrysander's collected edition. The revivalist movement, which began in Germany shortly after the First World War, has, not surprisingly, been strongest in England: especially influential was the foundation of the Handel Opera Society in 1955 and the annual productions in Abingdon during the 1960s and 1970s under the direction of Alan Kitching. Though some of Handel's works have appeared at major opera-houses, and many

have been recorded, none has yet gained a permanent place in the mainstream repertoire. Not least among the difficulties facing potential performers are the demanding vocal lines, the reassignment of the castrato parts, and the lack of resources to match the opulence of the original productions, on which much of the works' success depends. However, the most serious problem remains the inherent artificiality of opera seria itself, which Handel, for all his ingenuity and talent, was unable to overcome completely. Although he has fared much better in the modern environment than any other 18th-cent. composer of opera seria, it is likely to be many years yet before any single opera of Handel's will achieve the status of a repertory work.

WORKLIST (excluding pasticcios, staged oratorios, and masques; all first prod. London unless otherwise stated): *Almira* (Feustking, after Pancieri; Hamburg 1705, some music lost); *Nero* (Feustking; Hamburg 1705, music lost); *Vincer se stesso è la maggior vittoria (Rodrigo)* (after Silvani; Florence 1707); *Der beglückte Florindo* (Hinsch, after anon.; Hamburg 1708); *Die verwandelte Daphne* (Hinsch, after anon.; Hamburg 1708); *Agrippina* (Grimani; Venice 1709); *Rinaldo* (Rossi, after Hill, after Tasso; 1711, rev. 1712, 1713, 1714, 1717, 1731); *Il pastor fido* (Rossi, after Guarini; 1712, rev. twice 1734); *Teseo* (Haym, after Quinault; 1713); *Silla* (Rossi; 1713); *Amadigi di Gaula* (Haym, after La Motte; 1715, rev. 1716, 1717); *Radamisto* (Haym, after Lalli; 1720, rev. 1720, 1721, 1728); *Muzio Scevola* (Act I only: with Amadei & Bononcini) (Rolli, after Stampiglia; 1721, rev. 1722); *Floridante* (Rolli, after Silvani; 1721, rev. 1722, 1727, 1733); *Ottone, re di Germania* (Haym, after Pallavicino; 1723, rev. 1723, 1726, 1727, 1733); *Flavio* (Haym, after Noris; 1723, rev. 1732); *Giulio Cesare in Egitto* (Haym, after Bussani; 1724, rev. 1725, 1730, 1732); *Tamerlano* (Haym, after Piovene, after Pradon; 1724, rev. 1731); *Rodelinda, regina de' Langobardi* (Haym, after Salvi, after Corneille; 1725, rev. 1725, 1731); *Scipione* (Rolli, after Salvi; 1726, rev. 1730); *Alessandro* (Rolli, after Mauro; 1726, rev. 1727, 1732); *Admeto, re di Tessaglia* (?, after Mauro, after Aureli; 1727); *Riccardo Primo, re d'Inghilterra* (Rolli, after Briani; 1727); *Genserico* (?, after Beregani; some music for Act I only, later used in *Siroe* and *Tolomeo*); *Siroe, re di Persia* (Haym, after Metastasio; 1728); *Tolomeo, re di Egitto* (Haym, after Capece; 1728, rev. 1730); *Lotario* (Salvi; 1729); *Partenope* (?, after Stampiglia; 1730, rev. 1730, 1737); *Poro, re dell'Indie* (?, after Metastasio; 1731, rev. 1731, 1736); *Tito* (?, after Racine; music for Act I only, later used in *Ezio*); *Ezio* (Metastasio; 1732); *Sosarme, re di Media* (?, after Salvi; 1732, rev. 1734); *Orlando* (Capece, after Ariosto; 1733); *Arianna in Creta* (?, after Pariati; 1734, rev. 1734); *Ariodante* (?, after Salvi, after Ariosto; 1735, rev. 1736); *Alcina* (unknown, after Ariosto; 1735, rev. 1736, 1737); *Atalanta* (Valeriani; 1736, rev. 1736); *Arminio* (after Salvi; 1737); *Giustino* (?, after Pariati, after Beregani; 1737); *Berenice* (?, after Salvi; 1737); *Faramondo*

318

(Zeno; 1738); *Serse* (?, after Stampiglia, after Minato; 1738); *Imeneo* (?, after Stampiglia; 1740, rev. 1742); *Deidamia* (Rolli; 1741).

CATALOGUE: B. Baselt, *Händel-Handbuch* (Leipzig, 1978–86).

BIBL: W. Dean, *Handel and the Opera Seria* (Berkeley, CA, 1969); R. Strohm, *Essays on Handel and Italian Opera* (Cambridge, 1985); W. Dean and J. M. Knapp, *Handel's Operas* (Oxford, 1987); E. Harris, ed., *Handel: Opera Librettos* (New York, 1989).

**Handel Opera Society.** The English Handel Opera Society was founded in 1955 at the instigation of E. J. Dent, and began with *Deidamia*, in his translation, under Charles Farncombe in St Pancras TH. Notable subsequent productions have included *Alcina* with Joan Sutherland (1957), *Semele* and *Rodelinda* with Sutherland and Janet Baker (1959—the first annual season at London, SW), *Serse* and *Giulio Cesare* with Sutherland (1963), *Orlando* with Baker (1966). Many other distinguished singers have taken part, sometimes early in their careers, including Heather Harper, Geraint Evans, Richard Lewis, Benjamin Luxon, and Philip Langridge. The society has also performed in Liège, Göttingen, Halle, and Drottningholm.

**Handlung für Musik** (Ger.: 'action in music'). The term employed by Wagner to describe his libretto for *Lohengrin*. Following D'Indy's model it was applied by French Wagnerians, in translation as **action musicale**, to many of their own works, which they saw as standing apart from the mainstream operatic tradition of the 19th cent.

**Hann, Georg** (*b* Vienna, 30 Jan. 1897; *d* 9 Dec. 1950). Austrian bass-baritone. Studied Vienna with Lierhammer. Munich, S, 1927–50. Also London, CG; Paris; Berlin; Milan, S. Created La Roche (*Capriccio*). A distinguished buffo artist, famous for his Kecal, Ochs, Nicolai's Falstaff, Leporello; also sang Pizarro, Sarastro, Amfortas. Possessed a large, attractive voice. (R)

**Hanover** (Ger., Hannover). City in Lower Saxony, Germany. The first opera was probably P. *Cesti's Orontea* in 1649. Opera was given in the Ballhaus to 1678; in 1679 the Schlossflügel, or Kleines Schlosstheater, opened; then the Schlosstheater (cap. 1,200) opened 1689 (modelled on Vicenza) with Agostino *Steffani's Henrico Leone.* He composed seven more operas for Hanover, and was music director until 1698. Handel was Kapellmeister 1710. The city was dependent on visiting companies until 1773, when Friedrich Schroeder gave Singspiels and other works including Benda's *Ariadne*. Grossmann's company, with B.A. Weber as director, gave performances in the 1790s, including Gluck and Mozart. With the establishment of the Kingdom of Hanover in 1815, opera developed, especially under Wilhelm Sutor and the opening of a new theatre 1818. Heinrich *Marschner was appointed music director 1831, and the theatre rebuilt as the Hofoper 1837; five of his operas were premièred here. The new Hoftheater (cap. 1,300) opened 1 Sep. 1852 with Marschner's *Natur und Kunst*; he retired 1859. Bülow was music director 1877–9. The Hoftheater became the Städtische Bühnen 1921; Rudolf Krasselt was music director 1923–43, greatly raising standards. The theatre was bombed in 1943, and reopened as the Landestheater (later Niedersächsiches Staatstheater) (cap. 1,207) on 30 Nov. 1950 with *Rosenkavalier*. Franz Konwitschny was music director 1945–9, Johannes Schüler 1949–60, Günther Wich 1961–5, Christoph Perick from 1993. The theatre's centenary was celebrated in 1952 with the première of Henze's *Boulevard Solitude*, the Schlosstheater's tercentenary in 1989 with *Henrico Leone*.

The first post-war German opera performances were given in a temporary theatre, the Galeriegebäude, in nearby Herrenhausen in 1945. Festival performances are also given in the Baroque Heckentheater here (opened 1693) and in the open air.

**Hänsel und Gretel.** Opera in 3 acts by Humperdinck; text by Adelheid Wette (composer's sister), after the story in the Grimm brothers' *Kinder- und Hausmärchen* (1812–14). Prod. Weimar 23 Dec. 1893, with Kayser, Schubert, Tibelti, Finck, Wiedey, cond. R. Strauss; London, Daly's T, 26 Dec. 1894, with Douste, Elba, Miller, Lennox, Copland, cond. Arditi; New York, Daly's T, 8 Oct. 1895, with Douste, Elba, Meisslinger, Gordon, Brani, Johnston, Bars, cond. Seidl.

Hänsel (mez) and Gretel (sop) are sent by their angry mother (sop) to collect strawberries in the wood. Their father (bar), returning home drunk, is worried for their safety, knowing of a wicked Witch who lures children and eats them; the parents go in search of the children.

Tired out, but afraid to return home because they have eaten all the strawberries, Hänsel and Gretel fall asleep in the woods, aided by the Sandman (sop); they are protected by guardian angels.

On being woken by the Dew Fairy (sop) the following morning, Hänsel and Gretel see a house made entirely of marzipan and sweets. Hungry, they start to eat, and are captured by the wicked Witch (mez). She puts Gretel to work and fattens Hänsel up ready for eating; but Gretel manages to trick the Witch, pushing her head-first into the oven. As the Witch dies, the house collapses and previously imprisoned children re-emerge. The Witch herself is baked into a huge cake, which

# Hans Heiling

Hänsel and Gretel, now reunited with their parents, share with the other children.

Another opera on the tale is by Reichardt (1773).

**Hans Heiling.** Opera in a prologue and 3 acts by Marschner; text by Devrient, after an old folktale. Prod. Berlin, H, 24 May 1833, with Grünbaum, Valentini, Devrient, Bader, cond. Marschner; Oxford 2 Dec. 1953, with Blacker, Rawson, Wasserman, Townend, cond. Westrup.

The Bohemian mountains, 14th cent. A spirit, Hans Heiling (bar), has found a human girl he loves and, ignoring the warnings of his mother (con), decided to forgo the realm of spirits and marry her.

He brings to earth jewels for Anna, his bride (sop). He is, however, perturbed by the warmth of her affection for Konrad (ten).

At evening, Anna wanders distractedly in the forest; she loves Konrad and yet is married to Heiling. Suddenly the Queen of the Spirits, Heiling's mother, appears and beseeches Anna to relinquish him. Anna faints, and is found by Konrad, who carries her home. She rejects Heiling's gifts and reveals that she knows he is a spirit. Irate, Heiling stabs Konrad and flees.

Heiling returns to the spirit world; however, on learning that Konrad is not dead but plans to marry Anna, he resolves to hide among the wedding guests. He grasps Anna's hand; as Konrad hastens to his wife's aid, his hunting knife shatters. When Heiling summons a throng of spirits to annihilate humanity, his mother appears and leads Heiling back to the spirit world.

**Hanslick, Eduard** (*b* Prague, 11 Sep. 1825; *d* Baden, 6 Aug. 1904). German critic. His first work on aesthetics, *Vom Musikalisch-Schönen*, remains his most important book, arguing a case for music as non-representational, and for form expressed in sound as being the true nature of music. This, with some intent, placed him in opposition to Liszt and especially Wagner, whose early works he admired but whose later course he feared and deplored in direct proportion to his deep admiration for Wagner's genius. His stance led to his being pilloried by Wagner in *Meistersinger* as Beckmesser (the second prose sketch of the work names the Marker as Hanslich). They were never reconciled, but Hanslick's posthumous tribute to Wagner was generous and moving. He has passed into history as the epitome of the ultra-conservative critic; but his conservatism was based on deeply felt, thoroughly argued artistic criteria, and it was these that gave his writings, during his long years as a critic for the Vienna *Neue freie Presse* (1855–95), so much weight, and which make his essays still worth reading.

WRITINGS: *Vom Musikalisch-Schönen* (Leipzig, 1854, 16/1966, trans. 1891, R/1974); *Aus meinem Leben* (Berlin, 1894, 4/1911); H. Pleasants, ed., *E. Hanslick: Music Criticisms 1846–99* (New York, 1963).

**Hans Sachs.** See SACHS, HANS.

**Hanswurst** (Ger.: 'John Sausage'). A ubiquitous comic character of the 18th-cent. Viennese popular theatre, introduced into the repertoire by Stranitzky. Elements of his personality can be traced in many later Viennese operatic figures, including notably Papageno in Mozart's *Die Zauberflöte*.

**Hardy, Thomas** (*b* Upper Bockhampton, 2 June 1840; *d* Dorchester, 11 Jan. 1928). English poet and novelist. Operas on his works are as follows:

*Far From the Madding Crowd* (1873): E. Harper (*Fanny Robin*, 1976)
*The Mayor of Casterbridge* (1886): Tranchell (1951)
*The Woodlanders* (1887): Paulus (1985)
*Three Strangers* (1888): Bath (20th cent.); Gardiner (1936)
*Tess of the D'Urbervilles* (1891): D'Erlanger (1906)
*The Queen of Cornwall* (1923): Boughton (1924)

**Harewood, Earl of** (George Henry Hubert Lascelles) (London, 7 Feb. 1923). English critic and administrator. Son of the former Princess Royal and first cousin of Queen Elizabeth II. Founded magazine *Opera* 1950, editor to 1953. Board of directors, London, CG, 1951–3, 1969–72, controller of opera planning 1953–60. Director Edinburgh Festival 1961–5. Managing director, London, SW, later ENO, 1972–85; his years here saw the expansion of the company's range and enterprise. Compiled and wrote revised edition of *Kobbé's Complete Opera Book* (1954, 1987). Has championed the EOG, Opera School, and other native operatic enterprises.

**Harmes, Johann Oswald** (*b* Hamburg, *bapt.* 30 Apr. 1643; *d* ?Brunswick, 1708). German designer. Studied Hamburg and Rome, also observing theatrical work in Venice and Vienna. Chief designer Dresden 1677, then working in other North German cities including Hanover and Hamburg. Essentially a painter, he made, for Kusser, Steffani, Keiser, and Handel, designs of an expressive fluency and realism (often using local and familiar scenes) that set new standards and were influential on 18th-cent. German designers.

**Harmonie der Welt, Die** (The Harmony of the World). Opera in 5 scenes by Hindemith; text by the composer. Prod. Munich, P, 11 Aug. 1957, with Fölser, Töpper, Holm, Metternich, cond. Hindemith.

Set during the Thirty Years War, the opera

concerns the quest of the astronomer Johannes Kepler (bar) and General Wallenstein (ten) for harmony in the universe; each seeks a different form. At the end of the opera, Kepler is transformed into the Earth, and Wallenstein into Jupiter and they, together with other humans transformed into symbols of the zodiac, shine in the heavens. This work is, like *Mathis der Maler*, a study of the artist's or philosopher's relationship to the political and social movements of his times.

**Harper, Heather** (*b* Belfast, 8 May 1930). Irish soprano. Studied London with H. Isepp and Husler. Début Oxford, OUOC, 1954 (Lady Macbeth). Gly. 1957, 1960, 1963. EOG 1956–75; London, CG, from 1962. Bayreuth 1967, 1968; Buenos Aires, C, 1969–72; also New York, M, 1976–8; San Francisco. Roles include Eva, Micaëla, Arabella, Ellen Orford, Governess; created Mrs Coyle (*Owen Wingrave*) and Nadia (*The Ice-Break*). A musicianly singer with a well-controlled voice. (R)

**Harris,** (Sir) **Augustus** (*b* Paris, 1852; *d* Folkestone, 22 June 1896). English impresario. Stage manager and producer Mapleson Co., then manager Drury Lane 1879. Brought Carl Rosa OC to London 1883, managing their seasons until 1887, when he gave an Italian season. This led to him taking Covent Garden, where he was manager 1888–96, achieving great financial and artistic success with a company including Melba and the De Reszkes. He introduced opera in the original language to Covent Garden, and did much to popularize Wagner.

**Harrison, William** (*b* London, 15 June 1813; *d* London, 9 Nov. 1868). English tenor and impresario. Studied London, RAM. Début London, CG, 1839 (in the première of Rooke's *Henrique*). After seasons at Drury Lane and Her Majesty's, creating leading roles in *The Bohemian Girl*, *Maritana*, etc., he formed an English opera company in 1856, with Louisa *Pyne. This performed every autumn and winter at Covent Garden, 1858–64, and gave the first performances of works by Balfe, Wallace, Benedict, and others.

**Harsdörffer, Georg Philipp** (*b* Nuremberg, 1 Nov. 1607; *d* Nuremberg, ?22 Sep. 1658). German poet and librettist. In 1644 he founded the Pegnesischer Blumenorden, a group dedicated to the purification of the German language. Much of his work was influenced by the formative years he spent in Italy; his collection of musical texts *Frauenzimmer Gesprächspiele* contains the libretto for the oldest surviving German musical drama, Staden's *Seelewig* (*c*.1644).

BIBL: P. Keller, *Die Oper Seelewig von Sigmund Theo-phil Staden und Georg Philipp Harsdörffer* (Berne, 1977).

**Harshaw, Margaret** (*b* Narberth, PA, 12 May 1909). US soprano. Studied New York, Juilliard School. Début as mezzo-soprano, New York, M, 1942 (Second Norn). Sang there until 1950 as Amneris, etc., then until 1964 as soprano, establishing herself as a leading Wagnerian. London, CG, 1953–6; Gly. 1954. An earnest singer with a firm, even tone, successful as Brünnhilde, Kundry, and Isolde. (R)

**Hart, Fritz** (*b* London, 11 Feb. 1874; *d* Honolulu, 9 July 1949). English conductor and composer. Studied London with Stanford. Director of Melbourne Conservatory 1915; where Melba taught, and many of his operas were produced. A gifted writer, he provided most of his own texts, favouring Celtic twilight subjects; and his intense practical sense (he had appeared as an actor) helped to make his operas theatrically effective.

**Hartford.** Town in Connecticut, USA. Opera was given by visiting companies until 1941, when the Connecticut O Foundation was founded. Seasons with star casts and minimal productions have drawn audiences from Boston and New York to hear e.g. Stignani as Azucena, Sutherland as Mary Stuart, and Olivero as Adriana Lecouvreur. The Hartt Opera T in the University of Hartford stages rare works in often controversial productions.

**Hartmann, Karl Amadeus** (*b* Munich, 2 Aug. 1905; *d* Munich, 5 Dec. 1963). German composer. Studied Munich with Haas and Scherchen. His *Simplicius Simplicissimus* was composed in 1935 but, owing to his refusal to take part in musical life under the Nazis, not staged until after the war, when its vigorous and effective music won it a following in Germany.

SELECT WORKLIST: *Simplicius Simplicissimus* (Scherchen, Petzet, & Hartmann, after Grimmelshausen; comp. 1934–5, Cologne 1949).

**Hartmann, Rudolf** (*b* Ingolstadt, 11 Oct. 1900; *d* Munich, 26 Aug. 1988). German producer and manager. Studied stage design Munich, Bamberg with Berg-Ehlert. Producer Altenberg 1924; Nuremberg 1928–34, 1946–52; Berlin, S, 1934–8, where he began his collaboration with Clemens Krauss; Munich 1938–44, director, Munich, P, 1952–67. Staged premières of *Friedenstag* (1938) and *Capriccio* (1942), and revivals of most of Strauss's works, also première of *Der Liebe der Danae*, Salzburg 1952. London, CG: *Elektra* 1953, *Arabella* 1965, *Frau ohne Schatten* 1967; new *Ring* 1954. A producer with a strong and effective sense of tradition.

**Harwood, Elizabeth** (*b* Barton Seagrave, 27 May 1938; *d* Fryerning, 22 June 1990). English soprano. Studied RNCM. Début Gly. 1960 (Second Boy, *Zauberflöte*). London: SW from 1961; CG from 1968. Salzburg from 1970; Milan, S, 1972; New York, M, 1975. Roles included Lucia, Zerbinetta, Musetta, Tytania, Fiordiligi. An expressive and elegant singer with a bright tone and a warm stage personality. (R)

**Háry János** (John Háry). Opera in prologue, 5 parts, and epilogue by Kodály; text by B. Paulini and Z. Harsányi, after János Garay's poem. Prod. Budapest 16 Oct. 1926, with Nagy, Palló, cond. Rékai; New York, Juilliard School, 18 Mar. 1960, with Anker, Whitesides, cond. Waldmann; London, Camden TH, 28 Nov. 1967, with Temperley, Olegario, cond. Wales.

**Häser, Charlotte** (*b* Leipzig, 24 June 1784; *d* Rome, 1 May 1871). German soprano, daughter of the composer **Johann Georg Häser** (1729–1809), sister of the composer and singing teacher **August Ferdinand Häser** (1779–1844). Studied with her father, and with Ceccarelli at Dresden. Here she was taken up by Paer, appearing with success at T-Italien, Paris, 1803–6. After engagements in Prague and Vienna, she sang in Italy 1806–12, one of the first German singers to achieve international fame, and hailed as 'la divina tedesca'. After further successes in Germany, where she was much admired by Spohr, she married and retired to Rome. One of the first women to sing men's roles (e.g. Tamino). A contemporary account praised the simplicity and musicianship of her singing, and its absence of unnecessary flourishes.

**Hasse, Johann Adolf** (*b* Bergedorf, nr. Hamburg, 25 Mar. 1699; *d* Venice, 16 Dec. 1783). German composer. Sang as tenor at Hamburg under Keiser (1718) and at the court of Braunschweig-Wolfenbüttel (1719), where his first opera *Antioco* was produced. Sent to Italy, he studied with Alessandro Scarlatti and settled in Naples from 1724 to 1729, where his operas and dialect comedies, including *Sesostrate* and *La sorella amante* enjoyed great popularity, and made him a much sought-after social personality. In 1730 *Artaserse* was staged in Venice, with a cast including Farinelli and Cuzzoni: together with the setting of the same libretto by Vinci, made one month earlier, it proved a turning-point in the history of opera. Metastasio's reform of the opera seria libretto was matched in a musical style which, while appropriate to its stylized conventions, embodied sufficient contrast and expressive content to maintain the interest of the audience, while also providing brilliant opportunities for the display of the singers' talents. The libretto was immediately taken up by many other composers; Hasse's *Artaserse*, like Vinci's, was

performed in houses across Italy, and the composer immediately capitalized upon his success in another work in similar vein, *Dalisa*. For this the prima donna was Cuzzoni's great rival, Faustina Bordoni: one month after the première she married Hasse.

In the months immediately after their wedding the couple toured Europe: Hasse produced several new operas, while Faustina increased her fame in performances both of his and of other composers' work. Meanwhile, they had also attracted the attention of Prince Frederick Augustus II of Saxony, whose father, the Elector Augustus ('August the Strong'), presided at Dresden over a court increasingly Italianate in taste. Although the Elector himself preferred French ballet and theatre, his son was a great admirer of Vivaldi's work and had a burning ambition to establish an Italian opera company in Dresden. To this end he wanted first to recruit a composer of substance, and the success of *Artaserse*, which made Hasse the favoured candidate, fortuitously occurred with a vacancy on the court establishment following the death of the Kapellmeister. Appointed early in 1730, Hasse arrived in Dresden in 1731, when he produced *Cleofide*, a brilliant display piece for Bordoni's talents which had the added bonus of a heroic plot guaranteed to appeal to August the Strong, who still viewed Italian opera with some suspicion. A triumphant success, it secured Hasse's position and on the accession of Frederick Augustus II, an Italian opera company was finally established which Hasse directed.

While the Saxon court now provided the main base for Hasse's activities, he found the intrigues of the Italian musical retinue wearisome and, together with Faustina, took every opportunity to be absent. Most of these trips were connected with the many performances of his works throughout Italy during the 1730s. When the Opera of the Nobility attempted to fuel their conflict with Handel's Royal Academy of Music by securing Hasse's services as director in 1734 they were unsuccessful, but the production of *Artaserse* which was mounted that year in London was rapturously received. After 1740 Hasse spent more time at the Saxon court, whose supremacy in Germany as an operatic centre was rivalled only by that of Berlin. Over the next 20 years he produced at least one opera each year, most to Metastasian librettos, and many smaller dramatic works. Of particular importance was the second *Arminio*, performed to celebrate Frederick the Great's stay in Dresden after his victory at Kesseldorf. Impressed, the Emperor invited Hasse to Berlin in 1753, when several of his works were staged.

Though Hasse was held in high regard, his

Dresden years were not trouble-free. He found it hard to work with younger singers, such as Regina Mingotti, who gradually appeared to challenge Faustina's declining powers, while the appointment of Porpora as Kapellmeister in charge of sacred music in 1748 was far from welcome. However, his eventual departure was caused by political circumstances: following the Siege of Dresden in 1760, in which most of Hasse's possessions, including musical manuscripts, were destroyed, the Elector was forced to disband the opera company. In 1763 the Hasses moved to Vienna, where they had been well received in 1760–2, and where Metastasio was engaged in a bitter struggle with the advocates of operatic reform, Calzabigi and Gluck. Hasse naturally supported Metastasio's cause, which also enjoyed the backing of the conservative court, and the new works he wrote for Vienna were enthusiastically received: most important was *Egeria*, performed to celebrate the coronation of Joseph II. In 1773 he settled in Venice, participating in the active musical life of the city until his death.

Hasse's contribution to the development of opera seria was crucial. Together with Graun he was one of the last composers to work mainly in a court rather than in a commercial context, and it is impossible to separate the spirit of his achievements from the flourishing cultural life of the setting for which most of them were conceived. The broad lines on which his opera seria were designed, the majestic sweep of the music, and the heroic characters around whose actions and emotions the drama unfold, are an ideal reflection of the opulent court for which he worked. However, he was equally well regarded in Italy and it is significant that, unlike his contemporary Graun, he composed Italian opera from the beginning of his career and never showed any inclination towards the older German-Italian hybrid style which flourished in Hamburg, and which Graun imitated in his early works. Dresden, considered by many at this time as the Florence of Germany, was the ideal environment for Hasse, who in all major respects must be considered an 'Italian' composer. In addition to his many Italian triumphs, he was also successfully received at the courts of Munich (1746) and Paris (1750), while through the historic connection of the Saxon Electors with Poland he also had an important influence on the development of opera in Warsaw, where several of his works were performed.

Vinci's death shortly after his *Artaserse* in 1730 has enabled Hasse to gain the major credit for the establishment of the Metastasian *opera seria style. In fact, there are so many similarities of approach between the two composers, with much borrowing of particular devices and techniques, that it is impossible to give one pre-eminence over the other. But by virtue of his longevity and productivity, Hasse's name became inextricably linked with the foundation of the conventionalized opera seria idiom, with its stereotyped aria formulas, fast-moving recitative, and emphasis upon the primacy of the solo singer; and this basic formula, once devised, stood Hasse in good stead and was repeated throughout his operas, though he was not afraid to break the strict alternation of aria and recitative for dramatic effect. Although the most devoted admirer of Metastasio, he did not hesitate to follow contemporary practice and interpolate 'foreign' arias into his works: for example, roughly half of the numbers in *Cleofide* set text not found in Metastasio's libretto but were imported from earlier operas.

Although Hasse's music has yet to undergo any significant revival, a knowledge of it is crucial for an understanding of the development of opera seria. For most Italian and German composers of the 18th cent. he was the point of departure: Gluck's first opera, for example, was a setting of *Artaserse* (1741) in which Hasse's influence was particularly noticeable. Mozart's early opere serie also owe much to Hasse: interestingly, his *Ascanio in Alba* was performed in Milan in 1771 for the wedding of Archduke Ferdinand on the same occasion as the performance of Hasse's last opera, *Ruggiero*. Like Handel, he suffers from having composed in an era whose conventions and aesthetic values, particularly that of the *doctrine of affections, make appreciation of his work difficult for a modern audience. However, in its melodic invention, rhythmic vitality, and expressive treatment of the voice (especially in the accompanied recitative) Hasse's music is almost without parallel in the 18th cent. and is eminently worthy of serious rediscovery.

WORKLIST (excluding smaller dramatic pieces): *Antioco* (Feind, after Zeno & Pariati; Brunswick 1721); *Sesostrate* (Carasale after Zeno & Pariati; Naples 1726); *Astarto* (Zeno & Pariati; Naples 1726); *Gerone, tiranno di Siracusa* (Aureli; Naples 1727); *Attalo, re di Bitinia* (Silvani; Naples 1728); *Tigrane* (Silvani; Naples 1729); *Ulderica* (?; Naples 1729); *Ezio* (Metastasio; Venice 1730, rev. 1755); *Artaserse* (Metastasio; Venice 1730, rev. 1740, 1760); *Dalisa* (Lalli, after Minato; Venice 1730, later known as *La costanza vincitrice*); *Arminio* (Salvi; Milan 1730); *Cleofide* (Boccardi, after Metastasio; Dresden 1731, rev. 1736, later known as *Alessandro nell'Indie*); *Catone in Utica* (Metastasio; Turin 1731); *Demetrio* (Metastasio; Venice 1732, rev. 1740; later known as *Cleonice*); *Cajo Fabrizio* (Zeno; Rome 1732, rev. 1734); *Euristeo* (Lalli, after Zeno; Venice 1732); *Issipile* (Metastasio; Naples 1732, rev. 1742, 1763); *Siroe, re di Persia* (Metastasio; Bologna 1733, rev. 1747, 1763); *Tito Vespasiano* (Metastasio; Pesaro

1735, rev. 1738, 1759; later known as *La clemenza di Tito*); *Senocrita* (Pallavicino; Dresden 1737); *Atalanta* (Pallavicino; Dresden 1737); *Asteria* (Pallavicino; Dresden 1737); *Irene* (Pallavicino; Dresden 1738); *Alfonso* (Pallavicino; Dresden 1738); *Viriate* (Lalli, after Metastasio; Venice 1739); *Numa Pompilio* (Pallavicino; Hubertusburg 1741); *Lucio Papiro* (Zeno; Dresden 1742, rev. 1746, 1766); *L'asio d'amore* (Metastasio; Naples 1742); *Didone abbandonata* (Algorotti, after Metastasio; Hubertusburg 1742); *Antigono* (Metastasio; Hubertusburg 1743 or 1744, rev. 1744); *Ipermestra* (Metastasio; Vienna 1744, rev. 1746, 1751); *Semiramide riconosciuta* (Metastasio; Venice 1744, rev. 1747, 1760); *Arminio* (Pasquini; Dresden 1745, rev. 1753); *La spartana generosa* (Pasquini; Dresden 1747); *Leucippo* (Pasquini; Hubertusburg 1747, rev. 1749, 1751, 1765); *Demofoonte* (Metastasio; Dresden 1748, rev. 1749, 1758); *Il natale di Giove* (Metastasio; Hubertusburg 1749); *Attilio Regolo* (Metastasio; comp. 1740, Dresden 1750); *Ciro riconosciuto* (Metastasio; Dresden 1751); *Adriano in Siria* (Metastasio; Dresden 1752); *Solimano* (Migliavacca; Dresden 1753, rev. 1754); *L'eroe cinese* (Metastasio; Hubertusburg 1753, rev. 1773); *Artemisia* (Migliavacca; Dresden 1754); *Il re pastore* (Metastasio; Hubertusburg 1755, rev. ?1760 or 1762); *L'Olimpiade* (Metastasio; Dresden 1756, rev. 1761, 1764); *Nitteti* (Metastasio; Venice 1758); *Achille in Sciro* (Metastasio; Naples 1759); *Alcide al bivio* (Metastasio; Vienna 1760); *Zenobia* (Metastasio; Vienna ?1761); *Il trionfo di Clelia* (Metastasio; Vienna 1762, rev. 1763); *Egeria* (Metastasio; Vienna 1764); *Romolo ed Ersilia* (Metastasio; Innsbruck 1765); *Partenope* (Metastasio; Vienna 1767); *Piramo e Tisbe* (Coltellini; Vienna 1768, rev. Vienna 1790); *Ruggiero* (Metastasio; Milan 1771).

BIBL: R. Gerber, *Der Operntypus J. A. Hasses und seine textlichen Grundlagen* (Leipzig, 1925); F. Millner, *The Operas of Johann Adolf Hasse* (Ann Arbor, 1979).

**Hauer, Josef** (*b* Wiener Neustadt, 19 Mar. 1883; *d* Vienna, 22 Sep. 1959). Austrian composer and theorist. His theories of atonal music, which preceded and were respected by Schoenberg, found expression in two operas, *Salambo* and *Die schwarze Spinne*. The latter, staged largely as an act of homage, proved to have little theatrical feeling.

WORKLIST: *Salambo* (after Flaubert; part. prod. 1930); *Die schwarze Spinne* (The Black Spider) (Schlesinger, after Gotthelf; comp. 1932, prod. Vienna 1966).

**Hauk, Minnie** (*b* New York, 16 Nov. 1851; *d* Tribschen, 6 Feb. 1929). US soprano. Studied New York with Errani, later with Strakosch. Début Brooklyn 1866 (Amina). London, CG, 1868; Paris 1869; Moscow, St Petersburg 1869–70; Vienna 1870–4; Berlin 1874–7; London, HM, 1878–81. New York, M, 1890–1, then organized her own company, but retired after one season, and went to live in Wagner's villa at Tribschen with her husband. Her repertory encompassed 100 roles, including Juliette, Manon, Selika. A famous Carmen, she was a considerable singer-actress, possessing a forceful character, a penetrating tone, and scrupulous attentiveness to detail.

BIBL: M. Hauk, *Memories of a Singer* (London, 1925).

**Hauptmotiv.** See LEITMOTIV.

**Hauptprobe** (Ger.: 'chief rehearsal'). In German opera-houses, the final rehearsal before the dress rehearsal, or *Generalprobe.

**Hauptstimme** (Ger.). Principal voice or role.

**Häusliche Krieg, Der.** See VERSCHWORENEN, DIE.

**haute-contre.** Term generally accepted as meaning a high tenor (but not a falsettist or a castrato); the voice of most leading male roles in 17th- and 18th-cent. French opera, with a range normally about d–b'. However, there has been confusion about its definition. In early 18th-cent. Italy, it seems to have been synonymous with 'contralto' (a literal equivalent, 'against the high voice') or contratenor; it could also designate an alto castrato voice. In France, it denoted a particular natural male voice that went up to bb' in full (not falsetto) voice, exceptionally as high as d'', while 'contralto' referred specifically to women, and was not a voice type in vogue. The reign of the haute-contre was established in Lully's operas, especially with DuMesny and Boutelou, and reached a peak with Denis-François Tribou and Pierre *Jélyotte, for both of whom Rameau wrote many leading roles. Later, Gluck occasionally wrote for the haute-contre, e.g. Achille (*Iphigénie en Aulide*) and Pylade (*Iphigénie en Tauride*). In the 19th cent. the term came to mean a tenor with a high extension in falsetto (in Italy, a *tenore contraltino*). Rousseau did not regret the decline of the haute-contre, whose timbre he too often found harsh and forced, and Berlioz noted with relief that it was superseded by the deep female contralto, with her more satisfying sound.

**Havana.** Capital of Cuba. The first theatre, the T Coliseo, opened with an opera on Metastasio's *Didone abbandonata*. This was renamed the T Principal in 1803; it was severely damaged by storms in 1844 and 1846. Meanwhile there was opened the T Tacon, modelled on the T Real of Madrid and the T Liceo of Barcelona, and in its day said to be the largest opera-house in the world. It was for long the home of visiting European companies. In 1915 it was reinaugurated as the T Nacional under Serafin. A number of other theatres have housed operetta and zarzuela. A Havana OC, organized by Francesco Marty under Arditi with leading French and Italian singers, appeared in New Orleans 1842, New York 1847–50.

**'Ha! welch' ein Augenblick!'** Pizarro's (bs) aria in Act I of Beethoven's *Fidelio*, in which he determines to murder his prisoner Florestan.

**Hawes, William** (*b* London, 21 June 1785; *d* London, 18 Feb. 1846). English composer and conductor. From 1824 he was director of the London, L, where he introduced Weber's *Der Freischütz*, adapted and with some airs of his own (1824). Apart from staging his own operettas, he adapted many others for the English stage, among them *Così fan tutte*, *Don Giovanni*, and *Der Vampyr*.

**Haydn, Joseph** (*b* Rohrau, 31 Mar. 1732; *d* Vienna, 31 May 1809). Austrian composer. Despite a sustained attempt in recent years to revive interest in Haydn's dramatic music, his operas are still largely accepted as inferior in quality to his instrumental and symphonic music. Yet the composition of opera held a central place in his life: he himself considered the works he wrote for Eszterháza to be among the finest of his œuvre, and much of his time was taken up in supervising the production of opera. Despite the judgement of posterity, Haydn was arguably the most highly regarded opera composer of his day and placed by many above Mozart.

After leaving St Stephen's Cathedral, where he had been a choirboy, Haydn spent some time as an independent musician in Vienna. Around 1752 he made the acquaintance of Joseph von *Kurz, who recruited him to write the music for a comic opera, *Der krumme Teufel*. Though popular, the work is now lost, as are the four following operas. These were all written for performance in Eisenstadt, the country seat of Count Paul Anton Esterházy, whose service Haydn joined in 1761. The following year Paul Anton was succeeded by his brother Nikolaus and under his guidance the Esterházy court became one of the cultural centres of the Austro-Hungarian Empire. The palace at Eszterháza was to provide the setting for almost all of Haydn's operatic activities, and many of his works were first given as part of one of the splendid court entertainments which the prince organized there.

Although the first surviving work for Eisenstadt, *Acide*, was an opera seria, Haydn was predominantly attracted to opera buffa. In 1768 *Lo speziale*, a work featuring *janissary music, opened the new opera-house at Eszterháza and the following year *Le pescatrici* was the highlight of the marriage festivities for the Prince's niece. For the visit of the Empress Maria Theresa in 1773 the Prince mounted his most lavish festival, at which *L'infedeltà delusa* was given a rapturous second performance: the first had been for the name-day of Paul Anton's widow. *L'incontro improvviso* was first performed for a visit of the Archduke Ferdinand, while *Il mondo della luna* was given to celebrate the marriage of the prince's second son.

In the field of opera buffa comparison with Mozart is inevitable, and Haydn's works are undeniably inferior, lacking as they do his symphonic conception. But viewed against the background of his Italian contemporaries they have much to commend them, particularly in their handling of ensembles, in their musical characterization, in their subtle mixture of tragic and comic elements, and in their consummate orchestration. Indeed, such was Haydn's growing reputation in this field that in 1777 he was commissioned to write an opera for the Italian company in Vienna; unfortunately, musical intrigues caused him to withdraw his work and *La vera costanza* was first performed at Eszterháza in 1779 instead.

In addition to composing opera, Haydn was also kept busy writing incidental music for the plays of the wandering theatrical troupes which gravitated towards the Esterházy court, as well as Singspiels for the marionette theatre which opened at Eszterháza in 1773. Works performed there included *Der Götterrath*, *Philemon et Baucis*, *Das abgebrannte Haus*, and *Die bestrafte Rachbegierde*: not surprisingly, the relatively little surviving music for these puppet operas suggests that the emphasis was on catchy melodic writing, rather than expansive or dramatic gesture. In November 1779 the opera-house burned down and performances were shifted to the marionette theatre and to the main hall of the palace. Haydn's only opera seria for Eszterháza, *L'isola disabitata*, was given in these circumstances, but by 1781 the opera-house had been rebuilt and *La fedeltà premiata* was given in celebration. Both this work and its successor, *Orlando paladino*, were hugely successful, the latter being performed throughout Europe. Haydn's last opera for Eszterháza, *Armida*, took as its subject a heroic theme, and treated it in a correspondingly grandiose fashion: it proved hugely popular.

In 1790 Prince Nikolaus died and the theatre closed, bringing to an end a spectacular period in the annals of opera. In addition to his own works, Haydn had also been responsible for conducting over 100 operas by other composers, rearranging the scores and adding arias where necessary, and had presented for his patron a repertoire to rival that of any of the great Italian or German houses: the decade 1780–90 saw over 1,000 operatic performances at Eszterháza. No composer of note failed to appear in the repertoire during Haydn's reign and composers and performers were attracted to Eszterháza from across Europe.

His obligations to the Esterházy court now much reduced, Haydn was free to travel. In 1791 he accepted the invitation of the impresario Salomon to come to London where as part of his contract he was to compose an opera for Sir John Gallini's company. The chosen libretto was a treatment of the Orpheus story, but because it proved impossible to obtain a licence, the work was never performed. Whether the surviving four acts of *L'anima del filosofo* constitute the complete work is uncertain, since Haydn himself talks of a 5-act work: if there is no missing final act, it is the only operatic treatment of the Orpheus legend to end tragically, except for Monteverdi's *Orfeo* (1607). The most striking feature of *L'anima del filosofo* is its debt to Gluck, manifested in the tragic chorus writing and restraint of the vocal lines. One of the last great opere serie of the 18th cent., it is a fitting final testament to Haydn's lifelong involvement with the world of the stage.

Operas on his life are by Hetzel (*La jeunesse de Haydn*) and Suppé (*Joseph Haydn*, 1887).

WORKLIST (all first produced Eszterháza unless otherwise stated): *Der krumme Teufel* (The Crooked Devil) (Kurz-Bernadon, after Le Sage; Vienna 1753, lost, rev. 1758 as *Der neue krumme Teufel*); *La marchesa Nespola* (?; Eisenstadt 1762, lost); *La vedova* (The Widow); (?; Eisenstadt 1762, lost); *Il dottore* (?; Eisenstadt 1762, lost); *Il Sganarello* (?; Eisenstadt 1762); *Acide* (Migliavacca; Eisenstadt 1763); *La canterina* (The Poor Singer) (?; Pressburg 1767); *Lo speziale* (The Pharmacist) (Goldoni; 1768); *Le pescatrici* (The Fishermen) (Goldoni; 1770); *L'infedeltà delusa* (Faithlessness Deceived) (Coltellini; 1773); *L'incontro improvviso* (The Unexpected Encounter) (Friberth, after Dancourt; 1775); *Il mondo della luna* (The World of the Moon) (Goldoni; 1777); *La vera costanza* (True Constancy) (Travaglia, after Puttini; 1779); *L'isola disabitata* (Metastasio; 1779); *La fedeltà premiata* (?, after Lorenzi; 1781); *Orlando paladino* (Porta, after Ariosto; 1782); *Armida* (Durandi, after Tasso; 1784); *L'anima del filosofo* (The Philosopher's Spirit) (*Orfeo ed Euridice*) (Badini; comp. London 1791; prod. Florence 1951).

CATALOGUE: A. Hoboken: *Joseph Haydn: Thematisch-Bibliographisches Werkverzeichnis* (Mainz, 1957–78).

BIBL: H. C. Robbins Landon, 'The Operas of Haydn', in *The New Oxford History of Music*, vii (Oxford, 1973).

**Haydn,** (Johann) **Michael** (*b* Rohrau, *bapt.* 14 Sep. 1737; *d* Salzburg, 10 Aug. 1806). Austrian organist and composer, brother of JOSEPH HAYDN. Studied in Vienna as a choirboy at St Stephen's Cathedral; in 1762 Kapellmeister to Archbishop of Salzburg, where he spent his whole life, also becoming organist of several local churches.

Michael Haydn possessed a sure if unremarkable gift for composition: he is remembered today mainly for his sacred music although his achievements have inevitably been overshadowed by those of his older brother. In addition to an opera, *Andromeda e Perseo*, he composed a quantity of sacred dramatic music, including *Pietas christiana* (1770) and *Abels Tod* (1778), most of which took after the model of Singspiel. In 1767 he contributed the second part to *Die Schuldigkeit des ersten Gebotes*, a collaboration in which Mozart participated.

His wife **Maria Magdalena Lipp** (1745–1827) was a gifted soprano who created Rosina in the première of Mozart's *La finta semplice* and also sang in many of her husband's works.

SELECT WORKLIST: *Die Schuldigkeit des ersten Gebotes* (Wimmer; Salzburg 1767, lost, comp. with Mozart); *Andromeda e Perseo* (Salzburg; 1787).

BIBL: M. H. Schmid, *Mozart und die Salzburger Tradition* (Tutzing, 1976).

**Haym, Nicola Francesco** (*b* Rome, 6 July 1678; *d* London, 11 Aug. 1729). Italian librettist, composer, and cellist of German descent. Employed 1694–1700 by Cardinal Ottoboni in Rome; came to England *c.*1700 and entered the service of the 2nd Duke of Bedford as 'household musician'. Together with Clayton was involved with the earliest attempts to establish Italian opera in London, including the productions of Bononcini's *Camilla* (1706) and *Etearco* (1711) and Scarlatti's *Pirro e Demetrio* (1708), for which he composed additional music. After the success of Handel's *Rinaldo* (1711) Haym perceived that the future of opera in England lay with the newly arrived composer. Entering Handel's circle, Haym provided the libretto for his third London opera, *Teseo*, in 1713 beginning a collaboration which was to last until 1728. In 1722, following Rolli's departure, he became librettist and Italian secretary to the Royal Academy of Music, also providing librettos for Ariosti (e.g. *Vespasiano*, 1724) and Bononcini (e.g. *Calfurnia*, 1724).

Very few of Haym's texts were original: mostly he took pre-existing librettos, many of them by Salvi, and altered or adapted them. While his commercial acumen cannot be faulted, there is little to suggest that he was anything but a moderately gifted librettist; perhaps more significant was his, and Handel's, ability to select the stories which would most engage the attention of their audience. Operas by Handel with librettos by Haym include *Flavio, Giulio Cesare, Radamisto, Rodelinda,* and *Tamerlano*.

**Haymarket.** See KING'S THEATRE.

**head voice.** A method of tone production in the upper register, so called from the sensation experienced by the singer of the voice functioning at the top of the head. The tone is light and clear. See also CHEST VOICE.

**Hector.** In classical mythology, the greatest hero of Troy in the Trojan war, husband of *Andromache. He was slain by Achilles. He appears (ten) in Tippett's *King Priam*, and his ghost (bs) comes to urge Aeneas on to found Rome in Berlioz's *Les Troyens*.

**Heger, Robert** (*b* Strasburg, 19 Aug. 1886; *d* Munich, 14 Jan. 1978). German conductor and composer. Studied Strasburg, Zurich, and Berlin under Schillings. Strasburg 1907; Ulm 1908; Vienna, V, 1911; Nuremberg 1913; Munich 1920; Vienna, S, 1925; Berlin, S, 1933–4, and SO 1944–50; Zoppot (Sopot) regularly; music director Kassel 1935–44. London, CG, 1925–35, and with Munich Co. 1953, conducting first London *Capriccio*; Bayreuth 1964 (*Meistersinger*). A sound, reliable musician who had the confidence of more famous conductors in more prominent positions than he ever achieved. His five operas include *Der Bettler namenlos* (1932). (R)

**Heidegger, Johann Jakob** (*b* Zurich, 19 June 1666; *d* Richmond, 5 Sep. 1749). Swiss impresario. In London at DL and H, he helped to establish Italian opera in England, playing a central role in the foundation of the RAM 1719 and working closely with Handel especially 1729–34; remained in control of the opera until *c*.1745. Let the theatre to the Opera of the Nobility 1734; closed 1737.

**Heilbronn, Marie** (*b* Antwerp, 1851; *d* Nice, 31 Mar. 1886). Belgian soprano. Studied Paris with Duprez, London with Wartel. Début Paris, I (*Marie, Fille du régiment*). Paris, OC, 1878, 1884–6, and O, 1880; Milan, S, and London, CG, 1879. Despite faults, her attractive voice and lively singing brought success as Violetta, Juliette, and Manon, which she created (having sung in Massenet's first opera, *La grand'tante*). When the composer first played her *Manon*, she cried, saying 'It's the story of my own life'.

**Heine, Heinrich** (orig. Harry) (*b* Düsseldorf, 13 Dec. 1797; *d* Paris, 17 Feb. 1856). German poet, playwright, and novelist. A sometime friend of Meyerbeer, and critic of the Augsburg *Allgemeine Zeitung*: he wrote valuable reports on Paris musical life. His *Almansor* (1823) was the basis of an opera partly composed and later destroyed by Debussy. Operas on his works are as follows:
*Almansor* (1823): Behrend (1931)
*William Ratcliff* (1823): Cui (1869); Bavrinecz (1885); Pizzi (1889); Mascagni (1895); Leroux (1906); Dopper (1909); Andreae (1914); Ostendorf (1982)
*Der Schelm von Bergen* (1846): Atterberg (*Fanal*, 1834); Liebeswogen; Gerlach (1903)

*Aus den Memorien des Herrn von Schnabelewopski* (1831): Wagner (*Der fliegende Holländer*, 1843)

**Heinrich.** Heinrich der Vogler (Henry the Fowler) (bs), the King of Saxony in Wagner's *Lohengrin*.

**Heise, Peter** (*b* Copenhagen, 11 Feb. 1830; *d* Tårbæk, 12 Sep. 1879). Danish composer. Studied Copenhagen with Berggreen, Leipzig with Hauptmann. Though chiefly known for his songs, he became increasingly interested in the stage, and his *Drot og marsk* is regarded as one of the most important Danish national operas.

SELECT WORKLIST: *Drot og marsk* (King and Marshal) (Richardt; Copenhagen 1878).

**Heldy, Fanny** (*b* Ath, 29 Feb. 1888; *d* Passy, 13 Dec. 1973). Belgian soprano. Studied Liège. Début Brussels, M, 1910 (Elena, Gunsbourg's *Ivan the Terrible*); sang there until 1912. Paris, OC, 1917–20, and O, 1920–39; London, CG, 1926, 1928. Created roles in Hahn's *Le marchand de Venise*, and Honegger and Ibert's *L'aiglon*, among others. A prominent singer-actress of great character and beauty, she was an outstanding Violetta, Manon, and Concepción. (R)

**Helen.** In classical mythology, Helen of Troy was the most beautiful woman of the ancient world. Her elopement with Paris from her husband Menelaus caused him to lead the Greek armies to besiege Troy. She appears as the heroine (sop) of Offenbach's *La belle Hélène*, (as Elena) (sop) in Boito's *Mefistofele*, and also (mez) in Tippett's *King Priam*.

**Hélène.** The Duchess Hélène (sop), sister of Duke Frederick of Austria and lover of Henri, in Verdi's *Les vêpres siciliennes*.

**Helsinki** (Swed., Helsingfors). Capital of Finland. Opera was given by touring German companies in the early 19th cent. The first permanent opera company was founded in 1873. The Alexander T (a miniature version of St Petersburg, M; cap. 500) opened in 1880 and was used for opera after Finland achieved independence from Russia in 1917. In 1911 Edward Fazer and Aïno Ackté put opera on a more secure footing with their Domestic Opera; in 1914 this became the Suomalainen Ooppera (Finnish O). A new opera-house first planned in the 1970s finally opened Nov. 1993.

**Heming, Percy** (*b* Bristol, 6 Sep. 1883; *d* London, 11 Jan 1956). English baritone. Studied London with King and Blackburn, and Dresden with Grosse. Début London, Beecham OC, 1915 (Paris, *Roméo et Juliette*). Sang with this Co. until 1919, then joined BNOC, 1922. London, SW,

1933–40, 1942. Director London, CG English Co., 1937–9. A highly versatile and impressive artist with a full, mellow tone; roles included Amfortas, Scarpia, Mozart's Dr Bartolo, Macheath. (R)

**Hempel, Frieda** (*b* Leipzig, 26 June 1885; *d* Berlin, 7 Oct. 1955). German, later US soprano. Studied Berlin with Nicklass-Kempner. Début Breslau 1905 (Violetta, Pamina, Rosina); Schwerin, 1905–7; Berlin 1905, 1907–12; New York, M, 1912–19; Chicago 1914, 1920–1. Thereafter sang only in concert, giving Jenny Lind recitals in costume. An exceptionally gifted and refined singer, whose pure, rounded tone, brilliant technique, and dramatic sensitivity enabled her to shine in a wide repertory, including Queen of the Night, Lucia, Euryanthe, and Eva. (R)

**Hemsley, Thomas** (*b* Coalville, 12 Apr. 1927). English baritone. Studied with Manén. Début London, Mermaid T (Purcell's Aeneas). Gly. 1953–71; Zurich 1963–7; Bayreuth 1968–70; London, CG, from 1970. Created Demetrius (*Midsummer Night's Dream*), Mangus (*Knot Garden*), Caesar (Hamilton's *Catiline Conspiracy*). With his excellent enunciation and strong stage presence, a distinguished Beckmesser, Malatesta, and Count Almaviva (Mozart). (R)

**Henderson, Roy** (*b* Edinburgh, 4 July 1899). English baritone. Studied London, RAM. Début London, CG, 1929 (Donner). Almaviva (*Figaro*) in first Gly. season, and regularly until 1939 (Guglielmo, Masetto, Papageno). A sensitive, intelligent artist who also made a considerable reputation as a teacher, incl. of Kathleen Ferrier. (R)

**Hendricks, Barbara** (*b* Stephens, AR, 20 Nov. 1948). US soprano. Studied New York with Tourel. Salzburg 1981; Paris, O, and New York, M, 1982; also Berlin, Vienna, etc. Roles include Susanna, Nannetta, Gilda. Possesses a pure, clear, and expressive voice. (R)

**Henri.** A young Sicilian patriot (ten), lover of Hélène, in Verdi's *Les vêpres siciliennes*.

**Henri VIII.** Opera in 4 acts by Saint-Saëns; text by Détroyat and Silvestre. Prod. Paris, O, 5 Mar. 1883, with Krauss, Richard, Renaud, Lassalle, cond. E. Altes; London, CG, 14 July 1898, with Pacary, Helgon, Renaud, cond. Mancinelli.

England, *c.*1534. The opera concerns the love of the King (bar) for Anne Boleyn (sop), whom he marries despite her love for Gomez (ten), the Spanish ambassador, and in the face of the Pope's disapproval.

Operas on Henry VIII's wives include Donizetti's *Anna Bolena* (1830) and Lillo's *Caterina Howard* (1849).

**Henze, Hans Werner** (*b* Gütersloh, 1 July 1926). German composer. Initially self-taught, then studied Heidelberg with Fortner. His first opera was *Das Wundertheater*; this was followed by the radio opera *Ein Landarzt* and by *Boulevard Solitude*. The latter, a version of the Manon Lescaut story, was an instant success; it makes much use of ballet, in which Henze has shown great interest, and is constructed in separate numbers, as is the radio opera *Das Ende einer Welt*. This first revealed, in the form of a puncturing satire on snobbish arty society, Henze's abiding interest in the problem of the artist's relationship and duty to his fellows.

In 1956 *König Hirsch* revealed Henze as a major operatic talent: it is a very large-scale treatment of a Gozzi fable concerning the conflicting claims on a king's duty, and initially aroused violent enthusiasm and antagonism. Henze's particular gift for a somewhat fantastic vein of lyricism was first fully disclosed here, and this was perhaps brought to flower by his growing love of Italy, where he lived for many years. In emphasizing the poetic timelessness of the hero's predicament in his next opera, *Der Prinz von Homburg*, he further showed his distaste for German military values: these are subdued in his operatic treatment of Kleist's ambivalent soldier-dreamer. The score reveals a new power in absorbing the various influences and manners to which Henze has been dangerously subject by virtue of his sheer skill in mastering all forms of musical language and his extreme technical fluency, while the vein of fantasy is undiminished. *Elegy for Young Lovers* concerns a great poet who feeds his art with the lives of all those around him, callously destroying them in the process if need be; a further treatment of Henze's ideas about the relationship between the artist and society, here portrayed as predator and prey, it is more disciplined in manner without excluding an affecting lyricism. *Der junge Lord* is a tart little satire, by no means without a streak of cruelty, on the *kleinbürgerlich* element in German society which Henze despises.

*The Bassarids* returns to the voluptuous richness of *König Hirsch* without loss of disciplines meanwhile acquired. Instructed by his librettist, W. H. Auden, to 'make his peace with Wagner' before beginning work, it duly shows a stronger, less wary acknowledgement of the German tradition than previously, and includes some fine large-scale music. A number of music-theatre works followed. *We Come to the River* marked Henze's return to opera, after a 10-year absence, and to the theme of the individual in conflict with authority, with a work on a left-wing political subject using his familiar neo-Romantic language. (R)

WORKLIST: *Das Wundertheater* (The Magic Theatre) (Schack, after Cervantes; 'opera for actors' Heidelberg 1949, rev. for singers, Frankfurt 1965); *Boulevard Solitude* (Weil, after Jokisch; Hanover 1952); *Ein Landarzt* (A Country Doctor) (radio opera, after Kafka; Hamburg radio 1951, prod. Frankfurt 1965); *Das Ende einer Welt* (The End of a World) (radio opera, Hildesheimer; Hamburg 1953, prod. Frankfurt 1965); *König Hirsch* (King Stag) (Cramer; Berlin 1956, rev. as *Il re cervo*, Kassel 1963); *Der Prinz von Homburg* (The Prince of Homburg) (Bachmann, after Kleist; Hamburg 1960); *Elegy for Young Lovers* (Auden & Kallman; Schwetzingen 1961); *Der junge Lord* (The Young Lord) (Bachmann, after Hauff; Berlin 1965); *The Bassarids* (Auden & Kallman, after Euripides; Salzburg 1966); *We Come to the River* (Bond; London 1976). *Pollicino* (Leva, after Collodi, Grimm and Perrault; Montepulciano 1980); *The English Cat* (Bond, after Balzac; Schwetzingen 1983); *Das verratene Meer* (Treichel, after Michima; Berlin 1990).

**Herbert, Victor** (*b* Dublin, 1 Feb. 1859; *d* New York, 26 May 1924). US composer, conductor, and cellist of Irish birth. Studied Stuttgart with Cossmann and Seifriz. Married the soprano **Therese Foerster** (1861–1927) in 1886 and went with her to New York, where she sang at the Metropolitan and he played in the orchestra. He is remembered chiefly for his tuneful and highly successful operettas, of which he wrote over 40 in 30 years. Of his two serious operas, *Natoma* is somewhat Wagnerian in manner; neither this nor the lighter 1-act *Madeleine* succeeded in spite of carefully prepared performances.

SELECT WORKLIST: *The Wizard of the Nile* (Smith; Philadelphia 1895); *Cyrano de Bergerac* (Reed & Smith, after Rostand; Montreal 1899); *Babette* (Smith; Washington, DC, 1903); *Babes in Toyland* (MacDonough, after Baum (*The Wizard of Oz*); New York 1903); *Mme Modiste* (Blossom; Trenton 1905); *The Red Mill* (Blossom; New York 1906); *Naughty Marietta* (Young; New York 1910); *Natoma* (Redding; Philadelphia 1911). *Madeleine* (Stewart; New York 1914).

BIBL: E. Waters, *Victor Herbert* (New York, 1955).

**Herincx, Raimund** (*b* London, 23 Aug. 1927). English bass-baritone. Studied Belgium with Van Dyck, Milan with Valli. Début WNO 1950 (Mozart's Figaro). London, SW, 1956–66; subsequently with ENO; CG from 1968. Also Boston, Salzburg. Created Faber (*Knot Garden*), White Abbot (*Taverner*), Segura (Williamson's *Our Man in Havana*). A forceful singer with a powerful voice, successful as Wotan, Escamillo, and Macbeth. (R)

**Herklots, Carl Alexander** (*b* Dulzen, 19 Jan. 1759; *d* Berlin, 23 Mar. 1830). German librettist. In 1790 he moved to Berlin, working as lawyer at the Prussian court. He was also extremely active as a translator, adapting more than 70 French and Italian works, including *Così fan tutte* and

Salieri's *La cifra*. These demonstrated considerable sensitivity for the matching of words to music and did much to expand the repertory of the Berlin stage. His own librettos, which proved very popular among German composers, included *Die Geisterbeschwörung* (Cartellieri, 1792) and *Der Mädchenmarkt* (Kospoth, 1793).

**Her Majesty's Theatre.** The *King's Theatre in the Haymarket, London, was renamed Her Majesty's Royal Italian Opera House on the accession of Queen Victoria in 1837, during *Laporte's management. Two years later, after refusing to re-engage Tamburini in protest against a singer's clique known as *La vieille garde*, he was replaced by *Lumley. Mapleson followed 1862–7 and 1877–87. With the opening of Covent Garden in 1847 came the rivalry of a second Royal Italian O. The theatre burnt down in 1867; reopened 1869, it did not become an opera-house again until 1877. The Lumley and Mapleson managements saw the English premières of operas including later Donizetti, early and middle Verdi, Gounod's *Faust*, *Carmen*, and *Mefistofele*. The Carl Rosa London seasons were given there in 1879, 1880, and 1882. Seidl gave the first London *Ring* in 1882. Galli-Marié sang her original role of Carmen in a French season 1886, and Patti made her sole appearance 1887. The theatre was pulled down 1891, and the present building opened in 1897 under the management of Beerbohm Tree. Beecham gave the English première of *The Wreckers* in 1909, an opéra-comique season in 1910, and the first London *Ariadne* in 1913. The BNOC gave its London seasons there after leaving Covent Garden in 1924. The theatre has rarely housed opera since.

**Hermann.** A young officer (ten), loved by Lisa, in Tchaikovsky's *The Queen of Spades*.

**Herod.** Herod Antipas (4 BC–AD 39), the 'tetrarch' of the Gospels, husband of Herodias; he beheaded John the Baptist. He appears as Hérode (bar) as Massenet's *Hérodiade*, and (ten) in R. Strauss's *Salome*.

**Hérodiade.** Opera in 4 acts by Massenet; text by Paul Millet and 'H. Grémont' (Georges Hartmann), after Flaubert's story (1877). Prod. Brussels, M, 19 Dec. 1881, with Duvivier (as Salomé), Deschamps, Vergnet, Manoury, cond. Dupont; New Orleans, French OH, 13 Feb. 1892, with Caignart, Duvivier (as Hérodiade), Verhèes, Guillemot, Rey, Dulin, Rossi; London, CG, 6 July 1904 (as *Salome*), with Calvé, Lunn, Dalmorès, Renaud, Plançon, cond. Lohse.

Jerusalem, AD 30. Hérodiade (mez), married to Hérode (bar), abandons her daughter Salomé (sop), who loves the prophet Jean (ten). Hérode perceives Jean as a threat and throws him into

# Herodias

prison, encouraged by the discovery of Salomé's love for Jean; Hérode himself desires Salomé. When Salomé vows to die with Jean, he accepts her love for the first time, seeing it to be sincere. Salomé is taken to a banquet at Hérode's palace, and throws herself at his feet, begging forgiveness for Jean. However, he has already been beheaded and Salomé kills herself.

**Herodias.** The wife of Herod Antipas. She appears (mez) in Massenet's *Hérodiade* and Strauss's *Salome*.

**Hérold, Ferdinand** (*b* Paris, 28 Jan. 1791; *d* Paris, 19 Jan. 1833). French composer of Alsatian origin. Studied Paris with his father, François-Joseph Hérold, later with Adam, Catel, and Méhul. Winning the Prix de Rome, he went to Italy and had his first opera, *La gioventù di Enrico Quinto* successfully produced in Naples. Back in Paris, he became maestro al cembalo at the T-Italien. He collaborated with Boieldieu on *Charles de France*, but had his first great success with *Les rosières*, following this up with another success, *La clochette*. Other short, light operas were composed while he continued working at the T-Italien, 1820–7, including another great success, *Marie*; this combined elements of Italian vocal writing and Singspiel within an essentially French framework. In 1827 he moved to the Opéra as répétiteur. Among some ballets and less successful operas, he produced the first of his last two masterpieces, *Zampa*, a work which initially did not make its way chiefly because of the Opéra-Comique being forced to close three times during the year 1831. After an unsuccessful collaboration on *La marquise de Brinvilliers* with eight other composers, he wrote *Le pré aux clercs*, whose triumph he barely lived to see.

With *Zampa* and *Le pré aux clercs* alone, Hérold earns a place as an important minor figure in the history of French opera. There are suggestions in his music of Rossini and (especially in *Zampa*) of Weber, not least in the increased importance given to the orchestra he would also have observed in Méhul; Wagner's admiration of the work is reflected in some details in *The Ring*. *Le pré aux clercs* shows Hérold moving into the arena of Grand Opera with a work that in some respects anticipates Meyerbeer's *Les Huguenots*. But he was essentially a follower of Boieldieu, with an individual vein of melody that attempts greater depths of Romantic feeling without forfeiting lightness of manner. His early death was a tragedy for French opera: he himself remarked in his last illness, 'I was just beginning to understand the stage.'

SELECT WORKLIST (all first prod. Paris, OC, except the first): *La gioventù di Enrico Quinto* (The Youth of Henry V) (Landriani, after Hérold, after Duval;

Naples 1815); *Charles de France* (Rance, Théaulon, & Artois; 1816—Act 1 Boieldieu, Act 2 Hérold); *Les rosières* (Théaulon; 1817); *La clochette* (The Little Bell) (Théaulon; 1817); *Marie* (Planard; 1826); *Zampa* (Mélesville; 1831); *La marquise de Brinvilliers* (Scribe & Castil-Blaze, 1831; comp. with Auber, Batton, Berton, Blangini, Boieldieu, Carafa, Cherubini, & Paer); *Le pré aux clercs* (Planard, after Mérimée; 1832); *Ludovic* (Saint-Georges; 1833; completed Halévy).

BIBL: B. Jouvin, *Hérold: sa vie et ses œuvres* (Paris, 1868); A. Pougin, *Hérold* (Paris, 1906).

**Herrenhausen.** See HANOVER.

**Hertz, Alfred** (*b* Frankfurt, 15 July 1872; *d* San Francisco, 17 Apr. 1942). German, later US, conductor. Studied Frankfurt. Début Halle 1891; Altenburg 1892–5; Barmen-Elberfeld 1895–9; Breslau 1899–1902. Moved to USA; New York, M, 1902 as cond. of German opera until 1915, including the first US *Parsifal* (before the expiry of the Bayreuth copyright; he never cond. in Germany again). Cond. premières of *Königskinder* (1910) and W. Damrosch's *Cyrano de Bergerac* (1913) and several other US premières. London, CG, 1910. Latterly turned to concert conducting.

**Hervé** (orig. Ronger, Florimond) (*b* Houdain, 30 June 1825; *d* Paris, 3 Nov. 1892). French composer, singer, and conductor. Studied Paris with Elwart and Auber. Though working as an organist under his baptismal name, he used the name Hervé for his theatrical career, which included the composition of over 100 light operettas. His first success was with *Don Quichotte et Sancho Pança*. At the Odéon and Palais-Royal his experience in many roles, including as composer, conductor, producer and tenor, led to the satire *Les folies dramatiques* and also to the opening of his Folies-Concertantes in 1854 (later renamed Folies-Nouvelles). He continued to write and appear in a stream of light works of which the most famous was *Mam'zelle Nitouche*. He also worked in London, writing *Aladdin the Second* for the Gaiety and acting as music director at the Empire from 1886. At their best, his fluent, skilful pieces have an immediate charm and appeal.

SELECT WORKLIST: *Don Quichotte et Sancho Pança* (?; Paris 1848); *Les folies dramatiques* (Hervé; Paris 1853); *Chilpéric* (Hervé; Paris 1868); *Le petit Faust* (Crémieux & Jaime; Paris 1869); *Aladdin the Second* (Hervé & Thompson; London; 1870); *Mam'zelle Nitouche* (Meilhac, Millaud, & Blum; Paris 1882).

BIBL: L. Schneider, *Les maîtres de l'opérette française: Hervé, Charles Lecocq* (Paris, 1924).

**Herz, Joachim** (*b* Dresden, 14 June 1924). German producer and administrator. Studied Dresden, conducting with Hintze, producing with Arnold, also acting as répétiteur at O School.

Further studies Berlin 1949–51. Producer Dresden-Radebeul Touring O 1951–3; Berlin, K, assistant to Felsenstein 1953–6; Cologne 1956–7. Chief producer Leipzig 1957, director 1959–76: *Ring* 1976. London, C: *Salome* 1975, *Fidelio* 1980, *Parsifal* 1986. Directed film *Fliegender Holländer* 1963 and Danish TV *Serse* 1973. Director Berlin, K, 1976–81: first revival original 2-act *Butterfly* 1978, *Peter Grimes* 1981. WNO: original *Butterfly* 1979, *Forza* 1981. Munich: *Ägyptische Helena* 1981. Dresden: *Freischütz* 1985 (reopening of Semper O), Mayer *Goldene Topf* 1989. Helsinki: *Così* 1989. A disciple of Felsenstein with his own sharp ideas about music theatre, he has written (of *Fidelio*), 'the individual case has a significance that is generally applicable'.

**'Herzeleide'.** Kundry's (sop) narration in Act II of *Parsifal*, in which she tells Parsifal of his mother Herzeleide.

**Heuberger, Richard** (*b* Graz, 18 June 1850; *d* Vienna, 28 Oct. 1914). Austrian conductor, critic, and composer. After a mildly distinguished career as a choral conductor he turned to criticism, writing for Viennese and Munich newspapers and succeeding Hanslick on the *Neue freie Presse* in 1896. Here, as in his later editorship of the *Neue musikalische Presse*, he propagated lively if sometimes irascible views on modern music. He composed four operas, including *Mirjam* and *Barfüssele*, and five operettas. Of these the most famous was *Der Opernball*: although Heuberger's music lacks the inventiveness or immediate appeal of that of his successors Kálmán and Lehár, it was much admired in its day and *Der Opernball* was taken up abroad, notably in the USA, where it was given both in English and in German.

SELECT WORKLIST: *Mirjam* (Ganghofer; Vienna 1894); *Der Opernball* (Leon & Waldberg; Vienna 1898); *Barfüssele* (Leon; Dresden 1905).

**Heure Espagnole, L'** (The Spanish Hour). Opera in 1 act by Ravel; text by Franc-Nohain, after his own comedy. Prod. Paris, OC, 19 May 1911, with Vix, Périer, cond. Ruhlmann; London, CG, 24 July 1919, with Donalda, Maguénat, cond. Pitt; Chicago, Auditorium, 5 Jan. 1920, with Gall, Maguénat, cond. Hasselmans.

Toledo, 18th cent. The clockmaker Torquemada (ten) goes off to attend to the town clocks, leaving a customer, the muleteer Ramiro (bar), in the shop to await his return. Concepción (sop), who is accustomed to receive her lovers in her husband's absence, sets him carrying clocks about. Gonzalve (ten), a poet, enters and serenades her protractedly. On the arrival of a second lover, Don Inigo Gomez (bs), Ramiro is made to carry Gonzalve, hidden in a clock, up to the bed-room. The same happens to Don Inigo, while the first clock is brought down. Annoyed by their ineffectualness, Concepción eventually admiringly orders Ramiro upstairs again—without a clock. Torquemada returns, finds the two lovers inside the clocks 'examining' them, and effects quick sales.

**Hidalgo, Elvira de** (*b* Aragón, 27 Dec. 1892; *d* Milan, 21 Jan. 1980). Spanish soprano. Studied Barcelona with Bordalba, Milan with Vidal. Début Naples, C, 1908 (Rosina). Monte Carlo from 1910; Paris, O, from 1912; Milan, S, 1916; London, CG, 1924; New York, M, 1924–6; also Chicago, Buenos Aires, etc. Taught from 1933; the only teacher of Callas. A vivacious singer with a high, delicate soprano, she excelled as Rosina, Lakmé, Musetta, and Philine. (R)

**Hidalgo, Juan** (*b* Madrid, *c*.1612/16; *d* Madrid, 30 Mar. 1685). Spanish composer. Though he worked for most of his life as a church composer, his enthusiasm for Italian opera led him to incorporate an operatic style into his secular vocal settings and to collaborate with Calderón on the first Spanish opera, *La púrpura de la rosa*, then on the first to survive, *Celos aun del aire matan*. He also wrote the first surviving zarzuela, *Los celos hacen estrellas*.

SELECT WORKLIST: *La púrpura de la rosa* (Calderón; ?Madrid *c*.1660); *Celos aun del aire matan* (Jealousy even of Air Kills) (Calderón; Madrid 1660); *Los celos hacen estrellas* (The Skies have Stars) (Vélez; ?Madrid 1672).

**Hildegard of Bingen** (*b* Bemersheim, 1098; *d* Rupertsberg, 17 Sep. 1179). German abbess, mystic, writer, and composer. Founded two monasteries near Bingen. Famed for her visions, prophecies, and intellectual powers, and known as 'the Sybil of the Rhine', she was revered by the most powerful churchmen and rulers, as well as by the humblest laymen. Among her numerous medical, literary, and musical works is the musical drama *Ordo virtutum* (1158), depicting the fight for Anima (the Soul) between the Devil and 16 Virtues including Caritas (alto range) and Misericordia (soprano range). As well as being one of the first morality plays ever written, it also anticipates vocal categorization not otherwise encountered until the 18th cent.: the voice range for Caritas is a–b' and for Misericordia f'–a''. Opera *The Vision of Hildegard* (Alisdair Nicholson, 1988).

**Hill, Alfred** (*b* Melbourne, 16 Nov. 1870; *d* Sydney, 30 Oct. 1960). Australian composer and conductor. Studied Leipzig with Sitt, Schreck, and Paul. Returning to Australia in 1896, by way of New Zealand, he formed the Australian Opera League, which produced his *Giovanni, the Sculp-*

*tor*. His operas include works on Maori legend as well as on more familiar subjects.

SELECT WORKLIST: *Tapu* (Adams; New Zealand 1902–3); *Giovanni, the Sculptor* (?; Melbourne 1914).

BIBL: J. Thomson, *A Distant Music: The Life and Times of Alfred Hill* (n.d.).

**Hiller, Johann Adam** (*b* Wendisch-Ossig, 25 Dec. 1728; *d* Leipzig, 16 June 1804). German composer and writer. Studied in Dresden. At the request of the impresario Koch he contributed original music to Weisse's versions of the texts *Die verwandelten Weiber* and its sequel *Der lustige Schuster*. Hiller wrote numbers in a simple, popular vocal style, matching those from a previous version by Standfuss which were incorporated in this setting.

In an attempt to create a more operatic genre, after the manner of Hasse and Graun, whom he admired, Hiller approached Schiebeler. Their first collaboration was the 'romantisch-comische Oper' *Lisuart und Dariolette*. Comments by its creators in the *Wöchentliche Nachrichtungen*—one of the earliest musical periodicals, edited and largely written by Hiller—show that their aim was to emulate Italian models. Neither *Lisuart* nor a second collaboration, *Die Muse*, was well received, and Hiller returned to Weisse, following the librettist's preference for an undemanding form of entertainment in a rustic, sentimental mode. The first three of their five subsequent works played a vital role in the establishment of a modern form of Singspiel (although the term used was 'comische Oper'), and were widely imitated. The plots of *Lottchen am Hofe*, *Die Liebe auf dem Lande*, and *Die Jagd* were all drawn from French opéras-comiques, and followed Rousseau's ideology in glorifying the simplicity of country life in preference to the artificiality of the town. The music, too, draws on German songs and elements of opéra-comique, alongside the bel canto lines of opera seria and the chattering arias and ensembles of opera buffa. By his choice of style, Hiller achieved some measure of characterization. He was restricted by the vocal abilities of the actors at his disposal which were, by all reports, limited. Burney described Hiller's music as 'very natural and pleasing . . . and deserving of much better performers than the present Leipsic company can boast'.

Hiller was, more than any other single composer, the founder of *Singspiel. His widespread success throughout Germany (though not in Vienna) was aided by his initial publication of vocal scores of his works in simplified versions, aimed to encourage amateur performance. He believed the improvement of German opera depended on raising the standard of singing among Germans, and to this end he published several pedagogic works, as well as founding a successful song-school in Leipzig. In addition to composing, Hiller promoted many concerts in Leipzig, and was an important critic who wrote on Metastasio, Rousseau, and aesthetics. Autobiography, *Lebensbeschreibung* (Leipzig, 1784).

SELECT WORKLIST (all prod. Leipzig): *Die verwandelte Weiber* (The Transformed Wives), Pt1 of *Der Teufel ist los* (The Devil's Let Loose) (Weisse, after Coffey; 1766); *Die lustige Schuster* (The Jolly Cobblers) Pt2 of *Der Teufel ist los* (Weisse, after Coffey; 1766); *Lisuart und Dariolette* (Schiebeler, after Favart; 1767); *Lottchen am Hofe* (Weisse, after Favart; 1767); *Die Muse* (Schiebeler; 1767); *Die Liebe auf dem Lande* (Love in the Country) (Weisse, after Favart; 1768); *Die Jagd* (The Hunt) (Weisse, after Collé; 1770); *Der Dorfbalbier* (The Village Barber) (Weisse, after Sedaine; 1771); *Der Aerndtekranz* (The Harvest Wreath) (Weisse; 1771).

**Himmel, Friedrich Heinrich** (*b* Treuenbrietzen, 20 Nov. 1765; *d* Berlin, 8 June 1814). German composer. A protégé of Frederick William II of Prussia, he studied in Dresden (with Naumann) and in Italy. His first opera, *La morte di Semiramide*, was given in Naples 1795, following which he succeeded Reichardt as Kapellmeister to the Berlin court. Although he wrote two further Italian opere serie—*Alessandro* for St Petersburg and *Vasco di Gama* for Berlin—his German works brought him to greater prominence. After *Frohsinn und Schwärmerei*, a 1-act Liederspiel, he had his greatest success with *Fanchon, das Leiermädchen*, a Singspiel reworking by Kotzebue of an old French *vaudeville. Such was its simple charm and captivating music that it was one of the most frequently performed stage works of the 19th cent. and was heard throughout Europe. *Die Sylphen* was also well regarded and is seen by some historians as a forerunner to Weber's *Der Freischütz*. Himmel was essentially a conservative composer: his German works show clearly the influence of Hiller and Weisse at a time when the genuine Singspiel tradition had largely been absorbed into the operatic mainstream.

SELECT WORKLIST: *La morte di Semiramide* (Bendetto, after Voltaire; Naples 1795); *Alessandro* (Moretti; St Petersburg 1799); *Vasco di Gama* (Filistri; Berlin 1801); *Frohsinn und Schwärmerei* (Cheerfulness and Sentiment) (Herklots; Berlin 1801); *Fanchon, das Leiermädchen* (Fanchon the Hurdy-Gurdy Girl) (Kotzebue; Berlin 1804); *Die Sylphen* (Robert, after Gozzi's *La donna serpente*; Berlin 1806).

**Hindemith, Paul** (*b* Hanau, 16 Nov. 1895; *d* Frankfurt, 28 Dec. 1963). German composer, conductor, and violist. Studied Frankfurt, violin with Hegner and Rebner, composition with A. Mendelssohn and Sekles. His first three operas were somewhat ephemeral 1-act pieces, though the scandal over the outspokenness of the first two helped to draw attention to his exceptional gifts. The first opera to hold the stage to any degree was *Cardillac*, which treats Hoffmann's tragic subject of a master-jeweller's obsession with his creations, and murder on their behalf, in an ingenious reconciliation between set forms and Expressionist elements. To some extent, the next opera, *Hin und Zurück*, belongs with the earlier group, in that it is a 1-act chamber opera of considerable irreverence, drawing upon jazz. Written for the 1927 Baden-Baden Festival, it reverses the action (though not the music) halfway through its 12 minutes. *Neues vom Tage* is another so-called *Zeitoper in which the intensity of the style explored in *Cardillac* is brought together with the satirical sharpness of *Hin und Zurück*, not always happily. The work's determined modernity involves an aria sung by the heroine in her bath praising the merits of electric heating over gas: this brought an injunction from the local gas company. There was here the turning away from Romanticism and the embracing of Baroque ideals that increasingly marked his art. School works composed at this time included *Lehrstück*, a teaching piece with a political message on the insignificance of the individual written with Brecht, and the children's play, written with children, *Wir bauen eine Stadt*.

The political stance and the effectiveness in reaching new audiences shown by these last pieces contributed to Hindemith falling foul of the Nazis. He had also, in 1932, begun work on his most serious engagement with the theme of the artist's relationship with society, *Mathis der Maler*. Hindemith's own dilemma between accommodation with unacceptable political views, some form of compromise, and isolation from the society he should serve, found apt expression in the story of the painter Matthias Grünewald. The work draws upon old German sources, including folk-song, and is essentially cast in a series of linked set numbers; and in its language it reaffirms the belief in tonality which he held as something of a bastion against serialism. The opera was the source of the material for his *Mathis der Maler* symphony. The success of *Mathis* proved unwelcome to the authorities: in 1934 his music was subjected to a partial boycott (against which he was publicly defended by Furtwängler), and in 1935 he was dismissed from his teaching post at the Berlin Hochschule. These events impelled him to ponder further his views on the position of the artist in society: a later revision of *Cardillac* was made on the basis of rejecting the idea of the artist as above morality.

Hindemith's last full-length opera, *Die Harmonie der Welt*, was based on the life of the Renaissance astronomer Johannes Kepler, taking his work as an example of human intelligence discovering in the laws of planetary motion a 'harmony of the spheres' with direct musical relevance. There is here a further justification of his belief in music deriving from the natural order rather than artificial construction: the stance is once more against serialism. His only other opera, the 1-act *The Long Christmas Dinner*, is also strongly tonal.

WORKLIST: *Mörder, Hoffnung der Frauen* (Murderer, Women's Hope) (Kokoschka; Stuttgart 1921); *Das Nusch-Nuschi* (marionette play, Blei; Stuttgart 1921); *Sancta Susanna* (Stramm; Frankfurt 1922); *Cardillac* (Lion, after Hoffmann; Dresden 1926, rev. Hindemith, Zurich 1952); *Hin und Zurück* (There and Back) (Schiffer; Baden-Baden 1927); *Neues vom Tage* (News of the Day) (Schiffer; Berlin 1929); *Lehrstück* (Teaching Piece) (Brecht; Baden-Baden 1929); *Wir bauen eine Stadt* (We're Building a Town) (Seitz; Berlin 1930); *Mathis der Maler* (Mathis the Painter) (Hindemith; Zurich 1938); *Die Harmonie der Welt* (The Harmony of the World) (Hindemith; Munich 1957); *The Long Christmas Dinner* (Wilder; Mannheim 1961).

CATALOGUE: H. Rösner, *Paul Hindemith* (Frankfurt, 1970).

BIBL: D. Neumayer, *The Music of Paul Hindemith* (Yale, 1986).

**Hines, Jerome** (*b* Hollywood, 8 Nov. 1921). US bass. Studied Los Angeles with Curci. Début San Francisco 1941 (Monterone). New York, M, from 1946; Gly. 1953; Milan, S, and Bayreuth 1958; Moscow, B, 1962. Also Paris, Munich, Vienna, etc. A notable Boris, Wotan, and Philip, among a large repertory, with a smooth and sonorous voice. He celebrated the 50th year of his career in 1991. (R)

**Hin und Zurück** (There and Back). Opera in 1 act by Hindemith; text by Marcellus Schiffer, after an English revue sketch. Prod. (with works by Milhaud, Toch, and Weill) Baden-Baden 17 July 1927, with J. Klemperer, Mergler, Lothar, Giebel, Pechner, cond. Mehlich; Philadelphia, Broad St. T, 22 Apr. 1928, cond. Smallens; London, SW, 14 Feb. 1958, by Opera da Camera of Buenos Aires, with Chevaline, Valori, Feller, cond. Sivieri.

On discovering that she has a lover, Robert (ten) wounds his wife Helen (sop) fatally. Supernatural intervention then reverses the plot: Helen comes back to life and the original conversation takes place in reverse. The opera ends with Robert giving Helen a birthday present. The entire proceedings are witnessed by Helen's aunt, who takes no notice of any of the events around her.

**Hippolyte et Aricie.** Opera in prologue and 5 acts by Rameau; text by Abbé Pellegrin after Euripides and Racine (*Phèdre*). Prod. Paris, O, 1 Oct. 1733 with Antier, Pélissier, Tribou, Chassé de Chinais, Jélyotte (Amour), cond. Francœur; Birmingham, 13 May 1965, with Hickey, Baker, Tear, Shirley-Quirk, cond. Lewis; Boston, 6 Apr. 1966.

The plot concerns the incestuous love of Phèdre (sop), wife of Thésée (bar), for her stepson Hippolyte (ten); he, however, loves Aricie (sop). Thésée, persuaded that Hippolyte is guilty of assaulting his stepmother, banishes his son, who is then killed by a sea-monster while attempting to flee with Aricie. Phèdre kills herself; Hippolyte, exonerated, is resurrected, and reunited with Aricie.

**Hislop, Joseph** (b Edinburgh, 5 Apr. 1884; d Upper Largo, 6 May 1977). Scottish tenor. Studied Stockholm with Bratt. Début Stockholm 1916 (Gounod's Faust). Chicago 1920–1; London, CG, 1920–8; Milan, S, 1923; Paris, OC, 1926. Retired 1937; taught in Stockholm 1936–48, his pupils including J. Björling and B. Nilsson. Rich-voiced and handsome, he was called 'my ideal Rodolfo' by Puccini; also successful in Verdi and the French repertory. (R)

**Hockney, David** (b Bradford, 9 July 1937). English painter and designer. Début London, Royal Court T, 1966 (Ubu roi, Brecht). Gly. 1975 (Rake's Progress), 1978 (Magic Flute); New York, M, 1981 (the triple bills Parade and Stravinsky); also Los Angeles (Tristan), Dallas (Magic Flute).

BIBL: M. Friedman, Hockney Paints the Stage (London, 1983).

**Hoddinott, Alun** (b Bargoed, 11 Aug. 1929). Welsh composer. Studied Cardiff, London with Benjamin. The Beach of Falesá (1974) brought his lyrical, highly chromatic style to bear upon an exotic South Seas plot and included some fine orchestral writing. He followed this with The Magician (1976), What the Old Man Does is Always Right (1977, a Hans Andersen setting combining the talents of Sir Geraint Evans and local Fishguard schoolchildren), The Rajah's Diamond (TV, 1979), and The Trumpet Major (1981), a work of considerable dramatic strength.

**Hoengen, Elisabeth** (b Gevelsberg, 7 Dec. 1906). German mezzo-soprano. Studied Berlin with Weissenborn. Début Wuppertal 1933. Vienna, S, from 1943. London, CG, 1947, 1959–60; New York, M, 1951–2; Bayreuth 1951. An impressive singing actress; roles included Lucretia, Klytemnestra, Ortrud. (R)

**Hofer, Josepha.** See WEBER.

**Hoffman, François-Benoit** (b Nancy, 11 July 1760; d Paris, 25 Apr. 1828). French poet, critic, and librettist. His success with the libretto for Lemoyne's Phèdre (1786) was followed by a series

for Méhul, whose enterprising structures he encouraged in his texts and whose work he vigorously defended. He also wrote Médée for Cherubini.

**Hoffmann, E. T. A.** (Ernst Theodor Amadeus, orig. Wilhelm) (b Königsberg, 24 Jan. 1776; d Berlin, 25 June 1822). German novelist, critic, conductor, and composer. One of the most famous and influential figures of the Romantic movement in literature. Directed the Bamberg T Company 1808, and the Seconda Company in Leipzig and Dresden 1813–14. After a Singspiel, Die Maske, he wrote Scherz, List und Rache for Posen. There followed other number operas with dialogue, though Aurora and Undine have important freely composed finales.

Undine anticipates much in later Romantic opera. It was greatly admired by Weber, whose review is famous for its statement of Romantic ideals, as an example of 'the opera Germans want, a self-contained work of art in which all elements, contributed by the related arts in collaboration, merge into one another and are absorbed in various ways so as to create a new world'. Undine also contains another curious anticipation of Wagner in its final destruction of the lovers in what is referred to as a Liebestod (though the music at this point is quite un-Wagnerian). As a critic Hoffmann took an equivocal position about Romanticism: he had a wary friendship with Weber, whose Freischütz was attacked in some notorious anonymous articles possibly by him; in Berlin in 1821 he translated the rival opera Olympie for Spontini. His writings influenced generations of Romantics, and are still valuable. They include some perceptive critical articles, but much of his most interesting work is in his imaginative writings, especially in the stories contained in the Serapionsbrüder collection. Among these are Ritter Gluck, a study of the nature of inspiration, and Don Juan. This makes a mysterious imaginative connection between the author and Donna Anna, and first suggested Mozart's Don Giovanni as a Romantic work with a strong association between love and death. In Der Dichter und der Komponist there is a discussion on the priorities of words and music in opera, with the hope that they may one day be written by the same man.

SELECT WORKLIST: Die Maske (Hoffmann; Berlin 1799); Scherz, List und Rache (Jest, Deceit, and Revenge) (Hoffmann, after Goethe; Posen ?1801–2, lost); Die lustigen Musikanten (The Merry Musicians) (Brentano; Warsaw 1805); Liebe und Eifersucht (Love and Jealousy) (Hoffmann, after Calderón, trans. Schlegel; comp. 1807, unprod.); Der Trank der Unsterblichkeit (The Draught of Immortality) (Soden; comp. 1808, unprod.); Aurora (Holbein; comp. 1811–12, prod. Bamberg 1933); Undine (Fouqué; Berlin 1816); *Der Liebhaber nach dem Tode* (The Posthumous Lover) (Salice-Contessa; comp. 1818–22, unprod.).

Hoffmann himself appears in operas by Laccetti (*Hoffmann*, 1912), Kosa (*Anselmus diák*, 1944), and Besch (1945), and in Offenbach's setting of three of his stories, *Les *contes d'Hoffmann* (1881): these are *Der Sandmann*, *Geschichte vom verlorenen Spiegelbilde*, and *Rath Krespel*. Operas on his works are as follows:

*Fantasiestücke in Callots Manier* (1814): Malipiero (*I capricci di Callot*, 1942)
*Der goldene Topf* (1815): Braunfels (unfin.); Petersen (1941); Kosa (1945)
*Die Elixiere des Teufels* (1816): Rodwell (1829)
*Nussknacker und Mausekönig* (1816): Szeligowski (*Kratatuk*, 1955)
*Der Sandmann* (1817): Adam (*La poupée de Nuremberg*, 1852); Audran (*La poupée*, 1898)
*Das Majorat* (1817): Weigl (*Die eisene Pforte*, 1823)
*Klein Zaches* (1819): Hausegger (*Zinnober*, 1898)
*Meister Martin, der Küfer, und seine Gesellen* (1819): Weissheimer (1879); Blockx (*Maître Martin*, 1892); Lacombe (1897)
*Die Bergwerke von Falun* (1819): libretto by Wagner for Dessauer (unused, 1842); Holstein (*Der Haideschacht*, 1868)
*Rath Krespel* (1819): Cadaux (*Le violon de Crémone*, no date)
*Das Fräulein von Scuderi* (1819): Stern and Zamarra (*Der Goldschmied von Toledo*, 1919); Hindemith (*Cardillac*, 1926, rev. 1952)
*Die Brautwahl* (1820): Busoni (1912)
*Signor Formica* (1820): Rastrelli (1832); Schütt (1892)
*Die Königsbraut* (1821): Offenbach (*Le roi Carotte*, 1872)
*Prinzessin Brambilla* (1821): Braunfels (1909)

Also: Elsner (*Stary trzpiot i młody mędrzec*, 1805); Harsányi (*Illusion*, 1949).

CATALOGUE: G. Allroggen, *E. T. A. Hoffmanns Kompositionen* (Regensburg, 1970).

BIBL: F. Schnapp, ed., *E. T. A. Hoffmann: Schriften zur Musik* (Munich, 1963, 2/1978); F. Schnapp, *Der Musiker E. T. A. Hoffmann* (Hildesheim, 1981); D. Charlton, ed., *E. T. A. Hoffmann's Musical Writings* (Cambridge, 1989).

**Hoffmann, Grace** (*b* Cleveland, 14 Jan. 1925). US mezzo-soprano. Studied New York with Schorr, Stuttgart with Wetzelsberger. Début with US touring co., 1951 (Lola). Milan, S, 1955; Bayreuth 1957–70; London, CG, 1959; Paris, O, 1962; New York, M, 1958. A successful Kundry, Brangäne, Eboli, and Cassandra. (R)

**Hofkammersänger.** See KAMMERSÄNGER.

**Hofkapellmeister.** See KAPELLMEISTER.

**Hofmann, Peter** (*b* Marienbad, 12 Aug. 1944). German tenor. Began as rock singer. Studied

Karlsruhe with Seiberlich. Début Lübeck 1969 (Tamino). Stuttgart 1975; Paris, O, 1976; Bayreuth from 1976; London, CG, 1979; New York, M, 1980; also Vienna, Chicago, San Francisco, etc. His handsome physique contributed to his success as Siegmund; roles also incl. Lohengrin, Tristan, Loge, Parsifal. (R)

**Hofmannsthal, Hugo von** (*b* Vienna, 1 Feb. 1874; *d* Rodaun, 15 July 1929). Austrian poet, dramatist, and librettist. His first works appeared in print when he was only 16. He met Richard Strauss while on holiday in Paris in 1900, beginning a relationship which was to result in six operas, while many more projects were discussed and eventually discarded. Their first collaboration, *Elektra* (1909), is recognized as one of the seminal works of Expressionism, perhaps betraying Hofmannsthal's early acquaintance with Stefan George. Its successor, *Der Rosenkavalier* (1911), occupied quite different emotional ground, although this text shared the same concern for economy of language and precision of expression. The works which followed drew inspiration from a variety of sources: *Ariadne auf Naxos* (1st version, 1911–12; 2nd version, 1915–16), *Die Frau ohne Schatten* (1914–17), *Die ägyptische Helena* (1924–7; rev. 1933), and *Arabella* (1924–9). Hofmannsthal also drafted a sketch for *Die Liebe der Danae*: after his death this was worked into a libretto by Gregor.

The relationship between Strauss and Hofmannsthal was never easy: their artistic views differed widely, as did their personal taste regarding suitable operatic subject matter. But each recognized the talents of the other, and their collaboration resulted in an equality of librettist and composer not seen since the 18th cent. Their working relationship was conducted by post, resulting in a voluminous correspondence which throws much light on the gestation of each opera and reveals that Strauss played a major role in the shaping of each libretto. This is abundantly demonstrated in *Arabella*, where the action flags noticeably in Acts II and III: this text was left unaltered by Strauss as a tribute to Hofmannsthal, who died after finishing the revision of Act I, which has a tauter dramatic structure.

Hofmannsthal also wrote the libretto for Wellesz's *Alkestis* (1924); his play *Die Hochzeit der Sobeide* was the source of Cherepnin's opera (1933); and his *Das Bergwerk zu Falun* of Wagner-Régeny's opera (1961).

BIBL: F. and A. Strauss, eds., *Briefwechsel Richard Strauss und Hugo von Hofmannsthal* (1926, 2/1955, trans. 1961).

**Hofoper** (Ger.: 'court opera'). The name given in pre-1918 days to the court or royal opera-houses in a number of German and Austrian cities, including Munich and Vienna.

**'Ho-jo-to-ho!'** The cry of Brünnhilde (sop) in Act II, and of the other Valkyries in Act III, of Wagner's *Die Walküre*.

**Holbrooke, Joseph** (*b* Croydon, 5 July 1878; *d* London, 5 Aug. 1958). English composer. Studied London with Corder. His first opera was a light work, *Pierrot and Pierrette*. He then devoted himself to a massive commission to set Lord Howard de Walden's poem *The Cauldron of Annwyn* as a trilogy of operas, *The Children of Don*, *Dylan*, and *Bronwen*. These achieved performance and, in the case of the first work, even some popularity, in days when Wagner's example still seemed a viable one to some composers working in different national traditions. Holbrooke makes use of some superficially Wagnerian characteristics—a national myth, a vast orchestra, symphonic motivic methods—without approaching Wagner's creative power, though there are sections which suggest a feeling for mythic opera.

SELECT WORKLIST: *Pierrot and Pierrette* (Grogan; London 1909); *The Cauldron of Annwyn* (T. E. Ellis, i.e. Lord Howard de Walden): *The Children of Don* (London 1912); *Dylan* (London 1914); *Bronwen* (London 1929) ); *The Enchanter* (?; Chicago 1915).

BIBL: G. Lowe, *Josef Holbrooke* (London, 1920).

**Holland.** In 1634 De Musijk Kamer, a group modelled on the Florentine Camerata, met to discuss musical drama, though no original works seem to have resulted. *De triomfeerende Min* (1678, to celebrate the Peace of Nijmegen), by Carolus Hacquart (*c*.1640–?1701), is often claimed as the first Dutch opera but is really a pastoral play with music. Attempts at Dutch opera were made by Hendrik Anders (1657–1714) and Servaas de Konink (?–1717 or 1718), especially in the latter's *De vrijadje van Cloris en Roosje* (1688, lost). Bartholomeus Ruloffs (1741–1801) was conductor at the rebuilt theatre from 1774, translating foreign works and inserting his own music in opéras-comiques. Slow to develop an independent tradition, Holland was for long dependent on visiting Italian, French, and German companies. The many short-lived Dutch national opera companies were shared between Amsterdam and The Hague. An early attempt at Dutch opera was by J. B. van Bree (1801–57) with *Saffo* (1834). However, despite encouragement for opera, little was composed, though an enlightened welcome was given to much foreign opera. The Wagner-Vereniging (1883) had great success in stimulating interest in its subject. A new impulse came in modern times, especially with the example of Willem Pijper (1894–1947) and his *Halewijn* (1933).

Henk Badings (1907–87) had some success with *De Nachtwacht* (The Night Watch, 1942).

The Holland Festival, founded in 1947, is unique in being a nation-wide festival, centred on Amsterdam and The Hague (latterly, through budget cuts, more exclusively on Amsterdam). It embraces music, drama, and the visual arts as well as other entertainments. Opera has always played an important part, and the programmes have shown a high regard for quality as well as great enterprise. Opera Forum, based on Enschede, tours opera throughout the country.

See also AMSTERDAM.

**Holloway, Robin** (*b* Leamington Spa, 19 Oct. 1934). English composer and critic. Studied with Goehr, also at Oxford and Cambridge. His major operatic work is a sensitive and intelligently constructed setting of Richardson's *Clarissa*, a work, composed 1968–76, which made a powerful impression on its production in London, C, 1990. A thoughtful critic, he has written a perceptive and influential study *Wagner and Debussy* (London, 1979).

**Holm, Richard** (*b* Stuttgart, 3 Dec. 1912; *d* Munich, 20 July 1988). German tenor. Studied Stuttgart with Ritter. Début Kiel 1937. Munich, S, from 1948. Gly. 1950. New York, M, 1952–3. London, CG, 1953, 1958–60, 1964–6. Also Bayreuth, Vienna, Berlin. Created Kent in Reimann's *Lear*; also sang Belmonte, Flamand, David, Aschenbach. (R)

**Holst, Gustav** (*b* Cheltenham, 21 Sep. 1874; *d* London, 25 May 1934). English composer of Swedish origin. His first five operas represent attempts to build upon the slender traditions and imported conventions of 19th-cent. English opera. *Lansdowne Castle*, written for his home town and performed there with piano, is one of many Sullivan-derived operettas of the day, though it includes some enterprising concerted numbers and an individual handling of magic scenes. *The Revoke* (1895), to a text by his friend Fritz Hart, was written and rehearsed at the London RCM. It retains Sullivanesque features (though the text lacks Gilbertian wit); however, it is continuously composed and shows awareness of the more ambitious operas being written by Stanford and Ethel Smyth, not least in the enterprising handling of the orchestra, some use of *Leitmotiv, and a skilful overture. Of some short children's operettas which followed, the only survivor is *The Idea*, a return to Sullivanesque conventions.

Holst had meanwhile been introduced by Hart to Wagner. He had sketched part of a Wagnerian opera, *The Magic Mirror*, in 1896, and now began planning *Sita*. However, he broke off to write *The Youth's Choice*, 'a musical idyll' in one act, to his own text. The latter is as feebly Wagnerian as much of the music, attempting to cover lack of characterization with turgid symbolism in pursuit of a trite moral. Despite many Wagnerian clichés, there is nevertheless a more assured handling of the Leitmotiv technique that Holst deployed with his ever-increasing mastery of the orchestra.

*Sita*, later dismissed by Holst as 'good old Wagnerian bawling', was his first full-scale opera. Its weaknesses include lame efforts at an English equivalent of *Stabreim, and some two-dimensional characterization; but Holst has, although intermittently, built on experience especially in the development of vivid orchestral commentary and an intelligent handling of Leitmotiv.

Only the Hindu subject matter foretells the astonishing achievement of *Savitri*. Set for only three singers and off-stage chorus with a 12-piece orchestra, this is the first significant English chamber opera since Purcell's *Dido and Aeneas*. Not only does the text show an understanding of Hindu ideas far in advance of its time for a Western artist (Holst had by now studied Sanskrit seriously); it anticipates modern chamber opera in its simplicity and directness of means. Turning away from Wagnerian expansiveness, Holst draws on his more natural sense of musical economy, and makes skilful dramatic use of his characteristic mastery of bitonality to express the essence of his brief drama both swiftly and evocatively.

*The Perfect Fool* contrasts, less effectively, a somewhat remote, ethereal manner with an earthy directness that does not exclude the hearty and the bathetic; though, as the popular ballet excerpts reveal, there is some excellent music buried in a cumbersome structure. *At the Boar's Head* collects the Falstaff episodes from Shakespeare's Henry IV plays, and draws on Holst's love of early English dance music to make a technically ingenious and often touching score. *The Tale of the Wandering Scholar* (later retitled *The Wandering Scholar*) is a 1-act chamber opera which returns with success to the comic style sought and only intermittently grasped in the previous two operas. Laconic and deft, it has a certain dryness, but at its best the dryness of good wit.

WORKLIST: *Lansdowne Castle* (Canningham; Cheltenham 1892); *The Revoke* (Hart; comp. 1895, unprod.); *The Idea* (Hart; London c.1896); *The Youth's Choice* (Holst; comp. 1902, unprod.); *Sita* (Holst, after the Ramayana; comp. c.1900–6); *Savitri* (Holst, after the Mahabharata; comp. 1908, prod. London 1916), *The Perfect Fool* (Holst; London 1923); *At the Boar's Head* (Holst, after Shakespeare; Manchester 1925); *The Tale of the Wandering Scholar*

(C. Bax, after Waddell; Liverpool 1934, rev. Britten, 1951, ed. Britten & I. Holst, 1968).

BIBL: I. Holst, *The Music of Gustav Holst* (London, 1951, 2/1968); M. Short, *Gustav Holst* (Oxford, 1990).

**Holý, Ondřej František** (*b* ?, *c*.1747; *d* Breslau, 4 May 1783). Czech composer and conductor. He was a prominent member of several of the most important opera troupes working in Bohemia and Germany (including those of Koch and Wäser), and set a number of Singspiel texts in the manner of Hiller. Like Hiller, he set Weisse's *Die Jagd*, but his most successful opera was *Der Kaufmann von Smyrna* (The Merchant of Smyrna) (Berlin 1773).

**Holzbauer, Ignaz** (*b* Vienna, 17 Sep. 1711; *d* Mannheim, 7 Apr. 1783). Austrian composer. Studied briefly with Fux, and in Venice. Held posts as Kapellmeister in Vienna (1744–6), Stuttgart (1750), and Mannheim (1753), after the success of *Il figlio delle selve*. His most successful work, *Günther von Schwarzburg*, was his only German opera, the remaining 16 stage works being in Italian or French, and including settings of Metastasio's texts *La clemenza di Tito* (1757) and *Adriano in Siria* (1757).

The text of *Günther*, by Anton Klein, takes an historical German theme: the work is significant as the first effort at using a German subject in a full-length German opera, and prompted a re-awakening of interest in a national genre after a lengthy domination by French and Italian opera. Following the manner of Holzbauer's Italian operas, in *Günther* the spoken dialogue of Singspiel is replaced by recitative; several dramatic scenes are built by using accompanied recitative. Holzbauer also deviated from stereotyped forms in his arias, writing in a freer manner which added to the drama of the work. Mozart described Holzbauer's music as 'very beautiful', adding 'the poetry doesn't deserve such music'. Despite the work's immense success, encouraged by its overt patriotism, Holzbauer wrote no further German operas. Mozart planned, at one stage, to write a companion piece to *Günther*, setting Klein's text *König Rudolf von Hapsburg*. However, the project was never realized due to the 'insufficiently patriotic' frame of mind of the Viennese theatre management.

SELECT WORKLIST: *Il figlio delle selve* (The Son of the Woods) (Capece; Schwetzingen 1753); *Don Chisciotte* (?; Schwetzingen 1755); *Nitteti* (Metastasio; Turin 1758); *Alessandro nell'Indie* (Metastasio; Milan 1759); *Günther von Schwarzburg* (Klein; Mannheim 1777).

**Homer, Louise** (*b* Pittsburgh, 30 Apr. 1871; *d* Winter Park, FL, 6 May 1947). US contralto. Studied Paris with Koenig. Début Vichy 1898,

(Léonore, *Favorite*). London, CG, 1899; New York, M, 1900–19, 1927–29; Chicago 1920–5. Created Witch in *Königskinder*, and leading roles in Paderewski's *Manrù* and Converse's *The Pipe of Desire*; also sang Amneris, Gluck's Orfeo, Laura (*Gioconda*). A popular artist with a full-toned voice, she was particularly admired as Dalila, and in Wagner. Wife of the composer Sidney Homer, and aunt of composer Samuel Barber. (R)

BIBL: A. Homer, *Louise Homer and the Golden Age of Opera* (New York, 1974).

**'Home, Sweet Home'.** Originally an aria in Bishop's *Clari, or The Maid of Milan* (called by the composer a 'Sicilian Air'); also occurs in altered form in Donizetti's *Anna Bolena*, which resulted in Bishop's bringing an action for 'piracy and breach of copyright'. It used to be sung by Patti, Melba, and other prima donnas in the Lesson Scene in *Barbiere*; they often accompanied themselves on a piano wheeled on to the stage for the purpose.

**Honegger, Arthur** (*b* Le Havre, 10 Mar. 1892; *d* Paris, 27 Nov. 1955). Swiss composer. Studied Zurich, Paris with Widor. His large output includes a number of works which show an original approach to the stage. His most substantial opera, *Antigone*, is less remarkable than the stage oratorio *Jeanne d'Arc au bûcher*, in which the skilful blend of speech, song, orchestral narration, and other effects makes a powerful dramatic presentation of the Joan story. He also wrote a number of operettas, notably *Les aventures du roi Pausole*.

SELECT WORKLIST: *Antigone* (Cocteau, after Sophocles; Brussels 1927); *Les aventures du roi Pausole* (Willemetz, after Louÿs; Paris 1930); *Jeanne d'Arc au bûcher* (Claudel; Basle 1938); *L'aiglon* (Cain, after Rostand; Monte Carlo 1937; comp. with Ibert); *Charles le téméraire* (Morax; Mézières 1944).

WRITINGS: *Je suis compositeur* (Paris, 1951; trans. 1966).

BIBL: G. Spratt, *The Music of Arthur Honegger* (Cork, 1987).

**Höngen, Elisabeth.** See HOENGEN, ELISABETH.

**Hook, James** (*b* Norwich, ?3 June 1746; *d* Boulogne, 1827). English composer and organist. Taught by Burney among others, he wrote his first dramatic work, a ballad opera, aged only 8. Beginning as a music teacher, he moved to London as an organist, but soon became known as a composer of light, tuneful music for social entertainment. Composer and organist at Marylebone Gardens 1768; Vauxhall Gardens 1714. He composed a prodigious quantity of music, including more than 30 stage works, which were very popular in their day. Many of these, although called 'operas', were essentially plays with interpolated numbers, often based on folk or popular tunes.

His most successful work, the 'musical entertainment' *The Double Disguise* (1784), was to a libretto by his wife.

**Hopf, Hans** (*b* Nuremberg, 2 Aug. 1916; *d* June 1993). German tenor. Studied Munich with Bender, Oslo with Bjärne. Début Augsburg 1936 with Bayrisches Landestheater. Dresden and Oslo, 1942–4. Bayreuth 1951, 1961–6. New York, M, from 1951; London, CG, 1951–3, 1963. Also Milan, S; Buenos Aires; Moscow; etc. A solid singer, particularly successful in Wagner. (R)

**Horn, Charles Edward** (*b* London, 21 June 1786; *d* Boston, 21 Oct. 1849). English composer and singer. Studied with his father, Karl Friedrich Horn, and Bath with Rauzzini. Début London, L, 1809 (in King's *Up All Night*). Contributed to many English operatic ventures as a singer and conductor, and brought Balfe to England in 1823. His voice, though poor, was of extensive range: he could sing both tenor and baritone parts, including Macheath and Caspar. A prolific composer, although his operas are generally no more than plays with inserted songs (e.g. 'Cherry Ripe' in *Paul Pry*). His only full-scale opera, *Ahmed al Ramel*, is lost, as is *Dirce*, apart from one duet. In the USA, 1827–49, he successfully produced several operas, some his own, others adaptions, e.g. *Die Zauberflöte*, *Cenerentola*.

**Horne, Marilyn** (*b* Bradford, PA, 16 Jan. 1934). US mezzo-soprano. Studied with her father, and University of S California with Vennard. Début Los Angeles 1954 (Háta, *Bartered Bride*). Gelsenkirchen 1956–60; London, CG, from 1964; Milan, S, from 1969; New York, M, from 1970. A versatile singer with an opulent voice of enormous compass, and an excellent technique. A notable contributor to the revival of Rossini, Bellini, Donizetti, etc. Repertory includes Rinaldo, Orfeo (Gluck), Rosina, Arsace, Adalgisa, Eboli, Carmen, Marie (*Wozzeck*), Charlotte, Dalila. Also dubbed the voice of Dorothy Dandridge in the 1954 film of *Carmen Jones*. Autobiography, *My Life* (London, 1984). (R)

**Hosenrolle** (Ger.: 'trouser-role'). See TRAVESTI.

**Hotter, Hans** (*b* Offenbach-am-Main, 19 Jan. 1909). German bass-baritone. Studied Munich with Roemer. Début Opava (Troppau) 1929 (Speaker, *Zauberflöte*). Prague 1932–4; Hamburg 1934–7; joined Munich, S, 1937, and subsequently Vienna, S; London, CG, from 1947; New York, M, from 1950; Bayreuth from 1952. Also Milan, S; Paris, O; Chicago; etc. Produced *The Ring* at London, CG, 1961–4. Retired 1972. Created Kommandant (*Friedenstag*) and Olivier (*Capriccio*); also Jupiter in the abortive first production of *Die Liebe der Danae*, 1944. A famous

Pizarro, Count Almaviva (Mozart), Sarastro, Philip, and Amonasro, and the dominating Wagner Heldenbariton of his day, celebrated as Amfortas, Kurwenal, Sachs, and above all Wotan. Possessed a warm, eloquent voice and a powerful stage presence, and excelled in the noble, the tragic, and the contemplative. An artist of penetrating intelligence, he considerably influenced succeeding generations of singers. (R)

BIBL: P. Turing, *Hans Hotter* (London, 1984).

**House of the Dead, From the** (Cz.: *Z mrtvého domu*). Opera in 3 acts by Janáček; text by the composer, after Dostoyevsky's novel based on his prison reminiscences (1862). Prod. Brno 12 Apr. 1930, with Šíma, Žlábková, Fischer, Olšovský, Pelc, Šindler, Pribytkov, cond. Bakala; Edinburgh, by Prague, N, 28 Aug. 1964, with Jedlička, Tattermuschová, Blachut, Kočí, Zídek, Karpísek, Švorc, cond. Gregor; New York, Net TV, 3 Dec. 1969, with Rounseville, Lloyd, Jagel, Reardon, cond. H. Adler.

The opera presents scenes of life in a Siberian prison camp. A political prisoner, Alexander Petrovich Goryanshikov (bar), arrives, and immediately arouses the anger of the Commandant (bar). The convicts are ordered to work; a few remain behind, listening to a story told by Luka (ten).

A year later Goryanshikov is teaching Alyeya (sop) to read; Skeratov (ten) tells of how he shot a rival for a girl's love. In celebration of Easter, the convicts perform two plays, watched by the townspeople. Alyeya is injured in a fight with another prisoner.

Goryanshikov visits Alyeya in the prison hospital; he overhears two convicts recounting their pasts. Luka dies, and one of the story-tellers, Shishkov (bs), recognizes him as his former rival. As Goryanshikov leaves the camp, having been pardoned by the Commandant, the other prisoners release a caged eagle as a symbol of the inability to suppress the human spirit.

**Houston.** City in Texas, USA. Opera was first heard in 1867, but found little support until the New York, M, visited in 1901. The Houston Grand Opera Association was founded in 1955 by Walter Herbert, opening with *Salome*. Growing rapidly, it moved to the new Jesse H. Jones Hall (cap. 3,000) in 1966. New directions came with David Gockley in 1973, and the repertory has included, as well as star-studded occasions, novelties including Bernstein's *A Quiet Place* (1983), Adams's *Nixon in China* (1987), Glass's *Planet 8* (1988), and Tippett's *New Year* (1989), also an operetta series, a Spring Opera Festival of free performances, and the foundation of the touring Texas Opera T and the Houston Opera Studio. The company performs in the Wortham

Theater Complex, opened 1987, which includes two theatres (cap. 2,176 and 1,066).

**Hovhaness, Alan** (*b* Somerville, MA, 8 Mar. 1911). US composer of Armenian and Scottish descent. Studied Boston with Converse. His long list of works includes a number of short chamber operas whose nature reflects his interest in Japanese music and Noh drama.

**Howarth, Elgar** (*b* Cannock, 4 Nov. 1935). English conductor. Studied Manchester. Stockholm 1978, first *Grand Macabre* (Ligeti). Birtwistle premières: London, ENO, 1986 (*Mask of Orpheus*); Opera Factory 1986 (*Yan Tan Tethera*); London, CG, 1991 (*Gawain*). Guest cond. O North 1985–8. A specialist in contemporary music.

**Howell, Gwynne** (*b* Gorseinon, 13 June 1938). Welsh bass. Studied Manchester with Llewellyn, London with O. Kraus. Début London, SW, 1968 (Monterone). London, CG, and Gly., from 1969; ENO from 1986; Paris 1983; New York, M, 1985; also Chicago, San Francisco, Hamburg. Roles incl. Padre Guardiano, Landgrave, Sachs, Pogner, Gurnemanz, Pimen, Colline; created title-role of Maxwell Davies's *Taverner*. (R)

**Hughes, Arwel** (*b* Rhosllanerchrugog, 25 Aug. 1909; *d* Cardiff, 23 Sep. 1988). Welsh composer. Studied with Kitson and Vaughan Williams. Joined BBC in Wales 1935, which he headed 1965–71. His first opera *Menna* was one of the earliest Welsh operas on a national theme and was followed by a comic work, *Serch yw'r doctor*. Hughes also appeared as conductor with the WNO.

SELECT WORKLIST: *Menna* (Griffiths; Cardiff 1953); *Serch yw'r doctor* (Love's the Doctor) (Lewis, after Molière; Cardiff 1960).

**Hugh the Drover.** Opera in 2 acts by Vaughan Williams; text by Harold Child. Prod. London, RCM, 7 July 1924 (dress rehearsal, with De Foras, Benson, Trefor Jones, Leyland White, cond. Waddington); first public perf., London, HM, 14 July 1924, by BNOC, with Mary Lewis, Willis, Tudor Davies, Frederic Collier, cond. Sargent; rev. with added Act II, scene 1, RCM, 15 June 1933, cond. Beecham; Washington, Poli's T, 21 Feb. 1928, with Montana, Tudor Davies, cond. Goossens.

England, early 19th cent. Hugh the Drover (ten) and John the Butcher (bs-bar) are rivals for the hand of Mary (sop). She is in love with Hugh, but her father wishes her to marry the wealthy John. In a boxing match, Hugh wins but is accused by John of being a Napoleonic spy and put in the village stocks. He is released by the Sergeant, who turns out to be an old friend, and Hugh and Mary leave together happily.

**Hugo, Victor** (*b* Besançon, 26 Feb. 1802; *d* Paris, 22 May 1885). French poet, novelist, and playwright. His only attempt at a libretto was his adaptation of *Notre-Dame de Paris* as *Esmeralda* for Bertin (1836); but his vigorous Romantic polemics, both by precept and by example, exercised a great influence on librettists and hence upon Romantic opera. Operas on his works are as follows:

*Marion DeLorme* (1829): Bottesini (1862); Pedrotti (1865); Ponchielli (1885)

*Les orientales* (1829): J. Cohen (*Les Bluets*, 1867)

*Hernani* (1830): Bellini (unfin., sketched 1830); Gabussi (1834); Mazzucato (1843); Verdi (1844); Laudamo (1849); Hirschmann (1908)

*Notre-Dame de Paris* (1831): L. Bertin (1836); Birch-Pfeiffer (1836); Rodwell (*Quasimodo*, 1836); Prévost (*c.*1840); Mazzucato (1838); Dargomyzhsky (1847); Poniatowski (1847); Battista (censored title *Ermelinda*, 1851; rev. as *Esmeralda*, 1857); Lebeau (1857); W. Fry (1864); Wetterham (1866); Campana (1869); Pedrell (*Quasimodo*, 1875); G. Thomas (*Esmeralda*, 1883); Giro (1897); F. Schmidt (1914); Bosch y Humet (*Febo*, early 20th cent.)

*Le roi s'amuse* (1832): Verdi (*Rigoletto*, 1851)

*Lucrèce Borgia* (1833): Donizetti (1833)

*Marie Tudor* (1833): Balfe (*The Armourer of Nantes*, 1836); Schoberlechner (*Rossana*, 1839); Ferrari (*Maria d'Inghilterra*, 1840); Pacini (1843); Bognar (1856); Kashperov (1860); Chiaromonte (1862); Blaramberg (*Maria Burgundskaya*, 1878); Gomes (1879); Wagner-Régeny (*Der Günstling*, 1935)

*Angelo* (1835): Mercadante (*Il giuramento*, 1837); Ponchielli (*La Gioconda*, 1876); Cui (1876); Villate (1880); Bruneau (1928)

*Ruy Blas* (1838): Poniatowski (1843); Besanzoni (1843); G. Rota (1858); W. Glover (1861); Zenger (1868); Marchetti (1869); G. Braga (comp. 1868); F. Franchetti (1868); Pietri (1916). *Les Burgraves* (1843): Salvi (1845); Dobrzyński (1860); Orsini (1881); Podesta (1881); L. Nielsen (1917); Sachs (1926)

*Les misérables* (1862): Duniecki (1864); Bonsignore (1925); Michetti (*Vagabonda*, 1933); after pt1: Ratbert (*L'Italie*); pt2, *Les pauvres gens* after pt2: Gilson (*Zeewik*, 1904); C.-M. Schönberg (musical, 1985; lyrics Boulbil & Kretzmer)

*La grand-mère* (1865): Silver (1930)

*L'homme qui rit* (1869): Ronzi (1894); Enna (*Komedianter*, 1920); Pedrollo (1920)

*Torquemada* (1882): Rota (1943)

*La légende des siècles* (1859–85): Mancinelli (*Isora di Provenza*, 1884)

**Huguenots, Les.** Opera in 5 acts by Meyerbeer; text by Scribe and Deschamps. Prod. Paris, O, 29 Feb. 1836, with Falcon, Dorus-Gras, Nourrit, N. Levasseur; New Orleans, T d'Orléans, 30 Apr. 1839, with Julie Calvé, Bamberger, Heymann, Bailly; London, CG, 20 June 1842, with Stöckl-Heinefetter, Lutzer, Wettlaufer, Breiting, Mellinger, Schwener, Staudigl, cond. Lachner. At New York, M, in the 1890s the performances were known as the *nuits des sept étoiles*, when the cast included Nordica, Melba, Scalchi, Jean and Édouard De Reszke, Maurel, and Plançon.

Touraine, France, August 1572. A Huguenot nobleman, Raoul de Nangis (ten), is in love with a woman he assumes to be the mistress of the Catholic Count de Nevers (bar). Raoul is summoned to a meeting with an undisclosed lady.

In the gardens at Chenonceaux, Valentine (sop), daughter of Count de Saint-Bris (bar), awaits the arrival of Raoul, whom she loves and wishes to marry. Valentine recognizes her as the woman he loves but, believing her to be Nangis's mistress, refuses her hand. The Catholics are furious and a fight breaks out.

Valentine is to marry Nevers, to whom she was originally betrothed; Raoul challenges him to a duel. Valentine overhears a plan to ambush Raoul on his way to the fight and warns him to go well protected; a riot between the two parties ensues.

The Catholics, led by Saint-Bris, resolve to launch a fierce attack against the Huguenots; their plans are overheard by Raoul.

Raoul warns the Huguenots, who are gathered for the marriage of Marguerite de Valois to Henri IV, of the impending attack. In a Huguenot churchyard, Valentine finds Raoul and his servant. She is now free since Nevers has died in combat, and wishes Raoul to adopt her faith and marry her. When he refuses to relinquish his religion, she vows to die with him. The couple survive an attack by Catholic troops and return to Paris, but are shot by her father's men.

The opera was chosen to open the present Covent Garden theatre 15 May 1858.

**Hullah, John** (*b* Worcester, 27 June 1812; *d* London, 21 Feb. 1884). English composer, manager, and teacher. Studied London with Horsley and Crivelli. Came to notice as composer of music to Dickens's opera *The Village Coquettes*; *The Barbers of Bassora* and *The Outpost* were produced at London, CG, in 1837 and 1838. He subsequently turned to teaching and the training of teachers.

SELECT WORKLIST (all first prod. London): *The Village Coquettes* (Dickens; 1836); *The Barbers of Bassora* (Morton; 1837); *The Outpost* (Serle; 1838).

BIBL: [F. Hullah,] *Life of John Hullah* (London, 1886).

**Hummel, Johann Nepomuk** (*b* Pressburg, 14 Nov. 1778; *d* Weimar, 17 Oct. 1837). Austrian pianist, conductor, and composer, son of **Joseph Hummel**, who during the late 1780s conducted

Schikaneder's troupe in Vienna. A child prodigy whose talent was much admired by Mozart, Hummel made extensive concert tours through Europe, including Scotland and England in 1790–2, before returning to Vienna to study with Albrechtsberger, Haydn, and Salieri. Kapellmeister to the Esterházy family at Eisenstadt 1804; after 1811 worked independently in Vienna, before becoming Kapellmeister at Stuttgart in 1816. Took a similar post at Weimar 1819, where he became friendly with Goethe. During the 1833 season he conducted the German OC at London, H. In addition to several ballets he composed four operas, which owe much to Mozart in their dramatic method and musical language. He was also influenced to some extent by French rescue opera, especially in *Die Rückfahr des Kaisers*, while in *Mathilde von Guise* his approach is coloured by anticipations of the Romantic style. In 1836 he wrote a new finale to Auber's *Gustave III*.

SELECT WORKLIST: *Mathilde von Guise* (Mercier-Dupary; Vienna 1810); *Die Rückfahr des Kaisers* (Veith; Vienna 1814).

**Humming Chorus.** The name usually given to the hidden chorus which ends Act II of Puccini's *Madama Butterfly* as Butterfly, Suzuki, and the baby take up their stand at the *shosi* awaiting the return of Pinkerton.

**Humperdinck, Engelbert** (*b* Siegburg, 1 Sep. 1854; *d* Neustrelitz, 27 Sep. 1921). German composer. As a child he learnt the piano and began composing opera and Singspiels; studied Cologne with Hiller, Jensen, and Gernsheim. Meeting Wagner in Italy in 1880, he was asked to help prepare *Parsifal* for Bayreuth; his many tasks included the composition of some music to cover a scene change at the first performance. He also worked as an editor of music, critic, and teacher: he was partly responsible for turning Siegfried Wagner's interests from architecture to music. His own most famous opera (and the first to survive), *Hänsel und Gretel*, began life as song settings before being made into a Singspiel and then into an opera. Ever since its successful première under Richard Strauss, it has remained his most popular work. Its simple nursery tunes are used in a superficially Wagnerian manner, though there is no systematic attempt to develop an association between persons or ideas and a musical motive. A work of unique charm, it has been an enduring success with German children, perhaps still more with nostalgic German parents.

Humperdinck's next work, *Die sieben Geislein*, was also based on a fairy story; and this was followed by his second most popular work, *Königskinder*, which made constructive use of melodrama; it was later revised as an opera.

*Dornröschen*, a third attempt at fairy-tale opera, was less successful. Strauss was again the first conductor of *Die Heirat wider Willen*, a work that looked back to an earlier tradition of pre-Wagnerian Romantic opera. Weak librettos were partly responsible for the failure of two later Spielopern, *Die Marketenderin* and *Gaudeamus*. His son **Wolfram Humperdinck** (*b* Frankfurt, 29 Apr. 1893; *d* 16 Apr. 1985) worked as a producer in Detmold, Leipzig 1933–41, then was Intendant at Kiel 1941–5.

WORKLIST: *Hänsel und Gretel* (Wette, after J. & W. Grimm; Weimar 1893); *Die sieben Geislein* (The Seven Young Kids) (Wette, after J. & W. Grimm; Berlin 1895); *Königskinder* (The Royal Children) (melodrama, Rosmer, i.e. Bernstein-Porges; Munich 1897; rev. as opera, New York 1910); *Dornröschen* (The Sleeping Beauty) (Eberling & Filhès, after Perrault; Frankfurt 1902); *Die Heirat wider Willen* (The Forced Marriage) (H. Humperdinck, after Dumas; Berlin 1905); *Die Marketenderin* (The Camp Follower) (Misch; Cologne 1914); *Gaudeamus* (Misch; Darmstadt 1919).

BIBL: W. Humperdinck, *Engelbert Humperdinck* (Frankfurt, 1965).

**Hunding.** A Neiding (bs), Sieglinde's husband, in Wagner's *Die Walküre*. His name derives from Old Norse *hundr* and Middle High German *Hunt* (dog).

**Hungary.** The Turkish occupation from 1541 to 1686 inhibited the development of national music, and Hungary's first opera performances were given in Pozsony by visiting Italian companies (*Mingotti, 1740; Bon, 1760; Zamperini, 1764). The country was under *Habsburg rule 1711–1919. From 1762 Haydn directed his operas at Kismarton (Eisenstadt); and in 1768 his *Lo speziale* opened Prince Esterházy's theatre at Eszterháza, where, especially from 1776, many of his operas were performed. Other nobles' theatres included that of Count Erdődy at Pozsony, where the first Mozart opera in Hungary was given (*Entführung*, 1785), and that of Bishop Patachich at Nagyvárad, where Dittersdorf worked (1764–9). In Buda and Pest performances were given in German (from 1787) and in Hungarian (1793–6, 1807–15). In 1719 the theatre of the Piarist School in Pest opened; from 1790 it began working towards national opera by staging Singspiels and opere buffe, but the first Singspiel in Hungarian was *Pikkó Herceg* (Prince Pikkó, 1793) by József Chudy (1751– or 1754–1813) (music lost, but evidently Hungarian songs interpolated in a Viennese Zauberposse).

In the early years of the 19th cent., the German company performed works, especially by Mozart, Beethoven, and Weber, which keenly interested Hungarian musicians. The inclusion

341

of many Italian and French works (Rossini, Bellini, Donizetti, Verdi, Auber, Meyerbeer) soon after their premières gave an example to the tentatively emerging Hungarian national opera. Some composers began using Hungarian elements in their operas, drawing on the *verbunkos tradition; others incorporated sections of foreign works into them. The first true Hungarian opera (though it includes German and Italian elements) was *Béla futása* (Bela's Flight, 1822) by József Ruzitska (c.1755–1823), which made use of the national verbunkos music, and became very popular. Other operas followed, notably Ruzitska's *Kemény Simon* (1822); *Csel* (The Ruse, 1839), by András Bartay (1799–1854); *Visegrádi kincskeresők* (The Treasure-Seekers of Visegrád, 1839), by Mark Rózsavölgyi (1789–1848); and *Benyovszky* (1847), by Ferenc Doppler (1821–83). These combine large dramatic scenes based on Italian and Viennese models with lyrical and heroic episodes deriving their style from the verbunkos.

The true founder of Hungarian opera was Ferenc *Erkel (1810–93), the first composer gifted enough to fuse the above elements into a vivid and personal musical language. From 1838 he directed the opera company of the Nemzeti Színház (National T), which had opened in Pest on 22 Aug. 1837. Hitherto Hungarian opera companies had worked in provincial towns, from 1799 in Kolozsvár (Cluj) and Debrecen, from 1818 in Székesfehérvár, from 1828 in Kassa (*Košice). Erkel's most important fellow-composer was Mihály *Mosonyi (1815–70); others included Károly Thern (1817–86), Károly Doppler (1825–1900), György Császár (1813–50), Ignác Bognár (1811–1900), and Károly Huber (1828–85); but for all their useful experimental work, they could not match Erkel's example. A parallel development, to which Erkel contributed, was the *népszínmű*, or popular play, which had musical insertions similar to ballad opera.

In 1884 Erkel moved to the newly opened Operaház, where he directed a large repertory. His death opened the theatre to Wagner, whose music he had resisted; of a Wagnerian generation which followed, the most outstanding was Ödön Mihalovich (1842–1929), composer of Wagnerian operas including *Toldi szerelme* (Toldi's Love, 1893) and a founder of the Budapest Wagner Society. A number of composers found international careers, especially in Habsburg lands: those who continued to work for national opera at home included Ferenc Sárosi (1855–1913), Jenő Sztojanovics (1864–1919), and Emil *Ábrányi (1882–1970). Their work helped to pave the way for the major talents of Béla *Bartók and Zoltán *Kodály. Other successful opera composers of the inter-war years included Ferenc Farkas (*b*

1905), Leó Weiner (1885–1960), Albert Siklós (1878–1942), and Pál Kadosa (1903–83).

After the war, with massive state support, opera developed and spread further, and new companies were formed in a number of provincial and industrial centres. The composer to win greatest international fame has been György Ligeti (*b* 1923), whose *Le grand macabre* was produced in Stockholm 1978. Among other composers to emerge have been Emil Petrovics (*b* 1930), with *C'est la guerre* (1962), *Lysistrate* (comp. 1962, prod. 1971), and *Bűn és bűnhődés* (Crime and Punishment, 1969); Sándor *Szokolay (*b* 1931); Sándor Balassa (*b* 1935) with *Az ajtón kívül* (Outside the Door, 1978); and Attila Bozay (*b* 1939), with *Csongor és Tünde* (1984).

See also BUDAPEST, DEBRECEN, ESZTERHÁZA, GYŐR, PÉCS, SZEGED.

**Hunter, Rita** (*b* Wallasey, 15 Aug. 1933). English soprano. Studied Liverpool with E. Francis, London with R. Llewellyn; later with E. Turner. London, SW, chorus, 1954–6; début in *Figaro* (First Bridesmaid). CR 1956–8; London, SW, 1959–66; London, CG, from 1963. ENO from 1968; New York, M, from 1972. Also Munich, San Francisco, Sydney. A generous singer with an ample, brilliant voice, notable as Brünnhilde, Norma, Amelia, Elisabeth de Valois. Autobiography, *Wait Till the Sun Shines, Nellie* (London, 1986). (R)

**Huntsmen's Chorus.** The name usually given to the chorus in Act III of Weber's *Der Freischütz*, in which the huntsmen sing of the joys of the chase.

**Hunyadi László.** Opera in 4 acts by Erkel; text by Béni Egressy, after Lőrince Tóth's drama. Prod. Budapest 27 Jan. 1844.

Hungary, 15th cent. László Hunyadi (ten), leader of the Hungarian army, learns of a plan by the new king László V (ten) to massacre his troops. When confronted, the King feigns reconciliation. He is attracted to Maria (sop), László's beloved, and, with the support of her father, Gaga (bar), has László arrested in order that he may marry Maria. László is condemned to death and beheaded.

**Huon.** Sir Huon of Bordeaux (ten), lover of Reiza, in Weber's *Oberon*.

**Huron, Le.** Opéra-comique in 2 acts by Grétry; text by Jean François Marmontel, after Voltaire's story *L'ingénu* (1767). Prod. Paris, CI, 20 Aug. 1768.

A coastal village in Brittany. The daughter (sop) of Saint-Yves (bs) does not wish to marry Gilotin (ten), as their parents have arranged, since she has fallen in love with a Huron Indian (bar). However, Abbot Kerkabon (bs) and his sister (sop) recognize the Huron as the son of their brother, who died on an expedition to the Hurons. Kerkabon tries to instil socially acceptable behaviour in his nephew, but the latter is impatient to marry the girl he loves. In order to prevent what he sees as an unsuitable marriage, Saint-Yves decides to send his daughter to a convent, but she is forgiven and the couple are happily united.

**Hüsch, Gerhard** (*b* Hanover, 2 Feb. 1901; *d* Munich, 21 Nov. 1984). German baritone. Studied Hanover with Emge. Début Osnabrück 1923 (Liebenau in Lortzing's *Der Waffenschmied*). Berlin: SO, 1930–5; then S, 1937–44. London, CG, 1930; Bayreuth 1930–1. Roles included Papageno, Wolfram. Possessed a warm lyric baritone. (R)

**Hyde, Walter** (*b* Birmingham, 6 Feb. 1875; *d* London, 11 Nov. 1951). English tenor. Studied London with G. García. Début Terry's T 1905 (in *My Lady Molly*). London, CG, 1908–24, singing Siegmund in Richter's *Ring*. New York, M, 1909–10. Member of Beecham OC, and BNOC, of which he became a director. A distinguished

Wagner singer, especially as Parsifal. Also created title-role in Holst's *The Perfect Fool*. (R)

**Hylas.** A young sailor (ten) in Berlioz's *Les Troyens*.

**Hymn of the Sun.** The chorus in Act I (repeated in Act III) of Mascagni's *Iris*.

**Hymn to the Sun.** The aria with which the Queen of Shemakha (sop) introduces herself in Act II of Rimsky-Korsakov's *The Golden Cockerel*.

Also aria in Glass's *Akhnaten*, in which Akhnaten declares his ideas for reform.

**Hytner, Nicholas** (*b* Manchester, 7 May 1956). English producer. Studied Cambridge. Début Kent O 1979 (*Turn of the Screw*). Kent O 1984–5; London, ENO, from 1983; Paris, O, 1987; London, CG, 1987 (first British *King Goes Forth to France*, Sallinen), 1988; Houston and Geneva 1989; Gly. 1991 (*Idomeneo*). Prods. incl. *Giulio Cesare*, a much-praised *Xerxes*, *Figaro*, *Magic Flute*, *Clemenza di Tito*, a famous *Rienzi* on a nearly bare stage, *Knot Garden*, *King Priam*. A very self-critical producer, his intelligent, cool, yet vital productions combine a sharp originality with a sensitivity to the music.

# I

**Iago.** A Venetian soldier, ensign to Otello, (ten) in Rossini's and (bar) in Verdi's *Otello*.

**Iaşi.** (Ger., Jassy). City in Moldavia, Romania. Opera was given by Italian touring companies from St Petersburg and Odessa from 1795, and under Gaetano Madji in 1806. There were seasons in the Teatrul Naţional under Alexander Flechtenmacher (1823–98) in 1844–7. A company worked under Tozzolli in 1851. The present company was formed in 1956. In 1958 the theatre was modernized, and the company began touring Moldavia.

**Ibert, Jacques** (*b* Paris, 15 Aug. 1890; *d* Paris, 5 Feb. 1962). French composer. Studied Paris with Vidal, Pessard, and Gédalge. His wide cultural interests and his love of drama found expression in seven operas, two of them in collaboration with Honegger. Almost all are in light vein. The witty farce *Angélique*, his most popular stage work, successfully recreates the spirit of 19th-cent. French comic opera; he then turned with almost equal success to opéra-comique with *Le roi d'Yvetot*. The more ambitious *L'aiglon* won more respect than popularity. Director, Union des Théâtres Lyriques from 1955.

> WORKLIST: *Persée et Andromède* (Nino, after Laforgue; Paris 1929); *Angélique* (Nino; Paris 1927); *Le roi d'Yvetot* (Limozin & Tourrasse; Paris 1930); *Gonzague* (Kerdick, after Veber; Paris 1930); *L'aiglon* (Cain, after Rostand; Monte Carlo 1937, comp. with Honegger); *Les petites cardinal* ( Willemetz & Brach, after Halévy; Paris 1939, comp. with Honegger); *Barbe-bleue* (Aguet; French radio 1943).

> BIBL: G. Miche, *Jacques Ibert* (Paris, 1968).

**Ibsen, Henrik** (*b* Skien, 20 Mar. 1828; *d* Oslo, 23 May 1906). Norwegian playwright. Operas on his works are as follows:

*The Feast at Solhaug* (1856): Stenhammar (1899)
*Olav Liljekrans* (1856): A. Eggen (1940)
*The Warriors at Helgeland* (1858): Rerny (*Hjördis*, 1900); K. Moor (*Hjördis*, 1905); Sandby (1920)
*Peer Gynt* (1867): Ullmann (20th cent.); Heward (unfin., from 1922); Egk (1938)
*A Doll's House* (1879): Mack (1967). *Kongsemnerne* (1863): Bucht (1966)
*Hedda Gabler* (1890): E. Harper (1985)
*Terje Vigen*: Lie (1904)

**Ice Break, The.** Opera in 3 acts by Tippett; text by the composer. Prod. London, CG, 7 July 1977, with Harper, McDonnell, Barstow, Vaughn, C. Walker, Shirley-Quirk, R. Kennedy, J. Dobson, also Anne Wilkens and James Bowman representing a single character, cond. C. Davis; Boston, Savoy OH, 18 May 1979, cond. Caldwell.

N America, present-day. The subject of the opera is stereotypes—their imprisoning characteristics and the need for rebirth. After 20 years in prison camps, Lev (bs) arrives in a new world to join his wife Nadia (sop), who had emigrated with their baby son Yuri. Also at the airport are Yuri's girl-friend Gayle (sop) and her black friend Hannah (mez), there to meet the black 'champion' Olympion (ten) together with his fan club. Out of a series of violent tensions, individual and collective, there develops a riot in which Olympion and Gayle are killed and Yuri (bar) is near-fatally injured. Nadia dies peacefully. During an interlude in which a group attempts a psychedelic 'trip', the messenger Astron (mez and high ten or counter-ten) is mistaken for God, though he disclaims his divinity. Yuri is operated on by Luke (ten), a young doctor, and, released from the cracking plaster, he finds reconciliation with his father.

**Iceland.** Though opera had occasionally been given in Reykjavik early in the century, regular performances date from the opening of the National T (cap. 661) in 1950. Iceland Opera was founded in 1979, and in 1980 commercial sponsorship enabled the company to buy and renovate a cinema, the Gamla Bió (cap. 505); this opened in 1982 with a *Zigeunerbaron* that ran to 49 performances with full houses. In the same year the National T staged with great success *Silkitromman* (The Silk Drum) by Atli Heimir Sveinsson (*b* 1938). New operas by Jón Ásgeirsson and Karólina Eiríksdóttir have also been staged. Iceland Opera mounts three productions a year, sometimes sung in Icelandic, and tours the island with piano-accompanied productions.

**'Ich baue ganz auf deine Stärke'.** Belmonte's (ten) aria in Act III of Mozart's *Die Entführung aus dem Serail*, which he is instructed by Pedrillo to sing so as to conceal the placing of the escape ladders to the seraglio windows.

**'Ich sah das Kind'.** Kundry's (sop) narration in Act II of Wagner's *Parsifal*, in which she relates to Parsifal the grief of his mother Herzeleide when he left her never to return.

**Idamante.** Idomeneo's son (sop, cas, or ten), lover of Ilia, in Mozart's *Idomeneo*.

**Idomeneo, re di Creta.** Opera in 3 acts by Mozart; text by G. B. Varesco, after Danchet's text for Campra's *Idomenée* (1712) and the ancient legend. Prod. Munich 29 Jan. 1781, with D. and E. Wendling, Raaff, Del Prato, Panzacchi, Valesi; Glasgow 12 Mar. 1934; Tanglewood, Lenox, MA, 4 Aug. 1947, with Bollinger, Trickey, Lenchner, Laderoute, Guarrera, cond. Goldovsky.

Sidon, in Crete, shortly after the end of the Trojan wars. Idomeneo, King of Crete (ten), has sent home from Troy captives including Ilia (sop), daughter of King Priam. She and Idomeneo's son Idamante (sop cas or ten) are in love, though he has not declared himself: Elettra (sop) also loves him. The impending return of Idomeneo is the sign for an amnesty of prisoners. But a sudden storm causes the King to vow to the seagod Nettuno (Neptune, bs) a sacrifice of the first living thing he meets on shore. This is his son. Horrified, he hurries away without speaking; a joyful chorus welcomes the warriors.

The King tries to evade his vow by sending Idamante to escort Elettra home to Argos, much to the distress of Ilia. But a storm arises, and there appears a monster who ravages the island. The people hold that some unknown sinner has offended the gods, and Idomeneo admits his guilt and is ready to die.

Idamante and Ilia declare their love before he sets out to attack the monster. Elettra, mad with jealousy, interrupts, followed by the King, who is torn between guilt and anxiety for his son. He is forced, by the people's demand for a victim, to reveal the truth. On hearing that the monster has been killed by Idamante, the High Priest hesitates to make the sacrifice. Idamante nevertheless offers himself as a victim so as not to break his father's vow. But the voice of the god spares him, announcing that Idomeneo must abdicate. Idamante ascends the throne with Ilia at his side.

Mozart revived *Idomeneo* for Vienna in 1786, rewriting the part of Idamante for a tenor, and it is in this version that the work is usually performed.

Other operas by Campra (1712), Galuppi (1756), Gazzaniga (1790), Paisiello (1792), Paer (1794), Federici (1806), and Farinelli (1812).

**'I have attained the highest power'.** Boris's (bs) monologue in Act II of Musorgsky's *Boris Godunov*.

**'Il balen'.** Di Luna's (bar) aria in Act II of Verdi's *Il trovatore*, in which he sings of the tempest raging in his heart.

**Ilia.** A Trojan princess (sop), daughter of King Priam and lover of Idamante, in Mozart's *Idomeneo*.

**'Il lacerato spirito'.** Fiesco's (bs) aria in the prologue to Verdi's *Simon Boccanegra*, in which he sings of his tortured soul.

**Illica, Luigi** (*b* Piacenza, 9 May 1857; *d* Colombarone, 16 Dec. 1919). Italian playwright and librettist. He wrote or collaborated on about 80 librettos for some of the leading composers of his day, beginning with Smareglia's *Il vassallo di Szigeth* (1889). Others include Catalani's *La Wally* (1892), Giordano's *Andrea Chenier* (1896), Mascagni's *Iris* (1898) and *Le maschere* (1901), D'Erlanger's *Tess* (1906); but he is most famous for his collaboration on Puccini's *Manon Lescaut* and, with Giuseppe Giacosa, *La bohème* (1896), *Tosca* (1900), and *Madama Butterfly* (1904). A man of exuberant and violent passions, he was at his best at inventing strong characters and situations. He lacked Giacosa's sensibility, but complemented him well, and in Puccini's words 'he had plenty of imagination'. He reflected a French influence in his work, in the conflict of a powerful love element with social or political tensions, portrayed in plots of some accuracy in historical atmosphere and detail, and supported by thorough stage directions (a French practice which he introduced into Italian opera).

**'Il mio tesoro'.** Don Ottavio's (ten) love song to Donna Anna in Act II of Mozart's *Don Giovanni*.

**'Il segreto'.** The *brindisi sung by Maffeo (con) in Act II of Donizetti's *Lucrezia Borgia* as the banqueters drink the wine poisoned by Lucrezia.

**imbroglio** (It.: 'entanglement, intrigue', orig. from the Broglio, the arcade of the Doge's Palace in Venice in which such intrigues were conducted). A scene in which confusion is suggested by great diversity, both melodic and rhythmic, of the parts given to the singers or groups of singers. Though originating in 18th-cent. Italian opera, the imbroglio at its most advanced and intricate can be seen in the end of Act II of Wagner's *Die Meistersinger*.

**Imeneo.** Opera in 3 acts by Handel; text an almost unaltered revision of Stampiglia's libretto set by Porpora, 1723. Prod. London, Lincoln's Inn Fields, 22 Nov. 1740, with Francesina, Edwards. The work was a failure and not revived between 1742 and 1960.

**Immolation.** The name generally given to the scene of Brünnhilde's death in Act III of Wagner's *Götterdämmerung* as she rides (or in most modern productions walks) into the flames of Siegfried's funeral pyre.

**Immortal Hour, The.** Opera in 2 acts by Boughton; text by 'Fiona Macleod' (William Sharp). Prod. Glastonbury 26 Aug. 1914, with Lemon, Jordan, Austin, Boughton (replacing an unwell singer as Dalua), cond. Kennedy Scott; New York, Grove St. T, 6 Apr. 1926, with Borden, Kuschke, Mordhurst, Vining, Cox, Rothwell, Gurney, cond. Bimboni.

Etain (sop), princess of the fairies, is travelling in the human world in search of some new, unidentified experience. The mortal King Eochaidh (bar) is also seeking such an experience: the 'immortal hour'. Dalua (bar), the Lord of Shadow, engineers the meeting of the couple; they fall in love and marry. A year later Midir (ten), prince of the fairies, arrives to take Etain back to her spirit world. As she departs, Dalua touches Eochaidh, who falls dead at this fatal contact.

The opera had an outstandingly successful season of 216 nights at the Regent T, London, from 13 Oct. 1922; this was followed by another run of some 160 performances from 17 Nov. 1923. By 1932 the work had received over 500 London performances. A revival at Sadler's Wells in 1953 was, however, unsuccessful.

**impresario** (from It. *impresa*: 'undertaking'). Organizer and/or manager of an opera company. The Italian equivalent of Ger. *Schauspieldirektor*, both in general use and as title for Mozart's opera.

BIBL: J. Rosselli, *The Opera Industry in Italy from Cimarosa to Verdi: the Role of the Impresario* (Cambridge, 1984).

**'Improvviso'.** The name generally given to the aria 'Un dì all'azzurro spazio' in Act I of Giordano's *Andrea Chénier*, in which Chénier composes a poem on the beauty of the world and man's inhumanity to man, and exhorts Maddalena not to scorn love.

**in alt**. The term used to describe the notes in the octave immediately above the top line of the treble stave, running from g'' to f'''. The next octave, from g''' to f'''' is **in altissimo**.

**Incoronazione di Poppea, L'** (The Coronation of Poppea). Opera in a prologue and 3 acts, usually attrib. Monteverdi, though almost certainly composed with the assistance of others; text by G. F. Busenello, after Tacitus, *Annals*, xiii–xv. Prod. Venice, autumn 1643; Northampton, MA, Smith Coll., 27 Apr. 1926, with Fatman, Miliette, Donovan, Lyman, Pitts, Sinclair, McNamara, cond. Josten; Oxford, OUOC, 6 Dec. 1927, ed. and cond. Westrup; Gly. (Leppard version) 29 June 1962, with Laszlo, Bible, Marimpietri, Dominguez, Lewis, Cava, Alberti, cond. Pritchard.

In the prologue the allegorical figures of Fortune and Virtue argue over which is the stronger, though Love claims he is more powerful than either, as the action of the opera will show.

Rome, AD 65. Nerone, the Roman Emperor, has taken as his mistress the scheming Poppea, much to the consternation of Ottone, who is in love with her himself. Ottavia, Nerone's wife, is grief-stricken at her husband's actions, and the efforts of her nurse Arnalta and page to comfort her are in vain, as are those of the philosopher Seneca. When Nerone announces that he is to banish Ottavia and marry Poppea, Seneca accuses him of insanity. This, together with Poppea's insinuations that Seneca is trying to usurp the Emperor's position, persuades Nerone to sentence the philosopher to death. Ottone takes a last chance at reconciliation with Poppea; when this is unsuccessful he resolves to kill her, and begins to woo Drusilla, a courtier who has long sought his attention.

Told by Mercury that Nerone has ordered his death, Seneca commits suicide: the Emperor celebrates with the poet Lucan. Ottavia meets Ottone and persuades him, with some difficulty, that he must go ahead with his plan to kill Nerone. He borrows clothes for a disguise from Drusilla, who is overjoyed that he now desires her. Poppea prepares for bed, and will let only Arnalta sleep near her. After the nurse has sung a lullaby, Love appears, to keep watch over Poppea. When the disguised Ottone enters to carry out the murder, Love prevents him, singing a triumphant aria as he makes his escape.

Arrested for her apparent attempt on Poppea's life, Drusilla protests her innocence to Nerone, though when Ottone tries to admit his guilt she protects him by confessing to the plot. Nerone, not knowing who is speaking the truth, resolves to kill them both, but eventually relents and orders them to be exiled instead. Discovering Ottavia to have been involved in the plan, Nerone banishes her as well. After Ottavia's final distraught appearance, the coronation of Poppea concludes the opera, watched over by Venus and Love.

Neither of the two extant scores of *L'incoronazione di Poppea* mentions the composer by name, but the work was attributed to Monteverdi by *Ivanovich. In recent years this claim has been strongly challenged: critical examination of the sources suggests that several scenes (including the whole final scene) and most of Ottone's music were written by a younger composer. It has been argued very plausibly that this person was probably Francesco *Sacrati, while the final duet 'Pur ti miro' (which was long cited as an example of Monteverdi's lyrical genius) has been shown to be a re-working of a piece of the same title by Benedetto *Ferrari. For a full discussion of the issues of authenticity see the edn. by Alan Curtis (London, 1989).

**Indes galantes, Les** (Gallantry in the Indies). Ballet-héroïque in a prologue and 4 entrées by Rameau; text by Louis Fuzelier. Prod. Paris, O, 23 Aug. 1735 with prologue and 2 entrées—third added at third performance, fourth added 10 Mar. 1736, with Eremans, Petipas, Pélissier, Antier, Cuignier, Jélyotte, Dun, De Clusse, cond. Cheron; New York, TH, 1 Mar. 1961, with Raskin, Ferriero, Bressler, Shirley, Trehy, cond. Dunn.

Following war, Amor is abandoned by the youth of four European nations (France, Italy, Spain, and Poland). The entrées which follow tell four tales of love in different parts of the world.

Turkey. The pasha Osman (bs) loves his Provençal slave girl, Émilie (sop), whose lover, Valère, has also been captured. However, recognizing Valère as his rescuer, Osman permits the couple to be reunited.

Peru. Don Carlos (ten), a Spanish officer, loves an Inca princess, Phani (sop). Jealous of the emotions she returns, the high priest Huascar (bar) causes a volcano to erupt; however, it is Huascar who dies, while Don Carlos carries Phani to safety.

Persia. Tacmas (ten), a Persian prince, loves Zaïre (sop), one of Ali's (bar) slaves; Fatima (sop), Tacmas's slave, loves Ali. Confusions are resolved when the two exchange slaves and celebrate together.

America. A French officer, Damon (ten), and Spanish officer, Don Alvar (bar), both love a native American girl, Zima (sop), though she prefers her compatriot Adario (ten or bar). Finally, they all join in festivities of peace.

Rameau's third opera, and greatest success. It was revived at the Paris Opéra in 1952, and perf. 246 times in the following ten years.

**India.** Indian musical theatre is a mixed genre. The fusion of epic poetry with the music and dancing of the temple led to Sanskrit musical drama (*geya nataka*), of which classics included the *c.*4th-cent. Kalidasa's *The Recognition of Shakuntala* (a story well known in Europe and the subject of several operas, including one by Schubert). The situations were formal, revolving round the stereotypes of hero, villain, and clown. The accompaniment was by instruments from eight to 60 in number, and was designed to frame a mood shared between performers and audience rather than to set words or illustrate actions; the songs were sung by members of the orchestra. There was a minimum of scenery or props.

With the Moslem raids of the 12th cent. and the rise of the Mogul Empire, Sanskrit music drama loosened its ritual forms; more realistic acting was demanded, with singing as well as dancing by the actors. Culturally, North and South India were divided. In the North, the emphasis was on pageantry and formal drama which made great demands on performers and audience alike: one drama, the *Ramalila*, takes a

fortnight to perform, the *Krishnalila* a month, and both are still given on the sites of their legendary events. In the South, music drama is dependent more on the village storyteller's art, and tends to consist of a narration illustrated with song, dance, and mime, though the subjects are also mythological. A later development, *charitram*, is closer to Western opera, with less improvisation and a more disciplined balance between drama and music. In each, vocal standards are regarded as being as irrelevant to the details of the executions as handwriting is to a composer, though this has not prevented the recent emergence of popular singers. In the North, the wealthy 19th-cent. amateur Wajid Ali Shah developed a genre of music drama illustrating medieval paintings that depicted ragas by means of tableaux, with miming actors and an instrumental ensemble. Around 1900 the Parsees toured a kind of musical theatre which won popularity by virtue of its large numbers of performers and its accent on spectacle. Bengali theatre, inclined to similar ideas, was disciplined by the example of Rabindranath Tagore (1861–1941), who attached the highest importance to his music and insisted on simplicity of form and seriousness of theme. Familiar with Western music, he was not afraid to use European folk-song in a dramatic form he called *Rabindra-Sangeet*: one of his works, *Shyama* (1939), is a simple ballad opera on a tragic story. He was the first Indian to achieve a successful cross-fertilization of East and West in drama.

Performances of Western opera were sometimes staged during the British Raj. Scenes from *Don Giovanni* were given in 1833 and *Cenerentola* was given in 1834 in Calcutta; *Cavalleria rusticana* was given at Simla in 1901 and in Bombay in 1925. The so-called Opera House in Calcutta was really intended for drama and variety, though Pollard's Lilliputian Opera Company (probably a concert party group) performed there in 1896–7.

New Indian operas produced within the first decade of independence included *Meena Gurjari* by Nat Mandal; this makes use of Punjabi folk-songs, treated in Leitmotiv manner and demanding part-singing, even a vocal trio.

**Indian Queen, The.** Semi-opera by Purcell (last act by Daniel Purcell); text by Dryden and Robert Howard. Prod. London, DL, 1695.

In reward for his success aginst the Mexicans, Montezuma asks for the hand of Orazia, daughter of the Inca king. The King's fury forces Montezuma to flee and seek refuge with the Mexicans; Zempoalla, the Mexican queen, falls in love with him. When Orazia and her father are captured by the Mexicans, Montezuma tries to save them. In order to prevent their execution,

Acacis, the Mexican prince, commits suicide; he also loves Orazia. The Inca king is discovered to be the King of the Mexicans; Montezuma and Orzia are permitted to marry, and Zempoalla kills herself.

**'In diesen heil'gen Hallen'.** Sarastro's (bs) aria in Act II of Mozart's *Die Zauberflöte*, in which he tells Pamina that no thought of violence is entertained within the sacred walls of the Temple of Isis and Osiris.

**Indy, Vincent d'** (*b* Paris, 27 Mar. 1851; *d* 2 Dec. 1931). French composer. Studied Paris with Lavignac. Apart from a number of projects and the unimportant 1-act *Attendez-moi sous l'orme* (1882), D'Indy's first opera was *Le chant de la cloche*, heavily indebted to *Parsifal*. He further expressed his love of Wagner, his ancestral feeling for the Cévennes, and his discipleship of Franck in his most important stage work, *Fervaal*. *L'étranger* continues in the same line while discovering a greater refinement of texture. Nevertheless, it came to stand for virtues opposite to those expressed in *Pelléas*. *La légende de St Christophe* continued to oppose the new values in an extraordinary, ambitious mélange of Wagner and medieval mysteries. His last stage work, *Le rêve de Cinyras*, is an unambitious operetta.

SELECT WORKLIST: *Attendez-moi sous l'orme* (Wait for me under the elm) (Prével & Bonnières, after Régnard; Paris 1882); *Le chant de la cloche* (The Song of the Bell) (D'Indy, after Schiller; Brussels 1912); *Fervaal* (D'Indy; Brussels 1897); *L'étranger* (D'Indy; Brussels 1903); *La légende de St Christophe* (D'Indy, after Voragine; Paris 1920); *Le rêve de Cinyras* (Courville; Paris 1927).

**Inez.** The daughter (sop) of Don Diego in Meyerbeer's *L'Africaine*.

**'In fernem Land'.** Lohengrin's (ten) narration (Ger., *Gralserzählung*) in Act III of Wagner's *Lohengrin*, in which he discloses his identity and tells of the Holy Grail.

**ingénu(e)** (Fr.: 'artless, innocent'). The ingénue of 18th-cent. French opera, a development by Favart of the *damigella of 17th-cent. Italian opera, was a type of girl of 15 or 16 whose innocence and light approach to love had a pronounced sentimentality but was also marked by a latent sensuousness; her tendency to tears was part of her charm, and this characteristic helped her to develop the *comédie larmoyante*, or sentimental comedy, which was a part of French comic opera in the later 18th cent. The ingénu, a less common or typical character, was usually little more than a village simpleton. See also OPERA SEMISERIA.

**Inghilleri, Giovanni** (*b* Porto Empedocle, 9 Mar. 1894; *d* Milan, 10 Dec. 1959). Italian baritone and composer. Début Milan, T Carcano, 1919 (Valentin). Successful career in Italy, including at Milan, S, after 1945; London, CG, 1928–35; Chicago 1929–30. Created leading roles in works by Casella, Malipiero, and Cattozzo. Roles included Scarpia, Gérard (*Chenier*), Amfortas. He also wrote an opera (*La burla*), a ballet, and several songs. (R)

**Innsbruck.** City in Tyrol, Austria. The Comedi-Haus opened 1631, with a small orchestra and 41 singers performing opera mostly for the Archduke Leopold's Court. His son Ferdinand Karl commissioned Christoph Gumpp to build a larger Komödienhaus opposite the Imperial Palace, opened 1654 with *Cleopatra* by the most distinguished of the Italians he engaged, Pietro *Cesti (court composer for eight years). This was the first German theatre with a permanent company for opera and Singspiel. Cesti's *Argia* was prod. 1655 in honour of the visit of Queen Christina of Sweden on her way to Rome; it lasted over five hours. Two more Cesti works were prod., including *Venere cacciatrice*, 1659, the year in which the first permanent theatre and opera companies were formed. Bankruptcy forced the dismissal of the ensemble, though amateur performances were given (including of Cesti's *La magnanimità d'Alessandro* with six counts, eight countesses, one marchese, and one baron). Under Maria Theresa the orchestra was dismissed and the southern part of the theatre was converted into a public library. The last court performance was the première of Hasse's *Romolo ed Ersilia* (1765). Schikaneder was producer and actor 1775–6. The repertory expanded rapidly in the 19th cent. In 1827 the theatre was modernized; a new theatre opened 1846. From 1886 it belonged to the municipality as the Stadttheater, and with the German annexation in 1838 became a Landestheater, after the war the Tyroler Landestheater. Siegfried Nessler was music director, and Robert Nessler conductor. The theatre was rebuilt 1967 (cap. 793), opening with *Meistersinger*.

**'In quelle trine morbide'.** Manon's (sop) aria in Act II of Puccini's *Manon Lescaut*, in which she tells of the chill in the splendour among which she lives, and wishes she were back in the humble dwelling where she and Des Grieux were once so happy.

**'In questa reggia'.** Turandot's (sop) narration in Act II of Puccini's *Turandot*, in which she recounts how the rape of her ancestress Princess Lo-u-Ling made her resolve to be revenged upon all men through her vow to set riddles to those seeking her love, with death as the price for failure.

**Insanguine, Giacomo** (*b* Monopoli, 22 Mar. 1728; *d* Naples, 1 Feb. 1795). Italian composer. Studied Naples with Abos, Feo, and Durante. Although not well known outside Naples, he was one of the most frequently performed composers there in the mid-18th cent. Of his *c*.20 operas most successful were the comic *Lo funnaco revotato* and the serious *Didone abbandonata*. 1785 became director of the S Onofrio Cons.

SELECT WORKLIST: *Lo funnaco revotato* (?Mililotti, after Oliva; Naples 1756); *Didone abbandonata* (Metastasio; Naples 1770).

**Inszenierung** (Ger.). Production. Hence *Neuinszenierung*, for the new production of an opera.

**Intendant** (Ger.: 'superintendent'). The administrator of a German opera-house; not necessarily the artistic or musical director.

**interlude.** See INTERMEZZO.

**intermède** (Fr.: 'interlude'). A term current in France in the 17th and 18th cents. with a variety of applications, but always carrying the connotation of a self-contained musical entertainment, invariably featuring dance, inserted between the acts of another, larger work. Hence many of the Latin tragedies given at the Jesuit college of Louis-le-Grand were performed with interpolated intermèdes, while Molière used them as a means of introducing secondary plots into his *comédies-ballets. This latter practice, which began with *La princesse d'Élide*, influenced the eventual development of opéra-comique. Intermèdes were also introduced into the Italian operas performed in Paris during the 17th cent.: especially important were those provided for performances of Cavalli's *Serse* (1660) and *Ercole amante* (1662).

**intermedio** (It.: 'interlude'). Following the renaissance of secular drama in Italy, which began with Poliziano's *Orfeo* (1480), it became customary to include musical interludes between the acts of the spoken plays. These intermedi frequently had an allegorical connection with the substance of the main plot and in their combination of drama and music were an important precursor of opera, since they often told a continuous, if interrupted, story. At aristocratic courts, such as those of Mantua, Ferrara, and above all Florence, intermedi were usually given for a celebration, such as a birthday or wedding, and called for elaborate choral and instrumental forces. Particularly important intermedi included those performed at Florence in 1539 (for Landi's *Il commodo*; music by Francesco Corteccia), in 1565 (for D'Ambra's *La cofanaria*; music by Corteccia and Alessandro Striggio), and in 1589. This last occasion, at which the intermedi framed Bargagli's comedy *La pellegrina*, represented the high point of the genre, and among those who collaborated in its success were several of the important figures of early opera, such as Peri, Caccini, and Rinuccini, while Cavalieri also supervised the whole production.

Given the invariable 5-act structure of the plays, which were fashioned after the classical model, there were usually six intermedi, which preceded and followed the stage drama, as well as being interpolated between acts. Although the intermedi are cited as an important influence on the birth of opera, their grandiose scale and spectacular presentation were quite at variance with the Bardi *Camerata's ideas. While providing a model through their presentation of mythological themes and their union of music and drama, they had less direct influence than is sometimes claimed. For many the first operas seemed too scholastic and sterile compared to the more lavish display of the intermedi, and it is no surprise that the latter should have remained popular long after the first operatic experiments, only waning when the power and wealth of the Italian courts had begun to decline. The performance of Caccini's *Il rapimento di Cefalo* in 1600, in which elaborate intermedi were given between the acts, was an intriguing meeting of the old and new.

**intermezzo** (It.: 'interlude'). Originally, the intermezzo was a short, comic entertainment inserted between the acts of a serious opera, often including grotesque elements. The earliest instance probably occurred at Bologna in 1623, when intermezzi were interpolated into *L'amorosa innocenza*; despite their separation, they formed a continuous opera of their own, *La coronazione di Apollo*. This joining together of a serious and a comic entertainment was very popular throughout the 17th cent., and received impetus from the reforms of opera seria led by Zeno and Metastasio, in which comic scenes, a staple feature of Venetian opera, were excised. The intermezzo grew in importance and the appeal of the more realistic, popular genre in contrast to the gods, kings, and heroes of opera seria had become too strong to be contained. Frequently, the two intermezzi performed during a 3-act opera seria formed a continuous plot, so that the audience was effectively presented with two operas on the same evening; sometimes a third part was given before the penultimate scene of the main work.

Following the wave of interest in the genre occasioned by Pariati's librettos for *Pimpinone* (1708) and *Paspagnacco* (1708), Naples was established as the main centre for composition and performance of the intermezzo. By the time of the most famous example, Pergolesi's *La serva padrona* (1733)—which was originally given with his opera seria *Il prigionier superbo*—the genre was widely cultivated throughout Italy. Rather than the stilted da capo structure of opera seria, it employed a wide variety of aria forms, and made distinctive use of the bass voice. Classically, there would be two singing characters, each of whom would sing one or two arias and a duet, and a mute. If opera seria stood for the establishment of order and the importance of conformity to social and moral standards, the intermezzo satisfied the complementary wish not to conform and, drawing from the lively world of the *commedia dell'arte, so came to make the larger entertainment seem artificial and stilted. Eventually it was subsumed, together with the wider tradition of non-serious operas such as Abbatini's *Dal male il bene* and Scarlatti's *Trionfo d'onore*, into the mainstream *opera buffa tradition.

In opera the word 'intermezzo' is also used in the sense of an interlude, a short piece of music, or even a short scene interpolated in the course of an opera. An orchestral intermezzo may be virtually a miniature tone-poem, perhaps denoting the passing of time (as between scenes 8 and 9 of Mascagni's *Cavalleria rusticana*) or describing or summarizing events between scenes (as between Acts II and III of Puccini's *Manon Lescaut*). A dramatic intermezzo, or interlude, may take the form of a scene to some degree outside the main plot (as with the choral interlude between Acts I and II of Schoenberg's *Moses and Aron*). Henze's 'opera seria' *The Bassarids* (1966), includes a dramatic intermezzo, *The Judgement of Calliope*, which though relevant to the main plot is a conscious reversion to the earlier practice of the 18th cent.

**Intermezzo.** Opera in 2 acts by Richard Strauss; text by the composer. Prod. Dresden 4 Nov. 1924, with Lotte Lehmann, Correck, cond. Busch; New York, Philharmonic Hall, 11 Feb. 1963, with Curtin, Bell, cond. Scherman; Edinburgh, King's T, 2 Sep. 1965, with Steffek, Prey, cond. Zallinger.

While the famous conductor Robert Storch (bar) is away, his wife Christine (sop) is pursued by the young Baron Lummer (ten). However, she soon realizes that all he wants is her money. A letter arrives for her husband, a passionate love-letter, causing Christine to send her husband a telegram refusing to see him again.

Storch receives the telegram when in the company of another conductor, Stroh (ten), for whom the love-letter was intended. Stroh goes ahead to explain the situation to Christine, but she still receives her husband coldly. However,

the couple are reconciled and appreciate the true happiness of their marriage.

The opera is based on incidents in Strauss's own life, when his marriage was threatened by having received, by mistake, a love-letter from an unknown admirer.

Also opera by G. F. Majo (1752).

**Intolleranza 1960.** Opera in 2 acts by Nono; text by Nono, after Ripellino, Brecht, Sartre, Fučik, and Mayakovsky. Prod. Venice, F, 13 Apr. 1961, with Munteanu, Gayer, Rehfuss, Tajo, Henius, cond. Maderna; Boston 22 Feb. 1965. Rev. as *Intolleranza 1970*, Florence 1974.

A woman (sop) turns against her lover, a refugee miner (ten), not understanding his homesickness. After becoming involved accidentally in a political demonstration he is sent to a prison camp, where he finds love and humanity despite the harsh conditions. The opera ends with a purifying flood across the land, symbolizing a new sympathy between men.

**intonation.** The quality of singing or playing in tune.

**introduzione** (It.: 'introduction'). One of the main structural units of 19th-cent. Italian opera, comprising choral and solo material, invariably occuring at the beginning of an opera, after the overture or prelude. As it often featured one or more cavatine, it was sometimes also referred to as an **introduzione e cavatina**.

**intrusive H.** A common singer's fault found chiefly in long runs on one syllable: each new note is started with an unvocalized breath so that the effect is not 'a-a-a' but 'ha-ha-ha'. An accusation of this fault levelled against Steuart Wilson by a schoolmaster in a letter to the *Radio Times* in 1933 led to a libel action. Wilson won his case, in which a large number of musicians were called as witnesses, and was awarded £2,100 damages against the BBC and the schoolmaster. He spent it on a production of Boughton's *The Lily Maid*.

**'Invano, Alvaro'.** The duet between Carlo (bar) and Alvaro (ten) in Act IV of Verdi's *La forza del destino*, in which Alvaro (now Padre Raffaello) tries to dissuade Carlo from challenging him to a duel.

**Invisible City of Kitezh, The** (Full title: *The Legend of the Invisible City of Kitezh and of the Maid Fevronia*) (Russ.: *Skazaniye o nevidimom grade Kitezhe i deve Fevronii*). Opera in 4 acts by Rimsky-Korsakov; text by V. I. Belsky, after the legend. Prod. St Petersburg, M, 20 Feb. 1907, with Filippov, Labinsky, Kuznetsova-Benois, Ershov, Sharonov, Markovich, Zabela, cond. Blumenfeld; London, CG (concert), 30 Mar. 1926, with Smirnova, Davidova, Pozenkovksy,

Carewia, Kaidonov, cond. Coates; Philadelphia 4 Feb. 1936, with Palmer, cond. Smallens.

While out hunting, Vsevolod (ten), Prince of Kitezh, falls in love with Fevronia (sop), the sister of a woodcutter, who goes to his aid when he is injured. Not knowing his identity, she agrees to marry him; only when he sends a wedding procession to fetch her does she discover who he really is.

The wedding festivities in the city of Kitezh the Less are interrupted when the Tartars ransack the city; only Fevronia and the drunkard Kutierma (ten) survive. He is obliged to direct the Tartars to Great Kitezh.

Vsevolod gathers men to fight the Tartars, but they are killed, and their city vanishes completely.

Fevronia and Kutierma escape to the forest. She falls, exhausted, and as she dies, she sees a vision of Vsevolod guiding her to a rebuilt Kitezh.

In another world, the city stands again and its people celebrate the wedding of Vsevolod and Fevronia.

**Iolanta.** See YOLANTA.

**Iphigenia.** In Greek mythology, the daughter of Agamemnon and Clytemnestra. Agamemnon killed a hart in the sacred grove of Artemis, who then becalmed the Greek fleet waiting to sail from Aulis for Troy. Agamemnon decided to follow the priest Calchas's advice and sacrifice Iphigenia, but Artemis carried her off to Tauris to be a priestess. There she was discovered years later by her wandering brother Orestes, and brought home again to Mycenae.

The legend was given dramatic expression by Euripides in his *Iphigenia in Aulis* (?406–405 BC, unfin.) and *Iphigenia in Tauris* (414–412 BC). Most of the many operas on Iphigenia are based on these plays. Among at least 40 operas on the Aulis episode are those by Löwe (1661), D. Scarlatti (1713), Caldara (1718), Graun (1728), Orlandini (1732), Porpora (1735), Jommelli (1751), Franchi (1766), Guglielmi (1768), Gluck (1774), Pleyel (1785), Zingarelli (1787), and Cherubini (1788). Operas on the Tauris episode are by Campra and Desmarets (1704), Gluck (1779), Piccinni (1787), D. Scarlatti (1713), Traetta (1758), Araia (1758), Galuppi (1708), Jommelli (1771), and Carafa (1817).

Also opera by J. André (*Iphigenia in Rheinsberg*, 1853).

**Iphigénie en Aulide.** Opera in 3 acts by Gluck; text by Du Roullet, after Racine (1674), after Euripides' tragedy (?406–405 BC, unfin.). Prod. Paris, O, 19 Apr. 1774, with Arnould, Duplant, Legros, Larrivée, cond. Gluck; Oxford 20 Nov. 1933, with Green, Philips, Heseltine, Dance,

Wade, Downing, Douglas, cond. Harvey; Philadelphia, AM, 22 Feb. 1935, with Tentoni, Van Gordon, Bentonelli, Baklanov, cond. Smallens.

The island of Aulis. The oracle of Diana has stated that the Greeks' journey to Troy, which has been interrupted, may continue only if Iphigénie (sop), daughter of the Greek King Agamemnon (bs-bar), is sacrificed. Iphigénie and her mother Clytemnestre (mez) have been summoned to Aulis, supposedly for Iphigénie's marriage to Achille.

Reports of Achille's infidelity, intended to encourage Iphigénie to leave the island, prove to be unfounded.

The wedding celebrations are suspended when Arcas (bs) informs the company that Agamemnon intends to kill his daughter at the altar. After a violent argument between Agamemnon and Achille, the King orders his wife and daughter to quit Aulis.

Iphigénie decides to submit to the sacrifice, but the ceremony is interrupted by Achille. The high priest Calchas (bs) finally announces that Diana has been moved by the sincerity of emotion displayed, and will permit the army to move on without any sacrifice.

Also operas by D. Scarlatti (1713) and Cherubini (1788).

**Iphigénie en Tauride.** Opera in 4 acts by Gluck; text by Guillard and Du Roullet, after Euripides' drama (414–412 BC). Prod. Paris, O, 18 May 1779, with Levasseur, Larrivée, Legros, cond. Francœur; London, H, 7 Apr. 1796, with Giorgi-Banti, Roselli, Viganoni, Rodevino; New York, M, 25 Nov. 1916, with Kurt, Sembach, Weill, cond. Bodanzky.

Tauris. Iphigénie (sop) is unaware that her brother Oreste (bar) killed their mother Clytemnestre, after she had murdered her husband Agamemnon. As priestess to Diana, she is obliged by Thoas, King of Scythia (bs), to put all strangers to death. When Oreste and his friend Pylade (ten) arrive on Tauris, Iphigénie fails to recognize them.

Oreste is tormented by the thought that he is the cause of Pylade's death, and by memories of his crime. In a state of delirium, he tells Iphigénie the fate of the royal family, without disclosing his own identity.

Disconcerted by the stranger's resemblance to her brother, Iphigénie tries to save him by sending him with a message to her sister Electre (Fr.). Oreste refuses to leave Pylade, and when he vows to commit suicide if his friend is sacrificed, Pylade agrees to go.

At the moment of sacrifice, Iphigénie recognizes her brother. Thoas demands the sacrifice, and Iphigénie swears she will die with Oreste.

The situation is saved by a Greek party, led by Pylade. Thoas is killed; Diana appears, pardons Oreste, and returns to the Greeks her image, which had been stolen by the Scythians.

Also operas by D. Scarlatti (1713), Traetta (1763), and Piccinni (1781).

**Ippolitov-Ivanov, Mikhail** (b Gatchina, 19 Nov. 1859; d Moscow, 28 Jan. 1935). Russian composer and conductor. Studied St Petersburg with Rimsky-Korsakov. Conducted Tiflis O from 1884, also in Moscow at Mamontov O (1898–1906), Zimin O, and Bolshoy from 1925. He conducted a number of the premières of operas by Rimsky-Korsakov, whose music, as well as Tchaikovsky's, much influenced his own conservative idiom. Wrote text and music of Acts II–IV in completion of Musorgsky's *The Marriage*.

SELECT WORKLIST: *Ruth* (after the Bible; Tiflis 1887); *Azra* (?; Tiflis 1890, destroyed); *Asya* (?, after Turgenev; Moscow 1900); *Izmena* (Treachery) (?, after Sumbatov, after a Georgian folk-tale; ? 1910).

WRITINGS: *Pyatdesyat let russkoy muzyki v moikh vospominaniyakh* (50 Years of Russian Music in My Reminiscences) (Moscow, 1934).

BIBL: S. Bugoslavsky, *Ippolitov-Ivanov* (Moscow, 1936).

**Ireland.** The success of *The Beggar's Opera* in Dublin in 1728 led to many imitations using Irish songs as well as other folk and art music. Irish composers of opera, including Balfe, Wallace, and Stanford, mostly worked abroad. The first opera with a Gaelic (as well as English) libretto was *Muirgheis* (1903) by O'Brien Butler, followed by *Eithne* (1910) by Robert O'Dwyer.

See also BELFAST, DUBLIN, WEXFORD.

**Irene.** Rienzi's sister (sop), lover of Adriano, in Wagner's *Rienzi*.

**Iris.** Opera in 3 acts by Mascagni; text by Illica. Prod. Rome, C, 22 Nov. 1898 with Darclée, De Lucia, cond. Mascagni; rev. Milan, S, 19 Jan. 1899; Philadelphia, AM, 14 Oct. 1902, with Farneti, Schiavazzi, Bellati, cond. Mascagni; London, CG, 8 July 1919, with Sheridan, Capuzzo, Couzinou, cond. Mugnone.

The opera, set in 19th-cent. Japan, tells of the vain attempt by Osaka (ten) to win the love of the pure young Iris (sop). He arranges with Kyoto (bar), the keeper of a brothel, to have her abducted, and her blind father (Il Cieco, bs), thinking she has gone there voluntarily, curses her and flings mud at her. Iris drowns herself in a sewer.

**Irische Legende.** Opera in 5 scenes by Egk; text by the composer after W. B. Yeats's drama *The Countess Cathleen* (1892). Prod. Salzburg, Festspielhaus, 17 Aug. 1955, with Borkh, Klose, Lorenz, Böhme, Frick, cond. Szell.

In Ireland, some time in the future, Satan's powers have caused a famine and the people sell their souls to him in exchange for food. Countess Cathleen (sop) offers to sell her soul not only for this reason, but because her lover, the poet Aleen (bs), has been abducted by demons. Like Marguerite in *Faust*, she is saved and ascends to heaven.

**Irmelin.** Opera in 3 acts by Delius; text by the composer. Prod. Oxford, New T, 4 May 1953, with Graham, Copley, Round, Hancock, cond. Beecham.

The Princess Irmelin (sop) awaits her true love. Nils (ten), a prince disguised as a swineherd, is searching for the ideal woman; he is told he will find her at the end of the silver stream, and there he finds Irmelin.

**Isaac, Adèle** (b Calais, 8 Jan. 1854; d Paris 22 Oct. 1915). French soprano. Studied Paris with Duprez. Début Paris, T Montmartre, 1870 (Massé's *Les noces de Jeannette*). Brussels, M, 1872–3; Paris: OC, between 1873 and 1894; O, 1883–5. A brilliant coloratura singer, successful as Gounod's Juliette, Violetta, etc., she created Olympia and Antonia in the first *Tales of Hoffmann*, persuading Offenbach to rewrite the Doll's music to suit her skills. (He also altered the other roles for her, but died before completing a florid aria for Giulietta; this act was thus omitted in the première.) First Minka in Chabrier's *Le roi malgré lui*.

**Isabeau.** Opera in 3 acts by Mascagni; text by Illica. Prod. Buenos Aires, Coliseo, 2 June 1911, with Farneti, Saludas, Galeffi, cond. Mascagni; Chicago, Auditorium, 12 Nov. 1917, with Raisa, Crimi, Rimini, cond. Campanini.

Cavaggio, the Middle Ages. Unwilling to choose a husband, the Princess Isabeau (sop) is made to ride naked through the streets in the middle of the day; anyone who dares look at her will be condemned to death. This edict is disobeyed by Folco (ten), a young forester, and Isabeau falls in love with him; the crowd, however, lynch him, and Isabeau kills herself over his dying body.

**Isabella.** (1) An Italian lady (con), lover of Lindoro, in Rossini's *L'italiana in Algeri*.

(2) The Princess Isabelle (sop), later married to Robert, Duke of Normandy, in Meyerbeer's *Robert le diable*.

**Ismaele.** The nephew (ten) of Sedecia, King of Jerusalem, lover of Fenena and loved by Abigaille, in Verdi's *Nabucco*.

**Isolde.** The Irish princess (sop), wife of King Mark and lover of Tristan, in Wagner's *Tristan und Isolde*.

**Isolier.** The Count's page (mez) in Rossini's *Le Comte Ory*.

**Isouard, Nicolò** [Nicolò de Malte] (*b* Malta, 6 Dec. 1775; *d* Paris, 23 Mar. 1818). Maltese composer; known simply as 'Nicolò' because his family disapproved of his music. Studied Paris; abandoned commerce to study composition, notably with Sala and Guglielmi. His first opera, *L'avviso ai maritati*, had some success, but it was *Artaserse* which launched him. Held ecclesiastical appointments Malta 1794–8, also composing seven operas for the theatre in Valetta, including *Il barbiere di Siviglia* and *Ginevra di Scozia*. After the French occupation of Malta in 1798, he went to Paris, where he contributed numbers to his friend Rodolphe Kreutzer's *Le petit page*, 1800. Their unsuccessful *Flaminius à Corinthe* was followed by revivals of two of his Italian operas with rewritten librettos. These works, *Le tonnelier* and *L'impromptu de campagne*, established him as a leading composer for the French stage.

His first original works for Paris were *La statue*, initiating his collaboration with the librettist F. B. Hoffman, and *Michel-Ange*. During 18 years in Paris he wrote 30 operas, the most successful being *Les rendez-vous bourgeois*, *Cendrillon*, *Joconde*, and *Jeannot et Colin*. In his early works, with their instant melodic charm and light-hearted spirit, he reflects the Neapolitan style in which he was reared, showing a special talent for the handling of ensembles. Later, in France, he developed the lyrical aspects of the opéra-comique in a series of comic operas best characterized as escapist: a distraction from current events, rather than a reflection or witty distortion. His initial popularity in France lay partly in the absence of any serious competitor: after Boieldieu returned from Russia in 1811, and Auber's first Parisian work was performed in 1813, Isouard lost some of his following. He was, however, appreciated abroad, especially in Vienna, where *Le billet de loterie* and *Le magicien sans magie* were given shortly after their Paris premières.

SELECT WORKLIST (all first prod. Paris unless otherwise stated): *L'avviso ai maritati* (Gonella; Florence 1794); *Artaserse, re di Persia* (Metastasio; Livorno 1794); *Il barbiere di Siviglia* (?Petrosellini, after Beaumarchais; Valetta 1796); *Il tonneliere* (?; Valetta 1796; rev. with lib. by Delrieu and Quêtant as *Le tonnelier*, 1801); *L'improvisata in campagna* (Valetta 1797; rev. with lib. by Delrieu as *L'impromptu de campagne*, 1801); *Ginevra di Scozia* (?; Valetta c.1798); *Le petit page* (Pixérécourt; 1800, rev. 1804; comp. with R. Kreutzer); *Flaminius à Corinthe* (Pixérécourt & Lambert; 1801; comp. with R.

Kreutzer); *La statue* (Hoffman; 1802); *Michel-Ange* (Delrieu; 1802); *Les rendez-vous bourgeois* (Hoffman; 1807); *Un jour à Paris* (Étienne; 1808); *Cendrillon* (Étienne, after Perrault; 1810); *Le billet de loterie* (Creuzé de Lesser & Roger; 1811); *Le magicien sans magie* (Creuzé de Lesser & Roger; 1811); *Lully et Quinault* (Gaugiran-Nanteuil; 1812); *Joconde, ou Les coureurs d'aventures* (Étienne; 1814); *Jeannot et Colin* (Étienne; 1814).

BIBL: E. Wahl, *Nicolò Isouard: sein Leben und sein Schaffen auf dem Gebiet der Opéra Comique* (Munich, 1906).

**Israel.** In 1920 Mordecai Golinkin organized a benefit in Petrograd with Shalyapin, and the proceeds enabled him, when he emigrated to Palestine, to stage a performance of *La traviata* in a Tel Aviv cinema and launch the Palestine OC. By 1927, when the company disbanded, he had presented over 20 operas, all in Hebrew, including *Samson et Dalila*, *La Juive*, and Rubinstein's The Maccabees. He was later one of the founders of the Palestine Folk O in 1941: the repertory included M. Lavry's *Dan ha-Shomer* (Dan the Guard, 1945). In 1933 a Chamber Opera was formed under Benno Frankel.

In 1945 the soprano Edis de Philippe formed a national opera company to coincide with the foundation of the Jewish state. On 29 Nov. 1947, the day on which the UNO voted the partition of Palestine, she staged a gala of excerpts by the new Hebrew National O, followed by *Thaïs* with herself in the title role. In 1958 the company, now the Israel National OC, moved into the former Knesset (cap. *c*.900) in Tel Aviv. By the 25th season there had been 43 productions. Edis de Philippe died in 1978, and a difficult period followed, with the country largely dependent on visiting companies and conductors. In 1985 the New Israeli O was formed, and plays in the Noga Hall (cap. 830) and Duhl Auditorium. Operas are given in the vernacular, with Hebrew surtitles, and an international repertory includes operetta and musicals, finding a place for works with special associations (e.g. *Nabucco*, Goldmark's *The Queen of Sheba*). Wagner and Strauss are still excluded on the grounds of being associated with Nazism. The company also plays in Haifa and Jerusalem, where there is an Opera Society that also tours, and performances are given in the Roman amphitheatre at Caesarea.

Operas by native-born Israelis include *Alexandra* (1959) and *Independence Night Dream* (1973) by Menachem Avidom. Operas by composers settled in Israel include those by Josef *Tal (*b* 1910) and *The War of the Sons of Light* (1972) by Ami Maayani (*b* 1936).

**Istanbul.** City in Turkey, originally Byzantium (*c*.658 BC–AD 330), then Constantinople 330–1930, capital of the Ottoman Empire

353

1453–1924. The first theatre on Western models was the Naum T, opened by the Syrian Mihail Naum in 1844 with Donizetti's *Lucrezia Borgia*. An Italian company played 1845–6. The theatre burnt down, but reopened 1848 with *Macbeth*. The 1849–50 season included six operas by Bellini, Donizetti, and Verdi, given by an Italian company of 36. The 1850–1 season opened with *Robert le diable* and included *Attila*. Naum went on to stage *Il trovatore* (1858) and *Il barbiere* (1854), but despite the grant of a monopoly he needed subsidy. In 1864–5 another Italian company, which included Adelaide Ristori, opened with *I vespri siciliani*. In the rebuilt theatre, *La muette de Portici* was given in 1869, but the theatre burnt down in 1870. A private opera-house (cap. 300) was built in 1859 by Sultan Abdülmecit in his Dolmabahçe Palace, with Donizetti's brother Giuseppe as Director; this burnt down in 1863. Another theatre was built by Sultan Abdülhamit in the Yıldız Palace in 1889: opera was staged here until 1908. Despite other enterprises, including operetta at the Şehir Tiyatrosu T in the 1930s, opera was not established permanently until the Istanbul State O was formed in 1969. In that year was opened the Atatürk Kültür Mevrkezi (Atatürk Culture Centre), with an opera-house (cap. 1,307). The theatre burnt down in 1971. The first Istanbul Festival in 1973 included the first performance of any kind in the precincts of Topkapı Palace, of Mozart's *Entführung* (concluding on a reconciliatory gesture of released doves). The company tours to Ankara, İzmir, Bursa, and the Turkish part of Cyprus.

**'It ain't necessarily so'.** Sporting Life's (ten) cynical song in Act II of Gershwin's *Porgy and Bess*, casting doubt on the truths put forward by the Bible.

**Italiana in Algeri, L'** (The Italian Girl in Algiers). Opera in 2 acts by Rossini; text by Angelo Anelli, originally set by Luigi Mosca (1808). Prod. Venice, B, 22 May 1813, with Marcolini, Annibaldi, Berni Chelli, Gentili, Galli, Rosich; London, H, 26 Jan. 1819, with Giorgi-Belloc, Ambrogetti, García, Placci; New Orleans, T d'Orléans, 24 Apr. 1832, with Saccomano, Marozzi, Verducci, Manetti, Orlandi, Fornasari, Sapignoli, cond. Rapetti.

Mustafà, the Bey of Algiers (bs), wishes to find an Italian wife, and force his present wife Elvira (sop) to marry an Italian slave, Lindoro (ten). By chance an Italian ship is wrecked off the coast, and Isabella (con)—who has been travelling with an elderly admirer, Taddeo, whom she passes off as her uncle—comes ashore. She is in fact seeking her long-lost lover, Lindoro. She persuades Mustafà to give her Lindoro as her personal slave.

Lindoro convinces Mustafà that Isabella will marry him if he joins the noble order of the 'Pappataci', whose primary rules are to eat and be silent. During the initiation ceremony, organized by the Italians, Isabella and Lindoro make their escape, and Mustafà is obliged to return to Elvira.

**Italian overture.** In contrast to the earlier practice of Venetian opera, where the overture (usually called 'canzona' or 'sonata') took the form of a slow introduction in duple metre, followed by a faster movement in triple rhythm, A. Scarlatti introduced a new form *c.* 1680 (in *Tutto il mal non vien per nuocere*) in which the slow movement was framed by two fast outer movements. Usually known as the Italian overture, or 'sinfonia', this pattern suggests the later structure of the symphony and was in vogue, alongside the *French overture, up to the middle of the 18th cent.

**Italy.** The emergence of opera as a distinctive art form is widely accepted as dating from the meetings of the Bardi *camerata in Florence in the late 16th cent., although the influence of earlier forms in which music and drama combined, notably the *intermedio and *pastorale, must be acknowledged. The first opera produced by members of the Camerata was *Peri's *Dafne* (1598), but little of this survives. More significant therefore are the *Euridice* operas of Peri and Caccini (1600); also Cavalieri's *Rappresentazione di anima e di corpo* (1600), which introduced the new *stile recitativo to Rome, and Agazzari's *Eumelio* (1606).

One of the earliest composers to emerge was also to prove one of the art's greatest geniuses, Claudio *Monteverdi, then working at Mantua. Not only is his music of the highest inspiration, but his vision of opera as a drama in music has remained valid through all its history. The early operas had placed great emphasis on the stile recitativo, pursuing the intention of the Camerata to imitate (as they thought) ancient Greek drama, but Monteverdi greatly enriched the genre, both in form and in subtlety of characterization, giving the recitative expressive force, the orchestra eloquence, and the solo arioso numbers concentrated emotional power. His *Orfeo* (1607), represents a merging of the old tradition of the intermedio with the new practices of opera; particularly important was his skilful use of the choral or orchestral ritornello as a unifying element. Its success was repeated with *Arianna* (1608), of which only the famous lament survives.

Gradually opera spread to other Italian cities, notably Rome, where it flourished under the patronage of the *Barberini family up to their exile in 1644. Roman composers continued the process begun by Monteverdi, and increasingly differentiated more clearly between aria and recitative: in the operas of Stefano *Landi there is an increased emphasis on melody in the arias, while in his *Sant'Alessio* (1632) a sinfonia before Act II suggests the arrangement of the later *Italian overture in its 3-movement structure. These works also made increased use of ensemble and, beginning with Landi's *La morte d'Orfeo* (1619), comic scenes and characters were introduced.

With the opening to the public of the T San Cassiano in Venice in 1637, opera was made available to a wider audience than the aristocracy. At the same time, the formality of the manner had begun to demand comic relief. Venetian opera of this period, as exemplified above all in the late works of Monteverdi, and those of *Cesti, and *Cavalli, developed greater variety and expressiveness, with more complicated plots, livelier characterization, comic episodes, and numbers which displayed some of the favourite singers who had begun to make careers in opera and to cultivate the art of *bel canto. With firmer separation between recitative and aria, the latter developed greater formality, the *da capo form making its first appearance. Under the new commercial constraints, the role of both orchestra and chorus was much reduced. Opera continued to spread across Italy, back to Florence with Cesti, and to Naples with the work of Francesco *Provenzale. Among the first Italians to establish themselves in Germany were Agostino *Steffani, P. A. *Ziani, A. *Draghi, *Sances, and *Pallavicino.

By the beginning of the 18th cent. Italian opera had achieved a standardization that, though rigid, provided an example for native and now also for foreign composers. The genre *opera seria, frequently also called Neapolitan opera (though the validity of this connection has long been challenged) was established largely by the librettists Apostolo *Zeno and Pietro *Metastasio who, in organizing the free forms of 17th-cent. opera into a firm 3-act structure, dispensed with much that threatened dramatic logic, but provided a confining strictness. The subjects of opera seria were generally classical history and legend, with action and reflection scrupulously balanced in the course of recitative and aria: in general, the action took place in recitative, with the arias providing points of reflection and repose, also of musical expression and the chance of personal display by the singer. To this pattern the other operatic ingredients were subject; and great attention devolved upon the *aria. This form not only gave opera a strict framework in which composers could work; it encouraged the rise of a school of virtuoso singers, especially the

*castrato. The major figure in the transition to opera seria was Alessandro *Scarlatti and the genre was firmly established with Antonio *Vivaldi, Nicola *Porpora, Leonardo *Leo, and Johann Adolf *Hasse. But the composer who produced the greatest and most original work in the convention was *Handel. Subsequent composers of opera seria include Nicolò *Piccinni, Antonio *Sacchini, Antonio *Salieri, and the young *Gluck and *Mozart. It was the strictness and high seriousness of opera seria that encouraged the development of separate comedies, sometimes ending an act or inserted as *intermezzos, hence the growth of *opera buffa.

Reaction against so strict a form was inevitable, and Nicolò *Jommelli and Tommaso *Traetta helped to pave the way for Gluck. By now, Italian opera had begun to establish itself firmly in most countries. Germany was long dependent on the many travelling companies that crossed Europe, such as those of the brothers *Mingotti, for vernacular entertainment, but many courts also supported an Italian composer. France had developed its tragédie lyrique from the example of an Italian, Lully, and was to be frequently dependent upon great Italian composers such as Cherubini, Spontini, and Rossini. Though attempts to found a school of English opera in the late 17th cent. foundered, the influence of Italian works was equally strong here, and seen especially in Purcell's *Dido and Aeneas*. Russia cultivated its national tradition by the classic method of first importing Italians to show the way, then sending native composers to study in Italy, finally developing a national genre based on folk-song and folk legend grafted at first on to Italian stock. By the early 19th cent. Germany and France were becoming determinedly independent of Italian example, and the Paris Opéra was forming a style, *Grand Opera, that in turn embraced Italians including Cherubini, Spontini, Rossini, and Verdi.

However, Italian opera continued as a central tradition, and Italian singers and singing teachers continued to set such a powerful example that much of the desire for reform sprang from a wish to diminish their influence in favour of a more dramatic art. Rossini, though happy to write for great singers, restrained them by writing out most of his notes as they were to be sung. He revealed the comic possibilities of opera buffa to a matchless degree, and confirmed a number of conventions in what has been called the *Code Rossini; he was also a master of Romantic opera and Grand Opera. Italian Romantic opera reaches its fullest stature in *Bellini and *Donizetti; its greatest master was *Verdi. For all its advances, Italian opera remained grounded on its original principles—melody that was a stimulus to the finest vocal art, dramatic plots, clarity of presentation, and powerful emotions presented with a sense of theatrical effect.

Such was the strength of this tradition that little effective influence reached Italy, though there were composers (including Verdi) who felt the force of Wagner. But partly in reaction to the drama that dealt with grand and heightened emotions, there developed the manner known as *verismo, the so-called *squarcio di vita* or 'slice of life' referred to in an early classic of the genre, *Leoncavallo's *Pagliacci*. The emotions here, too, quickly tended to be extreme, as in *Mascagni, *Giordano, and *Puccini, whose work proved immensely popular and stimulating to other countries' composers. Puccini remains one of the last composers with a substantial body of work that features regularly in world repertories.

Italian opera continues to be vigorous, and to draw composers from Puccini's successors such as *Zandonai, *Malipiero, and many others, including Nono and Maderna, down to the youngest generation. In recent times the increasing diversity of musical drama has challenged the traditional Italian supremacy; but the art remains powerfully rooted in Italy, and the traditions that spring from much in Italian nature continue to be carefully cultivated in many cities and towns of the peninsula.

See also BARI, BERGAMO, BOLOGNA, BRESCIA, CAGLIARI, CATANIA, COMO, FERRARA, FLORENCE, GENOA, MACERATA, MANTUA, MILAN, NAPLES, NOVARA, PADUA, PALERMO, PARMA, PESARO, PIACENZA, RAVENNA, REGGIO EMILIA, ROME, SIENA, SPOLETO, TREVISO, TRIESTE, TURIN, UDINE, VENICE, VERONA, VICENZA.

**'Ite sull' colle'.** Oroveso's (bs) aria opening Act I of Bellini's *Norma*, urging the Druids to go upon the hills to see when it is the new moon.

**Ivanhoe.** Opera in 5 acts by Sullivan; text by Julian Sturgis, after Scott's novel (1818). Prod. for inauguration of short-lived Royal English Opera House (now Palace T) in Cambridge Circus, London, 31 Jan. 1891, with Macintyre, Ben Davies, ffrangcon-Davies, cond. Sullivan.

England, late 12th cent. Ivanhoe (ten) is in love with Lady Rowena (sop), but her guardian Cedric (bar) wishes her to marry a descendant of the crown. Two knights Templar, De Bracy (ten) and Gilbert (bar), attend a tournament and are so enamoured of Rowena and a Jewish girl, Rebecca (sop), that they capture them, along with Cedric and Rebecca's father Isaac (bar), and imprison them in the castle of Torquilstone. Ivanhoe besieges the castle, but the knights escape, taking Rebecca with them. In a duel between Ivanhoe and Gilbert, the latter is killed. King Richard the Lionheart (bs) orders the disbandment of the Templars, and permits the marriage of Ivanhoe and Rowena.

Sullivan's only Grand Opera. Originally had a continuous run of 160 performances.

Other operas on the subjects are by Rossini (1826), Marschner (1829), Pacini (1832), Savi (1863), and Ciardi (1886).

**Ivanov, Alexey** (*b* Chizhovo, 22 Sep. 1904; *d* 1988). Russian baritone. Studied Tver, and Leningrad with Bosse. Leningrad, Maly T, 1932–6; Saratov O and Gorky O, 1936–8; Moscow, B, 1938–67. Roles incl. Pizarro, Prince Igor, Onegin, Escamillo, Bolkonsky (*War and Peace*); sang Ryleyev in first *The Decembrists* (Shaporin, 1955). (R)

**Ivanov, Andrei** (*b* Samostye, 13 Dec. 1900; *d* Moscow, 1 Oct. 1970). Ukrainian baritone. Studied Kiev with Lund. Début with travelling opera co. 1925. Baku 1926; Odessa 1928–31; Sverdlovsk, 1931–4. Kiev O 1934–49; Moscow, B, 1950–6. His large repertory incl. Rigoletto, Prince Igor, Onegin, Mazeppa, Mizgir. (R)

**Ivanov, Nikolay** (*b* Poltava, 22 Oct. 1810; *d* Bologna, 7 July 1880). Russian tenor. Travelled with Glinka to Italy, where he studied Milan and Naples with Bianchi, Nozzari, and Fodor-Mainvielle. Début Naples, C, 1832 (Percy, *Anna Bolena*). Paris, I, 1833; London, CG, 1834; Milan, S, 1843. Subsequently sang mainly in Italy; never returned to Russia. Much admired in operas by Bellini, Donizetti, and Rossini (who was his close friend). Created roles in works by Pacini; Verdi added two arias, suggested by Rossini, to *Ernani* and *Attila* for him. Though uninspired on stage, he was an elegant singer, of whom Chorley said, 'Nothing could be more delicious as to tone'.

**Ivanovich, Cristoforo** (*b* Budva, 1628; *d* Venice, Jan. 1689). Italian librettist and theatre historian. Lived in Venice from 1657. Wrote librettos for Ziani and Cavalli, but is best remembered for the chronical of Italian opera, *Le memorie teatrali di Venezia*, which he first published 1681 in his *Minerva al tavolino*. In its second edition (1688), this is one of the most important sources for the early Venetian repertory, covering the period 1637–87.

**Ivan Susanin.** A peasant (bs), father of Antonida, in Glinka's *A Life for the Tsar*. Sometimes used as the title of the opera.

**Ivan the Terrible.** Ivan IV (1530–84) reigned as tsar 1547–84. He appears (bs) in Bizet's *Ivan IV* (comp. 1862–3) and Rimsky-Korsakov's *The Maid of Pskov* (1873), in a silent role in Rimsky-Korsakov's *The Tsar's Bride*, and as a crucial part of the plot of Tchaikovsky's *The Oprichnik*

(1874). The censor refused to allow his appearance in the latter work, though he is referred to by the word 'grozny', or 'terrible', here and also in Musorgsky's *Boris Godunov*. Other works on him are by D'Orgeval (1876) and Gunsbourg (1910).

**Ivogün, Maria** (*b* Budapest, 18 Nov. 1891; *d* Beatenberg, 2 Oct. 1987). Hungarian soprano. Studied Vienna with Schlemmer-Ambros, Munich with Schöner. Début Munich 1913 (Mimì). Sang there until 1925. London, CG, from 1924. Berlin, SO, 1925–34. Chicago 1922-3. Created Ighino (*Palestrina*). A musical singer with an exceptionally flexible voice, famous as Constanze, Norina, Gilda, etc. Strauss described her Zerbinetta as 'without rival'. (R)

**Ivrogne corrigé, L', ou Le mariage du diable** (The Drunkard Reformed, or The Devil's Marriage). Opera in 2 acts by Gluck; text by Louis Anseaume, after La Fontaine's fable *L'ivrogne en Enfer*. Prod. Vienna, B, Apr. 1760; London, Birkbeck Coll., 12 Mar. 1931 (as *The Devil's Wedding*); Hartford, CT, 26 Feb. 1945 (as *The Marriage of the Devil*).

A vineyard, *c.*1760. Colette (sop) and her lover Cléon (bar) devise a scheme to outwit her Uncle Mathurin's plan to marry her to his old friend Lucas. When Mathurin (ten) and Lucas (bs) are drunk one evening, the couple simulate a scene in hell, terrifying the men into allowing Colette to marry Cléon. When the deception is revealed, Mathurin resolves to give up drinking, and to abandon Lucas.

# J

**jácara.** A kind of *entremés performed after the second act of a serious play, usually featuring the escapades of a lovable rogue.

**Jackson.** City in Mississippi, USA. Opera/South was founded in 1971 to involve the local Black community as audience and performers, and opened with *Aida* in May in the Municipal Auditorium. It has also given the premières of William Grant Still's *Bayou Legend* (comp. 1941) and *Highway no. 1, USA* (comp. 1962), and Ulysses Kay's *Jubilee* (1976), as well as standard works including *L'elisir d'amore* set on a Mississippi steamboat and an all-Black *Otello* with, as late substitute, a White hero. Debria Brown and Faye Robinson are among artists who have gone on to successful careers. Walter Herbert was music director 1971–5.

**Jackson, William** (*b* Exeter, 29 May 1730; *d* Exeter, 5 July 1803). English composer, organist, essayist, and painter. Active in Exeter as teacher, and from 1777 as organist at the cathedral. Popular as a song composer, especially through his three sets of *Twelve Songs* (1755, *c*.1765, *c*.1770). His first opera, *The Lord of the Manor*, was in repertory for over 50 years and, like the comic *The Metamorphosis*, was characterized by simple and graceful music, with an immediate appeal. His *Observations on the Present State of Music in London* (1791) is an important record of contemporary musical life.

WORKLIST: *The Lord of the Manor* (Burgoyne, after Marmontel; London 1780); *The Metamorphosis* (Jackson; London 1783).

**Jacobin, The** (Cz.: *Jakobín*). Opera in 3 acts by Dvořák; text by Marie Červinková-Riegrová. Prod. Prague, C, 12 Feb. 1889, with Foerstrová-Lautererová, Cavallarová, Veselý, Heš, Kroessing, cond. Čech; London, St George's H, 22 July 1947 by Worker's Music Assoc., with Vowles, Taplin, Lensky, Dargavel, Laurie, Davidson, R. Davies, cond. Corbett; Washington, Lisner Auditorium, 28 Oct. 1979, cond. Husa.

The opera tells of how the Jacobin Bohuš (bar), returned from political exile, is helped to re-establish his position in the community by his friend Benda (ten), a musician.

**Jacovacci, Vincenzo** (*b* Rome, 14 Nov. 1811; *d* Rome, 30 Mar. 1881). Italian impresario. Began career as a fishmonger, then managing seasons in various Rome theatres. His enterprise led to his being arrested for selling too many tickets for the première of Donizetti's *Adelia* (1841, with Giuseppina Strepponi, godmother to his daughter). He went on to become one of the leading impresarios of mid-19th-cent. Italy. He championed Verdi, giving at T Apollo, Rome, the premières of *Trovatore* (1853) and *Ballo in maschera* (1859) and the Italian première of *Forza* (1863). However, his sharp business sense led him to engage cheap singers, to the irritation of Verdi (who called one 'a wax dummy'). His productions were nevertheless based on spectacle and a sense of show that attracted the public.

**Jacquino.** Rocco's assistant gaoler (ten) in Beethoven's *Fidelio*.

**Jadlowker, Hermann** (*b* Riga, 20 July 1877; *d* Tel Aviv, 13 May 1953). Latvian tenor. Studied Vienna with Gänsbacher. Début Cologne 1899 (Gomez in Kreutzer's *Nachtlager von Granada*). Karlsruhe 1906–9; Paris, C, 1910; New York, M, 1910–13; Berlin, H, 1909–21. Created King's Son (*Königskinder*), and Bacchus (*Ariadne auf Naxos*). Possessed a large, mellow voice and a superb technique; his repertory encompassed Rossini's Count Almaviva and Parsifal. (R)

**James, Henry** (*b* New York, 15 Apr. 1843; *d* London, 28 Feb. 1916). US, later English, novelist. Operas on his works are as follows:

*The Aspern Papers* (1888): Argento (1988)
*Owen Wingrave* (1892): Britten (1971)
*The Turn of the Screw* (1898): Britten (1954)

**Janáček, Leoš** (*b* Hukvaldy, 3 July 1854; *d* Moravská Ostrava, 12 Aug. 1928). Czech composer. Studied Brno with Křížkovský, Prague with Skuherský, Leipzig with Paul and Grill, Vienna with Krenn.

Janáček always regarded opera as lying at the centre of his creative life. The opening of the Czech Theatre in Brno in January 1885 encouraged him to sketch an operatic synopsis on Chateaubriand's *The Last Abencerage*, but his first completed opera was *Šárka*, to a text originally intended for Dvořák. When Dvořák turned it down, Janáček composed it without the author's permission. It is a remarkably assured and well-written work with some anticipations of his later style; and it is a stronger piece than its successor, *The Beginning of a Romance*. Here, in an attempt to compose a light folk comedy, he reverted to an unsuccessful version of opera based on folk-songs and dances.

Through this work, however, he came to know

Gabriela Preissová's story *Její pastorkyňa*, which provided the basis of his next opera. Better known outside Czechoslovakia as *Jenůfa*, it was his first operatic masterpiece and has remained perhaps the best-known of them. The work's gestation was long, and shows in the move away from number-orientated opera (which lies behind the first act) to a more freely composed manner depending largely on monologue or soliloquy. This also enabled him to develop a more personal idiom by the use of speech-melodies, melodic curves which owed their shapes to his observations of Czech (especially Moravian) vocal inflections but their character to his own lyrical gift. They also enabled him to place this gift at the service of the truthful yet compassionate observation of human nature which is first fully revealed in *Jenůfa*. Further, he was able to make use of these identifying melodic phrases motivically (though not in any manner approaching Leitmotiv) so as to construct whole scenes. It is his first opera to a prose text, though he was obliged to give the text some metrical symmetries so as to match his musical intentions. The first performance of *Jenůfa* caused little widespread attention, and it was not until the Prague première of a revised version in 1916 that the work achieved real success.

Meanwhile, Janáček turned his attention first to *Osud*, or *Fate*, a work which marks some musical advance but a reverse in its use of a clumsy, semi-autobiographical text. There are also problems with the text of *The Excursions of Mr Brouček*. The opera is divided into two parts (a third was initiated but not completed). The adventures of the stolid burger Brouček as he is precipitated first into a society of empty aesthetes on the moon, and then back into the Hussite wars, owe something to the Viennese magic comedy of Raimund, but are essentially local; yet the opera's musical inventiveness has earned it increasingly wide acceptance.

Among other subjects contemplated by Janáček, in his affection for all things Russian, was *Anna Karenina*; but he eventually turned to Ostrovsky for *Káťa Kabanová*. Even more than in *Jenůfa*, he here develops a motivic identification with characters, including the expression of the subtlety of Káťa herself by the use of an exceptionally wide range of motives for her, and the dependence of some of the subsidiary characters on her by their assumption of her motives. He can also characterize more freely than in earlier works by means of elegance and charm of line for the suffering Káťa, light superficiality for certain others, a grim abruptness for the mother-in-law responsible for Káťa's downfall and death.

Janáček was by now internationally famous, and his operas were taken up especially in Germany. More favourable operatic conditions at home gave him encouragement; and he was sustained by a passionate friendship with a younger woman, Kamila Stösslová, with whom he identified the heroines of his operas. His next work, however, showed his ability to draw on unlikely sources, in his use of a strip cartoon for an opera on the life and death of a fox, *The Cunning Little Vixen* (as it is known in English: a more correct translation is 'The Adventures of Vixen Sharpears'). The narrative is here somewhat disjunct, but the subject enabled Janáček further to refine and concentrate his idiom without loss of lyrical intensity. It is characteristic in its laconic yet passionate, as well as compassionate, invention, and in its nearly pantheistic devotion to nature and the forces binding men, animals, and the forest. (There is, however, no attempt to make human-animal equivalents in the plot, as in the German version by Max Brod).

*The Makropoulos Case* turns from the life of the woods to an urban world of legal wrangles and the unnatural figure of a singer whose use of an elixir of life has condemned her to a tragic immortality; yet the theme is related in that Janáček is pointing to the justness of a natural life span, and insisting that it is the certainty of death that gives beauty and meaning to life. The musical idiom is sharper and more rapid than in previous works, with—until the final monologue—an even greater detachment between voice and orchestra. His final work was on another unlikely subject, Dostoyevsky's prison diaries *From the House of the Dead*. There is no female part here, and the series of short narratives, deriving from multiple characterizations of the various prisoners, is made to build up an image of different conditions of men drawn together by suffering and rising above it. At the head of the score he set the words, 'In every human being there is a divine spark.'

WORKLIST (all first prod. Brno unless otherwise stated): *Šárka* (Zeyer; comp. 1887–8, rev. 1918–19, 1925, prod. 1925); *Počátek románu* (The Beginning of a Romance) (Tichý, after Preissová; 1894); *Její pastorkyňa* (Her Foster-daughter, known as *Jenůfa*) (Janáček, after Preissová; comp. 1894–1903, prod. 1904); *Osud* (Fate) (Bartošová & Janáček; comp. 1903–5, rev. 1906–7, prod. 1958); *Výlet pana Broučka do měsíce* (Mr Brouček's Excursion to the Moon) (Janáček, with Gellner, Dyk & Procházka, after Čech; comp. 1908–17, prod. Prague 1920); *Výlet pana Broučka do XV. století* (Mr Brouček's Excursion to the 15th Century) (Procházka, after Čech; comp. 1917, prod. Prague 1920); *Káťa Kabanová* (Janáček, after Ostrovsky; 1921); *Příhody Lišky Bystroušky* (The Adventures of Vixen Sharpears, known as *The Cunning Little Vixen*) (Janáček, after Těsnohlídek; 1924); *Věc Makropulos* (The Makropulos Affair) (Janáček, after Čapek; 1926); *Z mrtvého domu* (From the House of the Dead) (Janáček, after Dostoyevsky; 1930).

CATALOGUE: B. Štědroň, *Dílo Leoše Janáčka* (Prague, 1959; trans. London, 1959).

WRITINGS: B. Štědroň, ed., *Janáček ve vzpomínkách a dopisech* (Prague, 1946), (trans. as *Janáček: Letters and Reminiscences* (Prague, 1954)); V. & M. Tausky, *Leoš Janáček: Leaves from his Life* (New York, 1982); M. Zemanová, *Janáček's Uncollected Essays on Music* (London, 1989).

BIBL: J. Vogel, *Leoš Janáček* (Kassel, 1958, trans. 1962, 2/1982); E. Chisholm, *The Operas of Leoš Janáček* (Oxford, 1971); M. Ewans, *Janáček's Tragic Operas* (London, 1971).

**janissary music.** The *yeniçeri* (from Turk. *yeni*, new and *çeri*, soldiery) were originally raised during the 14th cent. to provide a bodyguard for the Sultan; gradually the term came to refer less specifically to any Turkish infantryman. As they marched, the janissaries were accompanied by the distinctive sound of a Turkish band, usually including crescent ('Jingling Johnny'), kettledrum, bass drum, cymbals, and triangle. These bands attracted widespread attention during the mid-18th cent.: following the gift of one from the Turkish Sultan to August II of Poland (1735) they were acquired by militia throughout Europe. Many opera composers of the period sought to recreate the unique sound, as well as some of the characteristic rhythmic features of their music, especially in those works which, following a vogue of the 1770s and 1780s, dealt with Turkish themes. Probably the earliest suggestion of 'Turkish' music in opera is the use of the cymbals in Strungk's *Esther* (1680); more famous examples are to be found in Haydn's *Lo speziale* (1769), Gluck's *La rencontre imprévue* (1764), Grétry's *La fausse magie* (1778), Süssmayr's *Soliman der Zweyte* (1799) and *Gülnare* (1800), and above all in Mozart's *Die Entführung aus dem Serail* (1782), where the influence is most strongly felt, and most explicitly acknowledged.

**Janowitz, Gundula** (*b* Berlin, 8 Feb. 1937). German soprano. Studied Graz with Thöny. Début Vienna, S, 1960 (Barbarina). Regular appearances there, and at Salzburg from 1963. Gly. 1964; New York, M, 1967; Paris, O, 1973; London, CG, 1976. Roles include Pamina, Donna Anna, Leonore, Eva, Sieglinde, Ariadne. Possesses a vibrant and expressive voice. (R)

**Janssen, Herbert** (*b* Cologne, 22 Sep. 1892; *d* New York, 3 June 1965). German, later US, baritone. Studied Berlin with Daniel. Début Berlin, S, 1922 (Schreker's *Der Schatzgraber*), S until 1938; London, CG, 1926–39; Paris, O, 1929; Bayreuth 1930–7; Buenos Aires 1938; New York, M, 1939–52. A fine singer with a warm tone and much dramatic intelligence, celebrated as Wolfram, Kurwenal, Kothner, Gunther, Amfortas; sang Sachs and Wotan less successfully. (R)

**Japan.** Japanese musical theatre is the most important form of Japanese drama, and centres on the Noh Play ('accomplishment' or 'art' play). Several elements, including the narrative courtly dance of Bugaku and the importation of Gagaku from Korea in AD 612, combined to bring Noh to its flowering in the 14th cent.; nothing of importance has been written since 1600.

The Noh Play is a fusion of song (*uta*) by a chorus (*ji*) of 8–10, recitation (*kotoba*), dance (*mai*), and instrumental music (*hayashi*), the latter provided by flute and three drums of variable, determinate pitch. An average evening Noh consists of several plays, each lasting up to about three-quarters of an hour, first a warrior play; second a female-wig play, in which a man impersonates a woman, often mad or in tragic circumstances; third a play of freer nature and subject, usually sensational; and a concluding dance play. The most famous author of Noh Plays was Zeami Motokiyo (1363–1443), who also composed, directed, acted, and danced; his son Jūro Motomasa (1395–1431) was the author of *Sumidagawa*, a celebrated female-wig play now also known in the West as the source of Britten's *\*Curlew River*. This quasi-operatic treatment of Noh was anticipated by Yoshiro Irino (*b* 1921) with *Aya no tsuzumi* (The Silken Drum, 1962), after Motokiyo, also the subject of Sveinsson's *Silkitromman* (1982), and by other Japanese composers.

Western opera was first given in Japan in 1894, but only established on a regular basis by the Fujiwara OC, founded in 1933 by the tenor Yoshie Fujiwara. The Niki-kai troupe, a kind of singers' co-operative founded in 1952, has reflected the Japanese enthusiasm for Wagner. Of many other groups, the Tokyo Chamber Opera Group has been outstanding; founded in 1970, it gives five productions a year. In spite of poor conditions, it has given operas ranging from *The Play of Daniel* to contemporary works. There have been many visits by Western companies, giving Japan a place on the international opera circuit. In general, local production and design standards have not matched those of performance. Opera is also given in Kyoto, Osaka, Fujisawa, Utsonomiya, and other cities and towns, and in universities.

An early opera by a Japanese was *Ochitaru tennyo* (The Depraved Heavenly Maiden) by Kosaku Yamada (1886–1965), written in 1912 while the composer was studying with Bruch in Berlin but not staged until 1929; he also wrote

*Ayame* (The Sweet Flag, 1931) for Paris. The most popular opera by a Japanese composer is *Yuzuru* (The Twilight Heron, 1952) by Ikuma Dan (*b* 1924), also composer of *Kikimimi zukin* (The Listening Cowl, 1955) and *Hikarigoke* (Shining Moss, 1972). Other successful works include *Aoki okami* (The Dark Blue Wolf, 1972) by Saburo Takata (*b* 1913); *Kesa to Morito* (1968) by Ishii Kan (*b* 1921); *Arima-no Miko* (Prince Arima, 1967) by Sadao Bekku (*b* 1922), a pupil of Milhaud and Messiaen; *Kurai kagami* (The Black Mirror, 1960) by Yasushi Akutagawa (1925–89); and *Iwai uta ga nagareru yoru ni* (The Night of the Wedding Song, 1985), by Kazuko Hara.

See also TOKYO.

**Jasager, Der** (The Yea-Sayer). Opera in 2 acts by Kurt Weill; text by Brecht, after a 15th-cent. Japanese Noh drama. Prod. Berlin 24 June 1930.

A boy plans to accompany his teacher and three older students on a strenuous expedition in search of a wise man, in order to find a cure for his ailing mother. Unable to complete the journey, he agrees to submit to a tradition which demands that anyone failing to reach the goal must be thrown into the valley; he does so on condition that the teacher takes the medicine back to his mother.

Weill intended this as a didactic opera concerned with intellectual and moral matters characterizing it as a *Lehrstück.

**Jason.** The leader of the Argonauts (ten), former husband of Medea, in Cherubini's *Médée*.

**Jassy.** See IAŞI.

**Jean de Paris** (John of Paris). Opera in 2 acts by Boieldieu; text by C. Godard d'Aucour de Saint-Just. Prod. Paris, OC, 4 Apr. 1812; London, CG, arr. Bishop with additions 12 Nov. 1814; New Orleans, T. d'Orléans, 26 Mar. 1816.

A Pyrenean village, 17th cent. The young widowed Princess of Navarre (sop) is destined to marry the Crown Prince of France (ten), but intends first to spend some time travelling. At a Pyrenean inn, preparations for her arrival are interrupted by the arrival of the suite of 'Jean de Paris' (the Prince in disguise); they take possession of the inn. The Princess's Seneschal (bar) is outraged, but Jean declares that the Princess may be his guest. She recognizes him, but conceals this, telling him as they dance that she has already chosen her husband; all is disclosed, and the Prince and Princess are united.

Other operas on the subject are by Morlacchi (text by Romani, after Saint-Just, 1818) and Donizetti (using Romani's text, 1839).

**'Je crois entendre encore'.** Nadir's (ten) aria in Act I of Bizet's *Les pêcheurs de perles*, in which he sings of his love for Leïla.

**'Je dis que rien ne m'épouvante'.** Micaëla's (sop) aria in Act III of Bizet's *Carmen*, in which she prays to God to deliver her from the fear that against her will has overcome her as she seeks Don José in the mountains.

**Jélyotte, Pierre** (*b* Lasseube, 13 Apr. 1713; *d* Oloron, 12 Oct. 1797). French tenor and composer. Studied Toulouse. Début Paris, O, 1733 (Blamont's *Les fêtes greques et romaines*). Remained at the Opéra until his retirement in 1765, replacing Denis-François Tribou in 1738 as leading singer. Sang many of the roles Tribou had created, e.g. Castor and Hippolyte, which Rameau revised for him. He also created several parts for Rameau, who was inspired by his high range (reaching d'') and flexibility to write the most brilliant and difficult music in the repertory of the haute-contre (e.g. in *Zaïs*), to which few subsequent singers could do justice. Lalande admired his voice for its 'beauty and large volume'; Rousseau, though disliking the voice type, was so impressed that he wrote Colin (*Le devin du village*) for him. Jélyotte's compositions include a comédie-ballet, *Zéliska* (1746).

**Jenifer.** A young girl (sop), bride of Mark, in Tippett's *The Midsummer Marriage*.

**Jeník.** A young villager (ten), son of Tobias Micha by his first marriage, in Smetana's *The Bartered Bride*.

**Jenůfa** (Cz.: *Její pastorkyňa*: Her Foster-Daughter). Opera in 3 acts by Janáček; text by the composer, after Gabriela Preissová's drama (1890). Prod. Brno 21 Jan. 1904, with Kabeláčová, Svobodová, Procházka, Staněk-Doubravský, cond. Hrazdira; New York, M, 6 Dec. 1924, with Jeritza, Matzenauer, Oehmann, Laubenthal, cond. Bodanzky; London, CG, 10 Dec. 1956, with Shuard, Fisher, Lanigan, E. Evans, cond. Kubelík.

Moravia, 1900. In the mill of Grandmother Buryjovka in the Moravian mountains lives the ne'er-do-well Števa Buryja (ten); his stepbrother Laca Klemeň (ten) is a farm-hand and their cousin Jenůfa (sop) helps in the house. She is the stepdaughter of Grandmother Buryjovka's daughter-in-law, who, from her position as sexton, is known as the Kostelnička (sop). Jenůfa is expecting Števa's child, and anxiously awaits the result of a conscription ballot to know whether he will return and marry her. He returns, unrecruited but also drunk, and the Kostelnička forbids him to marry Jenůfa until he has proved his worth by a year's total abstinence. When Števa, who is only physically attracted by Jenůfa, leaves, Laca offers her flowers and tries to kiss her; on being repelled, he slashes her face with a knife.

Jenůfa has secretly had a son. The Kostelnička, tormented by the disgrace, sends for Števa to marry Jenůfa; but he denies responsibility, and says that he is engaged to the mayor's daughter Karolka (mez). Laca appears, penitent and willing to marry Jenůfa though shocked by hearing of the birth of Števa's child. The Kostelnička, hoping to help the marriage on, tells Laca that the child has died; she takes it and drowns it in a brook. When Jenůfa awakes, she is also told that the child has died.

Jenůfa and Laca are about to be married, as are Števa and Karolka. As the ceremony begins, news comes that the body of a baby has been found under the ice. Jenůfa realizes the truth and reveals whose baby it is. To save her from being accused, the Kostelnička steps forward and confesses her guilt. Comforted by Jenůfa's forgiveness, she is led away. Jenůfa now turns to Laca and gives him his freedom; but he is faithful, and as the curtain falls they pledge their love.

**Jeremiáš, Otakar** (*b* Písek, 17 Oct. 1892; *d* Prague, 5 Mar. 1962). Czech composer and conductor. Studied with his father, **Bohuslav Jeremiáš** (1859–1918), founder of the South Bohemian Conservatory, and in Prague with Novák. Among other activities, director Prague, N, 1945–7. The most important of his operas, Romantic and post-Janáček in manner, is a setting of *The Brothers Karamazov* that has won much respect in his own country. His older brother **Jaroslav Jeremiáš** (*b* Písek, 14 Aug. 1889; *d* České Budějovice, 16 Jan. 1919) was a pianist and the composer of a number of works including an opera *Starý král* (The Old King) (Gourmont; Prague 1919).

SELECT WORKLIST: *Bratři Karamazovi* (O. Jeremiáš, after Dostoyevsky; Prague 1928); *Enšpígl* (Mařánek, after Coster; Prague 1949).

**Jerger, Alfred** (*b* Brno, 9 June 1889; *d* Vienna, 18 Nov. 1976). Austrian bass-baritone. Studied conducting, Vienna. Passau, Berlin, and Zurich as conductor. Singing début Zurich 1917 (Lothario, *Mignon*). Munich 1919; Vienna 1920–64; London, CG, 1924, 1934. A highly versatile artist; also a producer and libretto reviser. Large repertory included Leporello, Don Giovanni, Sachs, Beckmesser, Ochs, Barak, and Jonny (Krenek); created Mandryka (*Arabella*), and the Man (*Die glückliche Hand*). Appointed director of Vienna, S, 1945, and sang Count Almaviva (Mozart) in its reopening season. (R)

**Jeritza, Maria** (*b* Brno, 6 Oct. 1887; *d* Orange, NJ, 10 July 1982). Moravian soprano. Studied Brno, Prague with Auspitzer, later New York with Sembrich. Début Olomouc 1910 (Elsa). Vienna, S, 1912–35. New York, M, 1921–32; London, CG, 1925–6. Created title-role of

*Ariadne auf Naxos* (in both versions), and Empress (*Die Frau ohne Schatten*); also the inspiration for *Die ägyptische Helena*. A glamorous and gifted singing actress who used her resources to full effect, she was a celebrated Tosca, Turandot, Minnie, and Salome. Autobiography, *Sunlight and Song* (London, 1924). (R)

BIBL: R. Werba, *Maria Jeritza* (Vienna, 1987).

**'Jerum! jerum!'** Hans Sachs's (bar) cobbling song in Act II of Wagner's *Die Meistersinger von Nürnberg*.

**Jérusalem.** See LOMBARDI, I.

**Jerusalem, Siegfried** (*b* Oberhausen, 17 Apr. 1940). German tenor. Bassoonist until 1977. Studied Stuttgart with Kalcher. Début Stuttgart 1975 (Prisoner, *Fidelio*). Hamburg 1976. Bayreuth from 1977; Berlin, D, from 1978; New York from 1980; London, CG, from 1986; also Geneva, Munich, Zurich, etc. A gifted Wagner singer, combining a fine physique, dramatic conviction, and a well-projected, lyrical style. Roles incl. Erik, Lohengrin, Walther, Siegfried, Siegmund, Parsifal. (R)

**Jessonda.** Opera in 3 acts by Spohr; text by E. H. Gehe, after Antoine Lemierre's tragedy *La veuve de Malabar*. Prod. Kassel 28 July 1823, with Schröder-Devrient, Beltheim, Keller, Bergmann, Mayer, Tourny, cond. Spohr; London, Prince's T, 18 June 1840; Philadelphia 15 Feb. 1864.

Goa, early 16th cent. The Rajah's widow Jessonda (sop) must die, according to custom, on his funeral pyre. Nadori (ten), a young priest of Brahma, must announce this to her; but he falls in love with her sister Amazili (sop) and promises to help save Jessonda. Purifying herself for her ordeal, Jessonda is recognized by her former lover, the Portuguese general Tristan d'Acunha (bar), besieging Goa. A truce forbids him to attack and rescue her; but when he finds that the priest Dan-dau (bs) has broken the truce, he feels free to enter the temple by a secret passage and rescue Jessonda.

**'Je suis Titania'.** Philine's (sop) aria in Act II of Thomas's *Mignon*, sung as she remembers her role in the performance the strolling players have been giving of *A Midsummer Night's Dream*.

**Jesuit drama.** The involvement with dramatic music of the Society of Jesus (founded Paris, 1534, by Saint Ignatius of Loyola) had an important influence on the early development of opera: the staging of Agazzari's *Eumelio* at the Collegium Germanicum in Rome in 1606 was one of the earliest of all operatic performances. The most distinctive contribution of the Order came through their cultivation of the school drama, a morally or religiously didactic piece in which musical items separated passages of spoken dialogue, somewhat in the manner of a *Singspiel. Such works enjoyed particular popularity in Germany, where their main exponents were Ferdinand Tobias Richter (c.1649–1711), Johann Michael Zacher (1651–1712), and above all Johann Kaspar Kerll (1627–93).

**Jewels of the Madonna, The**. See GIOIELLI DELLA MADONNA, I.

**'Jewel Song'** (Fr., 'Air des bijoux') Marguerite's (sop) aria in Act III of Gounod's *Faust*, admiring the jewels in the casket which has been brought her by Faust and Méphistophélès.

**Ježibaba**. The Witch (con) in Dvořák's *Rusalka*.

**Joan of Arc** (Fr., Jeanne d'Arc) (b Domrémy, 6 Jan. 1412; d Rouen, 30 May 1431). The French martyr who led an army against the English besieging Orléans, and was burnt at the stake. Operas about her are by Andreozzi (1789), R. Kreutzer (1790), Carafa (1821), Vaccai (1827), Pacini (1830), Balfe (1837), Vesque von Püttlingen (1840), Verdi (1845), Langert (1862), Duprez (1865), Mermet (1876), Bruneau (1878), Tchaikovsky (1879), Chausson (1880), Reznicek (1886), Widor (1890), Wambach (1900), Morera (1907), Roze (1911), Marsh (1923), Anderson (1934), Honegger (1938), Bastide (1949), Dello Joio (*The Trial at Rouen*, 1955 TV, rev. for stage as *The Triumph of St. Joan*, 1959), Humphreys (1968), Klebe (1976), J. Joseph (1978).

**Jobin, Raoul** (b Quebec, 8 Apr. 1906; d Quebec, 13 Jan. 1974). Canadian tenor. Studied Paris. Début Paris, O, 1930 (Gounod's Tybalt). Paris, O, 1930–9, 1946–52, and OC from 1937; London, CG, 1937; New York, M, 1940–50; also Chicago, Buenos Aires, San Francisco, New Orleans. Roles incl. Lohengrin, Walther, Gounod's Faust, Don José, Werther, Pelléas, Radamès, Canio, Cavaradossi; created Fabrice (Sauguet's *Chartreuse de Parme*). (R)

**Jocasta**. In Greek mythology, the Queen of Thebes, mother and later wife of Oedipus; it was the discovery of his unwitting incest, coupled with his unwitting murder of his father, that caused him to put out his eyes and leave Thebes. She appears (mez) in Stravinsky's *Oedipus rex* and Enescu's *Oedip*.

**Jochanaan**. John the Baptist (bar) in Strauss's *Salome*.

**Jochum, Eugen** (b Babenhausen, 1 Nov. 1902; d Munich, 26 Mar. 1987). German conductor. Studied Augsburg, Munich with Waltershausen and Hausegger. Répétiteur Munich 1924–5, Mön-

chen-Gladbach 1925–6. Chief cond. Kiel 1926–9, Mannheim 1929–30; music director Duisburg 1930–2, Hamburg 1934–5. Bayreuth 1953 (*Tristan*), 1954 (*Tannhäuser*), 1971–3 (*Parsifal*). (R) His brother **Georg** (b Babenhausen, 10 Dec. 1909; d Mülheim, 1 Nov. 1970) was a conductor who worked at Frankfurt 1934–7, Plauen 1937–40, and as music director Linz 1940–5.

**John of Leyden**. The Anabaptist prophet and self-proclaimed Emperor (ten) in Meyerbeer's *Le prophète*.

**Johnson, Edward** (b Guelph, Ontario, 22 Aug. 1878; d Toronto, 20 Apr. 1959). Canadian tenor and manager. Studied New York with Feilitsch, Florence with Lombardi. Début (as Edoardo di Giovanni) Padua 1912 (Chénier). Milan, S, 1913–14, first Italian Parsifal. Between 1914 and 1919 he created leading tenor roles in works by Montemezzi, Pizzetti, and Alfano; was also the first Italian Rinuccio and Luigi, and an admirable Siegfried and Tannhäuser. Chicago 1919–22; New York, M, 1922–35, creating leading roles in *The King's Henchman*, *Peter Ibbetson*, and *Merry Mount*. London, CG, with BNOC, 1923 (Gounod's Faust). Was also a fine Pelléas. Manager New York, M, 1935–50. Then active in opera class, Toronto U. An intelligent and musicianly singer with a fine stage presence and a strong lyrical voice; latterly a skilful manager. (R)

BIBL: R. Mercer, *The Tenor of his Time* (Toronto, 1976).

**Johnston, James** (b Belfast 17 Aug. 1903; d Belfast, 17 Oct. 1991). Irish tenor. Début Dublin 1940 (Duke of Mantua). London: SW, 1945–50, as guest until 1957; CG, 1950–8. Created Hector in *The Olympians* (Bliss). Roles included Canio, Radamès, Manrico, first English Gabriele Adorno. Possessed a ringing, Italianate tenor. (R)

**Jolie fille de Perth, La** (The Fair Maid of Perth). Opera in 4 acts by Bizet; text by Saint-Georges and Adenis, loosely based on Scott's novel *The Fair Maid of Perth* (1832). Prod. Paris, L, 26 Dec. 1867, with J. Devriès, Ducasse, Lutz, Barré; Manchester 4 May 1917, with Nelis, Clegg, Hyde, Millar, cond. Beecham.

Perth, c.1500. Henry Smith (ten), an armourer, gives shelter to the gipsy Mab (sop or mez); when his beloved Catherine Glover (sop) arrives, Mab is forced to hide, but her subsequent appearance makes Catherine suspicious of Henry's behaviour. The Duke of Rothsay's (bar or ten) plan to abduct Catherine, whom he also loves, is upset when Mab takes Catherine's place. Mab gives the Duke a golden rose, a gift from Henry which Catherine threw away in her previous anger with him. On arriving to announce his formal betrothal to Catherine, Henry sees the rose and suspects her of infidelity with the Duke. Ralph (bs or bar), the apprentice to Catherine's father, swears her innocence, and agrees to fight a duel with Henry to prove it. The Duke prevents the duel, and Catherine's temporary loss of her senses is cured by Henry's singing.

**Jommelli, Nicolò** (*b* Aversa, 10 Sep. 1714; *d* Naples, 25 Aug. 1774). Italian composer. Studied Aversa and Naples with Feo and Fago. Achieved success with his first two operas, *L'errore amoroso* and *Odoardo*, and in 1740 travelled to Rome, where *Ricimero* and *Astianatte* were produced. Moved to Bologna 1741, where he composed *Ezio* (first version) and developed a close friendship with Martini; then Venice, where he was director of the Ospedale degli Incurabili 1743.

Now recognized as one of the leading composers of the day, he wrote works for most major Italian houses: particularly notable were *Semiramide*, *Tito Manlio*, *Demofoonte*, and *Sofonisba*. In 1747 he settled in Rome, becoming assistant maestro di cappella at St Peter's 1749: here his operas included *Ifigenia in Aulide* and *Talestri*. In Vienna the same year he composed two operas, *Achille in Sciro* and *Didone abbandonata*, both to texts by the court poet *Metastasio, who admired him and with whom he became friends. In 1753 he became Kapellmeister to the Duke of Württemberg and in Stuttgart composed nearly 40 operas, some for other German courts, as well as for Italy, which he visited regularly. Among significant works of this period are *Pelope*, *Fetonte* (first version), *La clemenza di Tito*, *Il re pastore*, and *Bajazette*. In 1769 he returned to Italy, where he composed *Armida abbandonata* and *Ifigenia in Tauride*. These were found too German in style; but the four dramatic works written for Lisbon, including *Le avventure di Cleomede* (a dramma serio-comico) and *Il trionfo di Clelia*, were very successful. The failure of his last operas for Italy brought on a fatal fit of apoplexy.

Jommelli composed around 80 operas, although only 53 of them survive. These include a number of comic operas and intermezzos, among them *Don Trastullo*, *Il matrimonio per concorso*, and *La critica*, but his most important works were his opere serie. Of the generation following Alessandro Scarlatti, he was reared in the conventional tradition of Metastasian opera seria, and was much influenced by Hasse, who provided crucial support in his early career. However, like Traetta, he possessed a radical streak nurtured by his long stay outside Italy which enabled him to enrich the opera seria tradition with French and German elements. But although even his early operas show signs of reform, his

friendship with Metastasio hindered an appreciation of more modern librettos, and the main developments in the Stuttgart operas were in the musical style. The most important was the replacement of the rigid secco recitative/da capo aria with a more fluid structure, in which elements of aria, arioso, and recitative combined. Increasingly, he preferred accompanied recitative to secco, heightening the possibility for dramatic expression, and allowing fuller use of the disciplined orchestra at his command, one of the finest in Germany. This was also heard to good effect in the many spectacular scenes which were included after the French model, and in Noverre's ballet sequences. The chorus, which had long disappeared from Italian opera, was reinstated in a prominent role in his later operas and made an active protagonist in the drama. Even the overture became more integral, notably in *Fetonte*.

Although Jommelli's 'reform' of opera seria was not as radical or as prominent as Gluck's or even Traetta's, he played an important role in developing a more realistic, less stereotyped approach to the genre. His works soon disappeared from the repertory and have still to be seriously revived, yet their expressive melodic style and sensitive characterization suggest that he is one of the most undervalued of 18th-cent. composers.

SELECT WORKLIST: *L'errore amoroso* (Palomba; Naples 1737, lost); *Odoardo* (Fiorentini; Naples 1738, lost); *Ricimero, re de' Goti* (Zeno & Pariati; Rome 1740); *Astianatte* (Salvini; Rome 1741); *Ezio* (Metastasio; Bologna 1741); *Semiramide* (Silvani; Venice 1742); *Tito Manlio* (Roccaforte; Turin 1743); *Demofoonte* (Metastasio; Padua 1743); *Sofonisba* (A. & G. Zanetti; Venice 1746); *Don Trastullo* (Valle; Rome 1749); *Achille in Sciro* (Metastasio; Vienna 1749); *Didone abbandonata* (Metastasio; Vienna 1749); *Ifigenia in Aulide* (Verazi; Rome 1751); *Talestri* (Roccaforte; Rome 1751); *La clemenza di Tito* (Metastasio; Stuttgart 1753); *Pelope* (Verazi; Stuttgart 1755); *Bajazette* (Piovene; Turin 1753); *Il re pastore* (after Metastasio; Ludwigsburg 1764, lost); *La critica* (Martinelli; Ludwigsburg 1766); *Il matrimonio per concorso* (Martinelli; Ludwigsburg 1766); *Vologeso* (Verazi; Ludwigsburg 1766); *Fetonte* (Verazi; Ludwigsburg 1768); *Armida abbandonata* (De Rogatis; Naples 1770); *Ifigenia in Tauride* (Verazi; Naples 1771); *Le avventure di Cleomede* (Martinelli; Lisbon 1772); *Il trionfo di Clelia* (Metastasio; Lisbon 1774).

**Jommellino.** See ANDREOZZI.

**Jones, Della** (*b* Neath, 13 Apr. 1946). Welsh mezzo-soprano. Studied London, RCM. Début Geneva 1970 (in *Boris*). London, ENO 1973, 1977–82, and subsequently; London, CG, 1990; WNO; Scottish O; also Paris, O; New York; Los Angeles; Venice; Geneva. Roles include Ottavia, Rosina, Cenerentola, Dorabella, Donna Elvira,

Despina, Dido (Purcell and Berlioz), Carmen, Baba the Turk. A versatile performer with a firm tone and a vivid, engaging stage presence. (R)

**Jones,** (Dame) **Gwyneth** (*b* Pontnewynydd, 7 Nov. 1936). Welsh soprano. Studied London with Packer; Siena, and Geneva with Carpi. Début (as mezzo-soprano) London 1963, RCM (Gluck's Orfeo). Engaged as mezzo-soprano for Zurich O 1962; soprano début as Amelia (*Ballo*). London, CG, from 1963; Vienna and Bayreuth from 1966; New York, M, from 1972. Roles include Senta, Kundry, Brünnhilde, Leonore, Salome, Marschallin, Verdi's Desdemona, Tosca, Turandot. Impressive in her dramatic commitment, and, at her peak, with a voice of golden tone. (R)

**Jones, Inigo** (*b* London, *bapt.* 19 July 1573; *d* London, 21 June 1652). English architect and designer. After travelling on the Continent, entered service of Queen Anne, later James I and Charles I, designing for royal masques. His elegant stage designs were based on a visual sense whose fluency and grace, making skilful use of illusory perspective, rest on a sound understanding of stage practicalities.

**Jones, Parry** (*b* Blaina, 14 Feb. 1891; *d* London, 26 Dec. 1963). Welsh tenor. Studied London, RCM, and with Colli, Scheidemantel, and John Coates. Début 1914. D'Oyly Carte O 1915; London, CR 1919–22; BNOC 1922–8; London, CG, 1949–53. Also Italy, Germany, Belgium. A musicianly, intelligent singer outstanding in character roles, e.g. Captain (*Wozzeck*), Shuisky, Monostatos, Bob Boles. (R)

**Jones, Richard** (*b* London, 7 June 1953). English producer. Studied Hull and London. Scottish O from 1982; Kent O; Wexford 1986; O North; London, ENO, 1989; Scottish O *Ring* from 1985; also Stuttgart, Amsterdam, Bregenz. Productions include *Così*, *Carmen*, *Rigoletto*, *Mazeppa*, *Mignon*, *Into The Woods*, première of Blake's *Plumber's Gift* and a much-admired *Love of the Three Oranges*. His iconoclastic and bizarre style is marked by a distinctive black humour.

**Jongleur de Notre-Dame, Le** (Our Lady's Juggler). Opera in 3 acts by Massenet; text by Maurice Léna, after a story by Anatole France in *L'étui de nacre* (1892). Prod. Monte Carlo 18 Feb. 1902, with Maréchal, Renaud, Soulacroix, cond. Jéhin; London, CG, 15 June 1906, with Laffitte, cond. Messager; New York, Manhattan OC, 27 Nov. 1908, with Mary Garden as Jean (orig. tenor role), cond. Campanini.

Cluny, 14th-cent. The juggler Jean (ten) is reprimanded by the Prior (bs) for singing a blasphemous song, and is told that he must abandon his juggling in order to gain the Virgin's forgive-

ness. He joins the monks in the abbey and, unable to sing in Latin, wishes to praise the Virgin through his juggling. The monks are divided as to the propriety of this, but when a statue of the Madonna moves and blesses Jean, they know his offering was acceptable. Unaware of the miracle, Jean dies.

Also musical drama by Maxwell Davies (1978).

**Jonny spielt auf** (Johnny Strikes Up). Opera in 2 parts (11 scenes) by Krenek; text by composer. Prod. Leipzig 10 Feb. 1927, with Cleve, Schulthess, Beinert, Horand, Spilcker, cond. Brecher; New York, M, 19 Jan. 1929, with Easton, Fleischer, Kirchhoff, Bohnen, Schorr, cond. Bodansky; Leeds 6 Oct. 1984, with Sprague, Mackay, Woollam, Sullivan, cond. Lloyd-Jones.

Jonny (bar), a jazz-band leader, steals a violin from Daniello (bar) and becomes so immensely successful that his performance from the North Pole sets the world dancing the Charleston.

**Jonson, Ben** (*b* London, *c.*11 June 1572; *d* London, 6 Aug. 1637). English poet and dramatist. Together with Inigo Jones, he was instrumental in the creation of the English *masque. Operas on his works are as follows:

*Volpone* (1606): Gruenberg (1945); Demuth (1949); Antheil (1953); Coombs (1957); Zillig (1957); Zimmermann (1957); Burt (1960)
*Epicoene* (1609): Salieri (*Angiolina*, 1800); Lothar (*Lord Spleen*, 1930); Strauss (*Die schweigsame Frau*, 1935)
*The Alchemist* (1610): Lang (1969)
*The Masque of Oberon* (1615): Arne (*The Fairy Prince*, 1771)

**Joplin, Scott** (*b* Marshall, TX, or Shreveport, LA, 24 Nov. 1868; *d* New York, 1 Apr. 1917). US composer. His first opera, *A Guest of Honor* (1903), was lost on its way to the copyright office. His only surviving opera, *Treemonisha* (*c.*1907), was to his distress never adequately performed during his lifetime, and was only revived in 1972. It uses ragtime tunes within a conventional framework, and includes substantial use of chorus.

**José.** See DON JOSÉ.

**Joseph.** Opera in 3 acts by Méhul; text by Duval. Prod. Paris, OC, 17 Feb. 1807, with Gavaudan, Elleviou, Solié; New Orleans, T St Philippe, 21 Apr. 1812; London, CG, 3 Feb. 1914, with Jonsson, Sembach, Van Hulst, Kiess, Plaschke, cond. Pitt.

Memphis, biblical times. Under the assumed name of Cleophas, Joseph (ten or sop) has saved Egypt from famine; his blind father Jacob (bar) and brothers arrive in Memphis to beg for food, but do not recognize him. When he reveals his identity, he forgives his brothers and begs his father to do likewise.

**Joubert, John** (*b* Cape Town, 20 Mar. 1927). South African composer. Studied Cape Town and London, RAM, where his teachers included Howard Ferguson, before taking university appointments at Hull and Birmingham. His dramatic writing embraces children's operas (*The Quarry* and *The Prisoner*), as well as a radio opera (*Antigone*) and a rare adaptation of George Eliot (*Silas Marner*).

SELECT WORKLIST: *Antigone* (Trickett, after Sophocles; Cape Town 1954); *Silas Marner* (Trickett, after Marner; Cape Town 1961); *The Prisoner* (Tunnicliffe, after Tolstoy; Barnet 1973).

**Journet, Marcel** (*b* Grasse, 25 July 1867; *d* Vittel, 7 Sep. 1933). French bass. Studied Paris with Obin and Seghettini. Début Béziers 1891 (in *La favorite*). Brussels, M, 1894–1900; London, CG, 1897–1909, 1927–8; New York, M, 1900–8; Paris, O, 1908–32; Chicago 1915–19; Milan, S, 1917, 1922–7. First Simon Mago (Boito's *Nerone*). A basso cantante with a voice of great beauty and versatility; his wide repertory included Guillaume Tell, Golaud, Gounod's Méphistophélès, Dosifey, Fafner, Wanderer, Sachs, Tonio, Scarpia. (R)

**Jouy, Étienne de** (*b* Jouy-en-Josas, 12 Sep. 1764; *d* Saint-Germain-en-Laye, 4 Sep. 1846). French playwright, librettist, and journalist. As librettist at the Paris Opéra, he worked for some of the most successful composers of the day, including Méhul, Boieldieu, Catel, and Dalayrac; but his greatest and most characteristic collaborations were in the field of *Grand Opera with Cherubini (*Les Abencérages*), Rossini (*Guillaume Tell*), and especially Spontini (*La vestale*, *Fernand Cortez*, and *Milton*). His vivid appreciation of the sensational could go to extremes of literalism: he introduced into *Cortez* the same number of horses with which the conquistador had subdued Mexico on the ground that the shock of the occasion to the audience matched that of the original Aztecs. At his finest, as in *La vestale*, he responded to the demands of the heroic and spectacular in Grand Opera with some imposing scenes and verses.

**Juch, Emma** (*b* Vienna, 4 July 1863; *d* New York, 6 Mar. 1939). US soprano. Studied Detroit with Murio Celli. Début London, HM, 1881 (Philine). American OC (later National OC) 1884. After its collapse, she founded the Juch Grand Opera Co., which toured USA, Canada, and Mexico until 1891, when she retired from the stage. A champion of opera in English, she possessed impeccable diction, while her vocal abilities encompassed the Queen of the Night and Senta. (R)

**Judith.** In the book of the biblical Apocrypha named after her, a Jewish girl who made her way to the tent of Nebuchadnezzar's general, Holofernes, and cut off his head, thus saving her native town of Bethulia. Operas on the subject (sometimes under her Italian name, Giuditta), and mostly based on Metastasio's *Betulia liberata* or Hebbel's tragedy *Judith* (1839), are by Gagliano (an azione sacra, 1626), Scacchi (1636), Demmler (1780), L. Kozeluch (1799), Fuss (1814), Levi (1844), Hebenstreit (1849, on Nestroy's parody of Hebbel), E. Naumann (1858), A. Peri (1860), Serov (1863), Doppler (1870), Eithner (*c*.1870), Hillemacher (1876), Pauline Thys (1883), Vietinghoff-Scheel (1884), Silveri (1885), Falchi (1887), Götze (1887), Moss (1891), Cadwich (1900), Polignac (1916), Ettinger (1921), Chishko (1923), Reznicek (1923), Honegger (1926), Goossens (1929), C. N. Berg (1936), Gnecchi (1953), Matthus (1985). See also GIUDITTA.

**Jugoslavia.** See CROATIA, SERBIA, SLOVENIA.

**Juive, La** (The Jewess). Opera in 5 acts by Halévy; text by Scribe. Prod. Paris, O, 23 Feb. 1835, with Falcon, Dorus-Gras, Nourrit, N. Levasseur; New Orleans, T d'Orléans, 13 Feb. 1844, with Fleury-Joly, Lecourt, Grosseth, Blès, cond. E. Prévost; London, DL, 29 July 1846, with Julien, Charton, Laborde, Zelger, cond. Henssens.

Constance, 15th cent. Samuel (Prince Léopold in disguise) (ten) works in the shop of the Jewish goldsmith Eléazar (ten), whose daughter Rachel (sop) he loves. Assuming Samuel to be a fellow Jew, Rachel invites him to their Passover meal.

They are interrupted by Prince Léopold's wife Eudoxia (sop), who requests a gold chain for her husband; Léopold is filled with remorse. Eléazar is furious when the truth emerges.

In Eudoxia's presence, Rachel claims that she and Léopold are lovers. The Cardinal de Brogni (bs), persecuter of the Jews in Constance, condemns them to death.

Eudoxia persuades Rachel to confess that she lied, and that Léopold is innocent. Eléazar refuses the Cardinal's offer that Rachel will be saved if he renounces his own faith, reminding him that it was a Jew who rescued the Cardinal's daughter from a fire in their home many years previously.

Rachel, too, refuses to abandon Judaism and, as she is executed, Eléazar reveals that she was the Cardinal's daughter.

**Julia.** The eponymous Vestal Virgin (sop), lover of Licinius, in Spontini's *La vestale*.

**Julien.** Opera in a prologue and 4 acts by Charpentier; text by the composer. Prod. Paris, OC, 4 June 1913, with Carré, Rousselière, cond. Wolff; New York, M, 26 Feb. 1914, with Farrar, Caruso, cond. Polacco.

The dead Louise (sop) appears to Julien (ten) in a vision and attempts to rekindle his faith in art and beauty. Julien, however, is frustrated in the attempt to find his soul by this means, and dies.

Composed as a sequel to *Louise*, the work was unsuccessful, receiving only 20 Paris performances.

**Julietta.** Opera in 3 acts by Martinů; text by the composer, after Georges Neveux's drama *Juliette, ou La clé des songes* (1930). Prod. Prague, N, 16 Mar. 1938, with Horáková, Podvalová, Gleich, Ludek, Mandaus, cond. Talich; London, C, 5 Apr. 1978, with Roberts, Kale, Du Plessis, Wicks, cond. Mackerras.

The opera tells of the efforts of a Paris bookseller, Michel (ten), to find Julietta (sop), whom he encountered three years previously. He discovers that the inhabitants of her home town, including Julietta herself, have lost their memory; Michel also begins to be afflicted. When, impulsively, Julietta struggles to tear herself away from him, Michel shoots her and is thereafter tormented by the sound of her voice. He eventually recedes into a dream-world, still in quest of Julietta.

**Juliette.** Roméo's lover (sop) in Gounod's *Roméo et Juliette*.

**Julius Caesar** (*b* Rome, 12 July 100 (?102) BC; *d* Rome, 15 Mar. 44 BC). Roman dictator. Most operas on Caesar, many of them drawing on Bussani's text for Sartorio's *Giulio Cesare*, treat the story of his love for *Cleopatra. The most celebrated of them is Handel's *Giulio Cesare*.

**Jullien, Louis** (*b* Sisteron, 23 Apr. 1812; *d* Paris, 14 May 1860). French conductor. Studied Paris. Conducted in London from 1840, where his flamboyant personality attracted much attention to his concerts. Composed the unsuccessful *Pietro il grande* (London, CG, 1852, with Tamberlik). Originator of the disastrous speculation at Drury Lane when operas were to be conducted by, among others, Berlioz (who gives a sardonic but tolerant account of Jullien in his *Mémoires*).

**Junge Lord, Der** (The Young Lord). Opera in 2 acts by Henze; text by Ingeborg Bachmann, after a parable in Hauff's *Der Scheik von Alexandria und seine Sklaven* (1827). Prod. Berlin, D, 7 Apr. 1965, with Mathis, Otto, Johnson, Graf, McDaniel, Driscoll, Hesse, Grobe, cond. Dohnányi; San Diego, Civic T, 17 Feb. 1967, with Lynn, Curatillo, Turner, Toscano, Fredericks, Cole, Remo, cond. Herbert; London, SW, by Cologne O, 14 Oct. 1969, with Fine, Ahlin, Harper, Mohler, Nicolai, cond. Janowski.

'Hülsdorf-Gotha', *c*.1830. Sir Edgar (silent role) introduces his nephew, Lord Barrat (ten) to the inhabitants of the small German provincial town of Grünwiesel. The young lord shocks the local snobbish establishment, but his behaviour is sycophantically condoned until it is revealed that he is a circus ape dressed in man's clothing.

**Jupiter.** In Roman mythology, the most powerful of the Olympian gods; also known in English as Jove (from Lat., Jovis pater; It., Giove), and in Greek mythology as Zeus. He appears in many Baroque operas, sometimes as lover (e.g. Handel's *Semele*), sometimes as a *deus ex machina.

**Jurinac, Sena** (*b* Travnik, 24 Oct. 1921). Bosnian soprano. Studied Zagreb with Kostrenčić. Début Zagreb 1942 (Mimì). Vienna from 1945, Salzburg from 1947; Milan, S, from 1948; London, CG, 1947, 1959–63; Gly. 1949–56. An exceptionally sensitive artist, with a sweet-toned yet powerful voice, outstanding in Mozart (as Ilia, Pamina, Donna Elvira, Cherubino, Countess, Donna Anna, Fiordiligi), and as Octavian and the Composer. Roles also include Marschallin, Verdi's Desdemona, Leonore, Butterfly. (R)

BIBL: U. Tamussino, *Sena Jurinac* (Augsburg, 1971).

# K

**Kabaivanska, Raina** (*b* Burgas, 15 Dec. 1934). Bulgarian soprano. Studied Sofia with Prokopova, Milan with Fumagalli-Riva, Vercelli with Tess. Début Sofia 1957 (Tatyana). Milan, S, from 1961; London, CG, and New York, M, from 1962; Paris, O, from 1975; also Vienna, Chicago, Buenos Aires, etc. An expressive singer and effective actress, notably in the verismo repertory, e.g. as Nedda, Adriana Lecouvreur, Butterfly, Tosca. (R)

**Kabalevsky, Dmitry** (*b* St Petersburg, 30 Dec. 1904; *d* Moscow, 14 Feb. 1987). Russian composer. Studied Moscow with Catoire and Myaskovsky. His successful career, including official positions in publishing and administration, also included composition in many genres, including opera. His first, *Colas Breugnon*, has become well known for its sparkling overture; others include works responding to contemporary events, such as *In the Fire*, on the defence of Moscow against the Germans. He also rewrote one work, *The Taras Family*, in the wake of the 1948 decree, which he supported, attacking formalism in Soviet music.

SELECT WORKLIST: *Colas Breugnon* (Bragin, after Rolland; Leningrad 1938); *V ogne* (In the Fire) (Solodar; Moscow 1943: withdrawn—music used in next opera); *Semya Tarasa* (The Taras Family) (Tsenin, after Gorbatov; Leningrad 1947, rev. 1950, 1967); *Nikita Vershinin* (Tsenin, after Ivanov; Moscow 1955).

BIBL: G. D. Abramovsky, *Kabalevsky* (Moscow, 1960).

**Kabanicha.** Marfa Kabanová (con), widow of the rich merchant Kabanov and hence known as Kabanicha, in Janáček's *Káťa Kabanová*.

**Kaiserslautern.** Town in Rhineland-Palatinate, Germany. A resident company was established 1787. The theatre (cap. 1,200) opened 1862 with Weber's *Preciosa*; burnt down 1867. A second theatre opened 1897; bombed 1944. The present theatre (cap. 750) opened 1950.

**Kaliningrad** (Ger., Königsberg). City in the Russian Federal Republic, until 1945 East Prussia. The first stage musical piece was the dramatic allegory *Cleomedes* (1635) by Heinrich Albert (1604–51). The first opera-house opened 1755, and Singspiels were given. After the Russian occupation of 1758–62, captured Austrian officers helped to foster Viennese music. Mozart was

first heard 1793 (*Don Giovanni*), and with French opera dominated the repertory in the early 19th cent. The theatre burnt down 1798; a new one was opened and took over the ensemble of the Schuch sisters. With Hiller's son Friedrich Adam as music director, the company played in Königsberg in winter, Danzig in summer. The Schauspielhaus opened 1806; burnt down 1808. The old theatre reopened, with an ill-assorted company and a repertory of French and German works. A new theatre opened (cap. *c*.750) in 1810, but standards fell and a chaotic situation ensued. Matters improved in 1812, and a subscription system was introduced. Wagner worked here 1836–7, but one of many economic crises led to the dismissal of Minna Planer, his first wife. A Wagner tradition developed in the late 19th cent., with ten Wagner works receiving over 200 perfs. 1889–1900. The Ostpreussisches Landestheater was rebuilt (cap. 1,435) in 1910.

**Kálmán, Emmerich** (orig. Imre) (*b* Siófok, 24 Oct. 1882; *d* Paris, 30 Oct. 1953). Hungarian composer. Studied Budapest with Koessler and first won attention as a composer of serious music. Music critic of the *Pesti Napló* 1904–8, before the success of a cabaret, written under a pseudonym, turned him towards lighter music. In 1908 his first operetta, *Tatárjárás*, was given in Budapest, and shortly afterwards in Vienna, to great acclaim and was followed by other successes, including *Der Zigeunerprimas* and *Der kleine König*.

Although Kálmán drew inspiration from stylized features of Hungarian life and music, his main model was the Viennese operetta of Johann Strauss and Lehár. Following his first successes he settled in Vienna and, while not as consummate a melodist as Lehár, secured a loyal and enthusiastic following. His greatest triumph and most enduring work was *Die Csárdásfürstin*, which was followed by eight more operettas for the Viennese stage, of which *Gräfin Mariza* and *Die Herzogin von Chicago* were best received. After the rise of Nazism Kálmán emigrated, first to Paris, and then to the USA, where he was already known through his music for *Golden Dawn* and where he found further favour with *Marinka*. His last work, *Arizona Lady*, was completed by his son **Charles Kálmán** (*b* Vienna, 17 Nov. 1929), himself a composer of musicals.

SELECT WORKLIST: *Tatárjárás* (Autumn Manœuvres) (Bakonyi; Budapest 1908); *Der Zigeunerprimas* (The Gypsy Primate) (Sari, Grünbaum, & Wilhelm; Vienna 1912); *Der kleine König* (Bodanzky, after Bakonyi; Vienna 1912); *Die Csárdásfürstin* (The Gypsy Princess) (Stein & Jenbach; Vienna 1915); *Gräfin Mariza* (Countess Mariza) (Brammer & Grünwald; Vienna 1924); *Golden Dawn* (Harbach & Hammerstein; New York 1927; with Stothart); *Die*

*Herzogin von Chicago* (The Duchess of Chicago) (Brammer & Grünwald; Vienna 1928); *Marinka* (Farkas & Marton; New Haven, CT, 1945); *Arizona Lady* (Grünwald & Beer; Berne 1954).

BIBL: R. Oesterreicher, *Emmerich Kálmán: der Weg eines Komponisten* (Vienna, 1954).

**Kalniņš, Jānis** (*b* Pernav, 3 Nov. 1904; *d* Riga, 23 Dec. 1951). Latvian composer. Studied with his father, the composer and organist **Alfrēds Kalniņš** (1879–1951), and Riga with Vītols. Music advisor and conductor Riga, National T, 1923–33; conductor National O 1933–44; then emigrated to USA.

**Kalomiris, Manolis** (*b* Smyrna, 14 Dec. 1883; *d* Athens, 3 Apr. 1962). Greek composer. Studied Athens with Xanthopoulos, Constantinople with Spanoudi, Vienna with Grädener. Very active and influential in the development of modern Greek music; founded Hellenic Conservatory 1919, director until 1926; founded National Conservatory, director until 1948; chairman National Opera 1950–2. His most successful operas, *The Shadowy Waters* and *The Mother's Ring*, show his warm Romantic idiom and his understanding of how the example of the Russian nationalists could help the cause of an independent Greek musical style.

WORKLIST (all first prod. Athens): *O protomastoras* (The Master Builder) (Kazantzakis; 1916); *To dakhtilidi tis manas* (The Mother's Ring) (Kambyssis; 1917); *Anatoli* (Sunrise) (Kambyssis; 1945); *Ta xotika nera* (The Shadowy Waters) (Yeats text; 1950); *Constantinos o Palaeologos* (Kazantzakis; 1962).

**Kamieński, Maciej** (*b* ?Sopron or Magyar-Ovar, 13 Oct. 1734; *d* Warsaw, 25 Jan. 1821). Polish composer of Slovak origin. Composed the first publicly performed Polish opera, *Misery Made Happy*, in 1778. His music reflects contemporary European influences, with some native ingredients.

SELECT WORKLIST: *Nędza uszczęśliwiona* (Misery Made Happy) (Bogusławski, after Bohomolec; Warsaw 1778).

**Kammersänger(in)** (Ger.: 'chamber singer'). High honorary title given by German and Austrian governments to distinguished singers. Originally the title was bestowed by the various courts hence the titles Hofkammersänger(in).

**Kanawa, Kiri Te.** See TE KANAWA, KIRI.

**Kanne, Friedrich August** (*b* Delitzsch, 8 Mar. 1778; *d* Vienna, 16 Dec. 1833). German composer and critic. Studied in Leipzig; moved to Vienna 1808; Kapellmeister at the Pressburg Opera 1809. Returned to Vienna in 1810, becoming editor of the Vienna *AMZ*, an important source of information for operatic activities. His ten stage works included heroic and magic opera,

Singspiel, and melodrama, and were very popular in their day. A critic of some influence, he was a close friend of Beethoven.

WORKLIST (all first prod. Vienna unless otherwise stated): *Orpheus* (Kanne; 1807); *Miranda oder das Schwert der Rache* (Miranda, or the Sword of Vengeance) (Kanne; 1811); *Die eiserne Jungfrau* (The Iron Maiden) (Biedenfeld; 1822); *Lindane, oder die Fee und der Haarbeutelschneider* (Lindane, or the Fairy and the Cutpurse) (Bäuerle; 1824); *Die Mainacht* (Kanne; Berlin 1834).

**Kansas City.** City in Missouri, USA. The Lyric Opera of Kansas City opened in Sep. 1958 with *Bohème* and has had a policy of giving opera in English by young US artists. It performs in the Lyric T (cap. 1,659).

**Kapellmeister** (Ger.: 'chapel-master'). Originally the choirmaster in a court chapel, and formerly generally used in Germany for a conductor. Those who held appointments in a court would be known as **Hofkapellmeister**; assistants were often called **Vizekapellmeister**. The equivalent of the French *maître de chapelle* and Italian *maestro di cappella*.

**Kapp, Julius** (*b* Steinbach, 1 Oct. 1883; *d* Sonthofen, 18 Mar. 1962). German critic. Production adviser Berlin, S, 1921–45, editing the *Blätter der Staatsoper*; Berlin, SO, 1948–54. His books include a substantial biography, *Richard Wagner* (Berlin 1910, 32/1929), an edition of Wagner's writings, and various useful studies of other composers intended for the general reader.

**Kappel, Gertrude** (*b* Halle, 1 Sep. 1884; *d* Pullach, 3 Apr. 1971). German soprano. Studied Leipzig Cons. Début Hanover 1907 (Leonore). Sang there until 1924. London, CG, 1912–14, 1924–6; Vienna 1921–7; Munich, S, 1927–31; New York, M, 1928–36. Retired 1937. A highly distinguished Wagner singer, with a warm, powerful tone, celebrated as Brünnhilde and Isolde. (R)

**Karajan, Herbert** (orig. Heribert) **von** (*b* Salzburg, 5 Apr. 1908; *d* Arif, nr. Salzburg, 16 July 1989). Austrian conductor. Studied Salzburg Mozarteum with Schalk and Vienna Conservatory. Début Salzburg, Landestheater, 1927 (*Fidelio*). Ulm 1929–34; Aachen 1934–8; Berlin, S, 1938–45; Vienna, S, as artistic director 1956–64; Milan, S, 1948–68; Bayreuth 1951–2; artistic director Salzburg Festival 1958–60 and from 1964. Established his Salzburg Easter Festival in 1967, giving one opera of *The Ring* each year, but never a complete cycle; and then *Parsifal, Tristan, Meistersinger, Trovatore*, and *Don Carlos*. These performances with the Berlin Philharmonic were subsequently recorded, and were intended to be filmed, as part of Karajan's skilful and assertive use of all forms of modern media to further his intentions. He rarely worked with other producers, preferring to control everything himself, and subjugating dramatic meaning to the beauty of sound he was a master at creating with his orchestra. Some singers admired him and owed much in their careers to his power; but others suffered from early over-exposure in roles for which they were not ready but found it difficult to resist. He was probably a member of the Nazi party from 1933; he admitted to the date 1935, and to career opportunism rather than political conviction. From musicians' viewpoint, the indictment was the use of exceptional gifts for performances that dealt in a self-indulgent, predictable gloss as part of a success mechanism and evaded the spontaneity, adventure, and risks of great music-making. (R).

BIBL: R. Osborne, *Conversations with Karajan* (Oxford, 1989).

**Karfreitagzauber.** See GOOD FRIDAY MUSIC.

**Karlsruhe.** Town in Baden-Württemberg, Germany. Opera was first given in the 17th cent. in the Residenz Durlach, before the foundation of Karlsruhe in 1715; a theatre had been opened in 1712, and opera was given. Visiting companies performed in the 18th cent. The Grand Ducal Theatre opened 1810 with Ferdinand *Paer's *Achille*; burnt down 1847. Franz Danzi was Kapellmeister 1812–24, Josef Strauss 1824–63. The Court Theatre opened 1851; bombed 1944. A strong Wagner tradition was initiated under Strauss and maintained under Herman Levi (1864–72) and Felix Mottl (1881–1903). The theatre (from 1918 Badisches Landestheater, from 1933 Staatstheater) was rebuilt 1954 (cap. 1,055; 330 in small theatre). Music directors have included Josef Krips (1926–33), Joseph Keilberth (1935–40), and Alexander Krannhals (1955–62).

**Kärntnertortheater.** Viennese theatre, the full title of which was 'Kaiserlich-königliches Hoftheater nächst dem Kärntnertor' (lit. 'Imperial-royal court theatre next to the Carinthian gate', a reference to its location). Built in 1761, it was burnt down following an accident during a performance of Gluck's *Don Juan* ballet. After rebuilding, it reopened in 1763; in 1776 it was taken under direct imperial control, with a management approved by the Emperor. Now that the general public was no longer freely admitted to the theatre, the Emperor encouraged the establishment of private theatres in the Viennese suburbs: this led to the opening of the *Theater in der Leopoldstadt, the *Theater auf der Wieden, and the *Theater in der Josefstadt. Originally the home of drama, from 1790 the Kärntnertortheater was also used for opera and operetta, but

with a repertory that was largely non-German. After 1810 the Kärntnertortheater concentrated exclusively on opera and ballet, balancing the simultaneous development of the *Burgtheater into a theatre for spoken drama. *Barbaia was director 1821–8 and invited Rossini, who had been popular in Vienna since his performance in the theatre of *L'inganno felice* in 1816, to give a season of opera there in 1822. Despite the efforts of Weber, and the success of *Freischütz* in 1822, the repertory became increasingly dominated by Italian works in the 1820s: this fundamental taste prevailed well into the century. Premières given in the Kärntnertortheater included the final version of *Fidelio*, *Euryanthe*, *Linda di Chamounix*, *Maria di Rohan*, and *Martha*.

**Kašlik, Václav** (*b* Poličná, 28 Sep. 1917; *d* Prague. 4 June 1989). Czech producer, composer, and conductor. Studied Prague with Talich. Début Brno 1941, cond. and prod. Gluck's *Orfeo*. Music director Brno 1943–5; Prague, T of 5 May, 1945–8, Chief producer Prague, N, from 1953: première of Martinů's *Mirandolina* 1959. Milan, S; Venice, F, première of Nono's *Intolleranza* (1961). London, CG, *Pelléas* 1969, *Nabucco* 1972, *Tannhäuser* 1973. Verona, *Boris* 1976. Berlin, S, Gounod's *Faust* 1977. Ottawa, *Idomeneo* 1981. Has worked with Josef *Svoboda on the 'laterna magica' technique using film, stereophonic sound, and other media. He has explored some original ideas in opera production. The most successful of his operas has been *Krakatit* (after Čapek; Ostrava 1961), which draws on mixed genres and media; *La strada* (Prague 1982) is an adaptation of Fellini's film.

**Kaspar.** See CASPAR.

**Kassel.** City in Hesse, Germany. The Landgrave Moritz the Learned, who was a composer, opened the Ottoneum 1605, the first permanent theatre in Germany. Ruggiero Fedeli gave Singspiel annually 1701–21. A new Italian operahouse opened in Prince Maximilian's palace in 1764 with Ignatio Fiorillo's *Diana e Endimione*. French opera was given at the Bauhaus 1776–85 and especially under Felice *Blangini 1809–13, when the conductor Legaye was observed directing the orchestra 'in the French manner, with a baton'. The Hoftheater opened Feb. 1814 with Winter's *Unterbrochene Opferfest*, beginning the tradition of German opera which continued under Spohr (1822–57) and Mahler (1883–5). Later directors have included Paul Bekker (1925–7, with Krenek as Dramaturg: his *Orpheus* was premièred 1926), Ernst Legal (1927–8), Robert Heger (1935–44), Karl Elmendorff (1948–51), Christoph von Dohnányi (1963–6), Gerd Albrecht (1966–72), James Lockhart (1972–80). A new theatre opened 1909; it was bombed 1943, and performances were given in the Stadthalle; the Staatstheater (cap. 1,010) opened Sep. 1959 with the première of Wagner-Régeny's *Prometheus*.

**Kastorsky, Vladimir** (*b* Bolshiye Soly, 14 Mar. 1871; *d* Leningrad, 2 July 1948). Russian bass. Studied St Petersburg with Gabel and Cotogni. Début 1894 with Champagner's touring co., St Petersburg, M, 1898–30; later member of Moscow, Z. Roles included Miller (Dargomyzhsky's *Rusalka*), Gremin, Pimen, Hagen, Wotan. Sang in Dyagilev's Russian seasons in the West, 1907–8.

**Káťa Kabanová.** Opera in 3 acts by Janáček; text by the composer, after Ostrovsky's tragedy *The Storm* (1859). Prod. Brno 23 Nov. 1921, with Veséla, Hladiková, Zavřel, Šindler, Pustinská, Jeral, cond. Neumann; London, SW, 10 Apr. 1951, with Shuard, R. Jones, cond. Mackerras; Cleveland, Karamu House, 26 Nov. 1957, with piano; Bear Mountain, NY, Empire State Music Festival, 2 Aug. 1960, with Shuard, Doree, Petrak, Gari, Frankel, cond. Halasz.

Kalinov, second half 19th cent. Boris (ten) is in love with Káťa (sop), who is married to Tichon (ten). Káťa's mother-in-law Kabanicha (con) is jealous because her son pays more attention to Káťa than to her. As Tichon is about to depart on a journey, his mother insists that he admonish Káťa not to take a lover during his absence.

Káťa obtains a key to the garden and, uncertainly, agrees to meet Boris. At length she returns his expression of love.

Tichon returns unexpectedly, causing Káťa to become hysterical. She interprets a violent storm as a divine signal of knowledge of her sin, and confesses her affair to Kabanicha and Tichon. Boris is to be sent to Siberia and Káťa, convinced that she has no future, throws herself into the Volga.

BIBL: J. Tyrrell, *Leoš Janáček: 'Káťa Kabanová'* (Cambridge, 1982).

**Katerina Ismaylova.** See LADY MACBETH OF MTSENSK, THE.

**Katowice.** City in Poland. The Silesian State O was founded in 1945 by Adam *Didur, and gives regular seasons.

**Kauer, Ferdinand August** (*b* Klein-Thaya, 18 Jan. 1751; *d* Vienna, 13 Apr. 1831). Austrian composer of Moravian birth. Studied Znaim, Tyrnau, and Vienna; in 1781 joined Marinelli's newly formed company at Vienna, L, as leader and conductor. From 1782 also composed music for the theatre, including Singspiels, operas, and incidental songs, mostly to texts by the house poet Karl Friedrich Hensler. Their first major success was *Das Faustrecht in Thüringen*, eclipsed

two years later by *Das Donauweibchen*, which was taken up throughout central and eastern Europe and followed by an even more popular sequel, *Das Waldweibchen*. He was Kapellmeister in Graz 1810–11; he returned to Vienna, L, but in 1814 succeeded Leopold Huber at Vienna, J. Shortly afterwards the company was wound up and he was unable to find another position; most of his manuscripts were lost in the Danube flooding of Mar. 1830.

Kauer made an important contribution to the development of the Singspiel in Vienna, composing music for at least 160 dramatic works. Like the works of Dittersdorf, Wenzel Müller, and Schenk, these range from mere pantomimes with the occasional song to continuous opera. Despite his significant development of the Zauberposse, no other works matched *Das Donauweibchen* and *Das Waldweibchen* in popularity: these fairy-tale operas were among the most frequently performed German stage works of the 19th cent. and of some importance for the eventual formulation of a national style of opera. A third part (with music by Bierey) followed 1801; Friedrich Adam Hiller provided music for an interlude 1802.

*Das Donauweibchen* was also important in the history of opera in Russia, where it was first given in 1803 in an adaptation by Stepan *Davydov as *Lesta, dneprovskaya rusalka*, with additional music by *Cavos.

SELECT WORKLIST (all first prod. Vienna): *Das Faustrecht in Thuringen* (The 'Law of the Jungle' in Thüringia) (Hensler; pt1 1796; pt2 1796; pt3 1797); *Das Donauweibchen* (The Danube Sprite) (Hensler; 1798; pt2 Feb. 1803); *Das Waldweibchen* (The Wood Sprite) (Hensler; 1800).

**Kazakhstan.** The Abay State Academic Opera and Ballet Theatre opened in Alma-Ata in 1934 with *Kyz-Zhibek*, by the founder of Kazakh professional musical culture, Evgeny Brusilovsky (1905–81), who spent much of his life in Kazakhstan. This, and his later operas such as *Er-Targyn* (1938), draw on Kazakh folk stories. The first opera by Kazakh composers was *Abay* (1944), a collaboration between Akhmet Zhubanov (1906–68) and Latif Khamidy (1906–83) on the life of the Kazakh national poet Abay Kunanbayev. Brusilovsky's pupil Mukan Tulebayev (1912–60) wrote *Birzhan men Sara* (1946), a love story of two singers based on the historical figures Birzhan Kozhagulov and Sara Zhienkulova; this is regarded as an outstanding example of the successful combination of national musical traditions with Western operatic forms. Kazakh music was also enriched by the various national minorities settled in the region: the first opera by an Uighur (a Central Asian Turkic people) was *Nazugum* (1956) by Kuddus Kuzhamyarov (*b* 1918). Further advances came in the work of

Sydykh Mukhamedjanov (1924–91), with the comedies *Aysulu* (1964) and *Zhumbak kyz* (1971), the opera *Akhan-sere Aktokty* (1982). Other important works include *Kamar-Sulu* (The Fair Kamar, 1963) and *Alpamys* (1972) by Erkegaly Rakhmadiyev (*b* 1932); *Enlik-Kebek* (1974), *Dvadsat vosem* (Twenty-eight, 1985) and *Kurmangazy* (1989) by Zhubanov's daughter Gaziza Zhubanova (*b* 1927); and *Makhambet* (1989) by Bazarbay Djumaniyazov (*b* 1936). Other new developments came with the rock opera-ballet *Brat moy, Maugly* (My Brother Mowgli, 1975), by Almas Serkebayev (*b* 1948), and the children's opera *Kanbak zhal* (Naughty Kanbal, 1982) by Zholan Dastenov (*b* 1943).

BIBL: P. Momynov, *Kazakhkoye opernoye iskusstvo* (Alma-Ata, 1963).

**Kecal.** The marriage broker (bs) in Smetana's *The Bartered Bride*.

**Keil, Alfredo** (*b* Lisbon, 3 July 1850; *d* Hamburg, 4 Oct. 1907). Portuguese composer of German and Alsatian descent. Studied Lisbon with Soares, Cinna, and Vieira, Nuremberg with Kaulbach, returning in 1870 and exhibiting as a painter. He began writing light music, including the 1-act *Susana*. This was followed by more serious operas, all on Portuguese subjects. The last of these, *Serrana*, was the first substantial opera to be written in Portuguese (though the première was in Italian).

WORKLIST: *Susana* (Lopes de Mendonça; Lisbon 1883); *Donna Bianca* (Fereal, after Garrett; Lisbon 1888); *Irene* (Fereal; Turin 1893); *Serrana* (Lopes de Mendonça, after Branco; Lisbon 1899).

**Keilberth, Joseph** (*b* Karlsruhe, 19 Apr. 1908; *d* Munich, 20 July 1968). German conductor. Studied Karlsruhe. Répétiteur Karlsruhe 1925; music director 1935. Cond. Dresden 1945–50; Munich 1951, music director 1959–68. Bayreuth 1952–6 (incl. *Ring*). Edinburgh Festival with Munich Company 1952. A fine conductor especially of Wagner, Richard Strauss, and Pfitzner. His declared wish to die as Mottl had, while conducting *Tristan*, was uncannily granted him. (R)

**Keiser, Reinhard** (*b* Teuchern, *bapt.* 12 Jan. 1674; *d* Hamburg, 12 Sep. 1739). German composer. His principal teacher was almost certainly Kuhnau; *c.*1693 became Kapellmeister at Brunswick in succession to Kusser, who moved to Hamburg and produced Keiser's first opera, *Basilius*. Further operas were written for Brunswick, but in 1696 Keiser followed his mentor to Hamburg, where he began a lengthy collaboration with the librettist Postel. *Adonis* ushered in a series of successful operas for the T am Gänsemarkt, sometimes five per season. In 1702 he took over the theatre with Drüsicke: shortly afterwards Handel arrived, several of whose early works were given under Keiser's régime, as were operas by Grünwald and Mattheson. Keiser was later sole director, and under him Hamburg became the most important commercial operatic centre in Germany. In 1707 he resigned through financial difficulties, but continued to write for the Hamburg stage. Among his most successful works were *Störtebecker und Jödge Michel*, *Claudius*, *Octavia*, and *Croesus*, which with their emotional and lyrical power for a while elevated German opera to a standard rivalling that of the Italian and French traditions. *Der Carnaval von Venedig* mirrored the contemporary Neapolitan tradition of dialect opera in its extensive use of Platt-Deutsch.

In 1718 Keiser left Hamburg, first for Stuttgart, where he failed to obtain a post with the Duke of Württemberg. In 1721–3 he divided his time between Hamburg and Copenhagen, where he directed a German company in two new operas, as well as in revivals. He was rewarded with the unsalaried title of Kapellmeister to the King of Denmark, but in 1723 returned to Hamburg. His late work here, which included *Der Hamburger Jahrmarkt* and *Circe*, was gradually supplanted by that of Telemann, and he turned to sacred music, succeeding Mattheson as Kantor of Hamburg Cathedral 1728.

The number of Keiser's operas is variously estimated at between 75 and 125: only *c.*30 survive. Like so many of the great composers of the Baroque, his main achievement lay in transcending the distinct national styles of German, Italian, and French music and elements of all three may be discerned in his operas, some of which include Italian arias within a mainly German text: one such, *La fedeltà coronata*, was the first Hamburg opera with an Italian title. From Franck and Kusser, Keiser inherited a style fundamentally German in the stolidity of its texture and the shaping of the melodic line; under the influence of French and Italian opera, this was gradually softened in works of a lighter conception which approached the new galant style. From French opera Keiser borrowed the choruses and incidental music, while the enhanced quality of German instrumental music enabled him to develop the accompaniment, which often employed prominent obbligato instruments. As in Italian opera, display pieces for the singers were essential: although these were provided by da capo arias after the style of Scarlatti, Keiser was no slave to convention and reverted to earlier Venetian opera in his use of a variety of shorter forms, such as arioso and Lied. The librettos he set show a similarly rich cosmopolitanism.

Keiser was highly regarded by his peers,

especially for his rich orchestral writing and his skilled handling of the voices. Mattheson called him the first dramatic composer in the world, while Handel learnt much from him, as is seen especially in his *Almira*. Many of Handel's later developments, such as his picturesque accompaniments, were derived from Keiser's models, and he borrowed extensively from Keiser several times at later stages of his career. His works were not widely performed outside Hamburg, however, and he had the misfortune to live at a time when German popular taste declined towards the vulgar and farcical. Most of his later operas, such as *Printz Jodelet*, followed this trend, eventually leading to the closure of the Hamburg Opera shortly before his death.

Keiser's allegedly dissolute character, largely a 19th-cent. invention, was the subject of an opera to his own melodies arranged by Benno Bardi, *Der tolle Kapellmeister* (Danzig 1931).

SELECT WORKLIST (all first prod. Hamburg unless otherwise stated): *Der königliche Schäfer, oder Basilius in Arkadien* (The Royal Shepherd, or Basilius in Arcadia) (Bressand, after Parisetti; ?1693); *Procris und Cephalus* (Bressand; Brunswick 1694); *Der geliebte Adonis* (Postel; 1697); *Die sterbende Eurydice, oder Orpheus* (The Dying Euridice, or Orpheus) (Bressand; Hamburg 1702; 2 parts); *Störtebecker und Jödge Michels* (?Hotter; 1701; 2 parts); *Die verdammte Staat-Sucht, oder der verführte Claudius* (The Cursed State-Search, or Claudius Seduced) (Hinsch; 1703, rev. 1726); *Der gestürzte und wieder erhöhte Nebucadnezar, König zu Babylon* (The Destroyed and Restored Nebuchadnezzar, King of Babylon) (Hunold; 1704, rev. 1728 with Telemann); *Die römische Unruhe, oder Die edelmüthige Octavia* (The Roman Disturbance, or The Aristocratic Octavia) (Feind; 1705); *La fedeltà coronata, oder Die gekrönte Treue* (Fidelity Crowned) (Hinsch; 1706) *Der angenehme Betrug, oder Der Carneval von Venedig* (The Pleasant Deception, or The Carnival of Venice) (Meister & Curro; 1707; with Graupner); *Der hochmüthige, gestürzte und wieder erhabene Croesus* (The Proud, Fallen, and Restored Croesus) (Postel, after Minato; 1710, rev. 1730); *Ulysses* (Lersner; Copenhagen 1722); *Der Armenier* (Lersner; Copenhagen 1722); *Der lächerliche Printz Jodelet* (The Comical Prince Jodelet) (Prätorius, after Franck; 1726); *Circe* (Maurizius & Prätorius; 1734; with Giacomelli, Hasse, & Vinci).

BIBL: K. Zelm, *Die Opern Reinhard Keisers* (Munich, 1975).

**Kelemen, Zoltan** (*b* Budapest, 12 Mar. 1926; *d* Zurich, 9 May 1979). Hungarian bass. Studied Budapest, and Rome with Pediconi. Début Augsburg 1959 (Kecal). Cologne 1961 until his death. Bayreuth from 1962; New York, M, from 1968; London, CG, from 1970. Made skilful use of his rich vocal palette in a variety of characters, e.g. Osmin, Verdi's Falstaff, Ochs, Rangoni, Pizarro, Klingsor. (R).

**Kellogg, Clara Louise** (*b* Sumterville, SC, 9 July 1842; *d* New Haven, CT, 13 May 1916). US soprano and impresario. Studied New York. Début New York, AM, 1861 (Gilda). London, HM, 1867, 1879; toured USA 1868–72. In 1872 formed the short-lived Lucca-Kellogg Co. with Pauline Lucca; director of English Opera Co. 1873–6, also singing and supervising the staging, chorus, and libretto translations. Vienna 1880; St Petersburg 1881. Intelligent, and with a fine technique, she was successful as Marguerite, Philine, Violetta, Aida. One of the first internationally famous American singers. Her memoirs contain acute observations on contemporary artists, e.g. Patti, Nordica.

BIBL: C. L. Kellogg, *Memoirs of an American Prima Donna* (New York, 1913, R/1978).

**Kelly, Michael** (*b* Dublin, 25 Dec. 1762; *d* Margate, 9 Oct. 1826). Irish tenor and composer. Studied Dublin with Passerini and Rauzzini, later Italy with Finaroli and Aprile. Début Dublin 1779 (Count in Piccinni's *La buona figliuola*). After singing in Italy 1779–83, he was engaged at the Court T, Vienna, 1783–7. Here he created Basilio and Curzio in *The Marriage of Figaro*, and became a close friend of Mozart, as well as meeting other famous composers of the time. London, DL, from 1787, H, from 1793, where besides being leading tenor he was acting manager for 30 years. Last appearance Dublin 1811 in *The Bard of Erin*, one of his many popular compositions. As a singer he had 'amazing power and steadiness' (James Boaden), and was among the first English tenors to sing high notes in full voice. With Nancy Storace, he considerably raised the standard of singing at Drury Lane. His *Reminiscences* (London, 1826) give an amusing and valuable picture not only of the operatic scene of his day, but also of theatrical life in London.

BIBL: S. Ellis, *The Life of Michael Kelly* (London, 1930, R/1968).

**Kemble, Adelaide** (*b* London, 1814; *d* Warsash, 4 Aug. 1879). English soprano. Younger daughter of actor Charles Kemble. Studied London with Braham, Italy with Pasta. Début Venice 1838 (Norma). After much success in Italy, she returned to England. Sang Norma, London CG, 1841; and as member of the English Co. there 1841–2, appeared in *Figaro*, *Sonnambula*, *Semiramide*, etc. Retired 1843 upon her marriage to Edward Sartoris. An impressive and intelligent artist, she was, according to Chorley, one of the greatest English singers of the century. Accounts of her career are to be found in her sister Fanny Kemble's *Record of a Girlhood* (London, 1878).

**Kemp, Barbara.** See SCHILLINGS, MAX VON.

**Kempe, Rudolf** (*b* Niederpoyritz, 14 June 1910; *d* Zurich, 11 May 1976). German conductor. Studied Dresden with Busch. First oboe Dortmund O 1928, Leipzig Gewandhaus 1929–36. Début as conductor Leipzig 1935; répétiteur 1935–9. Chemnitz 1942; music director 1946–8; Weimar 1946–9; Dresden 1949–52; Munich 1952–4. London, CG, with Munich company 1953; regularly cond. CG until 1974. New York, M, 1954–6. Bayreuth 1960 (new *Ring*), 1967 (*Lohengrin*). First made his name in Wagner and Strauss, but was also one of the few German conductors widely admired in Verdi and Puccini. A first-rate orchestra trainer, always obtaining beautiful, lucid playing. (R)

BIBL: C. Kempe-Öttinger, *Rudolf Kempe* (Munich, 1977).

**Kenny, Yvonne** (*b* Sydney, 25 Nov. 1950). Australian soprano. Studied Sydney Cons. and Milan, S, O school. Début in concert, London, QEH, 1975 (title-role of Donizetti's *Rosmonda d'Inghilterra*). London, CG, from 1975; ENO 1977, Gly. 1985, Scottish O; also Paris, O; Vienna; Munich, S; Aix; Salzburg. Roles include Semele, Ilia, Constanze, Pamina, Aennchen, Micaëla, Oscar. (R)

**Kent Opera.** Founded in 1969, with encouragement from Alfred Deller, by Norman and Johanna Platt, the company opened with Monteverdi's *L'incoronazione di Poppea* at Tunbridge Wells on 10 December. It was conducted in the Leppard edition by Roger Norrington, who as the company's music director made his own version for production in 1974. This was in support of Platt's concern to present operas as nearly as possible as the composers imagined them. Similar considerations were brought to bear on Mozart and Verdi, and the company quickly made a reputation not only for vigour, enterprise, and effectiveness, but for a style of performance based on authenticity as a means to the greatest dramatic impact. In 1970 the company followed up its initial success with the first modern performance of Handel's *Atalanta*. Next came *Figaro*, then *Dido and Aeneas* in 1971, both at Canterbury. That year also saw two new operas commissioned from Alan Ridout, *The Pardoner's Tale* and *Angelo*, to texts by Platt. Other important productions have included Telemann's *Patience of Socrates* (1974), *Così fan tutte* (1974), *Rigoletto* (1975), *Eugene Onegin* (1977), *Il ritorno d'Ulisse* (1978), *The Turn of the Screw* (1979), *Falstaff* and *Figaro* (1981), *Agrippina* (1982), *Fidelio* (1983), and the première of Judith Weir's *A Night at the Chinese Opera* (1987). Many of the productions were in the hands of Norman Platt; from 1974, visiting producers included principally Jonathan Miller, taking charge of one production each year

until 1981, also Elijah Moshinsky and Nicholas Hytner. Norrington left in 1982, having conducted over 30 operas. Iván Fischer, music director since 1984, succeeded Norman Platt as artistic director in 1989. Singers who appeared with the company include Sandra Browne, Janice Cairns, Eiddwen Harrhy, Felicity Lott, Felicity Palmer, Anne Pashley, Rosalind Plowright, Laura Sarti, Sarah Walker, Alfred and Mark Deller, Graham Allum, Neil Jenkins, Thomas Hemsley, Benjamin Luxon, and John Tomlinson. The company's tours chiefly covered the southern part of England. Arts Council cuts reduced activities in 1985, and though there were those who felt that the company had lost something with the withdrawal of Miller and Norrington, there was widespread dismay at the council's threat to remove its grant in 1987. This threat became reality in 1989, and the company went into liquidation.

**Kerman, Joseph** (*b* London, 3 Apr. 1924). US critic and musicologist. Studied London, New York, Princeton with Strunk. His concern to give criticism a sounder musicological basis, and musicology firmer critical attitudes, first found expression in *Opera as Drama* (New York, 1956, 2/1989). This influential book pioneered other serious critical responses to opera by discussing chosen works as complete dramatic entities, arguing that 'the imaginative function of music in drama and that of poetry in drama are fundamentally the same', each bearing responsibility for the success of the drama.

**Kern, Adele** (*b* Munich, 25 Nov. 1901; *d* Munich, 6 May 1980). German soprano. Studied Munich. Début Munich 1924 (Olympia). Vienna 1929–30; Salzburg 1927–35. London, CG, 1931, 1934; Munich 1937–46. Member of the celebrated Clemens Krauss ensembles, and admired as Susanna, Zerlina, Sophie, Zerbinetta. (R)

**Kern, Jerome** (*b* New York, 27 Jan. 1885; *d* New York, 11 Nov. 1945). US composer. Studied under Galico, Lambert, and Pierce. He first earned a living by writing songs for insertion in Broadway revues, which brought him to popular attention. His first successful musical was *The Red Petticoat*, followed by a string of works in similar vein, including most notably *Sally*, *Stepping Stones*, and *Sunny*. These works, in which words and music are closely integrated, are among the earliest in the general shift from operetta towards the musical. All were eclipsed by *Show Boat*, which rapidly became one of the most popular musicals of all time both in America and in England. Containing such hit numbers as 'Can't Stop Lovin' Dat Man' and 'Ol' Man River', it provided a model for many later composers with its direct yet touching expressions of emotion, and instantly memorable melodies, while also including some sophisticated use of motive. After 1939 Kern retired from the stage and devoted himself to song-writing in Hollywood.

SELECT WORKLIST (all first prod. New York): *The Red Petticoat* (West; 1912); *Sally* (Grey; 1920); *Stepping Stones* (Caldwell; 1923); *Sunny* (Harbach & Hammerstein; 1925); *Show Boat* (Hammerstein, after Ferber; 1927).

BIBL: H. Fordin, *Jerome Kern: The Man and his Music* (Santa Monica, 1975).

**Kern, Patricia** (*b* Swansea, 4 July 1927). Welsh mezzo-soprano. Studied London with Parry Jones. Début Opera for All 1952 (Cenerentola). London, SW, 1959–69; and CG, 1967–72; also Spoleto, Dallas, New York City O. With her excellent coloratura singing and lively characterization, she was particularly successful in Rossini (Isabella, Rosina, etc.). Created Josephine in Williamson's *Violins of St. Jacques*. (R)

**Kertész, István** (*b* Budapest, 28 Aug. 1929; *d* nr. Tel-Aviv, 16 Apr. 1973). Hungarian, later German, conductor. Studied Budapest with Kodály, Weiner, and Somogyi, Rome with Previtali. Cond. Györ 1953; Budapest 1954–6. Left Hungary 1956. Music director: Augsburg 1958–63; Cologne 1964–73. London, CG, 1966–8. Though he was notable for his Mozart, he covered a range that included Wagner, Bartók, Britten, and Prokofiev. (R)

**Khaikin, Boris** (*b* Minsk, 26 Oct. 1904; *d* Moscow, 10 May 1978). Byelorussian conductor. Studied Moscow with Malko and Saradzhev. Cond. Moscow, Stanislavsky T, 1928–35; Leningrad, Maly T, 1936–43, and Kirov, 1943–54; Moscow, B, 1954–78. Cond. many premières, incl. Prokofiev's *Duenna* (1946) and *Story of a Real Man* (1948). Florence 1953 (*Khovanshchina*). Leipzig 1964 (*Queen of Spades*). (R)

**Khanayev, Nikandr** (*b* Pesochnya, 8 June 1890; *d* Moscow, 23 July 1974). Russian tenor. Studied Moscow with Zvyagina. Moscow, B, 1926–54. A striking singer with a powerful voice, particularly successful as Hermann, Sadko; also Sobinin, Shuisky, Siegmund, Verdi's Otello. (R)

**Khan Konchak.** A Polovtsian Khan (bs) in Borodin's *Prince Igor*.

**Khokhlov, Pavel** (*b* Ustye, 2 Aug. 1854; *d* Moscow, 20 Sep. 1919). Russian baritone. Studied Moscow with Yury Arnold (as a bass) and Alexandrova-Kochetova. Début Moscow, B, 1879 (Valentin). Remained member of Moscow, B, until 1900. St Petersburg, M, 1881, 1887–8. Prague 1889. A master of the bel canto style, with a warm voice and personality, he was especially famous as Onegin (which he sang at the first pub-

lic performance, 1881, and thereafter 138 times in Moscow), and as Rubinstein's Demon. Also sang Don Giovanni, Germont, Telramund, Boris, Prince Igor. Though a conscientious artist (he continually restudied his roles), he suffered an early vocal decline, as even his friend and admirer Tchaikovsky admitted in 1886.

BIBL: V. Yakovlev, *P. A. Khokhlov* (Moscow, 1950).

**Khovanshchina** (The Khovansky Affair). Opera in 5 acts by Musorgsky; text by the composer and Stasov. Left unfinished, completed and orchestrated by Rimsky-Korsakov. Prod. St Petersburg, Kononov T, 21 Feb. 1886, cond. Goldstein; first major prod. St Petersburg, M, 7 Nov. 1911, with Zbruyeva, Ershov, Labinsky, P. Andreyev, Sharonov, Shalyapin, cond. Coates; London, DL, 1 July 1913, with Petrenko, Andreyev, Zaporozhets, Shalyapin, cond. Cooper; Philadelphia, Metropolitan OH, 18 Apr. 1928, with Fedorova, Criona, Windheim, Shvetz, Figaniak, cond. Grigaitis. Other versions by Stravinsky and Ravel (Paris 1930) and Shostakovich (Leningrad 1960).

The opera, based on the complicated political events of the time of the accession of Peter the Great in 1682, concerns the strife between various factions. The Streltsy, or Guards, are led by Prince Ivan Khovansky (bs), whose son Andrey (ten) in a sub-plot tries to rape a German girl Emma (sop), but is prevented by his former mistress Marfa (sop). The rival faction is led by Prince Golitsyn (ten). An important part is also played by the Old Believers, the ultra-orthodox group in the Church who had refused to accept the reforms of the 1650s: they are led by Dosifey (bs), and are eventually killed, together with the reconciled Marfa and Andrey.

**Khrennikov, Tikhon** (*b* Elets, 10 June 1913). Russian composer. Studied with Agarkov and Vargunina, Moscow with Gnesin and Shebalin. The success of his song-opera *Into the Storm*, especially with Stalin, led to this type of simple, melody-orientated opera in separate numbers being associated with the doctrine of Socialist Realism in opposition to more ambitiously, continuously composed works such as Prokofiev's *Semyon Kotko*. Khrennikov became a spokesman for Socialist Realism, and in 1948 headed the musical opposition at the Zhdanov tribunal attacking Prokofiev, Shostakovich, and others for 'formalism' and other deviant sins. Nevertheless, he succeeded in giving the Composers' Union which he led great influence in Soviet society, and there is evidence of private helpfulness. His bureaucratic activities limited his time for composition, but he had a further operatic success with *The Mother*.

SELECT WORKLIST (all first prod. Moscow): *V buryu*

375

(Into the Storm) (Faiko & Virta, after Virta; 1939, rev. 1952); *Frol Skobeyev* (Tsenin, after a 17th-cent. Russian tale; 1950, rev. 1966); *Mat* (The Mother) (Faiko, after Gorky; 1957).

BIBL: Y. Kremlev, *Tikhon Khrennikov* (Moscow, 1963).

**Khromchenko, Solomon** (*b* Zlatopol, 4 Dec. 1907). Ukrainian tenor. Studied Kiev with Engel-Krom. Moscow, B, 1934–57. With his pliant, soft-grained voice and clear diction, he was an admired Almaviva, Gérald (*Lakmé*), Lensky, Berendey, Boyan, Simpleton. (R)

**Kiel.** City in Schleswig-Holstein, Germany. Opera was first given in 1764 in the Opern- und Komödienhaus. The Stadttheater was built on its site 1841; rebuilt 1907 (cap. 960); bombed 1944. The summer Tivoli T opened 1845; burnt down 1870; re-established as Schillertheater 1890; as the Schauspielhaus, used for opera 1945–53. The Bühnen der Landeshauptstadt Kiel opened in the rebuilt Stadttheater (cap. 866) 1953. Directors have included Georg Hartmann 1924–32 and Rudolf Sellner 1950–5; music directors Georg Winkler 1950–9, Peter Ronnefeld 1963–5, and Hans Zender 1969–76.

**Kienzl, Wilhelm** (*b* Waizenkirchen, 17 Jan. 1857; *d* Vienna, 19 Oct. 1941). Austrian composer. Studied Graz, Prague, Leipzig, and Vienna. After extensive travels became first Kapellmeister at the German opera in Amsterdam; following posts in Hamburg and Munich he was director of the Steiermärkischer Musikverein, Graz 1897–1917, when he retired to Vienna.

His first opera, *Urvasi*, attracted some attention, despite its dramatic weaknesses: its successor, *Heilmar der Narr*, was not produced until 1892 because of staging difficulties. With *Der Evangelimann* Kienzl achieved his greatest success: its folk scenes proved particularly effective and it was favourably compared to Humperdinck's *Hänsel und Gretel*. Kienzl composed seven further operas, but none was as well received as *Der Evangelimann*, which was given in several countries.

Kienzl belonged to that group of composers, including Humperdinck, which was reared in the shadow of Wagner, and which sought to apply Wagnerian techniques of music drama, especially Leitmotiv, to non-heroic subjects. To the modern audience his works appear over-sentimental and lacking in dramatic sincerity: they are rarely seen today. In addition to his compositions, Kienzl edited Mozart's *La clemenza di Tito* and published a study of Wagner, with whose family he had once lived at Bayreuth.

SELECT WORKLIST: *Urvasi* (Gödel, after Kalidasa; Dresden 1886); *Heilmar der Narr* (Heilmar the Fool) (Kienzl; Munich 1892); *Der Evangelimann* (The

Evangelical) (Kienzl, after Meissner; Berlin 1895); *Don Quixote* (Kienzl, after Cervantes; Berlin 1898).

WRITINGS: W. Kienzl, *Meine Lebenswanderung: Erlebtes, Erschautes* (Stuttgart, 1926).

**Kiepura, Jan** (*b* Sosnowiec, 16 May 1902; *d* Harrison, NY, 15 Aug. 1966). Polish tenor. Studied Warsaw with Brzeziński and Leliva. Début Lwów 1925 (Gounod's Faust). Vienna, S, from 1926; London, CG, 1927; Paris, O, 1928; Chicago 1931–2; New York, M, 1938–41. A famous Calaf, Don José, Cavaradossi, and Des Grieux (Puccini), and, with his striking good looks, later a successful film star. Married the soprano and film actress **Marta Eggerth** (*b* 1912), with whom he sang in numerous tours of *The Merry Widow*. (R)

**Kindermann, August** (*b* Potsdam, 6 Feb. 1817; *d* Munich, 6 Mar. 1891). German bass-baritone. No formal training. Début from chorus of Berlin, H, 1837 (in Spontini's *Agnes von Hohenstaufen*). Leipzig 1839–46, becoming a close friend of Lortzing, who wrote the title-role of *Hans Sachs* and Eberhardt (*Der Wildschütz*) for him. Munich 1846–91, singing in a wide repertory (e.g. Don Giovanni, Mozart's and Rossini's Figaros, Gounod's Méphistophélès, Sachs), and creating Wotan in both *Das Rheingold* and *Die Walküre*. Also first Titurel, Bayreuth 1882. Possessed a powerful voice, musical sensitivity, and dramatic inspiration. His children **Marie, Franziska,** and **August** had good singing careers in Germany, while his daughter Hedwig *Reicher-Kindermann was a distinguished Wagner singer.

**King, James** (*b* Dodge City, KS, 22 May 1925). US tenor. Studied New York with Singher. Début 1961, as baritone; tenor début San Francisco 1961 (Don José). Berlin, D, from 1962; Bayreuth 1965–75; London, CG, 1966–76; New York, M, from 1966; Milan, S, 1968; Paris, O, 1972. Roles include Florestan, Manrico, Calaf; especially successful as Walther, Lohengrin, Parsifal, Bacchus. (R)

**King, Matthew** (*b* London, *c*.1733; *d* London, Jan. 1823). English composer. A popular song and instrumental composer, he also wrote a number of theoretical works. He composed numerous stage works 1804–19, some of which were collaborations. These ranged from popular farces, such as *Too Many Cooks*, to a so-called 'grand romantic opera', *One O'Clock*. King also composed melodramas, including *Ella Rosenburg* and *The Fisherman's Hut*.

SELECT WORKLIST (all prod. London): *Too Many Cooks* (Kenney; 1805); *Ella Rosenburg* (Kenney; 1807); *One O'Clock* (Lewis; 1811; with Kelly); *The Fisherman's Hut* (Tobin; 1819; with Davy).

**King Arthur, or The British Worthy.** 'A drama-tick opera' in prologue, 5 acts, and epilogue, by Purcell; text by Dryden. Prod. London, Dorset Gardens T, prob. early June 1691, with Better-ton, Williams, Hodgson, Kynaston, Sandford, Alexander, Bowen, Harris, Bracegirdle, Richardson, Butler, Bowman; New York, Park T, 25 Apr. 1800.

Arthur, King of the Britons, and Oswald, King of Kent, are rivals for the hand of Emme-line, and resolve to fight a battle on St George's day. Each is supported by different spirits, The armies meet, and when Arthur has the oppor-tunity of killing Oswald, he spares him, thus win-ning Emmeline's hand.

There are some 30 operas on Arthurian legends.

**King Fisher.** A business man (bar), Jenifer's father, in Tippett's *The Midsummer Marriage*.

**King Mark.** See MARK.

**King Priam.** Opera in 3 acts by Tippett; text by the composer, after Homer's *Iliad*. Prod. Coventry T, 29 May 1962, with Collier, Veasey, Elkins, Lewis, F. Robinson, Godfrey, Dobson, Lanigan, cond. Pritchard; Karlsruhe 26 Jan. 1963, with Moussa-Felderer, Wolf-Ramponi, Graf, Reynolds, Harper, Vandenburg, cond. Grüber.

Troy, antiquity. Priam (bs-bar), supported by Hecuba (sop), orders the death of his baby son Paris, who (it is prophesied by the Old Man (bs)) will cause his father's death. But he is relieved to find, when out hunting, that the boy (sop) has been spared; he and Hector (bar) take him to Troy. Here Paris (ten) and Hector quarrel; Paris fetches Helen (mez) from Sparta. In the last scene of the act, Hermes (ten) arranges the Judgement of Paris; Aphrodite (i.e. Helen) is chosen.

In the war that ensues, Achilles (ten) is sulking in his tent; the Trojans drive the Greeks back and fire their ships, but are weakened by Hector and Paris quarrelling. Patroclus (bar) puts on Achilles' armour but is killed by Hector. The Trojans' rejoicing is interrupted by Achilles' war-cry.

The women reflect upon their role in the war. News of Hector's death at the hands of Achilles is brought to Priam. He goes to beg his son's body, and rouses Achilles' pity. He then withdraws into himself, and is killed before the altar by Achilles' son.

**King Roger** (Pol.: *Król Roger*). Opera in 3 acts by Szymanowski; text by Jarosław Iwaszkiewicz and the composer. Prod. Warsaw 19 June 1926, with Mossakowski, Korwin-Szymanowska, Dobosz, Wraga, cond. Młynarski; London, SW, 14 May 1975, with Gail, Knapp, cond. Mackerras; Long Beach, CA, 24 Jan. 1988, cond. Sidlin.

Sicily, 12th cent. Queen Roxane (sop) falls in love with a shepherd-prophet (ten), from India, denounced as a heretic. King Roger (bar), after initially denouncing the shepherd as a blas-phemer, is finally converted. The opera ends with a bacchanal in the ruins of a Greek temple, at which the shepherd, now transformed into Dionysus, officiates.

Another opera on the subject is by H. Berton (1817).

**King's Henchman, The.** Opera in 3 acts by Deems Taylor; text by Edna St Vincent Millay. Prod. New York, M, 17 Feb. 1927, with Easton, Johnson, Tibbett, cond. Serafin.

Eadgar (bar), King of England, sends Æthel-wold (ten) to win the Princess Ælfrida (sop) for him; Æthelwold marries her, sending word that she is ugly, but when Eadgar comes to visit them and finds how beautiful she is, Æthelwold kills himself in remorse.

**King's Theatre.** Built by Vanbrugh in the Hay-market, London, to house the Lincoln's Inn Fields Theatre Company, and opened 9 Apr. 1705 as the Queen's T with Greber's *Gli amori d'Ergasto*, the first Italian opera in London. Opera alternated with drama until the end of 1707, when opera alone was given. At first works were sung some in English, some in Italian, and some mixed. After 1710 it was the London home of Italian opera. Renamed the King's T on the accession of George I in 1714. *Heidegger was the manager 1710–34, sharing with Handel 1729–34. The premières of more than 24 of Han-del's operas, seven of his pasticcios, and some of his secular choral works and oratorios were given here, also the English premières of works by Bononcini, Tarchi, and Cimarosa. The new King's T, then the largest in England (cap. 3,300), opened under the management of Storace and Kelly on 16 Jan. 1793 with Paisiello's *Il bar-biere di Siviglia*. During 1793–1830 many major works had their English premières there, with singers including Giorgi-Banti, Bellochi, Mrs Billington, Catalani, Fodor-Mainvielle, Grassini, Mara, Garcia, Kelly, and Naldi. During Ebers's régime, 1821–7, a number of works by Rossini had their English premières, also the first Meyer-beer opera in England, *Il crociato in Egitto*. Singers included Brambilla, Colbran, Caradori, De Begnis, Pasta, and Velluti. During the régime of *Laporte (1828–42) the theatre became *Her Majesty's.

BIBL: D. Nalbach, *The King's Theatre, 1704–1867* (London, 1972).

**Kipnis, Alexander** (*b* Zhitomir, 13 Feb. 1891; *d* Westport, CT, 14 May 1978). Ukrainian, later

US, bass. Studied Berlin with Grenzebach. Though interned in 1914 as an enemy alien, he was released through the influence of an officer who had heard him sing. Début Hamburg 1915. Wiesbaden 1916–18; Berlin, SO, 1919–30; Berlin, S, 1930–4, after which he settled in USA. London, CG, 1927, 1929–35; Gly. 1936; Chicago 1923–32; New York, M, 1940–6. Also Bayreuth, Salzburg, Buenos Aires. Outstanding in German opera (his Gurnemanz was especially celebrated), he was also notable as Arkel, Philip, and Boris. His voice possessed great flexibility and tonal variety, and a wide compass, while his dramatic versatility was equal to both Sarastro and Ochs, and also Hans Sachs. (R)

**Kirghizistan.** See KYRGYZSTAN.

**Kirkby Lunn, Louise.** See LUNN, LOUISE KIRKBY.

**Kirov Theatre.** See ST PETERSBURG.

**Kirsten, Dorothy** (*b* Montclair, NJ, 6 July 1910; *d* Los Angeles, 18 Nov. 1992). US soprano. Studied New York, Juilliard School, and Rome with Pescia. Début Chicago 1940 (Pousette, *Manon*). New York, M, 1945–57, 1959–79. Possessor of a clear, confident voice; roles included Butterfly, Tosca, Minnie, Louise. Appeared in several films, including *The Great Caruso*. (R)

**Kiss, The** (Cz.: *Hubička*). Opera in 3 acts by Smetana; text by Eliška Krásnohorská, after the story (1871) by 'Karolina Světlá' (Joanna Mužáková). Prod. Prague, P, 7 Nov. 1876, with Sittová, Cachová, Lausmannová, Vávra, Čech, Lev, Mareš, Šára, cond. Čech; Chicago, Blackstone T, 17 Apr. 1921; Liverpool 8 Dec. 1938 (amateur); London, King's T, Hammersmith, 18 Oct. 1948, cond. Tausky.

A North Bohemian village. The widower Lukaš (ten) is anxious to exchange a betrothal kiss with Vendulka (sop), but she stubbornly refuses him before they are married; such an act is said to rouse the ghostly wrath of a dead wife. After many tricks and much indignation, Vendulka agrees.

**Kitezh.** See INVISIBLE CITY OF KITEZH, THE.

**Klafsky, Katharina** (*b* Mosonszentjanós, 19 Sep. 1855; *d* Hamburg, 29 Sep. 1896). Hungarian soprano. Studied Vienna with Marchesi, later Leipzig with Sucher. Début as mez, Salzburg 1875. Discovered by Neumann; Leipzig 1876–82, graduating from small roles to Brangäne, Venus, and Alice (*Robert le diable*). Sang with Neumann's Wagner Co. 1882–3. London: HM, 1882; CG and DL, 1892; DL, 1894. Hamburg 1886–95; US tour 1895–6 with Damrosch Co. under her third husband, the conductor Otto Lohse. Returned 1896 to Hamburg, where she died suddenly at the height of her powers. She was buried in her *Tannhäuser* Elisabeth costume. One of the foremost dramatic sopranos of her day, much admired (including by Mahler) as Leonore, Agathe, Eglantine, Isolde, Elisabeth, Brünnhilde (incl. all three on consecutive nights).

BIBL: L. Ordemann, *Aus dem Leben und Wirken von Katharina Klafsky* (Hamelin, 1903).

**Klagenfurt.** Town in Carinthia, Austria. Opera is given in the Stadttheater (cap. 770), opened 1910.

**Klausenburg.** See CLUJ.

**Klebe, Giselher** (*b* Mannheim, 28 June 1925). German composer. Studied Berlin with Wolfurt, Rufer, and Blacher. He was one of the leading opera composers to emerge in Germany after the Second World War. His dramatic idiom was influenced both by Blacher and by Expressionist elements deriving from Berg and Schoenberg. Often the dense ingenuity of the dodecaphonic music has made comprehensibility difficult in the theatre. Latterly his style simplified and clarified, and he has sometimes turned to number opera. He has always shown an interest in operas based on distinguished literary originals, and his technique is often to let the story, or a reordered version of it, unfold in the dramatist's terms with musical commentary superimposed.

SELECT WORKLIST: *Die Räuber* (The Robbers) (Klebe, after Schiller; Düsseldorf 1957); *Die tödlichen Wünsche* (The Fatal Wishes) (Klebe, after Balzac; Düsseldorf 1959); *Die Ermordung Cäsars* (The Murder of Caesar) (Klebe, after Shakespeare; Essen 1959); *Alkmene* (Klebe, after Kleist; Berlin 1961); *Figaro lässt sich scheiden* (Figaro Gets Divorced) (Klebe, after Horvath; Hamburg 1963); *Jacobowsky und der Oberst* (Jacobowsky and the Colonel) (Klebe, after Werfel; Hamburg 1965); *Das Märchen von der schönen Lilie* (The Tale of the Beautiful Lily) (Klebe, after Goethe; Schwetzingen 1969); *Ein wahrer Held* (A True Hero) (Klebe, after Synge; Zurich 1975); *Das Mädchen aus Domrémy* (The Girl from Domrémy) (Klebe, after Schiller; Stuttgart 1976).

**Kleiber, Carlos** (*b* Berlin, 3 July 1930). Austrian, later Argentinian, conductor; son of below. Began musical studies Buenos Aires 1950. Début La Plata 1952. Studied chemistry Zurich, returning to music as répétiteur Munich, Gärntnerplatz, 1954–6. Cond. Potsdam, Deutsche O am Rhein, 1956–64; Zurich 1964–6; Stuttgart 1966. Also Munich (from 1968); Vienna, S (from 1973); Bayreuth 1974–6 (*Tristan*). Edinburgh Festival 1966 (*Wozzeck*). London, CG, since 1974 (*Rosenkavalier, Elektra, Bohème*). Milan, S, 1977 (Verdi's *Otello*); New York, M, 1988. An individualist and a perfectionist, he demands long rehearsal periods and will refuse to conduct if dissatisfied; his grasp of a score is always serious and original, his handling of it passionate and lyrical. (R)

**Kleiber, Erich** (*b* Vienna, 5 Aug. 1890; *d* Zurich, 27 Jan. 1956). Austrian, later Argentinian, conductor; father of above. Studied Vienna and Prague. Cond. Darmstadt 1912–18; Wuppertal 1919–21; Mannheim 1922–3. Music director Berlin, S, 1923–34. Guest cond. at many operahouses 1933–56, notably Buenos Aires, C, 1937–49; London, CG, 1938, 1950–3; Amsterdam 1933–8, 1949–50. His directorship in Berlin was one of the most brilliant in the theatre's history with premières of *Wozzeck* (1925) and Milhaud's *Christophe Colomb* (1930) and the introduction of *Jenůfa* and *Shvanda* into the repertory. Resigned 1934 at the time of the *Mathis der Maler* controversy and did not return to Germany until 1950. Reappointed to Berlin, S, 1954, resigning as protest against political interference. His contribution to the development of the English company at London, CG, 1950–3 was crucial. A fine, sensitive orchestra trainer, with a sharp intellect and quick sensitivity. (R).

BIBL: J. Russell, *Erich Kleiber* (London, 1958).

**Klein, Herman** (*b* Norwich, 23 July 1856; *d* London, 10 Mar. 1934). English critic and teacher. Studied singing with Manuel García. Critic of *Sunday Times* 1881–1901, also acting as unofficial adviser to the paper's owner, Augustus Harris, in his engagement of Covent Garden singers. Critic *New York Herald* 1902–9, also publishing many books on singers and singing.

**Klein, Peter** (*b* Zündorf, 25 Jan. 1907 *d* Vienna, 3 Oct. 1992). German tenor. Studied Cologne Cons. Début 1930 in small roles, Cologne. Hamburg 1937–42; Vienna from 1942; Salzburg 1946–56; London, CG, from 1947; New York, M, 1949–51. A highly successful character tenor, acclaimed as Pedrillo, Don Basilio (Mozart), M. Taupe (R. Strauss), David, Mime. (R)

**Klemperer, Otto** (*b* Breslau (Wrocław), 14 May 1885; *d* Zurich, 6 July 1973). German conductor. Studied Frankfurt with Knorr, Berlin with Pfitzner. Début Berlin 1906 (*Orpheus in the Underworld*). Recommended by Mahler to the German T, Prague, where he cond. 1907–10. Hamburg 1911–12; Bremen 1913–14; Strasburg 1914–17. Music director: Cologne 1917–24; Wiesbaden 1924–7; Berlin 1927–33, first at the Kroll 1927–31, then S: premières incl. Hindemith *Neues vom Tage* (1929) and many Berlin premières. His work at the Kroll was vital in advancing the cause of new ideas in opera which the theatre represented. Goethe Medal 1933; shortly afterwards forced to leave Germany. Apart from a period at the Budapest O 1947–50, conducted little opera after the war, but made a

belated London, CG, début conducting and producing *Fidelio* 1961; also *Magic Flute* 1962, *Lohengrin* 1963. His greatness as a concert conductor made the rarity of these distinguished late appearances the more regrettable. Operas and operatic projects include *Wehen* (Birth Pangs, 1915; rev. as *Das Ziel* (The Goal) 1929). (R)

BIBL: P. Heyworth, *Otto Klemperer*, i (Cambridge, 1983).

**Klenau, Paul von** (*b* Copenhagen, 11 Feb. 1883; *d* Copenhagen, 31 Aug. 1946). Danish composer and conductor. Studied Copenhagen with Malling, Berlin with Bruch, Munich with Thuille. Kapellmeister Freiburg O, 1907. Further study Stuttgart with Schillings, also working at the Hofoper. Freiburg 1912. His Icelandic *Kjarten und Gudrun* was premièred under Furtwängler. Still further study with Schoenberg, whose music influenced his later operas. His tastes led him to keep one foot in Germany, and he held conducting posts in Vienna as well as Copenhagen, whither he finally retired in 1940.

SELECT WORKLIST: *Sulamith* (Klenau, after the Song of Solomon; Munich 1913); *Kjarten und Gudrun* (Klenau; Mannheim 1918); *Die Lästerschule* (School for Scandal) (Hoffmann, after Sheridan; Frankfurt 1927); *Michael Kohlhaas* (Klenau, after Kleist; Stuttgart 1933); *Rembrandt von Rijn* (Klenau; Berlin 1937); *Elisabeth von England* (Klenau; Kassel 1939).

**Klimentova, Maria** (*b* ?, 1857; *d* Moscow, 1946). Russian soprano. Studied Moscow with Galvani. Moscow, B, 1880–9. Created Tatyana 1879, when Tchaikovsky, though admitting her lack of acting ability, liked her 'warmth and sincerity'. Also sang Oxana in the first *Cherevichki* (the revised *Vakula the Smith*). Later taught in Moscow.

**Klingsor.** The magician (bs) in Wagner's *Parsifal*.

**Klose, Margarete** (*b* Berlin, 6 Aug, 1902; *d* Berlin, 14 Dec. 1968). German mezzo-soprano. Studied Berlin with Marschalk and Bültemann. Début Ulm 1927 (in Kálmán's *Gräfin Mariza*). Mannheim 1928–31. Berlin: SO, 1931–49, 1958–61; S, 1949–55. Bayreuth 1936–42; London, CG, 1935, 1937; San Francisco 1953. Also S America, etc. A distinguished singer, particularly effective in Wagner and Verdi, but also as Dalila, Klytemnestra, and the Kostelnička. Created Oona in Egk's *Irische Legende*. (R)

**Kluge, Die** (The Clever Woman). Opera in 6 scenes by Carl Orff; text by the composer, after Grimm's story *Die Geschichte von dem König und der klugen Frau*. Prod. Frankfurt 20 Feb. 1943, with Wackers, Gonszar, Staudenmeyer, cond. Winkler; Cleveland, Karamu House, 7 Dec.

1949; London, SW, 27 July 1959 with Harper, cond. Priestman.

The King (bar), tired of the wisdom of the Clever Woman (sop), whom he married after she had successfully answered his riddles, sends her away, telling her she can take with her whatever she wants from the palace. She sets off down the road, and opens the trunk from which the King emerges, he being the one thing she wanted above all else. Delighted at her choice, he offers to take her back, if she promises not to be clever any longer.

**Klytemnestra.** See CLYTEMNESTRA.

**Knappertsbusch, Hans** (*b* Elberfeld, 12 Mar. 1888; *d* Munich, 25 Oct. 1965). German conductor. Studied philosophy at Bonn, then music at Cologne with Steinbach and Lohse. Mülheim 1910–12, also spending summers assisting Richter and Siegfried Wagner at Bayreuth. Cond. Elbefeld 1913–18; Leipzig 1918–19; Dessau 1919, music director 1920–2; Munich 1922–36. Dismissed by the Nazis. Vienna, S, 1936–50. Returned to cond. in Munich, and was regularly at Bayreuth 1951–64. Guest appearances Paris; Milan; Rome; Zurich; London, CG, 1937 (*Salome*); but never sought international career. An 'aristocratic master of the art of conducting' (Hans Hotter). His art came from deep in the German tradition he so fully exemplified, and his finest interpretations, sober, grandly conceived, and gravely moving, were of Bruckner and Wagner: no conductor has more sensitively conveyed the spirituality of *Parsifal* (the last work he conducted, Bayreuth 1964). (R)

BIBL: R. Betz and W. Panovsky, *Knappertsbusch* (Ingolstadt, 1958).

**Knight, Gillian** (*b* Redditch, 1 Nov. 1934). English mezzo-soprano. Studied London, RAM. Début with D'Oyly Carte Co. 1959 (in Gilbert and Sullivan). London: SW/ENO from 1968; CG from 1970. Paris, O, from 1978; also Pittsburgh, Tanglewood, Montreal, Vancouver, Nice. Roles incl. Ragonde (*Ory*), Ulrica, Olga, Carmen, Gertrude (Thomas's *Hamlet*). (R)

**Knipper, Lev** (*b* Tbilisi, 3 Dec. 1898; *d* Moscow, 30 July 1974). Georgian composer. After a brief army career, studied Moscow with Glier and Zhilyayev, Berlin with Jarnach. Also worked in Moscow Art T and Nemirovich-Danchenko Musical T. He maintained his army connections, also aligning himself at first with the Association of Contemporary Music in Moscow. His most important work from this period is the opera *The North Wind*, a work whose iconoclastic manner was advanced for its day. He later embraced a more orthodox party idiom, also interesting himself in the music of the Soviet republics and writing an opera, *On Lake Baykal*, on Mongolian themes.

SELECT WORKLIST: *Severniy veter* (The North Wind) (Knipper, after Kirshon; Moscow 1930); *Na Baykale* (On Lake Baykal) (Tsidinzhapov & Feinberg; Ulan-Ude 1948).

**Knote, Heinrich** (*b* Munich, 26 Nov. 1870; *d* Garmisch, 15 Jan. 1953). German tenor. Studied Munich with Kirschner. Début Munich 1892 (Georg, *Der Waffenschmied*). Sang there until 1931. London, CG, 1901, 1903, 1907–8, 1913; New York, M, 1904–8. A much-acclaimed Heldentenor, with a clear, resonant voice, excellent diction, and handsome appearance, he was extremely successful as Siegfried, Tristan, Lohengrin, and Tannhäuser. After retiring in 1931, he taught in Munich. (R)

**Knot Garden, The.** Opera in 3 acts by Tippett; text by the composer. Prod. London, CG, 2 Dec. 1970, with Barstow, Gomez, Minton, Tear, Herincx, Carey, Hemsley, cond. Davis; Evanston, IL, Northwestern U, 22 Feb. 1974, with Strauch, Jaffe, Evans, Kraus, Pollock, Cooper, Dickson, cond. Rubenstein (first Tippett opera in USA).

A walled garden, near a city. The analyst Mangus (bar) has been invited by Faber (bar) and Thea (mez) to treat their ward Flora (sop), but quickly perceives that the real trouble lies in the marriage: Faber has grown too far outward, Thea too far inward, and they no longer meet in a true marriage. A homosexual couple arrive: they are the black writer Mel (bar) and the musician Dov (ten). A further arrival is Denise (sop), Thea's sister and 'a dedicated freedom fighter' who has suffered torture. Mangus contrives a series of devices to set their difficulties to rights, including an elaborate charade based on *The Tempest*; the stage devices include the revolving of the symbolic knot garden of the title, in which they meet and play out their relationships. Eventually Mel leaves with Denise, and Faber and Thea find renewed marriage: Flora has found adulthood and independence, and Dov is left to set off upon a journey (recorded in Tippett's *Songs for Dov*, which arose out of the opera).

**Knussen, Oliver** (*b* Glasgow, 12 June 1952). English composer. Studied London with Lambert, Tanglewood with Schuller. His first opera, *Where the Wild Things Are*, revealed a strong stage instinct and the ability to reach both children and adults with witty and appealing music; this was joined in a double-bill by *Higglety Pigglety Pop!*, and with the benefit of Maurice Sendak's entertaining fantasy monsters and other beasts has proved widely successful.

WORKLIST: *Where the Wild Things Are* (Sendak & Knussen; Brussels 1980, rev. London, Gly. O,

1984); *Higglety Pigglety Pop!* (Sendak & Knussen; Gly. (incomplete) 1984, (complete) 1985).

**Koanga.** Opera in a prologue, 3 acts, and epilogue by Delius; text by C. F. Keary, after George Washington Cable's novel *The Grandissimes* (1880). Prod. Elberfeld 30 Mar. 1904, with Kaiser, Whitehill, cond. Cassirer. London, CG, 23 Sep. 1935, with Slobodskaya, Brownlee, cond. Beecham; Washington, DC, 18 Dec. 1970, with Lindsey, Holmes, cond. Callaway.

Louisiana, late 18th cent. On a Mississippi river plantation, Palmyra (sop) spurns the slave-driver Simon Perez (ten) and falls in love with Koanga (bar) a prince of her own tribe. The planter Don José Martínez allows the wedding, during which Palmyra is abducted by Perez. Koanga quarrels with the planter and flees into the forest, where he and a voodoo priest invoke a plague on their enemies. But Palmyra is also afflicted. Koanga arrives in time to save her from Perez, whom he kills before being himself killed. Palmyra stabs herself with Koanga's spear.

**Kobbé, Gustav** (*b* New York, 4 Mar. 1857; *d* Long Island, 27 July 1918). US critic. Studied Wiesbaden and New York. After editing *The Musical Review*, served as critic on several New York papers including *The World*, which sent him to Bayreuth for the *Parsifal* première 1882. His books include a 2-vol. study of Wagner (1890), but he is best known for *The Complete Opera Book* (1922), which, especially in its revisions by the Earl of Harewood (1954, 1976, 1987), remains a helpful collection of opera synopses.

**Kobelius, Johann Augustin** (*b* Wachlitz, 21 Feb. 1674; *d* Weissenfels, 17 Aug. 1731). German composer. He is known to have been an important and prolific composer of Singspiel for the Weissenfels court at the start of the 18th cent., but all his music is lost.

**Koch, Gottfried** (*b* ?, 1703; *d* ?, 1775). German impresario. After 21 years acting with the Neuber Company, founded his own in 1850, taking over Schönemann's in 1758 and settling as near-permanent company, Hamburg 1763, playing a mixed repertory. First to stage Weisse's *Der Teufel ist los*, music by Standfuss, and also gave revision with music by Hiller. Leipzig 1763, where he was so popular with the students that an irritated professor, facing an empty lecture hall, had his matinées reduced to two per week. Weimar 1768–71. Important among travelling theatre managers for seriously attempting higher standards and for maintaining a repertory in a single town for several years.

**Kochubey.** A rich Cossack (bs), Maria's father, in Tchaikovsky's *Mazeppa*.

**Kodály, Zoltan** (*b* Kecskemét, 16 Dec. 1882; *d* Budapest, 6 Mar. 1967). Hungarian composer. Initially self-taught, then studied Budapest with Koessler. His early and enduring friendship with Bartók led to researches into Hungarian folk music out of a shared wish to give their country musical maturity and independence. This is expressed in all his music, including his stage works: here he prefers number opera with dialogue to forms involving continuous development, in the belief that audiences must first be educated in Hungarian folk music before a more advanced language employing it could be used. *Háry János* is essentially a Hungarian Singspiel, drawing on the tales of an historical folk hero, a braggart soldier whose fantastic adventures celebrate Magyar virtues. At the centre of the concept is a contrast between natural Hungarian characters and the sophistication of the foreign court as it impresses them. *The Transylvanian Spinning-Room* brings together many of the songs and dances he had himself collected into seven scenes: there is no real plot, rather a celebration of village life in a complex situation involving the parting and reuniting of two lovers. Kodály uses a certain amount of counterpoint and can combine separate folk-songs, but keeps the orchestra in a subsidiary role. *Czinka Panna* is an historical piece written to celebrate the centenary of the 1848 revolution, and makes greater use of dialogue at the expense of musical substance.

WORKLIST (all first prod. Budapest): *Háry János* (John Háry) (Paulini & Harsányi; 1926); *Székely fonó* (The Transylvanian Spinning-Room) (Kodaly, after a folk tale; 1932); *Czinka Panna* (Balázs; 1948).

BIBL: L. Eősze, *Kodály Zoltán élete és munkássága* (Budapest, 1956; trans. 1962); P. Young, *Zoltan Kodály: A Hungarian Musician* (London, 1964).

**Kollo, René** (*b* Berlin, 20 Nov. 1937). German tenor. Grandson of operetta composer **Walter Kollo** (1878–1940). After period as pop singer and in operetta, studied Berlin with Varena. Début Brunswick 1965 (in *Oedipus Rex*). Düsseldorf 1967; Bayreuth from 1969; London, CG, and New York, M, from 1976. Roles include Tamino, Max, Lensky, Laca; also, despite an essentially light lyric voice, Lohengrin, Siegmund, Parsifal, Walther. A singer of great musicianship. (R)

**Köln.** See COLOGNE.

**Kolozsvár.** See CLUJ.

**Koltai, Ralph** (*b* Berlin, 31 July 1924). Hungarian, later British, designer. Studied London. London: CG, *Tannhäuser* 1954, *Taverner* 1972, première of *The Ice Break* 1977; SW, *Mahagonny* 1963; C, *Ring* 1973, *Seven Deadly Sins* (Weill) 1978, *Pacific Overtures* (Sondheim) 1987. Scottish O: *Elegy for Young Lovers* 1970, *Rake's Pro-*

gress 1971. WNO: *Midsummer Marriage* 1976, *Don Giovanni* 1982. Munich: *Fidelio* 1974. Lyons: *Die Soldaten* 1983. Has often worked closely and successfully with Michael *Geliot; a pioneering designer with an interest in novel materials and structures, including tubular scaffolding.

**Konchak.** The Khan (bs), chieftain of the Polovtsians, in Borodin's *Prince Igor*.

**Konetzni, Anny** (*b* Ungarisch-Weisskirchen, 12 Feb. 1902; *d* Vienna, 6 Sep. 1968). Austrian soprano. Sister of below. Studied Vienna with Schmedes, Berlin with Stückgold. Début as contralto, Vienna, V, 1925 (Adriano, *Rienzi*). Berlin, S, 1931–6; Vienna, S, from 1933; London, CG, 1935–9, 1951; New York, M, 1934–5. Specialized in Wagner and Richard Strauss repertory. (R)

**Konetzni, Hilde** (*b* Vienna, 21 Mar. 1905; *d* Vienna, 20 Apr. 1980). Austrian soprano. Sister of above. Studied Vienna Cons., Prague with Prochaska-Neumann. Début Chemnitz 1929 (Sieglinde, with her sister as Brünnhilde). Prague 1932–8; Vienna, S, from 1936; London, CG, 1938–9, 1947, 1955; Gly. 1938. Like her sister, sang the Marschallin; also Donna Elvira, Leonore, Chrysothemis, Gutrune. A popular singer with a warm, lyrical voice, she continued to appear in cameo roles in Vienna well into the 1970s. (R)

**König, Johann Ulrich von** (*b* Esslingen, 8 Oct. 1688; *d* Dresden, 14 Mar. 1744). German librettist, poet, and dramatist. Settled in Hamburg in 1710, where he was heavily involved with the direction of the T am Gänsemarkt. In 1720 he went for the first time to Dresden, as court poet; 1730–5 he was back in Hamburg, but returned to Dresden to become director of court ceremonies and librarian.

König played a major role in the history of German opera. His serious librettos took the model of Bostel, Bressand, and Postel, and turned it into a refined expression of German Baroque ideals, frequently drawing inspiration from Italian or French models. Often his texts embraced a moral message, as in *L'inganno fedele oder Der getreue Betrug*, set by Keiser in 1714 and also given in the same year in England for the coronation of George I as *Die gecrönte Tugend*. While in Hamburg he also wrote librettos for Schürmann (*Die getreue Alceste*, 1719, after Quinault) and Graun (*Die in ihrere Unschuld siegende Sinilde*, 1727, based on Silvani) at Brunswick. König also excelled in comic texts, of which the most famous was that for Telemann's *Der geduldige Socrates* (1721, based on Minato).

**König Hirsch** (King Stag). Opera in 3 acts by Henze; text by Heinz von Cramer, after Gozzi's *fiaba*, *Re Cervo* (1762). Prod. Berlin, S, 23 Sep. 1956 with Pilarczyk, Fischer-Dieskau, cond. Scherchen. Rev. and shortened as *Il re Cervo*, Kassel 1963, cond. Henze; Santa Fe 4 Aug. 1965, with Shirley, Allen, cond. Baustian.

Abandoned in the forest as a child by the Governor (bs-bar), the King (ten) has been cherished by wild beasts. Grown up, he returns to claim his throne and choose a bride, but the Governor's scheming contrives that he shall renounce the crown and return to the forest from what he now believes to be a world of lies. After various adventures he enters the body of a stag, while the Governor takes his shape and returns to the city to initiate a reign of terror. But, driven by human longings, the Stag King returns to the city; the Governor is killed by his own assassins, and the King regains human form.

**Königin von Saba, Die** (The Queen of Sheba). Opera in 4 acts by Goldmark; text by Salomon Mosenthal. Prod. Vienna, H, 10 Mar. 1875, with Materna, Wilt, Beck; New York, M, 2 Dec. 1885 with Lilli Lehmann, Brandt, Stritt, Fischer, cond. Seidl; London, Kennington T, 29 Aug. 1910, with Woodall, Wheatley, Winckworth, cond. Goossens.

Jerusalem and the Syrian desert, *c*.950 BC. The plot, based on a biblical source, relates the story of the Queen of Sheba's visit to King Solomon. Assad (ten), the King's favourite courtier, has fallen in love with the Queen (mez), which causes him to commit an act of profanation in the temple during his own pre-arranged wedding to Sulamith (sop). After his release, Assad seeks Sulamith out in the desert, but the couple can only pledge their love briefly before Assad dies. Also subject of an opera by Gounod. See REINE DE SABA, LA.

**Königsberg.** See KALININGRAD.

**Königskinder, Die** (The Royal Children). Opera in 3 acts by Humperdinck; text by 'Ernst Rosmer' (Else Bernstein-Porges). As melodrama, prod. Munich 23 Jan. 1897. Prod. New York, M, 28 Dec. 1910, with Farrar, Jadlowker, Goritz, cond. Hertz; Berlin, H, 14 Jan. 1911, with Artôt de Padilla, Kirchoff, Hoffmann; London, CG, 27 Nov. 1911, with Gura-Hummel, Langendorff, Wolf, cond. Schalk.

A Prince (ten) falls in love with the beautiful Goose-girl (sop), but a Wicked Witch (con), in whose power the girl is held, refuses to release her. The miracle of a star falling on to a particular lily enables the girl to escape. The Prince has now become King, but the people refuse to accept the Goose-girl as their Queen and, ostracized, the couple die in each other's arms in the snow.

**Konwitschny, Franz** (*b* Fulnek, 14 Aug. 1901; *d* Belgrade, 28 July 1962). German conductor. Studied Brno and Leipzig. Répétiteur, then cond., Stuttgart 1926–30; Freiburg cond., then music director, 1933–7; Frankfurt 1937–45; Hanover 1945–6; Dresden 1953–5; Berlin, S, 1955–62. London, CG, 1959 (*Ring*). His warm, expansive musicianship was heard at its finest in Richard Strauss and Wagner. (R)

**Kónya, Sándor** (*b* Sarkad, 23 Sep. 1923). Hungarian tenor. Studied Budapest with Székelyhidy, Detmold with Husler. Début Bielefeld 1951 (Turiddu). Berlin, SO, from 1955; Bayreuth 1958–60, 1967; Paris, O, 1959; Milan, S, 1960; New York, M, 1961–74; London, CG, from 1963. Roles incl. Max, Don Carlos, Calaf, Walther, Parsifal; particularly successful as Lohengrin. (R)

**Korngold, Erich** (*b* Brno, 29 May 1897; *d* Hollywood, 29 Nov. 1957). Austrian composer. Studied Vienna with Fuchs, Zemlinsky, and Grädener. Very precocious, he made his début as a stage composer with a ballet *Der Schneemann* (orch. Zemlinsky) aged 11. His first two (1-act) operas *Der Ring des Polykrates* and *Violanta* were produced when he was 19. However, his greatest success was with *Die tote Stadt*, a mildly Expressionist work in which his rich, fluent, late-Romantic music, succulently orchestrated and carrying echoes of Puccini and even Richard Strauss (both of them admirers), is shown to good effect. Later moved to Hollywood, working as a film composer; one of his films, *Give Us This Night*, included an original 1-act opera.

SELECT WORKLIST: *Der Ring des Polykrates* (after Teweles; Munich 1916); *Violanta* (Muller; Munich 1916); *Die tote Stadt* (The Dead City) (Schott, after Rodenbach; Hamburg & Cologne 1920).

**Korrepetitor** (Ger.: 'rehearser'). The term used in German-speaking opera-houses for the member of the music staff who coaches the singers in their roles and is responsible for other musical tasks such as sub-conducting. See RÉPÉTITEUR.

**Košice.** Town in Slovakia, Czechoslovakia. Opera is given in the State T usually 3–4 times a week by a company of about 30 soloists.

**Košler, Zdeněk** (*b* Prague, 25 Mar. 1928). Czech conductor. Studied Prague with Ančerl. Répétiteur Prague, N, 1948, cond. 1951–8; Music director Olomouc 1958–62; Ostrava 1962–6; Berlin, K, 1966-8; Bratislava 1971. (R)

**Kostelnička** (Cz.: 'sexton's wife'). The name by which the widow (sop) of Toma Buryja is known in Janáček's *Jenůfa*. She looks after Jenůfa, her stepdaughter (and also her foster-daughter, in Cz. 'její pastorkyňa', the true title of the opera).

**Kothner.** Fritz Kothner (bs), a baker and Mastersinger, in Wagner's *Die Meistersinger von Nürnberg*.

**Kotzebue, August von** (*b* Weimar, 3 May 1761; *d* Mannheim, 23 Mar. 1819). German dramatist. He travelled widely in his career as a diplomat, including as a Russian councillor. His vast output of second-rate plays was quarried by many composers who found their colourful, sometimes absurd, Romantic plots and settings sufficiently appealing; these included Boieldieu (*Béniowski*, 1800), Schubert (*Des Teufels Lustschloss*, 1814), Lortzing (*Der Wildschütz*, 1842), and Spohr (*Der Kreuzfahrer*, 1848). He was murdered by a student who suspected him of being a spy.

**Koussevitsky, Sergey** (*b* Vishny-Volochek, 26 July 1874; *d* Boston, 4 June 1951). Russian, later US, conductor. Studied Moscow as bass player and had career as soloist. Began conducting 1907. Though primarily known as a concert conductor, he was one of the pioneers in introducing Russian opera to Western Europe, in 1921 conducting *Boris*, *Khovanshchina*, *Igor*, and *The Queen of Spades* in Paris, and *Snow Maiden* and other works in Barcelona. Responsible for commissioning Britten's *Peter Grimes*.

**Kovalev.** Platon Kuzmich Kovalev (bar), a college assessor, in Shostakovich's *The Nose*.

**Kovařovic, Karel** (*b* Prague, 9 Dec. 1862; *d* Prague, 6 Dec. 1920). Czech conductor and composer. Studied Prague with Fibich. Conductor Brno 1885–6, Plzeň 1886–7. As opera director of the Prague, N, 1900–20 he greatly influenced the development of Czech national opera, partly by the catholic choice of his repertory: he championed Czech opera, but also loved French opera, and introduced Bizet, Massenet, and Charpentier to Prague audiences, as well as Musorgsky, Wagner, and Richard Strauss. He was also a brilliant if controversial conductor who did much to raise orchestral and singing standards in the ensemble. He was particularly successful with the operas of Smetana and Dvořák, but was less enthusiastic for new music. The first Prague performance of *Jenůfa* led to the work's wider fame, but the revisions were his own and not Janáček's. As a composer, he had his greatest success with *The Dogheads*, a patriotic work in the Smetana–Dvořák tradition and still popular.

SELECT WORKLIST (all first prod. Prague): *Ženichové* (The Bridegrooms) (Kovařovic, after Macháček; 1884); *Psohlavci* (The Dogheads) (Šípek, after Jirásek; 1898); *Na starém bělidle* (The Old Bleaching-House) (Šípek, after Babiček; 1901).

**Kozlovsky, Ivan** (*b* Maryanovka, 24 Mar. 1900) Ukrainian tenor. Studied Kiev with Muravyova. Début Poltava 1918. Kharkov 1924; Sverdlovsk

1925; Moscow, B, 1926–57. A perceptive artist with a clear tone and an excellent technique, he sang the French, German (including Lohengrin), and Italian repertory, but was most famous in Russian opera, especially as Lensky, Berendey, and the Simpleton (*Boris*). Founded his own ensemble (1938–41), producing and singing in *Werther*, Gluck's *Orfeo*, and Arka's *Katerina*. (R)

**Kraków.** See CRACOW.

**Krásnohorská, Eliška** (*b* Prague, 18 Nov. 1847; *d* Prague, 26 Nov. 1926). Czech poet, editor, and librettist. She wrote librettos for K. Bendl and Fibich, but her most famous collaboration was with Smetana. Her simple and appealing verses contributed much to *The Kiss* and *The Secret*, but there were differences over her more ambitious words for *The Devil's Wall*.

**Krásová, Marta** (*b* Protivín, 16 Mar. 1901; *d* Vráž, 20 Feb. 1970). Czech mezzo-soprano. Studied Prague with Borová-Valoušková, Vienna with Ullanovský. Début Bratislava 1924 (Julia, Dvořák's *Jakobín*). Prague, N, 1926, 1928–66. Appearances throughout Europe. Outstanding in Czech opera, e.g. as the Kostelnička, Isabella (Fibich's *Bride of Messina*), the Witch (*Rusalka*); also sang Eboli, Amneris, Carmen. Possessed a beautiful voice and dramatic flair. (R)

**Kraus, Alfredo** (*b* Las Palmas, 24 Sep. 1927). Spanish tenor of Austrian descent. Studied Barcelona, Madrid, and Milan with Llopart. Début Cairo 1956 (Duke of Mantua). Milan, S, and London, CG, from 1959; Chicago 1962; New York, M, from 1966. A highly regarded *tenore di grazia* with a warm, effortless voice and an elegant style. Roles include Arturo, Alfredo, Werther, Des Grieux (Massenet), Hoffmann, and Ernesto. (R)

**Kraus, Auguste.** See SEIDL, ANTON.

**Kraus, Ernst** (*b* Erlangen, 8 June 1863; *d* Wörthsee, 6 Sep. 1941). German tenor. Studied Munich with Schimon-Regan, Milan with Galliera. Début Mannheim 1893 (Tamino). Damrosch OC 1896–8; Berlin, H, 1898–1924; Bayreuth 1899–1909; New York, M, 1903–4; London, CG, 1900, 1907, 1910. Also Paris, O; Milan, S; etc. An outstanding Wagner singer, with a large and expressive voice. Sang Siegfried, Siegmund, Walther, Erik, Herod. (R) His son **Richard Kraus** (1902–78) was a conductor. He held posts at Hanover; Stuttgart; Cologne; and Berlin, SO. Particularly associated with Mahler and R. Strauss; conducted revival of Busoni's *Faust*, 1955.

**Kraus, Joseph Martin** (*b* Miltenberg am Main, 20 June 1756; *d* Stockholm, 15 Dec. 1792). Swedish composer and conductor of German origin. Studied with Vogler; went to Stockholm 1778, where he became conductor at the theatre. In 1781 he was promoted to second court Kapellmeister; in the same year his first opera, *Proserpin*, was given. In 1782–8 he travelled through Europe on a study visit at the expense· of King Gustav III, visiting London, Naples, Paris, Rome, and Vienna. On his return he succeeded Uttini as first court Kapellmeister, playing an important role in the establishment of Swedish opera. His operas show the influence of both Gluck, who like Haydn held him in high regard, and Hasse, and are distinguished by his expressive handling of the orchestra, which is reminiscent of the style of the Mannheim school.

SELECT WORKLIST: *Proserpin* (Kellgren, after Gustav III; Ulriksdal 1781); *Soliman II* (Oxenstjerna, after Favart; Stockholm 1789); *Aeneas i Carthago* (Kellgren; Stockholm 1799).

**Kraus, Otakar** (*b* Prague, 10 Dec. 1909; *d* London, 27 July 1980). Czech, later British, baritone. Studied Prague with Wallerstein, Milan with Carpi. Début Brno 1935 (Amonasro). Bratislava 1936–9. London: CR, 1943–5; EOG, 1946; CG, 1951–68. Bayreuth 1960–2. Created Nick Shadow (*Rake's Progress*), King Fisher (*Midsummer Marriage*), Tarquinius (*Rape of Lucretia*), Diomede (*Troilus and Cressida*). A forceful and daring actor-singer, memorable in dark roles, particularly as Pizarro, Alberich, Iago, Scarpia. Also an excellent teacher; pupils include Robert Lloyd, Gwynne Howell. (R)

**Krause, Tom** (*b* Helsinki, 5 July 1934). Finnish baritone. Studied Vienna and Hamburg. Début Berlin, SO, 1959 (Escamillo). Bayreuth 1962; Gly. 1963; New York, M, from 1967; Paris, O, from 1975; also London, CG; Milan, S; Salzburg; etc. Created Jason in Krenek's *Der goldene Bock*. Possesses a warm, virile tone; roles include Mozart's Count Almaviva, Don Giovanni, Pizarro, Germont, Amfortas. (R)

**Krauss, Clemens** (*b* Vienna, 31 Mar. 1893; *d* Mexico City, 16 May 1954). Austrian conductor. Studied Vienna with Graedener and Heuberger. Début Brno 1913 (*Zar und Zimmermann*). Riga 1913–14; Nuremberg 1915–16; Stettin (Sczeczin) 1916–22. Assistant cond. Vienna 1922–4. Music director Frankfurt 1924–9. Director Vienna, S, 1929–35; Berlin 1935–7; Munich 1937–44: some of his leading singers, including Adele Kern, Viorica *Ursuleac (whom he married), and Julius Patzak followed him from Vienna to Berlin to Munich. Vienna, W, 1947. A close friend of Strauss, and one of his most gifted interpreters, he conducted the premières of *Arabella* (1933), *Friedenstag* (1938), *Capriccio* (1942, also collaborating with Strauss on the text), and *Die Liebe der Danae* (1952). London: CG, 1934 (*Arabella*,

*Shvanda*), 1947 with Vienna, S (*Salome, Fidelio*); Stoll T 1949 (*Falstaff, Tosca*); CG 1951–3 (*Tristan, Fidelio, Meistersinger*); Bayreuth 1953. (R)

BIBL: J. Gregor, *Clemens Krauss* (Vienna, 1953).

**Krauss, Gabrielle** (*b* Vienna, 24 Mar. 1842; *d* Paris, 6 Jan. 1906). Austrian soprano. Studied Vienna with Marchesi. Début Vienna 1859 (Mathilde, *Guillaume Tell*). Sang there until 1867; Paris: I, 1867–70, 1873; O, 1875–87. Milan, S, 1872–3. An exceptional actress as well as a singer of quality, she was known in France, after the great tragic actress, as 'la Rachel chantante'. Repertory included Donna Anna, Leonore, Norma, Lucia, Elsa, Senta, Aida. Created roles in Gounod's *Polyeucte* and *Tribut de Zamora* and Saint-Saëns's *Henri VIII*.

**Kravattentenor** (Ger.: 'necktie tenor'). A tenor whose tone suggests that he is being strangled by his neckwear.

**Krefeld** and **Möndchen-Gladbach**. Neighbouring towns in North-Rhine Westphalia, Germany. The first opera in Krefeld was given in 1794. A new theatre opened 1825 with Boieldieu's *La dame blanche*; enlarged 1886; bombed 1943. Performances were given in the Aula of the Lyzeum until a new theatre (cap. 832) opened 1952 with *Lohengrin*; the present theatre opened 1963 with *Don Giovanni*. In Möndchen-Gladbach the first opera-house (cap. 753) opened 1959. The two companies joined forces 1966. Robert Satanowski was music director 1969–76.

**Krenek** (orig. Křenek), **Ernst** (*b* Vienna, 23 Aug. 1900; *d* Palm Springs, CA, 23 Dec. 1991). Austrian, later US, composer. Studied Vienna and Berlin with Schreker. Worked at Kassel and Wiesbaden 1925–7, where his scenic cantata *Die Zwingburg* and comic opera *Der Sprung über den Schatten*, were composed. These employed a basically atonal idiom, as did the Expressionist opera *Orpheus und Eurydike*. However, in the political climate of the day Krenek became increasingly concerned with the relationship between the individual and the community, as well as in the power of the state. This prompted a series of Zeitopern in which his musical language underwent radical reform.

The first of these, *Jonny spielt auf*, was one of the greatest operatic successes of the 20th cent., although it caused scandal with its portrayal of the relationships between a Black musician and White women. But with its fantastical plot, exuberant staging (which called for a real train), and exploitation of such technical devices as radio, it was quickly taken up by opera-houses throughout Europe. Much of its effect also derived from the skilled use of jazz, already employed to a lesser degree in *Der Sprung über den Schatten*.

Although Krenek considered it to have a serious message, it was much misunderstood at the time and seen as a simple comedy: none the less, its commercial success enabled him to retire to Vienna.

None of the works which followed repeated the impact of *Jonny spielt auf. Der Diktator, Das geheime Königreich*, and *Schwergewicht* all contained a more overt political message, while the catchy rhythms and melodies of their predecessor were replaced with a more restrained, neo-Romantic style, described by the composer as 'Schubertian': this was also employed in the self-styled Grand Opera *Leben des Orest*. For political reasons *Kehraus um St Stephan*, a satire on postwar pre-Depression Vienna, could not be performed at Dresden in 1930. Its rejection hastened the composer's turn to serialism, used in *Karl V.*, a 'play with music' uniting elements of opera, pantomime, and film: although commissioned by the Vienna, S, its première was banned by the Nazis. In 1938 Krenek emigrated to the USA, where he made a distinguished reputation as a teacher. His later operas, which include *Cefalo e Procri, Dark Waters*, and notably *Der goldene Bock*, return to themes which had already fascinated him, such as the modern treatment of mythic subjects. He also wrote two television operas, *Ausgerechnet und verspielt* and *Der Zauberspiegel*.

WORKLIST: *Zwingburg* (Dungeon Castle) (Werfel, after anon.; Berlin 1924); *Der Sprung über den Schatten* (The Leap over the Shadows) (Krenek; Frankfurt 1924); *Orpheus und Eurydike* (Kokoschka; Kassel 1926); *Bluff* (Gribble & Levetzow; unperf.); *Jonny spielt auf* (Jonny Strikes Up) (Krenek; Leipzig 1927); *Der Diktator* (Krenek; Wiesbaden 1928); *Das geheime Königreich* (The Secret Kingdom) (Krenek; Wiesbaden 1928); *Schwergewicht, oder Die Ehre der Nation* (Heavyweight, or The Honour of the Nation) (Krenek; Wiesbaden 1928); *Leben des Orest* (Krenek; Leipzig 1930); *Kehraus um St Stephan* (Round St Stephan's Cathedral) (Krenek; comp. 1930, perf. Vienna 1990; *Karl V.* (Krenek; Prague 1938); *Cefalo e Procri* (Küfferle; Venice 1934); *What Price Confidence* (Krenek; comp. 1945–6, Saarbrücken 1960); *Tarquin* (Lavery; Cologne 1950); *Dark Waters* (Krenek; Los Angeles 1950); *Pallas Athene weint* (Krenek; Hamburg 1955); *The Bell Tower* (Krenek, after Melville; Urbana 1957); *Ausgerechnet und verspielt* (Krenek; Vienna 1962); *Der goldene Bock* (The Golden Ram) (Krenek; Hamburg 1964); *Der Zauberspiegel* (The Magic Mirror) (Krenek; Munich 1966).

BIBL: J. Stewart, *Ernst Krenek: the Man and his Music* (Berkeley, 1991).

**Krenn, Fritz** (*b* Vienna, 11 Dec. 1897; *d* Vienna, 17 July 1964). Austrian bass. Studied Vienna with Iro. Début Vienna, V (in *Cavalleria rusticana*). Vienna, S, 1920–5, 1934–42, 1946–59;

Berlin, S, 1927–43; London, CG, 1935, 1938; New York, M, 1951. A famous Ochs; also distinguished in Verdi, Wagner, and R. Strauss. Created roles in Hindemith's *Neues vom Tage* and *Cardillac*. (R)

**Kreutzer, Conrad** (from 1799, Conradin) (*b* Messkirch, 22 Nov. 1780; *d* Riga, 14 Dec. 1849). German composer and conductor. His first stage work was an operetta, *Die lächerliche Werbung*, performed by students of the U of Freiburg im Breisgau where he studied law; he sang a leading tenor role. In 1804 went to Vienna, where he probably studied with Albrechtsberger and composed several operas, of which only *Jery und Bätely* was produced. The success of *Konradin von Schwaben* and *Feodora* in Stuttgart led to his appointment to succeed Danzi as court conductor, 1812–16. After working in Schaffhausen and Donaueschingen as court conductor to the Prince of Fürstenburg, he returned to Vienna in 1822, where *Libussa* was successfully performed. He conducted at Vienna, K, 1822–7, 1829–32, and 1835–40; from 1833–5 at Vienna, J, where his most successful works, *Das Nachtlager von Granada* and the incidental music to Raimund's *Der Verschwender*, were produced. He was music director in Cologne 1840–2, eventually settling in Riga with his daughter Marie, a singer. In 1823 he set Grillparzer's *Melusine* libretto (intended for Beethoven).

Kreutzer wrote almost 40 works for the stage, but is remembered today chiefly for *Libussa* and *Das Nachtlager von Granada*, which offer a *rapprochement* between the spirit of the 18th cent. Singspiel and that of German Romantic opera. Although in a light, popular style, with immediately appealing melodies, Kreutzer's works were thought undramatic even in their day and his death went largely unnoticed.

SELECT WORKLIST: *Die lächerliche Werbung* (The Comical Courtship) (?; Freiburg 1800); *Jery und Bätely* (Goethe; Vienna 1810); *Konradin von Schwaben* (Guseck; Stuttgart 1812); *Feodora* (Kotzebue; Stuttgart 1812); *Libussa* (Bernhard; Vienna 1822); *Melusine* (Grillparzer; Berlin 1833); *Das Nachtlager von Granada* (The Night-Camp at Granada) (Braun, after Kind; Vienna 1834).

**Kreutzer, Rodolphe** (*b* Versailles, 16 Nov. 1766; *d* Geneva, 6 Jan. 1831). French violinist and composer. Studied with his father and Stamitz. Best known as a violinist (and the dedicatee of Beethoven's Op. 47 violin sonata). From 1790 had works produced at Paris, I and OC, notably *Paul et Virginie* (notable in that Act III is virtually continuous) and *Lodoïska* (the latter was initially rated above Cherubini's work of the same year. Continued to write many other operas, the most successful being *Astyanax*, *Aristippe*, and *Abel*:

the latter was thought 'belle' by Berlioz. In 1810 a fracture of his arm ended his solo career; but in 1816 he became second conductor of the Opéra, chief conductor 1817–24, director 1824–6. His music became outdated in his own day, and his last opera, *Matilde*, was rejected by the Opéra.

SELECT WORKLIST (all first prod. Paris): *Paul et Virginie* (Favières; 1791); *Lodoïska* (Dejaure; 1791); *Astyanax* (Dejaure; 1801); *Aristippe* (Giraud & Leclercq; 1808); *Abel* (Hoffman; 1810).

**Krieger, Johann Philipp** (*b* Nuremberg, *bapt*. 27 Feb. 1649; *d* Weissenfels, 6 Feb. 1725). German composer and organist. Studied Nuremberg and Copenhagen; *c*.1670 Kapellmeister to the Bayreuth court, then travelled 1673 to Italy. 1677 organist at Halle court; in 1680 he moved with the court to Weissenfels, becoming Kapellmeister. In 1680–1725 he directed opera there, with a repertory including Strungk and later Telemann. Composed *c*.18 operas for various German courts; some were also taken into the repertory of the Hamburg O. All are lost except for two collections of airs.

**Krips, Josef** (*b* Vienna, 8 Apr. 1902; *d* Geneva, 13 Oct. 1974). Austrian conductor. Studied Vienna with Weingartner. Répétiteur Vienna, V, 1921–4. Ustí nad Labem (Aussig) 1924–5. Music director Dortmund 1925–6; Karlsruhe 1926–33; Vienna, S, 1933–8, when dismissed by the Nazis. Cond. first post-war operas in Vienna at the Volksoper and T an der Wien, and under his direction the Staatsoper was restored to its previous eminence, 1945–50. Salzburg 1946, helping to re-establish the festival. London début, CG, 1947 with Vienna, S, Co. (*Don Giovanni*); 1963, 1971–4. Chicago 1960, 1964; New York, M, 1966-8, 1969–71; Bayreuth 1961. A warm, energetic man who contributed much to his country's musical traditions, which he cherished in many affectionate performances. (R)

**Krivchenya, Alexey** (*b* Odessa, 12 Aug. 1910; *d* Moscow, 10 Mar. 1974). Ukrainian bass. Studied Odessa with Selyavina. Sang in various Ukrainian theatres 1938–44; Novosibirsk 1944–9; Moscow, B, 1949. Roles incl. Mozart's Don Basilio, Ivan Susanin, Varlaam, Khovansky, Farlaf, Konchak, Kutuzov, Tkachenko (*Semyon Kotko*). (R)

**Kroll Oper.** See BERLIN.

**Krombholc, Jaroslav** (*b* Prague, 30 Jan. 1918; *d* Prague, 16 July 1983). Czech conductor. Studied Prague with Novák and Talich. Répétiteur Prague, N, 1940, début as cond. 1942 (Bořkovec's *Satyr*); music director Ostrava 1945; Prague, N, from 1945. Vienna 1959. London: CG, 1959 (*Boris*), 1961 (*Bartered Bride*); C, 1978 (*Don Giovanni*). Budapest, Stuttgart, Holland, and Edinburgh Festivals with Prague, N, Co. A fine conductor especially of Czech music, for which he did much as an ambassador as well as at home. (R)

**Krull, Annie** (*b* Rostock, 12 Jan. 1876; *d* Schwerin, 14 June 1947). German soprano. Studied Berlin with Brämer. Début Plauen 1898. Dresden 1900–12, where she created Diemut (*Feuersnot*); subsequently chosen by Strauss, who wanted 'the most dramatic soprano possible', to be the first Elektra. Repeated the role when *Elektra*, the first Strauss opera to be heard in Britain, was given in London, CG, 1910. (R)

**Krushelnytska** (or Krusceniski), **Salomea** (*b* Bilavyntsy (Pol., Biała), 23 Sep. 1872; *d* Lvov, 16 Nov. 1952). Ukrainian, later Italian, soprano. Studied Lvov with Wysocki, Milan with Crespi. Début Lvov 1893 (Léonore, *Favorite*). Warsaw 1898–1902; Naples, C, 1903; Milan, S, 1907, 1909, 1915; Buenos Aires 1906–13. A beautiful woman who possessed great dramatic and vocal gifts, she was described by Strauss as 'perfect as both Salome and Elektra'. Also an acclaimed Selika, Aida, Butterfly (in the revised, successful version at Brescia after its Milan fiasco), Tatyana, Isolde, and Brünnhilde, among a large repertory. Retired from stage 1920. (R)

**Kubelík, Rafael** (*b* Býchory, 29 June 1914). Czech, later Swiss, conductor and composer. Studied Prague with Šín. After a period with Czech Philharmonic, Music director Brno 1939–41. Left Czechoslovakia 1948 when he went to Edinburgh to conduct *Don Giovanni*. London, SW, 1954 (*Káťa*), Music director CG 1955–8, when his successes included *Otello*, a superb first English *Jenůfa* (1956), and *Troyens*. Encouraged native singers, and tried to keep casts intact throughout the season. Guest appearances Hamburg, Munich. Music Director New York, M, 1973 (first ever), resigning over budget cuts 1974. San Francisco 1977. His operas include *Veronika* (Brno 1947) and *Cornelia Faroli* (Augsburg 1947). (R) His wife, the Australian soprano **Elsie Morison** (*b* Ballarat, 15 Aug. 1924), studied Melbourne with Carey, also with him in London. Début London, RAH, 1948 (Galatea). London: SW, 1948–54; CG, 1954–62. Gly. 1953, 1956–9. First English Anne Trulove, and Blanche (*Carmélites*). Also sang Fiordiligi, Susanna, Marzelline. (R)

**Kuhlau, Friedrich** (*b* Uelzen, 11 Sep. 1786; *d* Copenhagen, 12 Mar. 1832). Danish composer of German origin. After early success in Hamburg as a keyboard player, he fled to Copenhagen in 1810 to avoid conscription into Napoleon's army. Here he made frequent appearances as a pianist;

from 1813 he was *Kammermusikus* to the king, playing flute in the court orchestra.

Beginning with Oehlenschläger's *Røverborgen*, Kuhlau composed much incidental music. In some of these dramas the music is of a quantity and quality which renders them close to genuine opera, as in *Elverhøj*. With its use of folk-song, and an overture which ends with the national anthem, it was, not surprisingly, an instant success and played an important role in the foundation of Danish opera.

SELECT WORKLIST (all first prod. Copenhagen): *Røverborgen* (The Robbers' Castle) (Oehlenschläger; 1814); *Lulu* (Güntelberg, after Wieland; 1824); *William Shakespeare* (Boye; 1826); *Elverhøj* (The Fairies' Mound) (Heiberg; 1828).

BIBL: K. Graupner, *Friedrich Kuhlau* (Leipzig, 1930).

**Kühleborn.** A water spirit, father of Undine, (bs) in Hoffmann's and (bar) in Lortzing's *Undine*.

**Kundry.** The messenger (sop) of the Grail, also in Klingsor's power, in Wagner's *Parsifal*.

**Kunz, Erich** (*b* Vienna, 20 May 1909). Austrian bass-baritone. Studied Vienna with Lierhammer and Duhan. Début Troppau (Opava) 1933 (Osmin). Vienna, S, from 1940. Bayreuth 1943–4, 1951; New York, M, 1952–4; London, CG, 1947; Gly. 1948, 1950; Salzburg 1942–60. Roles included Leporello, Figaro, Guglielmo; especially popular as Beckmesser and Papageno. (R)

**Kunzen, Friedrich Ludwig Aemilius** (*b* Lübeck, 4 Sep. 1761; *d* Copenhagen, 28 Jan. 1817). German composer, pianist, and editor. Studied with his father, the organist **Adolph Karl Kunzen** (1720–81). In 1787 worked at the Copenhagen O, where his first opera, *Holger Danske*, was given. This had only six performances, so he moved to Berlin, and joined Reichardt in editing the *Musikalisches Wochenblatt*. He then held theatrical posts at Frankfurt and Prague; returned to the Copenhagen O, succeeding Schulz as director 1795.

Kunzen was one of the most important figures in the history of Danish opera and responsible for introducing Mozart's dramatic works to Denmark. His style betrays their influence, as well as that of Gluck, and of Schulz: his most successful works, which included *Hemmeligheden*, *Dragedukken*, *Erik Ejegod*, and *Kaerlighed paa Landet*, represent a Danish branch of German Singspiel and treat many of the same themes, sometimes in a nationalist context.

SELECT WORKLIST (all prod. Copenhagen unless otherwise stated): *Holger Danske* (Oger the Dane) (Baggesen, after Wieland; 1789); *Die Weinlese* (The Grape Harvest) (Ihlee; Frankfurt 1793; rev. as *Viinhøsten*, 1796); *Hemmeligheden* (The Secret) (Thoroup, after Quétant; 1796); *Dragedukken* (The Dragon Doll) (Falsen; 1797); *Erik Ejegod* (Baggesen; 1798); *Min bedste moder* (My Grandmother) (Falsen; 1800); *Kjaerlighed paa landet* (Love in the Country) (Bruun, after Wieland; 1810).

**Kupfer, Harry** (*b* Berlin, 12 Aug. 1935). German producer and designer. Studied Leipzig. Début Weimar 1967, director 1967–72, then Dresden (*Hoffmann, Tristan, Parsifal, Moses and Aaron*). Bayreuth 1978 (*Flying Dutchman* staged as Senta's dream), 1988–92 (*Ring*). Director Berlin, K, 1981, where his striking productions included *Meistersinger*, and the première of Matthus's *Judith*. His work has shown a remarkable ability to present operas of different periods in a fresh and revealing light.

**Kupper, Annelies** (*b* Glatz, 21 Jul. 1906; *d* Munich, 8 Dec. 1987). German soprano. Studied Breslau. Début Breslau 1935 (Second Boy, *Zauberflöte*). Hamburg 1940–6; Munich, S, 1946–61; Bayreuth 1944; London, CG, 1953. A sensitive singer and a good musician; roles included Aida, Tatyana, Eva, Chrysothemis, and Danae (promised to her by Strauss before his death) in the official première of *Die Liebe der Danae*. (R)

**Kurpiński, Karol** (*b* Włoszakowice, 6 Mar. 1785; *d* Warsaw, 18 Sep. 1857). Polish composer and conductor. Studied with his father, the local organist. After holding various appointments, became deputy conductor, and then principal conductor, of the Warsaw O (1824–40). Founded and taught at his own schools of singing and drama, and founded the first Polish music journal, *Tygodnik muzyczny* (1820). Played an important part in Warsaw's concert life (he conducted Chopin's first concerts); also made a vital contribution to the development of Polish opera. He wrote 24 stage works, of which nine survive complete: these include *Czorstyn Castle*, which has occasionally been revived. Most of them are on Polish themes, and graft nationalist ideas on to a more international and particularly Viennese idiom. He married (1815) the soprano **Zofia Brzowska** (*b* Warsaw, 19 Jan. 1800; *d* Warsaw, 28 June 1879), who made her début aged 14 and sang soubrette roles at the Warsaw O.

SELECT WORKLIST: *Zamek na Czorsztynie* (Czorstyn Castle) (Krasiński; Warsaw 1819).

BIBL: T. Przybylski, *Karol Kurpiński, 1785–1857* (Warsaw, 1975).

**Kurt, Melanie** (*b* Vienna, 8 Jan. 1880; *d* New York, 11 Mar. 1941). Austrian soprano. Studied Vienna with Müller, later Berlin with Lilli and M. Lehmann. Début Lübeck 1902 (Elisabeth). Brunswick 1905–8; Berlin, H, 1908–12; London, CG, 1910, 1914; New York, M, 1914–17; Berlin, V, 1920–5. Powerful of voice and presence, she was a successful Leonore, Isolde,

Kundry, Brünnhilde, Marschallin, and Iphigénie (*Tauride*). (R)

**Kurwenal.** Tristan's retainer (bar) in Wagner's *Tristan und Isolde*.

**Kurz, Joseph Felix von** (*b* Vienna, 22 Feb. 1717; *d* Vienna, 3 Feb. 1784). Austrian actor, singer, dramatist, and theatre manager. Though he made many trips abroad, notably to Prague, Dresden, Frankfurt, Munich, and Pressburg, Kurz's activities were firmly centred on Vienna, where he first appeared with the German troupe at the Kärntnerthortheater, 1737–40. From 1734 he organized lavishly staged burlesques in which he often participated himself; he usually took the part of a comic character whom he called 'Bernardon', hence his own nickname 'Kurz-Bernardon'. Though continuing in the uniquely Viennese tradition of popular comedy established by the previous generation, his productions were noted for their higher musical standards. With their blend of satire and wit deriving from the *Hanswurst tradition and deft handling of interpolated songs and other pieces, they were important in paving the way for the foundation of the National-Singspiel in 1778. Haydn was recruited by Kurz to write music for *Der krumme Teufel* (1753).

**Kurz, Selma** (*b* Bielitz, 15 Nov. 1874; *d* Vienna, 10 May 1933). Austrian soprano. Studied Vienna with Ress, Paris with Marchesi. Début Hamburg 1895 (Mignon). Engaged by Mahler for Vienna 1899; sang there until 1927. London 1904–5, 1907; then, owing to Melba's jealousy, not again until 1924. Repertory included Tosca, Eva, Sieglinde, Violetta, Philine, and Gilda. Though not a good actress, she was a virtuoso singer, with an outstanding voice, and a superb trill, which she could sustain for 20 seconds. Greatly admired by Mahler for 'the incomparable gentleness of her personality, her legato, and her vocal control', she was also Strauss's favourite Zerbinetta, which she sang in the first revised version of *Ariadne*, 1916. (R) Her daughter **Desi Halban-Kurz** (*b* Vienna, 1912) had a short career as a singer.

BIBL: H. Goldmann, *Selma Kurz* (Bielitz, 1933).

**Kusche, Benno** (*b* Freiburg, 30 Jan. 1916). German bass-baritone. Studied Karlsruhe, and Freiburg with Harlan. Début Koblenz 1938 (Fra Melitone, *Forza*). Augsburg 1939–44; Munich from 1946; London, CG, 1952–3; Gly. 1954, 1963–4; New York, M, 1971–2. A fine character and buffo singer, acclaimed as Papageno, Leporello, Beckmesser, Faninal, Schicchi. (R)

**Kusser, Johann Sigismund** (*b* Pressburg, *bapt.* 13 Feb. 1660; *d* Dublin, Nov. 1727). Austrian composer. Studied Paris with Lully between

1674–82; c.1690 became Kapellmeister at Brunswick, where he wrote seven German operas, including *Julia* and *Ariadne*. Following a quarrel with his librettist Bressand, he moved to Hamburg 1694, taking over direction of the T am Gänsemarkt. In two years he produced new operas, such as *Erindo*, *Scipio Africanus*, and revivals of earlier works. After touring Germany with an opera company, he settled in Stuttgart 1698, composing three operas before leaving in 1704. Having visited London, he arrived in Dublin in 1709, where he had considerable success, rising to be chief composer and music-master at Dublin Castle.

Kusser made a distinctive contribution to the development of opera in Germany by following Conradi in the cultivation of the French style acquired during his stay in Paris. Little of his music survives, but it was highly regarded by his contemporaries and had a considerable influence upon Keiser and Mattheson.

SELECT WORKLIST: *Julia* (Bressand; Brunswick 1690); *Ariadne* (Bressand; Brunswick 1692); *Porus* (Bressand; Brunswick 1693); *Erindo oder Die unsträfliche Liebe* (Erindo or Blameless Love) (Bressand; Hamburg 1694); *Der grossmüthige Scipio Africanus* (The Magnanimous Scipio Africanus) (Fiedler, after Minato; Hamburg 1694); *The Man of Mode* (Etherege; London 1705).

**Kutuzov.** Fieldmarshal Mikhail Kutuzov, the commander of the Russian armies confronting Napoleon's armies in 1812. He appears (bs-bar) in Prokofiev's *War and Peace*.

**Kuzhamyarov, Kuddus** (*b* Kaynazar, 21 May 1918). Uighur composer. Studied Alma-Ata with Brusilovsky. His people's first professional composer, he has composed operas based on the music and history of Kazakhstan, and given the country a distinctive musical lead. His idiom draws on local song and dance styles, and seeks to incorporate these into a simple operatic framework.

SELECT WORKLIST: *Nazugum* (Khasanov; Alma-Ata 1956); *Zolotye gorye* (The Golden Mountains) (Tlendeyev; Alma-Ata 1960).

BIBL: P. Aravin, *Kuddus Kuzhamyarov* (Moscow, 1962).

**Kuznetsova, Maria** (*b* Odessa, 1880; *d* Paris, 25 Apr. 1966). Russian soprano and dancer. Began career as ballerina, in St Petersburg. Studied singing St Petersburg with Tartakov. Début St Petersburg Cons. 1904 (Tatyana). St Petersburg, M, 1905–13; Paris, O, 1908, 1910, 1912; London: CG, 1909, 1910, 1920; DL, 1914. Chicago 1916. Founded her own Russian Co. 1927, with which she sang in London, Paris, Barcelona, and S America until 1933, introducing much Russian repertory to the West. Other parts included Norma, Violetta, Tosca, Thaïs, Carmen; also created Fevronia (*Invisible City of Kitezh*), and Massenet's Fausta (*Roma*) and Cléopâtre. A much-admired artist, with a strong, flexible, bell-like voice, she was committed equally to singing and acting. As a dancer she performed in Russia and Paris, and created Potiphar's Wife in Strauss's *Josephslegende*, Paris 1914. (R)

**Kyrgyzstan.** The first musical theatre in Kyrgyz was formed in Frunze (until 1926, and from 1991, Bishkek) in 1936 out of a theatre and opera studio. The operettas *Altyn kyz* (The Golden Girl, 1937) and *Adzhal orduna* (Life, not Death, 1938) were written for it, as well as the first Kyrgyz opera, *Ay-churek* (Fair as the Moon, 1939), all in collaboration between three composers, the Russians Vladimir Fere (1902–71) and Vladimir Vlasov (1903–86) and the Kyrgyz Abdylas Maldybayev (1906–78). Broadly, Maldybayev was responsible for the melodic material, the others for constructing the works. Vlasov and Fere also collaborated on *Za schastye naroda* (For the People's Happiness, 1941), all three again on *Patrioty* (1941). Three operas have been written on the founder of modern Kyrgyz literature, the singer and writer Toktogul Saltyganov (1864–1933): *Toktogul* in 1940 by Alexander Veprik (1899–1958), in 1956 by Maldybayev and Mukash Abdrayev (1920–79), and in 1958 by Maldybayev, Vlasov, and Fere. *Manas*, by the same three composers on the national epic, was prod. 1946, but withdrawn for revision after accusations of formalism. Other Kyrgyz operas include *Aysha i Aydar* (1952) by Sergey Germanov (*b* 1918) and Achmet Amanbayev (*b* 1920); *Molodye serdtse* (Young Hearts, 1953) by Abdrayev; and *Kokul* (The Golden Forelock, 1942) and *Djamilya* (1961) by the Russian Mikhail Rauchwerger (*b* 1901), the latter making copious use of folk-song in a classical context. The Kyrgyz bass Bulat Minshilkiyev has sung in Moscow and abroad (incl. Milan, S).

In 1942 the Frunze theatre was renamed the Kyrgyz State Academy Opera and Ballet T; it makes summer tours with a predominantly Russian, Italian, and Kyrgyz repertory.

BIBL: B. Alagushov and A. Kaplan, *Kirgizskiye opery* (Frunze, 1973).

# L

**Labia, Fausta** (*b* Verona, 3 Apr. 1870; *d* Rome, 6 Oct. 1935). Italian soprano. Sister of below. Studied Verona with Aldighieri. Début Verona 1893 (Alice, *Robert le diable*). Stockholm 1893–5; Spain and Italy 1895–1905; Buenos Aires 1912, when she retired. Specialized in Wagner. Set up singing school in Rome. Her daughter **Gianna Perea-Labia** (*b* 1908) sang in Italy during the 1930s and 1940s.

**Labia, Maria** (*b* Verona, 14 Feb 1880; *d* Malcesine, 11 Feb. 1953). Italian soprano. Sister of above. Studied Verona with her mother Cecilia Labia. Début Stockholm 1905 (Mimì). Berlin, H, 1907–11; New York, Manhattan OC, 1908; Milan, S, 1912; Paris, O, 1913. Rome 1919; Buenos Aires 1920. Retired 1936. From 1930 taught in Warsaw, Siena, and then Lake Garda. Roles included Tosca, Carmen, Salome, Thaïs, and Felice (Wolf-Ferrari's *Quattro rusteghi*). Possessed a warm voice and a pretty figure, and was notable for her intense, exuberant, and often sensual acting. (R)

**Lablache, Luigi** (*b* Naples, 6 Dec. 1794; *d* Naples, 23 Jan. 1858). Italian bass. Studied Naples with Valente. Début Naples, T San Carlino, 1812 (Palma's *Erede senza eredità*). Palermo 1813–17; Milan, S, 1817–24; Vienna 1824; Naples, C, 1824–30; London, H and HM, 1830–2, 1835–52. Paris, I, 1830, and regularly until 1851; St Petersburg 1852; London, CG, 1854–6. A highly influential figure in the history of opera; his enormous repertory included Leporello, Geronimo (*Matrimonio segreto*), Rossini's Doctor Bartolo, Dandini, Figaro, Assur; Donizetti's Henry VIII, Podestà, Baldassare; and the creation of numerous roles, including his celebrated Don Pasquale, Marin Faliero, Lord Walton (*Puritani*), Massimiliano (Verdi's *I masnadieri*), and leading parts in other works by Bellini, Donizetti, Mercadante, Pacini, Vaccai, Mosca, Balfe, etc. Magnificent of voice and person, he possessed a large range (Eb –eb '), great vocal flexibility, and a genius for both the comic and the tragic that led Lumley to declare him 'the greatest dramatic singer of our time'; Chorley also praised his sensitivity, and his 'entire avoidance of grossness'. Though becoming gigantically fat, he was still robust and vocally fresh at 60. Among his numerous admirers were Schu-

bert (who dedicated three Italian songs, Op. 83, to him), Wagner, and several monarchs, Tsar Alexander I, Ferdinand I, and Queen Victoria, who took lessons from him. Of his 13 children, a son, **Federico**, was also an operatic singer, and a daughter, **Francesca**, married the pianist Thalberg.

BIBL: F. Castil-Blaze, *Biographie de Luigi Lablache* (Paris, n.d.).

**La Borde, Jean-Benjamin de** (*b* Paris, 5 Sep. 1734; *d* Paris, 22 July 1794). French composer and violinist. Studied with Rameau and Dauvergne and wrote his first stage work, *La chercheuse*, aged only 14. His first success was *Gilles*: 1758–73 he composed *c*.30 dramatic works which embraced both opéra-comique and tragédie lyrique. Always an amateur musician, he was briefly governor of the Louvre; during the French Revolution his home was destroyed and he was guillotined.

In his serious works, which included *Ismène* and *Amadis de Gaule*, he demonstrated a clear debt to Lully, without approaching similar depths of expression. More successful were his opéras-comiques, such as *Candide*, *Le billet de mariage*, and *Jeannot et Collin*, where the influence of the Italian style is noteworthy. Though popular, his works did not enjoy a good critical reception. Today he is remembered more for his seminal *Essai sur la musique ancienne et moderne* (1780).

SELECT WORKLIST (all first prod. Paris unless otherwise stated): *La chercheuse d'oiseaux* (The Bird Catcher) (Rozée; Mons 1748); *Gilles* (Poinsinet; 1757); *Ismène et Isménias* (Laujon; Choisy 1763); *Candide* (Voltaire; 1768); *Jeannot et Collin* (Desfontaines; 1770); *Amadis de Gaule* (Quinault; 1771); *Le billet de mariage* (The Marriage Contract) (Desfontaines; 1772).

BIBL: J. de Visme, *Un favori des dieux* (Paris, 1953).

**Labroca, Mario** (*b* Rome, 22 Nov. 1896; *d* Rome, 1 July 1973). Italian composer, administrator, and critic. Studied Parma with Respighi and Malipiero. Artistic director Florence Maggio Musicale 1936–44. Organized seasons of contemporary opera, Milan, S, and Rome, R, 1942. Administrator Venice, F, 1946–7, 1959–73; Milan, S, 1947–9. Director RAI 1949–58, during which period he arranged for Furtwängler to conduct *The Ring*. His music includes two children's operas.

WRITINGS: *L'usignolo di Boboli: 50 anni di vita musicale italiana* (Venice, 1959).

**Laca.** Laca Klemeň (ten), stepbrother of Števa Burya, in Janáček's *Jenůfa*.

**La calinda.** Dance in Act II of Delius's *Koanga*. Sometimes also known as *calenda*, the original dance was regarded as obscene and was at times banned.

**'La calunnia'.** Don Basilio's (bs) aria in Act I of Rossini's *Il barbiere di Siviglia*, in which he describes the growth of slander from a tiny breeze to a gale that can blast a man's reputation.

**Lachner, Franz** (*b* Rain am Lech, 2 Apr. 1803; *d* Munich, 20 Jan. 1890). German conductor. Studied with his father, Vienna with Sechter. Assistant cond. Vienna, K, 1827, chief cond. 1829–34. Mannheim 1834–6. Munich 1836, music director 1852–68. The real fame of the Munich O dates from his directorship. At first a great opponent of Wagner and his music, he was persuaded to produce *Tannhäuser* 1855, *Lohengrin* 1858, and behaved generously to Wagner. He was also a prolific composer, his operas including *Catarina Cornaro* (1841) and *Benvenuto Cellini* (1849). His younger brother **Ignaz** (1807–95) held appointments in Vienna, Stuttgart, Munich, Hamburg, Stockholm, and Frankfurt. A third brother, **Vincenz** (1811–93), conducted at Vienna and Mannheim (1836–73), and conducted a German company at London, CG, 1842, including the first English *Huguenots*.

**Lachnith, Ludwig** (*b* Prague, 7 July 1746; *d* Paris, 3 Oct. 1820). Bohemian composer. Studied Paris with Philidor. He was notorious as an arranger of famous operas to suit them to the lowest common denominator of public taste. He made *The Magic Flute* begin with the finale and included Don Giovanni's 'Finch'han dal vino', arranged as a duet, as well as excerpts from other Mozart operas and from Haydn's symphonies: this entertainment was entitled *Les mystères d'Isis* (1801). Reichardt and Berlioz were among those who protested, to no avail at first: despite being known as 'les misères d'ici', the piece had 134 performances. Lachnith also wrote some operas unaided.

**'Là ci darem la mano'.** Duet in Act I of Mozart's *Don Giovanni* in which Don Giovanni (bar) woos the initially reluctant but then complaisant Zerlina (sop).

**Lacy, Michael** (*b* Bilbao, 19 July 1795; *d* 20 Sep. 1867). Irish violinist. He was a skilful if unprincipled adapter of many famous operas for the London stage, among them a pasticcio jumbling Handel's *Israel in Egypt* and Rossini's *Mosè in Egitto* as *The Israelites in Egypt* (London, CG, 1833).

**'La donna è mobile'.** The Duke of Mantua's (ten) aria in Act III of Verdi's *Rigoletto* declaring his philosophy of fickleness in women. Verdi shrewdly anticipated its great fame by keeping it secret until the day of the première. Also the title of an opera by R. Malipiero (1957).

**Lady Billows.** An elderly autocrat (sop) in Britten's *Albert Herring*.

**Lady Macbeth of Mtsensk, The** (or *The Lady Macbeth of the Mtsensk District* (Russ.: *Ledy Makbet Mtsenskovo uyezda*), later revised and retitled *Katerina Izmaylova*). Opera in 4 acts by Shostakovich; text by A. Preis and the composer, after Leskov's story (1865). Prod. Leningrad, Maly T, 22 Jan. 1934, with Sokolova, Modestov, Zasetsky, Adrianova, Orlov, Balashov, cond. Samosud; Cleveland, Severance H, 31 Jan. 1935, with Leskaya, cond. Rodzinski; London, C, 22 May 1987, with Barstow, Trussel, White, Kale, cond. Elder. *Katerina Izmaylova* prod. Moscow, Stanislavsky Teatr, 8 Jan. 1963, with Andreyeva, Yefimov, Bulavin, Shchavinsky, cond. Provatorov; London, CG, 2 Dec. 1963, with Collier, Craig, Evans, Kraus, Howitt, cond. Downes.

Mtsensk, 1865. Katerina (sop), wife of a wealthy merchant Zinovy Izmaylov (ten), is bored, and is treated harshly by her father-in-law Boris (bs). When Zinovy has to go away on business, he makes Katerina swear to be faithful. Despite Boris's watchfulness, Katerina has an affair with the new servant, Sergey (ten).

When Boris discovers the secret and has Sergey beaten, Katerina poisons him. Sergey persuades Katerina also to murder Zinovy on his return.

During the wedding festivities of Katerina and Sergey, a peasant discovers Zinovy's body in the cellar. They are arrested.

Katerina and Sergey join other convicts on the road to Siberia. Bored with Katerina, he seduces another convict, Sonyetka (con), by bribing her with Katerina's stockings. In despair, Katerina throws herself into the river, dragging Sonyetka with her.

This opera unleashed the first serious attack on Shostakovich and on Modernist art in general from the Russian Communist party. It originally had considerable success, including 83 performances in Leningrad and 97 in Moscow; then on 28 Jan. 1936 *Pravda* published a vitriolic attack on the opera entitled 'Confusion instead of Music'. In this, the anonymous writer protested against the opera's 'deliberately discordant, confused flood of sound', and abused 'the basic concept of the music as a complete negation of opera'. This was followed on 6 Feb. with a second article attacking Shostakovich's ballet *The Limpid Stream*. These articles had the standing of official policy pronouncements, and changed the course both of Soviet music and of Shostakovich's career.

**'La fleur que tu m'avais jetée'.** See FLOWER SONG.

**La Fontaine, Jean de** (*b* Chateau-Thierry, 8 July 1621; *d* Paris, 13 Apr. 1695). French poet, fabulist, and librettist. His *Fables choisies mises en vers*, his most characteristic works (pub. 1668 (bks 1–6), 1678–9 (bks 7–11), 1694 (bk 12) ) are drawn from many sources, including Aesop, Phaedrus, Horace, and Bidpai, and are the basis of several operas. In 1674 La Fontaine wrote a libretto, *Daphne*, at Lully's request, but the composer rejected it, considering it insufficiently heroic. La Fontaine wrote attacks on Lully (*Le florentin*, 1674) and on opera (verse letter to Pierre de Nyert, 1677), but despite this began another libretto, *Galatée* (1682), and wrote the dedicatory verses for Lully's *Amadis* and *Roland*.

**Lakmé.** Opera in 3 acts by Delibes; text by Gondinet and Gille, after the former's *Le mariage de Loti*. Prod. Paris, OC, 14 Apr. 1883, with Van Zandt, Talazac, and Cobalet, cond. Danbé; London, Gaiety T, 6 June 1885, with Van Zandt, Dupuy, and Carroul, cond. Bevignani; Chicago, Grand OH, 4 Oct. 1883, but first full-scale US prod. New York, AM, 1 Mar. 1886, with L'Allemand, Bartlett, Canddus, Stoddard, cond. Thomas.

India, mid.-19th cent. An Englishman, Gérald (ten), is captivated by Lakmé (sop), the daughter of a Brahmin priest Nilakantha (bs), but her father swears to kill the man who has violated his holy temple.

In order to identify the intruder, Nilakantha forces Lakmé to sing at the bazaar: when Gérald appears, she faints, indicating the depth of her feeling for him. The priest stabs Gérald.

Lakmé tends Gérald in her hut in the forest; she presses him to drink magical water which will ensure the couple eternal love, but Gérald hesitates, unable to decide between his love for her and his duty to his regiment, a claim which is pressed by his fellow officer Frédéric (bs). Lakmé makes the decision for him by eating a poisonous leaf.

**Lalande, Henriette.** See MÉRIC-LALANDE, HENRIETTE.

**Lalli, Domenico** (*b* Naples, 27 Mar. 1679; *d* Venice, 9 Oct. 1741). Italian librettist. After travels through Italy, including a spell as impresario at T San Giovanni Grisostomo, Venice, he became court poet to the Elector of Bavaria in 1727. A close acquaintance of both Metastasio and Goldoni, traces of whose style are to be found in his work, he is credited with the first comic Venetian libretto, *Elisa*, set by Ruggeri in 1711. Important original librettos include *Pisistrato* (Leo, 1714), *Cambise* (A. Scarlatti, 1719), and *La verità in cimento* (Vivaldi, 1720); Lalli also made important adaptations of texts by Metastasio and

Zeno for Hasse and Galuppi. Some of his librettos were written in collaboration with *Silvani.

**Lalo, Édouard** (*b* Lille, 27 Jan. 1823; *d* Paris, 22 Apr. 1892). French composer. Studied Lille, Paris with Schulhoff and Crèvecœur. Originally a violinist, he initially composed mostly chamber music. His first opera, *Fiesque*, failed to win the Paris Théâtre-Lyrique prize in 1866, which greatly discouraged him. However, the success of his instrumental works encouraged him to write another opera, and this proved to be his masterpiece, *Le roi d'Ys*. Initially rejected by theatres including the Opéra, the work was eventually produced at the Opéra-Comique to great acclaim. Though in the manner of Grand Opera, and sometimes falling victim to its mannerisms, it is marked by Lalo's own personality, especially his capacity for elegant and charming melody and his vigorous sense of movement. His third opera was left with only one act completed. He married the singer Julie de Maligny; their son **Pierre Lalo** (1866–1943) was a well-known music critic.

WORKLIST: *Fiesque* (Beauquier, after Schiller; comp. 1866–7); *Le roi d'Ys* (Blau, after a Breton legend; Paris 1888); *La jacquerie* (Act I only, completed by Coquard: Blau & Arnaud; Monte Carlo 1895).

BIBL: G. Servières, *Édouard Lalo* (Paris, 1925).

**'L'altra notte'.** Margherita's (sop) aria in Act III of Boito's *Mefistofele*, as she lies in prison and describes the drowning of her child.

**'La luce langue'.** Lady Macbeth's (sop) aria in Act II of Verdi's *Macbeth*, welcoming the darkness of night that will shelter the murder of Duncan. Composed for the 1865 Paris revival.

**'La maja y el ruiseñor'.** Rosario's (sop) aria, known in English as 'The Lover and the Nightingale', sung in Tableau III of Granados's *Goyescas* as she waits for her lover and reflects on the nature of love. Originally a piano piece, based on a painting by Goya.

**'La mamma morta'.** Madeleine de Coigny's (sop) aria in Act III of Giordano's *Andrea Chenier*, in which she tells Gérard of the terrible death of her mother when their house was burned by the Revolutionary mob.

**Lambert, Constant** (*b* London, 23 Aug. 1905; *d* London, 21 Aug. 1951). English conductor, composer, and critic. Studied London with Vaughan Williams and R. O. Morris. Although primarily associated with the growth of ballet in England, had a great love of opera and cond. *Manon Lescaut* and *Turandot* at London, CG, also *Fairy Queen* (in his arrangement) and *Turandot* (1946–7).

WRITINGS: *Music, Ho!* (London, 1933, 1966).

BIBL: R. Shead, *Constant Lambert* (London, 1973).

**lament** (It., *lamento*). In its strictest sense, an aria or other passage in which the death of a character is mourned. More widely, the lament can also embrace general expressions of regret or sorrow; in opera, a lament is often sung upon the departure of a character, or at the ending of a love affair.

From the beginning, opera plots provided many opportunities for laments, such as that of Orpheus for Euridice, or Ariadne for Theseus. The 'Lamento d'Arianna' is the sole surviving number from Monteverdi's *Arianna* (1608) and set a standard for later composers which was approached, but rarely excelled. At this time it was also common practice for laments to be composed, in the monodic style, as self-contained pieces which were published as such. Later, in Venetian opera, the lament became in effect the hit number of the opera, and was an opportunity for the principal singers to demonstrate their command of pathetic and moving expressions of emotion: great care was taken over its placing to ensure maximum effect. Especially popular were laments over a chromatic ground bass: there are many in Cavalli's operas, and under his influence Purcell composed the finest example of the genre, 'When I am laid in earth' in *Dido and Aeneas*. Although there are many arias in opera seria in which a death or the loss of a lover is mourned, after 1700 the lament ceased to be recognized as an independent genre.

**'Lamento d'Arianna'.** The title usually given to Ariadne's scena 'Lasciatemi morire', the final and only surviving part of Monteverdi's *Arianna*, mourning her desertion by Theseus.

**Lammers, Gerda** (*b* Berlin, 25 Sep. 1915). German soprano. Studied Berlin with Mysz-Gmeiner and Schwedler-Lohmann. Spent 15 years as a concert singer. Stage début 1955 Bayreuth (Ortlinde). Kassel O from 1955; London, CG, 1957, 1959; New York, M, 1961–2; Hamburg 1959, 1967. A successful Senta, Brünnhilde, Isolde, Marie (*Wozzeck*), and Elektra, as which she scored a triumph when she replaced Goltz at London, CG, 1957. (R)

**La Motte, Antoine Houdar de** (*b* Paris, 15 Jan. 1672; *d* Paris, 26 Dec. 1731). French dramatist, librettist, theorist, and poet. His earliest dramatic work, the farce *Les originaux* (1693), was a failure, but through providing the text for Campra's opéra-ballet *L'Europe galante* he achieved great fame and was in considerable demand as a librettist: in later life he turned away from the theatre. His most significant collaborations were with Destouches (*Amadis de Grèce* and *Omphale*), Colasse (*Canente*), Francœur (*Scanderbeg*), and Marais (*Sémélé*). Through the unabashed inclusion of more spectacular, even fantastical elements he widened the scope of the tragédie en musique: many of his contemporaries considered him the most worthy successor to Quinault, although posterity has not.

**'L'amour est un oiseau rebelle'.** Carmen's (mez) Habanera in Act I of Bizet's *Carmen*, singing of the capricious and dangerous nature of love.

**Lampe, John Frederick** (Johann Friedrich) (*b* Saxony, *c*.1703; *d* Edinburgh, 25 July 1751). German composer who spent most of his life in Britain. He arrived in London *c*.1724, but his compositions date mainly from the 1730s. When in 1732 Thomas Arne's father organized a series of productions of 'English Opera' to further his son's musical career, it was, in fact, Lampe who composed most of the works: *Amelia* (1732), *Britannia* (1733), and *Dione* (1733) had poor responses, and *The Opera of Operas* (1733)—a revision of Fielding's *Tragedy of Tragedies*, set by Thomas Arne six months previously—fared little better. Lampe was at his best in burlesque; his greatest triumph was *The Dragon of Wantley* (1737), a satire on Handel and the Italian opera currently popular in London. Such was the work's success that audiences began to stay away from Italian opera, but, as so often, a sequel, *Margery, or A Worse Plague than the Dragon* (1738), failed. Among Lampe's four remaining stage works, only *Pyramus and Thisbe* (1745)—a mock opera, after Shakespeare—is of note. It was an all-sung opera with an overture, recitatives, and 14 songs, presented within the framework of a dialogue mocking contemporary operatic conventions.

**Lamperti, Francesco** (*b* Savona, 11 Mar. 1811; *d* Cernobbio, 1 May 1892). Italian singing-teacher. Studied Lodi and Milan. Director Lodi, T Filodrammatico, whither students came from all over Europe. Professor of singing Milan Cons. 1850–75: pupils included Albani, Campanini, Artôt, Cruvelli, Sembrich, Stolz, and Waldmann; he was a friend of Pasta and Rubini. He based his teaching on the method of the old Italian Rossinian school, and wrote several vocal studies and a treatise on singing. Though he conveyed a precise understanding of vocal anatomy, he placed emphasis on expression as much as technique. His elder son **Giuseppe** (1834–98) was an impresario at Rome, Ap., and Naples, C. His younger son **Giovanni** (1839–1910) was also a teacher, whose pupils included Bispham, Sembrich, Schumann-Heink, and Stagno.

**Lampugnani, Giovanni Battista** (*b* Milan, 1706; *d* ?Milan, after 1786). Italian composer. Studied Milan. Having had some success in his native city, where he composed ten operas, he succeeded Galuppi at the London, H, 1743, where he composed *Alfonso* and *Alceste*. He returned to Milan *c*.1745, where he eventually became maestro al cembalo at the T Regio Ducale and helped in the production of Mozart's *Mitridate* 1770. He wrote works for many of the major Italian houses; he may have revisited London 1755, when *Siroe* was given.

Lampugnani's career began with the composition of opera seria in a style comparable to Hasse's, but after 1755 he concentrated on opera buffa after the model of Galuppi.

SELECT WORKLIST: *Candace* (Lalli, after Silvani; 1732); *Antigono* (Marizoli; Milan 1736); *Ezio* (Metastasio; Milan 1736); *Angelica* (Vedoa; Venice 1738, rev. Milan 1738); *Rossane* (Rolli; London 1743); *Alfonso* (Rolli, after Pallavicino; London 1744); *Alceste* (Rolli, after Metastasio; London 1744); *Il gran Tamerlano* (Piovene; Milan 1746); *Siroe re di Persia* (after Metastasio; London 1755); *La contessina* (Goldoni; Milan 1759); *L'illustre villanella* (?; Turin 1769).

**Lanari, Alessandro** (*b* S Marcello di Jesi, 1790; *d* Florence, 3 Oct. 1862). Italian impresario. Lucca 1821, and other Italian towns. Florence, P, 1823–8, 1830–5, 1839–48, 1860–2. Commissioned Bellini's *Norma* and *Beatrice di Tenda*, Donizetti's *Elisir d'amore* and *Parisina*, and Verdi's *Attila* and *Macbeth*, also operas by Pacini and Mercadante. Though he had a somewhat stormy career, his expertise and command earned him the title 'the Napoleon of Impresarios'.

**Lancaster,** (Sir) **Osbert** (*b* London, 4 Aug. 1908; *d* London, 27 July 1986). English designer. Studied Oxford and London. Already distinguished as artist, cartoonist, and author, he turned to opera in 1952, designing *Love in a Village* for EOG. For Glyndebourne he designed *The Rake's Progress* (British première, 1953), *Falstaff* 1955, *L'italiana in Algeri* 1957, *La pietra del paragone* 1964, *L'heure espagnole* 1966, and *The Rising of the Moon* 1970. Also *Don Pasquale* for Sadler's Wells and *The Sorcerer* for D'Oyly Carte. The intelligent observation in his architectural books, coupled with the wit of his cartoons, made him especially well suited to the comedies he designed for Glyndebourne.

**Landgrave.** Hermann (bs), Landgrave of Thuringia, in Wagner's *Tannhäuser*.

**Landi, Stefano** (*b* Rome, 1586 or 1587; *d* Rome, 28 Oct. 1639). Italian composer. Studied at the Collegio Germanico in Rome. By 1618 maestro di cappella to the Bishop of Padua; the next year his *La morte d'Orfeo* was performed in Rome, whither he had returned by 1620. In 1624 he became maestro di cappella at S Maria dei Monti; in 1629 he joined the papal choir. Three years later *Il Sant'Alessio* was produced for the opening of the opera-house in the Barberini palace.

*La morte d'Orfeo* was the first secular opera given in Rome, inaugurating a tradition culminating in the works of Mazzochi, Marazzoli, and Rossi. Although clearly influenced by the Bardi Camerata and by Monteverdi, Landi opened a new path by his inclusion of grand choral scenes reminiscent of the intermedio. The ensemble, particularly the spectacular finale, became a hallmark of Roman opera and was used to even greater effect in *Il Sant'Alessio*, which eschews mythology in favour of a plot centring on the life of a human character. This move towards greater realism on stage was heightened by the inclusion of comic as well as serious episodes, while the strict stile rappresentativo of earlier opera was handled in a more flexible fashion, suggestive of the later distinction between recitative and aria. The orchestral sinfonias which preceded each act were some of the earliest forerunners of the *overture proper.

WORKLIST: *La morte d'Orfeo* (Landi; Rome 1619); *Il Sant'Alessio* (Rospigliosi; Rome 1632, rev. 1634).

BIBL: S. Leopold, *Stefano Landi: Beiträge zur Biographie* (Hamburg, 1976).

**Landowski, Marcel** (*b* Pont-l'Abbé, 18 Feb. 1915). French composer of Polish descent. Studied Paris with Busser. Much influenced by Honegger, on whom he wrote a book. His operas, for which he wrote his own texts, generally treat serious subjects in an accessible modern language. The most successful has been *Le fou* (1956).

**Langdon, Michael** (*b* Wolverhampton, 12 Nov. 1920; *d* Hove, 12 Mar. 1991). English bass. Studied Vienna with Jerger, Geneva with Carpi, London with O. Kraus. Début London, CG, 1950, in small roles. Remained member of the company. Gly. 1961, 1963, 1965, 1979. New York, M, 1964; Paris, O, 1971; also Vienna, Buenos Aires, Edinburgh. A versatile actor with a rich voice, successful as the Grand Inquisitor, Moses, Varlaam, Landgrave, Claggart, and especially Ochs, among a wide repertory. Created Mr Ratcliffe (*Billy Budd*), Recorder of Norwich (*Gloriana*), He-Ancient (*Midsummer Marriage*), and title-role of Orr's *Hermiston*. Autobiography, *Notes from a Low Singer* (London, 1982). (R)

**Lange, Aloysia.** See WEBER.

**Langridge, Philip** (*b* Hawkhurst, 16 Dec. 1939). English tenor. Studied London with Boyce and Bizony. Début Gly. 1964 (Servant, *Capriccio*). London, SW, 1972; EOG 1974; Gly. 1977; Milan, S, from 1979; Chicago 1981; London, CG, from 1983; ENO and New York, M, from 1984. A highly versatile singer, who has used his exceptional intelligence, range, and projection to master a large repertory, e.g. Dardanus (Rameau), Scipio (Handel), Don Ottavio, Idomeneo, Otello (Rossini), Huon, Florestan, Aeneas (Berlioz), Shuisky, Laca, and Živny (Janáček), Vere, Grimes, and Aschenbach. (R)

**Laporte, Pierre François** (*b* ?, 1799; *d* nr. Paris, 1841). French actor and opera manager. Manager London, H and HM, 1828–31, 1833–41. His management included the London premières of a number of operas by Bellini and Donizetti, and singers he brought to London for the first time included Rubini, Grisi, Nourrit, Tamburini, Persiani, Mario, Lablache, and Pauline Viardot.

**'Largo al factotum'.** Figaro's (bar) aria introducing himself in all his versatility and popularity in Act I of Rossini's *Il barbiere di Siviglia*.

**Larrivée, Henri** (*b* Lyons, 9 Jan. 1737; *d* Vincennes, 7 Aug. 1802). French baritone. Joined chorus of Paris, O, making début 1755 (Priest, *Castor et Pollux*). Leading bass there until 1797. Roles included Pollux, Piccinni's Roland and Orestes (*Iphigénie*), and Gluck's Hercules (*Alceste*); created Agamemnon (*Iphigénie en Aulide*), Orestes (*Iphigénie en Tauride*), Ubalde (*Armide*), and Salieri's Danaus (*Les Danaïdes*). His voice was sonorous and flexible, and had a large compass. Much admired by Gluck, though not as an actor. His wife **Marie Jeanne Le Mière** (1733–86) was a soprano who sang at the Opéra 1750–77. She created the title-role of Philidor's *Ernelinde*, in which her husband created the role of Ricimer.

**Larsén-Todsen, Nanny** (*b* Hagby, 2 Aug. 1884; *d* Stockholm, 26 May 1982). Swedish soprano. Studied Stockholm with Ljedström, also Berlin and Milan. Début Stockholm 1906 (Agathe). Sang there until 1923. Milan, S, 1923–4; New York, M, 1925–7; Bayreuth 1927–8, 1930–1; London, CG, 1927, 1930. Possessed a very powerful voice; successful as Isolde, Brünnhilde, Kundry, etc.

**Laruette, Jean-Louis** (*b* Paris, 7 Mar. 1731; *d* Paris, 10 Jan. 1792). French composer and singer. Joined the Paris Opéra-Comique as actor and tenor in 1752; beginning in 1753 he contributed single airs to vaudeville comedies for the T de la Foire. Together with Blaise and Duni he developed the comédie mêlée d'ariettes: among his many popular works was *L'ivrogne corrigé*, whose libretto was later set by Gluck. Remained a member of the Comédie-Italienne (with which the Opéra-Comique amalgamated) until 1779. His wife **Marie-Thérèse Villette** (1744–1837) was a popular singer and actress on the Parisian stage.

SELECT WORKLIST (all first prod. Paris): *Les amans trompés* (Anseaume; 1756); *La fausse aventurière* (Anseaume & Marcouville; 1757); *Le docteur Sangrando* (Anseaume & De Santerre; 1758; comp. with Duni); *Cendrillon* (Anseaume; 1759); *L'ivrogne corrigé* (The Drunkard Reformed) (Anseaume & De Sancerre; 1759); *Le Guy de chesne* (De Junquières; 1763); *Les deux compères* (The Two Confederates) (De Santerre; 1772).

**La Scala** (rightly, Teatro alla Scala). Theatre in Milan, built 1778 (cap. 2,800), after the T Regio Ducale burnt down 1776, on the site of the 14th-cent. church of Santa Maria della Scala (named after Regina della Scala, one of the Visconti family); opened 3 Aug. with the première of Salieri's *L'Europa riconosciuta*; renovated 1838. The theatre's reputation was established in the early 19th cent. with the premières of important works by Rossini, Donizetti, Meyerbeer, and Mercadante: some 250 performances of Rossini works were given 1824–6 under the English impresario Joseph Glossop. Domenico Barbaia succeeded him, 1826–32, commissioning works from Bellini; and he was followed by Bartolomeo Merelli, 1836–50 and 1861–3, who commissioned Verdi's first surviving opera, *Oberto* (1839), and three more, including *Nabucco* (1842).

The Scala's greatest periods were under *Toscanini, who ruled the theatre despotically but brilliantly 1898–1903, 1906–8, and 1920–9, attracting the greatest singers of the day. During the first he brought Wagner into the repertory and gave the Italian premières of *Eugene Onegin* (1900), *Salome* (1906), and *Pelléas* (1908); during the third he gave the premières of works by Pizzetti, Boito, Respighi, Puccini (*Turandot*, 1926), Giordano, and Zandonai. Toscanini left over quarrels with the Fascists, and was succeeded by Victor *De Sabata, 1930–57. The theatre was bombed Aug. 1943; reopened (cap. 3,600) 11 May 1946, concert under Toscanini (with Stabile, Nessi, Pasero, and Tebaldi's début). Ghiringelli succeeded as director, 1946–72. De Sabata was succeeded by Gavazzeni, Sanzogno, and Votto; and the presence of *Callas, 1950–8, led to revivals including *Anna Bolena*, *Pirata*, and *Medea*, and Visconti's productions of *Sonnambula* and *Traviata* under Giulini. The Piccola Scala (cap. 600) opened 1955 with *Matrimonio segreto*. Paolo Grassi was administrator 1972–7, with Massimo *Bogianckino director, 1972–5, and Claudio *Abbado music director, 1968–86; together they tried to give the theatre greater accessibility to a younger and wider audience. Riccardo *Muti became music director 1986. He has further broadened the repertory, not always to the taste of an audience that remains, after that of Parma, one of the most demanding in Italy.

BIBL: C. Gatti, *Il Teatro alla Scala* (Milan, 1964).

**Lassalle, Jean** (*b* Lyons, 14 Dec. 1847; *d* Paris, 7 Sep. 1909). French baritone. Studied Paris Cons. and with Novelli. Début Liège 1868 (Saint-Bris). Paris, O, 1872–6, 1877–92. Milan, S, 1879; London, CG, regularly 1879–93; New York, M, 1891–4, 1896–7. Retired 1901. Repertory included Guillaume Tell, Hamlet, Rigoletto, Flying Dutchman, Sachs; created roles in Massenet's *Roi de Lahore*, Gounod's *Polyeucte* and *Tribut de Zamore*, and Reyer's *Sigurd*, among others. Possessed a smooth and ample voice, and was at his best in noble and dignified roles. (R)

**'Last Rose of Summer, 'Tis the'.** An old Irish air, *The Groves of Blarney*, for which Thomas Moore wrote new words; in this form it was used by Flotow for Lady Harriet's (sop) song in Act II of *Martha*.

**Laszló, Magda** (*b* Marosvásáhely, 1919). Hungarian soprano. Studied Budapest with Stowasser and Székelyhedi. Début Budapest 1943. Remained until 1946. Then settled in Italy, singing at Milan, S, Rome, etc. Gly. 1953–4, 1962–3; London, CG, 1954. Roles include Poppea, Alceste, Norma, Senta, Daphne. Created Mother in Dallapiccola's *Il prigioniero*, and Cressida in *Troilus and Cressida*. An intelligent and graceful singer. (R)

**Latvia.** In 1772 a German theatre was built in Riga, and the repertory included opera. This proved too small, and was replaced in 1782 by the Vitinghoff T, where Wagner was music director, 1837–9. In his first season he conducted 85 performances of 16 works, in his second 83 of 22 works; the scope of his activity strained the resources of Riga and the goodwill of the Intendant, Carl von Holtei. A larger theatre, proposed in 1829, was not built until 1860–3; burned down in 1882, it was rebuilt in 1887. A Russian theatre was built in 1902. In the same year the Latvian Jaunais Teatris opened with a largely French and Italian repertory; this closed in 1905 on suspicion of collaboration with the Russian revolutionaries. The first organized company was founded by Pavils Jurjaņs (1866–1947), and operated 1912–15. With national independence in 1918, the Nacionalā opera was founded on the basis of the old German theatre; this opened with *Tannhäuser* in 1919. The independent Liepajas O was founded in 1922, becoming national in 1927. The touring Celojoša O was formed in 1929. Under Soviet domination the opera became the Valsts Operas un Baleta Teatris. Riga also has an operetta theatre.

The first, primitive attempt at a Latvian opera was *Spoka stunda* (1890) by Jēkabs Ozols. However, the real beginning of Latvian opera dates from 1919, when Alfrēds Kalniņš (1879–1951) wrote *Baņjuta* and Jānis Mediņš (1890–1966)

wrote *Uguns un Nakts* (Fire and Night) on the successful drama by Jānis Rainis: they were prod. 1921 and 1922 respectively. Other operas included *Salieieki* (The Islanders, 1926) by Kalniņš, and *Lolitas brīnumputns* (Lolita's Magic Bird, 1934) and *Hamlets* (1936) by his son Jānis (*b* 1904); also the children's opera *Spriditis* (1927) by Mediņš and the Wagner-influenced *Vaidelote* (The Vestal, 1927) by his brother Jāzeps (1877–1947). Other successful opera composers have been Arvīds Žilinskis (*b* 1905) with *Zelta zirgs* (The Golden Horse, 1965); Adolfs Skulte (*b* 1909) with *Princese Gundega* (1972); Margers Zariņš (*b* 1910) with *Uz jauno krasto* (To the New Shores, 1955) and *Zālas dzirnavas* (The Green Mill, 1958). The latter work, which makes controversial use of various expressive means, was stormily received, and his opera-ballet *Svētā Maurijica brīnum darbs* (The Miracle of St Maurice), comp. 1967, was not prod. until 1974. Imants Kalniņš (*b* 1941) has had a popular success with his *Spēlēju, dancoju* (I Sang and Danced, 1977), which mixes folk, symphonic, and rock music.

BIBL: S. Vēriņa, *Muzykalny teatr Latviy* (Leningrad, 1973).

**Laubenthal, Horst** (*b* Eisfeld, 8 Mar. 1939). German tenor. Studied Munich and with Laubenthal (below), whose name he adopted. Début Würzburg 1967 (Don Ottavio). Stuttgart 1967–73; Bayreuth 1970; Gly. 1972; Berlin, D, from 1973. Also Paris, O; Munich; Vienna; Barcelona; etc. A successful Lensky, Palestrina, Belmonte. (R)

**Laubenthal, Rudolf** (*b* Düsseldorf, 10 Mar. 1886; *d* Pöcking, 2 Oct. 1971). German tenor. Studied Berlin with Lilli Lehmann. Début Berlin, Deutsches Opernhaus, 1913, and member until 1918. Munich, S, 1919–23. London, CG, 1922, 1926–30; New York, M, 1923–33. A handsome Heldentenor with a strong dramatic sense, he was much esteemed as Siegfried, Tristan, and Walther; also sang Arnold (*Guillaume Tell*), John of Leyden, Števa. (R)

**Lauretta.** Schicchi's daughter (sop), lover of Rinuccio, in Puccini's *Gianni Schicchi*.

**Lauri-Volpi, Giacomo** (*b* Rome, 11 Dec. 1892; *d* Valencia, 17 Mar. 1979). Italian tenor. Studied Rome with Cotogni, later with Rosati. Début (as Giacomo Rubini) Viterbo 1919 (Arturo). Rome 1920, as Lauri-Volpi. Milan, S, from 1922; New York, M, 1922–33; London, CG, 1925, 1936; Rome, R (singing at its opening), from 1928; Paris, O and OC, 1929. In 1959 was still making occasional appearances, and sang at a gala in Barcelona 1972. An extraordinarily gifted singer in the line of Tamberlik, he possessed a full, poised voice with a wide range, a superb legato, and the capacity for both lyrical, pianissimo singing, and ringing, dramatic incisiveness. He also had excellent diction. Successful as Rossini's Count Almaviva, Arturo, Nemorino, and Massenet's Des Grieux, he was particularly acclaimed as Radamès, Calaf, and Manrico, among a large repertory. His writings include *Cristalli viventi* (Rome, 1948), *Voci parallele* (Milan, 1955), and *Misteri della voce umana* (Milan, 1957). (R)

BIBL: J. Menéndez, *Giacomo Lauri-Volpi* (Madrid, 1990).

**'La vergine degli angeli'.** Leonora's (sop) scene with chorus in Act II of Verdi's *La forza del destino*, in which she and the monks pray for her protection.

**Lavrangas, Dionysios** (*b* Argostolion, 17 Oct. 1860 or 1864; *d* Razata, 18 July 1941). Greek composer and conductor. Studied Naples with Scarano, Rossi, and Serao, Paris with Delibes, Massenet, and Dubois. Began conducting on his native island of Cephalonia, also toured. Returned to Greece 1894 and founded Greek Opera; also taught Athens Conservatory, Hellenic Conservatory (director Opera School), and elsewhere, and worked as a publisher and critic. He was more influential in his work for Greek opera as an administrator, conductor, and teacher than as a composer; his works reflect his Italian and especially his French training. Autobiography, *T'apomnimonevmata mou* (My Reminiscences) (Athens, 1940).

SELECT WORKLIST: *O lytrotis* (The Redeemer) (Papantoniou; comp. *c*.1902, prod. Corfu 1934); *Dido* (Dimitrakopoulos; Athens 1909); *Fakanapas* (Lavrangas, after Scribe; comp. 1935, prod. Athens 1950).

**Lawrence, Marjorie** (*b* Dean's Marsh, 17 Feb. 1907; *d* Little Rock, AR, 13 Jan. 1979). Australian, later US, soprano. Studied Melbourne with Boustead, Paris with C. Gilly. Début Monte Carlo 1932 (Elisabeth). Paris, O, 1932–6; New York, M, 1935–41, when she contracted polio. Resumed career in 1943, appearing seated in specially staged performances. Retired 1952. Possessed a vibrant, distinctive voice, and was successful as Brünnhilde, Isolde, Ortrud, Salome, Rachel, Alceste. Her autobiography *Interrupted Melody* (New York, 1949) was filmed, 1955. (R)

**Lazaridis, Stephanos** (*b* Ethiopia, 1944). Greek designer. Studied Geneva and London. London, ENO, from 1970, in notable partnership with David Pountney: their productions incl. *Rusalka, Doctor Faust, Lady Macbeth of Mtsensk, Traviata, Hansel and Gretel, Macbeth, Wozzeck*. Also London (CG and Earl's Court), Houston, San Francisco, Berlin, Sydney, Bologna, Flor-

ence, Seville, etc. Other producers with whom he has collaborated incl. Copley (*Figaro, Seraglio, Don Giovanni*), Friedrich (*Idomeneo*), Miller (*Tosca*), Lyubimov (*Tristan, Rigoletto, Fidelio*). Début as producer Scottish O, 1990 (Bartók's *Bluebeard*). His designs are characterized by a dramatic simplicity, startling use of colour, and striking, sometimes violent imagery, closely connected to the producer's concepts.

**Lázaro, Hipólito** (*b* Barcelona, 14 Sep. 1887; *d* Barcelona, 15 May 1974). Spanish tenor. No formal training. Début Barcelona 1909 in operetta. Then studied Milan with Colli. London, C, 1912 (as Antonio Manuele). Milan, S, from 1913 (under own name); New York, M, 1918–20. Paris, O, 1929; also S America, Italy, Spain. Retired 1950. An exuberant singer with a brilliant tone, he was a highly acclaimed Arturo; also sang Radamès, Vasco da Gama, Duke of Mantua, etc. Much admired by Mascagni; created leading roles in *Parisina* and *Piccolo Marat*. Also first Giannetto in Giordano's *Cena delle beffe*. (R)

**lazzi** (It. of uncertain etymology, but probably deriving from *far azi*, abbr. of *fare azione*: 'to perform an act'; hence *l'azi* for theatrical acts or turns, the spurious singular *lazo* or *lazzo*). Improvised acts in the course of a theatrical performance, above all in *commedia dell'arte, when one of the troupe performs a turn by himself. These lazzi became greatly ritualized, and through the commedia dell'arte exercised a great influence on comic librettos and on absurd or farcical arias within them. See also FARSA.

**Lear.** Opera in two parts by Reimann; text by Henneberg, after Shakespeare. Prod. Munich 9 July 1978, with Fischer-Dieskau, Knutson, Varady, Dernesch, Lorand, Götz, cond. Albrecht; London, C, 24 Jan. 1989, with Jaffe, Shilling, Cannan, Mannion, Tierney, Robson, cond. Daniel; San Francisco, 12 June 1981, with Stewart, Knutson, Dernesch, Lloyd, cond. Albrecht.

For other operas on the subject see SHAKESPEARE.

**Lear, Evelyn** (*b* Brooklyn, 8 Jan. 1926). US soprano. Studied New York, Juilliard School, and Berlin. Début Berlin, SO, 1959 (Composer), remained with company when it became Berlin, D, 1961. London, CG, 1965; New York, M, from 1967; also Munich, Hamburg, Paris, O. Roles include Countess (Mozart), Fiordiligi, Marschallin, Tatyana, Emilia Marty, Lulu. Created roles in operas by Egk and Klebe. (R)

**Lebrun, Franziska** (*b* Mannheim, *bapt.* 24 Mar. 1756; *d* Berlin, 14 May 1791). German soprano. Daughter of violinist Innocenz Danzi; wife of the composer and oboist August Ludwig Lebrun. Début Schwetzingen 1772 (Sandrina, Sacchini's

*La contadina in corte*). Mannheim, Court O, from 1772; London, H, 1777, 1779–80; Milan, S, 1778, singing at its opening; Munich 1782–8; Berlin 1779–90. Described as a *virtuosa da camera*, she was a prima donna of some importance, creating many roles in operas by Salieri, Holzbauer, Sacchini, and J. C. Bach, among others. Her fees were high; her tone, according to Burney, resembled that of an oboe ('travelling with her husband, she seems to have listened to nothing else').

**Lebrun, Louis-Sébastien** (*b* Paris, 10 Dec. 1764; *d* Paris, 27 June 1829). French composer and tenor. Had an undistinguished singing career in Paris at the Opéra (1787–91; 1799-1807) and T Feydeau (1791–99), but as a composer was very popular. The most successful of his 16 works, mainly opéras-comiques, were *Marcelin* and *Le rossignol* (more than 200 performances). In 1807 Lebrun joined Napoleon's private chapel as a singer, becoming director in 1810.

SELECT WORKLIST: *Marcelin* (Bernard-Valville; Paris 1800); *Le rossignol* (The Nightingale) (Étienne; Paris 1816).

**Leclair, Jean-Marie** (L'aîné) (*b* Lyons, 10 May 1697; *d* Paris, 22 Oct. 1764). French composer and violinist. One of the greatest performers and composers for his instrument during the 18th cent., he was at the centre of Parisian musical life from 1728 until his murder in 1764. His only opera, *Scylla et Glaucus*, was given a spectacular première and enjoyed fleeting success: he also composed a quantity of ballet and incidental music.

WORKLIST: *Scylla et Glaucus* (D'Albaret; Paris 1746).

BIBL: M. Pincherle, *Jean-Marie Leclair l'aîné* (Paris, 1952).

**Lecocq, Charles** (*b* Paris, 3 June 1832; *d* Paris, 24 Oct. 1918). French composer. Studied Paris Conservatoire with Halévy; in 1856 was joint winner (with Bizet) of an operetta competition sponsored by Offenbach. Although his entry, *Le docteur Miracle*, was well received, he spent several fruitless years before achieving full recognition. This eventually came with the success of *Fleur-de-thé*, followed by a series of almost 40 operettas of a similar style: some of these were first given in Brussels, where he lived during the 1870s, before being seen in Paris. His most successful work was the opéra-comique *La fille de Mme Angot*, which ran for 500 nights; other triumphs included *Cent vierges*, *Giroflé-Girofla*, and *Le petit duc*. Although his operettas did not achieve the lasting popularity of Offenbach's, their skilled orchestration and and captivating melodies put them among the best of the Parisian tradition. His sole attempt at a more serious vein,

*Plutus*, was a spectacular failure, and survived only eight performances.

SELECT WORKLIST: *Le docteur Miracle* (Battu & Halévy; Paris 1857); *Fleur-de-thé* (Chivot & Duni; Paris 1868); *Les cent vierges* (Chivot, Duni, & Clairville; Brussels 1872); *La fille de Mme Angot* (Clairville, Koning, & Siraudin; Brussels 1872); *Giroflé-Girofla* (Letterier & Vanloo; Brussels 1874); *Le petit duc* (Meilhac & Halévy; Paris 1878); *Plutus* (Millaud & Jollivet; Paris 1886).

BIBL: L. Schneider, *Une heure de musique avec Charles Lecocq* (Paris, 1930).

**Leeds.** City in Yorkshire, England. The Grand Theatre and Opera House opened in 1878, and saw performances by visiting companies including the Carl Rosa OC. The first *Ring* in the English provinces was given by the Denhof OC in 1911, and other companies to visit regularly included the BNOC (from 1923) and CG (from 1955). The Grand was restored in the 1980s (cap. 1,550), and is now the home of ★Opera North.

**'Legend of Kleinzach, The'.** Hoffmann's (ten) aria in the Prologue to Offenbach's *Les contes d'Hoffmann*, telling the story of the dwarf at the court of Eisenach.

**Legend of the Invisible City of Kitezh, The.** See INVISIBLE CITY OF KITEZH, THE.

**Legrenzi, Giovanni** (*b* Clusone, *bapt.* 12 Aug. 1626; *d* Venice, 27 May 1690). Italian composer. His first appointment was in Bergamo, at S Maria Maggiore (*c*.1645); in 1656 he became maestro di cappella at the Accademia dello Spirito Santo in Ferrara, where his first operas were composed, beginning with *Nino il giusto*, *Achille in Sciro*, and *Zenobia e Radamisto*. Around 1665 he left Ferrara and after 12 years travelling through Europe settled in Venice as director of the Ospedali dei Mendicanti; vice maestro di cappella at St Mark's 1681. He also composed about 14 operas, many of them lost, for Venice, but after becoming first maestro di cappella at St Mark's in 1685 devoted himself entirely to church and instrumental music. He was very influential as a teacher: among his pupils were Caldara, Gasparini, Lotti, and Pollarolo.

Legrenzi's operas represent the culmination of the 17th-cent. Venetian tradition, and owe much to Cavalli and Monteverdi. He was particularly famed for his 'heroic-comic' works, notably *Totila*, *Giustino*, and *I due Cesari*, in which a serious historical or quasi-historical theme was given elaborate musical and scenic treatment, with relief provided through the inclusion of comic scenes. In some respects he also anticipated the developments of the so-called Neapolitan School, not least in the importance given to the accompaniment and through the extensive use of independent instrumental numbers. His reign at St Mark's was distinguished by his augmentation of the orchestra. Legrenzi is credited with being the first to make significant use of the *motto beginning in his arias, especially in *Eteocle e Polinice*, and his style was a formative influence upon the young Handel.

SELECT WORKLIST: *Nino il giusto* (Bentivoglio; Ferrara 1662, lost); *Achille in Sciro* (Bentivoglio; Ferrara 1663, lost); *Zenobia e Radamisto* (Bentivoglio; Ferrara 1665, rev. Brescia 1666, Venice 1668 by Minato as *Tiridate*, Macerata 1669); *Eteocle e Polinice* (Fattorini; Venice 1675, rev. Naples 1680, Milan 1684, Modena 1690); *La divisione del mondo* (The Discord of the World) (Corradi; Venice 1675); *Totila* (Noris; Venice 1677, rev. Palermo 1696); *Giustino* (Beregani; Venice 1683, rev. Naples 1684, Milan 1689, Genoa 1689, Bologna 1691, Rome 1695, Verona 1696, Modena 1697, Vicenza 1697); *I due Cesari* (Corradi; Venice 1683, rev. Milan 1687).

BIBL: P. Fogaccia, *Giovanni Legrenzi* (Bergamo, 1959).

**Legros, Joseph** (*b* Monampteuil, 7 Sep. 1739; *d* La Rochelle, 20 Dec. 1793). French tenor (haute-contre) and composer. Début Paris, O, 1764 (Titon, Mondeville's *Titon et Aurore*). Sang there until he retired, because of obesity, in 1783. Repertory included operas by Lully, Rameau, and Rousseau; also created leading roles in works by Grétry, Piccinni, Sacchini, etc., and several for Gluck: Achilles (*Iphigénie en Aulide*), Pylades (*Iphigénie en Tauride*), Admète (*Alceste*), Cynire (*Echo et Narcisse*), and the tenor Orphée revised for him for the Paris version of 1774. Though an excellent singer with an attractive, flexible voice, particularly brilliant in its high register, he was ungainly, and stiff on stage. However, as Orphée he was considered admirable, 'a true-to-life actor, full of passion' (Mercure de France); a transformation declared by F. M. Grimm, among others, to be 'one of the most prominent miracles wrought by the enchanter Gluck'. His compositions include an opera, *Anacréon*.

**Lehár, Franz** (*b* Komáron, 30 Apr. 1870; *d* Bad Ischl, 24 Oct. 1948). Austrian composer and conductor. Studied Prague Cons., then worked as a military band conductor in Losoncz, Pola, Trieste, Budapest, and Vienna. He had an early stage success with the serious opera, *Kukuška*; leaving the army 1902 turned to operetta. Already known in Vienna through his popular waltz 'Gold und Silber', he had intended to join the T an der Wien as a conductor: so triumphantly received were his first two operettas, *Wiener Frauen* and *Der Rastelbinder*, that he devoted himself to composition. After the failure of *Der Göttergatte* and *Die Juxheirat*, he scored his greatest success with *Die lustige Witwe*, which rejuvenated the apparently doomed Viennese operetta tradition and opened the way for Oscar

Strauss, Fall, and Kálmán. Some 30 works in similar vein followed, including *Das Fürstenkind*, *Der Graf von Luxemburg*, *Zigeunerliebe*, and *Frasquita*, although he gradually began to lose his hold over the Viennese public. His fortunes revived after 1925, thanks in part to the championship of Richard Tauber, but the focus of his activities shifted from Vienna to Berlin. Notable successes of the later years included *Paganini*, *Der Zarewitsch*, *Giuditta*, and especially *Das Land des Lächelns*, which rivalled *Die lustige Witwe* in its melodic wealth. His final work was a revision of *Zigeunerliebe* as a Hungarian opera, *Garabonciás diák*.

At its height, Lehár's following rivalled that of the earlier generation of the Strausses, Suppé, Millöcker, and Zeller. Although less sparkling than Johann Strauss II's, his operettas have comparable melodic charm, and his gentle waltzes are probably the ancestors of the modern dance-music slow waltz. His lavish use of dance, dominated always by the waltz, almost created a form of ballet-operetta, which was imitated, always less convincingly, by several later composers. During his lifetime, his operettas were staged throughout Europe and America: today only a handful of the more important receive regular performances outside Germany, although many of the most popular numbers have a firm place in the repertory of light music.

SELECT WORKLIST (all first prod. Vienna unless otherwise stated): *Kukuška* (Falzari; Leipzig, rev. Budapest 1899 as *Tatjana*); *Wiener Frauen* (Tann-Bergler & Norini; 1902); *Der Rastelbinder* (Léon; 1902); *Der Göttergatte* (Léon & Stein; 1904); *Die Juxheirat* (The Joke Wedding) (Bauer; 1904); *Die lustige Witwe* (The Merry Widow) (Léon & Stein, after Meilhac; 1905); *Das Fürstenkind* (Léon, after About; 1909); *Der Graf von Luxemburg* (Willner & Bodanzky; 1909); *Zigeunerliebe* (Gypsy Love) (Willner & Bodanzky; 1910); *Frasquita* (Willner & Reichert; 1922); *Paganini* (Knepler & Jenbach; 1925); *Der Zarewitsch* (Jenbach & Reichert, after Zapolska-Scharlitt; Berlin 1927); *Das Land des Lächelns* (The Land of Smiles) (Herzer & Löhner; Berlin 1929) (a rev. of *Die gelbe Jacke*: Léon; 1923); *Giuditta* (Knepler & Löhner; 1934); *Garabonciás diák* (Vincze; Budapest 1943).

BIBL: B. Grün, *Gold and Silver: The Life and Times of Franz Lehár* (London, 1970).

**Lehmann, Lilli** (*b* Würzburg, 24 Nov. 1848; *d* Berlin, 17 May 1929). German soprano. Studied Prague with her mother Marie Loewe. Début Prague 1865 (First Boy, *Zauberflöte*). Berlin, Court T, 1869, 1870–85. Bayreuth 1876 (in first *Ring*, as Woglinde, Helmwige, and the Woodbird), 1896; London: HM, 1880, 1882, and CG, 1884, 1899; New York, M, 1885–92, 1898–9. Also Paris, Vienna, and Salzburg, where she was both artistic director and a principal artist. Last appearance in opera 1909, replacing the suddenly indisposed Bahr-Mildenburg in *Tristan*, Act III; gave recitals until 1920. A singer of phenomenal versatility, technique, and stamina, she possessed a vast repertory (170 roles), including Pamina, Donna Anna, Leonore, Lucia, Norma, Violetta, Marguerite, Philine, Isolde, Fricka, Brünnhilde, and Carmen. Her voice, though not large or rich, was used with resourceful intelligence, and her artistry was supreme. Hanslick refers to her 'superior mind', and 'a personality predestined for tragic and noble roles'; Farrar, one of her pupils, called her 'a daughter of Wotan'. Her writings include *Meine Gesangkunst* (Berlin, 1902; Eng. trans. *How to Sing* (London, 1902)), and her autobiography, *Mein Weg* (Leipzig, 1913). (R)

Her sister **Marie Lehmann** (*b* Hamburg, 15 May 1851; *d* Berlin, 9 Dec. 1931) was also a soprano. Studied with their mother. Début Leipzig 1867 (Aennchen). Bayreuth 1876, in first *Ring* (Wellgunde, Ortlinde), 1896; Vienna 1882–96. Retired 1897. A rather cool singer, though with an excellent technique, she sang Donna Elvira, Adalgisa, Marguerite de Valois (*Huguenots*). (R)

**Lehmann, Lotte** (*b* Perleberg, 27 Feb. 1888; *d* Santa Barbara, CA, 26 Aug. 1976). German, later US, soprano. Studied Berlin with Jordan, Reinhold, and Mallinger. Début Hamburg 1910 (Third Boy, *Zauberflöte*). London, DL, 1914; Vienna 1916–38; London, CG, 1924–35, 1938; Salzburg 1927–37; New York, M, 1934–45; San Francisco 1946. Thereafter sang in concert until 1951. Her repertory included Leonore, Verdi's Desdemona, Charlotte, Tatyana, Sieglinde, Eva, Suor Angelica, Arabella, Ariadne, Octavian, and her most famous role, the Marschallin. Much admired by Strauss, she was the first Composer in the revised *Ariadne auf Naxos*, and created the Dyer's Wife (*Die Frau ohne Schatten*), and Christine (*Intermezzo*). A exceptional singer, possessing a warm and beautiful tone, musicality, generosity, and charm, she was also an actress who brought a telling psychology and humanity to her interpretations. (R)

BIBL: L. Lehmann, *Anfang und Aufstieg* (Vienna, 1937; Eng. trans. *On Wings of Song* (London, 1938)); *Singing with Richard Strauss* (London, 1964).

**Lehmann, Maurice** (*b* Paris, 14 May 1895; *d* Paris, 17 May 1974). French producer and manager. Studied Paris. Paris, C, 1928, director 1931–66, himself producing a series of operettas. Also produced opéra-comique at the T Porte-Saint-Martin, including works by Auber and the première of Pierné's *Fragonard*. Reorganized Paris, O and OC, after the war; director Paris, O, 1945–6, 1951–5; was responsible for sumptuous productions of *Oberon*, *Magic Flute*, *Indes galantes*.

**Lehrstück** (Ger.: 'teaching piece'). A musico-dramatic genre popular in Germany in the 1920s and 1930s. It was probably invented by Brecht, who wrote such works not to entertain an audience, but in an attempt to educate the German masses in the face of the rise of Nazism. Most Lehrstücke contained an overtly Marxist message and usually presented a story whose main theme was the submission (or death) of an individual for the sake of the collective good. Representative examples are *Der Lindberghflug* (Hindemith and Weill, 1929), *Der Jasager* (Weill, 1930), and *Die Massnahme* (Eisler, 1930). Although written for performers as diverse as school groups and workers' choirs, many Lehrstücke adopted operatic principles in the shaping of the musical score.

**Leider, Frida** (*b* Berlin, 18 Apr. 1888; *d* Berlin, 4 June 1975). German soprano. Studied Berlin with Schwarz. Début Halle 1915 (Venus). Hamburg 1919–23; Berlin, S, 1923–40; London, CG, 1924–8; Chicago 1928–32; New York, M, 1933–4. Bayreuth regularly 1928–38. A great artist with a large, attractive voice, a fine legato line, and intense dramatic conviction, she was the outstanding Isolde and Brünnhilde of her day. Repertory also included Armide, Donna Anna, Leonore, Kundry, Marschallin, Santuzza. Autobiography, *Das war mein Teil* (Berlin, 1959; Eng. trans., *Playing my Part* (London, 1966)). (R)

**Leiferkus, Sergey** (*b* Leningrad, 4 Apr. 1946). Russian baritone. Studied Leningrad Cons. with Shaposhnikov. Début Leningrad, Operetta T, 1970 (small roles). Leningrad: Maly T, 1972–8; Kirov O from 1978. Wexford 1983, 1986; London, CG, 1987 (with Kirov O), and from 1988; ENO from 1987; San Francisco 1989; also Toronto, Dallas, Buenos Aires, Barcelona, Salzburg, etc. One of the first Leningrad artists to sing extensively in the West. A musical singer, successful as Onegin, Tomsky, Rangoni, Prince Igor, Scarpia, Escamillo. (R)

**Leigh, Walter** (*b* London, 22 June 1905; *d* nr. Tobruk, 12 June 1942). English composer. Studied Cambridge and in Berlin with Hindemith, from whom he inherited a concern for practical music-making which influenced his many stage works. With his melodic gift and expert craftsmanship Leigh raised the standard of light opera in *The Pride of the Regiment*, a clever parodistic piece, and *Jolly Roger*, which originally ran for six months.

SELECT WORKLIST: *The Pride of the Regiment* (Clinton-Baddeley; London 1932); *Jolly Roger* (Mackenzie & Clinton-Baddeley; London 1933).

**Leïla.** The Brahmin priestess (sop), lover of Nadir, in Bizet's *Les pêcheurs de perles*.

**Leinsdorf, Erich** (*b* Vienna, 4 Feb. 1912; *d* Zurich, 11 Sep. 1993). Austrian, later US, conductor. Studied Vienna with Pisk. Salzburg 1934–7, assisting Walter and Toscanini. New York, M, début 1938 (*Walküre*); succeeded Bodanzky as chief cond. German repertory 1938–43. Music director New York, CC, 1956–7, when he tried to pursue enterprising policies; New York, M, 1944–5, 1956–62, 1965–6 and guest cond. from 1971. San Francisco 1938–41, 1948, 1951, 1955, 1957. Bayreuth 1959 (*Meistersinger*) and 1972 (*Tannhäuser*). A stimulating conductor of considerable nervous intensity. Autobiography, *Cadenza* (Boston, 1976). (R)

**Leipzig.** City in Saxony, Germany. The first opera-house opened 1693 with *Alceste* by Nicolaus Strungk. He was succeeded by G. P. *Telemann; taste inclined towards German music. The theatre closed 1720, but in 1752 *Standfuss's *Der Teufel ist los* was given, encouraging a Singspiel tradition. The Schauspielhaus opened 1766, and visiting companies performed, including under Hiller, Neefe, Benda, Danzi, and E. T. A. Hoffmann. The Schauspielhaus was rebuilt 1817; replaced by the Neues Stadttheater (cap. 1,900) in 1867. Lortzing was conductor 1844–5; the premières of many of his works given here, and in 1850 Schumann's *Genoveva* was premièred. He was succeeded by Julius Rietz 1847–54.

Under Angelo Neumann's managership 1876–80, Leipzig became the base of the Wagner company that gave the first *Ring* performances in various European cities. Artur Nikisch was conductor 1878–89, assisted by Mahler 1886–8. Works premièred before 1914 include Smyth's *The Wreckers* (1906). With Gustav Brecher as director 1923–33, works premièred include Krenek's *Jonny spielt auf* (1927) and *Leben des Orest* (1930) and Weill's *Mahagonny* (1930), the latter causing a riot. Two complete cycles of Wagner's works (including the *Hochzeit* fragments) were given 1938. The opera-house was bombed 1943, and performances were given in the Dreilinden T; rebuilt (cap. 1,682) and opened 8 Oct. 1960 with *Meistersinger*, with operetta given in the Kleines Haus. Helmut Seidelmann was music director 1951–62. Premières have included Boris *Blacher's *Die Nachtschwalbe* (1947), Alan *Bush's *Wat Tyler* (1953) and *Guyana Jonny* (1966), and many important first German performances.

**'Leise, leise'.** Agathe's (sop) aria in Act II of Weber's *Der Freischütz*, in which she prays for protection for her lover Max.

**Leitmotiv** (Ger.: 'leading motif'; the form Leitmotif is often used in English). A term first used by F. W. Jähns in his *Carl Maria von Weber in seinen Werken* (1871) to denote a short musical figure identifying a person, thing, event, or idea in music and above all in opera. The origins of the device are disputed, and its history is controversial. There are suggestions of it in Gluck and Mozart, and a more consistent use of a musical figure to represent a character or idea in various early German Romantic operas and especially those of Weber, e.g. Samiel's diminished 7ths in *Der Freischütz*, Eglantine's theme in *Euryanthe*. Though instances in Weber caused Jähns to coin the term, it has since been generally reserved for a rather more advanced use of the device, above all as first developed by Wagner. Taking the old *reminiscence motive, Wagner showed how a subtle and intelligent use of it could not only recall characters or objects to mind, and further serve as a powerful structural force, but could convey to the listener an intricate understanding of how they change with the course of the drama, the musical modification or development expressing a new psychological or dramatic state. What are in *Das Rheingold* essentially reminiscence motifs are by *Die Walküre*, and above all by *Götterdämmerung*, formed into a subtle expressive network of ideas working in the music. The Leitmotif became an obsession with some of Wagner's followers, as it never had with him: its first use is popularly ascribed to Hans von Wolzogen in an article about the form of *Götterdämmerung* (1878), and it was employed in a study of *Tristan* by Heinrich Porges, written in 1866–77, although not published until 1902. Wagner's own preferred term (in 1867) was **Hauptmotiv**; he made use of other terms, but only once mentioned 'so-called Leitmotiv'. Later composers who adapted the principle of Leitmotiv to their own purposes include Richard Strauss, Humperdinck, and Kienzl.

**Leitner, Ferdinand** (*b* Berlin, 4 Mar. 1912). German conductor. Studied Berlin with Schreker and Prüwer. Assisted Busch, Gly. 1935. Cond. Berlin, Nollendorf T, 1943; Hamburg 1945–6; Munich 1946–7; Stuttgart 1947, music director 1950–69, giving the company new stature; Zurich since 1969. Guest appearances in many European cities and S America (especially Buenos Aires, C). A versatile and efficient conductor, and an excellent accompanist for singers. (R)

**Le Maure, Cathérine-Nicole** (*b* Paris, 3 Aug. 1704; *d* Paris, Jan. 1787). French soprano. Paris, O, chorus 1719; début 1721 (Astrée, Lully's *Phaéton*). Her career there lasted until 1744, punctuated by various withdrawals (one of which landed her in prison) and reappearances. When invited to sing for the wedding of the Dauphin in 1745, she demanded (and got) a carriage from the

King to take her there. She married the Baron Montbruelle in 1762; in 1771 she sang for the opening of the Colisée, to great acclaim, despite her age. 'Ni jolie, ni spirituelle' (La Borde), and small in stature, she triumphed through her sweet and powerful voice, excellent trill, and ability to reduce her audience to tears. She was celebrated in the works of Lully and Rameau, and created parts in numerous operas by Monteclair, Déstouches, Campra, etc. There was great rivalry between her and Pélissier; a contemporary (possibly Voltaire) preferred 'Pélissier par son art, Lemaure par sa voix'.

**Lemberg.** See LVOV.

**Lemeshev, Sergey** (*b* Knyazevo, 10 July 1902; *d* Moscow, 26 June 1977). Russian tenor. Studied Moscow with Raysky and Stanislavsky. Début Sverdlovsk 1926. Moscow, B, 1931–61. Repertory included Mozart's Count Almaviva, Gounod's Faust, Duke of Mantua, Lensky, Berendey, Vladimir (*Prince Igor* and *Dubrovsky*). Admired for his lyrical singing and intelligent acting. (R)

**Lemnitz, Tiana** (*b* Metz, 26 Oct. 1897; *d* Berlin, 1994). German soprano. Studied Metz with Hoch, Frankfurt with Kohmann. Début Heilbronn 1921 (Lortzing's *Undine*). Aachen 1922–8; Hanover 1929–33; Berlin, S, 1934–57; London, CG, 1936, 1938. Repertory included Eurydice, Agathe, Mimì, Aida, Sieglinde, Jenůfa; especially acclaimed as Octavian, Pamina. A sensitive, refined artist, called 'Piana' Lemnitz by her admirers on account of her exquisite pianissimo. (R)

**Lemoyne, Jean-Baptiste** (*b* Eymet, 3 Apr. 1751; *d* Paris, 30 Dec. 1796). French composer. Travelled with a wandering theatrical troupe to Germany, where he studied with Graun, Kirnberger, and Schulz. Appointed second Kapellmeister at the Berlin court, he also worked in Warsaw, where *Le bouquet de Colette* was given, before returning to Paris. His first Parisian opera, *Electre*, was a failure; Gluck, whose style he had imitated, vigorously repudiated his claim to be a pupil. In revenge he attached himself to the camp of Sacchini and Piccinni and wrote his next work after their model. *Phèdre* was an outstanding success and was followed by about 12 further operas in similar vein, of which the most important were *Les prétendus* and *Nephté*: at the première of the latter he was called out to acknowledge the applause, the first recorded instance of this happening in Paris. His son **Gabriel Lemoyne** (1772–1815) was a celebrated instrumental composer.

SELECT WORKLIST: *Le bouquet de Colette* (?; Warsaw 1775); *Electre* (Guillard; Paris 1782); *Phèdre* (Hoffmann; Paris 1786); *Les prétendus* (Rochon de Cha-

bannes; Paris 1789); *Nephté* (Hoffmann, after Corneille; Paris 1789).

**Leningrad.** See ST PETERSBURG.

**Lensky.** Vladimir Lensky (ten), a young poet and Olga's betrothed, in Tchaikovsky's *Eugene Onegin*.

**Lenya, Lotte** (*b* Vienna, 18 Oct. 1898; *d* New York, 28 Nov. 1981). Austrian, then US, singing actress. Began career as dancer. Zurich 1914–20; Berlin 1920, where she met and then married Kurt Weill. Sang in first (concert) performance of *The Rise and Fall of the City of Mahagonny* (Baden-Baden 1927); created Jenny in *The Threepenny Opera* (Berlin 1928), also appearing in the film (1931). Forced to flee Germany in 1933, she and Weill settled first in Paris (where she created Anna in *The Seven Deadly Sins*), then USA. There she created roles in *The Eternal Road*, and *The Firebrand of Florence*; also sang in *Street Scene* and *Down in the Valley*. After Weill's death she devoted herself to the performance of his works. Her inimitable, gravelly voice, highly charged singing, and keen dramatic perception contributed much to Weill's reputation, as well as her own. Acting parts incl. Rosa Klebb in the James Bond film *From Russia with Love* (1963). (R)

**Leo, Leonardo** (orig. Leonardo Ortensio Salvatore de) (*b* S Vito degli Schiavi, 5 Aug. 1694; *d* Naples, 31 Oct. 1744). Italian composer. Studied Naples with Basso and Fago, where his first opera, *Pisistrato*, was performed in 1714. His reputation was properly established with *Sofonisba*, by which time he also held a post in the royal chapel. With the success of *La 'mpeca scoperto* he began a notable series of comic operas in Neapolitan dialect, including *L'ammore fedele*, and *Amor vuol sofferenze*.

Leo's fame gradually spread, and commissions from opera-houses outside Naples, mainly for opere serie, resulted in *Timocrate*, *Trionfo di Camilla*, *La clemenza di Tito*, and *Siface*. He continued to hold posts in Naples, rising to be maestro di cappella at the Conservatorio della Pietà dei Turchini, 1741. Among operas written for Naples were *Farnace* (the last work to be given at the T San Bartolomeo), *L'Olimpiade* (the second work given at the T San Carlo), and *Demofoonte*: he also composed a festa teatrale, *Le nozze di Psiche con Amore*, for the wedding of Charles III. His pupils included Jommelli and Piccinni.

Though many scores are lost, Leo may have written over 70 operas. In his serious works, he was much influenced by Alessandro Scarlatti, though never his pupil, and was not afraid to depart from his model, notably in *L'Olimpiade*. This, in contrast to much Neapolitan opera, includes a prominent part for the chorus. More original were his dialect comedies, to which he brought many of the techniques of opera seria: his contribution to the development of the *ensemble finale was very important and seen to especially good effect in *Semmeglianza*, *Lo matrimonio annascuso*, and *L'ambizione delusa*.

SELECT WORKLIST (all first prod. Naples unless otherwise stated): *Pisistrato* (Lalli; 1714, lost); *Sofonisba* (Silvani; 1718, lost); *Timocrate* (Lalli; Venice 1723, lost); *La 'mpeca scoperto* (Confusion Exposed) (Oliva; 1723, lost); *L'ammore fedele* (Oliva; 1724, lost); *Il trionfo di Camilla* (Stampiglia; Rome 1726); *Semmeglianza* (Senialbo; 1726, lost); *Lo matrimonio annascuso* (Maltrone; 1727); *Demofoonte* (Metastasio; 1735; comp. with Mancini & Sarri); *La clemenza di Tito* (Metastasio; Venice 1735, lost); *Farnace* (Biancardi; 1736); *Siface* (Metastasio; Bologna 1737, lost); *L'Olimpiade* (Metastasio; 1737); *Le nozze di Psiche con Amore* (Baldassare; 1738); *Amor vuol sofferenze* (Federico; 1739; rev. as *La finta frascatana*, 1744); *L'ambizione delusa* (Canicà; 1742).

BIBL: G. Pastore, *Leonardo Leo* (Galatina, 1957).

**Leoncavallo, Ruggero** (*b* Naples, 23 Apr. 1857; *d* Montecatini, 9 Aug. 1919). Italian composer. Studied Naples with Ruta and Rossi, and shortly after graduating composed his first opera, *Chatterton*. Its production was abandoned shortly before the première, forcing him to teach and perform popular music. After travelling in Europe and Egypt, he returned home 1887 with plans for a vast Wagnerian trilogy embracing the Renaissance in Italy, *Crepusculum*, which he offered to Ricordi. Encouraged, he completed the first part, *I Medici*, but when no performance materialized he offered his next work, *Pagliacci*, to Sonzogno.

Together with Mascagni's *Cavalleria rusticana*, this ushered in a new epoch in Italian operatic history, that of *verismo. Though *Pagliacci* made Leoncavallo famous overnight, it did not ensure the success of his earlier works. *I Medici*, although well received by the audience, was attacked by the critics, as was *Chatterton*. *La bohème* arose from a suggestion to Puccini that Murger's *Scènes de la vie bohème* would make a good libretto: when Puccini was unenthusiastic, Leoncavallo adapted it for himself. Though his work contains many excellent ideas, and was quite successful in its day, it has suffered from comparison with Puccini's opera, with which it eventually coincided.

Leoncavallo remained at the forefront of operatic life in Italy, and also achieved popularity in the UK and in the USA, but none of his subsequent works had lasting success. *Zazà* won some praise from Fauré and others, but *Der Roland von Berlin*, commissioned by Wilhelm II in imitation of *I Medici* as a celebration of the House of Hohenzollern, had no appeal beyond the German court. Most of Leoncavallo's later works, which included *Malbruck*, *Il primo bacio*, and *La candidata*, were operettas; his final operas, the patriotic *Goffredo Mameli* and the 1-act *Edipo re*, attempted unsuccessfully to recapture his youthful grand style.

The best of Leoncavallo's music is in *Pagliacci*, the only opera by which he is remembered. One of the seminal works of verismo, it has a strong dramatic flair, sharp characterization, and direct melodic appeal. He drew inspiration from a wide variety of sources, including Wagner, Bizet, and Massenet as well as his Italian forebears. Like Boito, he wrote skilful librettos for many of his own works. Despite his achievements, his work, like that of all his Italian contemporaries, was quickly overshadowed by the greater talents of Puccini, hastening the rapid eclipse of everything save *Pagliacci*.

SELECT WORKLIST: *Chatterton* (Leoncavallo; comp. 1877, rev. Rome 1896); *I Medici* (Leoncavallo; comp. 1892, prod. Milan 1893); *Pagliacci* (Clowns) (Leoncavallo; Milan 1892); *La bohème* (Leoncavallo, after Murger; Venice 1897); *Zazà* (Leoncavallo, after Berton & Simon; Milan 1900); *Der Roland von Berlin* (Leoncavallo, after Alexis; Berlin 1904); *Malbruck* (Nessi; Rome 1910); *Zingari* (Gypsies) (Cavacchioli & Emanuel; London 1912); *La candidata* (Forzano; Rome 1915); *Goffredo Mameli* (Belvederi; Genoa 1916); *Edipo re* (Forzano, after Sophocles, completed Pennacchio; Chicago 1920).

BIBL: R. de Rensis, *Per Umberto Giordano e Ruggiero Leoncavallo* (Siena, 1949).

**Léonor.** Léonor de Guzman (sop), mistress of King Alfonso XI of Castile, in Donizetti's *La favorite*.

**Leonora.** 1. Oberto's daughter, rejected lover of Riccardo, in Verdi's *Oberto*.

2. The lady-in-waiting (sop) to the Princess of Aragon, and lover of Manrico, in Verdi's *Il trovatore*.

3. Donna Leonora de Vargas (sop), lover of Don Alvaro, in Verdi's *La forza del destino*.

Also operas by Mercadante (1844) and Fry (1845).

**Leonore.** Florestan's wife (sop), known as Fidelio, in Beethoven's *Fidelio*; also the title of the opera's original version. Also Léonore (sop), lover of Alonze, in Grétry's *L'amant jaloux*.

**Léonore, ou L'amour conjugal.** Opera in 2 acts by Gaveaux; text by Jean Nicolas Bouilly. Prod. Paris, T Feydeau, 19 Feb. 1798; New Orleans, T St. Philippe, 12 Mar. 1812. Historically important as the first setting of the *Fidelio* story: others are by Paer (1804), Mayr (1805), and Beethoven (*Fidelio*, 1805), all based on Bouilly. Also opera by Champein (1781).

**Leonova, Darya** (*b* Vishny Volochuk, 21 Mar. 1829; *d* St Petersburg, 6 Feb. 1896). Russian contralto. Studied St Petersburg. Début St Petersburg 1852 (Vanya, *Life for the Tsar*). She was then taught by Glinka, and may have been his mistress; he certainly admired her and her 'beautiful voice'. Sang in Moscow and St Petersburg until 1874. Toured Europe 1857–8; toured Russia, China, USA, and W Europe 1875–9. A great champion of Russian music, she created the Princess in Dargomyzhsky's *Rusalka*, Serov's *Rogneda*, the Hostess in *Boris Godunov*, Vlasevnya in *The Maid of Pskov*, and Margaret in Cui's *Ratcliff*. Other roles included Gluck's Orfeo, Azucena, Ortrud. Her voice was clear and sonorous, and she was a good actress. As a friend, she gave much practical help to Musorgsky, who was also her accompanist. She sang at his memorial service.

BIBL: V. Yakovlev, *D. M. Leonova* (Moscow, 1950).

**L'Épine, Margherita de**. See ÉPINE, MARGHERITA DE L'.

**Leporello**. Don Giovanni's servant (bs) in Mozart's *Don Giovanni* and Dargomyzhsky's *The Stone Guest*.

**Leppard, Raymond** (*b* London, 11 Aug. 1927). English conductor and musicologist. Studied Cambridge. Répétiteur Gly. 1954–6. Returned to Cambridge 1957 as lecturer, also beginning his exploration and editing of 17th- and early 18th-cent. music, especially Monteverdi and Cavalli. London, CG, 1958 (*Samson*), 1972 (*Figaro*, *Così*); Gly. 1962 (*Poppea*), 1967 (*Ormindo*, *Calisto*), 1970 (première of Maw, *Rising of the Moon*), 1972 (*Ritorno d'Ulisse*), 1975 (*Cunning Little Vixen*); London, SW, 1965 (Monteverdi's *Orfeo*), 1971 (*Poppea*); Santa Fe 1974 (*Egisto*). His realizations, which include additions from other works and some original composition, have been attacked for their inauthenticity, but served to introduce once neglected music to a wide general audience, and were skilfully judged in theatrical effect. (R)

**Lermontov, Mikhail** (*b* Moscow, 15 Oct. 1814; *d* Pyatigorsk, 27 July 1841). Russian writer. Operas on his works are as follows:

*The Angel* (1831): Koreshchenko (1900)
*Vadim* (1832–4): Aksyuk (*Pugachevtsy*, 1937); Kreitner (1952)
*The Boyar Orsha* (1835): Kashperov (1880); Agrenev-Slavyansky (1910); Krotkov (1898); Fistulari (?)
*Hajji-Abrek* (1835): Rubinstein (1858)
*Masquerade* (1835–6): Kolesnikov (*c*.1890); Mosolov (1940); Denbsky (1941); Bunin (1944); Zeidman (1945); Y. Nikolayev (1946); Nersesov (1948); D. Tolstoy (1955); Artamov (1957)

*The Tambov Treasurer's Wife* (1837): Asafyev (1937)
*The Song of the Merchant Kalashnikov* (1837): Rubinstein (1880)
*A Hero of Our Time* (1840). Pt1 *Bela*: Gaygerova (1941); A. Alexandrov (1946). Pt2 *Princess Mary*: Dekhterev (1941)
*The Demon* (final version, 1841): Rubinstein (1875)
*The Fugitive* (1841): Avetisov (1943)
*Tamara* (1841): Bourgault-Ducoudray (1891); Wietinghoff-Scheel (1886); Rogowski (1918)

**Le Rochois, Marthe**. See ROCHOIS, MARTHE.

**Lescaut**. A sergeant of the King's Guards (bar), Manon's cousin in Massenet's *Manon* and her brother in Puccini's *Manon Lescaut*.

**Le Sueur, Jean-François** (*b* Drucat-Plessiel, 15 Feb. 1760; *d* Paris, 6 Oct. 1837). French composer. Studied Abbeville and Amiens choir schools, Paris with Roze. Worked for a time as a choirmaster, usually controversially owing to his advanced ideas for a more theatrical church music. Turning to opera, with help and advice from Spontini, he began work on *Télémaque*, but then caught the temper of the Revolutionary times with *La caverne*. This made him, with Cherubini and Méhul, the most prominent composer of the day in France. A *rescue opera, *La caverne* is a powerful, even harsh work, formally pioneering for its successful reconciliation of elements from opera seria, opera buffa, and opéra-comique. *Paul et Virginie* and *Télémaque* followed, the former adumbrating some of the manner of Grand Opera.

This is the form of his greatest success, *Ossian, ou Les bardes*. The presence of Napoleon as patron helped to win the work favour, as did the current fascination with all things to do with 'Ossian'. Le Sueur's old love of choral music (evident in *La caverne*) tends to dominate the work and give it the semblance of oratorio; nevertheless, the most famous section, the tableau 'Ossian's Dream', is finely conceived and boldly written. Le Sueur also makes use in the work of some originally composed 'barbaric' melodies and textures, a fluent movement between recitative and aria, and certain devices (such as tense off-beat rhythmic figures and some effects of sonority) which impressed his pupil Berlioz. Two works written with Persuis followed before *La mort d'Adam*, a wildly ambitious work that demands a huge cast (including the entire population of heaven and hell) but that not surprisingly fails to match its theme with music of comparable majesty. *Alexandre à Babylone* takes some of the fluency of *Ossian* further, and anticipates the continuous manner of later 19th-cent. opera. Le Sueur was a teacher well loved by his pupils, among them Gounod, Thomas, and above all Berlioz, who failed to persuade him to accept even Beethoven, let alone the Romantics whose music his own in fact often anticipates.

WORKLIST (all first prod. Paris): *La caverne* (Palat-Dercy, after Lesage; 1793); *Paul et Virginie* (Dubreuil; 1794); *Télémaque* (Palat-Dercy; 1796); *Ossian, ou Les bardes* (Palat-Dercy, rev. Deschamps; 1804); *L'inauguration du temple de la Victoire* (with Persuis: Baour-Lormian; 1807); *Le triomphe de Trajan* (with Persuis: Esménard; 1807); *La mort d'Adam* (Guillard, after Klopstock, Milton, and the Bible; 1809); *Alexandre à Babylone* (Baour-Lormian; comp. 1815, unprod.).

CATALOGUE: J. Mongrédien, ed., *J.-F. Le Sueur: A Thematic Catalogue of his Complete Works* (New York, 1980).

BIBL: F. Lamy, *Jean-François Le Sueur (1760–1837)* (Paris, 1912).

**Let's Make an Opera.** Opera in 2 parts by Britten; text by Eric Crozier. Prod. Aldeburgh, Jubilee H, 14 June 1949, with E. Parry, Parr, Worthley, Lumsden, cond. Del Mar; St Louis, Kiel Auditorium, 22 Mar. 1950.

In the first part the preparations for the opera are discussed by the children and grown-ups taking part, and the audience is rehearsed for its role in four songs. The second part, *The Little Sweep*, is set in 1810 and tells the story of the encounter of Sammy (treb), an 8-year-old apprentice sweep, with the children who live at Iken Hall. The children manage to free Sammy from Black Bob (bs), his master.

**Letter Duet.** The Duettino 'sull'aria' between the Countess and Susanna (sops) in Act III of Mozart's *Le nozze di Figaro*, in which the Countess dictates to Susanna the letter to be sent to the Count making the assignation that evening in the garden. Despite all printed scores, some doubt exists as to which part consistently belongs to which singer.

**Letter Scene.** A favourite scene in 17th-cent. opera in which a character reads out a letter brought to him. The device long survived (e.g. in Massenet's *Werther*, where Charlotte rereads letters from the hero in the 'Air des lettres'), but is nowadays usually taken to refer to the extended scene for Tatyana (sop) in Tchaikovsky's *Eugene Onegin*, in which she sits up all night writing to Onegin declaring her love.

**Levasseur, Nicolas** (*b* Bresles, 9 Mar. 1791; *d* Paris, 7 Dec. 1871). French bass. Studied Paris with Garat. Début Paris, O, 1813 (Grétry's *Caravane du Caire*). London, H, 1815–16; Paris, I, 1819–28; Milan, S, 1820; Paris, O, 1827–45, 1849. Retired 1853. His repertory included much Rossini, and he created numerous roles: Bertram (*Robert le diable*), Marcel (*Huguenots*), Zacharie (*Le prophète*), Governor (*Ory*), Fürst (*Guillaume Tell*), Balthazar (*Favorite*), and Brogni (*La Juive*), among others. He was also the first Mosè in the Paris version. He possessed a magnificent true bass, and a refined style; he was however, uninspired dramatically, and was at his best in austere, dignified roles.

**Levasseur, Rosalie** (*b* Valenciennes, 8 Oct. 1749; *d* Neuwied, 6 May 1826). French soprano. Début Paris, O, 1766 (Zaïde, Campra's *L'Europe galante*). Sang there until 1785. For ten years performed as Mme Rosalie in minor roles (including Amour in the première of *Orphée*). Her talents were recognized and promoted by the Austrian ambassador, Mercy-Argentau, who became her lover. Resuming her own name, she succeeded Arnould as principal soprano at the Opéra, singing the title-role of the first Paris *Alceste*, and creating the leading roles in *Armide* and *Iphigénie en Tauride* (rechristened by Arnould *Iphigénie en Champagne* when her rival one day appeared drunk on stage), as well as in many works by Piccinni, Sacchini, J. C. Bach, Grétry, etc. Though her voice was powerful, it lacked flexibility, and was considered at its best only in the music of Gluck; however, she was undisputedly the outstanding actress of the day. After the death of Gluck, her teacher and close friend, she lived with his widow in Vienna.

**'Le veau d'or'.** The song in praise of gold, sung by Méphistophélès (bs) in Act II of Gounod's *Faust*.

**Levi, Hermann** (*b* Giessen, 7 Nov. 1839; *d* Munich, 13 May 1900). German conductor. The son of a rabbi; studied Mannheim with Lachner, Leipzig with Hauptmann and Rietz. Music director Saarbrücken, 1859–61; cond. Mannheim 1861; Rotterdam 1861–4; Music director Karlsruhe 1864–72, where he cond. Schumann's *Genoveva* and impressed Wagner with his *Meistersinger* and *Rienzi*. Cond. Munich 1872–6. He was in Bayreuth as Wagner's supporter from early days; cond. première of *Parsifal* 1882. His choice was successfully urged on Wagner, who greatly admired him but nearly wrecked the collaboration by his crude behaviour especially over the notion of a Jew conducting a work that centres on the representation of Christian ideas. Upset, Levi withdrew with dignity but was persuaded back when Wagner (largely on King Ludwig's urging) made peace. Their friendship survived, and Levi became, in Wagner's words, 'the ideal Wagner conductor'. He conducted at Wagner's funeral. Accounts speak of his great personal qualities, and the 'spiritual' nature of his interpretations. He made new versions of *Così fan tutte*, *Don Giovanni*, and *Figaro*; though now discredited, they helped to win a following for Mozart at a time when his operas were little known and much misrepresented. He also translated *Les Troyens* and Chabrier's *Gwendoline*.

BIBL: E. Possart, *Erinnerungen an Hermann Levi* (Munich, 1901).

**Levine, James** (*b* Cincinnati, 23 June 1943). US conductor. Studied Cincinnati with Levin, New York with Lhévinne, Morel, Serkin. Assistant cond. Cleveland Orch. 1964. Début New York, M, 1971 (Tosca), principal conductor 1974, music director 1975. He has conducted widely abroad: WNO 1970, *Aida*; Hamburg 1975, *Otello*; Salzburg from 1976; Bayreuth 1982–93 (*Parsifal*) and 1994 (new *Ring*); etc. His most distinguished achievement has been to sustain and develop the position of the Met. as an opera-house of world standing, in part thanks to his own wide repertory and catholic tastes. At his best, an inspiring and dramatically exciting conductor. (R)

**Lewis, Richard** (*b* Manchester, 10 May 1914; *d* Old Willingdon, 13 Nov. 1990). English tenor. Studied Manchester and London with Allin. Début Gly. 1947 (Male Chorus, *Rape of Lucretia*). London, CG, from 1947, SW, from 1948; San Francisco from 1953. Also Paris, O; Berlin; Buenos Aires; etc. Created Walton's Troilus, Tippett's Mark (*Midsummer Marriage*) and Achilles (*King Priam*), and Klebe's Amphitryon (*Alkmene*). Roles include Ferrando, Idomeneo, Florestan, Pinkerton, Dmitri, Herod, Tom Rakewell. A versatile and intelligent artist. (R)

**Lexington Theatre.** New York theatre on Lexington Avenue and 51st St. built by Oscar *Hammerstein to evade the 1910 Metropolitan contract which prevented him from giving opera for ten years. Sold, then scene of Boston Nat. OC 1916 season, Chicago O visits 1917–19, German OC 1923–4.

**Lhérie, Paul** (*b* Paris, 7 or 8 Oct. 1844; *d* Paris, 17 Oct. 1937). French tenor and baritone. Studied Paris with Obin. Début as tenor, Paris, OC, 1866 (Reuben, Méhul's *Joseph*). Remained with company, creating Don José 1875. Became baritone 1882. Milan, S, 1884, first Posa in the revised *Don Carlos*; London, CG, 1887. Created Rabbi David in Mascagni's *L'amico Fritz*. Though not especially distinguished as a tenor, as a baritone he had much success, being praised for his persuasive singing and elegant acting. Roles included Rigoletto, Iago (Verdi), and Hamlet.

**Liberec** (Ger., Reichenberg). Town in Bohemia, Czechoslovakia. Opera is given in the F. X. Šalda T about 3–4 times a week by a company of about 30 soloists. The company also gives a weekly performance at Jablonec. Though one of the country's smaller theatres, it has staged *Rienzi*,

*L'Africaine*, and *War and Peace*. The theatre was renovated in 1971.

**'Libiamo, libiamo'.** The *brindisi sung by Alfredo (ten) and Violetta (sop) in Act I of Verdi's *La traviata*.

**libretto** (It.: 'little book') (Fr., *livret*; Ger., *Textbuch*). The name generally given to the book of the words of an opera, hence the text itself. Though the earliest were some $8\frac{1}{2}$ in. in height, the diminutive was always used and the term has been current in English since about the mid-18th cent.

Early librettos usually began with a title-page succeeded by a preface in which the writer made his dedication to his patron, then by a few words addressed to the reader. The *argomento was a summary of the events preceding the action of the opera: these tended to increase in complexity, and even to take their place in the opera itself as a prologue (e.g. Verdi's *Il trovatore*). After the list of the characters in the opera came a catalogue of the scene-changes, dances, perhaps also the scenic effects: this trait survives on to playbills of the English musical theatre of the 19th cent. Until about the end of the 18th cent. there would also be a protesta, in which the author affirmed his sound Roman Catholic faith despite the pagan references to *numi*, *fati*, etc.: this arose from the *censorship applied especially in the Papal States of Italy. Though some were published with care, most were hurriedly printed for practical use by the audience in the theatre so as to enable them to follow the plot and, especially with opera seria, to appreciate the poetry of an aria that the composer had set and the singer was performing. When auditoria were darkened during the performance, librettos were sold as cereni (from It.: *cero*, a wax candle); and grease spots on some examples show that the use of candles as an aid to reading was a familiar practice.

In the 17th and 18th cents. the established pattern of recitative, aria, and chorus made special demands upon the librettist, and determined the course of the action. Changing musical conventions have similarly shifted the demands, with the comparative precedence of words over music remaining a topic argued in many theoretical works (e.g. by *Algarotti) and dramatized in operas as diverse as *Prima la musica poi le parole* (Music first and then words) by Salieri (1786) and Strauss's *Capriccio* (1942). The vital element has remained dramatic potency as it charges a composer's imagination, and a strong literary content in a libretto may prove obtrusive.

The sources of successful librettos have ranged from great dramatic masterpieces (*Othello* for

Verdi) and great novels (*War and Peace* for Pro-kofiev) to sentimental novels (*Scènes de la vie de Bohème* for Puccini) and narrative poems (*Eugene Onegin* for Tchaikovsky), from classical and heroic legend (most of Zeno and Metastasio) or Romantic legend (Wagner's *The Flying Dutch-man*) to real-life incident (supposedly *Pagliacci*, Janáček's *Osud*), from questions of belief (Wagner's *Parsifal*) to farce (most of Offenbach), from great painting (Hindemith's *Mathis der Maler*) to comic strip (Janáček's *Cunning Little Vixen*), from historical or biographical incident (much Grand Opera) to fairy-tale (much French, German, and Russian opera). In days when the conventions were firm, masters of them such as Zeno and Metastasio provided composers with material that could be used repeatedly, some-times more than once by the same composer: 27 of Metastasio's works did duty for at least 1,000 settings by at least 50 composers. Collaboration between a librettist who understands, and will respond to the demands of, his composer has regularly proved successful, though it is no guar-antee of quality. However, particularly fruitful results have come from the careful mutual plan-ning of Quinault and Lully, Calzabigi and Gluck, Da Ponte and Mozart, Boito and Verdi, Gilbert and Sullivan, Hofmannsthal and Strauss. Among the most successful composers to fashion their own librettos have been Menotti, Tippett, Ber-lioz, and above all Wagner, who more than any other artist thought out creatively ideas about the interaction of words and music.

The popularity of published librettos is as steady as ever, and has been given a new perspec-tive by the spread of recorded music, with record companies now usually accompanying complete opera sets with the original text and translation into several languages. A number of opera com-panies also accompany important new produc-tions with booklets that include the complete text; and a series of opera guides including the full text and translation of each opera has been published by English National Opera.

BIBL: P. Smith, *The Tenth Muse* (New York, 1970).

**Libuše.** Opera in 3 acts by Smetana; text (orig. German) by Josef Wenzig, trans. into Czech by Ervín Špindler. Prod. Prague, for the inaugu-ration of the Czech National T, 11 June 1881, with Sittová, Reichová, Fibichová, Vávra, Lev, Čech, Stropnický, Hynek, cond. Smetana.

Libuše (sop), Queen of Bohemia, settles a dis-pute between two brothers, Chrudoš (bs) and Šťáhlav (ten), over the division of their father's inheritance. Chrudoš is unhappy and insults Libuše, saying she is unfit to rule. Distraught at this, she marries a wise peasant, Přemysl (ten), to whom she abdicates power. The two brothers are reconciled and Přemysl founds the first Bohe-mian dynasty the Přemyslids.

**licenza** (It.: 'licence'). Sometimes used in the 17th and 18th cents. for an inserted cadenza, but more commonly the epilogue, consisting of arias and often a choral ode to the dedicatee of an opera or to the ruler before whom and under whose patronage the opera was given.

**Licette, Miriam** (*b* Chester, 9 Sep. 1892; *d* Twy-ford, 11 Aug. 1969). English soprano. Studied Paris with Marchesi and J. De Reszke, and Milan with Sabbatini. Début Rome 1911 (Butterfly). Beecham OC 1916–20; BNOC 1922–8; London, CG, 1919–29. Possessed a pure and even voice, and a wide repertory including Donna Elvira, Marguerite, Mimì, Verdi's Desdemona, Gutrune, Louise. (R)

**Liebe der Danae, Der** (The Love of Danae). Opera in 3 acts by Richard Strauss; text by Josef Gregor, after a sketch by Hofmannsthal. Prod. Salzburg 14 Aug. 1952, with Kupper, Gostic, Schoeffler, cond. Krauss. The work reached public dress rehearsal at Salzburg in 1944 (16 Aug. with Ursuleac, Taubmann, and Hotter, cond. Krauss), but then the theatres were closed by a decree of the Nazi minister Goebbels. Lon-don, CG, 16 Sep. 1953, with Kupper, Vanden-burg, Frantz, cond. Kempe; Los Angeles, U of S California, 10 Apr. 1964, with Weide, Gibson Riffel, cond. Ducloux.

Danae (sop), daughter of Pollux (ten), King of Eos, is visited in her dreams by Jupiter (bar). He takes the form of her lover Midas (ten), and, deceiving Danae, successfully woos her.

Despite Jupiter's warnings that he will loose his godly golden touch if he disobeys him, Midas touches Danae, and she turns to gold. Danae is allowed to choose between Jupiter and the mortal Midas, and selects the latter.

Midas reveals to Danae how he lost his wealth through his agreement with Jupiter. In order to try to regain her, Jupiter disguises himself as a poor tramp, but Danae rejects his promises of wealth and, impressed by her loyalty to Midas, Jupiter gives the couple his blessing.

**Liebermann, Rolf** (*b* Zurich, 14 Sep. 1910). Swiss composer, critic, and opera manager. Stu-died Zurich, then Budapest and Vienna with Scherchen, later Ascona with Vogel. His style extends from serial music to jazz, and though freely experimental is based on knowledge of practical effect. He virtually abandoned compo-sition on taking up his appointment as Intendant of the Hamburg, S (1959–73), one significant for his championship of many new works. At the Paris Opéra, 1973–80, he greatly raised the stan-dard of performances.

SELECT WORKLIST: *Leonore 40/45* (Strobel; Basle 1952); *Penelope* (Strobel, after Molière; Salzburg 1954); *Die Schule der Frauen* (Strobel; Louisville 1955); *La forêt* (Vidal, after Ostrovsky; Geneva 1987).

BIBL: I. Scharberth and H. Paris, eds., *Rolf Liebermann zum 60. Geburtstag* (Hamburg, 1970).

**Liebestod** (Ger.: 'love-death'). The first appearance of the term is in Hoffmann's *Undine* (1816), where it is used near the end of the work by Heilmann to describe the lovers' union in death. However, it is now always taken to refer to Isolde's death scene at the end of Act III of Wagner's *Tristan und Isolde*, beginning 'Mild und leise', though Wagner himself used it of the love duet in Act II.

**Liebesverbot, Das** (The Ban on Love). Opera in 2 acts by Wagner; text by composer, after Shakespeare's drama *Measure for Measure* (1604–5). Prod. Magdeburg 29 Mar. 1836, with Pollert, Freimüller, cond. Wagner; London, University Coll., 15 Feb. 1965, with Davies, Jenkins, Bentley, Kallipetis, cond. Badacsonyi.

**Lied** (Ger.: 'song'). In German opera, the term has generally been used for a number rather simpler than an aria, sometimes interpolated by a character and outside the plot. The May song in Weber's *Euryanthe* is described as 'Lied mit Chor', though other numbers in the work are described as Aria, Romanze, etc. See also AUFTRITTSLIED.

**Liederspiel** (Ger.: 'song play'). A German dramatic musical form, deriving from the *Singspiel, and consisting of songs joined by dialogue. The composer who gave Liederspiel its most characteristic form was J. F. Reichardt, who used the manner of German popular music in his songs. Subsequent composers, including F. H. Himmel, Carl Eberwein, and B. A. Weber, were successful, though to a lesser degree. Mendelssohn described his *Die Heimkehr aus der Fremde* (1829) as a Liederspiel, as did Lortzing his *Der Pole und sein Kind* (1832).

**Liège.** City in Liège, Belgium. Opera was given by visiting Italian companies in the T Baraque in 1745. In 1757 there was given the first Walloon opera, *Li voedge di Chaudfontaine* by Jean-Noël Hamal (1709–78). From 1767 operas were given in the Douane (cap. 300), including some by *Grétry (born in Liège). When this burnt down in 1805, opera was given in a Benedictine monastery (Salle Saint-Jacques). The Grand T (cap. 1,246), whose architecture imitated the Paris Odéon, opened in 1820 with Grétry's *Zémire et Azor*, and saw an expansion of operatic activity. Until 1914 the theatre had a permanent company of French and Belgian singers, and all operas were performed in French. Between the wars opera was given by visiting companies. The O Royale de Wallonie was established in 1967 in the theatre (cap. 1,048), also touring widely.

BIBL: H. Hamal, *Annales de la musique et du théâtre à Liège de 1738 à 1806* (Liège, 1989).

**lieto fine** (It.: 'happy ending'). A term apparently first used by Giacinto Cicognini, librettist of several operas in the mid-17th cent., including Cavalli's *Giasone*, to describe the final turn of an opera plot towards a happy resolution. Often this took the form of the last-minute repentance of a tyrant, shown the error of his ways or cured of his 'madness' by divine intervention, so as to provide a joyful conclusion, and to emphasize to the audience the benefits of liberty or enlightened patronage.

In the earliest operas the lieto fine was rarely employed; the first finale of Monteverdi's *Orfeo* (1607) follows the conventional tragic ending of the myth, with Orfeo torn to pieces by the Bacchantes. Even the revised finale (1609), which includes a *deus ex machina in the form of Apollo, is hardly joyful, for Orfeo is told that he will never see his beloved Euridice again.

However, these early operas were usually written for private performance before an intellectual, cultured audience. The development of public opera in Venice after 1637 brought with it a wider audience, whom composer and librettist wished to see leave the opera-house happy and uplifted. Monteverdi had already acknowledged the need for a lieto fine in operas performed to a less intellectual audience by his treatment of the Arianna story (1608), and eventual rejection of the libretto for *Narciso ed Ecco* in the same year. Both his surviving Venetian operas end happily, a model followed in the works of Cavalli and Cesti, where there is invariably a satisfying pairing-off of male and female characters.

In later Metastasian opera seria the lieto fine retained an important role, even when this necessitated a contrived mutilation of a well-known tragic myth. For example, in the 18th cent. Haydn's version of the Orpheus story, *L'anima del filosofo* (1791), alone preserves its tragic ending: all other treatments, including Gluck's so-called reform opera *Orfeo ed Euridice*, end with the joyful reunion of the lovers. The suspension of belief frequently demanded by the lieto fine did not hinder its becoming a staple feature of French opera, although it later became an object of parody also. In Rameau's *Hippolyte et Aricie*, for example, the hero has to be resurrected as a spirit, in order to be able to marry the still-mortal Aricie. All Mozart's opere serie conclude with a lieto fine: in *Idomeneo* a divine intervention spares the sacrifice of Idamante, while a similarly contrived denouement is provided for *La clemenza di Tito*.

**Life for the Tsar, A** (Russ.: *Zhizn za tsarya*). Opera in 4 acts and epilogue by Glinka; text by Baron Georgy Fyodorovich Rosen. Prod. St Petersburg, B, 9 Dec. 1836, with Stepanova, Petrova-Vorobyova, Leonov, Petrov, cond. Cavos; London, CG, 12 July 1887, with Albani, Scalchi, Gayarré, Devoyod, cond. Bevignani; San Francisco 12 Dec. 1936.

Russia and Poland, winter 1612. Ivan Susanin (bs) refuses to let his daughter Antonida (sop) marry her betrothed, Bogdan Sobinin (ten), until the future of Russia is secure and a new Tsar has been elected.

News of the election of Mikhail Romanov reaches the enemy Polish camp; the officers decide to launch an immediate attack.

While preparations for Antonida's wedding are under way, Polish soldiers burst into her home and compel her father to lead them to the Tsar. Secretly, Susanin tells his adopted son Vanya (con) to run ahead and warn the Tsar.

Susanin leads the Poles into the depths of the enchanted forest. They guess his trick and he prepares himself for death.

In the square of the Kremlin, the new Tsar praises his supporters and rejoices in their victory over the Poles; all lament the death of Susanin.

Originally entitled *Ivan Susanin*, the work was renamed *A Life for the Tsar* by permission of Nicholas I following his visit to one of the rehearsals. Under Communism it was known as *Ivan Susanin* and performed in a revised and distorted text by Sergey Gorodetsky.

**Ligeti, György** (*b* Dicsöszentmárton (Tîrnăveni), 28 May 1923). Hungarian, later Austrian, composer. Studied Kolozsvár (Cluj) with Farkas, Budapest with Kadosa and Veress. His major stage work is *Le grand macabre* (1978), in which his wide-ranging and highly original language find brilliant dramatic expression.

WORKLIST: *Le grand macabre* (Meschke and Ligeti, after Ghelderode; Stockholm, 1978).

BIBL: P. Griffiths, *György Ligeti* (London, 1985).

**Lille.** City in Nord, France. Opera was first given in the Hôtel de Ville; burnt down 1700. Rebuilt with a 90,000 florin gift from Louis XIV, it reopened 1718. A larger auditorium opened 1787; burnt down 1903. It was replaced by the T Sébastopol (cap. 1,450), which then staged operette when the T de l'Opéra (cap. 1,500) opened 1919. In 1979 the Association de l'Opéra de Nord was founded, embracing Lille, Tourcoing, and Roubaix; dissolved 1984.

**Lily of Killarney, The.** Opera in 3 acts by Benedict; text by John Oxenford and Dion Bouci-

cault, after the latter's drama *Colleen Bawn* (1860). Prod. London, CG, 8 Feb. 1862, with Pyne, Santley, Thirlwall, Harrison, cond. Mellon; Philadelphia 20 Nov. 1867.

Ireland, 19th cent. In order to solve his family's financial problems, Hardress Cregan (ten) is encouraged to marry the heiress Ann Chute (sop); however, he is already secretly married to a peasant, Eily O'Connor (sop), the Colleen Bawn. Mrs Cregan (con), without realizing the consequences of her action, passes to Danny Mann (bar) a glove from Hardress: this is the signal for Danny to kill Eily. He takes her out in a boat and pushes her under the water, but she is rescued by Myles na Coppaleen (ten). Corrigan (bs), believing Eily dead, has Hardress arrested. However, Myles and Eily suddenly appear and the truth about her marriage is disclosed.

**Lincoln Center.** The New York arts complex which houses the State T (home of the New York City O) and the *Metropolitan Opera House, also Avery Fisher Hall and the Juilliard School.

**Lincoln's Inn Fields Theatre.** The first theatre on the London site was the Duke's T, opened 1661 with *The Siege of Rhodes*. The second, 1695–1705, saw the first public *Dido and Aeneas*, 1700. The third, built by Christopher Rich and opened by his son John in 1714, saw the first *Beggar's Opera*, 1728. It was the home of the Opera of the Nobility, the Italian opera company set up in rivalry to Handel with Porpora as composer and Senesino as singer, 1733–4. Handel's last opera, *Deidamia*, was given there 1741. Later became a barracks; site now occupied by the Royal College of Surgeons.

**Lind, Jenny** (*b* Stockholm, 6 Oct. 1820; *d* Wynds Point, 2 Nov. 1887). Swedish soprano. Studied Stockholm with I. Berg. After many appearances as a child, she made her operatic début Stockholm 1838 (Agathe). By 1841 she had sung Euryanthe, Lucia, Norma, Donna Anna, and Julia (*La vestale*), and was in vocal trouble. Further study Paris with M. García jun. Returned triumphantly to Stockholm 1842–4. At Berlin, 1844, she was to have created Vielke (written specially for her) in Meyerbeer's *Ein Feldlager in Schlesien*; it was given instead to the soprano Tuczec. However, her Norma won high acclaim, as did her subsequent performances as Vielke. After further triumphs in Germany and Vienna, she was engaged by Lumley for Her Majesty's, London, 1847, and sang there (creating Amalia in *I masnadieri*) until her retirement from the stage in 1849. Thereafter she performed only in concert. She married the pianist Otto Goldschmidt in 1852, and lived in England from 1858 until her death. One of the 19th cent.'s most famous prima donnas, and known as 'the Swedish nightingale', she was inordinately idolized by the general public, and much admired by Mendelssohn, Hans Andersen, the Schumanns, Prince Albert, and Hanslick. The quality of her voice was remarkably fresh, pure, and sympathetic. Its upper register (c''–g''') was especially brilliant and powerful, and though its lower half, according to Chorley, was 'husky, and apt to be out of tune', its deficiencies were disguised by her skill in vocal colouring, while her breath control was extraordinary. Inimitable in roles of pathos and innocence, she was a celebrated Amina, Marie (*Fille du régiment*), and Alice (*Robert le diable*); however, her Norma, lacking in fire and grit, was not a great success in England.

BIBL: J. Bulman, *Jenny Lind* (London, 1956).

**Linda di Chamounix.** Opera in 3 acts by Donizetti; text by G. Rossi. Prod. Vienna, K, 19 May 1842, with Tadolini, Brambilla, Moriani, Varesi, Dérivis, Rovere; London, HM, 1 June 1843, with Persiani, Brambilla, Mario, Lablache; New York, P, 4 Jan. 1847, with Barili, Pico, Benedetti, Sanquirico, Beneventano, Riese, cond. Barili.

France, c.1760. When a tenant farmer, Antonio (bar), and his wife Maddalena (sop) find themselves in financial difficulties, their landlord's brother, the Marchese de Boisfleury (bar), offers to help by educating their daughter Linda (sop) at his château. However, the Prefect (bs) warns against this scheme, for the Marchese has designs on Linda, and she and some friends go to Paris in search of work instead.

The painter Carlo (ten) wishes to marry Linda, but his mother disapproves of the match since he is, in fact, the Visconti di Sirval, and the nephew of the Marchese. Although she does not yield to him, Linda occupies an apartment owned by Carlo. Misunderstanding the situation, Antonio curses her. When Pierotto (con) tells Linda that Carlo is to marry a rich girl this causes her to lose her reason.

Linda has returned to Chamounix and Carlo arrives there to look for her, having rejected his mother's choice of a match, and persuaded her to give her blessing to his marriage to Linda.

**Lindoro.** 1. A young Italian (ten), lover of Isabella, in Rossini's *L'italiana in Algeri*.

2. Assumed name of Count Almaviva in Rossini's *Il barbiere di Siviglia*.

**Lindpaintner, Peter von** (*b* Koblenz, 9 Dec. 1791; *d* Nonnenhorn, 21 Aug. 1856). German composer. Studied Munich with Winter. He became conductor at the new Isartortheater in Munich, 1812–19, also renewing his studies, with Grätz. When the opening of the Munich Hoftheater put the Isartortheater into difficul-

ties, he was forced to resign, and moved to Stuttgart; here he remained until his death, much admired as a conductor, including by Mendelssohn and Berlioz.

Lindpaintner's early operas are in Singspiel manner; later he became interested in Grand Opera and especially in Romantic opera, to which he made an interesting contribution. *Der Bergkönig* shows the strong influence of Weber, especially of *Euryanthe* in the sensational use of tremolo and the appeal to the spirits of darkness. *Der Vampyr* (on the same subject as Marschner's opera, but with a different librettist) is still more Weberian, particularly in its use of the popular polacca rhythm, its chromatic harmony, a Cavatine in the manner of Aennchen, and a Bridesmaids' Chorus; but it also anticipates Wagner with its powerful use of running figures under tense triplets, its choral prayer (suggesting that in *Lohengrin*) when the heroine Isolde is abducted, and the Paris-influenced collapse of a palace in ruins. *Die Macht des Liedes* represents a reconciliation with Singspiel, and shows Lindpaintner's gift with simple song forms, where he was really most at home. In *Die sizilianischen Vesper*, described as a 'grosse romantische Oper', his talent and range fail to match his intelligent understanding of what Grand Opera and German Romantic opera had to offer each other.

SELECT WORKLIST (all first prod. Stuttgart): *Der Bergkönig* (The Mountain King) (Hanisch; 1825); *Der Vampyr* (Heigel; 1828); *Die Amazone* (Robert; 1831); *Die Macht des Liedes* (The Power of Song) (Castelli; 1836); *Die sizilianischen Vesper* (Rau; 1843).

**Linley, Thomas** (*b* Badminton, 17 Jan. 1733; *d* London, 19 Nov. 1795). English singing teacher and composer. Studied Bath with Chilcott. Wrote *The London Merchant* (London, CG, 1767). Joint director London, DL, 1774–94. Composed, mostly for DL, a large number of operas, ballad operas, pantomimes, and farces. The best-known was *The Duenna* (CG, 1775), a composition and compilation of music for his son-in-law Sheridan's play in collaboration with his eldest son **Thomas** (*b* Bath, 5 May 1756; *d* Grimsthorpe, 5 Aug. 1778). A friend of Mozart, Thomas jun. showed precocious musical gifts, but was drowned in a boating accident, leaving a 3-act opera *The Cady of Bagdad* (1776), an oratorio, much violin music, and some songs.

Thomas sen. had 12 children, several of them becoming musicians. His eldest daughter **Elizabeth** (*b* Bath, 5 Sep. 1754; *d* Bristol, 28 June 1792), was a gifted soprano; début London, CG, 1767. She eloped with Sheridan 1772 and married him 1773. **Mary** (1758–87) and **Maria** (1763–84) were also sopranos, and the youngest, **Jane**, was an amateur soprano who married

Charles Ward, secretary of DL. Of their brothers, **Ozias** (1765–1831) was an organist and **William** (1771–1835) wrote some unsuccessful operas and shared unsuccessfully the DL management with Sheridan in the late 1790s. Gainsborough painted portraits of several of the family.

BIBL: C. Black, *The Linleys of Bath* (London, 1911, 3/ 1971); M. Bor and L. Clelland, *Still the Lark* (London, 1962).

**Lionel and Clarissa.** Opera in 3 acts by Dibdin; text by Isaac Bickerstaffe. Prod. London, CG, 25 Feb. 1768; Philadelphia 14 Dec. 1772.

**Lipkowska, Lydia** (*b* Babino, 6 June 1882; *d* Beirut, 22 Mar. 1958). Bessarabian soprano. Studied St Petersburg with Iretskaya. Début St Petersburg, M, 1906; St Petersburg, M, until 1908, 1911–13; Paris: C, O, and OC, 1909; New York, M, 1909–11; Chicago 1910; London, CG, 1911–12; Petrograd, T of Musical Drama, 1913–15; Rome 1915; left Russia 1919, continuing appearances in Europe and USA; USSR tour 1927–9; Odessa 1941. Retired to the Lebanon. An appealing singer with a bright, pure tone. Roles incl. Rosina, Lucia, Gilda, Violetta, Mimì, Snow Maiden, Tatyana, Iolanta, Susanna (Wolf-Ferrari). (R)

**Lisa.** The Countess's granddaughter (sop), lover of Hermann, in Tchaikovsky's *The Queen of Spades*.

**Lisbon** (Port., Lisboa). Capital of Portugal. Opera was first given regularly in the 18th cent. in the Royal Opera di Tejo, also at the Academia da Trinidade. The Teatro dos Paços da Ribeira opened on 31 Mar. 1755 with *Alessandro nell'Indie* (1755) by David Pérez (1711–88), but was destroyed in the earthquake that year. Opera was given at the T d'Ajunda and T do Bairro Alto, until 1792, when a group of business men built a replica of the Naples San Carlo. The T São Carlos (cap. 1,100) opened on 30 June 1793 with Cimarosa's *La ballerina amante*. In Dec. 1794 *A vingança da cigana*, by António Leal Moreira (1758–1819), became the first opera sung in Portuguese. The history of the theatre was that of a first-class Italian opera-house outside Italy, with occasional excursions into the French and German repertories. Though on the international circuit for famous singers, it declined after about 1910. It reopened as the Teatro Nacional de São Carlos (cap. 1,148) in 1940 with Rui Coelho's *Dom João*, but really regained its reputation under João Freitas Branco from 1970. The Portuguese OC operated at the T da Trinidade until 1974. The 1974 Revolution led to reduced prices and greater informality at the Nacional. Since 1981 it has had its own resident company, and in

that year João Paes, who had upheld dramatic and musical values, was replaced by José Serra Formigal, whose interests have been more in great singers. Wolf-Siegfried Wagner has staged controversial productions of his great-grandfather's works. Opera is also given in the Coliseu dos Recreios (cap. 7,000).

**Lisenko, Nikolay.** See LYSENKO, MYKOLA.

**Lisinski, Vatroslav** (*b* Zagreb, *bapt.* 8 July 1819; *d* Zagreb, 31 May 1854). Croatian composer. Studied Zagreb with Sojka and Wisner von Morgenstern. His first opera, *Ljubav i zloba*, was also the first Croatian opera. Its success encouraged him to devote himself wholly to music, and he studied further in Prague with Kittl and Pitsch. Returning to Zagreb, he found it impossible to make a living in music, and became a clerk. However, he left a second opera, *Porin*, in which he not only makes use of Croatian folk music but shows an awareness of Glinka's example in giving such ingredients a wider context and further makes some use of Leitmotiv.

WORKLIST: *Ljubav i zloba* (Love and Malice) (Demeter; Zagreb 1846); *Porin* (Demeter; comp. 1851, prod. Zagreb 1897).
BIBL: L. Županović, *V. Lisinski (1819–54), život—djelo—značenje* (Zagreb, 1969).

**Lisitsyan, Pavel** (*b* Vladikavkaz, 6 Nov. 1911). Armenian baritone. Studied Leningrad Cons. Début Leningrad 1935. Moscow, B, from 1940. New York, M, 1960; the first Soviet singer to appear there. Also San Francisco, Prague, Budapest. An impressive artist with a particularly warm and beautiful voice. Repertory includes Escamillo, Amonasro; also Onegin and many other Russian roles. (R)

**List, Emanuel** (*b* Vienna, 22 Mar. 1888; *d* Vienna, 21 June 1967). Austrian, later US bass. Studied New York with Zuro. Début Vienna, V, 1922 (Gounod's Méphistophélès). Berlin: SO, 1924; S, 1925–33. London, CG, from 1925; Bayreuth 1933; New York, M, 1935–50; also Chicago, Buenos Aires, Salzburg. Possessed a deep, full tone and a strong stage presence. Roles included Osmin, Rocco, Pogner; especially distinguished as Hunding, Hagen, and Ochs. (R)

**Liszt, Franz** (Ferenc) (*b* Raiding, 22 Oct. 1811; *d* Bayreuth, 31 July 1886). Hungarian composer, pianist, and conductor. His sole completed opera was *Don Sanche*, a 1-act piece written perhaps with some help over orchestration from his teacher Paer. It was produced in Paris in 1825, with Nourrit in the title-role, and after four performances lay neglected until modern times. On a fashionable subject of emotional trials and magic confusions, the work shows, unsurprisingly, little sign of individuality but includes

# Lithuania

some effective music (including for a storm). Of other operatic projects, the most ambitious, and one which absorbed Liszt for a considerable time, was *Sardanapale*.

Among Liszt's services to opera was his popularizing of works by many of his contemporaries with the piano transcriptions, paraphrases, and fantasias he played all over Europe during his years as a travelling virtuoso. In 1848 he became Kapellmeister in Weimar, where he championed, with characteristic generosity, many composers, including Wagner, whose *Lohengrin* had its première under him in 1850, and Berlioz, whose *Benvenuto Cellini* he gave in the revised 3-act version in 1852. He also gave the first performance of Schubert's *Alfonso und Estrella* in 1854, backing it up with a critical article, and of Rubinstein's *The Siberian Huntsmen* in the same year. Other pioneering productions included Schumann's *Genoveva* (1855). A plan to stage Wagner's as yet uncompleted *Ring* and make Weimar into what Bayreuth eventually became never materialized. Liszt's position eventually became difficult, partly for personal reasons, and matters came to head in his confrontations with the reactionary elements in the town over Cornelius's *The Barber of Bagdad* in 1858: he felt compelled to resign. His daughter Cosima married first Hans von Bülow, then Wagner.

WORKLIST: *Don Sanche* (Théaulon de Lambert & Rancé, after Florian; Paris 1825).

BIBL: A. Walker, *Franz Liszt: The Virtuoso Years 1811–1847* (New York, 1983); *Franz Liszt: The Weimar Years, 1848–1861* (London, 1989).

**Lithuania.** After the Lublin Union of 1569, Lithuania and Poland formed a united federative state, with the ruler as King of Poland and Grand Duke of Lithuania. Opera was first given on 4 Sept. 1636 in the well-appointed Court T of Vilnius Castle (*Il ratto di Helena*, set to a text by Virgilio Puccitelli by an unknown composer). A company of 90 was directed by Marco Scacchi (1602–85), playing in both capitals, Vilnius and Warsaw. Some ten court theatres were active in the latter half of the 18th cent., giving Italian opera. Outstanding was the Oginskis' theatre in Slonim (1764–84), and especially that of the Radvila family (Pol.: Radzwiłł) in Nesvyžius (Pol., Nieswież), where opera was given by amateurs from 1736, professionals 1756–1809.

The first public theatres were opened at Vilnius 1785 (by Poles) and Klaipėda (by Germans); there was also a German theatre in Vilnius 1835–44. From 1864 there was no Vilnius opera, but Italian and Russian companies visited, the latter giving a wide repertory. Political oppression and other difficulties inhibited the development of Lithuanian national opera dur-

ing the 19th cent., though there was clandestine pressure and the first libretto, Maironis's *Kame išganymas ?* (Where is Salvation?), was written 1895. The first Lithuanian opera was *Birutė* by Mikas Petrauskas (1873–1937): he then emigrated to the USA, where his *Eglė žalčių karalienė* (Egle the Snake Queen) was staged Boston 1924 (rev. Kaunas 1939).

After Lithuanian independence (1918), the Creative Artists' Society organized opera and drama companies, the former opening 31 Dec. 1920 in Kaunas (capital, 1920–39), with *Traviata*. Petrauskas's brother Kipras (1885–1968) was director, and a fine tenor, and Juozas Tallat-Kelpša (1889–1949) was conductor. The companies became the Valstybės teatras (State Theatre) in 1925. Italian and French opera dominated, and the company made a good reputation, with artists including Shalyapin, Coates, and Malko. Lithuanian operas included *Gražina* (1933) and *Radvila Perkūnas* (Radvila the Thunderer, 1937) by Jurgis Karnavičius (1884–1941), *Trys talismanai* (The Three Talismans, 1936) by Antanas Račiūnas (1905–84), and *Pagirėnai* (1942) by Stasas Šimkus (1887–1943).

Under German and Russian occupation, activity continued. Opera was given in Vilnius from 1941, operetta in Kaunas 1940 and Šiauliai 1942; the Russians closed those in Vilnius and Šiauliai. More Lithuanians emigrated to the USA, and the Lithuanian O of Chicago opened 1957 under Darius Lapinskas (*b* 1934) with *Rigoletto*; singing in Lithuanian, this has included in its repertory new Lithuanian operas by Julius Gaidelis (1909–83) (*Dana*, 1969) and Kazimieras Banaitis (1896–1963) (*Jūratė and Kastytis*, 1972). In 1948 the Kaunas company moved to Vilnius; a third of the repertory was Russian, and political pressures hampered development. Račiūnas's *Marytė* was produced 1953, and a few other works satisfied these pressures. Composers who took a more independent line included (at Vilnius) Vytautas Klova (*b* 1926) with *Pilėnai* (1956), *Vaiva* (1958), and *Du kalavijai* (Two Swords, 1966); Vytautas Paltanavičius (*b* 1924) with *Kryžkelėje* (At the Crossroads, 1967); Vytautas Barkauskas (*b* 1931) with *Legenda apie meilę* (The Legend of Love, 1975); Julius Juzeliūnas (*b* 1916) with *Sukilėliai* (The Insurgents, comp. 1957; banned; prod. 1977); Boris Borisov (*b* 1937) with *Piršlybos* (Matchmaking, 1983); Algimantas Bražinskas (*b* 1937) with *Kristijonas* (1984); and at Kaunas, Vitolis Baumilas (*b* 1928) with *Paskenduolė* (The Drowned Girl, 1957) and Benjaminas Gorbulskis (1925–86) with *Frank Kruk* (1959). At Klaipėda a Folk O began in 1956, becoming the professional Klaipėdos Muzikinis Teatras in 1986, and opening with *Mažvydas* (1988) by Audronė Žigaitytė (*b* 1957).

The increased activity has also included children's, TV, and radio operas, and latterly some rock operas e.g. *Komunarų gatvė* (Communard Street, 1979), by Faustas Latėns (*b* 1956).

BIBL: J. Bruveris, *Lietuvos operos ir baleto teatras* (Vilnius, 1991).

**Litolff, Henry** (*b* London, 7 Aug. 1818; *d* Bois-Colombes, 5 Aug. 1891). French composer, pianist, conductor, and publisher. Studied London with Moscheles. Though best known for his four piano *concertos symphoniques*, he also wrote a number of operas, including *Héloïse et Abélard* (Clairville & Busnach; Paris 1872) and *König Lear* (after Shakespeare & Holinshed; unprod.).

**Litvinne, Felia** (*b* St Petersburg, 11 Oct. 1860; *d* Paris, 12 Oct. 1936). Russian, later French, soprano. Studied Paris with Barthe-Banderali and Viardot. Début Paris, I, 1883 (Amelia, *Boccanegra*). New York, Mapleson Co., 1885–6; Paris, O, from 1889; Milan, S, 1890; New York, M, 1896–7; London, CG, 1899, 1902, 1905, 1907, 1910. Also St Petersburg; Brussels, M; Monte Carlo. Retired from stage 1916. An impassioned singer with a large, brilliant, and supple voice, she was famous as Alceste, Donna Anna, Aida, Isolde, and Brünnhilde, among many other roles. Her sister Hélène was married to bass Edouard \*De Reszke. Autobiography, *Ma vie et mon art* (Paris, 1933). (R)

**Liù.** A slave girl (sop), in love with Calaf, in Puccini's *Turandot*.

**Liverpool.** City in Lancashire, England. Largely dependent on touring companies, the city has also fostered local amateur groups. The most notable has been the Liverpool Grand OC, which has staged the British premières of Bizet's *Ivan IV* (1956) and Gounod's original *Mireille* (1981), as well as productions of *Hérodiade*, *Macbeth*, *La Gioconda*, and *Eugene Onegin*. Beecham staged *The Bohemian Girl* as part of the Festival of Britain in 1951. In 1971 the university gave the première of Elaine Murdoch's *Tamburlaine*.

**Livigni, Filippo** (*fl.* 1773–85). Italian librettist. He was one of the most successful Venetian comic librettists. Of his 13 works, the most famous were *La molinara* (Fischietti, 1778), *Il convito* (Cimarosa, 1782), and *Il due castellani burlati* (Valentini, 1785): these were set many times in the 18th cent. and followed the model of Goldoni.

**Ljubljana** (Ger., Laibach). Capital of Slovenia. Italian companies visited from 1652, regularly from 1740; from 1768 Germany companies also came. Opera was given in Count Auersperg's palace, from 1765 in the Staleško Gledališče (Estates T) (cap. 800; burnt down 1887). Visitors included Schikaneder's company. German influence was increasingly strong, and opera was given both at the Landestheater and the Dramatično Društvo (Dramatic Society), the latter giving rise to the Slovene O. In 1892 there was built a new theatre, the Slovensko Deželno Gledališče (Slovene Regional Theatre), which the ensembles shared. The musical director 1886–95 was the energetic and talented Fran Gerbič; he was followed by Hilarije Benišek 1895–1910, and by Vaclav Talich 1909–12. The theatre (cap. 700) is now the Opera of the Slovene National T, giving a season from Sep. to June. Opera is also given in the summer in the Krizanke, a former monastery.

**Ljungberg, Göta** (*b* Sundsvall, 4 Oct. 1893; *d* Lidingö, 28 June 1955). Swedish soprano. Studied Stockholm with Bratt. Début Stockholm 1918 (Elsa). Berlin, S, 1926. London, CG, 1924–9; New York, M, 1932–5, when she retired. Created roles in operas by Goossens, D'Albert, and Hanson. With her great beauty, clear voice, and dramatic acting, a much admired Salome; also sang Elektra, Chrysothemis, Brünnhilde, Sieglinde, Kundry. (R)

**Lloyd, Robert** (*b* Southend-on-Sea, 2 Mar. 1940). English bass. Studied London with O. Kraus. Début London, University Coll. O, 1969 (Minister, Beethoven's *Leonore*). London, SW, 1969–72, CG, from 1972; also Paris, Boston, etc. Repertory includes Sarastro, Gurnemanz, Leporello, Don Giovanni, Boris. (R)

**Lloyd-Jones, David** (*b* London, 19 Nov. 1934). Welsh conductor, translator, and editor. Studied Oxford. Coached *Boris* in Russian, London, CG, 1959. New OC 1960–4. Scottish O 1967; WNO 1968; London, SW/ENO, from 1969; London, CG, 1971; first artistic director O North 1977–90, greatly contributing to its success and developing a wide repertory. Cond. first British staged *War and Peace*. A specialist in Russian opera, his translations incl. *Onegin* and Rakhmaninov's *Francesca da Rimini*. His authoritative edition of *Boris*, using the full score of the original versions, has been welcomed and performed in Russia. (R)

**Lloyd Webber, Andrew** (*b* London, 22 Mar. 1948). English composer. Studied London at the GSM and RCM. In 1968 he brought out *Joseph and the Amazing Technicolour Dreamcoat*, for which the lyrics were provided by his schoolfriend Tim Rice. An instant success, it encouraged Lloyd Webber and Rice to another musical based on a biblical theme, *Jesus Christ, Superstar*. This first became known through the recording of the hit song, which paved the way for a stage show and eventually the musical itself, running in London for 3,358 performances. Further successes were achieved with *Evita* (a dramatization of the life of Eva Perón) and *Cats*; for *Starlight Express* and *The Phantom of the Opera* Lloyd Webber was joined by different collaborators, but managed to repeat his earlier triumphs.

In contrast to earlier musicals, Lloyd Webber's works have a tendency to rely on spectacular staging, seen most notably in *Starlight Express*, rather than musical individuality, although his best numbers, such as 'Don't Cry for Me Argentina', have earned a place in the repertory of popular song. He has also written a mini-opera for television, *Tell Me on a Sunday*.

SELECT WORKLIST: *Joseph and his Amazing Technicolour Dreamcoat* (Rice; London 1968); *Jesus Christ Superstar* (Rice; New York 1971, rev. London 1972); *Evita* (Rice; London 1978); *Tell Me on a Sunday* (Don Black; London 1980); *Cats* (Rice, after T. S. Eliot; London 1981); *Starlight Express* (Stilgoe; London 1984); *The Phantom of the Opera* (Hart; London 1986).

BIBL: M. Walsh, *Andrew Lloyd Webber: His Life and Works* (London, 1989).

**Locke, Matthew** (*b* ?Devon, 1621 or 1622; *d* London, Aug. 1677). English composer. He contributed the fourth entry to Davenant's *The Siege of Rhodes* and sang the role of the Admiral of Rhodes in the first performance. *Cupid and Death* (1653, part written by Christopher Gibbons) was described by Locke as a masque, although it resembles a full opera in having songs, recitative, short choruses, and dances, all by Locke. The first production was a private performance in 1653. Locke revised the work, adding music, before it received its first public performance in 1659. The score of *Cupid and Death* is the only complete extant copy of a 17th-cent. English masque.

After the Restoration in 1660, contact with French theatre increased and performances were given in London in 1673–4. Locke's *Psyche* (1675) was modelled on Molière's play and clearly emulates the form of the tragédie-ballet. It contains more music than any of Locke's other stage works (although the act tunes are by Draghi), but it is largely dull. The recitative in particular lacks the interest and variety of that in *Cupid and Death*, when Locke may have been influenced by music heard while travelling in the continent in 1648. Nevertheless, *Psyche* was sufficiently successful to inspire a parody, *Psyche Debauch'd*, at the rival Theatre Royal in 1674, and for Locke to publish a collection of vocal numbers from the work, under the title 'The English Opera', in 1675. Locke also wrote incidental music for many plays, including songs for Shadwell's famous version of *The Tempest*. Roger North comments that Locke 'composed to the semioperas divers pieces of vocall and instrumentall entertainment, with very good success'.

**Lockhart, James** (*b* Edinburgh, 16 Oct. 1930). Scottish conductor and pianist. Studied London. London: SW, 1960–2; CG, 1962–8; EOG, 1967. Music director WNO 1968–73, contributing much to its establishment and progress; music director Kassel 1972–80, also Koblenz O. London, ENO from 1984 (incl. at New York, M); also Hamburg, Munich, Florence, etc. Repertory incl. *Ring*, *Trovatore*, *Boris*, first *Lulu* by British co. (WNO), *War and Peace*, *Grimes*, first *Bear* (Walton). (R)

**Loder, Edward** (*b* Bath, 1813; *d* London, 5 Apr. 1865). English composer. Studied Frankfurt with Ries. His first opera, *Nourjahad*, opened the English Opera House (Lyceum) in 1834, and was acclaimed as a founding work of English opera. Other stage works followed, some little more than separate songs linked with dialogue, some ballad operas or something approaching English Singspiel. *The Night Dancers* was a characteristic example of English fairy opera, and like others failed to survive the satire of Sullivan's *Iolanthe*. His most important work was *Raymond and Agnes*, which reveals his melodic gift and his command of vivid orchestration, as well as a certain dramatic power. In more favourable conditions, and with stronger librettos, he might have achieved considerable stature as a composer of English Romantic opera. His sister **Kate Loder** (*b* Bath, 21 Aug. 1825; *d* Headley, 30 Aug. 1904) was a pianist and composer: her works include an opera *L'elisir d'amore* (1844).

SELECT WORKLIST: *Nourjahad* (Arnold; London 1834); *The Night Dancers* (Soane; London 1846); *Raymond and Agnes* (Fitzball, after Lewis; London 1855).

**Lodoïska.** Opera in 3 acts by Cherubini; text by C. F. Fillette-Loraux, after Louvet de Couvray's *Les amours de Faublas*. Prod. Paris, T Feydeau, 18 July 1791 (a fortnight before R. Kreutzer's opera on the same subject, prod. Paris, CI, 1 Aug. 1791); New York 4 Dec. 1826; London, Univ. Coll., cond. Addison.

Poland, 1600. Count Floreski (ten) goes in search of Lodoïska (sop), whom he wishes to marry. She is the prisoner of Dourlinski (bar), who also plans to wed her. The Tartars, under Titzikan (ten), attack the castle where she is held, enabling Floreski to rescue her.

Other operas on the subject are by Storace (pasticcio using music by Cherubini, Kreutzer, and Andreozzi, 1794), Mayr (1796), Caruso (1796), Paer (1804), Bishop (adapting Storace, 1816), Curmi (1846), and Succo (1849).

**Lodoletta.** Opera in 3 acts by Mascagni; text by Forzano, after Ouida's novel *Two Little Wooden Shoes* (1874). Prod. Rome, C, 30 Apr. 1917, with Storchio, Campioni, Molinari, cond. Mascagni;

New York, M, 12 Jan. 1918, with Farrar, Caruso, Amato, cond. Moranzoni.

Holland and Paris, 19th cent. Antonio (bs) gives Lodoletta (sop) a pair of new red shoes. After his death she falls in love with a painter, Flammen (ten), and follows him from Holland to Paris. She is afraid to enter his house, where a party is in progress, and dies in the snow. Flammen finds her and laments that he has always loved her.

Also opera on the subject by Hubay (*Moharósza*, 1903).

**Łódź.** City in Poland. Opera was first heard in the 1860s, when Polish operas and lighter works were given by companies from Warsaw and abroad. An unsuccessful attempt to establish a permanent company was made in 1888. Even after 1918, operatic life was dependent on guest performances. An operetta theatre opened in 1946, and on 18 Oct. 1954 the Society of Friends of Opera inaugurated the Łódź Opera. The new T Wielki (Grand T) (cap. 1,300: the largest in Poland after Warsaw) opened on 19 Jan. 1867. The company has built up a good reputation; it has given premières of works by Romuald Twardowski (*Lord Jim*, 1976; *Mary Stuart*, 1981) and Henryk Czyż and revived *Filenis* (1897) by Roman Statkowski (1859–1925).

**Loewe, Carl** (*b* Loebjuen, 30 Nov. 1796; *d* Kiel, 20 Apr. 1869). German composer and singer. Though famous for his songs, especially his ballad settings, Loewe also wrote some sacred and instrumental music and six operas, two based on Scott (*Malekhadel* (1832) on *The Talisman*, *Emmy* (1842) on *Kenilworth*).

**Loewe, Sophie** (*b* Oldenburg, 24 Mar. 1815; *d* Budapest, 28 Nov. 1866). German soprano. Daughter of the actor Ferdinand Löwe. Studied Vienna with Ciccimarra, Milan with Lamperti. Début Vienna, K, 1832 (Elisabetta in Donizetti's *Otto mesi in due ore*). Berlin 1837; London 1841; Milan, S, 1841; Venice 1843–4. A highly strung, dramatic singer, she created the title-role of Donizetti's *Maria Padilla* (written for her), Elvira in *Ernani* (when she sang badly, possibly annoyed that Verdi, striking a blow for composers' independence, had refused to write the usual rondo-finale for her), and Odabella in *Attila*. She retired in 1848 on her marriage to Prince Ferdinand of Liechtenstein; by then her voice was already in decline.

**Loewenberg, Alfred** (*b* Berlin, 14 May 1902; *d* London, 29 Dec. 1949). German, later British, musical historian. Studied Berlin and Jena. Emigrated to England 1935. His *Annals of Opera: 1597–1940* (Cambridge, 1943, rev. 3/1978) is a painstaking and invaluable compilation of performance details of some 4,000 operas in chronological order.

**Loge.** The God of Fire (ten) in Wagner's *Das Rheingold*.

**Logroscino, Nicola** (*b* Bitonto, *bapt.* 22 Oct. 1698; *d* ?Palermo, 1765). Italian composer. Studied Naples with Veneziano, Perugino, and Mancini; expelled from his conservatory for vice 1727. Organist to the Bishop of Conza 1728–31, also becoming known as an opera composer: especially successful were *Quinto Fabbio* and *Inganno per inganno*.

Though records suggest that he was one of the most important and prolific composers of Neapolitan comic opera in the 1730s, 1740s, and 1750s, only the music for *Il governatore* survives. This suggests that his description as 'il dio dell'opera buffa' ('the god of opera buffa') was justified, and is important for the development of the *ensemble finale.

SELECT WORKLIST: *Quinto Fabbio* (Salvi; Rome 1738, lost); *Inganno per inganno* (Deceit for Deceit) (Federico; Naples 1738, lost); *Il governatore* (Canicà; Naples 1747).

**Lohengrin.** Opera in 3 acts by Wagner; text by the composer, after the anonymous German epic. Prod. Weimar, Court T, 28 Aug. 1850, with Agthe, Fastlinger, Beck, Von Milde, Höfer, cond. Liszt; New York, Stadt T, 3 Apr. 1871, with Lichtmay, Friderici, Habelmann, Vierling, Franosch, Formes, cond. Neuendorff; London, CG, 8 May 1875, with Albani, D'Angeri, Nicolini, Maurel, Seidemann, cond. Vianesi.

King Henry the Fowler (bs), who has been visiting Antwerp to raise an army, holds court. He asks Frederick of Telramund (bar) why the kingdom of Brabant is torn by strife. Telramund accuses his ward Elsa (sop) of having murdered her young brother Gottfried in order to obtain the throne. Elsa describes a dream in which a knight in shining armour has come to defend her. The King's Herald (bar) twice calls for a champion. A swan-drawn boat bearing a knight in shining armour arrives. The knight (Lohengrin, ten) bids the swan farewell, and agrees to champion Elsa, offering her his hand in marriage on condition that she will never ask him his name or origin. Lohengrin defeats Telramund, generously sparing his life.

In the courtyard of the castle in Antwerp, Telramund, who has been banned as a traitor by the King, and his wife Ortrud (sop), are brooding on the state of events. Elsa appears on a balcony and sings a song to the night breezes. She descends, and Ortrud, offering her friendship, begins to sow distrust of Lohengrin in her mind. Dawn breaks, and processions form for the marriage of Elsa and Lohengrin. On the steps of the Cathedral, Ortrud accuses Lohengrin of having defeated Telramund by evil means, and then Telramund repeats his wife's accusation. Elsa assures the knight that she trusts him; but the seeds of suspicion have taken root.

A brilliant orchestral prelude and the celebrated Wedding March open a scene set in Elsa's bridal chamber. Elsa's happiness gives way to hysteria and she demands to know her husband's name. Telramund and four of his followers break into the room to attack Lohengrin, who immediately kills Telramund. He bids the nobles to bear the body to the King, and tells Elsa that he will reveal his secret to them all. The scene changes to the banks of the Scheldt. The King and court assemble, and Lohengrin tells them that he has come from the Temple of the Holy Grail in Monsalvat; his father was Parsifal, and Lohengrin is his name. He bids Elsa a sad farewell, and then turns to greet the swan which has brought the boat for him. Ortrud rushes on and reveals that the swan is in reality Gottfried, Elsa's brother. Lohengrin falls on his knees and prays. The swan becomes Gottfried, and a white dove of the Grail flies down and draws the boat away.

**Lokalposse.** See POSSE.

**Lola.** Alfio's wife (sop), lover of Turiddu, in Mascagni's *Cavalleria rusticana*.

**Lombardi alla prima Crociata, I** (The Lombards at the First Crusade). Opera in 4 acts by Verdi; text by Solera. Prod. Milan, S, 11 Feb. 1843, with Frezzolini-Poggi, Ruggeri, Guasco, Severi, Dérivis, Rossi, Marconi, Vairo, Gandaglia; London, HM, 12 May 1846, with Grisi and Mario; New York, Palmo's OH, 3 Mar. 1847 (first Verdi opera in USA).

Milan, Antioch, Jerusalem, late 11th cent. Two brothers, Pagano (bs) and Arvino (ten), both loved Viclinda (sop). She preferred Arvino but on the day of her wedding to him, Pagano tried to kill him. At the start of the opera, Pagano returns from exile to Milan and seems to be reconciled with his brother and Viclinda, and their daughter Giselda (sop). Still seeking revenge on his brother, however, Pagano plans an attack but murders his father by mistake. He is exiled to the Holy Land where he lives as a hermit.

Giselda accompanies her father on a crusade and is captured by Acciano (bs), the tyrant of Antioch. His son Oronte (ten) is in love with Giselda and wants to espouse her Christian faith. In a battle between the Christians and Moslems, it appears that Oronte is killed. Giselda curses her people for allowing the war, so her enraged father tries to kill her. Pagano, whom neither of them now recognizes, prevents him.

Mortally wounded, Oronte escapes with Giselda. He is baptized by Pagano shortly before dying in Giselda's arms. Pagano leads an attack on Jerusalem and is also wounded. He discloses his identity to his brother who forgives him.

*I Lombardi* was adapted by Verdi for the Paris Opéra in 1847. The libretto was refashioned in French by Gustav Vaëz and Alphonse Royer as *Jérusalem*. In this version Roger (bs), the uncle of Hélène (sop), engages a mercenary to kill Gaston (ten), to whom she is betrothed. By mistake he kills her father, the Count of Toulouse. Gaston is blamed but he and Hélène are reunited at the end when Roger confesses his guilt. Prod. Paris, O, 26 Nov. 1847, with Van Gelder, Duprez, Portheaut, Alizard, Bremont; New Orleans, T d'Orléans, 24 Jan. 1850, with R. De Vriès, Duluc, Bessin, Corradi, Galinier, cond. E. Prévost.

**London.** Capital of Great Britain. Masques were popular in the 17th cent., but opera began with *The Siege of Rhodes*, at Rutland House in Sep. 1656. Blow's *Venus and Adonis* was given during the 1680s, and many other works followed at the city's many theatres. In 1689 Purcell's *Dido and Aeneas* was given in Chelsea. After his death, opera in London was for long almost entirely Italian. The first operatic work sung in Italian was the *pastorale *Gli amori d'Ergasto* (1705) by Jakob Greber (?–1731), and the first complete Italian opera given was Bononcini's *Camilla* (1706), although this was sung in English. In 1710 Handel arrived, producing almost 40 operas between *Rinaldo* (1711) and *Deidamia* (1741). The first Mozart opera heard was *La clemenza di Tito* (1806); and his operas, together with Rossini's, dominated repertories during the 1820s. The popularity of *Der Freischütz* (of which six versions were current in the 1820s) led to the invitation to Weber to compose and conduct *Oberon* (1826); this helped to encourage native composers to write for London theatres, whose taste remained largely centred on pasticcios or other light musical entertainments. Italian opera remained popular. Wagner was first heard regularly in the 1870s, the complete *Ring* in 1882. Opera developed rapidly after the turn of the century, and several new companies were formed. As London, with its long and powerful dramatic tradition, has always been well furnished with theatres, it is chiefly by way of their histories that the course of opera in the city is best followed.

See also ALMEIDA THEATRE, CAMDEN FESTIVAL, CHELSEA OPERA GROUP, COLISEUM THEATRE, COVENT GARDEN, DORSET GARDEN THEATRE, DRURY LANE THEATRE, HER MAJESTY'S THEATRE, KING'S THEATRE, LINCOLN'S INN FIELDS THEATRE, LONDON OPERA HOUSE, LYCEUM THEATRE, NEW OPERA COMPANY, OLD VIC THEATRE, PANTHEON,

ROYAL ITALIAN OPERA, SADLER'S WELLS, SAVOY THEATRE.

**London, George** (*b* Montreal, 30 May 1919; *d* Armonk, NY, 24 Mar. 1985). Canadian bass-baritone. Studied Los Angeles with Strelitzer and Stewart. Début (as George Burnson) Hollywood Bowl 1941 (Doctor, *Traviata*). Vienna 1949; Gly. 1950; Bayreuth 1951–64; New York, M, 1951–66; Milan, S, 1952; Moscow, B, 1960 (the first non-Russian to sing Boris there). Cologne 1962–4 (Wotan). Retired 1967. Opera producer from 1971, including *Ring*, Seattle and San Diego, 1973–5; Director Washington O Soc. from 1975. Prod. 1st US English-language *Ring*, Seattle 1975. Roles included Don Giovanni, Scarpia, Amfortas, Flying Dutchman, Mandryka. A dramatically intense performer with a rich and resonant voice. (R)

**London Opera House.** Built by Oscar *Hammerstein in 1911 and opened 13 Nov. with Nouguès's *Quo vadis?*, followed by premières of works by Massenet and Holbrooke. Closed 1912. As the Stoll, became a variety theatre and cinema. Short seasons of opera were given in the First World War, including the first English *Queen of Spades*. Seasons were resumed after the Second World War, from 1949, including by touring Italian and Yugoslav companies, also the first English *Porgy and Bess* (1951) and Honegger's *Jeanne d'Arc* with Ingrid Bergman (1954). Later demolished.

**Loreley.** Opera in 3 acts by Catalani; text by Zanardini and D'Ormeville. Prod. Turin, R, 16 Feb. 1890, with Ferni-Germano, Dexter, Durot, Stinco-Palermini, Pozzi, cond. Mascheroni; London, CG, 12 July 1907, with Scalar, Kurz, Bassi, Sammarco, Journet, cond. Campanini; Chicago, Auditorium, 17 Jan. 1919, with Fitziu, Macbeth, Dolci, Rimini, Lazzari, cond. Polacco. Orig. entitled *Elda*, in 4 acts, prod. Turin 31 Jan. 1880, with Garbini, Boulicioff, Barbacini, Athos, E. De Reszke, cond. Pedrotti.

The Rhine, *c.*1500. The orphan girl Loreley (sop) has fallen in love with Walter (ten), but he spurns her advances in favour of Anna (sop). Loreley makes a pact with the king of the Rhine, Alberich, promising herself to him if he will transform her into an irresistible seductress. As such she interrupts the nuptials of Walter and Anna; he falls for her and abandons Anna, who dies of grief. The Loreley is about to embrace Walter when spirits carry her away to her rock, thereafter to lure sailors to their deaths. In despair, Walter hurls himself into the Rhine.

Other operas on the subject are by Lachner (1846), Mendelssohn (unfin. 1847), Wallace (*Lurline*, 1860), Bruch (1863), Mohr (1884), Pacius (1887), Bartholdy (1887), E. Naumann (1889), and A. Becker (1898).

**Lorengar, Pilar** (*b* Saragossa, 16 Jan. 1928). Spanish soprano. Studied Barcelona Cons. Début Barcelona 1949 (in zarzuela). Aix 1955; London, CG, from 1955; Gly. 1956–60; Berlin, D, from 1958; New York, M, from 1966. Roles include Euridice (Gluck), Donna Elvira, Donna Anna, Countess (Mozart), Alice Ford, Eva, Mélisande. Particularly esteemed in the classical repertory. (R)

**Lorenz, Max** (*b* Düsseldorf, 17 May 1901; *d* Salzburg, 11 Jan. 1975). German tenor. Studied Berlin with Grenzebach. Début Dresden 1928 (Walther von der Vogelweide, *Tannhäuser*). New York, M, from 1931; Bayreuth, and Berlin, S, from 1933; London, CG, 1934; Vienna, S, from 1937; also Milan, S; Paris, O; etc. Created several roles, including Joseph K in Einem's *Der Prozess*. An expressive singer with exceptional declamation, famous for his Wagner roles, especially Siegfried, Walther; also an outstanding Otello (Verdi). (R)

**Lortzing, Albert** (*b* Berlin, 23 Oct. 1801; *d* Berlin, 21 Jan. 1851). German composer, conductor, librettist, singer, and actor. Son of an actor; studied Berlin with Rungenhagen, later by himself when obliged by his father's profession to travel extensively. He played children's parts on the stage as both actor and singer.

Lortzing's first opera was *Ali Pascha von Janina*, a conventional work which includes a strong vengeance aria making use of four trumpets. For this he wrote his own text, as he did throughout his career: he is the most important German composer before Wagner to do so. He followed this with some works that confirmed his attraction to Singspiel (he was a lifelong defender of opera with spoken dialogue) and his devotion to the language of Mozart: *Szenen aus Mozarts Leben* uses only Mozart's music. He then turned to comedy: *Die beiden Schützen* made a reputation later confirmed with *Zar und Zimmermann*. These two works, enduringly popular in Germany, display his characteristic vein of light tunefulness (deriving from the manner of German popular song), his easy control of the stage, his light-hearted, sentimental approach to character, and a distinct individuality in the handling of the conventional ingredients of German musical comedy. *Caramo* and *Hans Sachs* proved less durable. Nevertheless, *Hans Sachs* includes some attractive music, and is remarkable for some anticipations of Wagner's *Die Meistersinger von Nürnberg*: there are similarities in the handling of the characters and in the dramatic sequence of events which are not accounted for by the operas' common textual ancestry. His next work, *Casa-*

*nova*, contains suggestions of Leitmotiv. *Der Wildschütz*, which followed, was his greatest triumph, and remains popular in Germany; it is characteristic of Lortzing's easy, attractive musical manner in drawing on the German feeling for the countryside which Weber had first aroused in opera. It also involves a satirical treatment of the current devotion to the classics, which Lortzing always rejected as subjects for opera.

Lortzing was now also working as producer, singer, and conductor: he became Kapellmeister in Leipzig, 1844–5, but was forced to leave and suffered great difficulty in providing for his large family. His next work, *Undine*, was a magic opera on the popular Romantic legend. It includes some striking effects: the conclusion, with rising waters and a palace crashing in ruins, was a common inheritance of Parisian Grand Opera in German Romantic opera up to and including Wagner. The work also includes some early use of true Leitmotif.

In 1846 he moved to Vienna, where *Der Waffenschmied* was given. This strengthened his popularity, and at the same time confirmed the pattern of opera he had made his own—a lightly romantic comedy, to his own text, often turning on the disguise of an aristocrat as a workman, and expressed in engaging and tuneful songs culminating in rather stronger finales. *Zum Grossadmiral* followed; and then Lortzing attempted to capitalize on the revolutionary events of 1848 with *Regina*. Though the work contains some enterprising ideas, it is in no sense revolutionary musically but repeats Lortzing's well-tried formulas; and he discovered to his cost that the idea was in uneasy times found too subversive. He lost his post, but was nevertheless called to Leipzig to supervise his *Rolands Knappen*, an impressive work which has a heroine named Isolda and also contains some harmonic glimpses of the world of *Tristan*. However, Lortzing's chromatic excursions, though often advanced, are hardly ever more than decorative or incidental. An appointment in Leipzig fell through, and in 1850 he moved to Berlin as conductor of a small theatre. Here his family lived in poverty despite the frequent performance of his works, and he died when about to be dismissed from even this humble post. His last work was *Die Opernprobe*, which satirizes some of the stage and operatic conventions he knew so well, especially the Italianate recitative for which he retained a lifelong dislike.

Lortzing's music has a simple charm that fits it well to the kind of romantic comedy he preferred, though he can hardly be categorized as a Romantic: he was really always more concerned with effective singer types in his roles rather than subtlety of characterization. If not a sophisticated composer, and sometimes an over-sentimental one, he was capable of some remarkably enterprising strokes, and his works represent the most agreeable type of German comic opera.

WORKLIST (all texts by composer): *Ali Pascha von Janina* (comp. 1824, prod. Münster 1828); *Der Pole und sein Kind* (The Pole and his Child) (Osnabrück 1832); *Der Weihnachtsabend* (Christmas Eve) (Münster 1832); *Andreas Hofer* (after Immermann; comp. 1832, prod. Mainz 1887); *Szenen aus Mozarts Leben* (comp. 1832); *Die beiden Schützen* (The Two Marksmen) (after Cords; Leipzig 1837); *Zar und Zimmermann* (Tsar and Carpenter) (after Römers; Leipzig 1837); *Caramo* (after Saint-Hilaire & Duport; Leipzig 1839); *Hans Sachs* (after Deinhardstein; Leipzig 1840); *Casanova* (after Lebrun; Leipzig 1841); *Der Wildschütz* (The Poacher) (after Kotzebue; Leipzig 1842); *Undine* (after Fouqué; Magdeburg 1845); *Der Waffenschmied* (The Armourer) (after Ziegler; Vienna 1846); *Zum Grossadmiral* (At 'The Grossadmiral') (after Iffland; Leipzig 1847); *Regina* (comp. 1848, prod. Berlin 1899); *Rolands Knappen* (Roland's Squires) (after Musäus; Leipzig 1849); *Die Opernprobe* (The Opera Rehearsal) (after Jünger; Frankfurt 1851).

BIBL: M. Hoffmann, *Gustav Albert Lortzing* (Leipzig, 1956); H. Schirmag, *Albert Lortzing* (Berlin, 1982).

**Los Angeles.** City in California, USA. Visiting companies in the 1880s and 1890s included the Abbott, Juch, and Del Conti (1st US *Bohème*). The New York Metropolitan made its first West Coast appearance 1900. In 1906 the Lombardi (later San Carlo) Co. began visits, and the Chicago O came for 15 years from 1914. Other visitors included the Scotti Co. and a Russian company with Shalyapin. The Los Angeles Grand Opera Association was formed 1924 with Gaetano Merola as director, working with the San Francisco Opera. Merola formed the rival California Grand Opera Association 1925, rejoining 1927, and seasons were given at the Shrine Auditorium until 1932. The San Francisco O visited 1938–65, when the Dorothy Chandler Pavilion (cap. 3,098) was opened. Operatic activity increased during the 1980s, and the Los Angeles Music Center Opera Association was founded in 1985 with Peter Hemmings as director, opening in Oct. 1986 with Domingo (later artistic consultant) in Verdi's *Otello*. Since then operatic activity has put down strong and healthy roots. The Wiltern T (cap. 2,300) was reopened 1985, and productions have included Jonathan Miller's *The Mikado* with Dudley Moore. Opera has also been given in the Hollywood Bowl, beginning with Parker's *Fairyland* in 1915.

**Los Angeles, Victoria de** (*b* Barcelona, 1 Nov. 1923). Spanish soprano. Studied Barcelona Cons. Début Barcelona 1945 (Mozart's Countess). Paris, O, 1949; London, CG, 1950–61; New

York, M, 1951–61; Bayreuth 1961–2. Also Vienna; Milan, S; Buenos Aires; Scandinavia; Australia; etc. Madrid 1980 (Mélisande). A soprano held in much affection for her clear-toned, flexible voice, rich vocal palette, and the touching quality of her singing. Highly acclaimed as Rosina, Donna Anna, Mimì, Salud, Marguerite, Eva, Santuzza, Manon, Verdi's Desdemona, and Carmen. (R)

BIBL: P. Roberts, *Victoria de los Angeles* (London, 1982).

**Lott, Felicity** (*b* Cheltenham, 8 May 1947). English soprano. Studied London, RAM. Début London, ENO, 1975 (Pamina). London, CG, from 1976; Gly. Touring O 1976; Gly. Fest. O from 1977; Chicago 1984; New York, M, 1986. Also Paris, O; Brussels; M; etc. Roles include Fiordiligi, Mozart's and Strauss's Countesses, Arabella, Christine, Marschallin, Anne Trulove, Louise, Eva. Possesses a radiant tone and an elegant style. (R)

**Lotti, Antonio** (*b* Venice or Hanover, *c*.1667; *d* Venice, 5 Jan. 1740). Italian composer, organist, and singer: his father was Kapellmeister in Hanover. Studied Venice with Legrenzi, whose opera *Giustino* (1683) is often wrongly attributed to him. St Mark's as singer 1689, second organist 1692, first organist 1704, maestro di cappella 1736. He also composed 16 opere serie for Venice theatres 1706–17, of which *Porsenna* and *Alessandro Severo* were especially well received: for Vienna he composed *Costantino*. In Dresden 1717–19 with his wife, the soprano Santa Stella, to establish an Italian company for the Crown Prince, also producing *Giove in Argo*, *Ascanio*, and *Teofane*. Returning to Venice, he wrote no further operas, but was active as a teacher, his pupils including Bassani and Galuppi.

Lotti's operas fall into the period between the decline of the Venetian and the rise of the Neapolitan school, and he set librettos from both traditions (e.g. Silvani and Zeno). From Cavalli, Cesti, and late Monteverdi he inherited his basic understanding of dramatic procedures, although the use of da capo aria, especially with obbligato instrument, looked forward to the later style of Scarlatti. A skilled craftsman capable of deeply expressive writing, Lotti was much admired by Hasse, whose operas owe something to his example.

SELECT WORKLIST: *Il trionfo dell'innocenza* (Cialli; Venice 1692); *Porsenna* (Piovene; Venice 1712, rev. Naples 1713); *Alessandro Severo* (Zeno; Venice 1716); *Costantino* (Pariati; Vienna 1716; comp. with Fux & Caldara); *Giove in Argo* (Luchini; Dresden 1717); *Ascanio* (Luchini; Dresden 1718); *Teofane* (Pallavicino; Dresden 1719).

**Louise.** Opera in 4 acts by Gustave Charpentier; text by the composer. Prod. Paris, OC, 2 Feb. 1900, with Rioton, Deschamps-Jehin, Maréchal, Fugère, cond. Messager—950 perfs. by 1950; New York, Manhattan OC, 3 Jan. 1908, with Garden, Bressler, Dalmorès, Gilibert, cond. Campanini; London, CG, 18 June 1909, with Edvina, Bérat, Dalmorès, Gilibert, cond. Frigara.

Paris, *c*.1900. Louise (sop) is in love with Julien (ten), but her parents (mez and bs) refuse to allow them to marry. They set up house together in Montmartre, but when Louise's mother comes to tell her that her father is seriously ill she returns home to help nurse him back to health on condition that she be allowed to return eventually to Julien. But when her father has recovered, Louise's parents refuse to allow her to rejoin Julien; she quarrels with her father and he throws her out of the house, accusing the city of Paris of destroying his life.

**Love in a Village.** Ballad opera and pasticcio in 3 acts composed and partly compiled by Arne and Bickerstaffe. Prod. London, CG, 8 Dec. 1762; Charleston, Queen St. T, 10 Feb. 1766. Rev. Aldeburgh June 1952 in arr. by Arthur Oldham.

**Love of the Three Oranges, The** (Russ.: *Lyubov k tryem apelsinam*). Opera in 4 acts by Prokofiev; text (in Russian) by the composer, after Gozzi's comedy (1761). Prod. (in French version by V. Janacopulos) Chicago, Auditorium, 30 Dec. 1921, with Koshetz, Pavloska, Mojica, Dua, Beck, Dufranne, Cotreuil, cond. Prokofiev; Leningrad, 18 Feb. 1926; Edinburgh, by Belgrade OC, 23 Aug. 1962, with Miladinović, Čakarević, Andrasević, Paulik, Djokić, Hejbalova, cond. Danon.

A mythical land. An opera is being presented in which the King of Clubs (ten) instructs his Prime Minister, Leandro (bar), to organize an entertainment to cure the Prince's (ten) fatal melancholy by making him laugh. Leandro is secretly plotting the Prince's death.

When the Prince offends the wicked witch Fata Morgana (sop) with his laughter, he is sent to find three oranges from another witch, Creonte (bs).

Truffaldino the jester (ten) cuts open two, but two princesses contained within die of thirst. The Prince opens the third orange and the Princess Ninetta is saved by some of the stage audience producing a bucket of water. But Smeraldina (mez), Fata Morgana's servant, turns Ninetta into a rat and takes her place.

The good magician Celio (bs) restores Ninetta to her original form and, having disposed of the plotters, the couple are united.

'Love, too frequently betrayed'. Tom Rake-well's (ten) aria in Act I of Stravinsky's *The Rake's Progress*, calling on love to remain with him even as he betrays it in Mother Goose's brothel.

**Lualdi, Adriano** (*b* Larino, 22 Mar. 1885; *d* Milan, 8 Jan. 1971). Italian composer, conductor, and critic. Studied Rome with Falchi, Venice with Wolf-Ferrari. From 1909 very active as conductor, especially in Trieste, and then as critic and musical correspondent both nationally and internationally. His fervent nationalism led him to embrace Fascism, and he was elected to parliament in 1929. Director Naples Conservatory 1936–43, Florence Conservatory 1947–56. His political sympathies and the uneven nature of his advanced musical ideas led to his being first overrated as a composer and then rejected. The most significant of his operas, for which he mostly wrote his own texts, is perhaps *La figlia del re*, a version of the Antigone story set in India. For the rest, he tended to prefer comic or satirical subjects, with very erratic success: in a number of these, for instance the 1-act opera-ballet *La Grançeola*, he made ingenious use of 18th-cent. styles updated, but his ingenuity could also be self-defeating, as in the stylistic jumble of *Il diavolo nel campanile*. His son **Maner Lualdi** (*b* Milan, 23 Mar. 1912) directed many plays, films, and operas, including his father's.

SELECT WORKLIST: *La figlia del re* (The King's Daughter) (Lualdi; Turin 1922); *Il diavolo del campanile* (The Devil in the Belfry) (Lualdi, after Poe; Milan 1925, rev. Florence 1954); *La Grançeola* (Lualdi, after Bacchelli; Venice 1932).

WRITINGS: A. Lualdi, *La bilancia di Euripide* (Milan, 1969) (10 opera librettos and critical assessments).

**Lübeck.** City in Schleswig-Holstein, Germany. Opera was probably first heard in 1701. An opera-house was opened 1799, and a company formed; demolished 1857; new theatre opened 1858 with *Freischütz*; closed 1905. The Städtische Oper (cap. 1,012) opened 1908. Music directors have included Hermann Abendroth (1907–11), Wilhelm Furtwängler (1911–15), and Christoph von Dohnányi (1957–63).

**Lubin, Germaine** (*b* Paris, 1 Feb. 1890; *d* Paris, 27 Oct. 1979). French soprano. Studied Paris with Isnardon and Martini, later Litvinne and Lilli Lehmann. Début Paris, OC, 1912 (Antonia). Paris, O, 1914–44; London, CG, 1937; Bayreuth 1938–9; also Salzburg, Vienna, Berlin, etc. An excellent singer, with a full tone and great nobility of presence, highly acclaimed as Alceste, Iphigénie (*Aulide*), Donna Anna, Leonore, Aida, Isolde, Brünnhilde, Kundry, Elektra, Marschallin, Ariane (Dukas). Career ended prematurely when she was imprisoned for collaboration with the Nazis. (R)

**Lucan.** Marcus Annaeus Lucanus (*b* Corduba, 3 Nov. AD 39; *d* Rome, 15 Apr. AD 65). Roman poet, the nephew of the younger Seneca, and the author of the *Pharsalia*. As Lucano, he appears in Monteverdi's *L'incoronazione di Poppea* in his role as Nero's friend (though later he suffered Seneca's fate of being commanded to commit suicide). His observation, 'the more that a good act costs us, the dearer it is to us', is quoted at the start of Bouilly's libretto for Cherubini's *Les deux journées* as its moral. An opera on his work is Hamilton's *Pharsalia* (1968).

**Lucca, Pauline** (*b* Vienna, 25 Apr. 1841; *d* Vienna, 28 Feb. 1908). Austrian soprano. Studied Vienna with Uffmann and Levy. Début Vienna, K, 1859 (Second Boy, *Zauberflöte*). Berlin 1861; London, CG, between 1863 and 1882; Russia 1868–9; Paris, O, 1872; USA 1872–4; Vienna, K, 1874–89. Though not endowed with a beautiful voice, she had a wide range (g–d'''), and was a colourful, uninhibited actress, known as 'the demon wild-cat'. Her varied repertory included Donna Anna, Zerlina, Valentine, Eva (which she studied with Wagner), Selika (written for her, though she did not create the part); also Leonora (*Trovatore*) and Azucena (once singing both in the same performance). Her Carmen caused a sensation, as did her appearances off-stage with Bismarck: Cosima Wagner, lamenting their public familiarity, wrote, 'Such things do no honour to the crown or to art.'

BIBL: A. Jansen-Mara and D. Weisse Zahrer, *Die Wiener Nachtigall* (Berlin, 1935).

**Luchetti, Veriano** (*b* Viterbo, 12 Mar. 1939). Italian tenor. Studied Milan with Piazza and Capuano. Début Wexford 1965 (Alfredo). Spoleto 1967; London, CG, from 1973; Milan, S, 1975; New York, M, 1977; also Paris, O; Berlin, D; Brussels, M; Vienna; S America; Budapest; etc. Roles include Pinkerton, Rodolfo, Don Carlos, Don José. A sensitive singer with a firm, rich tone. (R)

**Lucia di Lammermoor.** Opera in 3 acts by Donizetti; text by Cammarano, after Scott's novel *The Bride of Lammermoor* (1819). Prod. Naples, C, 26 Sep. 1835, with Persiani, Duprez, Zappucci, Cosselli, Porto, Balestrieri, Rossi; London, HM, 5 Apr. 1838, with Persiani and Rubini; New Orleans, T d'Orléans, 28 May 1841, with Julie Calvé and Auguste Nourrit.

Scotland, late 16th cent. Lord Enrico Ashton (bar) wishes his sister Lucia (sop) to marry Arturo Bucklaw (ten) in order to strengthen his own political alliances. She, however, loves Sir Edgardo Ravenswood (ten), an enemy of the family. He proposes to ask for Lucia's hand as a gesture of reconciliation.

In an attempt to encourage her to abandon Edgardo, Enrico shows Lucia a letter supposedly from her lover to another woman. Lucia consequently agrees to marry Arturo. Edgardo interrupts the wedding ceremony, cursing Lucia for betraying him.

Edgardo and Enrico have agreed to fight a duel when news arrives that Lucia has lost her senses and has killed Arturo. Edgardo resigns himself to death at Enrico's hand, but then learns that the mad Lucia is now dead. In grief, Edgardo kills himself.

**Lucio Silla.** Opera in 3 acts by Mozart; text by G. de Gamerra, rev. Metastasio, poss. after Plutarch's *Parallel Lives* (AD c.100). Prod. Milan, RD, 26 Dec. 1772, with Morgnoni, De Amicis, Rauzzini, Suardi, Mienci, Onofrio; London, Camden TH, 7 Mar. 1967, with Curphey, Bruce, Jenkins, Conrad, cond. Farncombe; Baltimore, Peabody Concert H, 19 Jan. 1968, with Perret, Winburn, Riegel, Gerber, cond. Conlin.

Rome, c.80 BC. Cecilio (sop cas) has been banished by the dictator Silla (ten), but has returned illegally to Rome. He seeks news of his wife Giunia (sop) from his friend Cinna (sop cas). Cinna tells Cecilio that she mourns him, for Silla, who is enamoured of Giunia, has told her that her husband is dead. Cecilio meets Giunia at the family tombs.

Silla is discussing his plans to marry Giunia when Cecilio rushes in, sword drawn. Celia (sop), Silla's sister, joins a conspiracy to murder Silla. She loves Cinna but cannot tell him because his hatred of Silla is so great. Cinna tells Giunia that she is to be forced to marry Silla. Giunia resolves to plead with the Senate for her husband's life; if they refuse, she would prefer to die with him rather than marry Silla. As she rejects Silla before the Senate, Cecilio enters, again with his sword drawn. Silla orders both husband and wife to be arrested.

Celia and Cinna wish to save Cecilio and Giunia. In a sudden act of magnanimity, Silla pardons the couple and the people unite in singing his praises.

Other operas on the subject are by Anfossi (1774), J. C. Bach (1774), and Mortellari (1778).

**Lucrezia Borgia.** Opera in a prologue and 2 acts by Donizetti; text by Romani after Victor Hugo's tragedy (1833). Prod. Milan, S, 26 Dec. 1833, with Méric-Lalande, Brambilla, Pedrazzi, Mariani; London, HM, 6 June 1839, with Grisi, Mario; New Orleans, American T, 11 May 1843, with Castellan, Maiocchi, Perozzi, Valtellina, Thames, Guissiner, cond. Mueller.

Venice and Ferrara, early 16th cent. Lucrezia Borgia (sop) alone knows that Gennaro (ten) is her son. Her fourth husband Alfonso (bar) suspects the couple of having an affair and orders Gennaro's arrest. Lucrezia arranges his escape.

At a party at the Princess Negroni's, Gennaro and his friends are given wine which, Lucrezia announces as she enters to the accompaniment of funeral music, is poisoned. She is insulted because one of them has mutilated the name Borgia on her family crest to read 'Orgia'. Finding Gennaro among them, she offers him an antidote, which he refuses; he dies having learnt the identity of his mother.

When it was produced in Paris in 1840, Hugo protested and the work was withdrawn; the libretto was then rewritten, the title changed to *La rinnegata* and the action transferred to Turkey.

**Ludwig, Christa** (b Berlin, 16 Mar. 1928). German mezzo-soprano. Studied with her mother, Frankfurt with Hüni-Mihacsek, New York with Milanov. Début Frankfurt 1946 (Orlovsky). Salzburg from 1954; Vienna from 1955; New York, M, from 1959; London, CG, 1969, 1976; also Milan, Chicago, Berlin, etc. A highly expressive and intelligent artist with a ripe, firm voice, even throughout its wide compass. Large repertory includes Dorabella, Rosina, Eboli, Dido (Berlioz), Kundry, Brangäne, Octavian, Carmen; also Leonore, Dyer's Wife, Lady Macbeth, Marschallin. Created roles in operas by von Einem (*Der Besuch der alten Dame*), and Martin. (R)

**Ludwig II, King of Bavaria** (b Nymphenburg, 25 Aug. 1845; d Lake Starnberg, 13 June 1886). Son of Maximilian II, he succeeded his father in 1864. He took only a sporadic interest in the affairs of state, and lived a partly withdrawn life. Devoting himself to the patronage of Wagner, whose art provided a focus for his fantasies, he provided a home in Munich, appointed Bülow as court pianist, allotted Wagner the Villa Pellet on Lake Starnberg, and made plans for a festival theatre in Munich to house *The Ring*, engaging Semper for the purpose. Though the latter project was prevented by the government, Ludwig was instrumental in getting the Court O to stage the premières of *Tristan* (1865), *Meistersinger* (1868), *Rheingold* (1869), and *Walküre* (1870). Wagner's autobiography *Mein Leben* was undertaken at Ludwig's request. Dissension followed Wagner's deception over his affair with Cosima von Bülow, though Ludwig continued to provide support. Finally he helped Wagner with money to build the Bayreuth Festspielhaus (he lent 200,000 marks) and the Villa Wahnfried. A solitary, unhappy man, he took consolation in supporting Wagner, and delight in listening to the operas, sometimes alone in the theatre.

BIBL: W. Blunt, *The Dream King* (London, 1970).

**Luigi.** A young stevedore (ten), lover of Giorgetta, in Puccini's *Il tabarro*.

**Luisa Miller.** Opera in 3 acts by Verdi; text by Cammarano, after Schiller's tragedy *Cabale und Liebe* (1784). Prod. Naples, C, 8 Dec. 1849, with Gazzaniga, Salvetti, Malvezzi, De Bassini, Selva, Salandri, Arati, Rossi, cond. Verdi; Philadelphia, Walnut St. T, 27 Oct. 1852, with C. Richings, Bishop, P. Richings, Röhr, McKeon, cond. Cunningham; London, SW, 3 June 1858.

Tyrol, early 18th cent. Rodolfo (ten), the son of Count Walter (bs), is expected to marry Federica (mez), but is in love with Luisa (sop). In order to save her father, who threatened the Count, Luisa has to write to Wurm (bs), an employee of the Count, saying she loves him. Rodolfo makes Luisa admit that she wrote the letter. Rodolfo poisons himself and Luisa, and also succeeds in killing Wurm before dying.

**Lully, Jean-Baptiste** (Giovanni Battista Lulli) (*b* Florence, 28 Nov. 1632; *d* Paris, 22 Mar. 1687). French composer of Italian birth. Studied privately. Brought to France in 1646 to the court of Mlle de Montpensier. He soon made a reputation as a singer, dancer, and violinist, and entered the service of Louis XIV 1653. Appointed 'compositeur de la musique instrumentale du Roi', he produced a quantity of ballet music for the court; naturalized 1661 and, with Boësset, became 'surintendant de la musique et compositeur de la musique de la chambre'. 'Maître de la musique de la famille royale' in 1662, shortly before his first collaboration with Molière in *Le mariage forcé*.

Until now Lully had composed mainly *ballets de cour, such as the *Ballet des bienvenues* (1655) and *Ballet de l'amour malade* (1657), in which his Italian musical ancestry was apparent. The collaboration with Molière in a series of *comédies-ballets saw the increasing adoption of French elements in his style, notably in the handling of the vocal line and in the rhythmic verve of the instrumental parts. Many of these works approach genuine opera in the quantity and quality of their music. In such comédie-ballets as *L'amour médicin* (1665) and especially *Le bourgeois gentilhomme* (1670) Lully served an apprenticeship for his later career as an opera composer, experimenting with different styles of recitative and air, and perfecting his instrumental technique.

Since the exile of the Barberini family to Paris in 1644, attempts had been made to establish opera in France, largely by Cardinal *Mazarin. The opposition to this movement, which included Lully, felt that the French language was unsuitable for opera, and capitalized upon Louis XIV's preference for ballet. However, Lully followed the fortunes of Perrin's operatic experiments with great interest and the success of *Pomone* (1671) convinced him that the future lay in opera rather than ballet. When Perrin was imprisoned in 1672, the ever-devious Lully moved swiftly to purchase his opera privilege. Using his considerable influence at court, he achieved an almost complete monopoly of French stage music, recruited the theatre machinist Vigarani and the librettist Quinault, and founded the Académie Royale de Musique, which was soon offered the theatre in the Palais-Royal as its base.

The first production of the new group was a so-called pastorale-pastiche *Les fêtes de l'Amour et de Bacchus*, followed shortly by *Cadmus et Hermione*. With this work Lully and Quinault may be said finally to have established French opera; over the next 12 years they were to collaborate in a series of such *tragédies en musique in which the same basic formula was replicated with considerable success. At the heart of the tragédie en musique was the glorification of the monarch: the prologue, which only rarely had any connection with the tale following, was essentially intended as a homage to Louis XIV, while many of the stories—sometimes chosen by the King himself—were presented as an allegory of the royal character. The conflict between *amour* and *gloire*, with its attendant obligations, was explored through treatments of well-known mythological or chivalric plots in which the spirit of classical French tragedy, seen most notably through the intervention of the *merveilleux* and the *deus ex machina*, was tempered by an emphasis on spectacle and celebration reminiscent of the ballet de court.

Despite the triumphant reception of *Cadmus et Hermione* and its successors, and his shrewd manipulation of royal patronage, Lully's career was not without difficulties. His wresting of the operatic privilege led to a rift with Molière, who turned to Charpentier instead, while vigorous attempts were made to remove Quinault from his favoured position. Lully resisted, claiming him to be the only poet with whom he could collaborate, but was forced to capitulate for a brief period when Quinault was disgraced at court by allegedly caricaturing the king's mistress, Mme de Montespan, in *Isis*. Ironically, his most successful work, *Bellérophon*, was written without Quinault's assistance: its first run lasted for nine months.

The synthesis of Italian and French musical

styles which had already begun in the comédie-ballets was brought to full fruition in the tragé-dies en musique, which demonstrate a remark-able unity of approach. In particular, Lully must be credited with the invention of a style of recita-tive suitable for the French language, which soon subdivided into two branches: 'récitatif ordi-naire', a French equivalent of Italian secco recita-tive, apparently much influenced by the actress La Champmeslé's delivery of Racine's verses, and 'récitatif obbligé', equating to the Italian arioso, and reserved for moments of greatest emotional importance. Although, not unnatur-ally, many elements of Italian music were adapted by Lully, such as expressive harmonic writing and the use of chaconnes, in his develop-ment of the air in its several guises he confirmed the importance of a uniquely French form which was to influence the development of opera up to Berlioz. Most significant, however, was Lully's use of the chorus, at a time when it had all but disappeared in Italy, which was employed as pro-tagonist in its own right, as well as to emphasize the spectacular element of the opera. Finally, the French tradition of orchestral playing was mas-terfully exploited in the overture, dances, and dramatic instrumental interludes, such as the sleep scene in *Atys*.

Lully's claim to be the true founder of French opera is beyond dispute. A wily operator, he suc-ceeded in neutralizing any serious opposition to his plans and maintained the active support of his monarch despite a salacious private life. But his success was not effected by generous patronage and an absence of any real challenge to his supremacy. His operas demonstrate a consum-mate command of the vocal and instrumental resources at his disposal, as well as a sympathetic understanding of the native love for tragedy and ballet, genres with which the new French opera had to compromise. Under Lully's supervision the individual elements of music, poetry, drama, dance, and staging were welded together into an indivisable whole: he was no mere composer, but played a central role in the creation of each opera, particularly in the shaping of the libretto.

By the time of his death—from gangrene caused by hitting his foot with a cane while beat-ing time at a concert—Lully was acknowledged at home and abroad as the leading French com-poser of his generation. His operas were widely performed and disseminated abroad; in England, where *Cadmus et Hermione* was given as early as 1686, they were an important element in attempts to found a national operatic tradition, while three years later *Acis et Galatée* was the first non-German opera to be produced at the Ham-burg Opera. In France Lully's works retained a certain following up to the French Revolution—

*Thésée* was still being performed in 1779—and were not totally dislodged either by Rameau, or by the *Guerre des Bouffons, one side of which they partly represented.

Operas on Lully are by Isouard (*Lully et Qui-nault*, 1812), Larochejagu (*La jeunesse de Lully*, 1846), Berens (1859), H. Hoffmann (*Lully*, 1889), and Peau (*La jeunesse de Lully*, and he appears in Grétry's *Les trois âges de l'Opéra* (1778).

WORKLIST (tragédies en musique only, all other stage works excluded; librettist Quinault unless otherwise stated): *Cadmus et Hermione* (Paris 1673); *Alceste, ou Le triomphe d'Alcide* (Paris 1674); *Thésée* (Saint-Ger-main 1675); *Atys* (Saint-Germain 1676); *Isis* (Saint-Germain 1677); *Psyché* (with T. Corneille & Fon-telle; Paris 1678); *Bellérophon* (Corneille, Fontenelle, & Despreaux; Paris 1679); *Proserpine* (Saint-Ger-main 1680); *Persée* (Paris 1682); *Phaëton* (Versailles 1683); *Amadis* (Paris 1684); *Roland* (Versailles 1685); *Armide* (Paris 1686); *Achille et Polyxène* (Campistron; Paris 1687; Act I only, completed by Collasse).

CATALOGUE: H. Schneider, *Chronologisch-Thema-tisches Verzeichnis sämtlicher Werke von Jean-Baptiste Lully* (Tutzing, 1981).

BIBL: R. Scott, *J.-B. Lully: The Founder of French Opera* (London, 1973); J. E. W. Newman, *Jean-Baptiste de Lully and his Tragédies lyriques* (Ann Arbor, MI, 1979); J. Hajdu Heyer, ed., *Jean-Bap-tiste Lully and the Music of the French Baroque* (Cam-bridge, 1989).

**Lulu.** Opera (unfinished) by Alban Berg; text by the composer, after Wedekind's dramas *Erdgeist* (1895) and *Die Büchse der Pandora* (1901). Prod. Zurich 2 June 1937, with Hadzič, Bernhard, Feichtinger, Baxevanos, Feher, Melzer, Stig, Emmerich, Monische, Frank, cond. Denzler; London, SW, by Hamburg O, 1 Oct. 1962, with Lear, cond. Ludwig; Santa Fe 7 Aug. 1963, with Joan Carroll, cond. Robert Craft. Completed ver-sion (orchestrated by Cerha), Paris, O, 24 Feb. 1979; Santa Fe 28 July 1979; London, CG, 16 Feb. 1981.

Prague, c.1930. A ring-master (bs) announces his show to the audience, presenting Lulu as one of the acts.

Lulu's (sop) elderly husband, Dr Goll, dies of a heart attack when he finds her in a compromis-ing position with the Painter (ten). The Painter, now her husband, kills himself with a razor when Lulu's former lover, Dr Schön (bar), tells him of her past. Lulu forces Dr Schön to break off his engagement by threatening to run away with a Prince (ten), who loves her.

Lulu has married Schön, but continues to receive her many admirers, including Schön's son Alwa (ten). After an argument, Lulu kills Schön. In a filmed trial, Lulu is convicted of murder. She is helped to escape from prison by the lesbian Countess Geschwitz (mez).

In order to escape being sold as a prostitute to an Egyptian, Lulu flees to London with Alwa, the Countess Geschwitz, and the elderly Schigolch (bs). One of her clients is Jack the Ripper, who murders first Lulu and then the Countess.

Berg died without quite finishing the orchestration of Lulu. The original première, which had to be given outside Nazi Germany, was of a version with parts of the suite drawn from the opera given as background to the final scenes. Berg's widow then refused to release the unpublished material, and not until her death in 1976 was it possible to publish Erwin Stein's completion of the vocal score and for the work to be performed, in the orchestration by Friedrich Cerha.

BIBL: G. Perle, *The Operas of Alban Berg: 'Lulu'* (Berkeley, CA, 1985).

Also opera (after Wieland) by Mainzer (?1853).

**Lumley, Benjamin** (*b* London, 1811; *d* London, 17 Mar. 1875). English manager. Legal adviser to *Laporte, whom he succeeded as manager of London, HM, 1841–52; forced to close by success of London, CG. Reopened 1856; retired 1859. Also manager Paris, I, 1850–1. Gave British premières of many works by Donizetti and Verdi, commissioning *I masnadieri*, and brought to London singers incl. Lind, Frezzolini, Cruvelli, Piccolomini, Tietjens, Staudigl, Giuglini, and Ronconi. His company also included Grisi, Mario, Persiani, Tamburini, and Lablache; his quarrel with the first four led to the establishment of the Royal Italian O at London, CG. His *Reminiscences of the Opera* (London, 1864) gives a vivid, if one-sided, picture of London operatic life in the 1840s and 1850s.

**Lunn, Louise Kirkby** (*b* Manchester, 8 Nov. 1873; *d* London, 17 Feb. 1930). English mezzosoprano. Studied Manchester with Greenwood, London with Visetti. Début, as student, London, DL, 1893 (Margaret, *Genoveva*). London, CG, 1896, 1901–14, 1919–22; CR, 1897–9; New York, M, 1902–3, 1906–8. Her large, well-controlled voice ranged from g–b''. Though somewhat cool dramatically, she was a successful Fricka, Brangäne, Ortrud, Dalila, and Amneris. (R)

**Lussan, Zélie de.** See DE LUSSAN, ZÉLIE.

**Lustigen Weiber von Windsor, Die** (The Merry Wives of Windsor). Opera in 3 acts by Nicolai; text by Mosenthal, after Shakespeare's comedy (1600–1). Prod. Berlin, Court T, 9 Mar. 1849, with Tuczek, Zschiesche, cond. Nicolai; Philadelphia, AM, 16 Mar. 1863. London, HM, 3 May 1864 (as *Falstaff*), with Tietjens, Santley.

Very much the same plot as Verdi's *Falstaff*, but without Bardolph and Pistol, and with Slender and Master Page. The Wives are given German names: Frau Fluth (Alice), Frau Reich (Meg).

Nicolai's last and most important opera, which for some time exceeded even Verdi's *Falstaff* in popularity on German stages.

**Lustige Witwe, Die** (The Merry Widow). Operetta in 3 acts by Lehár; text by Viktor Léon and Leo Stein, after Meilhac's comedy *L'attaché*. Prod. Vienna, W, 30 Dec. 1905, with Günther, Treumann; London, Daly's T, 8 June 1907, with Lily Elsie; New York, New Amsterdam T, 21 Oct. 1907; Paris, T Apollo, 28 Apr. 1909. The complicated plot deals with the attempts of Baron Zeta (bs) to obtain the Merry Widow Hanna Glawari's (sop) fortune for his impoverished country of Pontevedria by getting his compatriot Danilo (ten or bar) to marry her. A sub-plot is concerned with the flirtation between Zeta's young wife Valencienne (sop) and an officer, Camille (ten).

**Lutyens, Elisabeth** (*b* London, 6 July 1906; *d* London, 14 April 1983). British composer, daughter of Edwin Lutyens. Studied Paris and London. She first turned to opera with two chamber works, *The Pit* and *Infidelio*, by which time her individual style was already well developed. An important trilogy of larger-scale works appeared in the 1960s, but in *The Waiting Game* and *The Goldfish Bowl* she returned to the more intimate character of her early operas. Autobiography, *A Goldfish Bowl*, 1972.

WORKLIST (all first prod. London): *The Pit* (Rodgers; 1947); *Infidelio* (Lutyens; 1954); *The Numbered* (Volonakis, after Cannetti; 1967); *Time Off!—not a Ghost of a Chance* (Lutyens; 1972); *Isis and Osiris* (Lutyens, after Plutarch; 1970); *The Waiting Game* (Lutyens; 1973); *The Goldfish bowl* (Lutyens; 1975).

BIBL: M. and S. Harries, *A Pilgrim Soul: The Life and Work of Elisabeth Lutyens* (London, 1989).

**Luxon, Benjamin** (*b* Redruth, 24 Mar. 1937). English baritone. Studied London with Grüner. Début in EOG's tour of USSR 1963 (Sid, *Albert Herring*, and Tarquinius). London, CG, and Gly. from 1972; London, ENO from 1974; Milan, S, 1986. Created title-role of *Owen Wingrave*, Jester and Death in Davies's *Taverner*. Roles include Count Almaviva (Mozart), Don Giovanni, Onegin, Posa, Ford, Wolfram. Possesses a warm, round tone, and a strong stage personality. (R)

**Luzzati, Emanuele** (*b* Genoa, 3 July 1921). Italian designer. Studied Lausanne. Began theatrical career 1947, and has designed for most leading Italian houses, especially with the producer Franco Enriquez, who brought him to Glyndebourne 1963 for *Magic Flute*. Also for Glyndebourne designed *Macbeth* 1964, *Magic Flute* 1966, *Don Giovanni* 1967, *Entführung* 1968, *Così* 1969 (and Munich), *Turco in Italia* 1970. Florence, *Gazza ladra* 1965. EOG, *Midsummer Night's Dream* 1967. Scottish O, *Cenerentola* 1970. Many Italian cities, esp. Genoa, *Don Giovanni* 1976, *Figaro* 1979, *Salome* 1980, Tutino's *Pinocchio* 1985, *Turco in Italia* 1987; Bologna, *Fledermaus* 1978; Venice, *Attila* 1986; Pesaro, *Turco in Italia* 1986, *Scala di seta* 1988; Florence, *Káťa Kabanová* 1989. St Louis, *Oberon* 1988. His ingenious sets for Glyndebourne lent a consistency to the Mozart series, but his colourful, sometimes fantastic, invention has been most appreciated in Italy, where he has successfully designed a wide range of works.

**Lvov** (Ukrainian, Lviv; Pol., Lwów; Ger., Lemberg). City in the Ukraine; part of Austria 1772–1919, Poland 1919–45, USSR 1945–91. A German opera company was founded 1776, giving mostly opera and Singspiel; a Polish company followed in 1780, giving drama and vaudevilles but little opera; these played until 1809. Józef Elsner worked here 1792–9, Karol Lipiński 1809–14. The Ukrainian T, founded 1864, gave opera and operetta. Henryk Jarecki was music director 1874–1900, giving opera new prominence, with singers including Sembrich, Krushelnytska, and Didur. The German company closed 1872, and performances of an international repertory were given in Polish. With Russian control in 1939, the Franko Opera and Ballet T opened 1940 in the old theatre founded in 1900 (cap. 1,080). After war and German occupation, operations were resumed with the ceding of Lvov to the USSR in 1945; the theatre reopened with Gulak-Artemovsky's *Zaporozhets za Dunayem* (The Don Cossack), and has given many Ukrainian works in its repertory. The theatre was restored and reopened, and the company reorganized, in 1985.

**Lvov, Alexey** (*b* Reval (Tallinn), 5 June 1798; *d* Kovno (Kaunas), 28 Dec. 1870). Russian composer and violinist. Great-nephew of **Nikolay Lvov** (1751–1803), author of the text of Fomin's *The Postdrivers*. Studied privately before joining the army. After holding various appointments, including adjutant to the Tsar, director of Imperial Court Chapel, 1837. He won an international career as a violinist, and wrote a large amount of sacred music as well as some not very successful operas. Tchaikovsky used the text of *Undina* for his own abortive opera. Composer of the Tsarist national anthem.

WORKLIST: *Bianca und Gualtiero* (Grünbaum, after Guillau; Dresden 1844); *Undina* (Sollogub, after Fouqué; St Petersburg 1848); *Starosta Boris* (Boris the Headman) (Kulikov; St Petersburg 1854).

**Lyatoshinsky, Boris** (*b* Zhitomir, 3 Jan. 1895; *d* Kiev, 15 Apr. 1968). Ukrainian composer and conductor. Studied Kiev with Glière. A prominent figure in Ukrainian music, his many and varied activities included the composition of two operas, both on heroic national themes and making use of Ukrainian folk music.

WORKLIST: *Zolotoy obruch* (The Golden Ring) (Mamontov, after Franko: Kiev 1930, rev. Lvov 1970); *Shchors* (Kochersky and Rylsky; Kiev 1938, rev. as *Polkovodets* (The Commander) Kiev 1970).

BIBL: V. Samokhvalov, *B. N. Lyatoshinsky* (Kiev, 1970, 2/1974).

**Lyceum Theatre.** Theatre in Wellington Street, London, built in 1772 for exhibitions and concerts; converted for opera 1792 by Samuel Arnold. He did not obtain a licence until 1809, when the theatre housed the English OC after Drury Lane burnt down. From 1812 known as the English OH. Rebuilt 1815. Staged works by Bishop, Braham, Loder etc. First English *Freischütz* 1824; English *Così* 1830 (as *Tit for Tat*). Burnt down 1830; rebuilt 1834. A long series of English Romantic operas began with Barnett's *The Mountain Sylph*. Balfe, who sang there 1839, became manager 1840. As the Royal Italian O, housed Gye's company during the rebuilding of Covent Garden 1856–7. Mapleson took the theatre 1861, wrongly believing he had Patti under contract, and gave the first English *Ballo in maschera*. Seasons by the Carl Rosa OC were given 1876–7, also in the 1920s and 1930s. The La Scala, Milan, company gave the first English *Otello* (Verdi) 1889. Beecham gave his last Russian opera season 1931 with Shalyapin. In 1945 the theatre became a dance hall.

**Lyons** (Fr., Lyon). City in Rhône, France. An Académie Royale de Musique opened in the Jeu de Paume in 1688, with Lully's *Phaëton*; burnt down 1688. Opera, especially Lully, Desmarets, and Campra, was then given in the stables of the Hôtel de Chaponay, later in other of the city's hôtels, but various fires and other vicissitudes seriously affected the development of operatic life. The Grand T was opened 1756, but proved too small; demolished 1827. The new Grand T opened on 1 July 1831 with *La dame blanche*; enlarged 1837 (cap. 1,350). The repertory was modelled on that of the Paris Opéra, with visits from Italian companies, and French premières sometimes anticipated Paris. A Wagner tradition developed at the turn of the century, including the first complete French *Ring*, 1904. The company maintained its high reputation between the wars, and during World War II under Roger Lalande, with André Cluytens as music director. In the 1950s the appeal of the repertory was widened, and in 1969 the producer Louis Erlo became director, further developing and strengthening the repertory with works of all periods. He gave a number of premières of French works, and revived operas by Cavalli and Charpentier, also extending the audience to students, and children, sometimes involving them in the writing and performance of new pieces. Serge Baudo was music director 1969–71, Theodor Guschlbauer 1971–75, John Eliot Gardiner (the first Englishman to hold a major French opera post) 1983–8. The emphasis moved more to the revival and exploration of neglected Baroque and Classical works. Opera is also given in the Roman theatre at Fourvière, in the Maurice Ravel Auditorium, in the Fabrique (an old factory), and in the gymnasium of the former École de Santé Militaire. The Berlioz Festival Lyon-La Côte-Saint-André has given performances of his operas.

**Lysenko, Mykola** (*b* Hrynky, 22 Mar. 1842; *d* Kiev, 6 Nov. 1912). Ukrainian composer, conductor, and pianist. Studied Kiev with Panochini, Dimitryev, and Wilczek, Leipzig with Reinecke and Richter. His operas and operettas were very popular in the Ukraine, and were admired by Rimsky-Korsakov (who gave him some orchestration lessons) and Tchaikovsky. Many of them are based on Gogol. Attempts by Tchaikovsky and others to have his operas staged in Moscow were frustrated by his refusal to have them translated into Russian.

SELECT WORKLIST: *Nich pid Risdvo* (Christmas Eve) (Starytsky, after Gogol; comp. 1874, rev. 1877–82, prod. Kharkov 1883); *Utoplennitsa* (The Drowned Woman) (Starytsky & Kukharenko, after Gogol; Kharkov 1885); *Taras Bulba* (Starytsky, after Gogol; comp. 1880–90, prod. Kiev 1903); *Aeneid* (Sodovsky, after Kotlyarevsky, after Virgil; Kiev 1911).

BIBL: V. Chagovets, *M. V. Lysenko* (Kiev, 1949).

**Lysiart.** The Count of Forest (bs), lover of Eglantine, in Weber's *Euryanthe*.

**Lyudmila.** Prince Svetozar of Kiev's daughter (sop), lover of Ruslan, in Glinka's *Ruslan and Lyudmila*.

# M

**Maag, Peter** (*b* St Gallen, 10 May 1919). Swiss conductor. Studied Zurich, Basle, and Geneva, with Hoesslin and Marek. Répétiteur and chorus-master Biel-Solothurn 1943, cond. 1945–6; Düsseldorf 1952–4; music director Bonn 1954–9, incl. *Rappresentazione di anima e di corpo* and *Genoveva*; Vienna, V, 1964. London, CG, 1958, 1977. New York, M, 1972. After successful appearances in various Italian theatres, including Venice and Parma, music director Turin, R, 1974–6. Especially successful in Mozart. (R)

**Maazel, Lorin** (*b* Neuilly, 6 Mar. 1930). US conductor. Studied Pittsburgh with Bakaleinikoff. Début as conductor aged 9. Bayreuth 1968–9 (first US conductor); New York, M, 1962; Rome 1965; Artistic director Berlin, D, 1965–71; London, CG, 1978; director Vienna, S, 1982–4, from which he resigned amid controversy. Repertory incl. *Don Giovanni*, *Lohengrin*, *Ring*, *Luisa Miller*, *Otello* (Verdi), *Rosenkavalier*, *Thaïs*, first *Ulisse* (Dallapiccola). A talented, efficient, highly successful conductor. (R)

**Mabellini, Teodulo** (*b* Pistoia, 2 Apr. 1817; *d* Florence, 10 Mar. 1897). Italian conductor and composer. Studied Florence with Pillotti and Gherardeschi. Following his early success with *Matilda e Toledo*, the Grand Duke Leopold II provided funds for him to study with Mercadante. His next opera, *Rolla*, confirmed his promise; settled in Florence 1843, where he conducted the Società Filarmonica, and became maestro di cappella to the Tuscan court (1847), and director of the T alla Pergola (1848). His nine operas, which include *I Veneziani a Costantinopoli* and *Fiametta*, owe much to Mercadante's style, and enjoyed some popularity. Professor of composition at the Istituto Reale Musicale 1859–87.

SELECT WORKLIST: *Matilda e Toledo* (?; Florence 1836); *Rolla* (Giachetti; Turin 1840); *I Veneziani a Costantinopoli* (?Ballon; Rome 1844); *Fiametta* (Canovai; Florence 1857).

BIBL: A. Simonatti, *Teodulo Mabellini* (Pistoia, 1923).

**Macbeth.** Opera in 4 acts by Verdi; text by Piave, after Shakespeare's tragedy (1605–6). Prod. Florence, P, 14 Mar. 1847, with Barbieri-Nini, Rossi, Brunacci, Benedetti, Varesi, cond. Verdi; New York, Niblo's Garden, 24 Apr. 1850,

with Bosio, Badiali, cond. Arditi; Gly. 21 May 1938, with Schwarz, Valentino, Lloyd, Franklin, cond. Busch (productions had previously been announced in London for 1861 and 1870). Prod. St Petersburg 1854, as *Sivardo il sassone*. For the Paris première a new version was made by Verdi and Piave from the French translation of Nuittier and Beaumont, which includes the addition of Lady Macbeth's 'La luce langue' in Act II, the opening and final choruses of Act IV, and the battle fugue. Prod. Paris, L, 21 Apr. 1865, with Rey-Balla, Montjaure, Ismael, Petit, cond. Deloffre.

Scotland, 11th cent. A chorus of witches greets Macbeth (bar) as Thane of Cawdor and King of Scotland, and Banquo (bs) as the father of kings. Macbeth is duly announced Thane of Cawdor. Lady Macbeth (sop) persuades her husband to murder King Duncan (silent), when he stays with them at Dunsinane that night, so as to gain the throne for himself.

Macbeth is named King of Scotland and orders that Banquo and his son Fleance (silent) be killed in order to prevent the remainder of the witches' prophecy coming true. However, Fleance escapes, and Macbeth is haunted, during a banquet, by images of Banquo.

The witches tell the anxious Macbeth that he will be safe until Birnam Wood comes to Dunsinane.

Lady Macbeth walks in her sleep, to the alarm of her maid and the doctor. The English army, under Duncan's son Malcolm (ten), advances, disguised with branches taken from Birnam Wood. Macduff (ten) kills Macbeth.

See also under SHAKESPEARE for other *Macbeth* operas.

BIBL: D. Rosen and A. Porter, eds., *Verdi's 'Macbeth': A Sourcebook* (New York, 1984).

**McCabe, John** (*b* Huyton, 21 Apr. 1939). English composer. Studied Manchester with Pitfield; Munich with Genzner. Career as a concert pianist, then director of the London College of Music. An eclectic composer, he has drawn inspiration from Stravinsky, Hindemith, and Bartók among others. His stage works, usually written with the abilities of specific performers in mind, include the children's opera *The Lion, the Witch, and the Wardrobe* and the chamber opera *The Play of Mother Courage*.

SELECT WORKLIST: *The Lion, the Witch, and the Wardrobe* (Larner, after Lewis; Manchester 1969); *The Play of Mother Courage* (Smith, after Grimmelshausen; Stonyhurst 1974).

**McCormack, John** (*b* Athlone, 14 June 1884; *d* Dublin, 16 Sep. 1945). Irish, later US, tenor.

Studied Milan with Sabatini. Début Savona 1906 (Fritz). London, CG, 1907–14; New York: Manhattan OC, 1909; M, 1910–18. On his own admission a poor actor; retired from opera 1923, devoting himself to concert work. An outstanding singer, with a sweet, limpid tone, formidable breath control, and exquisite phrasing. Celebrated as Don Ottavio, Elvino, Rodolfo, Edgardo, and Duke of Mantua. (R)

BIBL: G. T. Ledbetter, *The Great Irish Tenor* (London, 1977).

**McCracken, James** (*b* Gary, IN, 11 Dec. 1926; *d* New York, 30 Apr. 1988). US tenor. Studied New York with Ezekiel and Pagano, later Milan with Conati. Début Central City, CO, 1952 (Rodolfo). New York, M, 1953–7, and from 1963; London, CG, from 1964; also Vienna, Salzburg, Zurich, etc. A powerful performer with an imposing physique. Roles included Verdi's Otello, Don José, Samson, Calaf, Hermann, Florestan, Tannhäuser. (R)

**MacCunn, Hamish** (*b* Greenock, 22 Mar. 1868; *d* London, 2 Aug. 1916). Scottish composer and conductor. Studied London with Parry, Stanford, and Taylor. Director opera class London, GSM, from 1912. Later conducted for CR (including the first English-language *Tristan*), also with the Moody-Manners Co., Savoy T, and Beecham OC. As a composer, he was basically German-influenced; and his talent, though distinctive, was not profound enough to reconcile this with the Scottish musical cause dear to his heart. The *rapprochement* is most successfully achieved in *Jeanie Deans*, to a lesser extent in *Diarmid*. He later moved away from Scottish interests, and two later operas are unremarkable.

SELECT WORKLIST: *Jeanie Deans* (Bennett, after Scott; Edinburgh 1894); *Diarmid* (Argyle; London 1897).

**Macerata.** Town in the Marche, Italy. Opera is given in a summer festival in the Arena Sferisterio (cap. 6,000), originally a site for sports and various events. The first opera was the Count of Macerata's production of *Aida*, 1921. Large-scale productions continue to be given.

**Macfarren,** (Sir) **George** (*b* London, 2 Mar. 1813; *d* London, 31 Oct. 1887). English composer and conductor. Studied with his father and Lucas, later with Potter. Conducted at Covent Garden, London from 1845. His ambition to succeed as an opera composer was only partly realized, principally with *King Charles II* and *Robin Hood*. Knighted 1883. His wife **Natalia Macfarren** (orig. Clarina Andrae, *b* Lübeck, 1828; *d* Bakewell, 9 Apr. 1916) was a contralto (she sang in *King Charles II*) and a gifted translator of opera and Lieder.

SELECT WORKLIST: *King Charles II* (Ryan; London 1849); *Robin Hood* (Oxenford; London 1860).

BIBL: H. Banister, *George Alexander Macfarren* (London, 1891).

**Macheath.** The highwayman (ten) in Pepusch's *The Beggar's Opera* and in subsequent versions including Weill's *Die Dreigroschenoper* (where he is also known as Mackie Messer (Mack the Knife)).

**McIntyre,** (Sir) **Donald** (*b* Auckland, 22 Oct. 1934). English bass-baritone. Studied London with Keeler, Essen with Kaiser-Breme. Début WNO 1959 (Zaccaria, *Nabucco*). London, SW, 1960–7; London, CG, and Bayreuth from 1967; New York, M, 1974. Also Vienna; Paris, O; Milan, S. Repertory includes Guglielmo, Pizarro, Caspar, Attila, Amfortas, Wotan, Golaud. Possesses a rounded, substantial tone and a strong stage presence. (R)

**Mackenzie,** (Sir) **Alexander** (*b* Edinburgh, 22 Aug. 1847; *d* London, 28 Apr. 1935). Scottish composer. Studied Sondershausen with Stein, London with Lucas. His first opera, *Colomba* (1883), was criticized as being superficially Wagnerian; his other operas are *The Troubadour* (1886), *Phoebe* (unprod.), *The Cricket on the Hearth* (comp. 1901, prod. 1914), and an operetta, *The Knights of the Road* (1905).

**Mackerras,** (Sir) **Charles** (*b* Schenectady, 17 Nov. 1925). Australian, later British, conductor. Studied Sydney, and Prague with Talich. Début London, SW, 1948 (*Fledermaus*). Principal cond. Hamburg O 1966–9; music director London, SW, 1970–7; music director WNO, 1987–92. Also New York, M; Vienna, S; Chicago; San Francisco; Berlin, D; Paris; Geneva; Zurich. Wide repertory incl. Handel, Mozart (e.g. *Così*, *Figaro*, *Flute*), and Czech composers, notably Janáček (e.g. *Cunning Little Vixen*, *Káťa Kabanová*). His early reputation was largely made in Mozart, when his scrupulous approach to texts involved the restoration of appoggiaturas, and especially in Janáček. He did much to establish Janáček's stature in England. A versatile, intelligent conductor, with a strong sense of structure and phrasing and a careful regard for singers. (R)

BIBL: N. Phelan, *Charles Mackerras* (London, 1987).

**McLaughlin, Marie** (*b* Hamilton, 2 Nov. 1954). Scottish soprano. Studied Glasgow with Alexander, and London O Centre. Début as student (Susanna). Scottish O; WNO; ENO from 1978; London, CG, from 1980; Hamburg; Berlin, D; Paris, O; Munich Fest.; New York, M; Gly. Roles include Despina, Marzelline, Norina, Violetta, Nannetta, Micaëla, Tytania. (R)

**MacNeil, Cornell** (*b* Minneapolis, 24 Sep. 1922). US baritone. Studied Hartford with

Schorr, New York with Lazzari and Guth. Début Philadelphia 1950 (first John Sorel, *The Consul*). New York, CC, 1952–5; Chicago 1957; Milan, S, 1958; New York, M, from 1959; London, CG, 1964; also Vienna; Buenos Aires; Paris, O; etc. A technically accomplished singer with a smooth, mellow tone, especially successful in Verdi, e.g. as Nabucco, Macbeth, Rigoletto, Carlo (*Ernani*), Luna. (R) His son **Walter MacNeil** (*b* 1949) is a tenor; father and son have appeared together as the Germonts.

**'Ma dall'arido stelo divulsa'.** Amelia's (sop) aria opening Act II of Verdi's *Un ballo in maschera*, when she has come to a lonely spot to pick a herb to cure her love for King Gustavus (Riccardo).

**Madama Butterfly.** Opera in 2 acts by Puccini; text by Giacosa and Illica, after David Belasco's drama (1900) on the story by John Luther Long, based on real events. Prod. Milan, S, 17 Feb. 1904, with Storchio, Zenatello, De Luca, Pini-Corsi, cond. Campanini (when it was a fiasco); New version (in 3 acts) prod. Brescia, T Grande, 28 May 1904, with Krusceniski, Zenatello, cond. Campanini; London, CG, 10 July 1905, with Destinn, Caruso, Scotti, cond. Campanini; Washington, Columbia T, 15 Oct. 1906, with Szamosy, Sheehan, Goff, cond. Rothwell.

Nagasaki, early 20th cent. The geisha Cio-Cio-San, Madam Butterfly (sop), marries an American naval officer, Pinkerton (ten), who, despite the warnings of the American consul Sharpless (bar), treats the matter lightly.

Deserted by Pinkerton shortly after the wedding, Butterfly awaits his return; she now has a son, 'Trouble'. She remains insistent that Pinkerton will return, and Sharpless has not the heart to read her a letter from her husband saying he now has an American wife. The harbour cannon announces the arrival of Pinkerton's ship; Butterfly and her servant Suzuki (mez) decorate the house with petals, happily anticipating his return.

The following morning Pinkerton approaches the house with his wife Kate (mez) but, suddenly filled with guilt, cannot bring himself to face Butterfly. She tells Kate that she will hand over the child if Pinkerton will come and fetch him. Rather than face a life without honour, she kills herself with her father's ceremonial sword; Pinkerton arrives to find her dead.

**Madame Chrysanthème.** Opera in a prologue, 4 acts, and epilogue by Messager; text by Hartmann and André Alexandre, after Pierre Loti's novel (1888). Prod. Paris, T Renaissance, 26 Jan. 1893; Chicago, Auditorium, 19 Jan. 1920.

A French warship, Nagasaki, late 19th cent. The opera tells of the brief marriage of Pierre (ten), a French naval officer, to a Japanese geisha, Madame Chrysanthème (sop). He abandons her lightly, but is touched by the sincerity of a letter which he receives on board ship on his return voyage to France, and thinks with affection of the wife he has left behind.

**Madame Sans-Gêne.** Opera in 3 acts by Giordano; text by Simoni, after Sardou and Moreau's drama. Prod. New York, M, 25 Jan. 1915, with Farrar, Martinelli, Amato, cond. Toscanini; Turin 28 Feb. 1915, with Farneti, Grassi, Stracciari, cond. Panizza.

Also opera by Dłuski (1903).

**'Madamina'.** Leporello's (bs) *catalogue aria in Act I of Mozart's *Don Giovanni* enumerating Giovanni's conquests to Donna Elvira.

**Maddalena.** Sparafucile's sister (con) in Verdi's *Rigoletto*.

**Madeleine de Coigny.** The Comtesse de Coigny's daughter (sop), lover of Chénier, in Giordano's *Andrea Chenier*.

**Maderna, Bruno** (*b* Venice, 21 Apr. 1920; *d* Darmstadt, 13 Nov. 1973). Italian conductor and composer. Studied Venice with Malipiero, Milan with Pizzetti, Vienna with Scherchen. Cond. premières of Nono's *Intolleranza* (1961), Berio's *Passaggio* (1963), and his own *Satyricon* (1973). Florence, Maggio Musicale, 1964, 1970. Holland Festival 1965–8, 1973. Milan, S, 1967. New York, Juilliard School, 1970 (*Il giuramento*), 1971 (*Clemenza di Tito*); CO 1972 (*Don Giovanni*). As well as being expert at presenting difficult new scores to maximum dramatic effect, had a special interest in Monteverdi and Rameau. (R)

BIBL: R. Fearn, *Bruno Maderna* (Chur, 1990).

**'Madre, pietosa vergine'.** Leonora's (sop) aria in Act II of Verdi's *La forza del destino*, begging the Virgin Mary for forgiveness as she arrives at the Monastery of Hornachuelos.

**Madrid.** Capital of Spain. Opera was first given in the private royal theatres, initially in the Palacio Real Buen Retiro for Philip IV. At the Real Sitio del Buen Retiro, an island was used for open-air performances. An Italian company played in 1703 and 1708, then at the T de los Caños del Peral. Philip V arranged for Farinelli to come and sing to him as a cure for his melancholy. From 1740 the theatre gave performances in Spanish. The T del Real Palacio was opened in 1849, the T Real in 1850. Theatres that opened in the 19th cent. included the T del Liceo, T del Insituto (1839) where zarzuela was given, T del Circo (originally for equestrian performances but also used by Italian singers), T del Museo, T de Variedades (1842), Circo del Príncipe Alfonso (as an opera-house renamed the Circo de Rivas, 1863), and the T Rossini (1864). The important T la Zarzuela opened in 1856. The T Lírico opened in 1902 with the intention of paying special attention to Spanish opera. Many of these and other theatres later closed, but the principal theatres continued to house short seasons, often with Italian singers as well as Spanish. The T Real closed in 1925, to reopen in 1966. It is currently being renovated, due to reopen in 1992. The T de la Zarzuela (cap. 1,242) is meanwhile the company's home, and the country's second house after the Liceu, Barcelona. Madrid also holds an annual opera festival, extended in 1984 to include zarzuela and ballet.

**madrigal comedy** (sometimes misleadingly also known as **madrigal opera**). A dramatic sequence of madrigals, presenting a loosely knit plot. Although it has been claimed that such works were staged, and thus represent an important forerunner of opera, there is little evidence for this. The most famous early example, Vecchi's *L'Amfiparnasso* (1597), was certainly designed to appeal to the ear and not the eye, while Banchieri's *La saviezza giovanile* (1598) was given a dramatic performance, with actors who mimed to the accompaniment of singers and instrumentalists placed behind a screen.

**mad scene** (Fr., *folie*; Ger., *Wahnsinnsszene*). A scene in which the hero or heroine goes mad, generally with tragic results. Its first use was in 17th-cent. Venetian opera: there are many examples in Cavalli, notably in *Egisto*. The mad scene became a common feature of opera seria, as in Handel's *Orlando*, where the hero, *furioso* ('mad'), suffers a vision of Hell in which the disturbance of his reason is expressed by irregular rhythms including probably the first use of quintuple metre (which Burney thought would have been intolerable in any sane context). Handel later makes satirical use of the convention in *Imeneo* as the heroine feigns madness. The mad scene was a common element in 19th-cent. opera: the most famous is in Donizetti's *Lucia di Lammermoor* ('Ardon gl'incensi'), and others occur in his *Anna Bolena* and *Linda di Chamounix*, in Bellini's *I Puritani*, and in Thomas's *Hamlet*. This phase of the convention was satirized by Sullivan with Mad Meg in *Ruddigore*, and in Britten's *Midsummer Night's Dream*. Madness or mental derangement of one sort or another is also depicted in Strauss's *Elektra*, Berg's *Wozzeck*, Stravinsky's *The Rake's Progress*, and Britten's *Peter Grimes*.

**Madwoman, The.** The mother (ten) of the dead boy in Britten's *Curlew River*.

**maestro** (It.: 'master'). A courtesy title given to composers, conductors, and even impresarios in Italy, sometimes more colloquially in other countries. In the 17th-cent. the **maestro al cembalo** (at the keyboard) sat at the harpsichord, guiding the performance; by the end of the 18th cent. this function had been reduced to accompanying secco recitative, and it disappeared with the end of the basso continuo. The **maestro di cappella** (of the chapel) was originally the equivalent of the German *Kapellmeister, with duties that included presiding as maestro al cembalo. The **maestro sostituto** (deputy) is a coach, répétiteur, and general musical assistant, now usually known as **maestro collaboratore** (in collaboration); the prompter is sometimes known as the **maestro suggeritore**. The former term **maestro concertatore** (conductor) has now generally given way to **direttore (d'orchestra)**. The only other musical officials with the title maestro in an Italian opera-house are normally the **maestro di coro** (choirmaster) and **maestro di banda** (bandmaster).

**Maeterlinck, Maurice** (*b* Ghent, 29 Aug. 1862; *d* Nice, 6 May 1949). Belgian writer. Operas on his works are as follows:

*La Princesse Maleine* (1890): Lili Boulanger (unfin.)
*Les aveugles* (1890): Achron (1919)
*L'intruse* (1890): Pannain (1940)
*Les sept princesses* (1891): Nechayev (1923)
*Pelléas et Mélisande* (1894): Debussy (1902)
*Alladine et Palomides* (1894): E. Burian (1923); Chlubna (1925); Burghauser (1944)
*La mort de Tintagiles* (1894): Nouguès (1905); Collingwood (1950)
*Ariane et Barbe-bleue* (1901): Dukas (1907)
*Monna Vanna* (1902): E. Abrányi (1907); Février (1909); Pototsky (1926); Brânzeu (1934); Velasques (1900)
*Sœur Béatrice* (1902): Yanovsky (1907); Grechaninov (1912); Mitropoulos (1919); Laliberté (1920); Rasse (1944); Hoiby (1959)
*Joyselle* (1903): A. Cherepnin (1926)
*L'oiseau bleue* (1909): Wolff (1919).

**Magdalene.** Eva's nurse (sop), lover of David, in Wagner's *Die Meistersinger von Nürnberg*.

**Magda Sorel.** The wife (sop) of John Sorel in Menotti's *The Consul*.

**Magdeburg.** City in Saxony, Germany. The Magdeburger Nationaltheater opened 1795 with K. G. Döbbelin as music director. Standards were at first poor. A French company introduced the city to opéra-comique in 1810. New impetus came in 1825, and the repertory was enlarged. Wagner was conductor 1834–6, and his *Das Liebesverbot* was produced here for one night in 1836. Lortzing's *Undine* was premièred 1845. The Stadttheater (cap. 1,200) opened 1877;

bombed 1944. The Maxim-Gorki-Theater (cap. 1,109) opened 1959.

**Maggio Musicale.** See FLORENCE.

**Maggiore, Francesco** (*b* ?Naples, *c.*1715; *d* ?Holland, ?1782). Italian composer. Studied Naples with Durante. During an itinerant life was involved with opera composition in many European cities, including Verona and Frankfurt. Wrote *c.*12 operas, both comic and serious, but is remembered as the first Neapolitan opera buffa composer to write for Venice with *I rigiri delle cantarine*.

SELECT WORKLIST: *I rigiri delle cantarine* (Tricks of the Singer's Trade) (Vitturi; Venice 1745).

**Magic Fire Music.** The name generally given to the music accompanying the final scene of Wagner's *Die Walküre* as Wotan conjures up the fire which will defend Brünnhilde as she sleeps on her rock, and through which only a hero can pass.

**Magic Flute, The.** See ZAUBERFLÖTE, DIE.

**Magnard, Albéric** (*b* Paris, 9 June 1865; *d* Baron, 3 Sep. 1914). French composer. After hearing *Tristan* at Bayreuth, studied Paris with Dubois and Massenet, then D'Indy. His first opera, *Yolande*, met with little success. *Guercœur* concerns a liberating hero who returns from heaven to earth to find himself betrayed both by his widow and by the people he has liberated and who now, unable to make a success of a free society, are turning back to dictatorship. It reflects Magnard's exalted pessimism, as well as his admiration for Wagner (by way of D'Indy) in his use of Leitmotiv and his rich chromatic vein. In the long preface to *Bérénice*, he openly declared that Wagner's style suited his classical tastes and traditional musical culture; and there is in the work a feeling for what Wagner had admired in Gluck, another of Magnard's heroes. The scores of *Yolande* and of the outer two acts of *Guercœur* were destroyed when Magnard's house was burnt down by German invaders in riposte to his firing upon them; the latter work was reconstructed by Guy Ropartz.

WORKLIST: *Yolande* (Magnard; Brussels 1892); *Guercœur* (Magnard; comp. 1900, prod. Paris 1931); *Bérénice* (Magnard, after Racine; Paris 1911).

BIBL: B. Bardet, *Albéric Magnard, 1865–1914* (Paris, 1966).

**Mahagonny.** See AUFSTIEG UND FALL DER STADT MAHAGONNY.

**Mahler, Gustav** (*b* Kalischt (Kaliště), 7 July 1860; *d* Vienna, 18 May 1911). Austrian composer and conductor. Studied Vienna with Fuchs and Krenn. Début as conductor, Summer T, Bad Hall, 1880. Appointments followed at Ljubljana

1881; Olomouc 1882–3; Kassel 1884; Prague 1885 (where Seidl was first conductor and where Mahler gave notable performances of *The Ring*); Leipzig 1886–8 (under Nikisch); Budapest 1888–91 (as director—originally engaged for ten years, but resigning after two owing to insuperable difficulties); Hamburg 1891–7; Vienna, H, 1897–1907 (appointed conductor in May 1897, director July 1897, artistic director Oct. 1897). Also conducted London: CG, 1892, first *Ring* cycle there, and other German works; DL. New York, M, 1907–10, including first professional US performances of *The Bartered Bride* and *The Queen of Spades*.

It was during his ten years at the Vienna Hofoper (which later became the Staatsoper) that Mahler's true greatness as a conductor and director was revealed. He was an ardent perfectionist and innovator; he remarked at the outset, in famous words, 'Tradition ist Schlamperei' (Tradition is slovenliness). He built up an ensemble of singers who brought to the Hofoper some of its greatest glories; they included Gutheil-Schoder, Kurz, Mildenburg, Weidt, Mayr, Slezak, Schmedes, Winkelmann, and Weidemann; and his chief designer was Alfred Roller. Mahler's new productions in Vienna included *The Ring*, also *Figaro*, *Così fan tutte*, *Die Entführung*, *Don Giovanni*, and *Die Zauberflöte* (these five for the 150th anniversary of Mozart's birth), *Fidelio*, *Aida*, *Falstaff*, *Die lustigen Weiber*, *Louise*, *Der Corregidor*, *The Taming of the Shrew*, and *Iphigénie en Aulide*. He had many enemies during his period in Vienna—he was Jewish, dictatorial, and he spent lavishly on productions—yet he succeeded in wiping out the deficit which the Staatsoper had accumulated, and in raising the performances to a standard that has remained exemplary.

His early operas *Herzog Ernst von Schwaben* and *Die Argonauten* were destroyed, probably unfinished, and *Rübezahl* was abandoned, its music perhaps absorbed into other projects. He completed, from fragments and by adding other pieces, Weber's *Die drei Pintos* (1888), and made new performing versions of *Euryanthe* (1904) and *Oberon* (1906), and of Mozart's *Figaro*.

WORKLIST: *Herzog Ernst von Schwaben* (Steiner, ?after Uhland; comp ?1877-8); *Die Argonauten* (Mahler, ?after Grillparzer, after classical legend; comp. ?1880); *Rübezahl* (Mahler, after the Silesian legend; comp. ?1879–83).

BIBL: H. de la Grange, *Mahler* (New York, 1985).

**Maid of Orleans, The** (Russ.: *Orleanskaya deva*). Opera in 4 acts by Tchaikovsky; text by the composer, after Zhukovsky's version of Schiller's tragedy. Prod. Moscow, M, 25 Feb. 1881, with Kamenskaya, Vasilyev, Raab, F.

Stravinsky, Pryanishnikov, Koryakin, cond. Nápravník; New York, M, by Bolshoy, 2 July 1991.

The opera relates the story of Joan of Arc's (mez or sop) recognition of King Charles VII of France (ten) at Chinon, of her love for a Burgundian knight, Lionel (ten), of her denouncement by her father (bs) at the coronation at Rheims, and her eventual sentence to be burnt at the stake.

**Maid of Pskov, The** (sometimes also known as *Ivan the Terrible*) (Russ.: *Pskovityanka*). Opera in 4 acts by Rimsky-Korsakov; text by the composer, after Lev Mey's drama (1860). Prod. St Petersburg, M, 13 Jan. 1873, with Platonova, Leonova, Melnikov, Orlov, Solovev, Petrov, cond. Nápravník. Rev. 1877 and 1891–2; 3rd version prod. St Petersburg, Panayevsky T, 18 Apr. 1895, by members of St Petersburg Musical Society. New prologue, *Boyarynya Vera Sheloga*, for 2nd version, comp. 1876–7, prod. separately, Moscow, Solodovnikov T, 27 Dec. 1898; London, DL, 8 July 1913, with Brian, Nikolayeva, Petrenko, Andreyev, Alchevsky, Shalyapin, cond. Cooper.

Russia, late 16th cent. Tsar Ivan the Terrible (bs) has ransacked the city of Novgorod, whose citizens have rebelled against him; the people of Pskov fear a similar fate. He promises to leave the city in peace if he can take with him Olga (sop), niece of Prince Tokmakov (bs), Governor of Pskov. Olga plans to elope with her beloved Tucha (ten), but the couple are surprised, and Olga taken to the Tsar, who she learns is her father. As Tucha fights his captors, Olga is hit by a stray bullet, and dies as a sacrifice to Pskov.

**Maid of the Mill, The.** Pasticcio in 3 acts arranged by Samuel Arnold; text by Bickerstaffe, after Richardson's 'series of familiar letters', *Pamela* (1740). Prod. London, CG, 31 Jan. 1765, with Brent, Dibdin, Mattocks; New York, John St. T, 4 May 1769.

The first of Arnold's many stage pieces. It drew on music by 18 composers.

**Maillart, Aimé** (*b* Montpellier, 24 Mar. 1817; *d* Moulins-sur-Allier, 26 May 1871). French composer. Studied Paris Cons. with Halévy 1833–41; won Prix de Rome with cantata *Lionel Foscari*. Beginning with *Gastibelza*, contributed a series of works to the Paris stage, of which *Les dragons de Villars* was most popular. His success derived in part from a gift for writing light, captivating melodies; he was also helped by his choice of colourful librettos. Several works, including his last, *Lara*, were given abroad, including London.

SELECT WORKLIST (all first prod. Paris): *Gastibelza* (Cormon & D'Ennery; 1847); *Les dragons de Villars* (Lockroy & Cormon; 1856; known in Germany as *Das Glöckchen des Eremiten*); *Lara* (Cormon & Carré; 1864).

**Mainz.** City in Rhineland Palatinate, Germany. Singspiels were given 1665. A court theatre opened 1711, and French opera predominated. Italian opera was given in the Redoutensaal 1858. The Nationaltheater opened 1767. The Stadttheater (cap. 1,400) opened 1833 with *La clemenza di Tito*; bombed 1942. The present theatre (cap. 1,100) opened 1951.

**Maiorano, Gaetano.** See CAFFARELLI.

**Maître de Chapelle, Le.** Opera in 2 parts by Paer; text by Sophie Gay, after Duval's comedy *Le souper imprévu* (1796). Prod. Paris, OC, 29 Mar. 1821, with Boulanger, Martini, Féréol; London, CG, 13 June 1845; New Orleans, T d'Orléans, 2 Oct. 1823.

Barnabé (bs) tries to enlist the help of his French cook Gertrude (sop), in a performance of his opera *Cléopâtre*. At first she tricks him and his nephew Benedetto (ten) in order to delay the singing, but eventually succumbs and manages to sing the words and music correctly.

Paer's most successful opera.

**Majo, Gian Francesco di** (*b* Naples, 24 Mar. 1732; *d* Naples, 17 Nov. 1770). Italian composer. Studied with his father Giuseppe di *Majo, his uncle Manno, his great-uncle Feo, and later with Martini. Joined his father at the Neapolitan court as organist 1747. His first opera, *Ricimero*, was well received and his fame assured with *Cajo Fabricio*. This led to commissions from many Italian cities, from Vienna (for which *Alcide* was written), and from Mannheim (where he produced *Ifigenia in Tauride*). Returned to Naples 1767 to resume his courtly duties, but when not chosen to succeed his father as first maestro di cappella undertook another extensive tour of Italy, producing operas in many cities.

Majo wrote over 20 dramatic works, many of them opere serie to Metastasian librettos. These were popular and much praised in their day, employing a fluent and sensuous melodic line: Mozart spoke of their 'bellissima musica'. Majo's freer handling of the aria, loosening the distinctions between it and the recitative, his musical integration of the overture, and his enhancement of the orchestral accompaniment represented a significant departure from firmly established conventions. His most striking modification of contemporary practice came in *Ifigenia*, which is a pioneering adumbration of some of Gluck's later reforms. But like Jommelli's and Traetta's, Majo's contribution to the reform of opera seria in the mid-18th cent. has been largely overshadowed by that of Gluck, and despite their considerable originality his works have remained neglected.

SELECT WORKLIST: *Ricimero, re dei Goti* (?Silvani; Parma 1758); *Cajo Fabricio* (Zeno; Naples 1760); *Alcide negli Orti Esperidi* (Alcida in the Gardens of the Hesperides) (Coltellini; Vienna 1764); *Ifigenia in Tauride* (Verazi; Mannheim 1764).

**Majo, Giuseppe di** (*b* Naples, 5 Dec. 1697; *d* Naples, 18 Nov. 1771). Italian composer. Father of above. Studied Naples under Fago and Basso; a contemporary of Leo and Feo. Joined Neapolitan court, rising to become first maestro di cappella (1744). His earliest stage works, including *Lo vecchio avaro*, were opere buffe; like his two later opere serie, *Arianna e Teseo* and *Semiramide*, these were conventional in conception.

SELECT WORKLIST (all first prod. Naples): *Lo vecchio avaro* (The Old Miser) (Tullio; 1727); *Arianna e Teseo* (Pariati; 1747); *Semiramide riconosciuta* (Metastasio; 1751).

**Makropoulos Affair, The** (Cz.: *Věc Makropulos*). Opera in 3 acts by Janáček; text by the composer, after Karel Čapek's drama (1922). Prod. Brno 18 Dec. 1926, with Čvanová, Miřiovská, Otava, Pelc, Olšovský, Pour, Šindler, cond. Neumann; London, SW, 12 Feb. 1964, with Collier, Dempsey, Herincx, cond. Mackerras; San Francisco 19 Nov. 1966, with Collier, Dempsey, Ludgin, cond. Horenstein.

The famous singer Emilia Marty (sop) intervenes in the lawsuit of Gregor v. Prus, and shows first-hand knowledge of the actions, a hundred years previously, of Elian MacGregor, the mistress of the deceased Baron Prus.

It gradually emerges that she is not concerned with the outcome of the case, but only to recover an old Greek document which contains the formula for giving 300 years of life. Emilia was, as Elina Makropoulos, used as a guinea-pig for an elixir of eternal life 337 years previously, and is now longing for death; throughout many generations and changes of name she has always retained the same initials.

Prus (bar) reluctantly hands over the document in return for spending the night with her. Now in possession of the secret, she can end her life, which has become an intolerable burden. The others long to know the formula, but Marty wisely burns the parchment, and dies.

**Malanotte, Adelaide** (*b* Verona, 1785; *d* Salò, 31 Dec. 1832). Italian contralto. Début Verona 1806. After successes in Turin and Rome, created title-role in Rossini's *Tancredi*. Stendhal relates how, when she rejected her entrance aria the day before the première, Rossini produced the famous 'Di tanti palpiti'. Impressive on stage, especially in travesti roles, she sang with 'perfect intonation and refined taste', according to Hérold, though he disliked her voice—'too like a cor

anglais'. Her son **Giovan Battista Montresor** (*b* Salò, *c*.1800; *d* ?) was a tenor. Début Bologna 1824. Sang until 1860 in Italy and USA, with particular success as Rossini's Otello. Taught in Bucharest after retiring from stage.

**Malatesta.** Dr Malatesta (bar), Pasquale's friend, in Donizetti's *Don Pasquale*.

**Maldere, Pierre van** (*b* Brussels, 16 Oct. 1729; *d* Brussels, 1 Nov. 1768). Belgian composer. Studied Brussels with Fiocco. His *Le déguisement pastoral* (1759), though produced in Vienna, is an early Belgian opera. Later operas were produced in Brussels.

**Male Chorus.** The commentator (ten) who, with the *Female Chorus, introduces, comments upon, and sums up the action of Britten's *The Rape of Lucretia*.

**Malfitano, Catherine** (*b* New York, 18 Apr. 1948). US soprano. Studied New York with J. Malfitano. Début Central City, CO, 1972 (Nannetta). New York: City O, 1974–9, M, from 1979. Paris, OC, 1984. Also Salzburg; Vienna; Munich Fest.; Florence; Amsterdam; Brussels, M. Roles incl. Poppea, Zerlina, Amelia, Violetta, Juliette, Manon, Mimì, Liù, Sophie, Salome, Lulu. (R)

**Malherbe, Charles** (*b* Paris, 21 Apr. 1853; *d* Cormeilles, 5 Oct. 1911). French musicologist and composer. Studied Paris, first law, then music with Massenet. Assistant archivist Paris, O, 1896; archivist 1899. Collaborated with Albert Soubies on books on opera, especially *Histoire de l'Opéra-Comique* (Paris, 1892–3). Composed several opéras-comiques. Contributed invaluable material to the Rameau *Œuvres complètes* (with Saint-Saëns) from 1894, and initiated *Hector Berlioz: Werke* (with Weingartner), the first Berlioz edition (1900–10). His vast autograph collection was left to the library of the Paris Conservatoire.

**Malheurs d'Orphée, Les** (The Sorrows of Orpheus). Opera in 3 acts by Milhaud; text by Armand Lunel. Prod. Brussels, M, 7 May 1926, with Bianchini, Thomas, cond. De Thoran; New York, TH, 29 Jan. 1927 (concert), Hunter Coll. 22 May 1958; London, St Pancras TH, 8 Mar. 1960, with J. Sinclair, Cameron, cond. Fredman.

A modernized version of the Orpheus and Eurydice legend. Orphée (bar), an animal healer, plans to marry the gypsy girl Eurydice (sop). When she develops a serious illness, Orphée tries in vain to cure her, and Eurydice dies. He is killed by her sisters, who blame Orphée for Eurydice's death.

**Malibran, Maria** (*b* Paris, 24 Mar. 1808; *d* Manchester, 23 Sep. 1836). Spanish mezzo-soprano.

Elder daughter of Manuel *García, and sister of Pauline *Viardot. Studied with her father. Appeared aged 5 in Naples, 1814, in Paer's *Agnese*. Début London, H, 1825 (Rosina). New York, with her father's Italian OC, 1825–6. While there married Eugène Malibran, to escape from her father's dominance, but returned to Europe without him, 1827. Paris, I, 1828; London, H, 1829; performed at both until 1832. Bologna, Naples, 1832–3; Milan, S, 1834–5; London CG, 1835. In 1836 she married the violinist Charles de Bériot, by whom she had already had a son. She was the embodiment of Romanticism, in her brilliant yet melancholic personality, free and feverish life-style, and charismatic, impassioned performances. Even her early death was, as Rossini put it, an 'advantage'. Her genius was appreciated not only by an often hysterical public, but also by many great figures of the age, e.g. Lafayette, Vigny, Musset, Hans Andersen, Donizetti, Bellini, Chopin, Moscheles, and Paganini. Liszt and George Sand were both profoundly influenced by her. Fétis (like Verdi) found her 'sublime' if 'unequal', and declared her 'the most astonishing singer of her century'. She excelled and fascinated as an actress; her voice, though peculiar and uneven, had an exciting quality, while its weakness between f' and f'' was disguised with an incomparable technique. A wide range, g–e''', enabled her to sing a variety of roles, including Norma, Cenerentola, Amina, Rossini's Desdemona, Semiramide, Leonore, Maria Stuarda. She created no major roles, but she was, as Legouvé said, 'one of those artists who cause art to advance, because they are always seeking'. When she was thrown from a horse in June 1836, she refused to concede to her injuries and went on performing as usual; she died three months later.

BIBL: A. Fitzlyon, *Maria Malibran* (London, 1987).

**Malipiero, Gian Francesco** (*b* Venice, 18 Mar. 1882; *d* Treviso, 1 Aug. 1973). Italian composer and editor. Studied Venice with Bossi, Bologna, and Vienna. Taught composition at Parma Cons. 1921–3; 1932 joined Liceo Musicale in Venice, where he established a formidable teaching reputation; director 1939–52. One of the most eclectic and prolific 20th-cent. opera composers, with a style shaped by influences as diverse as verismo, Expressionism, and neoclassicism. Though he rejected serialism, Malipiero's outlook was fundamentally that of a spirited avant-gardist; his distinctive musical language owes most to modal harmony and to a dislike of counterpoint and thematic development, while his vocal lines exhibit great rhythmical freedom following syllabic declamation.

Malipiero destroyed or subsequently rejected most of his music written before 1913, including his earliest operatic attempts, such as the 1-act *Canossa* and *Sogno d'un tramonto d'autunno*. In the trilogies *L'Orfeide* and the *Tre commedie goldoniane* he reveals a love of old music, reflecting his studies of earlier Venetian masters: this eventually resulted in an important complete critical edition of Monteverdi, and significant work on Vivaldi. Elements of the bizarre and fantastic, which form an integral part of the Venetian dramatic tradition, also appear in many early works. With the rise of Mussolini, Malipiero turned to more heroic gestures: *Giulio Cesare* was in part a celebration of the Duce's achievements. After a series of works on a similar scale, including *Antonio e Cleopatra* and *Ecuba*, he returned to the fantastical style of his earlier works with *I capricci di Callot*, *L'allegra brigata*, and *Mondi celesti e infernali*. The latter works are both cyclic: *L'allegra brigata* is based on collection of six novels fashioned into a single drama, while the *Mondi celesti e infernali*, described as 'three acts with seven women', offers portraits of seven female dramatic figures, including Medea and Poppea. Of his later operas the most interesting are the *Rappresentazione e festa di Carnasciale e della Quaresima*, whose inclusion of dance episodes recalls the *intermedio tradition, and the Pushkin-based *Don Giovanni*.

SELECT WORKLIST (all texts by Malipiero unless otherwise stated): *Canossa* (Benco; Rome 1914); *Sogno d'un tramonto d'autunno* (Dream of an Autumn Sunset) (D'Annunzio; comp. 1913, RAI 1963); *L'Orfeide* (comprising *La morte delle maschere* (The Death of the Shades), *Sette canzoni*, and *Orfeo*) (*Sette canzoni* only, Paris 1920; complete, Düsseldorf 1925); *Tre commedie goldoniane* (comprising *La bottega da caffè* (The Coffee Shop), *Sior Todero Brontolon*, and *Le baruffe chiozotte*) (after Goldoni; Darmstadt 1926); *Giulio Cesare* (after Shakespeare; Genoa 1936); *Antonio e Cleopatra* (after Shakespeare; Florence 1938); *Ecuba* (after Euripides; Rome 1941); *I capricci di Callot* (after Hoffmann; Rome 1942); *L'allegra brigata* (after various 14th-, 15th-, and 16th-cents. Italian novels; Milan 1950); *Mondi celesti e infernali* (RAI 1951, Venice 1961); *Donna Urraca* (after Mérimée; Bergamo 1954); *Rappresentazione e festa di Carnasciale e della Quaresima* (after Florentine libretto of 1558: comp. 1961, prod. Venice 1970); *Don Giovanni* (after Pushkin; Naples 1963).

WRITINGS: *Scrittura e critica* (Florence, 1984).

BIBL: J. Waterhouse, *La musica di Gian Francesco Malipiero* (Turin, 1990).

**Malipiero, Riccardo** (*b* Milan, 24 July 1914). Italian composer, nephew of above. Studied Milan (with Dallapiccola) and Venice (with his uncle). Working largely in a serialist idiom, he has written three operas: *Minnie la candida*, the comic *La donna è mobile* (caricaturing traditional operatic forms), and a television opera, *Battono alla porta*. Also active as a critic and writer.

WORKLIST: *Minnie la candida* (Bontempelli; Parma 1942); *La donna è mobile* (Woman is Fickle) (Zucconi, after Bontempelli; Milan 1957); *Battono alla porta* (Buzzati; RAI 1962; prod. Genoa 1963).

**Mallinger, Mathilde** (*b* Zagreb, 17 Feb. 1847; *d* Berlin, 19 Apr. 1920). Croatian soprano. Studied Prague with Gordigiani and Vogl, Vienna with Loewy. Début Munich 1866 (Norma). Sang there until 1868, her roles including Elsa, Elisabeth, and the first Eva. Berlin 1869–82, when she retired. A great rivalry with Pauline Lucca culminated in a performance of *Figaro* (1872) in which they both appeared; Lucca was hissed on her entry, as a result of which she broke her contract. Taught singing in Prague and Berlin; one of her pupils was Lotte Lehmann.

**Malmö**. City in Malmöhus, Sweden. The City Theatre opened in 1944, and stages a mixed repertory.

**Malta**. In the mid-17th cent. masques were performed on carnival days, mostly by the Knights of St John and their followers, in one of the great halls of the Auberges (generally that of Italy). In 1732 the T Pubblico opened (later known as the T Manoel, after Antonio Manuel de Vilhema, Grand Master of the Knights), and is now the National T of Malta. It was modelled on the T Santa Cecilia of Palermo, and works by Jommelli, Piccinni, Galuppi, Paisiello, and Cimarosa were staged during the next 60 years. In 1796 the Maltese composer Nicolò Isouard returned home, and several of his works were given.

During the French occupation of 1798–1800, works by Dalayrac and other French composers were produced. In the succeeding British occupation opera was again given by Italian singers, generally from Naples and Palermo. These seasons lasted from Sep. until May, and the impresario was contracted to stage 12 new operas each season, of which five had to be new to Malta. By the 1880s 32 of Donizetti's had been staged, and almost all Verdi's. A new theatre, designed by the Covent Garden architect Edward Barry, opened on 9 Oct. 1866; it burnt down in 1873 and was rebuilt by 1877.

By the turn of the century many famous singers had been heard in Malta, including Albani, Bellincioni, and Scotti (who made his début here in 1890). From 1900 to 1939 the seasons were similar to those of the larger Italian theatres, with all works, including Richard Strauss, Wagner, and the Russian and French repertory, being sung in Italian. Unfamiliar works by Giordano, Mulé, Zandonai, and others were heard in Malta often within a few months of their premières. During the Second World War the Royal O was destroyed, and performances were given by local artists at the T Manoel and

elsewhere; these included *La figlia del sole* by Cervello (a sequel to *Madama Butterfly*). After the war, Italian singers were heard again, and in 1961 opera returned to the T Manoel, restored in 1986 (cap. 400).

Four operas by the Maltese composer Carmelo Pace have been staged, *Caterino Desguanez* (1970), *I martiri*, *Angelica* (1973), and *Ipogeana* (1976). Opera is also given in the Aurora Theatre Complex on Gozo.

**Malten, Therese** (*b* Insterburg, 21 June 1855; *d* Neuzschieren, Saxony, 2 Jan. 1930). German soprano. Studied Berlin with Engel. Début Dresden 1873 (Pamina). Sang there for 30 years. Admired by Wagner, who hoped to hear her as Isolde. He chose her as one of the three singers of Kundry (with Brandt and Materna) in the first performances of *Parsifal*. Bayreuth 1882–94; London, DL, 1882; Moscow 1889. Repertory included Armide, Leonore, Isolde, Eva, Sieglinde, Brünnhilde. A handsome and talented actress, she possessed a powerful voice with a wide range.

**Mamelles de Tirésias, Les** (The Breasts of Tiresias). Opera in 2 acts by Poulenc; text by Guillaume Apollinaire. Prod. Paris, OC, 3 June 1947, with Duval, Payen, cond. Wolff; Brandeis U, Waltham, MA, 13 June 1953; Aldeburgh 16 June 1958, with Vyvyan, Pears (acc. two pianos), cond. Mackerras.

When Thérèse (sop), fed up with her submissive life as a woman, casts aside her breasts and adopts a masculine life as Tirésias, her husband (ten) has to take her place as a woman, but soon tires of her arrogant behaviour. Declaring he will produce children without her, he proceeds to bring 40,000 children of varying ages into the world. A policeman (bar) consults a fortune-teller as to how the children are to be supported, but is nearly strangled by her, when he attacks her after having been accused of being sterile. The husband saves him, and it emerges that the fortune-teller is none other than Thérèse. She begs her husband's forgiveness and the couple are reunited.

**Manchester**. City in Lancashire, England. The first theatre in Manchester was built in 1753 and in 1775 the first T Royal was approved by Parliament. Opera was provided by touring companies and operatic excerpts in the 'Gentleman's Concerts' until Hallé conducted a season of operas at the T Royal in 1855, including *Fidelio*, *Don Giovanni*, *Der Freischütz*, *Robert le diable*, *Les Huguenots*, *Lucrezia Borgia*, and *La favorite*. He followed this with concert performances at the Free Trade Hall of *Fidelio*, *Die Zauberflöte*, and Gluck's *Armide*, *Iphigénie en Tauride*, which he also took to London, and *Orfeo ed Euridice*. The performances of *Iphigénie* and *Orfeo* were the first in English. Opera was also provided by the Rouseby and Parepa-Rosa Companies. The latter began its existence in England with a performance in Manchester of *Maritana* on 1 Sep. 1873. After Hallé's death opera was again provided only by touring companies and by ambitious amateurs, though in 1897 Cowen conducted a concert perf. of *Les Troyens à Carthage* after giving the first British concert performance at Liverpool earlier the same year. Between 1916 and 1919 Thomas Beecham presented an ambitious series of opera seasons at the New Queen's T, and it seemed as though Manchester would rival London as an operatic centre until Beecham's enforced withdrawal in 1920, when touring companies again took over. Since then amateur opera has flourished, and seasons have been given by London CG and SW, and Italian companies, at the Palace T, Opera House, and occasionally at the Hippodrome, Free Trade Hall, and suburban theatres.

**Mancia, Luigi** (*b* ?Venice, ?1660s; *d* ?, after 1708). Italian composer and librettist. A shadowy figure, whose activities have been traced in Hanover, Rome, Berlin, Naples, and Düsseldorf during the years *c*.1687–1708. Is credited with *c*.10 operas; the most successful, *Partenope*, enjoyed considerable fame in its day.

SELECT WORKLIST: *Partenope* (Stampiglia; Naples 1699, rev. Rovigo 1699).

**Mancinelli, Luigi** (*b* Orvieto, 6 Feb. 1848; *d* Rome, 2 Feb. 1921). Italian conductor and composer. Studied Orvieto with his brother Marino, Florence with Mabellini. Led cellos at Florence, P, and Rome, Ap., where in 1874 he was called upon to fill the place of a drunk conductor in *Aida*. After engagements in various Italian cities, conducted a concert in London in 1886, on the strength of which he was engaged by Harris as chief conductor Drury Lane 1887, and Covent Garden 1888–1905. Here he conducted many English premières, and many of the De Reszke Wagner performances in Italian. Music director Madrid, T Real, 1888–95. Chief Italian conductor New York, M, 1893–1903. His own operas included *Ero e Leandro* (Norwich 1896) and *Paolo e Francesca* (Milan 1907). The most important Italian conductor between Faccio and Toscanini, he anticipated the latter in his powerful authority, personal magnetism, and deep fidelity to the score. His Italian Wagner performances were controversial, but not without German admirers (including Weingartner). As a composer, he resisted verismo in favour of a classicism he associated with Boito, who in 1877 called him the ideal conductor of *Mefistofele*.

**Mancini, Francesco** (*b* Naples, 16 Jan. 1672; *d* Naples, 22 Sep. 1737). Italian composer. Studied Naples with Provenzale and Ursino; his first opera was *Ariovisto*. Joined the Neapolitan court as first organist 1704; became maestro di cappella 1707. Less than a year later was forced to cede the post to its former holder, A. Scarlatti: he became vice maestro di cappella, but was guaranteed the right to succeed Scarlatti, which he did in 1725. 1720–35 director of the Cons. S Maria di Loreto, where he was succeeded by Fischietti.

Mancini composed *c*.20 operas for Naples, and is generally regarded as Scarlatti's most important contemporary there. His works, which included *Lucio Silla*, and *Trajano*, enjoyed some popularity abroad. Especially important was the performance of *Idaspe fedele* in London in 1710 which, after *Almahide* of the same year, was the first opera to be given there entirely in Italian. His work was overshadowed by that of Scarlatti, and younger composers such as Hasse, Leo, and Vinci, whom he followed increasingly after *Trajano*, a work generally considered the climax of the Neapolitan baroque pageant opera.

SELECT WORKLIST: *Ariovisto* (?; Naples 1702); *Lucio Silla* (Rossini; Naples 1703); *Idaspe fedele* (?Candi; London 1710); *Trajano* (Noris; Naples 1723).

**Mandikian, Arda** (*b* Smyrna, 1 Sep. 1924). Greek-Armenian soprano. Studied Athens with De Hidalgo and Trianti, Oxford with Wellesz. Début Oxford, OUOC, 1950 (Berlioz's Dido). Created title-role *Incognita* (Wellesz) Oxford, OUOC, 1951. Paris, CE, 1952; London, CG, 1953; EOG 1954. Her strong and individual dramatic powers contributed to the character of Miss Jessel, which Britten wrote for her in *The Turn of the Screw*. This she created and often sang. Other roles incl. Savitri and Elettra. Assistant general director Athens O 1974–80. (R)

**Mandini.** Italian family of singers.

1. **Stefano** (*b* 1750; *d* *c*.1810). Bass-baritone. Début Venice 1775. Vienna 1783–8, singing in operas by Cimarosa, Paisiello, Salieri, Storace, etc., and creating Count Almaviva in *Figaro*. Appearances in Naples, Paris, Venice, Vienna again, and St Petersburg 1788–94; possibly Berlin 1804. A highly successful singer, who shone in comedy.

2. **Maria** (*fl* 1780s). Soprano, wife of above. Début Vienna 1783. Created Marcellina (*Figaro*), which was written for her. Popular in character parts, as which she often appeared with her husband and Nancy Storace in operas by Martín y Soler, Cimarosa, Sarti, and Paisiello.

3. **Paolo** (*b* Arezzo, 1757; *d* Bologna, 25 Jan. 1842). Tenor, the less talented brother of Stefano. Studied with Valente. Début Brescia 1777. Milan, S, 1781. Sang for Haydn at Eszterháza

1783–4, creating Idreno in *Armida*; Vienna 1785–6 (with Stefano and Maria), and 1789; St Petersburg 1796 (with Stefano).

**Mandryka.** A rich young landowner (bar) in Strauss's *Arabella*.

**Manelli, Francesco** (*b* Tivoli, after 13 Sep. 1594; *d* Parma, July 1667). Italian composer and bass. 1605–24 sang at Tivoli Cathedral; 1627–9 maestro di cappella there. During early 1630s in Rome, where his wife participated in opera productions. Settled in Rome 1637; at St Mark's (as singer) 1638–46, but 1645 went to Parma to sing at S Maria della Steccata. Subsequently became vice maestro di cappella there and from 1653 worked at the Parmese court, possibly as maestro di cappella.

In 1637 he collaborated with Ferrari on *Andromeda*; this was produced, with Manelli himself taking the parts of Nettuno and Astarco Mago, to inaugurate the first Venetian public opera-house in the T San Cassiano. It was followed by several more works, including *La maga fulminata* and *Alcate*, which consolidated the new commercial opera: *Delia* was performed for the opening of T Santi Giovanni e Paolo, Venice, in 1639. None of these works survives, but they probably followed the Roman style of Landi and Rossi, as did the lost works, including *Le vicende del tempo*, he wrote for the Parmese court.

SELECT WORKLIST: *Andromeda* (Ferrari; Venice 1637, lost); *La maga fulminata* (The Sorceress) (Ferrari; Venice 1638, lost); *Delia* (Strozzi; Venice 1639); *Alcate* (Tirabosco; Venice 1642, lost); *Le vicende del tempo* (Morando; Parma 1652, lost).

**Manfredini, Vincenzo** (*b* Pistoia, 22 Oct. 1737; *d* St Petersburg, 16 Aug. 1799). Italian composer. Brother of the castrato **Giuseppe Manfredini**, he studied with his father, the violinist **Francesco Manfredini** (1688–1744), and with Perti and Fiorino. Travelled to Russia with his brother 1758, succeeding Raupach as court composer; here he wrote five operas, including *Carlo Magno* and *L'Olimpiade*. Returned to Italy 1769, having been replaced at court by Galuppi. Living in Bologna and Venice, he worked for a time as a literary editor and wrote several important theoretical works. He returned to St Petersburg 1798 at the request of Tsar Paul I, whom he had earlier instructed on the harpsichord. His operas, though important contributions to the St Petersburg repertoire, were cast in a conventional mould.

SELECT WORKLIST: *L'Olimpiade* (Metastasio; Moscow 1762); *Carlo Magno* (Lazzaroni; St Petersburg 1763).

**Manfredini-Guarmani, Elisabetta** (*b* Bologna, ?1786; *d* ?). Italian soprano, daughter of Vincenzo *Manfredini. Début Bologna 1809. Admired by Rossini, for whom she created

several roles: Amira (*Ciro in Babilonia*), Amenaide (*Tancredi*), Aldimira (*Sigismondo*), Elisabetta (*Adelaide di Borgogna*). Successes in Turin, and at La Scala, Milan. Possessed a beautiful but inexpressive voice.

**Manhattan Opera Company.** New York company founded by Oscar *Hammerstein in 1906 at the Manhattan OH on W 34th St. Opened 3 Dec. 1906 with *Puritani* under the director *Campanini. The first season included appearances by Calvé and Melba. The 1907–8 season saw the first US *Thaïs*, *Louise*, and *Pelléas*, all with Mary Garden; the final season, 1909–10, saw the first US *Elektra*, *Sapho*, and *Grisélidis*. Artists included Nordica, Schumann-Heink, Tetrazzini, Cavalieri, Zenatello, McCormack, and Dalmorès. The company gave 463 performances of 49 operas in its four seasons, also appearing in Philadelphia in a theatre built by Hammerstein. From 1908 to 1910 the Philadelphia Co. had its own chorus, orchestra, and conductors, with singers commuting from New York. The company also visited Boston, Cincinnati, Pittsburgh, and Washington.

Such was its success that in 1910 the Metropolitan offered Hammerstein $2,000,000 if he refrained from giving opera in New York for ten years. He tried to break the agreement in 1913 by building the *Lexington Theatre, but was legally restrained. The Manhattan OH was used for the Chicago O's seasons, but later sold, becoming New York's leading opera recording studio for RCA and the Metropolitan recordings.

BIBL: J. F. Cone, *Oscar Hammerstein's Manhattan Opera Company* (New York, 1964).

**Mann, Thomas** (*b* Lübeck, 6 June 1875; *d* Zurich, 12 Aug. 1955). German writer. He was much influenced by Wagner, constructing his novels leitmotivically and including in them some fine descriptions of the music's effect, including its dangerous qualities as portrayed by Nietzsche. Of his writings on Wagner, the essay *Leiden und Grösse Richard Wagners* (The Sufferings and Greatness of Richard Wagner, 1933) is an important and perceptive argument for Wagner's independent stature, written at a time when the Nazis were trying to claim him as one of them. His *Der Tod in Venedig* (1912) was set by Britten as *Death in Venice* (1973).

**Manners, Charles** (*b* London, 27 Dec. 1857; *d* Dundrum, Co. Dublin, 3 May 1935). Irish bass and impresario. Studied Dublin, London, RAM, and Italy. Début London with D'Oyly Carte Co. 1882 (creating Pte. Willis, *Iolanthe*). Joined CR, then London, CG, in 1890. First English Gremin, 1892. USA 1893. Toured S Africa 1896–7 with his wife Fanny Moody. In 1898 they founded the Moody-Manners Co., which lasted until 1916. Manners promoted British opera by both producing it and awarding prizes to the best works. He was largely responsible for establishing the Glasgow Grand Opera Society, which later gave the first British *Idomeneo* and *Les Troyens*. Retired from singing 1913.

**Mannheim.** City in Baden-Württemberg, Germany. The Schlosstheater, designed by Bibiena, opened 1742 with *Meride* by Carlo Grua (*c.*1700–73), music director until 1753 and required to compose an opera a year. Visitors included Hasse, Galuppi, and Jommelli; other distinguished Italians came at the invitation of Ignaz *Holzbauer, music director 1753–78. Due partly to the famous orchestra, opera flowered 1760–70; and after a period in which J. C. *Bach's Italian operas were popular, there was a turn to German opera. Holzbauer's *Günther von Schwarzburg* triumphed in 1777. The Nationaltheater opened 1777, but the court move to Munich badly affected standards. Opera prospered under Vincenz Lachner 1837–72, with a large repertory, and more progressively under August Nassermann 1895–1900. Music directors have included Weingartner (1889–91), Bodanzky (1909–15), Furtwängler (1915–20), Kleiber (1922–3), Elmendorff (1935–42). Premières have included Goetz's *Taming of the Shrew* (1874), Wolf's *Corregidor* (1896), and Wellesz's *Alkestis* (1924). The Nationaltheater was bombed 1943; the new Nationaltheater (cap. 1,200) opened Jan. 1957 with *Der Freischütz*. Horst Stein was music director 1963–70.

**Manolov, Emanuil** (*b* Gabrovo, 7 Jan. 1860; *d* Kazanlŭk, 2 Feb. 1902). Bulgarian composer and conductor. Studied Moscow. Composer of the first Bulgarian opera, *Siromakhkinya* (The Poor Woman) to his own text, after Vasov. Though unfinished, amateur production under his direction Kazanlŭk, 1900; later completed by others and produced Sofia 1910. Italianate in style, it also makes use of Bulgarian urban folk-songs.

BIBL: A. Balareva, *Emanuil Manolov* (Sofia, 1961).

**Manon.** Opera in 5 acts by Massenet; text by Meilhac and Gille, after Prévost's novel *Manon Lescaut* (1731). Prod. Paris, OC, 19 Jan. 1884, with Heilbronn, Talazac, Taskin, Danbé; Liverpool 17 Jan. 1885, with Roze, M'Guckin, cond. Goossens I.; New York, AM, 23 Dec. 1885, with Hauk, Giannini, Del Puente, cond. Arditi.

France, second half 18th cent. The Chevalier Des Grieux (ten) falls in love with Manon (sop), whom he meets as she stops at an inn in Amiens with her cousin Lescaut (bar) on her way to a convent.

The young lovers elope to Paris. De Brétigny (bar), a friend of Lescaut, persuades Manon to go away with him.

Des Grieux, in despair, decides to enter the priesthood, despite pleas from his father, the Comte Des Grieux (bs). Manon comes and persuades Des Grieux to go off with her.

At a gambling house Des Grieux is accused of cheating and Manon is arrested as a prostitute and condemned to transportation.

Des Grieux bribes an officer for permission to speak to her and tries to persuade her to run away with him. She is too weak to do so and dies in his arms.

Ten years later Massenet wrote a sequel, *Le portrait de Manon* (1894). Other operas on the story are by Auber (1856), Kleinmichel (1887), Puccini (see below) and Henze (1952).

**Manon Lescaut.** Opera in 4 acts by Puccini; text by Illica and Oliva, after Prévost's novel *Manon Lescaut* (1731). Prod. Turin, R, 1 Feb. 1893, with Ferrani, Cremonini, Moro, cond. Pomé; London, CG, 14 May 1894, with Olghina, Beduschi, Pini-Corsi, cond. Seppilli; Philadelphia, Grand OH, 29 Aug. 1894, with Kört-Kronold, Montegriffo, cond. Hinrichs.

France, second half 18th cent. The young Chevalier Des Grieux (ten) sees Manon Lescaut (sop) in an inn on her way to a convent; they fall in love and elope, stealing the carriage of Géronte de Ravoir (bs).

Encouraged by her brother Lescaut (bar), Manon has abandoned Des Grieux and is living lavishly as Géronte's mistress. Des Grieux makes his fortune gambling and persuades Manon to escape with him again; she takes with her jewels she received from Géronte, who sends the police after her and has her arrested for theft and prostitution.

At Le Havre, Des Grieux and Lescaut try in vain to secure Manon's release; she has been sentenced to deportation. In desperation, Des Grieux pleads with the captain to be allowed to sail with her.

Manon and Des Grieux escape from New Orleans but Manon collapses from exhaustion and dies in her lover's arms.

**Manrico.** A chieftain (ten) under the Prince of Biscay, lover of Leonora and the eponymous troubadour of Verdi's *Il trovatore*.

**Manru.** Opera in 3 acts by Paderewski; text by Nossig, after Kraszewski's novel *The Cabin Behind the Wood* (1843). Prod. Dresden 29 May 1901; New York, M, 14 Feb. 1902, with Sembrich, Scheff, Homer, Bispham, cond. Damrosch.

Against her mother's wishes, Ulana (sop) marries the gypsy Manru (ten). She revives his love with a potion, but the gypsy girl Asa (sop) lures him back to his people. Ulana commits suicide and Manru, now chief of the tribe, is killed by Oros (bs), the man he had deposed who himself loves Asa.

**Mantua** (It., Mantova). City in Lombardy, Italy. The first city after Florence to stage opera, though lavish dramatic entertainments had been a feature at the *Gonzaga court during the 16th cent. Monteverdi's *Favola d'Orfeo* was produced 1607, *Arianna* 1608. Other early operas premièred include Gagliano's *Dafne* (1608). Singers then employed by the Duke, Vincenzo Gonzaga, included Claudia Cattaneo (Monteverdi's wife), Lucrezia Urbana, Caterinuccia Martinelli, and Adriana Basile. The Regio Ducale T Nuovo, designed by F. Galli-Bibiena and A. Galuzzi, opened 1732 with *Cajo Fabrizio*; burnt down 1780; reopened 1783 with Sarti's *Il trionfo della pace*. Vivaldi's *Candace* (1720) and *Semiramide* (1732) were premièred here. The T Sociale opened 1822 with Mercadante's *Alfonso ed Elisa*; the T Andreani opened 1862 with Verdi's *I masnadieri*.

**Man without a Country, The.** Opera in 2 acts by Damrosch; text by Arthur Guiterman, based on Edward Everett Hale's story (1863). Prod. New York, M, 12 May 1937, with Traubel, Carron, cond. W. Damrosch.

Philip Nolan (ten), a young US Army officer, is deported for his impulsive shouts of treason in a courtroom. For 50 years he travels without a country, and gradually becomes a fanatical patriot. His bravery in a battle secures his return to his homeland, but the official pardon never arrives and he dies clutching the Stars and Stripes.

**Manzuoli, Giovanni** (*b* Florence, *c.*1720; *d* Florence 1782). Italian soprano castrato. Début Florence 1731. After successes in Naples, Rome, and Venice, invited by Farinelli to Madrid 1749–52, and again 1755. Bologna 1763, in the première of Gluck's *Trionfo di Clelia*. London, H, 1764–5, to great acclaim. Here he also gave lessons to the 8-year-old Mozart, exerting a formative influence on his musical and vocal taste. Vienna 1765; Italy 1766–8, when he officially retired, and became chamber singer to the Grand Duke of Tuscany. However, he performed in Rome, 1770, and finally in Milan, 1771, creating the title-role of *Ascanio in Alba*, written for him by Mozart. Burney described his voice as 'powerful and voluminous', and his style as 'grand and full of taste and dignity'.

**Mapleson, James Henry** (*b* London, 4 May 1830; *d* London, 14 Nov. 1901). English singer, violinist, and impresario. Studied London. Sang Verona 1854 as Enrico Mariani; unsuccessfully under his own name DL, once. Played London, HM, 1848–9. London: manager DL 1858; L 1861, incl. British première *Ballo*; HM 1862–7; DL 1868; joined *Gye at CG 1869–70; DL 1871–6; HM 1877–81; CG 1885, 1887; HM 1887, 1889. New York, AM, 1878–96. 1896–7, also touring other US cities. Gave many British and US premières, and brought to London Di Murska, Gerster, C. Nilsson, Trebelli, Hauk, Nordica, Campanini, Jean De Reszke (as baritone), Pandolfini. Known as 'the Colonel'; *The Mapleson Memoirs* (London, 1888, 2/1966) give an amusing and colourful account of his activities.

**'M'apparì'.** The Italian translation of Lionel's (ten) 'Ach, so fromm' in Act III of Flotow's *Martha*, in which he sings of his hopeless love for Martha.

**Mara, Gertrud** (*b* Kassel, 23 Feb. 1749; *d* Reval, 20 Jan. 1833). German soprano. Studied London with Paradisi, Leipzig with Hiller. Début Dresden 1767. Engaged 1771 by Frederick the Great for Berlin O. Against his wishes, she married a dissolute cellist, Johann Baptist Mara, but he gave his approval on condition that she stayed at Berlin for life. After much subsequent friction, she was released in 1779. Munich and Vienna 1780–2; Paris 1782, enjoying a celebrated rivalry with Todi. Sang in London from 1784, with enormous success, notably at the Haymarket, 1789, as Cleopatra in Handel's *Giulio Cesare*, and at Covent Garden, 1790, in Nasolini's *Andromaca*. She was also famous for her *Messiah*. Having divorced Mara, she left London 1802 with a flautist, Florio, with whom she toured the Continent. After their separation in 1803, she lived in Moscow until the burning of 1812, when she settled in Reval (Tallinn). One of the first German singers to become internationally famous; though a bad actress, she possessed an expressive voice of great beauty and virtuosic powers, with a range of g–e'''. Mozart disliked both her singing and her arrogance; Goethe, however, celebrated her artistry in two poems.

BIBL: O. Anward, *Die Prima Donna Friedrichs des Grossen* (Berlin, 1931).

**Marais, Marin** (*b* Paris, 31 May 1656; *d* Paris, 15 Aug. 1728). French composer and viola da gamba player. Studied with Lully. A gifted instrumentalist, he joined the orchestra of the Académie Royale de Musique as solo gamba player, sharing the direction of the orchestra with Colasse. Titular director 1695–1710. Best known for his instrumental music and virtuosic performances, he also composed four highly regarded operas after the model of Lully's tragédies en musique. Most innovative was *Alcyone*, which contains one of the earliest operatic representations of a storm. All his operas were successfully revived in the 18th cent.

> WORKLIST (all first prod. Paris): *Alcide* (Lully; 1693, comp. with L. Lully); *Ariane et Bacchus* (Saint-Jean; 1695); *Alcyone* (La Motte; 1706); *Sémélé* (La Motte; 1709).

**Mařák, Otakar** (*b* Esztergom, Hungary, 5 Jan. 1872; *d* Prague, 2 July 1939). Czech tenor. Studied Prague with Paršova-Zikešová. Début Brno 1899 (Gounod's *Faust*). Prague, N, 1900–7, 1914–34; Vienna 1903; Berlin, H, 1906; London, CG, 1908, and HM, 1913; Chicago 1914. Dubbed 'the Czech Caruso', he was a distinguished Canio; also sang Bacchus, Parsifal, and Gennaro in the first *Gioielli della Madonna*. After retiring in 1934, he became destitute in Chicago; funds were raised in Prague to enable its former star to return. He died soon after. (R)

**Marazzoli, Marco** (*b* Parma, *c*.1602 or *c*.1608; *d* Rome, 26 Jan. 1662). Italian composer, singer, and harpist. Possibly studied with Allegri; 1637 joined the papal choir as tenor. A protégé of the *Barberini family, with whom he travelled throughout Italy. In 1643 he was brought by Mazarin to Paris, where *Il capriccio* was probably given; stayed until 1645 and was involved with efforts to establish Italian opera there.

In 1637 he produced, with Mazzochi, the first comic opera, *Il falcone* (usually known by its title upon revival 1639, *Chi soffre, speri*). This includes some of the earliest *recitativo secco, but otherwise follows closely the style of earlier operas. *Dal male il bene*, written jointly with Abbatini to celebrate the return of the exiled Barberini family 1653, is a more original work, and set a standard for later comic opera composers, especially in its use of ensembles. Marazzoli's other operas included works for Venice (*Gli amori di Giasone e d'Issifile*) and Ferrara (*Le pretensioni del Tebro e del Po*).

> SELECT WORKLIST: *Chi soffre, speri* (Who Suffers May Hope) (Rospigliosi; Rome 1637 (as *Il falcone*), rev. 1639; comp. with Mazzochi); *Gli amori di Giasone e d'Issifile* (Persiani; Venice 1642, lost); *Le pretensioni del Tebro e del Po* (Pio di Savoia; Ferrara 1642); *Il capriccio* (Buti; ?Rome 1643, ?rev. Paris 1645); *Dal male il bene* (Rospigliosi; Rome 1653; comp. with Abbatini).

**Marcellina.** A duenna (sop) in Mozart's *Le nozze di Figaro*.

**Marcello.** The Bohemian painter, lover of Musetta, (bar) in Puccini's and (ten) Leoncavallo's *La bohème*.

**Marcello, Benedetto** (*b* Venice, 1 Aug. 1686; *d* Brescia, 24 or 25 July 1739). Italian composer, writer, and theorist. Trained as a lawyer, but also studied composition with Gasparini and Lotti. Held civic positions in Venice (1711 made a member of the Council of Forty), Pola, and Brescia. Composed a quantity of vocal music, including a famous set of 50 psalm settings, many instrumental works, and possibly the opera *La fede riconosciuta*. He was a gifted poet and wrote several librettos, but is best remembered for his satirical pamphlet *Il teatro alla moda* (The Fashionable Theatre, Venice *c*.1720). This took the form of a series of satirical addresses to all those connected with operatic production, from composer down to stage-hand, mocking their shortcomings and shoddy methods, and is one of the most valuable sources for an appreciation of 18th-cent. theatrical practice and malpractice.

**Märchen** (Ger.: 'tale', esp. fairy-tale). A title occasionally used for a German fairy opera.

**Marchesi.** Family of singers.

1. **Salvatore** (Cavaliere de Castrone, Marchese della Rajata) (*b* Palermo, 15 Jan. 1822; *d* Paris, 20 Feb. 1908). Italian baritone, husband of below. Studied Palermo with Raimondi, Milan with Lamperti, London with García jun. Début New York 1848 (Carlo, *Ernani*). Berlin 1852; Ferrara 1853; London, HM, 1863–4. Roles included Leporello, Gounod's Méphistophélès, Rossini's Figaro. Taught Vienna Cons.; wrote a singing method, and translated several librettos into Italian, including *Tannhäuser* and *Lohengrin*.

2. **Mathilde** (née Graumann) (*b* Frankfurt, 24 Mar. 1821; *d* London, 17 Nov. 1913). German mezzo-soprano and teacher, wife of above. Studied Paris with García jun. Mainly a concert singer, appearing often with Salvatore, whom she married in 1852; only operatic appearance Bremen 1853 (Rosina). Taught Vienna Cons. 1854–61 and 1868–78; Paris 1861–4; Cologne 1865–8; established own studio in Paris 1881–1908. Her method, based on García's, drew on the traditions of bel canto, and aimed for evenness throughout the register, precise attack, excellent intonation, brilliance and ease at the top of the voice, and vocal longevity. Among her numerous prima donna pupils were Krauss, Di Murska, Calvé, Sanderson, Kurz, Eames, Garden, and Melba. Farrar, who declined to learn with her, preferred not to sacrifice 'all dramatic expression . . . to sound only'.

3. **Blanche** (*b* Paris, 4 Apr. 1863; *d* London, 15 Dec. 1940). French soprano, daughter of (1) and (2). Studied Paris with her mother. Début Prague 1900 (Brünnhilde, *Walküre*). Member of Moody-Manners Co. in London, where she settled as a teacher. Roles included Elsa, Isolde, Santuzza.

> BIBL: M. de Castrone Marchesi, *Erinnerungen aus meinem Leben* (Vienna, 1877); English version,

enlarged, *Marchesi and Music* (New York, 1897). B. Marchesi, *A Singer's Pilgrimage* (London, 1923, R/1977).

**Marchesi, Luigi** (*b* Milan, 8 Aug. 1755; *d* Milan, 14 Dec. 1829). Italian soprano castrato. His castration was apparently voluntary. Studied Modena with Caironi, Milan with Albuzzi. Début Rome, T delle Dame, 1773 (Anfossi's *L'incognita perseguitata*). Engaged for the Munich court, 1776–8. Naples 1778–81; Florence 1779–81; Milan from 1782; Turin 1782–98. Vienna 1785; Russia 1785–6 (inaugurating the Hermitage T, St Petersburg). London 1788–90. Thereafter he spent most of his time in Italy, retiring in 1805. Sang in numerous operas by Cimarosa, Bianchi, Sarti, Mayr, Mysliveček, etc., including many premières. One of the greatest (and vainest) castratos of his time, he was celebrated all over Europe for his strong, pure, and brilliant tone, his formidable technique, and his 'ravishing taste' (Kelly). Mount-Edgcumbe found him 'incomparable' in 'scenes of energy and passion', while criticizing his lavish ornamentation; Gerber admired his declamation and deportment, Stendhal his powers of improvisation. He was also adored for his good looks, one woman even leaving her husband and children in order to follow him round Europe.

**Marchetti, Filippo** (*b* Bolognola, 26 Feb. 1831; *d* Rome, 18 Jan. 1902). Italian composer. Studied Naples with Lillo and Conti; also helped by Mercadante. The success of his first opera, *Gentile da Varano*, in Naples was not matched by that of *La demente* in Turin and elsewhere, and the difficulties of finding a stage for his third opera, *Il paria*, coupled with the rise of Verdi, led him to withdraw from composition. He moved to Rome in 1862 and began work on *Romeo e Giulietta*, which was initially a failure in Trieste in 1865 but had a success in Milan two years later. He fully regained his popularity with *Ruy Blas* when it was staged in Florence after a poor initial reception in Milan; the work then swept Europe, and was even staged in Cairo, Sydney, and in North and South America. One of the only Italian operas apart from Verdi's to achieve real success in these years, it was followed by two more failures, *Gustavo Wasa* and *Giovanni d'Austria*. Though of the second rank, Marchetti's work was intermittently popular in its day for its melodic charm (the duet 'O dolce voluttà' from Act III of *Ruy Blas* was long famous); to some extent it anticipates verismo.

WORKLIST: *Gentile da Varano* (Marchetti; Turin 1856); *La demente* (The Madwoman) (Checchetelli; Turin 1856); *Il paria* (Checchetelli; comp. 1859, unprod.); *Romeo e Giulietta* (Marcello, after Shakespeare; Trieste 1865); *Ruy Blas* (D'Ormeville, after

Hugo; Milan 1869); *Gustavo Wasa* (D'Ormeville; Milan 1875); *Don Giovanni d'Austria* (D'Ormeville; Turin 1880).

**Marcolini, Marietta** (*b* Florence, *c*.1780; *d* ?). Italian mezzo-soprano. Début uncertain, but already singing in Venice 1800. Naples 1803–4; Rome 1808; Milan, S, 1809. A brilliant comic actress with a magnificent voice, she inspired Rossini (for whom she is said to have left Lucien Bonaparte) to write five roles for her: Ernestina (*L'equivoco stravagante*), the title-role of *Ciro in Babilonia*, Clarice (*La pietra del paragone*), Isabella (*L'italiana in Algeri*), and the title-role of *Sigismondo*. The finale of *L'italiana* is a tribute to her virtuosity and stamina. Sang in Milan until 1820, when she retired.

**Marcoux, Vanni** (also known as Vanni-Marcoux) (*b* Turin, 12 June 1877; *d* Paris, 22 Oct. 1962). French bass-baritone. Studied Turin with Collini, Paris with Boyer. Début Turin 1894 (Sparafucile). Bayonne 1899. London, CG, 1905–12, 1937. Paris: O, from 1908; OC, 1918–36. Milan, S, 1910; Boston 1911–12, Chicago 1913–14, 1926–32. His repertory of 240 roles included Don Giovanni, Don Basilio (Rossini), Hunding, Iago (Verdi), Scarpia, Golaud, Ochs, a much-acclaimed Boris, and many creations, e.g. Guido Colonna (*Monna Vanna*). He was also a celebrated Don Quichotte, a part which though created by Shalyapin was written for him by Massenet. An outstanding artist, he never allowed his moving and powerful dramatic sense to distort his superb singing. Retired from stage 1940, though sang Quichotte at Paris, OC, 1947. Taught Paris Cons. 1938–43; director Grand T, Bordeaux, 1948–51. (R)

**Maréchal, Adolphe** (*b* Liège, 26 Sep. 1867; *d* Brussels, 1 Feb. 1935). Belgian tenor. Studied Liège. Début Liège 1891. Paris, OC, 1895–1907. Monte Carlo and London, CG, 1902. Created several roles, e.g. Julien (*Louise*), Jean in *Le jongleur de Notre-Dame*. One of the Paris OC's foremost singers, he possessed a warm lyric tenor. Lost his voice in 1907. (R)

**Mařenka.** The daughter (sop) of Krušina and Ludmila, the eponymous intended bride of Jeník, in Smetana's *The Bartered Bride*.

**Marfa.** A young widow (sop), one of the Old Believers and lover of Andrey Khovansky, in Musorgsky's *Khovanshchina*.

**Marguerite.** A young girl (sop), lover of Faust, in Gounod's *Faust*.

**Marguerite de Valois.** Henri of Navarre's Queen (sop) in Meyerbeer's *Les Huguenots*.

**Maria.** The daughter (sop) of Kochubey and Lyubov, lover of Mazeppa, in Tchaikovsky's *Mazeppa*.

**Maria di Rohan.** Opera in 3 acts by Donizetti; text by Cammarano and Lillo, after Lockroy's *Un duel sous le cardinal de Richelieu* (orig. *Il Conte de Chalais*). Prod. Vienna, K, 5 June 1843, with Tadolini, Novarra, Guasco, G. Ronconi; London, CG, 8 May 1847, with E. Ronconi, Alboni, Ronconi, cond. Costa; New York, Astor Place OH, 10 Dec. 1849, with Bertucca, Perrini, Forti, Salvatore Patti, Beneventano, Giubilei, cond. Maretzek.

France, early 17th cent. Maria (sop) is secretly married to Henri, duc de Chevreuse (bar), but later falls in love with the comte de Chalais (ten), after interceding with him on behalf of her husband, who has been arrested for killing Richelieu's nephew. Chevreuse challenges Chalais to a duel and kills him; Maria too wants to die, but her husband decrees that she live a life of disgrace.

Also operas by Lillo (1839) and F. Ricci (1839); similar plot, with changed names, set by Giordano (*Regina Diaz*, 1892).

**Maria Egiziaca.** Opera in 1 act (3 episodes) by Respighi; text by C. Guastalla. Prod. (concert) New York, Carnegie H, 16 Mar. 1932, with Boerner, Eddy, cond. Respighi; (staged) Buenos Aires, C, 23 July 1933, with Dalla Rizza, Damiani, cond. Marinuzzi; London (concert), Hyde Park Hotel, 11 Apr. 1937.

**Mariani, Angelo** (*b* Ravenna, 11 Oct. 1821; *d* Genoa, 13 June 1873). Italian conductor. Studied Ravenna with Roberti. His music attracted the interest of Rossini, with whom he became friends. Cond. Messina 1844–5, where he tried to develop the orchestra in opera against some opposition. Milan, RD, 1846 (*Due Foscari*), so impressing Verdi that he was asked to conduct the première of *Macbeth*; this came to nothing, though he gave performances of *I Lombardi* and *Nabucco* so exciting that he was threatened with imprisonment for stirring up rebellion. Copenhagen 1847–8; Constantinople 1848–51; Genoa, CF, 1852 (*Robert le diable*), where he made the orchestra the best in Italy. First worked with Verdi, Rimini 1857 (*Aroldo*), when they became close friends. Bologna 1860, opening a long series of Verdi performances with *Ballo* and also giving the first Italian *Lohengrin* (1871) and *Tannhäuser* (1872). His great contribution was, by unremitting attention to detail and thoroughness of preparation, to make opera more genuinely integrated as a total theatrical experience. His refusal when seriously ill to conduct the première of *Aida* in Cairo caused a breach with Verdi.

BIBL: T. Mantovani, *Angelo Mariani* (Rome, 1921); F. Walker, *The Man Verdi* (London, 1962).

**Mariani, Luciano** (*b* Cremona, 1801; *d* Castell' Arquato, 10 June 1858). Italian bass. Milan, S, 1820–37. Created Oroe (*Semiramide*), Rodolfo

(*La sonnambula*), and Alfonso (*Lucrezia Borgia*). Possessed a voluminous voice—'too voluminous', said a Milanese critic—and had a long and distinguished career in Italy. His sister **Rosa** (*b* Cremona 1799; *d* ?) often sang with him. Début Cremona 1818. London, H, 1823. Created Arsace in *Semiramide*. Stendhal thought her 'remarkable', and 'the finest contralto now living'.

**Mariani-Masi, Maddalena** (*b* Florence 1850; *d* Erba, 25 Sep. 1916). Italian soprano. Studied Florence with Cortesi, Vienna Cons. Début Florence 1871. Milan, S, 1872. Created title-role in *La Gioconda*. Also sang Agathe, Lucrezia Borgia, Elena (*Vespri siciliani*). Capable of great variety in her singing, and imposing on stage. Retired in 1890, having forced her voice, and taught; her pupils included Lina Cavalieri. She appeared frequently with her sister **Flora Mariani de Angelis**, a talented mezzo-soprano who sang in Italy 1871–90.

**Maria Padilla.** The daughter (sop) of Don Ruiz de Padilla in Donizetti's *Maria Padilla*.

**Maria Stuarda.** Opera in 3 acts by Donizetti; text by Bardari after Schiller's tragedy (1800). Prod. (as *Buondelmonte*, with libretto changed by Salatino) Naples, C, 18 Oct. 1834, with Ronzi de Begnis, Del Serre, Pedrazzi, Crespi, cond. Donizetti; given in Donizetti's original form, Milan, S, 30 Dec. 1835, with Malibran, Puzzi-Toso, Reina, Marini, Novelli; New York, Carnegie H (concert), 16 Nov. 1964, with Hoffman, Jordan, Traxel, Michalski, Metcalf, cond. Scherman; London, St Pancras TH, 1 Mar. 1966, with Landis, Jolly, Hillmann, Drake, cond. Gover; New York, City O, 7 Mar. 1972, with Sills, Tinsley, Stewart, Fredericks, Devlin, cond. Rudel.

London and Fotheringhay, 1587. Elizabeth's (mez) favourite, Leicester (ten), and Talbot (bs) have received a request for help from her rival for the throne, Mary Stuart (sop). Leicester persuades Elizabeth to visit Mary in prison at Fotheringhay.

On Leicester's advice, Mary is submissive to the Queen, but nevertheless finds herself accused of treachery. She insults Elizabeth and is sentenced to death.

Learning of her sentence, Mary makes a full confession, denying any involvement in the murder of her husband. As a last wish, granted by Elizabeth, Mary asks that her maid may accompany her to the steps of the scaffold. She goes to her death, still protesting her innocence.

**Marie.** 1. A *vivandière* (sop), the eponymous 'daughter of the regiment' and lover of Tonio, in Donizetti's *La fille du régiment*.

2. Wozzeck's wife (sop) in Berg's *Wozzeck*.

**Marina.** Marina Mnishek (sop), daughter of a Sandomir nobleman, in Musorgsky's *Boris Godunov*.

**Marini, Ignazio** (*b* Tagliuno, 28 Nov. 1811; *d* Milan, 29 Apr. 1873). Italian bass. Début probably Brescia 1832. Milan, S, 1833–47; London, CG, 1847–9; New York 1850–2; St Petersburg 1856–63, succeeding Lablache. One of the most eminent singers of his time, he was a huge man with a deep and flexible basso cantante. Created many roles, including Guido (Donizetti's *Gemma di Vergy*), and title-roles of *Oberto* and *Attila*; he was also a good friend to the young Verdi. A new cabaletta (possibly not by Verdi) was written for him as Silva in *Ernani*, Vienna 1844. Acclaimed as Rossini's Mosè and Mustafà, and Oroveso in *Norma*. His wife **Antonietta Rainieri-Marini** was a talented mezzo-soprano; they often appeared together, e.g. in the first *Gianni di Parigi* (Donizetti), and in *L'italiana in Algeri*. Not at her best in soprano roles; but successfully created Leonora in *Oberto*. Verdi also wrote the Marchesa in *Un giorno di regno* for her, keeping both parts well within her range.

**Marino Faliero.** Opera in 3 acts by Donizetti; text by E. Bidèra, after Casimir Delavigne's tragedy (1829) and also Byron's tragedy (1821). Prod. Paris, I, 12 Mar. 1835, with Grisi, Rubini, Tamburini, Lablache, Santini, Ivanoff; London, CG, 14 May 1835; New Orleans, St Charles T, 22 Feb. 1842, with Salvatori, Ober-Rossi, Cecconi, Perozzi, cond. Rapetti. Has been performed in Germany as *Antonio Grimaldi*.

Venice, 1355. The Doge Marino Faliero (bs), is enraged at the leniency of the sentence imposed on Michele Steno (bs), who has accused his wife Elena (sop) of infidelity. She has, in fact, fallen in love with his nephew Fernando (ten). Marino's conspiracy against the Council is betrayed; as he is led to his death he forgives her.

**Marinuzzi, Gino** (*b* Palermo, 24 Mar. 1882; *d* Milan, 17 Aug. 1945). Italian conductor and composer. Studied Palermo with Zuelli. Palermo, T Massimo, conducting German as well as Italian repertory (first *Tristan* there, 1909), also in Italy and Buenos Aires (first *Parsifal* there, 1913). Cond. première of Puccini's *La rondine*, Monte Carlo 1917. Chicago 1919–21. Chief cond. Rome R, 1928–34. London, CG, 1934. Milan, S, with De Sabata 1934–44, director 1944. He wrote three operas, *Barberina* (Palermo 1903), *Jacquerie* (Buenos Aires 1918), and *Palla de' Mozzi* (Milan 1932). His son **Gino** (*b* New York, 7 Apr. 1920) was assistant cond. Rome R, 1946–51, and is a composer. (R)

**Mario, Giovanni** (*b* Cagliari, 17 Oct. 1810; *d* Rome, 11 Dec. 1883). Italian tenor. Of aristocratic birth, was an officer until he deserted the army for political reasons. Studied Paris with Bordogni and Ponchard. Début Paris, O, 1838 (title-role, *Robert le diable*). London, HM 1839–46, and CG 1847–71; Paris, I, from 1840; St Petersburg 1849–63, 1868–70; New York 1854; Madrid 1859, 1864. Retired from stage 1871; toured USA with Patti 1872–3. Then, having spent all his considerable earnings, lived in poverty until his death. Considered Rubini's successor, he was elegantly handsome, with a winning stage presence, and an extraordinarily sweet-toned voice. Chorley, however, wrote that he 'fascinated everyone . . . into forgetting incompleteness and deficiency which study might have remedied'. He was acclaimed in a wide repertory, e.g. Don Ottavio, Rossini's Otello and Count Almaviva, Arturo (*Puritani*), Raoul, John of Leyden, Eléazar, Duke of Mantua, Manrico, Edgardo, and first Ernesto in *Don Pasquale*. Verdi wrote an extra cabaletta for him in *I due Foscari*. His partnership with Grisi from 1839 was one of the most successful in operatic history; their personal relationship lasted until she died, but they were never able to marry.

BIBL: Mrs G. Pearse (his daughter) and F. Hird, *The Romance of a Great Singer* (London, 1910, R/1977); E. Forbes, *Mario and Grisi* (London, 1985).

**marionette opera-theatre.** See PUPPET OPERA.

**Maritana.** Opera in 3 acts by Wallace; text by Edward Fitzball, after the drama *Don César de Bazan* by D'Ennery and Dumanoir. Prod. London, DL, 15 Nov. 1845; Philadelphia, Walnut St. T, 9 Nov. 1846.

Madrid, early 19th cent. The opera tells of the complicated efforts of Don José (bar) to try to gain the attentions of the Queen of Spain. They revolve around the street-singer Maritana (sop), whom he persuades Don Caesar (ten) to marry on the eve of his execution; he subsequently implies to the Queen that the King (bs) is rather too fond of Maritana. However, plans go awry when Don Caesar's 'execution' fails to use real bullets and he returns to claim his bride. Don José is killed and Maritana and her husband live happily.

Also operetta *Don César* by Dellinger (1885).

**Mark.** 1. King Mark of Cornwall (bs), husband of Isolde, in Wagner's *Tristan und Isolde*.

2. A young man of unknown parentage (ten), bridegroom of Jenifer, in Tippett's *The Midsummer Marriage*.

**Marlowe, Christopher** (*b* Canterbury, 6 Feb. 1564; *d* Deptford, 30 May 1593). English poet and dramatist. Busoni's *Doktor Faust* (1925) is partly based on his *Tragedy of Dr Faustus* (?, entered on Stationers' Register, 1601); and he is the central figure of Mellers's *The Tragicall History of Christopher Marlowe* (comp. 1950–2).

**Marmontel, Jean François** (*b* Bort, 11 July 1723; *d* Abloville, 31 Dec. 1799). French dramatist, librettist, and critic. His many librettos include four for Rameau, seven for Grétry, and others for composers including Cherubini (*Démophon*, 1788) and Zingarelli (*Antigone*, 1790). His writings include important articles in the *Encyclopédie* and an *Essai sur les révolutions de la musique* (1777) in which he sided with Piccinni in the controversy against Gluck. His novel *Les Incas* (1773) was the basis of the operas *Cora* by Naumann (1782) and Méhul (1791). In his librettos, he initiated a number of ideas that were to prove attractive to his contemporaries and successors. His enthusiasm for Rousseau, whom he knew personally, marks his libretto *Le Huron* for Grétry (1768), with its adulation of the Noble Savage; he helped to give impetus to the interest in the medieval; and his story *Soliman Second*, made into a Favart libretto and then a verse play, helped to kindle interest in the oriental. It even directly influenced a number of more famous librettos, including that of Mozart's *Die Entführung aus dem Serail*, in its characters, which number among them a Chief Eunuch named Osmin. However, his tragic ambitions were not matched by a comparable talent, and for all the range of his subjects his handling of them is conventional and emotionally limited.

**Mârouf, Savetier du Caire** (Mârouf, the Cobbler of Cairo). Opera in 4 acts by Rabaud; text by Népoty, after a story in Mardrus's French version of the *1,001 Nights*. Prod. Paris, OC, 15 May 1914, with Davelli Jean Périer, Vieuille, cond. Ruhlmann; New York, M, 19 Dec. 1917, with Alda, De Luca, Rothier, cond. Monteux.

The cobbler Mârouf (ten) escapes his wife Fattoumah (sop) by going to sea. Shipwrecked, he is introduced by his friend Ali (bar) to the Sultan (bs) as a wealthy merchant; he marries the Sultan's daughter Saamcheddine (sop) and rifles his treasury, continually promising that his caravans will soon arrive. The princess loves him even when he confesses to her, and they flee. A magic ring brings Mârouf the service of a genie, and the pursuing Sultan forgives Mârouf when his rich caravans do indeed arrive.

**Marquise de Brinvilliers, La.** Opera in 3 acts by Auber, Batton, Berton, Blangini, Boieldieu, Carafa, Cherubini, Hérold, and Paer; text by Scribe and Castil-Blaze. Prod. Paris, OC, 31 Oct. 1831.

The most famous of several collective works of the period, treating the life and loves of the famous poisoner.

**Marriage, The** (Russ.: *Zhenitba*). Opera (unfinished) by Musorgsky; text after Gogol's comedy (1842). One act (four scenes) only completed in 1868; concert perf. at Rimsky-Korsakov's house, 1906; stage perf. with piano, rev. Rimsky-Korsakov, St Petersburg, Suvorin School, 1 Apr. 1909; first full prod., orch. Rimsky-Korsakov and Gauk, Petrograd 26 Oct. 1917. Completion (Acts II to IV) by Ippolitov-Ivanov, 1931; new completion in 1 act by Cherepnin, 1933. Ravel also had plans to orchestrate the work.

St Petersburg, 1830. Councillor Podkolesin (bar) decides to marry. The marriage-broker Fekla (con) tries to do a deal with him for Agafya's hand. His friend Kochkarev (ten) intervenes and tries to precipitate matters. (In Gogol's play, and in the revisions, Agafya (sop) is so alarmed by Podkolessin's attempts at proposal that she jumps out of the window. Podkolessin also vanishes leaving Kochkarev laughing triumphantly.)

Also opera in 1 act by Martinů; text by the composer, after Gogol. Prod. NBC TV, 7 Feb. 1953, with Stollin, Heidt, Gramm, cond. Adler; first stage prod. Hamburg 13 Mar. 1954, with Görner, Litz, Gura, Katona, Gollnitz, Marschner, Blankenheim, Roth, cond. Stein.

**Marriage of Figaro, The.** See NOZZE DI FIGARO, LE.

**Marschallin, Die.** The Princess von Werdenberg (sop), wife of the absent Field Marshal and lover of Octavian, in R. Strauss's *Der Rosenkavalier*.

**Marschner, Heinrich** (*b* Zittau, 16 Aug. 1795; *d* Hanover, 14 Dec. 1861). German composer. Studied Bautzen with Hering, Prague with Tomášek, Leipzig with Schicht. Apart from an early attempt, *Titus*, and a work, *Der Kyffhäuserberg*, in the manner of Viennese Singspiel, his first opera was *Heinrich IV und d'Aubigné*. This was staged by Weber, who backed the performance up with an appreciative article. By then Marschner had already written a Romantic opera, *Saidar und Zulima*, which Weber was unable also to perform in Dresden, whither Marschner moved in 1821. In 1824 he became Weber's assistant (as director of both Italian and German companies), though their initially cordial relationship had by now deteriorated, partly because Weber felt that Marschner was at once imitating his style and using it for commercial purposes rather than in the cause of German opera. *Der Holzdieb*, written in such time as he could find from his duties, is another Viennese-style Singspiel with anticipations of Lortzing. That year, 1826, Weber died, and on being passed over for the succession to his post, Marschner left Dresden.

His first new post was as music director in Danzig, where he produced an unsuccessful and stylistically jumbled attempt at Spontini's Grand Opera manner in *Lucretia* (with his wife Marianne Wohlbrück in the title-role). Moving to Magdeburg on their way to Leipzig, where Marianne was engaged, he began work with his brother-in-law on his first great success, *Der Vampyr*. The work takes much from *Der Freischütz*, both in the treatment of horror and in the form and melodic nature of the music, while lacking its predecessor's distinction especially in orchestration. It does, however, anticipate aspects of Wagner's *Der fliegende Holländer* in the handling of the accursed hero and his relationship with the pure heroine. Especially, Marschner here develops the idea of the character containing both good and evil, and divided against himself. He followed some of these ideas up in *Der Templer und die Jüdin*, which reflects the more elevated, courtly manner of *Euryanthe* and anticipates much in *Lohengrin* and *Tannhäuser* both in the functional use of orchestration and in the self-divided character of Bois-Guilbert.

In 1831 Marschner became Kapellmeister in Hanover, where he first wrote the inflated *Des Falkners Braut* before turning to the third of his great successes, *Hans Heiling*. Weber's influence, and that of E. T. A. Hoffmann, remain in the treatment of a character torn between the real and the magic world, and in such details as the use of melodrama, but the opera shares with *Der Vampyr* and *Der Templer und die Jüdin* an individual voice that absorbs the influences effectively. None of his five remaining operas matches these achievements. The dramatic weakness of *Der Schloss am Aetna* is only partly due to a poor libretto, though this affected the work and, more seriously, *Der Bäbu*, which contains some fine music in a cumbersome plot. *Kaiser Adolf von Nassau*, reluctantly rehearsed by Wagner and deplored by him as superficial, was Marschner's first attempt at a continuously composed opera. *Austin* again suffers from a confused plot and poor characterization, as does his last opera, *Sangeskönig Hiarne*.

Marschner did not really depart from Singspiel structure, and was more concerned with extending the range of set forms and finding links between them than in developing a continuous drama based on motive (of which he made inventive use within scenes). But he is capable of striking dramatic strokes, of the strong characterization of his hero/villains with divided souls, of some powerful orchestral writing and some expressive chromatic harmony. With his three most successful works, he makes an individual contribution to German Romantic opera, in the

history of which he earns a place as more than a link between Weber and Wagner.

WORKLIST: *Titus* (Mazzolà's text for Mozart, after Metastasio; comp. 1816, unprod.); *Der Kyffhäuserberg* (Kotzebue; comp. 1816, prod. Zittau 1822); *Heinrich IV und d'Aubigné* (Hornbostel; comp. 1817–18, prod. Dresden 1820); *Saidar und Zulima* (Hornbostel; Pressburg 1818); *Der Holzdieb* (The Wood-thief) (Kind; Dresden 1825); *Lucretia* (Eckschlager, after Livy's account; Danzig 1827); *Der Vampyr* (Wohlbrück, after Nodier (with Carmouche & Jouffroy), Planché, Ritter, & Polidori; Leipzig 1828); *Der Templer und die Jüdin* (Wohlbrück, after Scott & Lenz; Leipzig 1829); *Des Falkners Braut* (The Falconer's Bride) (Wohlbrück, after Spindler; Leipzig 1832); *Hans Heiling* (Devrient, after an old legend; Berlin 1833); *Das Schloss am Ätna* (The Castle at Etna) (Klingemann, after an old marionette play; Leipzig 1836); *Der Bäbu* (Wohlbrück, after an anon. tale; Hanover 1838); *Kaiser Adolf von Nassau* (Rau; Dresden 1845); *Austin* (Marianne Marschner; Hanover 1852); *Sangeskönig Hiarne* (Grothe, after Tegnér, after the saga; comp. 1857–8, prod. Frankfurt 1863).

BIBL: G. Münzer, *Heinrich Marschner* (Berlin, 1901); A. Dean Palmer, *Heinrich August Marschner, 1795–1861* (Ann Arbor, MI, 1980).

**Marseilles** (Fr., Marseille). City in Bouches-du-Rhône, France. The first opera given was Pierre Gautier's *Le triomphe de la paix* in the hall of the Jeu de Paume on 28 Apr. 1685. Opera was also given in a hall before the opening of the Grand T (Salle Beauvau) on 31 Oct. 1787 with Ponteuil's *Tartuffe* and Champein's *La mélomanie*. A strong tradition developed in the 19th cent. of enterprising repertory, especially Grand Opera, and distinguished guest singers. The theatre burnt down 1919; reopened 1924 as Opéra Municipal de Marseille (cap. 1,350); refurbished 1972. The theatre has seen the French premières of a number of important works, but under Jacques Karpo (since 1975) has placed greater emphasis on 19th-cent. revivals. Opera is also given in other theatres, and in open-air arenas, notably the T Silvair (cap. 4,000) and Parc Borély.

**'Martern aller Arten'.** Constanze's (sop) aria in Act II of Mozart's *Die Entführung aus dem Serail*, in which she declares that neither torture nor death itself will make her yield to the Pasha. Written for Katharina Cavalieri.

**Martha, oder Der Markt von Richmond** (Martha, or Richmond Market). Opera in 4 acts by Flotow; text by W. Friedrich (Friedrich Wilhelm Riese), after Vernoy de Saint-Georges's ballet-pantomime *Lady Harriette, ou La servante de Greenwich*, to which Flotow had contributed some music. Prod. Vienna, K, 25 Nov. 1847, with Anna Zerr, Schwarz, Erl, Formes, Just; London, DL, 4 July 1849; New York, Niblo's

Garden, 1 Nov. 1852, with Anna Bishop, Jacques, Guidi, Leach, Strini, Rudolph, cond. Bochsa.

Richmond, *c.*1710. Lady Harriet (sop), maid of honour to Queen Anne, is tired of court life, and so she and her maid Nancy (mez) disguise themselves as country girls. Under the names of Martha and Julia they go to Richmond Fair, where they are hired as servants by two young farmers, Lionel (ten) and Plunkett (bs). The two men fall in love with the girls; when they later return to court life, Lady Harriet realizes that she loves Lionel. A replica of Richmond Fair is set up in her own garden and there, once again, Lionel sees 'Martha' in her humble clothes and the two couples are happily united.

Suppé wrote a parody, *Martl* (1848).

**Martin, Frank** (*b* Geneva, 15 Sep. 1890; *d* Naarden, 21 Nov. 1974). Swiss composer. His two operas are *Der Sturm* (Vienna 1956), a word-for-word setting of Schlegel's translation of Shakespeare's *The Tempest*, and *Monsieur de Pourceaugnac* (Geneva 1963) after Molière, which moves from song by way of recitative and melodrama to unaccompanied speech in the interests of serving the drama.

**Martin, Jean-Blaise** (*b* Paris, 24 Feb. 1768; *d* Ronzières, 28 Oct. 1837). French baritone. Studied Paris with Dugazon. Début Paris, T de Monsieur, 1788. Paris: T Favart, 1794–1801; OC, 1801–23, 1826, 1833. Taught Paris Cons. 1816–18, 1832–7. With his range of E♭ –a', plus an octave falsetto extension, he gave his name to the voice type known as *baryton Martin* (see BARITONE) that became an important feature of opéra-comique thereafter. He created numerous roles in works by Dalayrac, Méhul, Boieldieu, Halévy, etc. and was an excellent comic actor.

**Martinelli, Giovanni** (*b* Montagnana, 22 Oct. 1885; *d* New York, 2 Feb. 1969). Italian tenor. Studied Milan with Mandolini. Début Milan, V, 1910 (Ernani). London, CG, 1912, 1913–14, 1919, 1937. Milan, S, 1912. New York, M, 1913–46. Also Chicago, San Francisco. Last appearance Seattle 1967 (Emperor, *Turandot*). Enormously popular, and an artist of high calibre; possessed a ringing tone, a superb legato and breath control, and impeccable style. Admired by Puccini, and by Barbirolli, who refers to his 'unfailing musicality'. Roles included Guillaume Tell, Faust (Gounod), Eléazar, Don José, Lensky, Manrico, Don Carlos, Rodolfo, Dick Johnson, Cavaradossi, Calaf, Otello (Verdi), and one Tristan. (R)

**Martinů, Bohuslav** (*b* Polička, 8 Dec. 1890; *d* Liestal, 28 Aug. 1959). Czech composer. Initially largely self-taught, then studied Prague with Suk

and Paris with Roussel. His first opera was a 3-act comedy, *The Soldier and the Dancer*. There followed *Les larmes du couteau*, a 1-act jazz opera, and *Trois souhaits*, a 3-act film opera also making use of jazz. *The Miracles of Mary* consists of a cycle of four mystery plays. Martinů's first real success on the stage came with *Julietta*, which concerns the relationship of dream and reality, and of reality and memory, in which his invariably fluent idiom achieves considerable lyrical intensity. *Mirandolina* was an attempt to update the spirit of Goldoni. A serious work followed, *The Greek Passion*, but the steadily cumulative intensity of the original proved difficult to translate to the stage in short scenes lacking true dramatic power. Martinů also wrote three radio operas, *The Voice of the Forest*, *Comedy on a Bridge*, and *What Men Live By*; the pungent wit of *Comedy on a Bridge*, one of his most attractive works, has successfully transferred to the stage.

SELECT WORKLIST: *Voják a tanečnice* (The Soldier and the Dancer) (Budín; Brno 1928); *Les larmes du couteau* (The Tears of the Knife) (Ribemont-Dessaignes; comp. 1928, prod. Brno 1968); *Tři přání* (Three Wishes) (Ribemont-Dessaignes; film opera, comp. 1929); *Hry o Marii* (The Miracle Plays of Mary) (Martinů, after Ghéon; Brno 1935); *Hlas lesa* (The Voice of the Forest) (Nezval; Czech Radio 1935); *Veselohra na mostě* (Comedy on a Bridge) (Martinů, after Klicpera; Czech Radio 1937); *Julietta* (Martinů, after Neveux; Prague 1938); *Čím člověk žije* (What Men Live By) (Martinů, after Tolstoy; TV opera, New York 1953); *Ženitba* (The Marriage) (Martinů, after Gogol; TV opera, New York 1953); *Mirandolina* (Martinů, after Goldoni; Prague 1959); *Řecké pašije* (The Greek Passion) (Martinů, after Kazantzakis; Zurich 1961).

BIBL: M. Šafránek, *Bohuslav Martinů* (Prague, 1961; trans. London, 1962); B. Large, *Martinů* (London, 1975).

**Martinucci, Nicola** (*b* Taranto, 28 Mar. 1941). Italian tenor. Studied Milan with Sforni. Début Milan, N, 1966 (Manrico). Milan, S; Venice, F; Florence; etc. Deutsche Oper am Rhein from 1973; Verona from 1982; London, CG, 1985; New York, M, 1988. Repertory incl. Verdi, Puccini, Mascagni, Bizet. (R).

**Martín y Soler, Vicente** (*b* Valencia, 2 May 1754; *d* St Petersburg, 11 Feb. 1806). Spanish composer. Trained as a choirboy; went to Madrid, where he made his début as an opera composer 1776, possibly with the zarzuela *La Madrileña*. Then established his reputation in Italy, producing both serious and comic works for Lucca, Parma, Turin, Venice, and above all Naples. Encouraged to Vienna by Nancy Storace 1785: here he collaborated with Da Ponte on three operas: *Il burbero di buon cuore, L'arbore di*

*Diana*, and his greatest triumph, *Una cosa rara*. Such was its success that it supplanted *Figaro* which had been premièred earlier in the year; Mozart later quoted an aria from it ('O quanto si bel giubilo') in the banquet scene of *Don Giovanni*. Invited to St Petersburg by Catherine the Great 1788, replacing Sarti as court composer. Here he wrote two operas, *Gore Bogatyr Kosometovich* and *Pesnolyubie*, directed many productions, and played a significant role in the early cultivation of Russian opera. Da Ponte engaged him as composer at the London Haymarket 1794–6 but *c*.1796 he returned to St Petersburg and was inspector of the Italian court theatre 1800–4.

Although his works have been largely neglected, Martín y Soler was one of the most gifted and popular opera buffa composers of the late 18th cent., rivalling Cimarosa, Paisiello, and even Mozart. He possessed a sure gift for captivating melody, dramatic pace, and sharp musical characterization, talents recognized by Da Ponte, who much admired his work. At his best, e.g. in *Una cosa rara*, he approaches Rossini in his perceptive and engaging treatment of comic situations. During his lifetime his works were translated into many languages, including Danish, English, Hungarian, Polish, and Russian.

SELECT WORKLIST: *La Madrileña* (?; Madrid ?1776); *Il burbero di buon cuore* (The Moaner with a Heart of Gold) (Da Ponte, after Goldoni; Vienna 1786); *Una cosa rara* (A Rarity) (Da Ponte, after Vélez de Guevara; Vienna 1786); *L'arbore di Diana* (Da Ponte; Vienna 1787); *Gore Bogatyr Kosometovich* (The Unfortunate Hero Kosometovich) (Catherine the Great & Khrapovitsky; St Petersburg 1789); *Pesnolyubie* (Beloved Songs) (Khrapovitsky; St Petersburg 1790).

**Marton, Eva** (*b* Budapest, 18 June 1943). Hungarian soprano. Studied Budapest with Rösler and Sipos. Début Budapest 1968 (Queen of Shemakha). Frankfurt 1974–7; Bayreuth 1977–8; Milan, S, from 1978; New York, M, from 1981; Chicago 1980; London, CG, 1987. San Francisco, Buenos Aires, Vienna, etc. Possesses an even, powerful, dramatic soprano. Roles include Donna Anna, Elsa, Brünnhilde, Ariadne, Salome, Elektra, Turandot, Tosca. (R)

**Martyrs, Les.** Opera in 4 acts by Donizetti; text by Scribe, after Corneille's tragedy *Polyeucte*, orig. composed to an Italian libretto by S. Cammarano, *Poliuto*, in 3 acts, and written for Nourrit in Naples (rehearsed 1838) but banned by the censor. Prod. Paris, O, 10 Apr. 1840, with Dorus-Gras, Duprez, Massol, Derivis, Wartel, Serda; New Orleans, T d'Orléans, 24 Mar. 1846, with J. Calvé, Arnaud, Garry, Douvry, Mordant, cond. E. Prévost; London, CG, 20 Apr. 1852, as

*I martiri*, with Jullienne, Tamberlik, Roncini, Marini, cond. Costa.
For plot see POLIUTO.

**Mary Stuart.** See MARIA STUARDA.

**Marzelline.** The daughter (sop) of the gaoler Rocco, in love with Fidelio, loved by Jacquino, in Beethoven's *Fidelio*.

**Masaniello.** Tommaso Aniello (ten), a Neapolitan fisherman and brother of Fenella, in Auber's *La muette de Portici*.

**Mascagni, Pietro** (*b* Livorno, 7 Dec. 1863; *d* Rome, 2 Aug. 1945). Italian composer and conductor. Studied Livorno with Soffredini and Milan Cons. with Ponchielli and Saladino, where his friends included Puccini. Despite early recognition of his talent, he found the training stultifying and left before completing his formal studies. For a while he travelled Italy with a touring opera company; joined the orchestra of the T dal Verme, Milan 1884; conducted an operetta season in Parma 1885; eventually settled at Cerignola as a piano teacher.

Mascagni's first operatic work, *Pinotta*, which he mislaid for 50 years, was an adaptation of a cantata; his next attempt, the romantic tragedy *Guglielmo Ratcliff*, remained unfinished when he was unable to find a theatre willing to stage it. But in 1890 he won first prize in a competition sponsored by the publishers Sonzogno with his 1-act opera *Cavalleria rusticana*. At its première this melodramatic opera of love and hatred in Sicily was given a tumultuous reception, making its composer famous overnight; it was quickly taken up across Europe (Berlin 1890; London 1891; Paris 1892) and began a vogue for similar 1-act works. With Leoncavallo's *Pagliacci* of 1892 it represented the beginning of *verismo opera and attracted numerous admirers, including Verdi and Puccini, partly because it was generally regarded as a brilliantly successful counter to the principles of Wagnerian music drama.

The rest of Mascagni's career was effectively spent trying to recapture the success of *Cavalleria rusticana*, but to no avail. *L'amico Fritz* was quickly produced, but despite some charming pastoral passages failed to convince; the works which followed, *I Rantzau*, the completed *Guglielmo Ratcliff*, and *Silvano* were all soon forgotten. By the time that *Zanetto* was performed Mascagni had moved to Pesaro as director of the conservatory; modest in conception it fared no better. With the orientally inspired *Iris*, Mascagni fleetingly regained some of his former popularity, not least because of his inventive orchestration. The short-lived success of *Isabeau* apart, this was the last of his works to achieve serious recognition. The bold concept of *Le mas-*

*chere*, with its six simultaneous premières, was fundamentally flawed and the opera had a brief and stormy career, while *Amica* was equally short-lived. *Il piccolo Marat*, which many consider Mascagni's best work, found few admirers; none of the later operas, which include *Parisina*, *Lodoletta*, and the rediscovered *Pinotta*, have had any following.

Although Mascagni enjoyed some fame as an opera conductor, he was not highly regarded by his peers. His precarious position foundered completely when he allowed himself to become too closely associated with Fascism, composing his *Nerone* for La Scala, Milan, with Mussolini in mind, as well as choral and orchestral works for various political occasions. As a result he was ostracized by many Italian musicians including Toscanini, and spent the last few years of his life in comparative poverty and disgrace in a Rome hotel. Of his works, only *Cavalleria rusticana* has found a permanent place in the present-day repertory. Although easily dismissed as a rather tawdry piece of melodrama, it was strikingly original and daring in its time, and shows Mascagni to have possessed, at his best, considerable dramatic skill. (R)

SELECT WORKLIST: *Pinotta* (Targioni-Tozzetti; comp. c.1880, prod. San Remo 1932); *Guglielmo Ratcliff* (Maffei, after Heine; Milan 1895); *Cavalleria rusticana* (Rustic Chivalry) (Targioni-Tozzetti & Menasci, after Verga; Rome 1890); *L'amico Fritz* (Suardon, after Erckmann-Chatrian; Rome 1891); *I Rantzau* (Targioni-Tozzetti & Menasci, after Verga; Florence 1892); *Zanetto* (Targioni-Tozzetti & Tenasci, after Coppée; Pesaro 1896); *Iris* (Illica; Rome 1898); *Le maschere* (The Masks) (Illica; Genoa, Milan, Rome, Turin, Venice, Verona 1901); *Amica* (Berel; Monte Carlo 1905); *Isabeau* (Illica; Buenos Aires 1911); *Parisina* (D'Annunzio; Milan 1913); *Lodoletta* (Forzano, after Ouida; Rome 1917); *Il piccolo Marat* (Forzano & Targioni-Tozzetti; Rome 1921); *Nerone* (Targioni-Tozzetti, after Cossa; Milan 1935).

BIBL: D. Cellmare, *Pietro Mascagni* (Rome, 1965).

**Maschere, Le** (The Masks). Opera in a prologue and 3 acts by Mascagni; text by Illica. Prod. simultaneously in six Italian cities, 17 Jan. 1901. In Milan (La Scala, with Brambilla, Carelli, Caruso, cond. Toscanini), Venice, Turin, and Verona it was hissed; in Genoa it was not allowed to be completed; only in Rome, where Mascagni conducted it, did it meet with any kind of favour. It has been revived a few times.

In a plot of commedia dell'arte characters, Florindo (ten) and Rosaura (sop), aided by Columbina (sop) and Arlecchino (ten), attempt to prevent the marriage which Rosaura's father, Pantalone (bs), has planned for her. They eventually succeed.

**Mascheroni, Edoardo** (*b* Milan, 4 Sep. 1892; *d* Como, 4 Mar. 1941). Italian conductor and composer. Studied Milan with Boucheron. Cond. Brescia 1880; Rome, Ap., 1884–8; Chief cond. Milan, S, 1891–4, conducting premières of *La Wally* (1892) and, at Verdi's request, *Falstaff* (1893), the latter with such effect that Verdi called him the 'third author'. Cond. Germany, Spain, S America. His *Lorenza* (text by Illica) was prod. Rome 1901, and *La perugina* (Illica) Naples 1909. His brother **Angelo** (1855–95) studied Paris with Delibes and was also a conductor who toured with Patti.

**Mascotte, La.** Operetta in 3 acts by Audran; text by Duru and Chivot. Prod. Paris, B, 28 Dec. 1880 (1,000 perfs. by 1885); Boston, Gaiety T, 11 Apr. 1881; Brighton 19 Sep. 1881.

Piombino, 17th cent. The goose-girl Bettina (sop), sent as a good luck mascot to the farmer Rocco, is loved by the shepherd Pippo (bar). Prince Laurent (ten) takes her away to his castle, leaving his daughter Fiametta (sop) to try to seduce Pippo. He agrees to marry Fiametta but, at the last moment, escapes with Bettina. After a war, in which Laurent is on the losing side, the Prince agrees to the marriage of Pippo and Bettina.

Audran's most popular work.

**Masetto.** A peasant (bar), betrothed to Zerlina, in Mozart's *Don Giovanni*.

**Masini, Angelo** (*b* Forlì, 28 Nov. 1844; *d* Forlì, 28 Sep. 1926). Italian tenor. Studied with Minguzzi. Début Finale Emilia 1867 (Pollione). Appearances in Venice, Rome, Florence, etc. until 1875; sang the first performances of Verdi's Requiem under his direction in Paris, London, and Vienna. Cairo 1875–6; St Petersburg 1879–1903; Madrid, 1881–2, 1884–5; Barcelona 1886; retired 1905. A virtuoso singer, he possessed what Tamagno called 'un voce di paradiso', a ravishing mezza voce, and superb breath control. Successful as Don Ottavio, Count Almaviva (Rossini), Nemorino, Duke of Mantua (having seven different endings for 'La donna è mobile'). Verdi thought there would have been 'no one better' as the first Fenton, but Masini accepted a more lucrative contract in Russia (where he became the Tsar's favourite tenor).

**Maskarade.** Opera in 3 acts by Nielsen; text by Vilhelm Andersen, after a play by Holberg. Prod. Copenhagen 11 Nov. 1906, with Ulrich, Møller, Knudsen, Neiiendam, Kierulf, Jerndoff, Mantzius, cond. Nielsen; St Paul, MN, 23 June 1972, with Neil, Williams, Atherton, Christeson, Beri, cond. Buketoff; London, Morley Coll., 9 May 1986, with Rivera, Dyer, Chard, and Tansky.

Copenhagen, 1723. Leander (ten) has fallen in love with an unknown woman at a masked ball, and vows to marry her, although his father Jeronimus (bs) has already arranged another marriage for him. Against Jeronimus's instructions, Leander goes to the masquerade; he is sighted by Mr Leonard (ten or bar), the father of his intended wife. Leander's beloved, Leonora (sop), turns out to be none other than Leonard's daughter, so the predestined couple are now happy to be united.

**Mask of Orpheus.** Opera in three acts by Birtwistle; text by Peter Zinovieff, after the classical legend. Prod. London, C, 21 May 1986, with Rigby, Dilke, Robinson, Angel, Langridge, Walters, Robson, McDonnell, cond. Howarth.

Each major role is taken by three performers, representing different aspects of the character as man (or woman), hero, and myth—Orpheus (ten, ten, dancer/mime), Eurydice (mez, mez, dancer/mime), and Aristaeus (bs-bar, bs-bar, dancer/mime)—and the three acts treat different aspects of the legend.

**Maslennikova, Leokadia** (b Saratov, 3 June 1918). Russian soprano. Studied Minsk, and Kiev with Yevtushenko. Kiev 1944–6; Moscow, B, 1946–59. Roles incl. Parasya (*Sorochintsy Fair*), Tatyana, Lisa, Yolanta, Marguerite, Micaëla, Nedda, Tosca. (R)

**Masnadieri, I** (The Robbers). Opera in 4 acts by Verdi; text by Maffei, after Schiller's drama *Die Räuber* (1781), after C. F. D. Schubart's story (1777). Prod. London, HM, 22 July 1847, with Lind, Gardoni, Coletti, Lablache, Corelli, Bouché, cond. Verdi; Rome, Ap., 12 Feb. 1848; New York, Winter Garden T, 31 May 1860, with Olivieri, Guerra, Luisia, Mirandola.

Germany, early 18th cent. Carlo (ten) is, through the machinations of his brother Francesco (bar), disinherited by their father Massimiliano Moor (bs) after leaving home and his lover Amalia (sop) and becoming chief of a robber band. Francesco announces Carlo's death, and also that of his father (whom he has imprisoned). But Amalia learns the truth, and also repels Francesco's advances. The lovers meet in a forest, and Carlo frees his father without, from loyalty to the robbers, disclosing his identity. He is later recognized by his father, but kills Amalia rather than bind her to his life of crime.

**masque** (sometimes also **mask**). A 16th- and 17th-cent. English stage entertainment combining poetry, music, singing, dancing, and acting, often setting mythological subjects in elaborate scenery. It grew out of the French *ballet de cour*, and especially the model of Beaujoyeulx, but in England was not restricted to courtly performance: masques were, for example, sometimes given in the Inns of Court. The most important early writer was Ben Jonson who, beginning with the *Masque of Blackness* (1605), provided a notable series of works for the court up to 1631. Although there was no rigid pattern for the masque, the action proceeded along broadly fixed lines: invariable was the presence of the three main dance sections (the 'entry', 'main dance', and 'going off') and the concluding 'revels', dances in which the audience participated. In *The Hue and Cry after Cupid* (1608) and *The Masque of Queens* (1609) Jonson introduced the *antimasque, a grotesque interlude comparable to the *intermezzo and, following its introduction to England through Dowland's lute songs, the *stile rappresentativo made a gradual appearance too. According to its preface, Jonson's *Lovers Made Men* (1617) was set throughout in this style by its composer Lanier: if true, this would lend it a good claim to be the first English opera.

As an art-form uniting music and drama, the masque was an important precursor of short-lived English opera. It reached its most lavish form under Queen Caroline (reigned 1625–49), in such spectacular works as Shirley's *The Triumph of Peace* (1634; music by Ives) and *Britannia trumphans* (1638; music by William Lawes). During the Protectorate performances of masques were so restricted that only one was performed, Shirley's *Cupid and Death* (1653), with music by Locke and Christopher Gibbons. Following the Restoration in 1660, there was a revival of interest in the genre, although many works so described are closer to genuine opera, among them Blow's *Venus and Adonis* (c. 1682).

The masque survived into the 18th cent.: important works were Congreve's *The Masque of Paris*, the subject of a famous competition in 1701 to find the best setting of its text, and Handel's *Semele*. The masque form has occasionally been revived in modern times, for example in Britten's *Gloriana* and Harrison Birtwistle's *The Mask of Orpheus*.

**Massé, Victor** (b Lorient, 7 Mar. 1822; d Paris, 5 July 1884). French composer. Studied Paris Conservatoire with Halévy, winning the Grand Prix de Rome. Began his career with *La chambre gothique*; this was followed by a string of popular comic works beginning with *La chanteuse voilée* and including *Galathée* and *Les noces de Jeannette* (his most successful opera: 1,000 performances by 1895). For a while Massé was seen as the successor to Auber, but his later works, which included *La reine Topaze* and *Les saisons*, did not prove enduring, though they are always charmingly tuneful.

# Massenet

He became chorus-master at the Opéra 1860, and professor of composition at the Conservatoire, 1866. After *Le fils du brigadier* he wrote no stage music for a while, but in 1876 produced a serious work, *Paul et Virginie*, which was well regarded. The same year he resigned from the Opéra because of illness; during his declining years he wrote another work in similar vein, *Une nuit de Cléopâtre*. Ambitious in scope, neither recaptured the success of his earlier comic operas, the best of which stayed in the repertoire long after his death.

SELECT WORKLIST (all first prod. Paris): *La chambre gothique* (Carmouche; 1849); *La chanteuse voilée* (Scribe & De Leuven; 1850); *Galathée* (Barbier & Carré; 1852); *Les noces de Jeannette* (Barbier & Carré; 1853); *Les saisons* (Barbier & Carré; 1855); *La reine Topaze* (Lockray & Battu; 1856); *Le fils du brigadier* (Labiche & Delacour; 1867); *Paul et Virginie* (Barbier & Carré, after Saint-Pierre; 1876); *Une nuit de Cléopâtre* (Barbier after Gautier; 1885).

BIBL: G. Ropartz, *Victor Massé* (Paris, 1887).

**Massenet, Jules** (*b* Montaud, 12 May 1842; *d* Paris, 13 Aug. 1912). French composer. Studied with his mother, Paris with Thomas. On his return from Italy, where he spent three years as a Prix de Rome winner, he had sufficient success with *La grand'tante* in 1867 for him to be commissioned, once the Franco-Prussian War was over, to write *Don César de Bazan*. However, his career was really set on its course by Pauline Viardot's championship of his 'sacred drama' *Marie-Magdeleine*, whose theme of the reformed courtesan touched off the vein of what D'Indy called 'a discreet and pseudo-religious eroticism' that was to mark much of his work. *Le roi de Lahore* was carefully tailored to the demands of the newly opened Opéra (Palais Garnier) and to an audience that had learnt to appreciate the orientalisms of Bizet's *Les pêcheurs de perles*. Its triumph made Massenet's name, and led to his establishment at the Conservatoire as the most influential teacher of his generation. He was as popular in this role as he saw to it that he remained with the public as a composer. Though *Hérodiade* shows some Wagnerian influences, and in its free melodic declamation anticipates aspects of Debussy's *Pelléas et Mélisande*, the style is essentially sweet and appealing whatever the dramatic situation. His capacity to charm and captivate under all circumstances, remarked upon by Fauré, was even more successfully deployed in *Manon*: though the orchestration and some use of motive were to lead to jibes about 'Mlle Wagner', the work put him in a commanding position as the most successful French opera composer of the day, with large and enthusiastic audiences.

Nevertheless, three comparative failures followed, *Le Cid*, *Esclarmonde*, and *Le mage*; Masse-net's preference for *Esclarmonde* above all his other works is perhaps in part connected to the young Sybil Sanderson, who sang the title-role and who played an important part in his life. In *Werther* the sentimental aspect of the novel touched his own, and his mastery of an elegant declamatory vocal line, subtly reinforced by the orchestra, is at the service of powerful and touching scenes handled with a sure dramatic instinct. With *Thaïs*, he returned to the contrast between sacred and profane love in *Hérodiade*, and to the treatment of the vulnerable courtesan of *Manon*; and though he does not portray in much depth either the sacred or the profane, or make Thaïs much larger than the touching Manon, the work has his now familiar skill and charm. *Le portrait de Manon* was a conscious return to the earlier theme with a sequel; a more original departure was *La navarraise*, a highly successful attempt to bring verismo back to its origins in France. *Sapho* again returns to formula in a work contrasting rural innocence with urban sophistication, once more depicting the two in elegant tones rather than entering seriously into their characterization. Not for the first time, there is here repaid a debt which Tchaikovsky had incurred to Massenet in some of his own operas (especially *The Maid of Orleans* and *Eugene Onegin*).

*Sapho* was Massenet's last major success, though *Cendrillon*, a skilful confection of tried devices written for Carré's new reign at the Opéra-Comique, proved popular; and Massenet's mastery of the pretty was re-invoked with *Grisélidis*, whose plot it suits less well. With *Le jongleur de Notre-Dame* he abandoned the female portraiture in which had excelled for an all-male opera, and with all his old skill brings off a subject that tests even his ability with the religiously sentimental. Few of his remaining works showed such command, and the times were at last turning against him; but there is a genuinely affecting quality in *Don Quichotte*, in part through a subtle refinement of his skills to a moving simplicity. Autobiography, *Mes souvenirs* (Paris, 1912, trans. 1919, R/1970).

WORKLIST (all first prod. Paris unless otherwise stated): *Esmeralda* (?, after Hugo; unprod., comp. c.1865, lost); *La coupe du roi de Thulé* (Gallet; comp. c.1866, unprod.); *La grand'tante* (The Great-Aunt) (Adénis & Granvallet; 1867); *Don César de Bazan* (D'Ennery, Dumanoir, & Chantepie, after Hugo's character in *Ruy Blas*; 1872); *L'adorable bel'-boul'* (Gallet; 1874, lost); *Bérangère et Anatole* (Meilhac & Poirson; 1876); *Le roi de Lahore* (Gallet; 1877); *Robert de France* (?; comp. c.1880, unprod., lost); *Les Girondins* (?; comp. 1881, unprod., lost); *Hérodiade* (Millet, Grémont, & Zamadini, after Flaubert; Brussels 1881); *Manon* (Meilhac & Gille, after Prévost; 1884); *Le Cid* (D'Ennery, Blau, & Gallet, after

Corneille; 1885); *Esclarmonde* (Gallet & Gramont; 1889); *Le mage* (Richepin; 1891); *Werther* (Blau, Milliet, & Hartmann, after Goethe; Vienna 1892); *Thaïs* (Gallet, after France; 1894); *Le portrait de Manon* (Boyer; 1894); *La navarraise* (Claretie & Cain; London 1894); *Amadis* (Claretie; comp. *c*.1895, prod. 1922); *Sapho* (Cain & Bernède, after Daudet; 1897); *Cendrillon* (Cain, after Perrault; 1899); *Grisélidis* (Silvestre & Morand, after Boccaccio; 1901); *Le jongleur de Notre-Dame* (Our Lady's Juggler) (Léna, after France, after a medieval legend; Monte Carlo 1902) *Chérubin* (Cain, after Croisset, after Beaumarchais's character in *Le mariage de Figaro*; Monte Carlo 1905); *Ariane* (Mendès, after the classical legend; 1906); *Thérèse* (Claretie; Monte Carlo 1907); *Bacchus* (Mendès, after a Sanskrit epic; 1909); *Don Quichotte* (Cain, after Le Lorrain, after Cervantes; Monte Carlo 1910); *Roma* (Cain, after Parodi: Monte Carlo 1912); *Panurge* (Spitzmüller & Boukay, after Rabelais; 1913); *Cléopâtre* (Payen; Monte Carlo 1914).

BIBL: J. Harding, *Massenet* (London, 1970).

**Massol, Eugène** (or **Jean-Etienne**) (*b* Lodève, 23 Aug. 1802; *d* Paris, 30 Oct. 1887). French baritone. Studied Paris Cons. Début as tenor, Paris, O, 1825 (Licinius, *La vestale*). Paris, O, until 1845 (by then a baritone), and 1850–8, when retired. Brussels, M, 1845; London: DL, 1846; CG, 1848–50; HM, 1851. Sang as tenor in the premières of *Guillaume Tell*, *La muette de Portici*, *Robert le diable*. As baritone created Sévère (Donizetti's *Les martyrs*), Reuben (Auber's *L'enfant prodigue*), De Nevers (*Les Huguenots*), and lead roles in Halévy's *La reine de Chypre* and *Le Juif errant*. Also successful as Alfonso (*Favorite*). Lumley praised him as 'a thorough artist'.

**Master Peter's Puppet Show.** See RETABLO DE MAESE PEDRO, EL.

**Mastersingers of Nuremberg, The.** See MEISTERSINGER VON NÜRNBERG, DIE.

**Masterson, Valerie** (*b* Birkenhead, 3 June 1937). English soprano. Studied London with Clinton and Asquez, Milan with Saraceni. Début Salzburg 1963 (Frasquita). D'Oyly Carte Co. 1966–70. London, SW/ENO from 1971, and CG from 1974; Paris, O, from 1978. Also Gly., Buenos Aires, Barcelona, Prague. Roles include Semele, Countess (Mozart), Constanze, Adèle (*Ory*), Marguerite, Violetta, Sophie. A singer of poise and distinction, especially successful in Handel, Mozart, and the French repertory. (R)

**Matačić, Lovro von** (*b* Sušak, 14 Feb. 1899; *d* Zagreb, 4 Jan. 1985). Croatian conductor and producer. Studied Vienna with Schalk, Herbst, and Nedbal. Chorus master Cologne 1917, cond. 1919. Cond. Novi Sad 1919; Ljubljana 1924–6; Belgrade 1926–32, 1938–42; Zagreb 1932–8; Vienna, V, 1942–5. Guest appearances Germany and Italy from 1953. Music director Dresden

1956–8, jointly with Konwitschny Berlin, S, 1956–8. Chicago 1959. Frankfurt 1961–6. Bayreuth 1959 (*Lohengrin*). Also produced many operas, including *Orfeo*, *Poppea*, *Parsifal*, and *Turandot*. (R)

**Materna, Amalie** (*b* St Georgen, 10 July 1844; *d* Vienna, 18 Jan. 1918). Austrian soprano. Studied Graz. Début Graz 1864 (Suppé's *Light Cavalry*). Vienna, H, 1869–94. Bayreuth 1876, 1882–91; New York, M, 1885. A leading Wagnerian of her time; sang Brünnhilde in the first complete *Ring* (Bayreuth 1876); also created Kundry. Much admired by Wagner, who declared her the one woman capable of singing Brünnhilde; he also found her 'sympathetic' and visually convincing. Other roles included Elisabeth, Rachel (*La Juive*), Valentine. Created title-role of Goldmark's *Königin von Saba*. Possessed a large, bright voice and immense stamina.

**Mathis, Edith** (*b* Lucerne, 11 Feb. 1938). Swiss soprano. Studied Lucerne, and Zurich with Basshart. Début Lucerne 1956 (Second Boy, *Zauberflöte*). Cologne 1959–63; Berlin, D, from 1963; Gly. 1962–3, 1965; Hamburg 1960–72; New York, M; 1970; London, CG, 1970, 1972; Salzburg and Munich Festivals. Successful as Zerlina, Pamina, Nannetta, Mélisande, Zdenka, and particularly as Sophie. (R)

**Mathis der Maler** (Mathis the Painter). Opera in 7 scenes by Hindemith; text by the composer, after Matthias Grünewald's life and his altarpiece at Colmar. Prod. Zurich 28 May 1938, with Hellwig, Funk, Stig, cond. Denzler; Edinburgh 29 Aug. 1952, with Wasserthal, Rothenberger, Ahlersmeyer, cond. Ludwig; Boston U, 17 Feb. 1956, with Gay, cond. Caldwell.

In the Peasants' War of 1542, Grünewald (bar) leads the peasants against the church. Losing faith in his cause, he escapes with Regina (sop). He renounces the outside world in favour of his art.

This portrayal of people rising against authority alarmed the Nazis, who despite protests by Furtwängler banned the scheduled 1934 Berlin première. Furtwängler was dismissed and Hindemith had to leave the country. A symphony drawing on music from the opera has become well known in the concert hall.

**Matinsky, Mikhail** (*b* Pokrovsk, 1750; *d* St Petersburg, *c*.1820). Russian librettist. Studied Moscow and in Italy. In 1782 he wrote the text for Pashkevich's *Sanktpeterburgsky gostiny dvor* (St Petersburg Market). This only survives in its revised version (1792) as *Kak pozhivyosh, tak i proslyvyosh* (As you live, so shall you be judged), and makes much use of folk music and of the chorus (including a remarkable and effective sequence of seven wedding choruses) to give one of the earliest operatic pictures of Russian life.

**Matrimonio segreto, Il** (The Secret Marriage). Opera in 2 acts by Cimarosa; text by Bertati, after the comedy *The Clandestine Marriage* (1766) by Colman and Garrick. Prod. Vienna, B, 7 Feb. 1792, with Bussani, Bosello, Mandini, when the whole work was encored at the request of Leopold II; London, H, 11 Jan. 1794, with Pastorelli, Casentini, Braghetti, Rovedino, Morelli; New York, Italian OH, 4 Jan. 1834.

Bologna, 1780. The opera tells of the attempts of Geronimo (bs), a wealthy citizen of Bologna, to marry off his daughter Elisetta (sop) to an English 'Milord', Count Robinson (bs). The latter prefers Geronimo's other daughter Carolina (sop), who is secretly married to Paolino (ten), a young lawyer and business associate of Geronimo. Geronimo's sister Fidalma (mez), who rules the household, is herself in love with Paolino. Carolina and Paolino plan an elopement, and after a bedroom scene of mistaken identities, all ends happily with the Count agreeing to marry Elisetta.

Cimarosa's most popular work, frequently revived. It was chosen to inaugurate La Piccola Scala, Milan, 26 Dec. 1955.

Operas on the same subject by Graffigna (1883) and Gast (*Der Löwe von Venedig*, prod. 1891 as *Die heimliche Ehe*).

**Mattei, Stanislao** (*b* Bologna, 10 Feb. 1750; *d* Bologna, 12 May 1825). Italian composer. Studied Bologna with Martini, whom he succeeded at San Francesco. A rigid conservative, he was the successful teacher of, among others, Rossini, Donizetti, Morlacchi, and Bertolotti.

**Matters, Arnold** (*b* Adelaide, 11 Apr. 1904; *d* Adelaide, 21 Sep. 1990). Australian baritone. Studied Adelaide with Bevan and Carey, London with Johnstone-Douglas. Début London, SW, 1932 (Valentin). London: SW, 1932–9, 1947–53; CG, 1935–9, 1946–54. Created Pilgrim (*Pilgrim's Progress*), Cecil (*Gloriana*). Roles included Don Giovanni, Sachs, Wotan, Falstaff (Verdi), Scarpia, and a memorable Boccanegra. Returned to Adelaide 1954 to teach. (R)

**Mattheson, Johann** (*b* Hamburg, 28 Sep. 1681; *d* Hamburg, 17 Apr. 1764). German composer, singer, harpsichordist, and organist. Joined the chorus of the Hamburg O while still a boy, singing soprano parts. He later graduated to tenor roles and 1699 his first opera *Die Plejades* was given. By 1700 he was established as one of the leading figures in Hamburg's operatic life, variously taking principal roles, directing performances from the harpsichord, and contributing

new works to the repertoire. In 1703 his friendship with Handel began: though on one occasion they fought a famous duel, the two composers maintained a generally cordial relationship; Mattheson's appearances in Handel's *Almira* and *Nero* were his last on the Hamburg stage. In 1704 he began a long and fruitful relationship with the British ambassador in Hamburg, acting first as tutor to his son (Cyrill Wich) and later as his secretary. His musical activities continued with the composition of further operas, and appointments as director of the Hamburg cathedral (1715), and Kapellmeister to the Duke of Holstein (1719).

Mattheson occupied an important pivotal position in the history of opera in N Germany, standing midway between the earlier tradition of Kusser and Keiser, an influence readily detectable in much of his vocal music, and the later developments of his near-contemporary Telemann, whose career in the city began somewhat later. He was a particularly important influence on Handel at a formative stage in his career, assisting with the composition of *Almira*; several of Handel's operas later borrowed from Mattheson's music. Full appreciation of his operatic talents is not possible: of his eight works, only one, *Die unglückselige Cleopatra*, survives. For many years it was believed that most of Mattheson's manuscripts had been destroyed in the bombing of Hamburg in the Second World War, but in 1983 evidence emerged to suggest that many were taken to Russia and have survived.

As a theorist and writer on music, Mattheson was unsurpassed in his generation, either in the quantity or importance of his work. His most important publications were *Der vollkommene Kapellmeister* (1739) and *Grundlage einer Ehrenpforte* (1740).

SELECT WORKLIST: *Die unglückselige Cleopatra* (Feustking; Hamburg 1704); *Boris Goudonow* (Mattheson; Hamburg 1710).

BIBL: H. Marx, *Johann Mattheson* (Hamburg, 1982); G. Buelow, ed., *New Mattheson Studies* (Cambridge, 1983).

**Matzenauer, Margarete** (*b* Temesvár, 1 June 1881; *d* Van Nuys, CA, 19 May 1963). Austro-Hungarian soprano and contralto. Studied Graz with Janushovsky, Berlin with Mielke and Emmrich. Début Strasburg 1901 (Fatima, *Oberon*). Munich 1904–11; Bayreuth 1911; New York, M, 1911–30; London, CG, 1914. Possessed a rich voice of extraordinary range, singing Isolde, Erda, Brünnhilde, Amneris, Aida, Fidès, Kostelnička, etc., with equal ease. (R)

**Maurel, Victor** (*b* Marseilles, 17 June 1848; *d* New York, 22 Oct. 1923). French baritone. Studied Paris with Vauthrot and Duvernoy. Début Marseilles 1867 (*Guillaume Tell*). Paris, O, 1868,

1879–94. Milan, S, from 1870; London, CG, 1873–9, 1891–5, 1904; New York, M, 1894–6, 1898–9. Roles included Don Giovanni, Rigoletto, Amonasro, Hamlet, Telramund; Boccanegra in the revised 1881 version (when Verdi lowered the tessitura for him); first Iago, Falstaff (described by W. J. Henderson as 'one of the great creations of the lyric stage'), and Tonio, persuading Leoncavallo to write the Prologue for him. One of the greatest singing actors of his time: intelligent, subtle, and cultivated, possessing a wide variety of vocal colour, and an exceptional dramatic instinct. Shaw, while admiring him, sometimes found him 'artificial' (e.g. as Don Giovanni), but Lilli Lehmann was once so moved by his Valentin that she was 'speechless for hours'. Designed sets for *Mireille* at Met. (1919). Taught New York 1909 until his death. Autobiography, *Dix ans de carrière* (Paris, 1897, R/1977). (R)

**Mauro.** Venetian family of designers. The first generation consisted of the brothers **Gaspare** (*fl* 1657–1719), **Pietro** (*fl* (1669–97), and **Domenico** (*fl* 1669–1707), who worked principally in various Venetian theatres. Influenced especially by *Burnacini, they emphasized the more realistic and popular aspects of opera, not least by their ingenious and even witty use of stage devices. The most important of their children was Domenico's son **Alessandro** (*fl* 1709–48), who had an international career that prospered especially in Dresden with Hasse. Of the next generation, the most important were his sons **Domenico** (*fl* 1733–80) and **Gerolamo** (1725–66), who worked in many Venetian theatres. Like all the family, they were efficient craftsmen and imaginative designers who served opera well, particularly in its popular aspects, without conferring a strong individual style on the art.

**Mavra.** Opera in 1 act by Stravinsky; text by B. Kochno, after Pushkin's poem *The Little House at Kolomna* (1830). Prod. Paris, O, 3 June 1922, with Slobodskaya, De Sadowen, Belina, cond. Fitelberg (after private perf. at the Hôtel Continental); Philadelphia, Orch. OC, 28 Dec. 1934, with Kurenko; Edinburgh, King's T, 21 Aug. 1956, by Hamburg S, with Muszely, Ast, Litz, Förster, cond. Ludwig.

A Russian village, *c*.1800. In order for her lover Vassily (ten) to gain entry to her house, Parasha (sop) persuades her mother (con) to accept him (in disguise) as a new maid, Mavra, to replace their recently deceased servant. When she returns slyly to check on the maid's work, the mother finds Mavra shaving; neighbours chase Vassily away.

Also opera by Solovyev (n.d.).

**Maw, Nicholas** (*b* Grantham, 5 Nov. 1935). English composer. Studied London with Berkeley and Steinitz, Paris with Deutsch. An early comedy, *One Man Show*, was notable for some engaging music whose promise was fulfilled in *The Rising of the Moon*, a well-crafted work, witty and lyrical, showing that an understanding of the strengths of Romantic opera (in particular Strauss) need not involve either pastiche or anachronism.

WORKLIST: *One Man Show* (Jacobs, after Saki; London 1964); *The Rising of the Moon* (Cross; Gly. 1970).

**Max.** A huntsman (ten), lover of Agathe, in Weber's *Der Freischütz*.

**Maxwell Davies, Peter.** See DAVIES, PETER MAXWELL.

**May Night** (Russ.: *Mayskaya noch*). Opera in 3 acts by Rimsky-Korsakov; text by the composer, after Gogol's story (1831–2). Prod. St Petersburg, Maryinsky T, 21 Jan. 1880, with Slavina, Bichurina, Velinskaya, F. Stravinsky, Lody, Melnikov, Solovyev, Ende, cond. Nápravník; also used to reopen the Maryinsky T, renamed the Russian State OH, Petrograd, 12 Mar. 1917; London, DL, 26 June 1914, with Petrenko, Smirnov, Andreyev, Belyanin, Ernst, cond. Steinberg.

Ukraine. Levko (ten) and Hanna (mez) are in love, but his father the Mayor (bs) disapproves. Levko tells her the legend of Pannochka (sop) who drowned herself to escape from her stepmother and become a rusalka (water sprite): the stepmother was also drowned, but cannot be distinguished now from the good rusalki. Levko is furious to find his father later serenading Hanna.

Great complications ensue at a drinking orgy of the Mayor, leading to his sister-in-law being mistaken for the Devil and nearly burnt.

Near the castle by the lake, Levko is singing in praise of Hanna when the rusalki appear; in return for his identifying the disguised stepmother, they provide him with evidence that leads to the Mayor being outwitted and his own wedding to Hanna being celebrated.

**Mayr, Johann Simon** (*b* Mendorf, 14 June 1763; *d* Bergamo, 2 Dec. 1845). German composer. Studied with his father, then Bergamo with Lenzi, Venice with Bertoni. On the death of the patron who supported him, he was urged by Piccinni to try his hand at opera, and the success of his *Saffo* in 1794 led to many commissions. His first opera for Milan was the 1799 revision of *Lodoiska*, though his national fame really dates from the success of *Ginevra di Scozia*, which opened the T Nuovo in Trieste in 1801. Thereafter he wrote operas for Naples and Rome as well as Milan and Venice, and his fame also spread internationally with productions in Germany and also in London, St Petersburg, New York, and elsewhere. As late as 1850 his masterpiece, *Medea in Corinto*, was revived in London for Pasta's last appearances.

Mayr's chief model in *Saffo* was Gluck, whose mastery of the chorus and orchestra the work reflects, as well as a dramatic fluency not then familiar in Italian opera. Most of his works are in buffo style, and he brought a new vividness of orchestration to the genre, with the addition of harps to the orchestra and a novel virtuosity of wind writing (as well as regular use of the stage band). He introduced orchestral numbers depicting storms, earthquakes, and other natural phenomena, and was one of various composers to anticipate the so-called 'Rossini crescendo'. He further developed the role of the chorus with music of a variety that extended from church polypony to double choruses and folk-like ensembles. He modified the exit aria convention in these dramatic interests, giving his characters a much wider expressive range in the use of different aria forms. His most celebrated work, *Medea in Corinto*, makes pioneering advances in continuity between numbers, and it is chiefly for this greater emphasis on dramatic content and fluency that he was influential on the succeeding generation of Italian composers, among them Rossini.

Mayr wrote on music, especially a treatise on Haydn, and in 1805 founded a conservatory in Bergamo, where his pupils included Donizetti (whom he taught without charge for ten years). He died blind, greatly honoured by musicians including Verdi, a mourner at his funeral; his influence on early 19th-cent. opera extended beyond Rossini, whose rise to some extent eclipsed him, to the composers of early Italian Romanticism, including Verdi.

SELECT WORKLIST: *Saffo* (Sografi; Venice 1794); *Lodoiska* (Gonella; Venice 1796); *Che originali* (Rossi; Venice 1798); *Ginevra di Scozia* (Rossi; Trieste 1801); *Le due giornate* (The Two Days) (Foppa, after Bouilly; Milan 1801); *Elisa* (Rossi; Malta 1801); *Alonso e Cora* (Bernardoni, after Marmontel; Milan 1803); *L'amor conjugale* (Married Love) (Rossi, after Bouilly; Padua 1805); *I Cherusci* (Rossi; Rome 1808); *Il sacrifizio d'Ifigenia* (Arici, after Du Roullet; Brescia 1811); *La rosa bianca e la rosa rossa* (The White Rose and the Red Rose) (Romani; Genoa 1813); *Medea in Corinto* (Romani; Naples 1813); *Tamerlano* (Romanelli; Milan 1813).

BIBL: C. Scotti, *Giovanni Simone Mayr* (Bergamo, 1903).

**Mayr, Richard** (*b* Henndorf, 18 Nov. 1877; *d* Vienna, 1 Dec. 1935). Austrian bass-baritone. Studied Vienna Cons. Début Bayreuth 1902 (Hagen). Vienna 1902–35. Frankfurt from 1905; London, CG, 1911–13, 1924–31; New York, M, 1927–30. With his round tone, sensitive singing, and versatile dramatic gifts, an outstanding artist; acclaimed as Sarastro, Leporello, Gurnemanz, etc. Created several roles, notably Barak (*Die Frau ohne Schatten*). Though not the first Ochs, as Strauss and Hofmannsthal had hoped, he later became the fulfilment of their ideal: 'Falstaffian, comfortable, laughable'. (R)

**Mazarin,** (Cardinal) **Jules** (*b* Pescina, 14 July 1602; *d* Vincennes, 9 Mar. 1661). Italian, later French, politician. His significance for opera lies in his advocacy in France of his native Italian opera, in part to aid his own political and ecclesiastical interests. His efforts were decisively aided by the exile of the *Barberini family from Rome in 1644, who brought with them a number of Italian musicians to Paris. Among them was Luigi Rossi, whose *Orfeo* (1647) impressed its public but also excited antagonism against Mazarin for the expense of an art foreign to France. Exiled by the first Fronde (the revolt against royal absolutism, 1648), he did not manage to return effectively until 1653 after the second Fronde, and was then unable to establish a permanent Italian opera. He commissioned *Ercole amante* from Cavalli for Louis XIV's wedding (1660), but neither the opera nor the special theatre was ready in time. His influence survived in the Italian elements in the operas of his countryman *Lully.

BIBL: C. Massip, *La vie des musiciens de Paris au temps de Mazarin* (Paris, 1976).

**Mazeppa.** Opera in 3 acts by Tchaikovsky; text by the composer and V. P. Burenin, after Pushkin's poem *Poltava* (1829). Prod. Moscow, B, 15 Feb. 1884, with Krutikova, Pavlovskaya, Korsov, Borisov, Usatov, cond. Altani; Liverpool 6 Aug. 1888, with Winogradoff, cond. Truffi; Boston, OH, 14 Dec. 1922, with Guseva, Valentinova, Radeyev, Karlash, Danilov, Alimov.

Ukraine, 18th cent. Maria (sop), daughter of Kochubey (bs), loves Mazeppa (bar) and rejects Andrey (ten). When Mazeppa is refused her hand by Kochubey, he abducts her; Andrey offers to approach the Tsar with information about Mazeppa's treachery in siding with the Swedes.

But the Tsar believes Mazeppa, and Kochubey is arrested; torture fails to break him or to part him from his treasure. Maria is told of this and pleads with Mazeppa, who is absorbed in his plans to set up a separate state under his rule. Kochubey is led to execution.

After the symphonic picture *The Battle of Poltava*, the last act shows Andrey pursuing the fleeing Swedes; however, he is shot by Mazeppa, and Maria, now gone mad, sings a lullaby over his body.

Other operas on the subject (those marked S based on Słowacki's drama, 1840) are by Maurer (1837), Campana (1850), Wietinghoff (1859), Pedrotti (1861), Pourny (1872), Jarecki (S, 1876), Pedrell (1881), Minchejmer (S, comp. 1885, prod. 1900), Grandval (1892), Koczalski (S, 1905), and Nerini (1925).

**Mazurok, Yury** (*b* Krasnik, 18 July 1931). Polish baritone. Studied Moscow with Sveshnikova. Moscow, B, from 1963; London, CG, from 1975; New York, M, from 1978; also Vienna, S; Berlin; Prague; Budapest etc. Possesses an attractive tone and stage presence. Repertory incl. Figaro (Rossini), Germont, Luna, Escamillo, Onegin, Yeletsky, Andrey (*War and Peace*). (R)

**Mazzinghi, Joseph** (*b* London, 25 Dec. 1765; *d* Downside, 15 Jan. 1844). English composer of Corsican origin. Studied London with J. C. Bach and Sacchini. At the age of 19 he was appointed music director at the H, for which he composed a number of popular operas of little substance. He was replaced by Storace in 1793, and dismissed from the theatre in 1798.

**Mazzocchi, Domenico** (*b* Civita Castellana, *bapt.* 8 Nov. 1592; *d* Rome, 21 Jan. 1665). Italian composer, brother of Virgilio *Mazzocchi. Trained for the priesthood; from *c*.1620 served various members of the Aldobrandini family.

His only surviving opera, *La catena d'Adone*, was one of the first to be performed in Rome. Both libretto and score adumbrate many later features of its tradition: in particular, the increased use of ensembles, a new emphasis on comic episodes and characterization, and a sharper distinction between monody and aria-type music. In his preface Mazzochi justifies his inclusion of the latter on the grounds that he wished to alleviate the 'tedium of the recitative'. The score is also noteworthy for including the first use in an operatic context of the word 'aria', although here it refers to ensemble as well as to solo numbers.

WORKLIST: *La catena d'Adone* (The Chain of Adonis) (Tronsarelli, after Marino; Rome 1626).

**Mazzocchi, Virgilio** (*b* Civita Castellana, *bapt.* 22 July 1597; *d* Civita Castellana, 3 Oct. 1646). Italian composer, brother of Domenico *Mazzocchi, with whom he studied in Rome. Maestro di cappella at St John Lateran, succeeding Abbatini 1628–9; Maestro di cappella at the Cappella Giulia, St Peter's 1629–46. Together with Marazzoli composed *Chi soffre, speri* for performance at the Barberini palace. One of the most characteristic works of the Roman tradition of the 1630s, this includes some of the earliest use of secco recitative.

WORKLIST: *Chi soffre, speri* (Who Suffers May Hope)

(Rospigliosi; Rome 1637 (with the title *Il falcone*); rev. 1639).

**Medea.** In Greek mythology, the sorceress who helped Jason to win the Golden Fleece, afterwards escaping with him. She prevented pursuit by casting the limbs of her brother behind her to delay the King, their father. Deserted by Jason, she killed his two children and poisoned his new wife: she then fled to Athens, where she married King Aegeus. In most operas she burns to death in Juno's Temple; see *Médée. Operas on her are by Cavalli (*Giasone*, 1649), Kerl (1662), Kusser (*Jason*, 1692), Collasse (1696), Charpentier (1693), Brusa (1726), Salomon (1713), Pérez (1744), Gebel (1752), Benda (1775), Andreozzi (1784), Vogel (*La toison d'or*, 1786), Naumann (1788), Cherubini (1797), Mayr (1813), Fontenelle (1813), Coccia (1816), Selli (1839), Pacini (1843), Mercadante (1851), Tommasini (1906), Bastide (1911), Milhaud (1939), and Wilson (1982).

**Médecin malgré lui, Le** (Doctor despite Himself). Opera in 3 acts by Gounod; text a slight alteration of Molière's comedy (1666) by the composer, Barbier, and Carré. Prod. Paris, L, 15 Jan. 1858, with Caye, Faivre, Lesage, Meillet, cond. Deloffre; London, CG, 27 Feb. 1865; Cincinnati, Odeon, 20 Mar. 1900.

France, 17th cent. The doctor Sganarelle (bar) is summoned to attend Lucinde (sop), daughter of a wealthy neighbour Géronte (bs). She is, her lover Léandre (ten) explains to the audience, pretending to be dumb in order to avoid having to marry a rich suitor. To help the couple, Sganarelle passes Léandre off as his apothecary; Lucinde is cured. The couple elope, and when news arrives that Léandre had inherited a large fortune, Géronte is happy for the marriage to proceed.

Gounod's first opéra-comique and first great success.

**Médée** (Medea). Opera in 3 acts by Cherubini; text by F. B. Hoffman, after Corneille's tragedy (1635). Prod. Paris, T Feydeau, 13 Mar. 1797, with Julie Legrand, Gaveaux, Dessaules; popular in Germany, especially in version with recits. by Lachner (Frankfurt 1855); London, HM (with recits. by Arditi), 6 June 1865, with Tietjens, Santley. Not perf. in Italy until 30 Dec. 1909 at Milan, S, with Mazzoleni, Frascani, Isalberti; New York, TH, 11 Nov. 1955, with Farrell, McCracken, Hurley, Scott, Lipton, cond. Gamson.

Corinth, mythological. After helping Jason (ten) to recover the Golden Fleece, Médée (sop) returned with him to Corinth, and there bore him two sons. Having abandoned her in order to marry Glauce (sop), Jason fears Médée's vengeance; she curses him.

Glauce's father Créon (bar), King of Corinth, orders Médée to leave the city; she feigns acceptance and sends Glauce a wedding gift of a nightgown and magic diadem, which she has covered with poison.

Médée's gifts kill Glauce and the people clamour for her life. She takes refuge in the Temple of Juno and, having killed her sons, sets fire to the temple and burns to death.

**Medici family.** Italian family of musical patrons and the most powerful family in Florence from the 15th cent. It was under the patronage of the Medicis that the earliest operas were given. During the reigns of Cosimo I (1537–74) and Francesco I (1574–87) musical life was greatly enhanced by the recruiting of Caccini and the number of elaborate court entertainments (such as the performance of lavish intermedi in 1539 to celebrate Cosimo's marriage) increased. For opera the most important member of the Medici family was Ferdinando I (1549–1609), whose marriage in 1589 to Catherine of Lorraine was celebrated with the most famous of the Florentine intermedi. The support of his court was crucial to those involved in the preparation of the first opera, Peri's *Dafne* (1598). Later, for the marriage of Maria de' Medici with Henri IV of France in 1600, two operas were given at the Florence court, Peri's *Euridice* and Caccini's *Il rapimento di Cefalo*. Later significant patrons in the Medici family included Cosimo II (1590–1620) and Cosimo III (1642–1723).

**medieval liturgical drama.** A form of church drama with continuous music which evolved in the 10th cent., originating as an expansion to the normal liturgy. Ornamental melismas known as tropes, inserted into the standard plainchant, were given words in order that they might more easily be memorized, as well as to add to their expressiveness. The tropes added to introits in services at Easter and at Christmas frequently took the form of a dialogue and were sung by different individual voices. One trope frequently cited as the basis of liturgical drama is the Easter trope beginning 'Quem quaeritis?', where three monks approach a fourth at the sepulchre (represented by the altar) seeking Christ's body. Originally only three sentences long, the trope was expanded, fulfilling the function stated by St Ethelwold in the *Regularis Concordia* (written AD 965–75) which was 'to fortify the faith of the ignorant multitude'.

From the 10th cent. on, dramatic episodes were often introduced into the liturgy. Tropes were expanded to depict entire biblical scenes and became too long to retain their original position as part of the service. The liturgical dramas

were therefore performed independently, often preceding or following a service.

The liturgical drama continued to expand in length and develop in complexity, reaching its peak in the 12th cent. A drama might have several scenes, given in different positions around the church, and contemporary directions indicate that increasingly elaborate costumes, props, and sets were employed: the clerical performers had originally relied on readily available ecclesiastical garments and objects. Directions also specified the movement of the performers using highly formalized gestures. Though conventions varied in different times and places, the liturgical drama was in general more concerned with representation than with characterization, and in this respect differs from the more humanitarian opera which evolved in the 17th cent.

Favourite subjects were the Easter sepulchre, 'Peregrinus' plays (i.e. the journey to Emmaus), Christmas plays (the Shepherds, the Visit of the Magi, the Massacre of the Innocents), Old Testament plays (Daniel, Joseph, Isaac), New Testament plays (the Raising of Lazarus), and the legends of the saints, especially St Nicholas. The most significant source of the dramas is the Fleury Play Book (13th cent.), which contains ten plays representing many of the above categories.

In literary form, the plays might use poetic or prose texts, the latter being biblical texts, or elaborated biblical phrases, or pre-existent antiphons. When old textual material was used, it frequently retained the melody which belonged to it in the normal course of the liturgy. Newly composed music for the prose text kept the same chant-like style. The nature of the music for soloists was generally syllabic with a few appropriate, short melismas. As with other sources of plainchant, extant notation does not give any indication of rhythm. The poetic texts tend to follow the literary pattern of the contemporary hymn and imply rhythmic performance, since they would have been sung by a chorus. They are often processionals, marking a change of scene in the drama, and may repeat a simple melodic formula many times over.

The notation of the music dramas was entirely monophonic, though counterpoint might have been improvised at climactic points, particularly from the 11th cent., when this practice was becoming common in many parts of the liturgy. Sources do not specify either what instrumental accompaniment, if any, was used in the plays. It seems likely that this was restricted to organ and bells during processionals, although the 12th-cent. *Play of Daniel* includes the lines 'let drums sound, let the harp players touch their strings; let the instruments of music sound'. The play also includes several short instrumental passages as bridge episodes.

Originating from a Beauvais manuscript, *The Play of Daniel* is one of the most varied and rewarding of the church music dramas. It contains over 50 different melodies and the musical styles are clearly chosen to reflect situations and emotions in a highly dramatic manner, e.g. sinewy vocal lines depict plotting against Daniel. The rubric shows the influence of contemporary secular drama, such as the convention of evil characters entering from the left.

Little liturgical drama was written after the 14th cent., though it continued to be performed into the 16th cent. Vernacular drama, frequently based on liturgical subjects (e.g. the mystery plays), took its place as a main creative outlet, and in these works the music had a far more incidental role.

BIBL: F. Collins jun., *The Production of Medieval Church Music-Drama* (Charlottesville, VA, 1972); W. L. Smoldon, *The Music of the Medieval Music Dramas* (Oxford, 1980).

**medieval vernacular drama.** A play in the vernacular, frequently on a religious subject, in which music was an important component. Developing concurrently with liturgical drama, the two forms shared many features although the vernacular plays were performed at first substantially, and later entirely, by non-clerics, outside the precincts of the church. In most countries, vernacular drama reached its peak in the 14th and early 15th cents.; in France and England in particular this coincided with the cultivation of cycles of plays known as miracle or mystery plays.

From the early Middle Ages, many minstrels combined music with acting, juggling, and puppetry to create a more substantial entertainment. Several forms of 12th-cent. troubadour song, such as the *jeu-parti* (a dialogue in the manner of a debate), the long epic *chanson de geste* (e.g. *The Song of Roland*), and the *planctus* (lament) also lent themselves to dramatic expansion. These secular sources combined with the influence of the many dramatic and semi-dramatic ceremonies which were an official part of the liturgy to create varied forms of vernacular drama throughout Europe. The only surviving early secular play with a significant amount of music is Adam de la Halle's *Le jeu de Robin et de Marion* (*c.*1283). Written in northern Italy, but as a Christmas entertainment for French soldiers, it contains short, simple melodies, many of which were popular pre-existing refrains. Many plays took religious themes: the lament of Mary at the Cross was the basis of the German form of *Marienklage*, of which some fifty examples survive. As with other extant plays of the 11th and 12th

cents., the *Marienklage* mix Latin and vernacular texts and also use both liturgical chant and popular song melodies.

The music of the vernacular dramas generally had one of three functions: to imitate other sounds (e.g. music for feasts, coronations); to further the stage business (e.g. a fanfare to mark the entrance or exit of an important character, or to summon the audience's attention after a change of scene); to act as a symbol (e.g. accompanying the appearance of God and the angels, or representing the din of hell). Very few manuscripts contain musical notation, but stage directions indicate that most of the music was monophonic. However, some numbers, including hymns and the music marking the changes of scene, called *silete* in French plays and performed by a group of angels, might well have been sung in written or improvised polyphony.

Vernacular drama used costumes and sets, although these were not very elaborate since performances were often out of doors and some was itinerant. Presentation of time and place were non-naturalistic; sets were emblematic, usually consisting of an unlocalized central acting 'place' in front of symbolic 'houses'. Costumes were more naturalistic than in the liturgical drama, but were entirely contemporary. The plays also imposed the conditions of medieval life on biblical characters, e.g. in one English play, Herod is assigned a minstrel.

Italian vernacular drama differed from that in England, France, and Germany in having a greater tendency towards spectacular visual effects and in showing a greater influence of the tradition of popular song. The *laudi spirituali*, sung by processing flagellants in the 12th cent., evolved into longer *laude drammatiche*, in which the music was almost exclusively a vehicle for a narrative. This form developed into the larger *rappresentazione sacra, which flourished in Tuscany in the 15th and 16th cents.

In Spain vernacular drama continued in a similar form into the 20th cent., whereas in the rest of Europe the place of the plays was taken in the late 15th and early 16th cents. by other forms of musico-dramatic entertainment. These may be very roughly divided into three groups—sumptuous court drama (including mumming, masques, and allegorical interludes); popular drama, which included morality plays and in which music played a less significant role; and educational dramas, performed at schools and colleges, designed for moral edification and to give a training in language.

Although music played an integral part in medieval vernacular drama, its secondary emotive role makes it closer to a play with substantial incidental music than a direct antecedent of opera.

**Medium, The.** Opera in 2 acts by Menotti; text by the composer. Prod. New York, Columbia U, 8 May 1946; London, Aldwych T, 29 Apr. 1948.

USA, today. The Medium, Madame Flora (con), helped by her daughter Monica (sop) and the mute Toby, cheats her clients. During a seance she feels a hand on her throat and confesses to her clients that she is a fraud. They refuse to believe her, but she loses her nerve, beats Toby and throws him out of the house; she then turns to the whisky bottle. Toby, in love with Monica, returns to find her and hides behind a puppet theatre. A noise awakens Madame Flora, and seeing the closed curtain move she shoots at it, killing Toby. 'I've killed the ghost!' she screams.

**Mefistofele.** Opera in a prologue, 4 acts, and epilogue by Boito; text by the composer. Prod. Milan, S, 5 Mar. 1868, with Reboux, Spallazzi, Junca; but not a success. Rev. and prod. Bologna 4 Oct. 1875, with Borghi-Mamo, I. Campanini, Nannetti; London, HM, 6 July 1880, with Nilsson, I. Campanini, Nannetti; Boston 16 Nov. 1880. Final rev., Milan, S, 25 May 1881, with Mariani-Masi, Marconi, Nannetti, cond. Faccio.

Unlike Gounod in his *Faust*, Boito based his opera on both parts of Goethe's work. Prologue: Mefistofele (bs) wagers God that he can win Faust's soul. Faust (ten) is followed by a sinister grey friar who reveals himself to be Mefistofele. He persuades Faust to sell his soul in return for one moment of perfect happiness.

Faust, under the name of Enrico, courts Margherita (sop). Mefistofele later takes him to witness the Witches' Sabbath, where Faust is distressed by an apparition of Margherita in chains.

Margherita, delirious, is in prison after poisoning her mother and murdering the son she had by Faust. Mefistofele offers to free her but she rejects his help. She regains her sanity briefly and dies.

In Greece, Faust admires Helen of Troy who, flattered, offers him her love. The epilogue depicts a disillusioned Faust near death. He dies repenting his bargain; God pities him and frustrates Mefistofele's attempts to claim his soul.

**Mehta, Zubin** (*b* Bombay, 29 Apr. 1936). Indian conductor. Studied Vienna with Swarowsky. Début Montreal 1964 (*Tosca*). New York, M, 1965–71; London, CG, 1977; also Vienna, Florence. Repertory incl. *Trovatore*, *Aida*, *Otello* (Verdi), *Tosca*, *Turandot*, *Carmen*, first *Mourning becomes Electra*. (R)

**Méhul, Étienne-Nicolas** (*b* Givet, 22 June 1763; *d* Paris, 18 Oct. 1817). French composer. Studied Givet with a local organist, Monthermé with Hanser, Paris with Edelmann. His first opera, *Cora*, was unperformed until after his next and more representative work, *Euphrosine*, an opéra-comique mixing serious and comic elements. It became immediately and enduringly famous for its duet 'Gardez-vous de la jalousie', in which a simple motive of alternating thirds identifies jealousy and its working both here and motivically elsewhere in the opera. The more severe *Stratonice* also contained a popular number, 'Versez tous vos chagrins'. Set in ancient times, and consciously simple in a Gluckian manner, it shows the care for achieving a vivid and individual setting that was to mark Méhul's operatic work: much of this he achieved by means of carefully devised orchestration. *Le jeune sage et le vieux fou* was a comedy, with an overture that like many others has had an independent life.

With the establishment of the new political régime, Méhul was appointed to the Institut National de Musique in 1793, and his next opera, *Horatius Coclès*, used a classical theme to make a political point in what Méhul himself drily called his 'style de fer', or 'iron style'. He was more successful with *Mélidore et Phrosine*, in which he takes still further his use of motive, making use of a motto theme associated with the guiding force of love, and also with his other major work of the decade, *Ariodant*. The latter's ambitious structures and knightly setting, as well as the opposition of a 'dark' and a 'light' couple, anticipates Weber's *Euryanthe*, which it also influenced in its remarkable use of dark-hued orchestration as a functional expressive device. Certainly Weber appreciated Méhul's pioneering determination to give each of his operas, through its orchestration, an individual character. In 1795 Méhul was granted a pension by the Comédie-Italienne, and in 1794 he was made a director of the Conservatoire. Possibly at the command of Napoleon, with whom he had many contacts, he wrote *La prise du Pont de Lodi*, celebrating the Austrian defeat of 1796.

From 1800 there followed a series of comedies, beginning with *Bion* and including *Une folie* and *Le trésor supposé*, both also popular in Germany, as well as *Héléna*, some of whose features (such as a dramatic trumpet call) influenced *Fidelio*. In the Ossianic *Uthal*, Méhul's care for an individual atmosphere shows in the banishing of violins and trumpets from the orchestra; the most original part is an overture that is virtually non-thematic and includes a mysterious voice calling through the mist. His last major success was *Joseph*, in which many of his gifts, including his melodic grace and his feeling for atmosphere and situation, are powerfully and movingly displayed. It was immediately successful, being as much admired in Germany and other European countries as in France, influencing a number of composers and surviving in repertories into the 20th cent. A few opéras-comiques followed, but Méhul never again gave himself so concentratedly to a dramatic subject. Though trained in the classical school, his Romantic impulses led him to explore the dramatic role of the orchestra, and with it of motive, and thereby to influence (especially through Weber's music and his advocacy) the course of German Romantic opera.

SELECT WORKLIST (all first prod. Paris): *Cora* (Valadier; comp. 1789, prod. 1791); *Euphrosine* (Hoffman; 1790); *Stratonice* (Hoffman; 1792); *Le jeune sage et le vieux fou* (The young sage and the old fool) (Hoffman; 1793); *Horatius Coclès* (Arnault; 1794); *Mélidore et Phrosine* (Arnault; 1794); *Le jeune Henri* (The Young Henry [IV] ) (Bouilly; 1797); *La prise du pont de Lodi* (The Taking of the Bridge of Lodi) (Delrieu; 1797); *Adrien* (Hoffman; 1799); *Ariodant* (Hoffman; 1799); *Bion* (Hoffman; 1800); *L'Irato* (Marsollier; 1801); *Une folie* (An Act of Folly) (Bouilly; 1802); *Le trésor supposé* (The Supposed Treasure) (Hoffman; 1802); *Héléna* (Bouilly; 1803); *Les deux aveugles de Tolède* (The Two Blind Men of Toledo) (Marsollier; 1806); *Uthal* (Saint-Victor; 1806); *Gabrielle d'Estrées* (Saint-Just; 1806); *Joseph* (Duval; 1807).

BIBL: A. Pougin, *Méhul: sa vie, son génie, son caractère* (Paris, 1889, 2/1893).

**Mei, Girolamo** (*b* Florence, 27 May 1519; *d* Rome, July 1594). Italian humanist and writer on music. Studied with the philosopher Piero Vettori, with whom he translated much Greek tragedy. Membership of several learned *academies, including that of the Accademia degli Alterati, and journeys through Italy and France enhanced an early literary and cultural awareness. Around 1551 Mei commenced an exhaustive study of Greek music theory, culminating in the treatise *De modis musicis antiquorum* (completed 1573). From 1559 he lived in Rome, where from 1572–81 he conducted his lengthy and celebrated correspondence with Vincenzo *Galilei (resident in Florence) concerning the nature of ancient Greek music, in particular its use within a dramatic context. Many of the ideas which Galilei later explored in his *Dialogo della musica antica et della moderna* owe their inspiration to Mei, whose cogent and sensitive reinterpretation of the culture of the Greeks may be counted among the most formative influences on the development of *stile rappresentativo and hence on the eventual emergence around 1600 of the new genre of opera.

WRITINGS: *Letters on Ancient and Modern Music to Vincenzo Galilei*, ed. C. Palisca (1960, 2/1977).

BIBL: D. Restani, *L'itinerario di Girolamo Mei dalla 'poetica' all musica* (Florence, 1990).

**Meier, Waltraud** (*b* Würzburg, 9 Jan. 1956). German mezzo-soprano. Studied Cologne with Thiesen and Jacob. Début Würzburg 1976 (Cherubino). Mannheim 1978–80; Dortmund 1980–3; Buenos Aires 1980; Cologne 1983; Bayreuth from 1983; London, CG, from 1985; New York, M, 1988; also Vienna, S; Paris, C; etc. A gifted singer and actress with a sensuous tone, successful as Venus, Brangäne, Fricka, Azucena, Eboli, Carmen, Santuzza, Octavian, and particularly Kundry. (R)

**Mei-Figner, Medea.** See FIGNER, MEDEA.

**Meilhac, Henri** (*b* Paris, 21 Feb. 1831; *d* Paris, 6 July 1897). French dramatist and librettist. Most of his librettos were written in collaboration, with Millaud for Hervé's *Mlle Nitouche*, with Gille for Massenet's *Manon* and Planquette's *Rip*; but by far the most important collaboration of his career was with Ludovic *Halévy. Their most famous libretto is certainly *Carmen*, for Bizet; but their most characteristic vein was operetta, especially as shown in Offenbach's *La belle Hélène*, *Barbe-bleue*, and *La vie parisienne*. Frequently by means of a classical setting, they satirized contemporary society in the guise of mocking an old myth, also parodying some of the conventions of opera. At a time when the Paris Opéra enshrined the pomp of the most elevated manner of Grand Opera, Meilhac and Halévy provided the necessary corrective of laughter at human foibles and the shortcomings of society.

**'Mein Herr Marquis'.** Adele's (sop) laughing song in Act II of Johann Strauss's *Die Fledermaus*, in which, disguised at Orlofsky's party, she flirts with her employer Eisenstein.

**Meistersinger von Nürnberg, Die** (The Mastersingers of Nuremberg). Opera in 3 acts by Wagner; text by the composer. Prod. Munich 21 June 1868, with Mallinger, Diez, Nachbaur, Schlosser, Betz, Hoelzel, Fischer, Bausewein, cond. Bülow; London, DL, 30 May 1882, with Sucher, Schefsky, Winkelmann, Landau, Gura, Ehrke, Kraus, Koegel, cond. Richter; New York, M, 4 Jan. 1886, with Seidl-Kraus, Brandt, Strift, Krämer, Fischer, Kemlitz, Lehmler, Staudigl, cond. Seidl.

Nuremberg, mid-16th cent. Walther von Stolzing (ten), falling in love with Eva (sop), daughter of the goldsmith Veit Pogner (bs), learns that she is to be betrothed next day to the winner of a singing contest held by the Guild of Mastersingers. The apprentice David (ten) explains the rules to Walther, who seeks to join the Mastersingers. He is invited to sing a trial song; Beckmesser (bs-bar), the town clerk, also in love with Eva, marks it harshly but Hans Sachs (bs-bar), the cobbler, discerns its originality and value.

On Midsummer Eve, Eva makes her way to Sachs's shop to question him about the competition. Despite being half in love with her himself, he resolves to help Eva and Walther, who have met and plan to elope. Beckmesser arrives to serenade Eva; Sachs, who is working on shoes for him, hammers every time he makes a mistake in his song. Sachs prevents Eva and Walther from escaping and takes the latter into his own house as a riot develops over the noise made by Beckmesser, who suffers a beating.

Sachs soliloquizes on the world's follies, as exemplified in the riot, and resolves to mend matters. When Walther relates the dream he had, Sachs writes down the words of two stanzas, sensing that it will be a prize song. The arrival of Eva, who visits on a pretext, inspires Walther to his final stanza. Beckmesser comes to see Sachs, finds the song, and is permitted by Sachs to make use of it. The company departs for the festal meadow. Beckmesser sings Walther's song disastrously, and Walther himself is summoned to show how it should be sung. He wins the prize and Eva's hand. He accepts membership of the Mastersingers after Sachs has reminded him of their purpose of preserving the art of German song.

BIBL: R. Rayner, *Wagner and 'Die Meistersinger'* (London, 1940).

**Melani**. Italian family of musicians. The most significant were the sons of Domenico Melani, bellringer at Pistoia Cathedral 1624–50:

1. **Jacopo** (*b* Pistoia, 6 July 1623; *d* Pistoia, 19 Aug. 1676). Composer. Member of the chapel of the Grand Duke of Tuscany; brought to Paris by Mazarin 1644 and 1647. Organist of Pistoia cathedral 1647; maestro di cappella 1657. Wrote *c*.10 operas, mostly for Florence; one of the most important composers of comic opera in the 17th cent. The most interesting is *La Tancia* (properly *Il potestà di Colognole*) which contains a famous parody of the incantation scene from Cavalli's *Giasone*: other works include *Il pazzo per forza* and *Il Girello*, whose performance in Rome in 1668 was responsible for rejuvenating comic opera there in the late 17th cent.

SELECT WORKLIST: *La Tancia* (Moniglia; Florence 1657); *Il pazzo per forza* (The Forced Madman) (Moniglia; Florence 1658); *Ercole in Tebe* (Moniglia; Florence 1661); *Il Girello* (Acciaiuoli; Rome 1668).

2. **Atto** (*b* Pistoia, 31 Mar. 1626; *d* Paris, 1714). Soprano castrato and composer. Studied Rome with Rossi and Pasqualini. Paris from 1644, singing in operas by Rossi and Cavalli. Secret agent of Mazarin, and gentleman of the chamber to Louis XIV. Exiled after Mazarin's death in 1661. Moved to Rome, singing for Cardinal Rospigliosi (later Pope Clement IX) till 1668. Returned to France 1679, remaining as politician and diplomat until his death. Composed several cantatas. His correspondence is important for its information on musical events 1644–61.

3. **Francesco Maria** (*b* Pistoia, 3 Nov. 1628; *d* ?Pistoia, 1663). Soprano castrato. In the service of Archduke Sigismund of Austria 1657–63, and of Grand Duke of Tuscany 1672–1700. Paris 1660 (Amastri, Cavalli's *Serse*).

4. **Bartolomeo** (*b* Pistoia, 6 Mar. 1634; *d* Pistoia, 1703). Mezzo-soprano castrato. Sang at Munich court 1657–60. Florence, P, 1661, creating roles in operas by his brother Jacopo.

5. **Vincentio Paolo** (*b* Pistoia, 15 Jan. 1637; *d* before 1667). Soprano castrato. Sang in Florence, P, 1661.

6. **Alessandro** (*b* Pistoia, 4 Feb. 1639; *d* Rome, Oct. 1703). Italian composer. After holding posts in Orvieto, Ferrara, and Pistoia, worked in Rome as maestro di cappella at various churches. Most important opera was *L'empio punito*, probably the first operatic treatment of the *Don Juan legend.

WORKLIST: *L'empio punito* (The Miscreant Punished) (Acciajuoli; Rome 1669).

**Melba**, (Dame) **Nellie** (*b* Richmond, Melbourne, 19 May 1861; *d* Sydney, 23 Feb. 1931). Australian soprano. Studied Paris with Marchesi. Début Brussels, M, 1887 (Gilda). London, CG, 1888–1914 (except 1909, 1912), 1919, 1922–4, 1926; Paris, O, 1888; Milan, S, 1892–4; New York, M, 1893–7, 1898–1901, 1904–5, 1910–11. Also Chicago, Sydney, Melbourne. Retired 1928. At first a high coloratura famous for her Lucia, Gilda, Violetta, etc. she later sang the lyric repertory: Marguerite, Ophélie, Verdi's Desdemona, and her most famous role, Mimì. Also Aida, Elsa, Nedda, and her one failure, Brünnhilde (*Siegfried*). One of the most consummate technicians of her era, she dazzled with her runs, trills, staccato singing, impeccable intonation, and ease of attack. Her voice, with a range of $b\flat$ –$f'''$ and absolute evenness throughout the scale, had a pure, incandescent quality, remaining fresh because never forced. She was an indifferent actress, but possessed a strong, queenly presence. At Covent Garden, the *prima donna assoluta* ruled off-stage as well as on it, deciding the castings of operas and the engaging (or not) of other artists. Her popularity with the public was enormous; in England she was considered an institution. Autobiography, *Melodies and Memories* (London, 1925, R/1970). (R)

BIBL: J. Hetherington, *Melba* (London, 1967).

**Melbourne.** City in Victoria, Australia. Melbourne was long dependent on touring companies, especially the Lyster 1861–80, with performances normally given in the Princess T. There were various visits by companies under *Melba (born in, and taking her professional name from, the city). The National Theatre OC of Melbourne was founded in 1950, merging with the New South Wales National O of Sydney in 1952. The Sidney Meyer Music Bowl (cap. 2,000, lawn seating for 30,000) opened 1959. The Victoria OC (Victoria State O since 1977) has also been very active in the face of financial and other difficulties. The Melba Memorial Centre opened 1973. The Globe OC was formed in 1979. The State Theatre (cap. 1,984) opened 1985.

**Melchior, Lauritz** (*b* Copenhagen, 20 Mar. 1890; *d* Santa Monica, CA, 18 Mar. 1973). Danish, later US, tenor. Studied Copenhagen with Bang. Début as baritone, Zwicki OC, 1912 (Germont). Official début Copenhagen 1913 (Silvio, *Pagliacci*). Retrained as tenor with Herold; second début Copenhagen 1918 (Tannhäuser). Further study Munich with Bahr-Mildenburg. London, CG, 1924 (Siegmund). Then Bayreuth 1924–31; London, CG, 1926–39; New York, M, 1925–7, 1928–50. Also Buenos Aires, Paris, Berlin, etc. Though a limited actor, and not always musically accurate, he was one of the most extraordinary Heldentenors in history, impressive as much for his eloquent singing as for his astonishing vocal endurance. His voice, plangent and dark, yet possessed of ringing top notes, had a moving quality, and his enunciation was exemplary. Famous as Siegmund, Siegfried, Parsifal, Tannhäuser, Lohengrin, and Tristan (which he sang more than 200 times); also sang Verdi's Otello. (R)

BIBL: S. Emmons, *Tristanissimo* (New York, 1990).

**Mélisande.** Golaud's second wife (sop), lover of Pelléas, in Debussy's *Pelléas et Mélisande*.

**melisma.** An expressive vocal line sung to one syllable.

**Melitone.** Fra Melitone (bar), a comic priest, in Verdi's *La forza del destino*.

**Melnikov, Ivan** (*b* St Petersburg, 4 Mar. 1832; *d* St Petersburg, 8 July 1906). Russian baritone. Studied St Petersburg with Lomakin, Italy with Repetto. Début St Petersburg, M, 1867 (Riccardo, *I Puritani*). Sang there until 1890; producer there 1890–2. Celebrated as Ruslan, the Miller (Dargomyzhsky's *Rusalka*), Susanin (*Life for the Tsar*). Created title-roles of *Boris Godunov* (as which he was much acclaimed), *Prince Igor*, and *The Demon*; Don Juan (*The Stone Guest*); roles in Rimsky-Korsakov's *May Night* and *Maid of Pskov*, and Tchaikovsky's *The Oprichnik*, Va-

kula the Smith, *The Sorceress*, and *The Queen of Spades*. Also sang Onegin, and Kochubey (*Mazeppa*). Tchaikovsky's favourite baritone, and a friend of Musorgsky; highly praised by Stasov for his dramatic interpretations, and by Modest Tchaikovsky for his declamatory and legato singing.

**melodrama** (Fr., *mélodrame*). A dramatic composition, or section of a composition, in which one or more actors recite to a musical commentary: if for one actor, the term 'monodrama' may be used, if two, 'duodrama' (as in the *duodramas of *Benda). The style became popular in the second half of the 18th cent., especially in opéracomique. The first full-scale melodrama was Rousseau's *Pygmalion* (1770), in which he tried to 'join the declamatory art with the art of music', alternating short spoken passages with instrumental music as a development of the *pantomime dialoguée*. The technique was enterprisingly used by *Fomin in *Orpheus*. The most successful examples of its power of heightening the dramatic tension are in the grave-digging scene in Beethoven's *Fidelio* and in the Wolf's Glen Scene in Weber's *Der Freischütz*. Mozart, who admired Benda's music, used melodrama in his *Zaide* and intended to write one himself, *Semiramis*. The form has retained some following in Czechoslovakia: Fibich wrote a trilogy *Hippodamia* (1890–1). It was also used by Humperdinck (e.g. the original version of *Königskinder*, 1897).

**melodramma** (It.). A dramatic text written to be set to music; hence **melodramma serio**, a synonym for *opera seria. Not to be confused with *melodrama, for which an alternative Italian word is *melologo*.

**Melot.** One of King Mark's courtiers (bar) in Wagner's *Tristan und Isolde*.

**Melusine.** In medieval French legend, a beautiful mermaid-like water sprite who is allowed to marry a mortal on condition that her nature is concealed; when her husband discovers it, she returns to her element. First given German form in 1486 by Thüring von Ringoltingen. Frequently reprinted; subsequent treatments by writers incl. Hans Sachs, Karl Hensler (*Das *Donauweibchen*), Grillparzer (prepared as libretto for Beethoven, 1823, set by Kreutzer), Tieck, and Goethe. Operas on her legend are by Kreutzer (1833), Schindelmeisser (1862), Grammann (1875), Reimann (1971). See also RUSALKA, UNDINE.

**Mendelssohn** (-Bartholdy), **Felix** (*b* Hamburg, 3 Feb. 1809; *d* Leipzig, 4 Nov. 1847). German composer. As a boy, Mendelssohn wrote a number of operas, mostly based on French vaudevilles and all in the *Liederspiel manner that

461

well suited his light, elegant gift. The first of these that can stand modern revival is *Die beiden Pädagogen*, a witty satire on the arguments between rival educational methods (Pestalozzi and Basedow) that must have been only too familiar to the young composer. Others of these early works include some delightful and effective numbers, sometimes well matched to dramatic situations, that do not, however, suggest a larger operatic gift. Mendelssohn's most important opera is *Die Hochzeit des Camacho* (1827), a setting of an episode from *Don Quixote*. Despite some charming and amusing numbers, and in the woodland scenes a manner suggesting that the example of Weber had not been lost on him, there is a tendency for the musical forms to dominate the drama rather than the other way about. The work's failure greatly depressed him, but he continued to search for operatic subjects. In his next work, *Die Heimkehr aus der Fremde*, he returned to his earlier Liederspiel manner, making effective use of his mature musical style and not least of his gifts as a Romantic orchestrator. His search for librettos led him to contemplate many subjects, and to embark upon various collaborations; but the fact that these all ran into the sand led his friends to suspect a deeper problem. The surviving fragments of his last operatic project, *Die Loreley*, do not suggest that this was being solved.

WORKLIST: [*Ich, J. Mendelssohn*] (Mendelssohn; unfin., 1820); *Die Soldatenliebschaft* (Military Love) (Casper; comp. 1820, prod. Wittenberg 1962); *Die beiden Pädagogen* (Casper, after Scribe; comp. 1821, prod. Berlin 1962); *Die wandernden Komödianten* (Casper; 1822, unprod.); *Der Onkel aus Boston* (Casper; Berlin 1824); *Die Hochzeit des Camacho* (Camacho's Wedding) (?, after Cervantes; Berlin 1827); *Die Heimkehr aus der Fremde* (The Return from Abroad; sometimes known in English as Son and Stranger) (Klingemann; Berlin 1829); *Die Loreley* (Geibel; unfin., 1847).

BIBL: J. Warrack, 'Mendelssohn's Operas', in *Music and Theatre: Essays for Winton Dean* (Oxford, 1987).

**Menorca.** See MINORCA.

**Menotti, Gian Carlo** (*b* Cadegliano, 7 July 1911). US composer of Italian birth. Having written two operas in childhood, he studied Milan, then Philadelphia with Scalero. His first surviving opera is *Amelia al ballo*, in which his light fertility of invention admirably serves the buffo plot. A more powerful dramatic vein is attempted in *The Old Man and the Thief* and *The Island God*; and these were followed by a thriller, *The Medium*, which has often shared a double bill with one of his most successful works, the comedy *The Telephone*. However, *The Consul* in 1950 showed Menotti at his most theatrically powerful and won him a world-wide audience, though it was noticeable that the power and sincerity of the dramatic plea was sharply at odds with the sub-Puccini lyricism of the score. Experience gained in filming *The Medium* stood Menotti in good stead in TV: *Amahl and the Night Visitors* has often been screened at Christmas. *The Saint of Bleecker Street* marked a return to blood-and-thunder verismo; but *The Unicorn, the Gorgon and the Manticore* was a satire on the vagaries of social fashion. In *Maria Golovin* Menotti tackled the deeper theme of personal imprisonment: his hero is blind and tormented by jealousy. A brilliantly skilled man of the theatre, quick and versatile, he has found it difficult to sustain the invention and theatrical power of his modern dramatic ideas with music that really belongs to a past verismo age.

He was librettist for Barber's *Vanessa*. In 1958 he founded the successful Festival of Two Worlds in *Spoleto, designed to give openings to young musicians. He has also produced opera in Scotland, which became his home.

SELECT WORKLIST (all texts by composer): *Amelia al ballo* (Amelia goes to the Ball) (Philadelphia 1937); *The Old Maid and the Thief* (NBC 1939; stage, Philadelphia 1941); *The Medium* (New York 1946); *The Telephone* (New York 1947); *The Consul* (Philadelphia 1950); *Amahl and the Night Visitors* (NBC-TV 1951; stage, Bloomington 1952); *The Saint of Bleecker Street* (New York 1954); *The Unicorn, the Gorgon and the Manticore* (Washington, DC, 1956); *Maria Golovin* (Brussels 1958); *Le dernier sauvage* (The Last Savage) (Paris 1963); *Help, Help, the Globolinks!* (Hamburg 1968); *The Most Important Man* (New York 1971); *Tamu-Tamu* (Chicago 1973) *The Hero* (Philadelphia 1976); *Goya* (Washington 1986).

**Méphistophélès.** The Devil (bs) in Gounod's *Faust* and in Berlioz's *La damnation de Faust*.

**Mercadante, Saverio** (*b* Altamura, *bapt.* 17 Sep. 1795; *d* Naples, 17 Dec. 1870). Italian composer. Studied Naples with Furno and Tritto, later Zingarelli. Here he attracted the attention of Rossini, who may have encouraged him to take up opera: *L'apoteosi d'Ercole* was well received in 1819. He followed this with others in quick succession, having his first great success with the seventh, *Elisa e Claudio*, in Milan and abroad. After failing to establish himself in Vienna, he renewed his success with *Caritea, regina di Spagna* in Venice. A proposed Spanish appointment having fallen through, he wrote *Ezio* for Turin, returned to Spain for *I due Figaro* (whose production was postponed from 1827 to 1835 on political grounds), then moved to Lisbon and back to Cadiz before settling in Italy in 1831. Here he had a triumph in Turin with *I Normanni a Parigi*. Invited to Paris by Rossini in 1835, he experienced a failure with *I briganti*, despite the presence in the cast of Grisi, Rubini, Tamburini, and Lablache. But his contact with Meyerbeer's *Les Huguenots* gave him decisive encouragement (though not a stylistic model) for his next and most famous work, *Il giuramento*, which triumphed in Milan in 1837. The striking expansion in his style, coupled with the withdrawal from the scene of Rossini and the death of Bellini, led to further successes with *Le due illustri rivali*, *Elena da Feltre*, *Il bravo*, and *La vestale*. Appointed director of the Naples Conservatory, he continued to compose busily, his later operas including notably *Il reggente* and *Orazi e Curiazi*.

Having begun in the heyday of Rossini, Mercadante's career extended well into the middle of Verdi's. His earliest operas are 18th cent. in manner, four of them deriving from Metastasio, and only later was he drawn to Cammarano and especially Romani as librettists, exchanging the manner of Rossini for lyrical melodrama and eventually, in *Il giuramento*, for music drama of considerable breadth and originality. Though this, and his succeeding works, are a clear bridge to Verdi, they have a strength of their own; structurally and harmonically they can be fluent and closely geared to the drama, while not losing a grasp of essentially Italian melody, and they impressed no less a judge of new music than Liszt. Verdi, who had been a victim of Mercadante's jealousy but forgave him, was fruitfully influenced (notably by *La vestale* in *Aida*). Though his later operas do not always maintain the ambition or quality of his best work, Mercadante is an important figure in 19th-cent. Italian music drama.

SELECT WORKLIST: *L'apoteosi d'Ercole* (The Apotheosis of Hercules) (Schmidt; Naples 1819); *Elisa e Claudio* (Romanelli, after Casari; Milan 1821); *Amleto* (Romani, after Shakespeare; Milan 1822); *Caritea, regina di Spagna* (Pola; Venice 1826); *Ezio* (Metastasio; Turin 1827); *I due Figaro* (Romani, after Martelly; comp. prob. 1827–9, prod. Madrid 1835); *I Normanni a Parigi* (Romani; Turin 1832); *I briganti* (The Robbers) (Crescini, after Schiller; Paris 1836); *Il giuramento* (The Oath) (Rossi, after Hugo; Milan 1837); *Le due illustri rivali* (Rossi; Venice 1838); *Elena da Feltre* (Cammarano; Naples 1838); *Il bravo* (Rossi and Marcello; Milan 1839); *La vestale* (The Vestal Virgin) (Cammarano; Naples 1840); *Il reggente* (Cammarano, after Scribe; Turin 1843); *Orazi e Curiazi* (The Horatii and the Curiatii) (Cammarano; Naples 1846).

BIBL: S. Palermo, *Saverio Mercadante* (Fasano, 1985).

**Merelli, Bartolomeo** (*b* Bergamo, 19 May 1794; *d* Milan, 3 Apr. 1879). Italian manager and librettist. Studied law, later music with Mayr. A fellow student was *Donizetti, for whom he later wrote the texts of five operas. Managed a theatrical agency, Milan 1826–30; managed seasons in Varese from 1830, Como and Cremona 1835; joint lessee Vienna, K, 1836–48, and engaged singers for Visconti at Milan, S, succeeding him 1836–50. Director, Milan, S, 1853–5, 1861–3. Also wrote librettos for Mayr, Vaccai, and Morlacchi, and suggested to Verdi the text of *Nabucco* that brought him back to opera. Though he treated Verdi generously, there were many complaints (including from Donizetti and later from Verdi) about his financial and artistic sharp practice. His popularity was not helped by possibly accurate rumours that he had spied for Radetzky. His son **Eugenio** (1825–82) organized tours throughout Europe, including to Edinburgh 1860; Vienna, K, 1864, and probably St Petersburg 1868 with Artôt.

**Méric-Lalande, Henriette** (*b* Dunkirk, 4 Nov. 1799; *d* Chantilly, 7 Sep. 1867). French soprano. Début Nantes 1814. Then studied Paris with García sen. and Talma; Milan with Bonfichi and Bandeali. Paris, T du Gymnase Dramatique, 1823. Venice, F, 1823–5; Naples, C, 1826; Milan, S, 1827–9, 1833; Paris, I, 1828; London, HM, 1830; Trieste 1838. One of Bellini's preferred singers; created Bianca (*Bianca e Gernando*), Imogene (*Il pirata*), title-role of *Zaira*, and Alaide (*La straniera*); also Palmide (Meyerbeer's *Crociato in Egitto*), and title-roles in Donizetti's *Elvida* (which demonstrates her virtuosity), and *Lucrezia Borgia* (when she demanded an inappropriate final bravura aria that he later cut). A remarkable dramatic soprano d'agilità, and an effective actress, she was successful in Rossini's *Semiramide*, *Donna del lago*, etc. By 1830, however, Chorley thought her appearance in London 'too late'.

**Merighi, Antonia** (*fl* 1717–40; *d* before 1764). Italian contralto. Sang in Italy (e.g. Venice, Bologna, Naples) 1717–35, in operas by Vivaldi, Gasparini etc. London, H, 1729–31, engaged by Handel. Sang in eight of his operas, creating Matilde (*Lotario*), Rosmira (*Partenope*), and Erissena (*Poro*) with particular success. Handel also rewrote other roles for her, including Armida (*Rinaldo*) and Elisa (*Tolomeo*). Considered a graceful and pleasant singer, and 'a perfect actress' (Rolli). Returned 1736 in works by Hasse, Veracini, etc., and 1737 for Heidegger's company, when she created Handel's Gernando (*Faramondo*) and Amastre (*Serse*). Her voice was by now declining. Munich 1740, then retired to Bologna.

**Merikanto, Aarre** (*b* Helsinki, 29 June 1893; *d* Helsinki, 29 Sep. 1958). Finnish composer and teacher, son of below. Merikanto's reputation as an operatic composer rests on *Juha*, a setting of a libretto rejected by Sibelius. It was not staged until long after its completion, largely owing to difficulties with the Finnish National Opera, but is gradually receiving recognition as seminal work in the country's opera. Merikanto's highly individual style was greatly influenced by Reger, with whom he studied, although the influence of atonality, serialism, and Finnish folk music may also be detected.

WORKLIST: *Helena* (Finne; 1912); *Juha* (Ackté & Aho; comp. 1920, rev. 1922, broadcast perf. 1958, staged perf. 1963).

**Merikanto, Oskar** (*b* Helsinki, 5 Aug. 1868; *d* Oitti, 17 Feb. 1924). Finnish composer and organist, father of above. Studied Leipzig and Berlin; Kapellmeister of Finnish Opera 1911–22. Although known mainly as a composer of salon music, Merikanto was responsible for the first opera to a Finnish text, *The Bothnian Maiden*. Like his two later operas, this work was composed in a popular, Romantic style, although the libretto is heavily coloured by national folk legend.

WORKLIST: *Pohjan neiti* (The Bothnian Maiden) (Ryktönen, after the Kalevala; comp. 1899, prod. Helsinki 1908); *Elinan surma* (Elina's Death) (Finne; Helsinki 1910); *Regina von Emmeritz* (Topelius & Leino; Helsinki 1920).

**Mérimée, Prosper** (*b* Paris, 28 Sep. 1803; *d* Cannes, 23 Sep. 1870). French author. Operas on his works are as follows:

*Inès Mendo* (1825): D'Erlanger (1897)
*Le carrosse du Saint-Sacrement* (1829): Offenbach (*La Périchole*, 1868); Berners (1924); Busser (1948)
*Matteo Falcone* (1829): Zöllner (1894); Cui (1907)
*L'occasion* (1830): Durey (1920)
*Le ciel et l'enfer* (1830): Malipiero (*Donna Urraca*, 1954)
*La Vénus d'Ille* (1837): Schoeck (1922); Wetzler (*Die baskische Venus*, 1928)
*Colomba* (1840): Grandjean (1882); Mackenzie (1883); Radeglia (1887); Büsser (1921)
*Carmen* (1845): Bizet (1875); Halffter (*La muerte de Carmen*, 1930)
*La dame de Pique* (translation of *Pushkin) (1849): Halévy (1850)
*La chambre bleu* (1869): Bouval (1902); Lazarus (1937)

**Merli, Francesco** (*b* Milan, 27 Jan. 1887; *d* Milan, 11 Dec 1976.) Italian tenor. Studied Milan with Negrini and Borghi. Début Milan, S, 1916 (Alvaro, *Fernand Cortez*). Sang there until 1942, sharing roles with Pertile. London, 1926–30; New York, M, 1931–2. Retired 1948. Roles included Don José, Calaf, Verdi's Otello, Walther; created roles in works by Respighi and Favara. (R)

**Merola, Gaetano** (*b* Naples, 4 Jan. 1881; *d* San Francisco, 30 Aug 1953). Italian conductor and manager. Studied Naples. Assistant cond. New York, M, 1899; Henry Savage and Hammerstein's Manhattan Co. 1906–10. London, OH, 1910–11. Toured USA with Naples San Carlo Co. Director San Francisco Co. 1923–53, raising it to rank second only to New York, M, and bringing many famous artists to the USA for the first time.

**Merrie England.** Opera in 2 acts by German; text by Basil Hood. Prod. London, Savoy T, 2 Apr. 1902, with Evett, Lytton, Brandram, Fraser, Passmore; New York, Hunter Coll., 13 Apr. 1956 (concert).

England, late 16th cent. The plot turns on the rivalry of the Earl of Essex (bar) and Sir Walter Raleigh (ten) for the favour of Queen Elizabeth I (mez), who discovers that Raleigh's true affections are with Bessie Throckmorton (sop), but whose attempts to have Bessie murdered are frustrated by a plot of Essex's.

**Merrill, Robert** (*b* Brooklyn, 4 June 1917). US baritone. Studied with his mother, and New York with Margolis. Début Trenton 1944 (Amonasro). New York, M, 1945–76, 1983–4; Milan, S, 1960; London, CG, 1967; also Chicago, S America, Italy. With his generous, Italianate sound, vocally very successful as Germont, Posa, Renato, Carlo (*Forza*), etc. (R)

**Merriman, Nan** (*b* Pittsburgh, 28 Apr. 1920). US mezzo-soprano. Studied Los Angeles with Bassian and Lotte Lehmann. Début Cincinnati 1942 (La Cieca). Aix from 1953; Milan, S, 1955; Gly. 1956; also Chicago; Vienna; Paris, O. Repertory included a highly acclaimed Dorabella, Laura (*The Stone Guest*), Baba the Turk. Chosen by Toscanini for several recordings, including Gluck's *Orfeo* (title-role). (R)

**Merritt, Chris** (*b* Oklahoma City, 27 Sep. 1952). US tenor. Studied Oklahoma. Début Santa Fe 1975 (*Fenton*) New York City O 1981; Paris, O, 1983; London, CG, 1985; Milan, S, 1988; New York, M, 1990. Also Vienna, S; Florence; Amsterdam; Naples; San Francisco; etc. Roles include Otello (Rossini), Rodrigo (*Donna del lago*), Arnold, Arturo, Nemorino, Cellini, Aeneas (Berlioz), Robert le diable, Leukippos. Particularly admired in Rossini. (R)

**Merry Widow, The.** See LUSTIGE WITWE, DIE.

**Merry Wives of Windsor, The.** See LUSTIGEN WEIBER VON WINDSOR, DIE.

**Mesplé, Mady** (*b* Toulouse, 7 Mar. 1931). French soprano. Studied Toulouse with Isar-Lasson, Paris with Micheau. Début Liège 1953 (Lakmé). Paris, OC from 1956, and O from 1958; Moscow, B, 1972; New York, M, 1973; also Naples, Aix, Holland Fest., etc. Roles incl. Queen of the Night, Rosina, Lucia, Norina, Philine, Sophie, Zerbinetta. (R)

**messa di voce** (It.: 'placing of the voice'; Ger., *Schwellton*). The art of swelling and diminishing on a single note. First mentioned by Caccini in the preface to *Le nuove musiche* (1602), it acquired great importance in the age of bel canto as a demonstration of vocal skill and control. The French term *son filé* means only diminishing the tone. See also FILAR IL TUONO.

**Messager, André** (*b* Montluçon, 30 Dec. 1853; *d* Paris, 24 Feb. 1929). French composer and conductor. Studied Paris with Fauré, Saint-Saëns, and others. Organist at St Sulpice, then began writing operettas and ballets (including *Les deux pigeons*). His first success came with *La béarnaise*, which was popular both in Paris and in London. After several more works in similar vein, he achieved his greatest hit with *La basoche*, which set a new standard at the Opéra-Comique; in London its performance followed that of Sullivan's *Ivanhoe* at the English OH and led to a commission for *Mirette*. During the 1890s he produced a string of operettas which were highly popular in their day, notably *Madame Chrysanthème*, *Les p'tites Michu*, and *Véronique*.

Messager's works soon left the repertoire, but his importance as a conductor was more lasting. Coming under Wagner's influence in the 1880s, he made a name as a Wagner conductor; he was also music director of the Opéra-Comique 1898–1903, during which time he conducted the première of *Pelléas et Mélisande* (1902). He was manager of the Grand Opera Syndicate at London, Covent Garden, 1901–7, director of the Opéra 1907–14, music director of the Opéra-Comique again 1919–20. After the First World War, when he toured N and S America, he gave more time again to composition: works of this period include *Monsieur Beaucaire* (prod. with Maggie Teyte) and *La petite functionnaire*. He did much to introduce Wagner into France, conducting the French premières of the *Ring*, *Parsifal*, and *Meistersinger*. He is also famous for his support of Russian music and Mozart; he revived Gluck and Rameau, as well as encouraging his contemporaries and introducing Charpentier's *Louise* and Massenet's *Grisélidis*. At Covent Garden he conducted a wide repertory that included premières of Saint-Saëns's *Hélène* and Massenet's *Le jongleur de Notre-Dame*. His music is unfailingly elegant and deft, with a theatrical flair and tunefulness that are the product of much musical learning lightly worn.

SELECT WORKLIST: *La béarnaise* (Leterrier & Vanloo; Paris 1855); *La basoche* (Carré; Paris 1890); *Madame Chrysanthème* (Hartmann & Alexandre, after Loti; Paris 1893); *Mirette* (Carré, Weatherly, & Ross; London 1894); *Les p'tites Michu* (Vanloo & Duval; Paris 1897); *Véronique* (Vanloo & Duval; Paris 1898); *Monsieur Beaucaire* (Lonsdale & Ross, after Tark-ington; Birmingham 1919); *La petite fonctionnaire* (Roux, after Capus; Mogador 1921).

BIBL: M. Augé-Laribé, *André Messager, musicien de théâtre* (Paris, 1952).

**Messel, Oliver** (*b* Cuckfield, 13 Jan. 1904; *d* Barbados, 13 July 1978). English designer. Studied London with Tonks. Began to work for theatre 1926. Designed ballets in London for Sadler's Wells, and for Covent Garden, where he also had a brilliant success with *The Magic Flute*, 1947. Here he also designed *The Queen of Spades* 1950, the première of *Gloriana* 1951, *The Golden Cockerel* 1954, *Samson* 1958. Edinburgh Fest. (Gly. Touring O) *Ariadne auf Naxos* 1950. Gly.: *Idomeneo* 1951, *Cenerentola* 1952, *Barbiere*, *Comte Ory* 1954, *Figaro* 1955, *Entführung*, *Magic Flute* 1956, *Rosenkavalier* 1959. Also designed postwar proscenium arch for Glyndebourne. New York, M, 1953. His beautiful, colourful sets and costumes, of which those for *The Magic Flute* were outstanding, were sensitively designed to give each opera an atmosphere in which its music could best flourish.

**Messiaen, Olivier** (*b* Avignon, 10 Dec. 1908; *d* Paris, 28 Apr. 1992). French composer. Messiaen did not consider writing an opera until commissioned by Rolf Liebermann, and the text and music of *St François d'Assise* (1983) took him seven years. Subtitled 'Franciscan scenes', it is an interior drama tracing the progress of the state of grace in Francis's soul, employing seven principal singers and a large chorus and orchestra; it lasts some five hours, and draws extensively on the bird-song (including the birds of Assisi) which was one of Messiaen's enduring preoccupations, and is here celebrated in some of his richest orchestral writing.

**metaphor aria** (or **simile aria**). A term used for arias in opera seria in which the singer takes a metaphor or simile, illustrated by the music, to illuminate a dramatic situation or emotional situation. Characteristic examples in Metastasio, in whose librettos they abound, are when the singer compares himself to a raging lion, a steersman in a storm, a turtle-dove awaiting his mate, or (in *Sosarme*) compares another singer to a deranged butterfly. The merit of the convention was to provide singers with a vivid excuse for virtuosity, and the composer with a chance to display his powers of illustration. A celebrated later example, containing in its genuine emotion a touch of parody of the convention, is Fiordiligi's 'Come scoglio' in Mozart's *Così fan tutte*.

**Metastasio** (orig. Trapassi), **Pietro** (*b* Rome, 3 Jan. 1698; *d* Vienna, 12 Apr. 1782). Italian librettist and poet. Adopted by a Roman literary dilettante, Gravina, who gave him the name

Metastasio and eventually bequeathed him his fortune, he received a rigorous classical education. He wrote his first tragedy at age 14 and later joined the Accademia degli Arcadi. He moved to Naples and trained for the law, but was soon drawn into theatrical life there, enjoying the protection of the singer Marianna Benti-Bulgarelli ('La Romanina'): he also studied with Porpora, and provided texts for some of his smaller dramatic works. With his first wholly original libretto, *Didone abbandonata* (Domenico Sarro, 1724), Metastasio rose to overnight fame; returning to Rome, he capitalized upon its success by producing a series of works in similar vein. In 1729 he was invited to Vienna, where he succeeded Zeno as *poeta cesareo to Charles VI, arriving there the following year. 1730 also saw the first settings of his *Artaserse* libretto, marking a watershed in operatic history. Although all of his work to date had included a certain element of reform, it was *Artaserse* which provided the clearest expression of the new ideals of opera seria, and in Hasse's setting, which followed on shortly from that of Vinci, Metastasio's libretto found its ideal musical match.

During his half-century at Vienna, Metastasio's reputation grew to make him not only the most famous Italian poet of his day, but also the most frequently set librettist of all time. Excluding the texts he provided for oratorios and smaller dramatic pieces, he wrote 27 large-scale opera librettos: these were set by composers throughout Europe, some being used 60 or 70 times. Despite Metastasio's best endeavours, many of his librettos were revised before being set, Mozart's *Il re pastore* and *La clemenza di Tito* being just two of many examples: if such works are included, the list becomes even more impressive. Among the most notable to set Metastasio's librettos were Caldara, Vinci, and Hasse, who later joined forces with Metastasio to resist the reforms of Gluck and Durazzo. Many librettos survived well into the 19th cent. (Meyerbeer set *Semiramide* in 1819, Mercadante *Didone abbandonata* in 1823), long after Metastasio's name had become a by-word for all that was supposedly undramatic, contrived, and over-conventionalized, while his verses also provided a treasure trove of song texts for composers as diverse as Beethoven, Rossini, and Weber.

To understand Metastasio's talent requires a knowledge of the circumstances in which he worked, the demands of contemporary audiences, and the aesthetic ideals of the day: when these factors are considered, it is clear that much of the posthumous criticism levelled at him was misplaced and has served to distort appreciation of his unique gifts as a librettist. Many of the so-called reforms with which he is credited were already in place when he first turned to opera seria: under Stampiglia, Zeno, and others, the 17th-cent. Venetian libretto had been largely shorn of its comic characters, complex sub-plots, and excessive use of spectacle and machinery. Building on these developments, Metastasio codified the libretto into a pure and elevated expression of Enlightenment ideals, while remaining true to some fundamental Baroque procedures. His work thus provides a bridge between the two eras, reflecting the changing mood of the times in which he himself lived. While observance of the *doctrine of affections led to a concentration within each aria on one basic emotion, effectively hindering any real character development, the resolution of the drama, featuring a *lieto fine, was almost invariably achieved by the placing of reason above sentiment.

The growing aesthetic concern for balance and proportion in art had already led Metastasio's predecessors towards a more regulated libretto structure; under him, this was developed into a highly conventionalized formula, which also accorded remarkably well with the demands of composers. Six singers were usually called for, spanning a balanced range of voice-types, to which arias were assigned in accordance with a strict pattern, so that the relative importance of the stage character, as well as his or her ability to express emotion, was maintained. The subject-matter was usually drawn from classical mythology or history and, while still intricate in design, the action was not encumbered with diverting sub-plots or comic episodes. At the centre of the drama was a conflict between love and duty, which was usually resolved by the triumph of conscience over the heart, thereby averting tragedy: often this denouement was achieved with the help of a magnanimous tyrant. The social structure and standard of conduct which Metastasio's librettos glorified was, of course, entirely appropriate in the context of an imperial court, and was often summed up in the *licenza. The Baroque influence was seen most clearly in the highly formalized arias, in which the expression of only one basic emotion was permissible, and which received an equally stereotyped musical setting through the da capo aria. The utterances of the stage figures were dependent on situation and effect, rather than character, permitting an interchangeability of arias between operas which was anathema to later generations of composers. But as vehicles for effective vocal display pieces, Metastasio's texts were unsurpassed. His elegant, charming verse, marked by a vivid gift for imagery, was seen at its best in the *metaphor arias, and was considered by many an artefact of beauty in its own right.

The main drawback to Metastasian opera seria, and the one which later drew most censure, was the role played by the exit aria. This interfered radically with the progress of the drama and enouraged an excess of vocal skill that was the main target of Gluck's reforms. What had begun as a refined, flexible response to changing aesthetic values turned gradually during the 18th cent. into a rigidity of approach. Hence the most inventive composers, of which Mozart was the greatest example, turned to opera seria only when forced by circumstances, as with *La clemenza di Tito*: opera buffa became the preferred mode of expression, and it was in this genre that the greatest advances were made. But to judge Metastasio's librettos by the criteria of a later age is unfair: viewed in the light of what preceded them, and the conventions of the day, they represented a pinnacle of poetic achievement unattained since the earliest experiments of Rinuccini, Striggio, and Chiabrera, as well as the last operatic works in which the power of the librettist triumphed over that of the composer.

An operetta on Metastasio is by L. Dall'Argine (1903). For a list of settings of his librettos, see *Enciclopedia dello spettacolo*, vii, columns 501–5.

The following is a list of his operatic librettos, with first composer and date of first setting, excluding his many smaller dramatic pieces and sacred works: *Siface* (Feo, 1723); *Didone abbandonata* (Sarro, 1724); *Siroe* (Vinci, 1725); *Catone in Utica* (Vinci, 1728); *Ezio* (Porpora, 1728); *Semiramide* (Porpora, 1729); *Alessandro nell'Indie* (Vinci, 1729); *Artaserse* (Vinci, 1730); *Demetrio* (Caldara, 1731); *Issipile* (Conti, 1732); *Adriano in Siria* (Giacomelli, 1733); *Demofoonte* (Caldara, 1733); *L'Olimpiade* (Caldara, 1733); *La clemenza di Tito* (Caldara, 1734); *Achille in Sciro* (Caldara, 1736); *Ciro riconsciuto* (Caldara, 1736); *Temistocle* (Caldara, 1736); *Zenobia* (Predieri, 1740); *Antigono* (Hasse, 1743); *Ipermestra* (Hasse, 1744); *Attilio Regolo* (Hasse, 1750); *Il re pastore* (Bonno, 1751); *L'eroe cinese* (Bonno, 1752); *Nitteti* (Conforti, 1756); *Il trionfo di Clelia* (Hasse, 1762); *Romolo ed Ersilia* (Hasse, 1765); *Ruggiero* (Hasse, 1771). Different titles for many of these works later gained common currency, e.g. *Poro* for *Alessandro nell'Indie*, and *Viriate* for *Siface*.

BIBL: J. Joly, *Les fêtes théâtrales de Métastase à la cour de Vienne (1731–1767)* (Clermont-Ferrand, 1978); M. Murano, *Metastasio e il mondo musicale* (Florence, 1986).

**Metropolitan Opera.** New York's leading opera-house formerly stood on Broadway between 39th and 40th Streets. It was opened 1883 when some New York business men, unable to get boxes at the *Academy of Music, subscribed $800,000; the Metropolitan (cap. 3,045) opened 22 Oct.

1883 with Gounod's *Faust*. Heavy losses under Henry Abbey led to stockholders appointing Leopold *Damrosch as director; dying before the end of his first season, he was succeeded by his son Walter *Damrosch (till 1891). Operas were sung in German, with US premières of much Wagner. The theatre was rebuilt after a fire in 1892 (cap. 3,849). Abbey returned 1891, with Maurice *Grau and Edward Schoeffel until 1897, with Grau as manager 1897–1903, Heinrich *Conried 1903–8. Singers included the De Reszkes, Eames, Lilli Lehmann, Nordica, Sembrich, Schumann-Heink, Ternina, Maurel, Plançon, and Van Dyck; then Caruso, Farrar, and Fremstad, with Mahler and Mottl as conductors. Conried enraged Bayreuth by giving the first US *Parsifal* before expiry of copyright, and public opinion in New York with *Salome*.

He was succeeded by *Gatti-Casazza (till 1935), whose shrewd and constructive policies included engaging Toscanini as chief conductor and staging the premières of *La fanciulla del West*, *Il trittico*, *Königskinder*, and *Madame Sans-Gêne*, *Goyescas* and, the first US opera here, Converse's *Pipe of Desire* (1910), and many US premières including *Queen of Spades*, *Boris Godunov*, and *Rosenkavalier*. Singers included Destinn, Alda, Tetrazzini, Bori, Hempel, Martinelli, Amato, and Renaud, with conductors including Hertz, Mahler, and Polacco; later virtually all the great singers of the century appeared (including the début of Rosa Ponselle, 1918). Edward *Johnson was manager 1935–50, and encouraged American singers including Warren, Peerce, Tucker, Traubel, Kirsten, Steber, and Harshaw, with guest conductors including Beecham, Walter, Busch, Stiedry, Szell, and Reiner; he also built up a Wagner tradition with the greatest Wagner singers of the day. He was succeeded by Rudolf *Bing, whose régime was conservative but who modernized stage techniques and brought in theatre producers including Alfred Lunt and Tyrone Guthrie, with operetta and a greater emphasis on the Italian repertory. The new Metropolitan (cap. 3,500) opened in the Lincoln Center on 16 Sep. 1966 with the première of Barber's *Antony and Cleopatra*. Kubelik became music director 1972, resigning 1974 and being succeeded by James Levine.

BIBL: I. Kolodin, *The Metropolitan Opera* (New York, 1966).

**Metz.** Town in Moselle, France. Opera is given in the T Municipal (cap. 760), opened 1752; renovated 1851. The company pursues a policy of a strong ensemble, and has also given the premières or French premières of a number of works.

# Mexico

**Mexico.** The first opera written in Mexico was *Partenope* (1711) by P. Manuel Zumaya (*c.*1678-1756). The T Coliseo opened in 1733, presenting various musical entertainments including zarzuela and Italian opera, including Paisiello's *Barbiere di Siviglia* in 1806. This was also the year of Manuel de Aremán's *El extranjero*. Operatic activity began to increase, and Italian opera achieved new popularity with the visit of García's company in 1830. Operas by Mexican composers followed. *Catalina de Guisa* (1859) by Cenobio Paniagua (1821–82) was popular, his *Piero d'Abano* (1863) less so. Melesio Morales (1838–1908) wrote *Romeo y Julieta* (1863) and the first Mexican opera to be staged abroad, *Ildegonda* (Florence 1866). The first true Romantic was Aniceto Ortega (1823–75), who was influenced by Beethoven and Weber, and whose *Guatimotzin* (1871) incorporates some native elements into its style. Gustavo Campo (1863–1934), an important teacher, organizer, and writer, encouraged a feeling for Wagner, and wrote *El rey poeta* (1901), influenced by his friends Massenet and Saint-Saëns.

Italian opera continued to be popular. In 1887 a pirated version was given of Verdi's *Otello*, orchestrated by P. Vallère from the vocal score. This was the first performance outside Italy; the genuine version followed in 1890. After the 1910 revolution, foreign companies virtually ceased to visit Mexico, though some open-air performances were given in bullrings by, among others, Caruso and Ruffo, to audiences of 25,000. During the 1930s opera was again in decline, though in 1934 the Palacio de las Bellas Artes, begun 1910, was finally opened (cap. 2,000). A local company performed there in 1935. In 1941 the first steps to establish a permanent company were taken by Franz Steiner, with his fellow Viennese refugees Karl Alwin and William Wymetal. The first production was *The Magic Flute*, with European and Mexican singers. Conductors who came during the war included Beecham, Kleiber, and Horenstein. In 1943 the contralto Fanny Anitúa founded the Opera Nacional, where Callas made her N American début in 1950; this went bankrupt in 1953. It was refounded in 1955. Opera, mostly 19th-cent. classics sung by Mexican casts in the original language, is given at irregular intervals during the year. Latterly there has been an attempt to encourage more native Mexican operas. Recent works have included *Orestes parte* (1988) by Federico Ibarra (*b* 1946). Seasons of about 50 annual performances are given with Mexican and international casts.

**Meyer, Kerstin** (*b* Stockholm, 3 Apr. 1928). Swedish mezzo-soprano. Studied Stockholm with Sunnegard, New York with Novikova. Début Stockholm 1952 (Azucena). New York, M, and London, CG, from 1960; Gly. 1961; Bayreuth 1962–5; also Salzburg, Edinburgh, etc. Roles included Dorabella, Carmen, Brangäne, Berlioz's Dido, Britten's Lucretia; created Carolina in first original *Elegy for Young Lovers*, Agave (*The Bassarids*), Mrs Arden (Goehr's *Arden Must Die*). A notable singing actress of great warmth, vitality, and perception. (R)

**Meyerbeer, Giacomo** (orig. Jakob Meyer Beer) (*b* Vogelsdorf, 5 Sep. 1791; *d* Paris, 2 May 1864). German composer of Jewish descent. Studied Berlin with Lauska, Zelter, and B. A. Weber, Darmstadt with Vogler. During his time with Vogler he wrote his first two operas, *Jephtas Gelübde* and *Wirth und Gast*: the former, more oratorio than opera, was a failure; and the latter, a comedy, also failed on its first performance and when it was revised for Vienna as *Die beyden Kalifen* in 1814. A Singspiel, *Das Brandenburger Tor*, missed its occasion, the return of the victorious Prussian army in 1814. He was at this stage better known as a piano virtuoso; but still nursing ambitions as an opera composer, he sought the advice of Salieri, who suggested he should go to Italy to study the voice. Here he wrote six Italian operas, with such success that he was even compared with Rossini. This brought him little approval in Germany: his former fellow-student Weber did produce some of his operas, but complained of his Italian 'aberration'. The success of *Il crociato in Egitto*, especially, encouraged him to turn his thoughts to Paris, where the work triumphed in 1825.

Though Meyerbeer travelled widely pursuing his career, this was centred henceforth on Paris, where he was to become the most important composer of French *Grand Opera. He spent some time assimilating French art, history, and literature, and, guided by his most important librettist, Scribe, planning his assault on the Opéra. His first conquest came in 1831 with *Robert le diable*, in which he successfully combined his German soundness of technique, Italianate melodies skilfully written for particular singers, and his shrewd understanding of the implications for a composer of the lavish staging of the Opéra. Even this triumph was overtaken in 1836 with the new commission from the Opéra's influential director Louis *Véron, *Les Huguenots*. Plans were laid for *L'Africaine*, but shelved when the intended prima donna, Marie *Falcon, lost her voice, in favour of *Le prophète* for Gilbert *Duprez. In 1842 Meyerbeer was appointed Prussian Generalmusikdirektor in Berlin. For the reopening of the theatre after a fire he wrote *Ein Feldlager in Schlesien*, with Jenny Lind. Controversies with the Intendant, Küstner, eventually led to his dismissal in 1848. He resumed work on *Le prophète*, remodelling it for Pauline *Viardot, and had another triumph with its première in 1849. With a new text by Scribe, he revised the music of *Ein Feldlager* for *L'étoile du nord*, successfully performed in 1854 at the Opéra-Comique, following this with another opéra-comique, *Le pardon de Ploërmel*, in 1859. He resumed work on *L'Africaine*, despite Scribe's death in 1851 and his own deteriorating health; the work went into rehearsal, but he died before preparations were complete.

Meyerbeer's contribution to Grand Opera was the most important of any single composer. His models were to a large extent Spontini, Auber, and Rossini's *Guillaume Tell*, though these were not really influences: what inspired him were the opportunities which Auber, Rossini, and others had found in the conventions developed at the Opéra in the 1820s by *Cicéri and *Daguerre and fostered by the astute management of Véron. His individual qualities included the mastery of orchestration which may be attributed in part to his German training and background, but which was extended to match particular dramatic demands. His inventiveness with the orchestra was much admired by Berlioz, and influenced Wagner (who bit the hand that had fed him help in Paris with some unpleasant references, including by implication in the anti-Semitic *Das Judentum in der Musik*). His collaboration with Scribe, who had a shrewd understanding of the Opéra and of the public, continued the tradition of Grand Opera in the use of historical subjects on a vast scale; it also gave him splendid musical opportunities, both for dazzling vocal writing carefully tailored for the greatest singers of the day and for grandiose scenes, or 'tableaux', in which they and massed choruses would conclude an act. He rarely made use of motive: his operas are constructed in a series of carefully planned numbers, and rely on individual and cumulative effect. Wagner described them as 'effects without causes'; and he has been by no means alone in finding in Meyerbeer's works too much calculation so as to achieve public success. That success can still be won by a production matching the works' nature.

WORKLIST: *Jephtas Gelübde* (Jephtha's Vow) (Schreiber; Munich 1812); *Wirth und Gast* (Host and Guest) (Wohlbrück, after the *1,001 Nights*; Stuttgart 1813, rev. as *Die beyden Kalifen*, Vienna 1814, rev. as *Alimelek*, Dresden 1820); *Das Brandenburger Tor* (Veith; comp. 1814, unprod.); *Romilda e Costanza* (Rossi; Padua 1817); *Semiramide riconosciuta* (Semiramide Rewarded) (Rossi, after Metastasio; Turin 1819); *Emma di Resburgo* (Rossi; Venice 1819); *Margherita d'Anjou* (Romani, after Pixérécourt; Milan 1820); *L'Almanzore* (Rossi; comp. 1821,

unprod.); *L'esule di Granata* (The Exile of Granata) (Romani; Milan 1821); *Il crociato in Egitto* (The Crusader in Egypt) (Rossi; Venice 1824); *Robert le diable* (Robert the Devil) (Scribe & Delavigne; Paris 1831); *Les Huguenots* (Scribe & Deschamps; Paris 1836); *Ein Feldlager in Schlesien* (A Camp in Silesia) (Rellstab, & Birch-Pfeiffer; Berlin 1844, rev. as *Vielka*, Vienna 1847); *Le prophète* (Scribe; Paris 1849); *L'étoile du nord* (The Northern Star) (Scribe; Paris 1854); *Le pardon de Ploërmel* (The Pilgrimage to Ploërmel; also known as *Dinorah*) (Barbier & Carré; Paris 1859); *L'Africaine* (The African Girl) (Scribe & Fétis; Paris 1865).

BIBL: H. and G. Becker, *Giacomo Meyerbeer: A Life in Letters* (Wilhelmshaven, 1983; trans. London, 1989).

**Meyerhold, Vsevolod** (*b* Pensa, 9 Feb. 1874; *d* Moscow, 2 Feb. 1940). Russian producer. His work in opera began at the Maryinsky T, St Petersburg in 1909 with enterprising productions including *Tristan and Isolde* 1909, *Boris Godunov* 1911, *Elektra* 1913, and *The Nightingale* 1917. One of his most famous productions attempted a reconciliation between Pushkin's and Tchaikovsky's *Queen of Spades* (1935). He worked closely with Shostakovich and with Prokofiev (whose *Love of the Three Oranges* took an example from Meyerhold's view of the *commedia dell'arte). He briefly took over Stanislavsky's company in 1938.

**mezza voce** (It.: 'half voice'). The direction to sing at half power.

**mezzo-contralto**. A term first applied to the singer Rosine Stoltz (1815–1903), whose voice was characterized by an especially strong lower register. More a description of timbre than of range.

**mezzo-soprano** (It., half-soprano). The middle category of female (or artificial male) voice. Though closer to soprano than to contralto, it is marked more by quality of tone (darker or richer than soprano) than by range, which may include many of the soprano's high or the contralto's low notes. (The normal mezzo range is approximately g–b''.) Many roles lying in the upper range, such as Poppea, Dido (Purcell and Berlioz), Cherubino, Charlotte, Eboli, Kundry, and the Kostelnička, are sung by both sopranos and mezzos. Parts written for castratos are now often taken by female mezzos, e.g. Handel's Serse and Gluck's Orfeo. Some mezzo-soprano roles given as such by composers in scores are as follows:
England: Hermia (*Midsummer Night's Dream*), Thea (*Knot Garden*).
France: Cassandre, Dalila, Fidès, Hérodiade, Carmen.
Germany: Ortrud, Octavian, Gräfin Geschwitz (*Lulu*).

Italy: Emilia (Rossini's and Verdi's *Otello*), Teresa (*Sonnambula*), Azucena, Amneris.
Russia: Hostess (*Boris*), Mme Larina and Filipyevna (*Onegin*).

**Miami.** City in Florida, USA. Opera is performed in the Miami Beach Auditorium (cap. 3,700) and the Dade County Auditorium (cap. 2,500) by the Greater Miami Opera, founded in 1941 by Arturo di Filippi, a tenor who took part in some of the early performances. The repertory is generally popular, with star guests in the leading roles, though Liszt's *Don Sanche* had its US première 1986 and Stephen Paulus's *The Postman Always Rings Twice* in 1986, and the company has developed a policy of a less experienced second cast performing at lower prices.

**Micaëla.** A peasant girl (sop), in love with Don José, in Bizet's *Carmen*.

**Micheau, Janine** (*b* Toulouse, 6 Jan. 1914; *d* Paris, 18 Oct. 1976). French soprano. Studied Toulouse, and Paris with Carré. Début Paris, OC, 1933 (La Plieuse, *Louise*). Paris, OC until 1956, and O 1940–56; London, CG, 1937; also Chicago, Buenos Aires, Florence. Roles included Violetta, Juliette, Mélisande, Sophie, Zerbinetta, Anne Trulove; created Milhaud's Creusa (*Medée*), and Manuela (*Bolivar*). (R)

**Michele.** A Seine bargee (bar), husband of Giorgetta, in Puccini's *Il tabarro*.

**'Mi chiamano Mimì'.** Mimì's (sop) aria in Act I of Puccini's *La bohème*, in which she describes herself to Rodolfo.

**Midsummer Marriage, The.** Opera in 3 acts by Tippett; text by the composer. Prod. London, CG, 27 Jan. 1955, with Sutherland, Leigh, Dominguez, Lewis, Lanigan, Kraus, cond. Pritchard; San Francisco 15 Oct. 1983, with Herincx, Johnson, Bailey, Greenwald, Nadler, cond. Agler.
England, the present. Morning. Midsummer Day, the wedding-day of Mark (ten) and Jenifer (sop). Her father King Fisher (bs), a business man, wishes to prevent the marriage. Before they can discover each other in marriage, each must first discover their true selves, so Jenifer goes in search of the physical reality and Mark in search of the imaginative intuition they each feel they lack.
Afternoon. In an acknowledged analogy with *The Magic Flute*, the couple are matched by an uncomplicated second pair, Jack (ten), a mechanic, and Bella (sop), secretary to King Fisher. They affirm their love. A series of ritual dances takes place.
Evening and Night. Madame Sosostris (con), a clairvoyante, arrives; King Fisher wishes her to return Jenifer to him. However, she sees a vision of Mark and Jenifer loving one another. When

King Fisher unveils Sosostris, since Jack has refused to do so, he discovers Mark and Jenifer hidden inside a huge flower-bud. He attempts to prevent their union, but dies. During a ritual fire dance the couple are transfigured and appear, reborn, ready to celebrate their wedding.

**Midsummer Night's Dream, A.** Opera in 3 acts by Britten; text by the composer and Peter Pears, after Shakespeare's comedy (*c*.1593–4). Prod. Aldeburgh 11 June 1960, with Deller, Vyvyan, Cantelo, Thomas, Hemsley (acting, sung by Joseph Ward), Maran, Pears, Brannigan, Lumsden, Kelly, Byles, Massine, cond. Britten; San Francisco, War Memorial OH, 10 Oct. 1961, with Oberlin, Costa, Evans, cond. Varviso.
The plot follows Shakespeare closely, with the characters separated musically into three groups: the fairies Oberon (counter-ten), Tytania (sop), and Puck (spoken acrobat); the lovers Lysander (ten) and Demetrius (bar), initially both in love with Hermia (mez), who is in love with Lysander, and Helena (sop), who is in love with Demetrius; and the rustics, Bottom (bs-bar), Quince (bs), Flute (ten), Snug (bs), Snout (ten), and Starveling (bar). The high sound of the fairies' voices (often associated by Britten with the exceptional or strange) is reinforced by the chorus of fairies, led by Cobweb, Peaseblossom, Mustardseed, and Moth (trebles). The playlet of *Pyramus and Thisbe* in Act III parodies various devices of Romantic opera.

**Migenes-Johnson, Julia** (*b* New York, 13 Mar. 1945). US soprano. Early career on Broadway. Studied Cologne with Ultman. Début New York, CO, 1965 (Musetta). Vienna, V, 1973–8; San Francisco 1978; New York, M, from 1979; London, CG, 1987. Also Los Angeles; Houston; Geneva; Vienna, S; Berlin. Repertory incl. Blonde, Susanna, *Hoffmann* roles, Carmen, Manon, Eurydice (Offenbach), Nedda, Salome, Lulu. (R)

**Mignon.** Opera in 3 acts by Thomas; text by Barbier and Carré, after Goethe's novel *Wilhelm Meisters Lehrjahre* (1795–6). Prod. Paris, OC, 17 Nov. 1866, with Galli-Marié, Cabel, Achard, Bataille; London, DL, 5 July 1870, with Christine Nilsson; New Orleans, French OH, 9 May 1871, with Dumestre, De Keghel, Naddi, Périé.
Germany and Italy, late 18th cent. Lothario (bs), a wandering minstrel, is searching for his long-lost daughter. A band of gypsies arrives; they try to make one of their number, Mignon (mez), dance. When she refuses they start to beat her; she is rescued by Wilhelm Meister, who engages her as his servant and then falls in love with her.
Mignon is saved from a burning castle by Wil-

helm who, with Lothario, nurses her back to health.

Lothario, who had lost his memory when Mignon was first kidnapped, now remembers that he is Count Lothario, and recognizes Mignon as his long-lost daughter Sperata.

**Mihalovich, Ödön** (*b* Feričance, 13 Sep. 1842; *d* Budapest, 22 Apr. 1929). Hungarian composer. Studied Pest with Mosonyi, Leipzig with Hauptmann, Munich with Cornelius. Much influenced by Wagner whose cause he promoted in Hungary: he even set a version of Wagner's libretto *Wieland der Schmied* after it had been rejected by Berlioz, among others. Director Budapest Music Academy from 1887, and responsible for engaging Mahler at the Royal Hungarian OH.

**Mikéli.** The water-carrier (bs) in Cherubini's *Les deux journées*.

**Mikhaylov, Maxim** (*b* Koltsovka, 25 Aug. 1893; *d* Moscow, 30 Mar. 1971). Russian bass. As child, sang in church choir; deacon in Omsk, Kazan, and Moscow. Moscow, B, 1932–56. Roles incl. Pimen, Varlaam, Konchak, Gremin, Miller, and his most famous, Ivan Susanin. (R)

**Mikhaylova, Maria** (*b* Kharkov, 3 June 1866; *d* Molotov (Perm), 18 Jan. 1943). Russian soprano. Studied St Petersburg Cons., Paris with Saint-Yves-Bax, Milan with Ronconi. Début St Petersburg, M, 1892 (Marguerite de Valois, *Huguenots*). Sang there until 1912; Prague 1903; Tokyo 1907; also Moscow, Kiev, etc. Repertory included Zerlina, Juliette, Nannetta, Lakmé, Lyudmila, Tamara (*Demon*), first Electra (Tanayev's *Oresteia*). A prolific recording artist, she achieved international fame though singing mainly in Russia. (R)

**Milan** (It., Milano). City in Lombardy, Italy. Staged musical works were given in the Salone Margherita of the Palazzo Ducale in the 16th cent. When this burnt down in 1708 it was replaced by the T Regio Ducale, 1717; burnt down 1776. Mozart visited there in the early 1770s and was much fêted. The 96 box-holders appealed directly to the Empress Maria Theresa, who gave permission for the building of *La Scala, opened 3 Aug. 1778 with Salieri's *L'Europa riconosciuta*. Opera has also been given at the T della Cannobiana, the T Carcano, the T Lirico, and the T dal Verme.

The Cannobiana (cap. 2,000) was built at the same time as La Scala on land given by Maria Theresa; opened 1779 with Salieri's *Fiera di Venezia* and *Il talismano*. *L'elisir d'amore* was premièred here 1832. Demolished 1894, replaced by T Lirico. Reopened 1894 with Samara's *Le martire*. Caruso made his Milan début here in the première of Cilea's *L'arlesiana*, 1897, and the theatre

saw other important premières by Cilea, Giordano, and Leoncavallo, and Italian premières of works by Massenet and Berlioz (*La prise de Troie*). The T Lirico was demolished 1938, rebuilt 1939.

The T Carcano opened Sep. 1803 with Federici's *Zaira*. Scene of the premières of *Anna Bolena* (1830) and *Sonnambula* (1831). The T dal Verme opened 1872, replacing the Politeama Ciniselli (1864), with *Les Huguenots*. Primarily the home of drama, but saw the premières of Puccini's *Le villi* (1884), Leoncavallo's *Pagliacci* (1892), and Zandonai's *Conchita* (1911).

**Milanov, Zinka** (*b* Zagreb, 17 May 1906; *d* New York, 30 May 1989). Croatian soprano. Studied Zagreb with Ternina, Prague with Carpi. Début Ljubljana 1927 (Leonora, *Trovatore*). Zagreb 1928–35; Prague, German T, 1936; New York, M, 1937–41, 1942–7, 1950–66; Buenos Aires, C, 1940–2; London, CG, 1956–7. Repertory included Donna Anna, Norma, Tosca; excelled in Verdi, e.g. as Leonora (*Forza* and *Trovatore*), Gilda, Amelia, Aida, and as Gioconda. Her voice possessed beauty of tone, power, and an unforgettable pianissimo. (R)

**Milde, Hans Feodor von** (*b* Petronell, 13 Apr. 1821; *d* Weimar, 10 Dec. 1899). German baritone. Studied Vienna with Hauser, Paris with García jun. Sang as guest, Weimar 1845; remained there for the rest of his career, staying because of Liszt. Retired 1884. Sang Flying Dutchman, Sachs, Kurwenal; created Telramund, and the Caliph (*Barber of Bagdad*). He married in 1851 the soprano **Rosa Agthe** (*b* Weimar, 25 May 1825; *d* Weimar, 25 Jan. 1906). Studied Weimar with Götze. Début Weimar 1845; sang there until 1867. Roles included Pamina, Lucia, Leonore, first Elsa, and first Margiana (*Barber of Bagdad*). Adelheid von Schorn praised husband and wife for their 'magnificent voices', 'wonderful artistry', and 'inspired acting'. Liszt held them both in high esteem as artists and friends, and wrote many of his songs for them. He was godfather to their son **Franz** (1855–1929), who sang at Weimar (1876–8) and Hanover (1878–1906). Another son, **Rudolf** (1859–1927), was also a singer.

**Mildenburg, Anna von.** See BAHR-MILDENBURG, ANNA.

**Milder-Hauptmann, Pauline Anna** (*b* Constantinople, 13 Dec. 1785; *d* Berlin, 29 May 1838). Austrian soprano. Studied Vienna with Tomaselli and Salieri (on Schikaneder's advice). Début Vienna 1803 (Juno, Süssmayr's *Der Spiegel von Arkadien*). Created Leonore in 1805 and in the two subsequent versions (persuading Beethoven to change 'unbeautiful, unsingable' passages in

Leonore's aria for 1814). Though at first not dramatically impressive, she later achieved many triumphs in the role. Berlin 1812, 1816–29. Further engagements in Vienna; retired 1836. One of the first of the German dramatic sopranos who propagated and influenced the ideas of Romanticism in the theatre, she also created leading roles in Cherubini's *Faniska* and operas by Spontini. Her nobility and emotional intensity were greatly admired in her Alceste, Armide, and especially Iphigénie (both *Aulide* and *Tauride*), as which she was praised by Goethe in verse. Her powerful voice ('like a house', as Haydn put it) was pure and expressive, and inspired Schubert's 'Shepherd on the Rock', and 'Suleika II'. Beethoven had hoped to write another opera for her.

**Mildmay, Audrey** (*b* Herstmonceux, 19 Dec. 1900; *d* London, 31 May 1953). English soprano. Studied London with Johnstone-Douglas, Vienna with Strasser. CR 1928–31. Married John *Christie, with whom she founded Gly. Fest. Sang there at its opening, 1934, then 1935–6, 1938–9. London, SW, 1939; Gly. 1940; Montreal 1943, then retired. Roles included Polly Peachum (*Beggar's Opera*), Zerlina, Susanna, Norina. Founded Edinburgh Fest. 1947 with Rudolf Bing. Vivacious and witty on stage, and an inspiration behind the scenes. (R)

**Milhaud, Darius** (*b* Aix-en-Provence, 4 Sep. 1892; *d* Geneva, 22 June 1974). French composer. Studied Aix-en-Provence, and at the Paris Conservatoire with Dukas, Gédalge, Leroux, Widor, and D'Indy. His studies interrupted by the First World War, he left the Conservatoire; made a formative visit to Brazil 1917–18. Returning to Paris, he became part of the avant-garde group known as 'Les Six'. Like his colleagues, he drew inspiration from a wide range of musical traditions, including jazz and folk music, particularly that of Latin America. Technically a very well equipped composer, Milhaud was soon recognized as the most original and resourceful of the group and, willing to produce music for almost any occasion, the prolificity of his operatic output was matched only by its inventiveness. In this, his musical eclecticism was brilliantly complemented by a sensitive reaction to sharply contrasting literary stimuli, ranging from the experimentalism of Jean Cocteau to a lively reinterpretation of the principles of classical theatre. On several occasions he was also attracted to large oratorio-type works (notably *Les Choéphores*) which, while only intended for semi-staged performance, at best evidence an approach similar to that of his operas.

For his first mature work he turned to the well-worn Orpheus theme, although *Les malheurs d'Orphée* treats the subject in the ironic fashion which was to characterize much of his later music: like the satirical work which followed it, *Le pauvre matelot*, it uses slender resources to maximum effect. Yet more concentrated expression is found in the three opéras-minute, *L'enlèvement d'Europe*, *L'abandon d'Ariane*, and *La délivrance de Thésée*, where the action is compressed into the space of 10–15 minutes. The ability to work on a larger canvas was demonstrated in the next work, *Christophe Colomb*, for which an impressive army of ten principal and 35 secondary soloists was marshalled with great skill; among other technical devices the opera calls for the use of cinema. In the works which followed, such as *Bolivar* and *David*, Milhaud returns to many of the same dramatic and musical techniques, while his three children's operas (*À propos de bottes*, *Un petit peu de musique*, and *Un petit peu d'exercice*) demonstrate a certain concern with *Gebrauchsmusik*. One of his last works is a setting of the little-known third play of Beaumarchais's Figaro trilogy, *La mère coupable*.

SELECT WORKLIST: *La brebis égarée* (The Lost Lamb) (Jammes; comp. 1910–15, perf. Paris 1923); *Les malheurs d'Orphée* (Lunel; Brussels 1925); *Le pauvre matelot* (The Poor Sailor) (Cocteau; Paris 1927); *L'enlèvement d'Europe* (Hoppenot; Baden-Baden 1927); *L'abandon d'Ariane* (Hoppenot; Wiesbaden 1928); *La délivrance de Thésée* (Hoppenot; Wiesbaden 1928); *Christophe Colomb* (Claudel; Berlin 1930); *À propos de bottes* (Chalupt; Paris 1932); *Un petit peu de musique* (Lunel; Paris 1932); *Un petit peu d'exercice* (Lunel; Paris 1934); *Bolivar* (Supervielle & Milhaud; Paris 1950); *David* (Lunel; Jerusalem 1954); *La mère coupable* (The Guilty Mother) (after Beaumarchais; Geneva 1966).

BIBL: C. Palmer, *Milhaud* (London, 1976); J. Drake, *The Operas of Darius Milhaud* (New York, 1989).

**Miller, Jonathan** (*b* London, 21 July 1934). English producer. First operatic production *Noyes Fludde*, London, Round House, then SW, New OC; Goehr's *Arden Must Die* 1974. Kent O 1974–82, in a series of enterprising productions working closely with Roger *Norrington. At London, C, his inventive stagings included a clever siting of *Rigoletto* in Mafia territory.

**Millico, Giuseppe** (*b* Terlizzi, 19 Jan. 1739; *d* Naples, 1 Oct. 1802). Italian soprano castrato and composer. Successful career throughout Europe; sang at court of Catherine the Great 1758–65. Heard in Parma, 1769, by Gluck, who took him to Vienna to create Paris in *Paride ed Elena*. Became Gluck's friend and companion, travelling with him to Paris, Mannheim, etc. Also taught singing to his niece Marianne. Vienna, H, 1772–80. London 1772–3, in operas by Sacchini; Berlin 1774; Naples from 1780 as maestro di cappella to the royal chapel. Here he composed a large quantity of music, including several operas (e.g. *La pietà d'amore*, *La Zelinda*, *Le cinesi*), and was a respected teacher; his pupils included Lady Hamilton. Kelly, who heard him at her house, found his singing 'enchanting'.

**Millo, Aprile** (*b* New York, 14 Apr. 1958). US soprano. Studied Juilliard School. Début Salt Lake City 1982. New York, M, 1984. Has made a mark especially as a Verdi singer: roles incl. Amelia (*Ballo*), Aida, Elvira (*Ernani*), Luisa Miller, Leonora (*Trovatore*), Elisabetta. (R)

**Millöcker, Karl** (*b* Vienna, 29 Apr. 1842; *d* Baden, nr. Vienna, 31 Dec. 1899). Austrian composer. Studied Vienna; joined the T in der Josephstadt as flautist 1858; became conductor at the Thalia T, Graz 1864, on Suppé's recommendation, where he composed his first operettas. Returned to Vienna 1866; was briefly conductor at the T an der Wien, then the Harmonie T. Conductor of the Deutsches T in Budapest 1868–9, before returning once again to Vienna, W. A succession of operettas, including *Abenteuer in Wien* and *Apajaune der Wassermann*, brought him to public attention and in 1882 he had his greatest triumph with *Der Bettelstudent*. Taken up throughout Europe and the USA, such was its success that he was able to retire from conducting. None of his subsequent works was as popular or highly regarded, although *Der arme Jonathan* had some following.

Together with Johann Strauss II and Suppé, Millöcker was one of the founding fathers of Viennese operetta, skilfully blending the style of Offenbach with the local tradition of light dramatic music. An immensely skilled craftsman, he possessed a surer theatrical instinct than most of his contemporaries, though his scores generally lack their characteristic Viennese verve and charm. Only *Der Bettelstudent* has survived the test of time; it is still popular in the German and Austrian operetta repertory today.

SELECT WORKLIST (all first prod. Vienna): *Abenteuer in Wien* (Adventure in Vienna) (Berla; 1873); *Apajune der Wassermann* (Apajune the Water-Carrier) (Zell & Genée; 1880); *Der Bettelstudent* (The Beggar Student) (Zell & Genée, after Sardou; 1882); *Der arme Jonathan* (Poor Jonathan) (Wittmann & Bauer; 1890).

BIBL: E. Nick, *Vom Wiener Walzer zur Wiener Operette* (Hamburg, 1954).

**Milnes, Sherrill** (*b* Downers Grove, IL, 10 Jan. 1935). US baritone. Studied Tanglewood with Goldovsky. Début Boston OC 1960 (Masetto). Milan, N, 1964; New York, M, from 1965; London, CG, from 1971; also Chicago, Buenos Aires, Vienna, etc. Roles include Don Giovanni, Figaro (Rossini), Escamillo, Jochanaan. His warm, incisive tone and excellent legato line particularly suit Verdi roles, e.g. Rigoletto, Di Luna, Posa, Macbeth, Germont. (R)

**Milton, John** (*b* London, 9 Dec. 1608; *d* London, 8 Nov. 1674). English poet. Operas on his works are as follows:

*Paradise Lost* (1667); Le Sueur (*Le mort d'Adam*, 1809—also based on Klopstock and the Book of Genesis); Spontini (unfin., begun 1838); Rubinstein (1856, prod. 1875); Penderecki (1978)

*Samson Agonistes* (1671): Handel (*Samson*, 1743)

Spontini's *Milton* (1804) is based on the poet's life and provided material for his *Das verlorene Paradies*; a projected *Miltons Tod* did not materialize.

**Mime.** A Nibelung (ten), brother of Alberich, in Wagner's *Siegfried*.

**Mimì.** A seamstress (sop), lover of Rodolfo, in Puccini's and Leoncavallo's *La bohème*.

**mimodrame.** An older name for *pantomime.

**Minato,** (Count) **Nicolò** (*b* Bergamo *c.*1630; *d* Vienna 1698). Italian librettist, poet, and impresario. During his early career he provided about 12 librettos for the Venetian public opera-houses, including those for Cavalli's *Pompeo Magna* (1666) and Antonio Sartorio's *La prosperità d'Elio Seiano* (1667). In 1669 he became court poet to Leopold I and while in Vienna wrote about 170 librettos, mostly for the court composer Antonio Draghi, which were staged by the designer Ludovico Burnacini. Like his Venetian librettos, these showed a fondness for historical subjects, such as *Temistocle in Persia* (1681) and *Scipione preservatore di Roma* (1690). *Gundeberga* (1672) is one of the earliest librettos based on a German historical subject. In adaptations and revisions his work enjoyed considerable longevity: several texts later inspired German librettos (e.g. König's *Der geduldige Socrates*, 1721, which was based on *La pazienza di Socrate*, 1680) while Handel's *Serse* (1738) was based on an adaptation of Minato's libretto of 1654.

**Mingotti.** Italian family of impresarios. **Angelo** (*b* ?, *c.*1700; *d* ?, after 1767) and his brother **Pietro** (*b* Venice, *c.*1702; *d* Copenhagen, 28 Apr. 1759) first formed companies in Dresden and Stuttgart, then travelled widely in Germany. They were among the most successful touring impresarios of the age, and did much to introduce Italian opera to other parts of Europe, especially Germany. They were in Brno 1732–6, Graz 1736 (where they built the first public theatre), then divided the company. Angelo was in Hamburg 1740; Pietro was in Pressburg 1741, returning to Graz and then going on to Prague. Both brothers recruited fine singers, including Cuzzoni. Failing to establish himself in Prague,

Pietro moved to Linz, but continued to tour. Angelo continued working in Graz, also visiting Prague, Leipzig, and Dresden. Pietro built a wooden theatre in the Zwinger in Dresden 1746. In 1747 **Caterina Valentini** (*b* Naples, 16 Feb. 1722; *d* Neuburg, 1 Oct. 1808), Italian soprano and sister of the composer Michelangelo Valentini, married Pietro, taking the professional name **Regina Mingotti**. She studed with Porpora; début Dresden 1747, Porpora's *Filandro*. At this stage Gluck worked with the company. Regina now became a fierce rival of Faustina *Bordoni. In 1748 the brothers visited Leipzig, and Pietro went on to Copenhagen, where, apart from further tours, he remained, living to see the decline of the Italian opera there. Angelo moved to Bonn, where he may have worked with Beethoven's father. Regina also appeared in England, where she sang in works by Jommelli and others. Her quarrels with Vaneschi at the Haymarket rivalled the Bordoni–Cuzzoni fights. Kitty Clive mimicked her in the musical satire *Lethe* (1775).

**Minnie.** The so-called 'Girl of the Golden West' (sop) and the owner of the Polka Saloon, lover of Dick Johnson, in Puccini's *La fanciulla del West*.

**Minorca.** *The Beggar's Opera* and other works were given in the 18th cent. during the British occupation (1708–1802). Seasons, mostly of Italian opera, are given in Maó in the T Principal (cap. 1,500), built 1829 and the oldest theatre in Spain.

**Minton, Yvonne** (*b* Sydney, 4 Dec. 1938). Australian mezzo-soprano. Studied Sydney with Walker, London with Cummings and Cross. Début London, City Literary Institute, 1964 (Lucretia). London, CG, from 1965; Milan, S, and Chicago 1970; New York, M, 1973; Paris, O, 1976; Australian O 1972–3; Bayreuth 1974–7. Created Thea (*The Knot Garden*). Repertory includes Sesto, Cherubino, Dorabella, Marina, Dalila, Brangäne, Fricka, Orfeo, and, with her handsome stage presence and smooth tone, a highly acclaimed Octavian. (R)

**Miolan-Carvalho, Marie** (*b* Marseilles, 31 Dec. 1827; *d* Château-Puys, 10 July 1895). French soprano. Studied with her father F. Félix-Miolan, then Paris with Duprez. Toured France with him 1848–9. Début Brest 1849 (Isabella, *Robert le diable*). Paris: OC, 1849–55, 1868–85 (when she retired); L, 1856–67; O, 1868–79. London, CG, 1856, 1859–64, 1871–2; also Berlin, St Petersburg. Married the impresario Léon *Carvalho. One of the most celebrated singers of the time, with an attractive voice covering nearly three octaves, she created Marguerite, Baucis, Juliette, and Mireille for Gounod, and roles in operas by Halévy, Thomas, Massé, etc. Though a dazzling virtuoso, she could also rise to the demands of Mozart, e.g. as Pamina, Zerlina, Countess, Cherubino. Halévy wrote of her: 'No words can express such talent, such charm, such artistry!'

BIBL: E. Accoyer-Spoll, *Mme Carvalho: notes et souvenirs* (Paris, 1885).

**'Mira, o Norma'.** The duet in Act II of Bellini's *Norma*, in which Adalgisa (sop) pleads with Norma (sop) not to give up her children.

**Mireille.** Opera in 5 acts by Gounod; text by Carré, after Mistral's poem *Mireio* (1859). Prod. Paris, L, 19 Mar. 1864, with Miolan-Carvalho, Faure, Lefèvre, Morini, Ismaël, Petit, cond. Deloffre; London, HM, 5 July 1864, with Tietjens, Trebelli, Giuglini, Santley, Junta, cond. Arditi; Philadelphia, AM, 17 Nov. 1864 (2 acts only), in full in Chicago 13 Sep. 1880. Rev. in 3 acts 1864.

Arles, 19th cent. The opera tells of the love of Mireille (sop) for Vincent (ten); this is opposed by Mireille's father Ramon (bs), and complicated by a rival for Mireille's affections in the bull-tender Ourrias (bar). The original 5-act version ended tragically with Mireille's death. The 1889 revised version ends with Vincent and Mireille happily united.

**Mir iskusstva** (Russ.: The World of Art). The fortnightly journal which, first appearing 10 Nov. 1898, edited by *Dyagilev and financed by *Mamontov, gave its name to an important group of artists. Its contents included discussions of music and opera, and among the contributors were Maeterlinck and Ruskin; its position was 'aesthetic' in opposition to the Russian Realist tendency. The operatic articles were generally pro-Wagnerian. In 1904 *Benois succeeded Dyagilev, but was forced to close the journal in the same year.

**'Mir ist so wunderbar'.** The canon quartet in Act I of Beethoven's *Fidelio*, in which Leonore (sop), Marzelline (sop), Jacquino (ten), and Rocco (bs) express outwardly similar emotions of wonder for inwardly different reasons.

**'Miserere'.** Properly the opening word ('Have mercy') of Psalm 51 (50 in the Vulgate). In opera, generally referring to the scene in Act IV of Verdi's *Il trovatore* in which a chorus of monks prays for the soul soon to depart as Leonora (sop) wanders beneath the tower in which her doomed lover Manrico (ten) is held.

**Mistress Quickly.** See QUICKLY, MISTRESS.

**'Mi tradì'.** Donna Elvira's (sop) aria in Act II of Mozart's *Don Giovanni*, in which she expresses her conflicting emotions of vengeance and pity towards Giovanni. Composed for Katharina

Cavalieri in answer to her protest that she had too little to sing for the Viennese première of the opera.

**Mitridate Eupatore.** Opera in 5 acts by A. Scarlatti; text by Girolamo Roberti. Prod. Venice, GG, carnival 1707.

Sinope, 120 BC. Mitridate Eupatore has taken refuge in the court of the Egyptian King Ptolemy after being banished following the murder of his father by his mother Stratonica (mez) and her lover Farnace (bar). Mitridate's death is called for but in a fight with Farnace, he kills his stepfather. Issicratea (mez), Mitridate's wife, kills Stratonica, and the couple are crowned rightful King and Queen of Pontus.

**Mitridate, re di Ponto.** Opera in 3 acts by Mozart; text by V. Cigna-Santi, after G. Parini's translation of Racine's *Mithridate* (1673). Prod. Milan, RD, 26 Dec. 1770, with Bernasconi, Benedetti, Varese, Cicognani, Muschietti, D'Ettore, Bassano; St. Louis, Loretto Hilton T, 1 June 1991, cond. Morgan.

The Crimea, c.135 BC. Sifare (sop cas) and his half-brother Farnace (con cas), both love Aspasia (sop), betrothed to their father Mitridate (ten). News arrives that reports of Mitridate's death were false and that his fleet is approaching. Mitridate returns, defeated, with Ismene (sop), a princess betrothed to Farnace. He is suspicious of his sons' feelings for Aspasia.

Farnace tells Ismene he no longer loves her. Mitridate questions Aspasia and is convinced that Farnace has designs both on his throne and on his wife. Alone, Aspasia confesses to Sifare that she loves him and he is torn between love for her and loyalty to his father. Mitridate imprisons Farnace for conspiring with the Romans. When he learns of Sifare and Aspasia's love, he curses them and imprisons them.

Ismene pleads in vain for Mitridate to be merciful, while Aspasia prepares to take poison. She is prevented by Sifare, who has been released by Ismene, and who now joins his father in a new fight against the Romans. A Roman helps Farnace to escape, but he repents of his treachery and joins his father's army. Mitridate is wounded but, before he dies, is reconciled with his sons. Begging Aspasia's forgiveness, he gives her hand to Sifare, and Farnace is united with Ismene.

Text first set by Q. Gasparini (1767). The subject, in the libretto by Zeno, has also been set by Caldara (1728), Porpora (1730), Terradellas (1746), Araia (1747), and Sarti (1779).

**Mitropoulos, Dmitri** (*b* Athens, 1 Mar. 1896; *d* Milan, 2 Nov. 1960). Greek, later US, conductor and composer. Studied Athens, Brussels, Berlin with Busoni. Répétiteur Berlin, S, 1921–5. Cond. some opera in Athens, but not again until the 1950s, when he gave concert performances of *Wozzeck* and *Elektra* with the New York Philharmonic. New York, M, 1954–60, including première of Barber's *Vanessa* (1958). His opera *Sœur Béatrice* was produced Athens Conservatory 1919.

**Miura, Tamaki** (*b* Tokyo, 22 Feb. 1884; *d* Tokyo, 26 May 1946). Japanese soprano. Studied Tokyo with Junker, Germany with Petzold and Sarcoli. Début Tokyo 1909 (Gluck's Euridice). London OH 1915; Chicago 1918, 1920; Italy, including Rome, C, 1921; 1924 toured USA with San Carlo OC; San Francisco 1924. Roles included Butterfly, Iris, Marguerite, Mimì, Santuzza. Created title-role of Aldo Franchetti's *Namiko-San* (written for her), Chicago Civic O 1925. The first Japanese-born singer to make an international career. (R)

**Mlada.** Opera-ballet in 4 acts by Rimsky-Korsakov; text by the composer, based on a text by Gedeonov for an earlier opera to be written with Borodin, Cui, and Musorgsky. Prod. St Petersburg, M, 1 Nov. 1892; New York, M, 28 June 1991, by Bolshoy company.

Slav lands, 9th–10th cent. Voyslava (sop) poisons Mlada (silent dance role) in order to replace her in Yaromir's (ten) affections. Morena, Queen of the Underworld (mez), causes Yaromir to fall in love with Voyslava, but the ghost of Mlada intervenes; she tells Yaromir what happened. A priest advises Yaromir to question spirits sent by Morena to taunt him; they confirm that Voyslava was the murderess. When Voyslava is condemned to death, Morena raises a furious storm in which Yaromir dies, thus uniting him forever with his beloved Mlada.

**Mödl, Martha** (*b* Nuremberg, 22 Mar. 1912) German mezzo-soprano, later soprano. Studied Nuremberg Cons. Milan with Mueller. Début Remscheid 1943 (Hänsel). Düsseldorf 1945–9 as mezzo-soprano. Hamburg from 1949; London, CG, from 1950; Berlin 1950–1 (by now singing soprano roles); Bayreuth 1951–67; Vienna from 1952; New York, M, 1957–60. Also Paris, O; Milan, S; Berlin; Salzburg; etc. Was still singing in 1984 (creating Mumie in Reimann's *Gespenstersonate*). Sang Carmen, Eboli, Marie, Octavian; also an outstanding Kundry, Isolde, Sieglinde, Gutrune, Brünnhilde, particularly in association with Bayreuth. An intensely dramatic performer, she used her warm, vibrant voice with artistry and intelligence. (R)

**Moïse.** See MOSÈ IN EGITTO.

**mojiganga** (Sp.: 'masquerade'). A short piece performed at the end of Spanish theatrical productions of the 17th and early 18th cents. featuring comic, satirical, or grotesque elements.

**Moldavia** (Rom., Moldova). A musical theatre was opened in Kishinev (Rom., Chişinău) in

475

1933; in 1955 this became the Pushkin National Opera and Ballet T. Though operas were written by Moldavian composers, notably *Zhar-ptitsa* (The Firebird, 1926) by Evgeny Koka (1893–1954), the first Moldavian national opera was *Grozovan* (1956) by David Herschfeld (*b* 1911). He also wrote *Aurelia* (1959). Other Moldavian operas include *Kaza mare* (The Big House, 1968) by Mark Kopitman (*b* 1929); *Klop* (The Bedbug, 1963) by Edward Lazarev (*b* 1935); an operetta *Na beregu Urala* (On the Edge of the Urals, 1943) by Solomon Lobel (1910–81); and a children's fairy-tale opera, *Koza s tremya kozlyatamy* (The Goat with the Three Kids, 1967) by Zlata Tkach (*b* 1928).

**Molière** (orig. Jean-Baptiste Poquelin) (*b* Paris, *bapt.* 15 Jan. 1622; *d* Paris, 17 Feb. 1673). French dramatist. Operas on his works are as follows:

*Les précieuses ridicules* (1659): Galuppi (1752); Blanchard (*c.*1830s); Mériel (1877); Galliera (1901); Goetzl (1905); Seymour (1920); Zich (1924); Behrend (1928); Lattuada (1929); Bush (*If the Cap Fits*, 1956)
*Sganarelle* (1660): Arcais (1871); Grosz (1925); Wagner-Régeny (1929); Kaufmann (1958); Zito (1972); Pasatieri (*Il signor deluso*, 1974); V. Archer (1974)
*L'école des maris* (1661): Alessandri (*Il vecchio geloso*, 1781); Mortari (1930); Bondeville (1936)
*L'école des femmes* (1662): Liebermann (1955); Mortari (1959)
*Le mariage forcé* (1664): F. Hart (1928)
*La Princesse d'Élide* (1664): Laverne (1706); Galuppi (*Alcimena*, 1749)
*Tartuffe* (1664): Haug (1937); Kosa (1952); Eidens (20th cent.); Benjamin (1964)
*L'amour médecin* (1665): Berton (1867); Poise (1880); Wolf-Ferrari (1913); Herberigs (1920); Bell (1930); Bentoiu (1964)
*Georges Dandin* (1666): Mathieu (1877); Sebastiani (1893); D'Ollone (1930)
*Le médecin malgré lui* (1666): Désaugiers (1792); Gounod (1858); Poise (1887); Veretti (1927); Behrend (1947); Kaufmann (1958)
*Le sicilien* (1667): Kospoth (*Adrast und Isidor*, 1799); Preu (1779); Levasseur (1780); Miča (1781); Mézeray (1825); Joncières (1859); Cadaux (mid-19th cent.); K. H. David (1924); Letorey (1930)
*La pastorale comique* (1667): Cabanet (1879)
*L'avare* (1668): Burghauser (1950)
*Amphytrion* (1668): Grétry (1786); Oboussier (1948)
*M de Pourceaugnac* (1669): Orlandini (1727); Hasse (1727); Jadin (1792); Mengozzi (1793); Alani (1851); Franchetti (1897); Bastide (1921); Martin (1963)
*Le bourgeois gentilhomme* (1670): Hasse (*Larinda e Vanesio*, 1726); Esposito (1905); Gargiulo (1947); Hlobil (1972). See also ARIADNE AUF NAXOS.

*Le malade imaginaire* (1673): Thern (*A képzett beteg*, mid-19th cent.); Napoli (1939); Dupérier (1943); Haug (*Le malade immortel*, 1946); L. Miller (1970); J. Paner (*Zdarý nemocný*, 1970).

Also Veretti (*Il medico volante*, comp. 1923–4, unprod.), Dibdin (*Dr Ballardo*, 1770), Galuppi (*Le virtuose ridicole*, 1752), Locke (*Psyche*, 1675), Isouard (*Les deux avares*, 1801), J. Kaffka (*So prellt man alte Füchse*, 1782).

**Molinara, La, ossia L'amor contrastato** (The Maid of the Mill, or Lovers' Rivalry). Opera in 3 acts by Paisiello; text by Giovanni Palomba. Prod. Naples, T dei Fiorentini, carnival 1789.

The lawyer Pistofolo (bs), the jilted Caloandro (ten), and the elderly Governor Rospolone (bs) all write letters proposing marriage to Rachelina (sop), a beautiful mill-owner. In order to avoid being pursued by the Governor, Pistofolo dresses as a miller and Caloandro as a gardener. When Rachelina chooses Pistofolo, Caloandro, in a rage of disappointment, tries to kill him. The Governor then implies that Pistofolo is also mad so, rather then be the wife of a lunatic, Rachelina decides not to marry at all.

**Molinari Pradelli, Francesco** (*b* Bologna, 4 July 1911). Italian conductor. Studied Bologna, Rome with Molinari. Cond. Bologna 1939, Bergamo, Brescia. Milan, S, 1946. London, CG, 1955, 1960. Vienna, S, from 1959. San Francisco 1957–66; New York, M, 1966–73. Vienna since 1959. (R)

**Moll, Kurt** (*b* Buir, 11 Apr. 1938). German bass. Studied Cologne, and Krefeld with E. Müller. Début Aachen 1961 (Lodovico, Verdi's *Otello*). Aachen until 1965. Bayreuth from 1968; Milan, S, and Paris, O, from 1972; New York, M, from 1976; London, CG, from 1977; Vienna, Berlin, Buenos Aires etc. Firm in tone and line; a successful Hunding, Fafner, King Mark, Pogner, Gurnemanz, also Osmin, Caspar. (R)

**Mombelli**. Italian family of singers.

1. **Domenico** (*b* Villanova, 17 Feb. 1751; *d* Bologna, 15 Mar. 1835). Tenor and composer. Sang mostly in Italy, especially Naples, 1780–1816, in operas by Paisiello, Cimarosa etc. Also appeared Madrid, Vienna, and Lisbon. Befriended the young Rossini, who considered him 'an excellent tenor'; created Demetrio in Rossini's first opera, *Demetrio e Polibio*. His own compositions included *Didone* (1776). His first wife

2. **Luisa Laschi**(-Mombelli) (*b* Florence, 1760s; *d* ?, *c.*1790) was a soprano. Début Vienna, H, 1784 (Cimarosa's *Giannina e Bernadone*). Naples 1785, meeting Mombelli; Vienna, H, as leading prima donna, 1785–8. Created Countess

(*Figaro*), and Zerlina in the first Vienna *Don Giovanni*, Mozart adding the duet 'Per queste due manine' for her and Benucci (Leporello). Described in the press as 'Grace personified', she was an unusually expressive singer, with a clear, appealing voice, and a charming figure. Difficulties with the management led to the Mombellis' dismissal in 1788.

In 1791 Domenico married the ballerina Vincenza Viganò, niece of Boccherini. She wrote the libretto for Rossini's *Demetrio e Polibio*, in which two of her and Domenico's daughters also appeared:

3. **Ester** (*b* Bologna, 1794; *d* ?), mezzo-soprano, and

4. **Anna** (Marianna) (*b* Milan, 1795; *d* ?), soprano. With their father and a bass, Lodovico Olivieri, they performed operas in Bologna, Milan, and elsewhere. Stendhal admired both, praising Anna's pure and simple style, and Ester's 'exquisite, crystalline voice' and technical skill. Anna retired after her marriage in 1817; Ester sang until 1827, becoming a celebrated Cenerentola, and creating Madama Cortese in *Il viaggio a Reims*; also Donizetti's Zoraide (described by Celli as 'ravishing . . . with her expressive pathos'), and Gilda (*L'ajo nell'imbarazzo*).

5. Their son **Alessandro** taught singing in Bologna.

**Monaco.** See MONTE CARLO. Also Italian for Munich.

**Mona Lisa.** Opera in 2 acts by Schillings; text by Beatrice Dovsky. Prod. Stuttgart 26 Sep. 1915 with Hedy Iracema-Brügelmann, Forsell; New York, M, 1 Mar. 1923, with Kemp, Bohnen, cond. Bodanzky.

Florence, late 15th cent. A Carthusian monk (ten) relates to a young honeymoon couple in Florence the story of Mona Lisa (sop). Her jealous husband Francesco del Giocondo (bar) shuts her lover Giovanni de Salviati (ten) in a cupboard, where he suffocates. In turn, Mona Lisa shuts Francesco in the cupboard. The opera's three characters are revealed to be the modern counterparts of the people in the story.

**'Mon cœur s'ouvre à ta voix'.** Dalila's (mez) aria in Act II of Saint-Saëns's *Samson et Dalila*, in which she almost overcomes Samson's resolve not to part with the secret of his strength.

**Mond, Der** (The Moon). Opera in 3 acts by Orff; text by the composer, after Grimm. Prod. Munich 5 Feb. 1939, with Patzak, cond. Krauss. New York, CC, 16 Oct. 1956, with Kelly, Treigle, cond. Rosenstock.

Fairyland. The narrator (ten) tells the story of four boys (ten, 2 bars, bs) who steal the moon, each taking a quarter to their graves. The world grows dark and the four stick the pieces of the moon together and hang it up as a lamp. This wakes up all the dead, who create such a tumult that it is heard in heaven; St Peter descends to the underworld, takes the moon away, and hangs it up on a star.

**Mondo della luna, Il** (The World on the Moon). Opera in 3 acts by Haydn; text by Goldoni, adapted by P. F. Pastor. Prod. Esterháza 3 Aug. 1777. Rev. in incomplete form, London Opera Club, Scala T, 8 Nov. 1951, cond. Pritchard; New York, Greenwich Mews Playhouse, 7 June 1949. Restored by Robbins Landon, perf. Holland Fest. 24 June 1959, with Adani, Rizzoli, Casoni, Alva, Cortis, cond. Giulini.

Venice, 1750. Bonafede has two daughters, Clarice (sop) who loves Ecclitico (ten) and Flaminia (sop) who loves Ernesto (sop); he opposes their marriage, also that of Ernesto's servant Cecco (ten) to Lisetta (mez). When Ecclitico tells Bonafede that he has received an invitation to visit the moon, Bonafede begs to go with him. Ecclitico and his friend Ernesto transform a garden into a lunar landscape and, waking from a sleeping potion, Bonafede believes himself to be on the moon. The others join in the charade, and marry their true loves 'on the moon'. Bonafede discovers the trick and is furious, but is eventually reconciled to the three marriages.

Also operas by Galuppi (1750), Piccinni (1770), and Paisiello (*Il credulo deluso*, 1774; *Il mondo della luna*, 1782).

**Mondonville, Jean-Joseph Cassanéa de** (*b* Narbonne, *bapt.* 25 Dec. 1711; *d* Belleville, 8 Oct. 1772). French violinist and composer. Paris début as violinist at the Concert Spirituel 1734; director 1755–62. Joined court retinue 1739; 'sous-maître' 1740; 'Surintendant de la Chapelle du Roi', 1744–58. Famed initially as a performer and composer of instrumental music, Mondonville's first opera, *Isbé*, was failure, though *Bacchus d'Erigone* and the ballet-héroïque *Le carnaval du Parnasse*, fared better. During the *Guerre des Bouffons he was chosen to represent the French tradition; the production of *Titon et l'Aurore* is said to have hastened the departure of the Italian troupe from Paris. His later operas included a popular work in Languedoc dialect, *Daphnis et Alcimadure*, and a controversial resetting of a libretto made famous by Lully, *Thésée*. Mondonville's rise to fame was aided by his court connections—he was a protégé of Mme de Pompadour—but at his best he was a worthy, if less skilful, imitator of the dramatic techniques of Lully and Rameau, with a gift for writing graceful and pleasing melodies.

SELECT WORKLIST: *Isbé* (La Rivière; Paris 1742); *Bacchus et Erigone* (La Bruère; Versailles 1747); *Le carnaval du Parnasse* (Fuzelier; Paris 1749); *Titon et l'Aurore* (La Marre & Voisenon; Paris 1753); *Daphnis et Alcimadure* (Mondonville; Paris 1754); *Thésée* (Quinault; Paris 1767).

BIBL: F. Hellouin, *Mondonville: sa vie et son œuvre* (Paris, 1903).

**Monelli, Raffaele** (*b* Fermo, 5 Mar. 1782; *d* S Benedetto del Tronto, 14 Sep. 1859). Italian tenor. Studied Fermo with Giordaniello. Sang with success in Italy 1808–20. Created Bertrando (*L'inganno felice*), and Dorvil (*La scala di seta*) for Rossini. His brother **Savino** (*b* Fermo, 9 May 1784; *d* Fermo, 5 June 1836) was a handome tenor, with a 'sweet but delicate voice' (Radiciotti). Studied Fermo with Giordaniello. Début Milan, S, 1816 (Tamino). Created Giannetto (*La gazza ladra*), Adalberto (*Adelaide di Borgogna*) for Rossini; Ramiro (*Chiara e Serafina*), Enrico (*L'ajo nell' imbarazzo*) for Donizetti.

**Mongini, Pietro** (*b* Rome, 29 Oct. 1828; *d* Milan, 27 Apr. 1874). Italian tenor. Began career as bass (Ascoli Piceno 1851) but was singing tenor by 1853 (Genoa). Paris, I, 1855; Milan, S, 1858, 1868–9; St Petersburg 1857–8, 1861–2, 1870–1; London: HM, 1858–67, DL and CG, 1868–73; Cairo, 1871, 1873–4. Created Radamès, and was a notable Alvaro, Don Carlos, Manrico; also Don Ottavio, Otello (Rossini), Arnold (*Guillaume Tell*). An ex-dragoon who cut an exciting figure on stage (and once nearly cut off Charles Santley's finger in *Forza*), he possessed a voice both powerful and flexible, if one not used with great subtlety.

**Moniglia, Giovanni Andrea** (*b* Florence, 22 Mar. 1624; *d* Florence, 21 Sep. 1700). Italian librettist. Wrote several texts for Melani, including *Il potestà di Colognole* (*La Tanchia*), probably the first comic opera libretto. His work was also set by Cavalli (*Ipermestra*, 1654), Cesti, Legrenzi, and Ziani.

**Moniuszko, Stanisław** (*b* Ubiel, 5 May 1819; *d* Warsaw, 4 June 1872). Polish composer. Studied Warsaw with Freyer, Minsk with Stefanowicz, Berlin with Rungenhagen. His first important opera, after some early operettas, was his best-known, *Halka*. On its delayed production in Warsaw, this was hailed as the first significant Polish national opera, and won international acclaim. Standing half-way between the old tradition of separate numbers and the new music drama, *Halka* makes use of an idiom that is recognizably Polish (without drawing on folk music). It was considerably influenced by Auber, whose music Moniusko admired and conducted often during his period at the Teatr Wielki (from 1859). His other most successful opera was *The Haunted Manor*, a less even work that suffers from an awkward plot but that contains some

excellent music: its setting and its powerful atmosphere, no less than its use of Polish dance rhythms, fostered nationalist loyalties to such an extent that it was banned from the Teatr Wielki by the Tsar.

SELECT WORKLIST: *Halka* (Wolski; comp. 1846–7, concert Vilnius 1848, rev. version prod. Warsaw 1858); *Hrabina* (The Countess) (Wolski; Warsaw 1860); *Straszny dwór* (The Haunted Manor) (Chęciński; Warsaw 1865).

BIBL: J. Prosnak, *Stanisław Moniuszko* (Cracow, 1964, 2/1969); B. Maciejewski, *Moniuszko, Father of Polish Opera* (London, 1979).

**Monnaie, Théâtre royale de la.** See BRUSSELS.

**Monna Vanna.** Opera in 4 acts by Février; text by Maeterlinck. Prod. Paris, O, 10 Jan. 1909, with Bréval, Muratore, Delmas, Marcoux, cond. Vidal; Boston, OH, 5 Dec. 1913, with Garden, Muratore, Marcoux, Ludikar, cond. Caplet.

Pisa, late 15th cent. Monna Vanna (sop), a childhood friend of Prinzivalle (ten), commander of the Florentine army, succeeds in getting him to lift the siege of Pisa. However, her husband Guido (bar), the Pisan commander, refuses to believe that she has achieved this innocently. When Prinzivalle is imprisoned, Monna Vanna sides with him, declaring her love for him, and the couple escape.

Also opera by Ábrányi (1907).

**Monnet, Jean** (*b* Condrieu, 7 Sep. 1703; *d* Paris, 1785). French impresario, writer, and composer. After various vicissitudes, director OC 1743–5, with Rameau as music director. Director Lyons 1745. London 1749; the company failed, and he was imprisoned. Paris, OC, 1758, achieving great success and contributing to the establishment of opéra-comique, to which genre he also contributed some unsuccessful examples.

**monodrama.** A *melodrama in which only one actor has any substantial role. A classic example is Benda's *Pygmalion*.

**monody** (from the It. *stile monodico*) See STILE RAPPRESENTATIVO.

**Monostatos.** A Moor (ten) in the service of Sarastro in Mozart's *Die Zauberflöte*.

**Monpou, Hippolyte** (*b* Paris, 12 Jan. 1804; *d* Orléans, 10 Aug. 1841). French composer and organist. Spent his early years as assistant at the Institution de Musique Religieuse, composing only sacred music. When this closed he turned to writing operas in order to support his family, quickly producing a series of comic works, including *Les deux reines*, *Le luthier de Vienne*, *Le Piquillo*, and *Le planteur*. Though popular in its day, his music soon passed out of fashion.

SELECT WORKLIST (all first prod. Paris): *Les deux*

*reines* (Soulié & Arnould; 1835); *Le luthier de Vienne* (Saint-Georges & De Leuven; 1836); *Le Piquillo* (Dumas; 1837); *Le planteur* (Saint-Georges; 1839).

**Monsigny, Pierre-Alexandre** (*b* Fauquembergues, 17 Oct. 1729; *d* Paris, 14 Jan. 1817). French composer. After working in Paris as a clerk and in a private theatre, studied with Gianotti and began composing operas. Encouraged by the success of the first, *Les aveux indiscrets*, he began collaborating with Sedaine with *On ne s'avise jamais de tout*. Anxious to give more substance to opéra-comique as it was practised by Philidor and Duni, he wrote *Le roy et le fermier*, which not only gave scope to his melodic charm and sensitive response to situations, but concerned itself with political and social issues and made use of more elaborate musical techniques. The popular *Rose et Colas* was a simple pastoral comedy; but with *Le déserteur*, his most famous and widely successful work, he used a dramatic subject to move away finally from the *comédie mêlée d'ariettes to what he described as opéra-comique larmoyant (or sentimental comedy). He strengthened his style still further in *La belle Arsène*, and again in *Félix*, his last opera. Impoverished by the removal of his aristocratic patrons in the Revolution, he withdrew from composition but survived on some pensions and lived to see his works recover their popularity. He had contributed to opéra-comique not only an individual melodic vein but a seriousness of purpose, some formal innovations (such as thematic relationships between final choruses and overtures), and an enhanced feeling for orchestration that anticipates Méhul.

SELECT WORKLIST (all first prod. Paris unless otherwise stated): *Les aveux indiscrets* (The Indiscreet Vows) (La Ribardière; 1759); *On ne s'avise jamais de tout* (One Never Thinks of Everything) (Sedaine; 1761); *Le roy et le fermier* (The King and the Farmer) (Sedaine, after Dodsley; 1762); *Rose et Colas* (Sedaine; 1764); *Le déserteur* (Sedaine; 1769); *La belle Arsène* (Fair Arsène) (Favart; Fontainebleau 1773); *Félix* (Sedaine; Fontainebleau 1777).

BIBL: P. Druilhe, *Monsigny* (Paris, 1955).

**Montagnana, Antonio** (*b* Venice; *fl* 1730–50). Italian bass. Possibly studied with Porpora. Rome 1730; Turin 1731; London, H, 1731–3, for Handel, creating Varo (*Ezio*), Altomaro (*Sosarme*), and Zoroastro (*Orlando*), and singing in several revivals (e.g. *Poro, Admeto, Flavio*) in which Handel wrote new music for him. Joined Porpora's Opera of the Nobility 1733–7, singing in works by Porpora, Bononcini, Hasse, etc.; Heidegger's company 1737–8, creating Handel's Gustavo (*Faramondo*), and Ariodate (*Serse*). Madrid 1740–50, at the royal chapel and in opera. A true bass (unlike *Boschi), his voice at its peak (before 1738) possessed a range of E to f',

particularly resonant low notes, and unusual agility and accuracy. Handel wrote many arias of considerable difficulty and virtuosity for him, but also found him effective in slow, sustained music.

**Montague, Diana** (*b* Winchester, 1954). English mezzo-soprano. Studied Manchester with Cox, Tunbridge Wells with Bruce-Lockhart. Début with Gly. Touring O 1977 (Zerlina). London, CG, from 1978; Bayreuth from 1983; Chicago 1984; Salzburg 1986; New York, M, and Frankfurt 1987; Gly. 1989; also London, ENO, Scottish O, Edinburgh. Roles incl. Idamante, Cherubino, Dorabella, Sesto, Nicklausse, Orlovsky, Wellgunde, Mélisande. (R)

**Monte Carlo**. Capital of the Principality of Monaco. The Grand T, or Salle Garnier (cap. 600), opened in 1879, and enjoyed its greatest days under the management of Raoul Gunsbourg from 1893 to 1951. It was the scene of many important premières, including Massenet's *Le jongleur de Notre-Dame* (1902) and *Don Quichotte* (1910), Fauré's *Pénélope* (1913), Puccini's *La rondine* (1917), and Ravel's *L'enfant et les sortilèges* (1925), of enterprising revivals, and also of famous performances by Patti, Melba, Caruso, and Shalyapin. Gunsbourg's successors attempted to continue the tradition, and under Renzo Rossellini (1972–6) a number of new works were introduced. Seasons are now somewhat reduced, and usually last from Jan. to Apr. and include works in the French and German repertories and Italian works sung by prominent artists.

BIBL: T. Walsh, *Monte Carlo Opera, 1879–1909* (London, 1975).

**Montéclair, Michel Pignolet de** (*b* Andelot, bapt. 4 Dec. 1667; *d* Aumont, 22 Sep. 1737). French composer. In service of Prince de Vaudémont, with whom he visited Italy; from 1699 in Paris; 1699–1737 double-bass player at the Opéra, where his opéra-ballet *Les festes de l'été* was given. His only opera, *Jephté*, was banned at first because of its unprecedented use of a biblical plot. A considerable success, it was revived many times up to 1761 and is reputed to have turned Rameau, whose use of the chorus it adumbrates, towards the tragédie lyrique.

WORKLIST: *Jephté* (Pellegrin; Paris 1732).

**Montemezzi, Italo** (*b* Vigasio, 4 Aug. 1875; *d* Vigasio, 15 May 1952). Italian composer. Studied Milan with Saladino and Ferroni. As a composer he stands closer to Boito than to his verismo contemporaries. His works have enjoyed more success abroad than in Italy, though his first opera, *Giovanni Gallurese*, was given 17 times in its first season. His masterpiece is *L'amore dei tre* *re*, a work which in Italy has been compared to *Pelléas*; however, the true influence is Wagner, and in *La nave* Richard Strauss. His eclectic late-Romantic style is also shown in his other operas.

SELECT WORKLIST: *Giovanni Gallurese* (D'Angelantonio; Turin 1905); *L'amore dei tre re* (The Love of the Three Kings) (Benelli; Milan 1913); *La nave* (The Ship) (D'Annunzio; Milan 1918).

BIBL: L. Tretti and L. Fiumi, *Omaggio a Italo Montemezzi* (Verona, 1952).

**Monterone.** Count Monterone (bs), the nobleman who curses Rigoletto in Verdi's *Rigoletto*.

**Monteux, Pierre** (*b* Paris, 4 Apr. 1876; *d* Hancock, ME, 1 July 1964). French, later US, conductor. Studied Paris. Began conducting aged 12; violist Paris, OC, 1890; then gained much experience as concert and ballet conductor (incl. for Dyagilev's Ballets Russes). Paris, O, 1913–14, 1931. New York, M, 1917–19, 1953–6, conducting US premières of *Golden Cockerel*, *Mârouf*, and between 1953 and 1956 authoritative performances of *Manon*, *Faust* and *Orfeo* (Gluck), *Hoffmann*, and *Samson et Dalila*. (R)

**Monteverdi, Claudio** (*b* Cremona, 15 May 1567; *d* Venice, 29 Nov. 1643). Italian composer. Though not the first opera, Monteverdi's *La favola d'Orfeo* is the earliest surviving work in the permanent repertory. Written for a private performance before the Accademia degli Invaghiti (Carnival 1607), it was inspired both by the *stile rappresentativo* of Peri, Caccini, and other members of the Bardi Camerata, as well as by the more elaborate musical and ballet entertainments of the Gonzaga court at Mantua. By turning to the Orpheus myth, Monteverdi and his librettist, Alessandro Striggio, chose a subject already used in many intermedi, as well as for the *Euridice* operas of Peri and Caccini (both 1600). Monteverdi almost certainly knew Peri's treatment of Rinuccini's libretto and clearly built upon the work of his predecessor, as for example in the recitative passages (notably the Messenger's report of Euridice's death), where he provides an expressive musical style to depict both events and emotions. But by including unifying instrumental ritornellos and homophonic choruses, modelled largely upon those of the intermedi, he transcended the dramatic limitations of the strict monodic style and demonstrated the true potential of the new art form.

Monteverdi's preface, and further instructions throughout the score, make apparently impressive instrumental demands: but the size of the probable venue for the first performance (the Galleria dei Fiumi in the ducal palace) meant that the orchestra must have been small, with performers doubling on several instruments. None the less, the instrumental requirements suggest a colourful array of orchestral timbres new to opera; sometimes Monteverdi even specified the precise instrumentation required, for example in Orfeo's lament 'Possente spirto'.

Despite the clear musical advances made by Monteverdi through his handling of monody, and expansion of instrumental and choral material and resources, *Orfeo* is clearly rooted in the spirit of the Renaissance, rather than being a true harbinger of Baroque opera. With an intellectual audience clearly in mind, Monteverdi and Striggio produced a 5-act work whose classical derivation and Neoplatonic ideals are far more evident than any appeal to human emotion, the lifeblood of later opera. Here Orfeo's love for Euridice, in most later treatments the substance of the emotional involvement, serves merely for the investigation of inner feelings. Significantly, neither libretto (1607) nor printed score (1609), with its revised ending, has a *lieto fine: in the former Orfeo withdraws in face of the Bacchic women, in the latter Apollo takes him up to heaven on a cloud machine with a Neoplatonic exhortation to pursue higher ideals.

Between *Orfeo* and *Il ritorno d'Ulisse in patria* Monteverdi was engaged on at least thirteen dramatic projects, yet only ballets such as *Il ballo delle ingrate* (1608) and *Tirsi e Clori* (1616), and cantatas such as the *Combattimento di Tancredi e Clorinda* (1624) have survived complete: of his operatic endeavours almost nothing is extant. *Arianna*, written for a Gonzaga wedding, was an appropriately festive piece, quite different from *Orfeo*. Only one fragment survives, the famous 'Lamento d'Arianna', whose popularity may be judged from the fact that it was printed as a monody (1623), and arranged as a madrigal (1614) and as a sacred parody (1638). The most widely known piece of Monteverdi's operatic output, it offers a distillation of his understanding of the monodic style and, like *Arianna* itself, served to strengthen his unassailable reputation.

In 1612 Monteverdi was dismissed from Mantua following the death of his patron Duke Vincenzo Gonzaga: in 1613 he became maestro di cappella at St Mark's, Venice, where he was to remain until his death. However, his duties were sufficiently light to allow him to accept commissions from elsewhere; ironically the Mantuan court, where the succession had changed once again, was to prove one of his best sources of income. Among important works written for Mantua were *Andromeda* (1620), whose recently discovered libretto shows it to have been an unusual, continuous drama, *Apollo*, and *La finta pazza Licori*, which was apparently never completed. *Proserpina rapita* was, however, written to celebrate a wedding between two Venetian aristocratic families.

It was in Venice that Monteverdi achieved the crowning glory of his operatic career with at least two works written for the new public opera-houses; *Arianna* was also revived here. With the successful establishment of the first public opera-house at the Teatro San Cassiano in 1637 the whole nature of the genre changed fundamentally: no longer aimed at exclusive, intellectual, or aristocratic audiences, it had to appeal to a wider cross-section of the public and be commercially viable without private patronage. In *Il ritorno d'Ulisse in patria* Monteverdi rose admirably to the new challenge of public opera; with its constant twists of plot and frequent use of disguise the pace of the drama never flags and the earlier preoccupation with allegorical figures is here supplanted by a real interest in human emotion. Penelope's love for Ulysses is movingly handled in the recitative and arioso passages, while in his ensembles, especially the duets, Monteverdi draws on his great skills as a madrigalist. Only the synopsis and scenario for *Le nozze d'Enea con Lavinia* survive, but they indicate that it too proceeded on much the same lines as *Ulisse* in being fundamentally tragic, yet with a happy ending. *L'incoronazione di Poppea* is cast on an even grander scale than *Ulisse*, requiring both a larger cast and better-endowed singers. Including such figures as nurses and servants, and provided with a simple, often comical style of music, it makes a direct appeal to the widest audience, while Poppea's conquest of Nero through the power of love moved Monteverdi to compose some of his most passionate and lyrical vocal music, in particular the laments. The sharp distinction between the various musical styles in *Poppea* reinforced a growing trend in operatic composition, whereby the characters' social status was reflected by an appropriately complex musical style. *Poppea* established the standards by which later Venetian opera was to be measured; but though long considered the crowning glory of Monteverdi's career, it is now commonly accepted that at least some of the music was not by him. The most likely collaborators, if not composers of the whole opera, were Sacrati and Ferrari.

Though Monteverdi's operatic abilities must be judged by a mere torso of his output, it is clear that he alone of the earliest operatic composers understood the problems of matching musical drama to its audience. Hence in *Orfeo* he provided an opera in which musical variety did not obscure a fundamentally scholastic work suitable for a learned academy; likewise, in *Arianna* he apparently matched the festive occasion for which it was required by an appropriately appealing musical style. That his powers were undiminished even in old age is proved by the hearty reception accorded him in Venice, while the many recent revivals of his operas have shown that his appeal has gone far beyond his own times.

WORKLIST: *La favola d'Orfeo* (Striggio; Mantua 1607); *Arianna* (Rinuccini; Mantua 1608, music lost); *Andromeda* (Marigliani; comp. 1618–20, music lost); *La finta pazza Licori* (Strozzi; Mantua 1627, lost); *Gli amori di Dafne e di Endimione* (Pio; Parma 1628, lost); *Prosperpina rapita* (Strozzi; Venice 1630, music lost except for one number); *Il ritorno d'Ulisse in patria* (Badoaro; Venice 1640); *Le nozze d'Enea con Lavinia* (?; Venice 1641, music lost); *L'incoronazione di Poppea* (Busenello; Venice 1643; ?collab.).

BIBL: D. Arnold, *Monteverdi* (London, 1963, rev. T. Carter, 3/1990); D. Stevens, ed., *The Letters of Claudio Monteverdi* (London, 1980); D. Arnold and N. Fortune, eds., *The New Monteverdi Companion* (London, 1985); S. Leopold, *Monteverdi* (Oxford, 1991).

**Monti, Anna Maria** (*b* Rome, 1704; *d* ?Naples, after 1727). Italian soprano. Début when only 13, Naples, T dei Fiorentini (Falco's *Lo mbruglio d'ammore*. Sang there regularly until 1727, mainly in comic servant roles. Her sister **Grazia** sang at Naples, N, 1728; another sister, **Laura** (*b* Rome, after 1704; *d* Naples, 1760), was also a soprano. Naples, N, 1726–32, in comic roles; Naples, B, 1733–5, singing with the bass Gioacchino Corrado in intermezzos. Created Serpina (*La serva padrona*), and parts written for her by Hasse and Leo. Naples, C, 1738, in first *La locandiera* (Auletta), when the new Queen of Carlos III was so charmed that she requested the revival of all the old intermezzos. Retired 1746, after further appearances in Naples.

**Monti, Marianna** (*b* Naples, 1730; *d* Naples, 1814). Italian soprano, cousin of Anna Maria *Monti and her sisters. Début Naples, T dei Fiorentini, 1746. Sang there and at T Nuovo until 1759, in comic servant roles. From 1760 an influential and popular singer, she created many roles written for her by Jommelli, Traetta, Paisiello, Cimarosa, etc. Also sang in operas by her brother **Gaetano Monti** (*b* Naples, *c*.1750; *d* ?Naples, ?1816), who was considered one of Naples's best composers, and was well known in Germany. His most popular stage works were *Le donne vendicate* (1781), and *Lo studente* (1783).

**Monticelli, Angelo Maria** (*b* Milan, *c*.1710; *d* Dresden, 1764). Italian soprano castrato. Début Rome *c*.1730. Venice, Milan, Florence, then London, H, 1741–4, 1746, with great success, including in Gluck's *Artamene*. Naples 1746; further appearances in Italy; Dresden 1756, working with Hasse. A distinguished singer, he was praised by Burney for his 'expression and gesture', and by Walpole, who pronounced him 'infinitely admired'.

**Montreal.** City in Quebec, Canada. The first opera to be heard was Dibdin's *The Padlock* in 1786. In the following years there were productions of Shield's *The Poor Soldier* (1787), Duni's *Les deux chasseurs et la laitière* (1789), and the première of Joseph Quesnel's *Colas et Colinette* (1790; the first Canadian opera). The T de Société opened in 1825, and saw opera from 1841. Throughout the 19th cent. Montreal was visited by touring companies giving a standard repertory of Italian opera, later also Wagner. The first attempt at a permanent company was the Société d'Opéra Français (1893-6). In 1910 the Montreal OC was formed and many famous singers were engaged, but it was forced to close in 1913. It was succeeded by the National OC of Canada, 1913-14. Operetta then took precedence (Société Canadienne d'Opérette, 1921-33; Variétés Lyriques, 1936-52).

The Montreal Festival began its summer performances in 1936 and for the following 30 years usually produced an annual opera, including many first Canadian performances. Beecham took part in the 1942 and 1943 festivals, conducting *Figaro*, *Roméo et Juliette*, and *Tristan*. In 1941 the Opera Guild of Montreal was formed by Pauline Donalda, and produced an opera a year until her death in 1969. In 1964 the Montreal Symphony Orchestra under Zubin Mehta began staging two or three operas each season in its new home, the Salle Wilfrid Pelletier (cap. 2,874) at the Place des Arts. Expo' 67, the 1967 World Fair, brought many major visiting companies to the city, including the Moscow Bolshoy, English Opera Group, Hamburg State Opera, La Scala, Royal Stockholm Opera, and Vienna State Opera. In 1971 the Provincial Government set up L'Opéra de Québec, intended not as a permanent company but to organize operatic activity with government support; it closed in 1975. L'Opéra de Montréal was founded in 1980.

**Moody, Fanny** (*b* Redruth, 23 Nov. 1866; *d* Dundrum, Co. Dublin, 21 July 1945). English soprano. Studied with Sainton-Dolby. Début Liverpool, with CR, 1887 (Arline, *The Bohemian Girl*). CR until 1898. With her husband Charles Manners formed the Moody-Manners Co. 1898-1916, also singing as leading soprano. London, CG, 1902-3, and DL, 1904. Repertory included Marguerite, Juliette, Elsa, Senta, Santuzza, Tatyana; created title-role of Pizzi's *Rosalba*, and Militza (McAlpin's *The Cross and the Crescent*).

**Moore, Douglas** (*b* Cutchogue, NY, 10 Aug. 1893; *d* Greenport, NY, 25 July 1969). US composer. Studied Yale with Parker, Paris with D'Indy and Boulanger. He was one of the few to resist the latter's influence, though his operas reveal fine craftsmanship as well as a manner affected by D'Indy's harmonic style. Of his operas, the most successful have been *The Devil and Daniel Webster* and *The Ballad of Baby Doe*, both of which have been regular items in US repertories. The latter, based on a true story of Colorado, recalls at times folk melodies and is a strong and effective work.

SELECT WORKLIST: *The Devil and Daniel Webster* (Benét; New York 1939); *The Ballad of Baby Doe* (Latouche; Central City, CO, 1956).

**Moore, Grace** (*b* Nough, TN, 5 Dec. 1898; *d* Copenhagen, 26 Jan 1947). US soprano. Studied New York with Marafioti. Début in operetta. Further study Antibes with Berthélemy. Opera début, Paris, OC, 1928. Paris, OC, 1929, 1938, 1946. New York, M, 1928-32, 1935-46; London, CG, 1935. Glamorous and full of personality, she was extremely popular as Louise, Manon, Mimì, Tosca, etc. Made several films, e.g. *One Night of Love*, *New Moon*. Died in an air crash. Autobiography, *You're only Human Once* (New York, 1944, R/1977). (R)

**Moore, Thomas** (*b* Dublin, 28 May 1779; *d* nr. Devizes, 28 Feb. 1852). Irish poet and musician. Active as a poet and arranger of songs which had an enormous popularity in their day, especially when sung by him. He wrote the libretto of a comic opera, *The Gypsy Prince* (1801), for Michael Kelly, and in 1811 produced an opera of his own, *M.P., or The Blue Stocking*. His *Lalla Rookh*, a story with four interpolated narrative poems, provided the texts for operas by C. E. Horn (1818), Kashin (*The One-Day Reign of Nourmahal*, Spontini (*Nurmahal*, 1822), F. David (1862), and Rubinstein (*The Veiled Prophet*, 1881). His *The Light of the Harem* was set as an opera by A. G. Thomas (1879).

**Moreira, António Leal** (*b* Abrantes, 1758; *d* Lisbon, 21 Nov. 1819). Portugese composer. Studied Lisbon; joined the Brotherhood of S Cecilia 1777 and became known as a composer of sacred music. Began to write Italian operas for the Lisbon court 1783, including a resetting of the text Mozart had used for *Ascanio in Alba*. Conductor at the T da Rua dos Condes 1790; at the T San Carlos 1793. Here he wrote the first two of the earliest operas to Portugese librettos, *A saloià namorada* and *A vingança da cigana*. A gifted melodist with a sure dramatic touch, Moreira made particularly effective use of the orchestra in his scores.

SELECT WORKLIST: *Ascanio in Alba* (Parini, after Stampa; Quelez 1785); *A saloià namorada* (Caldas Barbosa; Lisbon 1793); *A vingança da cigana* (Caldas Barbosa; Lisbon 1794).

**Morel, Marisa** (*b* Turin, 13 Dec. 1914). Italian soprano and producer. Studied Turin. Début

Milan, S, 1933 (Zerlina, Auber's *Fra Diavolo*). Further appearances at Milan, S, and elsewhere in Italy; New York, M, 1938–9. Roles included Amor (Gluck's *Orfeo*), Sophie. Formed (and sang with) Marisa Morel OC 1941 for the production of Mozart operas. This performed in Paris and elsewhere in Europe. Its singers included Danco, Simionato, Corena, Rothmüller; conductors included Ansermet, Böhm.

**Morena, Berthe** (*b* Mannheim, 27 Jan. 1878; *d* Rottach-Egern, 7 Oct. 1952). German soprano. Studied Munich with Röhr-Brajnin and De Sales. Début Munich, N, 1898 (Agathe). Sang there until 1924. New York, M, 1908–12, 1925; London, CG, 1914; retired 1927. A distinguished, intelligent singer of great beauty, she excelled in Wagner, e.g. as Isolde, Brünnhilde, Eva. (R)

BIBL: L. Vogl, *Berta Morena und ihre Kunst* (Munich, 1919).

**'Morgenlich leuchtend'.** The opening words of Walther's (ten) Prize Song in Act III of Wagner's *Die Meistersinger von Nürnberg*.

**Moriani, Napoleone** (*b* Florence, 10 Mar. 1806 or 1808; *d* Florence, 4 Mar. 1878). Italian tenor. Studied with Ruga. Début Pavia, 1833 (Pacini's *Arabi nelle Gallie*). Successes throughout Italy, including Milan, S, 1839–40; Vienna 1842; Prague and Dresden 1843; London, HM, 1844; Paris, I, 1845, 1849–50. His voice deteriorated from 1844; retired 1851. One of the leading singers of his time, dubbed 'il tenore della bella morte', he inflamed audiences with his intensely acted death-scenes. Famous for his Donizetti, e.g. in *Lucia di Lammermoor*, *Pia de' Tolomei*, *Lucrezia Borgia*; created Enrico (*Maria di Rudenz*) and Carlo (*Linda di Chamounix*). Also successful in Bellini and Verdi (who wrote an extra romanza for him in *Attila*, in 1847). Admired by Wagner, Mendelssohn, Fanny Mendelssohn ('He delighted me . . . he sings with such simplicity'), and Chorley, who appreciated his 'richly strong' voice even in its decline. He was very probably the father of his frequent partner ★Strepponi's illegitimate children.

**Morison, Elsie.** See KUBELÍK, RAFAEL.

**Morlacchi, Francesco** (*b* Perugia, 14 June 1784; *d* Innsbruck, 28 Oct. 1841). Italian composer. Studied Perugia with Mazzetti (his uncle) and Caruso, Naples with Zingarelli, Bologna with Mattei. The success of his first opera, *Il poeta in campagna*, and then of *Il ritratto* and especially *Il corradino*, led to commissions from Rome for *La principessa per ripiego* and *Le Danaidi*, and also from Milan for *Le avventure di una giornata*.

In 1810 Morlacchi succeeded Josef Schuster as Kapellmeister in Dresden, where he did much to improve standards, setting a fine example as a musician and as a virtuoso conductor and showing a capacity for selecting and training singers into an ensemble. His vanity and deviousness, and a tendency to indolence, were exacerbated by the rivalry with Weber, who arrived at the German opera in 1817. His first opera for Dresden was *Raoul di Créqui*, regarded by some as his masterpiece, and others included *Il nuovo barbiere di Siviglia*: of this, Weber remarked, 'The fellow has little musical knowledge, but he has talent, a flow of ideas, and especially a fund of good comic stuff in him.' The operas' greatest strength is in their handling of voices, especially in comic scenes; latterly Morlacchi showed an increasing response to German influences in his greater enterprise with orchestration, revealing some awareness of the work of Weber (with whom relations had improved). He also wrote a number of further operas for Italy. He was one of the last important Italian composers to work as Kapellmeister in a German theatre, and the closure of the Italian Opera in Dresden in 1832 was a step towards ending the supremacy of Italian opera in Germany.

SELECT WORKLIST: *Il poeta in campagna* (The Poet in the Country) (?; Florence 1807); *Il ritratto* (The Picture) (Romanelli; Verona 1807); *Il corradino* (Sografi; Parma 1808); *La principessa per ripiego* (The Substitute Princess) (Ferretti; Rome 1809); *Le avventure di una giornata* (A Day's Adventures) (Romanelli; Milan 1809); *Le Danaidi* (Scatizzi, after Metastasio; Rome 1810); *Raoul di Créqui* (Artusi; Dresden 1811); *Il nuovo barbiere di Siviglia* (The New Barber of Seville) (Petrosellini & Sterbini; Dresden 1816); *Gianni di Parigi* (Romani; Milan 1818); *Tebaldo e Isolina* (Dresden 1820).

BIBL: G. Ricci des Ferres-Cancani, *Francesco Morlacchi* (Florence, 1956).

**Morris, James** (*b* Baltimore, 10 Jan. 1947). US baritone. Studied Baltimore with Ponselle, New York with Valentino, Moscona, and with Hotter. Début New York City, CO, 1969 (Crespel, *Contes d'Hoffmann*). New York, M, from 1971; Salzburg from 1982; London, CG, from 1988. Possesses a fine legato and a rich, mellifluous tone. Roles include Count, Don Giovanni, Philip, Amonasro, Gounod's Méphistophélès, Scarpia, Flying Dutchman, and a much-acclaimed Wotan. (R)

**'Morrò, ma prima in grazia'.** Amelia's (sop) aria in Act III of Verdi's *Un ballo in maschera*, in which she accepts that she must die as the penalty for her accused adultery but begs to see her son first.

**Morselli, Adriano** (*fl.* Venice, 1679–83). Italian librettist. Worked as house poet at Venice, GG, producing *c*.20 librettos for composers including Pollarolo and Ziani. His texts reflected the gradual change in the nature of Venetian opera from the heroic-comic pattern of the early tradition to the more refined style of the immediate pre-Metastasian era.

**Mort d'Adam, La** (The Death of Adam). Opera in 3 acts by Le Sueur; text by Guillard, combining Klopstock's drama *Der Tod Adams* (1757) with material from the Book of Genesis and Milton's *Paradise Lost* (1667). Prod. Paris, O, 21 Mar. 1809, with Dérivis, Nourrit, Lainé, Lays.

Approaching death, Adam (bs) is worried because his sons, including Seth (ten) and Cain (ten), will have to pay for his sins. Cain curses his father, but Adam forgives him. Prayers for Adam drive back Satan, who had come to bear him to hell; Adam rises to heaven.

This extremely spectacular opera contains an elaborate motivic system; the death-scene recalls at least 12 earlier passages.

**Mosca, Giuseppe** (*b* Naples 1772; *d* Messina, 14 Sep. 1839). Italian composer. Studied Naples with Fenaroli; first opera, *Silvia e Nardone*, performed there 1791. Worked for various Italian theatres; maestro al cembalo at Paris, I, 1803. Returned to Italy 1810; director of the T Carolino in Palermo 1817–20; director at Messina OH 1827–39. Composed over 40 operas, reminiscent in style of Rossini. Chiefly remembered for his famous attack on the latter's *La pietra del paragone* (1812), claiming that its use of the *crescendo had been stolen from his *I pretendenti*.

SELECT WORKLIST: *I pretendenti* (The Pretend Madmen) (Prividali; Milan 1811); *Il finto Stanislao* (Rossi; Venice 1812).

His brother **Luigi Mosca** (*b* Naples 1775; *d* Naples, 13 or 30 Nov. 1824) was maestro al cembalo at Naples, C, and vice maestro di cappella at the Naples court. A famous singing teacher, his *L'italiana in Algeri* predates Rossini's by five years.

SELECT WORKLIST: *L'italiana in Algeri* (Anelli; Milan 1808).

**Moscow.** Capital of Russia, and of the USSR 1922–91. Though Italian troupes visited Moscow from 1731, the first opera-house in Russia was the Operny Dom (Opera House), opened in 1742 with *La clemenza di Tito*. In 1759 Locatelli opened another Operny Dom, which gave opera alternately with Russian drama. The opening of the Petrovsky T in 1780 by Prince Pyotr Urusov and the English ex-acrobat Michael Maddox was a major step: here most of the early Russian operas were given. When it burnt down in 1805, the independent theatres at which opera was also given absorbed the company.

In 1824 there opened the Maly T (Little T), so-called to distinguish it from the Bolshoy T

(Grand T), which was also built by Osip Bove (on a project by A. Mikhaylov) and opened in 1825. The Maly (cap. 560, later 900) was primarily a dramatic theatre, but in the Bolshoy a mixed repertory included works by Verstovsky and Glinka. Reconstructed by Alberto Cavos (son of Caterino) after a fire in 1853, the new theatre opened in 1856 (cap. 2,000) and was reserved for opera and ballet. This was the home of a rising generation of great singers, and saw the premières of many important Russian operas of the second half of the century. The production style (echoing French taste, as so much in Russian art) favoured an elaborate manner with large crowd scenes, histrionic gestures, and elaborately detailed sets, and this survives as a characteristic of the theatre. Among the most successful designers was Victor Vasnetsov (1848–1926), with his elaborate realistic sets. As all the Imperial Theatres were a Crown monopoly until a decree of Alexander III in 1882, no independent reform movement developed until the end of the century. In 1885 Savva Mamontov (1841–1918) opened his Moskovskaya Chastnaya Russkaya Opera (Moscow Private Russian Opera); when he was bankrupted and imprisoned in 1899, this was reorganized as the Tovarishchestvo Chastnoy Opera (Private Opera Society) under *Ippolitov-Ivanov. Among the singers it attracted was Shalyapin, and the enterprising production style served as a widely admired corrective to the official Bolshoy style. In 1904 this closed and was succeeded by the *Zimin Opera Theatre, which worked (from 1908 at the Solodovnikov T, later Filial T) under Ippolitov-Ivanov and Emil *Cooper, with Pyotr Olenin as director from 1907 and Fyodor Komisarzhevsky from 1915. This arrangement lasted until 1917, when a disastrous fire destroyed the company's property. At the Revolution, the Bolshoy was temporarily directed by the tenor Leonid *Sobinov. The theatre was reorganized, reopening in April 1918 (cap. now 2,155) and also taking over the Filial T. The theatre remained throughout the succeeding years the bastion of Moscow operatic life, preserving, though not without cost to theatrical originality, an essentially 19th-cent. Russian tradition. In the late 1980s productions (including by Elena *Obraztsova and the film director Sergey Bondarchuk) began to show more enterprise.

Also in 1919, Konstantin *Stanislavsky began working with the studio, which became the Stanislavsky Opera Studio in 1924 and was reorganized in 1926 and again, as the Stanislavsky OH, in 1928. The theatre known from 1926 as the Nemirovich-Danchenko Music T was founded in 1919; it specialized at first in opéra-comique and operetta, and later turned to the classics and to new Soviet works. Stanislavsky died in 1938, and the two groups were united in 1941, with *Nemirovich-Danchenko himself directing until his death in 1943. Though neither director had staged anything at the Bolshoy, their influence on the replacement of a fossilized style by vivid acting and staging was vital. The Stanislavsky-Nemirovich-Danchenko T is Moscow's second opera-house, where it still is that the more adventurous productions may be seen. From 1934 the Soviet Opera Ensemble gave concert performances of operas, including new ones (Prokofiev's *War and Peace*, 1944). Opera is also given, on a panoramic scale, in the vast Kremlin Palace of Congresses, built 1961 (cap. 6,000). The most advanced productions of all, however, are at the Kamerny Muzykalny T, opened in 1972 by Boris Pokrovsky and established in an old cinema in 1974, when Gennady Rozhdestvensky became music director. Early Russian opera and modern chamber works are given, including some pop opera; and works unsuccessful in larger theatres have found new careers (notably Shostakovich's *The Nose*).

See also RUSSIA.

**Mosè in Egitto** (Moses in Egypt). Opera in 3 acts by Rossini; text by Tottola, after Francesco Ringhieri's tragedy *L'Osiride* (1747). Prod. Naples, C, 5 Mar. 1818, with Colbran, Funk, Manzi, Benedetti, Nozzarri, Cicimarra, Remorini, Chizzola (the famous preghiera 'Dal tuo stellato soglio' was added for the 1819 revival); London, HM, 23 Apr. 1822, as *Pietro l'Eremita*, with Camporese, De Begnis, Zucchelli; New York, Masonic H, 22 Dec. 1832. Revised as Grand Opera in 4 acts, as *Moïse et Pharaon, ou Le passage de la Mer Rouge*, text by Balocchi and Jouy, Paris, O, 26 Mar. 1827, with Cinti-Damoreau, Mori, Nourrit, Dupont, Levasseur, L.-Z. and H.-B. Dabadie; London, CG, 20 Apr. 1850, as *Zora*, with Castellan, Vera, Tamberlik, Tamburini; New York, Italian OH, 2 Mar. 1835.

Near East, biblical times. Pharaon (bs) has decided to free the captive Hebrews, and Moïse (bs) hears a voice predicting that the Jews will at last reach their Promised Land. Pharaon's son Aménophis (ten) wishes to keep them in Egypt since he loves Moïse's niece Anaï (sop). Moïse obscures the sun to demonstrate his powers.

Aménophis's mother Sinaïde (sop) advises him to comply with his father's wishes and marry the King of Assyria's daughter.

Moïse and his people refuse to worship the god Isis; Moïse extinguishes the flame on the altar of the temple. Pharaon banishes the Jews.

The Hebrews, including Asaï, who was captured by Aménophis but has now been returned to her people, escape across the Red Sea, whose waves Moïse miraculously parts. The Egyptian troops in pursuit are drowned.

Other operas on Moses include Orefice's *Il Mosè* (1905); see also below.

**Moses und Aron.** Opera in 2 acts (3rd uncompleted) by Schoenberg; text by the composer, after the Book of Exodus. Prod. Zurich 6 June 1957, with Fiedler, Melchert, cond. Rosbaud; London, CG, 28 June 1965, with Robinson, Lewis, cond. Solti; Boston, Back Bay T, 30 Nov. 1966, with Gramm, Lewis, cond. McConathy.

Near East, biblical times. Moses (spoken) receives the word of God but lacks the gift of communication possessed by his less visionary brother Aron (ten). While he is on Mount Sinai receiving the Ten Commandments, Aron encourages the Hebrews to erect a Golden Calf, as a tangible object they, as a simple people, can worship. The subsequent orgy is interrupted by Moses's return from Sinai: appalled, he shatters the tablets of stone and resolves to be released from his mission. The second act ends with him sinking to the ground mourning, 'O word, thou word, that I lack'. The (unset) third act was to show Moses triumphant in the desert.

BIBL: K. Vörner, *Gotteswort and Magie, Die Oper 'Moses und Aron'* (Heidelberg, 1959, trans. 1963); P. White, *Schoenberg and the God-Idea* (Ann Arbor, 1985).

**Mosonyi, Mihály** (orig. Michael Brand) (*b* Boldogasszonyfalva, *bapt.* 4 Sep. 1815; *d* Pest, 31 Oct. 1870). Hungarian composer and writer on music. Studied Poszony with Turányi. During his first period he was entirely Germanic in his musical ideas: *Kaiser Max auf der Martinswand* was promised performance by Liszt but postponed and eventually abandoned; extensive revisions were planned, but Mosonyi had meanwhile turned his attention to Hungarian music. He Magyarized his name in 1859, but contined to champion Wagner and Liszt while also working by example and precept for Hungarian national music. *Szép Ilonka* was, in his words, 'written entirely in the Hungarian idiom to the exclusion of all foreign elements'. His last opera, *Álmos*, attempted to reconcile the two strains in his style; it is a Romantic opera incorporating Hungarian elements into a Germanic technique.

WORKLIST: *Kaiser Max auf der Martinswand* (Pasqué; comp. 1856–7, unprod.); *Szép Ilonka* (Fair Helen) (Fekete, after Vörösmarty; Pest 1861); *Álmos* (Szigligeti; comp. 1862, prod. Budapest 1934).

BIBL: F. Bonis, *Mosonyi Mihály* (Budapest, 1960).

**Mother, The** (Cz.: *Matka*). Opera in 10 scenes by Alois Hába; text by the composer. Prod. Munich 19 May 1931. Rev. Florence 1964.

Moravia, early 20th cent. A year after the death of his first wife, the smallholder Křen (ten) decides to take a new wife, Marusa (sop), to help him with his five children. When two years later she has still not had any children of her own, she blames her husband, and seeks fathers elsewhere. For ten successive years she produces children and is an excellent mother to them all. Even when they flee the nest, her marriage to Křen proves happy.

The first quarter-tone opera. Also the title of operas by Stanley Hollier (1954, after Hans Andersen) and Khrennikov (1957, after Gorky).

**Mother Goose.** A brothel-keeper (mez) in Stravinsky's *The Rake's Progress*.

**Mother of Us All, The.** Opera in 2 acts by Virgil Thomson; text by Gertrude Stein. Prod. New York, Columbia U, 7 May 1947 with Dow, Blakeslee, Mowland, Horne, Rowe. The opera tells the story of Susan B. Anthony, American feminist leader and leading suffragette. Other historical characters appear, including Ulysses S. Grant, Andrew Johnson, and Daniel Webster.

**motif** (sometimes also **motive**) (Fr., *motif*; Ger., *Motiv*; It., *motivo*). A short musical figure of characteristic design, used as a unifying device in a composition, or section of a composition. In opera it has more specific uses, notably in the *motto aria, the *reminiscence motive, and the *Leitmotiv.

**Mottl, Felix** (*b* Unter-Sankt-Veit, 24 Aug. 1856; *d* Munich, 2 July 1911). Austrian conductor and composer. Studied Vienna with Bruckner. One of Wagner's assistants at first Bayreuth Festival, 1876, and subsequently conductor there 1886–1902. One of the most important of the first Bayreuth conductors, with a particular clarity and flexibility. Conductor Karlsruhe 1881–1903, raising standards to great heights; gave the first near-complete performance of *Les Troyens* (on consecutive nights, Dec. 1890), as well as a revised version of Cornelius's *Barber of Bagdad*. London, CG, 1898–1900. New York, M, 1903–4. Munich 1904–11 (director from 1907). Collapsed while conducting *Tristan* and died shortly after. Composed three operas, including *Agnes Bernauer* (Weimar 1880), and edited vocal scores of all Wagner's operas. Married soprano Zdenka Fassbaender (1879–1954).

**motto aria** (Ger., *Devisenarie*). An aria type first used by Cesti and Legrenzi and subsequently by many later Baroque composers. Its name derives from the practice of starting the aria with the initial motif of the main melody—hence a kind of false beginning—which is then repeated and continued into the full melody. This initial 'motto' is sometimes echoed by the orchestra. Many examples are to be found in Legrenzi's *Eteocle e Polinice* (1675).

**Mount-Edgcumbe** (Richard Edgcumbe), **Earl of** (*b* Plymouth, 13 Sep. 1764; *d* Richmond, 26 Sep. 1839). English writer and composer. Having travelled in Italy, he wrote *Zenobia* (1800) for Banti. His *Musical Reminiscences* (London, 1824, 4/1834) give a vivid and readable account of Italian opera in London from 1773.

**Mouret, Jean-Joseph** (*b* Avignon, 11 Apr. 1682; *d* Charenton, 22 Dec. 1738). French composer. In service of Duchess of Maine 1708–36, for whom he composed one of the earliest pastoral operas, *Le mariage de Ragonde*, and the opéra-ballet *Le triomphe de Thalie*: its epilogue was revived 1722 as *La Provençale*, in which version it includes the first use in opera of Provençal dialect. Paris: director of the O 1714–18; director of the CI 1717–37, to whose repertoire he contributed. Director of the Concert Spirituel 1728–34. Mouret's later work included the tragédie lyrique *Ariane* and the opéra-ballet *Les amours des dieux*.

SELECT WORKLIST (all first prod. Paris): *Le mariage de Ragonde* (Destouches; 1714; rev. 1742 as *Les amours de Ragonde*); *Ariadne* (Roy & Lagrange; 1717); *La Provençale* (?; 1722).

**Mozart, Wolfgang Amadeus** (*b* Salzburg, 27 Jan. 1756; *d* Vienna, 5 Dec. 1791). Austrian composer. Mozart's musical talents manifested themselves at an early age, and much of his childhood was spent making lengthy concert tours around Europe, mostly accompanied by his father, who was employed by the Archbishop of Salzburg. An interest in composition soon became apparent, and while only ten he contemplated writing an opera for local children. In 1767 he made his first serious attempt at operatic style, contributing the first act of a dramatic Lenten cantata, *Die Schuldigkeit des ersten Gebotes*, which may have been staged. *Apollo et Hyacinthus*, his first true opera, was performed like an *intermezzo, between the acts of a play celebrating the end of the university year in Salzburg. In both works the music is already strikingly theatrical and couched in the grand style of opera seria. But both opera buffa and Singspiel were now claiming the attention of serious composers, and over the coming years he made his first attempts at these genres. *La finta semplice*, a setting of a well-worn opera buffa libretto, demonstrated for the first time his powers of musical characterization, while in *Bastien und Bastienne* (written for the Dr. Mesmer later satirized in *Così fan tutte*) he showed his skill in adopting some of the features of German Singspiel.

While touring Italy in 1770, Mozart was commissioned to write an opera seria for the T Regio Ducale in Milan. Despite difficulties with the singers, *Mitridate, re di Ponto* was an enormous success; in the complexities of its plot, which to some extent anticipates *Idomeneo*, it is entirely typical of Metastasian opera seria, including in its *lieto fine. The first opera in which Mozart wrote for the castrato voice, it depended heavily on the prevailing musical models of opera seria, particularly by its use of a wide range of sterotyped aria forms. The success of *Mitridate* led to a commission for an archducal wedding in Milan in the following year. *Ascanio in Alba* further demonstrated his careful craftsmanship and understanding of the conventions of opera seria. These features also mark *Lucio Silla*, in which he took his outstanding cast to the limits of its abilities, in music that is highly expressive and emotionally charged. The range of musical characterization is much increased, and he uses a distinctive vocabulary for each participant, while the role of the orchestra is also greatly enhanced.

The other early operas are less significant. *La finta giardiniera* returns to the world of opera buffa, though the work uncharacteristically treats the darker and irrational side of human experience; it also builds upon the techniques learnt in *Lucio Silla*. However, the opera seria *Il re pastore* and the serenata *Il sogno di Scipione* can only be viewed as retrograde steps. The Singspiel *Zaide* held much promise, but was abandoned when it became clear to Mozart that he would be unable to arrange its performance. It demonstrates none the less the interest in *melodrama which he had already displayed in connection with the works of Benda.

*Idomeneo* marks a clear landmark not only in Mozart's operatic output, but also in his whole career. Commissioned by the Elector of Bavaria, it is the first of his operas in which individuality outweighs dependence on pre-existing models or conventions. Following contemporary practice, he wrote only the recitatives before travelling to Munich, where he composed the arias while rehearsing the singers. His correspondence back to his father in Salzburg, through whom he collaborated with his librettist Varesco, about this process gives us an important insight into his creative processes. He was especially fortunate to have the services of one of the best orchestras in Europe, the famous Mannheim band, which had moved to Munich in 1778. His score contains a feast of instrumental music, including notably the concertato accompaniment to Ilia's aria 'Se il padre perdei' and the concluding ballet music. Although many arias have a static function, and such features of Metastasian opera as the lieto fine and castratos are retained, Gluck's influence is clearly discernible in the use of chorus, preference for accompanied recitative, and extensive instrumental writing. Most importantly, he begins to move away from the conventional chain of alternating arias and recitatives into more fluent structures, creating longer dramatic blocks (e.g. the shipwreck scene).

Following the success of *Idomeneo*, Mozart returned to Salzburg, where he was now in the service of the Archbishop. But disagreements with his employer, and a general wish to be master of his own destiny, led to his dismissal in 1781 while with the archiepiscopal retinue in Vienna. He now found himself in quite a different working environment, in which for a time he was fêted and financially successful. The accession as sole regent of Joseph II in 1780 had acted as a rejuvenating force, and the political and cultural walls which Empress Maria Theresa had erected to divide Austria from its neighbours were almost immediately pulled down. Under Joseph's enlightened despotism all forms of artistic expression flourished in Vienna, especially music and drama. Mozart's freelance status allowed him to take full advantage of the new artistic climate, although a new commercial element was now present in his working life. No longer were operas written at the request of employers or patrons: they were produced to fulfil commissions from opera-houses or companies.

One of Joseph's earliest achievements was the establishment of a German opera company in 1777, the National-Singspiel. At the request of its director, Gottlieb Stephanie, Mozart produced a Singspiel on a popular Turkish theme, *Die Entführung aus dem Serail*. The exotic plot was mirrored in stylized imitations of Turkish music, including the use of quasi-oriental percussion instruments for the overture and choruses (see JANISSARY MUSIC). Most of its numbers, such as Pedrillo's serenade and Osmin's drinking song, have counterparts in other Singspiels of the day, while the characters themselves were lifted from the stock figures of the genre. But Constanze's bravura aria 'Martern aller Arten', with its concertato accompaniment, and the moving *vaudeville finale surpass any contemporary models in their depth of expression.

The success of *Entführung* in 1782 strengthened Mozart's growing reputation, but the demands of his subscription concerts, for which he wrote many of his piano concertos, left him little time for opera. Two opere buffe, *L'oca del Cairo* and *Lo sposo deluso*, were begun, but left incomplete, and it was four years before the next work was staged. This was the 1-act *Der Schauspieldirektor*, a diverting trifle written for the small theatre in the grounds of the imperial summer palace at Schönbrunn, which pokes gentle fun at the world of opera and was first given with Salieri's *Prima la musica e poi le parole*.

Italian, rather than German, opera was now in vogue in Vienna, and it was in an optimistic and carefree mood that Mozart embarked upon *Le nozze di Figaro*. The choice of a text by Vienna's most gifted comic librettist, Lorenzo Da Ponte, was bold, for the Beaumarchais play upon which it was based had been recently banned owing to its inflammatory and subversive content. In the hands of Da Ponte and Mozart, the political message and social satire were almost totally excised, and replaced instead by a concern for the interrelationships of the sharply contrasting characters. In responding to the new, genuinely human figures which Da Ponte created, and to the opera's message of reconciliation, Mozart was moved to some of his most profound and perceptive music. *Figaro* does not merely revolve around a series of stock situations; Mozart allows us to penetrate the characters and appreciate their actions in terms not of stage farce, but of real human experience. Musically, the opera represented another stage in the development of opera buffa, providing the first union of the worlds of instrumental and operatic writing. In particular, the dramatic pace and tension of the finales are governed by a masterful adoption of symphonic principles, as are many of the arias.

*Figaro* became a success not only in Vienna, but also in nearby Prague, leading to a commission for the opera-house there in the following year. For his libretto Da Ponte turned to the familiar theme of Don Giovanni and the 'stone guest', well known both through morality plays and vulgar farces. He was undoubtedly influenced by many earlier stage treatments, especially Bertati's text for Gazzaniga. Librettist and composer again surpassed previous efforts, turning a hackneyed cautionary tale with the shallow characters of a morality play into a truly dramatic work. Technically a dramma giocoso, *Don Giovanni* oscillates between moods of high comedy and deep tragedy, strongly tinged with irony. Musically, it builds upon the new techniques of *Figaro*: ensemble writing occupies a crucial role in determining dramatic pace and progression, while the stock characters of the story are given more human identities. Although all main features of the legend, such as the cemetery and banquet scenes, and the \*catalogue aria detailing Don Giovanni's conquests, are present, they are here fashioned into a more dramatic framework. But although popular in Prague, the opera enjoyed only a mixed reception elsewhere. During the 19th cent. it was usually given in severely mutilated versions, with its buffo moralizing sextet omitted.

A revival of *Figaro* in Vienna in 1789 probably led to the commission for Mozart's next work, *Così fan tutte*. Da Ponte was naturally chosen as librettist, although he this time produced a more original work, yet dealing with the familiar oper-atic theme of fidelity in love. In outward structure the plot is conventional, with two balanced pairs of lovers, manipulated by two more worldly wise confidants. But Da Ponte and Mozart again penetrate beyond the superficiality of amorous intrigues, with their stereotyped representations of blind passion and revenge, to reveal the essential humanity of their characters. Perhaps of all the Da Ponte–Mozart collaborations it is *Così* which achieves the most complete union of words and music. The characterization and action of the libretto are matched by a similarly balanced musical style, with the pairing and re-pairing of the four lovers reflected in the distribution and organization of the individual numbers, while the cynicism of the old philosopher Don Alfonso is complemented by the scintillating \*soubrette performance of the maid Despina. Once again, it is in the ensemble writing that Mozart's genius appears, for the individuality of each character shines through in music of unsurpassed tenderness and perception. Probably because of its plot, which was considered somewhat risqué at the time, *Così* did not achieve the performances or popularity of the other Da Ponte operas.

*Die Zauberflöte* comes from a quite different world. Commissioned by Schikaneder for his troupe at the T auf der Wieden, it clearly had its origins in Viennese Singspiel. Although the diversity of its musical expression was without parallel in the contemporary tradition, its plot capitalized upon the vogue for magic plays and operas. Much has been written about the clear, if only partly understood, Masonic sub-plot, but whether the opera is regarded as mystic ritual or childish pantomime is irrelevant against the enlightened humanism of Mozart's score. The social and intellectual polarity of the stage figures is matched by an appropriate range of musical styles—Singspiel for Papageno and Papagena, opera seria for the higher characters, and opera buffa for the ensembles. Opera seria was indeed in Mozart's mind at this time, for he broke off work on *Die Zauberflöte* to compose an essentially conventional Metastasian work for Prague to celebrate the coronation of Leopold. But he had little sympathy with *La clemenza di Tito* and his motivation was clearly financial; his pupil Süssmayr composed all the recitatives, and probably some of the arias as well.

The amalgam of the three main strands of contemporary opera in *Die Zauberflöte* forms a fitting apotheosis of Mozart's craft. His letters reveal it to be a work with which he identified closely, and herein perhaps lies a clue to his own character, for throughout his operas he speaks at one and the same time with the wit and natural charm of Papageno, while placing before his audience the lofty and worthy aspirations of Tamino. But to understand Mozart's mature operas also requires recognition of the strong cross-currents between his stage works and other genres. Though he demonstrated an acute appreciation of the conventions of all contemporary styles of opera, he developed them by adapting the formal and textural principles of the sonata, symphony, and concerto, which were themselves in turn imbued with the spirit of vocal music in their melodic shaping and expressive content. But Mozart rose above his contemporaries not only because of his mastery of musical craft. He possessed the supreme skill of being able to transform stock stage events into reflections of human life, mocking the vanity and folly of the proud and ignorant, and commiserating with the unfortunate and wronged. A shrewd observer of human life, like Shakespeare he reshaped the familiar themes of love, loyalty, revenge, and hatred in a manner which deepened and enriched the experience of audiences not only in his own lifetime, but for all succeeding ages.

Operas on Mozart are by Riotte (*Mozarts Zauberflöte*, 1820), Lortzing (*Szenen aus Mozarts Leben*, comp. 1832), Flotow (*Die Musikanten*, 1887), M. Anzoletti (*La fine di Mozart*, 1898), Rimsky-Korsakov (*Mozart and Salieri*, 1898), and Reynaldo Hahn (1925).

WORKLIST: *Apollo et Hyacinthus* (Widl; Salzburg 1767); *La finta semplice* (The Feigned Simpleton) (Coltellini, after Goldoni; Salzburg 1769); *Bastien und Bastienne* (Weiskern, Müller, & Schachtner, after M. J. B. & C. S. Favart, and H. de Guerville, after Rousseau; Vienna 1768); *Mitridate, re di Ponto* (Cigna-Santi, after Parini, after Racine; Milan 1770); *Ascanio in Alba* (Parini, ?after Stampa; Milan 1771); *Il sogno di Scipione* (Scipio's Dream) (Metastasio, after Cicero; Salzburg 1772); *Lucio Silla* (Gamerra, rev. Metastasio, after Plutarch; Milan 1772); *La finta giardiniera* (The Feigned Gardener) (?Calzabigi, rev. Coltellini; Munich 1775); *Il re pastore* (The Shepherd King) (Metastasio, rev. Varesco; Salzburg 1775); *Zaide* (Schachtner, after Sebastiani; comp. 1779–80, incomplete); *Idomeneo, re di Creta* (Varesco, after Danchet; Munich 1781, rev. Vienna 1786); *Die Entführung aus dem Serail* (The Abduction from the Seraglio) (Stephanie jun., after Bretzner; Vienna 1782); *L'oca del Cairo* (The Goose of Cairo) (Varesco; comp. 1783, incomplete); *Lo sposo deluso* (The Deluded Bridegroom) (?Da Ponte; comp. ?1783, incomplete); *Der Schauspieldirektor* (The Impresario) (Stephanie jun.; Schönbrunn 1786); *Le nozze di Figaro* (The Marriage of Figaro) (Da Ponte, after Beaumarchais; Vienna 1786, rev. 1789); *Il dissoluto punito, ossia Il Don Giovanni* (The Degenerate Punished, or Don Giovanni) (Da Ponte, after Bertati; Prague 1787, rev. Vienna 1788); *Così fan tutte, ossia La scuola degli amanti* (Women are Like That, or The School for Lovers) (Da Ponte; Vienna 1790); *Die Zauberflöte* (The Magic Flute) (Schikaneder; Vienna 1791); *La*

*clemenza di Tito* (The Clemency of Titus) (Mazzolà, after Metastasio; Prague 1791).

CATALOGUE: L. Köchel, *Chronologisch-thematisches Verzeichnis* (Leipzig, 1862, 6/1964).

BIBL: E. J. Dent, *Mozart's Operas* (London, 1913, 2/1947); A. Einstein, *Mozart* (London, 1946); E. Anderson, ed., *The Letters of Mozart and his Family* (London, 1938, 3/1985); W. Mann, *The Operas of Mozart* (London, 1977); C. Gianturco, *Mozart's Early Operas* (London, 1981); W. Hildesheimer, *Mozart* (London, 1984); A. Steptoe, *The Mozart-Da Ponte Operas* (Oxford, 1988).

**Mozart and Salieri.** Opera in 2 acts by Rimsky-Korsakov; text a setting of Pushkin's 'little tragedy' (1830). Prod. Moscow, private opera co., 7 Dec. 1898, with Shkafer, Shalyapin, cond. Truffi; London, RAH, 11 Oct. 1927, with Ritch, Shalyapin, Lavretsky, cond. Coates; Forest Park, PA, Unity House, 6 Aug. 1933.

Vienna, end 18th cent. Salieri (bar) contemplates Mozart's (ten) genius in comparison to his own musical skills. He fears that such a gift is actually destructive to musical tradition, and decides that the only possible course of action is to poison Mozart's wine so as to ensure the future of music. Mozart dies, to Salieri's personal sadness but professional relief.

**Mravina, Evgeniya** (*b* St Petersburg, 16 Feb. 1864; *d* Yalta, 25 Oct. 1914). Russian soprano. Studied St Petersburg with Pryanishnikov, Berlin with Artôt. Début St Petersburg, M, 1886 (Gilda). Sang there until 1898. A highly attractive, intelligent woman, with a pure tone and exemplary diction, she was also influential as one of the first female singers in Russia to bring dramatic depth to her roles. These included Violetta, Elsa, Marguerite, Antonida (*A Life for the Tsar*), Lyudmila, Tatyana, first Fornarina (Arensky's *Raphael*), and first Oxana (*Christmas Eve*), when she 'sang and acted beautifully' (Yastrebtsev).

**Mshevelidze, Shalva** (*b* Tbilisi, 28 May 1904; *d* Tbilisi, 4 Mar. 1984). Georgian composer. An important ethnomusicologist and administrator, Mshevelidze's works for Georgian music included several operas based on national history and epic. The most successful has been *Ambavi Tarielisa* (The Legend of Tariel) (Pagava, after Rustaveli; Tbilisi 1946, rev. 1966).

**Muck, Carl** (*b* Darmstadt, 22 Oct. 1859; *d* Stuttgart, 3 Mar. 1940). German conductor. Studied classics Heidelberg and Leipzig. Chorus-master Zurich; cond. there and Salzburg, Graz, and Brno. Engaged by Neumann for Prague 1886. Cond. Neumann's Wagner Co. in first performances of *The Ring* in Moscow and St Petersburg 1889. Berlin, H, 1892–1912 (music director from 1908): cond. 1,071 perfs. of 103 operas, of which

489

35 were novelties. London, CG, 1899. Bayreuth 1901–30, where he was considered the greatest *Parsifal* conductor of his generation: his interpretations were praised for their clarity and fidelity in slow tempos, and their strict rhythmic and formal control. (R)

**Muette de Portici, La** (The Dumb Girl of Portici). Opera in 5 acts by Auber; text by Scribe and Delavigne. Prod. Paris, O, 29 Feb. 1828, with Damoreau, Nourrit, Dabadie, Dupont, Prévost; London, DL, 4 May 1829, as Masaniello; New York, Park T, 9 Nov. 1829.

Portici and Naples, 1647. When, during the wedding of the Spanish Princess Elvira (sop) and Alfonso (ten), son of the Duke of Arcos, the dumb girl Fenella (dancer) identifies Alfonso as her seducer, Elvira disowns him.

Fenella indicates to her brother, the fisherman Masaniello (ten), what has happened, though without naming the culprit; he and his friends vow revenge.

Elvira sends a messenger to fetch Fenella, and Masaniello realizes who his sister's seducer must have been.

Having incited the peasants of Portici to rebellion, Masaniello regrets their excessive action and helps Elvira and Alfonso to escape.

The revolutionaries gain control of the palace and declare Masaniello king. However, the Duke launches a counter-attack and Masaniello is killed while trying to save Elvira. After persuading Elvira to forgive Alfonso, Fenella hurls herself from the balcony of the castle as Vesuvius erupts.

A performance in Brussels on 25 Aug. 1830 sparked off a Belgian revolt. Other operas on the subject are by Carafa (1831), Pavesi (1831), and Napoli (1953).

**Mugnone, Leopoldo** (*b* Naples, 29 Sep. 1858; *d* Naples, 22 Dec. 1941). Italian conductor and composer. Studied Naples, beginning his career aged 12 with a comedy, *Il dottore Bartolo Sarsparilla*. Cond. Venice, F, aged 16. Cond. Rome, C, 1888, incl. première of *Cavalleria rusticana* (1890) and *Tosca* (1900); also did much to promote French opera in Italy. Milan, S, from 1880. London, CG, 1905, 1906, 1919, 1924, conducting first CG *Andrea Chenier* (1905), *Fedora* (1906), and *Iris* (1919). New York, Manhattan OC, 1922. His Verdi was much admired by Boito; and Beecham considered him the best Italian opera conductor of his period.

**Mulè, Giuseppe** (*b* Termini Imerese, 28 June 1885; *d* Rome, 10 Sep. 1951). Italian composer. Studied Palermo Conservatory, of which he was director 1922–5; then director S Cecilia Conservatory, Rome, till 1943. His operas were initially much influenced by verismo, in particular by

Mascagni (as in *La baronessa di Carini*, 1912), but he later developed a more advanced style. His culminating work was *La zolfara* (The Mines of Sulphur, 1939).

**Mulhouse** (Ger., Mühlhausen). Town in Haut-Rhin, France. An opera-house was opened 1868, becoming the T Municipal (cap. 867) in 1876. The company has placed a strong emphasis on ensemble, seldom inviting guest artists even for the most ambitious productions. Since 1972 part of *Opéra du Rhin,

**Müller, Adolf** (*b* Tolna, 7 Oct. 1801; *d* Vienna, 29 July 1886). Austrian composer. Trained first as an actor and performed in Brno, Prague, Lemberg. Came to Vienna 1823, where he composed his first operetta. Engaged as singer at the Kartnerthortheater 1826, but the success of *Die schwarze Frau* led to his becoming conductor instead. Conductor at the T an der Wien 1828–47, sometimes also performing at the T in der Leopoldstadt. He wrote over 600 works for the Viennese stage, including operettas, Singspiels, Possen, and much incidental music. His light-hearted approach and tuneful style helped to prepare the way for Johann Strauss and Suppé.

SELECT WORKLIST: *Die erste Zusammenkunft* (Meisl; Vienna 1827).

His son **Adolf Müller** (*b* Vienna, 15 Oct. 1839; *d* Vienna, 14 Dec. 1901) worked in Posen, Magdeburg, Düsseldorf, and Rotterdam, before joining the T an der Wien, Vienna, and was also known as an opera composer.

**Müller, Maria** (*b* Litoměřice, 29 Jan. 1898; *d* Bayreuth, 13 Mar. 1958). Bohemian soprano. Studied Vienna with Schmedes. Début Linz 1919 (Elsa). After engagements in Prague and Munich, New York, M, 1925–35. Berlin: SO, 1926, 1950–2 (when she retired); S, 1927–45. Bayreuth, 1930–44. London, CG, 1934, 1937. A warm-voiced, expressive performer with a wide repertory, e.g. Iphigénie (*Tauride*), Donna Elvira, Reiza, Marguerite, Amelia (*Boccanegra*), Jenůfa, and roles in works by Pizzetti and Weinberger. Particularly distinguished as Elsa, Elisabeth, Sieglinde, Eva. (R)

**Müller, Wenzel** (*b* Trnava, 26 Sep. 1767; *d* Baden, nr. Vienna, 3 Aug. 1835). Austrian composer and conductor. Studied Johannisberg with Dittersdorf; joined Brno T as violinist 1782, also composing a Singspiel, *Das verfehlte Rendezvous*; conductor 1785. Moved to Vienna, becoming conductor and composer at the T in der Leopoldstadt. Here he strove hard to improve musical and theatrical standards and with the demise of the National-Singspiel was able to give the theatre a position of great importance in Viennese musical life. During his career he produced c.250 works; his first major success came with *Das Sonnenfest der Braminen*, equalled by *Das Neusonntagskind*. Other important works included *Kaspar der Fagottist oder Die Zauberzither*, reflecting a vogue for magic operas seen also in Mozart's contemporaneous *Die Zauberflöte*, *Die Schwestern von Prague* (whose hit song 'Ich bin der Schneider Kakadu' forms the basis of a set of piano trio variations by Beethoven), *Die zwölf schlafenden Jungfrauen*, *Das lustige Beilager*, and the exceptionally successful *Die Teufelsmühle am Wienerberg*. Later works were less well received; in 1807 he moved to the German O in Prague, where his daughter the soprano Therese *Grünbaum was also engaged.

Müller left in 1813, having been less successful than in Vienna at raising standards. But back at the Leopoldstadt in Vienna, he resumed activities with much success: among his most popular works were *Tankredi* and *Aline*. He wrote several works with Raimund, including *Der Barometermacher auf der Zauberinsel*, *Die gefesselte Phantasie*, and notably *Der Alpenkönig und der Menschenfeind*, some of which interested the young Wagner.

Müller was the most successful of all early 19th-cent. Viennese popular composers, and a few of his works with Raimund remain in the local repertory. Though early in his career he showed greater operatic ambitions, writing extended finales, it was in simpler songs and ensembles that he was most at ease. Witty and satirical numbers were his strong point, but he could also write in a more tender, reflective vein that charmed his audiences.

SELECT WORKLIST (all first prod. Vienna unless otherwise stated): *Das verfehlte Rendezvous* (The Missed Rendezvous) (Zehnenmark; Brno 1783); *Das Sonnenfest der Braminen* (The Sun-Festivities of the Brahmins) (Hensler; 1790); *Kaspar der Fagottist oder Die Zauberzither* (Kaspar the Bassoonist or the Magic Zither) (Perinet; 1791); *Das Neusonntagskind* (The New Sunday-Child) (Perinet, after Hafner; 1793); *Die Schwestern von Prague* (Perinet; 1794); *Das lustige Beilager* (Perinet, after Hafner; 1797); *Die zwölf schlafenden Jungfrauen* (The Twelve Sleeping Maidens) (Hensler; 1797); *Die Teufelsmühle am Wienerberg* (The Devils' Mill on the Wienerberg) (Hensler; 1799); *Tankredi* (Bäuerle; 1817); *Aline* (Bäuerle; 1822); *Der Barometermacher auf der Zauberinsel* (The Barometer-Maker on the Magic Island) (Raimund; 1823); *Die gefesselte Phantasie* (Fantasy Chained) (Raimund; 1828); *Der Alpenkönig und der Menschenfeind* (The Alpine King and the Misanthrope) (Raimund; 1828).

BIBL: L. Raab, *Wenzel Müller: ein Tonkünstler Altwiens* (Baden, 1928).

**Mullings, Frank** (*b* Walsall, 10 May 1881; *d* Manchester, 19 May 1953). English tenor. Stu-

died Birmingham with Beard and Breedan. Début Birmingham 1913 (Tristan). Beecham OC 1916–20; BNOC 1922–9; sang in concert 1930–45. A notable artist, memorable for his moving Otello (Verdi), Tristan, and Radamès. (R)

**Munich** (Ger., München; It., Monaco). City in Bavaria, Germany. The first recorded operatic performance was of Giovanni Macchioni's dramatic cantata *L'arpa festante* in Aug. 1653 in the Herkules-Saal of the Residenz. Johann Kaspar Kerll (1627–93) inaugurated the theatre on the Salvatorplatz (the Haberkasten, an old granary) with his *Oronte* (1656), and gave Munich standards as high as any in Europe; he left after disagreements over his progressive policies 1673. Kerll was succeeded by Ercole Bernabei 1674–87.

An important period began with Agostino *Steffani and his *Marco Aurelio* (1681); this was interrupted by the court's transfer to Brussels 1692, but resumed in 1701 under Pietro Torri (c.1650–1737), especially after the court finally settled in Munich again in 1715. A French influence was now felt beside the Italian. The new Residenztheater (cap. 436) was built by François Cuvilliés 1753 (bombed; reconstructed 1958), and saw the premières of Mozart's *La finta giardiniera* (1775) and *Idomeneo* (1781). The Elector Carl Theodor forbade Italian opera 1787, and German opera was cultivated, with translations of foreign works, and Singspiels by composers including Franz Danzi, Ferdinand Fränzl (1767–1833), and especially Peter von *Winter. Joseph Babo became director 1799 and despite economic restraints further developed the repertory of French opera and Singspiel, with Italian opera returning 1805. Weber's *Abu Hassan* was premièred 1811.

The T am Isarthor opened 1812 under Peter von *Lindpaintner (closed 1825), the Hof- und Nationaltheater 1818 (burnt down, rebuilt 1823; bombed 1943; rebuilt 1963, cap. 1,773). But standards declined until Franz Lachner took over as music director 1836; he extended the repertory and built up a strong ensemble. The Staatstheater am Gärtnerplatz (cap. 932) opened 1865.

Since 1945 the Staatstheater has been Munich's second opera-house; rebuilt 1969. Music directors since have included Georg Solti 1947–52, Rudolf Kempe 1952–4, Ferenc Fricsay 1956–9, Joseph Keilberth 1959–68, Wolfgang Sawallisch 1971–92; Peter Jonas from 1992.

Wagner was in Munich at Ludwig II's invitation from 1864, but his hopes for a festival theatre dedicated to his works were frustrated by the court. Nevertheless Hans von *Bülow (music director 1867–9) conducted the premières of *Tristan und Isolde* (1865) and *Die Meistersinger* (1868), and many other Wagner performances (including a private one of *Parsifal* for the King). Franz Wüllner was music director 1869–71, Hermann Levi 1872–96. Karl von Perfall was an influential director 1867–93, as was his successor Ernst von Possart 1894–1905. Under Possart the Prinzregententheater (cap. 1,012) was built on the Bayreuth model; opened 1901 with *Meistersinger*. Felix Mottl was music director 1903–11, Bruno Walter 1913–22.

The Strauss tradition dates from 1919, when performances of his works became a feature of the season and of the summer festivals; Strauss had been Kapellmeister here 1886–9 and 1894–8, and *Friedenstag* (1938) and *Capriccio* (1942) were premièred here. The bombing of the opera-house in 1943 inspired his threnody *Metamorphosen*, and all 15 of his operas were given as a cycle 1988. Hans Knappertsbusch was music director 1922–35, Clemens Krauss 1937–44, both consolidating the company's reputation.

**Münster.** City in North-Rhine Westphalia, Germany. Long dependent on visiting companies and local school music dramas, the city opened the Komödienhaus 1775 and formed a company 1782; the repertory included Gluck, Grétry, and Mozart. Under Pichler 1818–41 the repertory developed further, and Lortzing's first opera, *Ali Pascha von Janina*, was premièred 1828. Lortzing settled here 1827–33, and the theatre opened in 1900 bore his name; bombed 1941. A new theatre (cap. 956) opened 1956 with *The Magic Flute*.

**Muratore, Lucien** (*b* Marseilles, 29 Aug. 1878; *d* Paris, 16 July 1954). French tenor. Studied Marseilles Cons., Paris with E. Calvé. Début Paris, OC, 1902 (King, Hahn's *Carmélite*). Paris: OC until 1904; O 1905–12. Chicago 1913–24. Paris: O, 1920; OC, 1923; retired 1931. Created many roles, e.g. title-role of Massenet's Bacchus, and Prinzivalle (Février's *Monna Vanna*). A skilful singer-actor, he was successful as Gounod's Faust, Puccini's Des Grieux, Strauss's Herod, and especially Don José. Married to soprano Lina *Cavalieri 1913–27. (R)

**Murray, Ann** (*b* Dublin, 27 Aug. 1949). Irish mezzo-soprano. Studied Manchester with Cox. Début Scottish O (in *Alceste*). London, ENO from 1975, and CG from 1976; Gly. 1979; Cologne, 1979–81; New York, M, 1984. Also Milan, S; Vienna; Salzburg; Moscow; Leningrad. Roles include Xerxes, Cherubino, Donna Elvira, Dorabella, Sesto, Rosina, Charlotte, Octavian, Composer. A poised singer of vocal

finesse, with a fine legato. Married to the tenor Philip *Langridge. (R)

**Murska, Ilma di.** See DI MURSKA, ILMA.

**Musetta.** A *grisette*, lover of Marcello and others, (sop) in Puccini's and (mez) in Leoncavallo's *La bohème*.

**Musgrave, Thea** (*b* Barnton, 27 May 1928). Scottish composer. Studied Edinburgh with Gál, Paris with Boulanger. Apart from a chamber opera and a work for children, her first opera was *The Decision*. This represented an expansion of already impressive creative powers, and brought a new warmth and humanity into her idiom and also, despite some Schoenbergian influences, a new individuality. *The Voice of Ariadne* accepts some of Britten's innovatory techniques in the church parables, as well as making use of prerecorded tape; though it extends a Henry James short story over three acts, it successfully explores a complex and mysterious set of relationships. However, her major work has been *Mary, Queen of Scots*, which reveals a more surefooted dramatic pacing that also reflects some of Britten's control if not his idiom, and a sympathetic, humane command of characterization in a historical context. It is, appropriately, one of the most lyrical works of a composer who has also shown an original sense of theatre in the quasi-dramatic presentation of some of her concertos.

WORKLIST: *The Abbot of Drimock* (Lindsay, after Wilson; comp. 1955, concert London 1958, prod. London 1962); *Marko the Miser* (Samson, after Afanasyev; Farnham 1963); *The Decision* (Lindsay, after Taylor, after a real incident; London 1967); *The Voice of Ariadne* (Elguera, after James; Aldeburgh 1974); *Mary, Queen of Scots* (Musgrave, after Elguera; Edinburgh 1977).

BIBL: D. Hixon, *Thea Musgrave* (Westport, 1984).

**Musi, Maria Maddalena** (*b* Bologna, 18 June 1669; *d* Bologna, 2 May 1751). Italian singer. In the service of the Duke of Mantua 1689; Bologna 1692, 1694–7; Naples 1696–1700, 1702, particularly in operas by A. Scarlatti. Retired 1726. A fine artist, much appreciated in travesti roles by her contemporaries, she was also a fascinating woman, who amassed both riches and lovers. Known as 'la Mignatta' (the leech).

**musical comedy.** A form of light opera in which romantic and comic elements predominate, set to music whose principal aim is to frame the plot with tuneful numbers. Characteristic composers of musical comedy in England, where the term and genre had a particular vogue between the wars, are Lionel Monckton, Montague Phillips, Noël Coward, Ivor Novello, and Vivian Ellis. In its abbreviation as musical, the term has been particularly applied to US works, especially those written for Broadway. The first successful US-written operetta was Willard Spencer's *The Little Tycoon* (1886); but the genre first began to develop in the 20th cent. with the work of George M. Cohan, and was taken forward with works by Friml, Romberg, and Jerome Kern (to books by P. G. Wodehouse, Guy Bolton, and Oscar Hammerstein II) among many others; the enormous success of the Hammerstein–Kern *Show Boat* (1927) really established the musical play, or musical in the modern sense. Outstanding composers of musicals have been George Gershwin, Richard Rodgers, Frederick Loewe, Frank Loesser, Irving Berlin, Cole Porter, Kurt Weill, and Leonard Bernstein. Latterly, the emphasis in the musical has been on the spectacular or fantastical, a movement led by Andrew Lloyd Webber, whose works have been extremely popular on both sides of the Atlantic. Though loosely applied to both musical play and musical comedy, each a distinct and thriving branch of the US musical theatre, the term musical is usually reserved for the play in which music serves the plot with song and dance.

**music drama.** The name given to works in which the musical and dramatic elements are (or are intended to be) entirely unified, with every other consideration (such as opportunities for display by singers) subjugated to this end. The term first came into general use with Wagner; he realized in his later works the ideal to which German opera had been moving for over a century.

**musico.** See CASTRATO.

**music theatre.** A term for works in which musical and dramatic elements are involved, not necessarily representationally. Thus instrumentalists may be costumed and on the stage, or a musician may use a mask, or a group of musicians operate in semi-dramatic conventions. Originally an attempt, deriving from Brecht, to infuse a new kind of dramatic immediacy into opera, it was taken up by a number of younger composers, especially Goehr, Birtwistle, Ligeti, and Maxwell Davies, with works in which a dramatic convention rather than a full staging is used, partly for reasons of economy but fundamentally in the wish to explore new relationships between music and drama.

**Musorgsky, Modest** (*b* Karevo, 21 Mar. 1839; *d* St Petersburg, 28 Mar. 1881). Russian composer. Studied with Herke, then entering the Cadet School of the Guards in St Petersburg, when he also began writing music. He came to know Balakirev, who gave him some lessons in form, and having resigned his Guards commission, took up music more seriously.

Though one of the great dramatic composers of history, Musorgsky completed only one opera, his masterpiece *Boris Godunov*; and all his operatic works present problems to editors and performers. His earliest projects were *Han d'Islande*, after Victor Hugo (1856) and *Noch na Ivanov den* (St John's Eve, after Gogol, 1858). Of *Oedipus in Athens* there survives only a single scene, later incorporated into *Salammbô*, *Mlada*, and *Sorochintsy Fair*, and some other choruses. *Salammbô* is also unfinished, and remains interesting chiefly for its lyrical elements and for a tough realism which hints at *Boris*. The latter quality is developed in the first and only completed act of *The Marriage*: this is composed as a continuous drama based on speech-melody. It is a lively comedy of manners that includes some characterization by way of Leitmotiv. Musorgsky had also closely associated with the group of five composers comprising Rimsky-Korsakov, Balakirev, Cui, and Borodin as well as himself, dubbed by Stasov the *moguchaya kuchka* (mighty handful).

Musorgsky was by now growing interested in an opera on Pushkin's *Boris Godunov*; the original version was completed by 1869. Though *Boris* has antecedents in Romantic opera and Grand Opera, it is a unique and original product of them. It is also the greatest operatic outcome of the theories advocating realism in art that occupied Russian artists in these years. Examples are the opening crowd scene, which is not the conventional opening chorus but a musical depiction of the many groups and voices that make up a corporate crowd entity, and the Death of Boris, which, instead of being a conventional death aria, gradually fragments his vocal line against steady tolling and chanting until no music is left for it. There is some use of motive, though it is not developed after the manner of Wagnerian Leitmotiv. The vocal writing is still speech-inflected, but also capable of strongly lyrical qualities; and there are some separate numbers within the continuous composition. When Musorgsky revised the opera, he added the so-called Polish acts and somewhat softened the original conception.

While negotiating the production of *Boris* in 1870, Musorgsky began a new opera, *Bobyl*, one scene of which was inserted in *Khovanshchina*; and in 1870 he was also considering an unidentified Gogol comic opera. He also collaborated with Rimsky-Korsakov, Borodin, and Cui on a projected opera-ballet, *Mlada*, using part of the *Oedipus* music and *A Night on the Bare Mountain*. He began planning another historical opera, *Khovanshchina* (1872–80). This suffers from a complex plot and a multiplicity of leading figures, but portrays with great power some clashing orthodoxies, such as the Slavophile–Westernizer opposition and that between the modern Church and the Old Believers; it also includes some powerful scenes between the main characters. One, the incantation scene, was taken over from the outline for *Bobyl*. Work on the opera was interrupted by yet another project destined to remain unrealized, the comedy *Sorochintsy Fair*. This draws heavily on Ukrainian folk music. For a possible *Pugachevshchina* (1877, on the Pugachev rebellion and partly based on Pushkin), Musorgsky had begun noting down Kirghiz, Transcaucasian, and other tunes; he also played his friends some ideas for another opera, *Byron*. However, his private life, always in a state of disarray, had now deteriorated into alcoholism, and he died with a number of his works incomplete.

Many attempts have been made to get Musorgsky's works into a performing condition. Rimsky-Korsakov made a standard version of *Boris Godunov*, rescoring, reharmonizing, and rewriting much of the music and reversing the order of the last two scenes, with the intention of making the work's originalities more generally acceptable; he claimed this to be a temporary measure, and it is Musorgsky's own music, sometimes with minor adjustments to the material included, which is now becoming accepted. Rimsky-Korsakov also made a performing version of *Khovanshchina*, but his smoothing out of Musorgsky's characteristics led both Boris Asafyev and Shostakovich to make new versions. *The Marriage* has been completed for performance by Ippolitov-Ivanov, and *Sorochintsy Fair* by Lyadov, Karatygin, and others.

WORKLIST (all texts by Musorgysky unless otherwise stated): *Edip v Afinakh* (Oedipus in Athens) (after Ozerov; unfin., 1858–60, unprod.); *Salammbô* (after Flaubert; unfin., 1863–6, prod. Naples 1983); *Zhenitba* (The Marriage) (after Gogol; Act I only, comp. 1868, prod. St Petersburg 1909); *Boris Godunov* (after Pushkin & Karamzin; 1st version, 7 scenes, comp. 1868–9, prod. Leningrad 1928, 2nd version, prologue and 4 acts, comp. 1871–2, rev. 1873, prod. St Petersburg 1874); *Bobyl* (The Landless Peasant) (after Spielhagen; outline for 1 scene, comp. 1870, unprod.); *Mlada* (Gedeonov, after Krylov; opera-ballet project comp. with Cui, Rimsky-Korsakov, and Borodin; part of Act II comp. (incl. reworking of *Night on the Bare Mountain*, 1867), unprod.); *Khovanshchina* (The Khovansky Affair) (& Stasov; unfin., comp. 1872–80, prod. St Petersburg 1886); *Sorochinskaya Yarmarka* (Sorochintsy Fair) (after Gogol; unfin., comp. 1874–80, prod. Moscow 1913).

WRITINGS: A. Orlova and M. Pekelis, eds., *M. P. Musorgsky: Literaturnoye Naslediye* (Moscow, 1971); anon., ed., *M. P. Musorgsky: Pisma* (Letters) (Moscow, 1981).

BIBL: M. Calvocoressi, *Modest Musorgsky* (London, 1956, R/1967); A. Orlova, ed., *Trudy i dny M. P. Musorgskovo* (Moscow, 1963; trans. 1983).

**Mustafà.** The Bey of Algiers (bs) in Rossini's *L'italiana in Algeri*.

**Muti, Riccardo** (*b* Naples, 28 July 1941). Italian conductor. Studied Milan with Votto. Début Naples 1970. Salzburg from 1971; Florence 1972; London, CG, 1977; music director Milan, S, from 1986, where he restored five Mozart operas to the repertory, also reviving Cherubini's *Lodoïska*. Also Vienna, Munich, Paris, Rome, etc. Repertory incl. *Puritani, Attila, Ballo, Aida*, etc., and revivals incl. *Agnes von Hohenstaufen* and the first uncut *Guillaume Tell*. One of the finest Italian opera conductors of the day, he has given some powerfully dramatic performances of Verdi in particular, controlling the music strongly but with sympathetic encouragement to his singers. (R)

**Muzio, Claudia** (*b* Pavia, 7 Feb. 1889; *d* Rome, 24 May 1936). Italian soprano. Studied Turin with Casaloni, Milan with Viviani. Début Arezzo 1910 (Manon). Milan, S, 1913–14, 1926–7, 1929–30; London, CG, 1914; New York, M, 1916–22, 1933–34; Chicago 1922–3, 1931–2; Paris, O, 1923–5; also Buenos Aires, Rome. Created several roles, including a powerfully acted Giorgetta (*Il tabarro*) whose scream froze the blood. An unusually warm and communicative artist, called 'the divine Claudia' by her public, she possessed a voice of great beauty, superb vocal mastery, and an excellent command of style. Esteemed as Norma, Aida, Mimì, Tosca, Nedda, and Madeleine de Coigny (*Chenier*), and particularly cherished for her moving Violetta. (R)

BIBL: E. Arnosi, *Claudia Muzio* (Buenos Aires, 1987).

**Mysliveček, Joseph** (*b* Prague, 9 Mar. 1737; *d* Rome, 4 Feb. 1781). Bohemian composer. Studied Prague with Habermann and Seeger, Venice with Pescetti. His first opera, *Medea*, was so successful that he was invited to Naples, where he had an even greater success with *Bellerofonte*. Some 30 operas followed, especially for Rome, Naples, Bologna, Milan, and Florence; his popularity was such that he became known to the Italians, unable to pronounce his name, by its translation as Venatorini. In 1773 he wrote *Erifile* for Munich, but he met with success only in Italy. Other triumphs included *Demofoonte*, *Ezio*, and *L'Olimpiade*. Though admired and liked by Mozart, and praised by singers for his understanding of the voice, he accepted the conventions of opera seria though in his best works he tried to do more than merely fill them elegantly.

SELECT WORKLIST: *Medea* (Golter; Parma 1764); *Bellerofonte* (Bonechi; Naples 1767); *Demofoonte* (Metastasio; Venice 1769); *Erifile* (De Gamerra; Munich 1773); *Ezio* (Metastasio; Naples 1775); *L'Olimpiade* (Metastasio; Naples 1778).

BIBL: R. Pečman, *Joseph Mysliveček und sein Opernepilog* (Brno, 1970).

# N

**Nabucco (Nabucodonosor)**. Opera in 4 acts by Verdi; text by Solera, after the drama by Anicet-Bourgeois and Francis Cornue, *Nabucodonosor* (1836). Prod. Milan, S, 9 Mar. 1842, with Strepponi, Bellinzaghi, Ruggeri, Miraglia, Marconi, G. Ronconi, Dérivis, Rossi; London, HM, as *Nino*, 3 Mar. 1846, with Sanchioli, Corbari, Fornasari, Botelli; New York, Astor Place OH, 4 Apr. 1848, with Truffi, Amalia Patti, Benventano, Rossi.

Originally entitled *Nabucodonosor* (Nebuchadnezzar), the opera was first billed as *Nabucco* at the T San Giacomo, Corfu, in 1844, and has since been generally known by this abbreviation.

Jerusalem and Babylon, 586 BC. The Jews lament their defeat by Nabucco, and implore Jehovah to spare the Temple. Ismaele (ten) is holding Nabucco's daughter Fenena (sop) hostage; they declare their love. Abigaille (sop), her supposed sister and also in love with Ismaele, appears with soldiers and offers to save him. Nabucco (bar) now arrives; Zaccaria (bs), the High Priest, threatens Fenena's life so as to deny him the Temple, but Ismaele frustrates him and Nabucco sacks the Temple.

The Jews are captive in Babylon, with Fenena acting as Regent in Nabucco's absence. Abigaille discovers that she is not Nabucco's daughter but an adopted slave. She plans to kill Fenena, who has converted to the Jewish faith, and seize the throne. But Nabucco appears and seizes the crown, proclaiming himself divine; he is immediately struck down by a thunderbolt and made mad.

Abigaille is now Regent, as Nabucco is forced to accept, and plans to kill the prisoners. By the Euphrates, they sing of their lost fatherland.

Nabucco, from prison, sees Fenena led to execution and prays to Jehovah. Freed by Abdallo (ten), Captain of the Guard, he rushes to the scaffold with his men, overthrows the false idols, and saves her. All join in prayer and thanks; Abigaille appears, having taken poison, and dies penitent.

Also opera by Ariosti (1706).

**Nachbaur, Franz** (*b* Giessen, 25 Mar. 1830; *d* Munich, 21 Mar. 1902). German tenor. Studied Milan with Lamperti, Stuttgart with Pišek. Début Passau 1857. Hanover 1859; Prague 1860–3; Munich, N, 1867–90, when he retired;

London, DL, 1882; Moscow 1887. A handsome, musical singer with a good technique, he pleased Wagner as the first Walther. Also first Froh; other roles included Rienzi, Lohengrin, Siegmund, Tannhäuser, Radamès, Gounod's Faust, Raoul, Chapelou (*Postillon de Lonjumeau*).

**Nacht in Venedig, Eine** (A Night in Venice). Operetta in 3 acts by Johann Strauss II; text by 'F. Zell' (Camillo Walzel) and Genée. Prod. Berlin, Friedrich-Wilhelm-Städtisches T, 3 Oct. 1883, for the opening of the theatre; New York 24 Apr. 1884; London, Cambridge T, 25 May 1944.

Venice, 18th cent. The old senator Delaqua (bs), decides to marry his ward Barbara (sop), who is being wooed by both Delaqua's nephew Enrico (ten or spoken) and the Duke of Urbino (ten or bar). The plot is further complicated by the fact that Annina, who had been brought up by the same nurse as Barbara, changes clothes with her during the Venetian Carnival, while Ciboletta (sop), engaged to Pappacoda (bar), pretends she is Barbara. By the end of the opera Caramello (ten), the Duke's barber, has married Annina, and Enrico has married Barbaro.

**Nachtlager von Granada, Das** (The Night Camp at Granada). Opera in 2 acts by Conradin Kreutzer; text by K. J. Braun von Braunthal, after Kind's drama. Prod. Vienna, J, 13 Jan. 1834; London, Prince's T, 13 May 1840; New York 15 Dec. 1862.

Spain, mid-16th cent. Disguised as a hunter, the Crown Prince of Spain (bar) is granted a night's shelter by shepherds, who decide, however, to kill and rob him when they find him kissing the shepherdess Gabriela (sop). She is loved by, and loves, Gomez (ten), but is also being pursued by Vasco (bs), whom her uncle Ambrosio (bs) wishes her to marry. She appeals to the 'hunter' for assistance, who promises to intercede for her with the Crown Prince. When she and Gomez, having found the Prince's followers, expose the plot, he reveals himself as the Prince and unites her and Gomez.

**'Nacqui all'affanno'**. Cenerentola's (mez) aria in Act II of Rossini's *Cenerentola*, rejoicing at the happy change in her fortunes and embracing her sisters and stepfather. One of the most difficult and brilliant of coloratura arias.

**Nadir**. A fisherman (ten), friend of Zurga and his rival for Leïla's love, in Bizet's *Les pêcheurs de perles*.

**Naldi, Giuseppe** (*b* Bologna, 2 Feb. 1770; *d* Paris, 14 Dec. 1820). Italian bass. Début Varese 1785. Milan, S, 1786; successes throughout Italy, then Lisbon, 1803; London 1806–18, where he was the first London Don Alfonso, Papageno,

Figaro (Mozart and Rossini), and Leporello; also sang in many works by other composers, e.g. Paisiello, Piccinni, Cimarosa. Though his voice was 'weak and uncertain' (Mount Edgcumbe), he was a gifted comic actor and an excellent musician. He was in García's apartment when a newly invented steam cooker burst and killed him. His daughter **Carolina Naldi** (1801–76) was a soprano. Début Paris 1819. Successful in Rossini and Bellini; sang in Madrid, Lisbon, and Italy. Retired when she married Count Sparri in 1824.

**Nancy.** Town in Meurthe-et-Moselle, France. The first theatre, built 1708, opened 1709 with Desmarets's *Le temple d'Astrée*. The Pavillon de la Comédie staged opera from 1755, regularly from 1850; burnt down 1906. Opera was then given in the Salle Poirel. The present Grand T (cap. 1,310), built behind the front of the Archbishop's Palace in the Place Stanislas, opened 14 Oct. 1919 with Reyer's *Sigurd*. The Opéra-Théâtre de Nancy plays in the renovated Grand T (cap. 950), and has given a number of important premières and French premières.

**Nannetta.** Ford's daughter (Anne) (sop), lover of Fenton, in Verdi's *Falstaff*.

**Nantes.** Town in Loire-Atlantique, France. The Grand T Graslin opened 23 Mar. 1788; burnt down 1796. Means for rebuilding were offered by Napoleon after a visit in 1808, and it opened 3 May 1813. Restored 1968 (cap. 980), and 1975–8. The Opéra de Nantes et des Pays de la Loire was formed 1973 and has given premières of various French operas.

**Nantier-Didiée, Constance** (*b* Saint-Denis, Île de Bourbon, 16 Nov. 1831; *d* Madrid, 4 Dec. 1867). French mezzo-soprano. Studied Paris with Duprez. Début Turin 1850 (Giulia, Mercadante's *La vestale*). Paris, I, 1852, 1856–8; London, CG, 1853–6, 1858 (at the opening of the new theatre), 1859, 1863–4; St Petersburg 1862. Roles included Ulrica, Siebel, Maddalena, Ascanio (*Cellini*), and the first Preziosilla (*Forza*). Chorley thought her better in supporting roles than as a principal.

**Naples** (It., Napoli). City in Campania, Italy. The first operas were given by visiting companies from Venice, probably invited by Count d'Oñate, Viceroy 1648–53, as part of a pacific policy after the revolt of Masaniello. These included Antonio Generoli's Accademia dei Febi Armonici, in a theatre in the grounds of the royal palace, 1651. *Ciro*, by Francesco *Provenzale, was probably given, 1653–4; then Francesco Cirillo's *Orontea*, 3 Apr. 1654. The T San Bartolomeo housed opera from 1654, with the Venetian repertory given by visitors. From 1676, under Filippo Coppola, operas were produced at the palace for royal occasions; he was succeeded by P. A. *Ziani, 1680, on whose death in 1684 Alessandro *Scarlatti was appointed. Under him, Naples became a great operatic centre, and the T San Bartolomeo was enlarged. Opera buffa also developed, the first known being Michelangelo *Faggioli's *Cilla*, 1706, the first in public Antonio *Orefice's *Patrò Calienno de la Costa*, 1709 at the T dei Fiorentini (closed 1820). An important consolidation came with Domenico *Sarro's Metastasio setting *Didone abbandonata*, 1724. New theatres opened including the T Pace (1724–49) and T Nuovo (1724–1828). With a new impetus to cultural life at the accession of the Bourbon Charles III, the T San Carlo (built in 270 days) was opened 4 Nov. 1737 with Sarro's *Achille in Sciro*; enlarged 1777 and 1812. The T del Fondo (now T Mercadante) opened 1779, the T San Ferdinando 1790.

Naples was ruled by the French 1806–15, and French opera was influential. Domenico *Barbaia became manager 1809, bringing *Rossini in 1815 (until 1822) to manage the T Fondo and write two operas a year. Several were for Barbaia's mistress Isabella *Colbran, whom Rossini later married. The T San Carlo burnt down Feb. 1816; rebuilt in six months and reopened (cap. 3,500) 12 Jan. 1817; rebuilt 1844. Barbaia also helped the careers of *Donizetti (director of the royal theatres 1827–38) and *Bellini. *Mercadante was active from 1838. *Verdi had his operas produced, with mixed success, and also had censorship troubles.

The T San Carlo went through a long period of decline in the late 19th and early 20th cents. Bombed 1943, but quickly restored. Under the British occupation, operatic life was encouraged, and the theatre became very popular with the Allied troops, introducing many to opera for the first time. With Pasquale di Costanzo as director 1946–75, a more adventurous policy was developed, though the theatre has retained its long-standing reputation of having the noisiest and worst-behaved audience in Italy, well able to compete with events on the stage.

BIBL: C. Roscioni, ed., *Il Teatro di San Carlo* (Naples, 1988).

**Napoli, Jacopo** (*b* Naples, 26 Aug. 1911). Italian composer. Studied Cons. San Pietro, Naples; held directorships of Naples, Milan, and Rome Cons. His comic operas, which use Neapolitan songs, include *Il malato immaginario*, *Miseria e nobiltà*, and *Un curioso accidente*, and have enjoyed a limited success. His *Mas'Aniello* was awarded one of the prizes in La Scala's Verdi competition.

SELECT WORKLIST: *Il malato immaginario* (Ghisalberti, after Molière; Naples 1939); *Miseria e nobiltà*

497

(Viviani, after Scarpetta; Naples 1946); *Un curioso accidente* (Ghisalberti; Bergamo 1950); *Mas'Aniello* (Viviani; Milan 1953).

**Nápravník, Eduard** (*b* Býšť, 24 Aug. 1839; *d* Petrograd, 23 Nov. 1916). Czech conductor and composer. Studied with Pŭhonný and Svoboda, then Prague with Blazek and Pitsch, also Kittl. Moved to Russia 1861 as director of Prince Yusupov's orchestra; when this was disbanded on the emancipation of the serfs in 1863, moved to St Petersburg, M, as répétiteur, becoming assistant conductor 1867, chief conductor 1869. Of his four operas, the most successful was *Dubrovsky*; but though praised by Tchaikovsky, the music is skilful rather than inspired. His reputation rests on the work he did in presenting other men's operas with an efficiency that greatly helped their success and that of the cause of Russian opera. He was a well-trained, hard-working conductor, unfailingly clear in detail. 'With all his pedantry, with all his inexorable dryness, he was an inspirer, he was a constructor . . . His hand moved like a metronome without volition, but under this seeming indifference there was an iron will' (Sergey Volkonsky). He conducted over 4,000 performances during his career, and among some 80 works he premièred were Cui's *William Ratcliffe* (1869); Serov's *The Power of Evil* (1871); Dargomyzhsky's *Stone Guest* (1872); Rimsky-Korsakov's *Maid of Pskov* (1873), *May Night* (1880), and *Snow Maiden* (1882); Musorgsky's complete *Boris Godunov* (1874); Tchaikovsky's *Oprichnik* (1874), *Vakula the Smith* (1876), *Maid of Orleans* (1881), *Queen of Spades* (1890), and *Yolanta* (1892); and Rubinstein's *Demon* (1875). He did much, almost single-handedly, to raise standards of performance and company discipline in Russia, also concerning himself with improving the lot of singers and players.

WORKLIST (all first prod. St Petersburg): *Nizhegorodtsy* (The People of Nizhegorod) (Kalashnikov; 1868); *Harold* (Weinburg, after Wildenbruch; 1886); *Dubrovsky* (M. Tchaikovsky, after Pushkin; 1895); *Francesca da Rimini* (Palecek & Ponomarev, after Philippe; 1902).

BIBL: Y. Keldysh, ed., *E. F. Nápravník* (documents and letters) (Leningrad, 1959).

**Narraboth.** A young Syrian (ten), the captain of the guard, in Strauss's *Salome*.

**Nash, Heddle** (*b* London, 14 June 1894; *d* London, 14 Aug. 1961). English tenor. Studied Milan with Borgatti. Début Milan, T Carcano, 1924 (Rossini's Count Almaviva). London, 1925, Old Vic, and with BNOC. London, CG, 1929–39, 1947–8; Gly. 1934–8; London, CR, during the war; New Opera Co. 1957–8. Highly gifted, with an excellent voice, technique, and

stage presence, he was a notable Don Ottavio, Ferrando, Duke of Mantua, David, Faust (Gounod), Rodolfo. (R) His son **John Heddle Nash** (*b* London, 30 Mar. 1928; *d* 29 Sep. 1994), a baritone, sang with London, SW and CR. (R)

**Nasolini, Sebastiano** (*b* Piacenza, *c*.1768; *d* ?Venice or Naples, *c*.1806). Italian composer. Studied Venice; maestro al cembalo San Pietro T, Trieste 1787, where his first opera *Nitteti* was given. Lived in Venice until 1799, but took many trips abroad, notably to London for the production of *Andromaca* 1790. Composed *c*.40 operas, both serious and comic, for various Italian houses. These works, which owe much to Paisiello's example, had considerable success in their day; many were also staged in London.

SELECT WORKLIST: *Nitteti* (Metastasio; Trieste 1788); *Andromaca* (Salvi; Venice 1790); *La morte di Cleopatra* (Sografi; Vicenza 1791); *Il medico di Lucca* (Bertati; Venice 1797); *Gli umori contrarii* (Contrasting Moods) (Bertati; Venice 1798); *Il ritorno di Serse* (Ferrari; Naples 1816).

BIBL: C. Anguissola, *Geminiano Giacomelli e Sebastiano Nasolini: musicisti piacentini* (Piacenza, 1935).

**Nathan, Isaac** (Canterbury, 1790; *d* Sydney, NSW, 15 Jan. 1864). Australian composer of Polish-Jewish descent. Studied singing with Corri in London; appeared at Covent Garden and received some recognition as a song composer, especially for his settings of Byron's *Hebrew Melodies* (1815–19). He composed and produced two comic operas, *The Alcaid* and *The Illustrious Stranger*, but financial problems eventually forced him to emigrate to Australia. Here his many activities culminated in the composition of the first two Australian operas, *Merry Freaks in Troublous Times*, a comic opera on Charles II, and *Don John of Austria*, a Spanish historical opera. Both are in the style of contemporary English ballad operas.

SELECT WORKLIST: *The Alcaid* (Kenney; London 1824); *The Illustrious Stranger* (Kenney & Millingen; London 1827); *Merry Freaks in Troublous Times* (Nagel; comp. 1843, never fully staged); *Don John of Austria* (Montefiore; Sydney 1847).

BIBL: C. Mackerras, *The Hebrew Melodist* (Sydney, 1963).

**National Broadcasting Company.** Formed in New York 1949–50 with Samuel Chotzinoff as producer and Peter Herbert Adler as music director, the NBC opera department was launched with Weill's *Down in the Valley* Jan. 1950. It has given US premières of many operas, commissioned works including Menotti's *Amahl*, and given the première of Martinů's *The Marriage*. Toured 47 US cities 1956, 55 cities 1957–8. In

Oct. 1953 *Carmen* was the first opera to be televised in colour.

**Naumann, Johann Gottlieb** (*b* Blasewitz, 17 Apr. 1741; *d* Dresden, 23 Oct. 1801). German composer. Studied Dresden Kreuzschule, later with Tartini and Martini in Italy, where he also met Hasse. His first opera, *Il tesoro insidiato*, was given in Venice 1762; in 1764 he was charged with composing sacred music for the Saxon court, becoming Kapellmeister 1776. He also wrote several operas for Italy, including the dramma giocoso *Li creduti spiriti* and the opere serie *Achille in Sciro* and *Solimano*, as well as works for Dresden, notably *La clemenza di Tito*. In 1777 he travelled to Stockholm to assist King Gustav III in his plans for the reform of Swedish musical life; he returned to Dresden 1778 to produce *Elisa*, but had a second sojourn in Sweden 1782–6. In Stockholm he composed two operas, *Cora och Alonzo*, and *Gustaf Wasa*: the latter, based on a sketch by Gustav III, was regarded in the 19th cent. as the first Swedish national opera. In 1785–6 he performed a similar function in Copenhagen, where *Orpheus og Eurydike* was produced: based on Calzabigi's libretto for Gluck, this was the first large-scale Danish opera. Such was Naumann's success that he was offered a prestigious position at court; this he declined, returning instead to Dresden, where he became Oberhofkapellmeister. During his latter years he composed several operas, including *Protesilao*, written for the Berlin court in conjunction with Reichardt, and the enormously successful dramma giocoso *La dama soldato*. His last opera was *Aci e Galatea*.

Naumann was one of the last great exponents of the Neapolitan style in Germany, although in his opere serie the Baroque procedures of Hasse and Graun were gradually tempered with a certain adherence to a new Classical spirit of sensibility and elegance: in his comic works the influence of Galuppi, Paisiello, and Piccinni is very evident. Most of his operas for the Dresden court were appropriately Italianate in character, with an emphasis upon vocal virtuosity; however, the flexibility of Naumann's approach is best demonstrated by his Scandinavian operas, where he appears closer to the 'reform' style of Gluck, reflecting a French influence. Here the approach is more magisterial, even pompous, with extensive employment of chorus and ballet, and some use of motivic technique; a similar approach was adopted in his work for Berlin. His orchestration was particularly enterprising and his later works anticipate some of the characteristics of Romantic opera. Naumann's international career, particularly his seminal influence on the establishment of opera in Scandinavia, was of great significance, and gave prestige to Dresden, where his chief loyalties lay; he was the most important opera composer in the city between Hasse and Weber.

SELECT WORKLIST: *Il tesoro insidiato* (?; Venice 1762); *Li creduti spiriti* (Bertati, after Kurz; Venice 1764); *Achille in Sciro* (Metastasio; Palermo 1767); *La clemenza di Tito* (Metastasio; Dresden 1769); *Solimano* (?, after Migliavacci; Venice 1773); *Elisa* (Mazzolà; Dresden 1781); *Cora och Alonzo* (Alderbeth, after Marmontel; Stockholm 1782); *Gustaf Wasa* (Kellgren, after King Gustav III; Stockholm 1786); *Orpheus og Eurydike* (Biehl & Lindemann, after Calzabigi; Copenhagen 1786); *Medea in Colchide* (Filistri; Berlin 1788); *Protesilao* (Sertor; Berlin 1789; comp. with Reichardt); *La dama soldato* (Mazzolà; Dresden 1791); *Aci e Galatea* (Foppa; Dresden 1801).

BIBL: R. Engländer, *Johann Gottlieb Naumann als Opernkomponist (1741–1801)* (Leipzig, 1922, R/ 1970).

**Navarraise, La** (The Girl from Navarra). Opera in 2 acts by Massenet; text by Jules Claretie and Henri Cain, after the former's story *La cigarette*. Prod. London, CG, 20 June 1894, with E. Calvé, Alvarez, Plançon, Gilibert, cond. Flon; Paris, OC, 3 Oct. 1895, with E. Calvé, Jérôme, Bouvet, Mondaud Carbonne, Belhomme, cond. Danbé; New York, M, 11 Dec. 1895, with E. Calvé, Lubert, Castelmary, Plançon, cond. Bevignani.

Nr. Bilbao, Spain, *c*.1860. Anita (sop) is in love with Araquil (ten), but his father Remigio (bs) opposes the match as Anita has no dowry. In order to get some money, Anita decides to help General Garrido (bar), who is leading the royalist troops against the Carlist enemy Zuccaraga. When Araquil learns of this he follows Anita to Zuccaraga's camp, but is killed; Anita goes mad.

**Neapolitan opera.** See OPERA SERIA.

**Neblett, Carol** (*b* Modesto, CA, 1 Feb. 1946). US soprano. Studied with Vennard, Lehmann, and Bernac. Début New York City O 1969 (Musetta). New York City O from 1969; Chicago from 1975; Vienna 1976; London, CG, 1977; New York, M, from 1979; also Leningrad, Salzburg, Sydney, etc. Roles incl. Mozart's Countess, Senta, Musetta, Minnie, Marietta (*Die tote Stadt*), Louise, Thaïs. (R)

**Nebra, José de** (*b* Calatayud, bapt. 6 Jan. 1702; *d* Madrid, 11 July 1768). Spanish composer. Held posts as first organist (1724) and vice maestro di cappella (1751) at the royal chapel in Madrid. Like his superior *Corselli, was also active as an opera composer: in 1725–47 he composed *c*.20 operas, of which only the librettos survive. His most important work was *Antes que celos*.

SELECT WORKLIST: *Antes que celos* (Before jealousy) (Martínez; Príncipe 1747).

**Nedda.** Canio's wife (sop), lover of Silvio, in Leoncavallo's *Pagliacci*.

**Neefe, Christian Gottlob** (*b* Chemnitz, 5 Feb. 1748; *d* Dessau, 26 Jan. 1798). German conductor and composer. Studied Chemnitz, Leipzig with Hiller. In Leipzig he wrote his first Singspiels; then succeeded Hiller as conductor of Abel Seyler's company. When this failed financially, he joined Grossmann's company, and became court organist at Bonn (also teaching Beethoven). Music director Dessau 1796–7. He won a considerable following in his day for his Singspiels, especially *Adelheit von Veltheim* (1780), which is an early and influential treatment of a Turkish subject. He was much admired for his melodic gifts and neatness of characterization, and for the witty touches with which he was able to enliven Singspiel conventions, sometimes by slyly satirizing them. He did much to give the genre considerable dramatic life, sometimes with use of melodrama, preparing the way for a more continuously composed and wide-ranging manner. He made vocal scores of five of Mozart's operas. Married the soprano **Maria Zink** (1751–1821). Three of their daughters were singers: **Louise** (1779–1846), **Felice** (1782–1826), who sang at Dessau and Vienna, and **Margarethe** (1787–1808), who married the famous actor Ludwig Devrient (uncle to Karl Devrient, the first husband of Wilhelmine Schröder-Devrient). All three sisters appeared together in Lichtenstein's *Bathmendi* in 1798. A son, **Hermann** (1790–1854), was a scene-painter, who provided the décor for Schubert's *Die Zauberharfe*.

SELECT WORKLIST: *Die Apotheke* (Engel, after Goldoni; Berlin 1771); *Amors Guckkasten* (Cupid's Peepshow) (Michaelis; Leipzig 1772); *Die Einsprüche* (The Protests) (Michaelis; Leipzig 1772); *Adelheit von Veltheim* (Grossmann; Frankfurt 1780).

**Negri, Maria Caterina** (*fl.* 1720–44). Italian contralto. Studied Bologna with Pasi. Début possibly Modena, 1720. Prague 1724–7 at Count von Sporck's theatre; Italy 1727–33; London, for Handel, 1733–7; Lisbon 1740–1; Bologna 1744. Created several roles for Handel, including Polinesso (*Ariodante*), Bradamante (*Alcina*), Arsace (*Berenice*). Her sister **Maria Rosa Negri** (*b* c.1715; *d* Dresden, 4 Aug. 1760) was a mezzo-soprano. Dresden 1730–56, in operas by Hasse; London 1733 with her sister, also for Handel, who wrote Euterpe (*Parnasso in festa*) for her.

**Negrini, Carlo** (*b* Piacenza, 24 July 1826; *d* Naples, 14 Mar. 1865). Italian tenor. Studied Milan. Début Milan, S, 1847 (*I due Foscari*). Milan, S, 1850–3, 1858, 1861–3; London, CG, 1852; Naples, C, 1861–3; also Venice, Rome, Trieste, etc. Extremely popular in Italy. Pos-

sessed a ringing, baritonal voice, and had much success in Verdi and Donizetti, especially as Poliuto, and Rodolfo (*Luisa Miller*). Created Gabriele Adorno (*Boccanegra*), and his warhorse, Glauco (Petrella's *Jone*).

**Neher, Caspar** (*b* Augsburg, 11 Apr. 1897; *d* Vienna, 30 June 1962). German designer and librettist. Studied Munich with Pasetti, Vienna with Roller. Worked with Brecht, then 1924–6 Berlin with Reinhardt. Essen 1927 (*Idomeneo*, 1927). Designer at Berlin: Kroll under Klemperer 1924–8 (*Carmen, House of the Dead*), SO with Ebert 1931–3 (*Macbeth* 1931, *Ballo* 1932, cond. Busch). Wrote text and designed première of Weill's *Die Bürgschaft* (1932), and worked with Brecht and Weill on various productions. Frankfurt 1934–41. Designed *Macbeth* for Gly. 1938, *Ballo* (based on Berlin sets) for Gly. and Edinburgh, 1949. Wrote four texts for Wagner-Régeny, incl. *Persische Episode* with Brecht (1949–50, prod. 1963). *Wozzeck* London, CG, 1952. Salzburg 1947–62, Milan, Hamburg, Vienna, Berlin, Stuttgart, Cologne. New York, M, 1959 (*Macbeth*). Brecht called him the greatest set designer of the day; though historically aware, including of his Baroque forebears, his designs were always planned as a functional and contemporary interpretation of the work.

WRITINGS: *Caspar Neher: Zeugnisse seiner Zeitgenossen* (Cologne, 1960).

BIBL: G. von Einem and S. Melchinger, eds., *Caspar Neher* (Hanover, 1966).

**Neiding.** In Wagner's *Die Walküre*, one of Hunding's tribe, enemies of the Wölfing tribe to which Siegmund and Sieglinde belong. From Middle High German *nît* = hatred, modern *Neid* = envy.

**Neidlinger, Gustav** (*b* Mainz, 21 Mar. 1912; *d* Bad Ems, 26 Dec. 1991). German bass-baritone. Studied Frankfurt with Rottsieper. Début Mainz 1931. Hamburg 1936–50; Bayreuth 1952–75; London, CG, 1963, 1965; New York, M, 1972; Vienna; Paris, O; Milan, S; etc. Roles included Pizarro, Lysiart, Amfortas, Barak, and an outstanding Alberich. (R)

**Nelepp, Georgy** (*b* Bobruiky, 20 Apr. 1904; *d* Moscow, 18 June 1957). Ukrainian tenor. Studied Leningrad with Tomars. Leningrad 1929–44; Moscow, B, 1941, 1944–57. Full of nervous energy, and possessing a powerful, intense tone, he was considered the most impressive Hermann since the role's creator, Figner; also sang Sadko, Golitsin, Dmitry, Florestan, Don José, Radamès, Manrico, and created Grigory Melekhov (*Quiet Don*), Kakhovsky (*The Decembrists*).

**Nelusko.** The slave (bar), in love with his fellow slave Selika, in Meyerbeer's *L'Africaine*.

**Németh, Maria** (*b* Körmend, 13 Mar. 1897; *d* Vienna, 28 Dec. 1967). Hungarian soprano. Studied Budapest with Anthes, Naples with De Lucia. Début Budapest 1923 (Sulamith, *Die Königin von Saba*). Vienna, S, 1924–46; London, CG, 1931; Milan, S; Berlin; Paris; etc. Her considerable abilities encompassed Constanze, Queen of the Night, Donna Anna, Amelia (*Ballo*), Turandot, Brünnhilde (*Siegfried*). (R)

**'Nemico della patria'.** Gérard's (bar) monologue in Act III of Giordano's *Andrea Chenier*, accusing Chénier of being an enemy of the Revolution.

**Nemirovich-Danchenko, Vladimir** (*b* Ozurgety, 23 Dec. 1858; *d* Moscow, 25 Apr. 1943). Georgian-Armenian producer. Studied Moscow. Together with *Stanislavsky, he founded the Moscow Arts T in 1898. Its revolutionary methods and choice of works influenced the staging of opera in Russia and eventually abroad. In 1919 he founded a musical studio, which became in 1926 the Nemirovich-Danchenko Musical T. Bolder in his opera productions even than in the straight theatre, he was one of the first producers to conceive of the 'singing actor' (his phrase). He resisted the star system and developed abstract productions in which acting, movement, settings, and lighting were subject to the music (which, however, he would sometimes radically rearrange). Though admired for his productions of operetta, e.g. of Offenbach and Lecocq, it was in his avant-garde productions that his work was at its most characteristic, e.g. a very popular version of *Carmen* known as *Carmencita and the Soldier* (1924), Knipper's *North Wind* (1930), and the Moscow version in 1934 of Shostakovich's *The Lady Macbeth of Mtsensk*. He also staged a famous production of *La traviata* (1934).

WRITINGS: *Iz proshlovo* (Moscow, 1936; trans. as *My Life in the Russian Theatre* (London, 1936)).

**Nemorino.** A young peasant (ten), lover of Adina, in Donizetti's *L'elisir d'amore*.

**Neri, Giulio** (*b* Siena 21 May 1909; *d* Rome, 21 Apr. 1958). Italian bass. Studied Florence with Ferraresi. Début Rome, T delle Quattro Fontane, 1935. Rome, R, 1937–58. London, CG, 1953; Milan, S; Munich; Buenos Aires, C; etc. Large repertory included Verdi and Wagner; famous as Grand Inquisitor, Rossini's Don Basilio, Boito's Mefistofele. (R)

**Nero.** Roman emperor (AD 37–68, reigned AD 54–68). Operas on him are by Strungk (1693), Pallavicino (1679), Handel (1705, lost), Orlandini (1708 and 1721), Vivaldi and others (1715, lost), Duni (1735), Szabo (1878), and Rubinstein (1879). Also operas by Boito and Mascagni; see below. He also appears in Monteverdi's *L'incoronazione di Poppea*. SEE CLAQUE.

**Nerone** (Nero). Opera in 4 acts by Boito; text by the composer. Music unfin. at composer's death and completed by Tommasini and Toscanini. Prod. Milan, S, 1 May 1924, with Raisa, Pertile, Galeffi, Journet, Pinza, cond. Toscanini; New York, Carnegie H (concert), 12 Apr. 1982, cond. Queler; London, Logan H (concert), 26 Mar. 1985, cond. Shelley.

The opera is a series of pictures of imperial Rome at the time of Nero (ten), in which the decadent civilization, corrupted by oriental influences represented by Simon Mago (bar) and Asteria (sop), is contrasted with the new world of Christianity, symbolized by Fanuel (bar) and Rubria (mez). The opera includes the scene of the burning of Rome.

Also opera in 3 acts by Mascagni; text by Targioni-Tozzetti, after Pietro Cossa's comedy (1872). Prod. Milan, S, 16 Jan. 1935, with Rasa, Carosio, Pertile, Granforte, cond. Mascagni.

**Nessler, Victor** (*b* Baldenheim, 28 Jan. 1841; *d* Strasburg, 28 May 1890). Alsatian composer. Studied theology in Strasburg, also music with Stern, but after the success of *Fleurette* he went to Leipzig to study with Hauptmann. Chorusmaster Leipzig 1870, later conductor Carlos T. He won popularity, and the means to give up conducting, with *Der Rattenfänger von Hameln*, though his most enduring success came with *Der Trompeter von Säckingen*. His operas reveal a gently nostalgic German Romanticism, and draw on old German tales and legends for their tuneful and highly effective scores.

SELECT WORKLIST: *Fleurette* (Feberel; Strasburg 1864); *Der Rattenfänger von Hameln* (The Pied Piper of Hamelin) (Hofmann, after Wolff; Leipzig 1879); *Der Trompeter von Säckingen* (Bunge, after Scheffel; Leipzig 1884).

**'Nessun dorma'.** Calaf's (ten) aria in Act III of Puccini's *Turandot*, in which he reflects that he alone can reveal his name, while Peking is searched all night to discover someone who can tell Turandot the secret.

**Nesterenko, Evgeny** (*b* Moscow, 8 Jan. 1938). Russian bass. Studied Leningrad with Lukanin. Début Leningrad, Maly T, 1963 (Gremin). Moscow, B, from 1971; Milan, S, from 1973; New York, M, 1975; London, CG, 1978. Possesses an expressive, rich voice, and a commanding presence; successful as Philip, Attila, Boris, Ruslan, Kutuzov, among many other roles. (R)

**Nestroy, Johann Nepomuk von** (*b* Vienna, 7 Dec. 1801; *d* Graz, 25 May 1862). Austrian playwright, actor, producer, and singer. Abandoned law studies for singing; appeared aged 17 as bass in Handel's *Alexander's Feast*. Stage début Vienna, K, aged 20 (Sarastro). Moved to

Amsterdam 1823, extending his repertory to include comic roles, in which he appeared in various cities. Settled Vienna 1831. A very popular actor-dramatist: his range included figures of the Viennese popular theatre (e.g. Kasperl) and roles in local and French operetta, though he was most associated with parts in his own works: he played 880 different roles. He wrote a series of popular opera parodies, incl. of Rossini, Meyerbeer, and Wagner (*Tannhäuser* 1857, *Lohengrin* 1859). He included songs in his plays, almost always for himself; his favourite composer was Adolf Müller (who wrote music for 41 of his plays). His name is forever associated with the witty, gay, irreverent, Viennese play that was almost operetta.

BIBL: W. Yates, *Nestroy* (Cambridge, 1972).

**Netherlands.** See HOLLAND.

**NET Opera.** See TELEVISION OPERA.

**Neues Deutsches Theater.** See PRAGUE.

**Neues vom Tage** (News of the Day). Opera in 3 acts by Hindemith; text by Schiffer. Prod. Berlin, Kroll, 8 June 1929, with Stückgold, Kalter, Cavara, Krenn, Ernster, cond. Klemperer; rev. version by Hindemith, prod. Naples 1954 and Cologne 1965; Santa Fe 12 Aug. 1961, with Willauer, Bonazzi, Driscoll, cond. Hindemith.

A modern city, present-day. A newly married couple, Edward (bar) and Laura (sop), decide to divorce after a fierce quarrel. Legal proceedings are so slow that they go to an agency to speed matters along. However, they are so irritated by the agency's methods that they are reconciled; but they are persuaded to act out their story in the agency's theatre, and are thus exploited by the gutter press.

**Neuinszenierung** (Ger.). New production.

**Neumann, Angelo** (*b* Vienna, 18 Aug. 1838; *d* Prague, 20 Dec. 1910). Austrian baritone and impresario. Studied Vienna with Stilke-Sessi. Début Cracow 1859. Vienna, H, 1862–76, when he gave up singing, owing to a heart condition. Roles included Don Giovanni; also sang in *Tannhäuser* and *Lohengrin* when produced by Wagner, whose dramatic talent made an indelible impression on him. Director, Leipzig O, 1876–82, opening with an unexpectedly successful *Lohengrin*; also gave other Wagner operas, including in 1878 the first complete *Ring* outside Bayreuth. After putting on another triumphant *Ring* in Berlin, 1881, and given exclusive rights until 1889 by Wagner, he formed his own company and toured with the *Ring* throughout Europe, 1882–3. Also gave it in St Petersburg and Moscow 1889. Director Bremen O 1882–5; Prague, Landestheater, 1885–1910. Wagner appreciated the indefatigable promoter of his work: 'You have done great

things for me, and procured me an income on which I would never have counted'. (1881).

BIBL: A. Neumann, *Erinnerungen an R. Wagner* (Leipzig, 1907, trans. 1908).

**Neumann, František** (*b* Přerov, 16 July 1874; *d* Brno, 25 Feb. 1929). Czech conductor and composer. Studied Leipzig. Prague 1906–9; Brno 1919–29. Conducted premières of Janáček's *Šárka*, *Káťa Kabanová*, *Cunning Little Vixen*, and *Makropoulos Case*, and Brno première of *The Excursions of Mr Brouček*. His compositions include eight operas, among them *Beatrice Caracci* (Brno 1922).

**Neumann, Václav** (*b* Prague, 29 Sep. 1920). Czech conductor. Studied Prague with Dědeček and Doležil. After many concert engagements, engaged by Felsenstein for Berlin, K, 1956–64, where he conducted with great success *The Cunning Little Vixen* 1956. Leipzig 1964–8. Music director Stuttgart 1969–73. A sensitive, very musical conductor with a great care for dramatic truth.

**Nevada** (orig. Wixom), **Emma** (*b* Alpha, CA, 7 Feb. 1859; *d* Liverpool, 20 June 1940). US soprano. Mother of below. Studied Vienna with Marchesi. Début London, HM, 1880 (Amina). Milan, S, 1881; Paris, OC, 1883–4; New York, AM, 1884–5; London, CG, 1887; Berlin, H, 1907, when she retired. An excellent Bellini singer; her medallion is alongside Pasta's and Malibran's on the composer's statue in Catania. Also sang Lucia, Mignon, Mireille.

**Nevada, Mignon** (*b* Paris, 14 Aug. 1886; *d* Long Melford, 25 June 1971). English soprano, daughter of above, with whom she studied. Début Rome, T Costanzi, 1907 (Rosina). London, CG, 1910, 1922; Milan, S, 1923; Paris, OC 1920, O 1932. Successful as Zerlina, Marguerite, Lakmé, Mimì; her Desdemona (Verdi) was much praised by Beecham. (R)

**Nevers.** The Comte de Nevers (bar), a Catholic nobleman in Meyerbeer's *Les Huguenots*. Also the title of Adolar (Graf von Nevers, ten) in Weber's *Euryanthe*.

**Neway, Patricia** (*b* Brooklyn, 30 Sep. 1919). US soprano. Studied New York with Gesell. Début Chautauqua 1946 (Fiordiligi). New York City O, 1948; Paris, OC, 1952–4; Aix 1952. Created several parts, e.g. Menotti's Magda Sorel (*The Consul*) and title-role of *Maria Golovin*. Also sang Iphigénie (*Tauride*), Tosca, Marie (*La favorite*). (R)

**Newman, Ernest** (*b* Everton, 30 Nov. 1868; *d* Tadworth, 6 July 1959). English author and critic. While still in commerce studied music and philosophy, writing his first books. Midland

Institute of Music, Birmingham, 1903–5; critic *Manchester Guardian* 1905–19; *Observer* 1919–20; *Sunday Times* 1920–58. Newman's first Wagner book, *A Study of Wagner*, is a sensitive but unchallenging response; and his finest work was expository, as in his thorough, elucidatory *Wagner Nights*, describing each mature opera, or biographical, as in his magisterial 4-volume life. Despite later amplification and correction, especially in the light of archival sources not easily available to him, this remains an authoritative and highly readable work based on clear historical perceptions. He never wrote the major critical study of Wagner for which his intelligence and acumen would also have equipped him, though he used his high standing as a journalist critic to maintain constant public interest. His study of Liszt, a by-product of his work on Wagner, is a piece of hostile pleading. He also translated most of Wagner's operas.

WRITINGS: *Gluck and the Opera* (London, 1895, R/1964, 1977); *A Study of Wagner* (London, 1899, R/1974); *Richard Strauss* (London, 1908); *Wagner as Man and Artist* (London, 1914, 2/1924); *The Life of Richard Wagner* (4 vols., London, 1933–47, R/1976); *The Man Liszt* (London, 1934, R/1971); *Opera Nights* (London, 1943); *Wagner Nights* (London, 1949, R/1977); *More Opera Nights* (London, 1954).

BIBL: V. Newman, *Ernest Newman: A Memoir by his Wife* (London, 1963).

**New Opera Company.** Founded in Cambridge in 1956, with Leon Lovett as music director, to stimulate interest in contemporary opera by promoting premières of British works and British premières of foreign works. In 1960–3 and from 1973 the company worked closely with Sadler's Wells, London. Premières include Benjamin's *A Tale of Two Cities* (1957) and *Tartuffe* (1964), Joubert's *In the Drought* (1959) and *Under Western Eyes* (1969), Crosse's *Purgatory* (1966), Musgrave's *The Decision* (1967), and Lutyens's *Time Off? Not a Ghost of a Chance* (1972) and *Infidelio* (1973). British premières included Egk's *Der Revisor* (1958), Dallapiccola's *Il prigioniero* (1959), Orff's *Die Kluge* (1959), Schoenberg's *Erwartung* (1960), Henze's *Boulevard Solitude* (1962), Prokofiev's *Fiery Angel* (1965), Hindemith's *Cardillac* (1970), Shostakovich's *The Nose* (1973), Goehr's *Arden Must Die* (1974), Szymanowski's *King Roger* (1975), Ginastera's *Bomarzo* (1976), and Martinů's *Julietta* (1978). Closed 1984.

**New Orleans.** City in Louisiana, USA. Capital of French colonial Louisiana until 1762, in Spanish possession until 1800, under American control from 1803. The T St Pierre, built by the Henry brothers, opened 4 Oct. 1792, and gave the first recorded opera in the city, Grétry's *Sylvain*, 22 May 1796. The T St Philippe opened 30 Jan.

1808 with Méhul's *Une folie*. The first T d'Orléans opened 1815, but burnt down 1816; the second was opened by John Davis 27 Nov. 1819, with Boieldieu's *Jean de Paris*, and remained the city's leading opera-house for 40 years, with annual seasons usually from October to early spring. The auditorium included *loges grillées* (latticed boxes) for those in mourning or for other reasons not wishing to be seen. The repertory initially comprised Grétry, Méhul, Isouard, and Dalayrac, but later expanded to include important works by Cherubini and Rossini. From 1827 Davis also toured with works that were to be local premières in New York, Philadelphia, and elsewhere. James Caldwell opened the rival Camp St T (cap. 1,100) on 1 Jan. 1824, with a repertory including opera; he then opened the lavish St Charles T (cap. 4,100) on 30 Nov. 1835, in further rivalry with Davis, importing Italian companies from Havana with US premières of works by Rossini, Bellini, and Donizetti, 1836–42. The T d'Orléans countered, appealing to Creole audiences wth French works by Adam, Auber, Halévy, and Meyerbeer, some of them also US premières. The St Charles T burnt down 1842, but reopened on a more modest scale 1843, surviving until another fire 1899. Caldwell also managed the American T (opened 1840, rebuilt after fire 1842) until 1855; the mixed repertory included opera. Guest artists included Sontag and Frezzolini.

On 1 Dec. 1859 Charles Boudousquié opened the French OH (cap. 1,800) with *Guillaume Tell*; this was superior to any US theatre of its time. In its second season the 17-year-old Patti appeared as Lucia, Gilda, Rosina, Valentine, Dinorah, Lady Harriet, and Leonora (*Trovatore*). Activity declined during the Civil War, and an attempt to establish a resident company failed when the ship carrying the singers and musicians was lost in a hurricane. However, by the 1870s there was a rich and formal operatic life, supplemented by visits from travelling companies. During the 19th cent. over 170 operas received their US premières in New Orleans (more than in any American city except New York), including 40 works by Bellini, Donizetti, Rossini, Spontini, Halévy, Auber, and Meyerbeer. The French OH, however, declined, closed 1913, was storm-damaged 1915, reopened 11 Nov. 1919, but burnt down 4 Dec. For over two decades the city was then dependent on visiting companies, including a Russian troupe. On 11 June 1943 the New Orleans O Association, founded by Walter Loubat, gave its first performance in City Park, moving to the unsuitable Municipal Auditorium (cap. 2,500) in the autumn to inaugurate 4- to 8-week autumn–winter seasons. Walter Herbert was Music Director 1943–54, Renato Cellini

1954–64, Knud Andersson 1964–70, Arthur Cosenza from 1970. In 1973 the Association began playing in the T for the Performing Arts (cap. 2,317).

**New Year.** Opera in three acts by Tippett; text by composer. Prod. Houston, Cullen T, 27 Oct. 1989, with Field, Kazaras, St Hill, Manager, Schiappa, Maddalena, Shaulis, cond. DeMain; Gly. 1 July 1990, with Field, Langridge, St Hill, Manager, Robson, Maddalena, Shaulis, cond. Andrew Davis.

The Presenter (ten) introduces the world of Somewhere and Today. Jo Ann (sop) is afraid of leaving her room to help the children of Terror Town; she is mocked by her delinquent foster-brother Donny (bar) and warned not to indulge him by their foster-mother Nan (mez). In the laboratory of Merlin (bar) in Nowhere and Tomorrow, Pelegrin (ten) is fascinated by the image of Jo Ann retrieved on a screen, but their boss Regan (sop), concerned with their voyage into the future, has it removed. However, Pelegrin leaves in his space ship to seek Jo Ann. When he finds her, she believes she may discover courage.

During New Year's Eve celebrations, Donny is symbolically beaten as representing the bad Old Year. Merlin's space ship arrives, Regan having been tricked into travelling into the past. When Donny tries to board the space ship, it takes off; he is rescued by Jo Ann from a real beating as New Year strikes.

In Jo Ann's room, Nan arrives to take Donny into her care. Pelegrin arrives to take Jo Ann away, and in a ritual exchange she tastes forgetfulness and memory, choosing the latter; they declare their love. Back in her room, he gives her a rose. Eventually Merlin, Regan, and Pelegrin return to the future; Jo Ann is able to leave her room.

**New York.** City in USA. Ballad operas were given during the early 18th cent. The first musical theatre was opened 1752, followed by the John St. T 1767. An early American opera, *Tammany*, by James Hewitt (1770–1827) was produced here 1794, followed in 1796 by *The Archers* by Benjamin Carr (1768–1831), the first American opera of which parts survive. The Park T opened 1798, and here García's company gave the first US *Don Giovanni*, 1826, and the first Italian opera in the US, *Barbiere*, 1828. Other New York theatres in which opera was given in the 19th cent. included the Italian OH (1833–9), the Astor Place OH (1847–57), the Broadway T, Richmond Hill T, Castle Gardens, Palmo's OH, and the Stadt T. See also ACADEMY OF MUSIC, CITY CENTER, DAMROSCH OPERA COMPANY, LEXINGTON THEATRE, LINCOLN CENTER, MANHATTAN OPERA COMPANY, METROPOLITAN OPERA, NATIONAL BROADCASTING COMPANY, NEW YORK CITY OPERA, NIBLO'S GARDEN.

**New York City Opera.** A company founded in Feb. 1944 by the city's Mayor, Fiorello La Guardia, and by Newbold Morris, originally housed at the City Center. The aim was 'to present opera with the highest artistic standards, while maintaining the civic and democratic ideas of moderate prices, practical business planning and modern methods.' Lászlo Hálász was music director 1944–51, but was then summarily dismissed; he was succeeded by Joseph Rosenstock, who resigned in 1956, then Erich Leinsdorf, Julius Rudel 1957–79, Beverly Sills 1979–89, and Christopher Keene from 1989. The first seasons were short, with a popular repertory, but these were later extended, eventually to two or three months.

The company moved to the State T, Lincoln Center, in 1966, opening with Ginastera's *Don Rodrigo*. It is a rival to the Metropolitan, in that it offers an interesting repertory of classical and contemporary works, often in controversial productions; it has also welcomed distinguished guests, and given an early chance to young US artists. Premières include Still's *The Troubled Island* (1949), Tamkin's *The Dybbuk* (1951), Copland's *The Tender Land* (1954), Kurka's *The Good Soldier Schweik* (1958), Weisgall's *Six Characters in Search of an Author* (1959), Floyd's *The Passion of Jonathan Wade* (1962), and Menotti's *The Most Important Man* (1971); it has also given many important foreign works, and continues to stage enterprising and original productions.

BIBL: M. Sokol, *The New York City Opera* (New York, 1982).

**New Zealand.** Opera was first heard in 1864 (Lyster's company), and the country was long dependent on touring companies, including Pollard's Lilliputian OC with children and teenagers. The New Zealand OC was founded as a touring company in 1954 by Donald Munro, with James Robertson as artistic director. After some short works (in English, with local choruses), the company gave its first full-length production, *The Consul*, in 1957. Other productions followed, with the company benefiting from New Zealanders making international careers but returning to sing in productions. It was forced to close in 1972. The National O of New Zealand opened in 1979, visiting Auckland, Christchurch, and Wellington, again under Robertson's musical direction, but financial problems in 1982 caused it to be effectively taken over by the Mercury T in 1984. However, in 1985 companies were formed in Dunedin, Wellington, and Canterbury, then some smaller ones including O Waikato, Perkel O, and O Boutique, leading to a healthy growth of regional opera. The International Festival of the Arts in Wellington, founded in 1987, included a successful *Meistersinger* in 1990. Works by New Zealand composers have included *Plague upon Eyam* (1985) and *The Birds* (1986) by John Drummond and *Green Leaf* (1985) by Dorothy Buchanan.

BIBL: A. Simpson, ed., *Opera in New Zealand* (Wellington, 1991).

**Nezhdanova, Antonina** (*b* Krivaya Balka, 16 June 1873; *d* Moscow, 26 June 1950). Ukrainian soprano. Studied Moscow with Masetti. Début Moscow, B, 1902 (Antonida, *Life for the Tsar*). Leading soprano at Moscow, B, up to and after the Revolution; taught from 1936. Only stage appearance abroad, Paris 1912 (as Gilda, with Caruso and Ruffo). Repertory included Queen of the Night, Snow Maiden, Queen of Shemakha, Gilda, Violetta, Tatyana, Lisa, Elsa, Tosca, Lakmé. Her singing of Rimsky-Korsakov's roles, in particular, helped to form a tradition of their interpretation. An outstanding artist, who combined a well-centred voice and brilliant technique with sensitive acting. Nemirovich-Danchenko, with whom she worked, praised her 'candid simplicity and profound nobility'. (R)

BIBL: G. Polyanovsky, *A. V. Nezhdanova* (Moscow, 1970).

**Niblo's Garden.** Originally a summer resort in New York, built by William Niblo at the corner of Broadway and Prince St. In 1828 Niblo built the Sans Souci T there, and it, and its two successors, later known as Niblo's Garden, were the scene of various opera seasons until 1895; these included Shireff's English Co. (1838), the New Orleans French Co. (1827–33), and the Havana Opera Co. (1848 and 1850). Important seasons were also given there in 1853 by a company that included Sontag, Badiali, and Pozzolini, and in 1858 by Strakosch's company, with Gazzaniga, Pauline Colson, and Marcel Junca. The theatre, which had a capacity of 1,700 with standing room for another 1,000, closed in 1895.

**Nice** (Ital., Nizza). City in Alpes-Maritimes, France. The wooden T Maccarani opened 1776 on the site of the present Opéra; renovated 1790 as T Royal, then T de la Montagne 1792–1815. The T Royal opened 1830 (renamed Impérial 1860, Municipal 1871), and gave largely Italian seasons (French being excluded during the Piedmontese domination 1814–60). Burnt down 1881. Rebuilt as T Municipal 1885 (O de Nice from 1902) (cap. 1,230). Seasons were at first Italian, but a French company was formally established 1887, until 1950 singing almost exclusively in French. This was a distinguished period in the opera's history, with many famous singers appearing as guests with a strong company that also introduced an impressive number of new works, including Nouguès's *Quo vadis?* (1909), as well as many works new to France. Opera was also given in other theatres, including the Cercle de la Méditerranée, Casino Municipal (first French *Otello* (Verdi) with Tamagno), and Casino de la Jetée-Promenade (première of Falla's *La vida breve*, 1913). Ferdinand Aymé was director 1950–82, and did much to re-establish the company after the war, also attracting international singers including Bergonzi, Corelli, and especially Caballé. He was succeeded in 1982 by Lucien Salles, who has developed a more innovatory policy towards production. Opera has also been given from 1985 in the Apollon auditorium (cap. 2,500).

**Nicholls, Agnes** (*b* Cheltenham, 14 July 1877; *d* London, 21 Sep. 1959). English soprano. Studied London with Visetti. Début London, L, 1895 (Dido). London, CG, 1901–24; also sang with Denhof and Beecham Companies, and BNOC. Roles included Donna Elvira, Venus, Sieglinde, Brünnhilde (*Siegfried*). Married the conductor Hamilton Harty (1879–1941). (R)

**Nicklausse.** Hoffmann's friend (mez) in Offenbach's *Les contes d'Hoffmann*.

**Nick Shadow.** The Devil (bar) in Stravinsky's *The Rake's Progress*.

**Nicolai, Otto** (*b* Königsberg, 9 June 1810; *d* Berlin, 11 May 1849). German composer and conductor. After running away from home, studied Berlin with Zelter. While holding various posts in Italy, he interested himself in opera composition and worked in Vienna at the Hoftheater before returning to Italy. Here he had some success with his first operas. Back in Vienna, he conducted from 1841 at the Hofoper, where he was the first to introduce Beethoven's *Leonore No. 3* overture as an entr'acte in *Fidelio*. When what was to prove his most successful work, *Die lustigen Weiber von Windsor*, was refused by Berlin as director of the Royal Opera, he moved to Berlin as director of the Royal Opera.

Nicolai's most famous opera has survived competition even from Verdi's *Falstaff*, on the same subject, to remain a staple item in German repertories and an occasional one in those of other countries. In it the German tradition of comic opera is fertilized by the grace and fluency Nicolai had learnt in Italy. An intelligent and literate man, concerned to make a contribution to German Romantic opera, he here brought together qualities which included a fine feeling for the voice and a sensitive command of the orchestra. He was greatly admired as a conductor.

# Nicolini

WORKLIST: *Enrico II* (Romani; Trieste 1839); *Il templario* (Marini, after Scott; Turin 1840; rev. as *Der Tempelritter*, Vienna 1845); *Gildippe ed Odoardo* (Solera; Genoa 1840, lost); *Il proscritto* (The Exile) (Rossi; Milan 1841, rev. as *Die Heimkehr des Verbannten*, Vienna 1846); *Die lustigen Weiber von Windsor* (The Merry Wives of Windsor) (Mosenthal, after Shakespeare; Berlin 1849).

BIBL: G. Kruse, *Otto Nicolai: ein Kunstlerleben* (Berlin, 1911).

**Nicolini** (Grimaldi, Nicolò) (*b* Naples, bapt. 5 Apr. 1673 ; *d* Naples, 1 Jan. 1732). Italian mezzo-soprano castrato. Studied Naples with Provenzale. Début Naples 1685 (Page, Provenzale's *Stellidaura vendicata*). Sang at San Bartolomeo T and at royal palace regularly between 1697 and 1724 in operas by Lotti, A. Scarlatti, Porpora, etc.; Venice regularly 1700–31; also Rome, Bologna. London, H, 1708–12, 1715–17, in works by Handel, Bononcini, Gasparini. Created title-roles of Handel's *Rinaldo* and *Amadigi*. A remarkable singer and actor, with a very flexible voice and excellent vocal control, he was described by Addison as 'the greatest performer in dramatic music that is now living'.

**Nicolini, Ernest.** See PATTI, ADELINA.

**Nicolò.** See ISOUARD, NICOLÒ.

**Nielsen, Carl** (*b* Sortelung, 9 June 1865; *d* Copenhagen, 3 Oct. 1931). Danish composer. Studied Copenhagen with Gade; worked as an orchestral violinist before making his reputation as a composer of symphonic music. In 1908 became conductor of the Royal O, resigning 1914. By the time of his death he was the most respected Danish composer of his generation.

His first opera, *Saul og David*, took a decisively modernist approach and was initially controversial: in it the legacy of his symphonic apprenticeship is clearly apparent. Nielsen's only other completed opera, *Maskarade*, was quite different in style: a light-hearted, whimsical comedy, it has won a position comparable to that of *Halka* or *The Bartered Bride* as a national opera. Another comic opera was left unfinished at his death.

WORKLIST: *Saul og David* (Christiansen; Copenhagen 1902); *Maskarade* (Andersen, after Holberg; Copenhagen 1906).

**Niemann, Albert** (*b* Erxleben, 15 Jan. 1831; *d* Berlin, 13 Jan. 1917). German tenor. Studied Dessau with Nusch, later Paris with Duprez. Début Dessau (1st Captain, *Le prophète*) 1849. Halle 1852; Hanover 1854–66; Berlin, H, 1866–88, when he retired. Paris, O, 1861; Munich, 1864; London, HM, 1882; New York, M, 1886–8. Possessed a huge physique and voice, and great powers of expression. Roles included Florestan, Gounod's Faust, Manrico, Rienzi, Lohengrin, and Tannhäuser in the 1861 Paris

fiasco, when he caused Wagner much trouble with his intransigent arrogance. Later (though still causing trouble), he was highly successful as Siegmund in the first *Ring*; also as Walther, Tristan, and Siegfried. Lilli Lehmann wrote in 1920, 'This Siegmund was unique; it will no more return than another Wagner will'; while his Tristan was 'the most sublime thing ever achieved in music drama'.

BIBL: E. Newman, *The Life of Richard Wagner* (London, 1933–47).

**Nietzsche, Friedrich Wilhelm** (*b* Röcken, 15 Oct. 1844; *d* Weimar, 25 Aug. 1900). German philosopher, poet, and amateur composer. His interest in music expressed itself chiefly in his writings on Wagner, whom he met in 1868 and who profoundly affected *Die Geburt der Tragödie aus dem Geiste der Musik* (The Birth of Tragedy from the Spirit of Music, 1870–1). In this seminal work he set forward the distinction between *Apollo and *Dionysus, the Apollonian principle representing order, balance, and reason as opposed to Dionysian irrationality, intoxication, and creative danger; to this he added a discussion of Wagner as exemplifying the new birth of tragedy. In the essay *Richard Wagner in Bayreuth* (1875–6) he further states, even overstates, the case for Wagner. This coincided with the first Bayreuth Festival, by when Nietzsche was suffering doubts about Wagner. These were first obliquely expressed in *Menschliches, Allzumenschliches* (Human, All Too Human, 1876–9), and were to come to a head in the frontal attack *Der Fall Wagner* (The Case of Wagner, 1888). He had meanwhile explained his change of attitude (which was coloured by strong personal revulsion) at the end of *Die fröhliche Wissenschaft* (The Joyous Science, 1882–7), where he claimed to have mistakenly identified Wagner's Romanticism with Dionysian values. To the 'sickness' of Wagner he now opposed the clarity and light of *Carmen* (privately admitting that this was partly ironic), demanding the 'Mediterraneanization' of music so that it should embody the truly Dionysian. Yet though he had in 1876 resisted Wagner's exposition to him of the Christian motifs and symbolism of *Parsifal*, he was still able in 1887 to regard the *Parsifal* prelude as Wagner's finest achievement. To 1888 (by when he was on the edge of the insanity which finally claimed him) also belong his autobiography *Ecce Homo* and *Nietzsche contra Wagner*. He championed the minor composer Peter Gast (1854–1918), his friend and secretary, whom he saw as 'a new Mozart'. He was self-taught musically, and a good enough pianist to annoy Wagner by playing Brahms well; his youthful compositions are fluent but unimportant.

**Nightingale, The** (Russ.: *Solovey*; Fr.: *Le rossignol*). Opera in 3 acts by Stravinsky; text by the composer and Stepan Mitusov, after Hans Andersen's story. Prod. Paris, O, 26 May 1914, with Dobrovolska, Petrentko, Andreyev, Brian, Varfolomeyev, Gulayer, Belialin, cond. Monteux; London, DL, 18 June 1914, same cast, cond. Cooper; New York, M, 6 Mar. 1926, with Talley, Bourskaya, Errolle, Didur, cond. Serafin.

Fairyland. The famous nightingale (sop), whose song thrills all, sings to a poor fisherman (ten). The imperial court arrives in search of the bird, who agrees to sing for the Emperor (bar).

The nightingale refuses any reward save the pleasure of seeing tears of emotion in the imperial eyes. A mechanical nightingale arrives from the Emperor of Japan, but a comparison is prevented by the real bird's disappearance. The mechanical bird is installed at the royal bedside.

Death (con) sits by the Emperor's bed, but promises to return the crown when the nightingale returns and sings. This it does, and the Emperor's strength returns.

Also operas by Lebrun (1816), Gallas-Montubran (1959).

**Nightwatchman.** The watchman (bs) who goes the rounds in Nuremberg in Wagner's *Die Meistersinger von Nürnberg*.

**Nile Scene.** The name generally given to Act III of Verdi's *Aida*, which takes place by moonlight on the banks of the Nile.

**Nilsson, Birgit** (*b* West Karup, 17 May 1918). Swedish soprano. Studied Stockholm with Hislop and Sunnegaard. Début Stockholm 1946 (Agathe). Sang there regularly thereafter. Gly. 1951; Bayreuth 1954–70; London, CG, from 1957; Milan, S, from 1958; New York, M, from 1959. Also Vienna, Rome, Munich, Chicago, etc. Retired 1984. One of the greatest Wagner singers of the century, she possessed an unusually ample, pure-toned, and effortless voice, and a powerful stage presence. Repertory included Leonore, Lady Macbeth, Aida, Elsa, Sieglinde, Dyer's Wife, Salome, Marschallin, Turandot; a famous Isolde, Brünnhilde, Elektra. (R)

**Nilsson, Christine** (*b* Sjöabol, 20 Aug. 1843; *d* Stockholm, 22 Nov. 1921). Swedish soprano. Studied Stockholm with Berwald, Paris with Wartel and Delle Sedie. Début Paris, L, 1864 (Violetta). Paris, L until 1867, and O 1868. London: HM, 1867, and then at CG and DL until 1882; New York, AM, 1871; also Moscow, St Petersburg, Brussels, Munich, etc. New York, M, at opening in 1883. Immensely popular for her beauty, charm, and vocal brilliance; her voice, which was sweet-toned and crystal-clear, extended to f'''. Was to have sung Catherine in the first *Jolie fille de Perth*, but broke her contract to create Ophélie (*Hamlet*), one of her best roles. Also successful as Queen of the Night, Lucia, Marguerite, Mignon, Martha, Valentine, Elsa.

BIBL: M. Leche-Löfgren, *Kristina Nilsson* (Stockholm, 1944).

**Nina, ossia La pazza per amore** (Nina, or The Girl Mad through Love). Opera in 2 acts by Paisiello; text (in French, *Nina, ou La folle par amour*) by Benoît Joseph Marsollier de Vivetières for Dalayrac, trans. by Giuseppe Carpani with additions by Giovanni Battista Lorenzi. Prod. Caserta, Royal Palace, 25 June 1789; London 27 Apr. 1797.

Italy, 18th cent. Nina (sop) goes mad after her lover Lindoro (ten) is shot in a duel by his rival; her father, the Count (bs), had decreed that the rival should have her hand and is consequently full of remorse.

Lindoro recovers from his wounds but is distraught to learn of Nina's madness. She gradually recognizes him and regains her sanity, happy that her father will now permit her marriage to Lindoro.

**Nissen, Hans Hermann** (*b* Zippnow, 20 May 1893; *d* Munich, 28 Mar. 1980). German bass-baritone. Studied Berlin with Raatz-Brockmann. Début Berlin, V, 1924 (Caliph, *Barber of Baghdad*). Munich, Bayerische Staatsoper, 1925–67. London, CG, 1928, 1934; Paris, O, 1930; Milan, S, 1936–8; New York, M, 1938–9; Bayreuth 1943; also Chicago, Vienna, Salzburg. Roles included Renato, Barak, Orestes; outstanding as Wotan and Sachs. (R)

**'Niun mi tema'.** Otello's death scene in Act IV of Verdi's *Otello*, in which he warns no one to fear him if they see him armed, for he is embarking upon his last journey, his own death.

**Noble, Dennis** (*b* Bristol, 25 Sep. 1899; *d* Javea, 14 Mar. 1966). English baritone. Studied London with Gilly. Début London, CG, 1924 (Marullo, *Rigoletto*). London, CG, until 1938, and in 1947; also CR and BNOC. Cleveland, USA, 1935–6. Created several roles, e.g. Sam Weller (*Pickwick*, Albert Coates), and was a successful Verdi and Puccini singer. (R)

**'Nobles seigneurs, salut!'** The page Urbain's (sop) salutation in Act I of Meyerbeer's *Les Huguenots*, introducing his aria 'Une dame noble et sage' in which he brings a note from his mistress (Marguerite de Valois) to Raoul de Nangis.

**Noh Play.** See JAPAN.

**Noni, Alda** (*b* Trieste, 30 Apr. 1916). Italian soprano. Studied Trieste and Vienna. Début Ljubljana 1937 (Rosina). Vienna, S, 1942–5; London, Cambridge T, 1946; Gly. 1949–54. Also

Milan, S; Paris, OC; Berlin S; etc. An excellent soubrette with great comic gifts, successful as Blonde, Despina, Clorinda (*Cenerentola*), Nannetta, Zerbinetta. (R)

**'Non lo diro col labro'.** Alessandro's (sop cas) aria in Act I of Handel's *Tolomeo*, known in English as 'Did you not see my lady?'

**'Non mi dir'.** Donna Anna's (sop) aria in Act II of Mozart's *Don Giovanni*, in which she bids Don Ottavio speak no more of his hopes of marriage so soon after her father's death.

**Nono, Luigi** (*b* Venice, 29 Jan. 1924; *d* Venice, 8 May 1990). Italian composer. Studied Venice with Malipiero, later with Maderna and Scherchen. His first opera, *Intolleranza*, drew on various 20th-cent. techniques, including serialism, in an intelligent dramatic plea, but ran into an organized demonstration from a Fascist group, Ordine Nuovo, at a stormy première whose notoriety has overshadowed the quality of the work. After various music theatre pieces, he wrote *Al gran sole carico d'amore*, a large-scale celebration of the Paris Commune, and then an 'opera without action', *Prometeo*, a more intimate piece performed in a Venetian church, San Lorenzo.

WORKLIST: *Intolleranza 1960* (Nono, after various texts; Venice 1961, rev. in 1 act as *Intolleranza 1970*, Florence 1974); *Al gran sole carico d'amore* (To the Great Sun Charged with Love) (Nono; Milan 1975); *Prometeo* (Cacciari; Venice 1984).

**'No, no, Turiddu'.** Santuzza's (sop) duet in Mascagni's *Cavalleria rusticana*, in which she begs Turiddu (ten) not to follow his new love Lola into the church but to return to her.

**'Non più andrai'.** Figaro's (bar) aria in Act I of Mozart's *Nozze di Figaro*, describing to the reluctant Cherubino his impending translation from civilian to military life. The melody is used for the last of the three numbers played by Giovanni's private band at supper in the penultimate scene of *Don Giovanni*, in recognition of the tune's popularity: Leporello exclaims, 'Questa poi la conosco purtroppo' ('I know this one only too well').

**'Non più di fiori'.** Vitellia's (sop) aria in Act II of Mozart's *La clemenza di Tito*, in which she reflects that the god of marriage she had hoped to welcome must now yield to death. A rondo with obbligato basset-horn, probably composed for Josepha Duschek some weeks before the commission for the opera.

**'Non so più'.** Cherubino's (sop) aria in Act I of Mozart's *Le nozze di Figaro*, in which he declares his bewilderment at the novel excitement of desire.

**'No! Pazzo son! Guardate!'** Des Grieux's (ten) appeal in Act III of Puccini's *Manon Lescaut*, in which he beseeches the ship's captain to allow him to accompany Manon on her journey into exile.

**Nordica** (Norton), **Lillian** (*b* Farmington, ME, 12 Dec. 1857; *d* Batavia, 10 May 1914). US soprano. Studied Boston with O'Neill, Milan with Sangiovanni, later Paris with Sbriglia. Début Milan, T Manzoni, 1879 (Donna Elvira). St Petersburg 1880; New York, AM, with Mapleson Co., 1883. London: CG, 1887–93, 1898–9, 1902; DL, 1888–93. New York, M, 1891–1909 (intermittently); Bayreuth 1894 (first US singer there); Paris, O, 1910. At first merely a skilful singer, she later became an artist of considerable dramatic expression, aided by both Cosima Wagner's tutelage and her own determined application. She possessed a pure style, a velvety tone, and a technique and versatility that enabled her to sing Brünnhilde and Violetta on consecutive nights. Other roles included Cherubino, Donna Anna, Norina, Marguerite, Philine, Selika, Aida, Elsa, Kundry, and a particularly triumphant Isolde with the De Reszke brothers. (R)

BIBL: I. Glackens, *Yankee Diva* (New York, 1963).

**Norena, Eidé** (*b* Horten, 26 Apr. 1884; *d* Lausanne, 19 Nov. 1968). Norwegian soprano. Studied Oslo with Gulbranson, London with R. von zur Mühlen. Début Oslo 1907 (Amor, Gluck's *Orfeo*). Oslo 1908–18; Milan, S, 1924; London, CG, 1924–5, 1930–1, 1934, 1937; Paris, O, between 1924 and 1937; Chicago 1926–8; New York, M, 1933–8. An accomplished and tasteful singer, especially successful as Gilda, Violetta, Desdemona, Mathilde, and the three sopranos of *Les contes d'Hoffmann*. (R)

**Nørgård, Per** (*b* Copenhagen, 13 July 1932). Danish composer. Studied Copenhagen and Paris with Boulanger. In 1961 won a prize at the Gaudeamus Festival (Holland), which brought him to international attention; in 1958–62 was music critic for the newspaper *Politiken*; from 1965 taught composition at the Conservatory in Århus.

SELECT WORKLIST (all texts by Nørgård): *Dommen* (The Judges) (Copenhagen 1962); *Labyrinten* (Copenhagen 1967); *Gilgamesh* (Århus 1973).

**Nørholm, Ib** (*b* Copenhagen, 24 Jan. 1931). Danish composer. Studied Copenhagen and Darmstadt; music critic for the newspaper *Information* 1956–64; has held positions as organist and teacher in Denmark; from 1973 has taught at the Royal Cons. He has written a television opera, *Invitation til skafottet*, and a chamber opera, *Den unge Park*.

SELECT WORKLIST: *Invitation til skafottet* (Invitation to a Beheading) (Borum, after Nabokov; Danish television 1967); *Den unge Park* (The Young Park) (Christensen; Århus 1970).

**Norina.** A young widow (sop), lover of Ernesto, in Donizetti's *Don Pasquale*.

**Norma.** Opera in 2 acts by Bellini; text by Romani, after Soumet's tragedy (1831). Prod. Milan, S, 26 Dec. 1831, with Pasta, Grisi, Donzelli, Negrini; London, H, 20 June 1833, with Pasta, De Méric, Donzelli, Galli; New Orleans, St Charles T, 1 Apr. 1836, with Pedrotti, Montressor, Rosa, Marozzi.

Gaul, during the Roman occupation. The Roman proconsul Pollione (ten) has abandoned the Druid high priestess Norma (sop), by whom he has two sons, in favour of another priestess, Adalgisa (sop). Norma protects Pollione by advising against war and, in 'Casta Diva', calls on the goddess of the moon for peace. Norma learns that Adalgisa's lover is none other than Pollione.

Norma intends to kill her sons, but cannot bring herself to do so. Adalgisa goes to Pollione to try to persuade him to return to Norma, but fails. In a rage, Norma incites the Gauls to war against the Romans. Pollione is captured near the Druids' temple and is sentenced to death. In a final attempt to save Pollione, whom she still loves, Norma offers herself as a replacement for him and, having confided her children to her father's care, she mounts the pyre, where she is joined in death by Pollione.

**Norman, Jessye** (*b* Augusta, GA, 15 Sep. 1945). US soprano. Studied Washington with Grant, Michigan U with E. Mannion and Bernac. Début Berlin, D, 1969 (Elisabeth, *Tannhäuser*). Milan, S, and London, CG, from 1972; New York, M, from 1983; Paris, OC, 1984. Roles include Countess (Mozart), Dido (Purcell and Berlioz), Cassandre, Ariadne, Aida, Sieglinde, Selika. Possesses a large, opulent, and expressive voice, and a statuesque stage presence. (R)

**Norrington, Roger** (*b* Oxford, 16 Mar. 1934). Studied London with Boult. COG as tenor, then conductor (*Rake's Progress*). First music director *Kent Opera 1969–82, with notable effect, conducting over 300 performances of 30 works, many in striking productions by Jonathan *Miller. London, SW, then ENO 1979, 1987; Milan, PS, 1982; Venice, F, 1983; Florence 1984; London, CG, from 1986. Formed London Classical Players and Early Music Project (a company committed to historical stagings); has given highly successful performances, with an emphasis on composers' original intentions as well as an 'authentic' sound. A conductor of great musical

subtlety, whose quick intelligence and infectious energy inform all his undertakings. (R)

**Nortsov, Panteleymon** (*b* Peskovshchino, 28 Mar. 1900). Ukrainian baritone. Studied Kiev with Tsvetkov. Kiev and Kharkov, 1926–7; Moscow, B, 1927–54, also working with Stanislavsky. Appeared with Nezhdanova and Sobinov. Roles incl. Figaro (Mozart), Don Giovanni, Luna, Eletsky, Mazeppa, and a much-acclaimed Onegin. (R)

**Norway.** Performances were given in 1749 during a state visit by Frederik V of Denmark (with which Norway was united 1380–1814); and opera was for long dependent on touring companies such as the *Mingotti troupe. During the union with Sweden (1814–1905), the first Norwegian opera was *Fjeldeventyret* (The Mountain Adventure, 1824) by Waldemar Thrane (1790–1828), which had a concert performance but was not staged until 1850: it is an opéra-comique with touches of Weber. The first professional opera performances were given in 1858, the year that also saw *Fredkulla* (The Peacemaker) by Martin Andreas Udbye (1820–99), a work also influenced by German Romanticism. A plan by Bjørn Bjørnson for a major national opera with *Grieg proved abortive: the three scenes of *Olaf Trygvason* were given in concert in 1889 and staged in 1908. Other opera composers to appear included Johannes Haarklou (1847–1925) with five operas of which the most successful was *Marisagnet* (Mary's Legend, 1910); Ole Olsen (1850–1927) with Wagner-influenced works to his own texts, including *Lajla* (1893); Christian Sinding (1856–1941), also influenced by Wagner in *Der heilige Berg* (The Holy Mountain, 1914); Gerhard Schjelderup (1859–1933), who tried to move away from Wagner in operas including *Vaarnat* (Spring Night, 1908); and Catharinus Elling (1858–1942) with *Kosakkerne* (The Cossacks, 1897). A later generation included Arne Eggen (1881–1955) with the folk-influenced Ibsen opera *Olav Liljenkrans* (1940); and Ludvig Irgens Jensen (1894–1969) with *Robin Hood* (1945) and a dramatic choral symphony, *Heimferd* (The Return, 1930) rev. as an opera (1947). More recent Norwegian operas include *Jeppe paa Bjerget* (1966) by Geirr Tveitt (1908–81) and *Anne Pedersdotter* (1971) by Edvard Fliflet Braein (1924–76). Five operas were commissioned in 1985 by Den Norske Opera. Outside Oslo, an early attempt was made at founding a Norse Opera in Bergen by the violinist Ole Bull: this opened in 1850, with Ibsen as resident dramatist, but failed to find official support. Opera is also given in Trondheim, and there are festivals in Kristiansand and Brummundal.

See also OSLO.

**Nose, The** (Russ.: *Nos*). Opera in 3 acts by Shostakovich; text by A. Preis, A. Zamyatin, G. Yonin, and the composer, after Gogol's story (1835), with extracts and suggestions from other works by Gogol and using Smerdyakov's song from Dostoyevsky's *The Brothers Karamazov*. Prod. Leningrad, Maly T, 18 Jan. 1930, with Zhuravlenko, Raykov, Zasetsky, Nechayev, cond. Samosud; Santa Fe 11 Aug. 1965, Toscano, Reardon, cond. Kunzel; London, SW, by New OC, 4 Apr. 1973, with Opie, Drake, Williams, Dickerson, cond. Lovett.

St Petersburg, 1850. An official, Major Kovalyev (bar), wakes one morning to find that his nose has disappeared. His barber, Ivan Jakovlevich (bs), is astonished to discover it in his bread. On his way to report his loss to the police, Kovalyev sights the nose, dressed as a state councillor, but it refuses to talk to him as a lower-ranking official.

Kovalyev tries to place an announcement in the newspaper, but is refused.

The nose is arrested by the police, but a surgeon is unable to sew it back in place. However, one day, the nose reappears on Kovalyev's face just as suddenly as it disappeared.

The events are a mixture of fantasy and reality, designed as a satire on bureaucracy and philistinism, and the roles include one of the highest tenor parts in the repertory, the Police Inspector, who is required to sing up to d'' and e''.

**'Nothung! Nothung!'** Siegfried's (ten) song in Act I of Wagner's *Siegfried*, forging the sword 'Needful' which he knows his father (Wotan) has promised him in his hour of need. Its name derives from *Noth* = both affliction and need.

**Nouguès, Jean** (*b* Bordeaux, 25 Apr. 1875; *d* Paris, 28 Aug. 1932). French composer. Studied Bordeaux with Sarreau. His early operas were produced in Bordeaux, but he made a wider name with *Quo vadis?*, which was championed by Battistini and chosen to open the London OH in 1911, though its shallowness may have contributed to the failure of this venture. Among later operettas, he had some success especially with *L'auberge rouge* and *L'aigle*.

SELECT WORKLIST: *Quo vadis?* (Cain, after Sienkiewicz; Nice 1909); *L'auberge rouge* (The Red Inn) (Basset, after Balzac; Nice 1910); *L'aigle* (The Eagle) (Cain & Payen; Rouen 1912).

**Nourrit, Adolphe** (*b* Montpellier, 3 Mar. 1802; *d* Naples, 8 Mar. 1839). French tenor, son of below. Studied Paris with García sen. Début Paris, O, 1821 (Pylade, *Iphigénie en Tauride*). Paris, O, until 1837. Other roles included Orphée, Don Giovanni, Otello (Rossini). After 1824, further training with Rossini, to whom he in turn gave valuable advice on French style. Created the tenor roles in *Le siège de Corinthe*,

*Moïse*, *Le Comte Ory*, and *Guillaume Tell*; also title-roles of *Robert le diable*, and Liszt's *Don Sanche*; Raoul in *Les Huguenots*, Masaniello in *La muette de Portici*, and Eléazar in *La Juive*, among others. When Duprez was engaged at the Opéra as leading tenor in 1837, he left, successfully touring Belgium and France, but was prone to depression and to vocal troubles. In Italy studied further with Donizetti, who agreed to write *Poliuto* for him. However, its performance was prohibited by the Neapolitan censors. (Duprez eventually created the title-role in Paris.) Despite subsequent triumphs in Naples, severe melancholia drove him to suicide in 1839. An exceptionally intelligent, cultivated man, he was a highly influential artist of the Romantic movement, admired by Malibran, Liszt (his frequent accompanist), Chopin (who played at his funeral), Rossini, Meyerbeer, and Halévy. Possessed a voice of great charm, flexibility, and expressiveness, subtlety and variety as a singer, and a genius for acting (Berlioz was 'electrified' by him). Other talents included a sure dramatic sense (Meyerbeer, taking much advice from him, called him 'the second father' of *Les Huguenots*), and a gift for writing, e.g. the words of 'Rachel, quand du seigneur' in *La Juive*, the scenarios of several ballets for Taglioni, notably *La sylphide*, and a march, 'La Parisienne', which he often sang with the Marseillaise at the time of the 1830 revolution, to wild enthusiasm. He helped to raise the artistic standards of not only the Opéra, but Paris itself, where he introduced the songs of Schubert and Berlioz, and trained large workingmen's choirs. He was Professor of Declamation at Paris Cons. 1828–37. His brother **Auguste** (1808–53), also a tenor, became director of theatres in Amsterdam, the Hague, and Brussels, and subsequently taught singing.

BIBL: M. Quicherat, *Adolphe Nourrit, sa vie* (Paris, 1867); E. de Monvel, *Adolphe Nourrit* (Paris, 1903).

**Nourrit, Louis** (*b* Montpellier, 4 Aug. 1780; *d* Brunoy, 23 Sep. 1831). French tenor, father of above. Studied Paris with Guichard and Garat. Début Paris, O, 1805 (Renaud, *Armide*). Leading tenor from 1812. A superior singer with an attractive voice, if rather cool. Sang in operas by Gluck, Grétry, and Spontini, and appeared with his son Adolphe, including at his début, and in the first *Siège de Corinthe*. They resembled each other physically and vocally so much that Daussoigne-Méhul's *Les deux Salem* was produced especially to exploit the fact. Retired 1826.

**Novák, Vítězslav** (*b* Kamenice, 5 Dec. 1870; *d* Skuteč, 18 July 1949). Czech composer. Studied Jindřichův Hradec with Pojman, Prague with Knittl and Stecker, later Dvořák. He was initially influenced by the German Romantics, and only made a reputation after encounters with Moravian and Slovak folk music in the late 1890s fertilized a more individual style. In this, and in his devotion to nature, he at one time invited comparisons with Janáček. He did not write his first opera, a comedy, until 1913–14. The sharp irony in this work softened with *Karlštejn*, which is a patriotic gesture that does not exclude comedy. His most successful opera, *The Lantern*, draws on both fairy-tale and a feeling for the rococo. *Grandfather's Legacy* suffers from a cumbersome libretto, but includes some fine intermezzos.

WORKLIST: *Zvíkovský rarášek* (The Imp of Zvíkov) (Stroupežnický, after a popular comedy; Prague 1915); *Karlštejn* (Fischer, after Vrchlický; Prague 1916); *Lucerna* (The Lantern) (Jelínek, after Jirásek; Prague 1923); *Dědův odkaz* (Grandfather's Legacy) (Klášterský, after Heyduk; Brno 1926).

BIBL: V. Lébl, *Vítězslav Novák: život a dílo* (Prague, 1964; trans. 1968).

**Novara.** Town in Piedmont, Italy. The first opera was *Antemio in Roma*, by Besozzi, Erba, and Battistini. The T Nuovo opened 1779 with Sarti's *Medonte*, 1779; rebuilt 1883; renamed T Coccia after Carlo *Coccia (the city's Maestro di Musica, 1840–73) in 1973.

**Novello, Clara** (*b* London, 10 June 1818; *d* Rome, 12 Mar. 1908). English soprano. Fourth daughter of Vincent Novello. Studied London, Paris, later Milan with Micheroux. After a career as a concert singer, admired by Mendelssohn, Schumann, and her lifelong friend Rossini, she made her stage début in Padua 1841 (Semiramide). Genoa 1841–2; Modena, Bologna, Rome, 1842–3; London, DL, 1843. Retired on her marriage to Count Gigliucci, but returned to Italian stage 1849–54. Roles included Lucrezia Borgia, Norma, Donna Elvira; however, her real success was as an oratorio singer in England, 1851–60. One of the best English artists of the 19th cent., she sang with a clear, sweet tone and great purity of style.

BIBL: A. Mackenzie-Grieve, *Clara Novello* (London, 1955).

**Noverre, Jean-Georges** (*b* Paris, 29 Apr. 1727; *d* Saint-Germain-en-Laye, 19 Oct. 1810). Franco-Swiss dancer and choreographer. Danced first in Paris, including in operas. Ballet-master Strasburg, then Lyons with Marie Camargo. London, DL, with Garrick 1755–6, 1756–7, also writing *Lettres sur la danse* (1760). Stuttgart 1760, influencing the musical entertainments of the court. Vienna 1767; Milan 1774; Vienna, K, 1776; Paris, O, 1776–9; London, H, 1781–2, 1787–9, 1793. His serious and intelligent attempts to make dance more functional, and a more central element in opera than the merely decorative, were largely frustrated, but his conscious attempt to match the reforms of Gluck with his own art were influential on succeeding generations.

**Novi Sad** (Ger., Neusatz). City in Serbia, Yugoslavia. It was an early centre of Hungarian opera, whose conductor introduced opera to Belgrade 1829, and of Serbian Singspiel. The Serbian National T was founded 1864. A private company, founded in Belgrade in 1900, settled here 1911. An opera was founded as part of the Serbian National T after the First World War, and remains an important and flourishing company.

**Novotná, Jarmila** (*b* Prague, 23 Sep. 1907; *d* NY, 9 Feb. 1994). Czech soprano. Studied Prague with Destinn. Début Prague 1926 (Violetta). Vienna, S, 1933–8; San Francisco 1939; New York, M, 1940–56. Also Milan, S; Buenos Aires, C; Paris, O. Successful as Pamina, Gilda, Butterfly, Octavian, Mélisande, and first Giuditta (Lehár). (R)

**'No word from Tom'.** Anne Trulove's (sop) aria in Act I of Stravinsky's *The Rake's Progress*, in which she grieves over Tom's absence in London and then, in a cabaletta, resolves to go to him.

**Noyes Fludde.** Miracle play in 1 act by Britten; text part of the Chester Miracle Play. Prod. Orford Parish church, 18 June 1958, with Brannigan, Parr, cond. Mackerras; New York, James Memorial Chapel, 16 Mar. 1959.

The Voice of God (spoken) tells Noye (bar) to build an ark and thus save his family, Mrs Noye (sop) and their children and wives (boy trebles and girl sops), from the Flood. Animals process in pairs into the ark, and the Flood rises. Eventually Noye releases a dove, which finds dry land; the ark's occupants leave.

The score, which is for a childrens' orchestra including recorders and bugles as well as a professional group, also involves the congregation with the use of three hymns: 'Lord Jesus, think on me', 'Eternal Father, strong to save', and 'The Spacious Firmament on High' (to Tallis's Canon).

**Nozzari, Andrea** (*b* Vertova, 1775; *d* Naples, 12 Dec. 1832). Italian tenor. Studied Bergamo with Petrobelli, Rome with Aprile. Probable début Pavia 1794. Milan, S, 1796–7, 1800, 1811–12; Paris, I, 1803, 1804; Naples, C, 1811–25. One of Rossini's greatest tenors, he created Leicester (*Elisabetta, regina d'Inghilterra*), title-role of *Otello*, Rinaldo (*Armida*), Osiride (*Mosè*), Agorante (*Ricciardo e Zoraide*), Pirro (*Ermione*), Rodrigo di Dhu (*La donna del lago*), Paolo Erisso (*Maometto II*), Antione (*Zelmira*). Also a famous

# Nozze di Figaro

Paolino (*Matrimonio segreto*). Possessed a strong voice, and an imposing, graceful stage presence. Stendhal thought him 'one of the finest singers in Europe'. Rubini was his pupil.

**Nozze di Figaro, Le** (The Marriage of Figaro). Opera in 4 acts by Mozart; text by Lorenzo da Ponte, after Beaumarchais's comedy *La folle journée, ou Le mariage de Figaro* (1778, prod. 1784). Prod. Vienna, B, 1 May 1786, with Laschi, Storace, D. Bussani, M. Mandini, Gottlieb, S. Mandini, Benucci, F. Bussani, Kelly, cond. Mozart; London, H, 18 June 1812, with Catalani, Fischer, Naldi, Righi, Miarteni, Di Giovanni, Bianchi, Pucitta, Luigia, Dickons (preceded by 1-act burletta at Pantheon, 2 May 1812); New York, Park T, 10 May 1824, arr. Bishop (1819), with Johnson, Holman, Jones, Pearman.

Count Almaviva's castle near Seville, mid-18th cent. Figaro (bar) and Susanna (sop), servants of Count Almaviva (bar), are preparing for their wedding. Susanna realizes that their room is too close to that of the Count, who is pursuing her. Dr Bartolo (bar), the Countess's ex-guardian, and Marcellina (mez) wish to prevent Figaro's marriage by holding him to an agreement he made to marry Marcellina if he could not repay a loan. While the young page Cherubino (sop, sometimes mez) is telling Susanna of his hopeless love for the Countess, the Count arrives. Cherubino hides and his master flirts with Susanna. Don Basilio (ten) is heard and the Count also hides, forcing Cherubino to move from his place. Basilio enters and after proclaiming the Count's love for Susanna, gossips about Cherubino's love for the Countess. At this, the Count emerges and tells Susanna how he discovered Cherubino with Barbarina (sop), the gardener's daughter. He pulls back a cover and discovers Cherubino. Realizing that the boy has overheard his own indiscretions with Susanna, he despatches him to the army.

The Countess laments that her husband no longer loves her. She, Figaro, and Susanna devise a plan to trap him: Susanna will write to him agreeing to a rendezvous, but Cherubino will go instead, dressed in her clothes, and the Countess will interrupt. While he is being fitted with his costume, the Count knocks and Cherubino is hurriedly pushed into the dressing-room. The Countess tells her husband that the noise he hears is only Susanna trying on her wedding dress, but he is suspicious. He departs to fetch tools to force the door open, locking the main door as he leaves. As soon as he has gone, Cherubino leaps out of the window and Susanna takes his place. When the Count returns, she steps out and he has to apologize for having doubted his wife. The drunken gardener Antonio (bs) enters bearing a smashed flowerpot and a document (Cherubino's

commission), dropped by the escapee. Figaro tries to cover up by taking the blame.

Figaro attempts to delay Marcellina's plans by saying that he cannot marry without the consent of his parents. As proof of his claim of noble birth, he shows a birthmark on his arm and Marcellina recognizes him as her long-lost son. Susanna enters and is furious to find Figaro embracing Marcellina. When the situation is explained, she joins in the rejoicing, and plans proceed for their wedding.

At night in the garden, Figaro comes across Barbarina, who unwittingly reveals that she is bearing a message from the Count to Susanna; Figaro believes his wife is already deceiving him. The Countess and Susanna appear in each other's clothes, and in the darkness there are many mistaken identities: Cherubino flirts with the Countess (thinking she is Susanna), and is succeeded by the Count, who does not realize he is paying court to his own wife. Figaro recognizes Susanna by her voice. When the Countess steps into the light, the Count realizes that he has been tricked and humbly begs his wife's forgiveness. All rejoice in the happy conclusion of the 'mad day'.

For a revival of *Figaro* in Vienna in 1789 Mozart made a number of small but significant revisions, and also provided two new arias for Susanna: 'Al desio, di chi t'adora' and 'Un moto di gioia', to replace 'Deh vieni non tardar' and 'Venite, inginocchiatevi' respectively.

BIBL: T. Carter, *W. A. Mozart: 'Le nozze di Figaro'* (Cambridge, 1987).

**Nozze di Teti e di Peleo, Le** (The Marriage of Thetis and Peleus). Opera in a prologue and 3 acts by Cavalli; text by Orazio Persiani. Prod. Venice, C, 24 Jan. 1639.

**Nucci, Leo** (*b* Castiglione dei Pepoli, 16 Apr. 1942). Italian baritone. Studied Milan with Marchesi, later Bizzarri. Début Spoleto 1967 (Rossini's Figaro). Padua 1975; Milan, S, from 1976; London, CG, from 1978; New York, M, 1980; Paris, O, 1981; also Chicago, Vienna, Tokyo, etc. Roles incl. Malatesta, Mercutio, Renato, Rigoletto, Posa, Onegin, Sharpless; particularly acclaimed as Rossini's Figaro. (R)

**number opera**. The name given to an opera written in separate numbers, that is, arias, duets, trios, ensembles, choruses etc., separated by spoken dialogue or recitative. The normal form until the early 19th cent., it survived especially in France and Italy, but in Germany was gradually replaced, after Mozart's tendency to link sections into a larger whole especially in his finales, by operas in which the music was increasingly continuous. Although Italian opera also moved towards a more continuous structure during the

19th cent., vestiges of such divisions are still apparent until late in the century. Wagner strongly opposed number opera, insisting on music forming a dramatic whole with the text. In the 20th cent. the number opera has been revived, e.g. by Hindemith, Stravinsky, and Weill.

**Nuremberg** (Ger., Nürnberg). City in Bavaria, Germany. Sigmund Staden's *Seelewig*, sometimes claimed as the first German opera, was privately performed 1644. The Komödienhaus opened 1668, and in 1798 there opened the Reichstädtische-Nürnbergische Nationalschaubühne, the first permanent company, but standards were poor. With the opening of the Stadttheater 1833, they improved, and many important singers visited the city. Wagner and Verdi entered the repertory from 1858. The Stadttheater (cap. 1,456) opened 1905, was damaged in the Second World War, reopened (cap. 1,061) in 1945.

# O

**obbligato** (It.: 'obligatory'). Strictly, a term for an instrumental part that cannot be omitted. (It is often used, paradoxically, for a part that may be omitted.) In vocal music it is essential but subordinate to the voice, e.g. the corno di bassetto obbligato in Vitellia's 'Non più di fiori' in Act II of Mozart's *La clemenza di Tito*. Arias with obbligato instrumental parts were an essential feature of much late 17th- and early 18th-cent. French opera, and played an important role in Metastasian opera seria.

**Oberlin, Russell** (*b* Akron, OH, 11 Oct. 1928). US counter-tenor. Studied New York, Juilliard School. Début as concert singer, New York 1951. Founder member of New York Pro Musica. New York 1956, sang title-role of Handel's *Giulio Cesare* (in concert). London, CG, 1961, Oberon (*A Midsummer Night's Dream*); first US Oberon, San Francisco 1961. Like Alfred Deller, a pioneer in the revival of the male alto voice and repertory. (R)

**Oberon.** Opera in 3 acts by Weber; text by James Robinson Planché, after William Sotheby's translation (1798) of Wieland's *Oberon* (1780), based in turn on a 13th-cent. *chanson de geste*, *Huon de Bordeaux*. Prod. London, CG, 12 Apr. 1826, with Paton, Goward, Cawse, Vestris, Braham, Fawcett, Bland, Isaacs, cond. Weber; New York, Park T, 20 Sep. 1826, with Austin, Wollack, Sharpe, Horn, Richings; Leipzig 23 Dec. 1826.

France, the Orient, and Fairyland, AD 800. Oberon (ten) has vowed not to meet Titania again until he has found a faithful pair of lovers. With some supervision and help from Puck (sop), and provision of a magic horn that can summon aid, he sends Sir Huon (ten), who has offended Charlemagne, to Baghdad to rescue Reiza (sop).

With the aid of the magic horn, Huon succeeds in releasing Reiza, while his squire Sherasmin (bar) carries off her attendant Fatima (sop). On their voyage to Greece, both couples are shipwrecked and captured by pirates.

Sherasmin and Fatima are sold as slaves to the Emir of Tunis. Huon appears and, when he refuses to concede to the Emir's wife's plans to murder the Emir and marry her, is sentenced to death. He is saved at the last moment by magic. Oberon accepts Huon's proof of his fidelity to Reiza, and the couple are restored to Charlemagne's court.

Other operas on the subject are by Kunzen (*Holger Danske*, 1789—the principal 18th-cent. Danish opera), Wranitzky (1789—very successful until replaced by Weber's opera), and Hanke (*Hüon und Amande*, 1794). Also opera by Grosheim (*Titania*, 1792) and Dessauer (19th cent.; unprod.). See also SHAKESPEARE.

**Oberspielleiter** (Ger.: 'senior producer'). The name given to the chief resident producer of an opera company. He might sometimes also be the *Generalintendant.

**Oberto, conte di San Bonifacio.** Opera in 2 acts by Verdi; text by Piazza, revised by Solera, probably from the original text for Verdi's uncomposed or adapted *Rocester* and perhaps *Lord Hamilton*. Prod. Milan, S, 17 Nov. 1839, with Rainieri-Marini, Shaw, Sacchi, Salvi, Marini; London, St Pancras TH, 8 Apr. 1965 (concert), with Edwards, Sinclair, Pilley, Robinson, Ruta, cond. Head; New York, Amato T, 18 Feb. 1978.

Ezzelino's castle, near Bassano, 1228. Cuniza (mez) is to marry Riccardo (ten). Oberto (bs), the exiled Count of Bonifacio, has disowned his daughter Leonora (sop) because she was seduced by Riccardo. Oberto and Leonora reunite so as to be avenged on Riccardo. Oberto persuades Leonora to disclose the whole story to Cuniza. She is horrified and confronts Riccardo, telling him he must now marry Leonora.

Not satisfied with this revenge, Oberto challenges Riccardo to a duel. Riccardo kills Oberto and, filled with remorse, flees Italy, leaving Leonora in despair. Libretto reset by Graffigna, prod. Venice as *I Bonifazi ed i Salinguerra* (1842).

Also opera by F. Ricci (*Corrado d'Altamura*, 1841).

**Obraztsova, Elena** (*b* Leningrad, 7 July 1937). Russian mezzo-soprano. Studied Leningrad Cons. with Grigoryeva. Début Moscow, B, 1963 (Marina). Moscow, B, from 1964; New York, M, 1975 (with Moscow B), and from 1977; Milan, S, 1964, 1973 (with Moscow B), 1976; also Barcelona, Vienna, San Francisco, etc. Roles include Rosina, Countess (*Queen of Spades*), Marfa (*Khovanshchina*), Eboli, Carmen, Britten's Oberon. Possesses a deep, vivid tone and a vigorous style. (R)

**O'Brien, Timothy** (*b* Shillong, 8 Mar. 1928). English designer. Studied Cambridge. Has worked in TV and in the theatre, especially at Stratford, but has also made a reputation as an exceptionally intelligent, inventive, and practical designer for opera (often in collaboration with Tazeena Firth). Productions include London, SW: *Flying Dutchman* (1958), *Girl of the Golden*

*West* (1962); London, CG: the première of Tippett's *Knot Garden* (1970), *Peter Grimes* (1975), *Rake's Progress* (1978), *Lulu* (1981), Verdi's *Otello* (1987); London, C: Henze's *Bassarids* (1974), *Meistersinger* (1984). *Wozzeck* (Adelaide 1976); *Turandot* (Vienna 1983); *Lucia* (Cologne 1985); *War and Peace* (Leningrad 1991).

**Oca del Cairo, L'** (The Goose of Cairo). Opera in 2 acts by Mozart; text by G. B. Varesco. Comp. 1783, unfin. Prod. Paris, FP, 6 June 1867, with Armand, Géraizer, Mathilde, Laurent, Bonnet, Masson; London, DL, 12 May 1870, with Lewitzky, Sinico, Gardoni, Gassier, cond. Arditi; Miami 2 Apr. 1967, cond. Csonka.

Celidora (sop), daughter of Don Pippo (bs), ruler of Ripasecca, has been promised in marriage to Lionetto, although she loves Biondello (ten). The opera deals with Biondello's attempts to rescue Celidora from a castle in which Don Pippo has imprisoned her, together with her companion Lavina (sop). At first the lovers try to build a bridge, but are discovered before they can complete it. Act II was to have featured the arrival of a mechanical toy goose in which Biondello was to hide.

Although a prose draft of the whole survives, only the first act of the libretto proper was completed. From Mozart's correspondence it seems that he would otherwise have finished the opera.

**'O Carlo, ascolta'.** The introduction to Posa's (bar) aria in Act IV of Verdi's *Don Carlos*, telling Carlos that the Queen will wait for him on the following day outside the Monastery of San Yuste to see him for the last time.

**'Ocean! thou mighty monster'.** Reiza's (sop) aria in Act II of Weber's *Oberon*, in which she first apostrophizes the ocean and then hails the boat she believes to be coming to her rescue.

**Ochs.** Baron Ochs auf Lerchenau (bs), the Marschallin's cousin, in Strauss's *Der Rosenkavalier*.

**Octavian.** Count Octavian Rofrano (sop or mez), the Knight of the Rose in Strauss's *Der Rosenkavalier*. His full names, recited by Sophie in Act II, are Octavian Maria Ehrenreich Bonaventura Fernand Hyacinth; his nickname Quinquin was taken from Count Franz 'Quinquin' Esterházy, brother of Haydn's patron.

**Odabella.** The daughter (sop) of the Lord of Aquileia in Verdi's *Attila*.

**'O don fatale'.** Eboli's (sop) aria in Act IV of Verdi's *Don Carlos*, in which she curses the fatal beauty that has blighted her life.

**'O du, mein holder Abendstern'.** Wolfram's (bar) song to the evening star in Act III of Wagner's *Tannhäuser*.

**Odysseus.** In classical mythology, as recounted in Homer's *Iliad* and *Odyssey*, the Greek hero, most nimble-witted of the leaders of the army besieging Troy, who on the fall of the city was condemned to long years of wandering, and after many adventures (including those on the island of *Circe) returned to his native Ithaca. Here he found his faithful wife *Penelope waiting for him; and after killing her importunate suitors with the help of his son *Telemachus, he claimed her back. Many operas on this legend deal with his sojourn with Circe or his return to Ithaca, and therefore overlap with those dealing with Circe or Penelope as their main subject. He is known in Latin, e.g. in Virgil's *Aeneid*, as Ulysses. Operas on the legend are by Branchi (1619), Monteverdi (*Il ritorno d'Ulisse in patria*, 1640), Sacrati (1644), Zamponi (*Circe*, 1650), Melani (1656), Rettenpacher (1680), Gandio (*Ulisse in Feazia*, 1681: staged using the carved figures of San Moisè in Venice), Acciajuoli (*Ulisse in Feaccia*, 1681), Keiser (1696), Pollarolo (*Ulisse sconosciuto in Itaca*, 1698, lost), Porsile (1707), Vogler (1721), Porta (1725), Treu (*Ulisse e Telemacco*, 1726), Sciroli (*Ulisse errante*, 1749), Majo (1769), Gazzaniga (1779), G. Giordani (1782), J. C. Smith (1783), Alessandri (1790), Basili (1798), Perrino (*Ulisse nell'isola di Circe*, 1805), Romberg (1807), Mayr (1809), L. Ricci (1828), Bungert (tetralogy *Die Homerische Welt*, consisting of *Kirke* 1898, *Nausikaa* 1901, *Odysseus' Heimkehr* 1896, and *Odysseus' Tod* 1903), Speer (*c*.1900), Cesek (1907), Tomasi (1965), and Dallapiccola (1968).

**Oedipus Rex** (King Oedipus). Opera-oratorio in 2 acts by Stravinsky; text by Cocteau, after Sophocles' tragedy (*c*.435–425 BC), trans. into Latin by Daniélou. Prod. Paris, SB, 30 May 1927, as oratorio; first stage prod. Vienna 23 Feb. 1928; Boston 24 Feb. 1928 (concert) and New York, M, 21 Apr. 1931 (stage), cond. Stokowski; London, QH, 12 Feb. 1936 (concert), with Slobodsakya, Widdop, Williams, Walker, cond. Ansermet; Edinburgh (by Hamburg Co.) 21 Aug. 1956, with Ilosvay, Melchert, Pease, Van Mill, cond. Ludwig.

Thebes, mythological times. The people of Thebes seek the murderer of their former king, Laius, in order that their plague may be ended. The present king, Oedipus (ten), is disturbed to learn from the soothsayer that the killer was himself a king.

Jocasta (mez), Oedipus' wife who was previously married to Laius, assures the people that her first husband was killed by an unknown traveller. It then emerges that Oedipus' father was not Polybus, but Laius. Oedipus has fulfilled the prophecy of the oracle that Laius would be killed by his own son, though without knowing his victim's identity and Jocasta would marry her son. In final realization of the situation, Jocasta commits suicide and Oedipus blinds himself and departs for Colonus.

The chorus of the work is static, and the dramatic metaphor somewhere between opera and oratorio.

Also operas by Desogier (1779), Méreaux (1791), Leoncavallo (unfin., prod. 1920), Enescu (1936), Orff (1959), Rihm (1987). Operas on *Oedipus at Colonus* (Sophocles, prod. *c*.402 BC) are by Sacchini (1786) and Zingarelli (1802).

**Oehmann, Carl Martin** (*b* Floda, 4 Sep. 1887; *d* Stockholm, 26 Dec. 1967). Swedish tenor. Studied Stockholm with Gentzel, Milan with Oxilia and Quadri. Début Livorno, 1914 (Chénier). Göteborg 1917–23; Stockholm 1919–23, 1927–37; New York, M, 1924–5; Berlin, SO, 1925–37; London, CG, 1928. Roles included Tannhäuser, Walther, Hermann, Verdi's Otello, Bacchus. Retired from stage 1937, and taught; pupils included J. Björling, Gedda, and Talvela. (R)

**Oestvig, Karl** (*b* Oslo, 17 May 1889; *d* Oslo, 21 July 1968). Norwegian tenor. Studied Cologne with Steinbach. Début Stuttgart 1914. Stuttgart until 1919; Vienna, S, 1919–27; Berlin, SO, 1927–30. Retired 1932. Director Oslo O 1941. Described by Lotte Lehmann as 'a dazzling meteor'. A distinguished Lohengrin, Parsifal, and Walther, and 'positively ideal' (Strauss) as Bacchus. Created Giovanni (*Monna Vanna*) and Emperor (*Die Frau ohne Schatten*). (R)

**Offenbach, Jacques** (*b* Cologne, 20 June 1819; *d* Paris, 5 Oct. 1880). German, later French, composer and conductor. Son of a Cologne synagogue cantor from Offenbach originally named Eberst; studied with Alexander and Breuer. Sent by his father to Paris 1833 because Jews enjoyed a more emancipated status there. Entered the Conservatoire—changing his forename from Jacob to Jacques—after Cherubini had heard him and overruled regulations forbidding foreigners to enrol. Studied cello, but forced to leave for financial reasons 1834; joined the orchestra of the T Ambigu Comique as cellist, moving to Opéra-Comique in 1837. In 1835 received instruction in composition from Halévy and began to write short instrumental pieces. In 1839 he composed his first dramatic work, the vaudeville *Pascal et Chambord*, but first came to popular attention as a salon musician. During the 1840s his reputation in this field increased, though he failed to make any headway on the stage and for a while contemplated emigrating to the USA. He became conductor of the T Français in 1850, and in 1855 rented the T Marigny in the Champs-Elysées which, renamed the Bouffes-Parisiens, was the

venue for a season of operettas during the Exhibition of 1855. This was the turning-point of his career: with works like *Les deux aveugles* and *Le violoneux*, performed as part of a repertoire including Adam and Delibes, he won overnight successes.

For the winter Offenbach gave up his position at the T Français and moved to the T Comte. Here he gave the mock-oriental *Ba-ta-clan* and organized a competition for the best operetta, won by Bizet and Lecocq with their setting of *Le docteur Miracle*. In 1857 his company made a successful visit to London, though he was soon forced to flee Paris for a while because of debt. He returned with *Orpheus in the Underworld*, whose première in 1858 was his greatest triumph, quickly making him famous throughout France, Europe, and the USA. In 1860 his first operetta was given at the Opéra-Comique, *Barkouf*, but in 1861 he gave up direction of the Bouffes-Parisiens in order to devote more time to composition. His greatest works date from this period, including *La belle Hélène*, *Barbe-bleue*, *La vie parisienne*, *La Grande-Duchesse de Gérolstein*, and *La Périchole*. For the Vienna Hofoper in 1864 he produced the 'grand romantic' (in fact, comic) opera *Die Rheinnixen* which was much admired by Hanslick. But after the war of 1871, during which he had had to flee France, the new mood in Paris was against his frivolous manner, while his German ancestry also provoked hostility. He took over the T de la Gaité 1873–5, but lost a considerable amount of money and was forced to take a trip to the USA to replenish his purse. During his last years he continued to write, and was still capable of securing a brilliant success, as with *La fille du tambour-major*. His last work, *The Tales of Hoffmann*, lay unfinished at his death and was completed by Ernest Guiraud.

A brilliant man of the theatre, with a flair for epitomizing the wit of the day, Offenbach discovered a style which appealed irresistibly to the French taste of the 1860s. His opéras bouffes satirize the classics or contemporary politics and society, sometimes by reference one to the other, and with especial allusion to the modes and manners of the Second Empire. Similarly, familiar music is often pressed into service in absurd situations or made ludicrous by association with unexpected works. The impeccable *boulevardier* manner and exhilarating high spirits conceal the skill of these captivating pieces as well as the composer's literary erudition; they also distract from the lack of a true lyrical gift such as Johann Strauss could display. That he never lacked melodic appeal is shown not only in the frothiest of the operettas but also in *The Tales of Hoffmann*, though structural slackness and an absence of real musical substance weaken this,

his one attempt at a larger-scale work with continuous music. At its best, its melodic manner could be touching (as with John Styx's aria in *Orpheus in the Underworld*, 'Quand j'étais roi de Béotie') as well as hilarious (the famous concluding Galop in the same work). There is, moreover, a recurrent cynicism and fatalism concealed beneath the Second Empire merriment which led Mauriac to write of *La Grande-Duchesse de Gérolstein*, 'The laughter I hear in Offenbach's music is that of the Empress Charlotte, gone mad.'

Offenbach's works proved especially popular in Vienna, where they inspired an Austrian operetta tradition led by Johann Strauss II, Suppé, and Millöcker.

SELECT WORKLIST (all first prod. Paris unless otherwise stated): *Pascal et Chambord* (Bourgeois & Brisebarre; 1839); *Les deux aveugles* (The Two Blind Men) (Moinaux; 1855); *Le violoneux* (Mestéphès & Chevalet; 1855); *Ba-ta-clan* (Halévy; 1855); *Orphée aux enfers* (Orpheus in the Underworld) (Crémieux & Halévy; 1858, rev. 1874 in 4 acts); *Barkouf* (Scribe & Boisseaux; 1860); *Die Rheinnixen* (The Rhine Maidens) (Wolzogen, after Nuitter & Tréfeu; Vienna 1864) *La belle Hélène* (Fair Helen) (Meilhac & Halévy; 1864); *Barbe-bleue* (Bluebeard) (Meilhac & Halévy; 1866); *La vie parisienne* (Paris Life) (Meilhac & Halévy; 1866); *La Grande-Duchesse de Gérolstein* (Meilhac & Halévy; 1867); *La Périchole* (Meilhac & Halévy; 1868, rev. 1874); *La fille du tambour-major* (The Drum Major's Daughter) (Chivot & Duru; 1879); *Les contes d'Hoffmann* (The Tales of Hoffmann) (Barbier & Carré, after Hoffmann; 1881).

**'O Isis und Osiris'.** Sarastro's (bs) prayer to the Egyptian gods in Act II of Mozart's *Die Zauberflöte*.

**Olczewska, Maria** (*b* Ludwigsschwaige, 12 Aug. 1892; *d* Klagenfurt, 17 May 1969). German mezzo-soprano. Studied Munich with Erler. Début Krefeld 1915 (Page, *Tannhäuser*). Hamburg 1917–20; Vienna 1921–3, 1925–30; London, CG, 1924–33; Chicago 1928–32; New York, M, 1933–5. A dramatically intelligent singer with a glowing tone, celebrated as Ortrud, Brangäne, Herodias, Octavian. Created Brigitte in Korngold's *Die tote Stadt*. (R)

**Oldenburg.** Town in Lower Saxony, Germany. The first permanent theatre opened 1833 with Auber's *La neige*. Until 1921 opera was generally given by the Bremen company. Johannes Schüler pursued an adventurous policy 1928–32, giving the first *Wozzeck* after the Berlin première. The theatre was renovated 1961.

**Old Vic, The.** Properly, The Royal Victoria Hall: the theatre in the Waterloo Road, South London, birthplace of the English opera company which later had its home at *Sadler's Wells. Built in 1818 as the Royal Coburg Hall by Joseph Glossop, who as Giuseppe Glossop was impresario at La Scala, Milan, and T San Carlo, Naples, where he married one of the singers. Their son Augustus Glossop Harris was the father of Augustus *Harris, manager of Covent Garden, London, 1888–96. Renamed the Royal Victoria Hall 1833. In 1880 its lease was acquired by Emma Cons, and it was renamed the Royal Victoria Coffee Hall: opera was first heard as excerpts in costume. She was succeeded in 1898 by her niece Lilian *Baylis, who as well as Shakespeare gave opera twice a week and on alternate Saturday matinées. Even works on the scale of *Tristan* were heard, in drastically reduced versions by the music director, Charles Corri. After the First World War, Baylis invited the soprano Muriel Gough to produce *Figaro*; she declined in favour of Clive Carey, who persuaded Baylis to use E. J. Dent's translation. This was the beginning of the Dent–Carey Mozart revival in England, and of a move towards new standards of translation. By 1931, when Sadler's Wells opened, the repertory had widened to include Verdi and Wagner, with singers including Joan Cross, Edith Coates, Sumner Austin, Tudor Davies, and Powell Lloyd, with Lawrance Collingwood as conductor. There were also ballet evenings under Ninette de Valois with Constant Lambert conducting. No opera was given at the Old Vic between 1935 and 1979, when the English Music T gave a short season; in 1991 *Carmen Jones* was performed there.

BIBL: E. Dent, *A Theatre for Everybody* (London, 1945).

**Olga.** Olga Larina (sop), lover of Lensky, in Tchaikovsky's *Eugene Onegin*.

**Olimpiade, L'** (The Olympic Games). Opera in 3 acts by Vivaldi; text by Metastasio. Prod. Venice, T San Angelo, carnival 1734.

Greece, mythological times. Megacles (bs-bar) is torn between helping his friend Lycidas (ten) and his love for Aristea (sop), daughter of King Clysthenes of Sicyon (bs). Lycidas has asked him to represent him in the Olympic Games, in which the champion will win Aristea's hand. When Clysthenes banishes Lycidas (having learnt of his betrayal of Argene (con), to whom he had sworn his love), Lycidas tries to kill the King and is sentenced to death. Argene offers to take her beloved's place, but it emerges that Lycidas is in fact Clysthenes' son, who was thrown into the sea as a baby after an oracle had declared that he would try to kill his father. The opera ends with a double wedding for Megacles and Aristea, Lycidas and Argene.

Also operas by Caldara (1733), Pergolesi (1735), Leo (1737), Cimarosa (1784, opening the T Eretenio, Vicenza), and Paisiello (1786).

**Oliver, Stephen** (*b* Liverpool, 10 Mar. 1950; *d* London, 29 Apr. 1992). English composer. Studied Oxford with Leighton and Sherlaw Johnson. An intelligent and fluent composer with a strong theatrical instinct, he wrote many operas. Some of the most succesful are brief, laconic pieces, e.g. the wordless squib *The Waiter* (1976); some approach the musical, e.g. *Jacko's Play* (1979) and *Blondel* (1983); others are more ambitious, e.g. *Tom Jones* and *Timon of Athens*. Has also written *Three Instant Operas* (1973) for children. Translated works by Sallinen, and made an arrangement of Peri's *Euridice*.

SELECT WORKLIST (all texts by Oliver): *The Duchess of Malfi* (after Webster; Oxford 1971); *Tom Jones* (after Fielding; Newcastle 1976); *Beauty and the Beast* (after Beaumont; Batignano 1984); *Timon of Athens* (after Shakespeare; London 1991).

**Olivero, Magda** (*b* Saluzzo, 25 Mar. 1910). Italian soprano. Studied Turin with Gerussi and Simonetto. Début Turin 1933 (Lauretta). Successful career in Italy until 1941, when she married and left the stage. Returned 1951, at Cilea's request, to sing Adriana Lecouvreur, of which he considered her the ideal interpreter. London, Stoll T, 1952; Edinburgh 1963; New York, M, 1975. A notable singing actress; her successes included Medea, Fedora, Tosca, Minnie, Mimì, Liù. (R)

**Olivier.** A poet (bar), rival of the musician Flamand for the Countess, in Strauss's *Capriccio*.

**Olmütz.** See OLOMOUC.

**Olomouc** (Ger., Olmütz). City in Moravia, Czechoslovakia. Opera is given in the Oldřich Stibor T (cap. 750) 3–4 times a week by a company of about 30 soloists.

**'O luce di quest'anima'.** Linda's (sop) aria in Act I of Donizetti's *Linda di Chamounix*, in which she sings of her love for Arthur.

**Olympia.** The doll (sop), Hoffmann's first love, in Offenbach's *Les contes d'Hoffmann*.

**Olympians, The.** Opera in 3 acts by Bliss; text by J. B. Priestley. Prod. London, CG, 29 Sep. 1949, with Grandi, Coates, Johnston, Franklin, Glynne, cond. Rankl.

**Olympie.** Opera in 3 acts by Spontini; text by M. Dieulafoy and C. Brifaut, after Voltaire's tragedy (1762). Prod. Paris, O, 22 Dec. 1819, with Albert, Branchu, Nourrit, Dérivis, cond. R. Kreutzer. Rev. Berlin (trans E. T. A. Hoffmann) 14 May 1821, with Schulz, Milder, Blume, Bader, cond. Spontini. Failed owing to rival popular support for Weber's *Der Freischütz*, though occasionally revived until 1870. Rev. Florence, 1950.

Ephesus, 332 BC. Antigone (bar) is reconciled to his former adversary Cassandre (ten); he has killed Alexander the Great, whose daughter Olympie (sop) he plans to marry. But she loves Cassandre, whom Arzana (mez) suspects of the murder. Arzana discloses herself to the priests as Alexander's widow Statyre. Mortally wounded, Antigone confesses to the murder. The original version ends with Olympie's death and Statyre's suicide; the revision has Olympie marrying Cassandre and Statyre resuming the throne.

Also opera by C. Conti (1829).

**O'Mara, Joseph** (*b* Limerick, 16 July 1864; *d* Dublin, 5 Aug. 1927). Irish tenor. Studied Milan with Moretti. Début London, Royal English OH, 1891 (Ivanhoe). London: CG and DL, 1894–5; Moody-Manners Co., 1902–8; CG, 1910 under Beecham. Formed O'Mara Co. 1912; sang with it until 1924. Large repertory included Walther, Lohengrin, Don José, and a famous Eléazar. Retired 1926. (R)

**'Ombra mai fù'.** Serse's (sop cas) aria in Act I of Handel's *Serse*, praising the tree that gives him shade. Though marked larghetto, it has become irrevocably known as 'Handel's Largo'.

**ombra scene.** In early opera, a scene taking place in Hades, or one in which a ghost or shade (ombra) is conjured up. Latterly, this was often allied to a *slumber scene, in which the sleeping hero was addressed, usually in warning or reproach, by a ghost. Though originating in 17th-cent. opera (and a regular feature of works portraying the legend of Orpheus), the effect has proved tenacious: among the most celebrated, and powerful, ombra scenes are those in Berlioz's *Les Troyens*.

**'O Mimì, tu più non torni'.** The duet between Rodolfo (ten) and Marcello (bar), mourning Mimì's fickleness, in Act IV of Puccini's *La bohème*.

**'O mio babbino caro'.** Lauretta's (sop) aria in Puccini's *Gianni Schicchi*, in which she appeals to her father to allow her to marry Rinuccio.

**'O mio Fernando'.** Léonor's (mez) aria in Act III of the Italian version of Donizetti's *La favorite*, in which she sings of her love for Fernando.

**'O my beloved father'.** See 'O MIO BABBINO CARO'.

**'O namenlose Freude'.** The duet of reunion for Leonore (sop) and Florestan (ten) in Act II of Beethoven's *Fidelio*.

**Oncina, Juan** (*b* Barcelona, 15 Apr. 1925). Spanish tenor. Studied Barcelona with Capsir. Début Barcelona 1946 (Massenet's Des Grieux). Sang widely in Italy 1946–52. Gly. 1952–61. A leading

Donizetti and Rossini singer; also sang Ferrando, Don José, Cavaradossi. (R)

**'One fine day'.** See 'UN BEL DÌ VEDREMO'.

**Onegin, Sigrid** (*b* Stockholm, 1 June 1889; *d* Magliaso, 16 June 1943). Swedish contralto. Studied Munich with Reiss, Milan with Di Ranieri. Début Stuttgart 1912 (Carmen). Munich 1919–22; New York, M, 1922–4; London, CG, 1927; Berlin 1926–31; Bayreuth 1933–4. Created Dryad in *Ariadne auf Naxos*; also successful as Gluck's Orfeo, Fidès, and especially, with her regal style, as Lady Macbeth, Amneris, and Fricka. Her excellent voice was of great range and richness. (R)

BIBL: F. Penzoldt, *Alt-Rhapsodie: Sigrid Onegin— Leben und Werk* (Magdeburg, 1933/1953).

**'Ô paradis'.** Vasco da Gama's (ten) apostrophe to the island of Madagascar in Act IV of Meyerbeer's *L'Africaine*.

**'O patria mia'.** Aida's (sop) aria in Act III (the Nile Scene) of Verdi's *Aida*, in which she mourns that she will never again see her beloved homeland.

**Opava** (Ger., Troppau). Town in Moravia, Czechoslovakia. Opera is given in the Silesian T 3–4 times a week by a company of about 20 soloists. The company also tours to Krnov.

**opera** (It.: 'work'; an abbreviation of *opera in musica*). A drama to be sung with instrumental accompaniment by one or more singers in costume; *recitative or spoken dialogue may separate musical numbers. See also GRAND OPERA, MUSIC DRAMA, OPERA BUFFA, OPÉRA-COMIQUE, OPERA SERIA, SINGSPIEL.

**Opéra** (Paris). The name by which the most important French operatic institution, and its building, have often been generally known, even when officially entitled Académie Royale de Musique, Théâtre des Arts, Théâtre de la Nation etc., recently Opéra-Bastille, and when occupying various of 15-odd theatres during its history. On 28 June 1669, Robert *Cambert and the Abbé Pierre *Perrin obtained a royal privilege from Louis XIV to perform 'académies d'opéra ou réprésentations en musique et en langue françoise, sur le pied de celles d'Italie'. They joined with the Marquis de Sourdéac as producer and Beauchamp as ballet-master to recruit singers in the Languedoc. In 1670 they rented the Salle du Jeu de Paume ('de la Bouteille') for five years, converted it into a theatre, and opened 3 Mar. 1671 with Cambert and Perrin's *Pomone*. *Lully secured the patent from Perrin in Mar. 1672, opening at the Salle du Jeu de Paume de Bel Air on 15 Nov. with *Les fêtes de l'Amour et de Bacchus*. He moved to the Palais-Royal 1674. Together with his librettist *Quinault he gave in 15 years some 20 works whose form and nature as 'tragédies mises en musique' (later described as *tragédie lyrique) set a style for the Opéra. The manner was given a new aspect by Lully's successors *Destouches and *Campra, later (in particular) *Rameau, who made his début at the Opéra with *Hippolyte et Aricie*, 1733, and had 24 of his operas produced by 1760. The theatre burnt down 6 Apr. 1763, and the company occupied the Salle des Tuileries 1764–9, returning to the renovated Palais-Royal (cap. 2,500) in 1770.

After Rameau's death in 1764, opera was principally in the hands of foreigners. The reign of *Gluck began with *Iphigénie en Aulide*, 1774; the arrival of *Piccinni, representing a more Italianate manner in contrast to Gluck's dramatic realism, precipitated a famous controversy. In 1781 the Palais-Royal again burnt down, and the Opéra transferred to the Salle des Menus-Plaisirs. The new theatre at the Salle de la Porte Saint-Martin (cap. *c*.2,000) was opened by Marie Antoinette Oct. 1781 with Piccinni's *Adèle de Ponthieu*. In 1794 the company moved again to the Salle Montansier (cap. 1,650: parterre audience seated for the first time). After various changes of name during and after the Revolution, it was established on a firmer footing with the title Académie Impériale de Musique. The repertory was widened, though the traditional emphasis on ballet remained.

In 1807 the era of *Spontini, Napoleon's favoured composer, opened with *La vestale*. Together with *Cherubini, he confirmed a style of *Grand Opera which marked the Opéra for many decades. In 1820 the company moved to the Salle Favart, opening 19 June with Kreutzer's *Clari*, briefly in 1821 to the Salle Louvois, and in 1822 to the Salle de la rue Peletier (cap. 1,954), opening 6 Feb. with *Isouard's *Aladin, ou La lampe merveilleuse*. In this work, suitably, gaslight was first used. Operas by *Rossini, *Donizetti, *Auber, *Meyerbeer, *Hérold, and *Halévy, in particular, were given in sumptuous stagings by *Cicéri and *Daguerre; and the conventions of the Opéra dictated a 5-act work, with multiple soloists and choruses, an Act 2 ballet, and no spoken dialogue. In 1849 Meyerbeer, the most successful of his day in meeting these requirements, had his *Le prophète* produced with electric lighting. This was the period in which the Opéra, under Louis Véron, dominated Europe, setting a grand style, rewarding success with fame and fortune, imposing powerful but demanding standards as an orthodoxy, and thereby compelling composers as diverse as Wagner, Verdi, and Tchaikovsky to attempt the challenge. Wagner failed to achieve performance with *Rienzi*, and met a notorious scandal when his rewriting of *Tannhäuser* did not satisfy in 1861. Berlioz never had *Les Troyens* produced here in his lifetime.

In 1873 the Opéra again burnt down, and the company moved to the Salle Ventadour. Charles Garnier's sumptuous theatre was planned in 1867, but delayed by the Franco-Prussian War. The 'Nouvel Opéra' in the Salle du Boulevard des Capucines (or Palais Garnier) (cap. 2,156) opened 5 Jan. 1875, and was one of the largest and grandest theatres in the world. The auditorium was electrically lit 1881, the stage 1887. The policy was virtually that of a museum of French opera. Under the directorship of Jacques Rouché, 1915–45, this policy was extended; he preserved tradition, resisting innovation in its name, welcomed only the most orthodox composers, and isolated his singers from international careers while also generally excluding even the most distinguished foreign artists.

In 1939 the Opéra and Opéra-Comique were united as the Réunion des Théâtres Lyriques Nationaux. Rouché continued in office, and his policies endured after his retirement aged 82, though there was a gradual relaxation of the rule against spoken dialogue (finally admitting *The Magic Flute*, 1954) and then foreign languages. This, and a more adventurous policy about modern opera, did not halt the Opéra's decline. Rolf Liebermann was director 1973–80, placing the Opéra back on the world stage, even if losing it some of the old character while retaining some of the abuses. Among important events were the premières of the complete *Lulu* (1979) and Messiaen's *St François d'Assise* (1983). After fierce public controversy, including the dismissal of the intended music director, Daniel Barenboim, in 1989, and amid technical troubles, the new Opéra-Bastille (cap. 2,716) opened 17 Mar. 1990 with the first almost complete French performance of *Les Troyens*, under the new music director, Myung-Whun Chung. It was intended as an 'opera for the people'. The building includes two smaller halls.

BIBL: J.-G. Prod'homme, *L'Opéra* (*1669–1925*) (Paris, 1925); C. Dupêchez, *Histoire de l'Opéra de Paris 1875–1980* (Paris, 1985); M. de Saint Pulgent, *Le syndrome de l'Opéra* (Paris, 1991).

**Opera 80.** See OPERA FOR ALL.

**opéra-ballet.** A dramatic entertainment popular in France in the early 18th cent. in which operatic elements were combined with ballet. Its characteristic feature was a lack of dramatic continuity; each *entrée presented a discrete plot, and usually included at least one *divertissement. Beginning with Campra's *L'Europe galante* in 1697, 18 opéras-ballets were given by the Académie Royale de Musique in the years up to 1735.

The structural model for the new genre was provided by Pascal Colasse's *Ballet des saisons* (1695), but this work differed from the opéra-ballet proper in its retention of a mythological plot. In contrast to the earlier traditions of *ballet de cour and *tragedie en musique, the opéra-ballet took as its subject-matter contemporary figures, so-called *petits maîtres* and their ladies, and through them elevated lyrical comedy to the high status formerly enjoyed only by tragedy on the French stage. In his most successful essay in the genre, *Les fêtes vénitiennes* (1710), Campra included parodies of some of the most famous operas of the day.

In part a reaction against the heroic, galant tradition of Lully, the opéra-ballet was eventually supplanted by the *ballet-héroïque, whose elevated tone, if not dramatic organization, approached more closely the world of the tragédie en musique.

**opéra-bouffe** (Fr.). A designation frequently given to satirical 19th-cent. French operetta: the classic example is Offenbach's *Orfée aux enfers*.

**opera buffa** (It.: 'comic opera'). Following their first appearance in Landi's *La morte d'Orfeo* (1619), comic scenes were included in most operas of the 17th-cent. Roman and Venetian repertory, where they provided relief from the prevailing serious tone of the drama. In the later works of Monteverdi (especially *Poppea*), and of Cavalli, humorous characters, who are usually pages, servants, or nurses, play a substantial role. Often portrayed as cripples or stutterers, they contrast markedly with the more refined main characters, and express their down-to-earth thoughts with a vigour which sometimes borders on vulgarity: parallels are often made between their treatment and that of the comic characters in Shakespeare's tragedies.

Beginning with Mazzocchi's *Chi soffre, speri* (1639) and Abbatini's *Dal male il bene* (1653), the exclusively comic opera was born, although it never took root in the 17th cent. With the libretto reforms of *Zeno and *Metastasio in the early 18th cent. comic elements were completely excised from opera seria and this led to an increased cultivation of the *intermezzo, the genre of 1-act comic opera which had first been seen as early as 1623. As the rigid conventions of opera seria came increasingly to be regarded as too restrictive, composers began to favour the intermezzo, which permitted more flexible handling of the aria and a more natural approach to the drama.

Alongside the independent production of intermezzos outside the context of opera seria there developed the new genre of opera buffa, which shared its essential features, but was larger in scale and constituted a full-length entertainment in its own right. It may be said to have begun with Orefice's *Patrò Calienno de la Costa*, whose first performance in Naples in 1709 established the city as the main centre for comic opera in the early and mid-18th cent.

Although technically still an intermezzo, Pergolesi's *La serva padrona* (1733) is often regarded as the first great example of opera buffa. Like most later works, it deals with common, everyday characters, rather than the heroes and kings of opera seria, eschews the *castrato in favour of natural voices, particularly the bass, and places some emphasis on the *ensemble. Pergolesi's work was supported by that of Logroscino, Leo, Vinci, Paisiello, Piccinni, Fischietti, and Cimarosa among others; particularly significant for the development of opera buffa were the librettos of *Goldoni, whose importance matches that of Metastasio for opera seria. At the heart of opera buffa lay an interest in sharp characterization which was achieved partly by witty and lively librettos, partly through the music itself, especially in the *ensemble finales. Rather than noble bearing, or formalized expressions of love, sentimentality and everyday emotions provided the spiritual essence of opera buffa. Characters were drawn from real life, or were extensions of the *commedia dell'arte tradition: particularly important was the role of the mute. During the mid-18th cent. the inclusion of *parte serie* (serious roles) as a foil to the antics of the comic characters proved effective in enhancing the emotional scope of opera buffa, and led to the development of the *dramma giocoso.

Although some composers enjoyed a special reputation for their opere buffe, many wrote them alongside serious works. With the rise of the Enlightenment, the previous supremacy of opera seria was overthrown and such works tended to remain in the province of courts and similar institutions, while opera buffa became the more popular Italian operatic genre: thus in Vienna in the 1780s it was the opere buffe of such composers as Martín y Soler and Sarti which provided the main challenge to Singspiel, not opera seria. Towards the end of the 18th cent. the distinction between comic and serious began to blur, and such hybrid genres as the dramma giocoso and *opera semiseria assumed greater importance. But this was not before Mozart had provided a fitting culmination to the development of opera buffa in his masterpieces *Le nozze di Figaro* and *Così fan tutte*, where broad adherence to the main conventions of the genre is tempered by a uniquely enquiring, yet humane, approach to outwardly comic situations, allowing him to uncover the real unhappiness, even tragedy, which they actually conceal.

**opéra-comique** (Fr.: 'comic opera'). Despite its literal meaning, 'comic opera' conveys a false impression of this vague but generally accepted term: indeed, the French themselves understand different things by it according to the date of its use. The origins of opéra-comique lay in the *comédies-ballets of Lully and Molière, which combined music and dance within a spoken drama, and in the activities of the Italian troupe resident in Paris from 1660. Their music consisted of *vaudevilles, parodies of serious opera, and some original works; although their productions at first involved interpolating such music into a spoken Italian text, on which the tradition of the *commedia dell'arte exerted a strong influence, French scenes were increasingly accommodated. Often farcical, if not downright coarse and vulgar in its productions, the Comédie-Italienne was banished from Paris in 1696, following an attack on the king's mistress Mme de Maintenon in the satire *La fausse prude*. Their repertoire, now exclusively French, was taken over in a somewhat modified form by the so-called Théâtres de la Foire, hence the derivation of the name by which it came to be known, *drame forain.

Favart, the first important librettist of opéra-comique, began his career writing for the fairs; though he continued the earlier tradition of musical parody, particularly of serious operas, he also encouraged the composition of entirely new vaudevilles. By now often referred to as opéra-comique, and already embracing the later essential features of the genre, the drame forain might have remained of secondary importance had it not been for the *Guerre des Bouffons. Out of the strife between the advocates of Italian comic opera and French tragédie lyrique, it gave rise to a new style of popular opera, represented by Rousseau's *Le devin du village*, which combined the Italian opera buffa structure (i.e. continuous music with recitative) with a typically French plot. Although this model had few direct imitators, it inspired a new generation of opéra-comique composers, including Duni, Monsigny, and Philidor among many others. For them opéra-comique was a genre in which musical numbers cast in a simple style were linked by spoken dialogue: this form was also known as the *comédie mêlée d'ariettes. At this stage the opéra-comique provided a sharp contrast with the earlier tradition of tragédie lyrique: the characters were not kings or mythological figures, but everyday people, often peasants, who lived in a state of natural contentment (reflective of Rousseau's philosophy).

With the next generation a certain emphasis on the heroic may sometimes be discerned, especially in such works as Grétry's *Richard Cœur* *de Lion* and Boieldieu's *Jean de Paris*. Certainly, the origins of opéra-comique in low farce had by now long been transcended, and the very word 'comique' seemed increasingly less appropriate for a genre in which serious themes or episodes played such a distinctive role. For a while opéra-comique developed along lines similar to those of the Italian *opera semiseria and there are many similarities between the works of Paisiello and Cimarosa, and the French tradition of the early 19th cent., represented above all by Dezède and Dalayrac. Gradually the worlds of serious opera and opéra-comique began to blend, both in their choice of subject-matter and general musical technique, so that the remaining distinctive feature of the latter genre was its retention of spoken dialogue rather than sung recitative.

In the 19th cent. the designation opéra-comique was thus applied to works which varied enormously in their physical and emotional scale. Serious, indeed tragic, themes were presented in such works as Méhul's *Joseph* and Cherubini's *Médée*, while Boieldieu, one of the most successful composers in the genre, showed a penchant for more Romantic themes, as in his masterpiece *La dame blanche*. A lighter strain was discernible in another of his successes, *Le petit chaperon rouge*, while the older style of opéra-comique, in which there was a greater emphasis upon intrigue and farce, continued in such works as Della Maria's *Le prisonnier* and Isouard's *Le rendez-vous bourgeois*. The performance of Rossini's *Le Comte Ory* in Paris in 1828 had a distinctive effect on the development of opéra-comique, introducing a more Italianate influence into the melodic line. This is seen especially in the works of Auber, above all in his *Fra Diavolo*. By the late 19th cent. the divide between serious opera and opéra-comique had all but disappeared, since spoken dialogue now made relatively rare appearances: one such was in Bizet's *Carmen*. However, with the strict traditions of the Paris *Opéra still largely observed, the rival house, the *Opéra-Comique, was the venue for a large number of more experimental works, which thus carried the generic description as well: among them may be included Delibes's *Lakmé* and Charpentier's *Louise*.

**Opéra-Comique.** Theatre in Paris, long the city's second opera-house; originally the home for French musical pieces with spoken dialogue. In 1714 an agreement between the Comédie-Française and the Académie Royale de Musique resulted in the setting up of the Opéra-Comique, inaugurating the T de la Foire Saint-Germain with the parody *Télémaque* (music by Gillier). Its success was so great that the Académie had it closed 1745; reopened 1752 at Saint-Germain,

joined the Comédie-Italienne, 1762, first at the Hôtel de Bourgogne (rue Mauconseil), then 1783 at the Salle Favart, later known as Comédie-Italienne, Théâtre-Italien, then Opéra-Comique. A rival 'Opéra-Comique' at the T Feydeau (which premièred Cherubini's *Médée*, 1797) bankrupted them both, and they merged to form the T National de l'Opéra-Comique, 1801. In 1805 this moved to the Salle Feydeau. The company gave important premières of works by *Spontini, *Méhul, *Dalayrac, *Boieldieu, *Isouard, and other composers whose opéras-comiques were pioneering examples of the genre and the most forward-looking operas of their day.

The company moved to the Salle Ventadour 1829, then to the T des Nouveautés, and in 1840 to the second Salle Favart. Here it maintained its reputation as a more open-minded, enterprising house than the convention-bound Opéra, welcoming many important new and foreign works, as well as being the home for opera with spoken dialogue when this was banned by the Opéra. Works by *Auber, *Hérold, *Massenet, and *Adam were prominent in the repertory, and premières included *Carmen* (1875), *Hoffmann* (1881), *Lakmé* (1883), and *Manon* (1884). The theatre burnt down 1887, and Carvalho (director from 1876) was briefly imprisoned for negligence. The company eventually moved into the rebuilt Salle Favart (cap. 1,750), inaugurated 7 Dec. 1898, with Albert Carré as director (till 1925). Under him, premières included *Louise* (1900), *Pelléas* (1902), and the Opéra-Comique became more of a rival to the Opéra. In 1939 the two were united as the Réunion des T Lyriques Nationaux. Closed Apr. 1972 by the government, the Opéra-Comique housed the Opéra-Studio under Louis Erlo, Jan. 1973. It reopened with the old name Salle Favart, under Liebermann, in Dec. 1976.

**opera di obbligo** (It.: 'obligatory opera'). In 19th-cent. Italy, an opera scheduled as one of the main attractions of the season.

**opera di ripiego** (It.: 'reserve opera'). In 19th-cent. Italy, an opera which was held in reserve, only to be performed if one of the opere di obbligo failed to materialize or was delayed in its production.

**Opéra du Rhin.** A syndicate formed in 1972 to co-ordinate performances in Alsace between Colmar, Mulhouse, and Strasburg.

**Opera Factory.** Company founded by David Freeman, Sydney 1973; established Zurich 1976; sister co., in conjunction with London Sinfonietta, established 1981 with Freeman's production of *Punch and Judy*. Other Opera Factory successes were *Calisto* and *Knot Garden* (1984), the 1986 revival of *Così fan tutte*, and the première of *Yan Tan Tethera* (London, QEH, 1986). Freeman's work at London, C, includes *Orfeo*, *Ritorno d'Ulisse*, *Akhnaten*, and the première of *The Mask of Orpheus*. His demanding rehearsals and emphasis on the physical can produce remarkable performances from singers, particularly in Opera Factory; his most striking productions can change a traditional view of the classical repertory (e.g. *Così*), though his strong ideas and inclination to the sensational can sometimes diminish the music.

**Opera for All.** A group founded in 1949 by the Arts Council of Great Britain to take opera to small towns, even villages, thoughout the country. The initial group of six members expanded by 1966–7 into three groups of 12, touring in a minibus, based on the London Opera Centre, WNO, and Scottish O. By 1974 the Scottish and Welsh groups had expanded into autonomous ensembles, and Opera for All's London group remained the only one with the original name. In 1980 it was replaced by Opera 80, touring with a small orchestra and chorus.

**opéra-lyrique** (Fr.: 'lyric opera'). A term used, mostly outside France, for a style of opera that flourished in 19th-cent. France in distinction to *Grand Opera and *opéra-comique. In general, it was shorter and less rich in choral, scenic, and other effects than the former, while dispensing with the spoken dialogue of the latter. Beginning in the late 18th and early 19th cents., it developed in the works of Gounod and Thomas, and later Massenet, as a genre in which, on the whole, romantic elements were handled with a clarity, simplicity, and elegance that was at once more intense than opéra-comique in its more literally comic manifestations, and yet lighter in treatment than the most profound serious opera. Though it was not a term in general use among composers, Massenet did designate some of his operas as *drame lyrique*, *épisode lyrique*, *pièce lyrique*, and *conte lyrique*.

**Opera North.** Formed in Leeds in 1975, the company opened in 1978 as English National Opera North with *Samson et Dalila* under its music director, David Lloyd-Jones. It became independent in 1981, assuming its present name. Among productions that have won it a name for enterprise and high standards have been the première of Josephs's *Rebecca* (1983), the British premières of Krenek's *Jonny spielt auf* (1984), Strauss's *Daphne* (1987), Verdi's *Jérusalem* (1990), and Nielsen's *Maskarade* (1990), and productions of *Oedipus rex* (1981), *La traviata* (1985), *The Trojans at Carthage* (1986), Kern's *Show Boat* (1989), Dukas's *Ariane et Barbe-bleue* (1990), and Tippett's *Midsummer Marriage*

(1985) and *King Priam* (1991). The company tours primarily in the north of England.

**Opera of the Nobility.** The opera company established in London in 1733–7 in opposition to the Royal Academy of Music. It was initially based at Lincoln's Inn Fields T, but later moved to the King's T. See HANDEL.

**opera semiseria** (It.: 'semi-serious opera'). The introduction of serious elements into opera buffa first occurred *c*.1750 in the librettos of *Goldoni. The style thus created, known as *dramma giocoso, gradually came to be more popular than the straight opera buffa, whose characters had always had a tendency to be mere comic stereotypes. During the later 18th cent., under the influence of the French comédie larmoyante, sentimental or melodramatic colouring became ever more popular, a characteristic which was formally recognized in the choice of such generic terms as **dramma di sentimento**, **dramma eroicomico**, and **dramma tragicomico**. From this tradition there developed opera semiseria, a term first used with reference to Paisiello's *Nina* (1789). Sharing their mixture of the comic and the sentimental, opera semiseria dealt with much of the same subject-matter, including the popular situation where the path of true love is thwarted by the conventions of society. A celebrated essay in the genre was Cimarosa's *Il matrimonio segreto* (1792).

**opera seria** (It.: 'serious opera'; sometimes also **melodramma serio**). The main operatic genre of the early- and mid-18th cent., opera seria was born out of the libretto reforms of *Zeno and *Metastasio, though some features of its style can be traced back to the late 17th cent. At the heart of the genre lay two main concerns: to deliver, in an effective and appropriate manner, elegant verses, carefully crafted according to the *doctrine of affections, and to provide a display vehicle for the talents of the solo singers whose presence now dominated opera. Unlike the Venetian style which it replaced, opera seria eschewed comic characters and scenes; these found their way into the *intermezzo, and thence eventually into *opera buffa.

The earliest phase of opera seria has often been described as **Neapolitan opera**, a misleading term in that neither the developments which led to the formulation of the genre, nor the activities of the composers associated with it, were restricted to the city. Of a large school, which also included Porpora, Vinci, and Leo, the most important was undoubtedly Alessandro Scarlatti. While his early works still owe a clear debt to the Venetian tradition, the later operas, such as *Griselda* (1721), adumbrate many of the precepts of opera seria in its classic form. Foremost among these was the presentation of a rationalist plot, frequently involving a conflict of love with duty, in which elevated characters, usually historical or quasi-historical, participated. The primacy of the solo singer, especially the *castrato, led to an emphasis upon the *da capo aria, which was a point of emotional repose whose composition and performance was hedged round by conventions and restrictions, while action on stage was advanced in the *recitative. The division of the arias among the singers, and the character in which they were to be cast, was also rigidly defined and largely adhered to (see ARIA).

The full flourishing of the opera seria was reached in the music of Hasse, who in works such as *Cleofide* raised it to a supreme and confident reflection of 18th-cent. ideals. Around the same time, Handel was engaged in London concert life, to which he contributed a notable succession of opere serie. Here the Metastasian conventions were subtly modified, in part under the influence of French opera. Together, the works of Hasse and Handel demonstrate the full potential of a genre which has often been described as over-conventionalized and sterile, largely because of constraints upon dramatic articulation placed by the constant alternation of aria and recitative. Though such criticisms have little validity if opera seria is judged on its own terms, several attempts were made to reform it during the later 18th cent. The most important of these were undertaken by Gluck, following earlier experiments by Jommelli and Traetta. Beginning in *Orfeo* and continuing in *Alceste*, whose famous preface provides a cogent explanation of his aims, he stripped opera seria of its musical excesses in an effort to provide a dramatically more satisfying work. This was mainly achieved by giving a functional role to the chorus, after the model of Rameau, as well as through a general simplification of the arias, which were reshaped along the lines of the French air and released from the confinement of the da capo structure.

Much has been made of Gluck's reforms, although their effect was short-lived. A certain influence can be traced in Mozart's *Idomeneo* (1781) and Haydn's *L'anima del filosofo* (1791), but the eventual overthrow of opera seria's supremacy came not through reform, but as a result of the challenges laid down by opera buffa and Singspiel. Gradually, these became the most widely performed genres on the public stage, while opera seria retreated into the world of court opera. One of the last, and finest, opere serie is Mozart's *La clemenza di Tito* (1791).

Also the title of an opera by Gassmann (1769).

**Opéras-minute.** The name for three short chamber operas by Milhaud, *L'enlèvement d'Europe*, *L'abandon d'Ariane*, and *La délivrance de Thésée*, to texts by H. Hoppenot, the first prod. Baden-Baden 17 Aug. 1927, the other two Wiesbaden 20 Apr. 1928. They really form a single entertainment, and were first prod. as such Budapest (in Hung.) 28 Feb 1932; Paris, T Récamier, 27 Mar. 1963. Each lasts no more than 10–15 minutes, and treats the classical legends of Europa, Ariadne, and Theseus with respect though not without humour.

**operetta** (It.: 'little opera') (Fr., *opérette*). Originally used in the 17th cent. for a short opera, the term became associated by the 19th cent. with comic opera, to describe a play with an overture, songs, interludes, and dances. Beyond this, there is no single definition which is applicable to all the works usually thus described. In the 20th cent. a further development of operetta resulted in the *musical comedy.

Two main strands of operetta may be identified. One was French, beginning with Offenbach, and continuing with such composers as Lecocq, and one Viennese. The latter, which drew inspiration from the established comic theatrical tradition in Vienna, may be traced back to Suppé and Johann Strauss II, although it achieved its most characteristic expression in the works of Lehár, Millöcker, Kálmán, and Ziehrer. In England the *Savoy operas of Gilbert and Sullivan are really operettas; in the USA an indigenous tradition was established with the works of Victor Herbert and Reginald de Koven.

BIBL: R. Traubner, *Operetta* (New York, 1990).

**opérette.** See OPERETTA.

**Opernball, Der** (The Opera Ball). Operetta in 3 acts by Heuberger; text by Léon and Waldberg, after Delacour and Hennequin's farce *Les dominos roses*. Prod. Vienna, W, 5 Jan. 1898, with Dirkens; New York 24 May 1909, with Morena, Bartet, Beria.

Heuberger's most successful work.

**Ophelia.** Polonius's daughter in Shakespeare's *Hamlet*. She appears as Ophélie (sop) in Thomas's *Hamlet*.

**'Ora e per sempre addio'.** Otello's (ten) farewell to his past glories in Act II of Verdi's *Otello*.

**Orange.** Town in Vaucluse, France. The Roman theatre was excavated 1835 (cap. 10,000). The first opera staged was Méhul's *Joseph*, 21 Aug. 1869, and operas have been given at various times since. The choice usually falls on familiar works with visiting star singers.

**oratorio.** Usually defined as a musical composition with a text of religious or contemplative nature. In its early stages of development the oratorio was sometimes staged, which has led to much confusion between it and related operatic genres. In essence the distinction which must be drawn is not whether a work was intended to be staged, or not, but whether the libretto is of a dramatic or contemplative character: thus Cavalieri's *Rappresentazione di anima e di corpo* is clearly a sacred opera, and not the first oratorio, as is often claimed. Further confusion is caused by a general similarity of musical style in the 18th cent., as well as changes in the use of terminology (see AZIONE SACRA). Handel's oratorios are particularly problematic. Mostly written after he had been forced to retire from operatic composition, the very first were indisputably staged; although the practice of performing sacred works in a dramatic context was soon banned, some of Handel's later oratorios contain stage directions, suggesting that he at least conceived of them in dramatic terms.

**orchestra pit** (Fr.: *fosse d'orchestre*, Ger.: *Orchestergraben*, It.: *fossa d'orchestra*). The space before and below the stage which contains the opera orchestra. The partly hooded pit in the Bayreuth Festspielhaus, designed by Wagner and described by him as the 'mystic gulf' (mystischer Abgrund), allows an exceptionally faithful balance and a tonal blend especially suitable for his works, as well as an absence of visual distraction for the audience. Most opera-houses are now built with pits in sections of variable heights to assist the balance of different operas.

**Orefice, Antonio** (*fl.* Naples, 1708–34). Italian composer. A lawyer by profession; first came to attention with the opera seria *Maurizio*. His *Patrò Calienno de la Costa* set a comic libretto in Neapolitan dialect; such was its success that several of Orefice's contemporaries, including Falco, Leo, and Vinci, were inspired to write similar operas. Composed *c*.14 further comic works for Naples; most of his music is now lost.

SELECT WORKLIST: *Maurizio* (Minati; Naples 1708, lost); *Patrò Calienno de la Costa* (Mercotellis; Naples 1709, lost).

**Orefice, Giacomo** (*b* Vicenza, 27 Aug. 1865; *d* Milan, 22 Dec. 1922). Italian composer and critic. Studied Bologna with Mancinelli, offering *L'oasi* as his graduation exercise. Performed widely as a piano virtuoso before becoming professor of composition at Milan Cons. 1909; from 1920 music critic of the Milan *Secolo*. He was one of the first in Italy to encourage interest in older music and raise editorial standards, himself preparing versions of Monteverdi's *Orfeo* and Rameau's *Platée*.

Orefice wrote 11 operas which met with a mixed reception: they include *Chopin*, based on the composer's life and using his music; *Mosè*, his most successful work; and two based on Russian authors, *Il pane altrui* and *Radda*. These technically proficient works belong to the verismo tradition.

SELECT WORKLIST: *L'oasi* (The Oasis) (Dal Monte; Bologna 1885); *Chopin* (Orvieto; Milan 1901); *Mosè* (Orvieto; Genoa 1905); *Il pane altrui* (The Bread of Others) (Orvieto, after Turgenev; Venice 1907); *Radda* (Vallini, after Gorky; Milan 1913).

**Orest.** Orestes (bs-bar), Elektra's brother, in Strauss's *Elektra*.

**Oresteia.** Opera-oratorios by Milhaud; texts by Paul Claudel, based closely on Aeschylus' tragedies. 1. *Agamemnon* comp. 1913; prod. Paris 16 Apr. 1927. Clytemnestre greets her husband on his return from the war, but then kills him, in revenge for his sacrificing their daughter Iphigénie in order to gain victory.

2. *Les Choéphores* prod. Paris 8 Mar. 1927 (concert); Brussels, M, 27 Mar. 1935 (staged); New York 16 Nov. 1950 (concert). This relates the story of Oreste, who kills his mother and her lover Aegisthe so as to avenge his father's death at his wife's hand. He is thereafter tormented by the Furies.

3. *Les Euménides* comp. 1917–22; prod. W Berlin Apr. 1963. This concerns Athene's and Apollon's pleas on behalf of Oreste and the eventual transformation of the Furies into the kindly Eumenides.

**Orestes.** In classical mythology, the son of Agamemnon and *Clytemnestra and brother of *Iphigenia and of *Elektra, at whose encouragement he avenged his father's murder by his mother. Pursued by the Furies for his matricide, he won expiation by going with his friend Pylades to Tauris, and with the help of Iphigenia carrying off a statue of Artemis. Operas on him are by Perti (1685), Pollarolo (1697), Micheli (1722), Perez (1744), Monza (1766), Agricola (1772), Cimarosa (1783), Horzicki (*c*.1795), Federici (1804), Morlacchi (1808), C. Kreutzer (1818), Asti (1820), and Alberti (1872). Taneyev's *Oresteia* (1895) is a three-part treatment of the full legend. Orestes also appears (bar) in Gluck's *Iphigénie en Tauride* and R. Strauss's *Elektra*, and other settings of that part of the legend.

**Orfeide, L'.** Operatic triptych by Malipiero; text by the composer, comprising (1) *La morte delle maschere*, (2) *Sette canzoni*, and (3) *Orfeo, ovvero L'ottava canzone*. (2) prod. Paris, O, 10 July 1920, with Lapeyrette, Noel, Duclos, cond. Grovlez. First complete prod. Düsseldorf 31 Oct. 1925, with Nettersheim, Schilp, Domgraf-Fassbänder, cond. Orthmann.

Each of the three parts is independent. In *La morte delle maschere* masked commedia dell'arte figures are replaced with characters from real life. *Sette canzoni* consists of seven separate dramatic scenes, the majority of which depict one individual behaving harshly to another, or celebrating in the face of sorrow. Each is an expansion of a brief moment from life. The texts are drawn from old, mostly Renaissance, Italian sources. The third opera of the triptych, *Orfeo*, continues from the scenes of the previous opera; Orfeo sings a song which lulls everyone in a theatre to sleep, except the Queen, with whom Orfeo departs. This opera, like the first of the triptych, deals with the conflict between the artificiality of the theatre and reality.

**Orfeo, La favola d'** (The Legend of Orpheus). Opera in a prologue and 5 acts by Monteverdi; text by Alessandro Striggio. Prod. (privately) Mantua, Accademia degli Invaghiti, Feb. 1607, and Mantua, Court T, with Gualberto, 24 Feb. 1607. Rev. Paris 25 Feb. 1904, concert version by Schola Cantorum (arr. D'Indy); first modern stage perf. Paris, T Réjane, 1911; New York, M (concert), 14 Apr. 1912, with Fornia, Weil, Witherspoon; first US prod. (Malipiero version) Northampton, Smith Coll., 14 May 1929, with Kullman. First English perf. London, Institut Français, in D'Indy version (concert), 1924; first English prod. Oxford 7 Dec. 1925, opening prod. of OUOC, cond. Westrup.

After a prologue, nymphs and shepherds are discovered rejoicing at the coming marriage of Orfeo and Euridice. In a wood, Orfeo sings to nature, but is interrupted by La Messaggera (the Messenger) bringing the news of Euridice's death. He descends to Hades to find her, and is comforted by Hope; lulling Charon with his song, he crosses the Styx and wins the agreement of Pluto and Proserpina that he may take her back to earth if he will not turn round to look at her on their journey. Unable to resist assuring himself that she is indeed there, he turns just as they reach the light, and she is snatched back to Hades. Orfeo laments his lot, but is consoled by Apollo with the promise of immortality with Euridice.

Of many editions, the most famous (in some cases notorious) are those by Eitner (1881), D'Indy (1904), Orefice (1909), Malipiero (1923), Orff (1925), Westrup (1925), Benvenuti (1934), Respighi (1935), Hindemith (1954), Wenzinger (1957), Maderna (1967), Stevens (1967), and Leppard (1965).

BIBL: J. Whenham, *Claudio Monteverdi: 'Orfeo'* (Cambridge, 1986).

**Orfeo ed Euridice.** Opera in 3 acts by Gluck; text by Calzabigi, after the classical legend. Prod. Vienna, B, 5 Oct. 1762, with Bianchi, Guadagni, Clavarau, cond. Gluck; London, H, 7 Apr. 1770, with Guadagni, Zamparini; New York, Winter Garden, 25 May 1863, with Vestvali, Rotter,

Geary. French version, trans. Moline, and with title-role transposed for tenor, prod. Paris 2 Aug. 1774, with Legros, Arnould, Levasseur, cond. Francœur. Revised by Berlioz and prod. Paris 1859, with Viardot, Sass, cond. Berlioz.

Greece, mythological times. Orfeo (con, in revision ten) mourns the death of his wife Euridice (sop). His grief is such that Zeus, the father of the gods, permits him to journey to Hades to try to reclaim her. If he can persuade Pluto through the power of his music, he may lead her home, but must not glance back at her until they have crossed the Styx.

The Furies try to block Orfeo's path and call on Cerberus, the triple-headed dog, to tear him to shreds. However, Orfeo charms them with his singing and continues on his way. In the Elysian fields he finds Euridice, and begins to guide her back to earth.

Euridice cannot understand her husband's strange behaviour, which he is forbidden to explain. Believing he no longer loves her, she says she would rather die. In desperation, Orfeo turns to face her; she falls dead. But Amor (sop), touched by the beauty of the lament Orfeo sings, restores Euridice to life.

BIBL: P. Howard, *C. W. Gluck: 'Orfeo ed Euridice'* (Cambridge, 1981).

For other operas on the subject, see ORPHEUS.

**Orff, Carl** (*b* Munich, 10 July 1895; *d* Munich, 29 Mar. 1982). German composer. Studied Munich, where he wrote his first opera, *Gisei*. Worked as répétiteur and then conductor at Munich, Mannheim, and Darmstadt. In 1920 he resumed study with Kaminski, who encouraged him in the study of old music, resulting in arrangements of Monteverdi, including his first of *Orfeo* (1925; other versions 1931 and 1940). In 1924 he founded (with Dorothée Günther) the Günther School in Munich; here a revolutionary approach was adopted, with musical gifts encouraged by their expression through gymnastics and dance. For this purpose Orff developed the concept of the *Schulwerk* (school-work), music dedicated to the pedagogical exercise, using recorders, percussion, and other technically straightforward instruments.

*Carmina Burana* (1937), called a scenic cantata since it is not staged but performed to the accompaniment of mime and dance, brought Orff immediate fame. Its methods were used, less successfully, in *Catulli carmina* (1943) and *Trionfo di Afrodite* (1953): all set Latin texts and appeal unashamedly to the most primitive levels of emotion in the musical setting. These were the first works to explore Orff's so-called 'total theatre', the fundamental principles of which underpin all his dramatic music. Seeking to free such works from what he considered the exaggerations that had accrued to them by the beginning of the 20th cent., he returned to using elementary rhythms and popular folk-song; counterpoint and thematic development were eschewed and opera was regarded as a total work of art. In practice this resulted in a series of works with such interrupted dramatic action and static settings that the description 'opera' may be but loosely applied. *Der Mond* (1939), a Bavarian fairy-tale, employs a narrator to tell the story, only three episodes of which are seen on the stage: it contains some of Orff's most lyrical music, as does the highly successful *Die Kluge* (1943). Similar techniques are used in *Die Bernauerin* (1947), *Oedipus der Tyrann* (1959), *Ludus de nato infante mirificus* (1960), *Prometheus* (1966), and *De temporum fine comoedia* (1973). Only in *Antigonae* does Orff approach the spirit of genuine opera in a sensitive and original treatment of Sophocles' tragedy which attempts to recreate the earliest operatic experiments.

In 1950–5 he taught composition at the Munich Hochschule.

SELECT WORKLIST: *Gisei* (after Florenz; Munich 1913); *Der Mond* (Orff, after Grimm; Munich 1939); *Die Kluge* (Orff, after Grimm; Frankfurt 1943); *Antigonae* (Sophocles, trans. Hölderlin; Salzburg 1949).

BIBL: R. Münster, ed., *Carl Orff: das Bühnenwerk* (Munich, 1970).

**Orgeni, Aglaja** (*b* Rimászombat, 17 Dec. 1841; *d* Vienna, 15 Mar. 1926). Hungarian soprano. Studied Baden-Baden with Viardot. Début Berlin, 1865 (Amina). London, CG, 1866. Extensive career in Europe 1867–86, celebrated in coloratura roles, and as Agathe, Leonora (*Trovatore*), etc. Taught 1886–1914 Dresden Cons., where she was the first female Royal Professor.

**Orlandini, Giuseppe Maria** (*b* Florence, 19 Mar. 1675; *d* Florence, 24 Oct. 1760). Italian composer. Probably studied Florence; *c*.1723 maestro di cappella to Grand Duke of Tuscany; 1732 maestro di cappella at Florence Cathedral. Composed *c*.44 operas which enjoyed considerable popularity throughout Italy. Although he also wrote opere serie, his most characteristic music is found in his comic works. Most famous is the intermezzo *Il marito giocatore*: possibly the most frequently performed operatic work of the 18th cent.—often as *Serpilla e Bacocco*—it reveals a light, witty touch and was important in the development of opera buffa.

SELECT WORKLIST: *Il marito giocatore* (The Gambling Husband) (Salvi; Venice 1719).

**Orlando.** Opera in 3 acts by Handel; text an anon. adaptation of Carlo Sigismondo Capece's libretto, after Ariosto's poem *Orlando furioso* (1516), set by Scarlatti (1711). Prod. London, H, 27 Jan. 1733, with La Strada, Senesino; St.

Louis, Washington U., 25 Feb. 1983, with Minter, Armistead, Solomon, cond. McGegan.

For other operas on the subject see ARIOSTO. See also ARIODANTE.

**Orlovsky.** The Russian prince (con, sometimes ten) in J. Strauss's *Die Fledermaus.*

**Ormindo.** Opera in 3 acts by Cavalli; text by Giovanni Faustini. Prod. Venice, C, 1644. New version by Leppard, prod. Gly. 16 June 1967, cond. Leppard; New York, Juilliard School, 24 Apr. 1968, cond. Mester.

**Oroveso.** The High Priest of the Druids (bs) in Bellini's *Norma.*

**Orphée aux enfers** (Orpheus in the Underworld). Operetta in 4 acts by Offenbach; text by Crémieux and L. Halévy, possibly from a German scenario by Cramer. Prod. Paris, BP, 21 Oct. 1858 (2 acts); Paris, Gaité, 7 Feb. 1874 (4 acts), with Tautin, Leonce, Tayan, Désire; New York, Stadt T, Mar. 1861 (in German); with Scheller, Frau and Herr Meaubert, Krilling, Klein; London, HM, 26 Dec. 1865, adapted by Planché as *Orpheus in the Haymarket.*

In this parody of the Orpheus legend, Orphée (ten) is a violin-teacher in Thebes; his wife Eurydice (sop) has a lover Pluton (ten), disguised as the shepherd and bee-keeper Aristée. Pluton carries her away to the Underworld, so she leaves Orphée a note saying she is dead. Opinion Publique (mez) demands that he try and recover Eurydice from Hades. At Mount Olympus the gods are complaining about Jupiter's (bar) unacceptable behaviour, and of their unvarying diet of nectar and ambrosia. Orphée and Opinion Publique arrive and, at Jupiter's command, Pluton is compelled to hand back Eurydice. The gods decide to go to Hades to look for her.

In Hades, Eurydice sings a love duet with Jupiter, who has taken the form of a fly. Resuming his true shape, he turns her into a Bacchante and together they dance the cancan. Orphée enters and is warned that he must not look at his wife as he leads her back to earth. But Jupiter lets loose a thunderbolt which causes Orphée to jump round in fright, and so he has to forgo Eurydice, who stays as a Bacchante.

**Orpheus.** In classical mythology, the poet and singer who could charm wild animals with the beauty of his song. When his wife Eurydice died, he followed her to Hades and won her back by his art, on the condition that he would not turn to look at her until he reached the upper world again. At the last moment his loving anxiety overcame him and, turning, he saw her snatched back to Hades. His grief turned him against all other women, and he was torn to pieces by Maenads in Thrace. The fragments of his body were collected by the Muses and buried at the foot of Olympus. The story first appears in early Greek writings; it is set down by Virgil in the 4th Georgic. The legend acquired religious significance, but it was principally the account of Eurydice's rescue, as told in Ovid's *Metamorphoses*, x. 1–85, that first drew opera composers, together with the theme of the musician imposing order upon the natural world by his art.

Operas on the subject (normally named after either or both the names Orpheus and Eurydice) are as follows: Peri (1600, the first true opera to survive), Caccini (1600), Monteverdi (*La favola d'Orfeo*, 1607), anon. (*Il pianto d'Orfeo*, 1608), Belli (1616), Landi (1619), anon. (*La morte d'Orfeo*, 1622), anon. (*La favola d'Orfeo*, 1633), Schütz (1638), Rossi (1647), D'Aquino (1654), Loewe (*Orpheus aus Thracien*, 1659), Sartorio (1672), Di Dia (1676), Della Torre (1677), Draghi (1683), Krieger (1683), Sabadini (*Amore spesso inganna*, 1689), Kuhnau (1689), L. and J. B. Lully (1690), anon. (1694), various (*Orfeo, ossia Amore spesso inganna*, 1695), anon. (?partly Sartorio) (*Orfeo a torto geloso*, 1697), various (*Le finezze d'amore*, 1698), ?R. Goodson (1698), Campra (1699), Dauvergne (18th cent., unperf.), anon. (1701), Keiser (pt 1, *Die sterbende Eurydike*, 1702; pt 2, *Die verwandelte Leyer des Orpheus*, 1702), anon. (*Orfeo a torto geloso*, 1706), Keiser (rev. and conflation of his operas above, 1709), Fux (1715), Telemann (1726), anon. (1729), Wagenseil (1740), Hill (1740), Ristori (1750), Graun (1752), Gluck (1762), Tozzi (1775), Bertoni (1776), Torelli (1781), Benda (1785), Naumann (1786), Bertoni (from opera of 1776) and Reichardt (1788), Amendola (1788), Winter (1789), Trento (1789), Haydn (*L'anima del filosofo*, comp. 1791, prod. 1951), Paer (1791), Lamberti (1796), Morolin (1796), Bachmann (1798), Cannabich (1802), Kanne (1807), Sampieri (1814), Godard (1887), De Azevedo (1907), Malipiero (trilogy, *L'Orfeide*, 1918–22), Krenek (1926), Milhaud (*Les malheurs d'Orphée*, 1926), Casella (*La favola d'Orfeo*, 1932), Damase (1972), Febel (1983). Bialinski (*Orpheus in the 20th century*, 1981); Birtwistle (1986).

Comic versions, parodies etc., are as follows: Barthélémon ('musical burletta' by Garrick, 1767); various (parody of Gluck, *Roger Bontemps et Javotte*, 1775), Dittersdorf (*Orpheus der Zweite*, from opera *Die Liebe im Narrenhaus*, 1788), Deshayes (*Le petit Orphée*, 1792), Kauer (parody, 1813), Offenbach (*Orphée aux enfers*, 1858), Michaelis ('Posse mit Gesang', *Orpheus auf der Oberwelt*, 1860), Konradin (*Orpheus im Dorfe*, 1867), Casiraghi and Offenbach (parody in Milanese dialect, after Offenbach, *Orfeo in Vioron*, 1871), Casiraghi and Offenbach (parody in Milanese dialect, reworking of *Orfeo in Vioron*, as

*Orfeo o La musica dell'avenire*, 1871), Selim ('commedia lirica', *Orphée et Pierrot*, 1902), Offenbach and Vicente (operetta, reduction of Offenbach, *Anda la diosa!*, 1907), and Coelho (*Orfeu em Lisboa*, 1966).

**Orpheus in the Underworld.** See ORPHÉE AUX ENFERS.

**'Or sai chi l'onore'.** Donna Anna's (sop) aria in Act I of Mozart's *Don Giovanni*, in which she reveals to Don Ottavio that it was Giovanni who tried to seduce her and upon him that he must take vengeance.

**Orsini.** Paolo Orsini (bs), a Roman patrician, in Wagner's *Rienzi*.

**Ortrud.** Telramund's wife (mez) in Wagner's *Lohengrin*.

**Osborne, Nigel** (*b* Manchester, 23 June 1948). English composer. Studied Oxford with Wellesz and Leighton, Warsaw with Rudziński. His first opera, *Hell's Angels* (1986) aroused little except distaste, but he drew much attention and admiration for *The Electrification of the Soviet Union* (the title is a quotation from Lenin, 'Communism is Soviet power plus the electrification of the whole country'). A Glyndebourne commission (Craig Raine after Boris Pasternak's *The Last Summer*, prod. 1987), this makes use of songs, a powerful orchestral narrative, some electronic music, and cinematic devices to articulate a drama whose action proceeds at different levels. The novelty of the work, and initially the difficulty of grasping the words, caused a good deal of puzzlement, but also widespread interest.

**Oscar.** The page (sop) to King Gustavus III (or Riccardo) in Verdi's *Un ballo in maschera*.

**Oslo.** Capital of Norway (Christiania 1624–1925). The first opera in Norway was given by an Italian company during a state visit by Frederik V of Denmark in 1749. The Mingotti company and others toured. The Christiania Dramatic Society, founded late 18th cent., performed opéras-comiques and Singspiels. Singspiels were given at the Strømberg T in 1827. Opera was given at the Christiania T from 1837, with visits from Italian, Swedish, and Danish companies, and some performances by Norwegian artists. The first professional performances were of Thrane's *Fjeldeventyret* (The Mountain Adventure, 1858). The theatre burnt down in 1877. In 1883–6 performances were given at the Tivoli T, and in 1890 Bjørn Bjørnson gave Gounod's *Faust* and *Carmen*, also Haarklou's *Fra gamle dage* (Of Olden Days) in 1870 and Olsen's *Lajla* in 1908. The National T opened in 1899 and gave some opera. The Opéra-Comique opened in 1918 but closed three years later having given only 26 works; opera was then given again in other theatres in mixed repertories.

In 1950 the Norsk Operaselskap was founded by Jonas and Gunnar Brunvoll with Istvan Pajor as music director. This became Den Norske Opera in Nov. 1957, with Kirsten *Flagstad as director; it opened in Feb. 1959 (cap. 1,050). Flagstad retired through ill health in 1960 and was succeeded by Odd Grüner-Hegge, who by the company's tenth year had built up a repertory of 40 operas and seven operettas, and had given 1,290 performances in Oslo and 329 on tour. The company was enlarged and improved under Lans Runsten (1969–73), then under Gunnar Brunvoll. The Czech conductor Martin Turnovský also improved standards during 1975–80 as did Heinz Fricke and then Antonio Papano. The repertory has continued to expand and includes contemporary Norwegian works: five were commissioned in 1985, from Johan Kvandal, Antonio Bibalo, Egil Hovland, Oddvar Kvam, and Trygve Madsen. Kvam's *I 13 Time* (In the 13th Hour) was given in 1987, Bibalo's *Macbeth* in 1989. The company plans to open a new opera-house by New Year's Eve 1999. Opera is also given by Summeroper in the Logen concert-hall. There is also a permanent chamber opera company, Opera Mobile, which has its own home in the Folkemuseum, Bydgøy, and tours.

**Osmin.** The overseer (bs) of the Pasha Selim's estates in Mozart's *Die Entführung aus dem Serail.*

**Osnabrück.** City in Lower Saxony, Germany. Opera was given by visiting companies in the 18th cent., together with local performances of Singspiel, opéra-comique, opera buffa, and Jesuit school drama. Through private subscription a permanent theatre was built in the Waisenhof; this was taken over by the city 1882. The T am Domhof opened 1909. Ernst Pabst was chief producer 1929–31 and 1950–5, and notable productions of works by Krenek, Korngold, and Hindemith were staged. The theatre was bombed 1945; reopened 1950.

**'O soave fanciulla'.** The love duet between Rodolfo (ten) and Mimì (sop) ending Act I of Puccini's *La bohème.*

**'O souverain! o juge!'** Rodrigue's (ten) prayer in Act II of Massenet's *Le Cid.*

**Ossian.** The name commonly given to Oisin, a near-legendary 4th-cent. Gaelic warrior and bard, son of Finn (Fingal). His 'rediscovered' works were a sensation of the 1760s: they were in fact substantially the work of James Macpherson, but despite the doubts of many, the denunciations of Dr Johnson, and the eventual exposure of the fraud, the fashion for them swept Europe and kindled an enthusiasm for Scotland as a

country of Romanticism with an alternative mythology to that of the Greek and Roman classics. Operas on Ossianic subjects are by Barthélémon (*Oithona*, 1768), Morandi (*Comala*, 1780), Shield (*Oscar and Malvina*, 1791), Kunzen (*Ossians Harfe*, 1799), Le Sueur (*Ossian*, 1804: written at Napoleon's suggestion and dedicated to him), Méhul (*Uthal*, 1806), Winter (*Colmal*, 1809), Sampieri (*Ossian e Malvina*, 1816), J. G. Kastner (*Oskars Tod*, *c.*1833, unprod.), Sobolewski (*Comala*, 1858 at Weimar, cond. Liszt), Hoffbauer (1872), Rododeato (1876), Corradi (1891), Carillo (*Ossian*, 1903), Bainton (*Oithona*, Glastonbury 1915), and I. Whyte (*Comala*, 1929). A version of Ossian provides the text for 'Pourquoi me réveiller', the aria which Werther sings in Massenet's opera as he remembers happier days when he had translated it for Charlotte.

**Ossian, ou Les bardes** (Ossian, or The Bards). Opera in 5 acts by Le Sueur; text by Palat-Dercy and Deschamps. Prod. Paris, O, 10 July 1804.

Scotland, mythical times. The Scandinavians, led by Duntalmo (bs), have captured Caledonia. Rosmala (sop) is to marry Duntalmo's son Morval (ten), although she loves the Caledonian warrior Ossian (ten). About to be sacrificed to Odin by the Scandinavians, Ossian (in a famous tableau) dreams of bards and heroes of the past. Rosmala and her father Rosmor (bs) are also captured as sacrificial victims. However, all three are rescued by the Caledonians, and the opera ends with the marriage of Ossian and Rosmala.

The fact that this was Napoleon's favourite opera (he presented Le Sueur with a gold snuffbox inscribed 'The Emperor of the French to the author of *Ossian*') undoubtedly contributed to its great popularity in the early 19th cent.

**Osten, Eva von der** (*b* Heligoland, 19 Aug. 1881; *d* Dresden, 5 May 1936). German soprano and producer. Studied Dresden with Iffert and Söhle. Début Dresden 1902 (Urbain). Sang there until 1927. Berlin, H, from 1906; London, CG, 1913–14, and HM, 1913; USA, with German OC, 1922–4. Chosen as first Octavian; also sang Ariadne, Dyer's Wife (as which Strauss thought her 'magnificent'), Brünnhilde, Isolde, Kundry, Tatyana, Louise. Married the baritone Friedrich *Plaschke. After retiring, became producer at Dresden, and was artistic adviser for the first *Arabella.* (R)

**ostinato** (It.: 'obstinate'). A melodic figure which is reiterated throughout the course of a composition, usually in the same part and at the same pitch. It makes its most characteristic appearance in opera as the **ground bass** (It., basso ostinato), where it is used as the foundation for a series of melodic variations in the upper parts, especially the voice. In opera this technique was used extensively in the works, particularly *laments, of 17th-cent. Venetian and French composers and those who imitated them, such as Purcell: Dido's 'When I am laid in earth' provides a classic example. This lament is built around a popular ostinato figure, the descending tetrachord G–F♯–F♮–E–E♭–D), usually known as a **Schmerzensbass**.

**Ostrava.** Town in Moravia, Czechoslovakia. Opera is given in the Antonín Dvořák T 3–4 times a week by a company of about 20 soloists. The theatre was reconstructed in 1945 and substantially rebuilt and enlarged in 1972.

**Ostrčil, Otakar** (*b* Prague, 25 Feb. 1879; *d* Prague, 20 Aug. 1935). Czech composer and conductor. Studied Prague with Fibich. Though initially a teacher of philology, he conducted at the National T from 1909 and was Dramaturg and conductor at the Vinohrady T 1914–18: here he introduced little-known operas by his contemporaries Foerster and Zich, also by Mozart, Weber, Berlioz, and Halévy. Director National T from 1920 (when he conducted the première of *The Excursions of Mr Brouček*). An effective and influential director, he gave cycles of operas by Smetana (1924, 1927, 1934), Fibich (1925, 1932), and Dvořák (1929, 1934). He also introduced many new works, including by Janáček, and conducted the Czech première of *Wozzeck*, among other important works. His own operas show him to be a follower of Smetana and Fibich. The most successful of them has been *Honzovo království* (Johnny's Kingdom) (Mařánek, after Tolstoy; Brno, 1934).

BIBL: J. Bartoš, *Otakar Ostrčil* (Prague, 1936).

**Ostrovsky, Alexander** (*b* Moscow, 12 Apr. 1823; *d* Shchelikovo, 14 June 1886). Russian playwright. Operas on his works are as follows:

*Don't Live as You Like* (1854): Serov (*The Power of Evil*, 1871, text partly by Ostrovsky)
*The Storm* (1859): Kashperov (1867, text by Ostrovsky); Janáček (*Káťa Kabanová*, 1921); Asafyev (1940); Trambitsky (1943); Rocca (1952); Dzerzhinsky (1955); Pushkov (1962)
*A Dream on the Volga* (1865): Blaramberg (1865); Tchaikovsky (*The Voyevoda*, 1869); Arensky (1891)
*Tushino* (1867): Blaramberg (1895)
*The Forest* (1871): Kogan (1854)
*The Snow Maiden* (1873): Rimsky-Korsakov (1882)
*An 18th century Comedian* (1873): Blaramberg (*Skomorokh*, 1887)
*The Girl without a Dowry* (1879): Novykov (comp. 1945); D. Frankel (1959); Asafyev (?)

**O'Sullivan, John** (*b* Cork, 1878; *d* Paris, 28 Apr. 1955). Irish tenor. Studied Paris Cons. Début

Geneva 1911 (Reyer's Sigurd). Paris, O, 1914, 1916–18, 1922, 1930–2; Chicago 1919–20; Milan 1924; London, CG, 1927. His fine top notes made him especially successful as Arnold, Raoul, and Manrico. Much admired by James Joyce. (R)

**Otello** 1. Opera in 4 acts by Verdi; text by Boito, after Shakespeare's tragedy. Prod. Milan, S, 5 Feb. 1887, with R. Pantaleoni, Tamagno, Maurel, cond. Faccio; New York, AM, 16 Apr. 1888, with Eva Tetrazzini, Marconi, Galassi, cond. Campanini; London, L, 5 July 1889, with Cataneo, Tamagno, Maurel, cond. Faccio.

Cyprus, 15th cent. The people of Cyprus anxiously await the safe return of the Moor Otello (ten), Venetian governor of the island. He lands unharmed, but Iago (bar) plots to kill him in revenge for Cassio's promotion over him. He obtains Cassio's (ten) dismissal by exaggerating reports of his involvement in a brawl.

Iago succeeds in making Otello believe that there is an affair between Desdemona (sop) and Cassio. So when, at Iago's instigation, she tries to secure Cassio's release, Otello refuses. Iago's wife Emilia (mez) helps him produce further 'evidence' which convinces Otello of his wife's guilt.

Otello again refuses Desdemona's plea for clemency for Cassio and, after further 'proof' supplied by Iago, decides to murder Desdemona that night. When a Venetian ambassador announces that Cassio is to succeed Otello as governor, Otello flies into a rage and flings his wife to the ground, cursing her. Prompted by Iago, Roderigo (ten), who was in love with Desdemona, attempts to kill Cassio.

Unable to understand her husband's behaviour, Desdemona recalls, in the 'Willow Song', one of her maid's tales of abandonment. Otello enters and, despite Desdemona's protests of innocence, suffocates her. Emilia reports Roderigo's death and exposes Iago's plot. Racked with guilt, Otello stabs himself, kissing Desdemona as he dies.

BIBL: J. Hepokoski, *Giuseppe Verdi: 'Otello'* (Cambridge, 1983); H. Busch, ed., *Verdi's 'Otello' and 'Simon Boccanegra'* (Oxford, 1988).

2. Opera in 3 acts by Rossini, text by Marchese Francesco Beria di Salsa, after Shakespeare's tragedy (1604–5). Prod. Naples, T del Fondo, 4 Dec. 1816, with Colbran, Nozzari, David, Manzi, Benedetti, Ciccimarra; London, H, 16 May 1822, with Camporese, Curioni; New York, Park T, 7 Feb. 1826, Malibran, García. Rossini's opera includes the Venetian first act from Shakespeare's play, in which Desdemona expresses her love for Otello, despite her betrothal to Roderigo. Iago convinces Otello of an affair between Desdemona and Roderigo but, after Otello has mur-

dered Desdemona in a jealous rage, confesses his plot and commits suicide. Plagued with his guilt and grief, Otello also kills himself.

For other operas see SHAKESPEARE.

**'O terra, addio'.** The closing duet between Aida (sop) and Radamès (ten) in Act IV of Verdi's *Aida*, in which they bid farewell to life on earth.

**Otter, Anne-Sofie von** (*b* Stockholm, 9 May 1955). Swedish mezzo. Studied Stockholm Cons., and London with Rozsa. Début Basle 1982 (Hänsel). London, CG, and Berlin, D, from 1985; also Geneva, Stockholm O, etc. New York, M, 1990. Roles include Cherubino, Dorabella, Sesto, Composer. Possesses an outstanding voice and technique. (R)

**Otto, Teo** (*b* Remscheid, 4 Feb. 1904; *d* Frankfurt, 9 June 1968). German designer. Studied Kassel, Paris, and the Bauhaus in Weimar. Engaged by Klemperer for Berlin, Kroll O, 1927. Chief designer Berlin State Theatres 1927. Left Germany 1933; mostly Zurich until 1945. Salzburg, Vienna, Zurich. New York, M, *Tristan* 1960. London, CG, *Figaro* 1963. Favoured real materials to give 'tactile excitements to the set' rather than the painted scenic object.

**Ottoboni, Pietro** (*b* Venice, 2 July 1667; *d* Rome, 28 Feb. 1740). Italian librettist and patron. Great-nephew of Pope Alexander VIII, he was elevated to cardinal in 1689 and held various ecclesiastical appointments, rising to be vice-chancellor of the Roman Catholic church. He used his wealth to support the arts, turning his Palazzo della Cancelleria into a centre of music-making. Among those who appeared there were Bononcini, Pasquini, and the Scarlattis: Handel also profited from his hospitality while in Rome. Ottoboni took a considerable interest in opera, opening a theatre in his palace, and writing several librettos, including that for Lanciani's *L'amante del suo nemico* (1688) and Scarlatti's *Statira* (1690), which were widely admired and reveal a sensitive literary spirit. He is sometimes incorrectly credited with the composition of the opera *Colomba* (1690), for which he provided only the libretto.

**Ottocento** (It.: lit, 'eight hundred') 'Nineteenth century'.

**Ottokar.** The Prince (bar) to whom the villagers owe allegiance in Weber's *Der Freischütz*.

**Ottone.** Opera in 3 acts by Handel; text by Nicola Francesco Haym (altered from Pallavicino's text for Lotti's *Teofane*, 1719). Prod. London, H, 12 Jan. 1723, with Cuzzoni (her début), Robinson, Senesino, Boschi, Berenstadt, Durastanti. Rev. Handel O Society, 1971.

**'O tu che in seno agli angeli'.** Alvaro's (ten) aria in Act III of Verdi's *La forza del destino*, in which he remembers Leonora, whom he believes dead.

**'O tu, Palermo'.** Procida's (bs) aria in Act II of the Italian version of Verdi's *Les vêpres siciliennes* (orig. 'O toi, Palerme'), invoking the spirit of his homeland after his long absence.

**overture.** From the French *ouverture*, opening, the word normally used for the instrumental prelude to an opera, oratorio, play, ballet, or other work; later a concert work unconnected with the stage.

The earliest operas usually began with little more than a flourish of instruments, e.g. the Toccata opening Monteverdi's *Orfeo* (1607), and sometimes had no overture at all. More elaborate introductions began to be used in Roman opera, such as the sinfonias which open the three acts of Landi's *Il sant'Alessio* (1632); the familiar form of later Venetian opera was the canzona overture, usually cast with an introductory slow movement in duple rhythm and a fast movement in triple rhythm. This model was adapted by Lully into the *French overture and supplanted in its native country by the *Italian overture around the turn of the 18th cent. This division was in practice one of nomenclature rather than nationality, for it was possible for an Italian opera to be prefaced with a French overture, as in the case of Handel.

The idea of using material from the opera in the overture is anticipated by the introductory sinfonia to Cesti's *Il pomo d'oro* (1668), where music from later scenes appears; the first important use of the technique occurs in Rameau's *Castor et Pollux* (1735); despite earlier suggestions, it did not become accepted practice until Gluck, whose overtures may prepare for the first scene or set the mood of the opera. Mozart continued the practice, setting the mood of *Don Giovanni* and *Die Zauberflöte* or using some elements that recur, and in *Così fan tutte* using a motto theme to represent the words of the title. Beethoven took this principle of thematic anticipation further in the three *Leonore* overtures he wrote for *Fidelio*; and it reached its climax in Weber, who (taking his cue from Spohr's *Faust*) made his later overtures virtually epitomes of the ensuing opera, while adhering to sonata form. In Italy the overture was still designed chiefly to hush talkers and admit late-comers, only incidentally attempting to compose the audience's mind for the opera, e.g. Rossini's overture to *Aureliano in Palmira*, which was re-used for the tragedy *Elisabetta, regina d'Inghilterra* and the comedy *Il barbiere di Siviglia*. In French Grand Opera, the overture usually took the form of a freely composed selection of the most important tunes in the opera. Light composers have generally remained faithful to this pot-pourri style of overture. Wagner, though he came to prefer the term Vorspiel (prelude), at first pursued Weber's methods; later he composed still more subtle introductions to his dramas, preparing the audience thematically and psychologically for what was to come. The verismo composers on the whole preferred the brief introduction that had latterly become Verdi's habit, while Strauss sometimes even raised the curtain directly on the drama.

**'O welche Lust'.** The prisoners' chorus in Act I of Beethoven's *Fidelio*, praising the sun and freedom as they emerge from their cells for exercise.

**Owen Wingrave.** Opera in 2 acts by Britten; text by Myfanwy Piper, after Henry James's story (1892). Written for TV and prod. BBC and NET (USA), 16 May 1971, with Harper, Vyvyan, Fisher, Baker, Pears, Luxon, Shirley-Quirk; London, CG, 10 May 1973, with almost same cast, cond. Bedford; Santa Fe 9 Aug. 1973, with Titus, Steber, Gramm, Kraft, cond. Nelson.

London and Paramore, family home of the Wingraves. Owen (bar), the last of a military family, rejects his Army studies with Spencer Coyle (bs-bar) because of his pacifism, thus affronting both his family, in particular his aunt Miss Wingrave (sop) and his grandfather Sir Philip (ten), and also his betrothed, Kate (mez). Despite the sympathy of Coyle, they reject him, and to prove his courage to Kate he agrees to spend a night in the haunted room of the family seat. He is found dead in the morning.

**'O, wie ängstlich'.** Belmonte's (ten) aria in Act I of Mozart's *Die Entführung aus dem Serail*, in which he reveals the anxiety in his heart as he hopes once again to find Constanze.

**Oxford.** City in England. Apart from Sir Hugh Allen's productions of *Fidelio* and *Der Freischütz* before the First World War, there was no opera until the formation of the Oxford University Opera Club after J. A. Westrup's production of Monteverdi's *Orfeo* (1925). In 1926 *L'incoronazione di Poppea* had its British première. Sumner Austin was adviser 1928–33, when productions included *The Bartered Bride*, *May Night*, and the first British *Devil and Kate*. Hans Strohbach supervised the productions 1931–3. Later productions have included *Idomeneo*, *Les Troyens*, *Khovanshchina*, *Ruslan and Lyudmila*, *Macbeth* (Heather Harper's début), Scarlatti's *Mitridate Eupatore*, Cavalli's *Rosinda*, Spohr's *Jessonda*, the premières of Wellesz's *Incognita* (1951) and Sherlaw Johnson's *The Lambton Worm* (1978), and the British premières of *Hans Heiling* (1953), *The Secret* (1956, Janet Baker's début), *L'enfant et les sortilèges* (1958), Schubert's *Fierrabras* (1986), Mendelssohn's *Hochzeit des Camacho* (1987), and (semi-staged) Hoffmann's *Undine* (1991).

**Ozawa, Seiji** (*b* Hoten, 1 Sep. 1935). Japanese conductor. Studied Tokyo with Saito, Tanglewood, Berlin with Karajan. Début Salzburg 1969 (*Così*); London, CG, 1974; Paris, O, 1983 (first *Saint François d'Assise*), 1987. New York, M, 1992. Repertory incl. *Oberon*, *Carmen*, *Onegin*, *Elektra*, *Wozzeck*. Though possessing a lucid style and sympathy with his singers, he is less at home with the dramatic demands of opera than on the concert platform. (R)

**'O zitt're nicht'.** The Queen of the Night's (sop) aria in Act I of *The Magic Flute*, charging Tamino to rescue her daughter Pamina from Sarastro.

# P

**Pacchierotti, Gaspare** (*b* Fabriano, *bapt.* 21 May 1740; *d* Padua, 28 Oct. 1821). Italian soprano castrato. Studied Venice with Bertoni; sang there 1765–70. Début Venice, GG, 1766 (Gassmann's *Achille in Sciro*). Naples, C, 1771–6; further successes in Italy, then regularly London, H, 1778–84, and 1791. Here he scored unparalleled triumphs, and was declared 'superior to any singer . . . since Farinelli' (*Public Advertiser*), and 'the most perfect singer it ever fell to my lot to hear' (Mount-Edgcumbe). He was accompanied in Haydn's cantata *Arianna a Nasso* by the much impressed composer. Repertory included Gluck's Orfeo, and operas by Jommelli, Bianchi, Bertoni, etc. He sang at the inauguration of Milan, S, 1778; also at that of Venice, F, in 1792, after which he retired to Padua. One of the greatest of the castratos, he was distinguished not only by his ravishing voice (with its wide range of B♭–c′′′), virtuoso technique, and dramatic accomplishment, but also for his culture, modesty, and rare sweetness of nature. He frequently reduced his listeners to tears, and was adored by women.

**'Pace, pace mio Dio'.** Leonora's (sop) aria in Act IV of Verdi's *La forza del destino*, in which she prays for peace of mind.

**Pacini, Giovanni** (*b* Catania, 17 Feb. 1796; *d* Pescia, 6 Dec. 1867). Italian composer. Son of Luigi Pacini, the first Geronio in *Il turco in Italia*; began to study singing with Marchesi at the age of 12, and also worked at composition with Mattei and Furnaletto. His first opera, *Don Pomponio* (1813), was not produced, but his second, *Annetta e Lucindo*, was given in the same year. He continued to pour out operas (a dozen comedies in four years); but his first real success was an opera semiseria, *Adelaide e Comingio*.

In 1820 he went to Rome, where his easy Rossinian manner enabled him to help Rossini himself by composing three numbers for *Matilde di Shabran*. Many works followed; his Naples début was with *Alessandro nell'Indie*, which had 70 consecutive performances, and after the success of *L'ultimo giorno di Pompei* in 1825 he became music director of the T San Carlo, Naples. Here he was contracted to compose two operas a year. By 1830 he was one of the most successful opera

composers in Italy: in 16 years he had produced almost 40 operas, including *Il corsaro* for the inauguration of the rebuilt T Apollo in Rome. Especially after some failures in 1833, he realized that he was threatened by the growing reputation of Bellini and Donizetti. Accordingly he withdrew to found a music school at Viareggio, to which he devoted himself after the failure of *Carlo di Borgogna* in 1835. He made a come-back after the death of Bellini and retirement of Rossini with *Saffo*, the most successful work in his new, more careful style and indeed his masterpiece. He continued to write successful works, especially *La fidanzata corsa, Maria, regina d'Inghilterra, Medea, Lorenzino de' Medici*, and *Bondelmonte*; but with the rise of Verdi he was once more eclipsed. His last real success was *Il saltimbanco*.

In his early works, Pacini took Rossini as his model, believing this to be the natural and successful way to write Italian opera, and he even copied Rossini's move towards a more serious manner. He was well aware of this, and of his carelessness; he relied on a fluency and dexterity which won him the respect of Rossini and Bellini, above all for his skill in writing singable melodies. He was known, from his energetic melodic vein, as *il maestro della cabaletta*, and it is in *cabaletta sections that his most characteristic music is to be found. In his middle works, he tried to improve on earlier weaknesses in harmony and orchestration; he helped to unify aria, ensemble, and chorus in the years before Verdi. For his pupils he wrote some theoretical treatises, and also left an entertaining autobiography, *Le mie memorie artistiche* (Florence, 1865, 2/1875).

SELECT WORKLIST: *Adelaide e Comingio* (Rossi; Milan, 1817); *Alessandro nell'Indie* (Tottola, after Metastasio; Naples 1824); *L'ultimo giorno di Pompei* (The Last Day of Pompeii) (Tottola; Naples 1825); *Il corsaro* (Ferretti, after Byron; Rome 1831); *Saffo* (Cammarano; Naples 1840); *La fidanzata corsa* (The Corsican Bride) (Cammarano, after Mérimée; Naples 1842); *Maria, regina d'Inghilterra* (Tarantini; Palermo 1843); *Medea* (Castiglia; Palermo 1843); *Lorenzino de' Medici* (Piave; Venice 1845); *Bondelmonte* (Cammarano; Florence 1845); *Il saltimbanco* (The Mountebank) (Checchetelli; Rome 1858).

BIBL: M. Davini, *Il maestro Giovanni Pacini* (Palermo, 1827).

**Padilla y Ramos, Mariano.** See ARTÔT, DÉSIRÉE.

**Padmâvatî.** Opéra-ballet in 2 acts by Roussel; text by Louis Laloy, after an event in 13th-cent. Indian history. Prod. Paris, O, 1 June 1923, with Lapeyrette, Laval, Franz, Rouard, Fabert, cond. Gaubert; London, C, 6 July 1969 (concert), with Gorr, Berbié, Chauvet, Souzay, cond. Martinon.

India, 1540. The Mogul sultan Alaouddin

(bar) proposes an alliance with Ratan-sen (ten), King of Tchitor. He is well received, but demands Ratan-sen's wife Padmâvatî (mez) as a condition; to this Ratan-sen reluctantly consents. A Brahmin (ten) who later asks for her to be handed over is stoned to death by the crowd riots. Alaouddin defeats Ratan-sen in battle, but rather than have the sin of betraying her rest on her husband's conscience, Padmâvatî stabs him. She therefore has to die on his funeral pyre.

**'Padre, germani'.** Ilia's (sop) aria in Act II of Mozart's *Idomeneo*, expressing her guilt for loving a Greek, Idomeneo's son Idamante, whom as a Trojan princess she should hate.

**Padre Guardiano.** The Abbot (bs) of the monastery of Hornachuelos in Verdi's *La forza del destino*.

**padre nobile** (It.: 'noble father'). The former term for a role, normally taken by a senior member of a company, representing a figure of dignity and authority, paternal either literally or metaphorically. Historically, it was a descendent of the *magnifico* of the *commedia dell'arte. A famous instance is the elder Germont in Verdi's *La traviata*. Despite the French usage *père sérieux* as virtually interchangeable with *père noble*, the role could include comic elements.

**Padua** (It., Padova). City in the Veneto, Italy. In 1636 a dramatic musical entertainment, *Ermiona*, with music by G. F. *Sances, was organized by Pio Enea degli Obizzi. The T dello Stallone opened 1642, the T degli Obizzi 1652. The T Nuovo opened 1751 with Galuppi's *Artaserse*. Opera was also given in the T del Recinto and in Count Pepoli's theatre. When the Obizzi family died out, their theatre was successively renamed T Già degli Obizzi, Vecchio 1797, Nuovissimo 1827, Dei Concordi 1845; demolished 1885. The T Nuovo was rebuilt 1847, renamed T Verdi 1884.

**Paer, Ferdinando** (*b* Parma, 1 June 1771; *d* Paris, 3 May 1839). Italian composer. Studied Parma with Fortunati and Ghiretti.He won early success with his operas, and was made honorary maestro di capella in 1792; his most important early work was *Griselda*, in the semiseria style that he particularly cultivated. As director of the Vienna, K, 1797–1801, he came to know Beethoven. The most striking successes of this period were *Camilla* and *Achille*, whose funeral march was not the only number in his music to impress Beethoven. He was in Dresden 1801–6: semiseria works composed here include *I fuorusciti di Firenze*, *Sargino*, and *Leonora*, to the plot used by Beethoven. In 1806 he was taken up by Napoleon, at whose behest he soon settled in Paris. He became singing teacher to Napoleon's empress, Marie-Louise of Austria, director of the Opéra-Comique (1807) and of the T-Italien (1812) in succession to Spontini. He continued as a successful teacher after the fall of Napoleon, his pupils in the 1820s including Liszt. In 1824 Rossini (who had made his stage début as the boy Adolfo in *Camilla*) joined him at the T-Italien in what was to prove a troublesome partnership.

Paer was expert at matching the taste of the day with fluent, singable melody. This gave him a dominating position in Italian music at the start of the 19th cent., and won him his many European appointments over the heads of local composers. His style was enriched by his contacts with Viennese music, and he even had some influence on Beethoven; but this did not spare him one of E. T. A. Hoffmann's most withering reviews, taking *Sofonisba* as an example of all that was unacceptable to Germans in Italian opera. Paer's real contribution is as a composer of *opera semiseria and *rescue opera, in which the mingling of a serious theme with often delightful elements of comic relief is agreeably expressed in music of great fluency and operatic efficiency. His originalities include a number of Romantic features, among them the Mad Scene and a concern for writing continuously composed opera, though his reach after these and other novelties was rather greater than his grasp of them. Together with Mayr, he was the most important Italian composer between the death of Cimarosa (who influenced his early operas) and the rise of Rossini (whom in turn he influenced, including in the use of the orchestral crescendo).

SELECT WORKLIST: *Griselda* (Anelli, after Boccaccio; Parma 1798); *Camilla* (Carpani, after Marsollier; Vienna 1799); *Achille* (De Gamerra; Vienna 1801); *I fuorusciti di Firenze* (The Exiles from Florence) (Anelli; Dresden 1802); *Sargino* (Foppa, after Monvel; Dresden 1803); *Leonora* (Schmidt; Dresden 1804); *Sofonisba* (Rossetti, after Zanetti; Bologna 1805); *Agnese* (Buonavoglia & Gianetti, after Opie; Parma 1809); *Le maître de chapelle* (Gay, after Duval; Paris 1821).

**Pagliacci** (Clowns). Opera in 2 acts (orig. 1 act) by Leoncavallo; text by the composer. Prod. Milan, V, 21 May 1892, with A. Stehle, Giraud, Maurel, Ancona, cond. Toscanini; London, CG, 19 May 1893, with Melba, De Lucia, Ancona, Green, cond. Bevignani; Philadelphia, 15 June 1893, with Kört-Kronold, Montegriffo, Campanari, Averill, cond. Hinrichs.

Montalto (Calabria), 15 Aug. 1865. In the Prologue, Tonio (bar) tells the audience that the story they are about to see is a real one about real people.

Canio (ten), the leader of a travelling troupe of players, warns that if he were to find his wife Nedda (sop) unfaithful, she would pay dearly for it. Tonio makes advances to Nedda but she repulses him; however, he overhears Nedda and her lover Silvio (bar) planning to run away together, and he hurries to bring Canio back from the village inn.

Canio gives vent to his grief; and in the play put on for the villagers (in which he takes the part of Pagliaccio, Nedda that of Colombine, and Beppe (ten) that of Harlequin), he finds the situation so like reality that he stabs first Nedda, and then Silvio, who rushes to her aid.

**Pagliardi, Giovanni Maria** (*b* Genoa, 1637; *d* Florence, 3 Dec. 1702). Italian composer. Worked as a church musician in Genoa and Rome; from *c*.1669 possibly at the Medici court in Florence. Wrote three operas for Venice, T Grimani, and four for Florence: these show him to possess a remarkable flair for dramatic writing, seen at best in *Caligula delirante*.

SELECT WORKLIST: *Caligula delirante* (Gioberti; Venice 1672).

**Pagliughi, Lina** (*b* Brooklyn, 27 May 1907; *d* Savignano, 1 Oct. 1980). US, later Italian, soprano. Studied San Francisco with Brescia, Milan with Bavagnoli. Début Milan, T Nazionale, 1927 (Gilda). Milan, S, 1937, 1947; London, CG, 1938; Australian tour 1932; Rome 1949. Retired 1957. One of the leading light sopranos in Italy, celebrated as Queen of the Night, Lucia, Violetta etc. (R)

**Paisiello, Giovanni** (*b* Taranto, 9 May 1740; *d* Naples, 5 June 1816). Italian composer. Studied Naples. His first successes were in opera buffa, beginning with *Il ciarlone*, *I francesi brillanti*, *Demetrio*, *Il negligente*, *La frascatana*, and above all *L'idolo cinese*. Performed in cities throughout Italy, these marked him out as a composer of rare talent: the series of works written for Naples, which also included opere serie (e.g. *Lucio Papirio dittatore*) set him up as a worthy rival to Piccinni, later Cimarosa, and Guglielmi. He went to St Petersburg 1776 at the invitation of Catherine the Great; here he helped to establish Italian opera at the court, and composed *Lucinda ed Armidoro* and *Nitteti*. A series of popular works followed, culminating in *Il barbiere di Siviglia*, which became so popular as to prejudice the public against the setting by Rossini that was to oust Paisiello's work. Returned to Italy 1784, passing through Vienna, where he composed *Il re Teodoro in Venezia*; became maestro di cappella to Ferdinand IV in Naples, where he wrote among other operas *L'amor contrastato* (or *La molinara*). This became one of his most popular works: it includes the famous aria 'Quant'è più bello' and the duet 'Nel cor più non mi sento' taken for variations by Beethoven (WoO 69 and 70), the latter also by Paganini. A private entertainment in

Vienna, influenced by *La molinara*, led Wilhelm Müller to write his cycle of poems *Die schöne Müllerin*, used by Schubert. For the opening of the Pantheon in London he wrote *La locanda*; for the opening of La Fenice, Venice, *I giuochi d'Agrigento*. Other popular works of these years included *Nina* and *I zingari in fiera*. He sided with Napoleon in the disturbances of 1799 and, out of favour at the Bourbon Restoration, went to Paris. He returned to Naples, but the Bourbons did not forgive his previous actions and he died in comparative poverty.

With some 80 operas to his credit, Paisiello was one of the most prolific composers of his generation. Though his works are distinguished more for their ease and accomplishment than for profundity, his elegant melodic manner is capable of much sensitivity, and his feeling for the orchestra is vivid. As a composer of opera buffa he was one of those most responsible for turning it from a piece of light entertainment into a more dramatically satisfying and emotionally charged genre: together with that of Cimarosa, his work represents the highest peak of its development before Mozart. He had a sure, light touch, and at his best was capable of sophisticated and tender characterization, seen nowhere better than in *Nina*, written under the influence of Dalayrac's version (1787): this became a seminal work for the development of *opera semiseria. His handling of the ensemble is particularly noteworthy; he wrote some of the most advanced examples of the *ensemble finale before Mozart.

SELECT WORKLIST: *Il ciarlone* (The Chatterbox) (Palomba; Bologna 1764); *I francesi brillanti* (The Brilliant Frenchmen) (Mililotti; Bologna 1764); *Demetrio* (Metastasio; Modena 1765); *Il negligente* (Goldoni; Parma 1765); *L'idolo cinese* (Lorenzo; Naples 1767); *Lucio Papirio dittatore* (Zeno; Naples 1767); *La frascatana* (Livigni; Venice 1774); *Nitteti* (Metastasio; St Petersburg 1777); *Lucinda ed Armidoro* (Coltellini; St Petersburg 1777); *Il barbiere di Siviglia* (Petrosellini; St Petersburg 1782); *Il re Teodoro in Venezia* (Casti; Vienna 1784); *L'amor contrastato* (or *La molinara*) (Love Defied, or The Maid of the Mill) (Palomba; Naples 1789); *Nina* (Lorenzi & Carpani, after Marsollier; Caserta 1789); *I zingari in fiera* (The Gypsies at the Fair) (Palomba; Naples 1789); *La locanda* (The Inn) (Tonioli, after Bertati; London 1791); *I giuochi d'Agrigento* (Pepoli; Venice 1792).

BIBL: J. Hunt, *Giovanni Paisiello: His Life as an Opera Composer* (New York, 1975).

**Palazzesi, Matilde** (*b* Montecarotto, 1 Mar. 1802; *d* Barcelona, 3 July 1842). Italian soprano. Studied Pesaro with Solarto. Début Dresden 1824 in title-role of Rossini's *Zelmira*. Remained in Dresden with Italian O until 1833; successful in Rossini and as Mozart's Countess and Donna Elvira. Milan, S, and other Italian theatres. Barcelona 1841–2, incl. in first Verdi opera there, *Oberto*. Repertory included works by Bellini, Donizetti, Spontini, Mercadante, and Morlacchi. Married flautist and composer Angelo Savinelli (1800–70).

**palco** (It.: 'stand or platform'; in theatrical usage 'box') (Fr. and Ger., *Loge*). The terms for special boxes in Italy are *palco reale* (or *ducale, governativo, elettorale*), royal box; *palco di proscenio*, stage box; *palco di pepiano*, a box on stage level; *palchettone*, a large central box, perhaps originally the royal box, used for general or special purposes; *palco della vedova*, 'widow's box', a box so placed as to afford a view of the stage while concealing its occupants from the rest of the audience.

In Italy especially, a certain theatrical etiquette grew up around the use of boxes. Beginning in the Venetian theatres of the mid-17th cent., boxes were either sold outright, or hired out for an annual subscription. Those who bought their boxes were still required to pay an admission charge, but enjoyed enhanced status in the eyes of their peers. Proximity to the royal box was of paramount importance and in some theatres the patrons were permitted to decorate their boxes on an individual basis. Within their intimate recesses it was possible to indulge in amorous liaisons, commercial transactions, or, especially during the Risorgimento, political intrigue. All of this served to confirm opera's social role while distancing the listener from the operatic experience.

**Palella, Antonio** (*b* Naples, 8 Oct. 1692; *d* Naples, 7 Mar. 1761). Italian composer. Worked at Naples, C, from 1737 as maestro al cembalo. His three opere buffe, which avoid the ubiquitous Neapolitan dialect, occupy an important position in the history of the genre: their emphasis upon serious characters helped to pave the way for the *dramma giocoso. His most successful work was *Origille*.

SELECT WORKLIST: *Origille* (Palomba; Naples 1740).

**Palermo.** City in Sicily. The first opera was Cavalli's *Giasone*, 1655, in the T della Misericordia; opera was also given in the T di Santa Cecilia, opened 28 Oct. 1693 with Ignazio Pollice's *L'innocenza penitente*, and the T Marmoreo (by the shore) 1682. The first opera by the Palermo-born Alessandro Scarlatti was *Pompeo*, 1690. The T della Corte del Pretore (cap. 700) opened 1726 for opera buffa; renamed T di Santa Lucia and T di Santa Caterina; rebuilt 1809 as Real T Carolino; after unification with Italy, 1860, renamed T Bellini; burnt down 1964. *Donizetti spent a difficult year conducting here 1825–6, and wrote his *Alahor di Granata*, 1826. *Balfe was principal

baritone 1829–30 and produced his first opera, *I rivali di se stessi*, here 1829. The Politeama Municipale opened 1874 with Bellini's *Capuleti*, renamed Politeama Garibaldi 1882: Toscanini's Sicilian début 1892. The T Massimo (cap. 3,200) opened, after much quarrelling and delay, 16 May 1897 with *Falstaff*; as its name ('greatest') implies, it is a huge theatre, one of Italy's largest. In the first season the 24-year-old Caruso appeared in *Gioconda*, and the theatre, unable to compete for top international stars, has a reputation for engaging promising young artists: those appearing very early in their careers include Giulini, Muti, and Abbado. In the 1950s the theatre set an example in reviving bel canto opera. Opera is also given in the open-air T di Verdura di Villa Castelnuovo, inaugurated 1957 with Verdi's *Otello*.

**Palestrina.** Opera in 3 acts by Pfitzner; text by the composer. Prod. Munich, N, 12 June 1917, with Karl Erb in the title-role and Ivogün, Krüger, Kuhn, Brodersen, Feinhals, Gustav Schützendorf, Bender, cond. Walter; London, Collegiate T, 10 June 1981.

Rome and Trent, 1563. The composer Palestrina (ten) refuses Cardinal Borromeo's (bar) commission to compose a mass for the Council of Trent and is imprisoned. However, urged by visions of past composers and by his wife Lucretia (con), he sets to work and completes the mass—the *Missa Papae Marcelli*—by the following morning.

The assembled Council of Trent discusses the role of music in the mass and learns that the Pope's approval depends on his receiving from Palestrina a work which conforms to his requirements.

Palestrina's mass is performed to great acclaim. Pope Pius IV (bs) congratulates him and the elderly Palestrina gives thanks to God.

**Pallavicino, Carlo** (*b* Salò, ?*c*.1630; *d* Dresden, 29 Jan. 1688). Italian composer. Worked first as a church musician in Padua. At the Dresden court, as Vizekapellmeister 1667–73; Kapellmeister 1672. Returned to Padua 1673; maestro di cappella at the Ospedali degli Incurabili in Venice 1674–85. Returned to Dresden 1687, as director of the first permanent opera troupe.

Beginning with *Demetrio*, Pallavicino wrote *c*.20 works for the Venetian houses, including *Enea in Italia*, *Bassiano*, and *Nerone*: *Vespasiano* was given to open the T San Giovanni Grisostomo 1678. These belong in all essential respects to the tradition of Cavalli and Cesti, as does his best-known opera, *Gierusalemme liberata*; though first seen in Venice, it was given a lavish performance at Dresden shortly afterwards to mark Pallavicino's return. Regarded by his contemporaries as an ingenious and capable composer, he showed a particular facility for effective orchestral writing; during his lifetime, several operas were revived in houses outside Venice.

SELECT WORKLIST (all first prod. Venice): *Demetrio* (Dall'Angelo; 1666); *Enea in Italia* (Bussani; 1675); *Vespasiano* (Corradi; 1678); *Nerone* (Corradi; 1679); *Bassiano* (Noris; 1682); *Gierusalemme liberata* (Corradi, after Tasso; 1687, rev. Dresden 1687, rev. Hamburg 1695 (in German) as *Armida*)).

His son **Stefano Benedetto Pallavicino** (*b* Padua, 21 Mar. 1672; *d* Dresden, 16 Apr. 1742) was court poet at Dresden (1688–1742); in Düsseldorf 1695–1716. His librettos were set by Steffani, Lotti, Hasse, and Ristori among others: his *Teofane* was adapted by Haym for Handel's *Ottone*.

**Palmer, Felicity** (*b* Cheltenham, 6 Apr. 1944). English soprano, later mezzo-soprano. Studied London, GSM. Début Kent O 1971 (Purcell's Dido). London, ENO from 1975; London, CG, from 1985; Gly. 1985, 1988; also Lausanne, Houston, Zurich, Frankfurt. Roles include Handel's Tamerlano, Agrippina, Pamina, Donna Elvira, Marcellina, Lyubov (*Mazeppa*), Kabanicha, Witch (Humperdinck), Quickly. A zestful, imaginative performer with an incisive tone. (R)

**Palomba, Giuseppe** (*fl.* Naples, 1765–1825). Italian librettist. One of the most prolific Neapolitan librettists, his earliest text was for Insanguine's *La vedova capricciosa* (1765). He wrote *c.*120 librettos, mostly comic, which were set by all leading Italian opera buffa composers of the late 18th cent.: Cimarosa set 11 of them, including *La villanella riconosciuta* (1783) and *Le astuzie feminili* (1794). Palomba's most famous librettos were those for Paisiello's *La molinara* (1788; first given under the title *L'amor contrastato*) and Fioravanti's *Le cantatrici villane* (1799), and he held sway over the Neapolitan stage until well into the 19th cent., his last work appearing in 1825, *L'ombra notturna*.

**Pamina.** The daughter (sop) of the Queen of the Night in Mozart's *Die Zauberflöte*.

**Pampanini, Rosetta** (*b* Milan, 2 Sep. 1896; *d* Corbola, 2 Aug. 1973). Italian soprano. Studied Milan with Molaioli. Début Rome, T Nazionale, 1920 (Micaëla). After further study, second début Naples, C, 1923. Milan, S, 1925–30, 1934–7; London, CG, 1928–9, 1933; Paris, O, 1935; also Chicago, Berlin, Vienna, and throughout Italy. Retired 1943. Her warm tone and direct simplicity made her a celebrated Desdemona (Verdi), Mimì, Butterfly, Liù, Manon Lescaut, and Iris. (R)

**Pandarus.** The brother (ten) of the High Priest Calkas in Walton's *Troilus and Cressida*.

**Pandolfini, Angelica** (*b* Spoleto, 21 Aug. 1871; *d* Lenno, 15 July 1959). Italian soprano, daughter of below. Studied Paris with Massart. Début Modena 1894 (Marguerite). Malta and Italian provinces 1895–6; Milan, S, 1897–9, 1906, and L, 1902; Naples, C, 1898, 1901; also Buenos Aires, Cairo, Lisbon, Madrid. Retired 1909, after her marriage. A full-blooded yet sensitive singer, celebrated as Violetta, Eva, Mimì, Tosca, Butterfly. Created Adriana Lecouvreur. (R)

**Pandolfini, Francesco** (*b* Termini Imerese, 22 Nov. 1833; *d* Milan, 15 Feb. 1916). Italian baritone, father of above. Studied Florence with Ronconi and Vannuccini. Début Pisa 1859 (Conte di Vergy, *Gemma di Vergy*). Milan, S, 1871, 1873–4; Lisbon 1872–88; Paris, I, 1876; London, CG, 1877, 1882. Retired 1890. His powerful, sweet tone and noble style made him a successful Macbeth, Amonasro, and Renato. Also famous in Bellini and Donizetti.

**Panerai, Rolando** (*b* Campo Bisenzio, 17 Oct. 1924). Italian baritone. Studied Florence with Frazzi, Milan with Armani and Tess. Début Naples, C, 1947 (Pharaoh, *Mosè*). Milan, S, from 1951; New York, M, 1955; Salzburg from 1957; London, CG, 1960. Also Berlin, D; Moscow, B; San Francisco; etc. Roles include Guglielmo, Henry Ashton (*Lucia*), title-role of *Mathis der Maler*; particularly acclaimed as Figaro (Rossini), Luna, Germont, Ford. (R)

**Panizza, Ettore** (*b* Buenos Aires, 12 Aug. 1875; *d* Buenos Aires, 27 Dec. 1967). Argentinian conductor and composer of Italian descent. Studied Milan with Ferroni. Début Rome, C, 1898, and had early success with his opera *Medio evo latino* (Illica; Genoa 1900). Milan, V, 1902; Genoa 1902–3. London, CG, 1907–14, 1924. Turin 1912–17; Milan, S, 1916–17, 1921–9 (as Toscanini's assistant, cond. *Ring*), 1930–2, 1946–8. Vienna, Berlin, 1932. Chicago 1922–4; New York, M, 1934–42. Buenos Aires, C, 1908 (his *Aurora* for reopening), 1921–67 (incl. his *Bisanzio*, 1939). One of the most highly regarded conductors of his day, especially of Italian opera, and admired by composers including Puccini and Richard Strauss. Introduced Wagner and Strauss into his repertories and gave the premières of many operas, including Zandonai's *Francesca da Rimini* (1914), Wolf-Ferrari's *Sly* (1927), and Menotti's *The Island God* (1942). Autobiography, *Medio sigla de vida musical* (Buenos Aires, 1952).

**Pantaleoni, Adriano** (*b* Sebenico (Šibenik), 7 Oct. 1837; *d* Udine, 18 Dec. 1908). Brother of below. Studied Udine and Milan. Leading baritone Milan, S, 1871–2, 1875, 1877. Also Vienna; London, HM; New York. Large repertory included Rigoletto, Germont, Escamillo, and Donizetti roles.

**Pantaleoni, Romilda** (*b* Udine, 1847; *d* Milan, 20 May 1917). Italian soprano, sister of above. Studied Milan with Prati, Rossi, and Lamperti. Début Milan, T Carcano, 1868 (Foroni's *Margherita*). Turin 1875; Milan, S, 1883–7; 1889, 1891. Also Rome, Naples, Vienna, etc. Created Desdemona (not entirely to Verdi's satisfaction); Tigrana in Puccini's *Edgar*, and title-role of Ponchielli's *Marion Delorme*. Intelligent and cultured, she was an exceptional actress. Sang Aida, Amelia, Santuzza; famous for her Gioconda, Margherita (*Mefistofele*).

**Pantheon.** Building in Oxford Street, London: a rotunda based on S Sophia and much admired, it was described by Burney as 'the most elegant structure in Europe'. Opened Jan. 1772 and used mostly for concerts and appearances by singers including Aguiari and Giorgi-Banti. After the destruction of the King's T in 1789, adapted as a theatre and used for Italian opera 1791–2. In 1812 it became the home of a company composed of singers from the King's who had quarrelled with the management. In 1813 there was a short-lived attempt to establish an English opera company.

**pantomime** (Gr., *pantomimos*: 'an actor miming everything'). In the Roman theatre, an actor who enacted the drama exclusively by dumb show, with music; hence a dramatic entertainment in which the artists express themselves thus. Roman authors who turned their hands to writing popular pantomimes included *Lucan. The Italian intermezzo often included a pantomime part since most of the singers were needed for the main opera, e.g. Vespone in Pergolesi's *La serva padrona*, 1733: the tradition survives with Sante in Wolf-Ferrari's *Il segreto di Susanna*, 1909. The 18th-cent. English pantomime derived largely from the mimetic elements in the *commedia dell'arte, and was established in London at Lincoln's Inn Fields T by John Rich, who often played Harlequin. At Drury Lane, Garrick developed pantomime, impressing the visiting *Noverre with techniques that were to influence the *ballet d'action*. The mixed genre, including vocal elements, spectacular scenery, instrumental music, dance, and mime, was highly popular in London as an approximation to opera in the 19th cent. One of the most successful pantomime writers, James Robinson *Planché, was commissioned by Kemble of Covent Garden as the author of *Oberon* (1826) for the unsuspecting Weber, who knew nothing of the genre. He had already written a mute role for the heroine of his *Silvana* (1810).

Pantomime was popular in 18th-cent. Vienna, with commedia dell'arte elements influencing the musical theatre. The most important author was Josef von *Kurz, who often played *Hanswurst in performances in which music played a part. In 1783 Mozart wrote the scenario and music for a pantomime (K446, fragments only surviving) including roles for commedia characters with himself as Harlequin: he declared that they performed 'really nicely' (*recht artig*).

*Pantomime dialoguée* was a kind of melodrama with a patchwork score, popular at the end of the 18th cent. in France, where ballet and mime have played an important part in the theatre. The tradition survives in the silent title-role of Auber's *La muette de Portici* (1828), and other works including the danced mime in the Olympia act of Offenbach's *Les contes d'Hoffmann* (1881).

**Papageno** and **Papagena.** The birdcatcher (bar) and his lover (sop) in Mozart's *Die Zauberflöte*.

**parable aria.** See METAPHOR ARIA.

**Parasha.** A village girl (sop), lover of the hussar Vasily, in Stravinsky's *Mavra*.

**Pardon de Ploërmel, Le.** See DINORAH.

**Parepa-Rosa, Euphrosyne.** See ROSA, CARL.

**Pareto, Graziella** (*b* Barcelona, 15 May 1888; *d* Rome, 1 Sep. 1973). Spanish soprano. Studied Milan with Vidal. Début Madrid 1908 (Amina). Buenos Aires 1907–9, 1926; Milan, S, from 1914; London, CG, 1920; Chicago 1921–2, 1923–5. Roles included Rosina, Norina, Gilda, Violetta, Leïla. Beecham admired her 'haunting pathos'. (R)

**Pariati, Pietro** (*b* Reggio Emilia, 26 or 27 Mar. 1665; *d* Vienna, 1733). Italian librettist and poet. From 1699 he worked in Venice, first collaborating with Zeno for Gasparini's *Antioco* (1705). They wrote many librettos together, including the first treatment of the Hamlet story (Gasparini's *Ambleto*, 1705). In 1714 Pariati went to Vienna, joining the court of Charles VI; Zeno joined him four years later and the collaboration continued successfully until 1729 and the appointment of Metastasio as *poeta cesareo. Pariati's function was to assist in turning Zeno's dramatic scenarios into verse: this he did with consummate ease, producing some of the most elegant, successful and frequently set librettos of the early 18th cent. Among their most famous collaborations were *Statira* (Gasparini, 1705) *Astarto* (Albinoni, 1708), and *Don Chisciotte* (1719). By himself Pariati produced several contrasting librettos, ranging from the pastoralism of Conti's *I satiri in Arcadia* (1714) and *Il finto Policare* (1716), to the grandeur of Fux's legendary *Costanza e Fortezza* (1723).

**Paride ed Elena** (Paris and Helen). Opera in 5 acts by Gluck; text by Calzabigi. Prod. Vienna, B, 3 Nov. 1770, with Schindler, Kurz, Millico; New York, TH, 15 Jan. 1954 (concert), by American Opera Society, cond. Gamson; Manchester, RMCM, 27 Nov. 1963, with Matthews, Howells, R. Davies, cond. Brierley.

Greece, mythological times. Paris (ten) goes to Sparta to claim the fairest woman in the country as reward for having chosen Venus in his famous Judgement. He woos Helen (sop), at first unsuccessfully, and is eventually ordered to leave; but as he does so she confesses love, and they flee together, braving the anger of Pallas Athene (sop), but comforted by the support of Cupid (sop)(who has helped them, disguised as Erasto).

**'Parigi, o cara'.** The duet in Act III of Verdi's *La traviata*, in which Alfredo (ten) proposes to Violetta (sop) that they resume their life together far from the bustle of Paris.

**Paris.** In classical mythology, the son of King Priam of Troy; he carried Helen off from her husband Menelaus, thus precipitating the Trojan war. He appears (ten) in Tippett's *King Priam* and in Gluck's *Paride ed Elena*.

**Paris.** Capital of France. At *Mazarin's invitation, Italian companies played here, at the Petit-Bourbon, and at the Tuileries: the first Italian work was probably Marazzoli's *Il giudizio* (1645), the first French work Dassoucy's *Andromède*, Feb. 1650. Operatic life developed with the establishment of the Académie de Musique (*Opéra), *Opéra-Comique, Comédie-Italienne (*Théâtre-Italien), and over 20 other theatres, including the Colisée and Odéon, in which established or visiting companies played, or occasional enterprises and performances took place. It is in the 'official' institutions that French operatic life developed; but the busy activity of many others is no less characteristic.

These included, at an early stage, the T du Palais-Royal (cap. 650), opened 1784, and particularly the T de Monsieur, opened in the Tuileries 29 June 1789 with Viotti as director; this sheltered various companies, especially the OC, and was demolished 1829. The T de l'Odéon opened 1793, and provided a home to various companies, notably the I. The T des Nouveautés (cap. 487) opened 1827, and was the home of the Opéra-Comique, 1839–69. The T des Variétés (cap. 1,100) was from 1855 a home of operetta. The T Marigny opened 1850, becoming later the *Bouffes-Parisiens. Various establishments at various times bore the name T de la Renaissance. The T de la Gaîté-Lyrique was an important centre of 19th-cent. operetta. The *Théâtre-Lyrique (cap. 1,242) opened 1862 opposite the T du *Châtelet. Other theatres giving opera in the late 19th cent. included the T du Château d'Eau, opened 1866, and the Éden-T, opened 1884. Early 20th cent. theatres giving opera included the Trianon-Lyrique, Ranelagh, Petit Scène, Ambigu, for operetta the Apollo, Bobino, Casino Montparnasse, Casino de Paris, Empire, and Mogador. More important than these was the T des *Champs-Élysées (opened 1913). The Maison de la Radio, opened 1964, has concentrated on 20th-cent. works. Opera has also been given at the huge Palais des Sports. In 1990 the Opéra moved into its new home at the Bastille, becoming the Opéra-Bastille.

See also CHÂTELET, THÉÂTRE DES NATIONS.

**'Pari siamo'.** Rigoletto's (bar) monologue in Act I of Verdi's *Rigoletto*, in which he reflects on the corrupt courtiers and the curse that he has drawn on himself.

**Parisina.** 1. Opera in 3 acts by Donizetti; text by Felice Romani, after Byron's verse tale *Parisina* (1816), 'grounded on a circumstance mentioned in Gibbon's *Antiquities of the House of Brunswick*'. Prod. Florence, P, 17 Mar. 1833, with Unger, Sacchi, Duprez, Cosselli, Porto; New Orleans, St. Charles T, 4 June 1837, with Pantanelli; London, HM, 1 June 1838, with Grisi.

Azzo d'Este (bar), Duke of Ferrara, marries Parisina (sop) after the death of his first wife. But he rightly suspects that she and his son Ugo are in love, and when she murmers Ugo's name in her sleep, he swears vengeance. The couple are caught trying to escape and Ugo is executed; Parisina collapses lifeless.

2. Opera in 4 acts (revised to 3 acts after first perf.) by Mascagni; text by Gabriele D'Annunzio, based on a historical event which took place in Ferrara at the time of Nicolò III. Prod. Milan, S, 15 Dec. 1913.

Ferrara, 1425. Nicolò d'Este (bar) abandons his mistress Stella de Tolomei (sop), by whom he has an illegitimate son, Ugo (ten), in order to marry Parisina Malatesta (sop).

On a pilgrimage, Parisina is rescued from pirates' attack by Ugo; she tends his wounds, and falls in love with him.

Parisina and Ugo are surprised by Nicolò, who sentences them to death.

Stella visits her son in prison and attempts to save him. Ugo rejects Parisina's pleas to leave her and be spared, and the couple are beheaded.

Also opera by Keurvels (1890).

**Parlamagni, Antonio** (*b* 1759; *d* Florence, 9 Oct. 1838). Italian bass. Rome 1789–90, 1818, 1821; Florence 1795–6; Milan, S, from 1797. A celebrated Neapolitan buffo, he created Macrobio in Rossini's *La pietra del paragone*, when he was

# parlando

'immensely successful' (Stendhal); also Isidoro in *Matilde di Shabran*, with his daughter **Anna** creating Edoardo.

**parlando** (It.: 'speaking'). A direction to let the tone of the voice approximate to that of speech. The term parlato (It.: 'spoken') is used synonymously, but also to distinguish the spoken sections in opera with dialogue.

**parlante** (It.: 'talking'). In 19th-cent. Italian opera, an orchestral theme in whose performance the singers participate intermittently. They may either contribute to the melodic line (*parlante melodico*), sing harmony notes (*parlante armonico*), or do both (*parlante misto*).

**Parma.** City in Emilia Romagna, Italy. In 1618 Duke Rinuccio Farnese commissioned a theatre from Gian Battista Aleotti, the wooden T Farnese (cap. 4,000) on the first floor of the Palazzo della Pilotta; opened 1628 with an entertainment *Mercurio e Marte*, intermedi by Monteverdi, for the wedding of Odoardo Farnese to Margherita de' Medici. Closed after performance to honour Carlo I of Bourbon 1732 and abandoned; bombed 1944; rebuilt in 1950s. Several other theatres were built in the 17th cent., including the T del Collegio dei Nobili (1600), the T della Racchetta (1674, demolished 1832), the Teatrino di Corte, opened 1689 with Tosi's *L'idea di tutte le perfezioni*, demolished 1822, and especially the T Ducale (cap. 1,200), opened 1688 with *Medea in Atene* by Antonio Giannettini (1648–1721). The scene of the premières of several of the Parma-born Paer's operas, including *Agnese*, 1809. Demolished 1829, by when the Parma audience had won its reputation of being the most difficult in Italy. In 1816 the tenor Curioni was arrested for insulting the audience who had protested at his poor performance, and a similar fate awaited the impresario of the 1818 season, who was gaoled for 'offending the public sensibilities' with his poor choice of artists and repertory.

In 1815 Napoleon's second wife, the former Empress Marie-Louise, now Duchess of Parma, demolished the Convent of Sant'Alessandro to build the Nuovo T Ducale (cap. 1,800), opened 16 May 1829 with Bellini's *Zaira*; after independence renamed T Regio 1849. The theatre was closely associated with Verdi, born at nearby Le Roncole, and though only staging one première (the first Italian *Vêpres siciliennes* as *Giovanna di Guzmann*, 1855), has produced all the others except *Un giorno di regno* and *Il corsaro*. Toscanini, born Parma 1867, played the cello in the orchestra, but never conducted. The Istituto di Studi Verdiani was founded here 1959. Tebaldi studied and began her career here.

**'Parmi veder le lagrime'.** The Duke of Mantua's (ten) aria in Act II of Verdi's *Rigoletto*, in which he regrets the loss of Gilda.

**parody** (Gr., *parodos*: 'beside the ode', i.e. singing a song in a different style, burlesquing a song). In operatic usage, a composition of specific or general satirical intent, mocking an individual work, composer, or genre. At times of controversy, such as the *Guerre des Bouffons in Paris, parody becomes a weapon in the battle. However, parody may also be constructive: Mozart's use of opera seria conventions in *Così fan tutte* with Fiordiligi's 'Come scoglio' skilfully suggests the note of exaggeration in her protests. Mozart was himself parodied by Wenzel Müller's farce *Der travestierte Zauberflöte* (1818). Such works are, of course, backhanded compliments to the success of the original. The triumph of *Der Freischütz* spawned parodies including *Samiel, oder die Wunderpille* (1824, translated into Danish and Swedish), and others in London. Nestroy's 11 parodies included *Zampa*, *Robert le diable*, *Martha*, *Tannhäuser*, and *Lohengrin*.

**Parsifal.** Opera in 3 acts by Wagner; text by the composer, principally after Wolfram von Eschenbach's poem, *Parzival* (early 13th cent.). Prod. Bayreuth 26 July 1882, with Materna, Winkelmann, Reichmann, Scaria, Hill, Kindermann, cond. Levi; New York, M, 24 Dec. 1903 (infringing the Bayreuth copyright, which did not expire until 31 Dec. 1913), with Ternina, Burgstaller, Van Rooy, Blass, cond. Hertz; London, CG, 2 Feb. 1914, with Eva von der Osten, Hensel, Bender, Knüpfer, cond. Bodanzky.

The Bayreuth copyright was also broken by performances in English in Boston and elsewhere in the USA 1904–5, in Amsterdam 1905, and in Buenos Aires and Rio de Janeiro 1913.

In a forest near a lake at Monsalvat, Gurnemanz (bs) and his two Esquires watch a procession taking Amfortas (bar), son of Titurel (bs), on a stretcher to a nearby lake to bathe his wounds. Amfortas is guardian of the Holy Grail, the sacred cup which caught the blood of Christ on the Cross; it is the Grail which unites and sustains the brotherhood. A wild woman, Kundry (sop), brings Gurnemanz balsam. He relates to his knights how Amfortas was seduced by Kundry and wounded by the magician Klingsor after entering Klingsor's garden. In the struggle he lost possession of the Sacred Spear (said to be that with which the Centurion pierced Christ on the Cross); his wounds need to be touched with the Spear in order to heal. Only a 'Pure Fool made wise through pity' can recover the Spear. A young man (Parsifal, ten) is dragged in, having killed a swan. Gurnemanz believes he may be the 'Pure Fool' and takes him to Amfortas's castle to witness the ceremony of the Holy Grail. Parsifal understands nothing of the ritual and is sent away by Gurnemanz.

Klingsor (bs) has recognized Parsifal as the only possible redeemer of Amfortas and instructs Kundry to seduce the youth. He remains indifferent to the charms of the Flower Maidens in Klingsor's magic garden but, when Kundry kisses him, is made aware of the nature of Amfortas's temptation, and becomes 'wise through pity'. Kundry realizes that her salvation, too, depends on Parsifal. Klingsor hurls the Sacred Spear at him, but it stops miraculously suspended in mid-air. When Parsifal seizes it and makes the sign of the Cross, Klingsor's domain falls in ruins.

Many years later, Gurnemanz is a hermit. On Good Friday Kundry goes to draw water for him and sees a knight in black armour approaching; it is Parsifal. He is recognized by her and later by Gurnemanz. After Kundry has bathed the knight's feet and dried them with her hair, Gurnemanz anoints Parsifal as the new King of the Holy Grail. He baptizes Kundry; the three go to the Hall of the Grail, where Amfortas is unable to celebrate the ritual uncovering of the Grail because of the pain of the wound. Parsifal heals it with the touch of the Sacred Spear. As the knights pay homage to their new King, Parsifal raises the Grail aloft; a white dove hovers over his head, and Kundry falls lifeless.

BIBL: L. Beckett, *Richard Wagner: 'Parsifal'* (Cambridge, 1981).

**Pasero, Tancredi** (*b* Turin, 11 Jan. 1893; *d* Milan, 17 Feb. 1983). Italian bass. Studied Turin with Pessina. Début Vicenza 1918 (Rodolfo, *Sonnambula*). Turin 1919; Milan, V, 1920, and S, 1926–43; New York, M, 1929–34; London, CG, 1931; Verona from 1933; Paris, O, 1935; Berlin, D, 1941. A moving artist with a rich, noble tone, and great dramatic authority. Wide repertory incl. Sarastro, Mosè, Philip, Mephistopheles (Gounod and Boito), Boris, Escamillo. (R)

**Pashchenko, Andrey** (*b* Rostov on Don, 15 Aug. 1885; *d* Moscow, 16 Nov. 1972). Russian composer. Studied St Petersburg with Witohl and Steinberg. His operas, conventional and straightforward in style, include *Orliny bunt* (The Eagles Revolt, on the Pugachev rebellion) (Spassky; Leningrad 1925) and *Tsar Maximilian* (Remizov; Leningrad 1930).

**Pashkevich, Vasily** (*b* ?, *c*.1742; *d* St Petersburg, 20 Mar. 1797). Russian composer. He played the violin in the opera at St Petersburg in 1763, then working in various capacities in the Imperial Theatres. An important pioneer of Russian opera, he became well known with *Misfortune from Owning a Carriage*: unlike many early na-

tionalist operas, this includes only one folk-song. *The Miser* looks forward to later Russian opera in its pioneering use of natural speech rhythms; and other developments in Russian opera are anticipated in *Fevey*, the first to make use of a fairy subject and to include exotic elements, also in one song using the technique of a repeated melody against a changing background which was to be developed by Glinka.

WORKLIST (all first prod. St Petersburg): *Neschastye ot karety* (Misfortune from Owning a Carriage) (Knyazhnin; 1779); *Skupoy* (The Miser) (Knyazhnin, after Molière; 1782); *Sanktpeterburgsky gostiny dvor* (The St Petersburg Bazaar) (Matinsky; 1782); *Pasha Tunissky* (The Pasha of Tunis) (Matinsky; 1783); *Fevey* (Catherine the Great & others; 1786); *Nachalnoye upravleniye Olega* (The Early Reign of Oleg) (Catherine the Great; 1790, comp. with Sarti & Canobbio); *Fedul s detmi* (Fedul with his Children) (Catherine the Great & Khrapovitsky; 1791, comp. with Martín y Soler).

BIBL: A. Rabinovich, *Russkaya opera do Glinki* (Russian Opera up to Glinka) (Moscow, 1948); D. Lehmann, *Russlands Oper und Singspiel in der zweiten Hälfte des 18. Jahrhunderts* (Leipzig, 1958).

**Pasini, Lina.** See VITALE, EDOARDO.

**Paskalis, Kostas** (*b* Levadia, 1 Sep. 1929). Greek baritone. Studied Athens Cons. Début Athens 1951 (Rigoletto). Vienna, S, from 1958; Gly. 1964–5, 1967; New York, M, 1965–7; Milan, S, from 1967; London, CG, 1969, 1971–2. With his warm tone and intense dramatic sense, a successful Macbeth, Renato, Iago (Verdi), and first Pentheus (*Bassarids*). (R)

**Pasta, Giuditta** (*b* Saronno, 28 Oct. 1797; *d* Blevio, 1 Apr. 1865). Italian soprano. Studied Como with Lotti, Milan with Asioli and Scappa. Début (under maiden name Negri) Milan, T Filodrammatici, 1815 (Scappa's *Le tre Eleonore*). Paris, I, 1816; London, H, 1817. By now married to the tenor Giuseppe Pasta. After further study with Scappa, engaged Venice 1818, then Padua, Rome, Brescia, Turin. First real success Paris, I, 1821. London, H, 1824; thereafter sang in both capitals until 1837. Naples, C, 1826; Vienna 1829; Milan, S, 1831; St Petersburg 1840. Her voice, always intractable, had by now deteriorated, but in 1850 she appeared twice in London. One of the great artists of the 19th cent., she was a 'consummate vocalist' (Chorley), and a dramatic genius who 'electrified the soul' (Stendhal). Her voice spanned a–d''' (though some considered her essentially a mezzo-soprano), and possessed a fascinating variety of colour. Though her passionate sincerity was unrestrained, her style was noble and majestic (in contrast with her rival Malibran, described by Legouvé as 'all fermentation'); she used embellishments with care and subtlety, and, unusually, kept the same ones

541

throughout her career. She inspired and created the title-roles of Pacini's *Niobe*, Donizetti's *Anna Bolena* and *Ugo, conte di Parigi*, and Bellini's *Beatrice di Tenda*; also his Amina, and Norma (a role she found difficult, insisting on transposing 'Casta diva' down a tone to F, where it has remained). She also created Corinna in *Il viaggio a Reims*, and was a famous Desdemona (Rossini), Ninetta (*La gazza ladra*), Tancredi, Elisabetta, and Elcia (*Mosè*) regretably, she had passed her peak before Verdi had the opportunity to write for her. Other celebrated roles were Mayr's Medea, and Bellini's and Zingarelli's Romeo.

BIBL: Stendhal, *Vie de Rossini* (Paris, 1824; Eng. trans. 1824, rev. 1971 by R. Coe); M. Ferranti-Giulini, *Giuditta Pasta e i suoi tempi* (Milan, 1935).

**pasticcio** (It.: 'pie', hence hotchpotch). In the 18th cent. the term was applied to a wide variety of music; more recently, writers have used it so loosely that a precise (and uncontroversial) definition is impossible. In its broadest sense, it may be applied to any 18th-cent. opera which is a collaborative effort on the part of more than one composer: a classic example is the *Muzio Scevola* opera produced in London in 1721, for which Mattei, Bononcini, and Handel each composed one act. Usually, however, the pasticcio was a looser structure, essentially a play with airs, ensembles, dances, and other movements assembled from one or more composers. It had often begun life as an opera, but on each successive production had had text and/or music added, frequently from other works, to the point where it bore only scant relationship to the original source; sometimes it simply consisted of a new libretto, to which pre-existing music was fitted. Popular in an age when it was considered entirely proper to transplant arias from one work to another, the pasticcio gave the audience an opportunity to hear the maximum number of its favourite tunes in the briefest space of time. A vintage example is *Thomyris* (1707), for which Pepusch wrote recitatives, and adapted and arranged airs by Bononcini, Scarlatti, Gasparini, and Albinoni. The pasticcio flourished especially in London in the mid-18th cent.: many important works, including Gluck's *Orfeo ed Euridice*, were first seen on the stage there as pasticcios rather than in their genuinely operatic version. Though used of paintings as early as 1706, the first musical use of the term probably occurs in Quantz's autobiography of 1725.

**pastorale.** A dramatic poem intended for reading or stage presentation, with a predominantly lyrical text featuring a pastoral, usually mythological, subject. The genre first flourished in Italy, under the influence of classical models. Poliziano's *Orfeo* (1480) was the earliest example and,

while no music survives, it is clear that parts of the play lend themselves to such accompaniment. After Agostino Beccari's *Sacrificio d'Abramo* (1554) the pastorale reached the height of its popularity, leading to the two most important works of the 16th cent., Guarini's *Il pastor fido* (1590) and Tasso's *Aminta* (1573). The earliest opera librettos of *Rinuccini and *Striggio owe much to the model of the pastorale, and its influence on Italian opera, although much reduced, can be detected right up to the 18th cent., e.g. in Handel.

In the 17th cent. the pastorale also flourished in France, having been introduced from Italy at the end of the preceding century. The first to be sung in its entirety was Charles de Bey's and Michel de La Guerre's *Les triomphes de l'Amour* (1655) and this influenced the operatic experiments of Cambert and Perrin, which included the *Pastorale d'Issy* (1659) and *Pomone* (1671). Later, when the pastorale had been supplanted by classical tragedy, many of its distinguishing features found their way into the new French opera. Most important of these was the *merveilleux*, or intervention of the supernatural, which provided opera with one of its most dramatic opportunities; likewise, the carefree Arcadian world of shepherds and shepherdesses reappears, often in the prologue. In Campra's *opéra-ballet *L'Europe galante* (1697) a pastoral is appropriately included to represent the French.

**Pastor fido, Il** (The Faithful Shepherd). Opera in 3 acts by Handel; text by G. Rossi after Guarini's pastoral play (1590); Prod. London, H, 22 Nov. 1712, with Valeriano, Pilotti-Schiavonetti, L'Épine, Barbier, Valentini, Leveridge; rev. London, H, 18 May 1734 with Carestini, La Strada, Durastanti, Scalzi, Negri, Waltz; further rev. London, CG, 9 Nov. 1734 (with new prologue *Terpsicore*, text by G. Rossi), with Carestini, Strada, Beard, M. & R. Negri, Waltz. Both 1734 revisions drew heavily on the serenata *Il Parnasso in festa*.

**Paton, Mary Ann** (*b* Edinburgh, Oct. 1802; *d* Chapelthorpe, 21 July 1864). Scottish soprano. Début London, H, 1822 (Susanna). London: CG, from 1822; L, 1824; also DL and H. Toured USA 1833–6, 1840–1, with her second husband, the tenor Joseph Wood. Roles included Amina, Agathe, Alice (*Robert le diable*). Created Reiza (*Oberon*) with acclaim. Though Weber found her stupid, and Kemble described her acting as 'like an inspired idiot', her great beauty (she was painted by Lawrence and Newton, among others) and sweet, brilliant singing brought her much success and popularity.

**patter song.** A comic song in which the greatest number of words, delivered rapidly in conversational style, is fitted into the shortest space of time, with the music supporting their inflection.

**Patti, Adelina** (*b* Madrid, 19 Feb. 1843; *d* Craig-y-Nos Castle, 27 Sep. 1919). Italian soprano, daughter of the tenor **Salvatore Patti** (1800–69), a Donizetti singer, later opera manager in New York, and the soprano **Caterina Chiesa Barilli-Patti** (*d* 1870), who reputedly sang Norma the night before Adelina was born. Studied New York with E. Valentini, and Ettore Barilli (her half-brother). Singing début aged 7; stage début New York, AM, 1859 (Lucia). New Orleans, French OH, 1860–1. London, CG, 1861–95; Berlin 1861; Paris, I, 1862 and O from 1872; Vienna 1863; Moscow, St Petersburg, 1869; Milan, S, from 1877; US tour 1881–4 with Mapleson Co.; New York, M, from 1887; Chicago 1889. Also Rome, Madrid, S America, etc. Retired 1906, but appeared occasionally until 1914. Roles included Zerlina, Rossini's Desdemona and Rosina; Amina, Lucia, Valentine, Violetta, Aida, Marguerite, Juliette, Lakmé. One of the greatest prima donnas in operatic history; her voice was astoundingly flexible, even, and sweet, and her technique superb. Verdi praised the balance between her singing and acting. Her contracts included exemption from rehearsals, as well as the highest fees of the day (after 1882, a minimum of $5,000 a performance), and a stipulation as to the size of her name on posters. Her jewel collection was so prodigious that she had to travel with several guards. But the beauty and charm of her singing won all hearts, including Hanslick's. Shaw admired her phrasing and her 'unerring ear'. She married three times; her second husband was the tenor Nicolini. (R) Her elder sisters were **Amalia** (1831–1915), who made her debut in 1848, later marrying Maurice *Strakosch, and **Carlotta** (1836–89), judged by some to be as good as Adelina (she sang the Queen of the Night's arias up a tone) but forced by her lameness to make a concert career.

BIBL: H. Klein, *The Reign of Patti* (London, 1921).

**Patzak, Julius** (*b* Vienna, 9 Apr. 1898; *d* Rottach-Egern, 26 Jan. 1974). Austrian tenor. Studied Vienna: conducting with Adler, composition with Schmidt. Self-taught as singer. Début Reichenberg (Liberec) 1926 (Radamès). Brno 1927–8; Munich 1928–45; London, CG, from 1938; Vienna from 1946; also Berlin, Salzburg, Barcelona, etc. A great artist of rare intelligence, penetration, and subtlety. Large repertory included Belmonte, Tamino, Lohengrin, Hoffmann, Dmitry, Herod; particularly acclaimed as Florestan and Palestrina. Created several roles, e.g. the Private (Strauss's *Friedenstag*), Desmoulins (Von Einem's *Dantons Tod*), and Tristan (Martin's *Le vin herbé*). His excellent diction and

care with words also made him a moving Lieder singer. (R)

**Pauer, Jiří** (*b* Kladno-Libušín, 22 Feb. 1919). Czech composer. Studied with Šín, then Prague with Hába and Bořkovec. Director Prague O, 1953–5. His most successful work has been his opera *Zuzana Vojířová* (after Bor; Prague 1958).

**Paul Bunyan.** Opera in a prologue and 2 acts by Britten; text by W. H. Auden. Prod. New York, Columbia U, 5 May 1941, with Warchoff, Woodward, Hess, cond. Ross; Aldeburgh Festival 4 Jan. 1976, with Saunders, Doghan, Jenkins, cond. Bedford.

America, pioneer days. Dissatisfied trees anticipate the arrival of the lumberjack Paul Bunyan (spoken) who, the narrator (ten-bar) tells them, never ages but grows taller every day, so that he is as big as the Empire State Building.

He arrives at the camp and meets four Swedish workers, whose leader is Hel Helson (bar). Paul Bunyan marries Carrie but the marriage soon disintegrates and she returns to her home, taking their daughter Tiny (sop). When Carrie dies shortly afterwards, Paul Bunyan fetches Tiny back and she helps the new camp cook Slim (ten). Helson wants to leave the camp and become a farmer.

Helson, irritated at the manner in which Bunyan has taken command before his own departure, decides to kill the giant. However, on attempting an attack, he realizes his mistake and reconciles himself to working under Bunyan. The narrator tells how the characters all move on to new lives as time passes and the new age of the railroad and telephone arrives. Paul Bunyan reminds the audience of what can be achieved in a free country.

**Paul et Virgine.** Opera in 3 acts by Le Sueur; text by Alphonse Du Congé Debreuil, after Bernadin de Saint-Pierre's novel. Prod. Paris, T Feydeau, 13 Jan. 1794.

The opera tells how Virginie is forced to leave the island paradise (Mauritius) she shares with Paul in order to go to France to be educated by a rich aunt. Unable to bear the rigours of her new life she decides to return to her island, but drowns during a storm on the journey. Paul finds her dead body washed ashore.

Also opera by Satie (unfin.), with text by Raymond Radiguet and Jean Cocteau.

**Pauly, Rosa** (*b* Eperjes, 15 Mar. 1894; *d* Kfar Shmaryahn, 14 Dec. 1975). Hungarian soprano. Studied Vienna with Papier. Début Vienna, S, 1918 (Verdi's Desdemona). Berlin, Kroll O, 1927–31, and S, 1929–38; Vienna, S, 1929–35; Milan, S, 1935–9; New York, M, and London, CG, 1938. Retired 1943, then taught in Israel.

Roles included Donna Anna, Leonore, Ortrud, Gutrune, Dyer's Wife, Marie (*Wozzeck*), and a famous Elektra. Admired by Strauss. (R)

**Pauvre matelot, Le** (The Poor Sailor). Opera in 3 acts by Milhaud; text by Cocteau, based on a newspaper report of an actual event. Prod. Paris, OC, 16 Dec. 1927, with Sibille, Legrand, Vieuille, Musy, cond. Lauweryns; Philadelphia, AM, 1 Apr. 1937; London, Fortune T, 9 Oct. 1950, with Vyvyan, Servent, Loring, Wallace, cond. Renton.

A sailor (ten) returns home after a long absence and is not recognized by his wife (sop). He tests her fidelity by pretending to be her husband's rich friend, and showing her jewels to prove his wealth, tries to make love to her. When he is asleep she steals the jewels in order to help pay for her husband's voyage back home, and murders him.

**Pavarotti, Luciano** (*b* Modena, 12 Oct. 1935). Italian tenor. Studied Modena with A. Pola, Mantua with Campogalliani. Début Reggio Emilia 1961 (Rodolfo). London, CG, from 1963; Gly. 1964; Milan, S, from 1965; Miami 1965; New York, M, from 1968; Paris, O, from 1974. Also Vienna, Barcelona, Verona, Montreal, etc. A hugely successful, vastly popular singer, with a ringing, high lyric tenor of great beauty, an excellent technique, and a conquering personality. Repertory includes Elvino, Nemorino, Arturo (*Puritani*), Duke of Mantua, Alfredo, Rodolfo, Cavaradossi, Calaf, Edgardo. (R)

**Pavesi, Stefano** (*b* Casaletto Vaprio, 22 Jan. 1779; *d* Crema, 28 July 1850). Italian composer. Studied Naples with Piccinni, but was expelled from the conservatory for his republican sympathies and joined Napoleon's army as a serpent player; later studed Crema with Gazzaniga. In Venice he met the librettist Foppa, who helped to arrange for the staging of his second opera. Director, Vienna Court O, for six months each year 1826–30. He was one of the most active and prolific Italian composers before the rise of Rossini, with a distinctive melodic vein and a stronger feel for the orchestra than many of his contemporaries. Among some 70 operas, the most successful was *Ser Marcantonio* (on a subject similar to that of Donizetti's *Don Pasquale*), which had an initial run of 54 consecutive performances.

SELECT WORKLIST: *Ser Marcantonio* (Anelli; Milan 1810); *Fenella* (Rossi, after Scribe; Venice 1831).

**Pearl Fishers, The**. See PÊCHEURS DE PERLES, LES.

**Pears, (Sir) Peter** (*b* Farnham, 22 June 1910; *d* Aldeburgh, 3 Apr. 1986). English tenor. Studied London with Gerhardt and Freer; USA, with T. Schnabel and Mundy; later with Manén. Début London, Strand T, 1942 (Hoffmann). London, SW, 1943–5; Gly. 1946–7; founder member of EOG 1946–76; London, CG, regularly 1948–78; Aldeburgh 1952–79; New York, M, 1974, 1978. Last stage appearance Edinburgh 1979. One of the most cultivated and intelligent artists of his day, who used an elegant and individual voice with sensitive musicianship and skilful technique. His artistry profoundly influenced the music of Britten, his long time companion, for whom he created the roles of Peter Grimes, Male Chorus, Albert Herring, Captain Vere, Essex, Quint, Flute, the Madwoman, Sir Philip Wingrave, Aschenbach. Also first Pandarus (*Troilus and Cressida*), and Boaz (Berkeley's *Ruth*). Other roles included Don Ottavio, Ferrando, Rossini's Count Almaviva, Alfredo, Vašek, David, Oedipus Rex. His perceptive response to words also showed in his affecting Lieder partnership with Britten. (R)

**Pêcheurs de perles, Les** (The Pearl Fishers). Opera in 3 acts by Bizet; text by Eugène Cormon (Pierre-Étienne Piestre) and Michel Carré. Prod. Paris, L, 30 Sep. 1863, with Léontine de Maësen, Morini, Ismaël, Guyot, cond. Deloffre; London, CG, 22 Apr. 1887 (as *Leïla*), with Föhström, Garulli, Lhérie, Miranda; Philadelphia, Grand OH, 23 Aug. 1893. Orig. version rev. WNO, 4 Oct. 1973.

Ancient Ceylon. Zurga (bar) is selected as king of the fishermen. His friend Nadir (ten) returns and, after a long estrangement caused by falling in love with the same priestess, Leïla (sop), they are reconciled. A new priestess arrives to pray to Brahma while the fishermen are at sea; she is sworn to a vow of chastity and is heavily veiled. However, when she removes her veil, Nadir recognizes her as Leïla, and their love is rekindled.

Leïla tells the high priest Nourabad (bs) how she risked death to save a fugitive and was rewarded with a gold chain. She hears Nadir singing a serenade; he joins her and the couple reaffirm their love. But Nourabad finds them, and the couple are accused of sacrilege. Zurga recognizes Leïla and vows revenge by refusing clemency.

Zurga refuses Leïla's pleas to spare Nadir's life; as she leaves, she hands Zurga her gold chain, asking that it may be sent to her mother. As Nadir and Leïla are about to mount the funeral pyre, Zurga rushes in shouting that the camp is ablaze; he recognized the chain Leïla gave him as his own and realized she was his rescuer. Zurga releases the couple, and they escape.

Berlioz admired the work, writing that 'it does M. Bizet the greatest honour'. However, it was subjected to various revisions, the most notorious but longest lasting being that of 1889; this ends with Zurga dying on the funeral pyre, and numbers among other musical mutilations the rewriting of a duet as a trio by Benjamin Godard.

**Pechkovsky, Nikolay** (*b* Moscow, 13 Jan. 1896; *d* Leningrad, 24 Nov. 1966). Russian tenor. Studied Moscow with Donsky. Début as actor, Moscow 1913; début there as tenor Sergyevsky T, 1918. Moscow, B (Stanislavsky's operatic studio), 1921–3; Leningrad 1924–41. A fine singing actor, whose best roles incl. Hermann, Golitsin, Manrico, Radamès, Otello (Verdi), Don José, Canio. (R)

**Pécs.** Town in Pécs, Hungary. Touring companies first performed in 1800, and the first permanent company was formed in 1881. The National T (cap. 820) opened in 1895, and tours Western Hungary; the present company was formed 1949.

**Pederzini, Gianna** (*b* Vò di Avio, 10 Feb. 1903; *d* Rome, 11 Mar. 1988). Italian mezzo-soprano. Studied Naples with De Lucia. Début Messina 1923 (Preziosilla). Milan, S, 1930–43, 1956–7; Rome 1939–52; London, CG, 1931 (Preziosilla, Maddalena); Buenos Aires 1937–9, 1946–7. One of the outstanding mezzo-sopranos of her day, her roles included Carmen, Mignon, and the Rossini heroines. She also sang the Countess in *The Queen of Spades* and Mistress Quickly, roles for which her vivid dramatic talents suited her admirably. Created the Prioress in *Les dialogues des Carmélites*, Milan 1957. (R)

**Pedrell, Felipe** (*b* Tortosa, 19 Feb. 1841; *d* Barcelona, 19 Aug. 1922). Spanish composer. Studied Tortosa with Nin y Serra. Assistant director at Barcelona, T Circo, 1873–4. His position as the father of Spanish music drama rests on his foundation of a Spanish national style. This took its inspiration only partly from folk music; his scholarly interests involved a profound study of earlier music, especially Victoria. Though he wrote some comic operas and zarzuelas, his major work was the consciously Wagnerian trilogy *Los Pirineos*. There is a note of studiousness in his operas which, though they contain much fine music, inhibits wide acceptance.

SELECT WORKLIST: *El último Abencerraje* (The Last Abencerage) (Altès y Alabert, after Chateaubriand; comp. 1868, 2nd rev., It. text Fors de Casamayor, prod. Barcelona 1874); *Los Pirineos* (*El conde de Foix*, *Rayo de luna*, *La jornada de Panissars*) (Balaguer; Barcelona 1902).

**Pedrillo.** Belmonte's servant (ten), lover of Blonde, in Mozart's *Die Entführung aus dem Serail*.

**Pedrotti, Carlo** (*b* Verona, 12 Nov. 1817; *d* Verona, 16 Oct. 1893). Italian composer and conductor. Studied Verona with Foroni. The success of his first performed opera, *Lina*, secured him the conductorship of the Italian O in Amsterdam. Worked there 1840–5; returned to direct the T Nuovo and the T Filarmonico in Verona. Here he wrote several operas, including his best-known work, *Tutti in maschera* (given in Paris, 1869, as *Les masques*). Turin 1868 appointed director of the Liceo Musicale and conductor at the T Regio; 1882 director of the Liceo Musicale, Pesaro. Composed *c*.18 stage works, mostly opere buffe or opere semiserie of a conventional kind. Wrote nothing after 1872, feeling himself to have been eclipsed by the younger generation; this lack of self-esteem eventually drove him to suicide, by drowning in the River Adige. Although his works enjoyed some popularity in their day, they were soon forgotten. His activities as a conductor had more influence on Italian operatic life: under his régime the T Regio rivalled La Scala, Milan in the number, quality, and importance of its productions.

SELECT WORKLIST: *Lina* (Marcello; Verona 1840); *Tutti in maschera* (Everyone Masked) (Marcello, after Goldoni; Verona 1856).

BIBL: T. Mantovani, *Carlo Pedrotti* (Pesaro, 1894).

**Peerce, Jan** (*b* New York, 3 June 1904; *d* New York, 17 Dec. 1984). American tenor. Began career as violinist. Studied with Borghetti. Début Philadelphia 1938 (Duke of Mantua). Chicago 1940; New York, M, 1941–66. Moscow, B, and Leningrad 1956. Also S America, Canada, Germany, Austria, Holland, etc. Roles included Don Ottavio, Edgardo, Gounod's Faust, Rodolfo, Cavaradossi. (R)

**Peking opera.** The form of opera now recognized as being most representative of the Chinese tradition. Peking opera originated at the end of the 18th cent., deriving elements mainly from the *Erhuang* style from Yihuang region and the *Xipi* style from Shaanxi region. Despite an official ban in 1798, the new form of opera flourished, reaching its peak between 1875 and 1926 with actors such as Mei Lanfang and Zhou Xinfang. It was performed in tea-houses, and also frequently at private parties given by the wealthy. Peking opera developed without imperial support and it was not until 1884 that the Empress Dowager Cixi initiated performances at court. Most of the actors came from provinces to the south of Peking, many beginning their training as young as 7 years.

The performance of Peking opera is governed by rigid conventions. Staging is extremely simple, without sets or a front curtain, and with very

# Pélissier

few props. Actors are concerned with the five harmonies of hand, eye, finger, mouth, and step. Movements are exaggerated and stylized and much is conveyed by symbolic gestures, e.g. the manner of walking indicates status. Performers in Peking opera are categorized not by their voices, but by the role type they play; an actor usually trains for only one of the four role categories. *Shen* is the main male character, subdivided into *laosheng* (older man in position of authority), *xiaosheng* (scholar-lover with high-pitched voice), and *wusheng* (warrior figure, who performs acrobatics). *Jing* is a male role of secondary importance, generally with a painted face. *Dan* is the female role, which would have been taken by boys. This category has two subdivisions: *gingyi* (the virtuous woman) and *huadan* (the flirtatious woman who wears a brighter costume and whose gestures and expressions are more lively). The clown character, *chou*, is allowed the most improvisation; he usually has a white oblong across his eyes and nose. Colours of face-paint and clothes have a symbolic meaning as well as helping to identify characters; red indicates loyalty, black represents strength and honesty.

In accompanying the singers, the *huqin*, a two-string bowed instrument, is predominant. The instrumental band, generally invisible to the audience, also includes percussion players and, least importantly, wind players. Spoken dialogue is traditionally highly stylized, not following normal speech inflexions. As in other forms of Chinese opera, plots usually fall either into the military or the civilian category, many being drawn from dramas of the Yuan or Ming Dynasties. Early Peking operas tended to be short, showing only one incident, but they later developed to enact a sequence of events. There are over 1,000 dramas in the repertory of classical Peking opera. Modern Peking opera also includes works inspired by the Revolution, often depicting class struggle. *Hongdengji* (The Red Lantern) relates how the Communist party delivered a secret code to guerrillas during the war against Japan. Many of these modern works disregard traditional conventions; they use scenery and are far longer. Although they no longer classify actors, and make far less use of symbolic mannerisms, the operas still employ complex acrobatics for battle scenes.

**Pélissier, Marie** (*b* 1707; *d* Paris, 21 Mar. 1749). French soprano. Début Paris, O, 1722. Rouen 1722–6, where she married the singer Pélissier. Paris, O, 1726–34, 1735–41, when she retired. Created roles in numerous operas, including Rebel and Francœur's *Pyramus et Thisbe*, Quinault's *Les amours des déesses*, and Rameau's *Hip-*

*polyte et Aricie, Les Indes galantes, Castor et Pollux, Les fêtes d'Hébé, Dardanus*. A highly successful singer, whose celebrated rivalry with Le Maure divided their public. Though her voice was unexceptional (unlike Le Maure's), she was a highly communicative and gifted actress. She also attracted much attention with her tempestuous private life.

**Pelléas et Mélisande.** Opera in 5 acts, 12 tableaux, by Debussy; text a slight alteration of Maeterlinck's tragedy (1892). Prod. Paris, OC, 30 Apr. 1902, with Mary Garden, Gerville-Réache, Périer, Dufranne, Vieuille, cond. Messager; New York, Manhattan OC, 19 Feb. 1908, with Garden, Gerville-Réache, Dufranne, Arimondi, Crabbé, cond. Campanini; London, CG, 21 May 1909, with Rose, Féart, Bourgeois, Warnery, Vanni-Marcoux, Bourbon, Crabbé, cond. Campanini.

Allemonde. Golaud (bar), grandson of King Arkel (bs), marries a mysterious girl, Mélisande (sop), he has found weeping in the forest. He is a widower and already has a son Yniold (sop) by his previous marriage. Geneviève (mez), Golaud's mother, reads to the almost blind Arkel a letter from Golaud to his half-brother Pelléas (ten or bar), telling of his marriage. Arkel instructs Pelléas to convey his acceptance to Golaud. The couple arrive in Allemonde.

Mélisande accidentally drops her wedding-ring into a fountain. Golaud, thrown from his horse at the moment the ring fell, is being tended by Mélisande. He notices the ring's absence, and tells her to go in search of it, taking Pelléas with her as it is night. Together they go to the gloomy grotto where she told Golaud she had lost the ring, and are frightened by three blind beggars.

At her window, Mélisande combs her hair and allows it to tumble around Pelléas, below; although they do not speak of love, it is clear that their emotions are closely entwined. In a sinister vault in the castle, Golaud warns Pelléas to let Mélisande alone. Later he holds his son Yniold (treb or sop) up to spy on the couple in a room: the child reports that they sit in silence, but Golaud's jealousy remains.

Pelléas has decided to leave the castle and arranges a final meeting with Mélisande. Golaud finds them and, enraged, drags Mélisande along the ground by her hair. Later, the couple meet by the fountain where Mélisande lost her ring; they confess their love for the first time. Golaud, who has been spying on them, kills Pelléas.

Arkel, Golaud, and the Physician wait by Mélisande's bed, where she is dying, having given birth to a child. Golaud, repentant but still jealous, questions her about her love for Pelléas—was it a 'forbidden' love? The castle servants enter, and fall on their knees as Mélisande dies without answering Golaud.

BIBL: R. Nichols and R. Langham Smith, *Claude Debussy: 'Pelléas et Mélisande'* (Cambridge, 1989).

**Pellegrin,** (Abbé) **Simon-Joseph** (*b* Marseilles, 1663; *d* Paris, 5 Sep. 1745). French poet, dramatist, and librettist. Published poems and plays before writing his first opera libretto, that for Desmarets's *Renaud* (1705), which appeared anonymously. Gradually his literary activities took over from his clerical duties: forced to choose between the two, he opted for the theatre and was excommunicated. He then plunged into theatrical life, writing more than 50 dramatic works for the Parisian stage. In addition to those for the Théâtres de la Foire, he wrote about 35 librettos for the Académie Royale, set by many of the most important composers of the day, including Destouches (*Télémaque et Calypso*), Lacouste (*Télégone, Orion*), and Montéclair (*Les fêtes d'été, Jepthé*).

Pellegrin's most important work was the libretto he provided for Rameau's *Hippolyte et Aricie* (1733). The first of Rameau's *tragédies lyriques*, its appearance marked the beginning of a new phase in French operatic history, and much of its success was due to the libretto. In his skilful treatment of the myth of Phaedra and Theseus, Pellegrin united the world of classical French tragedy with the more humanistic spirit of the Enlightenment, providing a model for later writers.

**Pelletier, Wilfred** (*b* Montreal, 20 June 1896; *d* New York, 9 Apr. 1982). Canadian conductor. Studied Paris with Philipp and Widor. New York, M, as répétiteur 1917. Cond. Met. Saturday Night Opera Concerts 1932, also directing the Met. French repertory. Also cond. Ravinia, San Francisco. Active in Montreal concert life, and director Montreal Cons. 1942–1961.

**Penderecki, Krzysztof** (*b* Dębica, 23 Nov. 1933). Polish composer. Studied Cracow with Malawski and Wiechowicz. The strong grasp of dramatic effect and skill with diverse musical effects shown in his religious concert works was first revealed in opera with *The Devils of Loudun*; here the effects are not always backed with sufficient musical substance and continuity. With the sacra rappresentazione *Paradise Lost* he stepped back from his earlier experimentation and drew upon influences that include Stravinsky, Berg, and even Wagner, though not their mastery of cumulative dramatic effect. *Die schwarze Maske*, set in the Thirty Years War, concerns the difficulty of transcending racial as well as political, national, and sexual types, in an idiom that returns to some of his most violent

early music, with a strong dissonant and percussive impulse.

WORKLIST: *Diabły z Loudun* (The Devils of Loudun) (Penderecki, after Whiting, after Huxley, after an historical event; Hamburg 1969); *Paradise Lost* (Fry, after Milton; Chicago 1978); *Die schwarze Maske* (The Black Mask) (Kupfer & Penderecki, after Hauptmann; Salzburg 1986); *Ubu rex* (Penderecki & Jerocki, after Jarry; Munich 1991).

BIBL: W. Schwinger, *Krzysztof Penderecki* (London, 1989).

**Penelope.** In Homer's *Odyssey*, the wife of *Odysseus and mother of *Telemachus. She waited faithfully for Odysseus' return from the Trojan war, resisting all advances from her suitors, and was finally reunited with him. Operas on her are by Draghi (1670), Pallavicino (1685), ?Nicolini (?1685), Perti (1696), A. Scarlatti (?1696), Keiser (1702), Chelleri (1716), Conti (1724), Gillier (1728), Galuppi (1741), Gazzaniga (1779), Carvalho (*Penelope nella partenza da Sparta*, 1782), Piccinni (1785), Cimarosa (1794), Rota (1866), Solomon (1889), Fauré (1913), and Liebermann (1954). She also appears in a number of operas primarily concerning Odysseus. Also Strobel (1954, a modern adaptation based on an incident of the Second World War).

**Pénélope.** Opera in 3 acts by Fauré; text by René Fauchois, after the *Odyssey*. Prod. Monte Carlo, 4 Mar. 1913, with Bréval, Raveau, cond. Jehin.

**Pepusch, John Christopher** (*b* Berlin, 1667; *d* London, 20 July 1752). English composer of German birth. After working at the Prussian court, he arrived in London *c.*1700, where he was employed as viola player and later harpsichordist at Drury Lane. His first theatrical composition was a masque *Venus and Adonis* (1715), described by the librettist Cibber in the preface as 'an Attempt to give the Town a little good Musick in a language they understand'. His comment refers to the current vogue for Italian and Italianate opera prevalent in London. Pepusch himself had contributed the recitatives to a pasticcio, *Thomyris, Queen of Scythia* (1707), the fourth and last all-sung Italianate work staged at Drury Lane, when the castrato Valentini had sung the hero's part in Italian, while the rest of the cast sang in English. He also collaborated on masques with John Hughes (*Apollo and Daphne*, 1716) and Barton Booth (*The Death of Dido*, 1716).

Pepusch is best remembered for his contribution to *The Beggar's Opera* (1728), for which he wrote the bass lines to melodies selected by Gay, and the overture. He also worked with Gay on the sequel, *Polly*, and with Hawker on another ballad opera, *The Wedding*. His setting of Betterton and Dryden's *Dioclesian* (1724) is described as an opera with spoken dialogue. Although his original contributions may seem minor, Pepusch was certainly of great significance in the successful establishment of ballad opera. He is also remembered as one of the earliest of musical antiquarians.

**Pepys, Samuel** (*b* London, 23 Feb. 1633; *d* London, 26 May 1703). English diarist. His diary, which he kept from 1 Jan. 1660 until 31 May 1669, was written in cipher and not decoded until 1825. In his entries he refers frequently to domestic music-making during the Restoration, and also records his impressions of Italian operatic singers. His comments on recitative—'in singing, the words are to be considered and how they are fitted with notes' (16 Feb. 1667)—show that he favours the expressive style.

Pepys himself has been the subject of operas: Shaw (*Mr Pepys*, 1926), Coates (*Samuel Pepys*, 1929).

**Pérez, David** (*b* Naples, 1711; *d* Lisbon, 30 Oct. 1778). Italian composer. Studied Naples with Mancini, Veneziano, and Galli. His first opera, *La nemica amante*, established his reputation, which he confirmed with *Li travestimenti amorosi* and especially *Siroe*. He also worked in Palermo until 1748, then moving to Lisbon. Here he was patronized by King José I, and his operas were sumptuously staged. His earlier operas are mostly in a strict opera seria convention, but his works for Lisbon are more freely composed: he drew for his texts mostly on Zeno and Metastasio. He introduced the style of Neapolitan opera to Portugal, and helped to encourage the growth of opera there.

SELECT WORKLIST: *La nemica amante* (The Loving Enemy) (?; Naples 1735); *Li travestimenti amorosi* (The Amorous Disguises) (Palombo; Naples 1740); *Siroe* (Metastasio; Naples 1740); *Demofoonte* (Metastasio; Lisbon 1752); *Solimano* (?Migliavacca; Lisbon 1757).

**Perfect Fool, The.** Opera in 1 act by Holst; text by the composer. Prod. London, CG, 14 May 1923, by BNOC, with Teyte, Thornton, Ellis, Parker, Hyde, Collier, cond. Goossens; Wichita 20 Mar. 1962.

The Princess's (sop) hand has been promised to any man 'who does the deed no other man can do'. Many suitors take part in a contest, among them a (Verdian) Troubadour (ten) and a (Wagnerian) Traveller (bar). However, the Princess's eye lights on the Perfect Fool (spoken), and asks to marry him. But he has no interest in her and refuses.

**Pergola, Teatro alla.** See FLORENCE.

**Pergolesi, Giovanni Battista** (*b* Jesi, 4 Jan. 1710; *d* Pozzuoli, 16 Mar. 1736). Italian composer. Studied Naples with Greco, Vinci, and Durante; employed there by various aristocrats as maestro di cappella. In 1732 his first stage work, the opera seria *Salustia*, was staged. It was unsuccessful but the warm reception accorded the comic *Lo frate 'nnamorato* later that year indicated where his real talents lay. While the opere serie which he wrote over the next few years, including *Il prigionier superbo* and *Adriano in Siria*, are largely forgotten, the comic *intermezzos with which they were performed—*La serva padrona* and *La contadina astuta* respectively—are regarded as seminal for the development of *opera buffa. Likewise, though now considered one of the best 18th-cent. opere serie, *L'Olimpiade* was regarded with indifference in its day. Pergolesi's final stage work, the opera buffa *Flaminio*, was given a triumphant reception.

Though a consummate master of opera seria, possessing a flair for expressive writing and showing some originality in his handling of the aria, it was in comic works that he excelled. *La serva padrona* is the classic example of the early 18th-cent. intermezzo and adumbrates many of the features of later opera buffa. Taken up by comic troupes and performed throughout Europe, it brought Pergolesi posthumous fame and soon came to be regarded as the epitome of Italian style: when given in Paris 1752 it sparked off the *Guerre des Bouffons. *Lo frate 'nnamorato* is musically an even more accomplished work and represents an important stage in the development of the *ensemble finale.

WORKLIST (all first prod. Naples unless otherwise stated): *Salustia* (Morelli, after Zeno; 1732); *Lo frate 'nnamorato* (The Beloved Brother) (Federico; 1732); *Il prigionier superbo* (?; 1732); *La serva padrona* (The Maid-Mistress) (Federico; 1733); *Adriano in Siria* (Metastasio; 1734); *La contadina astuta* (The Astute Peasant-Girl) (Mariani; 1734); *L'Olimpiade* (Metastasio; Rome 1735); *Flaminio* (Federico; 1735).

BIBL: F. Degrada, ed., *Studi Pergolesiani* (Naples, 1986).

**Peri, Jacopo** (*b* Rome, 20 Aug. 1561; *d* Florence, 12 Aug. 1633). Italian composer, singer, and harpist. Studied Florence with Malvezzi; *c.*1588 joined the Medici court, where he won fame as a singer, being nicknamed 'Il Zazzerino', from his long golden hair. Participated as a singer in lavish court entertainments, such as the famous 1589 intermedi for Bargagli's *La pellegrina*, for which he also composed some music. An important member of the Bardi *camerata, he was closely involved in the intellectual debate out of which opera evolved. Though not the first to write in the new *stile rappresentativo, he was the first to use it for a complete dramatic work. His (mostly lost) setting of Rinuccini's *Dafne* libretto, for which he collaborated with Corsi, is usually regarded as the first true opera; for its performance Peri took the part of Apollo. To celebrate the wedding of Henri IV of France to Maria de' Medici (1600) he was commissioned to write a work for public performance. This opera, *Euridice*, is the first for which complete music survives: it was published 1601. Peri again participated, taking the part of Orpheus. He planned two further operas, *Tetide* and *Adone*, but only the latter was written, and this was not performed. His later dramatic music included many small-scale pieces, three (lost) sacre rappresentazioni, and two collaborations with Gagliano, *Lo sposalizio di Medoro e Angelica* and *Flora*.

Together with his arch-rival Caccini, it was Peri who gave effect to the deliberations of the Bardi camerata and by turning theory into practice demonstrated the full potential of the new art-form. In this he was aided by his background as a singer, his literary sensitivity, and a streak of true dramatic flair. *Euridice* was soon recognized as a work of considerable emotional force and originality and had a notable effect on the early history of opera. With its affective handling of the stile rappresentativo, finely judged use of the chorus, and illuminating treatment of the mythological story, its impression is still as powerful in modern revivals as it must have been to the audience of the day.

WORKLIST (all first prod. Florence): *Dafne* (Rinuccini; 1598, mostly lost; comp. with Corsi); *Euridice* (Rinuccini; 1600); *Lo sposalizio di Medoro e Angelica* (Salvadori; 1619); *Flora* (Salvadori; 1628; comp. with Gagliano).

**Périchole, La.** Operetta in 3 acts by Offenbach; text by Meilhac and Halévy, after Mérimée's drama, *Le carrosse de Saint-Sacrement* (1829). Prod. Paris, T des Variétés, 6 Oct. 1868, with Schneider, Legrand, Dupuis, Grenier, Lecomte, Bondelet, Bac; New York, Pike's OH, 4 Jan. 1869, with Irma, Rose, Aujac, Leduc, Lagriddou; London, Princess's T, 27 June 1870.

Peru, mid-18th cent. A gypsy street-singer, La Périchole (mez), and her partner and lover Piquillo (ten), come to Lima. Don Andres (bar), the viceroy of Peru, is charmed by La Périchole and engages her as a lady-in-waiting. After many intrigues and complications, including an escape from the local prison, the lovers are reunited.

**Périer, Jean** (*b* Paris, 2 Feb. 1869; *d* Neuilly, 6 Nov. 1954). French baritone. Studied Paris with Taskin and Bussine. Début Paris, OC, 1892 (Monostatos). Paris, OC, 1892–4, 1900–20. New York, Manhattan OC, 1908; also Monte Carlo. Though possessing only a limited voice, he was a very fine actor, and had a successful career in films (1900–10, 1922–38). Created many leading

roles, including Ramiro (*L'heure espagnole*), Rabaud's Mârouf, and Pelléas, a role to which he gave 'admirable definition', with affecting simplicity (Inghelbrecht). Debussy, who had reservations, attributed his growing popularity to the fact that he got better and better because 'he has entirely given up singing what I wrote'. (R)

**Perlea, Jonel** (*b* Ograda, 13 Dec. 1900; *d* New York, 29 July 1970). Romanian conductor. Studied Munich with Beer-Waldbrunn, Leipzig with Lohse and Graener. Concert début Bucharest 1919. Répétiteur Leipzig 1922–3; Rostock 1923–5. Cond. Cluj 1927–8; Bucharest 1928–32, 1934–44, music director 1929–32, 1934–6. Gave Bucharest premières of many works, including *Meistersinger*, *Falstaff*, and *Rosenkavalier* (all in Romanian). Berlin, Vienna, Stuttgart from 1935; Milan, S, and other Italian houses 1945–58, giving many world and Italian premières. New York, M, 1949–50. Afflicted by a stroke in 1957, he took to conducting with his left arm. Taught cond. New York, Manhattan School, 1952–69. (R)

**Pernet, André** (*b* Rambervilliers, 6 Jan. 1894; *d* Paris, 23 June 1966). French bass. Studied Paris with Gresse. Début Nice 1921. Paris: O, 1928–31, 1932–45, 1948; OC, 1931–48. Created leading roles in many operas, e.g. Milhaud's *Maximilien*, Hahn's *Marchand de Venise*, Enescu's *Œdipe*. Played the Father in the film of *Louise* directed by Charpentier. (R)

**'Per pietà'.** Fiordiligi's (sop) aria in Act II of Mozart's *Così fan tutte*, her resistance weakening in the face of the advances of her sister Dorabella's disguised lover Ferrando.

**Perrault, Charles** (*b* Paris, 12 Jan. 1628; *d* Paris, 16 May 1703). French poet and critic. He was involved in a debate between the relative merits of the ancients and moderns, but is best known for his collection of fairy-tales, *Histoires et contes du temps passé, avec des moralités* (1697), subtitled *Contes de Ma Mère L'Oye* (Mother Goose's Tales). The collection included *La Belle au bois dormant* (Sleeping Beauty), *Le Barbe-bleue* (Blue-Beard), *Le Petit Poucet* (Tom Thumb), *La Belle et le Bête* (Beauty and the Beast), and *Cendrillon, ou La petite pantoufle de verre* (Cinderella, or The Little Glass Slipper).

Operas on the stories are:

*La Belle au bois dormant*: B. Rubinstein (1938); Faith (1971)
*La Belle et le Bête*: Giannini (1938); Murray (1974); Lasker (20th cent.)

See also BLUEBEARD and CINDERELLA.

**Perrin, Émile-César-Victor** (*b* Rouen, 19 Jan. 1814; *d* Paris, 8 Oct. 1885). French administrator. Director Paris, OC, 1848–57, 1862, and O, 1862–71. Persuaded Verdi to write *Don Carlos* for the Opéra. His managements saw the production of many new works, incl. Thomas's *Mignon* (1866), Massenet's first opera, *La grand' tante* (1867), Saint-Saëns's first opera, *La princesse jaune* (1872), *Carmen* (1875), and works by Halévy, Adam, Massé, Auber, and Offenbach.

**Perrin, Pierre** (*b* Lyons, *c*.1620; *d* Paris, 25 Apr. 1675). French librettist. Assumed the title of Abbé. The first important French librettist, and the founder of the *Académie Royale de Musique et de Danse (the Paris *Opéra). Previously, he had written an opera for Cambert for performance at the small royal theatre at Issy (*La pastorale d'Issy*, 1659), 'in an attempt to evolve a genre of opera simpler than the current forms of Italian opera'. He abandoned the alexandrines of the French classical theatre in favour of short lyric verses. His success led Louis XIV to consider the foundation of a national opera, and in 1669 letters patent were granted to Perrin for the formation of the Académie. In 1671 the work which has the best claim to be the first French opera, Cambert's *Pomone* to Perrin's text, was publicly given. But financial difficulties and malpractices led to his imprisonment, and he died in poverty, having sold his letters patent to Lully. Though a mediocre versifier, Perrin is remembered as a founder of French opera, and also for his influence, directly on Quinault and later on the pastoral operas and writings of Rousseau, in developing a simpler and more popular folk-influenced style.

**Perron, Karl** (*b* Frankenthal, 3 June 1858; *d* Dresden, 15 July 1928). German baritone. Studied Berlin with Hey, Frankfurt with Stockhausen. Début Leipzig 1884 (Wolfram). Leipzig till 1891; Dresden 1892–1924. Bayreuth 1889–1904. Also Munich, Amsterdam. Roles included Don Giovanni, Wotan, King Mark, Amfortas, Onegin. Created Jochanaan, Orestes; also Ochs, though not the composer's and Hofmannsthal's ideal choice (too thin, refined, and humourless). Strauss none the less considered him 'a first-class artist'.

**Persephone.** See PROSERPINE.

**Persiani, Fanny** (*b* Rome, 4 Oct. 1812; *d* Neuilly, 3 May 1867). Italian soprano. Studied with her father, the tenor Nicola *Tacchinardi. Début Livorno 1832 (Fournier's *Francesca da Rimini*). Successes in Venice; Milan; Naples; Paris, I, 1837–48. London: HM, 1838–46; CG, 1847–9. Vienna 1837, 1844; St Petersburg 1850–2, after which she retired. A considerable virtuoso, she possessed complete mastery of her clear and silvery voice (which, however, Doni-

zetti thought 'rather cold'). With her delicate, angelic appearance she epitomized the fragile heroines of contemporary opera. Created title-roles of Donizetti's *Lucia, Pia de' Tolomei*, and *Rosmonda d'Inghilterra*; also highly acclaimed as Zerlina, Amina, Adina, Linda di Chamounix, Lucrezia Borgia, etc. In 1830 she married the composer **Giuseppe Persiani** (1804–69). Her singing largely accounted for the success of his operas, though his most famous, *Inès de Castro* (1835), was written for Malibran. Husband and wife helped to establish London's Royal Italian O at Convent Garden, 1847, he putting up much of the money.

**Persiani, Orazio** (*fl* Venice, *c*.1640). Italian poet. Secretary of the Accademia degli Incogniti and closely involved with the establishment of the first Venetian public opera-house in the T San Cassiano. Librettist of Cavalli's *Le nozze di Teti e di Peleo* (1639), the first Venetian opera.

**Persichini, Venceslao** (*b* Rome, 1827; *d* Rome, 19 Sep. 1897). Italian composer and teacher. Taught at S Cecilia, Rome; his pupils included Battistini, Magini-Coletti, and De Luca.

**Perth.** City in Western Australia. The Western Australia OC performs at Her Majesty's T and tours the state.

**Perti, Giacomo Antonio** (*b* Bologna, 6 June 1661; *d* Bologna, 10 Apr. 1756). Italian composer. Studied Bologna with his uncle and with Franceschini. Maestro di cappella S Pietro, Bologna, 1690; S Petronio, 1696. Wrote *c*.20 operas, Bologna, Florence, Modena, Rome, and Venice: these show him to be a fluent exponent of the late 17th-cent. style. His pupils included Torelli and Martini.

SELECT WORKLIST: *Oreste in Argo* (Bergamori; Bologna 1681, lost); *Rosaura* (Arcoleo; Venice 1689, lost); *Venceslao* (Zeno; Bologna 1708, lost).

**pertichino** (It., in Tuscan dialect: 'an extra draught animal'). The term used for the character in opera who during a recitative or aria remains silent or makes occasional interjections. The common dramatic function is to provide a listener, reacting suitably, for a singer's narration: an example is Ines listening to Leonora's 'Tacea la notte' in Act I of Verdi's *Il trovatore*. The term was common in the 18th cent., but gradually fell into disuse (though it remained as a dramatic function) towards the end of the 19th cent.

**Pertile, Aureliano** (*b* Montagnana, 9 Nov. 1885; *d* Milan, 11 Jan. 1952). Italian tenor. Studied Padua with Orefice, Milan with Bavagnoli. Début Vicenza 1911 (Lionel, *Martha*). Milan, S, 1915, 1918, 1921–37; New York, M, 1921–2; Buenos Aires, C, 1923, 1925–6, 1929; London,

CG, 1927–9, 1931; also Verona, Berlin, Vienna etc. Created title-roles of both Boito's and Mascagni's *Nerone*, and Wolf-Ferrari's *Sly*. Roles included Edgardo, Fernando (*Favorita*), Manrico, Radamès, Verdi's Otello, Rodolfo, Lohengrin, Chenier, Canio. Toscanini's favourite tenor, he was described by the critic Eugenio Gara as a 'sculptor of melody'. Though his tone could be hard (and aroused controversy), he used it expressively, and gave unforgettable performances of exceptional artistry and fervour. (R)

**Peru.** Tomás de Torrejón y Velasco (1644–1728), the Lima cathedral organist, wrote a Calderón opera, *La púrpura de la rosa*. The first Peruvian opera was *Atahualpa* (1877) by an Italian, Carlo Enrico Pasta. *Ollanta* (1900) by José Maria Valle is on a national subject. Opera is given at the T Municipal in Lima.

**Pesaro.** Town in the Marche, Italy. Opera was first given at the T del Sole, opened 1637 with Hondedei's *Asmondo*; rebuilt 1682–94, 1723, 1790. Replaced by the T Nuovo, opened 10 June 1818 with the Pesaro-born Rossini directing his *Gazza ladra* from the keyboard. Renamed T Rossini 1855, and the scene of many Rossini seasons, also of the premières of Mascagni's *Zanetto* (1896) and Zandonai's *La via della finestra* (1919). The Centro di Studi Rossiniani was founded 1940. The T Rossini closed for renovation 1967; reopened 1980. Since then Pesaro has held a successful annual Rossini Festival, attracting distinguished artists and reviving neglected operas. Performances are given in the T Comunale G. Rossini (cap. 914) and the Auditorium Pedrotti (cap. 800).

**Pescetti, Giovanni Battista** (*b* Venice, *c*.1704; *d* Venice, 20 Mar. 1766). Italian composer. Studied with Lotti. His first opera, *Nerone detronato*, was for Venice; after some collaborations with Galuppi came to London where he replaced Porpora 1737. London: director CG 1739, H 1740. Operas produced here included a revival of *Demetrio*, and *Aristodemo*. Returned to Venice *c*.1745. His operas owe much to Galuppi in their elegance and poise.

SELECT WORKLIST: *Nerone detronato* (Nero Dethroned) (Pimbaloni; Venice 1725); *Demetrio* (Metastasio; Florence 1732); *Aristodemo* (Rolli; London 1744); *Zenobia* (Metastasio; Padua 1761).

**Peter Grimes.** Opera in a prologue and 3 acts by Britten; text by Montagu Slater, after Crabbe's poem *The Borough* (1810). Prod. London, SW, 7 June 1945, with Cross, Coates, Pears, R. Jones, Donlevy, Brannigan, cond. Goodall; Tanglewood, Lenox, MA, 6 Aug. 1946, with Manning, W. Horne, Pease, cond. Bernstein.

The Borough, a small Suffolk fishing village,

*c*.1830. At an inquest into the death of the apprentice of the fisherman Peter Grimes (ten) a verdict of accidental death is recorded, but Grimes is warned not to take on more boy apprentices.

Ellen Orford (sop), the schoolmistress, alone stands by Grimes, and goes to fetch his new apprentice. That evening, in The Boar, the village people accuse Grimes of being a murderer. As soon as Ellen arrives with the new boy, Grimes takes him off home through a raging storm.

Several weeks later Ellen discovers that Grimes has been ill-treating the boy; when she insists that he be allowed to rest as it is Sunday, she and Grimes quarrel. The villagers, coming out of Church, hear this, and set off for Grimes's hut to see what is happening. Grimes's efforts to comfort the boy are interrupted by the arrival of the village deputation. Grimes pushes him out through a back entrance on to the cliff where, missing his footing, he plunges to his death.

The people of the Borough have not seen Grimes and the boy for several days and assume they are out fishing. However, Ellen recognizes a pullover washed up on the beach as the one she knitted for the boy. Grimes's boat has returned to harbour and a posse of villagers sets off to find him. Grimes has lost control of his senses and fails to recognize even Ellen for a while. Captain Balstrode (bar) advises Grimes to take his boat out to sea and sink it. This he does as the village comes to life for another, ordinary day.

BIBL: P. Brett, *Benjamin Britten: 'Peter Grimes'* (Cambridge, 1983).

**Peter Ibbetson.** Opera in 3 acts by Deems Taylor; text by the composer and Constance Collier, after the latter's play founded on George du Maurier's novel (1892). Prod. New York, M, 7 Feb. 1931, with Bori, Johnson, Tibbett, cond. Serafin.

England and France, 1855–87, Peter Ibbetson (ten) murders his tyrannical uncle, Colonel Ibbetson (bar), and is imprisoned for life in Newgate. He succumbs to dreams and visions, and conjures up his past, including memories of his childhood sweetheart Mary, now the Duchess of Towers (sop). After nearly 40 years in prison, he learns that she has died, and losing the will to live he also dies; the prison walls disintegrate and Peter, young once more, finds Mary waiting for him.

One of the most successful American operas before Menotti.

**Peter the Great** (Tsar Peter I of Russia) (*b* Moscow, 9 June 1672; *d* St Petersburg, 8 Feb. 1725). Operas on him, usually taking as subject the period he spent working in the shipyards of Saar-

**Petrassi**

dam in Holland, are by Grétry (1790), Shield (*The Czar*, 1790), Rossi (1793), Weigl (*Die Jugend Peter des Grossen*, 1814), Lichtenstein (*Frauenort, oder Der Kaiser als Zimmermann*, 1814), Bishop (1818), Pacini (*Il falegname di Livonia*, 1819), Vaccai (*Pietro il grande*, 1824), Donizetti (*Il borgomastro di Saardam*, 1827), Flotow (*Pierre et Catherine*, 1829), Cooke (*The Battle of Pultawa*, 1829), Lortzing (*Zar und Zimmermann*, 1837), Frondoni (1839), Jullien (1852), Meyerbeer (*L'étoile du nord*, 1854), Arapov (*Frigate Victory*, 1959), and Lourié (1958).

**Petrassi, Goffredo** (*b* Zagarolo, 16 July 1904). Italian composer. Studied Rome with Bustini and Donato. His operas form a comparatively minor part of his output. *Il cordovano* was bitterly attacked for a supposedly pornographic libretto at its first performance; it uses the voices in a somewhat instrumental, even concertante, manner. *Morte dell'aria* has also attracted attention for its thoughtful treatment of a real-life incident, the death of a would-be birdman falling from the Eiffel Tower. Intendant Venice, F, 1937–40.

WORKLIST: *Il cordovano* (Cervantes, trans. Montale; Milan 1949, rev. Milan 1959); *Morte dell'aria* (Death of the Air) (Scialoja; Rome 1950).

**Petrella, Clara** (*b* Milan, 28 Mar 1914; *d* Milan, 19 Nov. 1987). Italian soprano. Studied Milan with her sister Micaëla, and Russ. Début Alexandria 1939 (Liù). Milan, S, from 1947. Also Naples, C; Rome; Mexico City; etc. Created roles in numerous operas, e.g. *Maria Golovin* (Menotti), *Cagliostro*, *La figlia di Jorio* (Pizzetti). Highly acclaimed as a singing actress, especially as Manon Lescaut, Iris, Magda Sorel. (R)

**Petrella, Errico** (*b* Palermo, 10 Dec. 1813; *d* Genoa, 7 Apr. 1877). Italian composer. Studied Naples Cons. with Bellini and Zingarelli. Aged only 15 he wrote his first opera, *Il diavolo color di rosa*: though a great success, it had been done against his teachers' will, so he was expelled from the Conservatoire. He wrote five operas for Naples 1831–9, including *I pirati spagnuoli*, but ceased composing following a quarrel. Resumed his career 1851 with the successful *Le precauzioni*, then with *Elena di Tolosa* and *Marco Visconti*, the latter rivalling even Verdi's *Il trovatore* in the same season. Taken up by the firm of Lucca to compete with Ricordi's Verdi, he wrote his first commission for La Scala, Milan, *Elnava*, in 1856. This ushered in a number of well-received works, of which the most widely publicized and successful was *Jone*.

Though his music had an appeal in its day, Petrella's success was ephemeral and his style remained firmly rooted in the 18th-cent. opera buffa tradition. He chose subjects that really demanded greater imaginative control than he could give them and, unlike his great contemporary Verdi, was unable to release himself from the restricting conventions and formulas of the day.

SELECT WORKLIST: *Il diavolo color di rosa* (The Roseate Devil) (Tottola; Naples 1829); *I pirati spagnuoli* (The Spanish Pirates) (Bidera; Naples 1838); *Le precauzioni* (D'Arienzo; Naples 1851); *Elena di Tolosa* (Bolognese; Naples 1852); *Marco Visconti* (Bolognese; Naples 1854); *I promessi sposi* (The Betrothed) (Ghislanzoni, after Manzoni; Lecco 1869).

BIBL: G. Cosenza, *Vita e opere di Errico Petrella* (Rome, 1909).

**Petrograd.** See ST PETERSBURG.

**Petrov, Ivan** (*b* Irkutsk, 29 Feb 1920). Russian bass. Studied Moscow with Mineyev. Début with Kozlovsky's Co. 1939 (small roles). Moscow, B, from 1943; Paris, O, from 1954. An authoritative singer with a wide range, renowned for his Boris, Dosifey, Kochubey, Méphistophélès (Gounod), and Philip. (R)

**Petrov, Osip** (*b* Elizavetgrad, 15 Nov. 1806; *d* St Petersburg, 12 Mar. 1878). Russian bass. Initially self-taught, later studying with Cavos. Début Elizavetgrad 1826 (Cavos's *Cossack Poet*). Joined I. F. Stein's troupe, and was much influenced by the actor Shchepkin. St Petersburg 1830–78. The first of the great Russian basses, he at once set an example for composers of the emergent nationalist school, and became one of their finest interpreters. He was accordingly sought out to create most of the important bass parts of new operas performed in St Petersburg for nearly half a century: Glinka's Ivan Susanin and Ruslan; Dargomyzhsky's Miller (*Rusalka*) and Leporello (*Stone Guest*); Ivan the Terrible (*Maid of Pskov*); the Mayor (*Vakula the Smith*); and Varlaam (*Boris Godunov*), as well as works by Rubinstein and Serov. His warm, dark voice (ranging from B♭ to f♯), versatility, and powerful dramatic perception were allied to artistic integrity and great generosity of spirit. At his funeral, Musorgsky mourned his 'Grandpa' as 'an irreplaceable instructor . . . who inspired me to creativity'. An unnamed critic wrote of him, 'He has shown us what nationality in art means.' His wife **Anna Vorobyeva** (*b* St Petersburg, 14 Feb. 1816; *d* St Petersburg, 26 Apr. 1901) was a contralto, the daughter of the bass **Yakov Vorobyev** (?1766–1809). Studied with Sapienza. Created Vanya (*A Life for the Tsar*). Though prevented by illness from singing the first Ratmir (*Ruslan*), she sang in the third performance, and Glinka considered that it was she who made the work a success. Her large voice was of 'irresistible charm' (Stasov), and she was a much admired performer, highly successful in the Italian repertory.

BIBL: E. Lastochkina, *Osip Petrov* (Moscow, 1950).

**Petrov, Vasily** (*b* Alexeyevka, 12 Mar. 1875; *d* Moscow, 4 May 1937). Ukrainian bass. Studied Moscow with Bartsal. Début Moscow, B, 1902 (Miller, Dargomyzhsky's *Rusalka*). Moscow, B, until 1936; Berlin 1909; London, DL, 1914. Large repertory incl. Sarastro, Mozart's Don Basilio, Ivan Susanin, Ruslan, Pimen, Konchak, Gremin, Dodon, Gounod's Méphistophélès, Wotan.

**pezzo concertato** (It.: 'concerted piece'). In 19th-cent. Italian opera, a large-scale ensemble forming the *cantabile of the finale: sometimes, although not always, followed by a *stretta.

**Pfitzner, Hans** (*b* Moscow, 5 May 1869; *d* Salzburg, 22 May 1949). German composer and conductor. Studied Frankfurt with Knorr and Kwast, Wiesbaden with Riemann. Conductor, Mainz Stadttheater, 1894–6, a post he took in order to speed the première of *Der arme Heinrich*; Berlin, T des Westens, 1903–6; Strasburg 1908–18 (director, Stadttheater, 1910–16, where his assistant from 1914 was Klemperer).

A great admirer of Wagner and Schopenhauer, Pfitzner expressed dislike for modernist tendencies in music and became a fervent nationalist. His early triumph with *Der arme Heinrich* was followed up in the no less successful *Die Rose vom Liebesgarten*: this was given in several cities and among its admirers was Mahler who staged it in Vienna. *Christelflein* made less impact, but with *Palestrina* Pfitzner produced his greatest work. His treatment of the apocryphal legend of Palestrina's dealings with the Council of Trent, for which he wrote his own libretto, elevates it to a Wagnerian exploration of the conflict between old and new artistic ideals. Though well regarded in its day, when it was even claimed as the successor to *Parsifal*, the score fails sufficiently to articulate the taut dramatic structure of the libretto. His last opera, *Das Herz*, found few supporters.

Pfitzner's operas have enjoyed little success outside Germany; even there, the general public has not been unanimous in its acceptance of them, though *Palestrina* has become a regular feature of the Munich Summer Festivals, and is given regularly at the Staatsoper, Vienna.

WORKLIST: *Der arme Heinrich* (Poor Heinrich) (Grun, after anon. medieval legend; Mainz 1895); *Die Rose vom Liebesgarten* (Grun; Elberfeld 1901); *Christelflein* (Stach; Munich 1906, rev. Dresden 1917); *Palestrina* (Pfitzner; Munich 1917); *Das Herz* (The Heart) (Mahner-Mons; Munich & Berlin 1931).

BIBL: J. Williams, *The Music of Hans Pfitzner* (Oxford, 1992).

**Pforzheim.** Town in Baden-Württemberg, Germany. Opera was given by travelling companies in the 18th cent. The present Stadttheater (cap 437) opened 1948, and has staged premières of works by Eastwood, Rivière, Chailly, and Sauguet.

**Philadelphia.** City in Pennsylvania, USA. The first opera heard was *Flora*, an anon. ballad opera, given by an English company on 7 May 1754 at Plumstead's Warehouse. The Southwark T opened 1766 with Arne's *Thomas and Sally*, and imported English comic opera. Most opera in the early 19th cent. was Italian, brought by Da Ponte to the Chestnut St. T ('Old Drury'). The first true opera was *Der Freischütz* in 1825. A French company from New Orleans appeared 1827–33, and the Havana OC gave the first Italian opera here, Mayr's *Che originali* in 1829. William Fry wrote 'the first publicly performed Grand Opera written by a native American', *Leonora*, libretto by his brother, given by the Seguin Co., 1845.

The Philadelphia Academy of Music, opened Jan. 1857, is the oldest opera-house in continuous use in America. Up to the end of the 19th cent. many works had their US premières there, including *Luisa Miller*, Gounod's *Faust*, *Fliegender Holländer*, *Pagliacci*, *Cavalleria*, *Amico Fritz*, *Pêcheurs de Perles*, and *Manon Lescaut*. Companies functioning included the Philadelphia O, which gave performances under Gustav Heinrichs 1891–6, under Damrosch 1896–7, and Damrosch and Ellis 1897–9. The New York Metropolitan appeared 1899–1961 at the Academy of Music. The Chicago Grand Opera Co. was known as the Chicago-Philadelphia Grand O, as financial support came from both cities. Works receiving US premières included Herbert's *Natoma* (1911), Goldmark's *Cricket on the Hearth* (1912), and Franchetti's *Cristoforo Colombo* (1913): seasons were given in the Metropolitan OH built by Hammerstein to house his Manhattan OC, originally known as the Philadelphia OH (cap. 4,000) (now abandoned).

The Philadelphia Civic OC, founded in 1923 with Alexander Smallens as music director, was renamed the Philadelphia Lyric OC in 1958; US premières included *Feuersnot* (1927) and *Ariadne* (1928). The Pennsylvania Grand Opera Co. was active 1927–30, and gave the first US *Khovanshchina* (1928); this became the Philadelphia La Scala Co., then the Philadelphia Grand Opera Co. The Lyric and Grand finally merged into the Philadelphia OC, March 1975, and performs at the Academy (cap. 2,818). New directions in the 1980s took the company into enterprises that included *Carmen* set in Fascist Spain and a revival of Sousa's *The Free Lance*, also the staging of opera for PBS TV.

Other short-lived enterprises included the Philadelphia Grand O, 1926–32, which launched the careers of singers including Rose Bampton,

# Philidor

and the Philadelphia Orchestra OC, 1930–5, which gave the first US *Glückliche Hand*, *Wozzeck*, and *Mavra*, with Stokowski, Goossens, and Reiner conducting. In 1938 the Philadelphia OC was founded to give opera in English; it gave the première of Menotti's *The Old Maid and the Thief*. The Curtis Institute has staged student and semi-professional performances, and gave the première of Menotti's *Amelia goes to the Ball* (1937).

**Philidor, François-André Danican** (*b* Dreux, 7 Sep. 1726; *d* London, 31 Aug. 1795). French composer. Studied Versailles with Campra. His proficiency as a chess-player—he toured as a master and published a study of the game—was one facet of intellectual gifts that made him the first truly learned composer of opéras-comiques. From 1756, after his Italian style debarred him from the Opéra, he developed his career as a successful composer of opéra-comique. With *Le sorcier*, he became the first composer to take a curtain call. He did much to develop the genre, by the greater character he gave to melody, by his richer harmonic palette (especially in modulation), and by novelties he introduced such as the unaccompanied canon quartet in *Tom Jones*, his masterpiece (though it was originally booed). These were, however, more skilful extensions of existing forms and techniques than real originalities, together with a capacity for handling the familiar with panache; his success here can be attributed in part to his Italianate training. He was also influenced by Gluck, to the point of probable plagiarism in *Le sorcier* and *Ernelinde*. He also had vivid imitative powers, and brought into his scores the sounds of a hammer, or a donkey, or other effects to underline the text wittily; and he gave the orchestra a more expressive role even in light works. He married the singer **Angélique Richer**. His half-brother **Anne Philidor** (1681–1728) founded the Concert Spirituel to provide music on days when for religious reasons the Opéra was closed, and wrote some pastorales.

SELECT WORKLIST (all first prod. Paris): *Blaise le savetier* (Blaise the Cobbler) (Sedaine, after La Fontaine; 1759); *Le sorcier* (Poinsinet; 1764); *Tom Jones* (Poinsinet, after Fielding; 1765); *Ernelinde* (Poinsinet, after Noris; 1767); *Le bon fils* (The Good Son) (Devaux; 1773); *Persée* (Marmontel, after Quinault; 1780).

**Philine.** An actress (sop) in the theatrical company joined by Wilhelm in Goethe's *Wilhelm Meister*.

**Philip.** Philip II, King of Spain (bs) in Verdi's *Don Carlos*.

**Piacenza.** City in Piacenza, Italy. Dramatic musical entertainments were popular in the 16th and 17th cents. Opera was first given at the T di

Palazzo Gotico (or T Nuovo), opened 1644 with *La finta pazza* by Francesco Sacrati (1605–50). Cavalli's *Coriolano* was premièred 1669. Opera was then staged at the T del Collegio dei Nobili, later at the T delle Saline (particularly opera buffa) and T Ducale della Citadella. The T Municipale (cap. 1,500) opened 10 Oct. 1804 with Mayr's *Zamori*; renovated 1830.

**Piave, Francesco Maria** (*b* Murano, 18 May 1810; *d* Milan, 5 Mar. 1876). Italian librettist. He wrote librettos for Balfe, Cagnoni, Mercadante, and Pacini, among others, but is most famous for his collaboration with Verdi. He provided the librettos for *Ernani*, *La forza del destino*, *Macbeth*, *Simone Boccanegra*, *Rigoletto*, *La traviata*, *Il corsaro*, *Stiffelio*, *Aroldo*, and *I due Foscari*. A close friend and complaisant colleague, he saw his role as servant to Verdi's demands, and was prepared to listen to, and answer, the most detailed instructions about versification.

BIBL: G. Quarti, *Francesco Maria Piave, poeta melodrammatico* (Rome, 1939).

**Piccaver, Alfred** (*b* Long Sutton, 24 Feb. 1884; *d* Vienna, 23 Sep. 1958). English tenor. Studied New York, Milan with Rosario, Prague with Prohaska-Neumann. Début Prague 1907 (Gounod's Roméo). Vienna 1910–37; Chicago 1923–5; London, CG, 1924. A much-loved artist with a sweet, clear tone and a distinguished stage presence, his roles included Florestan, Radamès, Gounod's Faust, Walther, Lensky, Canio, Werther. (R)

**Picchi, Mirto** (*b* S Mauro, 15 Mar. 1915; *d* Florence, 25 Sep. 1980). Italian tenor. Studied Florence with Armani and Tess. Début Milan, Palazzo dello Sport, 1946 (Radamès). London, Cambridge T, 1947–8; Gly. 1949; London, CG, 1952–3. Milan, S, from 1954; also Rome, Naples, etc. Notable for his acting, and for his advocacy of 20th-cent. music. Created roles in works by Pizzetti, Testi, Lizzi, Castro; acclaimed as Grimes, Vere, Oedipus, Tom Rakewell, Drum-Major, and Tiresias (*The Bassarids*). (R)

**Piccinni, Niccolò** (*b* Bari, 16 Jan. 1728; *d* Passy, 7 May 1800). Italian composer. Studied Naples with Leo and Durante. Made his début 1754 with the highly acclaimed *Le donne dispettose*, whose popularity placed him on a level with Logroscino: this was followed by a series of well-received works, culminating in *La Cecchina, ossia La buona figliuola*. One of the most successful opere buffe of its day, *La buona figliuola* (as it is usually known) found many admirers and imitators: it is still sometimes revived in Italy. While holding various positions in Naples, Piccinni wrote *c.*50 operas, including a masterful setting of *L'Olimpiade*: he was particularly well regarded in Rome, where *I viaggiatori* was received almost as enthusiastically as *La buona figliuola*, though his position was eventually usurped by Anfossi. In 1776 he was encouraged to travel to Paris by a faction that wished to counter the reforms of Gluck. Here he directed a company of Italian singers which performed on alternate nights at the Opéra. He made an attempt at a French opera (taking advice from Marmontel over the adaptation of the Quinault libretto) but *Roland* had not yet been staged when the feud, fostered not by the composers but by their supporters, became public. An enterprising director of the Opéra arranged for them both to compose an *Iphigénie en Tauride*, but Piccinni's setting, though successful, could not rival Gluck's. After Gluck's departure a new rival arose in Sacchini following the success of his *Renaud* in 1783. Despite the triumph of *Didon* (which ran until 1826), Piccinni's star was on the wane. At the outbreak of the Revolution (1789) he returned to Naples, where he was involved in political trouble. He was fêted on his eventual return to Paris 1798, but never regained his old position.

Among his *c*.120 works there are some fine opere serie, including *L'Olimpiade*, which count among the best in the Metastasian mould, but Piccinni is chiefly remembered as a composer of opera buffa: with Paisiello, Cimarosa, and Guglielmi he represents the last of the great Neapolitan tradition. He had a mastery of detail and a rare talent for writing engaging melodies; his handling of the orchestra also surpassed the general level of the times, as did his command of harmonic resources (especially major-minor relationships) which impressed many later composers, including Bellini. His masterpiece was *La buona figliuola*, which is a milestone in the history of Italian opera buffa; particularly noteworthy is its highly developed use of the *ensemble finale. With its expressive arias and fine dramatic sense it represents the apotheosis of the genre before Mozart, and suggests much of the potential he was later to explore, while maintaining greater subservience to the traditional framework. In essence, Piccinni did not, like Gluck, command a strong overall view of opera as drama: it has well been said that his art was of the kind which adapts itself to the age, where Gluck's is the art to which the age must adapt.

SELECT WORKLIST: *Le donne dispettose* (The Spiteful Women) (Palomba; Naples 1754); *La Cecchina, ossia La buona figliuola* (The Good Girl) (Goldoni, after Richardson; Rome 1760); *L'Olimpiade* (Metastasio; Rome 1761); *I viaggiatori* (The Travellers) (Mililotti; Rome 1775); *Roland* (Marmontel, after Quinault; Paris 1778); *Iphigénie en Tauride* (Dubreuil; Paris 1781); *Didon* (Marmontel; Paris 1783).

His son **Ludovico Piccinni** (*b* Rome, 1764; *d* Passy, 31 Jul. 1827) wrote a number of French

and Italian works for Paris and Naples. His grandson (illegitimate son of his elder son Giuseppe) **Louis Alexandre** (*b* Paris, 10 Sep. 1779; *d* Paris, 24 Apr. 1850) was accompanist at the T Feydeau and at the Opéra, 1802–6. He wrote over 200 stage works, including 25 comic operas (performed at various Paris theatres: *Alcibiade solitaire* was given at the Opéra, 1824).

BIBL: A. Della Corte, *Piccinni* (Bari, 1928).

**Piccolo Marat, Il.** Opera in 3 acts by Mascagni; text by Forzano and Targioni-Tozzetti. Prod. Rome, C, 2 May 1921, with Dalla Rizza, Lázaro, Franci, Badini, cond. Mascagni.

Paris, *c* 1793. In order to free his mother from prison, Prince Jean-Charles de Fleury (ten) joins the Revolution and becomes known as the Piccolo Marat ('little Marat'). With the help of his lover Mariella (sop) and the Carpenter (bar), they force the President of the Revolutionary council (bs) to sign a release order for the imprisoned princess, and all escape to freedom.

One of Mascagni's more successful works, it is still revived in Italy from time to time.

**Piccolomini, Marietta** (*b* Siena, 15 Mar. 1834; *d* Poggio Imperiale, 23 Dec. 1899). Italian soprano. Studied Florence with Mazzarelli and Raimondi. Début Florence, P, 1852 (Lucrezia). Paris, I, 1855; London, HM, 1856–63; also Rome, Dublin, USA. Despite her family's opposition and an inferior singing talent, she had a highly successful career, delighting the English particularly with her beauty and unabashed dramatic involvement (though her famous Violetta suffered from her enthusiastic coughing). Sang Zerlina, Norina, Lucia, Gilda, Luisa Miller. Retired on marriage to an Italian nobleman, 1863.

**Pierné, Gabriel** (*b* Metz, 16 Aug. 1863; *d* Ploujean, 17 July 1937). French composer. Studied Paris with Lavignac and Massenet. His long list of works includes eight operas, in which the influence of Massenet shows in the direct emotional appeal and in the elegance and fluency of the vocal writing. Among the most successful were *La coupe enchantée*, *On ne badine pas avec l'amour*, *Sophie *Arnould*, and his masterpiece, *Fragonard*. In this, he shows a light touch with rococo stylization but also the ability to write skilfully effective arias and set pieces.

SELECT WORKLIST: *La coupe enchantée* (The Enchanted Goblet) (Matrat, after La Fontaine; Royan 1895); *On ne badine pas avec l'amour* (One doesn't Jest with Love) (Nigond & Leloir, after Musset; Paris 1910); *Sophie Arnould* (Nigond; Paris 1927); *Fragonard* (Rivoire & Coolus; Paris 1934).

BIBL: H. Busser, *Notice sur la vie et les œuvres de Gabriel Pierné* (Paris, 1938); L. Davies, *César Franck and his Circle* (London, 1970).

**Pietra del Paragone, La** (The Touchstone). Opera in 2 acts by Rossini; text by Luigi Romanelli. Prod. Milan, S, 26 Sep. 1812, with Marcolini, Zerbini, Fei, Vasoli, Galli, Bonoldi, Parlamagni, Rossignoli; Hartford, Hartt Coll. of Music, 4 May 1955, with Kallisti, Dippe, De Vita, Verduce, Stuart, cond. Paranov; London, St Pancras TH, 19 Mar. 1963, with Clark, Bainbridge, Laura Sarti, Robertson, Hammond-Stroud, Mangin, Wicks, cond. Fredman.

In a Tuscan village, early 19th cent. The wealthy Count Asdrubale (bs) puts to the test three young widows who want to marry him—Aspasia (sop), Fulvia (mez), and Clarice (con). Disguising himself as a Turk, he arrives at his own castle claiming that Asdrubale owes him a large sum of money. Only Clarice and Giocondo (ten) stand by their friend. Clarice in turn resolves to test the Count's faithfulness and appears in the guise of her twin brother, a soldier, threatening to remove Clarice. When Asdrubale begs for Clarice's hand, she reveals her identity and the couple swear their love, leaving Fulvia and Aspasia to find other lovers.

Rossini's first opera for La Scala.

**Pilarczyk, Helga** (b Schöningen, 12 Mar. 1925). German soprano. Studied Hamburg with Dziobek, later Rennert. Début Brunswick 1951 (Irmentraud, *Der Waffenschmied*). Hamburg 1954–68; Berlin, SO, 1956–60; London, CG, and Gly., 1958; Milan, S, 1963; New York, M, 1965. A committed advocate of 20th-cent. opera, who brought intelligence and insight to the roles of Salome, Marie (*Wozzeck*), Lulu, the Woman (*Erwartung*), Jocasta, etc. (R)

**Pilgrims' Chorus.** The chorus of pilgrims, 'Zu dir wall ich' in Act I, and 'Beglückt darf nun dich' in Act III of Wagner's *Tannhäuser*.

**Pilgrim's Progress, The.** Opera in 4 acts, with prologue and epilogue, by Vaughan Williams; text by composer and Wood, after Bunyan's allegory (pt1, 1674–9; pt2, 1684) and the Bible. Prod. London, CG, 26 Apr. 1951, with Matters, Te Wiata, Walker, E. Evans, cond. Hancock.

John Bunyan (bas-bar) introduces the story of the journey of Pilgrim (bar) towards the Heavenly City. This is shown in a series of scenes depicting his most famous encounters, including that with Apollyon (bs). Vaughan Williams had already set the scene with the Shepherds of the Delectable Mountains as a 1-act 'pastoral episode' in 1922, and later incorporated it into the work.

**Pilou, Jeannette** (b Alexandria, July 1931). Greek, later Italian, soprano. Studied Milan with Castellani. Début Milan, T Smeraldo, 1958 (Violetta). Milan, S; Naples; Rome; Vienna 1965; Paris, O, 1973; New York, M, from 1967; London, CG, 1971; also Monte Carlo, Chicago, Buenos Aires, Wexford. Roles incl. Susanna, Nannetta, Marguerite, Manon, Manon Lescaut, Mélisande, Butterfly. (R)

**Pilsen.** See PLZEŇ.

**Pimen.** The old chronicler monk (bs) in Musorgsky's *Boris Godunov*.

**Pimmalione.** Opera in 1 act by Cherubini; text by Stefano Vestris, after Antonio Sografi's Italian version of Rousseau's *Pygmalion* (1770), after the classical legend. Prod. Paris, Tuileries, 30 Nov. 1809.

Greece, mythological times. The Cypriot sculptor Pygmalion (mez) creates an ivory statue so beautiful that he falls in love with it. In answer to his prayers, Aphrodite (sop) brings the statue Galatea (sop) to life.

The opera was composed as a vehicle for Napoleon's favourite singer, the mezzo-soprano castrato Crescentini.

**Pini-Corsi, Antonio** (b Zara, June 1858; d Milan, 22 Apr. 1918). Italian baritone. Studied with Ravasio. Début Cremona 1878 (Dandini). Berlin, H, 1889; Milan, S, from 1893; London, CG, 1894–6, 1902–3; New York, M, 1899, 1909–14. A born buffo singer with a powerful tone, he was a popular Leporello, Dr Bartolo (Rossini), and Pasquale; created Ford, Schaunard, Happy (*Fanciulla*), and the Innkeeper (*Königskinder*). Retired 1917. (R) His brother **Gaetano** (1860–after 1923) was a tenor. Milan, S, from 1898; roles included David, Mime, and first Goro (*Butterfly*). (R)

**Pinkerton.** The American naval lieutenant (ten), faithless husband of Butterfly, in Puccini's *Madama Butterfly*. Originally Sir Francis Blummy Pinkerton, but after the opera's disastrous première rechristened Benjamin Franklin Pinkerton.

**Pinza, Ezio** (b Rome, 18 May 1892; d Stamford, CT, 9 May 1957). Italian bass. Studied Bologna with Ruzza and Vezzani. Début Soncino 1914 (Oroveso). Milan, S, 1922–4; New York, M, 1926–48; London, CG, regularly 1930–9; also Buenos Aires, and all Europe's major operahouses. After 1948 sang in musical comedy (e.g. a highly successful Broadway *South Pacific*) and films. One of the noblest bassi cantanti of the century, he possessed not only a voice of great warmth, plasticity, and beauty, but a 'magnetic charm' (Gilman) and excellent dramatic skills. His large repertory (over 95 roles) included Figaro (Mozart), a famous Don Giovanni, Don Magnifico, Philip, Boris, Pogner, Amfortas, Golaud, and first Tigellino (Boito's *Nerone*). (R)

**Piovene, Agostino** (*fl.* 1709–26). Italian librettist. One of the principal Venetian librettists of the early 18th cent.; his work showed embryonic signs of the reforms later carried through by Zeno and Metastasio. His most important librettos were for Gasparini (*La principessa fedele*, 1709), Orlandini (*Nerone*, 1721), Vivaldi (*Cunegonda*, 1726), and Porpora (*Tamerlano*, 1730).

**Pipe of Desire, The.** Opera in 1 act by Converse; text by George Edward Burton. Prod. Boston, Jordan H, 31 Jan. 1906, with Cushingchild, Dean, Townsend, cond. Goodrich. The first US opera to be prod. at the New York, M, 18 Mar. 1910, with Homer, Martin, Whitehill, cond. Hertz.

A magic pipe belonging to the Elf King (bs) is selfishly used by Iola (ten) and brings disaster to himself and to his lover Naoia (sop). The King pipes, and they die.

**Piper, John** (*b* Epsom, Surrey, 13 Dec. 1903; *d* Fawley, 28 June 1992). English painter and critic. Studied Richmond and London. One of the founder members of the *English Opera Group, 1947. His sets and costumes for the revival of Vaughan Williams's ballet *Job*, London, CG, 1948, though not liked by the composer led to further stage designs. He has worked at Glyndebourne, Covent Garden, and Sadler's Wells; but his most distinctive stage work has been done for Britten's operas. His strong architectural sense, sometimes a use of tapestry effects in a semi-abstract manner, and his powerful feeling for light and for colour, have produced remarkable designs closely attuned to the music. His wife **Myfanwy** (*b* 1911) was librettist of Britten's *Turn of the Screw*, *Owen Wingrave*, and *Death in Venice*.

**Pirata, Il** (The Pirate). Opera in 2 acts by Bellini; text by Romani, after Raimond's drama *Bertram, ou Le pirate* (1826), after Maturin's *Bertram* (1816), possibly deriving from Scott's *Rokeby* (1813). Prod. Milan, S, 27 Oct. 1827, with Méric-Lalande, Sacchi, Rubini, Lombardi, Ansilioni, Tamburini, cond. Lavigna; London, HM, 17 Apr. 1830, same cast; New York, Richmond Hill T, 5 Dec. 1832, with Pedrotti, Verducci, Montressor, Fornasari, Placci, Sapignoli, cond. Rapetti.

Sicily, 13th cent. Imogene (sop), although in love with Gualtiero (ten), has been forced to marry Ernesto, Duke of Caldora (bar), in order to prevent her father's death. Now a pirate, Gualtiero is shipwrecked off the coast of Caldora and aided by Imogene. He is horrified to learn of her marriage to Ernesto and vows revenge.

When Ernesto accuses Imogene of infidelity, she admits she still loves Gualtiero. Ernesto secretly watches the lovers' assignment. Gualtiero kills Ernesto in a duel, but then gives himself up and is sentenced to death. As he mounts the scaffold, Imogene collapses from grief.

The first Bellini opera produced in London.

**Pirogov.** Family of Russian singers.

1. **Grigory** (*b* Novoselky, 24 Jan. 1885; *d* Leningrad, 20 Feb. 1931). Bass. Studied Moscow with Medvedyev and Donsky. Début Rostov-on-Don 1908. St Petersburg, M, 1909. Moscow, B, 1910–15, 1917–21. Also St Petersburg, Paris, Berlin, Prague. Retired 1930. His full-toned, even voice was of exceptional range; he sang the baritone roles of Tomsky, Escamillo, and Rubinstein's Demon, as well as Boris, Ruslan, Wotan, Dodon, and Don Basilio (Mozart).

2. **Alexander** (*b* Novoselky, 4 July 1899; *d* Moscow, 26 June 1964). Brother of (1), also a bass. Studied Moscow with Tyutyunnik. Début Moscow 1919. Moscow: Z, 1922–4; B, 1924–59. With his powerful voice and dramatic versatility, successful as Gounod's Méphistophélès, Ivan Susanin, Ruslan, Boris (which he sang in the film of the opera, 1955), Gremin etc.

Two other brothers, **Alexey** (1895–?) and **Mikhail** (1887–1933), were also successful basses who worked in Moscow.

**Pisaroni, Benedetta Rosamunda** (*b* Piacenza, 16 May 1793; *d* Piacenza, 6 Aug. 1872). Italian soprano, later contralto. Studied Milan with Marchesi, Velluti, and Pacchierotti. Début Bergamo 1811 (Mayr's *La rosa bianca e la rosa rossa*). Successes in Italy; then, having lost her top notes through smallpox, she developed her lower register so effectively that she became known as the first Italian leading contralto. Second début Padua 1814. Naples 1818–19; Milan, S, between 1821 and 1831; Paris, I, 1827–30; London 1829. Retired from stage 1832. Despite severe facial disfiguration (she always warned impresarios by sending them her picture), she won great acclaim with her superb singing and dramatic conviction. At her best in tragic roles, and in the Rossini repertory. Created Zomira (*Riccardo e Zoroaide*), Malcolm (*La donna del lago*), Andromaca (*Ermione*); also a celebrated Arsace and Tancredi.

**Pišek, Jan Křtitel** (*b* Mšeno, 13 Oct. 1814; *d* Sigmaringen, 16 Feb. 1873). Bohemian baritone. Studied Prague with Triebensee. Début Prague 1835 (Oroveso). Brno 1838–9; Vienna 1839–40; court singer to King of Württemberg, Stuttgart, 1844–63; London, DL, 1849; Prague 1865. Roles included Don Giovanni, Figaro (Mozart and Rossini), Rigoletto, Luna, Hans Heiling. Berlioz admired his rich, expressive voice and thought him one of the best singers in Europe, declaring, 'This man is Don Giovanni, Romeo and Cortez rolled into one!'

**Pistocchi, Francesco Antonio** (*b* Palermo, 1659; *d* Bologna, 13 May 1726). Italian contralto castrato and composer. Studied Bologna with Perti and Monari. First sang in public aged 3; wrote first composition aged 8. Employed as soprano at S Petronio, Bologna, 1670–5, until dismissal. Lost his voice, but re-emerged in 1686 with a particularly fine contralto. After a successful career in opera 1687–1705, singing in Italy, Berlin, and Vienna, he became one of the most important singing teachers in history. His pupils included Bernacchi, Fabri, and Gizzi, and the tradition he established continued well into the 19th cent. An excellent technician, he introduced an instrumental virtuosity to singing, but also refinement: he had 'inimitable taste' (Tosi). As a composer he was highly regarded by Burney. His surviving operas include *Pazzie d'amore e dell'interesse* (1699).

**Pitt, Percy** (*b* London, 4 Jan. 1870; *d* London, 23 Nov. 1932). English conductor. Studied Leipzig with Reinecke and Jadassohn, Munich with Rheinberger. Music adviser Grand O Syndicate at London, CG, 1902; director 1907–24, helping Richter with the English-language *Ring*, 1908–9. Beecham OC 1915–20; BNOC 1920–4. Music adviser BBC 1922–30, and cond. first complete broadcast opera, *Hänsel und Gretel*, 6 Jan. 1923.

BIBL: J. Chamier, *Percy Pitt of Covent Garden and the BBC* (London, 1938).

**Pittsburgh.** City in Pennsylvania, USA. Touring companies visited from 1838 to 1939, when the Pittsburgh Opera Society was founded. It gave occasional performances until 1942, when Richard Karp became director; the company moved to the Syria Mosque 1945, then (having established itself as a successful group able to attract top singers) to the restored Heinz Hall in 1971. Tito Capobianco was one of the leading producers, becoming director 1983; his developments included the use of Op-Trans, an early form of *surtitles. The company, which pursues conservative policies with star singers, now performs in the Benedum Center (cap. 2,800).

**Pixérécourt, Guilbert de** (*b* Nancy, 22 Jan. 1773; *d* Nancy, 25 July 1844). French playwright. The father of the French *mélodrame*, with its insistence on moral subjects, he also influenced the librettos of his time, both by his example and as director of the Opéra-Comique 1822–7. His *mélodrames* make skilful use of music to heighten dramatic tension, and were deliberately designed for popular appeal with their stereotyped innocent heroine, virtuous hero, evil villain etc.: 'I wrote for those who could not read.' Donizetti's *Chiara e Serafina* and *Otto mesi in due ore* and Meyerbeer's *Margherita d'Anjou* are based on his plays. He also had some influence on the later Romantic drama (Dumas and Hugo) and thence on Romantic opera.

**Pixis, Johann Peter** (*b* Mannheim, 10 Feb. 1788; *d* Baden-Baden, 22 Dec. 1874). German pianist and composer, son and pupil of the composer Friedrich Wilhelm Pixis. One of the most brilliant pianists of his time, and the composer of much virtuoso piano music, he also wrote four operas. The first, *Almazinde*, was said to have too noisy and busy an accompaniment for the singers. *Bibiana*, written for Schröder-Devrient, failed. His adopted daughter **Francilla** (1816–?) was a contralto who made her début Karlsruhe and sang widely in Germany, London, Paris, Milan, and Naples, where Pacini wrote *Saffo* for her.

WORKLIST: *Almazinde* (Schmidt; Vienna 1820); *Der Zauberspruch* (The Spell) (after Gozzi; Vienna 1822); *Bibiana* (Lax; Aachen 1829); *Die Sprache des Herzens* (The Language of the Heart) (Lyser; Berlin 1836).

**Pizarro.** 1. The prison governor (bar) in Beethoven's *Fidelio*.
2. Francisco Pizarro (*c*.1471–1541), the Spanish conqueror of Peru, appears in operas by Giordani (1783), Candeille (1785), and Bianchi (1787).

**Pizzetti, Ildebrando** (*b* Parma, 20 Sept. 1880; *d* Rome, 13 Feb. 1968). Italian composer. Studied Parma with Righi. Having written some operas when very young, including *Il Cid* (entered for the 1902 Sonzogno competition), he formed a friendship with D'Annunzio, who wrote *Fedra* with the intention of turning it into a libretto for him. It is, however, a wordy text, and the music lacks the freedom he later achieved. Conservative by nature, he evolved a free-flowing declamatory style that has certain obvious debts to Wagner and Debussy but that is wholly Italian in its contours. He also wrote powerfully for the chorus, notably in his finest opera, *Debora e Jaële*. Other works that have developed these methods successfully include *Lo straniero*, *Vanna Lupa*, *La figlia di Jorio*, and *Assassinio nella cattedrale*.

SELECT WORKLIST: *Fedra* (D'Annunzio; Milan 1915); *Debora e Jaele* (Pizzetti; Milan 1922); *Lo straniero* (The Stranger) (Pizzetti; Rome 1930); *Fra Gherardo* (Pizzetti; Milan 1928); *Vanna Lupa* (Pizzetti; Florence 1949); *La figlia di Jorio* (Pizzetti, after D'Annunzio; Naples 1954); *Assassinio nella cattedrale* (Murder in the Cathedral) (Pizzetti, after T. S. Eliot; Milan 1958).

BIBL: M. La Morgia, ed., *La città dannunziana a Ildebrando Pizzetti* (Pescara, 1958).

**Planché, James Robinson** (*b* London, 27 Feb. 1796; *d* London, 29 May 1880). English theatrical writer. Translated many operas for the English stage, including works by Rossini, Auber, Marschner, Bellini (*Norma*), Hérold, Offenbach, and Mozart (*Magic Flute* and *Marriage of Figaro*). His version of Weber's *Der Freischütz* led Kemble to engage him as the librettist for *Oberon*. The form this took reflects not only Planché's assessment of the contemporary English musical theatre, and what would be acceptable, but his own interest in pantomimes and in historical costume. His *Recollections and Reflections* (2 vols., 1872) give a detailed picture of the theatrical life of his day.

**Plançon, Pol** (*b* Fumay, 12 June 1851; *d* Paris, 11 Aug. 1914). French bass. Studied Paris with Duprez and Sbriglia. Début Lyons 1877 (Saint-Bris). Paris, T de la Gaieté, 1880, and O 1883–93; London, CG, 1891–1904; New York, M, 1893 regularly until 1908. Repertory included Sarastro, a highly acclaimed Méphistophélès (Gounod), Frère Laurent, Escamillo, Ramfis, Landgrave, Pogner; created several roles including Gormas (*Le Cid*). A highly polished artist, whose smooth, essentially profondo voice also possessed a baritonal high register and great flexibility. His florid singing and trill were exceptional. (R)

**Planquette, Robert** (*b* Paris, 31 July 1848; *d* Paris, 28 Jan. 1903). French composer. Studied Paris with Duprato. After trying to earn a living by making arrangements and by playing the piano in cafés, he had some 1-act operettas performed at the Eldorado Music Hall and the Délassements-Comiques; then achieved his greatest success with *Les cloches de Corneville*, which had 400 consecutive performances at the Folies Dramatiques, and was widely staged abroad. He wrote *Rip van Winkle* for London, following this with *Nell Gwynne*, and adapting other works for the English stage. Among his later works *Mam'zelle Quat'sous* is outstanding, but on the whole they were not well received. A careful and stylish composer, with a lively melodic vein, he lacked the ease and variety of invention to keep his name before a public notorious for tiring quickly of anything lacking novelty.

SELECT WORKLIST: *Les cloches de Corneville* (Clairville & Gabet; Paris 1877); *Rip van Winkle* (Irving; London 1882); *Nell Gwynne* (Farnie; London 1884); *Mam'zelle Quat'sous* (Mars & Desvallières; Paris 1897).

**Plaschke, Friedrich** (*b* Jaroměř, 7 Jan. 1875; *d* Prague, 4 Feb. 1952). Czech bass-baritone. Studied Prague with Dötscher and Sklenář-Mala, Dresden with Scheidemantel. Début Dresden 1900 (Herald, *Lohengrin*). Dresden until 1937; Bayreuth 1911; London, CG, 1914; toured USA 1922–4 with German O. Created several Strauss roles, including First Nazarene (*Salome*), Altair (*Die ägyptische Helena*), Waldner (*Arabella*), Morosus (*Die schweigsame Frau*); also sang Kur-

wenal, Amfortas, Sachs, Barak. Married the soprano Eva von der *Osten. (R)

**Platée.** Opera in a prologue and 3 acts by Rameau; text by Adrien Le Valois d'Orville, after Jacques Autreau's libretto after Pausanias' *Periegesis*, ch. 9 (*c*.AD 176). Prod. Versailles 31 Mar. 1745.

Greece, mythological times. The prologue depicts the birth of comedy: Thespis (ten) instigates the creation of an entertainment demonstrating the absurdities of the gods and thus reforming human behaviour. In the play itself, Jupiter (bar) feigns love for the grotesque nymph Platée (ten) in order to arouse the anger of Junon (sop). Duly furious at Jupiter's behaviour, she disguises herself and, as he is about to make his marriage vows, hurls herself at Platée and tears off her bridal veil, then collapsing in a fit of laughter. She and Jupiter are happily reconciled.

**Plishka, Paul** (*b* Old Forge, PA, 28 Aug. 1941). US bass. Studied New Jersey with Boyajian. Début New Jersey 1961. New York, M, from 1967; Milan, S, 1975; also Chicago, Paris, Berlin, Barcelona. Large repertory incl. Leporello, Oroveso, Silva, Philip, Varlaam, Boris, King Mark, Frère Laurent, Abimelech (*Samson et Dalila*), Falstaff. (R)

**Plovdiv.** City in Bulgaria. Long dependent on visiting Italian companies, which from the 1880s (De Lucia, 1889) played in the Luxemburg T and Bulgarian T. Interest led to the formation of a local society (1896). In 1920 the tenor Alexander Krayev (1885–1958) formed an operatic group which began with Hadjigeorgyev's *Takhirbegovitsa*. In 1922 the Plovdiv City O was organized: it began with *La Juive*. Seasons continued until the war. The Plovdiv National O was formed in 1953, and began with *The Bartered Bride*; it has developed a good international repertory.

**Plowright, Rosalind** (*b* Worksop, 21 May 1949). English soprano. Studied Manchester. Début Gly. Touring O 1975 (Agathe). WNO, ENO, Kent O 1975–8; London, CG, from 1980; Madrid, Hamburg 1982; Milan, S, 1983; Verona 1985; Paris, O, 1987; Vienna, S, 1990; also Munich, San Francisco, Rome, Buenos Aires, etc. Roles incl. Médée, Norma, Elizabeth (*Maria Stuarda*), Aida, Amelia, Rossini's Desdemona, Senta, Ariadne, Madeleine (*Andrea Chenier*). (R)

**Plzeň** (Ger., Pilsen). City in Bohemia, Czechoslovakia. Opera was given in Czech regularly from 1868. A new opera-house opened in 1902 (cap. 1,100) (from 1955 J. K. Tyl T); renovated 1986. Besides this so-called Velké Divadlo (Grand T) there is a Malé Divadlo (Small T) for operetta. Plzeň has traditionally been a useful training ground for artists.

**poeta cesareo** (It.: 'Caesarean poet'). The title given to the court poet in Vienna from 1701, first conferred on Bernadoni and held by Stampiglia, Zeno, and Metastasio among others. His duties were to provide one or two opera librettos each year, several texts for the smaller-scale sacred and secular entertainments of the court, and verses for imperial occasions, such as weddings and funerals. Normally offered to an Italian, and refused by Romani for patriotic reasons.

**Poggi, Antonio.** See FREZZOLINI, ERMINIA.

**Pogner.** Veit Pogner (bs), the goldsmith and Mastersinger, and Eva's father, in Wagner's *Die Meistersinger von Nürnberg*.

**Poisoned Kiss, The.** Opera in 3 acts by Vaughan Williams; text by Evelyn Sharp, after Richard Garnett's story *The Poison Maid* in the collection *The Twilight of the Gods* (1888). Prod. Cambridge 12 May 1936, with Field-Hyde, Ritchie, Jones, Dunn, cond. Rootham; New York, Juilliard School, 21 Apr. 1937.

Fairyland. The fantastic plot turns on the rivalry of a sorcerer and an empress; his daughter Tormentilla (sop) has been brought up on poisons so that when she meets the empress's son Amaryllus (ten) she will kill him with her kiss. In the end the sincerity of their love defeats the plot.

**Poissl, Johann Nepomuk von** (*b* Haukenzell, 15 Feb. 1783; *d* Munich, 17 Aug. 1865). German composer. Studied Munich with Danzi and Vogler. His first opera was the comic *Die Opernprobe*, but his first real successes were the serious *Antigonus* and *Ottaviano in Sicilia*. In 1811 he met Weber, from whose friendship and championship he greatly benefited. The success of *Athalia* and *Der Wettkampf zu Olympia* did not bring him much material reward, and his career was made difficult by his belonging to the nobility. He was further disappointed of hopes of an appointment to Darmstadt, for which he wrote the popular *Nittetis* and *Issipile*. In 1823 he gained a court appointment in Munich, and in 1825 became director of the court theatre; but the theatre lost money and he was forced to resign in 1832. Though he seldom staged his own operas, he did reopen the Nationaltheater after the 1823 fire with *Die Prinzessin von Provence*. He died in poverty.

Poissl is important as a transitional figure between Mozart and Weber, and was one of the first German composers to move away from Italian and French example in favour of a continuously composed German opera (as in *Antigonus*, which does, however, contain some Italian and French elements). Weber praised *Der Wettkampf*

and *Athalia* (which makes some use of motive) for their melodic qualities, commending the latter as a national achievement. Poissl was ahead of his time in writing most of his own librettos, and he published criticism and essays on theatre organization.

SELECT WORKLIST: *Die Opernprobe* (The Opera Rehearsal) (Poissl, after an Italian libretto; Munich 1806); *Antigonus* (Poissl, after Metastasio; Munich 1808); *Ottaviano in Sicilia* (Poissl, after Metastasio; Munich 1812); *Athalia* (Wohlbrück, after Racine; Munich 1814); *Der Wettkampf zu Olympia* (The Olympic Games) (Poissl, after Metastasio; Munich 1815); *Nittetis* (Poissl, after Metastasio; Darmstadt 1817); *Issipile* (Poissl, after Metastasio; comp. 1818, unprod.).

BIBL: E. Reipschläger, *Schubaur, Danzi und Poissl als Opernkomponisten* (Berlin, 1911).

**Pokrovsky, Boris** (*b* Moscow, 23 Jan. 1912). Russian producer. Worked in Gorky 1937–43, then in Moscow and Leningrad staging many distinguished productions. His most significant work has been at the Moscow Chamber Musical Theatre from 1972. Here he was able even in restrictive times to stage unfamiliar works in pioneering and highly imaginative productions, drawing his repertory from both classical pre-Glinka opera, neglected or officially disregarded contemporary works, and foreign opera. His revivals, with the support of Gennady Rozhdestvensky as music director, have included Pashkevich's *The Miser* and Shostakovich's *The Nose* (1974), as well as works by Bortyansky, Kholminov, and Stravinsky (first Russian *Rake's Progress*), among many others.

**Polacco, Giorgio** (*b* Venice, 12 Apr. 1874; *d* New York, 30 Apr. 1960). Italian conductor. Studied Venice, Milan, St Petersburg. Répétiteur London, CG, 1890, cond. Lago's Co., Shaftesbury T, 1891 (Gluck's *Orfeo*). After engagements in Italy (première of *Zazà*, Milan 1900), and in Warsaw, Lisbon, and Brussels, also Russia, cond. four seasons in Buenos Aires and seven in Rio de Janeiro. New York, M, 1912–17, conducting 327 performances in five seasons. Chicago 1918–19, music director 1921–30, directing many memorable performances of French operas with Mary Garden, also Italian and German repertory. London, CG, 1913–14, 1930 (*Pelléas* with Teyte, who found him 'the ideal interpreter'). Forced by ill health to retire 1930.

**Poland.** Opera was first given in Poland in 1613, when Prince Stanisław Lubomirski invited an Italian company to his residence at Wiśnicz. In 1625 Prince Władisław Zygmunt visited the Grand Duchess of Tuscany and heard, in his honour, Francesca Caccini's *La liberazione di*

*Ruggiero dall'isola d'Alcina*, which he may have brought home: a Polish libretto appeared in 1628. After he was crowned Władisław IV in 1632, he founded an Italian opera company (1634–48) with Margherita Cattaneo as prima donna; it included some young Poles. About 12 operas (probably adaptations) were performed and published, including one by a Pole, Piotr Elert (?–1653), *La fama reale, ovvero Il principe trionfante Ladislao IV* (1633). The librettos were mostly by Virgilio Puccitelli, the music to some operas by Marco Scacchi.

Frederick Augustus, Elector of Saxony, was crowned Augustus II in 1697, and brought in his retinue an ensemble directed by J. C. Schmidt and Jacek Różycki. In 1700 a company of 60, formed in Paris at his request by Angelo Costantini, was brought under Deschallières; other companies also toured. In 1725 a court theatre, the Operalnia, opened in Warsaw. Augustus III was crowned in 1733 and at once showed enthusiasm for opera; he provided large subsidies, and twice-weekly performances were given with an orchestra of over 100. Many Metastasian operas were given; the company included *Hasse and *Bordoni. From c.1725 private operas were also established in the residences of Polish nobles, e.g. in Nieświcz, Ołyka, Słuck, and Białystock, performing mostly French and Italian works.

The growth of Polish national opera dates from the reign (1764–95) of Poland's last King, the enlightened and intelligent Stanisław August Poniatowski, who summoned artists of every kind to Warsaw and encouraged the arts throughout the country. The opening of the first public theatre in Warsaw in 1765 under Karol Tomatis encouraged the popularity of opera; and from 1776 opera seria was added to opera buffa. The first Polish opera was *Nędza uszczęśliwiona* (Sorrow Turned to Joy, 1778) by Maciej Kamieński (1734–1821), consisting of 11 airs and two duets. In 1779 the T Narodowy opened, giving in its first year Kamieński's *Zośka* and *Prostota cnotliwa* (Virtuous Simplicity), also *Nie każdy śpi, co chrapi* (Not All who Snore, Sleep) by Gaetano (known as Kajetan Majer, ?–c.1793).

Despite the three partitions of Poland at the end of the 18th cent., Polish texts began superseding foreign ones, due largely to the 'father of the Polish theatre', Wojciech Bogusławski (1757–1829). An actor, producer, and singer (he was the first Antek as well as the librettist of *Nędza uszczęśliwiona*) who also directed the theatre 1782–4 and 1799–1814, he translated many librettos and wrote others, including *Krakowiacy i Górale* (The Krakowians and the Highlanders, 1794) by Jan Stefani (1746–1829). In four acts, this is the first full-scale Polish opera, and was based on peasant life. In the period of

national difficulties which followed, Bogusławski staged productions of foreign works, while Jószef Elsner (1769–1854) wrote a long series of Polish operas, including *Andromeda*: this was given on 14 Jan. 1807 before Napoleon, who followed the performance with a French translation. Other popular operas by Elsner, mostly on Polish historical themes, included *Leszek Biaty* (Leszek the White, 1809), *Król Tokietek* (King Lokietek, 1818), and *Jagiełło w Tenczynie* (Jagiełło at Tenczyn, 1820). In 1810 Karol *Kurpiński (1785–1857) was appointed to the opera, and with his work and music helped to maintain a Polish opera free from Russian influence. He introduced works by himself, Elsner, Mozart, Rossini, later Weber, Auber, Donizetti, Bellini, and eventually Verdi. The T Wielki opened in Warsaw on 24 Sep. 1833.

Stanisław Moniuszko (1819–72) first had *Halka* given in Wilno in 1848, but it was kept out of Warsaw by the director Tomasz Nidecki; when Giovanni Quattrini took over in 1852, he immediately staged the work hailed as Poland's first great national opera. In his period as music director (1858–72), many new Polish works were staged, including those by Ignacy Dobrzyński (1806–67), Gabriel Rożniecki (1815–87), Adam Minchejmer (1830–94), Stanisław Duniecki (1839–70), and Moniuszko himself. In the decade 1858–67 some 40 new operas appeared, the most important being Moniuszko's own *Straszny dwór* (The Haunted Castle, 1865). Wagner was first given under Cesare Trombini. New composers included Władisław Żeleński (1837–1921), whose four operas included *Wallenrod* (1885) and *Goplana* (1896); and Ludomir Różucki (1884–1953), conductor at Lvov, whose *Bolesław Śmiały* (Bolesław the Bold, 1909) and *Meduza* (1912) show an ability to reconcile post-Wagnerian and verismo methods within a Slavonic manner, and whose most important work was *Eros i Psyche* (1917). Dominating them, however, was Karol *Szymanowski, whose sumptuous post-Romantic manner finds expression in *Hagith* (comp. 1913; prod. 1922) and *Król Roger* (King Roger, 1926).

After the First World War, there came a period of intense activity especially in Warsaw, Poznań, Lvov, Cracow, and Katowice, with many famous singers appearing; in composition, it was a reactionary period, and in the 1930s only about 30 new Polish operas appeared. After the Second World War, activity was quickly renewed. Warsaw's T Wielki, burnt down in 1939 and bombed in 1944, was rebuilt in 1965, but opera continued meanwhile and outside the capital a number of new companies were formed. 19th-cent. Polish opera remains popular in repertories that concentrate on Italian and Russian opera but are increasingly catholic in taste. Successful post-war Polish operas have included *Bunt żaków* (The Goliards' Revolt, 1951) and *Krakatuk* (1955) by Tadeusz Szeligowski (1896–1963); *Chłopi* (The Peasants, 1974) by Witold Rudziński (*b* 1913); *Jutro* (Tomorrow, 1966) by Tadeusz Baird (*b* 1928); *Cyrano de Bergerac* (1962), *Lord Jim* (1976), and *Mary Stuart* (1981) by Romuald Twardowski (*b* 1930); *Diabły z Loudun* (The *Devils of Loudun, 1969) by Krzysztof Penderecki (*b* 1933); and works by Joanna Bruzdowicz (*Bramy raju* (The Gates of Paradise, 1988) ).

See also GDAŃSK, CRACOW (KRAKÓW), ŁÓDŹ, LVOV (LWÓW), POZNAŃ, SOPOT, WARSAW, WROCŁAW.

**Poliuto.** Opera in 3 acts by Donizetti; text by Cammarano, after Corneille's tragedy *Polyeucte* (1642). Comp. 1838 for Nourrit, for production in Naples, but not passed by censor; rehearsed by Ronzi de Begnis, Nourrit, Barroilhet, Finocchi. New 4-act version as Grand Opera, with French text by Scribe, prod. Paris, O, 10 Apr. 1840, as *Les martyrs*, with Dorus-Gras, Duprez, Massol, Dérivis. New Orleans, T d'Orléans, 24 Mar. 1846, with J. Calvé, Arnaud, cond. E. Prévost. Retranslated into Italian as *I martiri* by Bassi, prod. Lisbon 15 Feb. 1843; London, CG, 20 Apr. 1852, with Julienne, Tamberlik, Ronconi.

Mytilene, capital of Armenia, AD 257. Poliuto (ten) has become a secret convert to Christianity, and is arrested and condemned to death. His wife Paolina (sop), although still in love with Severo (bar), the Roman proconsul, decides to join her husband and die with him.

Also opera *Polyeucte* by Gounod (1878).

**Pollak, Anna** (*b* Manchester, 1 May 1912). English mezzo-soprano. Studied with Hislop, and London with Cross. Début London, SW, 1945 (Dorabella). London, SW, until 1961. London, EOG 1946; London, CG, 1952; Gly. 1952–3; also Holland Fest. A versatile and stylish actress; sang Cherubino, Fatima, Orlovsky, Siebel. Created Bianca (*Rape of Lucretia*), Berkeley's Lady Nelson (*Nelson*) and title-role of his *Ruth*. (R)

**Pollak, Egon** (*b* Prague, 3 May 1879; *d* Prague, 14 June 1933). Austrian conductor. Studied Prague with Knittl. Chorus master Prague, Landestheater, 1901. Cond. Bremen 1905–10; Leipzig 1910–12; Frankfurt 1912–17; music director Hamburg 1917–31. London, CG, 1914. Chicago 1915–16, 1929–32. Buenos Aires, C, 1928. Vienna, S, 1932–3. Admired for his Wagner and especially Richard Strauss (who thought him 'superb'). Collapsed and died while conducting *Fidelio*. (R)

**Pollarolo, Carlo Francesco** (*b* Brescia, *c*.1653; *d* Venice, 7 Feb. 1722). Italian composer. Studied Venice with Legrenzi. After early success as a church musician in Brescia came to Venice 1650. Venice, St Mark's: second organist 1690, vice maestro di cappella 1692. Cons. degli Incurabili *c*.1696–1722. Composed *c*.85 operas, many lost; 1691–1707 he was the most frequently performed composer at the T San Giovanni Grisostomo, but his works were given in other Venetian houses. Influenced by the generation of Venetian composers which followed Cavalli and Cesti, his style shares some similarity with Pallavicino's. His works are notable for their increased use of the orchestra, suggesting the influence of Lully: this is seen also in his occasional use of the *French overture, e.g. in *Faramondo*. Pollarolo's works represent an important transitory phase leading to Neapolitan opera seria: his settings of Zeno, which also include *Gl'inganni felice*, were among the first.

SELECT WORKLIST: *Gl'inganni felice* (Zeno; Venice 1696); *Faramondo* (Zeno; Venice 1699).

His son (Giovanni) **Antonio Pollarolo** (*b* Brescia, *bapt.* 12 Nov. 1676; *d* Venice, 30 May 1746), whom he taught, joined him at St Mark's 1702; vice maestro di cappella 1723; maestro di cappella 1740. Wrote *c*.11 operas for the Venetian houses, including *Aristeo* and *Griselda*.

SELECT WORKLIST: *Aristeo* (Corradi; Venice 1700, lost); *Griselda* (Zeno; Venice 1701, lost).

**Pollini, Bernhard** (*b* Cologne, 16 Dec. 1838; *d* Hamburg, 27 Nov. 1897). German tenor, baritone, and impresario. Début Cologne 1858 (Arturo). Later sang as baritone with an Italian co., and became its manager. Director Lvov O 1864; director Italian O, St Petersburg, and Moscow; director Hamburg O 1874–97, where he worked closely with Bülow, then Mahler. Important for his promotion of Wagner's operas (taking *The Ring* to London, CG, 1892) and other new works including *Carmen*, Verdi's *Otello*, *Manon Lescaut*. Engaged the best singers of the day, e.g. Niemann, Klafsky, the young Mildenburg.

**Pollione.** The Roman proconsul (ten), lover of Norma and of Adalgisa, in Bellini's *Norma*.

**Polly.** Polly Peachum (sop), Lucy Lockit's rival for Macheath's love, in Pepusch's *The Beggar's Opera*, and in Weill's version of it, *Die Dreigroschenoper*.

**Polovtsian Dances.** The dances in Act II of Borodin's *Prince Igor* with which the Khan Konchak entertains Igor.

**Pomo d'oro, Il** (The Golden Apple). Opera in 5 acts, with prologue, by Cesti; text by Sbarra;

designs by Ludovico Burnacini. Prod. Vienna 13 and 14 July 1668. Most of the music was composed for the wedding of Leopold I and the Infanta Margherita (12 Dec. 1666). A work of exceptionally large proportions, with 5 acts instead of the normal 3, and a much enlarged orchestra, it was perhaps the most elaborate opera production ever staged; the cost of the décor alone (there were 24 different sets) was estimated at 100,000 Reichsthaler. The greater part of the music for Act V is lost; the rest has been published in a modern edition. Two of Cesti's other operas had also been given in Vienna: *Nettuno e Flora*, text by Sbarra, July 1666 (containing an aria in Act II by Leopold), and *Le disgrazie d'amore*, text by Sbarra, 19 Feb. 1667; for this Leopold wrote the *licenza.

**Ponchard**. French family of singers and composers.

1. **Antoine** (*b* Péronne, 1758; *d* Paris, Sep. 1827). Composer, and director of Grand T, Lyons, 1803–13.

2. **Louis Antoine Éléonore** (*b* Paris, 31 Aug. 1787; *d* Paris, 6 Jan. 1866). Tenor and composer, son of (1). Studied Paris with Garat. Début Paris, OC, 1812 (Grétry's *L'ami de la maison*). Paris, OC, until 1837, singing in operas by Grétry, Dalayrac, Monsigny, Isouard; created several roles, notably George Brown in his friend Boieldieu's *La dame blanche*. Though his appearance was unappealing, and his voice unremarkable, his vocal artistry made him one of the most distinguished singers in France. Taught at Paris Cons. from 1816; pupils included Dabadie, Faure, Mario, R. Stoltz, Roger.

3. **Marie Sophie Carrault-Ponchard** (*b* Paris, 30 May 1792; *d* Paris, 19 Sep. 1873). Soprano, wife of (2). Studied Paris with Garat. Début Paris, O, 1814 (Iphigénie, *Aulide*). Paris, OC, 1818–36, creating roles in operas by Auber (Juliette, *Emma*; Tao Jin, *Cheval de bronze*), Hérold (Queen, *Le pré aux clercs*), and Carafa.

4. **Charles Marie Auguste** (*b* Paris, 17 Nov. 1824; *d* Paris, 26 Apr. 1891). Tenor, son of (2) and (3). Studied Paris with his father. Début Paris, O, 1846; Paris, OC, 1847–71, successful in *Trial roles, and 1875–90 as manager and producer, staging the premières of *Carmen*, *Manon*, and *Lakmé*. Taught at Paris Cons. from 1872.

**Ponchielli, Amilcare** (*b* Paderno Fasolaro, 31 Aug. 1834; *d* Milan, 17 Jan. 1886). Italian composer. Studied Milan with Frasi and Mazzucato. Composed his first operatic music, with fellow-students, for *Il sindaco babbeo*. Conducted at the T Carcano, Milan, 1860. His first opera was *I promessi sposi* in 1856, successful on its revival in 1872, as was the Ricordi commission *I lituani*. But his greatest, and only enduring, triumph was

*La Gioconda*. His dramatic instinct here finds its best outlet: Boito's text combines many features of French Grand Opera and Italian Romantic melodrama, in a manner that suited the composer's gift for agreeable, warm music. The work's musical and dramatic styles were already exhausted in opera; but it contains celebrated passages, including the arias 'Cielo e mar' and 'Suicidio', and the Dance of the Hours ballet. In his other operas Ponchielli toyed with exotic styles, Slavonic in *I lituani*, oriental in *Il figliuol prodigo*. His last work, *Marion Delorme*, is closer to opéra-comique in style. Married Teresa *Brambilla. His pupils included Puccini.

SELECT WORKLIST: *I promessi sposi* (The Betrothed) (?, after Manzoni; Cremona 1856, rev. Milan 1872); *La savoiarda* (Guidi; Cremona 1861); *Roderico, re dei goti* (Roderick, King of the Goths) (Guidi, after Southey; Piacenza 1863); *I lituani* (Ghislanzoni, after Mickiewicz; Milan 1874); *La Gioconda* ('Tobia Gorrio' [Boito], after Hugo; Milan 1876); *Il figliuol prodigo* (The Prodigal Son) (Zanardini; Milan 1880); *Marion Delorme* (Golisciani, after Hugo; Milan 1885).

BIBL: A. Damerini, *Amilcare Ponchielli* (Turin, 1940); N. Albarosa, *Amilcare Ponchielli* (Casalmorano, 1984).

**Poniatowski, Józef** (*b* Rome, 20 Feb. 1816; *d* Chislehurst, 3 July 1873). Polish composer and tenor. Great-nephew of Stanisław August, King of Poland 1764–95. Studied Rome, then Florence with Zanetti and Ceccherini. Début Florence in title-role of his *Giovanni di Procida* (to his own text, 1839). His *Don Desiderio* (1840) was also widely successful; other operas include *Ruy Blas* (1843), *Bonifazio de' Geremei* (1843), *Malek Adel* (1846), and *Esmeralda* (1847). Became Prince of Monte Rotondo, 1848; Ambassador in Brussels, London, and Paris, where he also became a senator. His *Pierre de Médicis* was given at the Opéra, 1860, *Au travers du mur* at the Opéra-Comique, 1861. He followed Napoleon III into exile to London, where his *Gelmina*, written for Patti, was given in 1872.

WRITINGS: *Le progrès de la musique dramatique* (Paris, 1859).

**Ponnelle, Jean-Pierre** (*b* Paris, 19 Feb. 1932; *d* Munich, 11 Aug. 1988). French designer and producer. Studied Paris. Invited by Henze to design *Boulevard Solitude* (1952) and *König Hirsch* (1956); then Germany, Italy, USA. Prod. *Tristan* Düsseldorf 1962. Rapidly in demand as producer in many theatres (also generally making his own designs) in a wide range of works, both classics, especially Mozart and Rossini, and new works, e.g. Reimann's *Lear*, Munich 1978. Bayreuth, *Tristan* 1981; Cologne, *Parsifal* 1983; Salzburg Mozart series and *Moses and Aaron* 1987; Zurich Monteverdi and Mozart series from 1979 with Harnoncourt, *Carmen* 1982. San Francisco over 20 productions from 1958, *Flying Dutchman*, *Carmen* 1981; Houston, *Arlecchino* and *Pagliacci* 1982; New York, M, *Flying Dutchman* 1979, *Titus* 1984. London, CG, designer *L'heure espagnole* 1962, designer-producer *Don Pasquale* 1973, *Aida* 1984, *Italiana* 1988. Gly., *Falstaff* 1977. Vienna, *Cavalleria* and *Pagliacci* 1985, *Italiana* 1987. Munich, *Lulu* 1985, Reimann *Troades* 1986. Also made many films for TV. One of the most gifted, hardworking, and stimulating producers of his day: many of his ideas excited controversy, e.g. his new angles on Mozart and his *Flying Dutchman* production seen as a dream by the Steersman, but his theatrical sense was potent and musically based, and he unfailingly stimulated audiences, as well as performers and younger producers, without losing his sense that the work's importance was greater than his own.

**Pons, Lily** (*b* Draguignan, 12 Apr. 1898; *d* Dallas, 13 Jan. 1976). French, later US soprano. Studied Paris with Gorostiaga. Début Mulhouse 1928 (Lakmé). New York, M, 1931–60; London, CG, 1935. Also Paris, O; Chicago; Buenos Aires; etc. Gifted with an unusually high and agile voice, and great vivacity, she had an extremely successful career on a very small repertory. This included Amina, Lucia, Gilda, Olympia, Philine. She also appeared in films. (R)

**Ponselle, Rosa** (*b* Meriden, CT, 22 Jan. 1897; *d* Green Spring Valley, MD, 25 May 1981). US soprano. Studied with Thorner and Romani. Early career in vaudeville. Discovered by Caruso; he sang at her début, New York, M, 1918 (Leonora, *Forza*), her first ever operatic performance. New York, M, until 1937; London, CG, 1929–31; Florence, Maggio Musicale, 1933. Described by Serafin as a 'miracle', and considered one of the most remarkable singers of the century, she possessed a phenomenal, powerful voice (Huneker called it 'vocal gold . . . dark, rich, and ductile'), a seamless legato and impeccable coloratura. Also striking were her intense musicality and dramatic authority. Her 26 roles included Julia (*La vestale*), Donna Anna, Reiza, Norma, Violetta, Aida, Gioconda, Santuzza, Romani's Fedra. Retired early, 1937; artistic director of Baltimore Civic O from 1954. (R) Her sister **Carmela** (1888–1977) was a mezzo-soprano who sang at New York, M, 1925–35 as Amneris, Azucena, etc. (R)

**ponticello** (It.: 'little bridge'). The term used by the bel-cantists for the join between the chest and head registers.

**Ponziani, Felice** (*b* ?; *d* ?, *c*.1826). Italian bass. Active in Parma 1784–5; Prague 1786, where he sang successfully as Figaro, and 1787, creating

Leporello, which Mozart wrote for him. Evidently a gifted performer, he was described in a Prague newspaper as 'the favourite of connoisseurs, and of all who have heard him'. Venice 1792; Vienna 1792–8.

**Popp, Lucia** (*b* Uhorská Ves, 12 Nov. 1939; *d* Munich, 16 Nov. 1993). Czech soprano. Studied Bratislava and Prague with Hrušovska-Prosenková. Début Bratislava 1963 (Queen of the Night). Vienna, S, from 1963; London, CG, from 1966; New York, M, from 1967; Paris, O, from 1976, and B, 1991; Cologne, Salzburg, Munich, etc. An artist of charm, with a well-projected, bright, and vibrant voice. Roles include Susanna, Countess (Mozart), Aennchen, Eva, Mařenka, Sophie, Manon. (R)

**Poppea.** See INCORONAZIONE DI POPPEA, L'.

**'Porgi amor'.** The Countess's (sop) aria in Act II of Mozart's *Le nozze di Figaro*, in which she laments the loss of her husband's love.

**Porgy and Bess.** Opera in 3 acts by Gershwin; text by Du Bose Heyward and Ira Gershwin, after the drama *Porgy* by Du Bose and Dorothy Heyward. Prod. Boston, Colonial T, 30 Sep. 1935, with Brown, Mitchell, Duncan, Elzy, Bubbles, Buck, Matthews, Harvey, Dowdy, Davis, Coleman, Johnson, cond. Smallens; London, Stoll T, 9 Oct. 1952, with L. Price, Warfield, Colbert, Dowdy, Calloway, McCurry, cond. Smallens.

Catfish Row, Charleston, 1920s. Crown (bs), a stevedore, is obliged to flee after killing his friend Robbins (ten) in a fight over a dice game. He leaves behind him his girl Bess (sop); she is taken care of by Porgy (bs-bar), who has long loved her.

Porgy is happy in his new life with Bess. But while on a picnic, she meets Crown again and is persuaded to return to him. After a few days, however, she runs home to Porgy seeking shelter and declaring she loves him. Crown sets out to rescue some fishermen during a storm. The women weep for the dead lost at sea.

As he promised, Crown goes to Porgy's house in search of Bess, but Porgy stabs him dead and is arrested. Alone again, Bess is persuaded by Sporting Life (ten) to follow him to New York. When Porgy is released through lack of evidence, he sets off to New York after her.

**Poro.** Opera in 3 acts by Handel; text an anon. adaptation, prob. by G. Rossi or S. Humphreys, of Metastasio's *Alessandro nell'Indie*, trans. Samuel Humphreys. Prod. London, H, 2 Feb. 1731, with Strada, Senesino, Merighi, Bertolli, Fabri, Commano.

**Porpora, Nicola** (*b* Naples, 17 Aug. 1686; *d* Naples, 3 Mar. 1768). Italian composer and singing teacher. Studied Naples with Grecos, Giordano, and Campaniles. His first operatic success came with *Agrippina*, followed by *Flavio Anicio Olibrio* and *Basilio*. Through Prince Philipp of Hessen-Darmstadt, whose service he had entered *c*.1707, his *Arianna e Teseo* was given to great acclaim at the Vienna court 1714. In 1715–21 he was maestro di cappella at the Cons. di S Onofrio, Naples, where he taught a number of celebrated singers, including Caffarelli, Farinelli, Porporino, Salimbeni, and (though unsuccessfully) Hasse. During the 1720s his fame as a composer began to increase both in Italy and elsewhere in Europe, especially Vienna. He began to set librettos by his pupil Metastasio: the first, *Didone abbandonata*, was for Metastasio's mistress, the soprano Marianna Benti-Bulgarelli. In 1726–33 he lived in Venice, writing *Siface*, *Siroe*, *Ezio*, *Semiramide riconosciuta*, and *Issipile*. Having failed to secure the post of maestro di cappella at St Mark's, he travelled to London in 1733, where he had been invited as principal composer of the *Opera of the Nobility. Here he became the main rival to Handel and produced five operas: *Arianna in Nasso*, *Enea nel Lazio*, *Polifemo*, *Ifigenia in Aulide*, and *Mitridate*. Following the success of Handel's *Atalanta*, which won the support of the Opera of the Nobility's main backer, the Prince of Wales, Porpora prudently left London, shortly before the collapse of the company. Returning to Italy, he held positions in Venice (1737), Naples (1739), and Venice (1742), but in 1747 went to Dresden as singing-teacher to Princess Maria Antonia Walpurgis. Appointed Kapellmeister in 1748, but shortly afterwards Hasse was appointed above him as Oberkapellmeister. In 1752 he moved to Vienna, where Haydn was his pupil, copyist, and accompanied his lessons. He finally returned to Italy, to Naples (1758), Venice (1759), and Naples again (1760), where he died in poverty.

Porpora composed *c*.50 operas which were highly regarded in their day: together with Hasse's, they represent Metastasian opera seria in its purest form. As befitted a famous singing-teacher, he displayed a consummate ability to write elegant melodic lines displaying the singer's abilities to best effect. But the emphasis on vocal virtuosity, excessive even by the standards of the day, hindered dramatic articulation. Though more talented than many of his contemporaries, Porpora was no match for Handel, beside whom his style is stilted and lacking in dramatic intensity. Even during his lifetime criticisms were levelled that his music was too florid and ornamented; and his works, though not his reputation, soon passed from the repertoire.

SELECT WORKLIST: *Agrippina* (Giuvo; Naples 1708); *Flavio Anicio Olibrio* (Zeno & Pariati; Naples 1711);

*Basilio* (?Neri; Naples 1713); *Arianna e Teseo* (Pariati; Vienna 1714); *Didone abbandonata* (Metastasio; Reggio Emilia 1725); *Siface* (Metastasio; Milan 1725); *Siroe* (Metastasio; Rome 1727); *Ezio* (Metastasio; Venice 1728); *Semiramide riconosciuta* (Metastasio; Venice 1729); *Issipile* (Metastasio; Rome 1733); *Arianna in Nasso* (Rolli; London 1733); *Enea nel Lazio* (Rolli; London 1734); *Polifemo* (Rolli; London 1735); *Ifigenia in Aulide* (Rolli; London 1735); *Mitridate* (?Cibber; London 1736).

**Porsile, Giuseppe** (*b* Naples, 5 May 1680; *d* Vienna, 29 May 1750). Italian composer and singing-teacher. Studied Naples with Ursino, Giordano, and Greco. Held positions at Spanish court and Naples, before becoming singing-teacher to the Dowager Empress Amalia in Vienna 1715. Court composer 1720–40, writing *c*.30 dramatic works, including six opere serie. His style owes much to Fux and Caldara and he showed an appropriate fondness for heroic and noble themes.

SELECT WORKLIST: *Il ritorno d'Ulisse alla patria* (Moniglia; Naples 1707); *Spartaco* (Pasquini; Vienna 1726); *La clemenza di Cesare* (Pasquini; Vienna 1727).

**portamento** (It.: 'carrying') The smooth carriage of the voice from one note to another. The Fr. *porte de voix* is generally reserved, from *c*.1800, for an appoggiatura, generally from below: Ger. *tragen der Stimme*.

**Porter, Cole** (*b* Peru, IN, 9 June 1891; *d* Santa Monica, CA, 15 Oct. 1964). US composer. Studied Yale and Harvard U; later Paris with D'Indy. Following the failure of his first Broadway show, *See America First*, he moved to Paris. After war service his songs were gradually heard as inserts in Broadway shows, but he came to attention with *Wake Up and Dream* and *Fifty Million Frenchmen*, for which he provided all the music. These were followed up by a string of popular successes, including *Anything Goes*, *Let's Face It*, and *Mexican Hayride*. A skilful and effective song-writer, his most convincing work was *Kiss Me, Kate*, an adaptation of *The Taming of the Shrew* which demonstrates a dramatic as well as a lyrical gift. He also had success as a film composer, notably with *High Society*.

SELECT WORKLIST: *Kiss Me, Kate* (B. & S. Spewack, after Shakespeare; New York 1948).

**Portland.** City in Oregon, USA. The first opera performance in the North West Pacific was in summer 1867 at the Oro Fino T when the Bianchi OC arrived by steamer from San Francisco, opening with *Trovatore*. The New Market T opened 1875, the Marquam Grand OH 1890, housing seasons by the Emma Juch Grand OC and other touring companies. The Chicago Grand OC visited in 1913, returning on several occasions until 1931. Portland's first resident company was founded 1917 under Roberto Corruccini. The San Carlo OC visited regularly 1918–48. The Portland Opera Association was founded 1964, playing in the Civic Auditorium from 1968.

**Portugal.** Though popular dramas with music were familiar in Portugal from the Middle Ages, Portuguese opera began in 1733 with performances of *A vida do grande dom Quixote da la Mancha* by A. J. da Silva and *La pazienza di Socrate* by Francisco António de Almeida (*c*.1702–55). The former was in Portuguese (music lost). Works by António Leal *Moreira (1758–1819) included opere serie (e.g. *Siface e Sofonisba*, 1783) and Portuguese operas (e.g. *A vingança da cigana*, 1794). The first important composer was M. A. *Portugal (or Portogallo) (1762–1830), who wrote many Italian operas but also Portuguese comedies beginning with *A casa de pasto* (1784); his most popular work was *A castanheira* (1787). Opera did not advance greatly during the 19th cent., despite the activity of Francisco de Sá Noronha (1820–81), José de Arnbeiro (1838–1903), Miguel Ángel Pereira, Francisco de Freitas Gazul, and Augusto Machado. The latter attempted to resist the prevalent Italian influence with elements of French opera. The father of Portuguese Romantic opera was Alfredo Keil (1850–1907): his *Serrana* (1899) drew on folk elements. His work was developed by Rui Coelho (1891–?) with a number of operas to Portuguese texts, including *Inês de Castro* (1925) and *Orfeu em Lisboa* (1966). There was also an attempt by various 19th-cent. composers to develop an indigenous strain of Portuguese operetta on the basis of popular music drama. Successful contemporary composers include J. B. Santos (*b* 1924).

See also LISBON.

**Portugal, Marcos António** (da Fonseca) (orig. Ascenção; also known as Portogallo) (*b* Lisbon, 24 Mar. 1762; *d* Rio de Janeiro, 7 Feb. 1830). Portuguese composer. Singer and organist at the Seminário da Patricial. Music director of the T Siltre 1785, where he wrote five Portuguese operettas, including the popular *A castanheira*. In 1792 he went to Naples to study; here he wrote 21 operas for various Italian cities. The first, *Le confusione della somiglianza*, was one of his greatest successes and was revived shortly afterwards in Germany. His work embraced both opere serie and opere buffe: highly regarded were *Lo spazzacamino principe*, *La vedova raggiratrice*, *Demofoonte*, *La donna di genio volubile*, *L'inganno poco dura*, *Il ritorno di Serse*, and *Fernando nel Messico*, which brought him fame throughout Europe. Director of the Royal O at S Carlos T,

Lisbon, 1800: here he engaged Catalani 1801–6. When the court fled to Brazil following French occupation he stayed first in Lisbon but then followed (1810); opened the T São João in Rio de Janiero in 1813. Illness prevented him from returning to Lisbon with the court in 1821.

Portugal composed *c*.35 Italian operas and 21 1-act Portuguese comic operas. His style in the Italian works, predominantly that of the Neapolitan school, rarely goes beyond conventional gestures, but as one of the earliest to set librettos in his native language, he represents, together with Moreira, the first generation of Portuguese opera composers. Equally important for the establishment of his country's national opera were his activities outside Portugal and the international reputation which he achieved both as composer and as conductor.

SELECT WORKLIST: *A castanheira* (The Chestnut Seller) (?; Lisbon 1787); *Le confusione della somiglianza* (The Problems of Looking Alike) (Mazzini, after Del Buono; Florence 1793); *Lo spazzacamino principe* (The Royal Chimney Sweep) (Foppa, after Popigny; Venice 1794); *La vedova raggiratrice* (The Cheating Widow) (?; Florence 1794); *Demofoonte* (Metastasio; Milan 1794); *La donna di genio volubile* (The Flighty Woman) (Bertati; Milan 1796); *L'inganno poco dura* (The Mean Trick) (Zini; Naples 1796); *Il ritorno di Serse* (Ferrari; Florence 1797); *Fernando nel Messico* (Tarducci; 1798).

**Posa.** Rodrigo, Marquis of Posa (bar), in Verdi's *Don Carlos*.

**Posen.** See POZNAŃ.

**Posse** (Ger.: 'buffoonery, farce'). A form of popular theatrical entertainment which flourished especially in the late 18th and early 19th cents. Above all a Viennese form, though also found in different aspects in a number of other German cities, including Hamburg, and also Berlin, where the manner tended to be more realistic than in the generally fantastic comedies of Vienna. Sometimes also described as a 'Posse mit Gesang', the genre was distinguished from that of *Singspiel by the comparatively smaller amount of music it contained. Local colour was important (**Lokalposse**), as was magic (**Zauberposse**); often the two were combined in plots that dealt with the intrusion of magic into normal life, perhaps by transferring an ordinary Viennese citizen to fairyland. This device was particularly popular with the most famous of all Viennese popular theatre authors, Ferdinand *Raimund (1790–1836), e.g. in *Der Alpenkönig und der Menschenfeind* (1828) and others. It was also typical of the Viennese Zauberposse to touch on moral and general issues while keeping to a comic and fantastic narrative. Other subdivisions of the genre were the **Charakterposse**, which presented a farce of characterization, and the **Situationsposse**, which exploited comedy of situation.

Some authors provided songs more or less of their own composition (adapted folk-melody, original song-melody); others drew on composers including Riotte, Drechsler, Wenzel Müller, and Conradin Kreutzer. The Zauberposse, especially in the hands of Raimund, was enormously popular and influenced a number of dramatists and composers, including Wagner.

**'Possente spirto'.** The arioso section in Act III of Monteverdi's *Orfeo* in which he charms Caronte (Charon) to sleep. Famous for its skilfully contrasting instrumental accompaniment.

**Postel, Christian Heinrich** (*b* Freiburg, nr. Stade, 11 Oct. 1658; *d* Hamburg, 22 Mar. 1705). German librettist and poet. Settled in Hamburg in 1688, practising as a lawyer. An acquaintance of Gerhard Schott, founder and first director of the Hamburg Opera, he wrote librettos for Conradi (*Die schöne und getreue Ariadne*, 1691), Förtsch (*Ancile Romanum*, 1690), Keiser (*Der geliebte Adonis*, 1697), and Kusser (*Gensericus*, 1694; also set by Telemann in 1722 as *Sieg der Schönheit*), and played an important role in the foundation of German opera. Like his contemporaries Bostel and Bressand, he was heavily influenced by Italian and French models, especially that of Minato and other Venetian librettists. However, the spirit and craftsmanship of German Baroque poetry is more evident in Postel's work, and he was widely recognized as the most gifted of the early Hamburg librettists.

**Postillon de Lonjumeau, Le.** Opera in 3 acts by Adam; text by De Leuven and Brunswick. Prod. Paris, OC, 13 Oct. 1836, with Prévost, Ray, Collet, Henri; London, St James T, 13 Mar. 1837; New Orleans, T d'Orléans, 19 Apr. 1838, with J. Calvé, Heymann, Bailey, Astruc.

The postilion Chapelou (ten) possesses a fine voice, and is engaged by De Courcy (bar), the manager of the royal amusements, to sing at Fontainebleau. Under the name of Saint-Phar he becomes a great singer, and promises marriage to the rich Madeleine (sop), whom he had previously married when he was a postilion and she the hostess of a village inn. All ends happily.

Adam's most popular opera outside France.

Adaptation by Oudrid y Segura, *El postillón de la Rioja* (1856). Also opera by P. A. Coppola (1838).

**Pougin, Arthur** (orig. François Auguste Arthur Paroisse-Pougin; pseudonym Pol Dax) (*b* Châteauroux, 6 Aug. 1834; *d* Paris, 8 Aug. 1921). French writer and critic. Published biographies of Rossini, Bellini, Meyerbeer, Verdi, Auber, and others, and of singers including Dugazon, Malibran, Favart, and Grassini.

**Poulenc, Francis** (Jean Marcel) (*b* Paris, 7 Jan. 1899; *d* Paris, 30 Jan. 1963). French composer. Studied with Koechlin. Like other members of the group Les Six, Poulenc was eclectic in his inspiration and his music is imbued with a healthy irreverence for tradition which places him among the most original of his generation. Most of his music was not for the theatre: *Le gendarme incompris*, a *comédie-bouffe*, which he later withdrew, was not followed for nearly a quarter-century, until the witty and surrealist satire *Les mamelles de Tirésias*. *Dialogues des Carmélites* is of an altogether different order, a deeply felt religious work written in a simple lyrical style. For its conception Poulenc owes much to Debussy and the model of *Pelléas*, seen not least in his handling of the orchestra. His final work, *La voix humaine*, is a 45-minute monologue for soprano, written for Denise Duval: ironically described by him as a 'tragédie lyrique', it is a vocal *tour de force* which fails to satisfy dramatically.

WORKLIST: *Les mamelles de Tirésias* (The Breasts of Tiresias) (Apollinaire; Paris 1947); *Dialogues des Carmélites* (Bernanos; Milan 1957); *La voix humaine* (Cocteau; Paris 1959).

**Pountney, David** (*b* Oxford, 10 Sept. 1947). English producer. Studied Cambridge: Scarlatti *Trionfo d'onore*, Cambridge Opera Society 1967 (cond. Elder), then Smetana *Kiss* 1969, British première Dessau *Lukullus* 1970, *Rake's Progress* 1971; latter also Scottish O 1971, Amsterdam 1972. Janáček cycle jointly with Scottish/Welsh O, some later transferred to London, C. Houston: *Macbeth* 1974. Aldeburgh: *Savitri* and *Wandering Scholar* 1974. London, EOG: Puccini *Rondine* 1974. Scottish O: *Macbeth* 1977, *Meistersinger* 1977, *Onegin* 1979, *Don Giovanni* 1979, *Street Scene* 1989. London, C: première of Blake *Toussaint*, designed Bjørnson, cond. Elder, 1977, initiating a collaboration which gave the work of the Coliseum a particular stamp, with innovative, provocative, sometimes brilliant productions; director of productions 1982; *Flying Dutchman*, *Makropoulos Case*, 1982; *Queen of Spades*, *Rusalka*, *Gambler*, 1983; *Walküre* 1983; *Osud* 1984; *Midsummer Marriage*, *Orpheus in the Underworld*, 1985; *Dr Faust* 1986; *Carmen*, *Lady Macbeth of Mtsensk*, *Hänsel und Gretel*, 1987; *Cunning Little Vixen*, *Traviata*, *Christmas Eve*, 1988; *Falstaff* 1989; *Macbeth*, première Holloway *Clarissa*, *Wozzeck*, 1990; *Königskinder* 1992. Berlin, K: *Iolanthe* 1984. Bregenz: *Flying Dutchman* 1989. A gifted and highly intelligent man of the theatre, with strong ideas articulated through his productions: these can be profoundly or wittily illuminating, often theatrically dazzling, but at their most wilful can cast only a selective spotlight and leave much of the work's quality in darkness.

**'Pourquoi me réveiller?'** Werther's (ten) aria in Act III of Massenet's *Werther*, in which he sings a song of tragic love from the verses of Ossian which in happier days he had translated with Charlotte.

**'Poveri fiori'.** Adriana's (sop) aria in Act IV of Cilea's *Adriana Lecouvreur*, in which she contemplates the faded bunch of violets (now poisoned by her rival, the Princesse de Bouillon) which she had given Maurizio.

**Poznań** (Ger., Posen). City in Poland. Opera was first given towards the end of the 18th cent. During the 19th cent. seasons were given by visiting companies, including the T Narodowy from Warsaw, from Cracow, and elsewhere. In 1875 the T Polski was founded as a permanent repertory theatre, which until 1918 was the principal encouragement to opera and which still exists. The T Wielki opened on 31 Aug. 1919, closing on 1 Sep. 1939, the only Polish company to perform during the period without interruption. During the war, a German company played in the theatre until 1944. The Poznań O (cap. 950) opened on 2 June 1945 (renamed the Stanisław Moniuszko OH in 1949). Between 1919 and 1969 435 stage works by 147 composers were given, including 22 premières of Polish works and many Polish premières of foreign classics. More recent premières have included *Ślepy* (The Blind, 1984) by Jan Astriab (*b* 1938) and the Polish première of Penderecki's *Die schwarze Maske* (1987).

**Pozsony.** See BRATISLAVA.

**Prague** (Cz., Praha). Capital of Czechoslovakia from 1918, formerly capital of Bohemia. Opera was first given in 1627, when Italian singers came from Mantua to perform at the coronation of Ferdinand III. In 1723 Fux's *Costanza e Fortezza* was performed for the 4,000 visitors to the coronation of Charles VI in an amphitheatre with room for 1,000 performers. Visiting Italian companies performed first in the Estates Riding School, then in the the home of Count František Špork, Na Poříčí. From 1739 opera was also given at the Kotzen (Kotce) T, the main public theatre until the 1780s. Pasquale Bondini's Italian company played in a theatre built by Count Thun at his palace in the Malá Strana; and Karel Wahr's German company played at the Gräflich-Nostitzsches Nationaltheater, built at his own expense by Count Nostic and opened in 1783 (cap. over 1,000). In 1784 Nostic replaced Wahr with Bondini, who after the sensational première of *Figaro* (said to be far superior to the original Vienna production) secured the commission for

*Don Giovanni* in 1787. The Kotce T closed in 1783, and the Thun T burnt down in 1794. The 1790s saw operatic performances in Czech by the Patriotic T company. The Nostic T became the Královské stavovské divadlo (Royal Estates T) in 1789. The director of the German company in 1803–13 was Wenzel Müller, and in 1813–16 it was Carl Maria von Weber, who did much to widen and develop the repertory and to improve theatrical standards. The playwright Jan Štěpánek revived the Czech professional company from 1824, and in the face of restrictions built up opera in Czech. The theatre became the Královské zemské divadlo (Royal Provincial T) in 1861. In 1837 František Škroup was appointed conductor, and the theatre developed under the management of Josef Tyl (1846–51: since 1949 the theatre has been known as the Tylovo divadlo). Opera was also given in summer theatres, including the Novoměstské divadlo (New Town T) from 1858 (cap. over 3,000) and the Nové české divadlo (New Czech T) from 1876 (cap. over 3,000).

Another theatre being urgently required, in part as a national gesture, a newly formed association built the Prozatímní divadlo (Provisional T) in 1862 (cap. 900). Its methods and resources were initially primitive, with most of its singers amateurs and a tiny scratch orchestra under Jan Maýr. Smetana was director 1866–74, and six of his operas were produced here, as were the first stage works of Dvořák and Fibich. Practical and political disputes delayed its replacement, and the laying of the foundation stone of the new theatre in 1868 was a national event. In 1881 the Národní divadlo (National T), incorporating the Provisional T, opened with *Libuše*; but shortly afterwards the theatre burnt down. Money was quickly found for another theatre, which opened in 1883 (cap. 1,598). This has remained at the centre of Prague operatic life. Chief conductors have included Adolf Čech (1883–1900), Karel Kovařovic (1900–20), Otakar Ostrčil (1920–35), Václav Talich (1935–44 and 1947–8), Zdeněk Chalabala (1953–62), Jaroslav Krombholc (1963–75), and Zdeněk Košler from 1980. Important premières included Schoenberg's *Erwartung* (1924). The theatre was rebuilt 1977–83 (cap. 986, with Chamber T cap. 500).

Meanwhile, replacing the Novoměstské divadlo, the German minority built the Neues Deutsches T (opened 1887, cap. 1,554). The company was directed by Angelo *Neumann and Alexander *Zemlinsky (1911–27), in a line of distinguished German musicians who have developed their careers in Prague, including Mahler (1885), Klemperer (1907–10), and Szell (1929–37). After 1945 the theatre became first the Velká opera 5 května (Grand O 5th May), then in

1948, as filial of the National T, the Smetanovo divadlo (Smetana T) (cap. 1,044). The theatre has won a name for enterprise, especially with the productions of Josef *Svoboda. Opera has also been given at the Městské divadlo na Král, Vinokradech (Vinokradské divadlo) opened in 1907; particularly under Ostrčil (1913–19), this was a pioneering centre of modern opera. Operetta has been given in various theatres but especially in the Variété in Karlín (built 1880–1), from 1962 known as the Hudebni divadlo v Karlíně, and the Velká Opereta. In 1934 Emil F. *Burian opened, in the Mozarteum, an avant-garde theatre called D34 (Divadlo (theatre) 1934) where, especially in the 1950s, chamber operas and plays with music were given. In 1983 an experimental theatre, the Nová Scéna (New Stage), was opened next to the National T.

**Prati, Alessio** (b Ferrara, 19 July 1750; d Ferrara, 17 Jan. 1788). Italian composer. Studied Ferrara with Marzola and Rome with Speranza; c.1775 travelled to Paris, where he had success with L'école de la jeunesse. In 1782 he went to St Petersburg at the invitation of Grand Prince Paul; other travels took him to Warsaw, Dresden, Vienna, and Munich. Returned to Italy 1784, where Ifigenia in Aulide was given. Though not prolific by the standards of the day, with only c.10 operas to his credit, Prati was one of the most skilled composers of the mid-18th cent. and his work was highly regarded.

SELECT WORKLIST: L'école de la jeunesse (Anseaume; Paris 1779); Ifigenia in Aulide (Serio; Florence 1784).

**Pré aux clercs, Le.** Opera in 3 acts by Hérold; text by Planard, after Mérimée's novel Chronique du règne de Charles IX (1829). Prod. Paris, OC, 15 Dec. 1832, with Ponchard, Casimir, Massy, Thénard, Lemonnier, Féréol, Gent; London, Adelphi T, 9 Sep. 1833; Baltimore, Holliday St. T, 14 Oct. 1833 by New Orleans Co.; London, John Lewis Partnership Music Society, 24 Apr. 1985. Highly popular in its time, it was given under various titles in England and the US, e.g. The Challenge, The Field of Honor.

Girot (bar), host of the 'Pré aux Clercs', celebrates his betrothal to Nicette (sop), goddaughter of Marguerite de Valois (sop or mez). The ambassador Mergy (ten), searching for the queen, meets his childhood love Isabelle (sop), who is promised to Comminges (bar). Aided by Cantarelli (ten—*Trial role), the queen helps Isabelle: Mergy challenges Comminges, kills him, and marries Isabelle.

**Predieri, Luca Antonio** (b Bologna, 13 Sept. 1688; d Bologna 1767). Italian composer. Studied with Vitali, Perti, and various members of his musical family. Member of the Accademia Filarmonica 1716; director 1723; also held ecclesiastical appointments in Bologna. Moved to Vienna 1737; appointed second Kapellmeister to Fux, in succession to Caldara 1739; first Kapellmeister 1746–51.

During his years in Italy Predieri composed for various Italian cities both opere serie and opere buffe, including Lucio Papirio and La serva padrona, while in Vienna he contributed to the long-established tradition of court opera. Though apparently not as skilled or distinguished a composer as his predecessors Fux and Caldara, his works, which include Astrea placata, share a similar fondness for grand themes and heroic gesture.

SELECT WORKLIST: Lucio Papirio (Salvi; Pratolino 1714; lost); La serva padrona (The Maid-Mistress) (Federico; Florence 1732; lost); Astrea placata (Metastasio; Vienna 1739, lost).

BIBL: R. Ortner, Luca Antonio Predieri und sein Wiener Opernschaffen (Vienna, 1971).

**Preetorius, Emil** (b Mainz, 21 June 1883; d Munich, 27 Jan. 1973). German designer. After working as a book illustrator, he was suggested to Bruno Walter by Thomas Mann in 1912 for the Munich Iphigénie en Aulide. Berlin, Dresden, Madrid in the 1920s. Bayreuth 1932–41, all productions except Parsifal. Also Tristan Paris 1936, Florence 1941, Rome 1943, Amsterdam 1948, Munich 1958, Vienna 1959. Ring Berlin, S, 1936; Milan, S, 1938; Rome 1953–4. His three-dimensional sets derived from Wagner's instructions but were more stylized; he insisted that, 'for his penetratingly descriptive music Wagner needs a certain nearness to nature which should, however, not be literally naturalistic'. He asked for emphasis on the things carrying symbolic weight, and sought to combine 'inner vision and outward nature'. He managed to resist Nazi pressures to make his designs more literal and nationalistic.

WRITINGS: Richard Wagner: Bild und Vision (Bad Godesberg, 1949).

**preghiera** (It.: 'prayer') (Fr., prière; Ger., Gebet). An aria or chorus of quiet or reverent character in which the singer or singers make a prayer to God or to supernatural powers. Designed to display mastery of soft vocalization and simple line, it became a very popular ingredient of Italian opera, especially in the 19th cent., and thence of French and German opera. An early example, if not the first, is in Salieri's Tarare. A famous instance is Desdemona's 'Ave Maria' in Verdi's Otello; others are Huon's 'Ruler of this awful hour' in Weber's Oberon and Rienzi's 'Allmächt'ger Vater' in Wagner's opera. The preghiera could be treated with considerable fluency, especially by Verdi: choral preghiere occur in Aroldo (the canonic 'Angiol di Dio'), La forza del destino ('Padre eterno Signor'), and Aida ('O tu che sei d'Osiride'), and a preghiera is built into the finales of *Luisa Miller* and Act II of *Aroldo*.

**prelude** (from Lat. *praeludium*: something played before another work) (Ger: *Vorspiel*). There is no clear distinction between the prelude and *overture, though in general the former may be shorter and may also run directly into the opera (or act of an opera) which it introduces.

**'Prendi, l'anel ti dono'.** Elvino's (ten) aria in the opening scene of Bellini's *La sonnambula* as he places the ring on Amina's finger.

**'Près des remparts de Séville'.** Carmen's (mez) aria in Act I of Bizet's *Carmen*, in which she tempts Don José to release her from arrest by proposing to spend the evening at Lillas Pastia's tavern drinking manzanilla and dancing the seguidilla with her lover.

**Presentation of the Rose.** The name generally given to the scene in Act II of Strauss's *Der Rosenkavalier* as Octavian comes to present Sophie with a silver rose as token of Baron Ochs's wooing.

**Pressburg.** See BRATISLAVA.

**Prêtre, Georges** (*b* Waziers, 14 Aug. 1924). French conductor. Studied Douai, Paris with Cluytens and Dervaux. Début Marseilles 1946 (*Samson*). Marseilles 1946–8; Lille 1948–50; Toulouse 1951–4; Lyons 1955; Paris: OC, 1956–9 (*Mignon, Capriccio,* première of *Voix humaine*); O, 1959–71, music director 1970–1. Chicago 1959; San Francisco 1963–4; New York, M, from 1964. London, CG, 1965, 1976. Milan, S, 1965. Composed operetta *Pour toi* (1951). (R)

**Prévost, Antoine-François** (Abbé) (*b* Hesdin, 1 Apr. 1697; *d* Chantilly, 23 Nov. 1763). French writer. His *L'histoire du Chevalier Des Grieux et de Manon Lescaut* (1731), vol. vii of his *Mémoires d'un homme de qualité*, was the source of the following operas: Auber (1856); Kleinmichel (*Das Schloss de l'Orme*, 1883); Massenet (1884; a sequel, *Le portrait de Manon*, appeared in 1894); Puccini (1893); more distantly, Henze (*Boulevard Solitude*, 1952).

**Prey, Claude** (*b* Fleury, 30 May 1925). French composer. Studied Paris with Messiaen and Milhaud. His music is essentially theatrical, and he has explored a wide range of forms and media in the interests of making a bridge between stage and concert-hall. His most important work, *Les liaisons dangereuses* (1974), follows the epistolatory form of Laclos's novel, and makes use of an instrumental 'double' for each character who takes part in the action as the recipient of the letters.

**Prey, Hermann** (*b* Berlin, 11 July 1929). German baritone. Studied Berlin with Prohaska, Baum,

and Gottschalk. Début Wiesbaden 1952 (2nd Prisoner, *Fidelio*). Hamburg 1953; Vienna, S, and Berlin, SO, from 1956; Munich and Cologne from 1959; Bayreuth 1965–7, 1981–2, 1986; New York, M, from 1960; London, CG, from 1973; also Chicago, Buenos Aires, etc. An engaging performer with a warm, lyrical tone. Roles include Don Giovanni, Papageno, Wolfram, Beckmesser, Storch, Marcello. Also a fine Lieder singer. Autobiography, *First Night Fever* (London, 1986). (R)

**Preziosilla.** A gypsy girl (mez) in Verdi's *La forza del destino*.

**Priam.** In classical mythology, the King of Troy, husband of Hecuba and father of Hector, Paris, and Cassandra. He appears (bs) in Berlioz's *Les Troyens* and (bar) in Tippett's *King Priam*.

**Přibyl, Vilém** (*b* Náchod, 10 Apr. 1925; *d* Brno, 21 July 1990). Czech tenor. Studied Hradec Králové with Jakoubkova. Début Ústí nad Labem 1959 (Prince, Dvořák's *Rusalka*). Brno O from 1961 throughout his career; also Prague Nat. O. Edinburgh 1964 (with Prague); London, CG, from 1964; also Vienna, Barcelona, Holland, Vancouver, Scandinavia. Retired 1984. Roles included Admetus (*Alceste*), Florestan, Lohengrin, Don José, Radamès, Verdi's Otello, Turiddu, Shuisky, Laca, Boris (*Káťa Kabanová*), Sergey (*Lady Macbeth of Mtsensk*), Dalibor. Created Claudius (Horký's *The Poison of Elsinore*). Possessed a voice of great power and beauty, and was a highly expressive performer, outstanding in noble roles. (R).

**Price, Leontyne** (*b* Laurel, MS, 10 Feb. 1927). US soprano. Studied New York, Juilliard School, and with Page. Début Broadway 1952 (in Thomson's *Four Saints in Three Acts*). Toured Europe and Moscow as Gershwin's Bess, 1952–4. Chicago 1957; London, CG, and Paris, O, from 1958; Milan, S, 1960; New York, M, 1961–85, when she retired. Repertory included Poppea, Donna Anna, Tatyana, Manon Lescaut. Created Barber's Cleopatra. Her fine spinto soprano, with a burnished glow, shone especially in Verdi roles, e.g. Aida, Leonora (*Trovatore* and *Forza*), Amelia (*Ballo in maschera*). (R)

**Price, Margaret** (*b* Blackwood, 13 Apr. 1941). Welsh soprano. Studied London, TCM. Début WNO 1962 (Cherubino). London, CG, 1963, 1968, 1973, 1980, 1988; EOG 1967; Gly. from 1968; Paris, O, from 1973; Cologne from 1971; Munich from 1972; Paris, O, and Vienna from 1973. New York, M, 1985. Also Chicago, Salzburg, etc. With her full, limpid tone, fine technique, and musical phrasing, a singer of great quality. Roles include Donna Anna, Pamina, Fiordiligi, Countess (Mozart), Norma, Tatyana,

Amelia (*Boccanegra*), Desdemona (Verdi), Ariadne. (R)

**prière.** See PREGHIERA.

**Prigioniero, Il** (The Prisoner). Opera in prologue and 1 act by Dallapiccola; text by the composer, after Villiers de l'Isle-Adam's *La torture par l'espérance* (1883) and Charles Coster's *La légende d'Ulenspiegel et de Lamme Goedzak*. Broadcast RAI 1 Dec. 1949, with Lászlo, Colombo, Renzi. Prod. Florence, C, 20 May 1950, with Lászlo, Colombo, Binci, cond. Scherchen; New York, Juilliard School, 15 Mar. 1951, cond. Waldmann; London, SW, 27 July 1959, with Raisbeck, Young, Cameron, cond. Lovett.

Saragossa, *c*.1570. The mother (sop) describes a recurrent dream in which she is haunted by the ghost of Philip II, who is gradually transformed into the symbol of death. A prisoner (bs-bar) of the Inquisition is treated kindly by his gaoler (ten), who calls him 'brother', but is subjected to harsh torture. Finding that the gaoler has left the door unlocked, the prisoner escapes, passing monks who seem not to notice him, and arrives in a garden. However, as he reaches out to embrace a cedar, the arms of his gaoler, the Grand Inquisitor, surround him. Approaching the execution pyre, he realizes that hope and the illusion of freedom have been used as the ultimate instruments of torture.

**prima donna** (It.: 'first lady'). The name given to the leading female singer in an opera, or the principal soprano of an opera company. In 19th-cent. Italy it carried specific connotations regarding the nature of the role and the amount of music to be sung. See CONVENIENZE.

**Prima Donna.** Opera in 1 act by Benjamin; text by Cedric Cliffe. Prod. London, Fortune T, 23 Feb. 1949 with Perilli, Maclean, Hughes, James, cond. Benjamin; Philadelphia, 5 Dec. 1953.

Set in 18th-cent. Venice, the opera includes a comic duet between rival prima donnas 'La Filomela' and Olimpia, who try to outdo one another in a scene from *Ariadne desolata* much in the manner of *Le cantatrici villane*.

**Prima la musica e poi le parole** (First the Music, then the words). Opera in 1 act by Salieri; text by Casti. Prod. Vienna, Schönbrunn Palace, 7 Feb. 1786. Prod. in English, Brooklyn Coll., 18 Nov. 1967; London, St John's, 11 June 1978.

Vienna, 18th cent. A composer (bar) and a poet (bar) argue about an opera commissioned by their patron. The composer wishes the poet to write new texts for an old score he has discovered. They also disagree violently over the choice of performer for the principal female part, the poet selecting an actress, Tonina (mez), while the composer favours a prima donna, Eleonora

(sop). The issue is resolved by the women sharing the role, taking the comic and tragic portions respectively.

The opera presents in witty, often sarcastic dialogue, the age-old argument of the respective importance of the various artistic components of opera. The subject is also treated in Strauss's *\*Capriccio*.

**primo musico** (It.: 'first musician'). A term synonymous during the 18th cent. with \*primo uomo.

**primo tempo** (It.: 'first speed'). In 19th-cent. Italian opera, an additional movement in an aria scene, inserted before the \*cantabile.

**primo uomo** (It.: 'first man'). The title given to the principal male singer: in the 18th cent. this would have been a \*castrato, hence the alternative term **primo musico**. See also PRIMA DONNA.

**primo violino direttore** (It.: 'first violin director'). In 19th-cent. Italian opera-houses, the first violinist who directed the performance. See CONDUCTOR.

**Prince Igor** (Russ.: *Knyaz Igor*). Opera in prologue and 4 acts by Borodin; text by composer after a sketch by Stasov. Music completed by Rimsky-Korsakov and Glazunov. Prod. St Petersburg, M, 4 Nov. 1890, with Olgina, Dolina, Slavina, Melnikov, Koryakin, Stravinsky, Ugrinovich, cond. Nápravník; London, DL, 8 June 1914, with Kuznetsov, Petrenko, Andreyev, Shalyapin, cond. Steinberg; New York, M, 30 Dec. 1915 (in It.), with Alda, Perini, Amato, Didur, cond. Polacco.

Polovtsia, 1185. The opera tells of the capture of Prince Igor (bar), with his son Vladimir (ten), by the Polovtsians. Their leader, Khan Konchak (bs), entertains his captive like a royal guest, treating him to a display of oriental dances. He offers to let Igor go free if he promises not to fight the Polovtsians again. Igor refuses, but manages to escape, and rejoins his wife Yaroslavna (sop). Vladimir remains behind in the Polovtsian camp, having fallen in love with the Khan's daughter Konchakovna (mez), whom he is allowed to marry.

**Prinzregententheater.** See MUNICH.

**Prinz von Homburg, Der.** Opera in 3 acts by Henze; text by Ingeborg Bachmann, after the drama by Heinrich von Kleist (1821). Prod. Hamburg, S, 22 May 1960, with Fölser, Aarden, Melchert, Ruzdak, Blankenheim, Fliether, cond. Ludwig; London, SW, 1962, cond. Henze; Paris, with Frankfurt O, 1962; Düsseldorf 1964 (title-role rewritten for tenor, sung by Kaposy).

Fehrbellin and Berlin, 1675. The opera is set

in the 17th cent. Prince Friedrich von Homburg (bar), the General of Cavalry, is a poet and dreamer. Confused by reality and imagination, he leads a nearly disastrous charge in a battle, and is court-martialled. His beloved, Princess Natalie (sop), pleads with her uncle, the Elector of Brandenburg (ten), for Friedrich's life. When Friedrich refuses to beg for mercy on the grounds that the judgement was just, the Elector magnanimously pardons him.

**Prise de Troie, La.** See TROYENS, LES.

**Prisoners' Chorus.** See 'O WELCHE LUST'.

**Pritchard, John** (*b* London, 5 Feb. 1921; *d* Daly City, CA, 5 Dec. 1989). English conductor. Studied with his father. Gly.: répétiteur 1947; chorus-master 1949; assistant cond. 1950, début 1951 (*Don Giovanni*); cond. 1951–78; music counsellor from 1963; music director from 1969. London, CG, from 1952, incl. premières of *Gloriana* (1953), *Midsummer Marriage* (1955), and *King Priam* (1962, at Coventry). Vienna, S, 1952–3, 1964–5; Salzburg 1966. New York, M, from 1971. Music director San Francisco 1970–89, Cologne 1978–89, joint music director Brussels, M, 1981–9. A highly talented, versatile conductor, whose gift centred on Mozart but also embraced the Italian and French repertory and Strauss. His quick professionalism always enabled him to grasp new scores and to give sensitive support to singers and players, though it was on occasion not matched by a comparable energy. A well-loved figure especially at Glyndebourne, which both shaped his career and which, as George Christie wrote, was in later years significantly shaped by him. (R)

**Prize Song.** See 'MORGENLICHT LEUCHTEND'.

**Probe** (Ger.: 'trial, rehearsal'). Thus *Beleuchtungsprobe* (lighting rehearsal), *Hauptprobe (main rehearsal), *Generalprobe (final [dress] rehearsal). For *Das Rheingold* some German opera-houses call a *Schwimmprobe* for the Rhinemaidens. See also PROVA, RÉPÉTITION, SITZPROBE.

**Prodaná nevěsta.** See BARTERED BRIDE, THE.

**Prodigal Son, The.** Church parable in 1 act by Britten; text by William Plomer, after Luke 15: 11–32. Prod. Orford Parish Church, 10 June 1968, with Pears, Tear, Shirley-Quirk, Drake, cond. Britten; Caramoor Fest., Katonah, NY, 29 June 1969, with Lankson, Velis, Clatworthy, Metcalf, cond. Rudel.

The Father (bs-bar), his Elder Son (bar), and Younger Son (ten), work in the fields, but the Younger Son is lured by the Tempter (ten: the Abbot in the dramatic fiction) to demand his inheritance and leave for the city. Here he is robbed by Parasites, and eventually returns home, where he is welcomed by his father and in the end reconciled to his brother.

**producer** (Fr., *metteur en scène*; Ger., *Spielleiter*; It., *regista*. In Germany and elsewhere the term *Regisseur* is used, but *régisseur* in France refers specifically to those working on stage matters, e.g. lighting. In the USA the term general stage director is normally used.) The term for the artist responsible for the dramatic presentation of a performance. The English term 'production' is in Ger. *Inszenierung*, in It. *regia*, in Fr. *mise-en-scène*. The terms are comparatively modern, and did not appear in programmes regularly until the 1920s.

When opera was in its infancy, the dramatic presentation was controlled by the composer; but with the growth of spectacle as an aspect of opera, the designer played an increasing role. Still more important was the role of the ballet-master, who placed the chorus, generally symmetrically on either side of the stage or across the rear, and instructed its members in a limited amount of formal movement. He also controlled the principals: the stereotyped gestures still sometimes observed, recalling the conventions of mime, may derive from this. With dance an essential ingredient, Lully formed a strong set of production conventions with his ensemble of singing actors. The expressive grouping and movement of dance also influenced Gluck by way of the ideas of *Noverre. With the development of opera buffa in the 18th cent., the principal bass often took over the stage direction, a convention that survived during the 19th (as with Tagliafico at Covent Garden, London) and even into the 20th cent. Mozart was one of the first musical stage directors who not only rehearsed his operas from the pit, but controlled his singers. He was followed in Germany by Spohr and especially Weber, who instructed his singers in the drama before rehearsing them in the music, and conducted the performance from a position close to the footlights. In Italy Verdi normally supervised every detail of the production. It was Wagner who worked most fully towards the *Gesamtkunstwerk*, or unified work of art, and used his considerable dramatic skills to rehearse his singers in every detail of gesture, movement, and expression. His influence increased the emphasis on opera as drama, whether by musicians (e.g. especially Mahler in Vienna, also Toscanini in Milan, later Klemperer in Germany and Karajan in Austria) or with men of the theatre establishing didactic principles and forming special ensembles (e.g. *Stanislavsky in Russia) or making individual contributions as producers. Wagner's pre-eminence reasserted itself in post-war Bayreuth with his grandson Wieland. Actors and men of the theatre who became distinguished opera producers include Ernst von Possart, Max Reinhardt, Carl Ebert, Günther Rennert, Walter Felsenstein, Jean Vilar, Jean-Louis Barrault, Jean Cocteau, Tyrone Guthrie, Peter Hall, and Sam Wanamaker; designers include Jean-Pierre Ponnelle and Filippo Sanjust; film-makers include Luchino Visconti and Franco Zeffirelli; ballet-masters include Margherita Wallmann and Maurice Béjart; singers include Tito Gobbi, Hans Hotter, Regina Resnik, and Geraint Evans.

The rise in importance of the producer has led to the widely held view that after the Age of the Singer, then the Age of the Conductor, opera entered the Age of the Producer. This in turn has led to the term 'producer's opera', used (generally pejoratively) when the producer's ideas are felt to have been wished upon an opera rather than to have arisen from it.

**Prohaska, Jaro** (*b* Vienna, 24 Jan. 1891; *d* Munich, 28 Sep. 1965). Austrian bass-baritone. Studied Vienna. Début Lübeck 1922. Nuremberg 1925–31; Berlin, S, 1931–53; Bayreuth 1933–44; Paris, O, 1936; Buenos Aires 1935, 1937. A notable Wotan, Sachs, Amfortas, Ochs. Pupils included Hermann Prey. (R)

**Prokofiev, Sergey** (*b* Sontsovka, 27 Apr. 1891; *d* Moscow, 5 Mar. 1953). Russian composer. Studied Sontsovka with Glière, by when he had already composed two operas, adding to these a third before studying St Petersburg with Lyadov and Rimsky-Korsakov; he was helped more, however, by Myaskovsky, and was also influenced by Tcherepnin, with whom he studied conducting. *Maddalena* to some extent anticipates both *The Gambler* and *The Fiery Angel* in its treatment of male obsession for a woman who exercises an influence both benign and malign. However, his first mature opera is in fact *The Gambler*, consciously anti-Wagnerian and both a comedy and a study of obsession: though conversational in technique, it first fully revealed Prokofiev's lyricism, his ability to characterize sharply, and his fascination with abnormal or extreme states of mind.

Leaving Russia after the Revolution, Prokofiev began work on *The Love of the Three Oranges* en route for America, where initial success was gradually replaced with frustration. The new opera, composed for Chicago, ran into difficulties, though not for its quality. Impressed by *Meyerhold's theories, which included a diminution of the role of the actor and a challenge to conventional audience relationships, Prokofiev based the work on a play by Gozzi (then popular in Russian literary circles). It represents a change of course: the treatment is anti-Romantic, and

the fantastic comedy provides the composer with the opportunity for some of his most pungent and wittily grotesque, also wry and tender, music, drawing much stimulus from the surrealist interaction of different planes of reality. Career difficulties impelled him to leave for Paris; he returned to the USA for the opera's première, retreating again when hopes faded for the production of his next opera, *The Fiery Angel*. Further difficulties surrounded the work, the opera Prokofiev not without reason thought his best. The fantastic strain wittily indulged in the previous opera is here turned to horrific ends. Based on a story of possession and sorcery, the work has its grotesque elements, but is openly Romantic and even Expressionist in its violence and its exploration of dark emotions and sinister experiences in music of unprecedented power. At its centre is the largest and most powerful role Prokofiev ever wrote, for the doomed heroine Renata. The problems over its production led him to develop some of its ideas in his Third Symphony.

By 1932 Prokofiev was considering returning to Russia, though he did not finally do so until 1936. This was the time of maximum Communist repression of artistic initiative. At first the constraints did not oppress him personally, though he had been attacked for his two previous operas' qualities of parody and Expressionism: he was not opposed to everything in Socialist Realism, but the official welcome given him as a returning prodigal was only partially extended to *Semyon Kotko*. Nor was the work more widely admired. It has the broad approach and epic ambitions that make it perhaps Prokofiev's nearest approach to Grand Opera, by way of Glinka (there is even a Meyerbeerian dedication of the partisans' weapons): the grandiose gestures attempting to express optimistic Communist emotions are predictably less successful than the lyrical love music and colourful ensembles that are also to be found.

However, Socialist Realism is abandoned in *The Duenna* (also known as *Betrothal in a Monastery*). The simplicity here is not artificial, and the composer responds unaffectedly with some of his most charmingly accomplished music. With *War and Peace* he returns to a grand manner that again acknowledges the example of Glinka (in particular, *A Life for the Tsar*), and includes some fine scenes of battle and desolation, sometimes recalling his expertise as a film composer; there is also some memorable handling of the 'peace' element in the plot in the detailed and lyrical portrayal of the Bolkonsky family and the love of Natasha and Andrey. Yet it was one of the works singled out by the notorious 1948 Zhdanov tribunal in the attack on Russia's leading composers for 'formalism' and other alleged errors. It is a remarkable achievement, with carefully opposed musical manners that succeed in contrasting the public and national with the private and lyrical in a way not representative of the novel. Its complete Russian rehabilitation was signalled by its choice, for the first time since 1842, as a work to open the Bolshoy season in place of Glinka's *A Life for the Tsar*. In the wake of the Zhdanov tribunal, Prokofiev wrote *The Story of a Real Man*, but failed to satisfy even the authorities it was designed to appease.

WORKLIST: *Velikan* (The Giant) (?Prokofiev; comp. 1900, unprod.); *Na pustynnykh ostrovakh* (Desert Islands) (?Prokofiev; incomplete, comp. 1900–2, unprod.); *Pir vo vremya chumy* (A Feast in Time of Plague) (?Prokofiev, after Pushkin; comp. 1903, unprod.); *Undina* (Kilstett, after Fouqué; comp. 1904–7, unprod.); *Maddalena* ('Baron Lieven' (Lieven-Orlova); comp. 1911–13, orch. Downes, perf. BBC 1979); *Igrok* (The Gambler) (Prokofiev, after Dostoyevsky; comp. 1917, rev. 1928, prod. Brussels, 1929); *Lyubov k tryem apelsinam* (The Love of the Three Oranges) (Prokofiev, after Gozzi; Chicago 1921); *Ognenniy angel* (The Fiery Angel) (Prokofiev, after Bryusov; comp. 1923, rev. 1927, prod. Paris, 1954); *Semyon Kotko* (Katayev & Prokofiev, after Katayev; Moscow 1940); *Obrucheniye v monastyre* (Betrothal in a Monastery: The Duenna) (Prokofiev & Mendelson, after Sheridan; comp. 1941, prod. Leningrad 1946); *Khan Buzay* (?; incomplete, comp. 1942); *Voyna i mir* (War and Peace) (Prokofiev, after Tolstoy; comp. 1943, rev. 1946–52, concert perf. Moscow 1944, part prod. Prague 1948, Leningrad 1946, fuller prod. Florence 1953, Leningrad 1955)); *Povest o nastoyashchem cheloveke* (The Story of a Real Man) (Prokofiev & Mendelson, after Polevoy; private concert perf. Moscow 1948, prod. Moscow 1960); *Dalyokiye morya* (Distant Seas) (Prokofiev, after Dikhovichny; incomplete, comp. from 1948, unprod.).

WRITINGS: *Avtobiografiya* (Moscow, 1973).

BIBL: I. Nestyev, *Prokofiev* (Moscow, 1957, 2/1973, trans. 1961).

**prologue.** The introductory scene of an opera, in which one or more characters often gives a brief summary of the action. A staple feature of Italian operas up to the reforms of *Metastasio, the prologue usually employed mythological or allegorical figures: classic examples are the appearance of 'Music' at the beginning of Monteverdi's *Orfeo*, and the figures of 'Fortune', 'Virtue', and 'Love' which introduce his *Poppea* as a contest between their conflicting influences on the actions of the main characters. In 17th-cent. French opera the prologue, which was a lavishly staged and substantial movement, also used allegory, but with the more important task of reflecting the *gloire* of the French court. To do this, it was often set in the present day rather than the mythological era of the opera proper, as in Lully's *Thésée*. In Rameau the prologue gradually assumed a more

generally allegorical function and, after *Zoroastre* (1749), was sometimes dropped completely. After the 18th cent. it makes only rare appearances: a famous example is the prologue to Leoncavallo's *Pagliacci* (1892), which had an important influence on the rise of *verismo.

**prompter** (Fr., *souffleur*; Ger., *Souffleur*; It., *maestro suggeritore*). Whereas in the theatre the prompter only intervenes when the actor forgets his words or cue, in the opera-house the prompter helps the performers by giving them the opening words of every phrase a few seconds in advance. Especially in Italy, he may also act as a sub-conductor, relaying the conductor's beat, which he picks up from either a mirror or nowadays more commonly a TV monitor. Extra prompters are sometimes concealed on the stage, e.g. lying under Tristan's couch. The prompter is normally a member of the theatre music staff. One makes a rare stage appearance in Strauss's *Capriccio* in the person of M. Taupe ('Mr Mole').

**prompter's box** (Fr., *trou de souffleur*; Ger., *Souffleurkasten*; It., *buca*). The compartment containing the prompter, usually situated immediately before the footlights and covered over so as to conceal its occupant from the audience. Normally unobtrusive, but on occasion made part of the décor, even of the 'business' of comic opera.

**Prophète, Le.** Opera in 5 acts by Meyerbeer; text by Scribe. Prod. Paris, O, 16 Apr. 1849, with Viardot, Castellan, Roger, N. Levasseur, cond. Girard; London, CG, 24 July 1849, with Viardot, Hayes, Mario, Tagliafico, cond. Costa; New Orleans, T d'Orléans, 1 Apr. 1850, with R. Devriès, Tabon-Bessin, Duluc, Scott, cond. E. Prévost. The opera is based on an historical episode during the Anabaptist rising in Holland in the 16th cent.; the real John of Leyden was Jan Neuckelzoon, born 1509, who had himself crowned in Münster in 1535.

The plot concerns the love of John of Leyden (ten) for Bertha (sop). Count Oberthal (bar), ruler of Dordrecht, desires her for himself and has her abducted; she escapes back to John, but Oberthal threatens to kill Fidès (mez), his mother, if Bertha is not returned to him. John's loyalty to his mother prevails, but in revenge he is now ready to join the Anabaptists, whose incitement to revolt he has previously rejected. He is proclaimed a prophet of God, leads a bloody rebellion, and is crowned Emperor. Fidès, now reduced to poverty, and believing her son murdered by the Prophet, recognizes him at the coronation; he compels her to deny that she is his mother, however, fearing the wrath of his followers, who consider him divine. He meets her in secret, and she persuades him to repent of his deeds. They are joined by Bertha; she, having escaped from Oberthal once more and learning of John's 'death', has set fire to the palace to destroy the Prophet who she believes has killed her lover. A captain arrives to tell John that he is betrayed by the Anabaptists. Discovering that the Prophet and John are the same man, Bertha kills herself. John and Fidès choose to die with the Anabaptists in the destruction of the palace.

**Proserpine** (Gr., Persephone). In classical mythology, the daughter of Jupiter (Zeus) and Ceres (Demeter) who was carried off by Pluto (Hades) to become his wife and rule over the shades. Ceres, goddess of agriculture, did not allow the earth to produce until Mercury (Hermes) had rescued Proserpine. Even after her rescue she was bound to return to the underworld for a third of the year: she thus represents the seed-corn, hidden in the ground and bursting forth to nourish men and animals. Operas on her are by Monteverdi (1630, lost), Sacrati (1644), Gallicano (1645), Ferrari (1645), Lully (1689), Stuck-Batistin (1714), Kraus (1781), Asioli (1784), Da Silva (1784), Cimador (1791), Paisiello (1803), Winter (1804), Berlyn (1842), and Saint-Saëns (1887).

**Protagonist, Der.** Opera in 1 act by Weill; text by Kaiser. Prod. Dresden 27 Mar. 1926, cond. Busch; Santa Fe, 3 July 1993. Weill's first significant work for the stage.

**protesta** (It.: 'protest, declaration'). The affirmation printed in many 17th- and 18th-cent. Italian librettos to the effect that, while employing such figurative terms as 'Fate' or other pagan usages, the author was a faithful Roman Catholic; this was often necessary, especially in the Papal States, so as to obtain the Church's imprimatur.

**prova** (It., 'rehearsal'). In normal Italian usage, a session of preparation preceding the performance of a work. In a 'prova all'italiana', performers and orchestra rehearse without scenery. A rehearsal with scenery, usually accompanied by piano, is the 'prova di scena'. The pre-dress rehearsal is the 'anti-generale'. The dress rehearsal is the 'prova generale'; this is often virtually a trial performance, without interruptions, sometimes before an invited audience or (with a new work) even members of the Press. See PROBE, REHEARSAL, RÉPÉTITION.

**Provenzale, Francesco** (b Naples, c.1626; d Naples, 6 Sep. 1704). Italian composer. Studied Naples with Sabino; held various appointments as maestro di cappella there, the last at the royal chapel. Wrote c. nine operas for Naples of which only two survive, *Il schiavo di sua moglie* and *La Stellidaura vendicata*. These reveal a gift for expressive melody combined with an effective use of harmonic resources. Though reared in the Venetian tradition, his style in part prefigures that of his pupil A. Scarlatti, which has led some to describe him, erroneously, as the father of Neapolitan opera.

SELECT WORKLIST: *Il schiavo di sua moglie* (His Wife's Slave) (Perrucci; Naples 1671); *La Stellidaura vendicata* (Perrucci; Naples 1674).

**Pryanishnikov, Ippolit** (b Kerch, 26 Aug. 1847; d Petrograd, 11 Nov. 1921). Russian baritone and producer. Studied St Petersburg with Corsi, Milan with Ronconi. Début Milan 1876 (*Maria di Rohan*). St Petersburg, M, 1878–86. Singer, producer, and manager, Tiflis O, 1886–9; Kiev 1889–92; Moscow 1892–3, with own company, the Private Opera Partnership. Tchaikovsky, who admired and wished to help him, conducted Gounod's *Faust*, *The Demon*, and *Onegin* for him. Created Lionel (*Maid of Orleans*) and Mizgir (*Snow Maiden*); also sang Mazeppa. Productions included *Stone Guest*, *May Night*, *Pagliacci*. Taught Figner, Mravina, and Slavina, among others.

**Puccini, Giacomo** (b Lucca, 22 Dec. 1858; d Brussels, 29 Nov. 1924). Italian composer. Studied Lucca with his father (a pupil of Donizetti and Mercadante) and his uncle, then formally with Angeloni, later Milan with Bazzini and Ponchielli.

In 1883 the latter urged him to enter the *Sonzogno competition: despite failure, *Le villi* sufficiently impressed Boito and Ricordi for them to arrange for a performance in 1884. Its success confirmed Ricordi's admiration and led to his commission for *Edgar*, which failed in Milan in 1889. For all their promise, both works suffer from awkward librettos and show unevennesses that include the intermittent influence of Ponchielli. For *Manon Lescaut* he chose his own subject, in spite of the success of Massenet's opera, and five librettists eventually contributed to what was his first mature opera and the work that led Shaw to proclaim him as the heir to Verdi's crown. Some infelicities, notably heavy scoring, are outweighed by the wealth of melodic ideas, the sensitive musical characterization, above all a flexible approach to musical and dramatic structure.

Puccini's closeness to *verismo, evident here, is confirmed in his next opera, *La bohème*. Despite his greater gifts as a melodist and his broader, symphonic conception of operatic form, much in Puccini's style here (such as choice of realistic subject-matter and exploitation of sharply conflicting emotions) sets him alongside the verismo composers in their appeal to the prevailing taste of Italian audiences. Initially the work had a lukewarm reception, many considering it a retrograde step; but the combination of a concise musical structure (making skilful personal use of motive) with a lyrical melodic style and a vivid sense of characterization, which reached full fruition in *La bohème*, is carried forward in the next two works.

Turning, after some hesitation and with numerous libretto revisions, to *Tosca*, Puccini employed many of his by now familiar technical devices in a darker, still naturalistic, melodrama. Similarly, while *Madama Butterfly* inhabits another radically different emotional world, it is closely allied to its two predecessors in general approach. Yet problems with the libretto led to many difficulties in the work's composition, and the first night in 1904 was a disaster; only on its third revision did the opera achieve proper dramatic integration, with the heroine firmly at the centre of the action. Considering and discarding many ideas for his next opera, Puccini eventually settled on *La fanciulla del West*. Progress was slow, however, not least through his traumatic marital problems in 1908, and the work was not ready until 1910. At first sight, he seemed merely to be supplanting the Japanese flavour of *Butterfly* with another exoticism: however, in its musical construction, *La fanciulla del West* is a major advance. It paves the way for the later operas in its absorption of aria into a more fluent style in which all elements are freely mixed in the interests of sharper dramatic characterization. A public which had expected a simple repeat of effective formulas might well have been disappointed; but the work triumphed when it became the first opera to receive its world première at the New York Metropolitan in 1910.

Once again considering and discarding many ideas for librettos, Puccini was next guided by a commission from Vienna for an operetta; agreeing on a comic opera, he wrote *La rondine*, which despite initial success has failed to hold its place in the repertory. He returned now to an earlier idea for a triptych of operas. The concision of the 1-act structure had always appealed to verismo composers for the enhancement of emotional conflict by its compression into a short time-scale. A visit to Paris in 1912 led to the source of the strongly veristic *Il tabarro*; this was quickly followed by the all-female *Suor Angelica*, finally a comic foil to these two contrasting tragedies with *Gianni Schicchi*. A chance conversation next led to *Turandot*, but the now familiar drawn-out problems with constructing the libretto delayed work, and Puccini died from cancer of the larynx before he could quite complete the score. Though there is a return to the orientalism of *Madama Butterfly*, his style had meanwhile undergone significant changes, absorbing a number of contemporary influences including

that of Debussy. At Toscanini's suggestion, Franco Alfano composed the final scene.

Puccini's immediate model was Verdi, especially the later operas in which the formal divisions between aria, recitative, and ensemble are dissolved. However, the influence of Wagner is also marked, since Puccini turned to his own account the practice of linking action to music by use of motive. Wagner also provides a point of departure for his rich harmonic language and vivid orchestration, which continues its development up to and including *Turandot*. While the bel canto tradition remains influential, and marks many of the famous set-piece arias, Puccini's ultimate goal was a flexible structure in which all the elements would be subordinated to the drama. In this respect he mirrors the development of his near-contemporary Richard Strauss.

Puccini's innate feel for dramatic situation was somewhat clouded by his fondness for absurd or trivial plots: of these, perhaps *Madama Butterfly*, with its faintly mawkish overtones, and *La fanciulla del West*, a somewhat sub-Wagnerian treatment of redemption in a mining camp, are the most dubious. Objections have also been levelled at his sentimental-cruel handling of the suffering heroine, and his inability to portray happy, fulfilled love. But such criticism must defer to his powers of musical characterization, at their most vivid in *Tosca*, where it is achieved in an intense, almost violent fashion, and *Turandot*, where the heroine's personality is passionately illuminated musically by comparison with the pale Gozzi original. His care for the text is seen in the extensive revisions he demanded of his librettists: Zangarini claimed that Puccini should really be credited among the authors of *La fanciulla del West*. His concern to evoke atmosphere, as in the orientalisms of *Madama Butterfly* and the quasi-Americanisms of *La fanciulla del West*, led him to much exploratory work, such as a visit to the Castel Sant'Angelo to listen to the bells in preparation for *Tosca*. Even in his own time, these conflicting elements aroused contradictory views of his talent, not least in his own country, where critics deplored his decadence even as the public flocked to the opera-house. He remains a controversial figure, whose evident faults are readily forgotten in the overwhelming appeal of his works in the theatre.

WORKLIST: *Le villi* (The Willis) (Fontana; Milan 1884, rev. Turin 1884); *Edgar* (Fontana, after Musset; Milan 1889, rev. Ferrara 1892, rev. Buenos Aires 1905); *Manon Lescaut* (Leoncavallo, Praga, Oliva, Illica, & Giacosa, after Prévost; Turin 1893); *La bohème* (Bohemian Life) (Giacosa & Illica, after Murger; Turin 1896); *Tosca* (Giacosa & Illica, after Sardou; Rome 1900); *Madama Butterfly* (Giacosa & Illica, after Belasco, after Long; Milan 1904, rev.

Brescia 1904, rev. London, 1905, rev. Paris 1906); *La fanciulla del West* (The Girl of the [Golden] West) (Civini & Zangarini, after Belasco; New York, 1910); *La rondine* (The Swallow) (Adami, after Willner & Reichert; Monte Carlo 1917); *Il Trittico* (The Triptych) (New York 1918), comprising *Il tabarro* (The Cloak) (Adami, after Gold), *Suor Angelica* (Sister Angelica) (Forzano), and *Gianni Schicchi* (Forzano, after Dante); *Turandot* (unfin., completed by Alfano) (Adami & Simoni, after Gozzi; Milan 1926).

BIBL: M. Carner, *Puccini* (London, 1958, 2/1974); C. Osborne, *The Complete Operas of Puccini* (London, 1981); W. Ashbrook, *The Operas of Puccini* (rev. edn. Ithaca, 1985).

**puntatura** (It.: 'pointing'). In 19th-cent. Italian opera, the practice whereby a singer's line would be refashioned to suit his or her register not by *transposition, but by altering the melody without changing the underlying harmony.

**puppet opera** (Fr., *opéra de marionettes*; Ger., *Puppenoper* or *Marionettenoper*; It., *opera delle marionette*). The popularity of puppet theatres, and the virtuosity of some puppeteers, has occasionally led managers to present opera. The first première given by puppets was probably Ziani's *Damira placata* (Venice 1680). Haydn wrote five puppet operas for Eszterháza, *Philemon und Baucis* (1773), *Hexenschabbas* (?1773, lost), *Dido* (?1776), *Die Feuerbrunst* (?1775–8), and *Die bestrafte Rachbegierde* (?1779). After his retirement, the castrato Guadagni set up a theatre in Padua with himself as puppet-master; his most famous performance was of Gluck's *Orfeo*, in which he sang the title-role he had once created. Falla's *El retablo de Maese Pedro* (1923) requires a puppet theatre which performs to live singers; among the latter is Don Quixote, who becomes confused about their reality and attacks them.

**Purcell, Henry** (*b* London, 1659; *d* London, 21 Nov. 1695). English composer. Although he wrote only one stage work which may properly be described as an opera, *Dido and Aeneas*, Purcell composed much dramatic music. He contributed music to 44 plays, mostly comedies, mainly between 1690–5. For some works, e.g. Shadwell's *Epsom Wells* (1693) and Dryden's *Love Triumphant* (1694), he composed only one song, though for Shadwell's adaptation of *Timon of Athens* (1694) and the first three parts of D'Urfey's *The Comic History of Don Quixote* (1694–5) his contribution was extensive, embracing a quantity of instrumental and vocal music.

Four of the stage works for which Purcell provided music use it in such an integrated or extensive way that they have come to be known as semi-operas, after first being so described by Roger North. Apart from *King Arthur*, this term is misleading, for usually the musical contribu-

tion merely forms a self-contained episode surrounded by dialogue, as with the song 'I Attempt from Love's Sickness to Fly' from *The Indian Queen*. However, in such passages as the Incantation Scene of the same work Purcell uses all the techniques of operatic music, including recitative and aria styles. Only in *King Arthur* is the musical material so extensive and continuous that the work approaches opera. The authenticity of the music for *The Tempest*, formerly counted among the semi-operas, is now questioned: it was probably written by John *Weldon.

*Dido and Aeneas* was composed for performance at a girls' school in Chelsea. Earlier attempts to found a tradition of English opera had been largely unsuccessful, and the fact that Purcell wrote no other opera after *Dido and Aeneas* suggests that it would be wrong to view it as anything other than an isolated masterpiece. Although it comprises three acts, *Dido and Aeneas* is cast on a small scale and makes modest demands on the performers, many of whom would probably have been pupils at the school. Its musical style betrays the influences of both French and Italian opera: the overture, use of the chorus, and intervention of the supernatural (in the shape of the witches) are strongly reminiscent of Lully, while the recitative, whose contours are uniquely moulded by the English language, owes much to the Italian monodic style. Even Dido's poignant final lament, 'When I am Laid in Earth', would be unthinkable without the model of Cavalli's chromatic ostinato bass laments.

With Purcell's early death and the invasion of Italian opera composers around the turn of the century, *Dido and Aeneas* came to occupy a unique place as the first surviving English opera. In its day it was largely without influence: it is as a rejuvenating force on modern English composers, especially Britten and Tippett, that it has made its greatest impact.

WORKLIST (Opera and Semi-opera: all first prod. London): *Dido and Aeneas* (Tate; 1689). *The Prophetess, or the History of Dioclesian* (Tate; 1690); *King Arthur, or the British Worthy* (Dryden; 1691); *The Fairy Queen* (Shakespeare, ad. ?Settle; 1692); *The Indian Queen* (Dryden, ad. Howard; 1695).

BIBL: J. A. Westrup, *Henry Purcell* (London, 1937, 4/1980); C. Price, *Henry Purcell and the London Stage* (London, 1984).

**Puritani, I.** Opera in 3 acts by Bellini; text by Pepoli, after the play *Têtes rondes et Cavaliers* (1833) by Ancelot and Saintine, in turn loosely derived from Scott's novel *Old Mortality* (1816). Prod. as *I Puritani e i Cavalieri* Paris, I, 24 Jan. 1835, with Grisi, Rubini, Tamburini, Lablache (who were then dubbed the 'Puritani' Quartet); London, H, 21 May 1835, with same cast; New

577

Orleans, American T, 17 Apr. 1843 with Corsini, Perozzi, Calvet, Valtellina.

The opera is set in Plymouth at the time of the Civil War. Queen Henrietta (mez), widow of Charles I, is held prisoner in a fortress whose warden is the Puritan Lord Walton (bs). His daughter Elvira (sop) is in love with Lord Arthur Talbot (ten), a Cavalier, and permission for their marriage has been obtained, though she is also loved by the Puritan Sir Richard Forth (bar). Arthur helps Henrietta to escape by dressing her in Elvira's bridal veil. Elvira, believing herself betrayed, loses her reason. Arthur, now in danger from the Puritans both for his political crime and as the agent of Elvira's madness, returns to her, refusing to desert her although his capture will mean certain death. He is seized by his pursuers, and about to be executed when news arrives of the defeat of the Stuarts, and a pardon for all prisoners. The joyful shock of being reunited to Arthur restores Elvira's sanity.

**Pushkin, Alexander** (*b* Moscow, 26 May 1799; *d* St Petersburg, 29 Jan. 1837). Russian poet and author. Operas on his works are as follows:

*The Triumph of Bacchus* (1818): Dargomyzhsky (1848)

*Ruslan and Lyudmila* (1820): Glinka (1842)

*The Captive of the Caucasus* (1821): Alyabyev (unfin., *c.*1820); Cui (1883); Popova (1929)

*The Fountain of Bakhchiserai* (1823): Měchura (*Marie Potocka*, 1871); Fyodorov (1895); Zubov (1898); Ilyinsky (1899); Parusinov (1912); Krylov (1912); Arkhangelsky (1915); Smetanin (*Khan Girey*, 1935); Shaposhnikov (1940)

*The Gypsies* (1824): Wielhorski (1838); Kashperov (1850); Lishin (1876, unfin.); Morosov (1892); Konius (1892); Rakhmaninov (*Aleko*, 1893); Zubov (1894); Juon (1896); Sacchi (1899); Ferretto (1899); Mironov (?1900); Shäfer (1901); Siks (1906); Galkauskas (1908); Leoncavallo (1912); Shostakovich (lost); Kalafaty (1941); Shakhmatov (1949); Snatokov (1952)

*The Bridegroom* (1825): Blüm (1899)

*Count Nulin* (1825): Lishin (1876); Zubov (1894); Strelnikov (1938); Koval (comp. 1949, unfin.)

*Boris Godunov* (1825): Musorgsky (1869, 1874); historical subject also set by Mattheson (1710)

*Scene from 'Faust'* (1825): Asafyev (comp. 1936)

*The Negro of Peter the Great* (1827, unfin.): Arapov (*The Frigate Pobeva*', comp. 1958); Lourié (1958)

*Eugene Onegin* (1828): Tchaikovsky (1879)

*Poltava* (1829): Wietinghoff-Scheel (*Mazeppa*, 1859); Sokalski (1859, unfin.); Du Bois (late 19th cent.); Tchaikovsky (*Mazeppa*, 1884)

*The Snowstorm* (1830): Dzerhinsky (*Winter Night*, 1946)

*Mistress into Maid* (1830): Larionov (1875); Zajc

(*Lizinka*, 1878); Ekkert (1911); Spassky (1923); Biryukov (1947); Kovner (1948); Dukelsky (1958)

*The Tale of the Parson and his Man Balda* (1830): Karagichev (1931); Bakalov (comp. 1937)

*The Little House at Kolomna* (1830): Solovyov (1899, unfin.); Stravinsky (*Mavra*, 1922)

*The Shot* (1830): Strassenburg (1936)

*The Post Master* (1830): Kryukov (1940); Reuter (*Postmeister Wyrin*, 1947)

*The Undertaker* (1830): Yanovsky (1923); Admoni-Krasny (1935)

*A Feast in Time of Plague* (1830): Cui (1901); Prokofiev (comp. 1903); Rechmensky (1927); Lourié (1933); Tarnopolsky (1937); Dodonov (1930); Asafyev (1940); Goldenweiser (1942)

*The Stone Guest* (1830): Dargomyzhsky (1872); Biryukov (1941); Malipiero (1963)

*Mozart and Salieri* (1830): Rimsky-Korsakov (1898)

*The Miserly Knight* (1830): Rakhmaninov (1906); Kryukov (1917)

*Roslavev* (1831): Bargrinovsky (1812, comp. 1927)

*The Tale of Tsar Saltan* (1831): Rimsky-Korsakov (1900); Nikolsky (1913); Nasedkin (1950s)

*Rusalka* (1832): Dargomyzhsky (1856); De Maistre (1870); Alexandrov (1913); Stepanov (1913)

*Angelo* (1833): Kuznetsov (1894)

*Dubrovsky* (1833): Nápravník (1895); Napoli (1973)

*The Tale of the Fisherman and the Fish* (1833): Polovinkin (1934); Gibalin (1936); Parchomenko (1936); Manukyan (1937); Croses (1942)

*The Bronze Horseman* (1833): Popov (1937); Asafyev (1942)

*The Dead Princess* (1833): Vessol (1909); Mittelstädt (1913); Krasev (1924); Weisberg (1936); Emelyanova (1939); Kotilko (1946); Chernyak (1947); Tsybin (1949)

*The Queen of Spades* (1834): Halévy (1850); Suppé (1865); Tchaikovsky (1890)

*The Captain's Daughter* (1834): Cui (1911); Katz (1941); Kryukov (1944)

*The Golden Cockerel* (1834): Drasev (1907); Rimsky-Korsakov (1909)

Also: Kreitner: *The Death of Pushkin* (1937, unfin.: uses settings of seven Pushkin poems); Shekhter: *Pushkin in Exile* (1958). Prokofiev planned an opera on Pushkin and Glinka. Opera on Pushkin and his poems by V. Kobekin (1988).

**Putnam, Ashley** (*b* New York, 10 Aug. 1952). US soprano. Studied Michigan with Mosher and Patterson. Début Virginia O 1976 (Lucia). New York City O 1978; Gly. from 1978; New York, tour 1983; London, CG, 1986; also Aix, Amsterdam, Cologne, Brussels, Santa Fe, St Louis, Philadelphia etc. Roles incl. Fiordiligi, Queen of the Night, Maria Stuarda, Gilda, Violetta, Musetta, Hanna Glawari (*Merry Widow*), Marschallin, Arabella, Danae. (R)

**Pyne, Louisa** (*b* ?, 27 Aug. 1832; *d* London, 20 Mar. 1904). English soprano. Studied London with Smart. Début Boulogne 1849 (Amina). London: DL, HM, CG, 1851; US tour 1854–6. In 1856, she and *Harrison formed the Pyne–Harrison Co., with which she appeared first at London, L and DL, then CG, 1858–64, creating many roles in operas by Balfe, Wallace, Benedict, and Glover. Known as 'the English Sontag', she also sang an impressive Queen of the Night and Arline (*Bohemian Girl*). She retired on her marriage to the baritone Bodda in 1868.

# Q

**Quadrio, Francesco Saverio** (*b* Ponte in Valtellina, 1 Dec. 1695; *d* Milan, 21 Nov. 1756). Italian author and theorist. Vols ii and iii of his 7-vol. *Della storia e della ragione d'ogni poesia* (1749) discuss opera, oratorio, and cantata. Probably the first theorist to recognize the relationship between poetry and music in the theatre. His ideas were put into practice by *Goldoni. Provided valuable comments on many of the singers active at the time, especially the castrati.

**Quaglio**. Italian, later German, family of designers active from the early 18th to the 19th cents. Generally working in pairs, they made an important contribution to the visual aspect of German Romantic opera. Of the six generations, 15 members of the family were active as designers.

1. **Giulio I** (*b* Laino, 1601; *d* Vienna, after 1668). A fresco painter who occasionally worked for the stage.

2. **Giulio II** (*b* Laino, 1668; *d* Laino, 3 July 1751). Son of (1). Not active in the theatre.

3. **Giulio III** or **Giovanni Maria I** (*b* Laino, *c.*1700; *d* Vienna, 1765). Known as G. Quaglio, he was the son or nephew of (2). Designed operas for Vienna, B and K, 1748–65, making considerable use of transparent backdrops, and responding to the reform movement in his designs for Gluck's *Cinesi* (1754), *Orfeo* (1762), and *Telemaco* (1765).

4. **Carlo** (*b* ?; *d* ?). Son of (3), with whom he worked in Vienna 1762–5. Warsaw 1765–7.

5. **Lorenzo I** (*b* Laino, 23 July 1730; *d* Munich, 2 May 1805). Son of (3). Worked in Mannheim and Schwetzingen from 1758, designing works by Salieri, Galuppi, Hasse, and Paisiello. Frankfurt, Dresden 1768–9; designed new Ducal T at Zweibrücken 1775–6, and new Schlosstheater at Mannheim (with Alessandro Bibiena) 1777. Munich from 1778; designed sets for première of *Idomeneo* (1781) and for Singspiels by Hiller, Schweitzer, and Holzbauer. His designs are preserved in the T Museum, Munich, and in the Uffizi Gallery, Florence.

6. **Martin** (*b* ?; *d* ?). Son of (3). Worked at Mannheim with his brother (5), later at Kassel.

7. **Domenico I** (*b* Laino, 1708; *d* Laino, 1773). Son of ? (2). Worked in Milan, Salzburg, Vienna.

8. **Giuseppe** (*b* Laino, 2 Dec. 1747; *d* Munich,

2 Mar. 1828). Son of (7). Second designer Mannheim and Schwetzingen, also worked at Ludwigsburg, Speyer, Frankfurt. Designed Redoutensaal in Mannheim. Munich from 1779: designs include those for *Don Giovanni*, *Magic Flute*, *Freischütz*, developing a more consciously Romantic approach.

9. **Giulio IV** (*b* Laino, 1764; *d* Munich, 21 Jan. 1801). Son of (7). Worked Munich, Mannheim, Zweibrücken, Dessau, where he collaborated with Pozzi and Koch in building the Court T. Returned Munich 1799; director of productions 1800.

10. **Giovanni Maria II** (*b* Mannheim, 1772; *d* Munich, 1813). Son of (5). Worked in Munich 1795–9. 1802–3, Mannheim 1800–2, but mostly engaged on military work.

11. **Angelo I** (*b* Munich, 13 Aug. 1784; *d* Munich, 2 Apr. 1815). Son of (8). Worked in Munich 1801–15, contributing to Romantic opera.

12. **Domenico II** (*b* Munich, 1 Jan. 1787; *d* Hohenschwangau, 9 Apr. 1837). Son of (8). Worked in Munich 1803–14.

13. **Lorenzo II** (*b* Munich, 19 Dec. 1793; *d* Munich, 15 Mar. 1869). Son of (8).

14. **Simon** (*b* Munich, 23 Oct. 1795; *d* Munich, 8 Mar. 1878). Son of (8). Succeeded his brother (12) 1814 and his father at the Court T 1828. One of the first designers to use built scenery, 1839; designed more than 100 productions 1828–60, incl. first Munich *Fidelio* and *Freischütz* 1822.

15. **Angelo II** (*b* Munich, 13 Dec. 1829; *d* Munich, 5 Jan. 1890). Worked with his father (14), Munich from 1849, designing Munich premières of *Tannhäuser* 1855, *Lohengrin* 1858, premières there of *Tristan* (1865), *Meistersinger* (1868), *Rheingold* (1869), *Walküre* (1870), and private productions of *Oberon* and Gluck's *Armida* for Ludwig II, all revealing a flair for epic architecture. The courtyard at Hohenschwangau is supposedly based on Act 2 of his 1867 Munich *Lohengrin*. Dresden 1865–80; also Prague, St Petersburg. Designs in the Munich T Museum and Ludwig II Museum at Herrenchiemsee.

16. **Franz** (*b* Munich, 22 Apr. 1844; *d* Wassenburg, 19 Feb. 1920). Son of (14).

17. **Eugen** (*b* Munich, 3 Apr. 1857; *d* Berlin, 24 Sep. 1942). Son of (15). Worked with his father Munich and later Berlin, Stuttgart, Prague until 1923.

18. **Angelo III** (*b* Munich, 11 June 1877; *d* Munich, 20 Mar. 1917). Son of (16).

**Quaisain, Adrien** (*b* Paris, 1766; *d* Paris, 15 May 1828). French composer and singer. Studied Paris with Berton. Début 1797 in his own *Sylvain et Lucette*. Cond. Ambigu-Comique, 1799–1819.

His many pieces were popular in their day, and even won a following in Germany: *Le jugement de Salomon* (Paris 1802) was intended to open the new Königsberg T, 1810.

**'Qual cor tradisti'.** The duet in Act II of Bellini's *Norma*, in which Norma (sop) reproaches the guilty Pollione (ten) for having betrayed her and now welcomes death with him.

**'Quand'ero paggio'.** Falstaff's (bar) arietta in Act II of Verdi's *Falstaff*, in which he recalls the days when he was a slender young page to the Duke of Norfolk.

**'Quando le sere al placido'.** Rodolfo's (ten) aria in Act II of Verdi's *Luisa Miller* grieving over the evidence in a letter that his wife Luisa has betrayed him with Wurm.

**'Quando m'en vo'.** Musetta's (sop) aria in Act II of Puccini's *La bohème*, hoping to recover Marcello's love as they meet in the crowd outside the Café Momus.

**'Quanto è bella!'** Nemorino's (ten) aria in the opening scene of Donizetti's *L'elisir d'amore*, which he sings as he gazes at the beautiful young Adina.

**quartet.** An ensemble for four singers. There are several quartets in 17th- and early 18th-cent. operas although, like most ensembles, they were not generally favoured until the time of opera buffa. Because of their vocal symmetry, they have been used to striking effect by composers as diverse as Mozart (e.g. in the last acts of *Idomeneo* and *Così fan tutte*), Beethoven (the canonic quartet in *Fidelio*), and Verdi, whose dramatic manipulation of the ensemble in *Rigoletto* set a standard rarely matched, and never surpassed, by later composers.

**Quattro rusteghi, I** (The Four Curmudgeons). Opera in 3 acts by Wolf-Ferrari; text by Sugana and Pizzolato, after Goldoni. Prod. in German as *Die vier Grobiane*, Munich 19 Mar. 1906, with Bosetti, Matzenauer, Tordek, Gever, Koppe, Walter, Sieglitz, Geis, Bender, Bauberger, cond. Mottl; London, SW, as *The School for Fathers*, 7 June 1946 with Gruhn, Jackson, Hill, Iacopi, Glynne, Franklin, cond. Robertson; New York, CC, as *The Four Ruffians*, 18 Oct. 1951, with Faull, Mayer, Pease, cond. Halasz.

Venice, end of the 18th cent. Lunardo, Maurizio, Simon, and Cancian (all bs) try to control their wive's excesses, as they see it. Lunardo's daughter Lucieta (sop) is to marry Maurizio's son Filipeto (ten), but custom forbids them even to meet until the wedding. Aided by Riccardo (ten), Filipeto gains entry to Lucieta's house disguised as a girl and the young couple meet, but they are discovered. Cancian's wife Felice (sop) takes a hand, and the old men are persuaded that their best interests lie in permitting the marriage. The Intermezzo before Act II is one of the composer's most delightful pieces.

**Queen of Shemakha, The.** The Queen (sop) who comes to seduce King Dodon in Rimsky-Korsakov's *The Golden Cockerel*.

**Queen of Spades, The** (Russ.: *Pikovaya dama*). Opera in 3 acts by Tchaikovsky; text by Modest Tchaikovsky, with suggestions and contributions from the composer, after Pushkin's story (1834). Prod. St Petersburg, M, 19 Dec. 1890, with M. Figner, Dolina, Piltz, Olgina, Slavina, N. Figner, Yakovlev, Melnikov, Vasilyev, Frey, Kondaraki, cond. Nápravník; New York, M, 5 Mar. 1910, with Destinn, Meitschik, Slezak, Forsell, Didur, cond. Mahler; London, London OH, 29 May 1915, with Nikitina, Krasavina, Rosing, Bonell, Kimbell, cond. Gurevich.

St Petersburg, 1790. Hermann, a young officer (ten), loves Lisa (sop), granddaughter of the old Countess (mez), once a gambler known as the Queen of Spades. She is described in a ballad sung by Hermann's friend Tomsky (bar) as possessing the secret of winning at cards. Hermann goes to the Countess's bedroom at night to obtain the secret from her so as to win enough money to marry, but so terrifies her that she dies without speaking. Her ghost appears to him and reveals the secret: 'Three, seven, ace'. As Hermann becomes obsessed with winning, Lisa despairs, and drowns herself. Staking all in a gambling session with Prince Eletsky (bar), Lisa's original betrothed, he wins on the first two cards, but finds to his horror that the third, which he thinks will be the ace, is the Queen of Spades. The Countess's ghost appears once more, and Hermann, losing his reason, stabs himself.

**Queen of the Night, The.** Pamina's mother (sop), Sarastro's adversary, in Mozart's *Die Zauberflöte*.

**Queen's Theatre.** See KING'S THEATRE.

**Querelle des Bouffons.** See GUERRE DES BOUFFONS.

**'Questa o quella'.** The Duke of Mantua's (ten) aria in the opening scene of Verdi's *Rigoletto*, in which he declares that all women attract him.

**Quickly, Mistress.** A lady of Windsor (mez) in Verdi's *Falstaff* and Vaughan Williams's *Sir John in Love*.

**'Qui la voce'.** Elvira's (sop) Mad Scene in Act II of Bellini's *I Puritani*.

**Quilico, Louis** (*b* Montreal, 14 Jan. 1929). Canadian baritone. Studied New York with Singher, Rome with Pizzolongo. Début New York, CC,

1953 (Germont). San Francisco 1956–9; London, CG, and Paris, O, 1962; New York, M, from 1971. Also Moscow, B; Paris, OC; Buenos Aires, etc. Roles incl. Enrico (*Lucia*), Rigoletto, Falstaff, Golaud. (R) His son **Gino Quilico** (*b* New York, 29 Apr. 1955) is also a baritone. Studied with his father, and at Toronto Cons. Début Toronto 1978 (TV in *The Medium*). Paris, O and OC, 1981, B 1991; London, CG, 1983; New York, M, 1987; also Frankfurt, Hamburg, Aix, Salzburg, etc. Roles incl. Oreste (Gluck), Don Giovanni, Figaro (Rossini), Malatesta, Valentin, Lescaut (Massenet), Posa, Marcello. (R)

BIBL: R. Mercer, *The Quilicos* (Toronto, 1991).

**Quinault, Philippe** (*b* Paris, 3 June 1635; *d* Paris, 26 Nov. 1688). French dramatist and librettist. Though an established playwright by the time he abandoned his career for opera, he was able to modify his technique with great skill to the requirements of his demanding (and sole) collaborator Lully. As servant of the composer who was directly subject to Louis XIV, he wrote formal scenes, expressing stereotyped noble sentiments, in lines whose stateliness and elegance reflected the court's expression of royal *gloire* (frequently remarked upon in a special Prologue). Thus, despite their often tremendous adventures, Quinault's characters, generally drawn from mythology or chivalry, seem to express above all the milder emotions. Heroic success is achieved without suffering; love is generally not tempestuous but idyllic. A good grasp of narrative and a flexible command of graceful verse (loosening the constraints of the alexandrine) help to give the texts their quality. The chief influences on Quinault were romantic novels and the melodramatic Spanish tragedies. He made much use of scenic effects and pantomime or dance sequences, all marked by the frequent intervention of the miraculous. Spectacle was therefore also a crucial ingredient, as was the inclusion of dance; and operas were planned with the detailed approval of the King. Quinault's development of tragédie lyrique for Lully was a crucial influence on later French opera, and had an important effect on the manner of *Grand Opera.

BIBL: E. Gros, *Philippe Quinault: sa vie et son œuvre* (Paris, 1926).

**quintet.** An ensemble for five singers. There are only a few examples of such ensembles in opera, of which the most famous is that in the last act of Wagner's *Die Meistersinger*.

**quodlibet** (Lat.: 'what you please'). A kind of musical game of the 15th, 16th, 17th, and early 18th cents. involving extempore juxtaposition of different melodies. The three dances in the Act I finale of Mozart's *Don Giovanni* thus form a quodlibet.

In the 19th-cent. German (and especially Viennese) theatre it came to have a variety of meanings, including that of a theatrical entertainment in which well-known artists performed excerpts of their favourite roles, as well as simply a potpourri of popular themes. Another interesting manifestation was as a kind of *pasticcio, setting pre-existing music to a new libretto: the most successful such work was Stegmayer's *Rochus Pumpernickel* (1809; music by Haibel & Seyfried).

# R

**Raaff, Anton** (*b* Gelsdorf, *bapt.* 6 May 1714; *d* Munich, 28 May 1797). German tenor. Studied Munich with Ferrandini, later Bologna with Bernacchi. Début Munich 1736 (in an opera by Ferrandini). Italy 1739; Bonn and Frankfurt 1742; Vienna 1749 (Metastasio commenting that he 'sang like an angel'); Italy 1752; Madrid 1755, singing for Farinelli, with whom he went to Naples 1759. Further successes there and in Florence. In 1770, he entered the service of the Elector Palatine, Carl Theodor, at Mannheim and from 1778 at Munich. In 1777 Mozart was at first dismissive of his singing and wooden acting, but later wrote (and 'tailored' for him) the aria K295 (1778), so impressing Raaff that he asked Carl Theodor to commission *Idomeneo*. Mozart, by now appreciative of the ageing tenor's still considerable artistry (though finding his style 'peculiar'), wrote the title-role with his qualities in mind. His vocal longevity was remarkable; Kelly notes that he was still in good voice in 1787.

**Rabaud, Henri** (*b* Paris, 10 Nov. 1873; *d* Paris, 11 Sep. 1949). French composer and conductor. Studied Paris with Gédalge and Massenet. His first opera was a historical piece, *La fille de Roland* (1904), but his greatest success came with *Mârouf*, an oriental comedy notable for its lightness of touch and its fine orchestration. None of his other works approached its success, though he continued to write works with an expert understanding of the stage and an indifference to fashion. Paris, O: conductor 1908–18, director 1914–18; director Paris Cons. 1920–40.

SELECT WORKLIST (all first prod. Paris): *La fille de Roland* (Ferrari, after Bornier; 1904); *Mârouf, savetier du Caire* (Mârouf, the Cobbler of Cairo) (Népoty; 1914); *L'appel de la mer* (The Call of the Sea) (Rabaud, after Synge; 1924); *Martine* (Bernard; 1947).

BIBL: M. d'Ollone, *Henri Rabaud* (Paris, 1958).

**Rabelais, François** (*b* Chinòn, *c*.1494; *d* ?Paris, ?9 Apr. 1553). French writer. Operas on his works are as follows:

*Pantagruel* (1532): Grétry (*Panurge dans l'île des lanternes*, 1785); Planquette (*Panurge*, 1895); Terrasse (*Pantagruel*, 1911); Massenet (*Panurge*, 1913); Sutermeister (*Séraphine*, 1959)

*Gargantua* (1534): Mariotte (1935). *L'isle sonnante* (1562–4): Monsigny (1767)

Also opera *Rabelais* by Ganne (1892).

**Rachel.** Éléazar's daughter (sop), lover of Léopold, the eponymous Jewess of Halévy's *La Juive*.

**'Rachel, quand du Seigneur'.** Éléazar's (ten) aria in Act IV of Halévy's *La Juive*, in which he struggles with his dilemma between letting Rachel die at the hands of the Christians or saving her life by telling Cardinal Brogni that she is in fact the latter's daughter.

**Rachmaninov, Sergey.** See RAKHMANINOV, SERGEY.

**Racine, Jean** (*b* La Ferté-Milon, 21 Dec. 1639; *d* Paris, 21 Apr. 1699). French poet and dramatist. Operas on his works are as follows:

*Andromaque* (1667): Grétry (1780); Rossini (*Ermione*, 1819)
*Britannicus* (1669): Graun (1751)
*Bérénice* (1670): Magnard (1911)
*Bajazet* (1672): Hervé (*Les Turcs*, parody of Racine, 1869); various 18th-cent. works deriving from the subject as *Bajazet, Tamerlano* etc.
*Mithridate* (1673): Graun (1750); Scheinpflug (1754); Q. Gasparini (1767); Mozart (1770)
*Iphigénie* (1675): Graun (1748); Gluck (1774)
*Phèdre* (1677): Lemoyne (1786); Rameau (1733); Bussotti (1986)
*Esther* (1688): Meyerowitz (1957)
*Athalie* (1691): Handel (1733); Poissl (1814); Weisgall (1964)

**Radamès.** The Captain of the Egyptian Guard (ten), lover of Aida, in Verdi's *Aida*.

**Radamisto.** Opera in 3 acts by Handel; text by Nicolo Haym, adapted from *L'amor tirannico* (attrib. D. Lalli and M. Noris, 1712) and *Zenobia* (Noris, 1710), both after Tacitus. Prod. London, H, 27 Apr. 1720, with Durastanti, Montagnana. Handel's first opera for Royal Academy of Music; also very popular in Hamburg, where it was given in an adaptation by Mattheson, 1722. Rev. Handel O Society, 1960.

**Radford, Robert** (*b* Nottingham, 13 May 1874; *d* London, 3 Mar. 1933). English bass. Studied London with Randegger, Haynes, and King. Début London, CG, 1904 (Commendatore). Sang Hunding and Hagen under Richter in English *Ring*; subsequently sang with Beecham OC, and with BNOC, of which he was a founder and director. Admired in Mozart and as Boris. (R) His daughter **Winifred** (*b* London, 2 Oct. 1901; *d* Cheltenham, 15 Apr. 1993), also his pupil, sang as a soprano at Gly. 1934–8. Intimate O 1936; director 1956–75. (R)

**radio opera** (Fr., *opéra radiophonique*; Ger., *Funkoper*; It., *opera radiofonica*). The term for works specially composed or mounted for broadcasting; the English term also covering the broadcasting of tapes provided by other organizations, relays from opera-houses, or records. A forerunner of radio opera was the telephonic relay of *Les Huguenots* from the Paris Opéra in 1881. An aria from *Carmen* sung by Mariette Mazarin was broadcast from the Manhattan OC in the early 1900s, and in 1910 parts of *Pagliacci* and *Cavalleria rusticana* were broadcast from the New York Metropolitan. Berlin Radio broadcast *Madama Butterfly* from the Städtische Oper on 8 June 1921, and in 1924 founded an Oper-Sendebühne. The London station 2LO broadcast *Hänsel und Gretel* from Covent Garden on 6 Jan. 1923. Czech Radio broadcast *The Two Widows* in 1925. In the same year a New York station began weekly opera broadcasts, and on 25 Dec. 1931 *Hansel and Gretel* was relayed from the Met as the first regular US opera performance to be broadcast. Italian Radio broadcast Perelli's *I dispettosi amanti* in 1926, and *Tosca* from La Scala in 1928; it and the New York M transmitted opera regularly from 1931. The first transatlantic opera broadcast was *Fidelio* from Dresden (16 Mar. 1930). French Radio made its first broadcast from the Paris Opéra in 1932 with Rabaud's *Mârouf*. The first opera written for broadcasting was Walter Goehr's *Malpopita* (Berlin 1930). Other early operas written for radio included Egk's *Columbus* (Bavarian Radio, 1933), Cadman's *The Willow Tree* (NBC, 1933), Martinů's *Comedy on a Bridge* (Czech Radio, 1937), and Sutermeister's *Die schwarze Spinne* (Swiss Radio, 1936). In 1939 NBC commissioned Menotti's *The Old Maid and the Thief*, and in 1943 Montemezzi's *L'incantesimo*. During the 1950s French Radio maintained a permanent opera company. Virtually all radio organizations now broadcast opera occasionally or regularly. See also BRITISH BROADCASTING CORPORATION, TELEVISION OPERA.

**Raff, Joachim** (*b* Lachen, 27 May 1822; *d* Frankfurt, 24 or 25 June 1882). German composer. Self-taught; given guidance by Liszt. He left several unproduced operas; the only two to reach the stage were *König Alfred* (Logau; Weimar 1851) and *Dame Kobold* (Reber, after Calderón; Weimar 1870), which even Liszt thought a hotch-potch.

**Raimondi, Ruggero** (*b* Bologna, 3 Oct. 1941). Italian bass. Studied Rome with Pediconi, Piervenanzi, and Ghibaudo. Début Spoleto 1964 (Colline). Rome 1964; Venice 1965–7; Gly. 1969; Milan, S, and New York, M, from 1970; London, CG, from 1972; Paris, O, from 1979. A handsome performer, particularly acclaimed for his Verdi roles, e.g. Attila, Philip; also as Count Almaviva (Mozart), Don Giovanni, Mosè, Boris. (R)

**Raimund, Ferdinand** (*b* Vienna, 1 June 1790; *d* Pottenstein, 5 Sep. 1836). Austrian dramatist, actor, and producer. Spent all his life in the theatre, working especially in the T in der Josephstadt and T in der Leopoldstadt, Vienna. A popular actor, but especially remembered for his brilliant series of magic comedies with music by Wenzel Müller, Drechsler, Riotte, and Conradin Kreutzer. He also wrote tunes for them himself. His subjects frequently treat the humour of situation and local allusion, such as the effect of transferring an ordinary Viennese citizen to fairyland and the ensuing comic contrast of the fantastic with the earthy. He frequently made use of comic musical quotation, in the tradition used by Mozart in the *Don Giovanni* supper scene. In his classic examples of the *Posse, he won wide popularity, first with *Der Barometermacher auf der Zauberinsel* (1823) and especially with *Der Alpenkönig und der Menschenfeind* (1828). Other great successes included *Der Diamant des Geisterkönigs* (1824) and *Das Mädchen aus der Feenwelt* (1826).

BIBL: D. Prochaska, *Raimund and Vienna* (Cambridge, 1970).

**Raisa, Rosa** (*b* Białystok, 23 May 1893; *d* Los Angeles, 28 Sep. 1963). Polish, later US, soprano. Fled from a pogrom in Poland, settling in Naples. Studied there with Marchisio and E. Tetrazzini. Début Parma 1913 (Leonora, *Oberto*). Chicago 1913–14, 1916–36; London, CG, 1914, 1933; Milan, S, 1924, 1926. Also Paris, O; Buenos Aires, etc. Roles included Donna Anna, Norma, Aida, Tosca; created Asteria (Boito's *Nerone*) and Turandot. With an opulent, well-controlled voice, statuesque beauty, and impressive acting ability, she was one of the leading singers of her time. Opened singing school in Chicago, 1937, with her husband, the baritone **Giacomo Rimini** (1887–1952); he made his début in 1910 (Albert, *Werther*), sang Falstaff under Toscanini, Milan 1915, also Pong in the first *Turandot* (1926). (R)

**Rake's Progress, The.** Opera in 3 acts and epilogue by Stravinsky; text by W. H. Auden and Chester Kallman, after Hogarth's eight engravings (1735). Prod. Venice, F, 11 Sep. 1951, with Schwarzkopf, Tourel, Rounseville, Kraus, Cuénod, cond. Stravinsky; New York, M, 14 Feb. 1953, with Gueden, Thebom, Conley, Harrell, cond. Reiner; Edinburgh 25 Aug. 1953, by Gly. Co., with Morison, Lewis, Hines, cond. Wallenstein.

England, 18th cent. Tom Rakewell (ten) declines the offer of a city position offered by Trulove (bs), father of his sweetheart Anne Trulove (sop), and is complaining of his lack of money when Nick Shadow (bar) appears with news of sudden wealth. Taking Shadow as his servant, to be paid after a year and a day, Tom leaves for London, where he quickly descends to a life of vice.

His pleasures palling, he is easily tempted by Shadow to marry the fantastic bearded lady Baba the Turk (mez), and then to place his trust in a fake machine for turning stones into bread. He goes bankrupt, and all his effects are sold, under Sellem (ten), the auctioneer.

The time comes to pay Nick Shadow, who reveals himself as the Devil, and claims Tom's soul. They gamble for it; Tom wins, and Shadow sinks into the ground, but leaves Tom mad. The final scene is in Bedlam; Tom thinks he is Adonis, and Anne now takes her last leave of him.

The characters point the moral in the Epilogue: 'For idle hands And hearts and minds The Devil finds A work to do.'

**Rakhmaninov, Sergey** (*b* Semyonovo, 1 Apr. 1873; *d* Beverley Hills, CA, 28 Mar. 1943). Russian composer and pianist. His first opera, *Aleko*, was the winner of a student competition, and on its production won some approval, including from Tchaikovsky (whose influence it reflects). Its lyrical qualities dominate its dramatic ones, though it includes some appealing genre music. *The Miserly Knight* has a stronger individuality, and adds the influence of Wagner while also suggesting the manner of Bartók's *Duke Bluebeard's Castle*. Written for an all-male cast, it is cast in three scenes, the central one consisting of a fine monologue for the Knight (intended for Shalyapin). For *Francesca da Rimini*, Rakhmaninov had a libretto by Modest Tchaikovsky whose weaknesses he was unable to overcome with music of sufficient power.

WORKLIST (all first prod. Moscow): *Aleko* (Nemirovich-Danchenko, after Pushkin; 1893); *Skupoy rytsar* (The Miserly Knight) (Pushkin original; 1906); *Francesca da Rimini* (M. Tchaikovsky, after Dante; 1906); *Monna Vanna* (Slonov, after Maeterlinck, Act I only, 1907).

BIBL: G. Norris, *Rakhmaninov* (London, 1976).

**Ralf, Torsten** (*b* Malmö, 2 Jan. 1901; *d* Stockholm, 27 Apr. 1954). Swedish tenor. Studied Stockholm with Forsell, Berlin with Dehmlow. Début Stettin 1930 (Cavaradossi). Dresden 1935–43; London, CG, 1935–9, 1948; Stockholm from 1941; New York, M, 1945–8. Roles included Lohengrin, Walther, Parsifal, Radamès, Otello (Verdi); created Apollo (*Daphne*). (R) His elder brother **Oscar** (1881–1964) was leading tenor of Stockholm O 1918–30; Bayreuth and Paris, OC, 1927. Sang Siegmund, Tristan, and Verdi roles. Translated over 40 operas and operettas. (R).

**Rameau, Jean-Philippe** (*b* Dijon, 25 Sep. 1683; *d* Paris, 12 Sep. 1764). French composer and theorist. Taught first by his father, Rameau studied briefly in Italy before becoming organist at Avignon in 1702. His early career centred around the church: he held posts in Clermont-Ferrand, Paris, and Dijon before coming to public attention with his *Traité de l'harmonie* (Paris, 1722), which established him as the foremost theorist of his day. Significantly, the *Traité* was not merely concerned with examining the scientific basis of harmony, but also dealt at length with its use as a means of musical expression.

Born in the shadow of Lully's triumphs, Rameau not surprisingly harboured operatic ambitions, though his first attempt only came in middle age. His activities as a theorist, organist, and composer of harpsichord suites in the tradition of Couperin and Daquin undoubtedly hindered his entry into the operatic world, but the main reason for delay was his inability to find a satisfactory libretto. After many abortive attempts, including a collaboration with Voltaire, an introduction to the Abbé Pellegrin in 1730 resulted in a libretto which Rameau considered worthy of his music.

In *Hippolyte et Aricie* Rameau established at a stroke a new formula for French opera which served to rouse it from the stylistic torpor into which it had declined following Lully's death. From the outset he was viewed as a controversial figure by critics and academicians, who unanimously condemned the work; fortunately, he was no less strongly acclaimed by the public, who saw in *Hippolyte* a way forward for French music. Though the Lullian model had provided a point of departure, Rameau's greater emotional sensitivity enabled him to develop a richer and more vivid interpretation of the plot, in which he was aided by a skilful librettist. The former emphasis on *gloire* and rich spectacle is supplanted by a heightened dramatic awareness, and the classical myth is transmuted into a more engaging story, in which human emotions play their part. Hence Aricie, who has only a minor role in Racine's *Phèdre*, the play on which Pellegrin based his libretto, participates fully in the action of opera, while the emphasis on Hippolyte's own heroism is now reduced so that the amorous side of his character, barely mentioned in *Phèdre*, can be fully explored. Phèdre's character is likewise reduced from that of tragic heroine to something approaching a caricature of a woman spurned. The insertion of totally new episodes, such as Thésée's struggle in Hades, provides an opportunity for Rameau to engage in some effective pictorial writing for the orchestra: there were to be many such episodes in subsequent works. Likewise, the dramatic handling of the chorus, which as well as forming a backcloth for the main characters also participates in the action, represents a notable advance on contemporary models.

*Hippolyte* was followed by a string of popular successes, in which many of the same techniques were employed: *Les Indes galantes, Castor et Pollux, Les fêtes d'Hébé* and *Dardanus*. After a five-year break from opera, during which time he founded a school of composition, Rameau returned in 1745 with the opulent *La princesse de Navarre*, written to celebrate the marriage of the Dauphin. This was followed by two more works in serious vein, *Zoroastre*, using parts of *Samson*, one of the abortive early collaborations with Voltaire, and *Linus*, in which his handling of the tragédie lyrique is further refined. By now Rameau was recognized as the most important composer of his generation, so that when a performance of Pergolesi's *La serva padrona* in 1752 sparked off the famous intellectual debate usually known as the *Guerre des Bouffons, it was only natural that he should become the main spokesman in defence of the conservative tradition. His lively promotion of French tragédie lyrique against the claims of lighter Italian opera eventually triumphed, and he went on to produce a final masterpiece in *Abaris, ou Les Boréades*. Throughout the latter years of his life he continued to compose music for ballets and divertissements, such as *La guirlande* (1751) and *Anacréon* (1754), and pursue his interests as a theorist. He died a national hero and, like Lully, his works remained in the repertory until the French Revolution, and are now being revived.

The influence of Lully on Rameau's musical style was frequently acknowledged by the composer himself and a comparison of their work reveals many similarities. For both, the production of opera involved a compromise with ballet, a necessary part of any French stage entertainment. Although some of Rameau's works, notably *Les Indes galantes*, straddle the two worlds in almost equal proportions, his understanding of dramatic writing is demonstrated by a more skilful interpolation of balletic sequences within the genuinely operatic structures of such works as *Hippolyte* and *Les Boréades*. His treatment of the air likewise represents a development of Lully's work. Although many are faithful to the traditional structures and external features of the Lullian model, the legacy of Rameau's brief stay in Italy is apparent in a more lyrical, melismatic vocal line. Undoubtedly Rameau's greatest contribution to the development of French opera lay in his handling of the recitative, however, where he reached new depths of expression in the setting of French words to music. In general he followed Lully's model in writing a vocal line in which speech inflections are closely mirrored, although he injected a more expressive note by the use of chromatic harmony and by a greater exploitation of instrumental resources. In his 'récitatif accompagné pathétique' and 'récitatif mesuré', which equate closely to Italian accompanied recitative and arioso, moments of poignancy, grief, terror, and fury are underpinned by dramatic and colourful instrumental writing.

Rameau's influence on the development of opera in Europe was profound and wide-ranging. For his own generation he provided an escape from the stasis of Lully's tragédie en musique by offering a genuinely dramatic entertainment focusing on human emotions in place of an outdated concentration on the glorification of royalty. Later he was to have an important rejuvenating influence on Italian opera seria, when Gluck took his work as a model for his first 'reform' opera, *Orfeo ed Euridice*, which itself shares several points of similarity with *Castor et Pollux*, including the opening choral scene. Finally, in the 19th cent. Berlioz found inspiration in Rameau for some of his grandest conceptions, including *Les Troyens*.

WORKLIST (excluding ballets and incidental music; all first prod. Paris): *Hippolyte et Aricie* (Pellegrin, after Racine; 1733, rev. 1742, 1757); *Samson* (Voltaire; c.1733, unperf., lost) *Les Indes galantes* (Gallantry in the Indies) (Fuzelier; 1735, rev. 1735, 1736, 1743, 1751, 1761); *Castor et Pollux* (Bernard; 1737, rev. 1754, 1764); *Dardanus* (Le Clerc de la Bruère; 1739, rev. 1744, 1760, 1770); *La princesse de Navarre* (Voltaire; 1745, rev. 1763); *Platée* (D'Orville; 1745); *Zoroastre* (Cahusac; 1749, rev. 1756, 1770); *Linus* (Le Clere de la Bruère; c.1752, unperf.); *Abaris, ou Les Boréades* (Cahusac; 1763; prod. Aix 1982.).

BIBL: C. Girdlestone, *Rameau* (London, 1957, 2/1969).

**Ramey, Samuel** (*b* Colby, KS, 28 Mar. 1942). US bass. Studied Wichita with Newman, New York with Boyajian. Début New York City O 1972 (Zuniga). New York City O for several seasons. Gly. 1976–8; Paris, O, from 1983; New York, M, from 1984; London, CG, from 1985. Also Milan, S; Salzburg; Florence; etc. Vocally gifted, and magnetic on stage. His unusually broad repertory includes Figaro (Mozart), Don Giovanni, Mustafà, Moïse, Gounod's and Boito's Mephistopheles, Don Quichotte, Attila, Philip, Pimen, Boris, Nick Shadow, Duke Bluebeard. One of the most prolific recording artists of his generation. (R)

**Ramfis.** The High Priest (bs) in Verdi's *Aida*.

**Ramiro.** The muleteer (bar), lover of Concepción, in Ravel's *L'heure espagnole*.

**rammentatore** (It.: 'prompter'). An alternative title for **maestro suggeritore**, or *prompter, in Italian opera-houses.

**Rance, Jack.** The sheriff (bar) in Puccini's *La fanciulla del West*.

**Randegger, Alberto** (*b* Trieste, 13 Apr. 1832; *d* London, 18 Dec. 1911). Italian/German, later British, conductor, teacher, and composer. Studied Trieste with Ricci. Composed operas and held appointments in Italy before moving to London 1854. Wrote *The Rival Beauties* (Leeds 1864). A very influential professor of singing London, RAM, 1868, later RCM. Cond. London: J, 1857, CR, 1879–85; CG and DL, 1887–98, helping to promote Wagner; also admired in Verdi.
WRITINGS: *Singing* (London, 1893).

**Rangoni.** The Jesuit (bs), Marina's confessor, in Musorgsky's *Boris Godunov*.

**Ránki, György** (*b* Budapest, 30 Nov. 1907). Hungarian composer. Studied Budapest with Kodály. His interest in folk music is shown in a number of theatrical works, of which the most successful has been the opera *Pomádé király új ruhája* (King Pomádé's New Clothes) (Károlyi, after Andersen; Budapest Radio 1950, Budapest 1953).

**Rankl, Karl** (*b* Gaaden, 1 Oct. 1898; *d* St Gilgen, 6 Sep. 1968). Austrian, later British, conductor and composer. Studied Vienna with Schoenberg and Webern. Vienna, V, 1924; Liberec 1925; Königsberg 1927; Berlin, Kroll T, 1928–31, as chorus-master and conductor under Klemperer; Wiesbaden 1931–2. Graz 1932–7. Prague, Landestheater, 1937–9 (première of Krenek's *Karl V*). Music director London, CG, 1946–51, building up the new company; this constructive work was more admired than his conducting. Music director, Elizabethan O Trust, Australia, 1958–60. His *Deirdre of the Sorrows* won a Festival of Britain prize but was unprod. (R)

**Raoul.** Raoul de Nangis (ten), a Huguenot nobleman, lover of Valentine, in Meyerbeer's *Les Huguenots*.

**Rape of Lucretia, The.** Opera in 2 acts by Britten; text by Ronald Duncan, based on André Obey's play *Le viol de Lucrèce* (1931), in turn based on Shakespeare's poem *The Rape of Lucrece* (1594) and on Livy's history *Ab Urbe Condita*, i. 57–9 (*c*.26 BC–AD 17). Prod. Gly. 12 July 1946, with Cross, Ferrier, Pears, Kraus, cond. Ansermet; Chicago, Schubert T, 1 June 1947, with Resnik, Kibler, Kane, Rogier, cond. Breisach.

Rome, 510 BC. With a male (ten) and female (sop) Chorus commenting and eventually drawing a Christian moral, the opera relates the story of the proud, destructive Tarquinius (bar). He rides from the camp where news has arrived of the Roman wives' infidelity to make an attempt on the virtue of the sole exception, Lucretia (mez), wife of Collatinus (bs). Claiming hospitality, he later enters her room and rapes her. Unable to bear the burden of her shame, she kills herself the next day in the presence of her urgently summoned husband.

**Rappresentazione di anima e di corpo, La** (The Representation of the Soul and the Body). Opera by Emilio de' Cavalieri; text by Agostino Manni. Prod. Rome, Oratorio del Crocifisso, Feb. 1600; Cambridge, Girton Coll., June 1949; U of N Dakota, 23 Feb. 1966. The characters represent not only man's soul and body, but also various human attributes.

**rappresentazione sacra** (It.: 'sacred representation'). A term used in the 15th and 16th cents. for religious plays with music: it lingered on as a description for some of the earliest operas on sacred subjects. See also AZIONE SACRA.

**Rasi, Francesco** (*b* Arezzo, 14 May 1574; *d* after 1620). Italian tenor and composer. Probably studied with Caccini. Success in Rome in the service of Ferdinando I of Tuscany in early 1590s. In Gesualdo's service 1594–6, and in the Gonzagas' from 1598. Florence 1600, singing in first performances of Peri's *Euridice*, and Caccini's *Rapimento di Cefalo*. Very probably created title-role of Monteverdi's *Orfeo*, Mantua 1607. Sang in first *Dafne* of Gagliano, Mantua 1608. The music of his opera, *Cibele ed Ati* (intended for the marriage of Ferdinando Gonzaga, 1617) is lost; his text survives.

**rataplan.** An onomatopoeic French word representing the sound of a drum. The title of solos and ensembles in operas by Donizetti (*Fille du régiment*), Meyerbeer (*Huguenots*), Verdi (*Forza*), and Sullivan (*Cox and Box*). Also Singspiel by Pillwitz (1830).

**Ratisbon.** See REGENSBURG.

**Rattle, Simon** (*b* Liverpool, 19 Jan. 1955). English conductor. Studied London. While a student, cond. *L'enfant et les sortilèges* (London, RAM, 1974). Début Gly. 1977 (*Cunning Little Vixen*). Gly. regularly; London, ENO, 1985; Los Angeles 1988; London, CG, from 1990. Repertory incl. *Idomeneo*, *Figaro*, *Cunning Little Vixen*, *Porgy and Bess*, *L'heure espagnole*. His brilliant, sensitive, and passionate musical direction has produced memorable performances, incl. concert version of *Idomeneo*, London, QEH, 1987, which had more dramatic vitality than many stage productions. (R)
BIBL: N. Kenyon, *Simon Rattle* (London, 1987).

**Rautavaara.** Family of Finnish musicians.

1. **Vaïnö** (*b* Ilmajoki, 21 June 1872; *d* Helsinki, 7 Aug. 1950) was a singer who taught in Helsinki.

2. **Eino** (*b* Ilmajoki, 14 Mar. 1876; *d* Helsinki, 7 Aug. 1939). Baritone, brother of (1). Studied in Italy, Paris, and Berlin. Leading baritone, Helsinki O, 1911–24.

3. **Aulikki** (*b* Vassa, 2 May 1906; *d* Helsinki, 29 Dec. 1990). Soprano, daughter of (1). Studied Helsinki with her father, Berlin with Eisner. Début Helsinki 1932. Gly. 1934–8, Countess (Mozart), Pamina. Salzburg 1937. Edinburgh 1950. (R).

4. **Einojuhani** (*b* Helsinki, 9 Oct. 1928). Composer, son of (2). Studied Helsinki with Merikanto, New York with Persichetti, Tanglewood with Copland and Sessions, Ascona with Vogel, Cologne with Petzold. He has been accused of eclecticism. His operas include the TV opera *Kaivos* ('The Mine') (Rautavaara; Finnish TV 1963) and two works based on the Kalevala.

**Rauzzini, Venanzio** (*b* Camerino, *bapt.* 19 Dec. 1746; *d* Bath, 8 Apr. 1810). Italian soprano castrato and composer. Studied Rome, and possibly Naples with Porpora. Début Rome, T della Valle, 1765 (Piccinni's *Il finto astrologo*). In the service of Max Joseph III, Munich 1766–72, singing in works by Hasse, Bernasconi, and himself. Italy 1772–4, creating Cecilio in Mozart's *Lucio Silla* (Milan 1772). London, H, 1774–7. In 1777 moved to Bath, managing concerts. Continued to sing in London, where he also staged operas, including some of his own, at Haymarket 1781–4, 1787. As a singer, he had great success with a sweet-toned voice and handsome looks (usually singing female roles). Mozart wrote 'Exsultate, jubilate', K165, for him. As a composer he was admired by Burney. His prolific compositions include the operas *L'ali d'amore*, *L'eroe cinese*, *Creusa in Delfo*, and the highly popular *Piramo e Tisbe*. Among his pupils were Nancy Storace, John Braham, Mrs Billington, and Michael Kelly. Haydn, one of his many guests in Bath, wrote a canon to the memory of his dog Turk.

**Ravel, Maurice** (*b* Ciboure, 7 Mar. 1875; *d* Paris, 28 Dec. 1937). French composer. His two operas are unique contributions to the art, calling for elaborate resources but focusing them on two themes that especially suited him. *L'heure espagnole* is a cynical, farcical romp with plenty of Spanish local colour, and treats with wit and poise the subject of a bored wife's afternoon infidelities. The sophistication of the means is, indeed, a necessary part of the worldly wise elegance of the treatment. With *L'enfant et les sortilèges*, the technical brilliance also helps to lend a detachment from the risk of sentimentality. The

story of the bad child punished and returned to innocence is told by way of first some brilliantly parodistic music for his revengeful tormentors, then some ravishingly lyrical nature music in a garden scene. Ravel is known to have contemplated at least three more serious subjects: *La cloche engloutie* (on which he was working until 1914), *Jeanne d'Arc*, and *Don Quixote*.

WORKLIST: *L'heure espagnole* (The Spanish Hour) (Franc-Nohain; Paris 1911); *L'enfant et les sortilèges* (The Child and the Enchantments) (Colette; Monte Carlo 1925).

WRITINGS: R. Chalupt and M. Gerar, eds., *Ravel au miroir de ses lettres* (Paris, 1956).

BIBL: R. Nichols, *Ravel* (London, 1977).

**Ravenna.** City in Emilia-Romagna, Italy. Opera was first performed in the 17th cent. The T dell'Industria opened, giving the première of Vivaldi's *Armida al campo d'Egitto*. Performances were also given at the T Comunitativo, later T Vecchio. The T Alighieri opened 1852; performances are also given at the T Angelo Mariani and T Luigi Rosi.

**Rebikov, Vladimir** (*b* Krasnoyarsk, 31 May 1866; *d* Yalta, 4 Aug. 1920). Russian composer. Studied Moscow with Klenovsky, Berlin with Meyerberger. His style moved from a 19th-cent. Russian Romanticism by way of Impressionism towards Expressionism, and in his operas, some of which he entitled 'musico-psycholographic dramas', he attempted to place the music entirely at the service of the action, though this did not exclude a sense of fantasy. In this he was much influenced by the Moscow Arts T. Admired as avant-garde in its day, his work was found forced and pretentious under Communism, despite its lyrical qualities.

SELECT WORKLIST: *V grozu* (Into the Storm) (Plakshin, after Korolenko; Odessa 1894); *Yolka* (Plakshin, after Dostoyevsky, Andersen, & Hauptmann; Moscow 1903); *Zhenshchina s kinzhalom* (The Woman with the Dagger) (Rebikov, after Schnitzler; comp. 1910, unprod.); *Alfa i omega* (Rebikov; comp. 1911, unprod.); *Narcissus* (Shchepkina-Kupernik, after Ovid; comp. 1912); *Dvoryanskoye gnezdo* (A Nest of Gentlefolk) (Rebikov, after Turgenev; comp. 1916).

**Re cervo, Il.** See KÖNIG HIRSCH.

**récit** (Fr.: 'narrative'; not to be confused with *récitatif*, which is the usual translation of 'recitative'). A term used in the 17th and 18th cents. for an independent piece for solo voice, with or without accompaniment, heard within the context of a larger dramatic work, such as a ballet de cour.

**recitative** (Fr., *récitative*, or *récitatif*; Ger., *Rezitativ*; It., *recitativo*). The name given to the declamatory portions of opera, in which the plot is generally advanced, as opposed to the more static or reflective lyrical settings; it developed during the 17th cent. from the branching of *stile rappresentativo into two distinct styles, the other of which became *aria.

In Italian opera this division is first formally recognized in Metastasian opera seria, where two kinds may be discerned. In the first, **recitativo secco** ('dry' recitative), the notes and rhythms of the vocal line follow the verbal accentuation; this recitative was accompanied by the basso continuo and was used to advance the plot. **Recitativo accompagnato** (sometimes also called **recitativo stromentato**) sets the text in a more lyrical fashion and calls for the accompaniment of instruments other than the continuo, sometimes strings alone, sometimes *obbligato wind. Because the style of its word-setting is often closer to aria than to recitative, it is sometimes also known as **recitativo arioso**. As the conventions of Italian opera seria were reformed around the middle of the 18th cent. by Gluck and others, increasing emphasis was placed upon the recitativo accompagnato.

In part these reforms drew upon the tradition of French opera, where from the beginning an extraordinary degree of interest had been taken in the recitative. Three main types were distinguished: **récitatif simple**, which equated closely to the Italian recitativo secco, although it was sometimes interspersed with short airs; **récitatif obligé**, or orchestrally accompanied recitative, and **récitatif mesuré**, a type not formally identified until Rameau's day, although present much earlier, in which récitatif obligé conforms to a more regular metrical pattern. In French opera, distinctions between the air and the récitatif were always more blurred than between the equivalents in Italian opera.

By the end of the 18th cent. there was a universal move to absorb the recitative into other portions of the music, which can be seen to early good effect in Mozart's *Idomeneo*, while in the 19th cent. it was assimilated as one element of the grand *scena in Italian opera. Although, at its most developed, accompanied recitative was a highly expressive and dramatic vehicle, secco recitative was invariably regarded as a merely routine aspect of opera. Many composers, Mozart among them on occasion, allowed their pupils to write the recitatives, thus saving creative energy for the composition of the arias.

**recitativo arioso.** See ARIOSO.

**'Recondita armonia'.** Cavaradossi's (ten) aria in Act I of Puccini's *Tosca*, in which he contrasts the dark beauty of his beloved Tosca with the fair beauty of the Marchesa Attavanti, the model for his painting of the Magdalen.

**recording.** Opera and opera singers have played an important part in the history of the recording; indeed, it used to be said with some truth that 'Caruso made the gramophone and the gramophone Caruso'. In the mid-1890s Gianni Bettini began to record the voices of singers in New York on cylinders, and he offered for sale items by Ancona, Saléza, Plançon, Van Dyck, and Van Rooy; it is also known that he possessed cylinders made by Arnoldson, E. Calvé, Melba, Nordica, Sembrich, Nicolini, Campanini, Tamagno, the De Reszkes, Lassalle, and Maurel. Lionel Mapleson recorded De Marchi, Nordica, Ternina, and others at the New York Metropolitan 1901–2. Meanwhile Emil Berliner had perfected his machine, which played a disc. In Europe, the Gramophone Company was established in 1898, and by 1902, under the artistic guidance of Fred Gaisberg, its artists included Caruso, Shalyapin, E. Calvé, and Battistini. Various other companies quickly established themselves, and virtually every singer of note began to record. The first 'complete' opera recording (though with many cuts) was *Il trovatore*, made sporadically 1903–6 with 16 different singers. There was also a virtually complete *Pagliacci* directed by the composer. In Germany there was a *Fledermaus* in 1907, Gounod's *Faust* (Destinn, Jörn, Knüpfer) in 1908, *Carmen* (Destinn, Jörn), *Cavalleria* (Hempel), and *Pagliacci*. In France nine complete operas were recorded by the Pathé Company before 1914, including *Roméo et Juliette* (Gall, Affre, Journet), *Faust* (Beyle, Campredon, Gresse), and *Carmen* (Affre, Albers, Merentié). In Italy from 1918 there were recordings of *Barbiere* (De Lucia), *Rigoletto*, and *Tosca*.

Electrical recording replaced the acoustic techniques in the 1920s. Sizeable extracts from *Parsifal*, *The Ring*, and *Tristan* were recorded at Bayreuth in 1927, and excerpts from a live performance in Berlin of *Meistersinger* (Schorr, List, Schützendorf, Hutt, Blech) in 1928. In the same year records were made of Shalyapin in *Boris* and *Faust* at London, Covent Garden, and complete recordings of operas by La Scala, Milan, began with *Rigoletto*; the popular Verdi–Puccini repertory was nearly all recorded twice, by HMV and Columbia. In Paris, performances of *Carmen*, *Pelléas*, *Werther*, *Manon*, *Faust*, and *Manon Lescaut* were recorded. In the 1930s and 1940s the Italian studios were able to record Gigli in complete performances of *Pagliacci*, *Cavalleria*, *Tosca*, *Bohème*, *Butterfly*, *Andrea Chenier*, *Ballo*, and *Aida*; and during the occupation of France there were complete recordings of *Pelléas* and *Damnation de Faust*. In America in 1947 Columbia signed a contract with the Metropolitan, and its first complete opera was *Hänsel und Gretel*. The introduction of the long-playing record in the 1950s opened up new horizons, especially with the invention of stereophonic sound. A huge repertory gradually became available, developing further with the success of the compact disc during the 1980s, later with opera on *film and video. The opera-lover now has access to a vast range of works he is unlikely ever to see in the opera-house. The benefits greatly outweigh the drawbacks, though the latter include the lack of the real theatrical impulse which must remain at the heart of opera, the danger that studio circumstances and editings condition listeners to false expectations of what singers and players may be able to achieve in reality, and not least the suggestion that a performance is something perfectible rather than adventurous.

**'Re dell'abisso'.** Mlle Arvidson's (con) aria in Act I of Verdi's *Un ballo in maschera*, invoking the spirits.

**Reeves, Sims** (*b* Shooters Hill, 26 Sep. 1818; *d* Worthing, 25 Oct. 1900). English tenor. Studied with his father, later Paris with Bordogni, Milan with Mazzucato. Début Newcastle 1839 as baritone (Rodolfo, *La sonnambula*). Début London, DL, as tenor, 1842–3; Milan, S, 1846; London, DL, 1847, and HM, between 1848 and 1878. Roles included Florestan, Huon, Edgardo, Faust (Gounod), and first Lyonnel (Balfe's *The Maid of Honour*). Berlioz, for whom he sang in *La damnation de Faust*, London, 1848, referred to him as 'the god-tenor' (somewhat ironically, since the singer deprecated Berlioz's own god, Gluck).

**Refice, Licinio** (*b* Patrica, Rome, 12 Feb. 1885; *d* Rio de Janeiro, 11 Sep. 1954). Italian composer and conductor. Studied Rome, S Cecilia, with Boenzi, Falchi, and Renzi. Ordained 1910; maestro di cappella at S Maria Maggiore, Rome 1911–47 . Most of his music is for the Church, but he also wrote two operas, *Cecilia* (for Muzio) and *Margherita da Cortona*. He died while conducting a performance of *Cecilia*, leaving incomplete a further opera, *Il mago*.

SELECT WORKLIST: *Cecilia* (Mucci; Rome 1934); *Margherita da Cortona* (Mucci; Milan 1938).

BIBL: T. Onofri and E. Mucci, *Le composizioni di Licinio Refice* (Assisi, 1966).

**Regensburg** (formerly Ratisbon). City in Bavaria, Germany. Travelling companies visited in the 17th cent. Under Theodor von Schacht, Italian opera was given 1774–8 and 1784–6. Emanuel *Schikaneder was director 1786–9, Ignaz Walter 1804-22. The theatre burnt down 1849; rebuilt 1852.

**Reggio Emilia.** Town in Emilia-Romagna, Italy. Opera is given at the T Municipale (cap. 1,600), opened 1857 with Achille Peri's *Vittore Pisani*. The auditorium, designed by Costa, has 106 boxes and is one of the most beautiful in Italy.

**Regie**. The term used in German opera-houses for 'production'.

**Regisseur** (Ger.) (Fr., *régisseur*). The *producer of an opera.

**register.** The tone quality in different parts of the voice: chest, middle, and head in women; chest, head, and falsetto in men. The difference in tone results from the variation of muscular tension in the larynx, arising from the vibration frequency of the note produced, but controllable by the singer. Chest tones have a weighty quality, head tones a light (and thus more flexible) nature. Ideal vocal production effects a seamless change between the registers (Melba succeeded brilliantly; Malibran had more difficulty), by combining the qualities of each. Different epochs placed different emphases on the various registers. Rossini favoured the use of head tone in his high passages (contemporary tenors used falsetto, until Donzelli introduced his A, and *Duprez his C, from the chest); the heavier orchestral textures of Romantic opera (as in Verdi and Wagner) demanded the development of the chest and middle registers. However, the use of too much chest voice in the higher registers, in an attempt to enlarge the tone, can lead to forcing, and consequent vocal deterioration.

Typical of most voices is the 'break' (It., *passaggio* or *ponticello*), an area between two registers where the vocal muscle, reaching the limit of tension, is in danger of disengaging, with resultant loss of tone and, frequently, pitch (e.g. around eb'–f♯' with sopranos; bb –d' with mezzo-sopranos). Phrases which cut across one or more registers can present problems with legato singing; much of the skill of good vocal composers lies in palliating this awkwardness, often by careful use of vowels.

**'Regnava nel silenzio'.** Lucia's (sop) aria in Act I of Donizetti's *Lucia di Lammermoor*, in which she recounts to Alisa the legend of the fountain.

**rehearsal.** A session of preparation preceding the performance of a work. In opera, first come the piano rehearsals, with a coach or *répétiteur first working privately with individual singers, then accompanying them, often under the direction of the conductor, in a studio. Then follow stage rehearsals, when the singers work on stage with the producer, sometimes with the conductor, accompanied by the piano. Orchestral rehearsals will meanwhile have been taken by the conductor, at first without the singers. The German term **Sitzprobe** is now widely used for the first (and any subsequent) unstaged rehearsal with seated singers and the orchestra, directed by the conductor. Stage and orchestra rehearsals follow, in which the work is rehearsed piecemeal. Technical rehearsals ('techs') are for the benefit of the lighting crew and stage staff, with the set and the producer, sometimes also with the singers in costume. However, the piano dress rehearsal may well be treated as the first stage and technical runthrough. The pre-dress rehearsal is the first complete stage run-through with orchestra, possibly also the last before the dress rehearsal. 'Dress rehearsal' remains the normal term for the final rehearsal, though this is often virtually a trial performance, without interruptions, sometimes before an invited audience or (with a new work) even members of the Press. See PROBE, PROVA, RÉPÉTITION.

The Russians use the word *repetitsiya*, and further distinguish *zastolnya repetitsiya* ('table rehearsal'), at which the intended presentation of the work is discussed, and *vygorodkaya repetitsiya* ('closed rehearsal'), taking place in a rehearsal room or perhaps the foyer. There is also the *montirovochnaya repetitsiya*, with scenery, the *progon* (*progonnaya repetitsiya*), or first full rehearsal, and finally the *generalnaya repetitsiya*.

**Reichardt, Johann Friedrich** (*b* Königsberg, 25 Nov. 1752; *d* Giebichenstein, 27 June 1814). German composer and writer on music. Trained Königsberg with local musicians; after university took an extended tour through Germany and Bohemia, written up as two important sets of *Briefe eines aufmerksamen Reisenden die Musik betreffend* (1774 and 1776). In 1776, on Agricola's death, obtained the post of Kapellmeister to Frederick the Great in Berlin by sending him a copy of his *Le feste galanti*. Here he produced several Italian operas, including *Andromeda*, *Protesilao*, and *L'Olimpiade* and conducted many works by Graun and Hasse; with the melodrama *Cephalus und Procris* he also contributed to the repertoire of the German company. In 1783 he founded the Concert Spirituel to promote the performance of neglected music. A visit to Paris in 1785 produced the French operas *Tamerlan* and *Panthée*, which were originally intended for the Opéra and were strongly influenced by Gluck. Returning to Berlin, he had one of his greatest successes with *Brenno*, which one critic compared to Gluck. In 1794 he was dismissed for his subversive views and retired to Giebichenstein, near Halle, which became a meeting-place for many luminaries of the nascent Romantic movement. He briefly took the post of Kapellmeister at Kassel in 1808; in the same year he made an important journey to Austria, where in the following year his *Bradamante* was produced in a concert version.

The importance of Reichardt's songs has overshadowed consideration of his other works. However, his Singspiels, which included settings of three Goethe librettos, *Claudine von Villa Bella*, *Erwin und Elmire*, and *Jery und Bätely*, were much admired at the time and their generally warm reception is indicative of the gradual shift in Berlin from Italian to German taste. Reichardt's German operas, of which *Die Geisterinsel*, *Das Zauberschloss*, and *Der Taucher* were the most successful, represent an important contribution towards the eventual formulation of a style of national opera. He also made an abortive attempt to establish a new genre, the *Liederspiel: his best examples are *Lieb' und Treue* and *Kunst und Liebe*. Of Reichardt's critical writings, the most important for their comments on opera are *Über die deutsche comische Oper* (Hamburg, 1774), concentrating on Hiller's *Die Jagd* as example, and *Vertraute Briefe geschrieben auf einer Reise nach Wien* (Amsterdam, 1810).

SELECT WORKLIST (all first prod. Berlin): *Le feste galanti* (Villati; 1775); *Cephalus und Procris* (Ramler; 1777); *Andromeda* (Filistri; 1788); *Protesilao* (Sertor; 1789; comp. with Naumann); *L'Olimpiade* (Metastasio; 1791); *Tamerlan* (Schaum, after de Mandenville; comp. 1786, perf. 1800); *Orfeo* (Calzabigi; 1788, comp. with Bertoni); *Brenno* (Filistri; 1789); *Claudine von Villa Bella* (Goethe; 1789); *Erwin und Elmire* (Goethe; 1793); *Die Geisterinsel* (Gotter, after Shakespeare; 1798); *Jery und Bätely* (Goethe; 1801) *Das Zauberschloss* (The Magic Castle) (Kotzebue; 1802); *Der Taucher* (The Diver) (Bürde, after Schiller; 1811).

BIBL: R. Pröpper, *Die Bühnenwerke Johann Friedrich Reichardts* (Bonn, 1965).

**Reichenberg.** See OLOMOUC.

**Reicher-Kindermann, Hedwig** (*b* Munich, 15 July 1853; *d* Trieste, 2 June 1883). German mezzo-soprano, then soprano. Studied with her father, the bs-bar August *Kindermann. Début Karlsruhe 1871. Munich, N, 1871–7. Bayreuth 1876, Grimgerde in first *Ring*. Vienna 1878–80; Leipzig 1880, then singing for Neumann in his European Wagner tour 1882–3. First English *Ring*, 1882. Ill for a part of this, she rejoined it in Italy, only to die a few weeks later. Thought by Neumann to be 'the greatest dramatic singer of the second half of the 19th century', she was a rare and gifted artist with a voice like 'a swordblade flashing in the sunlight' (Weingartner), and a striking presence. Roles included Gluck's Orfeo, Fidès, Leonore, Agathe, Ortrud, and a breathtaking Brünnhilde and Isolde.

**Reichmann, Theodor** (*b* Rostock, 15 Mar. 1849; *d* Marbach, 22 May 1903). German baritone. Studied Berlin with Elssler, Milan with Lamperti. Début Magdeburg 1869 (Ottokar). Hamburg 1873–5; Munich 1875–82; Neumann's Wagner tour, 1882–3; Vienna 1883–9, 1893–1903. Bayreuth 1882–94, 1902; London, CG, 1884, 1892;

New York, M, 1889-91. Roles included Don Giovanni, Pizarro, Guillaume Tell, Hamlet, Verdi's Iago, Flying Dutchman, Wolfram, Sachs, and first Amfortas. A highly regarded artist of great integrity and dramatic energy, he was notable for his carefully considered characterizations.

**Reimann, Aribert** (*b* Berlin, 4 Mar. 1936). German composer and pianist. Studied Berlin with Blacher. The early influences of Webern and particularly Berg remain in *Ein Traumspiel*, and are absorbed in *Melusine*. A modern treatment of the *Melusine myth, concerning a failure of two characters to reconcile civilized and natural order, with Nature wreaking final revenge, the work makes intelligent use of the Romantic elements now more prominent in Reimann's idiom; this includes some lyrical vocal writing for the spirits contrasting with a more declamatory manner for the mortals. Fischer-Dieskau, whom he often accompanied, suggested *Lear* to him, and has been identified with the title-role. The music is post-Bergian, both in atmosphere and in some of its technical resources, but finds its own structural strengths and its own vein of lyricism. After the bleak, claustrophic chamber work *Die Gespenstersonata*, he returned to full-scale opera with the more dramatic and approachable *Troades*.

SELECT WORKLIST: *Ein Traumspiel* (A Dream Play) (Henius, after Strindberg; Kiel 1965); *Melusine* (Henneberg, after Goll, after the legend; Schwetzingen 1971); *Lear* (Henneberg, after Shakespeare; Munich 1978); *Die Gespenstersonate* (The Ghost Sonata) (Schendel & Reimann, after Strindberg; Berlin 1984); *Troades* (The Trojan Women) (Albrecht & Reimann, after Werfel, after Euripides; Munich 1986).

**Reina, Domenico** (*b* Lugano, 1797; *d* Lugano, 29 July 1843). Italian tenor. Studied Milan with Boile. Début 1820. London, H, 1823; Milan, S, 1829, 1831, 1833-6. Created Arturo (*La straniera*), paying particular attention to Bellini's wishes; also impressed Bellini as Pollione (to Pasta's Norma) in 1832, when he sang 'with such fire', Bellini wrote, that people thought the music had been changed. Also first Leicester in the re-written *Maria Stuarda* with Malibran, and Tamas (*Gemma di Vergy*).

**Reinagle, Alexander** (*b* Portsmouth, *bapt.* 23 Apr. 1756; *d* Baltimore, 21 Sep. 1809). English composer of Austrian descent. Studied with his father, the trumpeter Joseph Reinagle, and Edinburgh with Taylor. In 1786 he travelled to the USA, settling in Philadelphia. For many years he was responsible for musical theatre in Philadelphia and Baltimore, directing and composing or arranging many pieces. Most of these are lost.

**Reine de Saba, La** (The Queen of Sheba). Opera in 4 acts by Gounod; text by Barbier and Carré, after Gérard de Nerval's *Les nuits de Ramazan* (in his *Le voyage en Orient*, 1851), in turn based on the Arab legend. (Nerval had originally submitted a libretto, now lost, to Meyerbeer.) Prod. Paris, O, 28 Feb. 1862, with P. Guéymard, Belval, L. Guéymard; Manchester 10 Mar. 1880; New Orleans, French OH, 12 Jan. 1899, with Fiérens, Bouxmann, Godefroy, and Darnaud.

Jerusalem, 950 BC. King Soliman's (bs) master architect Adoniram (ten) is in love with Balkis, the Queen of Sheba (sop); she leaves her betrothed, Soliman, but in the end Adoniram dies in her arms, killed by his own workmen and not, as she believes, by the vengeful Soliman.

Also subject of opera by Goldmark (see KÖNIGIN VON SABA, DIE).

**Reiner, Fritz** (*b* Budapest, 19 Dec. 1888; *d* New York, 15 Nov. 1963). Hungarian, later US, conductor. Studied Budapest with Kössler. Répétiteur Budapest, Comic O, 1909, cond. début 1909 (*Carmen*). Cond. Ljubljana 1910, Budapest 1911. Chief cond. Dresden 1914–21, where he was much influenced by Nikisch. Rome, Barcelona, 1921–2. In USA 1922–35, mostly concert conducting. London, CG, 1936–7, incl. Flagstad's London début as Isolde. San Francisco 1936–8. New York, M, 1949–53 (first US *Rake's Progress*, 1953). A strong orchestral disciplinarian with a gift for exceptionally lucid performances: Stravinsky had a high regard for '*l'amico* Fritz'. (R)

**Reinhardt, Delia** (*b* Elberfeld, 27 Apr. 1892; *d* Dornach, 3 Oct. 1974). German soprano. Studied Frankfurt with Strakosch and Schako. Début Breslau 1913. Munich, N, 1916–23; New York, M, 1922–4; Berlin, S, 1924–35; London, CG, 1924–7, 1929. Also Paris, O; Buenos Aires; etc. Sang Cherubino, Micaëla, Freia, Mimì, Verdi's Desdemona, Eva, Octavian, Empress, Christine. (R)

**Reinhardt, Max** (*b* Baden, 9 Sep. 1873; *d* New York, 31 Oct. 1943). Austrian producer, administrator, and actor. Although primarily a man of the theatre, he influenced operatic production in his collaboration with Roller on the première of *Der Rosenkavalier* (1911); he also staged the première of the original *Ariadne auf Naxos* (1912). He played a major part, with Strauss and Hofmannsthal, in the founding of the Salzburg Festival, 1920. His spectacular productions of *Hoffmann*, *La belle Hélène*, and *Die Fledermaus* were particularly remarkable for their close integration with the music.

**Reining, Maria** (*b* Vienna, 7 Aug. 1903; *d* Vienna, 11 Mar. 1991). Austrian soprano. Studied Vienna. Début Vienna, S, 1931 (soubrette roles). Vienna until 1933, 1937–58; Munich 1935–7; Salzburg from 1937; London, CG, and Chicago 1938. Milan, S; Paris, O; and New York City O, 1949. A refined and aristocratic singer with a warm, silvery tone, acclaimed as Countess (Mozart), Pamina, Euryanthe, Elsa, Eva, Tatyana, Butterfly, Marschallin, Ariadne, Arabella. (R)

**Reissiger, Karl** (*b* Belzig, 31 Jan. 1798; *d* Dresden, 7 Nov. 1859). German composer and conductor. Studied Leipzig with Schicht, Vienna with Salieri, Munich with Winter. Succeeded Weber as music director of the German Opera in Dresden, 1826; later he became Hofkapellmeister and also directed the Italian opera in place of Morlacchi. A vigorous supporter of Weber and Beethoven, he also championed Wagner and prepared the première of *Rienzi*; Wagner became his assistant, but they fell out over Reissiger's refusal to set Wagner's libretto *Die hohe Braut*, and Wagner then portrayed him as reactionary and indolent. Berlioz admired his work. Though a capable musician, he lacked distinctive talent as a composer: his operas mingle French and Italian styles with German Romantic elements. The most successful was *Die Felsenmühle zu Etalières* (Miltitz; Dresden 1831).

**Reiza.** Haroun al Raschid's daughter (sop), lover of Huon, in Weber's *Oberon*.

**Remedios, Alberto** (*b* Liverpool, 27 Feb. 1935). English tenor. Studied Liverpool with Francis, and London, RCM, with Carey. Début London, SW/ENO, 1955 (small roles). London, SW/ENO, then ENO, from 1955; Frankfurt 1968–70; London, CG, from 1965; Cologne, San Francisco from 1973; New York, M, from 1976. Roles include Don Ottavio, Don José, Dmitri, Bacchus, Verdi's Otello, Mark (*Midsummer Marriage*). Combines lyricism and stamina to great effect, especially as Walther, and Siegmund and Siegfried in the ENO *Ring*. (R)

**reminiscence motive.** A short theme identified with a person, place, object, or idea in an opera, which can then be reintroduced so as to recall its subject at a later stage of the drama. It is distinct from *Leitmotiv in that it does not normally change its shape very markedly, since its function is to represent and recall to mind, rather than to suggest dramatic progress. It was a feature of opera especially French opéra-comique, particularly at the end of the 18th cent. and in the early years of the 19th cent., before the development of Leitmotiv as a stronger structural and expressive force. An example is the chord sequence for the statue in Hérold's *Zampa*. In German **Reminiscenzmotiv** or **Erinnerungsmotiv**.

**Renard.** *Histoire burlesque chantée et jouée* in 2 parts by Stravinsky; text by composer, after Russian folk-tales, trans. into Fr. by C. F. Ramuz. Prod. Paris, O, 18 May 1922, by Dyagilev's Ballet Russe, with Fabert, Dubois, Narçon, Mahieux, cond. Ansermet; New York 2 Dec. 1923. Played by clowns, dancers, or acrobats, with singers (T.T.B.B.) placed in the orchestra.

The fox (ten) persuades the cock (ten) down from his perch by preaching to him; but the cat (bs) and the goat (bs) rescue him. Renewing his efforts, the fox again lures the cock into his reach, and he is rescued again, just in time, by the others suggesting that Mrs Fox is unfaithful. They strangle the fox with his own tail.

**Renaud, Maurice** (*b* Bordeaux, 24 July 1861; *d* Paris, 16 Oct. 1933). French baritone. Studied Paris and Brussels Cons. Début Brussels, M, 1883. Brussels until 1890, 1908–14; Paris, OC, 1890–1, and O, 1891–1902; London, CG, 1897–9, 1902–4; New York: Manhattan OC, 1906–7, 1909–10; M, 1910–12. Roles included Chorèbe, Telramund, Alberich, Beckmesser, Hérode (*Hérodiade*); created Reyer's High Priest (*Sigurd*) and Hamilcar (*Salammbô*). Retired 1919. A skilful singer and distinguished artist, he possessed great dramatic flair. (R)

**Re pastore, Il** (The Shepherd King). Opera in 2 acts by Mozart; text by Metastasio for Bonno (1751), rev. Varesco. Prod. Salzburg, Archbishop's Palace, 23 Apr. 1775, with Consoli; London, St Pancras TH, 8 Nov. 1954, with Ilse Wolf, J. Sinclair, Young, cond. Ucko; Yale Univ., Norwalk, CT, 7 July 1971, cond. Meyer.

Sidon, at the time of Alexander the Great (*c*.332 BC). After conquering Sidon, Alessandro (ten) decides to place on the throne a distant descendant of the royal house, Aminta (sop), who was brought up as a shepherd, and loves Elisa (sop), a shepherdess of noble Phoenician descent. Tamiri (sop), an exiled princess, is in love with Agenore (ten), a Sidonian noble and friend of Alessandro. Alessandro sends Agenore to offer Aminta the throne of Sidon and he is thus reunited with Tamiri.

But Elisa and Aminta fear that fate may ruin their happiness. Agenore persuades Aminta that as a king he cannot marry a shepherdess, and Alessandro, on hearing that Tamiri has been found, decides that Aminta shall marry her; Agenore renounces her. Rather than lose Elisa, Aminta gives up the throne, and Alessandro, impressed by their loyalty, makes them King and Queen of Sidon, promising to conquer new territory for Agenore and Tamiri.

Other operas on the subject are by Bonno (1751), Agricola (1752), Sarti (1753), Araia

(1755), Hasse (1755), Gluck (1756), Piccinni (1760), Galuppi (1762 and 1766), Jommelli (1764), Giardini (1765: seen by Mozart in London), and Guglielmi (1784).

**répétiteur** (Fr.: 'rehearser'). A pianist in an opera company who coaches and accompanies the singers. In France, however, the term normally used is *chef de chant*, in Italy *maestro collaboratore*, in Germany *Solo répétiteur* or sometimes *Korrepetitor*. A good 'rep' will be an excellent pianist and sight-reader, will correct musical and linguistic mistakes, and will be able to conduct a rehearsal.

**répétition** (Fr.: 'rehearsal'). In normal French usage, a session of preparation preceding the performance of a work. The early stages are taken by a 'répétiteur'. A 'répétition à l'italienne' is the equivalent of the German *Sitzprobe*. The predress rehearsal is known as the 'pré-générale', or 'colonelle', ranking one below the dress rehearsal or 'générale'; the latter is often virtually a trial performance, without interruptions, sometimes before an invited audience or (with a new work) even members of the Press. See PROBE, PROVA, REHEARSAL.

**rescue opera** (Fr, *pièce à sauvetage*, also in It. usage; Ger., *Rettungsoper* or *Schreckensoper*). The name given to an opera in which an essential part of the plot turns on the rescue of a hero or heroine from prison or some other threatening situation. Examples are to be found at various times in the 18th cent., but it developed into an identifiable genre with the French Revolution and the close involvement of opera with real-life situations, often highly dramatic. The first Revolutionary rescue opera was Berton's *Les rigueurs du cloître* (1790), concerning the repression of monasticism. Dalayrac's *Camille* (1791) involves the rescue of a girl from a haunted ruined castle; he also wrote *Léhéman, ou La tour de Neustadt* (1801). In Le Sueur's *La caverne* (1793) the heroine is held prisoner by brigands. Cherubini's *Les deux journées* (1800, to a libretto by Bouilly greatly prized by Beethoven) is set in the 17th cent. at the time of the Fronde but has contemporary allusions in its theme of the rescue of innocent aristocrats from arrest. Bouilly's libretto *Léonore* (1798) concerns the rescue of the hero more exclusively in its first setting by Gaveaux than in Beethoven's *Fidelio* (1805), in which the genre is raised to the level of the ideal of freedom.

**Residenztheater.** See MUNICH.

**Resnik, Regina** (*b* New York, 30 Aug. 1922). US soprano, later mezzo-soprano. Studied New York with R. Miller. Début New York, New OC, 1942 (Lady Macbeth). New York, M, from 1944; Bayreuth 1953. Roles included Donna Anna, Sieglinde, Ellen Orford, Alice Ford. Withdrew 1954 for further study with Danise. Mez début Cincinnati 1955 (Amneris). London, CG, 1957–72; also Chicago, Vienna, etc. A vivid dramatic imagination and firm, rich tone brought her much success as Carmen, Eboli, Klytemnestra, Mistress Quickly, Lucretia. From 1971 opera producer, e.g. *Carmen* (Hamburg), *Elektra* (Venice), *Falstaff* (Warsaw). (R)

**Respighi, Ottorino** (*b* Bologna, 9 July 1879; *d* Rome, 18 Apr. 1936). Italian composer. Studied with Torchi and Martucci in Bologna; in 1900 became violist in the opera orchestra at St Petersburg, where he studied with Rimsky-Korsakov; later attended lectures by Bruch in Berlin. His first operas, the comic *Re Enzo* and *Semirama*, made his name and led to an appointment at the conservatory in Rome in 1913, where he settled. Of the operas which followed, *Belfagor*, *Maria Egiziaca*, and *La fiamma* had some success, although none could match the popularity of *La bella dormente*.

Respighi did not have a natural talent for dramatic composition and relied instead for effect upon his sumptuous scoring and lyrical inventiveness. His style owes much to his teachers, especially Rimsky-Korsakov, as well as to Richard Strauss, whose *Salome* provided the model for *Semirama*, and to Ravel. He also had a strong interest in music of the past and made arrangements of works by Paisiello and Cimarosa. In 1934 he made a free transcription of Monteverdi's *Orfeo*, and in his last opera, *Lucrezia*, developed a principle of dramatic recitative based on that of monody. It was left unfinished at his death and completed by his wife **Elsa Respighi Olivieri-Sangiacomo** (*b* Rome, 24 Mar. 1894), a concert singer, who also composed two operas, *Alcesti* (1941) and *Samurai* (1945), in her own right.

WORKLIST: *Re Enzo* (Donini; Bologna 1905); *Semirama* (Cerè; Bologna 1910); *Marie-Victoire* (Guiraud; comp. 1913–14, unprod.); *La bella addormentata nel bosco* (later *La bella dormente nel bosco* (The Sleeping Beauty in the Wood) (Bistolfi, after Perrault; Rome 1922 as puppet opera, rev. for childrens' voices, Turin 1934, rev. for adult voices by Torchi, 1966); *Belfagor* (Guastalla; Milan 1923); *La campana sommersa* (The Sunken Bell) (Guastalla, after Hauptmann; Hamburg 1927); *Maria Egiziaca* (Guastalla; New York 1932); *La fiamma* (The Flame) (Guastalla; Rome 1934); *Lucrezia* (Guastalla; Milan 1937).

BIBL: R. de Renzis, *Respighi* (Rome, 1935); E. Respighi, *Ottorino Respighi: dati biografici ordinati* (Milan, 1954; trans. London, 1962).

**Reszke, de.** See DE RESZKE.

**Retablo de Maese Pedro, El** (Master Peter's Puppet Show). Opera in 1 act by Falla; text by composer after Cervantes's novel *Don Quixote* (1615). Prod. Seville, T San Fernando (concert), 23 Mar. 1923, with Redondo, Segura, Lledo, cond. Falla; Paris, Princesse Edmond de Polignac's house, 25 June 1923 (in French), with Dufranne, Salignac, Peris, cond. Golschmann, harpsichordist Landowska; Clifton 14 Oct. 1924, with Tannahill, Goody, Cranmer, cond. Boult; New York, TH, 29 Dec. 1925.

The piece was originally intended for performance with puppets taking all the parts, double-sized for the human beings. A boy narrator (treb) introduces the puppet show to an audience including Don Quixote (bar or bs) and Sancho Panza (silent). The interaction of reality and fiction characteristic of *Don Quixote* takes the form of the boy telling the story of Don Gayferos rescuing the fair Melisendra from the Moors, subject to interruptions from Don Quixote and Master Peter himself (ten). Eventually Don Quixote becomes confused about the reality of the puppets, and leaping forward, beheads the Moors and destroys the puppet show.

**Rethberg, Elisabeth** (*b* Schwarzenburg, 22 Sep. 1894; *d* New York, 6 June 1976). German, later US, soprano. Studied Dresden with Watrin. Début Dresden 1915 (Arsena, *Der Zigeunerbaron*). Dresden until 1922, then regularly; New York, M, 1922–6 (when she retired); London, CG, 1925, 1934–9. Also Milan, S; Salzburg, etc. Her enormous repertory included Constanze, Countess (Mozart), Pamina, Agathe, Marguerite, Selika, Eva, Sieglinde, Aida, Desdemona (Verdi), Marschallin, and first Helen (*Die ägyptische Helena*). An outstanding singer of impeccable style, she used her beautiful voice with consummate skill and musicality. Much admired by R. Strauss and Toscanini. (R)

**Reutter, Hermann** (*b* Stuttgart, 17 June 1900; *d* Heidenheim an der Brenz, 1 Jan. 1985). German composer and pianist. Studied with Courvoisier and Dorfmüller, before establishing a successful career as performer and teacher. Although his music is now seldom performed, in his day he was regarded as one of Germany's leading composers. His style owes much to the neo-Romanticism of Pfitzner, although in his early works he was also influenced both by the technique and spirit of Hindemith and traces of Orff's music are also to be detected.

Reutter has had a long involvement with opera, perhaps not surprisingly since he was himself a distinguished man of letters and through accompanying such singers as Karl Erb and Sigrid Onegin developed a sensitive insight into the handling of the human voice. Of his eight operas the most successful was *Dr Johannes Faustus*, which like much of Reutter's music at this time took inspiration from German folk-song and contrasts sharply with its successor, the epic *Odysseus*. Several works, including *Der verlorene Sohn*, *Der Weg nach Freudenstadt*, and *Der Lübecker Totentanz*, explore Expressionist themes.

WORKLIST: *Saul* (after Lernet-Holenia; Baden-Baden 1928, rev. Hamburg 1947); *Der verlorene Sohn* (The Prodigal Son) (Rilke, after Gide; Stuttgart 1929, rev. Dortmund 1952); *Dr Johannes Faustus* (Andersen; Frankfurt 1936, rev. Stuttgart 1955); *Die Prinzessin und der Schweinehirt* (The Princess and the Swineherd) (after Andersen; Mainz 1938); *Odysseus* (Bach, after Homer; Frankfurt 1942); *Der Weg nach Freudenstadt* (Reutter; Göttingen 1948); *Der Lübecker Totentanz* (The Lübeck Death-Dance) (Reutter, after Holbein; Göttingen 1948); *Don Juan und Faust* (Andersen, after Grabbe; 1949); *Die Witwe von Ephesus* (Reutter, after Petronius; Cologne 1954, Stuttgart 1950, rev. Schwetzingen 1966); *Die Brücke von San Luis Rey* (Wilder; Essen 1954).

BIBL: H. Lindlar, ed., *Hermann Reutter; Werk und Wirken: Festschrift der Freunde* (Mainz, 1965).

**Revisor, Der.** Opera in 5 acts by Egk; text by composer, after Gogol's story *The Government Inspector* (1836). Prod. Schwetzingen by Stuttgart O, 9 May 1957, with Sailer, Plümacher, Stolze, Wunderlich, Ollendorf, cond. Egk; London, SW, with New Opera Co., 25 July 1958, with Clarke, Peters, Young, Platt, cond. Lovett; New York, CO, 19 Oct. 1960, with Brooks, Kobart, Crain, Beattie, cond. Egk.

A small Russian town; *c.*1840. Khlestakov (ten), a penniless civil servant, is mistaken for the Government Inspector and lavishly entertained by the Mayor (bs), his wife (mez), and daughter (sop).

**Reyer, Ernest** (*b* Marseilles, 1 Dec. 1823; *d* Le Lavandou, 15 Jan. 1909). French composer. Studied Marseilles, Paris with Farrenc, but largely self-taught. Made an early reputation with his works, including two opéras-comiques and an opera, *Erostrate*. His most famous work was *Sigurd*, a success he almost equalled with *Salammbô*. Though *Sigurd* is Wagnerian in subject-matter, and Reyer was a great admirer of Wagner (as of Berlioz and Weber), his work is not greatly coloured by any of their influences. His music does not reveal great individuality, though it shows an appreciation of the strengths of Grand Opera and is excellently scored, but reflects the intelligence and independence of mind which made him a witty and forceful critic. He defended Wagner and Berlioz eloquently, and travelled widely in the course of his work, making a visit to Cairo for the première of *Aida*.

SELECT WORKLIST: *Erostrate* (Méry & Pacini; Baden-Baden 1862); *Sigurd* (Du Locle & Blau; Brussels

1884); *Salammbô* (Du Locle, after Flaubert; Brussels 1890).

WRITINGS: *Notes de musique* (Paris, 1875); *Quarante ans de musique* (Paris, 1909).

BIBL: H. de Curzon, *Ernest Reyer, sa vie et ses œuvres* (Paris, 1924).

**Reyzen, Mark** (*b* Zaytsevo, 3 July 1895 *d* Moscow, 25 Nov. 1992). Russian bass. Studied Kharkov with Bugamelli. Début Kharkov 1921 (Pimen). Leningrad, K, 1925–30; Moscow, B, 1930–54. Also Paris, O; Berlin, S; Dresden; Budapest; etc. Sang Gremin at Moscow, B, on his 90th birthday, still in good voice. A sensitive actor with a warm, solid tone, remarkable in a wide range of characters, e.g. Mozart's Don Basilio, Gounod's Méphistophélès, Philip, Susanin, Boris, Dosifey. (R)

**Rezia.** See REIZA.

**Rezniček, Emil von** (*b* Vienna, 4 May 1860; *d* Berlin, 2 Aug. 1945). Austrian composer and conductor. Studied Graz with Mayer, Leipzig with Reinecke and Jadassohn. He held many conducting posts, including at Mannheim (1896–9); Warsaw (1907–8); and Berlin (1909–11). Of his operas, the best known are *Donna Diana* (Rezniček, after Cavana; Prague 1894), especially for its sparkling overture, and *Till Eulenspiegel* (Rezniček, after Fischart; Karlsruhe 1902), though he also wrote a number of serious works.

BIBL: R. Sprecht, *Emil Nikolaus von Rezniček: eine vorläufige Studie* (Vienna, 1923).

**Rheingold, Das.** See RING DES NIBELUNGEN, DER.

**Riccardo Primo.** Opera in 3 acts by Handel; text chiefly by P. A. Rolli, after F. Briani's libretto *Isacio tiranno*, set by Lotti, 1710. Prod. London, K, 11 Nov. 1727, with Cuzzoni, Faustina, Senesino, Boschi. Given in Hamburg in an adaptation by Telemann, Feb. 1729. Rev. Handel O Society 1964.

**Ricci, Federico** (*b* Naples, 22 Oct. 1809; *d* Conegliano, 10 Dec. 1877). Italian composer. Studied Naples with Zingarelli and Raimondi, also with Bellini and his brother **Luigi** (*b* Naples, 8 July 1805; *d* Prague, 31 Dec. 1859), who studied Naples with Furno, Zingarelli, and Generali.

Apart from works written in collaboration with his brother, Federico had successes with *La prigione d'Edimburgo* (Rossi, after Scott; Trieste 1838), *Luigi Rolla* (Cammarano; Florence 1841), and *Corrado d'Altamura* (Sacchero; Milan 1841). The latter, based on the same plot as Verdi's *Oberto*, led to other commissions; but his serious operas fared less well than his comedies. *Il marito e l'amante* (The Husband and the Lover) (Rossi; Vienna 1852) was successful, but *Il paniere d'amore* (The Basket of Love) (?Ricci; Vienna

1853) failed, and he accepted a post in St Petersburg. In 1869 he moved to Paris, where he had some success with an opéra-bouffe, *Une folie à Rome* (An Escapade in Rome) (Wilder; Paris 1869). The most famous of the brothers' collaborations was *Crispino e la comare* (Crispino and the Fairy) (Piave, after Fabbrichesi; Venice 1850), to which he contributed the lighter numbers that reflect his elegant craftsmanship.

Luigi, the more imaginatively gifted brother, was also less successful as a serious composer than in comedy, though he first became popular with his *Chiara di Rosembergh* (Rossi; Milan 1831). He had a great success with *Un avventura di Scaramuccia* (Romani; Milan 1834). In Trieste from 1836, he unwisely attempted *Le nozze di Figaro* (Rossi; Milan 1838). He then moved to Odessa with his twin-sister mistresses, the sopranos **Francesca** and **Ludmilla Stolz** (*b* 8 May 1827; sisters of Teresa *Stolz). He wrote for them *La solitaria delle Asturie* (Romani; Odessa 1845). They travelled together to Copenhagen and back to Trieste. Ricci had further successes with comic operas, before dying insane. His music has a lively tunefulness, and he was probably the leading spirit in the collaborations with his brother. He married Ludmilla, and their daughter **Adelaide** (*b* Trieste, 1 Dec. 1850; *d* 1871) sang in Paris 1868–9; his son by Francesca, **Luigi** (*b* Trieste, 27 Dec. 1852; *d* Milan, 10 Feb. 1906), was a composer whose operas included *Frosina* (1870), *Cola di Rienzo* (1880), and *Don Chischiotte* (1887).

A younger brother, **Egidio**, became an impresario in Naples, then in Copenhagen (1850), where he married the soprano Amalie Luzio.

BIBL: L. de Rada, *I fratelli Ricci* (Florence, 1878).

**Ricciarelli, Katia** (*b* Rovigo, 18 Jan. 1946). Italian soprano. Studied Venice with Adami-Corradetti. Début Mantua 1969 (Mimì). Milan, S, from 1973; Chicago 1973; London, CG, and New York, M, from 1974. Also Moscow, B; Barcelona; Berlin, D; Verona; etc. An attractive artist with a warm, ripe, lucent tone, capable of much emotional depth. Large repertory includes Imogene, Giulietta (Bellini), Lucrezia Borgia, Micaëla, Elisabeth de Valois, Desdemona (Verdi), Mimì, Luisa Miller. (R)

**Rich, John** (*b* ?London, 1691 or 1692; *d* London, 26 Nov. 1761). English producer and manager. As manager of Lincoln's Inn Fields T, he introduced Gay's *The Beggar's Opera* to London; when this, in the famous mot of the time, 'made Gay rich, and Rich gay', he was able to build the first Covent Garden, 1732.

**Richard, Cœur-de-lion.** Opera in 3 acts by Grétry; text by Sedaine, after 13th cent. fable.

Prod. Paris, CI, 21 Oct. 1784, with Rosalie, Colombe, Dugazon, Philippe, Clairval; 2nd version 22 Dec. 1785; final version 29 Dec. 1785; London, CG, 16 Oct. 1786; Boston 23 Jan. 1797.

Austria, end 12th cent. Blondel (ten), Richard's (ten) minstrel, wanders in search of his King disguised as a blind singer. Helped by Sir Williams (bs), his daughter Laurette (sop), and Marguerite of Flanders (sop), he manages first to make contact with Richard by means of his romance, 'Une fièvre brûlante', and then to rescue him from the castle where he is held prisoner under Florestan (bs). Grétry consciously designed the romance, which appears nine times in the opera in various melodic and rhythmic transformations, as a form of motive; and it became a much-admired example in the later development of Leitmotiv.

Also opera by Shield (1786).

**Richter, Hans** (*b* Raab (Györ), 4 Apr. 1843; *d* Bayreuth, 5 Dec. 1916). Austro-Hungarian conductor. Studied Vienna with Sechter. Hornplayer, Vienna, K, 1862–6. Worked with Wagner at Tribschen 1866–7 on fair copy of *Meistersinger* and 1870 on *Ring*; played trumpet in first performance of *Siegfried Idyll*. Wagner recommended him to Bülow as chorus-master for Munich. Cond. Munich 1868–9 (début *Guillaume Tell*), also singing at short notice Kothner in sixth performance of *Meistersinger*. Cond. *Lohengrin*, Brussels 1870. Chief cond. Budapest 1871–5. Vienna 1875–1900, music director from 1893. Bayreuth 1876–1912, cond. first *Ring*. London: DL 1882, first English *Tristan* and *Meistersinger*; CG 1884, 1903–1910, where, after his production of *The Ring* in English, his attempts to found a permanent English national opera in association with Percy *Pitt, were thwarted by the Grand O Syndicate at HM.

One of the greatest German conductors of his time, Richter was (with *Levi, *Mottl, and *Muck) in the first generation of Wagner conductors who worked with the composer in Bayreuth and founded a tradition. Accounts of his interpretations suggest gravity, a deep understanding of structure, and a strong lyrical flow that did not exclude lightness of touch. (R)

**Ricordi.** Italian firm of music publishers.

1. **Giovanni** (1785–1853) began as a copyist and issued his first publication in 1808. Benefiting from his close association with La Scala, Milan, he built up the firm and secured himself a powerful commercial position. He was able to hire out works by Rossini and Bellini, and he published much Donizetti as well as works by Mercadante, Vaccai, Pacini, the Ricci brothers, and Meyerbeer.

2. **Tito** (1811–88), son of (1), was director 1853–88, greatly expanding the business by opening various branches and taking over other firms.

3. **Giulio** (1840–1912), son of (2), was director 1888–1912. He exercised a notable force on Italian opera. Himself a minor composer (his single opera, *La secchia rapita*, Turin 1910, was to a text by Renato Simoni, later librettist of *Turandot*), he combined shrewd musical judgement in his choice of composers with ruthless business acumen in furthering their and the firm's interests. Though he did attack Toscanini, and reject Bizet and Leoncavallo, he championed Verdi with a fervour based on personal friendship and belief in Verdi's supremely Italian genius (Wagner was an enemy so vile that Ricordi even continued the attacks after acquiring the composer's operas with Lucca's stock in 1887). Ricordi quickly recognized Puccini as the 'Crown Prince' to Verdi, and helped him incalculably in his early career. Librettos were bought in case Puccini might use them, and Ricordi was also able to persuade Franchetti, dissatisfied with Illica's *Tosca* text, to hand it over to Puccini. 'Don Giulio' became, for Puccini, 'The only person who inspires me with trust, and to whom I can confide what is going through my mind.' He was the first to advertise operas with illustrated posters (starting with *La bohème*), and these were later used for the early editions of the vocal scores.

4. **Tito** (1865–1933), son of (3), was impulsive and dictatorial; as producer of *Tosca* he showed the importance of good acting and realistic scenery in verismo opera, but his treatment of Puccini was less tolerant than his father's, and their relations deteriorated. Puccini even contemplated a London publisher for one projected opera. When Tito left the firm in 1919, Puccini patched up their quarrel. The firm retains many of Verdi's and Puccini's scores, and controversy over discrepancies between these and the published versions even reached the Italian Senate in 1961.

**Ridderbusch, Karl** (*b* Recklinghausen, 29 May 1932). German bass. Studied Essen. Début Münster 1961. Deutsche Oper am Rhein from 1965; New York, M, 1967–9, 1976–7; Bayreuth from 1967; London, CG, 1971, 1973. Also Paris, O; Buenos Aires; Milan, S; Chicago; Berlin, D. Powerful of voice and build; repertory includes Rocco, Daland, Hunding, Hagen, Sachs, Ochs, Boris, Doktor (*Wozzeck*). (R)

**Ride of the Valkyries.** The name generally given to the choral scene opening Act III of Wagner's *Die Walküre*, in which the warrior-maidens, Wotan's daughters, gather on a rocky peak.

**Riders to the Sea.** Opera in 1 act by Vaughan Williams; text, J. M. Synge's tragedy (1904). Prod. London, RCM, 30 Nov. 1937, with Olive Hall, Smith-Miller, Steventon, Coad, cond. Sargent; Cleveland, Western Reserve U, 26 Feb. 1950.

The setting is the west coast of Ireland, where Maurya (con) has lost her husband and four sons at sea. A fifth is also discovered to have been drowned when her daughters Cathleen (sop) and Nora (sop) identify some clothes brought in; and when her last son Bartley (bar) in turn is claimed by the sea as he is riding some horses to a fair, her grief finally turns to resignation and peace. Other operas on the subject are by Rabaud (*L'appel de la mer*, 1924), and Betts (1955).

**Rienzi** (orig. *Cola Rienzi, der letzte der Tribunen*). Opera in 5 acts by Wagner; text by composer, after Mary Russell Mitford's drama (1828) and Bulwer Lytton's novel (1835). Prod. Dresden, H, 20 Oct. 1842, with Wüst, Schröder-Devrient, Tichatschek, Dettmer, Wächter, cond. Reissiger; New York, AM, 4 Mar. 1878, with Pappenheim, Hüman, Adams, Wiegand, cond. Menetzek; London, HM, 27 Jan. 1879, with Crosmond, Vanzini, Maas, Olmi, cond. Rosa.

Set in 14th-cent. Rome, the opera tells of the struggle between the Orsinis and the Colonnas. Paolo Orsini (bs) attempts to abduct Irene (sop), sister of Cola Rienzi (ten), but is interrupted by Stefano Colonna (bs), whose son Adriano (mez) appears to defend Irene. Rienzi, the papal legate, appears, and encouraged by Cardinal Raimondi (bs) urges the people to resist the tyranny of the nobles. The nobles swear allegiance to Rienzi as tribune, but they plot to murder him, and almost succeed. Condemned to death, they are spared through Adriano's intercession; but when they break their oath of submission, the people rise and kill them. However, the people in turn prove disloyal to Rienzi, and Adriano also tries to kill him. Rienzi is excommunicated, and Adriano now warns Irene of her brother's danger. She finds him at prayer in the Capitol, and refusing to flee with Adriano, remains with him as the mob fires the Capitol; Adriano rushes into the flames to perish with them.

Other operas on the subject are by John Barnett (1828, based only on Mitford), Conrad (1839), A. Peri (1862), Lucilla (1872), Persichini (1874), and L. Ricci (1880).

BIBL: J. Deathridge, *Wagner's Rienzi* (Oxford, 1977).

**Riga.** See LATVIA.

**Righetti-Giorgi, Geltrude.** See GIORGI-RIGHETTI, GELTRUDE.

**Righini, Vincenzo** (*b* Bologna, 22 Jan. 1756; *d* Bologna, 19 Aug. 1812). Italian composer. Studied Bologna with Martini. His early career was as a singer, which took him to Parma and Prague. Here his first opera, *Il convitato di pietra*, was given, a treatment of the Don Juan legend which pre-dated that of Da Ponte and Mozart by ten years. In 1780 he was called to Vienna by Joseph II to become director of the Italian Opera and singing teacher to the Princess Elizabeth of Württemberg. After seven years he moved to Mainz, and quickly on to Berlin, where the successful performance of *Enea nel Lazio* secured for him the post of Hofkapellmeister and director of Italian Opera in 1793. His career here was highly successful and with the librettist Filistri he produced a number of popular triumphs, including *Atalante e Meleagro* and *Tigrane*.

He married first the contralto **Anna Maria Lehritter** (1762–83) and then the soprano **Henrietta Kneisel** (1767–1801). He also made a reputation as a singing-teacher.

SELECT WORKLIST: *Il convitato di pietra* (The Stone Guest) (Porta; Prague 1776); *Enea nel Lazio* (Filistri; Berlin 1793); *Atalante e Meleagro* (Filistri; Berlin 1797); *Tigrane* (Filistri; Berlin 1800).

**Rigoletto.** Opera in 3 acts by Verdi; text by Piave, after Hugo's drama *Le roi s'amuse* (1832). Originally entitled *La maledizione*. Prod. Venice, F, 11 Mar. 1851, with Brambilla, Saini, Mirate, Varesi, Ponz, Damini, cond. Marès; London, CG, 14 May 1853, with Bosio, Mario, Ronconi; New York, AM, 19 Feb. 1855, with Bertucca-Maretzek, Frezzolini Bolcioni, Barilli.

Mantua, 16th cent. Rigoletto (bar), the hunchbacked court jester, taunts Count Ceprano (bs), whose wife is the latest fancy of the licentious Duke. The courtiers resolve to abduct Rigoletto's 'mistress' in revenge. She is in fact his daughter Gilda (sop). Count Monterone (bar) bursts in to confront the Duke, who has seduced his daughter. He is arrested and curses Rigoletto, who is mocking him. On going home, Rigoletto is accosted by Sparafucile (bs), a professional assassin, who offers his services. The jester enjoins Gilda never to leave the house on her own. She does not tell him of the visits of the handsome young student she met in church (the Duke in disguise). The courtiers trick Rigoletto into helping them abduct Gilda.

In the ducal palace, the courtiers report their exploits with glee. The Duke is delighted to hear Gilda is there and rushes off to her. A despondent Rigoletto enters, and a series of lies leads him to realize that the Duke must be with Gilda. She runs in, having been seduced by the Duke, and throws herself into her father's arms. Even the courtiers are abashed to find that she is really Rigoletto's daughter. He swears vengeance.

Rigoletto and Gilda go to Sparafucile's inn outside the town, where Rigoletto has arranged for the Duke's murder. The Duke appears and flirts with Maddalena (con), Sparafucile's sister, singing to her. This distresses Gilda, who still loves him. Maddalena has grown fond of the Duke and tries to dissuade her brother from murdering him. They agree that any stranger coming to the inn before midnight should substitute for the Duke. Gilda, disguised in man's clothing, overhears this and resolves to sacrifice her life. Rigoletto returns and Sparafucile hands over a sack containing a body. As Rigoletto is about to throw it into the river, he hears the Duke singing his song again. Realizing that he has been deceived, Rigoletto opens the sack and a flash of lightning reveals Gilda's face. She is still alive, but dies in his arms, begging forgiveness. Monterone's curse is fulfilled.

**Rihm, Wolfgang** (*b* Karlsruhe, 13 Mar. 1952). German composer. Studied Karlsruhe, and with Stockhausen. After *Faust und Yorick* (1977), he had a striking success with *Jakob Lenz* (1979), in which he abandoned serialism for a freer manner (influenced by Berg) that included some extremes of vocal writing in an ultra-Expressionist manner. *Die Hamletmaschine* (1987) follows Stockhausen in seeking a total theatre of sound and non-narrative, ritualistic drama. *Oedipus* (1987) fragments Sophocles in order to reconstruct the essence of the drama through music. *Die Eroberung von Mexico* was produced 1992.

**Rijeka** (It., Fiume). City in Croatia. The first theatre was built 1765, closed by the government 1797. Ivan *Zajc was director 1895. The company re-formed in 1946.

**Rimsky-Korsakov, Nikolay** (*b* Tikhvin, 18 Mar. 1844; *d* Lyubensk, 21 June 1908). Russian composer. Studied piano with Canille, and had guidance from Balakirev.

With two exceptions, all Rimsky-Korsakov's operas are on Russian themes. Mythology attracted him, as did fantastic subjects, and his first opera, *The Maid of Pskov*, is unusual in being an historical costume drama that treats of human emotions. His gift for harmonic and orchestral colour, coupled with his love of legend and fairy-tale, drew him more to the type of opera of which *May Night* was the first. This is a Gogol-inspired work dealing with supernatural elements in a colourful folk setting, and characteristically includes a good number of set pieces such as dances for wedding guests, skomorokhi, and rusalki, choruses, and a song for a blind bard. These could release his imagination and indulge his virtuoso technique without demanding much of his feelings for human character.

In *The Snow Maiden*, fairy and human worlds impinge; but the heroine's dilemma is expressed more by Rimsky-Korsakov's demonstration that they are irreconcilable than by anything beyond the very simplest characterization of the heroine, who is touching but not truly sympathetic. This is a weakness underlined when he reverted to village comedy with *Christmas Eve*, which for all its entertaining brilliance lacks the warmth of Tchaikovsky's setting of the same subject (*Vakula the Smith*). This and *Mlada* were regarded as studies for *Sadko*, in which he responds vividly to the picturesqueness of a story that affords ample opportunity for spectacle and the exotic. His idiom was by now slightly tinged by a Wagnerian influence (*The Ring* had had its St Petersburg premiere in 1888–9) and yet also a wish to strengthen his feeling for Russia's past by a use of a special bardic type of declamation.

However, his next work was *Mozart and Salieri*, a neoclassical opera (the first) based on Pushkin's 'little tragedy' dramatizing the rumour that Mozart had been poisoned by his rival. It is a conscious descendant of Dargomyzhsky's *The Stone Guest*, though it is more openly melodic and includes some pastiche. But feeling that this was not a way forward, he turned in *The Tsar's Bride* back to the kind of historical costume drama with which he had begun his career. Less self-indulgently colourful, this represents a return also to a more lyrical manner, and is a bel canto opera with songs and ensembles; it is the peak of his achievement in melodic composition. *The Tale of Tsar Saltan* combines fantasy with satire, whereas *Kashchey the Immortal* is a short, magic opera in which the characters are represented as much by the orchestra as vocally, allowing the composer to indulge in some dazzling orchestration that had a distinct influence on his pupil Stravinsky's *The Firebird*. These qualities reappear in *The Legend of the Invisible City of Kitezh*, a work involving the familiar dances and set pieces, but also combining a feeling for Orthodox sanctity and Dostoyevskyan mysticism with a somewhat Tolstoyan moral fervour: it includes his best single character, the spiritually tormented, drunken scoundrel Grishka Kuterma. *The Golden Cockerel* mixes the old fantastic elements with a vein of satire: that much is left obscure is of little importance since the air of mystery is without meaning and is no more than the magician's deception of his audience.

Rimsky-Korsakov was also responsible for completing, revising, and orchestrating a number of works by other Russian composers which he believed should see as friendly a light of day as possible. His revisions of Musorgsky's *Boris Godunov* have been much criticized, and his version has now been generally replaced in the theatre by the original, but it should be remembered that this was his expressed intention: at the time, he did not think the world would understand Musorgsky's originalities. Autobiography, *Letopis moyey muzikalnoy zhizni* (Chronicle of my Musical Life) (St Petersburg, 1909; trans. 1924 & 1942).

WORKLIST (all first prod. St Petersburg unless otherwise stated): *Pskovityanka* (The Maid of Pskov) (Rimsky-Korsakov, after Mey; 1873, 3rd version 1895); *Mlada* (opera-ballet; Krylov, unfin., comp. with Borodin, Cui, Musorgsky, & Minkus); *Mayskaya Noch* (Rimsky-Korsakov, after Gogol; 1880); *Snegurochka* (The Snow Maiden) (Rimsky-Korsakov, after Ostrovsky; 1882, 2nd version 1898); *Mlada* (opera-ballet; Rimsky-Korsakov, after Krylov; 1892); *Noch pered Rozhdestvom* (Christmas Eve) (Rimsky-Korsakov, after Gogol; 1895); *Sadko* (Rimsky-Korsakov & Belsky; Moscow 1898); *Mozart i Salieri* (Pushkin original; Moscow 1898); *Boyarinya Vera Sheloga* (Rimsky-Korsakov, after Mey; Moscow 1898); *Tsarskaya nevesta* (Rimsky-Korsakov, after Mey; Moscow 1899); *Skazka o Tsare Saltane* (The Tale of Tsar Saltan) (Belsky, after Pushkin; Moscow 1900); *Servilia* (Rimsky-Korsakov, after Mey; 1902); *Kashchey bessmertny* (Kashchey the Immortal) (Rimsky-Korsakov, after Petrovsky; Moscow 1902); *Pan Voyevoda* (Tyumenev; 1904); *Skazaniye o nevidimom grade Kitezhe i deve Fevronii* (The Legend of the Invisible City of Kitezh and of the Maid Fevronia) (Belsky; 1907); *Zolotoy petushok* (The Golden Cockerel) (Belsky, after Pushkin; Moscow 1909).

BIBL: G. Abraham: *Rimsky-Korsakov* (London, 1945).

**Rinaldo.** Opera by Handel in 3 acts; text by G. Rossi, after a sketch by Aaron Hill from Tasso's first epic (1562). Prod. London, H, 24 Feb. 1711, with Boschi, Girardeau, Schiavonetti, Nicolini, Valentini; Houston, Jones H, 16 Oct. 1975, with Horne, Rogers, Mandac, Ramey, cond. Foster. The first of Handel's operas for England; one of the few original librettos he set.

**Rinaldo di Capua** (*b* Capua or Naples, *c*.1705; *d* ?Rome, *c*.1780). Italian composer. Music to only six of his 32 operas survives. His first success came in Rome in 1737 with an untitled comic opera and the opera seria *Ciro riconosciuto*. His most successful works were the drammi giocosi *La commedia in commedia* and *La libertà nociva*, which were given throughout Italy. The former was heard in an abridged version in Paris (1752) as *La donna superba*, where it was also parodied as *La femme orgueilleuse*. Together with performances of the intermezzo *La zingara* it played an important role in the *Guerre des Bouffons. He also composed many opere serie of which the most successful was *Vologeso, re de' Parti*. Burney characterized him as 'a composer of great genius and fire'.

SELECT WORKLIST (all first prod. Rome): *Ciro ricono-*

*sciuto* (Metastasio; 1737); *La commedia in commedia* (Barlocci; 1738); *La libertà nociva* (Barlocchi; 1740); *Vologeso, re de' Parti* (Luccarelli; 1739).

**Ring des Nibelungen, Der** (The Ring of the Nibelung). A stage-festival play for three days and a preliminary evening (*Ein Bühnenfestspiel für drei Tage und einen Vorabend*)—sometimes called a tetralogy—by Wagner; text by the composer, based on the Nibelung Saga. Prod. Bayreuth, Festspielhaus, 13, 14, 16, 17 Aug. 1876, with Betz (Wotan), Grün (Fricka), Elmblad (Donner), Engelhardt (Froh), H. Vogl (Loge), Haupt (Freia), Reichenberg (Fafner), Eilers (Fasolt), Hill (Alberich), Schlosser (Mime), Jaide (Erda), Niemann (Siegmund), Scheffzky (Sieglinde), Niering (Hunding), Unger (Siegfried), Materna (Brünnhilde), Lilli Lehmann (Woodbird), Siehr (Hagen), Gura (Gunther), Weckerlin (Gutrune), Jaide (Waltraute), Lilli and M. Lehmann, Lammert (Rhinemaidens), Jachmann-Wagner, Scheffzky, Grün (Norns), Haupt, Lilli and M. Lehmann, Weckerlin, Amann, Lammert, Reicher-Kindermann, Jachmann-Wagner (Valkyries), cond. Richter. The separate operas were produced as follows:

*Das Rheingold* (The Rhinegold). Prologue in 1 act to the trilogy *Der Ring des Nibelungen*. Prod. Munich, N, 22 Sep. 1869, with A. Stehle, H. Vogl, Schlosser, Nachbaur, Kindermann, Fischer, Baussewein, cond. Wüllner; London, HM, 5 May 1882, with Reicher-Kindermann, H. Vogl, Schlosser, Scaria, Burger, Wiegand, Schelper, Eilers, Biberti, cond. Seidl; New York, M, 4 Jan. 1889, with Moran-Olden, Alvary, Sedlmayer, Mittelhauser, Fischer, Beck, Grienaues, Mödlinger, Weiss, cond. Seidl.

*Die Walküre* (The Valkyrie). Music-drama in 3 acts. Prod. Munich, N, 26 June 1870, with A. Stehle, T. Vogl, Kaufmann, H. Vogl, Kindermann, Bausewein, cond. Wüllner; New York, AM, 2 Apr. 1877, with Pappenheim, Canissa, Listner, Bischoff, Preusser, Blum, cond. Neuendorf; London, HM, 6 May 1882, with Reicher-Kindermann, T. Vogl, Riegler, Niemann, Scaria, Wiegand, cond. Seidl.

*Siegfried.* Music-drama in 3 acts. Prod. Bayreuth, Festspielhaus, 16 Aug. 1876; London, HM, 8 May 1882, with T. Vogl, H. Vogl, Schlosser, Scaria, cond. Seidl; New York, M, 9 Nov. 1887, with Lilli Lehmann, Alvary, Ferenczy, Fischer, cond. Seidl.

*Götterdämmerung* (The Twilight of the Gods). Music-drama in 3 acts. Prod. Bayreuth, Festspielhaus, 17 Aug. 1876; London, HM, 9 May 1882, with T Vogl, Schreiber, Reicher-Kindermann, H. Vogl, Wiegand, Biberti, Schelper, cond. Seidl; New York, M, 25 Jan. 1888 (incomplete, without Norns or Waltraute scenes), with

Lilli Lehmann, Seidl-Kraus, Niemann, Robinson, Fischer, cond. Seidl.

In *Rheingold*, the Nibelung dwarf Alberich (bs-bar) renounces love so that he may steal the Rhinegold, guarded by the three Rhinemaidens (sop, sop, mez), and by forging a Ring from it become master of the world. Wotan (bs-bar), ruler of the gods, has engaged the giants Fafner and Fasolt (basses) to build Valhalla for the gods; unable to pay for it, he has promised them Freia (sop), goddess of youth. Loge (ten), the fire god, persuades Wotan to accompany him to Nibelheim where by a trick Wotan obtains the Ring and the Rhinegold from Alberich; he intends to pay the giants with the gold, and keep the Ring himself. Alberich curses the Ring. The giants see the Ring on Wotan's finger and demand it as well as a magic helmet, the Tarnhelm. Wotan at first refuses, and the giants prepare to drag Freia away. Wotan's wife Fricka (mez) urges Wotan to comply; Erda (con), the earth goddess, warns Wotan of the consequences of retaining the Ring. He adds it to the gold, whereupon Fasolt and Fafner quarrel. Fafner kills Fasolt, and takes away the gold, the Tarnhelm, and the Ring. The gods, watched cynically by Loge, enter Valhalla as the curtain falls.

(In order to safeguard Valhalla, Wotan begets with Erda nine warrior daughters, the Valkyries, who bear the bodies of dead heroes to Valhalla, where they are revived and help to defend the castle. But in order to restore the Ring to the Rhinemaidens and rid the gods of the curse, Wotan has to engender human children. He descends to earth and begets Siegmund and Sieglinde, hoping that the former will one day kill Fafner and return the Ring to the Rhinemaidens. The pair are separated, Sieglinde being forcibly married to Hunding, and Siegmund driven to lead a wandering life of hardship.)

In *Die Walküre*, Siegmund (ten) is driven to shelter in Hunding's hut. He and Sieglinde (sop) feel a mysterious attraction. Hunding (bs) returns, and realizing that Siegmund is the murderer of his kinsmen, challenges him to fight the next day. Later, after she has drugged Hunding's drink, Sieglinde shows Siegmund the sword Nothung that Wotan had left embedded in the tree growing in Hunding's hut to be withdrawn by a hero. He pulls out the sword, and flees with Sieglinde. Fricka, the guardian of marriage vows, forces Wotan to side with Hunding in the coming combat. But Brünnhilde (sop), Wotan's favourite Valkyrie daughter, disobeys him and sides with Siegmund. Wotan intervenes and Siegmund is killed, Nothung being shattered by Wotan's spear. Brünnhilde gathers up the fragments and entrusts them to Sieglinde, who will soon bear Siegmund's child—the hero Siegfried.

Wotan's punishment for Brünnhilde is to deprive her of her immortal status, putting her to sleep on a fire-girt rock until she is claimed by a man. He mitigates her fate, however, by granting her plea that only a true hero will win through the flames. The curtain falls on the magic fire.

(Sieglinde has died giving birth to Siegfried. The boy has been brought up by the dwarf Mime, brother of Alberich. Mime's dwelling is in the forest close to the cave where Fafner, who by means of the Tarnhelm has changed himself into a dragon, guards the Rhinegold. Mime hopes to weld the fragments of Nothung together so that Siegfried can kill Fafner; he means thereby to gain the Ring for himself.)

In *Siegfried*, Wotan, disguised as a Wanderer, visits Mime (ten) and prophesies that the sword will be forged by a hero. Mime recognizes Siegfried (ten) as this hero, and intends to kill him when his plan is accomplished. Siegfried successfully forges the sword Nothung, and with Mime sets out to seek Fafner. After Siegfried has aroused and killed Fafner, he burns his finger in the dragon's blood. Sucking it, he finds that he can understand the language of the birds, one of which (sop) warns him of Mime's treachery, and then tells him of the sleeping Brünnhilde. Siegfried kills Mime, and with the Ring and Tarnhelm follows the bird to the Valkyrie's rock. The Wanderer, although he has told Erda that he longs only for the end, tries to bar his path, but Siegfried shatters his spear with Nothung, and, making his way through the fire, awakens Brünnhilde and claims her as his bride.

In *Götterdämmerung*, the Three Norns (con, mez, sop) prophesy the end of the gods. Siegfried gives Brünnhilde the Ring, and leaving her, goes to seek adventure. He comes to the Hall of the Gibichungs, where Alberich's son Hagen (bs) lives with his half-brother Gunther (bar) and half-sister Gutrune (sop). Hagen, who knows all that has happened, devises a plot. He gives Siegfried a drug to make him forget Brünnhilde, and arranges for him to marry Gutrune, the price being that he will fetch Brünnhilde as Gunther's bride: thus Hagen will obtain the Ring. To Brünnhilde comes her sister Waltraute (mez), who urges her in vain to return the Ring to the Rhinemaidens. Siegfried, wearing the Tarnhelm and in the guise of Gunther, penetrates the fire again; he overcomes Brünnhilde, tears the Ring from her finger, and takes her back, an unwilling bride for Gunther. Hagen summons the Gibichungs for the double wedding. Gunther leads on Brünnhilde, unrecognized by Siegfried. Seeing the Ring on Siegfried's finger, she accuses him of treachery. With Hagen and the now suspicious Gunther she plans Siegfried's death in a hunt.

Siegfried is resting on the banks of the Rhine;

the Rhinemaidens appear and plead with him to return the Ring. Hagen, Gunther, and the huntsmen now arrive. Siegfried is asked to relate his adventures. Hagen gives him a second drug to restore his memory, and he speaks of his love for Brünnhilde. Hagen spears him in the back. Siegfried's body is carried back to the Gibichung Hall, and in a quarrel over the Ring, Gunther is killed by Hagen. When Hagen approaches the dead Siegfried to remove the Ring, Siegfried's hand rises in the air and stops him. Brünnhilde orders a funeral pyre to be built, and taking the Ring from Siegfried's finger, places it on her own. On her horse Grane, she plunges into the flames. The Hall collapses, the Rhine overflows, and when Hagen tries to snatch the Ring from Brünnhilde, he is dragged below the waters by the Rhinemaidens. Valhalla rises in flames, and as the kingdom of the gods is destroyed, a new era of love dawns.

**Rinuccini, Ottavio** (*b* Florence, 20 Jan. 1562; *d* Florence, 28 Mar. 1621). Italian librettist. Reared in the Renaissance spirit of Tasso and Guarini, his earliest work was for the lavish Florentine entertainments, including the famous 1589 intermedi. He may have been a member of the Bardi Camerata; he was actively involved in the earliest operatic experiments through the Accademia degli Alterati, where he was known as 'Il Sonnachioso' (Sleepy). His text for Peri's *Favola di Dafne* (1598) was the first opera libretto, later set by Gagliano (1608) and Schütz (trans. Opitz, 1627; music lost—probably the first German opera). He also provided the libretto for *Euridice* (set by both Peri and Caccini in 1600) and later, as a member of the Accademia degli Elevati in Mantua, wrote the libretto for Monteverdi's *Arianna*. This was his most famous work and led to a further planned collaboration on *Narciso ed Ecco* (1608); Monteverdi did not set the libretto, because he disliked its unhappy ending. Later the same year Rinuccini provided the text for Monteverdi's *Ballo delle ingrate*.

Rinuccini was one of the founding fathers of opera and the most talented of the early librettists. His texts were in one sense conservative, drawing inspiration from Ovid for their storyline and from the favola pastorale for their prevailingly lyrical approach; more radical was his development of a highly flexible style of verse, whose simplicity and suitability for the natural declamation of the words provided a model for many later librettists. It is significant that Monteverdi considered him his best librettist; it has indeed been suggested that it may have been Rinuccini, not Striggio, who provided the text for the second (Apollonian) finale to *Orfeo* in 1609.

**Rio de Janeiro.** City in Brazil, capital 1763–1960. Among various attempts to organize opera in the 18th cent. were in 1767 a Casa de Opera, which burnt down in 1770 or 1771, and a newly founded (but short-lived) Casa de Opera organized by Luiz de Ferreira. Visiting Italian companies played at the Real T de São João, opened in 1813, which burnt down in 1815 but was quickly rebuilt. Many other theatres were built in the first decades of the 19th cent., many of them housing occasional opera, including the T Provisório (opened 1852 with Verdi's *Macbeth*; demolished 1875), the T Lirico, and two operetta houses, the Fénix Dramática and the Alcazar Lyrique. Toscanini made his début at the Imperial T in 1886. The T Municipal opened in 1909, and was remodelled in 1934 and in the 1970s (cap. 2,357); it remains the centre of Rio's operatic life. Visiting Italian seasons were given by, among others, Mascagni. The tradition has remained almost exclusively Italian, with places found in the repertory for Brazilian works (including by Mignone, Nepumoceno, and especially Gomes). New enterprise was brought into a conventional tradition of staging by Serge Bitto in the 1970s, but after an excellent period in the theatre's history in 1980–4 standards declined (with quantity of performances preceding quality) until a new director was appointed in 1988. Opera is also given in a theatre opened in a cultural centre formerly the Bank of Brazil (cap. 300).

**Riotte, Philipp Jakob** (*b* St Wendel, 16 Aug. 1776; *d* Vienna, 20 Aug. 1856). German composer and conductor. Studied with André; *c.*1805 Kapellmeister at Gotha. After travels to Danzig, Magdeburg, and Erfurt arrived in Vienna in 1808, taking up a post at the Hofoper. Worked at the T an der Wien as music director 1818–20 and 1824–6. His reputation was achieved by many popular stage works, including the pantomimes *Staberl als Freischütz* (1826) and *Der Postillon von Stadt-Enzersdorf* (1840) and the Singspiel *Das Grenzstädtchen*. Of the *c.*50 works he produced, the most popular and lasting was the music for Raimund's *Moisasurs Zauberspruch* (1827).

SELECT WORKLIST: *Das Grenzstädtchen* (The Little Frontier Town) (Kotzebue; Vienna 1809).

**Rise and Fall of the City of Mahagonny.** See AUFSTIEG UND FALL DER STADT MAHAGONNY.

**Ristori, Giovanni Alberto** (*b* ?Bologna, 1692; *d* Dresden, 7 Feb. 1753). Italian composer. After early success in Italy, including *Pallide trionfante in Arcadia*, he went to Dresden in 1717 to become composer to the company directed by his father **Tommaso Ristori**. His first opera there was *Cleonice*; in the late 1720s he again produced Italian opera, notably *Calandro*. After the accession of Augustus II (1733) he was increasingly passed over in favour of Hasse and was demoted to chamber organist, though he did compose Augustus's coronation opera, *Le fate*. In 1731–2 he accompanied his father on a trip to Russia and was possibly Kapellmeister to the St Petersburg court briefly. The performance of *Calandro* in Moscow in Dec. 1731 was the first of an Italian opera in Russia. During Hasse's frequent absences from Dresden during the 1730s and 1740s, Ristori sought favour through composing feste teatrali for the court, including *I lamenti di Orfeo*. Though Hasse's more substantial works were closer to the tastes of the court than Ristori's pastoral offerings, he was eventually promoted to the post of Vizekapellmeister shortly before his death.

SELECT WORKLIST: *Pallide trionfante in Arcadia* (Mandelli; Padua 1713); *Cleonice* (Constantini; Dresden 1718); *Calandro* (Pallavicini; Dresden 1726); *Le fate* (Pallavicino; Dresden 1736).

**Risurrezione** (Resurrection). Opera in 4 acts by Alfano; text by Cesare Hanay, after Tolstoy's novel *Resurrection* (1900). Prod. Turin, V, 30 Nov. 1904, with Magliulo, Mieli, Scandiani, cond. Serafin; Chicago, Auditorium, 31 Dec. 1925, with Garden, Ansseau, Baklanoff, cond. Moranzoni.

Russia, *c*.1910. Katusha (sop) meets and falls in love with Prince Dimitri (ten), by whom she becomes pregnant. He has to join his regiment and fails to keep a rendezvous with her at a railway station. Katusha commits a murder and is sentenced to be deported to Siberia; Dimitri follows her there and obtains a free pardon for her; but she rejects him in favour of Simonson (bar), a fellow convict.

**Rita**. Opera in 1 act by Donizetti; text by Vaëz. Comp. 1841: prod. Paris, OC, 7 May 1860, with Faure-Lefèbvre, Warot, Barielle; New York, Hunter Coll., 14 May 1957; London, National School of O, 12 Dec. 1962.

**'Ritorna vincitor'**. Aida's (sop) aria in Act I of Verdi's *Aida*, in which she struggles with her conflicting emotions as she longs for Radamès to return victorious from a war against her own father and country.

**ritornello** (It.: 'little return', hence refrain). In the earliest operas the term was applied to opening instrumental sections, as with that to Orfeo's 'Vi ricorda' in Monteverdi's opera: apart from the sinfonia, it was the only instrumental piece in early opera. Later, in Metastasian opera seria, the ritornello became the structural linchpin of the *da capo aria. As its name implies, the ritornello was reprised several times during its course and also served the expressive function of summarizing the emotional content.

**Ritorno d'Ulisse in patria, Il** (The Return of Ulysses to his Homeland). Opera in a prologue and 5 acts by Monteverdi; text by Badoaro. Prod. Venice, C, Feb. 1640; London, St Pancras TH, 16 Mar. 1965, with Bainbridge, Dinoff, Sarti, Kentish, Dickerson, McKinney, McCue, cond. Marshall; Washington, DC, Kennedy Center, 18 Jan. 1974, with Stade, Allen, Stillwell, Gramm, cond. Gibson.

The opera relates the events of the closing book of Homer's *Odyssey*, with added comments from the gods and from allegorical figures (Human Frailty, Time, Fortune, and Love: these appear as a prologue). Penelope laments the continued absence of Ulisse to her nurse Ericlea. After a discussion of men's sins between Giove and Nettuno, Ulisse is put on shore at Ithaca and encouraged by Minerva to reclaim his palace, given over to the suitors of his wife Penelope. Eumete, his herdsman, is taunted by the jester Iro, and then welcomes the disguised Ulisse, who revives his hopes of his master's return. Eumete then welcomes Ulisse's son Telemaco, also returning home, and there is a joyful reunion, while Eumete tells Penelope that Ulisse may soon appear. The suitor Antinoo mocks Ulisse, disguised as a beggar, who first wins a contest with the suitors by being able to string his own bow, and then turns it upon them. Penelope's fear that he may not truly be Ulisse is overcome by Ericlea, and husband and wife join in a love duet.

**Ritter, Peter** (*b* Mannheim, 2 July 1763; *d* Mannheim, 1 Aug. 1846). German composer, conductor, and cellist. He came from a distinguished family of Mannheim musicians whose members included the virtuoso bassoonist **Georg Wenzel Ritter** (1748–1808) and the violinist **Heinrich Ritter** (*fl.* 1779–93). Studied composition with Vogler; entered the Mannheim orchestra as cellist 1783; Kapellmeister of the orchestra of the Grand Duchy of Baden 1803–23. He composed *c*.20 stage works, of which the most popular were *Der Zitherschläger* and *Der Sturm*.

SELECT WORKLIST: *Die lustigen Weiber von Windsor* (The Merry Wives of Windsor) (Römer, after Shakespeare; Mannheim 1794); *Der Sturm* (The Tempest) (Döring, after Shakespeare; 1799); *Der Zitherschläger* (Seidel; Stuttgart 1810).

**Robert le diable**. Opera in 5 acts by Meyerbeer; text by Scribe and Delavigne. Prod. Paris, O, 21 Nov. 1831, with Cinti-Damoreau, Dorus-Gras, Nourrit, N. Levasseur, and Taglioni dancing, cond. Habeneck; London, DL, 20 Feb. 1832; New York, Park T, 7 Apr. 1834 (Rophino Lacy's version), with Mrs Wood, Sharpe, Harrison, Wood, Clarke, Placide, Blakeley, Haydn.

In 13th-cent. Palermo, Robert, Duke of Normandy (ten), the son of a mortal and a devil, falls in love with the Princess Isabella (sop). Disguised, and under the name of Bertram (bs), the Devil tries to gain Robert's soul; he prevents Robert from winning Isabella in a tournament, and Robert is then willing to use diabolical means. At a midnight orgy with ghostly nuns, Robert acquires a magic branch with which he gains access to Isabella; but she persuades him to break it. Robert denounces his father the Devil, and marries Isabella.

Other operas on the subject are by J. Barnett (1829), Casimiro (1842). Parodies incl. A. Müller's *Robert der Teufel* (1833), Damse's *Robert Birbanduch* (1844). Also J. G. Kastner (*Les nonnes de Robert le diable*, comp. 1845).

**Roberto Devereux, ossia Il conte d'Essex.** Opera in 3 acts by Donizetti; text by Cammarano, after F. Ancelot's tragedy *Elisabeth d'Angleterre*. Prod. Naples, C, 29 Oct. 1837, with Ronzi de Begnis, Granchi, Basadonna, Barroilhet, Barrattini, Rossi, Benedetti; London 24 June 1841, with Grisi, Rubini, Tamburini; New York, Astor Place OH, 15 Jan. 1849, with Truffi, Amalia Patti, Benedetti, S. Patti, Rossi-Corsi, Giubilei, cond. Maretzek.

London, early 17th cent. The Earl of Essex, Robert Devereux (ten), although loved by Queen Elizabeth (sop), is in love with Sara (sop or mez), Countess of Nottingham. Essex is accused of treason and sentenced to death by the Queen, whose reprieve comes too late to save him.

**Robertson, James** (*b* Liverpool, 17 June 1912; *d* Ruabon, 18 May 1991). English conductor. Studied Cambridge, Leipzig, and London. Répétiteur Gly. 1937–9; cond. London, CR, 1938–9; music director London, SW, 1946–54. Director London O Centre 1964–78. (R)

**Robeson, Paul** (*b* Princeton, NJ, 9 Apr. 1898; *d* Philadelphia, 23 Jan. 1976). US bass and actor. Studied law, Columbia U. Main career as actor, notably as Othello, but gave unforgettable performances as Joe (*Show Boat*), London, DL, 1928, and in New York (also appearing in the film), and as Crown (*Porgy and Bess*). His huge frame, magnificent voice, and the moving quality of his singing (especially in Spirituals) left indelible impressions on all who heard him. (R)

BIBL: M. Duberman, *Paul Robeson* (New York, 1988).

**Robin, Mado** (*b* Yseures-sur-Creuse, 29 Dec. 1918; *d* Paris, 10 Dec. 1960). French soprano. Studied Paris with Podestà. Début Paris, O, 1945 (Gilda). Appeared at Paris, O and OC; Brussels; San Francisco; and in Russia. Chiefly remarkable for her extraordinarily high range, which reached c''''. Her most famous roles were Lucia and Lakmé. (R)

**Robin Hood.** Opera in 3 acts by Macfarren; text by Oxenford. Prod. London, HM, 11 Oct. 1860, with Lemens-Sherrington, S. Reeves, Santley. The earliest celebrations of the hero seem to date from 16th-cent. May Day feasts. Among many early masques and ballad operas on the subject are works by Watts, Mendez, and Shield. Also operas by Baumgarten (1786), Hewitt (1800), Dietrich (1879), Holmes, and De Koven (1890).

**Robinson, Anastasia** (*b* Italy, *c*.1692; *d* Southampton, Apr. 1755). English soprano, later contralto. Studied London with Sandoni and Lindelheim. Début London, H, 1714 (in the pasticcio *Creso*). Sang there until 1717, also 1718, 1719, as Almirena (*Rinaldo*), first Oriana (*Amadigi*), and in A. Scarlatti's *Pirro e Demetrio*. Her voice changing through illness, she sang as a contralto, London, DL, 1719–24, creating Matilda (*Ottone*), Teodata (*Flavio*), Cornelia (*Giulio Cesare*); also sang in operas by D. Scarlatti, Porta, and Bononcini. Retired 1724 after her secret marriage to the Earl of Peterborough. (He did not publicly acknowledge her until just before he died (1735), but thrashed Senesino for impugning her honour.) A singer of great personal charm and cultivation, she shone through expressiveness rather than virtuosity. After retiring she kept a musical salon, promoting the works of Greene, Tosi, Bononcini, etc.

**Robinson, Forbes** (*b* Macclesfield, 21 May 1926; *d* London, 13 May 1987). English bass. Studied at La Scala school, Milan. Début London, CG, 1954 (Monterone). CG from 1954. Appearances with WNO, Berlin, Zurich, Buenos Aires, etc. Repertory of over 70 roles includes Figaro (Mozart), Don Giovanni, Pizarro, Philip, Boris, Dodon, Kečal, Moses (*Moses und Aron*), Claggart, and first Priam (*King Priam*). An excellent singing actor with a dark timbre and striking powers of expression. (R)

**Robson, Christopher** (*b* Falkirk, 9 Dec. 1953). Scottish counter-tenor. Studied London with Gaddarn, Esswood, and Mott. Début Birmingham 1979 (in Handel's *Sosarme*). Kent O; O Factory (Zurich and London); London, ENO from 1985 (Moscow, Leningrad, Kiev 1990) and CG from 1988; New York City O 1985; also Frankfurt, Berlin, Houston, Aix, etc. Roles incl. Endymion, Sorceress (Purcell), Arsamene (*Serse*), Gluck's Orfeo, Akhnaten. (R)

**Rocca, Lodovico** (*b* Turin, 29 Nov. 1895; *d* Turin, 25 Jun. 1986). Italian composer. Studied Turin, and Milan with Orefice. His fame rests chiefly on his third opera, *Il dibuk* (Simon, after An-Ski; Milan 1934), a skilful and effective amalgam of various styles and techniques. Director Turin Cons., 1950–66.

**Rocco.** The gaoler (bs), Marzelline's father, in Beethoven's *Fidelio*.

**Rochois, Marthe** (*b* Caen, *c*.1658; *d* Sartrouville-sur-Seine, 9 Oct. 1728). French soprano. Possibly studied with Michel Lambert; then with his son-in-law Lully. Début Paris, O, 1678. Paris, O, until 1698. Much admired by Lully, for whom she created roles in *Proserpine*, *Persée*, *Amadis*, *Roland*; also title-role of *Armide*. Though ordinary in appearance, she possessed great vivacity and seductiveness on stage and outclassed other more beautiful actresses in all her parts.

**Rode, Wilhelm** (*b* Hanover, 17 Feb. 1887; *d* Icking, 2 Sep. 1959). German bass-baritone. Studied Hanover with Moest. Début Erfurt 1909 (Herald, *Lohengrin*). Munich 1922–30; Vienna 1930–2; Berlin, SO, 1932–45 (Intendant from 1934). Also Paris, O; London, CG; Madrid; Prague; etc. One of the leading Wagner baritones of his time, and much esteemed as Wotan and Sachs. (R)

**Rodelinda.** Opera in 3 acts by Handel; text by Nicola Haym, after libretto by Salvi (1710). Prod. London, H, 13 Feb. 1725, with Cuzzoni, Dotti, Senesino, Pacini, Boschi, Borosini; Northampton, MA, Smith Coll., 9 May 1931. Also operas by Perti (1710), Canuti (1724), Nelvi (1726), Cordans (1731), and Graun (1741).

**Roderigo.** 1. A Venetian gentleman (ten) in Verdi's *Otello*.

2. See POSA.

**Rodgers, Richard** (*b* Hammels Station, NY, 28 June 1902; *d* New York, 30 Dec. 1979). US composer. Self-taught, he began composing while a student at Columbia U. In 1918 he met his first collaborator, Lorenz Hart; in 1919 their song 'Any Old Place with You' appeared in Lew Fields's *A Lonely Romeo*. Beginning with revues, notably *The Garrick Gaieties* (1925), but soon branching out into full musical comedies, Rodgers and Hart produced more than 30 works before Hart's death in 1943. Of these the most important were *On Your Toes* (1936), which included the ballet sequence 'Slaughter on Tenth Avenue', and *Pal Joey* (1940). Rodgers's second partnership, with Oscar Hammerstein II, achieved even greater success, and resulted in a string of popular triumphs of which the most important were *Oklahoma!* (1943), *Carousel* (1945), *South Pacific* (1949), and *The King and I* (1951).

In the quality of his hit-songs, such as 'You'll Never Walk Alone' from *Carousel* and 'Some Enchanted Evening' from *South Pacific*, Rodgers matched the very best in the tradition of the musical. However, he showed greater concern for the unity of the drama and the music, which is sometimes emphasized by his welding the normal pattern of clearly separated numbers into a more continuous musical progression. Through the film versions which were made of many of the musicals, e.g. *The Sound of Music* (1959), Rodgers's work has reached a wider public. His daughter **Mary Rodgers** (*b* New York, 11 Jan. 1931) has also made a career as a composer of musicals: her works include *Once upon a Mattress* (1959) and *The Mad Show* (1973).

SELECT WORKLIST (all first prod. New York): *On Your Toes* (Hart; 1936); *Pal Joey* (Hart; 1940); *Oklahoma!* (Hammerstein II; 1943); *Carousel* (Hammerstein II; 1945); *South Pacific* (Hammerstein II; 1949); *The King and I* (Hammerstein II; 1951); *The Sound of Music* (Hammerstein II; 1959).

BIBL: D. Ewens, *With A Song in his Heart* (New York, 1963).

**Rodolfo.** 1. The Count (bs) in whose room Amina sleepwalks in Bellini's *La sonnambula*.

2. Count Walter's son (ten), lover of Luisa, in Verdi's *Luisa Miller*.

3. The Bohemian poet (ten), lover of Mimì, in Puccini's *La bohème*.

**Rodrigo** (correctly, *Vincer se stesso è la maggior vittoria*). Opera by Handel; text an anon. adaptation of F. Silvani's *Il duello d'amore e di vendetta*, 1700, set by Ziani. Prod. Florence, prob. Palazzo Pitti, end Oct./beginning Nov. 1707, with Frilli, Beccarina, Marcello, Guicciardi, Valentina, Perini (not Tarquini, 'La bombace',as often stated). Handel's first opera for Italy; some of the music is lost; that which survives draws heavily on *Almira* and works by others, esp. Kaiser.

**Rodrigo.** See POSA.

**Rodwell, George** (*b* London, 15 Nov. 1800; *d* London, 22 Jan. 1852). English composer and playwright. A pupil of Novello and Bishop, he was music director of the Adelphi T, where his brother J. T. G. Rodwell was manager. In 1825, on his brother's death, he took over his share in the theatre; in 1828 he became a professor at London, RAM. From 1836 he was music director at Covent Garden.

Rodwell composed about 20 works for the London stage, mostly farces and melodramas of which the music is now lost. The most successful was *Teddy the Tiler* (1830); other important works included *The Devil's Elixir* (1829), *Don Quixote* (1833), and *The Bronze Horse* (1835), a reworking of Auber's opera, described as a 'musical drama'.

BIBL: E. Fitzball, *Thirty-Five Years of a Dramatic Author's Life* (London, 1859).

**Rodzinski, Artur** (*b* Spalato (Split), 1 Jan. 1892; *d* Boston, 27 Nov. 1958). Polish, later US, conductor. Studied Lvov, Vienna with Marx, Schreker, and Schalk. Début Lvov 1920 (*Ernani*). Cond. Warsaw 1924–8. Moved to USA, conducting mostly concerts, but also US première of *Lady Macbeth of Mtsensk*, 1935, and Wagner and Richard Strauss in Cleveland and Chicago. Salzburg 1936, Vienna 1937. Florence 1953 (first non-Russian *War and Peace*), Milan, Rome, Naples. Chicago 1958 (*Tristan*). (R)

**Roger, Gustave-Hippolyte** (*b* Paris, 17 Dec. 1815; *d* Paris, 12 Sep. 1879). French tenor. Studied Paris with Martin. Début Paris, OC, 1838 (Georges, Halévy's *L'éclair*). Paris, OC until 1848, and O, 1849–59; London, CG, 1847; Berlin, H, 1851, 1859. Lost his right arm in an accident 1859, but continued to sing at Paris, OC, until 1862. A celebrated and popular singer; created John of Leyden (*Le prophète*), and Berlioz's *Faust*, as well as numerous roles in operas by Thomas, Halévy, and Auber. His circle of friends included Meyerbeer, Gounod, and Alexandre Dumas Fils. Wagner was highly impressed with his intelligence and knowledge of German, but collaboration over the French translation of *Tannhäuser* had to be abandoned. Autobiography, *Le carnet d'un ténor* (Paris, 1880).

**Roi de Lahore, Le.** Opera in 5 acts by Massenet; text by Gallet. Prod. Paris, O, 27 Apr. 1877 with Joséphine De Reszke, Fouquet, Salomon, Lassalle, cond. Deldevez; London, CG, 28 June 1879, with Turolla, Gayarre, Lassalle, cond. Vianesi.

India, 11th cent. Alim (ten), King of Lahore, and his minister Scindia (bar) love Sita (sop). Scindia kills Alim, who is allowed by a god to return as a beggar. Sita kills herself so as to join him in Paradise.

**'Roi de Thulé, Le'.** The ballad 'Il était un roi de Thulé' sung by Marguerite in Act III of Gounod's *Faust* as she reflects upon her meeting with Faust.

**Roi d'Ys, Le** (The King of Ys). Opera in 3 acts by Lalo; text by Blau. Prod. Paris, OC, 7 May 1888, with Deschamps-Jéhin, Simmonet, Talazac, Bouvet, Cobalet, cond. Danbé; New Orleans, French OH, 23 Jan. 1890, with Furst, Leavington, Beretta, Balleroy; London, CG, 17 July 1901, with Pacquot, Adams, Jérôme, Seveilhac, Plançon, cond. Flon.

Brittany, mythological times. So as to be reconciled to Karnac (bar), the king of Ys (bs) offers him the hand of his daughter Margared (sop). But she loves Mylio (ten), who loves the king's other daughter Rozenn (sop). Margared refuses Karnac. Mylio defeats Karnac, who conspires with Margared and they open the flood gates and submerge Ys. Mylio kills Karnac, and

Margared drowns herself. The waters recede, and the town is saved.

**Roi l'a dit, Le** (The King has commanded it). Opera in 3 acts by Delibes; text by Gondinet. Prod. Paris, OC, 24 May 1873, with Priola, Lhérie, Ismael, cond. Deloffre; London, Prince of Wales's T, 1 Dec. 1894; U of Iowa, 29 Apr. 1967.

**Roi malgré lui, Le** (The King despite himself). Opera in 3 acts by Chabrier; text by De Najac and Burani, after Ancelot. Prod. Paris, OC, 18 May 1887, with Isaac, Mézéray, Bouvet, Fugère, cond. Danbé. Revived 1929 in revised version by Albert Carré, with Brothier, Gyula, Bourdin, Musy, cond. Masson.

Poland, 1574. Henri de Valois (bar), about to be crowned King of France, learns from Minka (sop), his betrothed, that there is a plot to kill him. He disguises himself as his friend De Nangis (ten) and joins the plotters. De Nangis comes to the camp and is mistaken for Henri. The plot is foiled, and the lives of both men are spared. Henri is finally crowned King of France and Poland.

**Rolfe-Johnson, Anthony** (*b* Tackley, 5 Nov. 1940). English tenor. Studied London with Keeler. Début London, EOG; 1973 (Vaudémont, *Yolanta*). Gly. 1974–6; Milan, S, 1984, 1986; London, ENO, CG. Salzburg 1987; also Hamburg, Edinburgh. A fine performer with a warm, flexible voice and great vocal skill. Roles incl. Ulisse, Jupiter (*Semele*), Jason (*Médée*), Idomeneo, Ferrando, Don Ottavio, Tito, Lensky, Fenton, Aschenbach. (R)

**Roller, Alfred** (*b* Vienna, 2 Oct. 1864; *d* Vienna, 21 June 1935). Austrian designer. Studied Vienna. Worked with Mahler at the Vienna, H (later S); chief designer 1903–9, 1918–34, also Vienna, B, 1918–34. Designed premières of *Rosenkavalier* (Dresden 1911) and *Frau ohne Schatten* (Vienna 1919), also many Salzburg productions. His powerful sets, atmospheric and carefully attuned to the music, were highly influential.

**Rolli, Paolo Antonio** (*b* Rome, 13 June 1687; *d* Todi, 20 Mar. 1765). Italian librettist and poet. Came to England in 1715; while also teaching Italian to aristocratic families, he achieved fame through his literary activities, which included the first translation of Milton's *Paradise Lost*. Appointed secretary to the *Royal Academy of Music in 1719, he participated in the establishment of Italian opera in London through the librettos he provided for Handel (e.g. *Floridante*, *Scipione*) and Bononcini (*Astarto*, *Griselda*). After a quarrel he left the Royal Academy and eventually joined the rival *Opera of the Nobility, for whom he wrote nine librettos during 1733–7: some were set by Porpora, including *Polifemo* (1735) and *Ifigenia in Aulide* (1735). His later work includes a final libretto for Handel, *Deidamia* (1741), and that for Lampugnani's *Rossana* (1743).

Like Haym, who replaced him at the Royal Academy, Rolli's librettos were not original works, but adaptations of pre-existing texts by such authors as Salvi, Silvani, and Stampiglia. At his best, as in *Alessandro* and *Deidamia*, Rolli demonstrates a sure dramatic skill and some poetic talent, although many of his offerings were only faintly disguised hackwork.

BIBL: G. Dorris, *Paolo Rolli and the Italian Circle in London* (The Hague, 1967).

**romance** (Fr., *romance*; Ger., *Romanze*; It., *romanza*). In opera, an aria, generally amorous or soliloquizing, normally intended less for display than aria proper, and thus lacking ornamentation, cadenzas etc. However, the distinction is not precise: Verdi used the terms almost interchangeably. Examples are Pedrillo's 'Im Mohrenland gefangen war' in Mozart's *Entführung*, Matilde's 'Sombres forêts' in Rossini's *Guillaume Tell*, and Radamès's 'Celeste Aida' in Verdi's *Aida*.

**Romanelli, Felice** (*b* Rome, 21 July 1751; *d* Milan, 1 Mar. 1839). Italian librettist. At La Scala, Milan, for over 30 years from 1799, he was one of the most prolific librettists of his day, producing over 60 texts and collaborating with a large number of composers. He drew on a wide literary and linguistic knowledge to develop a set of operatic conventions which proved flexible to different demands; and many of what were to become standard operatic expressions (and hence clichés) originated with him. Basically a classicist, he also treated the medieval and exotic subjects that were becoming an important part of Romantic feeling. Composers who set his texts included Fioravanti, Mayr, Mercadante, Nicolini, Pacini, and Rossini (*La pietra del paragone*).

**Romani, Luigi** (*b* Genoa, 31 Jan. 1788; *d* Moneglia, 28 Jan. 1865). Italian librettist and critic. His first librettos were written in Milan for Mayr; moving to Turin to take up an editorship, he developed the skills which made him sought after by most important Italian composers: over 100 set his texts. Though he collaborated with Rossini (e.g. *Il turco in Italia*) and once with Verdi (*Un giorno di regno*), his most important work was done with Donizetti and Bellini. Classical in his training and in his sense of balance, he made use of Romantic ideas rather than being possessed by them. However, his quick dramatic instinct for the operatic essence of a subject and the musical elegance of his verse made him an outstanding

librettist for Donizetti and little short of an ideal one for Bellini. The haste with which he worked on his numerous commissions meant that his work is not free from cliché, and can lack profundity; but as well as the pastoral idylls of *L'elisir d'amore* and *La sonnambula* he could achieve the nobility of *Anna Bolena* and especially *Norma*.

**Romani, Pietro** (*b* Rome, 29 May 1791; *d* Florence, 11 Jan. 1877). Italian composer, conductor, and teacher. Studied with Fenaroli; singing-master at the Reale Istituto Musicale; later conductor at Florence. He composed two operas, *Il qui pro quo* and *Carlo Magno*. He is better known, however, for the aria 'Manca un foglio', which he wrote to replace Dr Bartolo's 'A un dottor della mia sorte' for a performance of *Il barbiere di Siviglia* in Rome in 1816 because it was beyond the capabilities of the singer Paolo Rosich. Romani's aria is still to be found in some modern scores.

SELECT WORKLIST: *Il qui pro quo* (?; Rome 1817) and *Carlo Magno* (Charlemagne) (?; Florence 1823).

**Romania.** Opera was first given at Sibiu in 1772 by Livio Cinti's Italian touring company; others played in Bucharest from 1784, and a German company visited Iaşi in 1795. French, German, and Italian companies toured in the early 19th cent. Early Romanian operas include *Braconierul* (The Poacher, 1833) and *Zamfira* (1834) by Ion Wachmann (1807–63); the first Romanian operetta, *Baba-Hîrca* (1848), by Alexandru Flechtenmacher (1823–98); and *Vîrful cu dor* (The Summit of Desire, 1879) by Liubicz Skibinski text by 'Carmen Sylva' (Queen Elizabeth of Romania). The first foreign opera given in Romanian was Boieldieu's *Jean de Paris*; the first libretto published in Romanian was *Norma*. The first opera season with performances in Romanian was in 1885–6 under George Stephănescu (1843–1925); he also translated librettos, wrote polemic articles, composed operettas (beginning with *Peste Dunăre*, 1880), and founded a singing school whose pupils included Hariclea Darclée. Other important works included *Candidatul Linte* (Candidate Linte, 1877) and *Crai nou* (New Moon, 1877) by Ciprian Porumbescu (1853–83); *Olteanca* (The Girl from Olt, 1880, with Gustav Otremba) and *Petru Rareş* by Eduard Caudella (1841–1924). Other successful works included the Tennyson opera *Enoch Arden* (1906) by Alexis Catargi (1876–1923), a pupil of D'Indy and Enescu who was also a diplomat; and *Şezătoare* (The Vigil, 1908) by Tiberiu Brediceanu (1877–1969), whose long career included posts conducting opera in Cluj (from 1920) and, in his 70s, directing the Bucharest O (1941–4).

French and German were languages in common use among the upper classes, slower than musicians to appreciate the musical qualities of Romanian. But with the development of a national opera in the 1920s, encouragement was given to new Romanian works. Some of the more important were *Năpasta* (The Plague, 1928) and *Constantin Brâncoveanu* (1935) by Sabin Drăgoi (1894–1968), who conducted opera in Cluj and Timişoara, and whose work reflects his studies in folk music; *Marin Pescarul* (1934) by Marţian Negrea (1893–1973), a pupil of Franz Schmidt; *Monna Vanna* (1934) by Nicolae Brânzeu (*b* 1907), a prominent opera conductor; *O noapte furtunoasă* (A Stormy Night, 1935), by Paul Constantinescu (1909–63); and settings of Chekhov, *La drumul mare* (The High Road, 1932) and Musset (*Cu dragostea nu se glumeşte* (On ne badine pas avec l'amour, 1941) by Constantin Nottara (1890–1951). George Enescu's only opera, *Œdip*, was comp. 1931 and prod. Paris 1936, Bucharest 1958.

See also BUCHAREST, CLUJ, IAŞI, TIMIŞOARA.

BIBL: V. Cosma, *Opera Romînească* (Bucharest, 1962).

**romanza.** See ROMANCE.

**Romanze.** See ROMANCE.

**Romberg, Sigmund** (*b* Nagykanizsa, 29 July 1887; *d* New York, 9 Nov. 1951). Hungarian, later US, composer. Studied Vienna with Heuberger. Moving to USA in 1909, he was soon writing songs and operettas, the latter with librettists including Ira Gershwin, Oscar Hammerstein II, and P. G. Wodehouse. *Blossom Time* (1921) was based on the life of Schubert, whose music he adapted for the score. Other popular successes in the vein of romantic operetta include *The Rose of Stamboul*, *The Student Prince*, and *The Desert Song*. Later he moved to Hollywood as a film composer, and also turned with success to American musical comedy. His stage shows included other composers' songs, including early ones by Gershwin and Rodgers.

SELECT WORKLIST: *Blossom Time* (Donnelly, after Berté's operetta *Das Dreimädlerhaus* (Vienna 1916), after Bartsch; New York 1921); *The Rose of Stamboul* (Atteridge, after Fall's operetta *Die Rose von Stambul* (Vienna 1916); New York 1922); *The Student Prince* (Donnelly, after Bleichmann, after Meyer-Forster; New York 1924); *The Desert Song* (Harbach, Hammerstein II, & Mandel; New York 1926).

**Rome** (It., Roma). Capital of Italy. The first opera produced in Rome was *Cavalieri's *Rappresentazione di anima e di corpo*: based on a religious theme, it is sometimes wrongly described as an oratorio. It introduced the new Florentine *stile rappresentativo, which was further exploited in *Agazzari's pastoral drama *Eumelio*, prod. by pupils of the Seminario Romano early 1606. Works by M. A. Rossi, Vit-

tori, Mazzocchi, and Marazzoli were given in various palaces and noble houses, in many cases circumventing the papal ban on women on the stage which encouraged the rise of the castrato. The hall of the Palazzo Barberini was inaugurated 1632 with *Landi's *Sant'Alessio*, text by Giulio Rospigliosi (later Pope Clement IX); repeated 1634 for the visit of Alexander Charles, brother of Władisław of Poland. Other works given there included Marazzoli's and Mazzocchi's *Chi soffre, speri* (1639), an early comic opera (text by Rospigliosi); the last was Marazzoli's *La vita umana* (1656) in honour of Queen Christina of Sweden, resident in Rome 1654–89 and herself a patron of opera at her palace.

Clement IX authorized the first public opera-house, the T Torre di Nona (or T Tordinona), opened 1670 with Cavalli's *Scipio Africano*; closed 1674 by Clement's less enthusiastic successors; reopened 1690; demolished by Innocent XII 1697; rebuilt 1733; burnt down 1781; rebuilt 1787; altered 1795 (22 rows of stalls and 174 boxes) and reopened as T Apollo. Staged premières of Rossini's *Mathilde de Shabran*, cond. Paganini, Verdi's *Trovatore* and *Ballo*, and Donizetti's *Duca d'Alba*; demolished 1889 for a Tiber embankment, the Lungotevere Torre di Nona.

Other Rome theatres included the T Capranica, opened as private theatre 1679 (189 private boxes), publicly 1695. Staged mostly opera seria 1711–47, opera buffa from 1754 till closure 1881. The T delle Dame opened 1717, giving mostly opera seria. The T Valle, opened 1727, was originally a wooden structure in the courtyard of the Palazzo Capranica; gave premières by Rossini (*Cenerentola*) and Donizetti. The T Argentina was built 1731–2 by the Duke Sforza-Cesarini, with 186 boxes; opened with *Sarro's *Berenice*, premièred Rossini's *Barbiere* and Verdi's *Due Foscari* and *Battaglia di Legnano*. A leading theatre until the building of the T Costanzi, and continued to give seasons; now a concert hall and playhouse.

The T Costanzi was built by Domenico Costanzi, a rich builder, designed by Achille Sfondrini (cap. 2,293); opened 27 Nov. 1880 with *Semiramide*. The first act was interrupted for the Royal March as King Umberto and the Queen arrived late; the King left early, but the Queen remained. In 1888 the publisher Edoardo Sonzogno became manager, and announced his second competition for a 1-act opera: *Cavalleria rusticana* won, premièred here 17 May 1890. Also scene of other *Mascagni operas, and premières of *Tosca* and Zandonai's *Giulietta e Romeo*. Directed by the soprano Emma Carelli, 1911–25. Taken over by the city, enlarged, renovated, reopened as T Reale dell'Opera (cap. 1,600) 28 Feb. 1928 with Boito's *Nerone*. Gino Marinuzzi was conductor 1928–34, Serafin 1934–43 (also artistic director). Mussolini hoped to challenge the supremacy of La Scala, Milan, and enticed singers back from the New York Metropolitan: Gigli, Lauri-Volpi, Caniglia, Cigna, and Stignani were in the company. Serafin's achievements included cycle of *The Ring* in Italian, largely with Italian singers; also Verdi cycles 1940–1, and a cycle of contemporary opera, including the first Italian performance of *Wozzeck*, given during the German occupation though banned in Germany.

The theatre was renamed T dell'Opera in 1946, and developed a policy of encouraging new Italian works, welcoming works new to Italy (e.g. *Mathis der Maler*, 1951) and inviting foreign companies. Massimo Bogianckino was director 1963–8, when the theatre gave some of the best opera in Italy; Bruno Bartoletti was conductor, with visits from Giulini and Gui, and from producers including Visconti. Alberto Antignani, director since 1988, has revived works by Spontini and Cherubini, and commissioned new works. The theatre also organizes the summer seasons at the 3rd-cent. Terme di Caracalla (cap. 6,000), where large-scale productions have been mounted since 1937.

**Romeo and Juliet.** For operas on the subject see SHAKESPEARE.

**Roméo et Juliette.** Opera in 5 acts by Gounod; text by Barbier and Carré, after Shakespeare's tragedy (1594–5). Prod. Paris, L, 27 Apr. 1867, with Carvalho, Michot, Barré, Cazaux, cond. Deloffre; London, CG, 11 July 1867, with A. Patti, Mario, Cotogni, Tagliafico, cond. Costa; New York, AM, 15 Nov. 1867, with Hauk, Pancani, Antonacci, Medini, cond. Bergman. Voices: Roméo (ten), Juliette (sop), Mercutio (bar), Frère Laurent (bs). Parody, *Rhum et Eau en Juilliet* by Dejazet (1867).

**Romer, Emma** (*b* 1814; *d* Margate, 11 Apr. 1868). English soprano. Studied London with Smart. Début London, CG, 1830 (Linley's *The Duenna*). Sang London at L, DL, and CG until 1848, in first of numerous operas including Balfe's *Joan of Arc*, *The Enchantress*, *The Bondsman*; Hatton's *Queen of the Thames*, Barnett's *Fair Rosamund*. Also sang (in English) a much admired Amina and Adina. Manager of Surrey T from 1852.

**'Romerzählung'.** Tannhäuser's (ten) long narration of his pilgrimage to Rome in the last act of Wagner's opera.

**Ronconi, Giorgio** (*b* Milan, 6 Aug. 1810; *d* Madrid, 8 Jan. 1890). Italian baritone. Studied with his father, the tenor **Domenico Ronconi** (1772–1839), who had sung at Milan, S, Vienna,

and Munich. Début Pavia 1831 (Valdeburgo, *La straniera*). Successes at Rome, Naples, and Venice, 1832–8; Milan, S, 1839, 1842; London: HM, 1842; CG, regularly 1847–66. Also Paris, I, 1843; St Petersburg 1852–5, 1858–9. Sang until 1870. Taught singing Madrid Cons. from 1874. Despite an early vocal decline, one of the great singers of his age; a superlative melodramatic performer who could 'chill the blood' and 'stop the breath' (Chorley, who 'owed some of his best opera evenings' to him), and led Donizetti to champion the baritone protagonist. Among the roles he created were Cardenio (*Il furioso all'isola di San Domingo*), title-role of *Torquato Tasso*, Enrico (*Il campanello*), Nello (*Pia de' Tolomei*), and a famous Chevreuse (*Maria di Rohan*). One of the earliest Verdi baritones; created Nabucco, and sang the Doge (*I due Foscari*), Don Carlo (*Ernani*), Rigoletto. His brother **Sebastiano** (1814–1900), was also a baritone. Début Lucca 1836. Sang in Italy, London, and USA in Donizetti and Verdi.

**Rondine, La** (The Swallow). Opera in 3 acts by Puccini; text by Adami, translated from a German libretto by A. M. Willner and H. Reichert. Prod. Monte Carlo 27 Mar. 1917, with Dalla Rizza, Ferraris, Schipa, Huberdeau, cond. Marinuzzi; New York, M, 10 Mar. 1928, with Bori, Fleischer, Gigli, Ludikar, cond. Bellezza; London, Fulham TH, 9 Dec. 1965, with Doyle, Morgan, Gloster, Allum, cond. Head. Intended as an operetta for Vienna, the work was kept from production by the war, and the composer decided to set the libretto rather differently.

Paris and the Riviera, 1850s. Magda de Civry (sop), mistress of the banker Rambaldo (bar), is giving a party in her Paris home. The poet-philosopher Prunier (bar, later ten) reads Magda's palm and predicts that, like the swallow, she will fly away from Paris, perhaps to find true love. A young man, Ruggero (ten), son of an old friend of Rambaldo's, arrives and Lisette (sop), Magda's maid, advises him to spend his first evening in Paris at Bullier's café. Magda decides to follow him, in disguise.

At Bullier's, Ruggero sits alone. When Magda arrives, however, he dances with her (though without recognizing her) and the couple are soon drinking a toast to their love. Suddenly Rambaldo storms in and demands that Magda return home. She refuses, saying she has found love.

Ruggero and Magda have been living together on the Riviera. He hopes that they may marry and has written to ask for his parents' consent. Magda admits her past and feels unfit to marry him. She departs for Paris, since Rambaldo has begged her to return, leaving behind the one love of her life.

**rondò.** A popular aria form in the 18th cent., which was somewhat related to the instrumental rondo. The operatic rondò usually involved just two sections, one slow and one fast, each repeated twice. The first important composer of the rondò was Piccinni. A classic example is 'Non temer, amato bene', written by Mozart as an inclusion for *Idomeneo*. In the 19th cent. the rondò developed into the cantabile-cabaletta structure, first made popular by Rossini.

**Ronzi, Giuseppina.** See DE BEGNIS, GIUSEPPINA.

**Rooy, Anton van** (*b* Rotterdam, 1 Jan. 1870; *d* Munich, 28 Nov. 1932). Dutch baritone. Studied Frankfurt with Stockhausen. Début Bayreuth 1897 (Wotan). Bayreuth until 1903; London, CG, 1898–1913; New York, M, 1898–1908. A sensitive and serious artist, he was impressive as Valentin, Escamillo, Jochanaan; above all as Wotan, Kurwenal, and Sachs. (R)

**Rosa, Carl** (*b* Hamburg, 22 Mar. 1842; *d* Paris, 30 Apr. 1889). German conductor and impresario, born Carl Rose. Studied Leipzig and Paris. After touring Europe and USA as a violinist, met his wife **Euphrosyne Parepa** (*b* Edinburgh, 7 May 1836; *d* London, 21 Jan. 1874) who was a soprano. Studied with her mother Elizabeth Seguin. Début Malta 1852 (Amina). London: L, 1857; CG and HM regularly, 1859–65; CG, 1872. Toured USA 1865 with Rosa, whom she married in 1867. They formed the Parepa-Rosa Co., 1867, with her as leading soprano; this had much success in USA until 1872. Terminal illness prevented her appearing with it in London, 1873; after her death it was renamed *Carl Rosa Opera Company. Her voice was strong and sweet-toned; roles included Donna Anna, Norma, Elvira (*Puritani*).

**Rosalinde.** Eisenstein's wife (sop) in J. Strauss's *Die Fledermaus*.

**Rosbaud, Hans** (*b* Graz, 22 July 1895; *d* Lugano, 29 Dec. 1962). Austrian conductor. Studied Frankfurt with Sekles. After a concert career in which he became a champion of new music, music director Münster 1937–41; Strasburg 1941–4. Chief cond. Aix Festival 1947–58. Cond. première of *Moses und Aron* Hamburg Radio 1954, then Zurich 1957; cond. *Erwartung* and *Von Heute auf Morgen* Holland Festival 1958. (R)

**Rosenberg, Hilding** (*b* Bosjökloster, 21 June 1892; *d* Stockholm, 19 May 1985). Swedish composer and conductor. Studied privately, Stockholm with Ellberg and Stenhammar. Worked with theatre director Per Lindberg from 1926. Studied conducting with Scherchen: assistant conductor Stockholm, Royal O, 1932–4, chief conductor 1934. His theatrical work led to much

incidental music, then to several successful operas that to some extent reflect his admiration for Schoenberg. Later he simplified his style, and *Marionetter* is more neo-classical.

SELECT WORKLIST: *Resa till Amerika* (Journey to America) (Henriksson; Stockholm 1932); *Marionetter* (Benavente; Stockholm 1939); *Josef och hans bröder* (opera-oratorio: Rosenberg, after Mann; Swedish radio 1946, 1947, 1948); *Porträttet* (Malmerg, after Gogol; Swedish radio 1956).

BIBL: M. Pergament, *Hilding Rosenberg* (Stockholm, 1956).

**Rosenkavalier, Der** (The Knight of the Rose). Opera in 3 acts by Richard Strauss; text by Hugo von Hofmannsthal. Prod. Dresden, H, 26 Jan. 1911, with Siems, Von der Osten, Nast, Perron, cond. Schuch; London, CG, 29 Jan. 1913, with Siems, Von der Osten, Dux, Knüpfer, cond. Beecham; New York, M, 9 Dec. 1913, with Hempel, Ober, Case, Goritz, cond. Hertz.

After a prelude depicting passionate love-making, the curtain rises on the Marschallin (sop) and her young lover Octavian (sop or mez) in her bedroom. Hearing voices, Octavian hides and disguises himself as a maid. The Marschallin's boorish cousin Baron Ochs (bs) enters; he wants her to nominate a Knight of the Rose to present the traditional silver rose to his intended bride, the very young Sophie von Faninal (sop). Octavian emerges, dressed as 'Mariandel', and is instantly pursued by the lascivious Ochs. After the Marschallin's customary levée, Octavian reappears as himself, only to find her sombrely fatalistic that he will soon leave her. He cannot accept this, but their parting is subdued. Regretting that she did not kiss him goodbye, she sends the silver rose after him.

In the house of the newly rich merchant Faninal (bar), Sophie and her duenna Marianne (sop) are awaiting the presentation of the silver rose. When Octavian arrives with it, Sophie and he are immediately entranced by each other. Ochs follows shortly, repelling Sophie and outraging Octavian with his coarse behaviour. Octavian provokes him to a duel, and wounds him lightly; in the ensuing uproar, Faninal threatens Sophie with a convent if she refuses Ochs. Octavian, wanting her himself, hires Ochs's two intriguers, Annina (sop) and Valzacchi (ten). Later, Ochs receives a note via Annina from 'Mariandel' arranging a rendezvous.

'Mariandel' meets Ochs at an inn, with the intention of discrediting him as Sophie's suitor. The room is fitted with numerous devices enabling Annina and Valzacchi to frighten the Baron with strange apparitions; then Annina appears surrounded by children, pretending to be his deserted wife. Amid increasing tumult, the innkeeper and servants arrive, then the police, then the Faninals and finally the Marschallin. She dispatches them all (including the discomfited Ochs). Gently yielding Octavian to Sophie, she leaves them alone in their happiness.

BIBL: A. Jefferson, *Richard Strauss: 'Der Rosenkavalier'* (Cambridge, 1985).

**Rosina.** Dr Bartolo's ward, lover of Count Almaviva, (sop) in Paisiello's and (mez) in Rossini's *Il barbiere di Siviglia*.

**Rosing, Vladimir** (*b* St Petersburg, 23 Jan. 1890; *d* Los Angeles, 24 Nov. 1963). Russian, later US, tenor. Studied St Petersburg with Tartakov, Paris with J. De Reszke and Sbriglia. Début St Petersburg 1910 (Lensky). London, OH, 1915, directing and singing; USA 1921; founded American OC, which toured USA till 1929. In 1936 organized British Music Drama OC with Albert Coates. Possessed a powerful tone, and was considered by G. B. Shaw the best Russian singer of his time after Shalyapin. Roles included Hermann, Cavaradossi. (R)

**Rospigliosi, Giulio** (Pope Clement IX) (*b* Pistoia, 28 Jan. 1600; *d* Rome, 9 Dec. 1669). As a young man he served under Urban VIII, the Barberini pope responsible for the early encouragement of opera in Rome. His poetic talents were demonstrated in a series of carefully crafted librettos, including those for Landi's *Sant'Alessio*—the work given for the opening of the Barberini opera-house in 1632—Michelangelo Rossi's *Erminia sul Giordano* (1633), Mazzocchi's *L'innocenza difesa* (1641), and Luigi Rossi's *Il palazzo incantato* (1642). These represented a distinct move away from the Florentine-Mantuan tradition of the previous generation in the direction of a libretto in which human interest played a fuller role. Most important was *Sant'Alessio*: although, appropriately for a Roman work, based on a religious story, it featured more realistic comic characters than had hitherto been seen on the operatic stage. This led Rospigliosi to develop the first genuinely comic opera libretto in *Chi soffre, speri* (1639, music by Mazzocchi and Marazzoli). For inspiration he turned to the commedia dell'arte, although he also drew on commonplace figures from Italian life, inaugurating a tradition of comic libretto which reached a culmination with Da Ponte.

Shortly before the death of Pope Urban VIII, which led indirectly to a hiatus in the development of Roman opera, Rospigliosi became papal nuncio in Madrid. Here he made the acquaintance of Calderón, who provided the idea for a further comic libretto, *Dal male il bene*, set by Abbatini and produced Rome 1653, shortly before Rospigliosi's return. Abbatini set another of his librettos, *La Baltesara*; also first given in Rome, 1668, one year after Rospigliosi had been elected pope.

**Rossellini, Renzo** (*b* Rome, 2 Feb. 1908; *d* Monte Carlo, 13 May 1982). Italian composer and critic. Studied Rome with Sallustio, Setacioli, and Molinari. Was a pungently conservative critic for *Il messagero*. His operas follow the examples of Zandonai, Respighi, and Alfano, and have been enjoyed by audiences content with that tradition.

SELECT WORKLIST: *La guerra* (War) (Rossellini; Naples 1956); *Uno sguardo dal ponte* (A View from the Bridge) (Rossellini, after Miller; Rome 1961); *Il linguaggio dei fiori* (The Language of Flowers) (Rossellini, after Lorca; Milan 1963); *La leggenda del ritorno* (Fabbri, after Dostoyevsky; Milan 1966).

**Rossi, Gaetano** (*b* Verona, 18 May 1774; *d* Verona, 25 Jan. 1855). Italian librettist. Worked at different times in Venice and Verona. He wrote over 120 librettos for composers including Carafa, Coccia, Donizetti (e.g. *Linda di Chamounix*), Mayr, Mercadante, Meyerbeer (four, incl. *Il crociato in Egitto*), Nicolai, Pacini, and Rossini (*La cambiale di matrimonio, Tancredi,* and *Semiramide*). Though of no great literary ambitions, he drew on a wide range of sources and put into operatic currency many of the themes of Romanticism. His plots were often taken from classical or historical drama, and embraced the fashionable Spanish, Nordic, and British subjects as well as more traditional material; and his techniques contributed to the loosening of set forms characteristic of the reforms of early 19th-cent. opera.

**Rossi, Lauro** (*b* Macerata, 19 Feb. 1812; *d* Cremona, 5 May 1885). Italian composer. Studied Naples with Zingarelli, Furno, and Crescentini. His first opera was a comedy, *Le contesse villane*, followed by the successful *Costanza e Oringaldo*. The further success of three more comedies attracted the attention of Donizetti, who recommended him to the T Valle, Rome (1831–3). *La casa disabitata* was so successful that Malibran arranged for Barbaia to commission *Amelia* for her; but, through Malibran's insistence on dancing a pas de deux with the ballerina Mathis, the work was hissed. Rossi then toured Mexico, producing *Giovanna Shore* there; he returned suffering from yellow fever in 1843.

Settling in Milan, he revived *La casa disabitata* in 1843 as *I falsi monetari*, restoring his fame, and confirming this with *Il domino nero*. Head of Milan Conservatory 1849–70; head of Naples Conservatory to 1880, where he was unpopular. His last opera, *Biorn*, was a version of *Macbeth* tranferred to Norway, written for London. Rossi was always more successful in comedy, where he was at his best lively and inventive if not innovative, and was considered by some to be Donizetti's natural successor. Married **Isabella Obermeyer**, who thereafter sang as Ober-Rossi.

SELECT WORKLIST: *Le contesse villane* (The Rustic Countesses) (Passaro; Naples 1829); *Costanza e Oringaldo* (Fortini; Naples 1830); *La casa disabitata* (The Deserted House) (Ferretti; Milan 1834); *Giovanna Shore* (Romani; Mexico City 1836); *Il domino nero* (The Black Domino) (Rubino; Milan 1849); *La contessa di Mons* (D'Arienzo, after Sardou; Turin 1874); *Biorn* (Marshall, after Shakespeare; London 1877).

**Rossi, Luigi** (*b* Torremaggiore, *c.*1597; *d* Rome, 20 Feb. 1653). Italian composer and singing-teacher. Probably studied Naples with Macque, for a while serving the Neapolitan court. In 1621 joined the retinue of the Borghese family in Rome; in 1633 became organist of S Luigi dei Francesi. In 1641 Rossi entered the service of the *Barberini family, thereby participating in Rome's emergent operatic life. His *Il palazzo incantato*, first given in the opera-house at their palace, was one of the most lavish of mid-17th-cent. Italian operas and one of the most highly regarded.

Following the exile of the Barberini family to Paris in 1644, Mazarin invited Rossi to the French court as part of his efforts to introduce Italian opera there. Rossi arrived in 1646 and spent one year, during which he composed, among other works, his second opera *Orfeo*. Regarded as the apotheosis of the spectacular 17th-cent. Roman tradition, it also provided the French with one of their first and most widely discussed experiences of Italian opera. Like *Il palazzo incantato*, it places great weight on visual and aural splendour, though in the handling of the vocal line Rossi reveals a talent for expressive, lyrical writing, seen also in his numerous cantatas, which was appreciated by and influenced many of his contemporaries, including Cesti.

WORKLIST: *Il palazzo incantato* (Rospigliosi; Rome 1642); *Orfeo* (Buti; Paris 1647).

BIBL: A. Ghislanzoni, *Luigi Rossi: biografica e analisi delle composizioni* (Milan, 1954).

**Rossi, Michelangelo** (*b* Genoa, 1601–2; *d* Rome, July 1656). Italian composer. Best known during his lifetime as a violinist; from 1624 his activities centred on Rome, where he enjoyed aristocratic patronage. In 1633 his *Erminia sul Giordano* was given at the Barberini theatre: its production was one of the most spectacular events in the history of 17th-cent. Roman opera. Rossi's lost *Andromeda* was written to celebrate a wedding in Ferrara.

WORKLIST: *Erminia sul Giordano* (Erminia at the Jordan) (Rospigliosi; Rome 1633); *Andromeda* (Pio; Ferrara 1638, lost).

**Rossignol, Le.** See NIGHTINGALE, THE.

**Rossi-Lemeni, Nicola** (*b* Istanbul, 6 Nov. 1920; *d* Bloomington, IN, 12 Mar. 1991). Italian bass. Studied with his mother Xenia Makadon. Début Venice, F, 1946 (Varlaam). Milan, S, 1947–60; London, CG, 1952; New York, M, 1953–4; Chicago 1954–6. Also Paris, O; Moscow; Leningrad; etc. Roles included Don Giovanni, Don Basilio (Mozart), Mephistopheles (Gounod and Boito), Philip, Boris, Claggart. Sang in Italian première of Gruenberg's *Emperor Jones*, and created Pizzetti's Becket. Though his voice declined early, he was notable for his intelligent and vivid characterizations. Married the soprano **Virginia Zeani** (*b* Solovastra, 21 Oct. 1928). (R)

**Rossini, Gioachino** (*b* Pesaro, 29 Feb. 1792; *d* Paris, 13 Nov. 1868). Italian composer. Rossini's parents were both musicians, his father a trumpeter and his mother a singer. By the age of 14 he had learnt the horn, violin, cello, and harpsichord, and had sung professionally as well as writing a cavatina in the buffo style. In 1806 entered the Bologna Conservatory, and during his student years wrote his first opera, *Demetrio e Polibio*, as well as working as a continuo player in local opera-houses. His first professional work for the stage was *La cambiale di matrimonio*, which already reveals characteristics of wit and sentiment that were to mark his work. Two semiseria works, *L'equivoco stravagante* and *L'-inganno felice*, show the emergent range and versatility of his work. The latter was a particularly popular and artistic success, more so than the biblical *Ciro in Babilonia* (a somewhat static work whose best music is to be found in the choruses) or the nimbly farcical *La scala di seta*.

Further success came with *La pietra del paragone*, written for the soprano Marcolini to a commission from La Scala, Milan. Its warm lyrical content, together with a new feeling for nature, at the service of an ingenious and amusing plot, greatly appealed to a public that had by now learnt to relish his verve and melodic grace. Four works for Venice followed between Nov. 1812 and May 1813: *L'occasione fa il ladro*, *Il signor Bruschino* (which includes some darker and more menacing elements in the farce), *Tancredi*, and *L'italiana in Algeri*. *Tancredi* first made his name known outside Italy, partly through the wise popularity of 'Di tanti palpiti', but more deservedly for the manner in which he was able to give heroic virtues a new sensibility and hence accessibility to contemporary audiences. There is also a fresh approach to the handling of the orchestra, especially in the use of woodwind to add expressiveness to the vocal line. In his first comic masterpiece, *L'italiana in Algeri*, Rossini displays his assurance not only in the music's lyrical eloquence but in the virtuosity with which he both uses and disturbs convention to comic effect. Having mastered the conventions inherited from an earlier generation, and transformed them, he was able to establish and further develop a range of formal procedures which subsequently became known as the *Code Rossini. The modest *Aureliano in Palmira* had little success, the routine use of convention proving inadequate to the classical theme; and *Il turco in Italia* (his first collaboration with Romani) at first suffered from comparison with *L'italiana in Algeri*, though its urbanity, reliance on ensembles, and use of one of the characters as a controller of the plot give the work its own special distinction. *Sigismondo*, by contrast, was a failure, and remains an improbable candidate for successful revival.

Rossini was now engaged by *Barbaia as music director of both the T San Carlo and the T del Fondo in Naples. His first opera to be written for Naples was the dramatically powerful *Elisabetta, regina d'Inghilterra*, in which he drew on the T San Carlo's superb roster of singers, including Nozzari, García, and Isabella Colbran, later to be his first wife and the singer whose mastery of tragic and serious style was profoundly to influence the course of his work during his remaining 16 years in Italy. The next opera was the slight, but sharp and consciously Neapolitan, *La gazzetta*, and this was followed by *Otello*. A public in 1816 unfamiliar with Shakespeare was disturbed by the latter's tragic ending, which was replaced for the Rome revival. In the first two acts, Rossini and his librettist reduce the tragic condition to convention, but the final act shows a touch of genius in introducing a gondolier singing Dante's lines about the misery of lost happiness, and Desdemona's 'Willow Song' and Prayer were sufficiently admired by Verdi to make him wary of undertaking a rival project.

With *Il barbiere di Siviglia*, Rossini achieved his most enduring masterpiece. It marks a radical break with a past symbolized by the work of Paisiello, whose own setting of Beaumarchais is decisively outmanœuvred, though it was at first (at a stormy première) unfavourably compared with the earlier work: Paisiello's supporters may well have noticed that Rossini was not above beating their hero at his own musical game in places, though the finest passages (not only Figaro's exuberant 'Largo al factotum' and the Calumny aria, but the incomparable Act I stretto) are the Rossinian genius at its purest. Here, pointed declamation, rhythmic verve, the witty use of musical forms, and sharp-eared orchestral commentaries help embody character and action with a degree of comic realism that can be as lethal as it is entertaining. *La Cenerentola*, by contrast, shows a Rossini capable of genuine tenderness and path-

os, as well as fine writing for the coloratura mezzo-soprano voice of which he was so fond. A degree of latent Romanticism again emerges both in the powerful and dramatically far-flung melodramma *La gazza ladra*, and in the exotic and, for Rossini, remarkably sensuous score of *Armida*, one of whose most remarkable sections is set in a haunted forest. In both works, there is an increased range and sophistication in the use of orchestral colour, something that preoccupied Rossini in his later Naples period. After the somewhat formal and monochrome *Adelaide di Borgogna*, he completed the masterly *Mosè in Egitto*, famous for Moses's Prayer in Act III, and memorable for the opening 'Scene of the Shadows', a piece of sustained music drama that makes the significant advance of dispensing entirely with the use of a curtain-raising overture. There followed the 1-act farce *Adina*, and two further Naples works, *Riccardo e Zoraide* and *Ermione*. The latter, later described by Rossini as 'my little *Guillaume Tell*', did not please the Naples public and was quickly dropped from the repertory, though it is now seen to be a powerful and impressive response to the Racine drama on which it is based. *Eduardo e Cristina*, written for Venice, is no more than a hastily cobbled together pasticcio.

Rossini's last six Italian operas exhibit no less variety. *La donna del lago* is the first major opera to exploit Europe's growing fascination with Scott. Though the work does not live up to the high promise of its first act, it is notable for its tender, Romantic scene-painting and for elements of continuous composition in the earlier scenes. In this respect, *Maometto II* is a more consistently and powerfully sustained musical structure, imposingly written for soloists, chorus, and orchestra. Though startlingly difficult to sing, *Bianca e Faliero* presents a striking picture of internecine strife within war-threatened Venice, Rossini directing the fierce coloratura writing towards the depiction of states of emotional excess. There is also a good deal of emotional belligerence in *Matilde di Shabran*, Rossini's final essay in the semiseria genre and a rumbustiously diverting piece (whose first performance was conducted by Paganini). Rossini's last Neapolitan opera, *Zelmira*, suffers from a dull libretto by Tottola, set in ancient Greece and drawing from the composer somewhat formal, only occasionally exciting, music. Rossini's Italian career did, none the less, reach a fitting climax with *Semiramide*, one of his longest and most ambitious works. It is an opera that effectively reworks on a more massive scale the plot and musical structure of *Tancredi*, even if the drama is more complex and challenging. Though the music can be static for what are dramatically dynamic situations, and the coloratura self-defeating, it is one of Rossini's most imposing scores. He was by now the most popular and prolific composer of his day, having in the ten years since *Tancredi* composed some 25 operas. Stendhal declared, 'The glory of the man is only limited by the limits of civilization itself, and he is not yet 32.'

After visits to Vienna, where he met Beethoven, and London, where he conducted and sang in concerts with his wife, he settled in Paris as director of the Théâtre-Italien. Under his influence, it underwent something of a renaissance. He retained Paer as maestro al cembalo, engaged the young Hérold as chorus-master, and introduced Meyerbeer to Paris with *I crociato in Egitto*. For the extravagant coronation of Charles X in 1825 he wrote for a star-studded cast of singers a long, sophisticated, and often brilliantly inventive entertainment, *Il viaggio a Reims*. Commissioned by the French government to · write a number of comic and serious operas for the Opéra, he bided his time, familiarizing himself with French prosody and training a hand-picked group of singers in the Italian bel canto manner. Initially, he provided theatrically elaborate revisions of two Neapolitan works, *Maometto II* and *Mosè in Egitto*, whose political overtones were a resonance of the interest of a public inflamed by news of the Greek War of Independence as surely as they met a new public taste for opera as spectacle. His first new work (though one that draws on *Il viaggio a Reims*) was *Le Comte Ory*, a piece unique in its warmth, sensuous elegance, and good humour. It is also remarkable for the irony that chastens even the most tender moments, lending them a touch of wryness. His culminating masterpiece was to be one of the masterpieces of French Grand Opera, *Guillaume Tell*. Though the scene setting in the first act is rather leisurely, and there are empty pages to be found in its great length, Rossini's sense of picturesque colour, first glimpsed in *La donna del lago*, reaches new heights in this score, while such scenes as the gathering of the Cantons, the oath-swearing, and the shooting of the apple are all dramatically thrilling. Act II was especially admired by composers as different as Donizetti and Berlioz, both of whom thought it sublime.

This was Rossini's last stage work, though he lived for another 39 years, composing two religious works and a host of engaging trifles (the so-called *péchés de vieillesse*, or Sins of Old Age). He died on Friday 13th Nov. 1868, and was buried in Père Lachaise near Cherubini, Chopin, and Bellini. In 1887 his body was handed over to the city of Florence for reburial in Santa Croce, with a procession of more than 6,000 mourners included four military bands, and a chorus of 300 which sang the Prayer from *Mosè* to such effect that the crowd in front of Santa Croce encored it.

WORKLIST: *Demetrio e Polibio* (Viganò-Mombelli, after Metastasio's *Demetrio*; comp. before 1809, prod. Rome 1812); *La cambiale di matrimonio* (The Bill of Marriage) (Rossi, after Federici's drama; Venice 1810); *L'equivoco stravagante* (Gasparri; Bologna 1811); *L'inganno felice* (The Happy Deceit) (Foppa, after Palomba's text for Paisiello; Venice 1812); *Ciro in Babilonia* (Aventi; Ferrara 1812); *La scala di seta* (The Silken Ladder) (Foppa, after Planard's text for Gaveaux; Venice 1812); *La pietra del paragone* (The Touchstone) (Romanelli; Milan 1812); *L'occasione fa il ladro* (Opportunity makes the Thief) (Prividali; Venice 1812); *Il Signor Bruschino* (Foppa, after Chazet & Ourry's drama *Le fils par hazard*; Venice 1813); *Tancredi* (Rossi, after Voltaire's drama *Tancrède*; Venice 1813); *L'italiana in Algeri* (after Anelli's text for Mosca; Venice 1813); *Aureliano in Palmira* (Romanelli; Milan 1813); *Il turco in Italia* (Romani, after Mazzolà's text; Milan 1814); *Sigismondo* (Foppa; Venice, 1814); *Elisabetta, regina d'Inghilterra* (Schmidt, after Federici's drama based on Lee's drama *The Recess*; Naples 1815); *Torvaldo e Dorliska* (Sterbini; Rome 1815); *Almaviva, ossia L'inutile precauzione*, later called *Il barbiere di Siviglia* (Sterbini, after Beaumarchais's drama *Le barbier de Séville* and Petrosellini's text for Paisiello; Rome 1816); *La gazzetta* (Palomba, after Goldoni's drama *Il matrimonio per concorso*; Naples 1816); *Otello* (Salsa, after Shakespeare's drama; Naples 1816); *La Cenerentola* (Ferretti, after Perrault's story, Étienne's text for Isouard, and Romani's text for Pavesi; Rome, 1817); *La gazza ladra* (The Thieving Magpie) (Gherardini, after Aubigny & Caigniez's drama; Milan 1817); *Armida* (Schmidt, after Tasso's poem *Gerusalemme liberata*; Naples 1817); *Adelaide di Borgogna* (Schmidt; Rome 1817); *Mosè in Egitto* (Tottola, after Ringhieri's drama *L'Osiride*; Naples 1818); *Adina* (Bevilacqua-Aldobrandini; comp. 1818, prod. Lisbon 1826); *Riccardo e Zoraide* (Salsa; Naples 1818); *Ermione* (Tottola, after Racine's drama *Andromaque*; Naples 1819); *Eduardo e Cristina* (Schmidt, rev. Bevilacqua-Aldobrandini & Tottola, after Pavesi's text; Venice 1819); *La donna del lago* (The Lady of the Lake) (Tottola, after Scott's poem; Naples 1819); *Bianca e Faliero* (Romani, after Arnhault's drama *Blanche et Montcassin*; Milan 1819); *Maometto II* (Della Valle, after his drama *Anna Erizo*; Naples 1820); *Matilde di Shabran* (Ferretti, after Hoffman's text for Méhul's *Euphrosine* and Boutet de Monvel's drama; Rome 1821); *Zelmira* (Tottola, after Dormont de Belloy's drama; Naples 1822); *Semiramide* (Rossi, after Voltaire's drama; Venice 1823); *Il viaggio a Reims* (The Journey to Rheims) (Balocchi, partly after Staël's drama *Corinne*; Paris 1825); *Le siège de Corinthe* (rev. of *Maometto II*) (Balocchi & Soumet, after earlier text; Paris 1826); *Moise et Pharon* (rev. of *Mosè in Egitto* (Balocchi & Jouy, after earlier text; Paris 1827); *Le Comte Ory* (Scribe & Delestre-Poirson, after their own drama; Paris 1828); *Guillaume Tell* (Jouy, Bis, & others, after Schiller's drama; Paris 1829).

BIBL: Stendhal, *Vie de Rossini* (Paris, 1824; trans. 1956, 2/1970); G. Radiciotti, *Gioacchino Rossini: vita documentata, opere, ed influenza su l'arte* (Tivoli, 1927–9); H. Weinstock, *Rossini* (New York, 1968); R. Osborne, *Rossini* (London, 1986).

**Rostand, Edmond** (*b* Marseilles, 1 Apr. 1868; *d* Paris, 2 Dec. 1918). French playwright. Operas on his works are as follows:

*Les Romanesques* (1894): F. Hart (1918)
*La princesse lointaine* (1895): Montemezzi (unfin.); Witkowski (1934); Barberis (*Domniţa din depărtări* 1948)
*Cyrano de Bergerac* (1897): Damrosch (1913); Alfano (1936)
*L'aiglon* (1900): Honegger & Ibert (1937)

Also opera by Yanovsky, *Colombine* (1907).

**Rostock.** City in Mecklenburg, Germany. Travelling companies visited from 1606, and Singspiel was popular during the 18th cent. The Hoftheater opened 1751, and Italian and German companies gave seasons. The new Schauspielhaus opened 1786. The theatre burnt down 1880; performances were given in the Thaliatheater. The Stadttheater opened 1895, and developed vigorously under Willibald Kähler 1897–9. Visiting artists included Nikisch and Lilli Lehmann. The theatre was bombed 1942; reopened after the war.

**Roswaenge, Helge** (*b* Copenhagen, 29 Aug. 1897; *d* Munich, 19 June 1972). Danish tenor. Début Neustrelitz 1921 (Don José). Berlin, S, 1929–45, 1949; Bayreuth, 1934, 1936; London, CG, 1938; Salzburg 1933–9; Vienna, S, 1936–58. Also Paris, O; Milan, S; etc. Large repertory included Belmonte, Tamino, Florestan, Huon, Duke of Mantua, Radamès, Parsifal, Calaf. A highly regarded singer, he possessed a warm, ringing tone and glorious top notes. Autobiography, *Skratte Pajazzo* (Copenhagen 1945; German trans. 1953 as *Lache Bajazzo*). (R)

**Rota, Nino** (*b* Milan, 3 Dec. 1911; *d* Rome, 10 Apr. 1979). Italian composer. Studied Milan with Orefice, Rome with Pizzetti and Casella. Held teaching posts in Taranto and Bari; director, Bari Conservatory, from 1950. Composed much music for the cinema and radio. His operas are unabashedly tuneful and direct in appeal, and he won a particularly wide following with *Il cappello di paglia di Firenze* (A Florentine Straw Hat) (N. & E. Rota, after Labiche; comp. 1946, prod. Palermo 1955) and *La visita meravigliosa* (The Visitation) (Rota, after Wells; Palermo 1970).

**Rothenberger, Anneliese** (*b* Mannheim, 19 June 1924). German soprano. Studied Mannheim with Müller. Début Koblenz 1943 (small roles). Hamburg 1946–56; Munich from 1955; Vienna from 1957; Salzburg 1954, 1957; Gly. 1959–60; New York, M, 1960. Roles incl. Ilia, Constanze, Susanna, Pamina, Sophie, Zdenka, Lulu; created Sutermeister's Madame Bovary, among other parts. An attractive and excellent actress, with a firm, light tone. (R)

**Rothmüller, Marko** (*b* Trnjani, 31 Dec. 1908; *d* Bloomington, IN, 20 Jan. 1993). Croatian baritone. Studied Zagreb, Vienna with Weiss and Steiner (composition with Berg). Début Hamburg 1932 (Ottokar). Zurich 1935–47; London, CG, 1939, 1947–52; Vienna 1946–9; Gly. 1949–55; New York: City O, 1948, M, 1958–61, 1964–5. A well-rounded and intelligent artist with a warm, flexible voice. His repertory included Don Giovanni, Papageno, Rigoletto, Iago (Verdi), Kurwenal, Amfortas, Prince Igor, Tomsky, Doktor (*Wozzeck*); first Truchsess (*Mathis der Maler*). (R)

**Rouen.** City in Seine-et-Marne, France. An Académie Royale de Musique opened 15 Sep. 1688 in the Salle des Deux-Maures with Lully's *Phaëton*; became the Comédie, 1696. The T de Rouen (cap. 1,600) opened 29 June 1776; T de la Montagne 1793; T des Arts from 1794. The first operas by Boieldieu (born Rouen) were given here, *La fille coupable* (1793) and *Rosalie et Myrza* (1795). The theatre burnt down 25 Apr. 1876; rebuilt (cap. 1,500), opened 30 Sep. 1882 with *Les Huguenots*. The policy of giving French premières attracted Parisian audiences on special trains. Bombed 1940, 1944; seasons given at the T-Cirque. Rebuilt (cap. 1,460), opened 11 Dec. 1962 with *Carmen*. The theatre continues to pursue an adventurous policy.

**Rousseau, Jean-Jacques** (*b* Geneva, 28 June 1712; *d* Ermenonville, 2 July 1778). Swiss philosopher, composer, author, and writer on music. Wrote his first opera, *Les muses galantes* (1745), in imitation of *Les Indes galantes* of Rameau, who had a poor regard for it. His most famous composition was *Le devin du village* (1752), where the use of simple melodies and accompaniments (all that Rousseau was really capable of) well expressed the victory of plain rustic virtue over aristocratic corruption. Siding with the Italians in the *Guerre des Bouffons, he published his important *Lettre sur la musique française* (1753), expounding his hostility to French opera and taking a strong stand for melody as a form of heightened speech. *Pygmalion* (1770) uses orchestral interludes between actual speech, and prompted a number of imitations. He also left parts of another opera, *Daphnis et Chloé* (1779), and six new arias for *Le devin*. His other writings on music include the important *Dictionnaire de musique* (1768).

Rousseau's influence on Romanticism was

crucial. His stand for personal sensibility and emotion as the most valuable guide to living (as evinced in the *Confessions*) spoke to the Romantics in their preoccupation with sensation; and his belief in natural virtues (the 'noble savage'), and the qualities to be found in simple people living close to nature, was still more potent for artists, including composers. The greatly increased role of the orchestra in French Revolutionary and early German Romantic opera, to describe natural events playing a part in the drama, derives in part from Rousseau; as does the increased importance given to simple people and their feelings.

**Roussel, Albert** (*b* Tourcoing, 5 Apr. 1869; *d* Royan, 23 Aug. 1937). French composer. After an early naval career, studied music Paris 1894 with Gigout. His major stage work is the opéra-ballet *Padmâvatî*, based on an Indian legend. In it he affirms his belief that opera was exhausted and that a renewed impulse could be found in the 17th- and 18th-cent. French tradition of the opéra-ballet. Accordingly he gives choruses, dances, and mime an integral rather than decorative role, finding thereby opportunity for the symphonic development of his musical ideas. The subject further allows him a sense of distance—of masking and ritualizing emotion—which suited his temperament and also his musical nature in his fondness for 'exotic' scales and rhythms. In his *conte lyrique La naissance de la lyre* the element of ritual dance is turned to a classical legend, with the mask consciously, and by virtue of the subject, Apollonian. These qualities hardly equipped him for the comically racy intrigues of the plot of *Le testament de la tante Caroline*.

WORKLIST: *Padmâvatî* (Laloy, after a Hindu poem based on a 13th-cent. historical event; Paris 1923); *La naissance de la lyre* (Reinach, after Sophocles; Paris 1925); *Le testament de la tante Caroline* (Nino; Olomouc 1936).

BIBL: B. Deane, *Albert Roussel* (London, 1961).

**Rousselière, Charles** (*b* Saint-Nazaire, 17 Jan. 1875; *d* Joué-lès-Tours, 11 May 1950). French tenor. Studied Paris with Vaguet. Début Paris, O, 1900 (Saint-Saëns's Samson). Paris, O, until 1912; Monte Carlo 1905–14; New York, M, 1906–7; Milan, S, 1909; Buenos Aires, C, 1910. Roles included Roméo, Manrico, Siegmund, Parsifal; created leading roles in operas by Mascagni (*Amica*), Fauré (*Pénélope*), Charpentier (*Julien*). (R)

**Rovetta, Giovanni** (*b* Venice, *c*.1595; *d* Venice, 23 Oct. 1668). Italian composer and singer. Joined St Mark's, Venice as a choirboy; 1623 bass; 1627 second, 1644 first maestro di cappella. He composed two operas for Venice, *Ercole in Lidia* (1645) and *Argiope* (1649). Writing about a

performance of the first, John Evelyn described it as 'one of the most magnificent and expensive diversions the wit of man can invent'.

**Royal Academy of Music.** The organization founded by a group of noblemen in London in 1718 as a vehicle for the performance of operas by Handel and others. See HANDEL.

**Royal Hunt and Storm.** The name given by Berlioz to the choral and orchestral symphonic interlude in Act III of his *Les Troyens* in which Dido and Aeneas take refuge in a cave from a storm while out hunting, and there consummate their love.

**Royal Italian Opera.** The name adopted by the annual subscription series of Italian opera given in London during the 19th cent.; these involved a variety of companies and theatres, including HM, CG, L and DL.

**Royal Opera House.** See COVENT GARDEN.

**Rozkošný, Josef** (*b* Prague, 21 Sep. 1833; *d* Prague, 3 June 1913). Czech composer and pianist. Studied Prague with Kittl. Became a banker, but was a significant figure in Prague cultural life, including as choral director. His first opera, *Mikuláš*, was to a text by Smetana's librettist Karel Sabina, and shows the influence of Smetana (with whom he had an equivocal relationship). Few of the remainder had any sucess, apart from the Romantic *Svatojanské proudy*, one of the few fairy-tale operas in a genre later to be distinguished by Dvořák, and *Stoja*, the first Czech verismo opera.

SELECT WORKLIST: *Mikuláš* (Sabina; Prague 1870); *Svatojanské proudy* (St John's Rapids) (Rüffer; Prague 1871); *Stoja* (Kučera, after Konrád; Prague 1894).

**rubato** (It.: 'robbed'). The art (sometimes the abuse) of hurrying or slowing the pace within a given tempo in varying degrees for expressive effect.

**Rubini, Giovanni Battista** (*b* Romano, 7 Apr. 1794; *d* Romano, 3 Mar. 1854). Italian tenor. Studied Bergamo with Rosio, later Naples with Nozzari. Début Pavia 1814 (Generali's *Lagrime di una vedova*). Naples 1816–29; Milan, S, 1818, 1825–31; Paris, I, 1825, 1831–43; London, H (later HM), 1831–43; St Petersburg 1843, 1844. A supreme singer, celebrated in Mozart and Rossini, whose vocal gifts also influenced Italian Romanticism through Bellini, for whom he created roles in *Bianca e Gernando*, *Il pirata* (Gualtiero, which Tamburini said 'placed him above all known tenors'), *La sonnambula* (Elvino), and *I Puritani* (Arturo); and through Donizetti, for whom he created seven roles, most importantly Percy (*Anna Bolena*) and Fernando (*Marin Faliero*). Also admired by Glinka,

Chopin, Liszt (with whom he sang on tour), and Rubinstein. Though lacking looks and dramatic presence, he was so brilliant, committed, and intense a performer that he 'entirely enchanted' his hearers (Chorley, who thought him a 'genius'). His range was high, extending to f' in falsetto, his voice sweet, powerful, flexible, and capable of subtle gradations of colour and volume. He was also famous for his expressive musical sob. He retired in 1845, later going to sing to the dying Donizetti, who was unable to respond.

BIBL: C. Traini, *Il cigno di Romano* (Bergamo, 1954).

His wife was the French soprano **Adelaide Comelli-Rubini** (*b* ?1798; *d* Romano, 30 Jan. 1874). Studied Paris. Début Paris; then Naples (Morlacchi's *Gianni di Parigi*), where she sang until 1829. Married Rubini 1819. Created Calbo (Rossini's *Maometto II*); sang Imogene in *Il pirata* (despite Bellini's aversion to her, and with great success), and for Donizetti created Matilde (*Gianni di Calais*), and Nina (*Il giovedì grasso*), both with her husband. After 1831 she went into semi-retirement.

**Rubinstein, Anton** (*b* Vikhvatinets, 28 Nov. 1829; *d* Peterhof, 20 Nov. 1894). Russian composer and pianist. Studied with Villoing, later Berlin with Dehn. His operas show his concern for developing various Russian national characteristics, but suffer from a certain emotional coolness allied to a self-defeating fluency. When his first two nationalist operas, *Dmitry Donskoy* and *The Siberian Huntsmen*, both failed, he attacked the whole concept of Russian opera and the amateur condition of Russian composers (he founded the Russian Musical Society, 1859, and the St Petersburg Conservatory, 1862). Even disregarding his own German-Jewish origins, this served to place him in the Westernizer camp against the more Slavophile Stasov and the 'Mighty Handful'. His most successful opera, *The Demon*, is strongly influenced by French music, and its harmonic idiom is not recognizably Russian, though his lyrical vocal writing was of a quality to influence Tchaikovsky (who greatly feared and admired him). Nevertheless, it has some strongly written scenes, and the title-role was for many years regarded as one of the great Russian bass parts. The most successful of his other operas was *The Merchant Kalashnikov*. He later turned to comedy and sacred opera. His brother **Nikolay** (*b* Moscow, 14 June 1835; *d* Paris, 23 Mar. 1881) was a pianist whose virtuosity rivalled that of his brother, and a similar force for musical and educational standards in Russia. Founded the Moscow Conservatory, 1864.

SELECT WORKLIST: *Dmitry Donskoy* (Sollogub & Zotov, after Ozerov; St Petersburg 1852); *Sibirskiye*

*okhotniki* (The Siberian Huntsmen) (Zherebtsov; Weimar 1854); *Feramors* (Rodenberg, after Moore; Dresden 1863); *Demon* (Viskovatov, after Lermontov; St Petersburg 1875); *Kupets Kalashnikov* (The Merchant Kalashnikov) (Kulikov, after Lermontov; St Petersburg 1880).

**Rudel, Julius** (*b* Vienna, 6 Mar. 1921). Austrian, later US, conductor. Studied Vienna and New York. New York, CC: répétiteur 1943, cond. 1944, music director 1957–79. Pursued an enterprising policy including seasons devoted to US opera. Music director Kennedy Center, Washington, DC, 1971–6 (incl. première of Ginastera's *Bomarzo*, 1967). Director Caramoor Festival 1963–76. Paris, O, 1972. New York, M, from 1978. (R)

**Ruffo, Titta** (*b* Pisa, 9 June 1877; *d* Florence, 5 July 1953). Italian baritone. Studied Rome with Persichini, Milan with Casini. Début Rome, T Costanzi, 1898 (Herald, *Lohengrin*). London, CG, 1903; Chicago, frequently, 1912–26; New York, M, 1922–9. Also Buenos Aires; Milan, S; and the major European opera-houses. Gifted with an extraordinarily noble, ringing, and extensive baritone (described by both Serafin and De Luca as 'a miracle'), he was impressive, as a singer, for his compelling vigour and incisiveness; his critics, however, found him too aggressive. Famous for his Figaro (Rossini), Rigoletto, Don Carlos (*Ernani*), Scarpia, Hamlet, Leoncavallo's Tonio and Cascart (*Zazà*). Autobiography, *La mia parabola* (Milan, 1937, R/1977). (R)

BIBL: A. Farkas, ed., *Titta Ruffo—an Anthology* (Westport, CT, 1984).

**Ruggieri, Giovanni Maria** (*fl.* *c*.1690–1720). Italian composer. Composed *c*.12 operas for Venice during 1696–1712; 1715 maestro di cappella in Pesaro. *Elisa* is sometimes claimed as the first comic opera to be produced in Venice.

SELECT WORKLIST: *Elisa* (Lalli; Venice 1711).

**Ruprecht, Martin** (*b* ?Vienna, *c*.1758; *d* Vienna, 7 June 1800). Austrian composer and tenor. He joined Schindler's troupe in 1776 and in 1778 took the part of Fritz in Umlauff's *Die Bergknappen* for the opening of the National-Singspiel in Vienna. During the 1780s he sang with both the German and Italian companies; in 1788 he joined the Hofkapelle ensemble, where he remained until his death.

Of his six Singspiels the most important was *Was erhält die Männer treu?* Among others, *Der Dorfhandel* and *Das wütende Heer* were also well regarded, and represent a significant contribution to the early development of the Viennese Singspiel. In addition to his stage music, Ruprecht was also known as a song composer.

SELECT WORKLIST (all first prod. Vienna): *Was erhält die Männer treu?* (What Keeps Men Faithful?) (Zehn-

617

mark; 1780); *Der Dorfhandel* (Weidmann; 1785); *Das wütende Heer* (?, after Bretzner; 1787).

**Rusalka.** In Slavonic mythology, the water spirit which a drowned girl became. Generally, in northern lands the rusalki were less benign than in the south, where they would bewitch the passer-by and where death in their arms was an agreeable translation into a new world. The legend overlaps with that of *Undine and the *Donauweibchen.

1. Opera in 4 acts by Dargomyzhsky; text by composer, after Pushkin's dramatic poem (1832). Prod. St Petersburg, Circus T, 16 May 1856, with Bulakhova, Leonova, Lileyeva, O. Petrov, Bulakhov, Gumbin, cond. Lyadov; Seattle, WA, Metropolitan T, 23 Dec. 1921; London, L, 18 May 1931, with Slobodskaya, Pozemkovsky, Shalyapin, cond. Steiman.

Natasha (sop), daughter of a miller (bs), is betrayed by the Prince (ten) and drowns herself in the mill-stream. She becomes a Rusalka who lures men to their death. When the Prince marries, he hears her cries every time he approaches his bride. Wandering by the stream, he meets a child who reveals herself as his daughter. The miller hurls the Prince into the stream, where he joins Natasha and their child.

2. Opera in 3 acts by Dvořák; text by Jaroslav Kvapil, based largely on Fouqué's *Undine*, with additions from Hans Andersen's *The Little Mermaid* and suggestions from Gerhard Hauptmann's *Die versunkene Glocke*. Prod. Prague, N, 31 Mar. 1901, with Maturová, Brandachová, Hájková, Ptak, Kliment, cond. Kovařovic; Chicago, Sokol Slav H, 10 Mar. 1935 with Valentinova, Mashir, Daen, Bujanovsky, Karlash, Tulchinov. London, John Lewis T, 9 May 1950; London, SW, 18 Feb. 1959, with Hammond, Pollak, Stuart, Denise, Craig, J. Ward, Glynne, cond. Tausky.

The water spirit Rusalka (sop) falls in love with the Prince (ten), and with the help of the witch Ježibaba (mez) becomes human so as to marry him. But she must remain silent, and when the Prince tires of her and is unfaithful, a condition of her becoming human is violated and she dies together with the remorseful Prince.

Other operas on the subject are by Czerwiński (1874), Blaramberg (1888), and A. Alexandrov (1913).

**Ruse.** City in Bulgaria. A Ruse Operatic Society was formed in 1914; re-formed in 1919, it gave Ivanov's *Kamen i tsena* and Hadjigeorgyev's *Takhirbegovitsa*. Regular professional opera began when the Ruse National Opera opened in 1949. In 1954 the company visited Sofia, and has toured Romania. The theatre was reconstructed in 1956 (cap. 670).

**Ruslan and Lyudmila** (Russ.: *Ruslan i Lyudmila*). Opera in 5 acts by Glinka; text by V. Shirkov, K. Bakhturin, and others, after Pushkin's poem (1820). Prod. St Petersburg, B, 9 Dec. 1842, with O. Petrova, Stepanova, A. Petrova, Lileyeva, Baikov, Tosi, Leonov, Marcel, Likhansky, cond. Albrecht; London, L, 4 June 1931, with Lissichkina, Yurenev, Kaydanov, cond. Steinman; New York TH 26 Dec. 1942 (in concert); Boston 5 Mar. 1977, with Scovotti, Evans, Moulson, Braun, Tozzi, cond. Caldwell.

Lyudmila (sop) disappears from a feast for her three suitors, the knight Ruslan (bar), the poet Prince Ratmir (con), and Farlaf (bs), a cowardly warrior. Her father Svyetozar (bs) promises her to the one who can find her. Ruslan learns that she has been stolen by the dwarf Chernomor (silent), and is warned by the Finn (ten) against the wicked fairy Naina (mez), who advises Farlaf to wait until Ruslan has rescued Lyudmila, and then abduct her. Ruslan meets a gigantic head on a battlefield, and subduing it, discovers a magic sword. In Naina's palace, Ruslan is saved from her sirens by the Finn. He comes to Chernomor's castle, and defeats the dwarf, who has put Lyudmila into a trance. Ruslan is only able to awaken her with the aid of a magic ring from the Finn.

**Russia.** Though various theatrical entertainments with songs and dances were given in the 17th cent., no true operatic influence was felt until the opening up of Russia to the West by Peter the Great (d 1725). Under the Empress Anna (reigned 1730–40), the first Italian opera troupes to arrive were those of Giovanni Ristori and Johann Keyser, and the first opera performed in Russia was the former's *Calandro* (Moscow 1731). Other Italian companies came in 1733 and 1734, and the excellent company under Francesco *Araia in 1735. His *La forza dell'amore e dell'odio* (in the 'new St Petersburg Imperial T', 1736) was a great success, and two more works followed. German Singspiel probably made its first appearance in Russia in 1740.

The Empress Elizabeth (reigned 1741–61), with her love of Italy, further encouraged opera, which soon began to put down Russian roots. The first opera in Russian was *Tsefal i Prokris* (Cephalus and Procris, 1755), an opera seria by Araia; this was followed, also to a libretto by Sumarokov, by Raupach's *Altsesta* (Alceste, 1858). Peter III (reigned 1762) sent for Galuppi and Tartini, perhaps under the influence of his wife and successor Catherine the Great (reigned 1762–96), who encouraged French opéra-comique: the invitation to Italy, renewed by her, continued to support the century-long dynasty of foreign, principally Italian, composers in Russia: Araia, 1735–40, 1742–59, 1762; the German

Raupach, 1755–62; G. B. Locatelli, 1757–62; Manfredini, 1758–69; Galuppi, 1765–8; Traetta, 1768–76; Paisiello, 1776–83; Sarti, 1784–6, 1792–6; Cimarosa, 1787–91; the Spaniard Martín y Soler, 1788–94; and Cavos, 1798–1840. French opéra-comique was first staged in the 1750s, English opera (Dibdin's *The Padlock*) in 1771. The late 18th cent. also saw the building of theatres in the provinces, and visits from companies whose repertories included opera.

Native Russian opera was slow to emerge with any true identity. An early piece was *Tanyusha* (1756), set by the actor and theatre director Fyodor Volkov (1729–63), who had also translated operas; but probably a more nearly operatic piece was *Anyuta* (1772: music lost, perhaps by Vasily *Pashkevich, c.1742–97). The first whose music has completely survived is *Pererozhdeniye* (Rebirth, 1777), by Zorin. This, and surviving librettos, indicate opéra-comique with spoken dialogue as the example, with ballad opera as the immediate model. The first significant Russian operas are *Melnik koldun* (The Miller-Magician) (1779) by Mikhail Sokolovsky (*fl.* 1750–80) (using folk-songs), and three by Pashkevich, *Neschastye ot karety* (Misfortune from Owning a Carriage, 1779), *Skupoy* (The Miser, 1782), and *Sankpeterburgsky gostiny dvor* (The St Petersburg Bazaar, 1782) (using town songs). Another important work of the period was *Sbitenshchik* (The Sbiten [a drink] Seller, 1784), by the Frenchman Anton Bullant (*d* 1821).

Pashkevich's *Fevey* (1786) was to a text at least in part by Catherine herself. As an admirer of the *Encyclopédistes*, she believed in opera's role as social commentary; and her librettos include the patriotic, anti-Turkish *Nachalnoye upravleniye Olega* (The Early Reign of Oleg, music by Pashkevich, Canobbio, and Sarti, 1790), also domestic homily and satire, as in her attack on Gustav III during the Swedish war, *Gore Bogatyr Kosometovich* (The Unfortunate Hero Kosometovich, music by Martín y Soler, 1789). Other successful works were by Mathias Stabinger (*c.*1750–*c.*1815), (*Shchastlivaya Tonya* (Happy Tonya) and *Baba Yaga*, both 1786), and especially by *Fomin and *Bortnyansky. Paul I severely censored and restricted opera, but with Alexander I's accession in 1801 interest was renewed, in a French direction. A fashion for German fairy opera followed, and successful examples were written by Stepan Davydov (1777–1825), part-composer of the popular *Lesta* (from 1803), by Alexander Titov (1769–1827) (at least 13), and especially by Caterino *Cavos and Alexey *Verstovsky. The influential Cavos's substantial reorganizations included the development of a native singing tradition.

It is with *Glinka that Russian opera attains

maturity. In *Zhizn za tsarya* (A Life for the Tsar, 1836) and *Ruslan i Lyudmila* he found an authentic Russian voice that proved to be enduringly influential, the former work setting the example of the national epic, the latter that of the fantastic fairy-tale. After Alexander II's accession in 1855, steps were taken to improve the training of composers and singers, with the foundation of the conservatories of St Petersburg (1862) and Moscow (1864); and the strong influence of their respective founders, the brothers Anton and Nikolay *Rubinstein, gave a lead to such diverse composers as the 'Westernizers' *Serov and *Tchaikovsky, and the 'Slavophile' *Dargomyzhsky and the group dubbed by the critic Vladimir Stasov the *moguchaya kuchka* ('mighty handful'), Balakirev, *Musorgsky, *Rimsky-Korsakov, *Borodin, and *Cui. An important development was the breaking of the monopoly of the Imperial Theatres in 1882, allowing the growth of the enterprising private opera-companies such as those of Mitrofan Belyayev (1836–1904), Savva Mamontov (1841–1918), and Sergey *Zimin. After this great flowering of Russian opera, and of a school of singers to meet and encourage its standards, there was a brief recession in the early 20th cent. Interest turned elsewhere, despite the contributions of *Rakhmaninov, *Arensky, Grechaninov, Rebikov, and others. The line of lyrical, Italian-inspired opera was consciously reinvoked, and closed, by *Stravinsky with *Mavra* (1922).

After the 1917 Revolution, opera was threatened by Lenin's doubts about its importance in the social priorities, and by its aristocratic associations; but largely through the able arguments of the critic Boris Asafyev, and with the support of the arts commissar Anatoly Lunacharsky, opera was able to be viewed as based on national music and folk images, thus forming a foundation for the development of a Soviet operatic style. A reluctance to take opera seriously on the part of producers was overcome, and men of the stature of *Stanislavsky, *Nemirovich-Danchenko, Kommisarzhevsky, and Meyerhold were persuaded to apply their talents to opera. Fruitful contact between the Moscow Arts T and the Bolshoy led to the evolution of true actor-singers, though both Stanislavsky and Nemirovich-Danchenko preferred to work in their own studios. New life was infused into old productions, with more attention given to intelligent acting, less to vocal display; and the more able-minded singers, including Shalyapin and Ershov, were encouraged to produce operas in which they appeared. Sometimes, for lack of original Revolutionary operas, new librettos were provided for old works (*Les Huguenots* became *The Decembrists*). New Western operas were also introduced, including works by Schreker and Berg (*Wozzeck*, Leningrad 1927). The first attempts at genuinely modern Soviet opera, vigorously advocated by Asafyev, were *Za krasny Petrograd* (For Red Petrograd) by Arseny Gladkovsky and E. Prussak, Andrey Pashchenko's *Orliny bunt* (Eagles in Revolt), and Vasily Zolotarev's *Dekabristi* (The Decembrists) (all 1925).

The quarrel between restraint and adventure came to a head, inevitably, not over the many mediocre works that followed but over a major talent. *Shostakovich's *Nos* (The Nose, 1930) aroused both enthusiasm and opposition; and the opposition intensified with *Ledy Macbeth Mtsenskovo uyezda* (The Lady Macbeth of Mtsensk District, 1934). This precipitated a notorious *Pravda* article, 'Chaos instead of Music' ('Sumbur vmesto muzyky') (28 Jan. 1936), and the establishment of a hard, probably Stalin-inspired line which preferred Dzerzhinsky's *Tikhy Don* (The Quiet Don, 1935). Even *Prokofiev, though interested in working for Soviet opera, found it virtually impossible to satisfy the new demands, and his operas were misunderstood by officials; the willingness of lesser composers to toe the line assured their works of careers, but usually short ones. Great difficulties confronted composers anxious to reconcile an ideology they often did not share with their artistic integrity. Initial success was sometimes followed by violent rejection, most notoriously with *Velikaya druzhba* (The Great Friendship, 1947) by Vano Muradeli (1907–70). The nadir of state intervention was reached at the notorious 1948 congress (repudiated ten years later) at which leading Soviet composers were abused by the minister Andrey Zhdanov.

Since then there has been cautious evolution; but despite the enthusiasm for opera in Russia, the excellence of the singers' training, and the generous facilities available, few modern Russian operas, apart from those by Prokofiev and Shostakovich, have found a secure place in world repertories. Among those wartime and post-war composers most successful within Russia are Tikhon Khrennikov (*b* 1913), for long secretary of the composers' union, with *V buryu* (Into the Storm, 1939), Vissarion Shebalin (1910–63) with *Ukroshcheniye stroptivoy* (The Taming of the Shrew, 1957), Dmitry Kabalevsky (1904–87) with *Colas Breugnon* (1938), and Yury Shaporin (1887–1966) with *Dekabristy* (The Decembrists, 1953). The 'thaw' in Khrushchev's régime (1953–64) allowed some avant-garde productions but was too brief to give new composers serious

encouragement; and even a composer with adventurous ideas such as Rodion Shchedrin (*b* 1932) settled for convention in *Mertviye dushi* (Dead Souls, 1977). With the reforms of Gorbachev in the interests of *glasnost* ('openness'), then of Yeltsin, new ideas and enterprises have developed, though the situation remains uncertain.

The principal opera-houses of Russia are in Chelyabinsk (Glinka Opera and Ballet T, opened 1956 (cap. 1205) ); Gorky (Nizhny Novogorod) (Pushkin Opera and Ballet T, opened 1935 (cap. 1200) ); Kuybyshev (Opera and Ballet T, opened 1931); *St Petersburg; *Moscow; Novosibirsk (Novonikolayevsk) (Opera and Ballet T opened 1931); Perm (first season, 1894; Tchaikovsky Opera and Ballet T, opened 1920); Saratov (opera given from 1860s); Chernyshevsky (Opera and Ballet T, opened 1962 (cap. 1200) ); and Ekaterinburg (opera given from 1879; Lunacharsky Opera and Ballet T opened 1924 (cap. 1200) ). An important and valuable feature of post-Revolutionary Russia has been the operatic encouragement given to the republics that became part of the USSR, later the CIS.

**Ruthven.** Lord Ruthven (bar), the eponymous vampire in Marschner's *Der Vampyr*.

**Ruzitska, József** (*b* ?Pápa, *c.*1775; *d* ?, after 1823). Hungarian composer and conductor. Little is known of his life, except for records of theatrical appointments at Nagyvárad (1821) and Kolozsvár (1822), where he was Kapellmeister of the Hungarian theatre. He wrote three stage works: *Arany idők* (Golden Ages) (Debrecen; 1821); *Béla futása* (Béla's Escape) (Kolozsvár 1822) and *Kemény Simon avagy dicsőség házáért meghalni* (Simon Kemény, or It is Glorious to Die for the Fatherland) (Kolozsvár 1822). The latter, which combines folk-music elements with traditional operatic procedures, is one of the earliest attempts at a national style of Hungarian opera.

**Rysanek, Leonie** (*b* Vienna, 12 Nov. 1926). Austrian soprano. Studied Vienna with Jerger, later Grossmann. Début Innsbruck 1949 (Agathe). Bayreuth regularly 1951–82; Munich from 1952; London, CG, 1953–5, 1959, 1963; New York, M, from 1959. Also Paris, O; Berlin, SO; Salzburg; etc. Particularly effective, with her dramatic talent and strong, opulent tone, as Sieglinde, Kundry, Tosca, Salome, Danae, Elektra, Empress (*Die Frau ohne Schatten*). Later career as mezzo, e.g. Kabanicha, Klytemnestra, Kostelnička, with great success. (R)

# S

**Saarbrücken.** City in Saarland, Germany. Three theatres, built in the Baroque palace of the last Prince of Nassau-Saarbrücken, were destroyed in the French Revolution. The Neue T (cap. 450, enlarged to 700 in 1912) opened 1897 with Thomas's *Mignon*. Run in conjunction with the Trier theatre under Heinz Tietjen, 1919–22. The Gau-Theater, Saarpfalz (cap. 1,056, enlarged to 1,132) opened 1938 with *The Flying Dutchman*; bombed 1945; rebuilt 1947.

**Sabata, Victor de.** See DE SABATA, VICTOR.

**Sacchini, Antonio** (*b* Florence, 14 June 1730; *d* Paris, 6 Oct. 1786). Italian composer. Studied Naples with Durante, where his intermezzos *Fra Donato* and *Il giocatore* brought him to early attention. Taught Naples (Cons. S Maria di Loreto) 1756–62; director of the Cons. dell'Ospedaletto, Venice 1768–72. During this time he established a glowing reputation as an opera composer, above all with *L'Olimpiade* and *Il finto pazzo per amore*, both in Italy and elsewhere. In 1772–81 he worked in London, where his operas *Il Cid* (a reworking of the earlier *ll gran Cidde*), *Tamerlano*, *Nitteti*, and *Montezuma* were well received; because of debt he fled to Paris in 1781, where he was quickly drawn into the controversy between Gluck and Piccinni. Though at first hailed as a new weapon for the Italian faction, fierce rivalry soon developed between him and Piccinni. Neither *Renaud* nor *Chimène*, reworkings of *Armida* and *Il gran Cidde*, were well received, nor was *Dardanus*, in which he tried more consciously to imitate the style of the tragédie lyrique. Only after the posthumous performance of *Œdipe à Colone*, a work which owes much to Gluck's example, were his talents recognized by the French public.

As a student, Sacchini was regarded by Durante as Italy's brightest operatic hope, a judgement which was borne out in part by the popularity and respect which his works later enjoyed. But though his opere serie represent the culmination of the 18th-cent. tradition, with finely spun melodic lines and an expressive use of harmonic resources, they largely suffer from the same dramatic weaknesses as those of his contemporaries. The height of his art is seen in the French operas, particularly *Œdipe*, where skilful use of the chorus and a more flexible approach to the aria permit more satisfactory reconciliation of dramatic and musical aims.

SELECT WORKLIST: *Fra Donato* (?Trinchera; Naples 1756); *Il giocatore* (?; Naples 1757); *L'Olimpiade* (Metastasio; Padua 1763); *Il finto pazzo per amore* (The Lover Feigned Madman) (Mariani; Rome 1765); *Il gran Cidde* (Pizzi; Rome 1764, rev. London 1773 as *Il Cid*, rev. Paris 1783 as *Chimène*); *Armida* (De Gamerra; Milan 1772, rev. London as *Rinaldo* 1780, Paris 1783 as *Renaud*); *Tamerlano* (Piovene; London 1773); *Nitteti* (Metastasio; London 1774); *Montezuma* (Botarelli; London 1775); *Œdipe à Colone* (Guillard; Paris 1786).

BIBL: U. Prota-Giurleo, *Sacchini a Napoli* (Naples, 1956).

**Sachs, Hans** (*b* Nuremberg, 5 Nov. 1494; *d* Nuremberg, 19 Jan. 1576). Poet and Mastersinger. He is the composer of 13 *Meistertöne* and the author of several thousand poems, prose dialogues, comedies, and tragedies. His poem *Die Wittenbergisch Nachtigall* hailing Luther and the Reformation, and beginning 'Wach auff! es nahent gen dem tag', is set in part by Wagner as a salute to Sachs himself in *Die Meistersinger von Nürnberg* (1868). Here Sachs appears as the wise old cobbler-poet (bs-bar). He is also, as a younger man, the hero of Lortzing's *Hans Sachs* (1840), which is based on the Deinhardstein play drawn on by Wagner. Also Singspiel by Gyrowetz, *Hans Sachs im vorgerückten Alter* (1834), and three *Hans-Sachs Spiele* by Behrend (1949).

**Sack, Erna** (*b* Berlin, 6 Feb. 1898; *d* Wiesbaden, 2 Mar. 1972). German soprano. Studied Berlin with Daniel. Début Berlin, SO, 1925 (as contralto). Coloratura soprano from 1930. Berlin, S, 1933; London, CG, 1936; Chicago 1937. Also Milan, S; Paris, O; and world-wide tours. Roles included Rosina, Lucia, first Isotta (*Die schweigsame Frau*). Had a voice of phenomenal range (up to c''''), for which Strauss wrote several cadenzas in Zerbinetta's aria. (R)

**sacra rappresentazione.** See RAPPRESENTAZIONE SACRA.

**Sacrati, Francesco** (*b* Parma, *bapt.* 17 Sep. 1605; *d* ?Modena, 20 May 1650). Italian composer. Worked in Venice from *c.*1640, where he composed five operas. 1648 moved to Bologna; 1649 maestro di cappella Modena Cathedral.

The rediscovery of Sacrati's *La finta pazza* brought to light one of the earliest works in the Venetian repertory, in which the style of Cavalli and Cesti is to some extent prefigured. Performed in Paris 1645, it was one of the first Italian operas to be seen in France. A case has been argued, with some conviction, that Sacrati was responsible for much of the *Poppea* score usually attributed to Monteverdi.

SELECT WORKLIST: *La finta pazza* (The Feigned Madwoman) (Strozzi; Venice 1641).

**Sadko.** Opera-legend in 7 scenes by Rimsky-Korsakov; text by composer and V. I. Belsky, after folk legends. Prod. Moscow, private theatre, by Solodovnikov Co., 7 Jan. 1898, with Sekar-Rozhansky, Rostovsteva, Strakova, Alexanov, Karklin, I. Petrov, Negrin-Schmidt, Bedlevich, cond. Esposito; New York, M, 25 Jan. 1930, with Bouraskaya, Swarthout, Johnson, Diaz, Basiola, Ludikar, cond. Serafin; London, L, 9 June 1931, with Lissichkina, Sadoven, Pozemkovsky, Ritch, Petrov, cond. Goossens (Eugene III).

The wanderings of the minstrel Sadko (ten) bring him to Volkhova (sop), Princess of the Sea, who promises that his net will be filled with golden fish. He wages his head against the wealth of the citizens of Novgorod that he can catch the golden fish; he succeeds, and sets sail with some merchants, the Viking (bs), the Indian (ten), and the Venetian (bar), to find their fortunes, leaving behind his wife Lyubava (mez).

Returning successfully 12 years later, they are becalmed, and throw gold overboard to pacify the King of the Sea (bs); but his daughter Volkhova tells Sadko that one of the company must be sacrificed. Sadko is set adrift, and sinks to the sea-bed, where his song wins Volkova's hand. At the wedding Sadko's singing so arouses the sea that many ships are sunk; this brings down the anger of St Nicholas, who orders Sadko back to land. On the shore of Lake Ilmen Sadko bids farewell to Volkhova, who is transformed into the river that bears her name. Lyubava, still waiting, is reunited with Sadko, now the richest man in Novgorod.

**Sadler's Wells.** Theatre in Rosebery Avenue, North London. Originally a pleasure garden belonging to Sadler, who in 1684 discovered a medicinal well there. The first theatre was built 1765, and saw performances by Bland, the young Braham, and the Dibdins. Opera was sometimes given in the 19th cent., including the English première of *Luisa Miller* (1858); it then became a music-hall before falling into disuse.

In Mar. 1925 a move began to turn it into an *Old Vic for north London, and £70,000 was raised. The theatre (cap. 1,550) was reopened under Lilian *Baylis on 6 Jan. 1931. Opera and ballet alternated with Shakespeare until 1934, when it became the home of the opera and ballet companies. The repertory was enlarged and included the English premières of *The Snow Maiden*, *Tsar Saltan*, and the original *Boris Godunov*, the first *Don Carlos* in English, and productions of Wagner and works by English composers including Benjamin, Smyth, and Stanford.

Singers included Joan *Cross, Edith Coates, Sumner Austin, Powell Lloyd, and Arnold Matters, with Warwick Braithwaite and Lawrance Collingwood as conductors and Clive Carey and Sumner Austin as producers; the company also invited guest artists including Florence Austral, Miriam Licette, Heddle Nash, and the conductors Beecham, Barbirolli, and Coates. After 1940 a reduced company toured, with short seasons at the New T.

The company was built up again by Joan Cross, with productions including *The Bartered Bride* and *Così fan tutte* and new artists including Owen Brannigan and Peter Pears. It reopened in Sadler's Wells on 7 June 1945 with the première of *Peter Grimes*. Joan Cross then resigned, to be succeeded by Clive Carey and in 1947 by Norman *Tucker: music directors were James Robertson 1946–54, Alexander Gibson 1957–9, Colin Davis 1961–5. Premières included Berkeley's *Nelson* (1954), Gardner's *The Moon and Sixpence* (1957), and Bennett's *The Mines of Sulphur* (1965); British premières included *Simon Boccanegra*, *Rusalka*, *Mahagonny*, and an influential Janáček series under Mackerras beginning in 1951 with *Káťa Kabanová*. Leading singers included Victoria Elliott, Amy Shuard, Charles Craig, James Johnston, Howell Glynne, David Ward, and Owen Brannigan. Producers included Denis Arundell, Basil Coleman, George Devine, Powell Lloyd, and Tyrone Guthrie; Glen Byam Shaw was director of productions from 1962. The strength of public feeling for the theatre and company was shown by the response to a financial crisis which threatened to close them in 1958, and it became recognized that a stable company providing sound but also enterprising repertory performances had at least as secure a claim on audiences as the more glamorous international fare offered at Covent Garden. Stephen Arlen became director 1966–72: his régime included the 1968 *Mastersingers* which marked the rediscovery of Reginald *Goodall as a great Wagner conductor.

Plans for a move to the South Bank site of the National T in 1961 were abandoned, and the company moved in 1968 to the *Coliseum. In 1974 it was renamed English National Opera. The theatre continued to play host to visiting opera companies from home and abroad, and has been the scene of the British premières of *Die Frau ohne Schatten* and *Lulu*.

BIBL: D. Arundell, *The Story of Sadler's Wells* (London, 1965, 2/1978).

**Saffo.** See SAPPHO.

**sainete.** A genre of Spanish comic opera, portraying scenes from everyday life, often in the form of low comedy; placed as intermezzo (*entremès) or conclusion to larger works. The earliest surviving example is *Il Mago* (1632), and composers included Pablo Esteve (*d* 1794), Antonio Soler (1729–83), Antonio Rosales (*c.* 1740–1801), Jacinto Valledor (1744–1809), Blas Laserna (1751–1816), and others associated with the *tonadilla.

**Saint-Aubin, Jeanne-Charlotte Schroeder** (*b* Paris, 9 Dec. 1764; *d* Paris, 11 Sep. 1850). French soprano. Début at Court aged 9 (Ninette, Favart's *Acajou*), enchanting Louis XV. Sang at Bordeaux 1778, Lyons 1781; Paris, O, 1786, and CI (later OC), 1786–1808, when she retired in disgust at backstage bickerings, and became a director. Sang there once more, 1818. Appearing in works by Dalayrac, Grétry, Boieldieu, etc., she charmed audiences with her fresh voice and artless style, excelling, according to Fétis, in all types of role. Her two daughters **Cécile Saint-Aubin Duret** (1785–1862), and **Alexandrine** (1793–?) were both singers; the former created roles written for her in several of Isouard's operas, e.g. *Le billet de loterie*, *Jeannot et Colin*.

**Saint-Bris.** The Comte de Saint-Bris (bar), a Catholic nobleman in Meyerbeer's *Les Huguenots*.

**Saint-Évremond, Charles de Saint-Denis,** Seigneur de (*b* St Denis-le-Gast, *bapt.* 5 Jan. 1614; *d* London, 29 Sep. 1703). French critic, his writings contributed significantly to the evolution of a philosophy of opera in the late 17th and early 18th cents. He was, on the whole, against entirely sung dramatic works, seeing music merely as an ornament to spoken drama. After leaving France in 1661, he spent most of the latter part of his life in England, where he was instrumental in guiding the cultural taste of the courts of Charles II, James II, and William III.

**Saint François d'Assise.** Opera in 8 scenes by Messiaen; text by composer. Prod. Paris, O, 28 Nov. 1984, with Van Dam, Riegel, Éda-Pierre, Sénéchal, Duminy, Gautier, cond. Ozawa. Written for a vast orchestra and a chorus of 150 as well as seven principal singers, the complete work lasts five hours, and contains an extended bird-symphony, for which Messiaen spent much time in Assisi notating the indigenous bird-song. The work describes various episodes in the life of St Francis, among them his healing of a leper, his preaching to the birds, his receiving of the stigmata, and his visitation by an angel.

**Saint-Huberty** (Antoinette-Cécile Clavel) (*b* Strasburg, 15 Dec. 1756; *d* London, 22 July 1812). French soprano. Studied Warsaw with Lemoyne. Début Warsaw *c.*1774 (Lemoyne's *Le bouquet de Colette*). Paris, O, 1777–90, also singing at OC, and in Marseilles. Much admired by

Gluck. Became the Opéra's reigning prima donna after Levasseur's departure. Her eloquent acting aroused wild enthusiasm, and she was the first singer to be crowned on stage after a performance. Created numerous parts, notably the title-roles of Piccinni's *Didon*, Sacchini's *Chimène*, Edelmann's *Ariane*, and Hypermnestre (Salieri's *Les Danaïdes*); also sang Alceste (Gluck), Angélique (Piccinni's *Roland*). Left Paris 1790 with her lover, the Comte d'Antraigues, whom she married; they were eventually murdered in England, possibly because of the Comte's activities as a spy.

**St Louis.** City in Missouri, USA. The first opera heard was *La muette de Portici* in 1830, staged by a local company, but for long the city was dependent on touring companies. Celebrations of the 150th anniversary in 1914 led to the formation of the St Louis Municipal Opera Association, which gave summer seasons of a mixed musical repertory, including opera since 1919 and the opening of the vast Municipal Opera T (cap. 11,745). The Opera T of St Louis was founded in 1976 by Richard Gaddes, who has pursued an enterprising policy, concentrating on good ensemble and intelligent choice, that has included a Mozart cycle largely directed by Jonathan Miller and the revival of overlooked works; he was succeeded by Charles Mackay and Colin Graham in 1989. Opera is given in the Loretto-Hilton Center (cap. 954).

**St Petersburg.** City in Russia, named St Petersburg 1703–1914, Petrograd 1914–24; Leningrad 1924–91; St Petersburg again from 1991: capital of Russia 1703–1917.

The first theatrical productions were given soon after Peter the Great's foundation of the city, but the first opera to be performed was *La forza dell'amore e dell'odio* (1736) by the company of Francesco *Araia; his *Cephalus and Procris* (1755) was the first opera in Russian. In 1757 the first public theatre was opened, and here a mixed repertory included opera: Karl Knipper's Volny T (Free T) of 1777 became the Bolshoy Kammeny T (Grand Stone T) in 1783, and was followed by the Maly T (Small T). Caterino *Cavos was conductor of the Bolshoy 1798–1840. Italian opera was given 1826–32, and 1840. In 1833 Nicholas I opened the Alexandrinsky T, built by Rossi. The Bolshoy was rebuilt in 1836, opening with Glinka's *A Life for the Tsar*. For many years this work opened every season, and by the end of the century had been given over 700 times. Giovanni *Rubini headed a company; government-supported opera then lasted until 1885. Opera was also given from 1855 at the T Tsirk (Circus T), opposite the Bolshoy: this burnt down in 1859 and was rebuilt by Alberto Cavos (son of

Caterino) as the Maryinsky T in 1860. The influential conductor from 1863 (chief from 1869) to 1913 was Eduard *Nápravník. Premières included *La forza del destino* (1862) as well as many of the operas of the leading Russian composers of the day. From 1860 visiting French operetta and Italian companies played at the T Mikhaylovsky (built by Bryullov 1831–3, rebuilt by Cavos 1859). The three Imperial Theatres of St Petersburg were the Maryinsky, Mikhaylovsky, and (for drama) Alexandrinsky.

At the Revolution there was more resistance to change in Petrograd than in Moscow, and the commissar, the enlightened Anatoly Lunacharsky, was obliged to bring pressure on the company to resume work. The Maryinsky reopened in 1919 as the State Academic T of Opera and Ballet (known as GATOB; cap. 1,780: some of the first Soviet operas were given here in 1925), the Mikhaylovsky in 1920 as the State Academic T of Comic Opera, and the Maly T in 1921 as the Little Petrograd Academic T, becoming in 1926 the State Academic Little Opera T (known as MALEGOT; cap. 1,212). The latter was the home of most of the more advanced operatic experiments, and of the more advanced foreign works of the 1920s; but many smaller experimental groups flourished, among them the T Komichesky Opery (from 1920), the T Muzykalny Komedy, and the Palas-T (for operetta). In 1922 an opera studio was established in the conservatory, and here much enterprising and successful work was done; in 1928–9 they toured clubs and factories, and even gave a season in Salzburg. In 1935 the State Academic T was renamed the Kirov Opera and Ballet T. While this theatre upheld a traditional style, experimental work continued at the MALEGOT, including new works and productions (e.g. a Meyerhold production of *The Queen of Spades*). This was sharply interrupted by the 1936 attack on *Shostakovich's *The *Lady Macbeth of Mtsensk*. During the war the Kirov was evacuated to Perm, the MALEGOT to Orenburg, both returning in 1944. A group of artists who remained in the city formed an Opernobalety Kollektiv. After the war the Kirov and Maly set about re-establishing their reputations, which have continued to be more traditional for the former, more experimental for the latter. Under the conductor Yury Temirkanov since 1977, standards at the Kirov have increased and the artistic enterprise widened. In the early 1960s the Leningrad Conservatory opened a section to train musicians as opera directors: there is also an opera training section at the State Institute of T Art. Following the dissolution of the USSR, many theatres have chosen to revert to their former names.

See also RUSSIA.

**Saint-Saëns, Camille** (*b* Paris, 9 Oct. 1835; *d* Algiers, 16 Dec. 1921). French composer and pianist. Studied Paris with Maleden and F. Halévy, and made a very early mark as a child prodigy pianist. His long list of works includes 13 operas. The first, *La princesse jaune*, showed a touch with operetta which he did not really pursue. The only one to survive at all regularly in repertories is *Samson et Dalila*. Here his unfailing craftsmanship is supplemented by some sense of characterization for the two principals and the ability to create a unique and consistent emotional world. It was these qualities which led Liszt to secure it for production at Weimar when it was (temporarily) banned from the Paris stage because of its biblical subject-matter. *Henri VIII* is a Grand Opera making some use of motive. Other of his operas include some attractive numbers and show his fluency, sometimes his skill in pastiche, seldom much in the way of dramatic flair.

SELECT WORKLIST: *La princesse jaune* (The Yellow Princess) (Gallet; Paris 1872); *Samson et Dalila* (Lemaire; Weimar 1877); *Henri VIII* (Détroyat & Silvestre; Paris 1883).

BIBL: J. Harding, *Saint-Saëns and his Circle* (London, 1965).

**Salammbô.** Opera in 5 acts by Reyer; text by Camille du Locle, after Flaubert's novel (1862). Prod. Brussels, M, 10 Feb. 1890, with Rose Caron, Sellier, Renaud, Sentain, cond. Barwolf; New Orleans, French OH, 25 Jan. 1900, with Pacary, Layolle, Bonnarde, Bouxmann, cond. Vianesi.

Carthage, 240 BC. Matho (ten) has stolen the sacred veil from the shrine of the Carthaginian goddess Tanit. He is condemned to die at the hands of the priestess Salammbô (sop), who has fallen in love with him. She kills herself in his place, and Matho stabs himself.

Other operas on the subject are by Fornari (1881), Massa (1886), Morawski (?early 20th cent.), Hauer (part perf. radio 1930), Cuscinà (1931), Stoyanov (1940), and Casavola (1948). Musorgsky's unfinished opera (1863–6), of which he wrote six scenes and orchestrated two, was prod. Naples, C, 29 Mar. 1983 (orchestration completed by Zoltan Peskó) with Bernard, Bakov, Stone, Petkov, cond. Peskó.

**'Salce, salce'.** See WILLOW SONG.

**Salieri, Antonio** (*b* Legnano, 18 Aug. 1750; *d* Vienna, 7 May 1825). Italian composer. Studied with his brother Francesco, Venice with Pescetti and Pacini; during a visit Gassmann recognized Salieri's talents and brought him to Vienna 1766. Here he taught him composition and gave him important introductions to the court. Salieri's first surviving opera, the dramma giocoso *Le donne letterate*, was well received. In 1774 he became court composer and took charge of Italian opera at the court; in 1788 he succeeded Bonno as Kapellmeister.

Vienna remained the focus of Salieri's activities until his death. He composed over 30 operas there and exerted a considerable influence as a teacher, his pupils including Beethoven, Hummel, Schubert, Weigl, and Winter. His work embraced opera seria (*Palmira*) and opera buffa (*La fiera di Venezia*, *La grotta di trionfo*, *Falstaff*): less characteristically, for the National-Singspiel he wrote *Der Rauchfangkehrer*. The most interesting of his Viennese works are the settings of Da Ponte, largely opere buffe including *Il ricco d'un giorno*, *Il talismano*, and *La cifra*. An effective writer in the comic style, Salieri turns these librettos into attractive and diverting works, though they lack Mozart's humane and sensitive insight. For a celebration at Schönbrunn, at which Mozart's *Der Schauspieldirektor* was also given, Salieri provided *Prima la musica e poi le parole*, a satire on the rival claims of text and music. Following the death of Joseph II (1790) his influence at court declined; he relinquished directorship of the opera (and was succeeded by Weigl), though he continued to compose for it up to 1804.

Outside Vienna, he scored notable successes in Italy during an extended visit 1778–80: *L'Europa riconosciuta* opened La Scala, Milan, and operas were also given in Rome and Venice. But his greatest triumph came in France, where *Les Danaïdes* proclaimed him Gluck's successor. Even more overwhelming was the reception of *Tarare*, also well regarded when it was revised by Da Ponte for Vienna as *Axur*. Written to a libretto by Beaumarchais (who also contributed a lively preface on the importance of words in opera) this was an ambitious and musically impressive work; its text was frequently revised in the succeeding years, chiefly in order to reflect changing social principles through Revolution and Empire. In 1801 the Trieste OH opened with *Annibale in Capua*, his last Italian opera to be performed; his last German opera was *Die Neger*.

Salieri was a composer of the old school who found it difficult to adapt to the changing climate of the late 18th cent., which he explicitly acknowledged by his retreat from the operatic stage after 1804. Whether in comic or serious works, he rarely moved beyond the conventions of the genre, though he showed sensitivity to the notion of reform, and partly acknowledged it in *Tarare*. Highly regarded in his day, not least by Gluck, who eased his passage into Parisian musical circles, Salieri manifestly fails when the inevitable comparison is made with his near-contemporary Mozart. The claim that he poisoned Mozart was dramatized first by Pushkin

and thence advanced in opera by Rimsky-Korsakov's *Mozart and Salieri*; the story also occurs in Lortzing's pasticcio *Szenen aus Mozarts Leben*, and was widely propagated in Peter Schaeffer's play and film *Amadeus* (which contains much factual error and speculation). It is without foundation. Though he failed to influence Joseph II in Mozart's favour, and even intrigued against him, he proved helpful to his son, as he did to Gassmann's family: his generosity to the families of impoverished musicians was well known. His misfortune was to live in a city and era whose achievements must be judged by posterity against the yardstick of an operatic genius. Considered more realistically, he may be seen as a skilled exponent of his craft whose works are not without charm, though they lack the ultimate quality of memorability.

SELECT WORKLIST (all first prod. Vienna unless otherwise stated): *Le donne letterate* (The Cultured Ladies) (Boccherini; 1770); *La fiera di Venezia* (Boccherini; 1772); *L'Europa riconosciuta* (Verazi; Milan 1778); *Der Rauchfangkehrer* (The Chimney Sweep) (Auenbrugger; 1781); *Il ricco d'un giorno* (A Rich Man for a Day) (Da Ponte; 1784); *Les Danaïdes* (Du Rollet & Tschudi; Paris 1784); *La grotta di trionfo* (Casti; 1785); *Prima la musica e poi le parole* (First the music, then the words) (Casti; 1786); *Tarare* (Beaumarchais; Paris 1787, rev. Vienna 1788 as *Axur, re d'Ormus*); *Il talismano* (Da Ponte; 1788); *La cifra* (The Cipher) (Da Ponte; 1789); *Palmira, regina di Persia* (De Gamerra; 1796); *Falstaff* (Defranchesi; 1799); *Annibale in Capua* (Sografi; Trieste 1801); *Die Neger* (Treitschke; 1804).

BIBL: R. Angermüller, *Salieri: sein Leben und seine weltlichen Werke* (Munich, 1971).

**Salignac, Thomas** (*b* Générac, 19 Mar. 1867; *d* Paris, 1945). French tenor. Studied Paris with Duvernoy. Début Paris, OC, 1893 (small roles). New York, M, 1896–1903; Paris, OC, 1905–13, 1923; London, CG, 1897–9, 1901–4; Brussels, M, 1919. Director Nice O, 1913–14. Toured Canada and USA 1926 with an opéra-comique co. Roles included Don Ottavio, Don José, Gounod's Faust, Canio, Mârouf, Salammbô (Reyer), and creations in operas by Milhaud, Missa, and Falla (title-role of *El retablo de Maese Pedro*). (R)

**Salimbeni, Felice** (*b* Milan, 1712; *d* Ljubljana, Aug. 1751). Italian soprano castrato. Studied Naples with Porpora. Début Rome 1731 (Hasse's *Cajo Fabrizio*). In the service of Charles VI, Vienna, 1733–9 and of Frederick the Great 1743–50. Dresden Royal T from 1750. Possessed an exceptionally beautiful voice and appearance, and was famous for his adagios.

**Sallinen, Aulis** (*b* Salmi, 9 April 1935). Finnish composer. Studied Helsinki with Merikanto and Kokkonen. His reputation as a concert composer was enhanced by the appearance of his first opera, *The Horseman*, a tight-knit and powerful nationalistic work. This dramatic control was demonstrated anew with *The Red Line*, which further developed an idiom owing something to Shostakovich in its direct, innovative traditionalism. *The King Goes Forth to France* was a part-commission from Covent Garden, London, where it was staged in 1987: it is a powerful though enigmatic work, centring on the growth of irrational power as the main character deteriorates morally, and builds on the previous operas with lighter textures and a more economic use of material. Sallinen was originally commissioned to write *Kullervo* for the opening of the new Helsinki Opera House.

WORKLIST: *Ratsumies* (The Horseman) (Haavikko; Savonlinna 1975); *Punainen viiva* (The Red Line) (Sallinen, after Kianto; Helsinki 1979); *Kuningas lähte Ranskaan* (The King Goes Forth to France) (Haavikko; Savonlinna 1984); *Kullervo* (Los Angeles 1992).

**Salmhofer, Franz** (*b* Vienna, 22 Jan. 1900; *d* Vienna, 22 Sep. 1975). Austrian composer and conductor. Studied Vienna with Schreker and Schmidt. Cond. Vienna: B, 1929–39; director S, 1945–55; V, 1955–63. He had successes in Vienna with *Dame im Traum* (1935), *Ivan Tarassenko* (1938, rev. 1946), and *Das Werbekleid* (1943).

**Salminen, Matti** (*b* Turku, 7 July 1945). Finnish bass. Studied Helsinki, and Rome with Ricci. Début Finnish National O 1966 (small roles). Cologne 1972–6; Milan, S, 1973; London, CG, 1974; Bayreuth 1976; Paris, O, 1977; New York, M, 1981; also Savonlinna, Budapest, Vienna, Munich, etc. Roles incl. Osmin, Sarastro, King Mark, Fafner, Hunding, Hagen, Philip, Gremin. (R)

**Salome.** Opera in 1 act by Richard Strauss; text, Oscar Wilde's tragedy (1893) in the Ger. translation of Hedwig Lachmann. Prod. Dresden, H, 9 Dec. 1905, with Wittich, Chavanne, Burrian, Perron, cond. Schuch; New York, M, 22 Jan. 1907, with Fremstad, Weed, Burrian, Van Rooy, cond. Hertz (public rehearsal); London, CG, 8 Dec. 1910, with Ackté, Metzger, E. Krauss, Whitehill, cond. Beecham.

Herod's Palace in Jerusalem, AD 30. During Herod's banquet, Jochanaan (John the Baptist) (bar), proclaims—from the cistern where he is imprisoned—the coming of the Messiah. He is brought out for Salome (sop) to see, and repels her fascinated advances: he urges her not to follow the ways of her mother Herodias (mez). He is taken back to the cistern. Herod (ten), her stepfather, asks Salome to dance; she agrees on condition that he will grant her what she desires. After her Dance of the Seven Veils she demands

the head of Jochanaan, which Herod is forced to have brought to her. She fondles and kisses it until the revolted Herod orders his soldiers to crush her with their shields.

BIBL: D. Puffett, *Richard Strauss: 'Salome'* (Cambridge, 1989).

**'Salut, demeure chaste et pure'.** Faust's (ten) apostrophe to Marguerite's house in Act II of Gounod's *Faust*.

**Salvadori, Andrea** (*b* ?, 1591; *d* Florence, 1635). Italian librettist. Poet at the Tuscan court, where he contributed to many smaller dramatic pieces. Chiefly remembered as librettist for Gagliano's *La flora* (1628) which, in translation, was one of the first and most influential Italian operas given in France. Salvadori was Francesca Caccini's first choice as librettist for her *Ruggiero*: his refusal weakened his influence at court.

**Salvi, Antonio** (*b* Florence, 1742; *d* ?). Italian librettist. Worked in Florence between 1701 and 1710, thereafter probably in Venice, producing about 25 librettos. These were set by many of the leading composers in the first half of the 18th cent., including Gasparini (e.g. *Il tartaro nella China*, 1715), Scarlatti (*Tamerlano*, 1706, and *Arminio*, 1703), and Vivaldi (*Scanderbeg*, 1718), and achieved great popularity.

**Salvini-Donatelli, Fanny** (*b* Florence, *c*.1815; *d* Milan, June 1891). Italian soprano. Début Venice, T Apollo, 1839 (Rosina). Vienna 1842; London, DL, 1858. Created Violetta, against Verdi's wishes; though she was a good singer, her comfortable nature and her plumpness contributed to the fiasco of the first *Traviata*. She was, however, praised in the role by some critics, and admired in *Macbeth*, *I due Foscari*, and *Ernani*.

**Salzburg.** City in Austria. On 10 Feb. 1614 a 'Hoftragicomoedia' was staged in the Archbishop's palace, probably the first opera performance outside Italy. Operas were then staged in the Steintheater in the gardens of Schloss Hellbrunn. The late 17th cent. saw operas and school dramas by Georg Muffat, Heinrich Biber, and Andreas Hofer, mostly lost. Antonio *Caldara was regularly in Salzburg 1716–27, and the Heckentheater in the gardens of Schloss Mirabell was probably opened with his *Dafne* (1719). Here Mozart (born in the city 1756) had his *La finta semplice* produced 1769, and *Sogno di Scipione* 1772. The Ballhaus was converted into the Hoftheater 1775, becoming the Nationaltheater 1806.

The first Mozart festival was held in 1842, with an important centenary festival in 1856 directed by Alois Taux, also an opera conductor and composer. The modern festival dates from 1877; concerts and operas were given under Richter,

Mottl, Mahler, Richard Strauss, Muck, Schalk, and Weingartner. Lilli Lehmann arranged two *Don Giovanni* performances in 1901, singing Donna Anna, and there were performances of *Don Giovanni* and *Figaro* under Mahler, in the Roller settings and with Mayr as Figaro, 1906. From 1922 opera was given in the Stadttheater. The Festspielhaus (former Winterreitschule) was converted 1924, rebuilt 1926 (cap. 1,200), first staged opera 1927 (*Fidelio* with Lotte Lehman). The 1930s saw opera under Walter, Toscanini, Krauss, Furtwängler, and Knappertsbusch, though the German annexation in 1938 caused some to leave. The festival resumed 1945. The most influential conductors were Furtwängler, Böhm, and Karajan, who was director 1956–60 and 1964-89, and in 1967 founded an Easter Festival annually including a new opera production. Opera has been given in the Felsenreitschule from 1949, altered 1970 (cap. 1,549); the Festspielhaus from 1927, rebuilt 1937 (renamed Kleines Festspielhaus 1960) (cap. 1,323); and the Grosses Festspielhaus (cap. 2,177), opened 1960 with *Rosenkavalier*. The festival has included premières of works by Strauss (*Liebe der Danae*, 1952), Von Einem, Orff, Egk, Liebermann, Henze, and Berio. Gerard Mortier and Georg Solti became director and artistic director 1992.

**Samaras, Spiridon** (*b* Corfu, 22 Nov. 1861; *d* Athens, 7 Apr. 1917). Greek composer. Studied Corfu with Xyndas, Athens with Stancabianco (with whom he wrote an opera), Paris with Delibes. His *Flora mirabilis* is a verismo work anticipating Mascagni's *Cavalleria rusticana* by four years, and was given throughout Greece with great success. A further verismo opera, *La martire*, confirmed his touch with the genre, and was chosen to open the Milan, TL, in 1894. Though he composed several other operas, his only comparable success was with *Rhea*.

SELECT WORKLIST: *Medgé* (Elzéar; Rome 1888); *Flora mirabilis* (Fontani 1886); *La martire* (Illica; Naples 1894); *La furia domata* (?, after Shakespeare; Milan 1895); *Rhea* (Milliet; Florence 1908).

**Samiel.** The devil (spoken) in Weber's *Der Freischütz*.

**Sammarco, Mario** (*b* Palermo, 13 Dec. 1868; *d* Milan, 24 Jan. 1930). Italian baritone. Studied Palermo with Cantelli, Milan with Emerich. Début Palermo 1888 (Valentin). Milan, S, 1895–6, 1902, 1905, 1913; Buenos Aires from 1897; London, CG, 1904–14, 1919; New York, Manhattan OC, 1907–10; Chicago 1909–13. Created Gérard (*Chenier*), Cascart (Leoncavallo's *Zazà*) among other roles; also sang Scarpia, Hamlet, Rigoletto, and Tonio, with a mellifluous tone and dramatic conviction. (R)

**Sammartini, Giovanni Battista** (*b* Milan, 1700 or 1701; *d* Milan, 15 Jan. 1775). Italian composer. The most important Milanese instrumental and church composer of his generation. Sammartini's music was widely disseminated throughout Europe and was crucial in the development of the Classical style. His three operas, of which *L'ambizione superata dalla virtù* was the most successful, are models of mid-18th-cent. opera seria. As a teacher he had a considerable influence on Gluck, whose adapted some of his symphonic music for the overtures to *Le nozze d'Ercole e d'Ebe* (1747) and *La contesa dei numi* (1749).

SELECT WORKLIST: *L'ambizione superata dalla virtù* (?; Milan 1734).

**Samson et Dalila.** Opera in 3 acts by Saint-Saëns; text by Lemaire, after Judges 14–16. Prod. Weimar (in German) 2 Dec. 1877, with Von Müller, Ferenczy, Dengler, cond. Lassen; New York 25 Mar. 1892 (concert); New Orleans, French OH, 4 Jan. 1893, with Mounier, Reynaud, Hourdin; London, CG, 25 Sep. 1893 (concert), 26 Apr. 1909, with Kirkby Lunn, Fontaine, Davey, cond. Frigara. First French perf. Rouen 3 Mar. 1890; Paris: T Eden, 31 Oct. 1890; O not until 23 Nov. 1892, then 500 performances in 30 years.

Israel, 1150 BC. Samson (ten), the Hebrew warrior, leads a revolt against the Philistines. Dalila (mez), a Philistine beauty, is urged by the High Priest of Dagon (bar) to seduce Samson and discover the secret of his strength. Samson reveals to her that it lies in his hair, which Dalila then cuts off, rendering him powerless. He is taken prisoner by the Philistines, who put out his eyes. Brought to the temple of Dagon, where he is mocked by his captors, he prays to God for the return of his strength, and pulls down the temple, killing himself and his enemies.

**San Carlo Company.** US touring company founded in 1919 by Fortune Gallo (1878–1970) which gave popular-priced opera throughout the USA. It attracted distinguished guest singers, and many of its artists went on to major careers, including Jean Madeira, Dorothy Kirsten, Eugene Conley, and Nicola Rescigno. Disbanded 1955.

**Sances, Giovanni Felice** (*b* Rome, *c*.1600; *d* Vienna, *c*.12 Nov. 1679). Italian composer and singer. Studied Collegio Germanico, Rome. Held positions in Rome, Bologna, Venice, and possibly Padua. Joined the Vienna court as singer 1636; assistant Kapellmeister 1649; Kapellmeister 1669.

He wrote five operas: for the first, *Ermiona*, he himself took the part of Cadmus. His other works were written for Vienna or Prague for imperial occasions and include the operas *Apollo deluso* and *Aristomene Messanio* and many \*sepolcri. Though conventional in conception these did much to establish the court opera in Vienna and laid the foundation for Draghi's later achievements.

SELECT WORKLIST: *Ermiona* (Obizzi; Padua 1636, lost); *Apollo deluso* (Draghi; Vienna 1669); *Aristomene Messanio* (Minato; Vienna 1670).

**Sanderson, Sibyl** (*b* Sacramento, CA, 7 Dec. 1865; *d* Paris, 15 May 1903). US soprano. Studied Paris with Sbriglia and Marchesi. Début The Hague 1888 (Manon). Paris, OC, 1889, and O, 1894; London, CG, 1891; New York, M, 1895, 1901. Adored by Massenet, who launched her career, and wrote the roles of Esclarmonde and Thaïs for her. Also created Saint-Saëns's Phryné; sang Juliette, Gilda, etc. Despite her beauty, dramatic talent, and impressive voice (rising to a g''' nicknamed the 'sol Eiffel'), few thought her as 'superbly gifted' as Massenet did.

**San Diego.** City in California, USA. Opera was given by touring companies from the late 19th cent., though the Civic Grand O performed 1919–32. The San Diego O Guild first organized performances by visiting companies in 1952, becoming one in its own right when the New Civic T opened 1965 with Walter Herbert as director. Tito Capobianco succeeded in 1976, exploring the Verdi repertory and expanding activities until he left in 1983; the policy has since been more conservative, with rising young singers encouraged partly out of financial limitations, though some new directions have also been sought. Opera is given in the Civic T (cap. 2,992).

**Sandunova, Elizaveta** (*b* St Petersburg, ?1772; *d* Moscow, 3 Dec. 1826). Russian soprano. Studied St Petersburg with Martín y Soler, Paisiello, and Sarti. Début St Petersburg, Hermitage T, 1790 (Amore, Martín y Soler's *Arbore di Diana*). From 1791 she had much success at the court of Catherine the Great, who commanded her to sing as Uranova (after the recently discovered planet Uranus). Moscow, Petrovsky T, 1794; St Petersburg 1813–23. A strongly expressive, versatile singer and actress with a fine, rich voice, she was greatly admired in a huge repertory (*c*.300 roles), e.g. as Queen of the Night and especially Cavos's Lesta.

**San Francisco.** City in California, USA. The first opera was *Sonnambula* at the Adelphi T by the Pellegrini Co. in 1851. Opera was performed in 11 theatres, including the Baldwin, Tivoli, Civic OH, and Auditorium. The largest was Wade's OH (cap. 2,500), built in 1876 and later expanded as the Grand OH (cap. 4,000). Visiting companies included those of Mapleson (from 1879, with Patti 1884 and 1889), Emma Juch, Del Conte, and Damrosch. The New York Metropolitan first visited in 1890 (with Tamagno in *Otello* (Verdi) and *Trovatore* and Patti in *Semiramide*, *Sonnambula*, *Traviata*, and *Lucia*), returning in 1900 with a complete *Ring*. The 1906 earthquake struck only hours after Caruso, Fremstad, and Journet had been singing in *Carmen*.

Not until 1909 was recovery sufficient for a permanent company to be proposed by Gaetano Merola, conductor of W. A. Edwards's International Grand OC. The Italian community having raised some funds, Merola returned 1919–20 with the San Carlo Co. and staged *Pagliacci*, *Carmen*, and Gounod's *Faust* in Stanford University's football stadium 1922. He founded San Francisco O 1923. Operas were staged in the Civic Auditorium (cap. 12,000) until 1932, with a largely Italian and French repertory (though *Tristan* was given in 1927 and *Salome* in 1931), and a few rarities such as *Mârouf* and *L'enfant et les sortilèges*. Singers who appeared included Jeritza, Muzio, Rethberg, De Luca, Didur, Gigli, Journet, Lauri-Volpi, Martinelli, Pinza, Schipa, and Scotti. The company opened the War Memorial T (cap. 3,176) in 1932 with *Tosca*. Merola continued as director until his death in 1953, staging a popular, mainly Italian repertory, and engaging many European singers before the Metropolitan. German opera fared better from the 1930s; singers included Flagstad, Melchior, Schorr, later Traubel and Lotte Lehmann.

Merola was succeeded by Kurt Herbert Adler, who initiated a more adventurous policy with the US premières of many works. Among singers making their US débuts were Tebaldi, Del Monaco, Simionato, Gobbi, Jurinac, Margaret Price, Nilsson, Rysanek, Schwarzkopf, Sciutti, Te Kanawa, Geraint Evans, and Richard Lewis. Artists singing roles for the first time included Martinelli (Otello) and Tibbett (Iago); Geraint Evans (Wozzeck); Björling (Riccardo, *Ballo*); Horne (Eboli); and Sutherland (Maria Stuarda). Terence McEwen succeeded Adler in 1982, restoring the company's fortunes with careful economies, but retired through illness in 1988 to be succeeded by Lotfi Mansouri.

In 1961 The Spring OC was founded to give popular operas with young US singers; gradually new US operas were added and new production styles encouraged. Western Opera T tours opera in English, in 1987 becoming the first US company to visit China.

**Sanjust, Filippo** (*b* Rome, 9 Sep. 1925; *d* Nov. 1993). Italian designer and producer. Collaborated with Visconti London, CG, *Don Carlos* 1958, and subsequently. Salzburg, première

*Bassarids* 1966. First production, Frankfurt, *Magic Flute* 1968. London, C, *Semele* 1970. Hamburg, Berlin, Amsterdam, producing works as diverse as *Poppea*, *Lohengrin*, *Lulu*. A sensitive designer with a strong feeling for atmosphere.

**Sanquirico, Alessandro** (*b* Milan, 27 July 1777; *d* Milan, 12 Mar. 1849). Italian designer. Milan, S, 1817–32, designing premières including *Gazza ladra* 1817, *Pirata* 1827, *Norma* 1831, also many works by Donizetti, Mercadante, and Pacini. Redecorated interior of La Scala 1829. His designs, most of which can be seen in the La Scala Museum and elsewhere in Milan, reveal an impressive neo-classical style, later a warmly Romantic one that included the architectural and historical realism associated with the comparable Grand Opera genre of the Paris Opéra.

**Santa Fe.** Town in New Mexico, USA. The Opera Association of New Mexico, later Santa Fe Opera, was founded in 1956 by John Crosby. Its opening 1957 season included *The Rake's Progress* (supervised by Stravinsky) and the première of Marvin David Levy's *The Tower*; and this commitment to 20th-cent. music has continued with the premières of Floyd's *Wuthering Heights* (1958), Berio's *Opera* (1970), and Villa-Lobos's *Yerma* (1971). For ten years the company performed at an open-air theatre, which burnt down in 1967 after the US première of *Cardillac*; the season continued at the local High School. A new theatre (cap. 1,773) opened on the site 1968, with stage and pit covered but auditorium partly covered. The company has maintained its distinguished record for world and US premières and its steadfast commitment to the operas—particularly the later ones—of Richard Strauss. Crosby continues as director and music director.

**Sant'Alessio, Il.** Opera in prologue and 3 acts by Landi; text by Rospigliosi. Prod. Rome, Palazzo Barberini, 21 Feb. 1632.

Alessio has renounced his wife and the world for the religious life. He returns home and lives there unrecognized as a beggar, remaining steadfast to his vocation despite temptation.

**Santi, Nello** (*b* Adria, 22 Nov. 1931). Italian conductor. Studied Padua. Début Padua 1951 (*Rigoletto*). Zurich from 1959. Salzburg 1960. London, CG, 1960 (*Traviata*). New York, M, from 1962. Salzburg 1960 (*Don Carlos*). Guest appearances Hamburg, Munich, Vienna. A good 'singers' conductor'. (R)

**Santiago.** Capital of Chile. Opera was first given in the 1830s, from the 1850s at the T de la República: the T Municipal opened in 1857, but burnt down in 1870. The new theatre opened in 1873 (cap. 1,420). The first permanent ensemble was formed in 1955. Though the season is short and

the repertory limited, visitors have included Cossutta, Scotto, and Plowright.

**Santini, Gabriele** (*b* Perugia, 20 Jan. 1886; *d* Rome, 13 Nov. 1964). Italian conductor. Studied Perugia and Bologna. Début Rome, C, 1906. Cond. Buenos Aires (eight seasons) and Rio de Janeiro. Chicago; New York, Manhattan OC. Assistant cond. to Toscanini at Milan, S, 1925–9. Rome O, 1929–33, music director 1944–7. Milan, S, 1960–1. A champion of new works, and one of the leading Verdi conductors of his day. (R)

**Santley, (Sir) Charles** (*b* Liverpool, 28 Feb. 1834; *d* London, 22 Sep. 1922). English baritone. Studied Milan with Nava, later London with García jun. Début Pavia 1857 (Doctor, *Traviata*). London, CG from 1859, Pyne-Harrison Co. 1859–63, Mapleson Co. 1862–70; Milan, S, 1866; sang with CR 1875–7, when he retired from the stage. Roles included Don Giovanni, Count Almaviva (Mozart), Nevers, Luna, Flying Dutchman, and Valentin (Gounod, impressed, wrote 'Avant de quitter ces lieux' for him). Also created roles in operas by Wallace (*Lurline*), Balfe (e.g. *The Puritan's Daughter*), and Benedict (*The Lily of Killarney*). A singer highly acclaimed for his inspired and polished performances, he also composed, sometimes under the name of Ralph Betterton. Autobiography, *Reminiscences of my Life* (London, 1909, R/1977). (R)

**Santuzza.** A village girl (sop), Turiddu's abandoned lover, in Mascagni's *Cavalleria rusticana*.

**Sanzogno, Nino** (*b* Venice, 13 Apr. 1911; *d* Milan, 4 May 1983). Italian conductor and composer. Studied Venice with Agostini and Malipiero, Brussels with Scherchen. Cond. Venice, F, 1937. Milan, S, 1939; opened Milan, PS, 1955 (*Matrimonio segreto*). Edinburgh Festival 1957 with Milan, PS. Cond. many premières and Italian premières Milan and Venice, and was an outstanding advocate of modern opera, also of 18th-cent. Italian opera. Cond. Milan, S, 1962–5. (R)

**São Paulo.** City in Brazil. Opera has been given regularly since 1874, first at the T Provisório, then (1876–97) at the T São José, with occasional performances at the Politeama Nacional, and 1901–9 at the T Santana. The T Municipal opened 1907 with *Hamlet*, and has been the home of opera ever since. The companies, mostly of Italian artists who had been appearing in Buenos Aires and Rio, generally went on to São Paulo every autumn, and most of the great names among Italian opera singers appeared there. As well as the popular Italian and French repertory, works by S American composers have been staged.

'**Saper vorreste**'. Oscar's (sop) aria in Act III of Verdi's *Un ballo in maschera*, refusing to divulge the costume the King is wearing at the masked ball.

**Sapho**. See SAPPHO.

**Saporiti, Teresa** (*b* 1763; *d* Milan, 17 Mar. 1869). Italian soprano. Engaged, with her elder sister **Antonia**, while still very young by Bondini's co., Leipzig, Dresden, and Prague. Despite adverse press coverage in 1782 and 1785 (for performances both on- and off-stage), she evidently sang well enough to create Donna Anna, Prague 1787. Her small, trim figure was also appreciated. Venice 1788; Milan, S, 1789; St Petersburg from 1795 as *prima buffa assoluta*; Moscow 1796. Successful in operas by Cimarosa, Paisiello, etc. Nothing is known of her later career; she died aged 105 or 106.

**Sappho**. (*b* Lesbos, *fl* *c*.mid-7th cent. BC). Greek poetess. The popular legend of her love for and rejection by Phaon, culminating in her death-leap from the Leucadian rock, seems to have no historical basis. Operas on her life are as follows.

1. *Saffo*. Opera in 3 acts by Pacini; text by S. Cammarano, based on the legend. Prod. Naples, C, 29 Nov. 1840; Boston, Howard Athenaeum, 4 May 1847.

2. *Sapho*. Opera in 3 acts by Gounod; text by Emile Augier, based on the legend. Prod. Paris, O, 16 Mar. 1851, with Viardot, Guéymard, Poinsot, Marié, Brémond, Aimès, cond. Girard; London, CG, 1851. Gounod's first opera, later rev. in 2, then 4, acts.

Other operas on the subject are Piccinni's *Phaon* (1778), Mayr's *Saffo* (1794), Rejcha's *Sapho* (1822).

3. *Sapho*. Opera in 5 acts by Massenet; text by Henri Cain and Arthur Bernède after Daudet's novel (1884). Prod. Paris, OC, 27 Nov. 1897, with E. Calvé, Wyns, Giraudon, Leprestre, Gresse, Marc-Nohel, cond. Danbé; New York, Manhattan OC, 17 Nov. 1909, with Garden, D'Alvarez, Dalmorès, Dufranne, Leroux, cond. De la Fuente; London, Camden TH, 14 Mar. 1967, with Domzolski, De Peyer, Christiansen, Lennox, Olegario, cond. Gover.

The opera tells of the unhappy love of Fanny Legrand (sop), an artist's model who has been posing as Sappho for the sculptor Caoudal (bar), and a country youth, Jean Gaussin (ten).

**Sarastro**. The High Priest (bs) of Isis and Osiris in Mozart's *Die Zauberflöte*.

**Sardou, Victorien** (*b* Paris, 5 Sept. 1831; *d* Paris, 8 Nov. 1908). French dramatist. He first made his name as a writer of comedies; later turned to historical dramas (e.g. *La Tosca*, 1887), in which he developed a characteristic theme of dark, passionate stories set against a background of war or political tension. He wrote a number of roles for Sarah Bernhardt, and two, *Robespierre* and *Dante*, for Henry Irving. He wrote the librettos for *Le roi Carotte* (1872) for Offenbach and *Les barbares* (1901) for Saint-Saëns; he is, however, best remembered for the operas based on his works, in which many different composers found his strong situations, vivid characters, and lavishly developed scenes a powerful stimulus to music. Operas on his works are as follows:

*Piccolino* (1861): Grandval (1869); J. Strauss II (*Karneval in Rom*, 1873); Guiraud (1876)
*Les Prés-St-Gervais* (1862): Lecocq (1874)
*La bataille d'amour* (1863): Vaucorbeil (1863)
*Don Quichotte* (1864): Renaud (1895)
*Le Capitaine Henriot* (1864): Gevaert (1864)
*Patrie!* (1869): L. Rossi (*La contessa di Mons*, 1874); Paladilhe (1880). The subject was also contemplated by Verdi
*Rabagas* (1872): De Giosa (1882)
*Les merveilleuses* (1873): Félix (1914)
*La haine* (1874): Solovyev (*Cordelia*, 1885)
*Les noces de Fernande* (1878): Deffès (1878); Millöcker (*Der Bettelstudent*, 1882)
*Fédora* (1882): Giordano (1898)
*Théodora* (1884, incidental music by Massenet): Leroux (1907)
*La Tosca* (1887): Puccini (1900)
*Madame Sans-Gêne* (1893): Caryll (*The Duchess of Dantzic*, 1903); Giordano (1915); Dluski (early 20th cent.); Pierre-Petit (1947)
*Gismonda* (1894): Février (1918)
*La fille de Tabarin* (1901): Pierné (1901)
*Les barbares* (1901): Saint-Saëns (1901)
*La sorcière* (1903): D'Erlanger (1912)
*Fiorella* (1905): Webber (1905)

**Sargent**, (Sir) **Malcolm** (*b* Ashford, 29 Apr. 1895; *d* London, 3 Oct. 1967). English conductor. Although primarily a concert conductor, he worked with the BNOC, conducting the premières of Vaughan Williams's *Hugh the Drover* and Holst's *At the Boar's Head* (1925). D'Oyly Carte OC, 1926–8, 1951. London, CG, 1936 (*Louise*), 1954 (première of Walton's *Troilus and Cressida*). (R)

BIBL: C. Reid, *Malcolm Sargent* (London, 1968).

**Šárka**. 1. Opera in 3 acts by Janáček; text by Julius Zeyer, after his own drama (1887). Orig. version 1887; rev. 1888, 1918; final version rev. Chlubna, prod. Brno 11 Nov. 1925, with Pirková, Flögl, Oslovský, V. Šindler, cond. Neumann.

Bohemia, mythological times. After the death of Libuše, her husband Přemysl (bs) plans to dispense with her council of women. They revolt, under Šárka (sop), who falls in love with the warrior hero Ctirad (ten); nevertheless, she brings about his death, eventually hurling herself onto his funeral pyre.

2. Opera in 3 acts by Fibich; text by Anežka

Schulzová, after the Czech legend. Prod. Prague, N, 28 Dec. 1897.

**Sarro, Domenico Natale** (*b* Trani, 24 Dec. 1679; *d* Naples, 25 Jan. 1744). Italian composer. Studied Naples; vice maestro di cappella to the Naples court, 1704–7, 1725–37; maestro di cappella 1737–44.

Sarro is particularly remembered for making the first setting of Metastasio's *Didone abbandonata*, his first wholly original libretto. Of the generation of Neapolitan composers which followed Scarlatti, his career was launched with a sacred opera, *L'opera d'amore*. Though he was to write several more works of this kind, he is seen at his most characteristic in the *c*.25 heroic operas which he wrote both for Naples and other Italian houses. These brought him great popularity during the 1720s, but he was gradually eclipsed by Vinci. *Achille in Sciro* was given 1737 for the opening of the T San Carlo, Naples.

SELECT WORKLIST (all first prod. Naples): *L'opera d'amore* (?; 1702); *Didone abbandonata* (Metastasio; 1724); *Achille in Sciro* (Metastasio; 1737).

**Sarti, Giuseppe** (*b* Faenza, *bapt.* 1 Dec. 1729; *d* Berlin, 28 July 1802). Italian composer. Studied Padua with Valotti and Bologna with Martini. Organist of Faenza Cathedral, 1748–52; director of Faenza T, 1752–5. His first opera, *Il re pastore*, brought him to attention; a visit to Copenhagen 1753 with *Mingotti's troupe resulted in his appointment as Kapellmeister at the Danish court and director of the Italian O, 1755; director of court music 1763; director of court T, 1770–5. Sarti at first concentrated on composing Metastasian opere serie during his years in Copenhagen, including *Achille in Sciro* and *Didone abbandonata*. Later he wrote some of the earliest Danish operas, including *Tronfølgen i Sidon* and *Aglae, eller Støtten*. Following political difficulties at court, Sarti returned to Italy 1775. He became maestro di cappella at Milan Cathedral 1779, but 1784 was appointed to direct the imperial chapel in St Petersburg, where he collaborated with Pashkevich and Canobbio on one of the first Russian operas *The Early Reign of Oleg*, to a libretto by Catherine the Great. Later he founded a school of singing in the Ukraine and became head of the St Petersburg Cons. Returned to Italy 1801, dying on the way.

An accomplished composer with dramatic flair and sure melodic gift, Sarti was one of the most important figures for the wider dissemination of Italian opera during the late 18th cent.: his influence was crucial for the establishment of opera in both Denmark and Russia. He was equally at home in comic or serious genres but the greatest successes of his *c*.70 works were the opere buffe *I finti eredi* and *Fra i due litiganti*. The latter, which was revived to great acclaim in Vienna 1784 as Sarti passed through on his way to St Petersburg, contains the aria 'Come un agnello', quoted by Mozart in the banquet scene of *Don Giovanni*.

SELECT WORKLIST: *Il re pastore* (The Shepherd King) (Metastasio; Venice 1753); *Achille in Sciro* (Metastasio; Copenhagen 1759); *Didone abbandonata* (Metastasio; Copenhagen 1762); *Tronfølgen i Sidon* (The Succession to the Throne in Sidon) (Bredal, after Metastasio; Copenhagen 1771); *Aglae, eller Støtten* (Aglae, or the Column) (Fasting & Carstens, after Poinsinet de Sivry; Copenhagen 1774); *Fra i due litiganti il terzo gode* (Between the two litigants, it's the third who rejoices) (Goldoni; Milan 1782); *I finti eredi* (Bertati; St Petersburg 1785); *Nachalnoye upravleniye* (The Early Reign of Oleg) (Catherine the Great; St Petersburg 1790; comp. with Pashkevich & Canobbio).

BIBL: M. Baroni and M. Tavoni, eds., *Giuseppe Sarti* (Modena, 1986).

**Sarti, Laura** (*b* Trieste, 1925). Italian, later British, mezzo-soprano. Studied with Lucie Manen and Emmy Heim. First major appearance with Gly.: has also performed with Scottish O, Kent O, London SW, and at Drottningholm and Aix-en-Provence. One of her most celebrated roles is as Ottavia in Monteverdi's *Poppea*; others include Dido (Purcell), Emilia (Verdi's and Rossini's *Otello*), Principessa (*Suor Angelica*), and Signora Fabien (*Volo di notte*). Possessing a rich mezzo voice and a powerful stage presence, Sarti has also proved an influential and highly regarded teacher. (R)

**Sartorio, Antonio** (*b* Venice, 1630; *d* Venice, 30 Dec. 1680). Italian composer. Nothing is known of his life before the appearance of his first opera *Gl'amori infruttuosi di Pirro* in Venice 1661. In 1666–75 he was Kapellmeister at the court of Duke Johann Friedrich of Brunswick-Lüneburg, though he made annual journeys back to Venice to recruit musicians and to arrange productions of his operas during the Carnival. In 1676 he was appointed vice maestro di cappella at St Mark's, Venice.

Together with the works of Pollarolo and Ziani, Sartorio's 15 operas represent the apotheosis of the 17th-cent. Venetian tradition and a crucial advance on the work of Cavalli and Cesti. Though he developed many of the elements which they had made an integral feature of opera, notably the lament, his use of heroic subjects reflected a decisive shift in popular taste. Among his successes were *La prosperità d'Elio Seiano*, *Ermengarda*, and above all *Adelaide*. *Orfeo*, when given at the T San Luca 1672, was so well received that Cavalli's *Massenzio*, scheduled for the same season, was cancelled because it was feared that it would prove dull in comparison: Sartorio was given the libretto instead.

SELECT WORKLIST (all first prod. Venice): *Gl'amori infruttuosi di Pirro* (The Fruitless Loves of Pyrrhus) (Aureli; 1661); *La prosperità d'Eliano Seiano* (Minato; 1667); *Ermengarda regina de' Longobardi* (Dolfin; 1669, lost); *Adelaide* (Dolfin; 1672); *Orfeo* (Aureli; 1672); *Massenzio* (Bussani; 1673).

**Sass, Marie Constance** (*b* Oudenaarde, 26 Jan. 1834; *d* Auteuil, 8 Nov. 1907). Belgian soprano. Studied Ghent with Gevaert, Paris with Ugalde, Milan with Lamperti. Début as Marie Sax (she was later forced by Adolphe Sax to change her name) Venice 1852 (Gilda). Paris, O, 1860–70; Milan, S, 1869–70; St Petersburg 1870–1. Also Brussels, M; Barcelona; Madrid; etc. Her voice was powerful, flexible, and appealing. As well as *falcon roles, her repertory included Donna Anna, Lucrezia Borgia, Elisabeth (first Paris *Tannhäuser*), Amelia, Valentine. Created Selika; also Elisabeth de Valois, with little conviction and much temperamental behaviour: Verdi refused to have her for either Aida or Amneris. Married to *Castelmary 1864–7; retired 1877; died in poverty.

**Satie, Erik** (*b* Honfleur, 17 May 1866; *d* 1 July 1925). French composer. Studied Paris with Taudou. His stage works include a puppet opera, *Geneviève de Brabant* (comp. 1899), and the operetta *Pousse l'amour* (comp. *c*.1905, prod. Paris, 1907; rev. as *Coco chéri*, Monte Carlo 1913, lost).

BIBL: J. Harding, *Erik Satie and his Circle* (London, 1965).

**Sauguet, Henri** (*b* Bordeaux, 18 May 1901; *d* Paris, 22 Jun. 1989). French composer. Studied Bordeaux with Vaubourgoin and Canteloube. His first stage piece, *Le plumet du colonel*, was a so-called 'opéra-bouffe militaire'; this was followed by another comedy displaying his urbane manner, *La contrebasse*. His most substantial opera is *La chartreuse de Parme*, showed his ability to handle a major subject, though it lacks the sharp focus of his lighter works. His dextrous and eminently civilized manner is shown to admirable effect in *Les caprices de Marianne*.

SELECT WORKLIST: *Le plumet du colonel* (The Colonel's Plumage) (Sauguet; Paris 1924); *La contrebasse* (Troyat, after Chekhov; Paris 1931); *La chartreuse de Parme* (Lunel, after Stendhal; Paris 1939); *Les caprices de Marianne* (Grédy, after Musset; Aix-en-Provence 1954).

BIBL: F.-Y. Bril, *Henri Sauguet* (Paris, 1967).

**Saul og David.** Opera in 4 acts by Nielsen; text by Einar Christiansen, after 1 Samuel. Prod. Copenhagen, Royal T, 28 Nov. 1902, with Dons, Lendrop, Hérold, Cornelius, Simonsen, Nissen, Müller, cond. Nielsen; London, Collegiate T, 23 Feb. 1977, with Gail, Hillman, Masterson-

Smith, Doghan, McDonnell, Francis, cond. Wolfenden.

Israel, *c*.1006 B.C. In this version of the biblical story, Saul (bs) is cast as a complex, questioning figure in contrast to the more straightforward David (ten). The events encompass David falling in love with Saul's daughter Michal (sop), the narrow defeat of Goliath, and the return of Jonathan (ten) with news of the victory over the Philistines, Saul's jealousy of David, and David's eventual accession as King.

**Saunders, Arlene** (*b* Cleveland, 5 Oct. 1930). US soprano. Studied New York with Barbour. Début National OC 1958 (Rosalinde). New York CO and Milan, N, 1961; Hamburg 1963; Gly. 1966; New York, M, 1976; London, CG, 1980. Also Paris, O; Vienna; Munich; Washington; Edinburgh; Rome. Roles incl. Pamina, Louise, Eva, Marschallin, Arabella, Countess (*Capriccio*), Minnie. Created Beatrix Cenci (Ginastera). (R)

**Savage, Henry** (*b* New Durham, NH, 21 Mar. 1859; *d* Boston, 29 Nov. 1927). US impresario. Originally an estate agent, he was forced to take over the Castle Square T, Boston, when a lessee failed in 1897. He proved a successful manager, and collaborated with Maurice Grau in forming the Grau–Savage Metropolitan English Grand OC 1900. Gave *Parsifal* throughout the USA 1904–5; similar tours with *Butterfly* and *Girl of the Golden West* 1911.

**Savitri.** Opera in 1 act by Holst; text by composer, after an episode in the Mahabharata. Prod. London, Wellington H, by London School of O, 5 Dec. 1916, with Corran, Pawlo, Cook, cond. Grunebaum; first public prod, London, Lyric T, Hammersmith, 23 June 1921, with Silk, Wilson, Carey, cond. Bliss; Chicago, Palmer House, 23 Jan. 1934, with Witwer, Colcaire, Schmidt, cond. Krueger.

India, legendary times. Death (bar) has come to claim Satyavan (ten), but agrees to grant his wife Savitri (sop or mez) any wish, excepting the life of Satyavan. She chooses that she herself shall have 'life in its fullness', thereby outwitting Death, because a full life will only be possible with her husband.

Also opera by Zumpe (unfin. 1903, completed Rössler 1907).

**Savonlinna.** Town in E Finland. The summer festival in the courtyard of Olavinlinna Castle (cap. 2,262) was founded in 1912 by Aïno *Ackte, who sang in Melartin's *Aïno*, and in the years up to 1916 included works by Pacius, Merikanto, Madetoja, and Hannikainen. Her attempt to revive the festival in 1930 was controversial, and it did not become annual until 1967.

A success then was August Everding's *Fidelio*, and his popular *Magic Flute* was in every festival 1972-89. These, as well as the charm of the setting, helped to assure the festival's survival, and the presence of Martti Talvela as artistic director 1972–80 further strengthened standards.

**Savoy operas.** The name given to the operettas by Gilbert and Sullivan, from the theatre at which many of them were first produced.

**Savoy Theatre.** Theatre in the Strand, London, built by Richard D'Oyly Carte to house the Gilbert and Sullivan operettas, consequently known as the Savoy Operas. Opened 1881 with *Patience*. Three seasons of opera were organized by Marie Brema 1910-11; she sang Gluck's Orfeo. The present theatre (cap. 1,122) opened 1930.

**Sawallisch, Wolfgang** (*b* Munich, 26 Aug. 1923). German conductor. Studied Munich with Haas. Répétiteur Augsburg 1947, later chief cond. until 1953. Music director: Aachen 1953–8; Wiesbaden 1958–60; Cologne 1960–3. Bayreuth 1957–62, conducting new productions of *Tristan* (1957), *Holländer* (1959), and *Tannhäuser* (1961). Music director Munich 1971–92. London, CG, with Munich co., 1972. An intelligent, judicious conductor with a particular feeling for Mozart and Strauss. (R)

**Sayão, Bidú** (*b* Rio de Janeiro, 11 May 1902). Brazilian soprano. Studied Nice with J. De Reszke. Début Rio de Janeiro 1924 (Rosina). Paris, O and OC, 1931; New York, M, 1937–51. Also Buenos Aires, Washington, Rome, etc. Retired 1958. Roles included Susanna, Adina, Juliette, Violetta, Mimì, Manon, Mélisande. An engaging performer with a refined style. (R)

**Sbarra, Francesco** (*b* Lucca, 19 Feb. 1611; *d* Vienna, 20 Mar. 1668). Italian poet and librettist. Of aristocratic lineage, Sbarra entered the priesthood in 1645. The favoured librettist of Cesti, for whom he wrote four works, including *Il pomo d'oro* (1668), and *Alessandro vincitor di se stesso* (1651), one of the most widely performed operas of its day.

**Sbriglia, Giovanni** (*b* Naples, 23 June 1829; *d* Paris, 20 Feb. 1916). Italian tenor and teacher. Studied Naples with Roxas. Début Naples, C, 1853. Successful in Italy; toured USA and S America. New York 1860. From 1875, taught in Paris, becoming one of the most celebrated teachers of his time. Pupils included the De Reszkes, Nordica, Sanderson, Plançon.

**Scala, Teatro alla.** See LA SCALA.

**Scala de seta, La** (The Silken Ladder). Opera in 1 act by Rossini; text by Foppa, after the play *L'échelle de soie* by François-Antoine-Eugène de Planard and possibly a libretto from it made for Gaveaux (1808). Prod. Venice, S Moisè, 9 May 1812, with (probably) M. Cantarelli, Nagher, T. Cantarelli, Monelli, Del Monte, De Grecis, Tacci; London, SW, by T dell'Opera Comica, Rome, 26 Apr. 1954, with Tuccari, Lollin, Bernardi, Catalani, Dulciotti, cond. Merelli; San Francisco O, 18 Feb 1966 with Lopez, Davis, cond. Samuel.

Venice, early 19th cent. The silken ladder of the title is used nightly by Dorvil (ten) to rejoin his wife Giulia (sop), whom he has secretly married, but who is living in the house of her tutor Dormont (bs); the intrigues are complicated by Giulia being promised by Dormont to Blansac (bar), who is loved by her cousin Lucilla (mez).

**Scalzi, Carlo** (*b* Voghera; *fl* 1719–38). Italian soprano castrato. Venice 1719–21, 1724–5, 1737–8; Naples 1726–7, 1730; London 1733–4. Also Parma, Modena, Genoa, etc. Very popular on the Continent; sang in operas by Vinci, Porpora, Hasse (creating title-role of *Ezio*), and Handel, for whom he created Alceste (*Arianna in Creta*), and who altered and expanded arias in several other works for him. Metastasio thought him (and Farinelli) 'incomparable'.

**Scapigliatura** (from It. *scapigliare*, to rumple or dishevel, hence *scapigliatura*, dishevelledness or Bohemianism). The name of a 19th-cent. Italian artistic movement, most active during the 1860s and 1870s, deriving from a novel by one of its members, Cletto Arrighi, *La scapigliatura e il 6 febbraio* (1862). Its members, under the guidance of the novelist Giuseppe Rovani, believed that the value of a work of art resided in its capacity to break down barriers between the arts. The most significant musicians in the group, which affected iconoclastic statements, disorderly behaviour, and casual dress, were *Boito and *Faccio. The influence shows mostly in Boito's liking for elaborate verbal sound-play in his texts, but more profoundly in the awareness of darkness and negation in his operatic characters from Mefistofele to Iago.

**Scarabelli, Diamante Maria** (*b* Bologna; *fl* 1695–1718). Italian soprano. Venice regularly 1695–1716; Bologna 1696–1718. Also Milan, Padua, Genoa, etc. A highly successful virtuoso singer, she created Poppea in Handel's *Agrippina*; also acclaimed in operas by Orlandini, Lotti, Caldara, and others.

**Scaria, Emil** (*b* Graz, 18 Sep. 1838; *d* Blasewitz, 22 July 1886). Austrian bass. Studied Vienna with Gentiluomo and Lewy. Later London with García. Début Budapest 1860 (Saint-Bris). Leipzig 1863; Dresden 1865. Vienna, H, 1873–86. Berlin 1881, and London, HM, 1882, with Neumann's co., when he had a mental collapse, and had to be replaced. But, recovering, he created Gurnemanz later that year and rejoined Neumann's co. touring Europe. USA tour 1883. Produced *Parsifal*, Bayreuth, 1883. Died insane. He was much admired by Wagner, also by Hanslick. An excellent and musical singer, with crystal-clear declamation. Roles included Rocco, Dulcamara, Falstaff (Nicolai), King Mark, Wotan.

**Scarlatti**, (Pietro) **Alessandro** (Gaspare) (*b* Palermo, 2 May 1660; *d* Naples, 22 Oct. 1725). Italian composer. Little is known about Scarlatti's early training, although it is often claimed that he took lessons in Rome (where he moved 1672) with Carissimi. In 1679 he came to attention with his first opera, *Gli equivoci nel sembiante*, which was quickly taken up by houses outside Rome. Around this time he became maestro di cappella to Queen Christina of Sweden, though he also enjoyed the patronage of Cardinal Pietro Ottoboni, himself a librettist of note. By the time that he left Rome for Naples in 1684, to become maestro di cappella to the Viceroy, he had already written six operas. The next eighteen years were even more productive: though Scarlatti's claim that he wrote over 80 operas while in Naples is almost certainly an exaggeration, at least 40 works dating from this period have survived. Many of these were for the T San Bartolomeo, where he was director from 1684, but a number were first given elsewhere, among them *Statira*, produced for the reopening of the T Tordinona in Rome, 1690. Gradually becoming disenchanted with life in Naples, Scarlatti resigned his position there in 1702. The next year he became assistant maestro di cappella at S Maria Maggiore in Rome, though he angled, unsuccessfully, for an appointment in Florence: in support of his supplication he sent at least four new operas to Prince Ferdinando Medici. In Rome he secured a position with Cardinal Ottoboni, but quickly found that the absence of opera in the city, following the closure of the theatres there in 1700, was a severe creative handicap. In 1708 he resumed his post in Naples, where he composed a further 11 operas, including two of his greatest successes, *Tigrane* and *Cambise*. His last operas, *Telemaco*, *Marco Attilio Regolo*, and *Griselda*, were written for Rome, where he made an extended visit 1718–22.

Scarlatti is often credited with being the founder of a new school of so-called Neapolitan opera, an exaggeration, if not a complete falsehood, which has done much to hinder appreciation of his historical significance. Rather than being the originator of a new tradition, he represented more the apotheosis of the first great age of opera during which activity had been focused on

Venice. Responsible more than any other single composer for the establishment of Naples as a centre of operatic activity at the beginning of the 18th cent., he gave his operas a dramatic and musical shape far closer to the Venetian tradition than to that of opera seria which followed. Though he did not share the Venetian fondness for mythology, Scarlatti made similar use of its characteristic comic characters and episodes and adopted many of its structural principles in his arias. Perception of him as a radical composer is suggested largely by those features which seem to adumbrate later elements of opera seria, mainly a significant use of the da capo aria (increasingly important from the 1690s onwards), and his choice, in later life, of librettos which bear traces of reform. Principal among these is *Griselda*, where the strong love-duty conflict and balanced dramatic structure looks forward to the age of Metastasio.

That Scarlatti himself realized his fundamentally conservative orientation is demonstrated by his gradual retreat from opera after his return to Naples in 1708. Though he made one attempt at a comic opera, *Il trionfo dell'onore*, he did not find the new genre to his liking: instead he concentrated more on instrumental music and smaller vocal forms. But his success, in his heyday, cannot be lightly discounted, as the numerous revivals of his operas attest, especially those of *Pirro e Demetrio* and *La caduta de' Decemviri*. His best work shows dramatic sensitivity of a rare order, coupled with a musical inventiveness and prolificacy which few others rivalled. His greatest contribution came, however, in his handling of the orchestra, where the skilful use of obbligato instruments in the aria and constant search for new textures expanded considerably the emotional and expressive potential of opera.

SELECT WORKLIST: *Gli equivoci nel sembiante* (Misunderstandings in Appearance) (Contini; Rome 1679); *Statira* (Ottoboni; Rome 1690); *Pirro e Demetrio* (Morselli; Naples 1694); *La caduta de' Decemviri* (The Fall of the Decemviri) (Stampiglia; Naples 1697); *Mitridate Eupatore* (Roberti; Venice, 1707); *Tigrane* (Lalli; Naples 1715); *Telemaco* (Capece; Rome 1718); *Il trionfo dell'onore* (Tullio; Naples 1718); *Cambise* (Lalli; Naples 1719); *Marco Attilio Regolo* (Noris; Rome 1719); *Griselda* (Zeno; Rome 1721).

BIBL: D. J. Grout, *Alessandro Scarlatti: an Introduction to his Operas* (Berkeley, 1979).

**Scarlatti, Domenico** (*b* Naples, 26 Oct. 1685; *d* Madrid, 23 July 1757). Italian composer, son of the above. Though he is remembered today principally for his keyboard works, his early career was dominated by his fruitless efforts to emulate his father's success as a composer of opera. Travels around Italy (including to Venice) and appointments elsewhere (notably in Rome, in the service of Maria Casimira, Queen of Poland) apart, Scarlatti's activities in this field centred on Naples; in 1719 he was called to Lisbon and thence to the Madrid court in 1728. His *c*.15 works reveal a certain dramatic flair, though fail entirely to live up to the high standards of originality set by his father: most important for their influence on later composers were *Orlando* and *Ambleto*.

His nephew **Giuseppe** (*c*.1720–77) composed *c*.32 operas, mostly for Italian opera-houses, though some were given at the Viennese court, where Gluck secured his appointment as composer of ballet music 1762–4.

WORKLIST (Domenico): *Giustino* (Convò, after Berengani; Naples 1703); *Orlando* (Capece, after Ariosto; Rome 1711); *Ambleto* (Zeno; Rome 1715).

BIBL: R. Kirkpatrick, *Domenico Scarlatti* (Princeton, 1953, 3/1968).

**Scarpia.** Baron Scarpia (bar), the Chief of Police, in Puccini's *Tosca*.

**scena** (from Gr. *skene*, lit. 'tent', or 'booth', hence 'stage'). (1) A solo operatic movement of primarily dramatic purpose, less lyrical or formally composed than an aria, though very similar, e.g. Leonore's 'Abscheulicher' in Beethoven's *Fidelio*.

(2) A sequence of music of different styles (recitative, aria etc.) which joined together in chain-like fashion form a dramatic block. The term has a particular connotation in 19th-cent. Italian opera, where it is a crucial element of the *Code Rossini and takes on a more regulated structure. At the heart of the scena is a two-part sequence, deriving from the exit aria convention of Metastasian opera seria, comprising a slow *cantabile followed by a fast and brilliant *cabaletta. This basic structure, which was used for duets as well as solo arias, could be expanded by the addition of further sections. In its fullest form it began with a long opening recitative which preceded the cantabile: this was sometimes also separated from the cabaletta by a transitional movement known as the *tempo di mezzo. See also CONVENIENZE.

**scenario** (It.: 'scenery', also used in the English sense of the word). An outline libretto indicating the characters and number and type of scenes. In German, *Scenarium* means a complete libretto with detailed indications of scenery and staging.

**scene.** The term used to denote the smaller subdivision of an *act.

**Schack, Benedikt** (*b* Mirotice, 7 Feb. 1758; *d* Munich, 10 Dec. 1826). Bohemian tenor and composer. Studied Vienna with Frieberth. Début with Schikaneder's co. 1786 (in Paisiello's *La frascatana*). Freihaus-T. auf der Wieden,

Vienna, 1789–93. Graz 1793; Munich, N, 1796–1814. Created Tamino (also playing the flute solos); first German-language Don Ottavio and Count Almaviva. Described by L. Mozart as an excellent singer with 'a beautiful voice' and 'a beautiful method'. As a composer (appreciated by Haydn, among others), he collaborated with F. X. Gerl on Schikaneder's 'Anton' Singspiels; also wrote several operas, in one of which, *Der Stein der Weisen*, Mozart assisted with the duet 'Nun liebes Weibchen'.

**Schalk, Franz** (*b* Vienna, 27 May 1863; *d* Edlach, 3 Sep. 1931). Austrian conductor. Studied Vienna with Bruckner. Cond. Reichenberg (Liberec) 1888; Graz 1890–5; Prague 1895–8; Berlin 1898–1900. Vienna, H, 1900, succeeding Gregor as director 1918, and shared first conductorship with Strauss 1919–24. Cond. first revised *Ariadne* (1910) and *Frau ohne Schatten* (1916). Differences arose, with the two eventually not on speaking terms, and the public generally siding with Schalk. Strauss resigned, leaving Schalk in control until 1929. London, CG, 1898, 1907, and (three highly successful *Ring* cycles) 1911. New York, M, 1898–9. A founder of the Salzburg Festival. Had the reputation of being a somewhat dour but extremely efficient conductor. (R)

**Schaunard.** The Bohemian musician (bar) in Puccini's *La bohème*.

**Schauspieldirektor, Der** (The Impresario). Opera in 1 act by Mozart; text by Gottlieb Stephanie the younger. Prod. Vienna, Schönbrunn Palace T, 7 Feb. 1786 (in same bill as Salieri's *Prima la musica e poi le parole*), with Lange, Cavalieri, Adamberger; London, St James's T, 30 May 1857; New York 9 Nov. 1870, with Lichtmay, Römer, Hölzel, Himmer, Rohbeck, Himmer.

Vienna, late 18th cent. The plot describes the rivalries between two prima donnas, Madame Silberklang (sop) and Madame Herz (sop): each offers a sample of her ability, the former in sentimental vein, the latter with a brighter rondo. They then come to blows over which of them deserves the higher salary, and try to outsing each other in a trio with M. Vogelsang (ten) in the presence of the long-suffering Impresario (spoken).

Various revisions have been made, including one that turns the singers into Aloysia Lange, Katharina Cavalieri, and Valentin Adamberger, and the Impresario into Mozart himself.

**Scheidemantel, Karl** (*b* Weimar, 21 Jan. 1859; *d* Weimar, 26 June 1923). German baritone. Studied Weimar with Borchers, Frankfurt with Stockhausen. Début Weimar 1878 (Wolfram).

Weimar until 1886; Dresden 1886–1911; London, CG, 1884, 1899; Bayreuth 1886–92; Milan, S, 1892. Roles included Pizarro, Kurwenal, Sachs, Amfortas, Scarpia; created Kunrad (*Feuersnot*), and Faninal, among other roles. Much admired and very popular, he was described by Strauss as 'wonderful'. Director Landestheater, Dresden, 1921–2; wrote two books on singing. (R)

**Schenk, Johann Baptist** (*b* Wiener Neustadt, 30 Nov. 1753; *d* Vienna, 29 Dec. 1836). Austrian composer. Studied Vienna with Wagenseil. From 1794 he was music director for Prince Auersperg, but never sought a career as a court musician. Five early works, written 1780–6, were given anonymously, but their success persuaded him to reveal his identity with *Im Finstern ist nicht gut tappen* (1789). In his rather unreliable autobiography, Schenk states that he decided in 1780 to devote himself exclusively to opera, having previously written symphonies, concertos, and chamber music.

His position in the court theatres of Vienna established by the mid 1790s with works such as *Achmet und Almanzine*, Schenk went on to produce his most successful work, *Der Dorfbarbier*, in 1796. The Singspiel remained popular for over 25 years, receiving performances as far afield as Moscow and New York; it is still revived occasionally. He wrote for the stage until 1802, after which he only revised previous works and composed some vocal and choral music. He abandoned his final work, in the manner of Gluck, through lack of progress and ill health. With *Der Dorfbarbier*, along with other successful works such as *Die Jagd* and *Der Fassbinder*, Schenk made a significant contribution to the establishment of an indigenous Viennese Singspiel, following in the line of Umlauff. He was highly regarded by Weber and Lorzting; Beethoven took some counterpoint lessons with him. He published an *Autobiographische Skizze* (1830, publ. 1924).

SELECT WORKLIST (all prod. Vienna): *Achmet und Almanzine* (?, after Lesage and Ormeville; 1795); *Der Dorfbarbier* (The Village Barber) (P. & J. Weidmann; 1796); *Die Jagd* (The Hunt) (?, after Weisse; 1799); *Der Fassbinder* (The Cooper) (?, after Audinot; 1802).

**Scherchen, Hermann** (*b* Berlin, 21 Jun. 1891; *d* Florence, 12 June 1966). German conductor. Largely self-taught. Gained early experience as viola player, then conducting various ensembles and orchestras. Though primarily a concert conductor, he made some operatic appearances, generally with the new music he always championed; cond. premières of Dallapiccola's *Il prigioniero* (Florence 1950), Dessau's *Das*

*Verurteilung des Lukullus* (Berlin 1951), Henze's *König Hirsch* (Berlin 1956). Also cond. successful prod. of *Moses und Aron*, Berlin 1959. Adapted Webern's Second Cantata for the stage as *Il cuore* (Naples 1958). (R)

WRITINGS: *Lehrbuch des Dirigierens* (Leipzig, 1929; trans. as *Handbook of Conducting*, London, 1933).

**Scherman, Thomas** (*b* New York, 12 Feb. 1917; *d* New York, 14 May 1979). US conductor. Studied New York with Bamberger and Rudolf. Début Mexico City 1947; in same year formed New York Little Orchestra Society, which has given many important concert performances of opera including *Ariadne auf Naxos*, *L'enfant et les sortilèges*, *Goyescas*, *Iphigénie en Tauride*, and *Euryanthe*. (R)

**Schikaneder, Emanuel** (*b* Straubing, 1 Sep. 1751; *d* Vienna, 21 Sept. 1812). German theatre manager, singer, actor, and playwright. Schikaneder's early life was spent as wandering musician, but in 1773 he fell in with a travelling theatrical troupe in Augsburg, married the daughter of the manager, and eventually succeeded to the directorship of the enterprise himself. The company appeared widely in Germany and Austria and Schikaneder acquired a reputation as one of the best comic actor-singers of his day. In 1780, the troupe appeared in Salzburg, and here Schikaneder made acquaintance with the young Mozart, who apparently provided an aria for one of his performances. After appearances in 1784 at Vienna, K and B, Schikaneder moved to Regensburg to take over the direction of the theatre there; he was soon recalled to Vienna by his wife to manage a small suburban theatre, the T auf der Wieden. For several years he encountered great difficulties in running it, but the success of Mozart's *Die Zauberflöte* in 1791 saved him from financial ruin. In 1800 he partnered a merchant in opening a new theatre, the T an der Wien, which he continued to manage until 1806. Although Schikaneder planned to build and manage other theatres in Vienna, his sudden decline into madness caused an abrupt departure from the Viennese operatic scene.

Schikaneder's reputation today rests upon the libretto which he provided for Mozart's *Die Zauberflöte*, in which he also took the part of Papageno. In recent years, however, the extent to which Schikaneder was involved with the project has been questioned, since he lacked the intimate knowledge of Freemasonry revealed in the libretto. In all probability his main contribution was restricted to the lighter scenes, in which his great skills as a comic actor would be revealed, while some other unknown person provided the serious Masonic text. In total Schikaneder wrote the librettos for about 50 operas or plays with music.

Composers who set his texts include Schack, Süssmayr, Hoffmeister, Paisiello, Seyfried, Winter, Teyber, and Weigl, as well as Mozart. Beethoven began but never completed a setting of *Vestas Feuer* in 1804; this same text was later set by Joseph Weigl.

BIBL: K. Honolka, *Papageno: Emanuel Schikaneder* (trans. Portland, OR, 1990).

**Schiller, Friedrich von** (*b* Marbach, 10 Nov. 1759; *d* Weimar, 9 May 1805). German poet and playwright. Operas on his works are as follows:

*Die Räuber* (1782): Mercadante (1836); Verdi (*I masnadieri*, 1847); Zajc (*Amelia*, 1860); Klebe (1957)
*Die Verschwörung des Fiesko* (1783): Hellmesberger (comp. 1848–9); Lalo (comp. 1866–7)
*Kabale und Liebe* (1784): Verdi (*Luisa Miller*, 1849); Von Einem (1976)
*Don Carlos* (1787): P. D. Deshayes (1799); Nordal (1843); M. Costa (1844); Bona (1847); Ferrari (1854); Moscuzza (1862); Ferrara (1863); Verdi (1867)
*Der Handschuh* (1798): Polgar (1973)
*Der Taucher* (1798): Reichardt (1811); Kreutzer (1813)
*Die Bürgschaft* (1798): Schubert (1816, unfin.); Lachner (1828); Hellmesberger (comp. 1851)
*Wallenstein* (1799): Seyfried (1813); Adelburg (*c.*1872); Musone (1873); Denza (1876); Ruiz (1877); Weinberger (1937); Shabelsky (1950); Zafred (1965)
*Das Lied von der Glocke* (1799): D'Indy (1912)
*Maria Stuart* (1800): Mercadante (1821); Donizetti (1834, as *Buondelmonte*); Lavello (1895)
*Die Jungfrau von Orleans* (1801): Vaccai (1827); Balfe (1837); Vesque von Püttlingen (1840); Verdi (1845); Langert (1861); Tchaikovsky (1881); Rezniček (1886); Klebe (1976)
*Die Braut von Messina* (1803): Vaccai (1839); Orzen (1840); Bonawitz (1874); Fibich (1884)
*Wilhelm Tell* (1804): Carr (*The Archers*, 1796); Rossini (1829); Van Overeen (1906)
*Demetrius* (1805, unfin.): Joncières (1876); Dvořák (1882)
*Der Gang nach dem Eisenhammer* (1804): B. A. Weber (1810); Schoenfeld (1832); C. Kreutzer (1837); Terry (1861)
*Turandot* (1804): Danzi (1816); Vesque von Püttlingen (1838); Busoni (1917); Puccini (1926)

**Schillings, Max von** (*b* Düren, 19 Apr. 1868; *d* Berlin, 24 July 1933). German composer, conductor, and manager. Studied Bonn with Brambach, and in Munich was influenced by Strauss. Assistant stage conductor, Bayreuth 1892, chorus-master 1902. His first opera, *Ingwelde*, was imitative of Wagner, and even the second, the comic *Der Pfeifertag*, did not escape this influence. Professor at Munich, 1903 (his pupils included Furtwängler and Heger). Wagner's influence continues in his third opera, the religious *Moloch*. Intendant, Stuttgart, 1908–18, Generalmusikdirektor 1911; was ennobled when the new opera-house opened in 1912. His régime was of great enterprise, and included the premières of the original *Ariadne auf Naxos*, of his version of *Les Troyens*, and of his most successful opera, *Mona Lisa*. Intendant, Berlin, 1919–25, when productions included *Palestrina* and *Die Frau ohne Schatten*, among much else. Toured USA as guest conductor, 1924, 1931. His wife was **Barbara Kemp** (*b* Kochem, 12 Dec. 1881; *d* Berlin, 17 Apr. 1959). German soprano. Studied Strasburg Cons. Début Strasburg 1903 (Priestess, *Aida*). Berlin, H (later S), 1913–32. New York, M, 1922–4; Bayreuth 1914–27. Sang Isolde, Senta, Kundry; created title-role of her husband's *Mona Lisa*, and produced this and *Ingwelde* in Berlin, 1938–9. (R)

WORKLIST: *Ingwelde* (Sporck; Karlsruhe, 1894); *Der Pfeifertag* (Day of the Piper) (Sporck; Schwerin 1899); *Moloch* (Gerhäuser, after Hebbel; Dresden 1906); *Mona Lisa* (Dovsky; Stuttgart 1915).

BIBL: J. Geuenich and K. Strahn, eds., *Gedenkschrift Max von Schillings zum 100. Geburtstag* (Düren, 1968).

**Schindelmeisser, Louis** (*b* Königsberg, 8 Dec. 1811; *d* Darmstadt, 30 Mar. 1864). German conductor and composer. Studied Berlin with Marx, Leipzig with Dorn. Kapellmeister Salzburg, 1832, then occupying various other posts until settling in Darmstadt 1853. An early defender of Wagner, and conductor of his works. His own operas, which include *Melusine* (Pasqué; Darmstadt 1861), are more directly Weberian.

**Schinkel, Karl Friedrich** (*b* Neuruppin, 13 Mar. 1781; *d* Berlin, 9 Oct. 1841). German architect and designer. Studied Berlin, Italy, and France. Worked as scene designer for Wilhelm Gropius 1807–15; chief designer at the Royal T 1815–28, initially for *Magic Flute*. These famous sets provided a panoramic background to the action, strongly atmospheric in mood and detail. He also designed some of Spontini's operas, in which his grasp of architectural design and ability to handle large structures were particularly appropriate (*La vestale*, 1818, *Olympie*, 1821).

**Schipa, Tito** (*b* Lecce, ?1888, birth cert. states 2 Jan. 1889: *d* New York, 16 Dec. 1965). Italian tenor. Studied Lecce with Gerunda, Milan with Piccoli. Début Vercelli 1910 (Alfredo). Milan, S, from 1915; Chicago 1919–22; New York, M, 1932–5, 1940–1. Also Buenos Aires, Monte Carlo, Rome, Italian provinces (never opera in England). Sang into the late 1950s. The outstanding *tenore di grazia* of his time; his sweet, slightly nasal, highly individual tone, superbly projected and controlled, was used with exceptional refinement and often with moving pathos. His roles included Don Ottavio, Elvino, Ernesto, Duke of Mantua, Werther, Mascagni's Fritz; created Ruggero (*La rondine*). Autobiography, *Si confessa* (Genoa, 1961). (R)

**Schippers, Thomas** (*b* Kalamazoo, MI, 9 Mar. 1930; *d* New York, 16 Dec. 1977). US conductor. Studied Philadelphia, and with Olga Samaroff. Début New York 1948 with Lemonade Co. New York, M, 1953. Cond. première of Copland's *The Tender Land* (1954), also many works by Menotti, including at Spoleto. Cond. première of Barber's *Antony and Cleopatra* opening New York, M, at Lincoln Center, 1966. Milan, S, 1967 (*Assedio di Corinto*). Bayreuth, 1963 (*Meistersinger*). London, CG, 1968 (*Elektra*). (R)

**Schira, Francesco** (*b* Malta, 21 Aug. 1809; *d* London, 15 Oct. 1883). Italian conductor and composer. Studied Milan with Basili and wrote his first opera, for La Scala, aged 23. Music Director Lisbon, San Carlos, 1834–42. Music director, London, Princess's T, 1842. Succeeded Benedict at DL, 1844–7; cond. CG, 1848; DL, 1852. The good reception given his operas in London was far exceeded by the success of *Selvaggia* in Venice (1875).

**Schlosser, Max Karl** (*b* Amberg, 17 Oct. 1835; *d* Utting, 2 Sep. 1916). German tenor. After unsuccessful early career in operetta, became a baker, but returned to singing 1868 to create David. Munich 1868–1904. Also created Mime in *Rheingold* and *Siegfried*, much to the satisfaction of Wagner, who (though ambiguously describing him as 'every inch a dwarf') found in him a rare performer who actually fulfilled his ideal of a role. Saint-Saëns and Grieg (who praised his declamation) both declared him one of the most convincing in the cast of the first *Ring*. Toured Europe with Neumann's co. as Mime, 1880–2. Other roles included Almaviva, Max, Lionel, and the baritone role of Beckmesser.

**Schlusnus, Heinrich** (*b* Braubach, 6 Aug. 1888; *d* Frankfurt am Main, 18 June 1952). German baritone. Studied Frankfurt, later Berlin with Bacher. Début Hamburg 1915 (Herald, *Lohengrin*). Berlin, S, 1917–50; Chicago 1927; Bayreuth 1933; Paris, O, 1937; Barcelona; etc. A singer of noble qualities, whose generous, Italianate sound, fine legato, and easy top notes made him an excellent Verdi baritone, e.g. as Rigoletto, Germont, Montfort (*Vêpres siciliennes*); also a successful Amfortas. (R)

**Schmedes, Erik** (*b* Gentofte, 27 Aug. 1866; *d* Vienna, 23 Mar. 1931). Danish tenor. Studied Berlin with Rothmühl, Paris with Artôt, later Vienna with Ress. Début Wiesbaden 1891 as baritone (Valentin). Nuremberg 1893–4; Dres-

den 1894–7, during when, after further study with Iffert, became tenor. Vienna 1898–1924; Bayreuth 1899–1902, 1906; New York, M, 1908–9. Also Paris, O; Berlin; Munich; Prague; etc. Roles included Florestan, Siegfried, Tristan, Parsifal, Palestrina. A blond giant with a huge voice, whose greatest musical achievements were reached under Mahler. (R)

**Schmidt-Isserstedt, Hans** (*b* Berlin, 5 May 1900; *d* Holm bei Wedel, 28 May 1973). German conductor. Studied Heidelberg, Münster, Berlin with Ertel and Schreker. Cond. Wuppertal 1928; Rostock 1928; Darmstadt 1931–3. Chief cond. Hamburg 1935–42; Berlin, Deutsches Opernhaus, 1942–5. Composed *Hassan gewinnt* (Elberfeld 1928). (R)

**Schneider, Hortense** (*b* Bordeaux, 30 Apr. 1833; *d* Paris, 6 May 1920). French soprano, later mezzo. Studied Bordeaux with Schaffner. Début Agen 1853 (Inès, *La favorite*). Paris from 1855 at various venues in works by Offenbach, Adam, etc. Created leading roles in Offenbach's *Belle Hélène*, *Grande-Duchesse de Gérolstein*, *Périchole*, with brilliant success. London 1867; St Petersburg 1872; retired 1878. An accomplished and captivating performer, with an exuberant temperament both on and off stage. Saint-Saëns once thought of her for Dalila.

**Schneider-Siemssen, Günther** (*b* Augsburg, 7 June 1926). German designer. Studied Munich with Preetorius. Bremen 1954–62; Vienna, S, from 1962; London, CG, *Ring* 1962–6; New York, M, from 1967, incl. *Ring*. Salzburg since 1965, *Boris*, *Don Giovanni*, *Frau ohne Schatten*, and all Karajan's Easter Festival productions. Adept at traditional settings and using modern techniques with sparse scenery, and what he has called 'painting with light' (or, in some of the Easter productions, darkness).

**Schnorr von Carolsfeld, Ludwig** (*b* Munich, 2 July 1836; *d* Dresden, 21 July 1865). German tenor. Studied Dresden with J. Otto, and Leipzig. Début Karlsruhe 1853 (in small roles). Karlsruhe until 1860; Dresden 1860–5; Munich 1865. Though corpulent, he was a heroic and affecting performer, intelligent, dramatically powerful, and vocally gifted. After triumphs as Lohengrin and Tannhäuser, created Tristan, with a dedication that contributed to his premature death within weeks. His loss profoundly affected Wagner. He had married 1860 the Danish soprano **Malvina Garrigues** (*b* Copenhagen, 7 Dec. 1825; *d* Karlsruhe, 8 Feb. 1904). She studied Paris with García. Début Breslau 1841 (in *Robert le diable*). Karlsruhe 1854–60, where she met her husband. Created Isolde with him in

Munich, to equal acclaim. She retired after his death, and taught; her pupils included Gudehus.

BIBL: C. H. Garrigues, *Ein ideales Sängerpaar* (Copenhagen, 1937).

**Schoeck, Othmar** (*b* Brunnen, 1 Sep. 1886; *d* Zurich, 8 Mar. 1957). Swiss composer. Studied Zurich with Attenhofer. His sensitivity to poetry and love of the human voice led him to express himself most fully in song and opera. His first opera, *Erwin und Elmire*, is written in somewhat Mozartian language and also shows the influence of Strauss's *Ariadne auf Naxos*, but its favourable reception gave Schoeck encouragement. He followed it with *Don Ranudo*, a comedy whose lack of love interest did not prevent its initial success; there ensued two works on a theme that absorbed Schoeck, *Das Wandbild*, a Hoffmannesque fantasy to a text by Busoni concerning a portrait that comes to life, and *Venus*, in which it is a statue that comes to life.

His most important opera, however, followed in 1927 with *Penthesilea*. Schoeck claimed to have learnt from Wagner, Strauss, and Ravel; its Expressionistic music is more advanced than that of *Elektra*, but does not go so far as *Lulu*. Despite his assertion that 'The artist must speak the language of his day', he reverted to a more tonally stable idiom with *Vom Fischer und syner Fru*. His dislike of German nationalism did not leave him unaware of the advantages of maintaining cultural links with Hitler's Germany. *Massimilla Doni* was accordingly premièred with success under Karl Böhm in Dresden, where the critics eagerly emphasized the work's supposed 'Germanness'. Similar success attended *Das Schloss Dürande* in Berlin, though its run was abandoned shortly after Goering made a furious attack on the libretto. None of Schoeck's operas found a place in the repertory during his lifetime, though *Penthesilea* has enjoyed a revival in Germany since the late 1970s. However, the best of his work has a subtlety and also a Romanticism that in varying ways and to varying degrees of success reconciles itself to more modern styles.

WORKLIST: *Erwin und Elmire* (after Goethe; Zurich 1916); *Don Ranudo* (Rüeger, after Holberg; Zurich 1919); *Das Wandbild* (The Picture on the Wall) (Busoni; Halle 1921); *Venus* (Rüeger, after Mérimée; Zurich 1922); *Penthesilea* (Schoeck, after Kleist; Dresden 1927); *Vom Fischer und syner Fru* (Schoeck, after Runge, after Grimm; Dresden 1930); *Massimilla Doni* (Rüeger, after Balzac; Dresden 1937); *Das Schloss Dürande* (Burte, after Eichendorff; Berlin 1943).

BIBL: H. Corrodi, *Othmar Schoeck* (Frauenfeld, 1956).

**Schoenberg, Arnold** (*b* Vienna 13 Sep. 1874; *d* Los Angeles, 13 July 1951). Austrian composer. Studied Vienna with Zemlinsky. Schoenberg first turned to opera in 1909 with the composition of *Erwartung*, though its première, like that of its successor *Die glückliche Hand*, was not given until 1924. Both are imbued with the spirit of the Expressionist movement and are among the earliest works in which tonality was definitely abandoned. *Erwartung*, which took him only 17 days to write, is a monodrama centring on the experiences of a woman who, wandering through the wood at night in search of her lost lover, discovers his dead body. To a generation which stood so much under the influence of Wagner's music dramas, the links between *Tristan* and *Erwartung*, which might be seen as an Expressionist version of its *Liebestod, were obvious. In his intense exploration of the woman's feelings and recollections, Schoenberg penetrates the depths of human experience, conveying the nightmare journey of her mind in music which eschews the use of obvious themes or their development. Despite its delayed production, *Erwartung* was known to the circle around Schoenberg and influenced a whole series of similarly Expressionist works by his fellow composers. He himself continued many of the same ideas in *Die glückliche Hand*, which again centres on one character, though it also calls for mimed dance and a chorus. Again, a re-interpretation of a characteristically Wagnerian theme is the basis for the action: the quest for truth by the artist, whose *glückliche Hand* is his individual, specially favoured touch.

Schoenberg's next opera, *Von Heute auf Morgen*, was the first in which twelve-note, serialist principles were used. But those who were acquainted with his instrumental music in this style and expected something arid, or uncompelling, were disappointed. A brilliantly observed comedy of a witty and determined woman keeping hold of her husband, it is satirical and farcical, yet manages nonetheless to mirror the symmetry of the music with a plot and action which is contrived in a similarly balanced fashion. Freer serialist principles were used for Schoenberg's last opera, *Moses und Aron*, which remained uncompleted on his death and was not staged until 1957. His largest dramatic work, it is also his fullest expression of the quest theme that absorbed him so deeply. Concerning communication between God and Man, its action hinges around the distortion suffered by pure truth (received by Moses) when it undergoes exposition (by Aron) in terms comprehensible to Man. Schoenberg's religious nature and his artist's concern over communication were profoundly touched by this theme, and the opera has by its emotional power done much to overcome prejudices against him. Of all Schoenberg's works composed after his discovery of serialism, it offers the closest *rap-*

*prochement* between new procedures and those of traditional tonal music.

WORKLIST: *Erwartung* (Expectation) (Pappenheim; comp. 1909, prod. Prague 1924); *Die glückliche Hand* (The Favoured Hand) (Schoenberg; comp. 1910–13, prod. Vienna 1924); *Von Heute auf Morgen* (From Day to Day) (G. Schoenberg, under the pseudonym Blonda; Frankfurt 1930); *Moses und Aron* (Schoenberg; comp. 1930–2, two acts only, prod. Zurich 1957).

WRITINGS: L. Stein, ed., *Style and Idea* (London, 1975, 2/1984).

BIBL: M. Macdonald, *Schoenberg* (London, 1976).

**Schöffler, Paul** (*b* Dresden, 15 Sep. 1897; *d* Amersham, 21 Nov. 1977). German, later Austrian, baritone. Studied Dresden with Staegemann, Berlin with Grenzebach, Milan with Sammarco. Début Dresden 1925 (Herald, *Lohengrin*). Dresden until 1938. Vienna, S, 1938–65; London, CG, 1934–9, 1949–53; New York, M, 1949–53, 1954–6, 1962–5; Bayreuth 1943–4; Salzburg 1938–41, 1947, 1949–65. Highly regarded as a Wagner interpreter, especially as Sachs, he also sang Figaro, Don Alfonso, Pizarro, Iago; created Jupiter (*Liebe der Danae*), and Danton (*Dantons Tod*). (R)

**School for Fathers, The.** See QUATTRO RUSTEGHI, I.

**Schopenhauer, Artur** (*b* Danzig, 22 Feb. 1788; *d* Frankfurt, 21 Sep. 1860). German philosopher. His greatest work is *Die Welt als Wille und Vorstellung* (The World as Will and Representation, 1818). Though ignored for many years, it came to be one of the most influential single works of philosophy in the 19th cent. for musicians, especially by way of its impact on Wagner. He claimed that his first reading in 1854 was the most important event of his life; Schopenhauerian turns of phrase in his previous writings may be evidence of closeness of mind. The concept of the 'will' as the urge to exist, involving hurt and destructiveness of self and fellow men, from which release into oblivion (particularly through renunciation) is the desired goal, coloured much in Wagner's subsequent works. The most remarkable is *Tristan und Isolde*, whose theme and text are strongly Schopenhauerian, particularly in the Act II duet in which the lovers seek release from the 'tückischer Tag' (malicious or deceitful day) into the bliss of night. However, Wotan's yearning for 'das Ende', even as his actions struggle against it, is also Schopenhauerian; and (especially in the original draft) Hans Sachs's consideration of 'Wahn' as 'destructive illusion', which may be healed or at any rate interpreted by the 'good illusion' of art, relates immediately to Schopenhauer's view of art,

especially music, as embodying a truer reality than that which is outwardly apprehensible.

**Schorr, Friedrich** (*b* Nagyvárád, 2 Sep. 1888; *d* Farmington, CT, 14 Aug. 1953). Hungarian, later US, bass-baritone. Studied Vienna with Robinson. Début Chicago 1912 (small roles). True début Graz 1912 (Wotan). Graz until 1916; Cologne 1918–23; Berlin, S, 1923–31; New York, M, 1924–43; Bayreuth 1925–31; London, CG, 1924–33. Settled in USA 1931. The dominating Wotan and Sachs of his generation, he could sing with warm, lyrical ease, yet also with arresting dramatic force, always with exemplary attention to words. Repertory also included Pizarro, Amonasro, Scarpia, Jochanaan, Orestes, Doktor Faust. (R)

**Schreier, Peter** (*b* Meissen, 29 July 1935). German tenor. Studied Leipzig with Polster, Dresden with Winkler. Début as student Dresden, S, 1957 (Paolino, *Matrimonio segreto*). Dresden, S, 1959–63; Berlin, S, from 1963; Vienna, Salzburg, and New York, M, from 1967; Milan, S, and Buenos Aires, C, from 1968. Also London, CG; Moscow, B; Budapest; Warsaw; etc. Large and varied repertory includes Belmonte, Ferrando, Tamino, Loge, David, Flamand, Lensky. An extremely intelligent singer and actor, with a fine legato line, and sympathetic feeling for the text. An excellent Lieder singer. (R)

**Schreker, Franz** (*b* Monaco, 23 Mar. 1878; *d* Berlin, 21 Mar. 1934). Austrian composer, teacher, and conductor. Studied Vienna with Fuchs; from 1912 taught composition at the Vienna Academy. He began work on his first opera, *Der ferne Klang*, as early as 1903, but temporarily abandoned it when friends criticized his libretto. After the successful concert performance of an interlude 1909, he was inspired to complete the work. Its performance in Frankfurt 1912 marked him down as the most gifted operatic composer of his generation and brought him instant recognition. Among its admirers was Schoenberg, who quoted passages of it in his *Harmonielehre*, and Berg, who made the piano score and was influenced (in *Wozzeck*) by its use of set forms. One of the first German works to acknowledge the impact of Debussy's music, it is an important essay in Expressionism and shares some of the same vision and musical language as Strauss's *Elektra* and Schoenberg's *Erwartung*.

Schreker's next work, *Das Spielwerk und die Prinzessin*, caused an outrage when performed before the conservative Viennese audience 1913; a more experimental work, making some use of pan-tonality, both its music and libretto offended. More successful were *Die Gezeichneten* and *Der Schatzgräber*, which confirmed his reputation as an avant-garde composer and led to the prestigious directorship of the Berlin Hochschule in 1920. During the 1920s and 1930s Schreker came increasingly under attack: from musicians, who began to revolt against the late Romantic style which he represented, and from the Nazis, because of his Jewish origins. The première of *Christophorus* was cancelled for fear of the Nazi demonstrations which later totally disrupted that of *Der Schmied von Gent*. In 1932 Schreker was removed from his position at the Berlin Hochschule for political reasons.

Though perhaps the most original German opera composer after Berg and Schoenberg during the first decades of the 20th cent., Schreker did not have a lasting influence, due in no small measure to the rise of the Nazis. But the important emancipating effect which his works had for later composers, opening up a world beyond that of post-Wagnerian music drama, cannot be underestimated, while at their best they have real dramatic power in their own right.

SELECT WORKLIST: *Der ferne Klang* (The Distant Sound) (Schreker; comp. 1901–10, prod. Frankfurt 1912); *Das Spielwerk und die Prinzessin* (The Clockwork and the Princess) (Schreker; Frankfurt & Vienna 1913, rev. Munich 1920); *Die Gezeichneten* (The Signified) (Schreker; Frankfurt 1918); *Der Schatzgräber* (The Treasure Seeker) (Schreker; Frankfurt 1920); *Christophorus* (Schreker; comp. 1924–7; prod. Freiburg 1978); *Der Schmied von Gent* (Schreker, after de Coster; Berlin 1932).

BIBL: G. Neuwirth, *Franz Schreker* (Vienna, 1959).

**Schröder-Devrient, Wilhelmine** (*b* Hamburg, 6 Dec. 1804; *d* Coburg, 26 Jan. 1860). Daughter of **Friedrich Schröder** (1744–1816), the first German Don Giovanni, and the actress **Sophie Bürger Schröder** (1781–1868), 'the German Mrs Siddons'. Appeared in ballet, Hamburg and Vienna, as child, later as actress. Studied with her mother, and Vienna with Mozatti. Début Vienna, K, 1821 (Pamina). Vienna, K, until 1822; Dresden 1822–47; Berlin 1828; Paris 1831–2; London, H, 1832–3, 1837. By now her voice was declining, but she successfully created Adriano (*Rienzi*), Senta, and Venus. Last appearance Riga 1847. Though deficient in vocal technique, she was a powerful and moving actress, dubbed 'the Queen of Tears' when seen really weeping on stage. She was famous for her Donna Anna, Euryanthe, Reiza, Norma, Valentine; and Weber thought her the best of all Agathes, who revealed more than he had believed the role to contain. Beethoven, who had rehearsed her, thanked her personally for her Leonore: this was by all accounts the greatest interpretation of the age; Moscheles preferred her to Malibran, and Wagner was aroused by it to his sense of vocation as a composer, as he describes in *Mein Leben*. Her style later disturbed

Berlioz with its exaggeration and vehemence, and she was criticized for declaiming rather than singing, but she was one of the principal performers to inspire and contribute to German Romantic opera with her committed and passionate approach. Her private life suffered: her first marriage to Karl Devrient ended with divorce; she remarried twice, and had a succession of lovers, defending her life as 'grist to her art'.

**Schröter, Corona** (*b* Guben, 14 Jan. 1751; *d* Ilmenau, 23 Aug. 1802). German actress, singer, and composer. Studied Leipzig with Hiller, where she established a reputation first as a singer and then as an actress. Goethe, who admired her greatly, brought her to the Weimar court 1776. Here she both participated in concerts and continued her acting career, creating roles in many of Goethe's plays. For *Die Fischerin* (1782) she composed incidental music, including a setting of the opening ballade, 'Der Erlkönig'.

**Schubert, Franz** (*b* Vienna, 31 Jan. 1797; *d* Vienna, 19 Nov. 1828). Austrian composer. Studied Vienna with Ruzička and Salieri. For the whole of his life, Schubert hoped for, and frequently attempted to write, the successful opera that would have helped to establish his wider renown. The operas he saw in Vienna which most influenced him were those by Weigl and Gyrowetz, as well as the many lighter pieces with music by Wenzel *Müller. His own completed operas fall broadly into three categories: most are Singspiels, one is a melodrama, and three are true operas.

For all the Romantic absurdities of the plot of *Des Teufels Lustschloss*, the music is said to have astonished Salieri with its quality and range, and does indeed contain some powerful invention, especially orchestral; but despite an impressive fluency between numbers, the work shows the lack of true dramatic instinct that was to plague Schubert. In the following year, 1815, he completed four Singspiels. *Der vierjährige Posten* is a slight, amusing piece in simpler style, with separate songs in a manner very different from the Lied, of which Schubert was already making himself a great master. *Fernando* is somewhat weightier, and in its scoring again reflects Schubert's ability to conceive of orchestral effects that belong purely in the theatre. Despite use of a more skilful libretto (by Goethe), *Claudine von Villa Bella* is not very distinctive; but by the end of the year, with *Die Freunde von Salamanka*, Schubert's technique had greatly matured, and there are signs that he was trying to make the music interact with the (lost) dialogue.

Apart from the unfinished *Die Bürgschaft* (which shows signs of growing out of hand), Schubert wrote no more theatre music until 1819 and the unfinished *Adrast* (which contains some splendid music) and *Die Zwillingsbrüder*; 1820 saw some delightful and well-planned music for a melodrama, *Die Zauberharfe*. In 1821 he was briefly appointed to a conducting post at the Court T, where his tasks included coaching Caroline Unger in *Così fan tutte*, but he soon resigned. In 1821–3 came the three works that crown his brief operatic career. Of all his stage works, Schubert set most store by *Alfonso und Estrella*, his most lyrical opera though one again weakened by a fatally static quality. The only one of his operas to have achieved much posthumous success came next, *Die Verschworenen* (or *Der häusliche Krieg*): here, the lyrical gifts are more successfully matched to deft dramatic strokes. With *Fierrabras*, there is not only orchestral mastery and some beautiful individual numbers, but a more expert use of *reminiscence motive and a freer dramatic construction. There are fewer set arias, more scenas that merge swiftly into one another, though Schubert's dramatic instincts are insufficiently strong to keep him from falling prey to banal situations and to numbers in which his genius for sustaining musical structures works against a true opera composer's sense of when these must be governed by dramatic considerations. At least some of the blame for Schubert's lack of operatic success can be blamed on the fact that only one of his operas was staged in his lifetime, *Die Zwillingsbrüder*, and this was withdrawn after about a dozen performances. *Fierrabras* contains enough achievement, and more promise, for his early death to seem tragic in the field of opera as in those of song, symphony, and chamber music where his genius is more fully revealed.

WORKLIST: *Der Spiegelritter* (The Knight of the Mirror) (Kotzebue; unfin., comp. 1811–12, broadcast Swiss Radio 1949); *Des Teufels Lustschloss* (The Devil's Pleasure Castle) (Kotzebue; comp. 1813–14, prod. Vienna 1879); *Adrast* (Mayrhofer; unfin., comp. ?1817–19, prod. Vienna 1868); *Der vierjährige Posten* (The Four-yearly Post) (Körner; comp. 1815, prod. Dresden 1896); *Fernando* (Stadler; comp. 1815, prod. Vienna 1907); *Claudine von Villa Bella* (Goethe; comp. 1815, prod. Vienna 1913); *Die Freunde von Salamanka* (The Friends of Salamanka) (Mayrhofer; comp. 1815, prod. Halle 1928); *Die Bürgschaft* (The Surety) (anon.; unfin., comp. 1816, prod. Vienna 1908); *Die Zauberharfe* (The Magic Harp) (Hofmann; Vienna 1820); *Die Zwillingsbrüder* (The Twin Brothers) (Hofmann; Vienna 1820); *Sakuntala* (Neumann, after Kalidasa; unfin., comp. 1820, prod. Vienna 1971); *Alfonso und Estrella* (Schober; comp. 1822, prod. Weimar 1854); *Die Verschworenen* (or *Der häusliche Krieg*) (The Conspirators, or War in the Home) (Castelli, after Aristophanes; comp. 1823, prod. Frankfurt 1861); *Rüdiger* (?Mosel; sketches, comp. 1823, perf. Vienna 1868);

*Fierrabras* (Kupelwieser; comp. 1823, prod. Karlsruhe 1897); *Der Graf von Gleichen* (The Count of Gleichen) (Bauernfeld; sketches, comp. 1827).

CATALOGUE: O. E. Deutsch (with D. Wakeling), *Schubert: Thematic Catalogue of all his Works* (London, 1951; rev. W. Dürr & others, Kassel, 1978).

BIBL: E. McKay, *Schubert's Music for the Theatre* (Tutzing, 1990).

**Schuch, Ernst von** (*b* Graz, 23 Nov. 1846; *d* Kötzschenbroda, 10 May 1914). Austrian conductor. Studied Graz with Stolz, Vienna with Dessoff. Music director, Lobe's T in Breslau 1867; répétiteur Würzburg 1868–70; Graz 1870–1; Basle 1871. Music director Dresden, Italian O, 1872; music director, Court O, with Rietz from 1873, with Wüllner from 1879; in sole charge 1882–1914. Schuch made this one of the great periods in the city's operatic history. He was a cultivated, enterprising, and highly practical musician, with a belief in opera as music drama which he had inherited from his predecessors Weber and Wagner and a close control over all aspects of performance. It was only in the interests of greater orchestral control that he felt obliged to move the conductor's rostrum back from the old position near the footlights to the orchestra rail. He gave 51 premières, including several of Strauss's most important works: *Feuersnot* (1901), *Salome* (1905), *Elektra* (1909), and *Rosenkavalier* (1911). Strauss negotiated hard with him, but declared, 'Schuch is a wonder'. Also conducted premières of Bungert's tetralogy *Die Odysee* (1896–1903), Paderewski's *Manru* (1901), Dohnányi's *Tante Simona* (1912), and Wolf-Ferrari's *L'amore medico* (1913), gave some celebrated Wagner performances, and introduced Puccini to Dresden. In 1875 married the Hungarian soprano Clementine Procházka, who then sang as **Klementine Schuch-Proska** (*b* Sopron, 12 Feb. 1850; *d* Kötzschenbroda, 8 June 1932). Studied Vienna with Marchesi. Dresden 1873–1904. Vienna 1879. London, CG, 1884 (Eva, Aennchen). Her voice was a light coloratura, and her repertory embraced Italian roles (Rosina, Amina) as well as Mozart (Susanna, Zerlina). Their daughter **Liesel von Schuch** (1891–1990) was a coloratura soprano who sang in Dresden and Vienna.

BIBL: F. von Schuch, *Richard Strauss, Ernst von Schuch und Dresdens Oper* (Dresden, 2/1952).

**Schuloper** (Ger.: 'school opera'). A German opera written expressly for schools, hence couched in an uncomplicated style and employing appropriate forces. Its function was to encourage community music-making and to convey a didactic message, usually moral, but sometimes political. It often approached the *Lehrstück and some works, notably Weill's *Der*

*Jasager*, are part of both traditions. The classic example is Hindemith's *Wir bauen eine Stadt*; Britten's *Let's Make an Opera* reflects in part the German influence.

**Schulz, Johann Abraham Peter** (*b* Lüneburg, 31 Mar. 1747; *d* Schwedt an der Oder, 10 June 1800). German composer and conductor. Studied Berlin with C. P. E. Bach and Kirnberger. In 1768–73 he served as teacher and accompanist to Princess Sapieha Woiwodin von Smolensk, with whom he travelled throughout Europe, meeting Haydn, Grétry, and Reichardt. After working as assistant to Kirnberger for a number of theoretical works 1773–6, Schulz became music director of the French O in Berlin, later taking an additional court appointment at Rheinsberg. His championship of new music, especially that of Gluck, brought him into disfavour, so he resigned 1787, taking a position in Copenhagen as Kapellmeister at court and director of the Royal Danish O. He retired 1795 and returned eventually to Berlin.

Schulz wrote several operas for Berlin and Rheinsberg, including *La fée Urgèle* and *Athalie*, but his most important stage works were given in Copenhagen. In his Singspiels *Indtoget*, *Høstgildet*, and *Peters bryllup*, which follow in essential outline the main N German style of Singspiel of the day, he made an important contribution to the foundation of national opera in Denmark.

SELECT WORKLIST: *La fée Urgèle* (Favart; ?Rheinsberg 1782); *Athalie* (Racine; Rheinsberg 1785); *Høstgildet* (Thaarup; Copenhagen 1790); *Indtoget* (Heiberg; Copenhagen 1793); *Peters bryllup* (Thaarup; Copenhagen 1793).

**Schuman, William** (*b* New York, 4 Aug. 1910; *d* New York, 15 Feb. 1992). US composer. Studied New York (Malkin Cons.) with Persin and Haubiel, Columbia U, Salzburg Mozarteum, Juilliard School. 1935–45 taught at Sarah Lawrence Coll.; 1945–62 president, Juilliard School; 1962–9 president of Lincoln Center.

His first major success came with the Third Symphony (1941), which placed him at the forefront of his generation. Though his reputation was established largely by his orchestral music, he wrote several ballets and the popular *The Mighty Casey*, whose plot centres around a baseball hero.

WORKLIST: *The Mighty Casey* (Gury, after Thayer; Hartford, CT, 1953).

BIBL: F. Schreiber and V. Persichetti, *William Schuman* (New York, 1954).

**Schumann, Elisabeth** (*b* Merseburg, 13 June 1888; *d* New York, 23 Apr. 1952). German soprano. Studied Dresden with Hänisch, Berlin with Dietrich, Hamburg with Schadow. Début Hamburg 1909 (Shepherd, *Tannhäuser*). Hamburg until 1919; Vienna 1919–38; London, CG, 1924–31; New York, M, 1914–15. Settled in USA 1938. Her fresh, silvery voice, used with unerring skill and intelligence, her immaculate phrasing, sensitivity, and purity of style, together with irresistible personal charm, made her a singer who inspired a rare affection in her listeners. Unforgettable as Blonde, Susanna, Zerlina, Eva, Adele, and especially Sophie, she was also treasured for her Lieder singing. Toured USA 1921 with R. Strauss, her great admirer and supporter, performing his songs. (R)

**Schumann, Robert** (*b* Zwickau, 8 June 1810; *d* Endenich, 29 July 1856). German composer. His one opera, *Genoveva* (Reinick, after Tieck and Hebbel; Leipzig 1850), includes some excellent music, but reveals his difficulty in thinking in dramatic terms: there is little characterization, and the effects tend to be muted, in part through his scorn of all he thought cheap in Italian opera. It reflects his admiration for *Euryanthe*, including in its more advanced use of motive, and harmonically approaches some of the practices of Wagner, who was bemused by Schumann's reluctance to take his advice about the work. He also left some fragments of a setting of Byron's *The Corsair*.

BIBL: G. Abraham, 'The Dramatic Music', in id., ed., *Schumann: a Symposium* (London, 1952).

**Schumann-Heink, Ernestine** (*b* Lieben, 15 June 1861; *d* Hollywood, 17 Nov. 1936). Austrian, later US, contralto. Studied Graz with Leclair and Wüllner. Début (as Tini Rössler) Dresden 1878 (Azucena). Dresden until 1883, and in 1909; Hamburg 1883–97; Bayreuth 1896 regularly until 1914; London, CG, 1897–1901; New York, M, 1899–1903, and frequently until 1932. Had three husbands, taking her stage name from her first and second (Ernest Heink and Paul Schumann). Possessed a deep, powerful voice and a compelling artistic personality. Her outstanding achievements were in Wagner, e.g. Ortrud, Fricka, and especially with her extraordinary Erda (as which Cosima Wagner always addressed her), still impressive at 70; but also in Donizetti, and as Amneris, Fidès, Orlovsky, Carmen, Witch (*Hänsel*). Created Klytemnestra, pronouncing the opera 'a fearful din'. (R)

BIBL: M. Lawton, *Schumann-Heink, the last of the Titans* (New York, 1928).

**Schürmann, Georg Caspar** (*b* Idensen, 1672 or 1673; *d* Wolfenbüttel, 25 Feb. 1751). German composer. Sang at Hamburg O from *c*.1692; 1697 joined Brunswick court as singer and conductor of opera. Studied in Italy 1701–2. Wrote *c*.40 operas (mostly lost), including *Salomon*, *Leonilde*, *Die getreue Alceste*, and *Ludovicus Pius*; though originally intended for the Brunswick court, many of these were eventually taken into the repertory of the Hamburg O. One of the most highly regarded early composers of German opera, his works show both French and Italian characteristics and were an important influence on succeeding generations, surpassed only by that of Keiser.

SELECT WORKLIST (all first prod. Brunswick): *Salomon* (?Knorr von Rosenroth; 1701); *Leonilde* (Fiedler; 1704 or 1705); *Die getreue Alceste* (König, after Quinault; 1719); *Ludovicus Pius* (Simonetti; 1726).

BIBL: C. Schmidt, *Die frühdeutsche Oper und die musikdramatische Kunst Georg Caspar Schürmanns* (Regensburg, 1933–4).

**Schuster, Ignaz** (*b* Vienna, 20 July 1779; *d* Vienna, 6 Nov. 1835). Austrian actor, singer, and composer. Studied Vienna with Eybler and Volkert. Became known as a comic actor and singer from 1801 at the T in der Leopoldstadt. From 1804 also wrote music for light theatrical pieces. In 1813 created the role of Staberl in Bäuerle's *Die Bürger von Wien*, making this one of the last standard comic figures of the Viennese popular theatre. In 1818 he appeared in the title-role of his own *Die falsche Primadonna* (a title censored from the original *Die falsche Catalani*); the work was long popular. The emergence of *Raimund did not diminish his popularity, despite his jealousy of the younger man. His many scores include parodies. Also a church musician; composed music for Beethoven's funeral and acted as coffin-bearer.

**Schütz, Heinrich** (*b* Köstritz, *bapt.* 9 Oct. 1585; *d* Dresden, 6 Nov. 1672). German composer. Studied in Italy with Giovanni Gabrieli (1609–12). Dresden from 1614, rising to the position of Electoral Kapellmeister. Although known today as a composer of sacred music, Schütz produced several dramatic entertainments for festivities at the Dresden court; one, a setting of Martin Opitz's translation of Rinuccini's *Dafne*, is generally regarded as the first German opera. Written to celebrate a royal marriage in 1627, the score of *Dafne*, like that of all his stage works, is lost. It has been suggested that Schütz wrote another opera in Italy in the following year, but there is no concrete evidence to support this claim.

**Schützendorf**. German family of singers, all brothers.

1. **Guido** (*b* Vught, 22 Apr. 1880; *d* Apr. 1967). Bass. Studied Cologne. Sang with German OC throughout Germany and USA. Title-role of silent film, *Der fliegende Holländer* (1919).

2. **Alfons** (*b* Vught, 25 May 1882; *d* Weimar, Aug. 1946). Bass-baritone. Studied Cologne with

Walter, Milan with Borgatti. Bayreuth 1910–12; London, CG, 1910; also Prague, Vienna etc. (R)

3. **Gustav** (*b* Cologne, 1883; *d* Berlin, 27 Apr. 1937). Baritone. Studied Cologne Cons. and Milan. Début Düsseldorf 1905 (Don Giovanni). Berlin, H; Munich, N, 1914–20; New York, M, 1922–35. Roles included Alberich, Beckmesser, Faninal, and first Luna (*Palestrina*). His first wife was Delia *Reinhardt. (R)

4. **Leo** (*b* Cologne, 7 May 1886: *d* Berlin, 18 Dec. 1931). Bass-baritone. Studied Cologne with D'Arnals. Début Düsseldorf 1908. Engagements at Krefeld, Darmstadt, Wiesbaden; Vienna 1919–20; Berlin, S, 1920–9. A versatile actor, he sang in a wide range of parts, including Gounod's Méphistophélès, Boris, Ochs. Created Wozzeck, when Berg called him 'superb'. (R)

The four brothers performed together only once, in *Die Meistersinger*, Bremen 1916: Alfons sang Sachs; Leo, Beckmesser; Gustav, Pogner; and Guido, Kothner.

**Schwanda the Bagpiper.** See SHVANDA THE BAGPIPER.

**Schwarzkopf,** (Dame) **Elisabeth** (*b* Jarocin, 9 Dec. 1915). German soprano. Studied Berlin with Mysz-Gmeiner, later Ivogün. Début Berlin, SO, 1938 (Flowermaiden). Berlin until 1942; Vienna 1942–7; London, CG, 1947, 1948–51, 1959; Milan, S, 1948–63; Chicago 1959; New York, M, 1964–6. Created Anne Trulove. Roles included Marzelline, Violetta, Mimì, Mélisande; particularly associated with Donna Elvira, Countess (Mozart), Marschallin. A beautiful woman with a fine voice, whose interpretations were always contrived to give maximum effect. (R)

**Schweigsame Frau, Die** (The Silent Woman). Opera in 3 acts by Richard Strauss; text by Stefan Zweig, after Ben Jonson's drama *Epicœne* (1609). Prod. Dresden 24 June 1935, with Cebotari, Sack, Kremer, Ahlersmeyer, Plaschke, cond. Böhm; New York, CC, 7 Oct. 1958, with Carroll, Moody, Alexander, Beattie, cond. Herrmann; London, CG, 20 Nov. 1961, with Holt, Vaughan, Macdonald, J. Ward, D. Ward, cond. Kempe.

London, 1780. Sir Morosus (bs), a retired English admiral, cannot abide noise of any kind. His nephew Henry (ten) is secretly married to Aminta (sop), and both are members of a group of travelling actors. With the help of the Barber (bar), they arrange a mock marriage between Morosus and 'Timida', the silent woman, actually Aminta in disguise. Once married, she emerges (like Norina in *Don Pasquale*) as a noisy termagant. Morosus gives the young couple his blessing, and adopts a more mellow attitude to life.

Other operas on the subject are by Salieri (*Angiolina*, 1800) and Lothar (*Lord Spleen*, 1930).

**Schwerin.** City in Mecklenburg, Germany. The first opera performances were given in the 18th cent. in the Schlosstheater. In 1792 Antonio Rosetti (Rösler) wrote *Das Winterfest der Hirten* for nearby Ludwigslust. The Ballhaustheater opened 1788; burnt down 1831, rebuilt in Schinkel style 1836. Performances of *Don Giovanni* were given 1790, and the repertory included opéra-comique. Under Friedrich von Flotow 1855–63, a high standard was maintained; among visitors was Jenny Lind in Bellini. This level was continued under Alois Schmitt, Alfred von Wolzogen 1867–83, and Karl von Ledebur 1894–1913, the latter organizing some distinguished Gluck performances. The theatre burnt down 1882, and the Staatstheater opened 1886. Herman Zumpe was music director 1897–1901; he and his successor Willibald Kaehler brought in Wagner and contemporary works. An operetta ensemble was formed in 1918.

**Scio, Julie Angélique** (*b* Lille, 1768; *d* Paris, 14 July 1807). French soprano. Début (under the name of Grécy) Lille 1786. Sang at Montpellier, Avignon, Marseilles. Married the violinist Étienne Scio (1766–96). Paris, OC, 1792–1807, becoming one of its leading singers. Much admired for her pure tone, intelligent acting, and musicality. Created leading parts in operas by Berton, Dalayrac, Le Sueur, and Cherubini (title-roles of *Médée*, *Elisa*, *Lodoïska*; Constance, *Les deux journées*).

**Scipione.** Opera in 3 acts by Handel; text by P. A. Rolli, after Zeno's *Scipione nelle Spagne* (orig. written in 1710, probably for Caldara) and other librettos on the same subject (Piovene, *Publio Cornelio Scipione*, 1712; Salvi, 1704: both after Livy). Prod. London, H, 12 Mar. 1726, with Cuzzoni, Baldi, Senesino, Boschi. Revived Handel O Society, 1967.

**Sciutti, Graziella** (*b* Turin, 17 Apr. 1927). Italian soprano. Studied Rome. Début Aix-en-Provence 1950 (Elisetta, *Matrimonio segreto*). Gly. 1954–9, 1970, 1977; London, CG, 1956–62; Milan, S, from 1956; New York, M, from 1961; also Salzburg, Vienna, Rome etc. An outstanding soubrette, vivacious, and with excellent diction. Roles included Susanna, Despina, Papagena, Rosina, Norina, Nannetta. Also producer (New York, Chicago, Coblenz, Gly.). (R)

**Scolari, Giuseppe** (*b* Vicenza, ?1720; *d* ?Lisbon, after 1774). Italian composer. A shadowy figure whose activities centred on Venice, where he was extremely successful. Composed *c*.30 operas,

both serious and comic. His most successful work, thoroughly representative of his style, was the dramma giocoso *La cascina*, which became popular throughout Europe. He apparently spent some time in Spain in the early 1750s and appears to have moved to Portugal in later life.

SELECT WORKLIST: *La cascina* (Goldoni; Venice 1755).

**Scotland.** See EDINBURGH, GLASGOW, SCOTTISH OPERA.

**Scott, (Sir) Walter** (*b* Edinburgh, 15 Aug. 1771; *d* Abbotsford, 21 Sep. 1832). Scottish novelist and poet. Operas on his works are as follows:

*The Lady of the Lake* (1810): Bishop (*The Knight of Snowdoun*, 1811); Rossini (*La donna del lago*, 1819); Lemière de Corvey (Rossini pasticcio, 1825); Vesque von Püttlingen (1829)

*Waverley* (1814): Holstein (*Die Gastfreunde*, 1852, rev. as *Die Hochländer*, 1876); Dulcken (*MacIvor*, ?1865)

*Old Mortality* (1817): Bishop (*The Battle of Bothwell Brigg*, 1820); Bellini (*I Puritani*, distantly, 1835)

*The Black Dwarf* (1817): C. E. Horn (*The Wizard*, 1817)

*Rob Roy* (1818): Blanchard (*Diane de Vernon*, 1831); Curmi (1833); Flotow (1836); De Koven (1894); Grieve (1950)

*The Heart of Midlothian* (1818): Bishop (1819); Carafa (*La prison d'Edimbourg*, 1833); Ricci (*La prigione d'Edimburgo*, 1838); Berlijn (*Le lutin de Culloden*, comp. *c*.1848); MacCunn (*Jeanie Deans*, 1894)

*The Bride of Lammermoor* (1819): Adam (*Le caleb de Walter Scott*, 1827); Carafa (*Le nozze di Lammermoor*, 1829); Rieschi (1831); Damse (1832); Bredal (*Bruden fra Lammermoor*, 1832); Mazzucato (*La fidanzata di Lammermoor*, 1834); Donizetti (*Lucia di Lammermoor*, 1835)

*The Abbot* (1820): Fétis (*Marie Stuart en Écosse*, 1823)

*The Legend of Montrose* (1820): Bishop (1820)

*Ivanhoe* (1820): Rossini (pasticcio, prod. 1826 in Scott's presence); Marschner (*Der Templer und die Jüdin*, 1829); Pacini (1832); Nicolai (*Il Templario*, 1840); Sari (1863); Pisani (*Rebecca*, 1865); Castegnier (*Rébecca*, *c*.1882); Ciardi (1888); Sullivan (1891); Lewis (1907)

*Kenilworth* (1821): Auber (*Leicester*, 1823); Donizetti (*Elisabetta al castello di Kenilworth*, 1829); Damse (1832); Weyse (*Festen på Kenilworth*, 1836); Seidelmann (1843); Schira (rehearsed 1848); Badia (*Il conte di Leicester*, 1851); Caiani (1878); De Lara (*Amy Robsart*, 1893); Klein (1895); Schiuma (1920)

*Peveril of the Peak* (1822): C. E. Horn (1826)

*Quentin Durward* (1823): Laurent (1848); Gevaert (1858); Maclean (comp. 1893, prod. 1920)

*Redgauntlet* (1824): Gomis (*Le revenant*, 1833)

*The Talisman* (1825): Bishop (1826); Pacini

(1829); Adam (*Richard en Palestine*, 1844); Balfe (1874)
*Woodstock* (1826): Flotow (*Alice*, 1837)
*The Highland Widow* (1827): Grisar (*Sarah*, 1836)
*Tales of a Grandfather* (1828): Rossini (pasticcio, *Robert Bruce*, 1846)
*The Fair Maid of Perth* (1828): Bizet (1867); Lucilla (1877).

BIBL: J. Mitchell: *The Walter Scott Operas* (Birmingham, AL, 1978).

**Scotti, Antonio** (*b* Naples, 25 Jan. 1866; *d* Naples, 26 Feb. 1936). Italian baritone. Studied Naples with Triffani Paganini. Début Naples 1889 (Cinna, *La vestale*). Appearances in Madrid, Buenos Aires, Moscow; Milan, S, 1898; London, CG, 1899–1910, 1913–14; New York, M, 1899–33. Repertory included Don Giovanni, Posa, Scarpia, Sharpless, Iago (Verdi), Falstaff. In his prime, possessed a smooth, fine tone; always sang and acted with great artistry though his voice declined latterly. (R)

**Scottish Opera.** Founded in Glasgow in 1962 by Alexander Gibson, with Peter Hemmings as general administrator, the company began with two operas in its first season, *Madama Butterfly* and *Pelléas*. By 1975 it had staged 50 operas. Among its many achievements have been the completion of a *Ring* in 1971, *The Trojans* with Janet Baker, and commissions from Scottish composers including Musgrave (*Mary, Queen of Scots*) and Hamilton (*The Catiline Conspiracy*). Until 1975 the company toured Scotland and Northern England from a base in the King's T (opened 1867), which was restored and made its permanent home in 1976 (cap. 1,560); this was the first British opera-house to open since the war. From 1967 the company appeared regularly at the Edinburgh Festival with a number of outstanding productions including *The Rake's Progress*. Its achievements also include a successful Janáček cycle. The faithful audience which had been built up found its loyalty tested by some productions in the late 1980s. Gibson was succeeded by John Mauceri in 1986.

BIBL: C. Wilson, *It is a Curious Story* (Edinburgh, 1988).

**Scotto, Renata** (*b* Savona, 24 Feb. 1933). Italian soprano. Studied Milan with Ghirardini and Llopart. Début Milan, T Nuovo, 1953 (Violetta). Milan, S, from 1953; London, Stoll T 1957, and CG from 1962; New York, M, from 1965. Also Vienna; Paris, O; Chicago; Buenos Aires; Moscow; etc. Roles include Amina (as which successfully replaced Callas, Edinburgh 1957), Norma, Lucia, Gilda, Amelia, Mimì, Butterfly, Gioconda. Possessing a full-toned, flexible lirico-spinto soprano, also very effective in her characterizations. (R)

BIBL: O. Roca, *Scotto, More than a Diva* (London, 1986).

**Scribe, Eugène** (*b* Paris, 14 Dec. 1791; *d* Paris, 20 Feb. 1861). French librettist. He began his theatrical career as a writer of comedies. His first great operatic success came with the text for Auber's opéra-comique *La dame blanche* (1825). However, by his skilful appreciation of the theatrical conditions of the Paris of his day, and of the sensibilities of the audiences, he reanimated the genre of French *Grand Opera as well as giving opéra-comique a new strength. He followed up his first success for Auber with *La muette de Portici* (1828), which gave Grand Opera a new stature and style: in all, Auber was to set 38 of his librettos. Among many other composers with whom he worked, Halévy and especially Meyerbeer stand out as central representatives of the genre of Grand Opera.

Scribe inherited the historical awareness of Jouy's librettos for Spontini, and capitalized on the opportunities for staging on the most elaborate scale afforded by the Opéra. If characterization was sacrificed to grandeur, the effectiveness of his texts assured him and his composers of the greatest success. His plots often draw on historical sources, but are thoroughly reworked rather than directly adapted from a literary original, and frequently deal in the clashing of religious, national, or political orthodoxies and the lives and loves of characters caught up in them: the chorus is thus made a more functional part of the drama. This did not preclude the use of collaborators, chiefly in order to polish or even write verses to suit the strong stage situations in which he specialized. He was also adept at accepting or inventing novel effects: examples in Meyerbeer are the scene of the ghostly nuns in *Robert le diable*, the bathing scene in *Les Huguenots*, and the skating scene in *Le prophète*. He gave Grand Opera a definitive form, and an influence that was potent even in those, such as Wagner, who most sharply attacked it.

**Scutta, Andreas** (*b* Vienna, 1806; *d* Prague, 24 Feb. 1863). Austrian composer. He took up music after acting in Rossini's *Mosè* at the T an der Wien, Vienna; studied singing and then performed in Graz, Linz, and Zagreb. Lost his voice 1829, so resumed his acting career. Joined T in der Leopoldstadt, Vienna, 1831, where he also composed music for many farces and parodies. His later years were spent in Prague, where he appeared in the German T. Of his *c*.30 scores written for Vienna, which include *Robert der Wau Wau*, a skilful parody of Meyerbeer's *Robert le diable*, the most successful was *Eisenbahnheiraten*. He married Therese Palmer, who had created Isouard's *Cendrillon* aged 13.

SELECT WORKLIST: *Robert der Wau Wau* (Schickh; Vienna 1833); *Eisenbahnheiraten* (Railway Marriages) (Nestroy; Vienna 1844).

**Sea Interludes.** The name generally given to the orchestral interludes in Britten's *Peter Grimes*. Though four are usually selected for concert performance, they are in fact six in number: 'Dawn' opening Act I, 'Storm' between scenes 1 and 2 of Act I, 'Sunday Morning' opening Act II, a passacaglia between scenes 1 and 2 of Act II, 'Moonlight' opening Act III, and a final interlude between scenes 1 and 2 of Act III.

**Searle, Humphrey** (*b* Oxford, 26 Aug. 1915; *d* London, 12 May 1982). English composer. Studied London with Jacob, Morris, and Ireland; Vienna with Webern. His first opera, *The Diary of a Madman*, was a Berlin Festival commission that made use of electronic music as well as a normal orchestra. *The Photo of the Colonel* confronts the question of making the serial technique of which Searle was one of the leading British exponents a real vehicle for comedy, solving it by extensive use of rapid parlando passages for the singers with the burden of the commentary in the orchestra. *Hamlet* matches contemporary operatic techniques to Shakespeare's tragedy, and provides some ingenious music (for instance in deriving all the material from the 'To be or not to be' series) that attempts to expose rather than interpret the issues at the centre of the play.

WORKLIST: (all texts by Searle) *The Diary of a Madman* (after Gogol; Berlin 1958); *The Photo of the Colonel* (after Ionesco; Frankfurt 1964); *Hamlet* (after Shakespeare; Hamburg 1968).

**Seattle.** City in Washington, USA. The first musical stage production was an operetta version of *Uncle Tom's Cabin* in the early 1870s. Touring companies provided opera until the 1963 foundation of the Seattle Opera Association under the enterprising Glynn Ross; this opened in the Civic Auditorium 1964. Built during the 1962 World Fair, it became the Opera House (cap. 3,017). The Seattle Opera Guild was founded 1965 to provide financial support. After a period when the emphasis was on popular works and stars (though it also saw the premières of Floyd's *Of Mice and Men*, 1970, and Pasatieri's *Black Widow*, 1972), new initiative came with *The Ring* produced by George *London. *The Ring* became a regular event. Speight Jenkins became director-general 1985, and has continued the adventurous policy, which includes that of a less experienced second cast performing at lower prices.

**Sébastian, Georges** (*b* Budapest, 17 Aug. 1903; *d* nr. Paris, 13 Apr. 1989). Hungarian, later French, conductor. Studied Budapest with Bartók, Kodály, and Weiner, Munich with Walter. Répétiteur Munich 1922; New York, M, 1923–4.

Cond. Hamburg 1924–5; chief cond. Berlin, SO, 1927–31. Moscow Radio 1931–7. San Francisco 1944–6. Chief cond. Paris, O, 1947–73, where he specialized in Wagner and Richard Strauss, also being much admired in Verdi and Puccini and for his *Carmen*; also Paris, OC. Frequent appearances in French provinces and Geneva. Mexico 1978. (R)

**Šebor, Karel** (*b* Brandýs nad Labem, 1843; *d* Prague, 17 May 1905). Czech composer. Studied Prague with Kittl. Kapellmeister, Erfurt 1863; chorus-master, Prague, P, 1865; returned after a period in Vienna and other cities to Prague, 1894. His operas are important as early attempts at a Czech national style.

SELECT WORKLIST (all first prod. Prague): *Templáři na Moravě* (The Templars in Moravia) (Sabina; 1865); *Drahomíra* (Böhm; 1867); *Husitská nevěsta* (The Hussite's Bride) (Rüffer; 1868); *Blanka* (Rüffer; 1869); *Zmařená svatba* (The Foiled Wedding) (Červinková-Riegrová; 1871).

**Secret, The** (Cz.: *Tajemství*). Opera in 3 acts by Smetana; text by Eliska Krásnohorská. Prod. Prague, C, 18 Aug. 1878, with Sittová, Fibichová, Vávra, Čech, Mareš, Lev; Oxford, OUOC, 7 Dec. 1956, with C. Hunter, Baker, D. Minton, Reynolds, N. Noble, cond. Westrup.

A small North Bohemian town, end 18th cent. The plot concerns the separation through pride and poverty of two lovers, Rose (con) and Councillor Kalina (bar), who after many vicissitudes discover one another when Kalina, seeking a promised treasure, follows an underground passage that leads him into Rose's house.

**Sedaine, Michel-Jean** (*b* Paris, 4 July 1719; *d* Paris, 17 May 1797). French librettist. Originally a stonemason, he began writing vaudevilles for the Opéra-Comique after the *Guerre des Bouffons. His first successes were comédies mêlées d'ariettes, in the style of Favart but, as in his text for Philidor's *Le diable à quatre* (1756) and *Blaise le savetier* (1759), with greater attention to the detail of the action and the setting. His texts for Monsigny were particularly successful, and *On ne s'avise jamais de tout* (1761) was an example for Beaumarchais in *Le barbier de Séville*. The realism of his plots and incidents (as opposed to the idealized pastoral convention) impressed his contemporaries, despite the poor quality of the actual verse: his greatest strength is his instinctive grasp of theatrical character and situation. His librettos for Monsigny's *Le déserteur* (1769) and Grétry's *Richard Cœur-de-Lion* (1784) transform the *larmoyant* element from being merely a touch of colour for the *ingénue into the atmosphere of the whole drama, and subdue the comic element: their mixture of comic and serious disconcerted audiences, but was influential on *rescue opera, and even on Romantic opera. Two more important librettos were for Grétry's *Raoul Barbe-bleue* (1789) and *Guillaume Tell* (1791). Sedaine helped to increase the importance of opéra-comique as a genre, and hence to give the Opéra-Comique a new stature vis-à-vis the Opéra.

BIBL: L. Arnoldson, *Sedaine et les musiciens de son temps* (Paris, 1934).

**Sedie, Enrico delle.** See DELLE SEDIE, ENRICO.

**Seedo** (Sydow) (*b* c.1700; *d* ?Prussia, c.1754). German musician who worked on ballad opera in England. Following the success of *The Beggar's Opera* (1728), Seedo was engaged in rival enterprises at Drury Lane, London. These included Coffey's *The Devil to Pay*—later to be of significance in the evolution of German Singspiel—and *The Boarding School* (1733) and Fielding's *The Mock Doctor* (1732). His tasks in these collaborations involved composing a few numbers in simple English ballad style and arranging the remainder of the music. In 1733 Seedo also composed an all-sung masque *Venus, Cupid and Hymen*, now lost, in which his Italian wife sang.

**Seefried, Irmgard** (*b* Köngetried, 9 Oct. 1919; *d* Vienna, 24 Nov. 1988). Austrian soprano. Studied Augsburg with Mayer. Début Aachen 1940 (Priestess, *Aida*). Vienna, S, from 1943. London, CG, 1947–9; New York, M, 1953–4. Also Milan, S; Salzburg; Edinburgh; etc. Her warm tone and appealing personality made her a successful Susanna, Fiordiligi, Eva, Octavian, Composer (Strauss commenting, 'I never knew what a good part that could be'); also sang Cleopatra (*Giulio Cesare*), Marie (*Wozzeck*), Blanche (*Carmélites*). An admired Lieder singer. (R)

**Segreto di Susanna, Il** (Susanna's Secret). Opera in 1 act by Wolf-Ferrari; text by Golisciani and Zangarini. Prod. Munich (as *Susannas Geheimnis*) 4 Dec. 1909, with Tordek, Brodersen, cond. Mottl; New York, M, by Philadelphia-Chicago Co., 14 Mar. 1911, with White, Sammarco, cond. Campanini; London, CG, 11 July 1911, with Lipkowska, Sammarco, cond. Campanini.

Piedmont, early 20th cent. A slight but charming curtain-raiser about a jealous husband, Count Gil (bar), who, smelling tobacco in the house, suspects his pretty wife Susanna (sop) of secretly entertaining a lover. Susanna's secret is that she herself smokes.

**Seguin, Arthur** (*b* London, 7 Apr. 1809; *d* New York, 13 Dec. 1852). English bass. Studied London, RAM. London 1831–8, at DL, CG, and HM. Formed Seguin OC, New York 1838, and toured USA. Became chief of an Indian tribe, being given the name 'the man with the deep, mellow voice'. He was the uncle of Euphrosyne Parepa, wife of *Carl Rosa.

**Seidl, Anton** (*b* Pest, 7 May 1850; *d* New York, 28 Mar. 1898). Hungarian conductor. Studied Leipzig. Much influenced by Richter, and went to Bayreuth 1872, helping Wagner with preparations for the first *Ring*. On Wagner's warm recommendation cond. Neumann's Wagner Co. in Leipzig 1879, then touring as chief cond. 1883. Cond. Bremen 1883–5. New York, M, 1885–92, 1893–7, US premières of *Meistersinger* and *Tristan* 1886, *Siegfried* 1887, *Götterdämmerung* 1888. London, HM, 1882, first English *Ring*; CG 1897. Bayreuth 1897, *Parsifal*. With *Richter, *Mottl, *Levi, and *Muck, one of the first generation of great Wagner conductors. In 1883 married the soprano **Auguste Kraus** (*b* Vienna, 28 Aug. 1853; *d* Kingston, NY, 17 July 1939), who then sang as Auguste Seidl-Kraus. Studied Vienna with Marchesi. Début Vienna, H, 1877 (small roles). Leipzig 1881–2. London, HM, 1882. Neumann's Wagner Co. 1883. Bremen 1883. New York, M, 1884–7. First US Eva, Woodbird, Gutrune.

**Seidl-Kraus, Auguste.** See SEIDL, ANTON.

**'Se il padre perdei'.** Ilia's (sop) aria in Act II of Mozart's *Idomeneo*, reflecting that if she had lost her father in Troy, Idomeneo would now be a father to her and Crete the place where she could be happy.

**'Selig wie die Sonne'.** The quintet in Act III of Wagner's *Die Meistersinger von Nürnberg*, in which Eva (sop) and Walther (ten) muse rapturously on the song which they hope will gain the prize in the coming contest, Magdalene (mez) and David (ten) reflect on the good fortune this is likely to bring them, and Hans Sachs (bs-bar) confides to himself his thoughts on the loss of Eva.

**Selika.** The eponymous African slave (sop), in love with Vasco da Gama and loved by Nelusko, in Meyerbeer's *L'Africaine*.

**Sellars, Peter** (*b* Pittsburgh, 27 Sep. 1957). US producer. Studied Harvard U, where his highly original theatrical productions earned him early notoriety. On graduating formed his own company, the Explosives B Cabaret; first came to wider attention with Gogol's *The Inspector General* (1980). In 1983 he became director of the Boston Shakespeare Co., in 1984 director of the American National Theater Co. at the Kennedy Center.

Sellars's first important production for the opera-house was Handel's *Orlando* (American Repertory Theater, 1981). With its up-dated setting (the most notable feature of which was Orlando's transformation into an astronaut) and generally loose interpretation, it established Sellars as one of the most controversial figures in US opera. A production of *The Mikado* (transferred to present-day Japan) began his relationship with the Chicago Lyric O (1984); his most notable production here was that of *Tannhäuser* in 1988. For this Sellars not only employed *Surtitles for the sung text, but also projected colourcoded lines of his own invention to convey the unsung thoughts of the characters. In John *Adams he found an ideal partner, and played a major role in the conception of both *Nixon in China* and *The Death of Klinghoffer*.

In the UK Sellars has been most closely associated with Gly., where his interpretations of the Mozart classics have aroused outrage and plaudits in equal measure. His decision, without prior consultation, to excise the spoken dialogue from a production of *The Magic Flute* (set in present-day California, with Sarastro as the leader of a hippie commune) was widely criticized and led to a management crisis. One of the most inventive and lively figures on the current scene, together with David Pountney he is the most striking practitioner of what may loosely be called 'producer opera'.

**Sellem.** The auctioneer (ten) in Stravinsky's *The Rake's Progress*.

**Sembrich, Marcella** (*b* Wiśniewczyk, 15 Feb. 1858; *d* New York, 11 Jan. 1935). Polish, later US, soprano. Taught piano and violin by her father, Kasimir Kochański (she later adopted her mother's maiden name for artistic purposes), and heard playing these, and singing, by Liszt. Though impressed by her talent in all three, he advised her to follow a singing career. Studied Milan with G. Lamperti, later with F. Lamperti. Début Athens 1877 (Elvira, *Puritani*). Dresden 1878–80; London, CG, 1880–5, 1895; New York, M, 1883, 1898–1909 (except one season); also Madrid, Hamburg, St Petersburg, where she so moved Alexander II with a Chopin song that he promised 'not to forget Poland' (but was dead within the week). A highly gifted artist, who gave concerts performing with equal virtuosity as violinist, pianist, and singer, she was universally admired, including by Patti, Clara Schumann, Brahms, Verdi, and Bernhardt; for Puccini she was 'the Mimì'. Repertory also included Zerlina, Rosina, Lucia, Marguerite, Violetta, Elsa, Eva. Using her warm, generous tone and technical brilliance with an exceptional musicality, she was mistress of every style. A pioneer of the song recital without operatic arias. (R)

**Semele.** Masque in 3 acts by Handel; text an anon. (? N. Hamilton) adaptation of Congreve's drama (1706) for John Eccles, after Ovid, *Metamorphoses* iii, 261f. Prod. London, CG, 10 Feb.

1744, with Duparc (La Francesina), Young, Avoglio, Beard, Sullivan, Reinhold; first stage prod. Cambridge 10 Feb. 1925 (Arundell version); Evanston, Northwestern U, Jan. 1959, cond. Thor Johnson.

**Semenoff, Ivan** (*b* Paris, 7 June 1917; *d* Paris, 14 June 1972). French composer of Russian descent. Studied Paris with Honegger. His operas, exploratory in nature and effectively written for the stage, have won a considerable following in France. They include a popular setting of Chekhov's *The Bear* (*L'ours*, 1958), *Évangéline* (1964), *Don Juan ou L'amour de la géométrie* (1969), and *Sire Halewyn* (1974).

**semi-opera.** A term used to describe a number of English works of the late 17th-cent. which approach genuine opera in the method of their union of drama and music, yet fall short of true opera since their main action is carried forward in speech. The earliest example is Thomas Betterton's *The Tempest* (1674; music by Locke and others); Purcell's *King Arthur* represents its most developed form. More typical is *The Indian Queen*, which contains lengthy passages of dialogue into which occasional songs and other pieces are inserted, some of which, e.g. the 'Incantation Scene', exploit operatic procedures to the full.

**Semiramide.** Opera in 2 acts by Rossini; text by G. Rossi, after Voltaire's tragedy *Sémiramis* (1748). Prod. Venice, F, 3 Feb. 1823, with Colbran, Mariani, Spagna, Galli, Sinclair; London, H, 15 July 1824, with Pasta; New Orleans, St Charles T, 19 May 1837, with Marozzi, Pantanelli, Badiali, Fornasari, Candi, cond. Gabici.

Babylon, 1200 BC Semiramide (sop), Queen of Babylon, and her lover Assur (bar) murder the King. She later falls in love with a young man who turns out to be her son Arsace (mez). She receives a mortal blow that Arsace intends for Assur. Arsace is made king.

There are some 65 operas on the subject, including those by Aldrovani (1701), Destouches (1718), Porpora (two: 1724 and 1729), Vinci (1729), Vivaldi (1732, lost), Hasse (1744), Gluck (1748), Pérez (1749), Galuppi (1749), C. H. Graun (1754), Fischietti (1759), Sacchini (1762), Bernasconi (1765), Bertoni (1767), Sarti (1768), Paisiello (1773), Mortellari (1784), Salieri (1784), Gyrowetz (1791), Borghi (1791), Cimarosa (1799), Catel (1802), Nicolini (early 19th cent.), Meyerbeer (1819), M. García (1828), Respighi (1910).

**'Sempre libera'.** The cabaletta to Violetta's (sop) aria 'Ah! fors' è lui' in Act I of Verdi's *La traviata*, in which she declares her intention of abandoning her dream of Alfredo's love and returning to her former life of pleasure.

**Semyon Kotko.** Opera in 5 acts by Prokofiev; text by V. Katayev and composer, after Katayev's story. Prod. Moscow, Stanislavsky T, 23 June 1940, with Voskresensky, Malkova, Panchekhin, cond. Zhukov (it was soon dropped, the subject-matter being deemed 'inappropriate'); Dresden 1960.

Ukraine, 1918. Semyon Kotko (ten) returns to the Ukraine after the First World War. He wishes to marry Sophia (sop), the daughter of á rich landowner, Tkachenko (bs-bar), who refuses his consent. Semyon's cottage is destroyed by a German detachment; he escapes, and leaves home to form a resistance movement. Returning several months later, he prevents Sophia's marriage to the rich Klembovsky (ten) by blowing up the church.

**Seneca.** Lucius Annaeus Seneca the Younger (*b* Corduba, *c*.4 BC; *d* Rome, 12 Apr. AD 65), the Stoic philosopher, appears (bs) in Monteverdi's *L'incoronazione di Poppea* in his historical role as Nero's mentor who obeys with dignity the command to commit suicide.

**Sénéchal, Michel** (*b* Paris, 11 Feb. 1927). French tenor. Studied Paris with Paulet. Début Brussels, M, 1950. Brussels, M, until 1952. Aix from 1954; Milan, S, 1960; Gly. 1966; Paris, O and OC regularly, B 1990–1; Salzburg 1972–80; New York, M, 1985; also Rome, Toulouse, Marseilles, etc. His many roles incl. Platée (Rameau), Don Basilio (*Figaro*), Don Ottavio, Tamino, Idiot (*Boris*), M. Triquet, Quint, Oberon (Britten), and much French repertoire. A character-actor of great vocal and dramatic finesse. (R)

**Senesino** (orig. Francesco Bernardi) (*b* Siena, *c*.1680; *d* ?Siena, before 27 Jan. 1759). Italian mezzo-soprano castrato. Venice 1707–8; then at Bologna, Genoa, Naples; Dresden 1717–20. London, H, 1720–8 for Handel, with enormous success. Though his moodiness and arrogance enraged Handel (who called him 'a damned fool', according to Rolli), he was re-engaged 1730-3, until he deserted to the Opera of the Nobility. Sang with this co. 1733–6, to further acclaim, rivalling even Farinelli in popularity. His departure from London was much mourned by the ladies. Florence 1737–9; Naples 1739–40, though here his style was considered old-fashioned. Created 17 parts for Handel, including Ottone, Giulio Cesare, Bertarido (*Rodelinda*), Siroe, Orlando. With a voice described by Quantz as 'powerful, clear, equal, and sweet', he was remarkable for his singing of recitative, his tastefully ornamented adagios, and fiery allegros; also for his acting and expressiveness.

**Senta.** Daland's daughter (sop) in Wagner's *Der fliegende Holländer*.

**Senta's Ballad.** The ballad relating the legend of the Flying Dutchman sung by Senta (sop) in Wagner's *Der fliegende Holländer*, beginning 'Johohoe!'

**'Senza mamma, o bimbo'.** Angelica's (sop) aria in Puccini's *Suor Angelica*, grieving over the death of her illegitimate baby son and longing to join him in Heaven.

**sepolcro** (It.: 'sepulcre'). A genre of sacred opera given in Vienna in the 17th cent. It was distinguished by its 1-act form and obligatory setting of the Passion story; its name derives from its invariable use of the Holy Sepulchre of Christ as the main setting, and it was performed on Maundy Thursday and Good Friday. Its most noted exponent was Draghi. Because of its markedly contemplative text and purpose, the sepolcro is perhaps closer to the tradition of *oratorio, although its musical similarities with the *rappresentazione sacra have caused it to be widely considered an operatic genre.

**Serafin, Tullio** (*b* Rottanova di Cavarzere, 8 Dec. 1878; *d* Rome, 2 Feb. 1968). Italian conductor. Studied Milan with Saladino and Coronaro. Played violin and viola in orchestra, Milan, S. Début (anagrammatically as Alfio Sulterni) Milan 1898 (*Elisir*). Répétiteur Milan, S, 1898, under Toscanini. Cond. Ferrara 1903; Turin 1904–5; also Brescia, Milan, Trieste, Venice, Palermo. London, CG, 1907. Chief cond. Milan, S, 1910–14, 1917–18, giving various premières (incl. Montemezzi's *L'amore dei tre re* (1913) and *La nave* (1918) and many Italian premières). Paris, O, 1912 (*Fanciulla del West* with Caruso); Madrid 1912. Inaugurated Verona Arena summer seasons 1913 (*Aida*). Buenos Aires, C, 1914–16. New York, M, 1924–34, giving various premières (incl. Taylor's *The King's Henchman* (1927) and *Peter Ibbetson* (1931), Gruenberg's *Emperor Jones* (1933), Hanson's *Merry Mount* (1934)) and other US premières. London, CG, 1931. Rome 1934–43 (first Italian *Wozzeck*, with Gobbi, Italian-language *Ring*). Milan, S, 1939–40, 1946–7 (first Italian *Peter Grimes*, 1947). Chicago 1956–8. London, CG, 1952, 1959, 1960. Rome 1962. A conductor of great authority and understanding, with a versatility that centred on the Italian tradition of which he became a distinguished representative, he was particularly admired for his care with singers; those who owed him much for his shrewdness and wisdom in coaching included Ponselle, Callas, Sutherland, and Gobbi (who thought him 'an infallible judge of voice and character' and 'the most complete man of the theatre of our time'). In no small part responsible for the revival of bel canto. (R) His wife, the Polish soprano **Elena Rakowska** (1878–1964), sang at the Milan, S, New York, M, and Buenos Aires, C. She was especially successful in Wagner. (R)

WRITINGS: (with A. Toni) *Stile, tradizioni e convenzioni del melodramma italiano del Settecento e dell'Ottocento* (Milan, 1958).

BIBL: T. Celli and G. Pugliese, *Tullio Serafin* (Venice, 1985).

**Seraglio, The.** See ENTFÜHRUNG AUS DEM SERAIL, DIE.

**Serban, Andrei** (*b* Bucharest, 1943). Romanian producer. Studied Theatre Institute, Bucharest. After working on theatrical productions in New York (with Peter Brook) and Paris, made his début as an opera producer with WNO 1981 (*Eugene Onegin*). Has worked with WNO on several subsequent occasions (*Puritani, Rodelinda, Norma*). London, CG (*Fidelio*); Nancy, 1979 (*Zauberflöte*); New York, City Opera (*Alcina*); première of Glass, *The Juniper Tree* (Cambridge, MA, 1985). Also Geneva, Los Angeles, Bologna.

**Serbia.** Serbia became independent of Ottoman rule in the first half of the 19th cent., and in 1831 Prince Miloš summoned Josif Šlesinger (1794–1870) to Kragujevac (then the capital). Šlesinger composed music for plays, of which the most important was *Ženidba cara Dušana* (Tsar Dušan's Wedding, 1840); described as 'in the manner of an Italian opera', this was closer to a play with extensive musical contributions. A form of Serbian Singspiel containing folk and oriental elements was also popular. The Slovene composer Davorin Jenko (1835–1914), director of the Belgrade T 1871–1902, was successful in the genre, e.g. with *Vračara* (The Sorceress, 1882, a local version of Zauberstück), and also wrote a Romantic opera, *Pribislav i Božana* (1894). The first important Serbian composer was Stanislav *Binički (1872–1942), who conducted the first operas at the National T and was director of the new ensemble 1920; he composed the successful *Na uranku* (At Daybreak, 1903). Of a group of musicians working in Novi Sad, the most important was Isidor Bajič (1878–1915), who wrote *Knez Ivo od Semberije* (Prince Ivo of Semberia, 1911), characterizing Turks and Serbs musically with some use of motive. Jenko's successor at the National T from 1903, Petar Krstić (1877–1957), wrote Singspiels including *Snežana i sedem patuljaka* (Snow White and the Seven Dwarfs, 1912) and the more ambitious *Zulumčar* (The Young Tyrant, 1928). Petar Stojanovič (1877–1957) wrote operas and operettas, but a stronger talent was Petar Konjovič (1883–1970),

who was influenced by Janáček and Musorgsky; his *Vilin veo* (The Fairy's Veil, 1917) is a Weberian fairy opera, but *Knez od Zete* (The Prince of Zeta, 1929) and especially *Koštana* (1931) are more advanced music dramas. Stevan Hristič (1885–1958), director of the Belgrade O 1924–34, wrote a psychological drama *Suton* (In the Dusk, 1925, rev. 1954), combining impressionist orchestration with the intimate atmosphere of chamber opera.

The 1930s proved a fallow period. After Konjovič's *Koštana*, the only Serbian opera before the war was *Đurađ Branković* (1940) by the amateur Svetomir Nastasijević (1902–79). After the war, interest in opera was shown only by Stanojlo Rajičić (*b* 1910) (with *Simonida*, 1957, a conventional work by a former avantgardist) and Mihovil Logar (*b* 1902) (with *1941*, 1961, a modern but accessible work on a war subject); but they had to look to Sarajevo for performances of their operas.

See also BELGRADE, NOVI SAD.

BIBL: S. Djurić-Klajn, *Serbian Music through the ages* (Belgrade, 1972).

**serenade** (It., *serenata* = evening song, from *sera* = evening). By origin, a song sung under his lady's window by a lover, with or without instrumental accompaniment, which might in opera be provided by himself (as in Don Giovanni's 'Deh vieni') or by a hired band (as in Almaviva's 'Ecco ridente in cielo' in Rossini's *Barbiere*.) The term was soon applied to any instrumental piece of a light nature. See also *SERENATA.

**serenata** (It.: *serenade). The term used for a serenade, but also used in the 18th cent. for a short operatic piece performed, probably in the evening, to celebrate royal birthdays and other occasions in a room with a small amount of scenery. It was similar in style to the *azione teatrale. A famous example is Handel's *Acis and Galatea* (1720).

**Sergey.** A workman (ten), Katerina Ismaylova's lover, in Shostakovich's *The Lady Macbeth of Mtsensk*.

**Serov, Alexander** (*b* St Petersburg, 23 Jan. 1820; *d* St Petersburg, 1 Feb. 1871). Russian composer and critic. Studied in St Petersburg, initially following a career in the Civil Service. He was greatly encouraged by Stasov, at whose suggestion he began an opera, *The Pagan*, in 1840; other abortive operatic projects included *The Merry Wives of Windsor* and Gogol's *May Night*. Frustrated by his lack of technique, he took correspondence lessons with Osip Hunke, but the problem dogged him all his life, as did his inability to get on with his librettists. His notor-

Serpette

iously sharp tongue, coupled with his intellectual acuteness, helped to make him a feared but respected critic. In 1858 he was greatly impressed by a performance of *Tannhäuser* in Dresden, and this not only made him into Wagner's main advocate in a Russia resistant to Wagnerian music drama, but confirmed him in his own creative aims. He contemplated various Wagnerian projects (including an *Undine*). Eventually he settled on *Judith*, paradoxically at first in an Italian treatment and then in a manner that recalls Parisian *Grand Opera rather than Wagner. He broke with Balakirev over *Judith*, and was also at odds with both the Slavophile and Westernizer groups, resisting what he saw as formlessness and subservience to the text (in Dargomyzhsky) in favour of expressive forms controlled by the composer. He took his Wagnerian precepts to the point of having a personal hand in the production of *Judith*.

The success of *Judith* established Serov artistically and materially, and enabled him to marry the pianist and composer **Valentina Bergmann** (1846–1924: she was the first Russian opera composer, and her *Uriel Acosta* was prod. Moscow 1885, *Ilya Muromets*, with Shalyapin in the title-role, Moscow 1899). He followed this with *Rogneda*, which initially achieved an even greater success than *Judith*. In it, he turned his attention to national legend, intending an opera that would appeal to some of the ideals of the Slavophiles and also be an organically satisfying entity. Ironically many preferred individual parts to the whole, in a work that is certainly very uneven. Once again, there is a spectacular element that suggests Grand Opera rather than Wagnerian ideals of any kind, and that vitiates Serov's own genuinely held ideals: the best of the work was to survive in its influence on Rimsky-Korsakov, Musorgsky, and Borodin.

Serov's last opera, *The Power of Evil*, began as a collaboration with Ostrovsky, whom Serov replaced with a more complaisant librettist. Here, the nationalism is achieved more through use of folk music, leading to a style now deeply imbued with Russian folk styles in its declamation, its harmony, and its rhythmic and metrical characteristics. Though highly controversial, dividing Serov's critics, the work can be defended for its serious, thoroughgoing attempt to unify Russian musical characteristics with music drama on a substantial scale. Serov died leaving the work not quite finished: the score was completed by his wife and Nikolay Solovev.

WORKLIST (all first prod. St Petersburg): *Judith* (Maykov & others; 1863); *Rogneda* (Serov & Averkyev; 1865); *Vrazhya sila* (The Power of Evil) (Serov, Ostrovsky, & Kalashnikov; 1871).

WRITINGS: N. Stoyanovsky and others, eds., *A. N.*

*Serov: kriticheskiye staty* (Critical Essays) (St Petersburg, 1892–5).

BIBL: G. Abraham, 'The Operas of Serov', in J. Westrup and F. Sternfeld, eds., *Essays presented to Egon Wellesz* (Oxford, 1966).

**Serpette, Gaston** (*b* Nantes, 4 Nov. 1846; *d* Paris, 3 Nov. 1904). French composer. Studied Paris Cons. with Thomas. Won the Prix de Rome 1871, but failed to gain acceptance as a composer of serious opera. The success of *La branche cassée* at the Bouffes-Parisiens 1874 turned him in the direction of operetta and vaudeville. For the Parisian stage he composed *c*.30 such works, whose deft charm won them great popularity.

SELECT WORKLIST: *La branche cassée* (The Broken Branch) (Jaime & Noriac; Paris 1874).

**Serpina.** Uberto's servant (sop), the eponymous maid-mistress of Pergolesi's *La serva padrona*.

**Serse** (Xerxes). Opera in 3 acts by Handel; text an anon. adaptation of Stampiglia's libretto for Bononcini (1694), itself a reworking of Minato's libretto (1654). Prod. London, H, 15 Apr. 1738, with Francesina, Lucchesina, Caffarelli, Montagnana; Northampton, MA, 12 May 1928, with Garrison, Ekberg, F. Martinelli, Kullman, Dickenson, Meyer, Marsh, cond. Josten. Revived Handel O Society, 1970.

Though set in ancient Persia, the work makes use of some London street songs. Handel's only opera containing a purely comic character (Elviro, bs), it also includes 'Ombra mai fù', which was satirical by intent, but is now invariably taken seriously as 'Handel's Largo'.

Other operas on the subject are by Cavalli (1655), Förtsch (1689), and Bononcini (1694).

**Serva padrona, La** (The Maid-Mistress). Intermezzo in 2 parts to the opera *Il prigionier superbo* by Pergolesi; text by Federico. Prod. Naples, B, 5 Sep. 1733; London, H, 27 Mar. 1750; Baltimore, New T, 14 June 1790.

The most famous of all *intermezzos describes how Serpina (sop) lures her master Uberto (bs) into marrying her by pretending to leave with a ferocious soldier, in fact another servant, Vespone (mute).

Text also set by Paisiello, prod. Tsarskoye Selo 10 Nov. 1781, with De Bernucci, Marchetti. Other operas on the same text are by Predieri (1732), Abos (1744), and, in modified form, P. A. Guglielmi (1790).

**Servilia.** 1. The sister (sop) of Sesto (Sextus), lover of Annio (Annius), in Mozart's *La clemenza di Tito*.

2. Soranus' daughter (sop), lover of Valerius Rusticus, in Rimsky-Korsakov's *Servilia*.

**Sessions, Roger** (*b* Brooklyn, 28 Dec. 1896; *d* Princeton, 16 Mar. 1985). US composer. Studied

652

Harvard and Yale U, also privately with Bloch. Subsequently held teaching positions at Cleveland Cons., Princeton, Berkeley, and Harvard U. Sessions's music owes most to the models of Stravinsky and the Second Viennese School, whose intellectual clarity he shares; while not as overtly lyrical as that of many of his US contemporaries, notably Copland, it also reflects Bloch's influence in its expansive phrase structure and rich expressive content. Though *The Trial of Lucullus* was well received, Sessions's operatic reputation rests on the epic *Montezuma*.

SELECT WORKLIST: *The Trial of Lucullus* (Brecht; Berkeley, 1947); *Montezuma* (Borghese; comp. 1941–63, prod. Berlin 1963).

BIBL: A. Olmstead, *Conversations with Roger Sessions* (Boston, 1987).

**Sesto.** Sextus (sop cas), a young Roman patrician, Servilia's brother and lover of Vitellia, in Mozart's *La clemenza di Tito*.

**Séverac, Déodat de** (*b* St-Felix-Lauragais, 20 July 1872; *d* Céret, 24 Mar. 1921). French composer. Studied Paris with D'Indy and Magnard. His first opera, *Le cœur du moulin*, indicated his deep love of his native countryside, and in *Héliogabale*, which had its first performance to a crowd of over 15,000 in an arena, he used some Catalan instruments. His style is restrained, and in some ways anticipates neoclassicism.

SELECT WORKLIST: *Le cœur du moulin* (The Heart of the Mill) (Magre; Paris 1909); *Héliogabale* (Sicard; Béziers 1910).

BIBL: B. Selva, *Déodat de Séverac* (Paris, 1930).

**'Se vuol ballare'.** Figaro's (bar) aria in Act I of Mozart's *Le nozze di Figaro*, in which he promises himself to pit his wits against those of Count Almaviva.

**Seydelmann, Franz** (*b* Dresden, 8 Oct. 1748; *d* Dresden, 23 Oct. 1806). German composer. Studied Dresden with Schürer and Naumann; toured Italy 1765–8. His career was spent entirely at the Dresden court, where he had an active involvement with opera: church composer 1772; Kapellmeister 1787. He wrote 12 stage works, including both Singspiel and opera buffa, the first being *La serva scaltra* and the most successful *Il mostro* and *Il turco in Italia*.

SELECT WORKLIST (all first prod. Dresden): *La serva scaltra* (?; 1773); *Il mostro* (Mazzolà; 1785); *Il turco in Italia* (Mazzolà; 1788).

**Seyfried, Ignaz** (*b* Vienna, 15 Aug. 1776; *d* 27 Aug. 1841). Austrian composer, conductor, and writer on music. Studied Vienna with Albrechtsberger, Kozeluch, Mozart, and Winter. He was conductor in Schikaneder's Freihaustheater, later at the T an der Wien, from 1797 to 1827; 1805 he directed the première of *Fidelio*. His

career as a composer began with *Der Friede*: of his later works most successful were *Der Wundermann am Rheinfall*, and the biblical music dramas, including *Saul* (1810), *Abraham* (1817), *Der Makabäer* (1818), and *Noah* (1819). He also made many arrangements, among them the popular *Rochus Pumpernickel*, and provided incidental music for many plays, including the première of Grillparzer's *Die Ahnfrau* (1817). He wrote a number of parodies, and his own *Idas und Marpissa* was parodied in 1818 by Perinet and Tuczek: both were successful.

One of the most influential figures in early 19th-cent. Viennese cultural life, Seyfried was also a prolific writer: his books and articles (many of which appeared anonymously in journals and newspapers) suggest an intelligent and sensitive musician. His pupils included Suppé.

SELECT WORKLIST (all first prod. Vienna): *Der Wundermann am Rheinfall* (Schikaneder; 1799); *Idas und Marpissa* (Stegmayer; 1807); *Rochus Pumpernickel* (Stegmayer; 1809, comp. with Haibel).

**Seyler, Abel** (*b* Liestal, 23 Aug. 1730; *d* Rellingen, 25 Apr. 1800). Swiss actor and manager. A founder member of the Hamburg theatre 1767, he was manager 1769 but was forced to move to Hanover, where he formed a company. This became one of the best troupes of the time, basing itself in various cities and acquiring a good reputation, partly through the excellence of the actors but also because of Anton Schweitzer's musical direction. Works performed included Singspiels by Benda and Neefe as well as Schweitzer himself. Their performance of Benda's duodrama *Medea* led Mozart to agree to write one of his own, *Semiramis*, for them (not eventually composed).

**'s Gravenhage** (Eng., The Hague). See HOLLAND.

**shake.** An old term for a trill.

**Shakespeare, William** (*b* Stratford-upon-Avon, *bapt.* 26 Apr. 1564; *d* Stratford-upon-Avon, 23 Apr. 1616). English poet and dramatist. Operas on his works are as follows (the dates of the first seasons are those suggested by Sir Edmund Chambers):

*Henry VI, pts2 and 3* (1590–1): None
*Henry VI, pt1* (1592–3): None
*Richard III* (1592–3): Meiners (1859); Canepa (1879); Salvayre (1883); Durme (c.1961); Turok (1980); Kuljević (1987); Testi (1987)
*The Comedy of Errors* (1592–3): Storace (*Gli equivoci*, 1786); Bishop (1819); Lorenz (c.1890); Krejči (*Pozdvižení v Efesu*, 1946); Wilson-Dickson (1980)
*Titus Andronicus* (1593–4): None
*The Taming of the Shrew* (1593–4): Braham and

others (1828); Goetz (1874); Samara (1895); Maclean (*Petruccio*, 1895); Chapí (1896); Le Rey (1896); Silver (1922); Bossi (*Volpino il calderaio*, 1925); Wolf-Ferrari (1927, the Induction only); Bottagisio (comp. *c.*1927); Persico (1931); Clapp (comp. 1945–8, unprod.); Porter (*Kiss Me, Kate*, 1948); Giannini (1954); Groth (1954); Shebalin (1957); Eastwood (1960); Argento (1963)

*Two Gentlemen of Verona* (1594–5): Seymour (comp. 1935); Frezza (1987)

*Love's Labour's Lost* (1594–5): Folprecht (1926); A. Beecham (comp. 1934); Nabokov (1973); G. Bush (1988)

*Romeo and Juliet* (1594–5): Schwanenberger (?1773); Benda (1776); Marescalchi (1789); Rumling (1790); Dalayrac (1792); Steibelt (1793); Maldonati (1796); Zingarelli (1796); B. Porta (1809); P. G. Guglielmi (1810); Vaccai (1825); Torriani (1828); Damrosch (1862); Storch (1863); Morales (1863); Marchetti (1865); Ivry (1867); Gounod (1867); Mercadal (1873); H. R. Shelley (pub. 1901); D'Ivry (*Les amants de Vérone*, 1878); Marshall-Hall (1912); Campo (1915); Barkworth (1916); Ferroni (early 20th cent.); Zandonai (1922); Sutermeister (1940); Blacher (1947); Malipiero (1950); Fribec (1954); Gaujac (1955); Bernstein (*West Side Story*, on the general theme, 1957); Gerber (comp. *c.*1961); Fischer (1962); Mullins (1965); Didam (before 1967); Liotta (1969); Zanon (1969); Matuszczak (1970); Kelterborn (1991)

*Richard II* (1595–6): None

*A Midsummer Night's Dream* (1595–6): Purcell (*The Fairy Queen*, 1692); Leveridge (*Pyramus and Thisbe*, masque, 1716); Lampe (*Pyramus and Thisbe*, 1745); J. C. Smith (*The Fairies*, 1755); E. W. Wolf (*Die Zauberirrungen*, 1785); Grosheim (1792); Alyabyev (*Volshebnaya noch*, comp. 1842); Manusardi (1842); Suppé (1844); Roti (1899); De Boeck (1902); Huë (1903); Mancinelli (comp. 1917); Vreuls (1925); Arundell (comp. 1930); Delannoy (*Puck*, 1949); Britten (1960); Neely (1967); Doubrava (1969); Convery (1982); Mosca (1982); Gerber (1984); Kurz (1984); Werle (1985); Urrows (1986)

*The Merchant of Venice* (1596): Just (1787); Pinsuti (1873); Deffès (*Jessica*, 1898); Foerster (*Jessika*, 1905); Saussine (1907); Alpaerts (*Shylock*, 1913); Radò (*Shylock*, comp. 1914); Taubmann (*Porzia*, 1916); Carlson (1920); A. Beecham (1920); Laufer (publ. 1929); La Violette (1930); Hahn (1935); Brumagne (1938); Castelnuovo-Tedesco (1961); Neely (1967); Nürnberg (1973); Carlson (comp. before 1976); Convery (1982); A. Tchaikovsky (comp. 1982); Gerber (1984); Magee (comp. before 1984)

*King John* (1596–7): None

*Henry IV*, pts 1 and 2 (1597–8): Pacini (*La gioventù di Enrico V*, 1820); Mercadante (*La gio-*

ventù di Enrico V, 1834); Holst (*At the Boar's Head*, 1925)

*Much Ado about Nothing* (1598–9): Berlioz (*Béatrice et Bénédict*, 1862); A. Doppler (1896); Puget (1899); Podestà (*Ero*, 1900); Stanford (1901); Mojsisovics (*c.*1930); Hahn (1936); Heinrich (1956); Khrennikov (1972); Chen Jing Gewn (1986)

*Henry V* (1598–9): Boughton (*Agincourt*, Dramatic Scene, 1924); Misterly (1964)

*Julius Caesar* (1599–1600): Seyfried (1811); Robles (late 19th cent.); Carl (comp. before 1911); Malipiero (1936); Consorti (1923); Berghorn (1940); Klebe (1959); Schmitz (1978); Rasmussen (1983)

*As You Like It* (1599–1600): Veracini (*Rosalinda*, 1744); F. Wickham (*Rosalind*, 1938); Jirko (1969); Hasquenoph (1982)

*Twelfth Night* (1599–1600): Steinkühler (*Cäsario*, 1848); Anger (comp. 1872); Rintel (1872); Taubert (*Cesario*, 1874); Gall (comp. *c.*1887); Weis (*Viola*, 1892); Arenson (comp. 1893); Kirchner (1904); Hart (*Malvolio*, 1913); Smetana (*Viola*, unfin., prod. 1924); Farina (1929); Kusterer (1932); Holenia (*Viola*, 1934); De Filippi (*Malvolio*, comp. 1937); Shenshin (1939); Hess (1941); Gibbs (comp. 1947); Falk (comp. before 1950); Jírko (1967); Amran (1968); Zeidman (1968); Wilson (1969); Butt (comp. 1974); Shen Li Quin (1986); Klusák (1987)

*Hamlet* (1600–1): Caruso (1789); Andreozzi (1792); Mercadante (1822); Mareček (1840); Thau (*c.*1841); Buzzolla (1848); Zanardini (1854); Stadtfeld (1857, prod. 1882); Moroni (1860); Faccio (1865); Thomas (1868); Hopp (1874); Hignard (1888); Keurvels (1891); Marescotti (1894); Grandi (1898); Heward (comp. 1916, unfin.); J. Kalniņš (1936); Lindsay (1952); Roters (*c.*1957); Kagen (1959); Zafred (1960); Machavariani (*c.*1963); Reif (before 1968); Searle (1968); Szokolay (1969); Horky (*Jed z Elsinoru*, 1969, distantly); Engelmann (1969); Barton-Armstrong (1972); Bentoiu (1972); Raphling (1974); Reutter (1980); Válek (1982); Murakata (1983); Kelterborn (1984); Rihm, (*Die Hamletmaschine*, 1987, distantly)

*The Merry Wives of Windsor* (1600–1): Papavoine (1761); Ritter (1794); Dittersdorf (1796); Salieri (*Falstaff*, 1799); Hoffmann (1800); Balfe (*Falstaff*, 1838); Nicolai (1849); Adam (*Falstaff*, 1856); Verdi (*Falstaff*, 1893); Vaughan Williams (*Sir John in Love*, 1929); Swier (1941); Rea (1982)

*Troilus and Cressida* (1601–2): Zillig (1951)

*All's Well that Ends Well* (1602–3): Lichtenstein (*c.*1800); Lacy (*c.*1832); Hamel (*Malvina*, 1857); F. David (*Le Saphir*, 1865); Audran (*Gillette de Narbonne*, 1882); Castelnuovo-Tedesco (*Giglietta di Narbona*, 1959)

*Measure for Measure* (1604–5): Wagner (*Das Liebesverbot*, 1836)

*Othello* (1604–5): Rossini (1816); Ducis (1821);

Verdi (1887); Scognamiglio (1894); Sztojano-vits (1917); Zavodsky (1945); Machavariani (?1963); A. Müller (*Othellerl, der Mohr von Wien*, parody, 1982); Zhang Baoyun (1983)

*King Lear* (1605–6): Kreuzer (1824); Séméladis (1854); Sauvage (before 1873); Shiwa (before 1873); Gobatti (1881); Raynaud (1888); Cagnoni (1890); Litolff (1890); G. Cottrau (1913); Ghislanzoni (1937); Frazzi (1939); Gerhard (comp. *c*.1955); Pogodin (1955); Durme (comp. 1957); Lackey (1977); Reimann (1978); Beck (before 1979); Božič (1986)

*Macbeth* (1605–6): Asplmayr (1777); Chelard (1827); Verdi (1847); Taubert (1857); L. Rossi (*Biorn*, 1877); E. Bloch (1910); Gatty (1920); Pless (comp. *c*.1925); Snyder (1929); Daffner (1930); Collingwood (1934); Goedicke (1947); Durme (1957); Fries (comp. *c*.1965); Halpern (comp. *c*.1965); Koppel (1970); Cagiarda (1972); Smith (comp. *c*.1976); Sciarrino (1981); Koblenz (1984); Bibalo (1989)

*Antony and Cleopatra* (1606–7): Kaffka (1779); Sacchi (1877); Sayn-Wittgenstein-Berleburg (1883); Yuferov (1900); Ardin (1919); Malipiero (1938); Rasheed (1942); Durme (1961); Barber (1966); Bondeville (1973); Soukup (1976); Glonti (1981); Nürnberg (1985)

*Coriolanus* (1607–8): Baeyens (1941); Šulek (1958); Cikker (1974); Hartwig-Steller (comp. 1984)

*Timon of Athens* (1607–8): ?Leopold I (1696); Nürnberg (comp. 1985); Oliver (1991)

*Pericles* (1608–9): Cottrau (comp. *c*.1915); Hovhaness (1975); Hecker (1981)

*Cymbeline* (1609–10): R. Kreutzer (*Imogène*, 1796); Sobolewski (*Imogene*, 1832); O. van Westerhout (1892); Missa (*Dinah*, 1894); A. Eggen (1951); Arrieu (1974)

*The Winter's Tale* (1610–11): C. E. Barbieri (*Perdita*, 1865); Bruch (*Hermione*, 1872); Nešvera (*Perdita*, 1897); Berény (comp. 1898); Zimmermann (1900); Goldmark (1908); Nürnberg (comp. 1966); Field (1975); Harbison (1979); Zhang Zhiliang (1986)

*The Tempest* (1611–12): ?Purcell (1695); J. C. Smith (1756); Asplmayr (1781); Rolle (1782); Fabrizi (1788); Hoffmeister (1792); Caruso (1798); Winter (1798); W. Müller (1798); Fleischmann (*Die Geisterinsel*, 1798); Reichardt (*Die Geisterinsel*, 1798); Zumsteeg (*Die Geisterinsel*, 1798); Haack (*Die Geisterinsel*, 1798); Ritter (1799); Hensel (1799); Schäffer (comp. 1805); Emmert (1806); Kanne (1808); Riotte (1833); Alyabyev (comp. *c*.1835); Raymond (comp. *c*.1840); Kunz (1847); Halévy (1850); Nápravník (1860); Duvernoy (1880); Kashperov (1867); D. Jenkins (1880); Frank (1887); Urspruch (1888); Fibich (1895); Angyal (*c*.1900); Delfante (1900); De Angelis (1905); Kazanly (1910); Beer-Waldmann (1914); Farwell (*Caliban*, masque, 1916); Hale (publ. 1917); Gatty (1920); Lattuada (1922); Brearley (1929); Canonica (*Miranda*, 1937); Sutermeister (*Die Zauberinsel*, 1942); Atter-

berg (1948); Martin (1956); Šulek (*Oluja*, 1969); Smart (comp. *c*.1974); Henderson (comp. *c*.1975); Berio (*Un re in ascolto*, 1984); Westgaard (?1984); Eaton (1985); Hoiby (1986)

*Henry VIII* (1612–13): None

*Two Noble Kinsmen* (1612–13): None

Also: *The Rape of Lucrece*: Respighi (comp. 1936); Britten (1946). *Venus and Adonis*: Weisgall (*The Gardens of Adonis*, comp. 1981).

Also: Logar (*Four Scenes from Shakespeare*) and Zelinka (*Spring with Shakespeare*, 1955).

Operas in which Shakespeare appears are by Lillo (*La gioventù di Shakespeare*, 1851), Benvenuti (*Guglielmo Shakespeare*, 1861); and Serpette (*Shakespeare*, 1899). In Thomas's *La songe d'une nuit d'été* (1850), Shakespeare, Queen Elizabeth, and Falstaff all appear.

BIBL: G. Schmidgall, *Shakespeare and Opera* (Oxford, 1990); B. Gooch and D. Thatcher, *A Shakespeare Music Catalogue* (Oxford, 1991).

**Shalyapin, Fyodor** (*b* Kazan, 13 Feb. 1873; *d* Paris, 12 Apr. 1938). Russian bass. Of humble peasant origins, he had little formal education, and was self-taught. After difficult years touring with several theatrical troupes, making formal début with Semyonov-Smarsky's Co., Ufa, 1890 (Steward, Moniuszko's *Halka*), he studied briefly Tiflis with his only teacher, Usatov, who taught him free, and gave him invaluable support. Tiflis O 1893, learning over 14 roles in five months. St Petersburg, Panayev's Co. 1894, and M 1895–6. Left, feeling artistically constricted, to work in Mamontov's Private OC, 1896–9. This was a formative period, when he sang 19 roles with the freedom and encouragement to explore and experiment. Moscow, B, 1899–1920, also, unusually, singing for St Petersburg, M. Milan, S, 1901, 1904, 1908, 1912, 1929–30, 1933; Monte Carlo 1905–37; New York, M, 1907–8, 1921–9; London: DL, 1913–14; CG, 1926, 1928–9; L, 1931. Left Russia 1922. Repertory included Leporello, Mozart's Don Basilio, Valentin, Gounod's and Boito's Mephistopheles, Philip, Ivan Susanin, Ivan the Terrible, Miller (Dargomyzhsky's *Rusalka*), Varlaam, Boris, Dosifey; created Salieri (Rimsky-Korsakov) and Massenet's Don Quichotte. His talent was colossal. His natural gifts comprised a beautiful voice with an instinctive control (his mezza voce was much praised), an imposing physique, and plastic features. Combining these with an untiring search for knowledge, acute perception of physical detail and pyschological truth, and painstaking costuming and make-up, he created characters with such depth that audiences were spellbound. His creativity extended not only to his own artistic expression (whether singing, acting, drawing, or writing), but also to any production in which he was involved, inspiring all concerned. The revelatory nature of his interpretations, particularly in his Russian roles, helped to bring Musorgsky to the world's attention. (R)

WRITINGS: F. Shalyapin, *Stranitsy iz moyey zhizny* (Leningrad, 1926; trans. as *Pages from my Life* (London, 1927) ); *Maska i dusha* (Mask and Soul) (Paris, 1932; trans. as *Man and Mask* (London, 1932) ).

BIBL: V. Borovsky, *Chaliapin* (London, 1988).

**Shaporin, Yury** (*b* Glukhov, 8 Nov. 1887; *d* Moscow, 9 Dec. 1966). Russian composer. Studied Kiev with Lyubomirsky, St Petersburg with Sokolov. Founded Grand Drama Theatre with Gorky, Blok, and Lunacharsky, worked as music director, and continued to work in experimental theatre music. A project for an opera with Gorky, *The Mother*, foundered with the writer's death. His only opera, *The Decembrists*, began life as *Pauline Goebbel*, and was eventually completed and performed in 1953. It was found very acceptable to Soviet officialdom for its grandiose treatment of the subject, but is nevertheless a work of some power and individuality, cast in the epic tradition of Glinka's *A Life for the Tsar*.

WORKLIST: *Pauline Goebbel* (2 scenes prod. Leningrad 1925), rev. as *Dekabristy* (The Decembrists) (Rozhdestvensky, after A. Tolstoy; Moscow 1953).

BIBL: I. Martynov, *Yury Shaporin* (Moscow, 1966).

**Sharpless.** The American consul (bar) in Puccini's *Madama Butterfly*.

**Shaw, (George) Bernard** (*b* Dublin, 26 July 1856; *d* Ayot St Lawrence, 2 Nov. 1950). Irish dramatist, novelist, and critic. Wrote regular music criticism for *The Hornet* (1876–7), under the pseudonym of 'Corno di Bassetto' for *The Star* (1880–90), and for *The World* (1890–4). He had an intimate knowledge of the subjects he wrote about, especially Italian opera. He was always highly personal in his approach, often devastatingly witty, and was allowed to make an essay of each occasion. Shaw remains the most intelligent and entertaining journalist music critic in English letters. The fundamental seriousness of his passion for music is openly displayed in *The Perfect Wagnerite* (1898, 4/1922), a closely argued thesis on a Socialist theory of *The Ring*; this is still one of the most important studies of Wagner's music to have appeared in English. Shaw's miscellaneous criticisms, well worth rereading today, were reprinted as *Music in London 1890–4* (3 vols., 1932), *London Music in 1888–9 as heard by Corno di Bassetto* (1937), and *How to Become a Musical Critic* (1960).

Music, especially opera, also permeates Shaw's plays. This extends from many scattered references, such as the prize-fighter Cashel Byron comparing his battles to those of Wagner ('a game sort of composer') and the love of music shown by many of his characters, to the basing of *Man and Superman* on *Don Giovanni*, also to Shaw's insistence that his plays were conceived as operas without music, with duets, trios etc., and with casting for actors with high or low voices to match their roles and provide effective vocal contrasts.

Operettas on his works are by O. Straus (*Der tapfere Soldat*, 1908, after *Arms and the Man*) and Lilien (*Die grosse Katharina*, 1932, after *Great Catherine*). *My Fair Lady* (1956), music by Loewe, is based on *Pygmalion*.

**Shaw, Glen Byam** (*b* London, 13 Dec. 1904; *d* Wargrave, 29 Apr. 1986). English producer. After a career as an actor and theatre producer, turned to opera in 1962 when appointed director of productions at London, SW. Staged many notable productions for SW/ENO, including *Idomeneo*, *Rake's Progress*, the 1968 *Meistersinger*, and, with John Blatchley, *The Ring*. His work is marked by an economy of movement and a concentration on the essentials of character.

**Shepherds of the Delectable Mountains.** See PILGRIM'S PROGRESS, THE.

**Sherasmin.** Huon's squire (bar), lover of Fatima, in Weber's *Oberon*.

**Sheridan, Richard Brinsley** (*b* Dublin, 30 Oct. 1751; *d* London, 7 July 1816). English dramatist. Operas on his works are as follows:

*St Patrick's Day* (1775): S. Hughes (1947)
*The Duenna* (1775): Linley sen. and jun., songs for original production (1775); Bertoni (*La governante*, 1779); Bell (1939); Prokofiev (1946); Gerhard (comp. 1948, prod. 1992)
*School for Scandal* (1777): Klenau (1926)
*The Critic* (1779): Stanford (1916)

**Shield, William** (*b* Swalwell, 5 Mar. 1748; *d* London, 25 Jan. 1829). English composer and librettist. Apprenticed first as a boat-builder, he later gained a reputation as a violinist in N England, before coming to London 1773, where he led the violas in the orchestra of the H. His first dramatic work, the afterpiece *The Flitch of Bacon*, was a considerable success and launched him on a career during which he wrote *c*.45 works, most of them for Covent Garden, where he was house composer for 15 years. Together with Dibdin and Storace, he dominated the London stage during the 1780s and 1790s and represent the apotheosis of 18th-cent. English comic opera. Ranging from full-length operas (*The Crusade*) to afterpieces (*Rosina*), their frequent use of fantastic and exotic plots (e.g. *The Crusade*, *The Magic Cavern*, *The Enchanted Castle*) reflects contemporary literary trends, sometimes even including hints of Gothic horror. Shield's most successful works were *Rosina* and *The Poor Soldier*; he also made adaptations of Grétry's *Richard Cœur de Lion* (1786) and Dalayrac's *Nina* (1787). His works were among the first operas to be given in the USA.

SELECT WORKLIST (all first prod. London): *The Flitch of Bacon* (Bate & Dudley; 1778); *Rosina* (Brooke; 1782); *The Poor Soldier* (O'Keeffe; 1783); *The Magic Cavern* (Pilon & Wewitzer; 1784); *The Enchanted Castle* (Andrews & others; 1786); *The Crusade* (Reynolds; 1790).

**Shirley, George** (*b* Indianapolis, 18 Apr. 1934). US tenor. Studied Detroit with Ebersole and Boatner, Washington with Georgi. Début Woodstock 1959 (Eisenstein). Spoleto 1961; New York, M, from 1961; London, CG, from 1967; Gly. 1967–74. Also New York City O; Buenos Aires, C; Santa Fe; Berlin, D; etc. Roles include Ferrando, Tamino, Alfredo, David, Loge, Pelléas, Apollo (*Daphne*), Alwa (*Lulu*). (R)

**Shirley-Quirk, John** (*b* Liverpool, 28 Aug. 1931). English baritone. Studied London with Henderson. Début Gly. 1962 (Doctor, *Pelléas*). London, EOG 1964–73; London, CG, from 1973; New York, M, 1973; also Holland, St Louis, Berlin, etc. Created leading roles in Britten's church parables, the baritone roles in *Death in Venice*, Coyle (*Owen Wingrave*); also Lev (*The Ice Break*). Other roles include Guglielmo, Mozart's Count Almaviva, Golaud, Duke Bluebeard, Wozzeck. Uses a firm, clean tone with intelligence and musicality. (R)

**Shostakovich, Dmitry** (*b* St Petersburg, 25 Sep. 1906; *d* Moscow, 9 Aug. 1975). Russian composer. Studied Petrograd with Steinberg. Shostakovich is known to have contemplated many operatic projects, including Gorky's *The Mother*, Sholokhov's *The Quiet Don*, Chekhov's *The Black Monk*, Lermontov's *A Hero of Our time* (to a libretto by Meyerhold), and Tolstoy's *Resurrection*. The latter was possibly to be the second part of his projected trilogy of operas about women, of which *The Lady Macbeth of Mtsensk* was the first.

Partly under the inspiration of Meyerhold, to whom he was close in the 1920s, Shostakovich composed *The Nose* in 1927–8. Gogol's satirical play provided him with a vein of barbed grotesquerie that closely matched the sharp, witty, iconoclastic musical manner that he was still able to indulge comically and openly. Influenced by Prokofiev and especially Stravinsky, the work also ingeniously uses a modern form of the kind of musical declamation explored by Shostakovich's admired predecessors Dargomyzhsky and Musorgsky. Some of this manner survives into *The Lady Macbeth of Mtsensk*, notably in the scene in the police station; but it is absorbed into a more lyrical and expansive style that can skilfully contrast the composer's warmly expressed sympathy for the heroine, Katerina Izmaylova, and his scorn for the surrounding characters, venial, shallow, and corrupt, who bring about her sins and downfall. The work's first enormous success was sharply curtailed by a notorious *Pravda* article, 'Confusion instead of Music' (28 Jan. 1936). Inspired by Stalin, this abused the work for its ugliness, tunelessness, noisiness, and especially its treating of sex and violence. It marked not only a tragic turning-point in Shostakovich's career, but showed composers the constricting limits set by the doctrine of Socialist Realism that was imposed in the wake of the permitted arguments of the 1920s.

Shostakovich never completed another opera, though he wrote a single act setting Gogol's *The Gamblers* in a manner renewing his devotion to Musorgsky. He is said to have abandoned this when he realized it was growing too long and would never be staged. He also completed a operetta based on a local joke, *Moscow, Cheremushky*, and made new versions of Musorgsky's *Boris Godunov* (1940) and *Khovanshchina* (1960, also filmed).

Shostakovich's own account of his sufferings at the hands of the Stalinist authorities is contained in *Testimony*, a book controversial for its unreliable editorial methods but accepted by his friends, and his son Maxim, as essentially true.

WORKLIST: *Nos* (The Nose) (Zamyatin, Yonin, Preis, & Shostakovich, after Gogol; Leningrad 1930); *Ledy Macbeth Mtsenskovo uyezda* (The Lady Macbeth of the Mtsensk District) (Preis & Shostakovich, after Leskov; Leningrad, 1934); second version as *Katerina Izmaylova* (comp. 1956–63; prod. Moscow, 1963); *Igroky* (The Gamblers) (Gogol; unfin., comp. 1941–2, perf. Leningrad 1978).

WRITINGS: S. Volkov (related to and ed. by), *Testimony* (New York, 1979). Also copious articles, variously published.

CATALOGUE: D. Hulme, *Dmitri Shostakovich* (Muir of Ord, 1983; Oxford, 2/1991).

BIBL: D. Rabinovich, *Dmitry Shostakovich* (Moscow, 1950; trans. 1959); G. Norris, 'The Operas', in C. Palmer, ed., *Shostakovich* (London, 1982).

**Show Boat.** Musical play in 2 acts by Jerome Kern; text by Oscar Hammerstein II, after Ferber. Prod. New York, Ziegfeld T, 27 Dec. 1927, with Terris, Marsh, Morgan, Bledsoe, Winninger; London, DL, 3 May 1928, with Day, Worster, Burke, Robeson, Hardwicke.

**Shuard, Amy** (*b* London, 19 July 1924; *d* London, 18 Apr. 1975). English soprano. Studied London with Warren, later with Sacher and Turner. Début Johannesburg 1949 (Aida). London, SW 1949–55, and CG 1954–74; Milan, S, 1962; Bayreuth 1968. Also Buenos Aires, C; Vienna; San Francisco, etc. Roles included Cassandra (Berlioz), Eboli, Lady Macbeth, Kundry, Brünnhilde, Elektra, Turandot, Jenůfa, Kátá. A sensitive, telling performer with an expressive, dramatic voice. (R)

**Shuisky.** Prince Vassily Ivanovich Shuisky (ten), a boyar, in Musorgsky's *Boris Godunov*.

**Shvanda the Bagpiper** (Cz.: *Švanda dudák*). Opera in 2 acts by Weinberger; text by Miloš Kareš and Max Brod, after the folk-tale by Tyl. Prod. Prague, N, 27 Apr. 1927, with Novák, Kejřová, Nordenová, Schütz, Koslíková, Munclingr, Pollert, Lebeda, Hruška, cond. Ostrčil; New York, M, 7 Nov. 1931, with Müller, Laubenthal, Schorr, cond. Bodanzky; London, CG, 11 May 1934, with Ursuleac, Kullmann, Schoeffler, cond. Krauss. Weinberger's most successful opera.
Bohemia, legendary times. Babinsky (ten) persuades Shvanda (bar) to try to win Queen Ice Heart (mez) by means of his piping. Finding Shvanda to be already married, she orders his execution, but he is saved by his music and by Babinsky. A rash promise lands him in Hell, whence he is again rescued by Babinsky, who cheats the Devil (bs) at cards.
Other operas on the subject are by Hřímaly (1896), Weis (1905), and Bendl (1906).

**Sibelius, Jean** (*b* Hämeenlinna, 8 Dec. 1865; *d* Järvenpää, 20 Sep. 1957). Finnish composer. His only opera is a youthful 1-act work, *Jungfrun i tornet* (The Maiden in the Tower) (Herzberg; prod. 1896), which owes much to the example of *rescue opera, allied to a nationalist musical style. Also wrote much incidental music.

**Siberia.** Opera in three acts by Giordano; text by Illica. Prod. Milan, S, 19 Dec. 1903, with Storchio, Zenatello, De Luca, A. Pini-Corsi, cond. Campanini; New Orleans, French OH, 13 Jan. 1906 (in French) with Galli-Sylva, Lucas, Mezy, Baer, Régis, Bourgeois, cond. Rey; London, Fulham TH, 8 Dec. 1972, by Hammersmith Municipal Opera, with Doyle, O'Neil, Metcalfe, Corner, cond. Vandernoot.
Russia, early 19th cent. Vassili (ten), in love with Stephana (sop), mistress of Prince Alexis, has wounded him in a duel, and for this is exiled to Siberia; there he is joined by Stephana, and then Gleby (bar), her manager and former lover. They attempt to escape together, but are shot at by guards; Stephana is fatally wounded, and dies in Vassili's arms.

**'Si colmi il calice'.** Lady Macbeth's *brindisi in Act II of Verdi's *Macbeth*.

**Siebel.** A village youth, in love with Marguerite, in Gounod's *Faust*.

**Siège de Corinthe, Le** (The Siege of Corinth). Opera in 3 acts by Rossini; text by Luigi Balocchi and Alexandre Soumet, after the Duca di Ventignano's original libretto for the 2-act *Maometto II*, of which *Le siège de Corinthe* is so radical a revision that it merits classification as a separate work. Prod. Paris, Salle Le Peletier, 9 Oct. 1826, with Cinti-Damoreau, Frémont, L. Nourrit, A. Nourrit, Dérivis, Prévost, Bonel; Parma (as *L'assedio di Corinto*), 26 Jan. 1828; Genoa, CF, 1828 (with an additional cabaletta to the Act II Pamira–Maometto duet by Donizetti, thereafter often included); Vienna July 1831; New York, Italian OH, 6 Feb. 1835. Rossini's first French opera.

**Siegfried.** 1. The third part of Wagner's *Ring* cycle. See RING DES NIBELUNGEN.
2. The young hero (ten), son of Siegmund and Sieglinde and lover of Brünnhilde, in Wagner's *Siegfried* and *Götterdämmerung*, in *Der Ring des Nibelungen*. His name derives from *Sieg* (victory) and *Friede* (peace or protection).
3. The Count Palatine (bar) in Schumann's *Genoveva*.

**Siegfried's Journey to the Rhine.** The name generally given to the orchestral interlude between the Prologue and Act I of Wagner's *Götterdämmerung* describing Siegfried's journey from Brünnhilde's fire-encircled rock to the castle of the Gibichungs.

**Sieglinde.** Hunding's wife, Siegmund's sister and lover (sop), in Wagner's *Die Walküre*. Her name derives from *Sieg* (victory) and *linde* (gentle, also the *Lindenbaum*, or lime tree, whose heart-shaped leaves symbolize love).

**Siegmund.** Sieglinde's brother and lover (ten) in Wagner's *Die Walküre*. His name derives from *Sieg* (victory), *Mund* (guardian).

**Siehr, Gustav** (*b* Arnsberg, 17 Sep. 1837; *d* Munich, 18 May 1896). German bass. Studied Berlin with Krause and Dorn. Début Neustrelitz 1863 (Oroveso). Prague 1865–70; Wiesbaden 1870–81; Bayreuth 1876, 1882–4, 1886, 1889; Munich 1881–96. Repertory included Sarastro, Don Basilio (Mozart), Kaspar, Daland, King Mark. Created Hagen; alternated with Scaria as Gurnemanz in first *Parsifal*. Wagner found his singing and characterization excellent in *The Ring*. He later considered him for Wotan, telling him he had conceived the part for a bass, rather than baritone, voice.

**Siems, Margarethe** (*b* Breslau, 30 Dec. 1879; *d* Dresden, 13 Apr. 1952). German soprano. Studied Dresden with Orgeni, Paris with Viardot. Début Prague 1902 (Marguerite, *Les Huguenots*). Dresden 1908–20; London, CG, 1913, DL, 1914; also Berlin, Milan, St Petersburg. A singer of extraordinary versatility, she created Chrysothemis, Marschallin (pronounced 'magnificent' by Strauss), and Zerbinetta. Also sang Queen of the Night, Bellini and Donizetti coloratura parts, Amelia, Aida, Venus, Elisabeth, Isolde. Retired 1925. (R)

**Siena.** City in Tuscany, Italy. The first opera was Cesti's *Argia*, given in 1670 by the Accademia dei Rinnovati in the T dei Rinnovati (in the Palazzo Pubblico), then renamed T Grande; burnt down 1742, 1751; rebuilt by Galli-Bibiena; restored 1950, renamed T dei Rinnovati. The Accademia Chigiana was founded 1932 by Count Chigi Saracini, and mounts a Settimana Chigiana in September: performances are given in the T dei Rinnovati and the T dei Rozzi. A number of neglected operas, particularly 18th cent., have been revived.

**Siepi, Cesare** (*b* Milan, 10 Feb. 1923). Italian bass. Self-taught. Début Schio 1941 (Sparafucile). Resumed career after the war; Milan, S, from 1946; London, CG, 1950, 1962–73; New York, M, 1950–73; also Salzburg, Verona, S America, etc. With a warm, even tone and handsome stage presence, successful as Figaro (Mozart), Don Giovanni, Philip, Mephistopheles (Gounod and Boito), Boris, Gurnemanz. (R)

**Siface** (orig. Giovanni Francesco Grossi) (*b* Chiesina Uzzanese, 12 Feb. 1653; *d* nr. Ferrara, 29 May 1697). Italian soprano castrato. Studied Rome with Redi. Début Rome 1672. Much early success in Italy, acquiring his nickname from his acclaimed performance of Syphax in Cavalli's *Scipione affricano*, Venice 1678. In the service of the Duke of Modena, 1679–97. Also taken up by Queen Christina, now living in Rome; further success at her theatre, 1679. England 1687, where he was much admired (Purcell writing a harpsichord piece, 'Sefauci's Farewell', lamenting his departure), though he did not sing in opera. Modena, Naples, Parma, Bologna 1688 until his murder by the family of his mistress, Contessa Elena Forni. A singer of great gifts, praised by Mary of Modena as 'the finest musico living', he was an arrogant, capricious, and insolent character whose indiscreet boasting undoubtedly hastened his own end: he was murdered by hirelings of the family of a girl with whom he had had an affair. As an artist, he was much mourned.

**'Sì, fui soldato'.** Chénier's (ten) defence of his actions in the revolutionary tribunal in Act III of Giordano's *Andrea Chénier*.

**Signor Bruschino, Il, ossia Il figlio per azzardo.** Opera in 1 act by Rossini; text by Foppa, after a French comedy by De Chazet and Ourry. Prod. Venice, S Moisè, late Jan. 1813, with Pontiggia,

Nagher, Raffanelli, Del Monte, Berti, De Grecis; New York, M, 9 Dec. 1932, with Fleischer, Tokatyan, Windheim, De Luca, Pinza, cond. Serafin; Orpington, Kent O Group, 14 July 1960, cond. Langford. Occasionally revived in Italy.

An Italian town, 1800. Sofia (sop), ward of Gaudenzio (bs), is being forced to marry Bruschino's (bar) son, whom she has never seen. Her lover, Florville (ten), passes himself off as the son; when Bruschino himself arrives, he helps the plot along for his own reasons.

**'Signore, ascolta'.** Liù's (sop) appeal to Calaf to give up his attempt to woo Turandot, in Act I of Puccini's *Turandot*.

**Sigurd.** Opera in 5 acts by Reyer; text by Du Locle and Blau, after the Nibelung legend. Prod. Brussels, M, 7 Jan. 1884, with Rose Caron, Deschamps-Jehin, Jourdain, M. Devriès, Renaud, cond. Dupont; London, CG, 15 July 1884, with Albani, Fursch-Madi, Jourdain, Devoyod, Soulacroix, E. De Reszke, cond. Bevignani; New Orleans, French OH, 24 Dec. 1891, with Baux, Duvivier, Priolaud, Paulin, Guillemot, Bordeneuve, cond. Warnots. Written by Reyer without knowing Wagner's *Ring*.

The Rhine, legendary times. Hilda (sop), sister of King Gunther (bar), asks Uta (con) for magic aid to win the hero Sigurd (ten). He appears, and having drunk a love potion given to him by Hilda, agrees to obtain the Valkyrie Brunehild (sop) from her fire-encircled rock for Gunther in return for Hilda's hand. Sigurd, in the guise of Gunther, rescues Brunehild and gives her to Gunther, while he marries Hilda; but Brunehild loves Sigurd, sensing him to be her real saviour. Hilda reveals the truth; Sigurd, now free from the effect of the potion, realizes that he loves Brunehild. He is killed by Hagen (bs), Gunther's companion, Brunehild dying beside him. They rise from their funeral pyre to Odin's paradise.

**Si j'étais roi** (If I were King). Opera in 3 acts by Adam; text by Ennery and Brésil. Prod. Paris, L, 4 Sep. 1852; New Orleans, T. d'Orléans, 10 Apr. 1856; Newcastle 20 Feb. 1893.

A kingdom by the sea, 16th cent. The fisherman, Zephoris (ten), longs to marry the Princess Nemea (sop), whom he rescued from drowning. As a joke, the King (bar) transports him to his castle to lead a courtly life, but evicts him when he becomes betrothed to Nemea. Back in his hut, Zephoris believes the whole episode a dream, but Kadoor (bs), a scheming Prince who intends to murder the King and marry Nemea, arrives to challenge his rival. The King appears and banishes Kadoor, giving his blessing to Zephoris and Nemea.

**Silja, Anja** (*b* Berlin, 17 Apr. 1935). German soprano. Studied with her grandfather Egon van Rijn. Gave public recital aged 10. Début Brunswick 1956 (Rosina). Stuttgart 1958, and from 1965; Frankfurt 1959 (singing Constanze, Turandot); Bayreuth 1960–6; London, CG, from 1967; also Vienna, Cologne, Chicago, etc. Closely associated with Wieland Wagner, she rapidly made a reputation as a stimulating singing actress. Repertory includes Leonore, Lady Macbeth, Senta, Venus, Isolde, Salome, Elektra, Marie, Lulu, Kostelnička. (R)

**Sills, Beverly** (*b* Brooklyn, 25 May 1929). US soprano. Studied New York with Liebling. Début Philadelphia 1947 (Micaëla). New York City O 1955–79; Vienna 1967; Milan, S, 1969; London, CG, 1970; New York, M, 1975. Director New York City O 1979–89. Roles included Cleopatra (Handel), Lucia, Anna Bolena, Queen of Shemakha, Manon. A highly popular singer, effective on stage. (R)

BIBL: B. Sills, *Beverly* (New York, 1987).

**Silva.** Don Ruy Gomez de Silva (bs), a Spanish grandee, Elvira's uncle and guardian, in Verdi's *Ernani*.

**Silvani, Francesco** (*b* Venice, *c.*1660; *d* 1st half 18th cent.). Italian librettist. Worked for the Gonzaga family in Mantua and appears to have written around 50 librettos, some in collaboration with Lalli. Many were set by Gasparini and Lotti: among his most important works were *Tigrane* (1729, Hasse), *Semiramide* (1742, Jommelli), and *Statira* (Porpora, 1742): many of his librettos received multiple settings during the 18th cent. and some were translated and adapted for the German stage.

**Silvio.** A villager (bar), lover of Nedda, in Leoncavallo's *Pagliacci*.

**simile aria.** See METAPHOR ARIA.

**Simionato, Giulietta** (*b* Forlì, 17 May 1910). Italian mezzo-soprano. Studied Rovigo with Lucatello, Milan with Palumbo. Début Florence 1938 (Pizzetti's *Orséolo*). Milan, S, 1939–66; London, CG, 1953, 1964; Chicago 1954; New York, M, 1959. Also Gly., Paris, Salzburg, etc. A versatile singer with a vibrant, rich-toned voice, excellent technique, and vivid stage personality. Highly acclaimed as Gluck's Orfeo, Cherubino, Servilia, Jane Seymour, Azucena, Eboli, Quickly; also the soprano role of Valentine. (R)

**Simon Boccanegra.** Opera in a prologue and 3 acts by Verdi; text by Piave, based on the drama by António García Gutiérrez. Prod. Venice, F, 12 Mar. 1857, with Bendazzi, Negrini, L. Giraldoni, Echeverria—libretto rev. Boito, prod. Milan, S, 24 Mar. 1881, with D'Angeri, Tamagno, Maurel, E. De Reszke, cond. Faccio; New York, M, 28 Jan. 1932, with Rethberg, Martinelli, Tibbett, Pinza, cond. Serafin; London, SW, 27 Oct. 1948, with Gartside, Johnston, Matters, Glynne, cond. Mudie.

Genoa, 14th cent. Simon Boccanegra (bar), a plebeian candidate, is elected Doge. He hopes this will enable him to marry Maria Fiesco, who already has a daughter by him, though the child is lost. Maria dies and her father Jacopo (bs) curses Boccanegra.

Twenty-five years later Simon's daughter Amelia (sop) has been brought up by his political enemies. In conversation with her, Simon discovers her identity. Paolo (bar), one of Simon's courtiers, loves Amelia, and when the Doge tells him that he must give up all hope of her, he arranges to have her abducted.

Gabriele (ten), who also loves Amelia, murders her abductor and a riot breaks out, interrupting a meeting of the Doge's Council. Gabriele believes Simon was behind the murder and tries to kill him. Amelia intervenes to save her father.

Paolo gives Simon a slow poison and tries to blackmail Gabriele and Fiesco (now living unrecognized under the name of Andrea) into killing the Doge. Gabriele agrees, on being told that Amelia is the Doge's mistress. Simon's life is once again saved by his daughter's intervention. When he discovers that Amelia is Simon's daughter, Gabriele begs forgiveness. He agrees to fight on behalf of the Doge, who agrees that he can marry Amelia.

Paolo joins the rebel army. The plebeians win, and on his way to his execution, Paolo admits to Andrea that he abducted Amelia and tried to kill Simon. Simon meets and recognizes Fiesco (Andrea), joyfully announcing that he can now restore his granddaughter to him. When Amelia enters, he reveals this news to her. As he dies from the effect of Paolo's poison, Simon names Gabriele as his successor.

BIBL: H. Busch, *Verdi's 'Otello' and 'Simon Boccanegra' in Letters and Documents* (Oxford, 1988).

**Simoneau, Léopold** (*b* Quebec, 3 May 1918). Canadian tenor. Studied Montreal with Issaurel, later New York with Althouse. Début Montreal 1941 (Hadji, *Lakmé*). Paris, OC, 1949–54; Gly. from 1951; Paris, O, 1948–51; Chicago 1954, 1959; New York, M, 1963; also Vienna, Buenos Aires, Salzburg. Director O du Québec 1971. Outstanding, with his poignant, sweet tone and lyrical singing, as Ferrando, Don Ottavio, Tamino, Nadir, Wilhelm Meister (*Mignon*). (R)

**Simpleton.** The name usually given in English to the *yurodivy* or holy fool (ten) in Musorgsky's *Boris Godunov*.

**sinfonia** (It.: 'symphony', from Gr. *sumphōnia*: 'an agreeing in sound'). In the earliest operas the term was used to denote an instrumental movement. Its later usage was as an alternative description for the *Italian overture.

**Singspiel** (Ger.: 'song-play'). A term for a German drama in which musical numbers are separated by dialogue. Though such works were written in the 17th cent., e.g. Staden's *Seelewig* (1644), the term (in various spellings and forms) was first regularly applied to dramas with music in the 18th cent. An early example was Keiser's *Croesus* (1710). The influence of French opéra-comique and English ballad opera helped to develop Singspiel as a genre of comic opera with spoken dialogue, especially with the translation *c*.1750 of Coffey's ballad opera *The Devil to Pay* (1731) and *The Merry Cobbler* (1735) by C. F. Weisse as *Der Teufel ist los* and *Der lustige Schuster*, with new music by J. C. Standfuss. Both texts were then set by J. A. Hiller, one of the most important Singspiel composers; he followed these with many more (e.g. *Die Jagd*, 1770), making the genre particularly distinctive in Leipzig. From here it spread to Berlin, where one of the most prominent composers was Benda (*Der Jahrmarkt*, 1775). The tradition thus established bore a strong relationship to opéra-comique in its emphasis upon sentimental, lyrical comedy and, with its frequent use of folk colour, was an important influence on the eventual development of German national opera in the early 19th cent. Other important composers of this tradition included André (*Das tartarische Gesetz*, 1789), Neefe (*Die Apotheke*, 1771), and Reichardt.

In Vienna, following the establishment of the National-Singspiel under Joseph II in 1778, the genre developed along markedly different lines. From Italian opera buffa there came a more lively, sometimes bravura melodic style, while the librettos were characterized by their sharp wit and not infrequent satire and farce. Magic and the supernatural were an important element in Viennese Singspiel, although they were often treated irreverentially. Established by Umlauff (*Die Bergknappen*, 1778), the Viennese Singspiel found its most consummate exponent in Dittersdorf (*Doktor und Apotheker*, 1786). Later composers included Schenk (*Der Dorfbarbier*, 1796), Wenzel Müller (*Die Schwestern von Prag*, 1794), and Schubert (*Die Zwillingsbrüder*, 1820). The greatest Singspiel composer was Mozart, who composed two mature samples, *Die Entführung aus dem Serail* (1782) follows more closely the Viennese tradition, and is characterized by its

Turkish flavour; *Die Zauberflöte* (1791), while having many points of contact with contemporary magic operas, is a musically more complex work, in which the traditions of Singspiel, opera buffa, and opera seria are all combined.

In the early 19th cent. aspects of Singspiel technique found their way into other genres, principally *rescue opera—the classic example is Beethoven's *Fidelio*—and German Romantic opera, notably Weber's *Der Freischütz*. A Danish branch of Singspiel, established in the late 18th cent. by Schulz and Kunzen, had an equally important influence on the development of national opera there in the 19th cent.

**'Si, pel ciel'.** The duet in Act II of Verdi's *Otello*, in which Otello (ten) is joined by Iago (bar) in swearing vengeance on Desdemona for her alleged infidelity.

**'Si può?'** The prologue to Leoncavallo's *Pagliacci*, in which the singer, generally Tonio (bar), asks the audience to listen to his exposition of the situation before the curtain goes up.

**Sir John in Love.** Opera in 4 acts by Vaughan Williams; text selected by composer from Shakespeare's *The Merry Wives of Windsor*, *Love's Labour's Lost*, and *Much Ado about Nothing*; also from Thomas Middleton, Ben Jonson, Thomas Campion, *Gammer Gurton's Needle*, Christopher Marlowe, John Fletcher, Psalm 137, George Peel, Nicholas Udall, Philip Sidney, Richard Edwards, Philip Rosseter, and the song 'Greensleeves'. Prod. London, RCM, 21 Mar. 1929, with Walmsley, Kennedy, Warde, Herbert, Leyland White, Hemming, Holmes, Hancock, Moore, Evers, Mansfield, Rickard, Bamfield Cooper (chorus of Fairies and Imps incl. Imogen Holst), cond. Sargent; New York, Columbia U, 20 Jan. 1949, with Kovey, Dettens, Symes, Witwer, Wheeler, Hester, Lalli.

The plot follows the story of *The Merry Wives of Windsor* and their discomfiture of Falstaff (bar). Several of the minor characters are given larger parts than in Verdi's or Nicolai's operas, as part of the composer's attempt to fill out the vivid English background to the comedy.

**Siroe, re di Persia.** Opera in 3 acts by Handel; text an adaptation by Nicola Haym of Metastasio's libretto for L. Vinci (1726). Prod. London, H, 17 Feb. 1728, with Cuzzoni, Faustina, Senesino, Boschi; rev. Halle 1962. The first of Handel's operas on a Metastasio-derived libretto: it was overshadowed by the success of *The Beggar's Opera*.

**Situationsposse.** See POSSE.

**Sitzprobe** (Ger.: 'seated rehearsal'). The term for the first complete rehearsal of an opera, when soloists and chorus join with the orchestra, generally in the auditorium, with the singers sitting either in the stalls or on chairs on the stage. The term has been adopted in English opera-houses. Known in Italian houses as *prova all'italiana*.

**Škroup, František** (*b* Osice, 3 June 1801; *d* Rotterdam, 7 Feb. 1862). Bohemian composer and conductor. Studied with his father Dominik, then after law studies in Prague was drawn into the movement for Czech national opera, first as an amateur and then as a professional répétiteur, conductor, and composer. With Josef Chmelenský as librettist, and basing his ideas on the established tradition of Czech Singspiel, he wrote the first Czech opera, *Dráteník* (The Tinker), and took the title-role at the very successful première (Prague 1826). Assistant Kapellmeister, Stavovské T, 1827, Kapellmeister, 1837. Here he introduced many new works, latterly including some by Verdi and Wagner. His own subsequent operas were unsuccessful, largely because he proved unwilling or unable to advance from the simple, charmingly naïve Singspiel manner of *Dráteník* to answer the demands of new expressive ideals. Conductor, Rotterdam O, 1861.

BIBL: J. Plavec, *František Škroup* (Prague, 1941).

**Slavina, Maria** (*b* St Petersburg, 17 June 1858; *d* Paris, 1951). Russian mezzo-soprano. Studied St Petersburg with Iretskaya, Everardi, Pryanishnikov. Début St Petersburg, M, 1877 (Amneris). St Petersburg, M, until 1917. Her repertory of over 50 roles included Ratmir (*Ruslan*), Lyubov (*Mazeppa*), Ortrud, Fricka, Fidès, Quickly, Dalila, and a famous Carmen; created Countess (*Queen of Spades*), Lyubasha (*Tsar's Bride*), and, alternating with Kaminskaya, Hanna (*May Night*). Possessd a deep, velvety voice and a forceful dramatic personality; at her best in heroic parts. Taught St Petersburg from 1917, Paris from the 1920s.

**Sleepwalking Scene.** The name usually given to the scene in Act IV of Verdi's *Macbeth*, beginning 'Vegliammo invan due notti', in which a doctor (bs) and a gentlewoman (mez) observe Lady Macbeth (sop) walking and talking in her sleep. See also SLUMBER SCENE.

**Slezak, Leo** (*b* Schönberg (Šumperk), 18 Aug. 1873; *d* Egern am Tegernsee, 1 June 1946). Austrian tenor. Studied Brno with Robinson, later Paris with J. De Reszke. Début Brno 1896 (Lohengrin). Breslau 1899; London, CG, 1900, 1909; Vienna 1901–12, 1917–27, 1933; New York, M, 1909–13; also Paris, Munich, Berlin. An immensely popular artist of generous stature; his brilliant tone and deeply felt singing were much admired in his Guillaume Tell, Manrico, Radamès, Otello (Verdi), Walther, Tann-

häuser, and Hermann, among a large repertory. His sense of humour was irrepressible: he once so convulsed the chorus at the Metropolitan during *Aida* that they were fined by the management (Slezak paid the fine). Later an effective character tenor; appeared in films after leaving the stage. (R) His son **Walter** (1902–83) and daughter **Margaret** (1901–53) were both singers.

BIBL: W. Slezak, *What Time's the Next Swan?* (New York, 1962).

**Slivinsky, Vladimir** (*b* Moscow, 4 June 1894; *d* Moscow, 7 Aug. 1949). Russian baritone. Studied Moscow with Raysky. Moscow, Z, 1922–4; Leningrad 1924–35; Moscow 1930–48. With his attractive stage presence, a successful Onegin, Mazeppa, Eletsky, Demon, Germont; created Sandy (Yurasovsky's *Trilby*, 1924). (R)

**Slobodskaya, Oda** (*b* Vilna (Vilnius), 28 Nov. 1888; *d* London, 29 July 1970). Russian soprano. Studied St Petersburg with Iretskaya. Début St Petersburg, M, 1919 (Lisa). St Petersburg, M, until 1922; Paris 1922; London, L, 1931, CG, 1932, 1935; Milan, S, 1933; Buenos Aires, C, 1936. Settled in England, where she continued to sing into the 1960s. Repertory included Marguerite, Elisabeth de Valois, Sieglinde, and many Russian roles, e.g. Tatyana, Natasha (Dargomizhsky's *Rusalka*), Fevronia (*Invisible City of Kitezh*); created Parasha (*Mavra*). A moving interpreter, in recital as well as on stage, with a well-placed and beautiful voice. (R)

**Slovenia.** The region formed part of Austro-Hungary until its incorporation into the state of Yugoslavia (1918–1992). Dramatic performances with music were a feature of the Ljubljana Jesuit theatre in the 17th cent. There were seasons of German and Italian opera after 1770. The first Slovenian opera composer was Jakob Zupan (1734–1810), with *Belin* (comp. 1780 or 1782, in Slovene but on a Metastasian model). In the 1770s and 1780s Zollner's and Schikaneder's companies visited Ljubljana and gave encouragement to the first Singspiel in Slovenian, Janez Novak's (1756–1833) *Figaro* (1790, after Beaumarchais), Franc Pollini (1762–1846) studied with Mozart and Zingarelli, and wrote *La casetta nei boschi* (The Hut in the Woods, 1798); Bellini dedicated *La sonnambula* to him. Slovenian Romantic opera was slow to develop because of the domination of the German *Ständetheater* (cond. Mahler 1881–2). The first examples came late in the century with the works of Benjamin Ipavec (1829–1908) (e.g. *Teharski plemiči*, 1862), Anton Foerster (1837–1926) (e.g. *Gorenjski slavč*, 1872), and Viktor Parma (1858–1924) (e.g. *Urh, grof celjski*, 1895). Risto Savin (pseud. of Friderik Širca) (1859–1948) moved away from a simple Romantic idiom and had success with

*Lepa Vida* (Fair Vida, 1907), much influenced by Wagner and Strauss.

In the 1920s Slovenian composers drew closer than those in the rest of the new state of Yugoslavia to the mainstream of Central European music. *Črne maske* (Black Masks, 1927), by Marij Kogoj (1895–1956) is a powerful, if uneven, Expressionist opera, freely serial in technique and influenced by Berg. Slavko Osterc (1895–1941) turned from his early historical *Krst pri Savici* (Baptism by the Savica, 1921) to Expressionist and neoclassical operas and opera-minutes influenced by Schoenberg and Hindemith, e.g. *Krog s kredo* (The Chalk Circle, 1929).

A fruitful period came after the war with operas by Danilo Švara (*b* 1902) (e.g. *Slovo od mladosti*, Farewell to Youth, 1954) and Marijan Kozina (1907–66). Important among more recent Slovenian operas is *Cortesova vrnitev* (Cortes's return, 1974), by Pavel Šivic (*b* 1908). Darjan Božič (*b* 1933) has combined singing, speech, and novel vocal sonorities in mixed genre works including *Lysistrata '75* (1980) and *King Lear* (1986).

There are permanent companies in Ljubljana and Maribor.

See also LJUBLJANA.

BIBL: D. Cvetko, *Musikgeschichte der Südslaven* (Kassel, 1975).

**slumber scene** (It., *sonno*). A stock scene in 17th-cent. Italian opera, in which a character is, or falls, asleep on stage. The character may fall asleep to a lullaby (e.g. Poppea, to Arnalta's lullaby, in Monteverdi's *Poppea*) or simply through exhaustion (e.g. Giustino, over his plough, in Legrenzi's *Giustino*, Gomatz in Mozart's *Zaide*). He or she may then risk murder (as in *Poppea*), or talk in his sleep revealing secret feelings, either of hatred for a tyrant or love for the astonished beloved, who has meanwhile arrived, or wake to overhear some plot, or be warned by a friendly shade of a course to pursue (a device that survives into the *ombra scene).

**Sly**. Opera in 2 acts by Wolf-Ferrari; text by Forzano, developed from an idea in the Induction of Shakespeare's comedy *The Taming of the Shrew* (1593–4). Prod. Milan, S, 29 Dec. 1927, with Llopart, Pertile, Rossi-Morelli, Badini, cond. Panizza; London, BBC broadcast, 11 Dec. 1955, with Sladen, Vandenburg, Hemsley, R. Jones, cond. Kempe.

London, 1603. Sly (ten), antagonizes the Duke of Westmoreland (bar) at an inn. He is taken to the Duke's castle, where he is told that he is a nobleman recovering from an illness; Dolly (sop), the Duke's mistress, is presented to him as his wife. When they fall in love, the Duke throws Sly into a dungeon, where he dies.

**Smallens, Alexander** (*b* St Petersburg, 1 Jan. 1889; *d* Tucson, AZ, 24 Nov. 1972). Russian, later US, conductor. Studied New York and Paris. Assistant cond. Boston 1911–14. Cond. Chicago 1919–23. Music director Philadelphia, Civic O, 1927–34. Cond. première of *Porgy and Bess*, Boston 1935, also on European tour 1952–3. Cond. US premières of R. Strauss's *Feuersnot* (1927) and *Ariadne* (1928). (R)

**'Smanie implacabili'**. Dorabella's (sop) aria in Act I of Mozart's *Così fan tutte*, in which she swears to give the Furies that torment her with longings an example of doomed love.

**Smareglia, Antonio** (*b* Pola, 5 May 1854; *d* Grado, 15 Apr. 1929). Italian composer. Studied mathematics in Graz, turning to music after hearings of Mozart, Beethoven, and especially Wagner: he came to blows with an opponent at the Milan première of *Lohengrin* 1873. Studied Milan with Faccio. His first opera, *Preziosa*, had a success which (despite failure with his third) led to sumptuous premières of *Il vassallo di Szigeth* in Vienna under Richter and *Cornil Schut* in Dresden under Schuch. Their suggestions of German Romantic opera and early Wagner in their reliance on the orchestra may have helped their popularity in other German cities, but militated against it in Italy. However, his finest work, *Nozze istriane*, also successfully given in a number of European cities, is closer to verismo, and remains popular in his native Istria. Poverty and encroaching blindness impeded his career, which was rescued by Toscanini, who had conducted the première of his *Oceàna*, and who arranged for the production of *Abisso* and commissioned Smareglia to complete Boito's *Nerone* (which he was unable to do).

SELECT WORKLIST: *Preziosa* (?, after Longfellow; Milan 1879); *Il vassallo di Szigeth* (Illica & Pozza; Vienna 1889); *Cornil Schut* (Illica; Dresden 1893); *Nozze istriane* (Istrian Nights) (Illica; Trieste 1895); *Oceàna* (Benco; Milan 1903); *Abisso* (Benco; Milan 1914).

**Smart, (Sir) George** (*b* London, 10 May 1776; *d* London, 23 Feb. 1867). English conductor and singing teacher. Although primarily a concert conductor he was closely associated with Covent Garden, London in the 1820s, and accompanied Kemble to Germany to engage Weber as the theatre's music director and to commission *Oberon*; Weber died in his house. Much sought after as a singing-teacher; his pupils included Sontag and Lind.

BIBL: H. and C. Cox, *Leaves from the Journals of Sir George Smart* (London, 1907).

**Smetana, Bedřich** (*b* Litomyšl, 2 Mar. 1824; *d* Prague, 12 May 1884). Czech composer. Studied privately, Prague with Proksch.

His first opera, *The Brandenburgers in Bohemia*, was written for Count Harrach's national opera competition, which it won. It already declares the interests and abilities that were to make him the first great nationalist composer of his country. Among these were not only a quick ear for the cadences of spoken Czech (of which he was not initially complete master), but a command of orchestral music which was at the service of a strong dramatic instinct. He was thus from the first able to write continuously composed opera, within which room could be found for arias and ensembles. However, the first version of *The Bartered Bride* was with spoken dialogue, partly, he said, out of a reaction to accusations of Wagnerism levelled at him over *The Brandenburgers*. Later versions moved towards a continuous texture, without losing hold of the essentially Czech lyricism that has kept it one of the most popular operas in the repertory. Smetana draws on folk dances and characteristic rhythms from Czech folk-song, as well as plenty of local colour, but the elegance and skill with which it is handled come from his own imaginative qualities. In 1866 he became conductor of the Provisional T, Prague.

Wagnerism was a charge Smetana had to face over other operas, including his next, *Dalibor*. Also often compared to *Fidelio* because of its theme of imprisonment and freedom, and similarly often staged at moments of national liberation, *Dalibor* is in fact a very different work. It uses motive, but in a non-Wagnerian way, and it continues to draw on Czech musical characteristics while forming a continuous symphonic texture. Its nationalism is more overt in *Libuše*, a festival work which celebrates the founding of the Czech nation in glowing terms; and Smetana ingeniously reconciles the personal with the heroic in his contrast of chromatic and diatonic harmony and in his discreet but expressive use of motive. Its production was delayed for the opening of the National Theatre in 1881.

The remaining operas are lighter. *The Two Widows* is a delightful sentimental comedy, and especially in its revision brings a touch of French opéra-comique to a Czech milieu. The gains of the previous works include mastery of Czech prosody now expressed in light, rapid conversational exchanges, which also mark the warm characterization of *The Kiss*. Smetana's sense of dramatic contrast is richer here, and is taken further forward in the more dramatic declamatory style and extended orchestral commentary of *The Secret*. However, the deafness (caused by encroaching syphilis) which had descended on him by the time of *The Kiss* made work on *The Devil's Wall* difficult. Despite this, and the problematic nature of the plot, the work includes some impressive and ingenious music. He did not manage to compose more than 365 bars of *Viola*.

WORKLIST (all prod. Prague): *Braniboři v Čechách* (The Brandenburgers in Bohemia) (Sabina; 1866); *Prodaná nevěsta* (The Bartered [lit. Sold] Bride) (Sabina; 1st version comp. 1863–6, prod. 1866; 2nd version comp. and prod. 1869; 3rd version comp. and prod. 1869; final version comp. 1869–70, prod. 1870); *Dalibor* (Wenzig, trans. Špindler; 1868); *Libuše* (Wenzig, trans. Špindler; comp. 1869–72, prod. 1881); *Dvě vdovy* (The Two Widows) (Züngel, after Mallefille; 1st version 1874; 2nd version 1878); *Hubička* (The Kiss) (Krásnohorská, after Světlá; 1876); *Tajemství* (The Secret) (Krásnohorská; 1878); *Čertova stěna* (The Devil's Wall) (Krásnohorská; 1882); *Viola* (Krásnohorská, after Shakespeare; comp. 1874, 1883–4, incomplete, unprod.).

BIBL: B. Large, *Smetana* (London, 1970); J. Clapham, *Smetana* (London, 1972).

**Smirnov, Dmitry** (*b* Moscow, 19 Nov. 1882; *d* Riga, 27 Apr. 1944). Russian tenor. Studied Moscow with Krzhizhanovsky and Dodonov, later Pavlovskaya. Début Moscow 1903 (Gigi, Esposito's *Camorra*). Moscow, B, 1904–10; St Petersburg, M, 1910–17; New York, M, 1910–12; London, DL, 1914; sang throughout Europe from 1919, including Paris with Dyagilev's co. A highly successful singer with a fine technique, and a warm, even, lyrical tone. Large repertory included Lensky, Hermann, Werther, Nadir, Lohengrin, Canio. (R)

**Smirnov, Dmitry** (*b* Minsk, 2 Nov. 1948). Belorussian composer. Studied Moscow with Sidelnikov and Denisov, privately with Gershkovich. His fascination with Blake's poetry and painting has found expression in two operas, *Tiriel* and its chamber companion *The Lamentations of Thel*. One of the most significant composers to have emerged in Russia since *glasnost*, he has drawn from European influences, especially French and German, in the formation of his own neo-Romantic style.

WORKLIST: *Tiriel* (Smirnov, after Blake; Freiburg 1989); *The Lamentations of Thel* (Smirnov, after Blake; London 1989).

**Smith, John Christopher** (*b* Ansbach, 1712; *d* Bath, 3 Oct. 1795). English composer of German origin. His father (also, Anglicized, John Christopher Smith) was brought from Germany by Handel to act as his treasurer and copyist; the younger Smith took some lessons with Handel, also with Pepusch and Roseingrave. With Arne he attempted to establish an English operatic tradition along Italian lines, but his first work, *Ulysses*, was so unsuccessful that he was discouraged in this venture. During the 1740s he made some settings of Metastasian librettos, but these were apparently not performed. In 1754 he was commissioned to write two full-length Shakespearian operas by Garrick, *The Fairies* and *The Tempest*: these were not well received, though his last dramatic work, the afterpiece *The Enchanter*, enjoyed some success.

SELECT WORKLIST (all first prod. London): *Ulysses* (Humphreys; 1733); *The Fairies* (?Smith, after Shakespeare; 1755); *The Tempest* (?Smith, after Shakespeare; 1756); *The Enchanter* (Garrick; 1760).

**Smyth, (Dame) Ethel** (*b* London, 22 Apr. 1858; *d* Woking, 9 May 1944). English composer. Studied Leipzig with Reinecke and Jadassohn, then with Herzogenberg. Difficulties in making her way as a woman composer, especially of opera, led to *Fantasio* being first performed in Weimar; it is a light piece which she later repudiated. Her second opera, *Der Wald*, had a wide success, and aroused admiration for its professionalism and its strong sense of atmosphere. However, her most important opera is *The Wreckers*, which shows remarkable stagecraft and an ability to write strong roles in effective dramatic situations. The idiom is basically Germanic but makes use of English ballads and has distinct individuality. There followed her most striking comedy, *The Boatswain's Mate*, which had considerable success in England. A series of autobiographical books, beginning with *Impressions that Remained* (London, 1919), give a vivid picture of her times no less than of the striking personality observing them: edited as *The Memoirs of Ethel Smyth* by R. Crichton (London, 1987).

WORKLIST: *Fantasio* (Smyth, after Musset; Weimar 1898); *Der Wald* (The Forest) (Smyth & Brewster; Berlin 1902); *The Wreckers* (Smyth & Leforestier (Brewster); Leipzig 1906); *The Boatswain's Mate* (Smyth, after Jacobs; London 1916); *Fête galante* (Shanks, after Baring; Birmingham 1923); *Entente cordiale* (Smyth; London 1925).

BIBL: C. St John, *Ethel Smyth* (London, 1959).

**Snegurochka.** See SNOW MAIDEN.

**Snow Maiden, The** (Russ.: *Snegurochka*). Opera in prologue and 4 acts by Rimsky-Korsakov; text by composer, after Ostrovsky's drama (1873) on a folk-tale. Prod. St Petersburg, M, 10 Feb. 1882, with Velinskaya, Kamenskaya, Stravinsky, Bichurina, Makarova, Pryanishnikov, Schröder, Vasilyev, Koryakin, Sobolov, cond. Nápravník; New York, M, 23 Jan. 1922, with Bori, D'Arle, Delaunois, Harrold, Laurenti, Rothier, cond. Bodanzky; London, SW, 12 Apr. 1933, with Dyer, Cross, Coates, Davies, Austin, Lelsey, cond. Collingwood.

The Snow Maiden (sop), who is safe from the sun only so long as she renounces love, begins the life of a mortal with Bobyl (ten) and Bobylikha (mez). She is attracted to the singer Lel (con), who resists her; but Mizgir (bar), come to marry Kupava (sop), falls in love with her, in vain. Tsar

Berendey (ten), asked to judge the Snow Maiden's actions, is impressed by her beauty, and promises a reward to anyone who can win her love. At a feast, Lel prefers Kupava to her, and she flees from Mizgir. She appeals to her mother, Spring (mez), and then greets Mizgir lovingly. But the warmth of love is fatal to her: she dies, and Mizgir throws himself into the lake.

**'Soave sia il vento'.** The trio in Act I of Mozart's *Così fan tutte*, in which Fiordiligi and Dorabella (sops) and Don Alfonso (bs) wish for a calm sea as Ferrando and Guglielmo pretend to set off for the wars.

**Sobinov, Leonid** (*b* Yaroslavl, 7 June 1872; *d* Riga, 14 Oct. 1934). Russian tenor. Studied Moscow with Dodonov and Santagano-Gorchakova. Début Moscow 1893, with a visiting Italian troupe (small roles). Moscow, B, 1897–1924; St Petersburg, M, from 1901; Milan, S, 1903, 1905, 1911; Berlin 1905. First elected director, Moscow, B, 1917. Retired from stage, 1924; later artistic consultant for *Stanislavsky's Opera T. One of the idols of his time, with a strong romantic appeal in voice and appearance, and intelligent sensitivity, he was famous for his Ernesto, Sinodal (*Demon*), Berendey, Gounod's Faust, Werther, Nadir, and especially Lensky. (R)

**Söderström, Elisabeth** (*b* Stockholm, 7 May 1927). Swedish soprano. Studied Stockholm with Skilondz. Début Stockholm, Drottningholm T, 1947 (Bastienne). Stockholm, Royal O, from 1949; Gly. from 1957; New York, M, from 1959; London, CG, from 1960, and the major European opera-houses. One of the leading singing actresses of the day, whose intense and penetrating interpretations have covered a wide range of roles, e.g. Susanna, Countess (Mozart and Strauss), Tatyana, Octavian, Jenůfa, Elena Makropoulos, Mélisande, Marie (*Wozzeck*). Autobiography, *In My Own Key* (1979). (R)

**Sofia.** Capital of Bulgaria. After the country's liberation from the Turks in 1878, efforts were made to establish opera. The moving spirits in the foundation of the Dramatichesko-Operna Trupa in 1890 were the baritone Dragomir Kazakov (1866–1948), the tenor Ivan Slavkov (?–?), and the conductor Angel Bukoreshchlyev (1870–1950). The first performance was given in the cultural society, Slavyanska Beseda, in 1891. Bulgarian singers returning from foreign training formed the Bulgarska Operna Druzhba (Bulgarian Operatic Society) in 1907: the leaders were Kazakov, the bass Ivan Vulpe (1876–1929) and his wife the soprano Begdana Gyuseleva-Vulpe (1878–1932), and especially the tenor Konstantin Mikhaylov-Stoyan (1853–1914), who had been a distinguished Bolshoy soloist and had sung with Shalyapin before settling in Bulgaria. Performances were given in the Naroden T (National T) in 1908. By 1910 a talented group of well-trained singers had gathered in Sofia, among them the tenor Panayot Dimitrov (1882–1941), the basses Ivan Vulpe and Georgy Donchev (1884–1936) (principal teacher of the next generation), the sopranos Christina Morfova (1887–1936) and Lyudmila Prokopova (1880–1959), the mezzo-soprano Ana Todorova (1892–?), the tenors Petr Raychev (1887–1960), and Stefan Makedonsky (1885–1952). Also in 1910 they gave the first important season of Bulgarian opera.

The Opera T survived with difficulty during the war; it received a subsidy for the first time in 1922, and was made the Darzhavna Narodna O (National State O). From 1929 it performed in the Ivan Vasov T. A strong company was built up: its repertory consisted chiefly of Russian and Italian opera, with an increasing number of native works. A new theatre was built after the Second World War and the repertory was further extended. In 1966 the company gave performances in W Europe. As well as achieving a high artistic standard, the Sofia Opera has produced a number of distinguished singers, notably Boris *Christoff and Nikolay *Ghiaurov.

The International Competition for Young Opera Singers, founded in 1961, has given a start to the careers of a number of artists, including Dimiter Petkov, Sylvia Sass, Rosalind Plowright, and Marie Slorach.

**'Softly awakes my heart'.** See 'MON COEUR S'OUVRE À TA VOIX'.

**Sogno di Scipione, Il** (Scipio's dream). Opera in 1 act by Mozart; text by Metastasio, based on Cicero's 'Somnium Scipionis' in De Republica (54 BC), orig. written for Predieri (1735). Prod. Salzburg, ? early May 1772 as part of the celebrations for the installation of Mozart's patron, the Archbishop of Salzburg.

North Africa, c.200 BC. Scipio (ten) is visited in his sleep by the Goddess of Fortune (sop) and the Goddess of Constancy (sop), who ask him to chose one of them to guide him though life. Constancy guides him to his ancestors in Paradise, where his uncle, Scipio Africanus the Elder (ten), lectures on the immortality of man's soul and his father Emilio (ten) emphasizes the insignificance of worldly pleasures. In order to join their heavenly company as he wishes, Scipio must save Rome. When Scipio chooses Constancy as his guide, Fortune summons a fierce storm. Scipio awakens, and the moral is made plain in a *licenza.

**Sokalsky, Pyotr** (b Kharkov, 26 Sep. 1832; d Odessa, 11 Apr. 1887). Ukrainian composer and critic. Studied science at Kharkov. Best known for his journalism (sometimes under the pseudonym 'Fagot') and for his work for music in Odessa, but also wrote three operas which reflect another of his interests, folk-song. They are Mazeppa (after Pushkin; 1858, unfin.), Mayskaya noch (May Night) (after Gogol; 1863), and Osada Dubno (The Siege of Dubno) (after Gogol; comp. 1878, unprod.).

**'Sola, perduta, abbandonata'.** Manon Lescaut's (sop) final aria in Act IV of Puccini's Manon Lescaut, in which she awaits her death near New Orleans.

**Soldaten, Die.** Opera in 4 acts (15 scenes) by Bernd Zimmermann; text by the composer from the drama by Jakob Michael Reinhold Lenz (1776). Prod. Cologne 15 Feb. 1965, with Gabry, De Ridder, Brokmeier, Nicolai, cond. Gielen; Edinburgh Fest., 21 Aug. 1972, by Deutsche Oper am Rhein, with Gayer, De Ridder, Runge, Rintzler, cond. Wich; Boston, Savoy OH, 22 Jan. 1982, with Hunter, Morgan, Evans, Meyer, Cochran, R. Freni, cond. Caldwell. Total theatre, with jazz, film, speech, electronic music, ballet etc., all used to the full.

Flanders, yesterday, today, and tomorrow. Marie (sop), who is engaged to Stolzius (bar), is seduced by the Baron Desportes (ten), a high-ranking army officer; rejected, degraded, she eventually becomes the soldiers' whore.

**Soldiers' Chorus.** The name generally given to the chorus of soldiers, 'Gloire immortelle', in Act IV of Gounod's Faust, in which they vow to perpetuate the glory of their ancestors.

**'Solenne in quest'ora'.** The duet in Act III of Verdi's La forza del destino, in which Alvaro (ten) and Carlo (bar) swear eternal friendship.

**Solera, Temistocle** (b Ferrara, 25 Dec. 1815; d Milan, 21 Apr. 1878). Italian librettist and composer. He ran away from boarding-school in Vienna and joined a travelling circus, being eventually arrested by the police in Hungary. Admired as a poet by *Merelli, he was given the libretto of Oberto to refashion and for three years remained Verdi's favourite librettist. His texts suited Verdi at this period, in their bold theatrical effects and vigorous, unsubtle verse. He worked on Nabucco, I Lombardi, Giovanna d'Arco, and Attila; with the latter not quite complete, he left for Spain to follow his wife, the singer Teresa Rosmini. His subsequent adventures include periods as a manager in Madrid, as Queen Isabella's counsellor and perhaps lover, as editor of a religous magazine in Milan, as secret courier between Napoleon III and Cavour, and organizer of the Khedive of Egypt's police force, before he fell on hard times. He died in poverty and neglect. He composed five operas to his own texts including *Ildegonda* (1840), *Genio e sventura* (Talent and misfortune) (1843), and *La hermana de Pelayo* (The Sister of Pelayo) (1845).

**Solié, Jean-Pierre** (*b* Nîmes, 1755; *d* Paris, 6 Aug. 1812). French composer and singer. Began his career as a cellist in theatre orchestras, until his success in replacing an indisposed tenor in a performance of Grétry's *La rosière de Salency* turned him to singing 1778. He joined the Comédie-Italienne, Paris, 1782, then sang in Nancy and Lyons before returning to Paris 1787. After a trimphant performance in Propiac's *La fausse paysanne*, 1789, he was established as one of the leading singers on the Paris stage. Having now developed into a baritone, he was a particular inspiration to Méhul, who wrote several parts for him, including that of Jakob in *Joseph*. Solié also achieved some popularity as a composer of opéras-comiques, though his works did not earn a permanent place in the repertory.

**Solti, (Sir) Georg** (*b* Budapest, 21 Oct. 1912). Hungarian, later British, conductor. Studied Budapest, piano with Bartók and Dohnányi, composition with Kodály. Répétiteur Budapest O, 1930–3. Assistant to Toscanini, Salzburg 1936–7. Début Budapest 1938 (*Figaro*). Munich 1946; music director Munich, S, 1946–52; director Frankfurt 1952–61; Gly. 1954. London, CG, 1959, director 1961–71, when he considerably raised the standards, as the house changed from the repertory system in English to the stagione system in the original language, facilitating the engagement of top international singers and conductors. CG subsequently; Bayreuth 1983. Also New York, M; Chicago; Paris, O; etc. A conductor of intense energy with a gift for generating dramatic excitement. Distinguished in Verdi, Wagner, and Richard Strauss. Repertory also incl. *Zauberflöte, Fidelio*, first British *Moses und Aron, Billy Budd, Midsummer Night's Dream*. (R)

**'Sombre forêt'.** Mathilde's (sop) romance in Act II of Rossini's *Guillaume Tell*, in which she chooses the wild countryside above splendid palaces and begs the evening star to guide her.

**Somers, Harry** (*b* Toronto, 11 Sep. 1925). Canadian composer. Studied Toronto Cons. 1942–9 with Godden and Kilburn. Rejected a career as a concert pianist in favour of composition, supporting himself by casual employment, including broadcasting. His first theatrical work, and one of the finest by a Canadian, was *Louis Riel*, commissioned for the Canadian OC in Toronto. It deals with the inflammatory situation which arose in Canada in 1869 when Riel, the visionary Métis leader in Manitoba, came into conflict with Sir John A. Macdonald, the Prime Minister. Riel was eventually executed for treason in 1885 but recent views have grown increasingly sympathetic towards him. The libretto is in both French and English as appropriate to the situation; Somers's score is eclectic but a personal and original blend of contemporary European techniques, including those of electronic music, and a distinctly Canadian musical heritage. Somers has also composed two 1-act operas, *The Fool* and *The Homeless One*.

WORKLIST (all first prod. Toronto): *The Fool* (Fram; 1953); *The Homeless One* (Fram; 1955); *Louis Riel* (Moore & Languirand; 1967); *Mario and the Magician* (Anderson; after Mann; 1992).

BIBL: B. Cherney, *Harry Somers* (Toronto, 1975).

**sommeil** (Fr.: 'slumber'). The French equivalent of the *slumber scene. One of its earliest appearances is in Lully's comédie-ballet *Les amants magnifiques* (1670); the most famous example is in his *Atys*.

**Sondheim, Stephen** (*b* New York, 22 Mar. 1930). US composer and librettist. Studied Princeton with Babbitt. Wrote lyrics for Bernstein's *West Side Story* (1957). His own scores, which treat the language of the Broadway musical in an advanced and sophisticated manner, show not only a wry, witty, and comprehensive mastery of timing and stage effect, but an interest in more continuous, near-operatic structures.

SELECT WORKLIST (all first prod. New York, all lyrics by Sondheim): *A Funny Thing Happened on the Way to the Forum* (Shevelove & Gelbart, after Plautus; 1962); *Company* (Furth; 1970); *Follies* (Goldman; 1971); *A Little Night Music* (Wheeler, after Bergman; 1973); *Pacific Overtures* (Wheeler, after Weidmann; 1976); *Sweeney Todd* (Wheeler; 1979); *Sunday in the Park with George* (Lapine; 1984); *Into the Woods* (Lapine; 1987).

BIBL: C. Zadan, *Sondheim & Co.* (New York, 1974).

**Songspiel** (Ger.: 'song play'). The description given by Weill to the first version of his *Aufstieg und Fall der Stadt Mahagonny*, the *Mahagonny-Gesänge*, in which a series of separate songs is connected by dialogue. In essence a modern interpretation of the *Singspiel tradition.

**Song to the Moon.** The name generally given to Rusalka's aria 'Měsíčku na nebi' in Act I of Dvořák's *Rusalka*, in which she begs the moon to stay and reveal her lover to her.

**'Son lo spirito che nega'.** Mefistofele's (bs) aria in Act I of Boito's *Mefistofele*, in which he declares himself as the spirit of eternal denial (Goethe's 'Geist der stets verneint').

**Sonnambula, La** (The Sleep-walker). Opera in 2 acts by Bellini; text by Felice Romani, based on the scenario for *La somnambule, ou L'arrivée d'un nouveau seigneur*, a ballet-pantomime by Scribe, the choreographer Jean-Pierre Aumer, and Hérold (1827), in turn derived from the comédie-vaudeville *La somnambule* by Scribe and Delavigne (1819). Prod. Milan, T Carcano, 6 Mar. 1831, with Pasta, Taccani, Baillou-Hillaret, Rubini, Crippa, L. Mariani, Biondi; London, H, 28 July 1831, with Pasta, Rubini, Santini; New York, Park T, 13 Nov. 1835, with Paton, Wood, Brough, cond. Penson.

In a quiet little Swiss village early in the 19th cent., Amina (sop), foster-daughter of Teresa (mez), owner of the mill, is to be betrothed to Elvino (ten), a young farmer. Lisa (sop), the proprietress of the local inn and herself in love with Elvino, gladly entertains Count Rodolfo (bs), the handsome lord of the castle recently returned to the village. Amina, unknown to her lover and friends, is a sleep-walker; she enters the Count's bedroom at night and is discovered asleep in his room.

The distraught Elvino is now ready to marry Lisa, but the Count tries to prevent this by explaining Amina's condition. The villagers scoff; but at that moment Amina is seen walking in her sleep along the edge of the mill roof (in some prods., the insecure bridge over the mill stream). Elvino gives her the ring he had taken back after she was discovered in the Count's room, and Amina awakens to find Elvino ready to marry her.

Also opera on the same text by Antoni (*Amina*, 1825). Other somnambulism operas are by Piccinni (*Il sonnambulo*, 1797), Paer (*La sonnambula*, 1800), L. Ricci (*Il sonnambulo*, 1829), and Miceli (*Somnambule*, 1870).

**Sonnleithner, Joseph** (*b* Vienna, 3 Mar. 1766; *d* Vienna, 25 Dec. 1835). Austrian writer, poet, editor, and librettist. Secretary to the court theatres 1804–14, he played an important role in Viennese musical life, founding the *Gesellschaft adeliger Frauen* in 1812 (the forerunner of the *Gesellschaft der Musikfreunde*). Active as a poet, editor, and translator and adaptor of librettos, he was a friend of Schubert and Grillparzer, and was briefly manager of the Vienna, W. Remembered today for his reworking of Bouilly's *Léonore, ou L'amour conjugal* for Beethoven's *Fidelio* (1805), and for his librettos for Cherubini's *Faniska* (1806) and Gyrowetz's *Agnes Sorel* (1806). One of those most responsible for popularizing French-style rescue opera in Vienna, his dramatic talents

were limited; only after his *Fidelio* libretto was revised (and shortened) by Treitschke in 1814 was Beethoven reasonably satisfied.

**sonno.** See SLUMBER SCENE.

**Sontag, Henriette** (*b* Koblenz, 3 Jan. 1806; *d* Mexico City, 17 June 1854). German soprano. Born into a theatrical family, she studied first with her mother. Earliest appearance Darmstadt aged 6. Studied Prague with Czegka. Début Prague 1821 (Princess, Boieldieu's *Jean de Paris*). Vienna 1822–3, creating Euryanthe, at Weber's request, after triumphing in *La donna del lago*. Berlin 1825, where her success led to 'Sontag-fever'. (Rellstab countered with a satirical article, and went to prison for libel.) Paris, I, 1826, 1828; London, H, 1828, again sweeping all before her, including the Duke of Devonshire, who proposed marriage; Berlin 1830, when she was forced to retire, having secretly married the diplomat Count Rossi in 1827. Known thereafter as the 'Rossi-gnol', she continued to sing privately, enchanting Tsar Nicholas I among others. Returned to the stage after Rossi's retirement, to undiminished acclaim in London, Paris, and Germany, 1849–51, and USA, 1852. Died of cholera while still at the height of her powers. One of the most naturally gifted singers of her time, she possessed golden beauty, great personal charm, and a sweet, highly flexible voice ranging from a to e''' that she used with consummate ease. She was admired for her 'brilliant, inventive, fresh . . . and pleasing' qualities (J. E. Cox) but also 'something more' (Berlioz). Goethe wrote his poem 'Neue Siren' for her. Roles included Susanna, Donna Anna, Semiramide, Isabella, Lucrezia Borgia, Agathe; also created Miranda (Halévy's *Tempestà*), and soprano parts in Beethoven's Ninth Symphony.

**Sonzogno, Edoardo** (*b* Milan, 21 Apr. 1836; *d* Milan, 14 Mar. 1920). Italian publisher. His firm was founded at the end of the 18th cent. by his grandfather G. B. Sonzogno. He began to publish French and Italian music in 1874, and also established a series of competitions for new works in 1883, the second contest of 1889 being won by Mascagni's *Cavalleria rusticana*; the firm was an important promoter of verismo. Opened the T Lirico Internazionale, Milan 1894.

**Soot, Fritz** (*b* Wellersweiler, 20 Aug. 1878; *d* Berlin, 9 June 1965). German tenor. Studied Dresden with Scheidemantel. Début Dresden 1908 (Tonio, *Fille du régiment*). Dresden until 1918. Stuttgart 1918–22; Berlin, S, 1922–4, 1946–52, and SO, 1946–8; London, CG, 1924–5. Created Italian Tenor (*Rosenkavalier*), Drum-Major (*Wozzeck*), and roles in operas by Pfitzner and Schreker. Repertory included Erik, Siegfried, Tristan, Otello (Verdi), Laca, Palestrina, Mephistopheles (*Doktor Faust*). (R)

**Sophie.** 1. Faninal's daughter (sop), lover of Octavian, in Strauss's *Der Rosenkavalier*.
2. Charlotte's sister (sop) in Massenet's *Werther*.

**Sopot** (Ger., Zoppot). Town in Poland. In 1909 the Zoppot Waldoper was founded (beginning with Kreutzer's *Das Nachtlager von Granada*) to give open-air performances for visitors to the resort in a forest arena with notable acoustics. The modest standards improved after the war, and many distinguished artists appeared there. Hermann Merz initiated a famous Wagner series in 1922 under Knappertsbusch, later Max von Schillings, Erich Kleiber, Furtwängler, and Pfitzner, with the great Wagner singers of the day. The performances were as faithful and naturalistic as possible, and included *Grane and the goats drawing Fricka's chariot. By 1937 annual visitors numbered nearly 35,000, and the festival was as popular with artists as with the public. The festival continued during the war but then lapsed, to be revived in 1984.

**soprano** (from It. *sopra*: 'above'). The highest category of female (or artificial male) voice. Many subdivisions exist within opera houses: the commonest in general use (though seldom specified by composers in scores) are given below, with examples of roles and their approximate tessitura. These divisions often overlap, and do not correspond exactly from country to country. In general, distinction is more by character than by tessitura, especially in France: thus, the examples of the roles give a more useful indication of the different voices' quality than any technical definition.
French: soprano dramatique (Valentine, Alceste: g–c'''); soprano lyrique (Lakmé: bb–c#'''); soubrette or soprano léger (Despina, Zerline in *Fra Diavolo*: bb–c'''); soprano demi-caractère (Manon, Cassandre: a–c#'''); *Dugazon, divided as jeune Dugazon (Bérénice in Thomas's *Psyché*), première Dugazon (Djelma in Auber's *Le premier jour de bonheur*), forte première Dugazon (La Comtesse in Thomas's *Raymond*), and mère Dugazon (Mistress Bentson in *Lakmé*); *Falcon (Alice in *Robert le diable*: bb–c#''').
German: dramatischer Sopran (Brünnhilde: g–c'''); lyrischer Sopran (Arabella: bb–c#'''); hoher Sopran or Koloratur Sopran (Zerbinetta, Queen of the Night: g–f'''); Soubrette (Blonde, Aennchen: bb–c''').
Italian: soprano drammatico (Tosca: g–c'''); soprano lirico (Mozart's Countess, Mimi: bb–c'''); soprano lirico spinto (Butterfly, Des-

demona: a–c#'''); soprano leggiero (Norina, Despina: g–f'').
See also CASTRATO, MEZZO-SOPRANO.

**Sorochintsy Fair** (Russ.: *Sorochinskaya yarmarka*). Opera (unfinished) by Musorgsky; text by composer, after Gogol's story *Evenings on a Farm near Dikanka* (1831–2). Musorgsky completed only the prelude, the market scene, and part of the sequel, most of Act II, a vision scene adapted from *A Night on the Bare Mountain*, an instrumental hopak, and two songs. The first editors were Lyadov (1904) and Karatygin (1912). A version from these editions, with Rimsky-Korsakov's version of *A Night on the Bare Mountain*, and additions from other hands, prod. St Petersburg, Comedy T, 30 Dec. 1911; Moscow, Free T, 21 Oct. 1913, with Makarova-Shevchenko, Milyavskaya, Monakhov, Draculi, Karatov, cond. Saradjev. Version by Cherepnin prod. Monte Carlo 17 Mar. 1923, with Luart, MacCormack, cond. Cherepnin; same version New York, M, 29 Nov. 1930, with Müller, Boursakaya, Jagel, Pinza, cond. Serafin; London, Fortune T, 17 Feb. 1934. In addition to these, a version by Shebalin was published 1933.
Sorochintsy, 19th cent. Cherevik (bs) has brought his daughter Parasya (sop) to the Fair, where she meets her lover Gritzko (ten), and obtains her father's consent to their betrothal. His wife Khivrya (mez) disapproves, but she is discovered to have taken a lover, the Priest's son (ten), and eventually all ends happily for the young couple.

**Sosarme, re di Media.** Opera in 3 acts by Handel; text an anon. ad. (? P. A. Rolli or S. Humphrey) of A. Salvi's libretto *Dionisio, re di Portogallo* (set by Perti, 1707). Prod. London, H, 15 Feb. 1732, with Strada, Senesino, Bagnolesi, Bertolli, Montagnana. Rev. Abingdon 1970.

**Sotin, Hans** (*b* Dortmund, 10 Sep. 1939). German bass. Studied with Hezel, Dortmund Cons. with Jacob. Début Essen 1962 (Police Commissioner, *Rosenkavalier*). Hamburg from 1964; Gly. 1968; Chicago 1971; New York, M, 1972; Bayreuth from 1972; London, CG, from 1974; Milan, S, 1976. Also Paris, O; Buenos Aires; Vienna, S; etc. Possesses a rotund, powerful tone and strong dramatic talent. Acclaimed as Sarastro, Hunding, Sachs, Gurnemanz, Grand Inquisitor, Ochs. (R)

**sotto voce** (It.: 'below the voice'). A direction to sing softly or 'aside'.

**soubrette** (Fr. from Old Fr., *soubret*: 'cunning or shrewd'). Used in opera for such roles as Serpina, Despina, Susanna etc.—the cunning servant-girl; then, more generally, to designate a light soprano comedienne, such as Marzelline (*Fide-

669

*lio*), Adele (*Fledermaus*). In Italian opera the term *servetta* is also used. See DAMIGELLA, SOPRANO.

**Souez, Ina** (*b* Windsor, CO, 3 June 1903 *d* Santa Monica 7 Dec. 1992). US soprano. Studied Denver with Hinrichs, Milan with Del Campo. Début Ivrea 1928 (Mimì). London, CG, 1929, 1935; Gly. 1934–9; New York, New OC, 1941, and City O 1945; then left opera for jazz. Highly praised for her Donna Anna and Fiordiligi. (R)

**souffleur.** See PROMPTER.

**Sousa, John Philip** (*b* Washington, DC, 6 Nov. 1854; *d* Reading, PA, 6 Mar. 1932). US composer. Studied Washington, DC (Esputa Cons.). Following military service with the US Marine Band worked in Philadelphia as a theatre violinist. Conductor of the US Marine Band 1880–92, then formed Sousa's Band, touring extensively in N America and Europe.

Sousa is remembered chiefly as a composer of military marches, including 'Liberty Bell' and 'The Stars and Stripes Forever', but his 11 operettas constitute an important contribution to the US genre. Including *El capitan* and *The American Maid*, they enjoyed some success during his lifetime, though they have been largely neglected since. He also wrote a considerable quantity of incidental music and songs.

SELECT WORKLIST: *El capitan* (Klein, Frost, & Sousa; Boston 1896); *The American Maid* (Liebling; Rochester 1913).

BIBL: P. Bierley, *John Philip Sousa: American Phenomenon* (New York, 1973).

**South Africa.** The first operas given in Cape Town were popular works by Dibdin, Storace etc. in the early 19th cent. The first major work given was *Der Freischütz* by amateurs (1831). The 1870s saw visits by touring companies, including the Carl Rosa, Moody-Manners, and Quinlan, and seasons were given in Johannesburg and elsewhere. An annual Johannesburg season grew out of the 'Music Fortnight' established in 1926, and visiting artists took part. Albert Coates conducted *Die Walküre* in the late 1940s in the City Hall. In Cape Town University, Erik Chisholm gave some outstanding opera performances; the first occasion on which a Black singer appeared in a leading role with White singers was here in 1980. In 1956 the Eoan Group, composed of black singers, gave its first season. The Nico Malan OH opened in 1970 (to all races); the Cape Performing Arts Board (CAPAB) Opera plays here and tours the country. After a period of low standards in the late 1970s, the appointment of Murray Dickie in 1982 led to substantial artistic improvements, and to co-operation with the Performing Acts Council, Transvaal (PACT), whose company performs in the Pretoria State T (cap. 1,322) and in the Johannesburg Civic T. The Sand du Plessis T opened in Bloemfontein in 1985, with opera staged by the Performing Arts Council of the Orange Free State (PACOFS); the Orange Free State Arts Foundation gives opera in the Civic T. The opera department of the Natal Performing Arts Council (NAPAC) performs in the Natal Playhouse, Durban.

**Souzay, Gérard** (*b* Angers, 8 Dec. 1918). French baritone. Studied Paris with Bernac, Croiza, Marcoux, and Lotte Lehmann. Distinguished principally as a concert-singer, he has appeared on the stage in Aix-en-Provence (Purcell's Aeneas, 1960) and Paris, OC, 1960 (Golaud), and O, 1963–6 (Don Giovanni). Outside France, he has sung Mozart's Count Almaviva at Gly. 1965 and at New York, M the same year.

**sovrintendente** (It.: 'superintendent'). The administrator of an Italian opera-house designated as *ente autonomo: his functions are distinct from those of the artistic or music director though he will normally work in close artistic contact with them.

**Spain.** Popular musical dramas were successful from the Middle Ages, especially with religious subjects. The first Spanish opera whose music survives is to Calderón's text *Celos aun del aire matan* (Jealousy even of Air Kills, 1660), with music by Juan *Hidalgo (1612/16–85). Calderón was also the author of the text for some of the first *zarzuelas, e.g. *El jardín de la Falerina* (1684). However, developments were halted by the restraint on native works in favour of Italian opera by Philip V. In the 18th cent. a number of Spanish composers wrote opera in Italian, notably *Martín y Soler and Terradellas. The history of 18th-cent. Spanish opera is largely that of struggle between zarzuela and Italian-modelled serious opera. Then a royal decree of 1800 forbade all operas except those sung in Spanish by Spaniards; translations of foreign works were still preferred, though these now included French opéra-comique. Ramón *Carnicer (1789–1855) wrote popular and pioneering operas; by contrast, in the 19th cent. there developed rivalry between two versions of zarzuela, the *género grande* and the *género chico*. Great influence was exercised by Felipe *Pedrell, and in the early years of the 20th cent. Spanish works were given at the Teatro Real (which closed in 1925). Despite difficulties, Spanish operas of European popularity were composed by Isaac *Albéniz (1860–1909), Enrique *Granados (1867–1916), and Manuel de *Falla (1876–1946).

See also BARCELONA, BILBAO, MADRID, MINORCA, VALÈNCIA, ZARZUELA.

**Sparafucile.** A professional assassin (bs) in Verdi's *Rigoletto*.

**Speziale, Lo** (The Apothecary). Opera in 3 acts by Haydn; text based by composer on a libretto by Goldoni. Prod. Eszterháza, autumn 1768; New York, Neighborhood Playhouse, 16 Mar. 1926.

Italy, 18th cent. Sempronio (bs), an old apothecary, wants to marry his ward Grilletta (sop). She is also loved by the frivolous Volpino (mez) and the serious young Mengone (ten), who eventually wins her hand.

**Spieloper** (Ger.: 'opera-play'). A type of 19th-cent. light opera, ressembling *Singspiel, with a comic subject and spoken dialogue, e.g. some of Lortzing's works.

**Spielplan** (Ger.: 'performance plan'). The published prospectus of the season's repertory. Also a monthly publication giving programmes in all German opera-houses.

**Spieltenor** (Ger.: 'acting tenor'). A light tenor in a German company who plays such character roles as Mime, David, and Pedrillo. See TENOR.

**Spinning Chorus.** The name generally given to the chorus 'Summ' und brumm" sung by the girls in Wagner's *Der fliegende Holländer*, wishing that the spinning-wheels could drive their lovers back from the sea.

**spinto** (It.: 'pushed'). The description given to a voice, almost always tenor, but sometimes also soprano, of particular vigour and attack; the description may also be qualified, e.g. tenore lirico spinto.

**'Spirto gentil'.** Fernando's (ten) aria in Act IV of Donizetti's *La favorita*, in which he reflects on his love for Leonora as he is about to take his monastic vows.

**Split** (It., Spalato). City in Croatia. The first theatre was built in the 17th cent. The Italians twice disbanded the company, and in 1859 built a theatre for their visiting companies. The first local opera was given in 1921, on the foundation of the National Dalmatian T, but the administration moved to Sarajevo. Outdoor summer festival performances are given in Diocletian's Palace.

**Spohr, Louis** (*b* Brunswick, 5 Apr. 1784; *d* Kassel, 22 Oct. 1859). German composer and violinist. Studied Seesen with Riemenschneider and Dufour, Brunswick with Kunisch and Hartung.

Spohr's first stage work was an operetta, *Die Prüfung*, which received a concert performance during his leadership of the ducal band at Gotha. Its successor, *Alruna*, was rehearsed at Weimar, when it won Goethe's admiration: it is a more advanced work, taking a supernatural subject and attempting continuous opera. The first production he achieved was of *Der Zweikampf mit der Geliebten*, something of a reversion to convention while containing attractive music. This was followed in 1816 by *Faust*.

*Faust* is a landmark in German Romantic opera. The overture's use of leading themes from the opera was one of many devices admired by Weber, who conducted the première and whose review of it writes of 'a few melodies, carefully and felicitously devised, which weave through the whole work like delicate threads, holding it together intellectually'. Spohr strengthened this aspect of the opera when he revised it in 1852, adding recitatives.

The work's popularity was exceeded by that of *Zemire und Azor*, given at the Frankfurt O during Spohr's two years as conductor, 1817–19. Here, the system of reminiscence motive and Leitmotiv is taken even further, and the theme of redemption through love achieves new significance. In other ways, too, the work anticipates Wagner, e.g. in the atmospheric use of harmony and orchestration.

Among Spohr's operatic plans was one for *Der Freischütz*; he abandoned this, without later regrets, on learning that it was being set by Weber, who recommended him for the post of Hofkapellmeister at Kassel, 1822. Here he remained for the rest of his life: Generalmusikdirektor 1847. His greatest operatic success was *Jessonda*. Spohr and his audiences were fascinated by the characteristically Romantic idea of an oriental subject: he was here also able to give his lyrical gift, and above all his mastery of chromatic music, their fullest effect. He also gave a new fluency to recitative and loosened the set forms of aria; some of the dramatic pacing of the sequence of arias is awkward, but the best of the work has an imagination ahead of its time.

*Der Berggeist*, which followed, suffers from a poor libretto, though Spohr again anticipates Wagner (in *Der fliegende Holländer*) by constructing the music in linked scenes rather than arias and thereby achieving considerable dramatic pace. *Pietro von Abano* aroused predictable distaste for its treatment of necrophily, and returns to spoken dialogue between larger scene-complexes in which there is an awkward mix of convention and originality. More enterprisingly constructed numbers built into scene-complexes mark *Der Alchymist*, together with a fashionable use of Spanish local colour. Like other of his operas, it was damned with faint praise as appealing chiefly to connoisseurs. His last opera, *Der Kreuzfahrer*, did, however, achieve success, and is indeed one of his most effective works in its cohesive and continuous structure.

In 1842 Spohr became the first musician of importance to support Wagner, when he produced *Der fliegende Holländer* in Kassel only five months after the Dresden première. He later staged *Tannhäuser* but was prevented from following this with *Lohengrin*. His *Autobiography* gives a vivid account of his life and times.

WORKLIST: *Die Prüfung* (The Test) (Henke; concert, Gotha 1806); *Alruna* (?, after Hensler; comp. 1808, unprod.); *Der Zweikampf mit der Geliebten* (The Duel with the Beloved) (Schink; Hamburg 1811); *Faust* (Bernhard; Prague 1816); *Zemire und Azor* (Ihlee, after Marmontel; Frankfurt 1819); *Jessonda* (Gehe, after Lemierre; Kassel 1823); *Der Berggeist* (The Mountain Spirit) (Döring; Kassel 1825); *Pietro von Abano* (Pfeiffer, after Tieck; Kassel 1827); *Der Alchymist* (Schmidt [Pfeiffer], after Irving; Kassel 1830); *Die Kreuzfahrer* (The Crusaders) (L. & M. Spohr, after Kotzebue; Kassel 1845).

WRITINGS: *Selbstbiographie* (Kassel, 1860–1; trans. 1865, R/1969, 2/1978).

BIBL: C. Brown, *Louis Spohr* (Cambridge, 1984).

**Spoleto.** Town in Umbria, Italy. Opera was given from 1667 in the T Nobile; rebuilt 1830 as T Caio Melisso (cap. 500). The T Nuovo (cap. 900) opened Aug. 1864 with works by Verdi and Petrella. In 1958 *Menotti established the Festival of Two Worlds, giving young European and US artists chances to appear in his and Visconti's productions and also involving local talent. Thomas *Schippers was music director 1958–70. The first production was of Verdi's *Macbeth*, by Visconti; others have included Donizetti's *Duca d'Alba* (1959, using the original 1882 designs), *Carmen* (1962, with Verrett and George Shirley), *Salome* (1961), *Manon Lescaut* (1973), and *Parsifal* (1987). Revivals have been of works by Graun, Gagliano, Mercadante, and Salieri; there was later more emphasis on modern works and radical productions, including Polanski's production of *Lulu* (1974). Menotti's own works have also featured. He has also founded Spoleto festivals near Charleston, SC and Melbourne.

**Spoletta.** The police agent (ten) in Puccini's *Tosca*.

**Spontini, Gaspare** (*b* Maiolati, 14 Nov. 1774; *d* Maiolati, 24 Jan. 1851). Italian composer and conductor. Studied Naples with Sala and Tritto.

An early commission led to the successful *Li puntigli delle donne*, immediately followed by five more operas. When the Neapolitan court moved to Palermo before the advancing French in 1798, Spontini took on Cimarosa's post as conductor and in 1800 produced three deft Neapolitan operas. Shortly afterwards he left for Paris, where his light style was unsuited to the special demands of opéra-comique; but with the carefully composed *Milton* he won a success that carried his name to Germany and Austria. Dealing with the poet in

his blind old age, its numbers include a fine aria to the sun and light. One of *Milton*'s librettists was Étienne de Jouy, who, understanding better than the composer himself where his gifts lay, provided for *La vestale* a text that fully released Spontini's talent and led to his recognition in 1807 as one of the leading opera composers of the day.

Spontini intensified selected features of Gluck, Cherubini, and Méhul, and put them to novel use in his serious operas. These are founding works of Parisian *Grand Opera in their generous use of processions, rituals, oath-takings, ceremonies, and other imposing stage effects, with extra bands, large choral groupings, and tableaux. Spontini loved grandiose effects and striking dramatic contrasts: his calculated *coups de théâtre* succeed because of his sense of dramatic continuity and timing and his grasp of large-scale dramatic effect. *La vestale* is Gluck's tragédie lyrique adapted to the taste of Empire audiences.

The Empress Josephine's patronage carried Spontini through all opposition; and after the constant rewriting that was his habit, *La vestale* triumphed. For *Fernand Cortez* the librettist was again Jouy, the patron now Napoleon (who felt that the subject might influence opinion more favourably in the Spanish Wars). A more polished work, it had a still greater success. *Cortez* is a grand historical pageant filled with tableaux, including a cavalry charge, the burning of the Spanish fleet, a heroine who plunges into a lake, and other sensational effects; the choruses of Spaniards and Aztecs are central characters, and the ballets are more functional than before in opera.

In 1810 Spontini took over the T-Italien, and at the Odéon formed a distinguished and enterprising ensemble before his dismissal in 1812; his restoration in 1814 was brief and again controversial, for his undoubted gifts were allied to an overbearing personality. A few minor pieces followed before *Olympie*, another tragédie lyrique which again included much spectacle—processions, a Bacchanal, a battle, an apotheosis. Though carefully composed, and including original elements (such as a Grand March that directly influenced Berlioz's Trojan March), it was not well received in its first form in 1819; but assiduous revision brought success in Paris and Berlin, where Spontini was now summoned. Much impressed, Frederick William III of Prussia engaged him for the Berlin Court Opera, where he worked in uneasy co-operation with the Intendant, Brühl. His temper and haughtiness quickly caused difficulties, and when in 1821 the imposing spectacle of *Olympie* was succeeded in a few weeks by the thrilling new popular experience of Weber's *Der Freischütz*, Berlin was divided, the court siding with Spontini, the public preferring the new German Romantic opera.

Spontini's painstaking methods of composition prevented him from producing as much as was now required of him. In *Nurmahal* and *Alcidor* he had to attempt German Romantic opera for Berlin; the works are actually closer in nature to French opéra-féerie. *Alcidor* in particular hardened the division of public opinion, and Spontini could not rally his scattering admirers even by a conscientious attempt at a German Grand Opera, *Agnes von Hohenstaufen*, in 1829. This was more continuously composed, building up large formal complexes and lengthening the finale to cover much of an act. In its nature and structure Spontini was attempting to regain his hold on German audiences by way of a historical drama.

The sympathies of the new king lay elsewhere, but Spontini also hastened his own downfall: though fairly, even generously treated, he felt obliged to leave. He made occasional return visits to Germany: on one of them, he conducted at Dresden a *Vestale* prepared by Wagner, who admired much in Spontini and especially his command as a conductor. The admiration was fully shared by Berlioz, especially for his orchestration. He spent most of his remaining years in Paris, before returning to his native village.

SELECT WORKLIST: *Milton* (Jouy & Dieulafoy; Paris 1804); *La vestale* (The Vestal Virgin) (Jouy; Paris 1807); *Fernand Cortez* (Jouy & D'Esménard; Paris 1809); *Olympie* (Dieulafoy & Briffaut, after Voltaire; Paris 1819, rev. trans. Hoffmann, Berlin 1821); *Nurmahal* (Herklots, after Moore; Berlin 1822); *Alcidor* (Lambert, after Chabannes; Berlin 1825); *Agnes von Hohenstaufen* (Raupach; Berlin 1827, rev. Berlin 1829, rev. Berlin 1837).

BIBL: P. Fragapone, *Spontini* (Florence, 1983).

**Sporting Life.** A dope-peddler (ten) in Gershwin's *Porgy and Bess*.

**Sposo deluso, Lo** (The Deceived Bridegroom). Overture and four numbers, intended for opera in 2 acts by Mozart; text possibly by Da Ponte. Comp. probably late 1783. The surviving libretto adumbrates some features of that for *Figaro*.

**Sprechgesang** (Ger.: 'speech-song'). A type of vocalization first used by Humperdinck in early versions of *Königskinder* (1897), describing a form of musical declamation in which the actual pitch of the notes is indicated but the voice falls or rises in a manner between true speech and true song. It is used by Schoenberg, who preferred the term *Sprechstimme*, in *Moses und Aron* and by Berg in *Wozzeck* and *Lulu*.

**sprezzatura** (in full *sprezzatura di canto*; It., 'scorn [of melody]'). Term first used by G. Caccini in the preface to *Le nuovo musiche* (1602), describing how *stile rappresentativo* should be performed: effectively, with a kind of rubato.

**Staatsoper** (Ger.). State opera.

**Stabile, Mariano** (*b* Palermo, 12 May 1885; *d* Milan, 11 Jan. 1968). Italian baritone. Studied Rome with *Cotogni. Début Palermo 1909 (Marcello). After modest successes in Italy and on foreign tours, Milan, S, 1921 (Falstaff: chosen and coached by Toscanini). Chicago 1924–9; London, CG, 1926–31; Gly. 1936–9; Salzburg 1934–9; London, Cambridge T and Stoll T, 1946–9; also Milan, S, until 1955. His extraordinary artistic achievements lay not in beauty of tone, but in subtlety of style and singing, and a psychological perception that made an individual mark on every role he interpreted. He was memorable in over 50 parts, e.g. Figaro (Mozart and Rossini), Don Giovanni, Malatesta, Beckmesser, Iago (Verdi), Scarpia, Gianni Schicchi. Of his greatest role it was said, 'La donna è mobile; Falstaff è Stabile.' (R)

**Stabinger, Mathias** (*b* ?, *c*.1750; *d* Venice, *c*.1815). Composer and conductor, probably of German origin. A shadowy figure, he worked first in France and then in Italy, where *Le astuzie di Bettina* was given 1780. This brought him considerable success: he travelled to Poland 1781, becoming maestro al cembalo at the Warsaw O, then on to Russia. He worked in St Petersburg with the Mattei-Orecia troupe and in Moscow at the Petrovsky T, staying until 1783. After a brief visit to Italy he was back in St Petersburg 1785–99, directing the orchestra at the Petrovsky T. He returned to Italy *c*.1800.

An important figure in late 18th-cent. Russian cultural life, he contributed four operas to the Moscow repertory, including *Baba Yaga* and *Pigmalion*.

SELECT WORKLIST: *Le astuzie di Bettina* (?; Genoa 1780); *Baba Yaga* (Gorchakov; Moscow 1786); *Pigmalion* (Maykov, after Rousseau; Moscow 1787).

**Stabreim** (Ger.: 'alliteration'). The system of giving cohesion to verse, and also suggesting contacts in meaning, by alliterating key words in a line and perhaps also an adjacent line. Originating in Old English and German verse, it was taken up again and cultivated for use in mythic opera by Wagner, in *Tristan* but with especial emphasis in *The Ring*. His need was not only to suggest the works' ancient subjects atmospherically, but to find a heightened poetic utterance that would avoid the cadential implications of end-rhyme at a time when he was seeking uncadenced, 'endless' melody. Wagner's Stabreim normally consisted of two or three alliterated *Hebungen* (a strong beat or arsis) with freely arranged, unalliterated *Senkungen* (a weak beat or thesis). An example (Wotan to the disobedient Brünnhilde, *Walküre* Act III) is:

| | |
|---|---|
| Was sonst du warst, | What once you were |
| Sagte dir Wotan. | Wotan decided. |
| Was jetzt du bist, | What now you are |
| das sage dir selbst. | you decide for yourself. |

Here the effect is not only assonant: a distinction is made between the 'w' sounds associated with Wotan and the roles and privileges formerly conferred upon Brünnhilde, and the 's' sounds associated with her disobedient separation from Wotan's will. The device was taken up by Wagner's followers. Attempts to transfer it to English have not been successful, whether in translations of Wagner (e.g. by H. and F. Corder) or in original librettos (e.g. by Holst in *Sita*).

**Stade, Frederica von** (*b* Somerville, NJ, 1 June 1945). US mezzo-soprano. Studied New York with Engelberg. Début New York, M, 1970 (Second Boy, *Zauberflöte*). New York, M; Paris, O, from 1973; Gly. 1973; London, CG, from 1975; Milan, S, 1976; Holland, Houston, San Francisco, etc. Her attractive presence and clear, sensuous tone have brought her much success as Cherubino (Mozart and Massenet), Sesto, Dorabella, Rosina, Cenerentola, Composer, Octavian, Charlotte, Mélisande. (R)

**Staden, Sigmund Theophil** (*b* Kulmbach, *bapt.* 6 Nov. 1607; *d* Nuremberg, *c*.30 July 1655). German composer. Studied Nuremberg with his father and Augsburg with Paumann, then returned to Nuremberg 1623, where he held municipal and ecclesiastical appointments for the remainder of his life. A gifted organist, instrumentalist, and composer, he wrote one of the earliest German musical dramas to have survived, *Seelewig*. This setting of a libretto by Harsdörffer was published in the journal *Frauenzimmer Gesprächspiele* and is a religious allegory concerning the spiritual conflict between evil and virtue. Staden's score mixes occasional attempts at *stile rappresentativo*—he claimed it to have been written 'after the Italian manner'—with strophic songs and short instrumental numbers. Though exaggerated claims have been made on its behalf, *Seelewig*'s importance derives mainly from its use of continous music in a dramatic context; it had little immediate impact and German opera cannot properly be said to have begun until much later.

WORKLIST: *Seelewig* (Harsdörffer; Nuremberg 1644).

**Städtische Oper** (Ger.). City opera.

**stage design**. Although the masques and other forms of elaborate entertainment which were forerunners of opera often required complex scenery, the first operas were simply staged. Not until the grandiose performances mounted in Cardinal Barberini's palace in Rome (*c*.1632–50), and the establishment of court theatres shortly after, did stage spectacle become the fashion.

It was in these court theatres, with their tiers of boxes, that stage became separated from audience by means of the proscenium frame. The perspective painting which characterized much Renaissance art was seized upon by theatre designers, and splendid two-dimensional scenery characterized the magnificent operatic productions of the Baroque. Even within this convention, the initial emphasis on central perspective soon gave way to angled perspectives, false perspective, *trompe-l'œil* effects, and other ingenuities. Pioneering work, particularly on machinery, was done by Giovanni *Burnacini in Venice, and by Giacomo *Torelli, especially with indoor sets and swifter scene changes. Greater realism was encouraged, with influential designers including the *Mauro family. In France, Torelli's successor Gaspare Vigarani created a formal beauty to match Lully's demands, in turn influencing Jean *Bérain. These influences spread to Germany, once the country began recovering its court operatic life in the mid-17th cent. after the Thirty Years War; while in England, Inigo *Jones produced beautiful and elegant décor for Stuart masques.

More subtle approaches to illusion and realism began to appear with a changing ethos in the 18th cent. A new range of stage presentation by way of more sophisticated use of such devices as diagonal perspective, *trompe-l'œil*, and views from above or below, together with great architectural sophistication, characterized the epoch-making work of Ferdinando *Galli-Bibiena and his dynasty. The reforms of Gluck encouraged a sharper realism represented by the *Galliari family (who increased the use of stage props), then by the important *Quaglio dynasty. Greater interest in Nature, decorative in the rococo, led to new methods, and with Revolutionary opera to a more functional use of scenery. In Germany, experiments in three-dimensional scenery were made by designers including Karl *Schinkel. In Paris, the emphasis on grandeur and a preoccupation with stage mechanics met in the work of *Cicéri and *Daguerre with their magnificent and ingenious sets. Their work also included sophisticated lighting and a greater concern for historicism, all reflecting the ethos of *Grand Opera. These interests influenced the Romantic designs of Alessandro *Sanquirico and Carlo Ferraio in Milan, and guided the ideas of artists working in countries where Romantic opera was associated with nationalism.

The developments in stage machinery in the 20th cent., together with new artistic movements, affected all countries. Among them was

Russia, where the dazzling designs of Alexandre *Benois captured European imagination, but where also Symbolist painters associated with the *Mir iskusstva group evolved new approaches to the visual. Painters, including Alfred *Roller, were more often brought into the opera-house, and new concepts of stage design were explored by Edward Gordon *Craig and Adolphe *Appia. Opera design was affected by virtually every artistic movement, including Expressionism, Futurism, Surrealism, Cubism, Constructivism, neoclassicism, and the work of particular artistic groups such as Walter Gropius's Bauhaus. Among important designers of the 1930s was Emil *Preetorius. Further developments after the Second World War focused initially on the work of Wieland *Wagner at Bayreuth, but also on that of many talented artists for whom opera has provided encouragement for technical and artistic experiment. In England original and successful designers have included Oliver *Messel, John *Piper, Osbert *Lancaster, and Ralph *Koltai, among many; those who have made an impression internationally include Emanuele *Luzzati, Jean-Pierre *Ponnelle, Josef *Svoboda, Filippo *Sanjust, Luchino *Visconti, and Franco *Zeffirelli. These and other designers have often worked regularly in a successful partnership with a producer, or as producers themselves.

**stagione** (It.: 'season'). The *stagione lirica* is the opera season at any Italian theatre. The stagione system (as opposed to the repertory system) has come to mean putting on one opera for a series of performances within a few weeks with the same cast throughout its run.

**Stagno, Roberto** (*b* Palermo, 11 Oct. 1840; *d* Genoa, 26 Apr. 1897). Italian tenor. Studied Milan with Lamperti. Début Lisbon 1862 (Rodrigo, Rossini's *Otello*). Madrid from 1865, replacing Tamberlik with great success; Moscow, B, 1868, partnering Artôt; Buenos Aires from 1879; opening season of New York, M, 1883–4; Paris, I, 1884. Last appearance 1897. Impressive for his vocal control, he encompassed florid Rossini and Donizetti roles, also Faust (Gounod), Lohengrin, Radamès, Manrico, and Otello (Rossini). Created Turiddu (with his wife Gemma *Bellincioni as Santuzza); famous as Robert le diable. His daughter **Bianca Stagno-Bellincioni** (1888–1980) was a singer and film actress.

**Stampiglia, Silvio** (*b* Civita Lavinia, 14 Mar. 1664; *d* Naples, 26 Jan. 1725). Italian librettist. A founder-member of the Accademia degli Arcadi, his first opera librettos were written while serving at the Naples court between 1696 and 1702: these included *Il trionfo di Camilla* (Bononcini, 1696),

*Partenope* (Mancia, 1699) and *Tito Sempronio Gracco* (Alessandro Scarlatti, 1702). He was briefly at the Medici court in Florence (1704–5); his most successful period was as *poeta cesareo in Vienna after 1706, where he provided several librettos for Bononcini, including *Il natale di Giunone* (1708) and *Li sacrifici di Romolo* (1708). He returned to Rome in 1718, and moved to Naples in 1722.

Stampiglia was one of the most popular and frequently set librettists of the early 18th cent. Many of his texts deal with historical themes: his most successful work was *Partenope* (later known under the title *Rosimira*), which was set by Caldara (1699), Handel (1730), Keiser (1733), Vivaldi (1738), and Porpora (1742) among many others. Together with several contemporaries, notably Zeno, he was responsible for a fundamental change in the Italian libretto, excising the comic scenes and refining the language of the prevailing Venetian model: this prepared for the later reforms of Metastasio.

**Standfuss, Johann** (*b* ?; *d* after *c*.1759). German composer. He composed the music for the first important German *Singspiel, *Der Teufel ist los* (1752), to Weisse's translation of Coffey's ballad opera, *The Devil to Pay*. New music was required to replace the original English airs because Weisse had difficulty fitting his newly translated text to the old melodies. The score for this, and for its sequel *Der lustige Schuster* (1759), is now lost but Hiller included many numbers in his own settings of the text (14 of Standfuss's songs in *Der Teufel*, and 33 in *Schuster*), and described Standfuss's music as having 'a not infelicitous expressiveness in the low comic vein'. The songs were folk-like in style, simple homophonic numbers with immediate popular appeal. The success of *Der Teufel ist los* precipitated a pamphlet war on the lines of the *Guerre des Bouffons, in which Gottsched and his supporters claimed that such popular forms of musical entertainment would lead to a decline in standards. Gottsched regarded opera as 'a nonsensical hotchpotch of poetry and music where the composer and poet violate each other and take great pains to bring about a very miserable work'.

Standfuss's talents as a composer were undeniably limited, but his style suited the light-hearted mood of the new Singspiel admirably, and the genre probably owes its establishment largely to the success of his works.

**Stanford, (Sir) Charles Villiers** (*b* Dublin, 30 Sep. 1852; *d* London, 29 Mar. 1924). Irish composer. Studied Dublin with Quarry, Leipzig with Reinecke, Berlin with Kiel. One of the few major British musical figures of his day to have faith in a national operatic revival, he wrote ten operas himself. It was symptomatic of the difficulties that the first two were produced in Germany. High claims have been made for a few of them, but the only one to achieve any success was *Shamus O'Brien*.

SELECT WORKLIST: *Shamus O'Brien* (Jessop, after Le Fanu; London 1896); *Much Ado about Nothing* (Sturgis, after Shakespeare; London 1901); *The Critic* (James, after Sheridan; London 1916); *The Travelling Companion* (Newbolt, after Andersen; Liverpool 1925).

**Stanislavsky, Konstantin** (*b* Moscow, 17 Jan. 1863; *d* Moscow, 7 Aug. 1938). Russian producer. Famous as actor and producer and director of the Moscow Arts T, and as developer of a new realism in acting, Stanislavsky also did much for opera. From 1885, with Tchaikovsky, Taneyev, Jürgenson, and Tretyakov, he was one of the directors of the Russian Musical Society. In 1898, with Vladimir *Nemirovich-Danchenko, he founded the Moscow Arts T, whose revolutionary methods and choice of works were also to influence the staging of opera in Russia and eventually abroad. He also sang in operetta with Savva Mamontov's company. In 1918 he founded the Opera Studio of the Bolshoy T, opening in 1919 with an act of *Eugene Onegin*; this separated from the Bolshoy in 1920.

Stanislavsky proposed to use the theatrical discoveries of the Moscow Arts T to refresh the tradition of staging opera in Russia, and hoped to develop a style of singing in which words and their meaning would be given new dramatic importance: to this he brought great system and powers of persuasion. His productions included *Werther* (1921), *Eugene Onegin* (1922: a famous 'workshop' production in everyday dress), *Il matrimonio segreto* (1925), works by Rimsky-Korsakov and Musorgsky, *The Barber of Seville*, and *Carmen*. In 1926 he moved his theatre to the Dmitrovsky T. The last production he supervised completely was *May Night* (1928). Becoming ill, he was obliged to conduct his rehearsals by summoning the cast to his bedside and even by use of the telephone. He organized a new Operatic-Dramatic Studio in 1938: the last production in which he took a personal part was *Madama Butterfly* (1938). After his death, *Meyerhold was director, 1938–9.

BIBL: E. Hapgood, ed., *Stanislavsky on Opera* (New York, 1975).

**Stara Zagora.** City in Bulgaria. Opera was first given under the auspices of the Kaval Music Society in 1919, but the first full-scale performance was of *Atanasov's *Gergana* in 1925. A new company was formed in 1928. The Stara Zagora National O opened in 1946 and has devel-oped an international repertory. A new opera-house opened in 1971 (cap. 900).

**Stasov, Vladimir** (*b* St Petersburg, 14 Jan. 1824; *d* St Petersburg, 23 Oct. 1906). Russian critic. Studied St Petersburg. After a period working as reviewer and archivist, he came to know Balakirev in 1856; the circle of five composers he championed, which also included Rimsky-Korsakov, Cui, Musorgsky, and Borodin, became known, through his coinage, as the Mighty Handful ('moguchaya kuchka'). He was the most important music critic to articulate the ideas of realist art, directly connected to humanity, which had been opposed to an abstract aestheticism by his literary predecessor Vissarion Belinsky. By temperament a Slavophile rather than a Westernizer, encouraging composers to seek out Russian subjects, he adopted a more liberal view than some of his colleagues.

WRITINGS: F. Jonas, ed., *V. V. Stasov: Selected Essays on Music* (London, 1968).

**Steber, Eleanor** (*b* Wheeling, WV, 17 July 1914; *d* Langhorne, PA, 3 Oct. 1990). US soprano. Studied Boston with Whitney, New York with Althouse. Unofficial début Boston, Commonwealth O, 1936 (Senta). New York, M, 1940–66; Gly. (Edinburgh) 1947; Bayreuth 1953; also Salzburg, Vienna, etc. Created Barber's Vanessa. A gifted and intelligent artist, highly regarded as Fiordiligi, Pamina, Violetta, Desdemona (Verdi), Elsa, Arabella, Marie (*Wozzeck*). (R)

**Steersman.** Daland's steersman (ten) in Wagner's *Der fliegende Holländer*.

**Steffani, Agostino** (*b* Castelfranco, 25 July 1654; *d* Frankfurt, 12 Feb. 1728). Italian composer, diplomat, and cleric. A protégé of Elector Ferdinand Maria of Bavaria, who brought him to Munich 1667, he studied with Kerll, and from 1672 in Rome with Bernabei. Returning to Munich 1674 he became court organist and was ordained 1680. From 1681 he was director of chamber music; he also wrote his first operas, beginning with *Marco Aurelio*. In 1688 he moved to the Hanover court as Kapellmeister, where he joined with the newly established Italian opera troupe in staging a number of productions. Eight of his own operas were given in Hanover, including *Orlando generoso* and *La libertà contenta*. By the later 1690s he was more actively engaged as a diplomat than as a musician, a function which he performed with some success. This led to a political appointment at the court of the Elector Palatine at Düsseldorf 1703; in 1709 he was appointed Apostolic Vicar of N Germany, basing himself at Hanover.

Steffani made an important contribution to the establishment of opera in N Germany, for though his works were essentially Italian in style, they were translated and taken into the repertory of the Hamburg O, where they influenced the following generation of composers. Many of the operas take historical subject-matter, sometimes Teutonic (e.g. *Alarico il Baltha*, *Henrico Leone*), and eschew the concentration on mythology which was typical at the time. Particularly interesting is the influence of French opera—Steffani made an extended visit to Paris 1678—notably in the overture and in the occasional use of a 5-part string texture, the prominent use of the motto aria, and the importance of contrapuntal writing. Among the composers whose operatic style was directly influenced by Steffani the most important was Handel.

SELECT WORKLIST: *Marco Aurelio* (Terzago; Munich 1681); *Alarico il Baltha* (Orlandi; Munich 1687); *Henrico Leone* (Mauro; Hanover 1689); *Orlando generoso* (Mauro; Hanover 1691); *La libertà contenta* (Mauro; Hanover 1693).

**Stehle, Sophie** (*b* Sigmaringen, 15 May 1838; *d* Hackerode, 4 Oct. 1921). German soprano. Studied Munich with Lachner. Début Munich 1860 (Emmeline, Weigl's *Schweizerfamilie*). Distinguished as Senta, Elsa, Elisabeth; created Fricka (*Rheingold*, 1869) and Brünnhilde (*Walküre*, 1870), but declined Brangäne. Also sang Pamina, Agathe, Selika, Marguerite. Hanslick admired her for her 'God-given talent', and for her persuasive, dark-hued voice.

**Stehle-Garbin, Adelina** (*b* Graz, 30 Jun. 1860; *d* Milan, 24 Dec. 1945). Austrian, later Italian soprano. Studied Milan Cons. Début Broni 1881 (Amina). Engagements in Bologna, Florence, Venice, S America; Milan, S, 1890–2, 1894; also appeared in Vienna, Berlin, Moscow, Madrid, etc. With her sweet tone, refined style, and dramatic skill, successful as Violetta, Manon, Ophelia; important as a verismo interpreter, e.g. as Adriana Lecouvreur. Created Nedda, Walter (*La Wally*), and first Nannetta, with her fiancé the tenor Edoardo Garbin as Fenton; they were later much admired in *La Bohème*. (R)

**Steibelt, Daniel** (*b* Berlin, 22 Oct. 1765; *d* St Petersburg, 20 Sep. 1823). German composer. Studied with Kirnberger. Following desertion from the Prussian army he visited several European cities, settling in Paris 1790: here his first opera, *Roméo et Juliette*, was successful only when revised as an opéra-comique. In 1810 he became director of the French O in St Petersburg, where he composed three operas, including *Cendrillon*. A colourful and rakish character, Steibelt was a competent composer with some dramatic flair, though his scores contain few memorable moments.

SELECT WORKLIST: *Roméo et Juliette* (Ségur, after Shakespeare; Paris 1793); *Cendrillon* (Étienne de Jouy; St Petersburg 1810).

**Stein, Horst** (*b* Elberfeld, 2 May 1928). German conductor. Studied Cologne with Wand. Répétiteur Wuppertal 1947–51. Cond. Hamburg 1951–5, 1961–3; Berlin, S, 1955–61; Mannheim 1963–70. San Francisco 1964–8. Chief cond. Vienna, S, 1970. Music director Hamburg from 1972. Bayreuth 1969–84, *Ring* and *Parsifal*. Paris, O, 1973, 1976. (R)

**Stendhal** (orig. Henri Beyle) (*b* Grenoble, 23 Jan. 1783; *d* Paris, 23 Mar. 1842). French writer. His musical biographies include essays on Mozart, Haydn, and Metastasio, though his best-known work is the entertaining *Life of Rossini* (1842). His novel *La chartreuse de Parme* (1839) was set as an opera by Sauguet (1939). His writings are an important source of information on performers of the time.

**Stenhammar, Wilhelm** (*b* Stockholm, 7 Feb. 1871; *d* Stockholm, 20 Nov. 1927). Swedish composer and conductor. Largely self-taught, Stenhammar was one of the central figures in Swedish music nationalism. As well as performing as a concert pianist, he had considerable success as a conductor and was briefly artistic director of the Royal Swedish O, 1900–1 and 1924–5. Though his musical style was forged, like that of many of his Scandinavian contemporaries, from the blending of nationalist folk influences into an essentially Romantic amalgam, Stenhammer showed greater interest than most in the more progressive music of Liszt and Wagner. This is seen nowhere better than in *Das Fest auf Solhaug* and *Tirfing*: both works were written early in his career and their conspicuous lack of success turned him away from opera.

WORKLIST: *Tirfing* (Boberg; Stockholm 1898); *Das Fest auf Solhaug* (Borch, after Ibsen; Stuttgart 1899; in Swed. as *Gildet på Solhaug*, trans. Elmblad (Stockholm 1902)).

**Stepanova, Elena** (*b* Moscow, 17 May 1891; *d* Moscow, 26 May 1978). Russian soprano. Studied Moscow with Polli. Début Moscow, B, 1912 (Antonida, *Life for the Tsar*). Moscow, B, until 1944, also Stanislavsky's opera studio. Sang with Shalyapin, Sobinov, Nezhdanova, etc; roles incl. Gilda, Tatyana, and was particularly associated with Rimsky-Korsakov roles, e.g. Snegurochka, Queen of Shemakha. (R)

**Stephanie, Gottlieb** (der Jüngerer) (*b* Breslau, 19 Feb. 1741; *d* Vienna, 23 Jan. 1800). Austrian dramatist, librettist, and theatre director. After military exploits he was encouraged into a stage career by Anton Mesmer, becoming director of the National-Singspiel in Vienna in 1779. His work here included the production of about 20

Singspiel librettos, notably those for Umlauff's *Das Irrlicht* (1782) and for Dittersdorf's *Doktor und Apotheker* (1786) and *Die Liebe im Narrenhaus* (1787); he also adapted many French and Italian works by composers including Grétry and Anfossi. He had a considerable influence on the development of the Singspiel in Vienna, consolidating the earliest experiments in the genre into a successful formula.

He is best remembered as the librettist of Mozart's *Die Entführung aus dem Serail* (1782), for which he adapted a libretto by Bretzner. Together with some changes of detail to make it closer to Viennese taste and custom, Stephanie also modified the ending, adding a new note of magnanimity reflecting the spirit of the Enlightenment. He completely rewrote five numbers and added four new ones: these revisions were mainly done at the behest of Mozart so as to tailor the arias to the singers at his disposal. Though he is not highly regarded as a librettist, his ready acquiescence to Mozart's suggestions resulted in the production of a Singspiel whose dramatic pace and close integration of text and music place it well above any contemporary works. Their later collaboration, *Der Schauspieldirektor* (1786), is an engaging satire of theatrical life, first given before a production of Salieri's *Prima la musica e poi le parole*.

**Stephens, Catherine** (*b* London, 18 Sep. 1794; *d* London, 22 Feb. 1882). English soprano and actress. Studied London with Lanza, later with Welsh. Début London, Pantheon T, 1812 (small roles). London: CG 1813–22, 1828; DL 1822–8. Retired 1835; married the aged Earl of Essex 1838 (he died 1839). 'The favourite of all', as Macready called her, she was admired for her attractive voice and gentle, appealing stage manner. Sang Polly Peachum, Clara (*The Duenna*), Susanna, Zerlina, Agnes (*Agathe*); also played Ophelia with Kemble. Weber wrote his last composition ('From Chindara's Warbling Fount') for her.

**Števa**. Števa Buryja (ten), half-brother of Laca Klemeň and faithless lover of his cousin Jenůfa, in Janáček's *Jenůfa*; also the name given to the baby he fathers on Jenůfa.

**Stevens, Risë** (*b* New York, 11 June 1913). US mezzo-soprano. Studied New York with Schoen-René, later Vienna with Gutheil-Schoder and Graf. Début New York, Little Theatre OC, 1931 (small roles). Prague and Vienna 1936–8; Gly. 1939, 1955; New York, M, 1938–61; Milan, S, 1954. A talented singing actress, much acclaimed as Cherubino, Mignon, Carmen, Dalila, Laura (*Gioconda*), Octavian. (R)

**Stewart, Thomas** (*b* San Saba, TX, 29 Aug. 1928). US baritone. Studied New York with Harrell. Début (as student) New York, Juilliard School, 1954 (La Roche in first US *Capriccio*). Appeared as bass with New York City O and Chicago O. Further study in Europe. Berlin, SO, 1958. London, CG, and Bayreuth from 1960; New York, M, from 1966; Paris, O, 1967; Orange 1974, 1980; San Francisco 1971, 1984–5. Roles include Don Giovanni, Escamillo, Kurwenal, Sachs, Wotan, Amfortas, Onegin. Married to the soprano Evelyn *Lear. (R)

**Stich-Randall, Teresa** (*b* West Hartford, CT, 24 Dec. 1927). US soprano. Studied Columbia U School of Music, singing Aida aged 15, and appearing in the première of Thomson's *Mother of Us All*. Sang in Toscanini's broadcasts of *Aida* (Priestess) and *Falstaff* (Nannetta) 1949–50. Vienna from 1952; Chicago from 1955; New York, M, from 1961; also Milan, S, Salzburg. Roles include Constanze, Donna Anna, Pamina, Violetta, Sophie, Ariadne. (R)

**Stiedry, Fritz** (*b* Vienna, 11 Oct. 1883; *d* Zurich, 8 Aug. 1968). Austrian, later US, conductor. Studied Vienna with Mandyczewski. Recommended by Mahler to Schuch, who engaged him for Dresden, 1907–8. After appointments at Teplice, Poznań, Prague, Nuremberg, and Kassel, chief cond. Berlin, H, later S, 1914–23. Vienna, V, 1924–8 (première of Schoenberg's *Die glückliche Hand*, 1924). Berlin, SO, 1928–33, succeeding Walter as chief cond. 1929 and working with Ebert on important revivals of *Macbeth* and *Boccanegra*; cond. première of Weill's *Die Bürgschaft* (1932). Forced to leave Germany by the Nazis, he cond. in Russia 1933–7. New York, New OC, 1941; Chicago 1945–6; New York, M, 1946–58, principal Wagner conductor and also cond. important Verdi revivals. Gly. 1947 (Gluck's *Orfeo*). London, CG, 1953–4 (*Ring*, *Fidelio*). (R)

**Stierhorn** (Ger.: 'bull horn' or 'cow horn'). A primitive horn of antiquity, in its modern form a set of three straight brass tubes with conical bore called for by Wagner in *Die Walküre*, Act II, and *Götterdämmerung*, Acts II and III; played by trombonists.

**Stiffelio**. Opera in 3 acts by Verdi; text by Piave, after the play by Émile Souvestre and Eugène Bourgeois, *Le pasteur, ou L'évangile et le foyer* (1849). Prod. Trieste, T Grande, 16 Nov. 1850, with Gazzaniga-Malaspina, De Silvestrini, Fraschini, Dei, Petrovich, Colini, Reduzzi, cond. Verdi; London, Collegiate T, 14 Feb 1973, with Conoley, Thomas, Kale, Lyons, Anderson, Seed, Dean, cond. Badacsonyi; Boston, Orpheum T, 17 Feb. 1978, with Moffo, Oostwoud, Ellis, Fleck, cond. Caldwell.

Germany, early 19th cent. Stiffelio (ten), an evangelical priest, does not detect the affair his wife Lina (sop) is having with Raffaele (ten). Her father Stankar (bar) suspects it and forbids her to confess to her husband as she wishes. Stankar challenges Raffaele to a duel.

The sound of the duel brings Stiffelio to the scene. When Stankar reveals that Raffaele is Lina's lover, Stiffelio starts to attack him himself.

Stiffelio offers Lina a divorce so that she may spend her life with Raffaele; he will dedicate himself to God. Stankar enters and announces that he has killed Raffaele. Following the example of the parable of Christ and the adulterous woman, Stiffelio is moved to forgive Lina.

The opera was unsuccessful and was revised by Verdi as *Aroldo*, first performed at the Teatro Nuovo, Rimini, 16 Aug. 1857; New York, AM, 4 May 1863; London, St. Pancras TH, 25 Feb. 1964.

England and Scotland, *c*.1200. Aroldo (ten) has just returned from a crusade. During his absence his wife Mina (sop) has been having an affair with Godvino (ten). Her father Egberto (bar) persuades her not to confess to Aroldo as the shock would kill him.

Egberto challenges Godvino to a duel. Aroldo interrupts the fight, but when he learns of Godvino's guilt attacks him himself. Briano (bs), Aroldo's spiritual mentor, reminds him of a Christian's duty to forgive.

Egberto, unable to live with the shame to his family's honour, is about to commit suicide when Briano enters and announces that Godvino has been recaptured. Aroldo has coerced Mina into signing a divorce agreement, when Egberto bursts in and announces that he has just killed Godvino.

Aroldo and Briano live in seclusion at Loch Lomond. During a storm on the loch, Egberto and Mina are washed ashore. Aroldo tries to drive Mina away but, after exhortations from both Briano and Egberto, forgives her.

BIBL: G. Morelli, ed., *Tornando a Stiffelio* (Florence, 1987).

**Stignani, Ebe** (*b* Naples, 11 July 1903; *d* Imola, 5 Oct. 1974). Italian mezzo-soprano. Studied Naples with Roche. Début Naples, C, 1925 (Amneris). Milan, S, 1926–56; London, CG 1937, 1939, 1952, 1955, 1957, and DL 1958; San Francisco 1948; also Verona, Gly., S America. A performer with a grand and fiery style, and a wide vocal range, famous for her Azucena, Ulrica, Eboli; also sang Orfeo (Gluck), Isabella (*L'italiana*), Adalgisa, Dalila, Ortrud. (R)

BIBL: B. De Franceschi, *Ebe Stignani* (Imola, 1982).

**stile monodico**. See STILE RAPPRESENTATIVO.

**stile rappresentativo** (It.: 'representative style'). The style of accompanied solo singing which developed around 1600 as a result of the deliberations of the Bardi *camerata. Apart from the experimental efforts of Galilei and others, the first sustained use of this new approach (sometimes also called **stile moderno** to distinguish it from the polyphony of the preceding generations) was in the operas of Peri, Caccini, and Cavalieri. Caccini's collection *Le nuove musiche* (1602) provides the most extensive early exploration of stile rappresentativo, and contains an illuminating preface. Over the following decades the style gradually split into two branches which later came to be recognized as *recitative and *aria. Despite many attempts by later scholars to distinguish between the terms **stile rappresentativo**, **stile recitativo**, and **stile monodico**, contemporary usage by composers and writers suggests that they were broadly interchangeable.

**stile recitativo**. See above.

**Still, William Grant** (*b* Woodville, MS, 11 May 1895; *d* Los Angeles, 3 Dec. 1978). US composer. Studied at Wilberforce Coll. and Oberlin Cons., then moved to New York where for several years he earned a living playing in theatre orchestras. The first Black American composer to receive widespread recognition, he composed seven operas beginning with *A Bayou Legend*. *Troubled Island*, given by the New York City Opera in 1949, was the first opera by a Black American to be staged by a major opera company.

SELECT WORKLIST (all texts by Arvey): *A Bayou Legend* (1941); *Troubled Island* (New York 1949); *Highway No. 1 USA* (1962).

BIBL: R. Haas, ed., *William Grant Still and the Fusion of Cultures in American Music* (Los Angeles, 1972).

**Stockhausen, Karlheinz** (*b* Burg Mödrath, 22 Aug. 1928). German composer and theorist. Studied Cologne Hochschule with Martin, and Cologne U; Paris with Messiaen. During the 1950s he was at the forefront of advances in electronic music; his influence spread through editorship of the journal *Die Reihe* (from 1954) and his directorship of summer composition courses at Darmstadt (from 1957). Taking Webern as a point of departure he developed the concept of 'total serialism', i.e. music in which all parameters, not just pitch, but duration, timbre, intensity as well, are subject to serial principles: many of his works also involve the use of aleatoric procedures.

Commencing in 1981, he has been involved with an operatic cycle of immense proportion, *Licht*, the first four parts of which, *Donnerstag aus Licht*, *Samstag*, *Montag*, and *Dienstag* have appeared to date.

# Stockholm

BIBL: M. Tannenbaum, *Conversations with Stockhausen* (Oxford, 1987).

**Stockholm.** Capital of Sweden. Opera was first given as a court entertainment in the mid-17th cent., and a theatre, the Confidencen, opened at the Palace of Ulriksdal in 1753; but the city was dependent on visiting French, German, and Italian companies until the accession of Gustav III in 1771. He converted the Bollhuset (or Ball House, for real tennis) for opera; the building opened in 1773 with the first opera in Swedish, the chief conductor Francesco Uttini's *Thetis och Pelée*. In 1782 the company moved to the New OH, opening with Johann *Naumann's *Cora och Alonzo*. The theatre was also used for masked balls, at one of which the King was assassinated (see BALLO IN MASCHERA). Franz *Berwald's cousin John was director 1823–49. The theatre was demolished in 1891 and reopened in 1898 as the Kungliga Teatern (cap. 1,090). Several former singers have been distinguished directors, among them John Forsell (1924–39), Joel Berglund (1949–56), and Set Svanholm (1956–63); an outstanding recent producer was Göran Gentele (1963–71). The theatre reopened after reconstruction in 1975. György Ligeti's *Le grand macabre* was commissioned in 1978. The theatre has an annexe, the Rotundan, opened in 1964 and used mainly for chamber opera. Opera is also given at the small Södra Teatern (Southern T), and operetta at the Oscar T. More experimental opera is also given by a lively young company in a converted cinema, the Folkoperan (cap. 585).

**Stollen.** See BAR.

**Stoll Theatre.** See LONDON OPERA HOUSE.

**Stoltz, Rosine** (*b* Paris, 13 Feb. 1815; *d* Paris, 28 July 1903). French soprano. Studied Paris with Ramier and L. Ponchard, Brussels with Cassel. Début (as Mme Ternaux) Brussels, M, 1832 (small roles); (as Heloïse Stoltz), 1835–7. Paris, O, from 1837, where she so much abused her influence as the director Pillet's mistress (ruthlessly eliminating rivals, e.g. Dorus-Gras) that they were both forced to resign 1847. Sang in French provinces, Lisbon, Vienna, Turin, and Brazil; returned to Opéra 1854–5; Lyons 1860, when she retired from the stage. Her subsequent private life was rich in aristocratic protectors, some of whom she married. Though vocally imperfect, she was a passionate, striking, and intelligent performer. Created many roles, e.g. title-role of Halévy's *Reine de Chypre*, Ascanio (*Benvenuto Cellini*), when Berlioz found her 'delightful', Marguerite (Auber's *Lac des fées*), and (her most famous) Léonore (Donizetti's *Favorite*); also sang Donna Anna, Rossini's Desdemona, Rachel, Fidès.

**Stolz, Robert** (*b* Graz, 25 Aug. 1880; *d* Berlin, 27 June 1975). Austrian composer and conductor, great-nephew of Teresa *Stolz. Studied with his parents, then Vienna Cons. with Fuchs and Berlin with Humperdinck. Joined Graz as répétiteur 1897; Marburg an der Donau as second conductor 1898; Salzburg as first conductor 1902; Brno, German T, 1903. Conductor, Vienna, W, from 1907, succeeding Bodanzky, where he gave the première of Straus's *Der tapfere Soldat* (1908). His first operetta was written in 1899, *Studentenulke*, but his first success was not until 1911 with *Servus Du!* This ushered in a series of light works which proved very popular in Vienna, but 1924 he moved to Berlin to work in a cabaret. Becoming well known as a film composer he moved to the USA 1940; he returned to Vienna 1946, where he wrote music for ice reviews 1952–71.

Stolz composed *c*.65 operettas of which the most popular were *Der Tanz ins Glück, Zwei Herzen im Dreivierteltakt*, and *Wenn die kleinen Veilchen blühen*. Though they fall firmly into the Viennese tradition, these have proved generally less lasting than the works of his contemporaries, above all Lehár. He also composed *c*.100 film scores and *c*.1,500 songs; one of these, 'Im Prater blüh'n wieder die Blumen', has proved his single most popular melody.

SELECT WORKLIST: *Der Tanz ins Glück* (The Dance into Happiness) (Bodanzky & Hardt-Warden; Vienna 1921); *Wenn die kleinen Veilchen blühen* (When Little Violets Bloom) (Hardt-Warden; The Hague 1932); *Zwei Herzen im Dreivierteltakt* (Two Hearts in Three-Four Time) (Knepler, Welleminsky, & Gilbert, after Reisch & Schulz; Zurich 1933).

BIBL: O. Herbrich, *Robert Stolz: König der Melodie* (Vienna, 1975).

**Stolz, Teresa** (Kostelec, 5 June 1834; *d* Milan, 23 Aug. 1902). Bohemian soprano. Studied Prague with Cabouna, Trieste with Ricci, Milan with Lamperti. Début Tiflis 1857. Appearances in Odessa, Constantinople; Bologna 1863; Milan, S, 1865–72; Cairo 1873–4; Vienna 1875; St Petersburg, with Paris I, 1876. Roles included Mathilde (*Guillaume Tell*), Lucrezia Borgia, Alice (*Robert le diable*), Norma, Giovanna d'Arco, Amelia, Gilda, Elisabeth de Valois, Leonora (*Trovatore* and *Forza*). A beautiful woman with a powerful, 'diamond-like' voice (Blanche Roosevelt), and impeccable, long-breathed phrasing, she was one of the greatest Verdi singers of her time, and created the soprano part in his Requiem. Much admired professionally and (to Giuseppina Verdi's distress) personally by the composer, who considered her the ideal Aida. Giuseppina bequeathed her a watch and a bracelet.

BIBL: F. Walker, *The Man Verdi* (London, 1962).

Her twin sisters (*b* 8 May 1827; *d* ?) **Francesca** (Fanny) and **Ludmilla** (Lidia) were both sopranos. See under FEDERICO *RICCI.

**Stolze, Gerhard** (*b* Dessau, 1 Oct. 1926; *d* Garmisch-Partenkirchen, 11 Mar. 1979). German tenor. Studied Dresden with Bader and Dittrich. Début Dresden 1949 (Moser, *Meistersinger*). Dresden until 1953; Bayreuth regularly 1951–69; Berlin, S, 1953–61; Vienna and Stuttgart from 1956; London, CG, 1960; New York, M, 1968. Also Paris, O; Naples, C; Stockholm, etc. An acclaimed David, Mime, Herod; also Captain (*Wozzeck*), Oberon (Britten), and first Oedipus (Orff). (R)

**Stone Guest, The** (Russ.: *Kamenny gost*). Opera in 3 acts by Dargomyzhsky; text Pushkin's 'little tragedy' (1830). Prod. St Petersburg, M, 28 Feb. 1872, with Platonova, Ilina, Komissarzhevsky, Melnikov, O. Petrov, cond. Nápravník; New York, Marymount Manhattan T, 25 Feb. 1986, with Stevens, Gentry, Messing, Cokorinos, Lau, cond. Kin; London, C, 23 Apr. 1987, with Harries, Burgess, Howlett, Connell, cond. Daniel.

Spain, 17th cent. Don Juan (ten) has killed the Commandant (bs), husband of Donna Anna (sop), and now returns to Seville with Leporello (bs). Seeing Anna arrive at a monastery, he resolves to seduce her. Meanwhile he dines at the home of Laura (mez), a former lover now bored with her present lover Don Carlos (bar). Juan kills Carlos in a duel and makes love to Laura.

Disguised, he gains promise of admittance to Anna's house, also inviting the Commandant's statue to dine.

Despite discovering Juan's identity, Anna welcomes Juan. They are interrupted by the arrival of the statue. Anna faints, and Juan is dragged down to Hell.

The opera was begun in 1866 but left unfinished: completed by Cui 1870, orchestrated by Rimsky-Korsakov 1870 (rev. 1902).

**Storace, Nancy** (*b* London, 27 Oct. 1765; *d* London, 24 Aug. 1817). English soprano, sister of below. Studied London with Sacchini and Rauzzini. Début aged 10 (Cupid, Rauzzini's *Le ali d'amore*). Florence 1780; out-sang the castrato Marchesi, and was obliged to leave. Vienna 1783–7, where her brilliant gift for comedy made her extremely popular. Here she married the composer John Abraham Fisher, and met Mozart, who began to write Emilia (*Lo sposo deluso*, unfinished) for her; also Susanna, which she created with enormous success, and the concert aria 'Ch'io mi scordi di te', K505. London, H, 1787–9, 1793; DL 1789–96, singing in her brother Stephen's operas. After his death in 1796, she toured the Continent with the tenor John Braham; they had a son, 1802, but could not marry with Fisher still living, and parted 1816.

**Storace, Stephen** (*b* London, 4 Apr. 1762; *d* London, 19 Mar. 1796). English composer, brother of above. His first two operas, *Gli sposi malcontenti* (1785) and *Gli equivoci* (1786), to a libretto by Da Ponte, were written in Vienna, where his sister Nancy and Michael Kelly sang in their premières. After returning to London in 1787 he wrote two further full-scale operas, one in Italian and one English opera seria, but they enjoyed little success. The remainder of his stage output consists of English dialogue operas, either full-length works or short afterpieces. Some were reworkings of operas by other composers, e.g. Cherubini's and Kreutzer's settings of *Lodoïska*, and Dittersdorf's *Doktor und Apotheker*; Storace rarely wrote an opera without some musical borrowings—even if they came from his own works.

The immediate success of *The Haunted Tower* (1789), which was one of the most popular full-length operas staged at Drury Lane during the entire 18th cent., established Storace as a leading theatre composer. His belief in the importance of collaboration between librettist and composer, demonstrated in *The Haunted Tower* and more strongly in *The Pirates* (1792), was, according to the *Thespian Dictionary* (1805), such that he thought it 'impossible for any author to procure a good opera without previously consulting his intended composer'. *The Pirates* consequently illustrates an increase in the importance of music in proportion to the words. Storace continued this practice in his later operas, and also adopted the long continuous finale movements his must have heard in Vienna. Storace's *The Cherokee* (1794) was the first English opera about the Wild West; Beethoven twice used what he assumed to be a Welsh folk tune taken from it.

**Storchio, Rosina** (*b* Venice, 19 May 1876; *d* Milan, 24 July 1945). Italian soprano. Studied Milan with Giovannini and Fatuo. Début Milan, V, 1892 (Micaëla). Milan: S, 1895–1904, and till 1918; L, 1900–6. Buenos Aires 1904–14; Rome 1917; New York, Manhattan OC, and Chicago 1921. Sang Mimì, Manon, and Sophie (Massenet); created Madama Butterfly, also Leoncavallo's Musetta (*Bohème*) and Zazà. A lyrical singer, and a sensitive exponent of vulnerable verismo heroines. (R)

**Story of a Real Man, The** (Russ.: *Povest o nastoyashchem cheloveke*). Opera in 3 acts by Prokofiev; text by composer and Mira Mendelson, after the story by Polevoy. Prod. Moscow, B, 7 Oct. 1960.

Russia, 1942. The airman Alexey (bar) is shot down and after wandering for 18 days in the forest has to have both legs amputated. When told of

681

another pilot who continued to fly after losing a leg, he resolves to return to service against doctors' orders. He is eventually reunited with his lover Olga (sop).

The opera, originally in 4 acts, was first performed privately (Leningrad, K, 3 Dec. 1948) but rejected after Zhdanov's attack on leading Russian composers.

**Stracciari, Riccardo** (*b* Casalecchio sul Reno, 26 June 1875; *d* Rome, 10 Oct. 1955). Italian baritone. Studied Bologna with Masetti. Début Bologna 1900 (Marcello). Milan, S, 1904–6, 1908–9; London, CG, 1905; New York, M, 1906–8; Paris, O, 1909; Chicago 1917–19. Thereafter mainly in Italy, Spain, S America. Last stage appearance Milan 1944. Possessed an exceptionally fine, velvety, and well-projected voice, and an aristocratic style. Large repertory included a celebrated Figaro (Rossini), Rigoletto, Amonasro, Germont. (R)

**Strada del Pò, Anna** (*fl* 1720–40). Italian soprano. Venice 1720–1; Naples 1724–6, marrying the theatre manager Aurelio del Pò (reputedly in exchange for a debt he owed her). Engaged by Handel for London 1729. She sang for him until 1737, and was the only singer to remain loyal when *Porpora's rival Opera of the Nobility drew others away. Sang more Handel roles than any other singer, and created many parts including Adelaide (*Lotario*), Partenope, Angelica (*Orlando*), Ginevra (*Ariodante*), Alcina, Atalanta, Berenice. Handel also wrote much new music for her in revivals. It was thanks to his patient help that she triumphed over the disadvantages of succeeding Faustina and Cuzzoni, and an appearance that led to the nickname 'Pig'. Her range encompassed c'–c''', and her virtuosic and expressive powers were considerable; she was also celebrated for her trill. She left England in 1738 after a quarrel with Heidegger; sang for Senesino, Naples 1739–40; retired 1741.

**Stradella, Alessandro** (*b* Rome, 1 Oct. 1644; *d* Genoa, 25 Feb. 1682). Italian composer. Sang as a boy at S Marcello del Crocifisso and was probably a pupil of Bernabei. Worked for Queen Christina of Sweden and Lorenzo Colonna, for whom his earliest dramatic works were written. During the early 1670s he was involved with Roman theatrical life, producing intermezzos and other interpolated pieces for revivals of Cavalli, Cesti, and others given at the newly opened Tordinona. In 1677 he was forced to leave Rome after quarrelling with the ecclesiastical authorities. Travelling via Venice and Turin he arrived in Genoa 1678. Here three of his operas were given at the T Falcone: *La forza dell'amor paterno*, *Le gare dell'amor eroico*, and, most innovatory, the comic *Trespolo tutore*, one of the

earliest operas in which the leading male part is taken by a bass. In many respects his work prefigures the early Neapolitan school epitomized by Alessandro Scarlatti.

Stradella's career was filled with amorous adventures, which nearly led to his murder for the abduction of a nobleman's mistress for whom he was engaged to teach music; he actually was murdered for an affair of which his mistress's brothers disapproved. His rakish behaviour is the subject of operas by Marchi (*Il cantore di Venezia*, 1835), Niedermeyer (1837), Flotow (1844), and Moscuzza (1850).

SELECT WORKLIST: *Trespolo tutore* (Villifranchi; Genoa c.1677); *La forza dell' amor paterno* (after Minato; Genoa 1678); *Le gare dell'amor eroico* (The Contests of Heroic Love) (after Minato; Genoa 1679).

**Strakosch, Maurice** (*b* Zidlochovice, 1825; *d* Paris, 9 Oct. 1887). Czech impresario. Studied Vienna; toured Europe and USA as pianist. Managed Adelina Patti's concert tours. Undertook first opera season, New York 1857; Chicago 1859; Paris 1873–4. Rome, Ap., 1884–5 with his brother **Max** (1834–92), who succeeded him as Manager of the New York Academy of Music 1887. Another brother, **Ferdinando** (?–1902), directed opera-houses in Rome, Florence, Barcelona, and Trieste. Autobiography, *Mémoire d'un impresario* (Paris, 1887).

**Straniera, La** (The Stranger). Opera in 2 acts by Bellini; text by Romani, after V.-C. Prévôt, Vicomte d'Arlincourt's novel *L'étrangère*, possibly from its dramatization by G. C. Cosenza (1827). Prod. Milan, S, 14 Feb. 1829, with Méric-Lalande, Unger, Reina, Tamburini; London, H, 23 Jun. 1832, with A. Tosi, Gioa-Tamburini, Donzelli, Tamburini; New York, Italian OH, 10 Nov. 1834, with Clementina and Rosina Fanti, Fabi, Porto, Monterasi, Sapignoli, cond. Boucher.

The Castle of Montolino, Brittany, c.1200. Arturo (ten), betrothed to Isoletta (sop or mez), falls in love with Alaide (sop), the stranger who is thought to be a witch. Discovering her with Valdeburgo (bar), in fact her brother, he suspects that she is betraying him and challenges his supposed rival to a duel. Valdeburgo is seen apparently drowning, and Alaide is accused of his murder although Arturo is ready to take the blame. Valdeburgo is found to be alive, and insists that Arturo marry Isoletta. Alaide is revealed as the morganatic wife of the French king, and in his grief Arturo kills himself.

**Strasburg** (Fr., Strasbourg; Ger., Strassburg). City in Bas-Rhin, France. The first opera was given by a visiting company, 1700. The first opera-house opened 19 June 1701; burnt down

31 May 1800; opera given in a smaller theatre, then in the church of Saint-Étienne. The T Municipal opened 23 May 1821 with Grétry's *La fausse magie*. Viardot sang Fidès and Orpheus here, and Galli-Marié made her début 1859. Destroyed in bombardment 10 Sep. 1870. Rebuilt (cap. 1,190) by German authorities, opened 4 Sep. 1893, remaining under German jurisdiction, and with Germanic repertory, until 1919. Pfitzner was director 1910–19; music directors included Otto Lohse 1897–1904, Furtwängler 1910–11, Klemperer 1910–14; Szell 1917-19. Under French control again, the theatre was directed by Paul Bastide 1919–38. Under German control again 1940–4: Rosbaud was music director 1941–4. Bastide returned 1945–8, and built up a new ensemble and repertory. In 1972 the city joined with Colmar and Mulhouse to form the *Opéra du Rhin. It has continued its enterprising policy of giving many premières and French premières.

**Stratas, Teresa** (*b* Toronto, 26 May 1938). Canadian soprano. Studied Toronto with Jessner. Début Toronto 1959 (Mimì). New York, M, from 1959; London, CG, from 1961; Milan, S, 1962; Moscow, B, 1963; Paris, O, from 1967; also Berlin, Salzburg, Lisbon. An impressive singing actress, displaying intense dramatic commitment. Roles include Cherubino, Violetta, Liù, Lisa, Salome, Lulu, Jenny (*Mahagonny*). (R)

BIBL: H. Rasky, *Stratas* (Toronto, 1988).

**Straus, Oscar** (*b* Vienna, 6 Mar. 1870; *d* Bad Ischl, 11 Jan. 1954). Austrian composer. Studied Berlin with Grädener and Bruch. In 1893 he held a succession of conducting posts in Pressburg, Brno, Teplitz, Mainz, and Hamburg, then engaged at Wolzogen's Überbrettl cabaret in Berlin 1900, where he wrote songs and stage works. Returning to Vienna, he had his first success with *Ein Walzertraum* 1907, which for a while threatened the popularity of Lehár's *The Merry Widow*. *Der tapfere Soldat* was equally well received, and enthusiastically taken up in the USA (as *The Chocolate Soldier*) but later works fared less well until *Der letzte Walzer* 1920. In 1927–48 Straus lived in France and the USA, writing several acclaimed film scores.

SELECT WORKLIST: *Ein Walzertraum* (Dörmann & Jacobson, after Müller; Vienna 1907); *Der tapfere Soldat* (The Brave Soldier) (Bernauer & Jacobson, after Shaw; Vienna 1908); *Der letzte Walzer* (The Last Waltz) (Brammer & Grünwald; Berlin 1920).

BIBL: B. Grün, *Prince of Vienna: The Life, Times, and Melodies of Oscar Straus* (London, 1955).

**Strauss, Johann (II)** (*b* Vienna, 25 Oct. 1825; *d* Vienna, 3 June 1899). Austrian composer, conductor, and violinist, eldest son of **Johann**

**Strauss I** (*b* Vienna, 14 Mar. 1804; *d* Vienna, 25 Sep. 1849). Studied with Drechsler; made his spectacular début as a composer and conductor of waltzes with his own orchestra 1844. Soon established as the most important composer of light music in Vienna, where he was known as the 'King of the Waltz', Strauss amalgated his orchestra with that of his father 1849. Appointed *Hofballmusikdirektor* 1863.

His success as a composer of dance music has tended to overshadow Strauss's importance as a composer of stage music. *Indigo und die vierzig Räuber* appeared at T an der Wien, Vienna, 1871, and ushered in a series of operettas of which *Die Fledermaus* and *Der Zigeunerbaron* were the most successful: his other works, which include *Cagliostro in Wien*, *Eine Nacht in Venedig*, and the more serious *Ritter Pázmán*, were more coolly received. But with *Die Fledermaus* he founded, with Suppé and Millöcker, a tradition of Viennese operetta which continued into the next century in the works of Straus, Stolz, Kálmán, and above all Lehár. Taking the exuberant and witty model of Offenbach's opéras bouffes, Strauss injected it with a peculiarly Viennese hedonism and sentimentality, seen not only in numbers of Offenbachian verve (such as the Champagne song, an idea which the librettist Haffner took from Lortzing's *Rolands Knappen*) but also in the ubiquitous Viennese waltz rhythm. *Der Zigeunerbaron* is a more carefully worked mixture of comic opera and operetta, and includes a strong Romantic element; and it was this which formed the manner of Viennese operetta in its first classic phase.

Strauss was at one time influenced by Liszt and Wagner, insisting on conducting the latter's music in Vienna, and in a few of his waltzes reflecting Lisztian and Wagnerian tendencies, to the grave displeasure of Hanslick.

SELECT WORKLIST (all first prod. Vienna unless otherwise stated): *Indigo und die vierzig Räuber* (Indigo and the Forty Thieves) (Steiner; 1871); *Die Fledermaus* (The Bat) (Haffner & Génée, after Meilhac & Halévy; 1874); *Cagliostro in Wien* (Zell & Génée; 1875); *Eine Nacht in Venedig* (Zell & Génée; Berlin 1883); *Der Zigeunerbaron* (Schnitzer, after Jókai; 1885); *Ritter Pázmán* (Dóczi, after Arany; 1892).

BIBL: M. Prawy, *Johann Strauss: Weltgeschichte im Walzertakt* (Vienna, 1975).

**Strauss, Richard** (*b* Munich, 11 June 1864; *d* Garmisch-Partenkirchen, 8 Sep. 1949). German composer and conductor. His father Franz (1822–1905) was a leading horn-player at the Munich Opera, and is credited with having rendered playable Wagner's original version of Siegfried's horn call. He was, however, an implacable opponent of Wagner's music and arranged for his prodigiously gifted son to be trained in the strictest classical manner. Private lessons with Meyer were combined with a traditional German schoolboy education, followed by study at the University of Munich in 1882 and 1883.

Though still relatively unadventurous in his style, Strauss had already made his mark as a composer when he left university in 1883. His early musical endeavours were greatly assisted by Hans von Bülow, who arranged for him to become assistant music director at Meiningen in 1885; he also became acquainted with the prominent Wagnerian Alexander Ritter, under whose guidance he rejected the conservative tradition in which his father had reared him and developed a more expressive, overtly Romantic style, in which traits of both Wagner and Liszt are strikingly apparent. Meanwhile, his conducting career was also advancing: in 1886 he was appointed sub-conductor at Munich, moving to Weimar in 1889. Strauss's new style was first exploited in a notable series of symphonic poems, whose extrovert emotional display and unprecedented orchestral demands earned him the reputation of an *enfant terrible*. His flamboyant and balletic demeanour as a conductor strengthened this perception, though it also brought him to the attention of theatres and concert-halls throughout Europe. By the time that he composed his first opera, *Guntram*, in 1892–3 he already had a consummate understanding of both instrumental and vocal technique, as well as a sure appreciation of the problems of the opera-house. In 1894 Strauss conducted at Bayreuth for the first time (*Tannhäuser*), while the unsuccessful première of *Guntram* marked out its composer as an avid disciple of Wagner. This was confirmed by his second opera, *Feuersnot*, produced in Dresden in 1901. Both works adhere strongly to the principles of music drama, particularly in their manipulation of *Leitmotivs, though Strauss's distinctive voice is already clearly discernible. Nowhere is this conflux of Wagner's influence and Strauss's inimitably sumptuous scoring more apparent than in the final scene of *Feuersnot*, which has enjoyed considerable popularity in the concert-hall and was the first of many memorable closing scenes.

By the time that Strauss produced his next opera, *Salome*, in 1905 his fame as a conductor had made him a household name. In 1895 he became conductor of the Berlin Philharmonic and in 1898 conductor of the Berlin Opera; in 1903 a Strauss Festival was held in London. The public outcry which followed the appearance of *Salome* thrust Strauss to the very forefront of musical avant-gardism. One of the earliest works to exhibit the Expressionist tendencies later associated with the Second Viennese School, it provoked violent opposition on account both of its libretto and its music. Wilde's horrifying, voyeuristic study of necrophilia was vividly illuminated by music whose expression ranges from that of brutal power to beguilingly sensuous beauty. Many aspects of the score seem to have been designed to cause maximum physical shock: the violent harmony, in which Wagnerian chromaticism is at times pushed to the verge of atonality; the deliberate garishness of the orchestration; the sheer loudness of the scoring; the sickly insidiousness of the melodic line: all serve to intensify the sensual horror of the story to stifling-point. Cast in one continuous act, *Salome* relies heavily upon Wagnerian principles in its construction, with the role of the orchestra assuming an even greater importance than in *Guntram* and *Feuersnot*. The view of some critics, that *Salome* is more a gargantuan orchestral tone-poem to which dramatic vocal parts have been added, is perhaps borne out in some measure by Strauss's famous exhortation to the orchestra, 'Louder! louder! I can still hear the singers.'

In *Elektra* Strauss continued along this revolutionary musical path and in setting his first libretto by Hugo von Hofmannsthal entered upon a long and active collaboration worthy of comparison with that of Mozart and Da Ponte. Though the opera contains the same emphasis on the darker side of human emotion, Hofmannsthal's treatment of the story of the murdered Agamemnon develops a strain of genuine tragedy. The horrific elements of *Salome* are still present in both libretto and music, but they are less an end in themselves, and more a means towards the articulation of a deeper message. Strauss's score is distinguished by a sharper awareness of the possibilities for musical characterization: compared to those of *Salome* the Leitmotivs are more clearly defined and generally reappear in a more recognizable form.

In 1910 Strauss resigned his post in Berlin; the following year *Der Rosenkavalier* received its first performance. Turning aside from the violence and horror of *Salome* and *Elektra*, Strauss and Hofmannsthal produced a comedy of manners concerning Viennese society of the 18th cent. Undoubtedly, this was Hofmannsthal's greatest libretto, and through the comic licentiousness of Baron Ochs, the ardent youthful passions of Octavian, and the tender resignation of the Marschallin, he created immensely sympathetic figures which have become some of the most widely loved of all operatic characters. Strauss responded with a score of unsurpassed warmth and tenderness, drawing noticeably upon the Viennese waltz tradition to provide an appropriate musical background for the amorous entanglements on stage. Despite its immediate appeal and charm, it would be wrong to think of *Der*

*Rosenkavalier* as being in any sense a simple work: it is one of Strauss's longest creations, and its touching warmth and whimsical humour belie a highly complex underlying structure.

The striking move away from Expressionism was continued in *Ariadne auf Naxos*. Intended originally as a 1-act opera for the concluding celebration at the end of Hofmannsthal's shortened German translation of Molière's *Le bourgeois gentilhomme*, for which Strauss had written the incidental music, it was first performed in 1912. The problems of assembling a cast of actors and singers on the same evening soon persuaded the authors to make their second, and better-known, version (1916), in which the events preparatory to the performance of *Ariadne* become a separate operatic Prologue. The opera itself, with its brilliant telescoping of the tragic story of Ariadne with the antics of characters from the commedia dell'arte, was one of Hofmannsthal's most inspired creations, while the new Prologue, in which the young Composer rails against the philistinistic attitudes of his master, brought forward one of Strauss's most moving characters.

In *Die Frau ohne Schatten* Strauss returned briefly to the Expressionist world of *Salome* and *Elektra*, though on this occasion the subject-matter was not concerned with physical horror, but intended as a complex psychological exploration of a kind later seen in Schoenberg's *Erwartung*. Hofmannsthal, who considered it his best work, produced a masterful libretto and if Strauss was unable to make the most of his carefully spun allegories, this was as much due to the inherent difficulties of the task itself, as to the composer's apparent inability to move beyond the realms of his own theatrical experience. Certainly, his score lacks for nothing in respect of musical invention, and there are many memorable passages and notable ideas.

It is generally considered that *Die Frau ohne Schatten* is the last work in which an overt strain of genuine modernism may be detected. By now Strauss's musical language was firmly established and already sounding conservative. Over the remaining 30 years of his life there is little sign of any serious musical development and against the backdrop of increasingly severe economic and political problems in Germany, Strauss's inspiration flagged noticeably, though he did not help himself by choosing to set a succession of weak librettos, beginning with his own *Intermezzo*. Based on a real-life incident, in which Strauss received an anonymous letter from a female admirer threatening the stability of his happy marriage, it continues his interest in drawing upon his own life's story for musical inspiration. The collaboration with Hofmannsthal, ended by his death in 1929, resulted in two further works:

the classically inspired *Die aegyptische Helena*, originally intended as an operetta, and the Viennese comedy *Arabella*. The former, like *Intermezzo*, contains some worthwhile ideas, but is generally lacking in inventiveness and memorability and suffers from an insufficiently taut structure. *Arabella*, which has enjoyed renewed popularity in recent years, was clearly designed to rekindle the success of *Der Rosenkavalier*, a task in which it fails, largely owing to its lack of the same melodic immediacy and appeal. That comparison apart, it is a charming dramatic piece, whose sensitive characterization is matched in a score of consummate craftsmanship.

For the libretto of *Die schweigsame Frau* Strauss turned to Stefan Zweig. In its handling of the well-worn themes of love and comic misunderstanding it stands together with *Der Rosenkavalier* and *Arabella*, and if less successful than either, it is only because Zweig, for all his other talents, lacked Hofmannsthal's superlative dramatic gifts and extensive theatrical experience. Strauss's score contains many fine moments, of which the prolonged ending to the second act is most noteworthy, suggesting that further collaboration between him and Zweig would have been fruitful. This was prevented by the intervention of the Nazis, who forbade any further work with Zweig, who was Jewish, and Strauss was forced to turn to the incomparably less gifted Josef Gregor, with whom he wrote three operas. For *Daphne* Strauss produced one of his most ravishing scores, while *Friedenstag*, though suffering from a noticeable paucity of dramatic and musical invention, received 100 performances. *Die Liebe der Danae* reached the stage of public dress rehearsal, but was not heard in Strauss's lifetime, owing to the sudden closing of all theatres by Goebbels in 1944.

Strauss's last opera, *Capriccio*, was based on an idea of his own, from which the conductor Clemens Krauss, who had made a notable contribution to the textual revision of *Daphne* and *Friedenstag*, fashioned the libretto. Its theme is the age-old question of whether words or music should take precedence in an opera, the conflicting arguments being placed in the mouths of a poet and a composer who are both suitors for the hand of the Countess. It is perhaps significant that *Capriccio* does not resolve the conflict—the work ends with the Countess still undecided in her choice—for Strauss himself had grappled with this problem from *Salome* onwards without coming down firmly on either side of the argument.

The last years of his life were unhappy and dogged with controversy. Having held public office during the Third Reich (he was briefly president of the Reichsmusikkammer), Strauss was forced to undergo de-Nazification after the Second World War. Despite formal acquittal, his apparently tacit support for the regime brought him hostility from many quarters and, like Wagner's, Strauss's music has been banned from public performance in Israel. This political controversy did nothing to help the career of a composer who, by the time of his death, was frankly anachronistic. Even today the prevailing view of Strauss emphasizes his fundamental conservatism, casting his music as the Liebestod of the Romantic era, while forgetting that at one point he stood at the very forefront of the avant-garde. Although the later works are regularly revived in Germany, and have been taken up in other countries, notably in the UK by the WNO, Strauss's operatic reputation rests upon *Salome, Elektra, Der Rosenkavalier*, and *Ariadne*. The influence of Wagner is obvious throughout his canon: less apparent is an equally strong debt to Mozart, a composer with whom Strauss had a great affinity. In Hofmannsthal's most successful librettos Strauss was provided with figures the equal of any that Da Ponte created for Mozart and he demonstrated the same wry humanity and sharp powers of musical characterization. (It must not be forgotten that Strauss took a large part in the shaping of the libretto: when he did not do so, as in Acts II and III of *Arabella*, there is a noticeable flagging of dramatic pace.) Despite his opulent and rich orchestration, Strauss was also capable of writing with a Mozartian delicacy and refinement. Like Mozart, he showed a particular talent for ensemble writing, which provides some of the most memorable passages in his works, including the trio at the end of *Der Rosenkavalier*. As befitted an experienced opera conductor, his handling of the vocal line was always skilled and assured, even when writing for singers at the very limits of their capabilities, as in Zerbinetta's coloratura aria in *Ariadne*, or the *Fiakermilli*'s song in *Arabella*.

Strauss once characterized himself not as a first-rate composer, but as a 'first-rate second-rate composer', and on balance this is perhaps the fairest assessment of his achievements. As a conductor, however, he had a reputation second to none in German opera, both in his own music and that of other composers. His repertoire was wide-ranging, but without doubt it was as an interpreter of Mozart that he achieved greatest fame: of particular note was his Munich production of *Don Giovanni* in 1896, for which he reinstated the buffo sextet, until that time customarily omitted. An almost demonic stage presence in his early years, with advancing age his gestures became discreet to the point of being almost invisible. The effects he achieved, however, were always

clear, lucid, and eloquent, even if frequently delivered at a tempo which seems alarmingly fast by modern standards. Having completed 1,192 performances (including 200 of his own works) when he left Berlin, Strauss became co-director, with Schalk, of the Vienna Staatsoper from 1919 to 1924, and conducted there regularly until 1942. He also conducted performances in Salzburg, Milan, London, Buenos Aires, and elsewhere, and after the end of the Second World War made an important visit to London at the encouragement of Thomas Beecham. (R) In 1894 he married the soprano **Pauline de Ahna** (1863–1950) who created Freihild in his *Guntram* (1894).

WORKLIST: *Guntram* (Strauss; Weimar 1894, rev. Weimar 1940); *Feuersnot* (Fire Famine) (Wolzogen; Dresden 1901); *Salome* (Wilde, trans. Lachmann; Dresden 1905); *Elektra* (Hofmannsthal, after Sophocles; Dresden 1909); *Der Rosenkavalier* (The Knight of the Rose) (Hofmannsthal; Dresden 1911); *Ariadne auf Naxos* (Ariadne on Naxos) (Hofmannsthal; Stuttgart 1912, rev. Vienna 1916); *Die Frau ohne Schatten* (The Woman Without A Shadow) (Hofmannsthal; Vienna 1919); *Intermezzo* (Strauss; Dresden 1924); *Die aegyptische Helena* (The Egyptian Helena) (Hofmannsthal; Dresden 1928); *Arabella* (Hofmannsthal; Dresden 1933); *Die schweigsame Frau* (The Silent Woman) (Zweig, after Jonson; Dresden 1935); *Friedenstag* (Armistice Day) (Gregor; Munich 1938); *Daphne* (Gregor; Dresden 1938); *Die Liebe der Danae* (The Love of Danae) (Gregor; rehearsed Salzburg 1944, prod. Salzburg 1952); *Capriccio* (Krauss; Munich 1942).

BIBL: W. Mann, *Richard Strauss: A Critical Study of his Operas* (London, 1964); N. Del Mar, *Richard Strauss* (3 vols., London, 1962–72); C. Osborne, *The Operas of Richard Strauss* (London, 1988).

**Stravinsky, Fyodor** (*b* Novy Dvor, 20 June 1843; *d* St Petersburg, 4 Dec. 1902). Russian bass. Father of below. Studied St Petersburg with Everardi. Début Kiev 1873 (Count Rodolfo, *Sonnambula*). Kiev until 1876; St Petersburg, M, 1876–1902. Repertory of 60 roles included Bartolo and Basilio (Rossini), Sparafucile, Henry the Fowler (*Lohengrin*), Mephistopheles (Gounod and Boito). Outstanding in the Russian repertory, e.g. as Farlaf (*Ruslan and Lyudmila*), the Miller (Dargomyzhsky's *Rusalka*), Rangoni, Varlaam (inspiring Rimsky-Korsakov to write the drinking song in *Sadko* for him). Sang in first performances of *May Night* (Golova), and *Snow Maiden* (Frost); also created roles in Tchaikovsky's *Vakula the Smith, Maid of Orleans* (Dunois), *The Enchantress* (Mamirov). Possessed a two-octave range, and a powerful, even tone. In the line of Petrov, he was, with Figner, a singer who approached his roles with an unusual care for detail and dramatic forcefulness, paving the way for Shalyapin. He was, however, also criticized for over-acting (Rimsky-Korsakov's biographer Yastrebtsev).

**Stravinsky, Igor** (*b* Oranienbaum, 17 June 1882; *d* New York, 6 Apr. 1971). Russian, then French, then US, composer. Studied St Petersburg with Kalafaty and Rimsky-Korsakov. Son of above.

Stravinsky's first opera, *The Nightingale*, both covers and concludes the period of his early spectacular, lavishly composed stage works. The first act was written in Russia in 1909; the last two were completed in Switzerland in 1914. In between came the three great early ballets, *The Firebird*, *Petrushka*, and *The Rite of Spring*. Stravinsky was the first to recognize the perils of returning to a half-finished work after such experiences, and agreed to do so only under pressure. The change of idiom is in theory justified by the contrast between the scenes in the forest with the child who loves the nightingale's song and the later ornate luxury of the Chinese court; but the stylistic and technical development is marked, and despite great beauties the opera as a whole cannot manage to bestride two separate worlds.

Most of Stravinsky's later stage works reflect a doubt about traditional opera and place the singers in a special position with regard to the action. *The Soldier's Tale* has speech and acting but no singing, though to some extent it anticipates modern chamber theatre. *The Wedding* is a ballet, or song-and-dance ritual, with songs and choruses, and Stravinsky's ambiguous feelings towards the stage were further revealed in his next dramatic work, *Renard*. Described as 'Une histoire burlesque chantée et jouée', this gives the action to dancers and places the singers (all male) in the orchestra. Written in 1915–16, it was not performed untill 1922, in the same programme as the conventionally operatic *Mavra*. Based on a Pushkin story (and dedicated to him, Glinka, and Tchaikovsky), it was written in deliberate reaction against Wagner's 'inflated arrogance' in the form of a modern opera buffa, 'because of a natural sympathy I have always felt for the melodic language, the vocal style and conventions of the old Russo-Italian opera'. There is no recitative, but a succession of arias and ensembles with simple accompaniment figures, reminiscent of classical opera buffa.

*Oedipus rex* again places the singers on the stage, but all are masked in the interest of preserving the impersonal and universal sense of tragedy that led to the choice of a dead language, Latin, for the libretto: the chorus is confined to one position and the soloists are allowed only formal entries and restricted gestures and movements. A speaker using the audience's language introduces each of the six episodes which make up the 'opera-oratorio'. Despite the wide variety of influences, the music has a powerful artistic unity, and the self-imposed constraints serve to depersonalize and ritualize the story, responding to its mythic stature.

Two ballets, *Apollon Musagète* and *Le baiser de la fée*, followed before Stravinsky resumed his highly individual relationship with the lyric stage in *Perséphone*; this is a 'melodrama' for reciter, tenor, chorus, dancers, and orchestra, closer to sung ballet than to danced opera. Two more ballets, *Jeu de cartes* and *Orpheus*, separate this from Stravinsky's next opera, *The Rake's Progress*. (A later opera, on the rebirth of the world after atomic disaster, with Dylan Thomas, was prevented by the poet's death.) Here the influences are again diverse, and in this culminating work of Stravinsky's neoclassical period the forms serve both to make contact with the world of Hogarth and to contain some of his most compassionate music. *The Flood* (1962), a setting of part of the York miracle play, is a slighter work in Stravinsky's neo-Webern manner. (R)

WORKLIST: *Solovey* (The Nightingale) (Stravinsky & Mitusov, after Andersen; comp. 1908–9, 1913–14, prod. Paris 1914); *Bayka* (Renard) (Stravinsky, after Russian folk-tales; comp. 1915–16, prod. Paris 1922); *Histoire du soldat* (The Soldier's Tale) (Ramuz; Lausanne 1918); *Mavra* (Kochno, after Pushkin; Paris 1922); *Svadebka* (The Wedding) (Stravinsky, after Russian folk texts; comp. 1914–17 and 1921–3, prod. 1923); *Oedipus rex* (Stravinsky & Cocteau, after Sophocles; Paris 1927); *Perséphone* (Gide, after the classical legend; Paris 1934); *The Rake's Progress* (Auden & Kallman, after Hogarth; Venice 1951).

BIBL: E. White, *Stravinsky: The Composer and his Works* (London, 1966, 2/1976); V. Borovsky and A. Schouvaloff, *Stravinsky on Stage* (London, 1982).

**Strehler, Giorgio** (*b* Trieste, 14 Aug. 1921). Italian producer. Studied Milan. Early career as actor. Founded Milan, PS, 1947 (with Paolo Grassi). Début Milan, S, 1947 (*Traviata*). Milan, S and PS subsequently, as the base of his operatic activities; also Paris, O, and Salzburg. Much appreciated by Brecht for productions of his work, he mounted two memorable *Die Dreigroschenoper* at PS (1956, 1973). Repertory incl. *Entführung*, *Don Giovanni*, *Zauberflöte*, *Macbeth*, *Boccanegra*, *Otello* (Verdi), *Falstaff*, *Love of Three Oranges*. Combines elements of the past (with commedia dell'arte) and present (social and political comments); and seriousness with comedy in comic roles (Cherubino, Falstaff).

**Streich, Rita** (*b* Barnaul, 18 Dec. 1920; *d* Vienna, 20 Mar. 1987). German soprano. Studied Augsburg, Berlin with Ivogün and Berger. Début Aussig (Ústi nad Labem) 1943 (Zerbinetta). Berlin, S, 1946–50, SO, 1950; Vienna

1953; London, with Vienna O, 1954; San Francisco 1957; Gly. 1958. An outstanding singer in Mozart and Strauss with coloratura roles (Queen of the Night, Zerbinetta), but also a fine, sensitive Constanze and Sophie. (R)

**Strepponi, Giuseppina** (*b* Lodi, 8 Sep. 1815; *d* Sant'Agata, 14 Nov. 1897). Italian soprano. Studied Milan Cons. Début possibly Adria 1834. Trieste and Vienna, K, 1835; Venice, Bologna, Rome, Florence, Verona, etc. 1836–42; Milan, S, 1839–42. With *Ronconi and *Moriani (the probable father of her two illegitimate children), she enjoyed a series of brilliant triumphs as Amina, Norma, Imogene, Elvira (*Puritani*), Lucia, Saffo (Pacini). She was highly praised for her attractive, smooth voice, and charming and spirited performances. However, overwork had led to a premature decline by the time she created Abigaille, 1842. She retired 1846, then taught. Her championing of Verdi's first opera, *Oberto*, contributed greatly to his recognition. She became his mistress, probably 1847, and in 1859 his second wife. A woman of extraordinary generosity and sensitivity, she loved Verdi with a devotion that survived even the difficult period of his relations with Teresa *Stolz.

BIBL: M. Mundula, *La moglie di Verdi* (Milan, 1938); F. Walker, *The Man Verdi* (London, 1962); E. Cazzulani, *Giuseppina Strepponi* (Lodi, 1984).

**stretta** (It.: 'tightening or squeezing'). The passage at the end of an act, ensemble, or aria, in which the tempo is accelerated to make a climax.

**'Stride la vampa'.** Azucena's (mez) aria in Act II of Verdi's *Il trovatore*, in which she contemplates the grim, mysterious memories reflected in the fire at which she stares.

**Striggio, Alessandro II** (*b* Mantua, ?1573; *d* Venice, 15 June 1630). Italian librettist, musician, and diplomat. The son of the composer **Alessandro Striggio I** (*c*.1540–92), he acquired a knowledge of dramatic music as a performer on the viol, participating in the famous 1589 intermedi. He served the Gonzaga court at Mantua during its artistically most creative phase, later becoming ambassador to Milan. A member of the Accademia degli Invaghiti, where he was known as 'Il Ritenuto' ('The Self-Possessed One'), Striggio provided librettos for small dramatic pieces by both Monteverdi and Gagliano. His most important work was the libretto written for Monteverdi's *Orfeo* in 1607, which represents the apotheosis of early opera. Drawing extensively on the dramatic traditions of classical tragedy and favola pastorale, it provided an admirable vehicle for the new *stile rappresentativo employed by the main characters, while the texts sung by the chorus are of a more madrigalian nature. With its

original tragic ending it stands apart from later librettos: for the printing of the score in 1609 a new ending was devised, with a *deus ex machina, but Striggio's authorship of this Neoplatonic conclusion is disputed.

**strophic.** The term used to describe arias or other vocal numbers in which all stanzas of the text are set to the same music.

**strophic variations.** A form of word-setting based upon the strophic principle, in which the bass remains unchanged, while the vocal melody is varied slightly for each stanza of the text. Widely used in early 17th-cent. opera: the classic example is 'Possente spirto' in Monteverdi's *Orfeo*.

**Strozzi, Giulio** (*b* Venice, 1583; *d* Venice, 31 Mar. 1652). Italian librettist, poet, and dramatist. Active in Venetian artistic circles from the early 1620s, he was a member of the Accademia degli Incogniti and founder of the famous Accademia degli Unisoni. With Busenello and Badoaro he played a major role in the establishment of commercial opera in Venice. Most settings of his librettos are lost, so his work must be judged on the texts themselves. These suggest a writer of considerable dramatic flair, although with their strong classical background they still reflect the earliest operatic experiments. For Monteverdi he provided *La finta pazza Licori* (1627) and *Proserpina rapita* (1630); in 1639 he collaborated with Manelli in *Delia*, which was given for the opening of the T Santi Giovanni e Paolo, Venice. His work became widely known outside Venice through Sacrati's setting of *La finta pazza* (1641) and through Cavalli's *Veremonda* (1653).

**Strungk, Nicolaus Adam** (*b* Brunswick, *bapt.* 15 Nov. 1640; *d* Dresden, 23 Sep. 1700). German composer. Engaged at courts of Wolfenbüttel, Vienna, Celle, and Hamburg before becoming music director to the cathedral and city of Hamburg. Court composer in Hanover 1682–6; Vizekapellmeister in Dresden 1688. In 1692 he founded an opera-house in Leipzig, where he lived from 1696: this opened 1693 with his *Alceste*. Strungk was one of the most important German opera composers of his generation: though most of his *c*.20 works are now lost, he made a significant contribution to the Hamburg O repertory in its earliest days and played a crucial role in the establishment of commercial opera elsewhere in Germany.

SELECT WORKLIST: *Semiramis* (Frank or Köler; Hamburg 1681); *Alceste* (Thymach, after Aureli; Leipzig 1693).

**Studer, Cheryl** (*b* Midland, MI, 24 Oct. 1955). US soprano. Studied Vienna with Hotter.

Munich, S, 1980 (Mařenka). Darmstadt 1983–5; Berlin, D, and Bayreuth from 1985; Paris, O, 1986; New York, M, 1988, also Milan, S; Vienna; Philadelphia. Roles include Pamina, Elettra, Donna Anna, Lucia, Odabella, Micaëla, Elsa, Elisabeth, Eva, Chrysothemis. (R)

**Stuttgart.** City in Baden-Württemberg, Germany. The Komödienhaus opened in the ducal Lustgarten 1674, and the first opera was probably *Amalthea* by the music director, Theodor Schwartzkopf, 1696. Under Johann Kusser opera developed 1699–1704 with French and Italian works. After a stagnant period, new interest grew after a visit by a company under Riccardo Broschi (Farinelli's brother). Singers who appeared at this time included Francesca Cuzzoni. Opera flourished under Duke Carl Eugen, from 1744. The Hoftheater opened 1750 with Graun's *Artaserse*, and was renovated 1758. Ignaz Holzbauer was Intendant from 1751, and gave works by Hasse and Galuppi. Jommelli developed operatic life 1753–69, and a number of smaller theatres were opened; the theatre was rebuilt 1759 to accommodate the ballet that, especially under J. G. Noverre 1760–7, became an important ingredient of Stuttgart opera. Antonio Sacchini succeeded Jommelli 1770, and under Johann Zumsteeg interest began to be shown in developing German opera 1793–1802, with Dittersdorf and Mozart added to the repertory. The Kleines T (cap. 1,200) opened 1812. Franz Danzi (who encouraged the young Carl Maria von Weber, the Duke's secretary 1807–10) was music director 1807–12; he was succeeded 1812 by Conradin Kreutzer, and then Hummel, who revised the repertory to include works by Méhul, Spontini, Boieldieu, and others. Peter von Lindpaintner was appointed 1819, raising standards, and remaining till his death in 1856. The Wilhelminatheater opened 1840, burnt down 1902.

Max Schillings was assistant 1908, then music director 1911–18; his *Mona Lisa* was premièred 1915. The Württembergisches Staatstheater (Grosses und Kleines Haus) (cap. 1,400) opened 1912 (when Schillings was ennobled as Von Schillings), and six weeks later saw the première of Strauss's *Ariadne auf Naxos* under the composer. Fritz Busch was music director 1918–22, conducting premières of works by Hindemith, also a Pfitzner cycle and works by Stravinsky and Schreker. He was succeeded by Carl Leonhardt 1922–37 (giving Wagner and Weber cycles) and Herbert Albert 1937–44. The company included Erb, Suthaus, and Fritz Windgassen. The theatre closed 1945, but after the war a new ensemble was quickly built up. In 1946 the German première was given of *Mathis der Maler*,

then the première of Orff's *Die Bernauerin*. Ferdinand Leitner was music director 1947–69 and Erich Schäfer Intendant 1949–72: he mounted 210 productions, including 52 contemporary works. Wieland Wagner (16 productions, more than in Bayreuth) and Günther Rennert (50 productions) worked regularly, and the company included Grace Hoffmann, Martha Mödl, Anja Silja, Gustav Neidlinger, Fritz Wunderlich, and Wolfgang Windgassen (director at the time of his death in 1974). Vaclav Neumann was music director 1969–72, Silvio Varviso 1972–80, Garcia Navarro 1987–91, Gabriele Ferro from 1992.

**Sucher, Rosa** (*b* Velburg, 23 Feb. 1847; *d* Eschweiler, 16 Apr. 1927). German soprano. Little formal vocal training. Début Munich, N, 1871 (Waltraute, *Walküre*). Berlin, H, 1875; Leipzig 1877–8; here she married the conductor Josef Sucher. Hamburg from 1878; London, DL, 1882 and CG, 1892; Vienna 1886; Bayreuth 1886, 1888, 1891, 1896; Berlin, Court O, 1888–98; New York, M, 1895, with Damrosch Co. Retired 1903. One of the finest of the first generation of Wagner singers, she was an intense performer with a splendid voice and stage presence. Her roles included Agathe, Euryanthe, Selika, Elsa, Eva, Brünnhilde, Isolde, Kundry, Desdemona (Verdi). Autobiography, *Aus meinem Leben* (Leipzig, 1914).

**Suchoň, Eugen** (*b* Pezinok, 25 Sep. 1908; *d* Bratislava, 5 Aug. 1993). Slovak composer. Studied Bratislava with Kafenda, Prague with Novák. His most successful work is *The Whirlpool*, which has been widely performed. Superficially close to Janáček in its drawing on folk inflections for its declamatory melodic lines, and in the plot of a village drama of murder, jealousy, and forgiveness, it is in fact very different in nature. The characterization is less vivid, but the work has its own powerful atmosphere: this derives partly from strongly composed singing parts that owe much to Slovak folk music, partly from excellent writing for an orchestra that plays a powerful narrative role, also from the original use of a chorus: this takes part (as in the wedding scene), comments, and acts as an abstract, somewhat pantheistic, exponent of the connection between Nature, music, and men's souls. Suchoň's other opera, *Svätopluk*, has not matched the success of this work: it makes use of serial techniques without relinquishing a lyrical manner.

WORKLIST: *Krútňava* (The Whirlpool) (Suchoň & Hoza, after Urban; Bratislava 1949); *Svätopluk* (Krčmeyrová, Stodola, & Suchoň; Bratislava 1960).

BIBL: E. Zavarský, *Eugen Suchoň* (Bratislava, 1955).

**suggeritore.** See RAMMENTATORE.

**'Suicidio!'** La Gioconda's (sop) dramatic monologue in Act IV of Ponchielli's *La Gioconda*, in which she contemplates suicide.

**Suitner, Otmar** (*b* Innsbruck, 19 May 1922). Austrian conductor. Studied Innsbruck, Salzburg with Krauss. Répétiteur Innsbruck 1942–4; music director Kaiserslautern 1957–60; chief cond. Dresden 1960–4; Berlin, S, from 1964. Bayreuth 1964 (*Tannhäuser*) and 1965 (*fliegende Holländer*). San Francisco from 1969. Admired in Mozart, Wagner, and Richard Strauss, also the Italian repertory. (R)

**Sullivan,** (Sir) **Arthur** (*b* London, 13 May 1842; *d* London, 22 Nov. 1900). English composer. Studied London RAM with Sterndale Bennett, and Leipzig (1858–61) with Hauptmann and Rietz. His first stage works were to librettos by F. C. Burnand, a 1-act farce *Cox and Box* and the 2-act *Contrabandista*. The first collaboration with W. S. Gilbert came in 1871 with *Thespis*, most of which is lost. In 1875 the impresario Richard D'Oyly Carte persuaded them to work together again on a piece to precede Offenbach's *La Périchole*: the result was the successful *Trial by Jury*, unique as being their only operetta without spoken dialogue. In the same year Sullivan collaborated with B. C. Stephenson on *The Zoo*. It had a few performances, but Sullivan soon decided to join forces with Carte and Gilbert, and a theatre was taken.

Their first production was *The Sorcerer*, which ran for 175 nights. This was followed by the immensely successful *HMS Pinafore* and *The Pirates of Penzance*, which established the pattern of what have become known, from the theatre to which Carte transferred in 1881, as the Savoy operas. Selecting venue and target carefully, Gilbert contrived successions of ingeniously whimsical or paradoxical situations in verse whose wit and brilliance of invention have rarely been matched: the chief reproaches against him are a certain mawkishness and his notoriously cruel handling of the middle-aged spinsters who recur as a fixation in many of the operettas. Sullivan brought to these admirable librettos many apt qualities. Above all, his quick melodic ear responded to the lilt and patter of Gilbert's rhythms, and gave them new and often unexpected twists. Harmonically he was less original, and occasionally matches Gilbert's sentimentality too well; but his melodic elegance extends to a contrapuntal ingenuity that enabled him to combine tunes of different character with satisfying musical and dramatic effect. The essence of his success as an operetta composer, however, was at least as much in his vivid sense of parody and the sure instinct with which he seized and both guyed and used his models. Sometimes the parody is direct: more often he fastens on a composer (e.g. Handel) or form (e.g. madrigal) and absorbs as much of it as is needed to start his own invention running. His technical skill and wide knowledge of music, especially of Romantic opera, stood him in good stead here.

*Patience* turns these gifts on the cult of the aesthetic, and in particular upon Oscar Wilde: the quality of the invention is shown by the work's continuing popularity long after the target had vanished. *Iolanthe* is a skilful mixture of fairy-tale and satire on the peerage; but in *Princess Ida* the combination of satire (on women's rights) and never-never-land court romp is less happily managed by Gilbert. Sullivan, however, provided for it some of his most ambitious and original music. A return to formula came with *The Mikado*, which was pure never-never-land, spiced as ever with topical allusions, though it took its impulse from the Victorian cult of *japonaiserie*. It was a brilliant success, running for 672 nights, and has remained the most popular of the series. *Ruddigore* followed, a comedy on country life with ghostly goings-on that reach their climax in one of Sullivan's finest extended scenes, 'The Ghosts' High Noon'. *The Yeomen of the Guard* was an attempt to break out into a more operatic manner; though it remains a vintage Savoy opera, it includes some of Sullivan's best numbers and successfully creates a pervading sense of grimness in the dominating presence of the Tower of London. *The Gondoliers* reverts brilliantly to the former type of success, and is further notable for the skilfully handled long introductions and finales without spoken dialogue.

During its run, Sullivan quarrelled with Gilbert, and having failed in his ambition to write a successful Grand Opera, *Ivanhoe*, collaborated with Sidney on the operetta *Haddon Hall*. Reconciliation with Gilbert followed, and the thirteenth of their operettas ensued, *Utopia Limited*, a satire which fails partly through trying to fire at too many Victorian targets. After a remodelling of the old *Contrabandista* as *The Chieftain*, the partners produced their last piece, *The Grand Duke*.

Sullivan went on to write *The Beauty Stone*, and with *The Rose of Persia* it was hoped that he had found his new Gilbert in Basil Hood. They began work on *The Emerald Isle*; after Sullivan's death it was completed by Edward German.

WORKLIST (texts by Gilbert and first prod. London unless otherwise stated): *The Sapphire Necklace* (Chorley (excerpts); 1867); *Cox and Box* (Burnand, after Morton; 1867); *The Contrabandista* (Burnand; 1867); *Thespis* (1871); *Trial by Jury* (1875); *The Zoo* (Rowe; 1875); *The Sorcerer* (1877); *HMS Pinafore* (1878); *The Pirates of Penzance* (Paignton; 1879); *Patience* (1881); *Iolanthe* (1882); *Princess Ida* (after Tennyson; 1884); *The Mikado* (1885); *Ruddigore*

(1887); *The Yeomen of the Guard* (1888); *The Gondoliers* (1889); *Ivanhoe* (Sturgis, after Scott; 1891); *Haddon Hall* (Grundy; 1892); *Utopia Limited* (1893); *The Chieftain* (Burnand; 1894); *The Grand Duke* (1896); *The Beauty Stone* (Pinero & Carr; 1898); *The Rose of Persia* (Hood; 1899); *The Emerald Isle* (Hood; 1901).

BIBL: G. Hughes, *The Music of Arthur Sullivan* (London, 1960); A. Jacobs, *Arthur Sullivan: A Victorian Musician* (London, 1984).

**Sulpice.** The sergeant (bs) of the 21st Grenadiers in Donizetti's *La fille du régiment*.

**'Summertime'.** Clara's (sop) lullaby to her baby in Act I of Gershwin's *Porgy and Bess*.

**Suor Angelica.** Opera in 1 act by Puccini; text by Forzano: pt2 of *Il trittico*. Prod. New York, M, 14 Dec. 1918, with Farrar, Perini, cond. Moranzoni; Rome, C, 11 Jan. 1919, with Dalla Rizza, Sadowen, cond. Marinuzzi; London, CG, 18 June 1920, with Dalla Rizza, Royer, cond. Bavagnoli.

A Tuscan convent, late 17th cent. Sister Angelica (sop) was forced by her noble family to take the veil after giving birth to a son out of wedlock. Seven years later her aunt, the Princess (con), arrives to ask her to sign away her share of the inheritance. Angelica asks for news of her son, only to be told that he has been dead for two years. She gathers herbs to make a potion to kill herself; but having drunk it, she realizes she has committed a deadly sin and prays to the Virgin for forgiveness. As she dies, she has a vision in which the Virgin brings her son to her.

**Supervia, Conchita** (*b* Barcelona, 9 Dec. 1895; *d* London, 30 Mar. 1936). Spanish mezzo-soprano. Little musical training. Début Buenos Aires 1910 (Bretón's *Los amantes de Teruel*). Rome 1911 (Octavian); Bologna 1912 (Carmen); Chicago 1915–16; Milan, S, from 1924; Turin 1925; Paris: CE, 1929; O, 1930; OC, 1933–4. London, CG, 1934–5. Died in childbirth at the height of her powers. An exceptional artist possessing an innate musicality and a *joie de vivre* that permeated her singing. Her voice was vibrant, very flexible, and had a range of g–b''. Famous as Carmen, and in Rossini, e.g. as Isabella, Rosina, Cenerentola. Also sang Cherubino, Adalgisa, Charlotte, Mignon. (R)

**Suppé, Franz (von)** (*b* Split, 18 Apr. 1819; *d* Vienna, 21 May 1895). Austrian composer of Belgian descent. Studied as a boy with a bandmaster, Ferrari, and with Cigalla, later Vienna with Seyfried and Sechter. Cond. Vienna: J, 1840–5 (also visiting Baden, Odenbürg, and Pressburg); W, 1845–62; Kaitheater, 1862–5; Carltheater from 1865. Inspired by the success of Offenbach's operettas, especially *Ba-Ta-Clan*, in Vienna, in 1860 he began, with *Das Pensionrat*,

the long series of works that was to give Viennese operetta its first classic form. His greatest successes in this manner were *Die schöne Galathea*, under the influence of Offenbach's classical parodies, and *Die leichte Kavallerie*, a piece with a contemporary Austrian military setting. In a rather more ambitious manner, he had further success with *Fatinitza* and especially with his masterpiece, *Boccaccio*. Here he comes closest to true opera, without sacrificing his typical melodic elegance and charm, his neat characterization, or his ability to develop musical ideas wittily and at some length. He had little success after *Boccaccio*; he was overshadowed by Johann Strauss II. Among his lesser-known works is a clever parody of Wagner's *Lohengrin*, *Lohengelb oder Die Jungfrau von Dragant*.

SELECT WORKLIST: *Das Pensionrat* (Suppé; Vienna 1860); *Die schöne Galathea* (Fair Galatea) (Henrion; Berlin 1865); *Die leichte Kavallerie* (Light Cavalry) (Costa; Vienna 1866); *Lohengelb oder Die Jungfrau von Dragant* (Costa & Grandjean; Vienna 1870); *Fatinitza* (Zell & Génée; Vienna 1876); *Boccaccio* (Zell & Génée; Vienna 1879).

**Surtitles** (proprietary term). Also supertitles, supratitles. A device for screening a summary translation of an opera in the course of its performance, normally projected above the proscenium arch. Invented in 1983 by John Leberg.

**Susanna.** 1. Figaro's betrothed (sop) in Mozart's *Le nozze di Figaro*.

2. An Old Believer (sop) in Musorgsky's *Khovanshchina*.

3. Count Gil's wife (sop) in Wolf-Ferrari's *Il segreto di Susanna*.

**Süssmayr, Franz Xaver** (*b* Schwanenstadt, 1766; *d* Vienna, 16 Sep. 1803). Austrian composer. He had some lessons with Mozart, and with Salieri in vocal composition. From 1795 he conducted at the Vienna, K, where he produced a number of his own operas. He is best remembered for his completion of Mozart's *Requiem*; he also probably wrote the secco recitatives for *La clemenza di Tito*.

Süssmayr's early works clearly follow the Singspiel vein of Hiller, and both composers made settings of the Rousseau-inspired *Die Liebe auf dem Lande*. Following the success of Cimarosa's *Il matrimonio segreto* (1792), Süssmayr adopted the same late Neapolitan buffo style, abandoning almost completely his previous model of Mozart; only in similarities of melody did his influence remain. Süssmayr went to Bertati, Cimarosa's librettist, for the text for his own next opera, *L'incanto superato*. Despite this enthusiasm, he completed only four Italian operas, in comparison with 21 German stage works, the most successful of which was the Singspiel

*Der Spiegel von Arkadien*. This has a text by Schikaneder, and follows in the tradition of *Die Zauberflöte*; it was immensely popular, receiving 113 performances at the T auf der Wieden alone before 1804. Encouraged by the success of this work, Süssmayr pursued the mixed form of Singspiel. Later works included some in the 'Turkish' mode of *Die Entführung aus dem Serail*—*Soliman der Zweite*, from which Beethoven took a theme for a set of piano variations, and *Gülnare, oder Die persische Sklavin*. All his stage works but two were first seen in Vienna; one of two operas staged in Prague was a setting of *Il turco in Italia*.

SELECT WORKLIST: *Die Liebe auf dem Lande* (Love in the Country) (Weisse; comp. 1785–9, unprod.); *L'incanto superato* (Bewitchment Overcome) (Bertati; Prague 1793); *Il turco in Italia* (Mazzolà; Prague 1794); *Der Spiegel von Arkadien* (Schikaneder; Weimar 1796); *Soliman der Zweite* (Huber, after Favart; Vienna 1799); *Gülnare* (Lippert, after Marsollier; Vienna 1800).

**Sutermeister, Heinrich** (*b* Feuerthalen, 12 Aug. 1910). Swiss composer. Studied Munich with Courvoisier. Worked as répétiteur, Berne, before devoting himself to composition. His intention to make strong audience appeal an essential of opera composition was rewarded with the success of the radio opera *Die schwarze Spinne*, and particularly with that of *Romeo und Julia*. Subsequent operas proved thinner in content, their effects not being commensurate with their effectiveness, and the acknowledged influence of Orff providing insufficient substance.

SELECT WORKLIST: *Die schwarze Spinne* (The Black Spider) (Rösler, after Gotthelf; Radio Berne 1936); *Romeo und Julia* (Sutermeister, after Shakespeare; Dresden 1940); *Raskolnikoff* (P. Sutermeister, after Dostoyevsky; Stockholm 1948); *Titus Feuerfuchs* (Sutermeister, after Nestroy; Basle 1958); *Madame Bovary* (Sutermeister, after Flaubert; Zurich 1967). BIBL: D. Larese, *Heinrich Sutermeister* (Amriswil, 1972).

**Suthaus, Ludwig** (*b* Cologne, 12 Dec. 1906; *d* Berlin, 7 Sep. 1971). German tenor. Studied Cologne. Début Aachen 1928 (Walther). Aachen until 1931; Stuttgart 1932–41; Berlin, S, 1941–8, and SO (later D), 1948–65; Vienna 1948–71; Gly. 1950; London, CG, 1952–3; Milan, S, 1954, 1958; also San Francisco, Buenos Aires, etc. Roles included Florestan, Erik, Tristan, Siegmund, Siegfried, Hermann, Otello (Verdi), Bacchus, Števa. A powerful, moving singer, much admired by Furtwängler. (R)

**Sutherland, (Dame) Joan** (*b* Sydney, 7 Nov. 1926). Australian soprano. Studied Sydney with her mother, later with John and Aida Dickens, London with Carey. Début Sydney 1951 (Goossens's Judith). Gly. 1959–60; London, CG, from 1952; New York, M, and Milan, S, from 1961. Also Vienna, S; Venice; Dallas; etc. Retired 1990. Gifted with a voice of outstanding flexibility, beauty, and ease in the upper register, and possessing a virtuosic technique (with a superb trill), she contributed greatly to the revival of the bel canto repertory. Though there were criticisms of her limited interpretative range and feeling for words, her extraordinary vocal abilities and attractive personality have made her one of the most popular prima donnas of her day. Roles included Alcina, Donna Anna, Gilda, Desdemona (Verdi), Eva, the *Hoffmann* heroines; particularly famous for her Rossini (e.g. Semiramide), Bellini (e.g. Elvira, Norma), and Donizetti (e.g. Lucia, Anna Bolena). Created Jenifer (*Midsummer Marriage*). Married to Richard *Bonynge, who became her preferred conductor. Awarded Order of Merit 1992. (R)

BIBL: N. Major, *Joan Sutherland* (London, 1987).

**Suzel**. A farmer's daughter (sop), lover of Fritz in Mascagni's *L'amico Fritz*.

**Suzuki**. Madam Butterfly's servant (mez) in Puccini's *Madama Butterfly*.

**Svanholm, Set** (*b* Västeras, 2 Sep. 1904; *d* Saltsjö-Duvnäs, 4 Oct. 1964). Swedish tenor. Studied Stockholm with Forsell. Début (as baritone) Swedish Royal O, 1930 (Silvio). Début there as tenor 1936 (Radamès); permanently engaged from 1937. Vienna 1938–42; New York, M, 1946–56; Milan, S, 1942, 1950; Bayreuth, 1942; London, CG, 1948–57. Director Swedish Royal O 1956–63. An intelligent, musical and robust singer, he was highly acclaimed as Tristan, Siegmund, Parsifal; also sang Florestan, Otello (Verdi), Herod, Peter Grimes. (R)

**Svoboda, Josef** (*b* Čáslav, 10 May 1920). Czech designer. Leading designer Prague O from 1948, especially Czech repertory. London: CG, *Frau ohne Schatten* 1967, *Pelléas* 1969, *Nabucco* 1972, *Tannhäuser* 1973, *Ring* 1974–6; C, *Vespri* 1984. Ottawa, *Idomeneo* 1981. Creator of the 'Lanterna magika' combining cinema projection on multiple screens. Makes inventive use of gauzes and ingenious lighting effects; and has often used a staircase as a central feature of his designs.

**Sweden**. Opera was first given during the reign of Queen Christina (1644–64), and for long was primarily a court entertainment. In the mid-18th cent. French and Italian companies were established, and German companies paid visits. Queen Lovisa Ulrika built the theatre at Drottningholm (1754). The Italian company that came in 1755 did not survive, but its director, Francesco Uttini, remained and was court music director from 1767. On his accession in 1771, Gustavus III dismissed the resident French troupe and set about encouraging Swedish opera. The old Boll-huset (Ball House for real tennis) was put in order and opened in 1773 with *Thetis och Pelée*, based on a text by the King with music by Uttini (formerly chief Drottningholm composer). This was later parodied in *Thetis och Pelée* (1779) by Stenborg, who also wrote the first Swedish historical opera, *Konung Gustaf Adolfs jakt* (1777). But also in 1773 the Swedish opera had given Gluck's *Orphée* before the Paris première, a much-prized achievement. Gluck was the model for the busy operatic activity which Gustaf stimulated in these years among composers, who included H. P. Johnsen (1717–79), with *Birger Jarl och Mechtild* (with Uttini, 1774); also the Germans Joseph Martin Kraus (1756–92), in Sweden from 1778, with *Proserpina* (1781) and *Aeneas i Carthago* (1799); Johann *Naumann (1741–1801), in Sweden from 1777 to assist Gustavus III's plans, with *Cora och Alonzo* (1782) and a national opera on an idea of the king, *Gustaf Wasa* (1786); and Georg Joseph *Vogler (1749–1814), in Sweden 1784–1806, with *Gustaf Adolph och Ebba Brahe* (1788). The first Royal OH opened in 1782. Gustav's assassination in his own opera-house in 1792 (see BALLO IN MASCHERA) led to the eclipse of the arts for some 20 years.

The various early 19th-cent. Swedish composers who attempted opera had little success. The only opera completed and staged was *Frondörerna* (The Rebels, 1835) by Adolf Lindblad (1801–78). Franz *Berwald (1796–1868) failed to win a performance for his *Drottningen av Golconda* (The Queen of Golconda, comp. 1864). Neither Ivar Hallstrøm (1826–1901), Andreas Hallén (1846–1925), Wilhelm *Stenhammar (1871–1927), nor Ture Rangström (1884–1947) had much success with their Wagnerian operas, though the former's *Waldemarsskatten* (1899) was written for the new Royal OH (opened 1898 with parts of Berwald's *Estrella di Soria*). Nevertheless, the encouragement given to singing helped the international fame of Jenny Lind, and the long tradition of fine Wagner singers included Christine Nilsson, Kerstin Thorborg, Nanny Larsén-Todsen, Joel Berglund, Sigurd Björling, Set Svanholm, and Birgit Nilsson. In composition, strong examples were set by Hilding *Rosenberg (1892–1985) and Natanael *Berg (1879–1957); and Swedish opera entered the international repertory with *Tranfjädrarna* (Crane Feathers, 1956) by Sven-Erik Bäck (b 1919) and especially *Aniara* (1959) by Karl-Birger *Blomdahl (1916–68). Among contemporary composers who have won international fame are Lars Johan *Werle (b 1926). The distinguished line of Swedish singers continued with Jussi Björling, Elisabeth Söderström, Nicolai Gedda, Kerstin Meyer, and Ingwar Wixell. In

1975 Ingmar Bergman staged a highly successful production of *The Magic Flute* for Swedish TV.

See also DROTTNINGHOLM, GÖTEBORG, MALMÖ, STOCKHOLM, UMEÅ.

**Switzerland** (Fr., Suisse; Ger., Schweiz; Ital., Svizzera). Opera did not develop in Switzerland until the composer Meyer von Schauensee (1720–89), who had served as a soldier in Italy and composed an opera for his colonel in Turin, produced some of his operas in Lucerne: the first was *L'ambassade de Parnasse* (1746). Though this was appreciated, little followed. Swiss-born composers including Rousseau, Du Puy, Schnyder von Wartensee, Stuntz, later Honegger and Liebermann, chose to work abroad. However, a number of Romantic composers, partly under the influence of *Rousseau as a philosopher, turned to Swiss subjects, including Cherubini (*Elisa*), Rossini (*Guillaume Tell*), Bellini (*La sonnambula*), and Catalani (*La Wally*). A number of foreign composers worked in Switzerland, notably Wagner during his years of exile (1849–61). Among the most important Swiss composers of the 20th cent. have been Ernest *Bloch (1880–1959), Frank *Martin (1890–1974), Willy *Burkhard (1900–55), and Rolf *Liebermann (b 1910). The most influential conductor has been Ernest *Ansermet. Opera is toured by Aargau Opera, based in Aarau.

See also BASLE, BERNE, GENEVA, ZURICH.

**Sydney.** City in New South Wales, Australia. Sydney was for many years dependent on touring companies, and the struggle to establish a permanent company in a properly equipped permanent home was long and difficult. In 1952 the New South Wales OC joined with the National Theatre OC of Melbourne; this merger was completed under the auspices of the Elizabethan Theatre Trust in 1954. Warwick Braithwaite was music director 1955–6, resigning over managerial interference. In 1955 the NSW State government announced a competition for the design for a new Sydney opera-house. This was won by Jørn Utzon, whose design immediately aroused controversy for its originality, problems of construction, and then soaring costs (finally $120m). Utzon resigned in 1966, and local architects completed the interior. In 1967 the large hall, intended for opera, was redesigned as a concert-hall (cap. 2,690) and the smaller hall became the opera theatre (cap. 1,547). In July 1970 the old opera theatre, Her Majesty's, burnt down, destroying most of the Opera's possessions. The new opera-house was opened on 20 Oct. 1973 by Queen Elizabeth II, but the first public performance was of *War and Peace*, conducted by the music director of Australian Opera, Edward Downes, on 28 Sep. He was succeeded in 1976 by

693

Richard Bonynge, whose policies led to a crisis over which the general manager, Peter Hemmings, was forced to resign: this led to an official enquiry by the Australia Council in 1986. Bonynge was obliged to step aside to the post of chief guest conductor in 1984. A crisis of confidence and cash ensued. Since then morale and standards have revived.

**Szeged.** City in Csongrád, Hungary. Touring companies first performed in 1800. The National T (cap. 1,029) opened in 1883. The present company was formed in 1945, and tours SE Hungary. Summer festivals began in 1932; interrupted by Second World War, then recommenced 1964, and have given particular prominence to Erkel's operas.

**Szell, George** (*b* Budapest, 7 June 1897; *d* Cleveland, 29 July 1970). Hungarian, later US, conductor. Studied Vienna with Mandyczewski, Foerster, and Reger. Encouraged to conduct by Richard Strauss, joined Berlin, H, 1915. Cond. Strasburg 1917; Darmstadt 1921; Düsseldorf 1922; Berlin, S, 1924–9. Music director Prague German O 1929–37. New York, M, 1946, 1953–4, much admired in Wagner and Strauss. Salzburg 1949–64, incl. premières of Liebermann's *Penelope* (1954) and Egk's *Irische Legende* (1955). A powerful and demanding conductor, who obtained performances of great energy and intensity. (R)

**Szirmai, Albert** (*b* Budapest, 2 July 1880; *d* New York, 15 Jan. 1967). Hungarian, later US, composer. Studied Budapest with Koessler. After working in a Budapest theatre as a composer, he moved to New York in 1926. His first operetta, *Sárga dominó* (1907), established him as a Hungarian composer of light opera to rank with Kálmán, and his works were in their day widely performed.

**Szokolay, Sándor** (*b* Kúnágota, 30 Mar. 1931). Hungarian composer. Studied Budapest with Szabó and Farkas. His first opera, *Blood Wedding*, drew attention to him as an opera composer of great dramatic command, and the work was quickly taken up by a number of European theatres. Its dark, somewhat Expressionist nature was modified for *Hamlet*, in which Szokolay found a more lucid and coherent style (including some skilful archaicism) upon which he was able to build with the colourful imagery of *Sámson*. *Ecce homo*, a setting of Kazantzakis's *Christ Recrucified*, develops this orchestral mastery and includes some fine choral writing, if sometimes at the expense of the solo vocal lines.

WORKLIST (all first prod. Budapest, texts by Szokolay): *Vérnász* (Blood Wedding) (after Lorca; 1964); *Hamlet* (after Shakespeare; 1968); *Sámson* (after Németh; 1973); *Ecce homo* (after Kazantzakis; 1987).

**Szymanowski, Karol** (*b* Tymoszówka, 6 Oct. 1882; *d* Lausanne, 29 Mar. 1937). Polish composer. Studied with his father and Neuhaus. Hopes of having his first opera, *Hagith*, produced in Vienna under his friend Fitelberg were not realized. It has made less mark than his most ambitious stage work, *King Roger*, an impressive attempt at treating the theme of the enticements and perils of Dionysus in a setting of the conflict between Christianity and paganism in 12th-cent. Sicily. Szymanowski's richly sensual orchestral style, and his skilful vocal writing, are at the service of an intelligent handling of a late-Romantic style.

SELECT WORKLIST: *Hagith* (Szymanowski, after Dörmann; Warsaw 1922); *Król Roger* (King Roger) (Szymanowski & Iwaszkiewicz; Warsaw 1926).

BIBL: C. Headington, *Szymanowski* (London, 1983); J. Samson, *The Music of Szymanowski* (London, 1980).

# T

**Tabarro, Il** (The Cloak). Opera in 1 act by Puccini; text by Adami, after Didier Gold's tragedy *La houppelande* (1910). Pt 1 of *Il trittico*. Prod. New York, M, 14 Dec. 1918, with Muzio, Crimi, Montesanto, cond. Moranzoni; Rome, C, 11 Jan. 1919, with M. Labia, Di Giovanni, Galeffi, cond. Marinuzzi; London, CG, 18 June 1920, with Dalla Rizza, Burke, Gilly, cond. Bavagnoli.

A barge on the Seine, Paris, early 20th cent. The bargee Michele (bar) suspects his wife Giorgetta (sop) of unfaithfulness, but tries to win her back by reminding her of how she used to shelter under his cloak. She has arranged a meeting with her lover, the barge-hand Luigi (ten), who mistakes Michele's lighting of his pipe for the signal that the coast is clear. Michele kills him and covers the body with his cloak. When Giorgetta appears, Michele invites her under the cloak again; he reveals Luigi's body and flings her down on top of it.

**tableau** (Fr.: 'picture'). In international usage, a stage 'picture' in which the action is frozen for a picturesque display. The *tableau générale*, popular in late 18th- and early 19th-cent. French theatre, was usually a statuesque representation of the dénouement or of a particularly dramatic scene. Tableaux were also a feature of some German theatrical traditions: they were popular in Dresden, and Weber originally planned to raise the curtain in the overture to *Euryanthe* on a tableau of the ghost scene.

**Tacchinardi, Nicola** (*b* Livorno, 3 Sep. 1772; *d* Florence, 14 Mar. 1859). Italian tenor. Studied Florence. Début Pisa or Livorno 1804. Milan, S, 1805; 1820–1. Successes in Rome: T Valle 1806–7, T Argentina 1809–10. Further acclaim Paris: Odéon 1811, I 1811–14. Spain 1815–17; Vienna 1816, 1823; Italy until 1831, when he retired and taught. Pupils included Frezzolini and his daughter Fanny *Persiani. Technically an accomplished singer with a generous, somewhat baritonal voice; his performances, much admired (including by Rossini), were enhanced by expressively mobile features despite his squat physique. Roles included Don Giovanni (transposed), Court Almaviva (Rossini), and a famous Otello (Rossini). See PERSIANI, FANNY.

**'Tacea la notte placida'.** Leonora's (sop) aria in Act I of Verdi's *Il trovatore*, in which she muses to Inez on the mysterious troubadour with whom she has fallen in love.

**Taddei, Giuseppe** (*b* Genoa, 26 June 1916). Italian baritone. Studied Rome. Début Rome, R, 1936 (Herald, *Lohengrin*). Rome from 1942; Vienna 1946–86; London, Cambridge T 1947, and CG between 1960 and 1967; Milan, S, 1948–51, 1955–61; also New York, M, 1985–7, Chicago, Salzburg, etc. A versatile singer impressive as Papageno, Figaro (Mozart and Rossini), Germont, Macbeth, Falstaff (Verdi), Alberich, Sachs, Scarpia. (R)

**Tadjikistan.** Not until the absorption of Tadjikistan into the USSR in 1924 was there any organized musical life. The foundation of the Opera and Ballet T (now the Ainy T) in Dushanbe in 1940 (on the basis of a musical theatre formed in 1936) was largely the work of an Armenian, Sergey Balasanyan (1902–82). He also wrote the first Tadjik national operas, *Shurishi Vose* (The Vose Revolt, 1939), *Kovay ochangar* (Kovay the Smith, 1941; with Sharif Bobokalanov (*b* 1910) ), and *Bakhtyor va Nisso* (1954). The Russian Alexander Lensky (1910–78), who lived in Tadjikistan 1937–56, wrote six operas on Tadjik subjects, including *Takhir i Sukhra* (1945) and *Arus* (The Bride, 1946). The first opera by a Tadjik composer was *Pulat i Gulru* (1957), by Sharafiddin Saifiddinov (*b* 1929), on a subject deriving much from *Fidelio* (including the critical arrival in a dungeon of rescuers heralded by a Revolutionary song); this was orchestrated by three Russian composers. Sayfiddinov also wrote *Ainy* (1978, on the national poet). Other Tadjik operas include *Komde i Madan* (1967) by Yankhel Sabzanov (*b* 1929). The subjects are almost invariably concerned with love caught up in heroic struggle; the first comedy was by a Russian composer, Samuil Urbach (1908–69), with *Bibi i Bobo* (1959).

BIBL: N. Nurjanov, *Tadzhiksky teatr* (Moscow, 1968).

**Tadolini, Eugenia** (*b* Forlì, 1809; *d* Paris, 1872). Italian soprano. Studied with Favi, Grilli, and her husband, the composer **Giovanni Tadolini** (1785–1872). Début Florence 1828. Paris, I, 1830–2; Milan, S, 1833, 1836, 1838–40, 1846; Vienna, K, 1842–4; Naples, C, and London, HM, 1848. Retired 1851. Roles included Donna Anna, Rosina, Amina, Norma, Lucia, Anna Bolena, Norina, Odabella; created Linda di Chamounix, Maria di Rohan, Verdi's Alzira. Donizetti thought highly of her ('She is a singer, she is an actress, she is everything') as did Strepponi, also Verdi, though he found her too beautiful, her voice too 'stupendous, clear, and pure', and her singing too 'perfect' for his conception of Lady Macbeth.

**Tagliafico, Joseph** (*b* Toulon, 1 Jan. 1821; *d* Nice, 27 Jan. 1900). French bass. Studied Paris with Piermarini and Lablache. Début Paris, I, 1844. London, CG, 1847–76 (1877–82 as stage-manager); also St Petersburg, New York, Paris, etc. An extremely versatile performer with a huge repertory, including Mozart, Rossini, Bellini, Verdi, Berlioz, Gounod.

**Tagliavini, Ferruccio** (*b* Reggio Emilia, 15 Aug. 1913). Italian tenor. Studied Parma with Brancucci, Florence with Bassi. Début Florence 1938 (Rodolfo). Milan, S, from 1940; New York, M, 1947–54, 1961–2; London, CG, 1950, 1955–6. Also Buenos Aires, C; Vienna; Paris; Chicago; etc. Retired 1965. A tenore di grazia outstanding as Arturo, Elvino, Edgardo, Nemorino, Werther, Fritz. (R) Married 1941 the Italian soprano **Pia Tassinari** (1903–90), with whom he often performed in *Bohème*, *L'amico Fritz*, etc. She studied Bologna with Vezzani. Début Casale Monferrato 1929 (Mimì). Milan, S, 1931–7, 1945–6; Rome 1933–43, 1951–2; New York, M, 1947–8. Later, her voice darkening, sang mez roles, e.g. Carmen, Charlotte. (R)

**taille.** The French term used for a tenor-range voice, between basse and *haute-contre, in opera from Lully to Rameau. It was also sometimes subdivided into *haute-taille* (or *première*) and *basse-taille* (or *concordant*), the latter approximating to a baritone.

**Tajo, Italo** (*b* Pinerolo, 25 Apr. 1915 *d* Cincinnati, 29 Mar. 1993). Italian bass. Studied Turin with Bertozzi. Début Turin 1935 (Fafner). Milan, S, 1940–56; Rome 1939–48; London, CG, 1950; San Francisco 1948–56; New York, M, 1948–50, 1975–90. Also Edinburgh; Paris, O; Naples, C; etc. A gifted buffo singer, with a large repertory including Mozart's Dr Bartolo, Leporello, Don Pasquale, Dulcamara, Doktor (*Wozzeck*), Ochs, and creations in works by Berio, Bucchi, Lualdi, Malipiero, Nono. (R)

**Taktakishvili, Otar** (*b* Tbilisi, 27 July 1924; *d* Tbilisi, 22 Feb. 1989). Georgian composer, conductor, and writer; son of below. Studied Tbilisi with Barkhudaryan. His operatic works include a number of historical and patriotic pieces, making use of an eclectic and popular style. His many official posts have included that of Georgian Minister of Culture.

SELECT WORKLIST: *Mindiya* (Tabukashvili, after Pshavela; Tbilisi 1961); *Sami novela* (Three Stories) (Taktakishvili, after Dzhavakhishvili & Tabidze; Tbilisi 1967).

BIBL: L. Polyakova, *Otar Taktakishvili* (Moscow, 1979).

**Taktakishvili, Shalva** (*b* Kvemo Khviti, 27 Aug. 1900; *d* Tbilisi, 18 July 1965). Georgian composer and conductor; father of above. Studied Tbilisi with Barkhudaryan. He directed opera studies at the Tbilisi Conservatory, and his own pioneering operas, which include works for children, helped to give stimulus to a younger generation.

SELECT WORKLIST: *Rassvet* (Dawn: for children) (Taktakishvili; comp. 1923, prod. Tbilisi 1926); *Deputat* (The Deputy) (Taktakishvili & Takayshvili; Tbilisi 1939).

BIBL: P. Huchua, *Shalva Taktakishvili* (Tbilisi, 1962).

**Tal, Josef** (*b* Pinne, 18 Sep. 1910). Israeli composer. Studied Berlin with Hindemith and others. Taught Palestine from 1934, director Israel Academy of Music 1948–52. Other important posts include with Hebrew University and Israel Broadcasting Authority. His most important opera is *Ashmedai*, a work in the form of a Jewish morality play; composed in free serial technique, it also employs jazz rhythms. *Massada 967* uses electronic techniques (of which Tal has been an expert teacher), organically and not merely atmospherically, but is dramatically somewhat halting. A similar lack of theatricality affects *Die Versuchung*.

SELECT WORKLIST: *Ashmedai* (Eliraz, after Buber, after the Talmud; Hamburg 1971); *Massada 967* (Eliraz, after Josephus; Jerusalem 1973); *Die Versuchung* (Eliraz; Munich 1976).

**Tale of Tsar Saltan, The.** See TSAR SALTAN.

**Talich, Václav** (*b* Kroměříž, 28 May 1883; *d* Beroun, 16 Mar. 1961). Czech conductor. Studied Prague with Mařák. Joined Berlin Philharmonic as violinist, becoming leader and thence being under the influence of Nikisch. After working in Odessa and Tbilisi as a violinist, conducted some opera in Ljubljana 1908–12, studied further Leipzig with Reger, Sitt, and Nikisch, Milan with Vigna. Cond. Plzeň 1912–15. Music director Prague, N, 1935–44, when the Nazis closed the theatre; conductor again 1947–8, again removed from post, and only restored to favour 1954. The greatest Czech conductor of his generation, he stood for much in his country's music both on the concert platform and in the opera-house, where his achievements included the advancement of Janáček's cause. His enterprising productions at the Prague, N, were frequently controversial. (R)

BIBL: H. Masaryk, ed., *Václav Talich: dokument života a díla* (Prague, 1967).

**Talvela, Martti** (*b* Hiitola, 4 Feb. 1935; *d* Juva, 22 July 1989). Finnish bass. Studied Lahti, and Stockholm with Oehmann. Début Swedish Royal O 1961 (Sparafucile). Bayreuth 1962–70; Berlin, D, from 1962; New York, M, from 1968; London, CG, 1970–3. Possessed a massive physique and voice, and a dramatic authority that could encompass both the frightening and the tender. Impressive as Osmin, the Grand Inquisitor, King Mark, Boris, Kecal, and Paavo Ruotsalainen (Kokkonen's *The Last Temptations*), written for him. Directed and performed in Savonlinna Opera Festival 1972–7, raising it to an international level, and greatly contributing to the appreciation of opera in Finland. (R)

**Tamagno, Francesco** (*b* Turin, 28 Dec. 1850; *d* Varese, 31 Aug. 1905). Italian tenor. Studied Turin with Pedrotti. Début Turin 1870 (Nearco, *Poliuto*). Palermo 1874 with great success; Venice, F, 1875; Milan, S, 1877–87, 1899, 1901; Chicago 1889; New York, M, 1890, 1894–5; London, L, 1889, and CG, 1895, 1901; also Buenos Aires, Madrid, St Petersburg. Last appearance 1904. Repertory included Arnold, Poliuto, Edgardo, Radamès, Don Carlos, Don José, Samson. Created Verdi's Otello, and roles in works by Ponchielli, Leoncavallo, and De Lara; also sang Gabriele Adorno in first revised *Boccanegra*, 1881. One of the most idolized tenors of his time, he possessed a thrilling, stentorian voice (described by Bispham as 'enduring brass') that reached a tireless top C. This, together with a natural, passionate temperament and instinctive dramatic perception (providing an ideal contrast with Maurel's Iago), made his Otello incomparable, earning even Verdi's approbation. Left his large collection of butterflies to Varese. (R)

BIBL: M. Corsi, *Francesco Tamagno* (Milan, 1937).

**Tamberlik, Enrico** (*b* Rome, 16 Mar. 1820; *d* Paris, 13 Mar. 1889). Italian tenor. Studied Naples with Zirilli and Borgna, Bologna with Guglielmi, Milan with De Abella. After a semi-public début Rome 1837, official début as Danieli, Naples 1841 (Tybalt, *I Capuleti e i Montecchi*). Naples, C, 1842–4 as Tamberlik; London, CG 1850–5, 1859–64, 1869, and HM 1877; St Petersburg, 1850–6, 1857–63 (creating Alvaro, *Forza del destino*, 1862); Paris, I, regularly 1858–77; toured Spain 1881–2; also Madrid, Buenos Aires, etc. An imposing man with a powerful, ringing, and virile tone, he was a highly sought-after artist who could sing lyrically as well as dramatically. His roles included Don Ottavio, Florestan, Arnold, Otello (Rossini), Poliuto, Cellini, Faust (Gounod). One of the first exponents of the high C♯ from the chest (to Rossini's dismay), he was also the first to interpolate the top Cs in 'Di quella pira' (with Verdi's approval). Retired to Madrid, and became an arms manufacturer.

**Tamburini, Antonio** (*b* Faenza, 28 Mar. 1800; *d* Nice, 8 Nov. 1876). Italian baritone. Studied with Boni and Osidi. Début Cento 1818 (Generali's *Contessa di colle erboso*). Milan, S, 1822,

1827–30; Naples, C, 1824, 1828–32; London, H, 1832–51, and CG 1847; Paris, I, regularly 1832–54. Also Vienna, K; Genoa; St Petersburg. He was 'singularly handsome', his voice 'rich, sweet, and equal' (Chorley), and he possessed a superb coloratura technique. Like Lablache, his frequent partner, he was a highly popular artist (his non-engagement in 1840 at London, HM, caused riots) with great comic gifts. His falsetto singing was, by all accounts, sensational. Created roles in Bellini's *Il pirata* (Ernesto), *La straniera* (Valdeburgo), and *I Puritani* (Sir Richard Forth), also in ten Donizetti operas, including *L'ajo nell'imbarazzo*, *Gianni di Calais*, *Marin Faliero*, *Don Pasquale* (Malatesta). Famous in Rossini's *L'italiana in Algeri*, *Barbiere*, *Tancredi*, *Cenerentola*, *Mosè*, *Semiramide*, *Otello*, and Donizetti's *Lucia di Lammermoor*, *Roberto Devereux*, *Lucrezia Borgia*, and as Mozart's Don Giovanni and Count Almaviva. Retired 1856, but sang Rossini's Figaro, Nice 1865.

BIBL: J. de Biez, *Tamburini et la musique italienne* (Paris, 1877).

**Tamerlano.** Opera in 3 acts by Handel; text by N. Haym, adapted from anon. libretto *Bajazet* (1719), itself based on Piovene's libretto (1711), both deriving from Pradon's play (1695), after Racine. Prod. London, H, 31 Oct. 1724, with Cuzzoni, Dotti, Senesino, Pacini, Boschi; Bloomington, Indiana U, 25 Jan. 1985, with Young, Medwetz, Patterson, Caldwell, Moore, cond. Bradshaw. Halle 1952.

Piovene's text was also set by Porpora (1730), and was used in an opera (*Tamerlano*, also called *Bajazet*) compiled from several composers by Vivaldi (1735). Other operas on the subject include Alessandro Scarlatti's *Il gran Tamerlano* (1706). See also RACINE.

**Tamino.** A Japanese prince (ten), lover of Pamina, in Mozart's *Die Zauberflöte*.

**Tancredi.** Opera in 2 acts by Rossini; text by G. Rossi, after Voltaire's *Tancrède* (1760). Prod. Venice, F, 6 Feb. 1813 with Malanotte-Montresor, Manfredini, Todràn, Bianchi, Marchesi; London 4 May 1820, with Belocchi, Corri, Torri, Angrisani; New York, Park T, 31 Dec. 1825, with Malibran, Barbieri, García, Crivelli, Angrisani, cond. Étienne.

Syracuse, 11th cent. Returning from exile in Sicily, Tancredi (con) prevents the marriage of his beloved Amenaide (sop) to Orbazzano (bs). The latter, however, intercepts a letter from her to Tancredi, which he alleges is being sent to the Saracen Chief, the enemy of the Sicilians. Amenaide is thrown in to prison and condemned to death unless a champion will fight for her honour.

Tancredi agrees to be her champion, although he believes her to be guilty of treason. He wins the duel, leads the Sicilians to victory, and finally learns that Amedaide was falsely accused. The lovers are reunited.

The work which established Rossini as an opera composer, and the first of his works to be translated. Tancredi's Act I aria *'Di tanti palpiti' became popular all over Europe. Rossini composed an alternative version of the opera with a tragic ending for Ferrara 1813.

**Taneyev, Sergey** (*b* Vladimir, 25 Nov. 1856; *d* Dyudkovo, 19 June 1919). Russian composer. Studied Moscow with Langer and Hubert, more importantly with Tchaikovsky and N. Rubinstein. His only operatic work is a trilogy based on the *Oresteia*, which he began in 1887 and finished in 1894. His painstaking approach is shown in a letter to Tchaikovsky describing how he devised the entire work in outline before planning acts, then scenes, then numbers, and only finally writing the actual music. The work reflects above all Taneyev's care, skill, and intelligence, as well as his independence: neither 'Slavophile' nor 'Westernizer', he attempted a setting that would keep aloof from Russian influences and concerns in the interests of a timeless classicism, especially by way of its elegant contrapuntal techniques.

WORKLIST: *Oresteya* (Venckstern, after Aeschylus; St Petersburg 1895).

BIBL: N. Bazhanov, *Taneyev* (Moscow 1971).

**Tanglewood.** See BERKSHIRE FESTIVAL.

**Tannhäuser und der Sängerkrieg auf Wartburg** (Tannhäuser and the Song Contest on the Wartburg). Opera in 3 acts by Wagner; text by the composer, based on a Middle High German poem, the *Wartburgkrieg* (*c*.1250), and on the *Tannhäuserlied* (*c*.1515), probably suggested by L. Bechstein's *Der Sagenschatz und die Sagenkreise des Thüringerlandes* (1835–8), with ideas from L. Tieck's *Der getreue Eckhart und der Tannenhäuser* (1799), E. T. A. Hoffmann's *Der Kampf der Sänger* (1818), and Heine's *Der Tannhäuser*. Prod. Dresden 19 Oct. 1845, with Johanna Wagner, Schröder-Devrient, Tichatschek, Mitterwurzer, Dettmer, cond. Wagner; New York, Stadt T, 4 Apr. 1859, with Siedenburg, Pikaneser, Graff, cond. Bergmann; London, CG, 6 May 1876, with Albani, D'Angeri, Carpi, Maurel, Capponi, cond. Vianesi. Rev. (Paris version), Paris, O, 13 Mar. 1861 (withdrawn by Wagner after three perfs.), with Saxe, Tedesco, Niemann, Morelli, Cazaux, cond. Dietsch; New York, M, 30 Jan. 1889, with Bettaque, Lilli Lehmann, Kalisch, Grienauer, Fischer, cond. Seidl (1st US Wagner); London, CG, 15 July 1895, with Eames, Adini, Alvarez, Maurel, Plançon, cond. Mancinelli.

The revision was dictated by the rule of the Opéra that foreign works must be in French (here, trans. Charles Truinet), and by the rigid convention that there must be an Act II ballet. Wagner insisted on opening with the new 'ballet' (the Venusberg music), for which the Opéra's resources were evidently inadequate. The main agents in whistling the work off the stage were the members of the Jockey Club.

The Wartburg, Thuringia, at the beginning of the 13th cent. Inside the Venusberg, Tannhäuser sings in praise of the pleasures offered him by Venus (mez). But sated, he longs to return to the world, and when he names the Virgin Mary, the Venusberg disappears and he finds himself in the valley of the Wartburg where a young shepherd (sop) is singing. A group of pilgrims passes on the way to Rome; then horns herald the Landgrave Hermann (bs), Tannhäuser's friend Wolfram (bar), and other knights. They welcome Tannhäuser after his year's absence, and he decides to return with them on hearing how sad the Landgrave's niece Elisabeth has been since his departure.

The Hall of Song in Wartburg Castle. Elisabeth (sop) greets the Hall, then welcomes Tannhäuser, who will not, however, reveal where he has been. The knights and their guests enter for the song contest; the Landgrave announces the theme as love. Wolfram sings of a pure, selfless love; but Tannhäuser follows with an outburst in praise of Venus. The knights threaten him, but Elisabeth intervenes. Tannhäuser promises atonement; he is banished to seek absolution from the Pope, and joins the pilgrims.

In the Valley of the Wartburg, several months later, Elisabeth is praying for Tannhäuser's forgiveness. When she sadly returns home, Wolfram prays to the evening star to guide and protect her. Tannhäuser returns, distraught at the Pope's refusal of absolution: he can now only return to Venus. A funeral procession approaches: Elisabeth has died of a broken heart. Tannhäuser sinks dead beside her. Pilgrims arrive from Rome bearing the Pope's staff, which has sprouted leaves in token that God has forgiven him.

Parodies by anon. (*Tannhäuser, oder Die Keilerei auf der Wartburg*, 1855), Binder (1857), Kalisch (1858). Other operas on the subject by Mangold (1846, reworked as *Der getreue Eckhart*, 1882: ends happily with Tannhäuser's marriage to Elisabeth), Giménez (zarzuelas, *Tannhäuser y el estanquero* and *Tannhäuser cesante*, 1890).

**Tarare.** Opera in prologue and 5 acts by Salieri; text by Beaumarchais, after a Persian tale (trans. Hamilton). Prod. Paris, O, 8 June 1787, with Maillard, Gavaudan, Chéron, Lainez, Rousseau, cond. Rey (33 consecutive performances); London, L, 15 Aug. 1825. Even more successful in Da Ponte's 4-act Italian version, with musical revisions, *Axur, re d'Ormus* (Vienna, B, 8 Jan. 1788, for wedding of Archduke Francis II, with Carmanini, Coldani, Costa).

Ormus, Persia, 1680. In the Prologue, the Genius of Nature (ten) and Genius of Fire (bar) conjure up a crowd of human shades. The popularity of Tarare (ten), a captain, with his troops arouses the envy of King Atar (bs), who tries unsuccessfully to murder him. Atar then abducts Tarare's beloved Astasie (sop). Aided by Calpigi (ten), Tarare eventually rescues her, and, after a rebellion during which Atar is killed, Tarare is acclaimed king (in a 1790 version he becomes a constitutional monarch, in 1795 an overthrower of tyranny, and with the 1815 Restoration the general in charge of the reformed Atar's army).

In the revised version as *Axur*, Astasie becomes Aspasia, and Calpigi becomes Biscroma; Taxare is now Axur (bar) and Atar is sung by a tenor.

Beaumarchais, who cast the work as an early example of *rescue opera, also contributed a preface that makes a lively and interesting defence of the librettist's position. Also opera by Mayr (*Atar*, 1812).

**Tarchi, Angelo** (*b* Naples, *c.*1760; *d* Paris, 19 Aug. 1814). Italian composer. Studied Naples with Fago and Sala. Had his first success there with the comic operas *L'archetiello*, *I viluppi amorosi*, and *Il barbiere d'Arpino*. Three operas written for Rome, including *Don Fallopio*, were also well received, as was the later opera seria *Ariarte*. In 1789 he was at London, H, producing *Il disertore francese* and *La generosità d'Alessandro*. From 1797 lived in Paris: wrote seven French comic operas for the Opéra-Comique of which the most successful was *D'auberge en auberge*. Tarchi was a fluent representative of the Neapolitan opera buffa style whose works enjoyed great popularity in their day: his opere serie were generally less highly regarded.

SELECT WORKLIST: *L'archetiello* (?; Naples 1778); *I viluppi amorosi* (The Entanglements of Love) (Mililotti; Naples 1778); *Il barbiere d'Arpino* (?; Naples 1779); *Don Fallopio* (?; Rome 1782); *Ariarte* (Moretti; Milan 1786); *Il disertore francese* (Benincasa; London 1789); *La generosità d'Alessandro* (Badini; London 1789, rev. of *Alessandro nell'Indie*, Milan 1788); *D'auberge en auberge* (Mercier-Dupaty; Paris 1800).

**tárogató.** A conical-bored wooden instrument, originally resembling a shawm and played with a double reed, brought by the Arabs to Hungary, where it became through its association with the Rákóczy movement a national symbol. It is used as such in Thern's *Svatopluc* (1839). In the 19th

cent. the double reed was replaced with a single reed and clarinet mouthpiece, and by keywork resembling that of a soprano saxophone. It is sometimes used for the second of the shepherd's tunes in Act III of *Tristan* where Wagner specifies a *Holztrompete*. Mahler's introduction of it at Budapest was followed at Bayreuth by Richter. See TRISTANSCHALMEI.

**Tarquinius.** Sextus Tarquinius, Prince of Rome, son of Lucius Tarquinius Superbus, appears (bar) in Britten's *The Rape of Lucretia*.

**Tassinari, Pia.** See FERUCCIO *TAGLIAVINI.

**Tasso, Torquato** (*b* Sorrento, 11 Mar. 1544; *d* Rome, 25 Apr. 1595). Italian poet. His persecution mania led to him being imprisoned 1579–86 by the Duke d'Este. Though there is no truth in the legend of his passion for Leonora d'Este, and the Duke's discovery leading to his imprisonment, it passed into writings by Milton, Byron, and Goethe, and is the subject of Donizetti's *Torquato Tasso* (1833). His major work is *Gerusalemme liberata* (1580–1), a poetic epic of the First Crusade which includes romantic and fabulous episodes; among the characters are Tancredi and Clorinda, Rinaldo and Armida, and Olindo and Sofronia. Operas on the work are as follows.

Generally: Zingarelli (*Gerusalemme distrutta*, 1794); Persuis (*Godefroy de Bouillon*, 1812).

The Rinaldo and Armida story: anon. (1633); Scacchi (1638); Ferrari (1639); Marazzoli (*L'amore triomfante dello sdegno*, 1641); anon. (1641); anon. (1664); anon. (1677); Lully (1686); Pallavicino (1687); Chiochiolo (1694); Orgiani (1695); anon. (1697); anon. (1698); Eccles (1699); Philip de Bourbon (1705); Boniventi (1707); Ruggieri (1707); Handel (1711); Rampini (1711); Buini (1716); Sarro (1718); Vivaldi (1718); Falco (1719); Buini (1720); Desmarets (1722); Bioni (1725); Albinoni (1726); Bioni (1726); anon. (1730); Bertoni (1746); Mele (1750–1); Graun (1751); Sarti (1759); Traetta (1761); G. Scarlatti (1766); Traetta (1767); Cannabich (?1768); Anfossi (1770); Jommelli (1770); Manfredini (1770); Salieri (1771); Sacchini (1772); Astarita (1773); Gazzaniga (1773); Naumann (1773); Gatti (1775); Tozzi (1775); Mortellari (1776); Cimarosa (1777); Gluck (1777); Mysliveček (1778); Reineck (1779); Bertoni (1780); Winter (1780); Apell (1782); Cherubini (1782); Righini (1782); Haydn (1784); Mortellari (1785); Prati (1785); Zumsteeg (1785); Sarti (1786); Zingarelli (1786); Skokoff (1788); Guglielmi (1789); Boutiny (1790); Alessandri (1794); Zingarelli and others (1794); Paradis (1797); André (1799); Mosca (1799); Haeffner (1801); Andreozzi (1802); Bianchi (1802); Righini (1803); Rossini (1817); Gläser (1825); Belisario (1828); Dabray (1851); Taducci (1868); Des

Roches (1888); Pellizona (1896); Zajc (1896); Dvořák (1904).

The Tancredi and Clorinda story: Monteverdi (1624); anon. (1633); Pollarolo (1693); Campra (1702); Bertoni (1766); Langlé (1782); Holzbauer (1783); Apell (1789); Gardi (1795); Méhul (1796); Zingarelli (1805); Pavesi (1812); Rossini (1813).

The Olindo and Sofronia story: Hertel (1767); Andreozzi (1788); Spontini (1800); Paer (?1824).

*Aminta* (1573) (the Erminia story): Rossi (1633); Moratelli (1687); anon. (1689); Pollarolo (1693); Bononcini (1719); A. Scarlatti (1723); Majo (1729); Calegari (1805); Gandini (1818); Lucantonio (1903).

Also Nicolai (*Gildippe ed Odoardo*, 1840).

**Tatar Republic.** The Musa Djalil Opera and Ballet T opened in Kazan in 1939 with *Kachkyn*, by the leading Tatar composer, Nazib *Zhiganov. Other composers of operas include Mikhail Yudin (1893–1948), with *Farida* (1944); Mansur Muzafarov (1902–66), with *Galiyabanu* (1940) and *Zulkhabire* (1943); the musical-comedy composer Jaudat Faisy (1910–73) with *Bashmagym* (The Cobblers, 1942) and *Altyn kez* (Golden Autumn, 1951); Allagiar Valiullin (1924–72) with *Samat* (1957); and Sofia Gubaidulina (*b* 1931) with *Djigangir* (1976).

**Tatyana.** Tatyana Larina (sop), in love with Onegin but later married to Prince Gremin, in Tchaikovsky's *Eugene Onegin*.

**Tauber, Richard** (*b* Linz, 16 May 1892; *d* London, 8 Jan. 1948). Austrian, later British, tenor. Studied Freiburg with Beines. Début Chemnitz (where his natural father, the actor Richard Anton Tauber, was director) 1913 (Tamino). Dresden 1913–18; Berlin, S, from 1919; Vienna, S, from 1925; Paris, O, 1928; London, CG, 1938–9, 1947; also Salzburg, Munich, Canada, S America, Australia. An exceptionally popular singer, with a lucent, mellow tone and an exquisite lyrical style, he lavished as much care on works by Lehár, who wrote several roles for him (e.g. in *Das Land des Lächelns*), as on Mozart (Belmonte, Don Ottavio, Tamino). Other roles included Florestan, Calaf, Bacchus. Also a gifted conductor. (R)

BIBL: C. Castle and D. Tauber, *This was Richard Tauber* (London, 1971).

**Taylor, Deems** (*b* New York, 22 Dec. 1885; *d* New York, 3 July 1966). US composer and critic. Studied New York, writing burlesques of opera and an original work, *The Echo*, before studying music seriously with Coon. Worked as a critic for the *New York World* (1921–5) and other journals, also as a broadcaster, especially on opera. Of his operas, both *The King's Henchman* and *Peter*

*Ibbetson* were highly successful in their day, having more performances at the Met. than those of any other American composer. Their idiom is essentially European and late-Romantic.

WORKLIST: *The King's Henchman* (Millay; New York 1927); *Peter Ibbetson* (Collier & Taylor, after Du Maurier; New York 1931); *Ramuntcho* (Taylor, after Loti; Philadelphia 1942); *The Dragon* (Taylor, after Gregory; New York 1958).

BIBL: J. Howard, *Deems Taylor* (New York, 1927, 2/1940).

**Tchaikovsky, Pyotr** (*b* Votkinsk, 7 May 1840; *d* St Petersburg, 6 Nov. 1893). Russian composer. Studied St Petersburg with Zaremba and A. Rubinstein.

Tchaikovsky declared that to refrain from writing opera was a heroism he did not possess. From his youth, the stage fascinated him: seeing a performance of *Don Giovanni* at the age of 10, he decided to devote himself to music, and throughout his life he returned to the writing of opera. His first serious attempts at composition came after the shock of his mother's death in 1854, when he wrote to a relation for a libretto for a 1-act opera to be called *Hyperbole*. Nothing came of this, but other early pieces included a scene of Pushkin's *Boris Godunov* (1863–4, lost).

Tchaikovsky's first completed opera was *The Voyevoda*, to a text partly by Ostrovsky. Though the pace is awkward and the situations are exaggerated, as a modern reconstruction shows, he was right to retain some of the best music for subsequent use when he abandoned the work (he passed the subject on to Arensky for his *A Dream on the Volga*). He then worked on *Undine*, but when it was rejected by the Imperial Theatres he destroyed it, again keeping some of the best music for re-use (e.g. a love duet for an Adagio in *Swan Lake*). His first surviving completed opera is *The Oprichnik*. Though he has not yet found his individual touch, there are some good numbers, some aptly handled motives, and passages in which convention gives way to genuine feeling. In *Vakula the Smith* he takes a much stronger operatic position, and develops a more personal vein of dramatic music. Though it lacks the comic realism and subtlety of character demanded by the plot, the work is well written, with some attractive lyrical love music and some delightful dances, and a neat contrast between peasant vigour and urban, aristocratic grace. It was later revised as *Cherevichki* ('The Slippers' or 'The Little Shoes').

His impetus to write opera was renewed in Paris in 1876 by a hearing of *Carmen*, which thrilled him especially for its association with love and a harsh fate, expressed in music of unfailing clarity and grace; and in the same year, his attendance as a special correspondent at the first Bayreuth Festival gave him an admiration for Wagner but no attraction to his ideas.

The virtues disclosed in his earlier operas find full expression in his operatic masterpiece, *Eugene Onegin*. Haunted by the predicament of Tatyana in her rejection by the subsequently remorseful Onegin, he began work on the famous *Letter Scene. He described the work as 'lyrical scenes', and hoped to get far from operatic convention as he knew it. His response to Pushkin's poem troubled some of his contemporaries (Turgenev was one who disliked the libretto), but though much has to be excluded, his response to some of its essential qualities is subtle and sensitive. Bringing together a story of frustrated love, scenes of Russian urban and country life, a feeling for the rococo and also for Russian folk music, *Onegin* touched on a great deal close to Tchaikovsky's nature, and the subject drew from him his freshest and most lyrical operatic music. Where the characters of *Vakula* are vivid types, those of *Onegin*, especially the tenderly drawn Tatyana, are real human characters.

Turning from Russian subjects, Tchaikovsky next set a version of *The Maid of Orleans*, making use of a text based on a version of Schiller. Again he was drawn above all to the heroine, and began by setting the scene of her narrative to the Dauphin and Count and her acclamation in Act II. His intention was to conquer the Paris Opéra, and the work is therefore cast in Grand Opera vein, with touches of Gounod and Massenet and with the pervading influence of Meyerbeer in its grandiose crowd scenes, processions, court ceremony, and battle scene. However, despite some fine strokes and effective passages, he is unable to give real emotional life to Joan: as he wrote in another context, 'Medieval dukes and knights captivate my imagination but not my *heart*, and where the heart isn't touched, there can't be any music.'

*Mazeppa* returns to a Russian subject. Again, Tchaikovsky began by composing a central, crucial scene so as to achieve immediate emotional identification with the characters: this was the Act II duet. The work exhibits some familiar Russian operatic characteristics, such as the suffering heroine, the intransigent old man, and the use of dances, and these are now fluently handled; but though there are delightful episodes and some striking dramatic ideas, Tchaikovsky fails to characterize the agonized Maria and the harsh, impressive Mazeppa very fully. Once again, he finds it difficult to enter into the depiction of figures so remote from his own emotional world, and so provides them with music that portrays them only partially. The finest parts of the opera are often in the minor episodes.

There was still less chance of success with *The Enchantress*, to a confused plot that drew on inconsistent musical characterization, with the tenor part French in manner, the bass Russian, and the central character of the 'Enchantress' or 'Sorceress' herself, the woman of overwhelming spirit and attraction, made by turns sentimental and dull; it is, again, in some of the dances and choruses and other incidental moments that the most successful music is to be found.

The influence of *Carmen* returns in *The Queen of Spades*, another Pushkin setting. Tchaikovsky's beloved Mozart, too, is evoked in some scenes recreating the rococo past of Catherine the Great's Petersburg and its French elegance. The opera is permeated with the sense of Fate which obsessed Tchaikovsky, operating, as in *Carmen*, within a convention of lyrical elegance; and though the work is inconsistent, above all in juxtaposing Mozartian grace with some almost Wagnerian chromatic intensity in the scenes of horror, it contains some of his finest operatic writing.

Tchaikovsky's last opera was the 1-act *Yolanta*, originally commissioned as a double bill with the ballet *The Nutcracker*. The remoteness of the medieval setting pleased him, and he was drawn to the touching idea of a blind heroine whose defect has been kept from her and is eventually cured through her love for the hero. It has much delicacy and grace, with some Wagnerian elements both harmonically and in the prominence of the orchestra in articulating the drama. Yet for all the excellence of the music in this work, and especially in *The Queen of Spades*, it is above all in the 'lyrical scenes' of *Eugene Onegin* that Tchaikovsky's individual gift for the operatic stage finds its fullest expression.

WORKLIST: *Voyevoda* (Ostrovsky & composer; Moscow 1869); *Undina* (Sollogub; unfin. 1869); *Oprichnik* (composer; St Petersburg 1874); *Kuznets Vakula* (Vakula the Smith) (Polonsky; St Petersburg 1876); *Eugeny Onegin* (composer & Shilovsky; Moscow 1879); *Orleanskaya deva* (The Maid of Orleans) (composer; St Petersburg 1881); *Mazeppa* (Burenin & composer; Moscow 1884); *Cherevichki* (The Slippers, rev. of *Vakula the Smith*) (Polonsky; Moscow 1887); *Charodeyka* (The Enchantress) (Shpazhinsky; St Petersburg 1887); *Pikovaya dama* (The Queen of Spades) (M. Tchaikovsky & composer; St Petersburg 1890); *Yolanta* (M. Tchaikovsky; St Petersburg 1892).

THEMATIC CATALOGUE: G. Dombayev, *Tvorchestvo Petra Ilyicha Tchaikovskovo* (The Works of Pyotr Ilyich Tchaikovsky) (Moscow, 1958).

WRITINGS: I. Tchaikovsky, ed., *Dnevniki Petra Ilyicha Tchaikovskovo* (The Diaries of P. I. Tchaikovsky) (Moscow, 1923; trans. New York, 1945); Various, eds., *Polnoye sobraniye sochineni* (Complete

Collected Writings), incl. letters and critical writings (Moscow, 17 vols., 1953–81).

BIBL: M. Tchaikovsky, *Zhizn Petra Ilyicha Tchaikovskovo* (The Life of P. I. Tchaikovsky) (Moscow, 1900–2; trans. London, 1906); V. Yakovlev, ed., *Dny i gody Petra Ilyicha Tchaikovskovo* (The Days and Years of P. I. Tchaikovsky) (Moscow, 1940); J. Warrack, *Tchaikovsky* (London 1973, 2/1989); D. Brown, *Tchaikovsky* (4 vols., London, 1979–91).

**Tear, Robert** (*b* Barry, 8 Mar. 1939). Welsh tenor. Studied Cambridge with Kimbell. Début Birmingham 1963 (Rameau's Hippolyte). London, EOG, 1963–71, creating roles in Britten's church parables, and 1973. Edinburgh 1968; London, CG, from 1970; Paris, O, from 1976. Also Munich; Milan, S; Berlin, D; San Francisco; Los Angeles; Geneva; etc. A singer of penetrating intelligence, vocal subtlety, and dramatic flair, distinguished as Loge, Lensky, Golitsyn, Herod, Tom Rakewell, Grimes, Aschenbach. Created Dov (*Knot Garden*), and title-role Penderecki's *Ubu rex*. (R)

**teatro di cartello.** See CARTELLONE.

**Tebaldi, Renata** (*b* Pesaro, 1 Feb. 1922). Italian soprano. Studied Parma with Campogalliani, Pesaro with Melis. Début Rovigo 1944 (Elena, Boito's *Mefistofele*). Milan, S, 1946–7, 1949–55, 1959–60; Naples, C, 1949–54; London, CG, 1950, 1955; Paris, O, 1951; New York, M, 1955–73; Chicago 1956–69. Also Buenos Aires, C; Vienna; Rome; etc. Her full, smooth, glowing tone was capable of both great power and a ravishing pianissimo; and her emphasis on the vocal and lyrical led her supporters to foster a rivalry with Callas. Roles included Donna Elvira, Elsa, Eva, Tatyana; particularly impressive in Verdi, Puccini, and as Madeleine (*Andrea Chenier*) and Adriana Lecouvreur. (R)

BIBL: V. Seroff, *Renata Tebaldi: The Woman or the Diva* (New York, 1975).

**Te Kanawa,** (Dame) **Kiri** (*b* Gisborne, 6 Mar. 1944). New Zealand soprano. Studied Auckland with Leo, London with Rozsa. Début London, Camden Fest., 1969 (Elena, *La donna del lago*). London, CG, from 1970; Gly. 1973; New York, M, from 1974; Paris, O, from 1974; also Vienna, Brussels, Santa Fe, San Francisco, Sydney, etc. One of the most successful divas of the day, whose chief asset is a sumptuous, creamy tone. Roles include Donna Elvira, Countess (Mozart and R. Strauss), Marguerite, Micaëla, Violetta, Elisabeth de Valois, Mimì, Arabella. (R)

BIBL: D. Fingleton, *Kiri* (London, 1982).

**Telemachus.** In Greek mythology, as recounted in Homer's *Odyssey*, the son of *Odysseus and *Penelope. After helping his father to kill his mother's suitors, he visited the island of *Circe

(in some versions of the legend, and hence in some operas, Calypso), marrying her or her daughter Cassiphone. He later killed Circe and fled to Italy, where he founded Clusium. Some operas on his story concern the Circe or Calypso episode, and therefore overlap with with those dealing with Circe as their main subject. Operas on him are by Campra (1704), Schürmann (1706), Graupner (1711), Destouches (1714), A. Scarlatti (1718), Gluck (1765), Meucci (1773), Bertoni (*Telemacco ed Euridice nell'isola di Calipso*, 1777), Grua (1780), Cipolla (1785), Da Silva (1787), Hoffmeister (*Der Königssohn von Ithaka*, 1795), Le Sueur (1796), S. Mayr (1797), Sor (1798), Kozeluch (1798), Kauer (1801), Boieldieu (1806), Bishop (ballad opera, 1816), and Triebensee (1824). Also zarzuela by Rogel (*Telemacco en la Albufera*, 1866).

**Telemaco.** Opera in 2 acts by Gluck; text by Coltellini, after Capece's text for A. Scarlatti (1718), in turn after the Odyssey. Prod. Vienna, B, 30 Jan. 1765.

Greece, antiquity. Ulisse (ten) is rescued from his capture on Circe's island by his son Telemaco (con), aided by Asteria (sop) and Merione (sop), son of King Idomeneo of Crete. In retribution, Circe (sop) conjures up before Telemaco's eyes images of his mother's death. However, all manage to escape from the island, which Circe then transforms to desert.

**Telemann, Georg Philipp** (*b* Magdeburg, 14 Mar. 1681; *d* Hamburg, 25 June 1767). German composer. Largely self-taught; his first opera, *Sigismundus* was written when aged 12. In 1700–4 he attended Leipzig U, where he founded the Collegium Musicum and wrote four operas. Held various positions in Sorau, Eisenach, Frankfurt, and Bayreuth before becoming music director of the principal churches in Hamburg 1721. Here he stayed for the rest of his career, though he made many journeys to Berlin and an important trip to France 1737.

Though roughly half of Telemann's 40 operas were composed for Leipzig, Eisenach, and Bayreuth, he is chiefly remembered for the *c*.20 works which he contributed to the repertory of the Hamburg Opera. His activities there, which began with *Der geduldige Socrates*, dominated the Opera until its closure in 1738 and represented the culmination of the first phase of Hamburg's operatic history. Though the influence of earlier German opera composers, especially Keiser, is apparent, Telemann was more cosmopolitan in style, drawing inspiration from France especially: as a boy he had studied the works of Lully and Campra. A more skilled dramatic talent than Mattheson or Graupner, he was much the most adventurous German composer of his day, and enjoyed greater fame even than J. S. Bach. His most successful work was the intermezzo *Pimpinone* (whose correct title is *Die ungleiche Heyrath*): an adaptation of a work by Albinoni, its mixture of German and Italian text recalls the earlier Hamburg tradition, though in its use of the buffo style it anticipates Pergolesi's *La serva padrona* by several years. Other important works were *Der neumodische Liebhaber Damon* and *Flavius Bertaridus*.

SELECT WORKLIST (all first prod. Hamburg): *Der geduldige Socrates* (Patient Socrates) (König, after Minato; 1721); *Der neumodische Liebhaber Damon* (?Telemann; 1724); *Die ungleiche Heyrath* (usually known as *Pimpinone*) (Praetorius, after Pariati; 1725); *Flavius Bertaridus* (Telemann & Wendt, after Ghigi; 1729).

**Telephone, The.** Opera in 1 act by Menotti; text by composer. Prod. New York, Heckscher T, 18 Feb. 1947, with Cotlow, Kwartin, cond. Barzin; London, Aldwych T, 29 Apr. 1948, with Cotlow, Rogier, cond. Balaban.

Lucy's apartment, present-day. Ben (bar) visits Lucy (sop) on the eve of his departure so as to propose marriage. His attempts are repeatedly frustrated by the telephone ringing. Eventually he finds the only recourse is to leave and ring her up himself; he is accepted.

**television opera** (Fr., *opéra en télévision*; Ger., *Fernseh-Oper*; It., *opera televisiva*). The first television experiments in opera took place in the BBC in 1936, when scenes from Albert Coates's *Pickwick* were broadcast. During the following three years the BBC gave 20 or more TV operas in full or shortened versions, including the English première of Busoni's *Arlecchino* and works by Handel, Falla, Méhul, Dibdin, and others. The first US TV opera was a shortened *Pagliacci* from Radio City Music Hall in March 1941. TV opera was resumed in England in 1946 with *The Beggar's Opera*. In November 1948 in New York the *Otello* opening of the Metropolitan season was televised live. Also in New York, NBC formed a Television Opera T, 1949–64, beginning with Weill's *Down in the Valley* and ending with *Lucia di Lammermoor*; it commissioned operas from Menotti (*Amahl and the Night Visitors*, the first TV commission, 1951) Martinů (*Ženitba* (The Marriage), 1953), Dello Joio, Foss, and others, and gave the US premières of *Billy Budd* and *War and Peace*. In France, transmissions began regularly in the 1950s, especially under the direction of Henri Spade. The first colour transmission of opera was NBC's *Carmen* (1953). In 1955 *The Magic Flute* was televised at the reopening of the Hamburg Opera, and *Don Giovanni* from the Berlin Deutsche Oper in 1961. The 1959 Salzburg prize for a TV opera was won by Sutermeis-

ter with *Das Gespenst von Canterville*. The first Eurovision broadcast was of *Fledermaus* from the Vienna Opera 1963. In 1969 the US National Education Television formed an opera theatre, in its first season giving the US première of *From the House of the Dead*. CBS commissioned Stravinsky's *The Flood*, 1962. In England, operas commissioned for TV include Benjamin's *Mañana* (1956), Bliss's *Tobias and the Angel* (1960), and Britten's *Owen Wingrave* (1971).

The many different techniques that have been employed include filming from a theatre, with several cameras, and studio performances, sometimes with actors or the singers miming and the sound-track added later, occasionally also making use of locations as well as the theatre. A remarkable example of a mixed solution was Bergman's version of *The Magic Flute* for Swedish TV (1975). The problem of the inadequate sound systems in most TV sets has sometimes been overcome by simultaneous stereo broadcasts.

**Telramund.** Friedrich von Telramund (bar), Count of Brabant, husband of Ortrud, in Wagner's *Lohengrin*.

**Temesvár.** See TIMIŞOARA.

**Templer und die Jüdin, Der** (The Templar and the Jewess). Opera in 3 acts by Marschner; text by W. A. Wohlbrück, after Scott's *Ivanhoe* (1820) and J. Lenz's *Das Gericht der Templer* (1824). Prod. Leipzig 22 Dec. 1829, rev. version with recit. replacing dialogue, Berlin 3 Aug. 1831, with Seidler, Devrient; London, Prince's T, 17 June 1840; New York, Stadt T, 29 Jan. 1872, with Fabbri-Mulder, Rosetti, Bernard, Formes. Rev. version by Mottl; also by Pfitzner, prod. Strasburg 1912.

York, 1194. In the nearby forest, the Norman Templars De Bracy (ten) and Bois-Guilbert (bar) meet in their pursuit of Rowena (sop), ward of the Saxon Cedric (bs), and Rebecca (sop), daughter of Isaac, the Jew of York. Cedric appears, angry that his disinherited son Wilfrid of Ivanhoe (ten) has been victor at a tournament, but he is reproached by Rowena, who loves Ivanhoe, and by his fool Wamba (ten). Isaac and Rebecca appear seeking protection for themselves and for a wounded companion (in fact, Ivanhoe, whom Rebecca also loves). In a hermit's cell, Friar Tuck (bs) entertains the Black Knight (bs), newly returned from the Crusades, with Locksley (bar) and his outlaws. In De Bracy's tower, Cedric, Wamba, and Rowena are prisoners. Rebecca, also a prisoner, is threatened by Bois-Guilbert, demanding her love. When the Saxons besiege the castle, he carries her off, and the Black Knight and Locksley rescue the others.

In a forest clearing, the Black Knight reveals

himself to the outlaws, Ivanhoe, Cedric, and the others as Richard Cœur de Lion. Isaac asks for Ivanhoe's help in rescuing Rebecca, accused by Bois-Guilbert of witchcraft and sentenced by the Grand Templar to death. At Templestowe, Bois-Guilbert is torn by conflicting emotions about his treatment of Rebecca, and plans to rescue her from being burnt alive.

In a cell at Templestowe, Rebecca is at prayer. Bois-Guilbert again tries to force her to accept him. In the tournament grounds, Rebecca's trial is held. To his horror, Bois-Guilbert is named as the Templars' champion; Ivanhoe is to be Rebecca's champion. Bois-Guilbert collapses, dead, Ivanhoe releases Rebecca, and the King arrives to dismiss the Templars and their courts.

Regarded in Marschner's lifetime as his finest work. For other Ivanhoe operas, see SCOTT.

**tempo d'attacco.** In 19th-cent. Italian opera, the fast opening section of a tripartite duet or ensemble.

**tempo di mezzo.** In 19th-cent. Italian opera, a free, transitional section which occurs between the cantabile and cabaletta of a duet or ensemble, or between the pezzo concertato and stretta of the finale.

**Tender Land, The.** Opera in 2 acts by Copland; text by Horace Everett, after Johns. Prod. New York, City O, 1 Apr. 1954, with Carter, Newton, Treigle, cond. Schippers. Rev. version, Berkshire Music Centre 2 Aug. 1954; Cambridge, Arts T, 26 Feb. 1962, by Cambridge University Opera Group, with Wells, Westwood, Ford.

USA, c.1950. Laurie Moss (sop), a farmer's daughter, graduates from high school and, despite being hampered by her over-protective mother (con), determines to embark fully upon adult life. She does so with the help of two strangers, Martin (ten) and Top (bar), from outside her close community. But Martin fails to keep a promise to elope with her, and the pain this causes her forces her at last to grow up.

**Tenducci, Giusti Ferdinando** (b Siena, c.1735; d Genoa, 25 Jan. 1790, or perhaps later). Italian soprano castrato. Earliest known appearance Venice 1753 (Bertoni's *Ginevra*). London, H, 1758, to great acclaim, particularly in Arne's *Artaxerxes*. Befriended Mozart there 1764, and gave him singing lessons. Dublin 1765, with much success. His scandalous marriage there to Dora Maunsell in 1766 resulted in his imprisonment. They were later reunited, and Casanova reports meeting their two children. (Her family had the marriage anulled 1775.) Worcester 1770 as Orfeo; he later established the popularity of 'Che farò' in London. Florence 1771; Venice and Naples 1772–5; Paris 1778, where Mozart composed an aria for him (now lost). His last stage appearance (London 1785) was not a success. After his wife died, he left England. Though not a great artist, he was much admired for his 'singing from the heart' (the publisher George Thomson).

**tenor** (It. *tenore*, 'holding': in early times the voice which held the plainsong). The highest category of natural male voice. Many subdivisions exist within opera-houses: the commonest in general use (though seldom by composers in scores) are given below, with examples of roles and their approximate tessitura. These divisions often overlap, and do not correspond exactly from country to country. In general, distinction is more by character than by tessitura, especially in France: thus, the examples of the roles give a more useful indication of the different voices' quality than any technical definition.

French: ténor (Don José: c–c''); ténor-bouffe (Paris in *La belle Hélène*; c–c''); Trial (Torquemada in *L'heure espagnole*); *taille.

German: Heldentenor (Huon, Bacchus: c–c''); Wagnerheldentenor (Siegfried: c–bb'); lyrischer Tenor (Max: c–c''); Spieltenor (Pedrillo, David: c–bb'); hoher Tenor (Brighella, *Rosenkavalier* tenor: c–c'').

Italian: tenore (Radamès: c–c''); tenore spinto (Rodolfo: c–c''); tenore di forza (Otello: c–c''); tenore di grazia (Nemorino: c–d''); tenor-buffo (Dickson, *La dame blanche*: c–bb').

See also CASTRATO, COUNTER-TENOR.

**tenuto** (It.: 'held'). A direction to hold notes for their full value, or very slightly longer. Consistently taken by singers as an invitation to hold them, especially if high ones, for as long as they please.

**Teresa.** The daughter (sop) of the papal treasurer Giacomo Balducci, lover of Cellini, in Berlioz's *Benvenuto Cellini*.

**Ternina, Milka** (b Vezišće, 19 Dec. 1863; d Zagreb, 18 May 1941). Croatian soprano. Studied Zagreb with Wimberger, Vienna with Gänzbacher. Début Zagreb 1882 (Amelia, *Ballo in maschera*). Leipzig 1883; Bremen 1886–90; Munich 1890–1906; Boston 1896; London, CG, 1898, 1900, 1906; Bayreuth 1899; New York, M, 1899–1904. Retired 1906, with a facial paralysis, and taught; her pupils included Zinka Milanov. A stupendous actress with a voice to match, she was triumphant in a wide range of roles, including Donna Anna, Reiza, Marguerite, Aida, Elsa, Ortrud, Sieglinde, Brünnhilde, Kundry, Tosca, and above all as Leonore and Isolde. Henry James found her singing 'a devastating experience'.

**terzett.** An ensemble for three singers. Although never as popular as the duet, the terzett was a feature of 17th-cent. opera, and also appeared occasionally in 18th-cent. opera seria. In opera buffa it became very popular towards the end of the 18th cent.: Mozart wrote one particularly fine example in *Figaro*, in which the characteristics and mannerisms of the three participants are brilliantly conveyed by rapid changes in the musical style. Several 19th- and early 20th-cent. composers showed considerable fondness for the terzett, notably Verdi and Strauss, who places one for the three principal female voices at the final climax of *Der Rosenkavalier*.

**Teschemacher, Margarete** (*b* Cologne, 3 Mar. 1903; *d* Bad Wiessee, 19 May 1959). German soprano. Studied Cologne. Début Cologne 1924 (Micaëla). Aachen 1925–7; London, CG, 1931, 1936; Stuttgart 1931–4; Dresden 1935–46; Düsseldorf 1947–52; Buenos Aires, C, 1934. Roles included Senta, Sieglinde, Countess (*Capriccio*), Arabella, and a much-acclaimed Jenůfa and Minnie. Created Daphne, as which she was praised by Strauss. (R)

**Teseo.** Opera in 5 acts by Handel; text by Nicola Haym, after Quinault's libretto *Thésée*, set by Lully (1675). Prod. London, Queen's T, 10 Jan. 1713, with Pilotti, Margarita, Barbier, Valeriano, Valentini, Leveridge, L'Épine.

**Tesi-Tramontini, Vittoria** (*b* Florence, 13 Feb. 1700; *d* Vienna, 9 May 1775). Italian contralto. Studied Florence with Redi, Bologna with Campeggi. Début Parma 1716 (in Astorga's *Dafni*). Dresden 1719; Florence 1721, and throughout Italy until 1747, with great success. Vienna 1748–54, after which she retired and taught, her pupils including Caterina Gabrielli. Sang in operas by Handel, Lotti, Jommelli, and Gluck. A somewhat masculine woman who held her own during a castrato-dominated period, she was highly regarded by Metastasio, Quantz, Dittersdorf, and Burney, among others, for her strong, extensive voice, imposing stage presence, and dramatic expression. She once prevented a duel between Cafarelli and the poet Migliavacca by stepping between their swords.

**Tess, Giulia** (*b* Milan, 19 Feb. 1889; *d* Milan, 17 Mar. 1976). Italian soprano. Studied Verona with Bottagisio. Début, as mezzo-soprano, Prato 1904. Venice and Prague 1909; also Vienna, St Petersburg, as Mignon, Amneris, Charlotte, etc. After further study, became soprano. Milan, S, 1922–36. Created Jaele (Pizzetti's *Deborah e Jaele*), Orsola (Wolf's *Il campiello*); also successful in verismo roles and as Salome, Elektra, and Honegger's Judith. Retired 1940, becoming a producer (e.g. Milan, S, 1946, *Cenerentola*) and

teacher, her pupils including Tagliavini, Barbieri. Married the Italian conductor **Giacomo Armani** (1868–1954).

**tessitura** (It.: 'texture'). A term indicating the prevailing range of a piece of music in relation to the voice for which it is written. Thus, Zerbinetta's aria in *Ariadne auf Naxos* has a particularly high tessitura.

**Tetrazzini, Eva** (*b* Milan, Mar. 1862; *d* Salsomaggiore, 27 Oct. 1938). Italian soprano, sister of below. Studied Florence with Ceccherini. Début Florence 1882 (Marguerite). New York, AM, 1888; Manhattan OC, 1908; London, CG, 1890. Roles included Aida, Valentine, Fedora, and first US Desdemona (Verdi). Married Cleofonte *Campanini. Recordings of 'E. Tetrazzini' refer to the third sister, **Elvira**, also a soprano.

**Tetrazzini, Luisa** (*b* Florence, 29 June 1871: *d* Milan, 28 Apr. 1940). Italian soprano, sister of above. Studied Florence, and with her sister Eva. Début Florence, T Pagliano, 1890 (Inez, *L'Africaine*). After much success in Italy, Russia, Spain, and especially S America, she made a belated but spectacular début London, CG, 1907. London, CG, 1908–12; New York, Manhattan OC, 1908–10, and M, 1911–12; Chicago 1912–13. Then sang in concert until 1934. A superb coloratura artist, whose upper register was warm and rounded, she dazzled audiences with her clear, brilliant, and spirited singing. Even a rotund figure and eccentric taste in costume did not detract from her Rosina, Amina, Lucia, Violetta, Gilda, Lakmé, and Ophélie. E. M. Forster, in *Where Angels Fear to Tread*, describes a singer 'stout and ugly' but with a beautiful voice, later admitting that this was the then unknown Tetrazzini. She died in poverty, her enormous fortune having been dissipated by her third husband. (R)

**Teyte, (Dame) Maggie** (*b* Wolverhampton, 17 Apr. 1888; *d* London, 26 May 1976). English soprano. Studied London, and Paris with J. De Reszke. Début Monte Carlo 1907 (Tyrcis, Offenbach's *Myriame et Daphné*). Paris, OC, 1907–10; London, HM, 1910, and CG, 1911, 1914, 1922–3, 1930, 1936–8; Chicago 1911–14; New York City Center 1948; London, Mermaid T, 1951, when she retired from the stage. Possessing a clear, firm tone and great interpretative gifts, she was admired as Zerlina, Blonde, Marguerite, Mimì, Butterfly, Cendrillon, and especially Mélisande—as which, however, Debussy thought her, while better vocally than Garden, 'an adorable doll' who sang the role with 'total incomprehension'. Created Princess, *The Perfect Fool*. Made celebrated recordings of French song. (R)

# Thaïs

BIBL: G. Connor, *The Pursuit of Perfection: A Life of Maggie Teyte* (London 1979).

**Thaïs.** Opera in 3 acts by Massenet; text by Louis Gallet, after Anatole France's novel (1890). Prod. Paris, O, 16 Mar. 1894, with Sybil Sanderson, Alvarez, Delmas, Beauvais, Delpouget, cond. Taffanel; New York, Manhattan OC, 25 Nov. 1907, with Garden, Dalmorès, Renaud, cond. Campanini; London, CG, 18 July 1911, with Edvina, Gilly, Darmel, cond. Panizza.

Egypt, 4th cent. Athanaël (bar), a young coenobite monk, resolves to persuade the courtesan Thaïs (sop) to return to a virtuous life. Eventually she allows him to take her to a convent. But unable to forget her, Athanaël returns to the convent, where Thaïs lies dying. As she expresses her gratitude to him for saving her soul, he collapses in anguish over the physical love for her that has overcome him.

The Méditation (properly Méditation réligieuse), the Act 2 intermezzo featuring violin and harp, has become well known as a concert piece.

**Theater an der Wien.** Viennese theatre, successor to the Theater auf der Wieden. Built by Schikaneder with funds provided by the merchant Zitterbarth, opened 13 June 1801 (cap. 1,232) with Teyber's *Alexander* (libretto by Schikaneder). Though seats were cheaper than at either of the 'imperial' theatres (the *Burgtheater and the *Kärntnertortheater), they were more expensive than at the other two suburban theatres (the *Theater in der Josefstadt and the *Theater in der Leopoldstadt). This reflected the more opulent scale of the building and its activities: praised at the time as the grandest and most beautiful theatre in Vienna, it had a stage which could accommodate up to 500 actors and 50 horses, and was particularly suited to the presentation of the popular *Zauberoper (magic operas). Here were given the premières of *Fidelio* (1805), *Der Waffenschmied* (1846), *Die Fledermaus* (1874), and other operettas by Johann Strauss II, Millöcker, and Lehár; the management of Barbaia (1821–2) saw the Viennese premières of many Rossini works. It was also the scene of Jenny Lind's Vienna triumphs (1846–7) in *Norma*, *Sonnambula*, *Les Huguenots*, *Freischütz*, and *Ein Feldlager in Schlesien*. The international fame of *The Bartered Bride* dates from its first performance in German at this theatre in 1893; and in Oct. 1897 *La bohème*, the first Puccini opera in Vienna, was produced there. From 6 Oct. 1945 to 1954 it was the home of the Vienna Staatsoper, while the opera-house was being rebuilt. Purchased by the City of Vienna in 1961, it was entirely renovated and reopened a year later on 30 May 1962 with *Zauberflöte*. Bernstein conducted a new production of *Fidelio* there on 24 May 1970 to commemorate the 200th anniversary of Beethoven's birth, and it is now the official theatre of the annual Vienna Festival. In recent years its repertory has increasingly focused on musical comedy.

**Theater auf der Wieden.** Viennese theatre, built in 1787 by *Schikaneder in one of the courts of the Freihaus and granted royal privilege by the emperor, hence sometimes known as the 'Kaiserlich-königliches priviligiertes Theater auf der Wieden', or the 'Freihaus Theater auf der Wieden'. Like the two other suburban Viennese theatres—the *Theater in der Josefstadt and the *Theater in der Leopoldstadt—the Theater auf der Wieden opened in the wake of the establishment of the *Burgtheater and *Kärntnertortheater as 'imperial' theatres. Since the general public was not freely allowed admission to these houses, the suburban theatres provided their main theatrical entertainment. The repertoire thus concentrated on German works, at first unsophisticated plays with incidental music, later Singspiels and genuine opera. The Theater auf der Wieden was the scene of the first production of Mozart's *Die Zauberflöte* (1791). It closed on 12 June 1801, and Schikaneder transferred his activities to the *Theater an der Wien.

**Theater in der Josefstadt.** Viennese theatre, opened in 1788, so called because of its location in the suburb of Josefstadt. It was the smallest of the suburban theatres which opened following the establishment of the *Burgtheater and *Kärntnertortheater as 'imperial' theatres in 1776. Like the *Theater auf der Wieden and the *Theater in der Leopoldstadt, its repertoire concentrated at first on unsophisticated plays with music, though Singspiels and genuine opera were also later given. It was enlarged in 1822: the opening work was Kotzebue's play *Die Weihe des Hauses*, for which Beethoven provided an overture and incidental music. For many years the Theater in der Josefstadt stood in the shadow of the other houses, but following the engagement of the talented comedian Wenzel Scholz in the 1830s it enjoyed greater prominence.

**Theater in der Leopoldstadt.** Viennese theatre, opened in 1781, so called because of its location in the suburb of Leopoldstadt. Also known as the 'Leopoldstädter-Theater'. Its opening followed the establishment of the *Burgtheater and *Kärntnertortheater as 'imperial' theatres in 1776 and a repertory specializing in Viennese popular drama was quickly established: from 1817 the company included *Raimund. Few of the works performed at the Theater in der Leopoldstadt can be described as 'opera'; they are rather plays with incidental music which was valued more for its facile charm than emotional profundity.

**Théâtre des Nations.** The name of various Paris theatrical enterprises, initially the theatre opposite the Châtelet. The name of the T Sarah Bernhardt 1879–99. Various operatic seasons were given there, e.g. an Italian season in 1905 and some Dyagilev seasons. From 1957 the name of a theatrical festival including operas, many of them French premières, often by distinguished visiting companies.

**Théâtre-Italien** (or Théâtre des Italiens). Paris theatrical institution that has performed in various establishments. The term came into particular use when the merger of the Opéra-Comique companies at the T Favart and T Feydeau in 1801 freed the use of the title and cleared the way for Italian performances that had a particular appeal to Napoleon. The company opened at the Société Olympique, renamed Salle de l'Opera Buffa, 1 May 1801 with Capua's *Furberia e Puntiglio*. It occupied various Paris theatres, and attracted the contribution of the most distinguished artists, including Spontini (1810–12), Paer (1812–27), and Rossini (1824–6), and the singers Catalani, Rubini, Ronconi, García, Pasta, Malibran, Viardot, Patti, and many more, a roll-call of almost all the great Italian singers of the 19th cent., who seldom if ever sang at the Opéra. The company was at the Salle Ventadour 1841–76. Perhaps the most famous première was *Don Pasquale* in 1843 with Grisi, Mario, Lablache, and Tamburini. But it was due to the Théâtre-Lyrique that Paris saw the French premières of Mozart's Italian operas, and of almost all the works of Mercadante, Bellini, Donizetti, Rossini, and Verdi, to a standard that regularly dwarfed the achievements of the official French establishments. The company declined in importance around 1850.

**Théâtre-Lyrique.** Paris operatic enterprise inaugurated, as the Opéra-National (originally opened 15 Nov. 1847), at the T Historique on 27 Sep. 1851 with Boisselot's *Mosquita et la sorcière*. Known as the Théâtre-Lyrique from 12 Apr. 1852. An excellent repertory of opéra-comique, Mozart, Rossini, even early Wagner developed, with distinguished singers including Viardot and Miolhan (later Carvalho). Léon *Carvalho was director 1856–60 and (having moved to the Châtelet, cap. 1,243) 1862–8, when the company enjoyed its greatest triumphs. These included many important premières of works by Gounod (*Faust*, 1859; *Roméo et Juliette*, 1867) and Bizet (*Pêcheurs de perles*, 1863; *Jolie fille de Perth*, 1867), also Berlioz's *Troyens à Carthage* (1863). Carvalho's enterprise was not always financially rewarded, but though he had to admit trivial works and unfortunately populist versions of the

classics, he did much to uphold standards of artistic enterprise. Pasdeloup succeeded Carvalho, 1860–70.

BIBL: T. Walsh, *Second Empire Opera: the Théâtre-Lyrique, 1851–1870* (London, 1981).

**Théâtres de la Foire.** See DRAME FORAIN.

**Thebom, Blanche** (*b* Monessen, PA, 19 Sep. 1918). US mezzo-soprano. Studied New York with Matzenauer and Walker. Début New York, M, 1944 (Fricka). New York, M, 1964–1967; Gly. 1950; London, CG, 1967. Paris, O; Milan; Brussels. Repertory included Mozart, Wagner, and an acclaimed Dido (Berlioz) and Amneris. (R)

**'There's a boat dat's leavin' soon for New York'.** Sporting Life's (ten) song in Act III of Gershwin's *Porgy and Bess*, trying to persuade Bess to flee with him.

**Thieving Magpie, The.** See GAZZA LADRA, LA.

**Thill, Georges** (*b* Paris, 17 Dec. 1897; *d* Draguignan, 15 Oct. 1984). French tenor. Studied Paris, and Naples with De Lucia. Début Paris, OC, 1918 (Don José). Paris, O, until after the war; London, CG, 1928, 1937; New York, M, 1931–2; Milan, S, 1928. Also Buenos Aires, C; Verona. Last stage appearance Paris, OC, 1953. Large repertory included Gluck, Cherubini, Méhul; also the roles of Aeneas (Berlioz), Don José, Werther, Lohengrin, Parsifal, Alfredo, Calaf. A considerable artist, impressive for his poised, full-toned singing and clarity of style. (R)

BIBL: A. Segond, *Georges Thill* (Lyons, 1980).

**Thoma, Therese.** See VOGL, HEINRICH.

**Thomas, Ambroise** (*b* Metz, 5 Aug. 1811; *d* Paris, 12 Feb. 1896). French composer. Studied privately, Paris with Zimmermann and Dourlen, also Barbereau, then Le Sueur. Between 1837 and 1843 he wrote nine stage works, most successfully *La double échelle*. There followed a series of opéras-comiques: in the tradition of Auber, these are both more lively and more consciously Romantic, if not very markedly individual. The best of them was *Le songe d'une nuit d'été* (non-Shakespearian, though it includes parts for Shakespeare, Falstaff, and Elizabeth I). Its success contributed to his election to the Académie in succession to Spontini (1851) and as professor at the Conservatoire (1856), where he was to exercise an influence on generations of students. He was director from 1871 until his death, maintaining a stoutly conservative régime in defiance of what he regarded as pernicious modern, especially Wagnerian, influences.

Himself always prone to the influence of more distinctive (or more successful) composers, he produced *Psyché* in imitation of Gounod's classically inspired *Sapho*, *Le carnaval de Venise* in imitation of Massé, and—after a failure with *Le roman d'Elvire* and then a pause—his greatest triumph, *Mignon*, in imitation of Gounod's *Faust*. He went on to repeat this with *Hamlet* in imitation of Gounod's *Roméo et Juliette*; but thereafter success largely eluded him. He was always willing to revise his works for changing audiences or conditions; and his careful eye for the main chance at the Opéra-Comique, supported by good craftsmanship and a certain gift for sentimental melodiousness, ensured his works popularity in their day, especially in France. These qualities were respected by many of his distinguished contemporaries, including Berlioz and Verdi.

SELECT WORKLIST (all first prod. Paris): *La double échelle* (The Double Staircase) (Planard; 1837); *Le songe d'une nuit d'été* (The Midsummer Night's Dream) (Rosier & De Leuven; 1850); *Raymond* (Rosier & De Leuven; 1851); *Psyché* (Barbier & Carré; 1857); *Le carnaval de Venise* (Sauvage; 1857); *Mignon* (Barbier & Carré, after Goethe; 1866); *Hamlet* (Barbier & Carré, after Shakespeare; 1868); *Françoise de Rimini* (Barbier & Carré, after Dante; 1882).

BIBL: H. Delaborde, *Notice sur la vie et les œuvres de M. Ambroise Thomas* (Paris, 1896).

**Thomas, Arthur Goring** (*b* Ratton Park, 20 Nov. 1850; *d* London, 20 Mar. 1892). English composer. Studied Paris with Durand, London with Sullivan and Prout, later Bruch. The success of *The Light of the Harem* brought a Carl Rosa commission, which produced his best-known and most successful work, *Esmeralda*. This was performed abroad, though its handling of the drama could not guarantee it long life once its easy charm had worn off. His talent did not prove up to the dramatic demands of *Nadeshda*, though there was admiration for his fluent recitative and for some of the lighter numbers.

SELECT WORKLIST: *The Light of the Harem* (Harrison, after Moore; London 1879); *Esmeralda* (Marzials & Randegger, after Hugo; London 1883); *Nadeshda* (Sturgis; London 1885).

**Thomas, Jess** (*b* Hot Springs, SD, 4 Aug. 1927; *d* San Francisco, 11 Oct. 1993). US tenor. Studied Stanford with Schulmann. Début San Francisco (Major-domo, *Rosenkavalier*). Karlsruhe 1958; Bayreuth 1961–9, 1976; Munich from 1961; New York, M, from 1962; London, CG, 1969–71; Paris, O, 1967, 1972; also Salzburg, Munich, Vienna. Large repertory includes Florestan, Tannhäuser, Walther, Tristan, Parsifal, and first Caesar (*Antony and Cleopatra*). (R)

**Thomson, Virgil** (*b* Kansas City, 25 Nov. 1896; *d* New York, 30 Sep. 1989). US composer and critic. Studied initially with Nadia Boulanger in Paris (1921–2), at which time he also met Les Six and Satie. Returning to Paris in 1925, he lived

there until 1940, and worked with Gertrude Stein on his first opera, *Four Saints in Three Acts*. Thomson and Stein began to plan an opera about Spanish saints set in a Spanish landscape in 1926; a piano score was completed in 1928, and the opera orchestrated in 1933. Thomson found that original libretto rambling and deleted about one-third of it, shaping the opera into a prologue and four acts; previously the text had been too reliant on word games and random remarks, not allowing for any realistic characterization.

The score of *Four Saints* includes elements ranging from Gregorian chant to American folk hymnody, and incorporates features characteristic of many of Thomson's works, notably the quotation of familiar tunes (e.g. 'God Save the King'), and the attention paid to word setting; Thomson takes care to ensure great clarity and fidelity to spoken English. Although it attracted much attention intitially, *Four Saints* has never retained a secure place in repertory.

Thomson composed his second opera, *The Mother of Us All*, while working as critic on the New York Herald Tribune, a post he held 1940–54, and in which he was noted for his lively, outspoken comments which often concentrated on features ignored by other critics. Based on the theme of women's suffrage, the opera includes real figures, including Stein and Thomson themselves. Again the music draws on a wide range of sources—revival hymnody, popular 19th-cent. melodies, gospel tunes—to create a distinctly American style.

For his third opera, *Lord Byron*, Thomson collaborated with librettist Jack Larson. This work, which took seven years to write, displays an emotional content entirely new to Thomson's generally muted language, expressed by increased lyrical expansiveness.

WORKLIST: *Four Saints in Three Acts* (Stein; Hartford 1934); *The Mother of Us All* (Stein; New York 1947); *Lord Byron* (Larson; New York 1972).

WRITINGS: *Music With Words* (Yale, 1989).

BIBL: K. Hoover and J. Cage, *Virgil Thomson* (New York, 1959).

**Thorborg, Kerstin** (*b* Venjan, 19 May 1896; *d* Falun, 12 Apr. 1970). Swedish mezzo-soprano. Studied Stockholm. Début Stockholm 1924 (Ortrud). Swedish Royal O 1924–30; Vienna, S, 1935–8; London, CG, 1936–9; New York, M, 1936–50; also Paris, Buenos Aires, Berlin, etc. Roles included Eglantine, Orfeo, Klytemnestra, Dalila, Fricka, Brangäne; Newman thought her 'the greatest Wagnerian actress' of the day. (R)

**Tibbett, Lawrence** (*b* Bakersfield, CA, 16 Nov. 1896; *d* New York, 15 July 1960). US baritone. Studied Los Angeles with Ruysdael, New York with La Forge. Début New York, M, 1923

(Lovitsky, *Boris*). New York, M, until 1950. London, CG, 1937; also Vienna, Prague, Australia. Roles included Valentin, Boccanegra, Ford, Verdi's Iago; created parts in *The Emperor Jones*, *The King's Henchman*, *Peter Ibbetson*, among others. Handsome, dramatically talented, and a fine singer with an attractive tone 'as solid as a piston rod' (Beckmesser of *Gramophone*), he was also successful in light opera and films. (R)

BIBL: A. Farkas, ed., *Lawrence Tibbett* (New York, 1990).

**Tichatschek, Joseph** (*b* Ober-Weckelsdorf (Teplice), 11 July 1807; *d* Blasewitz, 18 Jan. 1886). Bohemian tenor. Studied Vienna with Ciccimarra. Début Vienna, K, 1835 (Raimbaud, *Robert le diable*). Graz 1837; Dresden 1838–70; London, DL, 1841. Repertory included Idomeneo, Tamino, Adolar, Éléazar, Fernand Cortez (Spontini), Lohengrin; created Rienzi and Tannhäuser. His impassioned delivery and heroic presence, together with a fine and brilliant tone, brought him much admiration from Berlioz, Liszt, Nicolai, and Wagner (who appreciated his 'lively nature, glorious voice, and great musical ability' while deploring his lack of dramatic perception).

**Tiefland** (The Lowlands). Opera in a prologue and 3 acts by D'Albert; text by Rudolf Lothar, after the Catalan play *Terra Baixa* by Angel Guimeras. Prod. Prague, German T, 15 Nov. 1903, with Alfödý, Foerstel, Aranyi, Hunold, cond. Blech; rev. version New York, M, 23 Nov. 1908, with Destinn, L'Huillier, Schmedes, Feinhals, cond. Hertz; London, CG, 5 Oct. 1910, with Terry, Teyte, Coates, Radford, cond. Beecham.

The Pyrenees, 19th cent. The wealthy landowner Sebastiano (bar) has forced Marta (sop), whom he once found starving, to become his mistress. He now marries her off to a young shepherd, Pedro (ten), on condition that he abandon his mountain life for the lowlands; Sebastiano plans to continue his relationship with Marta under this cover. But Marta comes to love Pedro, and eventually she confesses the true situation to him. When Sebastiano tries to take her back, his marriage plans having collapsed when the story emerges, Pedro strangles Sebastiano and bears Marta off to the mountains.

D'Albert's most successful opera, and one of the best examples of German verismo; between the wars, it was given in Hamburg 264 times, more than *La bohème* or any Mozart opera.

**Tietjen, Heinz** (*b* Tangier, 24 June 1881; *d* Baden-Baden, 30 Nov. 1967). German conductor and producer. Producer Trier 1904–22, director 1907–22; Saarbrücken 1919–22; Breslau 1919–22. Director Berlin, SO, 1925–7, 1948–55, Prussian State T 1927–45. Artistic director Bayreuth 1931–44, producer from 1933, also cond. *Parsifal* (1933), *Ring* (1936, 1938, 1939, 1941), *Meistersinger* (1933–4), *Lohengrin* (1936–7). Director Hamburg 1955–9. London, CG, as producer 1950–1. His Wagner productions were traditional but not old-fashioned. Conducted Wieland Wagner production of *Lohengrin*, Hamburg 1958. Returned to Bayreuth to conduct 1959. (R)

**Tietjens, Therese** (*b* Hamburg, 17 July 1831; *d* London, 3 Oct. 1877). German, later British, soprano. Studied Vienna with Dellessie, Babing, and Proch. Début Hamburg 1848 (Erma, Auber's *Maçon*). Hamburg 1849, as Lucrezia. Vienna, 1853 and 1856. London, HM, 1858, making a great impression. London: DL, HM, L 1859 until her death; CG 1869. Paris, O, 1863; Naples, C, 1862–3, 1868–9. Possessed a generous, velvety tone, wide-ranging dramatic conviction, a magisterial presence, and great personal magnetism. Her superb technique encompassed the roles of Medea, Donna Anna, Elvira (*Puritani*), Norma, Lucia, Valentine, Fidès, Ortrud, and a particularly famous Leonore and Lucrezia; Wagner requested her in vain to create Isolde and to sing Sieglinde. She was adored by the British public; unusually for an international prima donna, she sang opera on stage in the provinces. Her last performance, as Lucrezia, was given in severe pain from cancer.

**Timişoara** (Hung., Temesvár). City in Banat, Romania. The region was long a centre of German culture, and opera (especially Mozart) was given in the early 19th cent., e.g. by Theodor Müller's company under Ion Wachmann (1807–63) in 1831–3. The present opera-house opened in 1947 and gave over 2,000 performances in its first five years.

**Tinsley, Pauline** (*b* Wigan, 23 Mar. 1928). English soprano. Studied Manchester with Dillon, London with Keeler, Cross, Henderson, and Turner. Début London 1961 (Rossini's Desdemona). London, SW/ENO from 1963; CG from 1965; New York City O 1971–2; Santa Fe 1969; also WNO, Boston, Hamburg, Amsterdam, Zurich, Verona, etc. An indomitable performer with a powerful, distinctive tone. Wide repertory includes Queen of the Night, Donna Anna, Zerlina, Anna Bolena, Leonora (*Trovatore* and *Forza*), Brünnhilde, Kundry, Elektra, Kostelnička, Lady Billows Ortrud, Lady Macbeth. (R)

**Tippett, (Sir) Michael** (*b* London, 2 Jan. 1905). English composer. Studied London with Wood and Morris. His deep humanity, richly stocked mind, and fascination with elaborate concepts and literary and other cultural allusions mark each of his operas, which are among the most original and important contributions to the genre in the 20th cent. He has always written his own texts, following early advice from T. S. Eliot, and there is a close interlocking between on the one hand his searching ideas and allusive, sometimes colloquial expression, and on the other the richness yet pungency and the lyrical exuberance of his music.

Through his operas there runs a thread of concern for reconciliation between opposing or sundered elements in human beings, epitomized in a line from his oratorio *A Child of Our Time*, 'I would know my shadow and my light, so shall I at last be whole.' *The Midsummer Marriage* takes as its metaphor a couple whose respective earthiness and airiness make them unfit for a full union until they have undergone various trials, conducted in semi-mythical English surroundings. Their search is expressed in some of Tippett's most ecstatic, haunting, and profusely lyrical music. Structurally, the work makes use of traditional arias, duets, and ensembles, and gives a particular importance to ballet (in the Ritual Dances). *King Priam* turns to an existing myth, that of the Trojan war, and, partly in response to criticisms of self-defeating prolixity in the earlier opera, now draws on a sparer, tauter idiom, a more declamatory vocal style, and sharper orchestral contrasts. The declared subject is 'the mysterious nature of choice'; and the plot is vividly dramatic, including some stunning *coups de théâtre* (such as the close of Act I on the wrath of Achilles) and some tenderly lyrical exchanges (as that between Achilles and Priam over Hector's body).

In *The Knot Garden*, proposing *Così fan tutte* as his example, Tippett returned to his reconciliation theme; here his central subject is a marriage which has gone wrong through the partners growing the one too mundane and the other too isolated from reality. They are set in a psychological (but dramatically and musically active) maze, from which they and the other characters emerge in new or renewed emotional bonds. The music draws on both the richness of Tippett's first opera and the spareness of his second, and finds in its confrontations between personalities a new concentration of expression. His fourth opera, *The Ice Break*, enlarges the reconciliation theme to wider issues, bringing into operatic context the divisions, acute but not unhealable, between East and West, Black and White, old and young, past and present. The attachment to the present (which includes some rather forced colloquialisms in the dialogue) shows in the enrichment of Tippett's lyrical language by his sympathy for blues and boogie-woogie, one that was already revealed in the previous opera.

Tippett's last opera, *New Year*, shows no slackening of contemporary awareness or lyrical impulse. The opera gathers up many of his long-held ideals and preoccupations, here in the context of contacts between a contemporary 'Terror Town' and a Utopian future, and of the characters' need to find reconciliation with their own pasts. In a work as rich as ever in layers of meaning, and in resonances that touch and stimulate, Tippett has again found his own reconciliation between his intense and personal sound world and various forms of urban popular music. Dance plays an organic part in a work that has more than a few links with the *musical. Not only does the orchestra use saxophones and electric guitars; jazz and pop styles, accepted as a genuine modern vernacular, are not merely imitated but fully composed in Tippett's own idiom. (R)

WORKLIST: *Robin Hood* (folk-song opera—Tippett, Ayerst & Pennyman; Boosbeck 1934); *The Midsummer Marriage* (Tippett; comp. 1946–52, prod. London 1955); *King Priam* (Tippett; Coventry 1962); *The Knot Garden* (Tippett; London 1970); *The Ice Break* (Tippett; London 1977); *New Year* (Tippett; Houston 1989).

BIBL: E. White, *Tippett and his Operas* (London, 1979); I. Kemp, *Tippett* (London, 1984).

**Tiresias.** In classical mythology, a Theban who was struck with blindness by Athene when he saw her bathing but given as a compensation the gift of prophecy. He was also said to have spent a part of his long life as a woman. He appears (bs) in Stravinsky's *Oedipus Rex*. He is also the male manifestation of Thérèse (sop) in Poulenc's *Les mamelles de Tirésias*.

**Titov, Alexey** (*b* St Petersburg, 23 July 1769; *d* St Petersburg, 20 Nov. 1827). Russian composer. Most of his works are for the stage, and based on a Mozartian idiom. About a dozen operas survive, including a 1-act comedy, *Yam*. This proved so popular that he wrote two sequels, to make up a trilogy. In them, he draws upon certain Russian musical characteristics and makes some use of folk-song (including the folk polyphony that was to be employed by Glinka and Tchaikovsky, among others). His brother **Sergey** (1770–1825) wrote an opera *Krestyane* (The Peasants) (1814).

SELECT WORKLIST (all first prod. St Petersburg; all texts by Knyazhnin): *Yam, ili Pochtovaya stantsiya* (Yam, or The Post Station) (1805); *Posidelky, ili Prodolzheniye Yama* (The Winter Party, or The Sequel to Yam) (1808); *Devichnik, ili Filatkina svadba* (The Wedding Eve Party, or Filatkin's Wedding) (1809).

**Titurel.** The former ruler (bs) of the Kingdom of the Grail, father of Amfortas, in Wagner's *Parsifal*.

**Todesverkündigung** (Ger.: 'announcement of death'). The name usually given to the scene in Act II of Wagner's *Die Walküre* in which Brünnhilde comes on Wotan's orders to tell Siegmund of his approaching death.

**Tofts, Catherine** (*b* *c.*1685; *d* Venice, 1756). English soprano. After singing in concert 1703–4, joined Drury Lane, London, 1705. Sang there until 1709. The first English-born singer to perform in Italian opera in England, and one of the earliest British prima donnas; sang in pasticcios (e.g. *Camilla, Thomyris, Pyrrhus and Demetrius*); also Clayton's *Arsinoe* and *Rosamond*. Her 'erect mien, charming voice, and graceful motion' (*Spectator*) won her much admiration, including Cibber's and Burney's. She also enjoyed a rivalry with Margherita de l'Épine (while swearing that she did not instigate some orange-throwing by her maid), and was far better paid, with a salary of £500. Leaving the stage with a fortune 1709, she settled in Venice, where she married the British consul, Joseph Smith. She seems to have died insane.

**Tokyo.** Capital of Japan. Opera was first heard in 1894, but not established on a regular basis until the formation of the Fujiwara OC by the tenor Yoshie Fujiwara in 1933. The company has no permanent home, but performs a selection of Western and Japanese opera with Japanese, sometimes Western, singers chiefly at the Bunka Kaikan concert-hall. The Niki-kai company, a singers' co-operative formed in 1952, performs mainly in the Nissei Theatre, opened in 1963 (cap. 1238). It gave *Parsifal* in 1967 and in 1973 began a *Ring* that got no further than the opening *Walküre*; *Siegfried* followed in 1984. Also important is the Tokyo Chamber Opera Group, founded in 1970 and performing works from all periods. Opera is also given in the NHK Broadcasting Corporation Hall (cap. 4,000), the Shinjuku Bunka Centre, the Bunka Kaikan, and the Globe T (opened 1988). A group entitled Tokyo Opera Produce was formed in 1980; others are the Nihon Opera Group, the Nagato Miha Opera, and the Creative Opera Series, all of which have a popular following.

**Tolomeo.** Opera in 3 acts by Handel; text by N. Haym, after Capece's libretto *Tolomeo ed Alessandro*, set by D. Scarlatti (1711). Prod. London, H, 30 Apr. 1728, with Faustina, Cuzzoni, Senesino, Baldi. Rev. Halle 1963; Abingdon 1973.

**Tolstoy,** (Count) **Lev** (*b* Yasnaya Polyana, 9 Sept. 1828; *d* Astapovo, 20 Nov. 1910). Russian writer. Although he played the piano, and could be deeply moved by music, Tolstoy was suspicious of music's power to affect men's moral judgement, as he suggested in *The Kreutzer Sonata*

(1890). His dislike of opera (apart from *Don Giovanni* and *Der Freischütz*) is evident from his mockery in *War and Peace*; he tried to persade Tchaikovsky not to persist with the form. He disliked Beethoven, and was unsympathetic to Russian music except for folk-songs, which he admired uncritically for their reflection of the simple peasant life he venerated. Operas on his works are as follows:

*War and Peace* (1863–9): Prokofiev (1946)
*Anna Karenina* (1876): Sassano (1905); Granelli (1906); Malherbe (1914); Hubay (1923); Robbiani (1924); Hlobil (1962; prod. 1972); Meylus (1970); Hamilton (1981)
*How Men Live* (1881): Martinů (1953)
*The Tale of Ivan the Jester* (1885): Ostrčil (*Honzovo království*, 1934); Fibich (*Bloud*, 1936)
*Ivan and the Drum* (1888): Reti (1933)
*Resurrection* (1899): Alfano (1904); Hristić (1912); Cikker (1962)
*The Living Dead* (1902): Leroux (1912)
*What For?* (1906): Strelnikov (*Beglets*, 1934).

**Tomášek, Václav** (*b* Skuteč, 17 Apr. 1774; *d* Prague, 3 Apr. 1850). Bohemian composer. Largely self-taught. His influence on Prague musical life in the first half of the 19th cent., especially as teacher, was powerful. The list of his works, in which the keyboard music and songs are most important, includes one completed opera, *Seraphine* (Dambek; Prague 1811). His autobiography gives a vivid picture of contemporary musical life.

WRITINGS: Z. Němec, ed., *Vlastní životopis V. J. Tomáška* (Prague, 1941).

**Tomasi, Henri** (*b* Marseilles, 17 Aug. 1901; *d* Avignon, 13 June 1978). French composer and conductor. Studied Marseilles, and Paris with Caussade, Vidal, and D'Indy. Cond. Paris radio 1930–6. Music director Monte Carlo O, 1946–50. An important and honoured figure in France, he was influential as a conductor with a strong feeling for the theatre. His operas, somewhat eclectic and on the whole conventional in style though theatrically effective, found a following in French opera-houses during the years after the Second World War.

SELECT WORKLIST: *Miguel Mañara* (or *Don Juan de Mañara*) (Tomasi, after Milosz; Paris radio 1952, Munich 1956); *L'Atlantide* (Didelot, after Benoit; Mulhouse 1954); *Sampiero Corso* (Cuttoli; Bordeaux 1956); *Le silence de la mer* (Tomasi, after Vercors; Toulouse 1964); *L'élixir du révérend père Gaucher* (Bancal, after Daudet; Toulouse 1964); *Ulysse* (Tomasi, after Giono; Mulhouse 1965).

**Tomlinson, John** (*b* Accrington, 22 Sep. 1946). English bass. Studied Manchester with McGuigan, London with O. Kraus. Début Gly. Touring O 1971 (Second Priest, *Zauberflöte*). Kent O; London: SW (New OC) 1972–4, ENO 1974–80,

CG from 1979; Bayreuth from 1988; New York, M, from 1991. Also O North; Berlin, D; San Francisco; Salzburg; Aix; etc. An impressively powerful performer with a large voice. Roles incl. Masetto, Padre Guardiano, Mosè, Attila, Hunding, Hagen, Wotan, Boris, Arkel; created Green Knight (Birtwistle's *Gawain*). (R)

**Tomowa-Sintow, Anna** (*b* Stara Zagora, 22 Sep. 1941). Bulgarian soprano. Studied Sofia with Zlatev-Cherkin. Début Stara Zagora 1965 (Tatyana). Leipzig 1967–72; Berlin, S, 1972–6; Salzburg from 1973; Paris, O, 1974; Milan, S, 1975; London, CG, from 1975; New York, M, from 1978; also Vienna, Munich, Moscow, Brussels; etc. Roles include Countess (Mozart and R. Strauss), Donna Anna, Elsa, Eva, Leonora (*Forza*), Desdemona (Verdi), Marschallin, Arabella. Possesses a strong, rich tone and a fine technique. (R)

**Tom Rakewell.** A young man (ten), lover of Anne, in Stravinsky's *The Rake's Progress*.

**tonadilla** (Sp.: 'a little tune'; dim. of *tonada*, deriving from *tono*, polyphonic song). A genre of Spanish comic opera which reached its greatest popularity in the 18th cent. Originally a simple song with guitar accompaniment, it grew out of the *sainete and developed into an operatic miniature (like the Italian intermezzo), ending with a lively dance, which was known as *a tonadilla escénica*. Its subject-matter was invariably satirical or political, its setting Spanish, and its emphasis upon lower-class life: sometimes folk dances were included. Though the tonadilla usually called for between two and four singers, some were written for a soloist, usually female, while the *tonadillas generales* might have up to 12 singers. The duration varied according to the number of characters; tonadillas for four characters seldom exceeded 20 minutes.

The earliest important composer of tonadillas was Antonio Guerrero (1700–76): the mature stage of the genre was initiated by Luis Misón (?–1766) in *La mesonera y el arriero*; he wrote many extremely popular tonadillas, as did José Palomino (1755–1810) and Manuel *García (1775–1832). Towards the end of the 18th cent. the genre took on some characteristics of Italian opera; it began to die out towards the mid-19th cent. and was supplanted by the *zarzuela. Other important tonadilla composers were Pablo Esteve (*d* 1794) and Blas Laserna (1751–1816). A singer of tonadillas is a *tonadillera*.

**Tonio.** 1. A Tyrolean peasant (ten), lover of Marie, in Donizetti's *La fille du régiment*.
2. A clown (bar) in Leoncavallo's *Pagliacci*.

**Toreador Song.** The name generally given to Escamillo's couplets in Act II of Bizet's *Carmen*, 'Votre toast, je peux vous le rendre', in which he acknowledges their toasts to him and sings of the toreador's life.

**Torelli, Giacomo** (*b* ?Fano, 1 Sep. 1608; *d* Fano, 17 June 1678). Italian designer and architect. Built T Novissimo, Venice's fourth public operahouse, including a new system for changing the sets in a single operation and thus increasing dramatic pace. Paris 1645, installing new stage machinery and working for the Italian troupe brought to Paris by *Mazarin.

**Toronto.** City in Ontario, Canada. The first operas given were Coleman's *The Mountaineers* and Storace's *No Song, No Supper* in 1825 by an American company, when Toronto, of fewer than 1,700 inhabitants, was mainly a garrison town and administrative capital of Upper Canada. During the years of growth to a commercial centre, visiting singers and small troupes appeared increasingly. In the second half of the century visiting singers included Sontag, Lind, and Patti (generally in concerts). Full-scale productions date from 1853, with Rosa Devriès in a *Norma* conducted by Luigi Arditi. Semi-staged performances were given, but little progress was made in resident production of opera until 1867, when the Holman English Opera Troupe leased the Royal Lyceum and installed itself as the city's permanent company for plays and opera until 1873. In 1867–80 the company gave many performances of about 35 operas and operettas. Visiting companies presenting major artists in leading roles included the Strakosch, Emma Abbott, Kellogg, English OC, New York, M, and troupes headed by Melba (1895) and Sembrich (1901).

Local productions were sporadic until 1929, when Sir Ernest Macmillan organized the Conservatory OC, playing at the Regent T; but owing to the financial crisis of the 1930s this never developed. In 1934 Harrison Gilmour, a wealthy financier married to a singer, organized the Opera Guild of Toronto, which gave opera until 1941. An Opera School was set up in 1946 as part of the Royal Conservatory of Music. The Royal Conservatory OC, formed in 1950, mounted three productions in the Royal Alexandra T (cap. 1,700), an opera festival so successful that it gradually severed ties with the Conservatory and grew into the Canadian OC in 1959 (with Herman Geiger-Torel as director). In 1960 the company moved to the O'Keefe Center (cap. 3,200), where it mounted six or seven operas each autumn. *Die Walküre* was the first Wagner opera to be staged, and gradually other Wagner and Strauss works entered the repertory. The 17 performances in 1961 and 1962 increased to 36 by 1973 and 1974, with many full houses. Since 1958 a touring group has toured Canada and the USA annually. From 1964 the Opera School has had its own theatre and workshops, and since 1969 has been a department of the university Faculty of Music, producing programmes of excerpts and two complete operas each year. Plans are in hand for a new theatre to replace the notoriously inadequate O'Keefe Center. Companies working locally include Opera Hamilton and Opera Atelier.

**Torquato Tasso.** Opera in 3 acts by Donizetti; text by Jacopo Ferretti, after Giovanni Rosini's drama *Torquato Tasso* (1832), Goldoni's drama *Tasso* (1755), Goethe's drama *Tasso* (1790), and Byron's poem *The Lament of Tasso* (1817). Prod. Rome, T Valle, 9 Sep. 1833, with Spech, Carocci, Lauretti, G. Ronconi, Rinaldi, Poggi; London, HM, 3 Mar. 1840, with De Varny, Coletti.

Ferrara, 16th cent. The poet Tasso (bar) has as rivals Roberto (ten) for his fame and Don Gherardo (bs) who believes that he loves Eleonora di Scandiani (mez): Tasso in fact loves another Eleonora (sop), sister of Duke Alfonso d'Este (bs). Gherardo steals a poem by Tasso in praise of his Eleonora. Plots and counterplots lead to the Duke declaring Tasso mad and having him confined in an asylum for seven years. When he is released, he learns that his Eleonora has died, and his mind now does give way; but he is urged to think of his future glory and to return to poetry.

**Torresella, Fanny** (*b* Tiflis, 1856; *d* Rome, 2 May 1914). Italian soprano. Studied with her father, the conductor Antonio Torresella. Début Trieste 1876 (Fenena). Milan, S, 1884, 1894, 1904; Naples, C, 1887; Buenos Aires, C, 1889, 1897, 1900; also much success in Spain. Retired 1909. Sang Zerlina, Lucia, Leïla, Philine, Musetta; created parts in works by Cilea and Mascagni. A virtuoso with a sweet-toned, agile voice, and an elegant style. (R)

**Tosca.** Opera in 3 acts by Puccini; text by Giacosa and Illica, after Sardou's drama *La Tosca* (1887). Prod. Rome, C, 14 Jan. 1900, with Darclée, De Marchi, E. Giraldoni, cond. Mugnone; London, CG, 12 July 1900, with Ternina, De Lucia, Scotti, cond. Mancinelli; New York, M, 4 Feb. 1901, with Ternina, Cremonini, Scotti, cond. Mancinelli.

Rome, June 1800. Angelotti (bs), a political prisoner, has escaped and is hiding in the Attavanti Chapel. He recognizes the painter Cavaradossi (ten) as an old friend and Cavaradossi gives him the keys to his villa as a refuge. Tosca (sop), a celebrated singer and Cavaradossi's lover, arrives and her jealousy is aroused when she recognizes the Marchesa Attavanti as the model for the portrait of Mary Magdalene which Cavaradossi is painting. A cannon signals Angelotti's escape and

when Tosca returns, the police chief Scarpia (bar) convinces her that Cavaradossi and the Marchesa have fled together. She rushes to the painter's villa, followed by the police.

Tosca is summoned to Scarpia, and while she is interrogated, Cavaradossi (who has been caught but denies all knowledge of Angelotti) undergoes torture in the next room. Unable to bear his agony, Tosca discloses Angelotti's hiding-place in the well of the villa. Tosca pleads for Cavaradossi and Scarpia agrees to release him if she will yield to him. She realizes this is the only hope and agrees. While he sits at his desk writing the safe-conduct pass which will enable her and Cavaradossi to leave the country, she stabs him.

Just before dawn Cavaradossi is led to his execution. Tosca tells him that this is to be a sham and that they can then escape; but he must pretend convincingly. When the squad fires, he falls. Tosca rushes to him but finds that Scarpia deceived her and that Cavaradossi is dead. In her grief, she flings herself from the parapet.

BIBL: M. Carner, *Giacomo Puccini: 'Tosca'* (Cambridge, 1985).

**Toscanini, Arturo** (*b* Parma, 25 Mar. 1867; *d* Riverdale, NY, 16 Jan. 1957). Italian conductor. Studied Parma. Began career as cellist, incl. in orchestra at première of Verdi's *Otello*. On the second night of an Italian opera season in Rio de Janeiro (30 Jun. 1886) he was called on by members of the company to replace the regular conductor Leopoldo Miguez, and two deputies against whom there were demonstrations, and conducted *Aida* from memory with great success. Back in Italy, he conducted Catalani's *Edmea* in Turin, 1886. In 1892 he conducted the première of *Pagliacci* at the T dal Verme, Milan. Music director Turin, R, 1895–8; opened with first Italian *Götterdämmerung* (in Italian) and gave première of *La bohème* (1896). Principal conductor La Scala, Milan, 1898–1903, opening with *Meistersinger* and including much Verdi and foreign works new to Italy. Also new to Italy was dramatic commitment of such intensity from a conductor in the interests of opera as an integrated art in the manner he admired in Wagner, and his departure was over demonstrations against him for refusing to allow Zenatello a disruptive encore in *Un ballo in maschera*.

From 1908 to 1915 Toscanini conducted at the other great centre of his operatic influence, the New York, M, whither he went with *Gatti-Casazza. Here he gave the premières of *La fanciulla del West* (1910) and Giordano's *Madame Sans-Gêne* (1915), and many first US performances including Gluck's *Armide* (1910), *Ariane et Barbe-bleue* (1911), and *Boris Godunov* (1913). Again he resigned over tensions between his ideals and what the management allowed. He conducted little until 1920, when he returned to La Scala, resuming his dominating role and bringing the ensemble to a high pitch of excellence. Resignation this time was over exhaustion and political tensions with the Fascists as much as managerial problems: at the first performance of *Turandot* (1926), at which he stopped the performance at the moment where Puccini had broken off, he had previously refused to conduct the Fascist anthem. He conducted at Bayreuth in 1930 (*Tristan* and *Tannhäuser*) and 1931 (*Parsifal* and *Tannhäuser*) (the first non-German), refusing to return in Nazi times; and at Salzburg 1934–7 (*Falstaff*, *Fidelio*, *Magic Flute*, *Fidelio*), refusing to return after the Nazi annexation of Austria. Between 1944 and 1954 he gave concert performances of a number of operas, including *Traviata*, *Aida*, *Falstaff*, and *Ballo in maschera*, for the NBC in New York; these were recorded. He returned to Milan to conduct the concert reopening La Scala in 1946, also the Boito memorial in 1948.

For parts of his life, Toscanini felt obliged to work outside the opera-house where his central interest lay. He won Verdi's friendship and admiration; he championed Puccini; he remained faithful to lesser Italian works, even naming a daughter Wally after Catalani's heroine. But he also introduced Italians to Gluck and to Wagner, as part of his dramatic vision; and it was the difficulty in realizing this, for all the scrupulousness of his preparation with singers and orchestra, that made his life in the theatre difficult. His genius rested on an uncanny memory for every detail of a score and an impassioned fidelity to it, as well as a compelling personal magnetism in making others share this. An autocrat in the opera-house, and fiercely intolerant of shortcomings, he spared himself least of all; and his intolerance of personal vanities, his intransigence with managements, and his notorious outbursts of rage were based on a love of music that seemed almost too intense for him to bear. (R)

BIBL: H. Sachs, *Toscanini* (New York, 1978).

**Tosi, Piero Francesco** (*b* ?Cesena, ?1653; *d* Faenza, 1732). Italian soprano castrato. Studied with his father, the composer **Giuseppe Tosi** (*fl* 1677–93). London 1692, with great success as singer and singing teacher. Vienna 1705–11 as court composer and diplomat. Dresden 1719; Bologna 1723, where he published his famous treatise on singing, *Opinioni de' cantori antichi e moderni*. Returned to London, remaining until 1727. Took holy orders Bologna 1730. His *Opinioni* provides information on and criticism of performing practices of the late 17th and early 18th cents., and his theories on bel canto are still

highly regarded. A singer of great expressiveness himself, he objected to the new 'instrumental' style of singing, and 'the Presumption of some Singers' with their inane ornamentation and casualness on stage; but also to the 'impudence' of composers who tried to dictate to singers by writing decorations into their music. He required of his pupils beauty of tone, agility, correct intonation, study of the text; above all, he declared, 'how great a master is the heart' which 'corrects the defects of nature'.

**Toten Augen, Die** (The Blind Eyes). Opera in a prologue and 1 act by D'Albert; text *Les yeux morts* by M. Henry, trans. H. Ewers. Prod. Dresden 5 Mar. 1916, with Forti, Taucher, Plaschke; Chicago, Gt Northern T, 1 Nov. 1923, with Gentner-Fischer, Hutt, Lattermann.

Jerusalem, the first Palm Sunday. Myrtocle (sop), the blind wife of the proconsul Arcesius (bar), has her sight given her by Jesus. When she discovers Arcesius's ugliness, she believes that the handsome Captain Galba (ten) is the man to whom she had been married. He is secretly in love with her, but when she discovers the truth, all her love for Arcesius returns, and she prays to be blind again. She gazes on the bright sun, and once again her eyes lose their sight.

**Tote Stadt, Die** (The Dead City). Opera in 3 acts by Korngold; text by 'Paul Schott' (Korngold and his father), after G.-R.-C. Rodenbach's novel *Bruges-la-morte* (1892). Prod. simultaneously 4 Dec. 1920 Hamburg with Münchow, Olczewska, Schubert, Degler, cond. Pollak, and Cologne with J. Klemperer, Rohr, Schröder, Renner, cond. Klemperer; New York, M, 19 Nov. 1921, with Jeritza, Telva, Harrold, cond. Bodansky.

Bruges, end 19th cent. Despite the remonstrances of Frank (bar), Paul (ten) cannot cease mourning his wife Marie (sop), and sees her once again in the person of the dancer Marietta (sop). In a nightmare sequence, he imagines Marie unfaithful to him with Frank, and his murder of her. Awake, he recognizes that he must distance himself from Bruges and Marie's memory.

Korngold's greatest success.

**Tottola, Leone** (*b* ?; *d* Naples, 15 Sep. 1831). Italian librettist. Working in the Neapolitan theatre at a time when opera buffa flourished, he placed his busy, not to say hasty, pen at the service of many different composers. Although his work was mocked for its lack of technique and taste, he sensed the impact of the novel ideas of Romanticism, and had some influence in impressing these on his composers. His subjects include many which became the stock in trade of Romantic opera, such as the biblical, oriental, and medieval, German legend, and English novels (including Scott). His librettos include *Adelson e Salvini* for Bellini, seven for Donizetti (including *Gabriella di Vergy* and *Il castello di Kenilworth*), six for Rossini (including *La donna del lago*), and many others for Pacini, Mercadante, Guglielmi, Fioravanti, Mayr, and others.

**Toulouse.** City in Haute-Garonne, France. An Académie Royale de Musique was opened by Lully's son-in-law in the Jeu de Paume in 1687. A new theatre belonging to the Capitouls (the elected consuls governing the city) opened in the Logis de l'Écu at the Capitole (Town Hall: hence later T du Capitole) on 11 May 1737. A new theatre opened 1 Oct. 1818; redesigned 1835, closed 1878. During the 19th cent. the theatre gained a high reputation partly through the excellence of the local singers (e.g. Merly, Gailhard, Capoul), also through the vigorously partisan popular application of the *trois débuts system. Rebuilt, reopened 7 Oct. 1880 with *La Juive*; burnt down 9 Aug. 1917; reopened (cap. 1,550) 6 Nov. 1923 with *Les Huguenots*. Renovated (cap. 1,300), reopened 10 Nov. 1950 with *Faust*. The tenor Louis Izar was director 1948–68, improving standards and extending the repertory, especially to cover Wagner. Renovated (cap. 1,200) in 1975. Michel Plasson was director 1973–82, strengthening and extending the theatre's activities. Opera has also been given in the Halle aux Grains (cap. 3,000).

**Tourel, Jennie** (*b* Vitebsk, 26 Jun. 1899; *d* New York, 23 Nov. 1973). US mezzo-soprano of Russian origin. Studied Paris with Hahn and El Tour (anagram of Tourel). Appeared Chicago 1930 (Second Scholar, Moret's *Lorenzaccio*); Paris: Opéra Russe, 1931, and OC, 1933–9. New York, M, 1937, 1943–7, and City Center, 1944; Montreal 1938–42. Also Holland, Israel, S America. A highly versatile and sensitive artist, acclaimed as Cherubino, Rosina, Mignon, Carmen. Created Baba the Turk. (R)

**tourney.** A term used to describe the music performed as an introduction to and during 16th- and 17th-cent. equestrian tournaments. Popular especially in N Italy (especially Turin and Florence) and S Germany, the tourney had much in common with early opera through its choice of mythological and allegorical plots, although little of its music survives.

**Tours.** City in Indre-et-Loire, France. Opera was first given in the Jeu de Paume from 1761. A theatre opened in the disused Couvent des Cordeliers on 7 Dec. 1796; rebuilt, reopened 8 Aug. 1872 with *Les noces de Jeannette*; burnt down 15 Aug. 1883. The T Français and Cirque de la Touraine opened 1884. A new theatre (cap. 1,200) opened 23 Nov. 1889. Jean-Jacques

Etcheverry was director of the Centre Lyrique de Tours 1973–83, extending and strengthening the company's artistic range. This was continued by his successor Michel Jarry.

**Traetta, Tommaso** (*b* Bitonto, 30 Mar. 1727; *d* Venice, 6 Apr. 1779). Italian composer. Studied Naples with Porpora and Durante. At Parma from 1758, he responded to the theatre director Du Tillot's wish to reconcile Italian and French operatic ideals with a setting of the text for Rameau's *Hippolyte et Aricie*. This goes some way in the direction of the ideals that were to be associated with Gluck, by bringing together dance, orchestration, choral scenes, and solo numbers in a functional manner. It also revealed the melodic grace in individual numbers that was one of Traetta's characteristics. With *Armida*, in Vienna, he further developed the fluency between numbers that was becoming another hallmark of his style, though he remained closely attached to the da capo aria. With *Sofonisba*, in Mannheim, he took this manner still further and was able to benefit from the availability of the famous ducal orchestra in developing his expressive symphonic scenes. *Ifigenia in Tauride* further benefits from knowledge of Gluck's *Orfeo*, including by way of collaboration with the castrato Gaetano Guadagni, the first Orfeo, and in the impressive use of the chorus. Moving to Venice in 1765, Traetta wrote two comic operas as well some more conventional Metastasian operas. In 1768 he succeeded Galuppi as Catherine the Great's opera director, also writing for the Russian court his *Antigone*. In this he returns with great effect to the manner he had been exploring in his earlier classical works, and further develops his most adventurous ideas especially in the field of building up large-scale dramatic scenes. He left Russia in poor health in 1775, and after an unsuccessful visit to London (where he was overshadowed by Sacchini), he returned to Venice. The best of his operas approach Gluck in manner and even at times in quality. His graceful cavatas or cavatinas can emerge from passages of very free and fluent arioso, as in Act II of *I Tintaridi*, and the forceful or elegant choral participations in his *Ifigenia*, if sometimes rather protracted, in turn influenced Gluck's setting.

SELECT WORKLIST: *Farnace* (Luchini, after Zeno; Naples 1751); *Ippolito ed Aricia* (Frugoni; Parma 1759); *I Tintaridi* (Frugoni; Parma 1760); *Armida* (Durazzo & Migliavacca, after Quinault; Vienna 1761); *Sofonisba* (Verazi, after Zeno; Mannheim 1762); *Ifigenia in Tauride* (Coltellini; Vienna 1763); *Antigone* (Coltellini; St Petersburg 1772).

BIBL: F. Casavola, *Tommaso Traetta di Bitonto (1727–1779): la vita e le opere* (Bari, 1957).

**tragédie-ballet** (Fr.: 'tragedy ballet'). A musico-dramatic form which emerged in 17th-cent. France, deriving from the *comédie-ballet, whose essential features it shared save for the character of the drama. The most famous example was *Psyché* (1671), for which Lully collaborated with Quinault, Molière, and Corneille: this had an important influence on the development of the tragédie en musique.

**tragédie en musique** (Fr.: 'tragedy in music'). The new style of French opera inaugurated by *Lully and *Quinault with *Cadmus* (1673). Although it more accurately describes the fundamental nature of the genre, the term tragédie en musique has been widely supplanted by that of *tragédie lyrique.

**tragédie lyrique** (Fr.: 'lyric tragedy'). The serious (though not necessarily tragic) French opera of the 17th and 18th cents. whose principal exponents were *Lully, *Campra, and *Rameau. Though the term tragédie en musique was preferred by composers, and is a more accurate description of the early stages of the genre, it was soon supplanted by that of tragédie lyrique, an invention of aestheticians. By the late 18th cent. it had such a wide application that any general stylistic definition is unrealistic.

**transposition**. The notation or performance of a piece of music in a key different from its original. Opera singers, particularly as they grow older, may often require an aria to be transposed down, usually to accommodate their top notes (a typical example is Manrico's 'Di quella pira'), even during a complete performance when it may make nonsense of the composer's key structure.

**Traubel, Helen** (*b* St Louis, 20 June 1899; *d* Santa Monica, CA, 28 July 1972). US soprano. Studied with Karst. Sang in concert 1926–37. Début New York, M, 1937 (creating Mary, Damrosch's *Man without a Country*). Sang with great success there 1939–53, replacing Flagstad 1941 as leading soprano. Also London, Buenos Aires, Mexico, Rio. Roles included Sieglinde, Brünnhilde, Isolde, Kundry, Marschallin. Her large, noble tone, stately presence, and interpretative depth evoked comparisons with Nordica. Left Metropolitan after quarrel with Bing over her night-club appearances; then worked in films, taught, and wrote detective novels. Autobiography, *St Louis Woman* (New York, 1959). (R)

**'Traurigkeit ward mir zum Lose'**. Constanze's (sop) aria in Act I of Mozart's *Die Entführung aus dem Serail*, in which she mourns her separation from Belmonte.

**Travelling Companion, The**. Opera in 4 acts by Stanford; text by Henry Newbolt, after Hans Andersen's fairy-tale. Prod. Liverpool 30 Apr. 1925.

**travesti** (Fr. past participle of *travestir*: 'to disguise'). The term used to describe such roles as Cherubino, Octavian, Orlofsky, Siebel, etc., which although sung by women are male characters. The English term is 'breeches part' or 'trouser-role', from the German *Hosenrolle*.

**Traviata, La** (The Strayed Woman). Opera in 3 acts by Verdi; text by Piave, after the drama *La dame aux camélias* (1852) by Dumas fils, after his novel (1848) based on his own experiences. Prod. Venice, F, 6 Mar. 1853, with Salvini-Donatelli, Graziani, Varesi, cond. Mares; London, HM, 24 May 1856, with Piccolomini, Calzolari, Benevenuto, cond. Bonetti; New York, AM, 3 Dec. 1856, with La Grange, Brignoli, Amodio, cond. Maretzek.

Paris, c.1850. The consumptive Violetta (sop) has won the love of Alfredo (ten). He declares himself to her but she is reluctant to enter into any serious attachment.

Violetta and Alfredo have been living together in the country for three months. While Alfredo is away, his father Giorgio Germont (bar) calls. He appeals to Violetta to break off this scandalous relationship since it is endangering the forthcoming marriage of his daughter. Because of her love for Alfredo, she agrees. As Violetta leaves, Alfredo is handed a letter from her declaring they must part. He notices an invitation to a party from Violetta's friend Flora and is sure he will find her there. Violetta has gone to the party with her former protector, Baron Douphol (bar). Alfredo arrives and Violetta pleads with him to leave before the Baron challenges him to a duel. Her unfaithfulness angers Alfredo and he flings at her money he has won at cards as repayment for what she has spent on him. Germont arrives and reprimands his son for his conduct.

Several weeks later Violetta is seriously ill. In a letter to her, Germont admits that he revealed her sacrifice to his son. Alfredo enters and begs forgiveness; he promises that they shall spend the rest of their lives together in happiness. He then realizes how ill Violetta is. She finds strength momentarily and rises to her feet, but falls dead.

**Trebelli, Zelia** (*b* Paris, 1838; *d* Étretat, 18 Aug. 1892). French contralto. Studied Paris with Wartel. Début Madrid 1859 (Azucena). Berlin, Court O, 1860; Paris, I, 1861; London: HM from 1862, DL, 1868–75, CG, 1868–82, 1888; New York, M, 1884. With her slim, graceful figure (preserved to the end of her career) and rich, extensive voice she was highly successful as Cherubino, Arsace, Maffeo Orsini, Siebel, etc; also as Zerlina, Fatima, Rosina, Léonore (*Favorite*). Married **Alessandro Bettini** (1825–98), a tenore di grazia famous for his sweet-toned Arturo and Almaviva.

**treble.** The term for the higher of the two categories of an unbroken boy's voice (the lower being alto).

**Treemonisha.** Opera in 3 acts by Scott Joplin; text by composer. Comp. *c*.1907; prod. (part only) Atlanta, Morehouse College, 11 Aug. 1972, with Balthrop, Allen, Rayam, White, cond. Hill.
  A plantation in Arkansas, 1884. The soothsayer Zodzetrick (ten) spreads superstition among the people, and is reproached by Treemonisha (con). In revenge, he and his companion Luddud (bar) kidnap her. She is rescued by Remus (ten), who scares Treemonisha's captors by disguising himself as the Devil. On her return to the village, Treemonisha agrees to become the planters' leader.
  Joplin's second and only surviving opera.

**Treigle, Norman** (*b* New Orleans, 6 Mar. 1927; *d* New Orleans, 16 Feb. 1975). US bass-baritone. Studied Louisiana with Wood. Début as student, New Orleans 1947 (Duke of Verona, *Roméo et Juliette*). New York City O 1953–72; London, CG, 1974; also San Francisco, Buenos Aires, Milan, Hamburg. A performer with a highly coloured dramatic style. Sang Don Giovanni, Boris, Gounod's and Boito's Mephistopheles, the *Hoffmann* baritone roles; created parts in *The Tender Land* (Copland) and *Susannah* and *Markheim* (Floyd). (R)

**Treptow, Günther** (*b* Berlin, 22 Oct. 1907; *d* Berlin, 28 Mar. 1981). German tenor. Studied Berlin, and Milan with Scarmeo. Début Berlin, SO, 1936 (Italian Tenor, *Rosenkavalier*). Berlin, SO (later D), until 1942, 1945–50, 1961–72; Munich 1942–7; Vienna 1947–55. Also Milan, S; London, CG; New York, M, 1951. A true Heldentenor, admired as Florestan, Siegmund, Siegfried, Tristan; also as Verdi's Otello, Števa. (R)

**Treviso.** City in the Veneto, Italy. The Bensi-Zecchini family's theatre opened 1670; closed 1719. Count Onigo's theatre opened 1691; closed 1714; reopened 1765 with Caldara's *Demofoonte*. The Delfin family's theatre opened 1721. In 1844 Count Onigo handed his theatre over to a group of box-holders; renamed T Sociale; burnt down 1868; rebuilt 1869; renamed T Comunale 1932; restored after period of neglect 1961. Since 1967 has staged the Autunno Trevigiano, in 1974 giving a complete Puccini cycle commemorating the 50th anniversary of his death.

**Trial.** The term, derived from Antoine *Trial, traditionally applied at the Paris OC to a tenor of dramatic rather than vocal excellence, specializing in comedy—e.g. Le petit Vieillard (Arithme-tic) in Ravel's *L'enfant et les sortilèges*. See also TENOR.

**Trial, Antoine** (*b* Avignon, 1737; *d* Paris, 5 Feb. 1795). Tenor and actor, brother of **Jean-Claude Trial** (1732–71), co-director (with Berton) of Paris, O, 1767–9, and one of the first French composers to write for the female contralto voice. Studied Avignon. First appeared in provincial theatre. Paris, CI, 1764 (Bastien, Philidor's *Le sorcier*). Possessing little vocal but much dramatic talent, he made a reputation as a character actor, giving his name to the type otherwise known as 'singer without voice'. (See below.) Appeared in works by Grétry, Monsigny, Champein, Kreutzer. Having supported Robespierre during the Terror, he killed himself when public opinion turned against him afterwards. His wife **Marie Jeanne Milon** (1746–1818) was his best pupil, and had much success in operas by Monsigny and Grétry. Their son **Armand-Emmanuel Trial** (1771–1803) was a singing-teacher and minor composer.

**Trier.** City in Rhineland-Palatinate, Germany. A theatre was opened in the former monastery in the Viehmarkt 1802. As well as opéra-comique, including works by Méhul and Boieldieu, German opera was given. Heinz Tietjen was an influential director until 1922. Theatre bombed 1944; rebuilt (cap. 622) 1964.

**Trieste.** City in Friuli Venezia-Giulia, Italy; originally Italian, then French, Austrian, since 1954 Italian. The first opera was Orlandini's *Serpilla e Bacocco*, given in 1730 in the Palazzo del Comune, where the T San Pietro (cap. 800) opened 1751; renamed by imperial decree Cesario Regio T San Pietro, 1760. Operas by Pergolesi, Piccinni, Cimarosa, Galuppi, Salieri, and Paisiello were given, with prominent singers. Da Ponte lived here for a time with his mistress Ferrarese, and persuaded leading Mozart singers from Vienna and Prague to appear. The T San Pietro closed 1800 with Zingarelli's *Giulietta e Romeo* with Catalani as Romeo. The T Nuovo opened 21 Apr. 1801 with Mayr's *Ginevra di Scozia*, commissioned for the occasion; renamed T Grande 1821, T Comunale 1861, T Comunale Giuseppe Verdi 1901. Verdi's *Corsaro* (1848) and *Stiffelio* (1850) both failed on their premières here. Many French works had their Italian premières, and a strong Wagner tradition developed: the Angelo Neumann co. performances of *The Ring* in 1883 at the newly opened T Politeama Rossini drew an average audience of 1,000 a night, a record for Wagner in Italy. The theatre became an *Ente Autonomo in 1936 under Giuseppe Antonicelli (1936–45, 1951–68), opening 9 Jan. 1937 with Verdi's *Otello*. Richard Strauss has also been popular, and the city has shown great loyalty to the locally born *Smareglia. Raffaello de Banfield, director since 1972, has encouraged Eastern European works. Open-air performances, mostly of operetta, are given at the Castello San Giusto.

**trillo caprino.** See BLEAT.

**trio.** See TERZETT.

**Triquet.** The Frenchman (ten) living near the Larins, who sings his couplets in Tatyana's praise at the dance, in Tchaikovsky's *Eugene Onegin*.

**Tristanschalmei** (Ger.: 'Tristan shawm'). A shawm in F built by Wilhelm Heckel to play the shepherd's tunes in Wagner's *Tristan*. Now normally replaced by the cor anglais.

**Tristan und Isolde.** Opera in 3 acts by Wagner; text by composer, after Gottfried von Strassburg's *Tristan* (c.1210), in turn based on Thomas von Britanje's *Roman de Tristan* (c.1150) and Eilhart von Oberge's *Tristrant und Isalde* (c.1170), after various versions of the old legend. Prod. Munich, N, 10 Jun. 1865, with Malvina and Ludwig Schnorr von Carolsfeld, Deinet, Zottmayer, Mitterwurzer, Heinrich, cond. Bülow; London, DL, 20 Jun. 1882, with Sucher, Brandt, Winkelmann, E. Kraus, Gura, Landau, cond. Richter; New York, M, 1 Dec. 1886, with Lilli Lehmann, Brandt, Niemann, Robinson, Fischer, cond. Seidl.

A ship at sea between Ireland and Cornwall, legendary times. Tristan (ten) is taking Isolde (sop) to be the bride of King Mark of Cornwall (bs). He refuses through his squire Kurwenal (bar) to see her on the ship. She describes to her woman Brangäne (mez) how he was wounded in slaying her betrothed, Morold, but healed by her. Refusing to admit her love for him, she orders Brangäne to prepare poison for them, but Brangäne substitutes a love potion. They drink it, and disclose their love for each other.

Outside King Mark's castle. Isolde takes advantage of her husband Mark's absence hunting with Melot (ten) to meet Tristan. The lovers declare their passion, and turn away from the harsh falsehoods of day to their own true world of night. Melot causes them to be surprised, but the King is too grief-stricken to show anger at Tristan's betrayal. Isolde answers Tristan that she will follow him wherever he goes. He is attacked by Melot and allows himself to be wounded.

Outside Tristan's castle of Kareol in Brittany. Kurwenal tries to cheer his sick master, who thinks only of Isolde. The repeated sad strain of a shepherd's pipe tells that Isolde, sent for by Kurwenal, is not in sight. When the shepherd's joyful tune announces her ship, Tristan excitedly tears off his bandages, and dies in her arms. A second ship brings Mark and Melot, and Kurwenal dies

killing Melot, unaware that they come to pardon Tristan. Isolde sings of the love which she can only now fulfil in the deeper night of death at Tristan's side.

Isolde's final soliloquy is now generally known as the 'Liebestod' (Love Death), but when Wagner first joined the Prelude and this finale for concert purposes, he entitled the former 'Liebestod' and the latter 'Verklärung' (Transfiguration).

**Trittico, Il.** The triptych of 1-act operas by Puccini, *Il *tabarro*, **Suor Angelica*, and **Gianni Schicchi*.

**Troilus and Cressida.** Opera in 3 acts by Walton; text by Christopher Hassall, after Chaucer's *Troilus and Criseyde* (c.1385) and Boccaccio's *Filostrato* (c.1350), in turn after Benoit de Saint-Maure's *Roman de Troie* (c.1160). Prod. London, CG, 3 Dec. 1954, with László, Lewis, Pears, Kraus, cond. Sargent; San Francisco, War Memorial OH, 7 Oct. 1955, with Kirsten, Lewis, McChesney, Weede, cond. Leinsdorf. Revised and shortened by Walton with Cressida's role transposed to mez range for Janet Baker, prod. London, CG, 12 Nov. 1976.

Troy, c.12th cent. BC. Cressida (sop or mez), daughter of the High Priest of Troy, Calkas (bs), is turned from her intention of becoming a priestess by the love of Troilus (ten), abetted by her uncle Pandarus (ten). When Calkas goes over to the Greeks, she is exchanged for their prisoner Antenor (bar), and agrees to marry Diomede (bar). Troilus comes to find her and is stabbed in the back by Calkas when fighting Diomede, who orders his body to be returned to Troy and Cressida to remain in the Greek camp as a harlot. She kills herself over Troilus's body.

**trois débuts.** The system operating in various French theatres, especially in the 19th cent., whereby before firm engagement for a season an artist had to appear in three different roles and be subject to three public votes on them.

**Trompeter von Säckingen, Der** (The Trumpeter of Säckingen). Opera in 4 acts by Nessler; text by Rudolf Bunge, after Scheffel's poem (1854). Prod. Leipzig 4 May 1884, with Jahns, Marion, Schelper, Grengg, cond. Nikisch; New York, M, 23 Nov. 1887, with Seidl-Kraus, Meisslinger, Ferenczy, Von Milde, Robinson, Fischer, cond. Seidl; London, DL, 8 July 1892, with Bettaque, Schumann-Heink, Landau, Litter, Reichmann, Wiegand, cond. Feld.

Säckingen, just after the Thirty Years War. Werner (bar) loves Maria (sop), whose parents want her to marry Damian (ten). Werner gives Maria trumpet lessons; their love is discovered. However, Damian proves to be a simpleton, and moreover a coward when the city is attacked. Werner is wounded in his heroic defence and then proves to be of noble birth; he is permitted to marry Maria.

Nessler's greatest success, formerly very popular in Germany. Also opera by Kaiser (1882).

**Troppau.** See OPAVA.

**trouser role.** See TRAVESTI.

**Trovatore, Il** (The Troubadour). Opera in 4 acts by Verdi; text by Cammarano and Bardare, after the drama *El trovador* (1836) by Gutiérrez. Prod. Rome, Ap., 19 Jan. 1853, with Penco, Goggi, Baucardé, Guicciardi, cond. Angelini; New York, AM, 2 May 1855, with Steffanone, Vestvali, Brignoli, Amodeo, cond. Maretzek; London, CG, 10 May 1855, with Ney, Viardot, Tamberlik, Graziani, cond. Costa.

Northern Spain, 15th cent. Ferrando (bs) tells how the Conte di Luna had a brother whose fortune was read by a gypsy when both boys were babies. The baby became ill and the gypsy was burned at the stake. In revenge, her daughter Azucena (mez) kidnapped the child, and charred bones were found at the site of the gypsy's execution. The baby's father could not believe his son dead and commanded that the brother, the present Conte di Luna, should continue the search. Leonora (sop), lady-in-waiting to the Princess of Aragon, is in love with Manrico (ten), a warrior troubadour. She hears his voice serenading her but finds the Conte di Luna (bar), who also loves her. Manrico reveals himself and Di Luna challenges him to a duel.

At the gypsy camp, Azucena relates to Manrico (her supposed son) the story of her mother's death. She confesses that in her desire to avenge it, she accidentally burnt her own son instead of the Di Luna's. When Manrico questions his own identity, Azucena hurriedly assures him that he is her son. News arrives that Leonora proposes to enter a convent since she believes Manrico dead, so he rushes off to prevent her. Di Luna is determined to abduct Leonora as she enters the convent, but Manrico is ahead of him.

The guards bring Azucena to Di Luna's camp. He is delighted when he realizes she is Manrico's mother and orders her to be tortured. As Manrico and Leonora are about to be married, they learn that Azucena has been captured. Manrico hurries to save her.

Manrico has failed to rescue Azucena and is a prisoner. Leonora offers herself to Di Luna if he will spare her beloved. He agrees, not knowing that her real intention is to poison herself rather than be unfaithful. She tells Manrico of his freedom but he, suspecting her action, denounces her. With Leonora dying in his arms, he realizes the sacrifice she has made. As Manrico is executed, Azucena reveals to Di Luna that he has killed his brother.

**Troyanos, Tatiana** (*b* New York, 12 Sep. 1938; *d* New York, 21 Aug. 1993). US mezzo-soprano. Studied New York with Heinz. Début New York City O 1963 (Hippolyta, *Midsummer Night's Dream*). Remained until 1965; Hamburg from 1965; Aix 1966; New York, M, from 1976; London, CG, from 1970; Milan, S, from 1977. Roles include Ariodante, Cherubino, Sesto, Adalgisa, Dido (Berlioz), Eboli, Carmen, Kundry, Octavian, Santuzza. Possesses a large, brilliant voice of extensive range, and a well-projected stage personality. (R)

**Troyens, Les** (The Trojans). Opera in 5 acts by Berlioz; text by composer, after Virgil's *Aeneid* (*c*.27–19 BC). Comp. 1856–8. To achieve a performance, Berlioz was obliged in 1863 to divide the work into 2 parts: Acts I and II into *La prise de Troie* (The Capture of Troy) (3 acts), prod. Karlsruhe 6 Dec. 1890, with Reuss, Harlacher, Oberländer, Cordes, Heller, Nebe, cond. Mottl; Acts III, IV, and V into *Les Troyens à Carthage* (The Trojans in Carthage) (5 acts, with added prologue), prod. Paris, L, 4 Nov. 1863, with Charton-Demeur, Monjauze, Cabel, De Quercy, Petit, Peront, cond. Deloffre. The whole work was first prod. Karlsruhe 6–7 Dec. 1890, with (in addition) Mailhac, Plank, Rosenberg, Guggenbühler, cond. Mottl; Glasgow (shortened) 18–19 Mar. 1935, with Black, Booth, Pugh, McCrone, Morrison, W. Dickie, Moir, W. Noble, Reid, Graham, cond. Chisholm; Boston, OH, 27 Mar. 1955 (9 scenes), with Moll, Schoep, Albert, McCollum, Joyce, cond. Goldovsky; San Francisco, OH, 4 Nov. 1966 (much shortened) with Crespin as Cassandra and Dido, Vickers, cond. Périsson; New York, M, 22 Oct. 1973, with Verrett (Cassandra and Dido), Blegen, Dunn, Vickers, Veasey, Riegel, Quilico, Macurdy, cond. Kubelik. First complete performance in French, London, CG, 17 Sep. 1969.

The abandoned Greek camp outside Troy, antiquity. The Greeks have departed, leaving behind the wooden horse. The forebodings of Cassandre (Cassandra) (mez), daughter of Priam (bs), are disbelieved even by her lover Chorèbe (Chorebus) (bar). The royal family and dignitaries of Troy give ceremonial thanks for their deliverance; the celebrations are interrupted first by the silent, grieving figure of Hector's widow Andromaque (Andromache) with their son. (At this point Berlioz originally included a scene in which the Greek spy Sinon explains the purpose of the horse.) The celebrations are further interrupted by Énée (Aeneas) (ten) with news that the priest Laocoön, who mistrusted the purpose of the horse, has been devoured by serpents; to the

Trojan March, the horse is dragged into the city so as to propitiate Athene.

A room in Énée's palace. Hector's ghost (bs) tells Énée to flee Troy and found a new city in Italy. He is joined by Panthée (Pantheus) (bs), his son Ascagne (Ascanius) (sop), and others; he leads an attempt to reach the citadel. Inside Priam's palace. Polyxène (Polyxenus) (sop) and a group of women are praying. Cassandre tells the women of Chorèbe's death and Énée's escape. As the Greeks rush in, the women kill themselves.

The Palace of Didon (Dido) in Carthage. Didon (mez) enters with Anna (con) and Narbal (bs), and the people celebrate peace. Iopas (ten) announces the arrival of the disguised Trojan wanderers. When invasion threatens, Énée reveals himself and leads his men in the defence of Carthage.

A forest outside Carthage. In the Royal Hunt and Storm, Didon and Énée seek shelter from the chase in a cave and consummate their love. Didon's gardens by the sea. An entertainment is being held; Didon is abstracted, and after the others have left she and Énée celebrate their love. Their idyll is interrupted by Mercure (bar or bs) summoning Énée to his duty to go on to Italy and the foundation of the new city.

The seashore. The young sailor Hylas (ten or con) sings of his homeland. Panthus summons the Trojans for departure; Énée is prevented by the Trojan ghosts from returning to Didon, but as they prepare to leave she appears and bitterly reproaches him. She calls on the gods for vengeance, and commands a funeral pyre to be built. Mounting the pyre, she dies with a vision of Rome before her eyes.

BIBL: I. Kemp, ed., *Hector Berlioz: 'Les Troyens'* (Cambridge, 1988).

**Tsar Saltan, The Tale of** (Russ.: *Skazka o Tsare Saltane*) (full title: The Legend of Tsar Saltan, of his son, the famous and mighty hero Prince Guidon Salanovich, and of the beautiful Swan Princess). Opera in a prologue and 4 acts by Rimsky-Korsakov; text by V. I. Belsky, after Pushkin's poem (1832). Prod. Moscow, Mamontov T, 3 Nov. 1900, with Mutin, Tsvetkova, Rostavtseva, Veretennikova, Strakhova, Shevelev, Zabela, Sekar-Rozhansky, Levandovsky, cond. Ippolitov-Ivanov; London, SW, 11 Oct. 1933, with Cross, Kennard, Palmer, Coates, Wendon, Kelsey, Austin, Hancock, cond. Collingwood; New York, St James's T, 27 Dec. 1937.

Tsar Saltan (bs) marries Militrisa (sop), whose sisters Tkachikha (mez) and Povarikha (sop) cause her and her son Prince Guidon (ten) to be thrown into the sea in a cask by telling the Tsar that she has produced a monster. Washed up on an island, Guidon rescues a swan fleeing from a

hawk: it turns out to be a disguised princess (sop). Turned by her into a bee, he goes to find the Tsar; when the sisters try to dissuade the Tsar from visiting the island, Guidon stings them mercilessly. Back on the island, he frees the princess from her enchantment, and she restores the Tsar's wife to him.

The orchestral 'Flight of the Bumble Bee' occurs in Act III.

**Tsar's Bride, The** (Russ.: *Tsarskaya nevesta*). Opera in 3 acts by Rimsky-Korsakov; text from L. A. Mey's drama (1849), additional scene by I. F. Tyumenev. Prod. Moscow, Mamontov T, 3 Nov. 1899, with Mutin, Zabela-Brubel, Shevelev, Tarasov, Sekar-Rozhansky, Rostovtseva, Shkafer, Gladkaya, Strakhova, cond. Ippolitov-Ivanov; Seattle 6 Jan. 1922; London, L, 19 May 1931, with Vechov, Antonovich, Sadoven, Victorov, Yurenev, Kaydanov, cond. Steinman.

Moscow, 1572. Marfa (sop), chosen by Ivan the Terrible as his bride, is loved by both Lykov (ten), whom she loves, and the oprichnik Gryaznoy (bar), who tries to win her with a love potion. Gryaznoy's mistress Lyubasha (mez) substitutes poison, and Marfa goes mad on learning, as she lies dying in the Kremlin, that the Tsar has beheaded Lykov for the crime. Gryaznoy kills Lyubasha.

**Tuček, Vincenc** (*b* Prague, 2 Feb. 1773; *d* Pest, 1821 or later). Bohemian composer, conductor, and singer. Son of **Jan Tuček** (*b* ?, *c*.1743; *d* Prague, 19 Sep. 1783), a composer and conductor who pioneered the introduction of Czech Singspiel into Prague theatres. Vincenc sang at the Patriotic T from 1793 (first Bohemian Tamino, 1794). Harpsichordist and composer there, 1794–6. Also worked in Wrocław (Breslau) and Vienna, L.: Kapellmeister 1807–10; then Budapest. Of his many operas and other stage pieces, largely ephemeral, the most successful were *Honza Kolohnát z Přelouče* (*Hans Klachl von Prelautsch*) (Steinsberg; Prague ?1796), (*Dämona* (Bullinger; Pest 1805), and *Lanassa* (Tuček; Pest 1805). A cousin, **Leopoldine Tuczek-Ehrenburg** (*b* Vienna, 11 Nov. 1821; *d* Baden, 20 Oct. 1883) was a soprano who sang in Vienna 1836–41, Berlin 1841–61.

**'Tu che di gel sei cinta'.** Liù's (sop) aria in Act III of Puccini's *Turandot*, assuring Turandot that one day she too will love the unknown prince (Calaf).

**'Tu che la vanità'.** Elisabetta's (sop) aria in Act V of Verdi's *Don Carlos*, praying before the tomb of Charles V and bidding farewell to her past and her illusions.

**Tucker, Norman** (*b* Wembley, 24 Apr. 1910; *d* London, 10 Aug. 1978). English administrator.

Studied Oxford and London. Began career as pianist. Joint director, London, SW, 1947–54, sole director 1954–66, surviving various crises but then resigning when the projected move of SW to the South Bank fell through. He presided over a period of steady development at SW, and also encouraged the introduction of Janáček's operas, some of which he translated.

**Tucker, Richard** (*b* Brooklyn, 28 Aug. 1913; *d* Kalamazoo, MI, 8 Jan. 1975). US tenor. Studied New York with Althouse. Début New York, Salmaggi Co., 1943 (Alfredo). New York, M, 1945–74; Verona 1947; London, CG, 1958; Milan, S, 1969; New Orleans 1948–73. Though a rudimentary actor, he sang with convincing fervour, excelling in Italian roles. Repertory included Alfredo, Manrico, Radamès, Enzo, Canio, Pinkerton, Eléazar, Lensky. His wife was the sister of tenor Jan *Peerce. (R)

**Turandot.** 1. Opera in 3 acts by Puccini, completed by Alfano; text by Adami and Simoni, after Gozzi's drama (1762), possibly after the *1,001 Nights*. Prod. Milan, S, 25 Apr. 1926, with Raisa, Zamboni, Fleta, cond. Toscanini; New York, M, 16 Nov. 1926, with Jeritza, Attwood, Lauri-Volpi, cond. Serafin; London, CG, 7 June 1927, with Scacciati, Schoene, Merli, cond. Bellezza. Unfin., completed by Alfano.

Peking, legendary times. The Princess Turandot (sop) is to marry whoever can answer her three riddles. If a candidate fails, he is beheaded. In the crowd awaiting the execution of a failed suitor is the deposed King of Tartary, Timur (bs); he recognizes his long-lost son Calaf (ten), who is obliged to keep his name and rank secret for fear of his enemies. Liù (sop), the slave-girl who has helped Timur since his escape, loves Calaf. He, however, falls in love with Turandot and resolves to win her, ignoring pleas from his father and Liù and from three Ministers of the Court, Ping (bar), Pang (bar), and Pong (ten).

The Emperor Altoum (ten) tries to persuade the stranger to give up the enterprise. Turandot proudly proclaims that no man will ever possess her. She poses the riddles and when Calaf answers each of them correctly, Turandot begs her father to release her from her oath. Calaf offers that if she can discover his name before dawn the following day, she may have him put to death.

Since Turandot has ordered mass execution if the stranger's name cannot be discovered, the people of Peking turn on him. Timur and Liù are captured and, in order to spare Timur torture, Liù confesses that she alone knows who the stranger is, but will not disclose his name. Her love for Calaf gives her courage. As she goes to her execution, she snatches a dagger and stabs

herself. Calaf declares his love to Turandot. She begs him to leave but he refuses and tells her his name. Turandot is softened by his sincerity and proclaims to her father that the stranger's name is Love.

BIBL: W. Ashbrook and H. Powers, *Turandot* (Princeton; 1991).

2. Opera in 2 acts by Busoni; text by composer, after Gozzi. Prod. Zurich, Stadttheater, 11 May 1917, with Enke, Smeikal, Richter, Peiroth, cond. Busoni; London, Cockpit T, 8 Mar. 1978; Stamford, 15 Nov. 1986, with Carmona, Ventriglia, Craig, Lancman, Craney, cond. Gilgore.

Other operas on the subject are by Blumenroeder (1810), Danzi (1817), Reissiger (1835), Vesque von Püttlingen (1838), Lövenskjow 1854), Jensen (*Die Erbin von Monfort*, 1864–5, adapted after his death to a *Turandot* libretto by his daughter), Bazzini (1867), Rehbaum (1888), Neumeister (1908).

**Turco in Italia, Il** (The Turk in Italy). Opera in 2 acts by Rossini; text by Romani, after Mazzolà. Prod. Milan, S, 14 Aug. 1814, with Maffei-Festa, Carpano, David, Galli, Pacini, Vasoli, Pozzi; London, HM, 19 May 1821, with Giuseppina and Giuseppe de Begnis; New York, Park T, 14 Mar. 1826, with Malibran, Barbieri, García, García jun., Crivelli, Rosich, Agrisani.

A gypsy camp near Naples, 1850. Zaida (sop) has fled Turkey, where her love for Selim (bs) was unrequited. The poet Prosdocimo (bar) is also in the camp, seeking material for a comedy less trite than the situation of the flirtatious Fiorilla (sop) and her ridiculous husband Geronio (bs), who now also arrive. At the quayside, a ship docks; Selim disembarks and immediately makes advances to Fiorilla, irritating her lover Narciso (ten) but interesting Prosdocimo. In her apartment, she entertains and reproves Geronio. Back in the camp, Selim has his fortune told by a 'gypsy' whom he recognizes as Zaida: their reconciliation is complicated by the arrival of Fiorilla.

Selim angers Geronio by suggesting that he sell his wife. Fiorilla and Zaida invite Selim to choose between them, but Zaida withdraws unhappily. Prosdocimo says that Selim will carry Fiorilla off during a masked ball, at which confusions deepen when Zaida appears as Fiorilla and Geronio and Narciso both appear as Selim. Everything is sorted out, with Selim and Zaida returning to Turkey and all pointing the moral that one should be content with one's lot.

At first coolly received as an imitation of *L'italiana in Algeri*, then very popular. Stendhal relates how at the first run Pacini, as Geronio, would change his cuckold representation nightly, once with an imitation of a famous member of the audience.

Other operas on the subject are by Seydelmann (1788), Süssmayr (1794). Irish burlesque by S. Lover, *Il Paddy Wack in Italia* (1841).

**Turgenev, Ivan** (*b* Orel, 9 Nov. 1818; *d* Bougival, 3 Sep. 1883). Russian writer. His association with Pauline Viardot led to collaboration in her operettas *Trop de femmes* (1867), *Le dernier sorcier* (1867), and *L'ogre* (1868). Operas based on his works are as follows:

*The Parasite* (1847): Dall'Olio (*Pasquino*, comp. *c*.1890, unprod.); Orefice (1907); Sauguet (comp. 1974, unprod.)
*A Sportsman's Sketches* (1847–51): *The Singers*: Goldenweiser (1945)
*Yermolay and the Miller's Wife*: Goldenweiser (1945)
*Asya* (1857): Ippolitov-Ivanov (1900)
*A Nest of Gentlefolk* (1858): Rebikov (comp. 1816); Bagadurov (1919)
*The Torrents of Spring* (1871): Goldenweiser (1955)
*The Song of Triumphant Love* (1881): Harteveld (1895); Simon (1897)
*Clara Milich* (1882): Kastalsky (1907, prod. 1916)
*The Dream* (1876): K. H. David (1928)

**Turiddu.** A young peasant (ten), former lover of Santuzza and lover of Lola, in Mascagni's *Cavalleria rusticana*.

**Turin** (It., Torino). City in Piedmont, Italy. The first dramatic work with music was Sigismondo d'India's pastoral *Zalizura*, 1611. From 1638 performances were given at the T delle Feste. The T Regio (cap. 2,500), Italy's finest opera-house until La Scala, opened 26 Dec. 1740 with Feo's *Arsace*, with Carestini in the title-role. The architect, Alfieri, also built the T Carignano (cap. 1,000), opened 1753. Many operas were written for the Regio during the 18th and early 19th cents., including works by Gluck (*Poro*), Jommelli, G. Scarlatti, Galuppi, Traetta, Paisiello, Martín y Soler, Cherubini, Cimarosa, Gazzaniga, Zingarelli, Meyerbeer, Mercadante, and Nicolai, with the greatest singers of the day.

The opening of La Scala in 1778, and various crises in the 1850s and 1860s, afflicted the Regio. Carlo Pedrotti became director 1865, and in 15 years made the Regio one of the leading Italian theatres. He conducted every production, and Verdi's operas were especially strongly cast; he also encouraged the Italian appreciation of Wagner. Seasons were then conducted by Faccio and Mascheroni. Toscanini was music director 1895–8, 1905–6, having already triumphed at the Carignano 1886. Premières included Catalani's *Loreley* (1890), Puccini's *Manon Lescaut* (1893)

and *Bohème* (1896), and Zandonai's *Francesca da Rimini* (1914). The Regio closed for modernizing 1901–5; reopened with Strauss conducting the first Italian *Salome*. The T di Torino gave important seasons of revivals and new works under G. M. Gatti, with Gui and Casella, 1925–31, during a time of the Regio's decline in the face of competition from Rome and Milan; notable events included Supervia in *L'italiana in Algeri*, Italian premières of *Ariadne auf Naxos*, *L'heure espagnole*, and Malipiero's *Sette canzoni* and premières of Gui's *La fata malerba* and Alfano's *Madonna imperia*, and an Italian *Ring* under Busch in Jan. 1936. The theatre burnt down 8 Feb. 1936. The new Regio (cap. 1,754) did not open until 10 Apr. 1973 with *Vespri siciliani*, with Callas (her first production), and Di Stefano also as producer. Italian Radio (formerly EIAR, now RAI) has broadcast outstanding performances of opera.

BIBL: A. Basso, ed., *Storia del Teatro Regio di Torino* (Turin, 1988).

**Turkey.** Turkey's European conquests, particularly in the 16th cent., left musical influences (such as individual scales and modes and the instruments and rhythms of *Janissary music) and also an 18th-cent. interest in Turkish plots, e.g. Gluck's *La rencontre imprévue* (1764), Haydn's *Lo speziale* (1768), Vogler's *Der Kaufmann von Smyrna* (1771), and Mozart's *Zaide* (1779) and *Entführung* (1782). Opera was first given in Turkey by a visiting company in 1797. Italian opera seasons were held in various theatres, especially the T Naum, 1841–70. Tigran *Chukhadjian (1837–98), though notable for his work for an independent Armenian opera, also helped by his presence to initiate operatic activity in Turkey; his works were given by Turkish operetta groups, especially that of Arshag Khachaturyan 'Benliyan' (1865–1923), active until 1910. A Turkish musical comedy, *Shaban*, by Vittorio Radeglia, was given at the Vienna Volksoper in 1918, and another by the Turkish composer Vedi Saba, *Kenan Çobaniari*. Operetta flourished in the 1930s at the Şehir Tiyatrosu T, especially with the works of Kemal Reşa Rey, composer of *Köyde bir Facia* (1929) and *Yan Marek* (1932), and of Muglis Sabahattin (1890–1947) with *Asaletmeap* (The Noble) and *Aşk mekteri* (The School for Love). The true founder of Turkish opera is Adnan Saygun, from whom Kemal Atatürk commissioned for the Shah of Persia's visit in 1934 *Özsoy*, on a Turkish-Persian legend. He also wrote *Taş bebek* (The Stone Doll) and *Kerem* (1952). Contemporary Turkish opera composers include Ferit Tüzün, with *Midas in kuluklari* (Midas's Ears, 1978); Cengis Tanç, with *Deli Dumrul* (1979);

Okan Demiris, with *Murat IV* (1979) and *Yusuf and Züleyha* (1989). See ANKARA, ISTANBUL.

**Turkmenistan.** The Makhtumkuly Opera and Ballet T was opened in Ashkabad in 1941, on the basis of an earlier musical theatre, with *Zokhre ve Takhir* by Adrian Shaposhnikov (1887–1967). The first Turkmen opera was Kakhiani's *Sudba bakhshi* (1941). A number of Turkmen composers followed with operas, sometimes in collaboration with Shaposhnikov or other composers from neighbouring republics; they included Dangatar Ovezov (*b* 1910), conductor at the theatre 1941–66, Ashir Kuliev (*b* 1918) (*Yusup i Akhmet*, with Boris Shekter (1900–61) ), and Veli Mukhatov (*b* 1916) with *Ganly saka* (The Bloody Watershed, 1967).

**Turnage, Mark Anthony** (*b* Grays, 10 June 1960). English composer. Studied London with Knussen and Lambert, Tanglewood with Henze and Schuller. His first opera, *Greek*, is a modern, polemical version of the Oedipus myth, vehemently scored and drawing on popular and music-hall song and the Blues as well as being influenced by Britten and Stravinsky.

WORKLIST: *Greek* (Turnage & Moore, after Berkoff; Munich 1988).

**Turner, (Dame) Eva** (*b* Oldham, 10 Mar. 1892; *d* London, 16 June 1990). English soprano. Studied London with M. Wilson, G. and E. Levi, later Broads. Début with London, CR, 1916 (Page, *Tannhäuser*). London, CR, until 1924; Milan, S, 1924–5, 1929; Buenos Aires 1927; London, CG, 1928–30, 1933, 1935, 1937–8, 1947–9; Rome 1930; Chicago 1928–30, 1938. One of the greatest British singers of the century, she possessed a clear, shining, full-toned voice of extraordinary resilience, and a cast-iron technique. Large repertory included Donna Anna, Leonore, Brünnhilde, Isolde, Aida, Butterfly, and an unforgettable Turandot, to which her somewhat cool artistic temperament was ideally suited. Taught Amy Shuard, Rita Hunter, Gwyneth Jones. (R)

**Turn of the Screw, The.** Opera in prologue and 2 acts by Britten; text by Myfanwy Piper, after Henry James's story (1898). Prod. Venice, F, 14 Sep. 1954 by EOG, with Vyvyan, Cross, Mandikian, Pears, Hemmings, Dyer, cond. Britten; London, SW, 6 Oct. 1954 by same cast; New York, Kaufman Concert Hall, by New York Coll. of Music, 19 Mar. 1958.

The narrator explains that a governess has been engaged to look after two orphans in the frequent absences of their guardian, with whom she must not communicate. Bly, a country house, mid-19th cent. Mrs Grose (sop), the housekeeper, introduces the Governess (sop) to her charges, Flora (sop) and Miles (treb). The Governess comes to realize that the children are in the power of two evil ghosts of former servants, Peter Quint (ten) and Miss Jessel (sop). In despair, the Governess writes to the guardian, but Quint persuades Miles to steal the letter. Flora denies all knowledge of Miss Jessel; but the Governess finally persuades Miles to utter Quint's name, only to find him dead in her arms.

The 'screw' of the title is represented by a theme that 'turns' through 15 variations as interludes between the eight scenes of each act.

BIBL: P. Howard, *Benjamin Britten, 'The Turn of the Screw'* (Cambridge, 1985).

**'Tutte le feste'.** Gilda's (sop) aria in Act II of Verdi's *Rigoletto*, telling her father of her seduction at the hands of the Duke.

**Two Widows, The** (Cz.: *Dvě vdovy*). Opera in 2 acts by Smetana; text by E. Züngel, after Pierre Mallefille's comedy *Les deux veuves* (1860). Prod. Prague, P, 27 Mar. 1874, with Boschetiová, Sittová, Vávra, Čech, Lausmannová, Šára, cond. Smetana. New version, with recitatives replacing dialogue, prod. Prague, P, 15 Mar. 1878; New York, Sokol H, 23 Oct. 1949; London, GSM, 17 Jun. 1963, with Reakes, Browning, Eales, Richard, cond. Thorne.

A Bohemian castle, 19th cent. The two widows Karolina (sop) and Anežka (sop) live on the former's estate. Their gamekeeper Mumlal (bs) brings in a poacher, Ladislav (ten), who proves to be the neighbouring landowner. Anežka is soon in love with him, but will not admit it until Karolina forces her hand by pretending rivalry.

# U

**Udine.** City in Friuli, Italy. The first public theatre opened 1681; demolished 1756 because of its embarrassing proximity to the cathedral. The T Nobile opened 1770; renamed T Sociale 1852, later T Puccini; closed 1963. Opera is now given in the T Comunale.

**'Udite, udite, o rustici'.** Dulcamara's (bs) aria in Act I of Donizetti's *L'elisir d'amore* in which he advertises his quack wares.

**Ugalde, Delphine** (*b* Paris, 3 Dec. 1829; *d* Paris, 19 July 1910). French soprano. First appearance aged 11 in a concert with Rubini, Tamburini, and Lablache. Studied Paris with Moreau-Sainti. Début Paris, OC, 1848 (in Auber's *Domino noir*). Paris: OC until 1853, 1870–1; L, 1854–60; B from 1861. A graceful, pert singer and actress, whose virtuoso technique was favourably compared with Cinti-Damoreau's by the tenor Roger. Created numerous roles in works by Massé, Thomas, Adam, etc.; also sang Mozart and Weber. Composed and took part in an operetta, *Halte au moulin* (1867). Her pupils included Marie Sass, and her daughter the soprano **Marguerite Ugalde** (1862–1940), who had a successful career in Paris and created Nicklausse.

**Uhde, Hermann** (*b* Bremen, 20 July 1914; *d* Copenhagen, 10 Oct. 1965). German bass-baritone. Studied Bremen with P. Kraus. Début Bremen 1936 (Titurel). Munich 1940–2, 1951–60; Hamburg 1947–50; Vienna, S, 1950–1; Bayreuth 1951–7, 1960; London, CG, 1953, 1954–60; New York, M, 1955–64. Large repertory included Mozart's Figaro, Pizarro, Caspar, Dappertutto, Philip, Wotan, Gunther, Klingsor Amfortas, Mandryka, Wozzeck, Tarquinius; created Creon (*Antigonae*). A highly cultivated, intelligent, and sensitive singer-actor of great integrity. (R)

**Ukraine.** From early days, Ukrainian composers made a crucial contribution to Russian operatic history. In 1739 a singing school opened in Glukhovo, the birthplace of Maxim Berezovsky (1745–77), who sang in opera in St Petersburg when only 14 and wrote the first 'Russian' opera to be produced in Italy (*Demofoonte*, 1773), and of Dmitry Bortnyansky (1751–1825), who also produced operas in Italy, later returning to Russia. A few small, privately run opera troupes were formed, and musical numbers were often inserted in popular dramas (e.g. *Natalka Poltavka*, 1819). But the first attempt at opera was *Zaporozhets za Dunayem* (The Danube Cossacks, 1863), by Semyon Gulak-Artemovsky (1813–73), a pupil of Glinka and a fine baritone. The first major Ukrainian composer was Mykola Lysenko (1842–1912). His operas, mostly based on Gogol, include *Risdivyana nich* (Christmas Eve, 1875), a new setting of *Natalka Poltavka* (1889), and the heroic opera *Taras Bulba* (comp. 1890, prod. 1924). Other Ukrainian operas of the period include *Mayskaya noch* (May Night, 1876) and *Osada Dubno* (The Siege of Dubno. comp. *c*.1884) by Pyotr Sokalsky (1832–87); *Kupalo* (comp. *c*.1892, prod. 1929) by Natal Vakhnyanin (1841–1908); *Katerina* (1899) by Nikolay Arkas (1853–1909); *Kupalna iskra* (The Kupalo Spark, 1901), by Boris Podgoretsky (1894–1919); and *Branka Roxolana* (1912) by Denis Sichinsky (1865–1912). The first Operatic Society in Greater Russia was organized at Kiev in 1889 by the singer Ippolit *Pryanishnikov: this gave Moscow seasons in 1892–9.

In Kiev opera was given in the Gorodsky (City) T (opened 1851); this was rebuilt in 1901, as the Liebknecht Opera T, and is now the Shevchenko Opera and Ballet T. Opera has been sung in Ukrainian since 1924. The company has toured widely, including to Barcelona (1979), Wiesbaden (1986), and Dresden (1987). In Kharkov an opera-house was opened in 1874, and here the first Ukrainian operas were given. This became the Lysenko Opera and Ballet T. The wide repertory includes foreign and well as Russian and Ukrainian works. There is also a flourishing operetta theatre. In Ekaterinburg (after 1926 Dnyepopetrovsk), an opera-house was opened in 1926; during the war this was destroyed, and the company evacuated to Krasnoyarsk. Besides this company, working in the reconstructed theatre, there are several important semi-professional groups. There is an opera-house in Odessa (1809), and some touring companies. The Franko T in Lvov (Lviv) was opened in 1940.

Post-Revolutionary operas included *Shchors* (1938) by Boris Lyatoshinsky (1895–1968); and *Perekop* (1939) by Mikhail Tits (1898–1978), Vsevolod Rybalchenko (*b* 1904), and Yuly Meytus (*b* 1903): the latter's ten operas include the widely performed *Molodaya Gvardiya* (Young Guard, 1947).

Since the Second World War there have been five opera-houses in the Ukraine. Successful operas have included *Bogdan Khmelnitsky* (1951) by Konstantin Dankevich (1905–84); *Gibel Eskadry* (Death of a Squadron, 1967) by Vitaly Gubarenko (*b* 1934); *Lisova pisnya* (Forest Song) (1958) and *Anna Karenina* (1970) by Vitaly Kireyko (*b* 1926); *Milana* (1957), *Taras Shevchenko*

(1964), and *Yaroslav Mudry* (1975) by Georgy Mayboroda (*b* 1913). A novel departure came in 1979 with *Tsvet paporotnika* (The Fern in Bloom) by Yevgeny Stankovich (*b* 1942), which uses folk tunes and national instruments but also a modern rock style. See also LVOV.

BIBL: A. Schreyer-Tkachenko, ed., *Istoriya ukrain-skoy muzyki* (Moscow, 1981).

**Ulfung, Ragnar** (*b* Oslo, 28 Feb. 1927). Norwegian tenor. Studied Oslo, and Milan with Minghetti. Début Oslo 1952 (Magician, *The Consul*). Swedish Royal O from 1958; London, CG, 1960, 1963, 1970–6; New York, M, 1972; Milan, S, 1973. A very gifted (and athletic) dramatic performer, impressive as Tamino, Don Carlos, Gustavus III, Lensky, Mime, Herod, Tom Rakewell, Števa, and in the parts he has created, notably Maxwell Davies's Taverner. (R)

**Ulm.** City in Baden-Württemberg, Germany. Ulm was an active centre of school music dramas, and had many visits from English touring companies. The Theater im Binderhof opened 1641, enlarged 1650. The Komödienhaus auf dem Kutschenhausplatz opened 1781, and developed a vigorous theatrical life under J. Dardenne 1834–47. Music directors have included Robert Heger 1908–9 and Karajan 1927–34.

**Ulrica.** See ARVIDSON, MLLE.

**Ulysses.** See ODYSSEUS.

**Umeå.** Town in Västerbotten, Sweden. Norrlands Opera, founded in 1975, is the world's most northern opera company and plays in a converted fire station.

**Umlauff, Ignaz** (*b* Vienna, 1746; *d* Vienna, 8 June 1796). Austrian composer, conductor, and viola player. He wrote his first Singspiel, *Die Insel der Liebe*, in 1772. When Josef II announced the opening of the new National-Singspiel at the Burgtheater in 1778, he was commissioned to write the work for the opening night and was appointed Kapellmeister. *Die Bergknappen* was an immediate success; one paper reported, 'at last the first German opera has arrived; it exceeds all the expectations of the public'. Umlauff remained the principal composer of the National-Singspiel and four further works were performed there during the following five years: *Die schöne Schusterin* (1779), *Das Irrlicht* (1782), *Die Apotheke* (1778), and *Welches ist die beste Nation?* (1782).

Umlauff was one of the principal founders of Viennese Singspiel. He and his successors combined folk-style music with elements of Italian opera, such as coloratura arias, frequently basing their works on the local theatrical tradition. The success of his works was due to this entertaining variety of material, and his ability to write humorous pieces with a strong sense of drama, as in the melodrama scene describing the avalanche in *Die Bergknappen*. His music contains some surprising modulations and sharp chromaticisms but it is above all attractive for its natural melodiousness.

Despite Mozart's description, in a letter to his father, of *Das Irrlicht* as 'that execrable opera by Umlauff', he suggested it or *Die schöne Schusterin* as works suitable for performance at the Salzburg court. Umlauff's later Singspiels, *Die glücklichen Jäger* (1785) and *Der Ring der Liebe* (1785, a sequel to Grétry's *Zémire et Azor*) never enjoyed the success of his earlier works.

His son **Michael Umlauff** (*b* Vienna, 9 Aug. 1781; *d* Vienna, 20 June 1842) was also a composer, conductor, and violinist. He studied with his father and later with Joseph Weigl. He was the composer of three Singspiels and many ballets, and held one of the Kapellmeister posts at the Viennese court theatres. However, he is best remembered for his association with Beethoven, the première of whose Ninth Symphony he conducted; he also directed the performance of the 1814 revival of *Fidelio*.

SELECT WORKLIST (all prod. Vienna): *Der Insel der Liebe* (The Island of Love) (Müller; ?1772); *Die Bergknappen* (The Miners) (Weidmann; 1778); *Die schöne Schusterin* (Stephanie, after Ferrières; 1779); *Das Irrlicht* (The Will o' the Wisp) (Stephanie, after Bretzaer; 1782).

**'Una bella serenata'.** The terzetto in Act I of Mozart's *Così fan tutte*, in which Ferrando (ten) plans a serenade for Dorabella (sop) and Guglielmo (bar) a banquet for Fiordiligi, while Don Alfonso (bs) hopes to be invited.

**'Una donna di quindici anni'.** Despina's (sop) aria in Act II of Mozart's *Così fan tutte*, describing the wiles of love to her mistresses.

**'Una furtiva lagrima'.** Nemorino's (ten) aria in Act II of Donizetti's *L'elisir d'amore*, in which he sees by the tear in Adina's eye that she loves him.

**'Un'aura amorosa'.** Ferrando's (ten) aria in Act I of Mozart's *Così fan tutte*, in which he sentimentally describes Dorabella's love.

**'Una voce poco fà'.** Rosina's (mez) aria in Act I of Rossini's *Il barbiere di Siviglia*, in which she recalls the serenade by Count Almaviva (Lindoro) that awoke her love, and promises to use all her powers to ensure that he will be hers.

**'Un bel dì vedremo'.** Butterfly's (sop) aria in Act II of Puccini's *Madama Butterfly*, in which she sings of Pinkerton's hoped-for return.

**'Un dì all' azzurro spazio'.** Chénier's (ten) aria in Act I of Giordano's *Andrea Chenier*, in which he denounces the selfishness of those in authority. Sometimes known as the 'Improvviso'.

**'Un dì felice'.** The love duet between Violetta (sop) and Alfredo (ten) in Act I of Verdi's *La traviata*.

**Undine.** The heroine of a popular Central European folk legend. Conceived by Paracelsus as one of the elemental spirits, she is the spirit of the waters; she has been created without a soul, and can only gain one by marrying a mortal and bearing him a child, though she must thereby take on all the penalties of humanity. The legend was given great popularity among the Romantics by Friedrich de la Motte Fouqué's tale (1811). She is the German equivalent of the Slavonic *Rusalka, and also appears as the *Donauweibchen.

(1) Opera in 3 acts by E. T. A. Hoffmann; text by Fouqué. Prod. Berlin, Royal OH, 3 Aug. 1816, with Friedrich, Therese and Johanna Eunicke, Leist, Willmann, Blume, Gern, Wauer, cond. Romberg; Oxford, Newman Rooms (semi-staged), 6 Dec. 1991, with Shaw, Sanchez, Arnold, Van Poznak, cond. Hinnells.

In a forest hut, Huldbrand (bar) and the Fisherman (bs) are anxious about the disappearance of Undine (sop). She has replaced the fisherfolk's drowned daughter, and Huldbrand has left Berthalda (sop) for her. Undine is in fact by a forest waterfall, where Kühleborn (bs) warns her of man's perfidy. Huldbrand arrives to find her. Their union is blessed by Heilmann (bs), and she confesses her nature to Huldbrand. In the town, Undine and Berthalda share the common bond of being foundlings. But when Undine reveals to the Count (ten) and Countess (mez) that Berthalda is really the fisherfolk's lost daughter, Berthalda is enraged. Huldbrand becomes increasingly uneasy at Undine's fairy nature, and turns to Berthalda, despite the menaces of Kühleborn. Heilmann tries to remind Huldbrand of Undine's fate if she is betrayed, but he insists on now marrying Berthalda. When at the wedding feast Berthalda has the well in the castle uncovered, Undine appears and gives Huldbrand her kiss of death. They are united in what Heilmann calls a Liebestod.

Hoffmann's most important opera, taken by Weber in his review as an example of the German ideal, 'namely a self-sufficient work of art in which every feature and every contribution by the related arts are moulded together in a certain way and dissolve, to form a new world'.

(2) Opera in 4 acts by Lortzing; text by composer after Fouqué. Prod. Magdeburg 21 Apr. 1845; New York 9 Oct. 1856, with Johanssen, Picaneser, Weinlich; London, John Lewis T, 30 Apr. 1973, with Bernardon, cond. Robertson.

Hugo (ten), accompanied by his squire Veit (ten), visits the forest and falls in love with Undine (sop); he returns to Court with her, though Veit has told Kühleborn (bar), her father and a powerful spirit, that he fears Hugo will desert her for Berthalda (sop). Undine discloses her fairy nature to Hugo, and Berthalda is revealed as being the child of Undine's foster-father Tobias (bs). Though protected by Undine, Berthalda seduces Hugo, and Undine returns to her kingdom; but he cannot forget her, and when Veit opens a sealed well, she returns from it to claim him as husband in her own realm.

Other operas on the legend are by Seyfried (1817), Girschner (1830), J. P. E. Hartmann (1842), Lvov (1848), Dargomyzhsky (*Rusalka*, 1856), Péan de la Roche-Jagu (*La reine de l'onde*, 1862), Semet (1863), Lecocq (operetta, *Les ondines au Champagne*, 1865), Mori (*The River Sprite*, 1865), Tchaikovsky (unfin., 1869), Pourny (operetta, *L'ondine de Plougastel*, 1871), Gothov-Grunecke (1874), Bottagisio (1893), Bucceri (1917), and Karzev (1923).

BIBL: F. Ferlan, *La thème d'Ondine dans la littérature et l'opéra allemande au XIX^{eme} siècle* (Berne, 1987).

**'Und ob die Wolke'.** Agathe's (sop) aria in Act III of Weber's *Der Freischütz*, in which she declares that even if the skies are veiled in cloud, God reigns above.

**Unger, Caroline** (*b* Stuhlweissenburg, 28 Oct. 1803; *d* Florence, 23 Mar. 1877). Austrian mezzo-soprano. Studied Vienna with Mozatti and Bassi, later Aloysia Lange and Vogl, and Milan with D. Ronconi. Appeared Vienna, H, 1819 (Cherubino); official début Vienna, H, 1821 (Dorabella). H until 1824. Naples 1825, C, 1826; Milan, S, 1827–30. Further successes in Italy, where her career reached its zenith; Paris, I, 1833; Vienna, H, 1839–40; Dresden 1843, then retired from stage, singing in concert under the name Unger-Sabatier (her husband was the writer François Sabatier). Combining a Teutonic dramatic intensity and an Italianate vocal facility, together with an immense range (a–d'''), she was one of the most outstanding and intelligent singers of her time. Sang Zerlina, Rosina, Isabella (*L'italiana*), Imogene (*Pirata*); created Isoletta (*La straniera*), Donizetti's Parisina, Maria di Rudenz, and Antonina (*Belisario*); also Pacini's Niobe. She sang the alto solo in the first performance of Beethoven's Ninth Symphony; at the end, it was she who turned the deaf composer round to see the rapturous applause.

**Unger, Georg** (*b* Leipzig, 6 Mar. 1837; *d* Leipzig, 2 Feb. 1887). German tenor. Studied Leipzig with Hey. Début Leipzig 1867. Returned there 1877–81. A singer with a huge physique and voice, but little artistic personality. Created Siegfried, with intensive coaching from Wagner, for whom the effort was worthwhile, but exhausting.

**United Kingdom.** See GREAT BRITAIN.

**United States of America.** For many years America saw little but ballad opera (*Flora, or The Hob in the Well* was given in Charleston, 1735, and *The Beggar's Opera* in New York in 1750) and Italian opera given by visiting companies. The first native American opera was *The Disappointment, or The Force of Credulity*, referred to in 1767 as 'a new American comic opera' (unprod.). A musical comedy *The Temple of Minerva* was given in 1781 in the presence of Washington; the music (lost) was by Francis Hopkinson, one of the signatories of the Declaration of Independence. The first American opera whose music survives was *The Archers* (1796), on the William Tell legend, by the English-born Benjamin Carr (1768–1831). Also in 1796 New York saw *Edwin and Angelina* by Victor Pellissier, a Frenchman who had settled in 1792 and went on to write other operas. John Bray wrote *The Indian Princess* (1808). A step forward came with *Enterprise* (1822), by Arthur Clifton (probably pseudonym of the Italian-born Philip Anthony Corri). The first opera written by a native American seems to have been *The Saw Mill, or A Yankee Trick* (1824), by Micah Hawkins (1777–1825): this is essentially a ballad opera. The arrangements of C. E. *Horn introduced Americans to operatic music in adaptations. The first true opera season was in *New Orleans (a US territory with the 1803 Louisiana Purchase).

Many famous Italian artists were now visiting America, some bringing their own companies, e.g. García and Da Ponte; and by the middle of the 19th cent. opera was virtually all Italian. In 1861 Lincoln attended a performance of *Un ballo in maschera*, in the Boston version. Its popularity set up a reaction among American composers, though William Henry Fry was unable to escape the influence in *Leonora* (1845). Others who attempted a more independent American national opera included George Frederick Bristow (1825–98) with *Rip Van Winkle* (1855), Dudley Buck (1839–1909) with *Deseret, or A Saint's Affliction* (1880), John Knowles Paine (1839–1906) with *Azara* (1901), and Silas Gamaliel Pratt (1846–1916) with *Zenobia* (1883). Despite the failure of this work, Pratt went on to campaign for native American opera, and in 1884 organized a Grand O Festival; but he met with little success, partly through his extravagant claims for his own work.

Towards the end of the 19th cent. a German influence was added as major German artists crossed the Atlantic. In 1911 the New York Metropolitan organized a prize, won by Horatio

Parker with *Mona* (1912). Many American operas now joined the German and Italian repertory at the Metropolitan and in other major American theatres. However, the fashion for European music and the difficulty of forming a true native tradition resulted in foreign works being long regarded as intrinsically superior. American composers either sought European training or settled in Europe in the hope of performance, e.g. Louis Adolf Coerne (1870–1922) with *Zenobia* (Bremen 1905).

A national tradition of operetta grew up early and developed into the classic American form, *musical comedy or musical. A number of musicians of European origin, first- or second-generation Americans in the great waves of immigration, or refugees from persecution, contributed to this genre: operetta composers (e.g. Friml, Romberg, and Victor Herbert) gave an example that was the basis for the original American achievements of Jerome *Kern (1885–1945), Irving *Berlin (1888–1989), Cole *Porter (1891–1964), Richard *Rodgers (1902–79), Frederick Loewe (1901–88), and above all George *Gershwin (1898–1937), whose *Porgy and Bess* (1935) remains the greatest example of an opera based on popular American musical idioms. Composers who returned from a European training (often Parisian) included Douglas *Moore (1893–1969) with *The Devil and Daniel Webster* (1938), Virgil *Thomson (1896–1989) with *Four Saints in Three Acts* (1934), Aaron *Copland (1900–90) with *The Second Hurricane* (1937), and Marc *Blitzstein (1905–64) with *The Cradle will Rock* (1937); of this generation, Samuel *Barber (1910–81) was the principal composer to be trained in America. Others who consolidated the position of American opera, as a characteristically active but various tradition, included Gian Carlo *Menotti (*b* 1911), Leonard *Bernstein (1918–90), Lukas *Foss (*b* 1922), and Carlisle *Floyd (*b* 1926). Among more recent composers who have further extended the range of opera are Hugo Weisgall (*b* 1912), Robert Ward (*b* 1917), Lee Hoiby (*b* 1926), John Eaton (*b* 1935), Philip *Glass (*b* 1937), and John *Adams (*b* 1947).

See also BALTIMORE, BERKSHIRE FESTIVAL, BOSTON, CENTRAL CITY, CHAUTAUQUA, CHICAGO, CINCINNATI, DALLAS, DETROIT, HARTFORD, HOUSTON, JACKSON, KANSAS CITY, LOS ANGELES, MIAMI, NEW ORLEANS, NEW YORK, PHILADELPHIA, PITTSBURGH, PORTLAND, ST LOUIS, SAN CARLO COMPANY, SAN DIEGO, SAN FRANCISCO, SANTA FE, SEATTLE, SPOLETO, WASHINGTON DC.

**Unterbrochene Opferfest, Das** (The Interrupted Sacrifice). Opera in 2 acts by Winter; text by F. X. Huber. Prod. Vienna, K, 14 June 1796;

London, H, 28 May 1834. Winter's most successful opera, very popular on German stages in the early 19th cent., and appearing regularly until the end of the century; also trans. into several other languages and given abroad (Eng. version, music arr. Hawes, London, L, 7 Aug. 1826).

**Uppmann, Theodor** (*b* Palo Alto, CA, 12 Jan. 1920). US baritone. Studied Curtis Institute with S. Wilson. Stage début New York City O 1948 (Pelléas). London, CG, 1951; New York, M, 1953–4, 1955–78. Also Santa Fe; Milan, S; Chicago. Roles include Guglielmo, Marcello, Sharpless; particularly acclaimed as Papageno, Pelléas, and Billy Budd, which he created. Also created parts in works by Villa-Lobos, Floyd, Pasatieri. (R)

**Urbain.** Marguerite de Valois's page (mez) in Meyerbeer's *Les Huguenots*.

**Urbani, Valentino** (*fl* 1690–1719). Italian mezzo-soprano castrato. Studied with Pistocchi. First known appearance Venice 1690; Bologna 1691, 1695; Berlin 1697–1700; Mantua 1703. London, DL, 1707–11, 1712–14, singing in pasticcios, and creating roles in Handel's *Rinaldo*, *Il pastor fido*, and *Teseo*. Venice 1717–19. The first castrato to sing regularly in London. Though limited in both range and execution, his 'chaste' style was appreciated, and he was an engaging actor.

**Urlus, Jacques** (*b* Hergenrath, 9 Jan. 1867; *d* Noordwijk, 6 July 1935). Dutch tenor. Studied Amsterdam with Averkamp, Nolthenius, and Van Zanten. Début Utrecht 1894 (Beppo, Pagliacci). Amsterdam 1894–9; Leipzig 1900–14; London, CG, 1910–14, 1924; Brussels, M, 1912–14; Bayreuth 1912; New York, M, 1912–17. Also Paris, O; Boston; Berlin. A notable Wagner singer, acclaimed as Siegmund and Tristan (which he was still singing in 1932); also sang Tamino, Samson, Verdi's Otello. (R)

**'Urna fatale'.** Carlo's (bar) aria in Act III of Verdi's *La forza del destino*, trying to reject the temptation to open the sealed envelope which contains the portrait of Leonora.

**Ursuleac, Viorica** (*b* Czernowitz (Černovcy, Cernăuţi), 26 Mar. 1894; *d* Ehrwald, 23 Oct. 1985). Romanian soprano. Studied Vienna with Forstén and Steiner. Début Agram (Zagreb) 1922 (Charlotte). Vienna, V, 1924–6, and S, 1930–5; Frankfurt 1926–30; London, CG, 1934; Berlin, S, 1935–7; Munich 1937–44. Repertory of over 80 roles included Fiordiligi, Verdi's Desdemona, Sieglinde, Tosca; for Strauss (who called her his 'Treueste aller Treuen') she created Arabella, the Countess, Maria (*Friedenstag*), and

Danae in the unofficial première. Also a celebrated Chrysothemis, Marschallin, Ariadne, Helena. Married Clemens *Krauss. (R)

**Uruguay.** Spanish light operas and tonadillas were given in Montevideo from the end of the 18th cent. A music school opened in 1831, followed by the Sociedad Filarmonica; the first complete opera given was Rossini's *L'inganno felice* in 1830 by the Tanni family. The T Solís opened in 1856, with a musical repertory based on the Spanish light tradition. Opera was mostly given by visiting Italians. The first Uruguayan opera was Giribaldi's *La parisina* (1878). National opera was more securely established by Eduoardo Fabini (1883–1950), who in 1903 founded the conservatory and other organizations before devoting himself to composition. Later operas include *La cruz del sur* (1920) by Alfonso Broqua (1867–1946), *Ardid de amor* (1917) and *La guitarra* (1924) by Carlos Pedrell (1878–1941), and *Paraná guazú* by Vincent Ascone (1897–1979).

**USSR** (Union of Soviet Socialist Republics). For countries of the former USSR (dissolved 1991, now Commonwealth of Independent States), see ARMENIA, AZERBAIJAN, BASHKIRIA, BELORUSSIA, BURYAT MONGOLIA, DAGESTAN, ESTONIA, GEORGIA, KAZAKHSTAN, KYRGYZSTAN, LATVIA, LITHUANIA, MOLDAVIA, RUSSIA, TADJIKISTAN, TATAR REPUBLIC, TURKMENISTAN, UKRAINE, UZBEKISTAN, YAKUTSK.

**Ústí nad Labem** (Ger., Aussig). Town in Bohemia, Czechoslovakia. Opera is given 3–4 times a week in the State T (cap. 900), modernized in 1947, by a company of about 20 soloists. The company also tours neighbouring industrial towns, including Karlovy Vary, Mariánské Lázně, and the Most district.

**Uthal.** Opera in 1 act by Méhul; text by J. M. B. Bins de Saint-Victor, 'imité d'Ossian'. Prod. Paris, OC, 17 May 1806. Famous for the original effect whereby, in order to create a dark and mysterious 'Ossianic' atmosphere, Méhul omitted violins and trumpets from the orchestra.

**Uzbekistan.** A Russian opera-house was opened in Tashkent in 1918. The first Uzbek musical theatre opened in 1929; the opera-house opened in 1939, but was reconstructed and opened as the Navi Opera and Ballet T. The first Uzbek opera was *Buran* (The Snowstorm, 1939) by Mukhtar Ashrafy (1912–75) with Sergey Vasilenko (1872–1956); this was modelled on Russian opera, using local folk-song and instruments. They collaborated again on *Ulug kanal* (The Great Canal, 1940); *Dilorom* (1958) was Ashrafy's own work. He was director and conductor of the Opera, 1930–70. The first Uzbek comic opera was *Prodelky Maysary* (Maisara's Tricks, 1958) by Suleiman Yudakov (*b* 1916). Other works of these years include *Leyla i Mejnun* (1940) by Reinhold Glier (1875–56) and Talib Sadykov (1907–57); *Ulugbek* (1942), on a 15th-cent. subject, by Alexey Kozlovsky (1905–77); and *Farkhad i Shirin* (1940) by Victor Uspensky (1879–1949) and Georgy Mushel (*b* 1909). The greatest popular success has been another collaboration between Glier and Sadykov, *Gulsara* (1949). Yudakov has also written *Pisma Zukhry* (Zukhry's Letters, 1964). Minasay Leviev (*b* 1912), Dany Zakirov (*b* 1914), and Boris Gienko (*b* 1917) have written musical comedies for the Tashkent Operetta T.

BIBL: J. Pekker, *Uzbekskaya opera* (Moscow, 1963).

# V

**Vaccai, Nicola** (*b* Tolentino, 15 Mar. 1790; *d* Pesaro, 5 or 6 Aug. 1848). Italian composer and singer. Studied Rome with Janacconi, Naples with Paisiello. Failing to follow up his successful Neapolitan début (with *I solitari di Scozia*) when he moved to Venice, he developed a career as a singing-teacher there and in Trieste. He sang in his own *Pietro il grande* in Parma, where he had his greatest operatic successes with *Zadig ed Astartea* and especially *Giulietta e Romeo*. These won the respect of Rossini, whose influence is clear. But the rise of Bellini eclipsed this moment of fame, especially when Bellini collaborated with Romani (with whom Vaccai had quarrelled) on his own Romeo and Juliet opera, *I Capuleti ed i Montecchi*. However, it became the custom (following a suggestion of Rossini) to incorporate the penultimate scene of Vaccai's opera into Bellini's.

In 1830 Vaccai went to Paris as a teacher, travelling on to England (1830–3) and publishing his still-respected singing method. *Giovanna Gray* was a failure, despite the presence of Malibran, but led to the post of *censore* at the Milan Conservatory: here he made a number of reforms. His last opera, *Virginia*, was an attempt to emulate the successes of French Grand Opera.

SELECT WORKLIST: *I solitari di Scozia* (The Hermits of Scotland) (Tottola, after De Gamerra; Naples 1815); *Pietro il grande* (Merelli; Parma 1824); *Zadig ed Astartea* (Tottola, after Voltaire; Naples 1825); *Giulietta e Romeo* (Romani, after Shakespeare; Milan 1825); *Giovanna Gray* (Pepoli; Milan 1836); *Virginia* (Giuliani; Rome 1845).

WRITINGS: *Metodo pratico di canto italiano per camera* (privately printed, 1832).

BIBL: G. Vaccai, *La vita di Nicola Vaccai* (Bologna, 1882).

**'V'adoro, pupille'.** Cleopatra's (sop) aria in Act II of Handel's *Giulio Cesare*, declaring her love to Caesar.

**Vakula the Smith**. Opera in 4 acts by Tchaikovsky; text by Yakov Polonsky, after Gogol's story *Christmas Eve* (1832). Prod. St Petersburg, M, 6 Dec. 1876, with Komissarzhevsky, Matchinsky, Petrov, Melnikov, Vasilyev, Bichurina, Raab, Ende, F. Stravinsky, Dyuzhikov, cond. Nápravník. Revised in 1885 as *Cherevichki* (The Little Boots—also known in the West as *Oxana's*

*Caprices*), prod. Moscow, B, 31 Jan. 1887, with Usatov, Streletsky, Dodonov, Matchinsky, Khokhlov, Vasilyevsky, Klimentova, Svyatlovskaya, cond. Tchaikovsky; New York, New Amsterdam T, 26 May 1922, with Mashir, Valentinova, Svetlov, Gorlenko, Koslov, Busanovsky.

The witch Solokha (mez) is approached by the amorous Devil (bar); as he flies off, he raises a storm and steals the moon so as to revenge himself on her son Vakula (ten), who has made an ugly painting of him; this hinders Vakula, who is trying to make his way to court Oxana (sop). When her father Chub (bs) comes home with his friend Panas (ten), both drunk, Vakula throws them out, but is himself thrown out by Oxana, though privately she admits that she loves him. In Solokha's hut, the Devil hides himself in a sack when the Mayor (bs), then the Schoolmaster (ten), and then Chub arrive; each in turn hides in a sack. Vakula staggers off with the sacks. Oxana demands a pair of *cherevichki* (high-heeled leather boots) belonging to the Tsaritsa, and Vakula flies off on the Devil's back to St Petersburg, where he is granted the *cherevichki*; he returns to claim Oxana.

Other operas on the subject are by Afansayev, written for the competition won by Tchaikovsky in 1875 (unprod.), and Rimsky-Korsakov: *Christmas Eve*, in 4 acts; text by composer after Gogol. Prod. St Petersburg, M, 10 Dec. 1895, with Mravina, Kamenskaya, Ershov, Chuprynikov, Ugrinovich, Koryakin, Mayboroda.

**Valdengo, Giuseppe** (*b* Turin, 24 May 1914). Italian baritone. Studied Turin with Accoriuti. Début Parma 1936 (Rossini's Figaro). Milan, S, 1941–3; New York, M, 1947–54; Gly. 1955. Also London, CG; Paris; Vienna; Buenos Aires. Roles included Don Giovanni, Tonio, Sharpless; his book *Ho cantato con Toscanini* (Como, 1962) describes his experiences when recording Verdi's Amonasro, Iago, and Falstaff for NBC. (R)

**València**. City in València, Spain. Opera is given in the Teatro Principal. Financial and other difficulties have made regular seasons precarious, and an ambitious attempt to improve matters in 1979 failed. Some improvement came in 1983, with a short season opening with the Valencian Martín y Soler's *L'arbore di Diana* (1787), and there was another new initiative in 1983 with a festival 'Opera i Solistes'.

**Valentin**. Marguerite's brother (bar) in Gounod's *Faust*.

**Valentine**. The Comte de Saint-Bris's daughter (sop), wife of the Comte de Nevers and lover of Raoul, in Meyerbeer's *Les Huguenots*.

**Valentini, Caterina**. See MINGOTTI.

**Valentini Terrani, Lucia** (*b* Padua, 28 Aug. 1946). Italian mezzo-soprano later contralto. Studied Padua. Début Brescia 1969 (Cenerentola). Milan, S, from 1973; New York, M, from 1974; Florence 1982. Also Paris, O; London, CG; Chicago; Moscow, B; Geneva; Munich; Prague; etc. Roles include Dido (Purcell), Tancredi, Isabella (*L'italiana*), Rosina, Marina (*Boris*), Quickly. One of the most successful contraltos of the day, with a particular sympathy for Rossini and his seria roles. (R)

**Valkyrie** (Ger., *Walküre*). In Teutonic mythology, the *Walküren* were the warrior maidens who took slain heroes from the battlefield to *Walhall* (Valhalla), there to feast and act as a bodyguard for Wotan. The word derives from *Wal* (battlefield), *küren* (to choose). In Wagner's *Ring*, they are nine in number (sop and con), *Brünnhilde, Gerhilde, Ortlinde, *Waltraute, Schwertleite, Helmwige, Siegrune, Grimgerde, and Rossweisse. Brünnhilde, Wotan's favourite among them, is the one referred to in the opera's title *Die Walküre*. See RIDE OF THE VALKYRIES.

**Vallin, Ninon** (*b* Montalieu-Vercieu, 9 Sep. 1886; *d* Lyons, 22 Nov. 1961). French soprano. Studied Lyons with Mauvarnay, Paris with Héglon. Début Paris, OC, 1912 (Micaëla). Paris: OC until 1916, 1926–46; O from 1920. Milan, S, and Buenos Aires, C, from 1916; also Rome, San Francisco. Created roles in operas by D'Erlanger and Leroux; also sang Iphigénie (*Tauride*), Mozart's Countess, Marguerite, Manon, Charlotte, Mignon, Mimì, Mélisande. A distinguished singer, whose vocal artistry and expressiveness won her many admirers, notably Debussy, with whom she worked and performed. (R)

**Valzacchi.** An Italian intriguer (ten), crony of Annina, in Strauss's *Der Rosenkavalier*.

**Vampyr, Der** (The Vampire). 1. Opera in 2 acts by Marschner; text by W. A. Wohlbrück, after the story *The Vampyre* (1819), attrib. Byron (actually by John Polidori), by way of a French *mélodrame* by Nodier, Carmouche, and De Jouffroy (1820), trans. L. Ritter (1822). Prod. Leipzig, Stadttheater, 29 Mar. 1828, with Röckert, Devrient, Streit, Höfler, Genast, Gay, Von Zieten, Vogt, Reinecke, Fischer, cond. Marschner; London, L, 25 Aug. 1829.

Scotland, 17th cent. Lord Ruthven (ten) has become a vampire, and must postpone the claiming of his soul by Satan with the sacrifice of three young girls. He destroys Janthe (sop), and is killed, but revives supernaturally in the moonlight through the help of Aubry (bar), to whom he entrusts his secret. He seduces and kills Emmy (sop), who is fascinated by vampires, and

nearly succeeds in claiming Aubry's bethrothed, Malvina (sop), as his third victim, but is denounced by Aubry just as the clock strikes.

2. Opera by Lindpaintner; text by C. M. Heigel on the same source as Marschner's opera. Prod. Stuttgart 21 Sep. 1828.

Other vampire operas are by Palma (*I vampiri*, 1812) and M. Mengal (*Le vampire*, 1826).

**Van Allan, Richard** (*b* Chipston, 28 May 1935). English bass. Studied Birmingham with Franklin. Début Gly. 1964 (Priest and Armed Man, *Zauberflöte*). London, ENO from 1969, CG from 1971. Also Paris, O; Boston; San Diego; Wexford, etc. Repertory includes Figaro (Mozart), Don Giovanni, Leporello, Padre Guardiano, Grand Inquisitor, Philip, Ochs, Claggart. An especially intelligent, versatile singing actor with a strong dramatic presence. (R)

**Vancouver.** City in British Columbia, Canada. A gala performance of *Lohengrin* was given in 1891, only five years after the city's establishment. This inaugurated the opera-house built by the Canadian Pacific Railway at its western terminus, and starred Emma Juch. The theatre thereafter rarely gave opera, and was sold to a vaudeville chain in 1912. Opera was given at the Theater under the Stars, 1940–63, but achieved no degree of permanence until 1958, when the Vancouver International Festival presented opera each summer until 1968. The festival included the North American première of Britten's *A Midsummer Night's Dream* and débuts of Günther Rennert and Joan Sutherland. In 1960 the Vancouver Opera Association gave its first performance in the new Queen Elizabeth T (cap. 2,821), and has continued to mount three or four operas each year. In addition to major Canadian singers, distinguished guests are also engaged, including Horne, Resnik, and Sutherland. Financial difficulties in the early 1980s were resolved in 1984, when the appointment of Brian McMaster of the WNO brought with it enterprising shared productions. In 1989 the International Opera Festival's travelling *Aida* drew 57,000 spectators to performances that included a 45-foot Sphinx floating in the bay.

**Van Dam, José** (*b* Brussels, 25 Aug. 1940). Belgian baritone. Studied Brussels with Anspach. Début Paris, O, 1961 (Escamillo). O from 1961; Berlin, D, from 1967; Vienna from 1970; London, CG, from 1973; New York, M, from 1975. Also Chicago, Salzburg; Venice, F, etc. Roles include Figaro (Mozart), Leporello, Méphistophélès (Gounod), Golaud, Flying Dutchman, Amfortas, Jochanaan, and first St François (Messiaen). An attractive and dramatically gifted artist, possessing a warm, extensive voice. (R)

**Van Dyck, Ernest** (*b* Antwerp, 2 Apr. 1861; *d* Berlaer-lez-Lierre, 31 Aug. 1923). Belgian tenor. Studied Paris with Bax. Début Paris, Eden T, 1887 (Lohengrin). Bayreuth 1888–1912 (as sole Parsifal, at Cosima's invitation); Vienna 1888–99; London, CG, 1891, 1897–8, 1901, 1907; Paris, O, 1891, 1908, 1914; Chicago and New York, M, 1898. Though strange in appearance, with a small head, spherical body, and round eyes and mouth, his magnificent voice and singing were much admired in Wagner, e.g. as Tannhäuser, Tristan, Loge, Siegfried; also as Cellini, Des Grieux (Massenet), and Werther, which he created. However, his vanity and sometimes ludicrously affected acting earned Mahler's dislike and his own subsequent dismissal from Vienna, despite his popularity. (R)

**Vaness, Carol** (*b* San Diego, CA, 27 July 1952). US soprano. Studied California State U. Début San Francisco 1976. Gly. from 1983; New York, M, from 1984; Chicago from 1985; London, CG, from 1982. Also Paris, O; Berlin, D; Vienna, etc. An incisive, lustrous-toned singer of ardent temperament, impressive as Alcina, Elettra, Donna Anna, Fiordiligi, Violetta, Nedda. (R)

**Vanzo, Alain** (*b* Monte Carlo, 2 Apr. 1928). French tenor and composer. Studied Aix-les-Bains. Début Paris, O, 1954 (A Pirate, *Oberon*). Paris, O from 1954, and OC from 1956; London, CG, 1961, 1963; New York, M, from 1973; also Vienna, Gly., Montreal, etc. Roles include Edgardo, Robert le diable, Cellini, Gounod's Faust, Werther, Gérald (*Lakmé*). A fine, sensitive, highly expressive artist with a sweet, firm tone and a subtle technique. His operas incl. *Pêcheurs d'étoiles* (1972) and *Chouan* (1982). (R)

**'Va, pensiero'.** The chorus of exiled Hebrews in Act III of Verdi's *Nabucco*, sending winged thoughts to their homeland.

**Varady, Julia** (*b* Oradea, 1 Sep. 1941). Romanian; later German, soprano. Studied Bucharest with Florescu. Début Cluj 1962. Cluj until 1972; Cologne 1972; Munich from 1973; New York, M, from 1977; Milan, S, 1984. Also London, CG; Berlin, D; Budapest; etc. Roles include Alceste, Donna Elvira, Vitellia, Violetta, Tatyana, Butterfly, Sieglinde, Arabella. A performer of great temperament and intensity. Married baritone Dietrich *Fischer-Dieskau 1977. (R)

**Varesco, (Abbate) Giambattista** (*fl.* Salzburg, 1775–83). Austrian librettist. Court chaplain to the Archbishop of Salzburg: librettist of Mozart's *Idomeneo* (1781), for which he adapted a text by Danchet. Mozart communicated with Varesco, whom he disliked, through his father: their correspondence gives an unparalleled insight into his creative process and thoughts on the nature of the libretto. Varesco also provided the text for the unfinished opera buffa *L'oca del Cairo* (1783); it is also possible that he adapted Metastasio's libretto for *Il re pastore* (1775).

**Varesi, Felice** (*b* Calais, 1813; *d* Milan, 13 Mar. 1889). Italian baritone. Début Varese 1834 (in Donizetti's *Furioso all'isola di San Domingo*). Sang throughout Italy until 1840; Milan, S, 1841; Vienna, K, 1842–7; Florence 1847; Naples, C, 1849–50; Venice, F, 1851, 1853; London, HM, 1864. Successful in Donizetti, and the first Antonio in *Linda di Chamounix*, he was also one of the earliest Verdi baritones, much influenced by *Ronconi. He excelled in the energetic and passionate, creating Macbeth (frequently consulted by the composer) and especially Rigoletto with enormous success. However, he was an unsympathetic Germont in the first disastrous *Traviata*; unable to do the role justice vocally or psychologically, he blamed the composer for 'misusing his resources'. Married the soprano **Cecilia Boccabadati** (1825–1906) (daughter of the soprano **Luigia Boccabadati** (*c.*1800–50)). She and her sister **Virginia Boccabadati** (1828–1922) were sopranos who excelled in the bel canto repertory. **Elena Boccabadati Varesi** (1854–1920), daughter of Felice and Cecilia, also sang this repertory, and was an especially successful Lucia; another daughter, Giulia, was a singing-teacher.

**Varlaam.** A vagabond friar (bs) in Musorgsky's *Boris Godunov*.

**Varna.** Town in Bulgaria (known as Stalin 1949–55). A Varna Operatic Society was founded in 1920 by Presiyan Dyukmedjev, out of which came a permanent company given in 1930 the title Varna Civic Opera, with Zlaty Atanasov as conductor. This closed in the following year. Another lasted from 1934 to 1935. A new civic company was formed in 1947 under the tenor Petar Baychev (1887–1960), performing in the National T.

**Várnay, Astrid** (*b* Stockholm, 25 Apr. 1918). Swedish, later US, soprano. Studied USA with her mother, the soprano Maria Yavor, later with Weigert. Début New York, M, 1941 (Sieglinde, replacing Lotte Lehmann at short notice). Replaced Traubel as Brünnhilde within the week. New York, M, until 1956. London, CG, 1948, 1951, 1958–9, 1968; Bayreuth 1951–67. Also Milan, S; Vienna; Chicago; Buenos Aires. Roles included Senta, Isolde, Kundry, Kostelnička; later Herodias, Klytemnestra, Von Einem's Old Lady (*Der Besuch der alten Dame*). An uneven singer, but a vivid and fervent performer who influenced the younger generation, e.g. Anja Silja. (R)

**Varvara.** The Kabanovs' foster-daughter (sop), lover of Kudrjaš, in Janáček's *Kářa Kabanová*.

**Vasco da Gama.** The Portuguese naval officer (ten), lover of Inez, in Meyerbeer's *L'Africaine*.

**Vašek.** A villager, Tobias Micha's son, in Smetana's *The Bartered Bride*.

**vaudeville** (Fr.: derived from the expression 'voix de ville', or 'vau de ville', i.e. 'voice of the town'). It originated in the 16th cent. as a short song, usually with an amorous text. By the 17th cent. it meant simply any song whose melody had become so popular that it was in the public domain; the vaudeville was identified by a *timbre*, a title which was usually the first line of the refrain from the original song. Its melodic style was invariably simple, using a restricted range and repeated rhythmic patterns: the tunes, called *fredons*, were sometimes also referred to as 'Pont Neuf tunes', after the wide bridge over the Seine which provided a meeting-place for local minstrels. These vaudevilles, some taking melodies from Lully's operas, were an important part of the repertoire of the Comédie-Italienne in Paris, who introduced them into comedies. The resulting mixture of music and drama was an important influence on the development of \*opéra-comique.

**vaudeville finale.** A kind of strophic finale, deriving from the \*vaudeville, in which each main character sings a stanza of the vaudeville in turn, with the full ensemble joining for the refrain. One of the first instances of its use is in Rousseau's *Le devin du village*; other early composers include Duni and Philidor. Later the style of the vaudeville finale was retained, although the music itself was newly composed. Classic examples may be found in Gluck's *Orfeo*, and Mozart's *Der Schauspieldirektor* and *Entführung*.

**Vaughan Williams, Ralph** (*b* Down Ampney, 12 Oct. 1872; *d* London, 26 Aug. 1958). English composer. Studied London and Cambridge with Parry, Wood, and Stanford.

Vaughan Williams turned to opera six times during his long life, and set considerable store by these works. The first, *Hugh the Drover*, is a ballad opera reflecting his earlier discovery of English folk-song as one of the bases for his musical language. It is in manner closer to the early nationalist operas of, for instance, Smetana than to its contemporaries, and was clearly intended to perform a similar service for the refoundation and emancipation of English opera. That it did not do so was largely due to its belatedness, in historical terms, and the inevitable artificiality in

1924 of a pastoral idyll set in the Napoleonic Wars. This does not vitiate the considerable charm and vigour of much of the score. In *Sir John in Love*, the folk influence is more fully absorbed, and, boldly challenging Verdi and Nicolai, Vaughan Williams attempts to give the character of Falstaff particular strength through the Englishness of the setting. However, by placing the characters in a 'real' Windsor, abandoning the conventions of the Elizabethan verse drama which Boito ingeniously replaced with those of opera buffa, Vaughan Williams makes the action seem less real, and cruder. For all the beauty of some of the settings, there is insufficient compensation in the use of much English poetry apart from Shakespeare and English folksong.

*Riders to the Sea* abandons these preoccupations in favour of a word-for-word setting of Synge's tragedy, drawing on a style that owes something to Wagner and to verismo, but that is also deeply characteristic of the composer. This is his operatic masterpiece. *The Poisoned Kiss* turns to a modern musical comedy, but suffers from a facetious text which has prevented the wider hearing of some of the composer's most delightful and satirical music.

Of all his operas, the work closest to Vaughan Williams's heart was *The Pilgrim's Progress*. Already in 1922 he had set an episode, *The Shepherds of the Delectable Mountains*: the completion of the whole work, which he always insisted was intended for the opera-house and not for cathedral performance, was an important event in his life. Inevitably uneven, given its long gestation, and suffering in places from an awkwardness in handling so large-scale a subject, it contains powerful and characteristic scenes, and is a moving statement of faith.

Together with \*Holst, Vaughan Williams did much to pioneer a revival of English opera; and quite apart from their individual achievements, it is certain that without them the way would have been less clear for the post-1945 revival given its impetus by Britten's *Peter Grimes*.

WORKLIST: *The Shepherds of the Delectable Mountains* (Vaughan Williams, after Bunyan: later incorporated in *The Pilgrim's Progress*; London 1922); *Hugh the Drover* (Child; London 1924); *Sir John in Love* (Vaughan Williams, after Shakespeare, with additions; London 1929); *The Poisoned Kiss* (Sharp, after Garnett; Cambridge 1936); *Riders to the Sea* (Synge's play; London 1937); *The Pilgrim's Progress* (Vaughan Williams & Wood, after Bunyan & the Bible; London 1951).

BIBL: M. Kennedy, *The Works of Ralph Vaughan Williams* (London, 1964); U. Vaughan Williams, 'Vaughan Williams and Opera', in *Composer*, 41 (1971), 25.

**Veasey, Josephine** (*b* Peckham, 10 July 1930). English mezzo-soprano. Studied London with Langford. Début London, CG, 1948 (small roles). London, CG, from 1954; Gly. 1957–9, 1964–5, 1969; Berlin, D, 1971; New York, M, from 1968; Paris, O, 1969, 1973. Also Milan, S; Vienna; Salzburg. Retired 1982, at the height of her powers. Large repertory included Dorabella, Amneris, Eboli, Dido (Berlioz), Fricka, Kundry, Charlotte; also first Andromache (*King Priam*) and Emperor (*We Come to the River*). An exceptionally gifted, musical artist with a warm, gleaming tone, and a sympathetic stage presence. (R)

**'Veau d'or, Le'.** Méphistophélès's (bs) mocking song to the crowd in Act II of Gounod's *Faust*.

**Vedernikov, Alexander** (*b* Mokino, 27 Dec. 1927). Russian bass. Studied Moscow with Alpert-Khasina, Milan with Barra. Début Leningrad, M, 1955. Moscow, B, from 1957; Milan, S, 1961. Also Paris, O; Vienna; Berlin; Finland. Repertory includes the Russian bass roles, Gounod's Méphistophélès, Daland, Philip. (R)

**'Vedrai, carino'.** Zerlina's (sop) aria in Act II of Mozart's *Don Giovanni*, in which she comforts her lover Masetto after his beating by Giovanni.

**Velluti, Giovanni Battista** (*b* Montolmo (Pausula), 28 Jan. 1780; *d* Sambruson di Dolo, 22 Jan. 1861). Italian soprano castrato. Studied Bologna with Mattei, Ravenna with Calpi. Début Forlì 1800. Naples, C, 1803; Rome 1805–7; Milan, S, 1808–9, 1810–11, 1813–14; Vienna and St Petersburg 1817–22; London, H, 1825–9, also manager 1826–9; Venice, F, 1822–3, 1824, 1826, 1830. Sang in operas by Cimarosa, Niccolini, Morlacchi. Created Rossini's only castrato role, Arsace (*Aureliano in Palmira*), enraging the composer with his lavish ornamentation (though they later became friends) and leading him to write out his own decorations. Also created Armando (Meyerbeer's *Crociato in Egitto*). The last great castrato, and singularly good-looking, he enjoyed enormous success on the Continent and in Russia, and had many mistresses. At his London début (with the young Malibran) he shocked and fascinated the English audience, by now unused to castrati, but soon gained popularity. At his peak he was a considerable singer, possessing a wide range, and both brilliance and taste. Stendhal admired his 'prodigious gifts'; Mount-Edgcumbe thought him effective in a 'grazioso' style, but limited emotionally and musically compared with *Pacchierotti. Mendelssohn found his singing distasteful. Retired from the stage in 1830 and farmed.

**vengeance aria.** An aria, common in the 18th and 19th cents., in which a character (often villainous) gives vent to his determination to avenge a perceived wrong. It served a useful purpose in giving a singer a powerful dramatic number, marked by forceful declamation and vigorous coloratura. Though there are many examples of female vengeance arias, of which the most famous is the Queen of the Night's 'Der Hölle Rache' in Mozart's *Die Zauberflöte*, it is a genre especially associated with male singers, and above all with baritones or basses, usually taking the parts of villains. The most celebrated is perhaps Pizarro's 'Ha! welch' ein Augenblick' in Beethoven's *Fidelio*, though the most influential example in early Romantic opera was Dourlinski's 'Oui! pour mon heureuse addresse' in Act III of Cherubini's *Lodoïska*. Mozart uses it in the context of comedy with Dr Bartolo's 'La vendetta' in *Figaro*.

**Venice** (It., Venezia). City in the Veneto, Italy. The first opera was probably Monteverdi's *Proserpina rapita*, given at the Palazzo Mocenigo Dandolo (now the Danieli Hotel) for the marriage of Giustiniana Mocenigo to Lorenzo Giustinian in 1630. The first public opera-house in the world opened 1637 with Manelli's *Andromeda*: this was the T San Cassiano, originally a private theatre built by the Tron family at the beginning of the 17th cent. Burnt down 1629; when it was rebuilt, the Trons decided to open it to the public, thus introducing a hitherto aristocratic art to a wider audience. Remained open until *c.*1800; staged the premières of Monteverdi's *Ulisse* and several works by Cavalli. The theatre's success led to the opening of other public opera-houses, as follows:

T Santi Giovanni e Paolo. Considered the finest theatre in Venice at the time, opened Jan. 1639 with Sacrati's *Delia*; rebuilt in stone 1654. Gave premières of Monteverdi's *Poppea* and works by Cavalli, Cesti, and Rovetta. Marco Faustini was director from 1658, and introduced operas by Luzzo and Ziani, and Cavalli's *Scipione*. After his death, directed by Grimani and Morich. Closed 1715, demolished 1748.

T San Moisè. Built by the San Bernaba branch of the Giustinian family, opened 1640 with Monteverdi's *Arianna*. Enlarged 1668 (cap. 800) and in 18th cent. Gave opera regularly until 1818, including premières of works by Vivaldi, Paisiello, Galuppi, and Anfossi. Commissioned Rossini's second opera, *La cambiale di matrimonio*, 1810, followed by *L'inganno felice*, *La scala di seta*, and *L'occasione fa il ladro* (all 1812) and *Il signor Bruschino* (1813). After 1818 a marionette theatre; rebuilt as the T Minerva. Now part shop and part block of flats.

T Novissimo. Opened 1641 with Sacrati's *La finta pazza*, and famous for the stage machinery of Giacomo *Torelli; gave opera until 1647. Mentioned by Evelyn in his diaries for 1645–6.

T Santi Apostoli (also San Aponal). Opened 1649 with Cesti's *Orontea*; gave opera until 1687.

T San Apollinare. Opened 1651 with Cavalli's *Calisto*; gave opera until 1669. Demolished 1690; today a poorhouse occupies the site.

T San Samuele. Opened 1656 for comedy; gave opera regularly 1710–1800, especially opera buffa. Gave premières of many operas by Porpora, Galuppi, and Piccinni, and saw triumphs of Carestini, Tenducci, Medici, and De Sanctis. Rebuilt 1748 after a fire; gave opera until 1894.

T San Salvatore. Built by the Vendramin family and the oldest theatre in Venice still standing; opened 1661 with Castrovillari's *Pasife*. Renamed T San Luca 1799, T Apollo 1833, T Goldoni 1875. Staged premières of Cavalli's last two operas. Goldoni wrote more than 60 plays for it. Pasta sang Norma here. First Venetian theatre with gaslight, 1826.

T Sant'Angelo. Built by the Capelli and Marcelli families on the Grand Canal; opened 1676 with Freschi's *Elena rapita*. Premièred several operas by Vivaldi, who was active as composer and impresario 1713–18 (possibly precipitating *Marcello's satire *Il teatro alla moda*), 1726–8, and 1733–5. Gave operas by Hasse, Albinoni, Lampugnani, and others until the end of the 18th cent.

T San Giovanni Grisostomo. Built by Grimani on the site of houses owned by Marco Polo, opened 1678 with Pallavicino's *Vespasiano*. Galuppi was at the theatre 1749–56. Renamed T Emoroniti, and entered period of decline. Malibran sang in *Sonnambula* 1835; at the end of Act I garlands were thrown on the stage and songbirds released from cages in the upper-tier boxes; Malibran refused her 4,000 Austrian lire fee and gave it to the impresario Giovanni Gallo for the theatre. Immediately renamed T Malibran. The most important Venetian theatre before the opening of La Fenice. Now a cinema, but still occasionally in operatic use.

T San Fantino. Opened 1699 with Pignatti's *Paolo Emilio*. Active until 1720.

T San Benedetto. Opened 1755 with Cocchi's *Zoe*. Damaged by fire 1773; reopened 1784, in 1787 renamed as T Venier (later renovated, cap. 1,500). By 1800 146 works had been premièred. Rossini's *Italiana in Algeri* was premièred 1813, *Eduardo e Cristina* 1819, and Meyerbeer's *Emma di Resburgo* 1819. Renamed T Rossini 1868. From 1925 a cinema.

Other theatres giving opera included the T San Festino (1699–1720), T Santa Margherita, T San Girolamo, and T Pepoli.

The T La Fenice, Venice's most famous opera-house, was planned as a successor to the burnt-down T San Benedetto, and was the phoenix (*fenice*) that arose from the ashes, though on another site. It was planned by a syndicate of Venetian patricians, citizens, and merchants, who chose Gianantonio Selva to design it. He was unpopular, and met with obstacles. On the theatre's façade he inscribed 'Societas', from which the Venetians formed the acrostic 'Sine Ordine Cum Irregularitate Erexit Theatrum Antonius Selva' ('without method and irregularly Antonio Selva built the theatre'). Opened 16 May 1792 with Paisiello's *I giuochi d'Agrigento*. The scene of the premières of Rossini's *Tancredi* (1813), *Sigismondo* (1814), and *Semiramide* (1823); Bellini's *Capuleti e Montecchi* (1830) and *Beatrice di Tenda* (1833); and Donizetti's *Belisario* (1836) and *Maria di Rudenz* (1838). Burnt down Dec. 1836; rebuilt on same site and plan, reopened 26 Dec. 1837. Staged premières of Verdi's *Ernani* (1844), *Attila* (1846), *Rigoletto* (1851), *Traviata* (1853), and *Boccanegra* (1857). However, the mediocrity of its chorus and orchestra in these years is reflected in some of the works written for it.

The theatre underwent alterations 1854 and 1936, when it opened as an *Ente Autonomo. Sometimes as part of the Biennale festival, it has given the premières of Stravinsky's *Rake's Progress* (1951), Britten's *Turn of the Screw* (1954), and Nono's *Intolleranza* (1961). Since 1968 it has staged important revivals and also productions of works by Rossellini, Rota, Malipiero, and Dallapiccola. Closed 1974–5 through financial crisis. The auditorium seats 1,500, and with its dazzling chandelier, 96 boxes, and blue, cream, and gold decorations is considered by many the most beautiful opera-house in the world. Music directors have included Ettore Gracis and Eliahu Iubal.

Opera has also been given in the courtyard of the Doge's Palace and on the island of San Giorgio.

BIBL: M. Gigardi and F. Rossi, *Il teatro La Fenice: cronologia degli spettacoli, 1792–1936* (Venice, 1989).

**Venus.** In Roman mythology, the goddess of love (in Greek, Aphrodite). Operas on the legend of Venus and Adonis are by Blow (*c*.1682) (see below), Desmarets (1697), Stölzel (1714), Bianchi (1781), De Pilles (1784), and Leroux (1905). Other Venus operas are by Sacrati (*Venere gelosa*, 1643), Bernabei (*Venere pronuba*, 1689), Molinari (*Venere travestita*, 1691), Pollarolo (*Venere travestita*, 1691, lost), Courcelle (*Venere placata*, 1731), Graun (*Venere e Cupido*, 1742), and Hubay (*Die Venus von Milo*, 1935). Also the title of an opera by Schoeck (1922). She appears (sop) in Wagner's *Tannhäuser* and Tippett's *King Priam*.

**Venus and Adonis.** Masque in prologue and 3 acts by Blow; text anon. Prod. London *c*.1682, when Venus was sung by Mary Davis, mistress of Charles II, and Cupid by their daughter Lady Mary Tudor; Cambridge, MA, 11 Mar. 1941.

**Venusberg Music.** The opening scene in Venus's realm in Wagner's *Tannhäuser*, substantially rewritten for the Paris version of 1861.

**Vêpres siciliennes, Les** (The Sicilian Vespers). Opera in 5 acts by Verdi; text by Scribe and Duveyrier. Prod. Paris, O, 13 June 1855, with Cruvelli, Guéymard, Bonnehée, Obin, cond. Dietsch; London, DL, 27 July 1859, with Tietjens, Mongini, Fagotti; New York, AM, 7 Nov. 1859, with Colson, Brignoli, Ferri, Junca, cond. Muzio.

Palermo, 1282. The French have occupied Sicily. The Duchess Hélène (sop) mourns her brother, executed by the French, and incites the Sicilians to rebel. Guy de Monteforte (bar), the French Governor, fails to persuade Henri, a Sicilian (ten), to join the French forces.

Hélène promises to marry Henri if he will avenge her brother's death. Procida (bs), a fanatical patriot, schemes to arouse the Sicilians' hatred of the invaders.

When Monteforte tells Henri that he is his son, Henri refuses to recognize his father, but realizing it will prevent him from marrying Hélène. At the Governor's Ball, Hélène and Procida plan to assassinate Monteforte; Henri defends him and the rebels are arrested.

Henri feels that he has now performed his filial duty and is free to join the conspirators again. However, he admits publicly his relationship to Monteforte in order to prevent the execution of Hélène and Procida. As an act of reconciliation, Hélène is to marry Henri.

Procida has planned that the wedding bells shall be the signal for the Sicilians to slaughter the French. Knowing this, Hélène tries, unsuccessfully, to delay the ceremony. As the bells ring out and the mob advances, the curtain falls.

**Verazi, Mattia** (*b* c.1725; *d* Munich, 20 Nov. 1794). Italian librettist. He wrote *c*.20 librettos, the earliest of them for Italy, including *Ifigenia in Aulide* (1751) for Jommelli. During the 1760s and 1770s he was at the Mannheim court, moving with it to Munich: his most important works there included *Sofonisba* (Traetta, 1763) and *Lucio Silla* (J. C. Bach, 1776). Verazi also wrote librettos for other courts, including that of the Duke of Württemberg, for whom he provided *Fetonte* (1768), the basis of Jommelli's most innovatory score.

**verbunkos** (Hung., from Ger., *Werbung*: 'recruiting'). A Hungarian dance used by recruiting-sergeants on their tours of towns and villages to try to persuade young men to enlist. The

music, based on the Hungarian scale including two augmented intervals, generally consisted of folk-tunes alternating slow and fast sections, later commonly an initial slow section (*lassu*) followed by a fast one (*friss*). This pattern survived into the csardás, but the verbunkos had already entered opera with József Ruzitska's *Béla futása* and *Kemény Simon* (both 1822), Ferenc *Erkel's *Bátori Mária* (1840) and *Hunyadi László* (1844), Ignác Bognár's *Tudor Mária* (comp. 1847, prod. 1856), and György Császár's *Kunok* (1848) and *Morsinai Erzsébet* (1850).

**Verdi, Giuseppe** (*b* Le Roncole, 10 Oct. 1813; *d* Milan, 27 Jan. 1901). Italian composer. Born of humble parents, Verdi owed his first musical education to the local organist and his first advancement to Antonio Barezzi, a wealthy Busseto music-lover. He studied with Ferdinando Provesi, also acting as deputy and sometimes conducting his own music. Having failed the entrance to the Milan Conservatory, he returned to Busseto and became maestro di musica; he also married Barezzi's daughter Margherita.

In the same year, 1836, Verdi completed his first opera, *Rocester*; it is lost, but is likely to have formed the substance of *Oberto*. This already shows Verdian traits, including an interest in the father–daughter relationship, alongside influences from Bellini, Rossini, and Mercadante. The work was staged perhaps partly through the good offices of the young soprano who was to have taken the lead, Giuseppina Strepponi. Verdi was promptly offered a contract for three more operas by La Scala's manager, Merelli; but the death of his wife, following the earlier death of their two children, as well as the failure of *Un giorno di regno*, so depressed him that he tried to cancel the contract. *Un giorno di regno* was Verdi's only comic opera before *Falstaff*; but despite some good ensembles, and an intelligent response to Donizetti's comic style, the work shows little gift for comedy.

Ever sympathetic, Merelli tempted Verdi to resume work with the libretto for *Nabucodonosor* (usually abbreviated as *Nabucco*); and the result was his first masterpiece. Though the influences of Rossini (especially of *Moïse*) and of Donizetti still show, there is a stronger and more dynamic individuality which dominates these; this is partly in the portrayal, especially in ensembles, of characters larger than life, but also in some rugged choral writing that is made a functional part of the drama. In the famous chorus 'Va, pensiero', sung by the exiled Hebrews, Risorgimento Italy found the first of many symbols in Verdi's music of an oppressed people suffering foreign domination.

*I Lombardi alla prima crociata* is an uneven

work consciously in *Nabucco* vein, but was originally a success with audiences on whose loyalties, musical and national, Verdi could now count. With *Ernani*, he turned away from his robust librettist Solera to Piave, who provided him with a first encounter with Hugo and a Romantic text concentrating on more personal emotions, among which the heroics of the opera are conducted. It also enabled Verdi to establish his own manner with vocal types, in particular the heroic, ardent tenor, the sturdy bass, and the emotionally complex baritone.

The success of *Ernani* led to a wealth of commissions, needed by Verdi financially but committing him to a period of labour often referred to as his 'anni di galera' ('prison years': Verdi himself used the phrase to cover his entire career up to *Un ballo in maschera*). Artistically he can here sometimes seem over-stretched. *I due Foscari* is a more intimate work than its predecessor, and makes some use of *reminiscence motive in a personal way, but suffers from the limitations of the subject. For *Giovanna d'Arco*, Verdi went back to Solera and the grand manner, without recapturing the actual grandeur of *Nabucco*; he then turned to Cammarano for *Alzira*, perhaps his least successful opera, and again to Solera for *Attila*, a work of infectious vigour that effectively closes the line of grandiose works begun with *Nabucco*.

With *Macbeth*, Verdi strikes out in a new direction, and makes his first encounter with Shakespeare. Even before the rewriting of the work for Paris in 1865, this 1847 version represents an original departure for Verdi and for Italian opera, in its serious attempt to reflect the spirit of Shakespeare and a Northern drama, but also in consequent expressive features such as the predominant use of minor tonalities, increased reliance on the orchestra, and the expectation that vocal drama should take precedence over vocal technique. *I masnadieri*, which followed, is by contrast a work whose beautiful parts do not add up to an individual or satisfying whole. Having revised, and improved, *I Lombardi* as *Jérusalem* (his first encounter with the conventions of the Paris Opéra), Verdi continued with *Il corsaro*, a work conventional in outward form but containing some subtle and delicate passages. *La battaglia di Legnano* also pays perceptive attention to detail, but does not succeed in accommodating this to the more spacious demands of Cammarano's politically angled plot. The same year, 1849, saw *Luisa Miller*, a work in which the more intimate study of emotions which had been increasingly absorbing Verdi becomes still more manifest. Not only is there a subtler understanding of the possibilities of vocal contrast, and with it a greatly increased range of melodic expressive-

ness, but there is also richer and more intricate use of the orchestra. Its success, and Verdi's consolidated position, gave him a new freedom to choose commissions as he wished.

*Stiffelio*, if not among Verdi's masterpieces, succeeds in dominating its unpromising subject to a remarkable degree, and in it he continues to expand his resources of characterization and his use of the orchestra for scene-painting. His personal life was also set on a new and happy course. In Paris, he had again met Giuseppina Strepponi: they became companions, and eventually (1859) man and wife. The mellowing of a character which always retained its tough and independent nature was largely due to her, and it is reflected in the greater breadth and assurance of his art. In 1851 came the first work of unquestioned genius, *Rigoletto*. Verdi declared it to be revolutionary, and though it does not dispense with the old forms of Italian opera, as he had distinguished them, he does now transcend them. His lifelong absorption in the father-daughter relationship here finds its most intense expression, with the baritone title-role dominating in expressive grandeur both the tender, touching soprano and the lightly charming tenor. Moreover, Verdi's increasing mastery of characterization in ensemble reaches new heights here in the famous quartet exploring four contrasting emotions in unified music. Even more than previously, it is in the orchestra that much of the drama, as well as the atmosphere, is enacted and developed with a new sense of continuity.

After *Rigoletto*, *Il trovatore* represents something of a regression to earlier simplicity and directness, with Cammarano's libretto enjoining upon the composer more static situations and conventional forms. Its superb tunes, bold characterization, and strong dramatic confrontations made it instantly and enduringly popular; but it is, essentially, the last early Verdi opera coming after the first of his middle period. The latter period is renewed with *La traviata*. At first a notorious fiasco, it has long been accepted as a masterpiece, one in which the old blunt divisions are swept away, conventions absorbed into music of a new subtlety of human understanding, and characterization as never before explored and developed in a continuous musical fabric.

With this central trilogy of works setting the seal on his achievement as the musical laureate of Italian Romanticism, Verdi now turned his attention to France. *Les vêpres siciliennes*, to a Scribe libretto, is (apart from *Jérusalem*) his first foray into French *Grand Opera. The demands of new styles and techniques show in the lack of spontaneity and a certain air of contrivance, though he does succeed, more than many experienced French composers, in giving some depth of character to the principals caught up in the drama. There is, however, greater range and feeling in *Simon Boccanegra*, a powerful, dark work whose sombre atmosphere Verdi attempted to lighten a little when he revised the original 1857 score in 1881. A less successful revision, forced on him by the censor, was that of *Stiffelio* as *Aroldo*. The beginning of 1859 saw *Un ballo in maschera* (q.v. for summary of its many censorship disputes). In it Verdi mixes into his tragic love-story elements of comedy in the manner he admired in Shakespeare, though for all the formal freedom, especially a new contrapuntal mastery, there is less development of the skilfully drawn characters than in other of his operas.

Verdi's name had long been associated with the Risorgimento as an artist of revolt. His operas were often read as political gestures for liberty, and his very name was used as an artistic slogan as an expression of hope for a united Italy: 'Viva VERDI' meant 'Viva Vittorio Emanuele, Re D'Italia'. In 1859 French and Italian armies drove the Austrians out of Lombardy; Verdi represented Busseto at the assembly in Parma and then, pressed by Cavour, became a member of the new Italian parliament. During his parliamentary career (1861–5) he found time for only one opera, *La forza del destino*. Written for St Petersburg, it is again a dark work, and though intended as an opera of ideas and to show an epic sweep, it suffers from a certain inability to rise above convention or to sustain a smooth dramatic flow. Individual scenes and arias are very fine; the pivotal role of the baritone is again magnificently asserted; and there is a strong central soprano role: these are the work's incidental strengths, not to mention a prophetic touch of comedy in the characterization of a tetchy friar.

*Macbeth* was revised in 1864–5 and presented in Paris, where Verdi agreed to write *Don Carlos*. It is his supreme work, if not the greatest of any composer, in the genre of Grand Opera. For all its problems, which include complicated issues concerning versions, it achieves an extraordinary degree of success in exploring both public and private emotions in a story of tragic grandeur. Verdi is here not only the master of individual and highly original portraits, but is able to relate them to outer worlds, as is shown nowhere more powerfully than in the scene when King Philip's grief at the absence of love is seen against the background of his royal confrontation with the Grand Inquisitor: here, as in the superb series of duets, formal demands are made the vehicle of the dramatic impetus and never impede the majestic flow.

*Aida* returns some of the French Grand Opera manner to Italy. If less subtle than its predecessor, it has always had a more immediate appeal for the clarity and boldness of its outlines and its simple, strong emotions, no less than its enticingly exotic setting. The spectacular elements are part of the plot, and so of the music, but Verdi places them firmly as the background against which his characters develop their tragedy. It is here, more than in *Don Carlos*, that the contrast between private emotion and patriotic pomp is explored and it was his last engagement with Grand Opera.

For his last two operas, Verdi turned to Shakespeare. He had long been haunted by the wish to set *King Lear*, and may have written for it some music that was absorbed into *Simon Boccanegra*; but it was his collaboration with Boito that led to the two closing Shakespeare masterpieces. Though there is a view which holds that Boito's sophistication operated to harmful effect on Verdi, it is one difficult to sustain in the face of the greatness of the closing tragedy and comedy. *Otello* responds to Boito's sensitively judged libretto with music that is now not only completely fluent, but that draws upon a greatly widened harmonic range and a heightened sense of orchestral tension, as with the storm which opens the whole work. Melody is still supreme, but it is melody subtly adaptable to situation and character, capable of running on in narration or bearing the weight of powerful expression. To discern the old formal divisions within certain sections of the music is possible, but is also to miss the expressive point.

By now an old man, Verdi took up *Falstaff* secretly from his curious public, and, he insisted, to amuse himself. More than *Otello*, it is an ensemble opera, with monologues or arias only playing an incidental part in the rapid flow. It is swift-paced and genial, but also sharp: Verdi's contrapuntal ingenuity plays a vital part in the pace and verve, as jealousy, tragic in *Otello*, becomes the source of laughter in *Falstaff*, and human fallibility the source of wit, not regret. The score flashes with brilliance, yet there is also a lyrical quality which gives it a peculiar grace. All Verdi's musical and dramatic skills are present, refined and sharpened; and this is never more so than when, at the end of this comedy that ends the whole great series of operas, Falstaff turns to the audience to lead a fugal chorus with the reminder that all the world's a stage.

Verdi died in a hotel in Milan in 1901. Despite his wish for a modest funeral, it became an occasion of national mourning; and as his coffin passed, the huge crowd burst spontaneously into 'Va, pensiero'.

WORKLIST: *Oberto, conte di San Bonifacio* (Solera, probably adapted from Piazza's *Rocester*; Milan 1839); *Un giorno di regno* (later *Il finto Stanislao*) (Romani, after Pineu-Duval; Milan 1840); *Nabuco-*

*donosor* (later *Nabucco*) (Solera, after Anicet-Bourgeois & Cornue; Milan 1842); *I Lombardi alla prima crociata* (Solera, after Grossi; Milan 1843); *Ernani* (Piave, after Hugo; Venice 1844); *I due Foscari* (Piave, after Byron; Rome 1844); *Giovanna d'Arco* (Solera, after Schiller; Milan 1845); *Alzira* (Cammarano, after Voltaire; Naples 1845); *Attila* (Solera, additions by Piave, after Werner; Venice 1846); *Macbeth* (Piave, additions by Maffei, after Shakespeare; Florence 1847, rev. version Paris 1865); *I masnadieri* (Maffei, after Schiller; London 1847); *Jérusalem* (Royer & Vaëz, adapted from *I Lombardi*; Paris 1847); *Il corsaro* (Piave, after Byron; Trieste 1848); *La battaglia di Legnano* (Cammarano, after Méry; Rome 1849); *Luisa Miller* (Cammarano, after Schiller; Naples 1849); *Stiffelio* (Piave, after Silvestre & Bourgeois; Trieste 1850); *Rigoletto* (Piave, after Hugo; Venice 1851); *Il trovatore* (Cammarano, additions by Bardare, after Gutiérrez; Rome 1853); *La traviata* (Piave, after Dumas fils; Venice 1853); *Les vêpres siciliennes* (Scribe & Duveyrier; Paris 1855); *Simon Boccanegra* (Piave, additions Montanelli, after Gutiérrez; Venice 1857, rev. with additions by Boito, Milan 1881); *Aroldo* (Piave, adapted from *Stiffelio*; Rimini 1857); *Un ballo in maschera* (Somma, after Scribe; Rome 1859); *La forza del destino* (Piave, after Saavedra, with Schiller scene trans. Maffei; St Petersburg 1862; rev. with additions Ghislanzoni, Milan 1869); *Don Carlos* (Méry & Du Locle, after Schiller; Paris 1867, rev. with additions Du Locle and trans. Zanardini, Milan 1884); *Aida* (Ghislanzoni, after Mariette; Cairo 1871); *Otello* (Boito, after Shakespeare; Milan 1887); *Falstaff* (Boito, after Shakespeare; Milan 1893).

BIBL: F. Walker, *The Man Verdi* (London, 1962); C. Osborne, ed., *Letters of Giuseppe Verdi* (London, 1971); J. Budden, *The Operas of Verdi*, 3 vols.: i. *From Oberto to Rigoletto* (London, 1973); ii. *From Il trovatore to La forza del destino* (London, 1978); iii: *From Don Carlos to Falstaff* (London, 1981); W. Weaver and M. Chusid, eds., *The Verdi Companion* (New York, 1979); D. Kimbell, *Verdi in the Age of Italian Romanticism* (Cambridge, 1981); J. Budden, *Verdi* (London, 1985).

**'Verdi prati'.** Ruggiero's (male sop) aria in Act II of Handel's *Alcina*, saying that the pleasant scene around him will soon be turned back to its former horror.

**Vere.** Capt. Vere (ten), commander of HMS *Indomitable*, in Ghedini's and in Britten's *Billy Budd*.

**verismo** (It.: 'realism'). The movement in Italian literature and opera that flourished briefly in the late 19th and early 20th cents. In literature, it is most fully typified in the works of the Sicilian writer Giovanni Verga, whose story *Cavalleria rusticana* (1880) was first dramatized (1884) and then provided the libretto for the parent opera of verismo, Mascagni's *Cavalleria rusticana* (1890). Its origins may be sought partly in the increased interest in naturalism in a post-unification Italy

whose regions sought to preserve their actuality and identity. Equally significant was the literary example deriving from such popular French authors as Dumas fils and especially Zola, acting under the positivism of Comte. In this philosophy, the observable fact was paramount; and the artistic outcome was for works to concentrate on areas previously regarded as better neglected, especially the criminal, the violent, low life, and wretched or squalid behaviour. Though verismo claimed to confront the 'slice of life' (the 'squarcio di vita' promised by Tonio in the Prologue to Leoncavallo's *Pagliacci*), the exceptional and horrifying as represented by the crime of passion, especially in a peasant or working-class milieu, were of particular fascination, and lie at the centre of many verismo operas.

The operatic impetus came from the second *Sonzogno competition, won by Mascagni with *Cavalleria rusticana*. Its success was immediate, impressing even the aged Verdi: 'he has invented a most effective genre: short operas without pointless longueurs'. The 1-act form was indeed to prove popular in the genre. Operas to appear in the wake of this triumph included 'low-life' Neapolitan works such as Tasca's *A Santa Lucia* (1892), Spinelli's *A Basso porto* (1894), Sebastiani's *A San Francesco* (1896), and Giordano's *Mala vita* (1892). In the same year Leoncavallo produced *Pagliacci*, on a plot drawn from one of his father's court cases. Puccini, who contemplated setting both Zola and Dickens, was to give the genre greater artistic range: *Il tabarro* (1918), the first part of *Il trittico*, is typical verismo in its crime of passion among Seine bargees, but verismo leaves its distinctive mark on *La bohème* (1896), *Tosca* (1900), and *La fanciulla del West* (1910), as well as much else in his work. It encouraged him to cultivate the accurate representation of different milieux, which become functional in the action, and to develop the particular type of aria in which a realistic parlando opening may lead to the most violent emotional climax. Puccini was also quick to respond to the erotic and the sadistic in which verismo dealt, chastened though this is by his own sentimentality. In France, verismo returned with several effective works. It marked the operas of, in particular, Massenet, whose *La navarraise* (1894) is the most vivid example of verismo. The most popular, however, was Charpentier's *Louise* (1900), even though sentimentality has here largely overcome realism. The movement has nothing to do with Russian traditions of realism, nor did it have any marked effect on German opera.

**Véron, Louis** (*b* Paris, 5 May 1798; *d* Paris, 27 Sep. 1867). French manager. Director of the

Paris Opéra 1831–5, when he confirmed its position as the most successful and influential theatre in Europe, including with the premières of *Robert le diable* (1831) and *La Juive* (1835) and the commissioning of *Les Huguenots* (1836). His business acumen, flair for publicity, and understanding of public taste combined to make the Opéra a great show-case of opera, though standards of artistry and execution were often criticized.

**Verona.** City in the Veneto, Italy. The T Filarmonico, designed by Francesco Galli Bibiena, opened 1732 with Vivaldi's *La fida ninfa*; burnt down 1749; reopened 1754 with Hasse's *Alessandro nell'Indie*. The 14-year-old Mozart played here on his first Italian visit in 1770; premières included Traetta's *L'Olimpiade* (1758) and Cimarosa's *Giunio Bruto* (1781). Bombed Feb. 1945. Opera has also been staged in the T dell'Accademia, T Nuovo (or Rena), T del Territorio, T Morando, T Leo di Castelnuovo, and T Ristori.

The Roman arena (cap. 16,663) staged operas by Rossini and Donizetti in 1856 and 1859; and in 1913 regular summer performances were initiated by Giovanni *Zenatello, his future wife Maria Gay, and the impresario Ottone Rovato. Serafin's advice was sought, and after acoustic tests, weather hazards, and other problems, *Aida* was staged on 10 Aug. 1913 to an audience including Puccini, Mascagni, Pizzetti, Zandonai, Montemezzi, and Illica. There have been summer seasons ever since, except 1915–18 and 1940–5. Although most operas are chosen for opportunities for the spectacular, including *Gioconda*, *Turandot*, *Mefistofele*, and *Forza del destino*, there have also been successful productions of more intimate works, including *Traviata* and *Bohème*. The atmosphere is naturally that of mass popular entertainment rather than artistic refinement; but it attracts distinguished artists, including Callas and Richard Tucker for their Italian débuts in *Gioconda*, 1947.

**Véronique.** Operetta in 3 acts by Messager; text by Vanloo and Duval. Prod. Paris, BP, 10 Dec. 1898, with Sully, Tariol-Bauge, Périer, Regnard; London, Coronet T, 5 May 1903; New Orleans, French OH, 21 Jan. 1900, with Rossil, Dambrine, Dupuis, Frasset, cond. Finance.

**Verrett, Shirley** (*b* New Orleans, 31 May 1931). US mezzo-soprano, later soprano. Studied Chicago with Fitziu, New York with Székely-Freschl. Début Yellow Springs 1957 (Lucretia). New York City O 1958; Moscow, B, 1963; London, CG, and Milan, S, from 1966; New York, M, from 1968; Paris, B, 1990 (at its opening). Her wide range, warm tone, and striking physical presence have made her a successful Amneris,

Eboli, Lady Macbeth, Dalila, Selika, Cassandra, Dido (Berlioz), Tosca. (R)

**Versailles.** Palace built by Louis XIV south-west of Paris. Although Lully wrote operas and ballets for the King, produced here, there was originally no permanent theatre, and temporary stages were erected either in the gardens or in various rooms and courtyards in the palace. Opera was given in a hall of the Grand Trianon 1686–1703, including premières of works by Lully and Destouches. Several of Rameau's operas were premièred, including *Platée* for the wedding of the Dauphin and Maria Teresa of Spain (1745). The opera-house, designed by Ange-Jacques Gabriel, was commissioned by Louis XV in 1748; after many delays due to lack of money, foreign wars etc., the theatre was completed for the celebrations of the marriage of the Dauphin (the future Louis XVI) and Marie Antoinette, and opened with Lully's *Persée*, 17 May 1770, with Sophie Arnould. Opera continued to be given in other places, including in the Petit Trianon and at the T du Jardin. It was also given from 1777 at the T Montansier: works premièred here included Grétry's *Céphale et Procris*, for the wedding of the Count of Artois and Marie-Thérèse of Savoy (1773), and Sacchini's *Œdipe à Colone* (1786).

Opera continued to be performed in Gabriel's theatre until the Revolution, when the theatre became the meeting-place of the Club des Jacobins; its furniture and decorations were plundered. Restored 1837 by King Louis-Philippe; reopened with scenes from *Robert le diable* with Falcon and Duprez, and a special ballet by Scribe with music by Auber, danced by the Elssler sisters. During the 1870s the theatre became the seat of the National Assembly, and of the Third Republic as proclaimed there Jan. 1875; then fell into disuse until 1952, when a campaign was launched for its restoration, and works (generally from the Opéra or Opéra-Comique) were staged. Rolf Liebermann began as director of the Opéra with a performance here of *Figaro* under Solti on 30 Mar. 1973. The theatre remains in use, chiefly for 18th-cent. revivals.

**Verschworenen, Die** (The Conspirators). Opera in 1 act by Schubert; text by Castelli, on the idea of Aristophanes' comedy *Lysistrata* (411 BC). Prod. Frankfurt 29 Aug. 1861; New York 16 June 1877; London, RCM, 20 June 1956.

The Crusaders' wives go on sexual strike until their husbands forswear war, as in Aristophanes' original. A touchy political censor insisted that the title be changed to *Der häusliche Krieg* (Domestic War).

**verso** (It.: 'verse'). In 19th-cent. Italian opera, lines of verse were classified according to their metre. Operatically, the most familiar are *verso* *endecasillabo*, 11-syllable verse normally in iambics; *decasillabo*, 10-syllable verse in anapaests; *ottonario*, 8-syllable verse in trochees; *settenario*, 7-syllable verse in iambics; *senario*, 6-syllable verse in broken anapaests; *quinario*, 5-syllable verse in iambics. Lines of Italian verse normally have feminine endings; if a line does end with a strong syllable, or is extended by means of an additional weak syllable, it may be described as *tronco* (truncated) or *sdrucciolo* (sliding). *Versi sciolti* are a free mixture of 7- and 11-syllable verse, or non-rhyming 11-syllable verse, often used in recitative. See also ACCENTO.

BIBL: W. Elwert, *Italienische Metrik* (Munich, 1968).

**Verstovsky, Alexey** (*b* Seliverstovo, 1 Mar. 1799; *d* Moscow, 17 Nov 1862). Russian composer. Abandoned engineering to study piano in St Petersburg with Steibelt and Field. Wrote music for various vaudevilles, then took up a full-time career in the theatre when appointed Inspector of Theatres in Moscow, 1825. Director of all the Moscow theatres, 1842, exercising a powerful and beneficial influence. Retired 1860.

Verstovsky's first opera, *Pan Twardowski*, met with little success. A Romantic Singspiel in the manner of *Der Freischütz*, it draws on some Weberian ingredients (a pact with the Devil, the *Preciosa* gypsy atmosphere) but is constructed of short, effective songs that show no influence of German Romantic opera. The work does perhaps owe to French opéra-comique the fascination with natural effects (a flood, an earthquake). *Vadim* includes some consciously Russian characteristics. His greatest success was *Askold's Tomb*, which was given some 600 times in Russia during the 19th cent. and in 1869 became the first Russian opera to be performed in America. Even this work, despite an incantation scene modelled on the Wolf's Glen scene in *Der Freischütz*, is essentially a collection of individual songs, which for all their appeal do not create much in the way of character or atmosphere except of the most rudimentary kind. It was the use of popular Russian elements in the songs and choruses, and strokes such as the introduction of folk instruments and players, that won the operas such an enthusiastic following.

WORKLIST (all first prod. Moscow): *Pan Twardowski* (Zagoskin; 1828); *Vadim* (Shevyryev, after Zhukovsky & Spiess; 1832); *Askoldova mogila* (Askold's Tomb) (Zagoskin; 1835); *Toska po rodine* (Longing for the Homeland) (Zagoskin; 1839); *Churova dolina* (The Chur Valley) (Shakhovskoy; 1841); *Gromoboy* (Lensky, after Zhukovsky; 1858).

BIBL: B. Dobrokhotov, *A. N. Verstovsky* (Moscow, 1949); G. Abraham, 'The Operas of Alexei Verstovsky', in *Nineteenth Century Music*, 7/3 (1984).

**Vespri siciliani, I.** See VÊPRES SICILIENNES, LES.

**Vestale, La** (The Vestal Virgin). Opera in 3 acts by Spontini; text by Étienne de Jouy. Prod. Paris, O, 16 Dec. 1807, with Branchu, Maillard, Laintz, Lays, Dérivis; London, H, 2 Dec. 1826; New Orleans, T d'Orléans, 17 Feb. 1828.

The opera tells of the love of the Roman captain Licinius (ten) for Julia (sop), who becomes a vestal virgin while her lover is in Gaul. He breaks into the temple to win her back, and she allows the holy fire to become extinguished. Condemned to death, she is being led to execution when a flash of lightning rekindles the flame. She is spared, and reunited with Licinius.

Other operas on the same libretto by Pucitta (1810) and Guhr (1814, new setting of Spontini text). Also operas by Vento (1776), Giordani (1786), Pacini (1823), Mercadante (1840); opera buffa, *Les petites vestales*, by Rey (1900).

**'Vesti la giubba'.** Canio's (ten) aria in Act I of Leoncavallo's *Pagliacci*, bewailing his duty to put on his motley and to clown, though his heart be breaking.

**Vestris, Lucia Elizabeth** (*b* London, 3 Jan. or 2 Mar. 1797; *d* London, 8 Aug. 1856). English contralto. Wife of Auguste Armand Vestris, ballet-master of London, H, 1809–17. Studied with Corri. Début London, H, 1815, at her husband's Benefit (Proserpina, Winter's *Il ratto di Proserpina*). London, H, 1816; Paris: I, and elsewhere, from 1816. After Vestris left her she returned to London: DL from 1820, H, 1821–5, CG, 1826; also Dublin 1824–7. Manager London: Olympic T, 1831–8, and, with her second husband, the actor Charles Mathews, of CG, 1839–42, L, 1847–55. Retired from stage 1854. An extremely beautiful woman with 'one of the most luscious of low voices' (Chorley), and much talent, she was a highly popular, though never a great, performer. Repertory included Arsace (*Semiramide*), Pippo (*La gazza ladra*), Emma (*Zelmira*), and even Don Giovanni and Macheath. Created a 'charming' Fatima (Planché) in *Oberon*, being the only singer completely adequate for her role.

**Viaggio a Reims, Il** (The Journey to Rheims). Opera in 2 acts by Rossini; text by Balocchi, after Mme. de Staël's novel *Corinne* (1807). Prod. Paris, I, for coronation of Charles X, 19 June 1825, with Pasta, Cinti-Damoreau, Levasseur, and Donzelli. Some of the music was re-used in *Le Comte Ory* in 1828; revived, without most of the *Ory* additions, Paris, I, 26 Oct. 1848 as *Andremo a Parigi*. London, CG, 4 July 1992.

**Vianesi, Auguste** (*b* Legnano, 2 Nov. 1837; *d* New York, 4 Nov. 1908). Italian, later French, conductor and composer. Studied Legnano with his father Giuseppe. Went to Paris 1857 with a letter of introduction to Rossini from Pasta. London, DL, 1858–9. Moscow 1863–4; St Petersburg 1867–9. Paris: I, 1873; O, 1887–91, giving premières of Saint-Saëns's *Ascanio* (1890) and Massenet's *Le mage* (1891). London, CG, 1870–80. Cond. London premières of *Lohengrin* (1875) and *Tannhäuser* (1876), also especially French and Italian repertory. New York, M, 1883–4 (inaugurating the house with Gounod's *Faust*), 1891–2. New Orleans, French OH, 1899–1900. Taught singing in New York from 1892.

**Viardot, Pauline** (*b* Paris, 18 July 1821; *d* Paris, 18 May 1910). Franco-Spanish mezzo-soprano. Daughter of Manuel *García, sister of Malibran. Early musical training with her father; studied piano with Liszt, composition with Rejcha, singing with her mother. Début London, HM, 1839 (Rossini's Desdemona). Paris, Odéon, 1839. Married Louis Viardot 1840; Paris, I, 1841–2; St Petersburg 1843–6, with great success; London, CG, 1848 regularly until 1855; Paris: O, 1849, 1851; L, 1856, 1859–61. An artist of unusual integrity and intelligence, and one of the greatest and most influential singers in operatic history. Unlike her rival Grisi, she possessed neither physical nor vocal beauty. Essentially a mezzo-soprano, she extended her range to cover g–f'''; her voice, like her sister's, was 'a rebel to subdue' (Chorley), but her masterly technique enabled her to sing Donna Anna, Leonore, Arsace, Amina, Bellini's Romeo, Lucia, Alice, Valentine, Azucena, Lady Macbeth. Her complete musical identification with her roles produced dramatic interpretations of great power. Created Fidès, written for her, with consummate success; Berlioz was overwhelmed by 'her almost deplorable ability to express profound grief' as Gluck's Orfeo (in Berlioz's own edition, 1859). She inspired Gounod's Sapho, Saint-Saëns's Dalila, and initially Berlioz's Dido; Schumann's Op. 24 Liederkreis and Brahms's Alto Rhapsody were written for her. She also did much to introduce Russian music to the West, and her influence on her lifelong admirer Turgenev was considerable, as on other literary figures including De Musset and Sand. She sight-read *Tristan*, with Wagner, at a private Paris concert for one of his patrons. A gifted pianist and linguist, she also composed (incl. operettas to words by Turgenev, e.g. *Le dernier sorcier*, 1869, which Tchaikovsky heard in 1889 and on which he made no comment). After an early vocal decline and retirement, she taught in Paris, her pupils including Artôt, Orgeni, and Brandt. Her long life included the acquaintance of both Da Ponte and Henry James. Her daughter **Louise Viardot** (1841–1918) was a contralto, teacher, and composer, whose works

include a comic opera, *Lindoro* (1879). Another daughter, Marianne, was briefly engaged to Fauré. Her son **Paul** (1857–1941) was a violinist and composer; also conducted Paris, O.

BIBL: A. Fitzlyon, *The Price of Genius* (London, 1964).

**vibrato** (It.: 'vibrated'). A fluctuation of pitch, intensity, and timbre in the voice. The rate of fluctuation increases with volume and dramatic intensity, thereby changing the colour of the tone produced. When the voice is under stress the vibrato may become too fast, producing a tremolando effect or 'flutter', or too slow, producing the unattractive 'wobble'. Some voices develop a permanent wobble through over-use.

**Vicenza.** City in the Veneto, Italy. The Accademia Olimpica was founded 1555 by leading figures including Palladio, who designed the T Olimpico (designs completed by Vincenzo Samozzi), opened 1585 and still open. Regular opera performances began in the T Castelli 1656. The T Eretenio opened 1784 with the première of Cimarosa's *L'Olimpiade*. A new theatre, built by Pamato 1828 and named after him, was later taken over by the city; renamed Politeama Comunale; renamed T Verdi 1901; bombed in Second World War.

**Vickers, Jon** (*b* Prince Albert, 29 Oct. 1926). Canadian tenor. Studied Toronto with Lambert and Geiger-Torel. Début Toronto, Canadian OC, 1954 (Duke of Mantua). Canadian OC, 1954–6; London, CG, from 1957; Bayreuth 1958, 1964; New York, M, from 1959. Also Milan, S; Paris, O; Vienna, S; Salzburg; San Francisco; etc. A moving dramatic singer with a robust and distinctive voice, clear articulation, and a magnetic stage presence. Successful as Jason (*Médée*), Florestan, Siegmund, Tristan, Cellini, Aeneas (Berlioz), Verdi's Otello, Canio, Grimes. (R)

**Vida Breve, La** (The Short Life). Opera in 2 acts by Falla; text by Carlos Fernandez Shaw. Prod. (in Fr.) Nice 1 Apr. 1913, with Lillianne Grenville, David Devriès, Eduard Cotreuil, cond. Miranne; New York, M, 6 Mar. 1926, with Bori, Howard, Alcock, Tokatyan, D'Angelo, cond. Serafin; Edinburgh 9 Sep. 1958, with Los Angeles, Gomez, Del Pozo, Martinez, Duis, cond. Toldrá.

Granada, 1900. Salud the gypsy (sop) loves Paco (ten), who is secretly about to marry Carmela (mez). At the wedding Salud denounces his perfidy, and dies broken-hearted at his feet.

**'Vieni, t'affretta'.** Lady Macbeth's (sop) aria in Act I of Verdi's *Macbeth*, in which she strengthens her husband's resolve to murder the King and gain the throne.

**Vienna** (Ger., Wien). Capital of Austria. Opera was first given by the Mantuan company, the Comici Fedeli, probably the anon. *Arcas* (1627). Initially encouraged by Ferdinand III, opera was firmly established at the Viennese court during the reign of Leopold I (1657–1705), largely following the Venetian model. Revivals of Cavalli's *Egisto* and *Giasone* were mounted (1643 and 1650 respectively), possibly that of Monteverdi's *Ulisse* also, and many composers and musicians, including *Bertali, *Sances, P. A. *Ziani, and *Cesti, were brought from Italy. To the popular Venetian style they added a sumptuousness and splendour appropriate to the imperial setting, creating a distinctively opulent style of Viennese court opera which survived well into the next century. The lavish production of Cesti's *Il pomo d'oro* in 1668, performed to celebrate the marriage of Leopold I to the Infanta Margherita, represented an early summit of Viennese operatic achievement. To house the elaborate event, Burnacini designed a theatre in the main square of the imperial palace, the Hofburg; this was the predecessor of the Burgtheater. During the mid-17th cent. Viennese ensembles were also responsible for introducing opera to several other cities, including Munich and Regensburg.

The tradition of court opera was continued by *Draghi who, working mainly with the librettist Minato, wrote some 175 works, many of them to heroic themes. Italian composers dominated into the next century; particularly important were M.A. *Ziani and *Bononcini, who continued to write in a predominantly Venetian style. The amazing productiveness of such composers placed Vienna at the forefront of European musical life: *c.*400 operas were produced during the second half of the 17th cent.; sometimes as many as 12 new works would be staged in one year. Work began on an opera-house in 1697, but this burnt down in 1699. In Joseph I's reign (1705–11) Giuseppe Galli-Bibiena was chief theatre architect; he built two theatres, one for smaller plays and musical works, another for the performance of the court opere serie. Under Charles VI (1711–40) Viennese Baroque opera reached its apotheosis, especially in the work of *Fux, which was characterized by a typically Teutonic emphasis on heavy counterpoint at the expense of dramatic expression. Despite public demand for a theatre, which eventually led to the building of the T am Kärntnertor in 1708, Charles VI insisted that production of opera should remain an imperial monopoly.

Around this time the influence of the new style of Neapolitan opera seria gradually began to be felt in Vienna; this was strengthened by the arrival of the librettists *Pariati and *Zeno in 1718, the latter holding the post of *poeta cesareo

to which *Metastasio succeeded in 1729. During the mid-18th cent. the Viennese operatic tradition was carried forward by *Conti, *Caldara, and L. A. *Predieri, but the accession of Empress Maria Theresa in 1740 ushered in a new era in which the magnificence of the High Baroque was replaced by a new emphasis upon the rococo and galant. In 1747 the Schönbrunn Schlosstheater opened; the following year Gluck's *Semiramide riconosciuta* was given to inaugurate the T bei der Hofburg. Court Kapellmeister from 1754, he produced several operas in traditional vein, before attempting an important reform of opera seria conventions in his *Orfeo ed Euridice* in 1762. This provoked a fierce backlash from Metastasio and *Hasse, but was of only passing significance, for the new genres of Singspiel and opera buffa were already claiming the attentions of composers and audiences alike.

From 1770 Joseph II ruled as co-regent with Maria Theresa, becoming emperor in 1780. The 20 years of his enlightened despotism were a watershed in Viennese operatic life during which the court's direct influence over opera was effectively renounced, allowing more adventurous experimentation: although such composers as the Kapellmeisters *Gassmann and *Bonno provided a link with earlier tradition, even they were somewhat attracted to lighter styles. Singspiel had been introduced to Vienna *c.*1760, but the establishment by Joseph II of the National-Singspiel in 1778, which opened with a performance of *Umlauff's *Die Bergknappen*, provided important encouragment, and irretrievably weakened the supremacy of Italian opera. The 1780s were golden years for Vienna, during which public taste alternated between German Singspiel and Italian opera buffa. The tensions caused by this rivalry are seen at best in *Mozart's mature works, which represent the pinnacle of Viennese operatic achievement; though the Burgtheater gradually yielded its position to the Kärntnertortheater, the former staged the premières of Mozart's *Entführung*, *Figaro*, and *Così*, and *Cimarosa's *Il matrimonio segreto*. On Bonno's death (1788), Salieri became court Kapellmeister, and his new operas were performed at the Kärntnertor.

The monopoly of the court theatres was challenged by the activities of those in the suburbs, principally Emanuel *Schikaneder's T auf der Wieden, where Mozart's *Zauberflöte* was given in 1791, and the T in der Leopoldstadt. With its more popular Singspiel style, which drew partly on a characteristic 18th-cent. Viennese tradition of satirical stage plays with interpolated music, the repertoire of these theatres proved highly popular and influenced the later work of *Raimund and *Nestroy. In 1801 Schikaneder moved to the T an der Wien, where he produced *Fidelio* (1805); although an important operatic venue during the first part of the 19th cent. (Lortzing was director 1846–8), it later became the home of operetta. The T in der Josefstadt opened 1822, with Beethoven conducting his music for it, *Die Weihe des Hauses*. Opera was also given in the summer Thaliatheater.

After the Napoleonic Wars, Rossini became the most popular composer in Vienna; and in 1821 an Italian, Domenico *Barbaia, became manager of the Kärntnertortheater and T an der Wien. He gave the première of *Euryanthe* and introduced the subscription system. Only three new operas were given 1823–42, but in 1842 *Donizetti became court composer and conductor, and wrote *Linda di Chamounix* and *Maria di Rohan* for the Kärntnertortheater. In 1847 the Carltheater opened on the site of the old T in der Leopoldstadt, giving musical comedy and operetta (demolished 1951). In 1861 Wagner conducted the local première of *Lohengrin*, which he was hearing for the first time in the theatre.

The 1857 plans for rebuilding the centre of Vienna included the new Oper am Ring (cap. 2,260), opened 25 May 1869 with *Don Giovanni*: public criticism led to the deaths of the architects Siccardsburg from a heart attack and Null by suicide before the opening. The first director was J. F. von Herbeck, 1870–5, followed by Franz Jauner, 1875–80, who brought Hans Richter to Vienna and staged *The Ring*, in turn followed by William Jahn: himself a conductor, he and Richter controlled the theatre 1880–96, gathering an ensemble that included Materna, Winkelmann, Reichmann, and Scaria (the four Bayreuth creators of Kundry, Parsifal, Amfortas, and Gurnemanz), also Ernest van Dyck and Marie Renard (creators of Werther and Charlotte). Standards were, typically, interpretative rather than innovative, with the première of *Werther* in 1892 an exceptional event. A further great period followed under Gustav *Mahler, 1897–1907, who brought in the designer Alfred *Roller, challenged the Opera's routine in the name of tradition with the strongest musical lead in the theatre's history, and formed an ensemble including Gutheil-Schoder, Mildenburg, Kurz, Weidt, Kittel, Schmedes, Slezak, Maikl, Demuth, and Mayr. Mahler left, a victim of the perennial Viennese hostility to challenge until it is comfortably in the past, but the singers continued with Weingartner's less distinguished directorship, 1907–11. Hans Gregor was more successful, 1911–18, engaging singers including Jeritza, Lotte Lehmann, and Piccaver, and developed an enterprising repertory. The Volksoper (cap. 1,473), originally the Kaiser Jubiläums Stadttheater (1898), opened 1904 with

*Freischütz* under Zemlinsky and was built up by the director, Rainer Simons, first with operetta. He then challenged the artistic lead of the Staatsoper with the first Viennese *Tosca* (1907) and *Salome* (1910).

In 1918 the Hofoper became the Staatsoper, under Franz Schalk 1918–29 (together with Richard Strauss 1920–4). Premières were few, apart from *Die Frau ohne Schatten*, but there were many Viennese premières; singers included Helletsgruber, Rajdl, Schumann, Jerger, Schipper, and Manowarda. At the Volksoper under Weingartner, 1919–24, the first Vienna *Boris* was given, also the première of Schoenberg's *Glückliche Hand*; it closed 1928, reopened 1929 as the Neues Wiener Schauspielhaus. Clemens Krauss was at the Staatsoper 1929–34, Weingartner 1934–6, Bruno Walter 1936–8, with singers including Németh, Anday, Novotná, Kern, Ursuleac, the Konetzni sisters, Olczewska, Dermota, Kiepura, Tauber, Schorr, and Rode, and the conductors Alwin and Krips. With the German annexation in 1938, many had to leave. The Volksoper became the Opernhaus der Stadt Wien. After a period of chaos, Karl *Böhm was appointed music director 1943 and began rebuilding the ensemble. He began with a Strauss cycle, and on 30 Nov. 1944 Knappertsbusch conducted the last performance in the old house, *Götterdämmerung*. In Sep. 1944 the theatre was closed by Goebbels; bombed 12 Mar. 1945.

The company resumed 1 May 1945 at the Volksoper with *Figaro* under Krips. Franz Salmhofer became director, and in Oct. the T an der Wien was reopened with *Fidelio*, the company playing in both houses. With Böhm, Krips, and Krauss, and an ensemble including Gueden, Konetzni, Jurinac, Reining, Schwarzkopf, Della Casa, Lipp, Welitsch, Seefried, Dermota, Lorenz, Hotter, Kunz, Klein, Patzak, Schoeffler, and Weber, it quickly regained its pre-war prestige. In 1954 Böhm became director to prepare for the reopening of the Staatsoper (cap. 1,709) on 5 Nov. 1955 with *Fidelio*. Böhm resigned in 1956 after public criticism, especially over a new production of *Wozzeck*, and was succeeded by Karajan, who made an agreement with La Scala for joint productions. Karajan gradually assumed the role of conductor-producer, becoming the most powerful figure at the Opera since Mahler. He introduced more contemporary music than had ever been Viennese taste, and opera in the original language. His ensemble included Ludwig, Nilsson, Popp, Rysanek, Berry, and Waechter. Various directors came and went, Egon Hilbert remaining in 1964 when Karajan resigned. Hilbert's successor Heinrich Reif-Grintl appointed Horst Stein as First Conductor, a new Vienna post. A slack period ensued, with standards restored by Lorin Maazel, 1982–4, who introduced a *stagione system but resigned in the face of public hostility. He brought in Ricciarelli, Marton, and Pavarotti. Claudio Abbado was music director 1986–91. The Volksoper has maintained its own traditions and standards.

**Vie parisienne, La.** Operetta in 5, later 4 acts, by Offenbach; text by Meilhac and Halévy. Prod. Paris, Palais-Royal, 31 Oct. 1866; New York, T. Français, 29 Mar. 1869; London, Holborn T, 30 Mar. 1872. The setting is the Paris of Haussman, to which tourists come from all over the world on the new railway system. The penniless Bobinet (spoken) and Gardefeu (ten) gull two Swedish aristocrats into thinking they are tourist guides; Gardefeu's house is used as a 'hotel', with Bobinet's servants dressed up as the 'guests'.

**Vieuille, Félix** (*b* Saugéon, 15 Oct. 1872; *d* Saugéon, 28 Feb. 1953). French bass. Studied Paris with Achard and Giraudet. Début Aix-les-Bains 1897 (Leporello). Paris, OC, 1898–1940; New York, Manhattan OC, 1908–9. Created numerous roles, including Father (*Louise*); Dukas's Barbe-bleue, Sultan (*Mârouf*), Eumée (*Pénélope*), Macduff (Bloch's *Macbeth*); also Arkel, praised by Debussy as 'the most profoundly human of grandfathers' and for the understanding of his artistic intentions. (R)

**Viglione-Borghese, Domenico** (*b* Mondovì, 3 July 1877; *d* Milan, 26 Oct. 1957). Italian baritone. Studied Pesaro with Leonese, Rome with Cotogni. Début Lodi 1899 (Herald, *Lohengrin*). Gave up singing, but was persuaded by Caruso to return to the stage. Toured S America 1905–6; Parma 1906; Milan, S, from 1910. Also Rome; Turin; Naples; Paris, O; Buenos Aires, C. Possessed a large, vigorous tone and a vivid stage personality. Particularly esteemed in dark roles, e.g. Amonasro, Iago (Verdi), Barnaba. His Jack Rance was dubbed 'the Prince of Sheriffs' by Puccini. (R)

**'Vilja, o Vilja'.** Aria in Act II of Lehár's *Die lustige Witwe*, in which she recounts a traditional national folk-tale of a maid of the woods and a huntsman's unrequited love for her.

**Village Romeo and Juliet, A.** Opera in prologue and 3 acts by Delius; text by composer, after the story by Gottfried Keller in his *Leute von Seldwyla* (vol. i, 1856; vol. ii, 1874). Prod. (in German) Berlin, 21 Feb. 1907, with L. Artôt de Padilla, Merkel, Egener, cond. Cassirer; London, CG, 22 Feb. 1910, with Vincent, Terry, Hyde, Dearth, cond. Beecham; Washington, DC, Kennedy Center, 26 Apr. 1972, with Wells, Stewart, Reardon, cond. Gallaway. The intermezzo before the final scene, *The Walk to the Paradise Garden*, is well known as a concert piece.

The theme is the love between two children of quarrelling farmers. Sali (treble, later ten) and Vrenchen (girl sop, later sop) leave their fathers Manz (bar) and Marti (bs), encouraged by the mysterious Dark Fiddler (bar), and from the fair in the Paradise Garden wander to the river and die together in a barge that sinks as it floats downstream.

**Villa-Lobos, Heitor** (*b* Rio de Janeiro, 5 Mar. 1887; *d* Rio de Janeiro, 17 Nov. 1959). Brazilian composer. Studied with his father. In a vast number of works he composed only two operas, *Izath* (with Azevedo and Júnio; comp. 1912–14, prod. 1958) and *Yerma* (Lorca; comp. 1955–6, prod. 1971). The latter is cast in a neo-Romantic idiom owing much to Puccini.

BIBL: S. Wright, *Villa-Lobos* (Oxford, 1992).

**Villati, Leopoldo di** (*b* ?, 1702; *d* Berlin, 9 July 1752). Italian librettist. His earliest librettos were for Vienna; remembered for his work at the court of Frederick the Great in Berlin, where he collaborated with Graun on 11 operas, beginning with *Le feste galanti* (1747). Many of these reflect the court taste by adapting French plays or librettos: *L'Europa galante* (1748, after La Motte), *Iphigénie in Aulide* (1748, after Racine), and *Armida* (1751, after Quinault). For *Coriolano* (1749) he expanded a draft by Frederick the Great.

**Villi, Le.** Opera in 2 acts by Puccini; text by Fontana after Alphonse Karr's *Les Willis* (and perhaps suggested by Adam's ballet *Giselle*). Prod. Milan, V, 31 May 1884, with Caponetti, A. d'Andrade, cond. A. Panizza; Manchester 24 Sep. 1897, with Rousney, Beaumont, Lord; New York, M, 17 Dec. 1908, with Alda, Bonci, Amato, cond. Toscanini. Rev. 2-act version prod. Turin, R, 26 Dec. 1884, with Filippi-Bresciani, E. Boronat, Gnaccarini.

The Black Forest, the Middle Ages. Villagers gather to celebrate the betrothal of Anna (sop) to Roberto (ten), who is about to leave on a journey to claim an inheritance. He promises Anna that he will return soon. Roberto returns many months later, having lived a carefree life. Anna has, unknown to him, died heartbroken; her father, Guglielmo (bar), calls on her spirit and on the Villi (the spirits of betrothed maidens deserted by their lovers) to take revenge on Roberto. Thinking it is his betrothed, Roberto then embraces Anna's ghost, and is forced to dance until he falls dead.

**Vilnius.** See LITHUANIA.

**Vinay, Ramón** (*b* Chillán, 31 Aug. 1912). Chilean baritone and tenor. Studied Mexico City with Pierson. Début, as baritone, Mexico City O 1931 (Alfonso, *La Favorite*). Sang baritone roles for several years. After further study, début as tenor Mexico City 1943 (Don José). New York, C, 1944, M, 1946–61; Milan, S, 1947–52; London, CG, 1950, 1953–60; Bayreuth 1952–7, 1962. Also Paris, O; Chicago; etc. Resumed baritone roles in 1962. Retired 1969. A highly musical, intelligent, and affecting singer actor. Roles included Lensky, Tristan, Parsifal, Don José, Canio, Herod, and a distinguished Otello (Verdi); also Doctor Bartolo (Mozart), Telramund, Rigoletto, Falstaff (Verdi), Scarpia. (R)

**Vinci, Leonardo** (*b* Strongoli, *c*.1690; *d* Naples, 27 or 28 May 1730). Italian composer. Studied Naples (Cons. dei Poveri di Gesù Cristo) with Greco; from 1719 maestro di cappella to the Prince of Sansevero; in 1725 became pro vice maestro di cappella to the royal chapel, Naples. In 1728 he was appointed maestro di cappella of his former conservatory in Naples, where his pupils included Pergolesi.

Vinci's earliest operatic works were simple comedies in Neapolitan dialect; though he composed 12 such works, all are lost except for *Li zite 'n galera*. Beginning with *Publio Cornelio Scipione*, he contributed *c*.24 works to the opera seria repertory, writing for Venice (e.g. *Ifigenia in Tauride*), Parma (e.g. *Il trionfo di Camilla*), and Rome (*Farnace*). His most important works were his settings of Metastasian librettos for Rome, including *Didone abbandonata*, *Catone in Utica*, and *Alessandro nell'Indie*. One of the first composers to set Metastasio, his *Artaserse* represents a watershed in the development of opera seria. In it signs of a break with the tradition of Alessandro Scarlatti are clearly apparent: in particular, the dramatic rigidity of the secco recitative–aria chain is given greater flexibility and the accompanied recitative reaches a new expressive peak. Though Vinci died shortly after the completion of *Artaserse*, it remained popular for many years and is viewed, together with Hasse's contemporaneous setting of the same libretto, as one of the seminal works for the establishment of Metastasian opera seria in the mid-18th cent.

SELECT WORKLIST: *Li zite 'n galera* (Saddumene; Naples 1722); *Publio Cornelio Scipione* (?; Naples 1722); *Ifigenia in Tauride* (Pasqueligo; Venice 1725); *Il trionfo di Camilla* (Frugoni; Parma 1725); *Didone abbandonata* (Metastasio; Rome 1726); *Catone in Utica* (Metastasio; Rome 1728); *Alessandro nell'Indie* (Metastasio; Rome 1729); *Farnace* (Lucchini; Rome 1729); *Artaserse* (Metastasio; Rome 1730).

**Violetta.** Violetta Valéry (sop), lover of Alfredo Germont, and the eponymous 'strayed woman', in Verdi's *La traviata*.

**Viotta, Henri** (*b* Amsterdam, 16 July 1848; *d* Montreux, 17 Feb. 1933). Dutch conductor of Italian descent. Studied Cologne with Hiller.

Founded the Wagnervereeniging, Amsterdam, and through it did much to propagate Wagner's music in Holland, staging the major works during the 1880s and in 1905 giving the first European *Parsifal* outside Bayreuth.

WRITINGS: *Richard Wagner* (Sneek, 1883).

**'Vi ravviso'.** Count Rodolfo's (bs) aria in Act I of Bellini's *La sonnambula*, in which he greets the village he last saw as a young man.

**Visconti, (Conte) Luchino** (*b* Milan, 2 Nov. 1906; *d* Rome, 17 Mar. 1976). Italian producer and designer. Originally a film- and theatre-producer, turning to opera in 1954 after seeing Callas's Norma. Worked with her Milan, S, producing *Vestale* 1954, *Sonnambula* 1955, *Traviata* 1956, *Anna Bolena* 1957, *Iphigénie en Tauride* 1958. Designed and produced *Don Carlos*, London, CG, 1958, returning for *Trovatore*, a 'black and white' *Traviata*, and an art-nouveau *Rosenkavalier*. Inaugurated Spoleto Festival with *Macbeth*, 1958, often returning, and ending with *Manon Lescaut* 1972. Callas's demonstration of the dramatic vitality of bel canto opera found a response in his historical sense, his feeling for tradition, and his eye for telling dramatic detail; his own revitalization of the staging of 19th-cent. Italian opera, previously thought largely the singers' prerogative, influenced a generation of designers including Zeffirelli and Sanjust. Opera on his ancestor by Amadei (1869).

**Vishnevskaya, Galina** (*b* Leningrad, 25 Oct. 1926). Russian soprano. Studied Leningrad with Garina. Début Leningrad 1944 (*Vogelhändler*). Moscow, B, 1952–74; New York, M, 1961–2, 1965, 1977; London, CG, from 1962; Milan, S, from 1964. A beautiful woman with a vibrant, typically Slavonic voice who brought her own highly coloured interpretations to a wide range of productions. Roles included Leonore, Aida, Tosca, Tatyana, Lisa, Natasha. Studied Katerina Ismaylova with Shostakovich for film of revised version of *Lady Macbeth of Mtsensk*; recorded original version with her husband, the cellist and conductor **Mstislav Rostropovich** (*b* 1926). They left USSR 1974. Her autobiography, *Galina* (London, 1984), is a fascinating account of artistic life in Soviet Russia. (R)

**'Vision fugitive'.** Hérode's (bs) aria in Act II of Massenet's *Hérodiade*, in which he sings of the vision of Salomé that haunts him day and night.

**'Vissi d'arte'.** Tosca's (sop) aria in Act II of Puccini's *Tosca*, in which she laments the unjustness of her fate.

**Vitale, Edoardo** (*b* Naples, 29 Nov. 1872; *d* Rome, 12 Dec. 1937). Italian conductor. Studied Rome with Terziani. Début aged 14 as operetta conductor. Rome, Arg., 1897 (first local *Götter-dämmerung*); Bologna 1897 (first local *Walküre*). Milan, S, 1908–10, giving first Italian *Medea* and *Boris* (1909). Music director Rome, C, 1913–26. Married **Lina Pasini** (*b* Rome, 8 Nov. 1872; *d* Rome, 23 Nov. 1959), a leading Wagner soprano who also had a success touring Italy with her husband. Their son **Riccardo** (*b* Rome, 1903) was artistic director of Rome O 1958–62.

**Vitali, Filippo** (*b* Florence, *c*.1590; *d* ?Florence, after 1 Apr. 1653). Italian composer and singer. Details of his early life are unclear; *c*.1631 joined the Barberini entourage in Rome as a priest. In 1633 he joined the Papal Chapel as a singer; 1642 became maestro di cappella to the Grand Duke of Tuscany in Florence. Vitali is chiefly remembered as the composer of *Aretusa*, the first opera to be staged in Rome. According to Vitali's own preface to the score, which shows the strong influence of the first generation of Florentine opera composers, he both composed and rehearsed the opera in 44 days.

WORKLIST: *Aretusa* (Corsini; Rome 1620).

**Vitellia.** The deposed Emperor Vitellius' daughter (sop) in Mozart's *La clemenza di Tito*.

**Vittori, Loreto** (*bapt.* Spoleto, 5 Sep. 1600; *d* Rome, 23 Apr. 1670). Italian singer, composer, and librettist. After early spells in both Rome and Florence, Vittori joined the entourage of Cardinal Ludovico Ludovisi in Rome in 1621. The next year he joined the Papal Choir and rapidly became recognized as one of the best singers of his generation. In addition to his ecclesiastical activities he made many appearances in opera, notably in the first performance of Mazzochi's *La catena d'Adone*. One of the first operatic castratos to achieve widespread fame, his own dramatic works included *Galatea*, an important addition to the Roman repertory.

SELECT WORKLIST: *Galatea* (Vittori; Rome 1639).

**'Viva il vino'.** Turiddu's (ten) *brindisi in Mascagni's *Cavalleria rusticana*.

**Vivaldi, Antonio** (*b* Venice, 4 Mar. 1678; *d* Vienna, 28 July 1741). Italian composer. Though Vivaldi is esteemed principally as an instrumental composer, and especially for his violin music, he made a notable contribution to the history of Venetian opera which is only now beginning to receive the critical attention it deserves. Perhaps the most significant result of recent researches, apart from the intrinsic value of discovering much unjustly neglected music of a high quality, is the light it has thrown on the general picture of opera composition in Italy in the early 17th cent. In particular, the firm Venetian focus of Vivaldi's activities destroys once and for all the justifica-

tion for describing this phase in the history of the genre as that of 'Neapolitan opera'.

Born the son of a violinist in the St Mark's orchestra, Vivaldi trained for the priesthood, being ordained in 1703, the year he became violin-master at the Pietà. Here his behaviour and personality brought him into conflict with the authorities and his career suffered many setbacks. After leaving the Pietà in 1717 he undertook extensive travels throughout both Italy and Europe, all the while composing and publishing instrumental music and staging operas. The first, *Ottone in Villa*, had been produced in Vicenza as early as 1713; during the 1720s and 1730s he became one of the most prolific and highly regarded Venetian opera composers, though his work soon lapsed into obscurity after his death.

Vivaldi's point of departure was the nascent genre of opera seria, to which he brought his own peculiar skills as an instrumental composer. In total he composed over 45 operas, of which 16 survive in their entirety. As is the case with many great opera composers of the Baroque, the high quality of Vivaldi's works does not lie for the most part in strikingly innovative gestures, but rather in the subtle adaption of conventional procedures and patterns. Like his contemporaries, Vivaldi showed a fondness in his librettos for historical or heroic subjects (e.g. *Griselda*, *Orlando*) which were sometimes also given exotic settings (e.g. *Montezuma*, *Teuzzone*). Together with Vinci's, his work carried forward that of Alessandro Scarlatti, leading to the solidification of the principles of Metastasian opera seria under Hasse. Unlike his near contemporary Handel, he showed less fluidity in his handling of the drama and abandoned the strict recitative–aria pattern far less readily. However, his sure melodic skill enabled him to rise above the more pedestrian efforts of most of his fellow Italians, while in some operas he directly utilized his unique ability to write effective descriptive music: for instance, the storm in *La fida ninfa* is paralleled in the *Tempesta di mare* violin concerto (Op. 8 No. 5).

SELECT WORKLIST: *Ottone in Villa* (Lalli; Vicenza 1713); *Teuzzone* (Zeno; Mantua 1719); *Orlando furioso* (Braccioli, after Ariosto; Venice 1727); *La fida ninfa* (The Faithful Nymph) (Maffei; Verona 1732); *Montezuma* (Giusti; Venice 1733, lost); *L'Olimpiade* (Metastasio; Venice 1734); *Griselda* (Zeno & Goldoni; Venice 1735); *Catone in Utica* (Metastasio; Verona 1737).

BIBL: M. Talbot, *Vivaldi* (London, 1978); M. Collins and E. Kirk, *Opera & Vivaldi* (Austin, TX, 1984).

**Vixen.** Bystrouška (sop), vixen bride of the fox Lišák, in Janáček's *The Cunning Little Vixen*.

**'Voce di donna'.** La Cieca's (mez) aria in Act I of Puccini's *La Gioconda*, voicing her thanks to

Laura for saving her from condemnation as a witch.

**Vogl, Heinrich** (*b* Au, 15 Jan. 1845; *d* Munich, 21 Apr. 1900). German tenor. Studied Munich with Lachner and Janke. Début Munich 1865 (Max). Munich 1865–1900; Bayreuth 1876–97; London, HM, 1882; Vienna, H, 1884–5; New York, M, 1890. Highly successful in the German repertory, and an outstanding Wagner singer. With his dark-hued and pleasing voice, fine technique and excellent projection, he was a successful Lohengrin, created Loge and Siegmund, and became the leading Tristan of his day. Also sang Siegfried and Parsifal. He was, however, considered 'prosaic' by Ludwig II, and Wagner rejected him as not 'aristocratic' enough for the first Walther. He wrote an unsuccessful opera, *Der Fremdling* (Munich 1899). His wife **Therese Thoma** (*b* Tutzing, 12 Nov. 1845; *d* Munich, 29 Sep. 1921), was a soprano. Admirable as Elsa, Isolde, and Brünnhilde, and created Sieglinde. Often performed with Vogl; however, his insistence that they be employed simultaneously at Bayreuth lost him the chance of the first Parsifal, as Wagner considered her wrong for Kundry.

**Vogler,** (Abbé) **Georg Joseph** (*b* Pleichach, 15 June 1749; *d* Darmstadt, 6 May 1814). German composer, pianist, and organist. Studied Würzburg. Widely known in his day as an organist, theorist, and teacher—including of Weber, Danzi, and Meyerbeer—he had little more than temporary success with his compositions. *La kermesse* (Paris 1783) did not even complete its première. Of his other operas, the most important was probably *Samori* (Vienna, 1804).

**'Voi, che sapete'.** The canzonetta written and sung by Cherubino (sop, sometimes mez) to the Countess in Act II of Mozart's *Le nozze di Figaro*.

**'Voi lo sapete, o mamma'.** Santuzza's (sop) account to Mamma Lucia of her betrayal by Turiddu in Mascagni's *Cavalleria rusticana*.

**Voix humaine, La** (The Human Voice). Opera (tragédie lyrique) in 1 act by Poulenc; text by Cocteau. Prod. Paris, OC, 6 Feb. 1959 with Duval, cond. Prêtre; New York, Carnegie H, 23 Feb. 1960, with Duval, cond. Prêtre (concert perf.); Edinburgh, King's T, 30 Aug. 1960, with Duval, cond. Pritchard. This is a 45-minute 'concerto' for soprano and orchestra—one side of a conversation between a young woman and her lover, who has jilted her. At the end she strangles herself with the telephone flex.

**voix sombrée** (Fr.: 'darkened voice'). A controversial vocal technique first used by the tenor Gilbert Duprez in which the larynx is kept at a low level, even when singing in the upper register. The tone thus produced is dark and sombre.

**Volo di Notte** (Night Flight). Opera in 1 act by Dallapiccola; text by composer after Saint-Éxupéry's novel *Vol de nuit* (1931). Prod. Florence, P, 18 May 1940, with Fiorenza, Danco, Pauli, Melandri, Valentino, Guicciardi, cond. Previtali; Glasgow, King's T, by Scottish O, 29 May 1963, with Collier, Laura Sarti, Andrew, Raymond Nilsson, Garrard, McCue, cond. Gibson; Stanford U, 1 Mar. 1962. Successfully given in many European theatres since the war.

The opera is set in the control room of an airport. It tells of the anxieties and relationships of Rivière, the flight controller (bs-bar), the radio-telephonist (ten), and Signora Fabien (sop), wife of one of the pilots, whose arrival on a night flight is anxiously awaited.

**Voltaire** (orig. François Marie Arouet) (*b* Paris, 24 Nov. 1694; *d* Paris, 30 May 1778). French writer. Operas on his works are as follows:

*Gertrude* (?): Grétry (1766)
*Samson* (1732): Rameau, with Voltaire as librettist (unprod.); Mercadante (1831); Champein (early 19th cent.)
*Zaïre* (1732): Nasolini (1797); Queiroz (1880); Portugal (1802); Federici (1803); Winter (1805); Lavigna (1809); García (1825); Bellini (1829); Manni (1845); Ernst II of Saxe-Coburg-Gotha (1846); Lefèbvre (1887); La Nux (1890)
*Alzire* (1736): Zingarelli (1794); Mayr (1806); Portugal (1810); Verdi (1845)
*Mérope* (1743): Graun (1756, text by Frederick II); Zandomenighi (1871)
*Zadig* (1748): Catrufo (1818); Vaccai (1825); Dupérier (1938)
*Sémiramis* (1748): Bianchi (1790); Catel (1802); Rossini (1823)
*L'orphelin de la Chine* (1755): Winter (*Tamerlan*, 1802)
*Candide* (1756): Knipper (comp. 1926–7); Bernstein (1956)
*Tancrède* (1760): Rossini (*Tancredi*, 1813)
*La Pucelle* (1762): Langlé (*Corisandre*, 1791)
*Ce qui plaît aux dames* (1764): Duni (*La fée Urgèle*, 1765); Schulz (1780–1); Fortia de Pilès (1784); Arquier (1804); Catrufo (1805); Blangini (1812)
*Olympie* (1764): Kalkbrenner (1798); Mosel (1813); Spontini (1819)
*Les scythes* (1767): Mayr (1800); Mercadante (1823)
*Charlot* (1767): Stuntz (*Heinrich IV zu Givry*, 1820)
*L'ingénu* (1767): Grétry (*Le Huron*, 1768)
*La bégueule* (1772): Monsigny (*La Belle Arsène*, 1773)

Operas on Voltaire are *Une matinée de Voltaire* by Solié (1800) and *Voltaire* by F. Müller (before 1845).

**Von der Osten, Eva.** See OSTEN, EVA VON DER.

**Von Heute auf Morgen** (From Day to Day). Opera in 1 act by Schoenberg; text by Max

motivic directness in *Das Rheingold* has given way by the end of the cycle to a dense conceptual network of ideas that, especially with all the accrued associations of four long evenings, speaks in music across word or image directly to the mind. The use of the orchestra, therefore, in *Das Rheingold* and *Die Walküre* accords more closely with Wagner's theories in *Oper und Drama* than is the case by *Siegfried*, when the expressive burden has passed more fully to the orchestra. Vocal lines that are often plain in *Das Rheingold* become lyrical in *Die Walküre*, and never again lose their expressive beauty: Wagner has evolved his so-called *unendliche Melodie* (that is, melody free of cadences or periods), in which details of text can be underlined, allusions made to previous or foretold events, and the emotions of the characters depicted in a freely flowing set of lines. It is the ideal technique for his method of unfolding events, feelings, and their inward meaning, largely in dialogues, especially when handled with such skill in conjunction with the enriched palette of orchestral sound, and with harmony that can range from the simple and diatonic to the subtlest chromatic intensity.

But a crucial element in the growth of *The Ring*, and of Wagner's whole art, was his reading in 1854 of Schopenhauer's *Die Welt als Wille und Vorstellung* (The World as Will and Representation), as he was working on *Die Walküre*. No single work was more important in enabling him to realize and understand ideas which he had intuitively and creatively glimpsed (and a part of his attraction to Schopenhauer was the book's insistence that understanding may be best reached by this route). The book was in the strictest sense a revelation, since it revealed to Wagner ideas which had long lain within him. Schopenhauer's reasoning that the exercise of latent energy (the will), whether for power, for sex, or for other achievements in living, is destructive, and that what should be sought is renunciation and annihilation, is but part (though an important part) of a complex philosophical argument that profoundly affected everything Wagner wrote after he encountered it. This is in spite of the fact that Schopenhauer's artistic theories, especially as regards opera, conflicted with Wagner's at almost every turn until he found ways of absorbing them into his own vision. Schopenhauer's ideas contributed fundamentally to the growth of Wotan from the grasping and deceitful tyrant of *Das Rheingold* through human suffering to the tragic god willing his own destruction in *Götterdämmerung*; but more crucially, Schopenhauer appeared to lend authority to Wagner's increasing reliance on music as the supreme art for articulating his drama.

*Tristan und Isolde* is deeply coloured by these experiences. Wagner's mastery of the orchestra enables him to place a crucially increased reliance on it, since the essence of the drama of the doomed lovers concerns their inward experience, articulated by the words but really expressed by the orchestra. This is achieved through Wagner's vastly expanded understanding of chromatic harmony (in which he was influenced by Liszt) but also by means of the subtly interweaving chromatic contrapuntal lines and not least through the orchestration, whose richness and density at once expresses these elements and is part of them. Tristan and Isolde turn their backs on the world's 'appearance' or outward manifestation (Schopenhauer's *Vorstellung*) for the mystic erotic union that must of necessity exclude the light of day in favour of the sympathetic, enfolding darkness of night; and thence it ensues that they can only truly find each other in the final darkness of annihilation. The combination of erotic tension and deeply pessimistic renunciation of the world gave *Tristan* a unique atmosphere which was immediately acknowledged, and the score's chromatic tension was but one element that left the course of European music permanently changed.

*Tristan* was in part an outcome of Wagner's reading of Schopenhauer and also of his love for Mathilde von Wesendonck, some of whose poems he set in Tristanesque vein during the time he was living in a house provided by her husband near Zurich. Once it was finished and due for production in 1861, he turned his thoughts to his 1845 scenario for *Die Meistersinger*. Originally intended as a comic afterpiece for *Tristan* it was transformed both in the redrafting of the scenarios and in the composition into a profound comedy. Though the language is necessarily more diatonic than that of *Tristan*, it is also subtly chromatic, and ingenious in re-creating a lost diatonic simplicity in 19th-cent. terms; and though suffused with warmth and tenderness, it is also a work that has at its centre the Schopenhauerian resignation of Hans Sachs, whose renunciation of his own 'will' is a sacrifice necessary for the happiness of the lovers and the restoration of order in Nuremberg. The work is also about tradition and renewal, the young artist Walther at once having to win Eva (in part a symbol of Paradisal innocence) away from the crabbed Beckmesser and even from Sachs himself, and to refresh a stultified tradition with his new art. This was of course close to Wagner's own preoccupations; but the work is again deeply imbued with Schopenhauer in Sachs's melancholy reflection on the *Wahn* (distorting illusion) that must be turned to the healing illusion which art gives the world. There is thus a tension between the robust, happy diatonic expression of order attached to the Mastersingers and citizens of Nuremberg, and the inner pessimism of the wise and noble Sachs. The opera is not only one of history's great comedies; it is the greatest of all works of art about art.

Wagner's fortunes in these years reached a low ebb, with his marriage to Minna in ruins and his creditors pressing him. Rescue came in 1864 in the form of support from King Ludwig II of Bavaria. Devoted to Wagner and obsessed by his music, Ludwig supported him personally and materially, though opposition in his Cabinet prevented him from building the theatre Wagner wanted. Meanwhile, Wagner had parted from Minna and was establishing himself with the wife of Hans von Bülow, Liszt's daughter Cosima. In exile again, he settled with her at Tribschen on Lake Lucerne; they were eventually married in 1870.

*Tristan* was finally produced in Munich in 1865 (with Bülow conducting), *Die Meistersinger* in 1868, and *Das Rheingold* and *Die Walküre* in 1869 and 1870. A long cherished plan for a special festival theatre, which accorded with Wagner's ideals of opera as a social ritual, was to be realized through the King's generosity. In 1871 Wagner and Cosima settled in *Bayreuth, and in the following year the foundation stone of the festival theatre was laid. Wagner and his family moved into a house built for them by the King, which he named Wahnfried ('peace won from illusion'). Elaborate preparations for the first festival culminated with the opening of the theatre in 1876. Such by now was Wagner's renown that the occasion drew a distinguished international audience from all corners of Europe.

Wagner had meanwhile begun work on what he knew would be his last opera, *Parsifal*, which was originally intended for Bayreuth alone. Described as a Bühnenweihfestspiel ('sacred festival drama'), it takes some of the chromatic language of *Tristan* to similar extremes but for entirely different expressive purposes. The theme of order won out of disarray, in the context of art and human society, which informs *Die Meistersinger*, is here turned to an ideal society, that of the Knights of the Grail as guardians of a system of belief by which men must live. The context is Christian, with a Eucharistic celebration at the heart of the work and the Holy Grail (the cup which caught the blood flowing from Christ on the Cross) its symbol. The corruption of the King, Amfortas, by the evil magician Klingsor through the seductress Kundry, damages the Grail's purity, which is eventually restored by the innocent Parsifal and his acquisition of understanding and compassion. There is again the contrast between diatonic purity and chromatic intensity of which Wagner made use in *Die Meistersinger*; but here the tension lies between the corroding evil of Klingsor and the agony of Amfortas on the one hand, on the other the suffusing radiance of the Grail to which Parsifal brings his redeeming innocence. Much of the opera has to be conducted in narrative, with a return to modified use of end-rhyme, and with the subtlest use of motive Wagner ever made; but the polarities of chromaticism and diatonicism also have in the context a subtle motivic force as good wars with evil for supremacy.

Wagner intended to write no more operas but to devote himself to one-movement symphonies, a plan that was never realized. In the autumn of 1882 he and his family left for Venice. Here he died the following February. His body was taken to Bayreuth and buried in the garden of Wahnfried. Cosima reluctantly survived him, an autocrat in her husband's artistic kingdom with their son Siegfried at her side until 1930, when she was laid to rest beside him.

Wagner's influence upon artists of all kinds was unique among composers, though in many cases his true nature was misunderstood. French Symbolist writing, for instance, derived from Baudelaire a confused vision of Wagner's methods. Versions of the technique of Leitmotiv were adopted by writers as different as Thomas Mann (author of an important essay *Leiden und Grösse Richard Wagners* (The Sufferings and Greatness of Richard Wagner) ), James Joyce, D. H. Lawrence, and T. S. Eliot. Many composers tried to emulate aspects of his achievement, among them August *Bungert in Germany and Joseph *Holbrooke and Rutland *Boughton in England. His more general musical influence was all-pervasive, in Germany especially on Strauss and Schoenberg, though it also left a strong mark in England, Italy, and Russia. French Wagnerians included Vincent d'*Indy; and *Debussy was more affected than he cared to admit. Further, the complex and far-ranging nature of Wagner's art led to interpretations, sometimes selective and incomplete, by admirers seeking justification of their own theories. The most notorious, and most damaging to Wagner's reputation, was Adolf Hitler, and Nazi apologists were not slow to find elements of Wagner's racialism and anti-Semitism in *Der Ring*. Wagner's own youthful revolutionary views as expressed in the cycle have also led to Socialist interpretations by critics (most famously, Bernard Shaw in *The Perfect Wagnerite*) and by some modern producers. Among other interpretations, that by Robert Donington in Jungian terms has aroused much interest.

WORKLIST (all texts by Wagner): *Die Hochzeit* (The Wedding) (fragment, comp. 1832–3, unprod.); *Die Feen* (The Fairies) (comp. 1833–4, prod. Munich 1888); *Das Liebesverbot* (The Ban on Love) (Magdeburg 1836); *Rienzi* (Dresden 1842); *Der fliegende*

*Holländer* (The Flying Dutchman) (Dresden 1843); *Tannhäuser* (Dresden 1845; Paris version 1861); *Lohengrin* (Weimar 1850); *Der Ring des Nibelungen: Das Rheingold* (Munich 1869); *Die Walküre* (Munich 1870); *Siegfried* and *Götterdämmerung* as part of complete cycle (Bayreuth 1876); *Tristan und Isolde* (Munich 1865); *Die Meistersinger von Nürnberg* (Munich 1868); *Parsifal* (Bayreuth 1882).

CATALOGUE: J. Deathridge, M. Geck, & E. Voss, eds., *Wagner Werk-Verzeichnis* (Mainz, 1985).

WRITINGS: R. Wagner, *Sämtliche Schriften und Dichtungen* (16 vols., Leipzig, 1914); M. Gregor-Dellin, ed., *Richard Wagner: Mein Leben* (Munich, 1963; trans. 1983); M. Gregor-Dellin and D. Mack, eds., *Richard Wagner: Die Tagebücher, 1869–1883* (2 vols., Munich, 1976–7; trans. as *Cosima Wagner's Diaries* 1978–80); R. Wagner, *Sämtliche Briefe* (Leipzig, from 1967); S. Spencer and B. Millington, eds., *Selected Letters of Richard Wagner* (London, 1987).

BIBL: E. Newman, *The Life of Richard Wagner* (London, 1933–47, R/1976); R. Donington, *Wagner's 'Ring' and its Symbols* (London, 1963); C. Dahlhaus, *Richard Wagners Musikdramen* (Munich, 1971; trans. 1979); P. Burbidge and R. Sutton, eds., *The Wagner Companion* (London, 1979); M. Gregor-Dellin, *Richard Wagner* (Munich, 1980; trans. London, 1983).

**Wagner, Siegfried** (*b* Tribschen, 6 June 1869; *d* Bayreuth, 4 Aug. 1930). German composer, conductor, and producer, son of Richard Wagner and Cosima von Bülow. Studied Frankfurt with Humperdinck, later Bayreuth with Richter and Kniese. Cond. Bayreuth, part of *Ring*, 1896; prod. *Der fliegende Holländer* Bayreuth 1906, when he became director of Bayreuth Festival. His attempts at scenic innovation were cautious, and could not go far because of his mother's loyalties to Wagner's memory. His own operas are mostly fairy-tale works in late-Romantic vein.

SELECT WORKLIST (all texts by S. Wagner): *Der Bärenhäuter* (The Bearskin Wearer) (Munich 1899); *Der Kobold* (The Goblin) (Hamburg 1904); *Der Schmied von Marienburg* (The Smith of Marienburg) (Rostock 1923).

WRITINGS: *Erinnerungen* (Stuttgart, 1923).

BIBL: Z. von Kraft, *Der Sohn* (Graz, 1963); P. Pachl, *Siegfried Wagners musikdramatisches Schaffen* (Tutzing, 1979).

**Wagner, Wieland** (*b* Bayreuth, 5 Jan. 1917; *d* Munich, 17 Oct. 1966). German producer and designer. Son of Siegfried and grandson of Richard Wagner. Studied privately, and for a short period with Roller in 1934. His first designs were for his father's *Der Bärenhäuter* (Lübeck 1936); this was followed by designs for *Parsifal* (Bayreuth 1937) and for two more of his father's operas in Antwerp 1937, and Düsseldorf 1938. Exempted from military service by Hitler so as, he said, to 'preserve and perpetuate the family name', he designed the 1943 Bayreuth *Meister-*

*singer* and produced *Walküre* and *Siegfried* in Nuremberg. Produced complete *Ring*, *Freischütz*, and his father's *An allem ist Hütchen schuld*, Altenburg 1943–4. Studied Wagner scores with Kurt Overhoff at Strauss's home in Garmisch 1947.

Cleared by a de-Nazification court, he assumed, with his brother Wolfgang, the artistic and business administration of post-war Bayreuth when it was removed from their mother Winifred. Together they completely revolutionized the style of Wagner production, abandoning pageantry and naturalism in favour of a more abstract approach designed to emphasize the works' more inward and mythic content. Less use was made of scenery, more of suggestive lighting, which in Wieland's skilled hands came almost to replace scenery at some points. He also produced regularly at Stuttgart, and occasionally in other German cities. His productions of *Fidelio*, *Carmen*, *Orphée*, and *Aida* aroused as much controversy as those of Wagner's works; however, his productions of *Salome*, *Elektra*, *Wozzeck*, and *Lulu* (generally with Anja Silja) were enthusiastically received. He hoped to 'replace the production ideas of a century ago, now grown sterile, by a creative intellectual approach which goes back to the origins of the work itself. Every new production is a step on the way to an unknown goal'.

BIBL: G. Skelton, *Wieland Wagner* (London, 1971).

**Wagner, Wolfgang** (*b* Bayreuth, 30 Aug. 1919). German producer, designer, and manager. Son of Siegfried and grandson of Richard Wagner. Worked as musical assistant Bayreuth 1942. Responsible for the business and administrative sides of the Bayreuth Festivals from 1951. Has produced and designed all the works in the Bayreuth canon, enterprisingly but with less of the imaginative flair shown by his brother Wieland.

**Wagner-Régeny, Rudolf** (*b* Szász-Régen, 28 Aug. 1903; *d* Berlin, 18 Sept. 1969). German composer and pianist of Romanian origin. Studied Leipzig and Berlin (with Rezniček, Koch, and Schreker). His earliest involvement with dramatic music came through the composition of the short theatre piece *Sganarelle* in which the influence of the French group Les Six is very clear. After several similar works, including *Der nackte König*, he scored a notable success with his first full-scale opera, *Der Günstling*; this prepared the way for two further collaborations with Neher, *Die Bürger von Calais* and *Johanna Balk*. The latter, in which he consciously adopted idioms calculated to offend the Nazi régime, was noisily received and Wagner-Régeny's career was momentarily halted. After war service and a nervous breakdown he took up a career as a teacher in the new German Democratic Republic, first in Rostock and later in Berlin at the Hochschule für Musik. Over the next 20 years he gradually developed a highly personal serialist idiom, elements of which are apparent in his later operas, *Persische Episode* and *Das Bergwerk zu Falun*.

SELECT WORKLIST: *Sganarelle* (Wagner-Régeny, after Molière; Essen 1929); *Der nackte König* (The Naked King) (Braun, after Andersen; Gera 1930); *Der Günstling* (Neher, after Hugo; Dresden 1935); *Die Bürger von Calais* (Neher, after Froissart; Berlin 1939); *Johanna Balk* (Neher, after Transylvanian folk-tales; Vienna 1941); *Persische Episode* (Neher; comp. 1940–50, prod. Rostock 1960); *Das Bergwerk zu Falun* (Wagner-Régeny, after Hofmannsthal; Salzburg 1961).

WRITINGS: R. Wagner-Régeny, *Begegnungen* (Berlin, 1968).

BIBL: D. Härtwig, *Rudolf Wagner-Régeny* (Berlin, 1965).

**Wagner tuba.** A valved brass instrument, with conical bore and funnel-shaped mouthpiece, built as Bb tenor and F bass at Wagner's suggestion for *The Ring* in 1876 to obtain a tone quality between the horns, which he regarded as lyrical and romantic, and the trombones, which he regarded as solemn, dignified, and heroic. The quartet used in *The Ring* consists of two tenors and two basses, played by the 5th–8th horns. Particularly associated with Hunding in *Walküre*. Also sometimes known as Waldhorntuba or Ringtuba. Used by Strauss in *Elektra* and *Frau ohne Schatten*. Rarely used elsewhere, though associated with mourning for Wagner's death in Bruckner's Seventh Symphony.

**Wagner-Vereniging.** See AMSTERDAM.

**'Wahn! Wahn!'** Hans Sachs's (bar) monologue in Act III of Wagner's *Die Meistersinger von Nürnberg*, in which he broods upon the follies of the world.

**Waldmann, Maria** (*b* Vienna 1842; *d* Ferrara, 6 Nov. 1920). Austrian mezzo-soprano. Studied Vienna with Passy-Cornet, Milan with Lamperti. Début St Petersburg 1865 (Pierotto, *Linda di Chamounix*). Milan, S, 1871–2; Paris, OC, 1874; also Moscow, Vienna, Trieste, Cairo. Retired on her marriage (when Verdi gave her away) in 1878. A very beautiful and dramatically gifted singer with a velvety tone, who during her short career became, in Verdi's view, his ideal mezzo-soprano. Sang Amneris in the first Italian *Aida*, and created the alto part in his Requiem.

**Wales.** See WELSH NATIONAL OPERA.

**Walker, Sarah** (*b* Cheltenham, 11 Mar. 1943). English mezzo-soprano. Studied London with Packer and Smith, later Rosza. Début Kent O 1969 (Ottavia), Gly. 1970; London, SW/ENO,

1971–8; Chicago 1977; London, CG, from 1979; Vienna, S, 1980; San Francisco 1981; Brussels, M, 1982; Geneva 1983; New York, M, 1986. Roles incl. Poppea, Penelope (*Ulisse*), Diana (*Calisto*), Dorabella, Maria Stuarda, Dido (Berlioz), Quickly, Countess (*Queen of Spades*), Fricka, Baba the Turk, Charlotte, Katisha, Gloriana. Created Agave (Buller's *The Bacchae*), Suzanne (*Toussaint*). A highly accomplished singer with a vivid presence. (R)

**Walk to the Paradise Garden.** The interlude between scenes 5 and 6 of Delius's *A Village Romeo and Juliet*, depicting the lovers Sali and Vreli journeying to the Paradise Garden (a dancing-place) and resting on the way.

**Walküre, Die.** See RING DER NIBELUNGEN, DER.

**Wallace, Ian** (*b* London, 10 July 1919). Scottish bass. Studied London with Allin and Borgioli. Début Cambridge T, 1946 (Schaunard). Edinburgh (Gly. co.) 1948; Gly. 1952–6, 1959–61; Parma and Venice, F, 1950; Berlin 1954; Rome 1955. A witty and engaging buffo singer, whose roles included Masetto, Bartolo (Rossini and Mozart), Don Magnifico, Sacristan. (R)

**Wallace, Vincent** (*b* Waterford, 11 Mar. 1812; *d* Vieuzos, 12 Oct. 1865). Irish composer. Studied with his father, a bandmaster, and with W. S. Conran and Haydn Corri. Played in the Dublin Theatre Royal; then emigrated to Tasmania. Later moved to Australia and made a name as a violin virtuoso. Many adventurous travels ensued, including visits to India and Chile. In Mexico City in 1841 he conducted the Italian opera season, moving on to New York to give concerts and then returning by way of Europe to England.

Soon after his return, he wrote *Maritana*, which had an immediate success. Wallace's travels had given him a musical sophistication: though the language is unadventurous, based on the simpler Romantic conventions of the day, he was able to introduce into the work effects ranging from the direct use of gypsy music and Spanish colouring to a feeling for the style of French and Italian opera. *Maritana* originally ran for 50 nights, and was quickly taken up by European and American theatres.

Wallace's next opera, *Matilda of Hungary*, was a failure. Its successor, in a manner closer to Grand Opera than *Maritana*, was *Lurline* (commissioned for Paris in 1848 and announced by Covent Garden but not given until 1860). Though it did not match the success of *Maritana*, *Lurline* won a considerable following, especially for its grandiose scenes and vigorous effects in a manner then unknown in English opera. After many more travels, concluded in time for *Lur-*

*line*'s London production, Wallace settled in London. His later operas were unsuccessful, and some operettas remained unperformed.

SELECT WORKLIST (all first prod. London): *Maritana* (Fitzball; 1845); *Matilda of Hungary* (Bunn; 1847); *Lurline* (Fitzball; 1860); *The Amber Witch* (Chorley; 1861); *Love's Triumph* (Planché; 1862); *The Desert Flower* (Harris & Williams; 1863).

**Wallerstein, Lothar** (*b* Prague, 6 Nov. 1882; *d* New Orleans, 13 Nov. 1949). Czech, later US, conductor and producer. Répétiteur Dresden 1909. Cond. and prod. Poznań 1910–14, chief prod. Breslau 1918–22. Duisburg 1922–4; Frankfurt 1924–6. Vienna 1927–38, where he prod. 65 operas including *Wozzeck*; Salzburg 1926–37. Guest prod. Milan, S, and Buenos Aires, C. Founded Opera School in The Hague 1939. New York, M, 1941–6.

**Wally, La.** Opera in 4 acts by Catalani; text by Illica, after Wilhelmine von Hillern's novel *Die Geyer-Wally* (1875; dram. 1880). Prod. Milan, S, 20 Jan. 1892, with Darclée (to whom the score is dedicated), Stehle, Guerrini, Suagnes, Pessina, cond. Mascheroni; New York, M, 6 Jan. 1909, with Destinn, L'Huilier, Martin, Amato, Rossi, cond. Toscanini; Manchester 29 Mar. 1919.

The Tyrol, *c*.1800. Catalani's last, and best, opera describes how la Wally (sop) is drawn still closer to her lover Hagenbach (ten) when her suitor Gellner (bar) tries to kill him; the lovers perish in an avalanche. Toscanini admired the opera so much that he named his daughter Wally.

**Walter, Bruno** (*b* Berlin, 15 Sep. 1876; *d* Beverly Hills, CA, 17 Feb. 1962). German, later French, later US, conductor. Studied Berlin with Ehrlich, Bussler, and Radeke. Répétiteur Cologne 1893–4, cond. début (*Waffenschmied*). Cond. Hamburg 1894–6, under Mahler; Breslau 1896–7; Pressburg 1897–8; Riga 1899–1900; Berlin 1900-1. Vienna 1901–12, first as Mahler's assistant. London, CG, 1910 (incl. Smyth, *The Wreckers*), 1924–31 (Mozart, Wagner, Richard Strauss). Music director Munich 1913–22, giving premières of Korngold's *Ring des Polycrates* and *Violanta* (1916), Pfitzner's *Palestrina* (1917), and Schreker's rev. *Das Spielwerk* (1920); Berlin, SO, 1925–9. Left Germany in 1933: guest cond. Vienna, S, 1935, artistic adviser 1936–8. New York, M, 1941–6, 1951, 1955–7, 1958–9. Also cond. Milan, Florence, Chicago. Walter's greatest interpretations were of Mozart, despite a certain sweetness, and Richard Strauss; he was also a fine Wagnerian. The warmth of his music-making was infectious, both to artists and audience, and was based on a strong lyrical impulse. Autobiography, *Theme and Variations*

(trans. New York, 1946; Ger. Frankfurt, 1947). (R)

**Walters, Jess** (*b* New York, 1906). US bass-baritone. Studied New York with Giuffrida. Début New York New OC 1941 (Macbeth). Appeared New Orleans, New York City Center; London, CG, 1947–59; Netherlands O 1960–4. Roles included Amonasro, Alfio. (R)

**Walther.** Walther von Stolzing (ten), a young Franconian knight, lover of Eva, in Wagner's *Die Meistersinger von Nürnberg*.

**Walton, (Sir) William** (*b* Oldham, 29 Mar. 1902; *d* Forio d'Ischia, 8 Mar. 1983). English composer. Sang as a boy chorister at Christ Church, Oxford; largely self-taught. Despite an almost lifelong love of Italian opera, connected to his devotion to the country and eventual residence there, Walton did not turn to writing an opera of his own until *Troilus and Cressida*. The work had a difficult gestation, and, despite its high reputation at the time of its première, was found by many to be cast in too conventional a Romantic vein. The best of the work (notably the characterization of Pandarus) transcends the limitations of the text and has some fine dramatic inventions. Walton's only other opera was *The Bear*, a vigorous Romantic comedy using only three characters.

WORKLIST: *Troilus and Cressida* (Hassall, after Chaucer; London, 1954); *The Bear* (Dehn, after Chekhov; Aldeburgh 1965).

BIBL: M. Kennedy, *Portrait of Walton* (London, 1989).

**Waltraute.** A Valkyrie (mez) in Wagner's *Die Walküre* and *Götterdämmerung*. Her name derives from the Old Norse *valr* (slain) and Middle High German *Wal* (battlefield) and *traut* (devoted).

**Waltz, Gustavus** (*fl* 1732–59). English bass of German origin. First known appearance London, Little Haymarket T, 1732 (in Lampe's *Amelia*). Sang for Handel 1733–6, at London H and CG and also in Oxford, creating roles in *Arianna in Creta*, *Ariodante*, *Alcina*, *Atalanta*. London: DL, 1733–4, 1736, 1743–5; CG, 1739–42, 1749–51. Also sang in Handel oratorios, and operas by Arne, Lampe, and Pescetti. Possibly cooked for Handel, but was in any case a good musician; this somewhat colours Handel's alleged comment, 'Gluck knows no more of counterpoint than my cook'. As a singer, he had a wide range and a dramatic ability that encompassed both the tragic and the comic.

**Wanderer.** The guise assumed by Wotan (bs-bar) in Wagner's *Siegfried*.

**Wandering Scholar, The.** Chamber opera in 1 act by Holst; text by Clifford Bax, founded on an incident in Helen Waddell's *The Wandering*

*Scholars* (1928). Originally entitled *The Tale of the Wandering Scholar*; rev. Britten (1951), ed. Britten and Imogen Holst (1968). Prod. Liverpool, David Lewis T, 31 Jan. 1934, with Pryce, Eastwood, Maher, J. Ward, cond. Wallace; Toronto, Royal Conservatoire of Music, 25 Mar. 1966; New York, State U, 1 Aug. 1968.

Father Philippe (bs) visits Alison (sop) in the absence of her husband Louis (bar), but is interrupted by the Scholar, Pierre (ten). They refuse him food and drink, and he leaves, but returns with Louis, and tells a pointed tale that leads to the discovery of unexpected food and drink, and then the priest hidden under a heap of straw. Another opera on the subject is by Bell (1935).

**War and Peace** (Russ.: *Voyna i mir*). Opera in 5 acts by Prokofiev; text by composer and M. Mendelson, after Tolstoy's novel (1869). 1st version, 11 scenes, comp. 1942, unprod. 2nd version, 11 scenes, Prague, N, 1948. 3rd version, 13 scenes, Leningrad, Maly T, 12 June 1946, with Lavrova, Petrov, Chishko, Zhuravlenko, Butyagin, cond. Samosud (8 scenes only). 4th version, 10 scenes, Florence 26 May 1953, with Carteri, Bastianini, Barbieri, Tajo, Corena, cond. Rodzinski. Final version, 13 scenes, Leningrad, Maly T, 1 Apr. 1955, with Lavrova, Shaposhnikov, Glebov, Matusov, Butyagin, cond. Grikurov (11 scenes only, cut); Moscow, Stanislavsky T, 8 Nov. 1957, with Kayevchenko, Morozov, Shchabinsky, Pirogov, cond. Shaverdov (cut); Moscow, B, 15 Dec. 1959, with Vishnevskaya, Kibkalo, V. Petrov, Krivchenya, cond. Melik-Pashayev; London, C, 11 Oct. 1972, with Barstow, McDonell, Collins, Shilling, Woollam, Brecknock, Bailey, cond. Lloyd-Jones; Boston, Savoy OH, 8 May 1974, with Saunders, Neill, Carlson, Gramm, cond. Caldwell.

**Ward, David** (*b* Dumbarton, 3 July 1922; *d* Dunedin, 16 July 1983). Scottish bass. Studied London with Carey, Munich with Hotter. Début London, SW, 1953 (Old Bard, *Immortal Hour*). London, SW, until 1959; Gly. 1958; London, CG, 1959–83; Bayreuth 1960–2; New York, M, 1963. Also Milan, S, Buenos Aires, Chicago, Vienna, etc. Roles included Don Basilio (Mozart), Rocco, Lord Walton, Philip, Flying Dutchman, King Mark, Hunding, Boris, Gremin, Arkel. An artist whose commanding stature and nobility of tone, allied to great sensitivity, lent particular distinction to his Wotan. (R)

**Warren, Leonard** (*b* New York, 21 Apr. 1911; *d* New York, 4 Mar. 1960). American baritone of Russian origin. Studied New York with Dietsch, Milan with Pais and Picozzi. Début New York, M, 1939 (Paolo, *Boccanegra*). New York, M, until 1960; Buenos Aires, C, 1942; Milan, S, 1953; toured USSR 1958. His voice was of an exceptional mellowness and range (reaching high C); he had an impeccable technique, and became a convincing actor. Famous for Verdi's Rigoletto, Macbeth, Germont, Boccanegra, Iago, Falstaff, and Puccini's Scarpia. Collapsed and died on Met stage during a performance of *Forza*, after singing 'Urna fatale'. (R)

**Warsaw** (Pol., Warszawa). Capital of Poland. The first opera to be given was *Galatea* (prob. by Santi Orlandi) in 1628, on the initiative of Prince Władysław Zygmunt. When crowned Władysław IV, he formed an Italian company (1634–48) with Margherita Cattaneo as prima donna and including some young Poles. In 1637 Bartolomeo Bolzoni built a provisional wooden theatre (cap. 1,000); and for his wedding celebrations Władysław built a permanent theatre in the castle (1637). When Augustus II was crowned in 1697, he brought with him a company under J. C. Schmidt and Jacek Różycki; in 1700 a French company under Deschallières spent five days in the city. Other companies made visits. In 1725 the court theatre, Operalnia, opened. Augustus III (crowned 1733) provided large subsidies and twice-weekly performances were given by a company with an orchestra of over 100; many Metastasian operas were given by a company from Dresden including *Hasse and *Bordoni.

More systematic encouragement came with the reign (1764–95) of Stanisław August Poniatowski. In 1765 the first public theatre opened, with Karol Tomatis (1739–1806) as director: under him French and Italian companies played until 1767. In 1772 a company under Kurz-Bernardon performed in the Radziwiłł Palace; the directorship passed in 1776 to Franciszek Ryx (1732–99), who gave the first opera seria with Italian singers and in 1778 gave the first Polish opera, *Nędza uszczęśliwiona* (Sorrow Turned to Joy) by Maciej Kamieński (1734–1821). The T Narodowy (National T) opened in 1779 with *Le tonnelier* (The Cooper) by Nicolas-Médard Audinot (1732–1801), sung in Polish. Ryx's directorship saw the début of the 'father of the Polish theatre', Wojciech Bogusławski (1757–1829), whose activities included directing the theatre (1782–4 and 1799–1814). French and Italian operas were given, from 1783 Mozart, in Polish. Opera was also staged at the Royal Lazienski Garden, in the T w Pomarańczarni (opened 1788), and in the open-air Amfiteatr on an island (also known as the T na Wyspie) (1790). After Bogusławski, the most important figure in the early 19th cent. was Józef *Elsner (1769–1854), who composed and conducted for the theatre. In 1810 Karol *Kurpiński (1785–1857) was appointed; he became director in 1823 and introduced a largely French and Italian repertory. The T Wielki (Grand T), built by Corazzi, opened on 24 Sep. 1833 with *Barbiere* in Polish (a language which under the Russians could only be heard publicly in opera, not drama); here the opera company reached a high standard under Kurpiński, maintained (1840–52) under the conductorship of Tomasz Nidecki and (1853–74) Jan Quattrini.

After the première of *Halka* (1858), Moniuszko became director, introducing many new Polish works. Cesare Trombini was music director 1874–81: works he introduced included *Lohengrin* (1879). Emil Młynarski was director 1898–1903 and 1919–29, introducing works by Szymanowski and other modern composers. A feature of the period from the 1890s to 1914 were the seasons of Italian opera with great singers, including Battistini (who sang in Warsaw every year from 1894 to 1912). At the T Nowości (opened 1901) operetta was given under the energetic Ludwik Śliwiński. Financial difficulties forced the T Wielki to interrupt its work repeatedly during the 1930s. The theatres closed in 1939, and the T Wielki was damaged by fire; in 1944 bombs destroyed all but the façade. Opera was resumed in 1945 first in a cinema and then in the Roma Hall, refitted in 1953 (cap. 1,000). The T Wielki reopened in 1965 (cap. 1,800). Founded in 1962, the Warsaw Chamber Opera is attached to the T Wielki, and as well as premièring new works has revived works by Hasse (for the first time since his visit): it moved into a theatre of its own in 1987. A State Operetta Co. was founded in 1949. The director and conductor 1981–91 was Robert Satanowski, whose forceful policy first concentrated on Italian and Slavonic opera, though Poland's first *Ring* was initiated in 1988 and completed in 1989. The theatre holds a ten-day annual opera festival.

**Washington, DC.** Capital of the USA. Visiting companies, especially from Philadelphia, gave seasons of ballad opera 1800–14. The first opera staged was *Barbiere*, by the Walton Co. in 1836. Opera was given at Albaugh's OH from 1884. The Washington National Opera Association performed in the 1920s, from 1924 in the new Auditorium. There were also Black groups, one formed in 1872; the National Negro OC formed by Mary Cardwell Dawson in 1943 gave performances until her death in 1962.

The Opera Society of Washington was founded in 1956, and despite early difficulties won a reputation for enterprise: the music director, Paul Callaway, conducted the first US *Erwartung* (1960), Hindemith conducted *The Long Christmas Dinner* and Stravinsky *Oedipus*

*Rex* and *The Nightingale*. The company also gave the première of Ginastera's *Bomarzo* (1967). Financial difficulties intervened, but the Kennedy Center (cap. 2,318) opened 1971 with the company in the commissioned première of Ginastera's *Beatrix Cenci*; performances are also given in the Eisenhower T (cap. 1,142) and the Terrace T (cap. 512). The company became Washington Opera in 1977, and has since expanded rapidly, maintaining its celebrated progressiveness under the general director, Martin Feinstein, through the production of US works, e.g. the world première of Menotti's *Goya* (1986).

**Watson, Claire** (*b* New York, 3 Feb. 1927; *d* Utting, 16 July 1986). US soprano. Studied New York with Schumann, Amsterdam with Lichtenstein. Début Graz 1951 (Verdi's Desdemona). Frankfurt 1956–8; Munich 1958; London, CG, 1958–64, 1970, 1972; Gly. 1960. Also Vienna; Milan, S; Chicago; Buenos Aires. Roles incl. Donna Elvira, Eva, Tatyana, Marschallin, Ariadne, Arabella. A musical, radiant toned, and eloquent performer. (R)

**Weber.** German singers, daughters of **Fridolin Weber** (1733–79), musician and uncle of Carl Maria von Weber.

1. **Josepha** (*b* Zell, 1758 or 1759; *d* Vienna, 29 Dec. 1819). Soprano. Sang from early 1780s. In 1788 married the violinist Franz Hofer (1755–96). Joined Schikaneder's co., Vienna, W, 1790. In 1797 married the actor-singer Sebastian Mayer (later the first Pizarro). Sang in Vienna until 1805. An accomplished singer, for whom Mozart wrote the Queen of the Night, which she created; also the bravura aria K580 to be sung in Paisiello's *Barbiere di Siviglia*.

2. **Aloysia** (*b* Zell or Mannheim, between 1759 and 1761; *d* Salzburg, 8 June 1839). Soprano. Studied Mannheim with Mozart, he being captivated by her 'pure, lovely voice' and 'superb cantabile singing'. They toured together 1778; he was much enamoured, but she soon rejected him. Munich 1778; Vienna 1779–92, singing first with German O, then with Italian O. In 1780 married the actor and painter **Joseph Lange** (1751–1831). Mozart wrote several arias for her, and the role of Mme Herz (*The Impresario*); she also sang Donna Anna in the first Vienna *Don Giovanni*, and Sesto (*Clemenza*) in concert, 1795.

3. **Constanze** (*b* Zell, 5 Jan. 1762; *d* Salzburg, 6 Mar. 1842). Soprano, wife of Mozart. A good amateur singer with an attractive voice, she sang one of the soprano parts in Mozart's C minor Mass, Salzburg 1783. After Mozart's death she devoted herself to having his last opera, *La clemenza di Tito*, performed. Married the diplomat Georg Nissen 1809, and completed the biography of Mozart he left unfinished at his death, 1826.

4. A fourth sister, **Sophie** (1763–1846), wrote a touching description of Mozart's death. She married the composer and singer Jakob *Haibel.

**Weber, Carl Maria von** (*b* Eutin, ?18 Nov. 1786; *d* London, 5 June 1826). German composer, conductor, pianist, and critic. Studied with various teachers on his youthful travels, including Heuschkel and M. Haydn, later Vogler.

Weber's first opera was a childhood essay, later destroyed. In 1800 came *Das Waldmädchen*, of which only fragments survive: this was later reworked as *Silvana*. Meanwhile, he had written *Peter Schmoll und seine Nachbarn*, which shows his interest in Romantic subjects and includes some original and charming music alongside much that is inexperienced. In 1804, not yet 18, he was appointed music director at Breslau: he instituted sweeping reforms anticipating those he was to pursue in Prague and Dresden, but largely through youth and inexperience failed to win sufficient support for them. While in Breslau, he began an opera, *Rübezahl*, of which three numbers survive. During a later (non-musical) post in Stuttgart he wrote *Silvana*: though still showing many signs of inexperience, it reveals an instinct for the theatre, and includes some woodland and chivalric music giving a foretaste of *Der Freischütz* and *Euryanthe*. Its title-role is silent, and the problems of writing an opera for a dumb heroine are ingeniously tackled by a composer with a developing interest in orchestral characterization. In Mannheim in 1810 he first read the story of *Der Freischütz*, and even planned an opera on it; the project fell through when the librettist defaulted. However, he did write the 1-act *Abu Hassan*, a witty little Singspiel in his freshest and most tuneful vein.

In 1816, in one of the reviews he had begun writing regularly, Weber referred in connection with Hoffmann's *Undine* to 'the German ideal, namely a self-sufficient work of art in which every feature and every contribution by the related arts are moulded together in a certain way and dissolve, to form a new world'. All his mature life, as Kapellmeister, critic, and composer, was dedicated to the aim of developing German opera along the lines of what Wagner was to call the Gesamtkunstwerk. Appointed to Prague as opera director in 1813, he set about reforming operatic practice there, recruiting singers for the capacity as members of an ensemble rather than as stars, revising rehearsal practice, re-forming the orchestra, and taking a practical interest in all theatrical matters from the scenery to the library catalogue. His repertory consisted of works that were the theoretical and creative basis of German Romantic opera, mostly French operas by Cherubini, Spontini, Isouard,

Boieldieu, Dalayrac, Méhul, Grétry, Catel, and others. He also prepared the way with essays on the operas he was to perform. He had no time for opera composition himself, though he contemplated a *Tannhäuser* to a text by Brentano.

In 1816 he resigned and was appointed Royal Saxon Kapellmeister in Dresden. Here his reforms were still more searching, and included the development of an elaborate rehearsal schedule for each work that began with a reading of the text so as to impress the drama upon the singers; he also re-formed the chorus, re-seated the orchestra, made important developments with scenery and lighting, and again planned his repertory so as to provide a background for the art of German opera, in competition to the Italian opera also flourishing in Dresden under Morlacchi.

In 1821 he went to Berlin to conduct his own contribution to German Romantic opera, *Der Freischütz*. The work triumphed; and especially in comparison with Spontini's *Olympie* in Berlin, it came at once to be seen as an opera that spoke for Germany. Though the work contains French elements, the plot, the characters, the legend, the music, all closely based on the daily country life of Germany, caught popular imagination in times when nationalism was emerging as a force. Basically a Singspiel, *Der Freischütz* transcends older examples by its imaginative strength and cunning structures. Some of the numbers, such as the Huntsmen's Chorus and Bridesmaids' Chorus, were to become almost folk-songs. The Wolf's Glen Scene, with its ingenious mixture of speech, song, orchestral description, and loosely constructed musical numbers, cleverly related to the powers of evil by use of motive, remained an unsurpassed example of Romantic horror in music.

The success of *Der Freischütz* led Barbaia to invite him to write a similar work for Vienna; but Weber was anxious to move on, and turned to a form of Grand Opera with *Euryanthe*. Despite inept writing by his librettist, Helmina von Chezy, he succeeded in constructing an opera, virtually continuous and without spoken dialogue, that anticipates much in Wagner and is itself a masterpiece. The weaknesses in the libretto have long inhibited the work's success, but the work has increasingly been recognized as an advanced and subtle music drama, giving example to its successors especially in its use of motive, its functional handling of the orchestra, and its dramatic momentum.

Weber did not build further on this new creative achievement because of the rapid advance of his tuberculosis. Invited to write an opera for Covent Garden, he was obliged to fall in with the suggestions of his English librettist, James

Robinson Planché, who pressed upon him the kind of musical pantomime that was current English taste. *Oberon* is consequently constructed in separate numbers, though there is some motivic linkage. The delicacy of the fairy music set new standards (noted by Mendelssohn), just as the storm and other elements were to influence Wagner in *Der fliegende Holländer* and beyond. However, this was, Weber sadly acknowledged, not what he meant by a German opera; and he planned to rewrite it, using recitative, in a more suitable manner. He did not live to do so, for having journeyed to London to conduct the première, he finally succumbed to his illness, dying in Sir George Smart's house the day before he was hoping to return to Germany.

A projected comedy, *Die drei Pintos*, was first entrusted to Meyerbeer for completion; later it was given to Mahler, who completed it with music from other of Weber's works and scored it.

Weber was also a pioneering opera conductor, experimenting with new seatings of the orchestra to improve the tonal blend. He normally conducted from close to the footlights, so as to maintain dramatic control, using a roll of paper held by the middle. As a critic he was influential, and his essays and reviews are valuable for their insight into the problems of Romantic opera and for their accounts of contemporary works.

WORKLIST: *Das Waldmädchen* (The Forest Girl) (Steinsberg; Freiberg 1800, fragments extant); *Peter Schmoll und seine Nachbarn* (Peter Schmoll and his Neighbours) (Türk, after Cramer; Augsburg 1803); *Rübezahl* (Rhode, after a folk legend; fragments, unprod.); *Silvana* (Hiemer, after text for *Das Waldmädchen*; Frankfurt 1810); *Abu Hassan* (Hiemer, after the *1,001 Nights*; Munich 1811); *Der Freischütz* (Kind, after Apel & Laun, after the folk legend; Berlin 1821); *Euryanthe* (Chezy, after a French romance; Vienna 1823); *Oberon* (Planché, after Wieland; London 1826). Also *Die drei Pintos* (The Three Pintos) (Hell, after Seidel; rev. C. Weber & Mahler, music completed by Mahler also using Weber's works; Leipzig 1888).

WRITINGS: G. Kaiser, ed., *Sämtliche Schriften von Carl Maria von Weber* (Berlin, 1908); J. Warrack, ed., trans. M. Cooper, *Carl Maria von Weber: Writings on Music* (Cambridge, 1981).

CATALOGUE: F. Jähns, *Carl Maria von Weber in seinen Werken* (Berlin, 1871).

BIBL: M. von Weber, *Carl Maria von Weber: ein Lebensbild* (Leipzig, 1864–6); J. Warrack, *Carl Maria von Weber* (London, 1968; Cambridge, 2/ 1976).

**Weber, Ludwig** (*b* Vienna, 29 July 1899; *d* Vienna, 9 Dec. 1974). Austrian bass. Studied Vienna with Borittau. Début Vienna, V, 1920 (Fiorello). Cologne 1930–3; Munich 1933–45, and subsequently; Milan, S, 1938–9, 1942, 1948, 1950; London, CG, 1936–9, 1950; Bayreuth

from 1951. Also Paris, O; Buenos Aires, C; etc. A tireless and outstanding Wagner singer with a rich, voluminous tone, impressive as King Mark, Fasolt, Hunding, Hagen, Gurnemanz; also as Osmin, Sarastro, Commendatore, Boris, Barak, Wozzeck. (R)

**Webster,** (Sir) **David** (*b* Dundee, 3 July 1903; *d* London, 11 May 1971). English manager. General administrator to the newly formed Covent Garden Opera Trust 1945–70, successfully establishing the Royal OH as the permanent home of the Sadler's Wells (later Royal) Ballet and the Covent Garden (later Royal) Opera. Helped to launch singers including Sutherland, Vickers, and Geraint Evans on their careers.

BIBL: M. Haltrecht, *The Reluctant Showman* (London, 1975).

**Weidt, Lucie** (*b* Troppau (Opava), 1876; *d* Vienna, 28 July 1940). Austrian soprano. Studied with her father, and Vienna with Papier. Début Leipzig 1900. Leipzig until 1902; Vienna 1902–26; New York, M, 1910–11; Buenos Aires 1912; also Paris, O, and London, CG. A celebrated Leonore, Brünnhilde, and Kundry, she was also a successful Marschallin. Created Nurse (*Die Frau ohne Schatten*). (R)

**Weigl, Joseph** (*b* Eisenstadt, 28 Mar. 1766; *d* Vienna, 3 Feb. 1846). Austrian composer and conductor. Studied with Albrechtsberger. His first dramatic work, the marionette opera *Die unnütze Vorsicht oder Die betrogene Arglist* (1783) so impressed Salieri that he introduced Weigl into the theatrical life of the court. Here in 1786 he aided in the preparation of *Le nozze di Figaro*, taking over from Mozart as conductor after the first two performances. Vizekapellmeister to the Viennese, 1827–38.

Weigl composed over 30 Italian and German operas in genres ranging from heroic opera (*Daniel in der Löwengrube*, 1820) to Italian *melodramma eroi-comico* (*Margeritta d'Anjou*, 1816). Some of his Italian works were composed especially for La Scala, Milan: *Cleopatra* and *Il rivale di se stesso* (1807–8), and *L'imbroscata* (1815); others were performed in Vienna. Weigl first achieved international recognition with *L'amor marinaro* (1797), from which Beethoven borrowed the theme of his Clarinet Trio Op. 11. In 1805 he set Schikaneder's text *Vestas Feuer*, which Beethoven had also begun, although not completed, in 1803. Widespread success came with his Singspiels *Das Waisenhaus*, and *Die Schweizerfamilie* (after *The Swiss Family Robinson*); the latter was performed throughout Europe well into the 19th cent. Weber regarded *Das Waisenhaus* as having initiated a vogue for 'grief and pain' operas—a short-lived genre that came to an end with Weigl's *Der Bergsturz* (1812). After

*Die eisene Pforte* he abandoned opera for sacred music. Zelter commented to Goethe that Weigl 'is most successful with modest situations'.

Weber considered his manner of writing 'typical of the Viennese school, founded on the solid qualities and strict attention to detail learnt from the examples of Haydn and Mozart'. Although *Die Schweizerfamilie* is occasionally revived, Weigl's significance today rests on his importance as a figure in pre-Romantic German opera.

SELECT WORKLIST (all prod. Vienna): *Vestas Feuer* (Schikaneder; 1805); *Das Waisenhaus* (The Orphanage) (Treitschke; 1808); *Die Schweizerfamilie* (Costelli; 1809); *Der Bergsturz* (The Avalanche) (Reils, after a real incident; 1813); *Die eisene Pforte* (The Iron Gate) (E.T.A. Hoffmann; 1823).

**Weill, Kurt** (*b* Dessau, 2 Mar. 1900; *d* New York, 3 Apr. 1950). German, later US, composer. Studied Dessau with Bing, Berlin with Krasselt and Humperdinck. Worked as Korrepetitor in Dessau 1919 under Bing and Knappertsbusch; director of opera, Lüdenscheid, 1920. Studied further in Berlin with Busoni and Jarnach. Through Fritz Busch he was introduced to the librettist Georg Kaiser with whom he wrote *Der Protagonist*. This was Weill's first major success and his next collaboration with Kaiser, *Der Zar lässt sich photographieren*, was also well received.

In 1927 Weill was invited by Hindemith to compose a work for his Deutsche Kammermusik Baden-Baden. For the Songspiel *Mahagonny* he collaborated for the first time with Brecht. This was a particularly inspired union, though the production of *Mahagonny*, with its overt political message, proved very controversial. The highly original approach which Weill had taken in *Mahagonny*, where elements of popular music from both modern and older traditions were used together to soften the more astringent avant-gardism of his earlier works, was taken further in his next collaboration with Brecht, *Die Dreigroschenoper*. Here their interest in artistic topicality was further signified (and largely misunderstood) in a transposition of *Gay's The Beggar's Opera* of 1728 to the Berlin of two centuries later. The international success of this work paralleled that of Krenek's *Jonny spielt auf*, given in Leipzig the previous year, with which it shares certain similarities both of dramatic substance and of musical style. In Weill as in Krenek, a strong political message is illuminated by the skilful exploitation of popular musical idioms: together these works offer a chilling insight into the *Zeitgeist* of pre-Nazi Germany. Together with Brecht, many of whose artistic aims he shared, Weill widened his range and effect in the moral and political *Zeitopern Aufstieg und Fall der Stadt Mahagonny* and *Der Jasager*. In the works which followed, notably *Die Bürgschaft* and the ballet *Die sieben Todesünden*, we find the fullest expression of his style: popular and up-to-the-minute, but uncompromising, basically Romantic, pointing with brilliant and wounding musical symbols the message of the text, direct in appeal but bearing a deeply felt political message.

Forced to leave Hitler's Germany in 1933, Weill spent some time in Paris and London, before moving to the USA in 1935, largely at the instigation of Max Reinhardt. Turning his back on his earlier career, he produced a number of musical comedies and light operas, including *Street Scene* and *Down in the Valley*. These enjoyed considerable success and, though cast in an unashamedly popular style, reveal skill and craftsmanship equal to that of his earlier works. In 1928 he married the actress **Lotte Lenya** (1898–1981).

SELECT WORKLIST: *Der Protagonist* (Kaiser; Dresden 1926); *Mahagonny-Gesänge* (Brecht; Baden-Baden 1927); *Der Zar lässt sich photographieren* (The Tsar Has Himself Photographed) (Kaiser; Leipzig 1928); *Die Dreigroschenoper* (Brecht & Hauptmann, after Gay; Berlin 1928); *Aufstieg und Fall der Stadt Mahagonny* (Rise and Fall of the City of Mahagonny) (Brecht; Leipzig 1930); *Der Jasager* (The Yea-Sayer) (Brecht; Berlin 1930); *Die Bürgschaft* (The Surety) (Neher; Berlin 1932); *Street Scene* (Rice & Hughes; New York 1947); *Down in the Valley* (Sundgaard; Bloomington 1948).

BIBL D. Drew, *Kurt Weill: a Handbook* (London, 1987).

**Weimar.** Town in Thuringia, Germany. Opera developed with the other arts under the Duchess Anna Amalia 1756–75. She encouraged Singspiel by establishing Abel Seyler in the castle theatre 1772–4, and bringing poets and composers together: Hiller's *Die Jagd* to Weisse's text was produced 1770, Schweitzer's *Alceste* to Wieland's text 1773; and she herself wrote music for *Erwin und Elmire* (1776) and *Das Jahrmarktsfest zu Plundersweilen* (1778) to texts by Goethe, under whose direction the theatre widened its range. A number of his Singspiels were performed. Hummel was music director 1818–37, extending the repertory to include works by Rossini, Auber, Meyerbeer, Spohr, and Bellini, but emphasizing German opera. The Hoftheater was built under Goethe's supervision 1825. Under Liszt, 1847–58, premières were given of *Lohengrin* (1850), Schubert's *Alfonso und Estrella* (1854), and Cornelius's *Barber of Baghdad* (1858), and productions of Berlioz's *Benvenuto Cellini* (1852 and 1856) and Schumann's *Genoveva* (1855). Liszt's advanced policies (and unorthodox private life) precipitated his departure, and he was succeeded by Eduard Lassen, under whom were given the second *Tristan* (1874), *Samson et Dalila*

(1877) and the first German *Mignon* (1868). *Hänsel und Gretel* was premièred 1893, also the first German *Werther*, under Richard Strauss, music director 1889–94. The Hoftheater was rebuilt as the Deutsches Nationaltheater (cap. 2,000) in 1907.

**Weinberger, Jaromír** (*b* Prague, 8 Jan. 1896; *d* St Petersburg, FL, 8 Aug. 1967). Czech composer. Studied Prague with Křička and Karel, then with Novák, and briefly (but significantly) Leipzig with Reger. *Shvanda the Bagpiper* was Weinberger's first performed opera, and had an immediate and widespread success. It followed in the Smetana tradition in its use of a folk-tale and of Czech dances (the Polka and Fugue survive as a popular concert item); Weinberger has also taken note of some German influences in his harmony, and writes well for the voice. There is nevertheless a somewhat manufactured note in the attempt to write a 19th-cent. nationalist opera in the 1920s, and this has meant that modern revivals are few. His only other important opera was *Waldštejn*.

SELECT WORKLIST: *Švanda dudák* (Shvanda the Bagpiper) (Kareš and Brod, after a folk-tale; Prague 1927); *Waldštejn* (Kareš, after Schiller; Vienna 1937).

**Weingartner, Felix** (*b* Zara, 2 June 1863; *d* Winterthur, 7 May 1942). Austrian conductor and composer. Studied Graz with Rémy, Leipzig, Weimar with Liszt. His *Sakuntala* was prod. Weimar 1884. Cond. Königsberg 1884–5; Danzig 1885–7; Hamburg 1887–9; Mannheim 1889–91. Chief cond. Berlin 1891–8. Vienna, H, 1908–11, succeeding Mahler. Hamburg 1912–14. Chief cond. Darmstadt 1915–19; Vienna, V, 1919–24, and S, 1935–6. Boston 1912–13. London, CG, 1939 (*Tannhäuser*, *Parsifal*). Wrote 11 other operas, including the quite successful *Genesius* (Berlin 1892) and an Aeschylean trilogy *Orestes* (Leipzig 1902), and edited versions of *Oberon*, *Fliegender Holländer*, and Méhul's *Joseph*. A conductor with a particular quality of lucidity and balance who found operatic conditions often at war with his ideals. Of his five wives, the third was the singer Lucille Marcel, the fifth his conducting pupil Carmen Studer. (R)

WRITINGS: *Weingartner on Music and Conducting* (New York, 1969).

**Weir, Judith** (*b* Cambridge, 11 May 1954). Scottish composer. Studied with Tavener, Cambridge with Holloway, Tanglewood with Schuller. *A Night at the Chinese Opera* was commissioned by the BBC for Kent O, and made an immediate impression for its theatrical originality and intelligent musical characterization. The work frames an old Chinese play, set with a stylization of oriental music, in a modern narrative using closed Western operatic forms, with a

beguiling inventive contact between the two dramatic layers and musical idioms. Weir's sharp, sometimes bleak wit and liking for interweaving different narrative levels were applied again to *The Vanishing Bridegroom*, deriving from three Scottish folk-tales; and the elegant vocal writing, subtle orchestration, and sure feeling for dramatic timing confirmed her as a major operatic talent.

WORKLIST: *The Black Spider*; *The Consolations of Scholarship*; *A Night at the Chinese Opera* (Weir, after an old Chinese play; Cheltenham 1987); *The Vanishing Bridegroom* (Weir, after Scottish folk-tales; Glasgow 1990).

**Weisgall, Hugo** (*b* Ivančice, 13 Oct. 1912). Czech, later US, composer and conductor. Studied with his father Adolf, an opera singer and synagogue cantor, then, after emigration, Baltimore with Sessions. His multifarious activities, diplomatic as well as musical, have included teaching and administration (founder Hilltop OC 1952), and much conducting and composition. His style is wide-ranging and eclectic, with various techniques and styles—neoclassical, serial, and pastiche popular music among them—being skilfully adapted to the creative task in hand.

SELECT WORKLIST: *The Tenor* (Shapiro & Lert, after Wedekind; Baltimore 1952); *The Stronger* (Hart, after Strindberg; Westport 1952); *Six Characters in Search of an Author* (Johnston, after Pirandello; New York, 1959); *Purgatory* (Yeats play; Washington, DC, 1961); *Athaliah* (Goldman, after Racine; New York 1964); *Nine Rivers from Jordan* (Johnston; New York 1968); *Jennie* (Hollander, after Mishima; New York 1976).

**Weisse, Christian Felix** (*b* Annaberg, 28 Jan. 1726; *d* Leipzig, 16 Dec. 1804). German librettist, dramatist, and poet. Active as a dramatist in Leipzig, where he was also a tax-collector. He translated Coffey's ballad opera *The Devil to Pay* when it was first performed with newly composed music in Germany. Set first by Standfuss (1752) and then by Hiller (1766), it served as the main model for German national Singspiel. Weisse subsequently wrote the texts for many of Hiller's Singspiels. These were mostly adaptations of French librettos (*Lottchen am Hofe*, 1767, after Favart and *Die Jagd*, 1770, after Collé) and frequently took as their subject the difference between town and country life, as treated by Rousseau in *Le \*devin du village* (1752). Gradually Weisse turned away from operatic work as Leipzig's importance in this field declined, preferring instead to concentrate on drama and poetry, for which he enjoyed a high reputation.

**Weissenfels**. Town in Germany, south of Halle, the site of the court of the Dukes of Saxe-Weissenfels. An opera theatre was opened 1685 by

Duke Johann Adolph I (1680–97); its most important period was 1680–1725 when it was directed by Krieger in a repertory including Strungk and Telemann. The importance of the court declined in the mid-18th cent.

**'Welche Wonne, welche Lust'.** Blonde's (sop) aria in Act II of Mozart's *Die Entführung aus dem Serail*, in which she delightedly receives the news of Belmonte's plans for escape.

**Weldon, John** (*b* Chichester, 19 Jan. 1676; *d* London, 7 May 1736). English composer and organist. A pupil of Purcell, he became a Gentleman of the Chapel Royal in 1701 and Organist in 1708. As a composer of dramatic music he is best remembered as winner of the first prize of the contest to set Congreve's masque *The Judgement of Paris* in 1700. A performance of *The Tempest* in July 1716 was advertised as having 'all the musick compos'd by Mr Weldon and perform'd compleat, as at the revival of the play': this statement, together with the evidence of the music itself, casts serious doubts on the authenticity of Purcell's music for the play.

**Welitsch, Ljuba** (*b* Borissovo, 10 July 1913). Bulgarian, later Austrian, soprano. Studied Sofia with Zlatev-Cherkin, Vienna with Lierhammer. Début Sofia 1934 (in *Louise*). Graz, Hamburg, Munich 1937–46; Vienna, S, 1947–64; London, CG, 1947–53; New York, M, 1948–52, 1972; Gly. 1948. Also Chicago; Milan, S; etc. A fullblooded and vital performer, with a large, bright, pure-toned voice. At her peak (till 1953) an astonishing Salome; also impressive as Donna Anna, Aida, Verdi's Desdemona, Tosca, Musetta, Lisa, Jenůfa. (R)

**Wellesz, Egon** (*b* Vienna, 21 Oct. 1885; *d* Oxford, 9 Nov. 1974). Austrian composer and musicologist. Studied musicology with Guido Adler, composition with Schoenberg. His early work as a musicologist brought him into contact with Venetian opera of the mid-17th cent. As well as being one of the first to write about this repertory, he took inspiration from the scores on which he worked and adopted some of their techniques in his own operas. Having already worked with him on the ballet *Achilles auf Skyros*, Wellesz collaborated with Hofmannsthal for his opera *Alkestis*. Later he sought inspiration in Goethe, in *Scherz, List und Rache*, and again in antiquity for *Die Bakchantinnen*. Forced to leave Austria in 1938, he settled in Oxford, where he became Reader in Byzantine Music in 1951. Here he wrote his last opera, *Incognita*, which was staged by the OUOC. Technically a highly accomplished composer, Wellesz brought to his work both intellectual clarity and literary sensitivity. Interested in expressing not his own des-

tiny but 'those things which betoken the link between the material and the spiritual world', he was drawn to myths, presenting them in modern interpretations that would stress their continuing validity. His highly personal musical style, which drew on the techniques of Baroque opera, the post-Romantic vocabulary of Schoenberg and his followers, and the textural fluidity of medieval and Byzantine music, found few followers; his neglected stage works are deserving candidates for modern revival.

SELECT WORKLIST: *Alkestis* (Hofmannsthal; Mannheim 1924); *Scherz, List und Rache* (Goethe; Stuttgart 1928); *Die Bakchantinnen* (Wellesz, after Euripides; Vienna 1931); *Incognita* (Mackenzie, after Congreve; Oxford 1950).

BIBL: C. Benser, *Egon Wellesz* (New York, 1985).

**Welsh National Opera.** Though there was an abortive Welsh National Opera Co. in 1890 and a Cardiff Grand Opera Society 1924–34, Wales was dependent on touring companies until after the war. The modern WNO was launched in Cardiff on 15 Apr. 1946 with *Cavalleria rusticana*. By 1956 the company was giving regular seasons in Wales and southern England with a repertory of 13 works, including Arwel Hughes's *Menna* (sung in Welsh at the National Eisteddfod); by 1966 this had increased to 34, by 1976 to over 70. In the early days exploration of the then lessknown early Verdi was a feature, and the fine amateur choral singing in *Nabucco* (1952), together with its nationalist fervour, identified the work closely with the company. Other successes included *Mefistofele* (1957), a moderndress *Battle of Legnano* (1960), *William Tell* (1961), *Fidelio* with Gwyneth Jones (1964), and *Moses* (1965). The regular visits to Sadler's Wells became very popular. Music directors included Idloes Owen (1946–52), Vilem Tausky (1955–6), Warwick Braithwaite (1956–61), Charles Groves (1961–3), and Bryan Balkwill (1963–8).

A more ambitious policy developed with the appointment of James Lockhart in 1968, with Michael Geliot in 1969 as director of productions. New productions included *Simon Boccanegra*, *Falstaff*, *Idomeneo*, *Don Carlos*, and the first British staging of *Lulu* (1971). Richard Armstrong became music director in 1973, and the repertory was further extended to include a Janáček series (with Scottish Opera) and a Britten series, Verdi's *Otello*, *Elektra*, and premières including Alun Hoddinott's *Beach at Falesá* (1974). The production of *The Midsummer Marriage* (1976) contributed to wider recognition of the work's stature. Brian McMaster, director from 1976, brought in foreign directors for innovative productions including Harry Kupfer for *Elektra* (1978), Lucian Pintilie for *Carmen* (1983) and *Rigoletto* (1985), Ruth Berghaus for *Don Giovanni* (1984), Andrei Serban for *Norma* (1985), Peter Stein for *Otello* (1986), and André Engel for *Der Freischütz* (1989). Distinguished Wagner productions have included *Tristan* (1979) and *Parsifal* (1983) under Reginald Goodall. *The Ring* was built up 1983–5, and *The Trojans* staged in 1987. Charles Mackerras became music director in 1986. The company's standing and popularity in Britain were shown in 1990 when the Arts Council was forced by public outcry to abandon its threat of closure. Welsh-born artists appearing with the company have included Anne Howells, Gwyneth Jones, Margaret Price, Elizabeth Vaughan, Eiddwen Harrhy, Geraint Evans, Delme Bryn-Jones, and Ryland Davies. The company plays in the New Theatre, Cardiff (cap. 1,140) and tours to Swansea and a number of smaller theatres which opened in Wales during the 1970s (e.g. the Theatr y Werin in Aberystwyth).

BIBL: R. Fawkes, *Welsh National Opera* (London, 1986).

**Werle, Lars Johan** (*b* Gävle, 23 June 1926). Swedish composer. Studied Uppsala with Moberg and Bäck. *Drömmen om Thérèse* (Dream about Thérèse, 1964) aroused much attention for its novel techniques of audience involvement and pre-recorded music, some of which were repeated in *Resan* (The Journey, 1969). *Tintomara* (1973), written for the bicentenary of the Royal OH in Stockholm, returns to the events at the time of the murder of Gustav III. Later, sometimes more conventional, operas include *Animalen* (The Animals, 1981) and *Leonardo* (1988).

**Werther.** Opera in 4 acts by Massenet; text by Blau, Milliet, and Hartmann, after Goethe's novel *Die Leiden des jungen Werthers* (1774). Prod. Vienna, O, 16 Feb. 1892, with Renard, Forster, Van Dyck, Neidl, cond. Jahn; Chicago, Auditorium, 29 Mar. 1894, by New York M Co, with Eames, Arnoldson, J. De Reszke, Martapoura, cond. Mancinelli; London, CG, 11 June 1894, with Eames, Arnoldson, J. De Reszke, Albers, cond. Mancinelli.

Wetzlar, July–Dec. 1772. Werther (ten) loves Charlotte (mez), who returns his love although betrothed to Albert (bar). Werther leaves, returning to find Charlotte married. She urges him to leave her again, but on finding that he has asked Albert for his pistols, rushes to him through a snowstorm to find that he has shot himself.

Other operas on the subject are by Kreutzer (1792), Puccita (1802), Schuster (1806), Benvenuti (1811), Müller (parody, 1830), Aspa (1849),

Conrady (1862), Gentili (1862), Randegger (1899), and Derozi (1906).

**Westrup,** (Sir) **Jack** (*b* London, 26 July 1904; *d* Headley, 21 Apr. 1975). English musicologist. One of the founder members of the Oxford University Opera Club, for which he prepared editions of *Orfeo* 1925, and *Poppea* 1927. As Professor of Music 1947–71, conducted the club's productions. His writings include translations for these productions, and valuable studies of Handel and Purcell.

**West Side Story.** Musical by Bernstein; text by Stephen Sondheim, after Arthur Laurents's book, based on Shakespeare's tragedy *Romeo and Juliet* (1594–5). Prod. Washington, National T, 19 Aug. 1957, with C. Lawrence, L. Kert; London, HM, 12 Dec. 1958; film 1961, with Natalie Wood and Richard Beymer (singing voices dubbed by Marni Nixon, Jim Bryant).

New York, 1950s. Tony and Maria, each belonging to rival gangs, the white Jets and the Puerto Rican Sharks, fall in love; Tony, trying to be a peacemaker, inadvertently kills Maria's brother, and then is himself killed.

**Wexford.** Town in Co. Wexford, Eire. An autumn festival was founded by Dr T. J. Walsh in 1951, when Balfe's *The Rose of Castile* was staged. Opera is given in the T Royal, and the repertory has been as notable for its adventurousness in reviving neglected works as the town has been for the conviviality of its welcome. The festival also has a reputation for giving an early chance to singers, including Mirella Freni, Ugo Benelli, and Christiane Éda-Pierre. The T Royal (opened 1832) was restored 1987 (cap. 550).

**Weyse, Christoph** (*b* Altona, 5 Mar. 1774; *d* Copenhagen, 8 Oct. 1842). Danish composer of German birth. After studying with J. A. P. Schultz in Copenhagen, he stayed in the city and became organist of the cathedral there. In 1819 he was appointed court composer and as such was required to compose either a Singspiel or cantata each year. Known mainly as a composer of sacred music and songs, Weyse wrote five Singspiels and a light opera *Et eventyr i Rosenborg Have* (An Adventure in Rosenborg Gardens, 1827). The only one of his Singspiels to gain an established place in the repertoire was *Sovedrikken* (The Sleeping Draught, 1809), which he began in 1800 but abandoned until a performance of *Don Giovanni* in 1807 inspired him to finish the work. His Singspiels combine examples of his individual fusion of simple, naïve song style with more sophisticated treatments of the text: his Romance settings are particularly attractive. Apart from *Sovedrikken*, his other Singspiels are *Faruk* (1812), *Ludlams hule* (1816), *Floribella* (1825), and *Festen på Kenilworth* (text by Hans Andersen, after Scott, 1836).

**'When I am laid in earth'.** Dido's (sop) lament before her death at the end of Purcell's *Dido and Aeneas*.

**'Where'er you walk'.** Jupiter's (ten) aria in Act II of Handel's *Semele*, promising Semele shade over her head and springing flowers at her feet.

**White, Willard** (*b* Jamaica, 10 Oct. 1946). Jamaican bass-baritone. Studied New York with Johnson, Tozzi, and Thorendahl. Début New York City Center 1974 (Colline). Successes in San Francisco, Washington, Tanglewood; London, ENO, from 1976; Gly. from 1978. (R)

**Whitehill, Clarence** (*b* Parnell, IA, 5 Nov. 1871; *d* New York, 18 Dec. 1932). US bass-baritone. Studied Paris with Sbriglia and Giraudet, later Frankfurt with Stockhausen. Début Brussels, M, 1898 (Donner). Brussels, M, 1899; Paris, OC, 1900, O, 1909. Lübeck 1901; Bayreuth 1904, 1908–9, studying with Cosima Wagner; London, CG, 1905–10, New York, M, 1909–10, 1914–32; Chicago 1911–16. Created Delius's Koanga. A singer of great distinction, particularly acclaimed as Sachs, Wotan, Gunther, Amfortas; also as Gounod's Méphistophélès, Capulet, Golaud. (R)

**Widdop, Walter** (*b* Norland, 19 Apr. 1892; *d* London, 6 Sep. 1949). English tenor. Studied London with Gilly. Début BNOC 1923 (Radamès). BNOC until 1929. London, CG, 1924, 1932–3, 1937–8; Barcelona 1928; also Holland, Germany. The leading English Wagner singer of his day, successful as Siegmund, Siegfried, Tristan; also sang Max, Aegisthus, Canio. (R)

**Widor, Charles-Marie** (*b* Lyons, 21 Feb. 1844; *d* Paris, 12 Mar. 1937). French composer and organist. Studied Brussels with Fétis and Lemmens. Best known for his organ music, Widor also wrote a number of stage works, including three operas. The most successful was *Les pêcheurs de Saint-Jean* (Cain; Paris 1905).

**Wieland, Christoph Martin** (*b* Oberholzheim, 5 Sep. 1733; *d* Weimar, 20 Jan. 1813). German librettist, dramatist, writer, and poet. One of the most prolific and influential figures of the rococo movement, Wieland played an important role in the development of German opera, not least through the critical opinions expressed in his journal *Teutscher Merkur*, where Gluck's operas found vociferous support. Librettist of the first original 5-act German opera, *Alceste* (Schweitzer, 1773), for which his model was Metastasio: its première was a watershed in German operatic development, the beginning of a gradual move away from predominantly comic plots to more serious drama.

**Wiesbaden.** City in Hesse, Germany. Opera was first given in 1765. In 1810 the theatre was named the Herzoglich-Nassauisches-Hoftheater; regular performances were given by a company from Mainz until the mid-19th cent. The theatre became the Königliches Hoftheater 1866. The Grosses Haus (cap. 1,325) opened Oct. 1894. Klemperer was music director 1924–7 with Carl Hagemann as Intendant; together they pursued a progressive policy, continued by Paul Bekker and Joseph Rosenstock until 1932, staging premières of works by Krenek, Milhaud, and others. Karl Elmendorff was music director 1932–5. The theatre reopened as the Hessisches Staatstheater (cap. 1,041, small theatre 328) in Sep. 1947. Wolfgang Sawallisch was music director 1957–9. The annual May Festival dates from 1896, and became well established under Bekker in the 1920s. In the 1960s it became a show-case for companies from Eastern Europe.

**Wildschütz, Der** (The Poacher). Opera in 3 acts by Lortzing; text by composer, after Kotzebue's comedy *Der Rehbock*. Prod. Leipzig, 31 Dec. 1842, with Günther, Lanz, Krüger, Düringer, Kindermann, Schmidt, Berthold, Wallmann; New York, Stadt T, 25 Mar. 1859, with Hübner, Heyde, Siedenberg, Lehmann, Graf, Fortner, Quint; London, DL, 3 July 1895.

A German village and castle, early 19th cent. Baculus (buffo bs), a schoolmaster on the estate of the Count of Eberbach (bar), has accidentally shot a buck. His betrothed, Gretchen (sop), is willing to intercede for him, but he is reluctant, and a young student offers to go instead, dressed as Gretchen. This is really Baroness Freimann (sop), the Count's sister, who accompanied by her maid Nanette (sop) sets off to keep an eye on her own betrothed, the Count's brother-in-law, Baron Kronthal (ten). The Count is much taken with 'Gretchen'; the Baron bribes Baculus with 5,000 thalers to give her up; meanwhile, 'Gretchen' is rescued from the situation by being allowed to spend the night in the room of the Countess (con) (whose unremitting Greek quotations are a satire on the recent success of Mendelssohn's *Antigone* music and the ensuing mania in Berlin for all things Greek). The unravelling of all these confusions still leaves Baculus guilty of poaching and of having sold his betrothed; but he is forgiven by all.

**Willan, Healey** (*b* Balham, 12 Oct. 1880; *d* Toronto, 16 Feb. 1968). English composer. Studied London with Hoyte. In 1913 emigrated to Canada, where he played an important role in the country's music. In addition to the church and organ music for which he was most famous, Willan also wrote incidental music for about twenty plays, arranged five ballad operas from folk material (including *The Beggar's Opera*), and composed two operas. *Transit through Fire* was the first opera to be commissioned by the Canadian Broadcasting Commission. A second radio commission followed for *Deirdre*: this, with its ancient story and richly Romantic music, is somewhat Wagnerian in manner, and had a considerable success.

SELECT WORKLIST: *Transit through Fire* (Coulter; Canadian radio 1942); *Deirdre* (Coulter; Canadian radio 1946, Toronto 1965).

**William Ratcliff.** Opera in 3 acts by Cui; text by A. Pleshcheyev, after Heine's drama (1823). Prod. St Petersburg, M, 26 Feb. 1869. Cui's most famous opera.

North of Scotland, 17th cent. Maria (sop) marries Douglas (ten), who is then told by her father Mac-Gregor (bs) how William Ratcliff (bar), whom Maria rejected, murdered her two subsequent fiancés. Douglas fights and defeats Ratcliff, but spares his life; Ratcliff appears in Maria's room, and she reveals her love for him. He kills her, her father, and himself.

Other operas on the subject are by Pizzi (1889), Bavrinecz (1895), Mascagni (1895), Leroux (1906), Dopper (1909), and Andreae (1914).

**Williams, Grace** (*b* Barry, 19 Feb. 1906; *d* Barry, 10 Feb. 1977). Welsh composer. Studied Cardiff, and London with Vaughan Williams and Jacob, Vienna with Wellesz. Her only opera, *The Parlour* (Williams, after Maupassant; 1961), is a comedy showing some influence of Britten.

BIBL: M. Boyd, *Grace Williams* (Cardiff, 1980).

**Williamson, Malcolm** (*b* Sydney, 21 Nov. 1931). Australian composer. Studied Sydney with Eugene Goossens III and London with Lutyens. Settled in London in 1953 where he established a successful career first as a pianist, later as a composer; appointed Master of the Queen's Music, 1975. His first important work for the stage was the opera *Our Man in Havana*, whose success was followed by *The Violins of Saint-Jacques*. Possessing a fluent, eclectic, and highly professional gift, he has shown a particular skill for writing shorter dramatic pieces, of which the children's operas *Julius Caesar Jones* and *The Happy Prince* are thoroughly representative. His later full-length operas, including *Lucky-Peter's Journey*, were coolly received and in recent years he has concentrated on smaller-scale works.

SELECT WORKLIST (all first prod. London): *Our Man in Havana* (Gilliat, after Greene; 1963); *The Violins of Saint-Jacques* (Chappell, after Leigh Fermor; 1966); *Julius Caesar Jones* (Williamson; 1966); *The Happy*

*Prince* (Williamson, after Wilde; 1965); *Lucky-Peter's Journey* (Tracey, after Strindberg; 1969).

**William Tell.** See GUILLAUME TELL.

**Willow Song.** (1) Desdemona's (sop) song 'Salce, salce' in Act IV of Verdi's *Otello*.
(2) Desdemona's (mez) song 'Assisa al piè d'un salice' in Act III of Rossini's *Otello*.

**Wilno.** See POLAND.

**Wilt, Marie** (*b* Vienna, 30 Jan. 1833; *d* Vienna, 24 Sep. 1891). Austrian soprano. Studied Vienna with Artôt and Gänsbacher. Début Graz 1865 (Donna Anna). London, CG, 1866–7, 1874–5; Vienna, H, 1867–77, 1886; Leipzig 1878–9. Though ungainly in appearance, she possessed 'the most beautiful voice of our time' (Artôt), and with her superb technique could sing Norma, Lucrezia Borgia, Valentine, Elsa, and Brünnhilde. Created Sulamith (Goldmark's *Königin von Saba*).

**Windgassen, Wolfgang** (*b* Annemasse, 26 May 1914; *d* Stuttgart, 8 Sep. 1974). German tenor. Studied Stuttgart with his father, the tenor **Fritz Windgassen** (1883–1963), Ranzow, and Fischer. Début Pforzheim 1941 (Alvaro). Stuttgart 1945–72; Bayreuth 1951–70; London, CG, 1955–66; New York, M, 1956–7; Paris, O, 1954–6, 1966. Also Milan, S, Vienna, etc. Roles included Tamino, Florestan, Duke of Mantua, Turiddu, Eisenstein; celebrated in Wagner, as Erik, Lohengrin, Tannhäuser, Tristan, Walther, Loge, Siegmund, Siegfried, Parsifal. With a lighter, more lyrical voice than the typical Heldentenor, he used his intelligence and technique to sing with consistent intensity and beauty of tone. He was particularly in sympathy with Wieland Wagner's productions. (R)

**Winkelmann, Hermann** (*b* Brunswick, 8 Mar. 1849; *d* Vienna, 18 Jan. 1912). German tenor. Studied Hanover with Koch. Début Sondershausen 1875 (Manrico). Hamburg 1878; London, DL, 1882; Bayreuth 1882–91; Vienna 1883–1906. The first Parsifal, praised by Wagner (and described as 'marvellous' by Neumann); also an acclaimed Tristan, Walther, Loge, Siegfried, Florestan, Verdi's Otello, Dalibor. He possessed a large, ringing tone and a lyrical style, though Wagner would have preferred a 'smoother mezza voce' in his Parsifal. (R)

**Winter, Peter von** (*b* Mannheim, *bapt.* 28 Aug. 1754; *d* Munich, 17 Oct. 1825). German composer. Winter received no training in composition until a short period of study with Vogler beginning in 1775; he later disowned his teacher.

He became musical director at Mannheim, and moved with the court to Munich in 1778, where he was responsible for conducting opéras-comiques. It was there that he began to compose music for the stage, including ballets, and melodramas modelled on those of Benda. He had written three melodramas by the time he composed his first opera, *Helena und Paris*, in 1782; the work was not well received, and he had no more success with his second opera, *Bellerophon* (1785). His career as a court musician progressed, with appointment to the post of Vizekapellmeister in 1787, and Kapellmeister in 1798, but he found it hard to establish a reputation as a successful composer. In quest of success, he travelled away from Munich to various operatic centres, writing opere serie and opere buffe for Naples and Venice in 1791–4, and tragédie lyrique (*Castor et Pollux*) for Paris in 1806.

Like many German composers of his time, Winter's musical style was firmly based on French and Italian operatic conventions. His most individual writing is to be found in his Singspiels, and the nine works produced in Vienna 1793–8 numbered some of his best. Foremost among these was *Das unterbrochene Opferfest*, which was his first great success, performed throughout Germany. The work demonstrates the influence of French *rescue opera in its plot, but also contains several notably Romantic features. Also in Vienna, he set *Das Labyrinth*, Schikaneder's sequel to *Die Zauberflöte*, in which he borrows themes from Mozart's Singspiel, including the triple chord; he also attempts to emulate the tonal unity of *Die Zauberflöte*. He returned to Munich but once again met with disappointment at the poor receptions of *Elisa* (1798), *Der Sturm* (1798, after *The Tempest*), and *Maria von Montalban* (1800). His heroic opera, *Colmal*, which he considered his masterpiece, had such a poor reception that he did not stage another work in the city for 11 years (his final Singspiel, *Der Sänger und die Sängerin*).

Winter's output of over 40 stage works spanned 40 years, and encompassed most of the operatic genres of the day. Described by Weber as 'an eminent master', he was generally disliked by his contemporaries, who found him a difficult man. The largely unadventurous style of his music, with its square rhythms and basic diatonic harmony, is compensated for in part by a profusion of charming melodies. His reputation rests largely on his operas, although he also wrote symphonies, concertos, chamber music, and sacred music, and his significance today is in his position of an immediate forerunner of the great composers of German Romantic opera.

SELECT WORKLIST: *Helena and Paris* (Förg, ? after Calzabigi; Munich 1782); *Das unterbrochene Opferfest* (The Interrupted Sacrifice) (Huber; Vienna 1796); *Das Labyrinth* (Schikaneder; Vienna 1798); *Colmal* (Colin, after Ossian; Munich 1809).

**'Winterstürme'.** Siegmund's (ten) song in Act I of Wagner's *Die Walküre* describing the yielding of winter to spring.

**Wittich, Marie** (*b* Giessen, 27 May 1868; *d* Dresden, 4 Aug. 1931). German soprano. Studied Würzburg with Ubrich. Début Magdeburg 1882 (aged 14) (Azucena). Schwerin 1886–9; Dresden 1889–1914; Bayreuth 1901–10; London, CG, 1905–6. Roles included Elsa, Isolde, Kundry, and a Brünnhilde that overwhelmed E. M. Forster: 'Force, weight, majesty! She seemed to make history.' Created Salome, while protesting, 'I won't do it, I'm a decent woman.' Her comfortable figure led Strauss to call her 'Auntie Wittich'.

**Wixell, Ingvar** (*b* Luleå, 7 May 1931). Swedish baritone. Studied Stockholm with Gustafsson. Début Stockholm 1955 (Papageno). Swedish Royal O from 1956; London, CG, from 1960; Gly. 1962; Berlin, D, from 1962; Salzburg 1966–9; Bayreuth 1971; New York, M, from 1973. Roles incl. Mozart's Count Almaviva, Pizarro, Posa, Boccanegra, Onegin, Mandryka. Possesses a strong tone and impressive stage personality. (R)

**Wolf, Hugo** (*b* Windischgraz (now Slovenj Gradec), 13 Mar. 1860; *d* Vienna, 22 Feb. 1903). Austrian composer. Studied Vienna with Fuchs and Krenn. His early opera-going drew him strongly to Wagner, whose influence shows in some aspects of his song-writing and hence in his only completed opera, *Der Corregidor*. He overcame his initial resistance to the libretto, whose detail nevertheless leads him into ignoring the demands of dramatic structure (his own version of Leitmotiv is insufficiently functional) in favour of lyrical illustration: in the individual scenes and songs, Wolf's genius dominates. There were many difficulties about securing a performance (Mahler was among those who turned it down, even after its première), and the work has never established itself in opera-houses. Only 600 bars survive of *Manuel Venegas*.

WORKLIST: *Der Corregidor* (Mayreder, after Alarcón; Mannheim 1896); *Manuel Venegas* (Hoernes, after Alarcon; 5 scenes completed 1897, staged Mannheim 1903).

BIBL: F. Walker, *Hugo Wolf: A Biography* (London, 1951, enlarged 2/1968).

**Wolfenbüttel.** Town in Germany, near Brunswick, seat of the Dukes of Brunswick-Lüneberg. The Wolfenbüttel court's importance for opera dates from the accession of Duke Anton Ulrich (reigned 1685 as co-regent with his brother; 1704–14 as sole regent), a talented poet and man of the theatre. He recruited a succession of gifted opera composers to the post of Kapellmeister, including Theile (1685), Kusser (1690), Keiser (1694), and Schürmann (1707); in 1690 an opera-house was opened. Under Dukes August Wilhelm (1714–31) and Ludwig Rudolph (1731–5) the musical retinue was increased and opera productions mounted on an even more lavish scale. The opera-house was closed 1753 when the court moved from Wolfenbüttel.

**Wolff, Albert** (*b* Paris, 19 Jan. 1884; *d* Paris, 21 Feb. 1970). French conductor. Studied Paris with Leroux, Gédalge, and Vidal. Paris, OC: chorus-master 1908, cond. from 1911, music director 1921–4, director 1945–6. Cond. premières of Laparra's *La jota* (1911), Charpentier's *Julien* (1913), Milhaud's *La brebis égarée* (1923), Ibert's *Angélique* (1927), Bruneau's *Angelo* (1928), and Poulenc's *Les mamelles de Tirésias* (1947), among many other works. Buenos Aires and Rio de Janeiro, 1912, 1940–4. New York, M, 1919–21, where his *L'oiseau bleu* was prod. 1919. London, CG, 1937 (*Pelléas*). His other operas included *Sœur Béatrice* (Nice 1948). Paris, O, 1949. A sensitive and scrupulous exponent of French music, who did much to sustain tradition at the Opéra-Comique. (R) His son **Stéphane Wolff** (1904–80) was a critic and historian of the Opéra and Opéra-Comique.

**Wolf-Ferrari, Ermanno** (*b* Venice, 12 Jan. 1876; *d* Venice, 21 Jan. 1948). Italian composer of German-Italian parentage. His original intention to be a painter was changed by a visit to Bayreuth. Studied Munich with Rheinberger. The failure of his first performed opera, *Cenerentola*, led him to return to Munich, where it had a success; thereafter he lived mostly near Munich. He had further success with *Le donne curiose*, *I quattro rusteghi*, and other Goldoni-based works, in which his affection for the world of 18th-cent. Venetian buffo and his German-trained skill with chromatic harmony combine to excellent theatrical effect. It was in comedy that Wolf-Ferrari was most at home, and where his elegant style, light touch, and skilful use of the orchestra are at their most effective. He wrote sympathetically for the voice in these, and in the trifle *Il segreto di Susanna*. He made one successful incursion into verismo with *I gioielli della Madonna*, but his nearest attempt at music drama, *Sly*, found his talent misdirected and overextended. He returned to comedy, re-creating with fair success the manner of his earlier works, especially in the popular *Il campiello*.

SELECT WORKLIST: *Cenerentola* (Cinderella) (Pezzè-Pascolato, after Perrault; Venice 1900); *Le donne curiose* (The Curious Ladies) (Sugana, after Goldoni; Munich 1903); *I quattro rusteghi* (The Four Curmudgeons) (Sugana & Pizzolato, after Goldoni; Munich 1906); *Il segreto di Susanna* (Susanna's Secret) (Goli-sciani and Zangarini; Munich 1909); *I gioielli della Madonna* (The Jewels of the Madonna) (Golisciani & Zangarini; Berlin 1911); *L'amore medico* (Love the Physician) (Golisciani, after Molière; Dresden 1913); *Sly* (Forzano, after Shakespeare; Milan 1927); *La vedova scaltra* (The Crafty Widow) (Ghisalberti, after Goldoni; Rome 1931); *Il campiello* (The Little Square) (Ghisalberti, after Goldoni; Milan 1936); *La dama boba* (The Confused Lady) (Ghisalberti, after De Vega; Milan 1939).

**Wolfram von Eschenbach** (*fl c.*1170–1220). German Minnesinger. He is the author of *Parzival* (?*c.*1200), upon which Wagner drew for his *Parsifal*, and *Titurel*, for which he composed a melody. He appears (bar) in Wagner's *Tannhäuser*.

**Wolf's Glen Scene** (Ger., *Wolfsschlucht*). The scene ending Act II of Weber's *Der Freischütz* in which seven magic bullets are forged, to the accompaniment of ghostly apparitions.

**Wood, (Sir) Henry** (*b* London, 3 Mar. 1869; *d* Hitchin, 19 Aug. 1944). English conductor. Studied London with Prout and García. Comp. four light operas 1890–2. Cond. Arthur Rouseby Co. 1889; helped Sullivan prepare *Ivanhoe* 1890; London, CR, 1891–2; Burns-Crotty Co. 1892; London, Olympic T, Lago's 1892 season, cond. English première of *Onegin*. Music adviser for Mottl Wagner concerts 1894. Cond. Stanford's *Shamus O'Brien* 1896–7 (100 nights), after which devoted his life to concert conducting. Autobiography, *My Life of Music* (London, 1938, R/1971). (R)

BIBL: R. Pound, *Sir Henry Wood* (London, 1969).

**Woodbird.** The Woodbird (sop) appears in Act II of Wagner's *Siegfried*; though Siegfried is at first unable to understand its song, this becomes comprehensible once he has tasted the blood of the giant Fafner whom he has slain; it tells him of the treasures to be found in Fafner's cave.

**Wotan.** The ruler (bs-bar) of the gods in Wagner's *Der Ring des Nibelungen*. He appears in *Das Rheingold* and *Die Walküre*, and as the Wanderer in *Siegfried*.

**Wotan's Farewell.** The name generally given to the scene closing Wagner's *Die Walküre*, in which Wotan takes leave of the sleeping Brünnhilde whom he has laid on her fire-girt rock.

**Wozzeck.** Opera in 3 acts by Berg; text by composer, after Büchner's dramatic fragment *Woyzeck* (1835–7, pub. 1879). Prod. Berlin, S, 14 Dec. 1925, with Johanson, Koettrick, Henke, Soot, Leo Schützendorf, Abendroth, cond. E. Kleiber; Philadelphia, AM (Orch. OC), 19 Mar. 1931, with Roselle, Eustis, Korell, Leonoff, Ivantzoff, Steshenko, cond. Stokowski; London, CG, 22 Jan.

1952, with Goltz, Sinclair, Hannesson, P. Jones, Rothmüller, Dalberg, cond. E. Kleiber.

A small Austrian garrison town, *c*.1820. Wozzeck (bar), an ordinary soldier, is lectured by the Captain (ten), who despises him. Gathering sticks with his friend Andres (ten), Wozzeck is alarmed by strange sounds and visions. Wozzeck's mistress Marie (sop) flirts from her window with a passing Drum-Major (ten). Wozzeck is examined by a half-crazed Doctor (bs) who is using him for experiments. The Drum-Major seduces Marie.

Wozzeck becomes suspicious of Marie. The Captain taunts him with suggestions of her infidelity, and Wozzeck accuses her. He finds her dancing with the Drum-Major, who boasts of his success. When Wozzeck refuses to drink with him, the Drum-Major beats him.

The repentant Marie is reading the Bible. As they walk by a pond, the distraught Wozzeck stabs her. Later, drinking in a tavern, he is seen to have blood on his hands, and rushes away. Wading into the pond to find the knife, he drowns. Their child is playing with his hobbyhorse, and does not understand when the other children tell him that his mother is dead.

BIBL: G. Perle, *The Operas of Alban Berg: 'Wozzeck'* (Berkeley, CA, 1985); D. Jarman, *Alban Berg: 'Wozzeck'* (Cambridge, 1989).

**Wranitzky, Paul** (*b* Nová Říše, 30 Dec. 1756; *d* Vienna, 26 Sep. 1808). Austrian composer of Moravian origin. He was an esteemed figure in Vienna, holding various positions at court, and being favoured by Haydn and Beethoven as a conductor; Beethoven chose Wranitzky to conduct the première of his First Symphony in 1800. He was also acquainted with Mozart; both were members of the same Freemasons' lodge. Of his ten stage works, all of which were first performed in Vienna, only *Oberon, König der Elfen* (1789) was a great success; it was performed during the festivities for the coronation of Leopold II in 1790, and throughout Germany. Described as a 'romantisches Singspiel', the work has a libretto by Giesecke, based on Wieland's poem; it is the first important fairy opera. The success of *Oberon* in Vienna encouraged Schikaneder to write the libretto for *Die Zauberflöte*.

Following a letter from Wranitzky requesting a libretto, Goethe replied offering him his sequel to *Die Zauberflöte*. However, the terms of collaboration did not meet with the approval of Wranitzky, who foresaw the project bringing him into competition with Mozart.

His brother **Anton** (*b* 13 June 1761; *d* 6 Aug. 1820) was also a composer, and had two daughters who became singers—**Anna Katherina** (married name Kraus) (1801–51) and **Karo-**

line (married name Seidler) (1794–1872), who was the first Agathe.

SELECT WORKLIST: *Oberon, König der Elfen* (Seyler, after Wieland; Vienna 1789).

**Wreckers, The.** Opera in 3 acts by Ethel Smyth; text by 'H. B. Laforestier' (Harry Brewster). Prod. Leipzig 11 Nov. 1906, cond. Nikisch; London, HM, 22 Jan. 1909, with Sapio, Amsden, Seiter, Coates, James, Winckworth, Byndon-Ayres, cond. Beecham; Newport Fest. 3 Aug. 1972.

The scene is an 18th-cent. Cornish village that lives by wrecking. Thirza (mez), the young wife of the preacher and headman Pascoe (bs-bar), is in love with Mark (ten), who is in turn loved by the lighthouse-keeper's daughter Avis (sop). Thirza and Mark, revolted by the *modus vivendi* of the villagers, light a bonfire to warn ships to keep clear of the rocky coast, but Pascoe is discovered near it by the suspicious villagers; if he is found to be guilty, he must die as a traitor. Mark admits that he lit the bonfire, and Thirza joins him in his death by drowning.

**Wrocław** (Ger., Breslau). City in Poland. In the late 17th cent. works by Johann Hallmann were given in which musical accompaniments and interludes played a part. The first operatic performances were organized by Daniel Treu (1695–1739), whose company gave *Orlando furioso* by Antonio Bioni (1698–after 1738) at the Ballhaus in 1725. Bioni had further successes in the city, which he dominated musically, before leaving in 1734. Operas given included Treu's own *Astarto* (1725), *Coriolano* (1726), *Ulisse e Telemacco* (1726), and *Don Chischiotte* (1727), and works by F. B. Conti, Albinoni, Porta, and Astorga (41 operas between 1725 and 1734). In 1782 the Opernhaus opened, with a repertory including much Mozart. In 1800 Vincenz Tuczek became director; when, on Vogler's recommendation, the 17-year-old Carl Maria von *Weber became music director in 1804, friction ensued, but outstanding performances were given of *Titus*, *Alceste*, and Weber's own *Peter Schmoll*. Weber started substantial reforms, but in his absence through an accident they were undone, and he resigned in 1806. Under Gottlob Bierey (1772–1840) opera flourished 1808–28; the company was enlarged, new singers engaged, and new works staged. Important Wagner performances were given 1852–88.

When the city became Polish after the Second World War, the Opera Dolnośląska (Lower Silesian O) from 1945 gave Polish works, especially by Moniuszko, as well as French and Italian, and later made a speciality of Czech opera. It gave the première of *Bunt żaków* (The Goliards' Revolt, 1951) by Tadeusz Szeligowski (1896–1963). In

1987 it staged *Un giorno di regno*, Verdi's comedy of Poland's King Stanisław Leszcyński.

**Wüllner, Franz** (*b* Münster, 28 Jan. 1832; *d* Braunfels, 7 Sep. 1902). German conductor and composer. Studied Münster with Arnold and Schindler. Cond. Munich Court T 1869 as Bülow's successor, having to face the difficulties surrounding the première of *Rheingold*. His success with this (1869) and *Walküre* (1870) led to his appointment as court conductor 1871. Dresden 1877, but he was forced to leave 1882. His son **Ludwig** (1858–1938) was a baritone who sang at Meiningen (1889–95) and then devoted himself to concerts.

BIBL: D. Kämper, ed., *Franz Wüllner* (Cologne, 1963).

**Wunderlich, Fritz** (*b* Kusel, 26 Sep. 1930; *d* Heidelberg, 17 Sep. 1966). German tenor. Studied Freiburg with Winterfeldt. Début Stuttgart 1955 (Tamino). Stuttgart until 1965; Frankfurt 1958; Munich from 1960; Vienna from 1962; London, CG, 1965. A lyrical singer remembered with great affection for his captivating tone and turn of phrase. Excellent as Belmonte, Don Ottavio, Lensky, Alfredo, Rodolfo, Palestrina; created Tiresias (Orff's *Oedipus der Tyrann*). Died at the peak of his powers. (R)

**Wuppertal**. City in North-Rhine Westphalia, Germany. The Wuppertaler Bühnen (cap. 845, small theatre 792) opened 1956 with *Mathis der Maler*. Under Grischa Barfuss, Intendant 1958–64, with Hans-Georg Ratjen as music director and Georg Reinhardt as designer, the company became one of the most progressive in Germany. Its productions of Monteverdi, Hindemith, and Busoni have won high praise.

**Wurm**. The servant (bs) of Count Walter in Verdi's *Luisa Miller*.

**Würzburg**. City in Bavaria, Germany. Jesuit school music dramas were given in the 17th cent., and travelling companies visited. Italian operas were given in the Residenz in the 1770s. Count Soden's theatre opened 1804 and became the Stadttheater 1843. In 1833 Wagner was invited as chorus-master by his brother Albert, who was singer and stage-manager. The theatre was bombed 1945; the Stadttheater (cap. 750) opened 1966 with *Meistersinger*.

# X

**Xenia.** Boris's daughter (sop) in Musorgsky's
*Boris Godunov*.

**Xerxes.** See SERSE.

**Xyndas, Spyridon** (*b* Corfu, 1814; *d* Athens, 12
Nov. 1896). Greek composer and guitarist. Studied Corfu with Mantzaros, and Naples. Of his
six completed operas, the most successful was
*The Parliamentary Candidate*, a political satire
that was the first opera composed to a Greek text.
His music is now lost.

SELECT WORKLIST: *Il conte Giuliano* (Marmora; Corfu
1857); *O ypopsiphios vouleftis* (The Parliamentary
Candidate) (Rinopoulos; Corfu 1867).

# Y

**Yakutsk.** The first moves towards the foundation of Yakut national opera were made on the basis of folk dramas with musical accompaniments. The national epic *Oloncho* was written down as *Nyurgun Bootur* and given music by Mark Zhirkov (1892–1951), then set by him as an opera (1947) with Heinrich Litinsky (1901–85). Other national operas include *Lookut i Nyurgusun* (1959) by the Armenian Grant Grigorian (1919–62); *Pesn o Manchary* (The Song of the Manchary, 1967), by Edward Alexeyev (*b* 1937) and Hermann Komrakov (*b* 1937); *Krasny Shaman* (The Handsome Shaman, 1967) by Litinsky; and *Neugasimoye plamya* (The Inextinguishable Flame, 1974) by N. Berestov.

**Yamada, Kōsaku** (*b* Tokyo, 9 June 1886; *d* Tokyo, 29 Dec. 1965). Japanese composer and conductor. Studied Tokyo with Werkmeister, Berlin with Bruch and Wolf. Production of his first opera, *Fallen Angel*, was delayed by the outbreak of war; its eventual production led to a commission for Paris which he fulfilled with *The Sweet Flag*. He was instrumental in the development of opera in Japan, and toured world-wide as a conductor. His most successful opera was *The Black Ships*. He remained throughout his life a strong advocate of Wagner and Richard Strauss, and sought reconciliation between a late German Romantic manner and Japanese musical characteristics.

SELECT WORKLIST (all texts by Yamada): *Ochitaru tennyo* (The Fallen Angel) (comp. 1912; prod. Tokyo 1929); *Ayame* (The Sweet Flag) (Paris 1931); *Kurofune* (The Black Ships) (Tokyo 1940).

**Yamadori.** A Japanese prince (bar, sometimes ten), suitor for Butterfly's hand, in Puccini's *Madama Butterfly*.

**Yaroslavna.** Princess Yaroslavna (sop), Igor's second wife, in Borodin's *Prince Igor*.

**Yeats, William Butler** (*b* Dublin, 13 June 1865; *d* Menton, 28 Jan. 1939). Irish poet and dramatist. Operas on his works are as follows:

*The Countess Cathleen* (1892): Egk (*Irische Legende*, 1955)
*The Land of Heart's Desire* (1894): F. Hart (1914)
*The Shadowy Waters* (1900): Kalomiris (1951); Swain (unfin.)

*The Only Jealousy of Emer* (1919): Harrison (1949)
*Purgatory* (1939): Weisgall (1961); Crosse (1966)

**Yeletsky.** See ELETSKY.

**Yniold.** The young son (sop) of Golaud, stepson of Mélisande, in Debussy's *Pelléas et Mélisande*.

**Yolanta.** Opera in 1 act (9 scenes) by Tchaikovsky; text by Modest Tchaikovsky, after Henrik Hertz's story *Kong Renés Datter* (trans. Zvantsev), after Hans Andersen. Prod. St Petersburg, M, 18 Dec. 1892, with Serebryakov, Yakovlev, N. Figner, Chernov, Karelin, Frey, M. Figner, Kamenskaya, Runge, Dolina, cond. Nápravník; Scarborough-on-Hudson, Garden T, 10 Sep. 1933; London, St Pancras TH, 20 Mar. 1968, with Barstow, De Peyer, Welsby, Raffell, N. Noble, cond. Lloyd-Jones.

Provence, Middle Ages. Yolanta (sop), daughter of King René of Provence (bs), has had knowledge of her blindness kept from her; the penalty for anyone revealing her disability to her is death. Ebn-Hakia (bar), a Moorish doctor, tells the King that only if she knows of her blindness and wills its removal can he cure her. Into her garden comes Robert, Duke of Burgundy (bsbar), with Count Vaudémont (ten), who discovers Yolanta's blindness and falls in love with her, as she with him. They are interrupted, and the King declares that Vaudémont must die unless she regains her sight. This ruse succeeds in curing her; Robert, who was engaged to Yolanta, but loved another, is released from his bond, and she is betrothed to Vaudémont.

Other operas on the subject are by Edwards (1893) and Behrend (1919). King René of Provence is also the subject of an opera by Hérold (1824).

**Young, Cecilia** (*b* ?London, 1711; *d* London, 6 Oct. 1789). English soprano. Studied with Geminiani. Probable first appearance London, DL, 1730 (concert). Sang for Handel's co. 1735 (in *Alcina* and *Ariodante*), and subsequently created parts in his oratorios *Saul* and *Alexander's Feast*. Married Thomas Arne 1737. Dublin 1742 (in his *Comus*, *Alfred*, and *Judgement of Paris*), 1748–9, 1755. London, DL, 1744. She and Arne separated 1756, but were reunited 1777. Possessed a light, flexible voice and was much admired by Burney (Arne's pupil), who thought her 'superior to any other Englishwoman of her time'. Her sister, **Isabella** (*fl.* 1730–53) sang at Covent Garden, and was married to the composer J. F. Lampe. Another sister, **Esther** (Mrs Jones) (*fl.* 1739–62), also a soprano, created Handel's Semele. Three nieces **Isabella** (Mrs Scott) (*d* 1791), **Elizabeth** (Mrs Dorman) (*fl.* 1756–65), and **Polly** (Mrs Barthélemon) (*c*.1749–99), were singers as well. All the above performed at some time under the name Miss Young. There has been much subsequent confusion.

**Young Lord, The.** See JUNGE LORD, DER.

**'Your tiny hand is frozen'.** See 'CHE GELIDA MANINA'.

**Ysaye, Eugène** (*b* Liège, 16 July 1858; *d* Brussels, 12 May 1931). Belgian violinist, conductor, and composer. Cond. London, CG, 1907 (*Fidelio*). Composed an opera to his own Walloon text, *Piér li houïeu* (Peter the Miner, 1931). Hoped to conduct this, but collapsed at the first rehearsal and only heard one performance.

**Yugoslavia.** See CROATIA, SERBIA, SLOVENIA.

**Yun, Isang** (*b* Tongyong, 17 Sep. 1917). Korean, later German, composer. Studied Osaka and Tokyo, later Paris with Revel, Berlin with Blacher and Rufer. Worked in Darmstadt summer schools; later taught Hanover and Berlin. In his operas, as elsewhere, he has sought a reconciliation between Western techniques and oriental styles; they are principally based on Taoist philosophy, and employ highly individual vocal techniques.

SELECT WORKLIST: *Der Traum des Liu-Tung* (Liu-Tung's Dream) (Bauernfeind, after Ma Chi-yuan; Berlin 1965); *Die Witwe des Schmetterlings* (Butterfly Widow) (Kunz, after anon. 16th-cent. Chinese story; Nuremberg 1969); *Geisterliebe* (Ghostly Love) (Kunz; Kiel 1971); *Sim Tjong* (Kunz, after a Korean legend; Munich 1972).

BIBL: H. Kunz, *Bewegtheit in Unbewegtheit: die Musik des Koreaners Isang Yun* (Munich, 1972).

# Z

**Zaccaria.** A Hebrew prophet (bs) in Verdi's *Nabucco*.

**Zafred, Mario** (*b* Trieste, 2 Mar. 1922; *d* Rome, 22 May 1987). Italian composer and critic. Studied Rome with Pizzetti, and in Paris. Critic of *Unità* (Rome) 1949–56. Artistic director, Trieste, T Giuseppe Verdi 1966, and Rome, T dell'Opera 1968. His operas include *Amleto* and *Wallenstein* and are very much in the vein of contemporary Soviet composers, in line with his own political philosophy.

SELECT WORKLIST: *Amleto* (Zafred, after Shakespeare; Rome 1961); *Wallenstein* (Zafred, after Schiller; Rome 1965).

**Zagreb** (Ger., Agram). Capital of Croatia. The first centre of operatic life was at Count Várkony's palace, where a German company played in 1797. For 37 years opera was performed here. In 1834 the first public theatre was built and was in use until 1895. From 1814 German companies' visits were organized by the German Opera and Drama Society, but performances in Croatian began about the same time, and in 1846 *Lisinski's *Ljubav i zloba* was premièred by amateurs. In 1861 official encouragement was given to Croatian opera. The Operetta T was founded in 1863. When *Zajc returned from Vienna in 1870, he reorganized the Zagreb Opera and Operetta Theatres. In 1889 the Hungarian government closed the theatre. Performances continued in the open air in summer, and in 1895 the present theatre (cap. 850) was opened with Zajc's *Nikola Šubić Zrinjski*, only to close again in 1902. It reopened in 1909 and maintained performances throughout two wars: it is in the house of the Opera of the Hrvatsko Narodno Kazalište (Croatian National T), and gives a season from Sep. to June. Operetta is given at the Comedy T.

**Zaide.** Opera in 2 acts (unfin.) by Mozart; text by Schachtner, after Sebastiani. Comp. 1779–80; prod. Frankfurt 27 Jan. 1866, compl. Gollmick with rewritten dialogue, overture and finale added by Anton André; London, Toynbee H, 10 Jan. 1953, with Barrett, cond. Dempster; Tanglewood 8 Aug. 1955, with Cardillo, Lussier, Boatwright, Zulick, cond. Sarah Caldwell.

Turkey, 16th cent. Gomatz (ten) has been taken prisoner by the Sultan Soliman (ten). He seeks refuge from his cares in sleep. Zaide (sop), the Sultan's favourite, declares her love for Gomatz while he sleeps and leaves a portrait of herself beside him. He awakes and declares his love for her. Allazim (bs), the Sultan's steward, offers to help them escape.

The Sultan is furious to find Zaide gone and, on learning of the couple's escape, plans a cruel revenge. They are recaptured, and neither the pleas of Allazim nor of Zaide can weaken the Sultan's resolve. The opera breaks off at this point. Various completions have been attempted in both staged and concert versions.

**Zajc, Ivan** (*b* Rijeka, 3 Aug. 1832; *d* Zagreb, 16 Dec. 1914). Croatian composer. Studied with his father, Rijeka, Milan with Ronchetti-Monteviti, Rossi, and Mazzucato. After working in Italy and Vienna, he returned home in 1870 and became director of the Zagreb Conservatory and conductor at the theatre. Here he exercised an important influence on the country's musical life. He was an important founder of Croatian opera, and was (with *Mislav*) the first to set Croatian texts. His most important work is *Nikola Šubić Zrinjski*. Based on the manner of Verdi's Risorgimento operas, the work includes some charming oriental music and some rousing tunes mostly built on the short phrases that were part of his idiom, and is well scored, especially for woodwind.

SELECT WORKLIST: *La tirolese* (The Tyrolean Girl) (Guidi; Milan 1855); *Die Hexe von Boissy* (The Witch of Boissy) (Costa; Vienna 1866); *Somnambule* (Young; Vienna 1868); *Mislav* (Markovič; Zagreb 1870); *Nikola Šubić Zrinjski* (Badalić, after Körner; Zagreb 1876); *Prvi grijeh* (Original Sin) (Kranjčević; comp. as oratorio 1907, prod. Zagreb 1912).

CATALOGUE: H. Pettan, *Popis skladbi Ivana Zajca* (Zagreb, 1956).

BIBL: H. Pettan, *Ivan Zajc* (Zagreb, 1971).

**Zamboni, Luigi** (*b* Bologna, *c*.1767; *d* Florence, 28 Feb. 1837). Italian bass. Début Bologna 1791 (in Cimarosa's *Fanatico in Berlina*). Rome, Naples, Venice; Milan, S, 1810–11, 1818. Retired 1825. A highly successful buffo singer, and childhood friend of Rossini, who wrote the title-role of *Barbiere di Siviglia* for him (describing him as 'valiant'). Also sang in operas by Fioravanti, Paisiello, Pavesi, etc.

**Zampa.** Opera in 3 acts by Hérold; text by Mélesville. Prod. Paris, OC, 3 May 1831, with Casimir, Boulanger, Chollet, Féréol, Moreau-Sainti; New Orleans, T d'Orléans, 16 Feb. 1833, with St Clair, Amédée. London, H, 19 Apr. 1833.

Sicily, 16th cent. At the head of his pirate band, Zampa (bar) invades the island of Castel Lugano and forces Camilla (mez), daughter of the Count, to abandon her betrothed Alfonso (ten) for him. As they celebrate this new betrothal, Zampa places a ring on the finger of the statue of his former bride Alice: the fingers close over the ring threateningly. Zampa remains unmoved by all pleas to release Camilla, and reveals that he is the Count of Monza, and thus Alfonso's brother. Zampa sends Alfonso to prison, and orders the statue to be thrown into the sea, while Camilla claims sanctuary at the altar; Zampa tears her away, but is dragged beneath the waves by the cold hand of the statue. Camilla is reunited with Alfonso.

With *Le pré aux clercs*, Hérold's masterpiece, it was immensely popular throughout the 19th cent.

**Zancanaro, Giorgio** (*b* Verona, 9 May 1939). Italian baritone. No formal training. Originally a policeman. Début Milan, N, 1971 (in *Puritani*). Bologna, Florence, Venice; Hamburg 1977; Milan, S, 1981; New York, M, 1982; London, CG, 1985. Roles incl. Rigoletto, Luna, Posa, Germont, Renato, Ford, Escamillo, Tonio. (R)

**Zandonai, Riccardo** (*b* Sacco, 30 May 1883; *d* Pesaro, 5 June 1944). Italian composer. Studied Rovereto with Gianferrari, Pesaro with Mascagni. He was taken up by Ricordi as a successor to Puccini in the verismo tradition after *Il grillo del focolare* in 1908, and went to Spain to absorb ideas for *Conchita* (a subject that had been rejected by Puccini). The popularity of this rested largely on Zandonai's ability to absorb some of the harmonic manner of, in particular, Richard Strauss into a bold gypsy plot with plenty of Spanish atmosphere. His most popular work, however, was *Francesca da Rimini*, which includes some finely written music but whose overstatements suggest an underlying uncertainty about the continuing validity of verismo in 1914. Similar considerations apply to *Giulietta e Romeo*, though he found a good stimulus to his sense of the stage and his harmonic adventurousness in *I cavalieri di Ekebù*. His wife was the soprano **Tarquinia Tarquini** (1883–1976), who created Conchita and whose roles also included Carmen, Salome, and Santuzza.

SELECT WORKLIST: *Il grillo del focolare* (The Cricket on the Hearth) (Hanau, after Dickens; Turin 1908); *Conchita* (Vaucaire & Zangarini, after Louÿs; Milan 1911); *Francesca da Rimini* (Ricordi, after D'Annunzio, after Dante; Turin 1914); *Giulietta e Romeo* (Rossato, after Shakespeare; Rome 1922); *I cavalieri di Ekebù* (The Knights of Ekebù) (Rossato, after Lagerlöf; Milan 1925).

BIBL: B. Cagnoli, *Riccardo Zandonai* (Trent, 1978); R. Chiesa, ed., *Riccardo Zandonai* (Milan, 1984).

**Zanelli, Renato** (*b* Valparaiso, 1 Apr. 1892; *d* Santiago, 25 Mar. 1935). Chilean baritone, later tenor. Studied Santiago with Querez. Début Santiago 1916 (Valentin). New York, M, 1919–23, as

baritone. After further study Milan with Lari and Tanara, tenor début Naples, C, 1924 (Raoul). Turin 1926; Parma 1927; London, CG, 1928–30; Rome 1928–32. Also Milan, S; Buenos Aires. Possessed a dark, attractive timbre, and much dramatic fire. Roles included Alvaro, and an outstanding Otello (Verdi). Also leading Wagner singer in Italy (Siegmund, Tristan). (R) His brother **Carlo Morelli** (1897–1970) was a baritone. Italy, and New York, M, 1935–40. (R)

**Zar und Zimmermann** (Tsar and Carpenter). Opera in 3 acts by Lortzing; text by the composer, after C. Römers, after Mélesville, Merle, and De Boirie (1818). Prod. Leipzig 22 Dec. 1837, with Günther, Frau Lortzing, A. Lortzing, Richter, Berthold; New York, Astor Place OH, 9 Dec. 1851, with Arming, Müller, Oehrlein, Beutler; London, Gaiety T, 15 Apr. 1871, with B. Cole, A. Tremaine, Lyall, Santley, Cook, cond. Meyer Lutz.

Saardam, 1698. Peter the Great (bar) is working in the Saardam shipyards under the name of Peter Michaelov so as to learn trades he cannot in Russia. He befriends a deserter, Peter Ivanov (ten), in love with Marie (sop), daughter of the Burgomaster Van Bett (bs). Asked by the English to discover if Peter the Great is indeed in the shipyard, Van Bett identifies the wrong Peter; the English Ambassador Lord Syndham (bs) and the Russian Ambassador Admiral Lefort (bs) are taken in, though the French Ambassador, the Marquis de Châteauneuf (ten), recognizes the real Tsar. After a festival, Van Bett prepares to send Peter Ivanov home with an anthem (whose rehearsal provides an entertaining number). Meanwhile, the real Tsar slips away. See also PETER THE GREAT.

**zarzuela**. A type of Spanish opera, mingling dialogue with music, usually popular in nature. The word derives from *zarza*, a bramble, which gave its name to the palace near Madrid, La Zarzuela, in which spectacular plays were mounted by Philip IV during the mid-17th cent. These lavish entertainments came to be known as 'fiestas de la Zarzuela': the earliest has been said to be Lope de Vega's *La selva sin amor* (1629). Important works included Calderón's *El jardín de Falerina* (for Philip's wedding to Mariana of Austria, 1648), *Fortunas de Andrómeda y Perseo* (1653), *El laurel de Apolo* (1657), and (the first described with the term zarzuela) *El golfo de las sirenas* (1657). Juan Vélez's *Los celos hacen estrellas* (1672) had music by Juan *Hidalgo, the most celebrated early zarzuela composer. The genre then resembled the French ballet de cour, with the formality of mythological subjects occasionally relieved by the introduction of popular songs, the musical style Italianate. In some early zarzuelas, the sung text is set to pre-existing music; even when original material is used, the composer is often unnamed. Gradually, more music came to be employed, and this was couched in the general style of contemporary Venetian opera.

During the first half of the 18th cent., while Spain was under Bourbon rule, the zarzuela went into a decline, and Italian opera was cultivated in its place, although isolated examples by Antonio Literes and Juan de Nebra were popular. After the accession of Charles V in 1759, and under the influence of the *sainete and *tonadilla, the zarzuela was gradually revitalized. The first important new works were Antonio Rodríguez de Hita's *Las segadoras de Valleca* (1768) and *Las labradoras de Murcia* (1769), which supplanted the earlier style of heroic plot with a new focus upon scenes of Spanish country life: the librettist, Ramón de los Cruz, largely shaped the late 18th-cent. zarzuela, collaborating with composers including Pablo Estere, Antonio Palomino, Fabián García Pachero, and Antonio Rosales, and even converting Italian operas into zarzuelas. On his death the genre once again languished, the victim of renewed interest in Italian opera and of such smaller entertaiments as the tonadilla.

By the early to mid-19th cent. a revival of the zarzuela came with the growth of nationalism. Manuel Bretón de los Herreros's *El novio y el concierto* (1839), with music by an Italian, Basilio Basili, parodying Italian operatic mannerisms was highly popular, and gave the genre new impetus: composers included Rafael Hernando and Joaquin Gaztambide. The fashion grew to the point at which the T de la Zarzuela was opened in Madrid (1856), followed by the T Apolo (1873). The most prominent composer of the period was Francisco Asenjo Barbieri (1823–94) with *Jugar con fuego* (1851), *Pan y toros* (1864) (in which Goya makes an appearance), and *El barberillo de Lavapiés*, among over 60. Also important were Pascual Arrieta (1823–94; *Marina*, 1871), Federico Chueca (1846–1908; *La canción de la Lola*, 1880, among almost 40). Different categories of zarzuela developed, principally *zarzuela grande*, normally in 3 acts and close to Romantic opera in style, and the *género chico* or *zarzuelita*, in 1 act and normally satirical in content. Other important composers of later 19th-cent. zarzuela included M. Fernández Caballero (1835–1906; *La viejecita*, 1897), Joaquin Valverde (1846–1910), Tomás Bretón y Hernández (1850–1923; *La verbena de la Paloma*, 1894), Ruperto Chapí y Lorente (1851–1909; *La revoltosa*, 1897), Jerónimo Giménez (1854–1923; *La tempránica*), Amadeo Vives (1871–1932; *Doña Francisquita*, 1923), Jesús Guridi (1886–1961; *El caserío*, 1926), Jacinto Guerrero (1895–1951; *La montería*, 1922), Federico Moreno Torroba (*b*

1891; *Luisa Fernanda*, 1932); José Maria Usandizaga (1887–1915; *Las golondrinas*, 1914).

**Zauberflöte, Die** (The Magic Flute). Opera in 2 acts by Mozart; text by Schikaneder, based on various sources, including *Lulu* in Wieland's collection of oriental fairy-tales *Dschinnistan* (1786) and the Abbé Terrasson's *Sethos* (1731). Prod. Vienna, W, 30 Sep. 1791, with Hofer, Gottlieb, Gerl, Schack, Nouseul, Schikaneder, Winter, J. Weber, cond. Mozart; London, H, 6 June 1811, with Griglietti, Bertinotti-Radicati, Cauvini, Sra. Cauvini, Righi, Naldi, Rovedino, cond. Weichsell (leader) and Pucilta (continuo); New York, P, 17 Apr. 1833, with Wallack, Austin, Sharpe, Jones, Placide, Horn.

Ancient Egypt. A young prince, Tamino (ten), is rescued from a serpent by the Three Ladies (sop, sop, mez) of the Queen of the Night. The birdcatcher Papageno (bar) boasts that it was he who killed the serpent. The Ladies place a padlock on Papageno's mouth as punishment and give Tamino a portrait of the Queen's daughter Pamina (sop), who has been abducted; falling in love with her picture, he vows to rescue her. The Queen (sop) announces that if Tamino can rescue Pamina, he may marry her. Papageno agrees to accompany him, and the Three Ladies give them magic bells and a magic flute to protect them. They will be led to Sarastro's palace, where Pamina is being held, by the Three Boys (trebs, or often sop, sop, mez). Monostatos (ten), a Moor in the service of Sarastro (bs), is about to rape Pamina when Papageno blunders in. He and Pamina go in search of Tamino, who has meanwhile entered a temple, where a priest has informed him that it is in fact the Queen and not Sarastro who is evil. Papageno and Pamina are surprised by Monostatos and his slaves, but bewitch them by playing the magic bells. Tamino and Pamina meet and immediately fall in love.

Tamino and Papageno are told they must undergo trials in preparation for initiation into the brotherhood; the first of these is to maintain a vow of silence. Pamina lies asleep and Monostatos approaches her once again. He is thwarted by the arrival of the Queen. She gives her daughter a dagger and explains that her only hope of regaining power is for Pamina to kill Sarastro. Monostatos is forcing himself on Pamina again when Sarastro enters. He says he is aware of the Queen's plans for revenge but that the brotherhood acts only by love. Papageno is taunted by a fleeting glimpse of his ideal young Papagena, but is deprived of her until the priests judge him worthy. Convinced that she has been deserted by Tamino, Pamina is about to stab herself but is prevented by the Three Boys. They take her to Tamino and she joins him in the trials of fire and water. They are brought through safely by Tamino playing his magic flute. Papageno is reunited with Papagena. The Queen, the Three Ladies, and Monostatos are engulfed in a clap of thunder before they can overthrow Sarastro and his followers. The others celebrate the victory of light over darkness, and the opera ends with a chorus of thanksgiving to the gods Isis and Osiris.

Goethe, who much admired *Die Zauberflöte* and was himself a Freemason, wrote part of a sequel to the opera, but this has never been set to music.

There is an opera *Mozarts Zauberflöte* (1820) by Riotte.

BIBL: J. Chailley, *La flûte enchantée, opéra maçonnique* (Paris, 1968, trans. 1972); E. Batley, *A Preface to the Magic Flute* (London, 1969); P. Branscombe, *W. A. Mozart: 'Die Zauberflöte'* (Cambridge, 1991).

**Zauberoper** (Ger.: 'magic opera'). A form of opera of greater substance than the Zauberposse, popular in Vienna in the late 18th and early 19th cents., in which a fairy story was told with sumptuous scenic effects, ribald comedy, and music by a distinguished composer. One of the earliest examples was Wranitzky's setting of the Oberon story (1789) and the latest Weber's (1828). Its greatest example, entirely transcending the form through its moral substance, is Mozart's *Die Zauberflöte*.

**Zauberposse.** See POSSE.

**Zazà**. Opera in 4 acts by Leoncavallo; text by composer, after the play by Simon and Berton. Prod. Milan, L, 10 Nov. 1900, with Storchio, Garbin, Sammarco, cond. Toscanini; San Francisco, Tivoli OH, 27 Nov. 1903; London, Coronet T, 30 Apr. 1909.

It tells of the affair between the Parisian music-hall singer Zazà (sop) and Milio Dufresne (ten), a married man about town, and Zazà's return to her lover Cascart (bar).

After *Pagliacci*, Leoncavallo's most successful opera.

**Zdenka**. Arabella's sister (sop) in Strauss's *Arabella*.

**Zeffirelli, Franco** (*b* Florence, 12 Feb. 1923). Italian producer and designer. After working in theatre and film, turned to opera in 1948. Produced and designed several works for Milan, S, notably *Turco in Italia, Cenerentola, Don Pasquale*. A disciple of *Visconti, he helped to give new dramatic validity to Italian Romantic opera with his production of *Lucia di Lammermoor* at London, CG, 1959, in the same year turning to verismo with *Cavalleria* and *Pagliacci*. Milan, S: *Bohème* 1963, *Ballo* 1972, Verdi's *Otello* 1976, *Turandot* 1983. New York, M: librettist and prod. Barber *Antony and Cleopatra* for opening of

new house 1966; *Otello* 1972, *Bohème* 1981, *Tosca* 1985, *Traviata* 1989, *Don Giovanni* 1990. Vienna: *Don Giovanni* 1972. Dallas, Chicago, San Francisco. Florence: *Traviata* 1985. His opera films have included a 1986 *Otello* that made some controversial cuts and rearrangements but that had considerable dramatic impact. A highly inventive producer with a quick eye for detail.

**'Zeffiretti lusinghieri'**. Ilia's (sop) aria in Act III of Mozart's *Idomeneo*, in which she bids the breezes fly to Idamante and tell him of her love.

**Zeitoper** (Ger.: 'opera of the time'). A term used to describe certain German operas of the 20th cent. in which topical themes are handled, often involving the use or imitation of contemporary popular music. Krenek wrote many such works, including *Jonny spielt auf* and *Der Diktator*.

BIBL: S. Cook, *Opera for a New Republic: the 'Zeitopern' of Krenek, Weil and Hindemith* (Ann Arbor, 1988).

**Żeleński, Władysław** (*b* Grodkowice, 6 July 1837; *d* Cracow, 23 Jan. 1921). Polish composer. Studied Cracow with Mirecki, Prague with Dreyschock and Krejči, Paris with Reber and Damcke. He was the most important Polish successor to Moniuszko, though his style is essentially conservative.

SELECT WORKLIST: *Konrad Wallenrod* (Sarnecki and Noskowski, after Mickiewicz; Lvov 1885); *Stara baśń* (Old Fable) (Bandrowski, after Kraszewski; Lvov 1907).

BIBL: Z. Jachimecki, *W. Żeleński* (Cracow, 1952).

**Zeller, Carl** (*b* St Peter in der Au, 19 June 1842; *d* Baden, nr. Vienna, 17 Aug. 1898). Austrian composer. He received a musical education in Vienna as one of the *Sängerknaben* at the imperial chapel, then studied law at Vienna U, while also taking composition lessons with Sechter. After practising as a lawyer, he joined the Ministry for Education and Culture, where a successful career was cut short by scandal.

Beginning with *Joconde*, Zeller made a notable, if limited contribution to the repertoire of Viennese operetta after the model of Johann Strauss II. He possessed a sure technique and a gift for writing appealing and instantly memorable tunes. His most successful work, and the only one by which he is today remembered, is *Der Vogelhändler*, which by 1900 had received nearly 3,000 performances on the German stage.

SELECT WORKLIST: *Joconde* (West & Moret; Vienna 1876); *Der Vogelhändler* (The Bird-Dealer) (West & Held; Vienna 1891).

BIBL: C. Zeller, *Mein Vater Carl Zeller* (St Pölten, 1942).

**Zémire et Azor.** Opera in 4 acts by Grétry, text by Marmontel, poss. after La Chaussée's comedy *Amour par amour* (1742). Prod. Fontainebleau (Court) 9 Nov. 1771, with Clavibal; London, DL, 5 Dec. 1776; New York 1 June 1787, cond. M. Huni. This version of the *Beauty and the Beast* fairy-tale was also set by Baumgarten (1775), Neefe (1776), Tozzi (1791), Spohr (1819), García (1827), and Chukhadjian (1891); sequel by Umlauff, *Der Ring der Liebe* (1786).

**Zemlinsky, Alexander** (*b* Vienna, 14 Oct. 1871; *d* Larchmont, NY, 15 Mar. 1942). Austrian composer and conductor. Studied Vienna with Fuchs. Conductor Vienna, K, 1899, and V, 1904 (Kapellmeister 1906–7, 1908–11)); Prague, Neues Deutsches T, 1911–27. Assistant to Klemperer in Berlin at the Kroll T, 1927–31. Emigrated to USA 1938. Zemlinsky's career included conducting many early performances of Berg, Hindemith, and Janáček. His most distinguished period was in Prague, where he conducted an extremely varied repertoire: Stravinsky commented ecstatically upon his performance of Mozart.

The composition of Zemlinsky's operas was spread out across his creative life and they offer an insight not only into his own development, but also into the main Austro-German musical trends of his day. The first three works, *Sarema*, *Es war einmal*, and *Der Traumgörge*, are all based on fairy-tales, suggesting a certain parallel with the contemporary *Jugendstil* movement. For Zemlinsky the fairy-tale opera was seen as a means of addressing the problems of the age through a subtle use of allegory, the simplicity and naïvety of the fairy-tale being used to mask a complex and subtle message. This is seen most clearly in *Der Traumgörge*, which in some respects prefigures Schreker's *Der ferne Klang*: the search of the central rustic character for a different world is here to be taken as an allegory of the favourite theme of the artist as outsider. After the comic *Kleider machen Leute*, he turned to two works which reflect a contemporary fascination for Renaissance settings, *Eine florentinische Tragödie* and *Der Zwerg*. His last completed opera, *Der Kreidekreis*, belongs to the genre of *Zeitoper* and has a clear didactic message. Though Zemlinsky's operas have occasionally been revived, they have been generally overshadowed by the works of Berg, Schoenberg, and Schreker. But in their day they were highly regarded; an enthusiastic champion was Mahler, who conducted the première of *Es war einmal* and whose wife Alma was a pupil of Zemlinsky. His other pupils included Korngold and, informally, Schoenberg, the première of whose *Erwartung* he conducted. Five further operas were planned but not completed.

WORKLIST: *Sarema* (Zemlinsky, after Gottschall; Munich 1897); *Es war einmal* (Once Upon a Time) (Singer, after Drachmann; Vienna 1900); *Der Traumgörge* (Wattke, after Feld; comp. 1905–6, prod. Nuremberg 1980); *Kleider machen Leute* (Clothes Make the Man) (Feld, after Kelter; Vienna 1910, rev. (in 2 acts) Prague 1922); *Eine florentinische Tragödie* (Meyerfeld, after Wilde; Stuttgart 1917); *Der Zwerg* (The Dwarf) (Klaren, after Wilde; Cologne 1922); *Der Kreidekreis* (The Chalk Circle) (Klabund; Zurich 1933).

**Zenatello, Giovanni** (*b* Verona, 22 Feb. 1876; *d* New York, 11 Feb. 1949). Italian tenor, orig. baritone. Studied Verona with Zannoni and Moretti. Début Belluno 1898 (Silvio). Tenor début Naples 1899 (Canio, having sung Silvio at previous performance). Lisbon 1902; Milan, S, 1902–7; London, CG, 1905–9, 1926; New York, Manhattan OC, 1907–10; New York, M, 1909; Chicago 1911–13. His extensive, even, rich-toned voice and impassioned characterizations produced a memorable Don José, Radamès, Alfredo, Faust (Gounod), Samson, first Pinkerton, and above all Otello (Verdi). Verona Arena 1913 (its inauguration), and subsequently manager; arranged Callas's momentous Gioconda there, 1947. After retiring in 1928, opened singing school New York, with his wife soprano Maria *Gay. (R)

**Zeno, Apostolo** (*b* Venice, 11 Dec. 1668; *d* Venice, 11 Nov. 1750). Italian librettist and literary scholar. A founder of the Accademia degli Animosi, he was an historian and critic, and editor of an important journal, regarding libretto-writing as incidental to his career. His first success in this field came in 1700 with *Lucio Vero* (set by Pollarolo) and was followed by further works in a similar vein, including *Griselda* (1701, for Pollarolo) and *Aminta* (1703, for Albinoni). In 1705 he collaborated for the first time with *Pariati in *Antioco* and this began a successful partnership which lasted for over 20 years. When Pariati went to the Viennese court in 1718, Zeno followed him shortly afterwards, becoming *poeta cesareo*, a post he held until replaced by Metastasio in 1729. In Vienna many of his librettos were set by Caldara, including *Ifigenia in Aulide* (1718) and *Mitridate* (1728).

Zeno is usually credited with being the first to 'reform' the Venetian-style libretto, paving the way for the more famous and radical changes of Metastasian opera seria. In truth, his work was the most successful embodiment of changes which were being more generally effected by Italian librettists (notably Stampiglia) at the beginning of the 18th cent. Foremost among these was

779

the paring of the cast to six, or at a maximum eight, characters, with just one main plot (and an occasional amorous sub-plot) presented in three, or less frequently five, acts, in place of the gross entanglements of Venetian opera. Comedy, and comic characters, were eschewed, and the main thrust of the drama increasingly concerned the conflict between love and duty, seen most clearly in *Griselda*. As befitted operas written for an imperial court, Zeno often portrayed this theme as a consecration of noble virtue, culminating in a triumphant final scene in which constancy and moral rectitude receive their just reward.

His scholarly bent ensured that the sources and background to his plots were carefully set out in his prefaces: authors upon whom he drew include Thucydides, Herodotus, Plutarch, and Livy, as well as Racine and Corneille. Once he had sketched the dramatic scenario, he and Pariati collaborated on the versification. The combination of Zeno's expertise at devising scenes of action, of confrontation, and of forceful emotion, and Pariati's talent for turning out elegant lyrics, ensured that his composers were provided with librettos of a dramatic and poetic quality unmatched in their day. Those who used his librettos included Araia, Ariosti, Bononcini, Cherubini, Ciampi, Duni, Fux, Galuppi, Gasparini, Guglielmi, Handel, Hasse, Pergolesi, Porpora, Sacchini, Scarlatti, Traetta, Vivaldi, Zingarelli, and many others.

BIBL: R. Freeman, 'Apostolo Zeno's Reform of the Libretto', *JAMS*, 21 (1968).

**Zerbinetta.** The leader (sop) of the Harlequinade in Strauss's *Ariadne auf Naxos*.

**Zerlina.** (1) A peasant girl (sop), betrothed to Masetto, in Mozart's *Don Giovanni*.

(2) The innkeeper Matteo's daughter (sop), lover of Lorenzo, in Auber's *Fra Diavolo*.

**Zhiganov, Nazib** (*b* Uralsk, 15 Jan. 1911; *d* Kazan, 2 June 1988). Tatar composer. Studied Kazan with Shevalina, Moscow with Litinsky, graduating with the opera *Kachkyn*. The works which followed did much to establish an independent genre of Tatar opera, using local folk styles and native historical or fantastic subjects within a conventional framework. His most powerful work, more Romantic in vein, is *Jalil*, on the life and poetry of Musa Jalil; this had wide success within the Soviet Union.

SELECT WORKLIST (all first prod. Kazan): *Kachkyn* (The Runaway) (Fayzy; 1939); *Irek* (Freedom) (Safin; 1940); *Altynchech* (Goldilocks) (Jalil; 1941); *Tyulyak* (Isanbet; 1945); *Shagyr* (The Poet) (Fayzy; 1947); *Jalil* (Fayzy; 1957).

BIBL: Y. Hirschman, *Nazib Zhiganov* (Moscow, 1957).

**Ziani, Marc'Antonio** (*b* Venice, *c*.1653; *d* Vienna, 22 Jan. 1715). Italian composer. Possibly studied with his uncle *Pietro Andrea Ziani. Details of his early Italian career are unclear, but in 1686 he became maestro di cappella at S Barbara, Mantua. Already he was involved with the activities of the Venetian opera-houses, making adaptations of Viennese works for performance there, including Cesti's *La schiava fortunata* (1676) and Draghi's *Leonida in Tegea* (1679). In 1679 he composed his own first opera, *Alessandro Magno in Sidone*, and by 1700 had contributed *c*.24 operas to the Venetian repertory, including *La finta pazzia d'Ulisse* and *Eumene*. That year he took up a position at the Viennese court as Vizekapellmeister, rising to Kapellmeister in 1712. Here he composed *c*.20 dramatic works, beginning with *Giordano Pio*. These represent a bridge between the earlier court-opera style of Draghi and the full flourishing of the Baroque tradition under Fux and Caldara.

SELECT WORKLIST: *Alessandro Magno in Sidone* (Aureli; Venice 1679); *La finta pazzia d'Ulisse* (Ulysses' Feigned Madness) (Noris; Venice 1696); *Eumene* (Zeno; Venice 1697); *Il Gordiano pio* (Cupeda; Vienna-Neustadt 1700); *Ercole vincitor dell'Invidia* (Mazza; Vienna 1706).

**Ziani, Pietro Andrea** (*b* Venice, ?before 21 Dec. 1616; *d* Naples, 12 Feb. 1684). Italian composer; uncle of above. After an early career as priest and organist in Venice he held posts in Bergamo and Innsbruck. His first opera, *La guerriera spartana*, was staged in 1654; in 1662 he moved to Vienna to become Vizekapellmeister to the Dowager Empress Eleonora. In 1666–7 he travelled to Dresden for the wedding of the Princess Anna Sophia: here *Teseo* was given. Returning to Venice in 1669, he became first organist at St Mark's; later passed over for the post of maestro di cappella there, he moved instead to the royal chapel in Naples.

Ziani was one of the most important successors to the Venetian tradition of Monteverdi and Cavalli and is of particular significance for his contribution to the establishment of court opera in Vienna. Much of his operatic writing is cast on similar lines to that of Cesti, with whom he shared a fondness for treating heroic themes in a humane fashion.

SELECT WORKLIST: *La guerriera spartana* (The Spartan Warrior Maid) (Castoreo; Venice 1654, lost); *Teseo* (Piccoli; Venice 1658, rev. Naples 1658, prob. by Provenzale, rev. Dresden 1667).

**Zigeunerbaron, Der** (The Gypsy Baron). Operetta in 3 acts by Johann Strauss II; text by Schnitzer, altered from a libretto by Jókai on his story *Saffi*. Prod. Vienna, W, 24 Oct. 1885, with Collin, Hartmann, Reisser, Streitmann, Girardi, cond. J. Strauss; New York, Casino T, 15 Feb. 1886, with Hall, Fritch, St John, Castle, Wilson,

Fitzgerald, cond. Williams; London, Rudolf Steiner T, 12 Feb. 1935, by amateurs.

The complicated plot tells of Sandor Barinkay (ten), who comes to claim his ancestral lands only to find them overrun by gypsies. He falls in love with one of them, Saffi (sop), who turns out to be a princess.

**Zimin, Sergey** (*b* Orekhovo-Zuyevo, 3 July 1875; *d* Moscow, 26 Aug. 1942). Russian impresario. Studied singing, then in 1903–4 began organizing opera seasons in his own name in Moscow, opening with *May Night* at the Aquarium T. He made one of the finest companies in Russia out of this, gathering excellent singers (many from the recently disbanded Mamontov Co.), directors, and designers. The company performed at the Solodovnikov T, which was taken out of Zimin's hands in the Revolution, though he continued working in opera as an adviser and producer, including 1919–20 at the Maly T and 1924–42 at the Bolshoy. In its day, the Zimin T did much to promote the works of Russian composers, giving the première of *The Golden Cockerel*, the first Moscow *Khovanshchina*, and a series of all Tchaikovsky's operas, 1904–17 (in the latter year a fire destroyed the company's property). It also gave the first Russian *Rienzi* and *Meistersinger*. Artists who worked with the company included the singers Ermolenko-Yuzhina, Matveyev, Shevelev, and Osipov, with as guests Shalyapin, Sobinov, and Battistini, the conductors Ippolitov-Ivanov and Cooper, and the producers Pyotr Olenin (chief 1907–15) and Kommisarzhevsky (chief 1915–17).

BIBL: V. Borovsky, *Moskovskaya Opera S. I. Zimina* (Leningrad, 1977).

**Zimmermann, Bernd Alois** (*b* Bliesheim, 20 Mar. 1918; *d* Königsdorf, 10 Aug. 1970). German composer. Studied Cologne with Lemacher and Jarnach. His sole opera was *Die Soldaten* (Zimmermann, after Lenz; Cologne 1965); this aroused considerable controversy, partly for its subject (which has affinities with *Wozzeck*) and partly for its use of so-called pluralism. This involves scenes playing simultaneously on various levels and in various periods, past, present, and future, and makes use of music, speech, electronic music, *musique concrète*, film, and other techniques and media.

**Zimmermann, Udo** (*b* Dresden, 6 Oct. 1943). German composer. Studied Dresden with Thilmann, Berlin with Kochan. Dramaturg and composer Dresden, S, from 1970. *Die weisse Rose* (The White Rose, 1967) is only marginally operatic, being scenes for two singers and chamber group, but in various stagings has had an international success. *Die zweite Entscheidung* (The Second Decision, Magdeburg 1970) and the suc-

cessful *Levins Mühle* (Levin's Mill, 1973) are influenced by Berg and also use lighter styles. He followed this with *Der Schuhu und die fliegende Prinzessin* (The Birdman and the Flying Princess, 1976), which again skilfully mixes styles, and *Die wundersame Schusterfrau* (The Wonderful Cobbler's Wife, 1982), the latter making some use of Spanish idioms in a tale based on Lorca. Intendant, Leipzig 1991.

**Zingarelli, Niccolò Antonio** (*b* Naples, 4 Apr. 1752; *d* Torre del Greco, 5 May 1837). Italian composer and teacher. Studied Naples with Fenaroli, Speranza, Anfossi, and Sacchini. His first dramatic work, the intermezzo *I quattro pazzi*, was produced while he was still a student, and shortly afterwards *Montezuma* was produced at Eszterháza and commended by Haydn. His greatest triumphs were in Milan, where he wrote popular works at incredible speed for La Scala. Returned to Milan after an unsuccessful venture in Paris and concentrated on comic opera, which spread his fame to Germany, before a church appointment turned his attention towards sacred music. In Rome from 1804, he composed comic operas, including the popular *Berenice*, his last opera. His resistance to Napoleon caused him to be arrested and taken to Paris, where the emperor, an admirer of his music, released him and granted him a pension. In 1813 he became director of Naples Cons., where his pupils included Bellini, Mercadante, and Morlacchi; in 1816 he became maestro di cappella at Milan Cathedral.

Zingarelli was the last major composer of opera seria, and most of his operas are on the conventional heroic subjects. The fluency of his technique enabled him to turn out large numbers of works that efficiently served the demands of the singers of the day, but this also encouraged him to write too much, too fast. He was a skilful orchestrator, and his tragic heroines are especially well drawn; but there is too much that is hasty or trivial in his work to encourage modern revival. His best-known opera, *Giulietta e Romeo*, loosely based on Shakespeare though with a *lieto fine, survived in repertories until the 1830s, largely through its adoption by Malibran.

SELECT WORKLIST: *I quattro pazzi* (The Four Madmen) (?; Naples 1768); *Montezuma* (Cigna-Santi; Naples 1781); *Giulietta e Romeo* (Foppa, after Shakespeare; Milan 1796); *Berenice* (Ferretti; Rome 1811).

**Zirra, Alexandru** (*b* Roman., 14 July 1883; *d* Sibiu, 23 Mar. 1946). Romanian composer. Studied Iaşi with Musicescu and Cerne, Milan with Gatti. Apart from his important activity as a teacher in Iaşi, he composed orchestral and especially stage music. In the latter he incorporates much indigenous Romanian musical material. His most important opera is *Alexandru Lăpuşneanu* (Zirra, after Neruzzi; comp. 1930, Bucharest 1941, rev. 1944).

**Zola, Emile** (*b* Paris, 2 Apr. 1840; *d* Paris, 29 Sep. 1902). French author and dramatist. Zola's wide human interests included a belief in the powers of the musical stage to express real and important issues; he collaborated on several operas with Alfred *Bruneau (or wrote texts later used by him): these were *Le rêve* (1890), *L'attaque du moulin* (1893), *Messidor* (1897), *L'ouragan* (1901), *L'enfant roi* (1902), *Naïs Micoulin* (1907), and *Les quatre journées* (1916). Other operas on his works are as follows:

*Nana* (1880): Gurlitt (1933)
*Germinal* (1885): Kaan z Albéstŭ (comp. 1902–8)
*L'attaque du moulin*: Bruneau (1893); K. Weis (*Útok na mlýn*, 1912).

**Zoppot.** See SOPOT.

**Zukunftmusik** (Ger.: 'music of the future'). The term coined by Wagner for his music in suggesting that he had discovered and opened up a new means of composition: it was much used.

**Zumpe, Hermann** (*b* Oppach, 9 Apr. 1850; *d* Munich, 4 Sep. 1903). German conductor and composer. Studied Leipzig with Tottmann. Helped Wagner to prepare first *Ring*, 1872–6. Cond. Hamburg (1884–6), Stuttgart (1891–5), Schwerin (1897), and Munich (1895), where he cond. the first Wagner at the Prinzregententheater. London, CG, 1898; he was compared to the greatest Wagner conductors of his day. He composed a number of operas and operettas.

**Zumsteeg, Johann Rudolf** (*b* Sachsenflur, 10 Jan. 1760; *d* Stuttgart, 27 Jan. 1802). German composer. Studied with Poli, whom he succeeded in 1792 as Kapellmeister and director of the Stuttgart Court Opera, where he spent his career. Though well known as a song composer, making a particular contribution to the development of the ballad, he also wrote a number of successful operas which showed an early interest in subjects that were to absorb the Romantics, including the exotic, the chivalric, and the fantas-

tic. His most successful and popular work was *Die Geisterinsel*.

SELECT WORKLIST: *Die Geisterinsel* (The Magic Island) (Gotter, after Shakespeare; Stuttgart 1798).

**Zurga.** The king of the fishermen (bar) in Bizet's *Les pêcheurs de perles*.

**Zurich.** City in Switzerland. Opera was first given at the Aktientheater, opened in 1833. This burnt down in 1890, and the Stadttheater (cap. 1,100) opened in 1891; renamed the Opernhaus 1964. It was in Zurich that the first 'legitimate' *Parsifal* outside Bayreuth was staged in 1913 (though still in the face of copyright argument from Cosima Wagner). Schoeck, Honegger, and Martin all worked here. Hans Zimmermann, the influential director 1937–56, established the June Festival, secured major artists (including Furtwängler and Knappertsbusch), and staged the premières of *Lulu* (1937), *Mathis der Maler* (1938), and Honegger's *Jeanne d'Arc au bûcher* (1942). Rosbaud conducted the stage première of *Moses und Aron* (1957). Under the directorship of Claus Helmut Drese (1975–86) an important Monteverdi cycle was staged.

BIBL: H. Erismann, *Das fing ja gut an . . . Geschichten und Geschichte des Opernhauses Zürich* (Zurich, 1984).

**Zweig, Stefan** (*b* Vienna, 28 Nov. 1881; *d* Petropolis, 22 Feb. 1942). German writer of Jewish descent. Studied Berlin and Vienna. His prolific output included the libretto for R. Strauss's *Die schweigsame Frau* (1935). Though the collaboration was itself cordial, the Nazi pressures on Zweig were distressing and angered Strauss, who successfully insisted that Zweig's name, omitted from the playbill, should be restored.

**Zwillingsbrüder, Die** (The Twin Brothers). Opera in 1 act by Schubert; text by G. von Hofmann, possibly after a French vaudeville, *Les deux Valentins*. Prod. Vienna, K, 14 June 1820, with Vogl, Rosenfeld, Gottdank, Meier; Lehigh Univ., Bethlehem, PA, 9 Mar. 1957.

**Zwischenspiel.** The usual German term for *intermezzo.